The Pesticide Manual

Eleventh Edition

Editor: C D S Tomlin

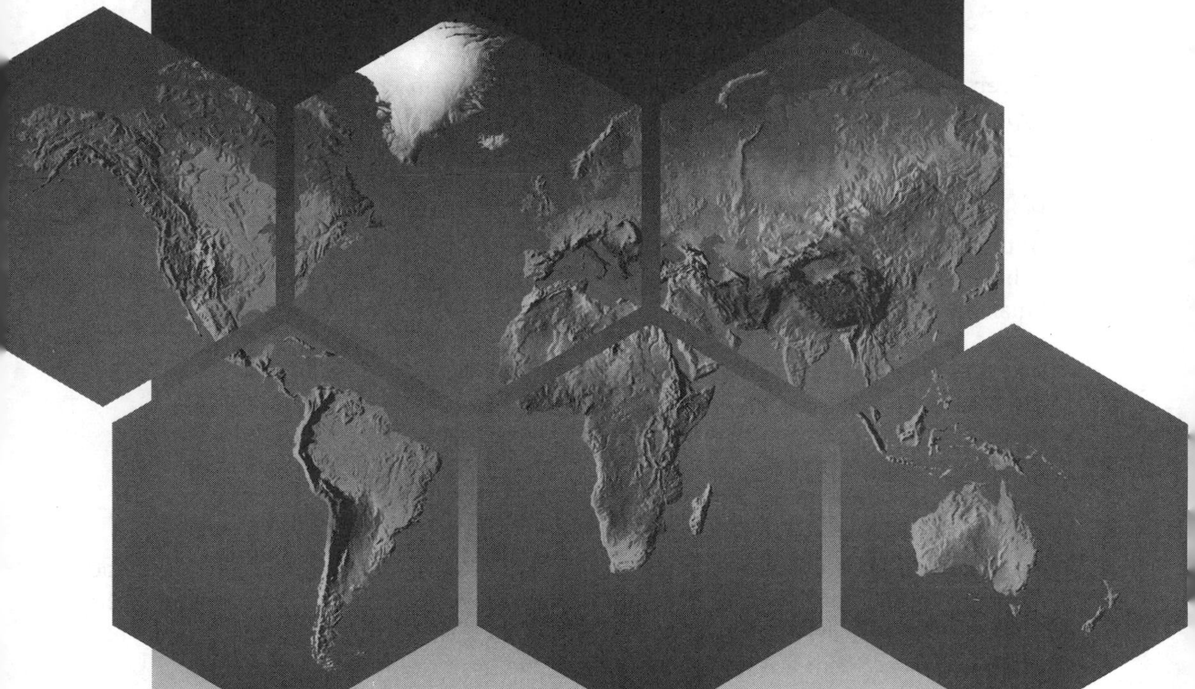

BRITISH
CROP
PROTECTION
COUNCIL

British Library Cataloguing in Publication Data.
A catalogue record of this book is available from the British Library.

ISBN 1 901396 11 8

First published	1968	Seventh edition	1983
Second edition	1971	Eighth edition	1987
Third edition	1972	Ninth edition	1991
Fourth edition	1974	Tenth edition*	1994
Fifth edition	1977	Eleventh edition	1997
Sixth edition	1979		

* The tenth edition incorporated *The Agrochemicals Handbook* previously published by The Royal Society of Chemistry.

Cover design by Major Design & Production, Nottingham
Typographic design by Alan Brannan Design, Stowmarket
Typeset and printed by Page Bros, Norwich

Published by:
British Crop Protection Council, 49 Downing Street, Farnham, Surrey GU9 7PH, UK
Tel: +44 (0)1252 733072   Fax: +44 (0)1252 727194
Email: md@BCPC.org   Internet: http://www.BCPC.org

All BCPC publications can be brought from:
BCPC Publications Sales, Bear Farm, Binfield, Bracknell, Berks RG42 5QE, UK
Tel: +44 (0)118 934 2727   Fax: +44 (0)118 934 1998
Email: publications@BCPC.org

Disclaimer
Every effort has been made to ensure that all information in this edition of *The Pesticide Manual* is correct at the time of going to press. However, the editor and the publisher do not accept liability for any error or omission in the content, or for any loss, damage or any other accident arising from the use of the products listed therein.

Before handling, storing or using any approved crop protection product, it is essential to follow the instructions on the label.

CONTENTS

FOREWORD

For nearly 30 years, *The Pesticide Manual* has served as a standard reference work on the active ingredients in products for the control of crop pests and diseases, weeds, animal ectoparasites and pests in public health. During that time it has gradually expanded its coverage to include plant growth regulators, repellents, synergists, herbicide safeners, and latterly beneficial microbial and invertebrate agents and pheromones. It has earned an enviable reputation for impartial, factual accuracy on the properties and uses of the many compounds that have been developed and marketed over the years as pesticides or enhancers of their effectiveness.

The agrochemical industry continues to make an enormous contribution to world food production and human welfare, and there is a continuing need for an up-to-date, authoritative and definitive source of information on all available pesticides, including their environmental impact. This eleventh edition of *The Pesticide Manual*, again produced and published solely by the British Crop Protection Council, provides an enormous amount of extremely useful data in a compact, concise and conveniently arranged form. It now includes improved coverage of ecotoxicological, degradation and environmental data to comply with the evolving EC and EPA regulatory requirements.

The editor and his colleagues at the British Crop Protection Council have further refined this scholarly work. It remains indispensable to all manufacturers, organisations, researchers, legislators, regulators, even journalists and private individuals with a professional interest in food and water quality and purity, in land use for forestry, farming, horticulture and recreation, and in conservation of terrestrial and aquatic flora and fauna. I strongly recommend this new edition to all previous buyers, and to new readers who need to be acquainted with the details of crop protection agents and their use.

Trevor Lewis
Lawes Trust Senior Fellow, Rothamsted
Formerly Director
Institute of Arable Crops Research, UK

THE PUBLISHER

This eleventh edition of *The Pesticide Manual* is published by the British Crop Protection Council (BCPC) - a registered charity. Formed in 1967, the principle objective of the BCPC is 'to promote and encourage the science and practice of crop proection for the benefit of all.'

BCPC brings together a wide range of organisations interested in the improvement of crop protection. The 43 members of the Board represent the interests of government departments, the agrochemical industry, farmers' organisations, the advisory services and independent consultants, distributors, the research councils, agricultural engineers, environment interests, training and overseas development.

The corporate members of the BCPC currently are:

ADAS
Agricultural Engineers Association
Association of Applied Biologists
Association of Independent Crop Consultants
ATB-Landbase
Biotechnology and Biological Sciences Reseach Council
British Agrochemicals Association
British Institute of Agricultural Consultants
British Society for Plant Pathology
British Society of Plant Breeders
Department of Agriculture for Northern Ireland
Department of the Environment
Ministry of Agriculture, Fisheries and Food
 - Pesticides Safety Directorate
National Association of Agricultural Contractors
National Farmers' Union
Natural Environment Research Council
Overseas Development Administration
Scottish Office Agriculture, Environment and Fisheries Department
Society of Chemical Industry - The Pesticides Group
United Kingdom Agricultural Supply Trade Association

To obtain further information about the BCPC, its activities, conferences or publications, please contact:

The General Secretary, British Crop Protection Council, 49 Downing Street, Farnham, Surrey GU9 7PH, UK

Tel: +44 (0)1252 733072   Fax: +44 (0)1252 727194
Email: gensec@BCPC.org   Internet: http://www.BCPC.org

EDITORIAL ADVISORY BOARD

An informal Editorial Advisory Board met to consider strategy and content for this edition of *The Pesticide Manual*. Thanks are due to them for many ideas and suggestions. Those attending were:

David Alford, ADAS
Duncan Batty, IACR-Rothamsted
Geoff Briggs, AgrEvo UK Ltd
Richard Bromilow, IACR-Rothamsted
David Cartwright, Zeneca Agrochemicals
John Caseley, IACR-Long Ashton
Catherine Eke, National Rivers Authority
Eric Ford, *Pesticide Science*
Terry Grayson
Keith Holly
Rob Jacobson, Horticulture Research International
Ann Kirkwood
Kevin Ledgerwood, Zeneca Agrochemicals
John Mumford, Imperial College, University of London
Roger Thornely, AgrEvo UK Ltd
Trevor Lewis, BCPC
Frances McKim, BCPC
Len Copping
Hamish Kidd, Royal Society of Chemistry

Written contributions were received from:
Keith Brent
John Dowsett, DowElanco Ltd
Andy Hart, Central Science Laboratory
Derek Hollomon, IACR-Long Ashton
Trevor Martin, Bayer plc
David Martin, MAFF
Richard Rees, AgrEvo GmbH
Dale Shaner, American Cyanamid Co.
Mike Upstone, Du Pont (UK) Ltd.

PREFACE

After three years of preparation, another edition of *The Pesticide Manual* is complete.

This eleventh edition contains over 30 more Main Entries than its predecessor. As well as many new molecules, I have also included for the first time some important pheromones, and more biological agents.

I have been able to include more detail on biochemistry and mode of action and more of the entries now have ecotoxicological data. An indication of application rates has been restored to this edition. Much toxicological and physico-chemical data has again been revised with new data supplied by manufacturers. New fields of information include Henry's constant (a measure of air/water partitioning), risk phrases and EEC numbers. I have changed the layout of some entries to give more emphasis to the ester or other form of commercial relevance.

The Guide to Main Entries has been further extended. The resistance surveys, introduced in the previous edition, have been updated and the Glossary of Species has been revised.

This edition would not have been possible without the help of many other people. I would like to thank Denise Ledgerwood and Douglas Hartley for their excellent and painstaking work checking the proofs, David Alford, George Cussans and David Yarham for their extensive revisions to the Glossary, Richard Bromilow for much useful advice on pKa and partition coefficients, Paul Lister for programming work, and my colleagues in BCPC for their support and advice throughout.

Finally, I would especially like to thank the many people in companies around the world who took so much trouble to provide the data which makes up the majority of this book.

Despite all these invaluable helpers, any errors in transcription, interpretation or abbreviation are of course my responsibility. I should be grateful if readers would let me know of errors or omissions, which can be rectified in future editions.

Clive Tomlin, *Editor*

GUIDE TO USING THE MAIN ENTRIES

Further details on many fields of information are given under the appropriate heading in the Notes to the Sample Entry, which follow.

The example entry illustrates a chemical substance which is used in its ester form, but which receives its approved common name for its parent (acid) form.

1. Sequential entry number.
2. Entry header. The preferred name for the active ingredient parent molecule. For details, see Notes to the Sample Entry.
3. Class. See Notes to the Sample Entry.
4. Chemical structure. See Notes to the Sample Entry.
5. Headers in italics refer to the different forms (ester/acid, salt/acid, etc.) for which data are given.
6. Common name. The data in parentheses list those standards organisations that have approved the preceding common name. For details, see Notes to the Sample Entry. See Abbreviations and Codes for the meaning of the acronyms.
7. IUPAC name. See Notes to the Sample Entry.
8. Chemical Abstracts name. See Notes to the Sample Entry.
9. Chemical Abstracts Service Registry Number. See Notes to the Sample Entry.
10. EEC no. See Notes to the Sample Entry.
11. Development codes. See Notes to the Sample Entry.
12. Official codes (fictional example). See Notes to the Sample Entry.
13. Composition. Purity of technical grade, isomer composition, etc. For details, see Notes to the Sample Entry.
14. Molecular weight. Given to one decimal place.
15. Molecular formula. Given in the order C,H, alphabetical.
16. Form. The physical form, appearance and odour (if any) of the active ingredient.
17. Melting point.
18. Boiling point (fictional example).
19. Vapour pressure. See Notes to the Sample Entry.
20. Partition coefficient between n-octanol and water (as the log value). For details, see Notes to the Sample Entry.
21. Henry's constant. See Notes to the Sample Entry.
22. Specific gravity, density, relative density or bulk density.
23. Solubility. Units are not repeated after every value, but may be found in parentheses at the end of a sentence, together with other data, where available, such as the temperature, the pH (for solubility in water) and the method used.
24. Flash point (fictional example).
25. Other physical properties (fictional example).

SAMPLE ENTRY

Note: This specimen entry has been deliberately shortened and modified (some items are fictitious, for example only) so that it illustrates the principles. It is **not** the same as the entry for diclofop-methyl.

❶ ❷ ❸
219 diclofop-methyl   *Herbicide*

2-(4-aryloxyphenoxy)propionic acid

NOMENCLATURE
❺ *diclofop-methyl*
❾ **CAS RN** *[51338-27-3]* unstated stereochemistry; *[71283–65–3]* (R)- isomer; *[75021-72-6]* (S)- isomer   ❿ **EEC no.** 257-141-8   ⓫ **Development codes** Hoe 023408; AE F023408

❺ *diclofop*
❻ **Common name** diclofop (BSI, E-ISO, (m) F ISO, ANSI, WSSA)
❼ **IUPAC name** (RS)-2-[4-(2,4-dichlorophenoxy)phenoxy]propionic acid
❽ **Chemical Abstracts name** (±)-2-[4-(2,4-dichlorophenoxy)phenoxy]propanoic acid
Other names Oname   ❾ **CAS RN** *[40843–25–2]* unstated stereochemistry
⓫ **Development codes** Hoe 021079   ⓬ **Official codes** OMS 9999

PHYSICAL CHEMISTRY
diclofop-methyl
⓭ **Composition** Tech. grade is ≥93% pure.   ⓮ **Mol. wt.** 341.2   ⓯ **M.f.** $C_{16}H_{14}Cl_2O_4$ ⓰ **Form** Colourless crystals.   ⓱ **M.p.** 39–41 °C   ⓲ **B.p.** 176 °C/0.5 mmHg
⓳ **V.p.** 0.25 mPa (20 °C) ...   ⓴ **K_{ow}** logP = 4.58   ㉑ **Henry** ...   ㉒ **S.g./density** 1.30 at 40 °C   ㉓ **Solubility** In water 0.8 mg/l (pH 5.7, 20 °C). In acetone, dichloromethane, dimethyl sulfoxide, ethyl acetate, toluene >500 g/l; in polyethylene glycol 148, methanol 120, isopropanol 51, n-hexane 50 (all in g/l, 20 °C).   **Stability** Stable to light. In water, DT_{50} (25 °C) 363 d (pH 5), 31.7 d (pH 7), 0.52 d (pH 9).   **Specific rotation** ...   ㉔ **F.p.** 165 °C (Tag open cup)
㉕ **Other properties** Tech. grade viscosity 15.7 mN/m (30 °C).

GUIDE TO USING THE MAIN ENTRIES

continued

26. pKa. Acid dissociation constant. See Notes to the Sample Entry.
27. No parent or derivative header. In these sections of an entry, it is usually not appropriate to distinguish whether data which follow apply to parent or derived form.
28. Patents. See Notes to the Sample Entry. WIPO country codes are used (see Abbreviations and Codes).
29. Manufacturer. The intention is that this should list only companies known to be manufacturing the active ingredient. In some cases, however, it may be that actual manufacture is carried out by a third party acting under contract to the named company.
30. Mode of action. The observed symptoms, and the physiology of the method by which the substance is effective.
31. Uses. See Notes to the Sample Entry.
32. Phytotoxicity. Any adverse effects on crops are listed. If the material is not phytotoxic, this is not normally noted.
33. Formulation type. For standard formulation types, the two-letter GCPF (formerly GIFAP) codes (see Abbreviations and Codes) are given. Some descriptions are too vague to allow conversion to a specific GCPF code, and have been left as full text.
34. Compatibility. Known incompatibilities with other pesticides, fertilisers or formulation adjuvants are given.
35. Selected tradenames. See Notes to the Sample Entry.
36. Analysis. See Notes to the Sample Entry.
37. IARC and EHC (fictional examples). See Notes to the Sample Entry.
38. Acute oral toxicity. Data are given in mg of active ingredient per kg of animal body weight.
39. Skin and eye. See Notes to the Sample Entry.
40. Acute inhalation toxicity. See Notes to the Sample Entry.
41. No Observable Effect Level. See Notes to the Sample Entry.
42. ADI. Acceptable daily intake (modified data). See Notes to the Sample Entry.
43. Toxicity class. See Notes to the Sample Entry.
44. EC Risk. See Notes to the Sample Entry.
45. PIC. See Notes to the Sample Entry.
46. Toxicology reviews. See Notes to the Sample Entry.

diclofop

⑭**Mol. wt.** 327.2   ⑮**M.f.** $C_{15}H_{12}Cl_2O_4$ ⑯**Form** Yellowish-white solid.
⑰**M.p.** 118-122 °C   ⑲**V.p.** 3.1 × 10^{-6} mPa (20 °C) ...   ⑳K_{ow} logP = 2.81
(pH 5), 1.61 (pH 7)   **S.g./density** 1.4 (20 °C)   ㉓**Solubility** In water 0.453
(pH 5), 122.7 (pH 7), 127.4 (pH 9) (all in g/l, 20 °C)   ㉖**pKa** 3.43

㉗**COMMERCIALISATION**
History Herbicidal activity of diclofop-methyl reported by ... Introduced by
Hoechst AG (now AgrEvo GmbH).   ㉘**Patent** DE 2136828; DE 2223894
㉙**Manufacturers** AgrEvo; Jingma

APPLICATIONS
diclofop-methyl
Biochemistry Fatty acid synthesis inhibitor. Destroys the cell membrane, ...
㉚**Mode of action** Diclofop-methyl is a selective systemic herbicide, also with
contact action ...   ㉛**Uses** Diclofop-methyl is used for post-emergence control of
wild oats ...   ㉜**Phytotoxicity** Phytotoxic to maize, sorghum, oats, sugar cane,
rice, and cotton.   ㉝**Formulation types** EC.   **Mixtures** *(diclofop-methyl +)*
fenoxaprop-P-ethyl.   ㉞**Compatibility** Incompatible with most other herbicides.
㉟**Selected tradenames** 'Hoegrass' (AgrEvo); 'Hoelon' (AgrEvo); 'Illoxan'
(AgrEvo)

㉗ ㊱**ANALYSIS**
Product analysis by glc (*CIPAC Handbook*, 1985, **1C**, 2096). Details of glc methods
for **product** and **residue** analysis are available from AgrEvo.

MAMMALIAN TOXICOLOGY
diclofop-methyl
㊲**IARC** 999   **EHC** 999   ㊳**Oral** Acute oral LD_{50} for rats 481-693 mg/kg (in
sesame oil), dogs 1600 mg/kg (highest dose without induction of vomiting).
㊴**Skin and eye** Acute percutaneous LD_{50} for rats >5000 mg/kg.
㊵**Inhalation** LC_{50} for rats >1.36 mg/l air.   ㊶**NOEL** (2 y) for rats 0.1 mg/kg b.w.;
(15 mo) for dogs 0.44 mg/kg b.w.   ㊷**ADI** 0.001 mg/kg b.w.   **Other** Non-
mutagenic in the Ames test.   ㊸**Toxicity class** WHO (a.i.) III; EPA (formulation) III
㊹**EC risk** Xn (R22, R43)   ㊺**PIC**

diclofop
㊻**Reviews** *Pesticide residues in food - 1995, FAO* ...   ㊳**Oral** Acute oral LD_{50} for
female rats 586 mg/kg.

GUIDE TO USING THE MAIN ENTRIES

continued

47. Other beneficial species (fictional example).
48. Daphnia. Where no species name is given, data are for *D. magna*. In this example, the EC_{50} for *Daphnia magna,* exposed for 48 hours, is 0.23 mg a.i. per litre of water.
49. Other aquatic species (fictional example)
50. Animals and Plants. The fate of material applied to test species. Data supplied by manufacturers have often been summarised.
51. Soil/Environment. See Notes to the Sample Entry.

ECOTOXICOLOGY

diclofop-methyl

Birds Acute oral LD_{50} for Japanese quail >10 000 mg/kg. Five-day dietary ... **Fish** LC_{50} (96 h) for rainbow trout 0.23 mg/l. **Bees** Non-toxic to bees under field conditions and application rate of 1.134 kg a.i./ha. **Worms** LC_{50} (14 d) for earthworms >1000 mg/kg soil, dry weight. ㊼**Other beneficial spp.** Not toxic to *Argsope argentata* (web-building spider), *Aleochara* sp. (staphylinid beetle) ... ㊽**Daphnia** LC_{50} (48 h) 0.23 mg/l. **Algae** EC_{50} (72 h) for *Scenedesmus subspicatus* 1.5 mg/l ㊾**Other aquatic spp.** LC_{50} (96 h) for mysid shrimp (*Mysidopsis bahia*) 5.7 mg/l ...

㉗ENVIRONMENTAL FATE

㊿**Animals** When fed to rats, diclofop-methyl is almost totally absorbed and then rapidly excreted ... ㊿**Plants** Diclofop-methyl is taken up rapidly and almost completely by plants, with little translocation ... (51)**Soil/Environment** In soil, diclofop-methyl is metabolised to diclofop, which then undergoes further degradation to ...

NOTES TO THE SAMPLE ENTRY

GENERAL POINTS

Active ingredient. Except where otherwise indicated, all data refer to the active ingredient.

Units. To avoid repetition, units are frequently omitted where a sequence of data is given; the relevant units are placed at the end of the sentence or phrase. For example:

"**Acute oral LD$_{50}$** for rats 250, rabbits 400, mice 300 mg/kg."

Company names. Company names are often given in a shortened form. The full name of the company, and an address, is given in the Directory of Companies, p. 1407.

Obsolete names. In the Main Entries, obsolete names are enclosed by brackets, e.g. [bencarbate].

HEADERS

Entry header. The preferred name for the active ingredient. For biological agents, the scientific name is used.

For chemicals, the ISO common name is used, where one exists. Failing that, the order of preference is:- a national approved common name (e.g. BSI, JMAF, etc), WSSA, BAN or ESA name, official code number, development code number, IUPAC chemical name.

Class. Also given in the heading are the field of use (herbicide, insecticide, etc), and the class of material.

Choice of the word used for a class name is somewhat arbitrary. Many of the names used here have common currency; in other cases, however, the name used is only one of several alternatives used by different groups of workers. The primary role of class names is to group chemicals according to their known, or assumed, biochemical mode of action, and by common chemical features; such grouping is, however, neither immutable nor infallible.

Often the type of chemical is also indicated by a stem in the common name (for example, 'uron' for ureas, and 'carb' for carbamates).

Chemical structure. Structures have been drawn giving first priority to clarity of presentation, the illustrator using artistic licence over the orientation and length of bonds, etc., in order to keep the structures within a reasonable area.

As far as possible, the structures have been drawn so as to show the three-dimensional aspect, when this is known, so that the reader can see that some groups are in the plane of the paper, with others above or below it. When compounds contain a mixture of stereoisomers, a single drawing may be given, with no representation of the stereochemistry of the individual components.

NOMENCLATURE

Common name and **Standard.** Most chemical pesticides have common names agreed by the International Organisation for Standardization through its Technical Committee 81 (ISO/TC 81) for which the secretariat is the British Standards Institution (BSI). The principles for coining these common names are explained in ISO 257: 1988 and in BS1831: Part 1: 1985.

If the molecule is, for example, an acid, it may be used in a salt or ester form. Usually the common name is assigned to the parent acid; from this name, the salt or ester name can be derived in a straightforward manner by adding the name of the ester or ion (e.g. diclofop, diclofop-methyl). Short form names are used for some ions and ester groups. These are indicated in the Abbreviations and Codes Section, p. 1425.

Names are not formally approved by ISO until they appear in a list issued by ISO. However, the last supplement to the list of ISO-approved names appeared in 1983. As names achieve draft ISO status rather more rapidly, a status earlier in the approval process has been accepted, that of Provisionally Approved ISO name; this is indicated in the Pesticide Manual as 'pa ISO'.

In practice, there are usually two ISO names; one in English (designated E-ISO) and one in French (designated F-ISO, with gender indicated as (*m*) or (*f*)). Note that the French ISO name is that used in French-speaking countries, with the significant exception of France itself. Names approved by the French standards organisation, AFNOR, are not necessarily the same as F-ISO ones.

IUPAC name, Chemical Abstracts name. The systematic chemical names are given, according to the rules of the International Union of Pure and Applied Chemistry, and of the 9th Collective Index period of the Chemical Abstracts Service, respectively.

The systems of nomenclature for molecules which can form stereochemically differing isomers can be complex. However, to appreciate the isomeric composition of such materials where described in the entries, an understanding of these systems is essential. For details, see the Stereochemistry Nomenclature Section, p. xxiii.

Scientific name. For biological agents, the Latin name of the organism. Synonyms may be included under **Other names**.

CAS RN. The Chemical Abstracts Service Registry Number, a useful unique reference number. Under the Chemical Abstracts system, differing isomers, including stereoisomers, are assigned different Registry Numbers. For example, the (R) and (S) optical isomers, as well as the (RS) racemic mixture, and material of undefined stereochemistry will, if they have been described in the chemical literature, all have different Registry Numbers. Further details are given in the introduction to Index 1.

EEC no. Number in the European Inventory of Existing Chemical Substances (EINECS) or in the European List of Notified (New) Chemicals (ELINCS). The EINECS Inventory lists and defines those chemical substances which were on the European Community market between 1 January 1971 and 18 September 1981, and to which the pre-marketing notification provisions of Amendment VI to the Directive 67/548/EEC do not apply. The final version was published in June 1990, in the Official Journal of the EC. ELINCS lists substances notified since EINECS was completed. The absence of information under this heading does not necessarily imply that the substance is not in one of these European lists.

Development codes. Code number usually assigned before the material was given a common name. In some cases, the name of the company is also given; this is the name of the company which actually developed the material, not necessarily the same as the name of the company which now markets it. For this reason, some of these names are of companies which are no longer involved in pesticides, and they may not appear in the Directory of Companies.

Official codes. These include numbers assigned by WHO (OMS series), the Entomological Society of America (ENT and AI3 series) and the American Society of Nematologists (AI4 series).

PHYSICAL CHEMISTRY

Composition. Technical grade (abbreviated 'tech.') material is used in practice. This is usually of a defined minimum purity, details of which are given here. Where a substance is capable of existing in different stereochemical forms (see **Chemical structure** and the Stereochemistry Nomenclature Section, p. xxiii), the material used may be a single stereochemical isomer, or a mixture of isomers. An attempt has been made to clarify what isomer or mixture of isomers is used; this does not always correspond to the isomers present in the common name as defined.

Vapour pressure. Values quoted in alternative sources can differ by several orders of magnitude; factors affecting determined vapour pressure values include temperature, purity and method of determination. Where differing values were available, the lowest figure has been chosen. In general, vapour pressures which were known to be those of technical material have not been quoted. In those few cases where a value without a temperature was supplied, it is left to the reader to decide whether or not to assume the determination was at room temperature. Manufacturers were also asked to supply the method of determination; where given, this is appended in parentheses. Most values have been converted to milliPascals (mPa).

1 mPa is equivalent to $c.$ 7.50×10^{-6} mmHg or torr, 10^{-8} bar or 9.87×10^{-9} atm.

Partition coefficient. A high value for the partition coefficient (P) between n-octanol and water is regarded as an indicator that a substance will bio-accumulate (unless other factors operate). Values are given as logP. Note that, for molecules that dissociate, logP values measured within 3 pH units of the pKa will represent an average value over two or more forms present.

Henry's constant. The tendency of a material to volatilise from aqueous solution to air. Sometimes measured, more usually calculated as the ratio of vapour pressure (in Pascals) × molecular weight / solubility (mg/l).

pKa. Strengths of acids and bases are indicated on a common scale (use of pKb for bases being no longer used widely). Where no pKa value is given, it may indicate that no significant dissociation occurs in the environmental range; values of pKa <4 for bases and pKa >8 for acids are not relevant in an environmental context.

COMMERCIALISATION

Patent. Usually only one patent or patent application is cited; patent applications or patents to the same invention in other countries will usually exist. Note that any protection afforded by any of the patents listed may have expired. A patent will not necessarily protect the active ingredient or its use as a pesticide; there may be a series of patent applications relating to the active ingredient, its manufacture, formulation or its use.

Company names refer to the original applicant; title to patents (as distinct from rights to it) may not be transferred when a product or company is bought by another company. For this reason, some company names given here are of ones no longer in the pesticides business.

Country codes are WIPO standard; see the Abbreviations and Codes Section, p. 1427.

APPLICATIONS

Uses. Crop names are in English; targets are given in either Latin or English. A Glossary of Latin and English names is included in *The Pesticide Manual*. Where British and American usage differs, alfalfa (lucerne), and peanuts (groundnuts) have been selected, but maize (corn) and aubergine (eggplant). Also, reference is made to couch, rather than to quackgrass, *Agropyron repens* = *Elytrigia repens*.

Selected tradenames. Up to three lead tradenames are given of the company (or its successor) inventing or introducing the product. For commodities, this is followed, in alphabetical order, by the main formulation name of other companies selling a product based on the a.i. The list of tradenames is therefore not comprehensive.

The name of the **company marketing** the formulation is given in parentheses after the tradename. Companies selling under licence may not be listed if the tradename owner retains, and exercises, marketing rights. Most company names are given in a shortened form. The full name of the company, and an address, is given in the Directory of Companies.

Names normally not included are those which represent a composite of the company name and the common name, names of mixtures (unless the a.i. is used only in mixture), and names used only in home and garden outlets. Suffixes are usually omitted.

ANALYSIS

Methods published in standard reference sources are given. References to AOAC methods which carry an asterisk are considered to be discontinued methods.

Note that many companies prefer workers to refer to them for advice as to the method most appropriate for a particular analytical problem.

MAMMALIAN TOXICOLOGY

IARC, EHC. The numbers given refer to the numbers of the relevant publication in the International Agency for Research on Cancer Monograph Series and Environmental Health Criteria Series, respectively. Both series are available from Distribution and Sales Service, World Health Organisation, 1211 Geneva 27, Switzerland.

Skin and eye. Manufacturers were asked to supply the acute percutaneous LD_{50}, skin and eye irritation effects, and skin sensitisation. LD_{50} data are given in mg of active ingredient per kg of animal body weight.

Inhalation. Usually, the acute inhalation LC_{50}. Data usually relate to rats, exposed to the material for 4 hours. However, these parameters are quoted where they are known; their absence does not necessarily imply that these standard conditions applied. Note that the units used differ between entries.

NOEL. No Observable Effect Level: the highest dose in an animal toxicology study at which no biologically significant increase in frequency or severity of an effect is observed. NOAEL 'No Observable Adverse Effect Level' is often used interchangeably with NOEL.

In its shortest form, an entry gives the duration of a trial (in parentheses), followed by the test species, value obtained, and the units. For example, in the sample entry, the no effect level in 2 year feeding trials for rats was 0.1 mg of active per kg of animal body weight; in 15 month feeding trials for dogs, the no effect level was 0.44 mg of active per kg of animal body weight.

ADI. Acceptable Daily Intake, defined as "The amount of a chemical which can be consumed every day for an individual's entire life span with the practical certainty, based on all the available evidence, that no harm will result".

The current value assigned by the JMPR is given first; any unattributed value has been issued by another body.

Toxicity class. Both WHO and EPA have issued standards by which pesticides may be classified according to toxicity hazard. EPA classes refer to formulated material. WHO also stress the importance of assessing the class for formulated material; their recommended classes, quoted in this Manual are, however, for the a.i. Details of the WHO and EPA classes are given in the Abbreviations and Codes Section, p. 1425; further details of the WHO standard are given in IPCS *The WHO Recommended Classification of Pesticides by Hazard and Guidelines to Classification 1996-1997*, WHO/PCS/96.3. Most of the WHO classes in this Manual are derived from this, or the previous edition, of this series. However, where a manufacturer has assigned their material to a more hazardous class than WHO, this value is given instead. Where toxicology data have been revised in this edition of the Manual, the assigned toxicity class may not reflect this change.

EC Risk. EC Directive 67/548/EEC and subsequent amendments provide a harmonised basis for the classification and handling of dangerous substances. They set out the criteria by which the risks in handling substances can be codified. *The Pesticide Manual* gives the risk symbol and risk codes that have been allocated under these guidelines.

The source used for the allocation was either the Health and Safety Executive of the UK under the CHIP2 regulations, or the manufacturer.

Environmental codes have been omitted, as these have so far been allocated to only a minority of substances. These data should not be used for purposes of labelling pesticides without referring to the detailed guides issued by EC member countries, in which additional information is given. A table of risk symbols and phrases appears in the Abbreviations and Codes Section, p. 1425.

PIC. Prior Informed Consent (PIC) was introduced in 1989 into the FAO Code of Conduct and into the UNEP London Guidelines For The Exchange of Information on Chemicals in International Trade. For pesticides, PIC is administered by FAO. Under PIC, importing countries are made aware of products which have been banned or severely restricted in other countries, on the basis of their unacceptability to human health and/or the environment, and are given the opportunity to object or consent to the continued import of such pesticides. For further details, see *Prior Informed Consent: A Guide to its Working*, GIFAP document D/1990/2537/13.

Where a pesticide is subject to PIC, this is noted in this edition of *The Pesticide Manual*.

Toxicity reviews. Review articles in various sources are given. Frequently, there is a report issued by the FAO Panel of Experts on Pesticide Residues in Food and the Environment, and the WHO Expert Groups on Pesticide Residues (JMPR). In many cases, an earlier review in these series is also available; these are not listed.

ENVIRONMENTAL FATE

Soil/Environment. Among the parameters commonly quoted are:

K_d - the ratio of sorbed to solution pesticide in a water-soil slurry. It is a measure of the relative affinities of the pesticide for water and a soil surface.

Freundlich K - used to define soil adsorption where it is non-linear with concentration.

K_{oc} - which adjusts K_d or Freundlich K for the proportion of organic carbon in the soil.

STEREOCHEMISTRY NOMENCLATURE

The shape of a pesticide molecule is often a vital factor influencing its biological activity.

The stereochemistry of organic compounds is indicated usually by employing the Cahn-Ingold-Prelog sequence rules (R. S. Cahn, C. Ingold & V. Prelog, *Experientia*, 1956, **12**, 81; *Angew. Chem., Int. Ed. Engl.*, 1966, **5**, 385, 511; see Section E of the IUPAC Rules of Nomenclature). By these rules, a substituent with the highest atomic number takes priority. For asymmetric molecules, the system uses the descriptors (R) and (S).

Nomenclature of Molecules with Two Chiral Centres (azoles)

Some molecules, including many azole fungicides, have two chiral (asymmetric) centres.

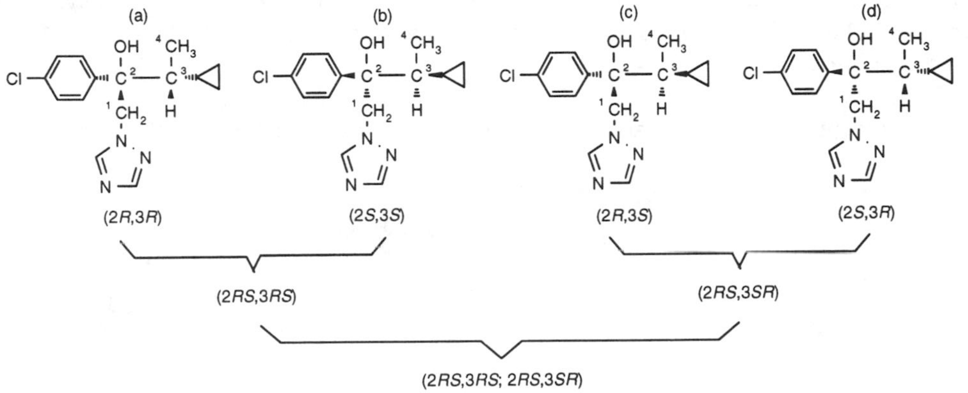

Four stereoisomers are possible in this case. In the IUPAC convention, assuming that the chiral atoms in the illustration are at positions 2 and 3, the combinations (a) (2R,3R), (b) (2S,3S), (c) (2R,3S) and (d) (2S,3R) are different. A racemate, comprising equal quantities of the first two, will be described as (2RS,3RS) - namely each position is considered in turn, and the first or second choice selected throughout. Similarly, a second racemate is possible comprising equal quantities of (c) and (d) and is represented by (2RS,3SR). A mixture of the two racemates contains all four possible stereoisomers and is shown as (2RS,3RS;2RS,3SR).

Chemical Abstracts also uses the (R) and (S) descriptors, but in a different way when there is more than one chiral centre. The *relative* configuration of the various centres is shown thus: (2R*,3R*) - this means that, if the 2-position has the (R) configuration, so does the 3-position, and so on. Finally the absolute position of the senior chiral position is given

[2R(2R*,3R*)]- or, in the case of a racemate (which has both stereoisomers), (±) is added, thus: (2R*,3R*)-(±).

Nomenclature of Pyrethroids

Whilst these rules refer to absolute configuration, an apparently minor change in a substituent remote from a chiral centre may reverse the ranking of a set of groups. Examples of these, taken from the class of pyrethroids, are shown below. If substituents of high atomic number, for example halogen atoms, are substituted for carbon atoms, the ranking may be reversed under the sequence rules and 3S (with a methyl substituent) becomes 3R (for bromine) even though the shape of the molecule is unaltered.

Hence, in the case of the pyrethrins and their synthetic analogues (which may have two or more asymmetric centres), a description proposed by the workers at Rothamsted Experimental Station, giving the relative stereochemistry, has been introduced (M. Elliott, N. F. Janes & D. A. Pulman, *J. Chem. Soc., Perkin Trans. I*, 1974, p. 2470 (first footnote)). The absolute configuration of the C-1 position of the cyclopropane ring is given as before, and the relative position (*cis* or *trans*) of the substituent at C-3 added, thus (1R)-*cis* or (1RS)-*cis-trans*, etc.

In the present entries, names according to this convention have been prefixed *Roth:*.

(1RS)-cis-trans-	(1R)-trans-	(1R)-cis-

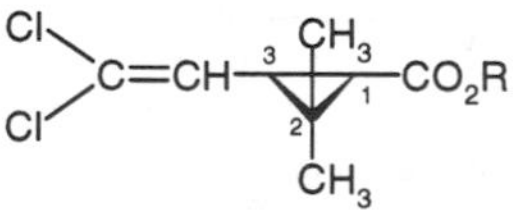

(1RS,3RS; 1RS,3SR)-	(1R,3R)-	(1R,3R)-

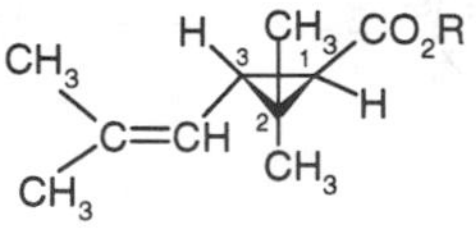

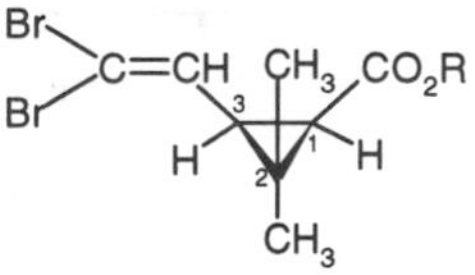

Several methods are used by *Chemical Abstracts* to indicate the stereochemistry of pyrethrins. In very simple cases, the absolute configuration (R or S) is given, a racemate being indicated by (±). The (1R-cis)-style notation is used for suitable cases, provided there is no additional chiral centre in the molecule. When more than two chiral centres are present, the absolute configuration (R or S) of the first is given with the descriptor α to indicate this group is above the plane of the molecule, subsequent chiral centres are described as α if the group with highest priority is above the plane and β if below it. Alternatively, the relative configurations may be given (1R*, 3R*) as mentioned previously.

Nomenclature of Geometrical Isomers

In both IUPAC and *Chemical Abstracts* style names, the shape of geometrical isomers is indicated by the descriptors (*E*) and (*Z*). Where the two senior groups under the sequence rules (see above) are on opposite sides of a double bond, the descriptor (*E*) is applied; where such groups are on the same side, the descriptor is (*Z*).

(*E*)   (*Z*)

Nomenclature of Amino Acids and Carbohydrates

The systems established for describing the stereochemistry of amino acids and carbohydrates (with D and L for the absolute configuration) are retained in IUPAC nomenclature and in *Chemical Abstracts*.

RESISTANCE TO PESTICIDES

The development of resistance to pesticides by target organisms has become an important issue. In order to understand and manage resistance and so minimise impact on crop production, four Resistance Action Committees exist as Specialist Technical Groups of the Global Crop Protection Federation, GCPF.

Reports were sought from these committees to provide up-to-date summaries on the present situation. They form the basis for the following statements. Users of individual entries in the manual should refer back to these statements for relevant information on resistance.

Further information on recommendations for pesticide use and the activities of the individual RACs can be obtained from GCPF, Avenue Louise 143, 1050 Brussels, Belgium.

FUNGICIDE RESISTANCE ACTION COMMITTEE (FRAC)

FRAC was founded in 1981 with the following objectives:

> To establish fungicide resistance monitoring surveys in order to detect or confirm cases of fungicide resistance.

> To harmonise recommendations for the use of fungicide groups marketed by more than one Company in order to prevent or manage fungicide resistance.

> To encourage research into fungicide use strategies aimed at preventing or managing fungicide resistance.

> To alert all interested parties to problems of resistance and educate them in the reasons why problems occur and how they can be avoided or managed.

To achieve these objectives, six working groups cover the major fungicide groups at most risk: benzimidazoles, dicarboximides, sterol biosynthesis inhibitors (SBIs), phenylamides, anilinopyrimidines and strobilurins. In addition, a special working group exists to cover fungicide use, mainly SBIs, on bananas. The working groups report into the Central Steering Committee. They are responsible, through their constituent companies, for collating data from monitoring surveys and recommending fungicide use strategies. These recommendations are ratified by the Central Steering Committee and published in a variety of formats.

Definition of Fungicide Resistance

FRAC is concerned that the term 'resistant' should be correctly used when applied to fungi and their response to fungicides. The following criteria are recommended:

The pathogen must at some time have been successfully controlled in commercial practice by the fungicide.

Resistance must show as a loss of efficacy against the pathogen when the product is used according to manufacturer's recommendations.

Evidence must be provided that the sensitivity of the pathogen to the fungicide has altered in order to render the product less effective.

Resistance Management Strategies

Strategies used or being evaluated differ between chemical groups and between crops. Current guidelines are available from FRAC or the manufacturer of the product.

Current Status of Resistance

The following statements are based on the results of surveys and information considered by the FRAC working groups. They indicate areas where resistance has been confirmed, but should not be used to suggest that a particular product should not be used in any particular locality. For more detailed information relating to specific uses and potential problem areas, reference must be made to the manufacturer of the product. '

Benzimidazoles

Resistance is widespread amongst Ascomycete and Deuteromycete fungi. Example genera are: *Fusarium, Venturia, Cercospora, Pseudocercosporella, Septoria, Botrytis, Mycosphaerella, Monilinia, Colletotrichum, Erysiphe, Penicillium, Pyrenophora.*

As a general policy, benzimidazoles should not be relied upon alone but are still extremely valuable as mixture partners or as part of an application programme alternating with other products.

Diethofencarb is a phenyl carbamate capable of controlling benzimidazole resistant *Botrytis cinerea* but with no activity against benzimidazole sensitive strains. Dual resistance to benzimidazoles and diethofencarb is known to occur in France and New Zealand.

Dicarboximides

Botrytis cinerea is the principal pathogen affected and resistance could occur in any area of use.

Resistance is known in other fungi, e.g. *Fusarium nivale*, *Sclerotinia homoeocarpa* in turf, *Monilinia fructicola* and *Alternaria spp.*, but has only been found sporadically.

Resistance management should be based on minimising use by restricting applications and the use of mixtures. Two way (dicarboximide + benzimidazole) or 3 way (dicarboximide + benzimidazole + diethofencarb) resistance is known in *Botrytis cinerea*.

Phenylamides

Phenylamides are active against Oomycete fungi. Resistance has been confirmed in *Phytophthora infestans* (potatoes), *Plasmopara viticola* (vines), *Peronospora destructor* (onions), *Bremia lactucae* (lettuce), *Pseudoperonospora humuli* (hops), *Pseudoperonospora cubensis* (cucurbits) and *Pythium* spp.

Resistance could occur in any area of intensive use. Used as part of a disease control programme in mixture and alternations with other fungicides, the phenylamides remain very useful compounds.

Sterol biosynthesis inhibitors (SBIs)

This group considers all SBIs, i.e. the demethylation inhibitors (DMIs) (azoles, pyrimidinyl carbinols) and morpholine and piperidine fungicides. Unlike the previous chemical groups, for which resistance can develop rapidly and is a very clear phenomenon, resistance to SBIs shows as a gradual reduction in the sensitivity of the target fungal population to the particular fungicide.

It is thus very difficult to define accurately when a target pathogen is 'resistant', as a decline in field activity may show without a total breakdown of control.

SBI fungicides are active against a wide range of Deuteromycete, Ascomycete and Basidiomycete genera. Individual compounds are individually selective against a range of species. In the list that follows, the genera and crops indicate areas where the sensitivity pattern of the target population has shifted sufficiently to DMI fungicides to cause concern in some locations. Where such shifts are confirmed, the margin for error in application rate and timing will be smaller, and it is thus imperative that the manufacturer's label recommendations are followed.

Cereals, wheat and barley	*Erysiphe graminis*
Vines	*Uncinula necator*
Banana	*Mycosphaerella fijiensis*
Apple	*Venturia inaequalis*
Cucumber	*Sphaerotheca fuliginea*

Recent surveys with morpholine fungicides and *E. graminis* have detected some shifts in sensitivity, but the products remain fully effective in the field. There is no cross resistance between morpholine fungicides and DMIs.

Resistance prevention and management strategies depend upon limiting the frequency of use and the use of mixtures, particularly DMI plus morpholine mixtures for cereals, alternating with non-DMI or morpholine products.

Anilinopyrimidines

The anilinopyrimidines, introduced to the market in 1994, are a new group of systemic, preventative and eradicant fungicides active against a wide range of Ascomycete and Deuteromycete fungi, but especially *Botrytis* spp. and *Venturia* spp. No resistance has been found in commercial practice. Resistance prevention recommendations are based on restricting use, alternations and mixtures.

Strobilurins

The strobilurins are a new group of broad spectrum, systemic, preventative and eradicant fungicides, active against Phycomycetes, Ascomycetes, Deuteromycetes and Basidiomycetes, introduced to the market in 1996. No resistance has been found in commercial practice. Resistance prevention recommendations are based on restricting use, alternations and mixtures.

General Statement

Fungicide resistance is a phenomenon that can seriously affect the success of disease control programmes. It can, however, be managed and none of the chemical groups considered by FRAC have gone out of use because of resistance. The success of resistance management depends upon the development and introduction of effective management strategies, followed by their use by all concerned with disease control. Prior to the commercial introduction of new fungicides, FRAC member companies ensure that:

- Relevant resistance risk assessment studies are carried out.

- Monitoring methodologies are developed.

- Field baseline studies on major target crop/pathogen combinations are carried out.

At and post introduction, they ensure that :

- Resistance management guidance is incorporated into use recommendations.

HERBICIDE RESISTANCE ACTION COMMITTEE (HRAC)

Organisation

HRAC is an industry-based group supported by GCPF. Its members are AgrEvo, American Cyanamid, BASF, Bayer, Novartis, DowElanco, DuPont, Monsanto, Rhône-Poulenc, Tomen, and Zeneca. In order to ensure effective cooperation and communication, its working groups are divided on a regional basis (Europe, North America, Australia, and Rest of World). These regional groups interact with an informal network of country committees which are often indirectly linked to HRAC and often led by government researchers. There is no set structure but our cooperation is fueled by our common aim -- to manage resistance.

Mission

The mission of HRAC is to facilitate the effective management of herbicide resistance by fostering understanding, cooperation, and communication between industry, government and farmers.

Aims

The aims of the HRAC to accomplish its mission are:

• To foster a responsible attitude to herbicide use.

• To support and participate in research, conferences and seminars which serve to increase our understanding of herbicide resistance.

• To promote a better understanding of the causes and results of herbicide resistance.

• To communicate herbicide resistance management strategies and support their implementation through practical guidelines.

- To seek active collaboration with public and private researchers, especially in the areas of problem identification and devising and implementing management strategies.

- To facilitate communication between industry representatives.

HRAC Publications:

HRAC has a webpage on the internet with current information on publications and other areas of interest to resistant weed management. The web page is **http://ipmww.ncsu.edu/orgs/hrac/hrac.html.**

HRAC has also supported the establishment of an international survey of herbicide resistant weeds. Access to this survey can be made through the internet at: **http://www.pioneer.net/~heapian/index.html.**

The following HRAC publications are available:

- *HRAC Overview - Partnership in the Management of Resistance*
- *HRAC Monograph 1 - A Review of Graminicide Resistance*
- *HRAC Monograph 2 - Herbicide Cross Resistance and Multiple Resistance in Plants*
- *HRAC Guidelines - How to Minimize Resistance and How to Respond to Cases of Suspected and Confirmed Resistance*
- *List of herbicide resistance testing facilities*
- HRAC Article - *Agrow* November, 1995
- *The Role of Industry in Herbicide Resistance Management* - HRAC Paper from the Brighton Crop Protection Conference, November, 1995
- *Classification of Herbicides According to Mode of Action.*

For further Information on these publications, please contact the address below:

HRAC Publicity Office
c/o David Nevill & Derek Cornes
Novartis Crop Protection AG
4002 Basel, Switzerland
HRAC Tel: 41-61-697-9146; Fax: 41-61-697-6855;
Email: david.nevill@cp.norvartis.com or
Email: derek.cornes@cp.novartis.com

Current Status of Herbicide Resistant Weeds

There are 185 documented cases of unique herbicide-resistant weed biotypes that have been reported worldwide in the 1996 International Survey of Herbicide-Resistant Weeds. Although sixty weed species have evolved resistance to triazine herbicides, triazine resistance now only accounts for one third of all documented cases of herbicide resistance, in contrast to a similar survey in 1990 where triazine resistant weeds accounted for over half of the documented cases of herbicide resistance. Many of the cases of herbicide resistance that were selected in the last decade have been to ALS inhibitor and ACCase inhibitor herbicides. Thirty-five weed species have evolved resistance to ALS inhibitor herbicides in eleven countries, and 13 species have evolved resistance to ACCase inhibitors also in eleven countries. See Figure 1.

Figure 1. The number of monocot and dicot weeds that have evolved herbicide-resistance to various classes of herbicides (I. M. Heap, The Occurrence of Herbicide-Resistant Weeds Worldwide, *Pestic. Sci.* (1997, in press)).

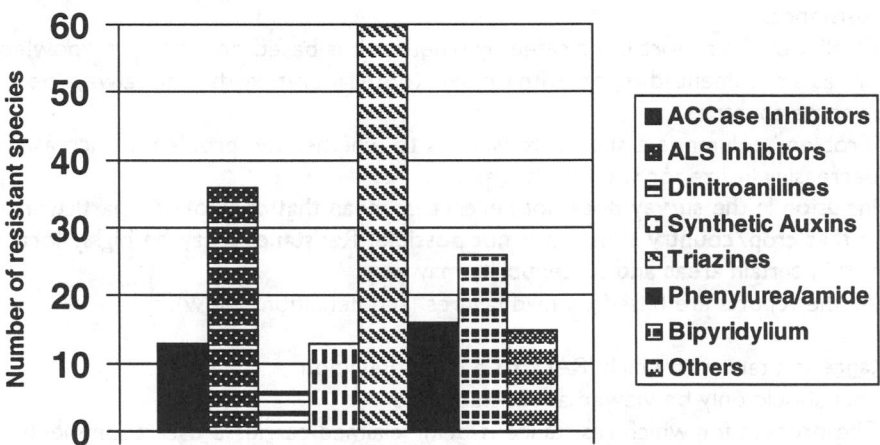

INSECTICIDE RESISTANCE ACTION COMMITTEE (IRAC)

The aim of IRAC is to define the size of the insecticide and acaricide resistance problems by surveys, funding research and advancing agreed techniques.

With the objective of advising or directing industry and non industry bodies on technically sound arthropod resistance management strategies.

We achieve this by having working groups which advise and direct work and strategy on a particular crop or group of crops. In addition, IRAC encourages local country organisations which work on resistance strategies and management in particular countries.

1996 Resistance Survey

The data for the present survey was collected during 1996. Obviously erroneous data were removed, as was any data which seemed doubtful. Data were then merged to give a single crop/country/pest entry. This survey should thus not be taken as inclusive, but represents IRAC's best efforts at this time.

- Columns refer to chemical class, even if only one member of the class shows resistance.
- Quality of the report is indicated by whether it is based on 'common knowledge' or has a documented report with lab data (note laboratory data not always possible in some countries).
- 'Problem' column is a subjective view as to whether the problem is increasing or decreasing in size, or whether it is stable.
- **Inclusion in the survey does not necessarily mean that control of a particular pest in that crop/country situation is not possible. Resistance may be highly localised within certain areas and susceptibility may vary.**
- All the reports are based on field failures (see definition below).

<u>Resistance</u> is a term for which IRAC has a clear definition.
An insect should only be viewed as resistant when:
- The product for which resistance is being claimed carries a use recommendation against the particular pest mentioned, and has a history of successful performance.
- Product failure is not a consequence of incorrect storage, dilution or application, and is not due to unusual climatic or environmental conditions.
- The recommended dosages fail to suppress the pest population below the level of economic threshold.
- Failure to control is due to a heritable change in susceptibility of the pest population to the product.

Note: Sensitivity to an insect or mite control agent must decrease significantly before field failure is experienced. The term 'resistance' should only be used once field failure has occurred and been confirmed.

Resistance Management

In order to combat the threat and spread of resistance, the working groups within IRAC have devised principles for specific pest/crop situations. The following are some generalised procedures for use management of insecticides and acaricides

1. Always include any efficient cultural/biological control practices in your pest control programme.
2. Time the application of compound against the most susceptible life stages based on local pest thresholds.
3. Do not rely on a single insecticide or acaricide class.
4. Use the insecticide/acaricide at recommended spray rates and intervals.
5. Where there is more than one generation of insects or mites, use different classes of insecticide/acaricide in alternation.
6. In the event of a control failure due to resistance, do not respray with an insecticide/acaricide of the same class.

IRAC Resistance Surveys - Tables:

Cotton
Field Crops and Vegetables
Ornamentals etc
Rice
Stored products
Top fruit

Cotton

Crop	Pest	Country	Exhibiting resistance to					Based on		Lab data		Problem		
			OP	Carb	Pyr	Others	Whole class	Rumour	Official publication	Yes	No	>	<	Stable
Cotton	*Alabama argillacea*	Argentina	X	X	X			X			X	X		
Cotton	*Alabama argillacea*	Brazil	X	X	X			X			X			
Cotton	*Amrasca bigutella*	India	X				X					X		
Cotton	*Anthonomus grandis*	Mexico	X					X		X		X		
Cotton	*Anthonomus grandis*	Guatemala	X		X		X			X		X		
Cotton	*Anthonomus grandis*	USA				CH	X		X	X			X	
Cotton	*Aphis gossypii*	Ethiopia	X					X		X			X	
Cotton	*Aphis gossypii*	China	X		X				X					
Cotton	*Aphis gossypii*	Zimbabwe	X						X					X
Cotton	*Aphis gossypii*	Greece	X		X		X	X		X				X
Cotton	*Aphis gossypii*	Brazil	X	X			X		X		X			X
Cotton	*Aphis gossypii*	India	X									X		
Cotton	*Aphis gossypii*	Peru	X	X					X	X				X
Cotton	*Aphis gossypii*	Egypt	X	X					X	X		X		
Cotton	*Aphis gossypii*	Turkey	X				X		X	X				X
Cotton	*Aphis gossypii*	RSA	X	X					X	X				X
Cotton	*Aphis gossypii*	Spain	X					X						X
Cotton	*Aphis gossypii*	USA	X	X	X	CH	X		X	X				X
Cotton	*Bemisia tabaci*	Israel	X	X			X		X	X				X
Cotton	*Bemisia tabaci*	USA	X						X	X				
Cotton	*Bemisia tabaci*	India	X											X
Cotton	*Bemisia tabaci*	Pakistan	X						X	X				
Cotton	*Bemisia tabaci*	Egypt	X	X	X				X	X				
Cotton	*Bemisia tabaci*	Turkey	X		X		X		X	X				X
Cotton	*Bucculatrix thurberiella*	Peru	X				X	X						X
Cotton	*Frankliniella occidentalis*	USA	X	X	X	CH	X		X		X			X
Cotton	*Helicoverpa armigera*	India	X	X	X	CH	X		X	X			X	
Cotton	*Helicoverpa armigera*	Turkey	X		X				X	X			X	
Cotton	*Helicoverpa armigera*	Australia	X	X	X	CH	X		X	X			X	
Cotton	*Helicoverpa armigera*	RSA			X				X	X			X	
Cotton	*Helicoverpa armigera*	China	X	X	X		X		X	X				X
Cotton	*Helicoverpa armigera*	Thailand	X	X			X		X	X			X	
Cotton	*Helicoverpa armigera*	Israel			X				X					
Cotton	*Helicoverpa armigera*	Pakistan	X		X	CH			X	X			X	
Cotton	*Helicoverpa armigera*	S.E. Asia			X			X	X	X			X	
Cotton	*Heliothis virescens*	Colombia	X		X		X		X	X				X
Cotton	*Heliothis virescens*	USA	X	X	X	CH	X		X	X			X	
Cotton	*Heliothis virescens*	Mexico	X	X			X	X		X			X	
Cotton	*Helicoverpa zea*	USA	X	X	X		X		X	X			X	
Cotton	*Lygus sp.*	Turkey	X					X				X	X	
Cotton	*Lygus sp.*	USA	X	X	X		X	X	X					
Cotton	*Myzus persicae*	Australia	X	X					X	X				
Cotton	*Myzus persicae*	India	X						X				X	
Cotton	*Pseudoplusia includens*	USA	X	X	X		X		X	X				X
Cotton	*Spodoptera exigua*	USA			X		X		X	X				X
Cotton	*Spodoptera exigua*	Mexico	X						X	X			X	
Cotton	*Spodoptera frugiperda*	USA	X	X	X	CH	X		X	X				X
Cotton	*Spodoptera littoralis*	Egypt	X						X	X				
Cotton	*Spodoptera littoralis*	Turkey	X		X		X		X	X				X
Cotton	*Spodoptera litura*	India	X	X					X	X				X
Cotton	*Tetranychus cinabarinnus*	Turkey	X		X	CH	X		X	X				X
Cotton	*Tetranychus sp.*	Greece	X				X				X			X
Cotton	*Tetranychus telarius*	Egypt	X						X					
Cotton	*Tetranychus urticae*	Brazil	X					X						X
Cotton	*Tetranychus urticae*	China	X		X	CH			X	X				X

CH = chlorinated hydrocarbons

Field Crops and Vegetables

Crop	Pest	Country	OP	Carb	Pyr	Others	Rumour	Official publication	Yes	No	>	<	Stable
Tea	Adoxophyes sp.	Japan		X			X		X				X
Pistachio	Agonoscena targionii	Turkey			X					X			
Potato	Agriotes sp.	Greece		X						X	X		
Brinjal	Amrasca biguttula	India	X					X	X		X		
Coffee	Antestia sp.	Zimbabwe	X										
Chillies	Anthonomus eugenii	Mexico	X	X			X					X	
Wheat	Aphis sp.	France			X		X			X		X	
Cucurbits	Aphis cucumeris	Sweden	X				X			X		X	
Chillies	Aphis gossypii	India	X									X	
Vegetables	Aphis gossypii	Greece	X				X		X				X
Vegetables	Aphis gossypii	Japan	X	X	X			X	X				X
Vegetables	Aphis gossypii	Korea	X	X	X			X	X				X
Vegetables	Aphis gossypii	Morocco	X	X	X			X	X				X
Watermelon	Aphis gossypii	Turkey	X						X		X		
Vegetables	Aphis gossypii	Turkey	X				X			X	X		
Cucurbits	Aphis gossypii	Bulgaria	X		X			X					X
Cucurbits	Aphis gossypii	Greece	X		X				X				X
Cucurbits	Aphis gossypii	Japan	X	X	X				X			X	
Cucurbits	Aphis gossypii	Netherlands		X									
Potato	Aphis nasturtii	Netherlands	X					X					X
Vegetables	Aphis sp.	Turkey	X	X			X			X			X
Beans	Bemisia tabaci	Brazil	X				X				X		
Brinjal	Bemisia tabaci	India	X										X
Okra	Bemisia tabaci	India	X										X
Tomato	Bemisia tabaci	Israel	X	X	X			X			X		
Vegetables	Bemisia tabaci	Japan	X	X				X		X	X		
Vegetables	Bemisia tabaci	Spain	X	X	X		X		X		X		
Vegetables	Bemisia tabaci	Netherlands				IGR		X	X				X
Rutabaga	Delia radicum	Canada		X			X			X		X	
Vegetables	Frankliniella intonsa	Japan	X		X		X			X			X
Vegetables	Frankliniella occidentalis	Spain	X	X			X			X	X		
Tomato	Frankliniella occidentalis	RSA	X	X	X		X			X	X		
Tomato	Frankliniella occidentalis	Spain	X	X			X			X			
Chickpeas	Helicoverpa armigera	Australia	X	X	X	CH				X			
Chickpeas	Helicoverpa armigera	India	X	X	X					X			
Chickpeas	Helicoverpa armigera	Spain	X	X	X					X			
Chillies	Helicoverpa armigera	India	X	X	X					X	X		
Groundnut	Helicoverpa armigera	India	X	X						X			
Sunflower	Helicoverpa armigera	India	X		X	CH							
Tomato	Helicoverpa armigera	India	X					X	X		X		
Tea	Homona magnanima	Japan				IGR		X					
Tomato	Keiferia lycopersicella	Mexico	X	X	X		X			X			
Potato	Leptinotarsa decemlineata	Canada	X	X	X	CH		X	X	X			
Potato	Leptinotarsa decemlineata	Eastern Europe	X		X	CH	X						X
Potato	Leptinotarsa decemlineata	Portugal			X								X
Potato	Leptinotarsa decemlineata	Spain	X										X
Potato	Leptinotarsa decemlineata	Turkey			X					X	X		
Potato	Leptinotarsa decemlineata	USA	X	X	X			X	X		X		
Eggplant	Leucinodes orbonalis	Bangladesh	X	X	X		X			X	X		
Eggplant	Leucinodes orbonalis	India	X	X	X		X			X	X		
Coffee	Leucoptera sp.	Tanzania	X					X	X				X
Mustard	Lipaphis erysimi	India	X	X	X	CH		X		X	X		
Musk melon	Liriomyza bryoniae	Taiwan		X			X			X	X		
Vegetables	Liriomyza cativae	Brazil	X	X	X		X				X		
Onion	Liriomyza chinensis	Japan	X				X			X			X
Potato	Liriomyza huidobrensis	Brazil	X	X	X		X			X	X		
Vegetables	Liriomyza huidobrensis	Brazil	X	X	X		X			X	X		
Tomato	Liriomyza strigata	Spain			X		X			X	X		
Eggplant	Liriomyza trifolii	Japan	X		X			X	X		X		
Vegetables	Liriomyza trifolii	Japan	X	X	X			X	X				X
Vegetables	Liriomyza trifolii	Kenya			X		X			X	X		
Vegetables	Liriomyza trifolii	Malaysia			X		X			X	X		
Vegetables	Liriomyza trifolii	USA	X				X			X	X		
Cabbage	Mamestra brassicae	Bulgaria	X					X					X
Pepper	Myzus persicae	Netherlands	X	X								X	
Tobacco	Myzus nicotianae	India								X	X		
Brinjal	Myzus persicae	India	X					X			X		
Cabbage	Myzus persicae	India	X					X			X		
Chillies	Myzus persicae	India	X	X			X	X		X			X
Eggplant	Myzus persicae	Japan	X	X	X			X	X		X		

Field Crops and Vegetables—continued

Crop	Pest	Country	OP	Carb	Pyr	Others	Rumour	Official publication	Yes	No	>	<	Stable
Lupin	*Myzus persicae*	Australia	X					X					
Pepper	*Myzus persicae*	Japan	X	X	X			X	X		X		
Sugarbeet	*Myzus persicae*	Germany	X					X				X	
Sugarbeet	*Myzus persicae*	Sweden	X					X	X				X
Tobacco	*Myzus persicae*	Greece	X	X	X		X	X	X				X
Tobacco	*Myzus persicae*	India	X	X			X			X			X
Vegetables	*Myzus persicae*	Greece	X		X			X	X				X
Vegetables	*Myzus persicae*	Japan	X	X	X			X	X				X
Vegetables	*Myzus persicae*	Japan	X	X	X			X	X				X
Vegetables	*Myzus persicae*	Japan	X	X	X			X	X				X
Vegetables	*Myzus persicae*	New Zealand	X					X	X		X		
Vegetables	*Myzus persicae*	UK	X	X	X			X	X		X		
Lettuce	*Nasonovia ribisnigri*	Spain	X					X			X		
Hop	*Phorodon humuli*	UK	X	X	X				X				X
Hop	*Phorodon humuli*	Germany	X	X	X			X					
Potato	*Phthorimaea operculella*	RSA	X	X				X			X	X	
Crucifer	*Phyllotreta sinuala*	Thailand	X	X	X			X				X	
Cabbage	*Pieris rapae*	China			X			X	X				X
Brassica	*Plutella xylostella*	Australia			X	CH		X	X				X
Brassica	*Plutella xylostella*	Malaysia		X		Bt, IGR		X	X				X
Brassica	*Plutella xylostella*	Indonesia						X	X				
Brassica	*Plutella xylostella*	Japan	X	X	X			X	X		X		
Brassica	*Plutella xylostella*	Kenya	X		X		X			X	X		
Brassica	*Plutella xylostella*	USA				Bt		X	X		X		
Brassica	*Plutella xylostella*	Korea	X					X	X		X		
Brassica	*Plutella xylostella*	Zambia	X		X					X	X		
Cabbage	*Plutella xylostella*	India	X	X	X			X	X		X		
Cabbage	*Plutella xylostella*	Indonesia			X			X	X		X		
Cabbage	*Plutella xylostella*	Japan	X	X	X	Bt, cartap		X	X		X		
Cabbage	*Plutella xylostella*	Philippines	X			CH, IGR, Bt		X					X
Cabbage	*Plutella xylostella*	Thailand	X	X	X			X	X				X
Cabbage	*Plutella xylostella*	Taiwan		X	X	IGR		X			X		
Cauliflower	*Plutella xylostella*	India	X		X	CH		X			X		X
Chinese kale	*Plutella xylostella*	Thailand	X	X	X		X			X	X		
Cole crops	*Plutella xylostella*	Canada	X	X	X			X	X		X		
Crucifer	*Plutella xylostella*	Thailand	X	X	X	IGR		X	X		X		
Wheat	*Schizaphis graminum*	USA	X	X				X	X		X		
Tea	*Scirtothrips dorsalis*	Japan	X	X	X			X	X		X		
Tomato	*Scrobipalpula absoluta*	Argentina	X		X		X						X
Tomato	*Scrobipalpula absoluta*	Brazil	X	X	X			X					X
Tomato	*Scrobipalpula absoluta*	Colombia	X		X		X						X
Chillies	*Spodoptera exigua*	Indonesia	X	X		IGR, CH				X			X
Crucifer	*Spodoptera exigua*	Thailand	X	X	X	IGR		X	X		X		
Scallion	*Spodoptera exigua*	Taiwan			X		X			X	X		
Shallot	*Spodoptera exigua*	Indonesia			X			X	X		X		
Vegetables	*Spodoptera exigua*	Japan	X	X	X			X	X				X
Vegetables	*Spodoptera exigua*	Spain	X	X	X		X			X			X
Welsh onion	*Spodoptera exigua*	Japan	X	X	X			X	X		X		
Maize	*Spodoptera frugiperda*	Brazil	X				X			X	X		
Maize	*Spodoptera frugiperda*	Mexico		X			X		X		X		
Sorghum	*Spodoptera frugiperda*	Colombia	X										X
Cabbage	*Spodoptera litura*	Japan	X	X	X			X	X				X
Cauliflower	*Spodoptera litura*	India				CH		X			X		
Groundnut	*Spodoptera litura*	India	X	X	X	CH		X	X		X		
Soy bean	*Spodoptera litura*	Indonesia	X					X	X		X		
Soy bean	*Spodoptera litura*	Japan	X	X	X			X	X		X		
Strawberry	*Spodoptera litura*	Japan	X	X	X			X	X				X
Sweet potato	*Spodoptera litura*	Japan	X	X	X			X	X		X		
Sweet potato	*Spodoptera litura*	Japan	X	X	X			X	X				X
Tobacco	*Spodoptera litura*	India	X	X	X	CH		X	X				
Chillies	*Spodoptera litura*	Indonesia	X	X		IGR, CH		X		X			X
Tea	*Tetranychus kanzawai*	Japan	X	X	X			X	X			X	
Tea	*Tetranychus kanzawai*	Taiwan				sulfur	X					X	
Vegetables	*Tetranychus kanzawai*	Japan	X					X		X		X	
Tomato	*Tetranychus urticae*	Australia				IGR			X			X	
Tomato	*Tetranychus urticae*	Bulgaria	X			formamidines			X	X		X	
Blackcurrant	*Tetranychus urticae*	New Zealand				propargite			X				X
Tomato	*Tetranychus urticae*	UK				Sn			X			X	
Tomato	*Tetranychus urticae*	Italy				IGR			X			X	
Vegetables	*Tetranychus urticae*	Japan	X	X	X			X	X			X	

Field Crops and Vegetables—continued

Crop	Pest	Country	Exhibiting resistance to				Based on		Lab data		Problem		
			OP	Carb	Pyr	Others	Rumour	Official publication	Yes	No	>	<	Stable
Lucerne	*Therioaphis trifolii*	Australia	X	X	X			X	X				X
Chillies	*Thrips sp.*	Indonesia	X	X		IGR, CH	X			X			X
Vegetables	*Thrips palmi*	Japan	X	X	X			X	X				X
Leek	*Thrips tabaci*	Netherlands	X		X								
Onion	*Thrips tabaci*	RSA	X	X	X		X			X	X		
Tobacco	*Thrips tabaci*	Greece	X	X			X			X	X		
Tomato	*Thrips tabaci*	Bulgaria	X					X			X		
Tea	*Tiomona magnaxima*	Japan	X	X	X	IGR		X	X		X		
Tomato	*Trialeurodes vaporariorum*	Belgium			X							X	
Tomato	*Trialeurodes vaporariorum*	Bulgaria		X	X			X				X	
Tomato	*Trialeurodes vaporariorum*	Sweden			X		X			X			X
Vegetables	*Trialeurodes vaporariorum*	Italy					X						X
Vegetables	*Trialeurodes vaporariorum*	Netherlands				IGR		X	X				X
Tomato	*Trialeurodes vaporariorum*	Belgium	X			IGR		X		X			
Vegetables	*Trialeurodes vaporariorum*	France				IGR	X			X	X		
Vegetables	*Trialeurodes vaporariorum*	Greece	X							X	X		
Vegetables	*Trialeurodes vaporariorum*	New Zealand	X		X			X	X		X		
Cucurbits	*Trialeurodes vaporariorum*	Spain	X		X	IGR	X				X		
Tomato	*Trialeurodes vaporariorum*	Japan						X			X		

Bt = *Bacillus thuringiensis*
IGR for whitefly is buprofezin
CH = chlorinated hydrocarbon
IGR for Lepidoptera is acylurea
Sn = organotin acaricides
IGR for mites is clofentezine/hexythiazox acaricides

Ornamentals etc.

Crop	Pest	Country	Exhibiting resistance to					Based on		Lab data		Problem		
			OP	Carb	Pyr	Others	Whole class	Rumour	Official publication	Yes	No	>	<	Stable
Ornamentals	*Aleurodes sp.*	Germany			X		X		X					X
Ornamentals	*Aphis cucumeris*	Sweden	X	X				X		X				
Ornamentals	*Aphis gossypii*	UK	X	X	X		X			X				
Ornamentals	*Aphis gossypii*	Japan	X	X	X				X	X		X		
Ornamentals	*Bemisia tabaci*	Israel				IGR		X		X				
Ornamentals	*Bemisia tabaci*	Sweden			X		X	X		X	X			
Ornamentals	*Bemisia tabaci*	USA	X	X	X				X	X	X			X
Ornamentals	*Bemisia tabaci*	UK	X	X	X	IGR	X		X					X
Turf	*Blissus insularis*	USA	X						X	X				
Ornamentals	*Frankliniella occidentalis*	New Zealand	X							X				
Ornamentals	*Frankliniella occidentalis*	Sweden	X		X		X	X		X				
Ornamentals	*Frankliniella occidentalis*	USA	X	X	X					X				
Ornamentals	*Frankliniella occidentalis*	Switzerland	X	X					X	X		X		
Ornamentals	*Frankliniella occidentalis*	Australia	X	X	X	CH				X		X		
Ornamentals	*Myzus persicae*	UK	X	X	X		X			X				
Ornamentals	*Myzus persicae*	Australia	X	X					X			X		
Lupins	*Myzus persicae*	Australia	X						X					X
Ornamentals	*Myzus persicae*	USA	X	X	X			X		X		X		X
Roses	*Phytoseiulus persimilis*	Australia	X						X		X	X		
Turf	*Schizaphis graminum*	USA	X				X	X				X		
Ornamentals	*Tetranychus cinnabarinus*	Bulgaria	X						X	X				
Rose	*Tetranychus cinnabarinus*	Taiwan				CH						X		X
Ornamentals	*Tetranychus urticae*	Australia			X	Sn			X		X			X
Roses	*Tetranychus urticae*	Australia	X	X					X					X
Roses	*Tetranychus urticae*	Brazil				MCL		X			X	X	X	X
Ornamentals	*Tetranychus urticae*	Sweden	X				X	X				X		X
Ornamentals	*Tetranychus urticae*	USA				MCL		X						
Ornamentals	*Trialeurodes vaporarium*	Sweden	X		X		X	X						

Bemisia tabaci includes *Bemisia argentafoli*

CH = chlorinated hydrocarbons
IGR for Lepidoptera is acylureas (benzoylureas)
IGR for mites is clofentezine/hexythiazox-type acaricides
IGR for whitefly is buprofezin
MCL = macrocyclic lactones
Sn = organotin acaricides

Rice

Crop	Pest	Country	OP	Carb	Pyr	Others	Whole class	Rumour	Official publication	Yes	No	>	<	Stable
Rice	*Chilo suppressalis*	Japan	X						X	X				X
Rice	*Chilo suppressalis*	Korea	X						X	X		X		
Rice	*Chilo suppressalis*	China	X						X	X				
Rice	*Cnaphalocrocis medinalis*	India	X		X			X				X	X		
Rice	*Dicladispa armigera*	Bangladesh			X	X		X							X
Rice	*Hydrella philippina*	Philippines	X	X						X					X
Rice	*Laodelphax stratellus*	Japan	X	X						X	X				X
Rice	*Laodelphax stratellus*	Korea	X	X						X	X		X		
Rice	*Marasmia patnalis*	India	X		X			X				X	X		
Rice	*Nephotettix cincticeps*	Japan	X	X						X	X				X
Rice	*Nephotettix cincticeps*	Korea	X	X						X	X		X		
Rice	*Nephotettix cincticeps*	Japan	X	X						X					X
Rice	*Nilaparvata lugens*	China		X		IGR		X							X
Rice	*Nilaparvata lugens*	Japan	X	X						X	X				X
Rice	*Nilaparvata lugens*	Korea	X	X						X	X		X		
Rice	*Oulema oryzae*	Japan	X	X		X				X	X		X		
Rice	*Oulema oryzae*	Japan	X	X						X	X		X		
Rice	*Schirpophaga innotata*	Indonesia		X						X	X				X
Rice	*Scotinophara coarctata*	Philippines	X	X						X		X			X
Rice	*Sogatella furcifera*	Japan	X	X						X	X				X
Rice	*Sogatella furcifera*	Korea	X						X	X		X		

IGR = buprofezin

Stored Products

Crop	Pest	Country	OP	Carb	Pyr	Others	Whole class	Official publication	Yes	No	>	<	Stable
Stored grains	*Cryptolestes sp.*	Australia	X		X	Pyrethroid synergist		X	X				
Stored grains	*Dermestes maculatus*	Australia	X					X	X				
Stored grains	*Oryzephilus surinamensis*	Australia	X		X	Pyrethroids		X	X				
Stored grains	*Oryzephilus surinamensis*	Australia	X					X	X				
Stored grains	*Rhyzopertha dominica*	Australia	X	X	X	Pyrethroid synergist		X	X				
Stored grains	*Sitophilus oryzae*	Australia	X		X	Pyrethroid synergist		X	X				
Stored grains	*Tribolium castaneum*	Australia	X		X	Pyrethroid synergist	X	X	X				
Stored grains	*Tribolium castaneum*	India	X			CH		X	X		X		
Stored grains	*Trogoderma granarium*	India				CH		X	X				

CH = chlorinated hydrocarbons

Top Fruit

Crop	Pest	Country	OP	Carb	Pyr	Others	Whole class	Rumour	Official publication	Yes	No	>	<	Stable
Citrus	Aculops pelekassi	Japan		X					X					
Pome	Adoxophyes orana	UK	X	X			X			X				X
Pome	Adoxophyes orana	RSA	X						X	X		X		
Pome	Adoxophyes orana	Japan	X						X	X		X		
Citrus	Aleurothrixus floccosus	Spain		X					X	X				X
Citrus	Aonidiella aurantii	RSA	X						X	X		X		
Citrus	Aonidiella aurantii	Australia	X					X	X					
Citrus	Aonidiella aurantii	USA	X	X					X	X		X		
Citrus	Aonidiella citrina	USA	X	X					X	X		X		
Fruit Trees	Aphis gossypii	Japan	X	X	X				X	X		X		
Citrus	Aphis gossypii	Greece	X	X			X			X				X
Citrus	Aphis gossypii	Spain	X		X					X				X
Pome	Aphis spiraecola	Japan	X	X	X					X		X		
Pome	Aphis spiraecola	Korea	X	X	X		X		X	X		X		
Citrus	Aspidiotus haederae	Spain	X											X
Citrus	Brevipalpus phoenicis	Brazil			X	CH, IGR			X	X				X
Banana	Cosmopolites sordidus	Australia	X	X					X	X				

Crop	Pest	Country	OP	Carb	Pyr	Others	Whole class	Rumour	Official publication	Yes	No	>	<	Stable
Pome	*Cydia pomonella*	Australia	X						X	X		X		
Pome	*Cydia pomonella*	RSA	X		X		X		X	X		X		
Pome	*Cydia pomonella*	France	X		X	IGR			X	X		X		
Pome	*Cydia pomonella*	Italy	X		X	IGR			X	X		X		
Pome	*Cydia pomonella*	Germany	X		X	IGR			X	X		X		
Pome	*Cydia pomonella*	Netherlands				IGR		X			X	X		
Pome	*Cydia pomonella*	Canada	X				X				X			X
Pome	*Dysaphis plantaginea*	Switzerland		X					X	X		X		
Citrus	*Frankliniella occidentalis*	Japan	X	X	X				X	X		X		
Peach	*Myzus persicae*	Argentina		X						X				X
Peach	*Myzus persicae*	Turkey	X	X			X	X		X		X		X
Peach	*Myzus persicae*	Greece	X		X		X		X	X				X
Peach	*Myzus persicae*	France	X	X	X		X		X	X		X		
Peach	*Myzus persicae*	Italy	X	X				X	X	X		X		
Peach	*Myzus persicae*	New Zealand	X				X		X	X		X		
Citrus	*Panonychus citri*	Spain				Sn				X				X
Citrus	*Panonychus citri*	Taiwan				IGR		X		X		X		
Citrus	*Panonychus citri*	Japan	X	X	X	METI, IGR	X		X	X		X		
Citrus	*Panonychus citri*	China	X			METI			X	X		X		
Citrus	*Panonychus citri*	Turkey				CH		X			X			X
Pome	*Panonychus ulmi*	Belgium				IGR	X			X		X		
Pome	*Panonychus ulmi*	Germany				IGR			X			X		
Grape	*Panonychus ulmi*	Germany				IGR		X	X			X		
Peach	*Panonychus ulmi*	Italy				IGR		X				X		
Pome	*Panonychus ulmi*	Italy			X	Sn, IGR		X				X		
Pome	*Panonychus ulmi*	Brazil				IGR			X	X				X
Pome/Stone	*Panonychus ulmi*	Japan	X	X	X	IGR	X		X	X				X
Apple	*Panonychus ulmi*	France				Sn								X
Citrus	*Panonychus ulmi*	Japan				METI			X		X			
Citrus	*Panonychus ulmi*	Korea				METI			X		X			
Citrus	*Panonychus ulmi*	China				METI	X		X		X			
Pome	*Panonychus ulmi*	Greece	X		X		X	X			X	X		
Pome	*Panonychus ulmi*	New Zealand				Sn, IGR, propargite				X				X
Pome	*Panonychus ulmi*	Korea	X						X	X				X
Citrus	*Parlatoria pergandei*	Spain	X											X
Citrus	*Phyllocnistis citrella*	Japan	X	X	X		X		X				X	
Citrus	*Phyllocoptruta oleivora*	Mexico	X					X					X	
Watermelon	*Plusia gamma*	Spain	X	X	X		X						X	
Pear	*Psylla piri*	Belgium			X				X					X
Pear	*Psylla piri*	Switzerland				IGR			X		X	X		
Pear	*Psylla piri*	Italy	X				X		X	X				X
Pear	*Psylla piri*	Spain		X	X			X	X	X				X
Pear	*Psylla piri*	Canada	X	X	X		X		X	X				X
Pear	*Psylla piri*	Greece	X		X			X			X			X
Pear	*Psylla piri*	Turkey			X			X			X			X
Citrus	*Scirtothrips aurantii*	RSA	X						X	X				X
Citrus	*Scirtothrips citri*	USA			X								X	
Pome/Stone	*Tetranychus cinnabarinus*	RSA				Sn, IGR			X	X			X	
Pome/Stone	*Tetranychus urticae*	Japan	X	X		IGR, METI			X	X			X	
Pome	*Tetranychus urticae*	Korea	X	X	X	METI, Sn			X	X			X	

CH = chlorinated hydrocarbons
IGR = acylureas (benzoylureas) for Lepidoptera
IGR = clofentezine/hexythiazox type for mites
Sn = organotin acaricides
METI = Metabolic Electron Transport Inhibitor acaricides

INTRODUCTION

RODENTICIDE RESISTANCE ACTION COMMITTEE (RRAC)

Background

RRAC is a group formed by members of the international rodenticide industry, within the framework of the Global Crop Protection Federation (GCPF). Participating companies are: AgrEvo, Bayer, Cyanamid, Lipha, LiphaTech, Rentokil, Rhône-Poulenc, Sorex and Zeneca. Senior technical specialists, with specific expertise in rodenticides, represent their companies on this Committee.

Purpose

The purpose of RRAC is to advise international agencies, government authorities, regulatory bodies and rodenticide users on technical matters relating to rodenticide resistance. It carries out this objective by:

- producing guidance leaflets for rodenticide users on the safe and effective use of rodenticide products;
- arranging seminars and conferences where members of industry can meet and exchange ideas with experts from universities, governments and international organisations;
- participating in trade shows and other similar events;
- sponsoring research projects on rodenticide resistance;
- developing and advocating the use of effective resistance management strategies.

Rodenticides are extremely valuable products in the prevention of human and animal diseases and the protection of property, agriculture and the environment. New molecules are difficult and expensive to bring to the market. In its activities, the overriding intention of RRAC is to develop our understanding of resistance and resistance management in an effort to maintain the effectiveness of currently available rodenticides.

Definition of Rodenticide Resistance

The term *rodenticide resistance* has several interpretations. The one adopted by RRAC is that formulated by Greaves, 1994 (Resistance to Anticoagulant Rodenticides, in A. P. Buckle & R. H. Smith eds., *Rodent Pests and their Control*, CABI, Wallingford, Oxon, UK, pp. 197-217):

Anticoagulant resistance is a major loss of efficacy in practical conditions where the anticoagulant has been applied correctly, the loss of efficacy being due to the presence of a strain of rodent with a heritable and commensurately reduced sensitivity to the anticoagulant.

This definition has three main elements. Firstly, that the phenomenon described should involve a significant change in susceptibility that brings about a practical effect. Thus, where the term resistance is applied to a compound and a pest species, practitioners should anticipate a real loss of efficacy at the resistance focus. Secondly, that the compound should have been applied correctly and, incidentally, be normally effective for the species involved. Often, when an anticoagulant treatment fails, the cause is attributed to resistance when the real reason is faulty application. Thirdly, that the resistance should have a genetic basis that makes it transmissible between rodent generations.

The detection of changes in the susceptibility to anticoagulants of rodent infestations that fall short of the above definition is important, however, so that appropriate remedial action can be taken to prevent the development of full-blown resistance.

Resistance that satisfies the definition above has only been detected in the case of the anticoagulant class of rodenticides.

Proper Use of Rodenticides

Because the incorrect use of rodenticides is often the cause of spurious reports of resistance to anticoagulants, RRAC has produced a leaflet entitled *Checklist for Rodenticide Users Experiencing Difficulties*. This leaflet is available in English, French, German and Spanish from the companies listed above.

Towards Improved Resistance Management

RRAC has sponsored a research programme, conducted in Germany by the Biologische Bundesanstalt für Land- und Forstwirtschaft (BBA), to investigate different practices for the management of anticoagulant resistance. The 3-year study on farms in the Münsterland focus of anticoagulant resistance is in its preliminary stages. Results will be published as they become available.

The Future

In order to fulfil its purpose, it is the intention of RRAC to participate actively in the field of rodenticide resistance, in particular to support studies to extend our understanding of the importance and distribution of rodenticide resistance, resistance management, resistance detection methods, and the biochemistry of resistance mechanisms.

While resistance to some anticoagulant compounds exists in certain localities, particularly in the countries of North America and northern Europe and in the species *Rattus norvegicus*, *Mus musculus/domesticus* and *Rattus rattus* (see Greaves, 1994, *loc. cit.* for a review), nowhere are resistant rodents impossible to control satisfactorily with currently available rodenticides. It is the opinion of RRAC that this will remain the case for the foreseeable future.

MAIN ENTRIES

Materials in Use or Being Developed

These entries include chemicals and biological agents used as active ingredients of products for the control of crop pests and diseases, animal ectoparasites and pests in public health. It also contains plant growth regulators, pest repellents, synergists, herbicide safeners and some timber preservatives.

The regulations governing the conditions under which a given pesticide may be used differ markedly between countries and in some cases between states. Mention of a particular use in *The Pesticide Manual* does not constitute a recommendation. It is essential to follow relevant local laws.

To fully appreciate the contents of these entries, read the Guide to using the Main Entries on p. x

abamectin

avermectin

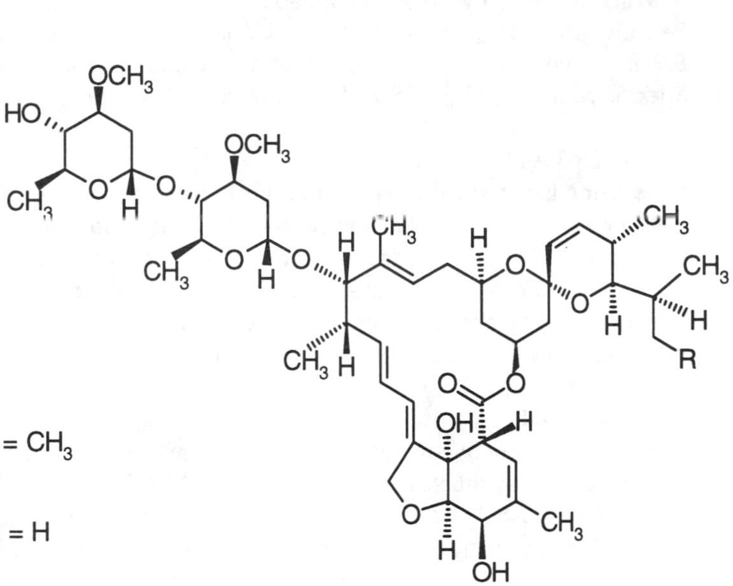

(i) R = CH$_3$

(ii) R = H

NOMENCLATURE

Common name abamectin (BSI, draft E-ISO, ANSI); abamectine ((*f*) draft F-ISO)
IUPAC name (10*E*,14*E*,16*E*,22*Z*)-(1*R*,4*S*,5'*S*,6*S*,6'*R*,8*R*,12*S*,13*S*,20*R*,21*R*,24*S*)-=
6'[(*S*)-*sec*-butyl]-21,24-dihydroxy-5',11,13,22-tetramethyl-2-oxo-3,7,19-=
trioxatetracyclo[15.6.1.1^{4,8}.0^{20,24}]pentacosa-10,14,16,22-tetraene-6-spiro-2'-=
(5',6'-dihydro-2'*H*-pyran)-12-yl 2,6-dideoxy-4-*O*-(2,6-dideoxy-3-*O*-methyl-α-L-=
arabino-hexopyranosyl)-3-*O*-methyl-α-L-*arabino*-hexopyranoside (i) mixture with
(10*E*,14*E*,16*E*,22*Z*)-(1*R*,4*S*,5'*S*,6*S*,6'*R*,8*R*,12*S*,13*S*,20*R*,21*R*,24*S*)-21,24-dihydroxy-=
6'-isopropyl-5',11,13,22-tetramethyl-2-oxo-3,7,19-trioxatetracyclo=
[15.6.1.1^{4,8}.0^{20,24}]pentacosa-10,14,16,22-tetraene-6-spiro-2'-(5',6'-dihydro-2'*H*-=
pyran)-12-yl 2,6-dideoxy-4-*O*-(2,6-dideoxy-3-*O*-methyl-α-L-*arabino*-=
hexopyranosyl)-3-*O*-methyl-α-L-*arabino*-hexopyranoside (ii) (4:1)
Chemical Abstracts name 5-*O*-demethylavermectin A$_{1a}$ (i) mixture with
5-*O*-demethyl-25-de(1-methylpropyl)-25-(1-methylethyl)avermectin A$_{1a}$ (ii)
Other names avermectin B1   **CAS RN** *[71751–41–2]* (abamectin);
[65195–55–3] (i); *[65195–56–4]* (ii)   **Development codes** MK-0936; C-076;
L-676,863

PHYSICAL CHEMISTRY

Composition A mixture containing ≥80% avermectin B$_{1a}$ (i) and ≤20% avermectin
B$_{1b}$ (ii).   **Mol. wt.** 873.1 (avermectin B$_{1a}$); 860.1 (avermectin B$_{1b}$)
M.f. C$_{48}$H$_{72}$O$_{14}$ (avermectin B$_{1a}$); C$_{47}$H$_{70}$O$_{14}$ (avermectin B$_{1b}$)   **Form** Colourless
to pale yellow crystals.   **M.p.** 150–155 °C   **V.p.** <2 × 10^{-4} mPa   **Solubility** In

water 7–10 µg/l (20 °C). In toluene 350, acetone 100, isopropanol 70, chloroform 25, ethanol 20, methanol 19.5, n-butanol 10, cyclohexane 6 (all in g/l, 21 °C).
Stability Stable to hydrolysis in aqueous solutions at pH 5, 7, and 9 (25 °C). Sensitive to stronger acid and base. U.V. irradiation causes conversion first to the 8,9-Z- isomer, then to unidentified decomposition products.
Specific rotation $[\alpha]_D^{22}$ +55.7° (c = 0.87 in $CHCl_3$)

COMMERCIALISATION
Production Isolated from fermentation of *Streptomyces avermitilis*.
History Anthelmintic and acaricidal activity of a group of chemically related compounds, the avermectins, reported by I. Putter et al. (*Experientia*, 1981, **37**, 963). A mixture of two of these, avermectin B_{1a} (i) and avermectin B_{1b} (ii) introduced as an acaricide and insecticide by Merck Sharp & Dohme Agvet.
Manufacturers Gilmore; Merck; Jingma; Sinon

APPLICATIONS
Biochemistry Acts by stimulating the release of γ-aminobutyric acid, an inhibitory neurotransmitter, thus causing paralysis. See M. J. Turner & J. M. Schaeffer in *Ivermectin and Abamectin*, W. C. Cambell ed., Springer-Verlag, New York (1989) p. 73.   **Mode of action** Insecticide and acaricide with contact and stomach action. Has limited plant systemic activity, but exhibits translaminar movement.
Uses Control of motile stages of mites, leaf miners, suckers, Colorado beetles, etc. on ornamentals, cotton, citrus fruit, pome fruit, nut crops, vegetables, potatoes, and other crops. Application rates are 5.6 to 28 g/ha for mite control, 11 to 22 g/ha for control of leaf miners. Also used for control of fire ants.
Formulation types EC; RB.   **Selected tradenames** 'Dynamec' (Merck); 'Abacide' (Mauget)

ANALYSIS
Product analysis by hplc with u.v. detection.   **Residues** by conversion to a fluorescent product followed by hplc with fluorescence detection. See T. Wehner et al. in *Comp. Anal. Profiles*, Chapt. 4.

MAMMALIAN TOXICOLOGY
Reviews *Pesticide residues in food – 1994*, FAO Plant Production and Protection Paper, 127, 1995. *Pesticide residues in food – 1994 evaluations. Part II – Toxicology.* World Health Organisation, WHO/PCS/95.2, 1995. *Pesticide residues in food – 1995*, FAO Plant Production and Protection Paper. G. Lankas & L. R. Gordon in *Toxicology in Ivermectin and Abamectin*, W. C. Campbell ed., Springer Verlag (1989) pp. 89–112.   **Oral** Acute oral LD_{50} for rats 10, mice 13.6 mg/kg (in sesame oil).
Skin and eye Acute percutaneous LD_{50} for rabbits >2000 mg/kg. Mild eye irritant; non-irritating to skin (rabbits).   **ADI** (JMPR) 0.0001 mg/kg b.w. [1992]; 0.0002 mg/kg b.w. [1994] (for mixture with Δ-8,9-isomer).   **Other** Non-mutagenic in the Ames test.   **Toxicity class** EPA (formulation) IV

ECOTOXICOLOGY
Birds Acute oral LD_{50} for mallard ducks 84.6, bobwhite quail >2000 mg/kg.
Fish LC_{50} (96 h) for rainbow trout 3.2, bluegill sunfish 9.6 µg/l. **Bees** Toxic to bees. **Daphnia** EC_{50} (48 h) 0.34 ppb. **Other aquatic spp.** LC_{50} (96 h) for pink shrimp (*Panaeus duorarum*) 1.6, mysid shrimp (*Mysidopsis bahia*) 0.022, blue crab (*Callinectes sapidus*) 153 ppb.

ENVIRONMENTAL FATE
Animals Metabolites include 3″-demethylavermectin B_1 and 24-hydroxymethyl= avermectin B_1. **Plants** 8,9-(Z)-avermectin B_1 has been identified. The polar degradates are the largest fraction; these are unidentified, but are non-toxic.
Soil/Environment Binds tightly to soil, with rapid degradation by soil micro-organisms. No bioaccumulation.

AC 263,222 *Herbicide*

imidazolinone

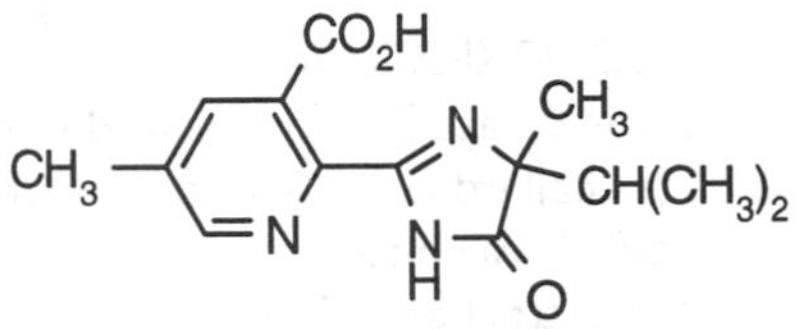

NOMENCLATURE
Common name imazapic (BSI proposed)
IUPAC name (RS)-2-(4,5-dihydro-4-isopropyl-4-methyl-5-oxoimidazol-2-yl)-5-= methylnicotinic acid
Chemical Abstracts name (±)-2-[4,5-dihydro-4-methyl-4-(1-methylethyl)-5-oxo-= 1H-imidazol-2-yl]-5-methyl-3-pyridinecarboxylic acid
Other names imazameth (rejected BSI name proposal) **CAS RN** *[104098–48–8]* acid, unstated stereochemistry (formerly *[81334–60–3]*); *[104098–49–9]* ammonium salt, unstated stereochemistry **Development codes** AC 263 222; CL 263 222

PHYSICAL CHEMISTRY
Mol. wt. 275.3 **M.f.** $C_{14}H_{17}N_3O_3$ **Form** Off-white to tan powder.
M.p. 204–206 °C **V.p.** $<1 \times 10^{-2}$ mPa (60 °C) **K_{ow}** logP = 0.393 (pH 4, 5, 6 buffer, 25 °C) **Solubility** In deionised water 2150 ppm (25 °C). In acetone 18.9 mg/ml (25 °C). **pKa** pKa_1 2.0, pKa_2 3.6, pKa_3 11.1

COMMERCIALISATION
History Reported by M. Wixson *et al.*, *Proc. South. Weed Sci. Soc.*, **45**, 341 (1992). Introduced by American Cyanamid Co. in 1996. **Manufacturers** Cyanamid

APPLICATIONS
Biochemistry Acetohydroxyacid synthase inhibitor. **Mode of action** Contact and residual herbicide. Weeds stop growing within 8 h of application and begin to turn yellow 1–3 d later. **Uses** For early post-emergence control of a wide range of weeds (including *Cassia obtusifolia*, *Desmodium tortuosum*, Cyperaceae, and *Panicum texanum*) in peanuts at 0.063 lb/a. **Formulation types** DG.
Selected tradenames 'Cadre' (Cyanamid)

MAMMALIAN TOXICOLOGY
Oral Acute oral LD_{50} for rats >5000 mg/kg. **Skin and eye** Acute percutaneous LD_{50} for rabbits >2000 mg/kg. Moderately irritating to eyes; slightly irritating to skin (rabbits). Not a skin sensitiser (guinea pigs). **Inhalation** LC_{50} for rats 4.83 mg/l air. **NOEL** (90 d) for rats 20 000 ppm (1625 mg/kg daily). Dermal NOEL (21 d) for rabbits 1000 mg/kg. Teratogenicity NOEL (maternal) for rats 1000, rabbits 350 mg/kg daily; (foetal) for rats 1000, rabbits 500 mg/kg daily.
ADI 0.50 mg/kg. **Other** Non-mutagenic, non-genotoxic, non-carcinogenic, non-teratogenic. **Toxicity class** EPA (formulation) III (70 DG)

ECOTOXICOLOGY
Birds Oral LD_{50} for mallard ducks and bobwhite quail >2150 mg/kg. LC_{50} (8 d) for mallard ducks and bobwhite quail >5000 ppm. **Fish** LC_{50} (96 h) for channel catfish, rainbow trout and bluegill sunfish >100 mg/l. **Bees** LD_{50} (contact) >100 µg/bee. **Daphnia** LC_{50} (48 h) >100 mg/l.

3   AC 94,377   *Plant growth regulator*

phthalimide

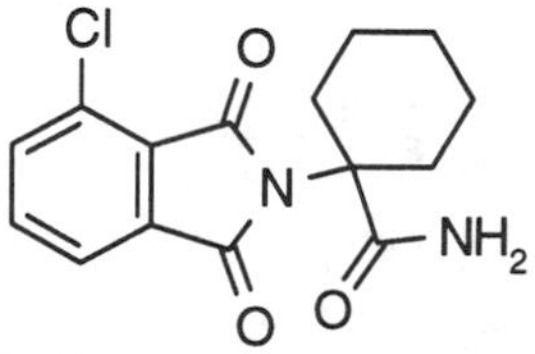

NOMENCLATURE
Chemical Abstracts name 1-(4-chloro-1,3-dihydro-1,3-dioxo-[2*H*]-isoindol-2-yl)-= cyclohexanecarboxamide

CAS RN *[51971–67–6]* **Development codes** AC 94,377; CL 94,377

PHYSICAL CHEMISTRY
Mol. wt. 306.7 **M.f.** $C_{15}H_{15}ClN_2O_3$ **Form** White crystalline solid.
M.p. 193–195 °C **Solubility** In water 30 ppm (25 °C). In acetone <2%,
N-methylpyrrolidone 35%, methylene chloride <2%, dimethyl sulfoxide >20% (all
at 20 °C).

COMMERCIALISATION
History Under evaluation by American Cyanamid Co.

APPLICATIONS
Biochemistry Not known. Activity mimics that of gibberellic acid in various
phytohormone bioassays (J. C. Suttle & J. F. Hultstrand, *Plant Physiology*, **80** (5),
115 (1986)). **Uses** Growth promoter, increasing the stem number, stem length
and commercial acceptability of cut stems in hybrid tea roses. Beneficial effects
have been reported in other crops. **Formulation types** SC.
Selected tradenames 'Surestem' (Cyanamid)

MAMMALIAN TOXICOLOGY
Oral Acute oral LD_{50} for male and female rats >5000 mg/kg. **Skin and eye** Acute
percutaneous LD_{50} for male and female rabbits >2000 mg/kg. Mildly irritating to
eyes; non-irritating to skin (male rabbits). **Other** Non-mutagenic in the Ames
test, microbial mutagenicity assay, and host-mediated assay.

4

acephate   *Insecticide*

organophosphorus

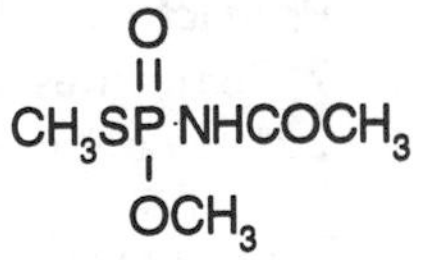

$$CH_3SP\!-\!NHCOCH_3$$

(structure: O,S-dimethyl acetylphosphoramidothioate)

NOMENCLATURE
Common name acephate (BSI, E-ISO, (*m*) F-ISO, ANSI, ESA, JMAF)
IUPAC name *O,S*-dimethyl acetylphosphoramidothioate
Chemical Abstracts name *N*-[methoxy(methylthio)phosphinoyl]acetamide
CAS RN *[30560–19–1]* **EEC no.** 250–241–2 **Development codes** Ortho 12 420
Official codes ENT 27 822

PHYSICAL CHEMISTRY
Composition Tech. grade is >97% pure. **Mol. wt.** 183.2 **M.f.** $C_4H_{10}NO_3PS$

Form Colourless crystals; (tech., a colourless solid). **M.p.** 88–90 °C; (tech. 82–89 °C) **V.p.** 0.226 mPa (24 °C) K_{ow} logP = −0.89 **S.g./density** 1.35 **Solubility** In water 790 g/l (20 °C). In acetone 151, ethanol >100, ethyl acetate 35, benzene 16, hexane 0.1 (all in g/l, 20 °C). **Stability** Relatively stable to hydrolysis; DT_{50} (40 °C) 60 h (pH 9), 710 h (pH 3).

COMMERCIALISATION
History Insecticide described by J. M. Grayson (*Pest Control*, 1972, **40**, 30). Chemical structure-biological activity relationships of analogues summarised by P. S. Magee (*Residue Rev.*, 1974, **53**, 3). Introduced by Chevron Chemical Co. **Patent** US 3716600; US 3845172 **Manufacturers** Aimco; Chevron; Jingma; Nagarjuna; Pesticides India; Productos OSA; Rallis; Shaw Wallace; Shenzhen Jiangshan; Sinon

APPLICATIONS
Biochemistry Cholinesterase inhibitor. **Mode of action** Systemic insecticide with contact and stomach action. **Uses** Control of a wide range of chewing and sucking insects, e.g. aphids, thrips, lepidopterous larvae, sawflies, leaf miners, leafhoppers, cutworms, etc. in fruit (including citrus), vines, hops, olives, cotton, soya beans, peanuts, macadamia nuts, beet, brassicas, celery, beans, potatoes, rice, tobacco, ornamentals, forestry, and other crops. Of moderate persistence, with residual activity lasting *c.* 10–21 d. **Phytotoxicity** Non-phytotoxic to most crops, but marginal leaf burn may occur on Red Delicious apples. **Formulation types** GR; SP; WP; CG; AE. **Selected tradenames** 'Orthene' (Tomen, Valent); 'Ortran' (Tomen, Valent); 'Acevol' (Voltas); 'Aimthene' (Aimco); 'Asataf' (Rallis); 'Cekucefate' (Cequisa); 'Lancer' (United Phosphorus); 'Racet' (Rotam); 'Rythane' (Ramcides); 'Saphate' (Sanonda); 'Starthene' (Shaw Wallace); 'Torpedo' (Searle India); 'Vital' (Productos OSA)

ANALYSIS
Product analysis by glc (J. B. Leary, *Anal. Methods Pestic. Plant Growth Regul.*, 1973, **7**, 363). **Residues** determined by glc (*idem, ibid.*; *Pestic. Anal. Man.*, 1979, **I**, 201-H, 201-I; *AOAC Methods*, 1995, 985.22).

MAMMALIAN TOXICOLOGY
Reviews *Pesticide residues in food – 1990*, FAO Plant Production and Protection Paper, 103, 1990. *Pesticide residues in food – 1990 evaluations. Toxicology*. World Health Organisation, WHO/PCS/91.47, 1991. **EHC** 63 (WHO, 1986; a general review of organophosphorus insecticides). **Oral** Acute oral LD_{50} for male rats 945, female rats 866, mice 361, dogs >681 mg/kg. **Skin and eye** Acute percutaneous LD_{50} for rabbits >2000 mg/kg. Non-irritating to skin (rabbits); non-sensitising to skin (guinea pigs). **Inhalation** LC_{50} >15 mg/l air. **NOEL** In 2 y feeding trials, dogs showed depression of cholinesterase at 100 mg/kg diet (maximum dose level) but no other significant effect; rats showed depression of

cholinesterase but no effect on weight gain nor pathological effect at 30 mg/kg diet. No teratogenic, mutagenic or carcinogenic effect observed. **ADI** (JMPR) 0.03 mg/kg [1990]. **Toxicity class** WHO (a.i.) III; EPA (formulation) III **EC risk** Xn (R22)

ECOTOXICOLOGY

Birds Acute oral LD_{50} for mallard ducks 350, chickens 568, ringneck pheasants 140 mg/kg. **Fish** LC_{50} (96 h) for bluegill sunfish 2050, rainbow trout >1000, channel catfish 2230, largemouth black bass 1725 mg/l. **Bees** LD_{50} 1.2 μg/bee. **Daphnia** (48 h) 67.2.

ENVIRONMENTAL FATE

Animals Metabolised to methamidophos (q.v.). **Plants** In plants, residual activity lasts for c. 10–15 d. The major metabolite is methamidophos (q.v.). **Soil/Environment** DT_{50} in soil 7–10 d; methamidophos (q.v.) has been identified as a metabolite.

acetamiprid   *Insecticide*

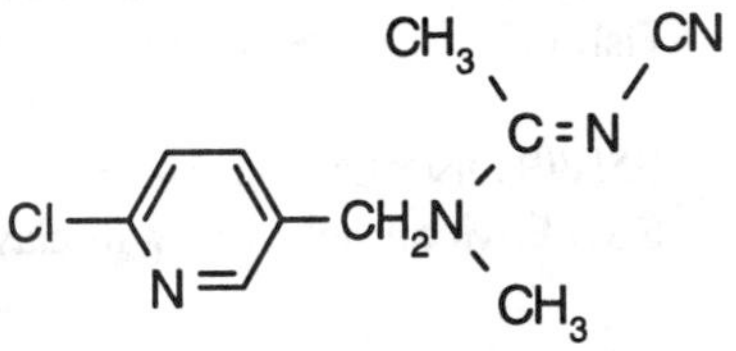

NOMENCLATURE
Common name acetamiprid (pa ISO)
IUPAC name (E)-N^1-[(6-chloro-3-pyridyl)methyl]-N^2-cyano-N^1-methylacetamidine
Chemical Abstracts name (E)-N-[(6-chloro-3-pyridinyl)methyl]-N'-cyano-N-=methylethanimidamide
CAS RN [135410–20–7]   **Development codes** NI-25 (Nippon Soda)

PHYSICAL CHEMISTRY
Mol. wt. 222.7   **M.f.** $C_{10}H_{11}ClN_4$ **Form** Colourless crystals.   **M.p.** 98.9 °C
V.p. <1 × 10^{-3} mPa (25 °C) K_{ow} logP = 0.80 (25 °C)   **S.g./density** 1.330 (20 °C)
Solubility In water at 25 °C, 4200 mg/l. Soluble in acetone, methanol, ethanol, dichloromethane, chloroform, acetonitrile and tetrahydrofuran.   **Stability** Stable in buffered solutions at pH 4, 5, 7. Degraded slowly at pH 9 and 45 °C. Stable under sunlight.   **pKa** 0.7, v. weak base

COMMERCIALISATION
History Reported by H. Takahashi et al. (*Proc. Br. Crop Prot. Conf. – Pests Dis.*, 1992, **1**, 89). Introduced by Nippon Soda Co., Ltd. **Patent** US 5304566 **Manufacturers** Nippon Soda

APPLICATIONS
Mode of action Systemic insecticide for soil and foliar application. **Uses** Control of Hemiptera, especially aphids, Thysanoptera and Lepidoptera on a wide range of crops, especially vegetables, fruit and tea. **Formulation types** SP; GR; WP; FU. **Selected tradenames** 'Mospilan' (Nippon Soda)

ANALYSIS
Tech. **product** by hplc; **residues** by gc.

MAMMALIAN TOXICOLOGY
Oral Acute oral LD_{50} for male rats 217, female rats 146, male mice 198, female mice 184 mg/kg. **Skin and eye** Acute percutaneous LD_{50} for male and female rats >2000 mg/kg. Non-irritating to skin and eyes (rabbits). **Inhalation** LC_{50} (4 h) for male and female rats >0.29 mg/l. **NOEL** (2 y) for rats 7.1 mg/kg b.w.; (18 mo) for mice 20.3 mg/kg b.w.; (1 y) for dogs 20 mg/kg b.w. **Other** Negative in the Ames test.

ECOTOXICOLOGY
Birds LD_{50} for bobwhite quail 180 mg/kg. LC_{50} for bobwhite quail >5000 ppm. **Fish** LC_{50} (24–96 h) for carp >100 mg/l. **Daphnia** LC_{50} (3–6 h) >1000 mg/l.

ENVIRONMENTAL FATE
Soil/Environment DT_{50} in clay loam 1 d; in light clay 1–2 d.

6   acetochlor   *Herbicide*

chloroacetanilide

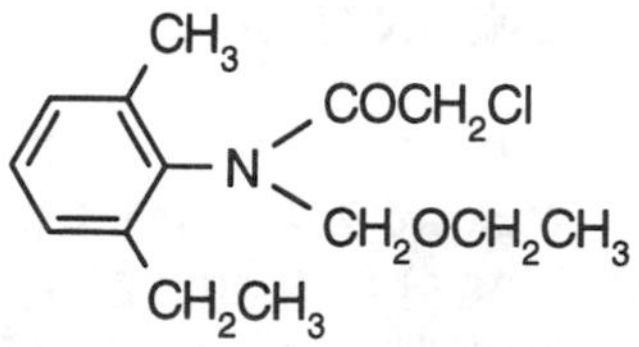

NOMENCLATURE
Common name acetochlor (BSI, E-ISO, ANSI, WSSA); acétochlore ((*m*) F-ISO)
IUPAC name 2-chloro-*N*-ethoxymethyl-6'-ethylaceto-*o*-toluidide

Chemical Abstracts name 2-chloro-*N*-(ethoxymethyl)-*N*-(2-ethyl-6-methylphenyl)= acetamide
CAS RN *[34256–82–1]* **Development codes** MON 097

PHYSICAL CHEMISTRY
Composition Tech. is 92% pure. **Mol. wt.** 269.8 **M.f.** $C_{14}H_{20}ClNO_2$ **Form** Tech. is a wine red to yellow or amber oil. **M.p.** <0 °C **B.p.** 162 °C/7 mmHg
V.p. 4.5×10^{-3} mPa (25 °C) **K_{ow}** logP = 3.03 **S.g./density** 1.110 (30 °C)
Solubility In water 223 mg/l (25 °C). Soluble in diethyl ether, acetone, benzene, chloroform, ethanol, ethyl acetate, and toluene. **Stability** Stable for over 2 years at 20 °C (EC formulation). **F.p.** >100 °C (Tag closed cup)

COMMERCIALISATION
History Herbicide discovered and introduced by Monsanto Co.
Manufacturers Monsanto; Nitrokémia; Sanachem; Sinon

APPLICATIONS
Biochemistry Inhibits protein synthesis; cell division inhibitor.
Mode of action Selective herbicide, absorbed mainly by the shoots of germinating plants. **Uses** Used pre-emergence or pre-plant to control annual grasses, certain annual broad-leaved weeds and yellow nutsedge in maize, peanuts, soya beans, cotton, potatoes and sugar cane. It is absorbed mainly by the shoots and secondarily by the roots of germinating plants. **Formulation types** EC.
Mixtures *(acetochlor +)* atrazine. **Selected tradenames** 'Harness' (Monsanto); 'Acenit' (Nitrokémia); 'Surpass' (Zeneca); 'Trophy' (Zeneca)

ANALYSIS
Product by gc. **Residues** by hplc. Details from Monsanto.

MAMMALIAN TOXICOLOGY
Oral Acute oral LD_{50} for rats 2148 mg/kg. **Skin and eye** Acute percutaneous LD_{50} for rabbits 4166 mg/kg. Contact sensitisation reactions observed in guinea pigs. Practically non-irritating to eyes and skin (rabbits). **Inhalation** LC_{50} (4 h) for rats >3.0 mg/l air. **NOEL** (2 y) for rats 10 mg/kg b.w. daily; (1 y) for dogs 12 mg/kg b.w. daily. **ADI** 0.01 mg/kg b.w. **Toxicity class** WHO (a.i.) III

ECOTOXICOLOGY
Birds Acute oral LD_{50} for bobwhite quail 1260 mg/kg. LC_{50} (5 d) for quail and mallard ducks >5620 mg/kg. **Fish** LC_{50} (96 h) for rainbow trout 0.45, bluegill sunfish 1.3 mg/l. **Bees** LD_{50} (96 h) 1.715 mg/bee. **Daphnia** LC_{50} (48 h) 16 mg/l.

ENVIRONMENTAL FATE
Animals The primary routes of metabolism are glutathione conjugation and

metabolism by cytochrome P450. **Plants** In maize and soya beans, rapidly absorbed and metabolised in the germinating plant. In maize, the first metabolite is glutathione, and in soya beans homoglutathione (E. J. Breaux, *J. Agric. Food Chem.*, 1986, **34**, 884). **Soil/Environment** Strongly absorbed by soil, with little leaching. Microbial degradation accounts for most loss from soil; DT_{50} 8–18 d.

7   acifluorfen   *Herbicide*

diphenyl ether

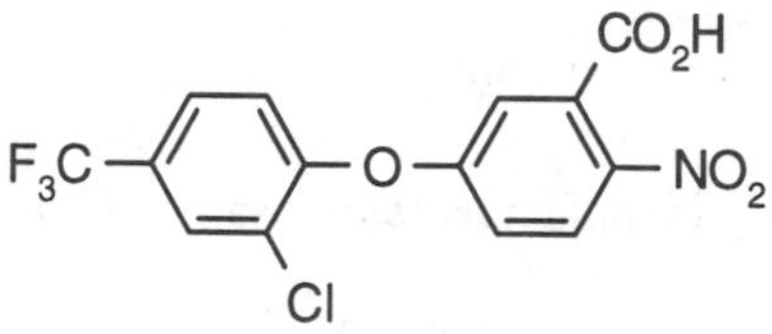

NOMENCLATURE
acifluorfen
Common name acifluorfen (BSI, E-ISO, ANSI, WSSA); acifluorfène ((*m*) F-ISO)
IUPAC name 5-(2-chloro-α,α,α-trifluoro-*p*-tolyloxy)-2-nitrobenzoic acid
Chemical Abstracts name 5-[2-chloro-4-(trifluoromethyl)phenoxy]-2-nitrobenzoic acid
CAS RN *[50594–66–6]*   **Development codes** MC 10978 (Mobil); RH-6201 (Rohm & Haas)

acifluorfen-sodium
Common name acifluorfen-sodium
IUPAC name sodium 5-(2-chloro-α,α,α-trifluoro-*p*-tolyloxy)-2-nitrobenzoate
CAS RN *[62476–59–9]*   **EEC no.** 263–570–7   **Development codes** RH 6201 (Rohm & Haas); MC 10978 (Mobil); BAS 9048.H (BASF)

PHYSICAL CHEMISTRY
acifluorfen
Mol. wt. 361.7   **M.f.** $C_{14}H_7ClF_3NO_5$ **Form** Light brown solid.   **M.p.** 142–160 °C
V.p. <0.01 mPa (20 °C) (evaporation rate method, CF/P006)   **S.g./density** 1.546
Solubility In water 120 mg/l (23–25 °C) (tech.). In acetone 600, ethanol 500, dichloromethane 50, xylene, kerosene <10 (all in g/kg, 25 °C).   **Stability** Decomposes at 235 °C. Stable in acid and alkaline media, pH 3 to pH 9 (40 °C). Decomposed by u.v. light, DT_{50} c. 110 h.

acifluorfen-sodium
Composition Tech. salt is not isolated in solid form, but as aqueous solution with 44% w/w a.i.   **Mol. wt.** 383.6   **M.f.** $C_{14}H_6ClF_3NNaO_5$ **M.p.** (dry) 274–278 °C

(decomp.)   **V.p.** <0.01 mPa (25 °C) (evaporation rate method, CF/P006)
K_{ow} logP = 1.19 (pH 5, 25 °C)   **Henry** <6.179 × 10^{-9} Pa m^3 mol^{-1} (calc.)
S.g./density Bulk density 0.4–0.5   **Solubility** In water (unbuffered) 62.07, (pH 7)
60.81, (pH 9) 60.71 (all in g/100 g, 25 °C). In octanol 5.37, methanol 64.15,
hexane <5 × 10^{-5} (all in g/100 ml, 25 °C).   **Stability** Stable >2 y at 20–25 °C in
aqueous solution.   **pKa** 3.86±0.12

COMMERCIALISATION
History Acifluorfen introduced as a herbicide independently by Mobil Chemical
Co. (now owned by Rhône-Poulenc Agrochimie, who no longer market it) and by
Rohm & Haas Co. (who transferred rights to BASF AG in 1987).   **Patent**
DE 2311638   **Manufacturers** Toll manufactured for BASF by Rohm & Haas

APPLICATIONS
Biochemistry Protoporphyrinogen oxidase inhibitor.   **Mode of action** Selective
contact herbicide, absorbed by the foliage and roots, with negligible translocation.
Activity is enhanced by sunlight.

acifluorfen
Uses Used post-emergence for the control of annual broad-leaved weeds
(*Abutilon, Amaranthus, Datura, Euphorbia, Polygonum, Ipomoea, Xanthium* spp.)
with some effects on grasses in soya beans, peanuts and rice. Applied at 0.2 to
0.6 kg/ha depending on crop.   **Phytotoxicity** Soya beans show good tolerance,
although some burning may be seen on new leaf growth which are readily
outgrown.   Phytotoxicity increased if mixed with fertilisers.
Compatibility Compatible with most other pesticides, but phytotoxicity to the
crop may be increased.

acifluorfen-sodium
Formulation types SL.   **Mixtures** (*acifluorfen-sodium* +) bentazone.
Selected tradenames 'Blazer' (BASF); 'Doble' (mixture) (BASF); 'Galaxy' (mixture)
(BASF); 'Storm' (mixture) (BASF)

ANALYSIS
Product analysis by glc. **Residue** analysis by glc or hplc (T. A. Ray *et al., J. Assoc.
Off. Anal. Chem.*, 1983, **66**, 1319).

MAMMALIAN TOXICOLOGY
acifluorfen-sodium
Oral Acute oral LD$_{50}$ for rats 1540, female mice 1370, rabbits 1590 mg/kg b.w;
(aqueous tech.).   **Skin and eye** Acute percutaneous LD$_{50}$ for rabbits >2000
mg/kg. Severe eye irritant; moderate skin irritant (rabbits) (aqueous tech.).
Inhalation LC$_{50}$ (4 h) for rats >6.91 mg/l air (aqueous formulation).   **NOEL** for
mice 7.5 ppm (mg/kg diet) (aqueous tech.).

ADI 10 µg/kg.   **Other** Non-mutagenic in Ames and mouse lymphoma assays.
Toxicity class WHO (a.i.) III; EPA (formulation) III

ECOTOXICOLOGY
acifluorfen-sodium
Birds Acute oral LD_{50} for bobwhite quail 325 mg/kg. LC_{50} (8 d) for bobwhite
quail and mallard ducks >5620 mg/kg diet.   **Fish** LC_{50} (96 h) for rainbow trout 17,
bluegill sunfish 62 mg/l.   **Bees** Use not expected to result in honey bee exposure;
tests not performed.   **Worms** LC_{50} (14 d) for *Eisenia foetida* >1800 mg/kg
substrate.   **Other beneficial spp.** Minimum inhibitory concentration for *Azotobacter
vinelandii* >1000 ppm, for *Bacillus subtilis* 1000 ppm.   **Daphnia** EC_{50} (48 h)
77 mg/l.   **Algae** EC_{50} for *Selenastrum capricornutum* >260, *Anabaena flos-aquae*
>350 µg/l.   **Other aquatic spp.** EC_{50} (96 h) for grass shrimps 189 mg/l.

ENVIRONMENTAL FATE
Animals After oral application in rats, a fast and almost complete absorption and
excretion occurs. Multiple application does not indicate a cumulative effect. The
dermal resorption is low. Acifluorfen-sodium is judged not to present a substantial
hazard to aquatic or terrestrial wildlife.   **Plants** Not transported within the plant;
degradation occurs at, or close to, the surface, DT_{50} c. 1 w. Metabolism is fast and
extensive, through amination, hydroxylation and carboxylation.
Soil/Environment The a.i. will be moderately quickly degraded, DT_{50} 108 d (silt
loam) – 200 d (clay loam), forming mainly bound residues and highly polar
metabolites. Degradation occurs through microbial activity; there is also photolytic
degradation on the soil surface. Accumulation in soil does not occur. Adsorption
K_{oc} 44–684, K_d 0.13–1.98; desorption K_{oc} 131–1955, K_d 0.39–4.6. In water,
acifluorfen is hydrolytically stable in the dark, but in light it is rapidly degraded,
DT_{50} c. 2 h, forming mainly CO_2. In extensive field trials under worst case
conditions, no case of leaching was recorded.

8   aclonifen   *Herbicide*

diphenyl ether

NOMENCLATURE
Common name aclonifen (BSI, draft E-ISO); aclonifène ((*m*) draft F-ISO)
IUPAC name 2-chloro-6-nitro-3-phenoxyaniline

Chemical Abstracts name 2-chloro-6-nitro-3-phenoxybenzenamine
CAS RN *[74070–46–5]* **EEC no.** 277–704–1 **Development codes** CME 127;
KUB 3359; LE84493

PHYSICAL CHEMISTRY
Composition ≥95% pure. **Mol. wt.** 264.7 **M.f.** $C_{12}H_9ClN_2O_3$ **Form** Yellow
crystals. **M.p.** 81–82 °C **V.p.** 1.6×10^{-2} mPa (20 °C) **K_{ow}** logP = 4.37
Henry 3.2×10^{-3} Pa m^3 mol^{-1} (20 °C) **S.g./density** 1.46 g/cm^3 **Solubility** In
water 1.4 mg/l (20 °C). In methanol 50, hexane 4.5, toluene 390 (all in g/kg,
20 °C). **Stability** Slowly decomposes when exposed to light.

COMMERCIALISATION
History Herbicide reported by W. Buck *et al.* (*Proc. Int. Congr., Plant Prot., 10th,*
1983, **1**, 307). Introduced by Celamerck GmbH & Co. (became Shell Agrar) and
later sold to Rhône-Poulenc Agrochimie. **Patent** US 4394159
Manufacturers Rhône-Poulenc

APPLICATIONS
Biochemistry Protoporphyrinogen oxidase inhibitor. **Mode of action** Systemic,
selective herbicide. **Uses** Pre-emergence control of grass and broad-leaved
weeds in winter wheat, potatoes, sunflowers, peas, carrots, maize, and other
crops. **Phytotoxicity** Non-phytotoxic to potatoes, sunflowers, and peas. May be
phytotoxic to cereals and maize at high dose rates. **Formulation types** SC.
Mixtures *(aclonifen +)* oxadiazon. **Selected tradenames** 'Challenge'
(Rhône-Poulenc)

ANALYSIS
Product and **residue** analysis by glc. Details from Rhône-Poulenc.

MAMMALIAN TOXICOLOGY
Oral Acute oral LD_{50} for rats and mice >5000 mg/kg. **Skin and eye** Acute
percutaneous LD_{50} for rats >5000 mg/kg. Slight skin irritant; non-irritating to eyes
(rabbits). **Inhalation** LC_{50} (4 h) for rats >5.06 mg/l air. **NOEL** (90 d) for rats
28 mg/kg b.w. daily; (180 d) for dogs 12.5 mg/kg b.w. daily. **ADI** 0.02 mg/kg
(France). **Other** Non-mutagenic in the Ames test. Not embryotoxic or
teratogenic in rats. Does not affect reproduction in rats at 2000 ppm over 2
generations. **Toxicity class** WHO (a.i.) III (Table 5)

ECOTOXICOLOGY
Birds Acute oral LD_{50} for Japanese quail and canaries >15 000 mg/kg b.w.
Fish LC_{50} (96 h) for rainbow trout 0.67, carp 1.7 mg/l. **Bees** Not toxic to bees.
LD_{50} (oral) >100 µg/bee. **Worms** LC_{50} (14 d) 300 mg/kg. **Daphnia** EC_{50} (48 h)
2.5 mg/l. **Algae** EC_{50} (96 h) 6.9 µg/l.

MAIN ENTRIES

ENVIRONMENTAL FATE

Animals In rats, following oral administration, 62–65% is excreted in the urine, principally in the form of polar compounds. No bioaccumulation. **Plants** In plants, hydroxylation occurs on both benzene rings. DT_{50} c. 14 d. **Soil/Environment** In sterile water, stable between pH 3 to 9. DT_{50} in water in presence of micro-organisms c. 1 mo. In soil, DT_{50} 36–80 d (22 °C). K_{oc} 5318–12 164. Unlikely to leach: $-0.13 <$ GUS < 0.64.

9   acrinathrin   *Acaricide, insecticide*

pyrethroid

NOMENCLATURE

Common name acrinathrin (BSI, draft E-ISO); acrinathrine (draft F-ISO)
IUPAC name (S)-α-cyano-3-phenoxybenzyl (Z)-(1R,3S)-2,2-dimethyl-3-[2-(2,2,2-= trifluoro-1-trifluoromethylethoxycarbonyl)vinyl]cyclopropanecarboxylate
Roth: (S)-α-cyano-3-phenoxybenzyl (Z)-(1R-*cis*)-2,2-dimethyl-3-[2-(2,2,2-trifluoro-= 1-trifluoromethylethoxycarbonyl)vinyl]cyclopropanecarboxylate
Chemical Abstracts name cyano(3-phenoxyphenyl)methyl 2,2-dimethyl-3-[3-oxo-= 3-[2,2,2-trifluoro-1-(trifluoromethyl)ethoxy]-1-propenyl]cyclopropanecarboxylate
CAS RN *[101007–06–1]* as defined; *[103833–18–7]* unstated stereochemistry
Development codes RU 38 702; HOE 076003; AE F 076003; NU 702

PHYSICAL CHEMISTRY

Composition Material is a single isomer.   **Mol. wt.** 541.4   **M.f.** $C_{26}H_{21}F_6NO_5$
Form Colourless crystals (tech. grade).   **M.p.** 81.5 °C (pure); 82 °C (tech.)
V.p. 4.4×10^{-5} mPa (20 °C)   **K$_{ow}$** logP = 5 (a.i., 25 °C)   **Henry** 4.8×10^{-2} Pa m^3 mol^{-1} (calc.).   **Solubility** In water ≤0.02 mg a.i./l (25 °C). In acetone, chloroform, dichloromethane, ethyl acetate, dimethylformamide >500, di-isopropyl ether 170, ethanol 40, hexane 10, *n*-octanol 10 (all in g a.i./l).   **Stability** Stable in acid but hydrolysis and epimerisation more important at pH >7. DT_{50} >1 y (pH 5, 50 °C), 30 d (pH 7, 30 °C); 15 d (pH 9, 20 °C), 1.6 d (pH 9, 37 °C). Stable for 7 d under 100 W light (tech.).   **Specific rotation** $[\alpha]_D^{20}$ +17.5°

COMMERCIALISATION

History Acaricide and insecticide reported by J. R. Tessier *et al.* (*IUPAC Pestic.*

Chem., 1983, **5**, 95). Introduced in France (1990) by Roussel Uclaf.
Patent EP 48186; FR 2486073   **Manufacturers** Roussel Uclaf

APPLICATIONS
Biochemistry Acts on the central nervous system.   **Mode of action** Contact and
stomach action.   **Uses** An ingested and contact acaricide effective against a wide
range of phytophagous mites on citrus, cotton, fruit, hops, ornamentals, soya
beans, tobacco, vegetables and vines. It also shows insecticidal properties, in
particular with high efficacy on thrips species on fruit trees, vines and vegetables.
Formulation types EC; SC; WP; EW.   **Mixtures** *(acrinathrin +)* propargite.
Compatibility May not be compatible with alkaline products.
Selected tradenames 'Rufast' (AgrEvo)

ANALYSIS
Product analysis by hplc.   **Residues** in plants, soil and water determined by glc
with ECD.

MAMMALIAN TOXICOLOGY
Oral Acute oral LD_{50} for rats and mice >5000 mg/kg (for tech. in corn oil).
Skin and eye Acute percutaneous LD_{50} for rats >2000 mg/kg. Non-irritating to
eyes and skin of rabbits. Not skin-sensitising to guinea pigs.   **Inhalation** LC_{50} (4 h)
for rats 1.6 mg/l air.   **NOEL** (90 d) for male rats 2.4 mg/kg b.w., for female rats
3.1 mg/kg b.w.; (1 y) for dogs 3 mg/kg b.w. Non-mutagenic and non-teratogenic
in rats (2 mg/kg b.w. daily) or rabbits (15 mg/kg b.w. daily).   **ADI** 0.02 mg/kg.
Other Low solubility in water and high adsorption on soil mean that low LC_{50} or
LD_{50} values under laboratory conditions do not present significant hazard under
practical field conditions.   **Toxicity class** WHO (a.i.) III (Table 5); EPA
(formulation) IV

ECOTOXICOLOGY
Birds Acute oral LD_{50} for bobwhite quail >2250, mallard ducks >1000 mg/kg.
LC_{50} (8 d) for bobwhite quail 3275, mallard ducks 4175 mg/kg diet.   **Fish** LC_{50} for
rainbow trout 5.66, mirror carp 0.12 mg/l.   **Bees** Oral LC_{50} (48 h) 150–200
ng/bee; contact (48 h) 200–500 ng/bee.   **Worms** LC_{50} (14 d) for earthworms
>1000 mg/kg; NOEC biomass 1.6 mg/kg.   **Daphnia** LC_{50} (48 h) 0.57 mg/l.
Algae EC_{50} (96 h) for green algae >0.82 mg/l.

ENVIRONMENTAL FATE
Animals No metabolites found representing >10% of parent compound.   **Plants**
The main residue is the parent compound.   **Soil/Environment** Strongly adsorbed
onto soil and immobile (irrespective of pH and o.m. content); K_d 2460–2780;
K_{oc} 127 500–319 610. Soil column leaching: <1% of applied acrinathrin found in
leachate. DT_{50} 5–100 d (4 soil types). DT_{50} under aerobic conditions (pH 6.2,
o.m. 3.1%) 52 d.

acrolein   *Herbicide*

$$H_2C{=}CH$$
$$\diagdown$$
$$CHO$$

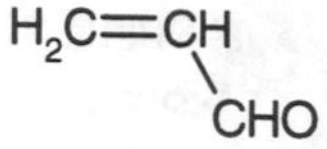

NOMENCLATURE
Common name acrolein (accepted in lieu of a common name: BSI, ANSI, WSSA)
IUPAC name prop-2-enal; acrylaldehyde
Chemical Abstracts name 2-propenal
CAS RN *[107–02–8]* **EEC no.** 203–453–4

PHYSICAL CHEMISTRY
Composition Tech. is 92–97%. **Mol. wt.** 56.1 **M.f.** C_3H_4O **Form** Colourless
mobile liquid with a pungent odour. **M.p.** –87 °C **B.p.** 52.5 °C **V.p.** 59 kPa
(38 °C) K_{ow} logP = 1.08 **Henry** 7.9 (calc.), 19.5 (measured) (both in kPa m^3
mol^{-1}) **S.g./density** 0.841 (4 °C–20 °C) **Solubility** In water 208 g/kg (20 °C).
Miscible with lower alcohols, ketones, benzene, diethyl ether and other common
organic solvents. **Stability** Stable ≤80 °C; highly reactive chemically. May
polymerise if exposed to light (stabiliser, e.g. hydroquinone). Polymerises slowly
on storage and violently in the presence of concentrated acids. Must be stored in
the dark, under nitrogen. Hydrolysis DT$_{50}$ 3.5 d (pH 5), 1.5 d (pH 7), 4 h (pH 10).
Transported in oxygen-free atmospheres in the presence of polymerisation
inhibitor. **F.p.** Closed cup <–17.8 °C

COMMERCIALISATION
History Algicide and aquatic herbicide introduced by Shell Chemical Co., now
marketed by Baker Performance Chemicals Inc. **Patent** US 2042220; US 2959476;
US 2978475 all to Shell

APPLICATIONS
Biochemistry Reaction with the sulfhydryl groups of enzymes.
Mode of action Contact herbicide which breaks down cell walls. **Uses** An aquatic
herbicide, injected below the water surface (1–15 mg/l), to control submersed
aquatic weeds and algae in irrigation canals and drainage ditches. Little effect on
emergent weeds at recommended rates. Floating weeds such as *Pistia*, *Eichornia*
and *Jussiaea* spp. controlled only if concentration is maintained for extended
period. **Formulation types** Liquid. **Compatibility** It is not customary to mix with
other preparations. **Selected tradenames** 'MAGNACIDE' (Baker Performance
Chemicals)

ANALYSIS
Residues in water determined by conversion to the 2,4-dinitrophenylhydrazone
and colorimetry of this derivative.

MAMMALIAN TOXICOLOGY

EHC 127 (WHO, 1992).   **Oral** Acute oral LD_{50} for rats 29, male mice 13.9, female mice 17.7 mg/kg.   **Skin and eye** Acute percutaneous LD_{50} for rabbits 231 mg/kg. Skin irritant; lachrymatory effect and irritant action on respiratory organs at <1 mg/kg.   **Inhalation** LC_{50} (4 h) for rats 8.3 mg/l air.   **NOEL** (90 d) for rats 5 mg/kg b.w. daily. Administration of 200 mg acrolein/l water to rats for 90 d causes no ill effects. No reproductive toxicity in 2-generation feeding study in rats at 7.2 mg/kg daily. No teratogenic effect in rabbits at levels causing maternal toxicity (maximum dose 2 mg/kg daily).   **Toxicity class** WHO (a.i.) Ia; EPA (formulation) I   **EC risk** F (R11); T+ (R26); T (R25); C (R34)

ECOTOXICOLOGY

Birds Acute oral LD_{50} for bobwhite quail 19, mallard ducks 30.2 mg/kg (studies were for tech. material; maximum rate in use is 15 mg/l water, so potential exposure is substantially reduced).   **Fish** Highly toxic to fish. Concentrations of 1–5 mg/l are deadly. LC_{50} (24 h) for rainbow trout 0.15, bluegill sunfish 0.079, shiners 0.04, and mosquito fish 0.39 mg/l.   **Algae** EC_{50} (5 d) for *Selenastrum capricornutum* 0.050, *Anabaena flos-aquae* 0.042, *Navicula pelliculosa* 0.07, *Skeletonema costatum* 0.03 mg/l.   **Other aquatic spp.** LC_{50} (48 h) for shrimps 0.10, oysters 0.46 mg/l. EC_{50} (14 d) for *Lemna gibba* 0.07 mg/l.

ENVIRONMENTAL FATE

Animals In goat and hen, no acrolein was detected in tissues or excreta, or in goat milk or hen eggs following administration of high doses. All residues identified are natural products. Data on naturally occurring metabolites found in aquatic species are also available.   **Plants** Following high application rates to lettuce, no acrolein was detected 1 day following last application. At harvest, 3 highly polar conjugates (representing in total <0.5 ppm) were detected.   **Soil/Environment** DT_{50} in water 150 h (pH 5), 120–180 h (pH 7), 5–40 h (pH 9). Acrolein is metabolised easily in soil, being mineralised to CO_2. Metabolic pathways involving oxidation, reduction and hydration have been proposed.

11   *Adoxophyes orana* GV *Biological agent*

granulosis virus

NOMENCLATURE

Scientific name *Adoxophyes orana* granulosis virus   **Strain** Swiss strain

PROPERTIES
Stability Stable for 4 weeks at room temperature; unlimited stability below 2 °C.

COMMERCIALISATION
History Discovered by A. Schmid in Switzerland 1975 (*Mitteilungen der Schweizerischen Entomologischen Gesellschaft*, **56**, 225–235 (1983)). 'Capex' registered in Switzerland since 1989. **Manufacturers** Andermatt

APPLICATIONS
Mode of action Acts by ingestion with following infection. **Uses** Used for control of summer fruit tortrix moths (*Adoxophyes orana*). **Formulation types** SC.
Compatibility Compatible with all non-copper fungicides, and all pesticides which do not have a repellent effect on *A. orana* and a pH between 6 and 8.
Selected tradenames 'Capex 2' (Andermatt)

ANALYSIS
By biotest with *Adoxophyes orana*.

MAMMALIAN TOXICOLOGY
Non-toxic.

12 AKD-2023 *Acaricide*

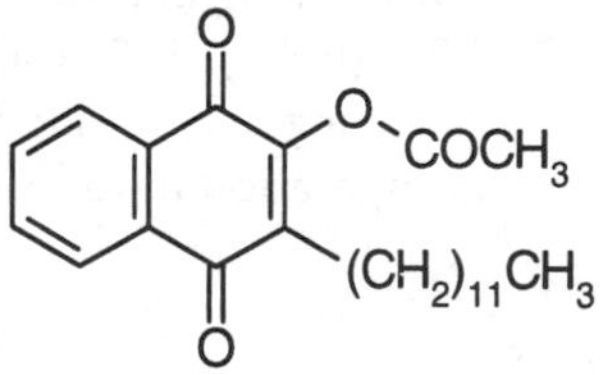

NOMENCLATURE
Common name acequinocyl (BSI proposed)
IUPAC name 3-dodecyl-1,4-dihydro-1,4-dioxo-2-naphthyl acetate
Chemical Abstracts name 3-(acetyloxy)-2-dodecyl-1,4-naphthalenedione
CAS RN [57960–19–7] **Development codes** AKD-2023; DPX-3792; DPX-T3792; AC-145

PHYSICAL CHEMISTRY
Mol. wt. 384.5 **M.f.** $C_{24}H_{32}O_4$ **Form** Yellowish fine powder. **M.p.** 59.6 °C
V.p. 5.1×10^{-2} mPa (40 °C) **K_{ow}** logP >6.2 **S.g./density** 1.15 (25 °C)

Solubility In water <10 µg/l (20 °C). In *n*-hexane 44, toluene 450, dichloromethane 620, acetone 220, methanol 7.8, DMF 190 (all in g/l).
Stability Hydrolysis DT_{50} 19 d (pH 1.2, 37 °C), 86 d (pH 4, 25 °C), 52 h (pH 7, 25 °C), 76 min (pH 9, 25 °C). Aqueous photolysis DT_{50} (pH 5) 6 d.

COMMERCIALISATION
History Licensed by du Pont to Agro-Kanesho Co. Ltd and under development by them. **Manufacturers** Agro-Kanesho

APPLICATIONS
Biochemistry Hydrolysed to 2-dodecyl-3-hydroxy-1,4-naphthoquinone which binds to the Qo centre (ubiquinol oxidation site) of complex III of the mitochondrial electron-transfer chain, which it inhibits. **Uses** Under development for control of *Panonychus* spp., *Tetranychus* spp., *Aculops* spp., *Polyphagatarsonemeus* spp. in citrus, apple, pear, peach, cherry, melon, cucumber, tea, ornamentals and vegetables.
Phytotoxicity None observed except in some varieties of strawberry.
Formulation types SC. **Compatibility** Not compatible with alkaline products.
Selected tradenames 'Kanemite' (Agro-Kanesho)

MAMMALIAN TOXICOLOGY
Oral Acute oral LD_{50} for rats and mice >5000 mg/kg. **Skin and eye** Acute percutaneous LD_{50} for rats >2000 mg/kg. Very mild skin irritant (rabbits). Not a skin sensitiser (guinea pigs). **Inhalation** LC_{50} for rats >0.84 mg/l. **NOEL** 9.0 mg/kg daily.

ECOTOXICOLOGY
Birds Acute oral LD_{50} for mallard duck and Japanese quail 2000 mg/kg. **Fish** LC_{50} for common carp and rainbow trout >100 ppm. **Daphnia** LC_{50} (24 h) 0.0039 ppm.

ENVIRONMENTAL FATE
Soil/Environment K_{oc} (four soil types) 31 800–123 000.

13 AKH-7088 *Herbicide*

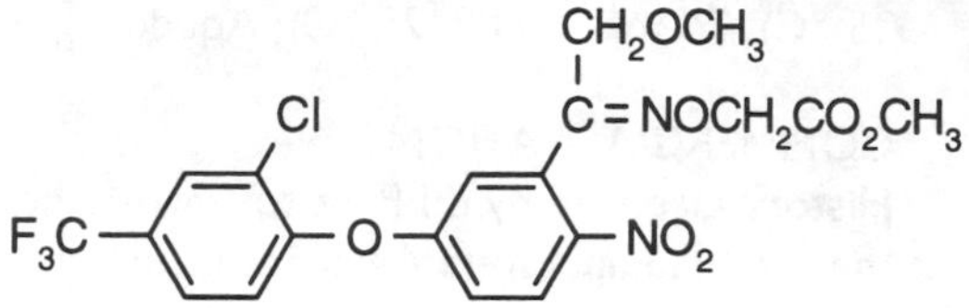
CH2OCH3
Cl
C = NOCH2CO2CH3
F3C
O
NO2

NOMENCLATURE

IUPAC name methyl (*EZ*)-1-[5-(2-chloro-α,α,α-trifluoro-*p*-tolyloxy)-= 2-nitrophenyl]-2-methoxyethylideneamino-oxyacetate

CAS RN *[104459–82–7]* **Development codes** AKH-7088

PHYSICAL CHEMISTRY

Mol. wt. 476.8 **M.f.** $C_{19}H_{16}ClF_3N_2O_7$ **Form** Colourless crystals.
M.p. 57.7–58.1 °C **Solubility** In water 1 mg/l (20 °C).

COMMERCIALISATION

Manufacturers Asahi

APPLICATIONS

Uses Post-emergence control of broad-leaved weeds, including velvetleaf, cocklebur, and jimsonweed, in soya beans.

ANALYSIS

Details from Asahi.

MAMMALIAN TOXICOLOGY

Oral Acute oral LD_{50} for rats 5000 mg/kg. **Skin and eye** Acute percutaneous LD_{50} for rats 2000 mg/kg. Moderate skin irritation, non-irritating to eyes (male rabbits). **Other** Non-mutagenic in the Ames test.

alachlor

chloroacetanilide

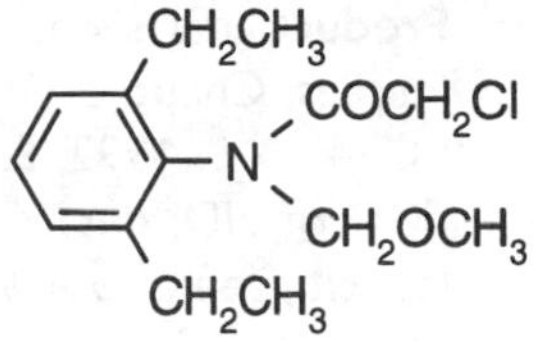

NOMENCLATURE
Common name alachlor (BSI, E-ISO, ANSI, WSSA, JMAF); alachlore ((*m*) F-ISO)
IUPAC name 2-chloro-2′,6′-diethyl-*N*-methoxymethylacetanilide
Chemical Abstracts name 2-chloro-*N*-(2,6-diethylphenyl)-*N*-(methoxymethyl)=
acetamide
CAS RN *[15972–60–8]*   **EEC no.** 240–110–8   **Development codes** CP 50 144;
MON 0144 (tech.)

PHYSICAL CHEMISTRY
Composition ≥93% pure.   **Mol. wt.** 269.8   **M.f.** $C_{14}H_{20}ClNO_2$ **Form** Yellow
white to wine red, odourless solid (room temperature); yellow to red liquid
(>40 °C).   **M.p.** 40.5–41.5 °C   **B.p.** 100 °C/0.0026 kPa   **V.p.** 2.1 mPa (25 °C)
K_{ow} logP = 3.09   **S.g./density** 1.125 (25 °C)   **Solubility** In water 242 mg/l
(25 °C). Soluble in diethyl ether, acetone, benzene, chloroform, ethanol, and
ethyl acetate. Slightly soluble in heptane.   **Stability** Hydrolysed by strong acids and
alkalis. Stable to u.v. light. Decomposes at 105 °C.   **F.p.** 137 °C (closed cup);
160 °C (open cup)

COMMERCIALISATION
History Herbicide reported by R. F. Husted *et al.* (*Proc. North Cent. Weed Control
Conf.*, 1966, **21**, 44). Introduced by Monsanto Co.   **Patent** US 3442945;
US 3547620   **Manufacturers** Comlets; Crystal; Krishi Rasayan; Makhteshim-Agan;
Monsanto; Pilarquim; Rallis; Sanachem; Sinon; Sundat

APPLICATIONS
Biochemistry Acts by inhibition of protein synthesis and root elongation. Maize
tolerance is attributed to rapid detoxification by glutathione transferases.   **Mode
of action** Selective systemic herbicide, absorbed principally by germinating shoots,
but also by the roots, with translocation throughout the plant, and accumulation
mainly in vegetative parts rather than in reproductive parts.   **Uses** Used pre-
emergence at 1.68–4.48 kg/ha to control annual grasses and many broad-leaved
weeds in cotton, brassicas, maize, oilseed rape, peanuts, radish, soya beans and
sugar cane.   **Formulation types** EC; GR; WG.   **Mixtures** *(alachlor +)* atrazine;
glyphosate; trifluralin; terbuthylazine; pendimethalin.   **Selected tradenames** 'Lasso'
(Monsanto); 'Alanex' (Makhteshim-Agan); 'Cattch' (Searle India); 'Lacorn'

(Efthymiadis); 'Satochlor' (Sagrochem); 'Sholay' (Rallis); 'Top 48' (Cequisa)

ANALYSIS
Product analysis by glc with FID (J. M. Warner & D. D. Arras in *Comp. Anal. Profiles,* Chapt. 5; *AOAC Methods,* 1995, 985.04, 988.03; *CIPAC Handbook,* 1985, **1C,** 4; *ibid.,* 1992, **E,** 4). **Residues** in water and in soil determined by glc with ECD or FID; details, and other methods reviewed by J. M. Warner & D. D. Arras, *loc. cit.;* see also *AOAC Methods,* 1995, 991.07. Details also available from Monsanto Co.

MAMMALIAN TOXICOLOGY
Oral Acute oral LD_{50} for rats 930–1350 mg/kg. **Skin and eye** Acute percutaneous LD_{50} for rabbits 13 300 mg/kg. Non-irritating to skin or eyes (rabbits). Contact sensitisation reactions observed in guinea pigs. **Inhalation** LC_{50} (4 h) for rats 1.04 mg/l air. **NOEL** (2 y) for rats ≤2.5 mg/kg b.w. daily; (1 y) for dogs ≤1 mg/kg b.w. daily. In 90 d feeding trials, rats and dogs receiving 200 mg/kg diet showed no ill-effects. **Other** Oncogenic in rats but not mice. **Toxicity class** WHO (a.i.) III; EPA (formulation) III **EC risk** (R40); Xn (R22); (R43)

ECOTOXICOLOGY
Birds Acute oral LD_{50} for bobwhite quail 1536 mg/kg. LC_{50} (5 d) for mallard ducks and bobwhite quail >5620 mg/kg diet. **Fish** LC_{50} (96 h) for rainbow trout 1.8, bluegill sunfish 2.8, fathead minnow 5.0, channel catfish 2.1 mg/l. **Bees** Not hazardous to bees when used as directed; LD_{50} 32 mg/bee. **Worms** LC_{50} (14 d) for earthworms 387 mg/kg dry soil. **Daphnia** EC_{50} (48 h) 10 mg/l. **Algae** TL_{50} (72 h) for *Selenastrum capricornutum* 12 µg/l. **Other aquatic spp.** EC_{50} (48 h) for crayfish >320 mg/l.

ENVIRONMENTAL FATE
Animals Rapidly oxidised by rat liver microsomal oxygenases to 2,6-diethylaniline (*Pestic. Biochem. Physiol.,* 1989, **33,** 16; *J. Agric. Food Chem.,* 1989, **37,** 1088; P. C. C. Feng et al., *Drug Metab. Dispos.,* 1990, **18,** 373). **Plants** Rapidly metabolised in plants to 2-chloro-2',6'-diethylacetanilide, with further degradation to the aniline derivative. **Soil/Environment** Rapidly degraded in soil by microbial action to 2-chloro-2',6'-diethylacetanilide, with further degradation to the aniline derivative; DT_{50} 1–30 d. Persists in soil for c. 6–10 w (J. Tiedie et al., *J. Agric. Food Chem.,* 1975, **23,** 77; J. K. Lee, *Hanguk Nanghwa Hakhoechi,* 1986, **29,** 182; *Rev. Environm. Contam. Toxicol.,* 1989, **110,** 110–114). In surface water, 55% degraded in 28 d.

alanycarb

Insecticide

oxime carbamate

NOMENCLATURE

Common name alanycarb (BSI, draft E-ISO)
IUPAC name ethyl (Z)-N-benzyl-N-[[methyl(1-methylthioethylideneamino-=
oxycarbonyl)amino]thio]-β-alaninate
Chemical Abstracts name (Z)-ethyl 3,7-dimethyl-6-oxo-9-(phenylmethyl)-5-oxa-=
2,8-dithia-4,7,9-triazadodec-3-en-12-oate
CAS RN *[83130–01–2]*   **Development codes** OK-135

PHYSICAL CHEMISTRY

Mol. wt. 399.5   **M.f.** $C_{17}H_{25}N_3O_4S_2$ **Form** Crystals; (tech., light yellow solid).
M.p. 46.8–47.2 °C   **V.p.** <0.0047 mPa (20 °C) K_{ow} logP = 3.43 (19–21 °C,
pH 7)   **S.g./density** 1.207 (20 °C)   **Solubility** In water 20 mg/l (20 °C). In
acetone, methanol, benzene, xylene, ethyl acetate, and dichloromethane, all >50%.
Stability Stable up to 100 °C; 0.2–1.0% decomposition at 54 °C for 30 days. Stable
in neutral and weakly basic media, but unstable in acidic and strongly basic media.
Degraded on glass plate under sunlight; DT_{50} 6 h.   **F.p.** 134 °C (closed cup)

COMMERCIALISATION

History Insecticide and nematicide reported by N. Umetsu (*J. Pesticide Science,*
1984, **9**, 169). Evaluated by Otsuka Chemical Co., Ltd.   **Patent** GB 2110206;
US 4444786; JP 8924144   **Manufacturers** Otsuka

APPLICATIONS

Biochemistry Cholinesterase inhibitor.   **Mode of action** Contact and stomach
action.   **Uses** A broad range insecticide applied mainly as a foliar spray. Effective
against Coleoptera, Hemiptera, Lepidoptera, and Thysanoptera in vines
(Aphididae, *Polychrosis botrana*), pome fruit (Aphididae), citrus (Aphididae,
Phyllocnistdidae, Diaspididae), tobacco (*Helicoverpa assulta*), and vegetables
(Aphididae, Lepidoptera).   **Formulation types** WP; EC.   **Mixtures** *(alanycarb +)*
fenpropathrin.   **Selected tradenames** 'Orion' (Otsuka)

ANALYSIS

Product analysis by hplc.   **Residues** of alanycarb in soil determined by hplc. Those
of alanycarb and its metabolite (methomyl) in crops by glc of a derivative with
FPD.

MAIN ENTRIES

MAMMALIAN TOXICOLOGY

EHC 64 (WHO, 1986; a review of carbamate insecticides in general). **Oral** Acute oral LD_{50} for male rats 440 mg/kg. **Skin and eye** Acute percutaneous LD_{50} for male rats >2000 mg/kg. Slightly irritating to eyes, not irritating to skin (rabbits). Non-sensitising to skin of guinea pigs. **Inhalation** LC_{50} (4 h) for rats >205 mg/m^3 air. **Other** Negative in the Ames test. Non-teratogenic in rats and rabbits. **Toxicity class** WHO (a.i.) II

ECOTOXICOLOGY

Birds LC_{50} (8 d) for bobwhite quail 3553, mallard ducks >5000 ppm. **Fish** LC_{50} (48 h) for carp 1.0 mg/l. **Bees** LD_{50} (topical) 0.8 µg/bee. **Daphnia** LC_{50} (3 h) >9.4 mg/l.

ENVIRONMENTAL FATE

Animals In rats, alanycarb is rapidly metabolised, either directly or via methomyl, to methomyl oxime, which is subsequently metabolised to unstable intermediates. They are converted to acetonitrile and CO_2, which are eliminated primarily via respiration and in the urine. **Plants** N-S bond cleavage is the initial metabolic step, giving rise to methomyl, which is further metabolised through methomyl oxime to acetic acid or acetonitrile, and eventually degraded to CO_2 (M. Kobayashi et al., IUPAC 7th Int. Congr. Pestic. Chem., 1990, **2**, 157). **Soil/Environment** DT_{50} in soil c. 1–2 d. Alanycarb is rapidly degraded to methomyl by chemical or microbial action. Methomyl formed is further degraded to methomyl oxime, which is eventually degraded to CO_2.

16 aldicarb   *Insecticide, acaricide, nematicide*

oxime carbamate

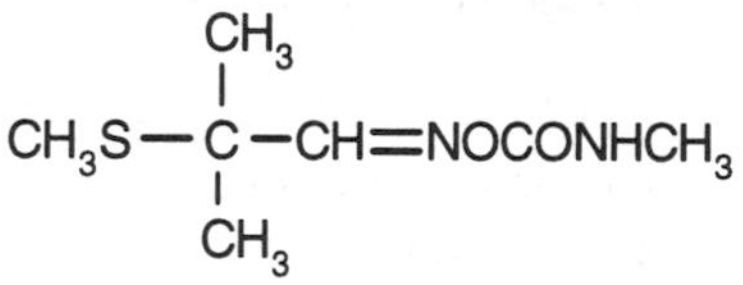

$$CH_3S-\underset{\underset{CH_3}{|}}{\overset{\overset{CH_3}{|}}{C}}-CH=NOCONHCH_3$$

NOMENCLATURE

Common name aldicarb (BSI, E-ISO, ANSI, ESA); aldicarbe ((*m*) F-ISO); no name (Germany)
IUPAC name 2-methyl-2-(methylthio)propionaldehyde O-methylcarbamoyloxime
Chemical Abstracts name 2-methyl-2-(methylthio)propanal O-[(methylamino)= carbonyl]oxime
CAS RN *[116–06–3]* **EEC no.** 204–123–2 **Development codes** UC 21 149
Official codes OMS 771; ENT 27 093; AI3–27 093

PHYSICAL CHEMISTRY

Mol. wt. 190.3   **M.f.** $C_7H_{14}N_2O_2S$ **Form** Colourless crystals.   **M.p.** 98–100 °C (tech.)   **V.p.** 13 mPa (20 °C)   **K$_{ow}$** logP = 0.053   **S.g./density** 1.195 (25 °C) **Solubility** In water 4.93 g/l (pH 7, 20 °C). Soluble in most organic solvents, e.g. in acetone 350, dichloromethane 300, benzene 150, xylene 50 (all in g/kg, 25 °C). Practically insoluble in heptane and in mineral oils. The sulfoxide has solubility >330 g/l in water.   **Stability** Stable in neutral, acidic, and weakly alkaline media. Hydrolysed by concentrated alkalis. Decomposes above 100 °C. Rapidly converted by oxidising agents to the sulfoxide, which is more slowly oxidised to the sulfone (aldoxycarb q.v.).   **F.p.** Non-flammable.

COMMERCIALISATION

History Insecticide reported by M. H. J. Weiden et al. (J. Econ. Entomol., 1965, **58**, 154). Introduced by the Union Carbide Corp. (now Rhône-Poulenc Agrochimie). **Patent** US 3217037   **Manufacturers** Rhône-Poulenc; Sanachem

APPLICATIONS

Biochemistry Cholinesterase inhibitor; specifically designed to resemble O-acetylcholine structurally (L. K. Payne & M. H. J. Weiden, J. Agric. Food Chem., 1966, **14**, 356). Metabolically activated to aldicarb sulfoxide.   **Mode of action** Systemic, with contact and stomach action. Absorbed rapidly through the roots, with translocation acropetally.   **Uses** Soil application for control of chewing and sucking insects (especially aphids, whitefly, leaf miners, and soil-dwelling insects), spider mites, and nematodes in glasshouse and outdoor ornamentals, sugar beet, fodder beet, strawberries, potatoes, onions, hops, vine nurseries, tree nurseries, peanuts, soya beans, citrus fruit, bananas, coffee, sorghum, pecans, cotton, sweet potatoes, sugar cane, and other crops.   **Formulation types** GR. **Mixtures** (aldicarb +) lindane; etridiazole + quintozene. **Compatibility** Incompatible with alkaline materials.   **Selected tradenames** 'Temik' (Rhône-Poulenc); 'Sanacarb' (Sanachem)

ANALYSIS

Product analysis by i.r. spectrophotometry (CIPAC Handbook, 1981, **1A**, 1094; AOAC Methods, 1995, 974.04).   **Residues** determined by rplc (ibid., 985.23), by glc with FPD after oxidation to aldoxycarb (R. T. Krause, J. Assoc. Off. Anal. Chem., 1980, **63**, 1114; R. R. Romine, Anal. Methods Pestic. Plant Growth Regul., 1973, **7**, 147; J. H. Smelt et al., loc. cit., pp. 279, 293). Methods based on colorimetry of a derivative were also used (D. P. Johnson et al., ibid., 1966, **49**, 399; D. F. Lee & J. A. Rougham, Analyst (London), 1971, **96**, 798). For methods in **drinking water**, see AOAC Methods, 1995, 991.06.

MAMMALIAN TOXICOLOGY

Reviews Pesticide residues in food – 1995, FAO Plant Production and Protection Paper, 133, 1996.   **IARC** 53   **EHC** 64 (WHO, 1986; a review of carbamate insecticides in general), 121 (WHO 1991).   **Oral** Acute oral LD_{50} for rats 0.93 mg/kg.   **Skin and eye** Acute percutaneous LD_{50} for male rabbits 20 mg/kg. **Inhalation** Rats were killed within 5 minutes by a dust concentration of 0.2 mg/l air.   **NOEL** In 2 y feeding trials, rats receiving 0.3 mg/kg daily were unaffected. **ADI** (JMPR) 0.003 mg/kg b.w. [1992].   **Toxicity class** WHO (a.i.) Ia; EPA (formulation) I   **EC risk** T+ (R27/28)

ECOTOXICOLOGY
Birds LC_{50} (8 d) for bobwhite quail 71 mg/kg diet.   **Fish** LC_{50} (96 h) for rainbow trout 0.88, bluegill sunfish 1.5 mg/l.   **Bees** Toxic to bees.

ENVIRONMENTAL FATE
Animals In rats, dogs and cows, absorbed rapidly and completely; >80% is excreted in the urine within 24 h, >96% within 3–4 d. Aldicarb is oxidised to the sulfoxide and sulfone, which undergo further metabolism. Metabolism of carbamate insecticides is reviewed (M. Cool & C. K. Jankowski in "*Insecticides*"). **Plants** In plants, the sulfur atom is oxidised to sulfoxide and sulfone groups. The highly soluble sulfoxide acts systemically on the plant, and is 10–20 times more active as a cholinesterase inhibitor than aldicarb itself. Further degradation leads to the formation of oximes, nitriles, amides, acids and alcohols which are present in the plant only in conjugated form.   **Soil/Environment** In soil, the sulfur atom is oxidised to sulfoxide and sulfone groups. Further degradation leads to the formation of oximes, nitriles, amides, acids and alcohols. Rapidly degraded in acid soils (pH >7.0), less so at pH ≤5.5. For degradation in soil and water, see H. Smelt *et al.*, *Pestic. Sci.*, 1995, **449**, 323, and refs. therein.

17 allethrin [(1*R*)- isomers]   *Insecticide*

pyrethroid

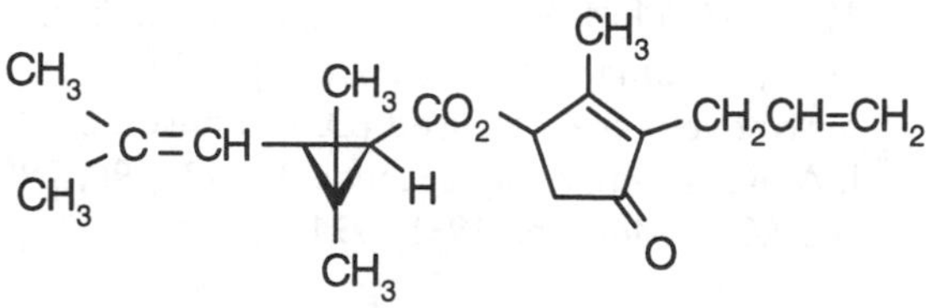

NOMENCLATURE
Common name allethrin (BSI, E-ISO, ESA, JMAF); alléthrine ((*f*) F-ISO); palléthrine ((*f*) France); no name (Germany) (all for (1*RS*)- isomers)

IUPAC name (RS)-3-allyl-2-methyl-4-oxocyclopent-2-enyl (1R,3R;1R,3S)-2,2-=
dimethyl-3-(2-methylprop-1-enyl)cyclopropanecarboxylate *Alt:* (RS)-3-allyl-2-=
methyl-4-oxocyclopent-2-enyl (+)-*cis-trans*-chrysanthemate *Roth:* (RS)-3-allyl-2-=
methyl-4-oxocyclopent-2-enyl (1R)-*cis-trans*-2,2-dimethyl-3-(2-methylprop-1-=
enyl)cyclopropanecarboxylate
Chemical Abstracts name 2-methyl-4-oxo-3-(2-propenyl)-2-cyclopenten-1-yl
2,2-dimethyl-3-(2-methyl-1-propenyl)cyclopropanecarboxylate
CAS RN *[584–79–2]* (unstated stereochemistry and also used for bioallethrin),
formerly *[137–98–4]* **EEC no.** 209–542–4 [for (1RS)- isomers]
Official codes OMS 468; ENT 17 510 [for (1RS)- isomers]

PHYSICAL CHEMISTRY
Composition d-Allethrin is ≥95% (1R)- isomers and ≥75% *trans* isomers; see also
bioallethrin. **Mol. wt.** 302.4 **M.f.** $C_{19}H_{26}O_3$ **Form** Tech. grade is a pale yellow
liquid. **B.p.** 281.5 °C/760 mmHg **V.p.** 0.16 mPa (21 °C) K_{ow} logP = 4.96
(room temperature) **S.g./density** 1.01 (20 °C) **Solubility** Practically insoluble in
water. In hexane 0.655 g/ml, methanol 72.0 ml/ml (both 20 °C). **Stability**
Decomposed by u.v. light. Hydrolysed in alkaline media. **F.p.** 87 °C

COMMERCIALISATION
History Insecticide reported by M. S. Schechter *et al.* (*J. Am. Chem. Soc.*, 1949, **71,**
3165). Development and properties reviewed by W. Barthel (*Wld. Rev. Pest
Control*, 1967, **6,** 59). Introduced by Sumitomo Chemical Co., Ltd.
Manufacturers Sumitomo

APPLICATIONS
Mode of action Non-systemic insecticide with contact, stomach, and respiratory
action. Gives rapid knockdown. Paralyses insects before killing them.
Uses Control of flies, mosquitoes, ants, and other household and public health
insect pests. Often used in combination with piperonyl butoxide or other
synergists, for control of chewing and sucking insects on ornamentals, vegetables,
and other crops; for household and public health insect control; for insect control
in animal houses; and as an animal ectoparasiticide. **Formulation types** AE; EC;
DP; WP; Coil. **Mixtures** *(allethrin [(1R)- isomers] +)* piperonyl butoxide.
Compatibility Incompatible with alkaline materials. **Selected tradenames** 'Pynamin
Forte' (Sumitomo)

ANALYSIS
Product analysis of pyrethroids reviewed by E. Papadopoulou-Mourkidou in *Comp.
Analyt. Profiles.* By glc (*CIPAC Handbook*, 1980, **1A,** 1097) or by titration of a
derivative (*AOAC Methods*, 1995, 953.05*). The proportions of stereoisomers
determined by glc of suitable derivatives (A. Murano, *Agric. Biol. Chem.*, 1972, **36,**
2203; F. E. Rickett, *Analyst (London)*, 1973, **98,** 687; A. Horiba *et al.*, *Agric. Biol.
Chem.*, 1977, **41,** 2003; 1978, **42,** 671) or by nmr (F. E. Rickett & P. B. Henry,

Analyst (London), 1974, **99**, 330). **Residues** determined by glc (*Anal. Methods Pestic. Plant Growth Regul.*, 1972, **6**, 283; J. Sherma, *ibid.*, 1976, **8**, 117). Details from McLaughlin Gormley King Co.

MAMMALIAN TOXICOLOGY
Reviews See A. J. Gray & D. M. Soderlund, Chapt. 5 in "*Insecticides*". See also *Evaluation of the toxicity of pesticide residues in food*, FAO Meeting Report, No. PL/1965/10; WHO/Food Add./26.65, 1965. *Evaluation of the toxicity of pesticide residues in food*. FAO Meeting Report, No. PL/1965/10/1; WHO/Food Add./27.65, 1965. **EHC** 87 (WHO, 1989). **Oral** Acute oral LD_{50} for male rats 2150, female rats 900 mg/kg. **Skin and eye** Acute percutaneous LD_{50} for male rabbits 2660, female rabbits 4390 mg/kg. **Inhalation** LC_{50} for rats >3875 mg/m^3 air. **ADI** (JMPR) No ADI [1965]. **Toxicity class** WHO (a.i.) III; EPA (formulation) III **EC risk** Xn (R22)

ECOTOXICOLOGY
Birds LC_{50} (8 d) for mallard ducks and bobwhite quail 5620 mg 'Pynamin Forte'/kg diet.

ENVIRONMENTAL FATE
Animals In mammals, following oral administration, one of the two terminal methyl groups of the chrysanthemic acid moiety is oxidised in the liver to an alcohol group, and further to a carboxyl group.

18   alloxydim   *Herbicide*

cyclohexanedione oxime

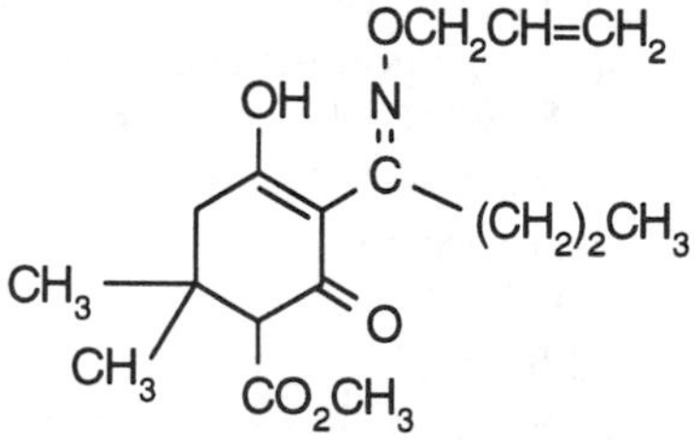

NOMENCLATURE
alloxydim
Common name alloxydim (BSI, draft E-ISO, (*m*) draft F-ISO)
IUPAC name methyl (*E*)-(*RS*)-3-[1-(allyloxyimino)butyl]-4-hydroxy-6,6-dimethyl-= 2-oxocyclohex-3-enecarboxylate

Chemical Abstracts name methyl 4-hydroxy-6,6-dimethyl-2-oxo-3-[1-[(2-= propenyloxy)imino]butyl]-3-cyclohexene-1-carboxylate (actual tautomer); methyl 2,2-dimethyl-4,6-dioxo-5-[1-[(2-propenyloxy)amino]butylidene]= cyclohexanecarboxylate (*C.A.* preferred tautomer)
CAS RN *[55634–91–8]* (*C.A.* preferred tautomer) **EEC no.** 259–733–1
Development codes NP-48 (Nippon Soda)

alloxydim-sodium
Common name alloxydim-sodium (BSI, E-ISO, *(m)* F-ISO); alloxydim (JMAF)
CAS RN *[66003–55–2]* **Development codes** NP-48 Na (Nippon Soda);
BAS 90 210H (BASF)

PHYSICAL CHEMISTRY
alloxydim
Mol. wt. 323.4 **M.f.** $C_{17}H_{25}NO_5$

alloxydim-sodium
Mol. wt. 345.4 **M.f.** $C_{17}H_{24}NNaO_5$ **Form** Colourless crystals. **M.p.** 185.5 °C (decomp.) **V.p.** <0.133 mPa (25 °C) K_{ow} logP = –0.20 **Solubility** In water >2 kg/l (30 °C). In dimethylformamide 1000, methanol 619, ethanol 50, methyl ethyl ketone 15, acetone 14, cyclohexanone 3, ethyl acetate 2, xylene 0.02 (all in g/kg, 30 °C). **Stability** Stable when dry, but very hygroscopic.

COMMERCIALISATION
History Herbicide reported by Y. Horono *et al.* (*Meet. Pestic. Sci. Soc. Jpn., 1st,* 1976) and A. Formigoni & Y. Horono (*Meded. Fac. Landbouwwet. Rijksuniv. Gent.,* 1977, **42,** 1597). Its sodium salt introduced by Nippon Soda Co., Ltd, and later by BASF AG and Rhône-Poulenc Agriculture. **Patent** JP 7795636 to Nippon Soda
Manufacturers Nippon Soda

APPLICATIONS
Biochemistry Mitosis inhibitor. Acetyl CoA carboxylase (fatty acid synthesis) inhibitor. **Mode of action** Selective systemic herbicide, absorbed predominantly by the leaves, and, to a lesser extent, by the roots.

alloxydim-sodium
Uses It is used post-emergence against grass weeds and volunteer cereals in sugar beet, vegetables and broad-leaved crops at 0.5–1.0 kg a.i./ha. Split applications with herbicides effective against broad-leaved weeds are recommended to increase the range of herbicidal activity. **Phytotoxicity** Non-phytotoxic to broad-leaved crops. **Formulation types** SP. **Selected tradenames** 'Clout' (Rhône-Poulenc); 'Kusagard' (Nippon Soda); 'Fervin' (AgrEvo)

ANALYSIS

Product analysis by u.v. spectrometry.   **Residues** determined by hplc (N. Watanabe et al., Mizu Shori Gijutsu, 1987, **28,** 729).

MAMMALIAN TOXICOLOGY

alloxydim-sodium

Oral Acute oral LD_{50} for male rats 2322, female rats 2260, mice 3000–4600 mg/kg.   **Skin and eye** Acute percutaneous LD_{50} for rats >5000, rabbits >2000 mg/kg.   **Inhalation** LC_{50} (4 h) for rats >4.3 mg/l air.   **NOEL** In 2 y feeding trials, rats and mice receiving 100 mg/kg diet showed no ill-effects.   **Other** Acute i.p. LD_{50} for rats 1700 mg/kg.   **Toxicity class** WHO (a.i.) III (Table 5); EPA (formulation) III

ECOTOXICOLOGY

alloxydim-sodium

Birds Acute oral LD_{50} for Japanese quail 2970 mg/kg.   **Fish** LC_{50} (48 h) for carp 3500 mg a.i./l. LC_{50} (96 h) for carp 2600, trout 2000 mg/l.   **Bees** Not toxic to bees.   **Other aquatic spp.** LC_{50} (3 h) for *Moina macrocopa* >4000 ppm.

ENVIRONMENTAL FATE

Soil/Environment In soil DT_{50} 2–10 d.

alpha-cypermethrin

See entry 184.

19   *Amblyseius* spp.   *Biological agent*

Acari

Thrips-consuming mite.

NOMENCLATURE

Amblyseius californicus
Scientific name *Amblyseius californicus*

Amblyseius cucumeris
Scientific name *Amblyseius cucumeris* (Oudemans)   **Strain** Diapause resistant
Other names *Typhlodromus thripsi* MacGill; *Amblyseius mckenziei* Sch. & Pr.;
Neoseiulus cucumeris

Amblyseius degenerans
Scientific name *Amblyseius degenerans* (Oudemans)   **Other names** *Ipheseius degenerans*

PROPERTIES
Amblyseius cucumeris
Stability Shelf life at 10 °C and complete darkness varies from 10 days down to 2 days, depending on formulation.

Amblyseius degenerans
Stability Shelf life at 10 °C and complete darkness, 2 days maximum.

COMMERCIALISATION
Amblyseius californicus
Manufacturers Arbico; Biobest; Koppert; Novartis BCM; Rincon-Vitova

Amblyseius cucumeris
Production A. *cucumeris* can be reared without using the pest target species. In commercial productions, an acarid mite is used as prey. This non-pest prey is fed on bran material.   **History** *Amblyseius cucumeris* was recognised as a predator of *Thrips tabaci* Lind, by MacGill in 1939. In 1973 it was re-discovered and associated with the Western Flower Thrips, *Frankliniella occidentalis*, Pergande 1895 (WFT), in Dutch glasshouses (Woetz 1973) and was then developed as a predator to control WFT (Ramakers 1978). First used on a commercial basis in the early 1980's, it is now the primary agent employed against this serious world-wide pest.
Manufacturers Applied Bio-nomics; Arbico; BCP; Biobest; Integrated Pest Management; Koppert; Nature's Alternative; Neudorff; Rincon-Vitova; Sautter & Stepper

Amblyseius degenerans
Production A. *degenerans* can be reared without using the target pest species. It is not amenable to high density rearings of the type used for A. *cucumeris*.
History A predator of *Tetranychus*. Tested as a predator of the Western Flower Thrips, *Frankliniella occidentalis* by van Houben et al. in 1992, due to its tolerance of low humidity.   **Manufacturers** Applied Bio-nomics; Biobest; Koppert; Novartis BCM

APPLICATIONS
Amblyseius spp.
Selected tradenames 'Spical' (Koppert)

Amblyseius californicus
Uses For control of spider mites.

Amblyseius cucumeris
Mode of action Because of its small size, all stages of A. *cucumeris* feed only upon thrips first instar larvae. The thrips egg is laid beneath the plant integument and is therefore not available. At 25 °C, development time from egg to adult on thrips larvae is 12 days. It will also complete its life cycle on various species of pollen grains when, at 25 °C, the development period averages 8.5 days. Predation occurs between 15 and 30 °C and 70 to 85 per cent relative humidity. Maximum performance relates to a temperature between 20–27 °C. It is an extremely mobile predator, actively searching for its prey. **Uses** A. *cucumeris* is used to control *Frankliniella occidentalis, Thrips tabaci* and *Thrips fuscipennis* on a wide range of vegetable crops and ornamentals under protected cultivation. Essentially the predator should be introduced before the pest invades. **Phytotoxicity** Non-phytotoxic and non-phytopathogenic. **Formulation types** Control Release System (CRS); Mixed with bran and vermiculite. **Compatibility** A. *cucumeris* is compatible for use with all other horticultural biological agents. Data on pesticides compatibility are provided by production companies. **Selected tradenames** 'Ambly-line cu CRS' (Novartis BCM); 'Thripex' (Koppert)

Amblyseius degenerans
Mode of action A. *degenerans* feeds on first instar and small second instar larvae of thrips, but not large larvae or adults. Predation occurs between 15–30 °C, and 60–85% r.h. **Uses** Used for control of the Western Flower Thrips, *Frankliniella occidentalis*, principally in peppers. It does not thrive on crops without excess pollen. **Selected tradenames** 'Ambly-line d' (Novartis BCM); 'Thripans' (Koppert)

MAMMALIAN TOXICOLOGY
Amblyseius spp.
Allergic response, such as rhinitis or skin irritation or both, may occur in production staff, more commonly due to the presence of the prey mite. Cases of allergy in workers applying the product to the crop have not been reported.

ametryn

1,3,5-triazine

$$CH_3S \diagdown \diagup N \diagdown \diagup NHCH_2CH_3$$
$$N \diagdown \diagup N$$
$$NHCH(CH_3)_2$$

NOMENCLATURE
Common name ametryn (BSI (since 1984), E-ISO, ANSI, WSSA, JMAF); ametryne (BSI (before 1984), (*f*) F-ISO)
IUPAC name N^2-ethyl-N^4-isopropyl-6-methylthio-1,3,5-triazine-2,4-diamine
Chemical Abstracts name N-ethyl-N'-(1-methylethyl)-6-(methylthio)-1,3,5-= triazine-2,4-diamine
CAS RN *[834–12–8]* **EEC no.** 212–634–7 **Development codes** G 34 162

PHYSICAL CHEMISTRY
Composition 96% pure. **Mol. wt.** 227.3 **M.f.** $C_9H_{17}N_5S$ **Form** White powder.
M.p. 84–86 °C **B.p.** 337 °C/98.6 kPa **V.p.** 0.365 mPa (25 °C) (OECD 104)
K_{ow} logP = 2.63 (25 °C) **Henry** 4.1×10^{-4} Pa m^3 mol^{-1} (calc.) **S.g./density** 1.19
(20 °C) **Solubility** In water 200 mg/l (20 °C). In acetone 610, methanol 510,
toluene 470, *n*-octanol 220, hexane 12 (all in g/l, 25 °C). **Stability** Stable in
neutral, weakly acidic, and weakly alkaline media. Hydrolysed by strong acids
(pH 1) and alkalis (pH 13) to the herbicidally-inactive 6-hydroxy derivative. Slowly
decomposed by u.v. light. **pKa** 4.1, weak base

COMMERCIALISATION
History Herbicide reported by H. Gysin & E. Knüsli (*Adv. Pest Control Res.*, 1960, **3,**
289). Introduced by J. R. Geigy S.A. (now Novartis Crop Protection AG).
Patent GB 814948; CH 337019 **Manufacturers** Crystal; Makhteshim-Agan;
Novartis; Oxon Italia

APPLICATIONS
Biochemistry Photosynthetic electron transport inhibitor.
Mode of action Selective systemic herbicide, absorbed by the leaves and roots,
with translocation acropetally in the xylem, and accumulation in the apical
meristems. **Uses** Pre- and post-emergence control of most annual grasses and
broad-leaved weeds in pineapples, sugar cane, bananas, citrus fruit, maize, cassava,
coffee, tea, sisal, cocoa, oil palms, and on non-crop land. Application rates are in
the range 2–4 kg/ha except when used as a directed spray on maize. Also used as
a potato haulm desiccant. **Phytotoxicity** Some sugar cane varieties show
temporary chlorosis and scorching of lower leaves. **Formulation types** FW; WP;
SC; EC; WG. **Mixtures** (*ametryn +*) amitrole; atrazine; 2,4-D; metolachlor;
prometryn; terbuthylazine; amitrole + atrazine; amitrole + 2,4-D.

Selected tradenames 'Evik' (USA) (Novartis); 'Gesapax' (Novartis); 'Amesip' (Sipcam); 'Ametrex' (Makhteshim-Agan); 'Mebatryne' (Rhône-Poulenc)

ANALYSIS
Product analysis by glc with FID (*CIPAC Handbook*, 1980, **1A**, 1102; FAO Specification CP/61; *AOAC Methods*, 1995, 971.08). **Residues** determined by glc (K. Ramsteiner et al., *J. Assoc. Off. Anal. Chem.*, 1974, **57**, 192; E. Knüsli, *Anal. Methods Pestic., Plant Growth Regul. Food Addit.*, 1964, **4**, 13; B. G. Tweedy & R. A. Kahrs, *Anal. Methods Pestic. Plant Growth Regul.*, 1978, **10**, 493). In **drinking water**, by gc with FID; *AOAC Methods*, 1995, 991.07.

MAMMALIAN TOXICOLOGY
Reviews *J Pest. Sci.*, **18** (4) 1993 (in Japanese). **Oral** Acute oral LD_{50} for rats 1160, mice 965 mg tech./kg. **Skin and eye** Acute percutaneous LD_{50} for rabbits >2020, rats >3100 mg/kg. Not a skin or eye irritant (rabbits). Not a skin sensitiser (guinea pigs). **Inhalation** LC_{50} (4 h) for rats >5170 mg/m^3 air. **NOEL** (2 y) for rats 50, mice 10 ppm; (1 y) for dogs 200 ppm. **ADI** 0.015 mg/kg. **Toxicity class** WHO (a.i.) III; EPA (formulation) III **EC risk** Xn (R22)

ECOTOXICOLOGY
Birds LC_{50} (5 d) for bobwhite quail and mallard ducks >5620 ppm. **Fish** LC_{50} (96 h) for rainbow trout 5, bluegill sunfish 19, channel catfish 25 mg/l. **Bees** Low toxicity to bees; LD_{50} (oral) >100 µg/bee. **Worms** LC_{50} (14 d) for earthworms 166 mg/kg soil. **Daphnia** LC_{50} (96 h) 28 mg/l. **Algae** EC_{50} (7 d) for *Selenastrum capricornutum* 0.0036 mg/l. **Other aquatic spp.** LC_{50} (96 h) for mysid shrimp (*Mysidopsis bahia*) 2.3 mg/l.

ENVIRONMENTAL FATE
Animals Irrespective of the dose or the dosing regime, most is excreted within 3 to 4 days. Conjugation with glutathione and dealkylation are the main metabolic pathways. **Plants** Metabolised by tolerant plants and, to a lesser extent, by sensitive plants, to non-toxic substances by replacement of the methylthio group by a hydroxy group, and by dealkylation of the amino groups.
Soil/Environment Loss from soil is principally by microbial degradation (H. O. Esser et al., *Herbicides: Chemistry, Degradation and Mode of Action*, 1975, **1**, 129). Median DT_{50} in soil 51 d (11–120 d). K_{oc} 300; however column leaching studies indicate ametryn does not leach significantly. Degradation in aquatic systems is caused by microbial processes, with photolysis also contributing. Adsorption to the sediment is the most efficient mechanism of elimination of ametryn from water.

amidosulfuron   *Herbicide*

---

sulfonylurea

$$CH_3SO_2-\underset{\underset{CH_3}{|}}{N}-SO_2NHCONH-\!\!\!\!\!\!\overset{\displaystyle OCH_3}{\underset{\displaystyle OCH_3}{\langle\!\!\!\langle\,}}$$

NOMENCLATURE
Common name amidosulfuron (BSI, pa E-ISO)
IUPAC name 1-(4,6-dimethoxypyrimidin-2-yl)-3-mesyl(methyl)sulfamoylurea
Chemical Abstracts name *N*-[[[[(4,6-dimethoxy-2-pyrimidinyl)amino]carbonyl]=
amino]sulfonyl]-*N*-methylmethanesulfonamide   **CAS RN** *[120923–37–7]*
Development codes AE F075032; Hoe 075032

PHYSICAL CHEMISTRY
Mol. wt. 369.4 **M.f.** $C_9H_{15}N_5O_7S_2$ **Form** White, crystalline powder.
M.p. 160–163 °C   **V.p.** 2.2×10^{-2} mPa (25 °C) K_{ow} logP = 1.63 (pH 2, 20 °C)
Henry 5.34×10^{-4} Pa m^3 mol^{-1} (20 °C)   **S.g./density** 1.5   **Solubility** In water, 3.3
(pH 3), 9 (pH 5.8), 13 500 (pH 10) mg/l (20 °C). In isopropanol 0.099, methanol
0.872, acetone 8.1 (all in g/l, 20 °C).   **Stability** Stable for 2 years at 25±5 °C in
unopened original containers. Abiotic hydrolysis, DT_{50} 33.9 d (pH 5), 365 d
(pH 7), 365 d (pH 9) (25 °C).   **pKa** 3.58

COMMERCIALISATION
Manufacturers AgrEvo

APPLICATIONS
Biochemistry Branched chain amino acid synthesis (acetolactate synthase or ALS)
inhibitor. Acts by inhibiting biosynthesis of the essential amino acids valine and
isoleucine, hence stopping cell division and plant growth. Selectivity derives from
rapid metabolism in the crop. Metabolic basis of selectivity in sulfonylureas
reviewed (M. K. Koeppe & H. M. Brown, *Agro-Food-Industry*, **6**, 9–14 (1995)).
Mode of action Selective, systemic herbicide, absorbed by the leaves and roots
and translocated throughout the plant. Plant growth is inhibited followed by the
development of chlorotic patches which spread acropetally and then basipetally.
Uses Post-emergence control of a wide range of broad-leaved weeds, especially
cleavers, in winter wheat, durum wheat, barley, rye, triticale and oats, at
30–60 g/ha; controls docks in pasture.   **Formulation types** WG.
Mixtures *(amidosulfuron +)* isoproturon.   **Selected tradenames** 'Gratil' (AgrEvo)

ANALYSIS
By reverse phase hplc. Methods for sulfonylurea residues in crops, soil and water

MAIN ENTRIES

reviewed (A. C. Barefoot *et al., Proc. Br. Crop Prot. Conf. – Weeds*, 1995, **2**, 707). Details available from AgrEvo.

MAMMALIAN TOXICOLOGY
Oral Acute oral LD_{50} for rats and mice ≥5000 mg/kg. **Skin and eye** Acute percutaneous LD_{50} for rats >5000 mg/kg. **Inhalation** LC_{50} (4 h) for rats >1.8 mg/l air. **NOEL** (2 y) for male rats 400 ppm diet (19.45 mg/kg b.w. daily). **ADI** 0.2 mg/kg b.w. **Other** Non-teratogenic and non-mutagenic.

ECOTOXICOLOGY
Birds LD_{50} for mallard ducks and bobwhite quail >2000 mg/kg. **Fish** LC_{50} (96 h) for rainbow trout >320 mg/l. **Bees** Acute oral LD_{50} >1000 µg/bee. **Worms** LC_{50} (14 d) for *Eisenia foetida* >1000 mg/kg. **Daphnia** LC_{50} (48 h) 36 mg/l. **Algae** E_bC_{50} (72 h) for *Scenedesmus subspicatus* 47 mg/l.

ENVIRONMENTAL FATE
Animals The major metabolic pathway in the rat is *O*-demethylation.
Soil/Environment In soil, amidosulfuron is degraded microbially, DT_{50} 3–29 d. Degradation is independent of the pH value, but dependent on the biological activity in the soil.

22 amitraz   *Acaricide, insecticide*

amidine

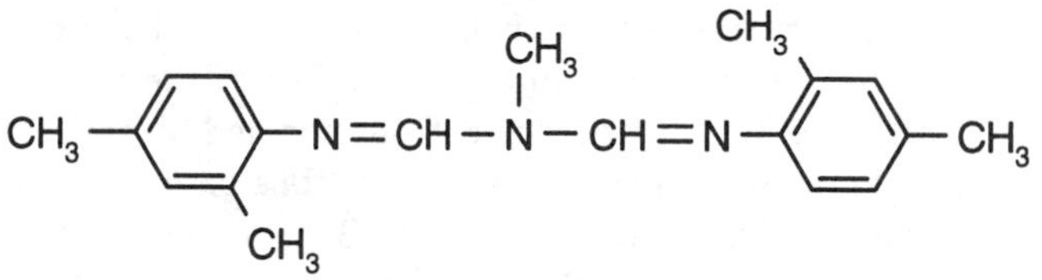

NOMENCLATURE
Common name amitraz (BSI, E-ISO, ANSI, ESA, BAN, JMAF); amitraze ((m) F-ISO)
IUPAC name *N*-methylbis(2,4-xylyliminomethyl)amine
Chemical Abstracts name *N'*-(2,4-dimethylphenyl)-*N*-[[(2,4-dimethylphenyl)= imino]methyl]-*N*-methylmethanimidamide
Other names *N,N*-bis(2,4-xylyliminomethyl)methylamine; *N*-methyl-*N'*-2,4-xylyl-= *N*-(*N*-2,4-xylylformimidoyl)-formamidine; 1,5-di-(2,4-dimethylphenyl)-3-methyl-= 1,3,5-triazapenta-1,4-diene **CAS RN** *[33089–61–1]* **EEC no.** 251–375–4
Development codes BTS 27 419 **Official codes** OMS 1820; ENT 27 967

PHYSICAL CHEMISTRY

Mol. wt. 293.4   **M.f.** $C_{19}H_{23}N_3$ **Form** White/pale yellow crystalline solid.
M.p. 86–88 °C   **V.p.** 0.34 mPa (25 °C) using dapeyron-clausius analysis
K_{ow} logP = 5.5 (25 °C, pH 5.8)   **Henry** 1.0 Pa m³ mol⁻¹ (measured)
S.g./density 1.128 (20 °C)   **Solubility** In water <0 .1 mg/l (20 °C). Soluble in
most organic solvents; in acetone, toluene, xylene >300 g/l.   **Stability** Hydrolysis
DT_{50} (25 °C) 2.1 h (pH 5), 22.1 h (pH 7), 25.5 h (pH 9). U.V. light appears to
have little effect on stability.   **pKa** 4.2, weak base

COMMERCIALISATION

History Acaricide reported by B. H. Palmer *et al.* (*Proc. Int. Congr. Acarol. 3rd,*
1971, p.687) for veterinary use and by D. M. Weighton *et al.* (*Meded. Fac.*
Landbouwwet. Rijksuniv. Gent, 1972, **37,** 765) for crop use. Introduced by The
Boots Co. Ltd.; (now marketed by AgrEvo GmbH for crop protection and by
Hoechst AG for veterinary use).   **Patent** GB 1327935 (Boots)
Manufacturers AgrEvo; Defensa; Rotam; Wujin

APPLICATIONS

Biochemistry Mode of action probably involves an interaction with octopamine
receptors in the tick nervous system, causing an increase in nervous activity.
Mode of action Non-systemic, with contact and respiratory action. Expellent
action causes ticks to withdraw mouthparts rapidly and fall off the host animal.
Uses Control of all stages of tetranychid and eriophyid mites, pear suckers, scale
insects, mealybugs, whitefly, aphids, and eggs and first instar larvae of Lepidoptera
on pome fruit, citrus fruit, cotton, stone fruit, bush fruit, strawberries, hops,
cucurbits, aubergines, capsicums, tomatoes, ornamentals, and some other crops.
Also used as an animal ectoparasiticide to control ticks, mites and lice on cattle,
dogs, goats, pigs and sheep.   **Phytotoxicity** At high temperatures, young capsicums
and pears may be injured.   **Formulation types** EC; WP; PO.   **Mixtures** *(amitraz +)*
bifenthrin; cypermethrin.   **Compatibility** Incompatible with alkaline materials,
parathion, and others.   **Selected tradenames** 'Mitac' (for crop protection)
(AgrEvo, NOR-AM); 'Taktic' (for veterinary) (Hoechst, NOR-AM); 'Byebye'
(Agriphar); 'Edrizar' (Siapa); 'Parsec' (Defensa); 'Racet' (Rotam); 'Sender'
(Sanonda)

ANALYSIS

Product analysis by glc (*CIPAC Handbook,* 1995, **G,** 5–10).   **Residue** analysis also
by glc. Details available from AgrEvo.

MAMMALIAN TOXICOLOGY

Reviews *Pesticide residues in food – 1990.* FAO Plant Production and Protection
Paper, 102, 1990. See also FAO/WHO reviews of 1980, 1984.   **Oral** Acute oral
LD_{50} for rats 650, mice >1600 mg/kg.   **Skin and eye** Acute percutaneous LD_{50}
for rabbits >200, rats >1600 mg/kg.   **Inhalation** LC_{50} (6 h) for rats 65 mg/l air.

NOEL In 2 y feeding trials, no adverse effect observed in rats receiving 50–200 ppm diet, or in dogs dosed 0.25 mg/kg daily. Human NOEL >0.125 mg/kg daily.   **ADI** (JMPR) 0.003 mg/kg b.w. [1990].   **Toxicity class** WHO (a.i.) III; EPA (formulation) III   **EC risk** Xn (R22)

ECOTOXICOLOGY
Birds LD_{50} for bobwhite quail 788 mg/kg. LC_{50} (8 d) for mallard ducks 7000, Japanese quail 1800 mg/kg.   **Fish** LC_{50} (96 h) for rainbow trout 0.74, bluegill sunfish 0.45 mg/l. Due to rapid hydrolysis, it is unlikely that this toxicity will be expressed in natural aquatic systems.   **Bees** Low toxicity to bees and predatory insects. LD_{50} (contact) 50 µg/bee (formulation).   **Worms** LC_{50} (14 d) for earthworms >1000 mg tech./kg.   **Daphnia** LC_{50} (48 h) 0.035 mg/l. Due to rapid hydrolysis, it is unlikely that this toxicity will be expressed in natural aquatic systems.   **Algae** EC_{50} for *Selenastrum capricornutum* >12 mg/l.

ENVIRONMENTAL FATE
Animals Rapid breakdown leading to excretion as a conjugate of 4-amino-3-=methylbenzoic acid and to a lesser extent to N-(2,4-dimethylphenyl)-N'-=methylformamidine.   **Plants** Rapidly degraded, mainly to N-(2,4-dimethylphenyl)-=N'-methylformamidine and to a smaller extent to 2,4-dimethyl-formanilide.
Soil/Environment Rapidly broken down in soil under aerobic conditions (*J. Appl. Bacteriol.*, 1977, **42**, 187; *ibid.* 1978, **44**, 383); DT_{50} in soil <1 d. Degradation occurs more rapidly in acid than in neutral or alkaline soils. Very strongly adsorbed to soil; K_{oc} c. 1000–2000.

23   amitrole   *Herbicide*

triazole

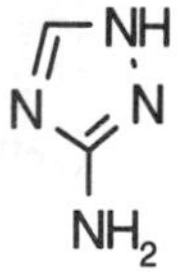

NOMENCLATURE
Common name amitrole (from 1984: BSI; E-ISO, (*m*) F-ISO, ANSI, WSSA); ATA (JMAF); aminotriazole (in lieu of a common name: (*m*) France; until 1984: BSI and USSR)
IUPAC name 1H-1,2,4-triazol-3-ylamine
Chemical Abstracts name 1H-1,2,4-triazol-3-amine
CAS RN *[61–82–5]*   **EEC no.** 200–521–5   **Official codes** ENT 25 445

PHYSICAL CHEMISTRY

Mol. wt. 84.1 **M.f.** $C_2H_4N_4$ **Form** Colourless crystals. **M.p.** 157–159 °C; (tech., 150–153 °C) **V.p.** 3.3×10^{-5} mPa (20 °C) **K_{ow}** logP = –0.97 **S.g./density** 1.138 (20 °C) **Solubility** In water 280 g/l (23 °C), 530 g/l (53 °C). In ethanol 26 g/100 g (75 °C). Moderately soluble in methanol, chloroform, dichloromethane, and acetonitrile. Sparingly soluble in ethyl acetate. Almost insoluble in diethyl ether, acetone, and non-polar solvents. **Stability** Stable in neutral, acidic and alkaline media. Powerful chelating agent. **pKa** pKa_1 4.2, pKa_2 10.7 (K. Chamberlain et al., Pestic Sci., **47**, 265 (1996) and refs. therein)

COMMERCIALISATION

History Herbicidal properties reported by R. Behrens (Proc. North Cent. Weed Control Conf., 1953, p. 61). Introduced by Amchem Products, Inc. (now Rhône-Poulenc Agrochimie). **Patent** US 2670282 **Manufacturers** Elf Atochem/CFPI; Makhteshim-Agan

APPLICATIONS

Biochemistry Inhibits chlorophyll formation and regrowth from buds. **Mode of action** Non-selective systemic herbicide, absorbed by the leaves and roots, with translocation in both the phloem and the xylem. **Uses** Total control of annual and perennial grasses and broad-leaved weeds around established fruit trees (including citrus) and bushes in orchards, in vineyards and olive groves, under ornamental trees and shrubs, in tree nurseries, in cereal stubble, and on fallow land and other non-crop areas such as paths, railway tracks, industrial areas, etc. Used for total herbicidal control pre-sowing or pre-planting of many crops. Also used for control of aquatic weeds in marshes, drainage ditches, etc. General reviews have been published (E. Kröller, Residue Rev., 1966, **12**, 163; M. C. Carter, Herbicides: Chemistry, Degradation and Mode of Action, 1975, **1**, p. 377).
Formulation types SP; SL; WP. **Mixtures** (amitrole +) ammonium thiocyanate; diuron; atrazine; simazine; atrazine + 2,4-D; atrazine + diuron; atrazine + simazine; bromacil + diuron; 2,4-D + diuron + simazine; diuron + simazine; bromacil; terbuthylazine; and many others. **Selected tradenames** 'Weedazol' (Rhône-Poulenc, CFPI); 'Azolan' (Makhteshim-Agan); 'Oriflam' (mixture) (CFPI); 'Superzol' (Productos OSA); 'Trinovin' (mixture) (Efthymiadis)

ANALYSIS

Product analysis by glc with FID (A. Jacques, Anal. Methods Pestic. Plant Growth Regul., 1984, **13**, 191), by acid-base titration (AOAC Methods, 1995, 967.06) or by formation of a silver complex during titration with silver salts (CIPAC Handbook, 1983, **1B**, 1720); other components of mixtures being determined by the appropriate methods (loc. cit.). **Residues** in soils determined by colorimetry (R. A. Herrett & A. J. Linck, J. Agric. Food Chem., 1961, **9**, 466; J. Burke & R. W. Storherr, J. Assoc. Off. Agric. Chem., 1961, **44**, 196; G. L. Sutherland, Anal. Methods Pestic., Plant Growth Regul. Food Addit., 1964, **4**, 17; Pestic. Anal. Man., 1979, II; H. Løkke, J. Chromatogr., 1980, **200**, 234).

MAIN ENTRIES

MAMMALIAN TOXICOLOGY

Reviews *Pesticide residues in food – 1993*, FAO Plant Production and Protection Paper, 122, 1993. *Pesticide residues in food – 1993 evaluations, Part II – Toxicology.* WHO, WHO/PCS/94.4, 1994. **IARC** 7, 41 **EHC** 158 (WHO, 1994). **Oral** Acute oral LD_{50} for rats 1100–24 600 mg/kg. **Skin and eye** Acute percutaneous LD_{50} for rats >2500, rabbits >10 000 mg/kg. Not a skin irritant; mild eye irritant (rabbits). Not a skin sensitiser. **NOEL** (24 mo) for rats 10 ppm; (18 mo) for mice 10 ppm. In 68 w feeding trials, rats receiving 50 mg/kg diet suffered no effect on growth or food intake, but male rats developed an enlarged thyroid after 13 w. But see T. H. Jukes & C. B. Schaeffer, *Science*, 1960, **132**, 296. **ADI** (JMPR) 0.0005 mg/kg b.w. [1993] (temporary). **Toxicity class** WHO (a.i.) III (Table 5); EPA (formulation) IV **EC risk** (R40); Xn (R48/22)

ECOTOXICOLOGY

Birds LD_{50} for bobwhite quail >2150 mg/kg. LC_{50} for bobwhite quail and mallard duck >5000 ppm diet. **Fish** LC_{50} (96 h) for rainbow trout and bluegill sunfish >1000 mg/l. **Bees** Non-toxic to bees; LD_{50} >10 µg/bee. **Other beneficial spp.** No adverse effect on soil microflora at 10× normal application rate. **Daphnia** LC_{50} (48 h) >10 mg/l. **Algae** NOEC for *Scenedesmus subspicatus* 8.96 mg/l.

ENVIRONMENTAL FATE

Plants Readily metabolised in plants, with formation of conjugates with endogenous plant constituents. **Soil/Environment** Persists in soil for *c.* 2–4 weeks; loss from soil is principally by microbial degradation.

24   ammonium sulfamate   *Herbicide*

$$NH_2SO_3^- \ NH_4^+$$

NOMENCLATURE

Common name AMS (WSSA); ammonium sulfamate (E-ISO, accepted in lieu of common name); ammonium sulphamate (JMAF, accepted in lieu of common name); sulfamate d'ammonium (F-ISO, accepted in lieu of common name)
IUPAC name ammonium sulfamidate
Chemical Abstracts name monoammonium sulfamate
CAS RN *[7773–06–0]* **EEC no.** (acid) 226–218–8

PHYSICAL CHEMISTRY

Mol. wt. 114.1 **M.f.** $H_6N_2O_3S$ **Form** Colourless, hygroscopic crystals. **M.p.** 131–132 °C **V.p.** Negligible at room temperature. **Solubility** In water 684 g/l (25 °C), 2300 g/l (30 °C). Moderately soluble in formamide, glycerol, and glycols. Slightly soluble in ethanol.   **Stability** Stable in neutral media at ordinary temperatures. Hydrolysed at higher temperatures and in acidic conditions. Decomposes at 160 °C.

COMMERCIALISATION
History Introduced as herbicide by E. I. du Pont de Nemours and Co. (who no longer manufacture or market it) and later by Albright and Wilson (Mfg) Ltd. **Patent** US 2277744 to Du Pont

APPLICATIONS
Mode of action Non-selective systemic herbicide, absorbed by the leaves, stems, and freshly-cut wood surfaces with translocation.   **Uses** A non-selective herbicide for control of woody plants, herbaceous perennials, and most annual broad-leaved weeds and grasses on non-crop land, land to be planted, and in forestry. Control of undesirable trees, and prevention of resprouting of freshly-cut stumps, by application to cut surfaces. Also used for control of poison ivy in pome fruit orchards.   **Formulation types** Crystals; SP.   **Selected tradenames** 'Sulfamate' (Mitsui Toatsu)

ANALYSIS
Product analysis by potentiometric titration with sodium nitrite (W. W. Bowler & E. A. Arnold, *Anal. Chem.*, 1947, **19**, 336).   **Residues** determined by colorimetry (H. L. Pease et al., *J. Agric. Food Chem.*, 1966, **14**, 140).

MAMMALIAN TOXICOLOGY
Oral Acute oral LD_{50} for rats 3900 mg/kg.   **Skin and eye** Acute percutaneous LD_{50} not available. Repeated applications of 50% aqueous solutions to the shaved skin of rats caused no irritation or systemic toxicity.   **NOEL** In 105 d feeding trials in rats, no adverse effect observed at 10 000 mg/kg diet; growth inhibition noted at 20 000 mg/kg diet.   **Toxicity class** WHO (a.i.) III (Table 5); EPA (formulation) III   **EC risk** (Acid) Xi (R36/38)

ECOTOXICOLOGY
Birds Acute oral LD_{50} for quail 3000 mg/kg.   **Fish** LC_{50} (48 h) for young carp 1000–2000 mg/l.

ENVIRONMENTAL FATE
Soil/Environment Microbially degraded in soil to ammonium sulfate within 6–8 weeks.

Ampelomyces quisqualis   *Biological agent*

fungus

NOMENCLATURE
Scientific name *Ampelomyces quisqualis* (AQ) isolate No. 10   **Other names** A.q.

PROPERTIES
Composition 'AQ10' formulation consists of 10^9 spores/g.   **Stability** The product has a shelf life of ≥ 6 months when stored in a cool, dry place, and ≥ 3 y under refrigeration.

COMMERCIALISATION
Production *Ampelomyces quisqualis* produces spores by semi-solid or submerged fermentation and the spores serve as the active ingredient of the WP formulation. **History** The fungus *Ampelomyces quisqualis* is a species of the order *Coleomycetes* subdivision *deuteromycotina* previously named *Cicinnobiolum cesatii* but renamed in 1959 with the current A.q. designation. The fungus is a well-known hyperparasite of the *Erysiphaceae* genus, a pathogenic fungus that causes powdery mildew diseases. In 1984, an isolate of A.q. was discovered in an arid zone of Israel and was thereafter designated AQ isolate No. 10 or AQ10. This isolate was licensed to and was developed by Ecogen Israel Partnership, Jerusalem (a subsidiary of Ecogen, Inc.). The WG formulation was introduced in 1995.   **Patent** US 5190754 **Manufacturers** Ecogen

APPLICATIONS
Mode of action Germinating spores suppress the development of powdery mildew by hyperparasitism. This process requires relative humidity of at least 60% in the microenvironment of the germinating spores. Once inside hyphae of the pathogen – following a process that takes 2–4 hours – the hyperparasite can propagate independently of the external environment; the end result is cessation of powdery mildew development.   **Uses** Selective fungal hyperparasite. Used to control powdery mildew in apples, cucurbits, grapes, ornamentals, strawberries, and tomatoes. Although each crop is attacked by a different species of the powdery mildew pathogen, *A. quisqualis* may hyperparasitise all of them to a similar extent. The product has been employed as part of integrated pest management programmes and in alternation with standard chemical mildewicides. It is commonly applied by standard spray application techniques in the presence of surfactants that are compatible with the viability of the organism.
Phytotoxicity Not phytotoxic or phytopathogenic.   **Formulation types** WG.
Compatibility Can be used concurrently with commercial biological insecticides such as *Bacillus thuringiensis*; however it cannot be comixed with currently used fungicides such as systemic sterol inhibitors.   **Selected tradenames** 'AQ10' (Ecogen)

ANALYSIS

The active ingredient of 'AQ10' is routinely subjected to two types of tests, these being a "germination test" and a "hyperparasitism test" to determine spore viability and performance, respectively. The germination test is conducted by plating of spores on agar and their incubation for 48 hours, during which time they start to germinate. The percent of germination indicates the viability within any population. The rate of hyperparasitism is determined in an *in-situ* test by the spraying of powdery-mildew-infested cucumber plants with an 'AQ10' spore suspension. Ten days incubating such plants in the greenhouse will lead to macroscopic symptoms of hyperparasitism; these can be assessed by semiquantitative parameters.

MAMMALIAN TOXICOLOGY

AQ_{10} has not demonstrated evidence of toxicity, infectivity, irritation or hypersensitivity to mammals. No allergic responses or health problems have been observed by research workers, manufacturing staff or users.

ECOTOXICOLOGY

Birds No demonstrated toxicity.   **Bees** No demonstrated toxicity.

Anagrus atomus   *Biological agent*

Hymenoptera

Wasp parasite of leafhoppers.

NOMENCLATURE
Scientific name *Anagrus atomus*

COMMERCIALISATION
Manufacturers English Woodlands

APPLICATIONS
Mode of action The adults are small (2 mm), short-lived (2–3 d) parasites which attack leafhopper eggs. Parasitised eggs are bright red in colour.   **Uses** For control of leafhoppers *Hauptidia maraccana* and *Empoasca decipiens* on tomatoes, peppers and ornamentals.   **Formulation types** Pupae in leaf pieces.

pyrimidinyl carbinol

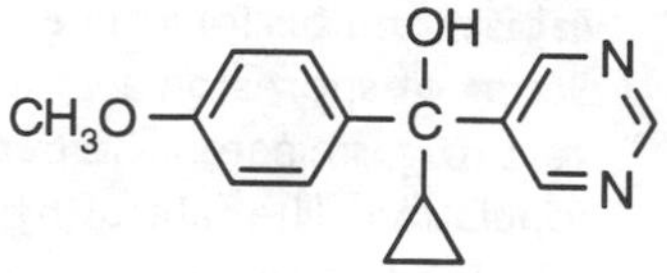

NOMENCLATURE
Common name ancymidol (BSI, E-ISO, ANSI); ancymidole ((*m*) F-ISO)
IUPAC name α-cyclopropyl-4-methoxy-α-(pyrimidin-5-yl)benzyl alcohol
Chemical Abstracts name α-cyclopropyl-α-(4-methoxyphenyl)-=
5-pyrimidinemethanol
CAS RN *[12771–68–5]*   **Development codes** EL-531; 69231

PHYSICAL CHEMISTRY
Mol. wt. 256.3   **M.f.** $C_{15}H_{16}N_2O_2$ **Form** White, crystalline solid.
M.p. 110–111 °C.   **V.p.** <0.13 mPa (50 °C) K_{ow} logP = 1.9 (pH 6.5, 25 °C)
Solubility In water *c.* 650 mg/l (25 °C). In acetone, methanol >250, hexane 37 (all in g/l). Readily soluble in ethanol, ethyl acetate, chloroform and acetonitrile. Moderately soluble in aromatic hydrocarbons. Slightly soluble in saturated hydrocarbons.   **Stability** Aqueous solutions are stable at pH 7 to 11. Decomposed in strongly acidic (pH <4) and strongly alkaline media. Stable to u.v. light. Stable to at least 52 °C.

COMMERCIALISATION
History Plant growth regulator reported by M. Snel & J. V. Gramlich (*Meded. Fac. Landbouwwet. Rijksuniv. Gent,* 1973, **38**, 1033). Introduced in USA (1973) by Eli Lilly & Co. (now DowElanco), and later sold to SePRO Corp.   **Patent** GB 1218623
Manufacturers SePRO

APPLICATIONS
Biochemistry Gibberellin synthesis inhibitor.   **Mode of action** Absorbed by the leaves and roots, with translocation in the phloem. Inhibits internode elongation. **Uses** Used to reduce internode elongation, to produce more compact plants. It has activity on a wide range of greenhouse plant species and is effective on most plants when applied to either foliage or soil.   **Formulation types** SL.   **Selected tradenames** 'A-Rest' (SePRO)

ANALYSIS
Product and **residue** analysis by glc with FID (R. Frank & E. W. Day, *Anal. Methods Pestic. Plant Growth Regul.,* 1976, **8**, 475; S. D. West & E. W. Day, *J. Assoc. Off. Anal. Chem.,* 1977, **60**, 904). Details available from DowElanco.

MAMMALIAN TOXICOLOGY
Oral Acute oral LD_{50} for rats 5000 mg/kg. **Skin and eye** Acute percutaneous LD_{50} for rabbits >200 mg/kg. Moderate eye irritant; mild skin irritant (rabbits). **Inhalation** LC_{50} (4 h) in rats no deaths (poor grooming) at 5.6 mg/l air. **NOEL** In 90 d feeding trials, rats and dogs receiving 8000 mg/kg diet suffered no ill-effect. Not carcinogenic. **Other** Not mutagenic or teratogenic. **Toxicity class** WHO (a.i.) III (Table 5); EPA (formulation) III

ECOTOXICOLOGY
Birds Acute oral LD_{50} for chickens >500 mg/kg. **Fish** LC_{50} for rainbow trout fingerlings 55, bluegill sunfish fingerlings 146, goldfish fingerlings >100 mg/l. **Bees** Not toxic to bees.

ENVIRONMENTAL FATE
Soil/Environment Undergoes microbial degradation in soil.

28 anilofos   *Herbicide*

organophosphorus herbicide

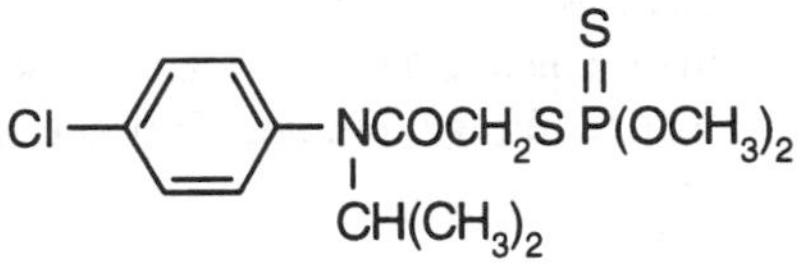

NOMENCLATURE
Common name anilofos (BSI, draft E-ISO, (m) draft F-ISO); no name (Japan)
IUPAC name S-4-chloro-N-isopropylcarbaniloylmethyl O,O-dimethyl phosphorodithioate
Chemical Abstracts name S-[2-[(4-chlorophenyl)(1-methylethyl)amino]-= 2-oxoethyl] O,O-dimethyl phosphorodithioate
CAS RN [64249–01–0] **Development codes** Hoe 30 374

PHYSICAL CHEMISTRY
Mol. wt. 367.8 **M.f.** $C_{13}H_{19}ClNO_3PS_2$ **Form** Crystalline solid. **M.p.** 50.5–52.5 °C **V.p.** 2.2 mPa (60 °C) K_{ow} logP = 3.81 **S.g./density** 1.27 (25 °C) **Solubility** In water 13.6 mg/l (20 °C). In acetone, chloroform, toluene >1000, benzene, ethanol, dichloromethane, ethyl acetate >200, hexane 12 (all in g/l). **Stability** Stable at pH 5–9 at 22 °C. Decomposes at 150 °C. Not sensitive to sunlight.

COMMERCIALISATION
History Herbicide reported by P. Langeluddeke *et al.* (*Proc. Asian Pacific Weed Sci.*

Soc. Conf., 8th, 1981, p. 449). Introduced by Hoechst AG (now AgrEvo GmbH).
Manufacturers AgrEvo; Gharda

APPLICATIONS
Mode of action Selective herbicide, absorbed through the roots, and, to some
extent, through the leaves.   **Uses** Control of annual grass weeds (e.g. *Echinochloa*,
Cyperus, Fimbristylis spp.) and sedges (e.g. *Scirpus* spp.) in transplanted rice.
Formulation types EC; GR.   **Mixtures** *(anilofos +)* 2,4-D esters.
Compatibility Compatible with herbicides which control broad-leaved weeds.
Selected tradenames 'Arozin' (AgrEvo); 'Rico' (AgrEvo); 'Aniloguard' (Gharda)

ANALYSIS
Product analysis by hplc.   **Residue** analysis by glc. Details available from AgrEvo.

MAMMALIAN TOXICOLOGY
Oral Acute oral LD_{50} for male rats 830, female rats 472 mg/kg.
Skin and eye Acute percutaneous LD_{50} for rats >2000 mg/kg. Slightly irritating to
skin and mucous membranes.   **Inhalation** LC_{50} (4 h) 26 mg/l air.
Toxicity class WHO (a.i.) II; EPA (formulation) III

ECOTOXICOLOGY
Birds Acute oral LD_{50} for male Japanese quail 3360, female Japanese quail 2339,
male chicken 1480, female chicken 1640 mg/kg.   **Fish** LC_{50} (96 h) for goldfish 4.6,
trout 2.8 mg/l.   **Bees** Contact LD_{50} 0.66 µg 'Arozin' 60 LC/bee.   **Daphnia** LC_{50}
(3 h) >56 mg/l.

ENVIRONMENTAL FATE
Soil/Environment Typical degradation for a phosphoric acid compound, with
chloroaniline and CO_2 as end-products. DT_{50} in soil 30–45 d at 23 °C.

29   anthraquinone   *Bird repellent*

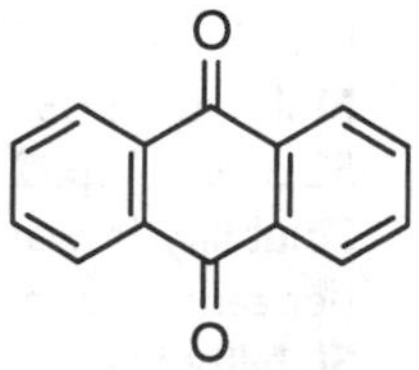

NOMENCLATURE
Common name anthraquinone (BSI, E-ISO, F-ISO, in lieu of common name)
IUPAC name anthraquinone

Chemical Abstracts name 9,10-anthracenedione
CAS RN [84–65–1]   EEC no. 201–549–0

PHYSICAL CHEMISTRY
Mol. wt. 208.2   M.f. $C_{14}H_8O_2$ Form Yellow-green crystals with an aromatic
odour (tech.).   M.p. 286 °C, subliming   B.p. 377–381 °C/760 mmHg (sublimes in
yellow needles)   V.p. 5 × 10^{-3} mPa (20 °C); 1 × 10^{-2} mPa (25 °C, OECD 104)
K_{ow} logP = 3.52 (22 °C)   S.g./density 1.44 (20 °C)   Solubility In water 0.084
mg/l (20 °C). In chloroform 6.1, benzene 2.6 (both in g/kg, 20 °C). In ethanol 4.4,
toluene 3.0, diethyl ether 1.1 (all in g/kg, 25 °C).   Stability Stable to acids and
alkalis.

COMMERCIALISATION
History Known for many years as a chemical, use as bird repellent reported by
F. Wenkel (Hoefchen-Briefe (Engl. Ed.), 1951, 4, 227). Marketed by Bayer AG.
Manufacturers Zeneca

APPLICATIONS
Mode of action Bird repellent which induces retching in birds.
Uses Anthraquinone is used as a seed treatment for cereals to deter attack by
birds, in particular rooks. Frequently formulated with insecticides and fungicides to
render treated seed unattractive to birds, so reducing hazard to wild life.
Formulation types DS; WP; FS.   Mixtures (anthraquinone +) bitertanol; triadimenol
+ triazoxide; bitertanol + fuberidazole; imidacloprid + bitertanol; tebuconazole +
captan.   Selected tradenames 'Corbit' (Bayer); 'Gaucho Ble' (mixture) (France)
(Bayer); 'Morkit' (Bayer); 'Sibutol' (mixture) (France, Germany, Belgium) (Bayer);
'Alpha Raxil CA' (mixture) (France) (Bayer); 'Brio' (Bayer)

ANALYSIS
Residues in plant material determined by Bayer (GLC method No. 00291,
W. Specht & M. Tillkes, Pflanzenschutz-Nachr. Bayer, 33(1), 61–85 (1980), Engl.
edition). Further information available upon request, from Bayer or Zeneca.

MAMMALIAN TOXICOLOGY
Oral Acute oral LD_{50} for rats >5000 mg/kg.   Skin and eye Acute percutaneous
LD_{50} for rats >5000 mg/kg. Not irritating to eyes and skin (rabbits).
Inhalation LC_{50} (4 h) for rats >1.3 mg/l air (dust).   NOEL (90 d) for rats 15
mg/kg diet.   Other Not mutagenic, not carcinogenic.   Toxicity class WHO (a.i.)
III (Table 5)

ECOTOXICOLOGY
Birds LD_{50} for Japanese quail >2000 mg/kg.   Fish LC_{50} (96 h) for rainbow trout
72, golden orfe 44 mg/l   Worms LC_{50} for Eisenia foetida >1000 mg/kg dry soil.
Daphnia LC_{50} (48 h) >10 mg/l.   Algae E_rC_{50} for Scenedesmus subspicatus
>10 mg/l.

ENVIRONMENTAL FATE

Animals Elimination is quick; almost 96% is excreted within 48 h in the urine and faeces. **Plants** Uptake by plants is negligible because anthraquinone is only used as a seed dressing and does not possess systemic properties. **Soil/Environment** Rapidly degraded in different soils; DT_{50} 7–10 d, DT_{90} 22–53 d (calc. from suitable laboratory trials). Degradation is apparently due to microbial activity in the first place. Laboratory trials with BBA standard soils did not reveal any leaching potential. In water, anthraquinone is stable to hydrolysis but extremely sensitive to light; DT_{50} in aqueous solution c. 9 min. On solid surface (silica gel), 80% disappeared within 1 d. Therefore, direct photolysis greatly contributes to dissipation in the environment. The low vapour pressure makes evaporation into the air unlikely.

30 *Aphelinus abdominalis*   *Biological agent*

Hymenoptera

Wasp parasite of aphids.

NOMENCLATURE
Scientific name *Aphelinus abdominalis*

PROPERTIES
Stability Shelf life 2–3 d at 8–10 °C in a cool, dark place.

COMMERCIALISATION
Manufacturers BCP; Biobest ; Koppert ; Neudorff; Novartis BCM; Sautter & Stepper

APPLICATIONS
Mode of action The parasitic wasp lays an egg inside the aphid. As the wasp develops, the aphid becomes a mummy, from which the adult wasp emerges. **Uses** For control of the potato aphid *Macrosiphum euphorbiae* and the greenhouse potato aphid *Aulacorthum solani* in protected crops. **Phytotoxicity** Non-phytotoxic and non-phytopathogenic. **Formulation types** Wasps without carrying material.

MAMMALIAN TOXICOLOGY
Non-toxic.

1   *Aphidius colemani*   *Biological agent*

Hymenoptera

Wasp parasite of aphids.

NOMENCLATURE
Scientific name *Aphidius colemani* (Vierick 1912)   **Other names** *Aphidius platensis*
(Bréthes 1913); *Aphidius transcaspicus* (Telenga 1958)

PROPERTIES
Stability For material reared at 20 °C and above, shelf life is 4 to 5 d under storage
conditions of 5 °C, 70–80% relative humidity and a 16:8 h L:D regime.

COMMERCIALISATION
Production In commercial productions, the parasites are reared on *Aphis gossypii*
or *Myzus persicae*; mummies of known age are collected, packaged in vials or
bottles, sent to the grower and opened in the crop close to the time of adult
emergence.   **History** According to Stary, 1974, this species possessed an original
discrete type locality of central India. However, in 1975, Stary placed two well-
known species *Aphidius platensis* and *Aphidius transcaspicus* as junior synonyms of
it, thereby extending its geographic range from Central Asia into the
Mediterranean basin, from which (under the name *A. platensis*) it has been
accidentally or intentionally introduced in Australia, Africa, Central America,
California, England, Norway and Holland. Historically *Aphidius matricariae* was the
principal agent in commercial production for use as an aphid control in
glasshouses, but recently strains from these productions, selected for their
superior performance against the aphid pest *Aphis gossypii*, have been identified as
Aphidius colemani contaminants. The majority of commercial producers have
abandoned *A. matricariae* production in favour of *A. colemani*.
Manufacturers BCP; Biobest; English Woodlands; Koppert; Neudorff; Novartis
BCM; Rincon-Vitova; Sautter & Stepper

APPLICATIONS
Mode of action The female stings and lays an egg within the body of the adult or
late nymphal stage of the aphid. Development takes place inside the body. At
pupal formation, the body of the host swells, the cuticle hardens, providing a
protective case (termed a 'mummy') from which the adult parasite emerges from
a hole it cuts in the dorsum of the host. Rate of development from egg to adult
varies with temperature; from 20 days at 15 °C reducing to 12 days at 24 °C. The
period of mummification encompasses one third of the development period. The
mean fecundity at 21 °C is about 45 mummies, with a 1:1 sex ratio.   **Uses** *A.
colemani* is employed in biological control programmes to control *Aphis gossypii*

(cotton aphid) and *Myzus persicae* (green peach aphid). It is used in all vegetable crops as well as strawberry and melon under protected cultivation. When *A. gossypii* is the pest, the first introduction of the parasite must be made into very low pest densities. Rates of introduction vary with aphid species, crop type and climatic conditions.   **Phytotoxicity** Non-phytotoxic and non-phytopathogenic. **Formulation types** Parasitised aphids mixed with sawdust; parasitised aphids on wheat (to produce a culture); pure parasite mummies.   **Selected tradenames** 'Aph-line c' (Novartis BCM); 'Aphipar' (Koppert); 'Aphibank' (Koppert)

MAMMALIAN TOXICOLOGY
There is no evidence of acute or chronic toxicity, eye or skin irritation or hypersensitivity to mammals. No allergic response or health problems have been observed in production staff or horticultural workers.

32 *Aphidoletes aphidimyza* Biological agent

Diptera

Predatory midge, consumer of aphids.

NOMENCLATURE
Scientific name *Aphidoletes aphidimyza* (Rondani)

PROPERTIES
Form Pupae.   **Stability** Storage/transportation life is about 7 d at 8 °C and 70–80% relative humidity, in dark or 16:8 L:D regime.

COMMERCIALISATION
Production *A. aphidimyza* is cultured commercially on sweet pepper, aubergine or beans, most often using green peach aphid (*Myzus persicae*) or the bean aphid (*Aphis fabae*) as the prey.   **History** The initial development of *A. aphidimyza* as an aphid control agent began in Finland and Russia in the early 1970's, culminating in commercial sales in Finland in 1978. Further trials and development of commercial rearing methods occurred simultaneously in northern European countries and in Canada through the first half of the 1980's, leading to commercial availability from the mid 1980's onwards.   **Manufacturers** Applied Bio-nomics; Arbico; BCP; Biobest; BioSafer; English Woodlands; Koppert ; Neudorff; Novartis BCM; Rincon-Vitova; Sautter & Stepper

APPLICATIONS

Mode of action In the glasshouse, the female midge seeks out aphid colonies close to where she will lay her eggs. Larvae are a distinctive orange colour. A larva will approach its aphid prey, disable it by biting through the genual joint of the legs (knee-caps), injecting a salivary paralysing toxin at the same time, after which it sucks out the body fluids. The larval stage lasts about five days, during which time it can kill up to 50 aphids. Life cycle, egg to adult, takes 21 days at 20+ °C and a female produces up to 200 eggs. *A. aphidimyza* enters diapause as the last larval instar in response to shortening autumn day lengths. However, if relatively low light levels are maintained within the glasshouse, diapause is precluded. **Uses** This predatory midge can use up to 60 species of aphids as food. Employed commercially to control major aphid pests of vegetable and ornamental crops under protected cultivation. It is frequently and most effectively used in a combined control programme with *Aphidius* parasites. Pupae can be distributed in the crop by placing small containers with moist peat. **Phytotoxicity** Non-phytotoxic and non-phytopathogenic. **Formulation types** Late development pupae mixed with vermiculite. **Selected tradenames** 'Aphido-line a' (Novartis BCM); 'Aphidend' (Koppert); AAP 2539 (BioSafer)

MAMMALIAN TOXICOLOGY

There is no evidence of acute or chronic toxicity, eye or skin irritation or hypersensitivity to mammals. No allergic response or health problems have been observed in production staff or horticultural workers.

3   asulam   *Herbicide*

$$H_2N - \langle\!=\!\rangle - SO_2NHCO_2CH_3$$

NOMENCLATURE

Common name asulam (BSI, E-ISO, ANSI, WSSA, JMAF); asulame ((m) F-ISO); no name (Germany)
IUPAC name methyl sulfanilylcarbamate
Chemical Abstracts name methyl [(4-aminophenyl)sulfonyl]carbamate
CAS RN *[3337–71–1]*; *[2302–17–2]* (sodium salt)   **Development codes** M&B 9057

PHYSICAL CHEMISTRY

Mol. wt. 230.2   **M.f.** $C_8H_{10}N_2O_4S$ **Form** Colourless crystals.   **M.p.** 142–144 °C (decomp.)   **V.p.** <1 mPa (20 °C)   **Solubility** In water 5 g/l (20–25 °C). In dimethylformamide >800, acetone 340, methanol 280, methyl ethyl ketone 280, ethanol 120, hydrocarbons and chlorinated hydrocarbons <20 (all in g/l, 20–25 °C). Other salts, in water, sodium >600, potassium >400, ammonium >400, calcium >200, magnesium >400 (all in g/l, 20–25 °C). **Stability** Stable in boiling water ≥ 6 h. Stable >4 y (pH 8.5, room temperature). **pKa** 4.82, forming water-soluble salts.

COMMERCIALISATION
History Herbicide reported by H. J. Cottrell & B. J. Heywood (*Nature (London)*, 1965, **207**, 655). Introduced by May & Baker Ltd (now Rhône-Poulenc Agrochimie). **Patent** GB 1040541 **Manufacturers** Rhône-Poulenc; Sanachem; Shionogi

APPLICATIONS
Biochemistry Cell division inhibitor. **Mode of action** Selective systemic herbicide, absorbed by the leaves, shoots, and roots, with translocation in both the symplastic and apoplastic systems to other parts of the plant. Causes a slow chlorosis in susceptible plants. **Uses** Control of annual and perennial grasses and broad-leaved weeds in spinach, oilseed poppies, alfalfa, some ornamentals, sugar cane, bananas, coffee, tea, cocoa, coconuts, rubber, etc.; wild oats in flax; docks (*Rumex* spp.) in grassland, fruit trees and bushes, and on non-crop land; and bracken (*Pteridium aquilinum*) in grassland, non-crop land, and forestry. Application rates from 1–10 kg/ha, depending on crop. **Formulation types** SL.
Mixtures (*asulam +*) atrazine; diuron; ioxynil; paraquat; MCPA + mecoprop.
Selected tradenames 'Asulox 40' (Rhône-Poulenc); 'Asilan' (Shionogi); 'Sanulam' (Sanachem)

ANALYSIS
Product analysis by hplc (A. Guardigli et al., *Anal. Methods Pestic. Plant Growth Regul.*, 1984, **13**, 197) or by hydrolysis with colorimetry of a derivative (C. H. Brockelsby & D. F. Muggleton, *ibid.*, 1973, **7**, 497). **Residues** determined by the latter method (*idem, ibid.*) or by hplc of a derivative (A. Guardigli et al., *loc. cit.*).

MAMMALIAN TOXICOLOGY
Reviews M. A. Gallo et al., *Effect of asulam in wildlife species, residues and toxicity in bobwhite quail after prolonged exposure*, Bull. Env. Contam. & Toxicol. 1975, **13(2)**, 200–205. B. Ingham & M. A. Gallo, *Effect of asulam in wildlife species – acute toxicity to birds and fish*, Proc. Northeast Weed Control Conf., 1975 **(1)** 194–199.
Oral Acute oral LD_{50} for rats, mice, rabbits, and dogs >4000 mg/kg.
Skin and eye Acute percutaneous LD_{50} for rats >1200 mg/kg. **Inhalation** LC_{50} (6 h) for rats >1.8 mg/l air. **NOEL** In 90 d feeding trials, rats receiving 400 mg/kg diet showed no significant ill-effects. No effect observed when fed to cows at 800 ppm over 8 w, or to sheep at 50 mg/kg over 10 d. Non-teratogenic.
Toxicity class WHO (a.i.) III (Table 5); EPA (formulation) IV

ECOTOXICOLOGY
Birds Acute oral LD_{50} for mallard ducks, pheasants, and pigeons >4000 mg/kg.
Fish LC_{50} (96 h) for rainbow trout, channel catfish, and goldfish >5000, bluegill
sunfish >3000, harlequin fish >1700 mg/l.   **Bees** Not toxic to bees at <2% w/v
by direct contact or ingestion.

ENVIRONMENTAL FATE
Animals In rats, following oral administration, 85–96% of the dose is eliminated,
predominantly in the urine, within 3 days.   **Soil/Environment** Has a short
persistence in soil, DT_{50} c. 6–14 d. Soil metabolism is by loss of amino group,
cleavage of carbamate group, or acetylation of amino group.

.4   atrazine   *Herbicide*

1,3,5-triazine

NOMENCLATURE
Common name atrazine (BSI, E-ISO, (f) F-ISO, ANSI, WSSA, JMAF)
IUPAC name 6-chloro-N^2-ethyl-N^4-isopropyl-1,3,5-triazine-2,4-diamine
Chemical Abstracts name 6-chloro-N-ethyl-N'-(1-methylethyl)-1,3,5-triazine-=
2,4-diamine
CAS RN *[1912–24–9]*   **EEC no.** 217–617–8   **Development codes** G 30 027

PHYSICAL CHEMISTRY
Composition Tech. is ≥96% pure.   **Mol. wt.** 215.7   **M.f.** $C_8H_{14}ClN_5$
Form Colourless powder.   **M.p.** 175.8 °C   **B.p.** 205.0 °C/101 kPa   **V.p.** 3.85 ×
10^{-2} mPa (25 °C) (OECD 104) K_{ow} logP = 2.5 (25 °C)   **Henry** 1.5 × 10^{-4} Pa m^3
mol^{-1} (calc.)   **S.g./density** 1.23 (22 °C)   **Solubility** In water 33 mg/l (pH 7,
22 °C). In ethyl acetate 24, acetone 31, dichloromethane 28, ethanol 15, toluene
4.0, n-hexane 0.11, n-octanol 8.7 (all in g/l, 25 °C).   **Stability** Relatively stable in
neutral, weakly acidic and weakly alkaline media. Rapidly hydrolysed to the
hydroxy derivative in strong acids and alkalis, and at 70 °C in neutral media; DT_{50}
(pH 1) 9.5, (pH 5) 86, (pH 13) 5.0 d.   **pKa** 1.7, v. weak base

COMMERCIALISATION
History Herbicide reported by H. Gysin & E. Knüsli (*Proc. Int. Congr. Crop Prot.,
4th,* Hamburg, 1957). Introduced by J. R. Geigy S.A. (now Novartis Crop

Protection AG).   **Patent** BE 540590; GB 814947   **Manufacturers** Crystal;
Du Pont; Makhteshim-Agan; Novartis; Oxon Italia; Rallis; Sanachem

APPLICATIONS
Biochemistry Inhibits photosynthetic electron transport. Maize tolerance is
attributed to rapid detoxification by glutathione transferases.   **Mode of action**
Selective systemic herbicide, absorbed principally through the roots, but also
through the foliage, with translocation acropetally in the xylem and accumulation
in the apical meristems and leaves.   **Uses** Pre- and post-emergence control of
annual broad-leaved weeds and annual grasses in maize, sorghum, sugar cane,
pineapples, chemical fallow, grassland, macadamia nuts, conifers, industrial weed
control. In Europe, use is increasingly being concentrated in maize and sorghum.
Used also in combinations with many other herbicides.   **Phytotoxicity** Phytotoxic
to many crops, including most vegetables, potatoes, soya beans, and peanuts.
Formulation types FW; SC; WP; GR; WG.   **Mixtures** *(atrazine +)* alachlor;
bentazone; bromoxynil; cyanazine; amitrole; imazapyr; pyridate; amitrole + 2,4-D;
amitrole + simazine; dichlobenil; diuron; diuron + simazine; metolachlor;
pendimethalin; simazine; dicamba; and many others.   **Selected tradenames**
'AAtrex' (USA) (Novartis); 'Bicep' (USA) (mixture) (Novartis); 'Gesaprim'
(Europe) (Novartis); 'Primagram' (mixture) (Novartis); 'Primextra' (Europe)
(mixture) (Novartis); 'Aktikon' (Nitrokémia); 'Atranex' (Makhteshim-Agan);
'Atrataf' (Rallis); 'Atratylone' (Agriphar); 'Atrazina' (Cequisa); 'Atrazol' (Sipcam);
'Coyote' (Defensa); 'Dhanuzine' (Dhanuka); 'Fogard' (Siapa); 'Hungazin'
(Budapesti Vegyimüvek); 'Mebazine' (Rhône-Poulenc); 'Sanazine' (Sanachem);
'Trinovin' (mixture) (Efthymiadis); 'Vectal' (AgrEvo)

ANALYSIS
Product analysis by glc with FID (*CIPAC Handbook*, 1980, **1A,** 1106; FAO
Specification (CP/61); *AOAC Methods*, 1995, 971.08).   **Residues** determined by glc
with ECD or FID (K. Ramsteiner et al., J. Assoc. Off. Anal. Chem., 1974, **57,** 92;
E. Knüsli, Anal. Methods Pestic. Plant Growth Regul., 1972, **6,** 600; B. G. Tweedy &
R. A. Kahrs, ibid., 1978, **10,** 493). In **drinking water**, by gc with FID (*AOAC
Methods*, 1995, 991.07); dealkylated atrazine can be determined by lc with u.v.
detection (*AOAC Methods*, 1995, 992.14, 10.7.01).

MAMMALIAN TOXICOLOGY
Oral Acute oral LD_{50} for rats 1869–3090 mg tech./kg, mice >1332–3992 mg/kg.
Skin and eye Acute percutaneous LD_{50} for rats >3100 mg/kg. Mild skin irritant;
non-irritating to eyes (rabbits). Skin sensitiser in guinea pigs, but not in humans.
Inhalation LC_{50} (4 h) for rats >5.8 mg/l air.   **NOEL** (2 y) for rats 10 mg/kg diet
(0.5 mg/kg daily), dogs 150 mg/kg diet (3.75 mg/kg daily), mice 10 mg/kg diet
(1.4 mg/kg daily).   **ADI** 0.005 mg/kg b.w.   **Toxicity class** WHO (a.i.) III (Table 5);
EPA (formulation) III   **EC risk** Xn (R22, R43)

ECOTOXICOLOGY
Birds Acute oral LD_{50} varies from 940 mg/kg for bobwhite quail to >2000 mg/kg for mallard ducks and 4237 for adult Japanese quail. Eight-day dietary LC_{50} for Japanese quail (chicks) >5000, (adults) >1000 mg/kg.   **Fish** LC_{50} (96 h) for rainbow trout 4.5–11.0, bluegill sunfish 16, carp 76, catfish 7.6, guppies 4.3 mg/l. **Bees** LD_{50} (oral) >97 µg/bee; (contact) >100 µg/bee.   **Worms** LC_{50} (14 d) for *Eisenia foetida* 78 mg/kg soil.   **Daphnia** LC_{50} (24 h) 87 mg/l.   **Algae** EC_{50} (72 h) for *Scenedesmus subspicatus* 0.043 mg/l, (96 h) for *Selenastrum capricornutum* 0.13 mg/l.

ENVIRONMENTAL FATE
Animals In mammals, following oral administration, atrazine is rapidly and completely metabolised, primarily by oxidative dealkylation of the amino groups (R. Ikonen et al., *Toxicol. Lett.*, 1988, **44**, 109; *Bull. Environ. Contam. Toxicol.*, 1989, **43**, 199; Y. Deng et al., *J. Agric. Food Chem.*, 1990, **38**, 1411), and by reaction of the chlorine atom with endogenous thiols. Diaminochlorotriazine is the main primary metabolite, which readily conjugates with glutathione. More than 50% of the dose is eliminated in the urine and around 33% in the faeces within 24 h.
Plants In tolerant plants, atrazine is readily metabolised to hydroxyatrazine and amino acid conjugates, with further decomposition of hydroxyatrazine by degradation of the side-chains and hydrolysis of the resulting amino acids on the ring, together with evolution of CO_2. In sensitive plants, unaltered atrazine accumulates, leading to chlorosis and death.   **Soil/Environment** Major metabolites under all conditions are desethylatrazine and hydroxyatrazine. Field DT_{50} 16–77 d (median 41 d), the longer values being from cold or dry conditions. In natural waters, DT_{50} 10–105 d (mean 55 d). DT_{50} under groundwater conditions 105->200 d, depending on test system (M. J. Wood et al. in: A Walker (Ed.), *Pesticides in soils and water: current perspectives* (BCPC Monograph 1991, **47**, 175–182)). K_d 0.2–18 ml/g, K_{oc} 39–173 ml/g; desalkylated metabolites had values similar to those of atrazine, while hydroxyatrazine was much more strongly adsorbed.

MAIN ENTRIES

azaconazole

Fungicide

azole

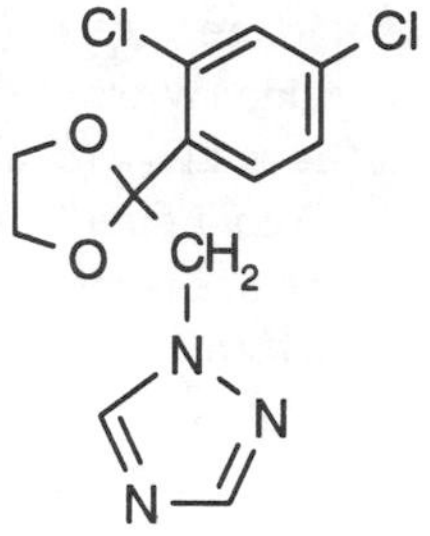

NOMENCLATURE
Common name azaconazole (BSI, draft E-ISI, (*m*) draft F-ISO)
IUPAC name 1-[[2-(2,4-dichlorophenyl)-1,3-dioxolan-2-yl]methyl]-1*H*-1,2,4-=
triazole
Chemical Abstracts name 1-[[2-(2,4-dichlorophenyl)-1,3-dioxolan-2-yl]methyl]-=
1*H*-1,2,4-triazole
CAS RN *[60207–31–0]* **EEC no.** 262–102–3 **Development codes** R028644

PHYSICAL CHEMISTRY
Mol. wt. 300.1 **M.f.** $C_{12}H_{11}Cl_2N_3O_2$ **Form** Beige to brown coloured powder.
M.p. 112.6 °C **V.p.** 0.0086 mPa (20 °C) **K_{ow}** logP = 2.17 (pH 6.4, 23±1 °C)
S.g./density 1.511 (23 °C) **Solubility** In water 0.3 g/l (20 °C). In acetone 160,
hexane 0.8, methanol 150, toluene 79 (all in g/l, 20 °C). **Stability** Stable ≤220 °C.
Stable to light under normal storage conditions, but not in ketone solvents. No
significant hydrolysis between pH 4 to pH 9. **pKa** <3, v. weak base **F.p.** 180 °C

COMMERCIALISATION
History Fungicide reported by J. van Gestel & E. Demoen (*Symp. Wood Pres.,*
Pretoria, March 1983). Introduced in Belgium (1983) by Janssen Pharmaceutica
N.V. **Manufacturers** Janssen

APPLICATIONS
Biochemistry Steroid demethylation inhibitor. **Uses** A fungicide particularly active
against wood-destroying and sapstain fungi. Also used as a disinfectant in
mushroom cultivation and on storage boxes for fruit and vegetables. Used in
combination with imazalil for wound healing in trees. **Formulation types** EC;
SL; OL. **Mixtures** *(azaconazole +)* imazalil; carbendazim; deltamethrin;
quaternary ammonium compounds; carbendazim + benomyl.
Selected tradenames 'Rodewod' (Janssen); 'Safetray SL' (Janssen)

ANALYSIS
Product analysis and **residues** in wood determined by glc or hplc. Details available from Janssen Pharmaceutica N.V.

MAMMALIAN TOXICOLOGY
Reviews *Evaluation of azaconazole*. MAFF, UK: No. 8, November 1988.
Oral Acute oral LD_{50} for rats 308, mice 1123, dogs 114–136 mg/kg.
Skin and eye Acute percutaneous LD_{50} for rats >2560 mg/kg. Slightly irritating to eyes and skin (rabbits). Non-sensitising to skin (guinea pigs). **Inhalation** LC_{50} (4 h) >0.64 mg/l air (5% and 1% formulations). **NOEL** for rats 2.5 mg/kg b.w. daily. **ADI** 0.03 mg/kg b.w. **Toxicity class** WHO (a.i.) II; **EC risk** Xn (R22)

ECOTOXICOLOGY
Birds LC_{50} (5 d) for ring-necked pheasants >5000 mg/kg. **Fish** LC_{50} (96 h) for rainbow trout 42 mg/l. **Daphnia** LC_{50} (96 h) 86 mg/l.

36 azadirachtin   *Insecticide*

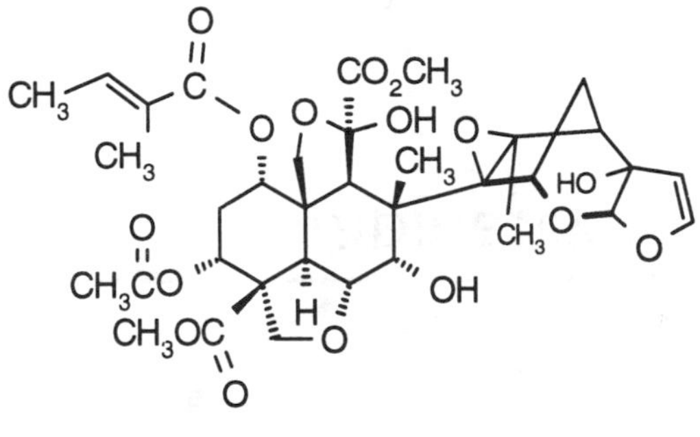

NOMENCLATURE
Chemical Abstracts name dimethyl [2aR-[2aα,3β,4β(1aR*,2S*,3aS*,6aS*,7S*,7aS*),=
4aβ,5α,7aS*,8β(E),10β,10aα,10bβ]]-10-(acetyloxy)octahydro-3,5-dihydroxy-4-=
methyl-8-[(2-methyl-1-oxo-2-butenyl)oxy]-4-(3a,6a,7,7a-tetrahydro-6a-hydroxy-=
7a-methyl-2,7-methanofuro[2,3-b]oxireno[e]oxepin-1a(2H)-yl)-1H,7H-=
naphtho[1,8-bc:4,4a-c′]difuran-5,10a(8H)-dicarboxylate
Other names azad **CAS RN** *[11141–17–6]*

PHYSICAL CHEMISTRY
Mol. wt. 720.7 **M.f.** $C_{35}H_{44}O_{16}$ **Form** Yellow-green powder with a strong garlic/sulfur odour. K_{ow} logP = 1.09 **F.p.** >140 °F (Tag closed cup)

COMMERCIALISATION
Production Extracted from the neem tree, *Azadirachta indica*.
History Extracts of the neem tree have long been known to have insect-controlling activity; azadirachtin is the principal active ingredient of such extracts.
Manufacturers Thermo Trilogy

APPLICATIONS
Biochemistry Ecdysone antagonist.   **Mode of action** Disrupts insect moulting.
Uses Neem tree extracts, and pure azadirachtin formulations are used for control of whitefly, leaf miners and other pests including pear psylla. Neem extracts also show anti-feedant and repellent properties, which have been shown to be due to other chemicals such as salannin. Extracts also show nematicidal and fungicidal activity. Dihydroazadirachtin is under development as an insecticide.   **Formulation types** EC.   **Selected tradenames** 'Azatin' (Thermo Trilogy); 'Kayneem' (neem oil) (Krishi Rasayan); 'Neemazad' (Thermo Trilogy); 'NeemAzal' (Andermatt); 'Neemix' (Thermo Trilogy); 'Neemolin' (seed extract) (Rallis)

MAMMALIAN TOXICOLOGY
Oral Acute oral LD_{50} for rats >5000 mg/kg.   **Skin and eye** Acute percutaneous LD_{50} for rabbits >2000 mg/kg.   **Toxicity class** EPA (formulation) IV

37   azafenidin   *Herbicide*

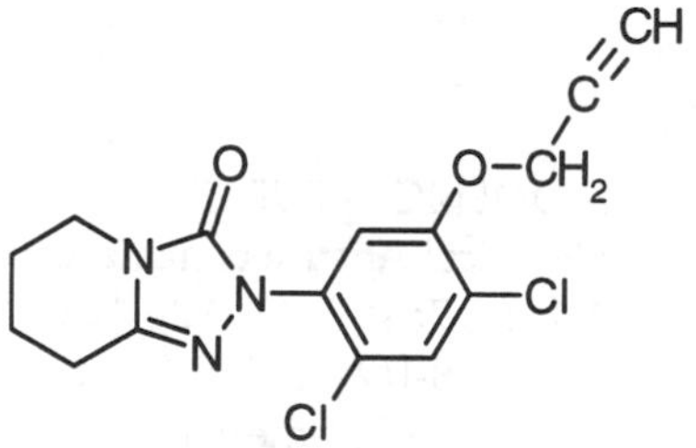

NOMENCLATURE
Common name azafenidin (BSI, pa ISO)
IUPAC name 2-(2,4-dichloro-5-prop-2-ynyloxyphenyl)-5,6,7,8-tetrahydro-=
1,2,4-triazolo[4,3-*a*]pyridin-3(2*H*)-one
Chemical Abstracts name 2-[2,4-dichloro-5-(2-propynyloxy)phenyl]-5,6,7,8-=
tetrahydro-1,2,4-triazolo[4,3-*a*]pyridin-3(2*H*)-one
CAS RN *[68049–83–2]*   **Development codes** R6447; DPX-R6447; IN-R6447

PHYSICAL CHEMISTRY
Composition Tech. is c. 97% pure. **Mol. wt.** 338.2 **M.f.** $C_{15}H_{13}Cl_2N_3O_2$
Form Rust coloured solid with a pungent odour. **M.p.** 168–168.5 °C
V.p. 10^{-6} mPa (20 °C) **K_{ow}** logP = 2.7 **Henry** 2.8×10^{-8} Pa m^3 mol^{-1} (calc.)
S.g./density 1.4 (20 °C) **Solubility** In water 16 ppm (pH 7). **Stability** Stable to hydrolysis; aqueous photolysis DT_{50} c. 12 h.

COMMERCIALISATION
History Under development by Du Pont Agricultural Products, with planned introduction in 1998–2000. **Manufacturers** Du Pont

APPLICATIONS
Biochemistry Inhibits porphyrin biosynthesis. **Mode of action** Primarily pre-emergence, shoot activity. **Uses** Intended for use as a pre-emergence herbicide in citrus, grapes, sugar cane and other perennial crops; active on both annual and perennial weeds. **Formulation types** WG.

MAMMALIAN TOXICOLOGY
Oral Acute oral LD_{50} for rats >5000 mg/kg. **Skin and eye** Acute percutaneous LD_{50} for rabbits >2000 mg/kg. Not an eye or skin irritant. Not a skin sensitiser. **Inhalation** LC_{50} for rats >5.3 mg/l. **NOEL** (90 d) for male and female rats 50, male mice 50, female mice 300, dogs 10 ppm.

ECOTOXICOLOGY
Birds Acute oral LD_{50} for bobwhite quail and mallard duck >2250 mg/kg. LC_{50} (8 d) for bobwhite quail and mallard duck >5620 ppm. **Fish** LC_{50} (96 h) for rainbow trout 33, bluegill sunfish 48 mg/l. **Bees** LD_{50} (oral) >20 µg/bee; (contact) >100 µg/bee. **Daphnia** EC_{50} (48 h) 38 mg/l. **Algae** EC_{50} 1.25 ppb.

38   azamethiphos   *Insecticide*

organophosphorus

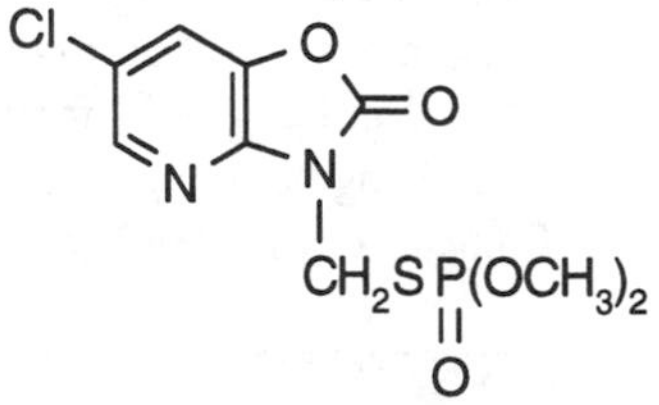

NOMENCLATURE
Common name azamethiphos (BSI, E-ISO, (*m*) F-ISO)

IUPAC name S-6-chloro-2,3-dihydro-2-oxo-1,3-oxazolo[4,5-b]pyridin-3-ylmethyl O,O-dimethyl phosphorothioate; 6-chloro-3-dimethoxyphosphinoylthiomethyl-= 1,3-oxazolo[4,5-b]pyridin-2(3H)-one
Chemical Abstracts name S-[(6-chloro-2-oxooxazolo[4,5-b]pyridin-3(2H)-yl)= methyl] O,O-dimethyl phosphorothioate
CAS RN [35575–96–3] **EEC no.** 252–626–0 **Development codes** CGA 18 809; GS 40 616 **Official codes** OMS 1825

PHYSICAL CHEMISTRY
Composition Tech. is >95%. **Mol. wt.** 324.7 **M.f.** $C_9H_{10}ClN_2O_5PS$
Form Colourless crystals; (tech. is a beige to grey powder). **M.p.** 89 °C
V.p. 0.0049 mPa (20 °C) **K_{ow}** logP = 1.05 **S.g./density** 1.60 (20 °C)
Solubility In water 1.1 g/l (pH 7, 20 °C). In dichloromethane 610, benzene 130, methanol 100, n-octanol 5.8 (all in g/kg, 20 °C). **Stability** Unstable in acids and alkalis; DT_{50} (20 °C) (calculated): 800 h (pH 5), 260 h (pH 7), 4.3 h (pH 9).
F.p. >150 °C (Marcusson, without stirring)

COMMERCIALISATION
History Insecticide and acaricide reported by R. Wyniger et al. (Proc. Br. Crop Prot. Conf. – Pests Dis., 1977, **3**, 1025). Introduced by Ciba-Geigy AG (now Novartis Crop Protection AG). **Patent** BE 769051; GB 1347373 **Manufacturers** Novartis

APPLICATIONS
Biochemistry Cholinesterase inhibitor. **Mode of action** Insecticide with contact and stomach action. Gives rapid knockdown, and has good residual activity.
Uses Control of flies and other insect pests in animal houses. Control of mosquitoes, tsetse flies, cockroaches, and other public hygiene pests.
Formulation types WP; AE. **Selected tradenames** 'Alfacron' (Novartis)

ANALYSIS
Product analysis by hplc. **Residues** determined by hplc. Details available from Novartis.

MAMMALIAN TOXICOLOGY
EHC 63 (WHO, 1986; a general review of organophosphorus insecticides).
Oral Acute oral LD_{50} for rats 1180 mg/kg. **Skin and eye** Acute percutaneous LD_{50} for rats >2150 mg/kg. Mild eye irritant; non-irritating to skin (rabbits).
Inhalation LC_{50} (4 h) for rats >560 mg/m³ air. **NOEL** (90 d) for rats 20 mg/kg diet (2 mg/kg daily), dogs 10 mg/kg diet (0.3 mg/kg daily). **ADI** 0.025 mg/kg.
Toxicity class WHO (a.i.) III; EPA (formulation) III

ECOTOXICOLOGY
Birds Acute oral LD_{50} for bobwhite quail 30.2, mallard ducks 48.4 mg/kg. Eight-day dietary LC_{50} for bobwhite quail 860, Japanese quail >1000, mallard

ducks 700 ppm. Based on the results of acute laboratory trials, azamethiphos is classified as highly toxic to birds. However, sublethal doses have a repellent effect on birds, so the risk for birds is significantly reduced. **Fish** LC_{50} (96 h) for channel catfish 3, crucian carp 6, guppy 8, rainbow trout 0.115–0.2, sheepshead minnow 2.22 mg/l. Based on the test with the most sensitive species, rainbow trout, azamethiphos is classified as highly toxic to fish. All other test species were less sensitive to this compound. **Bees** Toxic to bees; LD_{50} (24 h) (oral) <0.1 μg/bee; (contact) 10 μg/bee. **Daphnia** LC_{50} (48 h) 0.67 μg/l.

ENVIRONMENTAL FATE
Animals In rats and goats, the glucuronic acid conjugate of 2-amino-3-hydroxy-5-=chloro-pyridine represents the major metabolite, accounting for 27–48% of the dose, followed by the corresponding sulfuric acid conjugate accounting for 3–20% of the dose. **Soil/Environment** In loamy sand (aerobic) DT_{50} c. 6 h.

azimsulfuron   *Herbicide*

sulfonylurea

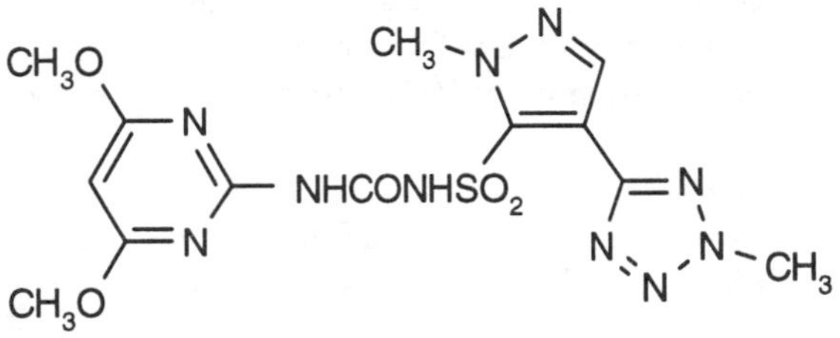

NOMENCLATURE
Common name azimsulfuron (BSI, pa E-ISO)
IUPAC name 1-(4,6-dimethoxypyrimidin-2-yl)-3-[1-methyl-4-(2-methyl-2*H*-=tetrazol-5-yl)pyrazol-5-ylsulfonyl]urea
Chemical Abstracts name *N*-[[(4,6-dimethoxy-2-pyrimidinyl)amino]carbonyl]-=1-methyl-4-(2-methyl-2*H*-tetrazol-5-yl)-1*H*-pyrazole-5-sulfonamide
CAS RN *[120162–55–2]*   **Development codes** IN-A8947; DPX A8947; A8947; JS-458

PHYSICAL CHEMISTRY
Composition Tech. is 98%.   **Mol. wt.** 424.4   **M.f.** $C_{13}H_{16}N_{10}O_5S$ **Form** White solid.   **M.p.** 170 °C   **V.p.** 4.0×10^{-6} mPa (25 °C) K_{ow} logP = 0.646 (pH 5), −1.37 (pH 7), −2.08 (pH 9) (25 °C)   **Henry** 8×10^{-9} (pH 5), 5×10^{-10} (pH 7), 9×10^{-11} (pH 9) (all in Pa m^3 mol^{-1}, calc.)   **Solubility** In water 72.3 (pH 5), 1050 (pH 7), 6536 (pH 9) (20 °C). In acetone 26.4, acetonitrile 13.9, ethyl acetate 13.0,

methanol 2.1, methylene chloride 65.9, toluene 1.8, hexane <0.2 mg/l.
Stability Hydrolysis DT_{50} 89 (pH 5), 124 (pH 7), 132 (pH 9) d (25 °C).
Degradation was through cleavage of the sulfonylurea bridge to give mainly the
tetrazolylpyrazole sulfonamide and aminodimethoxypyrimidine. DT_{50} in irradiated
aqueous solution, corrected for hydrolysis, was 103 (pH 5), 164 (pH 7), 225
(pH 9) d (25 °C). Indirect photodegradation has also been studied (A. C.
Barefoot, *Proc. Br. Crop Prot. Conf. – Weeds*, 1995, **2**, 713).   **pKa** 3.6

COMMERCIALISATION
History Reported by T. Marquez et al. (*Proc. Br. Crop Prot. Conf. – Weeds*, 1995, **1**,
65).   **Patent** US 4746353   **Manufacturers** Du Pont

APPLICATIONS
Biochemistry Branched chain amino acid synthesis (acetolactate synthase or ALS)
inhibitor. Acts by inhibiting biosynthesis of the essential amino acids valine and
isoleucine, hence stopping cell division and plant growth. Selectivity derives from
rapid metabolism in the crop. Metabolic basis of selectivity in sulfonylureas
reviewed (M. K. Koeppe & H. M. Brown, *Agro-Food-Industry*, **6**, 9–14 (1995)).
Mode of action Post-emergence herbicide with mainly foliar uptake; translocated
in xylem and phloem. Initial symptoms are cessation of growth and chlorosis,
followed by reddish colouration, then necrosis.   **Uses** Under development for
control of *Echinochloa* spp. and other annual and perennial broad-leaved and sedge
weeds in Southern European rice, at rates of 20–25 g/ha.
Formulation types WG.   **Selected tradenames** 'Gulliver' (Du Pont)

ANALYSIS
Residues Methods for sulfonylurea residues in crops, soil and water reviewed
(A. C. Barefoot et al., *Proc. Br. Crop Prot. Conf. – Weeds*, 1995, **2**, 707).

MAMMALIAN TOXICOLOGY
Oral Acute oral LD_{50} for rats >5000 mg/kg.   **Skin and eye** Acute percutaneous
LD_{50} for rats >2000 mg/kg. Not irritating to eyes or skin (rabbits). Not a skin
sensitiser (guinea pigs).   **Inhalation** LC_{50} (4 h) for rats >5.9 mg/l.   **NOEL** (2 y) for
rats 1000 mg/kg diet.   **ADI** 0.10 mg/kg b.w. (proposed).   **Other** Not mutagenic
(Ames).   **Toxicity class** WHO (a.i.) III (Table 5)

ECOTOXICOLOGY
Birds Acute oral LD_{50} for bobwhite quail and mallard duck >2250 mg/kg. Dietary
LC_{50} (8 d) for bobwhite quail and mallard duck >5260 mg/kg.   **Fish** LC_{50} (96 h)
for carp >300, bluegill sunfish >1000, rainbow trout 154 ppm.   **Bees** LD_{50} (48 h)
(oral) >25 μg/bee, (contact) >1000 μg/bee.   **Worms** LC_{50} >1000 mg/kg.
Daphnia LC_{50} (48 h) >1000 ppm.   **Algae** EC_{50} for *Selenastrum capricornutum*
12 μg/l.

ENVIRONMENTAL FATE

Animals Azimsulfuron was resistant to metabolism; >90% of dose was excreted unchanged. **Plants** Metabolism was rapid; little parent compound was found in any plant tissue at maturity. **Soil/Environment** Soil degradation in both flooded soils and under aerobic conditions has been studied. The most significant mechanisms are indirect photolysis and soil metabolism, together with chemical hydrolysis. Degradation products have been identified. No detectable residues in the field at harvest at depths of 0 to 50 cm. See A. C. Barefoot (*Proc. Br. Crop Prot. Conf. – Weeds*, 1995, **2**, 713).

azinphos-ethyl
Insecticide, acaricide

organophosphorus

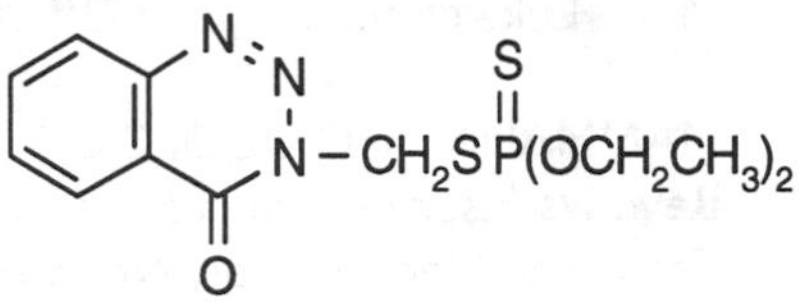

NOMENCLATURE

Common name azinphos-ethyl (BSI, E-ISO, (*m*) F-ISO); azinphosethyl (ESA); [triazotion] (former exception, USSR)
IUPAC name S-(3,4-dihydro-4-oxobenzo[*d*]-[1,2,3]-triazin-3-ylmethyl) *O,O*-diethyl phosphorodithioate
Chemical Abstracts name *O,O*-diethyl S-[(4-oxo-1,2,3-benzotriazin-3(4*H*)-yl)= methyl] phosphorodithioate
CAS RN [2642–71–9] **EEC no.** 220–147–6 **Development codes** Bayer 16 259; R 1513; E1513 **Official codes** ENT 22 014

PHYSICAL CHEMISTRY

Mol. wt. 345.4 **M.f.** $C_{12}H_{16}N_3O_3PS_2$ **Form** Colourless needles. **M.p.** 50 °C
B.p. 147 °C/1.3 Pa **V.p.** 0.32 mPa (20 °C) **K_{ow}** logP = 3.18 **S.g./density** 1.284 (20 °C) **Solubility** In water *c.* 4–5 mg/l (20 °C). In *n*-hexane 2–5, isopropanol 20–50, dichloromethane >1000, toluene >1000 (all in g/l, 20 °C).
Stability Rapidly hydrolysed in alkaline media. Relatively stable in acidic media. DT_{50} (22 °C) *c.* 3 h (pH 4), 270 d (pH 7), *c.* 11 d (pH 9).

COMMERCIALISATION

History Insecticide reported by E. E. Ivy et al. (*J. Econ. Entomol.*, 1955, **48**, 293). Developed by W. Lorenz and introduced by Bayer AG (who no longer manufacture or market it). **Patent** US 2758115; DE 927270
Manufacturers Makhteshim-Agan

APPLICATIONS
Biochemistry Cholinesterase inhibitor. **Mode of action** Non-systemic insecticide and acaricide with contact and stomach action. **Uses** Control of chewing and sucking insects and spider mites on fruit trees (including citrus), vines, hops, nuts, vegetables, potatoes, beet, oilseed rape, maize, cereals, tobacco, cotton, coffee, rice, ornamentals, forestry, and other crops. **Phytotoxicity** Non-phytotoxic when used as directed. Russetting is possible on some fruit varieties with the emulsifiable concentrate formulation. **Formulation types** EC; WP; UL.
Mixtures (*azinphos-ethyl +*) azinphos-methyl; dicofol; dimethoate.
Compatibility Incompatible with alkaline materials. **Selected tradenames** 'Azinugec E' (Sipcam Phyteurop); 'Cotnion-Ethyl' (Makhteshim-Agan)

ANALYSIS
Product analysis by colorimetric measurement of the phosphorodithioate moiety as a complex (*CIPAC Handbook*, 1970, **1**, 18; FAO Specification (CP/41)).
Residues determined by glc (D. C. Abbott *et al.*, *Pestic. Sci.*, 1970, **1**, 10; M. A. Luke *et al.*, *J. Assoc. Off. Anal. Chem.*, 1981, **64**, 1187).

MAMMALIAN TOXICOLOGY
Reviews *Pesticide residues in food.* FAO Agricultural Studies, No. 92; WHO Technical Report Series, No. 545, 1974. *1973 Evaluations of some pesticide residues in food.* FAO Agricultural Studies, No. 92; WHO Technical Report Series, No. 545, 1974. **EHC** 63 (WHO, 1986; a general review of organophosphorus insecticides). **Oral** Acute oral LD_{50} for rats c. 12 mg/kg. **Skin and eye** Acute percutaneous LD_{50} for rats c. 500 mg/kg (24 h). Not irritating to skin and eyes of rabbits. **Inhalation** LC_{50} (4 h) for rats c. 0.15 mg/l air. **NOEL** (2 y) for rats 2, dogs 0.1, mice 1.4 (all as mg/kg diet), monkeys 0.02 mg/kg b.w. **ADI** (JMPR) No ADI [1973]. **Other** Acute i.p. LD_{50} for rats >7.5 mg/kg. **Toxicity class** WHO (a.i.) Ib; EPA (formulation) I **EC risk** T+ (R28); T (R24)

ECOTOXICOLOGY
Birds Acute oral LD_{50} for Japanese quail 12.5–20 mg/kg. **Fish** LC_{50} (96 h) for golden orfe 0.03, rainbow trout 0.08 mg/l. **Bees** Not toxic to bees (depends on application method). **Daphnia** LC_{50} (48 h) 0.0002 mg/l.

ENVIRONMENTAL FATE
Animals In mammals, following oral administration, >90% is eliminated in the urine and faeces within 2 days. The major metabolites are the monodesethyl compound and benzazimide. **Plants** In plants, metabolites identified include azinphos-ethyl-oxon, benzazimide, dimethylbenzazimide sulfide and dimethylbenzazimide disulfide.
Soil/Environment Based on the K_{oc} value and leaching studies, azinphos-ethyl can be classified as a compound with very low mobility. The half-life is several weeks. Metabolites formed in soil under aerobic and anaerobic conditions are: desethyl azinphos-ethyl, sulfonmethylbenzazimid, bis(benzazimidmethyl)ether, methylthiomethylsulfoxide and methylthiomethylsulfone.

azinphos-methyl   *Insecticide*

organophosphorus

NOMENCLATURE
Common name azinphos-methyl (BSI, E-ISO, (*m*) F-ISO); azinphosmethyl (ESA);
[metiltriazotion] (former exception, USSR)
IUPAC name S-(3,4-dihydro-4-oxobenzo[*d*]-[1,2,3]-triazin-3-ylmethyl)
O,O-dimethyl phosphorodithioate
Chemical Abstracts name O,O-dimethyl S-[(4-oxo-1,2,3-benzotriazin-3(4*H*)-yl)=
methyl] phosphorodithioate
CAS RN *[86-50-0]* **EEC no.** 201-676-1 **Development codes** Bayer 17 147;
R 1582; E1582 **Official codes** ENT 23 233; OMS 186

PHYSICAL CHEMISTRY
Mol. wt. 317.3 **M.f.** $C_{10}H_{12}N_3O_3PS_2$ **Form** Yellowish crystals. **M.p.** 73 °C
V.p. 5×10^{-4} mPa (20 °C); 1×10^{-3} mPa (25 °C) **K_{ow}** logP = 2.96
Henry 5.7×10^{-6} Pa m^3 mol^{-1} (20 °C, calc.) **S.g./density** 1.518 (21 °C) **Solubility** In
water 28 mg/l (20 °C). In dichloroethane, acetone, acetonitrile, ethyl acetate,
dimethyl sulfoxide >250, *n*-heptane 1.2, xylene 170 (all in g/l, 20 °C).
Stability Rapidly hydrolysed in alkaline and acidic media; DT_{50} (22 °C) 87 d (pH 4),
50 d (pH 7), 4 d (pH 9). Photodegrades on soil surfaces and readily
photodegrades in water. Decomposes above 200 °C.

COMMERCIALISATION
History Insecticide and acaricide reported by E. E. Ivy *et al.* (*J. Econ. Entomol.*,
1955, **48**, 293). Developed by W. Lorenz and introduced by Bayer AG.
Patent US 2758115; DE 927270 **Manufacturers** Bayer; General Quimica;
Makhteshim-Agan

APPLICATIONS
Biochemistry Cholinesterase inhibitor. **Mode of action** Non-systemic, with
contact and stomach action. **Uses** Control of chewing and sucking insects of the
orders Coleoptera, Diptera, Homoptera, Hemiptera, and Lepidoptera, on fruit
trees (including citrus), vines, strawberries, nuts, vegetables, potatoes, cereals,
maize, cotton, ornamentals, beet, soya beans, tobacco, rice, coffee, sugar cane,
forestry, and other crops. **Phytotoxicity** Russetting is possible on some fruit
varieties with the emulsifiable concentrate formulation. **Formulation types** WP;
EC; DP; SC. **Compatibility** Incompatible with alkaline materials. **Selected
tradenames** 'Gusathion M' (Bayer); 'Acifon' (General Quimica); 'Azinugec' (Sipcam

MAIN ENTRIES

Phyteurop); 'Cotnion-Methyl' (Makhteshim-Agan); 'Valefos' (Productos OSA)

ANALYSIS
Product analysis by lc (*AOAC Methods*, 1995, 989.01, 7.7.02), by hplc (*AOAC Methods*, 1995, 989.01; *CIPAC Handbook*, 1992, **E**, 12) or by i.r. spectrophotometry (*ibid.*, 1995, 980.09; *CIPAC Handbook*, 1985, **1C**, 1970) or by colorimetric measurement of the phosphorodithioate moiety as a complex (*ibid.*, 1970, **1**, 25; FAO Specification CP/4). **Residues** determined by glc (*Analyst (London)*, 1977, **102,** 858; A. Ambrus *et al., J. Assoc. Off. Anal. Chem.*, 1981, **64,** 733; D. H. MacDougall, *Anal. Methods Pestic. Plant Growth Regul.*, 1972, **6,** 397). Methods for the determination of residues are available from Bayer.

MAMMALIAN TOXICOLOGY
Reviews *Pesticide residues in food – 1991.* FAO Plant Production and Protection Paper, 111, 1991. *Pesticide residues in food – 1991 evaluations. Part II – Toxicology.* World Health Organisation, WHO/PCS/92.52, 1992. **EHC** 63 (WHO, 1986; a general review of organophosphorus insecticides). **Oral** Acute oral LD_{50} for rats c. 9, male guinea pigs 80, mice 11–20, dogs >10 mg/kg. **Skin and eye** Acute percutaneous LD_{50} for rats 150–200 mg/kg (24 h); not a skin irritant, mild eye irritant (rabbits). **Inhalation** LC_{50} (4 h) for rats 0.15 mg/l air (aerosol). **NOEL** (2 y) for rats and mice 5 mg/kg diet; (1 y) for dogs 5 mg/kg diet. **ADI** (JMPR) 0.005 mg/kg b.w. [1991]. **Toxicity class** WHO (a.i.) Ib; EPA (formulation) I **EC risk** T+ (R24, R28)

ECOTOXICOLOGY
Birds Acute oral LD_{50} for bobwhite quail c. 32 mg/kg. Dietary LC_{50} (5d) for Japanese quail 935 mg/kg diet. **Fish** LC_{50} (96 h) for rainbow trout 0.02, golden orfe 0.12 mg/l. **Bees** Toxic to bees. **Worms** LC_{50} (14 d) 59 mg/kg. **Other beneficial spp.** Azinphos-methyl is an effective insecticide, therefore an effect on some non-target arthropods cannot be excluded, in particular, where those organisms are directly exposed to the spray treatment. **Daphnia** LC_{50} (48 h) 0.0011 mg/l. **Algae** E_rC_{50} (96h) for *Scenedesmus* 7.15 mg/l.

ENVIRONMENTAL FATE
Animals In mammals, following oral administration, >95% is eliminated in the urine and faeces within 2 days. The major metabolites are the monodesmethyl compound and benzazimide. **Plants** In plants, major metabolites identified include azinphos-methyl oxon, benzazimide, mercaptomethyl benzazimide and cysteinmethyl benzazimide. **Soil/Environment** Degradation involves oxidation, demethylation, and hydrolysis. Based on the K_{oc} values and leaching studies, azinphos-methyl can be classified as a compound with low mobility. The half-life in soil is several weeks.

azocyclotin

organotin

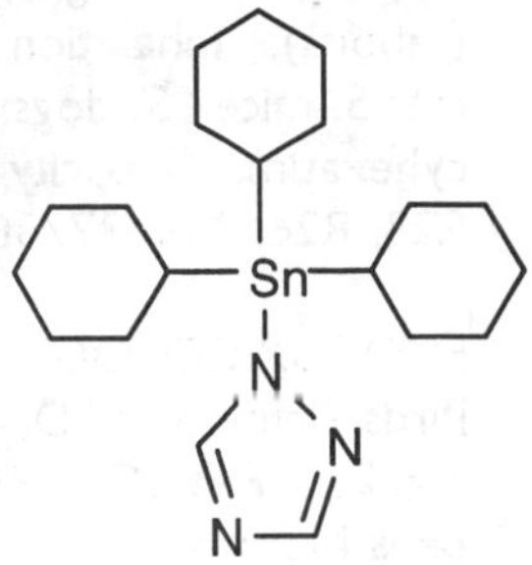

NOMENCLATURE
Common name azocyclotin (BSI, E-ISO, (*m*) F-ISO, ESA)
IUPAC name tri(cyclohexyl)-1*H*-1,2,4-triazol-1-yltin; 1-tricyclohexylstannanyl-1*H*-=
[1,2,4]triazole
Chemical Abstracts name 1-(tricyclohexylstannyl)-1*H*-1,2,4-triazole
CAS RN *[41083–11–8]* **EEC no.** 255–209–1 **Development codes** BAY BUE
1452

PHYSICAL CHEMISTRY
Mol. wt. 436.2 **M.f.** $C_{20}H_{35}N_3Sn$ **Form** Colourless crystals. **M.p.** 210 °C
(decomp.) **V.p.** 2×10^{-8} mPa (20°C); 6.0×10^{-8} mPa (25 °C) **K_{ow}** logP = 5.3
(20 °C) **Henry** 7×10^{-2} Pa m^3 mol^{-1} (20°C) (calc.) **S.g./density** 1.335 (21 °C)
Solubility In water 0.12 mg/l (20 °C). In dichloromethane 20–50, isopropanol
10–20, n-hexane 0.1–1, toluene 2–5 (all in g/l, 20 °C). **Stability** DT$_{50}$ (22 °C)
96 h (pH 4), 81 h (pH 7), 8 h (pH 9). **pKa** 5.36, weak base

COMMERCIALISATION
History Acaricide reported by W. Kolbe (*Pflanzenschutz-Nachr. (Engl. Ed.)*, 1977,
30, 325). Introduced by Bayer AG. **Patent** DE 2143252 **Manufacturers** Bayer

APPLICATIONS
Mode of action Long-acting acaricide with contact action. **Uses** Control of all
motile stages of spider mites on fruit (including citrus), vines, hops, cotton,
vegetables, and ornamentals. **Formulation types** WP.
Selected tradenames 'Peropal' (Bayer)

ANALYSIS
Product and **residue** analysis details available from Bayer AG. **Residue** analysis by
glc (E. Möllhoff, *ibid.*, 1977, **30**, 249).

MAMMALIAN TOXICOLOGY
Reviews *Pesticide residues in food – 1994*, FAO Plant Production and Protection

Paper, 127, 1995.   **Oral** Acute oral LD$_{50}$ for male rats 209, female rats 363, guinea pigs 261, mice 870–980 mg/kg.   **Skin and eye** Acute percutaneous LD$_{50}$ for rats >5000 mg/kg; strong dermal irritant; strong and corrosive eye irritant (rabbits).   **Inhalation** LC$_{50}$ (4 h) for rats c. 0.02 mg (as AE)/l air.   **NOEL** (2 y) for rats 5, mice 15, dogs 10 mg/kg diet.   **ADI** (JMPR) 0.007 mg/kg b.w. [1994] (with cyhexatin).   **Toxicity class** WHO (a.i.) II; EPA (formulation) I   **EC risk** T+ (R21, R23, R26; R36/37/38)

ECOTOXICOLOGY
Birds Acute oral LD$_{50}$ for male Japanese quail 144, female Japanese quail 195 mg/kg.   **Fish** LC$_{50}$ (96 h) for rainbow trout 0.004, golden orfe 0.0093 mg/l.
Bees Not toxic to bees; LD$_{50}$ >100 μg/bee (500 SC).   **Worms** LC$_{50}$ (28 h) 806 mg/kg (25 WP).   **Daphnia** LC$_{50}$ (48 h) 0.04 mg/l.   **Algae** EC$_{50}$ (96 h) for *Scenedesmus* 0.16 mg/l.

ENVIRONMENTAL FATE
Animals Metabolised by hydrolysis, forming 1,2,4-triazole and tricyclohexyl tin hydroxide, which is further oxidised to form dicyclohexyl tin oxide.
Plants Metabolites identified in plants include 1,2,4-triazole, tricyclohexyl tin hydroxide, and dicyclohexyl tin oxide.   **Soil/Environment** Half-life in soil ranges from a few days to many weeks, depending on soil type.

43   azoxystrobin   *Fungicide*

strobilurin analogue

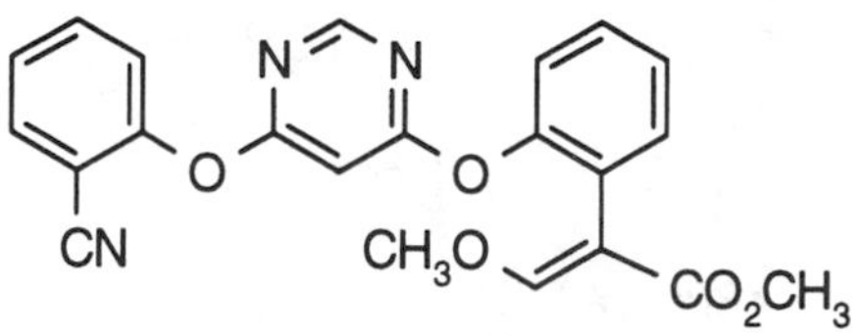

Synthetic analogue of naturally occurring fungal metabolites the strobilurins and oudemansins.

NOMENCLATURE
Common name azoxystrobin (BSI, pa ISO)
IUPAC name methyl (E)-2-{2-[6-(2-cyanophenoxy)pyrimidin-4-yloxy]phenyl}-=
3-methoxyacrylate
Chemical Abstracts name methyl (E)-2-[[6-(2-cyanophenoxy)-4-pyrimidinyl]oxy]-=
α-(methoxymethylene)benzeneacetate
CAS RN *[131860–33–8]*   **Development codes** ICIA5504

PHYSICAL CHEMISTRY
Mol. wt. 403.4   **M.f.** $C_{22}H_{17}N_3O_5$ **Form** White solid.   **M.p.** 116 °C (tech. 114–116 °C)   **V.p.** 1.1×10^{-7} mPa (25 °C)   **K_{ow}** logP = 2.5 (20 °C) **Henry** 7×10^{-9} Pa m^3 mol^{-1} **S.g./density** 1.34 (20 °C)   **Solubility** In water 6 mg/l (20 °C). Low solubility in hexane, *n*-octanol; moderate solubility in methanol, toluene, acetone; high solubility in ethyl acetate, acetonitrile, dichloromethane. **Stability** DT_{50} for aqueous photolysis 11–17 d.

COMMERCIALISATION
History Reported by J. R. Godwin et al. (*Proc. Br. Crop Prot. Conf. – Pests Dis.*, 1992, **1**, 435). Introduced by Zeneca Agrochemicals.   **Patent** EP 382375 **Manufacturers** Zeneca

APPLICATIONS
Biochemistry Inhibits mitochondrial respiration by blocking electron transfer between cytochrome b and cytochrome c_1. Controls pathogenic strains resistant to the 14-demethylase inhibitors, phenylamides, dicarboxamides or benzimidazoles.   **Mode of action** Fungicide with protectant, eradicant, translaminar and systemic properties. Inhibits spore germination and mycelial growth, and also shows antisporulant activity.   **Uses** Controls the following pathogens at application rates between 100 to 375 g a.i./ha: *Erysiphe graminis*, *Puccinia* spp., *Leptosphaeria nodorum*, *Septoria tritici* and *Pyrenophora teres* on temperate cereals; *Pyricularia oryzae* and *Rhizoctonia solani* on rice; *Plasmopara viticola* and *Uncinula necator* on vines; *Venturia inaequalis*, *Alternaria mali* and *Podosphaera leucotricha* on apples; *Sphaerotheca fuliginea* and *Pseudoperonospora cubensis* on cucurbitaceae; *Phytophthora infestans* and *Alternaria solani* on potato and tomato; *Mycosphaerella arachidis*, *Rhizoctonia solani* and *Sclerotium rolfsii* on peanut; *Monilinia* spp. and *Cladosporium carpophilum* on peach; *Pythium* spp. and *Rhizoctonia solani* on turf; *Mycosphaerella* spp. on banana; *Cladosporium caryigenum* on pecan; *Elsinoe fawcettii*, *Colletotrichum* spp. and *Guignardia citricarpa* on citrus; *Colletotrichum* spp. and *Hemileia vastatrix* on coffee.   **Phytotoxicity** Good crop safety, except on a few varieties of apple (e.g. McIntosh, Cox).
Formulation types SC, WG.   **Selected tradenames** 'Amistar' (not USA) (Zeneca); 'Heritage' (USA, Japan) (Zeneca); 'Quadris' (Zeneca)

MAMMALIAN TOXICOLOGY
Oral Acute oral LD_{50} for male and female rats and mice >5000 mg/kg.   **Skin and eye** Acute percutaneous LD_{50} for rats >2000 mg/kg. Slight eye and skin irritation (rabbits). Not a skin sensitiser (guinea pigs).   **NOEL** 18 mg/kg b.w. daily. **ADI** 0.2 mg/kg b.w.   **Other** Not oncogenic in rats or mice. No evidence of neurotoxicity, endocrine effects or teratogenicity.

ECOTOXICOLOGY
Birds Oral LD_{50} for bobwhite quail >2000 mg/kg b.w. Sub-acute dietary LC_{50} for

bobwhite quail and mallard duck >5200 mg/kg diet. **Other beneficial spp.**
Harmless to non-target organisms under field conditions at field application rates.

ENVIRONMENTAL FATE
Soil/Environment Rapidly degraded in soil, DT_{50} c. 1–4 w. On soil, photolysis
DT_{50} 11 d. Significant leaching has not been observed in field studies.

44 *Bacillus sphaericus*   *Biological agent*

bacterium

NOMENCLATURE
Scientific name *Bacillus sphaericus*   **Strain** Serotype H5a5b

COMMERCIALISATION
Manufacturers Abbott

APPLICATIONS
Uses Biological insecticide for use against mosquito larvae.
Formulation types CG.   **Selected tradenames** 'VectoLex' (Abbott)

MAMMALIAN TOXICOLOGY
Oral LD_{50} for rats >5000 mg/kg (tech.).   **Skin and eye** Acute percutaneous LD_{50}
for rabbits >2000 mg/kg (tech.). Mild skin irritant, eye irritant (rabbits).
Inhalation LC_{50} (4 h) c. 0.09 mg/l (tech.).

45 *Bacillus subtilis*   *Biological agent*

bacterium

NOMENCLATURE
Scientific name *Bacillus subtilis*   **Strain** GB03   **CAS RN** [68038–70–0]
Development codes GUS 2000

PROPERTIES
Form Dry off-white to tan solid.   **S.g./density** 26–29 lbs/ft^3 **Solubility** Insoluble
in water.   **Stability** Stable in dry, ambient conditions for at least two years.

COMMERCIALISATION
History Introduced in 1994. **Patent** US 5215747 **Manufacturers** Chris Hansen Biosystems

APPLICATIONS
Mode of action Bacteria colonise the plant root system, competing with disease organisms which attack root systems. **Uses** Seed treatment. **Formulation types** Seed treatment. **Mixtures** (*B. subtilis* +) metalaxyl + quintozene. **Selected tradenames** 'Kodiak' (Gustafson); 'System 3' (Uniroyal)

ANALYSIS
Fatty acid analysis, using gc.

MAMMALIAN TOXICOLOGY
Oral Not toxic or pathogenic to rats exposed to 10^8 cfu. **Skin and eye** Acute percutaneous LD_{50} for rabbits >2g/kg. **Inhalation** Not toxic to rats exposed to 10^8 cfu. **Toxicity class** EPA (formulation) IV

ECOTOXICOLOGY
Birds No pathogenicity or toxicity to young bobwhite quail by oral gavage over 5 days at 4×10^{11} spores/kg.

ENVIRONMENTAL FATE
Soil/Environment Naturally occurring in soil.

46   *Bacillus thuringiensis*   *Biological agent*

bacterium

NOMENCLATURE
Bacillus thuringiensis
Scientific name *Bacillus thuringiensis* Berliner. **Other names** Bt

subsp. kurstaki
Scientific name *Bacillus thuringiensis* subsp. *kurstaki*. **Var.** Serotype: 3a, 3b (Novartis) **Strain** EG2348, EG2349, EG2371, EG2424 (all Ecogen); Int. 15–313, SA-11, SA-12 (all Novartis); GC-91 (Novartis, a conjugate of *kurstaki* and *aizawai*). **Other names** Btk **Development codes** SAN 239 I; SAN 415 I; SAN 420 I (all Novartis); mixture with B.t. *aizawai*, CGA 237218 (Novartis)

subsp. aizawai
Scientific name *Bacillus thuringiensis* subsp. *aizawai*.   **Var.** Serotype: H-7 (Novartis)
Strain SA-2 (Novartis)   **Development codes** SAN 401 I (Novartis); mixture with
B.t. *kurstaki,* CGA 237218 (Novartis)

subsp. tenebrionis
Scientific name *Bacillus thuringiensis* subsp. *tenebrionis*   **Var.** Serotype: 8a, 8b
(Novartis)   **Strain** SA-10 (Novartis); NovoBtt (Abbott)   **Other names** *Bacillus
thuringiensis* subsp. *san diego*   **Development codes** SAN 418 I (Novartis)

subsp. israelensis
Scientific name *Bacillus thuringiensis* subsp. *israelensis*   **Var.** Serotype: H-14
(Novartis)   **Strain** SA-3 (Novartis)   **Development codes** SAN 402 I (Novartis)

PROPERTIES
Bacillus thuringiensis
Composition *Bacillus thuringiensis* is an aerobic spore-forming gram-positive,
rod-shaped bacterium, belonging to the family Bacillaceae. At sporulation, in
addition to spores, crystals of protein, the delta-endotoxin, are also formed.
Form Suspended solid in a fermentation broth or spray dried concentrate.
S.g./density Depends on fermentation materials and procedure.
Solubility Insoluble in water and organic solvents.   **Stability** Damaged by u.v. light;
dry powders stable up to 40 °C; useful shelf life of aqueous concentrates 0.5 y
(40 °C), 1.0 y (21–25 °C), >3 y (2–10 °C); stable at pH 4 to pH 7 (20 °C);
alkaline hydrolysis: 100% in 1 h at pH 11–12.

subsp. kurstaki
Form Crystals and bacterial spores.   **Stability** Shelf life of formulations is 1–3 y.

subsp. aizawai
Form Crystals and bacterial spores.

subsp. tenebrionis
Form Crystals and bacterial spores.

subsp. israelensis
Form Crystals and bacterial spores.   **Stability** Formulations are stable for 1–2 y in
a cool, dry, dark place.

COMMERCIALISATION
Production By fermentation under well-controlled conditions. **History** This
gram-positive bacterium was detected in 1902 in dying larvae of *Bombyx mori* Lin.
by S. Ishiwata (cited by K. Ishikawa, *Pathology of the Silkworm*) and was
characterised after isolation from larvae of *Ephestia kuehniella* Zell. by E. Berliner

(Z. Angew. Entomol., 1915, **2**, 29). First used as a microbial insecticide 'Sporeine' against lepidopterous larvae in 1938 (S. E. Jacobs, *Proc. Soc. Appl. Bacteriol.*, 1950, **13**, 83). Developed for the control of these pests by several companies, which currently market products. **Manufacturers** Abbott; Arbico; Becker Microbial; Ecogen; Gilmore; Novartis; Shionogi

subsp. kurstaki
History B. *thuringiensis* subsp. *kurstaki* strains EG2348, EG2349 and EG2371 were constructed at Ecogen Inc. (U.S. Patent 5080897) through conjugal mating between B. *thuringiensis* subsp. *kurstaki* strains (recipient) and B. *thuringiensis* subsp. *aizawai* strains (donor); they produce both CryI and CryII crystal proteins. Strain EG2424 was constructed through a multiple step conjugal mating between subsp. *kurstaki* strain EG2042 (recipient) and subspp. *morrisoni* and *kurstaki* strain EG2154 (donors) (see U.S. Patent 5024837 and C. Gawron-Burke & T. Johnson in *Advances in Potato Pest Biology and Management*, G. Zehnder, Ed.); it produces both CryIII and CryI crystal proteins. **Patent** US 5080897; US 5024837 (both to Ecogen) **Manufacturers** Abbott; Arbico; Novartis

subsp. aizawai
Manufacturers Novartis

subsp. tenebrionis
Manufacturers Arbico

subsp. israelensis
Manufacturers Novartis

APPLICATIONS
Bacillus thuringiensis
Mode of action Insecticide with stomach action. Following ingestion, the crystals of endotoxin are solubilised; the epithelial cells of the gut are damaged, insects stop feeding and eventually starve to death. **Mixtures** (*Bacillus thuringiensis* +) pyrethrins.

subsp. kurstaki
Mode of action See above. B. *thuringiensis* subsp. *kurstaki* cultures produce spores and CryI and CryII or CryIII protein crystals. Death occurs in 1–4 days. Products made from this isolate generally have field activity of less than 7 days. **Uses** Used for control of Lepidoptera larvae in agriculture, horticulture and forestry. Ecogen strain EG2424 is used for the control of Colorado potato beetle, *Leptinotarsa decemlineata*, and Lepidoptera. **Formulation types** WP; SC; WG; OF.
Compatibility Compatible with a number of acaricides, insecticides, fungicides, spreaders, stickers and wetters, but not compatible with alkaline products.
Selected tradenames 'Agree' (USA) [GC-91] (mixture) (Novartis); 'Bactospeine'

(Abbott); 'Bactucide' (Caffaro); 'Baturad' (Cequisa); 'Biobit' (Abbott); 'Bollgard' [EG2349] (Ecogen/Crop Care); 'Condor' [EG2348] (Ecogen); 'Cordalene' (Agrichem); 'Costar' [H3a, 3b] (Novartis); 'Crymax' [EG7841] (Ecogen); 'Cutlass' [EG2371] (Ecogen); 'Delfin' [H3a, 3b] (Novartis); 'Design' [GC-91] (Novartis); 'Dipel' (Abbott); 'Ecotech Bio' [EG2371] (Ecogen/Roussel-Uclaf); 'Ecotech Pro' [EG2348] (Ecogen/Roussel-Uclaf); 'Foil' [EG2424] (Ecogen); 'Foray' (for forestry) (Abbott); 'Halt' (Wockhardt); 'Jackpot' [EG2424] (Ecogen, Intrachem); 'Javelin' [H3a,3b] (Novartis); 'Larvo' (Troy); 'Lepinox' [EG7826] (Ecogen, Intrachem); 'Rapax' [EG2348] (Ecogen, Intrachem); 'Raven' [EG2424] (Ecogen); 'Steward' [H3a,3b] (Novartis); 'Thuricide' [H3a,3b] (Novartis); 'Turex' [GC-91] (mixture) (Novartis); 'Vault' [H3a,3b] (Novartis)

subsp. aizawai
Uses Used for the control of lepidopterous larvae (particularly for control of the diamond-back moth) in agriculture, horticulture and forestry, as well as wax moth control. **Selected tradenames** 'Certan' (Novartis); 'Florbac' (Abbott); 'XenTari' (Abbott)

subsp. tenebrionis
Uses Used for control of Coleoptera, especially the Colorado potato beetle (*Leptinotarsa decemlineata*). **Selected tradenames** 'Novodor' (Abbott)

subsp. israelensis
Uses Used for control of mosquito and blackfly larvae. Can be used by aerial application. **Formulation types** SC; WP. **Compatibility** Not normally used in mixture. **Selected tradenames** 'Acrobe' (Cyanamid); 'Bactimos' (public health) (Abbott); 'Bactis' (Caffaro); 'Gnatrol' (Abbott); 'Skeetal' (Abbott); 'Teknar' (Novartis); 'VectoBac' (Abbott)

ANALYSIS
Activity of *B. thuringiensis* is measured in i.u. relative to that of a standard product against *Trichoplusia ni, Spodoptera exigua, Leptinotarsa decemlineata, Aedes aegypti* or other appropriate susceptible species in standardised bioassays. Assays based on the number of spores are not satisfactory for potency determination. Photoimmunoassays are increasingly replacing bioassays as a more accurate and reliable method of determining potency.

subsp. kurstaki
The active ingredient of Ecogen's *B. thuringiensis* subsp. *kurstaki* strains is measured by a modified sodium dodecyl sulfate-polyacrylamide gel electrophoresis technique described by S. Brussock & T. Currier, in *Analytical chemistry of Bacillus thuringiensis*, Hickle and Fitch Eds. Insect bioassays can also be used to quantify the potency of formulations.

MAMMALIAN TOXICOLOGY
Bacillus thuringiensis
Extensive studies on *B. thuringiensis*-containing pesticides demonstrate that isolates
are not toxic or pathogenic. No adverse effects observed in body weight gain,
clinical effects, or on necropsy. Infectivity/pathogenicity studies show that rodents
gradually eliminate *B. thuringiensis* from the body after oral, inhalation or
intravenous application. Observed toxicity at high doses is attributed to the
vegetative growth stage, not to the insecticidal protein or to the spores. Early
formulations, produced from *B. thuringiensis* subsp. *thuringiensis* contained a toxic
β-exotoxin. See review by US EPA, J. T. McClintock *et al.*, *Pestic. Sci.*, **45**, 95
(1995). Some data are given under individual subspp.
Toxicity class EPA (formulation) III

subsp. kurstaki
Oral No infectivity or toxicity observed in rats at 4.7×10^{11} spores/kg. No
adverse effects at doses of 1×10^8 up to 7×10^{12} cfu per rat. **Skin and eye** No
infectivity or toxicity observed in rats at 3.4×10^{11} spores/kg. **Inhalation** No
infectivity or toxicity at 5.4 mg/l (2.6×10^7 spores/l). **NOEL** (2 y) for rats 8.4
g/kg b.w. daily; (13 w) for rats 1.3×10^9 spores/kg b.w. daily.

subsp. aizawai
Oral No adverse effects at $>1 \times 10^8$ cfu per rat.

subsp. tenebrionis
Oral No adverse effects at $>2 \times 10^8$ cfu per rat (tech.) or at >5 g WP/kg.

subsp. israelensis
Oral Acute oral LD_{50} for rats >2.67 g/kg, 1×10^{11} spores/kg; for rabbits
$>2.00 \times 10^9$ spores per rabbit. **Skin and eye** Acute percutaneous LD_{50} for rats
>2000 mg/kg (4.6×10^{10} spores/kg), rabbits >6.28 g/kg. **Inhalation** LC_{50}
8.0×10^7 spores/rat. **NOEL** for rats (3 mo) 4 g/kg b.w. daily.

ECOTOXICOLOGY
Bacillus thuringiensis
Fish LC_{50} $>12 \times 10^9$ spores/l.

subsp. kurstaki
Birds In 63-day feeding trials, chickens receiving up to 5.1×10^7 spores/g diet
showed no ill effects. **Fish** LC_{50} (96 h) for water gobie (*Pomatoschistus minutus*)
>400 mg/l (as 'Thuricide HP'). **Bees** Non-toxic to bees; LD_{50} (oral)
>0.1 mg/bee ('Delfin WG').

subsp. israelensis
Fish LC_{50} (96 h) for water feeder guppies (*Toecilia retriculata*) >156 mg/l (as 'Teknar'). **Daphnia** LC_{50} (96 h) >25 mg/l (tech.)

ENVIRONMENTAL FATE
Soil/Environment In clay loam of low nutrient status (pH 7.3, pF2, 25 °C), insecticidal activity declined rapidly in 20 d due to deterioration of the crystals; at pF3, it declined slowly for 500 d; at higher nutrient status there was a brief 10-fold increase of inoculum applied to soil. As a natural part of the ecosystem, it decays to complex and non-toxic organic compounds.

subsp. kurstaki
Soil/Environment *Bacillus thuringiensis* subsp. *kurstaki* spores have a very short persistence in the environment (DT_{50} 10 h), mainly due to their u.v.-light sensitivity.

47   *Bacillus thuringiensis* delta endotoxins

Insecticide

PHYSICAL CHEMISTRY
Composition 'MVP' (MYX7275), 'MVP II' (MYX104) and 'M-Peril' are based on *Bacillus thuringiensis* var *kurstaki* toxin Cry1A(c), as microcapsule (MVP, MVP II) and granular formulations. 'M-Trak' is based on B.t. var *san diego* toxin Cry3A. 'Mattch' (MYX 300) is a mixture of *kurstaki* Cry1A(c) and *aizawai* CryIC toxins. 'M/C' (MYX833) is based on *aizawai* toxin CryIC.

COMMERCIALISATION
Production Produced in cells of *Pseudomonas fluoresecens* which has been genetically modified to produce the *Bacillus thuringiensis* toxin. For all CellCap-based products, cells are then killed in such a way that they constitute a rigid microcapsule for the enclosed insecticidal protein.   **History** First introduced (as 'MVP') by Mycogen Corporation in 1991.   **Manufacturers** Mycogen

APPLICATIONS
Mode of action Stomach poison.   **Uses** Effective against Lepidoptera and other pests in vegetables, maize, tree fruit, vines and cotton. 'MVP' is recommended for diamond-back moth and other Lepidoptera, 'M/C' for armyworm species (*Spodoptera* spp.), 'M-Trak' for Colorado potato beetle and 'M-Peril' for corn borers. 'Guardjet' was developed for the Japanese crucifer market.
Formulation types CS (encapsulated in killed *Pseudomonas fluorescens*); GR.

Selected tradenames 'MVP', 'MVP II' (kurstaki, Cry1A(c)) (Mycogen); 'M/C' (aizawai, CryIC) (Mycogen); 'Mattch' (kurstaki + aizawai) (Mycogen); 'M-Trak' (san diego, Cry3A) (Mycogen); 'M-Peril' (kurstaki, Cry1A(c)) (Mycogen); 'Guardjet' (kurstaki, Cry1A(c)) (Kubota/Mycogen)

MAMMALIAN TOXICOLOGY
Oral Oral LD_{50} for rats >5050 mg/kg (formulation). **Skin and eye** Acute percutaneous LD_{50} for rabbits >2020 mg/kg (formulation). **Inhalation** All animals survived a dose of 9.98×10^{10} cells. **Other** See *Bacillus thuringiensis*.

BAS 480F   *Fungicide*

48

azole

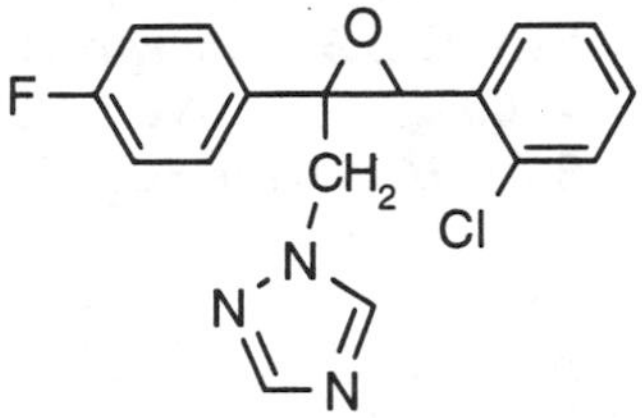

NOMENCLATURE
Common name epoxiconazole (BSI proposed)
IUPAC name (2RS,3SR)-1-[3-(2-chlorophenyl)-2,3-epoxy-2-(4-fluorophenyl)= propyl]-1H-1,2,4-triazole
Chemical Abstracts name *cis*-1-[[3-(2-chlorophenyl)-2-(4-fluorophenyl)= oxiranyl]methyl]-1H-1,2,4-triazole
CAS RN *[106325–08–0]* **EEC no.** 406–850–2 **Development codes** BAS 480F

PHYSICAL CHEMISTRY
Composition Material is the 2R,3S– 2S,3R– enantiomer pair. **Mol. wt.** 329.8
M.f. $C_{17}H_{13}ClFN_3O$ **Form** Colourless crystals. **M.p.** 136.2 °C. **V.p.** <0.01 mPa at 20 °C. **K_{ow}** logP = 3.44 (pH 7) **S.g./density** 1.384 (room temperature)
Solubility In water at 20 °C, 6.63×10^{-4} g/100 ml. In acetone 14.4, dichloromethane 29.1, heptane 0.04 (all in g/100 ml). **Stability** No hydrolysis at pH 5 and pH 7 within 12 days.

COMMERCIALISATION
History Developed and introduced by BASF AG; first registrations in 1993.
Patent EP 94564; US 4464381 **Manufacturers** BASF

APPLICATIONS

Biochemistry Inhibition of C-14-demethylase in sterol biosynthesis. **Mode of action** Preventive and curative fungicide. **Uses** Broad-spectrum fungicide, with preventive and curative action, for control of diseases caused by Ascomycetes, Basidiomycetes, and Deuteromycetes in cereals, sugar beet, peanuts, oilseed rape, and ornamentals. **Formulation types** SC; SE. **Mixtures** *(BAS 480F +)* tridemorph; fenpropimorph; carbendazim; thiophanate-methyl. **Compatibility** Compatible with morpholines and MBC-derivatives. **Selected tradenames** 'Opus' (BASF)

MAMMALIAN TOXICOLOGY

Oral Acute oral LD_{50} for rats >5000 mg/kg. **Skin and eye** Acute percutaneous LD_{50} for rats >2000 mg/kg. Non-irritating to eyes and skin of rabbits. **Inhalation** LC_{50} (4 h) for rats >5.3 mg/l air. **NOEL** (carcinogenicity) for mice 0.81 mg/kg b.w.

ECOTOXICOLOGY

Birds Acute oral LD_{50} for quail >2000 mg/kg. LC_{50} for quail 5000 mg/kg. **Fish** LC_{50} (96 h) for trout 2.2–4.6, bluegill sunfish 4.6–6.8 mg/kg. **Bees** LD_{50} >100 µg/bee. **Worms** EC_{50} (14 d) >1000 mg/kg soil. **Daphnia** LC_{50} (48 h) 8.7 mg/l. **Algae** EC_{50} (72 h) for green algae 2.3 mg/l

ENVIRONMENTAL FATE

Animals A.i. is readily excreted via faeces. There are no major metabolites, but a high number of minor metabolites was identified. The important metabolic reactions were cleavage of the oxirane ring, hydroxylation of the phenyl rings and conjugation. **Plants** There is extensive degradation. **Soil/Environment** Degradation in soil is by microbial activity, DT_{50} c. 2–3 mo. K_{oc} 957–2647.

---

49   BAS 620H   *Herbicide*

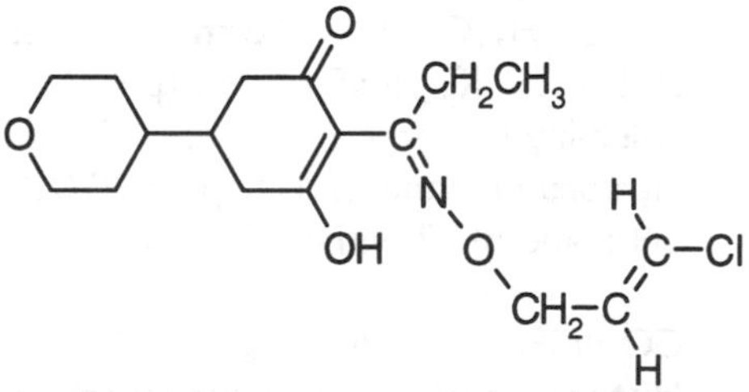

NOMENCLATURE
Common name tepraloxydim (BSI proposed)

IUPAC name (EZ)-(RS)-2-{1-[(2E)-3-chloroallyloxyimino]propyl}-3-hydroxy-=
5-perhydropyran-4-ylcyclohex-2-en-1-one
Chemical Abstracts name (E)-2-[1-[[(3-chloro-2-propenyl)oxy]imino]propyl]-=
3-hydroxy-5-(tetrahydro-2H-pyran-4-yl)-2-cyclohexen-1-one
Other names caloxydim (BSI rejected name)   **CAS RN** [149979–41–9]
Development codes BAS 620H

PHYSICAL CHEMISTRY
Mol. wt. 341.8   **M.f.** $C_{17}H_{24}ClNO_4$

COMMERCIALISATION
History Under development by Nisso BASF Agro Ltd, a joint venture company of
Nippon Soda Co., Ltd, BASF AG and Mitsui & Co.

APPLICATIONS
Uses Under development for soya beans and cotton.

BAS 654 00 H   *Herbicide*

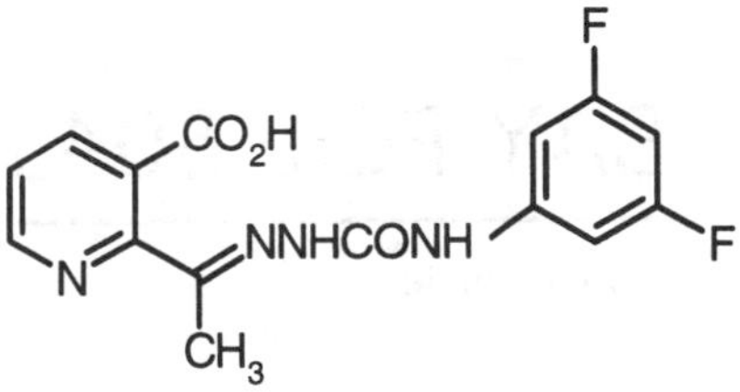

NOMENCLATURE
Common name diflufenzopyr (BSI proposed)
IUPAC name 2-{1-[4-(3,5-difluorophenyl)semicarbazono]ethyl}nicotinic acid
Chemical Abstracts name 2-[1-[[[(3,5-difluorophenyl)amino]carbonyl]=
hydrazono]ethyl]-3-pyridinecarboxylic acid
Other names [SAN 835 H]   **CAS RN** [109293–97–2]; (sodium salt
[109293–98–3])   **Development codes** BAS 654 00 H; [SAN 835 H] (acid);
[SAN 836 H] (sodium salt)

PHYSICAL CHEMISTRY
Mol. wt. 334.3   **M.f.** $C_{15}H_{12}F_2N_4O_3$ **Form** Tan coloured odourless solid.
M.p. 155 °C (dec.)   **V.p.** <10^{-2} mPa   **K$_{ow}$** logP = 0.037 (unstated pH)
S.g./density 0.24   **Solubility** In water 63 ppm.   **pKa** 3.18

COMMERCIALISATION
History Originated by Sandoz Agro (now Novartis Crop Protection AG); development and distribution rights acquired by BASF in December 1996.

APPLICATIONS
Biochemistry Polar auxin transport inhibitor. **Mode of action** Systemic, post-emergence herbicide. **Uses** For post-emergence control of annual broad-leaved and perennial weeds in maize. Initial commercialisation of a mixture with dicamba, both materials as sodium salts, is planned. To be applied at 0.2–0.4 kg a.e./ha. **Formulation types** WG.

MAMMALIAN TOXICOLOGY
Oral Acute oral LD_{50} >5.0 g/kg. **Skin and eye** Acute percutaneous LD_{50} >5.0 g/kg. **Inhalation** LC_{50} >3.14 mg/l.

ECOTOXICOLOGY
Birds LD_{50} for mallard duck >5620 ppm. **Fish** LD_{50} for bluegill sunfish >135, rainbow trout 106 mg/l. **Bees** LD_{50} >90 µg/bee.

ENVIRONMENTAL FATE
Soil/Environment Average DT_{50} in soil 4.5 d.

51   BAY FOE 5043   *Herbicide*

oxyacetamide

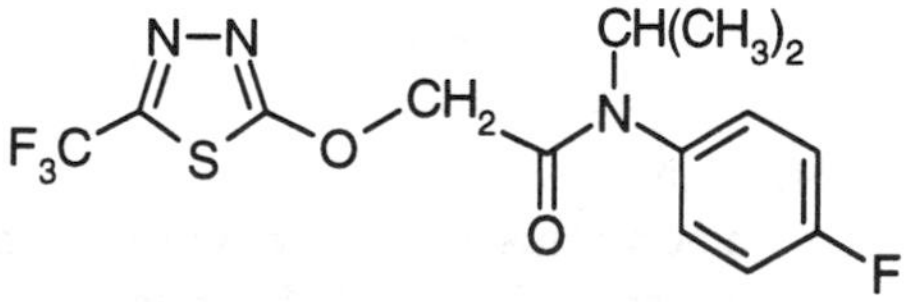

NOMENCLATURE
Common name fluthiamide (BSI proposed); thiadiazolamide (BSI alternative proposal)
IUPAC name 4'-fluoro-N-isopropyl-2-(5-trifluoromethyl-1,3,4-thiadiazol-=2-yloxy)acetanilide
Chemical Abstracts name N-(4-fluorophenyl)-N-(1-methylethyl)-=2-[(trifluoromethyl)-1,3,4-thiadiazol-2-yloxy]acetamide
CAS RN [142459–58–3]   **Development codes** BAY FOE 5043

PHYSICAL CHEMISTRY
Mol. wt. 363.34 **M.f.** $C_{14}H_{13}F_4N_3O_2S$ **Form** White to tan solid. **M.p.** 75–77 °C
V.p. 9×10^{-2} mPa (20 °C) **K$_{ow}$** logP = 3.2 **Solubility** In water 56 (pH 4), 56
(pH 7), 54 (pH 9) mg/l (25 °C). **Stability** Stable to hydrolysis at pH 5–9. Stable
to photolysis at pH 5.

COMMERCIALISATION
History Reported by R Deege *et al.* (*Proc. Br. Crop Prot. Conf. – Weeds*, 1995, **1**,
43). Under development by Bayer AG. **Manufacturers** Bayer

APPLICATIONS
Biochemistry Inhibits cell division and growth. **Mode of action** Pre- and early
post-emergence herbicide. **Uses** Under development for broad spectrum grass
control in a range of crops. **Mixtures** *(BAY FOE 5043 +)* metribuzin.
Selected tradenames 'Axiom' (mixture) (Bayer)

MAMMALIAN TOXICOLOGY
Oral Acute oral LD$_{50}$ for rats 589 mg/kg. **Skin and eye** Acute percutaneous LD$_{50}$
for rats >2000 mg/kg. Not an eye or skin irritant (rabbits). **Inhalation** LC$_{50}$ (4 h)
for rats >3740 mg/m^3 (aerosol). **NOEL** (2 y) for rats 25 mg/kg diet; (12 mo) for
dogs 40 mg/kg diet; (20 mo) for mice 50 mg/kg diet. **ADI** 0.011 mg/kg b.w.
(Bayer 1996) **Other** Non-mutagenic (Ames test); non-teratogenic (rabbit and
rat). **EC risk** Xn (R22)

ECOTOXICOLOGY
Fish LC$_{50}$ for bluegill sunfish 2.4, rainbow trout 3.5 mg/l. **Daphnia** LC$_{50}$
39.4 mg/l.

ENVIRONMENTAL FATE
Soil/Environment Aerobic DT$_{50}$ 34 d; K$_{oc}$ (sandy loam) 354 (range 233–613).
Stable to photolysis in soil.

52   *Beauveria bassiana*   *Biological agent*

entomopathogenic fungus

NOMENCLATURE
Scientific name *Beauveria bassiana* (Balsamo) Vuillemin **Strain** Bb 147 ('Ostrinil');
ATCC 74040 (= ARSEF 3097 = FCI 7744) ('Naturalis-L'); GHA ('BotaniGard' ES)
Development codes ESC 170 GH (Ecoscience); F-7744 (Troy)

PROPERTIES
Stability 'Ostrinil' has a shelf-life of 1 y when stored under 20 °C. It should not undergo thermal shock.

COMMERCIALISATION
Production B. bassiana is cultured by solid state fermentation on clay granules.
History An isolate of the Deuteromycete (Moniliales) fungus, *Beauveria bassiana* was obtained from a mycosed larva of *Ostrinia nubilalis* (*Lepidoptera: pyralidae*) or European corn borer, harvested in Beauce (North of France), by the Institut National de la Recherche Agronomique (France). This strain is particularly virulent (G. Riba, 1985. Thése de Doctorat d'Etat, Mention Sciences, Université Pierre et Marie Curie, France). A granular formulation was tested with success by A. Guerin (in *L'emploi d'ennemis naturels dans la protection des cultures*, Les Colloques de l'INRA No 34, Pub. INRA). A production process was developed by INRA and is now owned by Natural Plant Protection (NPP). Strain ATCC 74040, used in 'Naturalis-L', was isolated by Troy Biosciences from a coleopteran boll weevil in Texas (J. E. Wright & T. A. Knauf, *Proc. Br. Conf. Pests Dis.*, 1994, **1**, 45).
Patent EP 9040118330 **Manufacturers** Mycotech; NPP; Troy

APPLICATIONS
Mode of action Contact action, by attachment of fungal conidia to insect cuticle, followed by penetration and proliferation of hyphae in the insect body. The infection process requires freely available water and takes 24 to 48 hours. After the death of the insect, occurring between 3 to 5 days, fresh spores are produced on the outside of the cadaver. **Uses** 'Naturalis-L' controls various Coleoptera, Homoptera and Heteroptera. 'Ostrinil' is used to control European corn borer, at 25 kg/ha. 'BotaniGard' is used to control whitefly, thrips, aphids and mealybugs in vegetables and ornamentals. 'Back-Off' is under development for control of pests in cotton, vegetables and ornamentals. **Formulation types** 'Ostrinil' is MG (a clay microgranule formulation colonised by sporulating mycelia of "pyralidae-active" strain). 'Mycotrol' is a WP. 'Naturalis-L' is a SC. **Selected tradenames** 'BotaniGard' ES (Mycotech); 'Naturalis-L' [ATCC 74040] (Troy); 'Ostrinil' (NPP)

ANALYSIS
Activity of 'Ostrinil' is measured in terms of spore count (minimum of 5×10^8 spores/g) and efficacy against larvae (5th stage) of *Ostrinia nubilalis* (TL_{50} <4 days).

MAMMALIAN TOXICOLOGY
Oral Acute oral LD_{50} for rats >18 × 10^8 cfu/kg. No infectivity or pathogenicity was observed after 21 d. **Skin and eye** Acute percutaneous LD_{50} for rats >2000 mg/kg. **Inhalation** LD_{50} for rats >1.2 × 10^8 cfu/animal. **Other** Dermal, oral and inhalation studies with 'Naturalis-L' on rats indicate that the fungus is non-toxic and non-pathogenic.

ECOTOXICOLOGY

Birds Oral LD$_{50}$ (5 d) for quail >2000 mg/kg daily (by gavage). **Fish** 'Naturalis-L' does not affect fish embryos, larvae or adults. LC$_{50}$ (31 d) for rainbow trout 7300 mg/l. **Bees** 30-d Dietary and contact studies indicate that 'Naturalis-L' has no significant effect; LC$_{50}$ (23 d, ingestion) 9285 mg/kg. **Other beneficial spp.** No effect observed on beneficial spp., after field application. **Daphnia** EC$_{50}$ (14 d) 4100 mg/l.

ENVIRONMENTAL FATE

Soil/Environment Background levels of *B. bassiana* present in cotton fields were not significantly changed after repeated applications of 'Naturalis-L', indicating that the fungus does not accumulate.

MAIN ENTRIES

3 *Beauveria brongniartii*   Biological agent

entomopathogenic fungus

NOMENCLATURE

Scientific name *Beauveria brongniartii* **Var.** Bb 96 ('Betel'); Swiss isolates.

PROPERTIES

Stability Stable for 1 y at 2 °C.

COMMERCIALISATION

Production *B. brongniartii* is cultured by solid state fermentation on clay granules.
History An isolate of the Deuteromycete (Moniliales) fungus, *Beauveria brongniartii* was obtained from a mycosed larva of *Hoplochelus marginalis* (*Coleoptera: Melolonthinae*) or white grub, harvested in Madagascar, by the CIRAD/IRAT (France). This strain is particularly virulent (O. Goebel, 1989. Diplôme d'ingénieur en agronomie tropicale, CNEARC/ESAT, France). A fermented-rice preparation was tested with success (B. Vercambre *et al.*, 1991, in *Rencontres caraibes en lutte biologique. Les Colloques de l'INRA No 58*, p371–378, Pub. INRA). A production process was developed by INRA and is now owned by Natural Plant Protection (NPP). The formulation is marketed under the trade mark of 'Betel'.
'Engerlingspilz', owned by Andermatt, was registered in Switzerland in 1990.
Manufacturers Andermatt; NPP

APPLICATIONS
Mode of action Contact action by attachment of fungal conidia to insect cuticle, followed by penetration and proliferation of hyphae in the insect body. The infection process requires freely available water and takes 24 to 48 hours. After the death of the insect, occurring between 3 to 5 days, fresh spores are produced on the outside of the cadaver and other insects can be infected at their contact. **Uses** Used for control of cockchafers (*Melolontha melolontha*); inoculated barley is sown in the soil. 'Betel' is used specifically to control white grub (*Hoplochelus* spp.) on sugar cane. Application can be made either at planting time on the edges of the furrow, or at the foot of the ratoon canes. At the dose of 50 kg/ha, 'Betel' reduces the larval population below the damage threshold (3 larvae per sugar cane). 'Biolisa' is under development for control of *Longicornis*. **Formulation types** *B. brongniartii* from Andermatt is in the form of inoculated barley. 'Betel' is MG. **Selected tradenames** 'Betel' (NPP)

ANALYSIS
Activity of 'Betel' is measured in terms of spore count (minimum of 0.2×10^8 spores/g) and efficacy against larvae of *Hoplochelus marginalis*. Activity of 'Engerlingspilz' is measured by biotest with *Melolontha melolontha*.

MAMMALIAN TOXICOLOGY
Oral Acute oral LD_{50} for rats >5000 mg/kg. No toxicity, infectivity or pathogenicity from a single dose of 1.1×10^9 cfu/kg (rats). **Skin and eye** Acute percutaneous LD_{50} for rats >2000 mg/kg. Mildly irritant to skin (rabbit). **Other** Non-toxic.

ECOTOXICOLOGY
Birds Dietary LD_{50} (5 d) for quail and mallard ducks >4000 mg/kg. **Fish** LC_{50} (30 d) for rainbow trout 7200 mg/l, NOAEL (30 d) 3000 mg/l. **Daphnia** NOAEL (21 d) 500 mg/l.

benalaxyl

Fungicide

phenylamide (acylalanine type)

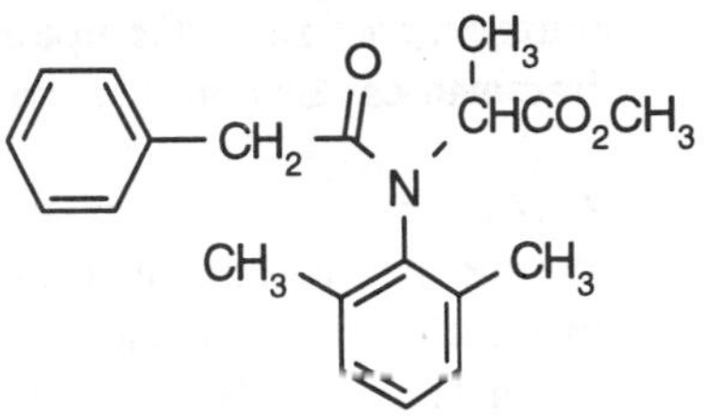

NOMENCLATURE
Common name benalaxyl (BSI, draft E-ISO, (*m*) draft F-ISO)
IUPAC name methyl *N*-phenylacetyl-*N*-2,6-xylyl-DL-alaninate
Chemical Abstracts name methyl *N*-(2,6-dimethylphenyl)-*N*-(phenylacetyl)-=
DL-alaninate
CAS RN *[71626–11–4]*   **EEC no.** 275–728–7   **Development codes** M 9834

PHYSICAL CHEMISTRY
Mol. wt. 325.4   **M.f.** $C_{20}H_{23}NO_3$ **Form** Colourless, almost odourless solid.
M.p. 78–80 °C   **V.p.** 0.66 mPa (25 °C, gas saturation method) K_{ow} logP = 3.54
(20 °C)   **Henry** 6.5 × 10^{-3} Pa m^3 mol^{-1} (20 °C, calc.)   **S.g./density** 1.181 (20 °C)
Solubility In water 28.6 mg/l (20 °C). In acetone, methanol and ethyl acetate 1.2,
dichloroethane and xylene 250, *n*-heptane <20 (all in g/kg, 22 °C).
Stability Hydrolysed in concentrated alkaline media. Stable in aqueous solutions at
pH 4–9. DT_{50} 86 d (pH 9, 25 °C). Thermally stable up to 250 °C under nitrogen.
Stable to sunlight in aqueous solution.

COMMERCIALISATION
History Fungicide reported by Garavaglia *et al.* (*Atti Simp. Chim. Antiparassitari, 3rd,
Piacenza,* 1981). Introduced by Farmoplant S.p.A. (now Isagro S.p.A.).
Patent BE 873908; DE 2903612; IT 19896/78   **Manufacturers** Isagro

APPLICATIONS
Biochemistry Nucleic RNA-polymerase inhibitor.   **Mode of action** Systemic
fungicide with protective, curative, and eradicant action. Absorbed by the roots,
stems, and leaves, with translocation acropetally to all parts of the plant, including
subsequent growth. Protective action is by inhibition of spore germination and of
mycelial growth, whilst curative action is by inhibition of mycelial growth, and
eradicant action is by inhibition of conidiophore formation.   **Uses** Control of
Oomycetes, particularly fungi of the family Peronosporaceae, *Phytophthora,
Plasmopara, Pseudoperonospora, Sclerospora, Bremia,* and *Pythium* spp. Particular
uses include control of late blights of potatoes and tomatoes; downy mildews of
hops, vines, lettuce, onions, soya beans, tobacco, and other crops; diseases caused

by *Phytophthora* spp. in strawberries; many diseases of flowers and ornamentals; and *Pythium* spp. on turf. Often used in combination with other fungicides. **Formulation types** GR; WP; EC. **Mixtures** *(benalaxyl +)* copper oxychloride; folpet; mancozeb. **Compatibility** Incompatible with alkaline materials. **Selected tradenames** 'Galben' (Isagro)

ANALYSIS

Product and **residue** analysis by hplc or glc. 'Gas Chromatographic Determination of Benalaxyl Residues in Different Crops and Water' in *Journal of AOAC International*, **76**, No. 3 (1993) 650–656. Details available from Isagro.

MAMMALIAN TOXICOLOGY

Reviews *Pesticide residues in food – 1987*. FAO Plant Production and Protection Paper 84, 1987. *Pesticide residues in food – 1987 evaluations. Part II – Toxicology*. FAO Plant Production and Protection Paper 86/2, 1988. **Oral** Acute oral LD_{50} for rats 4200, mice 680 mg/kg. **Skin and eye** Acute percutaneous LD_{50} for rats >5000 mg/kg. Non-irritating to skin and eyes (rabbits). Non-sensitising to skin (guinea pigs). **Inhalation** LC_{50} (4 h) for rats >10 mg/l air. **NOEL** (2 y) for rats 100 mg/kg diet; (1.5 y) for mice 250 mg/kg diet; (1 y) for dogs 200 mg/kg diet. **ADI** (JMPR) 0.05 mg/kg b.w. [1987]. **Other** Non-carcinogenic, non-mutagenic, and non-teratogenic. **Toxicity class** WHO (a.i.) III (Table 5); EPA (formulation) III

ECOTOXICOLOGY

Birds Acute oral LD_{50} for mallard ducks >4500, bobwhite quail >5000, chickens 4600 mg/kg. Five-day dietary LC_{50} for bobwhite quail and mallard ducks >5000 mg/kg. **Fish** LC_{50} (96 h) for rainbow trout 3.75, goldfish 7.6, guppies 7.0, carp 6.0 mg/l. **Bees** Non-toxic to bees; LD_{50} >100 µg/bee. **Worms** LC_{50} (48 h) for earthworms 0.0035 mg/cm^2. **Daphnia** LC_{50} (48 h) 0.59 mg/l. **Algae** EC_{50} (96 h) for *Selenastrum capricornutum* 2.4 mg/l.

ENVIRONMENTAL FATE

Animals In rats, following oral administration, rapidly metabolised, and eliminated in the urine (23%) and faeces (75%) within 2 days. **Plants** Slowly metabolised to glycosides in plants. **Soil/Environment** Slowly degraded by soil micro-organisms to various acidic metabolites. DT_{50} in soil is 20–71 d. K_{oc} 2728–7173 (3 soil types).

benazolin *Herbicide*

NOMENCLATURE
benazolin
Common name benazolin (BSI, E-ISO, WSSA); bénazoline ((*f*) F-ISO)
IUPAC name 4-chloro-2-oxobenzothiazolin-3-ylacetic acid;
4-chloro-2,3-dihydro-2-oxobenzothiazol-3-ylacetic acid
Chemical Abstracts name 4-chloro-2-oxo-3(2*H*)-benzothiazoleacetic acid
CAS RN *[3813–05–6]*   **EEC no.** 223–297–0   **Development codes** RD 7693

benazolin-ethyl
Common name benazolin-ethyl
CAS RN *[25059–80–7]*   **EEC no.** 246–591–0

PHYSICAL CHEMISTRY
benazolin
Composition Tech. grade is *c.* 90% pure.   **Mol. wt.** 243.7   **M.f.** $C_9H_6ClNO_3S$
Form Colourless, odourless crystals.   **M.p.** 193 °C; (tech. 189 °C)
V.p. 1×10^{-4} mPa (20 °C) K_{ow} logP = 1.34 (20 °C)   **Solubility** In water 500 mg/l
(pH 2.94, 20 °C). In acetone 100–120, ethanol 30–38, ethyl acetate 21–25,
isopropanol 25–30, dichloromethane 3.7, toluene 0.58, *p*-xylene 0.49, hexane
<0.002 (all in g/l, 20 °C).   **Stability** Stable in neutral, acidic, and weakly alkaline
media. Decomposed by concentrated alkalis.   **pKa** 3.04 (20 °C)

benazolin-ethyl
Mol. wt. 271.7   **M.f.** $C_{11}H_{10}ClNO_3S$ **Form** White crystalline solid; tech. is pale
yellow crystalline powder with a characteristic odour.   **M.p.** 79.2 °C;
(tech., 77.4 °C)   **V.p.** tech. 0.37 mPa (25 °C) K_{ow} logP = 2.50 (20 °C, in distilled
water)   **S.g./density** 1.45 (20 °C)   **Solubility** In water 47 mg/l (20 °C). In acetone
229, dichloromethane 603, ethyl acetate 148, methanol 28.5, toluene
198 (g/l, 20 °C).   **Stability** Stable to at least 300 °C. Stable in acid and in neutral
solution; at pH 9.18 DT_{50} 7.6 d (25 °C). No significant breakdown in water in
natural sunlight.

COMMERCIALISATION
History Herbicide reported by E. L. Leafe (*Proc. Br. Weed Control Conf.*, 7th, 1964,
p. 32). Introduced by The Boots Co. Ltd (now AgrEvo GmbH).

Patent GB 862226; GB 1243006   **Manufacturers** AgrEvo

APPLICATIONS
benazolin
Mode of action Selective, systemic, growth-regulator herbicide, absorbed principally by the leaves, and translocated readily throughout the plant in the phloem.   **Uses** Post-emergence control of many annual broad-leaved weeds, particularly black bindweed (*Bilderdykia convolvulus*), common chickweed (*Stellaria media*), cleavers (*Galium aparine*), and charlock (*Sinapis arvensis*), in oilseed rape. In combination with other herbicides, used for control of a wider range of broad-leaved weeds in oilseed rape, cereals (non-undersown, and undersown with grass or clover), grassland, clover, alfalfa, and flax. With dicamba, benazolin exhibits a strong synergistic effect, giving improved control of a number of species, including mayweeds.   **Phytotoxicity** Selective in cereals, oilseed rape, soya beans, maize, grassland.   **Mixtures** *(benazolin +)* clopyralid; 2,4-DB + MCPA; bromoxynil + ioxynil; bromoxynil + mecoprop.   **Selected tradenames** 'Keropur' (AgrEvo)

benazolin-ethyl
Formulation types SC; EC.   **Mixtures** As benazolin.   **Selected tradenames** 'Galtak' (AgrEvo)

ANALYSIS
Product analysis by glc or hplc. **Residue** analysis by glc.   Details available from AgrEvo.

MAMMALIAN TOXICOLOGY
benazolin
Oral Acute oral LD_{50} for rats >5000, mice >4000 mg/kg.   **Skin and eye** Acute percutaneous LD_{50} for rats >5000 mg/kg. Mild skin and eye irritant (rabbits). Not a skin sensitiser.   **Inhalation** LC_{50} (4 h) for rats 1.43 g/m^3 air.   **NOEL** (90 d) for rats 300–1000 mg/kg daily, for dogs *c.* 300 mg/kg daily.   **Toxicity class** WHO (a.i.) III (Table 5); EPA (formulation) III   **EC risk** Xi (R36/38)

benazolin-ethyl
Reviews *Informationen zum Wirkstoff: Benazolin-ethyl*, Industrieverband Agrar e.V., prepared by Schering AG (now AgrEvo). *Public Access to Information on Pesticides*, BAA Summary document: Benazolin-ethyl, prepared by Schering Agrochemicals (now AgrEvo).   **Oral** Acute oral LD_{50} for mice >4000, rats >6000, dogs >5000 mg/kg.   **Skin and eye** Acute percutaneous LD_{50} for rats >2100 mg/kg; not irritant to skin and eyes of rabbits. Not a sensitiser.   **Inhalation** LC_{50} (4 h) >5.5 mg/l. Low acute toxicity by inhalation route.   **NOEL** (2 y) for rats 12.5 mg/kg (0.61 mg/kg daily); (1 y) for dogs 500 mg/kg (18.6 mg/kg daily).   **ADI** 0.006 mg/kg (dogs); 0.36 mg for a 60 kg human.

ECOTOXICOLOGY
benazolin
Birds Acute oral LD_{50} for Japanese quail >10 200 mg/kg.   **Fish** LC_{50} (96 h) for trout 31.3, bluegill sunfish 27 mg/l.   **Bees** Non-toxic. LD_{50} 480 µg/bee.
Daphnia Low toxicity; LC_{50} (48 h) 233.4 mg/l (measured); 353.6 mg/l (nominal).

benazolin-ethyl
Birds Acute oral LD_{50} for bobwhite quail >6000, Japanese quail >9709, mallard ducks >3000 mg/kg. Dietary LC_{50} (5 d) for bobwhite quail and mallard ducks >20 000 mg/kg diet.   **Fish** LC_{50} (96 h) for bluegill sunfish 2.8, rainbow trout 5.4 mg/l.   **Bees** 10% EC formulation not harmful to bees.   **Worms** Low toxicity to earthworms: >1000 mg/kg dry soil.   **Daphnia** LC_{50} (48 h) 6.2 mg/l. 21 d NOEC (immobilisation) 0.05 mg/l, (reproduction) 0.158 mg/l.   **Algae** EC_{50} for *Selenastrum capricornutum* 16.0 mg/l;   NOEL 1.0 mg/l.

ENVIRONMENTAL FATE
benazolin
Animals In urine, major metabolites are *N*-[2-chloro-6-(methylsulfinyl)phenyl]= glycine and *N*-[*N*-[2-chloro-6-(methylthio)phenyl]glycinyl]aniline. There are also small amounts of other acid-labile polar conjugates of benazolin acid, and *N*-[2-chloro-6-(methylthio)phenylglycine]. Faecal metabolites are similar to those in urine.   **Plants** Benazolin acid is metabolised to form acid-labile conjugates. Minor pathways include hydroxylation of the aromatic ring of benazolin acid and loss of the acetic acid side-chain, followed by conjugation. Benazolin acid is translocated more rapidly and to a greater extent in sensitive species (e.g. mustard) than in resistant species (barley and rape). In barley, uptake and translocation are minimal and the majority of the applied material is recovered as parent compound at the site of application.   **Soil/Environment** Degrades to principally 'bound residues'. DT_{50} c. 14–28 d. K_d 1.0 (sandy loam), 0.4 (loam).

benazolin-ethyl
Animals Metabolism is by de-esterification of benazolin-ethyl to benazolin with a minor pathway involving opening of the thiazoline ring.   **Plants** The major metabolite is benazolin acid.   **Soil/Environment** On soil surface exposed to light, DT_{50} 3.5 d. In soil, DT_{50} 1–2 d. Benazolin-ethyl is degraded in soil by hydrolysis of the ester followed by a loss of the acetic acid side-chain, then ring opening and sulfonation. The major degradates are benazolin acid and 4-chloro-2-= oxobenzothiazolin. K_d 15 (sandy loam), 8 (loam).

MAIN ENTRIES

bendiocarb

carbamate

NOMENCLATURE

Common name bendiocarb (BSI, E-ISO, ANSI, ESA); bendiocarbe ((*m*) F-ISO)
IUPAC name 2,3-isopropylidenedioxyphenyl methylcarbamate;
2,2-dimethyl-1,3-benzodioxol-4-yl methylcarbamate
Chemical Abstracts name 2,2-dimethyl-1,3-benzodioxol-4-yl methylcarbamate
Other names [bencarbate]   **CAS RN** *[22781-23-3]*   **EEC no.** 245-216-8
Development codes NC 6897   **Official codes** OMS 1394

PHYSICAL CHEMISTRY

Mol. wt. 223.2   **M.f.** $C_{11}H_{13}NO_4$ **Form** Colourless, odourless, crystalline solid.
M.p. 124.6–128.7 °C   **V.p.** 4.6 mPa (25 °C) (gas saturation with glc detection)
K_{ow} logP = 1.72 (pH 6.55)   **S.g./density** 0.69 (20 °C)   **Solubility** In water 0.28 g/l
(pH 7, 20 °C). In dichloromethane 200–300, chloroform, dioxane 200, acetone
150–200, methanol 75–100, ethyl acetate 60–75, benzene, ethanol 40, *o*-xylene
10, *p*-xylene 11.7, *n*-hexane 0.225, kerosene <1 (all in g/l, 25 °C).   **Stability**
Hydrolysed rapidly in alkaline media, and more slowly in neutral and acidic media.
DT_{50} (tested under EPA guidelines) (25 °C) 4 d (pH 7), the products being
2,3-isopropylidenedioxyphenol, methylamine and carbon dioxide. Stable to light
and heat.   **pKa** 8.8, v. weak acid

COMMERCIALISATION

History Insecticide reported by R. W. Lemon (*Proc. Br. Insectic. Fungic. Conf. 6th*,
1971, **2**, 570) and P. J. Brooker *et al.* (*Pestic. Sci.*, 1972, **3**, 735). Introduced by
Fisons Ltd, Agrochemical Division. Formerly marketed by Schering Agrochemicals
for crop protection and by Cambridge Animal and Public Health Limited
(CAMCO) for public health use, now marketed for both uses by AgrEvo GmbH.
Patent GB 1220056   **Manufacturers** AgrEvo; Kuo Ching

APPLICATIONS

Biochemistry Cholinesterase inhibitor.   **Mode of action** Systemic insecticide with
contact and stomach action. Gives rapid knockdown, and has good residual
activity.   **Uses** Active against many public health, industrial and storage pests, such
as Formicidae, Blattodea, Culicidae, Muscidae, Siphonaptera. Particularly useful
inside buildings due to its low odour and lack of corrosive and staining properties.

Also used in turf and ornamentals for control of a broad spectrum of pests. In agriculture, it is used as a seed treatment and in granular formulations for the control of soil-dwelling pests and some foliar pests (*Agriotes* spp., *Atomaria linearis, Oscinella frit*), particularly in maize and sugar beet. Also used as a foliar spray on other crops to control Thysanoptera and other pests. **Phytotoxicity** Non-phytotoxic when used as directed. Not to be used on coleus.
Formulation types AE; DP; FS; GR; SC; UL; WP. **Mixtures** *(bendiocarb +)* piperonyl butoxide + pyrethrins. **Compatibility** Incompatible with alkaline materials. **Selected tradenames** 'Ficam' (mixture) (public health use) (AgrEvo); 'Garvox' (crop protection use) (AgrEvo); 'Seedox' (crop protection use) (AgrEvo); 'Dycarb' (Mallinckrodt); 'Multamat' (AgrEvo)

ANALYSIS
Product analysis by hplc (*CIPAC Handbook*, 1988, **D**, 10; *AOAC Methods, 1995,* 986.09). **Residue** analysis of foodstuffs by glc. Details available from AgrEvo.

MAMMALIAN TOXICOLOGY
Reviews *Pesticide residues in food – 1984.* FAO Plant Production and Protection Paper 62, 1985. *Pesticide residues in food – 1984 evaluations.* FAO Plant Production and Protection Paper 67, 1985. **EHC** 64 (WHO, 1986; a review of carbamate insecticides in general). **Oral** Acute oral LD_{50} for rats 40–156, mice 45, guinea pigs 35, rabbits 35–40 mg/kg. **Skin and eye** Acute percutaneous LD_{50} for rats 566–800 mg/kg. Non-irritating to skin and eyes. **Inhalation** LC_{50} (4 h) for rats 0.55 mg/l air. **NOEL** (90 d) and (2 y) for rats 10 mg/kg diet. In 90 d trials, rats receiving 250 mg/kg diet showed no ill-effects other than reversible inhibition of cholinesterase. **ADI** (JMPR) 0.004 mg/kg b.w. [1984]. **Toxicity class** WHO (a.i.) II; EPA (formulation) II **EC risk** T (R25); Xn (R21)

ECOTOXICOLOGY
Birds Acute oral LD_{50} for mallard ducks 3.1, bobwhite quail 19, domestic hens 137 mg/kg. **Fish** LC_{50} (96 h) for sheepshead minnow 0.86, bluegill sunfish 1.65, rainbow trout 1.55 mg/l. **Bees** Toxic to bees. Oral LD_{50} for honeybees 0.1 μg/bee. **Daphnia** LC_{50} (24 h) 0.33 mg/l; (48 h) 0.16 mg/l.

ENVIRONMENTAL FATE
Animals In rats and a range of other mammals, rapidly absorbed following oral administration or inhalation, but not following skin contact. Rapidly detoxified and eliminated almost completely within 24 hours as the sulfate and glucuronide conjugate of the principal metabolite 2,2-dimethyl-1,3-benzodioxol-4-ol (J. W. Adcock et al., *Pestic. Sci.,* 1981, **12**, 645). **Soil/Environment** Bendiocarb is rapidly degraded in soil, via hydrolysis of the methyl carbamate and heterocyclic rings, followed by oxidation to polar and soil bound residues, together with considerable mineralisation of the phenyl ring to CO_2. The rate of degradation of bendiocarb is pH-dependent, with slower breakdown under acidic conditions. DT_{50} values for

the rate of decline of bendiocarb in agricultural soils and standard Speyer 2.2 range from 0.5 to 10 d, depending upon soil type, moisture and temperature. In adsorption/desorption studies, bendiocarb was hydrolysed to 2,2-dimethyl-1,3-= benzodioxol-4-ol (NC 7312). Bendiocarb and NC 7312 are weakly adsorbed (K_{oc} 28–40); however because it is rapidly degraded, bendiocarb should not leach.

57 benfluralin   *Herbicide*

2,6-dinitroaniline

NOMENCLATURE
Common name benfluralin (BSI, E-ISO); benfluraline ((*f*) F-ISO); benefin (WSSA); bethrodine (JMAF)
IUPAC name N-butyl-N-ethyl-α,α,α-trifluoro-2,6-dinitro-*p*-toluidine
Chemical Abstracts name N-butyl-N-ethyl-2,6-dinitro-4-(trifluoromethyl)= benzenamine
CAS RN *[1861–40–1]* **EEC no.** 217–465–2 **Development codes** EL-110

PHYSICAL CHEMISTRY
Mol. wt. 335.3 **M.f.** $C_{13}H_{16}F_3N_3O_4$ **Form** Yellow-orange crystals.
M.p. 65–66.5 °C **B.p.** 121–122 °C/0.5 mmHg; 148–149 °C/7 mmHg
V.p. 8.7 mPa (25 °C). **K_{ow}** logP = 5.29 (20 °C, pH 7) **S.g./density** 1.28 (20 °C) (tech.) **Solubility** In water 0.1 mg/l (25 °C). In acetone, ethyl acetate, dichloromethane, chloroform >1000, toluene 330–500, acetonitrile 170–200, hexane 18–20, methanol 17–18 (all in g/l, 25 °C). **Stability** Decomposed by u.v. light. Stable for up to 30 days at pH 5–9 (26 °C).

COMMERCIALISATION
History Herbicide reported by J. F. Schwer (*Proc. North Cent. Weed Control Conf.*, 1965). Introduced in USA (1963) by Eli Lilly & Co (now DowElanco).
Patent US 3257190 **Manufacturers** Budapesti Vegyimüvek; DowElanco; Makhteshim-Agan

APPLICATIONS
Biochemistry Cell division inhibitor. **Mode of action** Selective soil herbicide,

absorbed by the roots. Affects seed germination and prevents weed growth by inhibition of root and shoot development.   **Uses** Control of annual grasses and some annual broad-leaved weeds in peanuts, lettuce, cucumbers, chicory, endive, field beans, french beans, lentils, alfalfa, clovers, trefoil, tobacco and established turf. Applied pre-emergence with soil incorporation, at 1.0–1.5 kg a.i./ha. **Formulation types** EC; GR; WG.   **Mixtures** *(benfluralin +)* oryzalin; trifluralin. **Selected tradenames** 'Balan' (DowElanco); 'Team' (mixture) (DowElanco); 'XI-2G' (mixture) (DowElanco); 'Benefex' (Makhteshim-Agan)

ANALYSIS
Product analysis by u.v. spectrometry *(AOAC Methods,* 1995, 973.13) or by glc with FID *(ibid.,* 973.14; *CIPAC Handbook,* 1983, **1B**, 1726).   **Residues** by glc with ECD (W. S. Johnson & R. Frank, *Anal. Methods Pestic. Plant Growth Regul.,* 1976, **8**, 335).   Details available from DowElanco.

MAMMALIAN TOXICOLOGY
Oral Acute oral LD_{50} for rats >10 000, mice >5000, dogs and rabbits >2000 mg/kg.   **Skin and eye** Acute percutaneous LD_{50} for rabbits >5000 mg/kg. Moderately irritating to skin and eyes (rabbits).   **Inhalation** LC_{50} (4 h) >2.16 mg/l air.   **NOEL** (2 y) for rats 1000 mg/kg diet, for mice 6.5 mg/kg b.w. daily. **ADI** 0.065 mg/kg.   **Toxicity class** WHO (a.i.) III (Table 5); EPA (formulation) II

ECOTOXICOLOGY
Birds Acute oral LD_{50} for mallard ducks, bobwhite quail, and chickens >2000 mg/kg.   **Fish** LC_{50} (96 h) for bluegill sunfish 6.0, rainbow trout 0.081 mg/l. **Bees** Low toxicity; 33% formulation had no effect up to 100 ppm in 60% sucrose; but at 1000 ppm mortality was increased.   **Daphnia** LC_{50} (48 h) >0.1 mg/l; using nominal concentrations, 50–100 mg/l.   **Algae** For *Selenastrum capricornutum* (7 d), 3.86 mg/l reduced specific growth rate and terminal biomass by 16.6% and 34.3% respectively.

ENVIRONMENTAL FATE
Soil/Environment In soil, chemical decomposition, evaporation and photochemical degradation take place. Duration of residual activity in soil is *c.* 4–8 months. T. Golab *et al.* *(J. Agr. Food Chem.,* 1970, **18**, 838); J. H. Miller *et al.* *(Weed Sci.,* 1975, **23**, 211); R. L. Zimdahl & S. M. Gwynn *(ibid.* 1977, **25**, 247). Adsorption Freundlich K 27 in sandy soil (pH 7.7) to 117 in clay loam (pH 6.9).

benfuracarb   *Insecticide*

---

carbamate

$$CH_3 \quad CH(CH_3)_2$$
$$OCON - S - NCH_2CH_2CO_2CH_2CH_3$$

(structure: 2,3-dihydro-2,2-dimethylbenzofuran ring with O and CH₃ CH₃ substituents)

NOMENCLATURE
Common name benfuracarb (BSI, draft E-ISO, (*m*) draft F-ISO)
IUPAC name ethyl *N*-[2,3-dihydro-2,2-dimethylbenzofuran-7-yloxycarbonyl=
(methyl)aminothio]-*N*-isopropyl-β-alaninate
Chemical Abstracts name 2,3-dihydro-2,2-dimethyl-7-benzofuranyl
2-methyl-4-(1-methylethyl)-7-oxo-8-oxa-3-thia-2,4-diazadecanoate; ethyl
N-[[[[(2,3-dihydro-2,2-dimethyl-7-benzofuranyl)oxy]carbonyl]methylamino]=
thio]-*N*-(1-methylethyl)-β-alaninate
CAS RN *[82560–54–1]*   **Development codes** OK-174

PHYSICAL CHEMISTRY
Mol. wt. 410.5   **M.f.** $C_{20}H_{30}N_2O_5S$ **Form** Viscous reddish brown liquid.
B.p. 110 °C/0.023 mmHg   **V.p.** 2.66×10^{-2} mPa (20 °C) K_{ow} logP = 4.3
(20–22 °C, pH 7)   **S.g./density** 1.142 (20 °C)   **Solubility** In water 8.1 mg/l
(20 °C). In benzene, dichloromethane, methanol, acetone, hexane, xylene and
ethyl acetate >50%.   **Stability** Stable in neutral and weakly basic media, but
unstable in acidic and strongly basic media. Decomposes at 225 °C.   **F.p.** 114 °C
(closed cup)

COMMERCIALISATION
History Insecticide reported by T. Goto *et al.* (*Proc. Int. Congr. Plant Prot. 10th,*
1983, **2**, 360). Introduced in UK, France and Spain (1984) by Otsuka Chemical
Co., Ltd.   **Patent** FR 2489329   **Manufacturers** Otsuka

APPLICATIONS
Biochemistry Cholinesterase inhibitor.   **Mode of action** Systemic and contact
insecticide with stomach and contact action.   **Uses** Used to control insect pests
(especially Chrysomelidae, Elateridae, *Lissorhoptrus oryzophilus*, *Plutella xylostella*
and Aphididae) in citrus, maize, rice, sugar beet and vegetables. Applied mainly as
soil treatment (0.5–2.0 kg a.i./ha for maize, 1.0–2.5 kg/ha for vegetables,
0.5–1.0 kg/ha for sugar beet), or seed treatment (0.4–1.5 kg/100 kg seed). Foliar
sprays (0.3-1.0 kg/ha) are also used on vegetables and fruit; rice is treated at
1.5–4.0 g/nursery box at transplanting.   **Formulation types** EC; GR; WP; SC.
Mixtures *(benfuracarb +)* propaphos; isoprothiolane; probenazole; diazinon;

disulfoton.   **Selected tradenames** 'Oncol' (Otsuka); 'Furacon' (Siapa); 'Nakar' (Makhteshim-Agan)

ANALYSIS
Product analysis by hplc or glc.   **Residues** of benfuracarb and its metabolites, carbofuran and analogues, determined by glc; those in soil by hplc (*J. Pesticide Science*, 1987, **12**, 491, 689).

MAMMALIAN TOXICOLOGY
Reviews *J. Pesticide Science*, **14**, 517–521 (1989).   **EHC** 64 (WHO, 1986; a review of carbamate insecticides in general).   **Oral** Acute oral LD_{50} for male rats 222.6, female rats 205.4, mice 175, dogs 300 mg/kg.   **Skin and eye** Acute percutaneous LD_{50} for rats >2000 mg/kg. Non-irritant to skin, slightly irritant to eyes (rabbits). Non-sensitising to skin (guinea pigs).   **Inhalation** LC_{50} (4 h) for rats 0.34 mg/l air. **NOEL** (2 y) for rats 25 mg/kg diet.   **Other** Non-mutagenic and non-teratogenic (rats and rabbits); non-carcinogenic (rats and mice).   **Toxicity class** WHO (a.i.) Ib

ECOTOXICOLOGY
Birds Acute oral LD_{50} for hens 92 mg/kg.   **Fish** LC_{50} (48 h) for carp 0.65 mg/l. **Bees** LD_{50} (topical) 0.29 µg/g.   **Daphnia** LC_{50} (3 h) >10 mg/l.

ENVIRONMENTAL FATE
Animals In rats, benfuracarb is metabolised rapidly and almost completely excreted in the urine and faeces within 7 days. Major metabolites in the faeces are carbofuran, carbofuran phenol, 3-hydroxycarbofuran, 3-hydroxyphenol, and 3-ketophenol. Urinary metabolites are β-glucuronide conjugates of these metabolites.   **Plants** In plants, *N-S* bond cleavage occurs as the initial step, giving rise to carbofuran, which is subsequently metabolised to 3-hydroxycarbofuran. The principal hydrolytic products are carbofuran phenol and 3-hydroxy- and 3-ketophenol, all present in the form of plant conjugates. See A. K. Tanaka *et al.*, *J. Agr. Food Chem.*, **33**, 1049 (1985) and N. Umetsu *et al.*, *J. Pesticide Science*, **10**, 501 (1985).   **Soil/Environment** DT_{50} in soil is *c.* 4–28 h. Under upland conditions, benfuracarb is decomposed to carbofuran, while under flooded conditions, carbofuran phenol is also found as a major degradation product.

MAIN ENTRIES

59   benfuresate   *Herbicide*

benzofuranyl alkanesulfonate

$CH_3CH_2SO_2O$ — (benzofuran ring with CH_3 and CH_3 substituents)

NOMENCLATURE
Common name benfuresate (BSI, draft E-ISO, (*m*) draft F-ISO)
IUPAC name 2,3-dihydro-3,3-dimethylbenzofuran-5-yl ethanesulfonate
Chemical Abstracts name 2,3-dihydro-3,3-dimethyl-5-benzofuranyl
ethanesulfonate
CAS RN *[68505–69–1]*   **Development codes** NC 20 484

PHYSICAL CHEMISTRY
Composition Tech. benfuresate is ≥95% pure.   **Mol. wt.** 256.3   **M.f.** $C_{12}H_{16}O_4S$
Form Off-white crystals; tech. is a light to dark brown highly viscous liquid. Slight
characteristic odour.   **M.p.** Stated melting points are: for pure a.i. 30.1 °C; for
tech. 32–35 °C.   **V.p.** 1.43 mPa (20 °C); 2.78 mPa (25 °C); (OECD 104).
K_{ow} logP = 2.41 (20 °C)   **S.g./density** 0.957   **Solubility** In water 261 mg/l
(25 °C). In dichloromethane >1220, acetone >1050, toluene >1040, methanol >980,
ethyl acetate >920, cyclohexane 51, n-hexane 15.3 (all in g/l).   **Stability** Stable to
hydrolysis at pH 5.0, 7.0, 9.2 for 31 d (37 °C); DT_{50} (25 °C) 12.5 d in 0.1 N
NaOH.   **F.p.** 37.5 °C (Abel Pensky). Non-flammable.

COMMERCIALISATION
History Herbicide reported by K. W. Chisholm *et al.* (*Proc. South. Weed Sci. Soc.*,
1980, **33**, 326) and S. D. Horne & S. D. van Hoogstraten (*Proc. Br. Weed Control
Conf.*, 1980, **1**, 201). Evaluated by FBC Limited (now AgrEvo GmbH).
Manufacturers AgrEvo

APPLICATIONS
Mode of action Main route of uptake of benfuresate by monocotyledonous plants
is by the emerging shoot when it grows through the treated soil layer. In broad-
leaved species, root uptake is more important.   **Uses** Post-emergence control of
grass and broad-leaved weeds in paddy rice, fruit, beans (*Phaseolus*), maize, sugar
cane, and perennial crops, at 450–600 g a.i./ha. Also used pre-plant incorporated
in cotton and tobacco at 2–3 kg a.i./ha.   **Phytotoxicity** The following crops are
known not to be sufficiently tolerant to required dose rates of benfuresate when
applied pre-plant incorporated or pre-emergence: radishes, tomatoes, swedes,
turnips, cabbages, Brussels sprouts, broccoli, carrots, cucurbits, lentils, kale,
oilseed rape, peanuts, millet, lettuce, spinach, soya beans, sorghum, and cereals.

Formulation types EC; GR; WG; Liquid.   **Selected tradenames** 'Cyperal' (AgrEvo)

ANALYSIS
Product analysis by hplc.   **Residue** analysis by glc. Details available from AgrEvo.

MAMMALIAN TOXICOLOGY
Oral Acute oral LD_{50} for male rats 3536, female rats 2031, male mice 1986, female mice 2809, dogs >1600 mg/kg.   **Skin and eye** Acute percutaneous LD_{50} for rats >5000 mg/kg. Not a skin or eye irritant, Non-sensitising to skin (guinea pigs).   **Inhalation** LC_{50} for rats >5.34 mg/l air; not harmful by inhalation.
NOEL (90 d) for mice 3000 mg/kg diet. In the rat combined chronic and oncogenicity study NOEL 60 ppm, corresponding to 3.07 mg/kg daily.
ADI 0.0307 mg/kg   **Other** Non-teratogenic. Non-mutagenic in the Ames test and in a cell transformation test, i.e. not genotoxic.   **Toxicity class** WHO (a.i.) III (Table 5)

ECOTOXICOLOGY
Birds Acute oral LD_{50} for bobwhite quail 32 272, mallard ducks >10 000 mg/kg.
Fish LC_{50} (96 h) for bluegill sunfish 22.3, carp 35, rainbow trout 12.28 mg/l.
Bees Data not required for registration, due to timing of application.
Worms LC_{50} (14 d) for earthworms 734.1 mg/kg.   **Daphnia** EC_{50} (48 h) 35.36 mg/l; NOEC 12.6 mg/l. EC_{50} (48 h) for *D. carinata* 42 mg/l.   **Algae** For algae (96 h), E_bC_{50} 3.8 mg/l; E_rC_{50} 15.1 mg/l; NOEC 0.6 mg/l.

ENVIRONMENTAL FATE
Animals Metabolism and elimination were rapid and complete; elimination was predominantly via the urine. Benfuresate is metabolised mainly to the ring open form of the lactone (2,3-dihydro-3,3-dimethyl-2-oxobenzofuran-5-yl ethanesulfonate); this material was eliminated in urine and faeces, along with small amounts of the lactone itself and both free and conjugated 2,3-dihydro-2-= hydroxy-3,3-dimethylbenzofuran-5-yl ethanesulfonate.   **Plants** The major route of metabolism was to conjugates which hydrolysed to 2,3-dihydro-3,3-dimethyl-2-= oxobenzofuran-5-yl ethanesulfonate in conjugated and fibre-bound forms. Under base reflux, the latter liberated the corresponding phenol. A minor metabolite in foliage was conjugated 2,3-dihydro-2-hydroxy-3,3-dimethylbenzofuran-5-yl ethanesulfonate.   **Soil/Environment** In soil, residue studies show that benfuresate is not readily leached and does not accumulate at depths lower than 7.5 cm under field conditions. K_{oc} 140–259 (mean 214), K_{des} 0.03–11.2. Laboratory DT_{50} 18–20 d (aerobic) and 300 d (anaerobic); field DT_{50} 7–29 d. Mainly degraded by soil microbial degradation.

MAIN ENTRIES

benomyl

benzimidazole

$CONH(CH_2)_3CH_3$

$NHCO_2CH_3$

NOMENCLATURE
Common name benomyl (BSI, E-ISO, (*m*) F-ISO, ANSI, JMAF)
IUPAC name methyl 1-(butylcarbamoyl)benzimidazol-2-ylcarbamate
Chemical Abstracts name methyl [1-[(butylamino)carbonyl]-1*H*-benzimidazol-=
2-yl]carbamate
CAS RN *[17804–35–2]* **EEC no.** 241–775–7 **Development codes** T 1991
(Du Pont)

PHYSICAL CHEMISTRY
Mol. wt. 290.3 **M.f.** $C_{14}H_{18}N_4O_3$ **Form** Colourless crystals. **M.p.** 140 °C
(decomp.) **V.p.** $<5.0 \times 10^{-3}$ mPa (25 °C) **K_{ow}** logP = 1.37 **Henry** $<4.0 \times 10^{-4}$
(pH 5), $<5.0 \times 10^{-4}$ (pH 7), $<7.7 \times 10^{-4}$ (pH 9) (all in Pa m^3 mol^{-1}, calc.)
S.g./density Bulk density 0.38 **Solubility** In water 3.6 (pH 5), 2.9 (pH 7),
1.9 (pH 9) (all in μg/l, room temperature). In chloroform 94, dimethylformamide
53, acetone 18, xylene 10, ethanol 4, heptane 0.4 (all in g/kg, 25 °C).
Stability Decomposed by strong acids and strong alkalis. Decomposes slowly in the
presence of moisture. In some solvents, dissociates to form carbendazim and butyl
isocyanate (M. Chiba & E. A. Cherniak, *J. Agric. Food Chem.*, 1978, **26**, 573).
Solubility in water and stability at various pH values investigated by R. P. Singh &
M. Chiba (*ibid.*, 1985, **33**, 63). Stable to light. Decomposed on storage in contact
with water and under moist conditions in soil. Mechanism of the acid-catalysed
decomposition in aqueous media, J. P. Calmon & D. R. Sayag (*ibid.*, 1976, **24**, 314,
317).

COMMERCIALISATION
History Fungicide reported by C. J. Delp & H. L. Klopping (*Plant Dis. Rep.*, 1968,
52, 95). Introduced by E. I. du Pont de Nemours and Co. **Patent** NL 6706331;
US 3631176 **Manufacturers** Aragonesas; Chinoin; Du Pont; Gilmore; Jiangsu;
Sinon

APPLICATIONS
Mode of action Systemic fungicide with protective and curative action. Absorbed
through the leaves and roots, with translocation principally acropetally.
Uses Effective against a wide range of Ascomycetes, and Fungi Imperfecti and
some Basidiomycetes in cereals, grapes, pome and stone fruit, rice and vegetables.
It is also effective against mites, primarily as an ovicide. Also used as pre-harvest

sprays or dips for the control of storage rots of fruit and vegetables. Typical rates are: on field and vegetable crops, 140–550 g a.i./ha; on tree crops 550–1100 g/ha; for post-harvest uses 25–200 g/hl. **Phytotoxicity** Non-phytotoxic if used as directed. Russetting is possible with Golden Delicious apples. **Formulation types** WP. **Mixtures** *(benomyl +)* thiram; mancozeb; folpet. **Compatibility** Incompatible with alkaline materials. **Selected tradenames** 'Benlate' (Du Pont); 'Benor' (Aragonesas); 'Fundozol' (Chinoin); 'Pilarben' (Pilarquim); 'Romyl' (Rotam)

ANALYSIS
Product analysis by i.r. spectrometry (F. J. Baude *et al., Anal. Methods Pestic. Plant Growth Regul.,* 1978, **10,** 157) or by hplc (*AOAC Methods,* 1995, 984.09; *CIPAC Handbook,* 1988, **D,** 14). **Residues** determined by cation-exchange hplc (J. J. Kirkland *et al., J. Agric. Food Chem.,* 1973, **21,** 368; N. Aharonson & A. Ben-Aziz, *J. Assoc. Off. Anal. Chem.,* 1973, **56,** 481; J. E. Farrow *et al., Analyst (London),* 1977, **102,** 752). Benomyl can be converted to a derivative and differentiated from carbendazim by hplc (M. Chiba & R. P. Singh, *J. Agric. Food Chem.,* 1986, **34,** 108).

MAMMALIAN TOXICOLOGY
Reviews *Pesticide residues in food – 1995,* FAO Plant Production and Protection Paper. *Pesticide residues in food: 1995 evaluations, Part II – Toxicology and Environment.* EHC 148 (WHO, 1993). **Oral** Acute oral LD_{50} for rats >5000 mg a.i./kg. **Skin and eye** Acute percutaneous LD_{50} for rabbits >5000 mg/kg; negligible irritant to skin, temporary to eyes (rabbits). **Inhalation** LC_{50} (4 h) for rats >2 mg/l air. **NOEL** (2 y) for rats >2500 mg/kg diet (maximum rate tested), no evidence of histopathological changes; for dogs 500 mg/kg diet. **ADI** (JMPR) 0.1 mg/kg b.w. [1995]. **Toxicity class** WHO (a.i.) III (Table 5); EPA (formulation) IV **EC risk** (R40)

ECOTOXICOLOGY
Birds Eight-day dietary LC_{50} for mallard ducks and bobwhite quail >10 000 mg/kg diet (using 50% formulation). **Fish** LC_{50} (96 h) for rainbow trout 0.27, goldfish 4.2 mg/l. LC_{50} (48 h) for guppy 3.4 mg/l. **Bees** Not toxic to bees. LD_{50} (contact) >50 μg/bee. **Worms** LC_{50} (14 d) 10.5 mg/kg. **Daphnia** LC_{50} (48 h) 640 μg/l. **Algae** E_{b50} (72 h) 2.0 mg/l, (120 h) 3.1 mg/l.

ENVIRONMENTAL FATE
EHC 148 concludes that although highly toxic to aquatic organisms, this effect is unlikely to be seen in the field due to the low bioavailability of sediment-bound residues. Earthworm populations may take more than two years to recover following field application. **Animals** In animals, the butylcarbamoyl group is removed to give the relatively stable carbendazim, followed by slow degradation to non-toxic 2-aminobenzimidazole. Hydroxylation also occurs, and the principal

metabolite 5-hydroxybenzimidazole carbamate is converted to the O– and N-conjugates; other possible metabolites include 4-hydroxy-2-benzimidazole= methylcarbamate. Benomyl and its metabolites are excreted in the urine and faeces within a few days, with no accumulation in animal tissue. **Plants** In plants, the butylcarbamoyl group is removed to give the relatively stable carbendazim, followed by slow degradation to non-toxic 2-aminobenzimidazole. Further degradation involves cleavage of the benzimidazole nucleus. Benomyl *per se* is stable on the surface of banana skins (J. Cox *et al.*, *Pestic. Sci.*, 1974, **5**, 135; J. Cox & J. A. Pinegar, *ibid.*, 1976, **7**, 193). **Soil/Environment** Benomyl is rapidly converted to carbendazim in the environment, DT_{50} 2 and 19 h in water and in soil, respectively. Data from studies on both benomyl and carbendazim are therefore relevant for the evaluation of environmental effects. K_{oc} = 1900.

61   benoxacor   *Herbicide safener*

benzoxazine

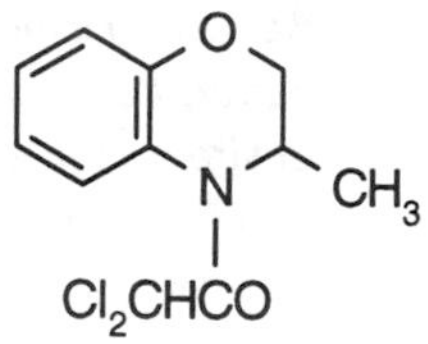

NOMENCLATURE
Common name benoxacor (BSI, draft E-ISO)
IUPAC name (±)-4-dichloroacetyl-3,4-dihydro-3-methyl-2*H*-1,4-benzoxazine
Chemical Abstracts name (±)-4-(dichloroacetyl)-3,4-dihydro-3-methyl-2*H*-= 1,4-benzoxazine
CAS RN [98730–04–2] unstated stereochemistry
Development codes CGA 154 281

PHYSICAL CHEMISTRY
Mol. wt. 260.1   **M.f.** $C_{11}H_{11}Cl_2NO_2$ **Form** Colourless solid.   **M.p.** 107.6 °C
V.p. 0.59 mPa (20 °C) K_{ow} logP = 2.6   **S.g./density** 1.52 (20 °C)   **Solubility** In water 20 mg/l (20 °C). In acetone 230, cyclohexanone 300, methylene chloride 400, methanol 30, *n*-octanol 11, isopropanol 13, toluene 90, hexane 3.2, and xylene 60 (all in g/kg).   **Stability** Stable up to 200 °C. Stable in acidic media, hydrolyses in alkaline media; DT_{50} c. 50 d (pH 7).   Undergoes rapid photolytic degradation (c. 0.5 h, pH 7, natural sunlight).   **F.p.** 185 °C (Pensky-Martens closed cup)

COMMERCIALISATION
History Herbicide safener reported by J. W. Peek *et al.* (*Abstr. Annu. Meeting Weed Sci. Soc. Am.*, 1988, **28**, 13). Introduced by Ciba-Geigy AG (now Novartis Crop Protection AG).   **Patent** US 4601745; EP 149974   **Manufacturers** Novartis

APPLICATIONS
Biochemistry Induces glutathione S-transferases.   **Mode of action** Absorbed mainly through the shoots of germinating seedlings; accelerates the detoxification of metolachlor in maize. Benoxacor has no herbicidal activity of its own.   **Uses** Used as a herbicide safener, benoxacor increases tolerance of maize to metolachlor under both normal and adverse environmental conditions. Does not affect activity of metolachlor on susceptible species. Available only in mixture with metolachlor. **Formulation types** SC; EC; FW.   **Mixtures** *(benoxacor +)* metolachlor; metolachlor + atrazine.

ANALYSIS
Residues determined by capillary glc. Details available from Novartis.

MAMMALIAN TOXICOLOGY
Oral Acute oral LD_{50} for rats >5000 mg/kg.   **Skin and eye** Acute percutaneous LD_{50} for rabbits >2010 mg/kg. Non-irritating to skin and eyes (rabbits). May cause skin sensitisation (guinea pigs).   **Inhalation** LC_{50} (4 h) for rats >2.0 mg/l air. **NOEL** (2 y) for rats 0.5 mg/kg b.w. daily.   **ADI** 0.005 mg/kg b.w. **Toxicity class** WHO (a.i.) III (Table V)   **EC risk** R43

ECOTOXICOLOGY
Birds Acute oral LD_{50} for mallard ducks >2150, bobwhite quail 2000 mg/kg. **Fish** LC_{50} (96 h) for rainbow trout 2.4, carp 10, bluegill sunfish 6.5 mg/l. **Bees** LD_{50} (48 h) (oral and contact) >100 µg/bee.   **Worms** LC_{50} (14 d) for earthworms >1000 mg/kg.   **Daphnia** LC_{50} (48 h) 4.8 mg/l.   **Algae** EC_{50} (72 h) for *Scenedesmus subspicatus* 0.63 mg/l.

ENVIRONMENTAL FATE
Animals Benoxacor is metabolised to water-soluble conjugation products following ring hydroxylation, deacetylation and reductive dechlorination.   **Plants** Metabolites are the same as in animals.   **Soil/Environment** DT_{50} in soil *c.* 5 d. K_{oc} 42–176.

sulfonylurea

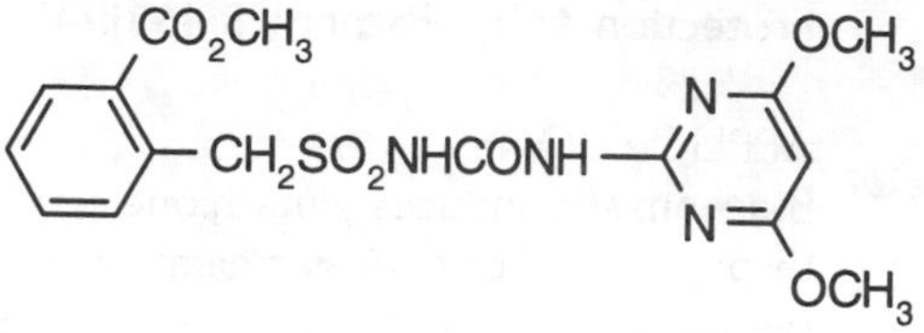

NOMENCLATURE

bensulfuron-methyl
Common name bensulfuron-methyl
Chemical Abstracts name methyl 2-[[[[[(4,6-dimethoxy-2-pyrimidinyl)amino]=
carbonyl]amino]sulfonyl]methyl]benzoate
CAS RN *[83055–99–6]*   **Development codes** DPX-F5384

bensulfuron
Common name bensulfuron (BSI, ANSI, WSSA, draft E-ISO, (*m*) draft F-ISO)
IUPAC name α-(4,6-dimethoxypyrimidin-2-ylcarbamoylsulfamoyl)-*o*-toluic acid
Chemical Abstracts name 2-[[[[[(4,6-dimethoxy-2-pyrimidinyl)amino]carbonyl]=
amino]sulfonyl]methyl]benzoic acid
CAS RN *[99283–01–9]*   **EEC no.** 401–340–6

PHYSICAL CHEMISTRY

bensulfuron-methyl
Composition Tech. is 96.5%.   **Mol. wt.** 410.4   **M.f.** $C_{16}H_{18}N_4O_7S$ **Form** White
to pale yellow, odourless solid.   **M.p.** 185–188 °C   **V.p.** 2.8×10^{-9} mPa (25 °C)
K_{ow} logP = 2.45 (pH 1.5), 0.62 (pH 7) (both at 30 °C)   **S.g./density** 1.41
Solubility In water 2.9 (pH 5), 12 (pH 6), 120 (pH 7), 1200 (pH 8) (all in mg/l,
25 °C). In acetone 1.38, acetonitrile 5.38, dichloromethane 11.7, ethyl acetate
1.66, hexane >0.01, xylene 0.280 (all in g/l, 20 °C).   **Stability** Aqueous solutions
are most stable under slightly alkaline conditions (pH 8), and slowly degrade under
acidic conditions; DT_{50} 11 d (pH 5), 143 d (pH 7) (25 °C).   **pKa** 5.2

bensulfuron
Mol. wt. 396.4   **M.f.** $C_{15}H_{16}N_4O_7S$

COMMERCIALISATION

History Herbicidal activity of bensulfuron-methyl reported by T. Yayama *et al.*
(*Proc. 9th Asian Pacific Weed Sci. Soc. Congr.*, 1983, p. 554). Introduced by E. I. du
Pont de Nemours and Co.   **Patent** US 4420325   **Manufacturers** Du Pont

APPLICATIONS
bensulfuron-methyl

Biochemistry Branched chain amino acid synthesis (acetolactate synthase or ALS) inhibitor. Acts by inhibiting biosynthesis of the essential amino acids valine and isoleucine, hence stopping cell division and plant growth. Selectivity is due to rapid metabolism in the crop. Metabolic basis of selectivity reviewed (M. K. Koeppe & H. M. Brown, *Agro-Food-Industry*, **6**, 9–14 (1995)).   **Mode of action** Selective systemic herbicide, absorbed by the foliage and roots, with rapid translocation to the meristematic tissues.   **Uses** Selective pre and post-emergence control of annual and perennial weeds and sedges (e.g. *Butomus umbellatus*, *Scirpus maritimus*, *Scirpus mucronatus*, *Alisma plantago-aquatica*, *Sparganium erectum*, *Cyperus* spp., *Typha* spp., etc.) in continuously flooded rice, at 30–100 g/ha.
Formulation types GR; WP; WG.   **Mixtures** *(bensulfuron-methyl +)* thiobencarb; dimepiperate; mefenacet; pretilachlor; esprocarb; butachlor; molinate; metsulfuron-methyl; pyributicarb; thenylchlor; thiobencarb + mefenacet.
Selected tradenames 'Fujigrass' (mixture) (Du Pont, Zeneca); 'Londax' (Du Pont); 'Zark' (mixture) (Du Pont); 'Wolf-Ace' (mixture) (Kumiai)

ANALYSIS
Product and **residue** analysis by hplc. Methods for sulfonylurea residues in crops, soil and water reviewed (A. C. Barefoot *et al.*, *Proc. Br. Crop Prot. Conf. – Weeds*, 1995, **2**, 707). Details from Du Pont.

MAMMALIAN TOXICOLOGY
bensulfuron-methyl

Oral Acute oral LD_{50} for rats >5000, mice >10 985 mg/kg.   **Skin and eye** Acute percutaneous LD_{50} for rabbits >2000 mg/kg. Non-irritant to skin and eyes.
Inhalation LC_{50} (4 h) for rats >7.5 mg/l air.   **NOEL** (2 y) for rats 750 mg/kg diet. Reproduction (2-generation) NOEL in rats 7500 mg/kg diet; teratogenicity NOEL in rats 500, rabbits 300 mg/kg. Not a mutagen.   **ADI** 0.2 mg/kg.
Toxicity class WHO (a.i.) III (Table 5); EPA (formulation) IV   **EC risk** Xi (R43)

ECOTOXICOLOGY
bensulfuron-methyl

Birds Acute oral LD_{50} for mallard ducks >2510 mg/kg. Eight-day dietary LC_{50} for bobwhite quail, mallard ducks >5620 mg/l.   **Fish** LC_{50} (96 h) for rainbow trout and bluegill sunfish >150 mg/l; LC_{50} (48 h) for carp >1000 ppm.   **Bees** LD_{50} >12.5 μg/bee.   **Daphnia** LC_{50} (48 h) >100 mg/l.

ENVIRONMENTAL FATE
Animals Almost completely biotransformed and rapidly excreted in urine and faeces of rats and goats.   **Plants** After uptake by rice, converted to non-herbicidal metabolite.   **Soil/Environment** DT_{50} 4–20 w on Flanagan and Keyport silt loam soils. In rice fields, DT_{50} in water averages 4–6 d.

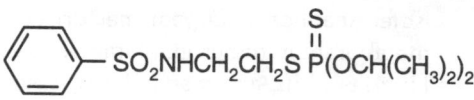

$$SO_2NHCH_2CH_2S\overset{\displaystyle S}{\overset{\|}{P}}(OCH(CH_3)_2)_2$$

NOMENCLATURE

Common name bensulide (BSI, E-ISO, (m) F-ISO, WSSA); SAP (JMAF)
IUPAC name S-2-benzenesulfonamidoethyl O,O-di-isopropyl phosphorodithioate; O,O-di-isopropyl S-2-phenylsulfonylaminoethyl phosphorodithioate
Chemical Abstracts name O,O-bis(1-methylethyl) S-[2-[(phenylsulfonyl)amino]= ethyl] phosphorodithioate
CAS RN [741–58–2] **EEC no.** 212–010–4 **Development codes** R-4461

PHYSICAL CHEMISTRY

Composition Tech. grade is 92% pure. **Mol. wt.** 397.5 **M.f.** $C_{14}H_{24}NO_4PS_3$
Form Colourless solid; (tech. grade is an amber solid or supercooled liquid with a characteristic camphor-like odour). **M.p.** 34.4 °C **V.p.** <0.133 mPa (20 °C)
K_{ow} logP = 4.2 **S.g./density** 1.25 (22 °C) **Solubility** In water 25 mg/l (25 °C). In kerosene 300 mg/l (20 °C). Miscible with acetone, ethanol, methyl isobutyl ketone, and xylene. **Stability** Relatively stable to acids and alkalis. Decomposes slowly in light. Tech. product decomposes at 155 °C; heating at 100 °C causes autocatalytic decomposition in 18–40 h. **F.p.** >104 °C

COMMERCIALISATION

History Herbicide reported by D. D. Hemphill (*Res. Rep. North Cent. Weed Control Conf.*, 1962, pp. 104, 111). Introduced by the Stauffer Chemical Co. (now Zeneca Agrochemicals). **Patent** US 3205253 **Manufacturers** Gowan

APPLICATIONS

Mode of action Selective herbicide, adsorbed on the root surfaces; a small amount is absorbed by the roots. Translocation of bensulide to the leaves does not occur, but metabolites are translocated. Acts by inhibition of germination.
Uses A pre-emergence herbicide suitable for pre-plant use at 2.3–5.0 kg a.i./ha, on crops such as brassicas, cotton, cucurbits, and lettuce; or for use on established turf at 11–22 kg/ha per season. **Formulation types** GR; EC.
Selected tradenames 'Betasan' (for turf use) (Gowan); 'Bensumac' (PBI/Gordon); 'Lawnpur' (Tomen); 'Prefar' (for crop use) (Gowan); 'Pre-San' (PBI/Gordon)

ANALYSIS

Product analysis by hplc (*Anal. Methods Pestic. Plant Growth Regul.*, 1984, **53**, 219).
Residues by glc with FID (*ibid.*). Analytical methods available from Zeneca. See

also R. W. Buxton et al., ibid., 1972, **6**, 672; G. Patchett et al., Anal. Methods Pestic., Plant Growth Regul. Food Addit., 1967, **5**, 483.

MAMMALIAN TOXICOLOGY
Oral Acute oral LD_{50} for male rats 360, female rats 270 mg/kg.   **Skin and eye** Acute percutaneous LD_{50} for rats >2000 mg/kg. Mild skin and eye irritant (rabbits). Not a skin sensitiser (guinea pigs).   **Inhalation** LC_{50} (4 h) for rats >1.75 mg/l   **NOEL** (90 d) for rats 25, dogs 2.5, mice 30 mg/kg daily. Not carcinogenic or teratogenic.   **Toxicity class** WHO (a.i.) II; EPA (formulation) III
EC risk Xn (R22)

ECOTOXICOLOGY
Birds Acute oral LD_{50} for bobwhite quail 1386 mg/kg. LC_{50} (5 d feed) for mallard ducks >5620 ppm; (21 d) for quail >1000 ppm.   **Fish** LC_{50} (96 h) for rainbow trout 1.1, goldfish 1–2, bluegill sunfish 1.4 mg/l.   **Bees** LD_{50} 0.0016 mg/bee. **Daphnia** LC_{50} (48 h) 0.58 ppm.

ENVIRONMENTAL FATE
Soil/Environment In soil, slowly degraded by microbial action. DT_{50} in soil 4–6 mo at 21–27 °C.

4   bensultap   *Insecticide*

2-dimethylaminopropane-1,3-dithiol

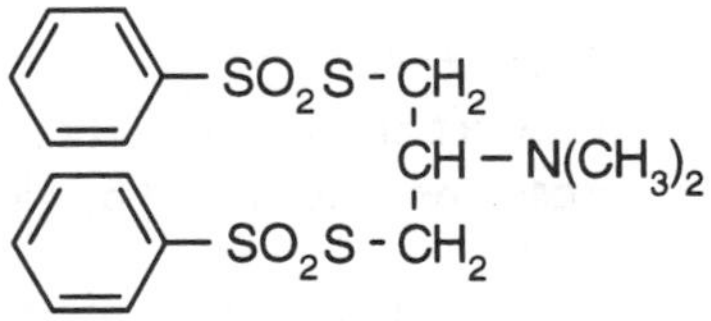

$$\text{(C}_6\text{H}_5\text{)}-SO_2S-CH_2$$
$$CH-N(CH_3)_2$$
$$\text{(C}_6\text{H}_5\text{)}-SO_2S-CH_2$$

NOMENCLATURE
Common name bensultap (BSI, draft E-ISO, (m) draft F-ISO)
IUPAC name S,S'-2-dimethylaminotrimethylene di(benzenethiosulfonate)
Chemical Abstracts name S,S'-[2-(dimethylamino)-1,3-propanediyl] di(benzenesulfothioate)
CAS RN [17606–31–4]   **Development codes** TI-78; TI-1671
Official codes OMS 3011

PHYSICAL CHEMISTRY
Mol. wt. 431.6   **M.f.** $C_{17}H_{21}NO_4S_4$ **Form** Pale yellow crystalline powder with slight characteristic odour.   **M.p.** 83–84 °C (decomposes at c. 150 °C)

V.p. 0.21 mPa (22 °C). **K$_{ow}$** logP = 2.28 (25 °C) **Solubility** In water 0.7–0.8 mg/kg (30 °C). In methanol 25, xylene 77, ethanol 13, hexane 0.17 (all in g/kg, 25 °C). In acetone, acetonitrile, *N,N*-dimethylformamide, chloroform >1000 (all in g/kg, 25 °C). **Stability** Stable ≤150 °C under dispersed light and at pH <5 (room temperature), but hydrolysed in neutral or alkaline solution.

COMMERCIALISATION
History Insecticide developed by Takeda Chemical Industries, Ltd.
Manufacturers Takeda

APPLICATIONS
Biochemistry Analogue or propesticide of the natural toxin nereistoxin. Acts as a synaptic blocking agent for the insect central nervous system. Inhibits the action of synaptic transmitter (acetylcholine) by occupying the post-synaptic membrane where the acetylcholine receptor exists. **Mode of action** Insecticide with contact and stomach action. **Uses** Control of major agricultural insect pests, particularly Coleoptera and Lepidoptera. In particular, control of Colorado beetles in potatoes and aubergines; tea leaf rollers and yellow tea thrips in tea; rice stem borers, rice leaf rollers, rice leaf miners and rice leaf beetles in rice; Oriental corn borers and *Tanymecus* spp. in maize; *Tanymecus* spp., root weevils, and tortoise beetles in beet; grape berry moths on vines; leaf beetles in cereals; boll weevils in cotton; codling moths, blister moths, and leaf miners in apples; small cabbage-white butterfly larvae, diamond-back moths, and blossom rape beetles on brassicas and oilseed rape; etc. Typical application rate is in the range 0.25–1.5 kg a.i./ha. **Phytotoxicity** May be phytotoxic to certain varieties of apple, peach and citrus. **Formulation types** WP; GR; DP. **Compatibility** Not compatible with alkaline materials. **Selected tradenames** 'Bancol' (Takeda); 'Victenon' (Takeda)

ANALYSIS
Product analysis by rp-hplc with u.v. detection (*CIPAC Handbook*, 1992, **E**, 17). **Residues** determined by glc with FPD.

MAMMALIAN TOXICOLOGY
Oral Acute oral LD$_{50}$ for male rats 1105, female rats 1120, male mice 516, female mice 484 mg/kg. **Skin and eye** Acute percutaneous LD$_{50}$ for rabbits >2000 mg/kg. Slight eye irritant; non-irritating to skin (rabbits). **Inhalation** LC$_{50}$ (4 h) for rats >0.47 mg/l air. **NOEL** (90 d) for rats 250, male mice 40, female mice 300 mg/kg diet; (2 y) for rats 10, mice 3.4–3.6 mg/kg b.w. daily. Non-oncogenic, non-teratogenic, and non-mutagenic. **Other** Acute i.p. LD$_{50}$ for male rats 503, female rats 438, male mice 442, female mice 343 mg/kg. **Toxicity class** WHO (a.i.) III; EPA (formulation) III

ECOTOXICOLOGY
Birds Acute oral LD$_{50}$ for bobwhite quail 311 mg/kg. Dietary LC$_{50}$ for bobwhite

quail 1784, mallard ducks 3112 mg/kg diet.   **Fish** LC_{50} (48 h) for carp 15, guppy 17, goldfish 11, rainbow trout 0.76 mg/l. LC_{50} (72 h) for carp 8.2, guppy 16, goldfish 7.4, rainbow trout 0.76 mg/l.   **Bees** Low toxicity to honey bees. LC_{50} (topical, 48 h) 25.9 µg/bee.   **Daphnia** LC_{50} (6 h) 40 mg a.i. (as EC)/l.

ENVIRONMENTAL FATE
Soil/Environment DT_{50} in soil varies from 3–35 d, depending on soil type. Soil DT_{50} ç. 7 d (upland conditions in laboratory).

5   bentazone   *Herbicide*

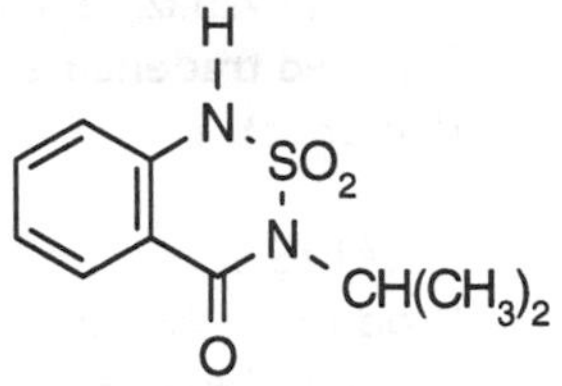

NOMENCLATURE
Common name bentazone (BSI, E-ISO, (*f*) F-ISO, JMAF); bentazon (ANSI, Canada, WSSA); bendioxide (Republic of South Africa)
IUPAC name 3-isopropyl-1*H*-2,1,3-benzothiadiazin-4(3*H*)-one 2,2-dioxide
Chemical Abstracts name 3-(1-methylethyl)-1*H*-2,1,3-benzothiadiazin-4(3*H*)-one 2,2-dioxide
CAS RN *[25057–89–0]*   **EEC no.** 246–585–8   **Development codes** BAS 35100H

PHYSICAL CHEMISTRY
Composition ≥96% pure.   **Mol. wt.** 240.3   **M.f.** $C_{10}H_{12}N_2O_3S$ **Form** Colourless crystals.   **M.p.** 139.4–141 °C   **V.p.** 0.17 mPa (20 °C) K_{ow} logP = 0.77 (pH 5), –0.46 (pH 7), –0.55 (pH 9)   **Henry** 7.167×10^{-11} Pa m³ mol⁻¹ (calc.)
S.g./density 1.41 (20 °C)   **Solubility** In water 570 mg/l (pH 7, 20 °C). In acetone 1507, ethanol 861, ethyl acetate 650, diethyl ether 616, chloroform 180, benzene 33, cyclohexane 0.2 (all in g/kg, 20 °C).   **Stability** Very resistant to hydrolysis in both acidic and alkaline media. Decomposed by sunlight.   **pKa** 3.3 (24 °C)

COMMERCIALISATION
History Herbicide reported by A. Fischer (*Proc. Br. Weed Control Conf., 9th*, 1968, **2**, 1042). Introduced by BASF AG in 1972.   **Patent** US 3708277; DE 1542836
Manufacturers BASF; High Kite; Q.E.A.C.A.

APPLICATIONS
Biochemistry Photosynthetic electron transport inhibitor.
Mode of action Selective contact herbicide, absorbed mainly by the foliage, with very little translocation, but also absorbed by the roots, with translocation acropetally in the xylem. **Uses** A contact herbicide controlling *Anthemis, Chamomilla* and *Matricaria* spp., *Chrysanthemum segetum, Galium aparine, Lapsana communis* and *Stellaria media* in winter and spring cereals at 1.0–2.2 kg a.i./ha. Other crops include peanuts, maize, peas, *Phaseolus* beans, rice (*Cyperus difformis, C. esculentus, C. serotinus, Monochoria vaginalis, Sagittaria pygmaea, S. sagittifolia, Alisma* and *Commelina* spp., *Scirpus maritimus* and *S. mucronatus*) and soya beans (*Abutilon theophrasti, Capsella bursa-pastoris, Cyperus esculentus, Datura stramonium, Helianthus* spp., *Polygonum* spp., *Portulaca* spp., *Sida spinosa, Ambrosia* spp., *Sinapis arvensis* and *Xanthium* spp.). **Formulation types** SL. **Mixtures** (bentazone +) acifluorfen-sodium; dichlorprop; mecoprop; MCPA; MCPB; atrazine; MCPA + MCPB; propanil; dichlorprop + ioxynil; dichlorprop + isoproturon; dichlorprop + MCPA; cyanazine + dichlorprop; dichlorprop-P; mecoprop-P.
Selected tradenames 'Basagran' (BASF); 'Galaxy' (BASF); 'Storm' (BASF); 'Benta' (Sanonda)

ANALYSIS
Product analysis by rplc with u.v. detection (*AOAC Methods*, 1995, 993.02, 7.4.02), by hplc (*CIPAC Handbook*, 1985, **1C**, 1973) or by glc of derivative.
Residues determined by glc of a derivative with ECD (BBA *Kurzfassungen zur Analytic von Pflanzenschutzmitteln in Wasser*, 1989, **2,** Auflage S. 42). In **drinking water**, by conversion to methyl ester with diazomethane, then gc with ECD (*AOAC Methods*, 1995, 992.32, 10.7.03). Details available from BASF.

MAMMALIAN TOXICOLOGY
Reviews *Pesticide residues in food – 1991*. FAO Plant Production and Protection Paper, 111, 1991. *Pesticide residues in food – 1991 evaluations, Part II – Toxicology*, WHO/PCS/92.52, 1992. **Oral** Acute oral LD_{50} for rats >1000, dogs >500, rabbits 750, cats 500 mg/kg. **Skin and eye** Acute percutaneous LD_{50} for rats >2500 mg/kg. Moderately irritating to skin and eyes (rabbits).
Inhalation LC_{50} (4 h) for rats >5.1 mg/l air. **NOEL** (1 y) for dogs 13.1 mg/kg b.w.; (2 y) for rats 10 mg/kg b.w.; (90 d) for rats 25, dogs 10 mg/kg b.w.; (78 w) for mice 12 mg/kg b.w. **ADI** (JMPR) 0.1 mg/kg [1991]. **Toxicity class** WHO (a.i.) III; EPA (formulation) III **EC risk** Xn (R22, R36, R43)

ECOTOXICOLOGY
Birds Acute oral LD_{50} for bobwhite quail 1140 mg/kg. Dietary LC_{50} for bobwhite quail and mallard ducks >5000 ppm. **Fish** LC_{50} (96 h) for rainbow trout and bluegill sunfish >100 mg/l. **Bees** Not toxic to bees; LD_{50} (oral) >100 μg/bee. **Worms** EC_{50} (14 d) >1000 mg/kg soil. **Daphnia** LC_{50} (48 h) 125 mg/l.
Algae EC_{50} (72 h) for green algae (*Ankistrodesmus*) 47.3 mg/l.

ENVIRONMENTAL FATE

Animals Metabolism studies in three different species showed that bentazone was only poorly metabolised, the parent compound being the predominant product. Only small amounts of hydroxylated bentazone metabolites could be detected. No conjugated products were found.   **Plants** In plants, rapidly metabolised to derivatives of anthranilic acid, with the principal metabolites being the 6- and 8-hydroxy derivatives.   **Soil/Environment** In soil, short-lived hydroxy compounds are first formed, which rapidly undergo further degradation (R. Huber & S. Otto in *Rev. Environ. Contam. & Toxicol.,*, **137**, 111–134). In sunlight, bentazone undergoes rapid degradation, ultimately to CO_2. It has low soil persistence: in freshly collected field soils, aerobic DT_{50} (lab., 20 °C) was 14 d. In lab. degradation studies, mostly with BBA-standard soils, DT_{50} (average) was 45 d; DT_{50} (field, average) was *c.* 12 d; corresponding DT_{90} was 44 d. K_{oc} 13.3–176 ml/g (average 42 ml/g), indicating mobility. However, when used according to good agricultural practice, bentazone is degraded more quickly than it can leach; in all lysimeter studies, average annual leachates contained <0.1 µg/l.

56 benzofenap   *Herbicide*

pyrazole (herbicide)

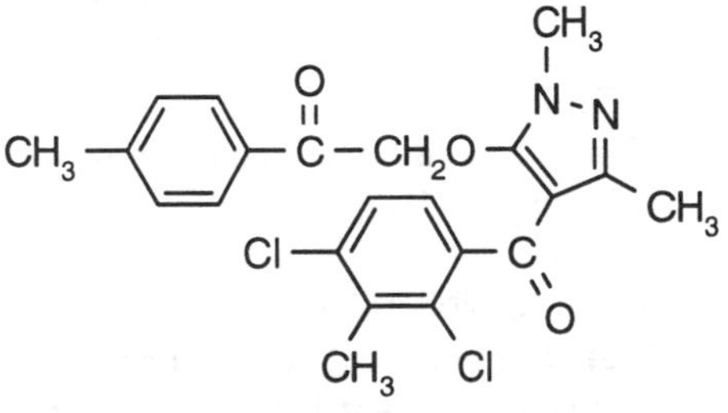

NOMENCLATURE
Common name benzofenap (BSI, draft E-ISO, (*m*) draft F-ISO)
IUPAC name 2-[4-(2,4-dichloro-*m*-toluoyl)-1,3-dimethylpyrazol-5-yloxy]-=4′-methylacetophenone
Chemical Abstracts name 2-[[4-(2,4-dichloro-3-methylbenzoyl)-1,3-dimethyl-=1*H*-pyrazol-5-yl]oxy]-1-(4-methylphenyl)ethanone
CAS RN *[82692–44–2]*   **Development codes** MY-71

PHYSICAL CHEMISTRY
Mol. wt. 431.3   **M.f.** $C_{22}H_{20}Cl_2N_2O_3$ **Form** White solid.   **M.p.** 133.1–133.5 °C
V.p. 0.013 mPa (30 °C) K_{ow} logP = 4.69   **Solubility** In water 0.13 mg/l (25 °C). In xylene 69, acetone 73, *n*-hexane 5.6, chloroform 920 (all in g/l, 20 °C).

COMMERCIALISATION
History Introduced by Mitsubishi Petrochemical Co. Ltd. (now Mitsubishi Chemical Corp.) and commercialised by Rhône-Poulenc Yuka Agro KK, a joint venture of Mitsubishi Chemical Corp. and Rhône-Poulenc Agro. **Manufacturers** Rhône-Poulenc Yuka Agro

APPLICATIONS
Biochemistry Inhibits chlorophyll synthesis by the same mechanism as pyrazolynate. **Mode of action** Systemic herbicide, absorbed principally through the roots and bases of target weeds, eventually causing death with typical pyrazole-like morphological changes, such as bleaching and yellowing. **Uses** In combination with pyributicarb and bromobutide, controls annual and perennial broad-leaved weeds in rice crops. **Phytotoxicity** Non-phytotoxic to rice plants. **Formulation types** SC; GR. **Mixtures** *(benzofenap +)* pretilachlor; pyributicarb + bromobutide. **Selected tradenames** 'Yukawide' (Mitsubishi)

ANALYSIS
Details from Mitsubishi.

MAMMALIAN TOXICOLOGY
Reviews *Pesticide Science Society of Japan*, **15**(1), 1990. **Oral** Acute oral LD_{50} for rats and mice >15 000 mg/kg. **Skin and eye** Acute percutaneous LD_{50} for rats >5000 mg/kg. **Inhalation** LC_{50} (4 h) for rats >1.93 mg/l. **NOEL** (2 y) for rats 0.15 mg/kg b.w. **ADI** 0.0015 mg/kg **Toxicity class** EPA (formulation) IV

ECOTOXICOLOGY
Fish LC_{50} (48 h) for rainbow trout, carp, and loach >10 mg/l. **Daphnia** LC_{50} (3 h) >10 mg/l.

ENVIRONMENTAL FATE
Plants No detectable residues in rice crops (detection limit 0.005 ppm).
Soil/Environment DT_{50} in paddy field soil 38 d. Movement of benzofenap in soil is only very slightly affected by soil type, the maximum movement being to 1 cm below the surface.

benzoximate

Acaricide

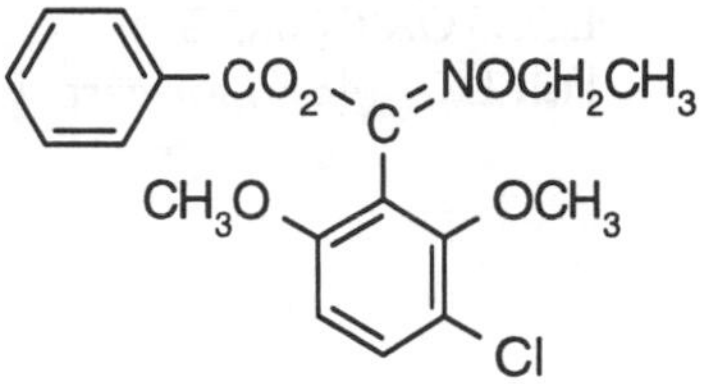

NOMENCLATURE
Common name benzoximate (BSI, E-ISO, (*m*) F-ISO); benzomate (JMAF)
IUPAC name 3-chloro-α-ethoxyimino-2,6-dimethoxybenzyl benzoate;
ethyl *O*-benzoyl-3-chloro-2,6-dimethoxybenzohydroximate
Chemical Abstracts name benzoic acid anhydride with 3-chloro-*N*-ethoxy-=
2,6-dimethoxybenzenecarboximidic acid
CAS RN *[29104–30–1]* **EEC no.** 249–439–1 **Development codes** NA-53M

PHYSICAL CHEMISTRY
Mol. wt. 363.8 **M.f.** $C_{18}H_{18}ClNO_5$ **Form** Colourless crystals. **M.p.** 73 °C
V.p. 0.45 mPa (25 °C). K_{ow} logP = 2.4 **S.g./density** 1.30 (20 °C) **Solubility** In
water 30 mg/l (25 °C). In dimethylformamide 1460, xylene 710, benzene 650,
hexane 80 (all in g/l, 20 °C). **Stability** Stable in acidic media, but decomposed by
concentrated alkalis.

COMMERCIALISATION
History Acaricide reported in *Jpn. Pestic. Inf.*, 1972, No. 9, p. 41. Introduced by
Nippon Soda Co., Ltd. **Patent** GB 1247817 **Manufacturers** Nippon Soda

APPLICATIONS
Mode of action Non-systemic acaricide with contact and stomach action.
Uses Control of all stages of spider mites, particularly *Panonychus* and
Eotetranychus spp., on pome fruit, stone fruit, citrus fruit, vines, and ornamentals.
Formulation types EC. **Compatibility** Incompatible with EPN and Bordeaux
mixture. **Selected tradenames** 'Citrazon' (Nippon Soda); 'Acarmate' (Sipcam);
'Artaban' (Procida)

ANALYSIS
Product analysis by u.v. spectrometry. **Residues** by colorimetry of a derivative.

MAMMALIAN TOXICOLOGY
Oral Acute oral LD_{50} for rats >15 000, Wistar rats >5000, male mice 12 000,
female mice 14 500 mg/kg. **Skin and eye** Acute percutaneous LD_{50} for rats and
mice >15 000 mg/kg. **NOEL** In 2 y feeding trials, rats receiving 400 mg/kg diet

showed no ill-effects.   **Other** Acute i.p. LD$_{50}$ for rats 4.2, male mice 4.6, female mice 4.3 mg/kg.   **Toxicity class** WHO (a.i.) III (Table 5); EPA (formulation) IV

ECOTOXICOLOGY
Fish LC$_{50}$ (48 h) for carp 1.75 mg/l.

68   6-benzylaminopurine   *Plant growth regulator*

cytokinin

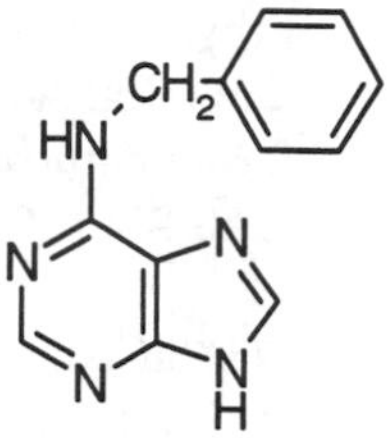

NOMENCLATURE
Common name 6-benzylaminopurine (JMAF)
IUPAC name 6-benzyladenine
Chemical Abstracts name N-phenylmethyl-1H-purin-6-amine
Other names 6-BAP; 6-BA; BAP   **CAS RN** *[1214–39–7]*   **EEC no.** 214–927–5

PHYSICAL CHEMISTRY
Composition Tech. is >99%.   **Mol. wt.** 225.3   **M.f.** C$_{12}$H$_{11}$N$_5$ **Form** Colourless, odourless, fine needles.   **M.p.** 234–235 °C   **V.p.** 2.373 × 10^{-6} mPa (20 °C)
K$_{ow}$ logP = 2.13 (unstated pH)   **Solubility** In water 60 mg/l (20 °C). Insoluble in most common organic solvents; soluble in dimethylformamide, dimethyl sulfoxide.
Stability Stable in acidic, alkaline and neutral aqueous solution. Stable to light and heat (8 h, 120 °C).

COMMERCIALISATION
History Reported by C. G. Skinner *et al.*, *Pl. Physiol.*, **33**, 190–194 (1958). Introduced in Japan by Kumiai Chemical Industry Co., Ltd in 1975.
Manufacturers Abbott; Chemol; Jingma; Kohjin

APPLICATIONS
Biochemistry Stimulates RNA, protein synthesis.   **Mode of action** Synthetic cytokinin; little translocated.   **Uses** Stimulates the following effects: cell division; lateral bud emergence (apples, oranges); basal shoot formation (roses, orchids); flowering (cyclamen, cacti); fruit set (grapes, oranges, melons). Inhibits senescence

of rice seedlings.   **Formulation types** PA; SL.   **Selected tradenames** 'Accel' (mixture) (Abbott); 'BA' (Kumiai); 'Beanin' (Riken Green); 'Paturyl' (Reanal); 'Promalin' (mixture) (Abbott)

ANALYSIS
Analysis by glc and hplc with spectophotometric detection.

MAMMALIAN TOXICOLOGY
Oral Acute oral LD_{50} for male rats 2125, female rats 2130, mice 1300 mg/kg.
Skin and eye Acute percutaneous LD_{50} for rats >5000 mg/kg. Not a skin or eye irritant (rabbits). Not a skin sensitiser.   **NOEL** (2 y) for male rats 5.2, female rats 6.5, male mice 11.6, female mice 15.1 mg/kg b.w. daily. **ADI** 0.05 mg/kg
Other Non-mutagenic in the Ames test. Non-teratogenic (rats and rabbits).
Toxicity class WHO (a.i.) III (Table 5)

ECOTOXICOLOGY
Fish LC_{50} (48 h) for carp >40 mg/l.   **Bees** LD_{50} (oral) 400 µg/bee, (contact) 57.8 l/ha (both 'Paturyl' 10 WSC formulation).   **Daphnia** LC_{50} (24 h) for *D. carinata* >40 mg/l.   **Algae** EC_{50} (96 h) 363.1 mg/l ('Paturyl' 10 WSC formulation).

ENVIRONMENTAL FATE
Animals Almost all of administered ^{14}C was excreted in urine and faeces. Three metabolites were identified.   **Plants** More than 9 metabolites were identified from metabolism studies in soya beans, grapes, maize and cocklebur.   **Soil/Environment** 16 Days after application to soil at 22 °C, 6-benzylaminopurine had degraded to 5.3% (sandy loam) and 7.85% (clay loam soil) of applied dose.

MAIN ENTRIES

beta-cyfluthrin

See entry 177.

beta-cypermethrin

See entry 185.

69   bifenox   *Herbicide*

diphenyl ether

CO$_2$CH$_3$

Cl—⟨benzene ring⟩—O—⟨benzene ring⟩—NO$_2$

Cl

NOMENCLATURE
Common name bifenox (BSI, draft E-ISO, (*m*) draft F-ISO, ANSI, WSSA)
IUPAC name methyl 5-(2,4-dichlorophenoxy)-2-nitrobenzoate
Chemical Abstracts name methyl 5-(2,4-dichlorophenoxy)-2-nitrobenzoate
CAS RN *[42576–02–3]*; formerly *[12680–11–4]*   **EEC no.** 255–894–7
Development codes MC-4379 (Mobil); MCTR-1–79; MCTR-12–79

PHYSICAL CHEMISTRY
Composition ≥97% pure.   **Mol. wt.** 342.1   **M.f.** C$_{14}$H$_9$Cl$_2$NO$_5$ **Form** Yellow
crystals with a slightly aromatic odour.   **M.p.** 84–86 °C   **V.p.** 0.32 mPa (30 °C)
K$_{ow}$ logP = 4.5   **Henry** 1.14 × 10^{-2} Pa m^3 mol^{-1} **S.g./density** 0.65 g/ml (bulk
density)   **Solubility** In water 0.35 mg/l (25 °C). In acetone 400, chlorobenzene
400, xylene 300, ethanol <50 (all in g/kg at 25 °C). Slightly soluble in aliphatic
hydrocarbons.   **Stability** Thermally stable up to 175 °C; total decomposition
occurs above 290 °C. Stable in slightly acidic or slightly alkaline media, but rapidly
hydrolysed above pH 9. DT$_{50}$ in saturated aqueous solution, 24 min at
250–400 nm; *c.* 5 h for a thin film on soil.

COMMERCIALISATION
History Herbicide reported by W. M. Dest *et al.* (*Proc. Northeast. Weed Sci. Conf.*,
1973, **27**, 31). Introduced by Mobil Chemical Co. Agrochemical Division (now
Rhône-Poulenc Agrochimie) and later by Eli Lilly & Co. (now DowElanco, who no
longer manufacture or market it).   **Patent** GB 1232368; US 3652645; US 3776715
all to Mobil   **Manufacturers** Rhône-Poulenc

APPLICATIONS
Biochemistry Protoporphyrinogen oxidase inhibitor. Acts by cellular membrane
disruption and by inhibition of photosynthesis.   **Mode of action** Selective
herbicide, absorbed by the foliage, emerging shoots and roots, with limited
translocation from roots and foliage to shoots.   **Uses** Control of annual broad-
leaved weeds and some grasses in cereals, maize, sorghum, soya beans, rice, and
some other crops at 0.75–1 kg a.i./ha. Applied pre-plant incorporated, pre-
emergence, or directed post-emergence. Often used in combination with other
herbicides to extend the spectrum of activity.   **Formulation types** SC; EC; GR;
WP.   **Mixtures** (*bifenox* +) mecoprop; linuron; chlorotoluron; propanil; clopyralid
+ mecoprop; isoproturon + mecoprop; ioxynil + mecoprop; isoproturon +

neburon.   **Selected tradenames** 'Modown' (Rhône-Poulenc); 'Tolkan Fox' (Rhône-Poulenc)

ANALYSIS
Product analysis is by rp hplc (*CIPAC Handbook,* 1995, **G**, 11–17).
Residues determined by glc with MCD. Details available from Rhône-Poulenc Agrochimie.

MAMMALIAN TOXICOLOGY
Reviews *Toxicological Evaluation of the Herbicide Bifenox,* J-O. Säfwenberg, Säfwenberg Consulting, Surbrunnsgatan 43 A, S-11348, Stockholm.   **Oral** Acute oral LD_{50} for rats >5000 mg tech./kg, mice 4556 mg/kg.   **Skin and eye** Acute percutaneous LD_{50} for rabbits >2000 mg/kg. Non-irritating to skin and eyes.
Inhalation LC_{50} for rats >0.91 mg/l air.   **NOEL** (2 y) for rats 80, dogs 145, mice 30 mg/kg b.w. daily.   **Other** Non-mutagenic in mouse lymphoma assay and Ames test. Non-teratogenic.   **Toxicity class** WHO (a.i.) III (Table 5); EPA (formulation) IV

ECOTOXICOLOGY
Birds Eight-day dietary LC_{50} for ducks and pheasants >5000 mg/kg diet.
Fish LC_{50} (96 h) for rainbow trout >0.67, bluegill sunfish >0.27 mg/l.   **Bees** LD_{50} (contact) >1000 μg/bee.   **Worms** EC_{50} and NOEC >1000 mg/kg.
Daphnia LC_{50} (48 h) 0.66 mg/l.   **Other aquatic spp.** LC_{50} (96 h) for *Palaemonetes pugio* (grass shrimp) 569 ppm, for *Mysidopsis bahia* (mysid shrimp) 0.065 mg/l; (48 h) for *Crassostrea virginica* (Eastern oyster) 210 mg/l.

ENVIRONMENTAL FATE
Animals Bifenox is relatively rapidly absorbed and eliminated from the body; 5-(2,4-dichlorophenyl)-2-nitrobenzoic acid was the major urinary metabolite, with no bifenox detected. Bifenox together with 5-(2,4-dichlorophenyl)anthranilate were detected in faeces.   **Soil/Environment** DT_{50} in soil c. 5–7 d. Duration of residual activity is c. 7–8 w. Degradation is mainly chemical and microbial, and the two principal metabolites are 5-(2,4-dichlorophenoxy)-2-nitrobenzoic acid and methyl 5-(2,4-dichlorophenoxy)anthranilate. K_d 50–300; K_{oc} 500–23 000.

MAIN ENTRIES

70 bifenthrin *Insecticide, acaricide*

pyrethroid

(*Z*)-(1*R*)-*cis*-

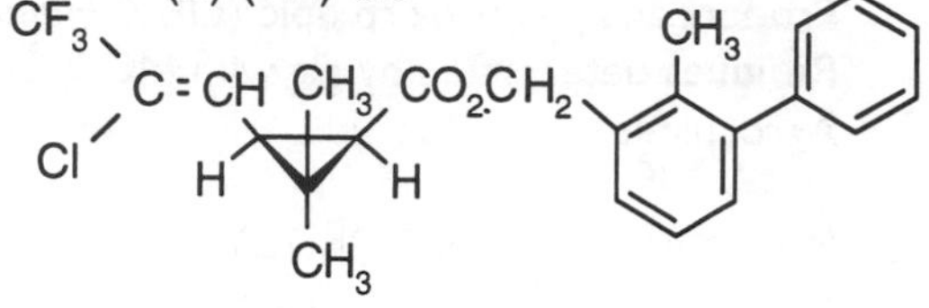

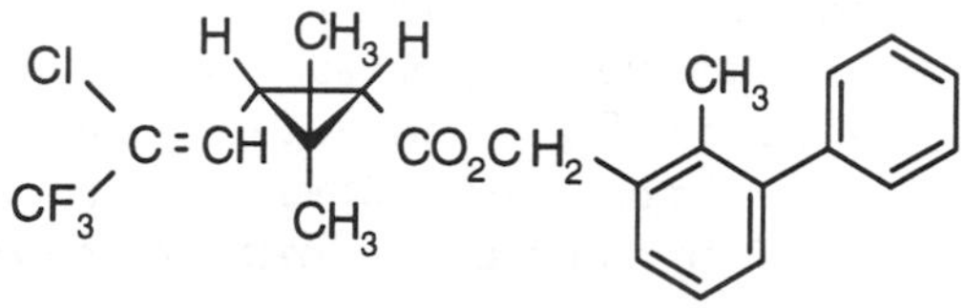

(*Z*)-(1*S*)-*cis*-

NOMENCLATURE

Common name bifenthrin (BSI, ANSI, draft E-ISO); bifenthrine ((*f*) draft F-ISO)
IUPAC name 2-methylbiphenyl-3-ylmethyl (*Z*)-(1*RS*,3*RS*)-3-(2-chloro-=3,3,3-trifluoroprop-1-enyl)-2,2-dimethylcyclopropanecarboxylate
Roth: 2-methylbiphenyl-3-ylmethyl (*Z*)-(1*RS*)-*cis*-3-(2-chloro-3,3,3-trifluoroprop-=1-enyl)-2,2-dimethylcyclopropanecarboxylate
Chemical Abstracts name (2-methyl[1,1′-biphenyl]-3-yl)methyl 3-(2-chloro-=3,3,3-trifluoro-1-propenyl)-2,2-dimethylcyclopropanecarboxylate
CAS RN *[82657–04–3]* **Development codes** FMC 54 800
Official codes OMS 3024

PHYSICAL CHEMISTRY

Composition Material contains ≥97% *cis-* isomer, ≤3% *trans-* isomer.
Mol. wt. 422.9 **M.f.** $C_{23}H_{22}ClF_3O_2$ **Form** Viscous liquid; crystalline or waxy solid. **M.p.** 51–66 °C **V.p.** 0.024 mPa (25 °C) **K_{ow}** logP >6 **S.g./density** 1.210 (25 °C) **Solubility** In water, 0.1 mg/l. Soluble in acetone, chloroform, dichloromethane, diethyl ether, and toluene. Slightly soluble in heptane and methanol. **Stability** Stable for 2 y at 25 °C and 50 °C (tech.). In natural daylight DT_{50} 255 d, stable 21 d at pH 5–9 (21 °C). **F.p.** 165 °C (Tag open cup); 151 °C (Pensky Martens closed cup)

COMMERCIALISATION

History Insecticide and acaricide reported by H. J. H. Doel *et al.* (*Meded. Fac. Landbouwwet. Rijksuniv. Gent,* 1984, **49,** 929) and A. R. Crossman *et al.* (*ibid.,* p. 939). Introduced by FMC Corp. **Manufacturers** FMC

APPLICATIONS

Mode of action Contact and stomach action.   **Uses** Effective against a broad range of foliar pests, including Coleoptera, Diptera, Heteroptera, Homoptera, Lepidoptera and Orthoptera; it also controls some species of Acarina. Crops include cereals, citrus, cotton, fruit, grapes, ornamentals and vegetables. Rates range from 5 g a.i./ha against Aphididae in cereals to 45 g/ha against Aphididae and Lepidoptera in top fruit.   **Formulation types** EC; SC; WP; UL; GR. **Compatibility** Not compatible with alkaline materials. **Selected tradenames** 'Talstar' (FMC)

ANALYSIS

Product analysis by glc; hplc for analysis of isomer content.   **Residue** analysis by glc with ECD. Details available from FMC Agricultural Chemicals Group. Analysis of pyrethroids reviewed by E. Papadopoulou-Mourkidou in *Comp. Analyt. Profiles*.

MAMMALIAN TOXICOLOGY

Reviews *Pesticide residues in food – 1992*, FAO Plant Production and Protection Paper, 116, 1993. *Pesticide residues in food – 1992 evaluations. Pt. II – Toxicology*. WHO, WHO/PCS/93.34 (1993).   **Oral** Acute oral LD_{50} for rats 54.5 mg/kg. **Skin and eye** Acute percutaneous LD_{50} for rabbits >2000 mg/kg. Non-irritant to skin; virtually non-irritating to eyes (rabbits). No skin sensitisation (guinea pigs). **NOEL** (1 y) for dogs 1.5 mg/kg daily. Non-teratogenic in rats (≤2 mg/kg daily) and rabbits (8 mg/kg daily).   **ADI** (JMPR) 0.02 mg/kg [1992]. **Toxicity class** WHO (a.i.) II; EPA (formulation) II

ECOTOXICOLOGY

Birds Acute oral LD_{50} for bobwhite quail 1800, mallard ducks 2150 mg/kg. Eight-day dietary LC_{50} for bobwhite quail 4450, mallard ducks 1280 mg/kg diet. **Fish** LC_{50} (96 h) for bluegill sunfish 0.00035, rainbow trout 0.00015 mg/l. **Bees** LD_{50} (oral) 0.1 µg/bee; (contact) 0.01462 µg/bee.   **Daphnia** LC_{50} (48 h) 0.00016 mg/l. Low solubility in water and high affinity for soil contribute to produce little impact in aquatic systems under field conditions.

ENVIRONMENTAL FATE

Soil/Environment Soil DT_{50} 65–125 d. K_{oc} 1.31–3.02 × 10^5.

phosphinico amino acid

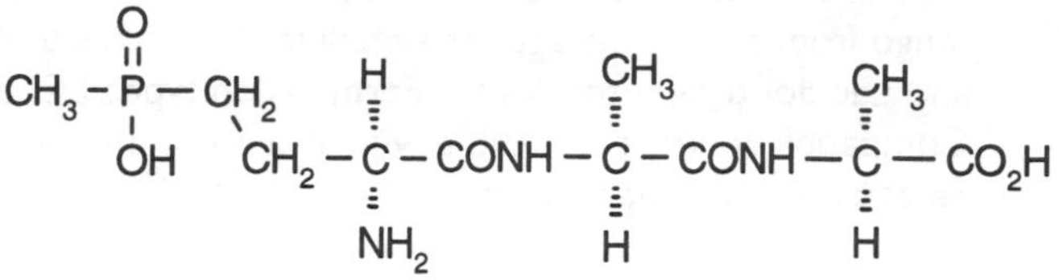

$$CH_3 - \overset{\overset{\displaystyle O}{\|}}{\underset{\underset{\displaystyle OH}{|}}{P}} - CH_2 \quad \underset{\underset{\displaystyle CH_2 - \overset{\overset{\displaystyle H}{|}}{\underset{\underset{\displaystyle NH_2}{|}}{C}} - CONH - \overset{\overset{\displaystyle CH_3}{|}}{\underset{\underset{\displaystyle H}{|}}{C}} - CONH - \overset{\overset{\displaystyle CH_3}{|}}{\underset{\underset{\displaystyle H}{|}}{C}} - CO_2H}{}$$

NOMENCLATURE
bilanafos
Common name bilanafos (BSI, draft E-ISO, (*m*) draft F-ISO); bialaphos (JMAF)
IUPAC name 4-[hydroxy(methyl)phosphinoyl]-L-homoalanyl-L-alanyl-L-alanine
Chemical Abstracts name 4-(hydroxymethylphosphinyl)-L-2-aminobutanoyl-=
L-alanyl-L-alanine
CAS RN *[35597–43–4]*   **Development codes** MW-801; SF-1293

bilanafos-sodium
Common name bilanafos-sodium
CAS RN *[71048–99–2]*

PHYSICAL CHEMISTRY
bilanafos
Mol. wt. 323.3   **M.f.** $C_{11}H_{22}N_3O_6P$ **Solubility** In water 1 kg/l. In methanol 500,
ethanol 250 (both in g/l). Practically insoluble in organic solvents such as acetone,
benzene, butan-1-ol, chloroform, diethyl ether, ethanol, hexane.   **Stability**
Unstable in strong acids and strong alkalis.   **Specific rotation** $[\alpha]_D^{25}$ –34° (10 g/l
water)

bilanafos-sodium
Mol. wt. 345.3   **M.f.** $C_{11}H_{21}N_3NaO_6P$ **Form** Colourless crystals.   **M.p.** *c.* 160 °C
(decomp.)

COMMERCIALISATION
Production Bilanafos-sodium is produced by *Streptomyces hygroscopicus* during
fermentation.   **History** Herbicidal activity of this fermentation product reported
by K. Tachibana *et al.* (*Abstr. 5th Int. Congr. Pestic. Chem.,* **IVa,** Abstract 19) and the
compound reviewed by S. Mase (*Jpn. Pestic. Inf.,* 1984, No. 45, p. 27). Introduced
by Meiji Seika Kaisha Ltd.   **Manufacturers** Meiji Seika

APPLICATIONS
Biochemistry Glutamine synthetase inhibitor, causing ammonia accumulation and
inhibition of photophosphorylation in photosynthesis.

Uses Post-emergence control of annual weeds in vines, apples, brassicas, cucurbits, mulberries, azaleas, rubber, etc.; and control of annual and perennial weeds in uncultivated land. **Formulation types** SL. **Selected tradenames** 'Meiji Herbiace' (Meiji Seika)

ANALYSIS
Product by nmr. **Residues** by gc. Details from Meiji Seika.

MAMMALIAN TOXICOLOGY
Oral Acute oral LD_{50} for male rats 268, female rats 404 mg sodium salt/kg.
Skin and eye Acute percutaneous LD_{50} for rats >5000 mg/kg. Non-irritating to skin and eyes (rabbits). **Other** In sub-acute and chronic toxicity studies, no ill-effects were observed. Non-carcinogenic, non-mutagenic and non-teratogenic. Not mutagenic in Ames and Rec assays. **Toxicity class** WHO (a.i.) II

ECOTOXICOLOGY
Birds Acute oral LD_{50} for chickens >5000 mg/kg. **Fish** LC_{50} (48 h) for carp 1000 mg/l. **Daphnia** LC_{50} (48 h) 1000 mg/l.

ENVIRONMENTAL FATE
Animals In the mouse, the main metabolite in the faeces following oral administration was 2-amino-4-[(hydroxy)(methyl)phosphinyl]butyric acid (A. Suzuki et al., J. Pestic. Sci., 1987, **12**, 105). **Plants** Metabolised in plants to the L-isomer of glufosinate, which has similar activity. **Soil/Environment** Inactivated in soil.

2 # bioallethrin *Insecticide*

pyrethroid

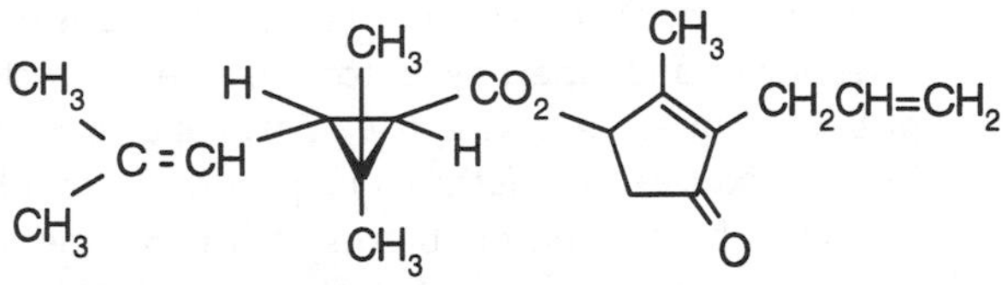

NOMENCLATURE
Common name bioallethrin (BSI, New Zealand); depalléthrine ((f) France (AFNOR)); d-trans-allethrin (ESA)
IUPAC name (RS)-3-allyl-2-methyl-4-oxocyclopent-2-enyl (1R,3R)-2,2-dimethyl-=
3-(2-methylprop-1-enyl)cyclopropanecarboxylate

Alt: (*RS*)-3-allyl-2-methyl-4-oxocyclopent-2-enyl (+)-*trans*-chrysanthemate
Roth: (*RS*)-3-allyl-2-methyl-4-oxocyclopent-2-enyl (1*R*)-*trans*-2,2-dimethyl-=
3-(2-methylprop-1-enyl)cyclopropanecarboxylate
Chemical Abstracts name 2-methyl-4-oxo-3-(2-propenyl)-2-cyclopenten-1-yl
2,2-dimethyl-3-(2-methyl-1-propenyl)cyclopropanecarboxylate
Other names *d*-allethrin   **CAS RN** *[584–79–2]* (same as for allethrin) formerly
[22431–63–6]   **EEC no.** 209–542–4   **Development codes** EA 3054; RU 11705
Official codes ENT 16 275; OMS 3044 (bioallethrin); OMS 3034 (*d*-allethrin)

PHYSICAL CHEMISTRY
Composition Bioallethrin contains ≥93% *m/m* total isomers of which there is
≥90% *trans*- isomer pair and ≤3% *cis*- isomer pair. Material in 'D-Trans' is
90% *m/m* bioallethrin. For *d*-allethrin see allethrin [(1*R*)- isomers].   **Mol. wt.** 302.4
M.f. $C_{19}H_{26}O_3$ **Form** Bioallethrin is an orange-yellow viscous liquid. The material
in 'D-Trans' is an amber viscous liquid.   **M.p.** Not applicable; no crystallisation
observed at −40 °C.   **B.p.** bioallethrin 165–170 °C/0.15 mmHg   **V.p.** 43.9 mPa
(25 °C)   **K_{ow}** logP = 4.68 (25 °C)   **S.g./density** 1.012 at 20 °C
Solubility Bioallethrin: in water 4.6 mg/l (25 °C). Completely miscible in acetone,
ethanol, chloroform, ethyl acetate, hexane, toluene and dichloromethane (20 °C).
Stability Degraded by u.v. light. In aqueous solution DT_{50} 1410.7 d (pH 5), 547.3 d
(pH 7), 4.3 d (pH 9).   **Specific rotation** $[\alpha]_D^{20}$ −18.5° to −22.5° (50 g/l toluene)
F.p. 87 °C (Pensky-Martens)

COMMERCIALISATION
History Improved insecticidal activity of this pair of esters of the (1*R*)-*trans* acid
described by J. Lhoste *et al.* (*C. R. Seances Acad. Agric. Fr.*, 1967, **53**, 686).
Introduced by Roussel Uclaf and by McLaughlin Gormley King Co.
Patent US 3159535 to Roussel Uclaf   **Manufacturers** AgrEvo

APPLICATIONS
Mode of action Non-systemic, non-residual insecticide with contact and stomach
action. Has rapid knockdown activity.   **Uses** Bioallethrin is a potent contact, non-
systemic, non-residual insecticide which produces a rapid knockdown and is used
against household insect pests (Blattodea, Culicidae and Muscidae) and in
insecticidal coils and electric thermal vapourisers. Used in mixtures with other
insecticides and piperonyl butoxide. The (*S*)-cyclopentenyl ester is the more
potent form − see next entry.   **Formulation types** EC; TC; AE; VP; OL; Oil; Coil;
Mat.   **Mixtures** (*bioallethrin +*) piperonyl butoxide; permethrin + piperonyl
butoxide; bioresmethrin; bioresmethrin + piperonyl butoxide; deltamethrin;
permethrin; resmethrin.   **Selected tradenames** 'Bioallethrine' (AgrEvo); 'D-Trans'
(MGK)

ANALYSIS
Product analysis of pyrethroids reviewed by E. Papadopoulou-Mourkidou in *Comp.
Analyt. Profiles.* By glc (*CIPAC Handbook*, 1980, **1A**, 1097; *AOAC Methods*, 1995,

973.12); proportions of stereoisomers determined by glc of suitable derivatives
(A. Murano, *Agric. Biol. Chem.*, 1972, **36,** 2203; A. Horiba et al., *ibid.*, 1977, **41,**
2003; 1978, **42,** 671; F. E. Rickett, *Analyst (London)*, 1973, **98,** 687) or by nmr
(F. E. Rickett & P. B. Henry, *ibid.*, 1974, **99,** 330). **Residues** determined by glc
(D. B. McClellan, *Anal. Methods Pestic., Plant Growth Regul. Food Addit.*, 1964, **2,** 25;
Anal. Methods Pestic. Plant Growth Regul., 1972, **6,** 283; J. Sherma, *ibid.*, 1976, **8,**
117).

MAMMALIAN TOXICOLOGY
Reviews A. J. Gray & D. M. Soderlund, Chapt. 5 in *"Insecticides"*. **EHC** 87
(WHO 1989). **Oral** Acute oral LD_{50} for male rats 709 mg bioallethrin/kg,
female rats 1042 mg/kg; for male rats 425–575 mg ('D-Trans')/kg, female rats
845–875 mg/kg. **Skin and eye** Acute percutaneous LD_{50} for rabbits >3000 mg
bioallethrin/kg. **Inhalation** LC_{50} (4 h) for rats 2.5 mg bioallethrin/l air.
NOEL (90 d) for rats 750 mg/kg diet. **Other** No mutagenic, carcinogenic,
embryotoxic or teratogenic effects have been observed. **Toxicity class**
WHO (a.i.) II; EPA II (tech.), III (formulations) **EC risk** Xn (R22)

ECOTOXICOLOGY
Birds Acute oral LD_{50} for bobwhite quail 2030 mg/kg. **Fish** Highly toxic. LC_{50} (96
h) (static test, flow-through test) for coho salmon 22.2, 9.40, steelhead trout 17.5,
9.70, channel catfish >30.1, 27.0, yellow perch –, 9.90 µg/l. **Daphnia** LC_{50} (96 h)
0.0356 mg/l.

ENVIRONMENTAL FATE
Animals [^{14}C-acid]-bioallethrin administered in rats with a single dose of either
200 or 100 mg/kg b.w. was readily eliminated in the urine and faeces of all dose
groups within 2–3 days after treatment. **Soil/Environment** [^{14}C-alcohol]-
bioallethrin degrades in pH 5 buffer to allethrolene, dihydroxyallethrolene, CO_2
and a number of small yield polar products as the result of photochemical
processes. *Cis/trans* isomerisation was not observed. The actual and extrapolated
DT_{50} of light-exposed and dark control samples is 48.8 h and 1447 h respectively.

73 bioallethrin S-cyclopentenyl isomer

Insecticide

pyrethroid

NOMENCLATURE
Common name esdepalléthrine (France (AFNOR))
IUPAC name (S)-3-allyl-2-methyl-4-oxocyclopent-2-enyl (1R,3R)-2,2-dimethyl-=
3-(2-methylprop-1-enyl)cyclopropanecarboxylate
Alt: (S)-3-allyl-2-methyl-4-oxocyclopent-2-enyl (+)-*trans*-chrysanthemate
Roth: (S)-3-allyl-2-methyl-4-oxocyclopent-2-enyl (1R)-*trans*-2,2-dimethyl-=
3-(2-methylprop-1-enyl)cyclopropanecarboxylate
Chemical Abstracts name [1R-[1α(S*),3β]]-2-methyl-4-oxo-3-(2-propenyl)-2-=
cyclopenten-1-yl 2,2-dimethyl-3-(2-methyl-1-propenyl)cyclopropanecarboxylate
Other names S-bioallethrin; Esbiol **CAS RN** *[28434–00–6]* **EEC no.** 249–013–5
Development codes RU 3054; RU 16 121 (for OMS 3046); RU 27 436 (for OMS
3045) **Official codes** AI3–29 024; OMS 3046 (for Esbiol); OMS 3045 (for
Esbiothrin).

PHYSICAL CHEMISTRY
Composition Esbiol (OMS 3046) contains ≥95% m/m total isomers of which ≥90%
is esdepalléthrine. Esbiothrin (OMS 3045) contains ≥93% m/m total isomers of
which ≥72% is esdepalléthrine, the remainder being essentially the (R)-cyclopentyl
isomer, with <3% *cis*-isomers. **Mol. wt.** 302.4 **M.f.** $C_{19}H_{26}O_3$ **Form** Orange-
yellow viscous liquid. **M.p.** Not applicable; no crystallisation observed at –40 °C.
B.p. 165–170 °C/0.15 mmHg (OMS 3046); 163–170 °C/0.15 mmHg (OMS 3045)
V.p. 44 mPa (25 °C) **S.g./density** 1.010 at 20 °C **Solubility** Completely miscible
at 20 °C with 100% ethanol, acetone, dichloromethane, di-isopropyl ether,
toluene, n-hexane, chloroform, ethyl acetate, methanol, n-octanol, petroleum
solvent. **Stability** Degraded by u.v. light. **Specific rotation** OMS 3046: $[\alpha]_D^{20}$
–47.5° to –55° (50 g/l toluene); OMS 3045: $[\alpha]_D^{20}$ ≥–37° (50 g/l toluene).
F.p. 113 °C (open cup)

COMMERCIALISATION
History Insecticide reported by F. Rauch *et al.* (*Meded. Fac. Landbouwwet. Rijksuniv.
Gent.*, 1972, **37,** 755). Introduced by Roussel Uclaf. **Manufacturers** Roussel Uclaf

APPLICATIONS

Mode of action Non-residual insecticide with rapid knockdown action. **Uses** Esdepalléthrine is by far the most potent constituent of allethrin. OMS 3046 is used in aerosols, coils, electric vapourisers, or sprays indoors against flying and crawling insects (Blattodea, Culicidae, Muscidae). Also used in aerosols and sprays in mixtures with other pyrethroids and piperonyl butoxide. OMS 3045 has similar uses to OMS 3046; both are non-residual products with rapid knockdown action. **Formulation types** AE; OL; VP; UL; Water-miscible liquid. **Mixtures** (bioallethrin S cyclopentenyl isomer +) piperonyl butoxide; deltamethrin; resmethrin; bioresmethrin; deltamethrin + piperonyl butoxide; permethrin + piperonyl butoxide; permethrin; permethrin + dichlorvos; RU 15525 + deltamethrin.

ANALYSIS

Product analysis of pyrethroids reviewed by E. Papadopoulou-Mourkidou in *Comp. Analyt. Profiles*. By glc (*CIPAC Handbook*, 1980, **1A**, 1097); proportions of stereoisomers determined by glc of suitable derivatives (A. Murano, *Agric. Biol. Chem.*, 1972, **36**, 2203; A. Horiba et al., *ibid.*, 1977, **41**, 2003; 1978, **42**, 671; F. E. Rickett, *Analyst (London)*, 1973, **98**, 687) or by nmr (F. E. Rickett & P. B. Henry, *ibid.*, 1974, **99**, 330). **Residues** determined by glc (D. B. McClellan, *Anal. Methods Pestic., Plant Growth Regul. Food Addit.*, 1964, **2**, 25; *Anal. Methods Pestic. Plant Growth Regul.*, 1972, **6**, 283; J. Sherma, *ibid.*, 1976, **8**, 117).

MAMMALIAN TOXICOLOGY

Reviews A. J. Gray & D. M. Soderlund, Chapt. 5 in "*Insecticides*". **EHC** 87 (WHO, 1989). **Oral** Acute oral LD_{50} esdepalléthrine:- for male rats 784, female rats 1545 mg/kg. OMS 3045:- for male rats 432, female rats 378 mg/kg. OMS 3046:- for male rats 574.5, female rats 412.9 mg/kg (in PEG 200). **Skin and eye** Acute percutaneous LD_{50} for rabbits:- 1545 mg esdepalléthrine/kg, >2000 mg OMS 3045/kg, >2000 mg OMS 3046/kg. **Inhalation** LC_{50} (4 h) for rats c. 1.26 mg esdepalléthrine/l air, c. 2.63 mg OMS 3045/l air. **NOEL** OMS 3045:- (2 y) for rats 500 ppm; not an oncogen up to 4500 mg/kg diet; (2 y) for mice 250 ppm; not an oncogen up to 1250 mg/kg diet; (1 y) for dogs 400 ppm. Esdepalléthrine:- (0.5 y) for rats 1000 mg/kg diet. **Other** No mutagenic, carcinogenic, embryotoxic or teratogenic effects have been observed. **Toxicity class** WHO (a.i.) II; EPA (formulation) III **EC risk** Xn (R21/22)

ECOTOXICOLOGY

Birds Acute oral LD_{50} for mallard ducks and bobwhite quail >5000 mg/kg; for chickens >5000 mg esdepalléthrine/kg. LC_{50} (8 d) for ducks and quail >5000 mg/kg diet. **Fish** LC_{50} (96 h) for rainbow trout 0.01, bluegill sunfish 0.033 mg/l. LC_{50} (96 h, static test) for fathead minnow 0.0800, yellow perch 0.0078 mg/l.

pyrethroid

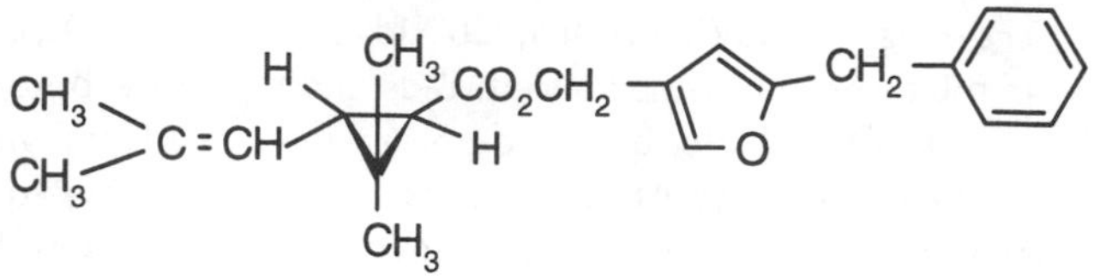

NOMENCLATURE
Common name bioresmethrin (BSI, E-ISO); bioresméthrine ((*f*) F-ISO); *d-trans*-resmethrin (ESA); cismethrin (BSI and E-ISO); cisméthrine ((*f*) F-ISO for the corresponding (1*R*)-*cis*- or (1*R*,3*S*)-stereoisomer)
IUPAC name 5-benzyl-3-furylmethyl (1*R*,3*R*)-2,2-dimethyl-3-(2-methylprop-= 1-enyl)cyclopropanecarboxylate
Alt: 5-benzyl-3-furylmethyl (+)-*trans*-chrysanthemate
Roth: 5-benzyl-3-furylmethyl (1*R*)-*trans*-2,2-dimethyl-3-(2-methylprop-= 1-enyl)cyclopropanecarboxylate
Chemical Abstracts name (1*R-trans*)-[5-(phenylmethyl)-3-furanyl]methyl 2,2-dimethyl-3-(2-methyl-1-propenyl)cyclopropanecarboxylate
Other names *d*-resmethrin   **CAS RN** *[28434–01–7]* formerly *[10453–54–0]* bioresmethrin; *[35764–59–1]* formerly *[31182–61–3]* cismethrin
EEC no. 249–014–0   **Development codes** NRDC 107; FMC 18 739; RU 11 484 (all bioresmethrin); NRDC 119 (cismethrin)   **Official codes** OMS 3043; ENT 27 622; AI3–27 622 (both bioresmethrin); OMS 1800 (cismethrin)

PHYSICAL CHEMISTRY
Composition Tech. grade contains ≥93% total isomers of which ≥90% is the (1*R*)-*trans*- isomer, and ≤3% is *cis*- isomers. *d*-Resmethrin contains ≥95% (1*R*)-isomers, ≥75% *trans*- isomer (bioresmethrin).   **Mol. wt.** 338.4   **M.f.** $C_{22}H_{26}O_3$
Form Tech. grade bioresmethrin is a viscous yellow to brown liquid which partially solidifies on standing. Tech. *d*-resmethrin is a colourless to yellowish liquid, sometimes partially crystalline at room temperature.   **M.p.** 32 °C   **B.p.** decomp. >180 °C   **V.p.** 18.6 mPa (25 °C) (gas saturation method). Tech. *d*-resmethrin 0.45 mPa (20 °C) K_{ow} logP >4.7   **S.g./density** 1.050 (20 °C); *d*-resmethrin 1.040 (25 °C)   **Solubility** In water <0.3 mg/l (25 °C). Soluble in ethanol, acetone, chloroform, dichloromethane, ethyl acetate, toluene and hexane. In ethylene glycol <10 g/l. *d*-Resmethrin: in water 1.2 mg/l (30 °C); in xylene 50% (25 °C).
Stability Decomposes above 180 °C, and under u.v. light. Readily hydrolysed in alkaline media. Sensitive to oxidation. *d*-Resmethrin decomposes under u.v. light. Readily hydrolysed in alkaline media.   **Specific rotation** $[\alpha]_D^{20}$ −5 to −9° (100 g/l ethanol)   **F.p.** *c*. 92 °C

COMMERCIALISATION
History Insecticide reported by M. Elliott *et al.* (*Nature (London)*, 1967, **213**, 493) and developed by Fisons Ltd, FMC Corp. (who no longer manufacture or market it), Roussel Uclaf (now AgrEvo GmbH) and the Wellcome Foundation. **Patent** GB 1168797; GB 1168798; GB 1168799 (all to NRDC) **Manufacturers** AgrEvo

APPLICATIONS
Mode of action Insecticide with contact action. **Uses** Bioresmethrin is a potent contact insecticide effective against a wide range of household and public health insects (Blattodea, Culicidae, Muscidae) and plant pests, pests of grain (*Rhyzopertha dominica*) and insects in animal housing. The bioresmethrin/cismethrin tech. has a similar range of activity, including Siphonaptera, *Cimex lectularius* and Acari. **Formulation types** AE; EC; OL; SL.
Mixtures (*bioresmethrin +*) cismethrin; piperonyl butoxide; bioallethrin + piperonyl butoxide; bioallethrin; pyrethrins + piperonyl butoxide.
Selected tradenames 'Isathrine' (AgrEvo)

ANALYSIS
Product analysis of pyrethroids reviewed by E. Papadopoulou-Mourkidou in *Comp. Analyt. Profiles*. By glc with FID (B. B. Brown, *Anal. Methods Pestic. Plant Growth Regul.*, 1973, **7**, 441) or by hplc (D. S. Gunew, *ibid.*, 1978, **10**, 19).
Residues determined by glc (*idem, ibid.*).

MAMMALIAN TOXICOLOGY
Reviews *Pesticide residues in food – 1991*. FAO Plant Production and Protection Paper, 11, 1991. *Pesticide residues in food – 1991 evaluations. Part II – Toxicology*. World Health Organisation, WHO/PCS/92.52, 1992. See also A. J. Gray & D. M. Soderlund, Chapt. 5 in "*Insecticides*". **EHC** 92 (WHO 1989). **Oral** Acute oral LD_{50} for rats 7070–8000 mg/kg; dissolved in corn oil ≥5000 mg/kg; for male rats 450, female rats 680 mg tech. *d*-resmethrin/kg. **Skin and eye** Acute percutaneous LD_{50} for female rats >10 000, rabbits >2000 mg/kg.
Inhalation LC_{50} (4 h) for rats 5.28 mg/l air, (24 h) 0.87 mg/l air; (4 h) for rats 1.56 mg tech. *d*-resmethrin/l air. **NOEL** (90 d) for rats 1200, dogs >500 mg/kg diet; (2 y) for rats 50 ppm diet (3 mg/kg b.w. daily). Rats tolerated 4000 mg/kg diet for 60 d. In rats dosed daily up to 200 mg/kg from day 6–15 of pregnancy, there was no teratogenic nor embryotoxic effect; nor in rabbits dosed daily up to 240 mg/kg from day 6–18 of pregnancy. Maternal NOEL were 100 and 60 mg/kg in rats and rabbits respectively. **ADI** (JMPR) 0.03 mg/kg [1991].
Other No carcinogenic, mutagenic or teratogenic effects have been demonstrated.
Toxicity class WHO (a.i.) III (Table 5)

ECOTOXICOLOGY
Birds Acute oral LD_{50} for chickens >10 000 mg/kg. **Fish** LC_{50} (96 h) for harlequin fish 0.014, guppies 0.5–1.0, rainbow trout 0.00062, bluegill sunfish

0.0024 mg/l; (48 h) for harlequins 0.018, guppies 0.5–1.0 mg/l. Despite high toxicity to fish in laboratory tests, bioresmethrin at the recommended rates does not present significant environmental hazard due to its rapid degradation under field conditions. **Bees** Very toxic to honeybees; LD_{50} (oral) 2 ng/bee, (topical) 6.2 ng/bee. **Daphnia** LC_{50} (48 h) 0.0008 mg/l.

ENVIRONMENTAL FATE
Animals Metabolism in rats was principally by oxidation reactions in the isobutenyl methyl groups, by ester cleavage and conjugation reactions. **Plants** In studies on tomatoes, very little material (entirely as degradation products) penetrated the skin. Similar results were obtained on cucumbers, but bioresmethrin on the skin of cucumbers is degraded more quickly than on tomatoes. **Soil/Environment** Strongly adsorbed on soil.

75   biphenyl   *Fungicide*

aromatic hydrocarbon

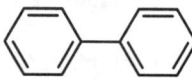

NOMENCLATURE
Common name biphenyl (BSI, E-ISO, F-ISO, accepted in lieu of common name)
IUPAC name biphenyl
Chemical Abstracts name 1,1'-biphenyl
Other names diphenyl   **CAS RN** *[92–52–4]*   **EEC no.** 202–163–5

PHYSICAL CHEMISTRY
Mol. wt. 154.2   **M.f.** $C_{12}H_{10}$ **Form** Colourless crystals.   **M.p.** 69–71 °C
B.p. 254–255 °C   **S.g./density** 1.041   **Solubility** Practically insoluble in water. Readily soluble in ethanol, diethyl ether, and other organic solvents.
Stability Very stable in acidic and alkaline media.   **F.p.** 106 °C

COMMERCIALISATION
History Fungicidal activity against citrus-rotting fungi described by G. B. Ramsey *et al.* (*Bot. Gaz.*, 1944, **106,** 74).

APPLICATIONS
Mode of action Fungistat which inhibits sporulation.   **Uses** Used to impregnate citrus fruit wraps to prevent fungal attack.   **Formulation types** Impregnated material.

ANALYSIS

Residues determined by u.v. spectrometry (F. A. Gunther *et al., Analyst (London)*, 1963, **88**, 36; F. A. Gunther & D. E. Ott, *ibid.*, 1966, **91**, 475; S. Norman *et al., J. Assoc. Off. Anal. Chem.*, 1966, **49**, 590; AOAC Methods, 1995, 968.25), by glc (H. Pyysalo *et al., J. Chromatogr.*, 1979, **168**, 512; *Pestic. Anal. Man.*, 1979, **I**, 2011; *Man. Pestic. Residue Anal.*, 1987, **I**, 6) or by hplc (J. E. Farrow *et al., Analyst (London)*, 1977, **102**, 752).

MAMMALIAN TOXICOLOGY

Reviews *Pesticide residues.* FAO Meeting Report, No. PL:1967/M/11; WHO Technical Report Series, No. 391, 1968. *1967 Evaluations of some pesticide residues in food.* FAO/PL:1967/M/11/1; WHO/Food Add./68.30, 1968.
Oral Acute oral LD_{50} for rats 3280, rabbits 2400, cats >2600 mg/kg. Depression of the central nervous system, paralysis, and convulsions have been observed in experimental animals. **Inhalation** Prolonged exposure of humans to vapour concentrations >0.005 mg/l air is considered dangerous (J. Deichmann *et al., J. Ind. Hyg. Toxicol.*, 1947, **29**, 1). **ADI** (JMPR) 0.125 mg/kg [1967].
Toxicity class WHO (a.i.) III (Table 5); EPA (formulation) III
EC risk Xi (R36/37/38)

ENVIRONMENTAL FATE

Animals In mammals, following oral administration, metabolites identified include 4-hydroxybiphenyl and other phenolic compounds (H. D. West *et al., Arch. Biochem. Biophys.*, 1956, **60**, 14–20). **Plants** Residues in fruit have been reviewed (A. Rajzman, *Residue Rev.*, 1964, **8**, 1: S. W. Souci & G. Maier-Haarländer, *ibid.*, 1966, **16**, 103).

76   bispyribac-sodium   *Herbicide*

pyrimidinyloxybenzoic

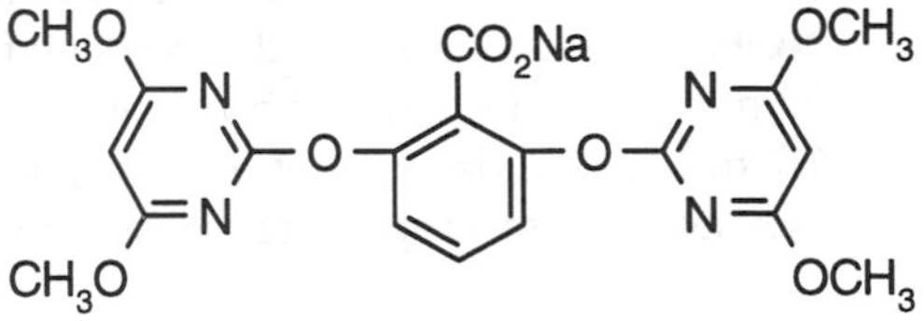

NOMENCLATURE

Common name bispyribac-sodium (BSI, pa ISO)
IUPAC name sodium 2,6-bis(4,6-dimethoxypyrimidin-2-yloxy)benzoate
Chemical Abstracts name sodium 2,6-bis[(4,6-dimethoxy-2-pyrimidinyl)oxy]=benzoate

CAS RN *[125401–75–4]* for acid; *[125401–92–5]* for sodium salt
Development codes KIH-2023 (for sodium salt)

PHYSICAL CHEMISTRY
Composition Tech. is >93%. **Mol. wt.** 452.4 **M.f.** $C_{19}H_{17}N_4NaO_8$; $C_{19}H_{18}N_4O_8$
for the acid **Form** White powder. **M.p.** 223–224 °C **V.p.** 5.05×10^{-6} mPa
(25 °C) K_{ow} logP = –1.03 (23 °C) **S.g./density** Bulk density 0.0737 (20 °C,
CIPAC MT 3) **Solubility** In water 73.3 g/l (25 °C). In methanol 26.3, acetone
0.043 g/l (25 °C). **Stability** Stable in water; DT_{50} >1 y (pH 7 & 9), 448 h (pH 4).
Stable to light. Not decomposed after 14 d at 55 °C. **pKa** 3.05

COMMERCIALISATION
History Reported by Kumiai Chemical Industry Co., Ltd (M. Yokoyama et al.,
Proc. Br. Crop Prot. Conf., Weeds, 1993, **1**, 61). Developed jointly by Kumiai and
Ihara Chemical Industry Co., Ltd. **Patent** US 4906285; EP 321846
Manufacturers Ihara/Kumiai

APPLICATIONS
Biochemistry Acts by inhibition of acetolactate synthase, blocking branched-chain
amino acid biosynthesis. **Mode of action** Selective, systemic post-emergence
herbicide, absorbed by foliage and roots. **Uses** Control of a wide range of
weeds, especially *Echinochloa* spp., in direct-seeded rice, at rates of 15–45 g/ha.
Also used to stunt growth of weeds in non-crop situations.
Formulation types SC; water-soluble liquid. **Selected tradenames** 'Nominee'
(Kumiai); 'Grass-short' (non-crop land) (Kumiai); 'Short-keep' (non-crop land)
(Riken Green)

ANALYSIS
By hplc. Details available from Kumiai.

MAMMALIAN TOXICOLOGY
Oral Acute oral LD_{50} for male rats 4111, female rats 2635 mg/kg.
Skin and eye Acute percutaneous LD_{50} for rats >2000 mg/kg. Non-irritating to
skin and slightly irritating to eyes (rabbits). **Inhalation** LC_{50} for rats >4.48 mg/l.
NOEL (2 y) for male rats 20 mg/kg diet (1.1 mg/kg b.w. daily), female rats
20 mg/kg diet (1.4 mg/kg b.w. daily), male mice 14.1 mg/kg daily, female mice
1.7 mg/kg b.w. daily. **ADI** 0.011 mg/kg. **Other** Non-mutagenic in the Ames
test, and non-teratogenic to rats and rabbits.

ECOTOXICOLOGY
Birds Acute oral LD_{50} for bobwhite quail >2250 mg/kg. Dietary LC_{50} (5 d) for
bobwhite quail and mallard duck >5620 mg/kg diet. **Fish** Acute LC_{50} (96 h) for
rainbow trout and bluegill sunfish >100 ppm. **Bees** LD_{50} (oral) >200 µg/bee;
LC_{50} (contact) >70 000 mg/l. **Worms** NOEL (14 d) for *Eisenia foetida*

>1000 mg/kg soil. **Daphnia** LC$_{50}$ (48 h) >100 ppm. **Algae** EC$_{50}$ (120 h) for *Selenastrum capricornutum* 3.4 mg/l, NOEC 0.625 mg/l.

ENVIRONMENTAL FATE
Animals >95% of dose applied to rats was excreted in urine and faeces within 7 d.
Plants Following application to foliage and soil at 5-leaf stage of rice plants, c. 10% of carbon label was distributed between straw and roots at harvest.
Soil/Environment In soil DT$_{50}$ <10 d (flooded and upland conditions).

7   bitertanol   *Fungicide*

azole

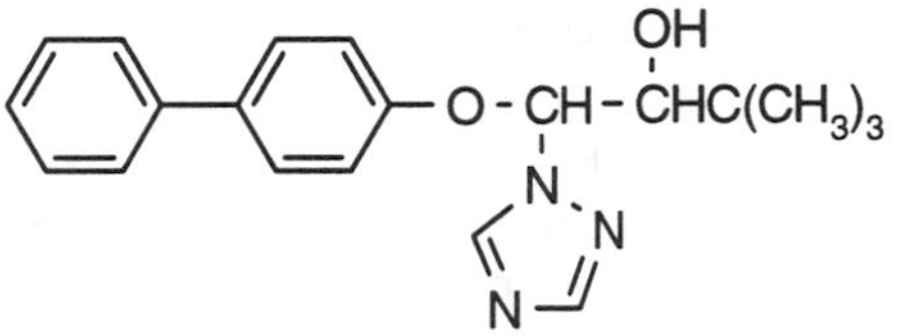

NOMENCLATURE
Common name bitertanol (BSI, draft E-ISO, (*m*) draft F-ISO)
IUPAC name 1-(biphenyl-4-yloxy)-3,3-dimethyl-1-(1*H*-1,2,4-triazol-1-yl)butan-2-ol (20:80 ratio of (1*RS,2RS*) and (1*RS,2SR*) isomers)
Chemical Abstracts name β-([1,1'-biphenyl]-4-yloxy)-α-(1,1-dimethylethyl)-1*H*-= 1,2,4-triazole-1-ethanol
CAS RN *[70585–36–3]* diastereoisomer A; *[70585–38–5]* distereoisomer B; *[55179–31–2]* unstated stereochemistry **Development codes** BAY KWG 0599

PHYSICAL CHEMISTRY
Composition Bitertanol is a mixture of 2 diastereoisomers.
Enantiomer A: (1*R,2S*) + (1*S,2R*); enantiomer B: (1*R,2R*) + (1*S,2S*) in the ratio A:B 7:3. **Mol. wt.** 337.4 **M.f.** C$_{20}$H$_{23}$N$_3$O$_2$ **Form** Colourless crystals; tech. white to tan crystals with a mild odour. **M.p.** 138.6 °C (A), 147.1 °C (B), 118 °C (eutectic of A and B) **V.p.** 2.2 × 10^{-7} mPa (A); 2.5 × 10^{-6} mPa (B) (both at 20 °C) **K$_{ow}$** logP = 4.1 (A), 4.4 (B) (20 °C) **Henry** 2 × 10^{-8} Pa m^3 mol^{-1} (A); 5 × 10^{-7} Pa m^3 mol^{-1} (B) (both 20 °C) **Solubility** In water 2.7 (A), 1.1 (B), 3.8 (eutectic) (all mg/l, 20 °C, not affected by pH). In dichloromethane >250, isopropanol 67, xylene 18, *n*-octanol 53 (all in g/l, 20 °C). **Stability** Stable in neutral, acidic, and alkaline media; DT$_{50}$ at 25 °C >1 y (pH 4, 7 and 9).

COMMERCIALISATION
History Fungicide reported by W. Brandes et al. (*Pflanzenschutz-Nachr. (Engl. Ed.)*, 1979, **32**, 1). Introduced by Bayer AG. **Patent** DE 2324010 **Manufacturers** Bayer

APPLICATIONS
Biochemistry Steroid demethylation inhibitor. **Mode of action** Foliar fungicide with protective and curative action. Inhibits ergosterol biosynthesis. Acts on spore germination, mycelium development, and sporulation. **Uses** Control of scab and *Monilinia* diseases on fruit; rusts and powdery mildews on ornamentals; black spot on roses; Sigatoka on bananas; and leaf spot and other diseases of vegetables, cucurbits, cereals, deciduous fruit, peanuts, soya beans, tea, etc. As a seed dressing, control of smuts and bunts of wheat and rye; in combination with other fungicides, also against seed-borne snow mould. **Phytotoxicity** Fruit crops are more tolerant to the wettable powder formulation than the emulsifiable concentrate formulation. **Formulation types** WP; EC; SC; PA; AE; FS; DS; WS; LS. **Mixtures** (*bitertanol +*) fuberidazole; triadimenol; captan; ziram; dodine. **Selected tradenames** 'Baycor' (for spray application) (Bayer); 'Sibutol' (for seed treatment) (Bayer)

ANALYSIS
Product analysis details available from Bayer AG. **Residue** analysis by glc (R. Brennecke, *ibid.*, 1985, **38**, 33; W. Specht & M. Tillkes, *ibid.*, 1980, **33**, 61) or by hplc (R. Brennecke, *ibid.*, 1988, **41**, 113). Methods for the determination of residues are also available from Bayer.

MAMMALIAN TOXICOLOGY
Reviews *Pesticide residues in food – 1988*. FAO Plant Production and Protection Paper 92, 1988. **Oral** Acute oral LD_{50} for rats >5000, mice c. 4300, dogs >5000 mg/kg. **Skin and eye** Acute percutaneous LD_{50} for rats >5000 mg/kg. Slightly irritating to skin and eyes (rabbits). Not a skin sensitiser. **Inhalation** LC_{50} (4 h) for rats >0.55 mg/l air (aerosol), >1.2 mg/l air (dust). **NOEL** (2 y) for rats and mice 100 mg/kg diet. In 12/20 mo trials, NOEL for dogs 25 mg/kg diet. **ADI** (JMPR) 0.01 mg/kg b.w. [1988]. **Toxicity class** WHO (a.i.) III (Table 5); EPA (formulation) III

ECOTOXICOLOGY
Birds Acute oral LD_{50} for Japanese quail >10 000, mallard ducks >2000 mg/kg. **Fish** LC_{50} (96 h) for rainbow trout 2.2–2.7 mg/l. **Bees** Not toxic to bees; LD_{50} >50 mg/bee. **Daphnia** LC_{50} (48 h) >1.8–7 mg/l. **Algae** E_rC_{50} for *Scenedesmus subspicatus* 0.31 mg/l.

ENVIRONMENTAL FATE
Animals After four days, elimination was almost complete; 98% was excreted with the faeces and urine. **Plants** The concentration of bitertanol in plant tissues can

be neglected. The active ingredient could be determined on the surface of fruits and leaves of treated plants. **Soil/Environment** Direct photodegradation in water contributes to overall degradation of bitertanol to only a low extent. Environmental DT_{50} in water is 1 mo to 1 y. Degradation in soil is rapid; CO_2 is the most important metabolite. Mobility in soil is low.

8 blasticidin-S   *Fungicide*

antibiotic

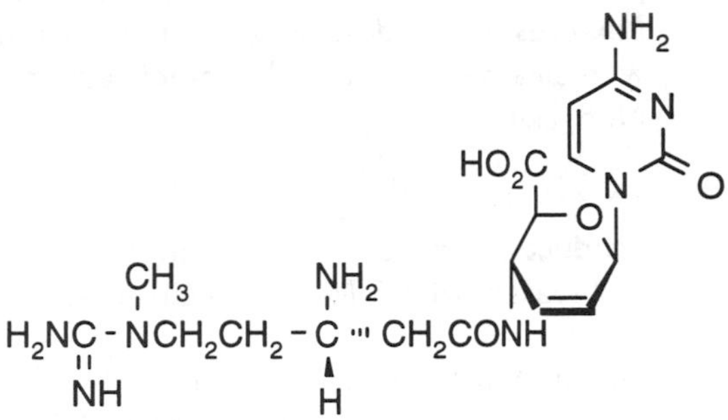

NOMENCLATURE
Common name blasticidin-S (JMAF)
IUPAC name 1-(4-amino-1,2-dihydro-2-oxopyrimidin-1-yl)-4-[(S)-3-amino-= 5-(1-methylguanidino)valeramido]-1,2,3,4-tetradeoxy-β-D-*erythro*-hex-= 2-enopyranuronic acid
Chemical Abstracts name (S)-4-[[3-amino-5-[(aminoiminomethyl)methylamino]-= 1-oxopentyl]amino]-1-[4-amino-2-oxo-1(2H)-pyrimidinyl]-1,2,3,4-tetradeoxy-= β-D-erythro-hex-2-enopyranuronic acid
CAS RN *[2079–00–7]* formerly *[11002–92–9]*, *[12767–55–4]*
Development codes BcS-3; -BAB; -BABS

PHYSICAL CHEMISTRY
Mol. wt. 422.4 **M.f.** $C_{17}H_{26}N_8O_5$ **Form** Colourless needles; (tech., light brown solid). **M.p.** 235–236 °C (decomp.) **Solubility** In water >30 g/l (20 °C). In acetic acid >30 g/l (20 °C). Practically insoluble in acetone, benzene, carbon tetrachloride, chloroform, cyclohexane, dioxane, ethanol, diethyl ether, ethyl acetate, methanol, pyridine, xylene. **Specific rotation** $[\alpha]_D$ +108.4°
pKa pKa_1 2.4 (carboxyl), pKa_2 4.6, pKa_3 8.0, pKa_4 >12.5 (three bases)

COMMERCIALISATION
Production By the fermentation of *Streptomyces griseochromogenes*.

History Antibiotic discovered by K. Fukunaga *et al.* (*Bull. Agric. Chem. Soc. Jpn.*, 1955, **19**, 18), its fungicidal properties reported by T. Misato *et al.*, (*Nippon Shokubutsu Byori Gakkaiho*, 1959, **24**, 302) and its structure elucidated by N. Otake (*Agric. Biol. Chem.*, 1967, **30**, 132). Its benzylaminobenzenesulfonate was introduced by Kaken Chemical Co., Ltd, Kumiai Chemical Industry Co., Ltd, and Nihon Nohyaku Co., Ltd. **Manufacturers** Kaken

APPLICATIONS
Biochemistry Protein synthesis inhibitor. **Mode of action** Contact fungicide with protective and curative action. **Uses** Control of rice blast (*Pyricularia oryzae*) by foliar application. **Phytotoxicity** Damage can be caused to alfalfa, aubergines, clover, potatoes, soya beans, tobacco, and tomatoes. Excessive application produces yellow spots on rice leaves. **Formulation types** DP; EC; WP. **Mixtures** *(blasticidin-S +)* fenitrothion + fenobucarb. **Compatibility** Incompatible with alkaline materials. **Selected tradenames** 'Bla-S' (Kaken, Kumiai, Nihon Nohyaku)

ANALYSIS
Product analysis by bioassay with *Bacillus cereus* IAM-1729. **Residues** determined by bioassay with *Fulvia fulva* ACI-1164.

MAMMALIAN TOXICOLOGY
Oral Acute oral LD_{50} for male rats 56.8, female rats 55.9, male mice 51.9, female mice 60.1 mg/kg. **Skin and eye** Acute percutaneous LD_{50} for rats >500 mg/kg. Severe eye irritant. **NOEL** (2 y) for rats 1 mg/kg diet. **Other** Non-mutagenic in bacterial reversion tests. **Toxicity class** WHO (a.i.) Ib; EPA (formulation) II **EC risk** T+ (R28)

ECOTOXICOLOGY
Fish LC_{50} (48 h) for carp >40 mg/l. **Daphnia** LC_{50} (3 h) for *D. pulex* >40 mg/l.

ENVIRONMENTAL FATE
Animals Almost all of ^{3}H-blasticidin-S administered to mice was excreted in the urine and faeces within 24 h. **Plants** In rice plants, cytomycin and deaminohydroxy blasticidin-S were identified as the main metabolites. **Soil/Environment** In soil DT_{50} <2 d (two soil types, o.c. 2.53%, 9.6%; moisture 42.6%, 87% respectively; pH 6.0, 25 °C).

borax   *Herbicide, fungicide, insecticide*

$$Na_2B_4O_7.10H_2O$$

NOMENCLATURE
Common name borax (E-ISO, F-ISO, accepted in lieu of common name)
IUPAC name disodium tetraborate decahydrate
Chemical Abstracts name borax
CAS RN borax: *[1303–96–4]* (formerly *[1344–90–7]*); sodium tetraborate
(anhydrous): *[1330–43–4]*; sodium octaborate (anhydrous): *[12008–41–2]*;
sodium metaborate: *[7775–19–1]*

PHYSICAL CHEMISTRY
Mol. wt. 381.4   **M.f.** $B_4H_{20}Na_2O_{17}$ **Form** Colourless crystalline solid, efflorescent
at low r.h.   **M.p.** 742 °C (anhydrous sodium tetraborate)   **S.g./density** 1.71
(20 °C)   **Solubility** In water 47.1 g/l (pH 9.3, 20 °C). Soluble in glycerol and
ethylene glycol. Insoluble in ethanol.   **Stability** Stable. When heated above
c. 62 °C, borax loses water of crystallisation, first forming the pentahydrate and
then anhydrous sodium tetraborate.   **F.p.** A fire retardant.

COMMERCIALISATION
History Borax has long been used as a mild antiseptic fungicide and as a herbicide.
Disodium tetraborate and sodium metaborate were introduced as herbicides by
the US Borax & Chemical Corp.   **Patent** US 2998310 (sodium octaborate);
US 3032405 (sodium metaborate)   **Manufacturers** Borax Français; Kerr-McGee;
US Borax

APPLICATIONS
Mode of action Non-selective herbicide, causing plants to yellow and desiccate.
Uses Total control of vegetation on non-crop land, such as paths, giving long-term
control. Also used in insect baits for food stores to control Formicidae.
Phytotoxicity Not to be used near desired plants.   **Formulation types** WP.
Mixtures *(borax +)* sulfur; bromacil; mancozeb + sulfur; benzalkonium chloride;
copper sulfate; aluminium ammonium sulfate + copper sulfate.   **Compatibility** Its
aqueous solution is alkaline and so incompatible with some herbicides.
Selected tradenames 'Dehybor' (US Borax); 'Neobor' (Borax Français); 'Pyrobor'
(Kerr-McGee)

ANALYSIS
Product analysis by acid-base titration (*AOAC Methods*, 1980, 20.056).
Residues determined by colorimetry, titration, atomic absorption spectroscopy
(*ibid.*, 1984, 20.057–20.067; 1980, 20.042–20.045).

MAIN ENTRIES

MAMMALIAN TOXICOLOGY
Reviews ECETOC Technical Report XXX, 'Toxicology and Risk Assessment of the Inorganic Borates' (1994).   **Oral** Acute oral LD_{50} for rats 4500–6000 mg/kg. **Skin and eye** Acute percutaneous LD_{50} for rabbits >10 000 mg/kg. Not an eye or skin irritant.   **NOEL** (2 y) for rats 154, dogs 78 mg/kg b.w. daily.   **Other** Not carcinogenic, not mutagenic. No reproductive or developmental toxicity noted. **Toxicity class** WHO (a.i.) III (Table 5); EPA (formulation) III

ECOTOXICOLOGY
Fish LC_{50} (24 d) for rainbow trout 1320 mg/l (disodium tetraborate decahydrate). **Bees** Not hazardous to bees.   **Daphnia** LC_{50} (48 h) 1170 mg/l (disodium tetraborate decahydrate).

ENVIRONMENTAL FATE
Animals Not metabolised in animals.   **Soil/Environment** No microbial degradation. Disappearance from soil is by washing out. Persistence in soil is ≤ 2 y, depending on rainfall and soil structure.

80   Bordeaux mixture   *Fungicide*

inorganic

NOMENCLATURE
Common name Bordeaux mixture (draft E-ISO, accepted in lieu of common name)
IUPAC name A mixture, with or without stabilising agents, of calcium hydroxide and copper(II) sulfate
Chemical Abstracts name Bordeaux mixture
Other names tribasic copper sulfate (in USA)   **CAS RN** *[8011–63–0]* Bordeaux mixture; *[11125–96–5]* Burgundy mixture

PHYSICAL CHEMISTRY
Form Fine, blue, precipitated powder, which contains 100% <40-μm particles and 80% <5-μm particles.   **Solubility** Insoluble in water. Very low solubility (<0.1%) in common solvents such as ketones, ethers, hydrocarbons, and chlorinated hydrocarbons. Dissolves in ammonium hydroxide, forming a cuprammonium complex.   **Stability** No observed deterioration in packs stored in cool, dry places for periods up to 2 years. Chemically not very reactive, but undergoes change if reacted with strong alkalis or acids. When heated, decomposes to black cupric oxide.

COMMERCIALISATION

History The mixture introduced by A. Millardet (*J. Agric. Prat. (Paris)*, 1885, **49,** 513); it may be used as a tank-mix or as pre-formed formulations. Burgundy mixture, introduced by E. Masson (*ibid.*, 1887, **51,** 814), is an analogous product produced from copper sulfate and sodium carbonate. **Manufacturers** Caffaro; Tomono

APPLICATIONS

Mode of action Foliar fungicide with protective action. **Uses** Bordeaux mixture is used as a protective fungicide for foliage applications, the freshly prepared precipitate having a high tenacity. Its use is limited to crop plants at stages of growth when its phytotoxicity is low. Major uses include the control of *Phytophthora infestans* on potatoes, *Venturia inaequalis* on apples, *Plasmopara viticola* on grapes and *Pseudoperonospora humuli* on hops. **Phytotoxicity** Plum and peach foliage may be injured, especially at low temperatures.
Formulation types WP; SC. **Mixtures** (*Bordeaux mixture +*) cufraneb; folpet; mancozeb; maneb; zineb; copper oxychloride + folpet.
Compatibility Incompatible with alkali-sensitive pesticides such as organophosphorus compounds and carbamates, and with strongly alkaline pesticides, such as lime sulfur. **Selected tradenames** 'Poltiglia' (Caffaro); 'Tri Cop Tri Basic' (Stoller); 'Z-Bordeaux' (Tomono)

ANALYSIS

Product analysis is by digestion with sulfuric acid and determination of the copper content by electrolytic, volumetric, gravimetric, colorimetric or atomic absorption spectroscopic methods (*AOAC Methods*, 1995, 920.27, 922.04, 922.05; *CIPAC Handbook*, 1970, **1,** 226; 1983, **1B,** 1915, 1918). **Residues** determined by digestion with sulfuric acid and perchloric acid, and colorimetry of a complex (*AOAC Methods*, 1980, 25.066–25.071).

MAMMALIAN TOXICOLOGY

Oral Acute oral LD_{50} for rats >4000 mg/kg (WP formulation).

ECOTOXICOLOGY

Fish Toxic to fish. **Bees** Not toxic to bees.

brodifacoum

coumarin anticoagulant

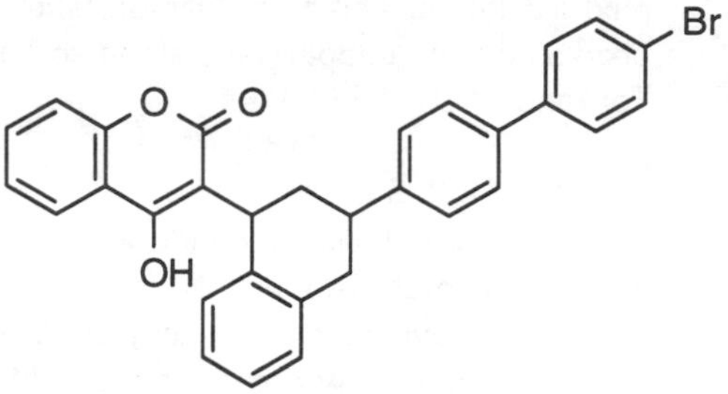

NOMENCLATURE
Common name brodifacoum (BSI, E-ISO, (*m*) F-ISO)
IUPAC name 3-[3-(4'-bromobiphenyl-4-yl)-1,2,3,4-tetrahydro-1-naphthyl]-=
4-hydroxycoumarin
Chemical Abstracts name 3-[3-(4'-bromo-[1,1'-biphenyl]-4-yl)-1,2,3,4-=
tetrahydro-1-naphthalenyl]-4-hydroxy-2*H*-1-benzopyran-2-one
CAS RN *[56073–10–0]* formerly *[66052–95–7]* **EEC no.** 259–980–5
Development codes WBA 8119 (Sorex); PP581 (Zeneca)

PHYSICAL CHEMISTRY
Composition Tech. material is >91% pure. **Mol. wt.** 523.4 **M.f.** $C_{31}H_{23}BrO_3$
Form White powder; (tech., off-white to buff or beige powder).
M.p. 228–232 °C (tech.) **V.p.** <<0.001 mPa (20 °C, gas saturation method)
K_{ow} logP = 8.5 **Henry** <10^{-1} (pH 5.2), <10^{-3} (pH 7.4), <10^{-5} (pH 9.3) (all in
Pa m^3 mol^{-1}, calc.) **S.g./density** 1.42 (25 °C, tech.) **Solubility** In water
3.8 × 10^{-3} (pH 5.2), 0.24 (pH 7.4), 10 (pH 9.3) (all in mg/l, 20 °C). In acetone 20,
chloroform 3, benzene <6 (all in mg/l, 20 °C). **Stability** Thermally (up to 50 °C)
and photolytically (30 d in direct sunlight) stable. Degraded by u.v. light when in
solution. **pKa** A very weak acid which is too lipophilic to form water-soluble
salts.

COMMERCIALISATION
History Rodenticide reported by R. Redfern *et al.* (*J. Hyg.*, 1976, **77**, 419).
Introduced by Sorex (London) Ltd and developed by ICI Agrochemicals (now
Zeneca Agrochemicals). **Patent** GB 1458670 to Sorex **Manufacturers** Sorex;
Zeneca

APPLICATIONS
Biochemistry Inhibits the vitamin K-dependent steps in synthesis of clotting factors
II, VII, IX and X. **Mode of action** Indirect anticoagulant. **Uses** Controls most
rodent pests, including *Rattus norvegicus*, *R. rattus* and *Mus musculus* at lower rates
than many other anticoagulants (e.g. warfarin and pindone); also *Cricetus cricetus*,

Mesocricetus auratus, Microtus pennsylvanicus, M. pinetorum, R. argentiventer, R. mindanensis and rodents, such as hamsters, that are difficult to control with other anticoagulants. Potency is such that a rodent may absorb a lethal dose by taking a 50 mg/kg bait as part of its food intake on only one occasion. For a review, see A. P. Buckle & R. H. Smith "Rodent Pests and their Control", CABI (1994). **Formulation types** RB; AB; BB. **Selected tradenames** 'Havoc' (Zeneca); 'Klerat' (Zeneca); 'Sorex Brodifacoum Rat & Mouse Bait' (Sorex); 'Talon' (Zeneca)

ANALYSIS
Product analysis by hplc (*CIPAC Handbook*, 1985, **1C**, 1981; *AOAC Methods*, 1995, 983.11; *Anal. Methods Pestic. Plant Growth Regul.*, 1988, **16,** 134).
Residues determined by hplc (*idem, ibid.*). Details from Zeneca.

MAMMALIAN TOXICOLOGY
EHC 175 (1995)   **Oral** Acute oral LD_{50} for male rats 0.27, male rabbits 0.3, male mice 0.4, female guinea pigs 2.8, cats c. 25, dogs 0.25–1.0 mg/kg.   **Skin and eye** Acute percutaneous LD_{50} for rabbits 0.25–0.63 mg/kg. Slight to mild skin and eye irritant (rabbits). Not a skin sensitiser (guinea pigs).   **Inhalation** LC_{50} (4 h) in range 0.0005 to 0.005 mg/l air.   **NOEL** 0.02 mg/kg diet daily.   **Toxicity class** WHO (a.i.) Ia; EPA (formulation) I   **EC risk** T+ (R27/28); T (R48/24/25)

ECOTOXICOLOGY
Birds Acute oral LD_{50} for chickens 4.5, mallard ducks 0.31 mg/kg. Dietary LC_{50} for mallard ducks 2.7, bobwhite quail 0.8 ppm.   **Fish** LC_{50} (96 h) for bluegill sunfish 0.165, rainbow trout 0.051 mg/l.   **Daphnia** LC_{50} (48 h) 0.064 mg/l.

ENVIRONMENTAL FATE
Animals In mammals, a number of hydroxycoumarins are formed.
Soil/Environment Degraded in soils (pH 5.5 to pH 8) under aerobic and flooded conditions. K_{oc} (average) 50 000, range 14 000–106 000; K_d 1040 (average), range 625–1320. DT_{50} >12 w. Unlikely to leach; <2% leached >2 cm in laboratory columns.

uracil

NOMENCLATURE

Common name bromacil (BSI, E-ISO, (*m*) F-ISO, ANSI, WSSA, JMAF)
IUPAC name 5-bromo-3-*sec*-butyl-6-methyluracil
Chemical Abstracts name 5-bromo-6-methyl-3-(1-methylpropyl)-2,4(1*H*,3*H*)-= pyrimidinedione
CAS RN *[314–40–9]*   **Development codes** DPX N976

PHYSICAL CHEMISTRY

Mol. wt. 261.1   **M.f.** $C_9H_{13}BrN_2O_2$ **Form** Colourless crystals.
M.p. 157.5–160 °C   **V.p.** 4.1×10^{-2} mPa (25 °C) K_{ow} logP = 1.88 (pH 5)
S.g./density 1.59 (23 °C)   **Solubility** In water 807 (pH 5), 700 (pH 7), 1287
(pH 9) (all in mg/l, 25 °C). In ethanol 134, acetone 167, acetonitrile 71, xylene 32,
3% aqueous sodium hydroxide 88 (all in g/l, 25 °C).   **Stability** Thermally stable up
to the melting point. Stable in aqueous bases. Slowly decomposed by
concentrated acids.   **pKa** 9.27, v. weak acid

COMMERCIALISATION

History Herbicide reported by H. C. Bucha *et al.* (*Science*, 1962, **137**, 537).
Introduced by E. I. du Pont de Nemours and Co.   **Patent** US 3325357;
US 3352862; BE 625897   **Manufacturers** Du Pont

APPLICATIONS

Biochemistry Photosynthetic electron transport inhibitor.   **Mode of action**
Herbicide, absorbed mainly through the roots, with slight absorption through the
leaves and stems.   **Uses** Total weed and brush control on non-crop land at
5–15 kg a.i./ha. Also for selective control of annual and perennial weeds and
grasses in citrus and pineapple plantations at rates in the range 1.5 to 8 kg/ha.
Formulation types WP; SL; GR.   **Mixtures** (*bromacil +*) picloram; diuron;
dichlobenil; amitrole + diuron; sodium chlorate + sodium metaborate;
2,4-D + dalapon + diuron; diuron + hexazinone; amitrole + dichlorprop + diuron.
Compatibility Water-soluble formulations are incompatible with ammonium
sulfamate and liquid preparations of amitrole. Herbicides containing soluble
calcium salts form precipitates when used with water-soluble formulations of
bromacil.   **Selected tradenames** 'Hyvar X' (Du Pont); 'Hyvar X-L' (lithium salt)
(Du Pont); 'Rokar X L' (Siapa); 'Uragan' (Makhteshim-Agan)

ANALYSIS

Product analysis by potentiometric titration (*CIPAC Handbook*, 1985, **1C**, 1986). **Residues** determined by glc (H. L. Pease, *J. Agric. Food Chem.*, 1966, **14**, 94; 1968, **16**, 54; H. Jarczyk, *Pflanzenschutz-Nachr. (Engl. Ed.)*, 1975, **28**, 319; H. L. Pease & J. F. Deye, *Anal. Methods Pestic., Plant Growth Regul. Food Addit.*, 1967, **5**, 335; *Anal. Methods Pestic. Plant Growth Regul.*, 1972, **6**, 663). In **drinking water**, by gc with FID (*AOAC Methods*, 1995, 991.07).

MAMMALIAN TOXICOLOGY

Oral Acute oral LD_{50} for male rats 2000, female rats 1300 mg/kg (for 80% formulation). **Skin and eye** Acute percutaneous LD_{50} for rabbits >5000 mg/kg. Mild eye and moderate skin irritant. Non-sensitising to skin. **Inhalation** LC_{50} (4 h) for rats >4.8 mg/l air. **NOEL** (2 y) for rats 50 ppm; (1 y) for dogs 625 ppm. **ADI** (EPA) 0.13 mg/kg [1993]. **Toxicity class** WHO (a.i.) III (Table 5); EPA (formulation) IV (dry); II (liquid) **PIC** Yes

ECOTOXICOLOGY

Birds Acute oral LD_{50} for bobwhite quail 2250 mg/kg. Eight-day dietary LC_{50} for mallard ducks and bobwhite quail >10 000 mg/kg diet. **Fish** LC_{50} (48 h) for rainbow trout 75, bluegill sunfish 71, carp 164 mg/l. **Bees** Not toxic to bees. **Daphnia** LC_{50} (48 h) 119 mg/l. **Other aquatic spp.** LC_{50} for mysid shrimp (*Mysidopsis bahia*) 112.9 mg/l; EC_{50} for oyster embryo larvae 130 mg/l.

ENVIRONMENTAL FATE

Animals The principal metabolite is 5-bromo-3-*sec*-butyl-6-hydroxymethyluracil (*J. Agric. Food Chem.*, 1969, **17**, 967–973). **Plants** See animals. **Soil/Environment** Duration of residual activity in soil is *c.* 5 mo.

33 # bromadiolone   *Rodenticide*

coumarin anticoagulant

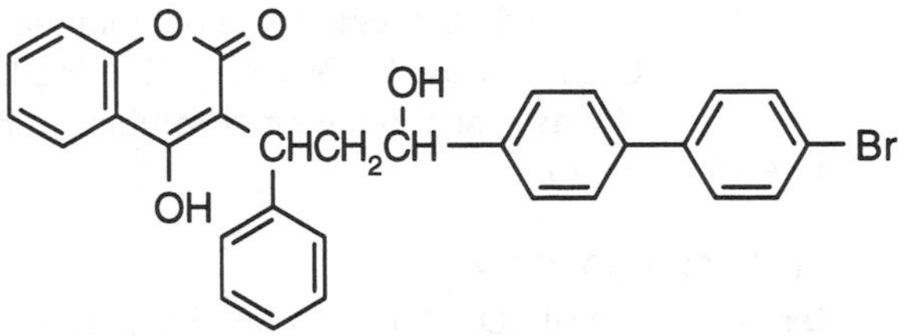

NOMENCLATURE

Common name bromadiolone (BSI, E-ISO, (*f*) F-ISO); broprodifacoum (Republic of South Africa)

IUPAC name 3-[3-(4'-bromobiphenyl-4-yl)-3-hydroxy-1-phenylpropyl]-=
4-hydroxycoumarin
Chemical Abstracts name 3-[3-(4'-bromo[1,1'-biphenyl]-4-yl)-3-hydroxy-=
1-phenylpropyl]-4-hydroxy-2H-1-benzopyran-2-one
CAS RN *[28772–56–7]* unstated stereochemistry **Development codes** LM 637

PHYSICAL CHEMISTRY
Composition Mixture of two diastereoisomers. Tech. grade bromadiolone is 97%
pure. **Mol. wt.** 527.4 **M.f.** $C_{30}H_{23}BrO_4$ **Form** Tech. grade is a yellowish
powder. **M.p.** 200–210 °C (mixture of two diastereoisomers) **V.p.** 0.002 mPa
(20 °C) **Solubility** In water 19 mg/l (20 °C). In dimethylformamide 730, ethyl
acetate 25, ethanol 8.2 (all in g/l, 20 °C). Soluble in acetone; slightly soluble in
chloroform; practically insoluble in diethyl ether and hexane. **Stability** Thermally
stable below 200 °C.

COMMERCIALISATION
History Rodenticide reported by M. Grand (*Phytiatr. Phytopharm.*, 1976, **25**, 69).
Introduced by Lipha S.A. **Patent** FR 96651; US 3764693; GB 1252088
Manufacturers Lipha; Rallis

APPLICATIONS
Biochemistry Second-generation anticoagulant rodenticide which also blocks
prothrombin formation. **Uses** Control of rats and mice (including those resistant
to warfarin) in areas containing stored products, household use, industrial
buildings, and other situations. **Formulation types** RB; CB; TP.
Mixtures (*bromadiolone* +) sulfaquinoxaline. **Selected tradenames** 'Maki' (Lipha);
'Super Caid' (Lipha); 'Lanirat' (Novartis); 'Ratoban' (Rallis)

ANALYSIS
Product and **residue** analysis by hplc (*CIPAC Handbook*, 1992, **E**, 23; *Anal. Methods
Pestic. Plant Growth Regul.*, 1988, **16**, 140). Details available from Lipha S.A.

MAMMALIAN TOXICOLOGY
Oral Acute oral LD_{50} for rats 1.125, mice 1.75, rabbits 1.00, dogs >10.0, cats
>25.0 mg/kg. **Skin and eye** Acute percutaneous LD_{50} for rabbits 1.71 mg/kg.
Inhalation LC_{50} 0.43 µg/l. **NOEL** In 90 d feeding trials on rats and dogs, the only
effect noted was reduction in prothrombin rating. **Toxicity class** WHO (a.i.) Ia;
EPA (formulation) I

ECOTOXICOLOGY
Birds Acute oral LD_{50} for bobwhite quail 138 mg/kg. LC_{50} (5 d) for mallard duck
110 ppm. **Fish** LC_{50} (96 h) for rainbow trout 1.4, bluegill sunfish 3.0 mg/l.
Bees Not hazardous to bees when used as directed. **Daphnia** LC_{50} (48 h)
2.0 mg/l.

ENVIRONMENTAL FATE

Soil/Environment Leaching behaviour is inversely related to clay and organic matter content of soils. In soil column and soil layer studies, the proportion of bromadiolone remaining in the top soil layer, using aged loamy sand soil (8.8% clay, 1.05% o.m.) was 97%, with 0.1% in leachate.

84 bromethalin *Rodenticide*

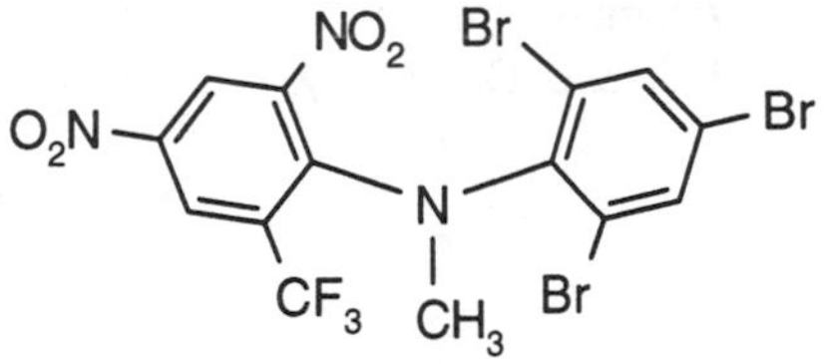

NOMENCLATURE

Common name bromethalin (BSI, draft E-ISO, ANSI); brométhaline ((f) draft F-ISO)

IUPAC name α,α,α-trifluoro-N-methyl-4,6-dinitro-N-(2,4,6-tribromophenyl)-= o-toluidine

Chemical Abstracts name N-methyl-2,4-dinitro-N-(2,4,6-tribromophenyl)-= 6-(trifluoromethyl)benzenamine

CAS RN *[63333–35–7]*   **Development codes** EL-614   **Official codes** OMS 3020

PHYSICAL CHEMISTRY

Mol. wt. 577.9   **M.f.** $C_{14}H_7Br_3F_3N_3O_4$ **Form** Pale yellow crystals.
M.p. 150–151 °C   **V.p.** 0.013 mPa (25 °C)   **Solubility** In water <0.01 mg/l. In dichloromethane 300–400, chloroform 200–300, methanol 2.3–3.4, heavy aromatic naphtha 1.2–1.3 (all in g/l).   **Stability** Good storage stability under normal conditions. Decomposed by u.v. light.

COMMERCIALISATION

History Rodenticide reported by B. A. Dreikorn et al. (*Proc. Br. Crop Prot. Conf.*, 1979, **2**, 491). Introduced by Eli Lilly & Co. (now DowElanco) and later by other companies.

APPLICATIONS

Biochemistry Uncoupler of oxidative phosphorylation.   **Mode of action** Acute rodenticide.   **Uses** Bait control of rats and mice, indoors and outdoors; effective against rodents resistant to anticoagulant rodenticides. Does not induce bait shyness.   **Formulation types** RB.

ANALYSIS
Product analysis by glc. Details from DowElanco.

MAMMALIAN TOXICOLOGY
Reviews For toxicology review, see B. A. Dreikorn et al., *Proc. Br. Crop Prot. Conf. – Pests Dis.*, 1979, 491. **Oral** Acute oral LD_{50} for rats and cats 2 mg tech. (in propane-1,2-diol)/kg; for mice and dogs 5 mg/kg. **Skin and eye** Acute percutaneous LD_{50} for male rabbits 1000 mg/kg. **Inhalation** LC_{50} (1 h) for rats 0.024 mg/l air. **NOEL** (90 d) for dogs and rats 0.025 mg/kg daily. **Toxicity class** WHO (a.i.) Ia; EPA (formulation) II

ENVIRONMENTAL FATE
Animals The major route of metabolism in the rat is *N*-demethylation (R. B. L. Van Lier & L. D. Cherry, *Fund. Appl. Toxicol.*, 1988, **11**, 664).

85 bromobutide   *Herbicide*

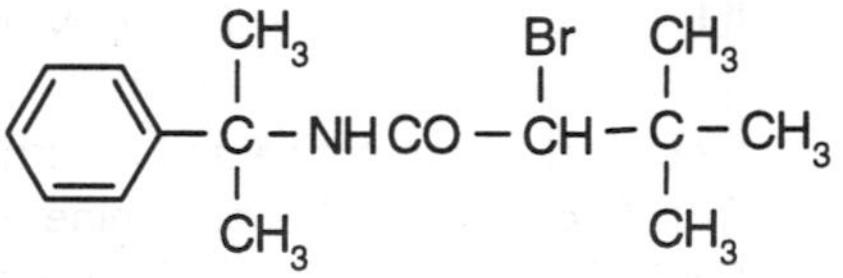

NOMENCLATURE
Common name bromobutide (BSI, draft E-ISO, (*m*) draft F-ISO)
IUPAC name 2-bromo-3,3-dimethyl-*N*-(1-methyl-1-phenylethyl)butyramide; 2-bromo-*N*-(α,α-dimethylbenzyl)-3,3-dimethylbutyramide
Chemical Abstracts name 2-bromo-3,3-dimethyl-*N*-(1-methyl-1-phenylethyl)= butanamide
CAS RN [74712–19–9]   **Development codes** S-4347

PHYSICAL CHEMISTRY
Mol. wt. 312.2   **M.f.** $C_{15}H_{22}BrNO$ **Form** Tech. grade forms colourless to yellow crystals.   **M.p.** 180.1 °C   **V.p.** 74 mPa (25 °C)   **Solubility** In water 3.54 mg/l (25 °C). In xylene 4.7, methanol 35, hexane 0.5 (all in g/l at 25 °C).
Stability Stable under normal storage conditions.

COMMERCIALISATION
History Herbicide reported by O. Kirino et al. (*Agric. Biol. Chem.*, 1981, **45**, 2669). Introduced by Sumitomo Chemical Co., Ltd.   **Patent** US 4288244; GB 2031420
Manufacturers Sumitomo

APPLICATIONS

Uses A selective herbicide used for paddy field rice, effective against sedges, especially *Echinochloa* spp., *Eleocharis acicularis* and *Scirpus juncoides*, and some broad-leaved weeds. It is also effective in upland rice. **Phytotoxicity** Rice injury may occur on sandy fields, or on rice which has been shallowly transplanted. **Formulation types** GR. **Mixtures** *(bromobutide +)* pyrazolynate; pyrazoxyfen. **Selected tradenames** 'Sumiherb' (Sumitomo)

ANALYSIS

Product analysis by glc with FID. **Residues** determined by glc. Details of methods available from Sumitomo Chemical Co. Ltd.

MAMMALIAN TOXICOLOGY

Oral Acute oral LD_{50} for rats >5000 mg/kg. **Skin and eye** Acute percutaneous LD_{50} for rats >5000 mg/kg. **Toxicity class** WHO (a.i.) III (Table 5); EPA (formulation) IV

ECOTOXICOLOGY

Fish LC_{50} (48 h) for carp >10 mg/l.

36 bromofenoxim *Herbicide*

hydroxybenzonitrile precursor

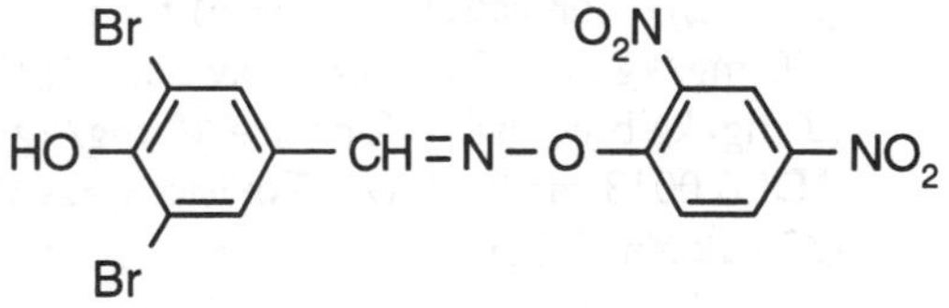

NOMENCLATURE

Common name bromofenoxim (BSI, E-ISO, WSSA); bromophénoxime ((f) F-ISO)
IUPAC name 3,5-dibromo-4-hydroxybenzaldehyde 2,4-dinitrophenyloxime
Chemical Abstracts name 3,5-dibromo-4-hydroxybenzaldehyde *O*-(2,4-dinitrophenyl)oxime
CAS RN *[13181–17–4]* formerly *[37273–85–1]* **EEC no.** 236–129–6
Development codes C 9122

PHYSICAL CHEMISTRY

Mol. wt. 461.0 **M.f.** $C_{13}H_7Br_2N_3O_6$ **Form** Cream-coloured crystals.
M.p. 196–197 °C **V.p.** $<10^{-5}$ mPa (20 °C) **K_{ow}** logP = 3.17 (pH 7.0)

S.g./density 2.15 (20 °C)   **Solubility** In water 0.6 (pH 3.8), 9.0 (pH 10) (both in mg/l, 20 °C). In hexane 200, isopropanol 400, acetone 9900, *n*-octanol 200 (all in mg/l, 20 °C).   **Stability** On hydrolysis at 70 °C, DT_{50} 41.4 h (pH 1), 9.6 h (pH 5), 0.76 h (pH 9).   **pKa** 5.46, weak acid

COMMERCIALISATION
History Herbicide reported by D. H. Green *et al.*, (*Symp. New Herbic., 3rd*, 1969, p. 77). Introduced by Ciba AG (now Novartis Crop Protection AG).
Patent BE 675444; GB 1096037   **Manufacturers** Novartis

APPLICATIONS
Biochemistry Generates bromoxynil which inhibits photosynthesis by uncoupling oxidative phosphorylation.   **Mode of action** Selective herbicide with contact action. Absorbed by the leaves and stems, with some translocation.   **Uses** Post-emergence control of many annual broad-leaved weeds in cereals, maize, grass-seed crops, grassland, and newly-seeded turf at rates in the range 1.0–2.5 kg a.i./ha. Used in combination with other herbicides to extend the spectrum and period of activity.   **Formulation types** WP; SC.   **Mixtures** (*bromofenoxim* +) terbuthylazine.   **Selected tradenames** 'Faneron' (Novartis)

ANALYSIS
Product analysis by hplc.   **Residues** determined by hplc with PCD. Particulars available from Novartis.

MAMMALIAN TOXICOLOGY
Oral Acute oral LD_{50} for rats 1217, dogs >1000, mice 940 mg/kg.   **Skin and eye** Acute percutaneous LD_{50} for rats >3000 mg/kg. Mild skin and eye irritant (rabbits).   **Inhalation** LC_{50} (6 h) for rats >0.242 mg/l air.   **NOEL** (2 y) for rats 100 mg/kg diet (3.8 mg/kg b.w. daily); (3 mo) for dogs 100 mg/kg diet (3 mg/kg b.w. daily), for mice 10 mg/kg diet (0.63 mg/kg b.w. daily).
ADI 0.0013 mg/kg b.w.   **Toxicity class** WHO (a.i.) III; EPA (formulation) III
EC risk Xn (R22)

ECOTOXICOLOGY
Birds Eight-day dietary LC_{50} for Japanese quail 5560, bobwhite quail 4000, mallard ducklings 2800 ppm.   **Fish** LC_{50} (96 h) for rainbow trout 0.18, carp 0.088 mg/l.
Bees WP non-toxic; SC toxic.   **Worms** LC_{50} (14 d) for earthworms c. 1300 mg/kg.   **Other beneficial spp.** No effect on soil respiration, ammonification or nitrification.   **Daphnia** LC_{50} (48 h) 1.2 mg/l.   **Algae** EC_{50} (96 h) for algae 2.2 mg/l.

ENVIRONMENTAL FATE
Animals In mammals, following oral administration, dinitrophenol and hydroxydibromobenzonitrile are formed, and are more slowly metabolised further

and excreted in the urine; unchanged parent is the primary excretion product in the faeces. **Plants** In plants, 3,5-dibromo-4-hydroxybenzoic acid has been identified as a metabolite. There is some conversion to bromoxynil (see Biochemistry). **Soil/Environment** Of low mobility with little or no persistence in the soil; DT_{50} 0.5–5 d. K_{oc} 1100.

7 bromopropylate *Acaricide*

benzilate

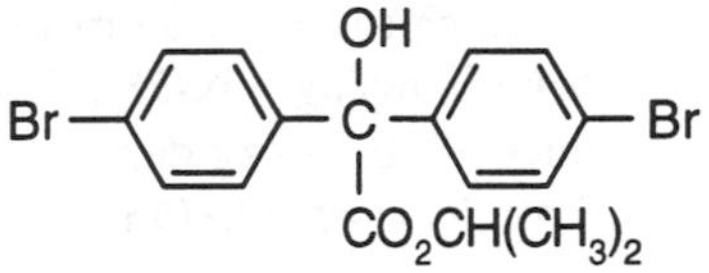

$$Br\text{---}\bigcirc\text{---}\overset{\overset{\text{OH}}{|}}{\underset{\underset{CO_2CH(CH_3)_2}{|}}{C}}\text{---}\bigcirc\text{---}Br$$

MAIN ENTRIES

NOMENCLATURE
Common name bromopropylate (BSI, E-ISO, (*m*) F-ISO, ANSI, ESA); phenisobromolate (JMAF)
IUPAC name isopropyl 4,4'-dibromobenzilate
Chemical Abstracts name 1-methylethyl 4-bromo-α-(4-bromophenyl)-= α-hydroxybenzeneacetate
CAS RN *[18181–80–1]* **EEC no.** 2420707 **Development codes** GS 19 851
Official codes ENT 27 552

PHYSICAL CHEMISTRY
Mol. wt. 428.1 **M.f.** $C_{17}H_{16}Br_2O_3$ **Form** White crystals. **M.p.** 77 °C
V.p. 1.1×10^{-2} mPa (20 °C) **K_{ow} logP** = 5.4 **S.g./density** 1.59 at 20 °C
Solubility In water <0.5 mg/l (20 °C). In acetone 850, dichloromethane 970, dioxane 870, benzene 750, methanol 280, xylene 530, isopropanol 90 (all in g/kg, 20 °C). **Stability** Fairly stable in neutral or slightly acidic media; DT_{50} 34 d (pH 9).

COMMERCIALISATION
History Acaricide reported by H. Grob *et al.* (*Abstr. Int. Congr. Plant Prot. 6th,* 1967, p. 198). Introduced by J. R. Geigy S.A. (now Novartis Crop Protection AG).
Patent GB 1178850; BE 691105; CH 471065 **Manufacturers** Novartis

APPLICATIONS
Mode of action Non-systemic acaricide with contact action, and long residual activity. **Uses** Control of all stages of tetranychid and eriophyid mites on pome fruit, stone fruit, citrus fruit, vines, strawberries, hops, cotton, soya beans, cucurbits, vegetables, and ornamentals. Also used to control parasitic mites in

beehives.   **Phytotoxicity** Slightly phytotoxic to certain varieties of apple, plum, and ornamentals.   **Formulation types** EC.   **Selected tradenames** 'Neoron' (Novartis)

ANALYSIS
Residues determined by glc with ECD (M. A. Luke et al., J. Assoc. Off. Anal. Chem., 1981, **64,** 1187). Methods reviewed by J. L. Daft in Comp. Analyt. Profiles, Chapter 11.

MAMMALIAN TOXICOLOGY
Reviews Pesticide residues in food – 1993, FAO Plant Production and Protection Paper, 122, 1993. Pesticide residues in food – 1993 evaluations, Part II – Toxicology. WHO, WHO/PCS/94.4, 1994.   **Oral** Acute oral LD_{50} for rats >5000 mg tech./kg.   **Skin and eye** Acute percutaneous LD_{50} for rats >4000 mg/kg. Slightly irritating to skin; non-irritating to eyes (rabbits).   **Inhalation** LC_{50} for rats >4000 mg/kg.   **NOEL** (2 y) for rats 500 mg/kg diet (c. 25 mg/kg daily); (1 y) for mice 1000 mg/kg diet (c. 143 mg/kg daily).   **ADI** (JMPR) 0.03 mg/kg b.w. [1993]. **Toxicity class** WHO (a.i.) III; EPA (formulation) IV

ECOTOXICOLOGY
Birds Acute oral LD_{50} for Japanese quail >2000 mg/kg. Eight-day dietary LC_{50} for Pekin ducks 600, Japanese quail 1000 mg/kg diet.   **Fish** LC_{50} (96 h) for rainbow trout 0.35, bluegill sunfish 0.5, carp 2.4 mg/l.   **Bees** Not toxic to bees; LC_{50} (24 h) 183 µg/bee.   **Daphnia** LC_{50} (48 h) 0.17 mg/l.

ENVIRONMENTAL FATE
Animals Bromopropylate is rapidly and efficiently eliminated in animals. Metabolism occurred by cleavage of the isopropyl ester and to a minor extent by oxidation. Metabolites formed after oxidation were 3-hydroxybenzilate and conjugates. **Plants** Studies with ^{14}C-labelled bromopropylate showed little penetration into leaves or fruit. Degradation was slow.   **Soil/Environment** The principal metabolite in soil is 4,4-dibromobenzilic acid. For details of persistence in soil, see J. Environ. Qual., 1973, **2,** 115. DT_{50} c. 40–70 d (lab. and field studies). Low mobility in soil.

bromoxynil   *Herbicide*

hydroxybenzonitrile

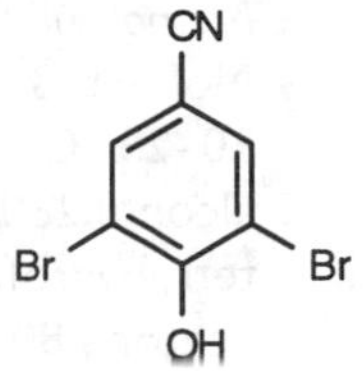

NOMENCLATURE
bromoxynil
Common name bromoxynil (BSI, E-ISO, (*m*) F-ISO, ANSI, WSSA)
IUPAC name 3,5-dibromo-4-hydroxybenzonitrile; 3,5-dibromo-4-hydroxyphenyl cyanide
Chemical Abstracts name 3,5-dibromo-4-hydroxybenzonitrile
CAS RN *[1689–84–5]* **EEC no.** 216–882–7 **Development codes** M&B 10 064
Official codes ENT 20852

bromoxynil octanoate
Common name bromoxynil octanoate
IUPAC name 2,6-dibromo-4-cyanophenyl octanoate
Chemical Abstracts name 2,6-dibromo-4-cyanophenyl octanoate
CAS RN *[1689–99–2]* **EEC no.** 216–885–3 **Development codes** M&B 10 731
(May & Baker); 16 272 RP (Rhône-Poulenc)

bromoxynil-potassium
Common name bromoxynil-potassium
CAS RN *[2961–68–4]*

PHYSICAL CHEMISTRY
bromoxynil
Composition Tech. grade is *c.* 95% pure. **Mol. wt.** 276.9 **M.f.** $C_7H_3Br_2NO$
Form Colourless solid. **M.p.** 194–195 °C (sublimes at 135 °C/0.15 mmHg);
(tech., 183–192 °C) **V.p.** 6.3×10^{-3} mPa (20 °C) **K$_{ow}$** logP = 2.8 (unionised)
Solubility In water 130 mg/l (20 °C). In dimethylformamide 610, tetrahydrofuran
410, acetone, cyclohexanone 170, methanol 90, ethanol 70, mineral oils <20,
benzene 10 (all in g/l, 25 °C). **Stability** Very stable to dilute alkalis and acids.
Stable to u.v. light. Thermally stable below the melting point. **pKa** 3.86

bromoxynil octanoate
Mol. wt. 403.0 **M.f.** $C_{15}H_{17}Br_2NO_2$ **Form** Cream, waxy solid. **M.p.** 45–46 °C
V.p. 1.9×10^{-1} mPa (25 °C) **K$_{ow}$** logP = 5.4 **Solubility** In water 3 mg/l (25 °C).
In chloroform 800, xylene, dimethylformamide 700, ethyl acetate 620,
cyclohexanone 550, carbon tetrachloride 500, *n*-propanol 120, acetone, ethanol

100 (all in g/l, 20–25 °C).   **Stability** Stable to sunlight and at its m.p.; readily hydrolysed to bromoxynil at pH >9.

bromoxynil-potassium
Mol. wt. 315.0   **M.f.** $C_7H_2Br_2KNO$ **M.p.** *c.* 360 °C   **Solubility** In water at 20–25 °C, 61 g/l. In acetone 70, 20% aqueous acetone 240, tetrahydrofurfuryl alcohol 260 (all in g/l, 20–25 °C). Sodium salt: In water 42 g/l (20–25 °C). In tetrahydrofurfuryl alcohol 430, methyl cellosolve 310, 20% aqueous acetone 150, acetone 80 (all in g/l, 20–25 °C).

COMMERCIALISATION
History Herbicidal properties of bromoxynil described independently by R. L. Wain (*Nature (London)*, 1963, **200**, 28), by K. Carpenter & B. J. Heywood (*ibid.*, p. 28), and by Amchem Products Inc. Development reviewed by B. J. Heywood (*Chem. Ind. (London)*, 1966, p. 1946). Introduced by May & Baker Ltd and by Amchem Products Inc. (both now within Rhône-Poulenc Group).
Patent GB 1067033 to May & Baker; US 3397054; US 4332613 both to Amchem
Manufacturers CFPI; Makhteshim-Agan; Rhône-Poulenc; Sanachem

APPLICATIONS
Biochemistry Inhibits photosynthesis, and uncouples oxidative phosphorylation.
Mode of action Selective contact herbicide with some systemic activity. Absorbed by the foliage, with limited translocation.   **Uses** Post-emergence control of annual broad-leaved weeds, especially young seedlings of the Polygonaceae, Compositae, and certain Boraginaceae, in cereals, maize, sorghum, flax, onions, garlic, mint, grass-seed crops, turf, and non-crop land. Often used in combination with other herbicides, to extend the spectrum of control.   **Formulation types** EC; SC; WP.
Mixtures *(bromoxynil +)* atrazine; ioxynil; dichlorprop; MCPA; clopyralid; 2,4-D; chlorsulfuron + ioxynil; clopyralid + mecoprop; dicamba + mecoprop; dichlorprop + ioxynil; ioxynil + mecoprop; ioxynil + isoproturon; ioxynil + linuron; MCPA + mecoprop; ioxynil + MCPA + mecoprop; ioxynil + linuron + mecoprop; dicamba + ioxynil + mecoprop; dichlorprop + ioxynil + linuron; dichlorprop + ioxynil + MCPA; and many more.   **Selected tradenames** 'Brominal' (octanoate) (Rhône-Poulenc); 'Buctril' (Rhône-Poulenc); 'Bromotril' (octanoate) (Makhteshim-Agan); 'Bromox' (octanoate) (Sanachem); 'Bromoxan' (octanoate) (Sanachem); 'Emblem' (octanoate) (CFPI); 'Kaleis' (mixture) (Novartis); 'Trionyl' (mixture) (Agriphar)

ANALYSIS
Product analysis (octanoate) by glc (*AOAC Methods*, 1995, 980.05, *CIPAC Handbook*, 1985, **1C**, 1989–1998) or by determination of bromine.
Residues determined by glc of a derivative (H. S. Segal & M. L. Sutherland, *Anal. Methods Pestic., Plant Growth Regul. Food Addit.*, 1967, **5**, 347; *Anal. Methods Pestic. Plant Growth Regul.*, 1972, **6**, 605); or by i.r. spectrometry.

MAMMALIAN TOXICOLOGY
bromoxynil
Oral Acute oral LD_{50} for rats 190, mice 110, rabbits 260, dogs c. 100 mg/kg.
Skin and eye Acute percutaneous LD_{50} for rats >2000, rabbits 3660 mg/kg. Mild
eye irritant; non-irritating to skin (rabbits). Not a skin sensitiser (guinea pigs).
Inhalation LC_{50} (4 h) for rats 0.41 mg/ml.   **NOEL** (2 y) for rats 100 ppm.
Toxicity class WHO (a.i.) II; EPA (formulation) II   **EC risk** (R63); T (R25)

bromoxynil octanoate
Oral Acute oral LD_{50} for rats 365 mg (formulated)/kg, mice 306 mg
(formulated)/kg, rabbits 325 mg/kg.   **Skin and eye** Acute percutaneous LD_{50} for
rats >2000, rabbits 1675 mg/kg.   **NOEL** (90 d) for rats 15.6, dogs 5 mg/kg daily.
EC risk (R63); Xn (R21/22)

bromoxynil-potassium
Oral Acute oral LD_{50} for rats 130, mice 100 mg (formulated)/kg.   **NOEL** (90 d)
for rats 16.6 mg/kg daily.

ECOTOXICOLOGY
bromoxynil
Birds Acute oral LD_{50} for pheasants 50, hens 240, quail 100–125, mallard ducks
200 mg/kg. Sub-acute oral LC_{50} (21 d) for pheasants 4000 ppm diet.   **Fish** LC_{50}
(48 h) for goldfish 0.46, catfish 0.063 mg/l.   **Bees** LD_{50} (48 h, oral) 4 μg/bee;
harmful to bees.   **Daphnia** LC_{50} (48 h) 12.5 mg/l.   **Algae** EC_{50} (72 h) for
Scenedesmus subspicatus 140 mg/l.

bromoxynil octanoate
Birds Acute oral LD_{50} for pheasants 175 mg (formulated)/kg.   **Fish** LC_{50} (48 h)
for rainbow trout 0.15 mg/l.   **Bees** Not toxic to bees.

bromoxynil-potassium
Birds Acute oral LD_{50} for pheasants 50 mg (formulated)/kg, hens 120 mg/kg.
Fish LC_{50} (48 h) for harlequin fish 5.0 mg/l.   **Bees** Not toxic to bees.

ENVIRONMENTAL FATE
Animals See plants.   **Plants** Metabolism in plants and animals is by hydrolysis of
the ester and nitrile groups with some debromination occurring (J. H. Buckland,
Pestic Sci., 1973, **4**, 149, 689).   **Soil/Environment** In soil, DT_{50} c. 10 d. Degraded
by hydrolysis and debromination to less toxic substances such as hydroxybenzoic
acid. For persistence in soil, see *Pestic Sci.,* 1980, **11**, 341.

bromuconazole

Fungicide

azole

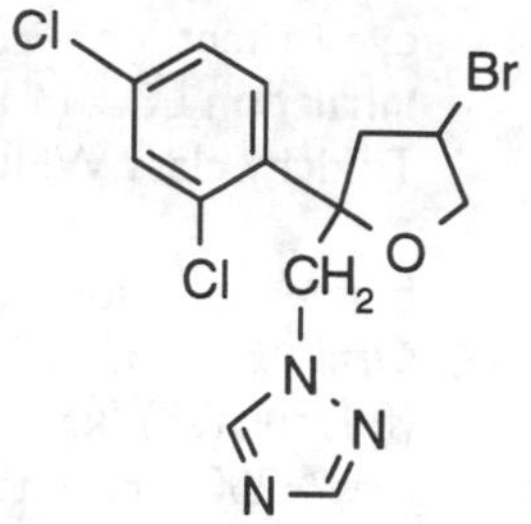

NOMENCLATURE
Common name bromuconazole (BSI, draft E-ISO)
IUPAC name 1-[(2RS,4RS:2RS,4SR)-4-bromo-2-(2,4-dichlorophenyl)=
tetrahydrofurfuryl]-1H-1,2,4-triazole
Chemical Abstracts name 1-[[4-bromo-2-(2,4-dichlorophenyl)tetrahydro-=
2-furanyl]methyl]-1H-1,2,4-triazole
Other names [bromoconazole] (rejected common name proposal)
CAS RN *[116255–48–2]* **EEC no.** 408–060–3 **Development codes** LS 860263

PHYSICAL CHEMISTRY
Composition Mixture of 2 diastereoisomers in proportion 54:46. Tech. is ≥96%
pure. **Mol. wt.** 377.1 **M.f.** $C_{13}H_{12}BrCl_2N_3O$ **Form** Colourless powder.
M.p. 84 °C **V.p.** 0.004 mPa (25 °C) **K_{ow} logP** = 3.24 (20 °C) **S.g./density** 1.72
(20 °C) **Solubility** In water 50 mg/l. Moderately soluble in organic solvents.
Stability In water, stable in the dark at acidic, basic or neutral pH values. Under
simulated sunlight, degradation is pH-dependent: bromuconazole is degraded
especially in acidic conditions, DT_{50} 18 d.

COMMERCIALISATION
History Fungicide reported by R. Pepin *et al.* (*Proc. 1990 Br. Crop Prot. Conf. –
Pests Dis., 1,* 459). Developed by Rhône-Poulenc Agrochimie.
Manufacturers Rhône-Poulenc

APPLICATIONS
Biochemistry Ergosterol biosynthesis inhibitor. **Mode of action** Systemic fungicide.
Uses Systemic fungicide generally applied by foliar spray at a max. rate of
300 g/ha against Ascomycetes, Basidiomycetes and Fungi Imperfecti including
Alternaria, Fusarium, black and yellow sigatoke, and *Pseudocercosporella* spp. in
cereals, fruit, vines, vegetables, tropical crops, turf and ornamentals.
Formulation types SC; EC; GR. **Mixtures** (*bromuconazole* +) iprodione;
fenpropimorph; prochloraz. **Selected tradenames** 'Granit' (Rhône-Poulenc);
'Vectra' (Rhône-Poulenc)

ANALYSIS

A.i. in tech. or formulation by hplc using u.v. detection.   **Residues** in crops by glc using ECD.

MAMMALIAN TOXICOLOGY

Oral Acute oral LD_{50} for rats 365, mice 1151 mg/kg.   **Skin and eye** Acute percutaneous LD_{50} for rats >2000 mg/kg. Non-irritating to eyes and skin of rabbits. Non-sensitising to skin of guinea pigs.   **Inhalation** LC_{50} (4 h) for rats >5 mg/l air.   **Other** Non-mutagenic in mouse micronucleus assay. Non-mutagenic in the Ames test.

ECOTOXICOLOGY

Birds Acute oral LD_{50} for bobwhite quail and mallard ducks >2150 mg/kg b.w. Eight-day LC_{50} for bobwhite quail and mallard ducks >5000 ppm.   **Fish** LC_{50} (96 h) for bluegill sunfish 3.1, rainbow trout 1.7 mg/l.   **Bees** Non-toxic by direct contact or ingestion at 500 µg/bee and 100 µg/bee respectively.
Worms Harmless.   **Other beneficial spp.** Harmless.   **Daphnia** LC_{50} (48 h) >5 mg/l.   **Algae** EC_{50} (96 h) for *Selenastrum capricornutum* 2.1 mg/l.

ENVIRONMENTAL FATE

Animals Metabolism in animals (rat, cow, hen) is extensive. Nearly 60 different metabolites were detected in the rat, 57 of which had structures assigned. No evidence was found to suggest that the compound and/or its metabolites accumulate in the organs and tissues of those species, even following repeated administration.   **Plants** Metabolism in cereals is characterised by the formation of bound residues of very polar metabolites and conjugates. The parent compound in the extractable residue has been identified as the major component. In the grain, no individual metabolite represents >0.01 mg/kg. In apples, about 23 different metabolites were detected; the most abundant was only present at 0.04 mg/kg.
Soil/Environment Under lab. and field conditions, bromuconazole exhibits very low mobility in soil. Field dissipation studies have shown that degradation in soil was much more rapid than predicted by lab. experiments.

$$HOCH_2 - \overset{\displaystyle Br}{\underset{\displaystyle NO_2}{C}} - CH_2OH$$

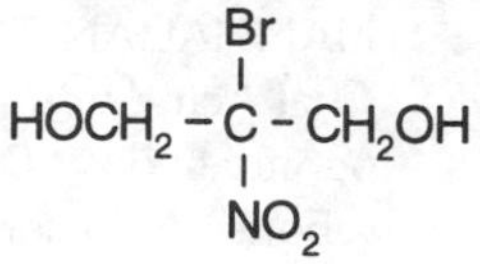

NOMENCLATURE
Common name bronopol (BAN)
IUPAC name 2-bromo-2-nitropropane-1,3-diol
Chemical Abstracts name 2-bromo-2-nitro-1,3-propanediol
CAS RN *[52–51–7]*   **EEC no.** 200–143–0

PHYSICAL CHEMISTRY
Mol. wt. 200.0   **M.f.** $C_3H_6BrNO_4$ **Form** Colourless to pale yellow-brown solid.
M.p. 130 °C   **V.p.** 1.68 mPa (20 °C)   **Solubility** In water 250 g/l (22 °C). In
ethanol 500, isopropanol 250, propylene glycol 143, glycerol 10, liquid paraffin <5
(all in g/l, 23–24 °C). Readily soluble in acetone and ethyl acetate; slightly soluble
in chloroform, diethyl ether, and benzene; insoluble in ligroin.   **Stability** Slightly
hygroscopic. Stable under normal storage conditions, but unstable in aluminium
containers.

COMMERCIALISATION
History Introduced in 1964 as a preservative for cosmetic and pharmaceutical
preparations, and later as an agricultural bacteriostat (D. F. Spooner &
S. B. Wakerley, *Proc. Br. Insectic. Fungic. Conf., 6th,* 1971, **1,** 201) by the Boots Co.
Ltd (now AgrEvo GmbH).   **Patent** GB 1193954   **Manufacturers** AgrEvo

APPLICATIONS
Biochemistry Oxidation of the mercapto group of bacterial enzymes. Inhibition of
dehydrogenase activity leads to irreversible membrane damage.   **Uses** Used as a
seed treatment for control of *Xanthomonas malvacearum* causing blackarm disease
of cotton.   **Phytotoxicity** Non-phytotoxic to a wide range of crops, including
cotton.   **Formulation types** WP; SP; DS.   **Selected tradenames** 'Bronotak'
(AgrEvo)

ANALYSIS
Product analysis by glc. Details available from AgrEvo.

MAMMALIAN TOXICOLOGY
Oral Acute oral LD_{50} for rats 180–400, mice 250–500, dogs 250 mg/kg.
Skin and eye Acute percutaneous LD_{50} for rats >1600 mg/kg. Moderate skin

irritant; mild eye irritant (rabbits).   **Inhalation** LC_{50} (6 h) for rats >5 mg/l air.
NOEL In 72 d feeding trials, rats receiving 1000 mg/kg diet showed no ill-effects.
Toxicity class WHO (a.i.) II; EPA (formulation) II   **EC risk** Xn (R21/22);
Xi (R37/38, R41)

ECOTOXICOLOGY
Birds Acute oral LD_{50} for mallard duck 510 mg/kg.   **Fish** LC_{50} (96 h) for trout
20.0 mg/l.   **Daphnia** LC_{50} (48 h) 1.4 mg/l.

ENVIRONMENTAL FATE
Animals Oral doses are rapidly absorbed and rapidly excreted, mainly in the
urine. The major metabolite has been identified as 2-nitropropane-1,3-diol.
Plants Following treatment of potatoes at a rate of 12 g bronopol/tonne, the
residues fall to <0.1 mg/kg after 6 months storage. Biochemical degradation in the
tuber leads to the metabolite 2-nitropropan-1,3-diol (Link *et al.*, *Nachrichtenbl.*
Pflanzenschutz DDR, 1985, **39**, 173).

▶1   bupirimate   *Fungicide*

pyrimidine

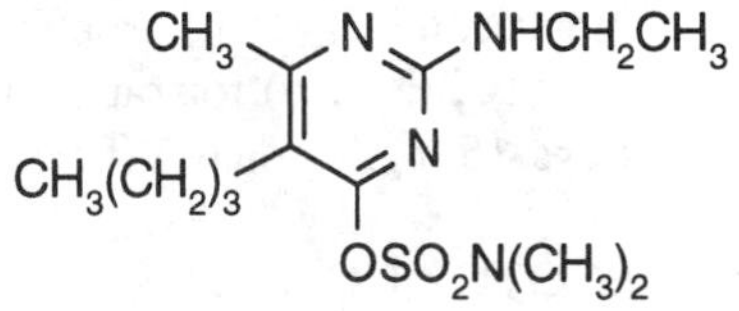

NOMENCLATURE
Common name bupirimate (BSI, E-ISO, (*m*) F-ISO, ANSI)
IUPAC name 5-butyl-2-ethylamino-6-methylpyrimidin-4-yl dimethylsulfamate
Chemical Abstracts name 5-butyl-2-(ethylamino)-6-methyl-4-pyrimidinyl
dimethylsulfamate
CAS RN *[41483–43–6]*   **Development codes** PP588

PHYSICAL CHEMISTRY
Composition Tech. is 90% pure.   **Mol. wt.** 316.4   **M.f.** $C_{13}H_{24}N_4O_3S$ **Form** Pale
tan, waxy solid.   **M.p.** 50–51 °C; (tech., c. 40–45 °C)   **V.p.** 0.1 mPa (25 °C)
K_{ow} logP = 3.9   **S.g./density** 1.2   **Solubility** In water 22 mg/l (pH 5.2, 25 °C).
Soluble in most common organic solvents except paraffins.   **Stability** Stable in
dilute alkalis, but readily hydrolysed by dilute acids. Rapidly decomposed by u.v.
irradiation in aqueous solution. Unstable on prolonged storage above 37 °C.
F.p. >50 °C

COMMERCIALISATION
History Fungicide reported by J. R. Finney *et al.* (*Proc. Br. Insectic. Fungic. Conf.*, *8th*, 1975, **2**, 667). Introduced by ICI Plant Protection Division (now Zeneca Agrochemicals). **Patent** GB 1400710 **Manufacturers** Zeneca

APPLICATIONS
Mode of action Systemic fungicide with protective and curative action. Absorbed by the leaves, with translocation in the xylem and translaminar action. Acts by inhibiting sporulation. **Uses** Control of powdery mildews of apples, pears, stone fruit, strawberries, gooseberries, currants, raspberries, vines, roses and other ornamentals, cucurbits, hops, beet, and other crops.
Phytotoxicity Chrysanthemums, roses, strawberries, and some varieties of apple may suffer slight injury. **Formulation types** EC; WP. **Mixtures** (*bupirimate +*) triforine; pirimicarb + triforine. **Selected tradenames** 'Nimrod' (Zeneca)

ANALYSIS
Product analysis by glc with FID (*CIPAC Handbook*, 1985, **1C**, 2007) or by hplc. Identity by glc, tlc, i.r. or nmr (*ibid.*, 1994, **F**, 406). **Residues** determined by glc of a derivative. Details available from Zeneca.

MAMMALIAN TOXICOLOGY
Oral Acute oral LD_{50} for rats, mice, rabbits, and guinea pigs >4000 mg/kg.
Skin and eye Acute percutaneous LD_{50} for rats 4800 mg/kg. Mild eye irritant; slight to mild irritant to skin (rabbits). Moderate skin sensitiser (guinea pigs).
Inhalation LC_{50} (4 h) no effect on rats exposed to 0.035 mg respirable particles/l air. **NOEL** (2 y) for rats 100 mg/kg diet; (90 d) for rats 1000 mg/kg diet, for dogs 15 mg/kg daily. **Toxicity class** WHO (a.i.) III (Table 5); EPA (formulation) III

ECOTOXICOLOGY
Birds Acute oral LD_{50} for quail >5200, pigeons >2700 mg/kg. **Fish** LC_{50} (96 h) for rainbow trout 1.0 mg/l. **Bees** NOEL: 0.20 mg/bee (oral), 0.050 mg/bee (contact). **Daphnia** LC_{50} (48 h) 7.3 mg/l.

ENVIRONMENTAL FATE
Animals In mammals, following oral administration, 68% of the dose is eliminated in the urine within 24 hours; 77% is eliminated in the urine and 21% in the faeces within 10 days. **Soil/Environment** In soil, the major degradation product is ethirimol. Soil DT_{50} 35–90 d (non-sterile flooded or non-flooded soils, pH 5.1 to pH 7.3).

buprofezin

Insecticide, acaricide

NOMENCLATURE
Common name buprofezin (BSI, draft E-ISO); buprofézine ((*f*) draft F-ISO)
IUPAC name 2-*tert*-butylimino-3-isopropyl-5-phenyl-1,3,5-thiadiazinan-4-one
Chemical Abstracts name 2-[(1,1-dimethylethyl)imino]tetrahydro-3-
(1-methylethyl)-5-phenyl-4*H*-1,3,5-thiadiazin-4-one
CAS RN *[69327–76–0]* **Development codes** NNI-750 (Nihon Nohyaku); PP618
(Zeneca)

PHYSICAL CHEMISTRY
Composition Tech. material is 98% pure. **Mol. wt.** 305.4 **M.f.** $C_{16}H_{23}N_3OS$
Form White crystals; (tech., white or pale yellow crystalline powder).
M.p. 104.5–105.5 °C **V.p.** 1.25 mPa (25 °C) **K_{ow}** logP = 4.3 **S.g./density** 1.18
(20 °C) **Solubility** In water 0.9 mg/l (20 °C). In chloroform 520, benzene 370,
toluene 320, acetone 240, ethanol 80, hexane 20 (all in g/l, 25 °C).
Stability Stable in acidic and alkaline media. Stable to heat and light.

COMMERCIALISATION
History Insecticide reported by H. Kanno *et al.* (*Proc. Br. Crop Prot. Conf. – Pests
Dis.*, 1981, **1**, 59). Introduced by Nihon Nohyaku Co., Ltd. **Patent** JP 1048643
Manufacturers Nihon Nohyaku

APPLICATIONS
Biochemistry Probable chitin synthesis and prostaglandin inhibitor. Hormone
disturbing effect, leading to suppression of ecdysis. **Mode of action** Persistent
insecticide and acaricide with contact and stomach action; not translocated in the
plant. Inhibits moulting of nymphs and larvae, leading to death. Also suppresses
oviposition by adults; treated insects lay sterile eggs. **Uses** Insecticide with
persistent larvicidal action against Homoptera, some Coleoptera and also Acarina.
Effective against Cicadellidae and Delphacidae (leafhoppers) in rice; Cicadellidae in
potatoes; Aleyrodidae (whitefly) in citrus, cotton and vegetables; Coccidae,
Diaspididae (scale insects) and Pseudococcidae (mealybugs) in citrus.
Phytotoxicity Slightly phytotoxic to Chinese cabbage. **Formulation types** WP; DP;
GR; SC. **Mixtures** *(buprofezin +)* deltamethrin; isoprocarb; fenobucarb; cartap;
lambda-cyhalothrin; pirimiphos-methyl; fenpropathrin. **Selected tradenames**
'Applaud' (Nihon Nohyaku)

ANALYSIS

Product by gc. **Residues** in soil and rice plants and water, by gc with ECD (M. Uchida et al., J. Pestic. Sci., **7**, 397 (1982)); in crops, by gc with NPD (H. Nishizawa et al., J. AOAC International, **77**, 1631 (1994)). Details from Nihon Nohyaku.

MAMMALIAN TOXICOLOGY

Reviews Pesticide residues in food – 1991. FAO Plant Production and Protection Paper, 111, 1991. Pesticide residues in food – 1991 evaluations. Part II – Toxicology. World Health Organisation, WHO/PCS/92.52, 1992. **Oral** Acute oral LD_{50} for male rats 2198, female rats 2355, male and female mice >10 000 mg/kg. **Skin and eye** Acute percutaneous LD_{50} for rats >5000 mg/kg. Mild skin irritant (guinea pigs), not irritating to skin and eyes (rabbits). **Inhalation** LC_{50} (4 h) for rats >4.57 mg/l air. **NOEL** for male rats 0.90, female rats 1.12 mg/kg daily. **ADI** (JMPR) 0.01 mg/kg [1991]. **Other** Non-carcinogenic, non-mutagenic. **Toxicity class** WHO (a.i.) III (Table 5); EPA (formulation) III

ECOTOXICOLOGY

Fish LC_{50} (48 h) for carp 2.7, rainbow trout >1.4 mg/l. **Bees** No direct effect at 2000 mg/l (WP formulation). **Other beneficial spp.** No effect on various predators (Euseius stipulatus 250 mg/l; Phytoseiulus persimilis 500 mg/l; Cyrtorhinus lividipennis, Microvelia atrolineata 250 mg/l; Lycosa pseudoannulata 2000 mg/l) or parasites (Aphytis linganensis 125 mg/l; Cales noacki, Encarsia formosa, Paracentrobia andoi 250 mg/l; Ephedrus japonicus 1000 mg/l). **Daphnia** LC_{50} (3 h) for D. pulex 50.6 mg/l.

ENVIRONMENTAL FATE

Soil/Environment DT_{50} (25 °C) 104 d (flooded conditions, silty clay loam, o.c. 3.8%, pH >6.4), 80 d (upland conditions, sandy loam, o.c. 2.4%, pH 7.0) (J. Pesticide Sci., **11**, 605–610 (1986)).

butachlor

Herbicide

chloroacetanilide

$$\text{CH}_2\text{CH}_3$$
$$\overset{\displaystyle\text{CH}_2\text{CH}_3}{\underset{\displaystyle\text{CH}_2\text{CH}_3}{\bigcirc}}-\text{N}\overset{\text{COCH}_2\text{Cl}}{\underset{\text{CH}_2\text{O(CH}_2)_3\text{CH}_3}{}}$$

NOMENCLATURE

Common name butachlor (BSI, draft E-ISO, (*m*) draft F-ISO, ANSI, WSSA, JMAF); no name (France)
IUPAC name *N*-butoxymethyl-2-chloro-2',6'-diethylacetanilide
Chemical Abstracts name *N*-(butoxymethyl)-2-chloro-*N*-(2,6-diethylphenyl)= acetamide
CAS RN *[23184–66–9]*   **Development codes** CP 53 619

PHYSICAL CHEMISTRY

Composition ≥93.5% pure.   **Mol. wt.** 311.9   **M.f.** $C_{17}H_{26}ClNO_2$ **Form** Light yellow to purple liquid with a faint, sweet odour.   **M.p.** –2.8 °C to 1.7 °C
B.p. 156 °C/0.5 mmHg   **V.p.** 2.4×10^{-1} mPa (25 °C)   **S.g./density** 1.076 (25 °C)
Solubility In water 20 mg/l (20 °C). Soluble in most organic solvents, including diethyl ether, acetone, benzene, ethanol, ethyl acetate, and hexane.
Stability Decomposes at ≥165 °C. Stable to u.v. light. Stable indefinitely ≤45 °C.
F.p. >135 °C (Tag closed cup)   **Other properties** Viscosity 37 cp (25 °C)

COMMERCIALISATION

History Herbicide reported by D. D. Baird & R. P. Upchurch (*Proc. South. Weed Control Conf., 23rd, 1970*, p. 101). Introduced by Monsanto Co.
Patent US 3442945; US 3547620   **Manufacturers** Comlets; Crystal; Hindustan ; Krishi Rasayan; Monsanto; Rallis; Searle India; Sinon; Sundat

APPLICATIONS

Biochemistry Protein synthesis inhibitor.   **Mode of action** Selective systemic herbicide, absorbed primarily by the germinating shoots, and secondarily by the roots, with translocation throughout the plant, giving higher concentrations in vegetative parts than in reproductive parts.   **Uses** Used pre-emergence for the control of annual grasses and certain broad-leaved weeds in rice, both seeded and transplanted. It shows selectivity in barley, cotton, peanuts, sugar beet, wheat and several brassica crops. Effective rates range from 1.0–4.5 kg a.i./ha. Activity is dependent on water availability such as rainfall following treatment, overhead irrigation or applications to standing water as in rice culture.
Phytotoxicity Non-phytotoxic to rice, cotton, barley, wheat, peanuts, sugar beet, and some brassicas.   **Formulation types** GR; EC.   **Mixtures** (*butachlor +*)

oxadiazon; pyrazolynate; propanil; 2,4-D. **Selected tradenames** 'Machete' (Monsanto); 'Butataf' (Rallis); 'Butanex' (Makhteshim-Agan); 'Dhanuchlor' (Dhanuka); 'Farmachlor' (Sanonda); 'Rasayanchlor' (Krishi Rasayan); 'Trapp' (Searle India)

ANALYSIS
Product analysis by glc with FID (*AOAC Methods*, 1995, 986.04; *CIPAC Handbook*, 1988, **D**, 17) or by i.r. spectrometry. **Residues** determined by glc. In **drinking water**, by gc with FID (*AOAC Methods*, 1995, 991.07). Details available from Monsanto Co.

MAMMALIAN TOXICOLOGY
Oral Acute oral LD_{50} for rats 2000, mice 4747, rabbits >5010 mg/kg.
Skin and eye Acute percutaneous LD_{50} for rabbits >13 000 mg/kg. Moderate skin irritant; practically non-irritating to eyes (rabbits). Contact sensitisation reactions observed in guinea pigs. **Inhalation** LC_{50} (4 h) for rats >3.34 mg/l air.
Other Oncogenic in rats but not mice. For detailed toxicology data, please contact Monsanto. **Toxicity class** WHO (a.i.) III (Table 5); EPA (formulation) III

ECOTOXICOLOGY
Birds Acute oral LD_{50} for mallard ducks >4640 mg/kg. Five-day dietary LC_{50} for mallard ducks >10 000, bobwhite quail 6597 mg/kg diet. **Fish** LC_{50} (96 h) for rainbow trout 0.52, bluegill sunfish 0.44, carp 0.32, channel catfish 0.10–0.14, fathead minnow 0.31 mg/l. **Bees** LD_{50} (contact) >100 μg/bee. **Daphnia** LC_{50} (48 h) 2.4 mg/l. **Other aquatic spp.** LC_{50} (96 h) for crayfish 26 mg/l.

ENVIRONMENTAL FATE
Animals Metabolised to water-soluble metabolites and excreted. **Plants** Rapidly metabolised in plants to water-soluble metabolites, leading eventually to mineralisation. **Soil/Environment** In soil, degradation is principally by microbial activity (Y.-L. Chen and T.-C. Wu, *Nippon Noyaku Gakkaishi*, 1978, **3**, 411). Persists for c. 6–10 weeks. Converted in soil or water to water-soluble derivatives, with a slow evolution of CO_2.

butamifos

phosphoramidate

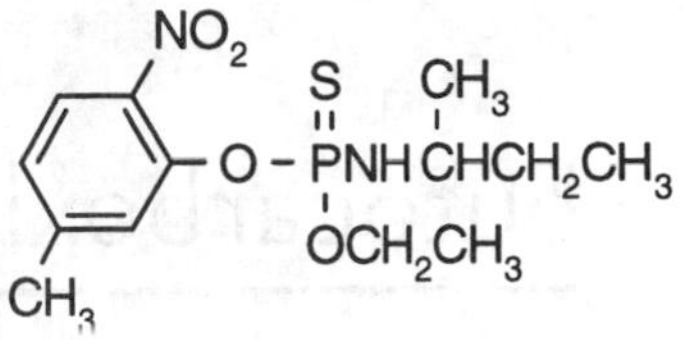

NOMENCLATURE
Common name butamifos (BSI, E-ISO, F-ISO, JMAF)
IUPAC name *O*-ethyl *O*-6-nitro-*m*-tolyl *sec*-butylphosphoramidothioate
Chemical Abstracts name *O*-ethyl *O*-(5-methyl-2-nitrophenyl) (1-methylpropyl)=
phosphoramidothioate
CAS RN *[36335–67–8]*   **Development codes** S-2846

PHYSICAL CHEMISTRY
Mol. wt. 332.4   **M.f.** $C_{13}H_{21}N_2O_4PS$ **Form** Tech. grade is a yellow-brown liquid.
V.p. 84 mPa (27 °C)   **K_{ow}** logP = 4.62 (25 °C)   **S.g./density** 1.188 (25 °C)
Solubility In water 6.19 mg/l (25 °C). Readily soluble in acetone, methanol, and
xylene at room temperature.

COMMERCIALISATION
History Herbicide reported by M. Ueda (*Jpn. Pestic. Inf.*, 1975, No. 23, p. 23).
Introduced by Sumitomo Chemical Co., Ltd.   **Patent** GB 1359727; US 3936433
Manufacturers Sumitomo

APPLICATIONS
Mode of action Selective non-systemic herbicide.   **Uses** A contact herbicide used
pre-emergence. Effective against annual and especially graminaceous weeds, in
beans, lawns, rice and vegetables.   **Formulation types** EC; WP.
Selected tradenames 'Cremart' (Sumitomo)

ANALYSIS
Product analysis by glc or by colorimetry (M. Horiba et *al.*, *Nippon Nogei Kagaku
Kaishi*, 1979, **53** 111).   Details of **residue** analysis available from Sumitomo
Chemical Co.

MAMMALIAN TOXICOLOGY
Oral Acute oral LD_{50} for male rats 1070, female rats 845 mg/kg.
Skin and eye Acute percutaneous LD_{50} for rats >5000 mg/kg. Non-irritant to skin
and eyes (rabbits).   **Inhalation** LC_{50} for rats >1200 mg/m^3 air.
Toxicity class WHO (a.i.) II

MAIN ENTRIES

95   butocarboxim   *Insecticide*

oxime carbamate

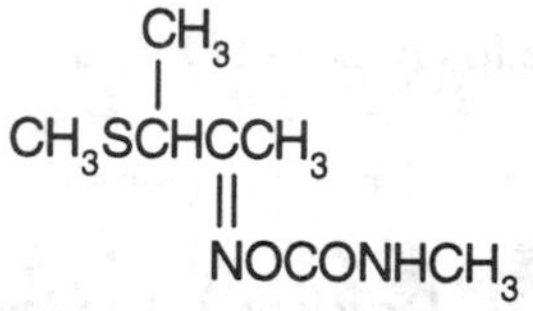

$$
\begin{array}{ccc}
CH_3 & & CH_3 \\
| & & | \\
CH_3SCHCCH_3 & & CH_3CCHSCH_3 \\
\| & & \| \\
NOCONHCH_3 & & NOCONHCH_3 \\
(E) & & (Z)
\end{array}
$$

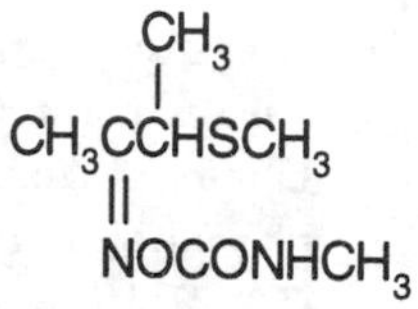

NOMENCLATURE
Common name butocarboxim (BSI, E-ISO); butocarboxime ((f) F-ISO)
IUPAC name 3-(methylthio)butanone O-methylcarbamoyloxime
Chemical Abstracts name 3-(methylthio)-2-butanone O-[(methylamino)carbonyl]=
oxime
CAS RN *[34681–10–2]* **Development codes** Co 755

PHYSICAL CHEMISTRY
Composition Tech. grade is liquid and obtained as an 85% tech. concentrate in
xylene containing the (E)- and (Z)- isomers in a ratio of 85–90:15–10.
Mol. wt. 190.3 **M.f.** $C_7H_{14}N_2O_2S$ **Form** Pale brown viscous liquid (tech.).
Crystalline at lower temperatures. **M.p.** 37 °C ((E)- isomer); ((Z)- isomer is an oil
at room temperature) **V.p.** 10.6 mPa (20 °C) (tech., without residual xylene)
K_{ow} logP = 1.1 **S.g./density** 1.12 (20 °C) **Solubility** In water 35 g/l (20 °C).
Completely miscible with aromatic hydrocarbons, esters, and ketones. Low
solubility in aliphatic hydrocarbons (11 g/l) and carbon tetrachloride.
Stability Stable at pH 5 to pH 7 (up to 50 °C), but hydrolysed by strong acids and
alkalis. Stable to sunlight and oxygen. Thermally stable up to 100 °C.

COMMERCIALISATION
History Insecticide reported by M. Vulic *et al.* (*Meded. Fac. Landbouwwet. Rijksuniv.
Gent*, 1973, **38**, 1175). Structure/activity of analogues discussed (T. A. Magee &
L. E. Limpel, *J. Agric. Food Chem.*, 1977, **25**, 1376). Introduced by Wacker Chemie
GmbH. **Patent** DE 2036491; US 3816532; GB 1353202 **Manufacturers** Wacker

APPLICATIONS
Biochemistry Cholinesterase inhibitor. **Mode of action** Systemic insecticide with
contact and stomach action. Absorbed by the leaves and roots. **Uses** Control of

aphids, thrips, whitefly, mealybugs, and other sucking insects (with secondary action against spider mites) on fruit trees (including citrus), vegetables, cereals, cotton, hops, tobacco, and ornamentals (including ornamentals cultivated by hydroponics). **Formulation types** EC; AE; SL. **Selected tradenames** 'Drawin' (Wacker)

ANALYSIS
Product analysis by i.r. spectrometry or by hplc. **Residues** of butocarboxim, its sulfoxide and sulfone (butoxycarboxim) determined by glc with TID. Details available from Wacker Chemie GmbH.

MAMMALIAN TOXICOLOGY
Reviews *Pesticide residues in food – 1985*. FAO Plant Production and Protection Paper 68, 1986. **EHC** 64 (WHO, 1986; a review of carbamate insecticides in general). **Oral** Acute oral LD_{50} for rats 153–215 mg/kg. **Skin and eye** Acute percutaneous LD_{50} for albino rabbits 360 mg/kg. Irritating to eyes. **Inhalation** LC_{50} (4 h) for rats 1 mg/l air. **NOEL** (2 y) for rats 100 mg/kg diet; (90 d) for dogs 100 mg/kg diet. No carcinogenic or mutagenic activity in routine tests at the highest dose (300 mg/kg diet) in rats during 2 y, nor any effect on fertility, growth rate or mortality. **ADI** (JMPR) No ADI [1985]; 0.02 mg/kg (Bundes Gesundheits Amt, Berlin). **Other** Not mutagenic in the Ames assay. **Toxicity class** WHO (a.i.) Ib; EPA (formulation) II

ECOTOXICOLOGY
Birds LD_{50} for mallard ducks 64 mg/kg. LC_{50} (8 d) for Japanese quail 1180 mg/kg diet. **Fish** LC_{50} (96 h) for rainbow trout 29 mg/l. LC_{50} (24 h) for rainbow trout 35, goldfish 55, guppy 70 mg/l. **Bees** Toxic to bees; LD_{50} 1 μg/bee. **Daphnia** LC_{50} (24 h) 3.2–5.6 mg/l. **Algae** NOEL for algae 62.5 mg/l.

ENVIRONMENTAL FATE
Animals In mammals, following oral administration, metabolised to butoxycarboxim, and excreted in the urine as butoxycarboxim and its degradation products. **Plants** In plants, degradation is the same as in soil. **Soil/Environment** In soil, the methylamine moiety is split off, and the sulfur atom is oxidised to sulfoxide and sulfone. DT_{50} in soil 1–8 d; DT_{50} for metabolites 16–44 d.

MAIN ENTRIES

butoxycarboxim *Insecticide, acaricide*

oxime carbamate

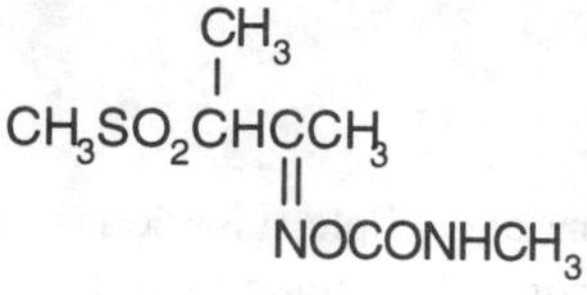

$$CH_3SO_2CHCCH_3$$ with CH_3 above and $NOCONHCH_3$ below

(*E*)

$$CH_3CCHSO_2CH_3$$ with CH_3 above and $NOCONHCH_3$ below

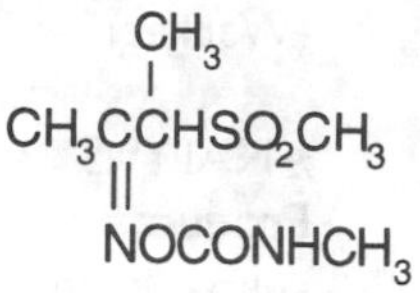

(*Z*)

NOMENCLATURE
Common name butoxycarboxim (BSI, E-ISO, JMAF); butoxycarboxime ((*f*) F-ISO)
IUPAC name 3-methylsulfonylbutanone *O*-methylcarbamoyloxime
Chemical Abstracts name 3-(methylsulfonyl)-2-butanone
O-[(methylamino)carbonyl]oxime
CAS RN *[34681–23–7]* **Development codes** Co 859

PHYSICAL CHEMISTRY
Composition Tech. butoxycarboxim contains the (*E*)- and (*Z*)- isomers in a ratio of
85–90:15–10. **Mol. wt.** 222.3 **M.f.** $C_7H_{14}N_2O_4S$ **Form** Colourless crystals.
M.p. 85–89 °C (tech.) **V.p.** 0.266 mPa (20 °C) (tech.) **S.g./density** 1.21 g/cm^3
(20 °C) **Solubility** In water 209 g/l (20 °C). Readily soluble in polar organic
solvents; slightly soluble in non-polar solvents. In chloroform 186, acetone 172,
isopropanol 101, toluene 29, carbon tetrachloride 5.3, cyclohexane 0.9,
heptane 0.1 (all in g/l at 20 °C). **Stability** Aqueous hydrolysis DT_{50} 501 d (pH 5),
18 d (pH 7), 16 d (pH 9). Stable to u.v. light. Thermally stable up to 100 °C.

COMMERCIALISATION
History Insecticide and special application method reported by M. Vulic &
H. Bräunling (*Meded. Fac. Landbouwwet. Rijksuniv. Gent*, 1974, **39**, 847).
Structure/activity of analogues reported (T. A. Magee & L. E. Limpel, *J. Agric. Food
Chem.*, 1977, **25**, 1376). Introduced by Wacker Chemie GmbH
Patent DE 2036491; US 3816532; GB 2353202 **Manufacturers** Wacker

APPLICATIONS
Biochemistry Cholinesterase inhibitor. **Mode of action** Systemic insecticide and
acaricide with contact and stomach action. Absorbed by the roots, with
translocation acropetally. **Uses** Control of aphids, spider mites, and other sucking
insects on potted ornamentals. **Formulation types** Cardboard sticks.
Selected tradenames 'Plant Pin' (Wacker)

ANALYSIS
Product analysis by i.r. spectrometry or by hplc. **Residues** determined by glc of a
derivative with TID. Details available from Wacker Chemie GmbH.

MAMMALIAN TOXICOLOGY
EHC 64 (WHO, 1986; a review of carbamate insecticides in general). **Oral** Acute oral LD_{50} for rats 458, rabbits 275 mg/kg. **Skin and eye** Acute percutaneous LD_{50} for rats >2000 mg/kg. **NOEL** (90 d) for rats 300 mg/kg diet; at 1000 mg/kg diet, only slight inhibition of erythrocyte and plasma cholinesterase noted. **Other** Oral LD_{50} for rats of the pasteboard stick formulation >5000 mg/kg. Acute subcutaneous LD_{50} for female rats 288 mg/kg. Butoxycarboxim is a metabolite of butocarboxim in plant and animal tissue, therefore the toxicological tests on the latter partly include butoxycarboxim. **Toxicity class** WHO (a.i.) Ib; EPA (formulation) II

ECOTOXICOLOGY
Birds Acute oral LD_{50} for hens 367 mg/kg. **Fish** LC_{50} (96 h) for carp 1750, rainbow trout 170 mg/l. **Bees** Not hazardous to bees (special test with 'Plant Pin'). **Daphnia** LC_{50} (96 h) 500 µg/l.

ENVIRONMENTAL FATE
Animals In mammals, following oral administration, eliminated in the urine both in the unchanged form and as metabolites. There is no accumulation in tissue. **Soil/Environment** DT_{50} for soil 41–44 d (20 °C). Butoxycarboxim diffuses from the pasteboard pin into the soil within 2 days, activity commences after c. 3 days, reaches its full extent in 7–14 days, and remains for 4–8 weeks, until the pasteboard pin rots.

7 butralin   *Herbicide, plant growth regulator*

2,6-dinitroaniline

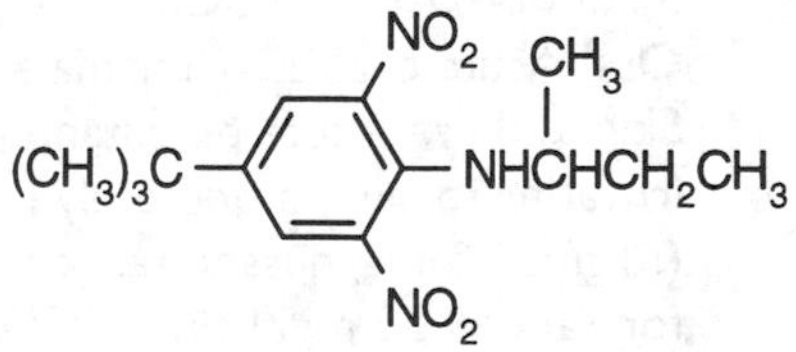

NOMENCLATURE
Common name butralin (BSI, E-ISO, ANSI, WSSA); butraline ((f) F-ISO); no name (Eire, Japan)
IUPAC name *N*-sec-butyl-4-*tert*-butyl-2,6-dinitroaniline
Chemical Abstracts name 4-(1,1-dimethylethyl)-*N*-(1-methylpropyl)-=2,6-dinitrobenzenamine
CAS RN *[33629–47–9]* **EEC no.** 251–607–4 **Development codes** Amchem 70–25; Amchem A-820

PHYSICAL CHEMISTRY
Composition Tech. is ≥98%. **Mol. wt.** 295.3 **M.f.** $C_{14}H_{21}N_3O_4$ **Form** Yellow-orange crystals with a slightly aromatic odour. **M.p.** 61 °C; tech. 59 °C
B.p. 134–136 °C/0.5 mmHg **V.p.** 0.77 mPa (25 °C) **K$_{ow}$** logP = 4.93
S.g./density 1.063 (25 °C) **Solubility** In water 0.3 mg/l (25 °C). In methanol 97.5, hexane 300 (both in g/l, 25 °C). **Stability** Decomposes at 265 °C. Hydrolytically and photochemically stable. Concentrates are stable on storage under dry conditions >3 y, but should not be stored <–5 °C or allowed to freeze.

COMMERCIALISATION
History Herbicide reported by S. R. McLane *et al.* (*Proc. South. Weed Sci. Soc.*, 1971, **24**, 58). Introduced by Amchem Products Inc. (now Rhône-Poulenc Agrochimie) and later acquired by CFPI-Agro. **Patent** US 3672866
Manufacturers CFPI

APPLICATIONS
Biochemistry Cell division inhibitor. **Mode of action** Selective herbicide, absorbed by germinating seedlings, with slow translocation acropetally. Also acts as a growth regulator, suppressing the growth of shoots, branches, and suckers.
Uses Pre-emergence control of annual broad-leaved weeds and grasses in cotton, soya beans, rice, barley, beans, alliums, vines, ornamentals, and orchards of fruit and nut trees; used at 1.12–3.4 kg a.i./ha (depending on soil type). Also used to control suckers on tobacco at 125 mg/plant. **Formulation types** EC.
Mixtures (*butralin* +) atrazine; linuron; monolinuron; neburon.
Compatibility Incompatible with strong oxidising agents.
Selected tradenames 'Amexine' (CFPI); 'Tamex' (CFPI, Rhône-Poulenc)

ANALYSIS
Product analysis by glc using an internal standard.

MAMMALIAN TOXICOLOGY
Oral Acute oral LD_{50} for male rats 1170, female rats 1049 mg/kg (tech.).
Skin and eye Acute percutaneous LD_{50} (tech.) for rabbits ≥2000 mg/kg. Slightly irritating to skin, moderately irritating to eyes (rabbits). Moderate skin sensitiser (Kligman & Magnusson test); not a skin sensitiser (Buehler test). **Inhalation** LC_{50} for rats >9.35 mg/l air. **NOEL** (2 y) for rats 500 ppm (20–30 mg/kg daily).
ADI 0.5 mg/kg (provisional). **Toxicity class** WHO (a.i.) III (Table 5); EPA (formulation) IV **EC risk** Xn (R22, R63)

ECOTOXICOLOGY
Birds Acute oral LD_{50} for bobwhite quail >>2250 mg/kg, for Japanese quail >>5000 mg/kg. Dietary LC_{50} (8 d) for bobwhite quail and mallard ducks >10 000 mg/kg diet. **Fish** LC_{50} (96 h) for bluegill sunfish 1.0, rainbow trout 0.37 mg/l. **Bees** Slightly toxic to bees. **Other beneficial spp.** Not toxic to soil

microflora at usual application rate. Application at 2–10 × recommended rate did not significantly alter microbial population.   **Daphnia** EC_{50} (48 h) 0.12 mg/l. **Algae** EC_{50} (5 d) for *Selenastrum capricornutum* 0.12 mg/l.

ENVIRONMENTAL FATE
Animals Extensively metabolised and excreted in urine and faeces. Metabolised in the rat by the primary metabolic processes of N-dealkylation, oxidation and nitro reduction, and by secondary processes of N-acetyl and glucuronic acid conjugation. The majority (85%) of applied butralin was excreted in the urine within 48 h. None was in the organs after 72 h. Butralin is ultimately metabolised to CO_2.   **Plants** In plants, metabolism leads to evolution of CO_2.
Soil/Environment In soil, microbial degradation occurs, with formation of the corresponding aniline, ring splitting and evolution of CO_2 (P. C. Kearney et al., J. Agric. Food Chem., 1974, **22**, 856). Persists in soil for c. 3 months. In sterile medium (pH 5, 7 and 9), no degradation after 21 d. Strongly adsorbed and not leached; lysimeter studies show little migration of butralin beyond the top 6 cm. DT_{50} for solution exposed to light 14 d. In rivers, c. 45% decomposed after 21 d.

8   # butroxydim   *Herbicide*

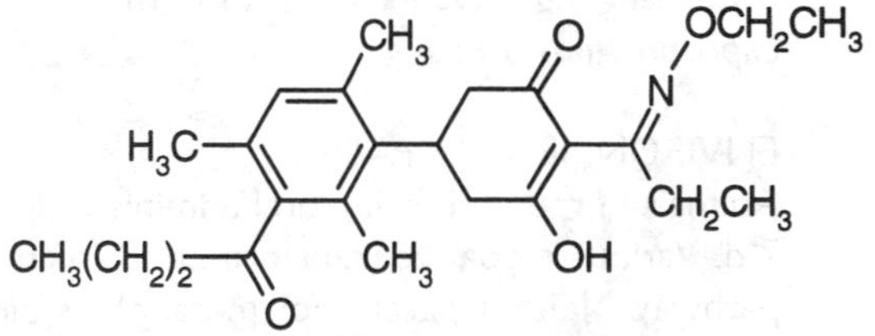

NOMENCLATURE
Common name butroxydim (BSI, pa E-ISO)
IUPAC name 5-(3-butyryl-2,4,6-trimethylphenyl)-2-[1-(ethoxyimino)propyl]-= 3-hydroxy-cyclohex-2-enone
Chemical Abstracts name 2-[1-(ethoxyimino)propyl]-3-hydroxy-5-[2,4,6-= trimethyl-3-(1-oxobutyl)phenyl]-cyclohexen-1-one
CAS RN *[138164–12–2]*   **Development codes** ICIA0500

PHYSICAL CHEMISTRY
Mol. wt. 399.5   **M.f.** $C_{24}H_{33}NO_4$ **Form** Off-white powdery solid.   **M.p.** 80.8 °C
V.p. 1×10^{-3} mPa (20 °C) K_{ow} logP = 1.90 (pH 7, 25 °C)   **Solubility** In water 6.9 mg/l (20 °C).   **Stability** Hydrolysis DT_{50} 10.5 d (pH 5), >240 d (pH 7), stable (pH 9) (all 25 °C).   **pKa** 4.36 (23 °C), weak acid

COMMERCIALISATION
Manufacturers Zeneca

APPLICATIONS
Uses A post-emergence grass herbicide for use on broad-leaved crops.
Selected tradenames 'Falcon' (Zeneca)

ANALYSIS
Residues determined by hplc with u.v. detection, or by hplc-ms.

MAMMALIAN TOXICOLOGY
Oral LD_{50} for male rats 3476, female rats 1635 mg/kg. **Skin and eye** Acute
percutaneous LD_{50} for rats >2000 mg/kg. Moderate skin irritant, mild eye irritant
(rabbits). Not a skin sensitiser (guinea pigs). **Inhalation** LC_{50} (4h) for rats
>2.99 mg/l. **NOEL** NOAEL (1 y) for dogs 5 mg/kg b.w. daily; NOEL (2 y) for
rats 2.5, mice 10 mg/kg b.w. daily (liver tumours only in males at high dose);
NOEL developmental for rats 5, rabbits 15 mg/kg b.w. daily. **Other** Genotoxicity
negative. Not oncogenic (rats).

ECOTOXICOLOGY
Birds Acute oral LD_{50} for mallard duck >2000, bobwhite quail 1221 mg/kg.
Sub-acute dietary LC_{50} (5 d) for mallard duck >5200, bobwhite quail 5200 mg/kg.
Fish LC_{50} (96 h) for rainbow trout >6.9, bluegill sunfish 8.8 mg/l. **Bees** LD_{50}
(contact, 24 h) >200 µg/bee. **Worms** LC_{50} (14 d) for *Eisenia foetida*
>1000 mg/kg. **Daphnia** LC_{50} (48 h) >3.7 mg/l. **Algae** E_bC_{50} for *Selenastrum
capricornutum* 0.71 mg/l.

ENVIRONMENTAL FATE
Animals In rats, following oral administration, >90% of the dose is excreted within
7 d. Various oxidative transformations of the butyryl chain dominate the metabolic
pathway. Neither parent nor metabolites bioaccumulate in tissues. **Plants** Rapidly
metabolised in plants. **Soil/Environment** Soil K_{oc} 6–1270 (stronger adsorption in
low pH soils). Rapid degradation in aerobic soil, lab. DT_{50} c. 9 d (20 °C,
40% MHC, pH 7.0, 4.09% o.m.); metabolites include 5-(3-butyryl-2,4,6-=
trimethylphenyl)-3-hydroxy-2-(1-iminopropyl)cyclohex-2-enone, 6-(3-butyryl-=
2,4,6-trimethylphenyl)-2-ethyl-4,5,6,7-tetrahydro-4-oxo-1,3-benzoxazole,
2-(3-butyryl-2,4,6-trimethylphenyl)glutaric acid and 5-(3-butyryl-2,4,6-=
trimethylphenyl)-3-hydroxy-2-propionylcyclohex-2-enone.

sec-butylamine

Fungicide

$$CH_3CH_2CH(CH_3)NH_2$$

NOMENCLATURE

Common name *(RS)-sec-butylamine* (proposed E-ISO, in lieu of a common name)
IUPAC name *(RS)-sec-butylamine; (RS)-2-aminobutane*
Chemical Abstracts name 2-butanamine
CAS RN *[13952–84–6]* (RS)-sec-butylamine; *[13250–12–9]* (R)-sec-butylamine
EEC no. 237–732–7

PHYSICAL CHEMISTRY

Mol. wt. 73.1 **M.f.** $C_4H_{11}N$ **Form** Colourless liquid, with an ammoniacal odour.
B.p. 63 °C **V.p.** 18 kPa (20 °C) **S.g./density** 0.724 (20 °C) **Solubility** Miscible
with water and with most organic solvents. With acids, forms water-soluble salts.
Stability Stable at room temperature. **pKa** Base, forming water-soluble salts with
acids.

COMMERCIALISATION

History Fungicidal properties of this fumigant reported by J. W. Eckert &
M. J. Kolbezen (*Nature (London)*, 1962, **194**, 188). Introduced by BASF AG
Manufacturers BASF

APPLICATIONS

Uses Fungicidal fumigation of seed and ware potatoes for control of skin spot and
gangrene. As fungicidal dips or sprays on harvested fruit, e.g. control of citrus blue
and green moulds and stem rot of oranges. Prevention of *Botrytis* on
chrysanthemums and gladioli during storage and transit. **Formulation types** SL.
Selected tradenames 'Deccotane' (Elf Atochem)

ANALYSIS

Product analysis by acid-base titration. **Residues** determined by glc of a
derivative (*Anal. Methods Pestic. Plant Growth Regul.*, 1976, **8**, 251; *Pestic. Anal.
Man.*, 1979, **II**; E. W. Day et al., *J. Assoc. Off. Anal. Chem.*, 1968, **51**, 39).

MAMMALIAN TOXICOLOGY

Reviews *Pesticide residues in food – 1984*. FAO Plant Production and Protection
Paper 62, 1985. **Oral** Acute oral LD_{50} for rats 380, dogs 225 mg/kg.
Skin and eye Acute percutaneous LD_{50} for rabbits >2500 mg/kg. Irritating to skin
and eyes. **Inhalation** Vapours dangerous if inhaled. **NOEL** In 2 y feeding trials,
teratogenicity studies, reproduction studies, rats and dogs receiving
2500 mg/kg diet suffered no ill-effect. In teratogenicity studies NOEL: for rats
2500 mg/kg diet; for rabbits 5000 mg/kg diet. In reproduction studies in rats

MAIN ENTRIES

NOEL 2500 mg/kg diet.   **ADI** (JMPR) Temporary ADI withdrawn [1984].
Toxicity class WHO (a.i.) II   **EC risk** F (R11); Xn (R20/22); C (R35)

ECOTOXICOLOGY
Birds Acute oral LD_{50} for hens 250 mg/kg.   **Fish** LC_{50} for young bluegill sunfish
>50 mg/l.

100 butylate   *Herbicide*

thiocarbamate

$$[(CH_3)_2CHCH_2]_2NC(O)SCH_2CH_3$$

NOMENCLATURE
Common name butylate (BSI, E-ISO, WSSA); butilate ((*m*) F-ISO); no name
(Germany)
IUPAC name S-ethyl di-isobutylthiocarbamate
Chemical Abstracts name S-ethyl bis(2-methylpropyl)carbamothioate
CAS RN *[2008–41–5]*   **Development codes** R-1910

PHYSICAL CHEMISTRY
Mol. wt. 217.4   **M.f.** $C_{11}H_{23}NOS$ **Form** Colourless liquid with an aromatic
odour; (tech., amber liquid).   **B.p.** 137.5–138 °C/21 mmHg; 71 °C/1 mmHg
V.p. 1.73×10^3 mPa (25 °C)   **S.g./density** 0.9402 (25 °C)   **Solubility** In water
36 mg/l (20 °C). Miscible with common organic solvents, e.g. acetone, ethanol,
xylene, methyl isobutyl ketone, kerosene.   **Stability** Thermally stable up to
200 °C. Hydrolysed by strong acids and alkalis, and in aqueous solution in sunlight.
F.p. 115 °C

COMMERCIALISATION
History Herbicide reported by R. A. Gray et al. (*Proc. North Cent. Weed Control
Conf.*, 1962, **19**, 19). Introduced by Stauffer Chemical Co. (now Zeneca
Agrochemicals).   **Patent** US 2913327   **Manufacturers** Sagrochem

APPLICATIONS
Biochemistry Inhibits lipid metabolism.   **Mode of action** Selective systemic
herbicide, absorbed by the roots and coleoptiles, with translocation acropetally.
Acts by inhibiting growth in the meristematic region. Germination inhibitor.
Uses Control of annual grass weeds and *Cyperus* spp. in maize and pineapples, by
pre-plant soil incorporation at 3 kg a.i./ha (4 kg for the control of broad-leaved
weeds in addition to grasses).   **Formulation types** EC; GR.   **Mixtures** *(butylate +)*

atrazine.   **Compatibility** Compatible with some other herbicides.
Selected tradenames 'Sutan Plus' (Zeneca); 'Anelda' (Sagrochem)

ANALYSIS
Product analysis by glc (*CIPAC Handbook*, 1983, **1B**, 1744; *AOAC Methods*, 1995, 974.05; *Anal. Methods Pestic. Plant Growth Regul.*, 1984, **13**, 295).   **Residues** in crops or soils determined by glc (*ibid.*) or by colorimetry of a derivative (J. E. Barney, *ibid.*, 1973, **7**, 641). In **drinking water**, by gc with FID (*AOAC Methods*, 1995, 991.07). Analytical methods available from Zeneca.

MAMMALIAN TOXICOLOGY
EHC 76 (WHO, 1988; general review of thiocarbamates).   **Oral** Acute oral LD_{50} for rats >3500 mg/kg.   **Skin and eye** Acute percutaneous LD_{50} for rabbits >5000 mg/kg. Slight/mild skin irritant, non-irritant to eyes (rabbit).
Inhalation LC_{50} (4 h) for rats 5.2 mg/l air.   **NOEL** (90 d) for rats 32, dogs 40 mg/kg daily. Not carcinogenic or teratogenic.   **Toxicity class** WHO (a.i.) III (Table 5); EPA (formulation) III

ECOTOXICOLOGY
Birds Seven-day dietary LC_{50} for bobwhite quail 40 000 mg/kg diet.   **Fish** LC_{50} (96 h) for rainbow trout 4.2, bluegill sunfish 6.9 mg/l.   **Bees** Not hazardous to bees when used as directed.

ENVIRONMENTAL FATE
Animals In mammals, following oral administration, the major metabolite is the *N,N*-dialkylcarbamoyl conjugate.   **Plants** Rapidly metabolised in plants to CO_2, diisobutylamine, fatty acids, conjugates of amines and fatty acids, and naturally-occurring plant constituents.   **Soil/Environment** In soil, microbial degradation involves hydrolysis to ethylmercaptan, diisobutylamine and CO_2. DT_{50} 1.5–10 w. Duration of residual activity in soil is *c.* 4 mo.

01   cadusafos   *Nematicide, insecticide*

organophosphorus

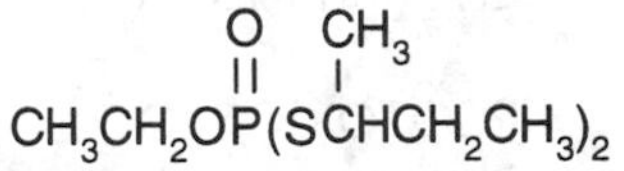

$$CH_3CH_2O\overset{\displaystyle O}{\overset{\displaystyle \|}{P}}(SCHCH_2CH_3)_2$$
$$\overset{CH_3}{\underset{}{|}}$$

NOMENCLATURE
Common name cadusafos (BSI, draft E-ISO)
IUPAC name *S,S*-di-*sec*-butyl *O*-ethyl phosphorodithioate

Chemical Abstracts name O-ethyl S,S-bis(1-methylpropyl) phosphorodithioate
Other names [ebufos] (rejected common name proposal)
CAS RN [95465–99–9] **Development codes** FMC 67 825

PHYSICAL CHEMISTRY
Mol. wt. 270.4 **M.f.** $C_{10}H_{23}O_2PS_2$ **Form** Colourless to yellow liquid.
B.p. 112–114 °C/0.8 mmHg **V.p.** 1.2×10^2 mPa (25 °C) **K$_{ow}$** logP = 3.9
S.g./density 1.054 (20 °C) **Solubility** In water 248 mg/l. Completely miscible with
acetone, acetonitrile, dichloromethane, ethyl acetate, toluene, methanol,
isopropanol, and heptane. **Stability** Stable up to 50 °C. Half-life in light <115 d.
F.p. 129.4 °C (Seta closed cup)

COMMERCIALISATION
History Nematicide and insecticide discovered by FMC Corp.
Manufacturers FMC

APPLICATIONS
Biochemistry Cholinesterase inhibitor. **Mode of action** Contact and stomach
action. **Uses** Controls various nematodes and larvae of Noctuidae, *Agriotes* spp.
and other soil insects in bananas, citrus, maize, potatoes, sugar cane, tobacco and
vegetables, at 3–10 kg/ha. **Formulation types** GR; EW.
Selected tradenames 'Apache' (FMC); 'Rugby' (FMC)

ANALYSIS
Product analysis by glc. **Residues** determined by glc with FID. Details available
from FMC Corp.

MAMMALIAN TOXICOLOGY
Reviews *Pesticide residues in food – 1991. FAO Plant Production and Protection
Paper, 111, 1991. Pesticide residues in food – 1991 evaluations. Part II – Toxicology.*
World Health Organisation, WHO/PCS/92.52, 1992. **EHC** 63 (WHO, 1986;
a general review of organophosphorus insecticides). **Oral** Acute oral LD$_{50}$ for
rats 37.1, mice 71.4 mg tech./kg. **Skin and eye** Acute percutaneous LD$_{50}$ for
male rabbits 24.4, female rabbits 41.8 mg/kg. Non-irritating to skin and practically
non-irritating to eyes of rabbits. **Inhalation** LC$_{50}$ (4 h) for rats 0.026 mg/l air.
NOEL (2 y) for rats 1 mg/kg diet; (1 y) for male dogs 0.001, female dogs
0.005 mg/kg daily. In oncogenicity tests (2 y), male mice 0.5, female mice 1 mg/kg
diet. **ADI** (JMPR) 0.0003 mg/kg [1991]. **Toxicity class** WHO (a.i.) Ib;
EPA (formulation) III

ECOTOXICOLOGY
Birds Acute oral LD$_{50}$ for bobwhite quail 16, mallard ducks 230 mg/kg. **Fish** LC$_{50}$
(96 h) for rainbow trout 0.13, bluegill sunfish 0.17 mg/l. **Worms** LC$_{50}$ (14 d) for
Eisenia foetida 72 mg/kg. **Daphnia** LC$_{50}$ (48 h) 1.6 μg/l. **Algae** EC$_{50}$ (96 h) 5.3
mg/l.

ENVIRONMENTAL FATE

Animals Readily absorbed, metabolised and eliminated in urine and faeces. Hydroxy sulfones were the major metabolites, followed by phosphorothioic and sulfonic acids. **Soil/Environment** DT_{50} in silty clay, sandy loam soils 11–55 d. K_{oc} 144–351.

cafenstrole   *Herbicide*

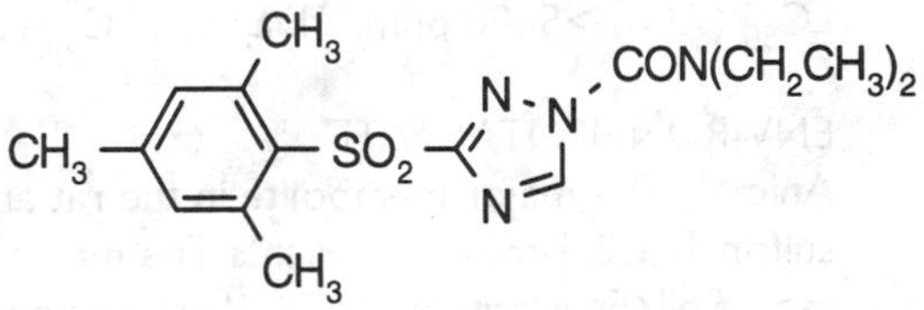

NOMENCLATURE

Common name cafenstrole (pa ISO)
IUPAC name *N,N*-diethyl-3-mesitylsulfonyl-1*H*-1,2,4-triazole-1-carboxamide
Chemical Abstracts name *N,N*-diethyl-3-[(2,4,6-trimethylphenyl)sulfonyl]-1*H*-=
1,2,4-triazole-1-carboxamide
CAS RN *[125306–83–4]*   **Development codes** CH-900 (Chugai)

PHYSICAL CHEMISTRY

Mol. wt. 350.5   **M.f.** $C_{16}H_{22}N_4O_3S$ **Form** Colourless crystals.   **M.p.** 114–116 °C
V.p. 2.99×10^{-6} mPa (20 °C)   **K_{ow}** logP = 3.21   **S.g./density** 1.30 (25 °C)
Solubility In water 2.5 mg/l (20 °C).   **Stability** Stable in neutral and weakly acidic media. Relatively stable to heat.

COMMERCIALISATION

History Discovered in 1987; reported by Chugai Pharmaceutical Co., Ltd (agrochemical interests now Eikou Kasei Co., Ltd), (M. Kanzaki et al., *Proc. Br. Crop Prot. Conf. – Weeds*, III, 923 (1991) (and refs. therein)). Introduced in Japan in 1997 by Eikou Kasei Co., Ltd.   **Patent** JP 1858798; US 5147445; EP 0332133
Manufacturers Eikou Kasei

APPLICATIONS

Uses Pre- and early post-emergence control of *Echinochloa oryzicola*, *Cyperus difformis* and other annual weeds in paddy rice.   **Formulation types** GR; WP; SC; WG.   **Mixtures** (cafenstrole +) bensulfuron-methyl + daimuron;

pyrazosulfuron-ethyl; imazosulfuron + daimuron; pyrazosulfuron-ethyl + cyhalofop-butyl; azimsulfuron + bensulfuron-methyl.

ANALYSIS
By hplc with u.v. detection.

MAMMALIAN TOXICOLOGY
Oral Acute oral LD_{50} for rats and mice >5000 mg/kg. **Skin and eye** Acute percutaneous LD_{50} for rats >2000 mg/kg. **Inhalation** LC_{50} (14 d) for rats >1.97 g/m^3. **ADI** 0.003 mg/kg. **Other** Non-mutagenic in the Ames test.

ECOTOXICOLOGY
Birds Acute oral LD_{50} for quail and mallard duck >2000 mg/kg. **Fish** LC_{50} (48 h) for carp >1.2 mg/l. **Bees** Acute oral LC_{50} (72 h) for bees >1000 ppm; contact LC_{50} (72 h) >5000 ppm. **Daphnia** LC_{50} (3 h) >500 mg/kg.

ENVIRONMENTAL FATE
Animals The major metabolite in the rat and dog is 3-(2,4,6-trimethylphenyl= sulfonyl)-1,2,4-triazole. **Plants** The major metabolite in rice is the same as in the rat. **Soil/Environment** DT_{50} (Japanese paddy) c. 7 d, (Japanese upland) c. 8 d.

103 calcium polysulfide

Fungicide, insecticide, acaricide

$$CaS_x$$

NOMENCLATURE
Common name lime sulfur (ESA, JMAF (which applies to an aqueous solution of calcium polysulfides)); calcium polysulfide (E-ISO, in lieu of a common name); polysulfure de calcium (F-ISO)
IUPAC name calcium polysulfide
Chemical Abstracts name calcium polysulfide
Other names Eau Grison (after Grison who used it as a fungicide in 1852).
CAS RN [1344–81–6] **EEC no.** 215–709–2

PHYSICAL CHEMISTRY
Composition It is an aqueous solution of calcium polysulfides, with traces of

calcium thiosulfate. **M.f.** CaS$_x$ **Form** Deep-orange liquid with unpleasant smell of hydrogen sulfide. **S.g./density** >1.28 (15.6 °C) **Solubility** Soluble in water. **Stability** Decomposed by carbon dioxide, acids, and soluble salts of metals which form insoluble sulfides; sulfur, hydrogen sulfide and the metal sulfide being formed.

COMMERCIALISATION
History Introduced in the 19th Century as a fungicide and for the control of scale insects.

APPLICATIONS
Mode of action Fungicide which acts directly and also by decomposition to elemental sulfur, which acts as a protective fungicide. Insecticide which acts by softening the wax of scale insects (G. D. Schafer, *Mich. Agric. Exp. Stn. Tech. Bull.*, 1911, No. 11; 1915, No. 21). **Uses** Control of powdery mildews, anthracnose, scab and other diseases on alfalfa, beans, clover, and fruit. Control of scale insects and spider mite eggs on fruit trees. **Phytotoxicity** Phytotoxic to sulfur-sensitive plants (apricots, raspberries, cucurbits, etc.). **Formulation types** Liquid. **Compatibility** Incompatible with many other pesticides. **Selected tradenames** 'Nevikén' (Chemol)

ANALYSIS
Product analysis for content of total polysulfide-sulfur, sulfide sulfur, thiosulfate, sulfate and calcium (*AOAC Methods*, 920.31–920.34; *MAFF Ref. Book (Tech. Bull.)*, 1958, No. 1; *CIPAC Handbook*, 1980, **1A,** 1279; J. R. Gray, *J. Sci. Food Agric.*, 1956, **7,** 3; FAO Specification (CP/58)).

MAMMALIAN TOXICOLOGY
Skin and eye Acute percutaneous LD$_{50}$ not available. Causes eye damage and skin irritation. **Toxicity class** EPA (formulation) I **EC risk** (R31); Xi (R36/37/38)

04 captafol   *Fungicide*

N-trihalomethylthio

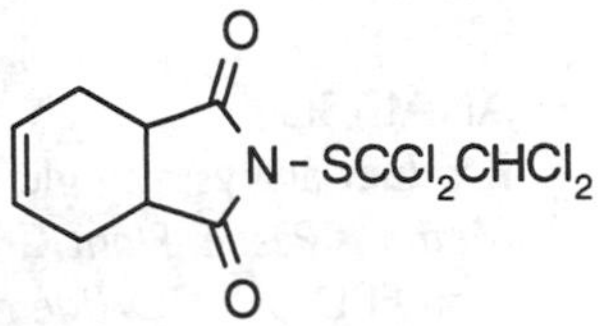

NOMENCLATURE
Common name captafol (BSI, E-ISO, (*m*) F-ISO, ANSI); difolatan (JMAF)

IUPAC name *N*-(1,1,2,2-tetrachloroethylthio)cyclohex-4-ene-1,2-dicarboximide; 3a,4,7,7a-tetrahydro-*N*-(1,1,2,2-tetrachloroethanesulfenyl)phthalimide
Chemical Abstracts name 3a,4,7,7a-tetrahydro-2-[(1,1,2,2-tetrachloroethyl)thio]-= 1*H*-isoindole-1,3(2*H*)-dione
CAS RN *[2425–06–1]*   **EEC no.** 219–363–3   **Development codes** Ortho-5865

PHYSICAL CHEMISTRY
Composition Tech. is ≥95%.   **Mol. wt.** 349.1   **M.f.** $C_{10}H_9Cl_4NO_2S$
Form Colourless to pale yellow crystals. The tech. grade is a light tan powder, with a characteristic odour.   **M.p.** 160–161 °C   **V.p.** Negligible at room temperature. K_{ow} logP = 3.8   **Solubility** In water 1.4 mg/l (20 °C). Slightly soluble in most organic solvents. In isopropanol 13, benzene 25, toluene 17, xylene 100, acetone 43, methyl ethyl ketone 44, dimethyl sulfoxide 170 (all in g/kg).   **Stability** Slowly hydrolysed in aqueous emulsion or suspension. Rapidly hydrolysed in acidic and alkaline media. Decomposes slowly at the melting point.

COMMERCIALISATION
History Fungicide reported by W. D. Thomas *et al.* (*Phytopathology*, 1962, **52**, 754). Introduced by Chevron Chemical Co.   **Manufacturers** Rallis

APPLICATIONS
Mode of action Non-systemic fungicide with protective and curative action. Acts by inhibiting germination of spores.   **Uses** Control of scab of pome fruit; shot-hole of stone fruit; peach leaf curl; downy mildew and black rot of vines; early and late blights of potatoes; *Alternaria* and mildew of carrots; celery leaf spot; *Septoria* of wheat; *Rhynchosporium* of barley; and various diseases of tomatoes, coffee, peanuts, citrus fruit, pineapples, macadamia nuts, onions, cucurbits, maize, sorghum, etc. Used as a seed treatment for control of *Pythium* and *Phoma* spp. and other emergence diseases of beet, cotton, peanuts, and rice. Also used as a wound protectant for grafting and pruning wounds and cankers on trees; and in the timber industry as a wood preservative.   **Phytotoxicity** Under certain conditions, grapes, apples, citrus fruit and roses may be injured.
Formulation types SC; WP; DP; LA; DS.   **Mixtures** *(captafol +)* propiconazole; pyrazophos; mancozeb + sulfur; oxadixyl; cyhalothrin + oxadixyl; copper oxychloride + cymoxanil; carbendazim + triadimenol.   **Compatibility** Incompatible with highly alkaline materials.   **Selected tradenames** 'Difoltan' (Rallis); 'Foltaf' (Rallis)

ANALYSIS
Product analysis by glc with FID or by hplc (A. A. Carlstrom & J. B. Leary, *Anal. Methods Pestic. Plant Growth Regul.*, 1978, **10**, 173).   **Residues** determined by glc with FPD or ECD (*idem, ibid.*; *Pestic. Anal. Man.*, 1979, **I**, 201-I; R. Mestres *et al.*, *Trav. Soc. Pharm. Montpellier*, 1979, **39**, 323; A. Ambrus *et al.*, *J. Assoc. Off. Anal. Chem.*, 1981, **64**, 733; M. A. Luke *et al.*, *ibid.*, p. 1187).

MAMMALIAN TOXICOLOGY

Reviews *Pesticide residues in food – 1985*. FAO Plant Production and Protection Paper 68, 1986. **IARC** 53 **Oral** Acute oral LD_{50} for rats 5000–6200 mg a.i./kg. **Skin and eye** Acute percutaneous LD_{50} for rabbits >15 400 mg/kg. Corrosive to eyes, mild skin irritant (rabbits). May cause allergic skin reactions in some people. **Inhalation** LC_{50} for male rats >0.72, female rats 0.87 mg/l (tech.). Dust may cause respiratory irritation. **NOEL** In 2 y feeding trials, no ill-effects observed in rats receiving 500 mg/kg diet and in dogs receiving 10 mg/kg daily. **ADI** (JMPR) Temporary ADI withdrawn [1985]. **Other** Carcinogenic in both rats and mice. **Toxicity class** WHO (a.i.) Ia; EPA (formulation) IV **EC risk** (R45); (R43) **PIC** Yes

ECOTOXICOLOGY

Birds Ten-day dietary LC_{50} for pheasants >23 070, mallard ducks >101 700 mg/kg diet. **Fish** LC_{50} (96 h) for rainbow trout 0.5, goldfish 3.0, bluegill sunfish 0.15 mg/l. **Bees** Not toxic to bees. **Daphnia** LC_{50} (96 h) 3.34 ppm. **Other aquatic spp.** Moderately to very highly toxic to freshwater invertebrates.

ENVIRONMENTAL FATE

Animals In mammals, following oral administration, captafol is hydrolysed to tetrahydrophthalimide (THPI) and dichloroacetic acid. THPI is degraded to tetrahydrophthalimidic acid and further to phthalic acid and ammonia. **Plants** In plants, metabolism is the same as in animals.

05 captan *Fungicide*

N-trihalomethylthio

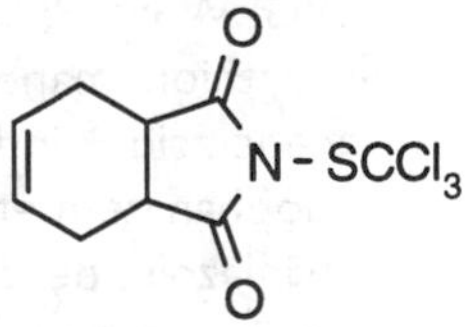

NOMENCLATURE

Common name captan (BSI, E-ISO, JMAF); captane ((m) F-ISO); captab (Republic of South Africa)
IUPAC name N-(trichloromethylthio)cyclohex-4-ene-1,2-dicarboximide
Chemical Abstracts name 3a,4,7,7a-tetrahydro-2-[(trichloromethyl)thio]-= 1H-isoindole-1,3(2H)-dione
CAS RN [133–06–2] **EEC no.** 205–087–0 **Development codes** SR 406
Official codes ENT 26 538

PHYSICAL CHEMISTRY
Composition Tech. is 90–95% pure. **Mol. wt.** 300.6 **M.f.** $C_9H_8Cl_3NO_2S$
Form Colourless crystals. Tech. is a colourless to beige amorphous solid, with a pungent odour. **M.p.** 178 °C (tech., 175–178°C) **V.p.** <1.3 mPa (25 °C)
K_{ow} logP = 2.8 (25 °C) **S.g./density** 1.74 (26 °C) **Solubility** In water 3.3 mg/l
(25 °C). In xylene 20, chloroform 70, acetone 21, cyclohexanone 23, dioxane 47, benzene 21, toluene 6.9, isopropanol 1.7, ethanol 2.9, diethyl ether 2.5
(all in g/kg, 26 °C). Insoluble in petroleum oils. **Stability** Hydrolysed slowly at neutral pH, rapidly in alkali. Thermal DT_{50} >4 y (80 °C), 14.2 d (120 °C).

COMMERCIALISATION
History Fungicide reported by A. R. Kittleston (*Science*, 1952, **115**, 84). Introduced by Standard Oil Development Co., later by Chevron Chemical Co.
Patent US 2553770; US 2553771; US 2553776 **Manufacturers** Crystal; Drexel; Makhteshim-Agan; Rallis; Tomen

APPLICATIONS
Mode of action Fungicide with protective and curative action. **Uses** Control of a wide range of fungal diseases, e.g. scab, *Gloeosporium* and other rots of pome fruit; shot-hole of stone fruit; peach leaf curl; brown rots of cherries, apricots, peaches, plums, and citrus fruit; downy mildew and black rot of vines; early and late blights of potatoes and tomatoes; *Alternaria* blight and leaf spot of carrots; anthracnose and downy mildew of cucurbits; leaf spot diseases of ornamentals; anthracnose and leaf spot diseases of tomatoes; brown patch on turf; *Botrytis* spp. on many crops; etc. Used on a large number of other crops. Used as a seed treatment or root dip for control of *Pythium*, *Phoma*, *Rhizoctonia* spp., etc. on maize, ornamentals, vegetables, oilseed rape, and other crops. **Phytotoxicity** Non-phytotoxic if used as directed, but some varieties of apple (e.g. Red Delicious, Winesap) and pear (e.g. D'Anjou, Bosc) may be injured, as also may lettuce seeds and, at higher dosages, celery and tomato seeds. **Formulation types** DS; WP; WG; WS; DP; Seed treatment. **Mixtures** (*captan* +) lindane; bromophos; malathion; maneb; carboxin; dicloran; 2-(1-naphthyl)acetic acid; 2-(1-naphthyl)= acetic acid + indol-3-ylbutyric acid; captafol + folpet; carbendazim; metiram; thiophanate-methyl; zineb; dinocap + zineb; mancozeb; ditalimfos; triadimefon; etridiazole; penconazole; fosetyl-aluminium; fosetyl-aluminium + thiabendazole; diazinon + lindane; hexaconazole; bronopol; benomyl; maneb + zineb; pyrifenox; anthraquinone + flutriafol; pencycuron. **Compatibility** Incompatible with alkaline materials, oil sprays, TEPP, and emulsifiable concentrate formulations of parathion. **Selected tradenames** 'Captaf' (Rallis); 'Dhanutan' (Dhanuka); 'Merpan' (Makhteshim-Agan); 'Phytocape' (Bayer)

ANALYSIS
Product analysis by glc (*CIPAC Handbook*, 1985, **1C**, p. 2013), by hplc (*AOAC Methods*, 1995, 980.06; *CIPAC Handbook*, 1985, **1C**, p. 2013), by i.r. spectrometry

or by total chlorine content after alkaline hydrolysis (*ibid.*, 1970, **1**, 171).
Residues determined by glc (A. Ambrus *et al.*, *J. Assoc. Off. Anal. Chem.*, 1981, **64**, 733; M. A. Luke, *et al.*, *ibid.*, p. 1187; J. N. Ospenson, *Anal. Methods Pestic.*, *Plant Growth Regul. Food Addit.*, 1964, **3**, 7; *Anal. Methods Pestic. Plant Growth Regul.*, 1972, **6**, 546) or by spectrophotometry of a derivative (*AOAC Methods*, 1995, 957.14).

MAMMALIAN TOXICOLOGY
Reviews *Pesticide residues in food – 1995*, FAO Plant Production and Protection Paper 133, 1996. *Pesticide residues in food – 1995 evaluations; Part II – Toxicology & Environment.* World Health Organisation, WHO/PCS/96.48, 1996.   **IARC** 30
Oral Acute oral LD_{50} for rats 9000 mg/kg.   **Skin and eye** Acute percutaneous LD_{50} for rabbits >4500 mg/kg. Corrosive to eyes, mild skin irritant (rabbits).
Inhalation LC_{50} for male rats >0.72, female rats 0.87 mg/l (tech.). Dust may cause respiratory irritation.   **NOEL** (2 y) for rats 2000, dogs 4000 mg/kg diet. No carcinogenic, teratogenic, or mutagenic effects observed.   **ADI** (JMPR) 0.1 mg/kg b.w. [1995].   **Toxicity class** WHO (a.i.) III (Table 5); EPA (formulation) IV
EC risk (R40); Xi (R36); (R43)

ECOTOXICOLOGY
Birds Acute oral LD_{50} for mallard ducks and pheasants >5000, bobwhite quail 2000–4000 mg/kg. Not toxic to starlings or red-winged blackbirds at 100 mg/kg.
Fish LC_{50} (96 h) for bluegill sunfish 0.072, harlequin fish 0.3, brook trout 0.034 mg/l.   **Bees** ED_{50} (oral) 91, (contact) 788 µg/bee.   **Daphnia** LC_{50} (48 h) 7–10 ppm.   **Other aquatic spp.** Moderately toxic to aquatic invertebrates.

ENVIRONMENTAL FATE
Animals In mammals, the three chlorine atoms are cleaved under the influence of cellular sulfhydryl compounds. Trithiocarbamate, thiophosgene, and tetrahydrophthalimide are formed (R. G. Owens, *Ann. New York Acad. Sci.*, 1969, **160**, 114–132).   **Soil/Environment** Soil K_d 3–8 (pH 4.5–7.2, o.m. 2.2–6.7%); DT_{50} c. 1 d (25 °C, pH 7.2, o.m. 1.2%).

Insecticide, plant growth regulator

carbamate

OCONHCH$_3$

NOMENCLATURE
Common name carbaryl (BSI, E-ISO, (*m*) F-ISO, ANSI, ESA, BAN); NAC (JMAF); [sevin] (former exception, USSR)
IUPAC name 1-naphthyl methylcarbamate
Chemical Abstracts name 1-naphthalenyl methylcarbamate
CAS RN *[63-25-2]* **EEC no.** 200-555-0 **Development codes** UC 7744
Official codes OMS 29; OMS 629; ENT 23 969

PHYSICAL CHEMISTRY
Composition Tech. grade is ≥99% pure. **Mol. wt.** 201.2 **M.f.** C$_{12}$H$_{11}$NO$_2$
Form Colourless to light tan crystals. **M.p.** 142 °C. **V.p.** 4.1 × 10^{-2} mPa
(23.5 °C) **K$_{ow}$** logP = 1.59 **S.g./density** 1.232 (20 °C) **Solubility** In water
120 mg/l (20 °C). Readily soluble in polar organic solvents. In dimethylformamide,
dimethyl sulfoxide 400–450, acetone 200–300, cyclohexanone 200–250,
isopropanol 100, xylene 100 (all in g/kg at 25 °C). **Stability** Stable under neutral
and weakly acidic conditions. Hydrolysed in alkaline media to 1-naphthol; DT$_{50}$
c. 12 d (pH 7), 3.2 h (pH 9). Stable to light and heat.

COMMERCIALISATION
History Insecticide reported by H. L. Haynes *et al.* (*Contrib. Boyce Thompson Inst.*,
1957, **18**, 507). Introduced by Union Carbide Corp. (now Rhône-Poulenc Ag Co.).
Patent US 2903478 **Manufacturers** Chunhu; Crystal; Drexel; Jin Hung; Kuo Ching;
Makhteshim-Agan; Rhône-Poulenc; Shenzhen Jiangshan

APPLICATIONS
Biochemistry Weak cholinesterase inhibitor. **Mode of action** Insecticide with
contact and stomach action, and slight systemic properties. **Uses** Control of
Lepidoptera, Coleoptera, and other chewing and sucking insects at 0.25–2.0 kg
a.i./ha on more than 120 different crops, including vegetables, tree fruit (including
citrus), mangoes, bananas, strawberries, nuts, vines, olives, okra, cucurbits,
peanuts, soya beans, cotton, rice, tobacco, cereals, beet, maize, sorghum, alfalfa,
potatoes, ornamentals, forestry, etc. Control of earthworms in turf. Used as a
growth regulator for fruit thinning of apples. Also used as an animal
ectoparasiticide. **Phytotoxicity** Non-phytotoxic if used as directed. Under certain
conditions, some varieties of apple and pear may be injured.
Formulation types TK; WP; DP; RB; SC; GR; OF. **Mixtures** (*carbaryl +*) rotenone;

propanil; tetradifon; diazinon; chlorfenson; lindane; sulfur; maneb; malathion; lindane + maneb; propaphos.   **Compatibility** Incompatible with alkaline materials such as Bordeaux mixture, lime, and lime sulfur.   **Selected tradenames** 'Sevin' (Rhône-Poulenc); 'Carbamec' (PBI/Gordon); 'Efaryl' (Efthymiadis); 'Karl' (Sanonda)

ANALYSIS

Product analysis by i.r. spectroscopy (*AOAC Methods*, 1995, 976.04; *CIPAC Handbook*, 1970, **1**, 185; 1980, **1A**, 1113; FAO Specification (CP/55)) or by hplc (G. W. Sheehan, *Anal. Methods Pestic. Plant Growth Regul.*, 1984, **13**, 157).
Residues determined by glc (*ibid.*, 1972, **6**, 478; *Man. Pestic. Residue Anal.*, 1987, **I**, 6; *Anal. Methods Residues Pestic.*, 1988, Part I, M2, M13; *AOAC Methods*, 1995, 975.40; A. Ambrus *et al.*, *J. Assoc. Off. Anal. Chem.*, 1981, **64**, 733), by rplc (*AOAC Methods*, 1995, 985.23), by colorimetry (*ibid.*, 964.18) or by tlc (*ibid.*, 968.26). For methods in **drinking water**, see *AOAC Methods*, 1995, 991.06.

MAMMALIAN TOXICOLOGY

Reviews *Pesticide residues in food*. FAO Agricultural Studies, No. 92; WHO Technical Report Series, No. 545, 1974. *1973 Evaluations of some pesticide residues in food*. FAO/AGP/1974/M/11; WHO Pesticide Residues Series, No. 4, 1975.   **IARC** 12   **EHC** 153 (WHO, 1994); 64 (WHO, 1986; a review of carbamate insecticides in general).   **Oral** Acute oral LD_{50} for male rats 850, female rats 500, rabbits 710 mg/kg.   **Skin and eye** Acute percutaneous LD_{50} for rats >4000, rabbits >2000 mg/kg. Non-irritating to skin and eyes (rabbits).
Inhalation LC_{50} for rats >206.1 mg/l air.   **NOEL** (2 y) for rats 200 mg/kg diet.
ADI (JMPR) 0.01 mg/kg b.w. [1973].   **Toxicity class** WHO (a.i.) II; EPA (formulation) I ('Tercyl' 85WP), II ('Sevin' 80S), III   **EC risk** Xn (R22)

ECOTOXICOLOGY

Birds Acute oral LD_{50} for young mallard ducks >2179, young pheasants >2000, Japanese quail 2230, pigeons 1000–3000 mg/kg.   **Fish** LC_{50} (96 h) for rainbow trout 1.3, sheepshead minnow 2.2, bluegill sunfish 10 mg/l.   **Bees** Toxic to bees; LD_{50} (topical) 1 µg/bee.   **Other beneficial spp.** Toxic to beneficial insects.
Daphnia LC_{50} (48 h) 0.006 mg/l.   **Other aquatic spp.** LC_{50} (96 h) for mysid shrimp (*Mysidopsis bahia*) 5.7 mg/l; LC_{50} (48 h) for Eastern oyster (*Crassostrea virginica*) 2.7 mg/l.

ENVIRONMENTAL FATE

Animals In mammals, carbaryl does not accumulate in body tissues, but is rapidly metabolised to non-toxic substances, particularly 1-naphthol. This, together with the glucuronic acid conjugate, is eliminated predominantly in the urine and faeces. Metabolism of carbamate insecticides is reviewed (M. Cool & C. K. Jankowski in "*Insecticides*").   **Plants** Metabolites are 4-hydroxycarbaryl, 5-hydroxycarbaryl and methylol-carbaryl.   **Soil/Environment** Under aerobic conditions, carbaryl at 1 ppm degraded with DT_{50} 7–14 d in a sandy loam and 14–28 d in a clay loam.

107 carbendazim   *Fungicide*

benzimidazole

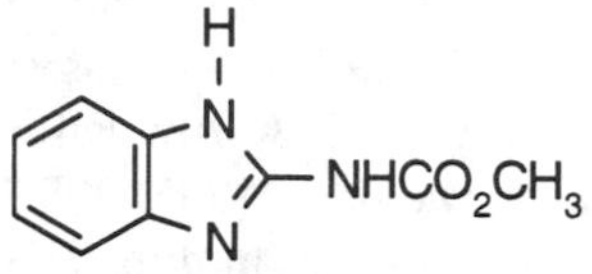

NOMENCLATURE
Common name carbendazim (BSI, E-ISO); carbendazime ((*f*) F-ISO); carbendazol (JMAF)
IUPAC name methyl benzimidazol-2-ylcarbamate
Chemical Abstracts name methyl 1*H*-benzimidazol-2-ylcarbamate
Other names MBC; BMC   **CAS RN** *[10605–21–7]*   **EEC no.** 234–232–0
Development codes BAS 346F (BASF); Hoe 017411 (Hoechst); DPX-E 965 (Du Pont)

PHYSICAL CHEMISTRY
Mol. wt. 191.2   **M.f.** $C_9H_9N_3O_2$ **Form** Crystalline powder.   **M.p.** 302–307 °C (decomp.)   **V.p.** 0.09 mPa (20 °C), 0.15 mPa (25 °C), 1.3 mPa (50 °C); separate study gives <0.0001 mPa (20 °C).   **K_{ow}** logP = 1.38 (pH 5), 1.51 (pH 7), 1.49 (pH 9)   **Henry** 3.6×10^{-3} Pa m^3 mol^{-1} (calc.)   **S.g./density** 1.45 (20 °C)
Solubility In water at 24 °C, 29 mg/l (pH 4), 8 mg/l (pH 7), 7 mg/l (pH 8). In dimethylformamide 5, acetone 0.3, ethanol 0.3, chloroform 0.1, ethyl acetate 0.135, dichloromethane 0.068, benzene 0.036, cyclohexane <0.01, diethyl ether <0.01, hexane 0.0005 (all in g/l, 24 °C).   **Stability** Decomposes at m.p.; stable for at least 2 y below 50 °C. Stable after 7 d at 20 000 lux. Slowly decomposed in alkaline solution (22 °C); DT_{50} >350 d (pH 5 and pH 7), 124 d (pH 9). Stable in acids, forming water-soluble salts.   **pKa** 4.2, weak base

COMMERCIALISATION
History Fungicide reported by H. Hampel & F. Löcher (*Proc. Br. Insectic. Fungic. Conf.*, 1973, **1**, 127, 301). Introduced (1974) by BASF AG, Hoechst AG (now AgrEvo GmbH) and E. I. du Pont de Nemours and Co.   **Patent** US 3657443; GB 1190614 to Du Pont   **Manufacturers** AgrEvo; Aimco; Atul; BASF; Du Pont; Gilmore; High Kite; Inquinosa; Jiangsu; New Chemi; Pilarquim; Sinon; Sundat

APPLICATIONS

Biochemistry Reported to inhibit beta-tubulin synthesis. **Mode of action** Systemic fungicide with protective and curative action. Absorbed through the roots and green tissues, with translocation acropetally. Acts by inhibiting development of the germ tubes, the formation of appressoria, and the growth of mycelia.

Uses Control of *Septoria, Fusarium, Erysiphe* and *Pseudocercosporella* in cereals, *Sclerotinia, Alternaria* and *Cylindrosporium* in oilseed rape, *Cercospora* and *Erysiphe* in sugar beet, *Uncinula* and *Botrytis* in grapes, *Cladosporium* and *Botrytis* in tomatoes, *Venturia* and *Podosphaera* in pome fruit and *Monilia* and *Sclerotinia* in stone fruit; application rates vary from 120–600 g/ha depending on crop. A seed treatment (0.6–0.8 g/kg) will control *Tilletia, Ustilago, Fusarium* and *Septoria* in cereals and *Rhizoctonia* in cotton. Also shows activity against storage diseases of fruit as a dip (0.3–0.5 g/l). **Formulation types** SC; WP; SL; OP; WG; Seed treatment. **Mixtures** *(carbendazim +)* propiconazole; triadimefon; chlorothalonil; maneb; mancozeb; imazalil; chlormequat chloride; fluquinconazole; flusilazole; maneb + tridemorph; maneb + sulfur; captan; sulfur; prochloraz; chlorothalonil + fenpropimorph; flutriafol; mancozeb + maneb; metalaxyl; fenarimol + maneb; cyproconazole; iprodione; fenpropimorph + mancozeb; tebuconazole; vinclozolin; diethofencarb; diniconazole + iprodione; fenbuconazole + prochloraz. **Compatibility** Incompatible with alkaline materials. **Selected tradenames** 'Bavistin' (BASF); 'Delsene' (Du Pont); 'Derosal' (AgrEvo); 'Troika' (mixture) (AgrEvo); 'Addstem' (Headland); 'Aimcozim' (Aimco); 'Arrest' (Searle India); 'Bencarb' (Productos OSA); 'Carbate' (PBI); 'Carezim' (Efthymiadis); 'Cekudazim' (Cequisa); 'Dhanustin' (Dhanuka); 'Fungy' (Ramcides); 'Hinge' (Quadrangle); 'Occidor' (Agriphar); 'Sabendazim' (Sanonda); 'Volzim' (Voltas)

ANALYSIS

Product analysis by titration against perchloric acid in acetic acid or by u.v. spectrophotometry. **Residues** in crops determined using methods for benomyl, hplc (J. J. Kirkland *et al., J. Agric. Food Chem.*, 1973, **21**, 368; *Pestic. Anal. Man.*, 1979, **II**; J. E. Farrow *et al., Analyst (London)*, 1977, **102**, 752) or fluorimetry or colorimetry of derivatives (H. L. Pease & J. A. Gardiner, *J. Agric. Food Chem.*, 1969, **17**, 267; N. Aharonson & A. Ben-Aziz, *J. Assoc. Off. Anal. Chem.*, 1973, **56**, 1330).

MAMMALIAN TOXICOLOGY

Reviews *Pesticide residues in food – 1995*, FAO Plant Production and Protection Paper 133, 1996. *Pesticide residues in food – 1995 evaluations; Part II – Toxicology & Environment.* World Health Organisation, WHO/PCS/96.48, 1996. **EHC** 149 (WHO, 1993). **Oral** Acute oral LD_{50} for rats >15 000, dogs >2500 mg/kg. **Skin and eye** Acute percutaneous LD_{50} for rabbits >10 000, rats >2000 mg/kg. Non-irritating to skin and eyes (rabbits). Not a skin sensitiser (guinea pigs). **Inhalation** LC_{50} (4 h) for rats, rabbits, guinea pigs or cats, no effect with suspension (10 g/l water). **NOEL** (2 y) for dogs 300 mg/kg diet, corresponding to

6–7 mg/kg b.w. **ADI** (JMPR) 0.03 mg/kg b.w. [1995]. **Other** Acute i.p. LD_{50} for male rats 7320, female rats 15 000 mg/kg. **Toxicity class** WHO (a.i.) III (Table 5) **EC risk** Xn (R40)

ECOTOXICOLOGY
Birds Acute oral LD_{50} for quail 5826–15 595 mg/kg. **Fish** LC_{50} (96 h) for carp 0.61, rainbow trout 0.83, bluegill sunfish >17.25, guppy >8 mg/l. **Bees** LD_{50} (contact) >50 μg/bee. **Worms** LC_{50} (4 w) for *Eisenia foetida* 6 mg/kg soil. **Daphnia** LC_{50} (48 h) 0.13–0.22 mg/l. **Algae** EC_{50} (72 h) for *Scenedesmus subspicatus* 419, *Selenastrum capricornutum* 1.3 mg/l.

ENVIRONMENTAL FATE
EHC 149 concludes that, although highly toxic to aquatic organisms, low bioavailability in surface waters makes it unlikely this toxicity will occur in the field. **Animals** In male rats, following a single oral administration of 3 mg/kg, 66% was eliminated in the urine within 6 hours. **Plants** Readily absorbed by plants. One degradation product is 2-aminobenzimidazole.
Soil/Environment 2-Aminobenzimidazole has been found as a minor metabolite. DT_{50} in soil 8–32 d under outdoor conditions. Carbendazim decomposes in the environment, DT_{50} 6 to 12 mo on bare soil, 3 to 6 mo on turf, and 2 and 25 mo in water under aerobic and anaerobic conditions, respectively. It is mainly decomposed by micro-organisms. K_{oc} 200–250.

108 carbetamide   *Herbicide*

carbamate

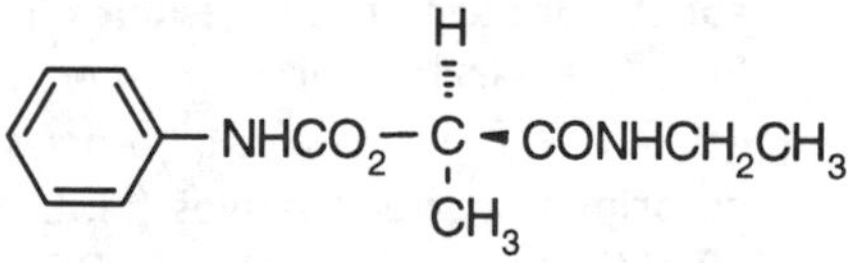

NOMENCLATURE
Common name carbetamide (BSI, E-ISO, (*m*) F-ISO, ANSI, WSSA); no name (Germany)
IUPAC name (*R*)-1-(ethylcarbamoyl)ethyl carbanilate
Chemical Abstracts name (*R*)-*N*-ethyl-2-[[(phenylamino)carbonyl]oxy]= propanamide
CAS RN *[16118–49–3]* **EEC no.** 240–286–6 **Development codes** 11 561 RP

PHYSICAL CHEMISTRY
Mol. wt. 236.3 **M.f.** $C_{12}H_{16}N_2O_3$ **Form** Colourless crystals. **M.p.** 119 °C;

(tech., >110 °C).   **V.p.** Negligible (20 °C)   **Solubility** In water *c.* 3.5 g/l (20 °C).
In acetone 900, dimethylformamide 1500, ethanol 850, methanol 1400,
cyclohexane 0.3 (all in g/l).   **Stability** Stable under normal storage conditions.

COMMERCIALISATION
History Herbicide reported by J. Desmoras *et al.* (*C. R. Journ. Etud. Herbic. Conf.
COLUMA, 2nd,* 1963, p. 14). Introduced by Rhône-Poulenc Agrochimie.
Patent GB 959204; BE 597035; US 3177061   **Manufacturers** Rhône-Poulenc

APPLICATIONS
Biochemistry Cell division inhibitor.   **Mode of action** Selective herbicide, absorbed
principally by the roots, and also by the leaves.   **Uses** Control of annual grasses
(including volunteer cereals) and some broad-leaved weeds at 2 kg/ha, in clover,
alfalfa, sainfoin, brassicas, field beans, peas, lentils, sugar beet, oilseed rape,
chicory, endive, sunflowers, caraway, strawberries, vines, and fruit orchards.
Formulation types WP; EC.   **Mixtures** *(carbetamide +)* dimefuron; oxadiazon.
Selected tradenames 'Legurame' (Rhône-Poulenc)

ANALYSIS
Product analysis by hplc (*CIPAC Handbook,* 1992, **E**, 28) or by titration of the
ethylamine liberated on hydrolysis (J. Desmoras *et al., Anal. Methods Pestic. Plant
Growth Regul.,* 1973, **7**, 509).   **Residues** determined by hydrolysis to aniline which
is measured by colorimetry of a derivative (*idem, ibid.*). Details of
chromatographic methods are available from Rhône-Poulenc Agrochimie.

MAMMALIAN TOXICOLOGY
Oral Acute oral LD_{50} for rats >2000, mice 1720, dogs 900 mg/kg.
Skin and eye Acute percutaneous LD_{50} for rabbits >500 mg/kg. Non-irritating to
eyes (rabbits).   **Inhalation** LC_{50} (4 h) for rats >0.13 mg/l air.   **NOEL** In 90 d
feeding trials, no effect observed with rats receiving 3200 mg/kg diet, or dogs
receiving 12 800 mg/kg diet.   **Toxicity class** WHO (a.i.) III (Table 5); EPA
(formulation) IV

ECOTOXICOLOGY
Birds Acute oral LD_{50} for bobwhite quail >2000 mg/kg.   **Fish** LC_{50} (96 h) for
rainbow trout and common carp >100 mg/l.   **Bees** Not hazardous to bees when
used as directed.   **Worms** LC_{50} 600 mg/kg soil.   **Daphnia** EC_{50} (48 h) 36.5 mg/l.

ENVIRONMENTAL FATE
Plants Rapidly metabolised, leaving no residues in the plant.   **Soil/Environment**
Microbially degraded in soil; DT_{50} *c.* 1 mo. Duration of activity at low
temperatures is *c.* 2–3 mo. K_d ranges from 0.10 (0.01% o.m., pH 6.6) to 7.92
(16.9% o.m., pH 6.8) (H. J. Pedersen *et al., Pestic. Sci.,* **44,** 131 (1995)).

109 carbofuran

Insecticide, nematicide

carbamate

NOMENCLATURE
Common name carbofuran (BSI, E-ISO, (*m*) F-ISO, ANSI, ESA)
IUPAC name 2,3-dihydro-2,2-dimethylbenzofuran-7-yl methylcarbamate
Chemical Abstracts name 2,3-dihydro-2,2-dimethyl-7-benzofuranyl
methylcarbamate
CAS RN *[1563–66–2]*; (carbofuran phenol *[1563–38–8]*; 3-ketocarbofuran
phenol *[17781–16–7]*) **EEC no.** 216–353–0 **Development codes** FMC 10 242;
BAY 70 143; D 1221 **Official codes** OMS 864; ENT 27 164

PHYSICAL CHEMISTRY
Mol. wt. 221.3 **M.f.** $C_{12}H_{15}NO_3$ **Form** Colourless crystals. **M.p.** 153–154 °C
(pure); 150–152 °C (tech.) **V.p.** 0.031 mPa (20 °C); 0.072 mPa (25 °C).
K_{ow} logP = 1.52 (20 °C) **S.g./density** 1.18 (20 °C) **Solubility** In water 320
(20 °C), 351 (25 °C) (both in mg/l). In dichloromethane >200, isopropanol
20–50, toluene 10–20 (all in g/l, 20 °C). **Stability** Unstable in alkaline media.
Stable in acidic and neutral media. Decomposes >150 °C. DT_{50} (22 °C) >>1 y
(pH 4), 121 d (pH 7), 31 h (pH 9).

COMMERCIALISATION
History Insecticide reported by F. L. McEwen & A. C. Davis (*J. Econ. Entomol.*,
1965, **58**, 369) and E. J. Armburst & G. C. Gyrisco (*ibid.*, p. 940). Introduced by
FMC Corp. and by Bayer AG. **Patent** US 3474170; US 3474171 both to FMC;
DE 1493646 to Bayer **Manufacturers** Bayer; Chunhu; FMC; Jin Hung; Kuo Ching;
Makhteshim-Agan; Mitsubishi Chemical; Pilarquim; Sanachem; Shenzhen Jiangshan;
Sinon; Sundat; Taiwan Tainan Giant

APPLICATIONS
Biochemistry Cholinesterase inhibitor. **Mode of action** Systemic, with
predominantly contact and stomach action. **Uses** Control of soil-dwelling and
foliar-feeding insects (including wireworms, white grubs, millipedes, symphylids, frit
flies, bean seed flies, root flies, flea beetles, weevils, sciarid flies, aphids, thrips,
etc.) and nematodes in vegetables, ornamentals, beet, maize, sorghum, sunflowers,
oilseed rape, potatoes, alfalfa, peanuts, soya beans, sugar cane, rice, cotton,
coffee, cucurbits, tobacco, lavender, citrus, vines, strawberries, bananas,
mushrooms, and other crops. **Formulation types** SC; GR; WP; FS.
Mixtures (*carbofuran +*) fenamiphos; isofenphos. **Compatibility** Incompatible with

alkaline materials. **Selected tradenames** 'Curaterr' (Bayer); 'Furadan' (FMC); 'Agrofuran' (Sanonda); 'Carbodan' (Makhteshim-Agan); 'Carbosip' (Sipcam); 'Cekufuran' (Cequisa); 'Chinufur' (Chinoin); 'Furacarb' (Aimco); 'Terrafuran' (Sanachem)

ANALYSIS
Product analysis by hplc (*CIPAC Handbook*, 1988, **D**, 20; *AOAC Methods*, 1990, 986.10) or by i.r. spectrometry. **Residues** determined by glc (*AOAC Methods*, 1995, 975.40; *Pestic. Anal. Man.*, 1979, **II**; *Man. Pestic. Residue Anal.*, 1987, **I**, 6; *Anal. Methods Residue Pestic.*, 1988, Part I, M113; R. F. Cooke, *Anal. Methods Pestic. Plant Growth Regul.*, 1973, **7**, 187; A. Ambrus et al., *J. Assoc. Off. Anal. Chem.*, 1981, **64**, 733) or by rplc (*AOAC Methods*, 1995, 985.23). Methods for determination of residues available from Bayer. For methods in **drinking water,** see *AOAC Methods*, 1995, 991.06, (includes its 3-hydroxy metabolite); for carbofuran phenol and 3-ketocarbofuran phenol, see also *AOAC Methods*, 1995, 992.14.

MAMMALIAN TOXICOLOGY
Reviews *Pesticide residues in food – 1982*. FAO Plant Production and Protection Paper 46, 1983. **EHC** 64 (WHO, 1986; a review of carbamate insecticides in general). **Oral** Acute oral LD_{50} for male and female rats c. 8, dogs 15, mice 14.4 mg/kg. **Skin and eye** Acute percutaneous LD_{50} for male and female rats >3000 mg/kg (24 h); mildly irritating to skin and eyes (rabbits). **Inhalation** LC_{50} (4 h) for male and female rats c. 0.075 mg/l air (aerosol). **NOEL** (2 y) for rats and mice 20 mg/kg diet; (1 y) for dogs 10 mg/kg diet. **ADI** (JMPR) 0.002 mg/kg b.w. [1996]. **Toxicity class** WHO (a.i.) Ib; EPA (formulation) I ('Furadan 4F'), II ('Furadan G') **EC risk** T+ (R26/28)

ECOTOXICOLOGY
Birds Acute oral LD_{50} for Japanese quail 2.5–5 mg/kg. LC_{50} for Japanese quail 60–240 mg (as GR5)/kg. Tech.: LC_{50} 0.7–8 mg/kg, depending on species.
Fish LC_{50} (96 h) for rainbow trout 22–29 mg (as GR5)/l, bluegill sunfish 1.75 mg (as GR3)/l, golden orfe 107–245 mg (as GR5)/l. Tech.; 7.3–362.5 μg/l, depending on species. **Bees** Toxic to bees (except for granular formulation). **Daphnia** LC_{50} (48 h) 38.6 μg/l.

ENVIRONMENTAL FATE
Animals Carbofuran is metabolised by hydrolytic and oxidative mechanisms in the rat. At 24 hours after treatment, 72% of the dose was eliminated in the urine, 2% in the faeces, and about 43% of the administered dose was hydrolysed. Over 95% of the material excreted in the urine was in the form of conjugated metabolites. The major metabolite was conjugated 3-ketocarbofuran phenol, while conjugated 3-hydroxycarbofuran was the predominant carbamate metabolite. Both metabolites were also present in the free form. Metabolism of carbamate insecticides is reviewed (M. Cool & C. K. Jankowski in "*Insecticides*").

Plants Carbofuran is quickly metabolised into 3-hydroxycarbofuran and ketocarbofuran. **Soil/Environment** DT_{50} in soil c. 30–60 d. Most important metabolite is CO_2 formed by microbiological degradation of the phenol compounds. $K_{oc} = 22$.

110 carbosulfan

Insecticide

carbamate

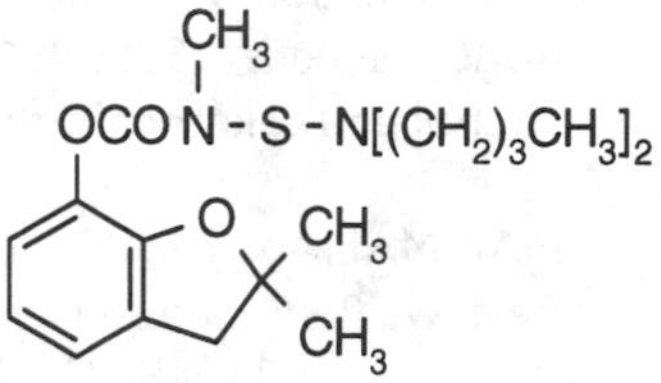

OCON - S - N[(CH$_2$)$_3$CH$_3$]$_2$
CH$_3$

CH$_3$
CH$_3$

NOMENCLATURE
Common name carbosulfan (BSI, ANSI, draft E-ISO, (m) draft F-ISO)
IUPAC name 2,3-dihydro-2,2-dimethylbenzofuran-7-yl (dibutylaminothio)=
methylcarbamate
Chemical Abstracts name 2,3-dihydro-2,2-dimethyl-7-benzofuranyl
[(dibutylamino)thio]methylcarbamate
CAS RN *[55285–14–8]* **Development codes** FMC 35 001
Official codes OMS 3022

PHYSICAL CHEMISTRY
Mol. wt. 380.5 **M.f.** $C_{20}H_{32}N_2O_3S$ **Form** Orange to brown clear viscous liquid.
B.p. 124–128 °C **V.p.** 0.041 mPa (25 °C) **S.g./density** 1.056 (20 °C)
Solubility In water 0.3 ppm (25 °C). Miscible with most organic solvents,
e.g. xylene, hexane, chloroform, dichloromethane, methanol, ethanol, acetone,
etc. **Stability** Hydrolysed in aqueous media; DT_{50} (25 °C) in pure water <1 h
(pH 4), 22 h (pH 6), 7.6 d (pH 7), 14.2 d (pH 8), >58.3 d (pH 9). **F.p.** 95 °C
(closed cup)

COMMERCIALISATION
History Insecticide reported by E. C. Maitlen & N. A. Sladen (*Proc. Br. Crop Prot. Conf.*, 1979, **2**, 557). Introduced by FMC Corp. **Manufacturers** FMC; Kuo Ching

APPLICATIONS
Biochemistry Cholinesterase inhibitor; activity is due to *in vivo* cleavage of the
N-S bond, resulting in conversion to carbofuran. **Mode of action** Systemic

insecticide with contact and stomach action.   **Uses** Control of a wide range of soil-dwelling and foliar insect pests. Examples of uses include control of millipedes, springtails, symphylids, wireworms, pygmy mangold beetles, frit flies, white grubs, aphids, caterpillars, flea beetles, Colorado beetles, stem borers, leafhoppers, planthoppers, codling moth, scales and free-living nematodes. The product is used in a wide range of crops, e.g. cotton, sugar beet, potato, rice, top fruit, citrus, maize, vegetables, sugar cane and coffee.   **Formulation types** GR; EC; WP; DP; UL; CS.   **Mixtures** *(carbosulfan +)* endosulfan; zeta-cypermethrin.   **Compatibility** Compatible with liquid fertilisers.   **Selected tradenames** 'Marshal' (FMC)

ANALYSIS
Product analysis by rp hplc (*CIPAC Handbook*, 1992, **E**, 35).   **Residues** determined by glc (B. Leppert et al., *J. Agric. Food Chem.*, 1983, **31**, 220; 1984, **32**, 1441) or by hplc. Details available from FMC Agricultural Chemicals Group.

MAMMALIAN TOXICOLOGY
Reviews *Pesticide residues in food – 1986*. FAO Plant Production and Protection Paper 77, 1986. *Pesticide residues in food – 1986 evaluations*. FAO Plant Production and Protection Paper 78/2, 1987.   **EHC** 64 (WHO, 1986; a review of carbamate insecticides in general).   **Oral** Acute oral LD_{50} for male rats 250, female rats 185 mg/kg.   **Skin and eye** Acute percutaneous LD_{50} for rats >2000 mg/kg. Slight eye irritant; moderate skin irritant.   **Inhalation** LC_{50} (1 h) for male rats 1.53, female rats 0.61 mg/l air.   **NOEL** (2 y) (oncogenic) for rats and mice 20 mg/kg diet.   **ADI** (JMPR) 0.01 mg/kg b.w. [1986].   **Toxicity class** WHO (a.i.) II; EPA (formulation) I (4 EC), II (2.5 EC)

ECOTOXICOLOGY
Birds Acute oral LD_{50} for mallard ducks 8.1, quail 82, pheasants 20 mg/kg.
Fish LC_{50} (96 h) for bluegill sunfish 0.015, trout 0.045 mg/l.   **Bees** Toxic to bees.
Daphnia LC_{50} (48 h) 1.5 μg/l.   **Algae** (96 h) 20 mg/l.

ENVIRONMENTAL FATE
Animals In rats, following oral administration, rapidly metabolised by hydrolysis, oxidation and conjugation, forming carbofuran methylol, carbofuran phenol, and their 3-hydroxy and 3-keto derivatives; the metabolites are rapidly excreted.
Plants Metabolites include carbofuran (*q.v.*) and 3-hydroxycarbofuran.
Soil/Environment In soil, rapidly degraded under both aerobic and anaerobic conditions, DT_{50} c. 2–5 d; DT_{90} <3–38 d; the principal metabolite is carbofuran (*q.v.*). Under field conditions, carbosulfan and carbofuran are unlikely to leach to groundwater.

carboxamide

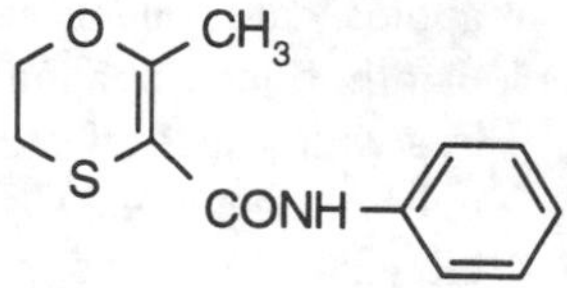

NOMENCLATURE
Common name carboxin (BSI, E-ISO, ANSI); carboxine ((f) F-ISO); carbathiin (Canada); no name (Denmark, Germany)
IUPAC name 5,6-dihydro-2-methyl-1,4-oxathi-ine-3-carboxanilide
Chemical Abstracts name 5,6-dihydro-2-methyl-*N*-phenyl-1,4-oxathiin-=3-carboxamide
CAS RN *[5234–68–4]*   **Development codes** D 735

PHYSICAL CHEMISTRY
Composition Tech. grade is >97% pure.   **Mol. wt.** 235.3   **M.f.** $C_{12}H_{13}NO_2S$
Form White crystals. Tech. is a white solid.   **M.p.** 91.5–92.5 °C and 98–100 °C depending on crystal structure   **V.p.** 0.025 mPa (25 °C)   **K$_{ow}$** logP = 2.2 (25 °C)
S.g./density 1.36   **Solubility** In water 199 mg/l (25 °C). In acetone 177, methylene chloride 353, methanol 88, ethyl acetate 93 mg/l (25 °C).   **Stability** Stable to hydrolysis (25 °C) at pH 5, pH 7 and pH 9. Aqueous solutions (pH 7) exposed to light DT$_{50}$ <3 h.   **pKa** <0.5

COMMERCIALISATION
History Fungicide reported by B. von Schmeling & M. Kulka (*Science*, 1966, **152,** 659). Introduced by Uniroyal Chemical Co., Inc.   **Patent** US 3249499; US 3393202; US 3454391   **Manufacturers** Jin Hung; Kemira FC; Sundat; Uniroyal

APPLICATIONS
Mode of action Systemic fungicide.   **Uses** Seed treatment for control of smuts and bunts (particularly loose smut) at 50–200 g a.i./100 kg seed, on barley, wheat, and oats; seedling diseases (particularly *Rhizoctonia* spp.) of barley, wheat, oats, rice, cotton, peanuts, soya beans, vegetables, maize, sorghum, and other crops; and fairy rings on turf. The dimorphic forms do not differ in fungicidal activity.
Formulation types WP; SC; FS; Seed treatment.   **Mixtures** (*carboxin* +) phenylmercury acetate; fenpiclonil + imazalil; lindane; captan; captan + maneb; thiram; imazalil; lindane + thiram; imazalil + thiabendazole; maneb; oxine-copper; thiabendazole; anthraquinone + oxine-copper; anthraquinone + lindane + oxine-copper.   **Compatibility** Not compatible with pesticides which are highly alkaline or acidic.   **Selected tradenames** 'Vitavax' (Uniroyal); 'Kemikar' (Kemira FC); 'Oxatin' (Diachem)

ANALYSIS
Product analysis by hplc or i.r. spectroscopy; details available from Uniroyal. **Residue** analysis by hydrolysis to aniline, which is determined by glc (H. R. Siskin & J. E. Newell, *ibid.*, 1971, **19**, 738; G. M. Stone, *Anal. Methods Pestic. Plant Growth Regul.*, 1976, **8**, 319). In **drinking water**, by gc with FID (*AOAC Methods*, 1995, 991.07).

MAMMALIAN TOXICOLOGY
Oral Acute oral LD_{50} for rats 3820 mg/kg. **Skin and eye** Acute dermal LD_{50} for rabbits >4000 mg/kg. Irritating to eyes (rabbits). **Inhalation** LC_{50} (1 h) for rats >20 mg/l air. **NOEL** In 2 y feeding trials, rats receiving 600 mg/kg diet showed no ill-effects. **ADI** 0.01 mg/kg. **Toxicity class** WHO (a.i.) III (Table 5); EPA (formulation) III

ECOTOXICOLOGY
Birds Eight-day dietary LC_{50} for mallard ducks >4640, bobwhite quail >10 000 mg/kg diet. **Fish** LC_{50} (96 h) for rainbow trout 2, bluegill sunfish 1.2 mg/l. **Bees** Not hazardous to bees when used as directed; LD_{50} >181 μg/bee. **Worms** LC_{50} (14 d) for earthworms 500–1000 ppm. **Daphnia** LC_{50} (48 h) 84.4 mg/l. **Algae** EC_{50} (96 h) for *Chlorella* 2.4 mg/l; (5 d) for *Selenastrum* 0.48 mg/l; (14 d) for *Lemna* 0.92 mg/l.

ENVIRONMENTAL FATE
Animals The principal metabolic pathway in rats and rabbits was *o*- or *p*-hydroxylation, followed by glucuronidation (R. N. Waring, *Xenobiotica*, 1973, **3**, 65–71). **Plants** In plants, undergoes oxidation to the sulfoxide, but not further to the sulfone. **Soil/Environment** DT_{50} in soil *c*. 24 h. K_{oc} 373.

12 carfentrazone-ethyl   *Herbicide*

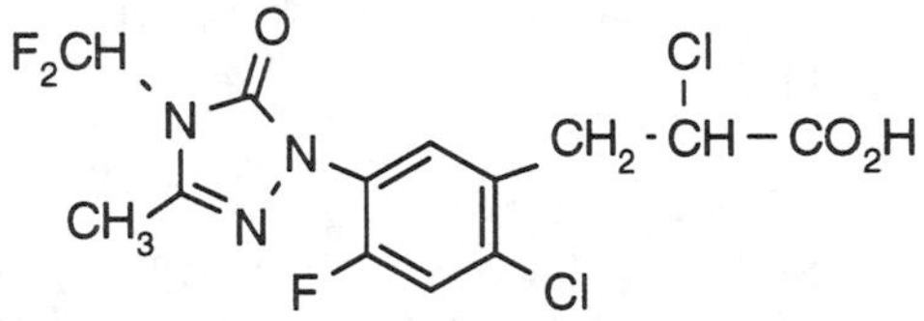

NOMENCLATURE
Common name carfentrazone-ethyl (BSI, pa ISO)

IUPAC name ethyl (RS)-2-chloro-3-[2-chloro-5-(4-difluoromethyl-4,5-dihydro-=
3-methyl-5-oxo-1H-1,2,4-triazol-1-yl)-4-fluorophenyl]propionate
Chemical Abstracts name ethyl α,2-dichloro-5-[4-(difluoromethyl)-4,5-dihydro-=
3-methyl-5-oxo-1H-1,2,4-triazol-1-yl]-4-fluorobenzenepropanoate
CAS RN [128621–72–7] (for the acid); [128639–02–1] (for the ethyl ester)
Development codes F8426; F116426

PHYSICAL CHEMISTRY
Mol. wt. 412.2   **M.f.** $C_{15}H_{14}Cl_2F_3N_3O_3$; $C_{13}H_{10}Cl_2F_3N_3O_3$ for acid
Form Viscous yellow liquid.   **M.p.** −22.1 °C   **B.p.** 350–355 °C/760 mmHg
V.p. 1.6×10^{-2} mPa (25 °C) K_{ow} logP = 3.36   **Henry** 2.47×10^{-4} Pa m^3 mol^{-1}
(20 °C, calc.)   **S.g./density** 1.457 (20 °C)   **Solubility** In water 12 µg/ml (20 °C),
22 µg/ml (25 °C), 23 µg/ml (30 °C).   **F.p.** >110 °C

COMMERCIALISATION
History Reported by W. A. van Saun et al. (Proc. Br. Crop Prot. Conf., Weeds, 1993,
1, 19).   **Manufacturers** FMC

APPLICATIONS
Biochemistry Acts by inhibition of protoporphyrinogen oxidase, leading to
membrane disruption.   **Mode of action** Absorbed by foliage with limited
translocation.   **Uses** Post-emergence control in cereals of a wide range of broad-
leaved weeds, especially Galium aparine, Abutilon theophrasti, Ipomoea hederacea
var. hederacea, Chenopodium album and several mustard species, at 9–35 g/ha.
Phytotoxicity Good tolerance in wheat, barley and rice.   **Formulation types** WG;
EC; SG.   **Mixtures** (carfentrazone-ethyl +) isoproturon; mecoprop-P.
Selected tradenames 'Affinity' (FMC); 'Aurora' (FMC)

ANALYSIS
Details from FMC.

MAMMALIAN TOXICOLOGY
Oral Acute oral LD$_{50}$ for female rats 5143 mg/kg.   **Skin and eye** Acute
percutaneous LD$_{50}$ for rats >4000 mg/kg. Minimally irritating to eyes and non-
irritating to skin (rabbits). No skin sensitisation (guinea pigs).   **Inhalation** LC$_{50}$
(4 h) for rats >5 mg/l.   **NOEL** (2 y) for rats 3 mg/kg daily.   **ADI** 0.03 mg/kg
(proposed).   **Other** Non-mutagenic in the Ames test.   **Toxicity class** WHO (a.i.)
III

ECOTOXICOLOGY
Birds LD$_{50}$ for quail >1000 mg/kg. LC$_{50}$ for quail and ducks >5000 ppm.
Fish LC$_{50}$ (96 h) 1.6–43 mg/l, depending on species.   **Bees** LD$_{50}$ (oral) >35,
(contact) >200 µg/bee.   **Worms** LC$_{50}$ >820 mg/kg soil.   **Daphnia** EC$_{50}$ (48 h)
9.8 mg/l.   **Algae** EC$_{50}$ 12–18 µg/l, depending on species.

ENVIRONMENTAL FATE

Soil/Environment Broken down in the soil by microbial action; not susceptible to photodecomposition nor volatility following soil application. Strongly absorbed to sterile soils (K_{oc} 750 ± 60 at 25 °C). In non-sterile soils, rapidly converted to the free acid, which has low soil binding (K_{oc} 15–35 at 25°C, pH 5.5). In the laboratory, soil DT_{50} is a few hours, degrading to the free acid, which in turn has DT_{50} 2.5–4.0 d.

13 cartap   *Insecticide*

2-dimethylaminopropane-1,3-dithiol

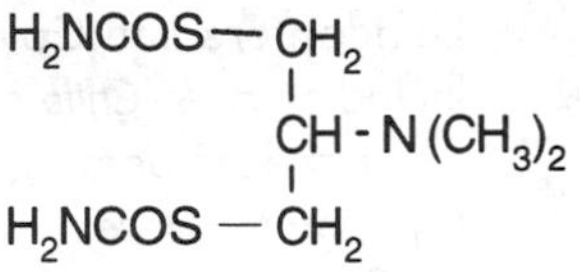

$$H_2NCOS—CH_2$$
$$CH-N(CH_3)_2$$
$$H_2NCOS — CH_2$$

NOMENCLATURE
cartap
Common name cartap (BSI, E-ISO, (*m*) F-ISO, JMAF)
IUPAC name *S,S'*-(2-dimethylaminotrimethylene) bis(thiocarbamate)
Chemical Abstracts name *S,S'*-[2-(dimethylamino)-1,3-propanediyl] dicarbamothioate
CAS RN *[15263–53–3]*   **Development codes** TI-1258

cartap hydrochloride
Common name cartap hydrochloride
CAS RN *[15263–52–2]* cartap monohydrochloride; *[22042–59–7]* cartap, unspecified hydrochloride   **EEC no.** 239–309–2

PHYSICAL CHEMISTRY
cartap
Mol. wt. 237.3   **M.f.** $C_7H_{15}N_3O_2S_2$

cartap hydrochloride
Mol. wt. 273.8   **M.f.** $C_7H_{16}ClN_3O_2S_2$ **Form** Colourless crystalline, slightly hygroscopic solid with slight odour.   **M.p.** 179–181 °C (decomp.)   **V.p.** Negligible.
Solubility In water *c.* 200 g/l (25 °C). Very slightly soluble in methanol and ethanol. Insoluble in acetone, diethyl ether, ethyl acetate, chloroform, benzene, and hexane.   **Stability** Stable in acidic conditions, but hydrolysed in neutral or alkaline media.

COMMERCIALISATION

History Insecticide reported by M. Sakai et al. (*Jpn. J. Appl. Entomol. Zool.*, 1967, **11**, 125), its action and structure-activity relationships reviewed (K. Konishi, *Pestic. Chem. [Congr. Pestic. Chem., 2nd, 1971]*, 1972, **1**, 179; M. Sakai & Y. Sato, *ibid.*, p. 445; M. Sakai, *Jpn. Pestic. Inf.*, 1971, No. 6, p. 15; 1978, No. 34, p. 22). The hydrochloride introduced by Takeda Chemical Industries, Ltd.
Patent GB 1126204; US 3332943; FR 1452338 **Manufacturers** Kuo Ching; Takeda

APPLICATIONS

cartap hydrochloride
Biochemistry Analogue or propesticide of the natural toxin nereistoxin. Causes paralysis by ganglionic blocking action on the central nervous system.
Mode of action Systemic insecticide with stomach and contact action. Insects discontinue feeding, and die of starvation. **Uses** Cartap hydrochloride is used, at c. 0.4–1.0 kg/ha, for control of chewing and sucking insects (particularly Lepidoptera and Coleoptera), at almost all stages of development, on many crops, including rice (*Chilo suppressalis*, *Cnaphalocrocis medinalis*, *Lissorhoptrus oryzophilus* and rice-leaf beetle), potatoes, cabbage and other vegetables (Agromyzidae, *Leptinotarsa decemlineata* and *Plutella xylostella*); also on soya beans, peanuts, sunflowers, maize, sugar beet, wheat, pearl barley, pome fruit, stone fruit, citrus fruit, vines, chestnuts, ginger, tea, cotton, and sugar cane. **Phytotoxicity** May be phytotoxic to cotton, tobacco, and apples, under certain soil and climatic conditions. **Formulation types** SP; DP; GR. **Mixtures** *(cartap hydrochloride +)* fenobucarb. **Compatibility** Not compatible with pesticides which are alkaline.
Selected tradenames 'Padan' (Takeda)

ANALYSIS

Product analysis by colorimetry (*CIPAC Handbook*, 1988, **D**, 24). **Residue** analysis by glc or by polarography (K. Nishi et al., *Anal. Methods Pestic. Plant Growth Regul.*, 1973, **7**, 371).

MAMMALIAN TOXICOLOGY

Reviews *Pesticide residues in food – 1995*, FAO Plant Production and Protection Paper 133, 1996. **EHC** 76 (1988).

cartap
ADI (JMPR) ADI withdrawn [1995].

cartap hydrochloride
Oral Acute oral LD_{50} for female rats 325, male rats 345, male mice 150, female mice 154 mg/kg. **Skin and eye** Acute percutaneous LD_{50} for mice >1000 mg/kg; no irritation to skin or eyes in rabbits. **Inhalation** LC_{50} (6 h) for rats >0.54 mg/l.
NOEL (2 y) for rats 10 mg/kg b.w. daily; (1.5 y) for mice 20 mg/kg b.w. daily.
Toxicity class WHO (a.i.) II; EPA (formulation) II **EC risk** Xn (R21/22)

ECOTOXICOLOGY
cartap hydrochloride
Fish LC_{50} for carp 1.6 mg/l (24 h) and 1.0 mg/l (48 h). **Bees** Moderately toxic to honey bees. **Other aquatic spp.** LC_{50} (24 h) for *Moina macrocopa* 12.5–25 mg/l.

ENVIRONMENTAL FATE
Animals In rats, the carbonyl carbon is hydrolysed, and the sulfur oxidised, with N-demethylation of thiomethyl derivatives. No accumulation occurs in tissues. Rapidly excreted in the urine. **Soil/Environment** DT_{50} in soil *c.* 3 d.

114 CGA 245704 *Plant activator*

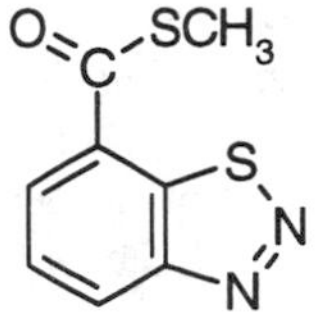

NOMENCLATURE
Common name acibenzolar (for thiol; BSI proposed)
IUPAC name *S*-methyl benzo[1,2,3]thiadiazole-7-carbothioate
Chemical Abstracts name 1,2,3-benzothiadiazole-7-carbothioic acid *S*-methyl ester
CAS RN *[135158–54–2]* **Development codes** CGA 245704

PHYSICAL CHEMISTRY
Mol. wt. 210.3 **M.f.** $C_8H_6N_2OS_2$ **Form** White to beige fine powder with a burnt-like odour. **M.p.** 132.9 °C **B.p.** *c.* 267 °C **V.p.** 4.4 × 10^{-1} mPa (25 °C) K_{ow} logP = 3.1 (25 °C) **Henry** 1.3 × 10^{-2} Pa m^3 mol^{-1} (calc.) **S.g./density** 1.54 (22 °C) **Solubility** In water 7.7 mg/l (25 °C). In methanol 4.2, ethyl acetate 25, *n*-hexane 1.3, toluene 36, *n*-octanol 5.4, acetone 28, dichloromethane 160 (all in g/l, 25 °C). **Stability** Hydrolysis DT_{50} (20 °C) 3.8 y (pH 5), 23 w (pH 7), 19.4 h (pH 9).

COMMERCIALISATION
History Reported by W. Ruess *et al.*, *XIII International Plant Prot. Congr.*, The Hague, Netherlands, July 2–7, 1995; *idem, Proc. Br. Crop Prot. Conf. – Pests Dis.*, 1996, **1**, 53. Introduced by Ciba-Geigy AG (now Novartis Crop Protection AG) in Germany and Switzerland in 1996. **Patent** EP 313512 **Manufacturers** Novartis

APPLICATIONS
Biochemistry Acts as a functional analogue of the natural signal molecule for systemic activated resistance, salicylic acid (H Kessmann et al. (Proc. Br. Crop Prot. Conf. – Pests Dis., 1996, **3**, 961). **Mode of action** Activates the plant's natural defence mechanism ("systemic activated resistance" (SAR)). Has no intrinsic fungicidal activity. **Uses** For use against a range of fungal infections of wheat. Must be applied as a protective treatment or early in the disease progress. Under development against a range of diseases in rice, bananas, vegetables and tobacco. **Formulation types** WG. **Selected tradenames** 'Bion' (Novartis)

ANALYSIS
Product analysis by gc with FID. **Residues** determined by hplc. Details available from Novartis.

MAMMALIAN TOXICOLOGY
Oral Acute oral LD_{50} for rats >2000 mg/kg. **Skin and eye** Acute percutaneous LD_{50} for rats >2000 mg/kg. Not irritating to skin or eye (rabbits). Skin sensitiser (guinea pigs). **Inhalation** LC_{50} (4 h) for rats >5000 mg/m^3 air. **NOEL** (2 y) for rats 8.5 mg/kg b.w. daily; (1.5 y) for mice 11 mg/kg b.w. daily; (1 y) for dogs 5 mg/kg b.w. daily. **ADI** 0.05 mg/kg b.w. **Other** Not oncogenic, not mutagenic; of no teratogenic relevance for humans. **Toxicity class** WHO (a.i.) III (Table 5) **EC risk** R43

ECOTOXICOLOGY
Birds LD_{50} (14 d) for mallard ducks and bobwhite quail >2000 mg/kg. Eight-day LC_{50} for mallard ducks and bobwhite quail >5200 mg/kg. **Fish** LC_{50} (96 h) for rainbow trout 0.4, bluegill sunfish 2.8 mg/l. **Bees** LD_{50} (48 h) (oral) 128.3 µg/bee; (contact) 100 µg/bee. **Worms** LC_{50} (14 d) for Eisenia foetida >1000 mg/kg soil. **Other beneficial spp.** Harmless to predator bug and mite, ground beetle and parasitic wasp. **Daphnia** LC_{50} (48 h) 2.4 mg/l. **Algae** E_bC_{50} (72 h) for Scenedesmus subspicatus 1.7 mg/l.

ENVIRONMENTAL FATE
Animals After oral administration, CGA 245704 is rapidly absorbed and also rapidly almost completely eliminated with urine and faeces. The metabolic pathways are independent of sex, pre-treatment or dose level administered. Residues in tissues were generally low and there was no evidence of accumulation or retention of CGA 245704 or its metabolites. **Plants** The metabolism proceeds via hydrolysis with subsequent conjugation with sugars or by oxidation of the phenyl ring followed by sugar conjugation. **Soil/Environment** In soil, the compound dissipates via hydrolysis; DT_{50} 0.3 d. The product further degrades, DT_{50} 20 d; metabolites become completely degraded and mineralised. Strong adsorption to soil, low mobility, K_{oc} 1394 ml/g. In water, for CGA 245704, DT_{50} <1 d; for the hydrolysis product, DT_{50} 8 d.

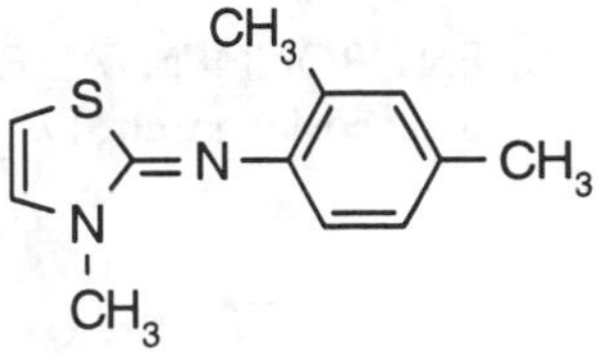

NOMENCLATURE

IUPAC name *N*-2,3-dihydro-3-methyl-1,3-thiazol-2-ylidene-2,4-xylidine;
N-3-methyl-4-thiazolin-2-ylidene-2,4-xylidine
Chemical Abstracts name 2,4-dimethyl-*N*-(3-methyl-2(3*H*)-thiazolylidene)=
benzenamine
CAS RN *[61676–87–7]* **Development codes** CGA 50 439

PHYSICAL CHEMISTRY

Mol. wt. 218.3 **M.f.** $C_{12}H_{14}N_2S$ **Form** Colourless crystals. **M.p.** 44 °C
V.p. 2.4 mPa (20 °C) **K**$_{ow}$ logP = c. 0.6 (unstated pH) **S.g./density** 1.19 (20 °C)
Solubility In water 150 mg/l (pH 9, 20 °C). In benzene 800, methanol 800,
dichloromethane 800, hexane 110 (all in g/kg, 20 °C). **Stability** Stable to
hydrolysis at ≤70 °C. **pKa** 5.2 (21 °C)

COMMERCIALISATION

History Ixodicide reported by R. M. Immler *et al.* (*Proc. Br. Crop Prot. Conf. – Pests
Dis.,* 1977, **2**, 383). Introduced by Ciba-Geigy AG (now Novartis Crop Protection
AG). **Patent** BE 841504; GB 1527807 **Manufacturers** Novartis

APPLICATIONS

Uses Used in dips or as a spray, it controls all Acarina species, including
strains resistant to organochlorine, organophosphorus and carbamate
ixodicides at concentrations >0.1 g a.i./l. **Formulation types** EC.
Selected tradenames 'Tifatol' (Novartis)

ANALYSIS

Product analysis by glc. **Residues** determined by glc using FPD or PND.
Particulars available from Novartis.

MAMMALIAN TOXICOLOGY

Oral Acute oral LD$_{50}$ for rats 725 mg/kg. **Skin and eye** Acute percutaneous LD$_{50}$
for rats >3100 mg/kg; slight irritant to skin and eyes of rabbits. **Inhalation** LC$_{50}$
(4 h) for rats >2800 mg/m^3 air. **NOEL** (2 y) for rats 10, mice 100 ppm; not
oncogenic, not carcinogenic. **Toxicity class** WHO (a.i.) II

ECOTOXICOLOGY

Birds Acute oral LD_{50} for Japanese quail 1212, Pekin ducklings 540 mg/kg.
Fish LC_{50} (96 h) for rainbow trout 12, carp 32 mg/l.

ENVIRONMENTAL FATE

Soil/Environment DT_{50} <14 d (aerobic soils).

116 chinomethionat   *Fungicide, acaricide*

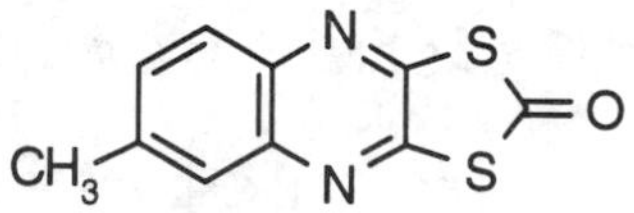

NOMENCLATURE

Common name chinomethionat (E-ISO); chinométhionate ((*m*) F-ISO);
quinomethionate (BSI); oxythioquinox (Australia, ESA); quinoxalines (JMAF);
no name (USA)
IUPAC name *S,S*-(6-methylquinoxaline-2,3-diyl) dithiocarbonate;
6-methyl-1,3-dithiolo[4,5-*b*]quinoxalin-2-one
Chemical Abstracts name 6-methyl-1,3-dithiolo[4,5-*b*]quinoxalin-2-one
CAS RN *[2439–01–2]* **EEC no.** 219–455–3 **Development codes** Bayer 36 205;
Bayer SAS 2074 **Official codes** ENT 25 606.

PHYSICAL CHEMISTRY

Mol. wt. 234.3 **M.f.** $C_{10}H_6N_2OS_2$ **Form** Yellow crystals. **M.p.** 170 °C
V.p. 0.026 mPa (20 °C) **K_{ow}** logP = 3.78 (20 °C) **S.g./density** 1.556 (20°C)
Solubility In water 1 mg/l (20 °C). In toluene 25, dichloromethane 40, hexane 1.8,
isopropanol 0.9, cyclohexanone 18, dimethylformamide 10, petroleum oils 4
(all in g/l, 20 °C). Soluble in hot benzene and dioxane. **Stability** Relatively stable
under normal conditions. Hydrolysed in alkaline media; DT_{50} (22 °C) 10 d (pH 4),
80 h (pH 7), 225 min (pH 9).

COMMERCIALISATION

History Acaricide and fungicide reported by K. Sasse (*Hoefchen-Briefe (Engl. Ed.)*,
1960, **13**, 197; K. Sasse *et al.*, *Angew. Chem.*, 1960, **72**, 973). Introduced by Bayer
AG. **Patent** DE 1100372; BE 580478 **Manufacturers** Bayer

APPLICATIONS

Mode of action Selective non-systemic contact fungicide with protective and eradicant action.   **Uses** Control of powdery mildews and spider mites on fruit (including citrus), ornamentals, cucurbits, cotton, coffee, tea, tobacco, walnuts, vegetables, and glasshouse crops; American gooseberry mildew on gooseberries and currants.   **Phytotoxicity** Phytotoxic to certain varieties of apple, pear, currant, rose, and ornamentals.   **Formulation types** WP; DP; FU; SC.   **Mixtures** *(chinomethionat +)* propineb; triadimenol.   **Compatibility** Incompatible with mineral oils (phytotoxicity may result), and with formulations based on thiram. **Selected tradenames** 'Morestan' (Bayer)

ANALYSIS
Product analysis by hplc (*CIPAC Handbook*, 1985, **1C**, 2019; *AOAC Methods*, 1995, 986.08) or by u.v. spectrometry (details available from Bayer AG).
Residues determined by glc (*Pestic. Anal. Man.*, 1979, **I**, 201-I, **II**; *Man. Pestic. Residue Anal.*, 1987, **I**, S13; A. Ambrus et al., *J. Assoc. Off. Anal. Chem.*, 1981, **64**, 733; R. T. Krause & E. M. August, *ibid.*, 1983, **66**, 1018), or by colorimetry after conversion to a derivative (H. Tietz et al., *ibid.*, 1962, **15**, 166; C. A. Anderson, *Anal. Methods Pestic. Plant Growth Regul. Food Addit.*, 1967, **5**, 277). Methods for the determination of residues are also available from Bayer.

MAMMALIAN TOXICOLOGY
Reviews *Pesticide residues in food – 1987*. FAO Plant Production and Protection Paper 84, 1987. *Pesticide residues in food – 1987 evaluations. Part II – Toxicology.* FAO Plant Production and Protection Paper 86/2, 1988.   **Oral** Acute oral LD_{50} for male rats 2541, female rats 1095 mg/kg.   **Skin and eye** Acute percutaneous LD_{50} for rats >5000 mg/kg. Slightly irritating to skin, severely irritating to eyes (rabbits).   **Inhalation** LC_{50} (4 h) for male rats >4.7, female rats 2.2 mg/l (dust). **NOEL** (2 y) for rats 40, male mice 270, female mice <90 mg/kg diet; (1 y) for dogs 25 mg/kg diet.   **ADI** (JMPR) 0.006 mg/kg b.w. [1987]. **Toxicity class** WHO (a.i.) III; EPA (formulation) III   **EC risk** Xn (R22); Xi (R36, R43)

ECOTOXICOLOGY
Birds Acute oral LD_{50} for bobwhite quail 196 mg/kg. Five-day dietary LC_{50} for bobwhite quail 2409, mallard ducks >5000 mg/kg diet.   **Fish** LC_{50} (96 h) for bluegill sunfish 0.0334, rainbow trout 0.131, golden orfe 0.24 mg/l.   **Bees** Not toxic to bees; LD_{50} >100 μg/bee.   **Worms** LC_{50} (14d) >1000 mg/kg. **Daphnia** LC_{50} (48 h) 0.12 mg/l.   **Algae** E_rC_{50} (96h) for *Scenedesmus* 0.14 mg/l.

ENVIRONMENTAL FATE
Animals In rats, following oral administration, chinomethionat is rapidly metabolised, and c. 90% is eliminated within 3 days in the faeces and urine. The main metabolite is chinomethionat acid (dimethylmercaptoquinoxaline-6-carboxylic acid) which also occurs in the conjugated form.   **Plants** After application to fruit, no penetration of the a.i. or metabolites in the fruit pulp was observed. The only

metabolite detected was dihydromethylquinoxalinedithiol. **Soil/Environment** K_{oc} 45–90 (3 soil types from sandy loam to high organic matter). DT_{50} in standard soil land 2 1–3 d.

117 chlomethoxyfen

Herbicide

diphenyl ether

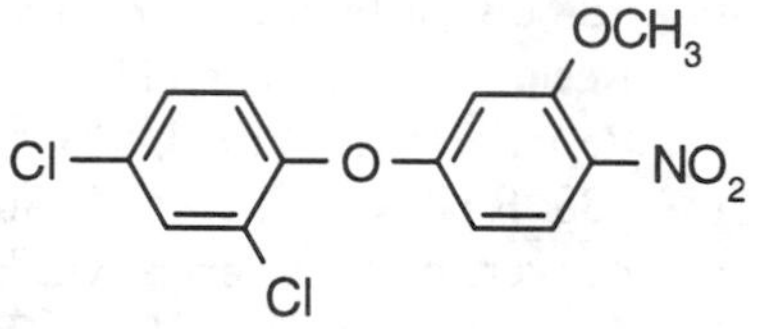

NOMENCLATURE
Common name chlomethoxyfen (BSI, draft E-ISO); chlométhoxyfène ((m) draft F-ISO); chlormethoxynil (JMAF)
IUPAC name 5-(2,4-dichlorophenoxy)-2-nitroanisole
Chemical Abstracts name 4-(2,4-dichlorophenoxy)-2-methoxy-1-nitrobenzene; 2,4-dichloro-1-(3-methoxy-4-nitrophenoxy)benzene
CAS RN *[32861–85–1]* **Development codes** X-52

PHYSICAL CHEMISTRY
Mol. wt. 314.1 **M.f.** $C_{13}H_9Cl_2NO_4$ **Form** Yellow crystals. **M.p.** 113–114 °C
B.p. 260 °C/25 mmHg **V.p.** 1.87 mPa (25 °C) **K_{ow}** logP = 3.34 (20 °C)
S.g./density 1.37 **Solubility** In water 0.3 mg/l (15 °C). In acetone 200, dimethyl sulfoxide 100, benzene 150 (all in g/kg, 15 °C). **Stability** Stable to acids, alkalis, and light.

COMMERCIALISATION
History Introduced as a herbicide by Nihon Nohyaku Co., Ltd. **Patent** JP 600441
Manufacturers Ishihara Sangyo

APPLICATIONS
Biochemistry Protoporphyrinogen oxidase inhibitor. **Mode of action** Selective contact herbicide, absorbed by the leaves and stems. **Uses** Pre-emergence control of *Scirpus* spp., *Echinochloa crus-galli*, and other annual weeds in transplanted rice at 1.5–2.5 kg a.i./ha. **Phytotoxicity** May scorch old leaves.
Formulation types WP; GR. **Mixtures** (chlomethoxyfen +) neburon.
Selected tradenames 'Diphenex' (Ishihara Sangyo); 'Ekkusugoni' (Nihon Nohyaku; Ishihara Sangyo)

ANALYSIS
Product analysis by glc with FID (F. Yamane & K. Tsuchiya, *Anal. Methods Pestic. Plant Growth Regul.*, 1978, **10**, 267). **Residues** determined by glc with ECD (*idem, ibid.*).

MAMMALIAN TOXICOLOGY
Oral Acute oral LD_{50} for rats and mice >10 000 mg/kg. **Skin and eye** Acute percutaneous LD_{50} for rats >5000 mg/kg. **Inhalation** LC_{50} for rats >1.767 mg/l. **Toxicity class** WHO (a.i.) III (Table 5); EPA (formulation) IV

ECOTOXICOLOGY
Fish LC_{50} (48 h) for carp 11.2 mg/l. **Daphnia** LC_{50} (24 h) >47.5 mg/l.

ENVIRONMENTAL FATE
Animals Metabolised in rats to form glucurone and glutathione conjugates.
Soil/Environment In soil, DT_{50} 7–43 d. K_d 525–1407.

18 chloralose   *Rodenticide*

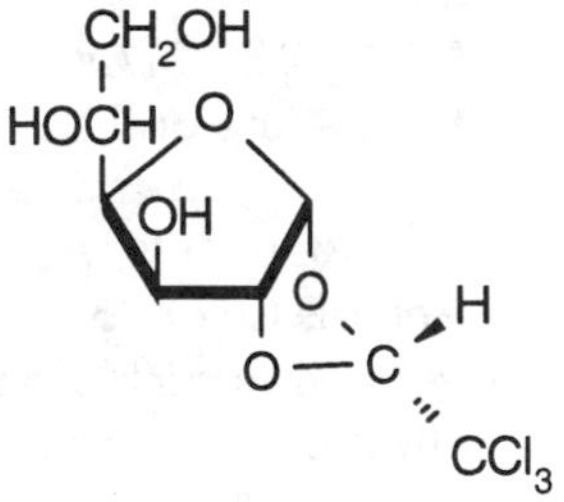

NOMENCLATURE
Common name chloralose (BSI, E-ISO, F-ISO); glucochloralose (BSI, E-ISO); glucochloral (F-ISO) are all accepted in lieu
IUPAC name (*R*)-1,2-*O*-(2,2,2-trichloroethylidene)-α-D-glucofuranose
Chemical Abstracts name (*R*)-1,2-*O*-(2,2,2-trichloroethylidene)-α-D-glucofuranose
Other names alphachloralose **CAS RN** *[15879–93–3]*, formerly *[39598–39–5]* or *[14798–36–8]*; *[16376–36–6]* beta-form **EEC no.** 240–016–7

PHYSICAL CHEMISTRY
Composition Chloralose also exists in a beta-form. **Mol. wt.** 309.5 **M.f.** $C_8H_{11}Cl_3O_6$ **Form** Crystalline powder. **M.p.** 187 °C (beta-form 227–230 °C) **V.p.** Negligible at room temperature. **Solubility** In water 4.44 g/l (15 °C). Soluble

in alcohols, diethyl ether and glacial acetic acid. Sparingly soluble in chloroform. Practically insoluble in petroleum ether. Beta-form is less soluble than the alpha-form in water, ethanol, diethyl ether. **Stability** Converted by acids and alkalis into glucose and chloral. **Specific rotation** $[\alpha]_D^{22}$ +19°

COMMERCIALISATION
Manufacturers Jewnin-Joffe

APPLICATIONS
Mode of action Narcotic, which renders birds easier to kill by other means. Rodenticide which acts by retardation of metabolism and lowering of body temperature to a fatal level. Rapidly metabolised and hence non-cumulative.
Uses Used as a rodenticide (particularly for mice), and also as a bird repellent and narcotic. **Formulation types** RB; CB.

ANALYSIS
Product analysis by hydrolysis and estimation of the trichloroacetaldehyde produced (K. C. Barrons & R. W. Hummer, *Agric. Chem.*, 1951, **6**(6), 48; G. I. Mills, *Anal. Chim. Acta*, 1952, **7**, 70).

MAMMALIAN TOXICOLOGY
Oral Acute oral LD_{50} for rats 400, mice 32 mg/kg. **Toxicity class** WHO (a.i.) II; EPA (formulation) II **EC risk** Xn (R20/22)

ECOTOXICOLOGY
Birds Acute oral LD_{50} for birds 32–178 mg/kg.

ENVIRONMENTAL FATE
Animals Chloralose is metabolised to chloral, oxidised to trichloroacetic acid, and reduced to trichloroethanol; the latter metabolite is responsible for the hypnotic effect of chloralose.

chloramben   *Herbicide*

benzoic acid (auxin)

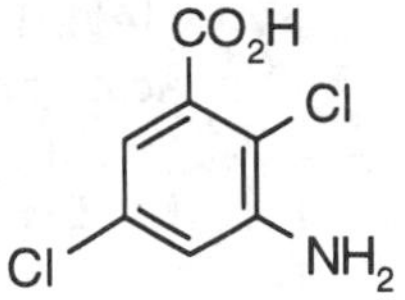

NOMENCLATURE
Common name chloramben (BSI, E-ISO, ANSI, WSSA); chlorambèn ((*m*) F-ISO);
[amben] (former WSSA name)
IUPAC name 3-amino-2,5-dichlorobenzoic acid
Chemical Abstracts name 3-amino-2,5-dichlorobenzoic acid
CAS RN *[133–90–4]* **Development codes** ACP M-629

PHYSICAL CHEMISTRY
Mol. wt. 206.0 **M.f.** $C_7H_5Cl_2NO_2$ **Form** Colourless crystals. **M.p.** 200–201 °C
V.p. 930 mPa (100 °C) **Solubility** In water 700 mg/l (25 °C). In
dimethylformamide 1206, acetone, methanol 223, ethanol 173, isopropanol 113,
diethyl ether 70, chloroform 0.9, benzene 0.2 (all in g/kg, room temperature).
Insoluble in carbon tetrachloride. **Stability** Thermally stable up to the boiling
point. Stable to oxidising agents, acids and alkalis. Decomposed by sodium
hypochlorite solutions. Sensitive to light.

COMMERCIALISATION
History Herbicide introduced by Amchem Products, Inc. (now Rhône-Poulenc
Agrochimie). **Patent** US 3014063; US 3174842 **Manufacturers** Rhône-Poulenc

APPLICATIONS
Biochemistry Auxin type. The *N*-glycoside was isolated from soya beans in
amounts equivalent to the chloramben applied, but little was recovered from
barley, a susceptible crop (S. R. Colby, *Science*, 1965, **150**, 619).
Mode of action Selective systemic herbicide, absorbed by the seeds and roots,
with limited translocation. Inhibits root development of seedling weeds.
Uses Used pre-planting incorporated and pre-emergence at 2–4 kg a.i./ha to
control grasses and broad-leaved weeds in seedling asparagus, navy beans,
peanuts, maize, sweet potatoes, pumpkins, soya beans, squash, sunflowers and
certain ornamentals. **Formulation types** DS; GR; SL; SP.
Selected tradenames 'Amiben' (ammonium salt) (Rhône-Poulenc)

ANALYSIS
Product analysis by u.v. spectrophotometry (*AOAC Methods*, 1995, 971.06; *CIPAC
Handbook*, 1983, **1B**, 1747) or volumetric methods (details available from Rhône-
Poulenc). See also: H. S. Segal & M. L. Sutherland, *Anal. Methods Pestic., Plant*

MAIN ENTRIES

Growth Regul. Food Addit., 1967, **5**, 321; *Anal. Methods Plant Growth Regul.*, 1972, **6**, 588. **Residue** analysis by glc of methyl ester; details available from Rhône-Poulenc.

MAMMALIAN TOXICOLOGY
Oral Acute oral LD_{50} for rats >5000 mg/kg. **Skin and eye** Acute percutaneous LD_{50} for rats >3160 mg/kg. Mild skin and eye irritant (rabbits). **NOEL** (2 y) for rats 10 000 mg/kg diet. **Toxicity class** WHO (a.i.) III (Table 5); EPA (formulation) IV

ECOTOXICOLOGY
Birds Eight-day dietary LC_{50} for mallard ducks 4640 mg/kg. **Fish** Not toxic to fish. **Bees** Not toxic to bees.

ENVIRONMENTAL FATE
Animals In rats, the major metabolites were 3-amino-5-chlorobenzoic acid, 3-aminobenzoic acid, 2,5-dihydroxybenzoic acid, 3,5-dihydroxybenzoic acid and 2,5-dichloroaniline. **Plants** In plants, chloramben is converted to a highly stable N-glucoside, which retards further degradation. **Soil/Environment** Undergoes microbial degradation in soil. Duration of activity is c. 6–8 w.

120 chlorbromuron   *Herbicide*

urea

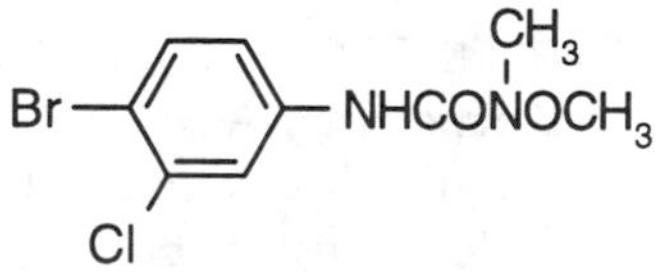

NOMENCLATURE
Common name chlorbromuron (BSI, E-ISO, WSSA, ex-ANSI); chlorobromuron ((m) F-ISO)
IUPAC name 3-(4-bromo-3-chlorophenyl)-1-methoxy-1-methylurea
Chemical Abstracts name N'-(4-bromo-3-chlorophenyl)-N-methoxy-N-methylurea
CAS RN [13360–45–7] **EEC no.** 236–411–9 **Development codes** C 6313

PHYSICAL CHEMISTRY
Mol. wt. 293.5 **M.f.** $C_9H_{10}BrClN_2O_2$ **Form** Colourless powder. **M.p.** 95–97 °C
V.p. 0.053 mPa (20 °C) **K_{ow}** logP = 2.9 **Henry** 4×10^{-4} Pa m^3 mol^{-1} (calc.)
S.g./density 1.69 (20 °C) **Solubility** In water 35 mg/l (20 °C). In acetone 460,

dichloromethane 170, hexane 89, benzene 72, isopropanol 12 (all in g/kg, 20 °C).
Stability Slowly hydrolysed in neutral, slightly acidic, and slightly alkaline media.

COMMERCIALISATION
History Herbicide reported by D. H. Green et al. (*Proc. Br. Weed Control Conf.*, *8th*, 1966, **2**, 363). Introduced by Ciba AG (now Novartis Crop Protection AG).
Patent BE 662268; GB 965313 **Manufacturers** Novartis

APPLICATIONS
Biochemistry Photosynthetic electron transport inhibitor.
Mode of action Selective herbicide, absorbed by the roots and leaves.
Uses Herbicide suitable for: pre-emergence use on carrots, peas, potatoes, soya beans, sunflowers at 1.0–2.5 kg/ha; post-emergence use on carrots and transplanted celery at 0.75–1.5 kg/ha. **Phytotoxicity** A number of crops may be injured by pre-emergence applications, e.g. rice, beet, okra, flax, cucurbits, tomatoes, strawberries, etc. **Formulation types** WP. **Mixtures** (*chlorbromuron +*) terbuthylazine; amitrole + simazine. **Selected tradenames** 'Maloran' (Novartis)

ANALYSIS
Product analysis by glc. **Residues** determined by hydrolysis to 4-bromo-3-chloroaniline, a derivative of which is determined by colorimetry, or by glc with ECD (G. Voss, *Anal. Methods Pestic. Plant Growth Regul.*, 1973, **7**, 569).

MAMMALIAN TOXICOLOGY
Oral Acute oral LD_{50} for rats >5000 mg tech./kg. **Skin and eye** Acute percutaneous LD_{50} for rats >2000, rabbits >10 000 mg/kg. Not a skin or eye irritant. **Inhalation** LC_{50} (6 h) for rats >1.05 mg/l air. **NOEL** (90 d) for rats 316 mg/kg diet (21.0 mg/kg daily), dogs >316 mg/kg diet (10.5 mg/kg daily).
Toxicity class WHO (a.i.) III; EPA (formulation) IV

ECOTOXICOLOGY
Birds Ten-day dietary LC_{50} for pheasants and mallard ducks >10 250 mg/kg diet.
Fish LC_{50} (96 h) for rainbow trout 5, bluegill sunfish 5, crucian carp 8 mg/l.
Bees Not toxic to bees.

ENVIRONMENTAL FATE
Plants In plants, metabolism involves demethylation and deamination-decarboxylation to the corresponding aniline molecule. **Soil/Environment** In biologically active soils, DT_{50} 8–28 w; strongly bound to soil (K_{oc} = 908), indicating low leaching potential. In water, stable against abiotic degradation, but in biologically active systems DT_{50} 4–11 w.

121 chlordane

Insecticide

cyclodiene organochlorine

NOMENCLATURE
Common name chlordane (BSI, E-ISO, (*m*) F-ISO, ESA, JMAF)
IUPAC name 1,2,4,5,6,7,8,8-octachloro-2,3,3a,4,7,7a-hexahydro-=
4,7-methanoindene
Chemical Abstracts name 1,2,4,5,6,7,8,8-octachloro-2,3,3a,4,7,7a-hexahydro-=
4,7-methano-1*H*-indene
CAS RN *[57–74–9]* chlordane; *[12789–03–6]* tech. grade; *[5103–71–9]*
cis- isomer (formerly *[22212–52–8]*); *[5103–74–2]* *trans*- isomer
EEC no. 200–349–0   **Development codes** Velsicol 1068; M 410 (Velsicol)
Official codes OMS 1437; ENT 9932

PHYSICAL CHEMISTRY
Composition Chlordane contains 60–75% of chlordane isomers, the major
components being two stereoisomers whose nomenclature at C(1) and C(2) has
been confused in the literature: the alpha or *cis*-isomer (1α,2α,3aα,4β,7β,7aα)
and the *trans*-isomer (1α,2β,3aα,4β,7β,7aα) usually known as gamma-,
occasionally as beta-chlordane. (The term gamma-chlordane has also been applied
to the 2,2,4,5,6,7,8,8-octachloro-isomer, CAS RN *[5566–34–7]*). The remainder
of the tech. grade comprises other stereoisomers (each ≤7%) and heptachlor.
Mol. wt. 409.8   **M.f.** $C_{10}H_6Cl_8$ **Form** Tech. grade is a viscous amber liquid.
M.p. *cis*- isomer 106–107 °C; *trans*- isomer 104–105 °C   **B.p.** 175 °C/1 mmHg
(pure)   **V.p.** 1.3 mPa (25 °C) (pure)   **S.g./density** 1.59–1.63 (25 °C, tech.)
Solubility In water 0.1 mg/l (25 °C). Miscible with most aliphatic and aromatic
organic solvents, including acetone, cyclohexanone, ethanol, deodorised kerosene,
isopropanol, trichloroethylene.   **Stability** Decomposed by alkalis, with the loss of
chlorine. Under u.v. irradiation, a change in the skeletal structure and of the
chlorine content occurs.

COMMERCIALISATION
History Insecticide, first described by C. W. Kearns *et al.* (*J. Econ. Entomol.*, 1945,
38, 661), discovered independently by R. Riemschneider (*Chim. Ind. (Paris)*, 1950,
64, 695). Introduced by Velsicol Chemical Corp.   **Patent** US 2598561 to Velsicol;
GB 618432 to J. Hyman   **Manufacturers** Velsicol

APPLICATIONS
Biochemistry Antagonist of the GABA receptor-chloride channel complex.
Mode of action Non-systemic insecticide with contact, stomach, and respiratory action. Long residual activity. **Uses** A persistent, non-systemic contact and ingested insecticide with some fumigant action. Used on land against Formicidae, Coleoptera, Noctuidae larvae, Saltatoria, subterranean termites (including *Coptotermes* spp.) and many other insect pests. It also controls household insects, pests of man and domestic animals, is used as a wood preservative, a protective treatment for underground cables and to reduce earthworm populations in lawns. It may be applied to soil, directly to foliage or as a seed treatment. Insecticidal activities of its components have been compared (S. J. Cristol, *Adv. Chem. Ser.*, 1950, No. 1, p. 541). **Phytotoxicity** Non-phytotoxic when used as directed. High concentrations may cause injury to some vegetables. Soil residues may depress germination. **Formulation types** EC; GR; DP; WP; Oil. **Mixtures** *(chlordane +)* heptachlor. **Compatibility** Incompatible with alkaline materials. **Selected tradenames** 'Octachlor' (Velsicol)

ANALYSIS
Product analysis by total chlorine content (*CIPAC Handbook*, 1970, **1**, 203; 1980, **1A**, 1119; *AOAC Methods*, 1995, 962.05). **Residues** determined by glc with ECD (*Pestic. Anal. Man.*, 1979, **I**, 201-A, 201-G, 201-I; *Anal. Methods Pestic. Plant Growth Regul.*, 1972, **6**, 315; *Analyst (London)*, 1979, **104**, 425; *Man. Pestic. Residue Anal.*, 1987, **I**, 5, 6, S9, S10, S12, S19; *Anal. Methods Residues Pestic.*, 1988, Part I, M1, M12). Chlordane isomers in **drinking water** determined by gc with ECD (*AOAC Methods*, 1995, 990.06).

MAMMALIAN TOXICOLOGY
Reviews *Pesticide residues in food – 1994*, FAO Plant Production and Protection Paper, 127, 1995. **IARC** 20, 42, 53 **EHC** 34 (WHO, 1984). **Oral** Acute oral LD_{50} for rats 133–649, mice 430, rabbits 300 mg/kg. **Skin and eye** Acute percutaneous LD_{50} for rabbits 200–2000, rats 217 mg/kg. Extremely irritating and corrosive to eyes; not a skin irritant (rabbits). Non-sensitising to skin (guinea pigs). **Inhalation** LC_{50} (4 h) for rats 0.56 mg/l (gravimetric) to >200 mg/l (nominal). **NOEL** (2 y) for dogs 3 mg/kg diet. Serious chronic and cumulative toxicity, including liver and kidney damage. Accumulates in body fat and lipid-containing organs. NOEL in a 3-generation study in rats 60 mg/kg diet. Not a teratogen in rabbits at 15 mg/kg daily. **ADI** (JMPR) 0.0005 mg/kg (PTDI) [1994]. **Other** Results of *in vivo* and *in vitro* studies indicate not mutagenic. Produced hepatocellular carcinomas in mice (S. S. Epstein, *Sci. Total Environ.*, 1976, **6**, 103–154). US National Academy of Sciences has ruled it is carcinogenic to certain strains of mice, but there is evidence that it may act as a promoter rather than as an initiator of carcinogenesis. **Toxicity class** WHO (a.i.) II; EPA (formulation) IIB **EC risk** (R40); Xn (R21/22) **PIC** Yes

MAIN ENTRIES

ECOTOXICOLOGY
Birds Acute oral LD_{50} for bobwhite quail 83 mg/kg. Eight-day dietary LC_{50} for bobwhite quail 421, mallard ducks 795 mg/kg diet. **Fish** LC_{50} (96 h) for rainbow trout 0.09, bluegill sunfish 0.07 mg/l. **Bees** Toxic to bees. **Daphnia** LC_{50} (48 h) 0.59 mg/l.

ENVIRONMENTAL FATE
Animals In rats, following oral administration, chlordane is metabolised via 1,2-dichlorochlordene and oxychlordane to 1-*exo*-hydroxy-2-chlorochlordene and 1-*exo*-hydroxy-2-*endo*-chloro-2,3-*exo*-epoxychlordene, and to various other hydroxylated products (S. Tashiro & F. Matsumura, *J. Agric. Food Chem.*, 1977, **25**, 872–880). **Soil/Environment** DT_{50} in soil *c.* 1 y.

122 chlorethoxyfos

Insecticide

organophosphorus

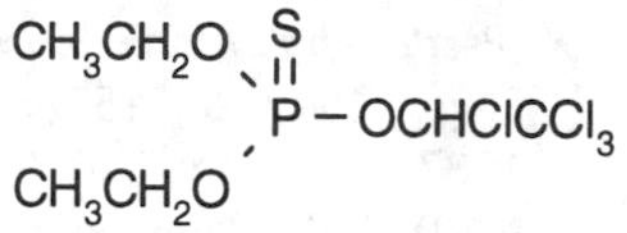

$$CH_3CH_2O \diagdown \underset{\diagup}{\overset{S}{\underset{\parallel}{P}}} - OCHClCCl_3$$
$$CH_3CH_2O$$

NOMENCLATURE
Common name chlorethoxyfos (BSI, pa E-ISO)
IUPAC name (±)-*O,O*-diethyl *O*-(1,2,2,2-tetrachloroethyl) phosphorothioate
Chemical Abstracts name *O,O*-diethyl *O*-(1,2,2,2-tetrachloroethyl) phosphorothioate
CAS RN *[54593–83–8]* **Development codes** SD 208 304; WL 208 304; DPX-43898

PHYSICAL CHEMISTRY
Composition Tech. is 88%. **Mol. wt.** 336.0 **M.f.** $C_6H_{11}Cl_4O_3PS$
B.p. 110–115 °C/0.8 mmHg **V.p.** *c.* 106 mPa (20 °C) **K_{ow}** logP = 4.59 (25 °C)
S.g./density 1.41 (20 °C) **Solubility** In water <1 mg/l; tech. 3 ppm (20 °C). Soluble in acetonitrile, chloroform, ethanol, hexane, xylene. **Stability** Stable at room temperature >18 months; stable at 55 °C for 2 weeks neat, in 316 stainless steel and in 504 ppm Fe_2O_3. Hydrolysis DT_{50} (25 °C) 4.3 d (pH 5), 59 d (pH 7), 72 d (pH 9). **F.p.** >230 °C; (tech., 38 °C)

COMMERCIALISATION
History Insecticide reported by I. A. Watkinson & D. W. Sherrod (*Proc. 1986 Br. Crop Prot. Conf. – Pests Dis.*, **1,** 107). Introduced by E. I. du Pont de Nemours and

Co. **Manufacturers** Du Pont

APPLICATIONS
Mode of action Works by vapour action. **Uses** Soil insecticide for use in controlling corn rootworm and other soil pests. **Formulation types** GR.
Selected tradenames 'Fortress' (Du Pont)

MAMMALIAN TOXICOLOGY
EHC 63 (WHO, 1986; a general review of organophosphorus insecticides). **Oral** Acute oral LD_{50} for female rats 1.8, male rats 4.8 mg/kg. **Skin and eye** Acute percutaneous LD_{50} for female rabbits 12.5, male rabbits 18.5 mg/kg. Moderate eye irritant but highly toxic by eye contact (rabbits). Not a skin irritant (rabbits); not a skin sensitiser (guinea pigs). **Inhalation** LC_{50} (4 h) for rats 0.58 ppm (8 mg/m^3), extremely toxic by inhalation. **NOEL** For male mice 0.18, female mice 0.21, male rats 0.18, female rats 0.25, male dogs 0.063, female dogs 0.065 mg/kg daily. **Other** Non-oncogenic, non-teratogenic, non-mutagenic.
Toxicity class EPA (formulation) I

ECOTOXICOLOGY
Birds Acute oral LD_{50} (gavage) for bobwhite quail 28 mg/kg. **Fish** LC_{50} (96 h) for rainbow trout 0.10, bluegill sunfish 0.0023, sheepshead minnow 0.00047 mg/l.

ENVIRONMENTAL FATE
Animals Major metabolites are CO_2 and biosynthetic intermediates such as serine, glycine and glycine conjugates. **Plants** The proposed pathway is degradation to trichloroacetic acid and thence to oxalic acid, etc. **Soil/Environment** Soil DT_{50} (25 °C) (lab. studies) 7 d (Hanford sandy loam), 20 d (Flanagan loam); field soil dissipation DT_{50} 2–3 d. K_d 33–98.

23 chlorfenapyr   *Insecticide, acaricide*

pyrazole (acaricide) analogue

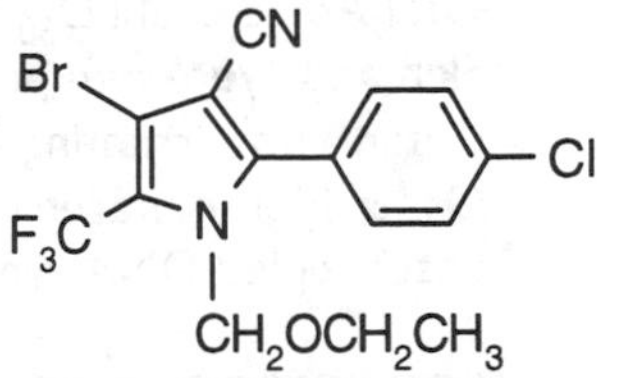

NOMENCLATURE
Common name chlorfenapyr (BSI, pa ISO)

IUPAC name 4-bromo-2-(4-chlorophenyl)-1-ethoxymethyl-=
5-trifluoromethylpyrrole-3-carbonitrile
Chemical Abstracts name 4-bromo-2-(4-chlorophenyl)-1-(ethoxymethyl)-=
5-(trifluoromethyl)-1*H*-pyrrole-3-carbonitrile
CAS RN *[122453–73–0]*   **Development codes** AC 303,630; CL 303,630
(Cyanamid)

PHYSICAL CHEMISTRY
Mol. wt. 407.6   **M.f.** $C_{15}H_{11}BrClF_3N_2O$ **Form** White solid.   **M.p.** 100–101 °C
K_{ow} logP = 4.83   **Solubility** Practically insoluble in water. Soluble in acetone,
diethyl ether, dimethyl sulfoxide, tetrahydrofuran, acetonitrile, and alcohols.

COMMERCIALISATION
History Under development by American Cyanamid Co.

APPLICATIONS
Biochemistry Oxidative removal *in vivo* of the *N*-ethoxymethyl group generates
the active species, which is a mitochondrial uncoupler.   **Mode of action** Insecticide
and acaricide with mainly stomach and some contact action. Exhibits good
translaminar but limited systemic activity in plants.   **Uses** Control of many species
of insects and mites, including those resistant to carbamate, organophosphate and
pyrethroid insecticides and also chitin-synthesis inhibitors, in cotton, vegetables,
citrus, top fruit, vines and soya beans. Among pests resistant to conventional
products which are controlled by chlorfenapyr are *Brevipalpus phoenicis* (leprosis
mite), *Leptinotarsa decemlineata* (Colorado potato beetle), *Helicoverpa/Heliothis*
spp., *Plutella xylostella* (diamond-back moth) and *Tetranychus spp*. Its use in
resistance management programmes for control of various cotton pests is under
evaluation.   **Phytotoxicity** No phytotoxicity observed at field use rates.
Formulation types EC; SC.   **Selected tradenames** 'Pirate' (cotton) (Cyanamid);
'Stalker' (Cyanamid)

ANALYSIS
Product by glc.   **Residues** by hplc. Details from Cyanamid.

MAMMALIAN TOXICOLOGY
Oral Acute oral LD_{50} for male rats 441, female rats 1152 mg tech./kg.
Skin and eye Acute percutaneous LD_{50} for rabbits >2000 mg/kg. Moderate eye
irritant; non-irritating to skin (rabbits).   **Inhalation** LC_{50} for rats 1.9 mg tech./l air.
Other Non-mutagenic in the Ames, CHO/HGPRT, mouse micronucleus and
unscheduled DNA synthesis tests.

ECOTOXICOLOGY
Birds Acute oral LD_{50} for mallard ducks 10, bobwhite quail 34 mg/kg. LC_{50} (8 d)
for mallard ducks 9.4, bobwhite quail 132 ppm.   **Fish** LC_{50} (48 h) for carp

500 µg/l. LC$_{50}$ (96 h) for rainbow trout 7.44, bluegill sunfish 11.6 µg/l.
Bees LD$_{50}$ 0.2 µg/bee. **Daphnia** LC$_{50}$ (96 h) 6.11 µg/l.

24 chlorfenvinphos *Insecticide, acaricide*

organophosphorus

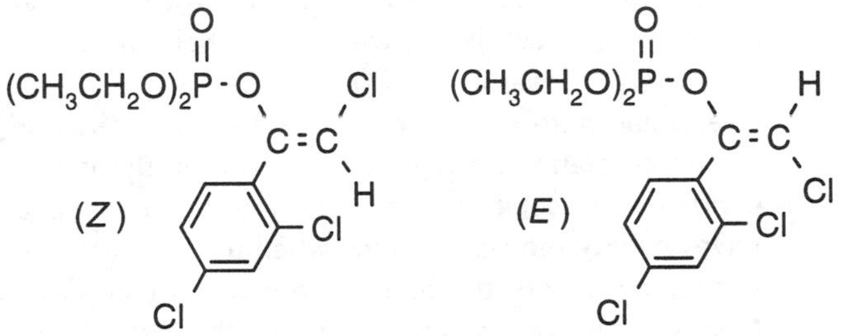

NOMENCLATURE
Common name chlorfenvinphos (BSI, E-ISO, (*m*) F-ISO, BAN); CVP (JMAF);
no name (USA)
IUPAC name 2-chloro-1-(2,4-dichlorophenyl)vinyl diethyl phosphate
Chemical Abstracts name 2-chloro-1-(2,4-dichlorophenyl)ethenyl diethyl
phosphate
CAS RN *[470–90–6]* (formerly *[2701–86–2]*) (*Z*)- + (*E*)-isomers; *[18708–87–7]*
(*Z*)-isomer; *[18708–86–6]* (*E*)-isomer **EEC no.** 207–432–0
Development codes SD 7859; C 8949 (Ciba-Geigy); GC 4072 (Allied);
CGA 26351 **Official codes** OMS 166; OMS 1328; ENT 24 969

PHYSICAL CHEMISTRY
Composition Tech. grade (≥90% (*Z*)- and (*E*)- isomers), typical ratio (*Z*):(*E*) 8.6:1.
The dimethyl ester analogue is chlorfenvinphos-methyl (*q.v.*). **Mol. wt.** 359.6
M.f. $C_{12}H_{14}Cl_3O_4P$ **Form** Colourless liquid; (tech., amber liquid).
M.p. –23 to –19 °C **B.p.** 167–170 °C/0.5 mmHg **V.p.** 1 mPa (25 °C); 0.53 mPa
(extrapolated to 20 °C) **K$_{ow}$** logP = 3.85 ((*Z*)-isomer), 4.22 ((*E*)-isomer)
S.g./density 1.36 at 20 °C **Solubility** In water 145 mg/l (23 °C). Miscible with
most common organic solvents, e.g. ethanol, acetone, dichloromethane, hexane,
xylene, propylene glycol, kerosene. **Stability** Hydrolysed slowly in neutral, acidic,
and slightly alkaline aqueous solutions. Hydrolysed rapidly in strongly alkaline
solutions. DT$_{50}$ (38 °C) >700 h (pH 1.1), >400 h (pH 9.1); (20 °C) 1.28 h
(pH 13).

COMMERCIALISATION

History Insecticide reported by W. F. Chamberlain et al. (*J. Econ. Entomol.*, 1962, **55**, 86). Introduced by the Shell International Chemical Company Ltd (now American Cyanamid Co.), Ciba AG (now Novartis Crop Protection AG) and by Allied Chemical Corp. (Novartis Crop Protection and Allied no longer produce or market it).   **Patent** US 2956075; US 3116201 to Shell

APPLICATIONS

Biochemistry Cholinesterase inhibitor.   **Mode of action** Insecticide and acaricide with contact and stomach action. Long residual activity.   **Uses** Soil application for control of root flies, rootworms, and other soil insects in vegetables; frit flies in maize; wheat bulb flies in wheat; bean seed flies; and phorid and sciarid flies in mushrooms. Foliar application for control of Colorado beetles on potatoes; scale insects and mite eggs on citrus fruit; stem borers and leafhoppers on rice; stem borers on maize and sugar cane; and whitefly on cotton. Used in public health for control of mosquito larvae. Also used as an animal ectoparasiticide.
Phytotoxicity Non-phytotoxic when used as directed. Some injury may occur if applied directly to the seeds of certain crops.   **Formulation types** GR; EC.
Mixtures (*chlorfenvinphos* +) cypermethrin; cypermethrin + methoxychlor; alpha-cypermethrin; dimethoate; methoxychlor; oxamyl; phenol; petroleum oils.
Compatibility Incompatible with alkaline materials.   **Selected tradenames** 'Birlane' (crop protection use) (Cyanamid); 'Supona' (veterinary use) (Cyanamid); 'Apachlor' (Rhône-Poulenc)

ANALYSIS

Product analysis by glc (*CIPAC Handbook*, 1980, **1A**, 1131; FAO Specification (CP/66)).   **Residues** determined by glc (*Pestic. Anal. Man.*, 1979, **I**, 201-I; 1979, **II**; D. C. Abbott et al., *Pestic. Sci.*, 1970, **1**, 10; A. Ambrus et al., *J. Assoc. Off. Anal. Chem.*, 1981, **64**, 733; *Man. Pestic. Residue Anal.*, 1987, **I**, 3, 5, 6, S8, S13, S17, S19; *Anal. Methods Residues Pestic.*, 1988, Part I, M2, M5, M12; *Analyst (London)*, 1980, **105**, 515; 1985, **110**, 765).

MAMMALIAN TOXICOLOGY

Reviews *Pesticide residues in food – 1994*, FAO Plant Production and Protection Paper, 127, 1995. *Pesticide residues in food – 1994 evaluations, Pt. II – Toxicology*, WHO, WHO/PCS/95.2, 1995.   **EHC** 63 (WHO, 1986; a general review of organophosphorus insecticides).   **Oral** Acute oral LD_{50} for rats 10, mice 117–200, rabbits 300–1000, dogs >12 000 mg/kg.   **Skin and eye** Acute percutaneous LD_{50} for rats 31–108, rabbits 400–4700 mg/kg. Non-irritating to skin and eyes (rabbits).   **Inhalation** LC_{50} (4 h) for rats c. 0.05 mg/l air.   **NOEL** (2 y) for rats and dogs 1 mg/kg diet (0.05 mg/kg daily).   **ADI** (JMPR) 0.0005 mg/kg b.w. [1994].
Toxicity class WHO (a.i.) Ia; EPA (formulation) I   **EC risk** T+ (R28); T (R24)

ECOTOXICOLOGY
Birds Acute oral LD_{50} for pheasants 107, pigeons 16 mg/kg.   **Fish** LC_{50} (96 h) for harlequin fish <0.32, guppies 0.3–1.6 mg/l.   **Bees** LD_{50} (24 h) (oral) 0.55 μg/bee, (topical) 4.1 μg/bee.   **Daphnia** LC_{50} (48 h) 0.3 μg/l.

ENVIRONMENTAL FATE
Animals Primary metabolic detoxification is by de-esterification to give 2-chloro-1-(2,4-dichlorophenyl)vinyl ethyl hydrogen phosphate. Ultimate metabolites include the glucuronides of 2,4-dichlorophenylethanediol and 1-(2,4-dichlorophenyl)=ethanol, and N-(2,4-dichlorobenzoyl)glycine; see D. H. Hutson in "*Prog. in Pesticide Biochem*", Vol. 1, D. H. Hutson & T. R. Roberts, eds., Wiley (1981).
Plants Metabolism is similar to that in mammals; see K. I. Beynon *et al., Residue Rev.*, 1973, **47**, 55.

25 chlorfluazuron   *Insecticide*

benzoylurea

NOMENCLATURE
Common name chlorfluazuron (BSI, draft E-ISO, (*m*) draft F-ISO)
IUPAC name 1-[3,5-dichloro-4-(3-chloro-5-trifluoromethyl-2-pyridyloxy)phenyl]-=3-(2,6-difluorobenzoyl)urea
Chemical Abstracts name *N*-[[[3,5-dichloro-4-[[3-chloro-5-(trifluoromethyl)-=2-pyridinyl]oxy]phenyl]amino]carbonyl]-2,6-difluorobenzamide
CAS RN *[71422–67–8]*   **Development codes** IKI-7899 (Ishihara); CGA 112 913 (Ciba-Geigy); PP145 (Zeneca); UC 64 644 (Rhône-Poulenc)

PHYSICAL CHEMISTRY
Composition Tech. is ≥94.0%.   **Mol. wt.** 540.7   **M.f.** $C_{20}H_9Cl_3F_5N_3O_3$
Form White crystals.   **M.p.** 226.5 °C (decomp.)   **V.p.** $<1 \times 10^{-5}$ mPa (20 °C)
K_{ow} logP = 5.8 (unionised)   **Henry** 1.82×10^{-7} Pa m^3 mol^{-1} (25 °C)
S.g./density 1.663 (20 °C)   **Solubility** In water <0.01 mg/l (20 °C). In hexane <0.01, *n*-octanol 1, xylene 2.5, methanol 2.5, toluene 6.6, isopropanol 7, dichloromethane 22, acetone 55, cyclohexanone 110 (all in g/l, 20 °C).
Stability Stable to heat and light, and to hydrolysis.   **pKa** 8.10, v. weak acid
F.p. 224.0–224.5 °C (OECD 102)

COMMERCIALISATION
History Insecticide reported by T. Haga et al. (Abstr. 5th IUPAC Congr. Pestic. Chem., 1982, IId-7). Discovered by Ishihara Sangyo Kaisha, Ltd, also evaluated by other companies. **Manufacturers** Ishihara Sangyo

APPLICATIONS
Biochemistry Chitin synthesis inhibitor (R. Neumann & W. Guyer, Proc. 10th Int. Congr. Plant Prot., 1983, **1**, 445). **Mode of action** Insect growth regulator which acts as an anti-moulting agent, leading to death of the larvae and pupae.
Uses Control of Heliothis, Spodoptera, Bemisia tabaci and other chewing insects on cotton, and Plutella, Thrips and other chewing insects on vegetables. Also used on fruit, potatoes, ornamentals and tea. **Formulation types** EC; SC; UL.
Mixtures (chlorfluazuron +) profenofos; ethiofencarb.
Selected tradenames 'Aim' (Ishihara Sangyo); 'Atabron' (Ishihara Sangyo, Zeneca); 'Helix' (ISK Biosciences, Crop Care); 'Jupiter' (Ishihara Sangyo)

ANALYSIS
Product analysis by hplc or glc. **Residues** by hplc with u.v. detection. Details from Ishihara Sangyo.

MAMMALIAN TOXICOLOGY
Oral Acute oral LD_{50} for rats >8500, mice 7000 mg/kg. **Skin and eye** Acute percutaneous LD_{50} for rats >1000, rabbits >2000 mg/kg. Non-irritating to skin; mild eye irritant (rabbits). Not a skin sensitiser. **Inhalation** LC_{50} (4 h) for rats >2.4 mg/l. **Other** Non-mutagenic in the Ames test. **Toxicity class** WHO (a.i.) III (Table 5)

ECOTOXICOLOGY
Birds Acute oral LD_{50} for quail and mallard ducks >2510 mg/kg. Eight-day dietary LC_{50} for quail and mallard ducks >5620 mg/kg diet. **Fish** LC_{50} (48 h) for carp >300 mg/l. **Bees** LD_{50} (oral) >100 µg/bee. **Worms** LC_{50} (28 d) for earthworms >1000 mg/kg soil. **Daphnia** LC_{50} (48 h) 0.908 µg/l. **Algae** EC_{50} (12 d) for green algae >1.8 mg/l.

ENVIRONMENTAL FATE
Animals Metabolised in rats by cleavage of the urea bridge. **Plants** Slowly degraded in plants. **Soil/Environment** Half-life in soil varies from c. 6 weeks to a few months. K_d 120–990. Slowly degraded in water, DT_{50} for aqueous photodegradation 20 h.

chloridazon   *Herbicide*

---

pyridazinone (PSII)

NOMENCLATURE
Common name chloridazon (BSI, E-ISO); chloridazone ((*f*) F-ISO); pyrazon, a
name formerly used by BSI and in many countries, is retained (ANSI, WSSA,
Canada, Denmark, Poland); PAC (JMAF)
IUPAC name 5-amino-4-chloro-2-phenylpyridazin-3(2*H*)-one
Chemical Abstracts name 5-amino-4-chloro-2-phenyl-3(2*H*)-pyridazinone
Other names PCA   **CAS RN** *[1698–60–8]* formerly *[58858–18–7]*
EEC no. 216–920–2   **Development codes** BAS 119H

PHYSICAL CHEMISTRY
Mol. wt. 221.6   **M.f.** $C_{10}H_8ClN_3O$ **Form** Colourless, odourless solid; (tech.,
brown, almost odourless, solid).   **M.p.** 206 °C (tech., 198–202 °C)
V.p. <0.01 mPa (20 °C)   **K$_{ow}$** logP = 1.19 (pH 7)   **S.g./density** 1.54 (20 °C)
Solubility In distilled water 0.34 g/l (20 °C). In methanol 15.1, ethyl acetate 3.7,
dichloromethane 1.9, toluene 0.1 (all in g/l, 20 °C). Practically insoluble in
n-heptane.   **Stability** Stable up to 50 °C for ≥2 years. Stable in aqueous media at
pH 3–9. DT_{50} in simulated sunlight (xenon lamp, 2100 µEinstein m^{-2} s^{-1}) 150 h
(pH 7, water).

COMMERCIALISATION
History Herbicide reported by A. Fischer (*Weed Res.*, 1962, **2**, 177). Introduced by
BASF AG.   **Patent** US 3222159; US 3210353; DE 1105232   **Manufacturers** BASF;
Oxon Italia

APPLICATIONS
Biochemistry Photosynthetic electron transport inhibitor at the photosystem II
receptor site.   **Mode of action** Selective systemic herbicide, rapidly absorbed by
the roots, with translocation acropetally to all plant parts.   **Uses** Control of
annual broad-leaved weeds in sugar beet, fodder beet and beetroot, by
application pre-plant incorporated, pre-emergence, or post-emergence at
1.3–3.25 kg a.i./ha. Often used in combination with other herbicides.
Phytotoxicity Phytotoxic to e.g. cabbage, carrots, cucumbers, lima beans and
tomatoes.   **Formulation types** FL; SC; WP; WG.   **Mixtures** *(chloridazon +)*
ethofumesate; lenacil; metolachlor; tri-allate; fenuron + propham;
chlorpropham + fenuron + propham; propachlor; chlorpropham; cycloate;
lenacil + metolachlor.   **Selected tradenames** 'Pyramin' (BASF); 'Rebell' (mixture)

(BASF); 'Flirt' (mixture) (BASF); 'Betozon' (Sipcam); 'Erbitox Bietole' (Siapa); 'Gladiator' (Tripart); 'Starter' (Truchem); 'Trojan' (AgrEvo)

ANALYSIS
Product analysis by i.r. spectrometry or by hplc (*CIPAC Handbook*, 1988, **D**, 31). **Residues** in plants determined by glc or hplc (*Rückstandanalytik von Pflanzenschutzmitteln*, 1987, Verlag Chemie, **9**, Lieferung 89-A-1).

MAMMALIAN TOXICOLOGY
Oral Acute oral LD_{50} for male rats 3830, female rats 2140, male mice 2860, female mice 3100 mg/kg. **Skin and eye** Acute percutaneous LD_{50} for rats >2000 mg/kg. Non-irritating to skin and eyes (rabbits). **Inhalation** LC_{50} (4 h) for rats >5.4 mg/l air. **NOEL** (2 y) for rats 16, mice 152 mg/kg b.w. daily. **ADI** 0.16 mg/kg. **Other** Non-mutagenic, non-teratogenic, non-carcinogenic. **Toxicity class** WHO (a.i.) III (Table 5); EPA (formulation) III **EC risk** Xi (R43)

ECOTOXICOLOGY
Birds Acute oral LD_{50} for bobwhite quail >2000 mg/kg b.w. Dietary LC_{50} for bobwhite quail >5000, mallard ducks 4260 mg/kg diet. **Fish** LC_{50} (96 h) for trout 32–46, bluegill sunfish 93 mg/l. **Bees** Not toxic to bees; LD_{50} (48 h) (oral and contact) >200 μg/bee. **Worms** LC_{50} (14 d) for *Eisenia foetida* 1050 ppm. **Daphnia** LC_{50} (48 h) 132 mg/l. **Algae** EC_{50} (120 h) for *Chlorella fusca* 1.9 mg/l.

ENVIRONMENTAL FATE
Animals In rats, following a single oral application, 85% is excreted in the urine within 1 hour, and 13% in the faeces. After repeated oral applications, excretion proceeds similarly, with 75% in the urine and 15% in the faeces. **Plants** In beet, conjugation to form the *N*-glucosyl metabolite occurs. **Soil/Environment** In soil, microbial degradation involves splitting off the phenyl group, to give 5-amino-= 4-chloropyridazin-3(2*H*)-one, which is not herbicidally-active (A. Fischer, *Weed Res.*, 1962, **2**, 177–184). Persists in soil for c. 6–8 weeks under sufficiently moist conditions. K_{oc} 89–340.

chlorimuron-ethyl *Herbicide*

sulfonylurea

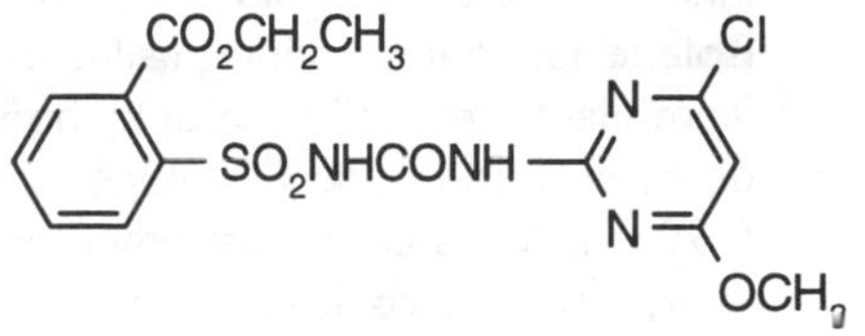

NOMENCLATURE
chlorimuron-ethyl
Common name chlorimuron-ethyl
IUPAC name ethyl 2-(4-chloro-6-methoxypyrimidin-2-ylcarbamoylsulfamoyl)=
benzoate
Chemical Abstracts name ethyl 2-[[[[(4-chloro-6-methoxy-2-pyrimidinyl)=
amino]carbonyl]amino]sulfonyl]benzoate
CAS RN *[90982–32–4]* **Development codes** DPX-F6025

chlorimuron
Common name chlorimuron (BSI, ANSI, WSSA, draft E-ISO, draft F-ISO)
IUPAC name 2-(4-chloro-6-methoxypyrimidin-2-ylcarbamoylsulfamoyl)benzoic
acid
Chemical Abstracts name 2-[[[[(4-chloro-6-methoxy-2-pyrimidinyl)amino]=
carbonyl]amino]sulfonyl]benzoic acid
CAS RN *[99283–00–8]*

PHYSICAL CHEMISTRY
chlorimuron-ethyl
Composition Tech. is >95%. **Mol. wt.** 414.8 **M.f.** $C_{15}H_{15}ClN_4O_6S$
Form Colourless crystals. **M.p.** 181 °C **V.p.** 4.9×10^{-7} mPa **K_{ow}** logP = 0.11
(pH 7) **S.g./density** 1.51 (25 °C) **Solubility** In water 9 (pH 5), 1200 (pH 7)
(both in mg/l, 25 °C). Low solubility in organic solvents. **Stability** In water DT_{50}
17–25 d (pH 5, 25 °C). **pKa** 4.2

chlorimuron
Mol. wt. 386.8 **M.f.** $C_{13}H_{11}ClN_4O_6S$

COMMERCIALISATION
History Chlorimuron-ethyl introduced as herbicide by E. I. du Pont de Nemours
and Co. First US registration in 1986. **Patent** US 4394506; US 4547215
Manufacturers Du Pont; IPESA

MAIN ENTRIES

APPLICATIONS
chlorimuron-ethyl
Biochemistry Branched chain amino acid synthesis (acetolactate synthase or ALS) inhibitor. Acts by inhibiting biosynthesis of the essential amino acids valine and isoleucine, hence stopping cell division and plant growth. Crop selectivity derives from plant metabolism both by homoglutathione conjugation and by de-esterification (M. K. Koeppe & H. M. Brown, *Agro-Food-Industry*, **6**, 9–14 (1995)).   **Uses** Used post-emergence for control of important broad-leaved weeds, such as cocklebur, pigweed, sunflower and annual morning glory, in soya beans and peanuts. Active at 9–13 g/ha.   **Formulation types** WG.
Mixtures *(chlorimuron-ethyl +)* metribuzin.   **Selected tradenames** 'Classic' (Du Pont); 'Darban' (IPESA)

ANALYSIS
Product analysis by hplc (R. A. Guinivan et al., *Anal. Methods Pestic. Plant Growth Regul.*, 1988, **16**, 37).   **Residues** determined by hplc (J. L. Prince & R. A. Guinivan, *J. Agric. Food Chem.*, 1988, **36**, 1). Methods for sulfonylurea residues in crops, soil and water reviewed (A. C. Barefoot et al., *Proc. Br. Crop Prot. Conf. – Weeds*, 1995, **2**, 707).

MAMMALIAN TOXICOLOGY
chlorimuron-ethyl
Oral Acute oral LD_{50} for rats 4102 mg/kg.   **Skin and eye** Acute percutaneous LD_{50} for rabbits >2000 mg/kg. Not a skin irritant or eye irritant (rabbits). Not a skin sensitiser (guinea pigs).   **Inhalation** LC_{50} (4 h) for rats >5 mg/l air.
NOEL (2 y) for rats 250 mg/kg diet (12.5 mg/kg daily); (1 y) for dogs 250 mg/kg diet (6.25 mg/kg daily). NOEL in: reproduction (2-generation) in rats 250 mg/kg diet; teratogenicity in rats 30, rabbits 15 mg/kg.   **ADI** 0.020 mg/kg b.w.
Toxicity class WHO (a.i.) III (Table 5); EPA (formulation) III

ECOTOXICOLOGY
chlorimuron-ethyl
Birds Acute oral LD_{50} (14 d) for mallard ducks >2510 mg/kg. Dietary LC_{50} for mallard ducks and bobwhite quail >5620 ppm.   **Fish** LC_{50} (96 h) for trout >1000, bluegill sunfish >100 mg/l.   **Bees** LD_{50} (48 h) >12.5 µg/bee.   **Worms** LC_{50} for earthworm (*Eisenia foetida*) >4050 mg/kg.   **Daphnia** LC_{50} (48 h) 1000 mg/l.
Other aquatic spp. *Lemna gibba* E_bC_{50} 0.45 ppb, E_rC_{50} 45 µg/l, EC_{50} (frond counts) 0.27 ppb. LC_{50} for crayfish >1000 ppm.

ENVIRONMENTAL FATE
Animals Chlorimuron-ethyl is rapidly and extensively metabolised in the hen; 18 metabolites in the excreta were resolved by hplc.   **Soil/Environment** In soil, K_d >1.60 (pH 4.5, 5.6% o.m.), 0.28 (pH 5.8, 4.3 % o.m.), <0.03 (pH 6.5, 2.1% o.m.), <0.03 (pH 6.6, 1.1 % o.m.).

chlormephos   *Insecticide*

organophosphorus

$$CICH_2SP(OCH_2CH_3)_2$$

with the structure showing $\overset{S}{\underset{\|}{}}$ above the P.

NOMENCLATURE
Common name chlormephos (BSI, E-ISO, (*m*) F-ISO)
IUPAC name S-chloromethyl O,O-diethyl phosphorodithioate
Chemical Abstracts name S-(chloromethyl) O,O-diethyl phosphorodithioate
CAS RN *[24934–91–6]* **EEC no.** 246–538–1 **Development codes** MC 2188
(Murphy)

PHYSICAL CHEMISTRY
Mol. wt. 234.7 **M.f.** $C_5H_{12}ClO_2PS_2$ **Form** Colourless liquid.
B.p. 81–85 °C/0.1 mmHg **V.p.** 7.6 × 10^3 mPa (30 °C) **S.g./density** 1.260
(20 °C) **Solubility** In water 60 mg/l (20 °C). Miscible with most organic solvents.
Stability Stable in neutral and weakly acidic media at room temperature, but
hydrolysed by dilute acids and alkalis at 80 °C. Rapidly hydrolysed in alkaline
media.

COMMERCIALISATION
History Insecticide reported by F. Colliot *et al.* (*Proc. Br. Insectic. Fungic. Conf.*, 7th,
1973, **2**, 557). Introduced by Murphy Chemical Ltd and developed under licence
by Rhône-Poulenc Agrochimie. **Patent** GB 1258922; GB 817360; GB 902795 to
Murphy **Manufacturers** Rhône-Poulenc

APPLICATIONS
Biochemistry Cholinesterase inhibitor. **Mode of action** Non-systemic insecticide
with contact and some stomach action. **Uses** Soil application for control of white
grubs, wireworms, millipedes, symphylids, and other soil-dwelling insects in maize,
sugar beet, fodder beet, sugar cane, potatoes, tobacco, cereals, and other crops.
Applied at 2–4 kg a.i./ha for overall application and 0.3–0.4 kg/ha for band
treatment. **Phytotoxicity** Non-phytotoxic when used as directed. Sorghum and
soya beans may be injured. **Formulation types** GR. **Compatibility** Can be
combined with various systemic organophosphorus insecticides.
Selected tradenames 'Dotan' (Rhône-Poulenc)

ANALYSIS
Product analysis by glc with FID (V. P. Lynch, *Anal. Methods Pestic. Plant Growth
Regul.*, 1978, **10**, 49). **Residues** determined by glc with TID (*idem, ibid.*).

MAMMALIAN TOXICOLOGY
EHC 63 (WHO, 1986; a general review of organophosphorus insecticides).

Oral Acute oral LD$_{50}$ for rats 7 mg/kg. **Skin and eye** Acute percutaneous LD$_{50}$ for rats 27 mg a.i./kg; for rats and rabbits >1600 mg (as 5% GR)/kg. **NOEL** (90 d) in rats 0.39 mg/kg diet. **Toxicity class** WHO (a.i.) Ia; EPA (formulation) I **EC risk** T+ (R27/28)

ECOTOXICOLOGY
Birds Acute oral LD$_{50}$ for quail 260 mg/kg. **Fish** Toxic to fish. LC$_{50}$ for harlequin fish 1.5 mg/l. **Bees** Toxic to bees.

ENVIRONMENTAL FATE
Animals In rats, following oral administration, there is almost complete elimination within 24 hours in the urine as diethyl phosphate and diethyl phosphorothioate. **Soil/Environment** In soil, chlormephos is converted to ethion.

129 chlormequat chloride *Plant growth regulator*

quaternary ammonium

$$ClCH_2CH_2N^+(CH_3)_3 \quad Cl^-$$

NOMENCLATURE
chlormequat chloride
Common name chlormequat chloride
Other names chlorocholine chloride; CCC **CAS RN** *[999–81–5]*
EEC no. 213–666–4 **Development codes** BAS 062 W; AC 38 555

chlormequat
Common name chlormequat (BSI, E-ISO, (*m*) F-ISO)
IUPAC name 2-chloroethyltrimethylammonium
Chemical Abstracts name 2-chloro-*N*,*N*,*N*-trimethylethanaminium
CAS RN *[7003–89–6]*

PHYSICAL CHEMISTRY
chlormequat chloride
Composition ≥96% pure; tech. concentrate 750 g/l. **Mol. wt.** 158.1
M.f. C$_5$H$_{13}$Cl$_2$N **Form** Colourless, extremely hygroscopic crystals, with a weak intrinsic odour (tech., pale yellow crystals with a fish-like odour). Usually produced as an aqueous solution. **M.p.** 235 °C **V.p.** <0.01 mPa (20 °C)
K$_{ow}$ logP = –1.59 (pH 7) **Solubility** In water >1 kg/kg (20 °C). In ethanol 320, dichloromethane, ethyl acetate, *n*-hexane <0.1, acetone 0.2, chloroform 0.3

(all in g/kg, 20 °C). **Stability** Extremely hygroscopic; aqueous solutions are stable. Decomposes at 245 °C.

chlormequat
Mol. wt. 122.6   **M.f.** $C_5H_{13}CIN$

COMMERCIALISATION
History Chlormequat chloride reported as plant growth regulator by N. E. Tolbert (*Plant Physiol.*, 1960, **35**, 380; *J. Biol. Chem.*, 1960, **235**, 475). Introduced, in collaboration with Michigan State University, by American Cyanamid Co. and by BASF AG in Germany (1966).   **Patent** US 3156554 to Cyanamid; US 3395009 to BASF; GB 1092138 to BASF; DE 1199048   **Manufacturers** Aimco; Allied Colloids; BASF; Cequisa; Cyanamid; Hico; Makhteshim-Agan; Nufarm GmbH; Sarabhai; UCB

APPLICATIONS
chlormequat chloride
Biochemistry Gibberellin biosynthesis inhibitor.   **Mode of action** Plant growth regulator which inhibits cell elongation, hence shortening and strengthening the stem and producing a sturdier plant. Also influences the developmental cycle, leading to increased flowering and harvest. May also increase chlorophyll formation and root development.   **Uses** Used to increase resistance to lodging (by shortening and strengthening the stem) and to increase yields in wheat, rye, oats, and triticale; to promote lateral branching and flowering in azaleas, fuchsias, begonias, poinsettias, geraniums, pelargoniums, and other ornamental plants; to promote flower formation and improve fruit setting in pears, almonds, vines, olives, and tomatoes; to prevent premature fruit drop in pears, apricots, and plums; etc. Also used on cotton, vegetables, tobacco, sugar cane, mangoes, and other crops.   **Formulation types** SL; DP.   **Mixtures** *(chlormequat chloride +)* choline chloride; carbendazim; ethephon; 9-(10-nor-*p*-menth-1-en-8-yl)-*p*-menth-= 1-ene; choline chloride + ethephon; ethephon + mepiquat chloride.
Compatibility Should not be combined with dinoseb, cyanazine, or other contact herbicides.   **Selected tradenames** 'Cycocel' (BASF, Cyanamid); 'CeCeCe' (BASF); 'CCC700' (FCC); 'Ceku-CCC' (Cequisa); 'Cropsafe 5C' (Hortichem); 'Cycogan' (Makhteshim-Agan); 'Hormocel' (Aimco); 'Atlas Quintacel' (Atlas Crop Protection); 'Titan' (AgrEvo)

ANALYSIS
Product analysis by potentiometric titration with silver nitrate (chlormequat chloride and choline chloride can thus be determined separately) (*CIPAC Handbook*, 1988, **D**, 39), by colorimetry (N. R. Pasarela & E. J. Orloski, *Anal. Methods Pestic. Plant Growth Regul.*, 1974, **7**, 523) or by ion chromatography (details available from BASF).   **Residues** determined by glc after reaction with sodium phenylsulfide (F. Tafuri *et al.*, *Analyst (London)*, 1970, **95**, 675;

R. P. Mooney & N. R. Pasarela, *J. Agric. Food Chem.*, 1967, **15**, 989) or by
colorimetry (N. R. Pasarela & E. J. Orloski, *loc. cit.*; *Anal. Methods
Residues Pestic.*, 1988, Part II; T. Stijve, *Deutsche Lebensm. Rundsch.*, 1980, **76**,
234). Details available from Cyanamid or BASF.

MAMMALIAN TOXICOLOGY
chlormequat chloride
Reviews *Pesticide residues in food – 1994*, FAO Plant Production and Protection
Paper, 127, 1995. *Pesticide residues in food – 1994 evaluations, Pt. II – Toxicology*,
WHO, WHO/PCS/95.2, 1995.   **Oral** Acute oral LD_{50} for male rats 966, female
rats 807 mg/kg.   **Skin and eye** Acute percutaneous LD_{50} for rats >4000, rabbits
>2000 mg/kg. Not irritating to skin and eyes; not a skin sensitiser.
Inhalation LC_{50} (4 h) for rats >5.2 mg/l air.   **NOEL** (2 y) for rats 50, male mice
336, female mice 23 mg/kg b.w.   **ADI** (JMPR) No ADI [1994].   **Other** Toxicity to
mammals is reduced by the addition of choline chloride (DEP 1215436).   **Toxicity
class** WHO (a.i.) III; EPA (formulation) III   **EC risk** Xn (R21/22)

ECOTOXICOLOGY
chlormequat chloride
Birds Acute oral LD_{50} for Japanese quail 555, pheasants 261, chickens 920 mg/kg.
Fish LC_{50} (96 h) for mirror carp and rainbow trout >100 mg/l.   **Bees** Non-toxic
to bees.   **Worms** LC_{50} (14 d) for *Eisenia foetida* 320 mg/kg soil.   **Daphnia** LC_{50}
(48 h) 16.9 mg/l.   **Algae** EC_{50} (72 h) for *Scenedesmus subspicatus* >100 mg/l; EC_{50}
(cell volume) for *Chlorella fusca* 5656 mg/l.   **Other aquatic spp.** LC_{50} (96 h) for
fiddler crab ≥ 1000, shrimp 804, oysters 67 mg/l.

ENVIRONMENTAL FATE
Animals In goats, 97% is eliminated within 24 h, principally unchanged.
Plants Converted to choline chloride in most plants studied.   **Soil/Environment** In
soil, rapidly degraded by microbial activity. It has no influence on soil microflora or
fauna. DT_{50} in 4 soils averaged 32 d at 10 °C; 1 d to 28 d at 22 °C. Low to
medium mobility. K_{oc} 203.

130 chloroacetic acid   *Herbicide*

$$ClCH_2CO_2H$$

NOMENCLATURE
chloroacetic acid
Common name chloroacetic acid (BSI, ISO-E, accepted in lieu of common name);
monochloroacetic acid (BSI, ISO-E, accepted in lieu of common name); acide

chloracétique (F-ISO, accepted in lieu of common name)
IUPAC name chloroacetic acid
Chemical Abstracts name chloroacetic acid
CAS RN *[79–11–8]* **EEC no.** 201–178–4

sodium chloroacetate
Common name sodium chloroacetate
Other names SMA; SMCA **CAS RN** *[3926–62–3]* **EEC no.** 223–498–3

PHYSICAL CHEMISTRY
chloroacetic acid
Composition Tech. grade is *c.* 90% pure. **Mol. wt.** 94.5 **M.f.** $C_2H_3ClO_2$
Form Chloroacetic acid forms a deliquescent solid and exists in three crystalline
forms. **M.p.** 63 °C (alpha), 55–56 °C (beta), 50 °C (gamma); (tech., 61–63 °C)
B.p. 189 °C **Solubility** Very soluble in water. Soluble in ethanol, benzene,
chloroform, and diethyl ether.

sodium chloroacetate
Mol. wt. 116.5 **M.f.** $C_2H_2ClNaO_2$ **Form** Colourless crystalline solid.
Solubility In water 850 g/l (20 °C).

COMMERCIALISATION
History Its herbicidal properties described by A. E. Hitchcock *et al.* (*Proc.
Northeast. Weed Control Conf.,* 1951, p. 105) and those of its sodium salt by
T. C. Breese & A. F. J. Wheeler (*Proc. Br. Weed Control Conf., 3rd,* 1956, p. 759).
Sodium chloroacetate was introduced by ICI Plant Protection Division (now
Zeneca Agrochemicals), who no longer manufacture or market it.

APPLICATIONS
sodium chloroacetate
Mode of action Selective contact herbicide. **Uses** Used post-emergence to
control a wide range of annual weeds at seedling stage in Brussels sprouts, kale,
leeks and onions at 20–25 kg a.i./ha. It is also used in combination with atrazine
for total weed control on industrial sites and other non-crop land.
Formulation types SP. **Mixtures** (*sodium chloroacetate +*) atrazine.
Selected tradenames 'Atlas Somon' (Atlas Crop Protection); 'Croptex Steel'
(Hortichem)

ANALYSIS
Product analysis by chlorine content determined by alkaline hydrolysis and
titration with silver nitrate, correcting for chloride ion and sodium dichloroacetate
originally present.

MAMMALIAN TOXICOLOGY

chloroacetic acid

Skin and eye Acute percutaneous LD_{50} not available. Irritating to eyes and skin (rabbits). **NOEL** Rats receiving 700 mg/kg diet for several months were not affected. **Toxicity class** WHO (a.i.) III **EC risk** T (R25); C (R34)

sodium chloroacetate

Oral Acute oral LD_{50} for rats 650, mice 165 mg/kg. **EC risk** T (R25); Xi (R38)

ECOTOXICOLOGY

chloroacetic acid

Birds Toxic to poultry. **Fish** LC_{50} (48 h) for rainbow trout 900 mg/l. **Bees** Toxic to bees.

131 chlorophacinone   *Rodenticide*

indandione anticoagulant

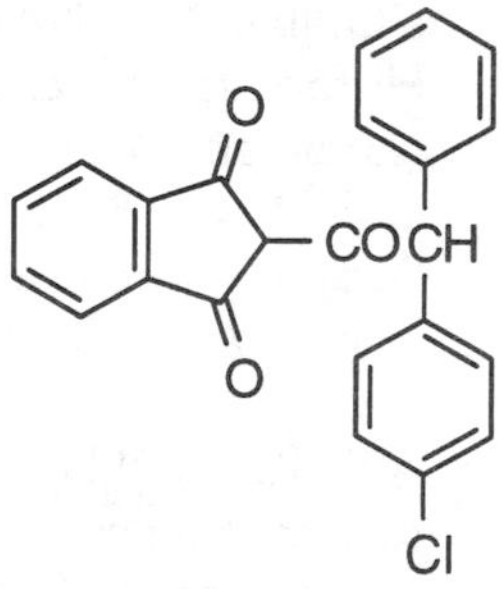

NOMENCLATURE

Common name chlorophacinone (BSI, E-ISO, (*f*) F-ISO, JMAF)
IUPAC name 2-[2-(4-chlorophenyl)-2-phenylacetyl]indan-1,3-dione
Chemical Abstracts name 2-[(4-chlorophenyl)phenylacetyl]-1*H*-indene-= 1,3(2*H*)dione
CAS RN *[3691–35–8]* **EEC no.** 223–003–0 **Development codes** LM 91

PHYSICAL CHEMISTRY

Mol. wt. 374.8 **M.f.** $C_{23}H_{15}ClO_3$ **Form** Pale yellow crystals. **M.p.** 140 °C
V.p. 1×10^{-4} mPa (25 °C) **S.g./density** Bulk density 0.38 g/cm^3 (20 °C)
Solubility In water 100 mg/l (20 °C). Readily soluble in methanol, ethanol, acetone, acetic acid, ethyl acetate, benzene and oil. Slightly soluble in hexane and diethyl ether. Soluble in aqueous alkalis with the formation of salts. **Stability** Very stable and resistant to weathering. **pKa** 3.40 (25 °C)

COMMERCIALISATION
History Rodenticide introduced by Lipha S.A. **Patent** US 3153612; FR 1269638
Manufacturers Lipha; Reanal

APPLICATIONS
Biochemistry Blocks prothrombin formation and uncouples oxidative phosphorylation. **Uses** Anticoagulant rodenticide, a single dose of a 50 mg/kg bait killing *Rattus norvegicus* from the 5th d. It is normally incorporated as 50–250 mg/kg bait. Does not induce 'bait-shyness'. **Formulation types** CB; RB; AB; TP; BB; PB. **Mixtures** *(chlorophacinone +)* sulfaquinoxaline. **Selected tradenames** 'Caid' (Lipha); 'Liphadione' (Lipha); 'Raviac' (Lipha); 'Redentin' (Reanal)

ANALYSIS
Product analysis by hplc (K. Hunter, *Anal. Methods Pestic. Plant Growth Regul.*, 1988, **16**, 119). **Residues** determined by hplc (*idem, ibid.*).

MAMMALIAN TOXICOLOGY
Oral Acute oral LD_{50} for rats 6.26 mg/kg. **Skin and eye** Acute percutaneous LD_{50} not available. Non-irritating to eyes and skin. Slight absorption through the skin. **Inhalation** LC_{50} (1 h) for rats 9.3 µg/l. **Other** In human volunteers, blood prothrombin levels fell to 35% within 2–4 d, following a single dose of 20 mg a.i., but recovered without treatment in 8 d. **Toxicity class** WHO (a.i.) Ia; EPA I (tech.); II (tracking powder) **EC risk** T+ (R27/28); T (R23, R48/24/25)

ECOTOXICOLOGY
Birds LC_{50} (30 d) for bobwhite quail 95, mallard duck 204 ppm. **Fish** LC_{50} (96 h) for rainbow trout 0.35, bluegill sunfish 0.62 mg/l. **Bees** Not hazardous to bees when used as recommended. **Daphnia** LC_{50} (48 h) 0.42 mg/l.

ENVIRONMENTAL FATE
Animals In mammals, following oral administration, 90% is eliminated in the faeces within 48 hours in the form of metabolites. **Soil/Environment** Parameter ranges for four soil types: K_d (adsorption) 80–1000, K_d(desorption) 57–579; K_{oc} 15 556–135 976.

Cl_3CNO_2

NOMENCLATURE
Common name chloropicrin (BSI, E-ISO, ESA, accepted in lieu of common name); chloropicrine (F-ISO, accepted in lieu of common name)
IUPAC name trichloronitromethane
Chemical Abstracts name trichloronitromethane
CAS RN *[76–06–2]* **EEC no.** 200–930–9

PHYSICAL CHEMISTRY
Mol. wt. 164.4 **M.f.** CCl_3NO_2 **Form** Colourless liquid with a lachrymatory action. **M.p.** –64 °C **B.p.** 112.4 °C/757 mmHg **V.p.** 3.2 kPa (25 °C) **S.g./density** 1.6558 (20 °C) **Solubility** In water 2.27 (0 °C), 1.62 (25 °C) (both in g/l). Miscible with most organic solvents, e.g. acetone, benzene, ethanol, methanol, carbon disulfide, diethyl ether, carbon tetrachloride. **Stability** Stable in acidic media, unstable in alkali. **F.p.** Non-flammable.

COMMERCIALISATION
History Used as an insecticide since 1908. **Patent** GB 2387 **Manufacturers** Great Lakes; Mitsui Toatsu; Niklor

APPLICATIONS
Mode of action Fumigant. **Uses** An insecticide used for the fumigation of stored grain. Also of soil to control nematodes and other soil-dwelling pests. **Phytotoxicity** Highly phytotoxic. **Formulation types** TC; TB; OL. **Mixtures** *(chloropicrin +)* methyl bromide; 1,3-dichloropropene; methyl isothiocyanate + 1,3-dichloropropene. **Compatibility** Compatible with other fumigants. **Selected tradenames** 'Chlor-O-Pic' (Great Lakes); 'Dorochlor' (Mitsui Toatsu)

ANALYSIS
Chloropicrin is determined by glc. For determination in air, collect in propan-2-ol and convert to a derivative which is measured colorimetrically (L. Feinsilver & F. W. Oberst, *Anal. Chem.*, 1953, **25**, 820). Methods reviewed by J. L. Daft in *Comp. Analyt. Profiles*, Chapter 11.

MAMMALIAN TOXICOLOGY
Reviews *Evaluation of the toxicity of pesticide residues in food.* FAO Meeting Report, No. PL/1965/10; WHO/Food Add./26.65, 1965. *Evaluation of the hazards to consumers resulting from the use of fumigants in the protection of food.* PL/1965/10/2; WHO/Food Add./28.65, 1965. **Oral** Acute oral LD_{50} for rats

250 mg/kg. **Skin and eye** Acute percutaneous LD$_{50}$ not available. Severe irritant to skin of rabbits. **Inhalation** At 0.008 mg/l air, it can be clearly detected; at 0.016 mg/l, it produces coughing and lachrymation; and at 0.12 mg/l, 30 to 60 minutes exposure can be fatal. Exposure of cats, guinea pigs and rabbits to concentrations of 0.8 mg/l air was lethal in 20 min. **ADI** (JMPR) No ADI [1965]. **Toxicity class** EPA (formulation) II **EC risk** Xn (R22); T+ (R26); Xi (R36/37/38)

ECOTOXICOLOGY
Fish TLm (48 h) for carp 0.168 mg/l. **Daphnia** LC$_{50}$ (3 h) 0.91 mg/l.

33 chlorothalonil *Fungicide*

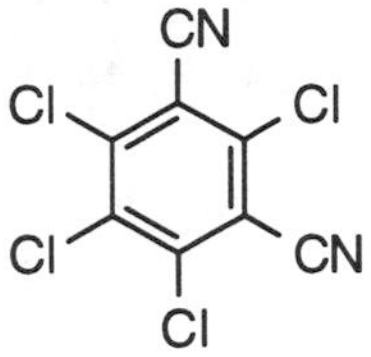

NOMENCLATURE
Common name chlorothalonil (BSI, E-ISO, (*m*) F-ISO, ANSI); TPN (JMAF)
IUPAC name tetrachloroisophthalonitrile
Chemical Abstracts name 2,4,5,6-tetrachloro-1,3-benzenedicarbonitrile
CAS RN *[1897–45–6]* **EEC no.** 217–588–1 **Development codes** DS-2787

PHYSICAL CHEMISTRY
Composition Tech. is *c.* 98% pure. **Mol. wt.** 265.9 **M.f.** C$_8$Cl$_4$N$_2$
Form Colourless, odourless crystals; tech. grade has a slightly pungent odour. **M.p.** 252.1 °C **B.p.** 350 °C/760 mmHg **V.p.** 0.076 mPa (25 °C)
K$_{ow}$ logP = 2.89 **Henry** 2.50 × 10^{-2} Pa m^3 mol^{-1} (25 °C) **S.g./density** 1.8
Solubility In water 0.81 mg/l (25 °C). In xylene 80, cyclohexanone, dimethylformamide 30, acetone, dimethyl sulfoxide 20, kerosene <10 (all in g/kg, 25 °C). **Stability** Thermally stable at ambient temperatures. Stable to u.v. light in aqueous media and in crystalline state. Stable in acidic and moderately alkaline aqueous solutions; slow hydrolysis at pH >9.

COMMERCIALISATION
History Fungicide reported by N. J. Turner *et al.* (*Contrib. Boyce Thompson Inst.,* 1964, **22**, 303). Introduced by the Diamond Alkali Co. (now ISK Biosciences Corp.). **Patent** US 3290353; US 3331735 **Manufacturers** Caffaro; Crystal; ISK Biosciences; Pilarquim; SDS Biotech; Sinon

APPLICATIONS

Biochemistry Conjugation with, and depletion of, thiols (particularly glutathione) from germinating fungal cells, leading to disruption of glycolysis and energy production, fungistasis and fungicidal action. **Mode of action** Non-systemic foliar fungicide with protective action. **Uses** Control of many fungal diseases in a wide range of crops, including pome fruit, stone fruit, citrus fruit, bush and cane fruit, cranberries, strawberries, pawpaws, bananas, mangoes, coconut palms, oil palms, rubber, pepper, vines, hops, vegetables, cucurbits, tobacco, coffee, tea, rice, soya beans, peanuts, potatoes, sugar beet, cotton, maize, ornamentals, mushrooms, and turf. **Phytotoxicity** Russetting is possible with flowering ornamentals, apples, and with grapes. Some varieties of flowering ornamentals may be injured. *Pittosporum* foliage is sensitive. Phytotoxicity may be increased with oils or oil-containing substances. **Formulation types** SC; WG; WP; Fogging concentrate.
Mixtures (*chlorothalonil +*) oxadixyl; carbendazim; vinclozolin; cymoxanil; cyproconazole; fenpropimorph; flusilazole; flutriafol; metalaxyl; nuarimol; propiconazole; carbendazim + propiconazole; sulfur; copper oxychloride; copper oxychloride + maneb; flutolanil. **Compatibility** Not compatible with oils.
Selected tradenames 'Bravo' (mixture) (ISK Biosciences); 'Daconil' (ISK Biosciences, SDS Biotech, Zeneca); 'Bombardier' (Unicrop); 'Clortocaffaro' (Caffaro); 'Clortosip' (Sipcam); 'Fungiless' (Sanonda); 'Repulse' (Hortichem); 'Rival' (Productos OSA); 'Teren' (Efthymiadis); 'Visclor' (Vischim)

ANALYSIS

Product analysis by glc (D. L. Ballee *et al., Anal. Methods Pestic. Plant Growth Regul.,* 1976, **8**, 263). **Residues** determined by glc (*idem, ibid.*; A. Ambrus *et al., J. Assoc. Off. Anal. Chem.,* 1981, **64**, 733; *Man. Pestic. Residue Anal.,* 1987, **I**, 6, S19; *Anal. Methods Residues Pestic.,* 1988, Part I, M1, M12). In **drinking water** by gc with ECD (*AOAC Methods,* 1995, 990.06). Details available from ISK Biosciences.

MAMMALIAN TOXICOLOGY

Reviews *Pesticide residues in food – 1992.* FAO Plant Production and Protection Paper, 116, 1993. *Pesticide residues in food – 1992 evaluations. Part II – Toxicology.* World Health Organisation, WHO/PCS/93.34. **IARC** 30 **EHC** 183 (1996).
Oral Acute oral LD_{50} for rats >10 000, dogs >5000 mg/kg. **Skin and eye** Acute percutaneous LD_{50} for albino rabbits >10 000 mg/kg. Severe eye irritant; mild skin irritant (rabbits). Evidence in humans of contact dermatitis. **Inhalation** LC_{50} (1 h) for rats >4.7 mg/l air; (4 h) for rats, (nominal concentration) 0.6 mg/l air; (4 h) for rats, (actual) 0.10 mg/l air. **NOEL** Chronic administration of chlorothalonil has been associated with tumour formation in the kidney and forestomach of rats and male mice. The mechanism has been demonstrated to be epigenetic with a NOEL of 1.8 in the rat and 1.6 in the mouse. In the dog, the pattern of toxicity is different from rodents, with a NOEL of at least 3 mg/kg b.w. (JMPR 1990). **ADI** (JMPR) 0.03 mg/kg [1992]. **Toxicity class** WHO (a.i.) III (Table 5); EPA (formulation) IV **EC risk** (R40)

ECOTOXICOLOGY

Birds Acute oral LD_{50} for mallard ducks >4640 mg/kg. Eight-day dietary LC_{50} for mallard ducks and bobwhite quail >10 000 mg/kg diet. **Fish** LC_{50} (96 h) for rainbow trout 49, bluegill sunfish 62, channel catfish 44 ppb. **Bees** Relatively non-toxic to bees. **Worms** LC_{50} (14 d) >1000 mg/kg. **Daphnia** LC_{50} (48 h) 70 ppb. **Algae** For *Selenastrum capricornutum*, EC_{50} (72 h) 0.13, (120 h) 0.21; NOEC (72 h) 0.05, (120 h) 0.1 (all in mg/l). **Other aquatic spp.** LC_{50} (48 h) for brown shrimp >1000 ppb; (96 h) for pink shrimp 165 ppb.

ENVIRONMENTAL FATE

Animals In non-ruminant mammals, chlorothalonil is largely unabsorbed. What is absorbed is conjugated with 1 to 3 glutathione molecules. The conjugation product undergoes further degradation to thiol and methylated thiol derivatives. In ruminant mammals, chlorothalonil is also largely unabsorbed. The major identifiable metabolite is the 4-hydroxy derivative with smaller quantities of very polar metabolites. In both non-ruminant and ruminant animals, chlorothalonil itself is not found in tissues or urine. **Plants** In plants, 4-hydroxy-2,5,6-trichloroiso= phthalonitrile is found to a limited extent as a metabolite. The majority of residue remains as parent compound. **Soil/Environment** K_{oc} 1600 (sand) to 14 000 (silt), indicating low mobility to immobile. In aerobic and anaerobic soil studies, DT_{50} is 5 to 36 d. In aerobic and anaerobic aquatic soil studies, DT_{50} is from a few hours to a few days. The principal metabolite is 4-hydroxy-2,5,6-trichloroiso= phthalonitrile, which has soil DT_{50} 6–43 d.

34 chlorotoluron *Herbicide*

urea

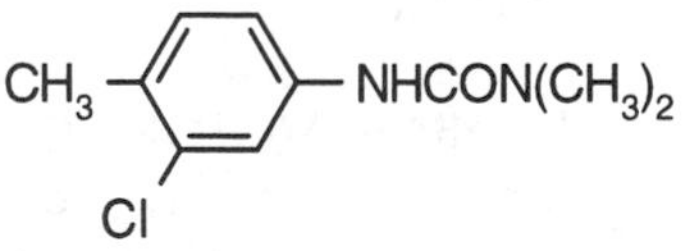

NOMENCLATURE

Common name chlorotoluron (BSI (from 1984), E-ISO, (*m*) F-ISO); chlortoluron (BSI (before 1984), (*m*) France, New Zealand)
IUPAC name 3-(3-chloro-*p*-tolyl)-1,1-dimethylurea
Chemical Abstracts name *N'*-(3-chloro-4-methylphenyl)-*N,N*-dimethylurea
CAS RN [15545–48–9] **EEC no.** 239–592–2 **Development codes** C 2242

PHYSICAL CHEMISTRY

Mol. wt. 212.7 **M.f.** $C_{10}H_{13}ClN_2O$ **Form** White powder. **M.p.** 148.1 °C

V.p. 0.005 mPa (25 °C)   **K$_{ow}$** logP = 2.5 (25 °C)   **S.g./density** 1.40 g/cm^3 (20 °C)
Solubility In water 74 mg/l (25 °C). In acetone 54, dichloromethane 51, ethanol 48, toluene 3.0, hexane 0.06, n-octanol 24, ethyl acetate 21 (all in g/l, 25 °C).
Stability Stable to heat and u.v. light. Slowly hydrolysed by strong acids and alkalis.
DT$_{50}$ (calculated) >200 d (pH 5, 7, 9; 30 °C).

COMMERCIALISATION
History Herbicide reported by Y. L'Hermite et al. (*C. R. Journ. Etud. Herbic. Conf.,* COLUMA, 5th, 1969, **II**, 349). Introduced by Ciba AG (now Novartis Crop Protection AG).   **Patent** BE 728267; GB 1255258   **Manufacturers** Jingma; Makhteshim-Agan; Novartis; Nufarm GmbH

APPLICATIONS
Biochemistry Photosynthetic electron transport inhibitor.
Mode of action Selective herbicide, absorbed by the roots and foliage.
Uses Effective at 1.5–3.0 kg a.i./ha, both as a residual soil-acting herbicide and a contact foliar-spray, against many broad-leaved and grass weeds of winter cereals, especially against *Alopecurus myosuroides*. It is combined with mecoprop to improve the control of *Galium*, *Papaver* and *Veronica* spp.   **Phytotoxicity** Some varieties of wheat and barley may be injured.   **Formulation types** FW; SC; WP; GR.   **Mixtures** (*chlorotoluron* +) bifenox; mecoprop; pendimethalin; terbutryn; trifluralin; terbuthylazine + terbutryn.   **Selected tradenames** 'Dicuran' (Novartis); 'Cekutoluron' (Cequisa); 'Chlortophyt' (Agriphar); 'Lentipur' (Nufarm GmbH); 'Tolurane' (Diachem); 'Tolurex' (Makhteshim-Agan)

ANALYSIS
Product analysis by determination of dimethylamine produced on hydrolysis, or by tlc (*AOAC Methods*, 1995, 977.06; *CIPAC Handbook*, 1980, **1A**, 1151).
Residues determined by hplc or by alkaline hydrolysis to 3-chloro-p-toluidine, a derivative of which is determined by glc with TID, MCD, or ECD. Details available from Novartis.

MAMMALIAN TOXICOLOGY
Oral Acute oral LD$_{50}$ for rats >5000 mg/kg.   **Skin and eye** Acute percutaneous LD$_{50}$ for rats >2000 mg/kg; not irritant to skin and eyes of rabbits. Not a skin sensitiser (guinea pigs).   **Inhalation** LC$_{50}$ (4 h) for rats >5300 mg/m^3.
NOEL (2 y) for rats 100 ppm (5 mg/kg daily), mice 100 ppm (11.3 mg/kg daily).
Toxicity class WHO (a.i.) III (Table 5)

ECOTOXICOLOGY
Birds Eight-day dietary LC$_{50}$ for mallard ducks >6800, Japanese quail >2150, pheasant >10 000 ppm.   **Fish** LC$_{50}$ (96 h) for rainbow trout 35, bluegill sunfish 50, crucian carp >100, catfish 60, guppy >49 mg/l.   **Bees** LD$_{50}$ (48 h) (contact) >20 µg/bee, (ingestion) >1000 ppm.   **Worms** LC$_{50}$ for earthworms

>1000 mg/kg.   **Daphnia** LC_{50} (48 h) 67 mg/l.   **Algae** EC_{50} (72 h) for *Scenedesmus subspicatus* 0.024 mg/l.

ENVIRONMENTAL FATE
Animals In mammals, following oral administration, >90% is eliminated in the urine and faeces within 24 hours. Main metabolism is via *N*-demethylation and stepwise oxidation of the ring methyl group to hydroxymethyl and carboxymethyl derivatives.   **Plants** Metabolites found in winter wheat include 3-chloro-= *p*-toluidine, 3-(3-chloro-4-methylphenyl)-1-methylurea, and 1-(3-chloro-= 4-methylphenyl)urea.   **Soil/Environment** DT_{50} in soil 30–40 d, in water >200 d.

35   chlorphonium chloride

Plant growth regulator

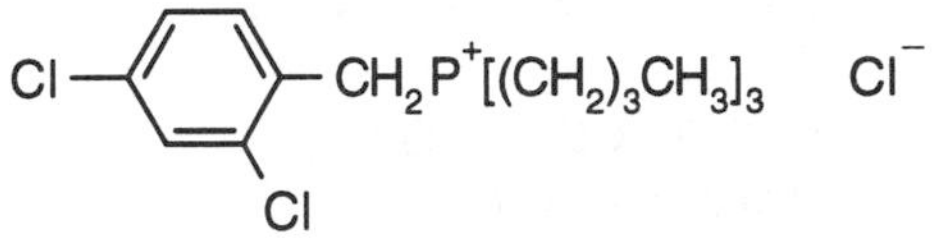

NOMENCLATURE
chlorphonium chloride
Common name chlorphonium chloride
CAS RN *[115–78–6]*   **EEC no.** 204–105–4

chlorphonium
Common name chlorphonium (BSI, E-ISO); chlorfonium ((*m*) F-ISO)
IUPAC name tributyl(2,4-dichlorobenzyl)phosphonium
Chemical Abstracts name tributyl[(2,4-dichlorophenyl)methyl]phosphonium

PHYSICAL CHEMISTRY
chlorphonium chloride
Mol. wt. 397.8   **M.f.** $C_{19}H_{32}Cl_3P$ **Form** Colourless crystals.   **M.p.** 114–120 °C
V.p. 9.33×10^{-2} mPa (20 °C)   **Solubility** In water 960 g/l (20 °C). In methanol 1030, acetone 200 (both in g/l, 20 °C). Insoluble in diethyl ether and hexane.
Stability Stable under normal storage conditions.

chlorphonium
Mol. wt. 362.3   **M.f.** $C_{19}H_{32}Cl_2P$

COMMERCIALISATION
History The chloride was introduced as a plant growth regulator by the Mobil Chemical Co. (who no longer manufacture or market it), currently marketed by Perifleur Products Ltd.   **Patent** US 3268323   **Manufacturers** Perifleur

APPLICATIONS
chlorphonium chloride
Uses Used to reduce internodal length in chrysanthemums, geraniums, petunias, and other responsive ornamentals; to increase the breaking of lateral shoots in geraniums and petunias; and to increase the number of buds and flowers in geraniums, azaleas, and rhododendrons. It is most effective as a soil treatment for potted ornamentals. In addition, treatment of stock plants enhances the uniformity of cuttings taken from them.   **Formulation types** GR; SL.
Selected tradenames 'Phosfleur' (Perifleur)

MAMMALIAN TOXICOLOGY
chlorphonium chloride
Oral Acute oral LD_{50} for rats 210 mg/kg.   **Skin and eye** Acute percutaneous LD_{50} for rabbits 750 mg/kg. Tech. product and formulations are irritating to eyes and skin.   **Toxicity class** WHO (a.i.) II; EPA (formulation) II   **EC risk** T (R25); Xn (R21); Xi (R36/38)

ECOTOXICOLOGY
chlorphonium chloride
Fish LC_{50} (96 h) for rainbow trout 115 mg/l.   **Bees** Not hazardous to bees.

ENVIRONMENTAL FATE
Soil/Environment DT_{50} of chlorphonium in standard German soil 2.2 & 2.3 >28 mo, when stored under BBA conditions.

chlorpropham

Herbicide, plant growth regulator

carbamate

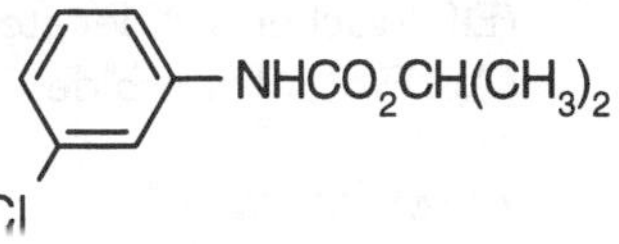

NOMENCLATURE
Common name chlorpropham (BSI, E-ISO, WSSA); chlorprophame ((*m*) F-ISO);
IPC (JMAF); chlor-IFC (USSR)
IUPAC name isopropyl 3-chlorocarbanilate
Chemical Abstracts name 1-methylethyl (3-chlorophenyl)carbamate
Other names CIPC; chloro-IPC   **CAS RN** *[101–21–3]*   **Official codes** ENT
18 060

PHYSICAL CHEMISTRY
Composition Tech. grade is 98.5% pure.   **Mol. wt.** 213.7   **M.f.** $C_{10}H_{12}CINO_2$
Form Colourless solid.   **M.p.** 41.4 °C (pure); 38.5–40 °C (tech.)
S.g./density 1.180 (30 °C)   **Solubility** In water 89 mg/l (25 °C). Readily soluble in
most organic solvents, e.g. alcohols, ketones, esters, chlorinated hydrocarbons,
aromatic hydrocarbons, etc. Moderately soluble in mineral oils (e.g. 100 g/kg in
kerosene).   **Stability** Stable to u.v. light. Decomposes above 150 °C. Hydrolysed
slowly in acidic and alkaline media.

COMMERCIALISATION
History Herbicide reported by E. D. Witman & W. F. Newton (*Proc. Northeast.
Weed Control Conf.*, 1951, p. 45). Introduced in 1951.   **Patent** US 2695225
Manufacturers Elf Atochem; Hodogaya; Kemira FC; United Phosphorus Ltd

APPLICATIONS
Biochemistry Inhibits root and epicotyl growth, normal cell division, protein and
RNA synthesis, suppresses transpiration and respiration, interferes with oxidative
phosphorylation and photosynthesis, and inhibits the activity of beta-amylase.
Mode of action Selective systemic herbicide and growth regulator, absorbed
predominantly by the roots and coleoptiles, and readily translocated acropetally.
Uses Pre-emergence control of many annual grasses and some broad-leaved
weeds in onions, leeks, garlic, shallots, alfalfa, clover, sugar beet, spinach, lettuce,
endive, chicory, herbs, peas, beans, carrots, celery, black salsify, fennel, soya beans,
cotton, rice, safflowers, sunflowers, cranberries, cane berries, ornamentals
(especially bulb flowers), ornamental trees and shrubs, perennial grass-seed crops,
etc. Also used as a sprouting inhibitor for ware potatoes, and as a sucker control
agent in tobacco. Often used in combination with propham as a potato sprouting

inhibitor, and in combination with other herbicides.   **Formulation types** EC; HN; GR; DP.   **Mixtures** (*chlorpropham* +) diuron; fenuron; linuron; propham; pentanochlor; propazine; propyzamide; fenuron + propham; diuron + propham; cetrimide; propham + thiabendazole; chloridazon + fenuron + propham. **Selected tradenames** 'Atlas Indigo' (Atlas Crop Protection); 'Decco Aerosol 273' (Elf Atochem); 'Neostop' (Agriphar); 'No Brot' (Productos OSA); 'Prevenol' (AgrEvo); 'Triherbide CIPC' (Elf Atochem); 'Warefog' (Mirfield)

ANALYSIS
Product analysis by glc or by i.r. spectroscopy (*Anal. Methods Pestic. Plant Growth Regul.*, 1972, **6**, 612; *CIPAC Handbook*, 1985, **1C**, 2025), or by titration of the carbon dioxide or 3-chloroaniline formed on hydrolysis (*CIPAC Handbook*, 1970, **1**, 223; 1983, **1B**, 1757; FAO Specification CP/73).   **Residues** determined by glc or hplc of derivatives (*Anal. Methods Pestic. Plant Growth Regul., loc. cit.*). In **drinking water**, by gc with FID (*AOAC Methods*, 1995, 991.07).

MAMMALIAN TOXICOLOGY
Reviews *Evaluation of the toxicity of pesticide residues in food.* FAO Meeting Report, No. PL/1965/10; WHO/Food Add./26.65, 1965. *Evaluation of the toxicity of pesticide residues in food.* PL/1965/10/1; WHO/Food Add./27.65, 1965.   **IARC** 12   **Oral** Acute oral LD_{50} for rats 5000–7500, rabbits 5000 mg/kg. **NOEL** In 2 y feeding trials, rats and dogs receiving 2000 mg/kg diet showed no ill-effects.   **ADI** (JMPR) No ADI [1965].   **Toxicity class** WHO (a.i.) III (Table 5); EPA (formulation) III

ECOTOXICOLOGY
Birds Acute oral LD_{50} for mallard ducks >2000 mg/kg.   **Fish** LC_{50} (48 h) for bluegill sunfish 12, bass 10 mg/l.   **Bees** Not hazardous to bees when used as recommended.

ENVIRONMENTAL FATE
Animals In mammals, following oral administration, the principal metabolic route is by hydroxylation at the para position and conjugation of the resultant 4-hydroxychloropropham with sulfate. There is also some hydroxylation of the isopropyl residue.   **Plants** In soya beans, three major metabolites have been identified, isopropyl N-4-hydroxy-3-chlorophenylcarbamate, isopropyl N-5-chloro-2-hydroxyphenylcarbamate, and 1-hydroxy-2-propyl-3'-chlorocarbanilate, whilst, in cucumbers, only the first of these has been identified. These aglycones are found in plants as water-soluble conjugates of glucose or other plant components. **Soil/Environment** In soil, microbial degradation leads to the production of 3-chloroaniline by an enzymic hydrolysis reaction, with liberation of CO_2. DT_{50} in soil *c.* 65 d (15 °C), 30 d (29 °C).

organophosphorus

$$Cl-\overset{N}{\underset{Cl}{}}-OP(OCH_2CH_3)_2$$

with structure: pyridine ring labeled Cl, N, Cl, Cl and $O\overset{S}{\underset{\|}{P}}(OCH_2CH_3)_2$

NOMENCLATURE
Common name chlorpyrifos (BSI, E-ISO, ANSI, ESA, BAN); chlorpyriphos
((*m*) F-ISO, JMAF); chlorpyriphos-éthyl ((*m*) France)
IUPAC name *O,O*-diethyl *O*-3,5,6-trichloro-2-pyridyl phosphorothioate
Chemical Abstracts name *O,O*-diethyl *O*-(3,5,6-trichloro-2-pyridinyl)
phosphorothioate
CAS RN *[2921–88–2]* **EEC no.** 220–864–4 **Development codes** Dowco 179
Official codes OMS 971; ENT 27 311

PHYSICAL CHEMISTRY
Mol. wt. 350.6 **M.f.** $C_9H_{11}Cl_3NO_3PS$ **Form** Colourless crystals with a mild
mercaptan odour. **M.p.** 42–43.5 °C **V.p.** 2.7 mPa (25 °C) **K_{ow}** logP = 4.7
Solubility In water *c.* 1.4 mg/l (25 °C). In benzene 7900, acetone 6500,
chloroform 6300, carbon disulfide 5900, diethyl ether 5100, xylene 5000,
iso-octanol 790, methanol 450 (all in g/kg, 25 °C). **Stability** Rate of hydrolysis
increases with pH, and in the presence of copper and possibly of other metals
that can form chelates; DT_{50} 1.5 d (water, pH 8, 25 °C) to 100 d (phosphate
buffer, pH 7, 15 °C).

COMMERCIALISATION
History Insecticide reported by E. E. Kenaga *et al.* (*J. Econ. Entomol.*, 1965, **58,**
1043). Introduced by Dow Chemical Co. (now DowElanco). **Patent** US 3244586
Manufacturers Aimco; Agriphar; Crystal; DowElanco; Excel; Ficom; Gharda;
Luxembourg; Makhteshim-Agan; Mitsu; Rallis; Sinon

APPLICATIONS
Biochemistry Cholinesterase inhibitor. **Mode of action** Non-systemic insecticide
with contact, stomach, and respiratory action. **Uses** Control of Coleoptera,
Diptera, Homoptera and Lepidoptera in soil or on foliage in a wide range of
crops, including pome fruit, stone fruit, citrus fruit, nut crops, strawberries, figs,
bananas, vines, vegetables, potatoes, beet, tobacco, soya beans, sunflowers, sweet
potatoes, peanuts, rice, cotton, alfalfa, cereals, maize, sorghum, asparagus,
glasshouse and outdoor ornamentals, mushrooms, turf, and in forestry. Also used
for control of household pests (Blattellidae, Muscidae, Isoptera), mosquitoes
(larvae and adults) and in animal houses. Also for stored products. **Phytotoxicity**
Non-phytotoxic to most plant species when used as recommended. Poinsettias,

MAIN ENTRIES

azaleas, camelias, and roses may be injured. **Formulation types** GR; EC; WP; DP; UL; Microcapsule. **Mixtures** (*chlorpyrifos* +) diflubenzuron; cypermethrin; dimethoate; disulfoton; lindane; pirimicarb; thiram. **Compatibility** Incompatible with alkaline materials. **Selected tradenames** 'Dursban' (DowElanco); 'Lorsban' (DowElanco); 'Agromil' (Westrade); 'Bullet' (Mitsu); 'Chlorfos' (Griffin); 'Destroyer' (Agriphar); 'Dhanvan' (Dhanuka); 'Dorsan' (Luxembourg); 'Omexan' (Delicia); 'Panda' (Cequisa); 'Pestan' (Efthymiadis); 'Piridane' (Diachem); 'Pyriban' (Aimco); 'Pyrifoz' (Sanonda); 'Pyrinex' (Makhteshim-Agan); 'Pyrivol' (Voltas); 'Radar' (Searle India); 'Robon' (Ramcides); 'Rochlop' (Rotam); 'Rodazim' (Rotam); 'Silrifos' (Siapa); 'Spannit' (PBI); 'Strike' (Wockhardt); 'Tafaban' (Rallis); 'Talon' (FCC); 'Terraguard' (Gharda); 'Tricel' (Excel)

ANALYSIS

Product analysis by hplc (*CIPAC Handbook*, 1985, **1C**, 2028; *AOAC Methods*, 1995, 981.03). **Residues** determined by glc (A. Ambrus et *al.*, *J. Assoc. Off. Anal. Chem.*, 1981, **64**, 733; *AOAC Methods*, 1995, 985.22; *Man. Pestic. Residue Anal.*, 1987, **I**, 6, S8, S9, S13, S19; *Anal. Methods Residues Pestic.*, 1988, Part I, M2, M5, M12). Details available from DowElanco.

MAMMALIAN TOXICOLOGY

Reviews *Pesticide residues in food – 1982*. FAO Plant Production and Protection Paper 46, 1983. *Pesticide residues in food – 1982 evaluations*. FAO Plant Production and Protection Paper 49, 1983. **EHC** 63 (WHO, 1986; a general review of organophosphorus insecticides). **Oral** Acute oral LD_{50} for rats 135–163, guinea pigs 504, rabbits 1000–2000 mg/kg. **Skin and eye** Acute percutaneous LD_{50} (in solutions) for rabbits c. 2000 mg/kg; (tech.) for rats >2000 mg/kg. **Inhalation** LC_{50} (4–6 h) for rats >0.2 mg/l (14 ppb). **NOEL** (2 y), based on blood plasma cholinesterase activity, for rats 0.03, dogs 0.01 mg/kg daily. **ADI** (JMPR) 0.01 mg/kg b.w. [1982]. **Other** Non-teratogenic. **Toxicity class** WHO (a.i.) II; EPA (formulation) II **EC risk** T (R24/25)

ECOTOXICOLOGY

Birds Acute oral LD_{50} for chickens 32–102 mg/kg. LC_{50} (8 d) for bobwhite quail 423 ppm. **Fish** LC_{50} (96 h) for rainbow trout 0.003, roach 0.25 mg/l. **Bees** Toxic to bees. LD_{50} (contact) 59 ng/bee, (oral) 250 ng/bee. **Other beneficial spp.** Toxic to Collembola (J. A. Wiles & G. K. Frampton, *Pestic. Sci.*, **47**, 273 (1996)). **Daphnia** LC_{50} (48 h) 1.7 µg/l.

ENVIRONMENTAL FATE

Animals In rats, dogs, and other mammals, following oral administration, rapid metabolism occurs, the principal metabolite being 3,5,6-trichloro-2-pyridinol. Excretion is principally in the urine. **Plants** Non-systemic in plants, not absorbed from soil via the roots. Residues taken up by plant tissues are metabolised to 3,5,6-trichloro-2-pyridinol which is conjugated and sequestered.

Soil/Environment In soil, chlorpyrifos is slowly degraded, DT_{50} c. 60–120 d, to 3,5,6-trichloropyridin-2-ol, which is subsequently degraded to organochlorine compounds and CO_2.

38 chlorpyrifos-methyl *Insecticide, acaricide*

organophosphorus

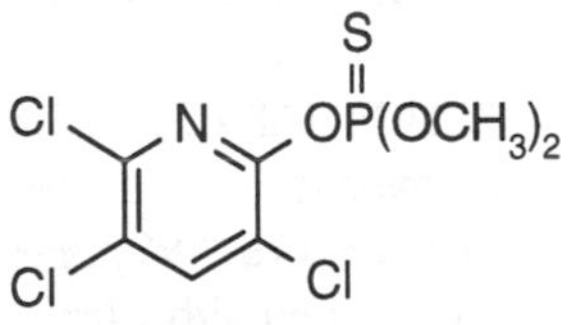

NOMENCLATURE
Common name chlorpyrifos-methyl (BSI, E-ISO, ANSI, ESA); chlorpyriphos-methyl ((*m*) F-ISO, JMAF)
IUPAC name *O,O*-dimethyl *O*-3,5,6-trichloro-2-pyridyl phosphorothioate
Chemical Abstracts name *O,O*-dimethyl *O*-(3,5,6-trichloro-2-pyridinyl) phosphorothioate
CAS RN *[5598–13–0]* **Development codes** Dowco 214 **Official codes** OMS 1155; ENT 27 520

PHYSICAL CHEMISTRY
Composition 97% pure. **Mol. wt.** 322.5 **M.f.** $C_7H_7Cl_3NO_3PS$ **Form** White crystals, with a slight mercaptan odour. **M.p.** 45.5–46.5 °C **V.p.** 3 mPa (25 °C) **K_{ow}** logP = 4.24 **S.g./density** 1.64 (23 °C) **Solubility** In water 2.6 mg/l (20 °C). In acetone >400, methanol 190, hexane 120 (all in g/kg, 20 °C).
Stability Hydrolysis DT_{50} 27 d (pH 4), 21 d (pH 7), 13 d (pH 9). Aqueous photolysis DT_{50} 1.8 d (June), 3.8 d (December). **F.p.** 182 °C (COC)

COMMERCIALISATION
History Insecticide reported by R. H. Rigterink & E. E. Kenaga (*J. Agric. Food Chem.*, 1966, **14**, 304). Introduced by Dow Chemical Co. (now DowElanco).
Patent US 3244586 **Manufacturers** Aimco; DowElanco

APPLICATIONS
Biochemistry Cholinesterase inhibitor. **Mode of action** Non-systemic insecticide and acaricide with contact, stomach, and respiratory action. **Uses** Control of Coleoptera, Diptera, Homoptera and Lepidoptera in cereals (including stored grains), and various foliar crop pests in pome fruit, stone fruit, citrus fruit, bush and cane fruit, vines, strawberries, vegetables, tomatoes, tea, rice, cotton, and

other crops. Industrial and public health uses include control of Muscidae and crawling insects. **Formulation types** EC; UL; Fogging concentrate. **Mixtures** (*chlorpyrifos-methyl* +) cypermethrin; fenobucarb; permethrin + piperonyl butoxide + pyrethrins. **Compatibility** Incompatible with alkaline and strongly acidic materials. **Selected tradenames** 'Reldan' (DowElanco); 'Pyriban-M' (Aimco)

ANALYSIS
Product analysis by hplc. **Residues** determined by glc (*Man. Pestic. Residue Anal.,* 1987, **I,** 6, S19; *Anal. Methods Residues Pestic.,* 1988, Part I, M2, M5; *Analyst (London),* 1979, **104,** 425; 1985, **110,** 765; P. Bottomley & P. G. Baker, ibid., 1984, **109,** 85). Details available from DowElanco.

MAMMALIAN TOXICOLOGY
Reviews *Pesticide residues in food – 1992.* FAO Plant Production and Protection Paper, 116, 1993. *Pesticide residues in food – 1992 evaluations. Part II – Toxicology.* World Health Organisation, WHO/PCS/93.34. **EHC** 63 (WHO, 1986; a general review of organophosphorus insecticides). **Oral** Acute oral LD_{50} for rats >3000, mice 1100–2250, guinea pigs 2250, rabbits 2000 mg/kg. **Skin and eye** Acute percutaneous LD_{50} for rabbits >2000, rats >3700 mg/kg. Non-irritating to skin and eyes. **Inhalation** LC_{50} (4 h) for rats >0.67 mg/l. **NOEL** In 2 y feeding trials NOEL, based on plasma cholinesterase levels, 0.1 mg/kg daily for dogs and rats. **ADI** (JMPR) 0.01 mg/kg b.w. [1992]. **Toxicity class** WHO (a.i.) III (Table 5); EPA (formulation) III

ECOTOXICOLOGY
Birds Acute oral LD_{50} for chickens >7950 mg/kg (capsule application). Eight-day dietary LC_{50} for mallard ducks 2500–5000 mg/kg. **Fish** LC_{50} (96 h) for rainbow trout 0.3 mg/l. **Bees** Very toxic to bees. LD_{50} (contact) 0.38 µg/bee. **Daphnia** LC_{50} (24 h) 0.016–0.025 ppm. **Other aquatic spp.** Toxic to crustaceans. LC_{50} (36 h) for crayfish 0.004 mg/l.

ENVIRONMENTAL FATE
Animals In rats and other mammals, following oral administration, rapid metabolism occurs, the principal metabolite being 3,5,6-trichloro-2-pyridinol. Excretion is principally in the urine. **Soil/Environment** In soil, undergoes microbial degradation to 3,5,6-trichloropyridin-2-ol, which is subsequently degraded to organochlorine compounds and CO_2; DT_{50} 1.5–33 d, DT_{90} 14–47 d, depending upon soil type and microbial activity. K_d 3.5–407 ml/g, depending on soil type. K_{oc} is more constant: 1190–8100 ml/g.

sulfonylurea

NOMENCLATURE
Common name chlorsulfuron (BSI, draft E-ISO, (*m*) draft F-ISO, ANSI, WSSA)
IUPAC name 1-(2-chlorophenylsulfonyl)-3-(4-methoxy-6-methyl-1,3,5-triazin-=
2-yl)urea
Chemical Abstracts name 2-chloro-*N*-[[(4-methoxy-6-methyl-1,3,5-triazin-2-yl)=
amino]carbonyl]benzenesulfonamide
CAS RN *[64902–72–3]* **Development codes** DPX 4189; W4189

PHYSICAL CHEMISTRY
Mol. wt. 357.8 **M.f.** $C_{12}H_{12}ClN_5O_4S$ **Form** White crystalline solid.
M.p. 170–173 °C **V.p.** 3×10^{-6} mPa (25 °C) **K$_{ow}$** logP = –0.99 (pH 7)
Henry 3.0×10^{-9} Pa m^3 mol^{-1} **S.g./density** 1.48 **Solubility** In water 587 ppm
(pH 5), 3.18 g/100g (pH 7) (both 25 °C). In dichloromethane 1.4, acetone 4,
methanol 15, toluene 3, hexane <0.01 (all in g/l, 20 °C). **Stability** Stable to light
when dry. Decomposes at 192 °C. In aqueous solutions, DT$_{50}$ 4–8 w
(pH 5.7–7.0, 20 °C); significant degradation in 24–48 h (pH <5). Hydrolysis is also
promoted by polar organic solvents such as methanol and acetone. **pKa** 3.6

COMMERCIALISATION
History Herbicide reported by P. G. Jensen (*Weeds Weed Control*, 1980, 21st, 24).
Introduced by E. I. du Pont de Nemours Co. **Patent** US 4127405
Manufacturers Du Pont

APPLICATIONS
Biochemistry Branched chain amino acid synthesis (acetolactate synthase or ALS)
inhibitor. Acts by inhibiting biosynthesis of the essential amino acids valine and
isoleucine, hence stopping cell division and plant growth. Selectivity derives from
rapid metabolism in the crop. Metabolic basis of selectivity in sulfonylureas
reviewed (M. K. Koeppe & H. M. Brown, *Agro-Food-Industry*, **6**, 9–14 (1995))
Mode of action Selective systemic herbicide, absorbed by the foliage and roots,
with rapid translocation acropetally and basipetally. **Uses** Control of most
broad-leaved weeds and some annual grasses in wheat, barley, oats, rye,
triticale, flax, and on non-crop land. Applied pre-emergence, early post-
emergence, pre-plant, or early post-plant incorporated at 9–25 g/ha.
Phytotoxicity Phytotoxic to many broad-leaved crops, particularly sugar beet

MAIN ENTRIES

and brassicas.   **Formulation types** WG.   **Mixtures** *(chlorsulfuron +)* methabenzthiazuron; flupyrsulfuron-methyl; metsulfuron-methyl; thifensulfuron-methyl; chlorotoluron.   **Selected tradenames** 'Glean' (Du Pont); 'Telar' (USA) (Du Pont); 'Lasher' (Sanonda)

ANALYSIS
Product analysis by hplc (R. V. Slates & M. W. Watson, *Anal. Methods Pestic. Plant Growth Regul.*, 1988, **16**, 53).   **Residues** determined by hplc *(idem, ibid.;* E. W. Zahnow, *J. Agric. Food Chem.*, 1982, **30**, 854) and by immunoassay (Kelly *et al., ibid.*, 1985, **33**, 962). Methods for sulfonylurea residues in crops, soil and water reviewed (A. C. Barefoot *et al., Proc. Br. Crop Prot. Conf. – Weeds*, 1995, **2**, 707).

MAMMALIAN TOXICOLOGY
Oral Acute oral LD_{50} for male rats 5545, female rats 6293 mg/kg.
Skin and eye Acute percutaneous LD_{50} for rabbits 2500 mg/kg. Mild eye irritant; non-irritating and non-sensitising to skin.   **Inhalation** LC_{50} (4 h) for rats >5.9 g/l air.   **NOEL** (2 y) in rats 100, mice 500 mg/kg diet; (1 y) in dogs 2000 mg/kg diet.   **ADI** 0.05 mg/kg.   **Other** No oncogenic, mutagenic or teratogenic activity detected in standard tests. Acute i.p. LD_{50} for rats 1450 mg/kg.   **Toxicity class** WHO (a.i.) III (Table 5); EPA (formulation) IV (75 WG)

ECOTOXICOLOGY
Birds Acute oral LD_{50} for mallard ducks and bobwhite quail >5000 mg/kg. Eight-day dietary LC_{50} for mallard ducks and bobwhite quail >5000 mg/kg diet.
Fish LC_{50} (96 h) for rainbow trout >250, bluegill sunfish >300 mg/l. LC_{50} for fathead minnow >300, catfish >50, sheepshead minnow >980 mg/l.   **Bees** LD_{50} (contact) >25 µg/bee.   **Worms** LC_{50} >2000 mg/kg.   **Daphnia** LC_{50} (48 h) 370 mg/l.

ENVIRONMENTAL FATE
Soil/Environment In soil, deactivation is through hydrolysis followed by complete metabolism to low-molecular-weight compounds through normal soil microbial processes. Rate of hydrolysis is increased by higher soil temperatures, low pH, and the presence of moisture. Average half-life under growing-season conditions is *c.* 4–6 w. K_{oc} 40 (pH 7).

benzoic acid

$$CO_2CH_3$$

(structure: benzene ring with Cl at positions, CO_2CH_3 top, CO_2CH_3 bottom, four Cl substituents)

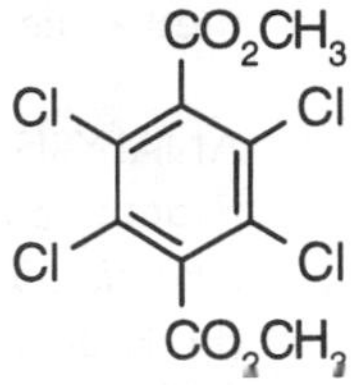

NOMENCLATURE
chlorthal-dimethyl
Common name chlorthal-dimethyl (pre-1990 BSI, E-ISO, *(m)* F-ISO); DCPA (WSSA); TCTP (JMAF)
Other names DCPA; dimethyl tetrachloroterephthalate **CAS RN** *[1861–32–1]*
Development codes DAC 893

chlorthal
Common name chlorthal (WSSA; since 1990 BSI, E-ISO, *(m)* F-ISO)
IUPAC name tetrachloroterephthalic acid
Chemical Abstracts name 2,3,5,6-tetrachloro-1,4-benzenedicarboxylic acid
CAS RN *[2136–79–0]*

PHYSICAL CHEMISTRY
chlorthal-dimethyl
Mol. wt. 332.0 **M.f.** $C_{10}H_6Cl_4O_4$ **Form** Colourless crystals. **M.p.** 156 °C
V.p. 0.21 mPa (25 °C, gas saturation method) **K_{ow}** logP = 4.28 (25 °C)
Solubility In water *c.* 0.5 mg/l (25 °C). In benzene 250, toluene 170, xylene 140, dioxane 120, acetone 100, carbon tetrachloride 70 (all in g/kg, 25 °C).
Stability Stable to heat and u.v. light. Decomposes at *c.* 360–370 °C.

chlorthal
Mol. wt. 303.9 **M.f.** $C_8H_2Cl_4O_4$

COMMERCIALISATION
History Herbicide reported by P. H. Schuldt *et al.* (*Proc. Northeast. Weed Control Conf.*, 1960, p. 42). Chlorthal-dimethyl introduced by Diamond Alkali Co. (now ISK Biosciences Corp.). **Patent** US 2923634

APPLICATIONS
chlorthal-dimethyl
Mode of action Selective non-systemic herbicide, absorbed by the coleoptiles (grasses) and hypocotyls. Kills germinating seeds. **Uses** Pre-emergence control of annual grasses and some annual broad-leaved weeds in onions, garlic, leeks, tomatoes, lettuce, cucurbits, capsicums, aubergines, brassicas, potatoes, sweet

potatoes, horseradish, field beans, soya beans, cotton, strawberries, ornamentals, established turf, and other crops. **Phytotoxicity** Phytotoxic to beet, spinach, flax, and trefoil. **Formulation types** WP; GR; SC. **Mixtures** *(chlorthal-dimethyl +)* methazole; propachlor. **Selected tradenames** 'Dacthal' (ISK Biosciences)

ANALYSIS
Product analysis by i.r. spectrophotometry (*AOAC Methods*, 1995, 970.06) or by glc (*ibid.*, 970.05, *CIPAC Handbook*, 1985, **1C**, 2032). **Residues** determined by glc (H. P. Burchfield & E. E. Storris, *Anal. Methods Pestic. Plant Growth Regul. Food Addit.*, 1964, **4**, 67; *Anal. Methods Pestic. Plant Growth Regul.*, 1972, **6**, 616). In **drinking water**, by gc with ECD (*AOAC Methods*, 1995, 990.06); the diacid may be estimated by gc with ECD (*AOAC Methods*, 1995, 992.32, 10.7.03).

MAMMALIAN TOXICOLOGY
chlorthal-dimethyl
Oral Acute oral LD_{50} for rats >12.5, dogs >10.0 g/kg. **Skin and eye** Acute percutaneous LD_{50} for albino rabbits >2 g/kg. Mild dermal irritant, slight eye irritant (rabbits). **Inhalation** LC_{50} (4 h) for rats >4.5 mg/l. **NOEL** (2 y) for mice 142, rats 1 mg/kg daily. **Toxicity class** WHO (a.i.) III (Table 5); EPA (formulation) IV

ECOTOXICOLOGY
chlorthal-dimethyl
Birds LD_{50} for bobwhite quail >2250 mg/kg. Five-day dietary LC_{50} for bobwhite quail and mallard ducks >5620 ppm. **Fish** Non-toxic to fish. LC_{50} (96 h) for rainbow trout >4.7, bluegill sunfish >5.4 mg/l. **Bees** Slightly toxic to bees. **Daphnia** LC_{50} (48 h) >4.6 mg/l.

ENVIRONMENTAL FATE
Animals In mammals, following oral administration, chlorthal-dimethyl is metabolised to monomethyl tetrachloroterephthalate and 2,3,5,6-tetrachloroterephthalic acid (chlorthal), which are eliminated in the urine. **Plants** Chlorthal-dimethyl is not metabolised by plants. **Soil/Environment** Soil DT_{50} 100 d. In soil, microbial degradation leads to monomethyl tetrachloroterephthalate and 2,3,5,6-tetrachloroterephthalic acid (chlorthal). Duration of residual activity in soil is *c.* 3 months.

chlorthiamid *Herbicide*

benzonitrile

S≈C–NH₂ structure

NOMENCLATURE
Common name chlorthiamid (BSI, E-ISO); chlortiamide ((*f*) F-ISO); DCBN (JMAF);
IUPAC name 2,6-dichlorothiobenzamide
Chemical Abstracts name 2,6-dichlorobenzenecarbothioamide
CAS RN *[1918–13–4]* **EEC no.** 217–637–7 **Development codes** WL 5792

PHYSICAL CHEMISTRY
Mol. wt. 206.1 **M.f.** $C_7H_5Cl_2NS$ **Form** Off-white solid. **M.p.** 151–152 °C
V.p. 0.13 mPa (20 °C) **Solubility** In water 950 mg/l (21 °C). In aromatic and
chlorinated hydrocarbons 50–100 g/kg. **Stability** Thermally stable up to 90 °C.
Stable in acidic solutions, but is converted to dichlobenil in alkaline solutions.

COMMERCIALISATION
History Herbicide reported by H. Stanford (*Proc. Br. Weed Control Conf., 7th*,
1964, p. 208). Introduced by Shell Research Ltd. **Patent** GB 987253

APPLICATIONS
Biochemistry Precursor of dichlobenil. **Mode of action** Systemic herbicide,
absorbed by the roots and, to some extent, by the leaves, with translocation
principally acropetally. Inhibits germination of seeds. **Uses** Selective control of
germinating weeds in pome fruit, currants, gooseberries, raspberries, citrus fruit,
vines, olives, lavender, and ornamental plants and shrubs. Control of bracken,
grass and broad-leaved weeds in forestry; and docks in meadows and pastures.
Total weed control on non-crop land, paths, industrial areas, etc.
Formulation types GR. **Compatibility** Incompatible with alkaline materials.

ANALYSIS
Product by potentiometric titration with silver nitrate (*CIPAC Handbook*, 1980, **1A**,
1158). **Residues** by glc with ECD (K. J. Benyon et al., *J. Sci. Food Agric.*, 1966, **17**,
151–155).

MAMMALIAN TOXICOLOGY
Oral Acute oral LD_{50} for rats 757, mice 500 mg/kg. **Skin and eye** Acute
percutaneous LD_{50} for rats >1000 mg/kg. **NOEL** (90 d) for rats 100 mg/kg diet.
Toxicity class WHO (a.i.) III; EPA (formulation) III **EC risk** Xn (R22)

MAIN ENTRIES

ECOTOXICOLOGY
Birds Acute oral LD_{50} for chickens 500 mg/kg.   **Fish** LC_{50} (24 h) for harlequin fish 41 mg/l.   **Bees** Not toxic to bees.

ENVIRONMENTAL FATE
Animals In rats, following oral administration, there is rapid metabolism and 70% elimination in the urine within 24 hours.   **Soil/Environment** In soil, chlorthiamid is converted to dichlobenil, which then gradually undergoes microbial degradation to 2,6-dichlorobenzamide and 2,6-dichlorobenzoic acid. Converted to dichlobenil in soil, DT_{50} for loss from soil (chlorthiamid + dichlobenil) is c. 35 d under dry and 14 d under wet conditions. Duration of residual activity in soil is c. 5–7 months. The fate in crops, soils and animals has been reviewed (K. I. Beynon & A. N. Wright, *Residue Rev.*, 1972, **43**, 23).

142 chlozolinate   *Fungicide*

dicarboximide

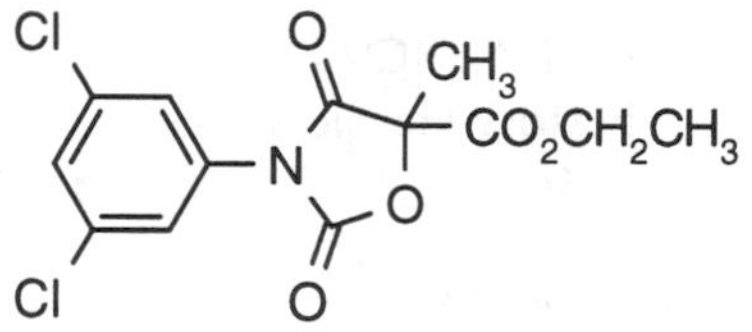

NOMENCLATURE
Common name chlozolinate (BSI, draft E-ISO, (*m*) draft F-ISO)
IUPAC name ethyl (±)-3-(3,5-dichlorophenyl)-5-methyl-2,4-dioxo-oxazolidine-=
5-carboxylate
Chemical Abstracts name (±)-ethyl 3-(3,5-dichlorophenyl)-5-methyl-2,4-dioxo-=
5-oxazolidinecarboxylate
CAS RN [84332–86–5] racemate; [72391–46–9] unstated stereochemistry
EEC no. 282–714–4   **Development codes** M 8164

PHYSICAL CHEMISTRY
Mol. wt. 332.1   **M.f.** $C_{13}H_{11}Cl_2NO_5$ **Form** Colourless, almost odourless solid.
M.p. 112.6 °C   **V.p.** 0.013 mPa (25 °C, gas saturation method) K_{ow} logP = 3.15
(22 °C)   **Henry** 2.29×10^{-3} Pa m^3 mol^{-1} (25 °C, calc.)   **S.g./density** 1.441 (20 °C)
Solubility In water c. 2 ppm (25 °C). In acetone, ethyl acetate and
1,2-dichloroethane >250, ethanol c. 13, *n*-heptane c. 2, xylene c. 60 (all in g/kg,
22 °C).   **Stability** Stable to 250 °C (under nitrogen). Stable to light. In solution,
hydrolysis occurs at pH 5–9.

COMMERCIALISATION

History Fungicide reported by Di Toro *et al.* (*Atti Giornate Fitopatol., Siusi*, 1980). Introduced by Agrimont (now Isagro S.p.A.). **Patent** BE 874406; DE 2906574; IT 20579/78 **Manufacturers** Isagro

APPLICATIONS

Biochemistry Lipid peroxidation in mitochondrial membranes.
Mode of action Contact fungicide with protective and curative action.
Uses Control of *Botrytis* spp., *Sclerotinia* spp., and *Monilia* spp. in ornamentals, vegetables, vines, pome fruit, stone fruit, sunflowers and strawberries. Can be applied as a foliar spray or as a soil drench. **Formulation types** WP; SC.
Mixtures (*chlozolinate* +) thiram. **Selected tradenames** 'Serinal' (Isagro)

ANALYSIS

Product and **residues** can be determined by glc. Details are available from Isagro.

MAMMALIAN TOXICOLOGY

Oral Acute oral LD_{50} for rats >4500, mice >10 000 mg/kg. **Skin and eye** Acute percutaneous LD_{50} for rats >5000 mg/kg. Non-irritating to skin (rabbits). Not a skin sensitiser (guinea pigs). **Inhalation** LC_{50} (4 h) for rats >10 mg/l air. **NOEL** (90 d) for rats 200 mg/kg diet; (1 y) for dogs 200 mg/kg diet. **Other** Not mutagenic, not teratogenic. **Toxicity class** WHO (a.i.) III (Table 5); EPA (formulation) III

ECOTOXICOLOGY

Birds Acute oral LD_{50} for quail and mallard duck >4500 mg/kg. **Fish** LC_{50} (96 h) for trout 27.5 mg/l. **Bees** LD_{50} (oral) >100 µg/bee. **Other beneficial spp.** Harmless to *Phytoseiulus persimilis* at up to 2 × recommended application rate. **Daphnia** LC_{50} (48 h) 1.18 mg/l. **Algae** EC_{50} (96 h) for *Selenastrum capricornutum* 30 mg/l.

ENVIRONMENTAL FATE

Animals Chlozolinate is readily absorbed, metabolised and excreted. Metabolites identified in urine of rats are: N-(3,5-dichlorophenyl)-5-methyloxazolidin-= 2,4-dione, N-(3,5-dichlorophenyl)-N-(α-hydroxypropionyl)carbamic acid, N-(α-hydroxypropionyl)-3,5-dichloroaniline and N-(3,5-dichlorophenyl)-= N-(α-hydroxy-α-carboxypropionyl)carbamic acid. **Plants** In plants, chlozolinate undergoes hydrolysis and decarboxylation processes, giving the same metabolites as those identified in animals. **Soil/Environment** In silt-loam soil, hydrolysis and decarboxylation occur; aerobic DT_{50} <7 d.

143 *Chrysoperla carnea*   *Biological agent*

Neuroptera

Entomophagous lacewing larva.

NOMENCLATURE
Scientific name *Chrysoperla carnea* (Stephens)   **Other names** *Chrysopa carnea*;
green lacewing

PROPERTIES
Form Eggs or larvae.   **Stability** Should be released on day of delivery.

COMMERCIALISATION
Manufacturers Arbico; BCP; Beneficial; Biobest; Biolab; Kunafin; Neudorff;
Novartis BCM; Rincon-Vitova; Sautter & Stepper

APPLICATIONS
Mode of action Lacewing larvae are voracious predators. They feed mainly on
aphids, but will eat a variety of soft-bodied, slow-moving prey including whitefly,
scales, mites, beetles and moth eggs. Larvae live for 2 to 3 weeks, eating *c.* 20
aphids a day.   **Uses** Introduced into protected and outdoor crops, mainly for the
control of aphids in strawberry, hops, and top fruit. Also used in cotton, potato
and maize.   **Formulation types** Various, on cards or in vials.

144 cinmethylin   *Herbicide*

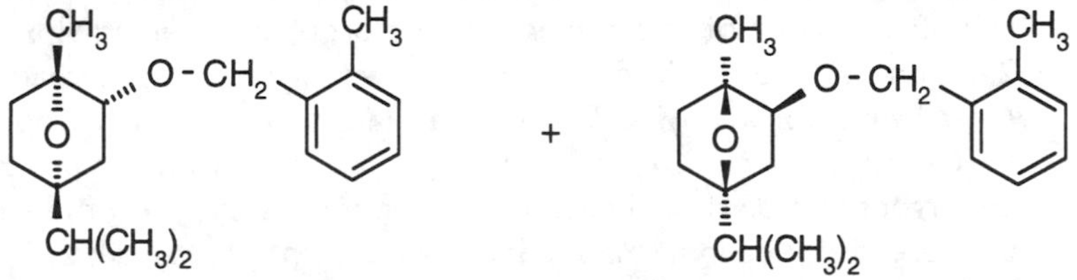

NOMENCLATURE
Common name cinmethylin (BSI, draft E-ISO, ANSI, WSSA); cinméthyline ((*f*) draft
F-ISO)
IUPAC name (1*RS*,2*SR*,4*SR*)-1,4-epoxy-*p*-menth-2-yl 2-methylbenzyl ether
Chemical Abstracts name *exo*-(±)-1-methyl-4-(1-methylethyl)-2-[(2-methylphenyl)=

methoxy]-7-oxabicyclo[2.2.1]heptane
CAS RN *[87818–31–3]* exo-(±)- isomers; *[87818–61–9]* exo-(+)- isomer;
[87819–60–1] exo-(–)- isomer   **Development codes** SD 95 481; WL 95 481

PHYSICAL CHEMISTRY
Mol. wt. 274.4   **M.f.** $C_{18}H_{26}O_2$ **Form** Dark amber liquid.   **B.p.** 313 °C/760
mmHg   **V.p.** 10.1 mPa (20 °C)   **K$_{ow}$** logP = 3.84   **S.g./density** 1.014 g/ml (20 °C)
Solubility In water 63 mg/l (20 °C). Miscible with a range of organic solvents.
Stability Thermally stable up to 145 °C. Hydrolytically stable from pH 5 to pH 9 at
25 °C. Light-catalysed decomposition occurs in the presence of air.   **F.p.** 147 °C
(ASTM D93)

COMMERCIALISATION
History Herbicide reported by J. W. Way *et al.* (*Proc. 1985 Br. Crop Prot. Conf. –
Weeds,* 1, 265). Introduced by Shell International Chemical Co. Ltd (now
American Cyanamid Co.) in China (1989) and by E. I. du Pont de Nemours Co.
(who no longer manufacture or market it).   **Manufacturers** Cyanamid

APPLICATIONS
Mode of action Absorbed from paddy water through the shoots and roots of
germinating or emerging weeds. Moves upwards through the plant and disrupts
meristematic development in growing points of roots and shoots.   **Uses** Control
of important weeds of rice, including *Echinochloa* spp., *Monochoria vaginalis*, and
Cyperus difformis, by post-transplanting application.   **Formulation types** EC; GR.
Mixtures (*cinmethylin* +) 2,4-D.   **Compatibility** Can be tank-mixed with broad-
leaved weed herbicides.   **Selected tradenames** 'Argold' (Cyanamid)

ANALYSIS
Product and **residue** analysis by glc. Details available from Cyanamid.

MAMMALIAN TOXICOLOGY
Oral Acute oral LD_{50} for rats 3960 mg/kg.   **Skin and eye** Acute percutaneous
LD_{50} for rats and rabbits >2000 mg/kg; moderate skin and mild eye irritant to
rabbits.   **Inhalation** LC_{50} (4 h) for rats >3.5 mg/l.   **NOEL** (2 y) for rats 100, mice
30 mg/kg diet.   **Toxicity class** WHO (a.i.) III (Table 5); EPA (formulation) III

ECOTOXICOLOGY
Birds Acute oral LD_{50} for bobwhite quail >2150 mg/kg. Five-day dietary LD_{50} for
bobwhite quail and mallard ducks >5620 mg/kg diet.   **Fish** LC_{50} (96 h) for
rainbow trout 6.6, bluegill sunfish 6.4, sheepshead minnow 1.6 mg/l.
Daphnia LC_{50} (48 h) 7.2 mg/l.   **Other aquatic spp.** LC_{50} for fiddler crab
>1000 mg/l.

ENVIRONMENTAL FATE

Animals For a study of the metabolism of cinmethylin in the goat, see *J. Agric. Food Chem.*, 1989, **37**, 787. **Soil/Environment** Strongly absorbed by soils (B. T. Grayson *et al.*, *Pestic. Sci.*, 1987, **21**, 143), but relatively short-lived in the environment, being broken down in the soil under aerobic conditions with DT_{50} 23–75 d, depending on soil texture. Under anaerobic conditions (as may be found in paddy rice), the rate of metabolism is reduced because of slower microbial breakdown.

145   cinosulfuron   *Herbicide*

sulfonylurea

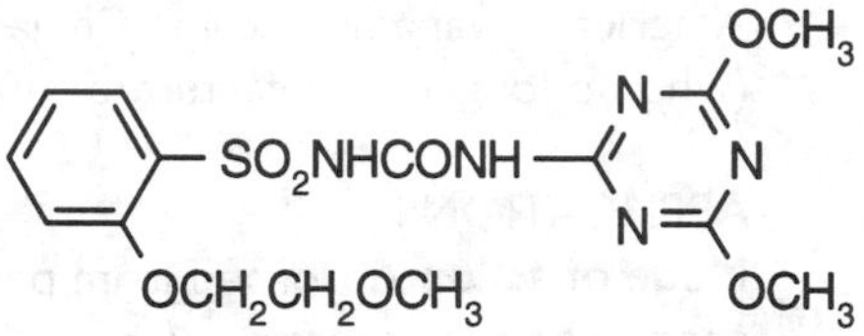

NOMENCLATURE
Common name cinosulfuron (BSI, draft E-ISO)
IUPAC name 1-(4,6-dimethoxy-1,3,5-triazin-2-yl)-3-[2-(2-methoxyethoxy)=phenylsulfonyl]urea
Chemical Abstracts name N-[[(4,6-dimethoxy-1,3,5-triazin-2-yl)amino]carbonyl]-=2-(2-methoxyethoxy)benzenesulfonamide
CAS RN *[94593–91–6]* **Development codes** CGA 142 464

PHYSICAL CHEMISTRY
Mol. wt. 413.4 **M.f.** $C_{15}H_{19}N_5O_7S$ **Form** Colourless crystalline powder.
M.p. 127.0–135.2 °C (pure) **V.p.** <0.01 mPa (25 °C) (OECD 104) **K_{ow} logP** = 2.04 (pH 2.1, 25 °C) **Henry** $<1 \times 10^{-6}$ Pa m^3 mol^{-1} (pH 6.7, calc.)
S.g./density 1.47 (20 °C) **Solubility** In water 120 (pH 5.0), 4000 (pH 6.7), 19 000 (pH 8.1) (all in mg/l, 25 °C). In acetone 36 000, ethanol 1900, toluene 540, *n*-octanol 260, *n*-hexane <1 (all in mg/l, 25 °C). **Stability** Decomposes above the melting point. No significant hydrolysis at pH 7–10; considerable hydrolysis at pH 3–5. **pKa** 4.72

COMMERCIALISATION
History Herbicide reported by M. Quadranti *et al.* (*Proc. 11th Conf. Asian Pacific Weed Sci. Soc.*, 1987, **1**, 117). Introduced by Ciba-Geigy AG (now Novartis Crop Protection AG). **Patent** US 4479821; EP 44807 **Manufacturers** Novartis

APPLICATIONS

Biochemistry Branched chain amino acid synthesis (acetolactate synthase or ALS) inhibitor. Acts by inhibiting biosynthesis of the essential amino acids valine and isoleucine, hence stopping cell division and plant growth. Selectivity derives from rapid metabolism in the crop. Metabolic basis of selectivity in sulfonylureas reviewed (M. K. Koeppe & H. M. Brown, *Agro-Food-Industry*, **6**, 9–14 (1995)). **Mode of action** Absorbed primarily by shoots and roots, and translocated to actively growing meristematic tissue. **Uses** Applied post-emergence to control many weeds including *Alisma,* annual *Cyperus, Eleocharis, Marsilea, Potamogeton* and *Sagittaria* spp., *Monochoria vaginalis* and *Sphenoclea zeylanica* in transplanted, direct seeded, wet-sown and water-sown, and dry-sown flooded rice crops, at 20–80 g/ha. Also used in tropical plantation crops. For full weed spectrum, may need to be combined with a grass herbicide in tank-mix or sequential treatment. **Phytotoxicity** No varietal restrictions for rice. **Formulation types** WG; GR. **Mixtures** *(cinosulfuron +)* quinclorac; pretilachlor + quinclorac; pretilachlor + fenclorim. **Selected tradenames** 'Setoff' (Novartis)

ANALYSIS

Product and **residue** analysis by hplc. Methods for sulfonylurea residues in crops, soil and water reviewed (A. C. Barefoot et al., *Proc. Br. Crop Prot. Conf. – Weeds,* 1995, **2**, 707). Details available from Novartis.

MAMMALIAN TOXICOLOGY

Oral Acute oral LD_{50} for rats and mice >5000 mg/kg. **Skin and eye** Acute percutaneous LD_{50} for rats >2000 mg/kg; not irritant to skin or eyes of rabbits. Non-sensitising to skin (guinea pigs). **Inhalation** LC_{50} (4 h) for rats >5 mg/l air. **NOEL** (2 y) for rats 400, mice 60 ppm; (1 y) for dogs 2500 ppm. **Toxicity class** WHO (a.i.) III (Table 5)

ECOTOXICOLOGY

Birds Acute oral LD_{50} for Japanese quail >2000 mg/kg. **Fish** LC_{50} (96 h) for rainbow trout >100 mg/l. **Bees** Non-toxic; LD_{50} (oral and contact) >100 μg/bee. **Worms** LC_{50} (14 d) for *Eisenia foetida* 1000 mg/kg. **Daphnia** LC_{50} (48 h) 2500 mg/l. **Algae** EC_{50} (72 h) for *Scenedesmus subspicatus* 4.8 mg/l.

ENVIRONMENTAL FATE

Animals Hydrolysis of methoxy groups and cleavage of the sulfonylurea bridge. It is rapidly excreted (80–100% within 24 h). **Plants** In plants, rapid degradation by cleavage of the sulfonylurea bridge. **Soil/Environment** In soil degrades a) by O-dealkylation in both ring systems, forming the corresponding phenol, mono- and di- hydroxytriazine and b) by cleavage of the sulfonylurea bridge; degradation proceeds further to form bound residues and CO_2; DT_{50} in soil (lab.) c. 20 d, in paddy fields c. 3 d. No potential for bioaccumulation. K_{oc} c. 20, indicating that cinosulfuron may leach; column data show however that rapid degradation prevents leaching. In paddy fields under natural conditions, cinosulfuron dissipates rapidly due to photolysis and to a lesser extent by adsorption to soil.

146 clethodim   *Herbicide*

cyclohexanedione oxime

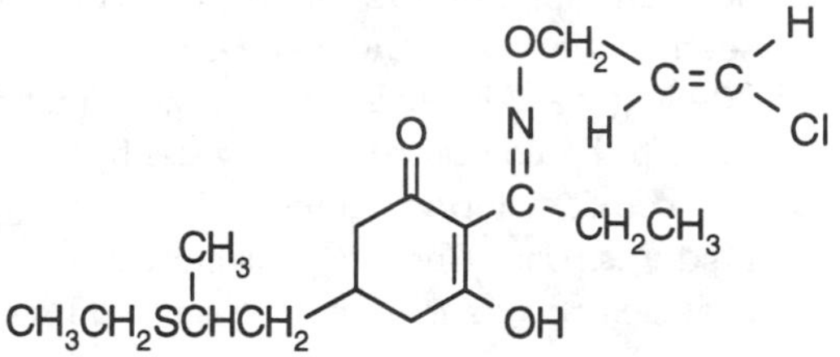

NOMENCLATURE
Common name clethodim (BSI, ANSI, draft E-ISO); clétodime ((f) draft F-ISO)
IUPAC name (±)-2-[(E)-1-[(E)-3-chloroallyloxyimino]propyl]-5-[2-(ethylthio)=propyl]-3-hydroxycyclohex-2-enone
Chemical Abstracts name (E,E)-(±)-2-[1-[[(3-chloro-2-propenyl)oxy]imino]=propyl]-5-[2-(ethylthio)propyl]-3-hydroxy-2-cyclohexen-1-one
CAS RN [99129–21–2]   **Development codes** RE-45601

PHYSICAL CHEMISTRY
Composition Tech. is >91% pure.   **Mol. wt.** 359.9   **M.f.** $C_{17}H_{26}ClNO_3S$
Form Clear, amber liquid.   **B.p.** Decomposes below boiling point.
V.p. $<1 \times 10^{-2}$ mPa (20 °C)   **S.g./density** 1.14 (20 °C)   **Solubility** Soluble in most organic solvents.   **Stability** Not stable to extremes of pH or temperature, or to exposure to u.v. light.

COMMERCIALISATION
History Herbicide reported by R. T. Kincade *et al.* (*Proc. 1987 Br. Crop Prot. Conf. – Weeds*, **1**, 49). Introduced by Chevron Chemical Co.   **Patent** GB 2090246
Manufacturers Tomen

APPLICATIONS
Biochemistry Acetyl CoA carboxylase (fatty acid synthesis) inhibitor.
Mode of action Systemic herbicide, rapidly absorbed and readily translocated from treated foliage to the root system and growing parts of the plant.   **Uses** Post-emergence control of annual and perennial grasses in a wide range of broad-leaved crops (including such field crops as soya beans, cotton, flax, sunflowers, alfalfa, peanuts, oilseed rape, sugar beet, tobacco, and potatoes), vegetable crops,

trees and vines. To be used with a non-phytotoxic crop oil concentrate.
Phytotoxicity Good tolerance in broad-leaved crops.   **Formulation types** EC.
Compatibility Antagonism has been observed when tank-mixed with Basagran or
Blazer.   **Selected tradenames** 'Select' (Tomen, Valent)

ANALYSIS
By hplc.

MAMMALIAN TOXICOLOGY
Reviews *Pesticide residues in food – 1994*, FAO Plant Production and Protection
Paper, 127, 1995. *Pesticide residues in food – 1994 evaluations, Pt. II – Toxicology*,
WHO, WHO/PCS/95.2, 1995.   **Oral** Acute oral LD_{50} for male rats 1630, female
rats 1360 mg/kg.   **Skin and eye** Acute percutaneous LD_{50} for rabbits
>5000 mg/kg.   **Inhalation** LC_{50} (4 h) >4.6 mg/l.   **NOEL** for mice 30, rats
16 mg/kg daily.   **ADI** (JMPR) 0.01 mg/kg b.w. [1994]; 0.16 mg/kg (Canada);
0.01 mg/kg (USA).   **Toxicity class** WHO (a.i.) III; EPA (formulation) III

ECOTOXICOLOGY
Birds LD_{50} for bobwhite quail and mallard ducks >2000 mg/kg. Eight-day dietary
LC_{50} for Japanese quail >6000 mg/kg.   **Fish** LC_{50} (96 h) for rainbow trout 56,
bluegill sunfish >120 mg/l.   **Bees** LD_{50} >100 μg/bee.   **Worms** LD_{50} for worms
454 mg/kg.   **Daphnia** LC_{50} (48 h) >120 mg/l.   **Algae** EC_{50} (5 d) for fresh water
algae 57.8 mg/l.

ENVIRONMENTAL FATE
Animals Metabolised to the sulfoxide.   **Plants** Metabolised to the sulfoxide.
Soil/Environment DT_{50} in soil 1–3 d. K_d 0.05–0.23 (5 soil types).

47   clodinafop-propargyl   *Herbicide*

2-(4-aryloxyphenoxy)propionic acid

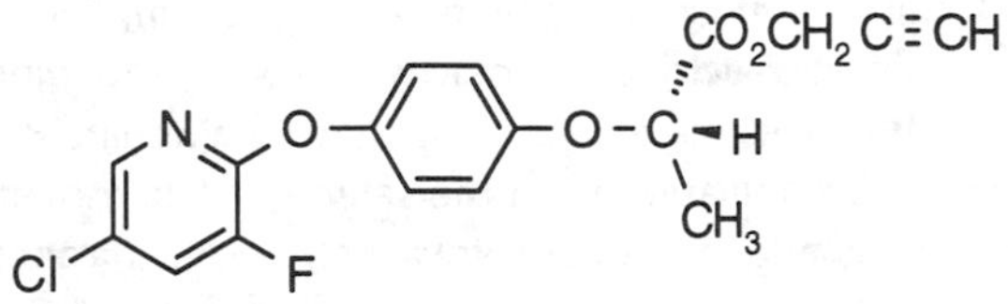

NOMENCLATURE
clodinafop-propargyl
Common name clodinafop-propargyl

IUPAC name prop-2-ynyl (*R*)-2-[4-(5-chloro-3-fluoropyridin-2-yloxy)phenoxy]=
propionate
Chemical Abstracts name propynyl (*R*)-2-[4-[(5-chloro-3-fluoro-2-pyridinyl)oxy]=
phenoxy]propanoate
CAS RN *[105512–06–9]*   **Development codes** CGA 184927

clodinafop
Common name clodinafop (BSI, pa E-ISO)
IUPAC name (*R*)-2-[4-(5-chloro-3-fluoro-2-pyridyloxy)phenoxy]propionic acid
Chemical Abstracts name (*R*)-2-[4-[(5-chloro-3-fluoro-2-pyridinyl)oxy]phenoxy]=
propanoic acid
CAS RN *[114420–56–3]*

PHYSICAL CHEMISTRY
clodinafop-propargyl
Composition (*R*)- isomer.   **Mol. wt.** 349.8   **M.f.** $C_{17}H_{13}ClFNO_4$ **Form** Colourless
crystals.   **M.p.** 59.5 °C; (tech., 48.2–57.1 °C)   **V.p.** 3.19×10^{-3} mPa (25 °C)
(OECD 104) K_{ow} logP = 3.9 (25 °C)   **S.g./density** 1.37 (20 °C)   **Solubility** In
water 4.0 mg/l (25 °C). In ethanol 97, acetone 880, toluene 690, *n*-hexane
0.0086, *n*-octanol 25 g/l (25 °C).   **Stability** Relatively stable in acidic media at
50 °C, hydrolyses in alkaline media; DT_{50} (25 °C) 64 h (pH 7), 2.2 h (pH 9).

clodinafop
Mol. wt. 311.7   **M.f.** $C_{14}H_{11}ClFNO_4$

COMMERCIALISATION
History Reported by J. Amrein *et al.* (*Proc. Br. Crop Prot. Conf. – Weeds*, 1989, **1**,
71–76). Introduced by Ciba-Geigy AG (now Novartis Crop Protection AG).
Patent EP B 0083556, US 4713109   **Manufacturers** Novartis

APPLICATIONS
clodinafop-propargyl
Biochemistry Inhibits lipid biosynthesis (ACCase).   **Mode of action** Post-
emergence, systemic grass herbicide. Phytotoxic symptoms appear within 1–3
weeks, affecting meristematic tissue.   **Uses** Used for post-emergence control of
annual grasses, including *Avena*, *Lolium*, *Setaria*, *Phalaris* and *Alopecurus*, in cereals.
Phytotoxicity Low toxicity to spring and winter wheat.   **Formulation types** EC.
Mixtures (*clodinafop-propargyl* +) cloquintocet-mexyl.   **Compatibility** Mainly used
in combination with the safener cloquintocet-mexyl.   **Selected tradenames** 'Topik'
(Canada, mixture) (Novartis); 'Celio' (France, mixture) (Novartis)

ANALYSIS
Details from Novartis.

MAMMALIAN TOXICOLOGY

clodinafop-propargyl

Oral Acute oral LD_{50} for rats 1829, mice >2000 mg/kg. **Skin and eye** Acute percutaneous LD_{50} for rats >2000 mg/kg. Non-irritating to eyes and skin (rabbits). May cause skin sensitisation (guinea pigs). **Inhalation** LC_{50} (4 h) for rats 2.325 mg/l air. **NOEL** (2 y) for rats 0.35, (18 mo) for mice 1.2, (1 y) for dogs 3.3 mg/kg b.w. daily. **ADI** 0.004 mg/kg b.w. **Toxicity class** WHO (a.i.) III **EC risk** R22, R43

ECOTOXICOLOGY

clodinafop-propargyl

Birds LD_{50} (8 d) for mallard ducks >2000, bobwhite quail >1455 mg/kg. **Fish** LC_{50} (96 h) for rainbow trout 0.39, carp 0.46, catfish 0.43 mg/l. **Bees** LD_{50} (48 h, oral and contact) >100 µg/bee. **Worms** LC_{50} for earthworms 210 mg/kg. **Daphnia** LC_{50} (48 h) >74 mg/l. **Algae** EC_{50} (96–120 h) for *Scenedesmus subspicatus* 25, *Microcystis* >65.5, *Navicula* 6.8 mg/l.

ENVIRONMENTAL FATE

Animals Hydrolysed to the corresponding acid. **Plants** In plants, rapidly degraded to the acid derivative as major metabolite. **Soil/Environment** In soil, undergoes rapid degradation to the free acid (DT_{50} <2 h) and then further to phenyl and pyridine moieties which are bound to the soil and mineralised. The free acid is mobile in soil but is further degraded with DT_{50} 5–20 d; in practice there is a negligible leaching potential.

48   clofencet   *Plant growth regulator*

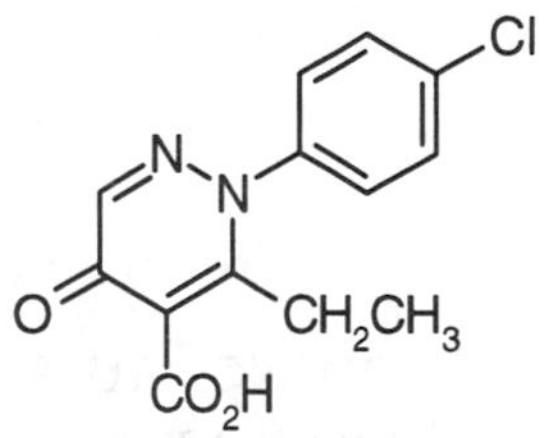

NOMENCLATURE

clofencet

Common name clofencet (pa ISO, ANSI)

IUPAC name 2-(4-chlorophenyl)-3-ethyl-2,5-dihydro-5-oxopyridazine-4-carboxylic acid

Chemical Abstracts name 2-(4-chlorophenyl)-3-ethyl-2,5-dihydro-5-oxo-=
4-pyridazinecarboxylic acid
CAS RN *[129025–54–3]*   **Development codes** MON 21200; RH 754; ICIA 0754;
FC 4001

clofencet-potassium
CAS RN *[82697–71–0]*

PHYSICAL CHEMISTRY
clofencet
Mol. wt. 278.7   **M.f.** $C_{13}H_{11}ClN_2O_3$

clofencet-potassium
Mol. wt. 316.8   **M.f.** $C_{13}H_{10}ClKN_2O_3$

COMMERCIALISATION
History Discovered by Rohm and Haas Co. Introduced by Monsanto Co. for use
in wheat.   **Manufacturers** Zeneca Life Science Molecules

APPLICATIONS
Mode of action Systemic. Inhibits pollen formation.   **Uses** Under development as
a cereal hybridising agent for wheat.   **Selected tradenames** 'Genesis' (potassium
salt) (Monsanto)

149 clofentezine   *Acaricide*

tetrazine

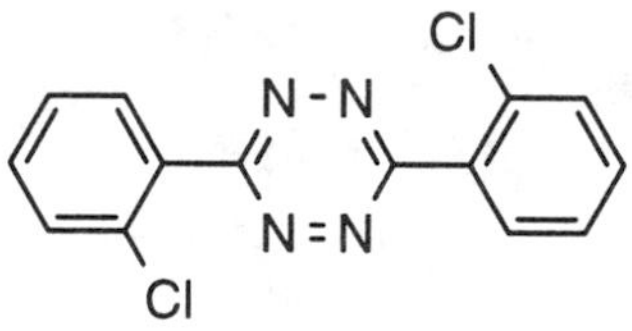

NOMENCLATURE
Common name clofentezine (BSI, ANSI, draft E-ISO, (f) draft F-ISO)
IUPAC name 3,6-bis(2-chlorophenyl)-1,2,4,5-tetrazine
Chemical Abstracts name 3,6-bis(2-chlorophenyl)-1,2,4,5-tetrazine
Other names [bisclofentezin] (rejected common name proposal)
CAS RN *[74115–24–5]*   **Development codes** NC 21 314

PHYSICAL CHEMISTRY
Mol. wt. 303.1   **M.f.** $C_{14}H_8Cl_2N_4$ **Form** Magenta crystals.   **M.p.** 182.3 °C
V.p. 1.3×10^{-4} mPa (25 °C) (gas saturation and glc) K_{ow} logP = 4.1 (25 °C)
S.g./density 1.51 (20 °C)   **Solubility** In water 2.5 µg/l (pH 5, 22 °C). In
dichloromethane 37, acetone 9.3, hexane 1, ethanol 0.5 (all in g/l, 25 °C).
Stability The a.i. and formulated products stable to light, heat and air; on
hydrolysis (22 °C) DT_{50} 248 h (pH 5), 34 h (pH 7), 4 h (pH 9).   **F.p.** Low
flammability.

COMMERCIALISATION
History Acaricide reported by K. M. G. Bryan et al. (*Proc. Br. Crop Prot. Conf. –
Pests Dis.*, 1981, **1**, 67) and relationships between chemical structure and biological
activity by P. J. Brooker et al. (*Pestic. Sci.*, 1987, **18**, 179). Introduced by FBC
Limited (now AgrEvo GmbH).   **Patent** EP 5912   **Manufacturers** AgrEvo

APPLICATIONS
Mode of action Specific acaricide with contact action, and long residual activity.
Inhibits embryo development.   **Uses** Control of eggs and young motile stages (but
not adults) of *Panonychus ulmi* and *Tetranychus* spp. on pome fruit, stone fruit,
citrus fruit, nuts, vines, hops, strawberries, cucurbits, cotton, and ornamentals.
Has no effect on predatory mites or beneficial insect species.   **Phytotoxicity** May
cause slight injury to glasshouse roses. May cause a slight pink deposit on petals of
white or pale flowers.   **Formulation types** WP; SC.   **Mixtures** (*clofentezine +*)
cyhexatin; fenpropathrin; bifenthrin; fenbutatin oxide; propargite.
Selected tradenames 'Apollo' (AgrEvo)

ANALYSIS
Product analysis by rp hplc (*CIPAC Handbook*, 1995, **G**, 18–23) or by visible
spectroscopy.   **Residues** in plants and soil determined by hplc. Details available
from AgrEvo.

MAMMALIAN TOXICOLOGY
Reviews *Pesticide residues in food – 1986*. FAO Plant Production and Protection
Paper 77, 1986. *Pesticide residues in food – 1986 evaluations. Part II – Toxicology.*
FAO Plant Production and Protection Paper 78/2, 1987.   **Oral** Acute oral LD_{50} for
rats >5200 mg/kg.   **Skin and eye** Acute percutaneous LD_{50} for rats >2100 mg/kg.
Mild eye and skin irritant.   **Inhalation** LC_{50} (4 h) for rats >9 mg/l air.   **NOEL** (2 y)
for rats 40 mg/kg diet; (1 y) for dogs 50 mg/kg diet.   **ADI** (JMPR) 0.02 mg/kg
b.w. [1986].   **Toxicity class** WHO (a.i.) III (Table 5); EPA (formulation) III

ECOTOXICOLOGY
Birds Acute oral LD_{50} for mallard ducks >3000, bobwhite quail >7500 mg/kg.
Eight-day dietary LC_{50} for mallard ducks and bobwhite quail >2000 mg/kg diet.
Fish LC_{50} (96 h) for rainbow trout >0.015, bluegill sunfish >0.25 mg/l (limits of

solubility).   **Bees** Acute LD_{50} (oral) >20 μg/bee; (contact) LC_{50} >1500 ppm.
Daphnia LC_{50} (48 h) >1.45 μg/l (limit of solubility).   **Algae** Not toxic to
Scenedesmus panonicus up to limit of solubility.

ENVIRONMENTAL FATE
Animals In mammals, undergoes metabolism by hydroxylation and exchange of the
chlorine atoms on the rings for methylthio groups. Following oral administration,
excretion occurs within 24–48 hours in the urine and faeces.   **Plants** In
metabolism studies, unchanged clofentezine was the major extractable residue.
Trace amounts (4%) of 2-chlorobenzonitrile, the major photodegradation product,
were also detected.   **Soil/Environment** In soil, the major degradation route leads
to 2-chlorobenzoic acid, and finally to CO_2. DT_{50} in soil 65 to 85 d (15 °C), 28 to
56 d (25 °C), depending upon soil type. However, in laboratory studies, no
leaching occurs. In water, 2-chlorobenzonitrile is the major product formed by
hydrolysis and photodegradation, with smaller amounts of other compounds. Low
solubility in water makes determination of soil adsorption constants difficult.

150   clomazone   *Herbicide*

NOMENCLATURE
Common name clomazone (BSI, ANSI, draft E-ISO, (f) draft F-ISO)
IUPAC name 2-(2-chlorobenzyl)-4,4-dimethyl-1,2-oxazolidin-3-one;
2-(2-chlorobenzyl)-4,4-dimethylisoxazolidin-3-one
Chemical Abstracts name 2-[(2-chlorophenyl)methyl]-4,4-dimethyl-=
3-isoxazolidinone
Other names [dimethazone]   **CAS RN** *[81777–89–1]*
Development codes FMC 57 020

PHYSICAL CHEMISTRY
Mol. wt. 239.7   **M.f.** $C_{12}H_{14}ClNO_2$ **Form** Clear colourless to light brown viscous
liquid.   **M.p.** 25 °C   **B.p.** 275 °C   **V.p.** 19.2 mPa (25 °C) K_{ow} logP = 2.5
Henry 4.19×10^{-3} Pa m³ mol⁻¹   **S.g./density** 1.192 (20 °C)   **Solubility** In water
1.1 g/l. Miscible with acetone, acetonitrile, chloroform, cyclohexanone,
dichloromethane, methanol, toluene, heptane, dimethylformamide.

Stability Stable at ambient temperatures for at least 2 y; stable at 50 °C for at least 3 mo. In sunlight DT_{50} >30 d in aqueous solution. **F.p.** 70–75 °C (closed cup)

COMMERCIALISATION
History Herbicide introduced by FMC Corp. **Patent** US 4405357
Manufacturers FMC

APPLICATIONS
Biochemistry Inhibits biosynthesis of photosynthetic pigments.
Mode of action Selective herbicide, absorbed by the roots and shoots and translocated upward. Susceptible species emerge but are devoid of pigmentation.
Uses Control of broad-leaved and grass weeds in soya beans, peas, maize, oilseed rape, sugar cane, cassava, pumpkins, and tobacco. Applied pre-emergence or pre-plant incorporated. **Phytotoxicity** Foliar contact or vapours may cause visual symptoms of chlorosis to nearby sensitive plants. **Formulation types** EC; WP; CS.
Mixtures *(clomazone +)* tebutam; trifluralin. **Compatibility** Compatible with many other herbicides, e.g. metribuzin, linuron, chloramben, alachlor, trifluralin, pendimethalin, metolachlor, oryzalin or ethalfluralin.
Selected tradenames 'Command' (FMC); 'Commerce' (mixture) (FMC)

ANALYSIS
Product by gc with FID or hplc with u.v. detection. Details and **residue** methods reviewed (A. W. Chen in *Comp. Anal. Profiles,* Chapt. 6).

MAMMALIAN TOXICOLOGY
Oral Acute oral LD_{50} for male rats 2077, female rats 1369 mg/kg.
Skin and eye Acute percutaneous LD_{50} for rabbits >2000 mg/kg. Practically non-irritating to eyes (rabbits). **Inhalation** LC_{50} (4 h) for rats 4.8 mg/l. **NOEL** (2 y) for rats 4.3 mg/kg daily. **ADI** 0.043 mg/kg (proposed). **Toxicity class** WHO (a.i.) II; EPA (formulation) III

ECOTOXICOLOGY
Birds Acute oral LD_{50} for bobwhite quail and mallard ducks >2510 mg/kg. LC_{50} (8 d) for bobwhite quail and mallard ducks >5620 ppm. **Fish** LC_{50} (96 h) for bluegill sunfish 34, Atlantic silverside 6.26, rainbow trout 19 mg/l. **Worms** LC_{50} (14 d) for *Eisenia foetida* 156 mg/kg. **Daphnia** LC_{50} (48 h) 5.2 mg/l. **Algae** EC_{50} (48 h) 2.10 mg/l. **Other aquatic spp.** LC_{50} for pink shrimp 8.9, eastern oyster 5.3 mg/l.

ENVIRONMENTAL FATE
Soil/Environment DT_{50} in soil *c.* 30–135 d.

aryloxyalkanoic acid

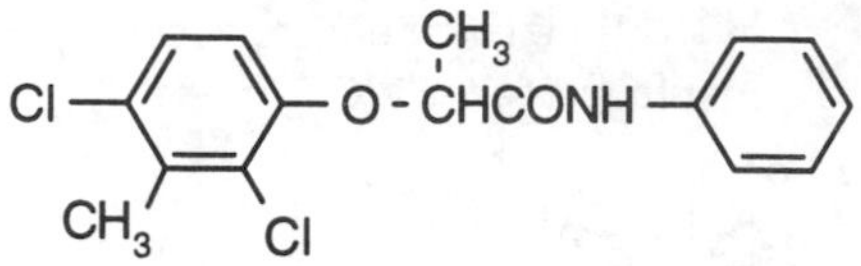

NOMENCLATURE
Common name clomeprop (BSI, draft E-ISO, (*m*) draft F-ISO)
IUPAC name (*RS*)-2-(2,4-dichloro-*m*-tolyloxy)propionanilide
Chemical Abstracts name (±)-2-(2,4-dichloro-3-methylphenoxy)-*N*-phenyl=
propanamide
CAS RN *[84496–56–0]* **Development codes** MY-15 (Mitsubishi)

PHYSICAL CHEMISTRY
Mol. wt. 324.2 **M.f.** $C_{16}H_{15}Cl_2NO_2$ **Form** Colourless crystals.
M.p. 146–147 °C **V.p.** <0.0133 mPa (30 °C) K_{ow} logP = 4.8 **Solubility** In water
0.032 mg/l (25 °C). In xylene 17, acetone 33, cyclohexane 9, dimethylformamide
20 (all in g/l, 20 °C).

COMMERCIALISATION
History Herbicide reported by K. Ikeda & A. Goh (*Jpn. Pestic. Inf.*, 1989, No. 55,
p. 15). Introduced by Mitsubishi Petrochemical Co. Ltd. (now Mitsubishi Chemical
Corp.) and commercialised by Rhône-Poulenc Yuka Agro KK, a joint venture of
Mitsubishi Chemical Corp. and Rhône-Poulenc Agro.
Manufacturers Rhône-Poulenc Yuka Agro

APPLICATIONS
Biochemistry Clomeprop shows auxin-type herbicidal activity, killing plants by
disturbing the hormone balance. It promotes RNA synthesis and influences protein
synthesis, cell division, and cell elongation. **Uses** Pre- to early post-emergence
control of broad-leaved and cyperaceous weeds in paddy rice (in combination
with pretilachlor). **Phytotoxicity** Growth of rice plants can be inhibited at higher
temperatures. **Formulation types** EC; GR. **Mixtures** (*clomeprop* +) pretilachlor.
Selected tradenames 'Yukahope' (Mitsubishi)

ANALYSIS
Details from Mitsubishi.

MAMMALIAN TOXICOLOGY
Reviews *Pesticide Science Society of Japan*, **16**(1), 1991. **Oral** Acute oral LD_{50} for
male rats >5000, female rats 3520, mice >5000 mg/kg. **Skin and eye** Acute

percutaneous LD$_{50}$ for rats and mice >5000 mg/kg.   **Inhalation** LC$_{50}$ (4 h) for rats >1.5 mg/l.   **NOEL** (2 y) for rats 0.62 mg/kg.   **ADI** 0.0062 mg/kg (Japan). **Toxicity class** WHO (a.i.) III (Table 5); EPA (formulation) IV

ECOTOXICOLOGY
Fish LC$_{50}$ (48 h) for carp, loach and rainbow trout >10 mg/l.   **Daphnia** LC$_{50}$ (3 h) >10 ppm.

ENVIRONMENTAL FATE
Plants Degrades quickly in the plant to the non-toxic glucosyl conjugate.
Soil/Environment Degrades quickly and disappears eventually as CO_2. Half-life in paddy-field soil is *c.* 3–7 d.

152 cloprop   *Plant growth regulator*

aryloxyalkanoic acid

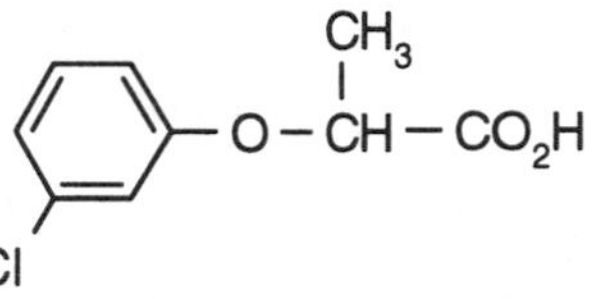

NOMENCLATURE
Common name cloprop (BSI, E-ISO, (*m*) F-ISO)
IUPAC name (±)-2-(3-chlorophenoxy)propionic acid
Chemical Abstracts name (±)-2-(3-chlorophenoxy)propanoic acid
Other names 3-CPA   **CAS RN** *[101–10–0]* unstated stereochemistry

PHYSICAL CHEMISTRY
Mol. wt. 200.9   **M.f.** $C_9H_9ClO_3$ **Form** Colourless crystals.   **Solubility** In water 1.2 g/l (22 °C). In acetone 790.9, dimethyl sulfoxide 2685, ethanol 710.8, methanol 716.5, iso-octanol 247.3 (all in g/l, 22 °C). In benzene 24.2, chlorobenzene 17.1, toluene 17.6 (all in g/l, 24 °C). In diethylene glycol 390.6, dimethyl formamide 2354.5, dioxane 789.2 (all in g/l, 24.5 °C).   **Stability** Highly stable.

COMMERCIALISATION
History Plant growth regulator introduced by Amchem Chemical Co. (now Rhône-Poulenc Agrochimie).   **Manufacturers** Rhône-Poulenc

APPLICATIONS
Mode of action Absorbed by the foliage and not readily translocated.
Uses Reduction of crown growth of pineapples, in order to increase the fruit size and weight in both plant and ratoon crops, and delay of maturation of fruit. Also used for fruit thinning in certain plum varieties.   **Formulation types** SL.
Selected tradenames 'Fruitone CPA' (Rhône-Poulenc)

ANALYSIS
By hplc with u.v. detection.

MAMMALIAN TOXICOLOGY
Oral Acute oral LD_{50} for male rats 3360, female rats 2140 mg/kg.
Skin and eye Acute percutaneous LD_{50} for rabbits >2000 mg/kg. Irritating to eyes, non-irritating to skin (rabbits).   **Inhalation** Non-toxic by inhalation at 200 mg/l air for 1 hour.   **NOEL** In feeding trials, no tumorigenic effect in mice (1.88 y) at 6000 mg/kg diet, in rats (2 y) at 8000 mg/kg diet.   **Toxicity class** EPA (formulation) III

ECOTOXICOLOGY
Birds Eight-day dietary LC_{50} for mallard ducks and bobwhite quail >5620 mg/kg.
Fish LC_{50} (96 h) for rainbow trout c. 21, bluegill sunfish c. 118 mg/l.

153 clopyralid   *Herbicide*

pyridinecarboxylic acid

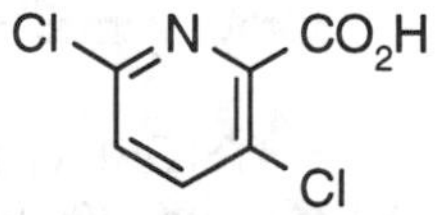

NOMENCLATURE
clopyralid
Common name clopyralid (BSI, ANSI, draft E-ISO, (*m*) draft F-ISO);
3,6-dichloropicolinic acid (WSSA, Canada, Finland and, until 1984, BSI; name accepted in lieu of a common name); acide dichloro-3,6 picolinique (France and, until 1984, F-ISO; name accepted in lieu of a common name)
IUPAC name 3,6-dichloropyridine-2-carboxylic acid
Chemical Abstracts name 3,6-dichloro-2-pyridinecarboxylic acid
Other names 3,6-DCP   **CAS RN** *[1702–17–6]*   **Development codes** Dowco 290

clopyralid-olamine
Common name clopyralid-olamine
CAS RN *[57754–85–5]*

PHYSICAL CHEMISTRY
clopyralid
Mol. wt. 192.0 **M.f.** $C_6H_3Cl_2NO_2$ **Form** Colourless crystals. **M.p.** 151–152 °C
V.p. 1.33 mPa (pure, 24 °C); 1.36 mPa (tech., 25°C) **K$_{ow}$** logP = –1.81 (pH 5),
–2.63 (pH 7), –2.55 (pH 9), 1.07 (unionised, 25 °C) **S g /density** 1.57 (20 °C)
Solubility Pure (99.2%) 7.85 (in distilled water), 118 (pH 5), 143 (pH 7), 157
(pH 9) (all in g/l, 20 °C). In acetonitrile 121, *n*-hexane 6, methanol 104 (all in
g/kg). Forms water-soluble salts, for example potassium, solubility >300 g/l
(25 °C). **Stability** Decomposes above m.p. Stable in acidic media and to light; on
hydrolysis DT$_{50}$ >30 d at pH range 5–9 (25 °C) in sterile water. **pKa** 2
F.p. No flashing exhibited in ignition test.

clopyralid-olamine
Mol. wt. 253.1 **M.f.** $C_8H_{10}Cl_2N_2O_3$ **Solubility** In water 560 g/l (25 °C).

COMMERCIALISATION
History Herbicide reported by T. Haagsma (*Down Earth,* 1975, **30**(4), 1).
Introduced in France (1977) by Dow Chemical Co. (now DowElanco).
Patent US 3317549 **Manufacturers** Agriphar; DowElanco

APPLICATIONS
Mode of action Selective systemic herbicide, absorbed by the leaves and roots,
with translocation both acropetally and basipetally, and accumulation in
meristematic tissue. Exhibits an auxin-type reaction. Acts on cell elongation and
respiration.

clopyralid
Uses Post-emergence control of many annual and perennial broad-leaved weeds
of the families Polygonaceae, Compositae, Leguminosae, and Umbelliferae, in
sugar beet, fodder beet, oilseed rape, maize, brassicas, onions, leeks, strawberries,
flax and grassland. Provides particularly good control of creeping thistle (*Cirsium
arvense*), perennial sow-thistle, coltsfoot, mayweeds, and *Polygonum* spp.
Phytotoxicity Good crop tolerance of graminaceous, cruciferous crops.
Formulation types SL. **Mixtures** *(clopyralid +)* benazolin; MCPA; mecoprop;
propyzamide; bromoxynil; fluroxypyr + MCPA; dichlorprop + MCPA;
fluroxypyr + ioxynil; MCPA + mecoprop; 2,4-D + MCPA; ioxynil + MCPA +
mecoprop; triclopyr; bifenox + mecoprop; cyanazine. **Selected tradenames** 'Clio'
(Agriphar); 'Cyronal' (AgrEvo)

MAIN ENTRIES

clopyralid-olamine
Selected tradenames 'Lontrel 35A' (DowElanco); 'MatriKerb' (PBI)

ANALYSIS
Product analysis by glc.   **Residues** determined by glc of a derivative (A. J. Pik & G. W. Hodgson, *J. Assoc. Off. Anal. Chem.*, 1976, **59**, 264). Details available from DowElanco.

MAMMALIAN TOXICOLOGY
clopyralid
Oral Acute oral LD_{50} for male rats 3738, female rats 2675 mg/kg.
Skin and eye Acute percutaneous LD_{50} for rabbits >2000 mg/kg; a severe eye irritant, not a skin irritant.   **Inhalation** LC_{50} (4 h) for rats >0.38 mg/l.
NOEL (2 y) for rats 15, male mice 500, female mice >2000 mg/kg b.w. daily.
ADI 0.15 mg/kg.   **Other** Non-carcinogenic, non-mutagenic, non-teratogenic and produces no significant toxicological effects on reproductive parameters.
Toxicity class WHO (a.i.) III (Table 5); EPA (formulation) IV

ECOTOXICOLOGY
clopyralid
Birds Acute oral LD_{50} for mallard ducks 1465, bobwhite quail >2000 mg/kg. Five-day dietary LC_{50} for mallard ducks and bobwhite quail >4640 mg/kg diet.
Fish LC_{50} (96 h) for rainbow trout 103.5, bluegill sunfish 125.4 mg/l.   **Bees** Non-toxic to bees. LD_{50} (48 h, oral and contact) >100 µg/bee.   **Worms** LC_{50} (14 d) for earthworms >1000 mg/kg soil.   **Other beneficial spp.** No effect on nitrification, nitrogen fixation or degradation of cellulose, starch, protein and leaf material at 1–10 ppm.   **Daphnia** EC_{50} (48 h) 225 mg/l. EC_{50} (21 d) immobilisation 69, reproduction 80 mg/l; NOEC 17 mg/l.   **Algae** EC_{50} (96 h) for *Selenastrum capricornutum*, cell count 6.9, cell volume 7.3 mg/l.   **Other aquatic spp.** EC_{50} (14 d) for *Lemna gibba* 89 mg/l.

ENVIRONMENTAL FATE
Animals In rats, following oral administration, there is rapid and almost quantitative unchanged elimination in the urine.   **Plants** Clopyralid is not metabolised in plants.   **Soil/Environment** In soil, microbial degradation occurs; slow degradation occurs in sterile soil. The major product is CO_2, only traces of one other metabolite have been recorded. Aerobic soil degradation depends on initial concentration (DT_{50} range 7 at 0.0025 ppm to 435 at 2.5 ppm, sandy loam), soil temperature and soil moisture; DT_{50} (BBA guidelines) 14–56 d; DT_{50} (USA guidelines) 2–94 d. Mean K_{oc} 4.64 (range 0.4–12.9), mean K_d 0.0412 (range 0.0094–0.0935); when aged in loam soil for 30 d, K_{oc} was c. 30 ml/g, suggesting clopyralid would be more tightly sorbed. Although data indicate potential for leaching, field dissipation and lysimeter studies demonstrate fairly rapid degradation and limited downward movement. Field dissipation DT_{50} was 8–66 d

(19 sites), with downward movement confined to 18 in. In lysimeter studies, the centre of mass movement ranged from 6–18 in. after 12 months, and cumulative (2 y) leachate concentrations were 0.002–0.14 ppb (0.1–0.6% of applied).

54 cloquintocet mexyl — *Herbicide safener*

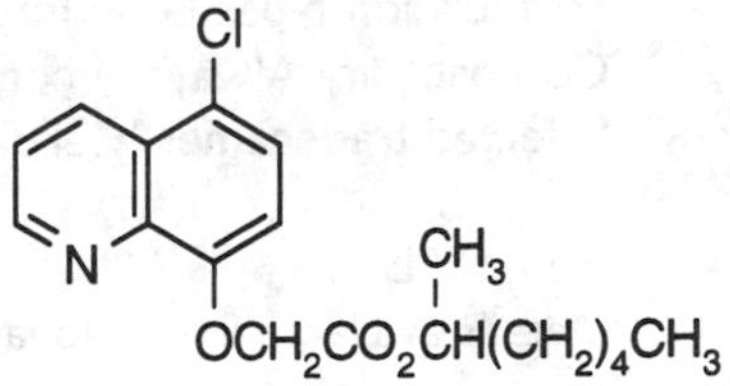

NOMENCLATURE

cloquintocet-mexyl
Common name cloquintocet-mexyl (BSI, pa E-ISO)
IUPAC name 1-methylhexyl (5-chloroquinolin-8-yloxy)acetate
Chemical Abstracts name 1-methylhexyl [(5-chloro-8-quinolinyl)oxy]acetate
CAS RN *[99607–70–2]*   **Development codes** CGA 185072

cloquintocet
Common name cloquintocet
CAS RN *[88349–88–6]*

PHYSICAL CHEMISTRY

cloquintocet-mexyl
Mol. wt. 335.8   **M.f.** $C_{18}H_{22}ClNO_3$ **Form** Colourless crystals.   **M.p.** 69.4 °C; (tech., 61.4–69.0 °C)   **V.p.** 5.31×10^{-3} mPa (25 °C) (OECD 104) K_{ow} logP = 5.03 (25 °C)   **Henry** 8.2×10^{-5} Pa m^3 mol^{-1} (calc.)   **S.g./density** 1.05 g/cm^3 (20 °C)
Solubility In water 0.59 mg/l (25 °C). In ethanol 190, acetone 340, toluene 360, *n*-hexane 0.14, *n*-octanol 11 (25 °C).   **Stability** Stable in acidic and neutral media, hydrolysed in alkaline media. DT$_{50}$ (25 °C) 133.7 d (pH 7).   **pKa** 3.5–4 (weak base, est.)

cloquintocet
Mol. wt. 237.6   **M.f.** $C_{11}H_8ClNO_3$ K_{ow} logP = –0.7 (25 °C) unstated pH

COMMERCIALISATION

History Reported by J. Amrein *et al.* (*Proc. Br. Crop Prot. Conf. – Weeds*, 1989, **1**,

71–76). Introduced by Ciba-Geigy AG (now Novartis Crop Protection AG).
Patent EP-B-0094349, US 4902340, US 5102445   **Manufacturers** Novartis

APPLICATIONS
cloquintocet-mexyl
Biochemistry Accelerates the detoxification process of clodinafop-propargyl in cereals (Kreuz *et al.*, *Z. Naturforsch.*, 1991, **46c**, 901–905).
Mode of action Improves cereal crop tolerance to the herbicide clodinafop-propargyl.   **Uses** Used as a herbicide safener in combination with clodinafop-propargyl for selective control of annual grasses (*Alopecurus myosuroides*, *Avena* spp., *Lolium* spp., *Phalaris* spp., *Poa trivialis*, *Setaria* spp.) in small grain cereals.
Formulation types EC.   **Mixtures** *(cloquintocet-mexyl +)* clodinafop-propargyl.
Compatibility Always in combination with clodinafop-propargyl.
Selected tradenames 'Celio' (mixture) (Novartis); 'Topik' (mixture) (Novartis)

ANALYSIS
Details available from Novartis.

MAMMALIAN TOXICOLOGY
cloquintocet-mexyl
Oral Acute oral LD_{50} for rats and mice >2000 mg/kg.   **Skin and eye** Acute percutaneous LD_{50} for rats >2000 mg/kg. Non-irritating to eyes and skin (rabbits). May cause skin sensitisation (guinea pigs).   **Inhalation** LC_{50} (4 h) for rats >0.935 mg/l air.   **NOEL** (2 y) for rats 4 mg/kg b.w. daily; (18 mo) for mice 106.5 mg/kg b.w. daily; (1 y) for dogs 44 mg/kg b.w. daily.   **ADI** 0.04 mg/kg b.w.
Toxicity class WHO (a.i.) III   **EC risk** R43

ECOTOXICOLOGY
cloquintocet-mexyl
Birds Non-toxic. LD_{50} for mallard ducks and bobwhite quail >2000 mg/kg.
Fish Non-toxic. LC_{50} (96 h) for rainbow trout and carp >76, bluegill sunfish >51, catfish 14 mg/l.   **Bees** Non-toxic. LC_{50} (48 h, oral and contact) >100 μg/bee.
Worms LC_{50} for earthworms >1000 mg/kg.   **Daphnia** LC_{50} (48 h) >100 mg/l.
Algae EC_{50} (96–120 h) for *Scenedesmus subspicatus* 0.63, *Microcystis* 2.5, *Navicula* 1.7 mg/l.

ENVIRONMENTAL FATE
Animals Hydrolysed to the free acid as the major metabolite.   **Plants** Rapidly degraded to the free acid as the major metabolite.   **Soil/Environment** In soil, rapid degradation to the free acid, DT_{50} 0.5–2.4 d. Further degradation and mineralisation of the acid occurs within a few weeks to a few months. Cloquintocet-mexyl and its major metabolites have low soil mobility, are strongly adsorbed to soil and have low leaching potential. DT_{50} (parent) <1 d in natural aquatic systems.

55 cloransulam-methyl

Herbicide

triazolopyrimidine sulfonanilide

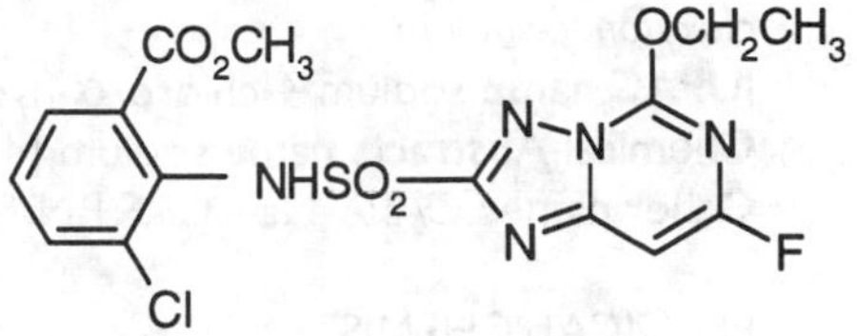

NOMENCLATURE
Common name cloransulam-methyl (BSI, pa ISO)
IUPAC name methyl 3-chloro-2-(5-ethoxy-7-fluoro[1,2,4]triazolo[1,5-c]=
pyrimidin-2-ylsulfonamido)benzoate; methyl 3-chloro-N-(5-ethoxy-7-fluoro[1,2,4]=
triazolo[1,5-c]pyrimidin-2-ylsulfonyl)anthranilate
Chemical Abstracts name methyl 3-chloro-2-[[(5-ethoxy-7-fluoro[1,2,4]triazolo=
[1,5-c]pyrimidin-2-yl)sulfonyl]amino]benzoate
CAS RN *[147150–35–4]*; (acid *[159518–97–5]*)   **Development codes** XDE-565

PHYSICAL CHEMISTRY
Mol. wt. 429.8 (acid 406.2)   **M.f.** $C_{15}H_{13}ClFN_5O_5S$ (acid $C_{14}H_{11}ClFN_5O_5S$)

COMMERCIALISATION
Manufacturers DowElanco

APPLICATIONS
Biochemistry Acetolactate synthase (ALS) inhibitor.   **Uses** Pre- and post-
emergence broad-leaved herbicide for use in soya beans at 17.5 to 44 g/ha.
Formulation types DG.   **Selected tradenames** 'First Rate' (DowElanco)

56 cloxyfonac

Plant growth regulator

aryloxyalkanoic acid

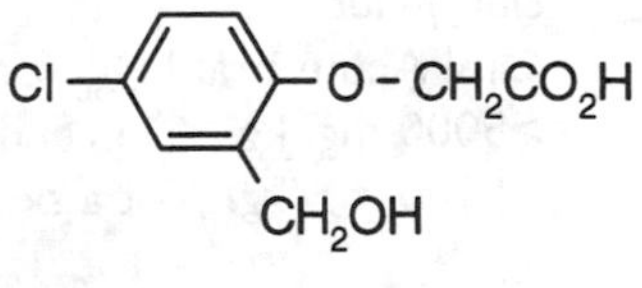

NOMENCLATURE
cloxyfonac
Common name cloxyfonac (BSI, draft E-ISO, (*m*) draft F-ISO)
IUPAC name 4-chloro-α-hydroxy-*o*-tolyloxyacetic acid

Chemical Abstracts name [4-chloro-2-(hydroxymethyl)phenoxy]acetic acid
Other names CHPA; PCHPA **CAS RN** [6386–63–6]

cloxyfonac-sodium
IUPAC name sodium 4-chloro-α-hydroxy-o-tolyloxyacetate
Chemical Abstracts name sodium [4-chloro-2-(hydroxymethyl)phenoxy]acetate
Other names CAPA-Na **CAS RN** [32791–87–0] **Development codes** RP-7194

PHYSICAL CHEMISTRY
cloxyfonac
Mol. wt. 216.6 **M.f.** $C_9H_9ClO_4$ **Form** Colourless crystals. **M.p.** 140.5–142.7 °C
V.p. 0.089 mPa (25 °C) (gas saturation method) **Solubility** In water 2 g/l. In
acetone 100, dioxane 125, ethanol 91, methanol 125 (all in g/l). Insoluble in
benzene and chloroform. **Stability** Stable at 40 °C, stable to light. Stable in
slightly acidic or slightly alkaline media.

cloxyfonac-sodium
Mol. wt. 238.6 **M.f.** $C_9H_8ClNaO_4$

COMMERCIALISATION
History Plant growth regulator evaluated by Shionogi and Co. Ltd.
Manufacturers Shionogi

APPLICATIONS
Biochemistry Auxin

cloxyfonac-sodium
Uses Fruit setting agent for tomatoes and aubergines at flowering, producing fruit
of uniform size. **Formulation types** SL. **Selected tradenames** 'Tomatlane'
(Shionogi)

ANALYSIS
Residues determined by glc.

MAMMALIAN TOXICOLOGY
cloxyfonac
Oral Acute oral LD_{50} for male and female rats and male and female mice
>5000 mg/kg. **Skin and eye** Acute percutaneous LD_{50} for male and female rats
>5000 mg/kg. Not a skin irritant (rats). **Toxicity class** WHO (a.i.) III (Table 5)

cloxyfonac-sodium
Oral Acute oral LD_{50} for male and female rats >5000 mg/kg. **Skin and eye** Acute
percutaneous LD_{50} for male and female rats >5000 mg/kg. Not a skin irritant
(rats).

cloxyfonac-sodium
Fish LC_{50} (48 h) for common carp 320 ppm as a.i. (9.8% liquid formulation).

ENVIRONMENTAL FATE
Soil/Environment Soil DT_{50} <7 d.

57 codlemone   *Insect pheromone*

pheromone

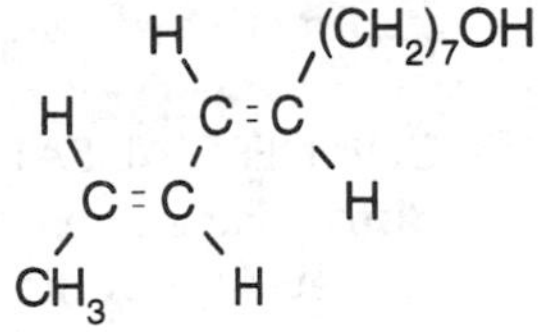

NOMENCLATURE
IUPAC name (*E,E*)-dodeca-8,10-dien-1-ol
Chemical Abstracts name (*E,E*)-8,10-dodecadien-1-ol
Other names codling moth pheromone; *Cydia pomonella* pheromone; E8E,
10–12OH (IOBC); EE8, 10–12OH   **CAS RN** *[33956–49–9]*

PHYSICAL CHEMISTRY
Composition The mating hormone of the codling moth (*Cydia pomonella*). Typical
composition for mating disruption contains 62.5% codlemone, with 31%
n-dodecanol and 6% *n*-tetradecanol.   **Mol. wt.** 182.3   **M.f.** $C_{12}H_{22}O$
M.p. *c.* 32 °C   **B.p.** 110–120 °C/2 mmHg   **V.p.** 69 mPa (20 °C) (evaporation
balance); 168 mPa (25 °C, calc.); 1.9×10^3 mPa (25 °C, calc., mixture for mating
disruption). K_{ow} logP >3.7   **Solubility** Insoluble in water. Readily soluble in
organic solvents.   **Stability** Stable in the dark; isomerises in light.   **F.p.** 91 °C
(closed cup)

COMMERCIALISATION
History In use since 1990.   **Manufacturers** Shin-Etsu

APPLICATIONS
Mode of action Acts by disruption of mating, either in trapping or in disorientation
mode.   **Uses** Used for control of codling moth.   **Formulation types** VP (coil).
Mixtures *(codlemone +)* *n*-dodecanol + *n*-tetradecanol; (*Z*)-11-tetradecenyl
acetate.   **Selected tradenames** 'Isomate-C' (mixture) (Shin-Etsu, Pacific
Biocontrol); 'NoMate CM' (Ecogen); 'RAK 3' (BASF)

158 copper hydroxide *Fungicide, bactericide*

inorganic

$Cu(OH)_2$

NOMENCLATURE
IUPAC name copper hydroxide; copper(II) hydroxide; cupric hydroxide
Chemical Abstracts name copper hydroxide ($Cu(OH)_2$)
CAS RN *[20427–59–2]*

PHYSICAL CHEMISTRY
Mol. wt. 97.6 **M.f.** CuH_2O_2 **Form** Blue-green solid. **Solubility** In water 2.9 mg/l (pH 7, 25 °C). Readily soluble in aqueous ammonia. Insoluble in organic solvents. **Stability** Dehydrated >50 °C for extended periods. Decomposes at 140 °C.

COMMERCIALISATION
History Fungicide introduced in USA (1968) by Kennecott Corp. and now marketed by Griffin Corp. **Manufacturers** Crystal; Griffin

APPLICATIONS
Mode of action Protectant fungicide and bactericide. **Uses** For control of Peronosporaceae in vines, hops, and brassicas, *Alternaria* and *Phytophthora* in potatoes, *Septoria* in celery, and *Septoria*, *Leptosphaeria*, and *Mycosphaerella* in cereals. **Formulation types** WP. **Compatibility** Not compatible with acids, dicloran or calcium polysulfide. **Selected tradenames** 'Cuproxyde' (La Cornubia); 'Kocide' (Griffin)

ANALYSIS
Product analysis by iodometric titration (*AOAC Methods*, 1984, 6.015–6.016; *CIPAC Handbook*, 1970, **1**, 226).

MAMMALIAN TOXICOLOGY
Oral Acute oral LD_{50} for rats 489 mg/kg (tech.). **Skin and eye** Acute percutaneous LD_{50} for rabbits >3160 mg/kg. Severely irritating and corrosive to eyes, mild skin irritant. **Inhalation** LC_{50} > 2 mg/l air. **Toxicity class** WHO (a.i.) III; EPA (formulation) III

ECOTOXICOLOGY
Birds Acute oral LD_{50} for bobwhite quail 3400, mallard ducks >5000 mg/kg. Dietary LD_{50} (8 d) for bobwhite quail and mallard ducks >10 000 ppm. **Fish** LC_{50} (24 h) for rainbow trout 0.08 mg/l; (96 h) for fathead minnow 0.023, bluegill sunfish >180 mg/l. **Bees** Non-toxic to honeybees. **Daphnia** LC_{50} 6.5 ppb.

59 copper oxychloride *Fungicide*

inorganic

$$Cu_2Cl(OH)_3$$

NOMENCLATURE
Common name copper oxychloride (E-ISO, accepted in lieu of common name); oxychlorure de cuivre (F-ISO)
IUPAC name dicopper chloride trihydroxide (approximate composition)
Chemical Abstracts name copper(II) chloride oxide hydrate
CAS RN *[1332–40–7]* (formerly *[1332–65–6]*, defined as copper(II) chloride hydroxide)

PHYSICAL CHEMISTRY
Composition Composition varies with the conditions of manufacture.
Mol. wt. 213.6   **M.f.** $ClCu_2H_3O_3$ **Form** Green to bluish-green powder.
V.p. Negligible at 20 °C   **Solubility** Practically insoluble in water, $<10^{-5}$ mg/l (pH 7, 20 °C). Insoluble in organic solvents. Soluble in dilute acids, forming Cu(II) salts; soluble in ammonium hydroxide, forming a complex ion.   **Stability** Very stable in neutral media. Decomposes on heating in alkaline media with the formation of copper oxides. Decomposes on heating above 220 °C, with the formation of copper oxides, and loss of hydrogen chloride.

COMMERCIALISATION
History Introduced as a fungicide in the early 1900s.   **Manufacturers** Aimco; Caffaro; Crystal; Novartis; Rallis

APPLICATIONS
Mode of action Foliar fungicide with protective action.   **Uses** Control of late blight of potatoes, tomatoes and other vegetables; leaf spot diseases of beet, celery, celeriac, parsley, olives, currants, and gooseberries; downy mildews of vines, hops, spinach, and ornamentals; canker and scab of pome fruit and stone fruit; scab, canker, and melanose of citrus fruit; asparagus rust; peach leaf curl; shot-hole of stone fruit; cane diseases of raspberries and blackberries; leaf spot and leaf scorch of strawberries; anthracnose and blister blight of tea; leaf spot and downy mildew of cucumbers and melons; bacterial diseases of lettuce; etc.   **Phytotoxicity** Non-phytotoxic at the recommended rates, except to carrots and potatoes under certain conditions. Russetting may occur with some varieties of apple.
Formulation types WP; WG; PA; DP; SC.   **Mixtures** (*copper oxychloride +*) benalaxyl; cymoxanil + oxadixyl; cymoxanil; dichlofluanid; folpet; metalaxyl; metiram; mancozeb; oxadixyl; propineb; sulfur; copper sulfate + folpet + oxadixyl; maneb + zineb; copper sulfate + mancozeb + oxadixyl; oxadixyl + propineb; copper carbonate (basic) + copper sulfate + mancozeb; captafol + cymoxanil; zineb; copper sulfate + cymoxanil + mancozeb.   **Compatibility** Incompatible with

mercury-containing compounds, thiram, DNOC, lime sulfur, and dithiocarbamates.
Selected tradenames 'Blitox' (Rallis); 'Cekucobre' (Cequisa); 'Cobox' (BASF);
'Coprantol' (Novartis); 'Coptox' (Aimco); 'Cupravit' (Bayer); 'Cuprenox'
(Diachem); 'Cuprokylt' (Unicrop); 'Dhanucop' (Dhanuka); 'Funguran' (Urania);
'Kapper' (Ramcides); 'Ossirame' (Sipcam); 'Pasta Caffaro' (Caffaro); 'Recop'
(Novartis); 'Styrocuivre' (La Cornubia)

ANALYSIS
Product: Copper content and formulation characteristics determined (*CIPAC Handbook*, 1970, **1**, 226; *ibid.*, 1992, **E**, 42). **Residues** determined by a colorimetric method (*AOAC Methods*, 1984, 3.020–3.028, 3.033–3.034) or by atomic absorption spectrophotometry (*ibid.*, 3.013–3.016).

MAMMALIAN TOXICOLOGY
Oral Acute oral LD_{50} for rats 700–800 mg/kg. **Skin and eye** Acute percutaneous LD_{50} for rats >2000 mg/kg. **Inhalation** LC_{50} (4 h) >30 mg/l.
Toxicity class WHO (a.i.) III; EPA (formulation) III

ECOTOXICOLOGY
Fish LC_{50} (48 h) for carp 2.2 mg/l. **Bees** Not toxic to bees. **Daphnia** LC_{50} (24 h) 3.5 mg/l.

ENVIRONMENTAL FATE
Animals Most of the copper oxychloride is excreted with the faeces. Small amounts may be incorporated in natural proteins, e.g. ceruloplasmin.
Soil/Environment Strongly absorbed by soils.

160   copper sulfate   *Algicide, fungicide*

$$CuSO_4.5H_2O$$

NOMENCLATURE
Common name copper sulfate (BSI; accepted in lieu of a common name)
IUPAC name copper sulfate
Chemical Abstracts name copper(2+) sulfate (1:1)
Other names blue vitriol; blue stone; blue copperas; copper sulphate; cupric sulphate **CAS RN** *[7758–98–7]* (anhydrous); *[7758–99–8]* (pentahydrate)
EEC no. 231–847–6

PHYSICAL CHEMISTRY
Mol. wt. 249.7 (pentahydrate) **M.f.** $CuH_{10}O_9S$ (pentahydrate) **Form** Blue crystals. **V.p.** Non-volatile. **S.g./density** 2.286 (15.6 °C) **Solubility** In water 148 (0 °C), 230.5 (25 °C), 335 (50 °C), 736 (100 °C) (all in g/kg). In methanol 156 g/l (18 °C). Practically insoluble in most other organic solvents. Soluble in glycerine to give an emerald green colour. **Stability** Slowly efflorescent in air. On heating, loses two molecules of water of crystallisation at 30 °C, two more molecules at 110 °C, and becomes anhydrous by 250 °C. By the reaction of alkalis on aqueous solutions, copper oxide is produced. With ammonia and amines, coloured complexes are formed. With numerous organic acids, sparingly-soluble salts are formed.

COMMERCIALISATION
Manufacturers Griffin; Ingenieria Industrial; Phelps-Dodge; Source Technology Biologicals

APPLICATIONS
Mode of action Aquatic algicide and foliar fungicide with protective action.
Uses Control of most species of algae in ponds, lakes, potable water, fish hatcheries, rice fields, streams, ditches, swimming pools, etc. Prevention and control of Dutch elm disease, by tree injection. Used as a general fungicide when mixed with lime to form Bordeaux mixture (*q.v.*). Also used as a wood preservative. **Phytotoxicity** Phytotoxic to most plants when not mixed with lime to form Bordeaux mixture (*q.v.*). **Formulation types** Crystals; WP; SC.
Mixtures (*copper sulfate +*) cymoxanil; cymoxanil + mancozeb; folpet; mancozeb; maneb; sulfur; zineb; copper oxychloride + cymoxanil + mancozeb; copper oxychloride + folpet + oxadixyl; copper oxychloride + mancozeb + oxadixyl; copper carbonate (basic) + copper oxychloride + mancozeb. **Compatibility** With several pesticides, salt or complex formation may take place.
Selected tradenames 'Blue Viking' (Griffin); 'Super Bouillie' (La Cornubia); 'Triangle Brand' (Phelps-Dodge)

ANALYSIS
Product analysis: copper content can be determined by electrolytic or volumetric thiosulfate methods (*CIPAC Handbook,* 1970, **1**, 226; *ibid.*, 1992, **E**, 42–48).
Residues by colorimetry or atomic absorption spectrophotometry (*AOAC Methods,* 1984, 3.013–3.034; *Methodensammlung Rückstandsanalytik,* 147).

MAMMALIAN TOXICOLOGY
Oral Acute oral LD_{50} difficult to determine since oral intake leads to nausea.
NOEL In feeding trials, rats receiving 500 mg/kg diet showed weight loss, whilst those receiving 1000 mg/kg diet exhibited damage to the liver, kidneys, and other organs. **Toxicity class** WHO (a.i.) II; EPA (formulation) I **EC risk** Xn (R22); Xi (R36/38)

ECOTOXICOLOGY
Fish Toxic to fish.

ENVIRONMENTAL FATE
Soil/Environment In soil, copper sulfate is partly washed down to lower levels, partly bound by soil components, and partly oxidatively transformed.

161 coumaphos   *Insecticide*

organophosphorus

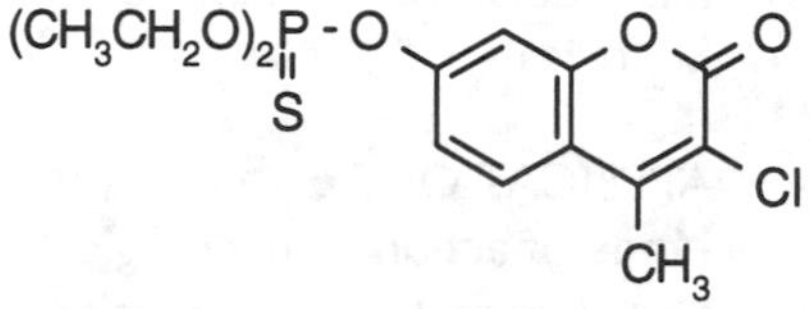

$(CH_3CH_2O)_2 \overset{\text{O}}{\underset{\text{S}}{P}} - O$

NOMENCLATURE
Common name coumaphos (BSI, E-ISO, (*m*) F-ISO, ESA, BAN); no name (Germany)
IUPAC name *O*-3-chloro-4-methyl-2-oxo-2*H*-chromen-7-yl *O,O*-diethyl phosphorothioate; 3-chloro-7-diethoxyphosphinothioyloxy-4-methylcoumarin
Chemical Abstracts name *O*-(3-chloro-4-methyl-2-oxo-2*H*-1-benzopyran-7-yl) *O,O*-diethyl phosphorothioate
CAS RN *[56–72–4]* **EEC no.** 200–285–3   **Development codes** Bayer 21/199
Official codes OMS 485; ENT 17 957

PHYSICAL CHEMISTRY
Mol. wt. 362.8   **M.f.** $C_{14}H_{16}ClO_5PS$ **Form** Colourless crystals. **M.p.** 95 °C; (tech., 90–92 °C)   **V.p.** 0.013 mPa (20 °C)   **K_{ow}** logP = 4.13   **S.g./density** 1.474 (20 °C)   **Solubility** In water 1.5 mg/l (20 °C). Limited solubility in organic solvents.
Stability Stable to hydrolysis in aqueous media, though in dilute alkali the pyrone is opened, reclosing on acidification.

COMMERCIALISATION
History Veterinary insecticide introduced by Bayer AG.   **Patent** DE 881194; US 2748146

APPLICATIONS
Biochemistry Cholinesterase inhibitor.   **Uses** Non-systemic insecticide for control of Diptera, and for control of Hypoderma on cattle. Also controls parasitic mites (*Varroa jacobsoni*) on bees.   **Formulation types** Powder; Spray; Liquid.

Selected tradenames 'Asuntol' (Bayer); 'Perizin' (Bayer)

ANALYSIS
Details available from Bayer AG.

MAMMALIAN TOXICOLOGY
Reviews *Pesticide residues in food – 1990.* FAO Plant Production and Protection
Paper, 102, 1990. *Pesticide residues in food – 1990 evaluations. Part II – Toxicology.*
World Health Organisation, WHO/PCS/91.47, 1991. **EHC** 63 (WHO, 1986; a
general review of organophosphorus insecticides). **Oral** Acute oral LD_{50} for male
rats 41, female rats 16 mg/kg (in peanut oil). **Skin and eye** Acute percutaneous
LD_{50} for male rats 860 mg/kg. **Inhalation** LC_{50} (1 h) for male rats >1081, female
rats 341 mg/m^3 air. **NOEL** In 2 y feeding trials, rats tolerated 100 mg/kg diet.
ADI (JMPR) No ADI [1960]; for 'Perizin' 0.035 mg/kg. **Toxicity class**
WHO (a.i.) Ia **EC risk** T+ (R28); Xn (R21)

ECOTOXICOLOGY
Birds LD_{50} for bobwhite quail 4.3, mallard ducks 29.8 mg/kg. **Fish** LC_{50} (96 h) for
bluegill sunfish 340, channel catfish 840 µg/l. **Daphnia** LC_{50} (48 h) 1.0 µg/l.

ENVIRONMENTAL FATE
Soil/Environment Photolytic DT_{50} on soil surface 23.8 d. K_d 61–298, K_{oc} 5778–
21 120 (4 soil types).

162 coumatetralyl   *Rodenticide*

coumarin anticoagulant

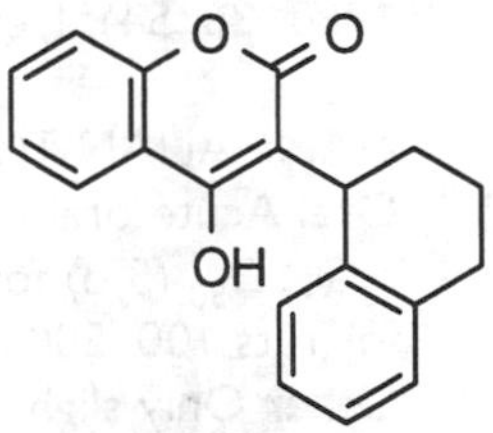

NOMENCLATURE
Common name coumatetralyl (BSI, E-ISO, (m) F-ISO); coumarins (JMAF – name
also applies to warfarin)
IUPAC name 4-hydroxy-3-(1,2,3,4-tetrahydro-1-naphthyl)coumarin

Chemical Abstracts name 4-hydroxy-3-(1,2,3,4-tetrahydro-1-naphthalenyl)-= 2H-1-benzopyran-2-one
CAS RN *[5836–29–3]* **EEC no.** 227–424–0

PHYSICAL CHEMISTRY

Mol. wt. 292.3 **M.f.** $C_{19}H_{16}O_3$ **Form** Colourless crystals; (tech., yellowish crystals). **M.p.** 172–176 °C; (tech., 166–172 °C) **V.p.** 8.5×10^{-6} mPa (20 °C) **K$_{ow}$** logP = 3.46 **Henry** 1×10^{-7} Pa m^3 mol^{-1} (pH 5, 20 °C) **Solubility** In water 4 (pH 4.2), 20 (pH 5), 425 (pH 7) (all in mg/l, 20 °C), 100–200 g/l (pH 9, 20 °C). Readily soluble in dimethylformamide. Soluble in alcohols and acetone. Slightly soluble in benzene, toluene and diethyl ether. In dichloromethane 50–100, isopropanol 20–50 (both in g/l, 20 °C). Readily soluble in alkalis with the formation of salts. **Stability** Thermally stable up to at least 150 °C. Not hydrolysed by water over 5 d (25 °C); DT$_{50}$ >1 y (pH 4–9). Rapidly decomposed in aqueous solutions exposed to sunlight or u.v. light; DT$_{50}$ c. 1 h. **pKa** 4.5–5.0

COMMERCIALISATION

History Rodenticide reviewed by G. Hermann & S. Hombrecher (*Pflanzenschutz-Nachr. (Engl. Ed.)*, 1962, **15**, 89). Introduced by Bayer AG. **Patent** DE 1079382; US 2952689 **Manufacturers** Bayer

APPLICATIONS

Biochemistry Anticoagulant rodenticide which inhibits blood coagulation by blocking prothrombin formation in the liver. **Mode of action** Requires multiple feedings to produce lethal effects. **Uses** Control of rats in a variety of situations, including areas containing stored products. **Formulation types** BB; RB; Liquid; Oil concentrate. **Mixtures** (*coumatetralyl +*) vitamin D$_3$. **Selected tradenames** 'Racumin' (Bayer)

ANALYSIS

Product analysis by hplc (*CIPAC Handbook*, 1995, **G**, 24–31; K. Hunter, *Analytical Methods Pestic. Plant Growth Regulat.*, 1988, **16**, 148). **Residues** in soil and water determined by hplc (*idem, ibid.*; E. Möllhoff, *Pflanzenschutz-Nachr. (Engl. Ed.)*, 1983, **36**, 54). Details available from Bayer AG.

MAMMALIAN TOXICOLOGY

Oral Acute oral LD$_{50}$ for rats 16.5, mice >1000, rabbits >500 mg/kg. Sub-chronic oral LD$_{50}$ (5 d) for rats 0.3 mg/kg/day. **Skin and eye** Acute percutaneous LD$_{50}$ for rats 100–500 mg/kg. **Inhalation** LC$_{50}$ (4 h) for rats 39, mice 54 mg/m^3. **Other** Only slightly dangerous to humans and domestic animals when used as directed, but young pigs are especially sensitive. **Toxicity class** WHO (a.i.) Ib; EPA (formulation) I **EC risk** T+ (R27/28); T (R48/24/25)

ECOTOXICOLOGY
Birds Acute oral LD_{50} for Japanese quail >2000 mg/kg b.w. Eight-day dietary LC_{50} for hens >50 mg/kg daily.   **Fish** LC_{50} (96 h) for guppies c. 1000, rainbow trout 48, golden orfe 67 mg/l.   **Daphnia** LC_{50} (48 h) >14 mg/l.   **Algae** E_rC_{50} >18 mg/l, E_bC_{50} (72 h) 15.2 mg/l.

ENVIRONMENTAL FATE
Soil/Environment BBA-standard soil 2.2 (aerobic conditions); 51% mineralisation in 90 days.

63 4-CPA *Plant growth regulator*

aryloxyalkanoic acid

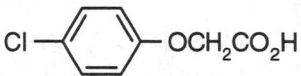

NOMENCLATURE
Common name 4-CPA (BSI, E-ISO, (*m*) F-ISO); 4-ChFU (USSR)
IUPAC name 4-chlorophenoxyacetic acid
Chemical Abstracts name (4-chlorophenoxy)acetic acid
CAS RN *[122–88–3]*   **EEC no.** 204–581–3

PHYSICAL CHEMISTRY
Mol. wt. 186.6   **M.f.** $C_8H_7ClO_3$ **Form** Colourless crystals.   **M.p.** 163–165 °C
Solubility Readily soluble in most organic solvents.

COMMERCIALISATION
History Plant growth regulator introduced by Dow Chemical Co. (now DowElanco), who no longer manufacture or market it.   **Manufacturers** Nissan

APPLICATIONS
Uses Used to improve fruit setting on tomato and inhibit sprout formation in mung beans. Also used for fruit thinning in peaches.   **Formulation types** SL; TB.
Selected tradenames 'Tomatotone' (Japan) (Nissan)

MAMMALIAN TOXICOLOGY
Oral Acute oral LD_{50} for rats 2200 mg/kg.   **Skin and eye** Acute percutaneous LD_{50} >2000 mg/kg. Not a skin irritant.   **Toxicity class** WHO (a.i.) III; EPA (formulation) III   **EC risk** Xn (R22)

NOMENCLATURE
Chemical Abstracts name trisodium hexafluoroaluminate
Other names sodium aminofluoride   **CAS RN** *[15096–52–3]*

PHYSICAL CHEMISTRY
Mol. wt. 209.9   **M.f.** AlF_6Na_3 **Form** Fine white odourless powder.   **M.p.** 1000 °C
V.p. No detectable vapour pressure.   **S.g./density** 0.890   **Solubility** In water
0.25 g/l (20 °C). Insoluble in organic solvents.   **Stability** Decomposed by hot
alkali.

APPLICATIONS
Mode of action Primarily a stomach poison.   **Uses** Applied at 5–30 kg/ha.
Formulation types WP.   **Selected tradenames** 'Kryocide' (Elf Atochem N America)

MAMMALIAN TOXICOLOGY
Oral Acute oral LD_{50} for rats >5000 mg/kg.   **Skin and eye** Acute percutaneous
LD_{50} for rabbits >2000 mg/kg. Not a skin irritant.   **Inhalation** LC_{50} for rats
>2 mg/l air.   **NOEL** (1 y) for dogs 3000 ppm. No reproductive or teratological
damage.   **Toxicity class** WHO (a.i.) III (Table 5)

ECOTOXICOLOGY
Birds Acute oral LD_{50} for bobwhite quail >2000 mg/kg. LD_{50} (8 d) for mallard
duck >10 000 mg/kg.

ENVIRONMENTAL FATE
Animals Metabolised as free fluoride ion.   **Soil/Environment** Not significantly
mobile in soil. In water, dissociates to free fluoride ion in a pH-dependent
equilibrium.

165 *Cryptolaemus montrouzieri* *Biological agent*

Coleoptera

Beetle, consumer of mealybugs.

NOMENCLATURE
Scientific name *Cryptolaemus montrouzieri*

PROPERTIES
Stability Shelf life 2–3 days at 15–20 °C in a dark place.

COMMERCIALISATION
Manufacturers BCP; Biobest; Bugs for Bugs; English Woodlands; Koppert; Nature's Alternative; Neudorff; Rincon-Vitova; Sautter & Stepper

APPLICATIONS
Mode of action The adult beetle and its larvae can kill and eat all stages of the mealybug.   **Uses** For control of mealybugs in protected crops.
Phytotoxicity Non-phytotoxic and non-phytopathogenic.   **Formulation types** Adult beetles without carrying material.   **Selected tradenames** 'Cryptobug' (Koppert)

MAMMALIAN TOXICOLOGY
Non-toxic.

66   cuprous oxide   *Fungicide*

inorganic

$$Cu_2O$$

NOMENCLATURE
Common name cuprous oxide (E-ISO, accepted in lieu of common name); oxyde cuivreux (F-ISO, accepted in lieu of common name)
IUPAC name copper(I) oxide; dicopper oxide
Chemical Abstracts name copper oxide (Cu_2O)
CAS RN *[1317–39–1]*

PHYSICAL CHEMISTRY
Mol. wt. 143.1   **M.f.** Cu_2O **Form** Yellow-to-red amorphous powder or octahedral crystals.   **M.p.** 1235 °C   **B.p.** 1800 °C   **V.p.** Negligible.
Solubility Practically insoluble in water and organic solvents. Soluble in aqueous solutions of ammonia and its salts. Soluble in dilute hydrochloric acid, forming cuprous chloride, which dissolves in excess acid. With dilute sulfuric or nitric acids, the cupric salt is formed and half the copper is precipitated as the metal.
Stability Stable under normal atmospheric (dry) conditions. Oxidises in the presence of moisture and at higher temperatures to copper(II) oxide and also undergoes conversion to copper carbonate.

COMMERCIALISATION

History Fungicidal properties as seed protectant reported by J. G. Horsfall (*N. Y. St. Agric. Exp. Stn. Bull.*, 1932, No. 615). Subsequently used for foliage protection.   **Manufacturers** Nordox; Novartis

APPLICATIONS

Biochemistry Because of its strong bonding affinity to amino and carboxyl groups, it reacts with proteins and acts as an enzyme inhibitor.   **Mode of action** Fungicide with protective action.   **Uses** Control of blights, downy mildews, rusts, and leaf spot diseases in a wide range of crops, including potatoes, tomatoes, vines, hops, olives, pome fruit, stone fruit, citrus fruit, beetroot, sugar beet, celery, carrots, coffee, cocoa, tea, bananas, etc.   **Phytotoxicity** Non-phytotoxic when used as directed, except to brassicas and copper-sensitive plants. Russetting is possible with some varieties of fruit.   **Formulation types** WG; WP.   **Mixtures** (*cuprous oxide +*) petroleum oils.   **Compatibility** Compatible with most other neutral pesticides. Incompatible with lime sulfur.   **Selected tradenames** 'Copper Nordox' (Nordox); 'Copper-Sandoz' (Novartis)

ANALYSIS

Product analysis: total copper determined by electrolytic or iodometric methods (*CIPAC Handbook*, 1992, **E**, 42–48; *AOAC Methods*, 1984, 6.015–6.016; *MAFF Ref. Book*, 1958, No. 1, p. 16); metallic copper in cuprous oxide may be determined (L. C. Hurd & A. R. Clark, *Ind. Eng. Chem. Anal. Ed.*, 1936, **8**, 380). Methods are available for total copper content of WP and DP (*CIPAC Handbook*, 1970, **1**, 226). **Residues** determined by reaction with concentrated sulfuric acid, and colorimetric estimation of derivatives or by atomic absorption spectroscopy (*AOAC Methods*, 1984, 3.020–3.028, 3.033–3.034, 3.013–3.016).

MAMMALIAN TOXICOLOGY

Oral Acute oral LD_{50} for rats 470 mg/kg.   **Skin and eye** Acute percutaneous LD_{50} for rats >2000 mg/kg.   **NOEL** No chronic toxic effects have been observed in rats receiving 500 mg/kg copper in their diet. No record of occupational diseases attributable to copper.   **Other** Sheep and calves are somewhat copper-sensitive, and livestock should not be allowed to graze on newly sprayed fields. **Toxicity class** WHO (a.i.) II

ECOTOXICOLOGY

Birds No significant history of ill-effect on birds.   **Fish** LC_{50} (48 h) for young goldfish 60, adult goldfish 150, young guppies 50 mg/l.   **Bees** Non-toxic to honeybees.   **Worms** Under conditions of moderate use and cultivation, the hazard to earthworms, and hence to soil structure, is insignificant.

67 cyanamide *Herbicide, plant growth regulator*

H_2NCN

NOMENCLATURE
Chemical Abstracts name cyanamide
Other names carbamic acid nitrile; amidocyanogen; hydrogen cyanamide; cyanoamine; cyanogenamide **CAS RN** [420–04–2] **EEC no.** 206–992–3

PHYSICAL CHEMISTRY
Mol. wt. 42.0 **M.f.** CH_2N_2 **Form** Colourless, hygroscopic crystals.
M.p. 45–46 °C **B.p.** 83 °C/0.5 mmHg **V.p.** 500 mPa (20 °C) K_{ow} logP = –0.82 (20 °C) **S.g./density** 1.282 (20 °C) **Solubility** In water 4.59 kg/l (20 °C). Soluble in alcohols, phenols and ethers. Sparingly soluble in benzene and halogenated hydrocarbons. Practically insoluble in cyclohexane. In methyl ethyl ketone 505, ethyl acetate 424, n-butanol 288, chloroform 2.4 (all in g/kg, 20 °C).
Stability Stable to light. Decomposed by alkalis (with the formation of dicyandiamide and polymerisation) and acids (with the formation of urea). On heating to 180 °C, dicyandiamide is formed, and polymerisation commences.
F.p. 207 °C

COMMERCIALISATION
Manufacturers SKW

APPLICATIONS
Mode of action Contact herbicide. Also acts as a growth regulator. **Uses** Early post-emergence control of broad-leaved weeds in alliums (onions, leeks, chives, shallots, and garlic) and bulb flowers. Also used to remove side shoots and inhibit sprouting in hops and vines. **Formulation types** EC. **Selected tradenames** 'Alzodef' (SKW)

ANALYSIS
Product by the Kjeldahl method or by argentometric determination. Details from SKW Trostberg. **Residues** (1) By colorimetric determination of a derivative. Details from SKW Trostberg. (2) By hplc (J. Prunonosa et al., *J. Chromatogr. Biomed. Appl.*, 1986, **50**, 253–269).

MAMMALIAN TOXICOLOGY
Oral Acute oral LD_{50} for rats 223 mg/kg (c. 300 mg/kg for 'Alzodef', 49% aqueous solution). **Skin and eye** Acute percutaneous LD_{50} for rabbits 848 mg/kg (c. 1700 mg/kg for 'Alzodef'). Severe skin and eye irritant. **Inhalation** LC_{50} (4 h) for rats >1 mg/l air (>13 mg/l air 'Alzodef'). **NOEL** (91 w) 1 mg/kg daily. **ADI** 0.01 mg/kg. **Other** A vasomotor reaction may occur when handling or using

cyanamide combined with alcohol consumption. **EC risk** T (R25); Xn (R21); Xi (R36/38); (R43)

ECOTOXICOLOGY
Birds Oral LD_{50} for bobwhite quail 350 mg/kg. Five-day dietary LC_{50} for bobwhite quail and mallard ducks >5000 ppm. **Fish** Toxic to fish; LC_{50} (96 h) for bluegill sunfish 44, carp 87, rainbow trout 90 mg/l. **Bees** Toxic to bees.
Other beneficial spp. EC_{10} (16 h) for *Pseudomonas putida* 24 mg/l. **Daphnia** LC_{50} (48 h) 3.2 mg/l. **Algae** EC_{50} (96 h) for *Selenastrum capricornutum* 13.5 mg/l.

ENVIRONMENTAL FATE
Animals The main urinary metabolite in the rat, rabbit and dog is acetylcyanamide.
Plants In plants, converted to nitrogenous compounds which are rapidly taken up.
Soil/Environment In soil, converted to nitrogenous compounds which are rapidly taken up. No residue problems.

168 cyanazine   *Herbicide*

1,3,5-triazine

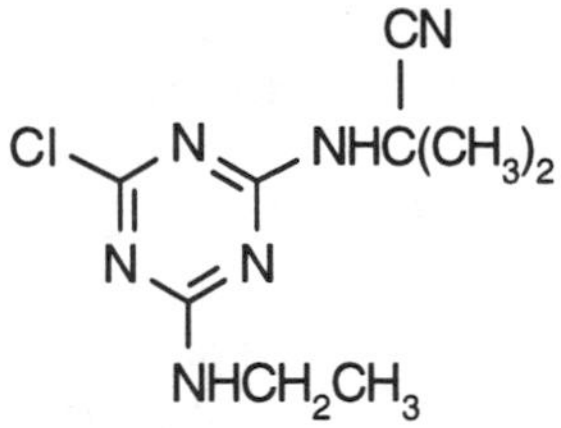

NOMENCLATURE
Common name cyanazine (BSI, E-ISO, (*f*) F-ISO, WSSA)
IUPAC name 2-(4-chloro-6-ethylamino-1,3,5-triazin-2-ylamino)-2-methyl= propiononitrile
Chemical Abstracts name 2-[[4-chloro-6-(ethylamino)-1,3,5-triazin-2-yl]amino]-= 2-methylpropanenitrile
CAS RN *[21725–46–2]* **EEC no.** 244–544–9 **Development codes** WL 19 805; SD 15 418; DW 3418

PHYSICAL CHEMISTRY
Composition Tech. grade ≥95% pure. **Mol. wt.** 240.7 **M.f.** $C_9H_{13}ClN_6$
Form Tech. cyanazine is a white crystalline solid. **M.p.** 167.5–169 °C (tech., 166.5–167 °C) **V.p.** 2×10^{-4} mPa (20 °C) K_{ow} logP = 2.1 **S.g./density** 1.29 kg/l (20 °C) **Solubility** In water 171 mg/l (25 °C). In methylcyclohexanone,

chloroform 210, acetone 195, ethanol 45, benzene, hexane 15, carbon tetrachloride <10 (all in g/l, 25 °C). **Stability** Stable to heat (1.8% decomposition after 100 h at 75 °C), and to light. Stable in solution between pH 5 and 9, hydrolysed by strong acids and alkalis. **pKa** 0.63, v. weak base

COMMERCIALISATION
History Herbicide reported by W. J. Hughes et al. (*Proc. North Cent. Weed Control Conf.,* 21st, 1967, p. 27). Introduced by Shell Research Ltd (now American Cyanamid Co.) **Patent** GB 1132306 **Manufacturers** Cyanamid; Griffin

APPLICATIONS
Biochemistry Photosynthetic electron transport inhibitor.
Mode of action Selective systemic herbicide, absorbed by the roots (with translocation acropetally to the leaves), and also by the foliage. **Uses** Used for general weed control (a) pre-emergence to the crop at 1–3 kg a.i./ha in broad beans, maize and peas; (b) post-emergence in barley and wheat during the early tillering stage at 0.26–0.33 kg/ha in combination with a variety of other herbicides for the control of broad-leaved weeds. Other crops for which it is used include: cotton, oilseed rape, forestry, potatoes, soya beans, sugar cane.
Phytotoxicity Selective if applied according to label recommendations.
Formulation types SC; WP; GR; DF. **Mixtures** (*cyanazine* +) linuron; MCPA; mecoprop; atrazine; pyridate; dichlorprop; isoproturon; clopyralid; fluroxypyr; bentazone + dichlorprop; bentazone + 2,4-DB. **Selected tradenames** 'Bladex' (Cyanamid, Du Pont); 'Cy-Pro' (Griffin)

ANALYSIS
Product analysis by i.r. (*CIPAC Handbook,* 1985, **1C,** 2039) or by glc or hplc (*AOAC Methods,* 1995, 991.32; *Anal. Methods Pestic. Plant Growth Regul.,* 1978, **10,** 275).
Residues in plants determined by glc with ECD or FID (*ibid.*), and in soils by glc or hplc (*ibid.*). **In water,** by lc with u.v. detection (*AOAC Methods,* 1995, 992.14, 10.7.01).

MAMMALIAN TOXICOLOGY
Oral Acute oral LD_{50} for rats 182–334, mice 380, rabbits 141 mg/kg.
Skin and eye Acute percutaneous LD_{50} for rats >1200, rabbits >2000 mg/kg.
Non-irritating to skin and eyes. **Inhalation** LC_{50} >2460 mg/m^3 air, as cyanazine dust. **NOEL** (2 y) for rats 12, dogs 25 mg/kg diet. **Toxicity class** WHO (a.i.) II; EPA II (DF, SC and WP formulations), III (GR formulation) **EC risk** Xn (R22)

ECOTOXICOLOGY
Birds Acute oral LD_{50} for mallard ducks >2000, quail 400 mg/kg. **Fish** LC_{50} (48 h) for harlequin fish 10 mg/l; LC_{50} (96 h) for fathead minnow 16 mg/l.
Bees Not toxic to bees. LD_{50} (topical) >100 μg/bee (tech. in acetone); (oral) >190 μg/bee (tech. as dust). **Daphnia** LC_{50} (48 h) 42–106 mg/l.

Algae EC_{50} (96 h) <0.1 mg/l.

ENVIRONMENTAL FATE
Animals In rats and dogs, following oral administration, cyanazine is rapidly metabolised and eliminated within c. 4 days. **Plants** In plants, the nitrile group is hydrolysed to a carboxylic acid group, and the chlorine atom is replaced by a hydroxy group (K. I. Beynon et al., *Pesticide Sci.*, 1972, **3**, 293–305).
Soil/Environment Microbial degradation in soil occurs within one growth period. Metabolism is similar to that in plants. DT_{50} in soil c. 2 w. (K. I. Beynon et al., *Pesticide Sci.*, 1972, **3**, 293–305, 379–401).

169 cyanophos   *Insecticide*

organophosphorus

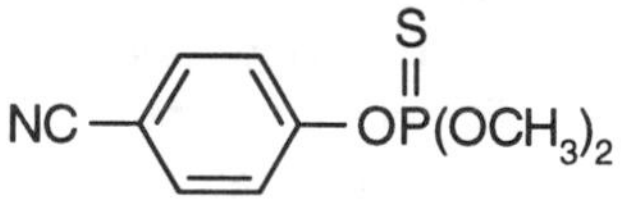

$$NC\text{—}C_6H_4\text{—}OP(OCH_3)_2 \ (\text{=S})$$

NOMENCLATURE
Common name cyanophos (BSI, E-ISO, (m) F-ISO); CYAP (JMAF)
IUPAC name O-4-cyanophenyl O,O-dimethyl phosphorothioate; 4-(dimethoxyphosphinothioyloxy)benzonitrile
Chemical Abstracts name O-(4-cyanophenyl) O,O-dimethyl phosphorothioate
CAS RN *[2636–26–2]* **EEC no.** 220–130–3 **Development codes** S-4084
Official codes OMS 226; OMS 869

PHYSICAL CHEMISTRY
Mol. wt. 243.2 **M.f.** $C_9H_{10}NO_3PS$ **Form** Yellow to reddish liquid with a faint characteristic odour. **B.p.** Decomposes at 119–120 °C **V.p.** 105 mPa (20 °C)
K_{ow} logP = 2.65 (room temperature) **S.g./density** 1.255–1.265 (25 °C)
Solubility In water 46 mg/l (30 °C). In methanol, acetone, chloroform >50% (20 °C). **F.p.** 104 °C

COMMERCIALISATION
History Insecticide reported by Y. Nishizawa (*Agric. Biol. Chem.*, 1961, **25**, 597). Introduced by Sumitomo Chemical Co., Ltd **Patent** JP 405852; JP 415199; US 3150040; US 3792132 **Manufacturers** Sumitomo

APPLICATIONS
Biochemistry Cholinesterase inhibitor. **Uses** Insecticide used at 25–50 g a.i./hl to

control Aphididae, Coccidae, Diaspididae, Lepidoptera and Margarodidae in cotton, fruit and vegetables. Used to control locusts and sanitary pests such as Blattodea, Culicidae and Muscidae. **Formulation types** DP; EC; UL; WP. **Selected tradenames** 'Cyanox' (Sumitomo)

ANALYSIS
Product analysis by glc or by u.v. spectrometry. **Residues** determined by glc with FTD. Details available from Sumitomo.

MAMMALIAN TOXICOLOGY
EHC 63 (WHO, 1986; a general review of organophosphorus insecticides).
Oral Acute oral LD_{50} for male rats 710, female rats 730 mg/kg.
Skin and eye Acute percutaneous LD_{50} for rats >2000 mg/kg.
Inhalation LC_{50} (4 h) for rats >1500 mg/m^3. **Toxicity class** WHO (a.i.) II; EPA (formulation) III **EC risk** Xn (R21/22)

ECOTOXICOLOGY
Fish LC_{50} (48 h) for carp 5 mg/l. **Bees** Toxic to honeybees.

70 cyclanilide *Plant growth regulator*

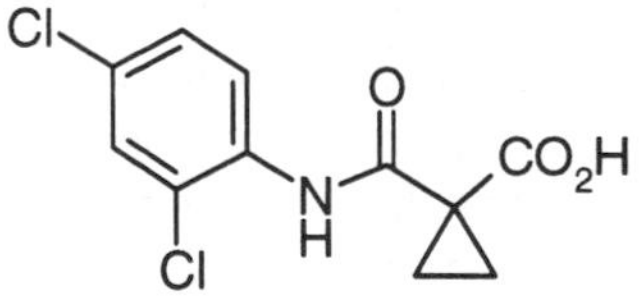

NOMENCLATURE
Common name cyclanilide (pa ISO)
IUPAC name 1-(2,4-dichloroanilinocarbonyl)cyclopropanecarboxylic acid
Chemical Abstracts name 1-[[(2,4-dichlorophenyl)amino]carbonyl]= cyclopropanecarboxylic acid
CAS RN [113136–77–9] **Development codes** RPA-090946

PHYSICAL CHEMISTRY
Mol. wt. 274.1 **M.f.** $C_{11}H_9Cl_2NO_3$ **Form** Powdery solid. **M.p.** 195.5 °C
V.p. <0.01 mPa (25 °C) **K_{ow}** logP = 3.25 **Henry** 7.41×10^{-5} Pa m^3 mol^{-1}
S.g./density 1.47 (20 °C) **Solubility** Low solubility in water. Soluble in organic solvents. **Stability** Relatively stable to photolysis. Stable to hydrolysis. Decomposes at c. 210 °C.

COMMERCIALISATION

History Reported by Fritz, *Proc. 21st Ann. Mtg. PGRSA* 1994, p. 154.
Patent US 4736056   **Manufacturers** Rhône-Poulenc

APPLICATIONS

Biochemistry Inhibits polar auxin transport.   **Uses** Plant growth regulator for cotton and other crops.   **Formulation types** SC.   **Mixtures** *(cyclanilide +)* ethephon.   **Selected tradenames** 'Finish' (mixture) (Rhône-Poulenc)

ANALYSIS

Product analysis by hplc.   **Residue** analysis by gc. Details available from Rhône-Poulenc.

MAMMALIAN TOXICOLOGY

Oral Acute oral LD_{50} for female rats 208, male rats 315 mg/kg.
Skin and eye Acute percutaneous LD_{50} for rabbits >2000 mg/kg. Slight skin irritant.   **Inhalation** LC_{50} (4 h) for rats >5.15 mg/l.   **NOEL** (2 y) for rats 7.5 mg/kg b.w.

ECOTOXICOLOGY

Birds Acute oral LD_{50} for bobwhite quail 216, mallard duck >215 mg/kg b.w. Dietary LC_{50} (8 d) for bobwhite quail 2849, mallard duck 1240 ppm diet.
Fish LC_{50} (96 h) for bluegill sunfish >16, rainbow trout >11 mg/l.   **Bees** Harmless to bees.   **Worms** Moderately toxic.   **Daphnia** EC_{50} (48 h) >13 mg/l.
Algae Moderately toxic.

ENVIRONMENTAL FATE

Animals Rapidly excreted, primarily as unchanged cyclanilide.   **Plants** Little degradation occurs in plants; cyclanilide is the major residue.
Soil/Environment Low to moderate persistence, DT_{50} c. 16 d under aerobic conditions. Degrades primarily by microbial activity. Medium to low mobility (average K_{oc} 346). Consequently little potential for leaching.

cycloate   *Herbicide*

thiocarbamate

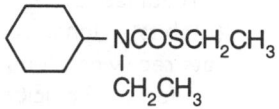

NOMENCLATURE
Common name cycloate (BSI, E-ISO, (m) Γ-ISO, WSSA), hexylthiocarbam (JMAΓ)
IUPAC name S-ethyl cyclohexyl(ethyl)thiocarbamate
Chemical Abstracts name S-ethyl cyclohexylethylcarbamothioate
CAS RN *[1134–23–2]* **Development codes** R-2063

PHYSICAL CHEMISTRY
Composition Tech. material is 95% pure. **Mol. wt.** 215.4 **M.f.** $C_{11}H_{21}NOS$
Form Colourless liquid with an aromatic odour. **B.p.** 145–146 °C/10 mmHg
V.p. 2.13 mPa (25 °C) **K$_{ow}$** logP = 3.88 **S.g./density** 1.024 g/ml **Solubility** In
water 75 mg/l (20 °C), Miscible with most organic solvents, e.g. acetone, benzene,
methanol, ethanol, xylene, kerosene, etc. **Stability** Hydrolysed by strong acids
and alkalis. Thermally stable (120 °C). **F.p.** >100 °C

COMMERCIALISATION
History Herbicide reported (*Nat. Weed Comm. Can. Dep. Agric. Res. Rep.*, 1963,
p. 51). Introduced by Stauffer Chemical Co. (now Zeneca Agrochemicals)
Patent US 3175897 **Manufacturers** Sagrochem; Zeneca

APPLICATIONS
Biochemistry Inhibits lipid metabolism. **Mode of action** Selective systemic
herbicide, absorbed by the roots and coleoptiles, with translocation acropetally to
the stems and leaves. Inhibits growth in the meristematic region of the leaves.
Uses Control of annual grass weeds, *Cyperus* spp., some perennial grasses, and
some annual broad-leaved weeds in sugar beet, fodder beet, beetroot and
spinach. Applied pre-planting, with soil incorporation at 3–4 kg a.i./ha.
Formulation types EC; GR. **Mixtures** *(cycloate +)* phenmedipham; lenacil.
Selected tradenames 'Ro-Neet' (Zeneca); 'Sabet' (Sagrochem)

ANALYSIS
Product analysis by glc (*CIPAC Handbook*, 1980, **1A**, 1190; *AOAC Methods*, 1995,
974.05). **Residues** in crops and soil determined by glc or colorimetry after
conversion to a derivative (J. R. Lane, *Anal. Methods Pestic. Plant Growth Regul.
Food Addit.*, 1967, **5**, 491; W. A. Ja *et al.*, *Anal. Methods Pestic. Plant Growth Regul.*,
1972, **6**, 686). In **drinking water**, by gc with FID (*AOAC Methods*, 1995, 991.07).
Details of methods available from Zeneca.

MAMMALIAN TOXICOLOGY

EHC 76 (WHO, 1988; general review of thiocarbamates). **Oral** Acute oral LD_{50} for male rats 2000–3190, female rats 3160–4100 mg/kg. **Skin and eye** Acute percutaneous LD_{50} for rabbits >5000 mg/kg. Non-irritating to skin and eyes (rabbits). **Inhalation** LC_{50} (4 h) for rats 4.7 mg/l **NOEL** In 90 d feeding trials, rats receiving 55 mg/kg daily and dogs receiving 240 mg/kg daily showed no ill-effects. **Toxicity class** WHO (a.i.) III; EPA (formulation) III

ECOTOXICOLOGY

Birds LD_{50} for Japanese quail >2000 mg/kg. Seven-day dietary LC_{50} for bobwhite quail >56 000 mg/kg diet (720 g/l for EC formulation). **Fish** LC_{50} (96 h) for rainbow trout 4.5 mg/l. **Bees** Not toxic to bees at 0.011 mg/bee. **Daphnia** LC_{50} (48 h) 5.6 mg/l.

ENVIRONMENTAL FATE

Animals Cycloate is converted to the sulfoxide by microsomal enzymes and then cleaved rapidly by the liver soluble-glutathione system. **Plants** Rapidly metabolised in the roots and leaves of beets. Metabolites include ethyl cyclohexylamine, CO_2, amino acids, sugars, and other natural plant constituents. **Soil/Environment** Microbial degradation is responsible for a large part of the disappearance of cycloate from soil. DT_{50} in soil c. 4–8 w.

172 cycloprothrin   *Insecticide*

pyrethroid

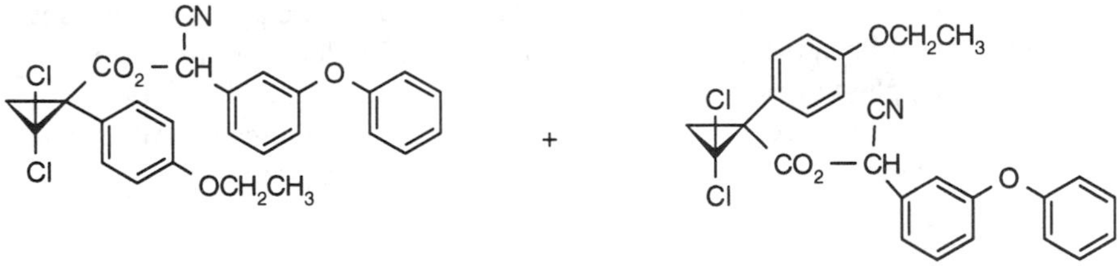

NOMENCLATURE

Common name cycloprothrin (BSI, draft E-ISO); cycloprothine ((f) draft F-ISO)
IUPAC name (RS)-α-cyano-3-phenoxybenzyl (RS)-2,2-dichloro-1-(4-ethoxy=phenyl)cyclopropanecarboxylate
Chemical Abstracts name cyano(3-phenoxyphenyl)methyl 2,2-dichloro-1-=(4-ethoxyphenyl)cyclopropanecarboxylate

CAS RN *[63935–38–6]* unstated stereochemistry   **Development codes** GH-414; NK-8116   **Official codes** OMS 3049

PHYSICAL CHEMISTRY
Mol. wt. 482.4   **M.f.** $C_{26}H_{21}Cl_2NO_4$ **Form** Yellow to brown viscous liquid (tech.).
B.p. 140–145 °C/0.001 mmHg   **V.p.** 2.13×10^{-3} mPa (20 °C) K_{ow} logP = 4.19
S.g./density 1.256 (25 °C)   **Solubility** In water 0.091 mg/l (25 °C). Readily soluble in most organic solvents, but only moderately soluble in aliphatic hydrocarbons.
Stability Stable ≤190 °C and to light.

COMMERCIALISATION
History Insecticide invented at CSIRO and reported by G. Nolan *et al. (Nature (London),* 1978, **272,** 734), and its development described by S. Kirihara & Y. Sakurai *(Jpn. Pestic. Inf.,* 1988, No. 53, 22). Introduced in Japan (1987) by Nippon Kayaku Co., Ltd.   **Manufacturers** Nippon Kayaku

APPLICATIONS
Mode of action Insecticide with contact and stomach action. Also exhibits antifeeding and repellent activities.   **Uses** Control of rice water weevils (*Lissorhoptrus oryzophilus*) and other insects in paddy rice, fruit, vegetables, tea, cotton, soya beans, and in forestry.   **Formulation types** GR; DP; EC.
Selected tradenames 'Cyclosal' (Nippon Kayaku)

ANALYSIS
Product and **residue** analysis by glc.

MAMMALIAN TOXICOLOGY
Reviews *J. Pesticide Science,* **16,** 697–702 (1991).   **Oral** Acute oral LD_{50} for rats and mice >5000 mg/kg.   **Skin and eye** Acute percutaneous LD_{50} for rats >2000 mg/kg; no skin or eye irritation by tech., moderate irritation by GR and DP.   **Inhalation** LC_{50} (4 h) for rats >1.5 mg/l air.   **NOEL** (101 w) for rats 20 ppm.
Other No teratogenic, carcinogenic or reproductive effect in rats in lifetime tests.
Toxicity class WHO (a.i.) III (Table 5)

ECOTOXICOLOGY
Birds Acute oral LD_{50} for Japanese quail >5000, hens >2000 mg/kg.   **Fish** LC_{50} (48 h) for carp >50, goldfish >10, rainbow trout 1.6 mg/l.   **Bees** Toxic to honeybees.   **Daphnia** LC_{50} (3 h) >10 mg/l.

ENVIRONMENTAL FATE
Animals In continual oral administration to rats, cycloprothrin is rapidly and completely excreted in urine and faeces (K. Seguchi *et al., J. Pesticide Sci.,* **16,** 591–598, (1991)).   **Plants** For metabolism in rice plants, see K. Seguchi *et al., J. Pesticide Sci.,* **16,** 599–607 (1991).   **Soil/Environment** When cycloprothrin was

applied to simulated paddy water, the concentration absorbed in rice plants increased with time to a maximum within 7 days. The concentration was very low and unchanged cycloprothrin was not found in the grain.

173 cyclosulfamuron

Herbicide

sulfamoylurea

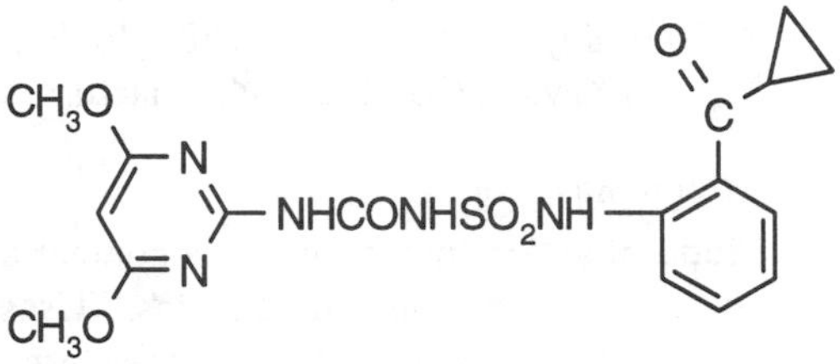

NOMENCLATURE

Common name cyclosulfamuron (pa ISO, ANSI)
IUPAC name 1-[2-(cyclopropylcarbonyl)phenylsulfamoyl]-3-(4,6-dimethoxy=pyrimidin-2-yl)urea
Chemical Abstracts name N-[[[2-(cyclopropylcarbonyl)phenyl]amino]sulfonyl]-=N^1-(4,6-dimethoxypyrimidin-2-yl)urea
CAS RN [136849–15–5]   **Development codes** AC 322,140

PHYSICAL CHEMISTRY

Mol. wt. 421.4   **M.f.** $C_{17}H_{19}N_5O_6S$ **Form** Off-white solid.   **M.p.** 160.9–162.9 °C; (tech., 149.6–153.2 °C)   **V.p.** *c.* 2.2×10^{-2} mPa (20 °C, gas saturation method) **K_{ow} logP** = 1.58 (pH 3), 2.045 (pH 5), 1.69 (pH 6), 1.41 (pH 7), 0.7 (pH 8) (all 25 °C)   **S.g./density** 0.624 (20 °C)   **Solubility** In water 0.17 (pH 5), 6.52 (pH 7), 549 (pH 9) (all in ppm, 25 °C).   **Stability** DT_{50} in 50mM phosphate buffer 2.2 (pH 3), 2.2 (pH 5), 5.1 (pH 6), 40 (pH 7), 91 (pH 8) (all in days). Stable 18 months (25 °C), 12 months (36 °C), 3 months (45 °C).   **pKa** 5.04

COMMERCIALISATION

History Reported by M. E. Condon *et al.* (*Proc. Br. Crop Prot. Conf., Weeds*, 1993, **1**, 41).   **Patent** US 5009699   **Manufacturers** Cyanamid

APPLICATIONS

Biochemistry Acts by inhibition of acetohydroxyacid synthase (AHAS), disrupting protein synthesis. Contributing factors for herbicide safety in rice, wheat and

barley are herbicide and crop placement, metabolic inactivation in the shoot, and the ability to translocate the herbicide to sites of potential metabolic inactivation (S. J. Rodaway et al., Proc. Br Crop Prot. Conf., 1993, **1**, 239).   **Uses** Control of a variety of dicotyledonous and sedge weeds, e.g. *Cyperus serotinus*, *Eleocharis kuroguwai* and *Sagittaria pygmaea*, in rice, and *Galium aparine*, *Matricaria* spp., *Veronica* spp., *Sinapis arvensis* and *Brassica napus* in wheat and barley at 25–50 g/ha, and in rice at 45–60 g/ha.   **Formulation types** GR; WP.
Mixtures *(cyclosulfamuron +)* pendimethalin; dithiopyr; cinmethylin; esprocarb.
Selected tradenames 'Jin-Qiu' (Cyanamid); 'Saviour' (Cyanamid)

MAMMALIAN TOXICOLOGY
Oral Acute oral LD_{50} for rats and mice >5000 mg/kg.   **Skin and eye** Acute percutaneous LD_{50} for rabbits >4000 mg/kg. Non-irritating to skin, and mildly irritating to eyes (rabbits).   **Inhalation** LC_{50} (4 h) for rats >5.2 mg/l.   **NOEL** (2 y) for rats 1000 mg/kg diet (50 mg/kg b.w. daily); (1 y) for dogs 100 mg/kg diet (3 mg/kg b.w. daily).   **Other** Non-mutagenic in the Ames test.

ECOTOXICOLOGY
Birds Acute oral LC_{50} for quail >1880 mg/kg. Dietary LC_{50} (8 d) for quail >5010 mg/kg.   **Fish** LC_{50} (72 h) for carp >50 ppm; (96 h) for trout >7.7 mg/l.
Bees Acute LD_{50} (24 h) (oral) >99 µg/bee; contact >106 µg/bee.   **Worms** No adverse effects at 892 mg/kg.   **Daphnia** LC_{50} (48 h) >9.1 mg/l.   **Algae** EC_{50} (72 h) 0.44 µg/l.

ENVIRONMENTAL FATE
Animals In rats, cyclosulfamuron is rapidly absorbed and readily excreted, mainly via the faeces.   **Plants** In crop plants, by hydrolysis of the urea bridge giving inactive compounds. See biochemistry.   **Soil/Environment** Soil adsorption (K_{oc}) 4.6 m³/kg in 3 Japanese paddy soils.

174 cycloxydim

cyclohexanedione oxime

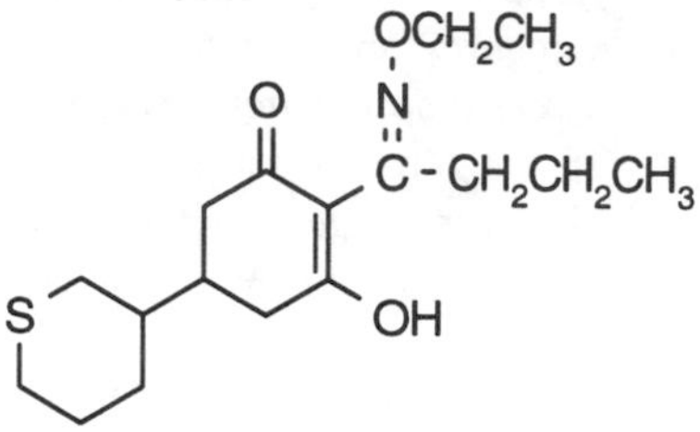

$$OCH_2CH_3$$

(structure: cyclohexenone ring with C=O, C=N-OCH₂CH₃, C-CH₂CH₂CH₃, OH, and a thian-3-yl (S-containing) ring substituent)

NOMENCLATURE
Common name cycloxydim (BSI, draft E-ISO); cycloxydime ((f) draft F-ISO)
IUPAC name (±)-2-[1-(ethoxyimino)butyl]-3-hydroxy-5-thian-3-ylcyclohex-2-=
enone
Chemical Abstracts name 2-[1-(ethoxyimino)butyl]-3-hydroxy-5-(tetrahydro-=
2H-thiopyran-3-yl)-2-cyclohexene-1-one; in an early publication the tautomer
2-[1-(ethoxyamino)butylidene]-5-(tetrahydro-2H-thiopyran-3-yl)-1,3-cyclo=
hexanedione was named
CAS RN [101205–02–1] cycloxydim; [99434–58–9] for tautomer
Development codes BAS 517H

PHYSICAL CHEMISTRY
Mol. wt. 325.5 **M.f.** $C_{17}H_{27}NO_3S$ **Form** Colourless odourless crystals; (tech.,
yellow beige crystals with a weak aromatic odour; dark brown oil above m.p.).
M.p. c. 41 °C **V.p.** <0.01 mPa (20 °C) **K$_{ow}$** logP c. 1.36 (pH 7, 25 °C).
S.g./density 1.12 (20 °C) **Solubility** In water 40 mg/l (20 °C). In acetone,
ethanol, dichloromethane, ethyl acetate, toluene >1000 (all in g/kg, 20 °C). In
n-hexane 29 g/kg (20 °C). **Stability** Stable at room temperature for ≥1 y;
unstable above 30 °C; decomposes c. 127 °C. DT$_{50}$ in argon atmosphere at
20 000 lux (simulated sunlight, 15–20 °C) >192 h; in air at 80 000 lux (simulated
sunlight, 20 °C) 0.8 h. **pKa** c. 4.17 **F.p.** 119 °C (tech.)

COMMERCIALISATION
History Herbicide reported by W. Zwick et al. (Proc. Br. Crop Prot. Conf. – Weeds,
1985, **1**, 85) and by N. Meyer et al. (ibid., p. 93). Introduced by BASF AG.
Patent EP 70370; EP 71707; US 4422864; DE 3121355 **Manufacturers** BASF

APPLICATIONS
Biochemistry Acetyl CoA carboxylase (fatty acid synthesis) inhibitor.
Mode of action Selective herbicide, absorbed rapidly by the foliage, with
translocation both acropetally and basipetally. **Uses** Post-emergence control of
annual and perennial grasses (except red fescue and Poa spp.) in broad-leaved
crops (e.g. oilseed rape, beans, potatoes, brassicas, cotton, celery, fennel,

beetroot, sugar beet, fodder beet, sunflowers, soya beans, flax, alfalfa, crucifers, alliums). **Phytotoxicity** Non-phytotoxic to all broad-leaved crops, when used as directed. Phytotoxic to graminaceous crops. **Formulation types** EC. **Compatibility** Should not be mixed with broad-leaved herbicides. **Selected tradenames** 'Focus' (BASF); 'Laser' (BASF); 'Stratos' (BASF)

ANALYSIS
Product analysis by hplc. **Residues** analysed by derivatisation to glutaric acid derivatives and measurement by glc with S-FPD. Details available from BASF.

MAMMALIAN TOXICOLOGY
Reviews *Pesticide residues in food – 1992. FAO Plant Production and Protection Paper, 116, 1993. Pesticide residues in food – 1992 evaluations. Part II – Toxicology. World Health Organisation, WHO/PCS/93.34.* **Oral** Acute oral LD_{50} for rats c. 5000 mg/kg. **Skin and eye** Acute percutaneous LD_{50} for rats >2000 mg/kg. Non-irritating to skin and eyes (rabbits). **Inhalation** LC_{50} (4 h) for rats >5.28 mg/l. **NOEL** (18 mo) for rats 7 mg/kg b.w. daily; (2 y) for mice 32 mg/kg b.w. daily. **ADI** (JMPR) 0.07 mg/kg [1992]. **Other** Non-mutagenic, non-carcinogenic, non-teratogenic. **Toxicity class** WHO (a.i.) III (Table 5); EPA (formulation) IV

ECOTOXICOLOGY
Birds Acute oral LD_{50} for quail >2000 mg/kg. **Fish** LC_{50} (96 h) for trout 220, bluegill sunfish >100 mg/l. **Bees** Non-toxic to bees; LD_{50} >100 µg/bee. **Daphnia** LC_{50} (48 h) 132 mg/l.

ENVIRONMENTAL FATE
Animals See Soil/Environment. **Plants** See Soil/Environment. **Soil/Environment** Rapidly broken down in plants, animals and in the soil by oxidation, conjugation, rearrangement, hydroxylation, and reductive ether-cleavage (*Br. Crop Prot. Conf. – Pests Dis.*, 1988, **1**, 335–341). DT_{50} (lab.) 0.5–12 d (20 °C). K_{oc} <10–183.

175 *Cydia pomonella* GV   *Biological agent*

granulosis virus

NOMENCLATURE
Scientific name *Cydia pomonella* granulosis virus   **Strain** Mexican strain
Other names Codling moth Granulosis Virus; CmGV; CpGV

PROPERTIES
Composition 'Carpovirusine' contains 6.7×10^{12} GV/l. **Stability** 'Carpovirusine' has a shelf-life of 2 y when stored at 4 °C. 'Madex' is stable for 4 weeks at room temperature, and has unlimited stability below 2 °C.

COMMERCIALISATION
Production CmGV is multiplied *in vivo*, on codling moth infested at the larval stage and isolated by centrifugation techniques. 'Carpovirusine' is then formulated as a suspension enclosing the entomopathogenic germ and specific additives.
History An isolate of this baculovirus was obtained from a virosed larva of *Cydia pomonella* L. (Lepidoptera: Tortricidae) or codling moth, collected in Mexico (*J. Insect Pathol.*, 1964, **6**, 373–386). **Manufacturers** Andermatt; NPP

APPLICATIONS
Mode of action Acts by ingestion with following infection. **Uses** For control of codling moth (*Cydia pomonella*) in pear, apple, apricot and walnut trees at 10^{13} GV/ha. Used in IPM programmes. **Formulation types** Liquid; SC.
Compatibility Compatible with all pesticides which do not have a repellent effect on the codling moth and a pH between 6 and 8. **Selected tradenames** 'Carposin' (Agrichem); 'Carpovirusine' (NPP); 'CYD-X' (Thermo Trilogy); 'Granupom' (AgrEvo); 'Madex 3' (Andermatt)

ANALYSIS
Biotest with codling moth.

MAMMALIAN TOXICOLOGY
Oral Oral LD_{50} for rats >2000 mg/kg. **Skin and eye** Does not penetrate the skin; mild skin irritant (rabbits). **Inhalation** LD_{50} (rats) $>2 \times 10^{12}$ GV/ml.
Other There is no evidence of acute or chronic toxicity, eye or skin irritation to mammals. No allergic symptoms or health problems have been observed by research workers, manufacturing staff or users.

ECOTOXICOLOGY
Birds No toxicity or pathogenicity at 10 000 mg/kg daily for 5 d. **Fish** LC_{50} (96 h) >250 mg/l. **Bees** LC_{50} (contact) $>1 \times 10^{10}$ GV ml^{-1} bee^{-1}. **Worms** LC_{50} >1000 mg/kg soil. **Other beneficial spp.** Harmless to mite predators *Typhlodromus pyri* and *Neoseiulus californicus*. **Daphnia** EC_{50} (48 h) >250 mg/l.

ENVIRONMENTAL FATE
CpGV is naturally occurring. **Soil/Environment** In soil, no viral activity found 4 months after soil application. In water, no real persistence (sedimentation of viral particles).

pyrethroid

NOMENCLATURE

Common name cyfluthrin (BSI, draft E-ISO, BAN); cyfluthrine ((f) draft F-ISO)
IUPAC name (RS)-α-cyano-4-fluoro-3-phenoxybenzyl (1RS,3RS;1RS,3SR)-3-= (2,2-dichlorovinyl)-2,2-dimethylcyclopropanecarboxylate
Roth: (RS)-α-cyano-4-fluoro-3-phenoxybenzyl (1RS)-cis-trans-3-(2,2-dichlorovinyl)-= 2,2-dimethylcyclopropanecarboxylate
Chemical Abstracts name cyano(4-fluoro-3-phenoxyphenyl)methyl 3-(2,2-= dichloroethenyl)-2,2-dimethylcyclopropanecarboxylate (unstated stereochemistry)
CAS RN [68359–37–5] unstated stereochemistry   **EEC no.** 269–855–7
Development codes BAY FCR 1272   **Official codes** OMS 2012

PHYSICAL CHEMISTRY

Composition Comprises a mixture of four diastereoisomeric pairs of enantiomers: I (R)-α-cyano-4-fluoro-3-phenoxybenzyl (1R)-cis-3-(2,2-dichlorovinyl)-2,2-= dimethylcyclopropanecarboxylate + (S)-α, (1S)-cis-; II (S)-α, (1R)-cis- + (R)-α, (1S)-cis-; III (R)-α, (1R)-trans- + (S)-α, (1S)-trans-; IV (S)-α, (1R)-trans- + (R)-α, (1S)-trans-. Tech. grade contains: 23–27% diastereoisomer I, 17–21% diastereoisomer II, 32–36% diastereoisomer III, and 21–25% diastereoisomer IV.
Mol. wt. 434.3   **M.f.** $C_{22}H_{18}Cl_2FNO_3$ **Form** Colourless crystals; (tech. is a brown, oily, viscous mass, with crystalline parts).   **M.p.** I 64 °C; II 81 °C; III 65 °C; IV 106 °C; tech. c. 60 °C   **V.p.** I 9.6×10^{-4}; II 1.4×10^{-5}; III 2.1×10^{-5}; IV 8.5×10^{-5} (all in mPa, 20 °C) K_{ow} logP: I 6.0; II 5.9; III 6.0; IV 5.9 (all 20 °C)
Henry I 1.9×10^{-1}; II 2.9×10^{-3}; III 4.2×10^{-3}; IV 1.3×10^{-2} (all in Pa m³ mol⁻¹, 20 °C)
S.g./density 1.28 (20 °C)   **Solubility** Diastereoisomer I: In water 2.5 (pH 3), 2.2 (pH 7) (both in µg/l, 20 °C). In dichloromethane, toluene >200, n-hexane 10–20, isopropanol 20–50 (all in g/l, 20 °C). Diastereoisomer II: In water 2.1 (pH 3), 1.9 (pH 7) (both in µg/l, 20 °C). In dichloromethane, toluene >200, n-hexane 10–20, isopropanol 5–10 (all in g/l, 20 °C). Diastereoisomer III: In water 3.2 (pH 3), 2.2 (pH 7) (both in µg/l, 20 °C). In dichloromethane, toluene >200, n-hexane, isopropanol 10–20 (all in g/l, 20 °C). Diastereoisomer IV: In water 4.3 (pH 3), 2.9 (pH 7) (both in µg/l, 20 °C). In dichloromethane >200, n-hexane 1–2, toluene 100–200, isopropanol 2–5 (all in g/l, 20 °C).   **Stability** Thermally stable at room temperature. In water DT_{50} for diastereoisomer I: 36, 17, 7; II 117, 20, 6; III 30, 11, 3; IV 25, 11, 5 (all in days, pH 4, 7, 9 respectively, 22 °C).   **F.p.** 107 °C (tech.)

COMMERCIALISATION

History Insecticide reported by I. Hammann & R. Fuchs (*Pflanzenschutz-Nachr. (Engl. Ed.)*, 1981, **34**, 121) and W. Behrenz et al. (*ibid.*, 1983, **36**, 127). Introduced by Bayer AG. **Patent** DE 2709264; DE 2852028 **Manufacturers** Bayer

APPLICATIONS

Mode of action Non-systemic insecticide with contact and stomach action. Acts on the nervous system, with rapid knockdown and long residual activity. **Uses** An insecticide effective against many pests, especially Lepidoptera, Coleoptera, Homoptera and Hemiptera on cereals, cotton, fruit and vegetables; also against migratory locusts and grasshoppers. Used against Blattellidae, Culicidae and Muscidae in public health situations, stored products, domestic use and animal health. It has a rapid knockdown effect and long-lasting residual activity. **Formulation types** AE; EC; EO; ES; EW; UL; WP; GR; Oilspray. **Mixtures** (*cyfluthrin +*) phoxim; methamidophos; omethoate; dichlorvos + propoxur; fenitrothion; profenofos; chlorpyrifos; tetramethrin. **Compatibility** Incompatible with azocyclotin. **Selected tradenames** 'Baythroid' (agricultural use) (Bayer); 'Baygon aerosol' (public health use) (Bayer); 'Solfac' (animal health use) (Bayer)

ANALYSIS

Residues determined by glc (*Anal. Methods Residues Pestic.*, 1988, Part I, M11). Details of methods for **product** and **residue** analysis available from Bayer AG. Analysis of pyrethroids reviewed by E. Papadopoulou-Mourkidou in *Comp. Analyt. Profiles.*

MAMMALIAN TOXICOLOGY

Reviews *Pesticide residues in food – 1987*. FAO Plant Production and Protection Paper 84, 1987. *Pesticide residues in food – 1987 evaluations. Part II – Toxicology.* FAO Plant Production and Protection Paper 86/2, 1988. **Oral** Acute oral LD_{50} for rats c. 500 mg/kg (in xylol), c. 900 mg/kg (PEG 400), c. 20 mg/kg (water/cremophor); for dogs >100 mg/kg. **Skin and eye** Acute percutaneous LD_{50} (24 h) for male and female rats >5000 mg/kg; non-irritating to skin, mildly irritating to eyes (rabbits). **Inhalation** LC_{50} (4 h) for male and female rats 0.5 mg/l air (aerosol). **NOEL** (2 y) for rats 50, mice 200 mg/kg diet; (1 y) for dogs 160 mg/kg diet. **ADI** (JMPR) 0.02 mg/kg b.w. [1987]. **Toxicity class** WHO (a.i.) II; EPA (formulation) II

ECOTOXICOLOGY

Birds Acute oral LD_{50} for bobwhite quail >2000 mg/kg. **Fish** LC_{50} (96 h) for golden orfe 0.0032, rainbow trout 0.0006–0.0029, carp 0.022, bluegill sunfish 0.0015 mg/l. **Bees** Toxic to honeybees. **Worms** LC_{50} for *Eisenia foetida* >1000 mg/kg dry soil. **Daphnia** LC_{50} (48 h) 0.00016–0.0027 mg/l. **Algae** E_rC_{50} for *Scenedesmus subspicatus* >10 mg/l.

ENVIRONMENTAL FATE

Animals Cyfluthrin was largely and very quickly eliminated; 98% of the administered amount was eliminated after 48 h via the urine and the faeces.
Plants Since cyfluthrin is not systemic, it penetrates only slightly into the plant tissues and is hardly translocated into other parts of the plant. The concentration is very low and can be neglected.　**Soil/Environment** Degradation in different soils is rapid. The leaching behaviour can be classified as immobile. The metabolites of cyfluthrin are subject to further microbial degradation to the point of mineralisation to CO_2.

77　beta-cyfluthrin　　　　　　　　　　*Insecticide*

pyrethroid

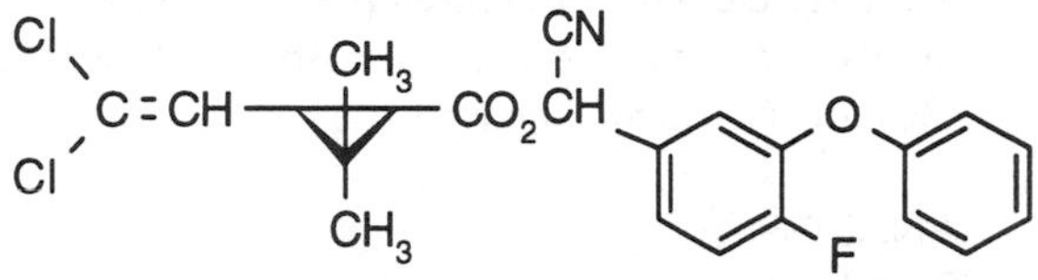

NOMENCLATURE

Common name beta-cyfluthrin (BSI, draft E-ISO); beta-cyfluthrine ((f) draft F-ISO); cyfluthrin-beta (Canada)
IUPAC name (RS)-α-cyano-4-fluoro-3-phenoxybenzyl (1RS,3RS;1RS,3SR)-3-= (2,2-dichlorovinyl)-2,2-dimethylcyclopropanecarboxylate
Chemical Abstracts name cyano(4-fluoro-3-phenoxyphenyl)methyl 3-(2,2-= dichloroethenyl)-2,2-dimethylcyclopropanecarboxylate
CAS RN *[68359–37–5]* unstated stereochemistry　**EEC no.** 269–855–7
Development codes FCR 4545 (Bayer)　**Official codes** OMS 3051

PHYSICAL CHEMISTRY

Composition Comprises a mixture of four diastereoisomeric pairs of enantiomers: I (R)-α-cyano-4-fluoro-3-phenoxybenzyl (1R)-cis-3-(2,2-dichlorovinyl)-2,2-= dimethylcyclopropanecarboxylate + (S)-α, (1S)-cis-; II (S)-α, (1R)-cis- + (R)-α, (1S)-cis-; III (R)-α, (1R)-trans- + (S)-α, (1S)-trans-; IV (S)-α, (1R)-trans- + (R)-α, (1S)-trans-. Tech. grade contains: <2% diastereoisomer I, 30–40% diastereoisomer II, <3% diastereoisomer III, and 53–67% diastereoisomer IV.　**Mol. wt.** 434.3
M.f. $C_{22}H_{18}Cl_2FNO_3$　**Form** Colourless crystals; (tech. is a white powder with slight characteristic smell).　**M.p.** (II) 81 °C; (IV) 106 °C　**V.p.** (II) 1.4×10^{-5} mPa; (IV) 8.5×10^{-5} mPa (both 20 °C)　**K$_{ow}$** (II) logP = 5.94; (IV) logP = 5.91　**Henry** (II) 2.9×10^{-3}; (IV) 3.1×10^{-2} (both Pa m^3 mol^{-1}, 20 °C)　**S.g./density** 1.34 g/cm^3

(22 °C)   **Solubility** In water: (II) 2.1, (IV) 1.2 (both $\mu g/l$, 20 °C). In dichloromethane and toluene, both pairs >200 g/l (20 °C). (II) In n-hexane 2–5, isopropanol 5–10 (both g/l, 20 °C). (IV) in n-hexane 1–2, isopropanol 2–5 g/l (20 °C).   **Stability** Thermally stable at room temperature. Hydrolysis DT_{50} (22 °C) for (II): 117 d (pH 4), 20 d (pH 7), 6 d (pH 9); for (IV): 25 d (pH 4), 11 d (pH 7), 5 d (pH 9).

COMMERCIALISATION
History Reported by M. C. Botte et al., 3rd ANPP Conf., 1993, **1**, 185. Introduced by Bayer AG.   **Manufacturers** Bayer

APPLICATIONS
Mode of action Non-systemic insecticide with contact and stomach action. Acts on the nervous system, with rapid knockdown and long residual activity.
Uses Insecticide effective against Lepidoptera, Coleoptera, Hemiptera and Homoptera on cotton, fruit, vegetables, cereals and other crops; also against migratory locusts and grasshoppers; and in animal health.   **Formulation types** EC; SC; EW; UL; GR.   **Mixtures** (beta-cyfluthrin +) methamidophos; oxydemeton-methyl; chinomethionat; fenitrothion.   **Selected tradenames** 'Bulldock' (Bayer); 'Responsar' (animal health) (Bayer)

ANALYSIS
Details of **product** and **residue** analysis available from Bayer AG. Analysis of pyrethroids reviewed by E. Papadopoulou-Mourkidou in Comp. Analyt. Profiles.

MAMMALIAN TOXICOLOGY
Oral Acute oral LD_{50} for rats c. 500 mg/kg (in polyethyleneglycol), c. 270 mg/kg (xylene), c. 15 mg/kg (cremophor/water); for mice c. 140 mg/kg.
Skin and eye Acute percutaneous LD_{50} (24 h) for rats >5000 mg/kg. No irritation to skin and slight primary irritation to the eye (rabbits).   **Inhalation** LC_{50} (4 h) for rats c. 0.1 mg/l (aerosol), 0.53 mg/l (dust).   **NOEL** (90 d) for rats 125, dogs 60 mg/kg diet.   **ADI** 0.02 mg/kg b.w.   **Toxicity class** WHO (a.i.) II; EPA (formulation) II

ECOTOXICOLOGY
Birds Acute oral LD_{50} for Japanese quail >2000 mg/kg.   **Fish** LC_{50} (96 h) for golden orfe 330.9, rainbow trout 89, bluegill sunfish 28 ng/l.   **Daphnia** LC_{50} (48 h) 0.00029–0.0018 mg/l.

ENVIRONMENTAL FATE
Animals Beta-cyfluthrin was largely and very quickly eliminated; 98% was eliminated after 48 h via the urine and the faeces.   **Plants** Since beta-cyfluthrin is not systemic, it penetrates only slightly into the plant tissues and is hardly translocated into other plant parts. The concentration is very low and can be

neglected. **Soil/Environment** Degradation in different soils is rapid. The leaching behaviour can be classified as immobile. The metabolites of beta-cyfluthrin are readily accessible to further microbial degradation to the point of mineralisation to CO_2 in the soil.

78 cyhalofop-butyl

Herbicide

2-(4-aryloxyphenoxy)propionic acid

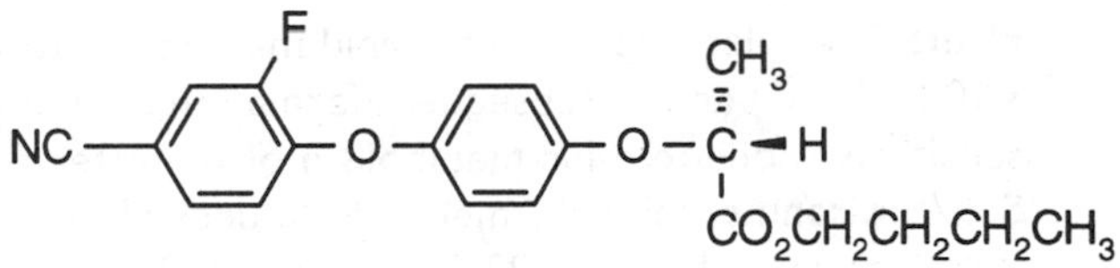

NOMENCLATURE
Common name cyhalofop-butyl (pa ISO, ANSI)
IUPAC name butyl (*R*)-2-[4-(4-cyano-2-fluorophenoxy)phenoxy]propionate
Chemical Abstracts name butyl (*R*)-2-[4-(4-cyano-2-fluorophenoxy)phenoxy]= propionate
CAS RN *[122008–85–9]; [122008–78–0]* (for acid)
Development codes XDE 537; DEH 112 (DowElanco)

PHYSICAL CHEMISTRY
Composition Material is the resolved (*R*)- isomer.   **Mol. wt.** 357.4
M.f. $C_{20}H_{20}FNO_4$ **Form** White crystalline solid.   **M.p.** 50 °C   **V.p.** 1.2×10^{-3} mPa (20 °C) K_{ow} logP = 3.31   **Solubility** In water 0.7 ppm (pH 7.0, 20 °C). In xylene 47.3, acetone 60.7 (wt. % a.i., 20 °C).   **Stability** Stable at pH 4, hydrolysed slowly at pH 7. At pH 1.2 or pH 9, decomposition is rapid.

COMMERCIALISATION
History Reported by DowElanco (P. G. Ray *et al., Proc. 10th Australian & 14th Asian-Pacific Weed Conference,* Brisbane, Australia, September 1993, p. 41).
Manufacturers DowElanco

APPLICATIONS
Biochemistry Inhibition of acetyl CoA carboxylase. Selectivity is due to differential metabolism of the molecule by rice and target grass weeds.   **Uses** For post-emergence control of grass weeds in rice.   **Selected tradenames** 'Clincher' (DowElanco)

MAMMALIAN TOXICOLOGY

Oral Acute oral LD_{50} for male and female rats, and for male and female mice >5000 mg/kg. **Skin and eye** Acute percutaneous LD_{50} for male and female rats >2000 mg/kg. **Other** Non-mutagenic in Ames, DNA repair and micronucleus tests. Not teratogenic. In *in vivo* cytogenetic studies, no induction of structural chromosomal aberration observed. Rat and rabbit studies indicate XDE 537 is not teratogenic.

ECOTOXICOLOGY

Fish LC_{50} for Japanese carp 1.54, rainbow trout 1.65 mg/l. **Daphnia** LC_{50} >100 mg/l.

ENVIRONMENTAL FATE

Plants Rice tolerance is due to rapid metabolism to the inactive diacid (DT_{50} <10 h, 30 °C) and to subsequent formation of non-polar metabolites. *Echinochloa* sensitivity is due to rapid metabolism of the ester to the active acid.
Soil/Environment Rapidly metabolised under both flooded and upland conditions. Relatively immobile, R_f 0.22 (sandy soil), 0.35 (loamy soil).

179 cyhalothrin   *Insecticide*

pyrethroid

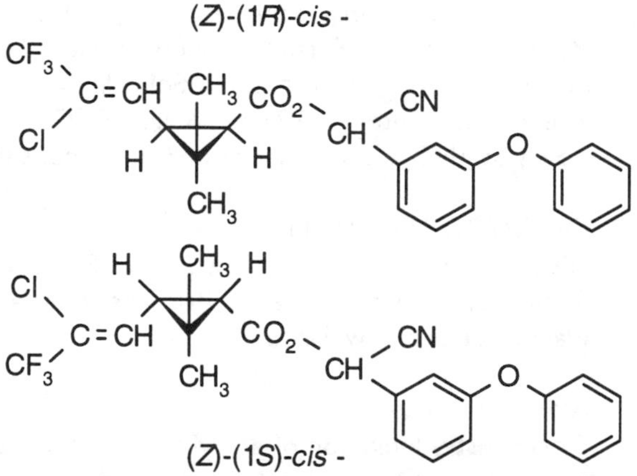

(Z)-(1R)-cis -

(Z)-(1S)-cis -

NOMENCLATURE

Common name cyhalothrin (BSI, draft E-ISO, BAN); cyhalothrine ((f) draft F-ISO)
IUPAC name (RS)-α-cyano-3-phenoxybenzyl (Z)-(1RS,3RS)-(2-chloro-=

3,3,3-trifluoropropenyl)-2,2-dimethylcyclopropanecarboxylate.
Roth: (*RS*)-α-cyano-3-phenoxybenzyl (*Z*)-(1*RS*)-*cis*-3-(2-chloro-3,3,3-trifluoro=
propenyl)-2,2-dimethylcyclopropanecarboxylate
Chemical Abstracts name [1α,3α(*Z*)]-(±)-cyano-(3-phenoxyphenyl)methyl
3-(2-chloro-3,3,3-trifluoro-1-propenyl)-2,2-dimethylcyclopropanecarboxylate
CAS RN *[68085–85–8]* **Development codes** PP563; ICI 146 814
Official codes OMS 2011

PHYSICAL CHEMISTRY
Composition Tech. grade cyhalothrin has purity ≥90% *m/m*, of which ≥95% is
cis-isomers. **Mol. wt.** 449.9 **M.f.** $C_{23}H_{19}ClF_3NO_3$ **Form** Yellow to brown
viscous oil (tech.). **B.p.** 187–190 °C/0.2 mmHg **V.p.** *c.* 0.001 mPa (20 °C)
K_{ow} logP = 6.8 (20 °C) **S.g./density** 1.25 (25 °C) **Solubility** In water 0.004 ppb
(20 °C). In acetone, dichloromethane, methanol, diethyl ether, ethyl acetate,
hexane, toluene, all >500 g/l (20 °C). **Stability** Stable to decomposition and
cis/trans isomerisation for at least 4 years in the dark at 50 °C. Stable to light;
loss on storage in the light is <10% in 20 months. Decomposes at 275 °C. Slowly
hydrolysed by water in sunlight at pH 7–9, more rapidly at pH >9. **F.p.** >100 °C

COMMERCIALISATION
History Insecticide reported by P. D. Bentley *et al.* (*Pestic. Sci.*, 1980, **11,** 156) and
its acaricidal properties by V. K. Stubbs (*Austral. Vet. J.*, 1982, **59,** 152). Introduced
by ICI Australia and ICI Agrochemicals (now Zeneca Agrochemicals).
Patent AU 521136; US 4183948; GB 2000764 **Manufacturers** Zeneca

APPLICATIONS
Mode of action Non-systemic insecticide with contact and stomach action, and
repellent properties. Gives rapid knockdown and long residual activity.
Uses Control of animal ectoparasites, especially *Boophilus microplus* or *Haematobia
irritans* on cattle, and *Bovicola ovis*, *Linognathus* spp. and *Melophagus ovinus* on
sheep. Applied as an animal dip or as a spray around animal houses.
Formulation types EC; WP. **Selected tradenames** 'Cyhalon' (Zeneca); 'Grenade'
(Zeneca)

ANALYSIS
Product analysis of pyrethroids reviewed by E. Papadopoulou-Mourkidou in *Comp.
Analyt. Profiles.* By hplc (A. Jacques *et al., Anal. Methods Pestic. Plant Growth Regul.,*
1984, **13,** 9). **Residues** determined by glc with ECD (*idem, ibid.*). Details available
from Zeneca.

MAMMALIAN TOXICOLOGY
Reviews *Pesticide residues in food – 1984.* FAO Plant Production and Protection
Paper 62, 1985. *Pesticide residues in food – 1984 evaluations.* FAO Plant Production
and Protection Paper 67, 1985. **EHC** 99 (WHO, 1990). **Oral** Acute oral LD_{50}

for male rats 166, female rats 114, guinea pigs >5000, rabbits >1000 mg/kg.
Skin and eye Acute percutaneous LD_{50} for male rats 1000–2500, female rats
200–2500, rabbits >2500 mg/kg. Moderate eye irritant; not a skin irritant
(rabbits). Moderate skin sensitiser (guinea pigs). **Inhalation** LC_{50} (4 h) for rats
>0.086 mg/l. **NOEL** No significant toxicological effect observed in feeding trials
with rats (2 y) or dogs (0.5 y) at 2.5 mg/kg daily. **ADI** (JMPR) 0.02 mg/kg b.w.
[1984]. **Other** No evidence of carcinogenicity, mutagenicity or disturbed
reproductive functions, no adverse effects on foetal development. A facial
sensation may be experienced by users of the chemical; this is transient, with
complete recovery. **Toxicity class** WHO (a.i.) II

ECOTOXICOLOGY
Birds Acute oral LD_{50} for mallard ducks >5000 mg/kg. **Fish** LC_{50} (96 h) for
rainbow trout 0.00054 mg/l. **Daphnia** LC_{50} (48 h) 0.38 µg/l.

ENVIRONMENTAL FATE
Animals In rats, following oral administration, cyhalothrin is rapidly eliminated in
urine and faeces. The ester group is hydrolysed, both moieties forming polar
conjugates. **Soil/Environment** In soil, DT_{50} c. 4–12 w. Leaching of cyhalothrin and
its degradation products through a range of soil types is negligible.

180 lambda-cyhalothrin   *Insecticide*

pyrethroid

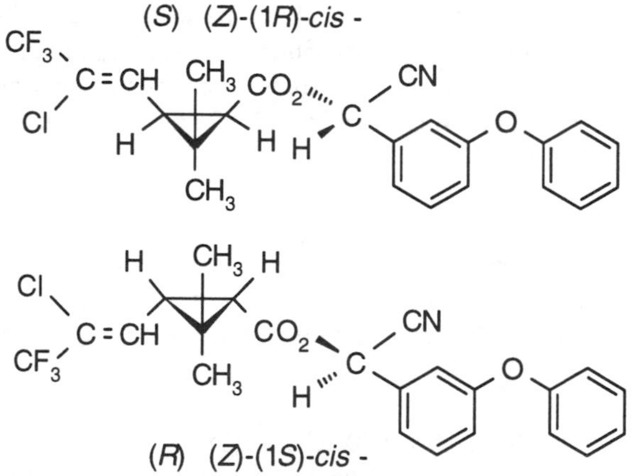

NOMENCLATURE
Common name lambda-cyhalothrin (BSI, draft E-ISO); lambda-cyhalothrine
((*f*) draft F-ISO)

IUPAC name A reaction product comprising equal quantities of (S)-α-cyano-3-phenoxybenzyl (Z)-(1R,3R)-3-(2-chloro-3,3,3-trifluoropropenyl)-2,2-= dimethylcyclopropanecarboxylate and (R)-α-cyano-3-phenoxybenzyl (Z)-(1S,3S)-= 3-(2-chloro-3,3,3-trifluoropropenyl)-2,2-dimethylcyclopropanecarboxylate. *Roth*: (S)-α-cyano-3-phenoxybenzyl (Z)-(1R)-*cis*-3-(2-chloro-3,3,3-trifluoro= propenyl)-2,2-dimethylcyclopropanecarboxylate and (R)-α-cyano-3-phenoxy= benzyl (Z)-(1S)-*cis*-3-(2-chloro-3,3,3-trifluoropropenyl)-2,2-dimethyl= cyclopropanecarboxylate (1:1)
Chemical Abstracts name [1α(S*),3α(Z)]-(±)-cyano(3-phenoxyphenyl)methyl 3-(2-chloro-3,3,3-trifluoro-1-propenyl)-2,2-dimethylcyclopropanecarboxylate
CAS RN *[91465–08–6]* **Development codes** PP321; ICIA0321
Official codes OMS 3021

PHYSICAL CHEMISTRY
Composition Tech. grade is *c.* 81% pure. **Mol. wt.** 449.9 **M.f.** $C_{23}H_{19}ClF_3NO_3$
Form Colourless solid; (tech., is a dark brown/green solidified melt).
M.p. 49.2 °C; (tech., 47.5–48.5 °C) **V.p.** 2×10^{-4} mPa (20 °C); 2×10^{-1} mPa
(60 °C) **K_{ow}** logP = 7 (20 °C) **S.g./density** 1.33 g/ml (25 °C) **Solubility** In
purified water 0.005 (pH 6.5), buffered water 0.004 (pH 5.0) (both in mg/l). In
acetone, methanol, toluene, hexane, ethyl acetate, *c.* 500 g/l.
Stability Stable to light. Stable on storage for more than 6 months at 15–25 °C.

COMMERCIALISATION
History Insecticide reported by A. R. Jutsum *et al.* (*Proc. Br. Crop Prot. Conf.,
Pests – Dis.*, 1984, **2**, 421). Introduced in Central America and Far East (1985) by
ICI Agrochemicals (now Zeneca Agrochemicals). **Patent** EP 107296; EP 106469
Manufacturers Zeneca

APPLICATIONS
Mode of action Non-systemic insecticide with contact and stomach action, and
repellent properties. Gives rapid knockdown and long residual activity.
Uses Control of a wide spectrum of insect pests, e.g. aphids, Colorado beetles,
thrips, Lepidoptera larvae, Coleoptera larvae and adults, etc., in cereals, hops,
ornamentals, potatoes, vegetables, cotton, and other crops. Provides good
control of insect-borne plant viruses. Also used for control of insect pests in
public health. **Formulation types** CS; WG; WP; EW; EC; ED; UL.
Mixtures *(lambda-cyhalothrin +)* buprofezin; pirimicarb; dimethoate; tetramethrin.
Selected tradenames 'Icon' (Zeneca); 'Karate' (Zeneca); 'Warrior' (Zeneca)

ANALYSIS
Product analysis of pyrethroids reviewed by E. Papadopoulou-Mourkidou in *Comp.
Analyt. Profiles*. By capillary glc (*CIPAC Handbook*, 1992, **E**, 49–57).
Residues determined by glc. Details available from Zeneca.

MAMMALIAN TOXICOLOGY

Oral Acute oral LD_{50} for male rats 79, female rats 56, male mice 19, female mice 31 mg/kg. **Skin and eye** Acute percutaneous LD_{50} for rats 1293–1507 mg/kg. Mild eye irritant; non-irritant to skin (rabbits). Not a skin sensitiser (guinea pigs). **Inhalation** LC_{50} (4 h) 0.06 mg/l air (total particulate). **NOEL** (1 y) for dogs 0.5 mg/kg daily. **Other** Non-mutagenic in the Ames test. **Toxicity class** WHO (a.i.) II

ECOTOXICOLOGY

Birds Acute oral LD_{50} for mallard ducks >3950 mg/kg. Dietary LC_{50} for quail >5000 mg/kg. No accumulation of residues in eggs or tissues. **Fish** LC_{50} for bluegill sunfish 0.21, rainbow trout 0.24 µg/l. **Bees** LD_{50} (oral) 38 ng/bee, (contact) 909 ng/bee. **Daphnia** LC_{50} (48 h) 0.36 µg/l.

ENVIRONMENTAL FATE

Animals In rats, following oral administration, rapidly eliminated in urine and faeces. The ester group is hydrolysed, both moieties forming polar conjugates. **Plants** For details of metabolism of lambda-cyhalothrin in cotton and soya leaves, see D. A. French & J. P. Leahey, *Br. Crop Prot. Conf. – Pests Dis.*, 1990, **3**, 1029–1034. **Soil/Environment** In water-sediment mixtures in sunlight DT_{50} c. 20 d. In soil, DT_{50} c. 4–12 w. Leaching of lambda-cyhalothrin and its degradation products through a range of soil types is negligible.

181 cyhexatin

Acaricide

organotin

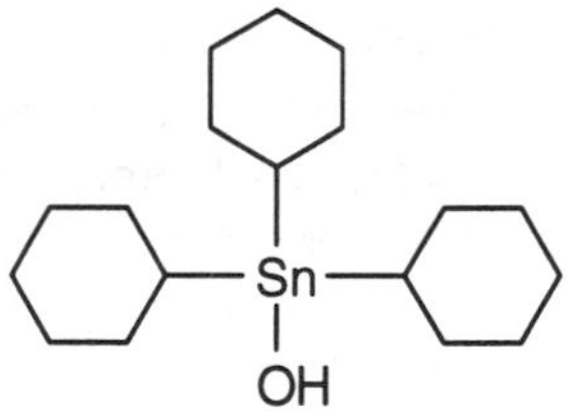

NOMENCLATURE

Common name cyhexatin (BSI, E-ISO, (*m*) F-ISO, ANSI, ESA); tricyclohexyltin hydroxide (JMAF)
IUPAC name tricyclohexyltin hydroxide
Chemical Abstracts name tricyclohexylhydroxystannane
CAS RN *[13121–70–5]* **EEC no.** 236–049–1 **Development codes** Dowco 213
Official codes ENT 27 395-X; OMS 3029

PHYSICAL CHEMISTRY
Mol. wt. 385.2 **M.f.** $C_{18}H_{34}OSn$ **Form** Colourless crystals. **V.p.** Negligible (25 °C). **Solubility** In water <1 mg/l (25 °C). In chloroform 216, methanol 37, dichloromethane 34, carbon tetrachloride 28, benzene 16, toluene 10, xylene 3.6, acetone 1.3 (all in g/kg, 25 °C). **Stability** Stable to 100 °C in aqueous suspensions from slightly acid (pH 6) to alkaline conditions; degraded by u.v. light.

COMMERCIALISATION
History Acaricide reported by W. E. Allison et al. (*J. Econ. Entomol.*, 1968, **61**, 1254). Developed from a joint project of Dow Chemical Co. and M & T Chemicals Inc. and introduced by Dow Chemical Co. (now DowElanco, who no longer manufacture or market it). **Patent** US 3264177; US 3389048
Manufacturers Chemia; Elf Atochem; Oxon Italia

APPLICATIONS
Mode of action Non-systemic acaricide with contact action. **Uses** Control of the motile stages of a wide range of phytophagous mites (adults and larvae) on pome fruit, stone fruit, vines, hops, nuts, bush fruit, strawberries, vegetables, tomatoes, cucurbits and ornamentals. **Phytotoxicity** Non-phytotoxic to deciduous fruit, vines, vegetables, and outdoor ornamentals. Slight injury (usually in the form of localised spotting) may occur on citrus fruit (particularly developing fruits and immature foliage), and glasshouse ornamentals and vegetables (seedlings and immature foliage). **Formulation types** WP; SC. **Mixtures** *(cyhexatin +)* clofentezine; tetradifon; dicofol + tetradifon; hexythiazox; benzoximate.
Compatibility Should not be applied in combination with wetting agents.
Selected tradenames 'Acarstin' (Sipcam); 'Aracnol F' (Diachem); 'Mitacid' (Sipcam); 'Oxotin' (Oxon Italia); 'Pennstyl' (Elf Atochem); 'Sipcatin' (Sipcam)

ANALYSIS
Product analysis by hplc (*CIPAC Handbook*, 1988, **D**, 46; *AOAC Methods*, 1995, 988.02). **Residues** determined by glc (*Pestic. Anal. Man.*, 1979, **II**; *Anal. Methods Residues Pestic.*, 1988, Part II; E. Möllhoff, *Pflanzenschutz-Nachr. (Engl. Ed.)*, 1977, **30**, 249) or by atomic absorption spectroscopy (J. L. Love & J. E. Patterson, *J. Assoc. Off. Anal. Chem.*, 1978, **61**, 627).

MAMMALIAN TOXICOLOGY
Reviews *Pesticide residues in food – 1994*, FAO Plant Production and Protection Paper, 127, 1995. *Pesticide residues in food – 1994 evaluations, Pt. II – Toxicology*, WHO, WHO/PCS/95.2, 1995. **EHC** 15 (WHO 1980). **Oral** Acute oral LD_{50} for rats 540, rabbits 500–1000, guinea pigs 780 mg/kg. **Skin and eye** Acute percutaneous LD_{50} for rabbits >2000 mg/kg. Irritating to eyes. **NOEL** (2 y) for dogs 0.75, mice 3, rats 1 mg/kg daily. **ADI** (JMPR) 0.007 mg/kg b.w. [1994] (with azocyclotin). **Toxicity class** WHO (a.i.) III; EPA (formulation) III
EC risk Xn (R20/21/22) **PIC** Yes

ECOTOXICOLOGY

Birds Acute oral LD_{50} for chickens 650 mg/kg. Eight-day dietary LC_{50} for mallard ducklings 3189, bobwhite quail 520 mg/kg diet. **Fish** LC_{50} (24 h) for large-mouth bass 0.06, goldfish 0.55 mg/l. **Bees** Virtually non-hazardous to honeybees (dermal LD_{50} 0.032 mg/bee) at the recommended rates of use.

Other beneficial spp. Virtually non-hazardous to most predacious mites and insects at the recommended rates of use.

ENVIRONMENTAL FATE

Soil/Environment Dicyclohexyl tin hydroxide, monocyclohexyl tin hydroxide, and inorganic tin compounds are formed as metabolites. Degradation is promoted by u.v. light.

182 cymoxanil

Fungicide

$$CH_3CH_2NHCONHCOC \overset{CN}{=} NOCH_3$$

NOMENCLATURE

Common name cymoxanil (BSI, ANSI, draft E-ISO, (*m*) draft F-ISO)
IUPAC name 1-(2-cyano-2-methoxyiminoacetyl)-3-ethylurea
Chemical Abstracts name 2-cyano-*N*-[(ethylamino)carbonyl]-2-(methoxyimino)= acetamide
CAS RN *[57966-95-7]* **Development codes** DPX-T3217

PHYSICAL CHEMISTRY

Composition Tech. is >95%. **Mol. wt.** 198.2 **M.f.** $C_7H_{10}N_4O_3$
Form Colourless, odourless crystals. **M.p.** 160–161 °C **V.p.** 0.15 mPa (20 °C)
K$_{ow}$ logP = 0.59 (pH 5), 0.67 (pH 7) **Henry** 3.8×10^{-10} Pa m^3 mol^{-1} (calc.)
S.g./density 1.31 (25 °C) **Solubility** In water 890 mg/kg (pH 5, 20 °C). In hexane 1.85, toluene 5.29, acetonitrile 57, ethyl acetate 28, *n*-octanol 1.43, methanol 22.9, acetone 62.4, methylene chloride 133.0 (all in g/l, 20 °C).
Stability Hydrolysis DT_{50} 148 d (pH 5), 34 h (pH 7), 31 min (pH 9). Sensitive to light. **pKa** 9.7 (decomp.)

COMMERCIALISATION

History Fungicide reported by J. M. Serres & G. A. Carraro (*Meded. Fac. Landbouwwet. Rijksuniv. Gent*, 1976. **41**, 645). Introduced by E. I. du Pont de Nemours Co. **Patent** US 3957847 **Manufacturers** Du Pont; Griffin

APPLICATIONS
Mode of action Foliar fungicide with protective and curative action. Has contact and local systemic activity, and also inhibits sporulation. **Uses** Control of Peronosporales, especially *Peronospora*, *Phytophthora*, and *Plasmopara* spp. Normally used in combination with protectant fungicides (to improve residual activity) on a range of crops, including vines, hops, potatoes, and tomatoes. **Formulation types** WP. **Mixtures** *(cymoxanil +)* oxadixyl; maneb; mancozeb; folpet; metiram; propineb; chlorothalonil; copper oxychloride; zineb; propineb + triadimefon; mancozeb + oxadixyl; copper oxychloride + oxadixyl; Bordeaux mixture + zineb; captafol + copper oxychloride; captafol + folpet; folpet + mancozeb; folpet + zineb; copper sulfate + mancozeb; copper sulfate + folpet; copper oxychloride + copper sulfate + mancozeb. **Compatibility** Incompatible with alkaline materials. **Selected tradenames** 'Curzate' (Du Pont); 'Manex C-8' (mixture) (Griffin); 'Pulsan' (mixture) (Novartis)

ANALYSIS
Product analysis by hplc. **Residues** determined by glc. Details available from Du Pont

MAMMALIAN TOXICOLOGY
Oral Acute oral LD_{50} for male rats 1196, female rats 1390, guinea pigs 1096 mg/kg. **Skin and eye** Acute percutaneous LD_{50} for male rabbits and dogs >3000 mg/kg. Mild eye irritant; non-irritating to skin. Non-sensitising to skin. **Inhalation** LC_{50} (4 h) for male and female rats >5.06 mg/l. **NOEL** for male rats 4.1, female rats 5.4, male mice 4.2, female mice 5.8, male dogs 3.0, female dogs 1.6 mg/kg b.w. daily. **ADI** 0.016 mg/kg. **Toxicity class** WHO (a.i.) III; EPA (formulation) III

ECOTOXICOLOGY
Birds Acute oral LD_{50} for bobwhite quail and mallard ducks >2250 mg/kg. Eight-day dietary LC_{50} for bobwhite quail and mallard ducks >5620 mg/kg diet. **Fish** LC_{50} (96 h) for rainbow trout 61, bluegill sunfish 29, common carp 91, sheepshead minnow >47.5 mg/l. **Bees** Not toxic to bees; LD_{50} (48 h, contact) >25 µg/bee; LC_{50} (48 h, oral) >1000 ppm. **Worms** LC_{50} (14 d) >2208 mg/kg soil. **Daphnia** LC_{50} (48 h) 27 mg/l. **Algae** 23% Inhibition of *Selenastrum capricornutum* at 1.05 mg/l after 5 d.

ENVIRONMENTAL FATE
Animals Radiolabelled cymoxanil was metabolised in the goat to natural products, including fatty acids, glycerol, glycine and other amino acids, lactose, and acid-hydrolysable formyl and acetyl groups. **Plants** Degraded to glycine with subsequent incorporation into natural products (proteins and starch). **Soil/Environment** In lab. soils, DT_{50} 0.75–1.5 d (5 soils, pH range 5.7–7.8, o.m. 0.8–3.5%). In the field, DT_{50} (bare soil) 0.9–9 d. In aquatic studies DT_{50} <1 d. K_{oc} 39–250.

pyrethroid

NOMENCLATURE
Common name cypermethrin (BSI, E-ISO, ANSI, BAN); cyperméthrine ((f) F-ISO)
IUPAC name (RS)-α-cyano-3-phenoxybenzyl (1RS,3RS;1RS,3SR)-3-(2,2-dichloro=
vinyl)-2,2-dimethylcyclopropanecarboxylate
Roth: (RS)-α-cyano-3-phenoxybenzyl (1RS)-*cis-trans*-3-(2,2-dichlorovinyl)-=
2,2-dimethylcyclopropanecarboxylate
Chemical Abstracts name cyano(3-phenoxyphenyl)methyl 3-(2,2-dichloro=
ethenyl)-2,2-dimethylcyclopropanecarboxylate
CAS RN [52315–07–8] (formerly [69865–47–0], [86752–99–0])
Development codes NRDC 149; PP383 (Zeneca); WL 43 467 (Shell); LE 79–600
(Rhône-Poulenc); FMC 30980   **Official codes** OMS 2002

PHYSICAL CHEMISTRY
Composition Tech. grade is 90% pure.   **Mol. wt.** 416.3   **M.f.** $C_{22}H_{19}Cl_2NO_3$
Form Odourless crystals (pure); yellow-brown viscous semi-solid at ambient
temperatures (tech.).   **M.p.** 80.5 °C; 60–80 °C (tech.)   **V.p.** 2.3×10^{-4} mPa
(20 °C) K_{ow} logP = 6.6   **S.g./density** 1.23 (20 °C)   **Solubility** In water
0.004 mg/l (pH 7). In acetone, chloroform, cyclohexanone, xylene >450, ethanol
337, hexane 103 (all in g/l, 20 °C).   **Stability** Relatively stable in neutral and
weakly acidic media, with optimum stability at pH 4. Hydrolysed in alkaline media.
Relatively stable to light in field situations. Thermally stable up to 220 °C.
F.p. Not auto-flammable; non-explosive.

COMMERCIALISATION
History Insecticide reported by M. Elliott *et al.* (*Pestic. Sci.*, 1975, **6**, 537).
Developed by Ciba-Geigy (now Novartis Crop Protection AG), ICI (now Zeneca
Agrochemicals), Mitchell Cotts and Shell International Chemical Co. (now
American Cyanamid Co.).   **Patent** GB 1413491 to NRDC   **Manufacturers** Aimco;
Ankur; Atabay; BASF; Cyanamid; Dhanuka; Ficom; FMC; Gharda; Krishi Rasayan;
Mitchell Cotts; Mitsu; New Chemi; Rallis; Rotam; Sanachem; Searle India; Tagros;
United Phosphorus; Zeneca

APPLICATIONS
Mode of action Non-systemic insecticide with contact and stomach action. Also
exhibits anti-feeding action. Good residual activity on treated plants.

Uses Control of a wide range of insects, especially Lepidoptera, but also Coleoptera, Diptera, Hemiptera, and other classes, in fruit (including citrus), vines, vegetables, potatoes, cucurbits, lettuce, capsicums, tomatoes, cereals, maize, soya beans, cotton, coffee, cocoa, rice, pecans, oilseed rape, beet, ornamentals, forestry, etc. Control of flies and other insects in animal houses; and mosquitoes, cockroaches, houseflies and other insect pests in public health. Also used as an animal ectoparasiticide. **Formulation types** EC; GR; WP; UL. **Mixtures** *(cypermethrin +)* monocrotophos; profenofos; sulfur; chlorfenvinphos. **Compatibility** Incompatible with alkaline materials. **Selected tradenames** 'Arrivo' (FMC); 'Cymbush' (Zeneca); 'Cymperator' (Zeneca); 'Cynoff' (FMC); 'Ripcord' (Cyanamid); 'Agrotrina' (Westrade); 'Basathrin' (BASF); 'Cekumetrin' (Cequisa); 'Cyperguard' (Gharda); 'Cypersan' (Sanachem); 'Cypersect' (Barclay); 'Cyproid' (Aimco); 'Cyrux' (United Phosphorus); 'Cythrine' (Agriphar); 'Demar' (Mitsu); 'Durin' (Dhanuka); 'Grand' (Sanonda); 'Lacer' (Searle India); 'Ralothrin' (Rallis); 'Ranjer' (Ramcides); 'Rasayanrin' (Krishi Rasayan); 'Rocyper' (Rotam); 'Sherpa' (Rhône-Poulenc); 'Starcyp' (Shaw Wallace); 'Volcyper' (Voltas)

ANALYSIS
Product Analysis of pyrethroids reviewed by E. Papadopoulou-Mourkidou in *Comp. Analyt. Profiles*. By glc with FID (*AOAC Methods*, 1995, 985.03, 986.02), by hplc or by glc with FID (*CIPAC Handbook*, 1994, **F**, 404; 1985, **1C**, 2047; A. Sapiets *et al.*, *Anal. Methods Pestic. Plant Growth Regul.*, 1984, **13**, 33). **Residues** determined by glc with ECD (*idem, ibid.*). Details available from Cyanamid or Zeneca.

MAMMALIAN TOXICOLOGY
Reviews See A. J. Gray & D. M. Soderlund, Chapt. 5 in "*Insecticides*". See also *Pesticide residues in food – 1981*. FAO Plant Production and Protection Paper 37, 1982. *Pesticide residues in food: 1981 evaluations*. FAO Plant Production and Protection Paper 42, 1982. **EHC** 82 (WHO, 1989). **Oral** Acute oral LD_{50} for rats 250–4150 (tech. 7180), mice 138 mg/kg. **Skin and eye** Acute percutaneous LD_{50} for rats >4920, rabbits >2460 mg/kg. Slight skin and eye irritant (rabbits). May be a weak skin sensitiser. **Inhalation** LC_{50} (4 h) for rats 2.5 mg/l. **NOEL** (2 y) for dogs 5, rats 7.5 mg/kg. **ADI** (JMPR) 0.05 mg/kg b.w. [1981]. **Other** Oral toxicity values for cypermethrin depend on such factors as: carrier, *cis:trans* ratio of the sample, species, sex, age and degree of fasting. Values reported sometimes differ markedly. **Toxicity class** WHO (a.i.) II; EPA (formulation) II **EC risk** R20/21/22; R36; R43

ECOTOXICOLOGY
Birds Acute oral LD_{50} for mallard ducks >10 000, chickens >2000 mg/kg. **Fish** LC_{50} (96 h) for rainbow trout 0.69, sheepshead minnow 2.37 µg/l; under field conditions fish are not at risk from normal agricultural usage. **Bees** Highly toxic to honeybees in laboratory tests, but field applications at recommended

dosages do not put hives at risk. LD_{50} (24 h) (oral) 0.035 µg/bee; (topical) 0.02 µg/bee. **Other beneficial spp.** Not toxic to Collembola (J. A. Wiles & G. K. Frampton, *Pestic. Sci.*, **47**, 273 (1996)). **Daphnia** LC_{50} (48 h) 0.15 µg/l.

ENVIRONMENTAL FATE
EHC review concludes that biological degradation is rapid, and consequently levels of cypermethrin and its degradation products in soil and surface waters are very low. **Animals** For studies of the biotransformation of cypermethrin in animal tissues, see *Pestic. Sci.*, 1987, **21**, 1 and *ibid.*, 1990, **30**, 159. **Soil/Environment** In soil, hydrolysis with cleavage of the ester bond occurs within *c.* 16 weeks. Further hydrolytic and oxidative degradation occurs. See T. R. Roberts & M. E. Standen, *Pestic. Sci.*, 1977, **8**, 305. In river water, rapid degradation occurs, DT_{50} *c.* 5 d.

184 alpha-cypermethrin   *Insecticide*

pyrethroid

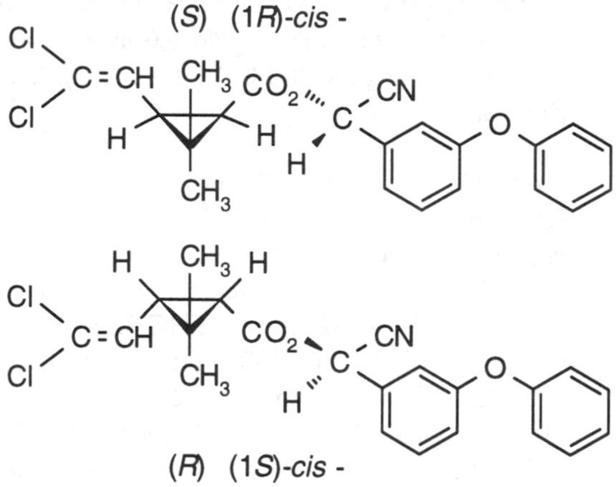

(S) (1R)-cis -

(R) (1S)-cis -

NOMENCLATURE
Common name alpha-cypermethrin (BSI, draft E-ISO); alpha-cyperméthrine ((*f*) draft F-ISO)
IUPAC name A racemate comprising (S)-α-cyano-3-phenoxybenzyl (1R,3R)-3-= (2,2-dichlorovinyl)-2,2-dimethylcyclopropanecarboxylate and (R)-α-cyano-3-= phenoxybenzyl (1S,3S)-3-(2,2-dichlorovinyl)-2,2-dimethylcyclopropanecarboxylate *Roth*: A racemate comprising (S)-α-cyano-3-phenoxybenzyl (1R)-*cis*-3-(2,2-= dichlorovinyl)-2,2-dimethylcyclopropanecarboxylate and (R)-α-cyano-3-= phenoxybenzyl (1S)-*cis*-3-(2,2-dichlorovinyl)-2,2-dimethylcyclopropanecarboxylate
Chemical Abstracts name [1α(S*),3α]-(±)-cyano-(3-phenoxyphenyl)methyl

3-(2,2-dichloroethenyl)-2,2-dimethylcyclopropanecarboxylate
Other names [alfoxylate]; [alphamethrin] (rejected common name proposals)
CAS RN *[67375–30–8]* correct stereochemistry; *[52315–07–8]* (formerly
[86752–99–0], *[86753–92–6]*) cypermethrin (no stereochemistry stated) were
sometimes used in *Chemical Abstracts* **Development codes** WL 85871;
FMC 63318; FMC 39391 **Official codes** OMS 3004

PHYSICAL CHEMISTRY
Composition Tech. grade alpha-cypermethrin is >90% pure *m/m*, typically >95%.
Mol. wt. 416.3 **M.f.** $C_{22}H_{19}Cl_2NO_3$ **Form** Colourless crystals; tech. is a white to
pale powder with a weak aromatic odour. **M.p.** 78–81 °C **B.p.** 200 °C/9.3 Pa
V.p. 2.3×10^{-2} mPa (20 °C) **K_{ow}** logP = 6.94 (pH 7) **S.g./density** 1.28 (22 °C)
Solubility In water *c.* 0.01 mg/l (25 °C). In acetone 620, dichloromethane 550,
cyclohexanone 515, ethyl acetate 440, chlorobenzene 420, acetophenone 390,
o-xylene 350, hexane 7 (all in g/l, 25 °C). In maize oil 19–20, ethylene glycol <1
(both in g/kg, 20 °C). **Stability** Very stable in neutral and acidic media.
Hydrolysed in strongly alkaline media. Thermally stable up to 220 °C. Field data
indicate that, in practice, it is stable to air and light. **F.p.** >80 °C (closed cup)

COMMERCIALISATION
History Insecticide reported by J. P. Fisher *et al.* (*Proc. Int. Congr. Plant Prot., 10th,*
1983, **1,** 452). Introduced by Shell International Chemical Co. (now American
Cyanamid Co.) as WL 85871. **Manufacturers** Cyanamid; FMC; Gharda; Mitsu;
Rotam; Tagros

APPLICATIONS
Mode of action Non-systemic insecticide with contact and stomach action. Acts on
the central and peripheral nervous system in very low doses. **Uses** Control of a
wide range of chewing and sucking insects (particularly Lepidoptera, Coleoptera,
and Hemiptera) in fruit (including citrus), vegetables, vines, cereals, maize, beet,
oilseed rape, potatoes, cotton, rice, soya beans, forestry, and other crops. Control
of cockroaches, mosquitoes, flies, and other insect pests in public health; and flies
in animal houses. Also used as an animal ectoparasiticide. **Formulation types** EC;
WP; SC; UL; TB. **Mixtures** *(alpha-cypermethrin +)* monocrotophos; dimethoate;
chlorfenvinphos. **Compatibility** Compatible with most organophosphorus
insecticides. **Selected tradenames** 'Fastac' (agronomic use) (Cyanamid); 'Fendona'
(public health use) (Cyanamid); 'Renegade' (veterinary use) (Cyanamid); 'Bestox'
(FMC); 'Dominex' (FMC); 'Alphadhan' (Dhanuka); 'Alphaguard' (Gharda);
'Bestseller' (Agriphar); 'Grander' (Sanonda); 'Stop' (Wockhardt)

ANALYSIS
Product Analysis of pyrethroids reviewed by E. Papadopoulou-Mourkidou in
Comp. Analyt. Profiles. **Residues** determined by glc with ECD. Details available
from Cyanamid.

MAMMALIAN TOXICOLOGY

Reviews See A. J. Gray & D. M. Soderlund, Chapt. 5 in "*Insecticides*". **EHC** 142 (WHO, 1992). **Oral** Acute oral LD_{50} for rats 79–400 mg/kg (in corn oil, value depending on concentration), 474 mg tech./kg. **Skin and eye** Acute percutaneous LD_{50} for rats and rabbits >2000 mg tech./kg; minimal irritant to the eyes of rabbits. **Inhalation** LC_{50} (4 h) for rats 0.32 mg/l air. **NOEL** In 90 d feeding trials, rats receiving 60 mg/kg diet showed no ill-effects. **Other** Non-mutagenic. **Toxicity class** WHO (a.i.) II; EPA (formulation) II

ECOTOXICOLOGY

Birds LD_{50} for quail and mallard ducks >10 000 mg/kg. **Fish** LC_{50} (96 h) for rainbow trout 0.0028 mg/l; due to rapid loss from water, no toxic effect to fish is observed under field conditions. **Bees** Toxic to bees. LD_{50} (24 h) 0.059 μg/bee. No toxic effect under field conditions. **Worms** LD_{50} (14 d) for earthworms >100 mg/kg artificial soil. **Daphnia** LC_{50} (48 h) 0.1–0.3 μg/l.

ENVIRONMENTAL FATE

EHC 142 notes that although highly toxic to fish, this is not realised under field conditions where rapid loss from water allows recovery of affected populations. **Animals** See cypermethrin. **Soil/Environment** Undergoes degradation in soil, DT_{50} c. 13 w in loamy soil.

185 beta-cypermethrin   *Insecticide*

pyrethroid

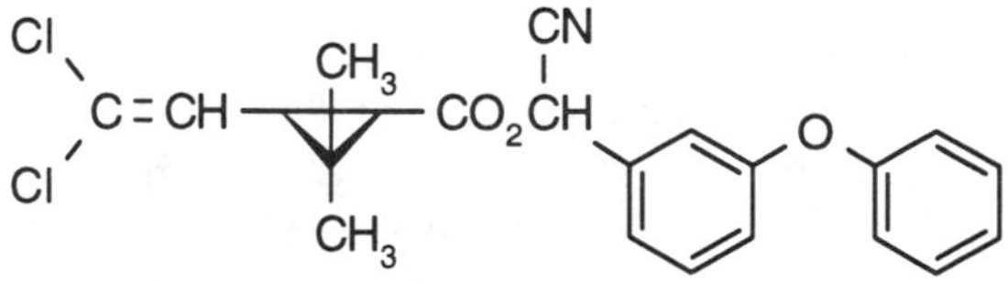

NOMENCLATURE

Common name beta-cypermethrin (BSI, draft E-ISO)
IUPAC name *Roth*: A reaction mixture comprising two enantiomeric pairs in ratio c. 2:3 (S)-α-cyano-3-phenoxybenzyl (1R)-*cis*-3-(2,2-dichlorovinyl)-2,2-dimethyl= cyclopropanecarboxylate and (R)-α-cyano-3-phenoxybenzyl (1S)-*cis*-3-= (2,2-dichlorovinyl)-2,2-dimethylcyclopropanecarboxylate with (S)-α-cyano-= 3-phenoxybenzyl (1R)-*trans*-3-(2,2-dichlorovinyl)-2,2-dimethylcyclopropane= carboxylate and (R)-α-cyano-3-phenoxybenzyl (1S)-*trans*-3-(2,2-dichlorovinyl)-= 2,2-dimethylcyclopropanecarboxylate

Chemical Abstracts name cyano(3-phenoxyphenyl)methyl 3-(2,2-dichloro-=
ethenyl)-2,2-dimethylcyclopropanecarboxylate; 2 parts of enantiomer pair
[(1R)-1α(S^*),3α] and [(1S)-1α(R^*),3α] with 3 parts of enantiomer pair
[(1R)-1α(S^*),3β] and [(1S)-1α(R^*),3β]
Other names [asymethrin] (rejected common name proposal)
CAS RN *[65731–84–2]* ([(1R)-1α(S^*),3α] isomer); *[72204–43–4]*
([(1S)-1α(R^*),3α] isomer); *[65732–07–2]* ([(1R)-1α(S^*),3β] isomer);
[83860–31–5] ([(1S)-1α(R^*),3β] isomer) **Development codes** CHINOIN
0619200

PHYSICAL CHEMISTRY
Composition A mixture consisting of two enantiomeric pairs in ratio 2:3:
(*S*) (1R)-*cis* and (*R*) (1S)-*cis* with (*S*) (1R)-*trans* and (*R*) (1S)-*trans*. Tech. grade beta-
cypermethrin contains ≥95% (normally >97%) relevant stereoisomers.
Mol. wt. 416.3 **M.f.** $C_{22}H_{19}Cl_2NO_3$ **Form** Tech. grade forms colourless or pale
yellow crystals. **M.p.** 64–71 °C (peak 67 °C), (varies with even small - 1% -
change in isomer ratio) **V.p.** 1.8×10^{-4} mPa (20 °C) **K_{ow}** logP >4.7 (pH 7)
S.g./density 1.32 g/ml (theoretical), 0.66 g/ml (crystal powder) (20 °C)
Solubility In water at pH 7, 51.5 (5 °C), 93.4 (25 °C), 276.0 (35 °C) (all in μg/l).
In isopropanol 11.5, xylene 349.8, dichloromethane 3878, acetone 2102, ethyl
acetate 1427, petroleum ether 13.1 (all in mg/ml, 20 °C). **Stability** Stable to
150 °C; to air and sunlight; in neutral and slightly acid media; epimerised in
presence of base, hydrolysed in strongly alkaline media. DT_{50} (extrapolated)
50 d (pH 3, 5, 6), 40 d (pH 7), 20 d (pH 8), 15 d (pH 9) (all at 25 °C).

COMMERCIALISATION
History Insecticide introduced in Hungary (1989) by Chinoin Pharmaceutical &
Chemical Works Co., Ltd. **Patent** HU 198612; EP 208758; US 4963584
Manufacturers Chinoin

APPLICATIONS
Biochemistry Acts on the nervous system of insects, disturbs the function of
neurons by interaction with the sodium channel. **Mode of action** Non-systemic
insecticide with contact and stomach action. **Uses** It can be used against a wide
range of insect pests in public health (e.g. flies, cockroaches, mosquitoes, fleas,
lice, bugs) and in veterinary applications (ectoparasitic ticks and mites). In plant
protection, it is effective against Coleoptera and Lepidoptera and gives good
protection against Orthoptera, Diptera, Hemiptera, and Homoptera. Mainly used
in alfalfa, cereals, cotton, grapes, maize, oilseed rape, pome fruit, potatoes, soya
beans, sugar beet, tobacco and vegetables. **Formulation types** EC; UL; ME; SC.
Selected tradenames 'Chinmix' (Chinoin); 'Cyperil' (Chinoin)

ANALYSIS
Product analysis by hplc or glc; analysis of pyrethroids reviewed by E.

Papadopoulou-Mourkidou in *Comp. Analyt. Profiles*. **Residues** determined by glc. Details available from Chinoin Agchem Business Unit.

MAMMALIAN TOXICOLOGY

Oral Acute oral LD_{50} for female rats 166, male rats 178, female mice 48, male mice 43 mg/kg. **Skin and eye** Acute percutaneous LD_{50} for rats >5000 mg/kg; mild skin and eye irritant (rabbits); non-sensitising to skin (guinea pigs). **Inhalation** LC_{50} for rats >1.97 mg/l per h. **NOEL** (2 y) for rats 250 mg/kg diet; (90 d) for rats 100 mg/kg diet. Not teratogenic in rats or rabbits. In 3-generation reproduction study, NOEL for rats 350 mg/kg. In 2 y carcinogenicity study, NOEL 500 mg/kg diet. **ADI** 0.05 mg/kg. **Other** Not mutagenic in Ames, SCE and micronucleus assays. **Toxicity class** WHO (a.i.) II; EPA (formulation) II (proposed)

ECOTOXICOLOGY

Birds Acute oral LD_{50} for quail 8030, pheasants 3515 mg 5% a.i. formulation/kg. LC_{50} (8 d) for pheasants and quail >5000 mg 5% a.i. formulation/kg diet. **Fish** LC_{50} (96 h) for carp 0.028, catfish 0.015, grass carp 0.035 mg 5% a.i. formulation/l. Fish are not harmed under normal field conditions. **Bees** LD_{50} (oral, 48 h) 0.0018 mg a.i./bee (as 5% formulation); (contact, 24 h) 0.085 l/ha (5% formulation); but not harmful at normal rate under field conditions. **Daphnia** LC_{50} (96 h) 0.00026 mg 5% a.i. formulation/l.

ENVIRONMENTAL FATE

Soil/Environment Soil DT_{50} 10 d (lime-furred chernozom, pH 6.96).

86 theta-cypermethrin *Insecticide*

pyrethroid

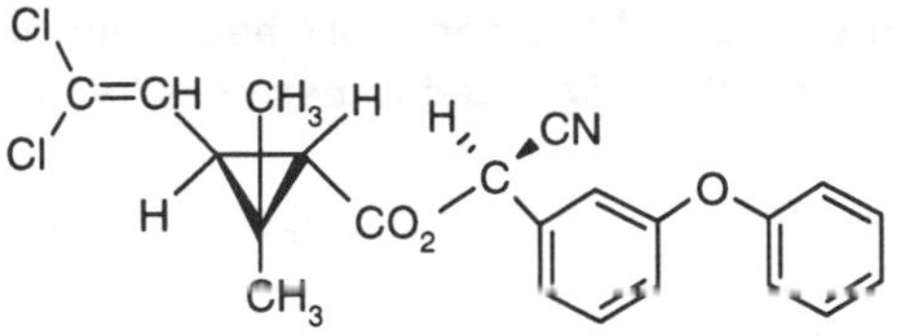

(*R*) (1*S*)-trans

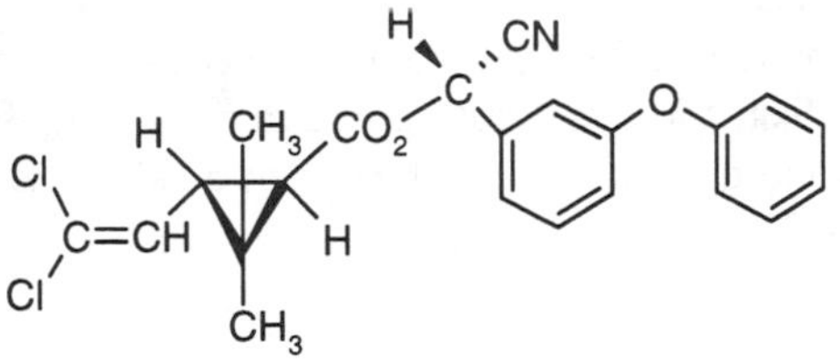

(*S*) (1*R*)-trans

NOMENCLATURE

Common name theta-cypermethrin (BSI, pa ISO)

IUPAC name A mixture of the enantiomers (*R*)-α-cyano-3-phenoxybenzyl (1*S*,3*R*)-3-(2,2-dichlorovinyl)-2,2-dimethylcyclopropanecarboxylate and (*S*)-α-cyano-3-phenoxybenzyl (1*R*,3*S*)-3-(2,2-dichlorovinyl)-2,2-dimethyl=cyclopropanecarboxylate in the ratio 1:1.

Roth: (*R*)-α-cyano-3-phenoxybenzyl (1*S*)-*trans*-3-(2,2-dichlorovinyl)-=2,2-dimethylcyclopropanecarboxylate and (*S*)-α-cyano-3-phenoxybenzyl (1*R*)-*trans*-3-(2,2-dichlorovinyl)-2,2-dimethylcyclopropanecarboxylate

Chemical Abstracts name [1α(*S**),3β]-(±)-cyano(3-phenoxyphenyl)methyl 3-(2,2-dichloroethenyl)-2,2-dimethylcyclopropanecarboxylate

Other names sigma-cypermethrin, tau-cypermethrin (rejected proposed names)

CAS RN *[71697–59–1]*; *[65732–07–2]* ([(1*R*)-1α(*S**),3β] isomer); *[83860–31–5]* ([(1*S*)-1α(*R**),3β] isomer) **Development codes** SK 80

PHYSICAL CHEMISTRY

Composition Racemic cypermethrin isomer pair consisting of (*S*) (1*R*)-*trans* and (*R*) (1*S*)-*trans* isomers in 1:1 ratio. Tech. contains >95% (normally >97%) relevant stereoisomers. **Mol. wt.** 416.3 **M.f.** $C_{22}H_{19}Cl_2NO_3$ **Form** White crystalline powder. **M.p.** 81–87 °C (peak 83.3 °C), by DSC **V.p.** 1.8×10^{-4} mPa (20 °C) **S.g./density** 1.33 g/ml (theoretical), 0.66 g/ml (crystal powder) (20 °C) **Solubility** In water 114.6 μg/l (pH 7, 25 °C). In isopropyl alcohol 18.0, di-isopropyl ether 55.0, hexane 8.5 (all in mg/ml, 20 °C). **Stability** Stable up to 150 °C. In water DT_{50} (extrapolated) 50 d (pH 3, 5, 6), 20 d (pH 7), 18 d (pH 8), 10 d (pH 9) (all at 25 °C).

COMMERCIALISATION
Patent US 4845126; EP 0215010; HU 198373. **Manufacturers** Chinoin

APPLICATIONS
Mode of action Contact insecticide. **Uses** For public health and animal pest control. **Formulation types** EC; PO. **Selected tradenames** 'Neostomosan' (Chinoin)

ANALYSIS
Analysis of pyrethroids reviewed by E. Papadopoulou-Mourkidou in *Comp. Analyt. Profiles*.

MAMMALIAN TOXICOLOGY
Oral Acute oral LD_{50} for male rats 7700, female rats 3200–7700, male mice 136, female mice 106 mg/kg. **Skin and eye** Acute percutaneous LD_{50} for rats >5000 mg/kg. Slight skin and eye irritant (rabbits); not a skin sensitiser (guinea pigs). **Other** Not mutagenic.

187 zeta-cypermethrin   *Insecticide*

pyrethroid

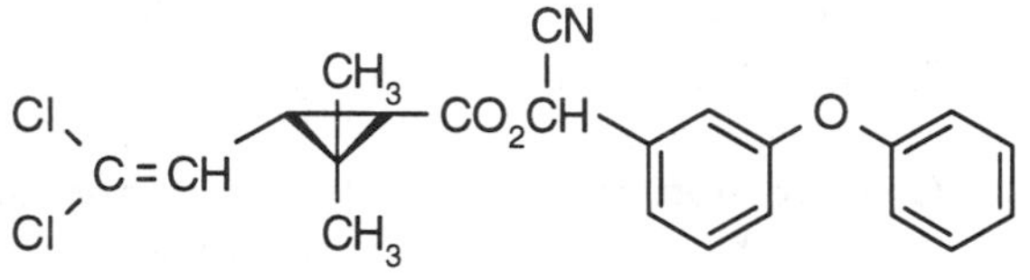

NOMENCLATURE
Common name zeta-cypermethrin (BSI, pa E-ISO, ANSI)
IUPAC name A mixture of the stereoisomers (S)-α-cyano-3-phenoxybenzyl (1RS,3RS;1RS,3SR)-3-(2,2-dichlorovinyl)-2,2-dimethylcyclopropanecarboxylate where the ratio of the (S);(1RS,3RS) isomeric pair to the (S);(1RS,3SR) isomeric pair lies in the ratio range 45–55 to 55–45 respectively
Roth: A mixture of the stereoisomers (S)-α-cyano-3-phenoxybenzyl (1RS)-*cis-trans*-3-(2,2-dichlorovinyl)-2,2-dimethylcyclopropanecarboxylate (where the ratio of isomeric pairs is as defined above)
Chemical Abstracts name (S)-cyano-(3-phenoxyphenyl)methyl ±cis-trans-3-= (2,2-dichloroethenyl)-2,2-dimethylcyclopropanecarboxylate
CAS RN [52315–07–8] (undefined stereochemistry) **Development codes** FMC 56701; F56701; F701

PHYSICAL CHEMISTRY
Composition The ratio of the isomer pairs is in the range 45–55 to 55–45.
Mol. wt. 416.3   **M.f.** $C_{22}H_{19}Cl_2NO_3$ **Form** Dark brown viscous liquid.
M.p. −22.4 °C   **B.p.** Decomposes before boiling at 760 mmHg
V.p. 2.5×10^{-4} mPa   **S.g./density** 1.219 g/cm^3 (25 °C)   **Solubility** In water
0.045 mg/l (25 °C). Miscible in most organic solvents.   **Stability** Stable at 50 °C
for 1 y. Photolysis DT_{50} (aqueous solution) 20.2–36.1 d (pH 7). Hydrolysis DT_{50}
508–769 d (pH 5), 188–635 d (pH 7), 2.9 d (pH 9).   **F.p.** >300 °C

COMMERCIALISATION
Manufacturers FMC

APPLICATIONS
Mode of action Insecticide acting by contact and by ingestion.   **Uses** Control of
beetles, aphids and Lepidoptera in cotton, fruit, vegetables, field crops and
ornamentals. Also used in forestry and in public health. Dosage rates vary from
7.5 g a.i./ha (beetles) to 30 g a.i./ha (cotton, *Heliothis*).   **Formulation types** EC;
WP; EW; EO.   **Selected tradenames** 'Fury' (FMC)

ANALYSIS
By hplc; details from FMC. Analysis of pyrethroids reviewed by E. Papadopoulou-
Mourkidou in *Comp. Analyt. Profiles*.

MAMMALIAN TOXICOLOGY
Oral Acute oral LD_{50} for rats 105.8 mg/kg.   **Skin and eye** Acute percutaneous
LD_{50} for rabbits >2000 mg/kg.   **Inhalation** LC_{50} (4 h) for rats 1 mg/l.
NOEL (1 y) for dogs 5 mg/kg b.w. daily.   **Toxicity class** WHO (a.i.) Ib

ECOTOXICOLOGY
Birds Acute oral LD_{50} for ducks >10 248 mg/kg.   **Fish** LC_{50} 0.69–2.37 µg/l
depending on species.   **Bees** Non-toxic under field conditions.   **Daphnia** LC_{50}
(48 h) 0.15 µg/l.

ENVIRONMENTAL FATE
Animals Elimination in mammals is rapid; 78% is eliminated via urine and faeces
after 24 h, 97% after 96 h.   **Soil/Environment** DT_{50} 14–28 d in a typical fertile
soil. Immobile; strongly adsorbs to organic material. K_{oc} 11 542–54 913.

188 cyphenothrin [(1R)-*trans*- isomers]

Insecticide

pyrethroid

(1R)-cis -

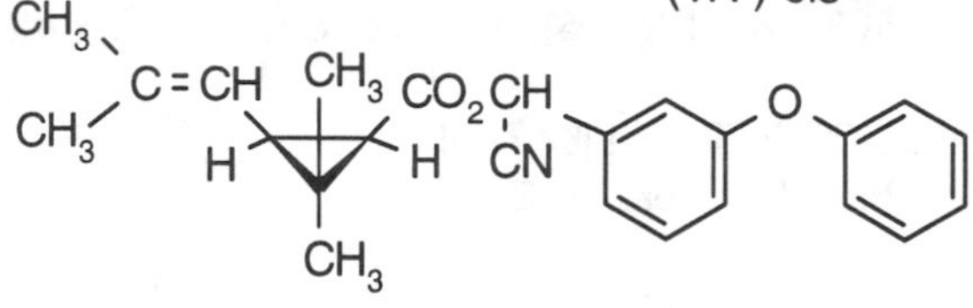

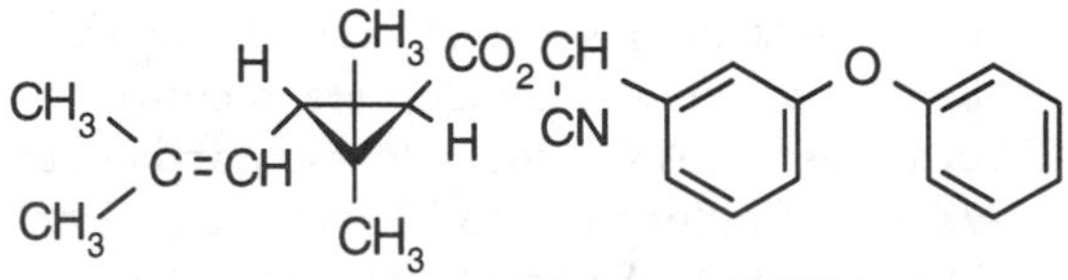

(1R)-trans -

NOMENCLATURE
Common name cyphenothrin (see composition) (BSI, draft E-ISO); cyphénothrine ((f) draft F-ISO)
IUPAC name (RS)-α-cyano-3-phenoxybenzyl (1RS,3RS;1RS,3SR)-2,2-dimethyl-= 3-(2-methylprop-1-enyl)cyclopropanecarboxylate. (See composition)
Roth: (RS)-α-cyano-3-phenoxybenzyl (1R)-*cis-trans*-2,2-dimethyl-= 3-(2-methylprop-1-enyl)cyclopropanecarboxylate
Chemical Abstracts name cyano(3-phenoxyphenyl)methyl 2,2-dimethyl-= 3-(2-methyl-1-propenyl)cyclopropanecarboxylate
CAS RN *[39515–40–7]* unstated stereochemistry **Development codes** S-2703 Forte (Sumitomo) **Official codes** OMS 3032

PHYSICAL CHEMISTRY
Composition The name cyphenothrin correctly applies to the (RS, 1RS)-*cis-trans*-mixed isomers. The material used is however rich (≥95%) in the (1R)- isomers, ≥75% *trans*-. **Mol. wt.** 375.5 **M.f.** $C_{24}H_{25}NO_3$ **Form** Viscous yellow liquid with a faint characteristic odour (tech.). **B.p.** 154 °C/0.1 mmHg **V.p.** 0.12 mPa (20 °C); 0.4 mPa (30 °C) **S.g./density** 1.08 (25 °C) **Solubility** In water <10 µg/l (25 °C). In hexane 4.84, methanol 9.27 g/100g (20 °C). **Stability** Stable for at least 2 years under normal storage conditions. Relatively stable to heat.
F.p. 130 °C

COMMERCIALISATION
History Insecticide reported by T. Matsuo *et al.* (*Agric. Biol. Chem.*, 1976, **40**, 247). Introduced by Sumitomo Chemical Co., Ltd. **Manufacturers** Sumitomo

APPLICATIONS

Mode of action Non-systemic insecticide with contact and stomach action. Acts on the nervous system of the insects, and has a rapid knock-down effect, in addition to a good residual effect. Also exhibits repellent properties. **Uses** Control of houseflies, midges and cockroaches in household use, public health, and industrial use. Also controls insects which attack wood and fabric. **Formulation types** AE; EC; UL; WP; HN. **Mixtures** *(cyphenothrin [(1R)-trans- isomers] +)* d-tetramethrin. **Compatibility** Incompatible with alkaline materials.
Selected tradenames 'Gokilaht' (Sumitomo)

ANALYSIS

Product analysis by glc. Details available from Sumitomo Chemical Co. Ltd.

MAMMALIAN TOXICOLOGY

Reviews See A. J. Gray & D. M. Soderlund, Chapt. 5 in *"Insecticides"*. **Oral** Acute oral LD_{50} for male rats 318, female rats 419 mg/kg (in corn oil).
Skin and eye Acute percutaneous LD_{50} for rats >5000 mg/kg. Non-irritating to skin and eyes. **Inhalation** LC_{50} (3 h) for rats >1850 mg/m^3.
Toxicity class WHO (a.i.) II; EPA (formulation) II

ENVIRONMENTAL FATE

Soil/Environment Degradation involves ester hydrolysis and oxidation.

89 cyproconazole *Fungicide*

azole

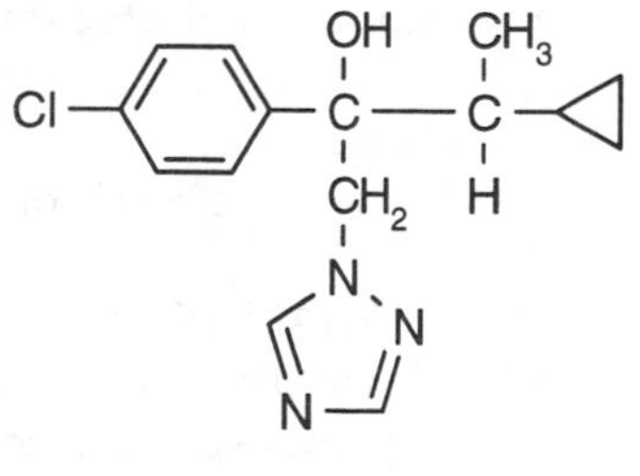

NOMENCLATURE

Common name cyproconazole (BSI, draft E-ISO, (m) draft F-ISO)
IUPAC name (2RS,3RS;2RS,3SR)-2-(4-chlorophenyl)-3-cyclopropyl-1-= (1H-1,2,4-triazol-1-yl)butan-2-ol
Chemical Abstracts name α-(4-chlorophenyl)-α-(1-cyclopropylethyl)-= 1H-1,2,4-triazol-1-ethanol

CAS RN *[94361–06–5]* for (2RS,3RS;2RS,3SR) isomers; *[113096–99–4]* for unstated stereochemistry; *[94361–07–6]* for (2RS,3SR) isomers
Development codes SAN 619 F

PHYSICAL CHEMISTRY
Composition It is a mixture (*c.* 1:1) of 2 diastereoisomers. **Mol. wt.** 291.8
M.f. $C_{15}H_{18}ClN_3O$ **Form** Colourless solid. **M.p.** 106–109 °C **B.p.** >250 °C
V.p. 3.46×10^{-2} mPa (20 °C) **K_{ow}** logP = 2.91 (pH 7) **S.g./density** 1.259 g/cm³
Solubility In water 140 mg/l (25 °C). In acetone 230, ethanol 230, dimethyl sulfoxide 180, xylene 120 (all in g/l, 25 °C). **Stability** Decomposition is <5% after storage for 2 years. Stable in aqueous solutions at pH 1 to 9 for a test period of 35 days (50 °C) or 14 days (80 °C). Slowly hydrolysed in 1N HCl and NaOH.
pKa No acid or base properties in range pH 3.5 to pH 10. **F.p.** No reaction up to 360 °C.

COMMERCIALISATION
History Fungicide reported by U. Gisi *et al.* (*Proc. Br. Crop Prot. Conf.*, 1986, **1**, 33; **2**, 857). Introduced in France and Switzerland (1989) by Sandoz AG (now Novartis Crop Protection AG). **Patent** US 4664696 **Manufacturers** Novartis

APPLICATIONS
Biochemistry Steroid demethylation inhibitor. **Mode of action** Systemic fungicide with protective, curative, and eradicant action. Absorbed rapidly by the plant, with translocation acropetally. **Uses** Systemic fungicide effective against a broad range of diseases. Cereal and sugar beet foliar diseases: *Septoria*, rust, powdery mildew, *Rhynchosporium*, *Cercospora*, *Ramularia* at 60–100 g a.i./ha. In fruit trees, vines, coffee, banana, turf, vegetables: *Venturia*, powdery mildew, rust, *Mycosphaerella*, *Monilia*, *Mycena*, *Sclerotinia*, *Rhizoctonia*. **Formulation types** SL; SC; WG; EW.
Mixtures (*cyproconazole +*) chlorothalonil; prochloraz; carbendazim; fentin acetate; sulfur; mancozeb; tridemorph; pyrazophos; captan; thiophanate-methyl; copper compounds. **Selected tradenames** 'Alto' (Novartis); 'Shandon' (Barclay)

ANALYSIS
Product analysis by glc or hplc. **Residues** determined by glc.

MAMMALIAN TOXICOLOGY
Oral Acute oral LD_{50} for male rats 1020, female rats 1333, male mice 200, female mice 218 mg/kg. **Skin and eye** Acute percutaneous LD_{50} for rats and rabbits >2000 mg/kg. Non-irritating to skin and eyes (rabbits). Non-irritating to skin and not a skin sensitiser (guinea pigs). **Inhalation** LC_{50} (4 h) for rats >5.65 mg/l air.
NOEL (1 y) for dogs 1 mg/kg b.w. daily; (2 y) for rats 1 mg/kg b.w. daily.
Other Not mutagenic in the Ames assay. **Toxicity class** WHO (a.i.) II
EC risk R22

ECOTOXICOLOGY

Birds Acute oral LD_{50} for Japanese quail 150 mg/kg. Eight-day dietary LC_{50} for Japanese quail 816, mallard ducks 1197 mg/kg diet. **Fish** LC_{50} (96 h) for carp 18.9, trout 19, bluegill sunfish 21 mg/l. **Bees** LD_{50} (contact) >0.1 mg/bee, (oral) >1 mg/bee. **Daphnia** LC_{50} (48 h) 26 mg/l.

ENVIRONMENTAL FATE

Animals In mammals, following oral administration, cyproconazole is rapidly absorbed, extensively metabolised and excreted, DT_{50} c. 30 h. No bioaccumulation. **Plants** In different crops, metabolic pathways are similar and the major residue is cyproconazole itself. **Soil/Environment** In soil, cyproconazole degrades moderately quickly, DT_{50} c. 3 mo. No accumulation and leaching potential. Soil adsorption, Freundlich K 4.1 (loam, pH 6.4, o.m. 2.3%), 16 (loamy sand, pH 5.1, o.m. 3.9%).

90   cyprodinil   *Fungicide*

anilinopyrimidine

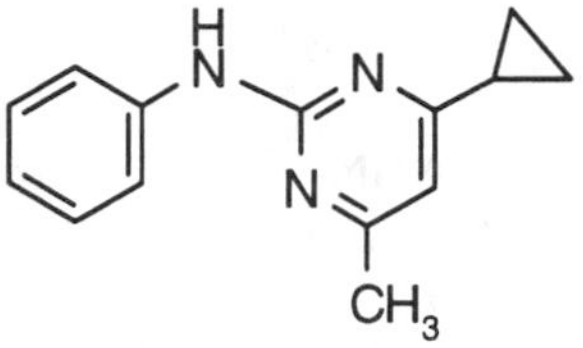

NOMENCLATURE

Common name cyprodinil (pa ISO)
IUPAC name 4-cyclopropyl-6-methyl-*N*-phenylpyrimidin-2-amine
Chemical Abstracts name 4-cyclopropyl-6-methyl-*N*-phenyl-2-pyrimidinamine
CAS RN *[121552–61–2]*   **Development codes** CGA 219417

PHYSICAL CHEMISTRY

Mol. wt. 225.3   **M.f.** $C_{14}H_{15}N_3$ **Form** Fine beige powder with a weak odour.
M.p. 75.9 °C   **V.p.** 5.1×10^{-1} mPa (crystal modification A); 4.7×10^{-1} mPa (crystal modification B) (both 25 °C) K_{ow} logP = 3.9 (pH 5.0), 4.0 (pH 7.0), 4.0 (pH 9.0) (25 °C)   **Henry** 6.6×10^{-3} to 7.2×10^{-3} Pa m^3 mol^{-1} (calc., depending on crystal modification)   **S.g./density** 1.21 (20 °C)   **Solubility** In water 20 (pH 5.0), 13 (pH 7.0), 15 (pH 9.0) (all mg/l, 25 °C). In ethanol 160, acetone 610, toluene 460, *n*-hexane 30, *n*-octanol 160 (all in g/l, 25 °C).   **Stability** Hydrolytically stable: DT_{50} in pH range 4–9 (25 °C) >>1 y. Photolysis DT_{50} in water 0.4–13.5 d.
pKa 4.44, weak base

COMMERCIALISATION
History Reported by U. J. Heye et al, (Proc. Br. Crop Prot. Conf., Pests Dis., 1994, **2**, 501). Developed and introduced by Ciba-Geigy AG (now Novartis Crop Protection AG). **Patent** EP 310550 **Manufacturers** Novartis

APPLICATIONS
Biochemistry Inhibitor of methionine biosynthesis. Cross-resistance with triazole, imidazole, morpholine, dicarboximide and phenylpyrrole fungicides is therefore unlikely. **Mode of action** Systemic product, with uptake into plants after foliar application and transport throughout the tissue and acropetally in the xylem. Inhibits penetration and mycelial growth both inside and on the leaf surface.
Uses As a foliar fungicide for use in cereals, grapes, pome fruit, stone fruit, strawberries, vegetables, field crops and ornamentals, and as a seed dressing on barley. Controls a wide range of pathogens such as *Pseudocercosporella herpotrichoides, Erysiphe* spp., *Pyrenophora teres, Rhynchosporium secalis, Septoria nodorum, Botrytis* spp., *Alternaria* spp., *Venturia* spp. *and Monilinia* spp.
Formulation types WG. **Mixtures** *(cyprodinil +)* fludioxonil; propiconazole; difenoconazole. **Selected tradenames** 'Chorus' (Novartis); 'Stereo' (Novartis); 'Switch' (mixture) (Novartis); 'Unix' (Novartis)

ANALYSIS
Product analysis by hplc. **Residues** determined by hplc. Details available from Novartis.

MAMMALIAN TOXICOLOGY
Oral Acute oral LD_{50} for rats >2000 mg/kg. **Skin and eye** Acute percutaneous LD_{50} for rats >2000 mg/kg. Non-irritant to eyes and skin (rabbits). Skin sensitiser (guinea pigs). **Inhalation** LC_{50} (4 h) for rats >1200 mg/m^3 air. **NOEL** (2 y) for rats 3 mg/kg b.w. daily; (1.5 y) for mice 196 mg/kg b.w. daily; (1 y) for dogs 65 mg/kg b.w. daily. **ADI** 0.03 mg/kg b.w. **Other** Non-mutagenic, non-teratogenic and non-oncogenic. **Toxicity class** WHO (a.i.) III (Table 5)

ECOTOXICOLOGY
Birds LD_{50} for mallard ducks and bobwhite quail >2000 mg/kg. Eight-day LC_{50} for mallard ducks and bobwhite quail >5200 ppm. **Fish** LC_{50} (96 h) for rainbow trout 0.98–2.41, carp 1.17, bluegill sunfish 1.07–2.17, sheepshead minnow 1.25, catfish 1.03 mg/l. **Bees** LD_{50} (48 h, oral) >316 μg/bee; LC_{50} (48 h, contact) >101 μg/bee. **Worms** LC_{50} (14 d) for *Eisenia foetida* 192 mg/kg.
Other beneficial spp. Harmless to *Poecillus cupreus* and beetles. **Daphnia** LC_{50} (48 h) 0.033–0.10 mg/l. **Algae** E_bC_{50} (72 h) for *Scenedesmus subspicatus* 0.75 mg/l. **Other aquatic spp.** EC_{50} for *Lemna gibba* 7.71 mg/l.

ENVIRONMENTAL FATE
Animals After oral administration, cyprodinil is rapidly absorbed and also rapidly

and almost completely eliminated with urine and faeces. The metabolic pathways are independent of sex, pre-treatment or dose level administered. Residues in tissues were generally low and there was no evidence for accumulation or retention of cyprodinil or its metabolites. **Plants** In tomatoes, the metabolism of cyprodinil proceeded mainly via hydroxylation of the 6-methyl group of the pyrimidine ring as well as hydroxylation of the phenyl and pyrimidine ring. **Soil/Environment** In soil, the compound dissipated with DT_{50} 20–60 d at normal soil humidities and soil temperatures, whereby formation of bound residues represents the major route for dissipation. In leaching and adsorption/desorption experiments, the compound proved to be immobile in soil, RMF ≤0.1. Photolytic DT_{50} in water 13.5 d.

91 cyromazine   *Insecticide*

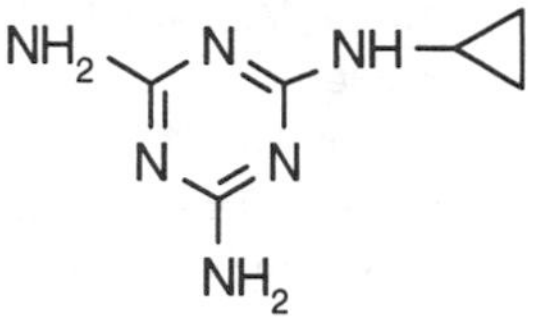

NOMENCLATURE
Common name cyromazine (BSI, draft E-ISO, (f) draft F-ISO, ANSI)
IUPAC name N-cyclopropyl-1,3,5-triazine-2,4,6-triamine
Chemical Abstracts name N-cyclopropyl-1,3,5-triazine-2,4,6-triamine
CAS RN [66215–27–8]   **Development codes** CGA 72 662
Official codes OMS 2014

PHYSICAL CHEMISTRY
Mol. wt. 166.2   **M.f.** $C_6H_{10}N_6$ **Form** Colourless crystals.   **M.p.** 224.9 °C
V.p. 4.48×10^{-4} mPa (25 °C) (OECD 104) K_{ow} logP = –0.061 (pH 7.0) (OECD 107)   **S.g./density** 1.35 (20 °C)   **Solubility** In water 13 g/l (OECD 105) (pH 7.1, 25 °C). In methanol 22, isopropanol 2.5, acetone 1.7, n-octanol 1.2, methylene chloride 0.25, toluene 0.015, hexane 0.0002 (all in g/kg, 20 °C).   **Stability** Stable up to 310 °C. Stable to hydrolysis for 28 days at up to 70 °C.   **pKa** 5.22, weak base

COMMERCIALISATION
History Insect growth regulator reported by R. D. Hall et al. (J. Econ. Entomol., 1980, **73**, 564) and by R. E. Williams et al. (Poultry Sci., 1980, **59**, 2207).

Introduced as an insecticide by Ciba-Geigy AG (now Novartis Crop Protection AG).   **Patent** GB 1587573; BE 857896   **Manufacturers** Novartis

APPLICATIONS
Mode of action Insect growth regulator with contact action, which interferes with moulting and pupation. When used on plants, action is systemic: applied to the leaves, it exhibits a strong translaminar effect; applied to the soil, it is taken up by the roots and translocated acropetally.   **Uses** Control of Diptera larvae in chicken manure by feeding to the poultry or treating the breeding sites. Used as a foliar spray to control leaf miners (*Liriomyza* spp.) in vegetables (e.g. celery, melons, tomatoes, lettuce), mushrooms, potatoes and ornamentals. Also used to control flies on animals.   **Formulation types** SC; SP; SG; SL; WP.   **Selected tradenames** 'Trigard' (for plant protection) (Novartis); 'Vetrazin' (for veterinary use) (Novartis); 'Neporex' (for veterinary use) (Novartis)

ANALYSIS
Product analysis by hplc.   **Residues** determined by glc or hplc.   Details available from Novartis.

MAMMALIAN TOXICOLOGY
Reviews *Pesticide residues in food – 1990. FAO Plant Production and Protection Paper, 102, 1990. Pesticide residues in food – 1990 evaluations. Toxicology.* World Health Organisation, WHO/PCS/91.47, 1991.   **Oral** Acute oral LD_{50} for rats 3387 mg/kg.   **Skin and eye** Acute percutaneous LD_{50} for rats >3100 mg/kg. Mild skin irritant; non-irritating to eyes (rabbits).   **Inhalation** LC_{50} (4 h) for rats >2.720 mg/l air.   **NOEL** (2 y) for rats 300, mice 1000 mg/kg diet.   **ADI** (JMPR) 0.02 mg/kg b.w. [1990].   **Toxicity class** WHO (a.i.) III (Table 5); EPA (formulation) III

ECOTOXICOLOGY
Birds Acute oral LD_{50} for bobwhite quail 1785, Japanese quail 2338, Pekin ducks >1000, mallard ducks >2510 mg/kg.   **Fish** LC_{50} (96 h) for bluegill sunfish >90, carp, catfish and rainbow trout >100 mg/l.   **Bees** Non-toxic to adult honeybees (no contact action up to 5 μg/bee).   **Daphnia** LC_{50} (48 h) >9.1 mg/l.

ENVIRONMENTAL FATE
Animals In rats, cyromazine is efficiently excreted, mainly as the parent compound.
Plants Rapidly metabolised in plants. The principal metabolite is melamine.
Soil/Environment Cyromazine and its main metabolite melamine are moderately mobile. Numerous studies conducted (laboratory and field) demonstrate that cyromazine is efficiently degraded by biological mechanisms.

aryloxyalkanoic acid

CI —⟨benzene ring⟩— OCH$_2$CO$_2$H
            |
            CI

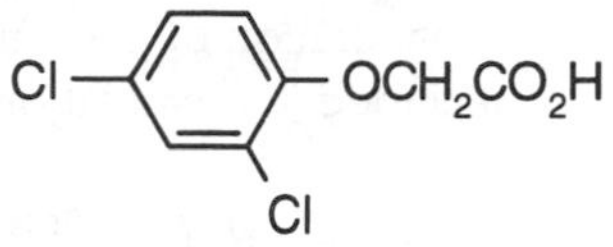

NOMENCLATURE
2,4-D
Common name 2,4-D (BSI, E-ISO, (m) F-ISO, WSSA); 2,4-PA (JMAF)
IUPAC name (2,4-dichlorophenoxy)acetic acid
Chemical Abstracts name (2,4-dichlorophenoxy)acetic acid
CAS RN [94–75–7] **EEC no.** 202–361–1

2,4-D-butotyl (2,4-D butoxyethyl ester)
CAS RN [1929–73–3]

2,4-D-butyl
CAS RN [94–80–4]

2,4-D-dimethylammonium
CAS RN [2008–39–1]

2,4-D-diolamine
CAS RN [5742–19–8]

2,4-D-2-ethylhexyl
CAS RN [1928–43–4]

2,4-D-isoctyl
CAS RN [25168–26–7] (formerly [1280–20–2])

2,4-D-isopropyl
CAS RN [94–11–1]

2,4-D-trolamine
CAS RN [2569–01–9]

PHYSICAL CHEMISTRY
2,4-D
Composition Tech. is ≥96% pure. **Mol. wt.** 221.0 **M.f.** C$_8$H$_6$Cl$_2$O$_3$
Form Colourless powder. **M.p.** 140.5 °C **V.p.** 1.1 × 10^{-2} mPa (20 °C)
K$_{ow}$ logP = 2.58–2.83 (pH 1) **S.g./density** 0.7–0.8 **Solubility** In water 311 mg/l

(pH 1, 25 °C); 0.6 g/l (20 °C). In ethanol 1250, diethyl ether 243, heptane 1.1, toluene 6.7, xylene 5.8 (all in g/kg, 20 °C). Insoluble in petroleum oils. Mono-*n*-butylamine salt: In water at 30 °C, 18 g/l.   **Stability** 2,4-D is a strong acid, and forms water-soluble salts with alkali metals and amines. Hard water leads to precipitation of the calcium and magnesium salts, but a sequestering agent is included in formulations to prevent this.   **pKa** 2.73

2,4-D-butotyl   (2,4-D butoxyethyl ester)
Mol. wt. 321.2   **M.f.** $C_{14}H_{18}Cl_2O_4$

2,4-D-butyl
Mol. wt. 277.1   **M.f.** $C_{12}H_{14}Cl_2O_3$

2,4-D-dimethylammonium
Mol. wt. 266.1   **M.f.** $C_{10}H_{13}Cl_2NO_3$ **M.p.** 85–87 °C   **Solubility** In water at 20 °C, 3 kg/l. Soluble in alcohols and acetone. Insoluble in kerosene and diesel oil.

2,4-D-diolamine
Mol. wt. 326.2   **M.f.** $C_{12}H_{17}Cl_2NO_5$

2,4-D-2-ethylhexyl
Mol. wt. 333.3   **M.f.** $C_{16}H_{22}Cl_2O_3$ **B.p.** >300 °C   **Solubility** In water <0.1 mg/l (25 °C).

2,4-D-isoctyl
Mol. wt. 333.3   **M.f.** $C_{16}H_{22}Cl_2O_3$ **Form** Yellowish-brown liquid with a phenolic odour.   **B.p.** 317 °C   **S.g./density** 1.14–1.17 g/ml (20 °C)   **Solubility** In water 10 mg/l.   **F.p.** 171 °C

2,4-D-isopropyl
Mol. wt. 263.1   **M.f.** $C_{11}H_{12}Cl_2O_3$ **Form** Colourless liquid.   **M.p.** 5–10 °C and 20–25 °C (two forms)   **B.p.** 130 °C/1 mmHg   **V.p.** 1.4 Pa (25 °C)
Solubility Practically insoluble in water; soluble in alcohols, most oils.

2,4-D-sodium
Mol. wt. 243.0   **M.f.** $C_8H_5Cl_2NaO_3$ **Solubility** In water 18 g/l (20 °C).

2,4-D-trolamine
Mol. wt. 370.2   **M.f.** $C_{14}H_{21}Cl_2NO_6$ **M.p.** 142–144 °C   **Solubility** In water 4.4 kg/l (30 °C).

COMMERCIALISATION
History The potent effects of its salts on plant growth were first described by P. W. Zimmerman & A. E. Hitchcock (*Contrib. Boyce Thompson Inst.*, 1942, **12**, 321),

and its early history is covered in *The Hormone Weedkillers*, C. Kirby (1980).
Manufacturers Aimco; Amvac; Ancom; Atul; Crystal; Defensa; Krishi Rasayan; Marks; Nissan; Nitrokémia; Nufarm Ltd; Rhône-Poulenc; Sanachem; SDS Biotech; Shenzhen Jiangshan; Uniroyal; United Phosphorus Ltd

APPLICATIONS
Mode of action Selective systemic herbicide. Salts are readily absorbed by the roots, whilst esters are readily absorbed by the foliage. Translocation occurs, with accumulation principally at the meristematic regions of shoots and roots. Acts as a growth inhibitor. **Uses** Post-emergence control of annual and perennial broad-leaved weeds in cereals, maize, sorghum, grassland, established turf, grass seed crops, orchards (pome fruit and stone fruit), cranberries, asparagus, sugar cane, rice, forestry, and on non-crop land (including areas adjacent to water) at 0.28–2.3 kg/ha. Control of broad-leaved aquatic weeds. The isopropyl ester can be also be used as a plant growth regulator to prevent premature fruit fall in citrus fruit. **Phytotoxicity** Phytotoxic to most broad-leaved crops, especially cotton, vines, tomatoes, ornamentals, fruit trees, oilseed rape and beet.

2,4-D
Formulation types EC; SL; SP; GR; SL. **Mixtures** *(2,4-D +)* amitrole; 2,4-DB; dichlorprop; MCPA; diuron; mecoprop; dalapon-sodium; simazine; ioxynil; bromoxynil; and many other herbicides. **Compatibility** Compatibility depends upon the particular formulation. **Selected tradenames** 'Agricorn D' (FCC); 'Capri' (Defensa); 'Dacamine' (ISK Biosciences); 'Damine' (Agriphar); 'Ded-Weed' (Uniroyal); 'Deferon' (Defensa); 'Desormone' (Rhône-Poulenc); 'Dikamin' (Nitrokémia); 'Dioweed' (United Phosphorus Ltd); 'Dymec' (PBI/Gordon); 'For-ester' (Vitax); 'Kay-D' (Krishi Rasayan); 'Novermone' (CFPI); 'Palormone' (Unicrop); 'Spritz-Hormin' (Nufarm B.V.); 'U 46 D' (BASF); 'Weedtox' (Aimco)

2,4-D-butotyl (2,4-D butoxyethyl ester)
Selected tradenames 'Erbitox LV-4' (Siapa)

2,4-D-dimethylammonium
Selected tradenames 'Erbitox Combi' (Siapa); 'Sanaphen D' (SL) (Sanachem); 'Staff' (Headland)

2,4-D-isoctyl
Formulation types EC. **Selected tradenames** 'Dicotox Extra' (Rhône-Poulenc); 'Supramone' (Rhône-Poulenc); 'Sanaphen D' (EC) (Sanachem)

ANALYSIS
Product analysis of 2,4-D, salts, esters and mixed combination products by acid-base titration, by glc (*CIPAC Handbook*, 1970, **1**, 241; 1980, **1A**, 1194; 1985, **1C**, 2060; 1994, **F**, 292–319; *Herbicides 1977*, pp. 6–21) or by hplc (*AOAC Methods*,

1995, 971.07, 976.03, 978.05, 984.07; *CIPAC Handbook*, 1985, **1C,** 2060; 1988, **D,** 51). FAO specification 1/TC/S/F (1992). Free phenol impurity determined by glc (*CIPAC Handbook*, 1994, **F,** 197), hplc (*ibid.*, 1994, **F,** 362) or electrochemically (*ibid.*, 1994, **F,** 368).   **Residues** determined by glc of derivatives (*Pestic Anal. Man.*, 1979, **I,** 201-D; *Anal. Methods Pestic. Plant Growth Regul.*, 1972, **6,** 630) or by hplc (M. Meier *et al.*, *Fresenius Z. Anal. Chem.*, 1989, **334,** 235). In **drinking water,** by conversion to methyl ester with diazomethane, then gc with ECD (*AOAC Methods*, 1995, 992.32, 10.7.03).

MAMMALIAN TOXICOLOGY
2,4-D

Reviews Pesticide residues in food. FAO Plant Production and Protection Series, No. 1; WHO Technical Report Series, No. 592, 1976. *1975 Evaluations of some pesticide residues in food.* AGP:1975/M/13; WHO Pesticide Residues Series no. 5, 1976. Toxicity and hazards to man, domestic animals and wildlife have been reviewed (J. M. Way, *Residue Rev.*, 1969, **26,** 37).   **IARC** 15   **EHC** 29 (WHO, 1984), 84 (WHO, 1989).   **Oral** Acute oral LD_{50} for rats 639–764, mice 138 mg/kg.   **Skin and eye** Acute percutaneous LD_{50} for rats >1600, rabbits >2400 mg/kg.   Skin and eye irritant (rabbits). Not a skin sensitiser (guinea pigs). **Inhalation** LC_{50} (24 h) for rats >1.79 mg/l   **NOEL** (2 y) for rats and mice 5 mg/kg b.w.; (1 y) for dogs 1 mg/kg b.w.   **ADI** (JMPR) 0.3 mg/kg b.w. [1975]. **Toxicity class** WHO (a.i.) II; EPA (formulation) II   **EC risk** Xn (R22); Xi (R36/37/38), (for salts and esters, Xn (R20/21/22))

2,4-D-2-ethylhexyl
ADI Due to toxicological equivalence to 2,4-D, ADI is assumed to be equivalent to that for 2,4-D.

2,4-D-isoctyl
Oral Acute oral LD_{50} for rats 650 mg/kg.   **Skin and eye** Acute percutaneous LD_{50} for rats >3000 mg/kg.   **NOEL** for rats 1250, dogs 500 mg/kg diet.

2,4-D-isopropyl
Oral Acute oral LD_{50} for rats 700 mg/kg.

2,4-D-sodium
Oral Acute oral LD_{50} for rats 666–805 mg/kg.

ECOTOXICOLOGY
2,4-D

Birds Acute oral LD_{50} for wild ducks >1000, Japanese quail 668, pigeons 668, pheasants 472 mg/kg. LC_{50} (96 h) for mallard ducks >5620 mg/l.   **Fish** Some formulations (e.g. esters) are toxic to fish, whilst others are not. LC_{50} (96 h) for rainbow trout >100 mg/l.   **Bees** Not toxic to bees; LD_{50} (oral) 104.5 µg/bee. **Daphnia** LC_{50} (21 d) 235 mg/l.

2,4-D-dimethylammonium
Fish LC_{50} (96 h) for rainbow trout 100 mg/l.

2,4-D-2-ethylhexyl
Fish LC_{50} greater than solubility in water.

2,4-D-isoctyl
Fish LC_{50} (96 h) for cut-throat trout 0.5–1.2 mg/l.

2,4-D-sodium
Birds Acute oral LD_{50} for wild ducks >2025 mg/kg.

ENVIRONMENTAL FATE
2,4-D
EHC 84 concludes that, when used as recommended, 2,4-D does not appear to produce direct toxic effects on any animal species. **Animals** In rats, following oral administration, elimination is rapid, and mainly as the unchanged substance. Following single doses of up to 10 mg/kg, excretion is almost complete after 24 hours, although, with higher doses, complete elimination takes longer. The maximum concentration in organs is reached after *c.* 12 hours. **Plants** In plants, metabolism involves hydroxylation, decarboxylation, cleavage of the acid side-chain, and ring opening. **Soil/Environment** In soil, microbial degradation involves hydroxylation, decarboxylation, cleavage of the acid side-chain, and ring opening. Half-life in soil <7 d. K_{oc} *c.* 60. For a review of environmental aspects of 2,4-D, see *Environmental Health Criteria 84* (WHO, 1989). Rapid degradation in the soil prevents significant downward movement under normal conditions.

2,4-D-2-ethylhexyl
Hydrolyses rapidly to the acid.

93 D2341 *Acaricide*

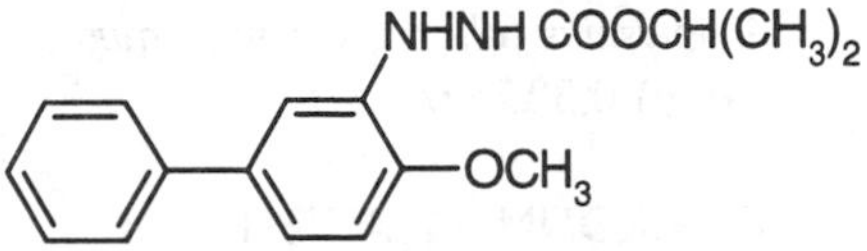

NOMENCLATURE
Common name bifenazate (BSI proposed)
IUPAC name isopropyl 3-(4-methoxybiphenyl-3-yl)carbazate

Chemical Abstracts name 1-methylethyl 2-(4-methoxy[1,1'-biphenyl]-3-yl)= hydrazinecarboxylate
CAS RN *[149877–41–8]*   **Development codes** D2341

PHYSICAL CHEMISTRY
Composition Tech. is >90% *m/m*.   **Mol. wt.** 300.4   **M.f.** $C_{17}H_{20}N_2O_3$
Form White, odourless crystals.   **M.p.** 120–124 °C   **V.p.** <1 × 10^{-2} mPa (25 °C)
K_{ow} logP = 3.4 (25 °C, pH 7)   **Henry** 1 × 10^{-3} Pa m^3 mol^{-1} **S.g./density** 1.31
Solubility In water 3.76 mg/l (20 °C).   **Stability** Stable for >1 y at 20 °C and
50% r.h. DT_{50} (hydrolysis) 20 h (25 °C, pH 7); DT_{50} (photolysis) 17 h (25 °C,
pH 5).   **F.p.** 110 °C

COMMERCIALISATION
History Reported by M. A. Dekeyser *et al.* (*Proc. Br. Crop Prot. Conf. – Pests Dis.*,
1996, **2**, 487). Introduced by Uniroyal Chemical Co., Inc in 1996.
Patent US 5367093; US 5438123   **Manufacturers** Uniroyal

APPLICATIONS
Mode of action Non-systemic acaricide with predominantly contact action and
long residual action.   **Uses** Under development by Uniroyal Chemical Co. for
control of phytophagous mites (both eggs and motile stages) on crops including
citrus, tree fruits, vines, hops, nuts, vegetables, ornamentals, cotton and maize.
Proposed use rates are 0.25–0.75 kg/ha.   **Formulation types** SC; WG; WP.

ANALYSIS
Product analysis by rp hplc.

MAMMALIAN TOXICOLOGY
Oral Acute oral LD_{50} for rats >5000 mg/kg.   **Skin and eye** Acute percutaneous
LD_{50} for rats >2000 mg/kg. Not a skin or eye irritant (rabbits).   **Inhalation** LC_{50}
for rats >4.4 mg/l.   **Other** Negative in Ames test, not teratogenic or mutagenic.

ECOTOXICOLOGY
Birds Acute oral LD_{50} for bobwhite quail 1142 mg/kg. Dietary LC_{50} (5 d) for
bobwhite quail 2298, mallard duck 726 mg/kg diet.   **Fish** LC_{50} (96 h) for bluegill
sunfish 0.58, rainbow trout 0.76 mg/l.   **Bees** LD_{50} (48 h, contact) 8.5 µg/bee.
Other beneficial spp. Harmless to predacious mites such as the phytoseiids
Amblyseius fallacis, Galendromus occidentalis and *Zetzellia mali*.   **Daphnia** EC_{50}
(48 h) 0.50 mg/l.

ENVIRONMENTAL FATE
Animals In animals, the product is considered of poor bioavailability and most of
the dose is excreted in the faeces. The small proportion that is absorbed
undergoes oxidation to the corresponding azo compound, and hydroxylation;

hydroxylated metabolites appear in the urine as sulfate or glucuronide conjugates. **Plants** Considered to be non-systemic; most residues stay on the surface and peel of the crops, where it is mostly not metabolised. Traces that penetrated the peel were metabolised as for animals.   **Soil/Environment** DT_{50} in aerobic soil is a few hours; DT_{50} (anaerobic) c. 80 d. Neither D2341 or its metabolites leached in a variety of soil types. DT_{50} in natural water 45 mins.

194 *Dacnusa sibirica*   *Biological agent*

Hymenoptera

Wasp parasite of leaf miners.

NOMENCLATURE
Scientific name *Dacnusa sibirica*

PROPERTIES
Stability Shelf life 2–3 d at 8–10 °C in a dark place.

COMMERCIALISATION
Manufacturers BCP; Biobest ; Koppert ; Neudorff; Novartis BCM

APPLICATIONS
Mode of action The parasitic wasp lays an egg inside the leaf miner larva; the egg develops and an adult wasp emerges from the leaf miner pupa.   **Uses** *Dacnusa sibirica* is used as an alternative to *Diglyphus sibirica* (q.v.), depending on the season. Koppert products 'Minex', 'Minusa' and 'Miglyphus' may consist of *Dacnusa isaea* and/or *Diglyphus isaea*.   **Phytotoxicity** Non-phytotoxic and non-phytopathogenic.   **Mixtures** *(D. sibirica +) Diglyphus isaea*.
Selected tradenames 'Dac-line s' (Novartis BCM); 'Minex' (Koppert); 'Minusa' (Koppert)

MAMMALIAN TOXICOLOGY
Non-toxic.

$$\text{Ph}-\underset{\underset{CH_3}{|}}{\overset{\overset{CH_3}{|}}{C}}-NHCONH-\text{C}_6\text{H}_4-CH_3$$

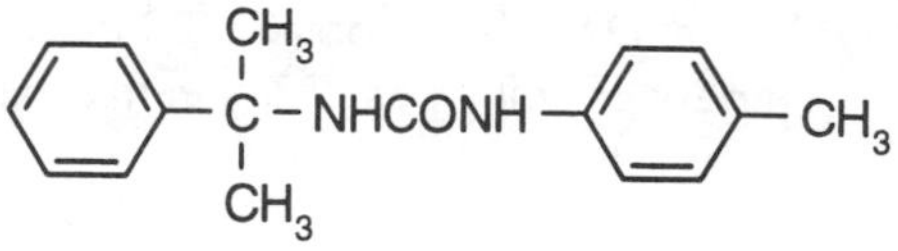

NOMENCLATURE

Common name daimuron (BSI, draft E-ISO, draft (*m*) F-ISO); dymron (JMAF)
IUPAC name 1-(1-methyl-1-phenylethyl)-3-*p*-tolylurea; 1-(α,α-dimethylbenzyl)-= 3-*p*-tolylurea
Chemical Abstracts name *N*-(4-methylphenyl)-*N*'-(1-methyl-1-phenylethyl)urea
CAS RN *[42609–52–9]* **Development codes** K-223; SK-23

PHYSICAL CHEMISTRY

Mol. wt. 268.4 **M.f.** $C_{17}H_{20}N_2O$ **Form** Colourless, odourless needles.
M.p. 203 °C **V.p.** 4.53×10^{-4} mPa (25 °C) **K_{ow}** logP = 2.7 **S.g./density** 1.108
(20 °C) **Solubility** In water 1.2 mg/l (20 °C). In acetone 16, methanol 10,
benzene 0.5, hexane 0.03 (all in g/l, 20 °C). **Stability** Stable to heat and light.
Stable between pH 4 and pH 9.

COMMERCIALISATION

History Herbicide reported by M. Tashiro (*Jpn. Pestic. Inf.*, 1979, No. 36, 40).
Introduced in Japan (1975) by Showa Denko KK (now SDS Biotech KK).
Patent JP 7335454 **Manufacturers** SDS Biotech

APPLICATIONS

Mode of action Selective herbicide, absorbed mainly by the roots, with rapid
translocation to the meristem. **Uses** Pre-emergence and early post-emergence
control of cyperaceous weeds and annual grass weeds in paddy rice at 0.45–2 kg
a.i./ha. **Phytotoxicity** Non-phytotoxic to rice. **Formulation types** WP; GR; SC.
Mixtures (*daimuron* +) imazosulfuron; bensulfuron-methyl; pretilachlor; mefenacet;
thenylchlor; cafenstrole; pyributicarb; pentoxazone.
Selected tradenames 'Showrone' (SDS Biotech)

ANALYSIS

Product and **residue** analysis by hplc and glc. Details are available from SDS
Biotech.

MAMMALIAN TOXICOLOGY

Oral Acute oral LD_{50} for rats and mice >5000 mg/kg. **Skin and eye** Acute
percutaneous LD_{50} for rats >2000 mg/kg. **Inhalation** LC_{50} (4 h) for rats 3250
mg/m^3. **NOEL** (1 y) for male dogs 30.6 mg/kg diet; (90 d) for male rats

3.118, female rats 3.430, mice 6.615 mg/kg daily. No ill-effects observed in a 3-generation study on rats. Non-teratogenic at 10 000 mg/kg diet. **ADI** 0.3 mg/kg. **Toxicity class** WHO (a.i.) III (Table 5)

ECOTOXICOLOGY
Birds Acute oral LD_{50} for bobwhite quail >2000 mg/kg. **Fish** LC_{50} (48 h) for carp >40 ppm. **Daphnia** LC_{50} (3 h) for *D. pulex* >40 ppm.

ENVIRONMENTAL FATE
Soil/Environment DT_{50} in paddy soil *c.* 50 d.

196 dalapon *Herbicide*

halogenated alkanoic acid

$$CH_3CCl_2CO_2H$$

NOMENCLATURE
dalapon
Common name dalapon (BSI, ANSI, WSSA, Canada, Germany, (*m*) France); proprop (Republic of South Africa); DPA (JMAF); 2,2-dichloropropionic acid (used in most other countries)
IUPAC name 2,2-dichloropropionic acid
Chemical Abstracts name 2,2-dichloropropanoic acid
CAS RN *[75–99–0]* **EEC no.** 200–923–0

dalapon-sodium
CAS RN *[127–20–8]*

PHYSICAL CHEMISTRY
dalapon
Mol. wt. 143.0 **M.f.** $C_3H_4Cl_2O_2$ **Form** Colourless liquid. **B.p.** 185–190 °C **V.p.** 0.01 mPa (20 °C) **Stability** Subject to hydrolysis; slight at 25 °C but comparatively rapid ≥50 °C; so aqueous solutions should not be kept for any length of time. Alkali causes dehydrochlorination above 120 °C. **pKa** 1.74–1.84, depending on the ionic strength of the solution.

dalapon-sodium
Mol. wt. 165.0 **M.f.** $C_3H_3Cl_2NaO_2$ **Form** Pale-coloured, hygroscopic powder.
M.p. Decomposes >191 °C **Solubility** In water 900 g/kg (25 °C). In ethanol 185, methanol 179, acetone 1.4, benzene 0.02, diethyl ether 0.16 (all in g/kg, 25 °C).

COMMERCIALISATION

History The sodium salt introduced as a herbicide by Dow Chemical Co. (now DowElanco) and by BASF AG (both no longer manufacture or market it). **Patent** US 2642354 to Dow

APPLICATIONS

Biochemistry Acts by precipitation of protein, which interferes with production of pantothenic acid. **Mode of action** Selective systemic herbicide, absorbed by the leaves and roots, with translocation via both the symplastic and apoplastic systems throughout the plant. **Uses** Dalapon-sodium is used for control of annual and perennial grasses (including couch grass) on non-crop land (embankments, roadside verges, industrial sites, railway tracks, irrigation channels, ditches, etc.), in orchards (pome fruit, stone fruit, bush fruit, citrus fruit, nuts), olive groves, vineyards, forestry, bananas, sugar cane, rhubarb, asparagus, potatoes, peas, soya beans, beet, oilseed rape, flax, maize, sorghum, coffee, tea, rubber, cotton, and ornamental shrubs. Control of reeds, sedges, rushes, halophytes, and semi-aquatic grass weeds in water courses. Application rates range from 37 kg/ha (non-crop areas) to 2–5 kg/ha (citrus). **Formulation types** SP; GR; WP. **Mixtures** (dalapon-sodium +) dichlobenil; atrazine; dalapon-magnesium; 2,4-D + diuron. **Compatibility** Should not be used in combination with oils or contact herbicides, as activity will be diminished due to reduction in translocation. **Selected tradenames** 'Dalacide' (Diachem); 'Diserbo Canali' (Siapa)

ANALYSIS

Product analysis (dalapon or salts) is by decomposition of a complex formed with mercury(II) nitrate and copper(II) nitrate and subsequent titration with acid (*CIPAC Handbook*, 1970, **1**, 274; 1980, **1A**, 1197; *AOAC Methods*, 1995, 962.06), by hplc (*ibid.*, 984.06; *CIPAC Handbook*, 1988, **1D**, 54) or by glc of a suitable ester (*Herbicides*, 1977, p. 45). **Residue** analysis is by glc of a suitable ester (G. N. Smith & E. H. Yonkers, *Anal. Methods Pestic., Plant Growth Regul. Food Addit.*, 1964, **4**, 79; *Anal. Methods Pestic. Plant Growth Regul.*, 1972, **6**, 621; *Rückstandsanal. Pflanzenschutzmitteln*, 28–1, 28–6).

MAMMALIAN TOXICOLOGY

dalapon-sodium

Reviews The toxic effects of dalapon have been reviewed (E. E. Kenaga, *Residue Rev.*, 1974, **53**, 104). **Oral** Acute oral LD_{50} for male rats 9330, female rats 7570, female mice >4600, female guinea pigs 3860, female rabbits 3860, cattle >4000 mg/kg. **Skin and eye** Acute percutaneous LD_{50} for rabbits >2000 mg/kg 85% formulation. Moderately irritating to skin and eyes (rabbits). **Inhalation** LC_{50} (8 h) >20 mg/l of 25% solution of formulated product. **NOEL** In 2 y feeding trials, no effect was observed in rats receiving 15 mg/kg daily, there was slight increase in kidney weight at 50 mg/kg daily. **Toxicity class** WHO (a.i.) III (Table 5); EPA (formulation) II **EC risk** Xn (R22); Xi (R38, R41)

ECOTOXICOLOGY
dalapon-sodium
Birds Acute oral LD_{50} for chickens 5660 mg/kg. Five-day dietary LC_{50} for mallard ducks, Japanese quail, and pheasants >5000 mg/kg diet. **Fish** LC_{50} (96 h) for rainbow trout, goldfish, and channel catfish >100, carp >500, guppies >1000 mg/l. **Bees** Non-toxic to honeybees.

ENVIRONMENTAL FATE
Animals In mammals, following oral administration, dalapon is rapidly eliminated. In dogs, following a single oral dose of 500 mg, 65–70% is excreted within 2 hours.
Plants In plants, dalapon-sodium does not undergo significant degradation.
Soil/Environment In soil, readily undergoes microbial degradation involving dechlorination and liberation of CO_2 (B. E. Day et al., *Soil Sci.*, 1963, **95**, 326). Following an application rate of 22 kg/ha, duration of residual activity in soil is *c.* 3–4 months.

97 daminozide *Plant growth regulator*

$$(CH_3)_2NNHCOCH_2CH_2CO_2H$$

NOMENCLATURE
Common name daminozide (BSI, E-ISO, (f) F-ISO, ANSI)
IUPAC name *N*-dimethylaminosuccinamic acid
Chemical Abstracts name butanedioic acid mono(2,2-dimethylhydrazide)
Other names SADH **CAS RN** *[1596–84–5]* **EEC no.** 216–485–9
Development codes B-995

PHYSICAL CHEMISTRY
Composition Tech. grade daminozide is >99% pure. **Mol. wt.** 160.2
M.f. $C_6H_{12}N_2O_3$ **Form** Tech. is a white powder with a faint amine-like odour.
M.p. 157–164 °C **V.p.** 22.7 mPa (23 °C) K_{ow} logP = –1.50 (pH 5, 7 and 9, 21 °C) **Solubility** In distilled water 100 g/kg (25 °C). In methanol 50, acetone 25 (both in g/kg, 25 °C). Insoluble in hydrocarbons. **Stability** No measurable hydrolysis over 30 d at pH 5, 7 and 9. Hydrolysed by acids and alkalis on heating. Solutions are slowly decomposed by light. **pKa** 4.68 (20 °C)

COMMERCIALISATION
History Plant growth regulator reported by J. A. Riddell et al. (*Science*, 1962, **136**, 391). Introduced by Uniroyal Chemical Co., Inc. **Patent** US 3240799; US 3334991
Manufacturers Fine; Uniroyal

APPLICATIONS
Biochemistry Interferes with gibberellic acid biosynthesis.
Mode of action Absorbed by the leaves, with translocation throughout the plant.
Uses To produce more compact plants (by inhibition of internodal elongation) of chrysanthemums, azaleas, hydrangeas, poinsettias, and other ornamentals.
Formulation types SP. **Compatibility** Incompatible with wetting agents, alkaline materials, oils, and copper-containing compounds. **Selected tradenames** 'Alar' (Uniroyal); 'B-Nine' (Uniroyal)

ANALYSIS
Product analysis by hplc. Details available from Uniroyal. **Residues** determined by glc of a derivative (J. R. Lane, *Anal. Methods Pestic., Plant Growth Regul. Food Addit.*, 1967, **5**, 499) or by alkaline hydrolysis to liberate 1,1-dimethylhydrazine which is determined colorimetrically (*Anal. Methods Pestic. Plant Growth Regul.*, 1972, **6**, 697; V. P. Lynch, *ibid.*, 1976, **8**, 491).

MAMMALIAN TOXICOLOGY
Reviews *Pesticide residues in food – 1991.* FAO Plant Production and Protection Paper, 111, 1991. *Pesticide residues in food – 1991 evaluations. Part II – Toxicology.* World Health Organisation, WHO/PCS/92.52, 1992. **Oral** Acute oral LD_{50} for rats >5000 mg/kg. **Skin and eye** Acute percutaneous LD_{50} for rabbits >5000 mg/kg. **Inhalation** LC_{50} (4 h) for rats >2.1 mg/l air. **NOEL** (1 y) for dogs 188 mg/kg b.w. daily, for rats 5 mg/kg b.w. daily. Non-oncogenic in 2 y feeding trials in rats and mice at 10 000 mg/kg diet. NOEL for teratogenicity and embryotoxicity in rabbits was 300 mg/kg diet. In 3-generation reproductive study, NOAEL for rats was 50 mg/kg b.w. daily. Non-mutagenic in *in-vivo* test. **ADI** (JMPR) 0.5 mg/kg b.w., for daminozide containing <30 mg 1,1-dimethyl= hydrazine/kg [1991]. **Toxicity class** WHO (a.i.) III (Table 5); EPA (formulation) IV **EC risk** (R40)

ECOTOXICOLOGY
Birds Eight-day dietary LC_{50} for mallard ducks and bobwhite quail >10 000 mg/kg diet. **Fish** LC_{50} (96 h) for rainbow trout 149, bluegill sunfish 423 mg/l. **Bees** Not toxic to bees; LD_{50} >100 µg/bee (85% formulation). **Worms** LC_{50} for earthworms >632 ppm. **Daphnia** LC_{50} (96 h) 76 mg/l. **Algae** EC_{50} for *Chlorella* 180 ppm.

ENVIRONMENTAL FATE
Plants Metabolites include 1,1-dimethylhydrazine. **Soil/Environment** Daminozide quickly dissipates in soil with most residues becoming bound, with some CO_2 formation. In aerobic soil, half of the daminozide is no longer recoverable after 17 h. No 1,1-dimethylhydrazine (UDMH) or *N*-nitrosodimethylamine were detected. In soil under anaerobic conditions, half of the daminozide had dissipated in 7.5 d. The major products were bound residues and CO_2. In a field study on

bare ground, about 90% of the daminozide had disappeared after 7 d. Neither hydrolysis nor photolysis are significant pathways for the degradation of daminozide.

98 dazomet

Nematicide, fungicide, herbicide, insecticide

methyl isothiocyanate precursor

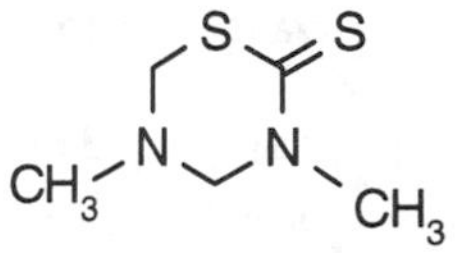

MAIN ENTRIES

NOMENCLATURE
Common name dazomet (BSI, E-ISO, (m) F-ISO, WSSA, JMAF); tiazon (USSR); [DMTT] (former WSSA name)
IUPAC name 3,5-dimethyl-1,3,5-thiadiazinane-2-thione; tetrahydro-3,5-dimethyl-= 1,3,5-thiadiazine-2-thione
Chemical Abstracts name tetrahydro-3,5-dimethyl-2H-1,3,5-thiadiazine-2-thione
CAS RN [533–74–4] **EEC no.** 208–576–7
Development codes N-521 (Stauffer Chemical Co.); Crag Fungicide 974 (Rhône-Poulenc); BAS 00 201N

PHYSICAL CHEMISTRY
Composition ≥94% pure. **Mol. wt.** 162.3 **M.f.** $C_5H_{10}N_2S_2$ **Form** Colourless crystals. **M.p.** 104–105 °C (decomp. - tech. grade). **V.p.** 0.37 mPa at 20 °C **K**$_{ow}$ logP = 0.15 (pH 7) **S.g./density** 1.37 (tech.) **Solubility** In water 3 g/kg (20 °C). In cyclohexane 400, chloroform 391, acetone 173, benzene 51, ethanol 15, diethyl ether 6 (all in g/kg, 20 °C). **Stability** Stable at temperatures up to 35 °C. Sensitive to temperatures >50 °C, and to moisture. Hydrolysed in acidic media to carbon disulfide, formaldehyde, and methylamine.

COMMERCIALISATION
History Originally prepared by M. Delépine (*Bull. Soc. Chim. Fr.,* 1897, **15**, 891), later introduced as a soil fumigant. **Manufacturers** BASF

APPLICATIONS
Biochemistry Non-selective inhibition of enzymes by degradation products.
Mode of action A pre-planting soil fumigant, acting by decomposition to methyl

isothiocyanate. **Uses** Soil sterilant, applied prior to planting out crops. Controls soil fungi (*Fusarium*, *Pythium*, *Rhizoctonia*, *Sclerotinia* and *Verticillium* spp. and *Colletotrichum coccodes (atramentarium)*), nematodes, germinating weed seeds, and soil insects. Also used as a slimicide in pulp and paper manufacture, and as a preservative in adhesives and glues. **Phytotoxicity** Highly phytotoxic to all green plants. Treated soils should not be planted until shown to be free of dazomet and its degradation products. **Formulation types** GR; DP; WP.
Selected tradenames 'Basamid' (BASF); 'Dazoberg' (Diachem)

ANALYSIS
Product analysis by acid hydrolysis, absorption of the carbon disulfide produced and iodometric titration (*CIPAC Handbook*, 1985, **1C**, 2074; H. A. Stansbury, *Anal. Methods Pestic., Plant Growth Regul. Food Addit.*, 1964, **3**, 119). **Residues** of parent compound in soil by hplc; of methyl isothiocyanate in crops by glc. Details available from BASF.

MAMMALIAN TOXICOLOGY
Oral Acute oral LD_{50} for rats 520, male mice 455, female mice 710, rabbits 320–620 mg/kg. **Skin and eye** Acute percutaneous LD_{50} for rats >2000 mg/kg; DP formulation is a skin and eye irritant to rabbits. **Inhalation** LC_{50} (4 h) for rats 8.4 mg/l air. **NOEL** in rats 20 mg/kg diet. **Other** Not oncogenic.
Toxicity class WHO (a.i.) III; EPA (formulation) III **EC risk** Xn (R22); Xi (R36)

ECOTOXICOLOGY
Birds LD_{50} for bobwhite quail 415 mg/kg b.w. LC_{50} for bobwhite quail 1850, mallard duck >5000 mg/kg diet. **Fish** LC_{50} (96 h) 0.16 mg/l. **Bees** Not toxic to bees when used as directed. **Daphnia** EC_{50} (48 h) 0.3 mg/l. **Algae** EC_{50} (96 h) for *Scenedesmus subspicatus* 1.0 mg/l. **Other aquatic spp.** EC_{10} (17 h) for *Pseudomonas putida* 1.8 mg/l.

ENVIRONMENTAL FATE
Soil/Environment In soil, in the presence of moisture, undergoes degradation to methyl(methylaminomethyl)dithiocarbamic acid, which then undergoes further degradation to methyl isothiocyanate, formaldehyde, hydrogen sulfide, and methylamine. See N. Drescher & S. Otto, *Residue Rev.*, 1968, **23**, 49.

Herbicide

aryloxyalkanoic acid

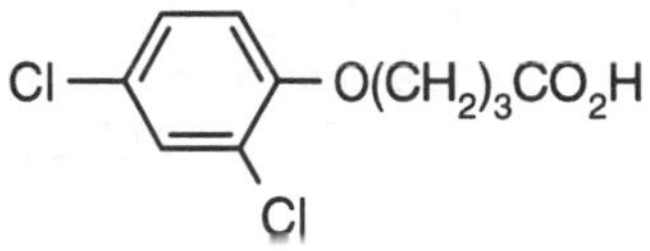

NOMENCLATURE
2,4-DB
Common name 2,4-DB (BSI, WSSA, draft E-ISO, (*m*) draft F-ISO)
IUPAC name 4-(2,4-dichlorophenoxy)butyric acid
Chemical Abstracts name 4-(2,4-dichlorophenoxy)butanoic acid
CAS RN *[94–82–6]*   **EEC no.** 202–366–9   **Development codes** M&B 2878

2,4-DB-butyl
CAS RN *[6753–24–8]*

2,4-DB-dimethylammonium
CAS RN *[2758–42–1]*

2,4-DB-isoctyl
CAS RN *[1320–15–6]*

2,4-DB-potassium
CAS RN *[19480–40–1]*

2,4-DB-sodium
CAS RN *[10433–59–7]*

PHYSICAL CHEMISTRY
2,4-DB
Mol. wt. 249.1   **M.f.** $C_{10}H_{10}Cl_2O_3$ **Form** Colourless crystals.   **M.p.** 117–119 °C
V.p. Negligible   **Solubility** In water at 25 °C, 46 mg/l. Readily soluble in acetone,
ethanol, and diethyl ether. Slightly soluble in benzene, toluene, and kerosene. The
alkali-metal and amine salts are readily soluble in water.   **Stability** Acidic in
reaction, and forms water-soluble alkali-metal and amine salts, but hard water
precipitates the calcium and magnesium salts. The acid and salts are very stable.
The esters are sensitive to acids and alkalis.   **pKa** 4.8

2,4-DB-butyl
Mol. wt. 305.2   **M.f.** $C_{14}H_{18}Cl_2O_3$ **V.p.** Substantial

2,4-DB-dimethylammonium
Mol. wt. 294.2   **M.f.** $C_{12}H_{17}Cl_2NO_3$

2,4-DB-isoctyl
Mol. wt. 361.3   **M.f.** $C_{18}H_{26}Cl_2O_3$

2,4-DB-potassium
Mol. wt. 287.2   **M.f.** $C_{10}H_9Cl_2KO_3$

2,4-DB-sodium
Mol. wt. 271.1   **M.f.** $C_{10}H_9Cl_2NaO_3$ **Solubility** In water >200 g/l (25 °C).

COMMERCIALISATION
History Its plant growth regulating properties reported by M. E. Synerholm & P. W. Zimmerman (*Contrib. Boyce Thompson Inst.*, 1947, **14**, 369). Introduced as a herbicide by May & Baker Ltd (now Rhône-Poulenc Agrochimie).
Patent CA 570065   **Manufacturers** Cedar; Marks; Rhône-Poulenc; United Phosphorus Ltd

APPLICATIONS
Biochemistry More selective than 2,4-D because its activity is dependent on β-oxidation to the latter within the plant.   **Mode of action** Selective, systemic, hormone-type herbicide, absorbed by the foliage, with translocation.   **Uses** Post-emergence control of many annual and perennial broad-leaved weeds in alfalfa, clovers, undersown cereals, grassland, forage legumes, soya beans, and peanuts at 1.5–3.0 kg a.i./ha.   **Phytotoxicity** Soya beans infested with *Phytophthora* should not be treated, as injury may occur.   **Formulation types** EC; SL.
Mixtures (*2,4-DB +*) MCPA; benazolin; mecoprop; benazolin + MCPA; linuron + MCPA; benazolin + cyanazine.   **Selected tradenames** 'Embutox' (Rhône-Poulenc); 'Butormone' (Unicrop); 'Butoxone' (Cedar); 'DB Straight' (United Phosphorus Ltd)

ANALYSIS
Product analysis by glc after conversion to suitable esters (*CIPAC Proc.*, 1981, **3**, 170; 1994, **F**, 292–319; W. H. Gutenmann, *Anal. Methods Pestic., Plant Growth Regul. Food Addit.*, 1967, **5**, 369). Free phenol impurity determined by glc (*CIPAC Handbook*, 1994, **F**, 197), or hplc (*ibid.*, 1994, **F**, 362).   **Residues** determined by glc of suitable esters (*Anal. Methods Pestic. Plant Growth Regul.*, 1972, **6**, 630, 636) or by hplc (Di-Corcia *et al.*, *Anal. Chem.*, 1989, **61**, 1363). In **drinking water**, by conversion to methyl ester with diazomethane, then gc with ECD (*AOAC Methods*, 1995, 992.32, 10.7.03).

MAMMALIAN TOXICOLOGY
2,4-DB
Oral Acute oral LD_{50} for rats 370–700 mg/kg.   **Toxicity class** WHO (a.i.) III; EPA (formulation) III   **EC risk** Xn (R21/22)

2,4-DB-potassium
EC risk Salts are Xn (R20/21/22)

2,4-DB-sodium
Oral Acute oral LD_{50} for rats 1500, mice 400 mg/kg.

ECOTOXICOLOGY
2,4-DB
Fish LC_{50} (96 h) for rainbow trout *c.* 4 mg/l (amine salt and esters). See also Statham *et al., Toxicol. Appl. Pharmacol.*, **36**, 281.   **Bees** Not toxic to bees.

ENVIRONMENTAL FATE
Plants In susceptible plants, rapidly undergoes β-oxidation to 2,4-D, which is then degraded to 2,4-dichlorophenol, followed by ring hydroxylation and ring opening. In tolerant plants, β-oxidation to 2,4-D is very slow.   **Soil/Environment** In soil, microbial degradation leads to the formation of 2,4-D, which subsequently undergoes further degradation. Half-life in soil <7 d (*Weed Res.* 1978, **18**, 275).

00 DCIP   *Nematicide*

$$[ClCH_2CH(CH_3)]_2O$$

NOMENCLATURE
Common name DCIP (JMAF)
IUPAC name bis(2-chloro-1-methylethyl) ether
Chemical Abstracts name 2,2'-oxybis[1-chloropropane]
CAS RN *[108–60–1]*   **Development codes** IK-141 (SDS Biotech)

PHYSICAL CHEMISTRY
Mol. wt. 171.1   **M.f.** $C_6H_{12}Cl_2O$ **Form** Light brown oily liquid.   **B.p.** 187 °C/760 mmHg   **V.p.** 74.6 Pa (20 °C)   **K_{ow}** logP = 1.86   **S.g./density** 1.1135   **Solubility** In water 1.7 g/l.   **Stability** Stable to heat, light and water.   **F.p.** 87 °C

COMMERCIALISATION
History Nematicide introduced by SDS Biotech.   **Manufacturers** SDS Biotech

APPLICATIONS
Uses Control of nematodes and citrus ground mealybug in vegetables, fruit, tobacco, ornamentals, tea, and mulberries. **Formulation types** GR; EC.
Mixtures *(DCIP +)* chloropicrin. **Selected tradenames** 'Nemamort' (SDS Biotech)

ANALYSIS
Product and **residue** analysis by glc.

MAMMALIAN TOXICOLOGY
Reviews Nouyaku Jihou (March 15, 1993). **Oral** Acute oral LD_{50} for male rats 503, male mice 599, female mice 536 mg/kg. **Skin and eye** Acute percutaneous LD_{50} for rats >2000 mg/kg. **Inhalation** LC_{50} (4 h) for mice and rats 12.8 mg/l.
NOEL (2 y) for male rats 13.4 mg/kg diet. **ADI** 0.13 mg/kg.

ECOTOXICOLOGY
Fish LC_{50} (48 h) for carp >40 mg/l. **Daphnia** LC_{50} (3 h) >40 mg/l.

---

201 DDT   *Insecticide*

---

organochlorine

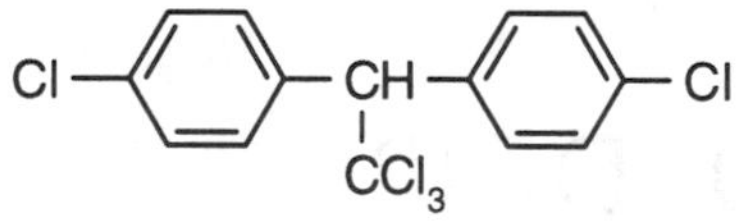

NOMENCLATURE
Common name DDT (BSI, E-ISO, *(m)* F-ISO, ESA, JMAF); zeidane (*(m)* France); dicophane (BAN); chlorophenothane (US Pharmacopoeia, for a mixture of isomers); pp'-DDT (BSI, draft E-ISO, *(m)* draft F-ISO, for the major component); pp' zeidane (France, for the major component); para,para'-DDT (Canada, for the major component)
IUPAC name of major component 1,1,1-trichloro-2,2-bis(4-chlorophenyl)ethane; 1,1,1-trichloro-di-(4-chlorophenyl)ethane; for mixture 1,1,1-trichloro-2,2-bis= (chlorophenyl)ethane
Chemical Abstracts name 1,1'-(2,2,2-trichloroethylidene)bis[4-chlorobenzene]
Other names dichlorodiphenyltrichloroethane   **CAS RN** *[50–29–3]*
EEC no. 200–024–3   **Official codes** OMS 16; ENT 1506

PHYSICAL CHEMISTRY
Composition Tech. product contains ≤ 30% *op'*-DDT [1,1,1-trichloro-2-(2-= chlorophenyl)-2-(4-chlorophenyl)ethane] which, being of insecticidal value, is not

usually removed. **Mol. wt.** 354.5 **M.f.** $C_{14}H_9Cl_5$ **Form** Colourless crystals
(*pp'*-DDT); tech., waxy solid. **M.p.** Indefinite (tech.); 108.5–109 °C (*pp'*-DDT)
B.p. 185–187 °C/0.05 mmHg (decomp.) (*pp'*-DDT) **V.p.** 0.025 mPa
(20 °C; *pp'*-DDT) **Solubility** Practically insoluble in water. Readily soluble in
aromatic and chlorinated solvents; moderately soluble in polar organic solvents
and petroleum oils. In cyclohexanone 1000, dioxane 1000, dichloromethane 850,
benzene 770, trichloroethylene 720, xylene 600, acetone 500, chloroform 310,
diethyl ether 270, ethanol 60, methanol 40 (all in g/l, 27 °C). **Stability** *pp'*-DDT
undergoes dehydrochlorination in alkaline solutions and at temperatures above the
melting point to the non-insecticidal 1,1'-(2,2-dichloroethenylidene)-bis(4-chloro=
benzene) (DDE). This reaction is catalysed by ferric chloride, aluminium chloride,
and u.v. light and, in solution, by alkali. Generally stable to oxidation.

COMMERCIALISATION
History Its insecticidal properties were discovered by P. Müller (P. Langer *et al.*,
Helv. Chim. Acta, 1944, **27**, 892). Introduced by J. R. Geigy S.A. (now Novartis
Crop Protection AG), who no longer manufacture or market it. **Patent** CH
226180; GB 547871

APPLICATIONS
Biochemistry Nerve poison, affecting sodium balance of nerve membranes. Insects
develop resistance through ability to dehydrochlorinate DDT, forming inactive
DDE. **Mode of action** Persistent non-systemic insecticide with contact and
stomach action. **Uses** Used as a mosquito vector control for the eradication of
malaria. Usage on crops has generally been displaced by less-persistent
insecticides. **Phytotoxicity** Phytotoxic to cucurbits, beans, young tomato plants,
and some varieties of barley. **Formulation types** EC; WP; DP; GR; Aerosol.
Compatibility Incompatible with alkaline substances.

ANALYSIS
Product analysis by glc with FID or i.r. spectrometry (*AOAC Methods*, 1995,
991.04, 7.6.28; 1995, 960.13; *CIPAC Handbook*, 1983, **1B**, 1760; *ibid.*, 1992, **E**,
58–64) or by potentiometric titration (*CIPAC Handbook*, 1994, **F**, 190).
Hydrolysable chlorine determined (*ibid.*, 183–189). Conversion to TDE
[1,1-dichloro-2,2-bis(4-chlorophenyl)ethane], which is no longer available
commercially, also occurs *in vivo* and in the environment, and must be considered
when DDT-derived residues are considered. **Residues** are normally determined
by glc or by tlc or paper chromatography (*AOAC Methods*, 1995, 970.52, 985.22;
Pestic. Anal. Man., 1979, **I**, 201-A, 201-G, 201-I; A. Ambrus *et al.*, *J. Assoc. Off.
Anal. Chem.*, 1981, **64**, 773; *Analyst (London)*, 1979, **104**, 425; *Man. Pestic. Residue
Anal.*, 1987, **I**, 4–6, S8-S10, S12, S19; *Anal. Methods Residues Pestic.*, 001988, Part I,
M1, M12).

MAMMALIAN TOXICOLOGY
Reviews *Pesticide residues in food – 1984.* FAO Plant Production and Protection

Paper 62, 1985. *Pesticide residues in food – 1984 evaluations*. FAO Plant Production and Protection Paper 67, 1985. The effects on reproduction of higher animals (G. W. Ware, *ibid.*, 1975, **59,** 119) have been discussed. Because of lipophilic properties of DDT and its breakdown products, it tends to be accumulated in food chains and in the environment. It has, therefore, largely been replaced by non-persistent insecticides. **IARC** 5, 42, 53 **EHC** 9 (WHO, 1979), 83 (WHO, 1989). **Oral** Acute oral LD_{50} for rats 113–118, mice 150–300, rabbits 300, dogs 500–750, sheep and goats >1000 mg/kg. **Skin and eye** Acute percutaneous LD_{50} for female rats 2510 mg/kg. **NOEL** (160 d) for rats 1 mg/kg diet. Though stored in body fat and excreted in milk, 17 humans who ate 35 mg/man daily *c*. 0.5 mg/kg daily) for 1.75 y suffered no ill-effect. **ADI** (JMPR) 0.02 mg/kg b.w. (PTDI) [1994]. **Toxicity class** WHO (a.i.) II; EPA (formulation) II **EC risk** T (R25, R48/25); (R40) **PIC** Yes

ECOTOXICOLOGY
Fish Toxic to fish and aquatic life.

ENVIRONMENTAL FATE
EHC 83 reviews resistance to degradation, widespread persistence in the environment, and high potential for bioaccumulation; this review concludes that DDT and its metabolites should be regarded as a major hazard to the environment. See also A. Bevenue, *Residue Rev.*, 1976, **61**, 37. Transformation of residues during cooking and food processing has been discussed (T. E. Archer, *Residue Rev.*, 1976, **61**, 29). The case for and against the compound was reviewed (K. Mellanby, *BCPC Monograph*, 1989, No. 43, p. 3). **Animals** In rats, following oral administration, DDT is metabolised to 1,1'-(2,2-dichloroethenylidene)-bis= (4-chlorobenzene) (DDE), 1,1'-(2,2-dichloroethylidene)-bis(4-chlorobenzene) (DDD), 1,1'-(2-chloroethenylidene)-bis(4-chlorobenzene), 1,1'-(2-chloro= ethylidene)-bis(4-chlorobenzene), 1,1'-bis(4-chlorophenyl)ethylene, 2,2-bis(4-chlorophenyl)ethanol, and 2,2-bis(4-chlorophenyl)acetic acid (DDA). Accumulates in the fatty tissues of mammals, and is excreted in milk.

debacarb

Fungicide

benzimidazole

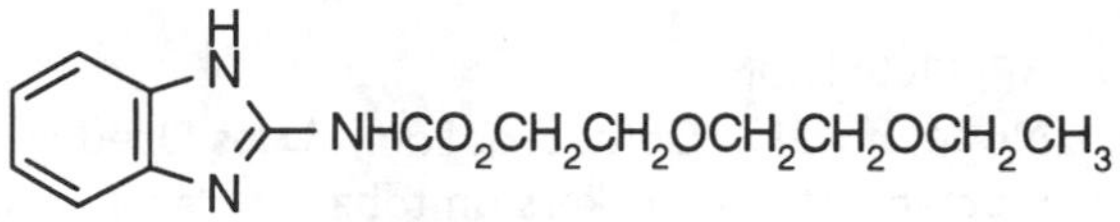

NOMENCLATURE
Common name debacarb (BSI, ANSI, pa E-ISO)
IUPAC name 2-(2-ethoxyethoxy)ethyl benzimidazol-2-ylcarbamate
Chemical Abstracts name 2-(2-ethoxyethoxy)ethyl 1*H*-benzimidazol-2-ylcarbamate
CAS RN *[62732–91–6]*

PHYSICAL CHEMISTRY
Mol. wt. 293.3   **M.f.** $C_{14}H_{19}N_3O_4$

COMMERCIALISATION
Manufacturers Mauget

APPLICATIONS
Uses Used by injection for control of various fungal diseases of trees.
Selected tradenames 'Abasol' (mixture) (Mauget); 'Fungisol' (mixture) (Mauget)

MAIN ENTRIES

decan-1-ol

Plant growth regulator

$CH_3(CH_2)_9OH$

NOMENCLATURE
Common name decan-1-ol (E-ISO accepted in lieu of common name)
IUPAC name decan-1-ol
Chemical Abstracts name 1-decanol
Other names *n*-decyl alcohol; *n*-decanol   **CAS RN** *[112–30–1]*

PHYSICAL CHEMISTRY
Composition Used alone, or in combination with *n*-octanol.   **Mol. wt.** 158.3
M.f. $C_{10}H_{22}O$ **Form** Viscous liquid.   **B.p.** 233 °C   **S.g./density** 1.4359 (20 °C)
Solubility Sparingly soluble in water; very soluble in most organic solvents.

COMMERCIALISATION
History Introduced as a plant growth regulator by Procter & Gamble Co. and by Panorama Chemicals (Pty) Ltd. **Manufacturers** Drexel

APPLICATIONS
Mode of action Contact action. **Uses** Used alone, or in combination with n-octanol, to kill suckers on tobacco plants. **Formulation types** EC.
Mixtures *(decan-1-ol +)* octan-1-ol. **Compatibility** Not to be mixed with other pesticides. **Selected tradenames** 'Antak' (Crystal, Drexel); 'Paranol' (Panorama); 'Royaltac' (Uniroyal)

MAMMALIAN TOXICOLOGY
Oral Acute oral LD_{50} for rats 18 000 mg/kg. **Skin and eye** Acute percutaneous LD_{50} not available. Mild skin and eye irritant.

204 deltamethrin   *Insecticide*

pyrethroid

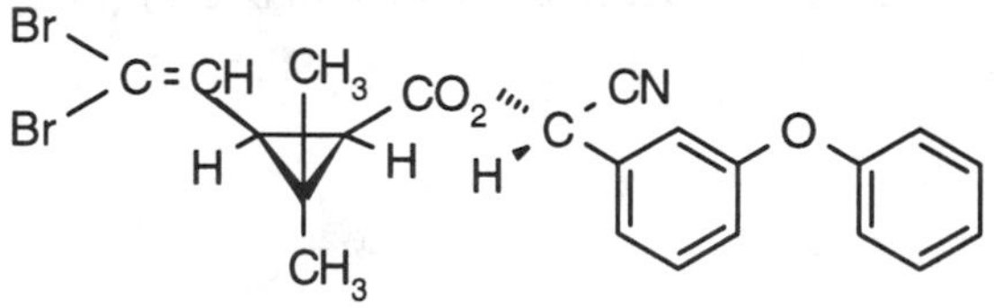

NOMENCLATURE
Common name deltamethrin (BSI, draft E-ISO); deltaméthrine ((f) draft F-ISO)
IUPAC name (S)-α-cyano-3-phenoxybenzyl (1R,3R)-3-(2,2-dibromovinyl)-= 2,2-dimethylcyclopropanecarboxylate. *Roth*: (S)-α-cyano-3-phenoxybenzyl (1R)-*cis*-3-(2,2-dibromovinyl)-2,2-dimethylcyclopropanecarboxylate
Chemical Abstracts name [1R-[1α(S*),3α]]-cyano(3-phenoxyphenyl)methyl 3-(2,2-dibromoethenyl)-2,2-dimethylcyclopropanecarboxylate
Other names [decamethrin] (rejected common name proposal)
CAS RN *[52918–63–5]*; *[52820–00–5]* ((RS)- (1R)-*cis*- isomer pair)
EEC no. 258–256–6 **Development codes** NRDC 161; RU 22 974 (Roussel Uclaf); AE F032640 **Official codes** OMS 1998

PHYSICAL CHEMISTRY
Composition Tech. produced industrially by Roussel Uclaf contains ≥98% deltamethrin (*i.e.* the single isomer). **Mol. wt.** 505.2 **M.f.** $C_{22}H_{19}Br_2NO_3$
Form Colourless crystals. **M.p.** 100–102 °C **V.p.** 1.24×10^{-5} mPa (25 °C, gas

saturation method) K_{ow} logP = 4.6 (25 °C) **Henry** 3.13 × 10^{-2} Pa m^3 mol^{-1} (calc.) **S.g./density** Bulk density 0.55 g/cm^3 (25 °C) **Solubility** In water <0.2 μg/l (25 °C). In dioxane 900, cyclohexanone 750, dichloromethane 700, acetone 500, benzene 450, dimethyl sulfoxide 450, xylene 250, ethanol 15, isopropanol 6 (all in g/l, 20 °C). **Stability** Extremely stable on exposure to air. Stable ≤190 °C. Under u.v. irradiation and in sunlight, a *cis-trans* isomerisation, splitting of the ester bond, and loss of bromine occur. More stable in acidic than in alkaline media; DT$_{50}$ 2.5 d (pH 9, 25 °C). **Specific rotation** $[\alpha]_D$ +61° (40 g/l benzene)

COMMERCIALISATION
History This single isomer, first described by M. Elliott et al. (*Nature (London)*, 1974, **248**, 710), and reviewed in *Deltamethrin Monograph* and by J. Tessier (*Chem. Ind.*, 1984, p. 199) was introduced by Roussel Uclaf. **Patent** GB 1413491 to NRDC **Manufacturers** Roussel Uclaf

APPLICATIONS
Biochemistry Like all pyrethroids, prevents the sodium channels from functioning, so that no transmission of nerve impulses can take place. **Mode of action** Non-systemic insecticide with contact and stomach action. Fast-acting. **Uses** A potent insecticide, effective by contact and ingestion against a wide range of pests. Crop protection uses include: Coleoptera (2.5–7.5 g/ha), Heteroptera (5.0–7.5 g/ha), Homoptera (6.2–12.5 g/ha), Lepidoptera (5.0–12.5 g/ha) and Thysanoptera (5–10 g/ha) in cereals, citrus, cotton, grapes, maize, oilseed rape, soya beans, top fruit and vegetables. It controls Acrididae (5.0–12.5 g/ha), and is recommended against locusts. Soil surface sprays (2.5–5.0 g/ha) control Noctuidae. It is used against indoor crawling and flying insects (12.5 mg/m^2) and pests of stored grain (0.25–0.5 g/t) and timber (Blattodea, Culicidae, Muscidae). Dip or spray (12.5–75 mg/l), and pour-on (0.75 mg/kg b.w.) applications give good control of Muscidae, Tabanidae, Ixodidae and other Acari on cattle, sheep and pigs, etc. **Formulation types** PO; SL; EC; WP; UL; SC; GR; DP; HN; EG. **Mixtures** (*deltamethrin* +) buprofezin; dimethoate; heptenophos; sulfur; triazophos; piperonyl butoxide; endosulfan; pirimicarb; chlorpyrifos-methyl; chlorpyrifos-ethyl; profenofos; bioallethrin S-cyclopentenyl isomer. **Selected tradenames** 'Butox' (veterinary uses) (Hoechst Roussel Vet.); 'Decis' (crop protection uses) (AgrEvo); 'K-Othrine' (environmental health uses) (AgrEvo); 'Kordon' (environmental health uses) (AgrEvo); 'Sadethrin' (Sanonda)

ANALYSIS
Product analysis by lc with u.v. detection (*AOAC Methods*, 1995, 991.03, 7.5.05) or by hplc (*CIPAC Handbook*, 1988, **D**, 57; M. Vaysse et al., *Anal. Methods Pestic. Plant Growth Regul.*, 1984, **13**, 53). **Residues** determined by glc with ECD (*Man. Pestic. Residue Anal.*, 1987, **I**, 6, S19; *Anal. Methods Residues Pestic.*, 1988, Part I, M11; A. Ambrus et al., *J. Assoc. Off. Anal. Chem.*, 1981, **64**, 733; P. G. Baker & P. Bottomley, *Analyst (London)*, 1982, **107**, 206; P. Bottomley & P. G. Baker, *ibid.*, 1984, **109**, 85).

MAMMALIAN TOXICOLOGY
Reviews See A. J. Gray & D. M. Soderlund, Chapt. 5 in "*Insecticides*". *Pesticide residues in food – 1982*. FAO Plant Production and Protection Paper 46, 1983. *Pesticide residues in food: 1982 evaluations*. FAO Plant Production and Protection Paper 49, 1983. **IARC** 53 **EHC** 97 (WHO, 1990). **Oral** Acute oral LD_{50} for rats ranges from 135–>5000 mg/kg depending upon carrier and conditions of the study; for dogs >300 mg/kg. Acute oral LD_{50} for formulations in rats: 1080 mg (of 15 g/l EC)/kg; 535 mg (of 25 g/l EC)/kg; >5000 mg (of 5 g/l UL)/kg; >16 000 mg (of 25 g/kg WP)/kg; >40 000 mg (of 25 g/l SC)/kg.
Skin and eye Acute percutaneous LD_{50} for rats and rabbits >2000 mg/kg. Non-irritating to skin; mild eye irritant (rabbits). **Inhalation** LC_{50} (4 h) for rats 2.2 mg/l air; (1 h) >4.6 mg/l air (micronised). **NOEL** (2 y) for mice 12, rats 1, dogs 1 mg/kg b.w. **ADI** (JMPR) 0.01 mg/kg b.w. [1982]. **Other** Non-mutagenic and non-teratogenic (mice, rats, rabbits). **Toxicity class** WHO (a.i.) II; EPA (formulation) II

ECOTOXICOLOGY
Birds Acute oral LD_{50} for mallard ducks >4640 mg/kg. Eight-day dietary LC_{50} for mallard ducks >8039, quail >5620 mg/kg diet. NOEL for reproduction for mallard ducks >70, bobwhite quail >55 mg/kg daily. **Fish** Toxic to fish under laboratory conditions; LC_{50} (96 h) for rainbow trout 0.91, bluegill sunfish 1.4 µg/l. Not toxic to fish under natural conditions. **Bees** Toxic to bees, LD_{50} (oral) 79 ng/bee, (contact) 51 ng/bee. Low LD_{50} and LC_{50} values under laboratory conditions do not represent a significant hazard to bees in normal field use.
Worms LC_{50} (14 d) for earthworms 28.57 mg/kg soil. **Daphnia** LC_{50} (48 h) 3.5 µg/l. **Algae** EC_{50} (96 h) for algae (*Selenastrum capricornutum*) >9.1 mg/l. Low LD_{50} and LC_{50} values under laboratory conditions do not represent a significant hazard to aquatic fauna in normal field use.

ENVIRONMENTAL FATE
Animals In rats, following oral administration, elimination occurs within 2–4 days. The phenyl ring is hydroxylated, the ester bond hydrolysed, and the acid moiety is eliminated as the glucuronide and glycine conjugate. **Plants** No uptake through leaves and roots – non-systemic compound. No major metabolites except in oily crops, where *trans*-deltamethrin is part of the residue definition.
Soil/Environment In soil, undergoes microbial degradation within 1–2 weeks. K_d 3790–30 000, K_{oc} 4.6×10^5–1.63×10^7 cm^3/g, confirms strong adsorption by soil colloids and no risk of leaching. DT_{50} (lab., aerobic) 21–25 d, (anaerobic) 31–36 d. In field DT_{50} <23 d. Soil photolysis DT_{50} 9 d. No incidence on soil microflora and nitrogen cycle.

demeton-S-methyl   *Insecticide, acaricide*

organophosphorus

$$CH_3CH_2SCH_2CH_2SP(OCH_3)_2$$

with O double-bonded to P.

NOMENCLATURE
Common name demeton-S-methyl (BSI, E-ISO, *(m)* F-ISO); methyl demeton (JMAF); methylmercaptofostiol (USSR); [methyl-mercaptofos teolovy] (former USSR name); no name (USA)
IUPAC name S-2-ethylthioethyl O,O-dimethyl phosphorothioate
Chemical Abstracts name S-[2-(ethylthio)ethyl] O,O-dimethyl phosphorothioate
Other names 2-ethylthioethyl dimethyl phosphorothiolate   **CAS RN** *[919–86–8]* demeton-S-methyl; *[867–27–6]* demeton-O-methyl; *[8022–00–2]* demeton-= methyl   **EEC no.** 213–052–6   **Development codes** Bayer 18 436; Bayer 25/154

PHYSICAL CHEMISTRY
Composition A reaction mixture containing demeton-S-methyl and demeton-O-= methyl is known as demeton-methyl (BSI) or methyl demeton (ESA).
Mol. wt. 230.3   **M.f.** $C_6H_{15}O_3PS_2$ **Form** Tech. grade is a pale yellow oil.
B.p. 74 °C/0.052 mmHg; 89 °C/0.15 mmHg; 118 °C/1 mmHg   **V.p.** 40 mPa
(20 °C) K_{ow} logP = 1.32 (20 °C)   **S.g./density** 1.207 (20 °C)   **Solubility** In water
22 g/l (20 °C). Readily soluble in common polar organic solvents (e.g. alcohols, ketones, chlorinated hydrocarbons, *c.* 200 g/l in dichloromethane, isopropanol, toluene).   **Stability** Hydrolysed rapidly in alkaline media, and more slowly in acidic and neutral aqueous media; DT_{50} (estimated) 63 d (pH 4), 56 d (pH 7), 8 d (pH 9) (22 °C). Oxidised to the sulfoxide (see oxydemeton-methyl) and sulfone (see demeton-S-methylsulphon).

COMMERCIALISATION
History In 1954 Farbenfabriken Bayer AG (now Bayer AG) introduced demeton-= methyl, a reaction mixture which contained demeton-S-methyl and demeton-O-= methyl (O-2-ethylthioethyl O,O-dimethyl phosphorothioate). An improved manufacturing process gave demeton-S-methyl, introduced by Bayer AG in 1957 (see G. Schrader, *Die Entwicklung neuer insektizider Phosphorsäure-Ester*).
Patent DE 836349; US 2571989   **Manufacturers** United Phosphorus Ltd

APPLICATIONS
Biochemistry It is metabolised in plants to the sulfoxide (oxydemeton-methyl) and sulfone (demeton-S-methylsulphon) *q.v.*. Cholinesterase inhibitor.
Mode of action Systemic insecticide and acaricide with contact and stomach action.   **Uses** Control of aphids (including virus vectors), other sucking insects, sawflies, and spider mites in fruit, vegetables, potatoes, cereals, beet, hops,

ornamentals. **Phytotoxicity** Non-phytotoxic when used as directed, except to some ornamentals (particularly certain chrysanthemum varieties).
Formulation types EC. **Compatibility** Incompatible with alkaline materials.
Selected tradenames 'Metaphor' (United Phosphorus Ltd); 'Mifatox' (FCC)

ANALYSIS
Product analysis based on alkaline hydrolysis determining the acid released (*CIPAC Handbook*, 1970, **1**, 312) or the thiol by titration with iodine (*ibid.*).
Residues determined by glc (*Man. Pestic. Residue Anal.*, 1987, **I**, 6, S13, S16, S19; *Anal. Methods Residues Pestic.*, 1988, Part I, M2, M5; A. Ambrus *et al.*, *J. Assoc. Off. Anal. Chem.*, 1981, **64**, 733). Methods for the determination of residues are also available from Bayer.

MAMMALIAN TOXICOLOGY
Reviews *Pesticide residues in food – 1989.* FAO Plant Production and Protection Paper 99, 1989. *Pesticide residues in food – 1989 evaluations. Part II – Toxicology.* FAO Plant Production and Protection Paper 100/2, 1990. **EHC** 63 (WHO, 1986; a general review of organophosphorus insecticides). **Oral** Acute oral LD_{50} for rats *c.* 30 mg/kg. **Skin and eye** Acute percutaneous LD_{50} for rats *c.* 30 mg/kg. Not irritating to eyes, mildly irritating to skin (rabbits).
Inhalation LC_{50} (4 h) for rats *c.* 0.13 mg/l air (as aerosol). **NOEL** (2 y) for rats and mice 1 mg/kg diet; (1 y) for dogs 1 mg/kg diet. **ADI** (JMPR) 0.0003 mg/kg b.w. for sum of demeton-S-methyl, demeton-S-methylsulphon and oxydemeton-methyl [1989]. **Toxicity class** WHO (a.i.) Ib; EPA (formulation) I
EC risk T (R24/25)

ECOTOXICOLOGY
Birds LD_{50} for male Japanese quail 50, female Japanese quail 44 mg/kg. **Fish** LC_{50} (96 h) for rainbow trout 6.4, golden orfe 23.2 mg/l (both 250 EC). **Bees** Toxic to bees. **Worms** LC_{50} for *Eisenia foetida* 250 mg/kg dry soil. **Daphnia** LC_{50} (48 h) 0.023 mg/l. **Algae** E_rC_{50} for *Scenedesmus subspicatus* 22.1 mg/l.

ENVIRONMENTAL FATE
Animals In mammals, following oral administration, metabolism involves oxidation of the thioethyl group to give the sulfoxide (oxydemeton-methyl) and sulfone (demeton-S-methylsulphon), and hydrolysis to dimethyl phosphate. Rapidly eliminated in the urine (97–99% within 24 h). **Plants** In plants, the thioethyl group is oxidised to give the sulfoxide (oxydemeton-methyl) and sulfone (demeton-S-= methylsulphon). Hydrolysis gives dimethyl phosphate.
Soil/Environment Degradation of demeton-S-methyl in soil is extremely rapid.

bis-carbamate

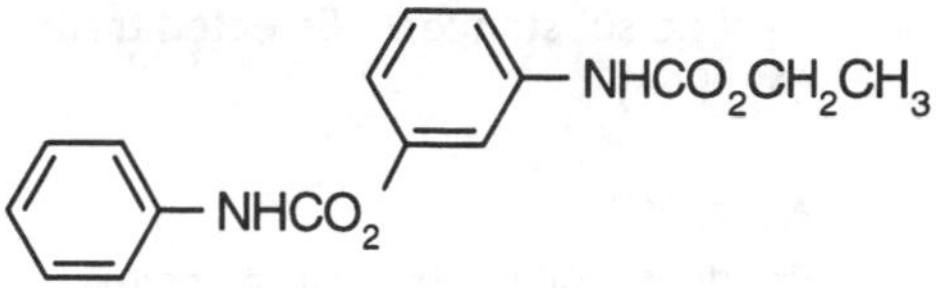

NOMENCLATURE

Common name desmedipham (BSI, E-ISO, ANSI, WSSA); desmédiphame ((*m*) F-ISO)
IUPAC name ethyl 3-phenylcarbamoyloxyphenylcarbamate; ethyl 3'-phenyl= carbamoyloxycarbanilate; 3-ethoxycarbonylaminophenyl phenylcarbamate
Chemical Abstracts name ethyl [3-[[(phenylamino)carbonyl]oxy]phenyl]carbamate
Other names DMP   **CAS RN** *[13684–56–5]*   **Development codes** SN 38 107; ZK 14 494; EP 475

PHYSICAL CHEMISTRY

Composition Tech. grade is *c.* 97.5% pure.   **Mol. wt.** 300.3   **M.f.** $C_{16}H_{16}N_2O_4$
Form Colourless crystals.   **M.p.** 120 °C   **V.p.** 4×10^{-5} mPa (25 °C)
K_{ow} logP = 3.39 (pH 5.9)   **Henry** 7.56×10^{-3} Pa m^3 mol^{-1} (calc.)
S.g./density Pour 0.536, tap 0.620   **Solubility** In water 7 mg/l (pH 7, 20 °C).
Readily soluble in polar organic solvents, e.g. acetone 400, isophorone 400, methanol 180, ethyl acetate 149, chloroform 80, dichloroethane 17.8, benzene 1.6, toluene 1.2, hexane 0.5 (all in g/l, 20 °C).   **Stability** Stable in aqueous acidic media, but hydrolysed in neutral and alkaline media. Stable for 2 years at 70 °C; DT_{50} 224 h in aq. solution at pH 3.8 exposed to light ≥ 280 nm; on hydrolysis DT_{50} *c.* 70 d at pH 5, *c.* 20 h at pH 7, 10 min at pH 9.

COMMERCIALISATION

History Herbicide reported by F. Arndt & G. Boroschewski (*Symp. New Herbic.*, *3rd*, 1969, p. 141). Introduced by Schering AG (now AgrEvo GmbH).
Patent GB 1127050   **Manufacturers** AgrEvo; Griffin; United Phosphorus

APPLICATIONS

Biochemistry Photosynthetic electron transport inhibitor.
Mode of action Selective systemic herbicide, absorbed through the leaves, with translocation primarily in the apoplast.   **Uses** Used post-emergence to control broad-leaved weeds, including *Amaranthus retroflexus,* in beet crops, in particular sugar beet. Usually sprayed in combination with phenmedipham and ethofumesate. Desmedipham acts through the leaves only, and does not depend on soil type and humidity under normal growing conditions. Due to its wide safety margin to the crop, spraying is merely timed according to the development stage of the weeds,

with optimum weed control when they are at the cotyledon stage.
Phytotoxicity Non-phytotoxic to beet crops.   **Formulation types** EC; SC.
Mixtures *(desmedipham +)* phenmedipham; phenmedipham + ethofumesate;
phenmedipham + ethofumesate + metamitron.   **Compatibility** Incompatible with
alkaline substances.   **Selected tradenames** 'Betanal AM' (AgrEvo); 'Kemifam'
(AgrEvo)

ANALYSIS
Product analysis by hplc or colorimetry of a derivative.   **Residues** determined by
hydrolysis to aniline using derivatives suitable for glc (C.-H. Roeder *et al.*, *Anal.*
Methods Pestic. Plant Growth Regul., 1978, **10**, 293) or hplc. Details available from
AgrEvo.

MAMMALIAN TOXICOLOGY
Oral Acute oral LD_{50} for rats >10 250, mice >5000 mg/kg.   **Skin and eye** Acute
percutaneous LD_{50} for rabbits >4000 mg/kg. Not a skin sensitiser.
Inhalation LC_{50} (4 h) for rats >7.4 mg/l.   **NOEL** (2 y) for rats 3.2 mg/kg daily, for
mice 22 mg/kg daily.   **ADI** 0.00125 mg/kg.   **Toxicity class** WHO (a.i.) III
(Table 5); EPA (formulation) III

ECOTOXICOLOGY
Birds LC_{50} (14 d) for bobwhite quail and mallard ducks >2000 mg/kg. Dietary
LC_{50} (8 d) for bobwhite quail and mallard ducks >5000 mg/kg diet.   **Fish** LC_{50}
(96 h) for rainbow trout 1.7, bluegill sunfish 3.2 mg/l.   **Bees** Not toxic to bees;
LD_{50} >50 µg/bee.   **Worms** LC_{50} (14 d) 466.5 mg/kg dry soil.   **Daphnia** LC_{50}
(48 h) 1.88 mg/l.   **Algae** IC_{50} (72 h) 0.061 mg/l.

ENVIRONMENTAL FATE
Animals In mammals, following oral administration, 80% of the parent compound
and its metabolites are eliminated in the urine within 24 hours. Hydrolysis to ethyl
N-(3-hydroxyphenyl)carbamate (EHPC) and conjugation (to glucuronides and
ethereal sulfates) are the major steps in metabolism.   **Plants** In sugar beet, EHPC
is the major metabolite, with *m*-aminophenol as a further metabolite.
Soil/Environment DT_{50} in soil *c.* 34 d, DT_{90} <115 d, with formation of ethyl
3-hydroxycarbanilate, which undergoes further degradation. Therefore,
desmedipham does not accumulate in soil, nor is there any relevant uptake by
following crops. Due to its favourable physico-chemical parameters, there is no
risk of groundwater contamination to be expected. K_{oc} = 1500.

desmetryn *Herbicide*

1,3,5-triazine

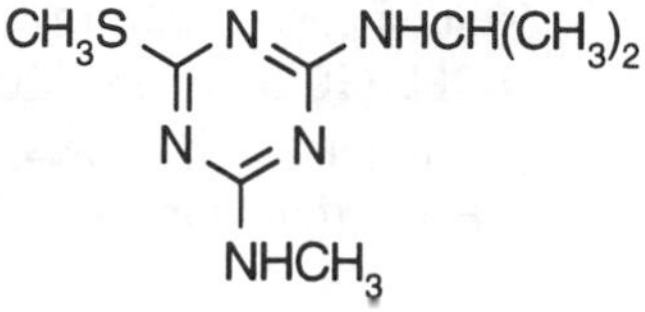

$$CH_3S \quad NHCH(CH_3)_2$$
$$NHCH_3$$

NOMENCLATURE
Common name desmetryn (BSI (since 1984), E-ISO, WSSA); desmetryne ((*f*) F-ISO, JMAF, BSI (before 1984)); no name (Portugal)
IUPAC name N^2-isopropyl-N^4-methyl-6-methylthio-1,3,5-triazine-2,4-diamine
Chemical Abstracts name N-methyl-N'-(1-methylethyl)-6-(methylthio)-1,3,5-= triazine-2,4-diamine
CAS RN *[1014–69–3]* **EEC no.** 213–800–1 **Development codes** G 34 360

PHYSICAL CHEMISTRY
Composition 95% pure. **Mol. wt.** 213.3 **M.f.** $C_8H_{15}N_5S$ **Form** White powder.
M.p. 84–86 °C **B.p.** 339 °C/98.4 kPa **V.p.** 0.133 mPa (20 °C)
K_{ow} logP = 2.38 **Henry** 4.8 × 10⁻⁵ Pa m³ mol⁻¹ (calc.) **S.g./density** 1.172 (20 °C)
Solubility In water 580 mg/l (20 °C). Readily soluble in most organic solvents, e.g. methanol 300, acetone 230, dichloromethane, toluene 200, hexane 2.6 (all in g/kg, 20 °C). **Stability** No significant hydrolysis at 70 °C detected at 5<pH<13.
pKa 4.0, weak base

COMMERCIALISATION
History Herbicide reported by J. G. Elliott & T. I. Cox (*Proc. Br. Weed Control Conf., 6th*, 1962, **2**, 759). Introduced by J. R. Geigy S.A. (now Novartis Crop Protection AG). **Patent** CH 337019; GB 814948 **Manufacturers** Novartis

APPLICATIONS
Biochemistry Photosynthetic electron transport inhibitor.
Mode of action Selective systemic herbicide, absorbed by the leaves and roots, with translocation acropetally through the xylem, and accumulation in the apical meristems. **Uses** Post-emergence control of fat-hen (*Chenopodium album*) and other annual broad-leaved weeds and some grasses at 0.25–0.5 kg/ha in brassicas (except cauliflowers and broccoli), herbs, onions, leeks, and conifer seedbeds.
Formulation types WP. **Compatibility** There may be incompatibility with some insecticides. **Selected tradenames** 'Semeron' (Novartis)

ANALYSIS
Product analysis by glc (*CIPAC Handbook*, 1983, **1B**, 1765). **Residues** determined by glc (K. Ramsteiner *et al., J. Assoc. Off. Anal. Chem.*, 1974, **57**, 192).

MAIN ENTRIES

MAMMALIAN TOXICOLOGY
Oral Acute oral LD_{50} for rats 1390, mice 1750 mg/kg. **Skin and eye** Acute percutaneous LD_{50} for rats 2000 mg/kg. Not a skin or eye irritant (rabbits). Not a skin sensitiser when applied epidermally; skin sensitiser when applied intradermally (guinea pigs). **Inhalation** LC_{50} (1 h) for rats >1563 mg/m^3 as WP 25.
NOEL (90 d) for rats 20 mg/kg diet (1.5 mg/kg daily); for dogs 200 mg/kg diet (6.6 mg/kg daily). **ADI** 0.0075 mg/kg. **Toxicity class** WHO (a.i.) III; EPA (formulation) III **EC risk** Xn (R22)

ECOTOXICOLOGY
Birds LC_{50} (8 d) for Japanese quail >10 000 mg/kg. **Fish** LC_{50} (96 h) for rainbow trout 2.2, common carp 37 mg/l. **Bees** Not toxic to bees; LD_{50} (oral) >197 μg/bee, (topical) >101 μg/bee. **Worms** LC_{50} (14 d) for earthworms 160 mg/kg soil. **Daphnia** LC_{50} (48 h) 45 mg/l. **Algae** EC_{50} (72 h) for *Scenedesmus subspicatus* 0.004 mg/l.

ENVIRONMENTAL FATE
Animals Major pathways in animal metabolism are hydrolysis of the methylthio group to form the hydroxy derivative, sulfoxidation of the methylthio group, and side-chain dealkylation. **Plants** Major pathways in plant metabolism are the hydrolysis of the methylthio group to form the hydroxy derivative, conjugation with glutathione and side-chain dealkylation. **Soil/Environment** Degradation involves oxidation of the methylthio group to sulfoxide and sulfone, hydrolysis with the introduction of a 2-hydroxy group, dealkylation at the substituted amino groups, cleavage of the amine moiety, and ring opening. Duration of residual activity in soil is *c.* 3 months, DT_{50} *c.* 50 d. K_{oc} 150, moderately mobile in soil. Degradation in aquatic systems is caused by microbial processes; desmetryn is also removed from water by adsorption to the sediment.

208 diafenthiuron   *Insecticide, acaricide*

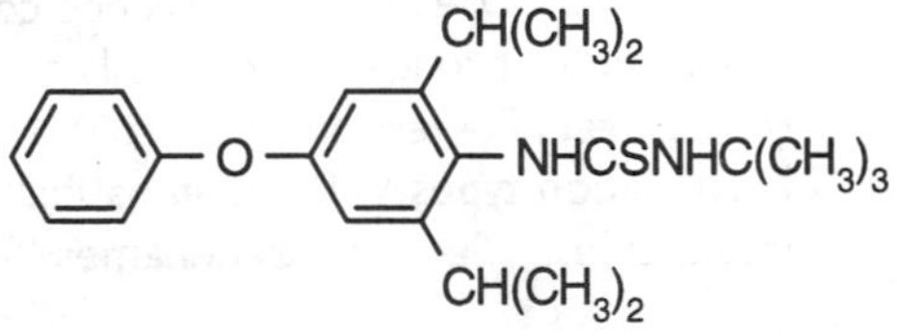

NOMENCLATURE
Common name diafenthiuron (BSI, draft E-ISO)

IUPAC name 1-*tert*-butyl-3-(2,6-di-isopropyl-4-phenoxyphenyl)thiourea
Chemical Abstracts name N-[2,6-bis(1-methylethyl)-4-phenoxyphenyl]-N'-=
(1,1-dimethylethyl)thiourea
CAS RN *[80060–09–9]* **Development codes** CGA 106 630

PHYSICAL CHEMISTRY

Mol. wt. 384.6 **M.f.** $C_{23}H_{32}N_2OS$ **Form** White powder. **M.p.** 144.6–147.7 °C
(OECD 102) **V.p.** <2 × 10^{-3} mPa (25 °C) (OECD 104) **K$_{ow}$** logP = 5.76
(OECD 107) **S.g./density** 1.09 (20 °C) (OECD 109) **Solubility** In water
0.06 mg/l (25 °C). In ethanol 43, acetone 320, toluene 330, *n*-hexane 9.6,
n-octanol 26 (all in g/l, 25 °C). **Stability** Stable in air and water and to light.

COMMERCIALISATION

History Insecticide and acaricide reported by H. P. Streibert *et al.* (*Proc. 1988 Br.
Crop Prot. Conf.*, *1*, 25). Introduced by Ciba-Geigy AG (now Novartis Crop
Protection AG). **Patent** GB 2060626; DE 3034905 **Manufacturers** Novartis

APPLICATIONS

Biochemistry Converted by light, or *in vivo* to the corresponding carbodiimide,
which is an inhibitor of mitochondrial respiration. **Mode of action** Insecticide and
acaricide which kills larvae, nymphs and adults by contact and/or stomach action;
also shows some ovicidal action. See J. Drabek *et al.* (*Recent Adv. Chem. Insect
Control II*, 1990, p. 170). **Uses** Insecticide and acaricide effective against
phytophagous mites (Tetranychidae, Tarsonemidae), Aleyrodidae, Aphididae and
Jassidae on cotton, various field and fruit crops, ornamentals and vegetables. Also
controls some leaf-feeding pests in cole crops (*Plutella xylostella*), soya beans
(*Anticarsia gemmatalis*) and cotton (*Alabama argillacea*). Is safe on adults of all
beneficial groups (Anthorcoridae, Coccinellidae, Miridae) and on adults and
immature stages of predatory mites (*Amblyseius andersoni*, *Typhlodromus pyri*),
spiders (Erigoridae, Lycosidae), *Chrysopa carnea*. Non-selective to immature stages
of Heteroptera (Anthocoridae, Miridae). Compatible with the biological control of
Aleyrodidae and mites in glasshouses. **Formulation types** SC; WP.
Selected tradenames 'Pegasus' (Novartis); 'Polo' (Novartis)

ANALYSIS

Product analysis by glc. **Residues** determined by glc. Details available from
Novartis.

MAMMALIAN TOXICOLOGY

Oral Acute oral LD$_{50}$ for rats 2068 mg/kg. **Skin and eye** Acute percutaneous
LD$_{50}$ for rats >2000 mg/kg. Non-irritant to eyes and skin (rats). **Inhalation** LC$_{50}$
(4 h) for rats 0.558 mg/l air. **NOEL** (90 d) for rats 4, dogs 1.5 mg/kg b.w. daily.
ADI 0.003 mg/kg b.w. **Toxicity class** WHO (a.i.) III **EC risk** R21/22, R23, R48

ECOTOXICOLOGY
Birds Acute oral LD_{50} for bobwhite quail and mallard ducks >1500 mg/kg. Eight-day dietary LC_{50} for bobwhite quail and mallard ducks >1500 mg/kg. No acute hazard under field conditions. **Fish** LC_{50} (96 h) for carp 0.0038, rainbow trout 0.0007, bluegill sunfish 0.0013 mg/l. Because of rapid degradation to non-toxic metabolites, there is no significant hazard under field conditions. **Bees** Toxic to honeybees; LD_{50} (48 h) (oral) 2.1 µg/bee; (contact) 1.5 µg/bee. No significant hazard under field conditions. **Daphnia** LC_{50} (48 h) <0.5 mg/l. **Algae** Non toxic.

ENVIRONMENTAL FATE
Animals Study of the adsorption, distribution and excretion in rats demonstrated that the major portion of the dose was excreted with the faeces. The compound is degraded to yield its corresponding carbodiimide, which in turn reacts with nucleophiles like water and fatty acids to form urea and fatty acid derivatives.
Plants In plants, diafenthiuron shows a complex metabolism pattern in all crops investigated, *i.e.* cotton, tomatoes and apples. Uptake of residue activity by plants from soil is low. **Soil/Environment** Diafenthiuron and its main metabolites show a strong sorptivity to soil particles. Degradation in soils proceeds rapidly: DT_{50} <1 h to 1.4 d.

209 diazinon   *Insecticide, acaricide*

organophosphorus

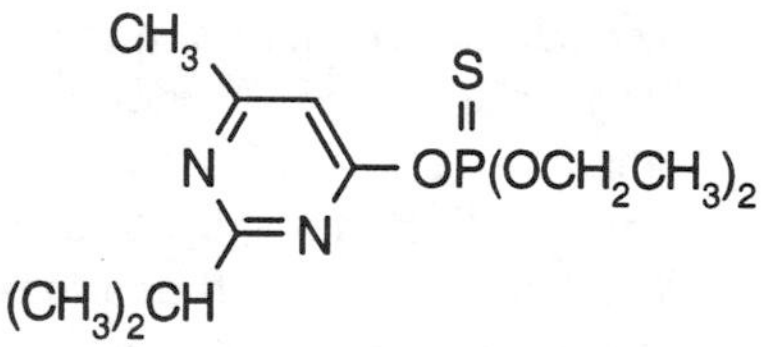

NOMENCLATURE
Common name diazinon (BSI, E-ISO, (*m*) F-ISO, ANSI, ESA, BAN, JMAF); dimpylate (INN)
IUPAC name *O,O*-diethyl *O*-2-isopropyl-6-methylpyrimidin-4-yl phosphorothioate
Chemical Abstracts name *O,O*-diethyl *O*-[6-methyl-2-(1-methylethyl)-4-pyrimidinyl] phosphorothioate
CAS RN *[333–41–5]* **EEC no.** 206–373–8 **Development codes** G 24 480
Official codes OMS 469; ENT 19 507

PHYSICAL CHEMISTRY
Composition Tech. is ≥95% pure. **Mol. wt.** 304.3 **M.f.** $C_{12}H_{21}N_2O_3PS$
Form Clear colourless liquid; (tech., yellow liquid). **B.p.** 83–84 °C/0.0002 mmHg;

125 °C/1 mmHg   **V.p.** 1.2 × 10^1 mPa (25 °C) (OECD 104)   **K$_{ow}$** logP = 3.30
(OECD 107)   **S.g./density** 1.11 (20 °C)   **Solubility** In water 60 mg/l (20 °C).
Completely miscible with common organic solvents, e.g. ethers, alcohols, benzene,
toluene, hexane, cyclohexane, dichloromethane, acetone, petroleum oils.
Stability Susceptible to oxidation above 100 °C. Stable in neutral media, but slowly
hydrolysed in alkaline media, and more rapidly in acidic media; DT$_{50}$ (20 °C)
11.77 h (pH 3.1), 185 d (pH 7.4), 6.0 d (pH 10.4). Decomposes above 120 °C.
F.p. ≥62 °C

COMMERCIALISATION
History Insecticide reported by R. Gasser (Z. *Naturforsch. Teil B,* 1953, **8,** 225).
Introduced by J. R. Geigy S.A. (now Novartis Crop Protection AG).
Patent BE 510817; GB 713278   **Manufacturers** Aimco; Elf Atochem; Drexel;
Makhteshim-Agan; Nippon Kayaku; Novartis

APPLICATIONS
Biochemistry Cholinesterase inhibitor.   **Mode of action** Non-systemic insecticide
and acaricide with contact, stomach, and respiratory action.   **Uses** Control of
sucking and chewing insects and mites on a very wide range of crops, including
deciduous fruit trees, citrus fruit, vines, olives, bananas, pineapples, vegetables,
potatoes, beet, sugar cane, coffee, cocoa, tea, tobacco, maize, sorghum, alfalfa,
flax, cotton, rice, ornamentals, glasshouse crops, forestry, etc.; soil insects (by soil
application); phorid and sciarid flies in mushroom cultivation; flies, lice, mites,
fleas, cockroaches, bedbugs, ants, and other insect pests in animal houses and
household use. Seed treatment for maize, for control of frit flies and also
conferring bird-repellent properties. Also used as a veterinary ectoparasiticide.
Phytotoxicity Non-phytotoxic when used as directed. Russetting may occur on
green and yellow apple varieties.   **Formulation types** SD; GR; WP; EC; DP; DS;
FT; CS; KN; Aerosol; Coating agent.   **Mixtures** *(diazinon +)* disulfoton; pyrethrins;
lindane; petroleum oils.   **Compatibility** Incompatible with copper-containing
compounds.   **Selected tradenames** 'Basudin' (Novartis); 'Cekuzinon' (Cequisa);
'Dianon' (Nippon Kayaku); 'Dianozyl' (Agriphar); 'Diazol' (Makhteshim-Agan);
'Ectoban' (for veterinary use) (Agropharm); 'Efdiazon' (Efthymiadis); 'Knox-out'
(Elf Atochem)

ANALYSIS
Product analysis by glc with FID (*CIPAC Handbook,* 1980, **1A,** 1199; *Anal. Methods
Pestic. Plant Growth Regul.,* 1972, **6,** 345; *AOAC Methods,* 1995, 971.08, 982.06).
Residues determined by glc with TID or MCD, or by tlc or paper chromatography,
or by single sweep oscillographic polarography (*ibid.,* 968.24, 970.33, 970.52,
970.53; *Analyst (London),* 1980, **105,** 515; *Man. Pestic. Residue Anal.,* 1987, **I,** 5, 6,
S8, S10, S13, S17, S19; *Anal. Methods Residues Pestic.,* 1988, Part I, M2, M5, M12).
In **drinking water,** by gc with FID (*AOAC Methods,* 1995, 991.07).

MAMMALIAN TOXICOLOGY

Reviews *Pesticide residues in food – 1993*, FAO Plant Production and Protection Paper, 122, 1993. *Pesticide residues in food – 1993 evaluations, Part II – Toxicology.* WHO, WHO/PCS/94.4, 1994. **EHC** 63 (WHO, 1986; a general review of organophosphorus insecticides). **Oral** Acute oral LD_{50} for rats 1250, mice 80–135, guinea pigs 250–355 mg/kg. **Skin and eye** Acute percutaneous LD_{50} for rats >2150, rabbits 540–650 mg/kg. Not an irritant (rabbits). **Inhalation** LC_{50} (4 h) for rats >2330 mg/m^3. **NOEL** (2 y) for rats 0.06 mg/kg b.w.; (1 y) for dogs 0.015 mg/kg b.w. daily, humans 0.02 mg/kg b.w. **ADI** (JMPR) 0.002 mg/kg b.w. [1993]. **Toxicity class** WHO (a.i.) II; EPA (formulation) II or III **EC risk** Xn (R22)

ECOTOXICOLOGY

Birds Acute oral LD_{50} for mallard ducklings 3.5, young pheasants 4.3 mg/kg. **Fish** LC_{50} (96 h) for bluegill sunfish 16, rainbow trout 2.6–3.2, carp 7.6–23.4 mg/l. **Bees** Highly toxic to bees. **Daphnia** LC_{50} (48 h) 0.96 µg/l.

ENVIRONMENTAL FATE

Animals The principal metabolites are diethyl thiophosphate and diethyl phosphate. **Plants** Studies with ^{14}C-labelled diazinon show a rapid absorption and translocation in plants. Metabolism proceeds via hydrolysis and subsequent transformation and degradation of the hydroxypyrimidine derivatives to CO_2. **Soil/Environment** Degradation involves oxidation to the phosphate (diazoxon) and hydrolysis (J. Pardue et al., J. Agric. Food Chem., 1970, **18**, 405–408). DT_{50} c. 11–21 d (laboratory). Diazinon is fairly strongly adsorbed onto soil, K_{om} 332 mg/g o.m. Mobility is low.

210 dicamba   *Herbicide*

benzoic acid (auxin)

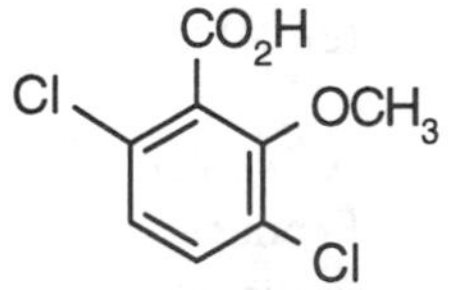

NOMENCLATURE
dicamba
Common name dicamba (BSI, E-ISO, (*m*) F-ISO, ANSI, WSSA); [dianat] (former exception, USSR); MDBA (JMAF)
IUPAC name 3,6-dichloro-*o*-anisic acid
Chemical Abstracts name 3,6-dichloro-2-methoxybenzoic acid

CAS RN *[1918–00–9]*; 5-hydroxy derivative *[7600–50–2]*
Development codes Velsicol 58-CS-11; SAN 837 H

dicamba-dimethylammonium
CAS RN *[2300–66–5]*

dicamba-potassium
CAS RN *[10007–85–9]*

dicamba sodium
CAS RN *[1982–69–0]*

PHYSICAL CHEMISTRY
dicamba
Composition Tech. grade purity is 80–90% *m/m*, remainder being mainly
3,5-dichloro-*o*-anisic acid.   **Mol. wt.** 221.0   **M.f.** $C_8H_6Cl_2O_3$ **Form** Colourless
crystals; tech. grade is a buff crystalline solid.   **M.p.** 114–116 °C   **B.p.** >200 °C
V.p. 4.5 mPa (25 °C, calc.)   **K$_{ow}$** logP = –0.15 (pH 7)   **S.g./density** 1.526 (25 °C)
Solubility In water 6.5 g/l (25 °C). In ethanol 922, cyclohexanone 916, acetone
810, dichloromethane 260, dioxane 1180, toluene 130, xylene 78 (all in g/l,
25 °C).   **Stability** Resistant to oxidation and hydrolysis under normal conditions.
Stable in acids and alkalis. Decomposes at *c.* 200 °C.   **pKa** 1.97

dicamba-dimethylammonium
Mol. wt. 266.1   **M.f.** $C_{10}H_{13}Cl_2NO_3$

dicamba-trolamine
Mol. wt. 346.2   **M.f.** $C_{12}H_{21}Cl_2NO_6$

dicamba-potassium
Mol. wt. 259.1   **M.f.** $C_8H_5Cl_2KO_3$

dicamba-sodium
Mol. wt. 243.0   **M.f.** $C_8H_5Cl_2NaO_3$

COMMERCIALISATION
History Herbicide reported by R. A. Darrow & R. H. Haas (*Proc. South. Weed
Conf., 14th*, 1961, p. 202). Introduced by Velsicol Chemical Corp., later
manufactured and marketed by Sandoz AG (now Novartis Crop Protection AG).
Now marketed in USA and Canada by BASF and elsewhere by Novartis.
Patent US 3013054   **Manufacturers** BASF

APPLICATIONS
Mode of action Selective systemic herbicide, absorbed by the leaves and roots,
with ready translocation throughout the plant via both the symplastic and
apoplastic systems. Acts as an auxin-like growth regulator.   **Uses** Control of

annual and perennial broad-leaved weeds and brush species in cereals, maize, sorghum, sugar cane, asparagus, perennial seed grasses, turf, pastures, rangeland, and non-crop land. Used in combinations with many other herbicides. Dosage varies with specific use and ranges from 0.1 to 11.2 kg a.i./ha. **Phytotoxicity** Most legumes are sensitive. **Formulation types** GR; SL. **Mixtures** (*dicamba* +) MCPA; mecoprop; atrazine; 2,4-D; dichlorprop; bromoxynil + mecoprop; 2,4-D + ioxynil; dichlorprop + MCPA; MCPA + mecoprop; mecoprop + triclopyr; and many more. **Compatibility** Precipitation of the free acid from water may occur if the dimethylammonium salt is combined with lime sulfur, heavy-metal salts, or strongly acidic materials. **Selected tradenames** 'Banvel' (dimethylammonium salt) (Novartis, BASF); 'Banvel SGF' (sodium salt) (Novartis, BASF); 'Marksman' (potassium salt, mixture) (Novartis, BASF); 'Diptyl' (Agriphar); 'Sivel' (dimethylammonium salt) (Siapa)

ANALYSIS
Product analysis by i.r. spectrometry (*AOAC Methods*, 1995, 969.07, 971.07; *CIPAC Handbook*, 1980, **1A**, 1204; M. A. Malina, *Anal. Methods Pestic. Plant Growth Regul.*, 1973, **7**, 545) or by hplc (*AOAC Methods*, 1995, 984.07; *CIPAC Handbook*, 1988, **D**, 51). **Residues** in plants and soil determined by glc of a suitable ester (*idem, ibid.*; H. K. Suzuki *et al., ibid.*, 1978, **10**, 305). In **drinking water**, dicamba, its 5-hydroxy, and des-methoxy derivatives may be determined by conversion to methyl ester with diazomethane, then gc with ECD (*AOAC Methods*, 1995, 992.32, 10.7.03).

MAMMALIAN TOXICOLOGY
dicamba
Oral Acute oral LD_{50} for rats 1707 mg/kg. **Skin and eye** Acute percutaneous LD_{50} for rabbits >2000 mg/kg. Extremely irritating and corrosive to eyes; moderately irritating to skin (rabbits). **Inhalation** LC_{50} (4 h) for rats >9.6 mg/l. **NOEL** (2 y) for rats 110 mg/kg b.w. daily; (1 y) for dogs 52 mg/kg b.w. daily. Developmental NOEL for rabbits 30 mg/kg b.w. daily, rats 160 mg/kg b.w. daily. Reproduction NOEL for rats 50 mg/kg b.w. daily. Not mutagenic. **Toxicity class** WHO (a.i.) III (Table 5); EPA (formulation) III **EC risk** R41

dicamba-dimethylammonium
Oral Acute oral LD_{50} for rats 1267 mg/kg (calc.).

dicamba-sodium
Oral Acute oral LD_{50} for female rats 4600 mg/kg.

ECOTOXICOLOGY
dicamba
Birds Acute oral LD_{50} for mallard ducks 2000 mg/kg. Eight-day dietary LC_{50} for mallard ducks and bobwhite quail >10 000 mg/kg diet. **Fish** LC_{50} (96 h) for

rainbow trout and bluegill sunfish 135 mg/l.   **Bees** Not toxic to bees; LD_{50} >100 µg/bee.   **Daphnia** LC_{50} (48 h) 110 mg/l.

ENVIRONMENTAL FATE

Animals In mammals, following oral administration, dicamba is rapidly eliminated in the urine, partly as a glycine conjugate.   **Plants** The degradation rate in plants varies greatly with species. In wheat, the major metabolite is 5-hydroxy-= 2-methoxy-3,6-dichlorobenzoic acid, whilst 3,6-dichlorosalicylic acid is also a metabolite.   **Soil/Environment** In soil, microbial degradation occurs, the principal metabolite being 3,6-dichlorosalicylic acid. Under conditions amenable to rapid metabolism, DT_{50} <14 d. K_{oc} = 2.

211 dichlobenil   *Herbicide*

benzonitrile

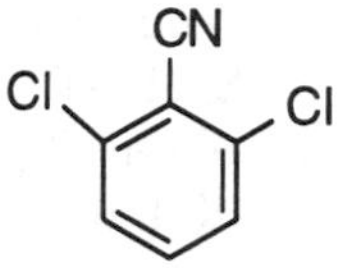

NOMENCLATURE

Common name dichlobenil (BSI, E-ISO, (*m*) F-ISO, ANSI, WSSA); DBN (JMAF)
IUPAC name 2,6-dichlorobenzonitrile
Chemical Abstracts name 2,6-dichlorobenzonitrile
CAS RN *[1194–65–6]*   **EEC no.** 214–787–5   **Development codes** H 133

PHYSICAL CHEMISTRY

Composition Tech. grade dichlobenil is ≥98% pure.   **Mol. wt.** 172.0
M.f. $C_7H_3Cl_2N$ **Form** Colourless crystals with an aromatic odour (tech., off-white crystals).   **M.p.** 145–146 °C (tech., 143.8–144.3 °C)   **B.p.** 270 °C/760 mmHg
V.p. 88 mPa (20 °C) (gas saturation method) K_{ow} logP = 2.70   **Henry** 1.10 Pa m^3 mol^{-1} (calc.)   **Solubility** In water 14.6 mg/l (20 °C). In acetone, dioxane, benzene 50 (all in g/l, 8 °C). In dichloromethane 100 g/l (20 °C). In xylene 53, ethanol 15, cyclohexane 3.7 (all in g/l, 25 °C). Very slightly soluble in non-polar solvents (<10 g/l).   **Stability** Stable to heat, <270 °C. Stable to acids, but rapidly hydrolysed by strong alkalis to 2,6-dichlorobenzamide. Photolytic DT_{50} in water 10.2 d (natural sunlight at 40° Northern latitude).

COMMERCIALISATION

History Herbicidal properties reported by H. Koopman & J. Daams (*Nature (London)*, 1960, **186**, 89). Introduced by Philips-Duphar B.V. (now Uniroyal Chemical Co., Inc).   **Patent** NL 572662; US 3027248   **Manufacturers** Uniroyal

APPLICATIONS

Biochemistry Cellulose biosynthesis inhibitor; has no effect on cell respiration or photosynthesis.   **Mode of action** Systemic herbicide. Inhibits actively dividing meristems, germination of seeds and damages rhizomes. Its selectivity can be ascribed to the fact that it is bound to the top 5–10 cm of the soil.   **Uses** For selective weed control of annual and many perennial weeds in woody ornamentals, fruit orchards, vineyards, bush fruit, forest plantations, public green areas at dosages between 2.7 and 5.4 kg a.i./ha. For total weed control in non-crop areas at dosages up to 8.1 kg a.i./ha. Control of floating, emergent or submerged aquatic plant growth in non-flowing water at 2.7–8.1 kg/ha, depending on water depth.   **Phytotoxicity** Some conifers are susceptible to dichlobenil vapour due to their bark structure.   **Formulation types** GR; WP.
Mixtures (*dichlobenil +*) diuron.   **Selected tradenames** 'Casoron' (Uniroyal); 'Barrier' (PBI/Gordon); 'Silbenil' (Siapa)

ANALYSIS

Product analysis by glc with FID (*CIPAC Handbook*, 1983, **1B**, 1769; *ibid.*, 1992, **E**, 65; *AOAC Methods*, 1995, 979.03; A. van Rossum, *Anal. Methods Pestic. Plant Growth Regul.*, 1978, **10**, 311) or by spectrometry.   **Residues** determined by glc (K. I. Beynon *et al.*, *J. Sci. Food Agric.*, 1966, **17**, 151).

MAMMALIAN TOXICOLOGY

Oral Acute oral LD_{50} for rats 4460, male mice 1014, female mice 1621 mg/kg.
Skin and eye Acute percutaneous LD_{50} for albino rabbits >2000 mg/kg. Non-irritating to skin or eyes (rabbits).   **Inhalation** LC_{50} (4 h) for rats >250 mg/m^3.
NOEL (2 y) for rats 50 mg/kg diet. In 2-generation feeding study for rats, NOEL 60 mg/kg diet.   **ADI** 0.025 mg/kg b.w.   **Other** Acute i.p. LD_{50} for female mice 603, rats 442 mg/kg.   **Toxicity class** WHO (a.i.) III (Table 5); EPA (formulation) III
EC risk Xn (R21)

ECOTOXICOLOGY

Birds Acute oral LD_{50} for bobwhite quail 683 mg/kg. Eight-day dietary LC_{50} for bobwhite quail *c.* 5200, mallard ducks >5200 mg/kg diet.   **Fish** LC_{50} (96 h) 5–13 mg/l (various fish species).   **Bees** Not toxic to bees; LD_{50} (contact) >11 µg/bee.   **Worms** LD_{50} for earthworms >1000 mg/kg substrate.
Other beneficial spp. Harmless to carabids. No effect on soil microflora.
Daphnia LC_{50} (48 h) 6.2 mg/l.   **Algae** EC_{50} (5 d) for *Selenastrum capricornutum* 2.0, *Anabaena flos-aquae* 2.7 mg/l.

ENVIRONMENTAL FATE

Animals Metabolised and excreted mainly as hydroxylated conjugates. For fate in animals, see K. I. Beynon & A. N. Wright, *Residue Rev.*, 1972, **43**, 23; A. Verloop, *ibid.*, 1972, **43**, 55.   **Plants** The soil metabolite 2,6-dichlorobenzamide can be taken up by plants via the roots. Plant metabolism involves ring hydroxylation (at the 3-position and, to a lesser extent, at the 4-position) of both dichlobenil and 2,6-dichlorobenzamide, and subsequently conjugation with a sugar. See K. I. Beynon & A. N. Wright, *Residue Rev.*, 1972, **43**, 23; A. Verloop, *ibid.*, 1972, **43**, 55.   **Soil/Environment** Has a low leaching potential. In soil, dichlobenil gradually undergoes microbial degradation to 2,6-dichlorobenzamide, which is slowly broken down to 2,6-dichlorobenzoic acid. Half-life of dichlobenil in soil may vary between 1 and 6 months depending on soil type. See K. I. Beynon & A. N. Wright, *Residue Rev.*, 1972, **43**, 23; A. Verloop, *ibid.*, 1972, **43**, 55.

212   dichlofluanid

Fungicide

N-trihalomethylthio

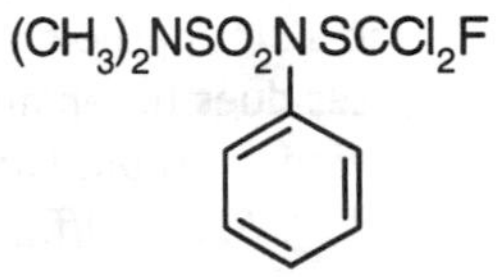

$(CH_3)_2NSO_2N\,SCCl_2F$

NOMENCLATURE

Common name dichlofluanid (BSI, E-ISO); dichlofluanide ((*f*) F-ISO); dichlorfluanid (JMAF); no name (USA)

IUPAC name *N*-dichlorofluoromethylthio-*N'*,*N'*-dimethyl-*N*-phenylsulfamide

Chemical Abstracts name 1,1-dichloro-*N*-[(dimethylamino)sulfonyl]-1-fluoro-*N*-=phenylmethanesulfenamide

CAS RN *[1085–98–9]*   **EEC no.** 214–118–7   **Development codes** Bayer 47 531; KUE 13032c

PHYSICAL CHEMISTRY

Mol. wt. 333.2   **M.f.** $C_9H_{11}Cl_2FN_2O_2S_2$ **Form** Colourless powder.   **M.p.** 106 °C   **V.p.** 0.015 mPa (20 °C)   **K_{ow}** logP = 3.7   **Solubility** In water 1.3 mg/l (20 °C). In dichloromethane >200, toluene 145, xylene 70, methanol 15, isopropanol 10.8, hexane 2.6 (all in g/l, 20 °C).   **Stability** Decomposed by alkaline media; DT_{50} (22 °C) >15 d (pH 4), >18 h (pH 7), <10 min (pH 9), and by polysulfides. Sensitive to light.

COMMERCIALISATION
History Fungicide reported by H. W. K. Müller (*Erwerbsobstbau*, 1964, **6,** 67); it is a member of the aryl(dichlorofluoromethylthio) compounds described by E. Kühle (*Angew. Chem.*, 1964, **76,** 807). Introduced by Bayer AG. **Patent** DE 1193498 **Manufacturers** Bayer

APPLICATIONS
Mode of action Basic contact foliar fungicide with protective action.
Uses Control of scab, brown rot and storage diseases in apples and pears; *Botrytis* spp., *Alternaria* spp., *Clasterosporium* spp., downy mildews, and other fungal diseases on vines, pome fruit, stone fruit, berry fruit, strawberries, hops, tomatoes, cucurbits, vegetables, ornamentals, and conifer seedbeds. Suppressive effect on spider and rust mites on fruit, grapes and other crops, with only slight effect on beneficial mites. Recommended in IPM programmes.
Phytotoxicity Some stone fruits and ornamentals may suffer slight injury.
Formulation types WP; DP; WG; Fumigant (smoke). **Mixtures** (*dichlofluanid* +) oxadixyl; copper oxychloride; cymoxanil; tebuconazole.
Compatibility Incompatible with liquid insecticides and with alkaline compounds. Should not be mixed with wetting agents or adhesives.
Selected tradenames 'Elvaron' (Bayer); 'Euparen' (Bayer)

ANALYSIS
Product analysis by hplc (*CIPAC Handbook*, 1985, **1C,** 2083) or by reaction with sodium methoxide and, ultimately, titration of the chloride (*ibid.*, 1983, **1B,** 1774).
Residues determined by glc (*Man. Pestic. Residue Anal.*, 1987, I; R. Brennecke, *Pflanzenschutz-Nachr. Bayer* (*Engl. Ed.*), 1988, **41,** 137; 1989, **42,** 223; A. Ambrus *et al.*, *J. Assoc. Off. Anal. Chem.*, 1981, **64,** 733). Details of methods available from Bayer AG.

MAMMALIAN TOXICOLOGY
Reviews *Pesticide residues in food – 1983*. FAO Plant Production and Protection Paper 56, 1984. *Pesticide residues in food: 1983 evaluations*. FAO Plant Production and Protection Paper 61, 1985. **Oral** Acute oral LD_{50} for rats >5000 mg/kg.
Skin and eye Acute percutaneous LD_{50} for rats >5000 mg/kg. Slight skin irritant; moderate eye irritant (rabbits). Skin sensitiser. **Inhalation** LC_{50} (4 h) for rats c. 1.2 mg/l air (dust), >0.3 mg/l (aerosol). **NOEL** (2 y) for rats <180 mg/kg diet; (2 y) for mice <200 ppm (threshold value); (1 y) for dogs 1.25 mg/kg b.w. **ADI** 0.0125 mg/kg b.w. (Bayer 1995). **Toxicity class** WHO (a.i.) III (Table 5); EPA (formulation) III **EC risk** Xi (R36); (R43)

ECOTOXICOLOGY
Birds Acute oral LD_{50} for Japanese quail >5000 mg/kg. **Fish** LC_{50} (96 h) for rainbow trout 0.01, golden orfe 0.12, bluegill sunfish 0.03 mg/l. **Bees** Not toxic

to bees. **Worms** LC_{50} for *Eisenia foetida* >890 mg/kg dry soil. **Daphnia** LC_{50} (48 h) >1.8 mg/l (90% pre-mix). **Algae** E_rC_{50} for *Scenedesmus subspicatus* 16 ml/l.

ENVIRONMENTAL FATE
Animals In rats, following oral administration, dichlofluanid is rapidly absorbed and excreted, mainly in the urine, with a small proportion in the faeces and via respiration. There is no accumulation in organs and tissues. Dichlofluanid is metabolised to dimethylsulfanilide which is further hydroxylated and/or demethylated. **Plants** In plants, dichlofluanid is metabolised to dimethylsulfanilide which is further demethylated and/or hydroxylated and conjugated.
Soil/Environment Due to its instability in soil, dichlofluanid is not leached into deeper soil layers. The main metabolite (dimethylsulfanilide) is further degraded and, according to parent and aged leaching studies, is unlikely to leach into deeper soil layers.

213 dichlormid *Herbicide safener*

chloroamide

$$Cl_2CHCON(CH_2CH=CH_2)_2$$

NOMENCLATURE
Common name dichlormid (WSSA)
IUPAC name *N,N*-diallyl-2,2-dichloroacetamide
Chemical Abstracts name 2,2-dichloro-*N,N*-di-2-propenylacetamide
CAS RN *[37764–25–3]* **Development codes** R-25788

PHYSICAL CHEMISTRY
Composition Tech. grade is *c.* 95% pure. **Mol. wt.** 208.1 **M.f.** $C_8H_{11}Cl_2NO$
Form Clear viscous liquid; (tech., amber-to-brown). **M.p.** 5.0–6.5 °C (tech.)
B.p. 130 °C/10 mmHg **V.p.** 800 mPa (25 °C) **K_{ow}** logP = 1.84±0.03 (25 °C)
S.g./density 1.202 (20 °C); tech., 1.192–1.204 **Solubility** In water *c.* 5 g/l. In kerosene *c.* 15 g/l. Miscible with acetone, ethanol, and xylene. **Stability** Unstable above 100 °C. Decomposes violently if heated with iron. Stable to light.

COMMERCIALISATION
History Its use to enhance herbicidal selectivity reported by F. Y. Chang *et al.* (*Can. J. Plant Sci.,* 1972, **52**, 707). G. R. Stephenson (*J. Agric. Food Chem.,* 1978, **26**, 137) compared the chemical structure/biological activity of analogues. Introduced by the Stauffer Chemical Co. (now Zeneca Agrochemicals). **Patent** US 4137070
Manufacturers Zeneca

APPLICATIONS

Mode of action See R. E. Wilkinson (*Chemistry and Mode of action of Herbicide Antidotes*, p. 85); uptake by plant roots studied (R. A. Gray & G. K. Joo, *ibid.*, p. 67). **Uses** Increases the tolerance of maize to thiocarbamate herbicides. **Formulation types** EC; GR; Liquid; CS; CG. **Mixtures** (*dichlormid +*) butylate; EPTC; vernolate; atrazine + butylate. **Selected tradenames** 'Eradicane' (mixture) (Zeneca)

ANALYSIS

Product and **residue** analysis by capillary glc, details available from Zeneca.

MAMMALIAN TOXICOLOGY

Oral Acute oral LD_{50} for male rats 2816, female rats 2146 mg/kg.
Skin and eye Acute percutaneous LD_{50} for rabbits >5000, rats >2000 mg/kg. Mild skin irritant; slightly irritating to eyes (rabbits). Mild skin sensitiser (guinea pigs). **Inhalation** LC_{50} (1 h) for rats >5.5 mg/l air. **NOEL** (90 d) for rats 20 mg/kg diet. **Toxicity class** WHO (a.i.) III

ECOTOXICOLOGY

Birds Five-day dietary LC_{50} for mallard ducks 14 500, bobwhite quail >10 000 mg/kg diet. **Fish** LC_{50} (96 h) for rainbow trout 141 mg/l. **Daphnia** LC_{50} (48 h) 161 mg/l.

ENVIRONMENTAL FATE

Animals For metabolism in rats, see J. B. Miallis *et al.*, *Chemistry and Mode of action of Herbicide Antidotes*, p. 109. **Plants** For metabolism in maize, see *idem, ibid.* **Soil/Environment** In soil DT_{50} *c.* 8 d (27–29 °C). For degradation in soil, see *idem, ibid.*

214 dichlorophen *Algicide, fungicide, bactericide*

chlorophenol

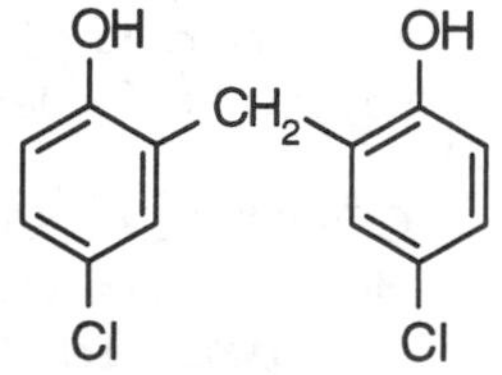

NOMENCLATURE

Common name dichlorophen (BSI, E-ISO, BAN); dichlorophène ((*m*) F-ISO)

IUPAC name 4,4'-dichloro-2,2'-methylenediphenol
Chemical Abstracts name 2,2'-methylenebis[4-chlorophenol]
Other names antiphen   **CAS RN** *[97–23–4]*   **EEC no.** 202–567–1

PHYSICAL CHEMISTRY
Mol. wt. 269.1   **M.f.** $C_{13}H_{10}Cl_2O_2$ **Form** Colourless, odourless crystals; (tech., light tan powder with a slight phenolic odour).   **M.p.** 177–178 °C; (tech., ≥164 °C)   **V.p.** 1.3×10^{-5} mPa (25 °C)   **Solubility** In water 30 mg/l (25 °C). In ethanol 530, isopropanol 540, acetone 800, propylene glycol 450 (all in g/l, 25 °C). Soluble in methanol, isopropyl ether, and petroleum ether. Sparingly soluble in toluene.   **Stability** Slowly oxidised in air. Acidic in reaction, and forms salts with aqueous alkalis. Photolysis in acidic solution in the absence of oxygen results in hydrolysis of one chlorine atom, to give the corresponding phenol; in the presence of oxygen, the corresponding benzoquinone is formed; the same products are formed at pH 9, together with 4-chloro-2,2'-methylenediphenol (E. Mansfield & C. Richard, *Pestic. Sci.*, **48**, 73 (1966)).   **pKa** pK_1 7.6; pK_2 11.6, weak acid

COMMERCIALISATION
History Its activity against mildew on cotton fabrics described by P. B. Marsh & M. L. Butler (*Ind. Eng. Chem.*, 1946, **38**, 701). Introduced by Sindar Corp. and by BDH Ltd.   **Patent** US 2334408

APPLICATIONS
Mode of action Contact action.   **Uses** Control of moss, red thread, *Fusarium* patch, and dollar spot in turf; and of moss on paths, walls, roofs, and non-crop land. Fungicidal, bactericidal and algicidal protection of horticultural benches and equipment, textiles, etc. for control of moulds and algae. Also used as an anthelmintic.   **Formulation types** EC; SL.   **Selected tradenames** 'Super Mosstox' (Rhône-Poulenc)

ANALYSIS
Product analysis by colorimetry (*AOAC Methods*, 1995, 972.48; J. R. Clements & S. H. Newburger, *J. Assoc. Off. Agric. Chem.*, 1954, **37**, 190).

MAMMALIAN TOXICOLOGY
Oral Acute oral LD_{50} for rats 2690, mice 1000, guinea pigs 1250, dogs 2000 mg/kg.   **NOEL** In 90 d feeding trials, rats receiving 2000 mg/kg diet showed no ill-effects.   **Toxicity class** WHO (a.i.) III; EPA (formulation) III   **EC risk** Xn (R22); Xi (R36)

ECOTOXICOLOGY
Fish Toxic to fish.

Nematicide

$$\underset{(E)}{\overset{ClCH_2}{\underset{H}{\diagup}}C=C\overset{H}{\underset{Cl}{\diagdown}}} \qquad \underset{(Z)}{\overset{ClCH_2}{\underset{H}{\diagup}}C=C\overset{Cl}{\underset{H}{\diagdown}}}$$

NOMENCLATURE
Common name 1,3-dichloropropene (BSI, E-ISO accepted in lieu of a common name); dichloro-1,3 propene (F-ISO)
IUPAC name (EZ)-1,3-dichloropropene
Chemical Abstracts name 1,3-dichloro-1-propene
CAS RN [542–75–6] (E)- + (Z)- isomers; [10061–02–6] (E)- isomer; [10061–01–5] (Z)- isomer **EEC no.** 208–826–5

PHYSICAL CHEMISTRY
Composition Tech. product is 92–97% pure. It is a mixture of approximately equal quantities of (E)- and (Z)- isomers. **Mol. wt.** 111.0 **M.f.** $C_3H_4Cl_2$
Form Colourless-to-amber liquid (tech.) with a sweet penetrating odour.
M.p. <–50 °C **B.p.** 108 °C; (Z)- isomer 104.3 °C; (E)- isomer 112 °C
V.p. 3.7 kPa; (E)- isomer 2.3 kPa; (Z)- isomer 3.5 kPa (all at 20 °C)
K_{ow} logP = 1.82 (20 °C); (E)- isomer logP = 2.03; (Z)- isomer logP = 2.06 (both 25 °C) **S.g./density** 1.214 (20 °C); (E)- isomer 1.224; (Z)- isomer 1.217
Solubility In water 2 g/l (20 °C); (E)- isomer 2.32, (Z)- isomer 2.18 (both in g/l, 25 °C). Miscible with hydrocarbons, halogenated solvents, esters, and ketones.
Stability Stable under normal conditions. DT_{50} 11.3 d (pH 5–9, 20 °C).
F.p. 25 °C (Abel closed cup)

COMMERCIALISATION
History Properties as a soil fumigant reported by Dow Chemical Co. (now DowElanco), who introduced it (*Down Earth*, 1956, **12**(2), 7).
Manufacturers DowElanco

APPLICATIONS
Mode of action Soil fumigant nematicide. The (Z)- isomer is more effective than the (E)- isomer to nematodes (C. R. Youngson & C. A. I. Goring, *Plant Dis. Rep.*, 1970, **54**, 196). **Uses** Pre-planting control of most species of nematode in deciduous fruit and nuts, citrus fruit, berry fruit, vines, strawberries, hops, field crops, vegetables, tobacco, beet, pineapples, peanuts, ornamental and flower crops, tree nurseries, etc. Also has secondary insecticidal (soil insects) and fungicidal activity, and, by controlling nematode virus vectors, control is obtained of virus diseases of strawberries, raspberries, tomatoes, hops, etc.
Phytotoxicity Phytotoxic, and should therefore not be applied near desired plants.
Formulation types Liquid. **Mixtures** *(1,3-dichloropropene +)* 1,2-dichloropropane.

Compatibility Compatible with 1,2-dichloropropane, forming the mixture D-D (this name now used for 1,2-dichloropropane itself).
Selected tradenames 'Telone' (DowElanco); 'D-D' (Cyanamid, Hokko); 'Nematrap' (Cyanamid); 'Nematox' (Siapa)

ANALYSIS
Product analysis by glc: details from DowElanco Ltd and Cyanamid.
Residues determined by glc (M. V. McKenry & I. J. Thomason, *loc. cit.*; T. R. Roberts & G. Stoydin, *loc. cit.*). Methods reviewed by J. L. Daft in *Comp. Analyt. Profiles*, Chapter 11.

MAMMALIAN TOXICOLOGY
EHC 146 (WHO, 1993). **Oral** Acute oral LD_{50} for rats 150 mg/kg.
Skin and eye Acute percutaneous LD_{50} for rats 1200 mg/kg; severely irritating to skin, eyes and mucous membranes. Prolonged contact with skin can cause severe burns. **Inhalation** LC_{50} (4 h) for rats 2.70–3.07 mg/l air. See also NOEL data. **NOEL** In 90 d inhalation studies, NOEL for rats and mice 0.05 mg/l air. In 2 y inhalation studies, NOEL for rats 0.099, mice 0.025 mg/l air. Not embryotoxic or teratogenic in rats and rabbits in inhalation studies at 0.59 mg/l. In 2-generation inhalation studies, NOEL on reproduction in rats 0.45 mg/l air. **ADI** Not appropriate. **Other** IARC (1987) Group 2B (Possibly carcinogenic in humans). **Toxicity class** EPA (formulation) III **EC risk** (R10); T (R25); Xn (R20/21); Xi (R36/37/38); (R43)

ECOTOXICOLOGY
Birds LD_{50} for bobwhite quail 152 mg/kg. Five-day dietary LC_{50} for mallard ducks and bobwhite quail >10 000 mg/kg diet. **Fish** LC_{50} (96 h) for rainbow trout 3.9, bluegill sunfish 7.1 mg/l. **Bees** LD_{50} (90 h) 6.6 µg/bee. **Worms** Initially toxic to earthworms; not persistent, therefore recolonisation is very quick. **Daphnia** LC_{50} (48 h) 6.2 mg/l. **Algae** Slightly toxic. **Other aquatic spp.** Moderate acute toxicity.

ENVIRONMENTAL FATE
Animals Excretion in rats occurred primarily as the mercapturic acid conjugate and its corresponding sulfoxide. **Plants** Metabolism is rapid, progressing through 3-chloroallyl alcohol, 3-chloroacrylic acid and 3-chloro-1-propanol, leading ultimately to normal plant constituents (D. L. Berry et al., *J Food Safety*, **4**, 247–55 (1980)); metabolic half-life of 1.5 hours for 1,3-D and 4.4 hours for 3-CAA was calculated. **Soil/Environment** Non-persistent in soil, undergoing hydrolysis to the corresponding 3-chloroallyl alcohols (C. E. Castro & N. O. Belser, *J. Agric. Food Chem.*, 1966, **14**, 69; T. R. Roberts & G. Stoydin, *Pestic. Sci.*, 1976, **7**, 325; M. V. McKenry & I. J. Thomason, *Hilgardia*, 1974, **42**, 393). Aerobic lab. DT_{50} (13 soils) 6–17 d (20 °C). Anaerobic DT_{50} (average) 8.4 d (15 °C), 2.4 d(25 °C). Hydrolysis DT_{50} 11 d (pH 5–9, 20 °C).

aryloxyalkanoic acid

(structure: CH₃ / OCHCO₂H group attached to 2,4-dichlorophenoxy ring)

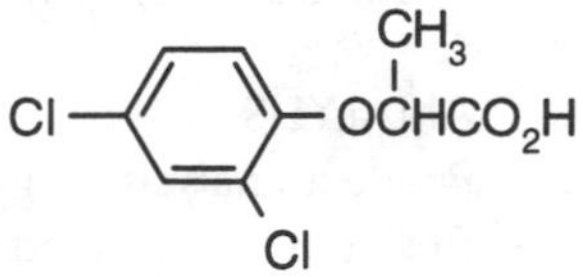

NOMENCLATURE

dichlorprop

Common name dichlorprop (BSI, E-ISO, (*m*) F-ISO, WSSA); [2,4-DP] (former exception, USSR)
IUPAC name (*RS*)-2-(2,4-dichlorophenoxy)propionic acid
Chemical Abstracts name (±)-2-(2,4-dichlorophenoxy)propanoic acid
Other names 2,4-DP   **CAS RN** *[7547–66–2]* racemic acid; *[120–36–5]* dichlorprop (unstated stereochemistry)   **EEC no.** 204–390–5
Development codes RD 406 (AgrEvo)

dichlorprop-isoctyl
CAS RN *[28631–35–8]*

PHYSICAL CHEMISTRY

dichlorprop

Composition Commercial dichlorprop is a racemate, only the (+)-isomer (*q.v.*) being herbicidally active.   **Mol. wt.** 235.1   **M.f.** $C_9H_8Cl_2O_3$ **Form** Colourless crystals; (tech. is a brown powder with a phenolic odour).   **M.p.** 116–117.5 °C; 114 °C (tech.)   **V.p.** $<1\times10^{-2}$ mPa (20 °C) K_{ow} logP = 1.77 (unstated pH)
S.g./density 1.42   **Solubility** In water at 20 °C, 350 mg/l. In acetone 595, isopropanol 510, benzene 85, toluene 69, xylene 51, kerosene 2.1 (all in g/l at 20 °C). Sodium salt: In water at 20 °C, 660 g acid equivalent/l. Diethanolamine salt: In water at 20 °C, 740 g acid equivalent/l.   **Stability** The acid is very stable, and forms sparingly-soluble, slightly-active salts with heavy metals.   **pKa** 3.00
F.p. 204 °C (open cup)

dichlorprop-butotyl (dichlorprop butoxyethyl ester)
Mol. wt. 335.2   **M.f.** $C_{15}H_{20}Cl_2O_4$ **Stability** The esters are hydrolysed on warming with acids or alkalis.

dichlorprop-dimethylammonium
Mol. wt. 280.2   **M.f.** $C_{11}H_{15}Cl_2NO_3$

dichlorprop-isoctyl
Mol. wt. 347.3   **M.f.** $C_{17}H_{24}Cl_2O_3$ **Form** Clear brown-orange liquid with a faint, slightly pungent odour.   **M.p.** <–37 °C   **V.p.** 0.45 mPa (20 °C)

S.g./density c. 1.12   **Stability** Stable after 14 d at 54 °C in the presence of metals. Some degradation over 14 d in sunlight.

dichlorprop-potassium
Mol. wt. 273.2   **M.f.** $C_9H_7Cl_2KO_3$ **Solubility** In water 900 g acid equivalent/l (20 °C).

COMMERCIALISATION
History Although its growth regulating properties were known earlier (P. W. Zimmerman & A. E. Hitchcock, *Proc. Am. Soc. Hortic. Sci.*, 1944, **45**, 353), it was not introduced as a herbicide commercially until 1961, by The Boots Co. Ltd (now AgrEvo GmbH, who do not manufacture or market it).   **Manufacturers** BASF; Marks; Nufarm GmbH; United Phosphorus Ltd

APPLICATIONS
Mode of action Selective, systemic, hormone-type herbicide, absorbed by the leaves, with translocation to the roots. Acts as a growth regulator by inhibiting formation of abscission zone.   **Uses** Post-emergence control of annual and perennial broad-leaved weeds (especially *Polygonum* spp., cleavers, chickweed, bindweed, etc.) in cereals and grassland; brush control on non-crop land; control of broad-leaved aquatic weeds; and chemical maintenance of embankments and roadside verges. Used at 2.7 kg a.e./ha, alone or in combination with other herbicides. Also used to prevent premature fruit fall in apples.
Formulation types SL; EC.   **Mixtures** (*dichlorprop* +) bentazone; bromoxynil; MCPA; 2,4-D; methabenzthiazuron; bentazone + cyanazine; bentazone + MCPA; bromoxynil + ioxynil; clopyralid + MCPA; dicamba + MCPA; and many others.

dichlorprop
Selected tradenames 'Dicopur DP' (Nufarm GmbH); 'Dromone' (Productos OSA); 'Polymone' (Unicrop); 'Redipon' (United Phosphorus Ltd); 'U 46 DP' (BASF); 'Weedone' (Rhône-Poulenc)

dichlorprop-dimethylammonium
Selected tradenames 'Seritone' (mixture) (Nufarm UK)

dichlorprop-potassium
Selected tradenames 'Seritox' (mixture) (Nufarm UK)

ANALYSIS
Product analysis by glc of the methyl ester or by titratable acid content (*CIPAC Handbook*, 1980, **1A**, 1211; 1994, **F**, 292–319). Free phenol impurity determined by glc (*ibid.*, 1994, **F**, 197), hplc (*ibid.*, 1994, **F**, 362) or electrochemically (*ibid.*, 1994, **F**, 368).   **Residue** analysis by glc or hplc (*CIPAC Handbook.*, 1985, **1C**, 2087). In **drinking water**, by conversion to methyl ester with diazomethane, then gc with ECD (*AOAC Methods*, 1995, 992.32, 10.7.03).

MAMMALIAN TOXICOLOGY
dichlorprop
Oral Acute oral LD_{50} for rats 825–1470, mice 400 mg/kg. **Skin and eye** Acute percutaneous LD_{50} for rats >4000, mice 1400 mg/kg; eye and skin irritant (rabbits). Not a skin sensitiser. **Inhalation** LC_{50} (4 h) for rats >0.65 mg/l air.
NOEL (3 mo) for rats 5 mg/kg daily; (2 y) for rats 3.6–4.2 mg/kg daily.
Toxicity class WHO (a.i.) III; EPA (formulation) III **EC risk** Xn (R21/22); Xi (R38, R41); (for salts, Xn (R20/21/22))

ECOTOXICOLOGY
dichlorprop
Birds LD_{50} for Japanese quail 504 mg/kg. **Fish** LC_{50} (96 h) for rainbow trout 521 mg/l. **Bees** Not toxic to bees. **Worms** (14 d) c. 1000 mg/kg dry soil.
Daphnia LOEC (survival) 100 mg/l. **Algae** E_rC_{50} for fresh-water green algae 1100 mg/l (as 600 g/l dimethylammonium salt).

dichlorprop-butotyl (dichlorprop butoxyethyl ester)
Fish LC_{50} (48 h) for bluegill sunfish 1.1 mg/l.

dichlorprop-isoctyl
Fish LC_{50} (48 h) for bluegill sunfish 16 mg/l.

ENVIRONMENTAL FATE
Animals Metabolism studies suggest that dichlorprop is absorbed and excreted essentially unchanged. **Plants** In plants, dichlorprop is metabolised in a similar manner to that in soil. For details of metabolites identified in barley, see G. Baerenwald *et al.*, *Z. Naturforsch. C: Biosci.*, 1987, **42**(4), 486.
Soil/Environment In soil, metabolism involves degradation of the side-chain to 2,4-dichlorophenol, ring hydroxylation, and ring opening; DT_{50} 21–25 d. K_{oc} c. 12–40. Due to the high rate of metabolic breakdown, traces of dichlorprop in leachate are very low.

217 dichlorprop-P   *Herbicide*

aryloxyalkanoic acid

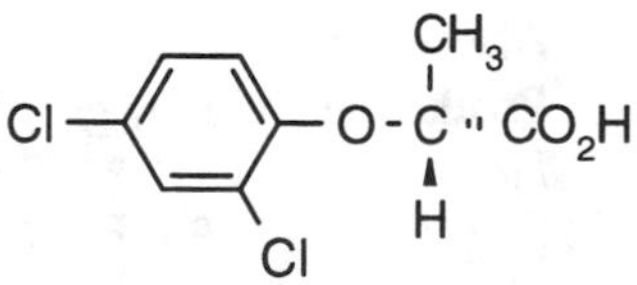

NOMENCLATURE
Common name dichlorprop-P (BSI, pa E-ISO)

IUPAC name (R)-2-(2,4-dichlorophenoxy)propionic acid
Chemical Abstracts name (+)-2-(2,4-dichlorophenoxy)propanoic acid
CAS RN *[15165–67–0]* **EEC no.** 403–980–1 **Development codes** BAS 044 H;
AHM867 (Marks & Co.)

PHYSICAL CHEMISTRY
Mol. wt. 235.1 **M.f.** $C_9H_8Cl_2O_3$ **Form** Colourless crystals with a weak intrinsic
odour. **M.p.** 121–123 °C; tech. 116–120 °C **V.p.** 0.062 mPa (20 °C)
K_{ow} logP = –0.25 (pH 7, 25 °C) **Henry** 2.471 × 10^{-5} Pa m^3 mol^{-1} (calc.)
S.g./density Tech. *c.* 1.47 (20 °C) **Solubility** In water 0.59 g/l (pH 7, 20 °C). In
acetone, ethanol >1000, ethyl acetate 560, toluene 46 (all in g/kg, 20 °C).
Stability Stable to heat and light. **Specific rotation** $[\alpha]_D^{21}$ +26.6° (c. 1.23 ethanol)
(Collins & Smith, *J. Sci. Food Agric.*, 1952, **3**, 248). **pKa** 3.67 (20 °C)

COMMERCIALISATION
History Herbicidal activity of this stereoisomer of dichlorprop reported by
S. T. Collins & F. E. Smith (*J. Sci. Food Agric.*, 1952, **3**, 248), by M. S. Smith *et al.*
(*Nature (London)*, 1952, **169**, 883) and later by W. O. G. Nuyken *et al.* (*Meded.
Fac. Landbouwwet. Rijksuniv. Gent*, 1987, **52**, 1139). Introduced in Germany (1987)
by BASF AG. **Manufacturers** BASF; Marks

APPLICATIONS
Mode of action Selective, systemic, hormone-type herbicide, absorbed by the
leaves, with translocation to the roots. **Uses** A post-emergence translocated
herbicide, particularly effective in controlling *Bilderdykia convolvulus*, *Polygonum
persicaria* and *P. lapathifolium*. Also controls *Galium aparine* and *Stellaria media*, but
is not consistently effective against *P. aviculare*. Used at 1.2–1.5 kg a.i./ha on
cereals alone or in combination with other herbicides. Also used to prevent
premature fruit fall in apples. **Formulation types** EC; DP; SL.
Mixtures (*dichlorprop-P +*) bentazone; 2,4-D; MCPA; bentazone + ioxynil;
bentazone + isoproturon; bentazone + MCPA; MCPA + mecoprop-P; and many
others. **Selected tradenames** 'Duplosan DP' (BASF)

ANALYSIS
Product analysis by rp hplc with u.v. detection (*CIPAC Handbook*, 1995, **G**, 32–38);
enantiomer ratio by chromatography of esters, with u.v. detection (*ibid., idem*).
Residues determined by gc-ms.

MAMMALIAN TOXICOLOGY
Oral Acute oral LD_{50} for rats 825–1470 mg/kg. **Skin and eye** Acute
percutaneous LD_{50} for rats >4000 mg/kg. **Inhalation** LC_{50} (4 h) for rats >7.4
mg/l air. **NOEL** (2 y) for rats 3.6 mg/kg b.w. daily.

MAIN ENTRIES

ECOTOXICOLOGY
Birds Acute oral LD_{50} for quail 250–500 mg/kg.   **Fish** LC_{50} (96 h) for trout
100–220 mg/l.   **Bees** Non-toxic to honeybees; LD_{50} (48 h) >25 µg/bee (tested
as dimethylammonium salt).   **Daphnia** EC_{50} (48 h) >100 mg/l.   **Algae** EC_{50} (72 h)
for *Pseudokirchneriella subcapitata* 676 mg/l.

218 dichlorvos   *Insecticide, acaricide*

organophosphorus

$$Cl_2C = CHO\overset{\text{O}}{\overset{\|}{P}}(OCH_3)_2$$

NOMENCLATURE
Common name dichlorvos (BSI, E-ISO, *(m)* F-ISO, BAN, ESA); DDVP (JMAF);
dichlorfos (USSR); [DDVF] (former exception, USSR)
IUPAC name 2,2-dichlorovinyl dimethyl phosphate
Chemical Abstracts name 2,2-dichloroethenyl dimethyl phosphate
CAS RN *[62–73–7]*   **EEC no.** 200–547–7   **Development codes** Bayer 19 149;
C 177   **Official codes** OMS 14; ENT 20 738

PHYSICAL CHEMISTRY
Mol. wt. 221.0   **M.f.** $C_4H_7Cl_2O_4P$ **Form** Colourless liquid; (tech., colourless-to-
amber liquid with an aromatic odour).   **B.p.** 234.1 °C/1×10^5 Pa (OECD 103);
74 °C/1.3×10^2 Pa   **V.p.** 2.1×10^3 mPa (25 °C) (OECD 104)   **K_{ow}** logP = 1.9
(OECD 117); 1.42 (separate study)   **Henry** 7×10^{-3} Pa m^3 mol^{-1}
S.g./density 1.425 (20 °C) (OECD 109)   **Solubility** In water *c.* 18 g/l (25 °C).
Completely miscible with aromatic hydrocarbons, chlorinated hydrocarbons and
alcohols; moderately soluble in diesel oil, kerosene, isoparaffinic hydrocarbons,
and mineral oils.   **Stability** Stable to heat. Slowly hydrolysed in water and in acidic
media, and rapidly hydrolysed by alkalis, to dimethyl hydrogen phosphate and
dichloroacetaldehyde; DT_{50} (estimated) 31.9 d (pH 4), 2.9 d (pH 7), 2.0 d (pH 9)
(22 °C).   **F.p.** >100 °C (DIN 51758); 172 °C (Pensky-Martens closed cup, 1×10^5
Pa)

COMMERCIALISATION
History Insecticide described by Ciba AG (now Novartis Crop Protection AG)
(GB 775085), but an incorrect structure given to the compound; later reported by
A. M. Martson *et al.* (*J. Agric. Food Chem.*, 1955, **3**, 319) as an insecticidal impurity
in trichlorfon. Introduced by Ciba-Geigy AG (now Novartis Crop Protection AG),
Shell Chemical Co. (now American Cyanamid Co.), and Bayer AG.   **Patent**
GB 775085 to Ciba-Geigy; US 2956073 to Shell   **Manufacturers** Aimco; Amvac;

Bharat; Denka; Jin Hung; Makhteshim-Agan; New Chemi; Nippon Soda; Novartis; Pesticides India; Q.E.A.C.A.; Rallis; Shenzhen Jiangshan; United Phosphorus

APPLICATIONS

Biochemistry Cholinesterase inhibitor. **Mode of action** Insecticide and acaricide with respiratory, contact, and stomach action. Gives rapid knockdown.
Uses Control of household and public health insect pests, e.g. flies, mosquitoes, cockroaches, bedbugs, ants, etc.; stored-product pests in warehouses, storerooms, etc.; flies and midges in animal houses; sciarid and phorid flies in mushrooms; sucking and chewing insects, and spider mites in a wide range of crops, including fruit, vines, vegetables, ornamentals, tea, rice, cotton, hops, glasshouse crops, etc. Also used as a veterinary anthelmintic. **Phytotoxicity** Non-phytotoxic when used as directed, except to some varieties of chrysanthemum.
Formulation types SL; EC; AE; GR; HN; KN; Impregnated strip; OL.
Mixtures (*dichlorvos +*) propoxur; hexythiazox; isoxathion; phosalone; jodfenphos; propetamphos; piperonyl butoxide + pyrethrins; cyfluthrin + propoxur; methoprene + propoxur. **Compatibility** Incompatible with alkaline materials, chinomethionat, and dichlofluanid. **Selected tradenames** 'Dedevap' (Bayer); 'Nuvan' (Novartis); 'Vapona' (agronomic uses only) (Cyanamid); 'Amidos' (Aimco); 'Charge' (Sanonda); 'Denkavepon' (Denka); 'Didivane' (Diachem); 'Divipan' (Makhteshim-Agan); 'Doom' (United Phosphorus); 'Phosvit' (Nippon Soda); 'Rupini' (Ramcides); 'Swing' (Siapa); 'Uniphos' (Florin); 'Vantaf' (Rallis)

ANALYSIS

Product analysis by reaction with excess of iodine which is estimated by titration (*CIPAC Handbook*, 1980, **1A**, 1214) or by glc (*CIPAC Proc.*, 1981, **3**, 173).
Residues determined by glc (*Pestic. Anal. Man.*, 1979, **I**, 201-H, 201-I; *Analyst (London)*, 1973, **98**, 19; 1977, **102**, 858; 1980, **105**, 515; *Man. Pestic. Residue Anal.*, 1987, **I**, 3, 6, S13, S17, S19; *Anal. Methods Residues Pestic.*, 1988, Part I, M2, M5). Sampling of atmospheres (*Anal. Methods Pestic. Plant Growth Regul.*, 1972, **6**, 529). In **drinking water**, by gc with FID (*AOAC Methods*, 1995, 991.07). Methods for the determination of residues are available from Bayer.

MAMMALIAN TOXICOLOGY

Reviews *Pesticide residues in food – 1993*, FAO Plant Production and Protection Paper, 122, 1993. *Pesticide residues in food – 1993 evaluations, Part II – Toxicology*. WHO, WHO/PCS/94.4, 1994. *Food Cosmet. Toxicol.*, 1974, **28**, 765–772 and A. S. Wright et al., *Arch. Toxikol.*, 1979, **42**, 1–18. **IARC** 20, 53 **EHC** 79 (WHO, 1989), 63 (WHO, 1986; a general review of organophosphorus insecticides).
Oral Acute oral LD_{50} for rats c. 50 mg/kg. **Skin and eye** Acute percutaneous LD_{50} for rats c. 90 mg/kg. Skin and eye irritant (rabbits). **Inhalation** LC_{50} (4 h) for rats 340 mg/m^3; (1 h) for rats 455 mg/m^3. **NOEL** (2 y) for rats 10 mg/kg diet. **ADI** (JMPR) 0.004 mg/kg b.w. [1993]. **Toxicity class** WHO (a.i.) Ib; EPA (formulation) I **EC risk** R23/24/25, R38, R43

MAIN ENTRIES

ECOTOXICOLOGY
Birds Acute oral LD_{50} for bobwhite quail 24 mg/kg. Sub-acute oral LD_{50} (8 d) for Japanese quail 300 mg/kg. **Fish** LC_{50} (96 h) for rainbow trout 200, golden orfe 450 µg/l (both 500 EC). **Bees** Acute oral LD_{50} 0.29 µg/bee. **Daphnia** LC_{50} (48 h) 0.19 µg/l. **Algae** EC_{50} (5 d) for *Scenedesmus subspicatus* 52.8 mg/l.

ENVIRONMENTAL FATE
EHC review concludes that, with the exception of gross spillage, recommended use does not constitute a hazard for aquatic or terrestrial organisms. The high toxicity to birds and bees necessitates caution in use. **Animals** In mammals, following oral administration, rapidly degraded in the liver by hydrolysis and O-demethylation, with a half-life of *c.* 25 minutes (L. Bull & R. L. Ridgway, *J. Agric. Food Chem.,* 1969, **17**, 837; D. H. Hutson & E. C. Hoadley, *Arch. Toxikol.,* 1972, **30**, 9–18; A. S. Wright *et al., Arch. Toxikol.,* 1979, **42**, 1–18). **Plants** Rapidly decomposed in plants (D. L. Bull & R. L. Ridgway, *J. Agric. Food Chem.,* 1969, **17**, 837). **Soil/Environment** Non-persistent in the environment, with rapid decomposition in the atmosphere. Undergoes hydrolysis in damp media, with the formation of phosphoric acid and CO_2. Half-lives <1 d in biologically active soils and water systems.

219 diclofop-methyl   *Herbicide*

2-(4-aryloxyphenoxy)propionic acid

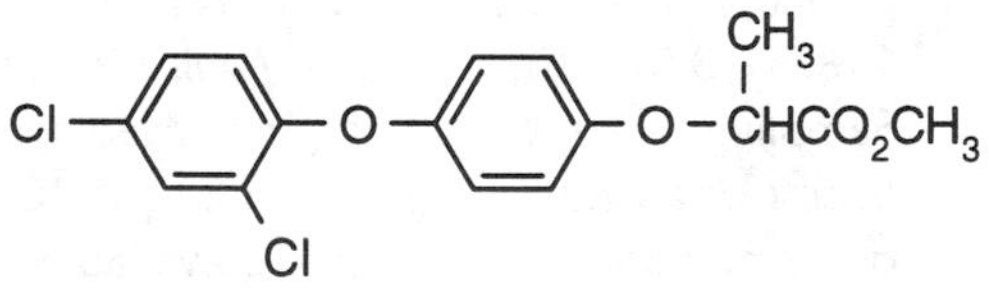

NOMENCLATURE
diclofop-methyl
CAS RN *[51338–27–3]* unstated stereochemistry; *[71283–65–3]* (R)-isomer; *[75021–72–6]* (S)-isomer **EEC no.** 257–141–8 **Development codes** Hoe 023408; AE F023408

diclofop
Common name diclofop (BSI, E-ISO, (*m*) F-ISO, ANSI, WSSA)
IUPAC name (*RS*)-2-[4-(2,4-dichlorophenoxy)phenoxy]propionic acid
Chemical Abstracts name (±)-2-[4-(2,4-dichlorophenoxy)phenoxy]propanoic acid
CAS RN *[40843–25–2]* unstated stereochemistry **Development codes** Hoe 021079

PHYSICAL CHEMISTRY
diclofop-methyl
Composition Tech. grade is ≥93% pure.   **Mol. wt.** 341.2   **M.f.** $C_{16}H_{14}Cl_2O_4$
Form Colourless crystals.   **M.p.** 39–41 °C   **V.p.** 0.25 mPa (20 °C); 7.7 mPa
(50 °C) (vapour pressure balance) K_{ow} logP = 4.58   **S.g./density** 1.30 at 40 °C
Solubility In water 0.8 mg/l (pH 5.7, 20 °C). In acetone, dichloromethane,
dimethyl sulfoxide, ethyl acetate, toluene >500 g/l; in polyethylene glycol 148,
methanol 120, isopropanol 51, *n*-hexane 50 (all in g/l, 20 °C).   **Stability** Stable to
light. In water, DT_{50} (25 °C) 363 d (pH 5), 31.7 d (pH 7), 0.52 d (pH 9).

diclofop
Mol. wt. 327.2   **M.f.** $C_{15}H_{12}Cl_2O_4$ **Form** Yellowish-white solid.
M.p. 118–122 °C   **V.p.** 3.1×10^{-6} mPa (20 °C), 9.7×10^{-6} mPa (25 °C),
1.7×10^{-3} mPa (50 °C) (vapour pressure balance) K_{ow} logP = 2.81 (pH 5), 1.61
(pH 7)   **S.g./density** 1.4 (20 °C)   **Solubility** In water 0.453 (pH 5), 122.7 (pH 7),
127.4 (pH 9) (all in g/l, 20 °C).   **pKa** 3.43

COMMERCIALISATION
History Herbicidal activity of diclofop-methyl reported by P. Langelüddeke *et al.*
(*Mitt. Biol. Bundesanst. Land.-Forstwirtsch. Berlin-Dahlem*, 1975, **165**, 169).
Introduced by Hoechst AG (now AgrEvo GmbH).   **Patent** DE 2136828;
DE 2223894   **Manufacturers** AgrEvo; Jingma

APPLICATIONS
diclofop-methyl
Biochemistry Fatty acid synthesis inhibitor. Destroys the cell membrane, prevents
the translocation of assimilates to the roots, reduces the chlorophyll content, and
inhibits photosynthesis and meristem activity.   **Mode of action** Diclofop-methyl is
a selective systemic herbicide, also with contact action, absorbed primarily by the
leaves, with some absorption by the roots in moist soil. Undergoes rapid
transformation to diclofop, which is translocated within the plant. Diclofop-methyl
(R)-(+)-enantiomer shows significantly greater herbicidal activity by foliar
application to 3 weed species than does the (S)-(–)-enantiomer, but there was less
difference by soil application (H. J. Nestler & H. Bieringer, *Z. Naturforsch. B. Anorg.
Chem. Org. Chem.*, 1980, **35B**, 366).   **Uses** Diclofop-methyl is used for post-
emergence control of wild oats, wild millets, and other annual grass weeds in
wheat, barley, rye, red fescue, and broad-leaved crops such as soya beans, sugar
beet, fodder beet, flax, legumes, oilseed rape, sunflowers, clover, alfalfa, peanuts,
brassicas, carrots, celery, beetroot, parsnips, lettuce, spinach, potatoes,
cucumbers, peas, beans, tomatoes, fennel, alliums, herbs, etc.
Phytotoxicity Phytotoxic to maize, sorghum, oats, sugar cane, rice, and cotton.
Formulation types EC.   **Mixtures** *(diclofop-methyl +)* fenoxaprop-P-ethyl.
Compatibility Incompatible with most other herbicides.
Selected tradenames 'Hoegrass' (AgrEvo); 'Hoelon' (AgrEvo); 'Illoxan' (AgrEvo)

ANALYSIS
Product analysis by glc (*CIPAC Handbook,* 1985, **1C,** 2096). Details of glc methods for **product** and **residue** analysis are available from AgrEvo.

MAMMALIAN TOXICOLOGY
diclofop-methyl
Oral Acute oral LD_{50} for rats 481–693 mg/kg (in sesame oil), dogs 1600 mg/kg (highest dose without induction of vomiting). **Skin and eye** Acute percutaneous LD_{50} for rats >5000 mg/kg. **Inhalation** LC_{50} for rats >1.36 mg/l air. **NOEL** (2 y) for rats 0.1 mg/kg b.w.; (15 mo) for dogs 0.44 mg/kg b.w. **ADI** 0.001 mg/kg b.w. (AgrEvo proposed value). **Other** Non-mutagenic in the Ames test. **Toxicity class** WHO (a.i.) III; EPA (formulation) III **EC risk** Xn (R22, R43)

diclofop
Oral Acute oral LD_{50} for female rats 586 mg/kg. **Skin and eye** Acute percutaneous LD_{50} for rats 1657 mg/kg.

ECOTOXICOLOGY
diclofop-methyl
Birds Acute oral LD_{50} for Japanese quail >10 000 mg/kg. Five-day dietary LD_{50} for bobwhite quail >1600, mallard ducks >1100 mg/kg b.w. **Fish** LC_{50} (96 h) for rainbow trout 0.23 mg/l. **Bees** Non-toxic to bees under field conditions and application rate of 1.134 kg a.i./ha. **Worms** LC_{50} (14 d) for earthworms >1000 mg/kg soil, dry weight. **Daphnia** LC_{50} (48 h) 0.23 mg/l. **Algae** EC_{50} (72 h) for *Scenedesmus subspicatus* 1.5 mg/l, (120 h) for *Selenastrum capricornutum* 0.53 mg/l.

diclofop
Fish LC_{50} (96 h) for rainbow trout 21.9, golden orfe 79.9 mg/l.

ENVIRONMENTAL FATE
Animals When fed to rats, diclofop-methyl is almost totally absorbed and then rapidly excreted; c. 90% is recovered unchanged in faeces and urine after 2 d, and 99% after 7 d. Accumulation of residues in the body is unlikely; low levels of total residues were found in organs and tissues 7 d after a single dose of 1.8 mg/kg b.w. Metabolites are identical with those in plants. **Plants** Diclofop-methyl is taken up rapidly and almost completely by plants, with little translocation. It is hydrolysed relatively quickly (DT_{50} in sugar beet 3 d), first to an isomeric mixture of hydrolysed free acids and their conjugates with glucuronic and sulfuric acid, and then to 4-(2,4-dichlorophenoxy)phenol. Total radioactive residues at the time of harvest in wheat, sugar beet and soya beans are generally low, below or around the determination limits (0.01 to 0.1 mg/kg). The same applies to rotation crops. **Soil/Environment** In soil, diclofop-methyl is metabolised to diclofop, which then undergoes further degradation to 4-(2,4-dichlorophenoxy)phenol, hydroxylated

free acids and CO_2. In various soils in field trials: DT_{50} 1–57 d, DT_{90} 30–281 d. Irrigation studies indicate low levels of leaching. From model calculations, a hazard to groundwater or to drinking water supplies can be excluded, even in sandy soil. Soil adsorption K_{oc} 14 000–24 400 mg/kg.

'20 diclomezine   *Fungicide*

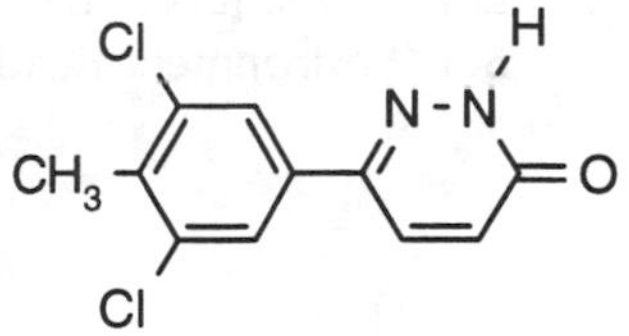

NOMENCLATURE
Common name diclomezine (BSI, draft E-ISO, (*f*) draft F-ISO)
IUPAC name 6-(3,5-dichloro-4-methylphenyl)pyridazin-3(2*H*)-one
Chemical Abstracts name 6-(3,5-dichloro-4-methylphenyl)-3(2*H*)-pyridazinone
CAS RN *[62865–36–5]* **Development codes** F-850 (Sankyo); SF-7531 (Sankyo)

PHYSICAL CHEMISTRY
Mol. wt. 255.1 **M.f.** $C_{11}H_8Cl_2N_2O$ **Form** Colourless crystals.
M.p. 250.5–253.5 °C **V.p.** <1.3 × 10^{-2} mPa (60 °C) **Solubility** In water 0.74 mg/l (25 °C). In methanol 2.0, acetone 3.4 (both in g/l, 23 °C).
Stability Slowly decomposed by sunlight. Stable under acidic, neutral, and alkaline conditions.

COMMERCIALISATION
History Fungicide reported by Y. Takahi (*Jpn. Pestic. Inf.*, 1988, No. 52, p. 31). Introduced in Japan (1988) by Sankyo Co., Ltd. **Patent** US 4052395; GB 1533010; JP 1170243 **Manufacturers** Sankyo

APPLICATIONS
Mode of action Fungicide with curative and protective action. Inhibits septum formation and mycelial growth. **Uses** Control of *Rhizoctonia* and *Sclerotinium* spp. in rice; white mould and twig rot on peanuts; and *Rhizoctonia* disease of turf.
Formulation types WP; DP; SC. **Selected tradenames** 'Monguard' (Sankyo)

ANALYSIS
Product analysis by hplc. **Residues** determined by glc of a derivative. Details available from Sankyo Co. Ltd.

MAMMALIAN TOXICOLOGY
Oral Acute oral LD$_{50}$ for rats >12 000 mg/kg.   **Skin and eye** Acute percutaneous
LD$_{50}$ for rats >5000 mg/kg; not a skin irritant.   **Inhalation** LC$_{50}$ (4 h) for rats
0.82 mg/l.   **Other** Non-mutagenic and non-teratogenic.
Toxicity class WHO (a.i.) III (Table 5)

ECOTOXICOLOGY
Birds Dietary LC$_{50}$ for bobwhite quail 7000 mg/kg diet.   **Fish** LC$_{50}$ (48 h) for carp
11.9 mg/l.   **Bees** LD$_{50}$ for honeybees 0.01 mg/bee.   **Daphnia** LC$_{50}$ (5 h)
>300 mg/l.

ENVIRONMENTAL FATE
Soil/Environment Readily adsorbed on soil particles.

221 dicloran   *Fungicide*

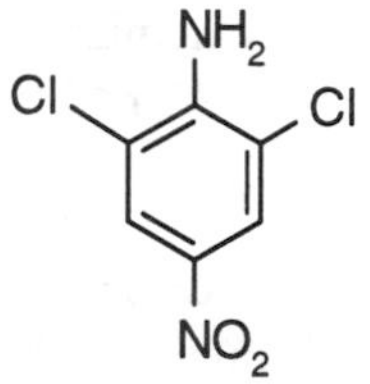

NOMENCLATURE
Common name dicloran (BSI, New Zealand); dichloran (Canada, South Africa);
CNA (JMAF); diclorane (France)
IUPAC name 2,6-dichloro-4-nitroaniline
Chemical Abstracts name 2,6-dichloro-4-nitrobenzenamine
Other names DCNA; ditranil   **CAS RN** *[99–30–9]*   **Development codes**
RD 6584 (Boots); U-2069 (Upjohn); SN 107 682 (AgrEvo)

PHYSICAL CHEMISTRY
Composition Tech. is ≥97% pure.   **Mol. wt.** 207.0   **M.f.** $C_6H_4Cl_2N_2O_2$
Form Yellow crystals.   **M.p.** 195 °C   **V.p.** 0.16 mPa (20 °C); 0.26 mPa (25 °C)
K$_{ow}$ logP = 2.8 (25 °C)   **Henry** 8.4 × 10^{-3} Pa m^3 mol^{-1} (calc.)   **S.g./density** 0.28
(bulk)   **Solubility** In water 6.3 mg/l (20 °C). In acetone 34, dioxane 40,
chloroform 12, ethyl acetate 19, benzene 4.6, xylene 3.6, cyclohexane 0.06 (all in
g/l, 20 °C).   **Stability** Stable to hydrolysis (pH range 5–9) and oxidation. Stable up
to 300 °C; in aqueous solution (pH 7.1) DT$_{50}$ 41 h (λ >290 nm).

COMMERCIALISATION

History Fungicide reported by N. G. Clark *et al.* (*Chem. Ind. (London)*, 1960, p. 572). Introduced by The Boots Co. Ltd (now AgrEvo GmbH, who no longer manufacture or market it). Marketed by Gowan since 1993.   **Patent** GB 845916
Manufacturers Wujiang

APPLICATIONS

Biochemistry Interferes with transfer of ribonucleic acid complex to the ribosome. Inhibits protein synthesis without affecting respiration.   **Mode of action** Protective fungicide which produces hyphal distortion but has little effect on spore germination.   **Uses** Control of *Botrytis, Monilinia, Rhizopus, Sclerotinia,* and *Sclerotium* spp. on stone fruit, berry fruit, vines, rhubarb, potatoes, ornamentals, lettuce, capsicums, cucurbits, glasshouse tomatoes, some vegetables, and other crops.   **Phytotoxicity** Injury is possible on glasshouse ornamentals and young lettuce plants.   **Formulation types** WP; DP; SC.   **Mixtures** (*dicloran* +) captan; imazalil.   **Compatibility** Mixing with some organophosphorus insecticides may lead to phytotoxicity.   **Selected tradenames** 'Allisan' (Gowan); 'Botran' (Gowan)

ANALYSIS

Product analysis by glc.   **Residues** determined by glc with ECD (M. A. Luke *et al., J. Assoc. Off. Anal. Chem.,* 1981, **64,** 1187; *Anal. Methods Pestic. Plant Growth Regul.,* 1972, **6,** 553).

MAMMALIAN TOXICOLOGY

Reviews *Pesticide residues in food – 1977.* FAO Plant Production and Protection Paper 10 Rev, 1978. *Pesticide residues in food: 1977 evaluations.* FAO Plant Production and Protection Paper 10 Sup, 1978.   **Oral** Acute oral LD_{50} for rats 4040, mice 1500–2500, guinea pigs 1450 mg/kg.   **Skin and eye** Acute percutaneous LD_{50} for mice >5000, rabbits >2000 mg/kg.   **Inhalation** LC_{50} (1 h) for rats >21.6 mg/l (for 75% WP).   **NOEL** (2 y) for rats 1000, dogs 100, mice 175 mg/kg diet.   **ADI** (JMPR) 0.03 mg/kg b.w. [1977].   **Toxicity class** WHO (a.i.) III (Table 5); EPA (formulation) III (WP, DP, SC)

ECOTOXICOLOGY

Birds Acute oral LD_{50} for bobwhite quail 900, mallard ducks >2000 mg/kg. Five-day dietary LC_{50} for mallard ducks 5960, bobwhite quail 1435 mg/kg diet.
Fish LC_{50} (96 h) for rainbow trout 1.6, bluegill sunfish 37, goldfish 32 mg/l.
Bees LD_{50} (contact) 0.18 mg/bee.   **Worms** LC_{50} (14 d) for *Eisenia foetida* 885 mg/kg.   **Daphnia** LC_{50} (48 h) 2.07 mg/l.

ENVIRONMENTAL FATE

Animals In rats, following oral administration, dicloran is rapidly metabolised, and excreted in the urine as the sulfate conjugate of 3,5-dichloro-4-aminophenol.
Plants In plants, the principal metabolites are 4-amino-3,5-dichlorophenol,

4-amino-2,6-dichloroacetanilide and 4-amino-2,6-dichloroaniline.
Soil/Environment Microbial degradation involves reduction to 4-amino-2,6-=
dichloroaniline. Soil DT_{50} 39–78 d (sandy loam, pH 6.3–7.1, o.m. 0.7–3.4%); K_{oc}
760 (sand, pH 6.0, 0.48% organic C), 1062 (sandy loam, pH 5.2, 1.45% organic C).

222 dicofol   *Acaricide*

organochlorine

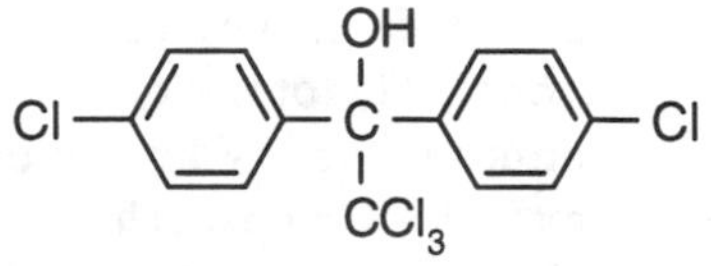

NOMENCLATURE
Common name dicofol (BSI, E-ISO, (m) F-ISO, ESA); kelthane (JMAF); no name
(Germany)
IUPAC name 2,2,2-trichloro-1,1-bis(4-chlorophenyl)ethanol
Chemical Abstracts name 4-chloro-α-(4-chlorophenyl)-α-(trichloromethyl)=
benzenemethanol
CAS RN *[115–32–2]* **EEC no.** 204–082–0 **Development codes** FW-293
Official codes ENT 23 648

PHYSICAL CHEMISTRY
Composition Tech. is 95% pure, composed of 80% dicofol and 20% its
-1-(2-chlorophenyl)-1-(4-chlorophenyl)- isomer ('*o,p*'-dicofol'). **Mol. wt.** 370.5
M.f. $C_{14}H_9Cl_5O$ **Form** Colourless solid; (tech. is a brown viscous oil).
M.p. 78.5–79.5 °C **B.p.** 193 °C/360 mmHg (tech.) **V.p.** 0.053 mPa (tech.)
K_{ow} logP = 4.30 **S.g./density** 1.45 (25 °C; tech.) **Solubility** In water 0.8 mg/l
(25 °C). In acetone, ethyl acetate, toluene 400, methanol 36, hexane, isopropanol
30 (all in g/l, 25 °C). **Stability** Stable to acids, but unstable in alkaline media,
being hydrolysed to 4,4'-dichlorobenzophenone and chloroform; DT_{50} (pH 5)
85 d, (pH 7) 64–99 h, (pH 9) 26 min. The 2,4'-isomer is hydrolysed more rapidly.
Degraded by light to 4,4'-dichlorobenzophenone. Stable ≤80 °C. WP formulations
are sensitive to solvents and surfactants, and these may affect acaricidal activity
and phytotoxicity. **F.p.** 193 °C (open cup)

COMMERCIALISATION
History Acaricide reported by J. S. Barker & F. B. Maugham (*J. Econ. Entomol.*,
1956, **49**, 458). Introduced by Rohm & Haas Co. **Patent** US 2812280;
US 2812362; US 3102070; US 3194730 **Manufacturers** Hindustan; Lainco;
Makhteshim-Agan; Rohm & Haas

APPLICATIONS

Mode of action Non-systemic acaricide with contact action.   **Uses** Non-systemic acaricide with little insecticidal activity, recommended (at 0.50–2.0 kg a.i./ha) for control of many species of phytophagous mite (including *Panonychus*, *Phyllocoptruta*, *Tetranychus*, and *Brevipalpus* spp.) on a wide range of crops, including fruit, vines, ornamentals, vegetables, and field crops.   **Phytotoxicity** Non-phytotoxic when used as directed, but aubergines and pears may be injured.
Formulation types EC; WP; DP; SC.   **Mixtures** *(dicofol +)* tetradifon; hexythiazox; dinocap; ethion; fenson; parathion-methyl; methomyl; sulfur; dimethoate; parathion-methyl + sulfur; ethion + parathion-methyl.
Compatibility Incompatible with highly alkaline materials. WP formulations are sensitive to solvents and surfactants, which may affect acaricidal activity and phytotoxicity.   **Selected tradenames** 'Kelthane' (Rohm & Haas); 'Acarin' (Makhteshim-Agan); 'Cekudifol' (Cequisa); 'Mitigan' (Makhteshim-Agan)

ANALYSIS

Product analysis by hplc (*CIPAC Handbook*, 1988, **D**, 67; *AOAC Methods*, 1995, 986.06), by potentiometric titration of the hydrolysable chlorine (*AOAC Methods*, 1995, 976.02**) or by glc.   **Residues** determined by glc (*Pestic. Anal. Man.*, 1979, 201-A, 201-G, 201-I; M. A. Luke et al., *J. Assoc. Off. Anal. Chem.*, 1981, **64**, 1187; *Anal. Methods Pestic. Plant Growth Regul.*, 1972, **6**, 415; *Manu. Pestic. Residue Anal.*, 1987, **I**, S8, S9, S12, S19).

MAMMALIAN TOXICOLOGY

Reviews *Pesticide residues in food – 1992*. FAO Plant Production and Protection Paper, 116, 1993. *Pesticide residues in food – 1992 evaluations. Part II – Toxicology.* World Health Organisation, WHO/PCS/93.34.   **IARC** 30 (Suppl. 7).   **Oral** Acute oral LD_{50} for male rats 595, female rats 578 mg/kg; rabbits 1810 mg tech./kg.
Skin and eye Acute percutaneous LD_{50} for rats >5000, rabbits >2500 mg/kg.
Inhalation LC_{50} (4 h) for rats >5 mg/l air.   **NOEL** In 2 y combined oncogenic and feeding trials, NOEL for rats 5 mg/kg diet (0.22 mg/kg daily for males, 0.27 mg/kg daily for females). In 2 generation reproduction study, NOEL for rats 5 mg/kg diet (0.5 mg/kg daily). In 1 y feeding study, NOEL for dogs 30 mg/kg diet (0.82 mg/kg daily); in 13 w trial for mice, 10 ppm (2.1 mg/kg b.w. daily).
ADI (JMPR) 0.002 mg/kg [1992].   **Toxicity class** WHO (a.i.) III; EPA (formulation) II or III   **EC risk** Xn (R22); Xi (R38); (R43)

ECOTOXICOLOGY

Birds LC_{50} (5 d) for bobwhite quail 3010, Japanese quail 1418, ring-necked pheasants 2126, mallard ducks 1651 ppm. In eggshell quality and reproduction studies, NOAEL for American kestrel 2, mallard 2.5, bobwhite 110 mg/kg diet.
Fish LC_{50} (96 h) for channel catfish 0.30, bluegill sunfish 0.51, largemouth bass 0.45, fathead minnow 0.183, sheepshead minnow 0.37 mg/l. LC_{50} (24 h) for rainbow trout 0.12 mg/l.   Lifecycle NOEC for fathead minnow 0.0045 mg/l; early

lifestage rainbow trout NOEC 0.0044 mg/l.   **Bees** Not toxic to bees; LD_{50} (contact) >50 µg tech./bee, (oral) >10 µg tech./bee.   **Worms** LC_{50} (7 d) 43.1 ppm, (14 d) 24.6 ppm. **Daphnia** LC_{50} (48 h) 0.14 mg/l.   **Algae** EC_{50} (96 h) for *Scenedesmus* 0.075 mg/l.   **Other aquatic spp.** LC_{50} (96 h) for mysid shrimp (*Mysidopsis bahia*) 0.06 mg/l; EC_{50} for oyster shell 0.15, fiddler crab 64, invertebrate early life stage (*Hyalella*) 0.19 mg/l.

ENVIRONMENTAL FATE

Animals In rats, following oral administration, 4,4'-dichlorobenzophenone and 2,2'-dichloro-1,1'-bis(chlorophenyl)ethanol are the principal metabolites. The same metabolites have been detected in laying hens and dairy goats.   **Plants** The principal metabolite in plants is 4,4'-dichlorobenzophenone.   **Soil/Environment** Soil photodegradation DT_{50} (silt loam) 30 d. Aqueous photodegradation DT_{50} (pH 5, sensitised conditions) 1–4 d; (unsensitised conditions) 15–93 d. Soil metabolism (aerobic) in silt loam 61 d (2,4'-isomer 7 d); (anaerobic) 16 d. Soil adsorption K_{oc} 8383 (sand), 8073 (sandy loam), 5868 (silty loam), 5917 (clay loam). Field dissipation DT_{50} 60–100 d. No mobility of parent or metabolites detected. Dichlorobenzophenone is a major degradate in all processes.

223 dicrotophos   *Insecticide, acaricide*

organophosphorus

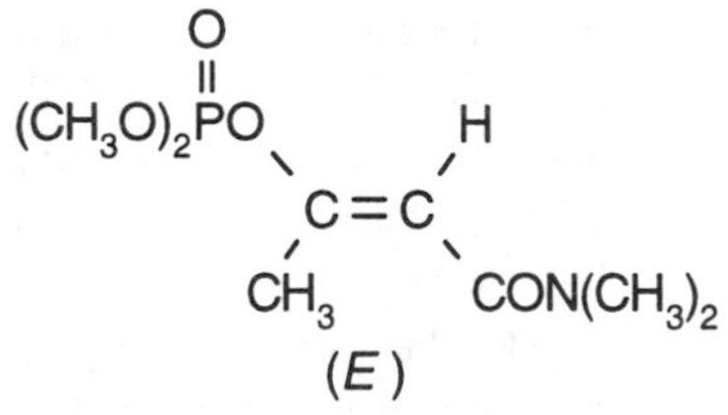

$$(CH_3O)_2\overset{\overset{O}{\|}}{P}O\underset{CH_3}{\diagdown}C=C\underset{CON(CH_3)_2}{\overset{H}{\diagup}}$$

(*E*)

NOMENCLATURE

Common name dicrotophos (BSI, E-ISO, F-ISO, ESA)
IUPAC name (*E*)-2-dimethylcarbamoyl-1-methylvinyl dimethyl phosphate; 3-dimethoxyphosphinoyloxy-*N*,*N*-dimethylisocrotonamide
Chemical Abstracts name (*E*)-3-(dimethylamino)-1-methyl-3-oxo-1-propenyl dimethyl phosphate
CAS RN [141–66–2] dicrotophos; [18250–63–0] the (*Z*)- analogue; [3735–78–2] (*E*)- + (*Z*)- compounds   **EEC no.** 205–494–3   **Development codes** C-709 (Ciba-Geigy); SD 3562 (Cyanamid)   **Official codes** OMS 253; ENT 24 482

PHYSICAL CHEMISTRY

Composition Tech. grade is 85% pure.   **Mol. wt.** 237.2   **M.f.** $C_8H_{16}NO_5P$

Form Yellowish liquid; (tech., amber liquid). **B.p.** 400 °C/760 mmHg; 130 °C/0.1 mmHg **V.p.** 9.3 mPa (20 °C) **S.g./density** 1.216 (20 °C) **Solubility** Totally miscible with water, acetone, alcohols, acetonitrile, chloroform, dichloromethane, xylene. Very slightly soluble (<10 g/kg) in diesel oil, kerosene. **Stability** Relatively stable to acids and alkalis. At 20 °C, 50% hydrolysis occurs in 88 days at pH 5, and in 23 days at pH 9. Subject to thermal decomposition.

COMMERCIALISATION
History Insecticide reported by R. A. Corey (*J. Econ. Entomol.*, 1965, **58**, 112). Introduced by Ciba AG (now Novartis Crop Protection AG, who no longer manufacture or market it) and later by Shell Chemical Co., USA (now DuPont Agricultural Products). **Patent** BE 552284, GB 829576 both to Ciba-Geigy; US 2956073, US 3068268 both to Shell **Manufacturers** Amvac; Cyanamid; Hui Kwang

APPLICATIONS
Biochemistry Cholinesterase inhibitor. **Mode of action** Systemic insecticide and acaricide with contact and stomach action. Moderately persistent. **Uses** Control of sucking, chewing, and boring insects and mites in cotton, coffee, rice, pecans, sugar cane, citrus fruit, tobacco, cereals, potatoes, palms, and other crops. Also used as an animal ectoparasiticide. **Phytotoxicity** Non-phytotoxic if used as directed. Phytotoxic to some varieties of apple and pear under certain conditions. **Formulation types** SL; UL; EC. **Selected tradenames** 'Bidrin' (Cyanamid); 'Dicron' (Hui Kwang)

ANALYSIS
Product analysis by i.r. spectrometry or glc; details from Cyanamid. **Residues** determined by glc (M.C. Bowman & M. Beroza, *J. Agric. Food Chem.*, 1967, **15**, 465; B. Y. Giang & H. F. Beckman, *ibid.*, 1968, **16**, 899). See also: P. E. Porter, *Anal. Methods Pestic. Plant Growth Regul.*, 1972, **6**, 287.

MAMMALIAN TOXICOLOGY
EHC 63 (WHO, 1986; a general review of organophosphorus insecticides). **Oral** Acute oral LD_{50} for rats 17–22, mice 15 mg/kg. **Skin and eye** Acute percutaneous LD_{50} for rats 110–180 mg/kg to 148–181 mg/kg, depending on the carrier and the conditions of the test; for rabbits 224 mg/kg; slight irritant to skin and eyes of rabbits. **Inhalation** LC_{50} (4 h) for rats c. 0.09 mg/l air. **NOEL** (2 y) for rats 1.0 mg/kg diet (0.05 mg/kg daily); for dogs 1.6 mg/kg diet (0.04 mg/kg daily). NOEL in a 3-generation reproduction study with rats 2 mg/kg daily. **Toxicity class** WHO (a.i.) Ib; EPA (formulation) I **EC risk** T+ (R28); T (R24)

ECOTOXICOLOGY
Birds Toxic to birds. Acute oral LD_{50} 1.2–12.5 mg/kg. Not neurotoxic to hens.

Fish LC$_{50}$ (24 h) for mosquito fish 200, harlequin fish >1000 mg/l.   **Bees** Very toxic to honeybees but, because surface residues rapidly decline, little effect seen in practice.

ENVIRONMENTAL FATE
Animals In rats and dogs, following oral administration, complete metabolism and elimination occur within a few days.   **Soil/Environment** The dimethylamino group is converted to an N-oxide and subsequently to -CH$_2$OH and -CHO groups, followed by demethylation and hydrolysis. See K. I. Beynon et al., Residue Rev., 1973, **47**, 55.

224 dicyclanil   *Insecticide*

NOMENCLATURE
Common name dicyclanil (pa ISO)
IUPAC name 4,6-diamino-2-cyclopropylaminopyrimidine-5-carbonitrile
Chemical Abstracts name 4,6-diamino-2-(cyclopropylamino)-5-pyrimidine=carbonitrile
CAS RN *[112636–83–6]*   **Development codes** CGA 183893

PHYSICAL CHEMISTRY
Mol. wt. 190.2   **M.f.** C$_8$H$_{10}$N$_6$

COMMERCIALISATION
History Reported by H. Kristinsson (*Proc. 8th IUPAC International Congress of Pesticide Chemistry*, Washington 1994). Under development by Ciba-Geigy AG (now Novartis Crop Protection AG).

APPLICATIONS
Mode of action When incorporated into the insect breeding substrate, prevents the development of larvae into pupae or adults.   **Uses** Insect growth regulator, with a high specificity against Diptera and Siphonaptera.

dienochlor   *Acaricide*

organochlorine

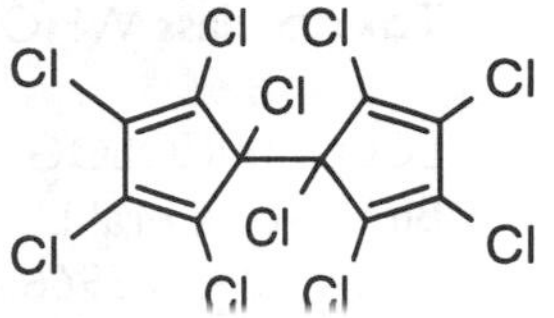

NOMENCLATURE
Common name dienochlor (BSI, E-ISO); diénochlore ((*m*) F-ISO)
IUPAC name perchloro-1,1'-bicyclopenta-2,4-diene
Chemical Abstracts name 1,1',2,2',3,3',4,4',5,5'-decachlorobi-2,4-cyclo=
pentadien-1-yl
CAS RN *[2227–17–0]* **Development codes** SAN 804 I **Official codes** ENT 25
718

PHYSICAL CHEMISTRY
Mol. wt. 474.6 **M.f.** $C_{10}Cl_{10}$ **Form** Grey crystals; (tech., light yellow powder,
sometimes with faint onion or garlic odour). **M.p.** 122–123 °C; (tech.,
111–128 °C) **B.p.** 250 °C (decomp.) **V.p.** 0.29 mPa (extrapolated to 25 °C)
K_{ow} logP = 3.23 (mean, 25 °C) **S.g./density** 1.923 (tech., 25 °C)
Solubility Practically insoluble in water (25 ppb). In iso-octane 7.89, toluene 59.00,
acetonitrile 1.16, *n*-octanol 4.77, tetrahydrofuran 98.32 (all in g tech./100 ml).
Stability Tech. dienochlor was stable when stored at 54 °C for 14 d, and at 42 °C
for 2 y. Hydrolysis DT_{50} (25 °C) 30.5 d (pH 9), 93 d (pH 7), 184 d (pH 5). DT_{50}
as thin film solution in simulated sunlight 1.6 min. **F.p.** 187.8 °C (open cup)

COMMERCIALISATION
History Acaricide reported by W. W. Allen *et al.* (*J. Econ. Entomol.*, 1964, **57**,
187). Introduced by Hooker Chemical Corp. and later by Zoecon Corp., followed
by Sandoz AG (now Novartis Crop Protection AG). **Patent** US 2732409;
US 2934470 to Hooker **Manufacturers** Novartis

APPLICATIONS
Mode of action Acaricide with predominantly contact action. Interferes with
oviposition. Long residual activity under glasshouse conditions. **Uses** Control of
mites (*Tetranychus* spp., *Panonychus ulmi*, and *Polyphagotarsonemus latus*) on roses,
chrysanthemums, and other ornamentals, normally under glass.
Formulation types WP; EW. **Selected tradenames** 'Pentac' (Novartis)

ANALYSIS
Product analysis by hplc spectroscopy. **Residues** determined by hplc.

MAMMALIAN TOXICOLOGY

Oral Acute oral LD_{50} for male and female rats >5000 mg/kg. **Skin and eye** Acute percutaneous LD_{50} for rabbits >3160, rats >2000 mg/kg. Mildly irritating to eyes and skin (rabbits); sensitising to skin. **Inhalation** LC_{50} (4 h) for rats 0.08 mg/l air. **Toxicity class** WHO (a.i.) III; EPA (formulation) I

ECOTOXICOLOGY

Birds Acute oral LD_{50} for bobwhite quail 4319 mg/kg. Eight-day dietary LC_{50} for mallard ducks 3966, bobwhite quail >5620 mg/kg diet. **Fish** LC_{50} (96 h) for rainbow trout 0.050, bluegill sunfish 0.6 mg/l. **Bees** LD_{50} (contact) >36 µg/bee. **Worms** LC_{50} (7d & 14 d) >1000 mg/kg dry soil. **Daphnia** LC_{50} (48 h) 1.2 mg/l. **Algae** NOEC (96 h) for *Scenedesmus subspicatus* >30 mg/l.

ENVIRONMENTAL FATE

Animals Rapid degradation was observed in rats. Within 4 days, following a single oral dose of 1 mg/kg, female rats excreted 2% and 88% of applied ^{14}C in urine and faeces respectively. **Plants** Major degradation products are perchloroketones. Degradation is mainly photochemical rather than metabolic; DT_{50} 2–3 d when plants are exposed to sunlight. **Soil/Environment** Soil adsorption K_d 384 (silty loam, pH 6.1, 3.6% o.m.), 196 (loamy sand, pH 7.5, 0.5% o.m.), 269 (silty clay loam, pH 6.8, 11.8% o.m.), 311 (sandy loam, pH 6.9, 1.3% o.m.). In soil DT_{50} 3.1 d.

226   diethofencarb   *Fungicide*

phenyl carbamate

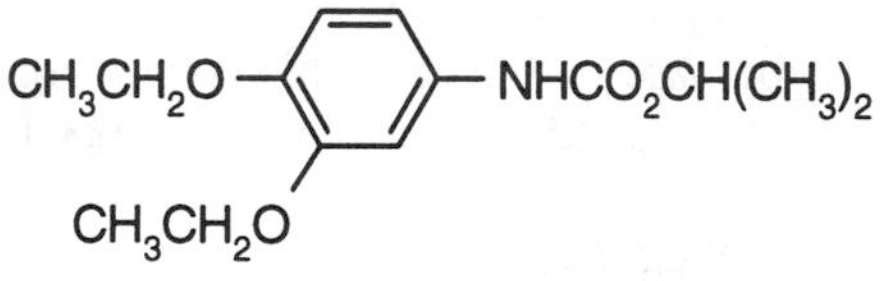

$$CH_3CH_2O{-}\underset{CH_3CH_2O}{\bigcirc}{-}NHCO_2CH(CH_3)_2$$

NOMENCLATURE

Common name diethofencarb (BSI, draft E-ISO); diéthofencarb ((m) draft F-ISO)
IUPAC name isopropyl 3,4-diethoxycarbanilate
Chemical Abstracts name 1-methylethyl (3,4-diethoxyphenyl)carbamate
CAS RN *[87130–20–9]* **Development codes** S-1605

PHYSICAL CHEMISTRY

Mol. wt. 267.3   **M.f.** $C_{14}H_{21}NO_4$ **Form** White crystalline solid. Tech. grade is a

colourless to light brown solid. **M.p.** 100.3 °C **V.p.** 8.4 mPa (20 °C)
K_{ow} logP = 3.02 (25 °C) **S.g./density** 1.19 (23 °C) **Solubility** In water 26.6 mg/l
(20 °C). In hexane 1.3, methanol 101, xylene 30 (all in g/kg, 20 °C). **F.p.** 140 °C

COMMERCIALISATION
History Fungicide introduced by Sumitomo Chemical Co., Ltd. **Patent** EP 78663
Manufacturers Sumitomo

APPLICATIONS
Mode of action Systemic fungicide with protective and curative action. Readily
absorbed though the leaves and roots and translocated throughout the plant.
Inhibits mitosis in grey mould germ tubes. **Uses** Control of benzimidazole-
resistant strains of *Botrytis* spp. on vines, cucumbers, aubergines, tomatoes,
strawberries, citrus fruit, lettuce, onions, and beans. Has secondary activity against
powdery mildews (*Cercospora, Venturia*). **Formulation types** WP.
Mixtures (*diethofencarb* +) carbendazim; procymidone; thiophanate-methyl.
Selected tradenames 'Sumico' (Sumitomo)

ANALYSIS
Product analysis details available from Sumitomo Chemical Co. Ltd.
Residues determined by glc with FTD.

MAMMALIAN TOXICOLOGY
Oral Acute oral LD_{50} for rats >5000 mg/kg. **Skin and eye** Acute percutaneous
LD_{50} for rats >5000 mg/kg. **Inhalation** LC_{50} (4 h) for rats >1050 mg/m^3.
Other Non-mutagenic in the Ames test. **Toxicity class** WHO (a.i.) III (Table 5)

ECOTOXICOLOGY
Birds Acute oral LD_{50} for bobwhite quail and mallard ducks >2250 mg/kg.
Fish LC_{50} (96 h) for rainbow trout >18 mg/l. **Bees** Contact LC_{50} 20 μg/bee.
Daphnia LC_{50} (3 h) (*D. pulex*) >10 mg/l.

ENVIRONMENTAL FATE
Animals After oral administration of ^{14}C to rats, 98.5–100 % of ^{14}C was excreted
within 7 days. The major metabolic routes in the rat were de-ethylation of the
4-ethoxy group, cleavage of the carbamate linkage, acetylation and finally
formation of the glucuronide and sulfate conjugates (*J. Pestic. Sci.*, 1990, **15**, 395).
Plants Readily degraded in plants. **Soil/Environment** Readily degraded in soil;
DT_{50} <1–6 d under aerobic conditions; only very slightly degraded under
anaerobic sterilised conditions.

coumarin anticoagulant

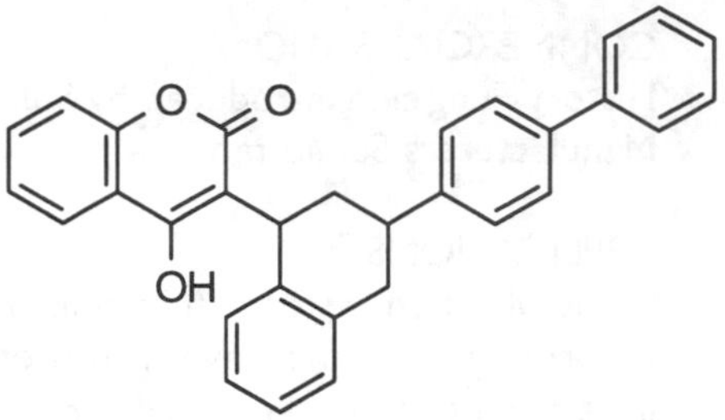

NOMENCLATURE
Common name difenacoum (BSI, E-ISO, (*m*) F-ISO)
IUPAC name 3-(3-biphenyl-4-yl-1,2,3,4-tetrahydro-1-naphthyl)-4-hydroxycoumarin
Chemical Abstracts name 3-[3-(1,1'-biphenyl)-4-yl-1,2,3,4-tetrahydro-=
1-naphthalenyl]-4-hydroxy-2*H*-1-benzopyran-2-one
CAS RN *[56073–07–5]* **EEC no.** 259–978–4

PHYSICAL CHEMISTRY
Composition Tech. material is >90% pure. **Mol. wt.** 444.5 **M.f.** $C_{31}H_{24}O_3$
Form Colourless, odourless crystals; (tech., buff/beige powder).
M.p. 215–217 °C **V.p.** 0.16 mPa (45 °C) **K_{ow}** logP >7 (calc., unionised)
Solubility In water 31×10^{-3} (pH 5.2), 2.5 (pH 7.3), 84 (pH 9.3) (all in mg/l, 20 °C).
Slightly soluble in alcohols. In acetone, chloroform >50, ethyl acetate 2, benzene
0.6 (all in g/l, 25 °C). **Stability** Stable to light, and to temperatures up to 100 °C.

COMMERCIALISATION
History Rodenticide reported by M. Hadler (*J. Hyg.*, 1975, **74**, 441). Introduced by
Sorex (London) Ltd and later by ICI Agrochemicals (now Zeneca Agrochemicals).
Patent GB 1458670 to Sorex **Manufacturers** Sorex; Zeneca

APPLICATIONS
Biochemistry Second-generation anticoagulant rodenticide. Inhibits the vitamin
K-dependent steps in the synthesis of clotting factors II, VII, IX and X.
Mode of action Indirect anticoagulant rodenticide. **Uses** Effective against rats and
most mice resistant to other anticoagulants. For a review, see A. P. Buckle & R. H.
Smith, "*Rodent Pests and their Control*", CABI. **Formulation types** RB; CB; AB; BB.
Mixtures *(difenacoum +)* ergocalciferol. **Selected tradenames** 'Neosorexa'
(Sorex); 'Ratak' (Zeneca); 'Ridak' (Zeneca)

ANALYSIS
Product analysis by hplc (K. Hunter, *Anal. Methods Pestic. Plant Growth Regulat.*,
1988, **16**, 151) or by u.v. spectrometry (details from Sorex (London) Ltd).
Residues determined by hplc (*idem, ibid.*).

MAMMALIAN TOXICOLOGY

EHC 175 (1995). **Oral** Acute oral LD_{50} for male rats 1.8, female rats 2.45, male mice 0.8, rabbits 2.0, female guinea pigs 50, dogs >50, cats >100, pigs >50 mg/kg. Sub-acute oral LD_{50} (5 d) for male rats 0.16 mg/kg/day. **Skin and eye** Acute percutaneous LD_{50} for male rats 27.4, female rats 17.2, rabbits 1000 mg/kg. Non-irritating to skin and eyes (rabbits). **Toxicity class** WHO (a.i.) Ia; EPA (formulation) I **EC risk** T+ (R28); T (R48/25)

ECOTOXICOLOGY

Birds Acute oral LD_{50} for chickens >50 mg/kg. **Fish** LC_{50} (96 h) for rainbow trout 0.10 mg/l. **Daphnia** LC_{50} (48 h) 0.52 mg/l.

ENVIRONMENTAL FATE

Soil/Environment DT_{50} (average) 290 d (range 146–439 d). Unlikely to leach; no leaching in 30 cm lab. columns.

28   difenoconazole   *Fungicide*

azole

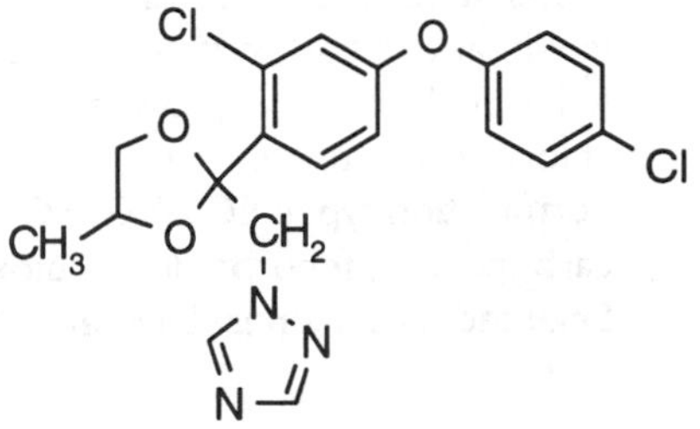

NOMENCLATURE

Common name difenoconazole (BSI, draft E-ISO)
IUPAC name *cis,trans*-3-chloro-4-[4-methyl-2-(1*H*-1,2,4-triazol-1-ylmethyl)-= 1,3-dioxolan-2-yl]phenyl 4-chlorophenyl ether
Chemical Abstracts name 1-[2-[4-(4-chlorophenoxy)-2-chlorophenyl]-4-methyl-= 1,3-dioxolan-2-ylmethyl]-1*H*-1,2,4-triazole
CAS RN *[119446–68–3]* unstated stereochemistry **Development codes** CGA 169 374

PHYSICAL CHEMISTRY

Composition Ratio of *cis*- to *trans*- isomers is in the range 0.7 to 1.5.
Mol. wt. 406.3 **M.f.** $C_{19}H_{17}Cl_2N_3O_3$ **Form** White to light beige crystals.
M.p. 78.6 °C **V.p.** 3.3×10^{-5} mPa (25 °C) **K_{ow}** logP = 4.20 (25 °C)

Henry 1.5×10^{-6} Pa m^3 mol^{-1} (calc.)   **S.g./density** 1.40 (20 °C)   **Solubility** In water 15 mg/l (25 °C). In ethanol 330, acetone 610, toluene 490, n-hexane 3.4, n-octanol 95 (all in g/l, 25 °C).   **Stability** Stable up to 150 °C. Hydrolytically stable.

COMMERCIALISATION

History Fungicide reported by W. Ruess *et al.* (*Proc. 1988 Br. Crop Prot. Conf.*, **2**, 543). Introduced in France (1989) by Ciba-Geigy (now Novartis Crop Protection AG).   **Patent** EP 65485   **Manufacturers** Novartis

APPLICATIONS

Biochemistry Sterol demethylation inhibitor. Inhibits cell membrane ergosterol biosynthesis, stopping development of the fungus.   **Mode of action** Systemic fungicide with preventive and curative action. Absorbed by the leaves with acropetal and strong translaminar translocation.   **Uses** Systemic fungicide with a novel broad-range activity protecting the yield and crop quality by foliar application or seed treatment. Provides long-lasting preventive and curative activity against Ascomycetes, Basidiomycetes and Deuteromycetes including *Alternaria, Ascochyta, Cercospora, Cercosporidium, Colletotrichum, Guignardia, Phoma, Ramularia, Septoria, Venturia* spp., Erysiphaceae, Uredinales and several seed-borne pathogens. Used against disease complexes in grapes, pome fruit, stone fruit, potatoes, sugar beet, oilseed rape, banana, ornamentals and various vegetable crops at 30–125 g a.i./ha. Used as a seed treatment against a range of pathogens in wheat and barley at rates of 3–24 g/100 kg seed.   **Phytotoxicity** In wheat, early foliar applications at growth stages 29–42 might cause, in certain circumstances, chlorotic spotting of leaves, but this has no effect on yield.
Formulation types EC; WG; SC; DS; FS.   **Mixtures** (difenoconazole +) carbendazim; fenpropidin; fludioxonil; metalaxyl-M; penconazole; propiconazole.
Selected tradenames 'Dividend' (Novartis); 'Score' (Novartis)

ANALYSIS

Residue determination by glc. Details from Novartis.

MAMMALIAN TOXICOLOGY

Oral Acute oral LD$_{50}$ for rats 1453, mice >2000 mg/kg.   **Skin and eye** Acute percutaneous LD$_{50}$ for rabbits >2010 mg/kg. Non-irritant to eyes and skin (rabbits). Non-sensitising to skin (guinea pigs).   **Inhalation** LC$_{50}$ (4 h) for rats ≥3300 mg/m^3 air.   **NOEL** (2 y) for rats 1.0 mg/kg b.w. daily; (1.5 y) for mice 4.7 mg/kg b.w. daily; (1 y) for dogs 3.4 mg/kg b.w. daily.   **ADI** 0.01 mg/kg b.w.
Other Not teratogenic or mutagenic.   **Toxicity class** WHO (a.i.) III   **EC risk** R22

ECOTOXICOLOGY

Birds 9–11 day LD$_{50}$ for mallard ducks >2150 mg/kg. LC$_{50}$ for bobwhite quail >4760, mallard ducks >5000 ppm.   **Fish** LC$_{50}$ (96 h) for rainbow trout 0.8,

bluegill sunfish 1.2, sheepshead minnow 0.82 mg/l. **Bees** Non-toxic to honeybees; LD_{50} (oral) >187 μg/bee; LC_{50} (contact) >100 μg/bee.
Worms LC_{50} for earthworms >610 mg/kg. **Daphnia** LC_{50} (48 h) 0.77 mg/l.
Algae EC_{50} (72 h) for *Scenedesmus subspicatus* 1.2 mg/l. **Other aquatic spp.** EC_{50} (96 h) for Eastern oysters 0.45 mg/l; (14 d) for *Lemna gibba* 18.5 mg/l.

ENVIRONMENTAL FATE
Animals After oral administration, difenoconazole was rapidly eliminated practically to entirety, with urine and faeces. Residues in tissues were not significant and there was no evidence for accumulation. **Plants** Two routes of metabolism: one by a triazole route to triazolylalanine and triazolylacetic acid; the other by hydroxylation of the phenyl ring followed by conjugation.
Soil/Environment Practically immobile in soil, strong adsorption to soil particles, low potential to leach below top soil layer. Soil dissipation rate is slow and dependent on application rate. DT_{50} for photolysis 145 d (natural sunlight).

229 difenzoquat metilsulfate

Herbicide, fungicide

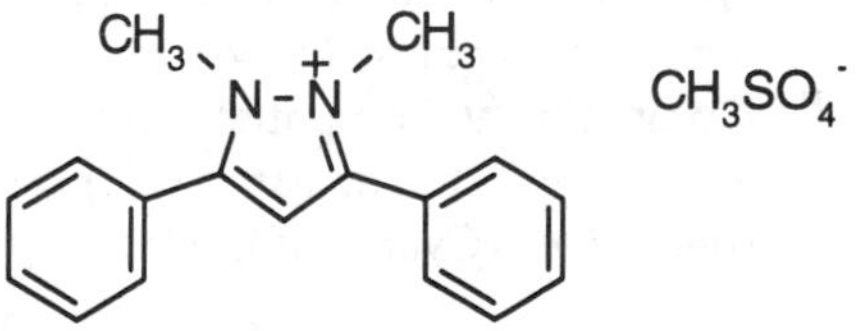

NOMENCLATURE
difenzoquat metilsulfate
Common name difenzoquat metilsulfate
CAS RN *[43222–48–6]* **EEC no.** 256–152–5 **Development codes** AC 84 777; CL 84 777

difenzoquat
Common name difenzoquat (BSI, E-ISO, (*m*) F-ISO, ANSI, WSSA)
IUPAC name 1,2-dimethyl-3,5-diphenylpyrazolium
Chemical Abstracts name 1,2-dimethyl-3,5-diphenyl-1*H*-pyrazolium
CAS RN *[49866–87–7]*

PHYSICAL CHEMISTRY

difenzoquat metilsulfate

Composition Tech. grade is 96% pure.   **Mol. wt.** 360.4   **M.f.** $C_{18}H_{20}N_2O_4S$
Form Colourless, hygroscopic crystals.   **M.p.** 150–160 °C (tech.)   **V.p.** $<10^{-2}$ mPa
K_{ow} logP = 0.648 (pH 5), –0.62 (pH 7), –0.32 (pH 9)   **S.g./density** 0.8 (25 °C)
Solubility In water 765 g/l (25 °C). In dichloromethane 360, chloroform 500,
methanol 588, 1,2-dichloroethane 71, isopropanol 23, acetone 9.8, xylene,
heptane <0.01 (all in g/l, 25 °C). Slightly soluble in petroleum ether, benzene, and
dioxane.   **Stability** Stable to light in aqueous media. Thermally stable. Stable in
weakly acidic media, but decomposed by strong acids and oxidants.   **pKa** *c.* 7
F.p. >82 °C (Tag open cup)

difenzoquat
Mol. wt. 249.3   **M.f.** $C_{17}H_{17}N_2$

COMMERCIALISATION

History Herbicidal properties of difenzoquat metilsulfate (the methyl sulfate)
reported by T. R. O'Hare & C. B. Wingfield (*Proc. North Cent. Weed Control Conf.*,
1973). This salt introduced by American Cyanamid Co; US registration in 1982.
Patent BE 792801; US 3882142   **Manufacturers** Cyanamid

APPLICATIONS

difenzoquat metilsulfate

Mode of action Selective herbicide, absorbed by the leaves, with translocation
mainly acropetally, and accumulation mostly near the treated area. Meristem
inhibitor.   **Uses** Post-emergence control of wild oats in barley, wheat, rye, maize,
ryegrass, and flax; applied at 750–1100 g/ha when used alone. Also used as a
fungicide to control powdery mildew in cereals.   **Formulation types** SL; SP.
Compatibility Compatible with many broad-leaved herbicides, cereal fungicides,
and the growth regulator chlormequat chloride.   **Selected tradenames** 'Avenge'
(herbicide) (Cyanamid); 'Match' (fungicide) (Cyanamid)

ANALYSIS

Product analysis by colorimetry (W. A. Steller, *Anal. Methods Pestic. Plant Growth
Regul.*, 1980, **11**, 291).   **Residues** determined by glc (W. A. Steller, *loc. cit.*).

MAMMALIAN TOXICOLOGY

difenzoquat metilsulfate

Oral Acute oral LD_{50} for male rats 617, female rats 373, male mice 31, female
mice 44 mg/kg.   **Skin and eye** Acute percutaneous LD_{50} for male rabbits
3540 mg/kg. Moderate skin and severe eye irritant (rabbits).   **Inhalation** LC_{50}
(4 h) 0.50 mg/l.   **NOEL** In 2 y feeding trials, rats receiving 500 mg/kg diet
showed no ill-effects.   **ADI** 0.2 mg/kg b.w. daily.   **Toxicity class** WHO (a.i.) II;
EPA (formulation) I (tech.)   **EC risk** Xn (R22)

ECOTOXICOLOGY
difenzoquat metilsulfate
Birds Eight-day dietary LC_{50} for bobwhite quail >4640, mallard ducks
>10 388 mg/kg diet. **Fish** LD_{50} (96 h) for bluegill sunfish 696, rainbow trout
694 mg/l. **Bees** Contact LD_{50} 36 µg/bee. **Daphnia** LC_{50} (48 h) 2.63 mg/l.

ENVIRONMENTAL FATE
Animals In rats, following oral administration, difenzoquat metilsulfate is excreted
unchanged in the urine and faeces. **Plants** No significant metabolism of
difenzoquat metilsulfate occurs in plants, removal being by photolytic
demethylation to the monomethyl pyrazole. **Soil/Environment** Strongly absorbed
by soil; K_d c. 400, K_{oc} c. 30 000. No significant microbial degradation occurs. Half-
life in soil is c. 3 months.

230 difethialone   *Rodenticide*

coumarin anticoagulant analogue

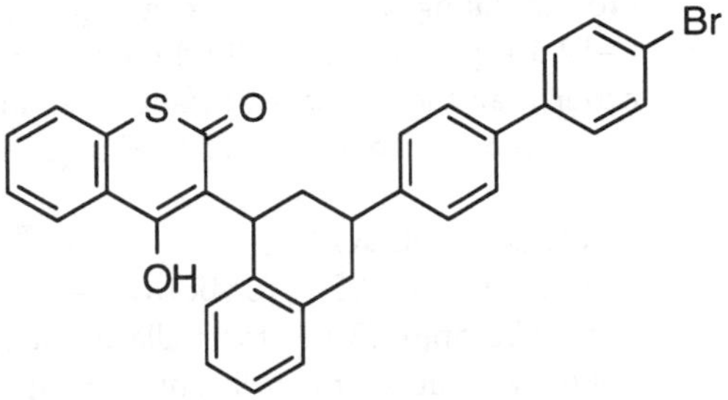

NOMENCLATURE
Common name difethialone (BSI, draft E-ISO, (f) draft F-ISO)
IUPAC name 3-[(1RS,3RS;1RS,3SR)-3-(4'-bromobiphenyl-4-yl)-1,2,3,4-tetrahydro-=
1-naphthyl]-4-hydroxy-1-benzothi-in-2-one where ratios of the racemates
(1RS,3RS) to (1RS,3SR) lie in the range 0–15 to 85–100
Chemical Abstracts name 3-[3-(4'-bromo[1,1'-biphenyl]-4-yl)-1,2,3,4-tetrahydro-=
1-naphthalenyl]-4-hydroxy-2H-1-benzothiopyran-2-one
CAS RN *[104653–34–1]* unstated stereochemistry **Development codes** LM 2219
Official codes OMS 3053

PHYSICAL CHEMISTRY
Composition Ratios of the racemates (1RS,3RS) to (1RS,3SR) lie in the range 0–15
to 85–100. **Mol. wt.** 539.5 **M.f.** $C_{31}H_{23}BrO_2S$ **Form** White, slightly yellowish

powder. **M.p.** 233–236 °C   **V.p.** 0.074 mPa (25 °C) K_{ow} logP = 5.17
S.g./density 1.3614 (25 °C)   **Solubility** In water 0.39 mg/l (25 °C). In ethanol 0.7,
methanol 0.47, hexane 0.2, chloroform 40.8, dimethylformamide 332.7, acetone
4.3 (all in g/l, 20–25 °C).

COMMERCIALISATION
History Rodenticide reported by J. C. Lechevin (*Confed. Europe Assoc. Pestic. Appl.
Parasitis, Geneva*, 1986). Introduced in France (1989) by Lipha.
Manufacturers Lipha

APPLICATIONS
Biochemistry Anticoagulant.   **Uses** Control of commensal rats and mice (including
warfarin-resistant strains).   **Formulation types** AB; BB.
Selected tradenames 'Baraki' (Lipha, Rhône-Poulenc)

ANALYSIS
Product by hplc and u.v. spectrophotometry.

MAMMALIAN TOXICOLOGY
Oral Acute oral LD_{50} for rats 0.56, mice 1.29, dogs 4, pigs 2–3 mg/kg.
Skin and eye Acute percutaneous LD_{50} for male rats 7.9, female rats 5.3 mg/kg.
Non-irritating to skin; mild eye irritant (rabbits).   **Inhalation** LC_{50} (4 h) for rats
5–19.3 µg/l.   **NOEL** In 90 d feeding trials, no toxicological effects were observed
other than those related to anti-vitamin K activity. Non-mutagenic and non-
teratogenic.   **Toxicity class** WHO (a.i.) Ia

ECOTOXICOLOGY
Birds Acute oral LD_{50} for bobwhite quail 0.264 mg/kg. LC_{50} (5 d) for bobwhite
quail 0.56 ppm; (30 d) for mallard ducklings 1.94 ppm.   **Fish** LC_{50} (96 h) for
rainbow trout 51, bluegill sunfish 75 µg/l.   **Daphnia** EC_{50} (48 h) 4.4 µg/l.

ENVIRONMENTAL FATE
Animals In rats, following oral administration, difethialone is characterised by a
short half-life in blood, a longer half-life in the liver, and an essentially faecal
elimination with almost complete absence of metabolism.
Soil/Environment Binds strongly to soil and would be classified as immobile; data
for a range of four soil types: (adsorption) K_d $2.3 \times 10^5 - 2.4 \times 10^7$;
K_{oc} $1.0 \times 10^8 - 5.3 \times 10^9$; (desorption) K_d $1.6 \times 10^5 - 1.8 \times 10^6$;
K_{oc} $5.4 \times 10^7 - 3.9 \times 10^8$.

benzoylurea

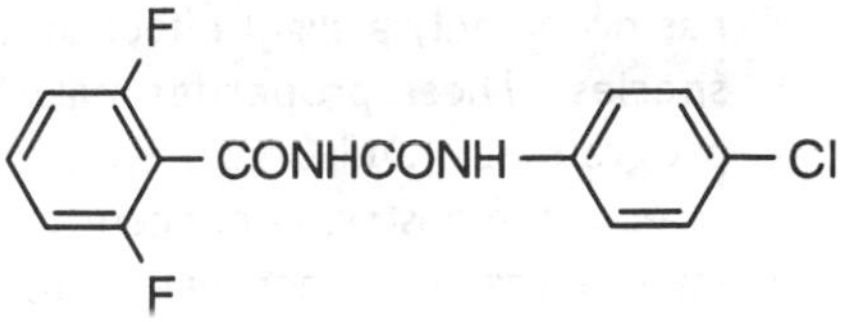

NOMENCLATURE
Common name diflubenzuron (BSI, E-ISO, (*m*) F-ISO, ANSI, ESA)
IUPAC name 1-(4-chlorophenyl)-3-(2,6-difluorobenzoyl)urea
Chemical Abstracts name *N*-[[(4-chlorophenyl)amino]carbonyl]-2,6-difluoro=
benzamide
CAS RN *[35367–38–5]*   **Development codes** DU 112307; PH 60–40; PDD
60–40-I (all Duphar B.V.); TH 6040 (T. H. Agriculture & Nutrition Co. Inc.)
Official codes OMS 1804; ENT 29 054

PHYSICAL CHEMISTRY
Composition Tech. grade diflubenzuron is ≥95% pure.   **Mol. wt.** 310.7
M.f. $C_{14}H_9ClF_2N_2O_2$ **Form** Colourless crystals; (tech., off-white to yellow
crystals).   **M.p.** 228 °C; (tech., 210–230 °C, decomp.)   **V.p.** 1.2×10^{-4} mPa
(25 °C) (gas saturation method) K_{ow} logP = 3.89   **Henry** 4.7×10^{-3} Pa m^3 mol^{-1}
(calc.)   **S.g./density** 1.56   **Solubility** In water 0.08 mg/l (pH 7, 25 °C). In
n-hexane 0.063, toluene 0.29, dichloromethane 1.8, methanol 1.1 (all in g/l,
20 °C).   **Stability** Light-sensitive when in solution, but stable to sunlight as a solid.
<0.5% decomposition after 1-d storage at 100 °C; <0.5% after 7-d at 50 °C.   In
aqueous solution (20 °C), stable at pH 5 and 7 (DT_{50} >150 d), at pH 9 DT_{50} 42 d.

COMMERCIALISATION
History Insecticide reported by J. J. van Daalen et al. (*Naturwissenschaften*, 1972,
59, 312) and reviewed by A. C. Grosscurt (*Pestic. Sci.,* 1978, **9,** 373) and W. Maas
et al. (*Chem. Pflanzenschutz-Schädlingsbekämpfungsmittel,* 1980, **6,** 423). Introduced
in 1975 by Philips-Duphar, B.V. (now Uniroyal Chemical Co., Inc).   **Patent** GB
1324293; US 3748356; US 3989842   **Manufacturers** Uniroyal

APPLICATIONS
Biochemistry Chitin synthesis inhibitor; and so interferes with the formation of the
insect cuticle. This action is quite specific; related biochemical processes, such as
chitin synthesis in fungi, and biosynthesis of hyaluronic acid and other
mucopolysaccharides in chickens, mice and rats are not affected.
Mode of action Non-systemic insect growth regulator with contact and stomach
action. Acts at time of insect moulting, or at hatching of eggs.   **Uses** For control
of a wide range of leaf-eating insects in forestry, woody ornamentals and fruit.

MAIN ENTRIES

Controls certain major pests in cotton, soya beans, citrus, tea, vegetables and mushrooms. Also controls larvae of flies, mosquitoes, grasshoppers and migratory locusts. Used as an ectoparasiticide on sheep for control of lice, fleas and blowfly larvae. Due to its selectivity and rapid degradation in soil and water, diflubenzuron has no or only a slight effect on the natural enemies of various harmful insect species. These properties make it suitable for inclusion in integrated control programmes. Diflubenzuron is effective at 25–75 g a.i./ha against most leaf-feeding insects in forestry; in concentrations of 0.01–0.015% a.i. against codling moth, leaf miners and other leaf-eating insects in top fruit; in concentrations of 0.0075–0.0125% a.i. against citrus rust mite in citrus; and at a dosage of 50–150 g a.i./ha against a number of pests in cotton (cotton boll weevil, armyworms, leafworms), soya beans (soya bean looper complex) and maize (armyworms). Also for control of larvae of mushroom flies in mushroom casing at 1 g a.i./m^2; mosquito larvae (25–100 g a.i./ha); fly larvae in animal housings (0.5–1 g a.i./m^2 surface); and locusts and grasshoppers at a dosage of 60–67.5 g a.i./ha. **Formulation types** WP; UL; GR; HN; OF; SC; WG. **Compatibility** Incompatible with strongly alkaline products. **Selected tradenames** 'Dimilin' (Uniroyal)

ANALYSIS
Product analysis by hplc (*CIPAC Proc.*, 1981, **3**, 185; *AOAC Methods*, 1995, 983.07; A. van Rossum et al., *Anal. Methods Pestic. Plant Growth Regul.*, 1984, **13**, 165). **Residues** determined by hplc or glc after hydrolysis to 4-chloroaniline which is converted to a derivative (B. Rabenort et al., ibid., 1978, **10**, 57).

MAMMALIAN TOXICOLOGY
Reviews *Pesticide residues in food – 1985*. FAO Plant Production and Protection Paper 68, 1986. *Pesticide residues in food – 1985 evaluations. Part II – Toxicology.* FAO Plant Production and Protection Paper 72/2, 1986. IPCS Health & Safety Guide no. 99 (WHO 1995). **EHC** 184 (1996). **Oral** Acute oral LD_{50} for rats and mice >4640 mg/kg. **Skin and eye** Acute percutaneous LD_{50} for rabbits >2000, rats >10 000 mg/kg. Not a skin irritant, slight eye irritant (rabbits); not a skin sensitiser (guinea-pigs). **Inhalation** LC_{50} for rats >2.88 mg/l. **NOEL** (2 y) for rats 40 mg/kg diet. No teratogenic, mutagenic, or oncogenic effect was observed. **ADI** (JMPR) 0.02 mg/kg b.w. [1985]. **Other** Acute i.p. LD_{50} for mice >2150 mg/kg. **Toxicity class** WHO (a.i.) III (Table 5); EPA (formulation) III

ECOTOXICOLOGY
Birds Eight-day dietary LC_{50} for bobwhite quail and mallard ducks >4640 mg/kg diet. **Fish** LC_{50} (96 h) for zebra fish (*Brachidanio rerio*) and rainbow trout >0.2 mg/l. **Bees** Not hazardous to bees and predatory insects; LD_{50} (oral and contact) >100 µg/bee. **Worms** NOEC for *Eisenia foetida* 780 mg/kg substrate. **Daphnia** LC_{50} (48 h) 7.1 µg/l. **Algae** LC_{50} for *Selenastrum capricornutum* >0.3 mg/l. **Other aquatic spp.** LC_{50} for molluscs >200 mg/l.

ENVIRONMENTAL FATE

Animals In rats, following oral administration, elimination is partly as the unchanged parent compound in the faeces, partly as hydroxylated metabolites (for c. 80%) and as 4-chlorophenylurea plus 2,6-difluorobenzoic acid (for c. 20%). The intestinal absorption is strongly related to the dosage administered – the higher the dosage, the more (relatively) is excreted unchanged in the faeces. See A. Verloop & C. D. Ferrell, *ACS Symp. Series*, 1977, No. 37, 237.

Plants Non-systemic. Not metabolised on plants (*idem, ibid.*).

Soil/Environment Diflubenzuron is strongly absorbed by soil/humic acid complex and is virtually immobile in soil. Rapidly degraded in soil, with a half-life of <7 days. The principal degradation products are 4-chlorophenylurea and 2,6-difluorobenzoic acid, (*idem, ibid.*).

232 diflufenican *Herbicide*

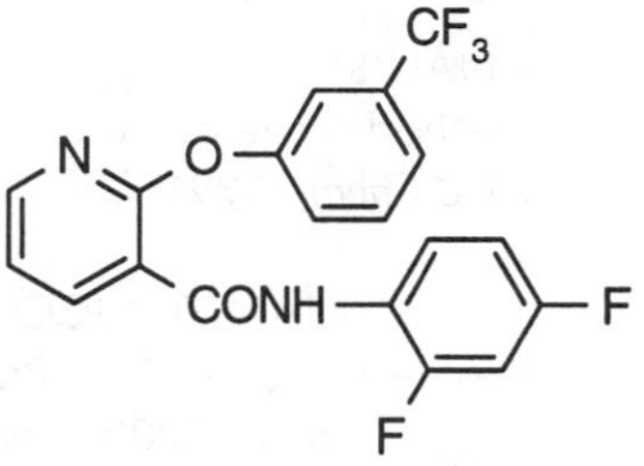

NOMENCLATURE

Common name diflufenican (BSI, draft E-ISO, (*m*) draft F-ISO); diflufenicanil ((*m*) France.)

IUPAC name 2',4'-difluoro-2-(α,α,α-trifluoro-*m*-tolyloxy)nicotinanilide

Chemical Abstracts name *N*-(2,4-difluorophenyl)-2-[3-(trifluoromethyl)= phenoxy]-3-pyridinecarboxamide

Other names DFF   **CAS RN** *[83164–33–4]*   **Development codes** M&B 38 544

PHYSICAL CHEMISTRY

Composition ≥97% pure.   **Mol. wt.** 394.3   **M.f.** $C_{19}H_{11}F_5N_2O_2$ **Form** Colourless crystals.   **M.p.** 159–161 °C   **V.p.** 4.25×10^{-3} mPa (25 °C, gas saturation method) K_{ow} logP = 4.9   **Henry** 0.033 Pa m^3 mol^{-1} **Solubility** In water <0.05 mg/l (25 °C). Soluble in most organic solvents, e.g. acetone, dimethylformamide 100, acetophenone, cyclohexanone 50, isophorone 35, xylene 20, cyclohexane, 2-ethoxyethanol, kerosene <10 (all in g/kg, 20 °C).   **Stability** Stable in air up to melting point. At 22 °C, very stable in aqueous solution at pH 5, 7 and 9. Fairly stable to photolysis.

COMMERCIALISATION

History Herbicide reported by M. C. C. Ramp et al. (*Proc. Br. Crop Prot. Conf. – Weeds*, 1985, **1**, 23) and by C. F. A. Kyndt et al. (*ibid.*, p. 29). Introduced by May & Baker Ltd (now Rhône-Poulenc Agrochimie).   **Patent** EP 53011   **Manufacturers** Rhône-Poulenc

APPLICATIONS

Biochemistry Carotenoid synthesis (phytoene desaturase) & photosynthetic electron flow inhibitor.   **Mode of action** Selective contact and residual herbicide, absorbed principally by the shoots of germinating seedlings, with limited translocation.   **Uses** Applied at 125–250 g/ha pre- or early post-emergence in autumn-sown wheat and barley to control grass and broad-leaved weeds, particularly *Galium*, *Veronica* and *Viola* spp. Normally used in combination with isoproturon or other cereal herbicides.   **Phytotoxicity** Any slight phytotoxicity is in the form of small transient patches on basal leaves, with no adverse effect on crop development.   **Formulation types** WP; SC.   **Mixtures** *(diflufenican +)* isoproturon; ioxynil + mecoprop; ioxynil + isoproturon + mecoprop; bromoxynil + diclofop-= methyl.   **Selected tradenames** 'Fenican' (Rhône-Poulenc); 'Javelin' (mixture) (Rhône-Poulenc)

ANALYSIS

Methods have been developed to determine **residues** in plants and soil (*J. Agric. Food Chem.*, 1991, **39**(5), 968–76).

MAMMALIAN TOXICOLOGY

Reviews *Proc. Br. Crop Prot. Conf. – Weeds* 1985, **(1)**, 23–28.   **Oral** Acute oral LD_{50} for rats >2000, mice >1000, rabbits >5000 mg/kg.   **Skin and eye** Acute percutaneous LD_{50} for rats >2000 mg/kg. Non-irritating to skin and eyes (rabbits).   **Inhalation** LC_{50} (4 h) for rats >2.34 mg/l air.   **NOEL** In 14 d sub-acute trials in rats, no adverse effect observed at 1600 mg/kg b.w. In 90 d feeding trials, NOEL for dogs was 1000 mg/kg b.w. daily, for rats 500 ppm diet.   **Other** Non-mutagenic in the Ames test.   **Toxicity class** WHO (a.i.) III (Table 5)

ECOTOXICOLOGY

Birds Acute oral LD_{50} for quail >2150, mallard ducks >4000 mg/kg.   **Fish** LC_{50} (96 h) for rainbow trout 56–100, carp 105 mg/l.   **Bees** Non-toxic by ingestion or contact.   **Worms** Non-toxic to earthworms.   **Daphnia** LC_{50} (48 h) – no effect at 10 mg/l.   **Algae** No growth inhibition of algae (96 h) at 10 mg/l.

ENVIRONMENTAL FATE

Plants In cereals, rapidly metabolised via the nicotinamide and nicotinic acid to CO_2. Following pre-emergence application in autumn, no residues are detectable in the grain and straw after c. 200–250 days.   **Soil/Environment** In soil, degradation proceeds via the metabolites 2-(3-trifluoromethylphenoxy)=

nicotinamide and 2-(3-trifluoromethylphenoxy)nicotinic acid to bound residues and CO_2. Half-life varies from 15 to 30 weeks, depending on soil type and water content.

.33 diflumetorim

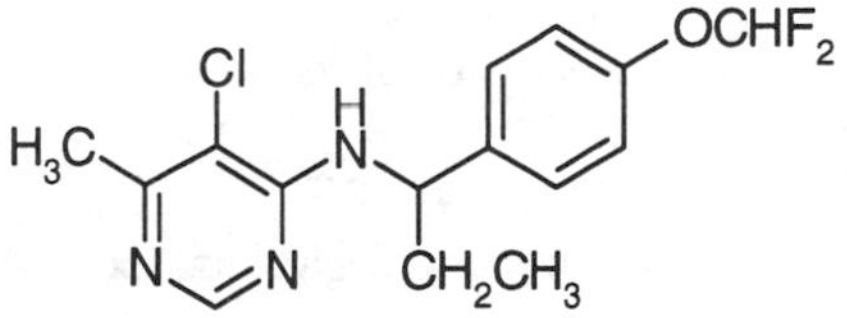

NOMENCLATURE
Common name diflumetorim (BSI, pa ISO)
IUPAC name (*RS*)-5-chloro-*N*-[1-(4-difluoromethoxyphenyl)propyl]-6-methyl= pyrimidin-4-ylamine
Chemical Abstracts name (±)-5-chloro-*N*-[1-[4-(difluoromethoxy)phenyl]= propyl]-6-methyl-4-pyrimidinamine
CAS RN *[130339–07–0]* **Development codes** UBF-002

PHYSICAL CHEMISTRY
Mol. wt. 327.8   **M.f.** $C_{15}H_{16}ClF_2N_3O$ **Form** Pale yellow crystals.
M.p. 46.9–48.7 °C   **V.p.** 3.21×10^{-1} mPa (25 °C)   **K_{ow}** logP = 0.62 (unstated pH)
S.g./density 0.490 (25 °C)   **Solubility** In water 33 mg/l.   **pKa** 4.5, weak base
F.p. 201.3 °C

COMMERCIALISATION
Patent US 5141941; EP 0370704   **Manufacturers** Ube

APPLICATIONS
Uses For control of powdery mildew and rust on cereals and ornamentals.
Formulation types EC.   **Selected tradenames** 'Pyricut' (Ube, Nissan)

MAMMALIAN TOXICOLOGY
Oral LD_{50} for male rats 448, female rats 534, male mice 468, female mice 378 mg/kg.   **Skin and eye** Acute percutaneous LD_{50} for male and female rats >2000 mg/kg.   **Inhalation** LC_{50} for male and female rats 0.61 mg/l.

ECOTOXICOLOGY
Birds LD_{50} for Japanese quail 881, mallard ducks 1979 mg/kg. **Fish** LC_{50} for rainbow trout 0.025, carp 0.098 mg/l. **Bees** LD_{50} (oral) >10 µg/bee. **Daphnia** LC_{50} 0.2 mg/l.

234 *Diglyphus isaea* Biological agent

Hymenoptera

Wasp parasite of leaf miners.

NOMENCLATURE
Scientific name *Diglyphus isaea* (Walker 1934)

PROPERTIES
Stability Shelf life 2–3 d at 8–10 °C in a dark place.

COMMERCIALISATION
History The initial work on the development of *D. isaea* as a biological agent to control leaf miner was completed in the Netherlands during the second half of the 1980's. The original population came from France, supplemented later by large collections of adults found as natural invaders of glasshouses.
Manufacturers Arbico; Biobest; BCP; Koppert; Neudorff; Novartis BCM; Rincon-Vitova

APPLICATIONS
Mode of action *D. isaea* is a larval parasite of leaf miners, although as many leaf miner larvae are killed through adult feeding as from larval feeding. The female *D. isaea* paralyses the host larva by stinging it through the leaf cuticle whilst it is feeding within the leaf. She then lays one or more eggs near the larval host, upon which the emerging larva will feed, eventually pupating within the mine. Development from egg to adult takes about 25 days at 15 °C, reducing to 10 days at 25 °C. The pupal stage represents half the development cycle. Female longevity is 30 days at 20 °C, during which over 250 eggs may be laid. **Uses** *D. isaea* is used as a biological control agent against the tomato leaf miner (*Liriomyza bryoniae*), American serpentine leaf miner (*Liriomyza trifolii*), the pea leaf miner (*Liriomyza huidobrensis*) and the chrysanthemum leaf miner (*Chromatomyia syngenesiae*). Main crops protected are tomato, chrysanthemum, cucumber, celery and certain ornamental species. **Phytotoxicity** Non-phytotoxic and non-phytopathogenic. **Formulation types** Adults without carrying material. **Mixtures** (*D. isaea* +) *Dacnusa sibirica*. **Selected tradenames** 'Dig-line i' (Novartis BCM); 'Miglyphus' (Koppert); 'Minex' (Koppert)

MAMMALIAN TOXICOLOGY

There is no evidence or acute or chronic toxicity, eye or skin irritation or hypersensitivity to mammals. No allergic response or health problems have been observed in production staff or horticultural workers.

.35 dikegulac   *Plant growth regulator*

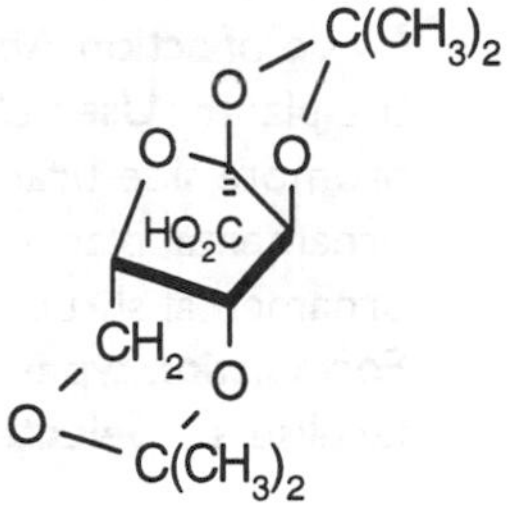

NOMENCLATURE
dikegulac
Common name dikegulac (BSI, E-ISO, (*m*) F-ISO, ANSI)
IUPAC name 2,3:4,6-di-*O*-isopropylidene-α-L-*xylo*-2-hexulofuranosonic acid
Chemical Abstracts name 2,3:4,6-bis-*O*-(1-methylethylidene)-α-L-*xylo*-=
2-hexulofuranosonic acid
CAS RN *[18467–77–1]*

dikegulac-sodium
CAS RN *[52508–35–7]*   **Development codes** Ro 07–6145/001 (Hoffmann-La Roche)

PHYSICAL CHEMISTRY
dikegulac
Mol. wt. 274.3   **M.f.** $C_{12}H_{18}O_7$ **M.p.** 74 °C   **V.p.** 4.0×10^3 mPa (25 °C) (OECD 104) K_{ow} logP = 1.35 (pH 1.9, 25 °C)   **Solubility** In ethanol, acetone >200, dichloromethane 100–200, toluene 2–5, *n*-hexane 0.1–1 (all in g/l, 25 °C).
F.p. >100 °C

dikegulac-sodium
Mol. wt. 296.3   **M.f.** $C_{12}H_{17}NaO_7$ **Form** Colourless crystals.   **M.p.** >300 °C
V.p. $<1.3 \times 10^{-3}$ mPa (25 °C)   **Solubility** In water 590 g/l (25 °C). In methanol 390, ethanol 230, chloroform 63, acetone, dimethylformamide, cyclohexanone, hexane, dioxane <10 (all in g/kg, 25 °C).   **Stability** Stable in the dry state. In

aqueous solution, stable in neutral and alkaline conditions, but unstable in acidic conditions. Decomposed when heated above 50 °C. Stable to light.

COMMERCIALISATION
History Plant growth regulating activity of the sodium salt described by P. Bocion et al. (*Nature (London)*, 1975, **258**, 142). Dikegulac introduced as a herbicide and plant growth regulator by F. Hoffman-La Roche & Co. and dikegulac-sodium by Dr R. Maag Ltd. All rights purchased in 1991 by PBI/Gordon.
Manufacturers PBI/Gordon

APPLICATIONS
dikegulac-sodium
Mode of action Absorbed by the leaves and roots, with translocation throughout the plant. **Uses** Dikegulac-sodium is used to reduce apical dominance and promote side-branching and flower-bud formation on azaleas, fuchsias, and other ornamental plants, and to retard temporarily longitudinal growth on hedges and ornamental shrubs. Also used to retard tree growth by trunk injection.
Formulation types SL. **Compatibility** Should not be mixed with pesticides or fertilisers. **Selected tradenames** 'Atrimmec' (PBI/Gordon)

ANALYSIS
Product and **residue** analysis is by glc after conversion to dikegulac-ethyl.

MAMMALIAN TOXICOLOGY
dikegulac-sodium
Oral Acute oral LD_{50} for male rats 31 000, female rats 18 000, mice 19 500 mg/kg. **Skin and eye** Acute percutaneous LD_{50} for rabbits >1000, rats >2000 mg/kg. Aqueous solutions are non-irritating to skin (guinea pigs) and eyes (rabbits). **NOEL** In 90 d feeding trials, rats receiving 2000 mg/kg daily, and dogs receiving 3000 mg/kg daily showed no ill-effects.
Toxicity class WHO (a.i.) III (Table 5)

ECOTOXICOLOGY
dikegulac-sodium
Birds Five-day dietary LC_{50} for mallard ducks, Japanese quail, and chickens >50 000 mg/kg diet. **Fish** LC_{50} (96 h) for bluegill sunfish >10 000, rainbow trout, goldfish, harlequin fish >5000 mg/l. **Bees** Not toxic to bees. LD_{50} (oral and topical) >0.1 mg/bee.

ENVIRONMENTAL FATE
Soil/Environment Hydrolysed to sodium 2,3-O-isopropylidene 2-keto-L-gulonate, the acid being formed in acidic media.

urea

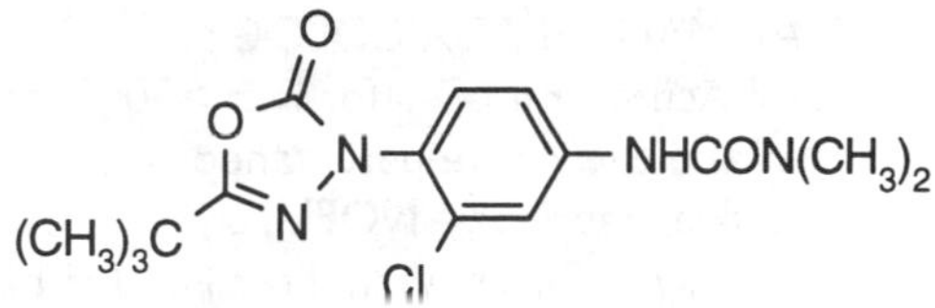

NOMENCLATURE
Common name dimefuron (BSI, E-ISO, (*m*) F-ISO)
IUPAC name 3-[4-(5-*tert*-butyl-2,3-dihydro-2-oxo-1,3,4-oxadiazol-3-yl)-=
3-chlorophenyl]-1,1-dimethylurea
Chemical Abstracts name *N*'-[3-chloro-4-[5-(1,1-dimethylethyl)-2-oxo-=
1,3,4-oxadiazol-3(2*H*)-yl]phenyl]-*N,N*-dimethylurea
CAS RN *[34205–21–5]* **EEC no.** 251–879–4 **Development codes** 23 465 RP

PHYSICAL CHEMISTRY
Mol. wt. 338.8 **M.f.** $C_{15}H_{19}ClN_4O_3$ **Form** Colourless, odourless crystals.
M.p. 193 °C **V.p.** 0.1 mPa (20 °C) **K_{ow}** logP = 2.51 **Solubility** In water 16 mg/l
(20 °C). Readily soluble in chloroform. Moderately soluble in acetone, acetonitrile,
chloroform, acetophenone and ethanol. Slightly soluble in benzene, toluene and
xylene. **Stability** Stable in aqueous solution in acidic conditions.

COMMERCIALISATION
History Herbicide reported by L. Burgaud *et al.* (*Proc. 12th Br. Weed Cont. Conf.*,
1974, **3**, 801). Introduced by Rhône-Poulenc Agrochimie. **Manufacturers** Rhône-
Poulenc

APPLICATIONS
Biochemistry Photosynthetic electron flow inhibitor. **Mode of action** Selective
herbicide, absorbed principally by the roots, but also, to some extent, by the
leaves. **Uses** Applied pre- and post-emergence at 0.2–2.0 kg a.i./ha in field
beans, certain cereals, cotton, peanuts, dormant alfalfa, oilseed rape and peas. It
has been developed for use in winter oilseed rape, where it is of particular value
for its activity against the difficult annual broad-leaved weeds, *Stellaria media*,
Chamomilla and *Matricaria* spp.; it also controls *Alopecurus myosuroides*, *Avena* spp.
and volunteer cereals. In mixture with carbetamide, it forms a useful broad-
spectrum herbicide for use in oilseed rape. It is moderately persistent in soil, a
period of up to 18 weeks being needed between application at rates selective in
oilseed rape and sowing certain subsequent crops. **Formulation types** WP.
Mixtures (*dimefuron +*) carbetamide; clopyralid; benazolin-ethyl; bentazone.
Selected tradenames 'Pradone' (mixture) (Rhône-Poulenc); 'Scorpion' (mixture)
(AgrEvo)

MAIN ENTRIES

ANALYSIS
Product and **residue** analysis by glc. Details available from Rhône-Poulenc Agrochimie.

MAMMALIAN TOXICOLOGY
Oral Acute oral LD_{50} for rats >5000, mice >10 000, dogs >2000 mg/kg.
Skin and eye Acute percutaneous LD_{50} for rats >2000 mg/kg. Non-irritant to eyes and skin (rabbits).   **NOEL** (2 y) for rats 150 mg/kg diet; (90 d) for dogs 20 mg/kg diet daily.   **Other** Acute i.p. LD_{50} for rats c. 4000 mg/kg.
Toxicity class WHO (a.i.) III (Table 5)

ECOTOXICOLOGY
Birds Non-toxic to birds.   **Fish** LC_{50} (96 h) for rainbow trout and bluegill sunfish >1000 mg/l.   **Bees** Non-toxic to bees. LD_{50} (48 h) (contact) >500 µg formulation/bee; (oral) >2563 µg formulation/bee.

ENVIRONMENTAL FATE
Animals Metabolites include desmethylaminodimefuron, monomethylamino-= *tert*-butylhydroxydimefuron and monomethylaminodimefuron.   **Plants** Metabolites include *N*-demethyldimefuron and *N*-didemethyldimefuron.
Soil/Environment Residual activity persists in soil for 3–6 months, following application in late autumn or winter. Microbial degradation increases rapidly in the spring as the temperature rises. K_{des} 19.0–44.9 (4 soil types), K_{oc} 1386–4490 (4 soil types).

237 dimepiperate *Herbicide*

thiocarbamate

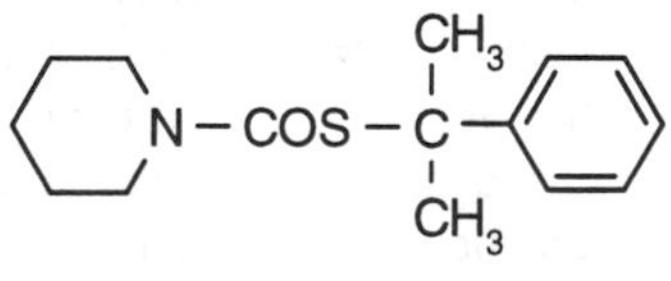

NOMENCLATURE
Common name dimepiperate (BSI, draft E-ISO, (*m*) draft F-ISO)
IUPAC name *S*-1-methyl-1-phenylethyl piperidine-1-carbothioate
Chemical Abstracts name *S*-(1-methyl-1-phenylethyl) 1-piperidinecarbothioate
CAS RN *[61432–55–1]*   **Development codes** MY-93; MUW-1193

PHYSICAL CHEMISTRY
Mol. wt. 263.4 **M.f.** $C_{15}H_{21}NOS$ **Form** Wax-like solid. **M.p.** 38.8–39.3 °C
B.p. 164–168 °C/0.75 mmHg **V.p.** 0.53 mPa (30 °C) K_{ow} logP = 4.02
S.g./density 1.08 (25 °C) **Solubility** In water 20 mg/l (25 °C). In acetone 6.2,
chloroform 5.8, cyclohexanone 4.9, ethanol 4.1, hexane 2.0 (all in kg/l, 25 °C).
Stability Stable >1 y (30 °C), and in light when dry. Aqueous solutions are stable at
pH 1 and pH 14.

COMMERCIALISATION
History Herbicide reported by M. Tanaka (*Jpn. Pestic. Inf.*, 1984, No. 45, p. 18)
and by K. Ikeda & A. Goh (*Jpn. Pestic. Inf.*, 1989, No. 55, p. 15). Introduced
and manufactured by Mitsubishi Petrochemical Co. Ltd. (now Mitsubishi
Chemical Corp.) and commercialised by Rhône-Poulenc Yuka Agro KK,
a joint venture of Mitsubishi Chemical Corp. and Rhône-Poulenc Agro.
Manufacturers Rhône-Poulenc Yuka Agro

APPLICATIONS
Biochemistry Inhibits lipid metabolism. **Mode of action** Systemic herbicide,
absorbed by the roots, leaves and stems, with translocation mainly upwards within
the plant. **Uses** Control of *Echinochloa crus-galli* in flooded rice.
Formulation types GR; EC. **Mixtures** (*dimepiperate +*) bensulfuron-methyl;
bensulfuron-methyl + benfuresate. **Selected tradenames** 'Yukamate' (Mitsubishi)

ANALYSIS
Residues determined by glc.

MAMMALIAN TOXICOLOGY
Reviews *Pesticide Science Society of Japan* **14**(1), 1989. **EHC** 76 (WHO, 1988;
general review of thiocarbamates). **Oral** Acute oral LD_{50} for male rats 946,
female rats 959, male mice 4677, female mice 4519 mg/kg. **Skin and eye** Acute
percutaneous LD_{50} for rats >5000 mg/kg. Non-irritating to eyes and skin
(rabbits). Non-sensitising to skin (guinea pigs). **Inhalation** LC_{50} (4 h) for rats
>1.66 mg/l. **NOEL** (2 y) for rats 0.104 mg/kg. **ADI** 0.001 mg/kg. **Other** No
teratogenic activity detected in rats or rabbits; 2-generation reproduction studies
in rats showed no abnormality. **Toxicity class** WHO (a.i.) III; EPA (formulation) III

ECOTOXICOLOGY
Birds Acute oral LD_{50} for male Japanese quail >2000, hens >5000 mg/kg.
Fish LC_{50} (48 h) for carp 5.8, rainbow trout 5.7 mg/l. **Daphnia** LC_{50} (3 h)
40 mg/l.

ENVIRONMENTAL FATE
Plants Both absorption and translocation are greater in cockspur than in rice.
Soil/Environment Rapidly degraded in paddy fields, DT_{50} <7 d.

chloroacetanilide

$$\begin{array}{c} CH_3 \\ \hline \\ N \\ CH_3 \end{array} \begin{array}{c} COCH_2Cl \\ \\ CH_2CH_2OCH_3 \end{array}$$

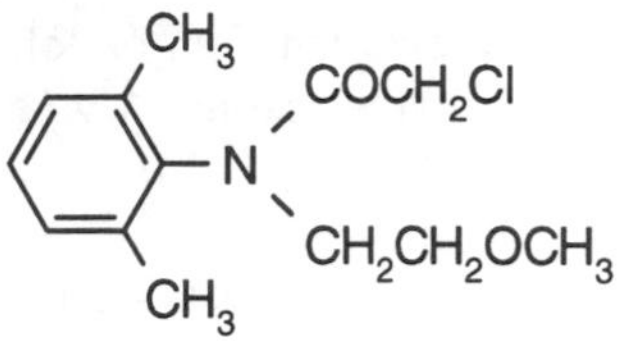

NOMENCLATURE
Common name dimethachlor (BSI, E-ISO, WSSA); diméthachlore ((*m*) F-ISO)
IUPAC name 2-chloro-*N*-(2-methoxyethyl)aceto-2',6'-xylidide
Chemical Abstracts name 2-chloro-*N*-(2,6-dimethylphenyl)-*N*-(2-methoxyethyl)=
acetamide
CAS RN *[50563–36–5]*   **EEC no.** 256–625–6   **Development codes** CGA 17 020

PHYSICAL CHEMISTRY
Mol. wt. 255.7   **M.f.** $C_{13}H_{18}ClNO_2$ **Form** Colourless crystals.
M.p. 45.8–46.7 °C   **B.p.** *c.* 320 °C (decomp. *c.* 300 °C)   **V.p.** 1.5 mPa (25 °C)
K_{ow} logP = 2.17   **Henry** 1.7×10^{-4} Pa m^3 mol^{-1} (calc.)   **S.g./density** 1.23 (20 °C)
Solubility In water 2.3 g/l (25 °C). Readily soluble in most organic solvents, e.g.
ketones, alcohols, chlorinated hydrocarbons, and aromatic hydrocarbons. In
methanol, benzene, dichloromethane >800, *n*-octanol 440 (all in g/kg, 25 °C).
Stability Very stable in neutral media and in dilute acids. Hydrolysed in alkaline
media (pH 13).

COMMERCIALISATION
History Herbicide reported by J. Cortier *et al.* (*C. R. Journ. Etud. Herbic. Conf.,*
COLUMA, 9th, 1977, **1,** 99). Introduced by Ciba-Geigy (now Novartis Crop
Protection AG).   **Patent** BE 795021; GB 1422473   **Manufacturers** Novartis

APPLICATIONS
Biochemistry Cell division inhibitor.   **Mode of action** Selective soil-herbicide,
absorbed mainly by the shoots of seedlings, and also by the roots of plants.
Uses Pre-emergence control of most annual grass (*Alopecurus myosuroides, Apera*
spica-venti and *Poa annua*) and broad-leaved weeds in oilseed rape at
1.25–2.00 kg a.i./ha.   **Phytotoxicity** Non-phytotoxic to oilseed rape.
Formulation types EC.   **Selected tradenames** 'Teridox' (Novartis)

ANALYSIS
Product analysis by glc.   **Residues** determined by glc with FID. Details available
from Novartis.

MAMMALIAN TOXICOLOGY

Oral Acute oral LD_{50} for rats 1600–>2000 mg/kg. **Skin and eye** Acute percutaneous LD_{50} for rats >3170 mg/kg; slight irritant to skin and eyes of rabbits. **Inhalation** LC_{50} (4 h) for rats >4450 mg/m^3. **NOEL** (2 y) for rats 300 mg/kg diet (12 mg/kg daily); (18 mo) for mice 300 mg/kg diet (31.8 mg/kg daily); (90 d) for dogs 350 mg/kg diet (10.4 mg/kg daily). **ADI** 0.12 mg/kg. **Toxicity class** WHO (a.i.) III **EC risk** R38, R43

ECOTOXICOLOGY

Birds LC_{50} for mallard ducks 200, Japanese quail 524 ppm. Five-day dietary toxicity for mallard ducks >10 000 ppm. **Fish** LC_{50} (96 h) for bluegill sunfish 15, rainbow trout 3.9, crucian carp 8 mg/l. **Bees** LD_{50} (oral and contact) >200 μg/bee. **Worms** LC_{50} (14 d) 130 mg/kg. **Daphnia** LC_{50} (48 h) 14.2–24 mg/l. **Algae** LC_{50} (72 h) for *Scenedesmus subspicatus* 0.053 mg/l.

ENVIRONMENTAL FATE

Animals The main metabolic reactions in rats are: O-dealkylation leading to O-desmethyl derivatives, subsitition of the chlorine by a hydroxy group or by glutathione, reduction of the methylene chloride moiety giving rise to acetyl derivatives, oxidation of a ring methyl resulting in hydroxymethyl derivatives. Most (88.7%–92.3%) of the administered dose was excreted within 7 d.

Plants Metabolism involves conjugation of the chloracetyl group, hydrolysis and sugar conjugation at the ether group, and formation of a heterocyclic ether-keto ring. **Soil/Environment** Mean K_{oc} 63 ml/g. Slightly mobile in soil. In soil, DT_{50} c. 4–15 d, DT_{90} <100 d; dissipation is by the formation of bound residues, and in parallel reactions, by successive oxidative degradation of the chloroacetyl side-chain to form polar metabolites and CO_2. In aquatic systems, DT_{50} 9–23 d.

239 dimethametryn *Herbicide*

1,3,5-triazine

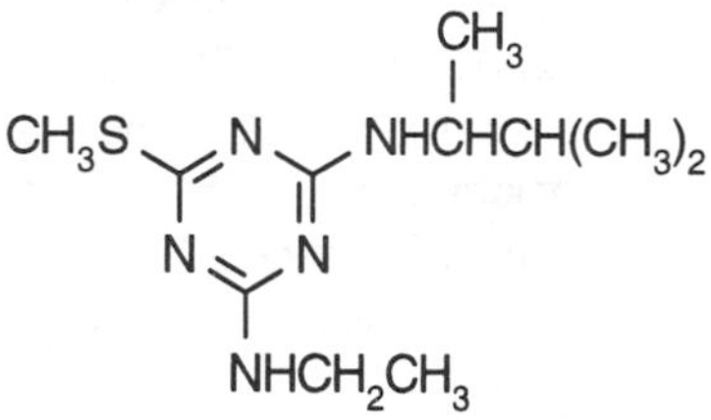

NOMENCLATURE

Common name dimethametryn (BSI, E-ISO, JMAF); diméthamétryne ((f) F-ISO)

IUPAC name N^2-(1,2-dimethylpropyl)-N^4-ethyl-6-methylthio-1,3,5-triazine-=
2,4-diamine
Chemical Abstracts name N-(1,2-dimethylpropyl)-N'-ethyl-6-(methylthio)-=
1,3,5-triazine-2,4-diamine
CAS RN [22936–75–0]   EEC no. 245–337–6   Development codes C 18 898

PHYSICAL CHEMISTRY
Mol. wt. 255.4   M.f. $C_{11}H_{21}N_5S$ Form Colourless crystals.   M.p. 65 °C
B.p. 151–153 °C/0.05 mmHg   V.p. 0.186 mPa (20 °C) K_{ow} logP = 3.8 (unstated
pH)   Henry 9.5×10^{-4} Pa m^3 mol^{-1} (calc.)   S.g./density 1.098 (20 °C)
Solubility In water 50 mg/l (20 °C). In acetone 650, dichloromethane 800, hexane
60, methanol 700, n-octanol 350, toluene 600 (all in g/l, 20 °C).   Stability No
significant hydrolysis occurred during 28 days at pH 5–9 and 70 °C.   pKa 4.1,
weak base

COMMERCIALISATION
History Herbicide reported by D. H. Green & L. Ebner (Proc. Br. Weed Control
Conf., 11th, 1972, p. 822). Introduced by Ciba-Geigy Ltd (now Novartis Crop
Protection AG).   Patent BE 714992; GB 1191585   Manufacturers Novartis

APPLICATIONS
Biochemistry Photosynthetic electron transport inhibitor.
Mode of action Selective herbicide, absorbed by the roots and leaves.
Uses Control of annual broad-leaved weeds in rice; normally used in combination
with other herbicides; 0.3–0.4 kg total a.i./ha used in transplanted rice and
0.2–0.3 kg/ha in seeded rice.   Formulation types EC; GR.
Mixtures (dimethametryn +) piperophos; pretilachlor + pyrazolynate; pretilachlor +
esprocarb + pyrazosulfuron-ethyl.   Selected tradenames 'Avirosan' (Novartis);
'Sparkstar' (mixture) (Nissan)

ANALYSIS
Product analysis by glc.   Residues determined by glc with TID. Details available
from Novartis.

MAMMALIAN TOXICOLOGY
Oral Acute oral LD_{50} for rats 3000 mg/kg.   Skin and eye Acute percutaneous
LD_{50} for rats >2150 mg/kg. Non-irritant to skin; slight irritant to eyes (rabbits).
Inhalation LC_{50} (4 h) for rats >5400 mg/m^3.   NOEL (2 y) for rats 25 ppm;
(23 mo) for mice 30 ppm.   ADI 0.01 mg/kg b.w.   Toxicity class WHO (a.i.) III;
EPA (formulation) II   EC risk R43

ECOTOXICOLOGY
Birds Eight-day dietary LC_{50} for Japanese quail >1000 ppm.   Fish LC_{50} (96 h) for

rainbow trout 5, crucian carp 8 mg/l. **Bees** LC$_{50}$ (48 h) (contact and ingestion) >100 μg/bee. **Daphnia** LC$_{50}$ 0.92 mg/l.

ENVIRONMENTAL FATE
Animals Metabolism of dimethametryn was found to proceed as for S-alkyl-= s-triazines, mainly via N-dealkylation, S-oxidation and conjugation with reduced glutathione, side-chain hydroxylation, and via conjugation leading to S-β-glucuronide conjugates. Metabolites found in faeces and urine are similar. **Plants** The leading reaction in the plant was the conjugation of the methylsulfinyl derivative with glutathione. After this conjugation, further side-chain transformations were also observed. In additional degradation or transformation steps, cysteine conjugates and triamino derivatives were formed.
Soil/Environment DT$_{50}$ in lab. soil c. 140 d.

.40 dimethenamid   *Herbicide*

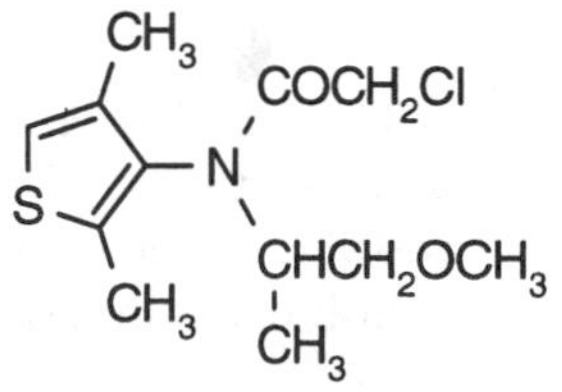

NOMENCLATURE
Common name dimethenamid (BSI, pa E-ISO)
IUPAC name (RS)-2-chloro-N-(2,4-dimethyl-3-thienyl)-N-(2-methoxy-= 1-methylethyl)acetamide
Chemical Abstracts name (RS)-2-chloro-N-(2,4-dimethyl-3-thienyl)-N-= (2-methoxy-1-methylethyl)acetamide
CAS RN [87674–68–8] **Development codes** SAN 582 H

PHYSICAL CHEMISTRY
Mol. wt. 275.8 **M.f.** C$_{12}$H$_{18}$ClNO$_2$S **Form** Yellowish-brown viscous liquid.
B.p. 127 °C/26.7 Pa **V.p.** 36.7 mPa (25 °C) **K$_{ow}$** logP = 2.15±0.02 (25 °C)
Henry 8.32 × 10^{-3} Pa m^3 mol^{-1} **S.g./density** 1.187 (25 °C) **Solubility** In water 1.2 g/l (pH 7, 25 °C). In heptane 282, iso-octane 220 (both in g/kg, 25 °C). In ether, kerosene, ethanol all >50% (25 °C). **Stability** Stable in storage at 54 °C for 4 weeks, and at 70 °C for 2 weeks. Estimated decomposition at 20 °C within

2 years is <5%. Stable at pH 5–9 (buffered, 25 °C) for 30 d. **F.p.** 91 °C (Pensky Martens closed cup)

COMMERCIALISATION
History Reported by J. Harr et al. (*Proc. Br. Crop Prot. Conf. – Weeds*, 1991, **1**, 87). Introduced by Sandoz AG (now Novartis Crop Protection AG); sold to BASF in 1996. **Manufacturers** BASF

APPLICATIONS
Biochemistry Alkylation of sulfhydryl groups of certain enzymes is probably involved. **Mode of action** Herbicide which is absorbed by the coleoptile, whereas practically no activity occurs after application to seeds, roots, or developed leaves. **Uses** Control of annual grasses and many broad-leaved weeds pre-emergence in maize, soya beans, and other crops. **Formulation types** EC.
Selected tradenames 'Frontier' (BASF)

MAMMALIAN TOXICOLOGY
Oral Acute oral LD_{50} for rats 1570 mg/kg. **Skin and eye** Acute percutaneous LD_{50} for rats and rabbits >2000 mg/kg. Non-irritating to skin; mild eye irritant (rabbits). Sensitising to skin. **Inhalation** LC_{50} (4 h) for rats >4990 mg/m^3 air. **NOEL** for rats 6.0, dogs 10, mice 4.3 mg/kg b.w. daily. **ADI** 0.04 mg/kg. **Other** Non-mutagenic in the Ames test and chromosome aberration assay. No carcinogenic or teratogenic potential.

ECOTOXICOLOGY
Birds Acute oral LD_{50} for bobwhite quail 1908 mg/kg b.w. Dietary LC_{50} for bobwhite quail and mallard ducks >5620 mg/kg b.w. **Fish** LC_{50} (96 h) for bluegill sunfish 6.4, rainbow trout 2.6 mg/l. LC_{50} for sheepshead minnow 7.2 mg/l. **Bees** Contact LD_{50} >1 mg/bee. **Worms** LC_{50} 459.8 mg/kg dry soil.
Other beneficial spp. Harmless to *Poecilus cupreus* and *Aleochara bilineata*.
Daphnia LC_{50} 16 mg/l. **Algae** LC_{50} for *Scenedesmus subspicatus* 0.062 mg/l.
Other aquatic spp. LC_{50} for mysid shrimp (*Mysidopsis bahia*) 4.8, eastern oyster (*Crassostrea virginica*) 5.0 mg/l.

ENVIRONMENTAL FATE
Animals Extensively broken down by animals; metabolites in rat, goat and hen include glutathione, cysteine and thioglycolic acid. **Plants** Metabolism in maize leads to glutathione, cysteine, thiolactic acid and thioglycolic acid.
Soil/Environment Rapidly degraded in soil, probably through microbial action, with DT_{50} 8–43 d, depending upon soil type and weather conditions. Photolysis DT_{50} on soil c. 7.8 d, in water 23–33 d. K_d (4 soils) 0.7–3.5; in water 23.9 d.

dimethipin   *Herbicide, plant growth regulator*

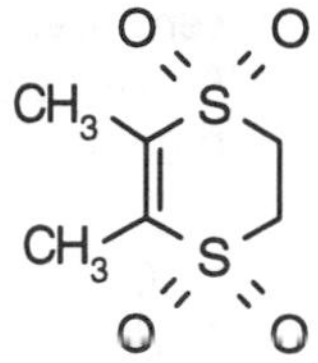

NOMENCLATURE
Common name dimethipin (BSI, ANSI, draft E-ISO, (*m*) draft F-ISO); no name (Japan)
IUPAC name 2,3-dihydro-5,6-dimethyl-1,4-dithi-ine 1,1,4,4-tetraoxide
Chemical Abstracts name 2,3-dihydro-5,6-dimethyl-1,4-dithiin 1,1,4,4-tetraoxide
CAS RN *[55290–64–7]*   **Development codes** N 252

PHYSICAL CHEMISTRY
Mol. wt. 210.3   **M.f.** $C_6H_{10}O_4S_2$ **Form** Tech. grade is a white crystalline solid.
M.p. 167–169 °C   **V.p.** 0.051 mPa (25 °C)   **K_{ow}** logP = –0.17 (24 °C)
S.g./density 1.59 g/cm³ (23 °C)   **Solubility** In water 4.6 g/l (25 °C). In acetonitrile
180, methanol 10.7, toluene 8.979 (all in g/l, 25 °C).   **Stability** Stable for 1 y at
20 °C, 14 d at 55 °C. Stable to light ≥7 d at 25 °C. Undergoes degradation in
aqueous solution.   **pKa** 10.88, v. weak acid

COMMERCIALISATION
History Plant growth regulator reported by R. B. Ames *et al.* (*Proc. Beltwide Cotton Produc. Res. Conf.*, 1974, p. 61). Introduced by Uniroyal Chemical Co., Inc.
Patent US 3920438   **Manufacturers** Uniroyal

APPLICATIONS
Biochemistry Interferes with protein synthesis in the plant epidermis.
Mode of action Defoliant and desiccant.   **Uses** Defoliation of cotton, nursery
stock, rubber trees, and vines. Potato haulm desiccant. Reduces seed moisture
content at harvest of maize, rice, oilseed rape, flax, and sunflowers.
Formulation types SC.   **Selected tradenames** 'Harvade' (Uniroyal)

ANALYSIS
Product analysis by glc or i.r. spectrometry.   **Residues** determined by glc using a
sulfur-specific detector. Details available from Uniroyal.

MAMMALIAN TOXICOLOGY
Reviews *Pesticide residues in food – 1988.* FAO Plant Production and Protection
Paper 92, 1988. *Pesticide residues in food – 1988 evaluations. Part II – Toxicology.*

MAIN ENTRIES

FAO Plant Production and Protection Paper 93/2, 1989. **Oral** Acute oral LD_{50} for rats 500 mg/kg. **Skin and eye** Acute percutaneous LD_{50} for rabbits >5000 mg/kg. Extreme eye irritant, not a dermal irritant (rabbits). Weak sensitiser (guinea pigs). **Inhalation** LC_{50} (4 h) for rats 1.2 mg/l. **NOEL** In chronic feeding studies, NOEL for rats 2, dogs 25 mg/kg b.w. daily. **ADI** (JMPR) 0.02 mg/kg b.w. [1988]. **Toxicity class** WHO (a.i.) III; EPA (formulation) I

ECOTOXICOLOGY
Birds Eight-day dietary LC_{50} for mallard ducks and bobwhite quail >5000 ppm. LC_{50} for mallard ducks 896 mg/kg. **Fish** LC_{50} (96 h) for rainbow trout 52.8, bluegill sunfish 20.9, sheepshead minnow 17.8 mg/l. **Bees** LD_{50} >100 μg/bee (25% formulation). **Worms** LC_{50} (14 d) for earthworms >39.4 ppm (25% formulation). **Daphnia** LC_{50} (48 h) 21.3 mg/l. **Algae** EC_{50} (5 d) for *Selenastrum capricornutum* 5.12 mg/l. **Other aquatic spp.** LC_{50} (96 h) in mysid shrimp (*Mysidopsis bahia*) 13.9 mg/l.

ENVIRONMENTAL FATE
Plants Unchanged. **Soil/Environment** DT_{50} in soil *c.* 104–149 d. K_d 0.092; K_{oc} 3.27.

242 dimethirimol *Fungicide*

pyrimidine

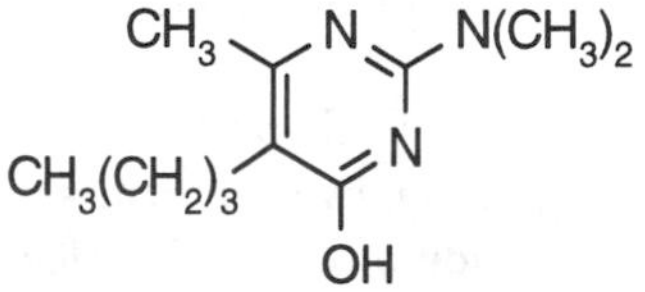

NOMENCLATURE
Common name dimethirimol (BSI, E-ISO, JMAF); diméthyrimol ((m) F-ISO)
IUPAC name 5-butyl-2-dimethylamino-6-methylpyrimidin-4-ol
Chemical Abstracts name 5-butyl-2-(dimethylamino)-6-methyl-4(1*H*)-pyrimidinone
CAS RN *[5221–53–4]* **Development codes** PP675

PHYSICAL CHEMISTRY
Mol. wt. 209.3 **M.f.** $C_{11}H_{19}N_3O$ **Form** Colourless needles. **M.p.** 102 °C
V.p. 1.46 mPa (30 °C) **K_{ow}** logP = 1.9 **Solubility** In water 1.2 g/l (25 °C). Readily soluble in aqueous solutions of strong acids. In chloroform 1200, xylene 360, ethanol 65, acetone 45 (all in g/l, 25 °C). **Stability** Stable in acidic and

alkaline solutions. Decomposed by sunlight when in aqueous solution; DT_{50} c. 7 d. Dissolves in aqueous solutions of strong acids to form water-soluble salts.

COMMERCIALISATION
History Systemic fungicide reported by R. S. Elias et al. (*Nature (London)*, 1968, **219**, 1160). Introduced by ICI Plant Protection Division (now Zeneca Agrochemicals).   **Patent** GB 1182584   **Manufacturers** Zeneca

APPLICATIONS
Mode of action Systemic fungicide with protective and curative action. Absorbed by the roots, with translocation in the xylem.   **Uses** Soil application for control of powdery mildews in cucurbits, tobacco, capsicums, tomatoes, and some ornamentals.   **Formulation types** SL.   **Selected tradenames** 'Milcurb' (Zeneca)

ANALYSIS
Product analysis by glc (J. E. Bagness & W. G. Sharples, *Analyst (London)*, 1974, **99**, 225).   **Residues** determined by hplc.   Details available from Zeneca.

MAMMALIAN TOXICOLOGY
Oral Acute oral LD_{50} for rats 2350, mice 800–1600, guinea pigs 500 mg/kg.
Skin and eye Acute percutaneous LD_{50} for rats >400 mg/kg. Non-irritating to skin and eyes (rabbits).   **NOEL** (2 y) for dogs 25 mg/kg, for rats 300 mg/kg diet.
Other Acute i.p. LD_{50} for rats 200–400 mg/kg.   **Toxicity class** WHO (a.i.) III (Table 5); EPA (formulation) III

ECOTOXICOLOGY
Birds Acute oral LD_{50} for hens 4000 mg/kg.   **Fish** LC_{50} for young brown trout 42 mg/l (24 h), 33 mg/l (48 h), 28 mg/l (96 h).   **Bees** Not toxic to bees; acute oral LD_{50} 4000 mg/kg.

ENVIRONMENTAL FATE
Plants In plants, metabolism involves demethylation of the dimethylamino group.
Soil/Environment Soil DT_{50} c. 120 d (sandy soil, greenhouse and John Innes composts).

243 dimethoate

organophosphorus

$$CH_3NHCOCH_2SP(OCH_3)_2$$
with S double-bonded to P

NOMENCLATURE
Common name dimethoate (BSI, E-ISO, (*m*) F-ISO, ANSI, ESA, JMAF); fosfamid (USSR)
IUPAC name *O,O*-dimethyl *S*-methylcarbamoylmethyl phosphorodithioate; 2-dimethoxyphosphinothioylthio-*N*-methylacetamide
Chemical Abstracts name *O,O*-dimethyl *S*-[2-(methylamino)-2-oxoethyl] phosphorodithioate
CAS RN *[60–51–5]* **EEC no.** 200–480–3 **Development codes** EI 12 880 (American Cyanamid); L 395 (Montedison); BAS 152J **Official codes** OMS 94; OMS 111; ENT 24 650

PHYSICAL CHEMISTRY
Composition Tech. grade is 95% pure. **Mol. wt.** 229.3 **M.f.** $C_5H_{12}NO_3PS_2$
Form Colourless crystals (tech., white to greyish crystals). **M.p.** 49 °C; (tech., 43–45 °C) **B.p.** 117 °C/0.1 mmHg **V.p.** 0.25 mPa (25 °C) **K$_{ow}$** logP = 0.704
Henry 1.2×10^{-6} Pa m^3 mol^{-1} **S.g./density** 1.277 at 65 °C **Solubility** In water 23.3 (pH 5), 23.8 (pH 7), 25.0 (pH 9) (all in g/l, 20 °C). Readily soluble in most organic solvents, e.g. in alcohols, ketones, benzene, toluene, chloroform, dichloromethane >300, carbon tetrachloride, saturated hydrocarbons, *n*-octanol >50 (all in g/kg, 20 °C). **Stability** Relatively stable in aqueous media at pH 2–7. Hydrolysed in alkaline solutions; DT$_{50}$ 12 d (pH 9). Decomposes on heating, forming the *O,S*-dimethyl analogue.

COMMERCIALISATION
History Insecticide reported by E. I. Hoegberg & J. T. Cassaday (*J. Am. Chem. Soc.,* 1951, **73**, 557; *Ital. Agric.,* 1955, **92**, 747). Introduced by American Cyanamid Co. (who have sold rights to Wilbur Ellis), by BASF AG, by Boehringer Sohn (now Cyanamid Agrar), and by Montecatini S.p.A. (now Isagro S.p.A.).
Patent US 2494283 to Cyanamid; DE 1076662 to Celamerck; GB 791824 to Montedison **Manufacturers** Aimco; Cheminova; Mico; Rallis; Shaw Wallace; Sinon

APPLICATIONS
Biochemistry Cholinesterase inhibitor. **Mode of action** Systemic insecticide and acaricide with contact and stomach action. **Uses** Control of a wide range of Acari, Aphididae, Aleyrodidae, Coccidae, Coleoptera, Collembola, Diptera, Lepidoptera, Pseudococcidae and Thysanoptera in cereals, citrus, coffee, cotton, fruit, grapes, olives, pastures, beetroot, potatoes, pulses, tea, tobacco, and vegetables. Also used for control of flies in animal houses. **Phytotoxicity** Non-

phytotoxic when used as directed, except to some varieties of lemon, peach, fig, olive, walnut, hop, tomato, bean, cotton, and pine. Russetting is possible with Red Delicious and Golden Delicious apples, and with some ornamentals.

Formulation types EC; WP; UL; GR; Aerosol. **Mixtures** *(dimethoate +)* permethrin; endosulfan; chlorpyrifos; fenitrothion; fenvalerate; flucythrinate; deltamethrin; prothoate; phenthoate; and many more.

Compatibility Incompatible with alkaline materials and with sulfur-based formulations. **Selected tradenames** 'Cygon' (Wilbur-Ellis); 'Perfekthion' (BASF); 'Rogor' (Isagro, Rallis); 'Roxion' (Wilbur-Ellis); 'Cekutoate' (Cequisa); 'Champ' (Searle India); 'Chimigor' (Diachem); 'Danadim' (Cheminova); 'Diadhan' (Dhanuka); 'Dicentra' (Sanonda); 'Dimezyl' (Agriphar); 'Efdacon' (Efthymiadis); 'Robgor' (Ramcides); 'Romethoate' (Rotam); 'Tara 909' (Shaw Wallace)

ANALYSIS

Product analysis by glc (*CIPAC Handbook*, 1980, **1A**, 1225; *CIPAC Proc.*, 1981, **3**, 204, 211; *CIPAC Handbook*, 1992, **E**, 69–72). **Residues** determined by glc (*Analyst (London)*, 1977, **102**, 858; 1980, **105**, 515; 1985, **110**, 765; J. E. Boyd, *Anal. Methods Pestic. Plant Growth Regul.*, 1972, **6**, 357; *Man. Pestic. Residue Anal.*, 1987, **I**, 3, 6, S8, S13, S17, S19; *Anal. Methods Residues Pestic.*, 1988, Part I, M5, M12).

MAMMALIAN TOXICOLOGY

Reviews *Pesticide residues in food – 1996*. FAO Plant Production and Protection Paper, in preparation. *Pesticide residues in food – 1996. Part II – Toxicology*. WHO/PCS/97.1. **EHC** 90 (WHO, 1989), 63 (WHO, 1986; a general review of organophosphorus insecticides). **Oral** Acute oral LD_{50} for rats 387, mice 160, rabbits 300, guinea pigs 350 mg/kg. **Skin and eye** Acute percutaneous LD_{50} for rats >2000 mg/kg. Non-irritating to skin (rabbits). **Inhalation** LC_{50} (4 h) for rats >1.6 mg/l air. **NOEL** (2 y) for rats 5.0 mg/kg diet (0.2 mg/kg daily).
ADI (JMPR) 0.002 mg/kg b.w. (sum of dimethoate and omethoate, expressed as dimethoate) [1996]. **Toxicity class** WHO (a.i.) II; EPA (formulation) II
EC risk Xn (R21/22)

ECOTOXICOLOGY

Birds Acute oral LD_{50} for male pheasants 15, quail 84, chickens 108, female mallard ducks 40 mg/kg. **Fish** LC_{50} (96 h) for mosquito fish 40–60, rainbow trout 6.2, bluegill sunfish 6 mg/l. **Bees** Toxic to bees. LD_{50} (oral and topical) 0.1–0.2 µg/bee. **Daphnia** LC_{50} (24 h) 4.7 mg/l; NOEC (24 h) 1 mg/l. **Algae** EC_{50} (72 h) for *Selenastrum capricornutum* 282.3 mg/l; NOEC (72 h) 30.5 mg/l.

ENVIRONMENTAL FATE

EHC review notes low risk to farm animals, moderate toxicity for birds, fish and aquatics, and very high toxicity for honeybees. This review concludes that, when used under proper conditions, exposure of the population through air, food or water is negligible. **Animals** In mammals, metabolism is the same as for plants.

Plants In plants, as well as oxidation to the phosphorothioate, there is also hydrolysis to O,O-dimethyl-phosphorodithioate, -phosphorothioate, and -phosphate. The ester group is demethylated and the methylamino group is hydrolytically cleaved. Oxidation of the phosphorothioate gives the corresponding oxone (omethoate q.v.), which is classified as toxic and a strong cholinesterase inhibitor, and which appears to show similar rapid degradation in environmental compartments as dimethoate. **Soil/Environment** Adsorption and desorption constants have been shown to be a linear function of soil silt content. K_{oc} ranges from 16.25 (sandy loam) to 51.88 (sand/loamy sand). Aerobic DT_{50} 2–4.1 d. Photolytic DT_{50} on soil surface 7–16 d.

244 dimethomorph   *Fungicide*

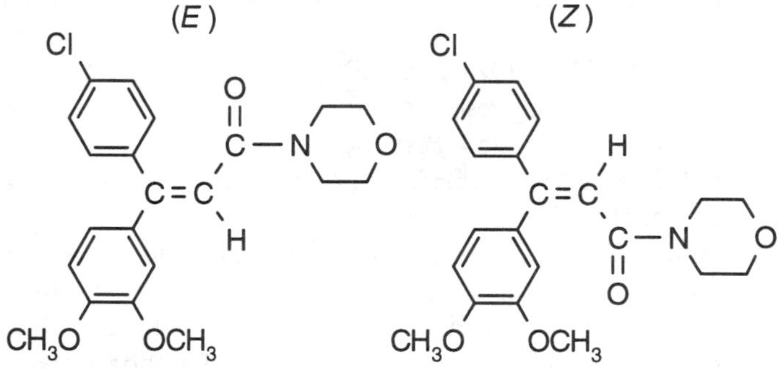

NOMENCLATURE
Common name dimethomorph (BSI, draft E-ISO); diméthomorphe ((m) draft F-ISO)
IUPAC name (E,Z)-4-[3-(4-chlorophenyl)-3-(3,4-dimethoxyphenyl)acryloyl]= morpholine
Chemical Abstracts name (E,Z)-4-[3-(4-chlorophenyl)-3-(3,4-dimethoxyphenyl)-= 1-oxo-2-propenyl]morpholine
CAS RN [110488–70–5] **EEC no.** 404–200–2 **Development codes** CME 151; WL 127 294; AC 336379; CL 336379

PHYSICAL CHEMISTRY
Composition (E)- to (Z)- ratio is c. 1:1. **Mol. wt.** 387.9 **M.f.** $C_{21}H_{22}ClNO_4$
Form Colourless crystals. **M.p.** 127–148 °C; (E)-isomer 135.7–137.5 °C; (Z)-isomer 169.2–170.2 °C **V.p.** (E)-isomer 9.7×10^{-4} mPa; (Z)-isomer 1.0×10^{-3} mPa (both 25 °C) **K_{ow}** logP = 2.63 (E)-isomer; 2.73 (Z)-isomer (both

20 °C)   **S.g./density** Bulk density 1318 kg/m^3 (20 °C).   **Solubility** In water
<50 mg/l (20–23 °C). In acetone 88 (*E*), 15 (*Z*), cyclohexanone 27 (*Z*),
dichloromethane 315 (*Z*), dimethylformamide 272 (*E*), 40 (*Z*), hexane 0.04 (*E*),
0.02 (*Z*), methanol 7 (*Z*), toluene 7 (*Z*) (all in g/l, 20–23 °C).   **Stability**
Hydrolytically and thermally stable under normal conditions. Stable for >5 years in
the dark. The (*E*)- and (*Z*)-isomers are interconverted in sunlight.

COMMERCIALISATION
History Fungicide reported by G. Albert *et al* (*Proc. 1988 Br. Crop Prot. Conf.
Pests Dis.*, **1**, 17). Introduced in 1993 by Shell (now American Cyanamid Co.).
Patent EP 120321   **Manufacturers** Cyanamid

APPLICATIONS
Biochemistry Inhibits the formation of the oomycete fungal cell wall.
Mode of action Local systemic fungicide with good protectant, curative and
antisporulant activity. Only the (*Z*)-isomer is intrinsically active, but, because of
rapid interconversion of isomers in the light, it has no advantage over the
(*E*)-isomer in practice.   **Uses** Fungicide effective against Oomycetes, especially
Peronosporaceae and *Phytophthora* spp. (but not *Pythium* spp.) in vines, potatoes,
tomatoes and other crops. Used in combination with contact fungicides (dithianon,
mancozeb, fentin hydroxide or copper compounds).   **Formulation types** EC; DC;
WP; WG.   **Mixtures** (*dimethomorph +)* mancozeb; dithianon; copper compounds;
fentin hydroxide.   **Selected tradenames** 'Acrobat' (Cyanamid); 'Forum'
(Cyanamid)

ANALYSIS
Product analysis by rp hplc (*CIPAC Handbook*, 1995, **G**, 39–46) or capillary gc.
Details available from Cyanamid.

MAMMALIAN TOXICOLOGY
Oral Acute oral LD_{50} for male rats 4300, female rats 3500, male mice >5000,
female mice 3700 mg/kg b.w.   **Skin and eye** Acute percutaneous LD_{50} for rats
>5000 mg/kg b.w.; not irritant to skin or eyes of rabbits. Not a skin sensitiser
(guinea pigs).   **Inhalation** LC_{50} (4 h) for rats >4.2 mg/l air.   **NOEL** (2 y) for rats
200 mg/kg diet; (1 y) for dogs 450 mg/kg diet. Non-oncogenic in 2 y studies in
rats and mice.   **ADI** 0.09 mg/kg b.w.   **Other** Acute i.p. LD_{50} for male rats 327,
female rats 297 mg/kg b.w.   **Toxicity class** WHO (a.i.) III (Table 5)

ECOTOXICOLOGY
Birds Acute oral LD_{50} for mallard ducks >2000 mg/kg.   **Fish** LC_{50} (96 h) for
bluegill sunfish >25, carp 14 mg/l.   **Bees** Non-toxic to honeybees at
0.1 mg/bee (contact or oral, highest dose tested).   **Worms** EC_{50} for earthworms
>1000 mg/kg soil.   **Daphnia** EC_{50} (48 h) 49 mg/l.   **Algae** EC_{50} (96 h) >29 mg/l.

ENVIRONMENTAL FATE
Animals In rats, the major route of metabolism is demethylation of one of the dimethoxy groups or by oxidation of one of the CH_2 groups (*ortho-* or *meta-* position) of the morpholine ring.   **Plants** The only significant component of the residue, when present, is the parent compound.

245   dimethylarsinic acid   *Herbicide*

organoarsenic

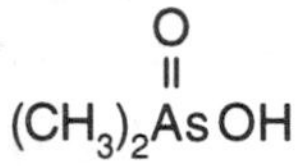

$$(CH_3)_2 \overset{\displaystyle O}{\overset{\displaystyle \|}{As}} OH$$

NOMENCLATURE
dimethylarsinic acid
Common name dimethylarsinic acid (draft BSI, draft E-ISO); cacodylic acid (WSSA, draft E-ISO)
IUPAC name dimethylarsinic acid
Chemical Abstracts name dimethylarsinic acid
CAS RN *[75–60–5]*

sodium dimethylarsinate
CAS RN *[124–65–2]*

PHYSICAL CHEMISTRY
dimethylarsinic acid
Composition Tech. grade is 65% pure, sodium chloride being one impurity.
Mol. wt. 138.0   **M.f.** $C_2H_7AsO_2$ **Form** Colourless crystals.   **M.p.** 192–198 °C
Solubility In water 2 kg/kg (25 °C). Soluble in short-chain alcohols; insoluble in diethyl ether.   **Stability** Decomposed by powerful oxidising or reducing agents. At pH 7, dimethylarsinic acid forms a water-soluble sodium salt (sodium dimethylarsinate), which is deliquescent.   **pKa** 6.29 (D. Wauchope, *J. Agric. Food Chem.*, 1976, **24,** 717)

sodium dimethylarsinate
Mol. wt. 159.9   **M.f.** $C_2H_6AsNaO_2$

COMMERCIALISATION
History Herbicidal properties reported by R. G. Mowev & J. F. Cornman (*Proc. Northeast. Weed Control Conf.*, 1959, **13,** 62). Introduced by Ansul Chemical Co. (which no longer exists) and later by Crystal Chemical Co.   **Patent** US 3056668
Manufacturers Luxembourg

APPLICATIONS
Uses Non-selective herbicide, used for post-emergence control of weeds on non-crop land, for lawn renovation, as a defoliant and desiccant for cotton, and for killing unwanted trees by injection.   **Formulation types** SL.
Mixtures *(dimethylarsinic acid +)* sodium dimethylarsinate; sodium dimethylarsinate + sodium hydrogen methylarsinate.   **Selected tradenames** 'Phytar' (Crystal, Vertac); 'Leaf-All' (Luxembourg)

ANALYSIS
Product analysis by acid/base titration of the oxidised product (E. A. Dietz & L. O. Moore, *Anal. Methods Pestic. Plant Growth Regul.,* 1978, **10**, 385).
Residues determined by oxidation followed by reduction to arsine which is determined by colorimetry *(idem, ibid.).*

MAMMALIAN TOXICOLOGY
dimethylarsinic acid
Oral Acute oral LD_{50} for rats 700 mg/kg.   **Skin and eye** Acute percutaneous LD_{50} not available. Non-irritating to skin and eyes (rabbits).   **Toxicity class** WHO (a.i.) III   **EC risk** Not specified, but arsenic compounds in general (with exceptions) are assigned T (R23/25); (preparations, 0.1%≤ concn. ≤0.2%, Xn (R20/22))

ENVIRONMENTAL FATE
Soil/Environment Inactivated on contact with soil.

246 dimethylvinphos   *Insecticide*

organophosphorus

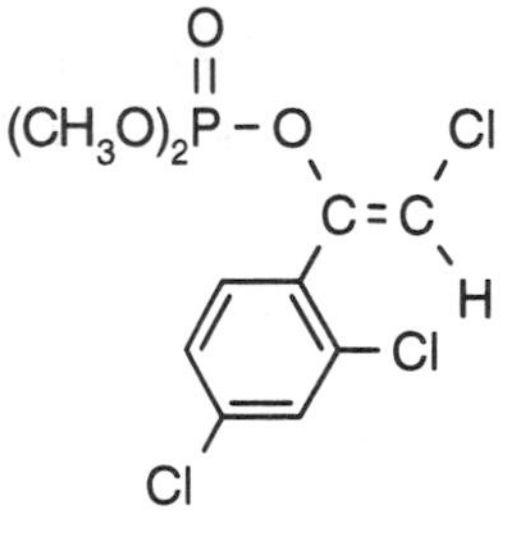

NOMENCLATURE
Common name dimethylvinphos (JMAF)
IUPAC name (Z)-2-chloro-1-(2,4-dichlorophenyl)vinyl dimethyl phosphate
Chemical Abstracts name 2-chloro-1-(2,4-dichlorophenyl)ethenyl dimethyl phosphate

Other names chlorfenvinphos-methyl   **CAS RN** *[2274–67–1]*
Development codes SD 8280; SKI-13

PHYSICAL CHEMISTRY
Composition Comprises >95.0% (*Z*)- isomer, <2.0% (*E*)- isomer. The diethyl ester
analogue (mixture of (*E*)- and (*Z*)- isomers) is chlorfenvinphos (*q.v.*).
Mol. wt. 331.5   **M.f.** $C_{10}H_{10}Cl_3O_4P$ **Form** Pale white crystalline solid.
M.p. 69–70 °C   **V.p.** 1.3 mPa (25 °C) K_{ow} logP = 3.12 (25 °C)
S.g./density 1.26 (25 °C)   **Solubility** In water 130 ppm (20 °C). In xylene
300–350, acetone 350–400, cyclohexanone 450–500 g/l (20 °C).
Stability Hydrolysis DT_{50} 40 d (pH 7.0, 25 °C). Unstable in sunlight.

COMMERCIALISATION
History Insecticide introduced by Shell Kagaku KK (now American Cyanamid Co.).
Patent JP-7720 (1966)   **Manufacturers** Cyanamid

APPLICATIONS
Biochemistry Cholinesterase inhibitor.   **Uses** Contact and stomach insecticide with
moderate persistence. Effective in rice against leaf roller, stem borer at
0.6–0.8 kg a.i./ha.   **Phytotoxicity** Non-phytotoxic to most crops at recommended
doses. Spotting seen on peach leaves.   **Formulation types** 2% dust.
Mixtures *(dimethylvinphos +)* iprobenfos + carbaryl; fenobucarb + diazinon;
etofenprox; isoprocarb; fenobucarb; metolcarb.   **Selected tradenames** 'Rangado'
(Cyanamid)

MAMMALIAN TOXICOLOGY
EHC 63 (WHO, 1986; a general review of organophosphorus insecticides).
Oral Acute oral LD_{50} for rats 155–210, mice 200–220 mg/kg.
Skin and eye Acute percutaneous LD_{50} for rats 1360–2300 mg/kg.
Inhalation LC_{50} (4 h) for male rats 970–1186, female rats >4900 mg/m^3.

ECOTOXICOLOGY
Fish LC_{50} (24 h) for carp 2.3 ppm.   **Daphnia** LC_{50} (24 h) 0.002 ppm.

azole

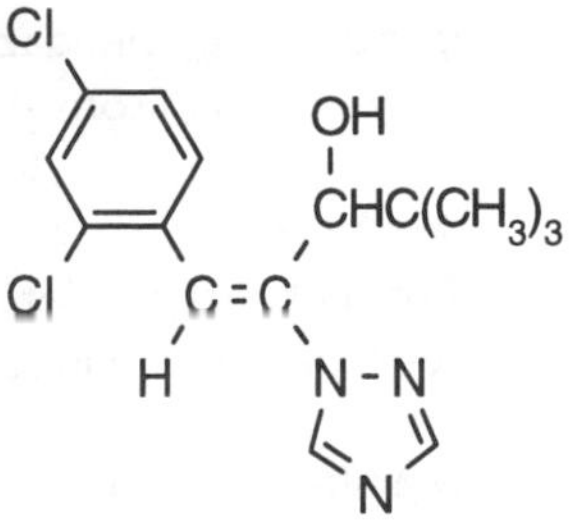

NOMENCLATURE
diniconazole
Common name diniconazole (BSI, ANSI, draft E-ISO, (m) draft F-ISO)
IUPAC name (E)-(RS)-1-(2,4-dichlorophenyl)-4,4-dimethyl-2-(1H-1,2,4-triazol-=
1-yl)pent-1-en-3-ol
Chemical Abstracts name (E)-(±)-β-[(2,4-dichlorophenyl)methylene]-α-=
(1,1-dimethylethyl)-1H-1,2,4-triazole-1-ethanol
CAS RN *[83657–24–3]*; *[76714–88–0]* (E)-isomers **Development codes** S-3308 L
(Sumitomo); XE-779 (Chevron)

diniconazole-M
IUPAC name (E)-(R)-1-(2,4-dichlorophenyl)-4,4-dimethyl-2-(1H-1,2,4-triazol-=
1-yl)pent-1-en-3-ol
CAS RN *[83657–18–5]*; *[83657–19–6]* (E)-(S)-isomer

PHYSICAL CHEMISTRY
diniconazole
Mol. wt. 326.2 **M.f.** $C_{15}H_{17}Cl_2N_3O$ **Form** Colourless crystals.
M.p. *c.* 134–156 °C **V.p.** 2.93 mPa (20 °C); 4.9 mPa (25 °C) **K_{ow}** logP = 4.3
(25 °C) **S.g./density** 1.32 (20 °C) **Solubility** In water 4 mg/l (25 °C). In acetone,
methanol 95, xylene 14, hexane 0.7 (all in g/kg, 25 °C). **Stability** Stable to heat,
light, and moisture.

COMMERCIALISATION
History Fungicide reported by H. Takano *et al.* (*Noyaku Kagaku*, 1983, **8**, 575);
diniconazole-M is more potent than diniconazole. Introduced by Sumitomo
Chemical Co., Ltd and Valent. **Manufacturers** Sumitomo

APPLICATIONS
Biochemistry Steroid demethylation (ergosterol biosynthesis) inhibitor.
Mode of action Systemic fungicide with protective and curative action.
Uses Control of leaf and ear diseases (e.g. powdery mildew, *Septoria*, *Fusarium*,

MAIN ENTRIES

smuts, bunt, rusts, scab, etc.) in cereals; powdery mildew in vines; powdery mildew, rust, and black spot in roses; leaf spot in peanuts; Sigatoka disease in bananas; and Uredinales in coffee. Also used on fruit, vegetables, and other ornamentals. **Formulation types** SC; WP; WG; EC. **Mixtures** *(diniconazole +)* iprodione; carbendazim + iprodione. **Selected tradenames** 'Spotless' (Sumitomo); 'Sumi-8' (Sumitomo)

ANALYSIS
Product analysis details available from Sumitomo Chemical Co. Ltd.
Residues determined by glc with FTD.

MAMMALIAN TOXICOLOGY
diniconazole
Oral Acute oral LD_{50} for male rats 639, female rats 474 mg/kg.
Skin and eye Acute percutaneous LD_{50} for rats >5000 mg/kg. Mild eye irritant; non-irritating to skin (rabbits). Not a skin sensitiser (guinea pigs).
Inhalation LC_{50} (4 h) for rats >2770 mg/m^3. **Toxicity class** WHO (a.i.) III; EPA (formulation) III

ECOTOXICOLOGY
diniconazole
Birds Acute oral LD_{50} for bobwhite quail 1490, mallard ducks >2000 mg/kg. Eight-day dietary LC_{50} for mallard ducks 5075 mg/kg diet. **Fish** LC_{50} (96 h) for rainbow trout 1.58, Japanese killifish 6.84, carp 4.0 mg/l. **Bees** Acute contact LD_{50} for bees >20 μg/bee.

ENVIRONMENTAL FATE
Animals In rats, following oral administration, diniconazole is rapidly metabolised by hydroxylation of the *tert*-butyl methyl groups. Within 7 days, 52–87% is excreted in the faeces and 13–46% in the urine. **Plants** Half-life in cereals is a few weeks.

2,6-dinitroaniline

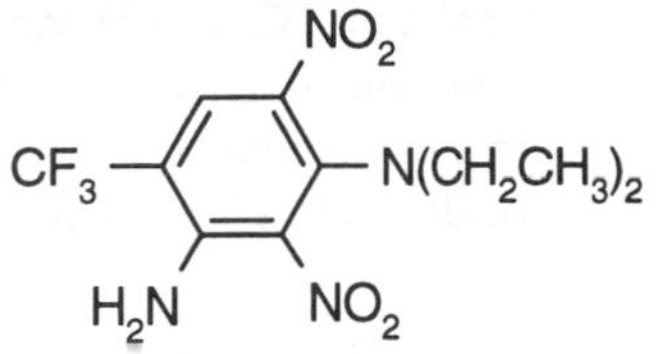

NOMENCLATURE
Common name dinitramine (BSI, E-ISO, (f) F-ISO, ANSI, WSSA)
IUPAC name N^1,N^1-diethyl-2,6-dinitro-4-trifluoromethyl-*m*-phenylenediamine
Chemical Abstracts name N^3,N^3-diethyl-2,4-dinitro-6-(trifluoromethyl)-1,3-=
benzenediamine
CAS RN *[29091–05–2]*   **Development codes** USB 3584

PHYSICAL CHEMISTRY
Composition Tech. grade is >83% pure *m/m*.   **Mol. wt.** 322.2
M.f. $C_{11}H_{13}F_3N_4O_4$ **Form** Yellow crystals.   **M.p.** 98–99 °C   **V.p.** 0.479 mPa
(25 °C) K_{ow} logP = c. 4.3   **Solubility** In water 1 mg/l (20 °C). In acetone 1040,
chloroform 670, benzene 473, xylene 227, ethanol 107, *n*-hexane 6.7 (all in g/l,
20 °C).   **Stability** Both a.i. and tech. showed no significant decomposition after 2-y
storage at ambient temperatures. Decomposes above 200 °C. Decomposed by
sunlight.

COMMERCIALISATION
History Introduced in USA (1973) by US Borax & Chemical Corp. (who no longer
manufacture or market it) and sold under licence since 1982 by Wacker GmbH.
Patent US 3617252 to US Borax   **Manufacturers** Wacker

APPLICATIONS
Mode of action Selective soil-herbicide, absorbed by the roots and shoots, with
very little translocation to the stems and leaves, and slightly more to the roots.
Prevents germination of seeds and inhibits root growth.   **Uses** Selective pre-plant
soil-incorporated control of many annual grass and broad-leaved weeds in cotton,
soya beans, peanuts, peas, beans, safflowers, sunflowers, carrots, turnips, fennel,
chicory, etc.; and in transplanted tomatoes, capsicums, aubergines, and brassicas at
0.4–0.8 kg a.i./ha.   **Formulation types** EC.   **Compatibility** Incompatible with
chlorthal-dimethyl.   **Selected tradenames** 'Cobex' (Wacker)

ANALYSIS
Product analysis by glc, details from Wacker Chemie GmbH.
Residues determined by glc with ECD (H. C. Newsom & E. M. Mitchell, *J. Agric.
Food Chem.*, 1972, **20**, 1222; H. C. Newsom, *Anal. Methods Pestic. Plant Growth
Regul.*, 1976, **8**, 359).

MAIN ENTRIES

MAMMALIAN TOXICOLOGY
Oral Acute oral LD_{50} for rats 3000 mg/kg. **Skin and eye** Acute percutaneous LD_{50} for rabbits >6800 mg/kg. Mild skin and eye irritant (rabbits).
Inhalation LC_{50} (4 h) for rats >0.16 mg/l air. **NOEL** In 90 d feeding trials, no ill-effect observed for rats and beagle dogs receiving 2000 mg/kg diet. In 2 y trials, no carcinogenic response observed in rats receiving 100 or 300 mg/kg diet.
Toxicity class WHO (a.i.) III (Table 5)

ECOTOXICOLOGY
Birds Acute oral LD_{50} for bobwhite quail >1200, mallard ducks >10 000 mg/kg. Eight-day dietary LC_{50} for bobwhite quail >5000, mallard ducks >10 000 mg/kg diet (25% EC formulation). **Fish** LC_{50} (96 h) for catfish 3.7, rainbow trout 6.6, bluegill sunfish 11.0 mg/l.

ENVIRONMENTAL FATE
Animals Degraded into small, polar fragments. **Plants** Metabolised rapidly by plants, and extensively degraded into small, polar fragments.
Soil/Environment Strongly adsorbed on soil, with no leaching. Metabolised by soil micro-organisms. In soils, DT_{50} 10–66 d. Less than 10% remains in most soils 90–120 days after application.

249 dinobuton
Acaricide, fungicide

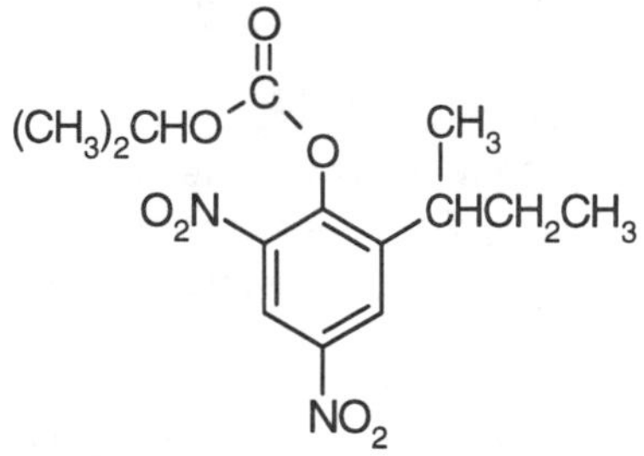

NOMENCLATURE
Common name dinobuton (BSI, E-ISO, (*m*) F-ISO)
IUPAC name 2-*sec*-butyl-4,6-dinitrophenyl isopropyl carbonate
Chemical Abstracts name 1-methylethyl 2-(1-methylpropyl)-4,6-dinitrophenyl carbonate
CAS RN [973–21–7] **Development codes** MC 1053 (Murphy)
Official codes OMS 1056; ENT 27 244

PHYSICAL CHEMISTRY
Composition Tech. is 97%.   **Mol. wt.** 326.3   **M.f.** $C_{14}H_{18}N_2O_7$ **Form** Pale yellow crystals.   **M.p.** 61–62 °C; (tech., 58–60 °C)   **V.p.** negligible (room temperature)
Solubility Practically insoluble in water; soluble in ethanol, aliphatic hydrocarbons, fatty oils; highly soluble in aromatic hydrocarbons, lower aliphatic ketones.

COMMERCIALISATION
History Acaricide and fungicide reported by M. Pianka & C. B. F. Smith (*Chem. Ind. (London)*, 1956, p. 1216). Introduced by Murphy Chemical Ltd and later manufactured by KenoGard AB (now owned by Rhône-Poulenc Agrochimie).
Patent GB 1019451

APPLICATIONS
Uses A non-systemic acaricide and a fungicide active against powdery mildews; recommended for glasshouse and field use against red spider mites and powdery mildews of apples, cotton and vegetables at 50 g a.i./100 l.
Formulation types EC.   **Selected tradenames** 'Acrex' (Efthymiadis)

ANALYSIS
Product and **residue** analysis by hydrolysis to dinoseb which is measured colorimetrically (D. S. Farrington et al., *Analyst (London)*, 1983, **103**, 353; *CIPAC Proc.*, 1980, **3**, 199; V. P. Lynch, *Anal. Methods Pestic. Plant Growth Regul.*, 1976, **8**, 275).

MAMMALIAN TOXICOLOGY
Oral Acute oral LD_{50} for mice 2540, rats 140 mg/kg.   **Skin and eye** Acute percutaneous LD_{50} for rats >5000 mg/kg.   **NOEL** for dogs 4.5, rats 3–6 mg/kg daily.   **Other** It acts as a metabolic stimulant, high doses causing loss of body weight.   **Toxicity class** WHO (a.i.) II

ECOTOXICOLOGY
Birds Acute oral LD_{50} for hens 150 mg/kg.

ENVIRONMENTAL FATE
Soil/Environment Persistence in soils is brief.

MAIN ENTRIES

dinitrophenol derivative

(i)

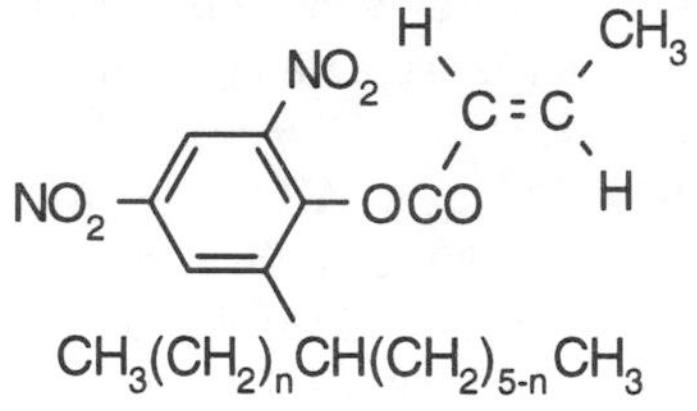

$CH_3(CH_2)_nCH(CH_2)_{5-n}CH_3$

n = 0,1,2

(ii)

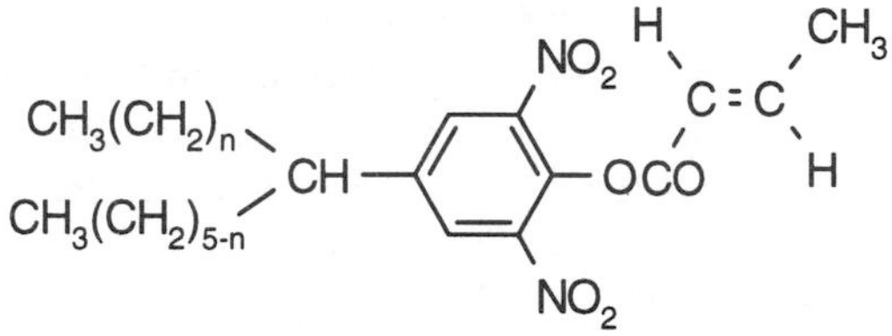

n = 0,1,2

NOMENCLATURE
Common name dinocap (BSI, E-ISO, (*m*) F-ISO); DPC (JMAF); no name (Republic of South Africa)
IUPAC name 2,6-dinitro-4-octylphenyl crotonates and 2,4-dinitro-6-octylphenyl crotonates in which 'octyl' is a mixture of 1-methylheptyl, 1-ethylhexyl and 1-propylpentyl groups
Chemical Abstracts name (*E*)-2-(1-methylheptyl)-4,6-dinitrophenyl 2-butenoate
CAS RN *[131–72–6]* single compound; *[39300–45–3]* mixed isomers but unstated stereochemistry **EEC no.** 254–408–0 **Development codes** CR-1693
Official codes ENT 24 727

PHYSICAL CHEMISTRY
Composition Originally thought to be 2-(1-methylheptyl)-4,6-dinitrophenyl crotonate (i, n=0). Now known that the commercial material is a mixture comprising 2.0–2.5 parts of 6-octyl isomers (i) to 1.0 part of 4-octyl isomers (ii). See IUPAC definition for specification of 'octyl'. **Mol. wt.** 364.3
M.f. $C_{18}H_{24}N_2O_6$ **Form** Dark brown liquid. **B.p.** 138–140 °C/0.05 mmHg
V.p. 5.3×10^{-6} mPa (20 °C); 7.5×10^{-5} mPa (25 °C) **K$_{ow}$** logP = 4.54
S.g./density 1.10 (20 °C) **Solubility** Almost insoluble in water (<0.1 mg/l). Soluble in most organic solvents, e.g. acetone, methanol, heptane.
Stability Rapidly decomposed by light. Decomposes >32 °C. Stable in acidic media. Hydrolysis of the ester group occurs in alkaline media.
2-(1-methylheptyl)-4,6-dinitrophenyl crotonate hydrolysed in aqueous methanol; (25 °C) DT$_{50}$ 229 d (pH 5), 56 h (pH 7), 17 h (pH 9).

COMMERCIALISATION
History Introduced by Rohm & Haas Co.　**Patent** US 2526660; US 2810767
Manufacturers Rohm & Haas

APPLICATIONS
Mode of action Contact fungicide with protective and curative action. Non-systemic acaricide. Compounds of type (i) are more effective as acaricides, and those of type (ii) as fungicides (A. H. M. Kirby & L. D. Hunter, *Nature (London),* 1965, **208**, 189; A. H. M. Kirby *et al., Ann. Appl. Biol.,* 1966, **57**, 21; R. J. W. Byrde *et al., ibid.,* p. 223). **Uses** Control of powdery mildews in pome fruit, stone fruit, citrus fruit, soft fruit, vines, cucurbits, ornamentals, tobacco, hops, and some vegetables; and American gooseberry mildew in gooseberries and currants. Secondary control of mites (*Panonychus, Tetranychus,* and *Aculus* spp.) in fruit trees and vines. **Phytotoxicity** Non-phytotoxic, except for glasshouse roses under certain conditions. Phytotoxicity is increased at higher temperatures.
Formulation types WP; EC; DP. **Mixtures** *(dinocap +)* myclobutanil; sulfur; dicofol; dicofol + tetradifon; mancozeb; and many more.
Compatibility Incompatible with oils and oil-based sprays, and alkaline materials.
Selected tradenames 'Karathane' (Rohm & Haas); 'Crotothane' (Rhône-Poulenc); 'Sialite' (Siapa)

ANALYSIS
Product analysis by spectrophotometry (R. F. Black *et al., J. Assoc. Off. Anal. Chem.,* 1974, **57**, 653). **Residues** determined by glc (A. Ambrus *et al., ibid.,* 1981, **64**, 733; *Pestic. Anal. Man.,* 1979, **I**, 201-A, 201-G, 201-I; *Manu. Pestic. Residue Anal.,* 1987, **I**, S19; *Anal. Methods Residues Pestic.,* 1988, Part II; *Anal. Methods Pestic. Plant Growth Regul.,* 1972, **6**, 568).

MAMMALIAN TOXICOLOGY
Reviews *Pesticide residues in food – 1989.* FAO Plant Production and Protection Paper 99, 1989. *Pesticide residues in food – 1989 evaluations. Part II – Toxicology.* FAO Plant Production and Protection Paper 100/2, 1990. **Oral** Acute oral LD_{50} for male rats 980, female rats 1190, dogs 100 mg/kg (F. S. Larson *et al., Arch. Int. Pharmacodyn. Ther.,* 1959, **119**, 31). **Skin and eye** Acute percutaneous LD_{50} for rabbits >4700 mg tech./kg, >2350 mg a.i./kg (as 250 g/kg WP). Irritating to skin and eyes (rabbits). **Inhalation** LC_{50} (4 h) for rats 0.36 mg a.i./l air (as EC).
NOEL (2 y) for rats 6–8 mg/kg daily (there was no carcinogenic effect); for dogs 0.4 mg/kg daily; cataracts were produced at toxic levels in several species. Teratogenic effects were produced in rabbits by oral gavage at 3 mg/kg daily, but not in rabbits treated dermally at 100 mg/kg daily. **ADI** (JMPR) 0.001 mg/kg b.w. [1989]. **Toxicity class** WHO (a.i.) III; EPA (formulation) III **EC risk** Xn (R22); Xi (R38)

MAIN ENTRIES

ECOTOXICOLOGY
Fish Toxic to fish. **Bees** Not toxic to bees.

ENVIRONMENTAL FATE
Animals In rats, following oral administration, dinocap is almost entirely eliminated rapidly in the urine and faeces. In dairy cows, dinocap and its metabolites are excreted almost completely in the faeces, with only a small amount in the urine. The nitro groups are enzymically reduced to amino groups, and ester hydrolysis also takes place. **Plants** Metabolism follows the same course as in animals.
Soil/Environment In soil, the principal metabolite is 2,4-dinitro-6-(2-octyl)phenol. Details are given in *J. Agric. Food Chem.*, 1984, **32**, 1151. Soil DT_{50} 4.5–6.1 d (silt loam and fine sandy loam) by microbial degradation.

251 dinoterb   *Herbicide*

dinitrophenol

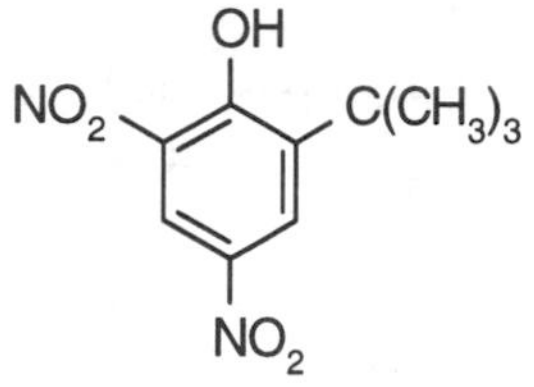

NOMENCLATURE
dinoterb
Common name dinoterb (BSI, E-ISO); dinoterbe ((*m*) F-ISO)
IUPAC name 2-*tert*-butyl-4,6-dinitrophenol
Chemical Abstracts name 2-(1,1-dimethylethyl)-4,6-dinitrophenol
CAS RN *[1420–07–1]* **EEC no.** 215–813–8 **Development codes** LS63 133 (Rhône-Poulenc); P 1108 (Murphy)

dinoterb-ammonium
CAS RN *[6365–83–9]*

dinoterb acetate
CAS RN *[3204–27–1]*

PHYSICAL CHEMISTRY
dinoterb
Mol. wt. 240.2 **M.f.** $C_{10}H_{12}N_2O_5$ **Form** Pale yellow solid with a phenol-like

odour. **M.p.** 125.5–126.5 °C **V.p.** 20 mPa (20 °C) **Solubility** In water 4.5 mg/l (pH 5, 20 °C). In cyclohexanone, ethyl acetate, dimethyl sulfoxide *c.* 200 (all in g/kg). In alcohols, glycols, aliphatic hydrocarbons *c.* 100 (all in g/kg). Soluble in aqueous alkalis with the formation of salts. **Stability** Stable below the melting point. Decomposes above 220 °C. Stable at least 34 d at pH 5–9 (22 °C).

dinoterb-ammonium
Mol. wt. 257.2 **M.f.** $C_{10}H_{15}N_3O_5$

dinoterb-diolamine
Mol. wt. 345.4 **M.f.** $C_{14}H_{23}N_3O_7$ **Solubility** In water 32.8 g/l.

dinoterb acetate
Mol. wt. 282.3 **M.f.** $C_{12}H_{14}N_2O_6$

COMMERCIALISATION
History Herbicide reported by G. A. Emery *et al.* (*Proc. Conf. EWRC/COLUMA, 2nd,* 1965, p. 41 for acetate) and by P. Poignant & P. Crisinel (*C. R. Journ. Etud. Herbic. Conf. COLUMA, 4th,* 1967, p. 196). Introduced by Pépro (now a subsidiary of Rhône-Poulenc Agrochimie) and by Murphy Chemical Ltd (who no longer manufacture or market it). **Patent** FR 1475686; FR 1532332; GB 1126658; US 3565601 all to Pépro **Manufacturers** Rhône-Poulenc

APPLICATIONS
Biochemistry Oxidative phosphorylation uncoupler. **Mode of action** Selective non-systemic herbicide with contact action. **Uses** Control of annual broad-leaved weeds post-emergence in cereals, maize, alfalfa, and beet; and pre-emergence in peas and beans. Also used for destruction of potato haulms. **Formulation types** SL; EC. **Mixtures** (*dinoterb* +) isoproturon; mecoprop. **Selected tradenames** 'Herbogil' (Rhône-Poulenc)

ANALYSIS
Product analysis by glc of a derivative (*CIPAC Handbook,* 1983, **1B,** 1797).
Residues by glc, details available from Rhône-Poulenc Agrochimie.

MAMMALIAN TOXICOLOGY
dinoterb
Oral Acute oral LD_{50} for rats 62, mice 25, rabbits 28 mg/kg. **Skin and eye** Acute percutaneous LD_{50} for guinea pigs 150 mg/kg. **NOEL** (2 y) for rats 0.375 mg/kg diet. **Toxicity class** WHO (a.i.) Ib **EC risk** (R61); T (also R24/25); Xi (R36); (R44); (salts and esters are (R61)); T (also R23/24/25)

ECOTOXICOLOGY
dinoterb
Fish LC_{50} (96 h) for rainbow trout 0.0034 mg/l. **Bees** Toxic to bees.

ENVIRONMENTAL FATE
Animals In rats, following oral administration, 98% is excreted in the faeces and urine within 7 days.

252 diofenolan   *Insecticide*

diphenyl ether

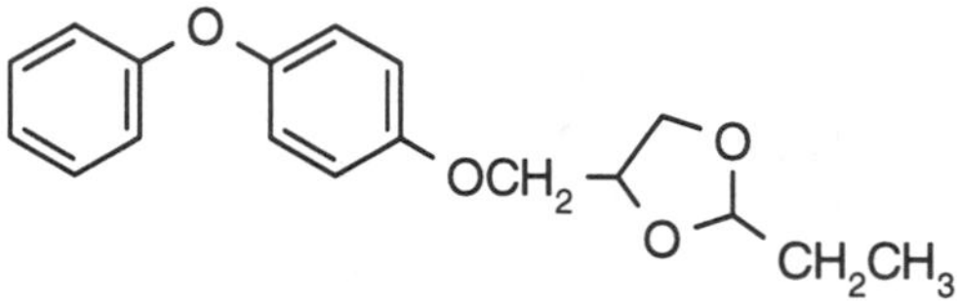

NOMENCLATURE
Common name diofenolan (BSI, pa E-ISO); diofenolane ((m) pa F-ISO)
IUPAC name A mixture of: (2RS,4SR)-4-(2-ethyl-1,3-dioxolan-4-ylmethoxy)phenyl phenyl ether (50%–80%) and (2RS,4RS)-4-(2-ethyl-1,3-dioxolan-4-ylmethoxy)= phenyl phenyl ether (50%–20%)
Chemical Abstracts name A mixture of (2RS,4SR)-2-ethyl-4-[(4-phenoxy= phenoxy)methyl]-1,3-dioxolane and (2RS,4RS)-2-ethyl-4-[(4-phenoxyphenoxy)= methyl]-1,3-dioxolane
CAS RN [63837–33–2]   **Development codes** CGA 059 205

PHYSICAL CHEMISTRY
Composition A mixture of 50–80% (2RS,4SR)- isomer with 50–20% (2RS,4RS)-= isomer.   **Mol. wt.** 300.3   **M.f.** $C_{18}H_{20}O_4$ **Form** Clear, pale yellow viscous liquid.
B.p. >250 °C/101.325 kPa (OECD 103)   **V.p.** 1.1×10^{-1} mPa (25 °C) (OECD 104)   **K_{ow}** logP = 4.8; 4.4 (*cis*- isomer); 4.3 (*trans*- isomer) (25 °C) (OECD 117)
S.g./density 1.141 (20 °C)   **Solubility** In water 4.9 mg/l (25 °C) (OECD 105).
Miscible with methanol, acetone, toluene, hexane, *n*-octanol.
Stability Hydrolytically stable at pH ≥7. Rapidly degraded by light.   **pKa** No dissociation in range pH 1–11

COMMERCIALISATION
History Reported by H. P. Streibert *et al.* (*Proc. Br. Crop Protection Conf. – Pests and Diseases*, 1994, **1**, 23). Introduced by Ciba-Geigy AG (now Novartis Crop Protection AG).   **Manufacturers** Novartis

APPLICATIONS
Mode of action Insect growth regulator with juvenile hormone activity. Inhibits development of first and second instar larvae of scales.   **Uses** Active against most

scale insects and eggs of some Lepidoptera in citrus, pome fruit, stone fruit, grapes, mango, olives, nuts, tea and ornamentals.   **Phytotoxicity** None observed. **Formulation types** EC.   **Selected tradenames** 'Arbor' (Novartis)

ANALYSIS
By hplc with u.v. detection.

MAMMALIAN TOXICOLOGY
Oral Acute oral LD_{50} for rats >5000 mg/kg.   **Skin and eye** Acute percutaneous LD_{50} for rats (24 h) >2000 mg/kg; not irritant to skin or eyes (rabbits); not a skin sensitiser.   **Inhalation** LC_{50} (4 h) for rats >3100 mg/m^3.   **NOEL** (3 mo) for rats 12 mg/kg b.w.; NOAEL (3 mo) for dogs 12 mg/kg b.w.   **ADI** 0.006 mg/kg (provisional).   **Toxicity class** WHO (a.i.) III

ECOTOXICOLOGY
Birds Acute oral LC_{50} for quail and ducks >2000 mg/kg.   **Fish** LC_{50} (96 h) for bluegill sunfish, catfish, carp and trout 1.0–1.7 mg/l.   **Bees** Acute oral LD_{50} and contact LC_{50} (48 h) >96 mg/bee. Tech. material has a high toxicity to the brood of the honeybee.   **Worms** LC_{50} (14 d) 204 mg/kg soil.   **Daphnia** LC_{50} (48 h) 0.5 mg/l.   **Algae** LC_{50} (72 h) 0.072 mg/l.

ENVIRONMENTAL FATE
Animals Intestinal absorption 80%. Rapid elimination. Tissue residues low.
Soil/Environment Rapid dissipation: field trials DT_{50} c. 6 d; laboratory DT_{50} <2 d. No potential for mobility (RMF 0.15).

253 # diphacinone   *Rodenticide*

indandione anticoagulant

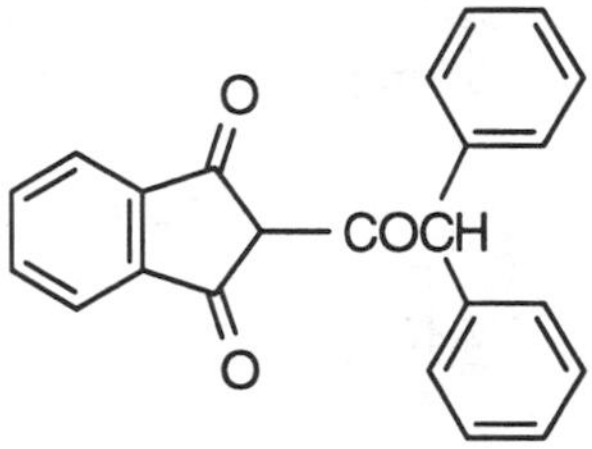

NOMENCLATURE
Common name diphacinone (BSI, E-ISO, (*m*) F-ISO, ANSI); diphacins (JMAF);

diphenadione (BAN); [diphacin] (former exception, Turkey); no name (Italy)
IUPAC name 2-(diphenylacetyl)indan-1,3-dione
Chemical Abstracts name 2-(diphenylacetyl)-1H-indene-1,3(2H)-dione
CAS RN *[82–66–6]*   **EEC no.** 201–434–5

PHYSICAL CHEMISTRY

Composition Tech. is 95% pure.   **Mol. wt.** 340.4   **M.f.** $C_{23}H_{16}O_3$ **Form** Yellow crystals; (tech., yellow powder).   **M.p.** 145–147 °C   **V.p.** 1.37×10^{-5} mPa (25 °C, tech.)   **S.g./density** 1.281   **Solubility** Practically insoluble in water (c. 0.3 mg/kg). In chloroform 204, toluene 73, xylene 50, acetone 29, ethanol 2.1, heptane 1.8 (all in g/kg). Soluble in alkalis with the formation of salts.   **Stability** Stable for 14 d (pH 6–9); hydrolysed in <24 h (pH 4). Rapidly decomposed in water by sunlight. Decomposes at 338 °C (without boiling).   **pKa** Acidic, forming water-soluble alkali metal salts.

COMMERCIALISATION

History Rodenticide reported by J. T. Correll *et al.* (*Proc. Soc. Exp. Biol. Med.,* 1952, **80,** 139). Introduced by Velsicol Chemical Corp. (now Novartis Crop Protection AG) and Upjohn Co. (neither company now manufactures or markets it).   **Patent** US 2672483

APPLICATIONS

Biochemistry Anticoagulant; inhibits the vitamin K-dependent steps in the synthesis of coagulation factors II, VII and X.   **Mode of action** Requires multiple feedings to produce lethal effects. Duration of action much longer than that of warfarin.   **Uses** Control of rats, mice, voles, prairie dogs (*Cynomys* spp.), ground squirrels, and other rodents.   **Formulation types** RB; CB.   **Compatibility** Chemically and physically compatible with other rodenticides.

ANALYSIS

Product analysis by hplc (K. Hunter, *Anal. Methods Pestic. Plant Growth Regul.,* 1988, **16,** 119, 128).   **Residues** determined by hplc or glc (*idem, ibid.*).

MAMMALIAN TOXICOLOGY

Oral Acute oral LD_{50} for rats 2.3, mice 340, rabbits 35, cats 14.7, dogs 3–7.5, pigs 150 mg/kg.   **Skin and eye** Acute percutaneous LD_{50} for rats <200 mg/kg. Non-irritating to skin and eyes; non-sensitising to skin (guinea pigs).   **Inhalation** LC_{50} (4 h) for rats <2 mg/l air (dust).   **NOEL** Chronic LD_{50} for albino rats is 0.1 mg/kg daily.   **Other** Non-mutagenic in the Ames test.   **Toxicity class** WHO (a.i.) Ia; EPA (formulation) I   **EC risk** T+ (R28); T (R48/23/24/25)

ECOTOXICOLOGY

Birds Acute oral LD_{50} for mallard ducks 3158 mg/kg. A 56-day secondary poisoning trial with bait (50 mg a.i./kg) revealed no hazard to sparrow hawks under conditions likely to be encountered in nature.   **Fish** LC$_{50}$ (96 h) for rainbow trout 2.8, bluegill sunfish 7.6, channel catfish 2.1 mg/l.

ENVIRONMENTAL FATE
Animals Not extensively metabolised in the rat; any metabolism mainly involves hydroxylation and conjugation.

254 diphenamid   *Herbicide*

alkanamide

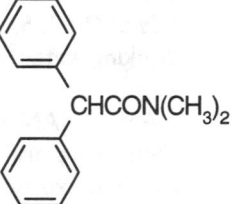

NOMENCLATURE
Common name diphenamid (BSI, E-ISO, ANSI, WSSA, JMAF); difénamide ((*f*) F-ISO); no name (Germany)
IUPAC name *N,N*-dimethyldiphenylacetamide
Chemical Abstracts name *N,N*-dimethyl-α-phenylbenzeneacetamide
CAS RN *[957–51–7]*   **EEC no.** 213–482–4   **Development codes** L-34 314 (to Lilly)

PHYSICAL CHEMISTRY
Mol. wt. 239.3   **M.f.** C$_{16}$H$_{17}$NO   **Form** Colourless crystals; (tech., off-white solid).   **M.p.** 134.5–135.5 °C; (tech., 132–134 °C)   **V.p.** Negligible at 20 °C
S.g./density 1.17 (23.3 °C)   **Solubility** In water 260 mg/l (27 °C). Moderately soluble in polar organic solvents. In acetone 189, dimethylformamide 165, xylene 50 (all in g/l, 27 °C).   **Stability** Relatively resistant to degradation by u.v. light. Moderately stable to heat (some decomposition occurs at 210 °C). Decomposed by strong acids and alkalis.

COMMERCIALISATION
History Herbicide reported by E. F. Alder *et al.* (*Proc. North Cent. Weed Control Conf.*, 1960, p. 55). Introduced by Eli Lilly & Co. and by The Upjohn Co. (neither now manufacture or market it).   **Patent** US 3120434

APPLICATIONS

Mode of action Selective systemic herbicide, absorbed by the roots, with translocation throughout the plant. Inhibits root elongation. **Uses** Pre-emergence control of annual grasses and some broad-leaved weeds in a wide range of crops, including cotton, potatoes, sweet potatoes, tomatoes, vegetables, capsicums, okra, soya beans, peanuts, tobacco, pome fruit, stone fruit, citrus fruit, bush fruit, strawberries, forestry nurseries, and ornamental plants, shrubs, and trees. Generally used at 4–6 kg a.i./ha. **Phytotoxicity** Phytotoxic to many crops, including cereals, maize, sorghum, sugar beet, beetroot, carrots, and spinach. **Formulation types** WP. **Compatibility** Incompatible with highly alkaline materials. **Selected tradenames** 'Enide' (Zeneca)

ANALYSIS

Residues determined by glc with FID (G. A. Boyack et al., J. Agric. Food Chem., 1966, **14**, 312; J. B. Tepe et al., Anal. Methods Pestic., Plant Growth Regul. Food Addit., 1967, **5**, 375; Anal. Methods Pestic. Plant Growth Regul., 1972, **6**, 637). In **drinking water**, by gc with FID (AOAC Methods, 1995, 991.07).

MAMMALIAN TOXICOLOGY

Oral Acute oral LD_{50} for rats 1050, mice 600, rabbits 1500, dogs 1000 mg/kg. **Skin and eye** Acute percutaneous LD_{50} for rats >225 mg/kg. Non-irritating to skin and eyes (rabbits). Not a skin sensitiser (guinea pigs). **NOEL** In 2 y feeding trials, rats and dogs receiving 2000 mg/kg diet showed no unusual effects on physiology and fertility. **Other** Acute i.p. LD_{50} for mice 569 mg/kg. **Toxicity class** WHO (a.i.) III; EPA (formulation) III **EC risk** Xn (R22)

ECOTOXICOLOGY

Fish Slightly toxic to fish.

ENVIRONMENTAL FATE

Animals In mammals, following oral administration, N-demethylation occurs, followed by elimination as the O-glucuronide; p-hydroxylation is reported as a minor route of metabolism (R. E. McMahon & H. R. Sullivan, Biochem. Pharmacol., 1965, **14**, 1085–1092). **Plants** In plants, metabolism involves N-demethylation. **Soil/Environment** In soil, diphenamid undergoes microbial degradation. Persistence under warm damp conditions is c. 3–6 months.

diphenylamine   *Fungicide*

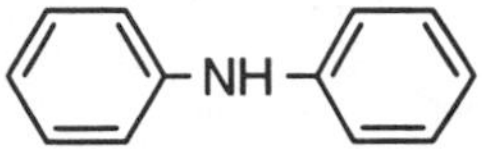

NOMENCLATURE
IUPAC name diphenylamine
Chemical Abstracts name *N*-phenylbenzenamine
Other names DPA   **CAS RN** *[122–39–4]*

PHYSICAL CHEMISTRY
Mol. wt. 169.2   **M.f.** $C_{12}H_{11}N$ **Form** Colourless crystals.   **M.p.** 53–54 °C
B.p. 302 °C   **Solubility** Sparingly soluble in water. Readily soluble in organic
solvents.   **Stability** Darkens on exposure to sunlight.   **pKa** Diphenylamine is a
weak base, and forms salts with concentrated acids.

COMMERCIALISATION
History Long known as a chemical, also used as fungicide.   **Manufacturers** Elf
Atochem; Productos OSA

APPLICATIONS
Uses Post-harvest fungicide protectant and scald inhibitor for pome fruit.
Formulation types EC.   **Selected tradenames** 'Coraza' (Productos OSA);
'No Scald' (Elf Atochem)

ANALYSIS
Residues determined by glc (*Man. Pestic. Residue Anal.*, 1987, **I**, 6; J. G. Allen &
K. J. Hall, *J. Agric. Food Chem.*, 1980, **28**, 255).

MAMMALIAN TOXICOLOGY
Reviews *Pesticide residues in food – 1984.* FAO Plant Production and Protection
Paper 62, 1985. *Pesticide residues in food – 1984 evaluations.* FAO Plant Production
and Protection Paper 67, 1985.   **ADI** (JMPR) 0.02 mg/kg b.w. (for 99.9% pure
material) [1984].

MAIN ENTRIES

256 diquat dibromide

bipyridylium

$2 Br^-$

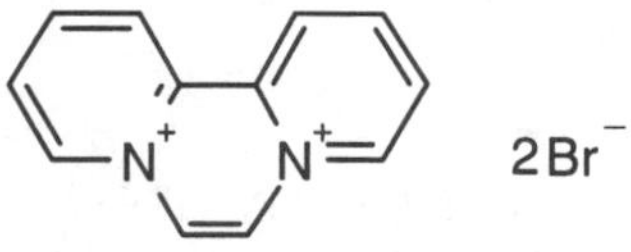

NOMENCLATURE
diquat dibromide
CAS RN *[85–00–7]*; *[6385–62–2]* for diquat dibromide monohydrate

diquat
Common name diquat (BSI, E-ISO, (*m*) F-ISO, ANSI, WSSA, JMAF); deiquat
(Germany); [reglon] (former exception, USSR)
IUPAC name 9,10-dihydro-8a,10a-diazoniaphenanthrene; 6,7-dihydrodipyrido-=
[1,2-*a*:2',1'-*c*]pyrazine-5,8-di-ium; 1,1'-ethylene-2,2'–bipyridyldiylium
Chemical Abstracts name 6,7-dihydrodipyrido[1,2-*a*:2',1'-*c*]pyrazinediium
CAS RN *[2764–72–9]* **EEC no.** 220–433–0

PHYSICAL CHEMISTRY
diquat dibromide
Mol. wt. 344.1 **M.f.** $C_{12}H_{12}Br_2N_2$ **Form** Colourless to yellow crystals.
M.p. (Monohydrate) Decomposes above 300 °C **V.p.** (Monohydrate)
<0.013 mPa **K**$_{ow}$ logP = –4.60 (20 °C) **S.g./density** 1.22–1.27 (20 °C)
Solubility In water 700 g/l (20 °C). Slightly soluble in alcohols and hydroxylic
solvents. Insoluble in non-polar organic solvents. **Stability** Stable in neutral and
acidic solutions, but readily hydrolysed in alkaline solutions. DT$_{50}$ at pH 7 in
simulated sunlight *c*. 74 d. Photochemically decomposed by u.v. irradiation.

diquat
Mol. wt. 184.2 **M.f.** $C_{12}H_{12}N_2$

COMMERCIALISATION
History Herbicide reported by R. C. Brian *et al.* (*Nature (London)*, 1958, **181**, 446).
Introduced by ICI Plant Protection Division (now Zeneca Agrochemicals).
Patent GB 785732 **Manufacturers** Zeneca

APPLICATIONS
diquat dibromide
Biochemistry During photosynthesis, superoxide is generated, which damages cell
membranes and cytoplasm. **Mode of action** Non-selective contact herbicide and
desiccant, absorbed by the foliage, with some translocation in the xylem.
Uses Pre-harvest desiccation of cotton, flax, alfalfa, clover, lupins, oilseed rape,

poppies, soya beans, peas, beans, sunflowers, cereals, maize, rice, sugar beet, and other seed crops; destruction of potato haulms; and stripping of hops. Control of annual broad-leaved weeds in vines, pome fruit, stone fruit, bush fruit, strawberries (also control of runners), citrus fruit, olives, hops, vegetables, ornamental plants and shrubs, and other crops. Control of emergent and submerged aquatic weeds. Weed control on non-crop land. Weed control and tassel inhibition in sugar cane. **Formulation types** SL; Gel. **Mixtures** *(diquat dibromide +)* paraquat dichloride; paraquat dichloride + simazine; simazine + TCA; amitrole + paraquat dichloride + simazine. **Compatibility** Incompatible with alkaline materials, anionic surfactants (e.g. alkyl sulfonates or alkyl aryl sulfonates), and alkali-metal salts of hormone-type herbicides. **Selected tradenames** 'Reglone' (Zeneca); 'Reglex' (Siapa); 'Desiquat' (Barclay)

ANALYSIS
Product analysis by u.v. spectrophotometry (*AOAC Methods*, 1995, 969.08; *CIPAC Handbook*, 1970, **1**, 342; 1992, **E**, 73–78; 1995, **G**, 47); impurities measured by glc (*ibid.*, 1980, **1A**, 1245; *Herbicides 1977*, p. 48). **Residues** determined by colorimetry (M. G. Ashley, *Pestic. Sci.*, 1970, **1**, 101; A. Calderbank *et al.*, *Analyst (London)*, 1961, **86**, 569; A. Calderbank & S. H. Yuen, *ibid.*, 1965, **90**, 95; 1966, **91**, 625; J. E. Pack, *Anal. Methods Pestic., Plant Growth Regul. Food Addit.*, 1967, **5**, 397; J. B. Leary, *Anal. Methods Pestic. Plant Growth Regul.*, 1978, **10**, 321; *Pestic. Anal. Man.*, 1979, **II**). Residues in potatoes by rplc with dual channel u.v. detection (*AOAC Methods*, 1995, 992.17). Details available from Zeneca.

MAMMALIAN TOXICOLOGY
diquat dibromide
Reviews *Pesticide residues in food – 1993*, FAO Plant Production and Protection Paper, 122, 1993. *Pesticide residues in food – 1993 evaluations, Part II – Toxicology.* WHO, WHO/PCS/94.4, 1994. **EHC** 39 (WHO, 1984). **Oral** Acute oral LD_{50} for rats 231, mice 125, rabbits 187, dogs 100–200, cows 37 mg/kg.
Skin and eye Acute percutaneous LD_{50} for rabbits >750, rats >2000 mg/kg. Irritating to skin and eyes (rabbits). Diquat is absorbed through human skin only after prolonged exposure; shorter exposure can cause irritation and a delay in the healing of cuts and wounds. Can cause temporary damage to nails. Irritant to eyes.
Inhalation Can cause nose bleeding if inhaled. **NOEL** (2 y) for rats 0.25 mg/kg b.w. daily; (4 y) for dogs 50 mg/kg diet. **ADI** (JMPR) 0.002 mg/kg b.w. as diquat ion [1993]. **Toxicity class** WHO (a.i.) II; EPA (formulation) II
EC risk T (R24/25); Xi (R36/37/38) (applies to all salts)

ECOTOXICOLOGY
diquat dibromide
Birds Acute oral LD_{50} for hens 200–400, partridges 295 mg/kg. **Fish** LC_{50} (96 h) for rainbow trout 21, mirror carp 67 mg/l.

ENVIRONMENTAL FATE
Animals In rats, following oral administration of diquat dibromide, the dose is completely eliminated in the urine and faeces within 4 days. **Plants** Metabolic breakdown of diquat dibromide does not occur in plants. On plant surfaces, photochemical degradation occurs. **Soil/Environment** Diquat dibromide is rapidly and completely inactivated on contact with soil. Adsorption capacity of soils 0.1–50 mg/kg.

257 disulfoton   *Insecticide, acaricide*

organophosphorus

$$CH_3CH_2SCH_2CH_2S\overset{\overset{\displaystyle S}{\|}}{P}(OCH_2CH_3)_2$$

NOMENCLATURE
Common name disulfoton (BSI, E-ISO, (m) F-ISO, ESA); ethylthiodemeton (JMAF); [M-74] (former exception, USSR)
IUPAC name O,O-diethyl S-2-ethylthioethyl phosphorodithioate
Chemical Abstracts name O,O-diethyl S-[2-(ethylthio)ethyl] phosphorodithioate
Other names [thiodemeton] (rejected common name proposal)
CAS RN [298-04-4]; (disulfoton sulfoxide [2497-07-6]; disulfoton sulfone [2497-06-5]) **EEC no.** 206-054-3 **Development codes** Bayer 19 639; S 276 (Bayer) **Official codes** ENT 23 347

PHYSICAL CHEMISTRY
Mol. wt. 274.4 **M.f.** $C_8H_{19}O_2PS_3$ **Form** Colourless oil with a characteristic odour (tech., pale yellow oil). **M.p.** <−25 °C **B.p.** 128 °C/1 mmHg
V.p. 7.2 mPa (20 °C), 13 mPa (25 °C), 22 mPa (30 °C) **K$_{ow}$** logP = 3.95
Henry 0.24 Pa m^3 mol^{-1} (calc.) **S.g./density** 1.144 (20 °C) **Solubility** In water 25 mg/l (20 °C). Readily miscible with *n*-hexane, dichloromethane, isopropanol, toluene. **Stability** Stable under normal storage conditions. Relatively stable in acidic and neutral media. Hydrolysed in alkaline media: DT$_{50}$ (22 °C) 133 d (pH 4), 169 d (pH 7), 131 d (pH 9). Photolysis DT$_{50}$ 1–4 d. **F.p.** 133 °C (tech.).

COMMERCIALISATION
History Described by G. Schrader (*Die Entwicklung neuer insektizider Phosphorsäure*). Introduced by Bayer AG and later by Sandoz AG (now Novartis Crop Protection AG). **Patent** DE 917668; DE 947369; US 2759010 all to Bayer
Manufacturers Bayer Corp.; Novartis; United Phosphorus Ltd

APPLICATIONS

Biochemistry Cholinesterase inhibitor. **Mode of action** Systemic insecticide and acaricide, absorbed by the roots, with translocation to all parts of the plant, giving long-lasting control. **Uses** Control of aphids, thrips, mealybugs, and other sucking insects, and spider mites in potatoes, vegetables, cereals, maize, sorghum, rice, tobacco, ornamentals, fruit and nut crops, and other crops. Also prevents cucumber mosaic and potato leaf roll viruses by controlling the virus vectors. **Phytotoxicity** Non-phytotoxic when used as directed. Treated seed may be injured by high doses. **Formulation types** GR; EC; DS. **Mixtures** (disulfoton +) chlorpyrifos; fonofos; quinalphos; diazinon; fenamiphos; triadimenol; isofenphos; cyproconazole. **Selected tradenames** 'Disyston' (Bayer); 'Frumin AL' (seed treatment) (Novartis); 'Solvirex' (granular) (Novartis)

ANALYSIS

Product analysis by glc with FID (*CIPAC Handbook*, 1985, **1C**, 2101; *AOAC Methods*, 1995, 980.10). **Residues** determined by glc (*Man. Pestic. Residue Anal.*, 1987, **I**, 3, 6, S8, S13, S16, S17, S19; *Anal. Methods Residues Pestic.*, 1988, Part I, M2, M5; *Analyst (London)*, 1980, **105**, 515). In **drinking water**, by gc with FID (*AOAC Methods*, 1995, 991.07). Details available from Novartis or Bayer.

MAMMALIAN TOXICOLOGY

Reviews *Pesticide residues in food – 1991*. FAO Plant Production and Protection Paper, 111, 1991. *Pesticide residues in food – 1991 evaluations. Part II – Toxicology*. World Health Organisation, WHO/PCS/92.52, 1992. **EHC** 63 (WHO, 1986; a general review of organophosphorus insecticides). **Oral** Acute oral LD_{50} for male and female rats 2–12, male and female mice 7.5, female dogs *c.* 5 mg/kg. **Skin and eye** Acute percutaneous LD_{50} for male rats 15.9, female rats 3.6 mg/kg; not irritating to skin and eyes (rabbit). **Inhalation** LC_{50} (4 h) for male rats *c.* 0.06, female rats *c.* 0.015 mg/l (for aerosol). **NOEL** (2 y) for rats and dogs 1, mice 4 mg/kg diet. **ADI** (JMPR) 0.0003 mg/kg b.w. [1991]. **Toxicity class** WHO (a.i.) Ia; EPA (formulation) I

ECOTOXICOLOGY

Birds Acute oral LD_{50} for bobwhite quail 39 mg/kg. Five-day dietary LC_{50} for mallard ducks 692, bobwhite quail 544 mg/kg diet. **Fish** LC_{50} (96 h) for bluegill sunfish 0.039, rainbow trout 3 mg/l. **Bees** Toxic to bees, depending on the mode of application. **Daphnia** LC_{50} (48 h) 0.013–0.064 mg/l.

ENVIRONMENTAL FATE

Animals In rats, following oral administration, ^{14}C-disulfoton is rapidly absorbed, metabolised, and the radioactivity excreted in the urine. The main metabolites are disulfoton sulfoxide and sulfone, their corresponding oxygen analogues and diethylthiophosphate. **Plants** In plants, disulfoton is very rapidly metabolised. The metabolism is the same as in animals. **Soil/Environment** In soil, disulfoton is very rapidly degraded. The metabolism is similar to that in animals and plants. It exhibits medium to low mobility in soil.

258 dithianon   *Fungicide*

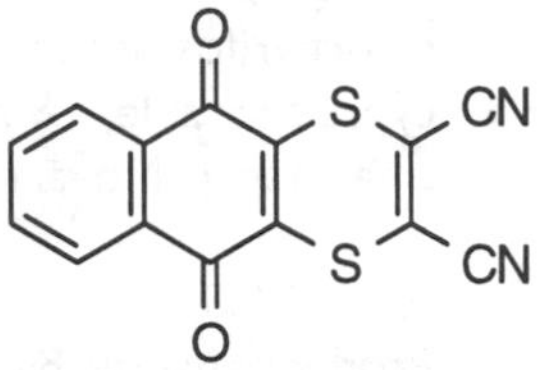

NOMENCLATURE
Common name dithianon (BSI, E-ISO, (*m*) F-ISO, JMAF); no name (Italy)
IUPAC name 5,10-dihydro-5,10-dioxonaphtho[2,3-*b*]-1,4-dithiine-2,3-dicarbonitrile
Chemical Abstracts name 5,10-dihydro-5,10-dioxonaphtho[2,3-*b*]-1,4-dithi-in-= 2,3-dicarbonitrile
CAS RN *[3347–22–6]* **EEC no.** 222–098–6 **Development codes** IT-931; MV 119A; CME107; SAG 107

PHYSICAL CHEMISTRY
Mol. wt. 296.3 **M.f.** $C_{14}H_4N_2O_2S_2$ **Form** Dark brown crystals with a coppery lustre; (tech., lighter brown). **M.p.** 225 °C; (tech., *c.* 217 °C)
V.p. 2.7×10^{-6} mPa (25 °C) **K_{ow}** logP = 3.2 **S.g./density** 1576 kg/m³ (20 °C)
Solubility In water 0.14 mg/l (pH 7, 20 °C). In chloroform 12, acetone 10, benzene 8 (all in g/l, 20 °C). Sparingly soluble in methanol and dichloromethane.
Stability Decomposed by alkaline media, by concentrated acids, and by prolonged heating; DT_{50} 12.2 h (pH 7, 25 °C). Stable up to 80 °C. Aqueous solution (0.1 mg/l) exposed to artificial sunlight had DT_{50} 19 h. **F.p.** Ignition temperature >300 °C

COMMERCIALISATION
History Fungicide reported by J. Berker *et al.* (*Proc. Br. Insectic. Fungic. Conf., 2nd,* 1963, p. 351). Introduced by E. Merck and later Shell Agrar GmbH (now American Cyanamid Co.). **Patent** GB 857383 **Manufacturers** Cyanamid

APPLICATIONS
Biochemistry Affects a range of fungal enzymes by reacting with thiol groups and interfering with cellular respiration. Multi-site mode of action.
Mode of action Foliar fungicide with protective and, to some extent, curative action. **Uses** Control of many foliar diseases (but not powdery mildews),

including scab on pome fruit; *Stigmina carpophila*, *Coccomyces hiemalis* and scab on cherries; *Monilia* spp., rust and leaf curl on peaches and apricots; leaf spot and rust on currants; *Didymella applanata* on raspberries; *Mycosphaerella fragariae* and *Diplocarpon earliana* on strawberries; *Plasmopara viticola* on vines; downy mildew on hops; scab and *Phomopsis citri* on citrus fruit; *Ascochyta chrysanthemi* on chrysanthemums under glass; *Glomerella cingulata* on coffee; *Marssonina* leaf spot on poplars; etc.   **Phytotoxicity** Slight russetting may occur on some apple varieties, such as Golden Delicious.   **Formulation types** SC; WG; WP. **Mixtures** *(dithianon +)* cymoxanil; copper oxychloride; dimethomorph; mancozeb. **Compatibility** Incompatible with petroleum oils, alkaline materials, and sulfur-containing compounds.   **Selected tradenames** 'Aktuan' (mixture) (Cyanamid); 'Delan' (Cyanamid); 'Cluster' (Barclay)

ANALYSIS
Product analysis by hplc or by colorimetry (*CIPAC Handbook*, 1985, **1C**, 2105; *ibid.*, 1995, **G**, 50–55; E. Amodori & W. Heupt, *Anal. Methods Pestic. Plant Growth Regul.*, 1978, **10**, 181).   **Residues** determined by hplc or by colorimetry (*idem, ibid.*). Details from Cyanamid.

MAMMALIAN TOXICOLOGY
Reviews *Pesticide residues in food – 1992*. FAO Plant Production and Protection Paper, 116, 1993. *Pesticide residues in food – 1992 evaluations. Part II – Toxicology*. World Health Organisation, WHO/PCS/93.34.   **Oral** Acute oral LD_{50} for rats 638, guinea pigs 115 mg/kg.   **Skin and eye** Acute percutaneous LD_{50} for rats >2000 mg/kg. Mild skin and eye irritant.   **Inhalation** LC_{50} (4 h) for rats 2.1 mg/l air.   **NOEL** (2 y) for rats 20, dogs 40 mg/kg diet; for mice 2.8 mg/kg daily. **ADI** (JMPR) 0.01 mg/kg b.w. [1992].   **Toxicity class** WHO (a.i.) III; EPA (formulation) III   **EC risk** Xn (R22)

ECOTOXICOLOGY
Birds Acute oral LD_{50} for male quail 280, female quail 430 mg/kg.   **Fish** Toxic to fish; LC_{50} (96 h) for common carp 0.1 mg/l.   **Bees** Contact LD_{50} >0.1 mg/bee. **Worms** LC_{50} (7 d) 588.4, (14 d) 578.4.   **Other beneficial spp.** Low hazard to predatory mites.   **Daphnia** LC_{50} (24 h) 2.4 mg/l.   **Algae** EC_{50} (96 h) 12 mg/l.

ENVIRONMENTAL FATE
Soil/Environment Freundlich K 18–56 mg/kg in soils (o.c. 0.7–2.6%, pH 4.8–6.6).

pyridine

$$F_3C \diagdown \quad N \quad \diagdown CHF_2$$
$$CH_3SOC \diagdown \qquad \diagdown COSCH_3$$
$$CH_2CH(CH_3)_2$$

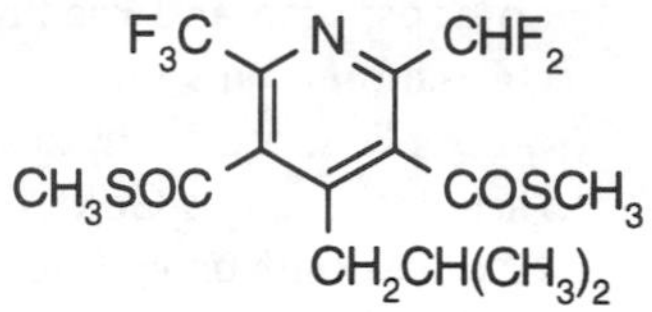

NOMENCLATURE
Common name dithiopyr (BSI, ANSI, draft E-ISO)
IUPAC name S,S'-dimethyl 2-difluoromethyl-4-isobutyl-6-trifluoromethylpyridine-=
3,5-dicarbothioate
Chemical Abstracts name S,S'-dimethyl 2-(difluoromethyl)-4-(2-methylpropyl)-=
6-(trifluoromethyl)-3,5-pyridinedicarbothioate
CAS RN *[97886–45–8]* **Development codes** MON 7200; MON 15 100;
RH 101664

PHYSICAL CHEMISTRY
Mol. wt. 401.4 **M.f.** $C_{15}H_{16}F_5NO_2S_2$ **Form** Colourless crystals. **M.p.** 65 °C
V.p. 0.53 mPa (25 °C) **Henry** 0.2 Pa m^3 mol^{-1} **Solubility** In water 1.4 mg/l
(20 °C).

COMMERCIALISATION
History Herbicide introduced by Monsanto Co. and subsequently sold to Rohm
and Haas Co. in 1994. **Patent** US 4692184 **Manufacturers** Rohm & Haas

APPLICATIONS
Biochemistry Inhibits cell division by disrupting spindle microtubule formation.
Mode of action Pre-emergence herbicide. **Uses** Pre-emergence and early post-
emergence control of annual grass and broad-leaved weeds in turf at 0.25 to
1.0 lb/a, and of annual grasses in rice. **Formulation types** EC; GR.
Selected tradenames 'Dictran' (Japan) (Rohm & Haas); 'Dimension' (Rohm &
Haas)

MAMMALIAN TOXICOLOGY
Oral Acute oral LD$_{50}$ for rats and mice >5000 mg/kg. **Skin and eye** Acute
percutaneous LD$_{50}$ for rabbits and rats >5000 mg/kg. Non-irritating to skin,
slightly irritating to eyes (rabbits). Non-sensitising to skin (guinea pigs).
Inhalation LC$_{50}$ (4 h) for rats >6 mg/l. **NOEL** (2 y) for rats ≤10 mg/kg; (1 y) for
dogs ≤0.5 mg/kg daily. **Other** Chronic oral exposure of rats and mice to
dithiopyr did not result in tumour formation. **Toxicity class** WHO (a.i.) III
(Table 5)

ECOTOXICOLOGY
Birds Acute oral LD$_{50}$ for bobwhite quail >2250 mg/kg. Five-day dietary LC$_{50}$ for bobwhite quail and mallard ducks >5260 mg/kg. **Fish** LC$_{50}$ (96 h) for rainbow trout 0.5, bluegill sunfish and common carp 0.7 mg/l. In a trout early life stage study, the maximum acceptable toxicant concentration was determined to be 0.082 mg/l. **Bees** LD$_{50}$ (topical) 0.08 mg/bee. **Worms** LC$_{50}$ (14 d) >1000 mg/kg. **Daphnia** LC$_{50}$ (48 h) >1.1 mg/l.

ENVIRONMENTAL FATE
Soil/Environment DT$_{50}$ in soil 17–61 d, depending on the formulation type. The major soil metabolites are the di-acid, the normal mono-acid and the reverse mono-acid; these metabolites, themselves, dissipate almost completely within 1 year.

260 diuron *Herbicide*

urea

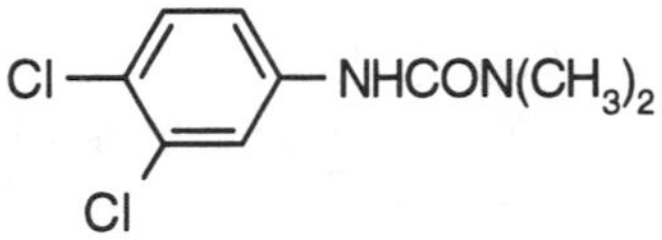

Cl —⟨ ⟩— NHCON(CH$_3$)$_2$
  Cl

NOMENCLATURE
Common name diuron (BSI, E-ISO, (*m*) F-ISO, ANSI, WSSA); DCMU (JMAF); [dichlorfenidim] (former exception, USSR)
IUPAC name 3-(3,4-dichlorophenyl)-1,1-dimethylurea
Chemical Abstracts name *N*'-(3,4-dichlorophenyl)-*N*,*N*-dimethylurea
CAS RN *[330–54–1]* **EEC no.** 206–354–4 **Development codes** DPX 14740

PHYSICAL CHEMISTRY
Mol. wt. 233.1 **M.f.** C$_9$H$_{10}$Cl$_2$N$_2$O **Form** Colourless crystals. **M.p.** 158–159 °C
V.p. 1.1 × 10^{-3} mPa (25 °C) **K$_{ow}$** logP = 2.85±0.03 (25 °C) **S.g./density** 1.48
Solubility In water 36.4 mg/l (25 °C). In acetone 53, butyl stearate 1.4, benzene 1.2 (all in g/kg, 27 °C). Sparingly soluble in hydrocarbons. **Stability** Stable in neutral media at normal temperatures, but hydrolysed at elevated temperatures. Hydrolysed by acids and alkalis. Decomposes at 180–190 °C.

COMMERCIALISATION
History Herbicide reported by H. C. Bucha & C. W. Todd (*Science*, 1951, **114**, 493). Introduced by E. I. du Pont de Nemours Co. **Patent** US 2655445

Manufacturers Ancom; Bayer; Cedar; Crystal; Defensa; Drexel; Du Pont; Griffin; Hodogaya; Makhteshim-Agan; Nufarm Ltd; Rhône-Poulenc; Sagrochem; Sanachem; United Phosphorus Ltd

APPLICATIONS
Biochemistry Inhibits photosynthesis (Hill reaction). **Mode of action** Systemic herbicide, absorbed principally by the roots, with translocation acropetally in the xylem. **Uses** Total control of weeds and mosses on non-crop areas at 10–30 kg a.i./ha. Selective control of germinating grass and broad-leaved weeds in many crops, including asparagus, tree fruit, bush fruit, citrus fruit, vines, olives, pineapples, bananas, sugar cane, cotton, peppermint, alfalfa, forage legumes, cereals, maize, sorghum, and perennial grass-seed crops at 0.6–4.8 kg/ha. Phytotoxic residues in soil disappear within 1 season at these lower rates. **Formulation types** WP; SC. **Mixtures** (*diuron +*) amitrole; bromacil; chlorpropham; methabenzthiazuron; hexazinone; paraquat dichloride; propham; linuron + terbacil; oryzalin; dalapon-sodium + MCPA; amitrole + bromacil; 2,4-D + dalapon-sodium; amitrole + atrazine; bromacil + hexazinone; amitrole + 2,4-D + simazine; propyzamide; and many more. **Selected tradenames** 'Karmex' (Du Pont); 'Cekuron' (Cequisa); 'Diurex' (Makhteshim-Agan); 'Direx' (Griffin); 'Dynex' (Crystal/Cedar); 'Sanuron' (Sanachem); 'Seduron' (Rhône-Poulenc); 'Toterbane' (Diachem); 'Unidron' (Unicrop)

ANALYSIS
Product analysis by hplc (G. W. Sheehan, *Anal. Methods Pestic. Plant Growth Regul.*, 1984, **13**, 227); or by hydrolysis and titration of the amine liberated (*CIPAC Handbook*, 1980, **1A**, 1251), details available from Du Pont. Impurities by tlc (*ibid.*, 1994, **F**, 336). **Residues** determined by glc (*Anal. Methods Pestic. Plant Growth Regul.*, 1972, **6**, 664; W. E. Bleidner, *J. Agric. Food Chem.*, 1954, **2**, 682). Hydrolysis to 3,4-dichloroaniline, a derivative of which is measured colorimetrically, may also be used (H. L. Pease, *ibid.*, 1962, **10**, 279; R. L. Dalton & H. L. Pease, *J. Assoc. Off. Anal. Chem.*, 1962, **45**, 377). In **water**, by lc with u.v. detection (*AOAC Methods*, 1995, 992.14, 10.7.01).

MAMMALIAN TOXICOLOGY
Oral Acute oral LD_{50} for rats 3400 mg/kg. **Skin and eye** Acute percutaneous LD_{50} for rabbits >2000 mg/kg for 80% DF. Mild eye irritant (WP formulation) (rabbits); non-irritating to intact skin (50% aqueous paste) (guinea pigs). Non-sensitising to skin (guinea pigs). **Inhalation** LC_{50} (4 h) for rats >5 mg/l.
NOEL (2 y) for rats 250, dogs 125 mg/kg diet (H. C. Hodge *et al.*, *Food Cosmet. Toxicol.*, 1967, **5**, 513). **ADI** (EPA) 0.002 mg/kg. **Toxicity class** WHO (a.i.) III (Table 5); EPA (formulation) III **EC risk** Xn (R48/22)

ECOTOXICOLOGY
Birds Eight-day dietary LC_{50} for bobwhite quail 1730, Japanese quail >5000,

mallard ducklings >5000, pheasant chicks >5000 mg/kg diet.   **Fish** LC_{50} (96 h) for rainbow trout 5.6, bluegill sunfish 5.9, guppies 25 mg/l.   **Bees** Not toxic to bees. **Daphnia** LC_{50} (48 h) 12 mg/l.

ENVIRONMENTAL FATE

Animals In mammals, metabolism is principally by hydroxylation and dealkylation (C. Boehme & W. Ernst, *Food Cosmet. Toxicol.*, 1965, **3**, 797–802).   **Plants** In plants, diuron undergoes demethylation of the nitrogen atom and hydroxylation at position 2 of the benzene ring.   **Soil/Environment** In soil, enzymic and microbial demethylation of the nitrogen atom and hydroxylation at position 2 of the benzene ring occur. Duration of activity in soil is *c.* 4–8 months, depending on soil type and humidity; DT_{50} 90–180 d (G. D. Hill et al., *Agron. J.*, 1955, **47**, 93; T. J. Sheets, *J. Agric. Food Chem.*, 1964, **12**, 30). K_{oc} 400.

:61   DNOC   *Insecticide, acaricide, herbicide*

dinitrophenol

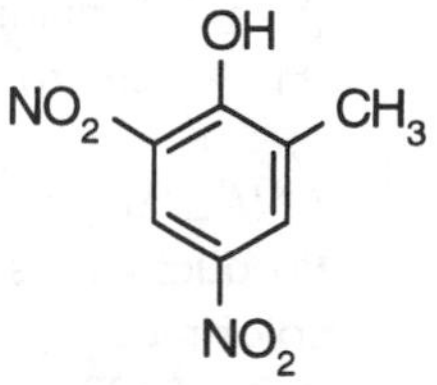

NOMENCLATURE

Common name DNOC (BSI, E-ISO, (*m*) F-ISO, WSSA, JMAF)
IUPAC name 4,6-dinitro-*o*-cresol
Chemical Abstracts name 2-methyl-4,6-dinitrophenol
Other names DNC   **CAS RN** *[534–52–1]*   **EEC no.** 208–601–1; (potassium salt, 219–007–7; ammonium salt, 221–037–0)   **Official codes** ENT 154

PHYSICAL CHEMISTRY

Composition Tech. is 95–98% pure.   **Mol. wt.** 198.1   **M.f.** $C_7H_6N_2O_5$
Form Yellow crystals (pure DNOC, but not tech., is explosive when dry).
M.p. 88.2–89.9 °C; (tech., 83–85 °C)   **V.p.** 16 mPa (25 °C) K_{ow} logP = 0.08
(pH 7)   **Henry** 2.41×10^{-7} Pa m^3 mol^{-1} (calc.)   **S.g./density** 1.58 (20 °C)
Solubility In water 6.94 g/l (20 °C, pH 7). In toluene 251, methanol 58.4, hexane

4.03, ethyl acetate 338, acetone 514, dichloromethane 503 (all in g/l, 20 °C). The sodium, potassium, calcium, and ammonium salts are readily soluble in water. **Stability** In water, DNOC degrades very slowly, DT_{50} >1 y. Photolysis DT_{50} c. 253 h (20 °C). In dry conditions, salts tend to detonate, and so they are usually moistened with up to 10% water in order to reduce the risk of explosion. **pKa** 4.48 (20 °C)

COMMERCIALISATION
History Introduced as an insecticide by Fr Bayer & Co. (now Bayer AG, who no longer manufacture or market it) in 1892, and as a herbicide by G. Truffaut et Cie in 1932.   **Patent** GB 3301 to Bayer; GB 425295 to Truffaut   **Manufacturers** Elf Atochem

APPLICATIONS
Biochemistry Respiratory electron transport inhibitor.   **Mode of action** Non-systemic insecticide and acaricide with contact and stomach action; contact herbicide.   **Uses** Control of overwintering stages of aphids, suckers, ermine moths, winter moths, tortrix moths, cherry blossom moths, scale insects, and spider mites on pome fruit trees, stone fruit trees, and soft fruit bushes. Control of annual broad-leaved weeds in cereals, maize, legumes, flax, tree fruit, bush fruit, hops, and grass-seed crops. Also used as a desiccant for leguminous seed crops; for destruction of potato haulms; and for chemical stripping of hops. **Phytotoxicity** Very phytotoxic.   **Formulation types** SC; PA; SL; WP. **Mixtures** (*DNOC +*) petroleum oils; metoxuron.   **Selected tradenames** 'Ibertox' (Siapa); 'Trifanex' (mixture) (Elf Atochem); 'Trifocide' (Elf Atochem); 'Trifrina' (Elf Atochem)

ANALYSIS
Product analysis by conversion to the ammonium salt which is measured by colorimetry (D. S. Farrington, *Analyst (London)*, 1983, **108**, 353; *CIPAC Proc.*, 1980, **2**, 206; 1981, **3**, 246) or by titration with titanium trichloride in an inert atmosphere (W. Fischer, *Fresenius' Z. Anal. Chem.*, 1938, **112**, 91; *MAFF Ref. Book.* No.1, 1958; *CIPAC Handbook*, 1970, **1**, 348).   **Residues** determined by colorimetry in alkaline solution (A. W. Avens *et al., J. Econ. Entomol.*, 1948, **41**, 432; W. H. Parker, *Analyst (London)*, 1949, **74**, 646).

MAMMALIAN TOXICOLOGY
Reviews *Evaluation of the toxicity of pesticide residues in food.* FAO Meeting Report, No. PL/1965/10/1; WHO/Food Add./27.65, 1965.   **Oral** Acute oral LD_{50} for rats 25–40, mice 16–47, goats 100, cats 50 mg/kg; for sheep 200 mg DNOC-sodium/kg.   **Skin and eye** Acute percutaneous LD_{50} for rats 200-600, rabbits 1000, mice 187 mg/kg. Skin irritant. Dangerous amounts can be absorbed through the skin.   **NOEL** (6 mo) for rats and rabbits >100, dogs 20 mg/kg diet; (28 d) for rats 13 mg/kg diet. In man, DNOC acts as a powerful cumulative

metabolic poison. There is a danger of chronic poisoning with repeated uptake (D. G. Harvey et al., Br. Med. J., 1951, **2**, 13; E. King & D. G. Harvey, Biochem. J., 1953, **53**, 185, 196). **ADI** (JMPR) No ADI [1965]. **Toxicity class** WHO (a.i.) Ib; EPA (formulation) I **EC risk** (R44); T+ (R27/28); (R40); Xi (R36); (R33); (potassium salt, T (R23/24/25); (R33); ammonium salt, T+ (R26/27/28); (R33))

ECOTOXICOLOGY
Birds LD_{50} for Japanese quail 15.7 mg/kg (14 d), ducks 23, partridges 20–25, pheasants 6–85 mg/kg. LC_{50} for Japanese quail 637 mg/kg diet. **Fish** LC_{50} for carp 6–13, trout 0.45, bluegill sunfish 0.95 mg/l. **Bees** LD_{50} 1.79–2.29 mg/bee; moderate to low toxicity under field conditions. **Worms** LC_{50} (14 d) for earthworms 15 mg/kg soil. **Daphnia** LC_{50} (24 h) 5.7 mg/l. **Algae** LC_{50} (96 h) 6 mg/l.

ENVIRONMENTAL FATE
Animals In mammals, following oral administration, DNOC undergoes metabolism, and is eliminated as the glucuronide conjugate and as 2-methyl-4,6-diaminophenol. DT_{50} in humans c. 150 h (estimated, Jastroch et al., Z. Ges. Hyg., 1978, **24**, 234). **Plants** In plants, the nitro groups are reduced to amino groups. **Soil/Environment** In soil, the nitro groups are reduced to amino groups. DT_{50} in soil 0.1–12 d (20 °C), 15 d (5 °C). In water, DT_{50} 3–5 w (20 °C).

262 dodec-8-en-1-yl acetate *Insect pheromone*

pheromone

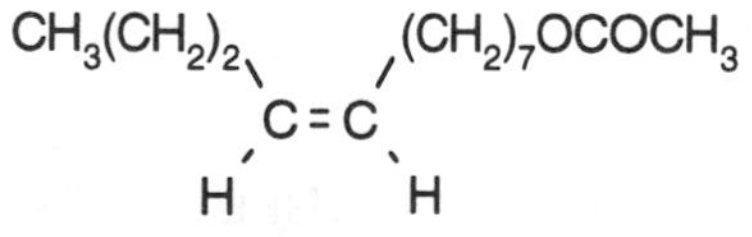

$$CH_3(CH_2)_2 \diagdown \qquad \diagup (CH_2)_7OCOCH_3$$
$$C = C$$
$$\diagup \qquad \diagdown$$
$$H \qquad H$$

(Z)- isomer

NOMENCLATURE
IUPAC name (Z)-dodec-8-en-1-yl acetate
Chemical Abstracts name (Z)-8-dodecen-1-ol acetate
Other names Oriental fruit moth pheromone; *Grapholitha molesta* pheromone; *Cydia molesta* pheromone; (Z8)-12Ac; E8-12Ac; (Z8/E8)-12Ac (IOBC codes)
CAS RN [28079–04–1]; (E)- isomer [38363–29–0]; (Z/E)- mixture [37338–40–2]; (Z)- alcohol [40642–40–8] **EEC no.** 248–823–6; 253–904–4

PHYSICAL CHEMISTRY
Composition Principal component of the sex pheromone of the Oriental fruit moth (*Grapholitha (Cydia) molesta*). Maximum attraction has been observed with a

mixture of (Z)- and (E)- isomers, with (Z)-dodec-8-en-1-ol, in the proportions 100:7:30. 'Isomate-M' contains these components in the ratio 93:6:1.
Mol. wt. 226.4 **M.f.** $C_{14}H_{26}O_2$ **Form** Yellow liquid with a characteristic odour. **B.p.** 115–125 °C/3 mmHg **V.p.** c. 70 mPa (20 °C, evaporation balance); 688 mPa (20 °C, calc.); 708 mPa (20 °C, calc., mixture for mating disruption). K_{ow} logP >4 **S.g./density** 0.879 (20 °C) **Solubility** Insoluble in water. Readily soluble in organic solvents. **Stability** Stable in range pH 5–7. **F.p.** 108 °C (closed cup)

COMMERCIALISATION
History In use since 1985. **Manufacturers** Shin-Etsu

APPLICATIONS
Mode of action Acts by disruption of mating, either in trapping or in disorientation mode. **Uses** Used for control of the Oriental fruit moth. **Formulation types** VP; Tube. **Selected tradenames** 'Confusalin' (AgrEvo); 'Isomate-M' (Shin-Etsu, Biocontrol); 'RAK 5' (BASF)

263 dodeca-7,9-dien-1-yl acetate

Insect pheromone

pheromone

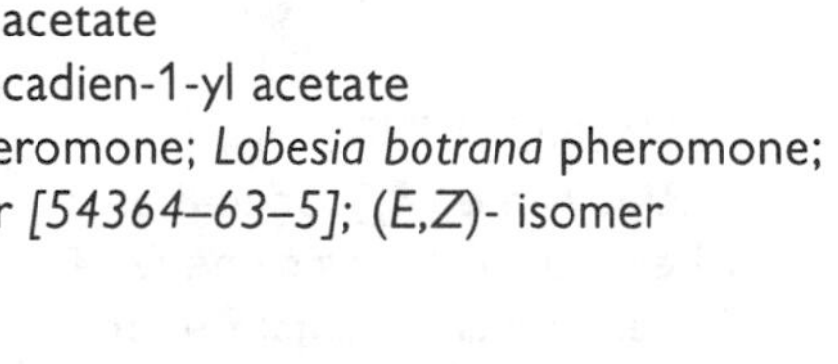

NOMENCLATURE
IUPAC name (7E,9Z)-dodeca-7,9-dien-1-yl acetate
Chemical Abstracts name (7E,9Z)-7,9-dodecadien-1-yl acetate
Other names European grapevine moth pheromone; *Lobesia botrana* pheromone; E7Z9-12Ac (IOBC) **CAS RN** (E,E)- isomer *[54364–63–5]*; (E,Z)- isomer *[55774–32–8]* **EEC no.** 259–127–7

PHYSICAL CHEMISTRY
Composition Mating hormone of the European grapevine moth (*Lobesia botrana*). The main component of the natural pheromone is the (7E,9Z)- isomer; the (7E,9E)- isomer is a second component. **Mol. wt.** 224.3 **M.f.** $C_{14}H_{24}O_2$ **Form** Oily, colourless liquid with a characteristic odour. **V.p.** 16×10^1 mPa (20 °C, evaporation balance) **Solubility** Insoluble in water. Readily soluble in organic solvents.

APPLICATIONS

Mode of action Acts as an attractant and by disruption of mating in the disorientation mode. **Uses** For control of the European grapevine moth. **Formulation types** VP. **Selected tradenames** 'RAK 2' (BASF)

264 dodemorph
Fungicide

morpholine

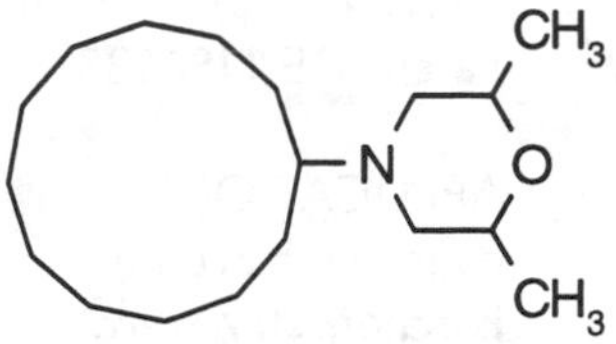

NOMENCLATURE

dodemorph

Common name dodemorph (BSI, E-ISO); dodémorphe ((m) F-ISO)
IUPAC name 4-cyclododecyl-2,6-dimethylmorpholine
Chemical Abstracts name 4-cyclododecyl-2,6-dimethylmorpholine
CAS RN *[1593–77–7]*; dodemorph was originally reported in the literature with an incorrect structure *[35842–17–2]* **EEC no.** 216–474–9
Development codes BAS 238F

dodemorph acetate

Common name dodemorph acetate (BSI, E-ISO); acétate de dodémorphe (F-ISO)
IUPAC name *N*-cyclododecyl-2,6-dimethylmorpholinium acetate
CAS RN *[31717–87–0]* **Development codes** BAS 238F

PHYSICAL CHEMISTRY

dodemorph

Composition Dodemorph contains *cis*- (*c.* 60% m/m) and *trans*- (*c.* 40%) isomers. **Mol. wt.** 281.5 **M.f.** $C_{18}H_{35}NO$ **Form** The *trans*- isomer is a colourless oil. The *cis*- isomer is a colourless solid with characteristic odour. **M.p.** 71 °C
B.p. *c.* 190 °C/1 mmHg **V.p.** *cis*- isomer: 0.48 mPa (20 °C) **K_{ow}** logP = 4.14 (pH 7) **Solubility** *cis*- isomer: In water <100 mg/kg. In chloroform >1000, ethanol 50, acetone 57, ethyl acetate 185 (all in g/l, 20 °C). **Stability** Stable to heat, light and water. **pKa** 8.08

dodemorph acetate

Mol. wt. 341.5 **M.f.** $C_{20}H_{39}NO_3$ **Form** Colourless solid (tech., viscous yellow liquid). **M.p.** 63–64 °C; *cis*- isomer 72 °C **B.p.** 315 °C (101.3 kPa)

V.p. 2.5 mPa at 20 °C   **K$_{ow}$** logP = 2.52 (pH 5), 4.23 (pH 9)   **Henry** 0.77 Pa m^3 mol^{-1} **S.g./density** c. 0.93   **Solubility** In water 1.1 mg/kg (20 °C). In benzene, chloroform >1000, cyclohexane 846, ethyl acetate 205, ethanol 66, acetone 22 (all in g/kg, 20 °C).   **Stability** Stable in unopened containers for >1 y. Stable at 50 °C for ≥2 y. Stable in neutral, moderate alkaline or acidic media.   **F.p.** Forms flammable decomposition products at high temperature.

COMMERCIALISATION

History Fungicidal properties of morpholines with large groups attached to the nitrogen described by K. H. König et al. (Angew. Chem., 1965, **77**, 327); those of dodemorph by J. Kradel & E. H. Pommer (Proc. Br. Insectic. Fungic. Conf., 4th, 1967, **1**, 170). Dodemorph acetate introduced in Germany (1968) by BASF AG. **Patent** DE 1198125   **Manufacturers** BASF

APPLICATIONS

dodemorph acetate
Biochemistry Steroid reduction (ergosterol biosynthesis) inhibitor.
Mode of action Dodemorph acetate is a systemic fungicide with protective and curative action. Absorbed through the leaves and roots, with translocation.
Uses Control of powdery mildews on roses and other ornamentals (outdoors and under glass).   **Phytotoxicity** Cinerarias, begonias, and some varieties of rose may be injured.   **Formulation types** EC.   **Mixtures** *(dodemorph acetate +)* dodine.
Selected tradenames 'Meltatox' (BASF)

ANALYSIS

Product analysis by potentiometric titration with perchloric acid in glacial acetic acid.   **Residues** in crops analysed by gc.

MAMMALIAN TOXICOLOGY

dodemorph
EC risk Xi (R36/37/38)

dodemorph acetate
Oral Acute oral LD$_{50}$ for male rats 3944, female rats 2465 mg/kg.
Skin and eye Acute percutaneous LD$_{50}$ for rats >4000 mg/kg (for 42.6% EC formulation). Severe skin and eye irritant (rabbits).   **Inhalation** LC$_{50}$ (4 h) for rats 5 mg/l air (for EC formulation).   **Toxicity class** WHO (a.i.) III (Table 5)

ECOTOXICOLOGY

dodemorph acetate
Fish LC$_{50}$ (96 h) for guppies c. 40 mg/l.   **Bees** Non-toxic to bees.   **Daphnia** LC$_{50}$ (48 h) 3.34 mg/l.

ENVIRONMENTAL FATE
Soil/Environment DT_{50} 73 d. K_{oc} 4200–48 000, indicating high adsorption and therefore no risk of leaching.

265 dodine   *Fungicide*

guanidine

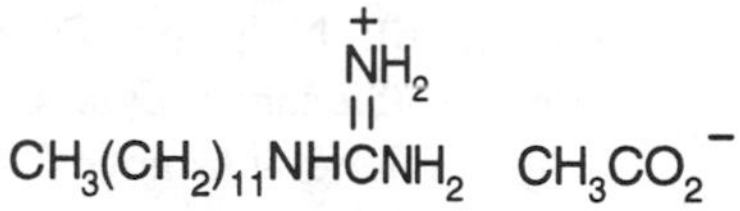

$$CH_3(CH_2)_{11}NHC\overset{\overset{+}{N}H_2}{\underset{}{||}}NH_2 \quad CH_3CO_2^-$$

NOMENCLATURE
dodine
Common name dodine (BSI, E-ISO, *(f)* F-ISO, ANSI); doguadine (France); [tsitrex] (former exception, USSR); [dodine acetate] (BSI, before 1969)
IUPAC name 1-dodecylguanidinium acetate
Chemical Abstracts name dodecylguanidine monoacetate
CAS RN *[2439–10–3]*   **EEC no.** 219–459–5   **Development codes** CL 7521; AC 5223

dodine free base
Common name guanidine (JMAF); [dodine] (BSI, before 1969)
CAS RN *[112–65–2]*

PHYSICAL CHEMISTRY
dodine
Mol. wt. 287.4   **M.f.** $C_{15}H_{33}N_3O_2$ **Form** Colourless crystals.   **M.p.** 136 °C
V.p. $<10^{-2}$ mPa (20 °C)   **Solubility** In water 630 mg/l (25 °C). Readily soluble in mineral acids, soluble in hot water and in alcohols: in butane-1,4-diol, *n*-butanol, cyclohexanol, *N*-methylpyrrolidone, *n*-propanol, tetrahydrofurfuryl alcohol >250 (all in g/l). Insoluble in most organic solvents.   **Stability** Stable in neutral, and moderately alkaline or acidic media, but the free base is liberated by concentrated alkali.

dodine free base
Mol. wt. 227.4   **M.f.** $C_{13}H_{29}N_3$

COMMERCIALISATION
History Fungicide reported by B. Cation (*Plant Dis. Rep.*, 1957, **41**, 1029). Introduced by American Cyanamid Co.   **Patent** US 2867562
Manufacturers Agrokémia; Cyanamid; General Quimica

APPLICATIONS

Mode of action Foliar fungicide with protective and some curative action.
Uses Dodine is used for control of scab on apples, pears, and pecans; leaf spot diseases of cherries, olives, blackcurrants, celery, and other crops; and foliar diseases of strawberries. Also used on other fruit, vegetable, nut, and ornamental crops, and on shade trees. Applied at 30–80 g a.i./100 l (0.25–1.5 kg/ha).
Phytotoxicity Peaches, plums, vines, and raspberries may be injured. Russetting may occur on some varieties of apple (including Golden Delicious).
Formulation types WP; SC; SL. **Mixtures** (dodine +) dodemorph acetate; fenarimol. **Compatibility** Incompatible with anionic surfactants, chlorobenzilate, lime sulfur, oils, oil emulsions, and hard water. **Selected tradenames** 'Cyprex' (Cyanamid); 'Melprex' (Cyanamid, Rhône-Poulenc); 'Venturol' (Cyanamid); 'Dodene' (Sipcam); 'Dodex' (Barclay); 'Efuzin' (Agrokémia); 'Guanidol' (Siapa); 'Sulgen' (General Quimica)

ANALYSIS

Product analysis by titration in non-aqueous media (*CIPAC Handbook*, 1983, **1B**, 1802; *AOAC Methods*, 1995, 970.07). **Residues** determined by colorimetry of a complex (*ibid.*, 964.19; *Pestic. Anal. Man.*, 1979, **II**; G. L. Sutherland, *Anal. Methods Pestic., Plant Growth Regul. Food Addit.*, 1964, **3**, 41).

MAMMALIAN TOXICOLOGY

dodine
Reviews *Pesticide residues in food*. FAO Food and Nutrition Series, No. 9; FAO Plant Production and Protection Series, No. 8; WHO Technical Report Series, No. 612, 1977. *1976 Evaluations of some pesticide residues in food*. **Oral** Acute oral LD_{50} for male rats c. 1000 mg/kg. **Skin and eye** Acute percutaneous LD_{50} for rabbits >1500, rats >6000 mg/kg. Irritating to skin. **NOEL** In 2 y feeding trials, rats receiving 800 mg/kg diet suffered a slight retardation of growth with no effect on reproduction or lactation. **ADI** (JMPR) 0.01 mg/kg b.w. [1976].
Toxicity class WHO (a.i.) III; EPA (formulation) I **EC risk** Xn (R22); Xi (R36/38)

ECOTOXICOLOGY

dodine
Birds Acute oral LD_{50} for mallard ducks 1142, Japanese quail 788 mg/kg.
Fish LC_{50} for harlequin fish (48 h) 0.53 mg/l; (96 h) 0.6 mg/l. **Bees** Topical LD_{50} for honeybees >0.011 mg/bee.

ENVIRONMENTAL FATE

Animals In rats, 95% of the applied dose is eliminated in urine and faeces within 8 d; c. 74% of excreted material is unchanged dodine; metabolites include creatine and guanidine derivatives. **Plants** In plants, dodine is converted to creatine by the action of a methyltransferase and simultaneous oxidative cleavage of the dodecyl moiety.

Insecticide

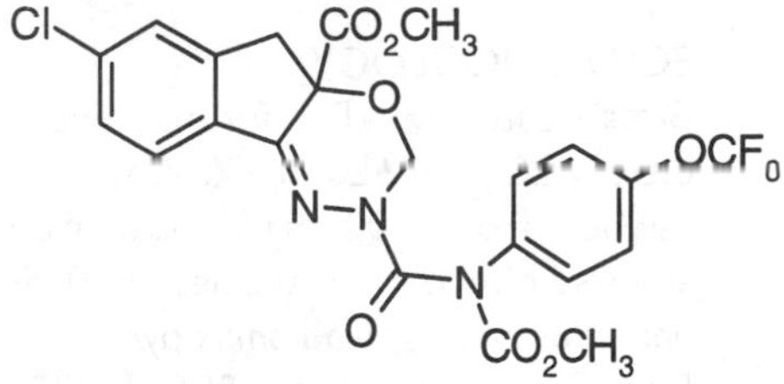

NOMENCLATURE
Common name indoxacarb (BSI proposed)
IUPAC name methyl 7-chloro-2,3,4*a*,5-tetrahydro-2-[methoxycarbonyl(4-trifluoro=
methoxyphenyl)carbamoyl]indeno[1,2-e][1,3,4]oxadiazine-4*a*-carboxylate
Chemical Abstracts name methyl 7-chloro-2,5-dihydro-2-[[(methoxycarbonyl)=
[4-(trifluoromethyl)phenyl]amino]carbonyl]-indeno[1,2-e][1,3,4]oxadiazine-=
4*a*(3*H*)-carboxylate
CAS RN *[144171–61–9]* (DPX-JW062); *[173584–44–6]* (DPX-KN128)
Development codes DPX-JW062; DPX-MP062; DPX-KN128; DPX-KN127 (see
Composition)

PHYSICAL CHEMISTRY
Composition DPX-JW062 is a 1:1 mixture of the (active) (*S*)- and (inactive)
(*R*)- isomers; DPX-MP062 is a 3:1 mixture of these isomers. The pure (*S*)- isomer
is referred to as DPX-KN128, the pure (*R*)- isomer as DPX-KN127. Unless
otherwise stated, data refers to DPX-JW062. **Mol. wt.** 527.8
M.f. $C_{22}H_{17}ClF_3N_3O_7$ **M.p.** 140–141 °C **V.p.** $<10^{-2}$ mPa (20–25 °C)
K_{ow} logP = *c.* 4.6 **Solubility** In water <0.5 mg/l. In *n*-octanol 0.48, methanol 0.39,
acetonitrile 76, acetone 140 g/l. **Stability** Aqueous hydrolysis DT_{50} >30 d (pH 5),
c. 30 d (pH 7), *c.* 2 d (pH 9).

COMMERCIALISATION
History Reported by H. H. Harder *et al.* (*Proc. Br. Crop Prot. Conf. – Pests Dis.,*
1996, **2**, 449). Under development by DuPont Agricultural Products.
Manufacturers Du Pont

APPLICATIONS
Biochemistry DPX-KN128, the active component, blocks sodium channels in nerve
cells. **Mode of action** Insecticide active by contact and ingestion. Affected insects
cease feeding, with poor co-ordination, paralysis and ultimately death.
Uses DPX-MP062 is under development for broad spectrum control of
Lepidoptera in cotton, vegetables and fruit, at 12.5–70 g/ha.

MAMMALIAN TOXICOLOGY

Oral Acute oral LD_{50} for rats >5000 mg/kg.   **Skin and eye** Acute percutaneous LD_{50} for rabbits >2000 mg/kg. No eye or skin irritation (rabbit). No skin sensitisation (guinea pig).   **Inhalation** LC_{50} for rats >2 mg/l.   **Other** Negative in the Ames test.

ECOTOXICOLOGY

Birds Acute oral LD_{50} for bobwhite quail and mallard duck >2250 mg/kg; 5-d dietary LC_{50} >5620 mg/kg diet.   **Fish** LC_{50} (96 h) for bluegill sunfish >1.0, rainbow trout >0.5 mg/l.   **Other beneficial spp.** DPX-KN128 had little or no adverse effects on 4 species at 30–50 g/ha; more extended studies on *Episyrphus balteatus* and *Typhlodromus pyri* described (M. Mead-Briggs *et al.*, *Proc. Br. Crop Prot. Conf. – Pests Dis.*, 1996, **1**, 307).

ENVIRONMENTAL FATE

Soil/Environment DT_{50} 4–5 d in tama silt loam soil. DT_{50} for aquatic photolysis 1–2 d (pH 5.0).

267   edifenphos   *Fungicide*

organophosphate ester

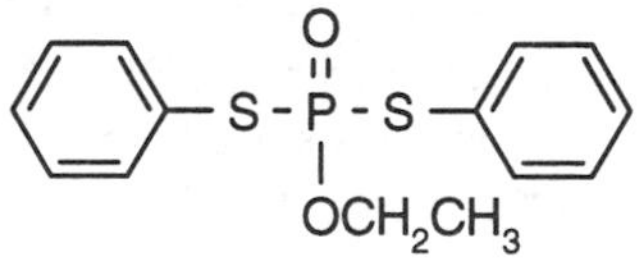

NOMENCLATURE

Common name edifenphos (BSI, E-ISO, (*m*) F-ISO); EDDP (JMAF)
IUPAC name *O*-ethyl *S,S*-diphenyl phosphorodithioate
Chemical Abstracts name *O*-ethyl *S,S*-diphenyl phosphorodithioate
CAS RN *[17109–49–8]*   **EEC no.** 241–178–1   **Development codes** Bayer 78 418; SRA 7847

PHYSICAL CHEMISTRY

Mol. wt. 310.4   **M.f.** $C_{14}H_{15}O_2PS_2$ **Form** Yellow to light brown liquid with a characteristic odour.   **M.p.** −25 °C   **B.p.** 154 °C/1 Pa   **V.p.** 0.032 mPa (20 °C)
K_{ow} logP = 3.83 (20°C)   **Henry** 2×10^{-4} Pa m^3 mol^{-1} (20 °C)
S.g./density 1.251 g/cm^3 (20 °C)   **Solubility** In water 56 mg/l (20 °C). In *n*-hexane 20–50, dichloromethane, isopropanol, toluene 200 (all in g/l, 20 °C).

Readily soluble in methanol, acetone, benzene, xylene, carbon tetrachloride and dioxane. Sparingly soluble in heptane.   **Stability** Stable in neutral media. Hydrolysed by strong acids and alkalis; at 25 °C, DT_{50} 19 d (pH 7), 2 d (pH 9). Degradable in the presence of light.   **F.p.** 115 °C (tech., DIN 51755)

COMMERCIALISATION
History Fungicide reported by H. Scheinpflug & H. F. Jung (*Pflanzenschutz-Nachr. (Engl. Ed.)*, 1968, **21**, 79). Introduced by Bayer AG.   **Patent** BE 686048; DE 1493736   **Manufacturers** Bayer; Hanwha

APPLICATIONS
Mode of action Foliar fungicide with protective and curative action (O. Kodama et al., *Agric. Biol. Chem.*, 1980, **44**, 1015).   **Uses** Control of rice blast. Also controls ear blight and stem rot in rice.   **Formulation types** EC; DP.
Mixtures (*edifenphos* +) phthalide; pencycuron; various insecticides.
Compatibility Incompatible with alkaline materials.   **Selected tradenames** 'Hinosan' (Bayer, Nihon Bayer); 'Hinorabcide' (Bayer)

ANALYSIS
Product analysis by glc with FID (*CIPAC Handbook*, 1992, **E**, 79). Details of methods for product and residue analysis available from Bayer AG.   **Residues** in rice determined by glc with TID (K. Vogeler, *ibid.*, p. 317; *Man. Pestic. Residue Anal.*, 1987, **I**, 6, S19).

MAMMALIAN TOXICOLOGY
Reviews *Pesticide residues in food – 1981*. FAO Plant Production and Protection Paper 37, 1982. *Pesticide residues in food: 1981 evaluations*. FAO Plant Production and Protection Paper 42, 1982.   **Oral** Acute oral LD_{50} for rats 100–260, mice 220–670, guinea pigs and rabbits 350–400 mg/kg.   **Skin and eye** Acute percutaneous LD_{50} for rats 700–800 mg/kg; slightly irritating to skin, not irritating to eyes (rabbits).   **Inhalation** LC_{50} (4 h) for rats 0.32–0.36 mg/l air (aerosol).
NOEL (2 y) for male rats 5, female rats 15, dogs 20 mg/kg diet; (18 mo) for mice 2 mg/kg diet.   **ADI** (JMPR) 0.003 mg/kg b.w. [1981].   **Toxicity class** WHO (a.i.) Ib   **EC risk** T (R23/24/25)

ECOTOXICOLOGY
Birds Acute oral LD_{50} for bobwhite quail 290, mallard ducks 2700 mg/kg.
Fish LC_{50} (96 h) for rainbow trout 0.43, bluegill sunfish 0.49, carp 2.5 mg a.i./l (using WP15).   **Bees** Not hazardous to bees when used as directed.
Daphnia LC_{50} (48 h) 0.032 µg/l.

ENVIRONMENTAL FATE
Animals Rapidly absorbed, metabolised and eliminated by rats and mice following oral administration. From 98.6 to 101.1% of the dose was recovered within

72 hours, at which time 4% was recovered from the tissues. Edifenphos was degraded primarily by loss of the phenyl, thiophenyl and ethyl groups to give organophosphorus acids. The ultimate metabolites were phosphoric and sulfuric acids. Furthermore, oxidation and methylation were apparent in the later stage of metabolism and conjugated metabolites were formed. **Plants** ^{14}C-Edifenphos was degraded in rice primarily by loss of phenyl, thiophenyl and/or ethyl groups to give organophosphorus acids, the ultimate products being benzenesulfonic acid and presumably phosphoric acid. **Soil/Environment** Leaching was very slight in soil columns, and inversely correlated with soil adsorption and soil o.m. content, indicating that edifenphos will be nearly immobile following application to soil. DT_{50} in soil was in the range of days to a few weeks. Edifenphos was degradable in sterile aqueous solutions; DT_{50} was in the range of minutes to a few days, depending on pH. DT_{50} in natural water was several hours. Metabolism in both soil and water was by loss of the phenyl, thiophenyl and ethyl groups to organophosphorus acids.

268 empenthrin [(*EZ*)- (1*R*)- isomers]

Insecticide

pyrethroid

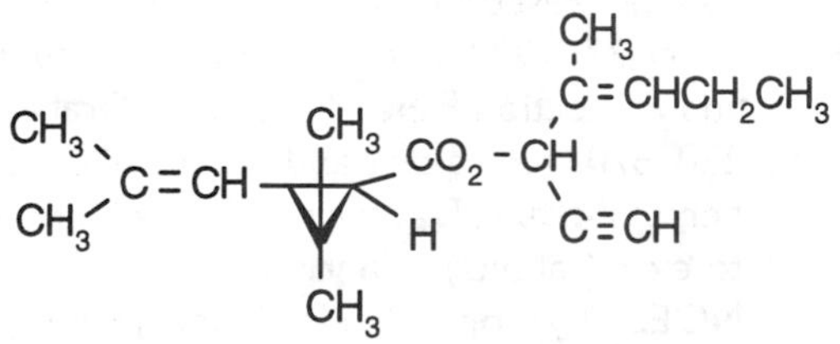

NOMENCLATURE
Common name empenthrin (BSI, draft E-ISO (see Composition)); empenthrine ((*f*) draft F-ISO)
IUPAC name (*E*)-(*RS*)-1-ethynyl-2-methylpent-2-enyl (1*R*,3*RS*;1*R*,3*SR*)-2,2-= dimethyl-3-(2-methylprop-1-enyl)cyclopropanecarboxylate (see Composition)
Roth: (*E*)-(*RS*)-1-ethynyl-2-methylpent-2-enyl (1*R*)-*cis-trans*-2,2-dimethyl-= 3-(2-methylprop-1-enyl)cyclopropanecarboxylate
Chemical Abstracts name 1-ethynyl-2-methyl-2-pentenyl (1*R*)-*cis-trans*-= 2,2-dimethyl-3-(2-methyl-1-propenyl)cyclopropanecarboxylate
CAS RN *[54406–48–3]* unstated stereochemistry **Development codes** S-2852 Forte (Sumitomo)

PHYSICAL CHEMISTRY

Composition The name empenthrin is defined for the (E)-(RS) (1RS)-cis-trans-mixed isomers. However, the material in use is the (EZ)-(RS) (1R)-rich cis-trans-isomer mixture. **Mol. wt.** 274.4 **M.f.** $C_{18}H_{26}O_2$ **Form** Yellow liquid.
B.p. 295.5 °C/760 mmHg **V.p.** 14 mPa (23.6 °C) **S.g./density** 0.927 (20 °C)
Solubility In water 0.111 mg/l (25 °C). Totally miscible in hexane, acetone and methanol (20 °C). **Stability** Stable under normal conditions for at least 2 years.
F.p. 107 °C

COMMERCIALISATION

History Insecticide introduced as (1R)-rich grade by Sumitomo Chemical Co. Ltd.
Manufacturers Sumitomo

APPLICATIONS

Mode of action Insecticide with contact action. Compared with other pyrethroids, has a high vapour pressure and hence is active against flying insects. **Uses** Used as a domestic insecticide, especially against moths, woolly bears and other insects attacking fabrics. **Formulation types** Aerosol; Impregnated material.
Compatibility Not compatible with alkaline materials.
Selected tradenames 'Vaporthrin' (Sumitomo)

ANALYSIS

Product analysis by glc. Details available from Sumitomo Chemical Co. Ltd.

MAMMALIAN TOXICOLOGY

Oral Acute oral LD_{50} for male rats >5000, female rats >3500 mg/kg.
Skin and eye Acute percutaneous LD_{50} for rats >2000 mg/kg. Not irritating to skin and eyes (rabbits). **Inhalation** LC_{50} (4 h) for rats >4610 mg/m^3.
Toxicity class WHO (a.i.) III

269 *Encarsia formosa*   *Biological agent*

Hymenoptera

Wasp parasite of glasshouse whitefly.

NOMENCLATURE

Scientific name *Encarsia formosa*   **Other names** Glasshouse whitefly parasite

PROPERTIES
Stability Recommended storage conditions are 8 °C and 70 to 80% r.h. The product should be used within 24 h.

COMMERCIALISATION
History In 1926, an English tomato grower noticed on his crop black pupae (sometimes referred to as scales) among the normal white pupae of the glasshouse whitefly (*Trialeurodes vaporariorum*). The emergents from these black pupae were identified as the parasitic wasp *Encarsia formosa* (Gahan). The parasite was produced and distributed during the 1930's and 1940's, but this ceased in 1956. Following the onset of resistance and very heavy whitefly infestations in the 1970's, commercial production of the parasite followed, so that today *E. formosa* is one of the most widely used and successful agents employed in protected crops. **Manufacturers** Applied Bio-nomics; Arbico; BCP; Biobest; Biological Services; English Woodlands; Koppert; M&R Durango; Nature's Alternative; Neudorff; Novartis BCM; Rincon-Vitova; Sautter & Stepper

APPLICATIONS
Mode of action The parasite reduces host numbers by stinging and feeding upon body fluids of early stage whitefly nymphs, and by parasitising the third instar stage, laying its egg in the immobile nymph. The female also feeds upon honeydew, to which it is attracted up to a range of 10 metres. Duration of life cycle is 34 days at 18 °C, down to 15 at 27 °C. Maximum oviposition occurs at 25 °C, up to a maximum of 300 eggs, and ceases below 12 °C. Dispersal flights are considerably curtailed below 18 °C and in low light levels. **Uses** The parasite is introduced into a crop towards the end of its developmental period within the nymph host. Parasitised pupae in determined numbers are glued to cards about 50 × 70 mm, incorporating a device for attachment to a plant. *E. formosa* is used to control the glasshouse whitefly *T. vaporariorum* and, less successfully, against the sweet potato whitefly, *Bemisia tabaci*. These whiteflies are important pests of tomatoes, cucurbits and a wide range of vegetables and ornamentals world-wide. With rare exceptions, control is limited to use on protected crops. The pupal parasite must be introduced when the pest is at a very low density. Several introductions are made until one month after the first pupal parasites are seen. **Phytotoxicity** Non-phytotoxic and non-phytopathogenic. **Formulation types** Parasitised pupae of whitefly on cards. **Compatibility** Use is entirely compatible with the use of other biological agents, but they are susceptible to a wide range of insecticides and some fungicides. **Selected tradenames** 'Encar-line' (Novartis BCM); 'En-Strip' (Koppert)

MAMMALIAN TOXICOLOGY
There is no evidence of acute or chronic toxicity, eye or skin irritation, or hypersensitivity to mammals. No allergic response or health problems have been observed in production staff or horticultural workers.

cyclodiene organochlorine

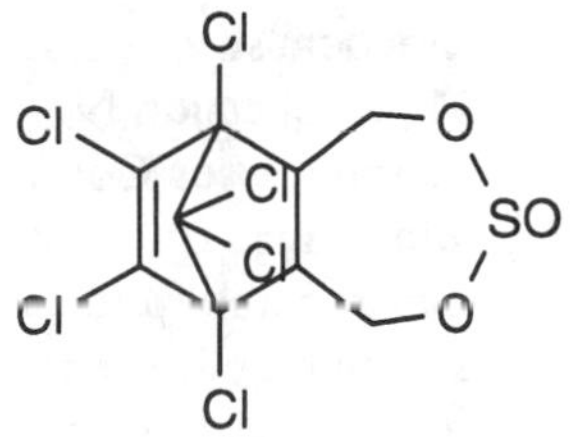

NOMENCLATURE
Common name endosulfan (BSO, E-ISO, (m) F-ISO, ANSI, ESA); thiodan (Iran, USSR); benzoepin (JMAF); no name (Italy)
IUPAC name (1,4,5,6,7,7-hexachloro-8,9,10-trinorborn-5-en-2,3-ylene= bismethylene) sulfite; 6,7,8,9,10,10-hexachloro-1,5,5a,6,9,9a-hexahydro-= 6,9-methano-2,4,3-benzodioxathiepine 3-oxide
Chemical Abstracts name 6,7,8,9,10,10-hexachloro-1,5,5a,6,9,9a-hexahydro-= 6,9-methano-2,4,3-benzodioxathiepine 3-oxide
CAS RN *[115–29–7]* endosulfan; *[959–98–8]* formerly *[33213–66–0]*, alpha-= endosulfan; *[33213–65–9]* formerly *[891–86–1]* and *[19670–15–6]*, beta-= endosulfan **EEC no.** 204–079–9 **Development codes** Hoe 02 671; FMC 5462
Official codes OMS 204 (α); OMS 205 (β); OMS 570; ENT 23979

PHYSICAL CHEMISTRY
Composition Endosulfan is a mixture of two stereoisomers: alpha-endosulfan, endosulfan (I), stereochemistry 3α,5aβ,6α,9α,9aβ-, comprises 64–67% of the tech. grade; beta-endosulfan, endosulfan (II), stereochemistry 3α,5aα,6β,9β,9aα-, 29–32%. Earlier reports on the stereochemistry of these isomers gave conflicting reports (W. Riemschneider, *World Rev. Pest Control*, 1963, **2**(4), 29).
Mol. wt. 406.9 **M.f.** $C_9H_6Cl_6O_3S$ **Form** Colourless crystals (tech., cream to brown, mostly beige). **M.p.** ≥80 °C (tech.); α- 109.2 °C; β- 213.3 °C
V.p. 0.83 mPa (20 °C) for 2:1 mixture of α- and β- isomers. **K_{ow}** logP for α- = 4.74; β- = 4.79 (both at pH 5) **Henry** α- 1.48, β- 0.07 (both Pa m^3 mol^{-1}, 22 °C, calc.) **S.g./density** c. 1.8 (20 °C) (tech.) **Solubility** In water alpha-= endosulfan 0.32, beta-endosulfan 0.33 (both in mg/l, 22 °C). In ethyl acetate, dichloromethane, toluene 200, ethanol c. 65, hexane c. 24 (all in g/l, 20 °C).
Stability Stable to sunlight. Slowly hydrolysed in aqueous acids and alkalis, with the formation of the diol and sulfur dioxide.

COMMERCIALISATION
History Insecticide reported by W. Finkenbrink (*Nachrichtenbl. Dtsch. Pflanzenschutzdienstes (Braunschweig)*, 1956, **8**, 183). Introduced by Hoechst AG (now AgrEvo GmbH) and, in the USA, by FMC Corp. **Patent** DE 1015797;

MAIN ENTRIES

US 2799685; GB 810602 all to Hoechst   **Manufacturers** AgrEvo; Aimco; Crystal; Excel; Hindustan; Makhteshim-Agan

APPLICATIONS
Biochemistry Antagonist of the GABA receptor-chloride channel complex.
Mode of action Non-systemic insecticide and acaricide with contact and stomach action.   **Uses** Control of sucking, chewing, and boring insects and mites on a very wide range of crops, including fruit (including citrus), vines, olives, vegetables, ornamentals, potatoes, cucurbits, cotton, tea, coffee, rice, cereals, maize, sorghum, oilseed crops, hops, hazels, sugar cane, tobacco, alfalfa, mushrooms, forestry, glasshouse crops, etc. Also controls tsetse flies.   **Phytotoxicity** Glasshouse geraniums and chrysanthemums, alfalfa, and lima beans may be injured.
Formulation types EC; WP; DP; GR; UL; SC; Powder concentrate.
Mixtures (endosulfan +) dimethoate; malathion; methomyl; monocrotophos; pirimicarb; triazophos; fenoprop; parathion; parathion-methyl; amitraz; deltamethrin; heptenophos; lindane + oxine-copper; petroleum oils; thiometon; anthraquinone + lindane + oxine-copper.   **Compatibility** Incompatible with strongly alkaline materials.   **Selected tradenames** 'Fan' (FMC); 'Thiodan' (AgrEvo); 'Afidanil' (Efthymiadis); 'Cekulfan' (Cequisa); 'Endocel' (Excel); 'Endodhan' (Dhanuka); 'Endosol' (Aimco); 'Endostar' (Shaw Wallace); 'Thionex' (Makhteshim-Agan)

ANALYSIS
Product analysis by i.r. spectrometry (*CIPAC Handbook*, 1970, **1**, 360) or by glc (*ibid.*, 1985, **1C**, 2110; *AOAC Methods*, 1995, 983.08).   **Residues** determined by glc with MCD (*ibid.*, 976.23; *Pestic. Anal. Man.*, 1979, 201-A, 201-G, 201-I; 405; A. Ambrus et al., *J. Assoc. Off. Anal. Chem.*, 1981, **64**, 773; *Man. Pestic. Residue Anal.*, 1987, **I**, 5, 6, S19; *Anal. Methods Residues Pestic.*, 1988, Part I, M1, M12). In **drinking water** by gc with ECD (*AOAC Methods*, 1995, 990.06). Further methods available on request from AgrEvo.

MAMMALIAN TOXICOLOGY
Reviews *Pesticide residues in food – 1989*. FAO Plant Production and Protection Paper 99, 1989. *Pesticide residues in food – 1989 evaluations. Part II – Toxicology*. FAO Plant Production and Protection Paper 100/2, 1990.   **EHC** 40 (WHO, 1984).   **Oral** Acute oral LD_{50} for rats 70 mg (in aqueous suspension)/kg, 110 mg tech. (in oil)/kg, 76 mg alpha-isomer/kg, 240 g beta-isomer/kg; for dogs 77 mg tech./kg.   **Skin and eye** Acute percutaneous LD_{50} for rabbits 359 mg (in oil)/kg; for male rats >4000, female rats 500 mg/kg.   **Inhalation** LC_{50} (4 h) for male rats 0.0345, female rats 0.0126 mg/l.   **NOEL** (2 y) for rats 15 ppm diet; (1 y) for dogs 10 ppm diet.   **ADI** (JMPR) 0.006 mg/kg b.w. [1989].   **Toxicity class** WHO (a.i.) II; EPA (formulation) I (tech.)   **EC risk** T (R24/25); Xi (R36)

ECOTOXICOLOGY

Birds Acute oral LD_{50} for mallard ducks 205–245, ring-necked pheasants 620–1000 mg/kg.   **Fish** Highly toxic (LC_{50} (96 h) for golden orfe 0.002 mg/l water) but, in practical use, should be harmless to wildlife.   **Bees** Not toxic to bees under field conditions at an application rate of 1.6 l/ha (560 g endosulfan/ha).   **Worms** NOEC 0.1 mg/kg dry weight.   **Daphnia** LC_{50} (48 h) 75–750 µg/l.   **Algae** EC_{50} (72 h) for green algae >0.56 mg/l.

ENVIRONMENTAL FATE

Animals The principal route of elimination is faeces; most of the radioactivity is excreted within the first 48 hours. The amounts excreted are independent of dose level, number of dosages and isomerism. There are indications of species-specificity. Residues of endosulfan accumulate in the kidneys rather than in fat. Elimination from the kidneys takes place with DT_{50} 7 d, but there is no sign of accumulation in the kidneys even after long-term feeding. Endosulfan is metabolised rapidly in mammalian organisms to less-toxic metabolites and to polar conjugates.   **Plants** The plant metabolites (mainly endosulfan sulfate) were also found in animals and have thus been investigated from a toxicological point of view. 50% of residues are lost in 3–7 days (depending on plant species).
Soil/Environment Endosulfan (alpha- and beta-) is degraded in soil with DT_{50} 30 to 70 d. The main metabolite usually found was endosulfan sulfate, which is degraded more slowly and is, for this reason, the most important metabolite. DT_{50} for total endosulfan (alpha- and beta- endosulfan and endosulfan sulfate) in the field is 5–8 mo. No leaching tendency was observed. K_{oc} 3000–20 000; K_d <3%.

271 endothal

Herbicide, algicide, plant growth regulator

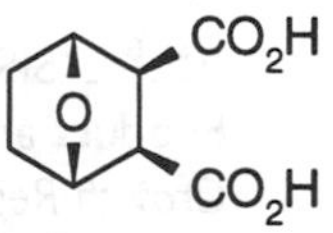

NOMENCLATURE

Common name endothal (BSI, France, New Zealand; since 1990 E-ISO, *(m)* F-ISO); endothall (ANSI, Canada, WSSA)
IUPAC name 7-oxabicyclo[2.2.1]heptane-2,3-dicarboxylic acid
Chemical Abstracts name 7-oxabicyclo[2.2.1]heptane-2,3-dicarboxylic acid

Other names 1,2-dicarboxy-3,6-*endo*-cyclohexane; 3,6-endoxohexahydrophthalic acid   **CAS RN** [145–73–3] endothal, unstated stereochemistry; [28874–46–6] endothal *rel*-(1R,2S,3R,4S)- isomer; [125–67–9] endothal-sodium, unstated stereochemistry; [17439–94–0] endothal-diammonium, unstated stereochemistry   **EEC no.** 205–660–5; 204–959–8 (sodium salt)

PHYSICAL CHEMISTRY

Composition Of the 4 theoretical stereoisomers of endothal, the *rel*-(1R,2S,3R,4S)- isomer is the most effective herbicide (US 2550494).
Mol. wt. 186.2   **M.f.** $C_8H_{10}O_5$ **Form** Colourless crystals (monohydrate).
M.p. 144 °C (monohydrate)   **V.p.** 2.09 × 10^{-5} mPa (24.3 °C) K_{ow} logP = −2.09 (unstated pH)   **Henry** 3.8 × 10^{-13} Pa m^3 mol^{-1} (calc.)   **S.g./density** 1.431 at 20 °C   **Solubility** In water 100 g/kg (20 °C). In methanol 280, dioxane 76, acetone 70, isopropanol 17, diethyl ether 1, benzene 0.1 (all in g/kg, 20 °C).   **Stability** Stable to light. Stable up to c. 90 °C, above which it undergoes a slow conversion to the anhydride. Endothal is a dibasic acid, and forms water-soluble amine and alkali-metal salts.   **pKa** A dibasic acid (pKa_1 3.4, pKa_2 6.7)
F.p. Non-flammable.

COMMERCIALISATION

History Herbicide reported by N. Tischler *et al.* (*Proc. Northeast. Weed Control Conf.*, 1951, p. 51). Introduced by Sharples Chemical Corp. (now Elf Atochem).
Patent US 2550494; US 2576080; US 2576081   **Manufacturers** Elf Atochem

APPLICATIONS

Mode of action Selective contact herbicide, absorbed by the leaves and roots, with limited acropetal translocation in the xylem. Also has algicidal properties, and acts as a defoliant and desiccant.   **Uses** Amine, ammonium and alkali-metal salts of endothal are used pre- and post-emergence for control of annual grass and broad-leaved weeds in sugar beet, fodder beet, beetroot, spinach, and turf at 2–6 kg a.i./ha. Control of algae and aquatic weeds (including use in rice). Also used as a desiccant for alfalfa, clover, and hops; for defoliation of cotton (as a harvest aid); and for destruction of potato haulms.   **Formulation types** GR; SL.
Selected tradenames 'Accelerate' (amine salt) (Elf Atochem); 'Herbicide 273' (dipotassium salt) (Elf Atochem)

ANALYSIS

Product analysis by glc with FID (R. Carlson *et al.*, *Anal. Methods Pestic. Plant Growth Regul.*, 1978, **10**, 327).   **Residues** determined by glc of a derivative, which is measured by MCD (*idem, ibid.*). Details available from Elf Atochem.

MAMMALIAN TOXICOLOGY

Oral Acute oral LD_{50} for rats 38–54 mg/kg (acid), 206 mg/kg (66.7% formulation of the amine salt).   **Skin and eye** Acute percutaneous LD_{50} for rabbits

>2000 mg/l (acid). **Inhalation** LC_{50} (14 d) 0.68 mg/l (acid). **NOEL** In 2 y feeding trials, rats receiving 1000 mg/kg diet showed no ill-effects. **Toxicity class** EPA (formulation) II **EC risk** T (R25); Xn (R21); Xi (R36/37/38) (also apply to sodium salt)

ECOTOXICOLOGY
Birds Acute LD_{50} for duck 111 mg/kg b.w. **Fish** LC_{50} (96 h) for trout 49, bluegill sunfish 77 mg/l. **Bees** Not toxic to bees. **Daphnia** LC_{50} (48 h) 92 mg/l. **Algae** Toxic to algae.

ENVIRONMENTAL FATE
Animals Rapidly absorbed. Elimination DT_{50} 1.8–2.5 h. **Plants** Residues in the plant were mainly endothal. **Soil/Environment** DT_{50} in aerobic soil 8.5 d. K_d 1.3–37.1. For a review, see Simsiman et al. (*Residue Rev.*, 1976, **62**, 131–174).

272 ENT 8184   *Insecticide synergist*

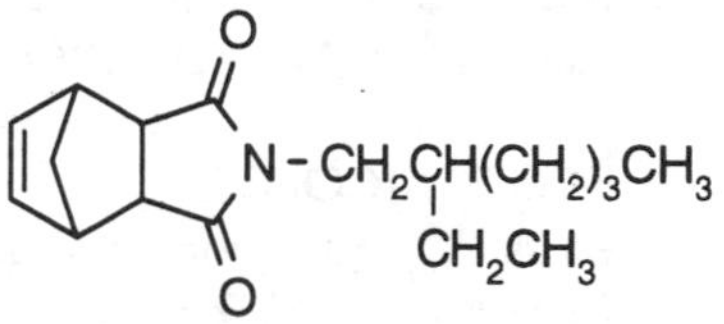

NOMENCLATURE
IUPAC name N-(2-ethylhexyl)-8,9,10-trinorborn-5-ene-2,3-dicarboximide; N-(2-ethylhexyl)bicyclo[2.2.1]hept-5-ene-2,3-dicarboximide
Chemical Abstracts name 2-(2-ethylhexyl)-3a,4,7,7a-tetrahydro-4,7-methano-= 1H-isoindole-1,3(2H)-dione
Other names N-octylbicycloheptenedicarboximide **CAS RN** [113–48–4]
Development codes MGK 264 **Official codes** ENT 8184

PHYSICAL CHEMISTRY
Composition The active ingredient is a 55:45 mixture of *cis*- and *trans*- isomers, both of which are biologically active. Tech. grade is 98% pure, main impurity 2-ethylhexanol. **Mol. wt.** 275.4 **M.f.** $C_{17}H_{25}NO_2$ **Form** Liquid. **M.p.** <–20 °C **B.p.** 150 °C/2 mmHg **V.p.** 2.4 mPa (25 °C) **K_{ow}** logP = 3.61 (*cis*- isomer), 3.80 (*trans*- isomer) **S.g./density** 1.046 (20 °C) **Solubility** Practically insoluble in water. Miscible with most organic solvents including petroleum oils, fluorinated hydrocarbons, also DDT and HCH. **Stability** Stable to light and heat. No hydrolysis after 30 d at pH 5–9.

COMMERCIALISATION
History Activity as synergist reported by R. H. Nelson *et al.* (*Soap Chem. Spec.*, 1949, **25**(1), 120). Introduced by Van Dyke Co. and later by McLaughlin Gormley King Co. **Patent** US 2476512 to McLaughlin Gormley King

APPLICATIONS
Uses A synergist for pyrethroids in aerosol sprays for household and veterinary use. **Formulation types** Aerosol concentrates; DP. **Mixtures** (*ENT 8184 +*) allethrin; allethrin + piperonyl butoxide; piperonyl butoxide + pyrethrins.

ANALYSIS
Product analysis by glc (*CIPAC Handbook*, 1985, **1C**, 2166; *AOAC Methods*, 1995, 980.04). **Residues** determined by glc. In **drinking water**, by gc with FID (*AOAC Methods*, 1995, 991.07). Details available from McLaughlin Gormley King.

MAMMALIAN TOXICOLOGY
Reviews *Pesticide residues*, FAO Rept. No. PL:1967/M/11; WHO Tech. Rept. Series, No. 391, 1968. *1967 Evaluations of some pesticide residues in food*, FAO/PL:1967/M 11/1; WHO/Food Add./68.30, 1968. **Oral** Acute oral LD_{50} for male rats 4990, female rats 4220 mg/kg. **Skin and eye** Acute percutaneous LD_{50} for rabbits 470 mg/kg. **Inhalation** LC_{50} (4 h) for rats >4.08 mg/l. **NOEL** for rats 50 mg/kg daily. NOEL for oncogenicity 450 mg/kg daily. **ADI** (JMPR) No ADI [1967]. **Toxicity class** WHO (a.i.) III

ECOTOXICOLOGY
Birds LC_{50} (8 d) for mallard ducks and bobwhite quail >5620 mg/kg diet. **Fish** LC_{50} (96 h) for rainbow trout 1.4, bluegill sunfish 2.4 mg/l. **Daphnia** LC_{50} (48 h) 2.3 mg/l.

273 EPN   *Insecticide, acaricide*

organophosphorus

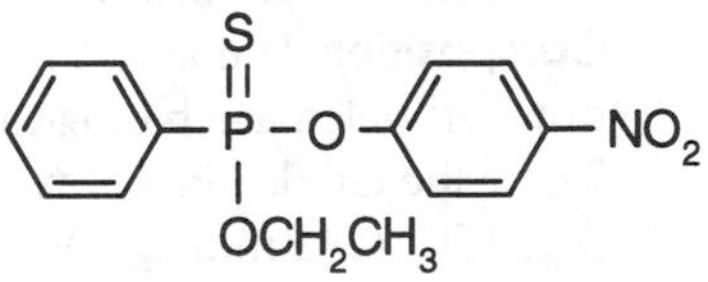

NOMENCLATURE
Common name EPN (JMAF)
IUPAC name *O*-ethyl *O*-4-nitrophenyl phenylphosphonothioate

Chemical Abstracts name O-ethyl O-(4-nitrophenyl) phenylphosphonothioate
CAS RN [2104–64–5]   **EEC no.** 218–276–8   **Official codes** OMS 219; ENT
17298

PHYSICAL CHEMISTRY
Mol. wt. 323.3   **M.f.** $C_{14}H_{14}NO_4PS$ **Form** Yellow crystals; (tech., dark amber
liquid).   **M.p.** 34.5 °C   **B.p.** 215 °C/5 mmHg   **V.p.** $<4.1 \times 10^{-2}$ mPa
K_{ow} logP >5.02   **S.g./density** 1.270 (20 °C)   **Solubility** Practically insoluble in
water. Soluble in many organic solvents, e.g. benzene, toluene, xylene, acetone,
isopropanol, methanol.   **Stability** Stable in neutral and acidic media, but
hydrolysed by alkali to liberate p-nitrophenol; DT_{50} 70 d (pH 4), 22 d (pH 7),
3.5 d (alkaline). On heating in a sealed tube, it is converted to the S-ethyl
isomer (R. L. Metcalf & R. B. March, *J. Econ. Entomol.*, 1953, **46**, 288).

COMMERCIALISATION
History Insecticide introduced by E. I. du Pont de Nemours Co. (who no longer
manufacture or market it) and subsequently by Nissan Chemical Industries, Ltd.
Patent US 2503390 to Du Pont   **Manufacturers** Nissan

APPLICATIONS
Biochemistry Cholinesterase inhibitor.   **Mode of action** Non-systemic insecticide
and acaricide with contact and stomach action.   **Uses** Effective at 0.5–1.0 kg
a.i./ha against a wide range of Lepidoptera larvae, especially bollworms (*Heliothis*
spp. and *Pectinophora gossypiella*) and *Alabama argillacea* in cotton, *Chilo* spp. in rice,
and other leaf-eating larvae in fruit and vegetables.   **Phytotoxicity** Non-phytotoxic,
except to some varieties of apple.   **Formulation types** DP; EC.   **Compatibility**
Incompatible with pesticides which are alkaline.   **Selected tradenames** 'EPN'
(Nissan)

ANALYSIS
Product analysis by glc.   **Residues** determined by glc (*AOAC Methods*, 1995,
974.22; H. L. Pease & J. J. Kirkland, *J. Agric. Food Chem.*, 1967, **15**, 187).

MAMMALIAN TOXICOLOGY
EHC 63 (WHO, 1986; a general review of organophosphorus insecticides).
Oral Acute oral LD_{50} for male rats 36, female rats 24, male mice 94.8, female
mice 59.4 mg/kg.   **Skin and eye** Acute percutaneous LD_{50} for male rats 2850,
female rats 538 mg/kg.   **NOEL** (104 w) for rats 0.67 mg/kg daily.   **Other** Causes
delayed neurotoxicity in hens.   **Toxicity class** WHO (a.i.) Ia   **EC risk** T+ (R27/28)

ECOTOXICOLOGY
Birds Acute oral LD_{50} for ring-necked pheasants >165, bobwhite quail 220 mg/kg.
Fish LC_{50} (48 h) for carp 0.20 ppm. LC_{50} for bluegill sunfish 0.37, rainbow trout
0.21 mg/l.   **Daphnia** LC_{50} (3 h) 0.0071 ppm.

ENVIRONMENTAL FATE

Animals Metabolised in mammals by desulfuration and removal of *p*-nitrophenol, and also by reduction of the nitro group to an amino group (M. Hitchcock & S. D. Murphy, *Biochem. Pharmacol.*, 1967, **16**, 1801–1811; R. L. Chrzanowski & A. G. Jelinek, *J. Agric. Food Chem.*, 1981, **29**, 580). **Plants** The major metabolite is ethyl phenylphosphonic acid. **Soil/Environment** In paddy soil, DT_{50} <15 d.

epoxiconazole

See under BAS 480F (entry 48).

274 EPTC   *Herbicide*

thiocarbamate

$$[CH_3(CH_2)_2]_2NC(O)SCH_2CH_3$$

NOMENCLATURE

Common name EPTC (BSI, E-ISO, (*m*) F-ISO, WSSA, JMAF)
IUPAC name *S*-ethyl dipropylthiocarbamate
Chemical Abstracts name *S*-ethyl dipropylcarbamothioate
CAS RN *[759–94–4]* **EEC no.** 212–073–8 **Development codes** R-1608

PHYSICAL CHEMISTRY

Composition Tech. material is 95% pure. **Mol. wt.** 189.3 **M.f.** $C_9H_{19}NOS$
Form Colourless liquid with an aromatic odour; (tech. is a yellow liquid).
B.p. 127 °C/20 mmHg **V.p.** 0.01 mPa (25 °C) **K_{ow}** logP = 3.2
S.g./density 0.9546 (30 °C) **Solubility** In water 375 mg/l (25 °C). Miscible with common organic solvents, e.g. acetone, ethanol, isopropanol, benzene, xylene, kerosene. **Stability** Stable up to 200 °C. Hydrolysed by strong acids on heating.
F.p. 110 °C

COMMERCIALISATION

History Herbicide reported by J. Antognini *et al.* (*Proc. Northeast. Weed Control Conf.*, 1957, p. 2). Introduced by Stauffer Chemical Co. (now Zeneca Agrochemicals). **Patent** US 2913327 **Manufacturers** Sagrochem; Zeneca

APPLICATIONS
Biochemistry Inhibits lipid metabolism. **Mode of action** Selective systemic herbicide, absorbed by the roots and shoots, with translocation acropetally to the leaves and stems. Kills germinating weed seeds and inhibits bud development from underground portions of some perennial weeds. **Uses** Control of annual and perennial grasses (especially couch grass), *Cyperus* spp., and some broad-leaved weeds in potatoes, beans, peas, forage legumes, beetroot, sugar beet, alfalfa, trefoil, clover, cotton, maize, flax, sweet potatoes, safflowers, sunflowers, strawberries, citrus, almonds, walnuts, ornamentals, pineapples, pine nurseries, and other crops. Applied pre-planting with soil incorporation. Combinations with the safener dichlormid are used for herbicidal control in maize.
Formulation types EC; GR. **Mixtures** *(EPTC +)* dichlormid.
Selected tradenames 'Eptam' (Zeneca); 'Eradicane' (mixture) (Zeneca); 'Alirox' (mixture) (Sagrochem)

ANALYSIS
Product analysis by glc (*AOAC Methods*, 1995, 974.05; *CIPAC Handbook*, 1983, **1B**, 1823; *Anal. Methods Pestic. Plant Growth Regul.*, 1984, **13**, 233).
Residues determined by glc: for crops (W. Y. Ja et al., ibid., 1972, **6**, 644; G. G. Patchett et al., *Anal. Methods Pestic. Plant Growth Regul. Food Addit.*, 1964, **4**, 117). In **drinking water**, by gc with FID (*AOAC Methods*, 1995, 991.07). Analytical methods available from Zeneca.

MAMMALIAN TOXICOLOGY
EHC 76 (WHO, 1988; general review of thiocarbamates). **Oral** Acute oral LD_{50} for rats >2000 mg/kg. **Skin and eye** Acute percutaneous LD_{50} for rats >2000, rabbits c. 10 000 mg/kg. Slight to mild eye irritant (rabbits). Not a skin sensitiser (guinea pigs). **Inhalation** LC_{50} (4 h) for male rats 4.3, female rats 3.8 mg/l.
NOEL (2 y) for mice 20 mg/kg daily; (90 d) for rats 16, dogs 20 mg/kg daily. Rats fed 326 mg/kg daily for 21 d showed no symptoms other than excitability and weight loss. Not teratogenic. **Toxicity class** WHO (a.i.) II; EPA (formulation) III
EC risk Xn (R22)

ECOTOXICOLOGY
Birds Seven-day dietary LC_{50} for bobwhite quail 20 000 mg/kg diet. **Fish** LC_{50} (96 h) for rainbow trout 19, bluegill sunfish 14 mg/l. **Bees** LD_{50} 0.011 mg/bee.
Daphnia LC_{50} (48 h) 14 mg/l.

ENVIRONMENTAL FATE
Plants In plants, EPTC is rapidly metabolised to CO_2 and other metabolites.
Soil/Environment In soil, EPTC rapidly undergoes microbial degradation to a mercaptan residue, an amino residue, and CO_2 (J. P. Hubbell & J. E. Casida, *J. Agric. Food Chem.*, 1977, **25**, 404). Decomposes in 4–6 weeks in warm, moist soils.

MAIN ENTRIES

275 *Eretmocerus californicus*   *Biological agent*

Hymenoptera

Wasp parasite of whitefly.

NOMENCLATURE
Scientific name *Eretmocerus* sp nr *californicus* (Howard)

COMMERCIALISATION
History Identification of species is difficult. The species now known as *Eretmocerus* near *californicus* has been recognised since at least 1980 and is the dominant species in the South Western United States, where it occurs in desert regions of Arizona and California.   **Manufacturers** Arbico; Koppert; Novartis BCM

APPLICATIONS
Mode of action All *Eretmocerus* species are obligate parasitoids of whitefly. Adult females lay single eggs beneath the immobile second or third instar larvae of the host, and the first instar parasite larva burrows into the host, where it completes its development.   **Uses** For control of whitefly *Bemisia tabaci*; it is also capable of parasitising *Trialeurodes vaporariorum*.   **Selected tradenames** 'Ercal' (Koppert); 'Eretmo-line cal' (Novartis BCM)

276 ergocalciferol   *Rodenticide*

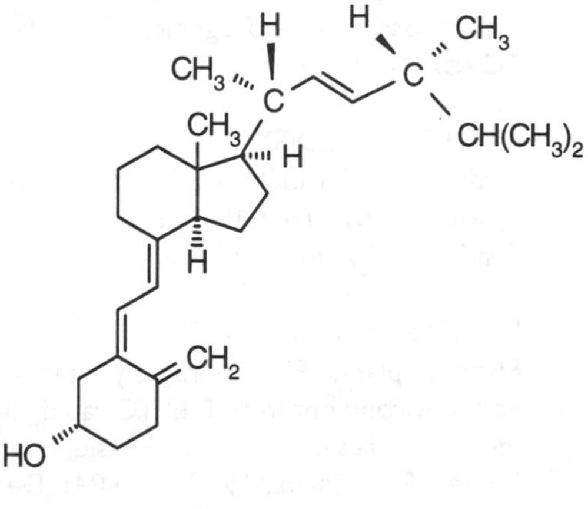

NOMENCLATURE
Common name ergocalciferol (INN, BSI proposed); calciferol (BAN)

IUPAC name (5Z,7E,22E)-(3S)-9,10-secoergosta-5,7,10(19),22-tetraen-3-ol
Chemical Abstracts name (3β,5Z,7E,22E)-9,10-secoergosta-5,7,10(19),22-= tetraen-3-ol
Other names Vitamin D_2; ercalciol   **CAS RN** [50–14–6]

PHYSICAL CHEMISTRY
Mol. wt. 396.7   **M.f.** $C_{28}H_{44}O$ **Form** Colourless crystals.   **M.p.** 115–118 °C
Solubility In water 50 mg/l (room temperature). In acetone 69.5, benzene 10, hexane 1 (all in g/l, room temperature).   **Stability** Stable in alkaline media. Unstable to light and air, and in acidic media. Above 120 °C, an irreversible reaction takes place, with conversion to the 10α- and 9β- isomers of the provitamin.   **Specific rotation** $[\alpha]_D$ +103° to +107°

COMMERCIALISATION
History Properties were reviewed by H. H. Inhoffen (*Angew. Chem.*, 1960, **72**, 875). Rodenticidal properties reported by M. Hadler (*Proc. Br. Pest Control Conf., 4th*). Introduced by Sorex Ltd.   **Patent** GB 1371135   **Manufacturers** Sorex

APPLICATIONS
Biochemistry Essential natural vitamin which, in high doses, produces a lethal hypervitaminosis, characterised by hypercalcaemia and increase in serum cholesterol.   **Uses** Control of rodents (by multiple feeding). Often used in admixture with warfarin to increase the efficacy.   **Formulation types** BB; AB; RB; SL.   **Mixtures** (*ergocalciferol +*) warfarin; difenacoum.   **Compatibility** Compatible with other rodenticides.   **Selected tradenames** 'Sorexa CD' (Sorex)

ANALYSIS
Product analysis by colorimetry or by u.v. spectrometry: details available from Sorex Ltd.

MAMMALIAN TOXICOLOGY
Oral Acute oral LD_{50} for rats 56, mice 23.7 mg/kg. Sub-acute oral LD_{50} (5 d) for rats 7 mg/kg daily.   **Other** Relatively safe to domestic animals.

ENVIRONMENTAL FATE
Animals The D vitamins, owing to their metabolic conversion products, play a central role in the regulation of calcium and phosphate metabolism in mammals, which is also affected by the peptide hormones calcitonin and parathormone. These promote uptake of calcium from the intestine, and are involved in intestinal transport of calcium and phosphate, as well as in mobilisation of calcium in the bones. Calciferol is hydroxylated first to the 25-hydroxy derivative in the liver, and this is then further hydroxylated to the 1α,25-dihydroxy derivative, as well as the 24(R),25-dihydroxy derivative, in the kidneys.

pyrethroid

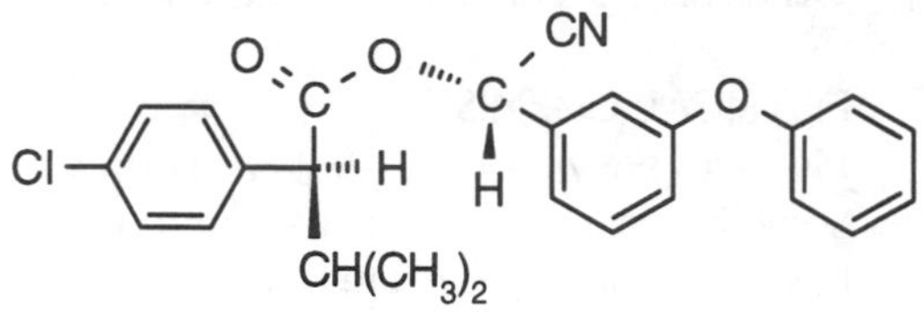

NOMENCLATURE
Common name esfenvalerate (BSI, E-ISO, (m) F-ISO)
IUPAC name (S)-α-cyano-3-phenoxybenzyl (S)-2-(4-chlorophenyl)-3-methyl= butyrate
Chemical Abstracts name [S-(R*,R*)]-cyano(3-phenoxyphenyl)methyl 4-chloro-= 2-(1-methylethyl)benzeneacetate
CAS RN *[66230–04–4]* **Development codes** S-1844; S-5602 Aα (both to Sumitomo); DPX-YB656 (Du Pont) **Official codes** OMS 3023

PHYSICAL CHEMISTRY
Composition Tech. is ≥98% total isomers and ≥75% resolved (S,S)- isomers.
Mol. wt. 419.9 **M.f.** $C_{25}H_{22}ClNO_3$ **Form** Colourless crystals; (tech., yellow-brown viscous liquid or solid at 23 °C). **M.p.** 59.0–60.2 °C; tech. 43.3–54 °C
B.p. 151–167 °C (tech.) **V.p.** 2×10^{-4} mPa (25 °C) **K**$_{ow}$ logP = 6.22 (25 °C)
S.g./density 1.26 (4–26 °C) **Solubility** In water 0.002 mg/l (25 °C). In xylene, acetone, chloroform, ethyl acetate, dimethylformamide, dimethyl sulfoxide >600, hexane 10–50, methanol 70–100 (all in g/kg, 25 °C). **Stability** Relatively stable to heat and light. **Specific rotation** $[\alpha]_D^{25}$ –15.0° (c 2.0 in methanol) **F.p.** 256 °C (Pensky-Martens)

COMMERCIALISATION
History The enhanced insecticidal activity of esfenvalerate over fenvalerate reported by I. Nakayama et al. (*Advances in Pestic. Sci.,* 1979, Part 2, p. 174), and its properties reviewed by H. Oo'uchi (*Jpn. Pestic. Inf.,* 1985, No. 46, p. 21). Introduced by Sumitomo Chemical Co., Ltd and developed under licence by Shell International Chemical Co., Ltd and E. I. du Pont de Nemours and Co.
Manufacturers Du Pont; Sumitomo

APPLICATIONS
Biochemistry Voltage dependent sodium channel agonist.
Mode of action Insecticide with contact and stomach action. **Uses** A potent contact and ingested insecticide with a very broad range of activity, especially effective against Coleoptera, Diptera, Hemiptera, Lepidoptera and Orthoptera on cotton, fruit, vegetables and other crops at 5–25 g a.i./ha. It is effective against strains resistant to organochlorine, organophosphorus and carbamate insecticides.

Phytotoxicity Some injury has been noted on crucifers, cucumbers, aubergines, tomatoes, pears and mandarin oranges. **Formulation types** EC; SC; UL.
Selected tradenames 'Sumi-alpha' (Sumitomo); 'Sumi-alfa' (Sumitomo); 'Asana' (Du Pont)

ANALYSIS
Product and **residue** analysis commonly by glc or by hplc. Details available from Sumitomo Chemical Co. Analysis of pyrethroids reviewed by E. Papadopoulou-Mourkidou in *Comp. Analyt. Profiles.*

MAMMALIAN TOXICOLOGY
Oral Acute oral LD_{50} for rats 75–88 mg/kg. **Skin and eye** Acute percutaneous LD_{50} for rats >5000, rabbits >2000 mg/kg. Slight skin irritant, mild eye irritant. Not a skin sensitiser. **NOEL** No effect on sub-chronic studies at ≥ 2 mg/kg daily. **ADI** 0.02 mg/kg. **Other** Acute LD_{50} values vary with the vehicle, concentration, route and species, etc.; values reported sometimes differ markedly. No carcinogenic, developmental or reproductive toxicity in animal tests.
Toxicity class WHO (a.i.) II; EPA (formulation) II

ECOTOXICOLOGY
Birds Acute oral LD_{50} for bobwhite quail 381 mg/kg. LC_{50} (8 d) for bobwhite quail >5620, mallard ducks 5247 ppm. **Fish** Extremely toxic to aquatic animals. LC_{50} (96 h) for fathead minnows 0.690, bluegill sunfish 0.26, rainbow trout 0.26 μg/l. **Bees** LD_{50} (contact) 0.017 μg/bee. **Daphnia** LC_{50} (48 h) 0.24 μg/l.

ENVIRONMENTAL FATE
Animals Rapid metabolism and elimination occurs in rats and other animals. Primary metabolism involves hydroxylation of 2'- and 4'- hydroxyl moieties, ester cleavage, hydroxylation and oxidation of the alcohol derivatives, oxidation of the cyano moiety and conjugation of the acidic metabolites with sulfate, glycine and glucuronic acid. **Plants** The major metabolite was decarboxylated fenvalerate. Ester cleavage, hydration of the cyano group to carboxamide and carboxylic acid, hydroxylation of the 2'- and 4'- phenoxy positions, conversion of the alcohol moiety to 3-phenoxybenzyl alcohol and 3-phenoxybenzoic acid, and conjugation of the resulting carboxylic acids and alcohols with sugars, also occur.
Soil/Environment In sand (0.38% o.m.) K_d (25 °C) = 4.4; in sandy loam (pH 7.3, 1.1% o.m.) K_d (25 °C) = 6.4, DT_{50} 88 d; in silty loam (pH 5.3, 2.0% o.m.) K_d (25 °C) = 71, DT_{50} 114 d; in clay loam (pH 5.7, 0.2% o.m.) DT_{50} 287 d; in clay loam (pH 6.4, 1.5% o.m.) K_d (25 °C) = 105. K_{oc} 5300.

MAIN ENTRIES

thiocarbamate

$$CH_3CH_2 \diagdown$$
$$\diagup NCOSCH_2 - \bigcirc$$
$$(CH_3)_2CHCH$$
$$|$$
$$CH_3$$

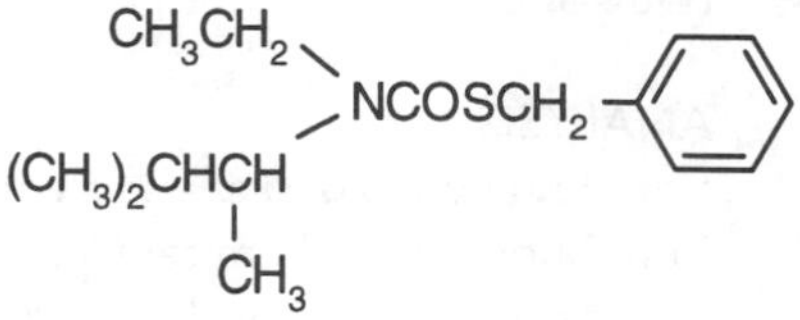

NOMENCLATURE
Common name esprocarb (BSI, draft E-ISO, (*m*) draft F-ISO)
IUPAC name *S*-benzyl 1,2-dimethylpropyl(ethyl)thiocarbamate
Chemical Abstracts name *S*-(phenylmethyl) (1,2-dimethylpropyl)=
ethylcarbamothioate
CAS RN *[85785–20–2]*   **Development codes** ICIA2957; SC-2957

PHYSICAL CHEMISTRY
Mol. wt. 265.4   **M.f.** $C_{15}H_{23}NOS$ **Form** Liquid.   **B.p.** 135 °C/35 mmHg
V.p. 10.1 mPa (25 °C)   **K**$_{ow}$ logP = 4.6   **S.g./density** 1.0353   **Solubility** In water
4.9 mg/l (20 °C). In ethanol, chlorobenzene, xylene, acetone, acetonitrile >1 (all
in g/kg, 25 °C).   **Stability** Stable at 120 °C. Photolysed in water, DT$_{50}$ 21 d (pH 7,
25 °C).

COMMERCIALISATION
History Herbicide discovered by Stauffer Chemical Co. (now Zeneca
Agrochemicals) and introduced in Japan (1988). Its herbicidal properties were
described in *J. Pestic. Sci.* (Japan), 1990, **15**, 1.   **Manufacturers** Zeneca

APPLICATIONS
Biochemistry Inhibits lipid metabolism.   **Uses** Pre- and post-emergence control of
annual weeds and *Echinochloa* spp. in paddy rice.   **Formulation types** GR.
Mixtures *(esprocarb +)* bensulfuron-methyl; pyrazosulfuron-ethyl + pretilachlor +
dimethametryn.   **Selected tradenames** 'Fujigrass' (mixture) (Du Pont, Zeneca);
'Sparkstar' (mixture) (Nissan, Zeneca)

ANALYSIS
Product and **residue** analysis by glc. Details available from Zeneca.

MAMMALIAN TOXICOLOGY
EHC 76 (WHO, 1988; general review of thiocarbamates).   **Oral** Acute oral LD$_{50}$
for female rats 3700 mg/kg.   **Skin and eye** Acute percutaneous LD$_{50}$ for rats
>2000 mg/kg. Mild skin and eye irritant (rabbits). Not a skin sensitiser (guinea
pigs).   **Inhalation** LC$_{50}$ (4 h) >4.0 mg/l air.   **NOEL** (2 y) for rats 1.1 mg/kg daily;
(1 y) for dogs 1.0 mg/kg daily. Non-teratogenic and non-oncogenic.
Toxicity class WHO (a.i.) III

Birds Acute oral LD_{50} for Japanese quail >2000 mg/kg. **Fish** LC_{50} for carp 1.52 mg/l.

ENVIRONMENTAL FATE
Soil/Environment In soil DT_{50} 30–70 d.

79 ethalfluralin

Herbicide

2,6-dinitroaniline

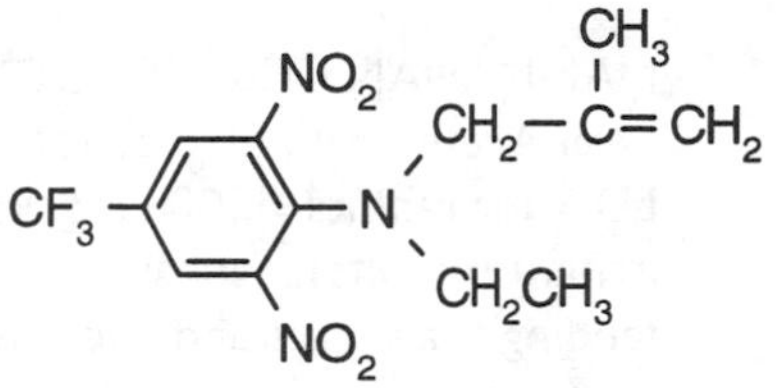

NOMENCLATURE
Common name ethalfluralin (BSI, E-ISO, ANSI, WSSA); éthalfluraline ((f) F-ISO)
IUPAC name N-ethyl-α,α,α-trifluoro-N-(2-methylallyl)-2,6-dinitro-p-toluidine
Chemical Abstracts name N-ethyl-N-(2-methyl-2-propenyl)-2,6-dinitro-= 4-(trifluoromethyl)benzenamine
CAS RN *[55283–68–6]* **Development codes** EL-161

PHYSICAL CHEMISTRY
Mol. wt. 333.3 **M.f.** $C_{13}H_{14}F_3N_3O_4$ **Form** Yellow-orange crystals, with a faint amine odour. **M.p.** 55–56 °C **B.p.** Decomposes at 256 °C **V.p.** 11.7 mPa (25 °C) **K_{ow}** logP = 5.11 (pH 7, 25 °C) **Solubility** In water 0.3 mg/l (pH 7, 25 °C). In acetone, acetonitrile, benzene, chloroform, dichloromethane, xylene >500, methanol 82–100 (all in g/l, 25 °C). **Stability** Tech. stable at 52 °C (highest storage temperature tested). No hydrolysis after 33 d at pH 3, 6 and 9 (51 °C). Photolysis DT_{50} (aqueous) 6.3 h; (vapour phase) 2 h.

COMMERCIALISATION
History Herbicide reported by G. Skylakakis *et al.* (*Proc. Br. Weed Control Conf.,* *12th*, 1974, **2**, 795). Introduced in Turkey (1974) by Eli Lilly & Co. (now DowElanco). **Patent** US 3257190 **Manufacturers** DowElanco

APPLICATIONS
Biochemistry Cell division inhibitor. **Mode of action** Selective soil-herbicide,

which affects seed germination and associated physiological growth processes. No significant absorption or translocation occurs in crops grown in soil treated with ethalfluralin. **Uses** Control of most germinating annual grasses and broad-leaved weeds in cotton, soya beans, dry beans, lentils, peanuts, cucurbits, safflowers, and sunflowers. Applied pre-plant soil-incorporated at 1.0–1.25 kg a.i./ha or, for peanuts and cucurbits, by post-planting surface application. Used in combination with atrazine for herbicidal control in maize and sorghum. **Formulation types** EC; GR; WG. **Selected tradenames** 'Sonalan' (USA) (DowElanco); 'Edge' (Canada) (DowElanco)

ANALYSIS
Product analysis by glc with FID or by spectrophotometry (E. W. Day, *Anal. Methods Pestic. Plant Growth Regul.*, 1978, **10**, 341). **Residues** determined by glc with ECD (*idem, ibid.*). Details available from DowElanco.

MAMMALIAN TOXICOLOGY
Oral Acute oral LD_{50} for rats >5000 mg/kg. **Skin and eye** Acute percutaneous LD_{50} for rabbits >5000 mg/kg. Moderate to severe irritant to skin; moderate eye irritant (rabbits). **Inhalation** LC_{50} (1 h) for rats >0.94 mg/m^3. **NOEL** In 2 y feeding trials, rats and mice receiving 100 mg/kg diet (for rats 4.2, mice 10.3 mg/kg b.w. daily) showed no ill-effects. **ADI** 0.042 mg/kg.
Toxicity class WHO (a.i.) III (Table 5); EPA (formulation) II

ECOTOXICOLOGY
Birds Acute oral LD_{50} for bobwhite quail >2000 mg/kg. Five-day dietary LC_{50} for bobwhite quail and mallard ducks >5000 mg/kg diet. **Fish** LC_{50} (96 h) for bluegill sunfish 0.102, rainbow trout 0.136 mg/l. **Bees** Not toxic to bees. LD_{50} (contact) 51 μg/bee (by linear regression). **Daphnia** LC_{50} (48 h) >0.365 mg/l; NOEC (21 d) 0.068 mg/l. **Algae** For *Selenastrum capricornutum* NOEL 0.004 mg/l; EC_{50} specific growth rate 0.009 mg/l.

ENVIRONMENTAL FATE
Animals In rats, following oral administration, 86% is excreted within 48 hours (64% in the faeces and 22% in the urine) and 95% within 7 days. In the urine, 3 metabolites have been identified, resulting from oxidation of the *N*-alkyl side-chain and/or by dealkylation. Glucuronide conjugates have been found as metabolites in the bile. **Plants** The same pattern of metabolism is observed as for trifluralin.
Soil/Environment Ethalfluralin is strongly adsorbed on soil, with negligible leaching; half-life in soils 25–46 days. Photolytic and microbial degradation takes place (B. J. Hayden & A. E. Smith, *Bull. Environ. Contam. Toxicol.*, 1983, **25**, 508). Aerobic metabolism in sandy loam soil (EPA laboratory studies) DT_{50} 45 d; more rapid metabolism anaerobically (DT_{50} 14 d) in same soil. K_{oc} 4000–8000; K_d 11.9 to 97 with o.m. 0.5 to 2.0%.

sulfonylurea

NOMENCLATURE
Common name ethametsulfuron (BSI, pa E-ISO – both for parent acid)
IUPAC name methyl 2-[(4-ethoxy-6-methylamino-1,3,5-triazin-2-yl)carbamoyl=
sulfamoyl]benzoate
Chemical Abstracts name methyl 2-[[[[[4-ethoxy-6-(methylamino)-1,3,5-triazin-=
2-yl]amino]carbonyl]amino]sulfonyl]benzoate
CAS RN *[97780–06–8]* methyl ester; *[111353–84–5]* parent acid
Development codes DPX-A7881

PHYSICAL CHEMISTRY
Mol. wt. 410.4   **M.f.** $C_{15}H_{18}N_6O_6S$ **Form** Colourless to light tan, odourless
crystals.   **M.p.** 194 °C   **V.p.** 7.73×10^{-10} mPa (25 °C) K_{ow} logP = 0.89 (pH 7),
1.588 (pH 5)   **S.g./density** 1.6   **Solubility** In water 50 mg/l (pH 7, 25 °C). In
acetone 1.6, acetonitrile 0.8, ethanol 0.17, methanol 0.35, methylene chloride 3.9,
ethyl acetate 0.68 (all in g/l).   **Stability** Stable at pH 7 and pH 9. Hydrolysis occurs
more rapidly at pH 5, DT_{50} 45 d. Photolysis is not a major degradation pathway.
Acidic in reaction.   **pKa** 4.6

COMMERCIALISATION
History Herbicide reported by J. R. Stone *et al.* (*Proc. N. Cent. Weed Control Conf.*,
1985, **40**, 17). Introduced by E. I. du Pont de Nemours and Co. in 1989.
Manufacturers Du Pont

APPLICATIONS
Biochemistry Branched chain amino acid synthesis (acetolactate synthase or ALS)
inhibitor. Acts by inhibiting biosynthesis of the essential amino acids valine and
isoleucine, hence stopping cell division and plant growth. Selectivity derives from
rapid metabolism in the crop. Metabolic basis of selectivity in sulfonylureas
reviewed (M. K. Koeppe & H. M. Brown, *Agro-Food-Industry*, **6**, 9–14 (1995)).
Uses Post-emergence control of wild mustard, hempnettle and other broad-leaved
weeds in oilseed rape at 15–20 g/ha.   **Formulation types** WG.
Selected tradenames 'Muster' (Du Pont)

ANALYSIS
By hplc. Methods for sulfonylurea residues in crops, soil and water reviewed
(A. C. Barefoot *et al., Proc. Br. Crop Prot. Conf. – Weeds*, 1995, **2**, 707).

MAMMALIAN TOXICOLOGY

Oral Acute oral LD_{50} for rats >5000, rabbits >5000 mg/kg. **Skin and eye** Acute percutaneous LD_{50} for rabbits >2000 mg/kg. Non-irritating to skin, slightly irritating to eyes of rabbits. Non-sensitising to skin (guinea pigs). **Inhalation** LC_{50} (4 h) for rats >5.7 mg/l air. **NOEL** (90 d) for rats and mice 5000 ppm; (1 y) for rats 500, dogs 3000 ppm; (18 mo) for mice 5000 ppm. Non-oncogenic and non-mutagenic in rats. Non-teratogenic in rats and rabbits. **Toxicity class** WHO (a.i.) Table 5; EPA Not registered in US

ECOTOXICOLOGY

Birds Acute oral LD_{50} for bobwhite quail and mallard ducks >2250 mg/kg. Five-day dietary LC_{50} for bobwhite quail and mallard ducks >5620 mg/kg diet. **Fish** LC_{50} (96 h) for bluegill sunfish and rainbow trout >600 mg/l. **Bees** Acute toxicity to honeybees >12.5 µg/bee. **Worms** Contact LD_{50} (14 d) for earthworms >1000 mg/kg soil. **Daphnia** LC_{50} (48 h) 34 mg/l.

ENVIRONMENTAL FATE

Animals Ethametsulfuron-methyl administered to male and female rats was rapidly metabolised and excreted in the urine and faeces. Half-lives for excretion range from 12 hours in male rats to 21–26 hours in female rats. Less than 0.2% of the dose remains in the tissues five days after dosing at the highest dose level. No preferential accumulation of ethametsulfuron-methyl or its metabolites.
Plants Oilseed rape was treated with 30 g/ha ethametsulfuron-methyl in a glasshouse: total radiolabelled residues in the foliage decreased rapidly from c. 1.0 ppm immediately after treatment to 0.02 ppm after 31 days; DT_{50} 1–3 h. Two primary metabolites were identified, formed by successive dealkylation; firstly of the ethoxy group, to give the corresponding hydroxytriazine, then of the methylamino substituent. Total radioactive residue in mature oilseed rape was very low (0.008 to 0.012 ppm). No ethametsulfuron-methyl was detected in the seed.
Soil/Environment Soil metabolism (aerobic, lab.) DT_{50} 9 w; three major metabolites were identified. In soil photolysis studies, sunlight accelerated degradation three-fold relative to dark controls. Aquatic metabolism (aerobic) DT_{50} 6 mo, (anaerobic) DT_{50} 2–9 mo, depending on sediment pH. In laboratory soil mobility studies based on soil TLC, soil column leaching, and adsorption/desorption studies, mobility is highly dependent on soil characteristics, primary organic matter content and soil pH. Mobility ranges from very mobile in sandy loam soil to very low mobility in loam soil.

ethylene generator

$$CICH_2CH_2\overset{\displaystyle O}{\overset{\|}{P}}(OH)_2$$

NOMENCLATURE
Common name ethephon (ANSI, Canada); chorethephon (New Zealand)
IUPAC name 2-chloroethylphosphonic acid
Chemical Abstracts name (2-chloroethyl)phosphonic acid
CAS RN *[16672–87–0]*

PHYSICAL CHEMISTRY
Mol. wt. 144.5   **M.f.** $C_2H_6ClO_3P$ **Form** Colourless solid; (tech. is a clear liquid).
M.p. 74–75 °C   **B.p.** *c.* 265 °C (decomp.)   **V.p.** <0.01 mPa (20 °C)
K_{ow} logP <–2.20 (25 °C, unstated pH)   **Henry** $<1.55 \times 10^{-9}$ Pa m^3 mol^{-1} (calc.)
S.g./density 1.409±0.02 (20 °C) (tech.)   **Solubility** Readily soluble in water,
c. 1 kg/l (23 °C). Readily soluble in methanol, ethanol, isopropanol, acetone,
diethyl ether, and other polar organic solvents. Sparingly soluble in non-polar
organic solvents such as benzene and toluene. Insoluble in kerosene and diesel oil.
Stability Stable in aqueous solutions having pH <5. At higher pH, decomposition
occurs with the liberation of ethylene. Sensitive to u.v. irradiation.
pKa pKa$_1$ 2.5, pKa$_2$ 7.2

COMMERCIALISATION
History Plant growth regulator introduced by Amchem Products Inc. (now Rhône-
Poulenc Agrochimie).   **Patent** US 3879188; US 3896163; US 3897486
Manufacturers Cedar; CFPI; Griffin; Rallis; Rhône-Poulenc

APPLICATIONS
Mode of action Plant growth regulator with systemic properties. Penetrates into
the plant tissues, and is decomposed to ethylene, which affects the growth
processes.   **Uses** Used to promote pre-harvest ripening in apples, currants,
blackberries, blueberries, cranberries, morello cherries, citrus fruit, figs, tomatoes,
sugar beet and fodder beet seed crops, coffee, capsicums, etc.; to accelerate post-
harvest ripening in bananas, mangoes, and citrus fruit; to facilitate harvesting by
loosening of the fruit in currants, gooseberries, cherries, and apples; to increase
flower bud development in young apple trees; to prevent lodging in cereals, maize,
and flax; to induce flowering of Bromeliads; to stimulate lateral branching in
azaleas, geraniums, and roses; to shorten the stem length in forced daffodils; to
induce flowering and regulate ripening in pineapples; to accelerate boll opening in
cotton; to modify sex expression in cucumbers and squash; to increase fruit setting
and yield in cucumbers; to improve the sturdiness of onion seed crops; to hasten
the yellowing of mature tobacco leaves; to stimulate latex flow in rubber trees,

and resin flow in pine trees; to stimulate early uniform hull split in walnuts; etc. **Formulation types** SL; EC. **Mixtures** (*ethephon* +) cyclanilide; mepiquat chloride; chlormequat chloride. **Compatibility** Incompatible with alkaline materials and with solutions containing metal ions, e.g. iron-, zinc-, copper-, and manganese-containing fungicides. **Selected tradenames** 'Ethrel' (Rhône-Poulenc); 'Cerone' (Rhône-Poulenc); 'Etherfon' (Productos OSA); 'Pluck' (Cedar/Crystal); 'Sierra' (CFPI); 'Super Boll' (Griffin); 'Terpal' (mixture) (BASF)

ANALYSIS
Product analysis by measuring the ethylene produced on treatment with concentrated alkali. **Residues** determined by conversion to the dimethyl ester, measured by glc with NPD or FPD (*Pestic. Anal. Man.*, Vol. II; *Anal. Methods Residues Pestic.*, 1988, Part II; W. P. Cochrane, *J. Assoc. Off. Anal. Chem.*, 1976, **59**, 617).

MAMMALIAN TOXICOLOGY
Reviews *Pesticide residues in food – 1995*, FAO Plant Production and Protection Paper 133, 1996. G. Hennighausen *et al.*, *Pharmazie*, **32**, 181 (1977). **Oral** Acute oral LD_{50} for rats 3030 mg/kg (tech.). **Skin and eye** Acute percutaneous LD_{50} for rabbits 1560 mg/kg (tech.). Irritating to skin and eyes. **Inhalation** LC_{50} (4 h) for rats 6.26 mg/l (tech.). **NOEL** (2 y) for rats 3000 ppm diet. **ADI** (JMPR) 0.05 mg/kg b.w. [1995]. **Toxicity class** WHO (a.i.) III (Table 5); EPA I (tech.) **EC risk** C (R21, R34, R41)

ECOTOXICOLOGY
Birds Acute oral LD_{50} for bobwhite quail 1072 mg/kg (tech.). Eight-day dietary LC_{50} for bobwhite quail >7000 ppm in diet (tech.). **Fish** LC_{50} (96 h) for carp >140, rainbow trout 720 mg/l (tech.). **Bees** Not toxic to bees. **Worms** Not toxic to earthworms. **Daphnia** EC_{50} (48 h) 577.4 mg/l. **Algae** EC_{50} (24–48 h) for *Chlorella vulgaris* 32 mg/l. **Other aquatic spp.** Low toxicity.

ENVIRONMENTAL FATE
Animals In animals, ethephon is rapidly excreted intact via the urine, and as ethylene via the expired air. **Plants** In plants, ethephon rapidly undergoes degradation to ethylene. **Soil/Environment** Rapidly degraded in soil, and strongly adsorbed; unlikely to leach.

ethiofencarb *Insecticide*

carbamate

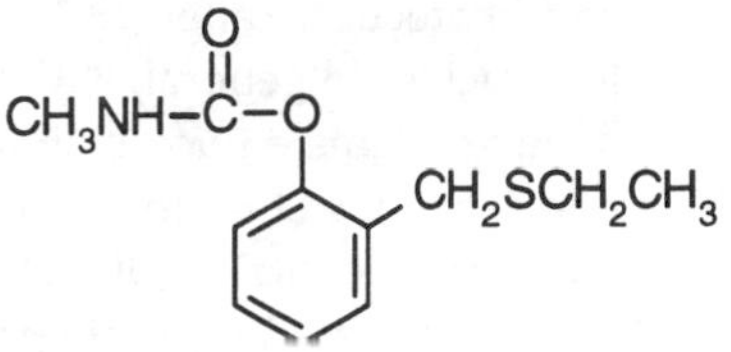

$$CH_3NH-\overset{\overset{\textstyle O}{\|}}{C}-O$$

CH_2SCH_2CH_3

NOMENCLATURE
Common name ethiofencarb (BSI, E-ISO); éthiophencarbe ((m) F-ISO)
IUPAC name α-ethylthio-o-tolyl methylcarbamate
Chemical Abstracts name 2-[(ethylthio)methyl]phenyl methylcarbamate
CAS RN *[29973–13–5]* **EEC no.** 249–981–9 **Development codes** HOX 1901;
BAY 108594

PHYSICAL CHEMISTRY
Mol. wt. 225.3 **M.f.** $C_{11}H_{15}NO_2S$ **Form** Colourless crystals. Tech. grade is a
yellow oil with a mercaptan-like odour. **M.p.** 33.4 °C **B.p.** Decomposes on
distillation **V.p.** 0.45 mPa (20 °C), 0.94 mPa (25 °C), 26 mPa (50 °C)
K_{ow} logP = 2.04 **S.g./density** 1.231 (20 °C) **Solubility** In water 1.8 g/l (20 °C).
In dichloromethane, isopropanol and toluene >200, hexane 5–10 (all in g/l,
20 °C). **Stability** Stable in neutral and acidic media, but hydrolysed in alkaline
solution. In isopropanol/water (1:1), DT_{50} (37–40 °C) 330 d (pH 2), 450 h
(pH 7), 5 minutes (pH 11.4). In aqueous solution, photodegradation in sunlight is
very rapid. **F.p.** 123 °C

COMMERCIALISATION
History Insecticide reported by J. Hammann & H. Hoffmann (*Pflanzenschutz-Nachr.
(Engl. Ed.)*, 1974, **27**, 267). Introduced by Bayer AG. **Patent** DE 1910588;
BE 746649 **Manufacturers** Bayer

APPLICATIONS
Biochemistry Cholinesterase inhibitor. **Mode of action** Systemic insecticide with
contact and stomach action. Absorbed by the leaves and roots. **Uses** Control of
aphids on pome fruit, stone fruit, soft fruit, vegetables, ornamentals and sugar
beet. **Phytotoxicity** Phytotoxic to anthurium and begonias.
Formulation types EW; GR; EC. **Selected tradenames** 'Croneton' (Bayer)

ANALYSIS
Product analysis by hplc with u.v. determination (*CIPAC Handbook*, 1992, **E**,
84–88). **Residues** determined by glc (G. Drager, *ibid.*, 1974, **27**, 144; *Anal.
Methods Residues Pestic.*, Part I, M13). Methods for the determination of residues
are available from Bayer.

ethiofencarb 479

MAMMALIAN TOXICOLOGY

Reviews *Pesticide residues in food – 1982*. FAO Plant Production and Protection Paper 46, 1983. *Pesticide residues in food: 1982 evaluations*. FAO Plant Production and Protection Paper 49, 1983. **EHC** 64 (WHO, 1986; a review of carbamate insecticides in general). **Oral** Acute oral LD_{50} for male and female rats *c.* 200, male and female mice *c.* 240, female dogs >50 mg/kg. **Skin and eye** Acute percutaneous LD_{50} for male and female rats >1000 mg/kg. Non-irritating to skin and eyes (rabbits). Not a skin sensitiser. **Inhalation** LC_{50} (4 h) for male and female rats >0.2 mg/l air (for aerosol). **NOEL** (2 y) for rats 330, mice 600, dogs 1000 mg/kg diet. **ADI** (JMPR) 0.1 mg/kg b.w. [1982]. **Toxicity class** WHO (a.i.) II; EPA (formulation) II **EC risk** Xn (R22)

ECOTOXICOLOGY

Birds Acute oral LD_{50} for Japanese quail 155, mallard ducks 140–275 mg/kg. **Fish** LC_{50} (96 h) for rainbow trout 12.8, golden orfe 61.8 mg/l. **Bees** Not toxic to bees. **Worms** LC_{50} for *Eisenia foetida* 262 mg/kg dry soil. **Daphnia** LC_{50} (48 h) 0.22 mg/l. **Algae** E_rC_{50} for *Scenedesmus subspicatus* 43 mg/l.

ENVIRONMENTAL FATE

Animals In animals, ^{14}C-ethiofencarb is rapidly excreted. The main metabolites are ethiofencarb sulfoxide and sulfone, ethiofencarb phenol and the corresponding sulfoxide and sulfone. Metabolism of carbamate insecticides is reviewed (M. Cool & C. K. Jankowski in "*Insecticides*"). **Plants** In plants, metabolites include ethiofencarb sulfoxide and sulfone and the hydrolysed products ethiofencarb phenol-sulfoxide and sulfone which form conjugates.

Soil/Environment Ethiofencarb has a relatively high mobility in soil, but is very rapidly degraded to its sulfoxide and sulfone and hydrolysed to the corresponding phenolic metabolites.

283 ethion   *Acaricide, insecticide*

organophosphorus

$$(CH_3CH_2O)_2\overset{\displaystyle S}{\overset{\displaystyle \|}{P}}SCH_2S\overset{\displaystyle S}{\overset{\displaystyle \|}{P}}(OCH_2CH_3)_2$$

NOMENCLATURE

Common name ethion (BSI, E-ISO, (*m*) F-ISO, ANSI, ESA, JMAF); diethion ((*m*) France, Republic of South Africa, formerly India); no name (Italy, Portugal)
IUPAC name *O,O,O',O'*-tetraethyl *S,S'*-methylene bis(phosphorodithioate)
Chemical Abstracts name *S,S'*-methylene bis(*O,O*-diethyl) phosphorodithioate

CAS RN *[563-12-2]*   EEC no. 209-242-3   Development codes FMC 1240
Official codes ENT 24 105

PHYSICAL CHEMISTRY
Mol. wt. 384.5   M.f. $C_9H_{22}O_4P_2S_4$ Form Water-white to amber-coloured liquid.
M.p. -15 to -12 °C   B.p. 164-165 °C/0.3 mmHg   V.p. 0.20 mPa (25 °C)
S.g./density 1.22 (20 °C); (tech., 1.215-1.230)   Solubility In water 2 ppm (25 °C).
Miscible with most organic solvents, e.g. acetone, methanol, ethanol, xylene,
kerosene, petroleum oils.   Stability Hydrolysed by aqueous acids and alkalis;
DT_{50} 390 d (pH 9). Slowly oxidised by air.

COMMERCIALISATION
History Insecticide reported (*Chem. Eng. News*, 1957, **35,** 87). Introduced by FMC
Corp.   Patent GB 872221; US 2873228   Manufacturers Cheminova; Krishi
Rasayan; Pesticides India; Rallis; Rhône-Poulenc; Shaw Wallace; Voltas

APPLICATIONS
Biochemistry Cholinesterase inhibitor.   Mode of action Non-systemic acaricide
and insecticide with predominantly contact action.   Uses Control of spider mites,
aphids, scale insects, thrips, lepidopterous larvae, leafhoppers, suckers, soil-
dwelling and other insects in pome fruit, stone fruit, citrus fruit, vines, vegetables,
ornamentals, cotton, maize, sorghum, cucurbits, strawberries, turf, and other
crops. Active against motile stages of mites when applied as a dormant spray, also
against overwintering forms of many species.   Phytotoxicity Non-phytotoxic when
used as directed, except to some varieties of apple (particularly early-maturing
varieties, like Early McIntosh).   Formulation types WP; EC; DP; GR; Seed
treatment.   Mixtures (*ethion +*) dicofol; dicofol + parathion-methyl; petroleum
oils; tetradifon; parathion.   Compatibility Incompatible with alkaline materials.
Selected tradenames 'Cekuetion' (Cequisa); 'Cethion' (Cheminova); 'Dhanumit'
(Dhanuka); 'Ethiol' (Rhône-Poulenc); 'MIT 505' (Shaw Wallace); 'Rayethion'
(Krishi Rasayan); 'Rhodocide' (Rhône-Poulenc); 'Tafethion' (Rallis)

ANALYSIS
Product analysis by hplc (*CIPAC Handbook*, 1983, **1B,** 1826; *AOAC Methods*, 1995,
979.04).   Residues determined by glc or by tlc or paper chromatography (*ibid.*,
968.24, 970.52; *Pestic. Anal. Man.*, 1979, **I,** 201-A, 201-G, 201-H, 201-I; D. C.
Abbott *et al.*, *Pestic. Sci.*, 1970, **1,** 10; *Anal. Methods Pestic. Plant Growth Regul.*,
1972, **6,** 396).

MAMMALIAN TOXICOLOGY
Reviews *Pesticide residues in food – 1990*. FAO Plant Production and Protection
Paper, 102, 1990. *Pesticide residues in food – 1990 evaluations. Toxicology.* World
Health Organisation, WHO/PCS/91.47, 1991.   EHC 63 (WHO, 1986; a
general review of organophosphorus insecticides).   Oral Acute oral LD_{50} for rats

208 mg/kg (pure), 47 mg/kg (tech.); for mice and guinea pigs 40–45 mg/kg.
Skin and eye Acute percutaneous LD_{50} for guinea pigs and rabbits 915 mg/kg;
(tech., for rabbits 1084 mg/kg).   **Inhalation** LC_{50} (4 h) for rats 0.45 mg tech./l.
NOEL (2 y) for rats 6 mg/kg diet (0.3 mg/kg daily), for dogs 2 mg/kg diet
(0.05 mg/kg daily).   **ADI** (JMPR) 0.002 mg/kg b.w. [1990].   **Toxicity class**
WHO (a.i.) II; EPA (formulation) II   **EC risk** T (R25); Xn (R21)

ECOTOXICOLOGY
Birds LD_{50} for quail 128, ducks >2000 mg tech./kg.   **Fish** Toxic to fish. Average
lethal concentration 0.72 mg/l (24 h); 0.52 mg/l (48 h).   **Bees** Toxic to bees.

ENVIRONMENTAL FATE
Animals Metabolised in animals by oxidation to a phosphorothioate, followed by
dealkylation and hydrolysis.   **Soil/Environment** Typical DT_{50} in soil 90 d.

284   ethirimol   *Fungicide*

pyrimidine

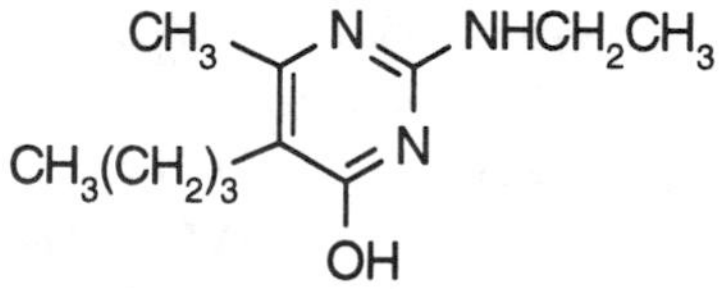

NOMENCLATURE
Common name ethirimol (BSI, E-ISO); éthyrimol ((*m*) F-ISO)
IUPAC name 5-butyl-2-ethylamino-6-methylpyrimidin-4-ol
Chemical Abstracts name 5-butyl-2-(ethylamino)-6-methyl-4(1*H*)-pyrimidinone
CAS RN *[23947–60–6]*   **EEC no.** 245–949–3   **Development codes** PP149

PHYSICAL CHEMISTRY
Composition Tech. material is 97% pure.   **Mol. wt.** 209.3   **M.f.** $C_{11}H_{19}N_3O$
Form Colourless crystals.   **M.p.** 159–160 °C (phase change at *c.* 140 °C)
V.p. 0.267 mPa (25 °C)   **K_{ow}** logP = 2.3 (pH 7, 20 °C)   **S.g./density** 1.21 (25 °C)
Solubility In water 253 (pH 5.2), 150 (pH 7.3), 153 (pH 9.3) (all in mg/l, 20 °C).
In chloroform 150, ethanol 24, acetone 5 (all in g/kg, 20 °C).   **Stability** Stable to
heat, and in acidic and alkaline solutions in the dark. Aqueous solution exposed to
sunlight and air had DT_{50} *c.* 21 d.   **pKa** 5, weak base (K. Chamberlain *et al.*, *Pestic.
Sci.*, **47**, 265 (1996) and ref. therein).

COMMERCIALISATION
History Fungicide reported by R. M. Bebbington *et al.* (*Chem. Ind. (London),* 1969, p. 1512). Introduced by ICI Plant Protection Division (now Zeneca Agrochemicals). **Patent** GB 1182584 **Manufacturers** Zeneca

APPLICATIONS
Mode of action Systemic fungicide with protective and curative action. Absorbed by the roots and leaves, with translocation in the xylem, but not in the phloem. **Uses** Control of powdery mildew on barley, wheat, and oats. Applied as a seed dressing, a foliar spray, or to the soil in the root zone. Also used for control of powdery mildews on cucurbits. **Formulation types** LS; SC; EC.
Mixtures *(ethirimol +)* flutriafol + thiabendazole; anthraquinone + flutriafol + oxine-copper; fuberidazole + imazalil + triadimenol. **Selected tradenames** 'Milgo' (Zeneca)

ANALYSIS
Product analysis by glc of a derivative (*CIPAC Handbook,* 1988, **D,** 72; J. E. Bagness & W. G. Sharples, *Analyst (London),* 1974, **99,** 225). Identity also by glc, tlc, i.r. and nmr (*CIPAC Handbook,* 1994, **F,** 404). **Residues** in crops determined by glc and in water by hplc or by spectrometry (M. J. Edwards, *Anal. Methods Pestic. Plant Growth Regul.,* 1976, **8,** 285). Details available from Zeneca.

MAMMALIAN TOXICOLOGY
Oral Acute oral LD_{50} for female rats 6340, mice 4000, female guinea pigs 500–1000, male rabbits 1000–2000 mg/kg. **Skin and eye** Acute percutaneous LD_{50} for rats >2000 mg/kg. Mild eye irritant; non-irritating to skin (rats). Not a skin sensitiser (guinea pigs). **Inhalation** LC_{50} for rats >4.92 mg/l. **NOEL** (2 y) for rats 200 mg/kg diet, for dogs 30 mg/kg daily. Not carcinogenic or teratogenic. **Toxicity class** WHO (a.i.) III (Table 5); EPA (formulation) IV **EC risk** Xn (R21)

ECOTOXICOLOGY
Birds Acute oral LD_{50} for hens 4000 mg/kg. **Fish** LC_{50} (96 h) for young brown trout 20, rainbow trout 66 mg/l. **Bees** LD_{50} (oral) 1.6 mg/bee; (contact) 20 g/l. **Daphnia** LC_{50} (48 h) 53 mg/l.

ENVIRONMENTAL FATE
Animals In rats, following oral administration, metabolism involves hydroxylation of the butyl group. Elimination is in the urine. **Plants** In plants, ethirimol undergoes rapid degradation, 2-amino-5-butyl-6-hydroxy-4-methylpyrimidine being found as a metabolite. DT_{50} in plants *c.* 3 d. **Soil/Environment** Soil DT_{50} 14–140 d (o.m. 1.0–10.1%, pH 7.8–8.1).

MAIN ENTRIES

benzofuranyl alkanesulfonate

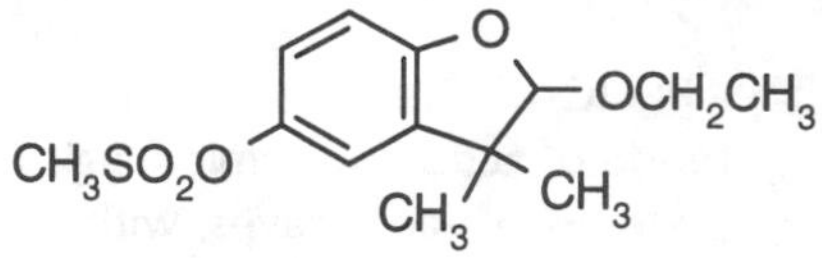

NOMENCLATURE

Common name ethofumesate (BSI, E-ISO, (*m*) F-ISO, ANSI, WSSA)
IUPAC name (±)-2-ethoxy-2,3-dihydro-3,3-dimethylbenzofuran-5-yl
methanesulfonate
Chemical Abstracts name (±)-2-ethoxy-2,3-dihydro-3,3-dimethyl-5-benzofuranyl
methanesulfonate
CAS RN *[26225–79–6]* **EEC no.** 247–525–3 **Development codes** AE BO49913;
NC 8438; SN 49913; ZK 49913

PHYSICAL CHEMISTRY

Mol. wt. 286.3 **M.f.** $C_{13}H_{18}O_5S$ **Form** White, crystalline solid; tech. is a light
brown crystalline solid; mild aromatic odour. **M.p.** 70–72 °C; tech., 69–71 °C
V.p. 0.12–0.65 mPa (25 °C) **K$_{ow}$** logP = 2.7 (pH 6.5–7.6, 25 °C)
Henry 3.7–6.8 × 10^{-3} Pa m^3 mol^{-1} **S.g./density** 1.29 (20 °C, tech.) **Solubility** In
water 50 mg/l (25 °C). In acetone, dichloromethane, dimethyl sulfoxide, ethyl
acetate >600, toluene, *p*-xylene 300–600, methanol 120–150, ethanol 60–75,
isopropanol 25–30, hexane 4.67 (all in g/l, 25 °C). **Stability** Stable to hydrolysis in
water at pH 7 and 9. At pH 5.0, DT_{50} 940 d forming the hydroxy analogue.
Phototransformed in water, DT_{50} 31 h. Degraded in air, DT_{50} 4.1 h.

COMMERCIALISATION

History Herbicide reported by R. K. Pfeiffer (*Symp. New Herbic., 3rd*, 1969, p. 1).
Introduced by Fisons Ltd Agrochemical Division (now AgrEvo GmbH) in 1974.
Patent GB 1271659 **Manufacturers** AgrEvo; Feinchemie Schwebda; Griffin;
United Phosphorus

APPLICATIONS

Biochemistry Not fully known. Inhibitor of cell division and lipid metabolism.
Mode of action Selective systemic herbicide, absorbed by the emerging shoots
(grasses) and roots (broad-leaved plants), with translocation to the foliage. Not
readily absorbed by leaves after the plant has generated a mature cuticle. Inhibits
the growth of meristems, retards cellular division, and limits formation of waxy
cuticle. **Uses** Used pre- and/or post-emergence in sugar and other beet crops,
turf, ryegrass and the other pasture grasses at 0.3–2.0 kg a.i./ha. It is effective in
controlling a wide range of important grasses and broad-leaved weeds, with a
good persistence of activity in the soil. In beet crops 1.0–2.0 kg/ha can be used,

but ethofumesate is normally recommended at 0.2–2.0 kg/ha in tank-mixtures or co-formulations with other residual or contact herbicides for use in beet. A high degree of tolerance is also shown by strawberries, sunflowers, Phaseolus beans and tobacco, depending on the time of application.   **Phytotoxicity** Onions, peas, beans, carrots, and cotton are tolerant to some extent.   **Formulation types** EC; SC.   **Mixtures** *(ethofumesate +)* chloridazon; lenacil; phenmedipham; metamitron; bromoxynil + ioxynil; phenmedipham + desmedipham; phenmedipham + metamitron; phenmedipham + desmedipham + metamitron.   **Selected tradenames** 'Betanal Tandem' (AgrEvo); 'Betanal Progress' (AgrEvo); 'Nortron' (AgrEvo, NOR-AM); 'Prograss' (AgrEvo), 'Tramat' (AgrEvo); 'Ethosat' (Feinchemie Schwebda); 'Ethosin' (Agriphar); 'Keeper' (Barclay); 'Primasan' (Mirfield)

ANALYSIS

Product and **residue** analysis by glc with FPD (R. J. Whiteoak *et al., Anal. Methods Pestic. Plant Growth Regul.*, 1978, **10**, 353). Details available from AgrEvo.

MAMMALIAN TOXICOLOGY

Oral Acute oral LD_{50} for rats >5000, mice >5000 mg/kg.   **Skin and eye** Acute percutaneous LD_{50} for rats >2000 mg/kg. Not a skin or eye irritant. Not a skin sensitiser.   **Inhalation** LC_{50} (4 h) for rats >3.97 mg/l air.   **NOEL** (2 y) for rats >1000 mg/kg diet (37.6 mg/kg b.w. daily).   **ADI** 0.4 mg/kg.   **Toxicity class** WHO (a.i.) III (Table 5); EPA (formulation) IV

ECOTOXICOLOGY

Birds Acute oral LD_{50} for mallard ducks >3552, bobwhite quail >8743 mg/kg. Eight-day dietary LC_{50} for mallard ducks >1082, bobwhite quail >839 mg/kg b.w. daily.   **Fish** LC_{50} (96 h) for rainbow trout 11.91–20.2, bluegill sunfish 12.37–21.2, mirror carp 10.92 mg/l.   **Bees** LC_{50} (contact and oral) >50 μg/bee.   **Worms** LC_{50} 134 mg/kg soil.   **Other beneficial spp.** LD_{50} for *Aleochara bilineata* >1250, *Poecilus cupreus* and *Chrysoperla carnea* >2000 g/ha.   **Daphnia** EC_{50} (48 h) 13.52–22.0 mg/l.   **Algae** EC_{50} 3.9 mg/l.   **Other aquatic spp.** EC_{50} (growth) (96 h) for *Crassostrea virginica* (Eastern oyster)1.7 mg/l; LC_{50} (96 h) for *Mysidopsis bahia* (mysid shrimp) 5.4 mg/l.

ENVIRONMENTAL FATE

Animals Major metabolite is the lactone or free acid form of the respective 2-oxo compound.   **Plants** In plants, ethofumesate is metabolised to the 2-hydroxy and 2-oxo derivatives, methanesulfonic acid, and CO_2.   **Soil/Environment** Ethofumesate is biologically degraded in soil to transient degradates which are rapidly converted to soil-bound residues and mineralised to CO_2. Photodegradation also occurs. DT_{50} ranges from 10–122 d (lab.) and 84–407 d (field). It has been demonstrated under field conditions that ethofumesate does not accumulate in soil and that uptake by succeeding crops is negligible. It is weakly/moderately adsorbed to soil (mean K_{oc} 203) but field lysimeter studies have demonstrated only low mobility, most residues being located in the top 30 cm. It does not leach into groundwater.

286 ethoprophos
Nematicide, insecticide

organophosphorus

$$CH_3CH_2OP(SCH_2CH_2CH_3)_2$$

with O double-bonded above P.

NOMENCLATURE
Common name ethoprophos (BSI, E-ISO, (*m*) F-ISO); ethoprop (ANSI, ESA, Society of Nematologists (USA))
IUPAC name *O*-ethyl *S,S*-dipropyl phosphorodithioate
Chemical Abstracts name *O*-ethyl *S,S*-dipropyl phosphorodithioate
CAS RN *[13194–48–4]* **EEC no.** 236–152–1 **Development codes** VC9–104 (Mobil Chemical Co.)

PHYSICAL CHEMISTRY
Mol. wt. 242.3 **M.f.** $C_8H_{19}O_2PS_2$ **Form** Pale yellow liquid. **B.p.** 86–91 °C/0.2 mmHg **V.p.** 46.5 mPa (26 °C) **K_{ow}** logP = 3.59 (21 °C) **S.g./density** 1.094 (20 °C) **Solubility** In water 700 mg/l (20 °C). In acetone, ethanol, xylene, 1,2-dichloroethane, diethyl ether, ethyl acetate, petroleum spirit, cyclohexane >300 (all in g/kg, 20 °C). **Stability** Very stable in neutral and weakly acidic media. Rapidly hydrolysed in alkaline media. Stable in water up to 100 °C at pH 7.
F.p. 140 °C (closed cup)

COMMERCIALISATION
History Nematicide reported by S. J. Locascio (*Proc. Fla. St. Hortic. Soc.*, 1966, **79**, 170). Introduced by Mobil Chemical Co. (who no longer manufacture or market it) and later by Rhône-Poulenc Agrochimie. **Patent** US 3112244; US 3268393 to Mobil **Manufacturers** Rhône-Poulenc

APPLICATIONS
Biochemistry Cholinesterase inhibitor. **Mode of action** Non-systemic nematicide and soil insecticide with contact action. **Uses** Control of plant-parasitic nematodes and soil insects in ornamentals, potatoes, sweet potatoes, tomatoes, vegetables, maize, soya beans, peanuts, cucurbits, strawberries, citrus, tobacco, bananas, pineapples, sugar cane, turf, and other crops at 1.6–6.6 kg a.i./ha.
Formulation types GR; EC. **Compatibility** Incompatible with alkaline materials.
Selected tradenames 'Mocap' (Rhône-Poulenc)

ANALYSIS
Product analysis by glc with TCD (F. A. Norris et al., Anal. Methods Pestic. Plant Growth Regul., 1988, **16**, 3).   **Residues** determined by glc with MCD or FPD (idem, ibid.; Man. Pestic. Residue Anal., 1987, **I**, 6, S19; Anal. Methods Residues Pestic., 1988, Part I, M2, M5; A. Ambrus et al., J. Assoc. Off. Anal. Chem., 1981, **64**, 733). In **drinking water**, by gc with FID (AOAC Methods, 1995, 991.07).

MAMMALIAN TOXICOLOGY
Reviews Pesticide residues in food – 1987. FAO Plant Production and Protection Paper 84, 1987. Pesticide residues in food – 1987 evaluations. Part II – Toxicology. FAO Plant Production and Protection Paper 86/2, 1988.   **EHC** 63 (WHO, 1986; a general review of organophosphorus insecticides).   **Oral** Acute oral LD_{50} for rats 62, rabbits 55 mg/kg.   **Skin and eye** Acute percutaneous LD_{50} for rabbits 26 mg/kg. May cause skin and eye irritation.   **Inhalation** LC_{50} for rats 123 mg/m^3. **NOEL** In 90 d feeding trials, rats and dogs receiving 100 mg/kg diet showed depression of cholinesterase levels but no other effect on pathology or histology. **ADI** (JMPR) 0.0003 mg/kg b.w. [1987].   **Toxicity class** WHO (a.i.) Ia; EPA (formulation) II   **EC risk** T+ (R27); T (R25)

ECOTOXICOLOGY
Birds Acute oral LD_{50} for mallard ducks 61, hens 5.6 mg/kg.   **Fish** LC_{50} (96 h) for rainbow trout 13.8, bluegill sunfish 2.1, goldfish 13.6 mg/l.   **Bees** Not hazardous to bees when used as directed.

ENVIRONMENTAL FATE
Animals In the rat, the principal metabolite is O-ethyl-S-propylphosphothioic acid, which is no more toxic than ethoprophos itself.   **Plants** In plants such as haricot bean and maize, ethoprophos is rapidly broken down into non-toxic metabolites such as methyl propyl sulfide, methyl propyl sulfoxide and methyl propyl sulfone. Although it enters the root system of plants, ethoprophos cannot be considered to be systemic, as it is not carried into the aerial parts. Therefore it usually leaves no detectable residues.   **Soil/Environment** DT_{50} in humus-containing soil (pH 4.5) c. 87 d; in sandy loam (pH 7.2–7.3) c. 14–28 d (J. H. Smelt et al., Pestic. Sci., 1977, **8**, 147–151). Freundlich K 1.08 (sandy loam, o.m. 1.0%), 1.24 (sandy loam, o.m. 1.98%), 2.10 (silt loam, o.m. 2.3%), 3.78 (silty clay loam, o.m. 4.1%).

MAIN ENTRIES

sulfonylurea

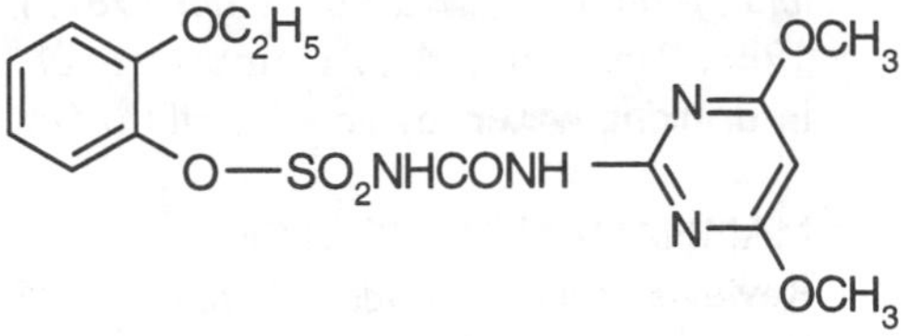

NOMENCLATURE
Common name ethoxysulfuron (BSI, pa ISO)
IUPAC name 1-(4,6-dimethoxypyrimidin-2-yl)-3-(2-ethoxyphenoxysulfonyl)urea
CAS RN *[126801–58–9]* **Development codes** Hoe 095404; Hoe-404

PHYSICAL CHEMISTRY
Mol. wt. 398.4 **M.f.** $C_{15}H_{18}N_4O_7S$ **Form** White to beige powder.
M.p. 144–147 °C **V.p.** 6.6×10^{-2} mPa K_{ow} logP = 2.89 (pH 3), 0.004 (pH 7),
–1.2 (pH 9) (20 °C) **Henry** (calc.) 1.00×10^{-3} (pH 5), 1.94×10^{-5} (pH 7),
2.73×10^{-6} (pH 9) Pa m^3 mol^{-1} (20 °C) **Solubility** In water 26 (pH 5), 1353
(pH 7), 9628 (pH 9) ppm (20 °C). **Stability** Hydrolytic DT_{50} 65 (pH 5), 259
(pH 7), 331 (pH 9) days.

COMMERCIALISATION
History Reported by E. Hacker *et al.* (*Proc. Br. Crop Prot. Conf. – Weeds*, 1995, **1**,
73). **Manufacturers** AgrEvo

APPLICATIONS
Biochemistry Branched chain amino acid synthesis (acetolactate synthase or ALS)
inhibitor. Acts by inhibiting biosynthesis of the essential amino acids valine and
isoleucine, hence stopping cell division and plant growth. Selectivity is due to
differential metabolism in crop and weed (H Köcher & G Dickerhof, *Proc. Br. Crop
Prot. Conf. – Weeds*, 1995, **1**, 249). Metabolic basis of crop selectivity in
sulfonylureas reviewed (M. K. Koeppe & H. M. Brown, *Agro-Food-Industry*, **6**, 9–14
(1995)). **Uses** Under development for broad-leaved and sedge weed control in
cereals, rice and sugar cane at 10–120 g/ha. **Formulation types** WG.

ANALYSIS
Methods for sulfonylurea **residues** in crops, soil and water reviewed (A. C.
Barefoot *et al., Proc. Br. Crop Prot. Conf. – Weeds*, 1995, **2**, 707).

MAMMALIAN TOXICOLOGY
Oral Acute oral LD_{50} for rats >3270 mg/kg. **Skin and eye** Acute percutaneous
LD_{50} for rats <4000 mg/kg. Not irritating to eyes or skin (rats). **Other** Not
mutagenic (Ames).

ENVIRONMENTAL FATE
Soil/Environment In lab. tests, DT_{50} in biologically active soil is c. 18–20 d. Under paddy conditions, DT_{50} is 10–60 d.

288 ethychlozate

Plant growth regulator

CH$_2$CO$_2$CH$_2$CH$_3$

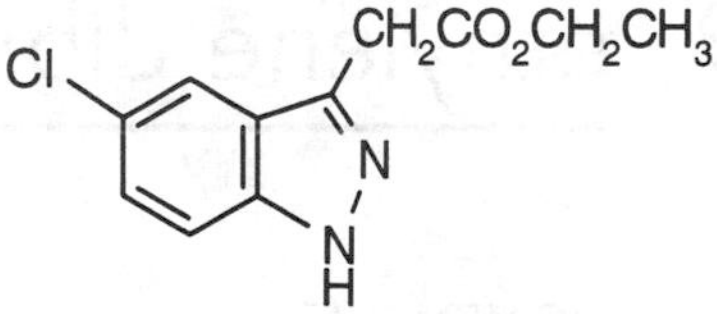

NOMENCLATURE
Common name ethychlozate (JMAF)
IUPAC name ethyl 5-chloro-3(1*H*)-indazolylacetate
Chemical Abstracts name ethyl 5-chloro-1*H*-3-indazole-3-acetate
CAS RN *[27512–72–7]* **Development codes** J-455 (Nissan)

PHYSICAL CHEMISTRY
Mol. wt. 238.7 **M.f.** $C_{11}H_{11}ClN_2O_2$ **Form** Yellow crystals. **M.p.** 76.6–78.1 °C
Solubility In water 0.225 g/l (24 °C). In acetone 673, ethyl acetate 496, ethanol 512, hexane 0.213, kerosene 2.19, methanol 691, isopropanol 381 (all in g/l, 24 °C). **Stability** Stable at 250 °C.

COMMERCIALISATION
History Plant growth regulator introduced by Nissan Chemical Industries, Ltd and distributed in Japan. **Manufacturers** Fujisawa

APPLICATIONS
Mode of action Has auxin-like activity, stimulating ethylene production with the formation of the abscission layer in young fruit. Rapidly translocates to the root system, and promotes root activity. **Uses** For thinning of young fruit, enhancement of colour, and improvement of quality in citrus fruit, especially in satsuma mandarin fruit. **Formulation types** EC. **Selected tradenames** 'Figaron' (Nissan)

MAMMALIAN TOXICOLOGY
Oral Acute oral LD_{50} for male rats 4800, female rats 5210, male mice 1580, female mice 2740 mg/kg. **Skin and eye** Acute percutaneous LD_{50} for rats
>10 000 mg/kg. Non-irritant to skin and eyes (rabbits). **Inhalation** LC_{50} (4 h) for

rats >1508 mg/m^3. **NOEL** (life-span) for mice 265 mg/kg daily. There was no evidence of teratogenicity or mutagenicity.

ENVIRONMENTAL FATE
Animals Rapidly excreted via the urine of rats within almost 24 hours.
Soil/Environment Decomposition in the soil takes place rapidly; DT$_{50}$ 1–4 d.

289 ethylene dibromide *Nematicide, insecticide*

$$BrCH_2CH_2Br$$

NOMENCLATURE
Common name EDB (JMAF); ethylene dibromide (BSI, E-ISO, ESA, accepted in lieu of a common name); dibromure d'éthylène (F-ISO, accepted in lieu of a common name)
IUPAC name 1,2-dibromoethane
Chemical Abstracts name 1,2-dibromoethane
CAS RN *[106–93–4]* **EEC no.** 203–444–5 **Official codes** ENT 15 349

PHYSICAL CHEMISTRY
Mol. wt. 187.9 **M.f.** C$_2$H$_4$Br$_2$ **Form** Colourless liquid. **M.p.** 9.3 °C
B.p. 131.5 °C **V.p.** 1.5 kPa (25 °C); 5.2 kPa (48 °C) **S.g./density** 2.172 (25 °C)
Solubility In water 4.3 g/kg (30 °C). Soluble in diethyl ether, ethanol and most common organic solvents. **Stability** Decomposed by alkalis and by light.
F.p. Non-flammable.

COMMERCIALISATION
History Its fumigant insecticidal properties described by I. E. Neifert *et al.* (*U.S. Dep. Agric. Bull.*, 1925, No. 1313). Introduced by Dow Chemical Co. (now DowElanco, who no longer manufacture or market it). **Patent** US 2448265; US 2473984

APPLICATIONS
Uses Control of nematodes, wireworms, and other soil pests. Fumigation of mills, warehouses, and households. **Phytotoxicity** Phytotoxic for green plants and germinating seed. **Formulation types** Oil. **Mixtures** *(ethylene dibromide +)* chloropicrin; methyl bromide. **Compatibility** Miscible with non-alkaline organic liquids. **Selected tradenames** 'Dibrome' (United Phosphorus)

ANALYSIS
Product analysis by glc (*CIPAC Handbook*, 1985, **1C**, 2115; *AOAC Methods*, 1995, 966.05). **Residues** determined by glc (*Analyst (London)*, 1974, **99,** 570; K. A. Scudamore, *Anal. Methods Pestic. Plant Growth Regul.*, 1988, **16,** 207, 226; *Anal. Methods Residues Pestic.*, 1988, Part I, M8). Atmospheres can be sampled by passing through 2-aminoethanol and determining bromide by Volhard titration (C. E. Castro, *Anal. Methods Pestic., Plant Growth Regul. Food Addit.*, 1964, **3,** 155; *Anal. Methods Pestic. Plant Growth Regul.*, 1972, **6,** 711). Methods reviewed by J. I. Daft in *Comp. Analyt. Profiles*, Chapter 11. Residues in **drinking water** by gc with linearised ECD (*AOAC Methods*, 1995, 993.15, 10.7.04).

MAMMALIAN TOXICOLOGY
Reviews *Pesticide residues in food.* FAO Agricultural Studies, No. 73; WHO Technical Report Series, No. 370, 1967. *Evaluation of some pesticide residues in food.* FAO/PL:CP/15; WHO/Food Add./67.32, 1967. **IARC** 15 **Oral** Acute oral LD_{50} for rats 146–420 mg/kg. **Skin and eye** Acute percutaneous LD_{50} not available. Dermal applications, if confined, cause severe burning of the skin. **Inhalation** Prolonged inhalation can lead to necrosis of the liver; acute inhalation toxicity 200 mg/l air. Rats tolerated 7 h exposures, 5 d/w, for 0.5 y at rates ≤0.21 mg/l. **ADI** (JMPR) 1.0 mg/kg b.w. as bromide ion [1966].
EC risk (R45); T (R23/24/25); Xi (R36/37/38): (for preparations, concn. ≥20%, T (R45; R23/24/25; R36/37/38), 1%≤ concn. <20%, T (R45; R23/24/25), 0.1%≤ concn. <1%, (R45; R20/21/22)) **PIC** Yes

290   ethylene dichloride   *Insecticide*

$$ClCH_2CH_2Cl$$

NOMENCLATURE
Common name EDC (JMAF); ethylene dichloride (BSI, E-ISO, ESA, accepted in lieu of a common name); dichlorure d'éthylène (F-ISO, accepted in lieu of a common name)
IUPAC name 1,2-dichloroethane
Chemical Abstracts name 1,2-dichloroethane
CAS RN *[107–06–2]* **EEC no.** 203–458–1 **Official codes** ENT 1656

PHYSICAL CHEMISTRY
Mol. wt. 99.0 **M.f.** $C_2H_4Cl_2$ **Form** Colourless liquid, with a chloroform-like odour. **M.p.** −36 °C **B.p.** 83.5 °C **V.p.** 10.4 kPa (20 °C) **S.g./density** 1.2569 (20 °C) **Solubility** In water 4.3 g/l (room temperature). Soluble in most organic

solvents.   **F.p.** 12–15 °C (Abel Pensky); lower and upper limits of flammability in air 275 and 700 mg/l.

COMMERCIALISATION
History Reported as a component of insecticidal fumigants by R. T. Cotton & R. C. Roark (*J. Econ. Entomol.,* 1927, **20,** 636).

APPLICATIONS
Mode of action Fumigant insecticide.   **Uses** An insecticidal fumigant used mainly in stored products.   **Mixtures** *(ethylene dichloride +)* carbon tetrachloride.

ANALYSIS
Product analysis by glc (*AOAC Methods,* 1995, 966.05).   **Residues** determined by glc (*ibid.,* 1984, 20.215–20.218; K. A. Scudamore, *Anal. Methods Pestic. Plant Growth Regul.,* 1988, **16,** 207, 231; *Analyst (London),* 1974, **99,** 570; *Anal. Methods Residues Pestic.,* 1988, Part I). Methods reviewed by J. L. Daft in *Comp. Analyt. Profiles,* Chapter 11.

MAMMALIAN TOXICOLOGY
Reviews *Evaluation of the toxicity of pesticide residues in food.* FAO Meeting Report, No. PL/1965/10; WHO/Food Add./26.65, 1965. *Evaluation of the hazards to consumers resulting from the use of fumigants in the protection of food.* PL/1965/10/2; WHO/Food Add./28.65, 1965.   **IARC** 20   **EHC** 62 (WHO, 1987).   **Oral** Acute oral LD_{50} for rats 670–890, mice 870–950, rabbits 860–970 mg/kg.   **ADI** (JMPR) No ADI [1965].   **EC risk** F (R11); (R45); Xn (also R22); Xi (R36/37/38): (for preparations, concn. ≥25%, T (R45; R22; R36/37/38); 20%≤ concn. <25%, T (R45; R36/37/38); 0.1%≤ concn. <20%, T (R45))

291   etobenzanid

Herbicide

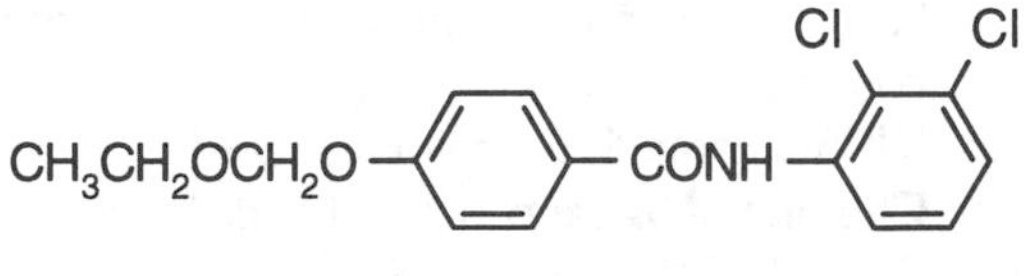

NOMENCLATURE
Common name etobenzanid (BSI, pa E-ISO); étobenzanide (pa F-ISO)
IUPAC name 2',3'-dichloro-4-ethoxymethoxybenzanilide
Chemical Abstracts name *N*-(2,3-dichlorophenyl)-4-(ethoxymethoxy)benzamide
CAS RN *[79540–50–4]*   **Development codes** HW-52

PHYSICAL CHEMISTRY
Mol. wt. 340.2 **M.f.** $C_{16}H_{15}Cl_2NO_3$ **M.p.** 92–93 °C **V.p.** 2.1×10^{-2} mPa (40 °C) **K_{ow}** logP = 4.3 (25 °C) **Solubility** In water 0.92 ppm (25 °C). In methanol 22.4, acetone >100, n-hexane 2.42 (all in g/l).

COMMERCIALISATION
Manufacturers Hodogaya

APPLICATIONS
Uses Pre- and post-emergence use in rice at 150 g/ha. **Formulation types** GR; WP. **Selected tradenames** 'Hodocide' (Hodogaya)

MAMMALIAN TOXICOLOGY
Reviews *Noyaku Jiho*, **462**, May 10, 1996 (issued by Japan Plant Protection Association). **Oral** LD_{50} for mice >5000 mg/kg. **Skin and eye** Slightly irritating to eyes and skin. **Inhalation** LC_{50} for rats 1503 mg/m³. **NOEL** for rats 4.4 mg/kg b.w. daily. **ADI** 0.044 mg/kg b.w.

ECOTOXICOLOGY
Birds Acute oral LD_{50} for quail >2000 mg/kg. **Fish** LC_{50} (72 h) for carp >1000 ppm. **Bees** LD_{50} >160 ppm diet. **Worms** LC_{50} >1000 ppm soil. **Daphnia** LC_{50} (3 h) >1000 ppm.

292 etofenprox   *Insecticide*

non-ester pyrethroid

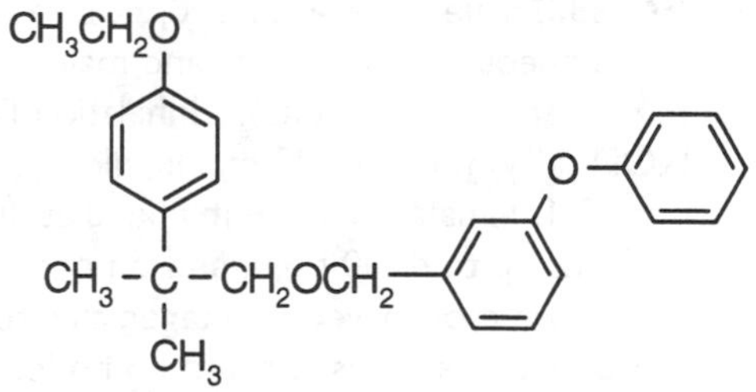

NOMENCLATURE
Common name etofenprox ((m) draft F-ISO, INN; BSI, draft E-ISO both since 1988); [ethofenprox] (before 1988)
IUPAC name 2-(4-ethoxyphenyl)-2-methylpropyl 3-phenoxybenzyl ether
Chemical Abstracts name 1-[[2-(4-ethoxyphenyl)-2-methylpropoxy]methyl]-=3-phenoxybenzene
CAS RN *[80844–07–1]* **Development codes** MTI-500 **Official codes** OMS 3002

PHYSICAL CHEMISTRY
Mol. wt. 376.5   **M.f.** $C_{25}H_{28}O_3$ **Form** White crystals.   **M.p.** 36.4–38.0 °C
B.p. 200 °C/0.18 mmHg   **V.p.** 32 mPa (100 °C) K_{ow} logP = 7.05 (25 °C)
S.g./density 1.157 (23 °C, solid); 1.067 (40.1 °C, liquid)   **Solubility** In water
<1 µg/l (25 °C). In chloroform 9, acetone 7.8, ethyl acetate 6, xylene 4.8,
methanol 0.066 (all in kg/l, 25 °C).   **Stability** Stable in acidic and alkaline media, at
80 °C for >90 days, and to light.

COMMERCIALISATION
History Insecticide introduced in Japan (1987) by Mitsui Toatsu Chemicals, Inc.
Patent GB 2118167; US 4570005   **Manufacturers** Mitsui Toatsu

APPLICATIONS
Mode of action Insecticide with contact and stomach action.   **Uses** Control of rice
water weevils, skippers, leaf beetles, leafhoppers, planthoppers, and bugs on
paddy rice; and aphids, moths, butterflies, whitefly, leaf miners, leaf rollers,
leafhoppers, thrips, borers, etc. on pome fruit, stone fruit, citrus fruit, tea, soya
beans, sugar beet, brassicas, cucumbers, aubergines, and other crops. Also used to
control public health pests, and on livestock.   **Formulation types** EC; WP; GR;
DP; EW; CS; SL; UL.   **Selected tradenames** 'Trebon' (Mitsui Toatsu)

ANALYSIS
Product analysis by glc with FID (*CIPAC Handbook*, 1995, **G**, 56–61).
Residues determined by glc with ECD.

MAMMALIAN TOXICOLOGY
Reviews *Pesticide residues in food – 1993*, FAO Plant Production and Protection
Paper, 122, 1993. *Pesticide residues in food – 1993 evaluations, Part II – Toxicology.*
WHO, WHO/PCS/94.4, 1994.   **Oral** Acute oral LD_{50} for male and female rats
>42 880, male mice >107 200, dogs >5000 mg/kg.   **Skin and eye** Acute
percutaneous LD_{50} for rats and male and female mice >2140 mg/kg; non-irritating
to skin and eyes (rabbits).   **Inhalation** LC_{50} (4 h) for rats 5900 mg/m^3.
NOEL (1 y) for dogs 32 mg/kg diet; (2 y) for male rats 3.7, female rats 4.8, male
mice 3.1, female mice 3.6 mg/kg diet. In life-span feeding studies on rats and mice,
at doses up to 4900 ppm, and on dogs, at doses up to 10 000 ppm, no adverse
effects were observed.   Mutagenicity, teratogenicity, and three-generation
reproduction studies showed no noticeable abnormalities.   **ADI** (JMPR)
0.03 mg/kg b.w. [1993].   **Toxicity class** WHO (a.i.) III (Table 5);
EPA (formulation) IV

ECOTOXICOLOGY
Birds Acute oral LD_{50} for mallards >2000 mg/kg.   **Fish** TLm (48 h) for carp
5.0 ppm.   **Worms** LC_{50} for earthworms (7 d) 43.1 ppm, (14 d) 24.6 ppm.
Daphnia LC_{50} (3 h) >40 mg/l.

ENVIRONMENTAL FATE

Animals In rats, parent compound, 2-(4-hydroxyphenyl)-2-methylpropyl 3-(phenoxybenzyl) ether, and 2-(4-ethoxyphenyl)-2-methylpropyl 3-(4-hydroxy= benzyl) ether are found.   **Plants** In rice, the main metabolite is 2-(4-ethoxy= phenyl)-2-methylpropyl 3-phenoxybenzoate.   **Soil/Environment** DT_{50} in soil *c.* 6 d.

---

293 etoxazole   *Acaricide*

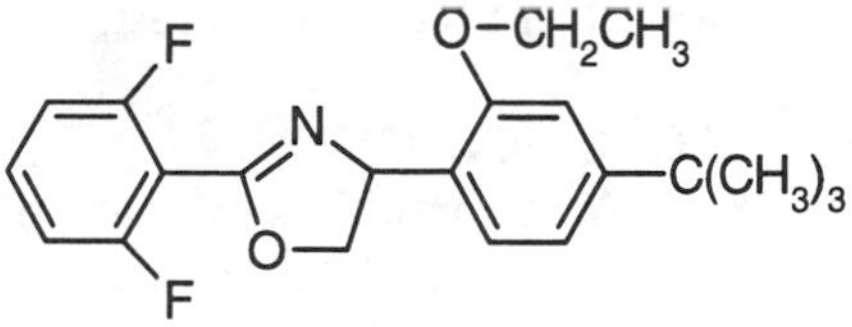

NOMENCLATURE
Common name etoxazole (BSI, pa ISO)
IUPAC name (*RS*)-5-*tert*-butyl-2-[2-(2,6-difluorophenyl)-4,5-dihydro-1,3-oxazol-= 4-yl]phenetole
Chemical Abstracts name 2-(2,6-difluorophenyl)-4-[4-(1,1-dimethylethyl)-2-= ethoxyphenyl]-4,5-dihydrooxazole
CAS RN *[153233–91–1]*   **Development codes** YI-5301 (Yashima); S-1283 (Sumitomo)

PHYSICAL CHEMISTRY
Composition Tech. is 93–98%.   **Mol. wt.** 359.4   **M.f.** $C_{21}H_{23}F_2NO_2$ **Form** White crystalline powder.   **M.p.** 101–102 °C   **V.p.** 2.18×10^{-3} mPa (25 °C)
K_{ow} logP = 5.59 (25 °C)   **S.g./density** 1.24 (20 °C)   **Solubility** In water 75.4 μg/l (20 °C). In acetone 300, methanol 90, ethanol 90, cyclohexanone 500, tetrahydrofuran 750, acetonitrile 80, ethyl acetate 250, xylene 250, *n*-hexane 13, *n*-heptane 13 (all in g/l, 20 °C).   **Stability** No decomposition after 30 d (50 °C). Stable in alkali.

COMMERCIALISATION
History Reported by T. Ishida *et al.* (*Proc. Br. Crop Prot. Conf., Pests Dis.*, 1994, **1**, 37).   **Manufacturers** Yashima

APPLICATIONS
Mode of action Contact acaricide. Appears to inhibit moulting of mites and aphids.
Uses Non-systemic acaricide with effect on eggs, larvae and nymphs, with no effect on adults. Controls many phytophagous mites (*Tetranychus, Panonychus* spp.)

in citrus, pome fruit, vegetables and strawberries at 50 g/ha, and tea at 100 g/ha.
Formulation types SC **Mixtures** *(etoxazole +)* fenpropathrin. **Compatibility**
Cannot be mixed with Bordeaux mixture. **Selected tradenames** 'Baroque
Flowable' (Yashima, Sumitomo); 'Biruku' (mixture) (Yashima, Sumitomo)

ANALYSIS
Product and **residue** analysis by hplc.

MAMMALIAN TOXICOLOGY
Oral Acute oral LD_{50} for male and female rats and mice >5000 mg/kg. **Skin and
eye** Acute percutaneous LD_{50} for male and female rats >2000 mg/kg; no skin or
eye irritation (rabbits). **Inhalation** LC_{50} for male and female rats >1.09 mg/l.
Other Negative in Ames test.

ECOTOXICOLOGY
Birds Acute oral LD_{50} for mallard duck >2000 mg/kg. Sub-acute oral LD_{50} for
bobwhite quail >5200 ppm diet. **Fish** LC_{50} (96 h) for Japanese carp 0.89 mg/l;
(48 h) for Japanese carp >20, rainbow trout >40 ppm. **Bees** LD_{50} (oral and
contact) >200 µg/bee. **Worms** NOEL (14 d) for *Eisenia foetida* >1000 ppm.
Daphnia LC_{50} (48 h) >40 ppm. **Algae** NOEL for *Selenastrum capricornutum*
>1.0 mg/l. **Other aquatic spp.** Disruption of moulting was observed in aquatic
arthropods.

ENVIRONMENTAL FATE
Soil/Environment In Japanese alluvial soil, DT_{50} 19 d, DT_{90} 90 d.

294 etridiazole   *Fungicide*

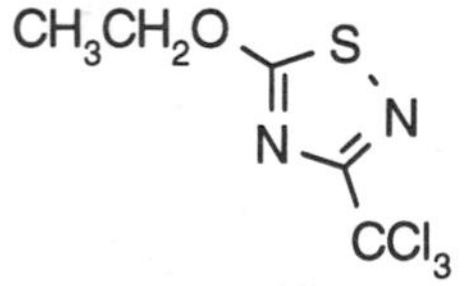

CH₃CH₂O — S
N — N
CCl₃

NOMENCLATURE
Common name etridiazole (BSI, E-ISO, *(m)* F-ISO); echlomezol (JMAF)
IUPAC name ethyl 3-trichloromethyl-1,2,4-thiadiazol-5-yl ether
Chemical Abstracts name 5-ethoxy-3-(trichloromethyl)-1,2,4-thiadiazole
Other names ethazol; ethazole; ETCMTD **CAS RN** *[2593–15–9]*
Development codes OM 2424

PHYSICAL CHEMISTRY

Composition Tech. grade is ≥96% pure. **Mol. wt.** 247.5 **M.f.** $C_5H_5Cl_3N_2OS$
Form Pale yellow liquid with a persistent odour. Tech. grade is a reddish-brown
liquid. **M.p.** 19.9 °C **B.p.** 95 °C/1 mmHg **V.p.** 1430 mPa (25 °C)
K_{ow} logP = 3.37 **S.g./density** 1.503 (25 °C) **Solubility** In water 117 mg/l
(25 °C). Miscible with ethanol, methanol, aromatic hydrocarbons, acetonitrile,
hexane, xylene. **Stability** Stable for 14 d at 55 °C. After continuous exposure to
sunlight for 7 d at 20 °C, 5.5–7.5% decomposition. Hydrolysis DT_{50} 12 d (pH 6,
45 °C), 103 d (pH 6, 25 °C). **pKa** 2.77, v. weak base **F.p.** 66 °C

COMMERCIALISATION

History Fungicide introduced by Olin Chemicals (who no longer manufacture or
market it) and by Uniroyal Chemical Co., Inc. **Patent** US 3260588; US 3260725
Manufacturers Uniroyal

APPLICATIONS

Mode of action Contact fungicide with protective and curative action.
Uses Control of *Phytophthora* and *Pythium* spp. in ornamentals, cotton, peanuts,
vegetables, turf, and other crops. Applied at 0.15–0.40 lb/a or 0.02–0.04 lb/cwt
seed. **Formulation types** DP; WP; EC; Seed treatment. **Mixtures** (*etridiazole* +)
quintozene; thiophanate-methyl; disulfoton. **Selected tradenames** 'Terrazole'
(Uniroyal)

ANALYSIS

Product analysis by glc. **Residues** determined by glc with ECD (*Pestic. Anal. Man.,
II, Pestic. Reg. Sec., 180, 370*). In **drinking water** by gc with ECD (*AOAC Methods,
1995, 990.06*). Details available from Uniroyal.

MAMMALIAN TOXICOLOGY

Oral Acute oral LD_{50} for rats 1100 mg/kg. **Skin and eye** Acute percutaneous
LD_{50} for rabbits >5000 mg/kg. Not irritating to skin, mild eye irritant (rabbits).
Inhalation LC_{50} (4 h) for rats >5.7 mg/l. **NOEL** (2 y oncogenic) for rats 4 mg/kg
b.w. daily; (2 y feeding) for dogs 2.5 mg/kg b.w. daily. **ADI** 0.025 mg/kg.
Toxicity class WHO (a.i.) III; EPA (formulation) III

ECOTOXICOLOGY

Birds Acute oral LD_{50} for bobwhite quail 560 mg/kg. Eight-day dietary LC_{50} for
bobwhite quail >5000 ppm, for mallard ducks 1650 mg/kg. **Fish** LC_{50} (216 h) for
rainbow trout 1.21, bluegill sunfish 3.27 mg/l. **Daphnia** LC_{50} (48 h) 4.9 mg/l.
Algae EC_{50} (5 d) for *Selenastrum* 0.072, *Anabaena* 0.29, *Navicula* 0.43, *Skeletonema*
0.38 mg/l. **Other aquatic spp.** LC_{50} (96 h) for mysid shrimp (*Mysidopsis bahia*)
2.5, oyster (*Crassostrea*) 3.0 mg/l.

ENVIRONMENTAL FATE

Animals In mammals, following oral administration, the water-soluble metabolite 3-carboxy-5-ethoxy-1,2,4-thiadiazole has been found. In rat urine, this compound was found as a major metabolite and N-acetyl-S-(5-ethoxy-1,2,4-thiadiazol-3-yl-= methyl)-L-cysteine was identified as a minor metabolite (R. T. H. Van Welie *et al.*, *Arch. Toxicol.*, 1991, **65**, 625).   **Plants** The trichloromethyl group is readily converted to the acid or alcohol and the ethoxy group is hydroxylated to form a hydroxyethyl derivative. Some plants convert etridiazole more extensively to form natural products.   **Soil/Environment** Soil DT_{50} in silt loam in laboratory at 25 °C (aerobic) 9.5 d, (anaerobic) 3 d. Field dissipation, in sandy clay loam DT_{50} 1 w. Soil adsorption K 5.31 (sandy soil), 1.41 (silt loam).

295 etrimfos   *Insecticide, acaricide*

organophosphorus

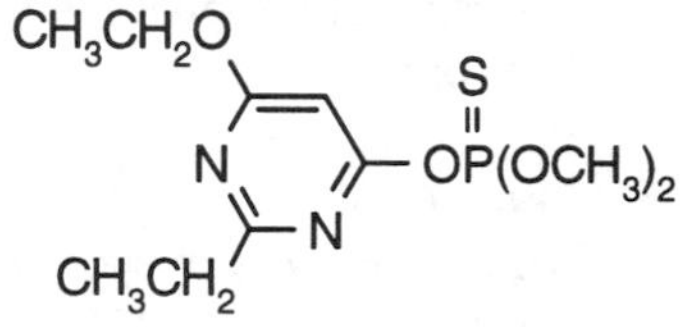

NOMENCLATURE

Common name etrimfos (BSI, E-ISO, (m) F-ISO, ANSI, ESA)
IUPAC name O-6-ethoxy-2-ethylpyrimidin-4-yl O,O-dimethyl phosphorothioate
Chemical Abstracts name O-(6-ethoxy-2-ethyl-4-pyrimidinyl) O,O-dimethyl phosphorothioate
CAS RN *[38260–54–7]*   **Development codes** SAN 1971   **Official codes** OMS 1806

PHYSICAL CHEMISTRY

Mol. wt. 292.3   **M.f.** $C_{10}H_{17}N_2O_4PS$ **Form** Colourless oil.   **M.p.** –3.35 °C
V.p. 6.5 mPa (20 °C); 18 mPa (30 °C); 49 mPa (40 °C) K_{ow} logP >3.3
S.g./density 1.195 at 20 °C   **Solubility** In water 40 mg/l (23–24 °C). Completely miscible with acetone, acetonitrile, chloroform, dimethyl sulfoxide, ethanol, diethyl ether, ethyl acetate, hexane, kerosene, methanol, toluene, xylene.
Stability Unstable when pure (c. 40% degradation in 28 d at 48–52 °C), but dilute solutions in non-polar solvents and its formulations are stable (c. 5% loss in 1 y at 18–22 °C). When stored in the dark at 22–24 °C for 28 d, no decomposition occurred in hexane, toluene, chloroform, acetone; c. 3% decomposition in

methanol; c. 12% decomposition in ethyl acetate. Hydrolysis (25 °C) DT_{50} 0.4 d (pH 3), 16 d (pH 6), 14 d (pH 9). High stability to light.

COMMERCIALISATION
History Insecticide reported by H. J. Knutti & F. W. Reisser (*Proc. Br. Insectic. Fungic. Conf., 5th*, 1975, **2**, 675). Introduced by Sandoz Agro AG (now Novartis Crop Protection AG).   **Manufacturers** Novartis

APPLICATIONS
Biochemistry Cholinesterase inhibitor.   **Mode of action** Non-systemic insecticide and acaricide with contact and stomach action.   **Uses** Control of Lepidoptera, Coleoptera, Diptera, and some Hemiptera at 250–750 g a.i./ha on fruit trees (including citrus), vines, olives, potatoes, vegetables, maize, alfalfa, ornamentals, tobacco. Control of Pyralidae in paddy rice at 1.0–1.5 kg a.i./ha. It is also used at 3–10 mg/kg for control of Lepidoptera, Coleoptera, mites, and lice in stored products.   **Phytotoxicity** Non-phytotoxic when used as directed, except to begonias, and, for EC formulations, cherries and some varieties of apple.
Formulation types DP; EC; GR; LS; UL.   **Compatibility** Not compatible with pesticides which are strongly alkaline.   **Selected tradenames** 'Ekamet' (Novartis); 'Satisfar' (Novartis)

ANALYSIS
Product analysis by glc.   **Residues** determined by glc (*Man. Pestic. Residue Anal.,* 1987, **I**, 6, S19; *Anal. Methods Residues Pestic.*, 1988, Part I, M2, M5; J. C. Karpally et al., *Anal. Methods Pestic. Plant Growth Regul.*, 1980, **11**, 125; *Analyst (London)*, 1985, **110**, 765).

MAMMALIAN TOXICOLOGY
EHC 63 (WHO, 1986; a general review of organophosphorus insecticides).
Oral Acute oral LD_{50} for rats 1600–1800, mice 470–620 mg/kg.
Skin and eye Acute percutaneous LD_{50} for rats >5000, male rabbits >500 mg/kg.
Inhalation LC_{50} (1 h) for rats >200 mg a.i. (as 50% EC)/l air.   **NOEL** (90 d) for rats 9 mg/kg diet; (6 mo) for dogs 12 mg/kg diet; (2 y) for rats 6, dogs 10 mg/kg diet.   **ADI** (JMPR) 0.003 mg/kg b.w. [1986].   **Toxicity class** WHO (a.i.) II; EPA (formulation) III

ECOTOXICOLOGY
Birds Eight-day dietary LC_{50} for quail 740 mg/kg diet.   **Fish** LC_{50} (96 h) for carp and guppies 5.5 mg/l, rainbow trout 24 μg/l.   **Bees** Toxic to bees by contact.
Daphnia LC_{50} (48 h) 0.0173 mg/l.   **Algae** EC_{50} (96 h) for *Scenedesmus subspicatus* 2.9 mg/l.

ENVIRONMENTAL FATE
Animals In mammals, following oral administration, etrimfos is rapidly metabolised

and excreted, mainly in the urine. Metabolites include 6-ethoxy-4-hydroxy-=
2-ethylpyrimidine, 2-ethyl-4,6-dihydroxypyrimidine, 6-ethoxy-4-hydroxy-=
2-(1-hydroxyethyl)pyrimidine, and other isomers.   **Plants** In plants, etrimfos
undergoes rapid metabolism to 6-ethoxy-4-hydroxy-2-ethylpyrimidine, and small
amounts of 2-ethyl-4,6-dihydroxypyrimidine and other hydroxylated derivatives.
Soil/Environment DT_{50} 3–10 d (o.c. 2.8%, pH 6.8). The principal degradation
products are 6-ethoxy-4-hydroxy-2-ethylpyrimidine, small amounts of 2-ethyl-=
4,6-dihydroxypyrimidine and other hydroxylated derivatives, and CO_2.

296 famoxadone   *Fungicide*

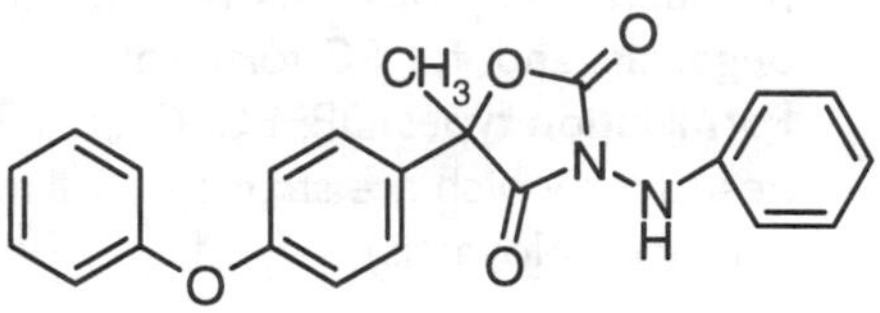

NOMENCLATURE
Common name famoxadone (BSI, pa ISO)
IUPAC name 3-anilino-5-methyl-5-(4-phenoxyphenyl)-1,3-oxazolidine-2,4-dione
Chemical Abstracts name 5-methyl-5-(4-phenoxyphenyl)-3-(phenylamino)-=
2,4-oxazolidinedione
CAS RN *[131807–57–3]*   **Development codes** JE874; DPX-JE874; IN-JE874

PHYSICAL CHEMISTRY
Mol. wt. 374.4   **M.f.** $C_{22}H_{18}N_2O_4$ **M.p.** 140.3–141.8 °C   **V.p.** 6.4 $\times 10^{-4}$ mPa
K_{ow} logP = 4.65 (pH 7)   **Solubility** In water, 52 ppb (20 °C).

COMMERCIALISATION
History Reported by M. M. Joshi & J. A. Sternberg (*Proc. Br. Crop Prot. Conf. – Pests
Dis.*, 1996, **1**, 21). Under development by DuPont Agricultural Products.
Manufacturers Du Pont

APPLICATIONS
Biochemistry Inhibits mitochondrial electron transport, by blocking
ubiquinol:cytochrome c oxidoreductase at complex III.
Mode of action Protectant, translaminar and residual fungicide.   **Uses** Under
development for control of a broad spectrum of plant pathogenic fungi at
50–200 g/ha. Particularly effective against grape downy mildew, potato and
tomato late and early blights, wheat leaf and glume blotch, and barley net blotch.

MAMMALIAN TOXICOLOGY
Oral Acute oral LD_{50} for rats >5000 mg/kg. **Skin and eye** Acute percutaneous LD_{50} for rats >2000 mg/kg. Not a skin or eye irritant (rabbits). **Other** Negative in Ames test; non-teratogenic.

297 famphur

Insecticide

organophosphorus

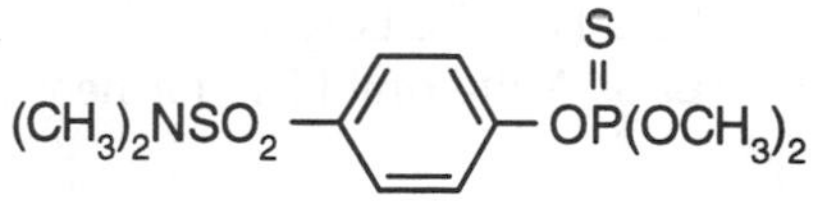

NOMENCLATURE
Common name famphur (ESA)
IUPAC name *O*-4-dimethylsulfamoylphenyl *O,O*-dimethyl phosphorothioate; 4-dimethoxyphosphinothioyloxy-*N,N*-dimethylbenzenesulfonamide
Chemical Abstracts name *O*-[4-[(dimethylamino)sulfonyl]phenyl] *O,O*-dimethyl phosphorothioate
Other names famophos **CAS RN** *[52–85–7]* **Development codes** CL 38 023; AC 38 023 **Official codes** OMS 584

PHYSICAL CHEMISTRY
Mol. wt. 325.3 **M.f.** $C_{10}H_{16}NO_5PS_2$ **Form** Colourless crystalline powder.
M.p. 52.5–53.5 °C **Solubility** Sparingly soluble in water. In aqueous isopropanol (45%) 23 g/kg (20 °C); in xylene 300 g/kg (5 °C); soluble in acetone, carbon tetrachloride, chloroform, cyclohexanone, dichloromethane, toluene; sparingly soluble in aliphatic hydrocarbons. **Stability** Stable at ambient temperatures >19 months.

COMMERCIALISATION
History Introduced by American Cyanamid Co. **Patent** US 3005004
Manufacturers Cyanamid

APPLICATIONS
Biochemistry Cholinesterase inhibitor. **Uses** Controls horn flies (*Haematobia irritans*), grubs (*Hypoderma* spp.) and lice (*Damalinia bovis*, *Bovicola bovis*, *Linognathus vituli*, *Solenopotes capillatus*, *Haematopinus eurysternus*) in cattle; also reindeer grub fly and reindeer nostril fly, and reduces lice infestations.
Formulation types PO; DP. **Compatibility** Incompatible with strong alkali.
Selected tradenames 'Bo-Ana' (Cyanamid); 'Warbexol' (AgrEvo)

ANALYSIS
Product analysis by glc (N. R. Pasarela et al., *J. Agric. Food Chem.*, 1967, **15**, 920).
Residues determined by glc (P. E. Gatterdam et al., ibid., p. 845). Details available from American Cyanamid Co.

MAMMALIAN TOXICOLOGY
EHC 63 (WHO, 1986; a general review of organophosphorus insecticides).
Oral Acute oral LD_{50} for male rats 35, female rats 62, male mice 27 mg tech./kg.
Skin and eye Acute percutaneous LD_{50} for albino rabbits 2730 mg/kg; the PO formulation produces eye and skin irritation. **Inhalation** No deaths occurred in rats exposed 7.5 h at 24 mg/l air. **Toxicity class** WHO (a.i.) Ib

ECOTOXICOLOGY
Birds Acute oral LD_{50} for hens 30 mg/kg.

298 farnesol with nerolidol   *Bioirritant*

pheromone

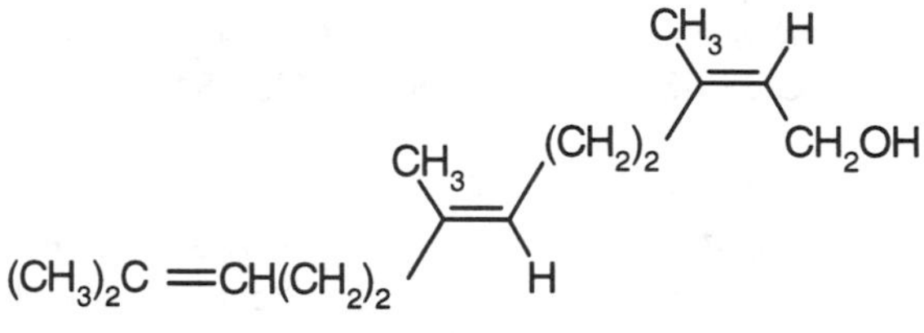

farnesol

NOMENCLATURE
Chemical Abstracts name (*Z,E*)-3,7,11-trimethyl-2,6,10-dodecatrien-1-ol (farnesol); 3,7,11-trimethyl-1,6,10-dodecatrien-3-ol (nerolidol)
CAS RN farnesol [4602–84–0]; nerolidol [7212–44–4], *cis*- isomer [3790–78–1], *trans*- isomer [40716–66–3]

PHYSICAL CHEMISTRY
Composition Mixture of structural isomers. **Mol. wt.** 222.4 **M.f.** $C_{15}H_{26}O$

APPLICATIONS
Mode of action Alarm pheromone of *Tetranychus urticae*. Increases activity of mites with consequent greater exposure to miticide. **Uses** Tank mixed with a miticide, to increase activity. **Formulation types** Controlled release liquid concentrate. **Selected tradenames** 'Stirrup M' (Troy)

fenamiphos   *Nematicide*

organophosphorus

$$CH_3S \longrightarrow \underset{OCH_2CH_3}{\overset{CH_3}{\underset{|}{\overset{|}{\bigcirc}}}} \longrightarrow \underset{OCH_2CH_3}{\overset{O}{\underset{|}{\overset{||}{OPNHCH(CH_3)_2}}}}$$

NOMENCLATURE
Common name fenamiphos (BSI, E-ISO); phénamiphos ((m) F-ISO)
IUPAC name ethyl 4-methylthio-*m*-tolyl isopropylphosphoramidate
Chemical Abstracts name ethyl 3-methyl-4-(methylthio)phenyl (1-methylethyl)=
phosphoramidate
Other names methaphenamiphos   **CAS RN** *[22224–92–6]*   **EEC no.** 244–848–1
Development codes BAY 68 138; SRA 3886

PHYSICAL CHEMISTRY
Mol. wt. 303.4   **M.f.** $C_{13}H_{22}NO_3PS$ **Form** Colourless crystals; (tech., tan waxy
solid).   **M.p.** 49.2 °C; (tech., 46 °C)   **V.p.** 0.12 mPa (20 °C); 4.8 mPa (50 °C)
K_{ow} logP = 3.30 (20 °C)   **Henry** 9.1×10^{-5} Pa m^3 mol^{-1} (20 °C)
S.g./density 1.191 (23 °C)   **Solubility** In water 0.4 g/l (20 °C). In
dichloromethane, isopropanol, toluene >200, hexane 10–20 (all in g/l, 20 °C).
Stability Hydrolysis DT_{50} 1 y (pH 4), 8 y (pH 7), 3 y (pH 9) (22 °C).   **F.p.** c. 200 °C

COMMERCIALISATION
History Nematicide reported by J. H. O'Bannon & A. L. Taylor (*Plant Dis. Rep.*,
1967, **51**, 995) and B. Homeyer (*Pflanzenschutz-Nachr. (Engl. Ed.)*, 1971, **24**, 48).
Introduced by Baychem Corp., Chemagro Division.   **Patent** DE 1121882;
US 2978479   **Manufacturers** Bayer

APPLICATIONS
Biochemistry Cholinesterase inhibitor.   **Mode of action** Systemic nematicide with
contact action. Absorbed by the roots, with translocation to the leaves. Against
soil populations of ectoparasitic nematodes, fenamiphos is more persistent and
more effective than the sulfoxide metabolite; the sulfone metabolite is least
effective. Against plant endoparasitic nematodes, the sulfone is more effective than
the sulfoxide (T. B. Waggoner & A. M. Khasawinah (*Residue Rev.*, 1974, **53**, 79)).
Uses Control of ectoparasitic, endoparasitic, free-living, cyst-forming, and root-
knot nematodes. Used in bananas, pineapples, citrus fruit, pome fruit, stone fruit,
vines, hops, cotton, cocoa, coffee, okra, peanuts, soya beans, cucurbits, tomatoes,
potatoes, vegetables, sugar beet, ornamentals, tobacco, and turf. Has secondary
activity against sucking insects and spider mites.   **Phytotoxicity** Non-phytotoxic
when applied to the soil.   **Formulation types** EC; GR; EW.

MAIN ENTRIES

Mixtures *(fenamiphos +)* isofenphos; carbofuran; disulfoton.
Compatibility Incompatible with alkaline materials.   **Selected tradenames**
'Nemacur' (Bayer)

ANALYSIS
Product analysis by glc; details available from Bayer AG.   **Residue** analysis by glc
(M. A. Luke *et al.*, *J. Assoc. Off. Anal. Chem.*, 1981, **64**, 1187; *Man. Pestic. Residue
Anal.*, 1987, **I**, 6, S16, S19; *Anal. Methods Residues Pestic.*, 1988, Part I, M5, M12; A.
R. C. Hill *et al.*, *Analyst (London)*, 1984, **109**, 483). In **drinking water**, by gc with
FID (*AOAC Methods*, 1995, 991.07); fenamiphos sulfoxide and sulfone can be
determined by lc with u.v. detection (*AOAC Methods*, 1995, 992.14, 10.7.01).
Details available from Bayer AG.

MAMMALIAN TOXICOLOGY
Reviews *Pesticide residues in food – 1987*. FAO Plant Production and Protection
Paper 84, 1987. *Pesticide residues in food – 1987 evaluations. Part II – Toxicology.*
FAO Plant Production and Protection Paper 86/2, 1988.   **EHC** 63 (general
review of organophosphorus insecticides, pub. 1986).   **Oral** Acute oral LD_{50} for
male and female rats *c.* 6, mice, dogs and cats *c.* 10 mg/kg.   **Skin and eye** Acute
percutaneous LD_{50} for rats *c.* 80 mg/kg. Slightly irritant to skin and eyes (rabbits).
Inhalation LC_{50} (4 h) for rats *c.* 0.12 mg/l air (aerosol).   **NOEL** (2 y) for rats 1,
mice 10 mg/kg diet; (12 mo) for dogs 1 mg/kg diet.   **ADI** (JMPR) 0.0005 mg/kg
b.w. [1987].   **Toxicity class** WHO (a.i.) Ia; EPA (formulation) I
EC risk T+ (R24, R28)

ECOTOXICOLOGY
Birds Acute oral LD_{50} for bobwhite quail 0.7–1.6, mallard ducks 0.9–1.2 mg/kg.
Dietary LC_{50} (5 d) for mallard ducks 316, bobwhite quail 38 mg/kg diet.
Fish LC_{50} (96 h) for bluegill sunfish 0.0096, rainbow trout 0.0721 mg/l.
Worms LC_{50} for *Eisenia foetida* 795 mg/kg dry soil (400 EC).   **Daphnia** LC_{50}
(48 h) 0.0019 mg/l.   **Algae** E_rC_{50} for *Scenedesmus subspicatus* 11 mg/l.

ENVIRONMENTAL FATE
Animals In mammals, following oral administration, there is rapid metabolism
involving oxidation to the sulfoxide and sulfone analogues, followed by subsequent
hydrolysis, conjugation and excretion via the urine; some *N*-dealkylation also
occurs (T. B. Waggoner & A. M. Khasawinah, *Residue Rev.*, 1974, **53**, 79).
Plants Degradation is by thiooxidation and hydrolysis. The major metabolites are
fenamiphos sulfoxide and fenamiphos sulfone, *idem, ibid.*   **Soil/Environment** No
effect on soil bacteria. Readily degradable in water, degradable on soil surfaces.
Duration of activity in soil is *c.* 4 months. Based on K_{oc} values and leaching studies,
fenamiphos can be classified as a compound with low mobility. Soil DT_{50} (aerobic
and anaerobic) several weeks. The major degradation products are fenamiphos
sulfoxide and fenamiphos sulfone and their phenols.

pyrimidinyl carbinol; DMI fungicide

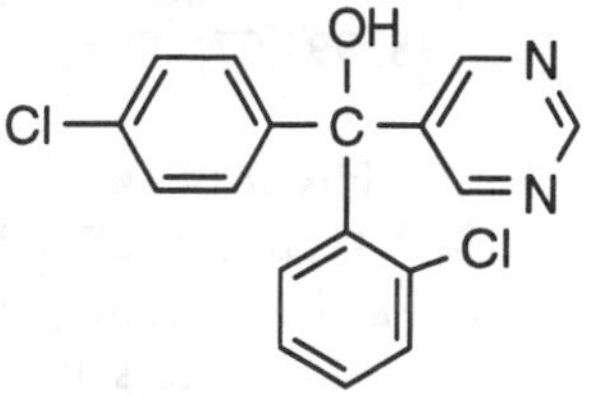

NOMENCLATURE
Common name fenarimol (BSI, E-ISO, (m) F-ISO, ANSI)
IUPAC name (±)-2,4'-dichloro-α-(pyrimidin-5-yl)benzhydryl alcohol
Chemical Abstracts name (±)-α-(2-chlorophenyl)-α-(4-chlorophenyl)-=
5-pyrimidinemethanol
CAS RN *[60168–88–9]* unstated stereochemistry   **Development codes** EL-222

PHYSICAL CHEMISTRY
Mol. wt. 331.2   **M.f.** $C_{17}H_{12}Cl_2N_2O$ **Form** Off-white crystals.   **M.p.** 117–119 °C
V.p. 0.065 mPa (25 °C) by vapour pressure balance   **K_{ow}** logP = 3.69 (pH 7,
25 °C)   **Solubility** In water 13.7 mg/l (pH 7, 25 °C). In acetone >250, methanol
125, xylene 50 (all in g/l, 25 °C). Readily soluble in most organic solvents, but only
slightly soluble in hexane.   **Stability** Decomposed rapidly by sunlight. Stable at
52 °C (highest storage temperature tested), and for 28 d at 52 °C and pH 3,
6 and 9.

COMMERCIALISATION
History Fungicide reported by I. F. Brown et al. (*Proc. Am. Phytopathol. Soc.,* 1975,
2, 31). Introduced in Lebanon (1977) by Eli Lilly & Co. (now DowElanco).
Patent GB 1218623   **Manufacturers** DowElanco

APPLICATIONS
Biochemistry Ergosterol biosynthesis inhibitor (demethylation inhibitor).
Mode of action Systemic fungicide with protective, curative, and eradicant action.
Translocated acropetally within the plant.   **Uses** Control of powdery mildews in
pome fruit, stone fruit, strawberries, vines, cucurbits, aubergines, peppers,
tomatoes, roses and other ornamentals, and beet; American gooseberry mildew in
gooseberries and currants; scab on pome fruit; and dollar spot, brown patch and
snow mould of turf.   **Phytotoxicity** Non-phytotoxic when used as directed. If
used in excessive amounts, abnormal leaf development and darker green
colouration may result.   **Formulation types** WP; EC; SC.   **Mixtures** *(fenarimol +)*
dodine; sulfur.   **Selected tradenames** 'Rubigan' (DowElanco)

ANALYSIS
Product analysis by glc with FID (E. W. Day & O. D. Decker, *Anal. Methods Pestic. Plant Growth Regul.*, 1984, **13**, 173). **Residues** in soil and plant tissue determined by glc with ECD (*idem, ibid.*). In **drinking water**, by gc with FID (*AOAC Methods*, 1995, 991.07). Details available from DowElanco.

MAMMALIAN TOXICOLOGY
Oral Acute oral LD_{50} for rats 2500, mice 4500, dogs >200 mg/kg.
Skin and eye Acute percutaneous LD_{50} for rabbits >2000 mg/kg. Mild eye irritant; non-irritating to skin (rabbits). **Inhalation** LC_{50} (1 h) – no adverse effect seen when rats exposed to 2.04 mg tech./l air. **NOEL** In 2 y feeding trials, rats receiving 25 mg/kg diet and mice receiving 600 mg/kg diet showed no ill-effects. **ADI** (JMPR) 0.01 mg/kg b.w. [1995]. **Toxicity class** WHO (a.i.) III (Table 5); EPA (formulation) III

ECOTOXICOLOGY
Birds Acute oral LD_{50} for bobwhite quail >2000 mg/kg. **Fish** LC_{50} (96 h) for rainbow trout 1.8, bluegill sunfish 3.1 mg/l. **Bees** Not toxic to bees; LC_{50} (contact) >1 g/l. **Worms** Not toxic to earthworms. **Daphnia** LC_{50} (48 h) 6.8 μg/l. **Algae** NOEC for *Scenedesmus subspicatus* 0.59 mg/l.

ENVIRONMENTAL FATE
Animals In mammals, following oral administration, fenarimol is rapidly excreted.
Plants Forms numerous photodegradation products. **Soil/Environment** DT_{50} >365 d under aerobic conditions in soil (28% sand, 14.7% clay, 57.3% silt, 2.3% o.m., pH 6.1). Microbial degradation is accelerated by light. K_{oc} 1.5–11.9 depending on soil type. K_d not determinable because of strong binding.

301   fenazaquin   *Acaricide*

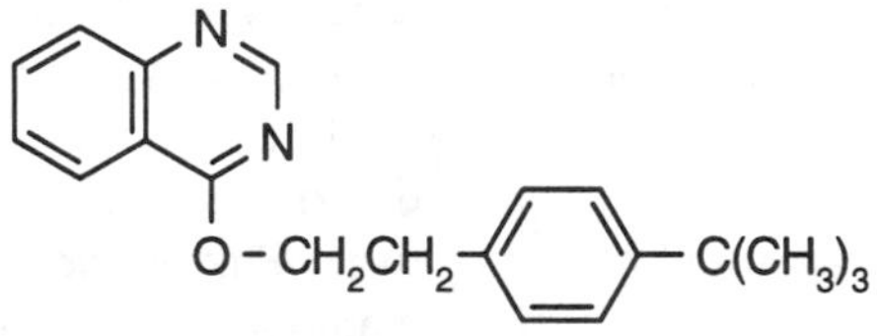

NOMENCLATURE
Common name fenazaquin (BSI, ANSI, draft E-ISO)
IUPAC name 4-*tert*-butylphenethyl quinazolin-4-yl ether

Chemical Abstracts name 4-[[4-(1,1-dimethylethyl)phenyl]ethoxy]quinazoline
CAS RN *[120928–09–8]* **Development codes** EL-436; 193136; XDE 436; DE 436

PHYSICAL CHEMISTRY
Mol. wt. 306.4 **M.f.** $C_{20}H_{22}N_2O$ **Form** Colourless crystals. **M.p.** 77.5–80 °C
V.p. 3.4×10^{-3} mPa (25 °C) **K_{ow}** logP = 5.51 **S.g./density** 1.16 **Solubility** In
water 0.22 mg/l (20 °C). In chloroform >500, toluene 500, acetone 400,
methanol 50, isopropanol 50, acetonitrile, hexane 33 (all in g/l, 20 °C).
Stability DT_{50} of aqueous solution exposed to sunlight 15 d (pH 7, 25 °C).

COMMERCIALISATION
History Reported by C. Longhurst *et al.* (*Proc. Br. Crop Prot. Conf. – Pests Dis.,*
1992, **1**, 51). Introduced by DowElanco. **Manufacturers** DowElanco

APPLICATIONS
Biochemistry Affects metabolism, inhibiting the mitochondrial electron transport
chain by binding with Complex I at Co-enzyme site Q. **Mode of action** Contact
acaricide with good knockdown activity on motile forms, as well as true ovicidal
activity, preventing eclosion of mite eggs. **Uses** Acaricide effective at 10–25 g/hl
against *Eutetranychus*, *Panonychus* and *Tetranychus* spp. and *Brevipalpus phoenicis* in
almonds, apples, citrus, cotton, grapes and ornamentals. **Formulation types** EC;
SC. **Selected tradenames** 'Magister' (DowElanco)

ANALYSIS
Details from DowElanco.

MAMMALIAN TOXICOLOGY
Oral Acute oral LD_{50} for male rats 134, female rats 138, male mice 2449, female
mice 1480 mg/kg. **Skin and eye** Acute percutaneous LD_{50} for rabbits >5000
mg/kg. Slightly irritating to eyes, non-irritating to skin. Not a skin sensitiser.
Inhalation LC_{50} (4 h) for male and female rats 1.9 mg/l air. **NOEL** 0.5 mg/kg b.w.
ADI 0.005 mg/kg. **Other** No evidence of mutagenicity, teratogenicity or
carcinogenicity. **Toxicity class** WHO (a.i.) II

ECOTOXICOLOGY
Birds Acute oral LD_{50} for bobwhite quail 1747, mallard ducks >2000 mg/kg.
Acute dietary LD_{50} for bobwhite quail and mallard ducks >5000 ppm. **Fish** LC_{50}
(96 h) for bluegill sunfish 34.1, trout 3.8 µg/l. **Bees** LD_{50} (contact) 8.18 µg/bee.
Daphnia LC_{50} (48 h) 4.1 µg/l.

ENVIRONMENTAL FATE
Soil/Environment Soil DT_{50} c. 45 d. K_{oc} 15 800 (sandy loam), 42 100 (clay loam);
K_d 54 (sand), 487 (clay loam).

azole

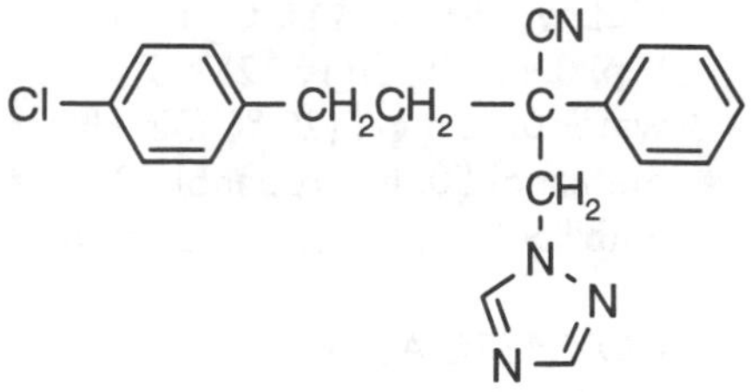

NOMENCLATURE
Common name fenbuconazole (BSI, draft E-ISO, ANSI)
IUPAC name 4-(4-chlorophenyl)-2-phenyl-2-(1H-1,2,4-triazol-1-ylmethyl)=
butyronitrile
Chemical Abstracts name α-[2-(4-chlorophenyl)ethyl]-α-phenyl-1H-=
1,2,4-triazole-1-propanenitrile
CAS RN *[114369–43–6]* unstated stereochemistry
Development codes RH-7592; RH-57592

PHYSICAL CHEMISTRY
Mol. wt. 336.8 **M.f.** $C_{19}H_{17}ClN_4$ **Form** Colourless crystals. **M.p.** 124–126 °C
V.p. 0.005 mPa (20 °C) **K$_{ow}$** logP = 3.23 **Solubility** In water at 25 °C, 0.2 mg/l.
Soluble in common organic solvents such as ketones, esters, alcohols, and
aromatic hydrocarbons. Insoluble in aliphatic hydrocarbons. **Stability** Stable to
hydrolysis in the dark, DT$_{50}$ >2210 d (pH 5), 3740 d (pH 7), 1370 d (pH 9).
Thermostable up to 300 °C.

COMMERCIALISATION
History Fungicide reported by D. Briant *et al.* (*Proc. 1988 Br. Crop Prot. Conf. –
Pests Dis.,* **1**, 33). Introduced by Rohm & Haas Co. **Manufacturers** Rohm & Haas

APPLICATIONS
Biochemistry Steroid demethylation inhibitor. **Mode of action** Systemic fungicide
with protectant, curative and eradicant properties. Translocated upwards in the
plant. **Uses** Control of *Septoria, Puccinia* rusts, bunt, smut and *Rhynchosporium
secalis* on cereals; powdery mildew and scab on pome fruit; brown rot and
powdery mildew on stone fruit; powdery mildew, black rot and grey mould on
vines; rust on beans; and beet leaf spot on sugar beet. Also for the control of a
wide range of diseases on field crops, rice, bananas, tree nuts, vegetables and
ornamentals. As well as its use by foliar treatment, it can also be used in seed and
post-harvest treatments. **Formulation types** EC; SC. **Mixtures** *(fenbuconazole +)*
fenpropidin; prochloraz; carbendazim + prochloraz. **Selected tradenames** 'Indar'
(Rohm & Haas); 'Troika' (mixture) (AgrEvo)

ANALYSIS
Details from Rohm & Haas.

MAMMALIAN TOXICOLOGY
Oral Acute oral LD_{50} for rats >2000 mg/kg. **Skin and eye** Acute percutaneous LD_{50} for rats >5000 mg/kg. Non-irritating to eyes and skin (tech.); severe irritant to eyes and skin (EC formulation) (rabbits). **Inhalation** LC_{50} (4 h) for rats >2.1 mg/l air (for tech.). **NOEL** (3 mo) for rats 20, mice 60, dogs 100 mg/kg daily. In rats, there were no observable teratogenic or embryotoxic effects at 30 mg/kg b.w. daily. **Other** Non-mutagenic in a variety of tests. **Toxicity class** EPA (formulation) II (tech., III)

ECOTOXICOLOGY
Birds Eight-day dietary LC_{50} for bobwhite quail 4050, mallard ducks 2110 mg/kg diet. LC_{50} (21 d) for bobwhite quail 2150 mg/kg daily. **Fish** LC_{50} (96 h) for bluegill sunfish 0.68 mg/l (tech.). **Bees** LC_{50} (96 h, dust exposure) >0.29 mg/bee.

ENVIRONMENTAL FATE
Soil/Environment Soil adsorption K_{oc} 2100–9000 (clay, loam, sand, sandy loam, silty clay loam).

303 fenbutatin oxide   *Acaricide*

organotin

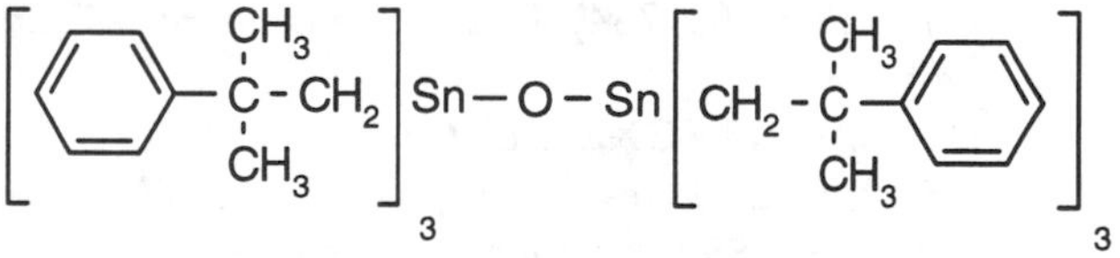

NOMENCLATURE
Common name fenbutatin oxide (BSI, E-ISO, JMAF); fenbutatin-oxyde ((m) F-ISO)
IUPAC name bis[tris(2-methyl-2-phenylpropyl)tin] oxide
Chemical Abstracts name hexakis(2-methyl-2-phenylpropyl)distannoxane
CAS RN [13356–08–6] **EEC no.** 236–407–7 **Development codes** SD 14 114
Official codes ENT 27 738

PHYSICAL CHEMISTRY
Mol. wt. 1052.7 **M.f.** $C_{60}H_{78}OSn_2$ **Form** Colourless crystals. **M.p.** 138–139 °C

(tech.)   **V.p.** 8.5×10^{-5} mPa (extrapolated to 20 °C) K_{ow} logP = 5.2
S.g./density 1290–1330 kg/m^3 (20 °C)   **Solubility** In water 0.005 mg/l (23 °C). In acetone 6, benzene 140, dichloromethane 380 (all in g/l, 23 °C). Very slightly soluble in aliphatic hydrocarbons and mineral oils.   **Stability** Extremely stable to heat, light, and atmospheric oxygen. Water causes conversion of fenbutatin oxide to tris(2-methyl-2-phenylpropyl)tin hydroxide which is reconverted to the parent compound slowly at room temperature and rapidly at 98 °C.   **F.p.** Not autoflammable but tech. will explode in a dust cloud if ignited.

COMMERCIALISATION
History Introduced in USA by Shell Chemical Co. (now DuPont Agricultural Products) and elsewhere by Shell International Chemical Company Ltd (now American Cyanamid Co.).   **Patent** US 3657451   **Manufacturers** Cyanamid; Du Pont

APPLICATIONS
Mode of action Non-systemic acaricide with contact and stomach action.
Uses Control of all motile stages of a wide range of phytophagous mites on pome fruit, stone fruit, citrus fruit, soft fruit, vines, bananas, cucurbits, ornamentals, and glasshouse crops. Gives long residual control.   **Phytotoxicity** Phytotoxic to tangerines, tangelos, and some varieties of grapefruit. Otherwise, non-phytotoxic when used as directed.   **Formulation types** WP; SC.
Selected tradenames 'Osadan' (Japan only) (Cyanamid); 'Torque' (Cyanamid); 'Vendex' (Du Pont, Cyanamid)

ANALYSIS
Product analysis by potentiometric non-aqueous titrimetry (*CIPAC Handbook,* 1988, **D**, 77).   **Residues** determined by glc with ECD or FPD of suitable derivative (*Anal. Methods Pestic. Plant Growth Regul.,* 1978, **10**, 139; *Methodensammlung Rückstandsanal. Pflanzenschutzmitteln,* 1987, S24; M. Sano et al., *J. Assoc. Off. Anal. Chem.,* 1979. **62,** 764). Details available from Cyanamid.

MAMMALIAN TOXICOLOGY
Reviews *Pesticide residues in food – 1992.* FAO Plant Production and Protection Paper, 116, 1993. *Pesticide residues in food – 1992 evaluations. Part II – Toxicology.* World Health Organisation, WHO/PCS/93.34.   **Oral** Acute oral LD_{50} for rats 2631, mice 1450, dogs >1500 mg/kg.   **Skin and eye** Acute percutaneous LD_{50} for rabbits >2000, rats >1000 mg/kg. Irritating to skin and severely irritating to eyes.
Inhalation LC_{50} 0.23 mg a.i. as dust/l.   **NOEL** (2 y) for rats 100 mg/kg diet, for dogs 15 mg/kg daily.   **ADI** (JMPR) 0.03 mg/kg b.w. [1992].   **Toxicity class** WHO (a.i.) III (Table 5); EPA (formulation) III   **EC risk** Xn (R21); Xi (R36/38)

ECOTOXICOLOGY
Birds Eight-day dietary LC_{50} for bobwhite quail 5065 mg/kg diet.   **Fish** LC_{50} (48 h)

for rainbow trout 0.27 mg a.i./l (as WP formulation). **Bees** Acute oral LD_{50} >0.1 mg/bee. **Other beneficial spp.** No adverse effects to a wide range of predatory and parasitic arthropods. **Daphnia** LC_{50} (24 h) 0.05–0.08 mg/l.

ENVIRONMENTAL FATE
Soil/Environment In soil, it is metabolised to dihydroxy-bis(2-methyl-2-phenyl= propyl)stannane and 2-methyl-2-phenylpropyl stannonic acid, presumably ultimately forming tin oxide/hydroxide. In tests in commercial use, there was minimal movement of fenbutatin oxide or its metabolites out of the top 30 cm of soil (A. Gray et al., Pestic. Sci., **43**, 295–302 (1995)).

304 fenclorim *Herbicide safener*

pyrimidine

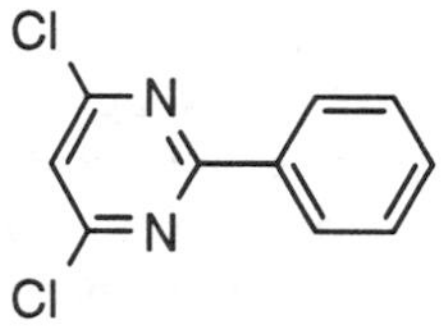

NOMENCLATURE
Common name fenclorim (BSI, draft E-ISO); fenclorime ((f) draft F-ISO)
IUPAC name 4,6-dichloro-2-phenylpyrimidine
Chemical Abstracts name 4,6-dichloro-2-phenylpyrimidine
CAS RN [3740–92–9]   **Development codes** CGA 123 407

PHYSICAL CHEMISTRY
Mol. wt. 225.1   **M.f.** $C_{10}H_6Cl_2N_2$ **Form** Colourless crystals.   **M.p.** 96.9 °C
V.p. 12 mPa (20 °C) K_{ow} logP = 4.17   **Henry** 1.1 Pa m^3 mol^{-1} (calc.)
S.g./density 1.5 g/cm^3 (20 °C)   **Solubility** In water 2.5 mg/l (20 °C). In acetone 14%, cyclohexanone 28%, dichloromethane 40%, toluene 35%, xylene 30%, hexane 4%, methanol 1.9%, n-octanol 4.2%, and isopropanol 1.8%.   **Stability** Stable in neutral, acidic and weakly alkaline media. Stable up to 400 °C.

COMMERCIALISATION
History Herbicide safener reported by J. Rufener & M. Quadranti (*Proc. 10th Int. Congr. Plant Prot.*, 1983, **1**, 332). Introduced by Ciba-Geigy Ltd. (now Novartis Crop Protection AG).   **Patent** EP 55693; US 4493726   **Manufacturers** Novartis

APPLICATIONS
Biochemistry Fenclorim influences the metabolism of pretilachlor in rice, which

proceeds via glutathione conjugation. **Mode of action** Herbicide safener. Taken up rapidly by the roots of germinating rice seeds and protects them from potential damage caused by pretilachlor. **Uses** Used as a safener for pretilachlor in direct-seeded rice. **Formulation types** EC. **Mixtures** *(fenclorim +)* pretilachlor. **Selected tradenames** 'Sofit' (mixture) (Novartis)

ANALYSIS
Residues determined by glc.

MAMMALIAN TOXICOLOGY
Oral Acute oral LD_{50} for rats >5000 mg/kg. **Skin and eye** Acute percutaneous LD_{50} for rats >2000 mg/kg. Non-irritating to skin and eyes (rabbits). Sensitising to skin (guinea pigs). **Inhalation** LC_{50} (4 h) for rats 2.9 mg/l air. **NOEL** (2 y) for rats 10.4, mice 113 mg/kg b.w. daily; (1 y) for dogs 10.0 mg/kg b.w. daily; (90 d) for rats 100 mg/kg b.w. daily. **Toxicity class** WHO (a.i.) III (Table 5) **EC risk** R20; R43

ECOTOXICOLOGY
Birds Acute oral LD_{50} for Japanese quail >500 mg/kg. LC_{50} for Japanese quail >10 000 ppm. **Fish** LC_{50} (96 h) for rainbow trout 0.6, catfish 1.5 mg/l. **Bees** Non-toxic to honeybees; LD_{50} (ingestion) >20 μg/bee, (contact) >1000 ppm. **Worms** LC_{50} (14 d) for earthworms >62.5 mg/kg. **Daphnia** LC_{50} (48 h) 2.2 ppm. **Algae** IC_{50} (5 d) for *Scenedesmus subspicatus* 20.9 ppm.

ENVIRONMENTAL FATE
Animals Rapidly absorbed, metabolised to polar compounds and excreted. No accumulation of residues in tissues. **Plants** Readily metabolised to polar compounds. At harvest time, residues are negligible. **Soil/Environment** In soil, DT_{50} 17–35 d. Fenclorim and its metabolites are strongly adsorbed onto soil.

305 fenfuram   *Fungicide*

carboxamide

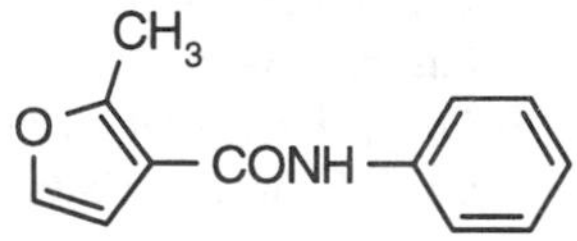

NOMENCLATURE
Common name fenfuram (BSI, E-ISO); fenfurame ((m) F-ISO)

IUPAC name 2-methyl-3-furanilide
Chemical Abstracts name 2-methyl-*N*-phenyl-3-furancarboxamide
CAS RN *[24691–80–3]*   **Development codes** WL 22 361

PHYSICAL CHEMISTRY
Mol. wt. 201.2   **M.f.** $C_{12}H_{11}NO_2$ **Form** Colourless crystals; (tech., cream solid).
M.p. 109–110 °C (tech.)   **V.p.** 0.020 mPa (extrapolated to 20 °C)   **Solubility** In
water 0.1 g/l (20 °C). In acetone 300, cyclohexanone 340, methanol 145, xylene
20 (all in g/l, 20 °C).   **Stability** Stable to heat and light. Stable in neutral media,
but hydrolysed by strong acids and alkalis.

COMMERCIALISATION
History Fungicide discovered by Shell Research Ltd and developed by KenoGard
VT AB (now Rhône-Poulenc Agrochimie).   **Patent** GB 1215066 to Shell
Manufacturers Rhône-Poulenc

APPLICATIONS
Uses Control of bunts and smuts (*Tilletia* and *Ustilago* spp.) in cereals, when
applied as a seed treatment.   **Formulation types** LS; DS.   **Mixtures** *(fenfuram +)*
guazatine acetates; guazatine acetates + imazalil; anthraquinone + imazalil +
lindane; guazatine acetates + imazalil + lindane.   **Selected tradenames** 'Pano-ram'
(Rhône-Poulenc)

ANALYSIS
Product analysis by glc with FID.   **Residues** in barley and wheat determined by glc.

MAMMALIAN TOXICOLOGY
Oral Acute oral LD_{50} for rats 12 900, cats 2450 mg/kg.   **Inhalation** LC_{50} (4 h) for
rats >10.3 mg/l air.   **NOEL** (2 y) for rats 10 mg/kg daily; (90 d) for dogs
300 mg/kg diet.   **Other** Acute i.p. LD_{50} for rats 1490 mg/kg. Non-mutagenic in
the Ames test.   **Toxicity class** WHO (a.i.) III (Table 5); EPA (formulation) IV

ECOTOXICOLOGY
Fish LC_{50} for guppies 11.0 mg/l.   **Bees** Not toxic to bees when used as directed.

ENVIRONMENTAL FATE
Animals In rats, following oral administration, up to 83% is eliminated within
16 hours, principally in the urine.   **Soil/Environment** DT_{50} in soil *c.* 42 d.

organophosphorus

O_2N —⟨CH₃ ring⟩— $OP(OCH_3)_2$ with S double bond

NOMENCLATURE
Common name fenitrothion (BSI, E-ISO, (*m*) F-ISO, ESA, BAN); MEP (JMAF)
IUPAC name *O,O*-dimethyl *O*-4-nitro-*m*-tolyl phosphorothioate
Chemical Abstracts name *O,O*-dimethyl *O*-(3-methyl-4-nitrophenyl) phosphorothioate
CAS RN *[122–14–5]*   **EEC no.** 204–524–2   **Development codes** Bayer 41831; S-5660; S-1102A (all to Bayer AG); AC 47 300 (American Cyanamid Co.)
Official codes OMS 43; OMS 223; ENT 25 715

PHYSICAL CHEMISTRY
Mol. wt. 277.2   **M.f.** $C_9H_{12}NO_5PS$ **Form** Yellow-brown liquid with a phenolic odour.   **M.p.** 3.4 °C   **B.p.** 164 °C/1.3×10^2 Pa (pure a.i.)   **V.p.** 15 mPa (20 °C)
K_{ow} logP = 3.5 (20 °C)   **S.g./density** 1.328 (20 °C)   **Solubility** In water 21 mg/l (20 °C). Readily soluble in alcohols, esters, ketones, aromatic hydrocarbons, and chlorinated hydrocarbons. In hexane 24, isopropanol 138 (both in g/l, 20 °C).
Stability Relatively stable to hydrolysis under normal conditions: DT_{50} (estimated) 108.8 d (pH 4), 84.3 d (pH 7), 75 d (pH 9) (22 °C).   **F.p.** 157 °C

COMMERCIALISATION
History Insecticide reported by Y. Nishizawa *et al.* (*Bull. Agric. Chem. Soc. Jpn.*, 1960, **24**, 744; *Agric. Biol. Chem.*, 1961, **25**, 605). Introduced by Sumitomo Chemical Co., Ltd and, independently, by Bayer AG and later by American Cyanamid Co. (who no longer manufacture or market it).   **Patent** BE 594669 to Sumitomo; BE 596091 to Bayer   **Manufacturers** Cheminova; Rallis; Shenzhen Jiangshan; Sumitomo

APPLICATIONS
Biochemistry Cholinesterase inhibitor.   **Mode of action** Non-systemic insecticide with contact and stomach action.   **Uses** Control of chewing, sucking, and boring insects in cereals, soft fruit, tropical fruit, vines, rice, sugar cane, vegetables, turf, and forestry. Also used as a public health insecticide for control of household insects (flies, cockroaches, and other insects) by application to breeding sites; for control of flies in animal houses; for control of stored product insect pests; for control of mosquito larvae (as a vector control agent for malaria); and for control of locusts.   **Phytotoxicity** Non-phytotoxic when used as recommended. Cotton, brassicas, and some fruits may be injured by high rates of application. Russetting is

possible with some apple varieties.   **Formulation types** EC; WP; GR; DP; UL; AE.
Mixtures *(fenitrothion +)* beta-cyfluthrin; trichlorfon; formothion; tetramethrin;
fenpropathrin; fenobucarb; fenvalerate; malathion; piperonyl butoxide +
tetramethrin; permethrin + resmethrin; deltamethrin + endosulfan + profenofos.
Compatibility Incompatible with alkaline compounds.
Selected tradenames 'Folithion' (Bayer); 'Sumithion' (Sumitomo); 'Cekutrotion'
(Cequisa); 'Dicofen' (PBI); 'Farmathion' (Sanonda); 'Fentron' (Efthymiadis);
'Shaminliulin' (Shenzhen Jiangshan)

ANALYSIS
Product analysis by glc with FID (*CIPAC Handbook*, 1985, **1C**, 2117; *AOAC Methods*,
1995, 989.02) or by the Averell & Norris method (*CIPAC Handbook*, 1980, **1A**,
1255; *FAO Specification* (CP/62)).   **Residues** determined by glc (*Pestic. Anal. Man.*,
1979, **I**, 201-A, 201-G, 201-H, 201-I; *Anal. Methods Residues Pestic.*, 1988, Part I,
M2, M5; *Analyst (London)*, 1980, **105**, 515; 1985, **110**, 765). Methods for the
determination of residues are available from Bayer.

MAMMALIAN TOXICOLOGY
Reviews *Pesticide residues in food – 1988*. FAO Plant Production and Protection
Paper 92, 1988. *Pesticide residues in food – 1988 evaluations. Part II – Toxicology*.
FAO Plant Production and Protection Paper 93/2, 1989.   **EHC** 133 (WHO,
1992), 63 (WHO, 1986; a general review of organophosphorus insecticides).
Oral Acute oral LD_{50} for rats c. 250 mg/kg.   **Skin and eye** Acute percutaneous
LD_{50} for rats c. 2500 mg/kg. Not a skin or eye irritant (rabbits).   **Inhalation** LC_{50}
(4h) for rats >1.2 mg/l air (aerosol).   **NOEL** (2 y) for rats and mice 10 mg/kg
diet; (1 y) for dogs 50 mg/kg diet.   **ADI** (JMPR) 0.005 mg/kg b.w. [1988].
Other Not carcinogenic, mutagenic, embryotoxic or teratogenic.
Toxicity class WHO (a.i.) II; EPA (formulation) II   **EC risk** Xn (R22)

ECOTOXICOLOGY
Birds Acute oral LD_{50} for quail 23.6, mallard ducks 1190 mg/kg.   **Fish** LC_{50} (48 h)
for carp 4.1 mg/l. LC_{50} (96 h) for bluegill sunfish 3.8, brook trout 1.7 mg/l.
Bees Toxic to bees.   **Other beneficial spp.** Highly toxic to non-target arthropods.

ENVIRONMENTAL FATE
Animals Fenitrothion is rapidly excreted in the urine and faeces. After 3 d, c. 90%
has been excreted by rats, mice and rabbits. The most important metabolites are
dimethylfenitrooxon and 3-methyl-4-nitrophenol.   **Plants** Fenitrothion applied in
the forest (balsam fir and spruce foliage) degraded with DT_{50} 4 d; 70–85% was
degraded within 2 w. Similar results were obtained in conifer foliage and cocoa
trees. Major metabolites are 3-methyl-4-nitrophenol, the oxygen analogue and
their decomposition products desmethylfenitrothion, dimethylphosphorothionic
acid and phosphorothionic acid.
Soil/Environment DT_{50} 12–28 d under upland conditions, 4–20 d under

submerged conditions. The major metabolites under upland conditions are 3-methyl-4-nitrophenol and CO_2, whereas, under submerged conditions, the major decomposition product was aminofenitrothion.

307 fenobucarb

Insecticide

carbamate

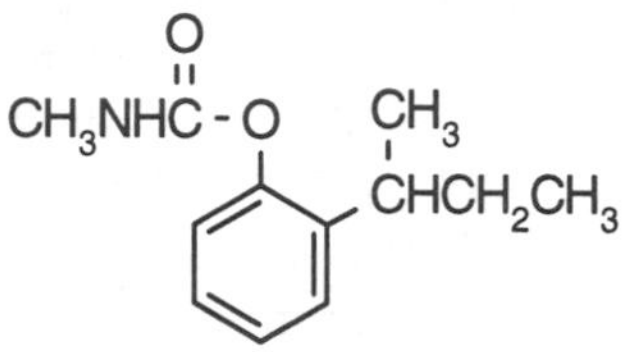

NOMENCLATURE
Common name fenobucarb (BSI, draft E-ISO, (*m*) draft F-ISO); BPMC (JMAF, Thailand)
IUPAC name 2-*sec*-butylphenyl methylcarbamate
Chemical Abstracts name 2-(1-methylpropyl)phenyl methylcarbamate
CAS RN *[3766–81–2]*   **Development codes** Bayer 41 367c   **Official codes** OMS 313

PHYSICAL CHEMISTRY
Composition Tech. is >97%.   **Mol. wt.** 207.3   **M.f.** $C_{12}H_{17}NO_2$ **Form** Colourless solid; (tech., colourless to yellow-brown liquid or solid).   **M.p.** 31–32 °C; tech., 26.5–31 °C   **B.p.** 112–113 °C/0.02 mmHg   **V.p.** 1.6 mPa (20 °C)
K_{ow} logP = 2.79   **S.g./density** 1.035 at 30 °C   **Solubility** In water 420 mg/l (20 °C), 610 mg/l (30 °C). In acetone, benzene, chloroform, xylene, toluene >1 (all in kg/kg at room temperature).   **Stability** Stable under normal storage conditions. Stable to light. Hydrolysed by acids and alkalis; (20 °C) DT_{50} >28 d (buffer at pH 2), 16.9 d (pH 9), 2.06 d (pH 10).   **F.p.** 142 °C (closed system)

COMMERCIALISATION
History Insecticide reported by R. L. Metcalf *et al.* (*J. Econ. Entomol.*, 1962, **55**, 889). Introduced by Sumitomo Chemical Co., Ltd, by Kumiai Chemical Industry Co., Ltd and Mitsubishi Chemical Industries (now Mitsubishi Chemical Corp.) jointly, and by Bayer AG (who no longer manufacture or market it).
Manufacturers Chunhu; Jin Hung; Mitsubishi Chemical; Kuo Ching; Shenzhen Jiangshan; Sinon; Sumitomo; Taiwan Tainan Giant

APPLICATIONS
Biochemistry Cholinesterase inhibitor.   **Mode of action** Non-systemic insecticide

with contact action. **Uses** Control of leafhoppers, planthoppers, thrips, and weevils on rice, tea, sugar cane, wheat, cucurbits, aubergines, and capsicums. Control of bollworms and aphids on cotton. **Formulation types** EC; DP; UL. **Mixtures** *(fenobucarb +)* fenitrothion; fenthion; cartap hydrochloride; chlorpyrifos-methyl. **Selected tradenames** 'Bassa' (Mitsubishi Kagaku, Kumiai); 'Osbac' (Sumitomo)

ANALYSIS
Product analysis by u.v. spectrometry of a derivative or by hplc (*CIPAC Handbook*, 1988, **D,** 86). Details available from Sumitomo Chemical Co. Ltd.

MAMMALIAN TOXICOLOGY
EHC 64 (WHO, 1986; a review of carbamate insecticides in general). **Oral** Acute oral LD_{50} for male rats 623, female rats 657 mg/kg. **Skin and eye** Acute percutaneous LD_{50} for rabbits 10 250 mg/kg. **Inhalation** LC_{50} (4 h) for rats >0.366 mg tech./l air. **NOEL** (2 y) for rats 4.1 mg/kg b.w. daily (100 mg/kg diet). **Toxicity class** WHO (a.i.) II; EPA (formulation) II

ECOTOXICOLOGY
Birds Acute oral LD_{50} for mallard ducks 323 mg/kg. Five-day dietary LC_{50} for mallard ducks >5500, bobwhite quail 5417 mg/kg diet. **Fish** LC_{50} (48 h) for carp 16 mg/l. **Daphnia** LC_{50} (3 h) 0.32 mg/l.

ENVIRONMENTAL FATE
Animals 2-(2-Hydroxy-1-methylpropyl)-phenyl *N*-methylcarbamate is a metabolite. Metabolism of carbamate insecticides is reviewed (M. Cool & C. K. Jankowski in "*Insecticides*"). **Plants** As for animals. **Soil/Environment** Soil K_{om} 125 (Utsunomia soil, 5.2% o.m.), 661 (Niigata soil, 1.8% o.m.); DT_{50} 6–30 d and 6–14 d under paddy and upland conditions, respectively.

308 fenothiocarb — *Acaricide*

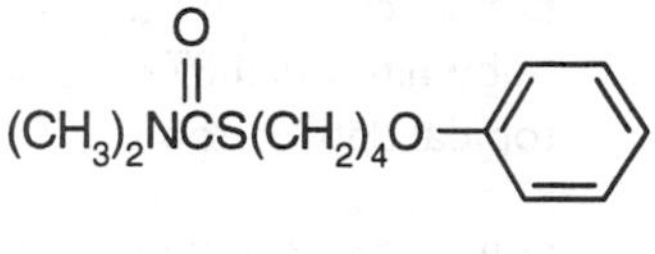

$$\text{(CH}_3)_2\overset{\displaystyle\text{O}}{\overset{\displaystyle\|}{\text{N}}}\text{CS(CH}_2)_4\text{O}-\!\!\bigcirc$$

NOMENCLATURE
Common name fenothiocarb (BSI, draft E-ISO, (*m*) draft F-ISO, JMAF)
IUPAC name *S*-4-phenoxybutyl dimethylthiocarbamate

Chemical Abstracts name S-(4-phenoxybutyl) dimethylcarbamothioate
CAS RN *[62850–32–2]*   **Development codes** KCO-3001; B1–5452 (both Kumiai)

PHYSICAL CHEMISTRY
Composition Tech. is >96%.   **Mol. wt.** 253.4   **M.f.** $C_{13}H_{19}NO_2S$
Form Colourless crystals.   **M.p.** 40–41 °C   **B.p.** 155 °C/0.02 mmHg
V.p. 0.166 mPa (23 °C)   **K$_{ow}$** logP = 3.28 (20 °C)   **S.g./density** 1.211 (20 °C)
Solubility In water 30 mg/l (20 °C). In cyclohexanone 3800, acetonitrile 3120,
acetone 2530, xylene 2464, methanol 1426, kerosene 80, hexane 66 (all in g/l,
20 °C).   **Stability** Slowly decomposed by sunlight (>10 d). Stable to hydrolysis for
5 d (pH 5–9, 40 °C). Stable to heat; <1% decomposition after 60 d at 55 °C.

COMMERCIALISATION
History Acaricide reported by Kumiai Chemical Industry Co., Ltd (H. Ogawa,
Jpn. Pestic. Inf., 1985, No. 46, 11). Introduced in Japan (1987) by Kumiai.
Patent GB 1508250; US 4101670; JP 1192876   **Manufacturers** Ihara/Kumiai

APPLICATIONS
Mode of action Non-systemic acaricide.   **Uses** Control of eggs and young stages
of *Panonychus citri*, *Panonychus ulmi*, and other *Panonychus* spp.
Phytotoxicity Certain apple varieties, cotton, peaches, melons, legumes, brassicas
and other crops may be injured at high rates of application.
Formulation types EC.   **Compatibility** Incompatible with lime sulfur which may
cause phytotoxicity.   **Selected tradenames** 'Panocon' (Kumiai)

ANALYSIS
Product and **residue** analysis by glc. Details available from Kumiai.

MAMMALIAN TOXICOLOGY
Oral Acute oral LD$_{50}$ for male rats 1150, female rats 1200, male mice 7000,
female mice 4875 mg/kg.   **Skin and eye** Acute percutaneous LD$_{50}$ for male rats
2425, female rats 2075, mice >8000 mg/kg.   **Inhalation** LC$_{50}$ (4 h) for rats
>1.79 mg/l.   **NOEL** (2 y) for male rats 1.86, female rats 1.94 mg/kg b.w. daily;
(1 y) for male dogs 1.5, female dogs 3.0 mg/kg b.w. daily.   **ADI** 0.0075 mg/kg.
Toxicity class WHO (a.i.) III

ECOTOXICOLOGY
Birds Acute oral LD$_{50}$ for mallard ducks >2000, male bobwhite quail 1013, female
bobwhite quail 878 mg/kg. **Fish** LC$_{50}$ (48 h) for carp 7.9 mg/l.   **Bees** LD$_{50}$ by
topical application 0.2–0.4 mg/bee.   **Daphnia** LC$_{50}$ (24 h) for *D. carinata* 6.7 mg/l.

ENVIRONMENTAL FATE
Animals Following dosing to rats, >90% of radio-label was excreted within 48 h in
the urine and faeces. Six metabolites were identified.   **Plants** Foliar applied

fenothiocarb rapidly disappeared from leaves and fruit of mandarin oranges. Nine metabolites were identified. **Soil/Environment** In the field DT_{50} 8 d (sandy loam soil), 15 d (sandy soil). K_{oc} 7380.

309 fenoxaprop-P-ethyl   *Herbicide*

2-(4-aryloxyphenoxy)propionic acid

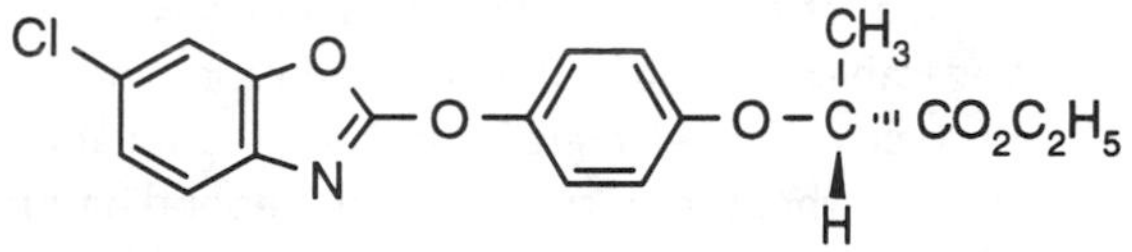

NOMENCLATURE
fenoxaprop-P-ethyl
Chemical Abstracts name ethyl (R)-2-[4-[(6-chloro-2-benzoxazolyl)oxy]phenoxy]= propanoate
CAS RN *[71283–80–2]* **Development codes** Hoe 046360; AE F046360

fenoxaprop-P
Common name fenoxaprop-P (BSI, draft E-ISO, (m) draft F-ISO)
IUPAC name (R)-2-[4-(6-chloro-1,3-benzoxazol-2-yloxy)phenoxy]propionic acid; (R)-2-[4-(6-chlorobenzoxazol-2-yloxy)phenoxy]propionic acid
Chemical Abstracts name (R)-2-[4-[(6-chloro-2-benzoxazolyl)oxy]phenoxy]= propanoic acid
CAS RN *[113158–40–0]* **Development codes** Hoe 088406; AE F088406

PHYSICAL CHEMISTRY
fenoxaprop-P-ethyl
Mol. wt. 361.8 **M.f.** $C_{18}H_{16}ClNO_5$ **Form** White odourless solid. **M.p.** 89–91 °C
V.p. 5.3×10^{-4} mPa (20 °C) **K_{ow}** logP = 4.58 **S.g./density** 1.3 (20 °C)
Solubility In water 0.7 mg/l (pH 5.8, 20 °C). In acetone 200, toluene 200, ethyl acetate >200, ethanol c. 24 (all in g/l, 20 °C). **Stability** Fenoxaprop-P-ethyl is stable for 90 d at 50 °C. Not sensitive to light. Decomposed by acids and alkalis. DT_{50} >1000 d (pH 5), 100 d (pH 7), 2.4 d (pH 9) (20 °C).

fenoxaprop-P
Mol. wt. 333.7 **M.f.** $C_{16}H_{12}ClNO_5$ **Form** Light beige, weakly pungent, fine powder. **M.p.** 155–161 °C **V.p.** 1.8×10^{-1} mPa (20 °C) **K_{ow}** logP = 1.83–0.24 (pH 5 – pH 9) **S.g./density** c. 1.5 (20 °C) **Solubility** In water 0.27 (pH 5.1), 61 (pH 7.0) (both in g/l, 20 °C). In acetone 80, toluene 0.5, ethyl acetate 36, methanol 34 (all in g/l, 20°C).

COMMERCIALISATION
History The herbicidal enantiomer of fenoxaprop was reported by H. P. Huff *et al.* (*Proc. Br. Crop Prot. Conf. – Weeds*, 1989, **2**, 717). Introduced by Hoechst AG (now AgrEvo GmbH).   **Manufacturers** AgrEvo

APPLICATIONS
fenoxaprop-P-ethyl
Biochemistry Fatty acid synthesis inhibition in grasses.
Mode of action Fenoxaprop-P-ethyl is a selective herbicide with contact and systemic action, absorbed principally by the leaves, with translocation both acropetally and basipetally to the roots or rhizomes.   **Uses** Post-emergence control of annual and perennial grass weeds in potatoes, beans, soya beans, beets, vegetables, peanuts, flax, oilseed rape, and cotton; and (when applied with the herbicide safener mefenpyr-diethyl) annual and perennial grass weeds and wild oats in wheat, rye, triticale and, depending on ratio, in some varieties of barley. **Phytotoxicity** Non-phytotoxic to broad-leaved crops.   **Formulation types** EC; EW; SE.   **Mixtures** *(fenoxaprop-P-ethyl +)* MCPA; bromoxynil; ioxynil; mecoprop-P; 2,4-D; diclofop-methyl; isoproturon.   **Selected tradenames** 'Puma' (mixture) (AgrEvo); 'Whip Super' (AgrEvo)

ANALYSIS
Details on hplc method are available from AgrEvo.

MAMMALIAN TOXICOLOGY
fenoxaprop-P-ethyl
Oral Acute oral LD_{50} for rats 3150–4000, mice >5000 mg/kg.
Skin and eye Acute percutaneous LD_{50} for rats >2000 mg/kg.   **Inhalation** LC_{50} (4 h) for rats >1.224 mg/l air.   **NOEL** (90 d) for rats 0.75 mg/kg b.w. daily (10 ppm), mice 1.4 mg/kg b.w. daily (10 ppm), dogs 15.9 mg/kg b.w. daily (400 ppm).

ECOTOXICOLOGY
fenoxaprop-P-ethyl
Birds Acute oral LD_{50} for bobwhite quail >2000 mg/kg.   **Fish** LC_{50} (96 h) for bluegill sunfish 0.58, rainbow trout 0.46 mg/l.   **Bees** LC_{50} (contact) >300 µg/bee, (feed) >1000 µg/bee.   **Worms** LC_{50} (14 d) for *Eisenia foetida* >1000 mg/kg soil.   **Daphnia** LC_{50} (48 h) 0.56 (pH 8.0–8.4), 2.7 (pH 7.7–7.8) (both mg/l). **Algae** LC_{50} (72 h) for *Scenedesmus subspicatus* 0.51 mg/l.

$$\langle\!\rangle - O - \langle\!\rangle - OCH_2CH_2NHCO_2CH_2CH_3$$

NOMENCLATURE

Common name fenoxycarb (BSI, draft E-ISO, (*m*) draft F-ISO, ANSI)
IUPAC name ethyl 2-(4-phenoxyphenoxy)ethylcarbamate
Chemical Abstracts name ethyl [2-(4-phenoxyphenoxy)ethyl]carbamate
CAS RN *[79127–80–3]*, formerly *[72490–01–8]* **Development codes**
Ro 13–5223/000 (to Roche) **Official codes** OMS 3010

PHYSICAL CHEMISTRY

Mol. wt. 301.3 **M.f.** $C_{17}H_{19}NO_4$ **Form** Colourless to white crystals.
M.p. 53–54 °C (OECD 102) **V.p.** 8.67×10^{-4} mPa (25 °C) (OECD 104)
K_{ow} logP = 4.07 (25 °C) (OECD 107) **S.g./density** 1.23 (20 °C) (OECD 109)
Solubility In water 6 mg/l (25 °C). In ethanol 510, acetone 770, toluene 630,
n-hexane 5.3, *n*-octanol 130 (all in g/l). **Stability** Stable to light. Stable to
hydrolysis in aqueous solutions at pH 3, 7, and 9 at 50 °C. **F.p.** *c.* 224 °C

COMMERCIALISATION

History Insecticide reported by S. Dorn et al. (*Z. Pflanzenkr. Pflanzenschutz*, 1981,
88, 269). Introduced by R. Maag Ltd (now Novartis Crop Protection AG).
Patent EP 4334; US 4215139 to Roche **Manufacturers** Novartis

APPLICATIONS

Mode of action Non-neurotoxic insect growth regulator with contact and stomach
action. Exhibits a strong juvenile hormone activity, inhibiting metamorphosis to the
adult stage and interfering with the moulting of early instar larvae. **Uses** Control
of Lepidoptera, scale insects, and suckers on fruit (including citrus), cotton, olives,
vines and ornamentals; Coleoptera and Lepidoptera in stored products; and also
cockroaches, fleas, mosquito larvae and fire ants in public health situations.
Phytotoxicity Some varieties of pears, vines, citrus fruit and ornamentals may be
injured. **Formulation types** WP; RB; EC. **Selected tradenames** 'Insegar'
(Novartis)

ANALYSIS

Product analysis by hplc. **Residues** determined by glc or preferably by hplc
(R. Haenni & P. E. Muller, *Anal. Methods Pestic. Plant Growth Regul.*, 1988, **16**, 21).

MAMMALIAN TOXICOLOGY

EHC 64 (WHO, 1986; a review of carbamate insecticides in general). **Oral** Acute
oral LD_{50} for rats >10 000 mg/kg. **Skin and eye** Acute percutaneous LD_{50} for

MAIN ENTRIES

rats >2000 mg/kg. Non-irritating and non-sensitising to skin (guinea pigs); mild eye irritant (rabbits). **Inhalation** LC_{50} for rats >0.48 mg/l air. **NOEL** (18 mo) for mice 4 mg/kg b.w.; (2 y) for rats 8 mg/kg b.w. **ADI** 0.04 mg/kg b.w. **Toxicity class** WHO (a.i.) III

ECOTOXICOLOGY
Birds Acute oral LD_{50} for Japanese quail >7000 mg/kg. LC_{50} (8 d) for bobwhite quail >25 000 ppm. **Fish** LC_{50} (96 h) for carp 10.3, rainbow trout 1.6 mg/l. **Bees** Non-toxic to bees; oral LC_{50} (24 h) >1000 ppm. **Daphnia** LC_{50} (48 h) 0.4 mg/l.

ENVIRONMENTAL FATE
Animals In rats, the major metabolic path is ring hydroxylation to form ethyl [2-[p-(p-hydroxyphenoxy)phenoxy]ethyl]carbamate. **Plants** Rapidly degraded in plants. **Soil/Environment** Low mobility in soil, no bioaccumulation. Relatively fast degradation in soil and water: DT_{50} 1.7–2.5 months (lab.), few to 31 days (field).

311 fenpiclonil   *Fungicide*

phenylpyrrole

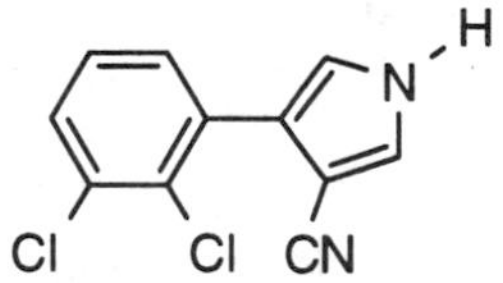

NOMENCLATURE
Common name fenpiclonil (BSI, draft E-ISO)
IUPAC name 4-(2,3-dichlorophenyl)pyrrole-3-carbonitrile
Chemical Abstracts name 4-(2,3-dichlorophenyl)-1H-pyrrole-3-carbonitrile
CAS RN [74738–17–3] **Development codes** CGA 142 705

PHYSICAL CHEMISTRY
Mol. wt. 237.1 **M.f.** $C_{11}H_6Cl_2N_2$ **Form** White crystals. **M.p.** 144.9–151.1 °C
V.p. 1.1×10^{-2} mPa (25 °C) K_{ow} logP = 3.86 (25 °C) **Henry** 5.4×10^{-4} Pa m^3 mol^{-1} (calc.) **S.g./density** 1.51 **Solubility** In water 4.8 mg/l (25 °C). In ethanol 73, acetone 360, toluene 7.2, n-hexane 0.026, n-octanol 41 (all in g/l, 25 °C).
Stability Not hydrolysed after 6 h at 100 °C between pH 3 and 9. Stable up to 250 °C.

COMMERCIALISATION
History Fungicide reported by D. Nevill *et al.* (*Proc. 1988 Br. Crop Prot. Conf. –*

Pests Dis., **1**, 65). Introduced in Switzerland (1988) by Ciba-Geigy AG (now Novartis Crop Protection AG). **Patent** EP 236272 **Manufacturers** Novartis

APPLICATIONS
Biochemistry Possibly acts by inhibition of transport-associated phosphorylation of glucose (A. B. K. Jespers & M. A. de Waard, *Pestic. Sci.*, **44,** 167 (1995)).
Mode of action Slightly systemic, contact fungicide with long-lasting activity.
Uses Effective against seed-borne pathogens of cereals, especially *Fusarium nivale* and *Tilletia caries*. On non-cereal crops, controls a wide range of seed- and soil-borne fungi (*Alternaria, Ascochyta, Aspergillus, Fusarium, Helminthosporium, Rhizoctonia* and *Penicillium* spp.). Having a different mode of action, fenpiclonil gives good control of all isolates of *F. nivale* including those resistant to carbendazim and related fungicides. On cereals and peas, used at 20 g a.i./100 kg seed; on potatoes, at 20–50 g/tonne (depending on target pathogens).
Formulation types WS; FS; DS. **Mixtures** *(fenpiclonil +)* imazalil; imazalil + carboxin. **Selected tradenames** 'Beret' (Novartis); 'Gambit' (Novartis)

ANALYSIS
Residue determination by hplc. Details available from Novartis.

MAMMALIAN TOXICOLOGY
Oral Acute oral LD_{50} for rats, mice, and rabbits >5000 mg/kg.
Skin and eye Acute percutaneous LD_{50} for rats >2000 mg/kg. Non-irritating to eyes and skin (rabbits); non-sensitising (guinea pig). **Inhalation** LC_{50} (4 h) for rats >1.5 mg/l air. **NOEL** for rats 1.25, mice 20, dogs 100 mg/kg b.w. daily.
ADI 0.0125 mg/kg b.w. **Other** Non-teratogenic and non-mutagenic.
Toxicity class WHO (a.i.) III (Table 5)

ECOTOXICOLOGY
Birds Acute oral LD_{50} for bobwhite quail >2510 mg/kg. LC_{50} for mallard ducks >5620, bobwhite quail 3976 ppm. **Fish** LC_{50} (96 h) for rainbow trout 0.8, carp 1.2, bluegill sunfish 0.76, catfish 1.3 mg/l. **Bees** Non-toxic to honeybees; LD_{50} (oral and contact) >5 µg/bee. **Worms** LC_{50} (14 d) for *Eisenia foetida* 67 mg/kg soil. **Other beneficial spp.** Harmless to carabid beetles. **Daphnia** LC_{50} (48 h) 1.3 mg/l. **Algae** LC_{50} (5 d) for *Scenedesmus subspicatus* 0.22 mg/l.

ENVIRONMENTAL FATE
Animals Rapidly absorbed from the gastrointestinal tract into the general circulation; rapidly excreted and almost completely eliminated, mostly in the faeces. The dominant metabolic pathway is oxidation of the pyrrole ring at the 2-position. A minor pathway is hydroxylation of the phenyl ring. All metabolites are excreted as conjugates, mainly as glucuronides. **Plants** Degradation proceeds via oxidation of the pyrrole ring followed by hydrolysis of the nitrile group. Opening of the pyrrole ring and hydroxylation of the phenyl ring are further

degradation steps. The parent is however the relevant residue; between 13–15 minor metabolites were also observed.   **Soil/Environment** Relatively persistent in soil; formation of bound residues represents the major route for dissipation. In leaching and adsorption/desorption experiments, the compound proved to be immobile in soil, RMF 0.3. Photolytic DT_{50} in water 70 min.

312 fenpropathrin *Acaricide, insecticide*

pyrethroid

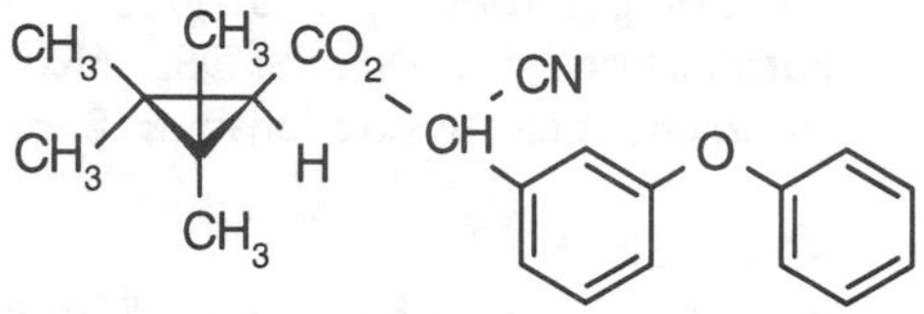

NOMENCLATURE
Common name fenpropathrin (BSI, E-ISO, ANSI); fenpropathrine ((m) F-ISO)
IUPAC name (*RS*)-α-cyano-3-phenoxybenzyl 2,2,3,3-tetramethylcyclopropane= carboxylate
Chemical Abstracts name cyano(3-phenoxyphenyl)methyl 2,2,3,3-tetramethyl= cyclopropanecarboxylate
CAS RN [64257–84–7] (racemate); [39515–41–8] (unstated stereochemistry)
Development codes S-3206   **Official codes** OMS 1999

PHYSICAL CHEMISTRY
Mol. wt. 349.4   **M.f.** $C_{22}H_{23}NO_3$ **Form** Yellow-brown solid (tech.).
M.p. 45–50 °C   **V.p.** 0.730 mPa (20 °C)   **K_{ow}** logP = 6 (20 °C)   **S.g./density** 1.15 (25 °C)   **Solubility** In water 14.1 µg/l (25 °C). In xylene, cyclohexanone 1000, methanol 337 (all in g/kg, 25 °C).   **Stability** Decomposed in alkaline solutions. Exposure to light and air leads to oxidation and loss of activity.

COMMERCIALISATION
History Insecticide reported by Y. Fujita (*Jpn. Pestic. Inf.*, 1981, No. 38, p. 21). Introduced by Sumitomo Chemical Co., Ltd and, in some countries, by Shell International Chemical Co. (now American Cyanamid Co.).   **Patent** GB 1356087; US 3835176   **Manufacturers** Cyanamid; Sumitomo

APPLICATIONS
Mode of action Acaricide and insecticide with repellent, and contact and stomach action.   **Uses** Control of many species of mites (except rust mites) and insects

(e.g. whitefly, lepidopterous larvae, leaf miners, leafworms, bollworms, etc.) on pome fruit, citrus fruit, vines, hops, vegetables, ornamentals (including ornamental trees), cotton, field crops, and glasshouse crops (cucurbits, tomatoes, ornamentals, etc.). **Formulation types** EC; SC; WP; UL.
Mixtures *(fenpropathrin +)* clofentezine; fenitrothion. **Compatibility** Incompatible with alkaline materials. **Selected tradenames** 'Danitol' (Valent/Sumitomo); 'Herald' (Cyanamid); 'Meothrin' (Sumitomo); 'Rody' (Sumitomo, Cyanamid); 'Digital' (Sanonda)

ANALYSIS
Product analysis of pyrethroids reviewed by E. Papadopoulou-Mourkidou in *Comp. Analyt. Profiles*. By hplc (S. Sakaue *et al.*, *Agric. Biol. Chem.*, 1982, **46**, 2165) or by glc with FID (*idem, ibid.*; Y. Takimoto *et al.*, *Anal. Methods Pestic. Plant Growth Regul.*, 1984, **13**, 69). **Residues** determined by glc with ECD (*idem, ibid.*); details available from Sumitomo Chemical Co.

MAMMALIAN TOXICOLOGY
Reviews *Pesticide residues in food – 1993*, FAO Plant Production and Protection Paper, 122, 1993. *Pesticide residues in food – 1993 evaluations, Part II – Toxicology.* WHO, WHO/PCS/94.4, 1994. See also A. J. Gray & D. M. Soderlund, Chapt. 5 in "*Insecticides*". **Oral** Acute oral LD_{50} for male rats 70.6, female rats 66.7 mg/kg (in corn oil). **Skin and eye** Acute percutaneous LD_{50} for male rats 1000, female rats 870, rabbits >2000 mg/kg. Not a skin irritant; mild eye irritant (rabbits). Non-sensitising to skin. **Inhalation** LC_{50} (4 h) for rats >96 mg/m^3. **ADI** (JMPR) 0.03 mg/kg b.w. [1993]. **Other** Non-mutagenic. **Toxicity class** WHO (a.i.) II; EPA (formulation) II

ECOTOXICOLOGY
Birds Acute oral LD_{50} for mallard ducks 1089 mg/kg. Eight-day dietary LC_{50} for bobwhite quail and mallard ducks >10 000 mg/kg diet. **Fish** LC_{50} (48 h) for bluegill sunfish 1.95 µg/l.

ENVIRONMENTAL FATE
Soil/Environment Degraded principally by photolysis, DT_{50} 2.7 w in river water. Duration of activity in soil is *c.* 1–5 d. Metabolites are listed in *Pestic. Sci.*, 1985, **16**, 119.

313 fenpropidin   *Fungicide*

morpholine analogue (piperidine derivative)

$$(CH_3)_3C \text{—} \bigcirc \text{—} CH_2 \text{ } \underset{\displaystyle CH_2}{\overset{\displaystyle \underset{\displaystyle CH}{\overset{\displaystyle CH_3}{|}}}{|}} \text{—N} \bigcirc$$

NOMENCLATURE
Common name fenpropidin (BSI, draft E-ISO); fenpropidine ((*f*) draft F-ISO)
IUPAC name (*RS*)-1-[3-(4-*tert*-butylphenyl)-2-methylpropyl]piperidine
Chemical Abstracts name (±)-1-[3-[4-(1,1-dimethylethyl)phenyl]-2-methyl=
propyl]piperidine
CAS RN *[67306–00–7]* unstated stereochemistry   **Development codes**
Ro 12–3049/000 (Maag); CGA 114900 (Ciba-Geigy)

PHYSICAL CHEMISTRY
Mol. wt. 273.5   **M.f.** $C_{19}H_{31}N$ **Form** Pale yellow, odourless, slightly viscous liquid.
B.p. >250 °C   **V.p.** 17 mPa (25 °C) K_{ow} logP = 2.59 (pH 7, 22 °C)
Henry 10 Pa m^3 mol^{-1} (calc.)   **S.g./density** 0.91 (20 °C)   **Solubility** In water
0.53 g/l (pH 7, 25 °C). Completely miscible in acetone, ethanol, toluene,
n-octanol, *n*-hexane (25 °C).   **Stability** Stable at room temperature in closed
container for at least 3 years. Stable to u.v. light when in aqueous solution. Stable
to hydrolysis at 80 °C at pH 4, 7 and 10.   **pKa** 10.1, strong base   **F.p.** 156 °C

COMMERCIALISATION
History Fungicide reported by K. Bohnen *et al.* (*Proc. Br. Crop Prot. Conf. – Pests
Dis.*, 1986, **1**, 27). Introduced by Dr R. Maag Ltd and developed world-wide by
Ciba-Geigy AG (now Novartis Crop Protection AG).   **Patent** GB 1584290;
DE 2752135; US 4241058   **Manufacturers** Novartis

APPLICATIONS
Biochemistry Inhibits ergosterol biosynthesis, but by a different mechanism
(inhibition of steroid reduction) from the triazole fungicides.
Mode of action Systemic foliar fungicide with protective and curative action, with
translocation acropetally in the xylem.   **Uses** Control of powdery mildew
(*Erysiphe graminis*), rusts (*Puccinia* spp.), and *Rhynchosporium secalis* in barley and
wheat.   **Formulation types** EC.   **Mixtures** *(fenpropidin +)* fenpropimorph;
prochloraz; propiconazole; fenbuconazole; difenoconazole; propiconazole +
fenpropimorph; fenpropimorph + prochloraz.   **Selected tradenames** 'Tern'
(Novartis); 'Columbia' (mixture) (Zeneca); 'Sponsor' (mixture) (AgrEvo)

ANALYSIS
Product analysis by glc with FID.   **Residues** determined by glc with FID or by hplc. Details available from Novartis.

MAMMALIAN TOXICOLOGY
Oral Acute oral LD_{50} for rats >1447 mg/kg.   **Skin and eye** Acute percutaneous LD_{50} for rats >4000 mg/kg.   Skin and eye irritant (rabbits). Not a skin sensitiser (guinea pigs).   **Inhalation** LC_{50} (4 h) for rats 1.22 mg/l air.   **NOEL** (2 y) for rats 0.5 mg/kg b.w. daily; (1.5 y) for mice 4.5 mg/kg b.w. daily; (1 y) for dogs 2 mg/kg b.w. daily.   **ADI** 0.005 mg/kg b.w.   **Other** Acute i.p. LD_{50} for rats 346 mg/kg. Non-mutagenic, non-carcinogenic and non-teratogenic.   **Toxicity class** WHO (a.i.) II   **EC risk** R20/22, R36/38, R43

ECOTOXICOLOGY
Birds Acute oral LD_{50} for mallard ducks 1900, pheasants 370 mg/kg.   **Fish** LC_{50} (96 h) for rainbow trout 2.6, mirror carp 3.6, bluegill sunfish 1.9 mg/l.
Bees Relatively non-toxic to bees; LD_{50} (48 h) (oral) >0.01 mg/bee; (contact) 0.046 mg/bee.   **Worms** LC_{50} (14 d) for earthworms >1000 µg/kg soil.
Other beneficial spp. Moderately harmful to the predatory mite *Phytoseiulus persimilis*.   **Daphnia** LC_{50} (48 h) 0.5 mg/l.   **Algae** E_bC_{50} (96 h) for *Microcystis aeruginosa* 4.4, *Scenedesmus subspicatus* 0.0057, *Navicula pelliculosa* 0.0025 mg/l.

ENVIRONMENTAL FATE
Animals In rats, following oral administration, fenpropidin is rapidly absorbed, distributed, metabolised and excreted in the urine and faeces. No bioretention potential.   **Plants** Relatively rapid and extensive degradation. In wheat, the principal metabolic pathway involves hydroxylation of the piperidine ring and oxidation of the tertiary butyl group. DT_{50} in wheat and barley plants *c.* 4–11 d.
Soil/Environment Strongly adsorbed, K_d 43.5 (sandy loam) – 117.1 (sandy clay loam), and extensively degraded, DT_{50} 58 (loam) – 95 (sandy loam) in soil. Fenpropidin and its metabolites have little or no tendency to leach.

morpholine

$(CH_3)_3C$ —⟨benzene ring⟩— CH_2CHCH_2 —N (with CH_3 substituent on the CH) — morpholine ring with CH_3, O, CH_3

NOMENCLATURE
Common name fenpropimorph (BSI, draft E-ISO); fenpropimorphe
((*m*) draft F-ISO)
IUPAC name (±)-*cis*-4-[3-(4-*tert*-butylphenyl)-2-methylpropyl]-2,6-dimethyl=
morpholine
Chemical Abstracts name *cis*-4-[3-[4-(1,1-dimethylethyl)phenyl]-2-methylpropyl]-=
2,6-dimethylmorpholine
CAS RN *[67564–91–4]* *cis*- isomer; *[67306–03–0]* unstated stereochemistry
EEC no. 266–719–9 **Development codes** Ro 14–3169 (Maag); ACR-3320 (Maag);
BAS 421F (BASF); CGA 101031 (Ciba-Geigy)

PHYSICAL CHEMISTRY
Mol. wt. 303.5 **M.f.** $C_{20}H_{33}NO$ **Form** Colourless, odourless oil (tech., yellowish
oil with an aromatic odour). **B.p.** >300 °C (101.3 kPa) **V.p.** 3.5 mPa (20 °C)
K_{ow} logP = 2.6 (pH 5), 4.1 (pH 7), 4.4 (pH 9) (all at 20 °C) **Henry** 0.3 Pa m^3
mol^{-1} (calc.) **S.g./density** 0.933 (20 °C) **Solubility** In water 4.3 mg/l (pH 7,
20 °C). In acetone, chloroform, ethyl acetate, cyclohexane, toluene, diethyl ether,
ethanol >1 (all in kg/kg, 20 °C). **Stability** Stable at room temperature in closed
container for at least 3 years. Stable to light. Stable to hydrolysis at pH 3, 7 and 9
(50 °C). **pKa** 6.98 (20 °C), base **F.p.** *c.* 105 °C (Pensky-Martens); 157 °C
(CIPAC MT12)

COMMERCIALISATION
History Fungicide reported by K. Bohnen & A. Pfiffner (*Meded. Fac. Landbouwwet.
Rijksuniv. Gent,* 1979, **44**, 487) and by E. H. Pommer & W. Himmele (*ibid.*, p. 499).
Introduced in Germany (1983) by BASF AG, and by Dr R. Maag Ltd and Rhône-
Poulenc Agriculture. **Patent** DE 2752135; DE 2656747; GB 1584290; US 4241058
Manufacturers BASF; Novartis

APPLICATIONS
Biochemistry Ergosterol biosynthesis inhibitor, by inhibition of steroid reduction
(sterol-Δ^{14}-reductase) and isomerisation (Δ^8 to Δ^7-isomerase).
Mode of action Systemic foliar fungicide with protective and curative action, with
translocation acropetally in the xylem. **Uses** Control of *Erysiphe graminis,
Rhynchosporium secalis,* and *Puccinia* spp. in cereals; *Cercospora beticola, Uromyces*

betae, and *Erysiphe betae* in sugar beet; *Uromyces* and *Puccinia* spp. in beans and leeks, and *Phomopsis* in sunflower. **Formulation types** EC; SC.
Mixtures *(fenpropimorph +)* carbendazim; carbendazim + mancozeb; chlorothalonil; carbendazim + chlorothalonil; fenbuconazole; prochloraz; fenpropidin; fenpropidin + prochloraz; propiconazole; propiconazole + fenpropidin; kresoxim-methyl; BAS 480F. **Selected tradenames** 'Corbel' (Novartis, BASF)

ANALYSIS
Product analysis by glc with FID (*CIPAC Handbook*, 1995, **G**, 56–61).
Residues determined by glc with FID or by hplc. Details available from Novartis or BASF.

MAMMALIAN TOXICOLOGY
Reviews *Pesticide residues in food – 1994*, FAO Plant Production and Protection Paper, 127, 1995. *Pesticide residues in food – 1994 evaluations, Pt. II – Toxicology*, WHO, WHO/PCS/95.2, 1995. **Oral** Acute oral LD_{50} for rats >1467 mg/kg.
Skin and eye Acute percutaneous LD_{50} for rats >4000 mg/kg. Skin irritant, not an eye irritant (rabbits). No skin sensitisation (guinea pigs). **Inhalation** LC_{50} (4 h) for rats >3580 mg/m^3 air, with moderate irritation of the respiratory organs.
NOEL for rats 0.3, mice 3.0, dogs 3.2 mg/kg b.w. daily. **ADI** (JMPR) 0.003 mg/kg b.w. [1994]. **Other** Not mutagenic, not carcinogenic, of no teratogenic relevance for humans. **Toxicity class** WHO (a.i.) III (Table 5)

ECOTOXICOLOGY
Birds Acute oral LD_{50} for mallard ducks >17 776, pheasants 3900 mg/kg. LC_{50} (5 d) for mallard ducks 5000, bobwhite quail >5000 mg/kg. **Fish** LC_{50} (96 h) for rainbow trout 9.5, bluegill sunfish 3.2–4.2, carp 3.2 mg/l. **Bees** Acute oral LD_{50} >100 µg/bee. **Worms** LD_{50} (14 d) for earthworms ≥520 mg/kg soil.
Other beneficial spp. Not dangerous to various beneficial insect species.
Daphnia LC_{50} (48 h) 2.4 mg/l. **Algae** EC_{50} (96 h) for *Chlorella fusca* 2.21 mg/l.
Other aquatic spp. EC_{10} (17 h) for *Pseudomonas putida* >1874 mg/l.

ENVIRONMENTAL FATE
Animals After oral administration to rats, fenpropimorph is rapidly absorbed and almost completely eliminated with urine and faeces. Residues in tissues were generally low. **Plants** Degradation in plants is similar to that in soil. Half-life in cereal plants is 3–7 days. **Soil/Environment** Degraded in soil by oxidation of the tertiary butyl group, in addition to oxidation and opening of the dimethylmorpholine ring. Half-life in soil ranges from 15 days in moderately humus loamy sand to 93 days in very humus loamy sand. K_{oc} 2772–5943. Strongly adsorbed to soil (low potential to leach), K_d 22.6 (sand) – 73.7 (loamy sand).

pyrazole (acaricide)

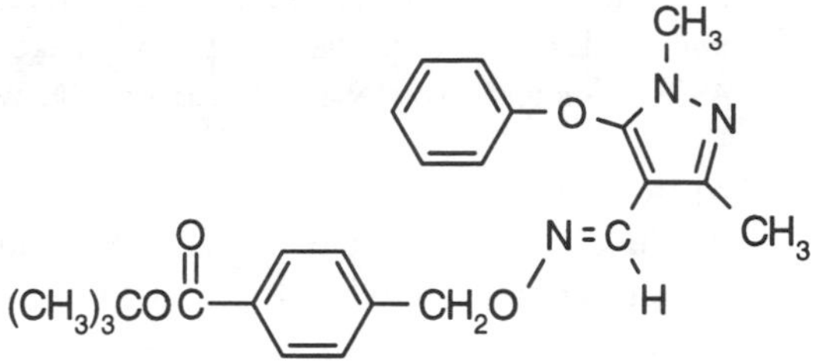

NOMENCLATURE
Common name fenpyroximate (BSI, draft E-ISO)
IUPAC name *tert*-butyl (*E*)-α-(1,3-dimethyl-5-phenoxypyrazol-4-ylmethylene=
amino-oxy)-*p*-toluate
Chemical Abstracts name (*E*)-1,1-dimethylethyl 4-[[[[(1,3-dimethyl-5-phenoxy-=
1*H*-pyrazol-4-yl)methylene]amino]oxy]methyl]benzoate
CAS RN *[111812–58–9]*   **Development codes** NNI-850; HOE 555–02A

PHYSICAL CHEMISTRY
Mol. wt. 421.5   **M.f.** $C_{24}H_{27}N_3O_4$ **Form** White crystalline powder.
M.p. 101.1–102.4 °C   **V.p.** 0.0075 mPa (25 °C)   **K_{ow}** logP = 5.01 (20 °C)
S.g./density 1.25 g/cm^3 **Solubility** In water 1.46×10^{-2} mg/l (20 °C). In methanol
15, acetone 150, dichloromethane 1307, chloroform 1197, tetrahydrofuran 737
(all in g/l, 25 °C).   **Stability** Stable in acid and alkali.

COMMERCIALISATION
History Acaricide reported by T. Konno *et al.* (*Proc. 1990 Br. Crop Prot. Conf. –
Pests Dis., 1,* 71). Introduced in 1991 by Nihon Nohyaku Co., Ltd.
Manufacturers Nihon Nohyaku

APPLICATIONS
Biochemistry Inhibitor of mitochondrial electron transport at complex I.
Mode of action Quick knockdown activity against larvae, nymphs and adults,
mainly by contact and ingestion. Also some moulting inhibitory activity on nymphs.
Uses Control of important phytophagous mites such as Tetranychidae,
Tarsonemidae, Tenuipalpidae, Eriophyidae, *Empoasca onukii* Matsuda and *Caloptilia
theivora* on citrus, apple, pear, peach, grape, tea, cherry, melon, etc., typically at
25–50 ppm.   **Phytotoxicity** Not phytotoxic to top fruit, citrus, tea, vegetables and
ornamentals.   **Formulation types** SC.   **Selected tradenames** 'Danitoron',
'Danitron' (Japan) (Nihon Nohyaku); 'Ortus' (Nihon Nohyaku); 'Pamanrin'
(China) (Nihon Nohyaku)

ANALYSIS
Product analysis by hplc; **residues** in soil by glc, in water by hplc, in plants by glc/hplc.

MAMMALIAN TOXICOLOGY
Reviews *Pesticide residues in food – 1995*, FAO Plant Production and Protection Paper 133, 1996. *Pesticide residues in food – 1995 evaluations; Part II – Toxicology & Environment*. World Health Organisation, WHO/PCS/96.48, 1996. See also *J. Pestic. Sci.*, **17**, S261 267 (1992). **Oral** Acute oral LD_{50} for male rats 400, female rats 245 mg/kg. **Skin and eye** Acute percutaneous LD_{50} for male and female rats >2000 mg/kg. Non-irritating to skin; slightly irritating to eyes (rabbits). **Inhalation** LC_{50} for male rats 0.33 mg/l. **NOEL** for male rats 0.97, female rats 1.21 mg/kg b.w. **ADI** (JMPR) 0.01 mg/kg b.w. [1995]. **Other** Not carcinogenic, teratogenic or mutagenic in long term studies. **Toxicity class** WHO (a.i.) II

ECOTOXICOLOGY
Birds LD_{50} for bobwhite quail and mallard ducks >2000 mg tech./kg. Dietary LD_{50} (8 d) for bobwhite quail and mallard ducks >5000 ppm. **Fish** LC_{50} (96 h) for rainbow trout 0.079, carp 0.29 mg/l. **Bees** No adverse effect on honeybees at 250 ppm (5 × recommended dose). **Other beneficial spp.** Relatively non-toxic to predacious mites. Little adverse effect at 25–50 ppm on *Chrysopa nipponensis*, *Harmonia axyridis*, *Ephedrus japonicus*, *Misumenops tricuspidatus*, *Licosa pseudoannulate*, *Orius* sp., *Scolothrips* sp. **Daphnia** LC_{50} (24 h) 0.204 mg/l.

ENVIRONMENTAL FATE
Soil/Environment DT_{50} 26.3–49.7 d (*J. Pestic. Sci.*, **18**, 67–75 (1993)).

316 fenthion   *Insecticide*

organophosphorus

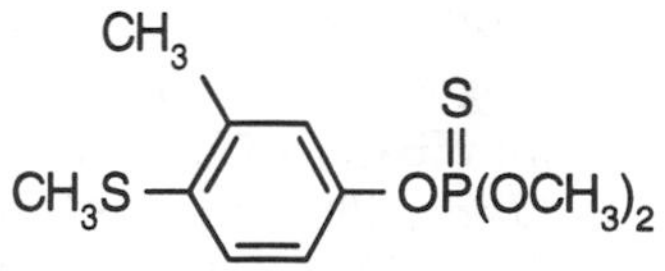

NOMENCLATURE
Common name fenthion (BSI, E-ISO, (*m*) F-ISO, ESA, BAN); MPP (JMAF)
IUPAC name *O,O*-dimethyl *O*-4-methylthio-*m*-tolyl phosphorothioate
Chemical Abstracts name *O,O*-dimethyl *O*-[3-methyl-4-(methylthio)phenyl] phosphorothioate

CAS RN *[55–38–9]* **EEC no.** 200–231–9 **Development codes** Bayer 29 493; S 1752; E 1752 **Official codes** OMS 2; ENT 25 540

PHYSICAL CHEMISTRY
Mol. wt. 278.3 **M.f.** $C_{10}H_{15}O_3PS_2$ **Form** Colourless oily liquid; (tech., brown oily liquid with a mercaptan-like odour). **M.p.** No solidification point down to −80 °C **B.p.** 90 °C/1 Pa (calc.), 117 °C/10 Pa (calc.), 284 °C (calc.) **V.p.** 0.74 mPa (20 °C); 1.4 mPa (25 °C) K_{ow} logP = 4.84 **Henry** 5×10^{-2} Pa m^3 mol^{-1} (20 °C) **S.g./density** 1.25 (20 °C) **Solubility** In water 4.2 mg/l (20 °C). In dichloromethane, toluene, isopropanol >250, hexane 100 (all in g/l, 20 °C). **Stability** Stable to light and up to 210 °C. Relatively stable in acidic conditions, and moderately stable in alkaline conditions; DT_{50} (22 °C) 223 d (pH 4), 200 d (pH 7), 151 d (pH 9). **F.p.** 170 °C (tech.)

COMMERCIALISATION
History Developed by G. Schrader & E. Schegk (G. Schrader, *Hoefchen-Briefe (Engl. Ed.)*, 1960, **13**, 1) and introduced by Bayer AG. **Patent** DE 1116656; US 3042703 **Manufacturers** Bayer; Shenzhen Jiangshan

APPLICATIONS
Biochemistry Cholinesterase inhibitor. **Mode of action** Insecticide with contact, stomach, and respiratory action. **Uses** Control of fruit flies, leafhoppers, leaf miners, leaf-eating larvae, stem borers, cereal bugs, and other insect pests in fruit (including citrus), vines, olives, vegetables, cotton, tea, sugar cane, rice, beet, tobacco, ornamentals, etc. Control of insect pests (flies, mosquitoes, cockroaches, fleas, ants, ticks, lice, etc.) in public health situations and animal houses, and control of animal ectoparasites. **Phytotoxicity** Non-phytotoxic when used as recommended. Some varieties of apples and cotton may be injured. **Formulation types** EC; WP; UL; GR; DP; HN; PO. **Compatibility** Incompatible with insecticides and fungicides which are highly alkaline. **Selected tradenames** 'Lebaycid' (Bayer); 'Beiliulin' (Shenzhen Jiangshan); 'Faster' (Sanonda); 'Pilartex' (Pilarquim)

ANALYSIS
Product analysis by colorimetry of a derivative (F. B. Ibrahim & J. C. Cavagnol, *J. Agric. Food Chem.*, 1966, **44**, 369; *CIPAC Handbook*, 1983, **1B**, 1830) – details from Bayer AG. **Residues** determined by glc (M. A. Luke *et al.*, *J. Assoc. Off. Anal. Chem.*, 1981, **64**, 1187; A. Ambrus *et al.*, *ibid.*, p. 733; *Anal. Methods Pestic. Plant Growth Regul.*, 1972, **6**, 301). Methods for the determination of residues are available from Bayer.

MAMMALIAN TOXICOLOGY
Reviews *Pesticide residues in food – 1995*, FAO Plant Production and Protection Paper 133, 1996. *Pesticide residues in food – 1995 evaluations; Part II – Toxicology &*

Environment. World Health Organisation, WHO/PCS/96.48, 1996. **EHC** 63 (WHO, 1986; a general review of organophosphorus insecticides). **Oral** Acute oral LD_{50} for male and female rats *c*. 250 mg/kg. **Skin and eye** Acute percutaneous LD_{50} for male and female rats *c*. 700 mg/kg (24 h); not irritating to skin and eyes (rabbit). **Inhalation** LC_{50} (4 h) for male and female rats *c*. 0.5 mg/l air (aerosol). **NOEL** (2 y) for rats <5, mice 0.1 mg/kg diet; (1 y) for dogs 2 mg/kg diet. **ADI** (JMPR) 0.007 mg/kg b.w. [1995]. **Toxicity class** WHO (a.i.) II; EPA (formulation) II **EC risk** T (R25); Xn (R21)

ECOTOXICOLOGY

Birds Acute oral LD_{50} for bobwhite quail 7.2 mg/kg. Five-day dietary LC_{50} for bobwhite quail 60, mallard ducks 1259 mg/kg. **Fish** LC_{50} (96 h) for bluegill sunfish 1.7, rainbow trout 0.8–1.0, golden orfe 2.7 mg/l. **Bees** Toxic to bees. **Worms** LC_{50} for *Eisenia foetida* 562 mg/kg dry soil. **Daphnia** LC_{50} (48 h) 0.0057 mg/l. **Algae** E_rC_{50} for *Scenedesmus subspicatus* 1.79 mg/l.

ENVIRONMENTAL FATE

Animals In mammals, following oral administration, elimination is mainly in the form of hydrolysis products in the urine. Major metabolites are fenthion sulfoxide, fenthion sulfone and their oxygen analogues. These metabolites are hydrolysed in further degradation, forming the corresponding phenols. **Plants** In plants, fenthion is oxidised to the sulfoxide and sulfone, both of which possess insecticidal properties, and to the monodesmethyl compound of fenthion sulfoxide. Further degradation occurs to the sulfone phosphate, which undergoes hydrolysis. **Soil/Environment** K_{oc} is 1500 (Ware, *Rev. Environ. Contam. & Toxicol.*, **123** (1992)). In sediment/water system with and without plants DT_{50} is *c*. 1.5 d (O'Neill *et al.*, *Environ. Toxicol. & Chem.*, **8**, 759–768 (1989)). The degradation of fenthion occurs rapidly under aerobic conditions, forming the metabolites fenthion sulfoxide, fenthion sulfone, followed by the analogous phenol compounds.

317 fentin *Fungicide, algicide, molluscicide*

organotin

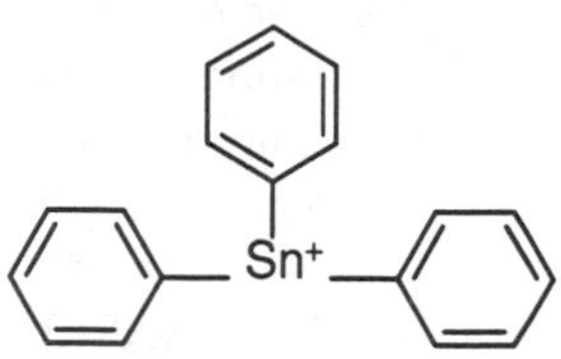

NOMENCLATURE
fentin
Common name fentin (BSI, E-ISO); fentine ((*f*) F-ISO); [fenolovo] (former

exception, USSR); chemical name is used (Republic of South Africa, USA)
IUPAC name triphenyltin
Chemical Abstracts name triphenylstannylium
CAS RN [668–34–8] **Development codes** VP 1940; Hoe 02824

fentin acetate
IUPAC name triphenyltin(IV) acetate; triphenyltin acetate
Chemical Abstracts name (acetyloxy)triphenylstannane
CAS RN [900–95–8] **EEC no.** 212–984–0 **Development codes** Hoe 002782
Official codes OMS 1020; ENT 25208

fentin hydroxide
IUPAC name triphenyltin(IV) hydroxide; triphenyltin hydroxide
Chemical Abstracts name hydroxytriphenylstannane
CAS RN [76–87–9] **EEC no.** 200–990–6 **Official codes** OMS 1017; ENT 28009

PHYSICAL CHEMISTRY
fentin
Mol. wt. 350.0 **M.f.** $C_{18}H_{15}Sn$

fentin acetate
Composition Tech. grade is ≥94% pure. **Mol. wt.** 409.0 **M.f.** $C_{20}H_{18}O_2Sn$
Form Colourless crystals. **M.p.** 121–123 °C; (tech., 118–125 °C) **V.p.** 1.9 mPa
(60 °C) **K$_{ow}$** logP = 3.43 **S.g./density** 1.5 (20 °C) **Solubility** In water at 20 °C
and pH 5, *c.* 9 mg/l. In ethanol 22, ethyl acetate 82, hexane 5, dichloromethane
460, toluene 89 (all in g/l, 20 °C). **Stability** Stable when dry. Converted to fentin
hydroxide in the presence of water. Unstable in acids and alkalis (22 °C), DT$_{50}$
<3 h (pH 5, 7 or 9). Decomposed by sunlight and by atmospheric oxygen.

fentin hydroxide
Composition Tech. grade is ≥95% pure. **Mol. wt.** 367.0 **M.f.** $C_{18}H_{16}OSn$
Form Colourless crystals. **M.p.** 118–120 °C **V.p.** 0.047 mPa at 50 °C
K$_{ow}$ logP = 3.43 **S.g./density** 1.54 at 20 °C **Solubility** In water at 20 °C and
pH 7, *c.* 1 mg/l (greater at lower pH-values). In ethanol *c.* 10, dichloromethane
171, diethyl ether 28, acetone *c.* 50 (all in g/l, 20 °C); in benzene 41 g – as
bis(triphenyltin) oxide/l. **Stability** Stable in the dark at room temperature.
Dehydration occurs on heating above 45 °C, yielding bis(triphenyltin) oxide, which
is stable up to *c.* 250 °C. Slowly decomposed by sunlight, and more rapidly by u.v.
light, to give inorganic tin via di- and mono- phenyltin compounds.

COMMERCIALISATION
History Fungicidal properties of organotin compounds investigated by G. J. M. van
der Kerk & J. G. A. Luijten (*J. Appl. Chem.*, 1954, **4**, 314; 1956, **6**, 56) and reviewed
by H. Bock (*Residue Rev.*, 1981, **79**, 1). Fentin acetate was introduced by Hoechst

AG (now AgrEvo GmbH) and fentin hydroxide by N.V. Philips-Duphar (now Uniroyal Chemical Co., Inc).

fentin acetate
Patent US 3499086 (Hoechst)   **Manufacturers** AgrEvo

fentin hydroxide
Manufacturers AgrEvo

APPLICATIONS
fentin acetate
Biochemistry Multi-site inhibitor, preventing spore germination, and inhibiting metabolism of the fungal organism, in particular respiration.
Mode of action Non-systemic fungicide with mainly protective action, but also some curative action. Also acts as an algicide and molluscicide. **Uses** Control of early and late blights of potatoes; *Septoria* leaf spot on celery and celeriac; leaf spot diseases of onions, sugar beet and peanuts; anthracnose of beans; rust and leaf blotch on wheat; rust on coffee; scab on pecans; *Phytophthora palmivora* and *Monilia roreri* on cocoa; *Pyricularia oryzae*, *Pellicularia sasakii* and *Helminthosporium oryzae* on rice; etc. Also used for control of algae and snails in rice fields, and for control of water snails in fish ponds. **Phytotoxicity** Vines, ornamentals, some fruits, and glasshouse crops may be injured. **Formulation types** WP.
Mixtures *(fentin acetate +)* maneb.   **Compatibility** Incompatible with oil emulsions and EC formulations.   **Selected tradenames** 'Brestan' (mixture) (AgrEvo); 'Radar' (Productos OSA); 'Suzu' (Nihon Nohyaku)

fentin hydroxide
Biochemistry Multi-site inhibitor, preventing spore germination, and inhibiting metabolism of the fungal organism, in particular respiration.
Mode of action Non-systemic fungicide with protective and curative action.
Uses Control of early and late blights of potatoes; *Septoria* leaf spot on celery and celeriac; leaf spot diseases of beet and peanuts; scab and other diseases of pecans; blast diseases of rice; and fungal diseases of cocoa, coffee, and soya beans. Exhibits an anti-feeding action on leaf-eating larvae, including *Spodoptera* spp.   **Phytotoxicity** Non-phytotoxic when used as directed. Tomatoes and apples may be injured. Surfactants, spreaders or stickers should not be used, as this can lead to phytotoxicity. **Formulation types** WP; SC.   **Mixtures** *(fentin hydroxide +)* maneb; metoxuron; sulfur.   **Compatibility** Not compatible with strongly acidic compounds. Incompatible with oils and liquid formulations.
Selected tradenames 'Brestan Flow' (AgrEvo); 'Super-Tin' (Chiltern, Griffin); 'Suzu-H' (Nihon Nohyaku); 'Tubotin' (Rhône-Poulenc)

ANALYSIS
Product analysis: fentin acetate and hydroxide may be determined by conversion

to fentin chloride, followed by rp hplc with u.v. detection (*CIPAC Handbook*, 1992, **E**, 89–94), or by hydrolysis of fentin acetate to the hydroxide which is measured by potentiometric titration (*CIPAC Handbook*, 1980, **1A**, 1263, 1266; *AOAC Methods*, 1995, 979.02) or by glc of a derivative (*ibid.*, 984.04; *CIPAC Handbook*, 1983, **1B**, 1837; Van Rossum *et al., Anal. Methods Pest. Plant Growth Regul.*, 1980, **11**, 227).   **Residues** determined by atomic-adsorption spectrophotometry of total tin (W. H. Evans *et al., Analyst (London)*, 1979, **104**, 16; T. Ferri *et al., Talanta*, 1989, **36**, 513) or by glc of a derivative (H. H. van den Broek *et al., Analyst (London)*, 1988, **113**, 1237; Van Rossum, *loc. cit.*; P. G. Baker *et al., Analyst (London)*, 1980, **105**, 282).

MAMMALIAN TOXICOLOGY
fentin
Reviews *Pesticide residues in food – 1991*. FAO Plant Production and Protection Paper, 111, 1991. *Pesticide residues in food – 1991 evaluations. Part II – Toxicology*, World Health Organisation, WHO/PCS/92.52, 1992. *Pesticide residues in food*. AGP:1970/M/12/1; WHO/Food Add./71.42, 1971. See also *Pesticide residues in food*, FAO Agr. Studies, No. 87; WHO Tech. Rept. Series, No. 474, 1971. *1970 Evaluations of some pesticide residues in food*, AGP:1970M/12/1; WHO/Food Add./71.42, 1971.   **EHC** 15 (WHO, 1980).   **ADI** (JMPR) 0.0005 mg/kg b.w. (for fentin compounds) [1991].

fentin acetate
Oral Acute oral LD_{50} for rats 140–298, guinea pigs 20, rabbits 30–50 mg/kg.
Skin and eye Acute percutaneous LD_{50} for rats *c.* 450, mice 350 mg/kg. Irritating to skin and mucous membranes.   **Inhalation** LC_{50} (4 h) for male rats 0.044, female rats 0.069 mg/l air.   **NOEL** (2 y) for dogs 5 mg/kg diet.   **Toxicity class** WHO (a.i.) II; EPA (formulation) II   **EC risk** T+ (R26); T (R24/25); Xi (R36/38); (R43)

fentin hydroxide
Oral Acute oral LD_{50} for male rats 171, female rats 110, male mice 245, female mice 209, male guinea pigs 27.1, female guinea pigs 31.1 mg/kg.
Skin and eye Acute percutaneous LD_{50} for rats 1600 mg/kg.   **Inhalation** LC_{50} (4 h) for rats 60.3 mg/m^3 air.   **NOEL** (2 y) for rats 2 mg/kg diet.
Toxicity class WHO (a.i.) II; EPA (formulation) II   **EC risk** T+ (R26); T (R24/25); Xi (R36/38)

ECOTOXICOLOGY
fentin acetate
Birds LD_{50} for quail 77.4 mg/kg.   **Fish** LC_{50} (48 h) for carp 0.32 mg/l.
Bees Formulation not toxic to bees.   **Daphnia** LC_{50} (48 h) 0.32–32 µg/l.

fentin hydroxide
Birds Eight-day dietary LC_{50} for bobwhite quail 38.5 mg/kg diet.   **Fish** LC_{50} (48 h)

for guppies 0.054, carp 0.05, golden orfe 0.11, Japanese killifish 0.072, harlequin fish 0.042 mg/l. **Bees** Not toxic to bees. **Daphnia** LC_{50} (48 h) 16.5 μg/l.

ENVIRONMENTAL FATE
fentin acetate
Soil/Environment In soil, degraded to inorganic tin via di- and mono- phenyltin compounds. Soil DT_{50} c. 140 d (biological degradation in agricultural loam, 1.39–1.69% organic carbon, pH 7.6, room temperature).

fenuron

Herbicide

urea

NOMENCLATURE
fenuron
Common name fenuron (BSI, E-ISO, F-ISO, ANSI, WSSA); [fenidim] (former exception, USSR); no name (Portugal, Sweden)
IUPAC name 1,1-dimethyl-3-phenylurea
Chemical Abstracts name *N,N*-dimethyl-*N'*-phenylurea
Other names PDU   **CAS RN** *[101–42–8]*

fenuron-TCA
Common name fenuron-TCA (WSSA)
IUPAC name 1,1-dimethyl-3-phenyluronium trichloroacetate
Chemical Abstracts name trichloroacetic acid compound with *N,N*-dimethyl-= *N*-phenylurea (1:1)
CAS RN *[4482–55–7]*   **Development codes** GC-2603

PHYSICAL CHEMISTRY
fenuron
Mol. wt. 164.2   **M.f.** $C_9H_{12}N_2O$ **Form** Colourless crystals.   **M.p.** 133–134 °C
V.p. 21 mPa (60 °C)   **S.g./density** 1.08 (20 °C)   **Solubility** In water 3.85 g/l (25 °C). In ethanol 108.8, diethyl ether 5.5, acetone 80.2, benzene 3.1, chloroform 125, hexane 0.2, groundnut oil 1.0 (all in g/kg, 20–25 °C).
Stability Stable to light and to oxidation. Stable in neutral media, but hydrolysed by strong acids and alkalis.

fenuron-TCA
Mol. wt. 327.6   **M.f.** $C_{11}H_{13}Cl_3N_2O_3$ **Form** Colourless crystals.   **M.p.** 65–68 °C

Solubility In water 4.8 g/l (room temperature). In 1,2-dichloroethane 666, trichloroethylene 567 (both in g/kg, room temperature). Sparingly soluble in petroleum oils.

COMMERCIALISATION
fenuron
History Herbicidal properties of fenuron described by H. C. Bucha & C. W. Todd (*Science*, 1951, **114**, 493). Introduced by E. I. du Pont de Nemours and Co. (who no longer manufacture or market it).   **Patent** US 2655447, GB 691403, GB 692589 all to Du Pont   **Manufacturers** United Phosphorus Ltd

fenuron-TCA
History Fenuron-TCA introduced by Allied Chemical Corp. Agricultural Division (now HACCO Inc.).   **Patent** US 2782112, US 2801911 (both to Allied)

APPLICATIONS
Biochemistry Inhibits photosynthesis.   **Mode of action** Non-selective systemic herbicide, absorbed predominantly by the roots, with translocation acropetally in the xylem.

fenuron
Uses Control of woody plants and deep-rooted perennial weeds, particularly on non-crop land. Often used in combination with chlorpropham to extend the weed spectrum and range of crops.   **Formulation types** WP.   **Mixtures** *(fenuron +)* chlorpropham; chlorpropham + propham; chloridazon + chlorpropham + propham; chloridazon + propham.   **Selected tradenames** 'Croptex Chrome' (mixture) (Hortichem); 'Kayron' (Krishi Rasayan)

fenuron-TCA
Uses Fenuron-TCA combines the herbicidal actions of fenuron and trichloroacetic acid, and is used for total weed control on non-crop areas and against woody plants.   **Formulation types** WP.   **Selected tradenames** 'Dozer' (HACCO); 'Urab' (HACCO)

ANALYSIS
Product analysis of fenuron based on hydrolysis (W. K. Lowen & H. M. Baker, *Anal. Chem.*, 1952, **24**, 1476).   **Residues** determined by colorimetry (R. L. Dalton & H. L. Pease, *J. Assoc. Off. Agric. Chem.*, 1962, **45**, 377).

MAMMALIAN TOXICOLOGY
fenuron
Oral Acute oral LD_{50} for rats 6400 mg/kg.   **Skin and eye** Acute percutaneous LD_{50} not available. Non-irritating to skin.   **NOEL** In 90 d feeding trials, no apparent effect observed with rats receiving 500 mg fenuron/kg diet.
Toxicity class WHO (a.i.) III (Table 5)

fenuron-TCA

Oral Acute oral LD_{50} for rats 4000–5700 mg/kg.   **Skin and eye** Acute percutaneous LD_{50} not available. Skin, eye, and respiratory tract irritant. **Toxicity class** WHO (a.i.) III (Table 5)   **EC risk** Xi (R38)

ECOTOXICOLOGY
fenuron
Fish LC_{50} (48 h) for guppies 610 mg/l.

ENVIRONMENTAL FATE
Plants In plants, *N*-demethylation occurs step-by-step.   **Soil/Environment** In soil, enzymic and microbial degradation involves step-by-step *N*-demethylation.

319 fenvalerate   *Insecticide, acaricide, ixodicide*

pyrethroid

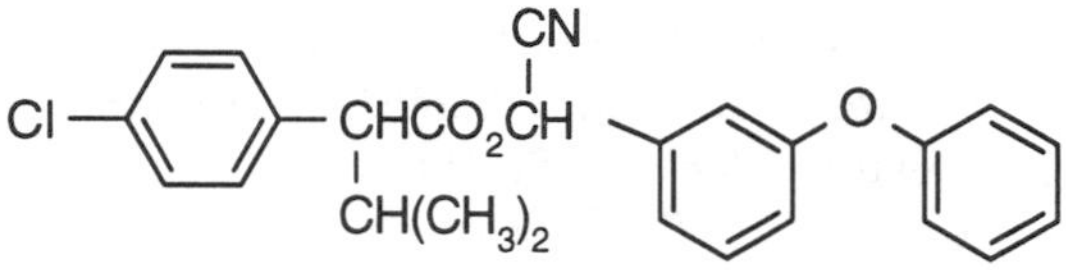

NOMENCLATURE
Common name fenvalerate (BSI, E-ISO, (*m*) F-ISO, ESA)
IUPAC name (*RS*)-α-cyano-3-phenoxybenzyl (*RS*)-2-(4-chlorophenyl)-= 3-methylbutyrate
Chemical Abstracts name cyano(3-phenoxyphenyl)methyl 4-chloro-α-= (1-methylethyl)benzeneacetate
CAS RN *[51630–58–1]* unstated stereochemistry   **Development codes** S-5602 (Sumitomo); WL 43 775 (Shell)   **Official codes** OMS 2000

PHYSICAL CHEMISTRY
Composition Tech. grade fenvalerate is 92.1% pure.   **Mol. wt.** 419.9
M.f. $C_{25}H_{22}ClNO_3$ **Form** Tech. grade fenvalerate is a viscous yellow or brown liquid, sometimes partly crystalline at room temperature.   **M.p.** 39.5–53.7 °C (pure)   **B.p.** Decomposes on distillation.   **V.p.** 1.92×10^{-2} mPa (20 °C)
K_{ow} logP = 5.01 (23 °C)   **S.g./density** 1.175 (25 °C)   **Solubility** In water <10 µg/l (25 °C). In *n*-hexane 53, xylene ≥200, methanol 84 (all in g/l, 20 °C).
Stability Stable to heat and moisture. Relatively stable in acidic media, but rapidly hydrolysed in alkaline media.   **F.p.** 230 °C

COMMERCIALISATION

History Insecticide reported by N. Ohno et al. (*Agric. Biol. Chem.*, 1974, **38**, 881) and results of field trials by M. D. Mowlam et al. (*Proc. Br. Crop Prot. Conf. – Pests Dis.*, 1977, **2**, 649). Introduced by Sumitomo Chemical Co., Ltd and, in some countries, by Shell International Chemical Co. (who no longer manufacture or market it). **Patent** GB 1439615; US 4062968 both to Sumitomo
Manufacturers Aimco; Ankur; Dhanuka; Ficom; Krishi Rasayan; New Chemi; Rallis; Sanachem; Searle India; Sumitomo; United Phosphorus

APPLICATIONS

Mode of action Non-systemic insecticide and acaricide with contact and stomach action. **Uses** Control of a wide range of pests, including those resistant to organochlorine, organophosphorus, and carbamate insecticides. Uses include control of chewing, sucking, and boring insects (particularly Lepidoptera, Diptera, Orthoptera, Hemiptera, and Coleoptera) in fruit, vines, olives, hops, nuts, vegetables, cucurbits, cotton, oilseed rape, sunflowers, alfalfa, cereals, maize, sorghum, potatoes, beet, peanuts, soya beans, tobacco, sugar cane, ornamentals, forestry, and on non-crop land. Used for control of flying and crawling insects in public health situations and in animal houses. Also used as an animal ectoparasiticide. **Formulation types** EC; UL; SC; WP. **Mixtures** *(fenvalerate +)* fenitrothion; dimethoate. **Compatibility** Incompatible with alkaline materials. **Selected tradenames** 'Sumicidin' (Sumitomo); 'Arfen' (Ramcides); 'Cekufenvalerato' (Cequisa); 'Dufen' (Dhanuka); 'Fenval' (Searle India); 'Kayvalerate' (Krishi Rasayan); 'Sanvalerate' (Sanachem); 'Shasicidin' (Sanonda); 'Starfen' (Shaw Wallace); 'Sumitox' (Aimco)

ANALYSIS

Product analysis of pyrethroids reviewed by E. Papadopoulou-Mourkidou in *Comp. Analyt. Profiles*. By hplc (*ibid., idem*; E. Papadopoulou-Mourkidou, *Anal. Methods Pestic. Plant Growth Regul.*, 1988, **16**, 31; 1984, **13**, 121) or by glc (*CIPAC Handbook*, 1988, **D**, 100; *ibid.*, 1995, **G**, 68–70). **Residues** determined by glc with ECD (*ibid.*; A. Ambrus et al., *J. Assoc. Off. Anal. Chem.*, 1981, **64**, 733; *Man. Pestic. Residue Anal.*, 1987, **I**, 6, S19; *Anal. Methods Residues Pestic.*, 1988, Part I, M11). Details available from Sumitomo Chemical Co.

MAMMALIAN TOXICOLOGY

Reviews *Pesticide residues in food – 1986*. FAO Plant Production and Protection Paper 77, 1986. See also A. J. Gray & D. M. Soderlund, Chapt. 5 in "*Insecticides*". **IARC** 53 **EHC** 95 (WHO, 1990). **Oral** Acute oral LD_{50} for rats 451 mg/kg. **Skin and eye** Acute percutaneous LD_{50} for rabbits 1000–3200, rats >5000 mg/kg. Slightly irritating to skin and eyes (rabbits). **Inhalation** LC_{50} for rats >101 mg/m^3. **NOEL** (2 y) for rats 250 mg/kg diet. **ADI** (JMPR) 0.02 mg/kg b.w. [1986]. **Toxicity class** WHO (a.i.) II; EPA (formulation) II

ECOTOXICOLOGY

Birds Acute oral LD_{50} for domestic fowl >1600, mallard ducks 9932 mg/kg. Dietary LC_{50} for quail >10 000, mallard ducks 5500 mg/kg. **Fish** LC_{50} (96 h) for rainbow trout 0.0036 mg/l. **Bees** Toxic to bees. Contact LD_{50} 0.23 µg/bee.

ENVIRONMENTAL FATE

EHC 95 cites rapid degradation and decomposition, reduced toxicity of its degradation products and absence of leaching in soil. Although laboratory tests show high toxicity for fish and bees, these effects are markedly reduced under field conditions because of adsorption to sediments, and the compound's strong repellent effect. The report concludes that risks to the environment are unlikely when it is applied as recommended. **Animals** In mammals, following oral administration, fenvalerate is rapidly metabolised. Up to 96% is excreted in the faeces within 6–14 days. For a study of the metabolism of fenvalerate in the laying hen, see *J. Agric. Food Chem.*, 1989, **37**, 190. **Plants** In plants, fenvalerate is split into two parts by cleavage of the ether group, followed by further hydroxylation in the 2- and 4-positions of the phenoxy ring, and hydrolysis of the nitrile group to amide and carboxyl groups. The majority of the acids and phenols thus formed are converted into glucosides. **Soil/Environment** In aqueous media, the ester bond is hydrolysed. In light, decarboxylation occurs, with recombination of the cleaved moieties (R. L. Holmstead *et al.*, *J. Agric. Food Chem.*, 1977, **25**, 56–58). DT_{50} in soil *c.* 75–80 d.

MAIN ENTRIES

320 ferbam *Fungicide*

dimethyldithiocarbamate

$$(CH_3)_2NCSS \diagdown \atop (CH_3)_2NCSS \diagup Fe-SCSN(CH_3)_2$$

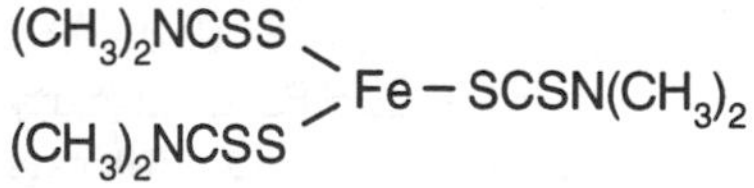

NOMENCLATURE

Common name ferbam (BSI, E-ISO, JMAF); ferbame ((*m*) F-ISO); no name (Germany)
IUPAC name iron tris(dimethyldithiocarbamate); iron(III) dimethyldithiocarbamate; ferric dimethyldithiocarbamate
Chemical Abstracts name (*OC*-6-11)-tris(dimethylcarbamodithioato-*S,S'*)iron
CAS RN *[14484–64–1]* **EEC no.** 238–484–2 **Official codes** ENT 14 689

PHYSICAL CHEMISTRY

Mol. wt. 416.5 **M.f.** $C_9H_{18}FeN_3S_6$ **Form** Black powder. **M.p.** Decomposes above 180 °C **V.p.** Negligible at 20 °C K_{ow} logP = –1.6 **S.g./density** Bulk density c. 0.6 kg/l.   **Solubility** In water 130 mg/l (room temperature). Soluble in organic solvents with high dielectric constant, e.g. chloroform, pyridine, acetonitrile, acetone.   **Stability** Stable to storage in closed containers. Tends to decompose on exposure to moisture and heat, and on prolonged storage.

COMMERCIALISATION
History Introduced by E. I. du Pont de Nemours and Co. (who no longer manufacture or market it).   **Patent** US 1972961   **Manufacturers** UCB

APPLICATIONS
Mode of action Foliar fungicide with protective action and secondary action against chlorosis.   **Uses** Control of scab on pome fruit, peach leaf curl (*Taphrina deformans*) on peaches, blue mould on tobacco, and other diseases on many other crops. Often used in combinations with dithiocarbamates to extend the spectrum of activity.   **Phytotoxicity** Non-phytotoxic, but may cause russetting in Golden Delicious apples.   **Formulation types** WP.   **Mixtures** *(ferbam +)* maneb + zineb.   **Compatibility** Incompatible with copper- or mercury-containing compounds, and with alkaline materials.   **Selected tradenames** 'Ferbam Granuflo' (UCB)

ANALYSIS
Product analysis by colorimetry (*CIPAC Handbook*, 1970, **1**, 397) or by acid hydrolysis, the liberated carbon disulfide being converted to dithiocarbonate which is estimated by titration with iodine (*AOAC Methods*, 1995, 965.15).
Residues determined by acid hydrolysis and suitable colorimetry of a derivative of the liberated carbon disulfide (H. L. Pease, *J. Assoc. Off. Anal. Chem.*, 1957, **40**, 1113; G. E. Keppel, *ibid.*, 1971, **54**, 528; *Analyst (London)*, 1981, **106**, 782).

MAMMALIAN TOXICOLOGY
Reviews *Pesticide residues in food – 1980*. FAO Plant Production and Protection Paper 26, 1981. *Pesticide residues in food: 1980 evaluations*. FAO Plant Production and Protection Paper 26 Sup, 1981.   **IARC** 12, 42   **EHC** 78 (WHO, 1988).
Oral Acute oral LD_{50} for rats >4000 mg/kg.   **Skin and eye** Acute percutaneous LD_{50} for rabbits >4000 mg/kg. Slightly irritating to eyes (rabbits). Irritating to skin and mucous membranes. Not a skin sensitiser (guinea pigs).   **Inhalation** LC_{50} (4 h) for rats 0.4 mg/l.   **NOEL** (2 y) for rats 250 mg/kg diet; for dogs 5 mg/kg daily.
ADI (JMPR) 0.02 mg/kg b.w. [1980].   **Toxicity class** WHO (a.i.) III (Table 5); EPA (formulation) IV   **EC risk** Xi (R36/37/38)

ECOTOXICOLOGY
Fish Moderately toxic to fish.   **Bees** Not toxic to bees.   **Daphnia** LC_{50} (48 h) 0.09 mg/l.

ENVIRONMENTAL FATE
Animals Not stored in the body tissues (H. C. Hodge *et al.*, *J. Pharmacol. Exp. Ther.*, 1956, **118**, 174).   **Plants** In plants, the principal metabolite is dimethylamine salt of dimethyldithiocarbamic acid; tetramethylthiuram disulfide, tetramethylthiuram monosulfide, tetramethylthiourea, carbon disulfide and sulfur can also be formed. Dimethyldithiocarbamic acid can be present as the free acid or as the metabolic conversion products DDC-β-glucoside, DDC-α-aminobutyric acid and DDC-α-alanine.

321 ferimzone   *Fungicide*

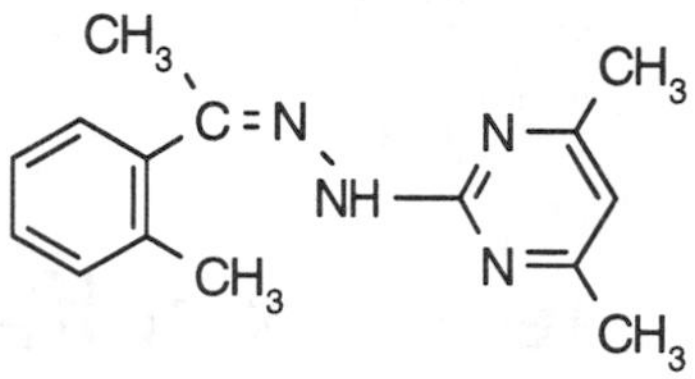

NOMENCLATURE
Common name ferimzone (BSI, draft E-ISO)
IUPAC name (Z)-2'-methylacetophenone 4,6-dimethylpyrimidin-2-ylhydrazone
Chemical Abstracts name (Z)-4,6-dimethyl-2(1H)-pyrimidinone [1-(2-methyl=phenyl)ethylidene]hydrazone
CAS RN [89269-64-7]   **Development codes** TF-164

PHYSICAL CHEMISTRY
Mol. wt. 254.3   **M.f.** $C_{15}H_{18}N_4$ **Form** Colourless crystals.   **M.p.** 175–176 °C
V.p. 4.11×10^{-3} mPa (20 °C)   **K_{ow}** logP = 2.89 (25 °C)   **S.g./density** 1.185
Solubility In water 162 mg/l (30 °C). Soluble in ethanol, ethyl acetate, chloroform, xylene, and acetonitrile.   **Stability** Stable to sunlight. Stable in neutral and alkaline aqueous solutions.

COMMERCIALISATION
History Fungicide introduced by Takeda Chemical Industries Ltd.
Manufacturers Takeda

APPLICATIONS
Uses Control of *Helminthosporium oryzae*, *Cercospora oryzae*, and *Pyricularia oryzae* in rice.   **Formulation types** WP.   **Mixtures** *(ferimzone +)* phthalide.
Selected tradenames 'Blasin' (mixture) (Takeda)

ANALYSIS
Product and **residue** analysis by hplc.

MAMMALIAN TOXICOLOGY
Oral Acute oral LD_{50} for male rats 725, female rats 642, male mice 590, female mice 542 mg/kg.   **Skin and eye** Acute percutaneous LD_{50} for rats >2000 mg/kg.
Inhalation LC_{50} (4 h) for rats >3.8 mg/l.   **Toxicity class** WHO (a.i.) III

ECOTOXICOLOGY
Birds Acute oral LD_{50} for bobwhite quail >2250, mallard ducks >292 mg/kg.
Fish LC_{50} (72 h) for carp 10 mg/l.   **Bees** LD_{50} (oral) >140 µg/bee.
Other aquatic spp. LC_{50} (24 h) for *Moina macrocopa* >40 mg/l.

ENVIRONMENTAL FATE
Soil/Environment DT_{50} 3–14 d, depending on soil type.

322 ferrous sulfate   *Herbicide*

$FeSO_4$

NOMENCLATURE
Common name ferrous sulfate (E-ISO, accepted in lieu of a common name); sulfate ferreux (F-ISO)
IUPAC name iron(II) sulfate
Chemical Abstracts name iron(2+) sulfate (1:1)
Other names green vitriol   **CAS RN** *[7782–63–0]* heptahydrate; *[7720–78–7]* anhydrous

PHYSICAL CHEMISTRY
Mol. wt. 151.9   **M.f.** FeO_4S **Form** Crystalline solid.   **Solubility** In water 266 g/l (10 °C).

COMMERCIALISATION
History Recommended by H. L. Bolley (*N. Dakota Agric. Exp. Sta. Rep.*, 1901, **11**, 48).

APPLICATIONS
Uses Used mainly as a constituent of lawn sand to kill moss. Higher plants may also be scorched.   **Formulation types** GR; SP; MG; DP.
Mixtures *(ferrous sulfate +)* chloroxuron; dichlorophen; dichlorprop + MCPA.

fipronil *Insecticide*

phenylpyrazole

NOMENCLATURE
Common name fipronil (BSI, pa E-ISO)
IUPAC name (±)-5-amino-1-(2,6-dichloro-α,α,α-trifluoro-p-tolyl)-4-trifluoro=
methylsulfinylpyrazole-3-carbonitrile
Chemical Abstracts name 5-amino-[2,6-dichloro-4-(trifluoromethyl)phenyl]-=
4-[(1R,S)-(trifluoromethyl)sulfinyl]-1H-pyrazole-3-carbonitrile
CAS RN *[120068–37–3]* **Development codes** MB 46030 (Rhône-Poulenc)

PHYSICAL CHEMISTRY
Mol. wt. 437.2 **M.f.** $C_{12}H_4Cl_2F_6N_4OS$ **Form** White solid. **M.p.** 200–201 °C;
(tech., 195.5–203 °C) **V.p.** 3.7×10^{-4} mPa (25 °C) **K$_{ow}$** logP = 4.0 (shake flask
method) **Henry** 3.7×10^{-5} Pa m^3 mol^{-1} (calc.) **S.g./density** 1.477–1.626 (20 °C)
Solubility In water 1.9 (pH 5), 2.4 (pH 9), 1.9 (distilled) (all mg/l, 20 °C). In
acetone 545.9, dichloromethane 22.3, hexane 0.028, toluene 3.0 (all g/l, 20 °C).
Stability Stable in water at pH 5 and 7; slowly hydrolysed at pH 9 (DT$_{50}$ c. 28 d).
Stable to heat. Slowly degrades in sunlight (c. 3% loss after 12 d continuous
irradiation); rapidly photolysed in aqueous solution (DT$_{50}$ c. 0.33 d).

COMMERCIALISATION
History Discovered by Rhône-Poulenc in 1987. Reported by F. Colliot et al., Proc.
Br. Conf. – Pests Dis., 1992, **1**, 29. Introduced by Rhône-Poulenc Agrochimie in
1993. **Manufacturers** Rhône-Poulenc

APPLICATIONS
Biochemistry Insecticide which acts as a potent blocker of the GABA-regulated
chloride channel. Insects resistant or tolerant to pyrethroid, cyclodiene,
organophosphorus and/or carbamate insecticides are susceptible to fipronil.
Mode of action Broad-spectrum insecticide, toxic by contact and ingestion.
Moderately systemic and, in some crops, can be used to control insects when
applied as a soil or seed treatment. Good to excellent residual control following
foliar application. **Uses** Control of multiple species of thrips on a broad range of
crops by foliar, soil or seed treatment. Control of corn rootworm, wireworms
and termites by soil treatment in maize. Control of boll weevil and plant bugs on
cotton, diamond-back moth on crucifers, Colorado potato beetle on potatoes by
foliar application. Control of stem borers, leaf miners, planthoppers, leaf

MAIN ENTRIES

folder/rollers and weevils in rice. Foliar application rates range from 10–80 g/ha; soil treatment rates 100–200 g/ha. **Formulation types** SC; GR; FS; EC; WG; UL. **Selected tradenames** 'Prince' (Nissan); 'Regent' (Rhône-Poulenc)

ANALYSIS
Product by hplc with u.v. detection. **Residues** by gc with ECD.

MAMMALIAN TOXICOLOGY
Oral Acute oral LD_{50} for rats 97, mice 95 mg/kg. **Skin and eye** Acute percutaneous LD_{50} for rats >2000, rabbits 354 mg/kg. Not a skin or eye irritant (OECD criteria). Not a skin sensitiser. **Inhalation** LC_{50} (4 h) for rats 0.682 mg/l (tech.; nose only exposure). **NOEL** (2 y) for rats 0.5 mg/kg diet; (18 mo) for mice 0.5 mg/kg diet; (52 w) for dogs 0.2 mg/kg b.w. daily (combined sexes). **Other** Non-mutagenic, non-teratogenic; no adverse effect on reproductive performance. Clinical signs of toxicity consistent with the interaction of the molecule at a neurotransmitter receptor were observed in all species tested but were completely reversible. **Toxicity class** WHO (a.i.) II; EPA (formulation) II

ECOTOXICOLOGY
Birds Acute oral LD_{50} for bobwhite quail 11.3, mallard ducks >2000, pheasants 31, red-legged partridges 34, house sparrows 1120, pigeons >2000 mg/kg. Dietary LC_{50} (5 d) for bobwhite quail 49, mallard ducks >5000 mg/kg diet. **Fish** Acute LC_{50} (96 h) for bluegill sunfish 85, rainbow trout 248, European carp 430 µg/l. **Bees** Highly toxic to honeybees both by direct contact and by ingestion. However no risk to bees when used as a soil or seed treatment. **Worms** Non-toxic. **Daphnia** LC_{50} (48 h) 0.19 mg/l; for *D. carinata* (48 h) 3.8 mg/l. **Algae** EC_{50} (96 h) for *Scenedesmus subspicatus* 0.068 mg/l; (120 h) for *Selenastrum capricornutum* >0.16, *Anabaena flos-aquae* >0.17 mg/l.

ENVIRONMENTAL FATE
In plants, animals and the environment, fipronil is metabolised via reduction to the sulfide, oxidation to the sulfone, and hydrolysis to the amide. In the presence of sunlight, a photodegradate also forms via sulfoxide extrusion. The sulfide, sulfone and photodegradate are known to act at the GABA receptor site, whereas the amide does not. **Animals** In the rat, once absorbed, the distribution and metabolism of fipronil is rapid. Elimination is mainly via the faeces as fipronil and its sulfone. The two major urinary metabolites in the rat were identified as conjugates of ring-opened pyrazole products. The distribution of radioactive residues in tissues was extensive after seven days. In goat and hen, the sulfone was the only metabolite identified in tissues. **Plants** When applied as an incorporated soil treatment to cotton, maize, sugar beet or sunflower, uptake of fipronil into plants in all cases was low (c. 5%). At crop maturity, the major residue components observed in all plants were fipronil, the sulfone, and the amide. Following foliar application to cotton, cabbage, rice and potatoes, at crop maturity

fipronil and the photodegradate were the major residue components.
Soil/Environment Results of lab. and field studies: Readily degraded: major
degradates in soil (aerobic) are sulfone and amide, (anaerobic) are sulfide and
amide. Photolysis of soil-applied fipronil gives the photodegradate together with
sulfone and amide. K_{oc} 427 (Speyer 2.2) to 1248 (sandy loam). Both fresh and
aged column leaching studies (5 soils) indicate that fipronil and its metabolites
present a low risk of downward movement in soil; this is supported by field
dissipation studies. Following soil incorporated in-furrow granular applications,
quantifiable residues were confined to the top 30 cm of soil, with no significant
lateral movement or residues.

324 flamprop-M   *Herbicide*

arylalanine

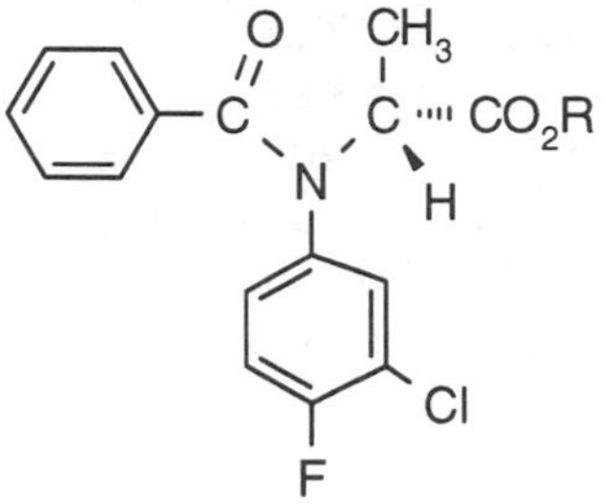

flamprop-M-methyl R = CH$_3$-

flamprop-M-isopropyl R = (CH$_3$)$_2$CH-

NOMENCLATURE
flamprop-M
Common name flamprop-M (BSI, draft E-ISO, (*m*) draft F-ISO)
IUPAC name *N*-benzoyl-*N*-(3-chloro-4-fluorophenyl)-D-alanine (previously
referred to erroneously in the literature as the L-alanine)
Chemical Abstracts name *N*-benzoyl-*N*-(3-chloro-4-fluorophenyl)-D-alanine
(formerly stated to be the L-alanine)
CAS RN *[90134–59–1]* D- acid; *[57353–42–1]* L- acid

flamprop-M-isopropyl
IUPAC name isopropyl *N*-benzoyl-*N*-(3-chloro-4-fluorophenyl)-D-alaninate
(previously referred to erroneously in the literature as the L-alaninate)
Chemical Abstracts name 1-methylethyl *N*-benzoyl-*N*-(3-chloro-4-fluorophenyl)-=
D-alaninate (formerly stated to be the L-alaninate)
Other names L-flamprop-isopropyl (rejected common name)
CAS RN *[63782–90–1]* D- form; *[57973–67–8]* L- form

Development codes WL 43425 (Shell); CL 901445 (Cyanamid); AC 901445 (Cyanamid)

flamprop-M-methyl
IUPAC name methyl *N*-benzoyl-*N*-(3-chloro-4-fluorophenyl)-D-alaninate
Chemical Abstracts name methyl *N*-benzoyl-*N*-(3-chloro-4-fluorophenyl)-=
D-alaninate
CAS RN *[63729–98–6]* D- form   **Development codes** WL 43423 (Shell);
CL 901444 (Cyanamid); AC 901444 (Cyanamid)

PHYSICAL CHEMISTRY
flamprop-M
Composition Tech. grade is ≥93%.   **Mol. wt.** 321.7   **M.f.** $C_{16}H_{13}CIFNO_3$
K_{ow} logP = 3.09 (25 °C, for flamprop, K. Chamberlain et *al.*, *Pestic Sci.*, **47**, 265
(1996))   **pKa** 3.7 (for flamprop, K. Chamberlain et *al.*, *Pestic Sci.*, **47**, 265 (1996)
and refs therein)

flamprop-M-isopropyl
Composition Tech. grade is >96%.   **Mol. wt.** 363.8   **M.f.** $C_{19}H_{19}CIFNO_3$
Form White crystals; (tech., off-white crystals).   **M.p.** 72.5–74.5 °C; (tech.,
70–71 °C)   **V.p.** 8.5×10^{-2} mPa (25 °C)   **K_{ow}** logP = 3.69
S.g./density 1315 kg/m^3 **Solubility** In water 12 mg/l (20 °C). In acetone 1560,
cyclohexanone 677, ethanol 147, hexane 16, xylene *c.* 500 (all in g/l, 20 °C).
Stability Stable to light and to heat, and at pH 2–8; DT_{50} (pH 7) 9140 d.
Hydrolysed at pH >8 to flamprop-M and isopropanol.   **F.p.** Non-flammable.

flamprop-M-methyl
Composition >96% pure.   **Mol. wt.** 335.8   **M.f.** $C_{17}H_{15}CIFNO_3$ **Form** White to
light grey crystals.   **M.p.** 84–86 °C; (tech., 81–82 °C)   **V.p.** 1.0 mPa (20 °C)
K_{ow} logP = 3.0   **S.g./density** 1.311 kg/l (22 °C)   **Solubility** In water 0.016 g/l
(25 °C). In acetone 406, *n*-hexane 2.3 (both in g/l, 25 °C).   **Stability** Stable to light
and to heat, and at pH 2–7. Hydrolysed in alkaline media (pH >7) to parent acid
and methanol.

COMMERCIALISATION
History Herbicidal properties of the isopropyl ester of the D- acid described by
R. M. Scott et *al.* (*Proc. Br. Crop Prot. Conf. – Weeds*, 1976, **2,** 723), design
discussed by M. A. Venis (*Pestic. Sci.*, 1982, **13,** 309) and development by D.
Jordan (*Span*, 1977, **20,** 21). Introduced by Shell Research Ltd (now American
Cyanamid Co.).   **Patent** GB 1437711; GB 1563201   **Manufacturers** Cyanamid

APPLICATIONS
Biochemistry Fatty acid synthesis inhibitor. Inhibits cell elongation and cell division,
and hence inhibits plant growth. Selectivity depends on differential rates of

hydrolysis to the free acid. In tolerant plants, the acid is further de-toxified by formation of conjugates.   **Mode of action** Flamprop-M-isopropyl and -M-methyl are selective systemic herbicides, absorbed by the leaves. Undergo hydrolysis to flamprop-M, which is the herbicidally active compound; in sensitive species, this is transported to meristems.

flamprop-M-isopropyl
Uses Post-emergence control of wild oats (*Avena* spp.) in barley and wheat, including those undersown with clover or ryegrass. Also controls *Alopecurus myosuroides* and *Arrhenatherum elatius*.   **Phytotoxicity** Some varieties of wheat and barley may be injured.   **Formulation types** EC.   **Compatibility** Antagonism with broad-leaved herbicides can be expected.   **Selected tradenames** 'Suffix BW' (Cyanamid)

flamprop-M-methyl
Uses Post-emergence control of wild oats (*Avena* spp.) in wheat, including crops undersown with clover or grass. Also controls *Alopecurus myosuroides*.
Phytotoxicity Non-phytotoxic to all spring and winter varieties of wheat.
Formulation types EC.   **Compatibility** Miscible with fungicides, chlormequat chloride and foliar nutrients. If applied together with phenoxy herbicides, the activity of flamprop-M-methyl may be reduced.   **Selected tradenames** 'Mataven L' (Cyanamid)

ANALYSIS
Product analysis for esters is by optical rotation and glc.   **Residues** determined by glc with ECD. Details available from Cyanamid.

MAMMALIAN TOXICOLOGY
flamprop-M
Toxicity class WHO (a.i.) III (Table 5)

flamprop-M-isopropyl
Oral Acute oral LD_{50} for rats and mice >4000 mg/kg.   **Skin and eye** Acute percutaneous LD_{50} for rats >1600 mg/kg. Not a skin or eye irritant.
Inhalation No effect (rats).   **NOEL** In 90 d feeding trials, rats receiving 50 mg/kg diet and dogs receiving 30 mg/kg diet showed no ill-effects.
Other Acute i.p. LD_{50} for rats >1200 mg/kg.

flamprop-M-methyl
Oral Acute oral LD_{50} for rats 1210, mice 720 mg/kg.   **Skin and eye** Acute percutaneous LD_{50} for rats >1800 mg/kg (as EC formulation). Non-irritating to skin and eyes; non-sensitising to skin.   **Inhalation** No effect (rats).   **NOEL** In 90 d feeding trials, rats receiving 2.5 mg/kg daily and dogs receiving 0.5 mg/kg daily showed no ill-effects.   **Other** Acute i.p. LD_{50} for rats 350–500 mg/kg.

ECOTOXICOLOGY

flamprop-M-isopropyl

Birds Acute oral LD_{50} for domestic fowl >2000 mg/kg. **Fish** LC_{50} (96 h) for rainbow trout 2.4 mg/l. **Bees** Non-toxic to bees. **Worms** Non-toxic. **Other beneficial spp.** Non-toxic to soil arthropods. **Daphnia** Slightly to moderately toxic. **Algae** EC_{50} (96 h) 6.8 mg/l. **Other aquatic spp.** Moderately toxic to freshwater and marine crustacea.

flamprop-M-methyl

Birds Acute oral LD_{50} for bobwhite quail 4640, pheasants, mallard duck, domestic fowl, partridges, pigeons all >1000 mg/kg. **Fish** LC_{50} (96 h) for rainbow trout 4.0 mg/l. **Bees** Non-toxic to bees. **Worms** Non-toxic. **Other beneficial spp.** Non-toxic to soil arthropods. **Daphnia** Slightly to moderately toxic. **Algae** EC_{50} (96 h) 5.1 mg/l. **Other aquatic spp.** Moderately toxic to freshwater and marine crustacea.

ENVIRONMENTAL FATE

Animals In mammals, following oral administration of flamprop-M-methyl or flamprop-M-isopropyl, complete metabolism and excretion occurs within 4 days. **Plants** In plants, flamprop-M-methyl and flamprop-M-isopropyl are hydrolysed to the biologically-active flamprop acid, which then undergoes conversion to a biologically-inactive conjugate. **Soil/Environment** The major soil degradate from both esters is flamprop free acid.

325 flazasulfuron   *Herbicide*

sulfonylurea

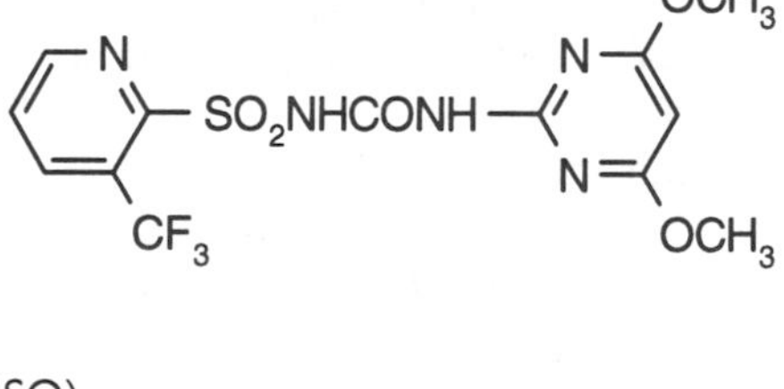

NOMENCLATURE

Common name flazasulfuron (BSI, draft E-ISO)
IUPAC name 1-(4,6-dimethoxypyrimidin-2-yl)-3-(3-trifluoromethyl-2-pyridyl= sulfonyl)urea
Chemical Abstracts name N-[[(4,6-dimethoxy-2-pyrimidinyl)amino]carbonyl]-= 3-(trifluoromethyl)-2-pyridinesulfonamide
CAS RN [104040–78–0]   **Development codes** SL-160; OK-1166

PHYSICAL CHEMISTRY
Composition Tech. is ≥92.0%.   **Mol. wt.** 407.3   **M.f.** $C_{13}H_{12}F_3N_5O_5S$
Form Odourless, white crystalline powder.   **M.p.** 166–170 °C   **V.p.** <0.013 mPa
K_{ow} logP = –0.06 (unstated pH)   **Henry** <2.58 × 10^{-6} Pa m^3 mol^{-1}
S.g./density 1.6055 (20 °C)   **Solubility** In water 2.1 g/l (pH 7, 25 °C). In octanol
0.2, methanol 4.2, acetone 22.7, toluene 0.56, acetonitrile 8.7 (all g/l, 25 °C); in
hexane 0.5 mg/l (25 °C).   **Stability** DT_{50} in water 11 d (25 °C).   **pKa** 4.37
(20 °C)   **F.p.** Non-flammable.

COMMERCIALISATION
History Herbicide introduced by Ishihara Sangyo Kaisha, Ltd.
Manufacturers Ishihara Sangyo

APPLICATIONS
Biochemistry Branched chain amino acid synthesis (acetolactate synthase or ALS)
inhibitor. Acts by inhibiting biosynthesis of the essential amino acids valine and
isoleucine, hence stopping cell division and plant growth.
Mode of action Systemic herbicide, rapidly absorbed through the leaves and
translocated throughout the plant.   **Uses** Post-emergence control of grass and
broad-leaved weeds and sedges (especially *Cyperus brevifolius* and *Cyperus
rotundus*) in warm-season turf (*Zoysia* and *Cynodon* spp.) at 25–100 g/ha.
Phytotoxicity Transient discolouration of newly developing leaves and stunting
may sometimes occur after treatment of turf, but recovery is rapid and complete.
Formulation types WP; WG.   **Selected tradenames** 'Katana' (Ishihara Sangyo);
'Shibagen' (Ishihara Sangyo)

ANALYSIS
Methods for sulfonylurea residues in crops, soil and water reviewed (A. C.
Barefoot *et al., Proc. Br. Crop Prot. Conf. – Weeds,* 1995, **2**, 707). Details from
Ishihara Sangyo.

MAMMALIAN TOXICOLOGY
Oral Acute oral LD_{50} for rats and mice >5000 mg/kg.   **Skin and eye** Acute
percutaneous LD_{50} for rats >2000 mg/kg. Moderately irritating to eyes; non-
irritating to skin (rabbits). Non-sensitising to skin (guinea pigs).   **Inhalation** LC_{50}
(4 h) for rats 5.99 mg/l.   **NOEL** for rats 1.313 mg/kg daily.   **Other** Non-
mutagenic in the Ames test, DNA repair test, and chromosomal aberration test.

ECOTOXICOLOGY
Birds Acute oral LD_{50} for Japanese quail >2000 mg/kg.   **Fish** LC_{50} (48 h) for carp
>20 mg/l.   **Bees** LD_{50} >100 µg/bee.   **Daphnia** LC_{50} (48 h) >20 mg/l.

ENVIRONMENTAL FATE
Soil/Environment In soil, DT_{50} <7 d under field conditions. DT_{90} 80–100 d
(15 °C) in several soils.

coumarin anticoagulant

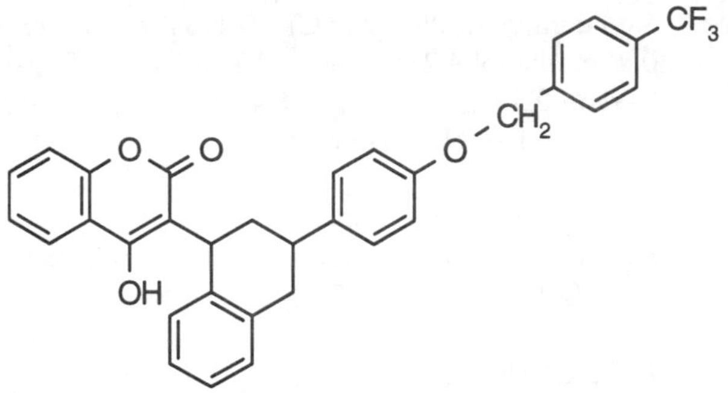

NOMENCLATURE

Common name flocoumafen (BSI, draft E-ISO); flocoumafène ((m) (draft F-ISO))
IUPAC name 4-hydroxy-3-[1,2,3,4-tetrahydro-3-[4-(4-trifluoromethylbenzyl=
oxy)phenyl]-1-naphthyl]coumarin (mixture of *cis*- to *trans*- isomers in the ratio
range 60:40 to 40:60)
Chemical Abstracts name 4-hydroxy-3-[1,2,3,4-tetrahydro-3-[4-[[4-(trifluoro=
methyl)phenyl]methoxy]phenyl]-1-naphthalenyl]-2*H*-1-benzopyran-2-one
CAS RN *[90035–08–8]* unstated stereochemistry **Development codes**
WL 108 366 **Official codes** OMS 3047

PHYSICAL CHEMISTRY

Mol. wt. 542.6 **M.f.** $C_{33}H_{25}F_3O_4$ **Form** Off-white solid. **M.p.** *cis*- isomer
181–191 °C, *trans*- isomer 163–166 °C **V.p.** 1.33×10^{-7} mPa (25 °C)
K_{ow} logP = 4.7 **Solubility** In water 1.1 mg/l. In acetone >600, ethanol 34, xylene
33, octanol 44 (all in g/l). **Stability** Stable to hydrolysis; at 50 °C and pH 7–9, no
detectable degradation occurred in 4 weeks. Thermally stable up to 250 °C.

COMMERCIALISATION

History Rodenticide reported by D. J. Bowler *et al.* (*Proc. Br. Crop Prot. Conf. –
Pests Dis.*, 1984, **2**, 397). Introduced by Shell International Chemical Co. Ltd (now
American Cyanamid Co.). **Manufacturers** Cyanamid

APPLICATIONS

Biochemistry Second-generation indirect anticoagulant. Inhibits the metabolism of
vitamin K_1, and thus depletes vitamin K_1-dependent clotting factors in plasma.
Blocks formation of prothrombin. **Uses** Control of rodents including *Mus
musculus, Rattus norvegicus, R. rattus, Arvicanthis niloticus, Bandicota* spp., *Praeomys
natalensis, R. argiventer, R. rattus diardii, R. losea, R. tiomanicus* and *Sigmodon
hispidus*. Effective against rodents which have become resistant to other
anticoagulant rodenticides. In addition to use around buildings, it is effective in

controlling rodents in field and plantation crops including cocoa, cotton, oilpalm, rice and sugar cane. **Formulation types** RB; BB; GB. **Selected tradenames** 'Storm' (Cyanamid)

ANALYSIS
Product analysis by hplc, details available from Cyanamid.

MAMMALIAN TOXICOLOGY
Oral Acute oral LD_{50} for rats 0.25, mice 0.8, dogs 0.075–0.25, rabbits 0.2, hamsters >50, gerbils 0.18, cats >10, goats >10, sheep >5, pigs c. 60 mg/kg.
Skin and eye Acute percutaneous LD_{50} for rats <3 mg tech./kg. **Inhalation** LC_{50} (4 h) for rats 0.16–1.4 mg/l (using 0.5% manufacturing master mix, as dust).
Toxicity class WHO (a.i.) Ia; EPA (formulation) I

ECOTOXICOLOGY
Birds Acute oral LD_{50} for chickens >100, Japanese quail >300, mallard ducks c. 24 mg/kg. Five-day dietary LC_{50} for Japanese quail 37, mallard ducks 1.7 mg/kg diet.
Fish Baits (50 mg a.i./kg) are non-toxic to aquatic species. LC_{50} (96 h) for carp 0.15 mg/l. **Daphnia** LC_{50} (48 h) immobilisation 0.66 mg/l nominal.

ENVIRONMENTAL FATE
Animals In rats (i.v.) (*Xenobiotica*, 1989, **19**, 51) and Japanese quail (oral, i.p.) (*Xenobiotica*, 1989, **19**, 63), flocoumafen is metabolised to a number of hydroxycoumarins.

327 fluazifop-butyl *Herbicide*

2-(4-aryloxyphenoxy)propionic acid

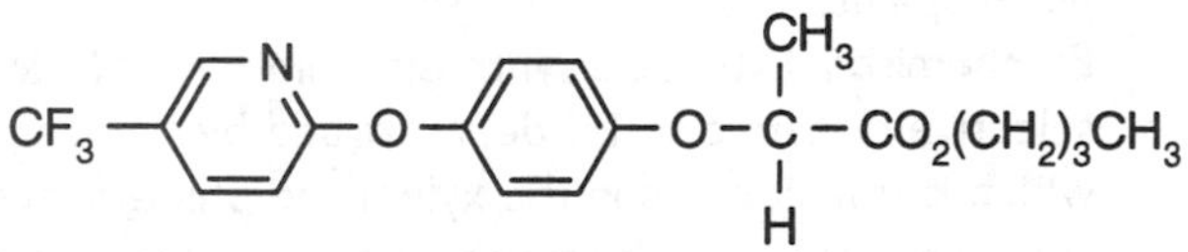

NOMENCLATURE
fluazifop-butyl
IUPAC name butyl (*RS*)-2-[4-(5-trifluoromethyl-2-pyridyloxy)phenoxy]propionate
Chemical Abstracts name butyl (±)-2-[4-[[5-(trifluoromethyl)-2-pyridinyl]oxy]= phenoxy]propanoate
CAS RN *[69806–50–4]* unstated stereochemistry **Development codes** SL-236 (to Ishihara); PP009; ICIA0009 (both to Zeneca)

fluazifop
Common name fluazifop (BSI, draft E-ISO, (*m*) draft F-ISO, ANSI)
IUPAC name (*RS*)-2-[4-(5-trifluoromethyl-2-pyridyloxy)phenoxy]propionic acid
Chemical Abstracts name (±)-2-[4-[[5-(trifluoromethyl)-2-pyridinyl]oxy]=
phenoxy]propanoic acid
CAS RN *[69335–91–7]* unstated stereochemistry
Development codes H-7236; SL-236

PHYSICAL CHEMISTRY
fluazifop-butyl
Composition Tech. material is 91% pure.　**Mol. wt.** 383.4　**M.f.** $C_{19}H_{20}F_3NO_4$
Form Pale straw-coloured liquid.　**M.p.** 13 °C; (tech., 10 °C)
B.p. 165 °C/0.02 mmHg　**V.p.** 0.055 mPa (20 °C)　K_{ow} logP = 4.5
S.g./density 1.21 (20 °C)　**Solubility** In water 1 mg/l (pH 6.5). Miscible with
acetone, cyclohexanone, hexane, methanol, dichloromethane and xylene. In
propylene glycol 24 g/l (20 °C).　**Stability** Stable for 3 years at 25 °C, and for
6 months at 37 °C. Reasonably stable in acidic and neutral conditions, but rapidly
hydrolysed in alkaline media (pH 9).　**F.p.** ≥37.8 °C

fluazifop
Mol. wt. 327.3　**M.f.** $C_{15}H_{12}F_3NO_4$　K_{ow} logP = 3.18 (25 °C, unionised,
K. Chamberlain *et al., Pestic. Sci.,* **47,** 265 (1996))　**pKa** 3.2 (K. Chamberlain *et al.,*
Pestic. Sci., **47,** 265 (1996))

COMMERCIALISATION
History Herbicidal properties of its butyl ester reported by R. E. Plowman *et al.*
(*Proc. Br. Crop Prot. Conf. – Weeds,* 1980, **1,** 29). Discovered by Ishihara Sangyo
Kaisha, Ltd, developed jointly with ICI Plant Protection Division (now Zeneca
Agrochemicals).　**Patent** GB 1599121 to Ishihara　**Manufacturers** Ishihara Sangyo;
Zeneca

APPLICATIONS
fluazifop-butyl
Biochemistry Fatty acid synthesis inhibitor.　**Mode of action** Fluazifop-butyl is a
selective systemic herbicide, absorbed by the leaves. Hydrolysed to fluazifop,
which is translocated in the xylem and phloem, accumulating in the meristems of
annual grasses, and the meristems, rhizomes, and stolons of perennial grasses.
Uses Post-emergence control of annual and perennial grass weeds in broad-leaved
crops. Particular uses include control of volunteer cereals and other grass weeds
in oilseed rape, sugar beet, fodder beet, potatoes, cotton, soya beans, peanuts,
pome fruit, bush fruit, vines, citrus fruit, pineapples, bananas, strawberries,
sunflowers, alfalfa, coffee, ornamentals, and many vegetables.　**Phytotoxicity** Non-
phytotoxic to broad-leaved crops.　**Formulation types** EC.　**Compatibility** With
some herbicides, crop damage may be increased and the effectiveness of

fluazifop-butyl may be reduced.   **Selected tradenames** 'Fusilade' (Zeneca); 'Hache Uno Super' (Ishihara Sangyo); 'Onecide' (Ishihara Sangyo)

ANALYSIS
Product analysis of fluazifop-butyl by glc with FID (*AOAC Methods*, 1995, 984.08; *CIPAC Handbook*, 1988, **1D**, 106).   **Residues** (fluazifop-butyl, fluazifop, free and conjugates) determined by hplc of the free acid. Details available from Zeneca.

MAMMALIAN TOXICOLOGY
fluazifop-butyl
Oral Acute oral LD_{50} for male rats >3030, female rats 3600, male mice 1600, female mice 1900, male guinea pigs 2659, rabbits 621 mg/kg.   **Skin and eye** Acute percutaneous LD_{50} for rats >6050, rabbits >2420 mg/kg. Mild skin irritant; practically non-irritating to eyes (rabbits). Mild skin sensitiser (guinea pigs). **Inhalation** LC_{50} (4 h) for rats >5.24 mg/l.   **NOEL** No toxicological effect noted in feeding trials: in dogs (1 y) at 5 mg/kg daily; in rats (90 d) at 100 mg/kg diet; in mice (2 y) at 5 mg/kg diet.   **Other** Acute i.p. LD_{50} for rats 1761 mg/kg.   **Toxicity class** WHO (a.i.) III (Table 5); EPA (formulation) II, III

ECOTOXICOLOGY
fluazifop-butyl
Birds Acute oral LD_{50} for mallard ducks >17 000 mg/kg. Five-day dietary LC_{50} for mallard ducks >25 000, ring-necked pheasants >18 500 mg/kg diet.   **Fish** LC_{50} (96 h) for rainbow trout 1.37, mirror carp 1.31, bluegill sunfish 0.53 mg/l. Low toxicity to aquatic invertebrates.   **Bees** Very low toxicity to bees, both orally and by contact.   **Daphnia** LC_{50} (24 h) >316 ppm.

ENVIRONMENTAL FATE
Animals Metabolised in rats to form fluazifop, which is not further metabolised.
Plants In plants, fluazifop-butyl is rapidly hydrolysed to fluazifop.
Soil/Environment In soil, rapid degradation of fluazifop-butyl occurs, DT_{50} <1 w. The major degradation product is fluazifop, which has DT_{50} in soil <3 w. Persistence in soil is lengthened by cold and dry conditions.

2-(4-aryloxyphenoxy)propionic acid

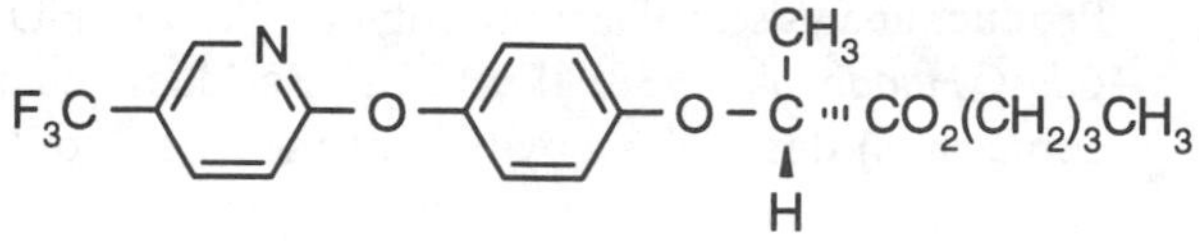

NOMENCLATURE
fluazifop-P-butyl
CAS RN [79241–46–6]

fluazifop-P
Common name fluazifop-P (BSI, draft E-ISO, (*m*) draft F-ISO, ANSI)
IUPAC name (*R*)-2-[4-(5-trifluoromethyl-2-pyridyloxy)phenoxy]propionic acid
Chemical Abstracts name (*R*)-2-[4-[[5-(trifluoromethyl)-2-pyridinyl]oxy]=
phenoxy]propanoic acid
CAS RN [83066–88–0]   **Development codes** PP005; ICIA0005 (both Zeneca);
SL-118 (Ishihara)

PHYSICAL CHEMISTRY
fluazifop-P-butyl
Composition Tech. material is ≥83% pure.   **Mol. wt.** 383.4   **M.f.** $C_{19}H_{20}F_3NO_4$
Form Colourless liquid.   **M.p.** –20 °C (glass-like)   **B.p.** 154 °C/0.02 mmHg
V.p. 0.033 mPa (20 °C)   **K_{ow} logP** = 4.5 (20 °C)   **S.g./density** 1.22 (20 °C)
Solubility In water 1.1 mg/l (20 °C). Miscible with acetone, hexane, methanol,
dichloromethane, ethyl acetate, toluene, and xylene.   **Stability** Fluazifop-P-butyl is
stable to u.v. light. Stable ≥8 y (25 °C), 12 w (50 °C); decomposes at 210 °C.
Hydrolysis DT_{50} >30 d (pH 5), 78 d (pH 7), 29 h (pH 9). Aqueous photolysis
DT_{50} 6 d (pH 5).   **pKa** –3.1 (calc.)   **F.p.** >50 °C

fluazifop-P
Mol. wt. 327.3   **M.f.** $C_{15}H_{12}F_3NO_4$ **Form** Pale yellow, glass-like material.
V.p. 7.9×10^{-4} mPa (20 °C)   **K_{ow} logP** = 3.1 (pH 2.6, 20 °C); –0.8 (pH 7, 20 °C,
est.)   **Solubility** In water 780 mg/l (20 °C).   **pKa** 2.98

COMMERCIALISATION
History The butyl ester, fluazifop-P-butyl, introduced as a herbicide by ICI Plant
Protection Division (now Zeneca Agrochemicals).   **Manufacturers** Zeneca

APPLICATIONS
fluazifop-P-butyl
Biochemistry Fatty acid synthesis inhibitor.   **Mode of action** Fluazifop-P-butyl is
quickly absorbed through the leaf surface, hydrolysed to fluazifop-P and

translocated through the phloem and xylem, accumulating in the rhizomes and stolons of perennial grasses and the meristems of annual and perennial grasses. **Uses** Post-emergence control of wild oats, volunteer cereals, and annual and perennial grass weeds in oilseed rape, sugar beet, fodder beet, potatoes, vegetables, cotton, soya beans, pome fruit, stone fruit, bush fruit, vines, citrus fruit, pineapples, bananas, strawberries, sunflowers, alfalfa, ornamentals, and other broad-leaved crops. **Phytotoxicity** Non-phytotoxic to broad-leaved crops. **Formulation types** EC; EW. **Mixtures** *(fluazifop-P-butyl +)* fomesafen. **Selected tradenames** 'Fusilade' (Zeneca), 'Venture' (Zeneca); 'Winner' (Barclay)

ANALYSIS
Product analysis for total fluazifop-butyl ((R)- and (S)- isomers) by gc with FID; hplc with u.v. detection is then used to determine the ratio of (R)- to (S)- isomers (*CIPAC Handbook*, 1995, **G**, 71–81). **Residues** (fluazifop-P-butyl, fluazifop-P, free and conjugates) as total acid by hplc. For details, see G. S. Davy et al. in *Comp. Anal. Profiles*, Chapt. 7.

MAMMALIAN TOXICOLOGY
fluazifop-P-butyl
Oral Acute oral LD_{50} for male rats 3680, female rats 2451 mg/kg.
Skin and eye Acute percutaneous LD_{50} for rabbits >2000 mg/kg. Mild skin and eye irritant (rabbits). Not a skin sensitiser (guinea pigs). **Inhalation** LC_{50} (4 h) for rats >5.24 mg/l. **NOEL** (2 y) for rats 10 mg/kg diet; (1 y) for dogs 5 mg/kg b.w. daily; (90 d) for rats 100 mg/kg diet. Multi-generation study (rats) 10 mg/kg diet; developmental toxicity for rats 2, rabbits 30 mg/kg b.w. daily. **ADI** 0.01 mg/kg (EPA). **Other** Genotoxicity negative.

ECOTOXICOLOGY
fluazifop-P-butyl
Birds Acute oral LD_{50} for mallard ducks >3528 mg/kg. **Fish** LC_{50} (96 h) for rainbow trout 1.07 mg/l. **Bees** Very low toxicity to bees. LD_{50} (oral and contact) >0.2 mg/bee.

ENVIRONMENTAL FATE
Animals In mammals, fluazifop-P-butyl is metabolised to fluazifop-P, which is rapidly excreted. **Plants** In plants, fluazifop-P-butyl is rapidly hydrolysed to fluazifop-P, which is then partly conjugated. Ether cleavage gives the pyridone and propionic acid metabolites, which may both be further metabolised or conjugated. **Soil/Environment** K_{oc} 5800 (for the acid, 30). In moist soils, rapid degradation of fluazifop-P-butyl occurs, DT_{50} 24 h. The major degradation product is fluazifop-P, which is hydrolysed to 5-trifluoromethylpyrid-2-one, and 2-(4-hydroxyphenoxy)= propionic acid, both of which are further degraded, ultimately to CO_2. In laboratory studies, DT_{50} for fluazifop-P acid ranges from <3 w (loamy soil) to >24 w (coarse sand). In the field, fluazifop-P degradation is more rapid, DT_{50} <4 w.

2,6-dinitroaniline

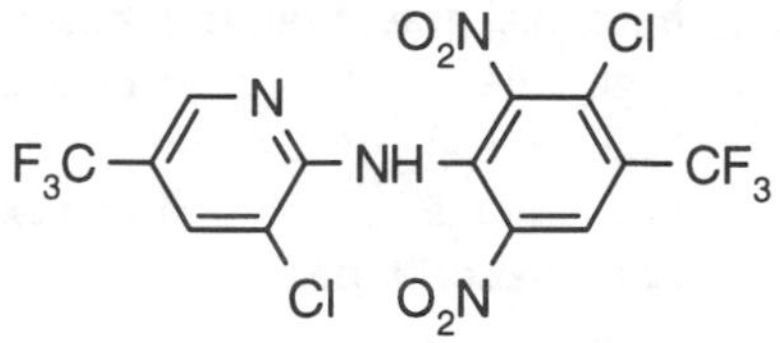

NOMENCLATURE
Common name fluazinam (BSI, draft E-ISO); fluaziname ((*m*) draft F-ISO)
IUPAC name 3-chloro-*N*-(3-chloro-5-trifluoromethyl-2-pyridyl)-α,α,α-trifluoro-=
2,6-dinitro-*p*-toluidine
Chemical Abstracts name 3-chloro-*N*-[3-chloro-2,6-dinitro-4-(trifluoromethyl)=
phenyl]-5-(trifluoromethyl)-2-pyridinamine
CAS RN *[79622–59–6]* **Development codes** B-1216; IKF-1216; ICIA0192

PHYSICAL CHEMISTRY
Mol. wt. 465.1 **M.f.** $C_{13}H_4Cl_2F_6N_4O_4$ **Form** Light yellow crystals.
M.p. 115–117 °C **V.p.** 1.5 mPa (25 °C) **K_{ow}** logP = 3.56 **S.g./density** 0.366
(25 °C, bulk) **Solubility** In water 1.7 mg/l (pH 6.8, 25 °C). In *n*-hexane 12,
acetone 470, toluene 410, diethyl ether 320, dichloromethane 330, ethanol 150
(all in g/l, 20 °C). **Stability** Stable to acid, alkali and heat.

COMMERCIALISATION
History Fungicide introduced by Ishihara Sangyo Kaisha, Ltd.
Manufacturers Ishihara Sangyo

APPLICATIONS
Biochemistry Uncouples mitochondrial oxidative phosphorylation.
Mode of action Fungicide with protective action. Has little curative and systemic
activity, but good residual effect and rain fastness. **Uses** Control of grey mould
and downy mildew on vines; apple scab; southern blight and white mould on
peanuts; and *Phytophthora infestans* and tuber blight on potatoes. Control of
clubroot on crucifers and rhizomania on sugar beet. Also for control of mites in
apples. **Formulation types** SC; WP; DP. **Selected tradenames** 'Shirlan' (Zeneca);
'Frowncide' (Ishihara Sangyo)

ANALYSIS
Details from Ishihara Sangyo.

MAMMALIAN TOXICOLOGY
Oral Acute oral LD_{50} for rats >5000 mg/kg. **Skin and eye** Acute percutaneous
LD_{50} for rats >2000 mg/kg. **Inhalation** LC_{50} for rats 0.463 mg/l.

ECOTOXICOLOGY
Birds Acute oral LD_{50} for bobwhite quail 1782, mallard duck 4190 mg/kg.
Fish LC_{50} (96 h) for carp 0.15, rainbow trout 0.11 mg/l. **Bees** LD_{50} (oral)
>100 µg/bee, (contact) >200 µg/bee. **Worms** LC_{50} (28 d) >1000 mg/kg.
Daphnia LC_{50} (48 h) 0.22 mg/l.

ENVIRONMENTAL FATE
Soil/Environment In soil, DT_{50} 33–62 d. K_d 143–820.

30 fluazuron

Ixodicide

benzoylurea

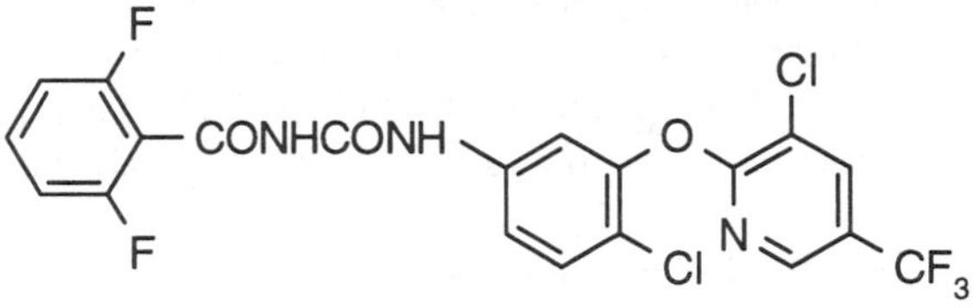

NOMENCLATURE
Common name fluazuron (BSI, pa E-ISO, INN)
IUPAC name 1-[4-chloro-3-(3-chloro-5-trifluoromethyl-2-pyridyloxy)phenyl]-=
3-(2,6-difluorobenzoyl)urea
Chemical Abstracts name N-[[[4-chloro-3-[[3-chloro-5-(trifluoromethyl)-=
2-pyridinyl]oxy]phenyl]amino]carbonyl]-2,6-difluorobenzamide
CAS RN *[86811–58–7]* **Development codes** CGA 157 419

PHYSICAL CHEMISTRY
Mol. wt. 506.2 **M.f.** $C_{20}H_{10}Cl_2F_5N_3O_3$ **Form** White to pink, odourless, fine
crystalline powder. **M.p.** 219 °C (decomp.) **V.p.** 1.2×10^{-7} mPa (20 °C)
K_{ow} logP = 5.1 (rp-tlc method) **S.g./density** 1.59 (20 °C) **Solubility** In water
<0.02 ppm (20 °C). In methanol 2.4, isopropanol 0.9 g/l (20 °C). **Stability** Stable
up to 219 °C. Hydrolysis DT_{50} (25 °C) 14 d (pH 3), 7 d (pH 5), 20 h (pH 7), 0.5 h
(pH 9).

COMMERCIALISATION
Manufacturers Novartis

APPLICATIONS
Biochemistry Interaction with chitin formation. **Uses** Strategic control of the
cattle tick *Boophilus microplus* (including all known resistant strains) on beef cattle.
Formulation types PO. **Selected tradenames** 'Acatak' (Novartis)

ANALYSIS
By hplc.

MAMMALIAN TOXICOLOGY
Oral Acute oral LD_{50} for rats >5000 mg/kg (OECD 401). **Skin and eye** Acute percutaneous LD_{50} for rats >2000 mg/kg (OECD 402). **Inhalation** LC_{50} (4 h) for rats >5994 mg/m^3 (OECD 403). **NOEL** (1 y) for dogs 7.5 mg/kg b.w. daily; (lifetime) for rats 400, mice (males and females) 4.5 mg/kg b.w. daily. In 2-generation reproduction tests (rats), NOEL 100 ppm. No carcinogenic, mutagenic, teratogenic or reprotoxicity potential. **ADI** (proposed) 0.45 mg/kg b.w. **Toxicity class** WHO (a.i.) III

ECOTOXICOLOGY
Birds Acute oral LD_{50} for bobwhite quail, mallard ducks >2000 mg/kg. Eight-day LC_{50} for bobwhite quail, mallard ducks >5200 ppm. **Fish** LC_{50} for rainbow trout >15, carp >9.1 mg/l (OECD class 'harmful'). **Bees** OECD class 'not toxic'. **Worms** LC_{50} (14 d) for earthworms >1000 mg/kg (not toxic) (OECD 207). **Daphnia** LC_{50} 0.0006 mg/l (OECD class 'very toxic'). **Algae** NOEC for freshwater green algae 27.9 mg/l (virtually non-toxic).

ENVIRONMENTAL FATE
Animals Virtually no metabolism. **Plants** Not relevant. **Soil/Environment** Not investigated; environmental contamination is unlikely because of pour-on application.

331 fluchloralin   *Herbicide*

2,6-dinitroaniline

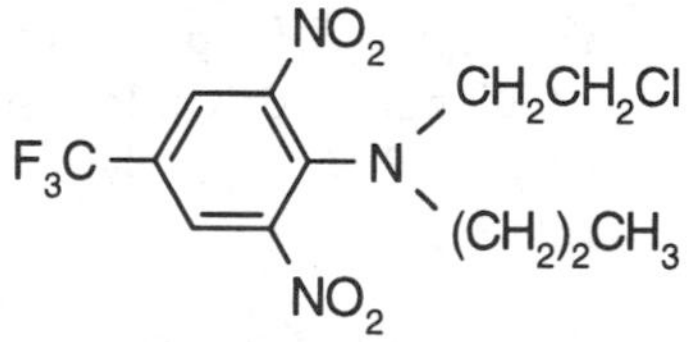

NOMENCLATURE
Common name fluchloralin (BSI, E-ISO, ANSI, WSSA); fluchloraline ((f) F-ISO)
IUPAC name N-(2-chloroethyl)-2,6-dinitro-N-propyl-4-(trifluoromethyl)aniline; N-(2-chloroethyl)-α,α,α-trifluoro-2,6-dinitro-N-propyl-p-toluidine
Chemical Abstracts name N-(2-chloroethyl)-2,6-dinitro-N-propyl-4-(trifluoro= methyl)benzenamine

CAS RN *[33245–39–5]*   **Development codes** BAS 392H; BAS 3920; BAS 3921; BAS 3922

PHYSICAL CHEMISTRY
Composition Tech. grade is ≥97% pure.   **Mol. wt.** 355.7   **M.f.** $C_{12}H_{13}ClF_3N_3O_4$
Form Orange-yellow solid.   **M.p.** 42–43 °C   **V.p.** 4 mPa at 20 °C; 3.3 mPa at
30 °C; 13 mPa at 40 °C; 53 mPa at 50 °C   **Solubility** In water <1 mg/l (20 °C). In
acetone, benzene, chloroform, diethyl ether, ethyl acetate >1000, cyclohexane
251, ethanol 177, olive oil c. 260 (all in g/kg, 20 °C).   **Stability** Decomposed by
u.v. light. Stable for at least 2 years when stored in original unopened container at
normal room temperatures.

COMMERCIALISATION
History Herbicide reported by A. Fischer (*Pestic. Chem. Proc. Int. Congr. Pestic.
Chem., 2nd,* 1972, **5**, 189). Introduced by BASF AG.   **Patent** DE 1643719;
BE 725877; GB 1241458 all to BASF   **Manufacturers** BASF

APPLICATIONS
Biochemistry Cell division inhibitor.   **Mode of action** Selective herbicide absorbed
via shoots and roots, with acropetal movement throughout the entire plant.
Affects seed germination and other physiological growth processes, especially in
the radicle.   **Uses** Control of annual grasses and some broad-leaved weeds in
cotton, rice (transplanted), soya beans, peanuts, snap beans, lima beans, okra,
jute, sunflowers, potatoes, and several vegetable crops. Applied pre-plant or pre-
emergence with soil incorporation.   **Phytotoxicity** Beet, sorghum, spinach and
oats are sensitive.   **Formulation types** EC.   **Selected tradenames** 'Basalin' (BASF)

ANALYSIS
Product and **residue** analysis by glc. Details available from BASF.

MAMMALIAN TOXICOLOGY
Oral Acute oral LD_{50} for rats 1550, rabbits 8000, mice 730, dogs 6400 mg/kg.
Skin and eye Acute percutaneous LD_{50} for rabbits >10 000 mg/kg. Mild skin and
eye irritant.   **Inhalation** LC_{50} (4 h) for rats 8.4 mg/l.   **NOEL** (90 d) for rats
250, dogs <750 mg/kg.   **Toxicity class** WHO (a.i.) III; EPA (formulation) III

ECOTOXICOLOGY
Birds Acute oral LD_{50} for mallard ducks 13 000, bobwhite quail 7000 mg/kg.
Fish LC_{50} (96 h) for bluegill sunfish 0.016, rainbow trout 0.012 mg/l.   **Bees** Not
toxic to bees.

ENVIRONMENTAL FATE
Plants In plants, rapidly metabolised, with some metabolites being incorporated
into plant structures.   **Soil/Environment** In soil, strongly adsorbed by clay colloids

and organic matter, with no leaching. Dealkylation occurs to N-(2-chloroethyl)-, N-propyl-, and unsubstituted trifluoro-2,6-dinitro-p-toluidine and to cyclised products. In light, there is also reduction of the nitro group to an amino group. Low-level soil residues may persist for more than one season after application, and phytotoxicity to very sensitive crops may occur.

332 flucycloxuron

Acaricide, insecticide

benzoylurea

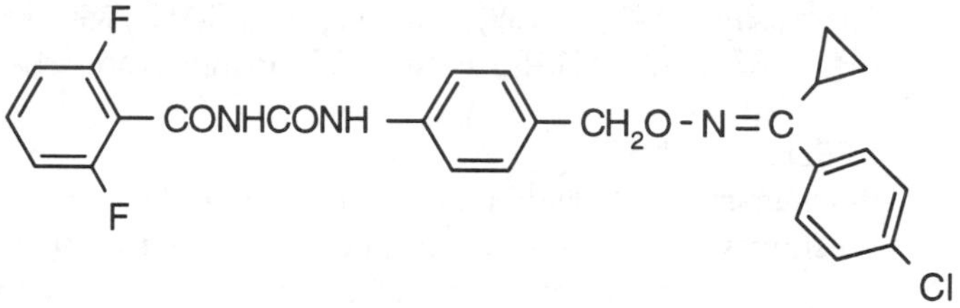

NOMENCLATURE
Common name flucycloxuron (BSI, draft E-ISO)
IUPAC name 1-[α-(4-chloro-α-cyclopropylbenzylideneamino-oxy)-p-tolyl]-= 3-(2,6-difluorobenzoyl)urea (ratio 50–80% (E)- and 50–20% (Z)- isomers)
Chemical Abstracts name N-[[[4-[[[[(4-chlorophenyl)cyclopropylmethylene]= amino]oxy]methyl]phenyl]amino]carbonyl]-2,6-difluorobenzamide
CAS RN [94050–52–9] (E)- isomer; [94050–53–0] (Z)- isomer; [113036–88–7] unstated stereochemistry   **Development codes** PH 70–23; DU 319722; UBI-A1335   **Official codes** OMS 3041

PHYSICAL CHEMISTRY
Composition Ratio 50 to 80% (E)- and 50 to 20% (Z)- isomers.   **Mol. wt.** 483.9
M.f. $C_{25}H_{20}ClF_2N_3O_3$ **Form** Off-white to yellow crystals (tech.).   **M.p.** 143.6 °C (decomp.) (for (E)-, (Z)- isomeric mixture)   **V.p.** 5.4×10^{-5} mPa (25 °C) (for (E)-, (Z)- isomeric mixture) K_{ow} logP = 6.97 (E)- isomer; 6.90 (Z)- isomer
Henry 2.6×10^{-2} Pa m^3 mol^{-1} (25 °C)   **Solubility** In water <1 µg/l (20 °C). In cyclohexane 0.2, xylene 3.3, ethanol 3.8, N-methylpyrrolidone 940 (all in g/l, 20 °C) for (E)-, (Z)- isomeric mixture.   **Stability** Thermally stable, <2% decomposition after 24-h storage at 50 °C. No degradation at pH 5, 7 or 9. Photolytically degraded, DT_{50} c. 18 d.

COMMERCIALISATION
History Acaricide and insecticide reported by A. C. Grosscurt *et al.* (*Pestic. Sci.*,

1988, **22,** 51). Introduced by Duphar B.V. in 1988 (now Uniroyal Chemical Co., Inc).   **Patent** EP 117320; US 4550202; US 4609676   **Manufacturers** Uniroyal

APPLICATIONS
Biochemistry Chitin synthesis inhibitor.   **Mode of action** Non-systemic acaricide and insecticide which inhibits the moulting process in mites and insects. It is only active against eggs and larval stages. Adult mites and insects are not affected.
Uses Control of eggs and larval stages of Eriophyid and Tetranychid mite species on a variety of crops, including fruit crops, vegetables and ornamentals. Also controls larvae of a number of insects. Because of its relative selectivity, flucycloxuron is well-suited for integrated control programmes. Recommended concentration for control of mites on fruit crops is 0.01 – 0.015% a.i. On insects, good activity has been found against codling moth, leaf miners and some leaf rollers in pome fruit at the same concentration. In ornamentals grown under glass/plastic, lower concentrations can be used. In grapes, the recommended dosage is 125 – 150 g a.i./ha.   **Formulation types** DC.
Selected tradenames 'Andalin' (Uniroyal)

ANALYSIS
Product and **residue** analysis by hplc; details available from Uniroyal.

MAMMALIAN TOXICOLOGY
Oral Acute oral LD_{50} for rats >5000 mg/kg.   **Skin and eye** Acute percutaneous LD_{50} for rats >2000 mg/kg. Non-irritating to skin; mild eye irritant (rabbits).
Inhalation LC_{50} (4 h) for rats >3.3 mg/l air.   **NOEL** In 2 y feeding trials, NOAEL for rats 120 mg/kg diet. In a 2-generation reproduction study, NOAEL for rats 200 mg/kg diet. Non-mutagenic and non-teratogenic. No evidence of carcinogenicity.   **Other** Acute i.p. LD_{50} for rats 2000 mg/kg.
Toxicity class WHO (a.i.) III (Table 5)

ECOTOXICOLOGY
Birds Acute oral LD_{50} for mallard ducks >2000 mg/kg. Eight-day dietary LC_{50} for mallard ducks and bobwhite quail >6000 mg/kg diet.   **Fish** LC_{50} (96 h) for rainbow trout and bluegill sunfish >100 mg/l.   **Bees** Low toxicity to bees. Contact LD_{50} for honeybees >100 µg/bee.   **Worms** EC_{50} (14 d) for earthworms >1000 mg/kg soil.   **Other beneficial spp.** Slightly to moderately harmful to predatory mites (IOBC).   **Daphnia** LC_{50} (48 h) 0.27 µg/l.   **Algae** NOEC for *Selenastrum capricornutum* >2.2 µg/l.   **Other aquatic spp.** LC_{50} (96 h) for mysid shrimp (*Mysidopsis bahia*) 340 ng/l.

ENVIRONMENTAL FATE
Animals Poorly absorbed from gastrointestinal tract; mainly excreted as unchanged parent compound in the faeces. Flucycloxuron absorbed is metabolised and further excreted in the urine, mainly as 2,6-difluorobenzoic acid and

N-(4-carboxyphenyl)-*N'*-(2,6-difluorobenzoyl)urea.   **Plants** Non-systemic. Not metabolised on plants.   **Soil/Environment** Strongly and irreversibly bound by soil/humic acid complex and virtually immobile in soil. Soil DT_{50} 0.25–0.5 y.

333 flucythrinate   *Insecticide*

pyrethroid

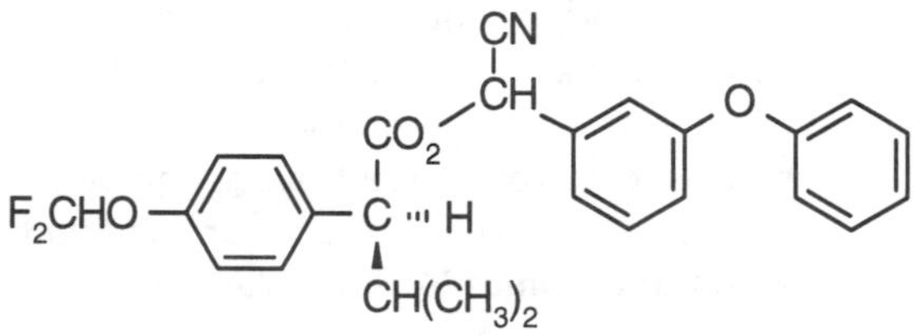

NOMENCLATURE
Common name flucythrinate (BSI, ANSI, draft E-ISO, (*m*) draft F-ISO)
IUPAC name (*RS*)-α-cyano-3-phenoxybenzyl (*S*)-2-(4-difluoromethoxyphenyl)-= 3-methylbutyrate
Chemical Abstracts name cyano(3-phenoxyphenyl)methyl 4-(difluoromethoxy)-= α-(1-methylethyl)benzeneacetate
CAS RN [70124–77–5], formerly [71611–31–9], unstated stereochemistry
Development codes AC 222 705; CL 222 705 (both to Cyanamid)
Official codes OMS 2007; AI3–29 391

PHYSICAL CHEMISTRY
Mol. wt. 451.4   **M.f.** $C_{26}H_{23}F_2NO_4$ **Form** Dark amber viscous liquid, with a faint ester-like odour (tech.).   **B.p.** 108 °C/0.35 mmHg   **V.p.** 0.0012 mPa (25 °C)
S.g./density 1.189 (22 °C)   **Solubility** In water 0.5 mg/l (21 °C). In acetone >820, xylene 1810, *n*-propanol >780, corn oil >560, cottonseed oil >300, soya bean oil >300, hexane 90 (all in g/l, 21 °C).   **Stability** Rapidly hydrolysed in water under alkaline conditions, but more slowly under neutral or acidic conditions; DT_{50} (27 °C) *c.* 40 d (pH 3), 52 d (pH 5), 6.3 d (pH 9). Stable for >1 y at 37 °C, >2 y at 25 °C. Degraded on soil plates by simulated sunlight, DT_{50} *c.* 21 d; aqueous solutions, DT_{50} *c.* 4 d.

COMMERCIALISATION
History Insecticide reported by W. K. Whitney & K. Wettstein (*Proc. Br. Crop Prot. Conf. – Pests Dis.*, 1979, **2**, 387). Introduced by American Cyanamid Co.
Patent US 4178308; GB 1582775   **Manufacturers** Cyanamid; Searle India

APPLICATIONS

Mode of action Non-systemic insecticide with contact and stomach action.
Uses Control of a wide range of insect pests, particularly on cotton (bollworms, leafworms, sucking insects, whitefly, beetles, etc.) and pome fruit and stone fruit trees (Lepidoptera, Homoptera, Coleoptera, etc.). Also used on vines, strawberries, citrus fruit, bananas, pineapples, olives, coffee, cocoa, hops, vegetables, soya beans, cereals, maize, alfalfa, sugar beet, sunflowers, tobacco, and ornamentals. Efficacy at higher temperatures is greater than for several other pyrethroid insecticides. Also suppresses phytophagous mites, and acts as a repellent to many pests. **Phytotoxicity** Non-phytotoxic when used as directed. In certain treated crops, e.g. cotton and fruit trees, the green colouration of the foliage is intensified, thereby improving appearance. **Formulation types** EC; WP; WG. **Mixtures** (flucythrinate +) chlorpyrifos; dimethoate; methomyl; phenthoate. **Selected tradenames** 'Cybolt' (Cyanamid); 'Cythrin' (Cyanamid); 'Pay-off' (Cyanamid); 'Fluent' (Sanonda)

ANALYSIS

Product analysis by hplc or glc; analysis of pyrethroids reviewed by E. Papadopoulou-Mourkidou in *Comp. Analyt. Profiles.* **Residues** determined by glc (*Pestic. Anal. Man.,* **I**, 201-I, **II**; *Anal. Methods Residues Pestic.,* 1988). Details available from American Cyanamid Co.

MAMMALIAN TOXICOLOGY

Reviews *Pesticide residues in food – 1985.* FAO Plant Production and Protection Paper 68, 1986. *Pesticide residues in food – 1985 evaluations. Part II – Toxicology.* FAO Plant Production and Protection Paper 72/2, 1986. **Oral** Acute oral LD_{50} for male rats 81, female rats 67, female mice 76 mg/kg. **Skin and eye** Acute percutaneous LD_{50} (24 h) for rabbits >1000 mg/kg; not irritating to skin and eyes of rabbits, but undiluted formulations may be irritating. **Inhalation** LC_{50} (4 h) for rats 4.85 mg/l air (aerosol). **NOEL** (2 y) for rats 60 mg/kg diet. **ADI** (JMPR) 0.02 mg/kg b.w. [1985]. **Other** In a 3-generation reproduction test on rats, no adverse effects were observed at 30 mg/kg diet. Not teratogenic or foetotoxic in tests on rats and rabbits. Not mutagenic in tests on rats. **Toxicity class** WHO (a.i.) Ib; EPA (formulation) I

ECOTOXICOLOGY

Birds Acute oral LD_{50} for mallard ducks >2510, bobwhite quail 2708 mg/kg. Eight-day dietary LC_{50} for mallard ducks 4885, bobwhite quail 3443 mg/kg diet.
Fish LC_{50} (96 h) for bluegill sunfish 0.71, channel catfish 0.51, rainbow trout 0.32, sheepshead minnow 1.6 µg/l. Low insecticidal dosages and immobility in soils should minimise hazards to fish. **Bees** Toxic to bees, but also has a repellent effect. LD_{50} (topical, as a dust) 0.078 µg/bee. **Daphnia** LC_{50} (48 h) 0.0083 mg/l.

ENVIRONMENTAL FATE

Animals In rats, following oral administration, 60–70% is eliminated within 24 hours, and >95% within 8 days, in the faeces and urine. In the faeces, the parent compound makes up most of the material excreted, but in the urine and in tissue, several metabolites are present. The major route of degradation is through hydrolysis, with subsequent hydroxylation of the hydrolysis products.

Soil/Environment In soil, there is low mobility and no leaching, with a half-life of c. 2 months.

334 fludioxonil   *Fungicide*

phenylpyrrole

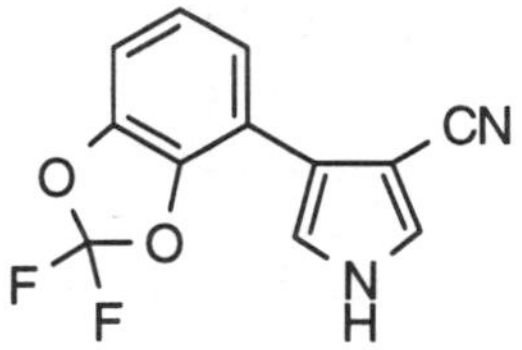

NOMENCLATURE

Common name fludioxonil (BSI, pa E-ISO)

IUPAC name 4-(2,2-difluoro-1,3-benzodioxol-4-yl)pyrrole-3-carbonitrile

Chemical Abstracts name 4-(2,2-difluoro-1,3-benzodioxol-4-yl)-1*H*-pyrrole-= 3-carbonitrile

CAS RN *[131341–86–1]*   **Development codes** CGA 173506

PHYSICAL CHEMISTRY

Mol. wt. 248.2   **M.f.** $C_{12}H_6F_2N_2O_2$ **Form** Colourless crystals.   **M.p.** 199.8 °C   **V.p.** 3.9×10^{-4} mPa (25 °C)   **K$_{ow}$** logP = 4.12 (25 °C)   **Henry** 5.4×10^{-5} Pa m^3 mol^{-1} (calc.)   **S.g./density** 1.54 (20 °C)   **Solubility** In water 1.8 mg/l (25 °C). In acetone 190, ethanol 44, toluene 2.7, *n*-octanol 20, hexane 0.0078 g/l (25 °C). **Stability** Practically no hydrolysis at 70 °C between pH 5 and 9.   **pKa** pKa_1 <0; pKa_2 c. 14.1

COMMERCIALISATION

History Developed and introduced by Ciba-Geigy AG (now Novartis Crop Protection AG).   **Patent** EP-206999; US 4705800   **Manufacturers** Novartis

APPLICATIONS

Biochemistry Mode of action is believed to be the same as for fenpiclonil. Possibly by inhibition of transport-associated phosphorylation of glucose

(A. B. K. Jespers & M. A. de Waard, *Pestic. Sci.*, **44**, 167 (1995)).
Mode of action Non-systemic foliar fungicide. Inhibits mycelial growth. **Uses** Seed treatment for control of *Gibberella* in rice and to control *Fusarium*, *Rhizoctonia*, *Tilletia*, *Helminthosporium*, and *Septoria* in both cereal and non-cereal crops. Also used as a foliar fungicide for control of *Botrytis*, *Monilia*, *Sclerotinia*, *Rhizoctonia*, and *Alternaria* in grapes, stone fruit, vegetables, field crops, turf and ornamentals.
Formulation types WP; WG; SC; DS; FS. **Mixtures** (*fludioxonil* +) cyprodinil; difenoconazole; metalaxyl; metalaxyl-M; furathiocarb; oxolinic acid; pefurazoate.
Selected tradenames 'Celest' (Novartis); 'Maxim' (Novartis)

ANALYSIS
Product by gc. **Residues** by hplc with u.v. detection. Details from Novartis.

MAMMALIAN TOXICOLOGY
Oral Acute oral LD_{50} for rats and mice >5000 mg/kg. **Skin and eye** Acute percutaneous LD_{50} for rats >2000 mg/kg. Non-irritating to eyes and skin (rabbits). Non-sensitising to skin (guinea pigs). **Inhalation** LC_{50} (4 h) for rats >2.6 mg/m^3 air. **NOEL** (2 y) for rats 40 mg/kg b.w. daily; (1.5 y) for mice 112 mg/kg b.w. daily; (1 y) for dogs 3.3 mg/kg b.w. daily. **ADI** 0.033 mg/kg b.w.
Other Not teratogenic, not mutagenic, not oncogenic. **Toxicity class** WHO (a.i.) III (Table 5)

ECOTOXICOLOGY
Birds Acute oral LD_{50} for mallard ducks and bobwhite quail >2000 mg/kg. LC_{50} for mallard ducks and bobwhite quail >5200 ppm. **Fish** LC_{50} (96 h) for bluegill sunfish 0.31, catfish 0.63, common carp 1.5, rainbow trout 0.5 mg/l. **Bees** Non-toxic; LD_{50} (48 h, oral) >329 µg/bee; LC_{50} (48 h, contact) >101 µg/bee.
Worms LC_{50} (14 d) for *Eisenia foetida* >1000 mg/kg soil.
Other beneficial spp. No long-term substantial reduction in major groups of zooplankton, bentic macro-invertebrates, emergent insects, peryphyton or phytoplankton. **Daphnia** LC_{50} (48 h) 1.1 mg/l. **Algae** EC_{50} (72 h) for *Scenedesmus subspicatus* 0.93 mg/l; (120 h) for *Selenastrum capricornutum* 0.092 mg/l.

ENVIRONMENTAL FATE
Animals Rapidly absorbed from the gastrointestinal tract into the general circulation and rapidly and almost completely excreted, via the faeces. The dominant metabolic pathway is oxidation of the pyrrole ring at the 2-position. A minor pathway is hydroxylation of the phenyl ring. All metabolites are excreted as conjugates, mainly as glucuronides. **Plants** Metabolism proceeds via the oxidation of the pyrrole ring, opening of the pyrrole ring and pyrrolidine carboxylic acid metabolite. In general, the compound is extensively metabolised to more than 10–15 minor metabolites. **Soil/Environment** Formation of bound residues is the major route for dissipation in soil; DT_{50} (lab.) 140–350 d, (field) 10–25 d. In

leaching and adsorption/desorption experiments, the compound proved to be immobile in soil. Photolytic DT_{50} in water 9–10 d (natural sunlight).

335 flufenoxuron   *Insecticide, acaricide*

benzoylurea

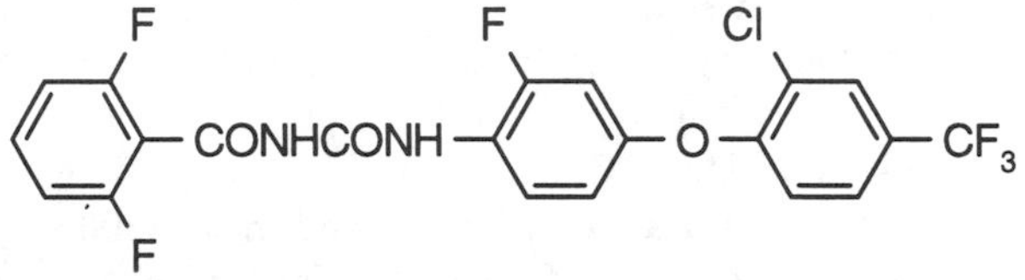

NOMENCLATURE
Common name flufenoxuron (BSI, draft E-ISO, (*m*) draft F-ISO)
IUPAC name 1-[4-(2-chloro-α,α,α-trifluoro-*p*-tolyloxy)-2-fluorophenyl]-=
3-(2,6-difluorobenzoyl)urea
Chemical Abstracts name *N*-[[[4-[2-chloro-4-(trifluoromethyl)phenoxy]-=
2-fluorophenyl]amino]carbonyl]-2,6-difluorobenzamide
CAS RN *[101463–69–8]*   **Development codes** WL 115 110

PHYSICAL CHEMISTRY
Composition Tech. grade is 98–100% pure.   **Mol. wt.** 488.8
M.f. $C_{21}H_{11}ClF_6N_2O_3$ **Form** Tech. is a white crystalline solid.   **M.p.** 169–172 °C
(decomp.)   **V.p.** 6.52×10^{-9} mPa (20 °C) K_{ow} logP = 4.0 (pH 7)
S.g./density 1.57 kg/l (20 °C) (typical value)   **Solubility** In water 7×10^{-11} g/l
(pH 7, 15 °C), 4 µg/l (25 °C). In acetone 82, xylene 6, dichloromethane 24 (all in g/l, 25 °C).   **Stability** Stable ≤190 °C; to natural sunlight; thin film on glass stable >100 h simulated sunlight. DT_{50} 11 d (ambient in water). On hydrolysis (25 °C) DT_{50} 206 d (pH 5), 267 d (pH 7), 36.7 d (pH 9), 2.7 d (pH 12).   **F.p.** Not autoflammable; no explosive properties.

COMMERCIALISATION
History Insecticide reported by M. Anderson *et al.* (*Proc. Br. Crop Prot. Conf. – Pests Dis.*, 1986, **1**, 89). Evaluated by Shell Research Ltd. Product now marketed by American Cyanamid Co.   **Patent** EP 161019   **Manufacturers** Cyanamid

APPLICATIONS
Biochemistry Chitin synthesis inhibitor   **Mode of action** Insect and acarid growth regulator with contact and stomach action. Treated larvae die at the next moult or during the ensuing instar. Treated adults lay non-viable eggs.   **Uses** Control of

immature stages of many phytophagous mites (*Aculus*, *Brevipalpus*, *Panonychus*, *Phyllocoptruta*, *Tetranychus* spp.) and insect pests on pome fruit, vines, citrus fruit, tea, cotton, maize, soya beans, vegetables and ornamentals.   **Phytotoxicity** None recorded.   **Formulation types** DG; EC; DC.   **Selected tradenames** 'Cascade' (Cyanamid)

ANALYSIS
Product by lc.   **Residues** by hplc. Details from Cyanamid.

MAMMALIAN TOXICOLOGY
Oral Acute oral LD_{50} for rats >3000 mg tech./kg.   **Skin and eye** Acute percutaneous LD_{50} for rats and mice >2000 mg/kg. Non-irritating to eyes and skin (rabbits).   **Inhalation** LC_{50} (4 h) for rats 5 mg/l air.   **NOEL** (1 y) for dogs 100 mg/kg diet; (90 d) for rats and mice 50 mg/kg diet.   **Toxicity class** WHO (a.i.) III (Table 5); EPA (formulation) III

ECOTOXICOLOGY
Birds Acute oral LD_{50} for bobwhite quail >2000 mg/kg.   **Fish** LC_{50} (96 h) for rainbow trout >100 mg/l.   **Other beneficial spp.** Low hazard to predatory mites and other predatory and parasitic arthropods.

ENVIRONMENTAL FATE
Soil/Environment Strongly adsorbed in soil, DT_{50} 42 d (clay loam). Use in orchards over 3 years at a dose of 97.5 g/ha gave rise to low residues in soil which did not significantly affect soil organisms, including earthworms (J. M. Gilbert et al., *Proc. Br. Crop Prot. Conf. – Pests Dis.*, 1992, **2**, 805–810).

336 flufenprox   *Insecticide*

non-ester pyrethroid

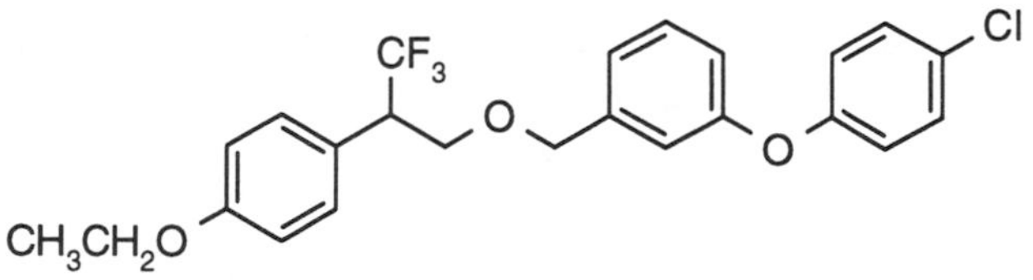

NOMENCLATURE
Common name flufenprox (BSI, draft ISO)
IUPAC name 3-(4-chlorophenoxy)benzyl (*RS*)-2-(4-ethoxyphenyl)-= 3,3,3-trifluoropropyl ether

Chemical Abstracts name 1-(4-chlorophenoxy)-3-[[2-(4-ethoxyphenyl)-=3,3,3-trifluoropropoxy]methyl]benzene
CAS RN *[107713–58–6]*   **Development codes** ICIA5682

PHYSICAL CHEMISTRY
Mol. wt. 450.9   **M.f.** $C_{24}H_{22}ClF_3O_3$ **Form** Odourless, transparent pale yellow-green liquid.   **B.p.** 204 °C/0.2 mmHg   **V.p.** 1.3×10^{-4} mPa (20 °C)
S.g./density 1.25 (25 °C)   **Solubility** In water at pH 7, 2.5 µg/l. Soluble (>500 g/l) in hexane, toluene, acetone, dichloromethane, ethyl acetate, *n*-octanol, acetonitrile and methanol.

COMMERCIALISATION
History Reported by R. F. S. Gordon *et al.* (*Proc. Br. Crop Prot. Conf. – Pests Dis.*, 1992, **1**, 81).   **Manufacturers** Zeneca

APPLICATIONS
Mode of action Insecticide with fast action and residual efficacy.   **Uses** Control of Homoptera, Heteroptera, Coleoptera and Lepidoptera on rice, including rice hoppers.   **Formulation types** DL; GR; EW; EC.   **Compatibility** Compatible with other insecticides, e.g. buprofezin, fenitrothion, and with fungicides, e.g. fthalide, isoprothiolane.

MAMMALIAN TOXICOLOGY
Oral Acute oral LD_{50} for rats >5000 mg/kg.   **Skin and eye** Acute percutaneous LD_{50} for rats >2000 mg/kg. Mild skin and eye irritant (rabbits); skin sensitiser (guinea pigs).

ECOTOXICOLOGY
Fish LC_{50} (96 h) for carp >10 mg tech./l; >25 mg as 20% EC/l; as 0.5% dust or granule, for red killifish and Asian pond loach >10 mg/l.   **Bees** LD_{50} 0.03 µg/bee; NOEL 100–1000 mg/kg.   **Daphnia** LC_{50} (48 h) 0.00035 mg/l.

337 flumethrin

Insecticide

pyrethroid

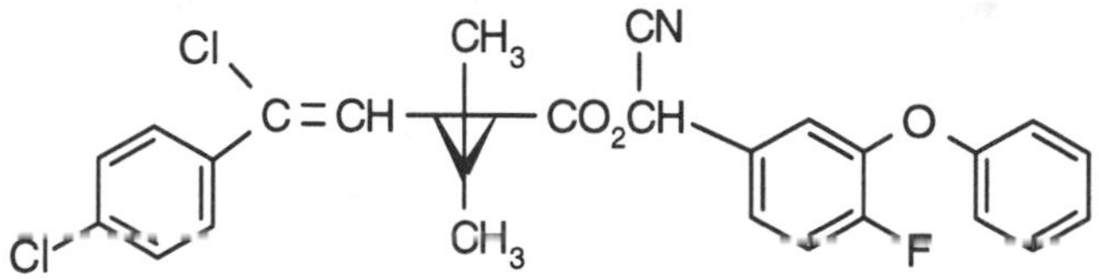

NOMENCLATURE

Common name flumethrin (BAN)
IUPAC name α-cyano-4-fluoro-3-phenoxybenzyl 3-(β,4-dichlorostyryl)-=
2,2-dimethylcyclopropanecarboxylate
Chemical Abstracts name cyano(4-fluoro-3-phenoxyphenyl)methyl 3-[2-chloro-=
2-(4-chlorophenyl)ethenyl]-2,2-dimethylcyclopropanecarboxylate
CAS RN *[69770–45–2]* unstated stereochemistry **Development codes** BAY V1 6045; BAY Vq1950

PHYSICAL CHEMISTRY

Mol. wt. 510.4 **M.f.** $C_{28}H_{22}Cl_2FNO_3$ **Form** Yellowish, highly viscous oil.
B.p. >250 °C

COMMERCIALISATION

History Insecticide developed by Bayer AG. **Patent** DE 2932920
Manufacturers Bayer

APPLICATIONS

Mode of action Non-systemic. **Uses** For the control of ticks, biting and sucking lice, psoroptic, chorioptic and sarcoptic munge; for diagnosis and control of varroatosis. **Selected tradenames** 'Bayticol' (Bayer); 'Bayvarol' (Bayer)

MAIN ENTRIES

338 flumetralin *Plant growth regulator*

2,6-dinitroaniline

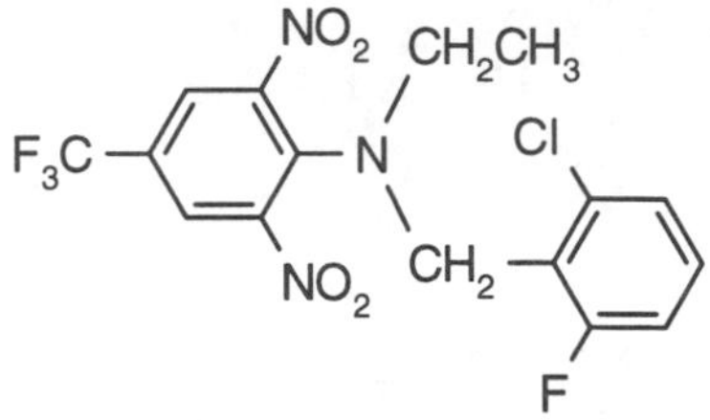

NOMENCLATURE

Common name flumetralin (BSI, draft E-ISO); flumétraline ((*f*) draft F-ISO)
IUPAC name *N*-(2-chloro-6-fluorobenzyl)-*N*-ethyl-α,α,α-trifluoro-2,6-dinitro-=*p*-toluidine
Chemical Abstracts name 2-chloro-*N*-[2,6-dinitro-4-(trifluoromethyl)phenyl]-=*N*-ethyl-6-fluorobenzenemethanamine
CAS RN *[62924–70–3]* **Development codes** CGA 41 065

PHYSICAL CHEMISTRY

Mol. wt. 421.7 **M.f.** $C_{16}H_{12}ClF_4N_3O_4$ **Form** Yellow to orange odourless crystals. **M.p.** 101.0–103.0 °C; (tech., 92.4–103.8 °C) **V.p.** 3.2×10^{-2} mPa (25 °C) (OECD 104) K_{ow} logP = 5.45 (25 °C) **Henry** 0.19 Pa m^3 mol^{-1} (calc.) **S.g./density** 1.54 **Solubility** In water 0.07 mg/l (25 °C). In acetone 560, toluene 400, ethanol 18, *n*-hexane 14, *n*-octanol 6.8 (all in g/l, 25 °C). **Stability** Decomposes exothermically above 250 °C. Stable between pH 5 and pH 9.

COMMERCIALISATION

History Plant growth regulator reported by M. Wilcox *et al.* (*Proc. Plant Growth Regul. Working Group*, 1977, **4**, 194). Introduced by Ciba-Geigy AG (now Novartis Crop Protection AG). **Patent** BE 891327; GB 1531260 **Manufacturers** Novartis

APPLICATIONS

Mode of action Plant growth regulator, which has a local systemic effect.
Uses Control of sucker growth in tobacco. **Phytotoxicity** May cause some injury to immature plants. **Formulation types** EC. **Selected tradenames** 'Prime' (Novartis)

ANALYSIS

Residues determined by glc using ECD or FID. Details available from Novartis.

MAMMALIAN TOXICOLOGY

Oral Acute oral LD_{50} for rats >5000 mg/kg. **Skin and eye** Acute percutaneous LD_{50} for rats >2000 mg/kg. EC formulation (150 g/l) is a moderate skin irritant and an extreme eye irritant. **Inhalation** LC_{50} for rats >2130 mg/m^3. **NOEL** (2 y) for rats and mice 300 ppm. **Toxicity class** WHO (a.i.) III (Table 5) **EC risk** R36/38, R43

ECOTOXICOLOGY
Birds LD_{50} for bobwhite quail and mallard ducklings >2000 mg/kg. LC_{50} for mallard ducklings >5000 ppm. **Fish** Toxic to fish; LC_{50} for bluegill sunfish and trout >3.2 µg/l. **Bees** Non-toxic. **Worms** LC_{50} for earthworms >1000 mg/kg. **Daphnia** LC_{50} (48 h) >2.8 µg/l. **Algae** EC_{50} for *Selenastrum* >0.85 mg/l.

ENVIRONMENTAL FATE
Animals Metabolic reactions include reduction of nitro groups, acetylation of the amino groups and hydroxylation of the phenyl ring. **Plants** Metabolism in tobacco is rapid and extensive; reduction or complete removal of substituents occurs. **Soil/Environment** Strongly adsorbed and therefore immobile in soil; degradation is by photolysis but not hydrolysis at pH 5, 7 and 9.

339 flumetsulam   *Herbicide*

triazolopyrimidine sulfonanilide

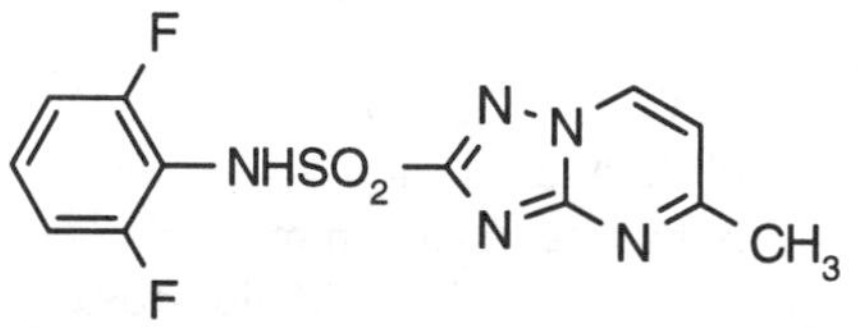

NOMENCLATURE
Common name flumetsulam (BSI, pa E-ISO, ANSI)
IUPAC name 2',6'-difluoro-5-methyl[1,2,4]triazolo[1,5-a]pyrimidine-2-sulfonanilide
Chemical Abstracts name N-(2,6-difluorophenyl)-5-methyl[1,2,4]triazolo[1,5-a]= pyrimidine-2-sulfonamide
CAS RN [98967–40–9] **Development codes** DE-498; XRD-498 (both DowElanco)

PHYSICAL CHEMISTRY
Mol. wt. 325.3 **M.f.** $C_{12}H_9F_2N_5O_2S$ **Form** Off-white odourless solid. **M.p.** 251–253 °C **V.p.** 3.7 × 10^{-7} mPa (25 °C) **K_{ow}** logP = –0.68 (25 °C, unstated pH) **S.g./density** 1.77 (21 °C) **Solubility** In water 49 mg/l (pH 2.5; solubility increases with pH). Very slightly soluble in acetone and methanol. Insoluble in hexane and xylene. **Stability** Aqueous photolysis DT_{50} 6–12 mo. Soil photolysis DT_{50} 3 mo. **pKa** 4.6 **F.p.** >93 °C.

COMMERCIALISATION
Manufacturers DowElanco

APPLICATIONS
Biochemistry Inhibits acetolactate synthase (ALS) enzyme, essential for synthesis of amino acids leucine, isoleucine and valine. Selectivity in soya is due to rapid metabolic deactivation. **Mode of action** Systemic herbicide, absorbed by roots and leaves of plants and translocated to growth points. **Uses** Used alone and in combination with trifluralin or metolachlor for control of broad-leaved weeds and grasses in soya beans, field peas and maize. **Phytotoxicity** Crops damaged by soil application of flumetsulam include sugar beet, cotton, oilseed rape, grain sorghum, tomatoes, and sunflowers. **Formulation types** OF; WG; SC.
Mixtures *(flumetsulam +)* trifluralin; metolachlor.
Selected tradenames 'Broadstrike' (DowElanco)

ANALYSIS
By gc with mass selection detection.

MAMMALIAN TOXICOLOGY
Oral Acute oral LD_{50} for rats >5000 mg/kg. **Skin and eye** Acute percutaneous LD_{50} for rabbits >2000. Non-sensitising to skin (guinea pigs). Slightly irritating to eyes (rabbits). **Inhalation** LC_{50} (4 h) >1.2 mg/l. **NOEL** for mice >1000, female rats 500, male rats 1000, dogs 1000 mg/kg. **Other** Non-teratogenic (dietary) in rats. Non-mutagenic in the Ames test.

ECOTOXICOLOGY
Birds Acute oral LD_{50} for bobwhite quail >2250 mg/l. Eight-day dietary LC_{50} for bobwhite quail and mallard duck >5620 mg/l. **Fish** LC_{50} (96 h) for silverside minnow >379 mg/l. Non-toxic to fathead minnow and bluegill sunfish. **Bees** LC_{50} >100 μg/bee. NOEL 36 μg/bee. **Daphnia** Non-toxic. **Other aquatic spp.** LC_{50} for shrimp >349.

ENVIRONMENTAL FATE
Animals Rapidly cleared via urine and faeces with no metabolites. 5-Hydroxy metabolite found in the hen. **Plants** DT_{50} in maize 2 h, soya beans 18 h, *Chenopodium* 131 h. Metabolites depend on the species; 5-hydroxy or 5-methoxy derivatives are common. **Soil/Environment** Availability of flumetsulam in soil is principally dependent upon soil pH and organic matter. Herbicidal activity increases as pH increases and organic matter decreases. DT_{50} in soil (25 °C, pH ⩾7, o.m. content <4%; or pH 6–7, o.m. content *c.* 1%), ⩽1 mo. DT_{50} in soil (pH 6–7, o.m. content 2–4%), 1–2 mo. K_{oc} 5–182; K_d 0.05–2.4.

340 flumiclorac-pentyl *Herbicide*

phthalimide analogue

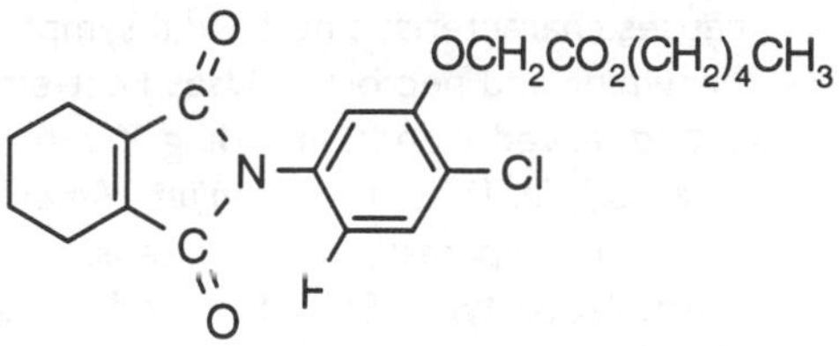
O
O
C
C
N
OCH2CO2(CH2)4CH3
Cl
F

NOMENCLATURE

flumiclorac-pentyl

Common name flumiclorac-pentyl (BSI, pa ISO, ANSI)
CAS RN *[87546–18–7]* **Development codes** S-23031 (Sumitomo); V-23031 (Valent)

flumiclorac

Common name flumiclorac (BSI, pa E-ISO, ANSI)
IUPAC name [2-chloro-5-(cyclohex-1-ene-1,2-dicarboximido)-4-fluorophenoxy]=acetic acid
Chemical Abstracts name [2-chloro-4-fluoro-5-(1,3,4,5,6,7-hexahydro-1,3-dioxo-=2*H*-isoindol-2-yl)phenoxy]acetic acid
CAS RN *[87547–04–4]*

PHYSICAL CHEMISTRY

flumiclorac-pentyl

Mol. wt. 423.9 **M.f.** $C_{21}H_{23}ClFNO_5$ **Form** Beige solid with a halide odour. **M.p.** 88.9–90.1 °C **V.p.** <0.01 mPa (22.4 °C) K_{ow} logP = 4.99 (20 °C) **S.g./density** 1.33 g/ml (20 °C) **Solubility** In water 0.189 mg/l (25 °C). In methanol 47.8, hexane 3.28, *n*-octanol 16.0, acetone 590 g/l. **Stability** Hydrolytic DT_{50} 4.2 d (pH 5), 19 h (pH 7), 6 min. (pH 9). Aqueous photolysis DT_{50} are similar. **F.p.** >88 °C

flumiclorac

Mol. wt. 353.7 **M.f.** $C_{16}H_{13}ClFNO_5$

APPLICATIONS

flumiclorac-pentyl

Biochemistry Protoporphyrinogen oxidase inhibitor. It is suggested that flumiclorac-pentyl induces accumulation of porphyrins in susceptible plants. The photosensitising action of accumulated porphyrins is likely to be one of the causes of membrane lipid peroxidation. The peroxidation of membrane lipids leads to irreversible damage of membrane function and structure in susceptible plants, and results in symptoms outlined below. Differences in metabolism of flumiclorac-pentyl seem to be the basis of crop selectivity. Experiments with

^{14}C-flumiclorac-pentyl suggest more rapid degradation in soya beans than in *Abutilon*. **Mode of action** Fast-acting, contact post-emergence herbicide. When applied to foliage of susceptible plants, it is readily absorbed into plant tissue and causes characteristic herbicidal symptoms such as desiccation, wilting, bleaching, browning and necrosis. **Uses** Post-emergence herbicide for control of problem broad-leaved weeds, including *Xanthium strumarium*, *Chenopodium album*, *Ambrosia artemisifolia*, *Datura stramonium*, *Amaranthus* spp., *Sida spinosa*, *Euphorbia maculata*, *Abutilon theophrasti*, in soya beans, and maize at 0.027–0.054 lb/a.
Formulation types EC. **Selected tradenames** 'Resource' (Sumitomo, Valent); 'Sumiverde' (Sumitomo)

MAMMALIAN TOXICOLOGY
flumiclorac-pentyl
Oral Acute oral LD_{50} for rats 3.6 g 0.86EC/kg. **Skin and eye** Acute percutaneous LD_{50} for rabbits >2.0 g 0.86EC/kg. Moderately irritating to skin and eyes; not a skin sensitiser. **Inhalation** LC_{50} for rats 5.51 mg 0.86EC/l.

ECOTOXICOLOGY
flumiclorac-pentyl
Birds Acute oral LD_{50} for bobwhite quail >2250 mg/kg. Dietary LC_{50} for bobwhite quail and mallard ducks >5620 ppm. **Fish** LC_{50} for rainbow trout 1.1, bluegill sunfish 13–21 mg/l.

ENVIRONMENTAL FATE
Soil/Environment Rapidly degraded in soil: DT_{50} 0.48–4.4 d in loamy-sand soil (pH 7); degradates have DT_{50} c. 2–30 d. The a.i. is immobile in soil; degradates have low to medium mobility. No residues of parent or degradates were observed below the upper 3 inches of soil.

341 flumioxazin   *Herbicide*

phthalimide

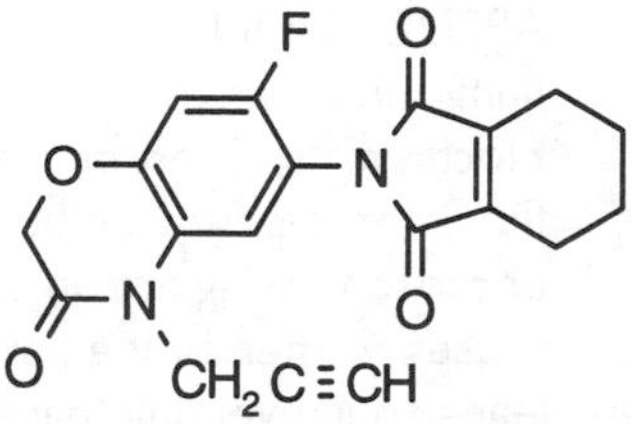

NOMENCLATURE
Common name flumioxazin (BSI, draft ISO, ANSI)

IUPAC name *N*-(7-fluoro-3,4-dihydro-3-oxo-4-prop-2-ynyl-2*H*-1,4-benzoxazin-= 6-yl)cyclohex-1-ene-1,2-dicarboxamide
Chemical Abstracts name 2-[7-fluoro-3,4-dihydro-3-oxo-4-(2-propynyl)-2*H*-= 1,4-benzoxazin-6-yl]-4,5,6,7-tetrahydro-1*H*-isoindole-1,3(2*H*)-dione
CAS RN *[103361–09–7]*   **Development codes** S-53482 (Sumitomo); V-53482 (Valent)

PHYSICAL CHEMISTRY
Mol. Wt. 354.3   **M.f.** $C_{19}H_{15}FN_2O_4$ **Form** Yellow-brown powder.
M.p. 201–204 °C   **V.p.** 0.32 mPa (22 °C)   **S.g./density** 1.5136 (20 °C)
Solubility In water 17.8 g/l (25 °C). Soluble in common organic solvents.
Stability Stable under normal storage conditions.

COMMERCIALISATION
History Reported by R. Yoshida *et al.* (*Proc. Br. Crop Prot. Conf. – Weeds*, 1991, **1**, 69). Introduced by Sumitomo.   **Manufacturers** Sumitomo

APPLICATIONS
Biochemistry Protoporphyrinogen oxidase inhibitor. Acts, in the presence of light and oxygen, by inducing massive accumulation of porphyrins, and enhancing peroxidation of membrane lipids, which leads to irreversible damage of the membrane function and structure of susceptible plants.
Mode of action Herbicide, absorbed by foliage and germinating seedlings.
Uses Control of many annual broad-leaved weeds and some annual grasses pre- and post-emergence in soya beans, peanuts, orchards and other crops.
Phytotoxicity Soya beans and peanuts are tolerant. Maize, wheat, barley and rice are moderately tolerant.   **Formulation types** DG; WP.
Selected tradenames 'Sumisoya' (Sumitomo)

MAMMALIAN TOXICOLOGY
Oral Acute oral LD_{50} for rats >5000 mg/kg.   **Skin and eye** Acute percutaneous LD_{50} for rats >2000 mg/kg. Non-irritating to skin; mild eye irritant (rabbits). Non-sensitising to skin (guinea pigs).   **Inhalation** LC_{50} (4 h) for rats >3930 mg/m^3 air.   **Other** Non-mutagenic in the Ames test, *in vivo* chromosomal aberration, and *in vivo/vitro* UDS.

ECOTOXICOLOGY
Fish LC_{50} (96 h) for rainbow trout 2.3, bluegill sunfish >21 mg/l.

342 fluometuron

Herbicide

urea

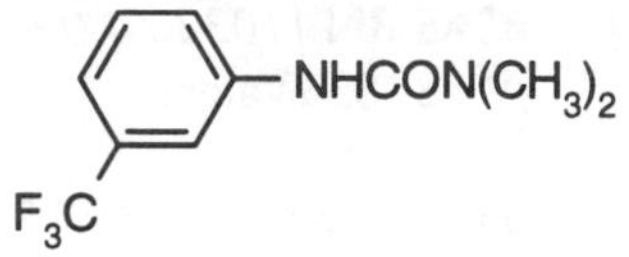

$$NHCON(CH_3)_2$$

F_3C

NOMENCLATURE
Common name fluometuron (BSI, E-ISO, (m) F-ISO, ANSI, WSSA)
IUPAC name 1,1-dimethyl-3-(α,α,α-trifluoro-*m*-tolyl)urea
Chemical Abstracts name *N,N*-dimethyl-*N'*-[3-(trifluoromethyl)phenyl]urea
CAS RN *[2164–17–2]* **EEC no.** 218–500–4 **Development codes** C 2059 (Ciba)

PHYSICAL CHEMISTRY
Mol. wt. 232.2 **M.f.** $C_{10}H_{11}F_3N_2O$ **Form** Colourless crystals.
M.p. 163–164.5 °C **V.p.** 0.125 mPa (25 °C), 0.33 mPa (30 °C) (OECD 104)
K_{ow} logP = 2.38 **S.g./density** 1.39 (20 °C) **Solubility** In water 110 mg/l
(20 °C). In methanol 110, acetone 105, dichloromethane 23, *n*-octanol 22, hexane
0.17 (all in g/l, 20 °C). **Stability** Stable in acidic, neutral, and alkaline media at
20 °C. Decomposed by u.v. irradiation.

COMMERCIALISATION
History Herbicide reported by C. J. Counselman *et al.* (*Proc. South Weed Conf.*,
17th, 1964, p. 189). Introduced by Ciba AG (now Novartis Crop Protection AG).
Patent BE 594227; GB 914779 **Manufacturers** Griffin; Makhteshim-Agan;
Novartis; Nufarm GmbH

APPLICATIONS
Biochemistry Photosynthetic electron transport inhibitor.
Mode of action Selective systemic herbicide, absorbed more readily by the roots
than by the foliage, with translocation acropetally. **Uses** Control of annual broad-
leaved weeds and grasses in cotton and sugar cane at 1.0–1.5 kg a.i./ha.
Phytotoxicity Phytotoxic to sugar beet, beetroot, soya beans, french beans,
brassicas, tomatoes, legumes, cucurbits, aubergines, and several other crops.
Formulation types WP; SC. **Mixtures** *(fluometuron +)* prometryn; metolachlor.
Selected tradenames 'Cotogard' (Novartis); 'Cotoran' (Novartis); 'Cottonex'
(Makhteshim-Agan); 'Meturon' (Griffin)

ANALYSIS
Product analysis by glc (*AOAC Methods*, 1995, 977.07; *CIPAC Handbook*, 1983, **1B**,
1842) or by hplc (G. W. Sheehan & A. Sobolewski, *Anal. Methods Pestic. Plant
Growth Regul.*, 1984, **13**, 241). **Residues** determined by hydrolysis to
α,α,α-trifluoro-*m*-toluidine, a derivative of which is measured by glc with ECD

(G. Voss et al., ibid., 1973, **7**, 569). **In water**, by lc with u.v. detection (AOAC Methods, 1995, 992.14, 10.7.01).

MAMMALIAN TOXICOLOGY
Oral Acute oral LD_{50} for rats >6000 mg/kg. **Skin and eye** Acute percutaneous LD_{50} for rats >2000, rabbits >10 000 mg/kg. Mild skin and eye irritant (rabbits). Not a skin sensitiser. **NOEL** (2 y) for rats 19, mice 1.3 mg/kg b.w. daily; (1 y) for dogs 10 mg/kg b.w. daily. **ADI** 0.013 mg/kg b.w. **Toxicity class** WHO (a.i.) III (Table 5); EPA (formulation) II

ECOTOXICOLOGY
Birds LD_{50} for mallard ducks 2974 mg/kg. Eight-day dietary LC_{50} for Japanese quail 4620, mallard ducks 4500, ringneck pheasants 3150 mg/kg diet. **Fish** LC_{50} (96 h) for rainbow trout 47, bluegill sunfish 96, catfish 55, crucian carp 170 mg/l. **Bees** LD_{50} (oral) >155 μg/bee, (topical) >190 μg/bee. **Worms** LC_{50} (14 d) for earthworms >1000 mg/kg soil. **Daphnia** LC_{50} (48 h) 10 mg/l. **Algae** EC_{50} (3 d) 0.16 mg/l.

ENVIRONMENTAL FATE
Animals In rats, mainly formation of demethylated metabolites and some conjugation with gluconuronic acid. Elimination, mainly in the urine, was 96% within one week. **Plants** In plants, degraded by a three-step process, first demethylation to a monomethyl, then to a demethylated intermediate, and finally deamination-decarboxylation to an aniline derivative. **Soil/Environment** In soil, microbial degradation is significant and rapid, with continual liberation of CO_2. Photodecomposition and volatilisation are insignificant. It is moderately leachable in all but very sandy soils: K_{oc} 31–117 (8 soil types); K_d 0.15–1.13. Median DT_{50} in soil c. 30 d; depending on environmental factors, range was 10 to almost 100 d; dry conditions reduce breakdown.

343 fluoroacetamide *Rodenticide*

$$FCH_2CONH_2$$

NOMENCLATURE
Common name fluoroacetamide (BSI, E-ISO, F-ISO, JMAF, accepted in lieu of a common name)
IUPAC name 2-fluoroacetamide
Chemical Abstracts name 2-fluoroacetamide
CAS RN [640–19–7] **EEC no.** 211–363–1

PHYSICAL CHEMISTRY
Mol. wt. 77.1 **M.f.** C_2H_4FNO **Form** Colourless crystalline powder.
M.p. 108 °C **Solubility** Very soluble in water. Soluble in acetone, moderately soluble in ethanol, and sparingly soluble in aliphatic and aromatic hydrocarbons.

COMMERCIALISATION
History Rodenticidal properties reported by C. Chapman & M. A. Phillips (*J. Sci. Food Agric.*, 1955, **6**, 231). Introduced as a rodenticide.
Manufacturers Jewnin-Joffe

APPLICATIONS
Mode of action Moderately fast-acting rodenticide which is less likely to lead to poison shyness because of sublethal dosing. It acts chiefly on the heart, with secondary effects on the central nervous system. **Uses** Control of rats and mice. Use restricted to sewers, locked warehouses, and other areas to which the public have no access. **Formulation types** GB. **Selected tradenames** 'Rodex' (Jewnin-Joffe)

ANALYSIS
Product analysis by titration (*AOAC Methods*, 1984, 6.023–6.024) or by reaction with sodium and precipitation as lead chloride fluoride (*cf. WHO Manual*, (2nd Ed.), SRT/5; *AOAC Methods*, 1984, 6.019–6.020).

MAMMALIAN TOXICOLOGY
Oral Acute oral LD_{50} for *Rattus norvegicus* c. 13 mg/kg (W. Bentley & J. H. Greaves, *J. Hyg.*, 1960, **58**, 125; E. W. Bentley *et al.*, *ibid.*, 1961, **59**, 413).
Toxicity class WHO (a.i.) Ib **EC risk** T+ (R28); T (R24) **PIC** Yes

344 fluoroglycofen-ethyl   *Herbicide*

diphenyl ether

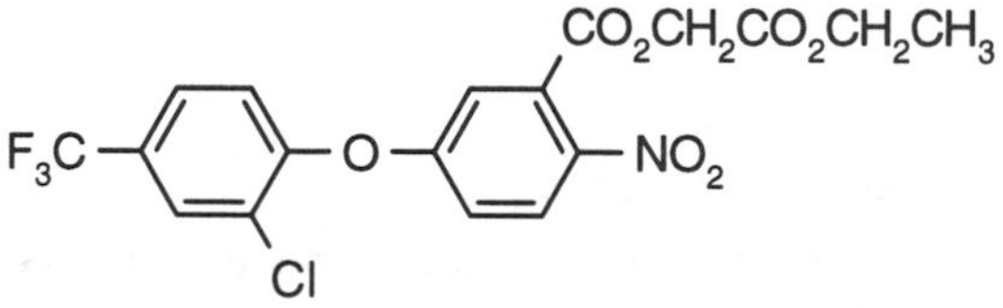

NOMENCLATURE
fluoroglycofen-ethyl
Common name fluoroglycofen-ethyl
CAS RN *[77501–90–7]* **Development codes** RH-0265

fluoroglycofen

Common name fluoroglycofen (BSI, ANSI, draft E-ISO); fluoroglycofène ((*m*) draft F-ISO)
IUPAC name *O*-[5-(2-chloro-α,α,α-trifluoro-*p*-tolyloxy)-2-nitrobenzoyl]glycolic acid
Chemical Abstracts name carboxymethyl 5-[2-chloro-4-(trifluoromethyl)= phenoxy]-2-nitrobenzoate
CAS RN *[77501–60–1]*

PHYSICAL CHEMISTRY
fluoroglycofen-ethyl
Mol. wt. 447.8   **M.f.** $C_{18}H_{13}ClF_3NO_7$ **Form** Dark amber solid.   **M.p.** 65 °C
V.p. <1.33 × 10^5 mPa (25 °C)   **K$_{ow}$** logP = 3.65   **S.g./density** 1.01 (25 °C)
Solubility In water 0.6 mg/l (25 °C). Readily soluble in most organic solvents (except hexane).   **Stability** In aqueous solution, 0.25 mg/l, at 22 °C, DT_{50} *c.* 231 d (pH 5), 15 d (pH 7), 0.15 d (pH 9); aqueous suspensions are rapidly decomposed by u.v. light.

fluoroglycofen
Mol. wt. 419.7   **M.f.** $C_{16}H_9ClF_3NO_7$

COMMERCIALISATION
History Herbicide developed by Rohm & Haas Co. as the ethyl ester.
Manufacturers Rohm & Haas

APPLICATIONS
fluoroglycofen-ethyl
Biochemistry Protoporphyrinogen oxidase inhibitor.
Mode of action Fluoroglycofen-ethyl is a selective leaf- and root- herbicide.
Uses Control of broad-leaved weeds and grasses (particularly *Galium*, *Viola*, and *Veronica* spp.) in wheat, barley, oats, peanuts, rice, and soya beans.
Formulation types EC; WP.   **Mixtures** *(fluoroglycofen-ethyl +)* mecoprop; triasulfuron; isoproturon; isoproturon + triasulfuron.   **Compatibility** Incompatible with suspension concentrate formulations.   **Selected tradenames** 'Compete' (Rohm & Haas); 'Satis' (mixture) (Novartis)

ANALYSIS
Product by gc.   **Residues** by glc with halogen-specific MCD.   Details from Rohm & Haas.

MAMMALIAN TOXICOLOGY
fluoroglycofen-ethyl
Reviews UK MAFF Scientific sub-committee on Pesticides, Disclosure Document for Fluoroglycofen-ethyl, Feb. 1992.   **Oral** Acute oral LD_{50} for rats 1500 mg/kg.

Skin and eye Acute percutaneous LD_{50} for rabbits >5000 mg/kg; slight irritant to their skin and eyes. Inhalation LC_{50} (4 h) for rats >7.5 mg EC formulation/l. NOEL (1 y) for dogs 320 mg/kg diet. ADI 0.009 mg/kg. Other Non-mutagenic in the Ames test. Toxicity class WHO (a.i.) III

ECOTOXICOLOGY
fluoroglycofen-ethyl
Birds Acute oral LD_{50} for bobwhite quail >3160 mg/kg. Eight-day dietary LC_{50} for mallard ducks and bobwhite quail >5000 mg a.i./kg. Fish LC_{50} (96 h) for bluegill sunfish 1.6, trout 23 mg/l. Bees LD_{50} (contact) >100 μg/bee. Daphnia LC_{50} (48 h) 30 mg/l.

ENVIRONMENTAL FATE
Animals Hydrolysis of the ester and reduction of the nitro group. Plants As for animals. Soil/Environment In soil and water, fluoroglycofen-ethyl is very rapidly hydrolysed to the corresponding acid. Microbial degradation follows in soil. In soil DT_{50} (to acifluorfen) c. 11 h.

345 fluoroimide

Fungicide

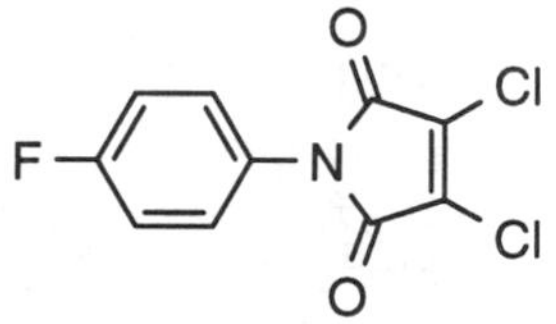

NOMENCLATURE
Common name fluoroimide (JMAF)
IUPAC name 2,3-dichloro-N-4-fluorophenylmaleimide
Chemical Abstracts name 3,4-dichloro-1-(4-fluorophenyl)-1H-pyrrole-2,5-dione
CAS RN [41205-21-4] Development codes MK-23 (Mitsubishi)

PHYSICAL CHEMISTRY
Mol. wt. 260.1 M.f. $C_{10}H_4Cl_2FNO_2$ Form Pale yellow crystals.
M.p. 240.5–241.8 °C V.p. 3.4 mPa (25 °C), 8.1 mPa (40 °C) K_{ow} logP = 2.3
S.g./density 1.59 Solubility In water 5.9 mg/l (20 °C). In acetone 19.2 g/l (20 °C). Stability Stable to 120 °C; under u.v. and sunlight; on hydrolysis DT_{50} 52.9 min (pH 3), 7.5 min (pH 7) and 1.4 min (pH 8).

COMMERCIALISATION
History Fungicide reported (*Jpn. Pestic. Inf.*, 1978, No. 34, p. 26). Introduced by

582 *fluoroimide*

Mitsubishi Chemical Industries Ltd (now Mitsubishi Chemical Corp.) and Kumiai Chemical Industry Co., Ltd. **Patent** JP 712681 **Manufacturers** Mitsubishi Chemical

APPLICATIONS
Biochemistry Inhibits sulfhydryl enzymes in the pathogen. **Mode of action** Foliar fungicide with protective action. Inhibits spore germination. **Uses** Control of *Monilinia mali* and *Mycosphaerella pomi* in apples, *Diaporthe citri* and *Elsinoe fawcetti* in citrus, *Corticium* sp. in rubber trees, *Botrytis cinerea* and *Peronospora destractor* in onions, *Colletotrichum theae-sinensis* and *Exobasidium vexans* in tea, and *Phytophthora infestans* in potatoes at 2–5 kg a.i./ha, and *Cercospora kaki* and *Mycosphaerella nawae* in persimmons at 0.5–2 kg/ha. **Phytotoxicity** May be phytotoxic to some varieties of pear. **Formulation types** WP. **Compatibility** Incompatible with alkaline materials. **Selected tradenames** 'Spartcide' (Mitsubishi Chemical, Kumiai)

ANALYSIS
Product and **residue** analysis by glc.

MAMMALIAN TOXICOLOGY
Oral Acute oral LD_{50} for rats and mice >15 000 mg/kg. **Skin and eye** Acute percutaneous LD_{50} for mice >5000 mg/kg. **Inhalation** LC_{50} (4 h) for male rats 0.57 mg tech./l air, for female rats 0.72 mg/l air. **NOEL** (2 y) for rats 600–2000 mg/kg diet. **Toxicity class** WHO (a.i.) III (Table 5)

ECOTOXICOLOGY
Birds Five-day dietary LC_{50} for pheasants >27 000 mg/kg diet. **Fish** LC_{50} (48 h) for carp 5.6 mg/l. **Daphnia** LC_{50} (3 h) 13.5 mg/l.

346 flupoxam   *Herbicide*

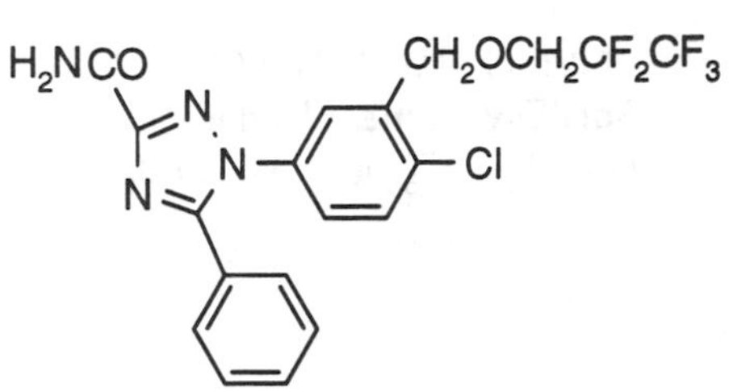

NOMENCLATURE
Common name flupoxam (BSI, pa E-ISO)

IUPAC name 1-[4-chloro-3-(2,2,3,3,3-pentafluoropropoxymethyl)phenyl]-=
5-phenyl-1*H*-1,2,4-triazole-3-carboxamide
Chemical Abstracts name 1-[4-chloro-3-[(2,2,3,3,3-pentafluoropropoxy)=
methyl]phenyl]-5-phenyl-1*H*-1,2,4-triazole-3-carboxamide
CAS RN *[119126–15–7]* **Development codes** KNW-739 (Kureha); MON-18500

PHYSICAL CHEMISTRY
Mol. wt. 460.8 **M.f.** $C_{19}H_{14}ClF_5N_4O_2$ **Form** Light beige, odourless crystals.
M.p. 144–148 °C **V.p.** 2.0×10^{-2} mPa (25 °C) **K_{ow}** logP = 3.27
S.g./density 1.433 (20.5 °C) **Solubility** In water 5.0 (pH 5.1), 1.0 (pH 7.4) mg/l.
In hexane 0.003, toluene 5.6, methanol 133, acetone 267, ethyl acetate 102 (all in
g/l at 20 °C).

COMMERCIALISATION
History Discovered and introduced by Kureha; also under development jointly
with Monsanto Co. for the European cereal market since 1987. Reported by
M. G. O'Keeffe *et al.* (*Proc. Br. Crop Prot. Conf. – Weeds*, 1991, **1**, 63) and by
B. Cagnac (ANPP, *15éme Conférence du Columa*, 1992). **Patent** EP 0282303;
US 4973353 **Manufacturers** Synexus (Kureha/Monsanto j.v.)

APPLICATIONS
Mode of action Cell elongation inhibitor which acts most effectively in actively
growing meristematic areas (both root and foliar). Activity is mainly through
contact with meristematic tissue, and there is little translocation within the plant.
Uses Control of annual broad-leaved weeds, pre- or early post-emergence, in
winter wheat and winter barley. Mixture with isoproturon extends the spectrum to
include control of grasses. **Phytotoxicity** Non-phytotoxic to winter wheat, durum
wheat and winter barley, when applied as directed. **Formulation types** SC.
Mixtures *(flupoxam +)* isoproturon. **Selected tradenames** 'Ovation' (Synexus)

MAMMALIAN TOXICOLOGY
Oral Acute oral LD_{50} for rats >5000 mg/kg. **Skin and eye** Acute percutaneous
LD_{50} for rabbits >2000 mg/kg. Non-irritating to skin; mild eye irritant (rabbits).
Other Non-teratogenic. Non-mutagenic in the Ames and micronucleus tests.

ENVIRONMENTAL FATE
Soil/Environment Undergoes microbial degradation in soil; does not move
laterally. DT_{50} in soil *c.* 69 d.

flupropanate   *Herbicide*

halogenated alkanoic acid

$$F_2CHCF_2CO_2H$$

NOMENCLATURE
flupropanate
Common name flupropanate (BSI, draft E-ISO, draft F ISO)
IUPAC name 2,2,3,3-tetrafluoropropionic acid
Chemical Abstracts name 2,2,3,3-tetrafluoropropanoic acid
CAS RN *[756–09–2]*

flupropanate-sodium
Common name flupropanate-sodium (BSI, draft E-ISO, draft F-ISO); tetrapion (JMAF)
CAS RN *[22898–01–7]*   **Development codes** Orga 3045; TOF(SW-6508)

PHYSICAL CHEMISTRY
flupropanate
Mol. wt. 146.0   **M.f.** $C_3H_2F_4O_2$

flupropanate-sodium
Composition Tech. is 60 wt% solution.   **Mol. wt.** 168.0   **M.f.** $C_3HF_4NaO_2$
Form Colourless crystals.   **M.p.** 165–167 °C (decomp.)   **V.p.** <40 mPa (138 °C)
K_{ow} logP <–1.9   **S.g./density** 1.45 (tech.)   **Solubility** In water 3.9 kg/l (20 °C).
Sparingly soluble in non-polar solvents.   **Stability** Stable to heat, light and water.

COMMERCIALISATION
History Herbicidal activity of flupropanate-sodium reported by N. V. Orgachemia
(E. Aelbers *et al., Symp. New Herbicides*, 1969, **1**, 17) and by Daikin Kogyo Co.
(now Daikin Industries Ltd) (*Jpn. Pestic. Inf.*, 1972, No. 13, p. 10). Now marketed
by ICI Australia and by Japanese agrochemical companies, mainly for non-crop
land applications.   **Manufacturers** Daikin

APPLICATIONS
flupropanate-sodium
Biochemistry Inhibition of the decarboxylation of aspartic acid to β-alanine.
Mode of action Systemic herbicide with low contact activity. Mainly taken up by
the roots.   **Uses** Control of annual and perennial grasses in pastures and in
uncultivated land.   **Formulation types** SL; GR.   **Selected tradenames** 'Frenock'
(Daikin, Sankyo, Hodogaya, ICI Australia, Sumitomo)

ANALYSIS
By ion chromatography. Column: SAM3–125; eluent: 4.4 mM Na_2CO_3, 1.2 mM $NaHCO_3$; flow rate: 2 ml/min.; injection: 100 μl; detection limit: 2 ppb.

MAMMALIAN TOXICOLOGY
flupropanate-sodium
Oral Acute oral LD_{50} for rats 11 900 mg tech./kg, for mice 9600 mg/kg.
Skin and eye Acute percutaneous LD_{50} for rats >5500, rabbits >4000 mg/kg.
Inhalation LC_{50} (14 d) for rats >1740 mg/m³. **NOEL** (3 mo) for rats
5 mg/kg diet; (12 mo) for mice 6.6 mg/kg diet. **Toxicity class** WHO (a.i.) III
(Table 5)

ECOTOXICOLOGY
flupropanate-sodium
Birds Acute oral LD_{50} for male quail 6750, female quail 11 000 mg/kg.
Fish TLm (48 h) for carp and rainbow trout >10 000 mg/l.

ENVIRONMENTAL FATE
Animals Flupropanate-sodium administered to rats was not detected in internal
organs after 14 d. **Plants** Flupropanate-sodium applied at 14.4 kg a.i./ha to white
clover was taken up rapidly by the plant. After 4 months, residues in the plant had
reduced to 256 ppm. **Soil/Environment** Field tests indicate that flupropanate is
immobile but persistent in the soil.

348 flupyrsulfuron-methyl-sodium *Herbicide*

sulfonylurea

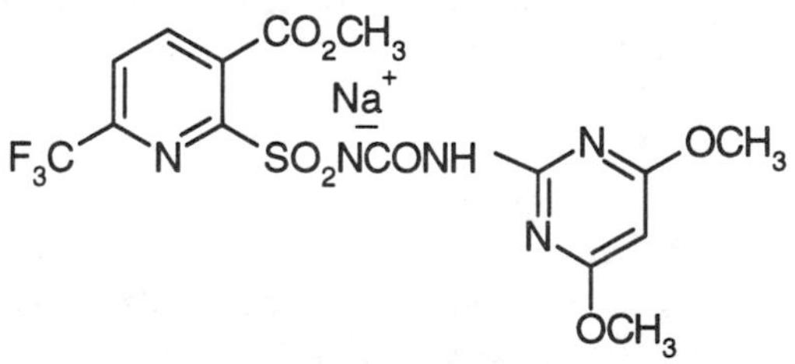

NOMENCLATURE
Common name flupyrsulfuron-methyl-sodium (BSI, pa ISO)
IUPAC name methyl 2-(4,6-dimethoxypyrimidin-2-ylcarbamoylsulfamoyl)-=
6-trifluoromethylnicotinate monosodium salt

Chemical Abstracts name methyl 2-[[[[(4,6-dimethoxy-2-pyrimidinyl)amino]= carbonyl]amino]sulfonyl]-6-(trifluoromethyl)-3-pyridinecarboxylate monosodium salt
CAS RN [144740–54–5]; (acid [150315–10–9])
Development codes DPX-KE459; JE138; IN-KE459

PHYSICAL CHEMISTRY
Mol. wt. 487.3   **M.f.** $C_{15}H_{13}F_3N_5NaO_7S$; (acid $C_{14}H_{12}F_3N_5O_7S$)   **Form** White powder with a pungent odour.   **M.p.** 165–170 °C   **V.p.** 10^6 mPa (20 °C)
K_{ow} logP = 0.96 (pH 5), 0.10 (pH 6) (both 20 °C)   **Henry** 10^{-8} Pa m^3 mol^{-1}
S.g./density 1.48   **Solubility** In water 63 mg/l (pH 5, 25 °C). In dichloromethane 600, acetone 3094, ethyl acetate 490, acetonitrile 4332, n-hexane >1.0 (all mg/l, 20 °C).   **Stability** DT_{50} 44 d (pH 5), 12 d (pH 7), 0.4 d (pH 9) (20 °C). Stable in most organic solvents.   **pKa** 4.9

COMMERCIALISATION
History Reported by S. R. Teaney et al. (Proc. Br. Crop Prot. Conf. – Weeds, 1995, 1, 49)   **Manufacturers** Du Pont

APPLICATIONS
Biochemistry Branched chain amino acid synthesis (acetolactate synthase or ALS) inhibitor. Acts by inhibiting biosynthesis of the essential amino acids valine and isoleucine, hence stopping cell division and plant growth. Selectivity derives from rapid metabolism in the crop. Metabolic basis of selectivity in sulfonylureas reviewed (M. K. Koeppe & H. M. Brown, Agro-Food-Industry, **6**, 9–14 (1995)).
Mode of action Active by foliar and root uptake.   **Uses** Under development for post-emergence, selective control of blackgrass (Alopecurus myosuroides) and other weeds in cereals at 10 g/ha.   **Formulation types** WG.   **Mixtures** (flupyrsulfuron-methyl-sodium +) thifensulfuron-methyl; metsulfuron-methyl; carfentrazone.

ANALYSIS
Methods for sulfonylurea **residues** in crops, soil and water reviewed (A. C. Barefoot et al., Proc. Br. Crop Prot. Conf. – Weeds, 1995, **2**, 707).

MAMMALIAN TOXICOLOGY
Oral Acute oral LD_{50} for rats and mice >5000 mg/kg.   **Skin and eye** Acute percutaneous LD_{50} for rabbits >2000 mg/kg. Not a skin or eye irritant (rabbits); not a skin sensitiser (guinea pigs).   **Inhalation** LC_{50} (4 h) for rats >2.0 mg/l.
NOEL (2 y) for rats 350 ppm; (1 y) for male dogs 16.4, female dogs 13.6 mg/kg.
ADI 0.035 mg/kg.   **Other** Non-mutagenic (Ames test).   **Toxicity class** WHO (a.i.) III (Table 5)

ECOTOXICOLOGY
Birds Oral LD_{50} for mallard duck >2250 mg/kg. Dietary LC_{50} for bobwhite quail

and mallard duck >5620 ppm.   **Fish** LC_{50} (96 h) for carp 820, rainbow trout 470 mg/l.   **Bees** LD_{50} (contact) >25, (dietary) >30 µg/bee.   **Daphnia** LC_{50} (48 h) 721 mg/l.   **Algae** EC_{50} (5 d) for green algae >20 µg/l (growth rate), 13 µg/l (cell density).   **Other aquatic spp.** EC_{50} (14 d) for *Lemna gibba*, frond 5.0 µg/l, biomass 2.5 µg/l.

ENVIRONMENTAL FATE
Low K_{ow} suggests that bioaccumulation should not occur.   **Soil/Environment** DT_{50} 8 to 25 d (20 °C), 58 d (10 °C) (lab. studies). Rapidly degraded to non-herbicidal materials.

349 fluquinconazole   *Fungicide*

azole

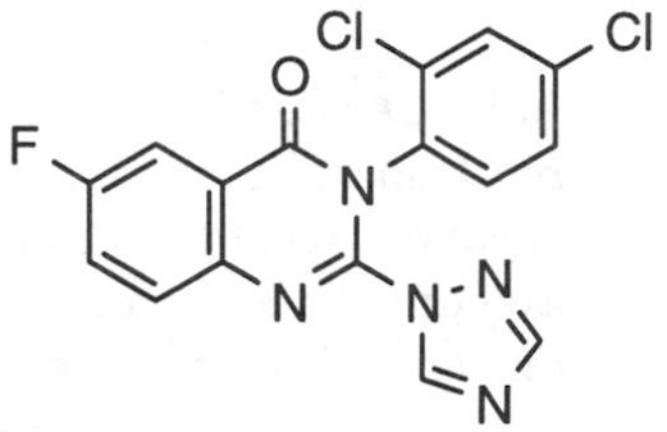

NOMENCLATURE
Common name fluquinconazole (BSI, draft ISO)
IUPAC name 3-(2,4-dichlorophenyl)-6-fluoro-2-(1*H*-1,2,4-triazol-1-yl)quinazolin-=4(3*H*)-one
Chemical Abstracts name 3-(2,4-dichlorophenyl)-6-fluoro-2-((1*H*)-1,2,4-triazol-=1-yl)-4(3*H*)-quinazolinone
CAS RN *[136426–54–5]*   **Development codes** SN 597265 (Schering)

PHYSICAL CHEMISTRY
Mol. wt. 376.2   **M.f.** $C_{16}H_8Cl_2FN_5O$ **Form** Cream/pale brown crystalline solid.
M.p. 191.9–193 °C (tech., 184–192 °C)   **V.p.** 6.4×10^{-6} mPa (20 °C)
K_{ow} logP = 3.24 (pH 5.6)   **Henry** 2.09×10^{-6} Pa m^3 mol^{-1} **S.g./density** 1.58 (20 °C)   **Solubility** In water at 20 °C, 1 mg/l (pH 6.6). In acetone 50, xylene 10, ethanol 3, dimethyl sulfoxide 200 g/l.   **Stability** In water DT_{50} (25 °C) 21.8 d (pH 7). Stable to light.

COMMERCIALISATION
History Reported by P. E. Russell *et al.* (*Proc. Br. Crop Prot. Conf. – Pests Dis.*, 1992, **1**, 411). Under development by AgrEvo GmbH.   **Manufacturers** AgrEvo

APPLICATIONS

Biochemistry Inhibitor of ergosterol biosynthesis. **Mode of action** Fungicide with protectant, eradicant and systemic properties. Is transported by the xylem, and there is also translaminar flow. **Uses** Control of a wide range of Ascomycetes, Deuteromycetes and Basidiomycetes, by foliar application, on broad-leaved and cereal crops, e.g. *Venturia inaequalis* and *Podosphaera leucotricha* in apple, *Uncinula necator* in vines, *Puccinia* spp. and *Septoria* spp. in wheat, *Cercospora*, *Erysiphe* and others in sugar beet, and other economically important diseases in crops such as oilseed rape, stone fruit and other crops. Application rates vary from c. 4–8 g/ha on fruit and vines to 125–190 g/ha on field crops. **Formulation types** SC; SE; WG; WP. **Mixtures** *(fluquinconazole +)* chlorothalonil; prochloraz. **Selected tradenames** 'Castellan' (AgrEvo); 'Vista' (mixture) (AgrEvo)

ANALYSIS

Product analysis by reverse-phase hplc. Crop **residues** analysis by solvent extraction followed by gc.

MAMMALIAN TOXICOLOGY

Oral Acute oral LD_{50} for male and female rats 112, male mice 325, female mice 180 mg/kg. **Skin and eye** Acute percutaneous LD_{50} for male rats 2679, female rats 625 mg/kg. Non-irritating to skin and eyes (rabbits), and not a skin sensitiser (guinea pigs). **Inhalation** LC_{50} (4 h) for rats 0.754 mg/l. **NOEL** (1 y) for rats 0.31, mice 1.1, dogs 0.5 mg/kg b.w. daily. **ADI** (proposed) 0.005 mg/kg. **Other** Negative in the Ames and other mutagenicity tests. **EC risk** T (R21, R23/25, R48/22)

ECOTOXICOLOGY

Birds Acute oral LD_{50} for bobwhite quail and mallard ducks >2000 mg/kg. **Fish** LC_{50} (96 h) for bluegill sunfish 1.34, rainbow trout 1.90 mg/l. **Bees** Oral and contact LD_{50} >100 µg/l. **Worms** LC_{50} (14 d) for *Eisenia foetida* >1000 mg/kg soil. **Other beneficial spp.** Harmless to a range of representative species in the field. **Daphnia** LC_{50} (48 h) >5.0 mg/l. **Algae** For *Selenastrum capricornutum*, E_rC_{50} 46, E_bC_{50} 14 µg/l.

ENVIRONMENTAL FATE

Animals Metabolised to 1,2,4-triazole. **Plants** Metabolised to 1,2,4-triazole. **Soil/Environment** Slowly degraded in soil via hydrolysis to the dione. Degradation rates depend on temperature, soil moisture, soil pH and organic matter content. Typical field DT_{50} 50–300 d. Strongly adsorbed to soils, K_{oc} >740. Soil column leaching studies and leaching models demonstrated that fluquinconazole will not leach to deeper soil layers and contaminate groundwater.

Cl—⟨thiazole ring with S, N⟩—CO$_2$CH$_2$—⟨phenyl⟩
CF$_3$

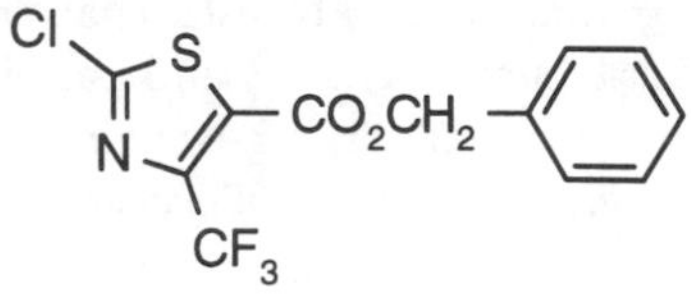

NOMENCLATURE
Common name flurazole (WSSA)
IUPAC name benzyl 2-chloro-4-trifluoromethyl-1,3-thiazole-5-carboxylate
Chemical Abstracts name phenylmethyl 2-chloro-4-(trifluoromethyl)-=
5-thiazolecarboxylate
CAS RN *[72850–64–7]*   **Development codes** MON 4606

PHYSICAL CHEMISTRY
Composition Tech. is 98% pure.   **Mol. wt.** 321.7   **M.f.** C$_{12}$H$_7$ClF$_3$NO$_2$S
Form Colourless crystals with a slight sweet odour; tech. is a yellow to tan solid.
M.p. 51–53 °C   **V.p.** 3.9 × 10^{-2} mPa (25 °C)   **S.g./density** 0.96 (tech.)
Solubility In water 0.5 mg/l (25 °C). Soluble in most organic solvents, including
alcohols, ketones, and xylene.   **Stability** Stable below 93 °C.   **F.p.** 392 °C (tech.,
Tag closed cup)

COMMERCIALISATION
History Herbicide safener introduced by Monsanto Agriculture Co.
Manufacturers Monsanto

APPLICATIONS
Uses Used as a herbicide safener. When applied as a seed protectant to grain
sorghum, acts as a safener against the phytotoxic effects of herbicides containing
alachlor or metolachlor.   **Formulation types** Seed treatment; WG.
Selected tradenames 'Screen' (Monsanto)

ANALYSIS
Details from Monsanto.

MAMMALIAN TOXICOLOGY
Oral Acute oral LD$_{50}$ for rats >5000 mg/kg.   **Skin and eye** Acute percutaneous
LD$_{50}$ for rabbits >5010 mg/kg. Non-irritating to skin; slightly irritating to eyes
(rabbits). Not a skin sensitiser (guinea pigs).   **Inhalation** No significant adverse
health effects reported through occupational exposure.   **NOEL** (90 d) for dogs
≤300 mg/kg daily, for rats ≤5000 mg/kg diet.

ECOTOXICOLOGY
Birds Acute oral LD_{50} for bobwhite quail >2510 mg/kg. Five-day dietary LC_{50} for mallard ducks and bobwhite quail >5620 ppm. **Fish** LC_{50} (96 h) for common carp 1.7, rainbow trout 8.5, bluegill sunfish 11 mg/l. **Daphnia** LC_{50} (48 h) 6.3 mg/l.

ENVIRONMENTAL FATE
Soil/Environment Rapidly degraded in soil, with the formation of water-soluble metabolites. Breakdown is primarily microbial, with chemical breakdown of lesser importance.

flurenol

Herbicide

morphactin

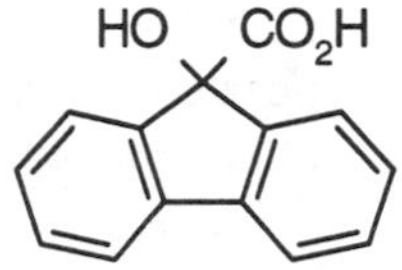

NOMENCLATURE
flurenol
Common name flurenol (BSI since 1990, E-ISO, (*m*) F-ISO); flurecol (BSI before 1990, Canada, Denmark, USA)
IUPAC name 9-hydroxyfluorene-9-carboxylic acid
Chemical Abstracts name 9-hydroxy-9*H*-fluorene-9-carboxylic acid
CAS RN *[467–69–6]*

flurenol-butyl
CAS RN *[2314–09–2]* **Development codes** EMD-IT-3233

PHYSICAL CHEMISTRY
flurenol
Mol. wt. 226.2 **M.f.** $C_{14}H_{10}O_3$ **K_{ow}** logP = 1.32 (unstated pH) **pKa** 1.09

flurenol-butyl
Mol. wt. 282.3 **M.f.** $C_{18}H_{18}O_3$ **Form** Colourless odourless crystals. **M.p.** 71 °C
V.p. 3.1×10^2 mPa (25 °C) **K_{ow}** logP = 3.7 **S.g./density** 1.15 (20 °C)
Solubility In water 36.5 mg/l (20 °C). In methanol 1500, acetone 1450, benzene 950, ethanol 700, carbon tetrachloride 550, isopropanol 250, cyclohexane 35, petroleum ether (b.p. 50–70 °C) *c.* 7 (all in g/l, 20 °C). **Stability** Decomposes in simulated daylight. Hydrolysed in acidic and alkaline media.

COMMERCIALISATION
History Effects of fluorene-9-carboxylic acids on plant growth reported by G. Schneider (*Naturwissenschaften*, 1964, **51**, 416) who proposed they be called morphactins (G. Schneider *et al.*, *Nature (London)*, 1965, **208**, 1013). Butyl ester introduced by E. Merck (now American Cyanamid Co.). **Patent** GB 1051652; GB 1051653 **Manufacturers** Cyanamid

APPLICATIONS
flurenol
Mode of action Synthetic growth regulator. Systemically transported acropetally and basipetally. Symptoms include inhibition of growth, dwarfing, inhibition of branching and of auxin transport. **Formulation types** EC.

flurenol-butyl
Mode of action Flurenol-butyl is a herbicide synergist, absorbed by the leaves and roots, with translocation acropetally and basipetally, and accumulation in growing tips of shoots and roots. Inhibits growth and development of growing tips and buds. **Uses** Flurenol-butyl is used in combination with phenoxy herbicides for post-emergence control of broad-leaved weeds in cereals, established grasses, and rice. **Formulation types** EC. **Mixtures** *(flurenol-butyl +)* MCPA esters and salts; 2,4-D esters and salts. **Selected tradenames** 'Aniten' (mixture) (Cyanamid)

ANALYSIS
Product analysis by glc of an ester (E. Amadori & W. Hempt, *Anal. Methods Pestic. Plant Growth Regul.*, 1980, **11**, 319). **Residues** determined by glc of an ester (*idem, ibid.*). Details of methods for product and residue analysis available from Cyanamid.

MAMMALIAN TOXICOLOGY
flurenol
Oral Acute oral for rats >6400, mice >6315 mg/kg. **Skin and eye** Acute percutaneous LD_{50} for rats >10 000 mg/kg. **NOEL** (117 d) for rats >10 000 ppm diet daily; (119 d) for dogs >10 000 ppm diet daily. **Toxicity class** WHO (a.i.) III (Table 5)

flurenol-butyl
Oral Acute oral LD_{50} for rats >10 000, mice >5000 mg/kg. **Skin and eye** Acute percutaneous LD_{50} for rats >10 000 mg/kg. Non-irritating to skin and eyes. **NOEL** (78 d) for rats >1000 mg/kg.

ECOTOXICOLOGY
flurenol
Fish LC_{50} (96 h) for trout 318 ppm. **Daphnia** LC_{50} (24 h) 86.7 mg/ml.

flurenol-butyl
Fish LC_{50} (96 h) for carp *c.* 18.2, rainbow trout *c.* 12.5 mg/l.   **Bees** Contact LD_{50}
for honeybees *c.* 0.10 mg/bee.   **Worms** LC_{50} >2000 mg/kg.   **Daphnia** LC_{50}
(24 h) 11.63 ppm.

ENVIRONMENTAL FATE
Animals In rats, following oral administration, 70–90% is eliminated within
24 hours, principally in the urine.   **Plants** Complete microbial degradation occurs
in plants.   **Soil/Environment** Rapid and complete microbial degradation occurs
in soil and water. DT_{50} in soil *c.* 1.5 d, in water *c.* 1–4 d. Adsorption Freundlich
K 1.6–5.0 mg/kg in soils with 0.5–2.6% o.c. and pH 6–7.6.

352 fluridone   *Herbicide*

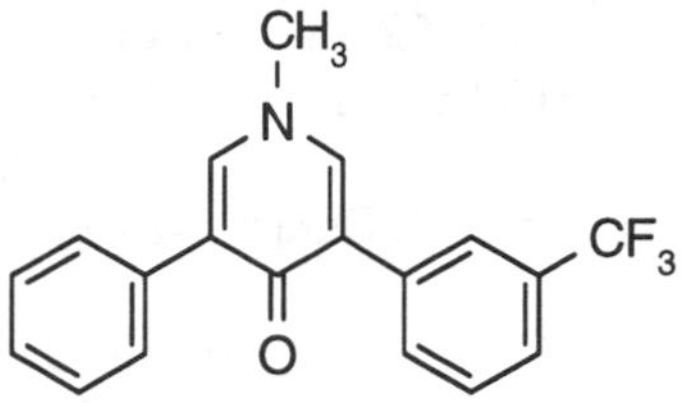

NOMENCLATURE
Common name fluridone (BSI, E-ISO, (*m*) F-ISO, ANSI, WSSA)
IUPAC name 1-methyl-3-phenyl-5-(α,α,α-trifluoro-*m*-tolyl)-4-pyridone
Chemical Abstracts name 1-methyl-3-phenyl-5-[3-(trifluoromethyl)phenyl]-=
4(1*H*)-pyridinone
CAS RN *[59756–60–4]*   **Development codes** EL-171

PHYSICAL CHEMISTRY
Mol. wt. 329.3   **M.f.** $C_{19}H_{14}F_3NO$ **Form** White crystalline solid; (tech. is a white
to tan crystalline solid).   **M.p.** 154–155 °C   **V.p.** 0.013 mPa (25 °C)
K_{ow} logP = 1.87 (pH 7, 25 °C)   **S.g./density** Bulk density, loose 0.358 g/cm^3,
packed 0.515 g/cm^3 **Solubility** In water *c.* 12 mg/l (pH 7, 25 °C). In methanol,
chloroform, diethyl ether >10, ethyl acetate >5, hexane <0.5 (all in g/l).
Stability Decomposes at 200–219 °C. Stable to hydrolysis at pH 3–9.
Decomposed by u.v. irradiation, DT_{50} 23 h in water.

COMMERCIALISATION
History Herbicide reported by T. W. Waldrep & H. M. Taylor (*J. Agric. Food Chem.*,

1976, **24,** 1250). Introduced in Syria (1977) by Eli Lilly & Co. (now DowElanco), and later sold to SePRO. **Patent** GB 1521092 **Manufacturers** SePRO

APPLICATIONS
Biochemistry Reduces carotenoid biosynthesis, by inhibition of phytoene desaturase, which causes chlorophyll depletion and hence inhibition of photosynthesis. **Mode of action** Selective systemic herbicide. In aquatic plants, absorbed by the foliage and roots. In terrestrial plants, absorbed predominantly by the roots, with translocation to the foliage (in susceptible species). In resistant species such as cotton, little or no translocation of root-absorbed fluridone occurs. **Uses** Used as an aquatic herbicide for control of most submerged and emerged aquatic plants (including *Utricularia* spp., *Ceratophyllum demersum*, *Elodea canadensis*, *Myriophyllum* spp., *Najas guadalupensis*, *Potamogeton* sp., *Hydrilla verticillata* and *Panicum purpurascens*) in ponds, lakes, reservoirs, irrigation ditches, etc. **Phytotoxicity** Cotton is the only crop which has been found to be tolerant. **Formulation types** SC; Pellets. **Selected tradenames** 'Sonar' (SePRO)

ANALYSIS
Product analysis by hplc (S. D. West, *Anal. Methods Pestic. Plant Growth Regul.*, 1984, **13,** 247). **Residues** in soil and plant tissue determined by hplc or glc with ECD of a suitable derivative (*idem, ibid.*; S. D. West & S. J. Parka, *J. Agric. Food Chem.*, 1989, **37,**). In **drinking water**, by gc with FID (*AOAC Methods*, 1995, 991.07). Details available from DowElanco.

MAMMALIAN TOXICOLOGY
Oral Acute oral LD_{50} for rats and mice >10 000, dogs >500, cats >250 mg/kg. **Skin and eye** Acute percutaneous LD_{50} for rabbits >5000 mg/kg. Non-irritating to skin; moderate eye irritant (rabbits). **Inhalation** LC_{50} (4 h) for rats >4.12 mg/l air. **NOEL** (24 mo) for mice 11.6, rats 8.5 mg/kg daily; (12 mo) for mice 11.4, rats 9.4, dogs 150 mg/kg daily; (90 d) for mice 9.3, rats 53, dogs 200 mg/kg daily. In 3-generation rat reproduction test, NOEL 121 mg/kg daily. Not carcinogenic. **Other** Not mutagenic. **Toxicity class** WHO (a.i.) III (Table 5)

ECOTOXICOLOGY
Birds Acute oral LD_{50} for bobwhite quail >2000 mg/kg. Eight-day dietary LC_{50} for bobwhite quail and mallard ducks >5000 mg/kg diet. **Fish** LC_{50} (96 h) for rainbow trout 11.7, bluegill sunfish 14.3 mg/l. **Bees** LD_{50} (oral) >362.6 μg/bee. **Worms** LC_{50} (14 d) for earthworms >102.6 ppm. **Daphnia** LC_{50} (48 h) 6.3 ppm. **Other aquatic spp.** EC_{50} (96 h) for pink shrimp 4.6, blue crab 34.0 ppm; (48 h) for oyster embryos 16.8 ppm.

ENVIRONMENTAL FATE
Plants Fluridone is not metabolised appreciably in terrestrial plants; see D. F. Berard *et al.*, *Weed Sci.*, **26,** 252 (1978). **Soil/Environment** Microbial degradation

is the principal cause of disappearance from soil. In a silt loam DT_{50} >343 d (pH 7.3, 2.6% o.m.). K_d 3–16, K_{oc} 350–1100 (5 soil types). In aquatic environments, degradation is principally by photolytic processes, but micro-organisms and aquatic vegetation are also factors. DT_{50} in water (anaerobic) 9 mo, (aerobic) c. 20 d. DT_{50} in the hydrosoil c. 90 d; fluridone gradually desorbs from the hydrosoil into the water column where it photodegrades.

353 flurochloridone   *Herbicide*

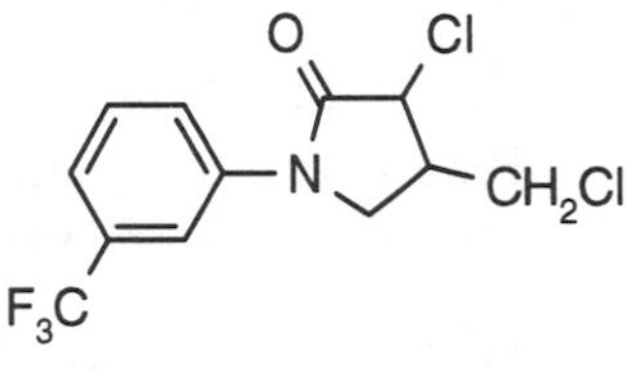

NOMENCLATURE
Common name flurochloridone (BSI, draft E-ISO, draft F-ISO); fluorochloridone (WSSA)
IUPAC name (3RS,4RS;3RS,4SR)-3-chloro-4-chloromethyl-1-(α,α,α-trifluoro-= m-tolyl)-2-pyrrolidinone (in ratio 3:1)
Chemical Abstracts name 3-chloro-4-(chloromethyl)-1-[3-(trifluoromethyl)= phenyl]-2-pyrrolidinone
CAS RN [61213–25–0] unstated stereochemistry   **Development codes** R-40244 (Zeneca)

PHYSICAL CHEMISTRY
Composition A mixture of *cis-* and *trans-* isomers in the ratio 1:3.   **Mol. wt.** 312.1
M.f. $C_{12}H_{10}Cl_2F_3NO$ **Form** Brown-beige waxy solid.   **M.p.** 55.6–79.7 °C (1:3 *cis:trans*); 55–57 °C (*cis-* isomer); 81–83 °C (*trans-* isomer)
B.p. 212.5 °C/10 mmHg   **V.p.** 0.44 mPa (25 °C) K_{ow} logP = 3.36
S.g./density 1.524 g/ml (20 °C)   **Solubility** In water 35.1 (distilled), 20.4 (pH 9) (both mg/l, 25 °C). In ethanol 100, kerosene <5 (both in g/l, 20 °C). Readily soluble in acetone, chlorobenzene, and xylene.   **Stability** Stable to hydrolysis at pH 5, 7 and 9 (25 °C); in acidic media and at elevated temperatures, decomposition occurs; DT_{50} 138 d (100 °C), 15 d (120 °C). DT_{50} (60 °C) 7 d (pH 4), 18 d (pH 7). Aqueous photolysis DT_{50} 4.3 d (*cis:trans*), 2.4 d (*cis*), 4.4 d (*trans*) (all pH 7, 25 °C).

COMMERCIALISATION
History Herbicide reported by F. Pereiro *et al.* (*Proc. 1982 Br. Crop Prot. Conf.* – *Weeds*, 1982, **1,** 225). Introduced by Stauffer Chemical Co. (now Zeneca Agrochemicals). **Manufacturers** Zeneca

APPLICATIONS
Biochemistry Inhibits synthesis of carotenoids (which prevent chlorophyll from undergoing photo-oxidation) by inhibition of phytoene desaturase.
Mode of action Selective herbicide, absorbed by the roots, stems, and coleoptiles.
Uses Applied pre-emergence to control *Stellaria media*, *Veronica hederifolia* and *Viola arvensis* in winter wheat and winter rye; *Amaranthus retroflexus*, *Portulaca oleracea* and *Solanum nigrum* in cotton; *Galium aparine*, *S. nigrum* and *Veronica persica* in potatoes; and a wide range of weeds in sunflowers.
Formulation types EC; CS. **Mixtures** *(flurochloridone +)* neburon.
Selected tradenames 'Racer' (mixture) (Zeneca)

ANALYSIS
Product analysis by glc. **Residues** in crops or soils determined by glc. Details of analytical methods available from Zeneca.

MAMMALIAN TOXICOLOGY
Oral Acute oral LD_{50} for male rats 4000, female rats 3650 mg/kg.
Skin and eye Acute percutaneous LD_{50} for rabbits >5000 mg/kg; practically non-irritating to skin, not irritant to eyes (rabbits). Not a skin sensitiser (guinea pigs).
Inhalation LC_{50} (4 h) for rats >0.121 mg/l air. **NOEL** (2 y) for male rats 100 mg/kg diet (3.9 mg/kg b.w. daily), female rats 400 mg/kg diet (19.3 mg/kg b.w. daily). **Other** Genotoxicity negative. **Toxicity class** WHO (a.i.) III (Table 5)

ECOTOXICOLOGY
Birds Acute oral LD_{50} for Japanese quail >2150 mg/kg. Eight-day dietary LC_{50} for mallard ducks and Japanese quail >5000 mg/kg diet. **Fish** LC_{50} (96 h) for rainbow trout 3.0, bluegill sunfish 5 mg/l. **Bees** Not hazardous to bees. LD_{50} (contact and oral) >100 μg/bee. **Daphnia** LC_{50} (48 h) 5.1 mg/l.

ENVIRONMENTAL FATE
Animals Flurochloridone is extensively metabolised and rapidly excreted in rats; >95% of orally applied dose was excreted in 90 h. Metabolism by oxidation, hydrolysis and glutathione conjugation yielded numerous metabolites in urine and faeces. **Plants** Flurochloridone is rapidly metabolised in plants. Numerous minor metabolites are formed through oxidation and conjugation. Residue levels of flurochloridone in crops are generally <0.05 mg/kg. **Soil/Environment** In the lab., flurochloridone is readily degraded in soil, mostly forming CO_2 and a bound residue; DT_{50} (3 soil types, aerobic, 28 °C) 4, 5 and 27 d; two metabolites were

formed which were readily degraded further. In aerobic sediment, flurochloridone degraded with DT_{50} 3–18 d (2 soils). In the field, DT_{50} 9–70 d. K_{oc} 680–1300, K_d 8–19, indicating low potential mobility on the McCall classification scale; flurochloridone does not leach because it is adsorbed and readily degraded in soil. For water, see Stability.

354 fluroxypyr *Herbicide*

aryloxyalkanoic acid

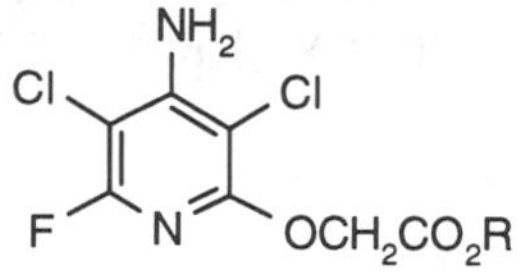

2-butoxy-1-methylethyl $\quad$ R = $CH_3(CH_2)_3OCH_2CH(CH_3)$-

meptyl (1-methylheptyl) $\quad$ R = $CH_3(CH_2)_5CH(CH_3)$-

NOMENCLATURE
fluroxypyr
Common name fluroxypyr (BSI, draft E-ISO, (*m*) draft F-ISO)
IUPAC name 4-amino-3,5-dichloro-6-fluoro-2-pyridyloxyacetic acid
Chemical Abstracts name [(4-amino-3,5-dichloro-6-fluoro-2-pyridinyl)oxy]acetic acid
CAS RN *[69377–81–7]* **Development codes** Dowco 433

fluroxypyr-meptyl
Other names fluroxypyr MHE **CAS RN** *[81406–37–3]* **EEC no.** 279–752–9
Development codes Dowco 433 MHE; XRD-433 1MHE; DOW-43300-H

fluroxypyr-2-butoxy-1-methylethyl
Other names fluroxypyr BPE **CAS RN** *[154486–27–8]*
Development codes DOW-43304-H; DOE-81680-H

PHYSICAL CHEMISTRY
fluroxypyr
Mol. wt. 255.0 **M.f.** $C_7H_5Cl_2FN_2O_3$ **Form** White crystalline solid.
M.p. 232–233 °C **V.p.** 3.784×10^{-6} mPa (20 °C, Knudsen effusion)
K_{ow} logP = –1.24 (unstated pH) **S.g./density** 1.09 (24 °C) **Solubility** In water 91 mg/l (20 °C). In acetone 51.0, methanol 34.6, ethyl acetate 10.6, isopropanol 9.2, dichloromethane 0.1, toluene 0.8, xylene 0.3 (all in g/l, 20 °C).
Stability Stable in acidic media. Fluroxypyr is acidic, and reacts with alkalis to form salts. DT_{50} in water 185 d (pH 9, 20 °C). Stable at temperatures up to melting point. Stable in visible light. **pKa** 2.94

fluroxypyr-meptyl
Mol. wt. 367.2   **M.f.** $C_{15}H_{21}Cl_2FN_2O_3$ **Form** Off-white solid.   **M.p.** 58.2–60 °C
V.p. 1.349×10^{-3} mPa (20 °C, Knudsen effusion); 1×10^{-2} mPa (20 °C, method
unspecified) K_{ow} logP = 4.53   **S.g./density** 1.322   **Solubility** In water 0.09 mg/l.
In acetone 867, methanol 469, ethyl acetate 792, dichloromethane 896, toluene
735, xylene 642, hexane 45 (all in g/l, 20 °C).   **Stability** Stable under normal
storage conditions; decomposes above m.p. Stable in visible light. Hydrolytic DT_{50}
454 d (pH 7), 3.2 d (pH 9); stable at pH 5. In natural waters, DT_{50} 1–3 d.

fluroxypyr-2-butoxy-1-methylethyl
Mol. wt. 369.2   **M.f.** $C_{14}H_{19}Cl_2FN_2O_4$ **Form** Viscous dark brown liquid.
B.p. decomp. 280 °C   **V.p.** 6×10^{-3} mPa (20 °C) K_{ow} logP = 4.17   **Henry** (calc.)
1.8×10^{-4} Pa m^3 mol^{-1} **S.g./density** 1.294 (22 °C)   **Solubility** In water 12.6
(purified water), 10.8 (pH 5), 11.7 (pH 7), 11.5 (pH 9) mg/l (20 °C). In toluene,
methanol, acetone, ethyl acetate >4000, hexane 68 g/l (20 °C).   **F.p.** 195.5 °C
(Pensky Marten closed cup)

COMMERCIALISATION
History Herbicide reported by O. Visbecq *et al.* (COLUMA Conf., 1983, p. 257).
Fluroxypyr-meptyl (the 1-methylheptyl ester) introduced in UK (1985) by Dow
Chemical Co. (now DowElanco).   **Manufacturers** DowElanco

APPLICATIONS
Mode of action Fluroxypyr is applied as fluroxypyr-meptyl. After predominantly
foliar uptake, the ester is hydrolysed to the parent acid, which is the herbicidally
active form, and translocated rapidly to other parts of the plants. Acts by inducing
characteristic auxin-type responses, e.g. leaf curling.   **Uses** Fluroxypyr is readily
translocated, effective by post-emergence foliar application, controlling a range of
economically important broad-leaved weeds (including *Galium aparine*) in all small
grain crops, and *Rumex* spp. and *Urtica dioica* in pastures. Directed applications are
used against herbaceous and woody broad-leaved weeds in orchards (apple only)
and plantation crops (rubber and oilpalm), and broad-leaved brush spp. in conifer
forests. Post-emergence, broadcast applications of fluroxypyr in maize up to the
6-leaf stage of the crop are used for control of *Calystegia sepium*, *Convolvulus
arvensis* and *Solanum nigrum*. The meptyl and 2-butoxy-1-methyl ethyl esters have
similar activity, the advantage of the latter being the wider range of formulation
options that are available.   **Phytotoxicity** Non-phytotoxic to the crops for which
its use is recommended.

fluroxypyr-meptyl
Formulation types EC.   **Mixtures** *(fluroxypyr-meptyl +)* bromoxynil; isoproturon;
bromoxynil + ioxynil; clopyralid + MCPA; dichlorprop + MCPA; clopyralid +
ioxynil; glyphosate; bromoxynil + ioxynil + clopyralid; metosulam.
Selected tradenames 'Starane' (DowElanco); 'Hurler' (Barclay)

ANALYSIS

Product analysis by hplc.   **Residues** of fluroxypyr and fluroxypyr-meptyl determined by hplc. Details available from DowElanco.

MAMMALIAN TOXICOLOGY

fluroxypyr

Oral Acute oral LD_{50} for rats 2405 mg/kg.   **Skin and eye** Acute percutaneous LD_{50} for rabbits >5000 mg/kg. Mild eye irritant; non-irritating to skin (rabbits). **Inhalation** LC_{50} (4 h) for rats >0.296 mg/l air.   **NOEL** (2 y) for rats 80 mg/kg b.w. daily; (1.5 y) for mice 320 mg/kg b.w. daily. No indication of carcinogenicity, teratogenicity or mutagenicity.   **ADI** 0.8 mg/kg b.w.   **Toxicity class** WHO (a.i.) III (Table 5)

fluroxypyr-meptyl

Oral Acute oral LD_{50} for rats >5000 mg/kg.   **Skin and eye** Acute percutaneous LD_{50} for rats >2000 mg/kg. Mild eye irritant, non-irritating to skin (rabbits). Not a skin sensitiser (guinea pigs).   **Inhalation** LC_{50} (4 h) for rats >1 mg/l.

fluroxypyr-2-butoxy-1-methylethyl

Oral Acute oral LD_{50} for rats >2000 mg/kg.   **Skin and eye** Acute percutaneous LD_{50} for rats >2000 mg/kg. Not a skin or eye irritant (rabbits); not a skin sensitiser (guinea pigs).   **NOEL** NOAEL (rats) 463 mg/kg b.w. daily.   **Other** Not mutagenic.

ECOTOXICOLOGY

fluroxypyr

Birds Acute oral LD_{50} for mallard ducks and bobwhite quail >2000 mg/kg. **Fish** LC_{50} (96 h) for rainbow trout and golden orfe >100 mg/l.   **Bees** Not toxic to bees. LD_{50} (contact, 48 h) >25 µg/bee.   **Daphnia** LC_{50} (48 h) >100 mg/l. **Algae** EC_{50} (96 h) for green algae >100 mg/l. EC_{50} (14 d) for *Lemna gibba* 12.3 mg/l.

fluroxypyr-meptyl

Birds Acute oral LD_{50} for mallard ducks and bobwhite quail >2000 mg/kg. **Fish** LC_{50} (96 h) for rainbow trout and golden orfe >solubility limit.   **Bees** Not toxic to bees. LD_{50} (oral and contact, 48 h) >100 µg/bee.   **Worms** LC_{50} (14 d) for earthworms >1000 mg/kg.   **Daphnia** LC_{50} (48 h) >solubility limit. **Algae** EC_{50} (96 h) for green algae >solubility limit.

ENVIRONMENTAL FATE

fluroxypyr

Animals In rats, following oral administration, fluroxypyr is not metabolised, but is rapidly excreted unchanged, principally in the urine.   **Plants** In plants, fluroxypyr is not metabolised, but biotransformation to conjugates occurs.

Soil/Environment In soil, fluroxypyr is rapidly degraded by micro-organisms in aerobic conditions to 4-amino-3,5-dichloro-6-fluoropyridin-2-ol, 4-amino-= 3,5-dichloro-6-fluoro-2-methoxypyridine, and CO_2. DT_{50} in laboratory soil studies 5–9 d (*c.* 23 °C). Lysimeter and field studies demonstrate there is no evidence of any significant leaching; M. Snel *et al.*, 15th Columa Conf., (1992) **1,** 125.

fluroxypyr-meptyl
Animals Hydrolysed to the parent acid, fluroxypyr. **Plants** Hydrolysed to the parent acid, fluroxypyr. **Soil/Environment** In laboratory soils, the ester is rapidly converted to fluroxypyr in all soil types with DT_{50} <7 d. In soil/water slurries DT_{50} 2–5 h (pH 6–7, 22–24 °C).

355 flurprimidol   *Plant growth regulator*

pyrimidinyl carbinol

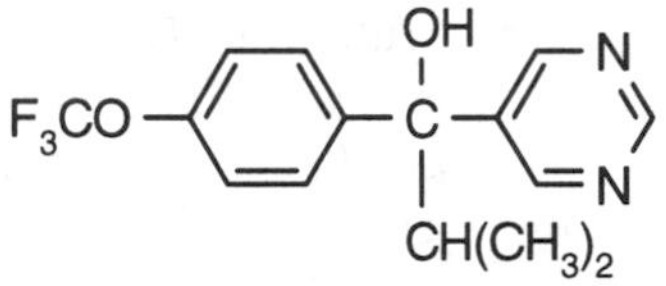

NOMENCLATURE
Common name flurprimidol (BSI, draft E-ISO, (*m*) draft F-ISO, ANSI)
IUPAC name (*RS*)-2-methyl-1-pyrimidin-5-yl-1-(4-trifluoromethoxyphenyl)= propan-1-ol
Chemical Abstracts name α-(1-methylethyl)-α-[4-(trifluoromethoxy)phenyl]-= 5-pyrimidinemethanol
CAS RN *[56425–91–3]* unstated stereochemistry   **Development codes** EL-500

PHYSICAL CHEMISTRY
Mol. wt. 312.3   **M.f.** $C_{15}H_{15}F_3N_2O_2$ **Form** Colourless crystals.
M.p. 93.5–97 °C   **B.p.** 264 °C   **V.p.** 4.85×10^{-2} mPa (25 °C) K_{ow} logP = 3.34 (pH 7, 20 °C)   **S.g./density** 1.34 (24 °C)   **Solubility** In water 114 (distilled), 104 (pH 5), 114 (pH 7), 102 (pH 9) (all in mg/l, 20 °C, OECD 105). In hexane 1.26, toluene 144, dichloromethane 1810, methanol 1990, acetone 1530, ethyl acetate 1200 (all in g/l).   **Stability** After 5 d at pH 4, 7 and 9 (50 °C), <10% hydrolysis had occurred. Stable for at least 14 mo at room temperature. Photolytically decomposed in water, DT_{50} *c.* 3 h.

COMMERCIALISATION
History Plant growth regulator reported by G. E. Brown *et al.* (*Abstr. Weed Sci.*

Soc. Am., No. 84, 1981). Introduced in USA (1989) by Eli Lilly & Co. (now DowElanco). **Patent** US 4002628 **Manufacturers** DowElanco

APPLICATIONS
Biochemistry Gibberellin synthesis inhibitor. **Mode of action** Plant growth regulator, absorbed by the leaves and roots, with translocation in the xylem and phloem. Reduces internode elongation. **Uses** Used to decrease the rate of growth in a wide range of mono- and dicotyledonous species, including perennial turf grasses, ornamental cover species, herbaceous and woody ornamentals, and deciduous and coniferous trees. **Formulation types** WP; EC; SC. **Selected tradenames** 'Cutless' (DowElanco)

ANALYSIS
Product analysis by glc with FID, or by hplc with u.v. detection. **Residues** in water or soils determined by glc with ECD, or by gc-ms (S. D. West in *Comp. Anal. Profiles*, Chapt. 9; S. D. West & B. S. Rutherford, *J. Assoc. Offic. Anal. Chem.*, 1986, **69**, 572).

MAMMALIAN TOXICOLOGY
Oral Acute oral LD_{50} for male rats 914, female rats 709, male mice 602, female mice 702 mg/kg. **Skin and eye** Acute percutaneous LD_{50} for rabbits >5000 mg/kg. Slight to moderate irritant to skin and eyes (rabbits). Not a skin sensitiser (guinea pigs). **Inhalation** LC_{50} (4 h) for rats >5 mg/l. **NOEL** (1 y) for dogs 7 mg/kg b.w. daily; (2 y) for rats 4, mice 1.4 mg/kg b.w. daily. **ADI** Not used on food crops. **Other** No teratogenic effects observed at doses of 200 mg/kg daily (rats) or 45 mg/kg daily (rabbits). Negative in Ames, DNA repair, primary rat hepatocyte and other *in vitro* assays. **Toxicity class** WHO (a.i.) III

ECOTOXICOLOGY
Birds Acute oral LD_{50} for quail and mallard ducks >2000 mg/kg. Five-day dietary LC_{50} for quail 560, mallard ducks 1800 mg/kg diet. **Fish** LC_{50} (96 h) for bluegill sunfish 17.2, rainbow trout 18.3 ppm. **Bees** LD_{50} (contact, 48 h) >100 µg/bee. **Daphnia** LC_{50} (48 h) 11.8 ppm. **Algae** EC_{50} for *Selenastrum capricornutum* 0.84 mg/l.

ENVIRONMENTAL FATE
Animals In mammals, the skin forms a significant barrier to absorption. Following oral administration, excretion follows in the urine and faeces within 48 hours, and more than 30 metabolites have been identified. No accumulation potential.
Soil/Environment Degradation in soil under aerobic conditions leads to more than 30 metabolites. Soil K_d 1.7 on sandy loam.

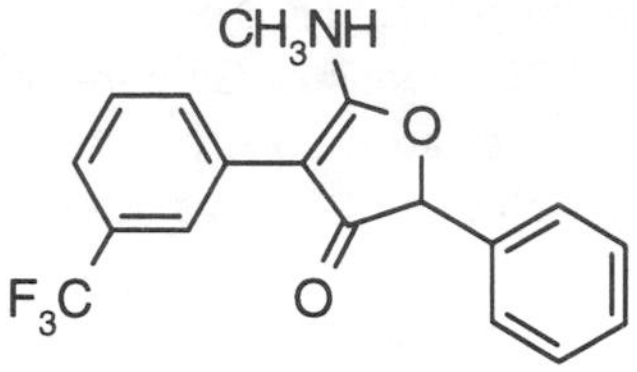

NOMENCLATURE

Common name flurtamone (BSI, ANSI, draft E-ISO)
IUPAC name (*RS*)-5-methylamino-2-phenyl-4-(α,α,α-trifluoro-*m*-tolyl)furan-=3(2*H*)-one
Chemical Abstracts name ($\pm$)-5-(methylamino)-2-phenyl-4-[3-(trifluoromethyl)=phenyl]-3(2*H*)-furanone
CAS RN *[96525–23–4]*   **Development codes** RE-40 885 (Rhône-Poulenc)

PHYSICAL CHEMISTRY

Mol. wt. 333.3   **M.f.** $C_{18}H_{14}F_3NO_2$ **Form** Ivory powder.   **M.p.** 152–155 °C
Solubility In water at 20 °C, 35 mg/l. Soluble in acetone, dichloromethane, methanol; slightly soluble in isopropanol.   **Stability** Stable but avoid concentrated acids or bases.

COMMERCIALISATION

History Herbicide reported by D. D. Rogers *et al.* (*Proc. 1987 Br. Crop Prot. Conf. – Weeds*, **1**, 69). Introduced by Chevron Chemical Co. and later acquired by Rhône-Poulenc Agrochimie.   **Patent** US 4568376; GB 2142629

APPLICATIONS

Biochemistry Carotenoid synthesis inhibitor.   **Uses** Pre-plant incorporated, pre-emergence or post-emergence control of broad-leaved and some grass weeds in small grains, peanuts, cotton, peas and sunflowers at 250–375 g a.i./ha..
Phytotoxicity Transitory bleaching effect.   **Formulation types** WP; WG; SC.
Selected tradenames 'Bacara' (Rhône-Poulenc)

MAMMALIAN TOXICOLOGY

Oral Acute oral LD_{50} for rats 500 mg/kg.   **Skin and eye** Acute percutaneous LD_{50} for rabbits 500, rats >5000 mg/kg. Non-irritating to skin and transient eye irritant (rabbits).   **NOEL** *c.* 50 mg/kg daily.   **Other** Not mutagenic in the Ames assay.

ECOTOXICOLOGY

Birds LC_{50} for bobwhite quail >6000, mallard ducks 2000 mg/kg diet.   **Fish** LC_{50} (96 h) for bluegill sunfish 11, rainbow trout 7 mg/l.   **Bees** LD_{50} (48 h, contact) >100 µg/bee.

ENVIRONMENTAL FATE
Plants No residues in peanuts and in cereals at harvest.   **Soil/Environment** Half-life in soil 4–8 w. Moderately adsorbed on soil colloids.

357 flusilazole   *Fungicide*

azole

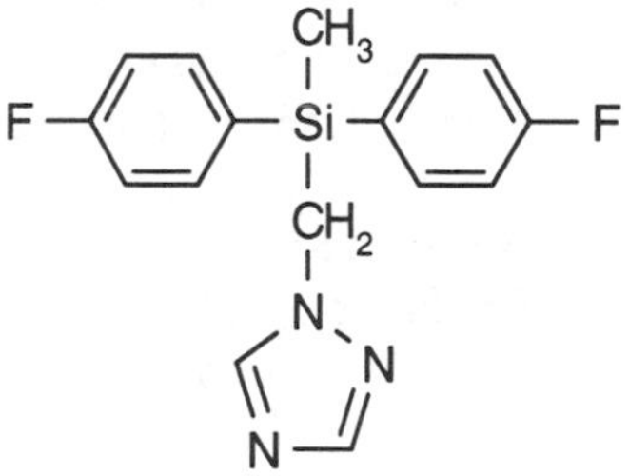

NOMENCLATURE
Common name flusilazole (BSI, ANSI, draft E-ISO, (*m*) draft F-ISO)
IUPAC name bis(4-fluorophenyl)(methyl)(1*H*-1,2,4-triazol-1-ylmethyl)silane;
1-[[bis(4-fluorophenyl)(methyl)silyl]methyl]-1*H*-1,2,4-triazole
Chemical Abstracts name 1-[[bis(4-fluorophenyl)methylsilyl]methyl]-1*H*-=
1,2,4-triazole
CAS RN *[85509–19–9]*   **Development codes** DPX-H 6573

PHYSICAL CHEMISTRY
Mol. wt. 315.4   **M.f.** $C_{16}H_{15}F_2N_3Si$ **Form** White, odourless crystals.
M.p. 53–55 °C   **V.p.** 3.9×10^{-2} mPa (25 °C, gas saturation method)
K_{ow} logP = 3.74 (pH 7, 25 °C)   **Henry** 2.7×10^{-4} Pa m^3 mol^{-1} (pH 8, 25 °C, calc.)
S.g./density 1.30   **Solubility** In water 45 (pH 7.8), 54 (pH 7.2), 900 (pH 1.1)
(all in mg/l, 20 °C). Readily soluble (>2 kg/l) in many organic solvents.
Stability Stable for more than 2 years under normal storage conditions. Stable to
light, and to temperatures up to 310 °C.   **pKa** 2.5, v. weak base

COMMERCIALISATION
History Fungicide reported by T. M. Fort & W. K. Moberg (*Proc. Br. Crop Prot.
Conf. – Pests Dis.*, 1984, **1**, 413). Introduced in France (1985) by E. I. du Pont de
Nemours and Co.   **Manufacturers** Du Pont

APPLICATIONS
Biochemistry Inhibits ergosterol biosynthesis (steroid demethylation inhibitor).
Mode of action Systemic fungicide with protective and curative action. Its

MAIN ENTRIES

resistance to wash-off, redistribution by rainfall and vapour phase activity are important components in its biological activity. **Uses** Broad spectrum, systemic, preventive and curative fungicide effective against many pathogens (Ascomycetes, Basidiomycetes and Deuteromycetes). It is recommended for many uses such as: – apples (*Venturia inaequalis, Podosphaera leucotricha*), – peaches (*Sphaerotheca pannosa, Monilia laxa*), – all major diseases damaging cereals, – grapes (*Uncinula necator, Guignardia bidwellii*), – sugar beet (*Cercospora beticola, Erysiphe betae*), – maize (*Helminthosporium turcicum*), – sunflowers (*Phomopsis helianthi*), – oilseed rape (*Pseudocercosporella capsellae, Pyrenopeziza brassicae*), – bananas (*Mycosphaerella* spp). **Formulation types** EC; WG; SE; EW; SC. **Mixtures** (*flusilazole +)* carbendazim; fenpropimorph; tridemorph. **Selected tradenames** 'Capitan' (cereals) (Du Pont); 'Nustar' (fruit) (Du Pont); 'Olymp' (fruit) (Du Pont); 'Punch' (cereals) (Du Pont); 'Sanction' (cereals) (Du Pont)

ANALYSIS
Product by gc. Details from Du Pont. **Residues** determined by gc (R. A. Guinivan & M. R. Gagnon, JAOAC Intl., **77**, 728 (1994)).

MAMMALIAN TOXICOLOGY
Reviews *Pesticide residues in food – 1995,* FAO Plant Production and Protection Paper 133, 1996. *Pesticide residues in food – 1995 evaluations; Part II – Toxicology & Environment.* World Health Organisation, WHO/PCS/96.48, 1996. **Oral** Acute oral LD_{50} for male rats 1100, female rats 674 mg/kg. **Skin and eye** Acute percutaneous LD_{50} for rabbits >2000 mg/kg; a mild irritant to skin and eyes, but not a skin sensitiser. **Inhalation** ALC_{50} for male rats 27, female rats 3.7 mg/l air. **NOEL** (2 y) for rats 10 mg/kg diet; (1 y) for dogs 5 mg/kg diet; (1.5 y) for mice 25 mg/kg diet. **ADI** (JMPR) 0.001 mg/kg b.w. [1995]. **Other** Not mutagenic. **Toxicity class** WHO (a.i.) III

ECOTOXICOLOGY
Birds Acute oral LD_{50} for mallard ducks >1590 mg/kg. **Fish** LC_{50} (96 h) for rainbow trout 1.2, bluegill sunfish 1.7 mg/l. **Bees** Not toxic to bees. LD_{50} >150 µg/bee. **Daphnia** LC_{50} (48 h) 3.4 mg/l.

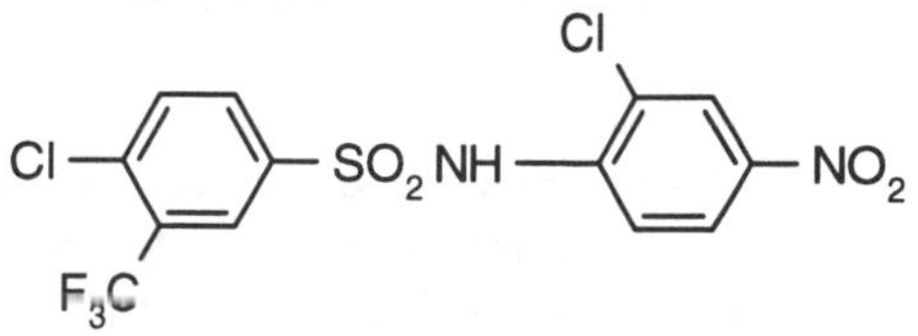

NOMENCLATURE
Common name flusulfamide (BSI, draft E-ISO)
IUPAC name 2',4-dichloro-α,α,α-trifluoro-4'-nitro-*m*-toluenesulfonanilide
Chemical Abstracts name 4-chloro-*N*-(2-chloro-4-nitrophenyl)-3-(trifluoro= methyl)benzenesulfonamide
CAS RN *[106917–52–6]*   **Development codes** MTF-651

PHYSICAL CHEMISTRY
Mol. wt. 415.2   **M.f.** $C_{13}H_7Cl_2F_3N_2O_4S$ **Form** Light yellow crystals.
M.p. 169.7–171.0 °C   **V.p.** 9.9×10^{-7} mPa (40 °C) K_{ow} logP = 2.8±0.5
S.g./density 1.739 gm/cm^3 (23 °C)   **Solubility** In water 2.9 mg/kg (25 °C). In methanol 24, acetone 314, xylene 14, tetrahydrofuran 592 (all in g/kg, 25 °C).
Stability Stable between 35 °C and 80 °C in the dark for 90 d. Stable in acidic media, stable in alkaline media.

COMMERCIALISATION
History Fungicide developed by Mitsui Toatsu Chemicals, Inc.   **Patent** EP 199433
Manufacturers Mitsui Toatsu

APPLICATIONS
Mode of action Inhibits spore germination.   **Uses** Used as a soil treatment for control of *Plasmodiophora brassicae* in brassicas and *Polymyxa betae* (fungal vector of beet yellow vein virus causing rhizomania disease) in sugar beet. Also controls *Pythium*, *Rhizoctonia*, *Phytophthora*, and *Fusarium* spp.   **Formulation types** DP; SC.
Selected tradenames 'Nebijin' (Mitsui Toatsu)

ANALYSIS
Product analysis by hplc.

MAMMALIAN TOXICOLOGY
Oral Acute oral LD_{50} for male rats 180, female rats 132 mg/kg.
Skin and eye Acute percutaneous LD_{50} for rats >2000 mg/kg.   **Inhalation** LC_{50} (4 h) for rats 0.47 mg/l

ECOTOXICOLOGY
Birds Acute oral LD_{50} for bobwhite quail 66 mg/kg. **Fish** TLm (96 h) for bluegill sunfish 1.2 mg/l. **Bees** LD_{50} >200 µg/bee. **Worms** >9 kg/ha. **Daphnia** EC_{50} (48 h) 0.29 mg/l. **Algae** E_rC_{50} (72 h) 4.2 mg/l.

ENVIRONMENTAL FATE
Animals In rats, the parent compound, 4-chloro-N-(2-chloro-4-hydroxyphenyl)-= α,α,α-trifluoro-m-toluenesulfonamide and 4-chloro-α,α,α-trifluoro-m-toluene= sulfonamide are found. **Plants** In cabbage, the parent compound, 4-chloro-= N-(2-chloro-4-formylaminophenyl)-α,α,α-trifluoro-m-toluenesulfonamide, 4-chloro-α,α,α-trifluoro-m-toluenesulfonamide and 4-chloro-α,α,α-trifluoro-= m-toluenesulfonate are found.

359 fluthiacet-methyl   *Herbicide*

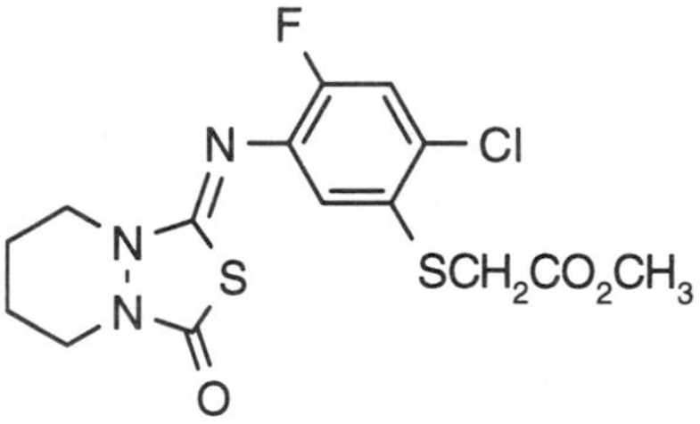

NOMENCLATURE
Common name fluthiacet-methyl (BSI, pa ISO)
IUPAC name methyl [2-chloro-4-fluoro-5-(5,6,7,8-tetrahydro-3-oxo-1H,3H-= [1,3,4]thiadiazolo[3,4-a]pyridazin-1-ylideneamino)phenylthio]acetate
Chemical Abstracts name methyl [[2-chloro-4-fluoro-5-[(tetrahydro-3-oxo-= 1H,3H-[1,3,4]thiadiazolo[3,4-a]pyridazin-1-ylidene)amino]phenyl]thio]acetate
CAS RN [117337–16–6]; ([149253–65–6] for the acid)
Development codes KIH-9201; CGA-248,757

PHYSICAL CHEMISTRY
Mol. wt. 403.9   **M.f.** $C_{15}H_{15}ClFN_3O_3S_2$ **Form** White powder.
M.p. 105.0–106.5 °C (OECD 102)   **V.p.** 4.41×10^{-4} mPa K_{ow} logP = 3.77 (25 °C)   **Henry** 2.1×10^{-4} Pa m^3 mol^{-1} (calc.)   **S.g./density** 0.43 (bulk, 20 °C)
Solubility In water 0.85 (distilled), 0.78 (pH 5 & 7) mg/l (25 °C). In methanol 4.41, acetone 101, toluene 84, n-octanol 1.86, acetonitrile 68.7, ethyl acetate 73.5, dichloromethane 9 (all in g/l, 25 °C); in n-hexane 0.232 g/l, (20 °C).   **Stability** In water DT_{50} 484.8 d (pH 5), 17.7 d (pH 7), 0.2 d (pH 9). In light DT_{50} 4.92 d.

COMMERCIALISATION
History Reported by Kumiai Chemical Industry Co., Ltd and Ciba-Geigy AG (now Novartis Crop Protection AG) (T. Miyazawa *et al.*, *Proc. Br. Crop Prot. Conf., Weeds*, 1993, **1**, 23). **Patent** US 4885023; US 4906279; EP-B-0273417 **Manufacturers** Ihara/Kumiai; Novartis

APPLICATIONS
Biochemistry Protoporphyrinogen oxidase inhibitor, causing accumulation of protoporphyrins, enhancing peroxidation of membrane lipids, and leading to irreversible damage to membrane structure and cellular function. Converted to the active substance in sensitive plants. **Mode of action** Selective, contact herbicide, requiring light for activity. **Uses** Post-emergence herbicide for control of broad-leaved weeds (e.g. *Abutilon theophrasti, Chenopodium album, Amaranthus retroflexus* and *Xanthium strumarium*) in maize and soya beans at 4–15 g/ha. **Formulation types** WP. **Selected tradenames** 'Action' (Novartis)

ANALYSIS
Product and **residue** analysis by glc with NPD.

MAMMALIAN TOXICOLOGY
Oral Acute oral LD_{50} for rats >5000 mg/kg. **Skin and eye** Acute percutaneous LD_{50} for rabbits >2000 mg/kg. Non-irritant to skin and slightly irritant to eyes (rabbits). **Inhalation** LC_{50} (4 h, nose exposure) for rats >5.048 mg/l air. **NOEL** (2 y) for male rats 2.1, female rats 2.5 mg/kg b.w. daily; (18 mo) for male and female mice 0.1 mg/kg b.w. daily. **ADI** 0.001 mg/kg (preliminary value). **Other** Non-mutagenic, non-teratogenic in rats and rabbits. **Toxicity class** WHO (a.i.) III (Table 5)

ECOTOXICOLOGY
Birds Acute oral LD_{50} for bobwhite quail and mallard ducks >2250 mg/kg. LC_{50} for blue quail >5620 ppm. Dietary LC_{50} (5 d) for bobwhite quail and mallard ducks >5620 mg/kg diet. **Fish** LC_{50} (96 h) for trout 0.004, carp 0.63, bluegill sunfish 0.14, sheepshead minnow 0.16 mg/l. **Bees** LD_{50} (contact) >100 µg/bee. **Worms** LC_{50} for earthworms >948 mg/kg dry soil. **Daphnia** LC_{50} >2.3 mg/l. **Algae** EC_{50} for *Selenastrum capricornutum* 2.86 µg/l. NOEL (5 d) for *Anabaena flos-aquae* 18.4 µg/l.

ENVIRONMENTAL FATE
Animals In the rat, within 48 h, 80% is eliminated via the faeces, 14% via urine. Metabolism proceeds via hydrolysis of the methyl ester, isomerisation at the thiadiazole ring and hydroxylation of the tetrahydropyridazine moiety. **Plants** Field residues in beans are >0.01 ppm. In glasshouse studies, negligible residues are found at 10 × rate; organosoluble metabolites are similar to those in the rat. **Soil/Environment** DT_{50} (hydrolysis, pH 7) 18 d, (photolysis on soil) 21 d,

(u.v. light) 2 h. In loam soil DT_{50} 1.2 d (25 °C, 75% of max. water capacity). K_{oc} (adsorption) 448–1883; K_{oc} (desorption) 1445–2782.

360 flutolanil   *Fungicide*

carboxamide

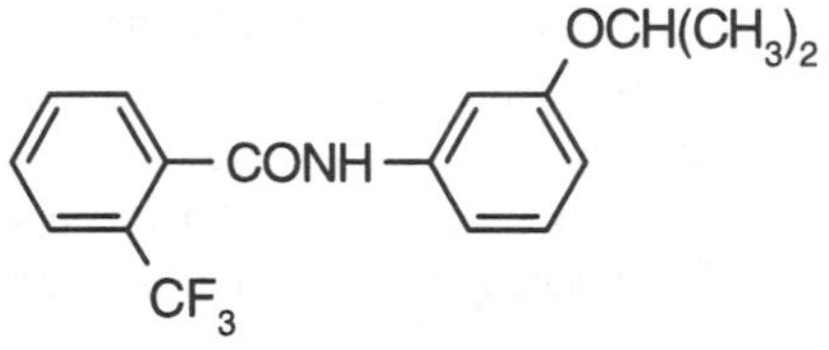

NOMENCLATURE
Common name flutolanil (BSI, draft E-ISO, (*m*) draft F-ISO); no name (Hungary)
IUPAC name α,α,α-trifluoro-3'-isopropoxy-*o*-toluanilide
Chemical Abstracts name *N*-[3-(1-methylethoxy)phenyl]-2-(trifluoromethyl)= benzamide
CAS RN *[66332–96–5]*   **Development codes** NNF-136 (Nihon Nohyaku)

PHYSICAL CHEMISTRY
Mol. wt. 323.3   **M.f.** $C_{17}H_{16}F_3NO_2$ **Form** Colourless odourless crystals.
M.p. 104–105 °C   **V.p.** 6.5×10^{-3} mPa (25 °C) K_{ow} logP = 3.7
S.g./density 1.32 (20 °C)   **Solubility** In water 6.53 mg/l (20 °C). In acetone 1439, methanol 832, ethanol 374, chloroform 674, benzene 135, xylene 29 g/l (20 °C).
Stability Stable in acidic and alkaline media (pH 3–11). Stable to heat and sunlight.

COMMERCIALISATION
History Fungicide reported by F. Araki & K. Yabutani (*Proc. Br. Crop Prot. Conf. – Pests Dis.*, 1981, **1**, 3). Introduced by Nihon Nohyaku Co., Ltd.
Patent JP 1104514   **Manufacturers** Nihon Nohyaku

APPLICATIONS
Biochemistry Inhibitor of succinate dehydrogenase complex, in the respiratory electron transport chain, leading to inhibition of aspartate and glutamate synthesis.
Mode of action Systemic fungicide with protective and curative action. Prevents fungal growth and penetration from infection cushions; causes collapse of hyphae and infection cushions.   **Uses** Control of Basidiomycetes, e.g. sheath blight (*Rhizoctonia solani*), white mould (*Corticium rolfsii*), snow blight (*Typhula* spp.) in rice, cereals, sugar beet and other crops.   **Phytotoxicity** Non-phytotoxic to cereals, rice, vegetables and fruits, when used as directed.

Formulation types WP; UL; SC; DP; GR.   **Selected tradenames** 'Moncut' (Nihon Nohyaku); 'Prostar' (AgrEvo, Nihon Nohyaku)

ANALYSIS
Product by glc; **residues** in soil, plants and water by glc. Details from Nihon Nohyaku.

MAMMALIAN TOXICOLOGY
Oral Acute oral LD_{50} for rats and mice >10 000 mg/kg.   **Skin and eye** Acute percutaneous LD_{50} for rats and mice >5000 mg/kg. Non-irritating to skin and slightly irritating to eyes (rabbits).   **Inhalation** LC_{50} for rats >5.98 mg/l.
Other Non-mutagenic in the Ames test, Rec assay.   **Toxicity class** WHO (a.i.) III (Table 5); EPA (formulation) IV

ECOTOXICOLOGY
Birds Acute oral LD_{50} for bobwhite quail and mallard ducks >2000 mg/kg.
Fish LC_{50} (96 h) for bluegill sunfish 5.4, rainbow trout 5.4, striped mullet 4.65 ppm; (72 h) for carp 2.3 ppm.   **Bees** No effect even with direct spray on insect.   **Daphnia** LC_{50} (6 h) 50 ppm.

ENVIRONMENTAL FATE
Plants In peanuts, the principal metabolites include both free and conjugated flutolanil and N-(3-hydroxyphenyl)-2-(trifluoromethyl)benzamide; see S. M. Smith *et al.*, *Proc 8th IUPAC International Congress of Pesticide Chemistry*, Washington 1994.   **Soil/Environment** DT_{50} 190–320 d (flooded soil), 160–300 d (upland) (*J. Pestic. Sci.*, **8**, 529–535 (1983)).

361 flutriafol   *Fungicide*

azole

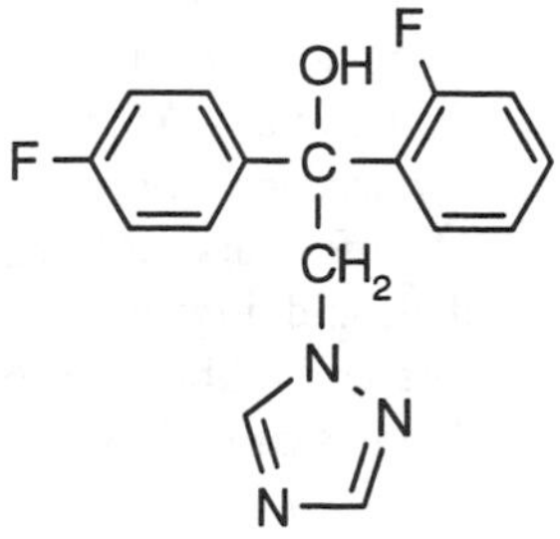

NOMENCLATURE
Common name flutriafol (BSI, ANSI, draft E-ISO, (*m*) draft F-ISO)

IUPAC name (RS)-2,4'-difluoro-α-(1H-1,2,4-triazol-1-ylmethyl)benzhydryl alcohol
Chemical Abstracts name (±)-α-(2-fluorophenyl)-α-(4-fluorophenyl)-1H-= 1,2,4-triazole-1-ethanol
CAS RN *[76674–21–0]* **Development codes** PP450

PHYSICAL CHEMISTRY
Mol. wt. 301.3 **M.f.** $C_{16}H_{13}F_2N_3O$ **Form** White crystalline solid. **M.p.** 130 °C
V.p. 7.1×10^{-6} mPa (20 °C) **K_{ow}** logP = 2.3 (20 °C) **S.g./density** 1.41 g/ml
(25 °C) **Solubility** In water 130 mg/l (pH 7, 20 °C). In acetone 190,
dichloromethane 150, methanol 69, xylene 12, hexane 0.3 (all in g/l, 20 °C).

COMMERCIALISATION
History Fungicide reported by A. M. Skidmore *et al.* (*Proc. 10th Int. Congr. Plant Prot.*, 1983, **1**, 368) and P. J. Northwood *et al.* (*ibid.*, 1983, **3**, 930). Introduced by ICI Plant Protection Division (now Zeneca Agrochemicals). **Patent** EP 15756
Manufacturers Zeneca

APPLICATIONS
Biochemistry Inhibits ergosterol biosynthesis (steroid demethylation inhibitor), causing fungal cell wall collapse and inhibition of hyphal growth.
Mode of action Contact and systemic fungicide with eradicant and protective action. Absorbed by the foliage, with translocation acropetally in the xylem.
Uses Control of a broad spectrum of leaf and ear diseases (including *Erysiphe graminis*, *Rhynchosporium secalis*, and *Septoria*, *Puccinia*, and *Helminthosporium* spp.) in cereals at 125 g a.i./ha. Also used in non-mercurial seed treatment formulations to control the major soil-borne and seed-borne diseases of cereals.
Formulation types SC. **Mixtures** *(flutriafol +)* carbendazim; chlorothalonil; ethirimol + thiabendazole. **Selected tradenames** 'Impact' (Zeneca); 'Vincit' (mixture) (Zeneca)

ANALYSIS
Residues by glc; details from Zeneca.

MAMMALIAN TOXICOLOGY
Oral Acute oral LD_{50} for male rats 1140, female rats 1480 mg/kg.
Skin and eye Acute percutaneous LD_{50} for rats >1000, rabbits >2000 mg/kg. Mild eye irritant; non-irritating to skin (rabbits). Not a skin sensitiser. **Inhalation** LC_{50} (4 h) for rats 1.65 mg/l. **NOEL** In 90 d feeding trials, rats receiving 20 mg/kg diet, and dogs receiving 5 mg/kg daily showed no ill-effects. Non-teratogenic in rats and rabbits. **Other** Non-cytogenic in *in vivo* studies. Non-mutagenic in the Ames assay. **Toxicity class** WHO (a.i.) III; EPA (formulation) III

ECOTOXICOLOGY
Birds Acute oral LD_{50} for female mallard ducks >5000 mg/kg. Five-day dietary

LC$_{50}$ for mallard ducks 3940, Japanese quail 17 100 ppm. **Fish** LC$_{50}$ (96 h) for rainbow trout 61, mirror carp 77 mg/l. NOEC 6.2 mg/l. **Bees** Low toxicity to honeybees. Acute oral LD$_{50}$ >5 μg/bee. **Daphnia** LC$_{50}$ (48 h) 78 mg/l.

ENVIRONMENTAL FATE
Soil/Environment Flutriafol has no effect on microbial populations, or on the carbon or nitrogen transformations occurring in the soil.

tau-fluvalinate   *Insecticide, acaricide*

pyrethroid

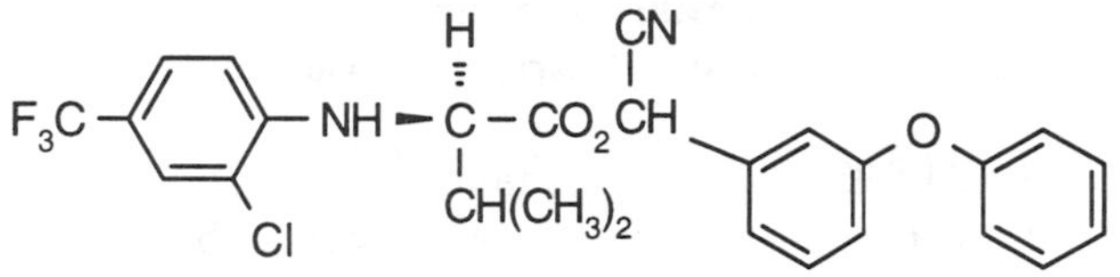

NOMENCLATURE
Common name tau-fluvalinate (BSI, draft E-ISO)
IUPAC name (RS)-α-cyano-3-phenoxybenzyl N-(2-chloro-α,α,α-trifluoro-p-tolyl)-= D-valinate.
Chemical Abstracts name cyano(3-phenoxyphenyl)methyl N-[2-chloro-= 4-(trifluoromethyl)phenyl]-D-valinate
CAS RN [102851–06–9] tau-fluvalinate; [69409–94–5] fluvalinate (DL-isomers) – many references in the literature have been made, erroneously, to the latter rather than the former **Development codes** SAN 527 I

PHYSICAL CHEMISTRY
Composition Material is a 1:1 mixture of (R)-α-cyano-, 2-(R)- and (S)-α-cyano-, 2-(R)- diastereoisomers. **Mol. wt.** 502.9 **M.f.** C$_{26}$H$_{22}$ClF$_3$N$_2$O$_3$ **Form** Viscous amber oil with a moderate or weak sweetish odour (tech.). **B.p.** 164 °C/0.07 mmHg (tech.) **V.p.** 9 × 10^{-8} mPa **K$_{ow}$** logP = 4.26 (25 °C) **Henry** 4.04 × 10^{-5} Pa m^3 mol^{-1} (calc.) **S.g./density** 1.262 (25 °C) **Solubility** In water, 1.03 ppb (pH 7, 20 °C). Soluble in toluene, acetonitrile, isopropanol, dimethylformamide, n-octanol; in iso-octane 108 g/l. **Stability** Tech. material is stable for 2 years at room temperature (20–28 °C). Undergoes decomposition on exposure to sunlight; DT$_{50}$ 9.3–10.7 min (aqueous solution buffered pH 5), c. 1 d (thin film on glass), 13 d (on soil surface). On hydrolysis (25 °C) DT$_{50}$ for a 9 ppb aqueous solution was 48 d (pH 5), 38.5 d (pH 7), 1.1 d (pH 9). **F.p.** 90 °C (tech.; Pensky-Martens closed cup)

COMMERCIALISATION

History Insecticidal and acaricidal activity of the various stereoisomers reported by C. A. Henrick et al. (*Pestic. Sci.*, 1980, **11**, 224) and by R. J. Anderson (*J. Agric. Food Chem.*, 1985, **33**, 508). Tau-fluvalinate has replaced fluvalinate; both introduced by Zoecon Corp. and purchased by Sandoz Agro AG (now Novartis Crop Protection AG).   **Patent** US 4243819   **Manufacturers** Novartis

APPLICATIONS

Mode of action Insecticide and acaricide with contact and stomach action.
Uses Control of a wide range of insects (including Lepidoptera, aphids, thrips, leafhoppers, whitefly, etc.) and spider mites on indoor and outdoor ornamentals, trees, apples, pears, peaches, citrus fruit, vines, cereals, vegetables, cotton, tea, tobacco, and turf. Also used (as 'Apistan') for control of *Varroa jacobsoni* in beehives.   **Formulation types** EC; EW; ULV; Controlled release strips; VP.
Mixtures *(tau-fluvalinate +)* thiometon; clofentezine; dicofol; acephate; methomyl; carbaryl.   **Compatibility** Incompatible with alkaline materials.
Selected tradenames 'Klartan' (Novartis); 'Mavrik' (Novartis)

ANALYSIS

Product analysis of pyrethroids reviewed by E. Papadopoulou-Mourkidou in *Comp. Analyt. Profiles*. By hplc (W. L. Fitch et al., *Anal. Methods Pestic. Plant Growth Regul.*, 1984, **13**, 79) or by glc with FID.   **Residues** determined by glc with ECD (*idem, ibid.*).

MAMMALIAN TOXICOLOGY

Oral Acute oral LD_{50} for female rats 261, male rats 282 mg a.i. (in corn oil)/kg.
Skin and eye Acute percutaneous LD_{50} for rabbits >2000 mg/kg; slight skin irritant, mild eye irritant (rabbits).   **Inhalation** LC_{50} (4 h) for rats >0.56 mg/l air (as 240 g/l EC formulation).   **NOEL** for rats 1 mg/kg b.w. daily.   **ADI** 0.01 mg/kg b.w.   **Toxicity class** WHO (a.i.) II; EPA (formulation) II

ECOTOXICOLOGY

Birds Acute oral LD_{50} for bobwhite quail >2510 mg/kg. Eight-day dietary LC_{50} for bobwhite quail and mallard ducks >5620 mg/kg diet.   **Fish** LC_{50} (96 h) for bluegill sunfish 0.0062, rainbow trout 0.0027, carp 0.0048 mg/l.   **Bees** Not hazardous to bees at recommended doses; LD_{50} (72 h, topical) 6 μg a.i. (tech.)/bee.
Worms LC_{50} (14 d) >1000 ppm.   **Daphnia** LC_{50} (48 h) 0.001 mg/l.   **Algae** LC_{50} for *Scenedesmus subspicatus* >2.2 mg/l.

ENVIRONMENTAL FATE

Animals In rats, following oral administration, c. 90% is excreted within 4 days, of which 20–40% is in the urine and 60–80% in the faeces. The principal metabolites are 'anilino acid', 3-phenoxybenzoic acid (3-PBA) and 4'-hydroxy-3-PBA as faecal metabolites and 4'-hydroxy-3-PBA, 3-PBA, and 3-phenoxybenzyl alcohol as urinary metabolites. **Plants** The parent compound itself accounts for >90% of the residue following treatment with tau-fluvalinate. Minor residues include decarboxyfluvalinate, 'anilino acid', haloaniline, 3-PBA and 4'-hydroxy-3-PBA. DT_{50} 2–6 w. **Soil/Environment** In soil under aerobic conditions (lab.) DT_{50} 12–92 d. Primary metabolites include the corresponding anilino acids and the parent aniline. K_{oc} (adsorption) >110 000, (desorption) >39 000.

363 fluxofenim *Herbicide safener*

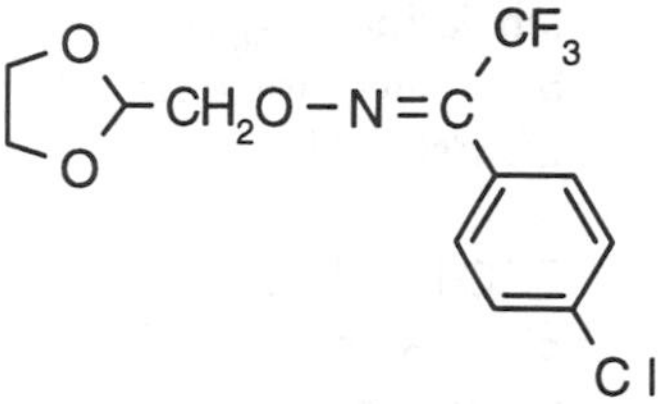

NOMENCLATURE
Common name fluxofenim (BSI, draft E-ISO)
IUPAC name 4'-chloro-2,2,2-trifluoroacetophenone O-1,3-dioxolan-2-ylmethyl= oxime
Chemical Abstracts name 1-(4-chlorophenyl)-2,2,2-trifluoro-1-ethanone O-(1,3-dioxolan-2-ylmethyl)oxime
CAS RN *[88485–37–4]* unstated stereochemistry
Development codes CGA 133 205

PHYSICAL CHEMISTRY
Composition Material comprises both (*Z*)- and (*E*)- isomers. **Mol. wt.** 309.7
M.f. $C_{12}H_{11}ClF_3NO_3$ **Form** Colourless oily liquid. **B.p.** 94 °C/0.1 mmHg
V.p. 38 mPa (20 °C) **K_{ow}** logP = 2.9 (rp-tlc method) **S.g./density** 1.36 (20 °C)
Solubility In water 30 mg/l (20 °C). Miscible with common organic solvents (acetone, methanol, toluene, hexane, octanol). **Stability** Stable ≥200 °C. Hydrolytically stable (>300 d, 50 °C, pH 5–9). **F.p.** >93 °C

COMMERCIALISATION
History Herbicide safener reported by T. R. Dill (*Abstr. Annu. Southern Weed Sci. Soc. Meeting, 39th, 1986*). Introduced by Ciba-Geigy AG (now Novartis Crop Protection AG). **Patent** US 4530716; EP 89313 **Manufacturers** Novartis

APPLICATIONS
Mode of action Used as a seed treatment, fluxofenim rapidly penetrates the seed and probably acts by enhancing metabolism of metolachlor. **Uses** Used as a herbicide safener to protect sorghum from injury by metolachlor; applied as a seed treatment at 0.3–0.4 g a.i./kg. Tolerance of sorghum to metolachlor is maintained in the presence of 1,3,5-triazines used in mixtures for additional control of broad-leaved weeds. **Formulation types** EC.
Selected tradenames 'Concep III' (Novartis)

ANALYSIS
By gc. Details from Novartis.

MAMMALIAN TOXICOLOGY
Oral Acute oral LD_{50} for rats 670 mg/kg. **Skin and eye** Acute percutaneous LD_{50} for rats 1540 mg/kg. Not a skin or eye irritant. Not a skin sensitiser.
Inhalation LC_{50} (4 h) for rats >1.2 mg/l. **NOEL** (90 d) for rats 10 ppm (1 mg/kg b.w. daily), for dogs 20 mg/kg b.w. daily. **Toxicity class** WHO (a.i.) II
EC risk R21/22

ECOTOXICOLOGY
Birds Acute oral LD_{50} for bobwhite quail >2000 mg/kg. Eight-day oral LC_{50} for bobwhite quail >5000 ppm. **Fish** LC_{50} for trout 0.86, bluegill sunfish 2.5 mg/l.
Daphnia LC_{50} (48 h) 0.22 mg/l.

ENVIRONMENTAL FATE
Animals In the rat, fluxofenim is rapidly absorbed and rapidly excreted via urine and faeces, with low tissue residues. Metabolism proceeds via hydrolysis of the dioxolane ring and subsequent oxidation steps, followed by cleavage of the oxime ether.

364 folpet   *Fungicide*

N-trihalomethylthio

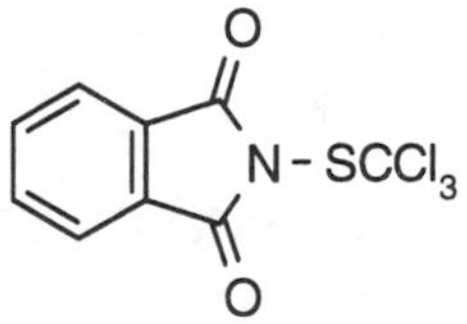

NOMENCLATURE
Common name folpet (BSI, draft E-ISO, (*m*) draft F-ISO, ANSI, JMAF); folpel ((*m*) France)

IUPAC name *N*-(trichloromethylthio)phthalimide; *N*-(trichloromethanesulfenyl)=
phthalimide
Chemical Abstracts name 2-[(trichloromethyl)thio]-1*H*-isoindole-1,3(2*H*)-dione
CAS RN *[133–07–3]* **EEC no.** 205–088–6

PHYSICAL CHEMISTRY
Composition Tech. grade is 90–95% pure. **Mol. wt.** 296.6 **M.f.** $C_9H_4Cl_3NO_2S$
Form Colourless crystals; (tech., yellow powder). **M.p.** 177 °C (decomp.)
V.p. 2.1 × 10^{-7} mPa (25 °C) **K$_{ow}$** log$P$ = 3.11 **Henry** 7.8 × 10^{-3} Pa m^3 mol^{-1}
(calc.) **S.g./density** 1.72 (20 °C) **Solubility** In water 0.8 mg/l (room
temperature). In carbon tetrachloride 6, toluene 26, methanol 3 (all in g/l, 25 °C).
Stability Stable in the dry state. Slowly hydrolysed by moisture at room
temperature. Rapidly hydrolysed in concentrated alkalis, and at elevated
temperatures.

COMMERCIALISATION
History Fungicide reported by A. R. Kittleson (*Science*, 1952, **115**, 84). Introduced
by the Standard Oil Development Co. and later by Chevron Chemical Co.
Patent US 2553770; US 2553771; US 2553776 **Manufacturers** Makhteshim-Agan

APPLICATIONS
Mode of action Foliar fungicide with protective action. **Uses** Control of downy
mildews, powdery mildews, leaf spot diseases, scab, *Gloeosporium* rots, *Botrytis*,
Alternaria, *Pythium*, and *Rhizoctonia* spp. in pome fruit, stone fruit, soft fruit, citrus
fruit, vines, olives, hops, potatoes, lettuce, cucurbits, onions, leeks, celery,
tomatoes and ornamentals. **Phytotoxicity** Non-phytotoxic, except to sweet
cherries and the D'Anjou variety of pears. Russetting is possible in sensitive apple
varieties if applied early in cropping. **Formulation types** WP; DP; SC; WG.
Mixtures *(folpet +)* fosetyl-aluminium; Bordeaux mixture; copper oxychloride;
cymoxanil; mancozeb; metalaxyl; cyprofuram; oxadixyl; benalaxyl; ofurace;
captafol + ofurace; copper oxychloride + copper sulfate + oxadixyl; copper
oxychloride + cymoxanil. **Compatibility** Incompatible with strongly alkaline
materials. **Selected tradenames** 'Folpan' (Makhteshim-Agan, Philagro)

ANALYSIS
Product analysis by i.r. spectroscopy or by hplc (*AOAC Methods*, 1995, 977.03;
CIPAC Handbook, 1983, **1B**, 1845). **Residues** determined by glc (*Pestic. Anal. Man.*,
1979, **I**, 201-A, 201-G, 201-I; M. A. Luke et al., *J. Assoc. Off. Anal. Chem.*, 1981, **64**,
1187; *Anal. Methods Pestic. Plant Growth Regul.*, 1972, **6**, 546).

MAMMALIAN TOXICOLOGY
Reviews *Pesticide residues in food – 1995*, FAO Plant Production and Protection
Paper 133, 1996. *Pesticide residues in food – 1995 evaluations; Part II – Toxicology &
Environment*. World Health Organisation, WHO/PCS/96.48, 1996. **Oral** Acute

oral LD_{50} for rats >9000 mg/kg.   **Skin and eye** Acute percutaneous LD_{50} for albino rabbits >4500 mg/kg; can cause irritation of mucous membranes, contact with eyes and skin or inhalation of dust or spray mist can result in local irritation (rabbits). Skin sensitiser (guinea pigs).   **Inhalation** LC_{50} (4 h) for rats >1.89 mg/l. **NOEL** In 1 y feeding trials, no ill-effect or tumour incidence was noted in albino rats receiving 800 ppm/kg diet, nor in dogs receiving 325 ppm/kg for 5 d/w. NOAEL for mice oncogenicity 450 ppm. No effect was observed on reproductive performance in rats over 3 generations at 1000 mg/kg diet; no teratogenic effect was noted in hamsters, monkeys or rats.   **ADI** (JMPR) 0.1 mg/kg b.w. [1995]. **Toxicity class** WHO (a.i.) III (Table 5); EPA (formulation) IV **EC risk** (R40); Xi (R36); (R43)

ECOTOXICOLOGY
Birds Acute oral LD_{50} for mallard ducks >2000 mg/kg.   **Fish** Toxic to fish. **Bees** Non-toxic to bees; LD_{50} (oral) >236 μg/bee; (contact) >200 μg/bee. **Worms** Non-toxic.   **Other beneficial spp.** Non-toxic to beneficial insects. **Daphnia** Toxic.   **Algae** Toxic.   **Other aquatic spp.** Non-toxic to aquatic organisms under practical conditions because of its instability in water.

ENVIRONMENTAL FATE
Animals Major metabolites are phthalimide, phthalic acid and phthalamic acid. **Plants** As for animals.   **Soil/Environment** DT_{50} (soil) 4.3 d; DT_{50} (water) <0.7 h. Strongly adsorbed to soil: K_d (adsorption) 0.126–0.216; (desorption) 0.042–0.112; unlikely to leach.

365 fomesafen   *Herbicide*

diphenyl ether

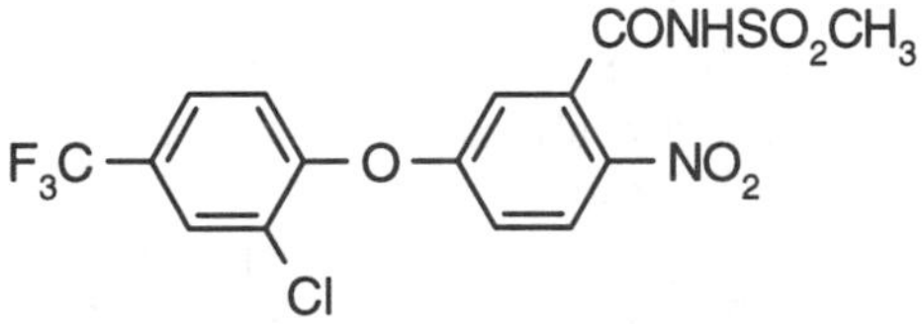

NOMENCLATURE
fomesafen
Common name fomesafen (BSI, draft E-ISO, ANSI); fomésafène ((m) draft F-ISO) **IUPAC name** 5-(2-chloro-α,α,α-trifluoro-p-tolyloxy)-N-methylsulfonyl-2-nitro= benzamide; 5-(2-chloro-α,α,α-trifluoro-p-tolyloxy)-N-mesyl-2-nitrobenzamide

Chemical Abstracts name 5-[2-chloro-4-(trifluoromethyl)phenoxy]-N-(methyl= sulfonyl)-2-nitrobenzamide
CAS RN *[72178–02–0]*   **Development codes** PP021

PHYSICAL CHEMISTRY
fomesafen
Mol. wt. 438.8   **M.f.** $C_{15}H_{10}ClF_3N_2O_6S$ **Form** White crystals.   **M.p.** 220–221 °C
V.p. <0.1 mPa (50 °C)   **K_{ow}** logP = 2.9 (pH 1)   **S.g./density** 1.28 g/ml (20 °C)
Solubility In water c. 50, <10 (pH 1–2), >600 (pH 7) (all in mg/l, 20 °C).
Stability Stable in storage for at least 6 months at 50 °C. Decomposed by light.
Stable at pH 3 and 11 (40 °C). Stable to aqueous photolysis for 32 d (pH 7,
25 °C).   **pKa** 2.7 (20 °C); forms water-soluble salts (e.g. fomesafen-sodium)

fomesafen-sodium
Mol. wt. 460.7   **M.f.** $C_{15}H_9ClF_3N_2NaO_6S$

COMMERCIALISATION
History Herbicide reported by S. R. Colby et al. (*Proc. Int. Congr. Plant Prot., 10th*,
1983, **1**, 295). Introduced by ICI Plant Protection Division (now Zeneca
Agrochemicals).   **Patent** EP 3416   **Manufacturers** Zeneca

APPLICATIONS
Biochemistry Protoporphyrinogen oxidase inhibitor.   **Mode of action** Selective
herbicide, absorbed by both leaves and roots, with very limited translocation in
the phloem.   **Uses** Early post-emergence control of broad-leaved weeds in soya
beans.   **Phytotoxicity** Non-phytotoxic to soya beans and to other crops such as
beans of the genus *Phaseolus* and the leguminous cover crops *Pueraria* and
Calapogonium.   **Formulation types** SL.   **Mixtures** (*fomesafen* +) fluazifop-P-butyl;
terbutryn.   **Selected tradenames** 'Flex' (Zeneca); 'Reflex' (Zeneca)

ANALYSIS
Product analysis by hplc with u.v. detection.   **Residue** analysis in soil by hplc; in
crops by tlc, hplc or nmr. Details available from Zeneca.

MAMMALIAN TOXICOLOGY
fomesafen
Oral Acute oral LD_{50} for male rats 1250–2000, female rats 1600 mg/kg.
Skin and eye Acute percutaneous LD_{50} for rabbits >1000 mg/kg. Mild skin irritant;
mild to moderate eye irritant (rabbits). Extreme skin sensitiser (Magnusson &
Kligman test), not a skin sensitiser (ear/flank Stevens test, guinea pigs).
Inhalation LC_{50} (4 h) for male rats 4.97 mg/l.   **NOEL** (2 y) for rats 0.25 mg/kg
b.w. daily; (18 mo) for mice 1 mg/kg b.w. daily (liver tumours in mouse due to
peroxisome proliferation – not relevant to man); (6 mo) for dogs 1 mg/kg b.w.
daily. NOEL developmental toxicity for rabbits 2.5 mg/kg b.w. daily.

ADI 0.0025 mg/kg.   **Other** Genotoxicity negative. Not oncogenic.
Toxicity class WHO (a.i.) III; EPA (formulation) III

fomesafen-sodium
Oral Acute oral LD_{50} for male rats 1860, female rats 1500 mg/kg.
Skin and eye Acute percutaneous LD_{50} for rabbits >780 mg/kg.   **Other** Not
oncogenic.

ECOTOXICOLOGY
fomesafen
Birds Acute oral LD_{50} for mallard ducks >5000 mg/kg. Five-day dietary LC_{50} for
mallard ducks and bobwhite quail >20 000 mg/kg.   **Fish** LC_{50} (96 h) for rainbow
trout 680, bluegill sunfish 6030 mg/l.   **Bees** Low oral and contact toxicity to bees.
LD_{50} (oral) $\geq$50 μg/bee, (contact) $\geq$100 μg/bee.

fomesafen-sodium
Fish LC_{50} (96 h) for rainbow trout 170 mg/l.

ENVIRONMENTAL FATE
Plants In soya beans, the diphenyl ether bond is rapidly cleaved to give inactive
metabolites.   **Soil/Environment** In soil, degrades slowly under aerobic conditions,
DT_{50} >6 mo, but degrades more rapidly under anaerobic conditions, DT_{50}
<1–2 mo. K_{oc} 34–164. Photodegradation occurs on the soil surface, DT_{50}
100–104 d. In the field, mean DT_{50} c. 15 w. No accumulation in lower soil levels;
the majority of fomesafen residues are present in the top six inches of soil.

366 fonofos   *Insecticide*

organophosphorus

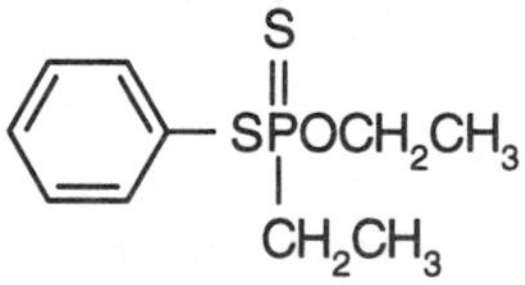

NOMENCLATURE
Common name fonofos (BSI, E-ISO, (*m*) F-ISO, ESA)
IUPAC name *O*-ethyl *S*-phenyl (*RS*)-ethylphosphonodithioate
Chemical Abstracts name (±)-*O*-ethyl *S*-phenyl ethylphosphonodithioate
CAS RN *[944–22–9]* unstated stereochemistry; *[66767–39–3]* racemate;
[62705–71–9] (*R*)-isomer; *[62680–03–9]* (*S*)-isomer   **EEC no.** 213–408–0
Development codes N-2790 (Zeneca)   **Official codes** OMS 410

PHYSICAL CHEMISTRY

Composition The chiral forms have been isolated (R. Allahyari *et al., J. Agric. Food Chem.*, 1977, **25**, 471; P. W. Lee *et al., Pestic. Biochem. Physiol.*, 1978, **8**, 146, 158). The (*R*)-isomer is more toxic to insects and mice and a more potent inhibitor of cholinesterase than the (*S*)-isomer.   **Mol. wt.** 246.3   **M.f.** $C_{10}H_{15}OPS_2$
Form Fonofos (99.5% pure) is a clear colourless liquid, with an aromatic odour.
B.p. *c.* 130 °C/0.1 mmHg   **V.p.** 28 mPa (25 °C) K_{ow} logP = 3.94
S.g./density 1.16 (25 °C)   **Solubility** In water 13 mg/l (22 °C). Miscible with organic solvents, e.g. acetone, ethanol, methyl isobutyl ketone, xylene, kerosene.
Stability Stable below 100 °C. Hydrolysed in acidic and alkaline media; DT_{50} 101 d (pH 4), 74–127 d (pH 7, depending on buffer), 1.8 d (pH 10) (40 °C). In light DT_{50} 12 d (pH 5, 25 °C).   **Specific rotation** Optical rotations of the chiral forms are reversed on solution in carbon tetrachloride, cyclohexane, methanol.
F.p. 179 °C

COMMERCIALISATION

History Insecticide reported by J. J. Menn and K. Szabo (*J. Econ. Entomol.*, 1965, **58**, 734). Introduced by Stauffer Chemical Co. (now Zeneca Agrochemicals).
Manufacturers Zeneca

APPLICATIONS

Biochemistry Cholinesterase inhibitor; proinsecticide, activated by metabolic oxidative desulfuration to the corresponding oxon.   **Mode of action** Soil-insecticide with contact and stomach action.   **Uses** Control of soil insects (especially mole crickets, corn rootworms, symphylids, wireworms, vine weevil larvae, cockchafer larvae, cabbage root fly, onion fly, carrot fly, wheat bulb fly, etc.) in cereals, maize, sorghum, vegetables, ornamentals, fruit (including citrus and bananas), vines, olives, potatoes, sugar beet, sugar cane, peanuts, tobacco, and turf at 1.0–1.5 kg a.i./ha.   **Formulation types** GR; MG; EC; SC; CS; Seed treatment.   **Mixtures** (*fonofos +*) disulfoton; lindane; pebulate; thiram.
Selected tradenames 'Dyfonate' (Zeneca); 'Capfos' (Zeneca)

ANALYSIS

Product analysis by capillary glc with N-PTD (J. E. Barney *et al., Anal. Methods Pestic. Plant Growth Regul.*, 1973, **7**, 269).   **Residues** in crops and soil determined similarly (*idem, ibid.*).   Details available from Zeneca.

MAMMALIAN TOXICOLOGY

EHC 63 (WHO, 1986; a general review of organophosphorus insecticides).   **Oral** Acute oral LD_{50} for male rats 11.5, female rats 5.5 mg/kg.   **Skin and eye** Acute percutaneous LD_{50} for rats 147, rabbits 32–261, guinea pigs 278 mg/kg. Non-irritating to skin and eyes (rabbits). Weak skin sensitiser (guinea pigs).
Inhalation LC_{50} (4 h) for male rats 51, female rats 17 µg/l.   **NOEL** (2 y) for rats 10 mg/kg diet (0.5 mg/kg daily); for dogs 0.2 mg/kg daily. Not carcinogenic or

teratogenic.   **Toxicity class** WHO (a.i.) Ia; EPA (formulation) I or II
EC risk T+ (R27/28)

ECOTOXICOLOGY
Birds Acute oral LD_{50} for mallard ducks 128 mg/kg.   **Fish** LC_{50} (96 h) for rainbow
trout 0.05, bluegill sunfish 0.028 mg/l.   **Bees** Toxic to bees; LD_{50} 0.0087 mg/bee.
Daphnia LC_{50} (48 h) 1 µg/l.

ENVIRONMENTAL FATE
Animals Metabolism in animals is similar to that in plants.   **Plants** In plants,
metabolism involves oxidation to the phosphonothioate, and hydrolytic cleavage
of the thiophenol residue.   **Soil/Environment** Under aerobic conditions, fonofos
at 10 ppm was degraded at a moderate rate, DT_{50} 3–16 w in soils ranging from
loamy sand to clay loam to peat, adsorption K 15.3 (loam soil). The major
degradate was O-ethyl ethanephosphonothioic acid; other degradates included
fonofos oxon, O-ethyl ethanephosphonic acid, O-ethyl O-methylethyl
phosphonate, diphenyl sulfide, methylphenyl sulfoxide and methylphenyl sulfone.

367 forchlorfenuron   *Plant growth regulator*

phenylurea

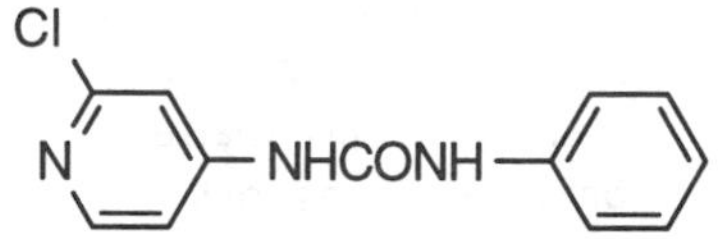

NOMENCLATURE
Common name forchlorfenuron (BSI, draft E-ISO, ANSI)
IUPAC name 1-(2-chloro-4-pyridyl)-3-phenylurea
Chemical Abstracts name N-(2-chloro-4-pyridinyl)-N'-phenylurea
CAS RN *[68157–60–8]*   **Development codes** KT-30; CN-11–3183; 4PU-30;
SKW 20010

PHYSICAL CHEMISTRY
Mol. wt. 247.7   **M.f.** $C_{12}H_{10}ClN_3O$ **Form** White to off-white crystalline powder.
M.p. 165–170 °C   **V.p.** 4.6×10^{-5} mPa (25 °C, gas saturation) K_{ow} logP = 3.2
(20 °C)   **Solubility** In water 39 mg/l (pH 6.4, 21 °C).   **Stability** Stable to heat,
light and water.

COMMERCIALISATION
History Plant growth regulator evaluated initially by Sandoz Crop Protection
Corp., USA (now Novartis Crop Protection AG).   **Manufacturers** Novartis

APPLICATIONS
Uses Used to increase the size of kiwi fruit and table grapes.
Selected tradenames 'Sitofex' (SKW)

ANALYSIS
By hplc with u.v. detection; method validated at the 0.01 ppm level (unpublished).

MAMMALIAN TOXICOLOGY
Oral Acute oral LD_{50} for rats 4918 mg/kg.   **Skin and eye** Acute percutaneous LD_{50} for rabbits >2000 mg/kg.   **Inhalation** LC_{50} (4 h) for rats: no mortality in saturated atmosphere.   **NOEL** 7.5 mg/kg b.w. daily.

ECOTOXICOLOGY
Birds Acute oral LD_{50} for bobwhite quail >2250 ng/kg. Five-day LC_{50} for bobwhite quail >5600 ppm.   **Fish** LC_{50} (96 h) for rainbow trout 9.2 mg/l.
Daphnia LC_{50} (48 h) 8.0 mg/l.

368 formaldehyde   *Fungicide, bactericide*

HCHO

NOMENCLATURE
Common name formaldehyde (BSI, E-ISO, JMAF, accepted in lieu of common name); aldéhyde formique (F-ISO, accepted in lieu of common name)
IUPAC name formaldehyde
Chemical Abstracts name formaldehyde
Other names formalin (37–41% solution)   **CAS RN** *[50–00–0]*
EEC no. 200–001–8

PHYSICAL CHEMISTRY
Mol. wt. 30.0   **M.f.** CH_2O **Form** Colourless, flammable gas with a pungent odour.   Formalin is a colourless liquid, with a pungent odour.   **M.p.** –92 °C
B.p. –19.5 °C   **S.g./density** 0.815 (–20 °C); 1.081–1.085 (25 °C, formalin)
Solubility Very soluble in water (up to 55%). Miscible with acetone, alcohols, and ethers. Formalin miscible with water, acetone, ethanol.   **Stability** Highly chemically-reactive. Polymerised in aqueous solution to trioxymethylene (retarded by the addition of methanol). Sensitive to light. Powerful reducing agent.
F.p. Flammable.

COMMERCIALISATION
History Introduced as a disinfectant by Loew in 1888 and was first used as a seed disinfectant by Geuther in 1896.

APPLICATIONS
Mode of action Fumigant fungicide and bactericide. **Uses** Soil sterilant with bactericidal and fungicidal properties used in glasshouses (after cropping) and mushroom houses. Also used as a silage preservative. **Phytotoxicity** Extremely phytotoxic. **Formulation types** SL; HN.

ANALYSIS
Product analysis by oxidation with hydrogen peroxide to form formic acid which is determined by titration (*MAFF Ref. Book.*, 1958, No. 1, p. 36; *CIPAC Handbook*, 1985, **1C,** 2130; *AOAC Methods*, 1995, 898.01) or by titration with thiocyanate (*ibid.*, 897.01, 931.03). Atmospheres may be measured by reaction with chromotropic acid and colorimetry (P. R. Ludlam & J. G. King, *Analyst (London)*, 1981, **106,** 488). Methods reviewed by J. L. Daft in *Comp. Analyt. Profiles*, Chapter 11. See also general review by E. R. Kennedy *et al.* (*Formaldehyde: Analytical Chemistry and Toxicology, ACS Series*, 1985, No. 210, pp. 3–12).

MAMMALIAN TOXICOLOGY
Reviews J. W. Goode *Formaldehyde: Analytical Chemistry and Toxicology, ACS Series*, 1985, No. 210, pp. 217–277. **EHC** 89 (1989). **Oral** Acute oral LD_{50} for rats 550–800 mg/kg (formalin). **Skin and eye** Acute percutaneous LD_{50} for rabbits 270 mg/kg (formalin). Absorbed through the skin. **Inhalation** LC_{50} for rats (0.5 h) 0.82 mg/l, (4 h) 0.48 mg/l; for mice (4 h) 0.414 mg/l; formaldehyde vapour is very irritating to the mucous membranes and toxic to animals, including man. **EC risk** T (R23/24/25); C (R34); (R40); (R43), (applies to concn. >25%); for 5%≤ concn. <25%, Xn (R20/21/22, R36/37/38, R40, R43); for 1%≤ concn. <5%, Xn (R40, R43)

ECOTOXICOLOGY
Fish Toxic to fish.

ENVIRONMENTAL FATE
Soil/Environment In the presence of air, formaldehyde is oxidised to formic acid.

carbamate

$$CH_3NH-\overset{\overset{O}{\|}}{C}-O-\underset{}{\bigcirc}-N=CHN(CH_3)_2$$

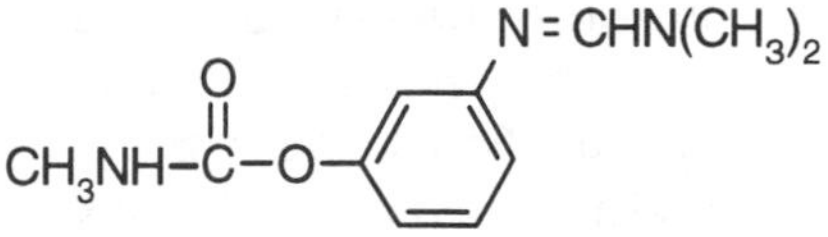

NOMENCLATURE
formetanate
Common name formetanate (BSI, E-ISO, (m) F-ISO, ANSI, ESA)
IUPAC name 3-dimethylaminomethyleneaminophenyl methylcarbamate
Chemical Abstracts name N,N-dimethyl-N'-[3-[[(methylamino)carbonyl]oxy]=
phenyl]methanimidamide
CAS RN [22259–30–9] **EEC no.** 244–879–0 **Development codes** SN 36 056;
ZK 10 970; EP-332 (AgrEvo)

formetanate hydrochloride
CAS RN [23422–53–9] **EEC no.** 245–656–0 **Official codes** ENT 27566

PHYSICAL CHEMISTRY
formetanate
Mol. wt. 221.3 **M.f.** $C_{11}H_{15}N_3O_2$ **pKa** 8.0 (25 °C), weak base

formetanate hydrochloride
Mol. wt. 257.8 **M.f.** $C_{11}H_{16}ClN_3O_2$ **Form** Colourless crystalline powder.
M.p. 200–202 °C (decomp) **V.p.** 0.0016 mPa (25 °C) **K_{ow}** logP ≤–2.7 (pH 7–9)
S.g./density Bulk density *c.* 0.5 g/ml. **Solubility** In water 822 g/l (25 °C). In
methanol 283, acetone 0.074, toluene 0.01, dichloromethane 0.303, ethyl acetate
0.001, *n*-hexane <0.0005 (all in g/l, 20 °C). **Stability** Stable for at least 8 years at
room temperature. Decomposes at *c.* 200 °C. Hydrolysis DT_{50} (22 °C) 62.5 d
(pH 5), 23 h (pH 7), 2 h (pH 9). Subject to photolytic degradation in aqueous
solution with DT_{50} 1333 h (pH 5), 17 h (pH 7), 2.9 h (pH 9) corrected for dark
controls.

COMMERCIALISATION
History Acaricidal properties of hydrochloride described by W. R. Steinhausen
(*Z. Angew. Zool.*, 1968, **55**, 107). Introduced by Schering AG (now AgrEvo
GmbH). **Patent** DE 1169194; GB 987381 **Manufacturers** AgrEvo

APPLICATIONS
Biochemistry Cholinesterase inhibitor. **Mode of action** Acaricide and insecticide
with contact and stomach action.

MAIN ENTRIES

formetanate hydrochloride

Uses Control of spider mites (motile stages) and some insects (Diptera, Hemiptera, Lepidoptera, Thysanoptera especially *T. occidentalis*) on ornamentals, pome fruit, stone fruit, citrus fruit, vegetables and alfalfa. **Phytotoxicity** Non-phytotoxic when used as recommended. At high rates of application, peas, beans, soya beans, peanuts, aubergines, cucumbers, and some varieties of rose may be injured. **Formulation types** SP. **Mixtures** None. **Compatibility** Incompatible with alkaline materials. **Selected tradenames** 'Carzol' (NOR-AM); 'Dicarzol' (AgrEvo)

ANALYSIS

Product analysis by hplc. **Residues** determined by hydrolysis to 3-aminophenol measured as derivatives by glc. Residues of parent compound determined by reverse-phase hplc (J. F. Lawrence, *J. Agric. Food Chem.*, 1981, **29,** 722) or colorimetry (N. A. Jenny & K. Kossmann, *Anal. Methods Pestic. Plant Growth Regul.*, 1973, **7,** 279). Details available from AgrEvo.

MAMMALIAN TOXICOLOGY

formetanate

EHC 64 (WHO, 1986; a review of carbamate insecticides in general).

formetanate hydrochloride

Reviews *Pesticides Studied in Man*, W Hayes, Williams and Wilkins (Baltimore) (1982). **Oral** Acute oral LD_{50} for rats 14.8–26.4, mice 13–25, dogs 19 mg/kg. **Skin and eye** Acute percutaneous LD_{50} for rats >5600, rabbits >10 200 mg/kg. Irritating to eyes. Sensitising in guinea pigs. **Inhalation** LC_{50} (4 h) 2.8 mg/l air (aerosol), 0.29 mg/l (dust). **NOEL** (2 y) for rats 10 mg/kg diet (0.52 mg/kg daily), mice 50 mg/kg diet (8.2 mg/kg daily); (1 y) for dogs 10 mg/kg diet (0.37 mg/kg daily). **ADI** 0.037 mg/kg (taking safety factor of 10 for cholinesterase inhibition). **Other** Not carcinogenic, teratogenic or mutagenic. **Toxicity class** WHO (a.i.) Ib; EPA (formulation) I **EC risk** T+ (R28)

ECOTOXICOLOGY

formetanate hydrochloride

Birds Acute oral LD_{50} for hens 21.5, mallard ducks 12, bobwhite quail 42 mg/kg. Five-day dietary LC_{50} for bobwhite quail 3963, mallard ducks 2086 mg/kg diet. **Fish** LC_{50} (96 h) for rainbow trout 4.42, bluegill sunfish 2.76 mg/l. **Bees** LD_{50} (contact) 14, (oral) 9.21 μg/bee. **Worms** Earthworm LC_{50} (14 d) 1048 mg/kg soil. **Daphnia** LC_{50} (48 h) 0.093 mg/l. **Algae** $E_b C_{50}$ (96 h) 1.5 mg/l. **Other aquatic spp.** Eastern oyster (*Crassostrea virginica*) EC_{50} (96 h) 2.5 mg/l.

ENVIRONMENTAL FATE

Animals Cleavage of *N*-methylcarbamate group with conjugation of resulting 3-dimethylamino-methylenequinophenol, also cleavage at amino nitrogen to give 3-formamido-*N*-methylcarbamate followed by loss of *N*-methylcarbamate to produce 3-formamidophenol which undergoes conjugation and further cleavage to 3-aminophenol, which is then acetylated to give 3-acetamidophenol and detoxified by conjugation. Metabolism of carbamate insecticides is reviewed (M. Cool & C. K. Jankowski in "*Insecticides*"). **Plants** Hydrolytic mechanisms provide the main route of metabolism to give a metabolite spectrum similar to that found in animals, water and soil. **Soil/Environment** Hydrolysis occurs rapidly in soil, DT_{50} 1–9 d (aerobic and anaerobic) under both laboratory and field conditions. Photolytic degradation on soil surface DT_{50} 16.3 h, involving cleavage of the carbamate group. K_d values for four soil types 1.49–3.00 (K_{oc} 140–620).

370 formothion   *Insecticide, acaricide*

organophosphorus

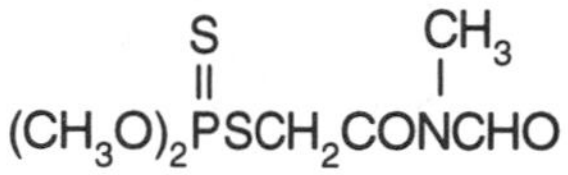

$$\underset{(CH_3O)_2PSCH_2CONCHO}{\overset{\displaystyle S \qquad\qquad CH_3}{\overset{\displaystyle \|\qquad\qquad\quad |}{}}}$$

NOMENCLATURE
Common name formothion (BSI, E-ISO, (*m*) F-ISO, ESA, JMAF)
IUPAC name *S*-[formyl(methyl)carbamoylmethyl] *O,O*-dimethyl phosphoro= dithioate; 2-dimethoxyphosphinothioylthio-*N*-formyl-*N*-methylacetamide
Chemical Abstracts name *S*-[2-(formylmethylamino)-2-oxoethyl] *O,O*-dimethyl phosphorodithioate
CAS RN *[2540–82–1]* **EEC no.** 219–818–6 **Development codes** J-38; SAN 69131 **Official codes** OMS 968; ENT 27 257

PHYSICAL CHEMISTRY
Mol. wt. 257.3 **M.f.** $C_6H_{12}NO_4PS_2$ **Form** Odourless, pale yellow, viscous liquid or crystalline mass. **M.p.** 25–26 °C **B.p.** Decomposes on distillation.
V.p. 0.113 mPa (20 °C) **S.g./density** 1.361 (20 °C) **Solubility** In water 2.6 g/l (24 °C). Completely miscible with common organic solvents, e.g. alcohols, ethers, ketones, chloroform, benzene, toluene, xylene. Slightly soluble in hexane. Very slightly soluble in ligroin and paraffin oil. **Stability** Hydrolysed by water to dimethoate and (dimethoxyphosphinothioylthio)acetic acid; DT_{50} ≤1 d (pH 3–9, 23 °C). Hydrolysed more rapidly in alkaline media. Dilute solutions in non-polar organic solvents are stable.

COMMERCIALISATION
History Insecticide reported by C. Klotzsche (*Mitt. Geb. Lebensmittelunters. Hyg.,* 1961, **52,** 341). Introduced by Sandoz AG (now Novartis Crop Protection AG). **Patent** US 3176035; US 3178337 **Manufacturers** Novartis

APPLICATIONS
Biochemistry Cholinesterase inhibitor. **Mode of action** Systemic insecticide and acaricide with contact and stomach action. **Uses** A contact and systemic insecticide and acaricide, effective at 250–500 g a.i./ha against a wide range of sucking and mining insects such as Aphididae, bugs, Cicadellidae, Coccidae, Diaspididae, Drosophilidae, Margarodidae, Psyllidae, Tephritidae and Thysanoptera, as well as against some chewing insects (*Cydia pomonella* and *Epilachna* spp.) and spider mites on a variety of field crops, fruit trees, citrus and other tropical fruit, cotton, grapes, hops, ornamentals, rice, sugar cane, tobacco and vegetables. **Phytotoxicity** Non-phytotoxic, except to some varieties of peach, apricot and cherry. **Formulation types** EC; UL. **Mixtures** *(formothion +)* fenitrothion. **Compatibility** Incompatible with alkaline materials. **Selected tradenames** 'Anthio' (Novartis)

ANALYSIS
Product analysis by glc (*CIPAC Handbook,* 1980, **1A,** 1270; *AOAC Methods,* 1995, 974.03). **Residues** determined by glc (A. Ambrus *et al., J. Assoc. Off. Anal. Chem.,* 1981, **64,** 733; M. Wisson *et al., Anal. Methods Pestic. Plant Growth Regul.,* 1976, **8,** 123).

MAMMALIAN TOXICOLOGY
Reviews *Pesticide residues in food.* FAO Agricultural Studies, No. 92; WHO Technical Report Series, No. 545, 1974. *1973 Evaluations of some pesticide residues in food.* FAO/AGP/1973/M/9/1; WHO Pesticide Residues Series, No. 3, 1974. **EHC** 63 (WHO, 1986; a general review of organophosphorus insecticides). **Oral** Acute oral LD_{50} for rats 365–500, mice 190–195, rabbits 570, cats 213 mg/kg. **Skin and eye** Acute percutaneous LD_{50} for male rats >1000 mg/kg. Slightly irritating to skin (rabbits). **Inhalation** LC_{50} (4 h) for rats 3.2 mg/l air (as 'Anthio' 50ZP). **NOEL** In 2 y feeding trials, rats and dogs receiving 80 mg/kg diet showed no adverse effects. **ADI** (JMPR) 0.02 mg/kg b.w. [1973]. **Toxicity class** WHO (a.i.) II; EPA (formulation) II

ECOTOXICOLOGY
Birds Acute oral LD_{50} for pigeons 630 mg/kg. **Fish** LC_{50} (72 h) for carp 10 mg/l; (96 h) for rainbow trout 38.3 mg/l. **Bees** Toxic to bees. LD_{50} (oral) 1.537, (topical) 1.789, (contact, deposit) 28.447 μg/g b.w. **Worms** LC_{50} (14 d) for earthworms 157.7 mg/kg soil. **Daphnia** LC_{50} (24 h) 16.1 mg/l. **Algae** EC_{50} (96 h) for *Scenedesmus subspicatus* 42.3 mg/l.

ENVIRONMENTAL FATE
Animals In mammals, following oral administration, formothion is metabolised to (dimethoxyphosphinthioylthio)acetic acid and polar metabolites, and elimination occurs within 24 hours.   **Plants** In plants, formothion is metabolised to dimethoate (*q.v.*), (dimethoxyphosphinthioylthio)acetic acid, bis(dimethylthiophosphoryl) disulfide, omethoate, and *O,O*-dimethyl hydrogen phosphorodithioate.
Soil/Environment DT_{50} in loamy soil <1 d, forming dimethoate (*q.v.*) and (dimethoxyphosphinthioylthio)acetic acid, which ultimately dissipates.

371 fosamine   *Herbicide*

organophosphorus herbicide

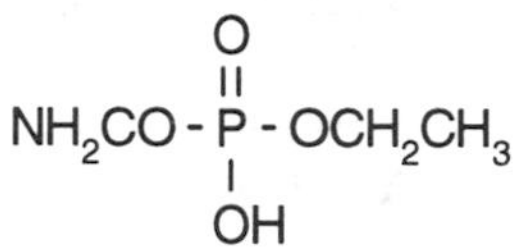

$$NH_2CO - \overset{\overset{\displaystyle O}{\|}}{\underset{\underset{\displaystyle OH}{|}}{P}} - OCH_2CH_3$$

NOMENCLATURE
fosamine
Common name fosamine (BSI, E-ISO, (*f*) F-ISO, ANSI, WSSA)
IUPAC name ethyl hydrogen carbamoylphosphonate
Chemical Abstracts name ethyl hydrogen (aminocarbonyl)phosphonate
CAS RN *[59682–52–9]*   **Development codes** DPX 1108

fosamine-ammonium
CAS RN *[25954–13–6]*   **Development codes** DPX R1108

PHYSICAL CHEMISTRY
fosamine
Mol. wt. 153.1   **M.f.** $C_3H_8NO_4P$

fosamine-ammonium
Composition Tech. is >95%.   **Mol. wt.** 170.1   **M.f.** $C_3H_{11}N_2O_4P$ **Form** White crystals.   **M.p.** 173–175 °C   **V.p.** 0.53 mPa (25 °C)   **S.g./density** 1.24
Solubility In water >2.5 kg/l (25 °C). In methanol 158, ethanol 12, dimethylformamide 1.4, benzene 0.4, chloroform 0.04, acetone 0.001, hexane <0.001 (all in g/kg, 25 °C).   **Stability** Stable in neutral and alkaline media. Decomposes in dilute solutions of acids.   **pKa** 9.25

COMMERCIALISATION
History Herbicide reported by O. C. Zoebisch et al. (*Proc. Northeast. Weed Sci. Soc.*, 1974, **28**, 347). The ammonium salt, fosamine-ammonium, introduced by E. I. du Pont de Nemours and Co.   **Patent** US 3627507; US 3846512
Manufacturers Du Pont

APPLICATIONS
Mode of action Contact herbicide, with slight systemic activity, absorbed by the foliage, stems, and buds. Inhibits bud development.

fosamine-ammonium
Uses Control of unwanted woody plants (including deciduous trees and shrubs) on non-crop areas, and in meadows and pastures. Also used to control field bindweed and bracken; and for selective control of deciduous species in conifer plantations.   **Phytotoxicity** Conifers are tolerant after the end of shoot growth and formation of terminal buds.   **Formulation types** SL.   **Compatibility** Not normally used in combination with other pesticides.   **Selected tradenames** 'Krenite' (Du Pont)

ANALYSIS
Residues in plant tissues and soil determined by glc of a derivative; details available from Du Pont.

MAMMALIAN TOXICOLOGY
fosamine-ammonium
Oral Acute oral LD_{50} for rats >5000 mg/kg.   **Skin and eye** Acute percutaneous LD_{50} for rabbits >1683 mg/kg. Non-irritating to skin and eyes (rabbits) (EC with or without surfactant). Non-sensitising to skin (guinea pigs).   **Inhalation** LC_{50} (1 h) for male rats >56 mg/l air (formulated product).   **NOEL** In 90 d feeding trials, rats receiving 1000 mg/kg diet showed no ill-effects.   **Toxicity class** WHO (a.i.) III (Table 5); EPA (formulation) IV

ECOTOXICOLOGY
fosamine-ammonium
Birds Acute oral LD_{50} for mallard ducks and bobwhite quail >10 000 mg/kg. LC_{50} for mallard ducks and bobwhite quail 5620 ppm.   **Fish** LC_{50} (96 h) for bluegill sunfish 590, rainbow trout 300, fathead minnow >1000 mg/l.   **Bees** Non-toxic to bees; acute topical LD_{50} >200 µg/bee.   **Daphnia** LC_{50} (48 h) 1524 ppm.

ENVIRONMENTAL FATE
Soil/Environment Fosamine-ammonium is rapidly decomposed by soil micro-organisms. Half-life in soil is c. 7–10 d.

$$\left(CH_3CH_2O - \overset{\displaystyle O}{\underset{\displaystyle H}{\overset{\|}{\underset{|}{P}}}} - O \right)_3 Al$$

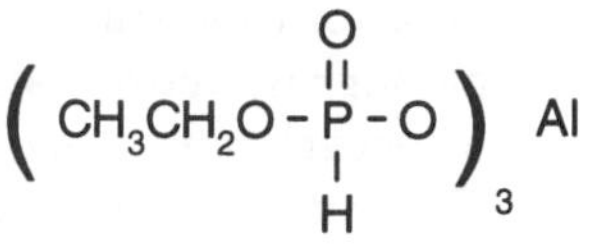

NOMENCLATURE
fosetyl-aluminium
Common name fosetyl-aluminium   **Other names** [efosite-Al] (rejected common name proposal); EPAL   **CAS RN** *[39148–24–8]*   **EEC no.** 254–320–2
Development codes LS 74 783; RP 32545 (Rhône-Poulenc)

fosetyl
Common name fosetyl (BSI, draft E-ISO, (*m*) draft F-ISO (since 1984)); [phosethyl] ((*m*) draft F-ISO)
IUPAC name ethyl hydrogen phosphonate
Chemical Abstracts name ethyl hydrogen phosphonate
Other names [efosite] (rejected common name proposal)
CAS RN *[15845–66–6]*

PHYSICAL CHEMISTRY
fosetyl-aluminium
Composition Tech. is ≥95% pure.   **Mol. wt.** 354.1   **M.f.** $C_6H_{18}AlO_9P_3$
Form Colourless powder; tech. is a white to yellowish powder.   **M.p.** >200 °C
V.p. <0.013 mPa (25 °C)   **K_{ow}** logP = –2.7 (pH 4)   **Henry** <3.93 × 10^{-8} Pa m^3 mol^{-1} (calc.)   **Solubility** In water 120 g/l (20 °C). In methanol 920, acetone 13, propylene glycol 80, ethyl acetate 5, acetonitrile 5, hexane 5 (all in mg/l, 20 °C).
Stability Hydrolysis of fosetyl-aluminium occurs under extreme acid or alkaline conditions. DT$_{50}$ 5 d (pH 3), 13.4 d (pH 13). Decomposes above 200 °C.

fosetyl
Mol. wt. 110.0   **M.f.** $C_2H_7O_3P$ **pKa** 0.8

COMMERCIALISATION
History Fungicidal activity of aluminium salt reported by D. Horrière *et al.* (*Phytiatr.-Phytopharm.*, 1977, **26**, 3). The salt was introduced by Rhône-Poulenc Agrochimie.   **Patent** FR 2254276   **Manufacturers** Rhône-Poulenc

APPLICATIONS
fosetyl-aluminium
Biochemistry Acts by inhibiting germination of spores or by blocking development of mycelium and sporulation.   **Mode of action** Systemic fungicide, rapidly absorbed through the plant leaves or roots, with translocation both acropetally

MAIN ENTRIES

and basipetally. **Uses** Control of diseases caused by Phycomycetes (*Phytophthora, Pythium, Plasmopara, Bremia* spp., etc.) on a variety of crops including vines, fruit (citrus, pineapples, avocados, stone fruit and pome fruit), berries, vegetables, hops, ornamentals and turf. Also useful activity against several bacterial plant pathogens. **Formulation types** WP; WG. **Mixtures** *(fosetyl-aluminium +)* folpet; mancozeb; iprodione; captan + thiabendazole; folpet + cymoxanil. **Compatibility** Incompatible with foliar fertilisers. **Selected tradenames** 'Aliette' (Rhône-Poulenc)

ANALYSIS

Product analysis by iodometric potentiometric titration (*CIPAC Handbook*, 1995, **G**, 82–88). **Residues** determined by glc of the parent compound and its metabolite. Details of methods are available from Rhône-Poulenc Agrochimie.

MAMMALIAN TOXICOLOGY

fosetyl-aluminium
Oral Acute oral LD_{50} for rats and mice >2000 mg/kg. **Skin and eye** Acute percutaneous LD_{50} for rats and rabbits >2000 mg/kg. Non-irritating to skin. **Inhalation** LC_{50} (4 h) for rats >1.73 mg/l air. **NOEL** (90 d) for rats 5000, dogs 50 000 mg/kg diet. Non-teratogenic and non-mutagenic. **ADI** 0–3.0 mg/kg b.w. **Other** Non-carcinogenic. **Toxicity class** WHO (a.i.) III (Table 5); EPA (formulation) III

ECOTOXICOLOGY

fosetyl-aluminium
Birds Acute oral LD_{50} for bobwhite quail >8000 mg/kg. **Fish** LC_{50} (96 h) for rainbow trout 94.3–428 mg/l. **Bees** No mortality to honeybees at 0.2 mg/bee (contact). **Worms** Harmless. **Other beneficial spp.** Harmless. **Daphnia** LC_{50} (96 h) 189 mg/l. **Algae** EC_{50} (96 h) for *Scenedesmus panonicus* 21.9 mg/l. **Other aquatic spp.** LC_{50} (96 h) for fiddler crab 145 mg/l.

ENVIRONMENTAL FATE

Animals Fosetyl-aluminium is almost completely absorbed and undergoes extensive metabolic transformation, the major end-products being CO_2 and phosphorous acid. **Plants** The metabolism of fosetyl-aluminium in plants proceeds through the hydrolytic cleavage of the ethyl ester bond. Phosphorous acid is detected as the major metabolite. **Soil/Environment** In soil, fosetyl-aluminium has an extremely short half-life under both aerobic and anaerobic conditions, with rapid dissipation and metabolism; DT_{50} (aerobic) 20 min. to 1.5 h. In water, its stability is pH-dependent.

organophosphorus

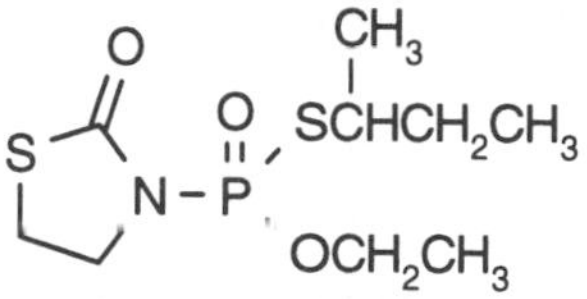

NOMENCLATURE
Common name fosthiazate (BSI, draft E-ISO)
IUPAC name (RS)-S-sec-butyl O-ethyl 2-oxo-1,3-thiazolidin-3-ylphosphonothioate;
(RS)-3-[sec-butylthio(ethoxy)phosphinoyl]-1,3-thiazolidin-2-one
Chemical Abstracts name O-ethyl S-(1-methylpropyl) (2-oxo-3-thiazolidinyl)=
phosphonothioate
CAS RN *[98886–44–3]* **Development codes** IKI-1145 (Ishihara);
ASC-66824 (ISK Biosciences)

PHYSICAL CHEMISTRY
Composition Tech. is ≥93.0%. **Mol. wt.** 283.3 **M.f.** $C_9H_{18}NO_3PS_2$ **Form** Light
brown liquid (tech.). **B.p.** 198 °C/0.5 mmHg **V.p.** 5.6×10^{-1} mPa (25 °C)
K_{ow} logP = 1.68 **Henry** 1.76×10^{-5} Pa m^3 mol^{-1} **S.g./density** 1.240 (20 °C)
Solubility In water 9.85 g/l (20 °C). In n-hexane 15.14 g/l (20 °C). **Stability** In
water, DT_{50} 3 d (pH 9, 25 °C). **F.p.** 127.0 °C (Pensky-Martens closed tester)

COMMERCIALISATION
History Nematicide discovered by Ishihara Sangyo Kaisha, Ltd.
Manufacturers Ishihara Sangyo

APPLICATIONS
Biochemistry Cholinesterase inhibitor. **Uses** Control of nematodes, aphids, thrips
etc. **Formulation types** GR; EC. **Selected tradenames** 'Nemathorin' (Ishihara
Sangyo)

ANALYSIS
Details from Ishihara Sangyo.

MAMMALIAN TOXICOLOGY
EHC 63 (general review of organophosphorus insecticides, pub. 1986).
Oral Acute oral LD_{50} for male rats 73, female rats 57 mg/kg. **Skin and eye** Acute
percutaneous LD_{50} for male rats 2396, female rats 861 mg/kg. Irritating to eyes,
not irritating to skin (rabbits). Sensitising to skin (guinea pigs). **Inhalation** LC_{50}
(4 h) for male rats 0.832, female rats 0.558 mg/l.

MAIN ENTRIES

ECOTOXICOLOGY
Birds Acute oral LD$_{50}$ for ducks 20, quail 10 mg/kg.   **Fish** LC$_{50}$ (96 h) for trout
114, bluegill sunfish 171 mg/l.   **Bees** LD$_{50}$ (48 h) 0.256 µg/bee.   **Daphnia** EC$_{50}$
(48 h) 279 µg/l.

374 fuberidazole   *Fungicide*

benzimidazole

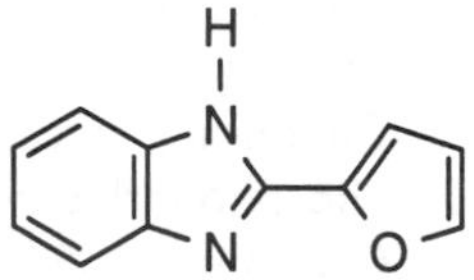

NOMENCLATURE
Common name fuberidazole (BSI, E-ISO, (*m*) F-ISO); no name (Canada)
IUPAC name 2-(2-furyl)benzimidazole
Chemical Abstracts name 2-(2-furanyl)-1*H*-benzimidazole
Other names furidazol; furidazole   **CAS RN** *[3878–19–1]*   **EEC no.** 223–404–0
Development codes Bayer 33 172; W VII/117

PHYSICAL CHEMISTRY
Mol. wt. 184.2   **M.f.** C$_{11}$H$_8$N$_2$O   **Form** Fine crystalline, light brown, odourless
powder (tech.).   **M.p.** 292 °C (decomp.)   **V.p.** 9 × 10^{-4} mPa (20 °C);
2 × 10^{-3} mPa (25 °C)   **K$_{ow}$** logP = 2.67 (22 °C)   **Henry** 2 × 10^{-6} Pa m^3 mol^{-1}
(20 °C)   **Solubility** In water 220 mg/l (pH 4), 71 mg/l (pH 7) (both 20 °C). In
isopropanol 31, toluene 0.35, dichloromethane 6.6 (all in g/l, 20 °C).
Stability Stable to hydrolysis in pure water, under sterile conditions, but sensitive
to light: DT$_{50}$ 15 min. (aqueous solution), 6 h (on silica gel).   **pKa** *c.* 4, weak base

COMMERCIALISATION
History Its preparation described by R. Weidenhagen (*Ber. Dtsch. Chem. Ges.,*
1936, **69**, 2271) and its fungicidal activity by G. Schuhmann (*Nachrichtenbl. Dtsch.
Pflanzenschutzdienstes (Braunschweig),* 1968, **20**, 1). Introduced by Bayer AG.
Patent DE 1209799   **Manufacturers** Bayer

APPLICATIONS
Biochemistry Mitosis inhibitor.   **Mode of action** Systemic fungicide with specific
action against *Fusarium*.   **Uses** Seed treatment for control of *Fusarium* spp. in
cereals, only in combination with other fungicides.   **Formulation types** DS; FS; LS;
WS.   **Mixtures** *(fuberidazole +)* bitertanol; triadimenol; imazalil + triadimenol;

bitertanol + triadimenol; anthraquinone + bitertanol.
Selected tradenames 'Baytan' (mixture) (Bayer); 'Sibutol' (mixture) (Bayer)

ANALYSIS
Product analysis by u.v. spectroscopy; details available from Bayer AG.
Residues determined by glc (W. Specht, *Pflanzenschutz-Nachr. (Engl. Ed.)*, 1977, **30**, 55; W. Specht & M. Tillkes, *ibid.*, 1980, **33**, 61). Methods for the determination of residues are also available from Bayer.

MAMMALIAN TOXICOLOGY
Oral Acute oral LD_{50} for male rats *c.* 336, mice *c.* 650 mg/kg. **Skin and eye**
Acute percutaneous LD_{50} for rats >5000 mg/kg; not irritating to skin and eyes
(rabbit). **Inhalation** LC_{50} (4 h) for rats >0.3 mg/l air (aerosol). **NOEL** (2 y) for
male rats 80, female rats 400, dogs 20, mice 100 mg/kg diet. **ADI** 0.006 mg/kg
b.w. (Bayer 1996). **Toxicity class** WHO (a.i.) II **EC risk** Xn (R22)

ECOTOXICOLOGY
Birds Acute oral LD_{50} for Japanese quail 500 mg/kg. **Fish** LC_{50} (96 h) for golden
orfe 16.2, rainbow trout 2.5 mg/l. **Bees** Not harmful to bees when used as
directed (seed treatment). **Worms** LC_{50} for *Eisenia foetida* >1000 mg/kg dry soil.
Daphnia LC_{50} (48 h) 5.6 mg/l. **Algae** E_rC_{50} for *Scenedesmus subspicatus*
0.49 mg/l.

ENVIRONMENTAL FATE
Animals In mammals, following oral administration, fuberidazole is rapidly
metabolised by hydrolysis and hydroxylation. Elimination is mainly in the urine.
Plants According to a balance study with labelled a.i. using seed-treated spring
wheat planted in a lysimeter, young plants absorbed about 10% of the applied
radioactivity which however decreased to 1–2% in later development stages.
Soil/Environment Mobility in different soils is low. Degradation is rapid; DT_{50}
0.6–11 d, DT_{90} 5–64 d measured in laboratory trials, using predominantly BBA
standard soils 1 and 2. Direct photodegradation in water contributes to overall
elimination of fuberidazole from the environment to a great extent; DT_{50} 0.5–3 d.
Volatilisation into the air of the parent compound or degradation products could
not be observed in the balance study described under 'Plants'.

375 furalaxyl

Fungicide

phenylamide (acylalanine type)

NOMENCLATURE
Common name furalaxyl (BSI, E-ISO, (*m*) F-ISO)
IUPAC name methyl *N*-(2-furoyl)-*N*-(2,6-xylyl)-DL-alaninate
Chemical Abstracts name methyl *N*-(2,6-dimethylphenyl)-*N*-(2-furanylcarbonyl)-=
DL-alaninate; *N*-(2,6-dimethylphenyl)-*N*-(2-furanylcarbonyl)-DL-alanine methyl
ester
CAS RN [57646–30–7]   **EEC no.** 260–875–1   **Development codes** CGA 38 140

PHYSICAL CHEMISTRY
Mol. wt. 301.3   **M.f.** $C_{17}H_{19}NO_4$ **Form** White, odourless crystals.   **M.p.** 70 °C
and 84 °C (dimorphic)   **V.p.** 0.07 mPa (20 °C)   **K_{ow}** logP = 2.7
Henry 9.3×10^{-5} Pa m^3 mol^{-1} (calc.)   **S.g./density** 1.22 (20 °C)   **Solubility** In
water 230 mg/l (20 °C). In dichloromethane 600, acetone 520, methanol 500,
benzene 480, hexane 4 (all in g/kg, 20 °C).   **Stability** Relatively stable in neutral or
weakly acidic media. Less stable in alkaline media; on hydrolysis DT_{50} (20 °C)
(calculated) >200 d at pH 1 and pH 9, 22 d at pH 10. Stable up to 300 °C.

COMMERCIALISATION
History Fungicide reported by F. J. Schwinn *et al.* (*Meded. Fac. Landbouwwet.
Rijksuniv. Gent*, 1977, **42**, 1181). Introduced by Ciba-Geigy AG (now Novartis
Crop Protection AG).   **Patent** BE 827419; GB 1448810   **Manufacturers** Novartis

APPLICATIONS
Mode of action Systemic fungicide with protective and curative action. Absorbed
through the roots, stalk, and leaves, with translocation.   **Uses** Control of soil-
borne diseases caused by *Phytophthora* and *Pythium* spp., and other Oomycetes, on
ornamentals.   **Formulation types** WP; GR.   **Compatibility** No incompatibility
known.   **Selected tradenames** 'Fongarid' (Novartis)

ANALYSIS
Product analysis by glc.   **Residues** in plants and soils determined by glc with TID
(D. J. Caverley & J. Unwin, *Analyst (London)*, 1980, **106**, 389). Details available
from Novartis.

MAMMALIAN TOXICOLOGY
Oral Acute oral LD_{50} for rats 940, mice 603 mg/kg. **Skin and eye** Acute percutaneous LD_{50} for rats >3100, rabbits 5508 mg/kg. Weak skin and eye irritant (rabbits). Non-sensitising to skin (guinea pigs). **NOEL** (90 d) for rats 82 mg/kg daily; for dogs 1.8 mg/kg b.w. daily. **Other** Not teratogenic, not mutagenic. **Toxicity class** WHO (a.i.) III **EC risk** R22

ECOTOXICOLOGY
Birds Acute oral LD_{50} (8 d) for penguin ducks and Japanese quail >6000 mg/kg. LC_{50} (8 d) for Japanese quail >6000 ppm. **Fish** LC_{50} (96 h) for rainbow trout 32.5, crucian carp 38.4, guppy 8.7, catfish 60.0 mg/l. **Bees** Not toxic to bees when used as directed; LD_{50} (24 h, oral) 5->20 μg/bee. **Worms** LC_{50} (14 d) for earthworms 510 mg/kg. **Daphnia** LC_{50} (48 h) 39.0 mg/l. **Algae** EC_{50} for *Scenedesmus subspicatus 27 mg/l.*

ENVIRONMENTAL FATE
Animals After oral administration, furalaxyl is rapidly metabolised and eliminated via urine and faeces. The metabolic pathways are independent of sex or dose level administered. Residues in tissues were generally low and there was no evidence for accumulation or retention of furalaxyl or its metabolites. **Plants** Furalaxyl is metabolised to polar, water-soluble, partially acidic, probably conjugated degradation products. **Soil/Environment** In soil, the compound dissipated with DT_{50} 31–65 d (20–25 °C). Degradation is by cleavage of the ester bond and by N-dealkylation.

376 furathiocarb *Insecticide*

carbamate

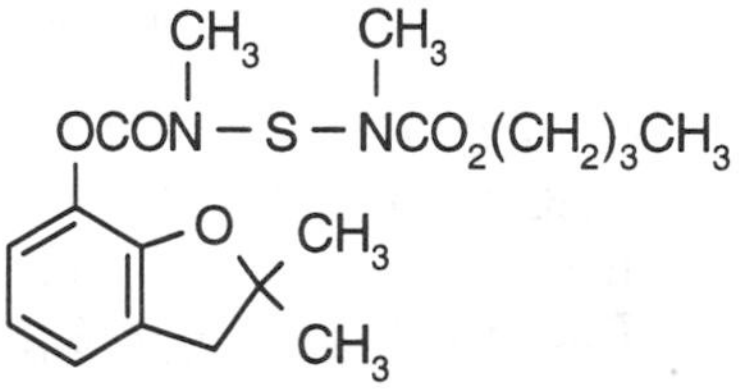

NOMENCLATURE
Common name furathiocarb (BSI, draft E-ISO, (m) draft F-ISO)
IUPAC name butyl 2,3-dihydro-2,2-dimethylbenzofuran-7-yl N,N'-dimethyl-=N,N'-thiodicarbamate

Chemical Abstracts name 2,3-dihydro-2,2-dimethyl-7-benzofuranyl 2,4-dimethyl-=
5-oxo-6-oxa-3-thia-2,4-diazadecanoate
CAS RN *[65907–30–4]* **Development codes** CGA 73 102

PHYSICAL CHEMISTRY
Mol. wt. 382.5 **M.f.** $C_{18}H_{26}N_2O_5S$ **Form** Yellow viscous liquid. **B.p.** >250 °C
(OECD 13) **V.p.** 3.9×10^{-3} mPa (25 °C) (OECD 104) K_{ow} logP = 4.6 (25 °C)
(OECD 117) **S.g./density** 1.148 (20 °C) (OECD 109) **Solubility** In water
11 mg/l (25 °C). Readily miscible with most common organic solvents, including
acetone, methanol, isopropanol, hexane, and toluene. **Stability** Stable up to
400 °C. In water DT_{50} 4 d (pH 9).

COMMERCIALISATION
History Insecticide reported by F. Bachmann & J. Drabek (*Proc. Br. Crop Prot. Conf.
– Pests Dis.*, 1981, **1**, 51). Introduced by Ciba-Geigy AG (now Novartis Crop
Protection). **Patent** BE 865290; GB 1583713 **Manufacturers** Novartis

APPLICATIONS
Biochemistry Cholinesterase inhibitor. **Mode of action** Systemic insecticide with
contact and stomach action. **Uses** Control of soil-dwelling insects in maize,
oilseed rape, sorghum, sugar beet, sunflowers, and vegetables. Applied as a foliar,
soil, or seed treatment. **Formulation types** DS; GR; EC; Liquid seed treatment.
Selected tradenames 'Deltanet' (Novartis); 'Promet' (Novartis)

ANALYSIS
Product analysis by glc. **Residues** determined by glc. Details available from
Novartis.

MAMMALIAN TOXICOLOGY
EHC 64 (WHO, 1986; a review of carbamate insecticides in general). **Oral** Acute
oral LD_{50} for rats 53, mice 327 mg/kg. **Skin and eye** Acute percutaneous LD_{50}
for rats >2000 mg/kg. Mild skin and eye irritant (rabbits). **Inhalation** LC_{50} (4 h)
for rats 0.214 mg/l air. **NOEL** for rats 0.35 mg/kg b.w. daily. **Toxicity class**
WHO (a.i.) Ib; EPA (formulation) II **EC risk** R25, R26, R36/38, R43

ECOTOXICOLOGY
Birds Acute oral LD_{50} for quail and mallard ducks <25 mg/kg. **Fish** LC_{50} (96 h)
for rainbow trout, carp, and bluegill sunfish 0.03–0.12 mg/l. **Bees** Toxic to bees.
Daphnia LC_{50} (48 h) 0.18 µg/l.

ENVIRONMENTAL FATE
Animals Metabolic transformation in rats proceeds via rapid and complete
hydrolysis, followed by oxidation and conjugation. Excretion occurs mainly via
kidney. **Plants** In plants, metabolised to carbofuran and its hydroxy and keto

derivatives. **Soil/Environment** In soil, rapidly decomposed to carbofuran, followed by complete metabolism. Under laboratory and field conditions, furathiocarb is not considered to be mobile. Main degradation product, carbofuran, is classified as mobile.

877 furilazole

Herbicide safener

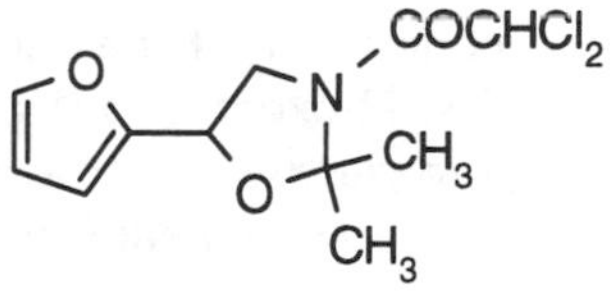

NOMENCLATURE
Common name furilazole (BSI, pa E-ISO)
IUPAC name (RS)-3-dichloroacetyl-5-(2-furyl)-2,2-dimethyloxazolidine
Chemical Abstracts name (±)-3-(dichloroacetyl)-5-(2-furanyl)-2,2-dimethyl=
oxazolidine
CAS RN [121776–33–8]; ((R)- isomer [121776–57–6])
Development codes MON 13900

PHYSICAL CHEMISTRY
Mol. wt. 278.1   **M.f.** $C_{11}H_{13}Cl_2NO_3$ **Form** Light brown powder (tech.)
M.p. 96.6–97.6 °C   **V.p.** 8.84×10^2 mPa (25 °C) K_{ow} logP = 2.12 (23 °C)
Solubility In water 0.0197 g/100 ml (20 °C).   **F.p.** 135 °C

COMMERCIALISATION
History Reported by B. H. Bussler *et al.* (*Proc. Br. Crop Prot. Conf. – Weeds*, 1991, **1**, 39). Introduced by Monsanto Co.   **Manufacturers** Monsanto

APPLICATIONS
Mode of action Safener which acts by enhancing herbicide metabolism.
Uses Reduces injury from several herbicides in certain grass crops. Especially effective for minimising growth retardation from sulfonylurea and imidazolinone herbicides in maize.   **Mixtures** *(furilazole +)* halosulfuron-methyl.
Selected tradenames 'Battalion' (mixture) (Monsanto)

MAMMALIAN TOXICOLOGY
Oral Acute oral LD_{50} for rats 869 mg/kg.   **Skin and eye** Acute percutaneous LD_{50} for rats >5000 mg/kg. Non-irritating to skin; slight eye irritant (rabbits).   Non-

sensitising to skin (guinea pigs).   **Inhalation** LC_{50} for rats >2.3 mg/l air.
NOEL (90 d) for rats 100 ppm (5 mg/kg), dogs 15 mg/kg.

ECOTOXICOLOGY
Birds LD_{50} for bobwhite quail >2000 mg/kg. LC_{50} for bobwhite quail and mallard
duck >5620 ppm.   **Fish** LC_{50} (96 h) for bluegill sunfish 4.6, rainbow trout
6.2 mg/l.   **Bees** LD_{50} >100 µg/bee.   **Daphnia** LC_{50} (48 h) 26 mg/l.

ENVIRONMENTAL FATE
Plants The metabolic pathway in maize and sorghum appears to involve
conversion to oxamic acid, (±)-2-[5-(2-furyl)-2,2-dimethyl-1,3-oxazolidin-3-yl]-=
2-oxoacetic acid, and/or an alcohol, interconversion of the acid and alcohol, and
conjugation of the alcohol to give 2-[5-(2-furyl)-2,2-dimethyl-1,3-oxazolidin-=
3-yl]-2-oxoethyl β-D-glucopyranoside. Further metabolism results in incorporation
into glucose/fructose and other natural plant components.
Soil/Environment Estimated half-life in aerobic soils 33–53 d, in anaerobic soils
13–15 d.

378 *Fusarium oxysporum*   *Fungicide*

fungus

NOMENCLATURE
Var. Fo 47

COMMERCIALISATION
Production By solid fermentation to produce clay microgranules; liquid
fermentation to produce liquid formulation.   **History** This strain was discovered in
suppressive soil of Chateaurenard, S.E. France, by INRA researchers.
Manufacturers NPP

APPLICATIONS
Mode of action Protects crops against pathogenic *Fusaria* spp. by: i) soil
competition at root level; *F. oxysporum* strain Fo 47 shows a strong colonisation
aptitude and high competivity for nutrients with other micro-organisms, ii)
competition at the surface of the root system, at the infection site, iii) elicitation;
the plant reacts to the inoculation, becomes auto-immune and produces
phytoalexins which inhibit the production of *Fusarium* enzymes and detoxify the
fusaric acid.   **Uses** A naturally-occurring mutant, antagonistic strain, for control of
Fusarium oxysporum and *F. moniliforme*. 'Fusaclean L' is under development for

treating rock-wool blocks, 'Fusaclean G' for use on field and glasshouse crops. **Formulation types** SL (spores, 10^{11} cfu/l); MG (mycelium and spores, 2.5×10^8 cfu/g). **Selected tradenames** 'Fusaclean L'; 'Fusaclean G' (NPP)

MAMMALIAN TOXICOLOGY
Oral Acute oral LD_{50} for rats >5000 mg/kg. **Skin and eye** Acute percutaneous LD_{50} for rats >2000 mg/kg. **Inhalation** Not toxic; not infective or pathogenic to rats after an intratracheal instillation. **Other** *F. oxysporum* strain Fo 47 is not pathogenic or infective to mammals.

ECOTOXICOLOGY
Birds Oral LD_{50} (5 d) for quail >9400 mg/kg (by gavage), equivalent to 1.2×10^3 cfu/g. **Fish** LC_{50} (96 h) >100 mg/l.

gamma-HCH

See entry 392.

379 gibberellic acid *Plant growth regulator*

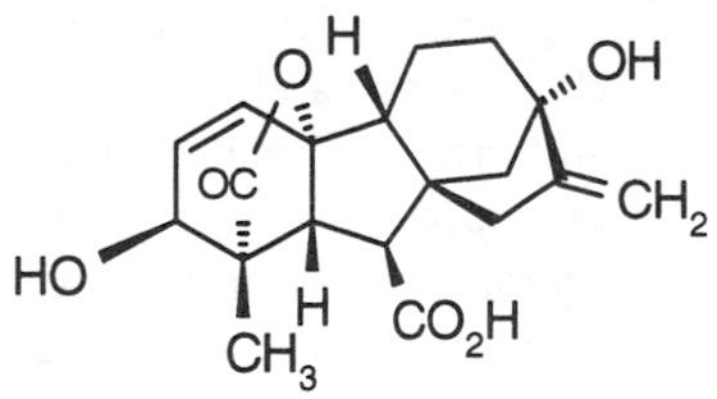

NOMENCLATURE
Common name gibberellic acid (BSI, draft E-ISO, accepted in lieu of a common name); acide gibbérellique (draft F-ISO)
IUPAC name (3S,3aS,4S,4aS,7S,9aR,9bR,12S)-7,12-dihydroxy-3-methyl-6-methyl=ene-2-oxoperhydro-4a,7-methano-9b,3-propeno[1,2-*b*]furan-4-carboxylic acid. *Alt:* (3S,3aR,4S,4aS,6S,8aR,8bR,11S)-6,11-dihydroxy-3-methyl-12-methylene-=2-oxo-4a,6-ethano-3,8b-prop-1-enoperhydroindeno[1,2-*b*]furan-4-carboxylic acid
Chemical Abstracts name (1α,2β,4aα,4bβ,10β)-2,4a,7-trihydroxy-1-methyl-8-=

methylenegibb-3-ene-1,10-dicarboxylic acid 1,4a-lactone
Other names gibberellin A$_3$; GA$_3$ (ambiguous)   **CAS RN** *[77–06–5]*

PHYSICAL CHEMISTRY
Mol. wt. 346.4   **M.f.** C$_{19}$H$_{22}$O$_6$ **Form** Crystalline solid.   **M.p.** 223–225 °C
(decomp.)   **Solubility** In water 5 g/l (room temperature). Soluble in methanol,
ethanol, acetone, and aqueous alkalis; slightly soluble in diethyl ether and ethyl
acetate. Insoluble in chloroform. Potassium, sodium, and ammonium salts: Readily
soluble in water (potassium salt 50 g/l).   **Stability** Dry gibberellic acid is stable at
room temperature, but slowly undergoes hydrolysis in aqueous or aqueous-
alcoholic solutions, DT$_{50}$ (20 °C) c. 14 d (pH 3–4), 14 d (pH 7). In alkalis,
undergoes a rearrangement to less biologically-active compounds. Decomposed
by heat.   **pKa** 4.0

COMMERCIALISATION
History Discovered by E. Kurosawa (*Trans. Nat. Hist. Soc. (Formosa)*, 1926, **16**,
213), who called it gibberellin A. Later ICI Plant Protection Ltd (now Zeneca
Agrochemicals) isolated a compound with similar biological properties and
chemical structure, and this was called gibberellic acid. It and other members of
the gibberellin group (over 70 are known) occur naturally in a wide variety of
plant species. The establishment of its chemical structure and stereochemistry has
been reviewed (J. F. Grove, *Q. Rev. Chem. Soc.*, 1961, **15**, 56; B. E. Cross *et al.*,
Adv. Chem. Series, 1961, No. 28, p. 13). Introduced by ICI (now Zeneca
Agrochemicals).   **Manufacturers** Abbott; Jingma; Zeneca

APPLICATIONS
Mode of action Acts as a plant growth regulator on account of its physiological
and morphological effects in extremely low concentrations. Translocated.
Generally affects only the plant parts above the soil surface.   **Uses** Plant growth
regulator, used in a variety of applications, e.g. to improve fruit setting of
clementines and pears (especially William pears); to loosen and elongate clusters
and increase berry size in grapes; to control fruit maturity by delaying
development of the yellow colour in lemons; to reduce rind stain and retard rind
ageing in navel oranges; to counteract the effects of cherry yellows virus diseases
in sour cherries; to produce uniform seedling growth in rice; to promote
elongation of winter celery crop; to induce uniform bolting and increase seed
production in lettuce for seed; to break dormancy and stimulate sprouting in seed
potatoes; to extend the picking season by hastening maturity in artichokes; to
increase the yield in forced rhubarb; to increase the malting quality of barley; to
produce brighter-coloured, firmer fruit, and to increase the size of sweet cherries;
to increase yields and aid harvesting of hops; to reduce internal browning and
increase yields of Italian prunes; to increase fruit set and yields of tangelos and
tangerines; to improve fruit setting in blueberries; to advance flowering and
increase the yield of strawberries; and also a variety of applications on

ornamentals.   **Formulation types** SP; Crystals; SG; EC; TB.
Compatibility Incompatible with alkaline materials and solutions containing
chlorine.   **Selected tradenames** 'Activol' (Zeneca); 'Berelex' (Zeneca); 'Ceku-Gib'
(Cequisa); 'GIB' (Burlington); 'Gibbex' (Griffin); 'Gibrel' (Thermo Trilogy);
'Kri-Gibb' (Krishi Rasayan); 'ProGibb' (Abbott); 'Release' (Abbott); 'RyzUp'
(Abbott); 'Strong' (Sanonda); 'Uvex' (Productos OSA)

ANALYSIS
Product analysis by hplc; details available from Zeneca.   **Residues** determined by
tlc (V. W. Winkler, *Anal. Methods Pestic. Plant Growth Regul.*, 1978, **10**, 545).

MAMMALIAN TOXICOLOGY
Oral Acute oral LD_{50} for rats and mice >15 000 mg/kg.   **Skin and eye** Acute
percutaneous LD_{50} not available. Non-irritating to skin and eyes.   **Inhalation** No
ill-effect on rats subjected to 400 mg/l for 2 h/d for 21 d.   **NOEL** (90 d) for rats
and dogs >1000 mg/kg diet (6 d/w).   **Toxicity class** WHO (a.i.) III (Table 5); EPA
(formulation) III

380   gibberellin A$_4$ with gibberellin A$_7$

Plant growth regulator

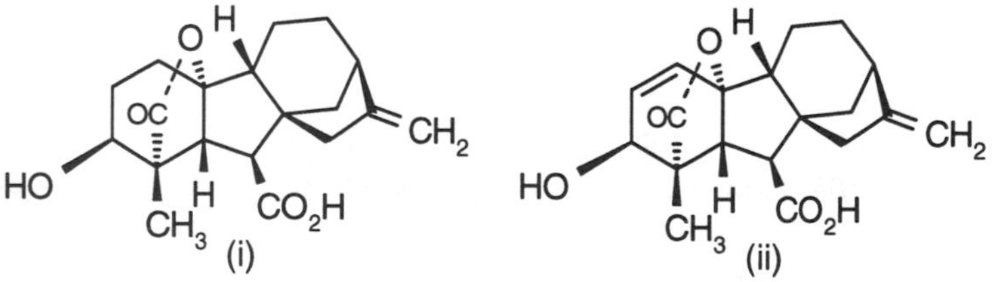

NOMENCLATURE
gibberellin A$_4$
IUPAC name (3S,3aR,4S,4aR,7R,9aR,9bR,12S)-12-hydroxy-3-methyl-6-methylene-
2-oxoperhydro-4a,7-methano-3,9b-propanoazuleno[1,2-*b*]furan-4-carboxylic acid
Chemical Abstracts name (1α,2β,4aα,4bβ,10β)-2,4a-dihydroxy-1-methyl-=
8-methylenegibbane-1,10-dicarboxylic acid 1,4a-lactone
CAS RN *[468–44–0]*

gibberellin A$_7$
IUPAC name (3S,3aR,4S,4aR,7R,9aR,9bR,12S)-12-hydroxy-3-methyl-6-methylene-=

2-oxoperhydro-4a,7-methano-9b,3-propenoazuleno[1,2-*b*]furan-4-carboxylic acid
Chemical Abstracts name (1α,2β,4aα,4bβ,10β)-2,4a-dihydroxy-1-methyl-=
8-methylenegibb-3-ene-1,10-dicarboxylic acid 1,4a-lactone
CAS RN *[510–75–8]*

PHYSICAL CHEMISTRY
gibberellin A$_4$
Mol. wt. 332.4 **M.f.** $C_{19}H_{24}O_5$

gibberellin A$_7$
Mol. wt. 330.4 **M.f.** $C_{19}H_{22}O_5$

COMMERCIALISATION
History Of the numerous gibberellins, a mixture of gibberellin A$_4$ (i) and
gibberellin A$_7$ (ii) was introduced as a plant growth regulator by ICI Plant
Protection Division (now Zeneca Agrochemicals). **Manufacturers** Abbott

APPLICATIONS
Uses Plant growth regulator used to reduce russetting in apples, increase fruit set
in pears, and germination and yield in celery. **Selected tradenames** 'Regulex'
(Zeneca); 'ProVide' (Abbott)

381 *Gliocladium virens* *Biological agent*

fungus

NOMENCLATURE
Scientific name *Gliocladium virens* **Strain** GL-21 **Other names** *Trichoderma virens*

PROPERTIES
Form Brown granules with an odour of molasses. **Stability** The product is stable
for 1 y in cool, dry conditions.

COMMERCIALISATION
History Discovered by the USDA and now marketed by Thermo Trilogy Corp.
Manufacturers Thermo Trilogy

APPLICATIONS
Mode of action Microbial fungicide with preventative rather than curative action.

Generates an antibiotic that kills the plant pathogens. Also parasitises soil pathogens and competes with them for nutrients. **Uses** Fungicide applied by soil incorporation, for control of soil pathogens such as Rhizoctonia, Pythium, Fusarium, Theilaviopsis, Sclerotinia and Sclerotium. For use in nurseries and glasshouses. **Formulation types** GR. **Compatibility** Fungicides should not be applied at time of incorporation. **Selected tradenames** 'SoilGard' (Thermo Trilogy)

MAMMALIAN TOXICOLOGY
Skin and eye May cause mild, reversible irritation to the eyes and skin.
Inhalation No acute toxicity, infectivity or pathogenicity.

ENVIRONMENTAL FATE
Occurs naturally in soil in low concentrations.

382 glufosinate-ammonium *Herbicide*

phosphinico amino acid

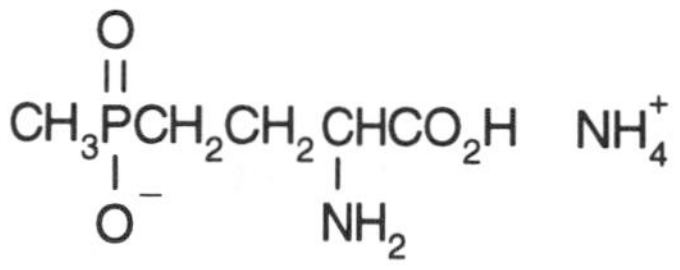

$$CH_3\overset{\overset{\displaystyle O}{\|}}{\underset{\underset{\displaystyle O^-}{|}}{P}}CH_2CH_2\underset{\underset{\displaystyle NH_2}{|}}{C}HCO_2H \quad NH_4^+$$

NOMENCLATURE
glufosinate-ammonium
IUPAC name ammonium 4-[hydroxy(methyl)phosphinoyl]-DL-homoalaninate; ammonium DL-homoalanin-4-yl(methyl)phosphinate
Chemical Abstracts name ammonium (±)-2-amino-4-(hydroxymethylphosphinyl)= butanoate
CAS RN *[77182–82–2]* unstated stereochemistry **Development codes** AE F039866; Hoe 039866

glufosinate
Common name glufosinate (BSI, E-ISO, (*m*) F-ISO); no name (Republic of South Africa)
IUPAC name 4-[hydroxy(methyl)phosphinoyl]-DL-homoalanine; DL-homoalanin-4-yl(methyl)phosphinic acid
Chemical Abstracts name (±)-2-amino-4-(hydroxymethylphosphinyl)butanoic acid
Other names phosphinothricin **CAS RN** *[53369–07–6]* racemic acid; *[51276–47–2]* acid, unstated stereochemistry; *[35597–44–5]* acid, D-isomer
Development codes AE F035956

PHYSICAL CHEMISTRY
glufosinate-ammonium
Mol. wt. 198.2 **M.f.** $C_5H_{15}N_2O_4P$ **Form** Crystalline solid with a slightly pungent odour. **M.p.** 215 °C. **V.p.** <0.1 mPa (20 °C). **K$_{ow}$** logP <0.1 (pH 7, 22 °C) **Henry** 4.48×10^{-9} Pa m^3 mol^{-1} (calc.) **S.g./density** 1.4 (20 °C) **Solubility** In water 1370 g/l (22 °C). In acetone 0.16, ethanol 0.65, ethyl acetate 0.14, toluene 0.14, hexane 0.2 (all in g/l, 20 °C). **Stability** Stable to light.

glufosinate
Mol. wt. 181.1 **M.f.** $C_5H_{12}NO_4P$ **V.p.** Low **pKa** pKa$_1$ <2, pKa$_2$ 2.9, pKa$_3$ 9.8

COMMERCIALISATION
History The ammonium salt was reported as a herbicide by F. Schwerdtle *et al.* (*Z. Pflanzenkr. Pflanzenschutz.*, 1981, Sonderheft IX, p. 431) and was introduced by Hoechst AG (now AgrEvo GmbH). **Manufacturers** AgrEvo

APPLICATIONS
glufosinate-ammonium
Biochemistry Glutamine synthetase inhibitor; leads to accumulation of ammonium ions, and inhibition of photosynthesis. **Mode of action** Non-selective contact herbicide with some systemic action. Translocation occurs only within leaves, predominantly from the leaf base to the leaf tip. **Uses** Glufosinate-ammonium is used for control of a wide range of annual and perennial broad-leaved weeds and grasses in fruit orchards, vineyards, rubber and oil palm plantations, ornamental trees and bushes, non-crop land, and pre-emergence in vegetables. Also used as a desiccant in potatoes, sunflowers etc. For control of annual and perennial weeds and grasses in glufosinate-tolerant crops (oilseed rape, maize, soya bean, sugar beet) developed through gene technology. **Formulation types** SL.
Compatibility Compatible with diuron, simazine, MCPA, and some other herbicides. **Selected tradenames** 'Basta' (AgrEvo); 'Liberty' (AgrEvo)

ANALYSIS
Product analysis by hplc with u.v. determination (*CIPAC Handbook*, 1995, **G**, 89–93). Details available from AgrEvo. **Residue** analysis by gc after derivatisation (*Man. Pestic. Residue Anal.* DFG651).

MAMMALIAN TOXICOLOGY
glufosinate-ammonium
Reviews *Pesticide residues in food – 1991*. FAO Plant Production and Protection Paper, 111, 1991. *Pesticide residues in food – 1991 evaluations*. World Health Organisation, WHO/PCS/92.52, 1992. E. Ebert *et al.*, *Fd. Chem. Toxic.*, **28** (5), 339–349 (1990); R. Hack *et al.*, *Fd. Chem. Toxic.*, **32** (5), 461–470 (1994).
Oral Acute oral LD$_{50}$ for male rats 2000, female rats 1620, male mice 431, female mice 416, dogs 200–400 mg/kg. **Skin and eye** Acute percutaneous LD$_{50}$ for male

rats >4000, female rats c. 4000 mg/kg. Not a skin or eye irritant. **Inhalation** LC_{50} (4 h) for male rats 1.26, female rats 2.60 mg/l air (dust); for rats >0.62 mg/l air (aerosol). **NOEL** (2 y) for rats 2 mg/kg b.w. daily. **ADI** (JMPR) 0.02 mg/kg b.w. [1991]. **Other** No teratogenic, carcinogenic, mutagenic or neurotoxic effects have been observed. **Toxicity class** WHO (a.i.) III; EPA (formulation) III

ECOTOXICOLOGY

glufosinate-ammonium

Birds Eight-day dietary LC_{50} for Japanese quail >5000 mg/kg. **Fish** LC_{50} (96 h) for rainbow trout 710, carp, bluegill sunfish, golden orfe >1000 mg/l. **Bees** Not hazardous to bees; LD_{50} >100 µg/bee. **Worms** LD_{50} for earthworms >1000 mg/kg soil. **Other beneficial spp.** Not toxic to beneficial arthropods. **Daphnia** LC_{50} (48 h) 560–1000 mg/l. **Algae** LD_{50} for *Scenedesmus subspicatus* ≥1000, *Selenastrum capricornutum* 37 mg/l.

ENVIRONMENTAL FATE

Animals Rapidly excreted, predominantly via faeces (90%). The principal metabolite is 3-methylphosphinico-propionic acid. A further faecal metabolite is *N*-acetylglufosinate, formed by intestinal micro-organisms. **Plants** Non-selective use: only the metabolite, 3-methylphosphinico-propionic acid (3-MPP), is taken up in traces from the soil. Desiccation: most of the residues consist of parent glufosinate-ammonium with minor amounts of metabolite 3-MPP. Selective use: the principal metabolite is *N*-acetylglufosinate with lesser amounts of parent and 3-MPP. **Soil/Environment** Rapidly degraded in surface levels of soil, and in water. Because of polarity, it and its metabolites do not bioaccumulate. Metabolism in soil and water reviewed (E. Dorn et al., Z. Pflanzenkr. Plflanzenschutz., Sonderheft XIII, 459–468 (1992)). Degraded to 3-methylphosphinico-propionic acid (3-MPP) and 2-methylphosphinico-acetic acid, and ultimately to CO_2 and bound residues. DT_{50} in soil c. 3–20 d, DT_{50} of metabolites c. 30 d, DT_{50} in water c. 2–30 d. Lysimeter studies and model calculations show that neither a.i. nor metabolites leach into groundwater. Adsorption is more correlated with clay content than organic matter, K_{clay} 2–115, K_{oc} 10–1230.

MAIN ENTRIES

$$HO_2CCH_2NHCH_2\overset{\overset{O}{\|}}{P}(OH)_2$$

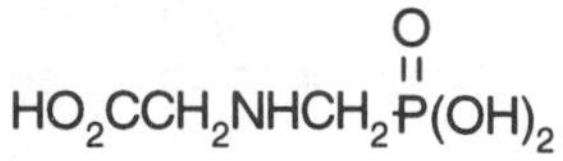

For a general review, see *"Glyphosate, A Unique Global Herbicide"*.

NOMENCLATURE
glyphosate
Common name glyphosate (BSI, E-ISO, (*m*) F-ISO, ANSI, WSSA, JMAF)
IUPAC name *N*-(phosphonomethyl)glycine
Chemical Abstracts name *N*-(phosphonomethyl)glycine
CAS RN *[1071–83–6]*   **EEC no.** 213–997–4   **Development codes** MON-0573;
CP 67573

glyphosate-ammonium
CAS RN *[40465–66–5]*   **Development codes** MON-8750

glyphosate-isopropylammonium
CAS RN *[38641–94–0]*   **EEC no.** 254–056–8   **Development codes** MON-0139;
MON 77209

glyphosate-sodium
CAS RN *[34494–03–6]*   **Development codes** MON-8722

glyphosate-trimesium
Other names sulfosate   **CAS RN** *[81591–81–3]*   **Development codes** SC 0224;
ICIA0224

PHYSICAL CHEMISTRY
glyphosate
Composition Tech. is ≥95% pure. Zwitterion structure (P. Knuuttila & H. Knuuttila, *Acta Chem. Scand.*, 1979, **33**, 623).   **Mol. wt.** 169.1   **M.f.** $C_3H_8NO_5P$
Form Colourless crystals.   **M.p.** 189.5±0.5 °C   **V.p.** Negligible K_{ow} logP <–3.4
(20 °C, OECD 107; EEC A8; unstated pH)   **S.g./density** 1.704 (20 °C)
Solubility In water 11.6 g/l (25 °C). Insoluble in common organic solvents, e.g.
acetone, ethanol and xylene. The alkali metal and amine salts are readily soluble in
water.   **Stability** Glyphosate and all its salts are non-volatile, do not
photochemically degrade and are stable in air.   **pKa** pKa_1 0.8, pKa_2 3, pKa_3 6.0,
pKa_4 11 (K. Chamberlain *et al.*, *Pestic. Sci.*, **47**, 265 (1996))

glyphosate-ammonium
Composition 95.2% pure.   **Mol. wt.** 186.1   **M.f.** $C_3H_{11}N_2O_5P$ **Form** Odourless

white powder. **M.p.** Decomp. >190 °C **V.p.** 9×10^{-3} mPa (25 °C) **Henry** 1.16×10^{-8} Pa m^3 mol^{-1} (calc.) **S.g./density** 1.433 (20 °C) **Solubility** In water 1445±19 g/l (20 °C, pH 3). Essentially insoluble in organic solvents.

glyphosate-isopropylammonium
Composition As a wet cake contains c. 62% w/w isopropylamine salt, c. 35% water. **Mol. wt.** 228.2 **M.f.** $C_6H_{17}N_2O_5P$ **Form** Odourless white powder. **M.p.** Occurs in 2 steps, 143–164 °C and 189–223 °C. **V.p.** 2.1×10^{-3} mPa (25 °C) **Henry** 4.6×10^{-10} Pa m^3 mol^{-1} (calc.) **S.g./density** 1.23 (20 °C) **Solubility** In water 1050 g/l (25 °C, pH 4.3). In dichloromethane 0.184, methanol 15.88 g/l (20 °C).

glyphosate-sodium
Mol. wt. 191.1 **M.f.** $C_3H_7NNaO_5P$ **Form** Odourless white powder. **M.p.** Decomp. >260 °C **V.p.** 7.56×10^{-3} mPa (25 °C) **Henry** 4.27×10^{-9} Pa m^3 mol^{-1} (calc.) **S.g./density** 1.622 (20 °C) **Solubility** Completely miscible in water.

glyphosate-trimesium
Mol. wt. 245.2 **M.f.** $C_6H_{16}NO_5PS$ **V.p.** 0.04 mPa (25 °C) **S.g./density** Bulk density 1.23g /cm^3 **Solubility** Very soluble in water. In acetone, chlorobenzene, ethanol, kerosene, xylene <5 (all in g tech./l). **Stability** Trimesium cation DT_{50} 6.7 d (100 °C); in aqueous solution exposed to light, trimesium cation DT_{50} >30 d (pH 9, 25 °C).

COMMERCIALISATION
History Herbicidal activity reported by D. D. Baird et al. (*Proc. North Cent. Weed Control Conf.*, 1971, **26**, 64). The isopropylammonium, sodium and ammonium salts introduced by Monsanto Co.; the trimesium (trimethylsulfonium) salt introduced in Spain (1989) by ICI Agrochemicals (now Zeneca Agrochemicals). **Patent** US 3799758 to Monsanto; EP 53871; US 4315765 both to ICI **Manufacturers** Aimco; Alkaloida; Aragonesas; Crystal; Excel; Feinchemie Schwebda; Herbex; High Kite; Kuo Ching; Makhteshim-Agan; Monsanto; Nufarm Ltd; Pilarquim; Productos OSA; Sanachem; Shenzhen Jiangshan; Sinon; Zeneca

APPLICATIONS
Biochemistry Inhibits 5-enolpyruvylshikimate-3-phosphate synthase (EPSPS), an enzyme of the aromatic acid biosynthetic pathway. This prevents synthesis of essential aromatic amino acids needed for protein biosynthesis.
Mode of action Non-selective systemic herbicide, absorbed by the foliage, with rapid translocation throughout the plant. Inactivated on contact with soil.
Uses Control of annual and perennial grasses and broad-leaved weeds, pre-harvest, post-planting/pre-emergence and in stubble, in cereals, peas, beans, oilseed rape, flax and mustard at c 1.5–2 kg/ha; as a directed spray in vines and olives at c. 4.3 kg/ha; in orchards, pasture, forestry and industrial weed control at

MAIN ENTRIES

c. 4.3 kg/ha. As an aquatic herbicide at *c.* 2 kg/ha.   **Formulation types** SL;
SG.   **Mixtures** *(glyphosate +)* simazine; terbuthylazine; 2,4-D; alachlor.
Compatibility Mixing with other herbicides may reduce the activity of glyphosate.
Selected tradenames 'Rodeo' (isopropylammonium salt) (Monsanto); 'Roundup'
(isopropylammonium salt) (Monsanto); 'Sting' (isopropylammonium salt)
(Monsanto); 'Asset' (isopropylammonium salt) (Ancom); 'Barbarian' (Barclay);
'Caoganlin' (Shenzhen Jiangshan); 'Glion' (Defensa); 'Glistar' (isopropylammonium
salt) (Alkaloida); 'Glycel' (isopropylammonium salt) (Excel); 'Glyphogan'
(isopropylammonium salt) (Makhteshim-Agan); 'Glyphotox' (Aimco); 'Mamba'
(isopropylammonium salt) (Sanachem); 'Pilarsato' (Pilarquim); 'Rondo'
(isopropylammonium salt) (Productos OSA); 'Rophosate' (Rotam); 'Sanos'
(isopropylammonium salt) (Sanonda); 'Taifun' (isopropylammonium salt)
(Feinchemie Schwebda); 'Touchdown' (trimesium salt) (Zeneca)

ANALYSIS
Product analysis by hplc *(AOAC Methods,* 1995, 983.10; *CIPAC Handbook,* 1985,
1C, 2130).   **Residues** determined by hplc *(Pestic. Anal. Man.,* 1979, **II**). In
environmental water, by lc, determination by postcolumn reactions specific for
primary amines *(AOAC Methods,* 1995, 991.08, 10.6.18).

MAMMALIAN TOXICOLOGY
glyphosate
Reviews *Pesticide residues in food – 1986.* FAO Plant Production and Protection
Paper 77, 1986. *Pesticide residues in food – 1986 evaluations. Part II – Toxicology.*
FAO Plant Production and Protection Paper 78/2, 1987.   **EHC** 159 (WHO).
Oral Acute oral LD_{50} for rats 5600, mice 11 300, goats 3530 mg/kg.
Skin and eye Acute percutaneous LD_{50} for rabbits >5000, rats >5010, mice
7030 mg/kg. Eye irritant; non-irritating to skin (rabbits).   **NOEL** In 2 y feeding
trials, no ill-effects were observed in rats receiving 410 mg/kg diet daily (average);
and in 1 y feeding trials, no ill-effects were observed in dogs receiving 500 mg/kg
daily (highest dose treated).   **ADI** 1.75 mg [1994] (EHC 159).   **Other** Not
mutagenic, not carcinogenic, not teratogenic, not neurotoxic. No adverse effects
on reproduction.   **Toxicity class** WHO (a.i.) III (Table 5); EPA (formulation) III
EC risk Xi (R41)

glyphosate-ammonium
Oral Acute oral LD_{50} for rats 4613 mg/kg.   **Skin and eye** Acute percutaneous
LD_{50} for rabbits >5000 mg/kg. Slight eye irritant, not a skin irritant (rabbits).
Inhalation LC_{50} >1.9 mg/l.

glyphosate-isopropylammonium
Oral Acute oral LD_{50} for rats >5000 mg/kg.   **Skin and eye** Acute percutaneous
LD_{50} for rabbits >5000 mg/kg. Slight eye irritant, not a skin irritant (rabbits).

glyphosate-trimesium
Oral Acute oral LD_{50} for male rats 748, female rats 755, mice 1250 mg/kg.
Skin and eye Acute percutaneous LD_{50} for rabbits >2000 mg/kg. **Inhalation** LC_{50}
(4 h) for rats >0.81 mg/l air. **NOEL** 100 mg/kg daily. No teratogenic effect
observed.

ECOTOXICOLOGY
glyphosate
Birds Acute oral LD_{50} for bobwhite quail >3851 mg/kg. Eight-day dietary LC_{50} for
quail and ducks >4640 mg/kg diet. **Fish** LC_{50} (96 h) for trout 86, bluegill sunfish
120, harlequin fish 168, sheepshead minnow >1000, fathead minnow 97 mg/l.
Bees LD_{50} (contact and oral) >100 μg/bee. **Daphnia** LC_{50} (48 h) 780 mg/l.
Algae EC_{50} (72 h) for *Selenastrum capricornutum* 485 mg/l, (7 d) 13.8 mg/l; (96 h)
for *Skeletonema costatum* 1.2 mg/l, (7 d) 0.64 mg/l; (7 d) for *Navicula pelliculosa*
42, *Anabaena flos-aquae* 15 mg/l. **Other aquatic spp.** LC_{50} (96 h) for mysid
shrimp (*Mysidopsis bahia*) >1000, grass shrimp 281, fiddler crab 934 mg/l. EC_{50}
(96 h) for sea urchin >1000 mg/l; (14 d) for *Lemna gibba* 25.5 mg/l.

glyphosate-isopropylammonium
Fish LC_{50} (96 h) for trout and bluegill sunfish >1000 mg/l. **Worms** LC_{50} (14 d)
for *Eisenia foetida* >5000 mg/kg soil. **Other beneficial spp.** No effects on carabid
beetle; harmless to slightly harmful to green lacewing, parasite species,
mites/spiders and insects, except moderately harmful to *Bembidion lampros*.
Daphnia LC_{50} (48 h) 930 mg/l. **Algae** EC_{50} (72 h) for *Scenedesmus subspicatus*
72.9 mg/l. **Other aquatic spp.** EC_{50} (48 h) for midge larvae 5600 mg/l.

glyphosate-trimesium
Birds Acute oral LD_{50} for bobwhite quail >2050, mallard ducks 950 mg tech./kg.
Fish LC_{50} (96 h) for trout 1800, bluegill sunfish >3500 mg/l. **Bees** LD_{50} (contact)
0.39 mg/bee, (oral) >0.4 mg/bee.

ENVIRONMENTAL FATE
Animals In mammals, following oral administration, glyphosate is very rapidly
excreted unchanged and does not bioaccumulate. **Plants** Slowly metabolised to
aminomethylphosphonic acid [1066–51–9], which is the major plant metabolite.
Soil/Environment In soil (field), DT_{50} 3–174 d, depending on edaphic and climatic
conditions. In water, DT_{50} varies from a few to 91 d. Photodegradation in water
occurs under natural conditions, DT_{50} ≤28 d; no substantial photodegradaton in
soil was recorded over 31 d. In a lab. whole system with water and sediment,
DT_{50} ≤14 d (aerobic), 14–22 d (anaerobic). The major metabolite in soil and
water is aminomethylphosphonic acid.

384 gossyplure *Insect pheromone*

pheromone

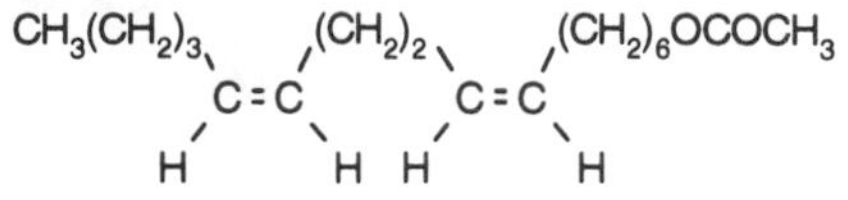

CH3(CH2)3 (CH2)2 (CH2)6OCOCH3
C=C C=C
H H H H

(Z,Z)-

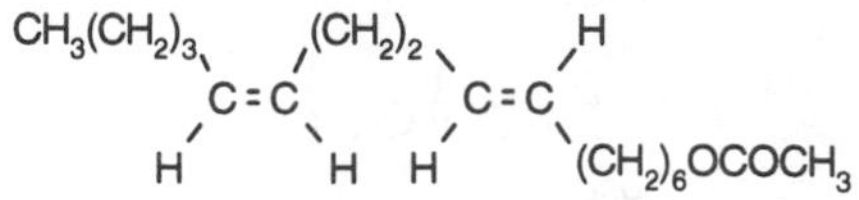

CH3(CH2)3 (CH2)2 H
C=C C=C
H H H (CH2)6OCOCH3

(Z,E)-

NOMENCLATURE

Common name gossyplure (name in common use)
IUPAC name a 1:1 mixture of (*Z*,*Z*)- and (*Z*,*E*)-hexadeca-7,11-dien-1-yl acetate
Chemical Abstracts name a 1:1 mixture of (*Z*,*Z*)- and (*Z*,*E*)-7,11-hexadecadien-= 1-ol acetate
Other names pink bollworm pheromone; *Pectinophora gossypiella* pheromone; PBW; hexadecadienyl acetate **CAS RN** *[53042–79–8]*; *[51607–94–4]* ((*Z*,*E*)- isomer); *[52207–99–5]* ((*Z*,*Z*)- isomer); *[122616–64–2]* (7-*Z*,11-unspecified stereochemistry)- isomer; *[50933–33–0]* (unspecified stereochemistry) **Development codes** PP761 (Zeneca)

PHYSICAL CHEMISTRY

Composition The mating pheromone of the pink bollworm moth (*Pectinophora gossypiella*) is a 1:1 mixture of (*Z*,*Z*)- and (*Z*,*E*)- isomers. **Mol. wt.** 280.4 **M.f.** $C_{18}H_{32}O_2$ **Form** Liquid, with a sweet odour. **B.p.** 170–175 °C/3 mmHg **V.p.** 11 mPa **K_{ow}** logP >4 **S.g./density** 0.885 (20 °C) **Solubility** Insoluble in water; soluble in organic solvents. **Stability** Stable in the range pH 5–7. **F.p.** 167 °C

COMMERCIALISATION

History In use since 1985. **Manufacturers** Ecogen; Shin-Etsu

APPLICATIONS

Mode of action Acts as an attractant and by disruption of mating. **Uses** For control of the pink bollworm moth in cotton. Used alone as a mating disruptant, and in combination with an insecticide to attract and kill.
Formulation types Granules, beads, flakes, hollow fibre, rope, PVC strips; all are slow-release. **Selected tradenames** 'Frustrate PBW' (Thermo Trilogy); 'NoMate PBW' (Ecogen); 'PB Rope', 'PB Rope-L' (Shin-Etsu); 'Stirrup PBW' (Troy)

guanidine

$$R\,NH - (CH_2)_8 - \overset{\overset{\textstyle R}{|}}{N} - [(CH_2)_8 - \overset{\overset{\textstyle R}{|}}{N}]_n H$$

n may be 0 or 1 or 2 etc.
and any R substituent may be

$$- H \ (17 - 23\%) \quad \text{or} \quad - \overset{\overset{\textstyle NH_2}{|}}{C} = NH \ (77 - 83\%)$$

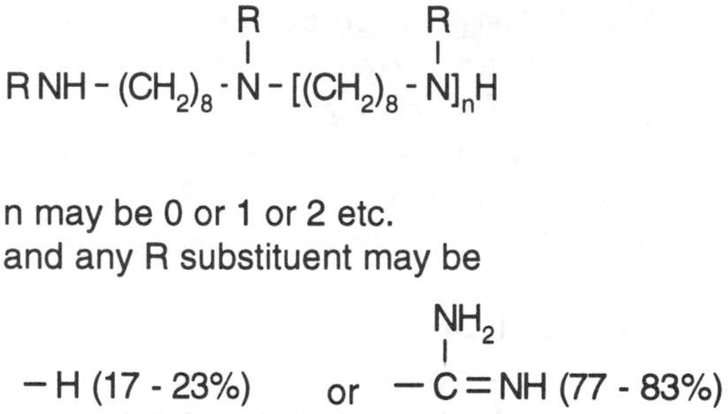

NOMENCLATURE

guazatine

Common name guazatine (BSI, draft E-ISO, *(f)* draft F-ISO — but see notes on composition)

IUPAC name 'A mixture of the reaction products from polyamines, comprising mainly octamethylenediamine, iminodi(octamethylene)diamine and octamethylenebis(imino-octamethylene)diamine, and carbamonitrile'

Chemical Abstracts name guazatine

CAS RN *[108173–90–6]*

guazatine acetates

Other names GTA   **CAS RN** *[115044–19–4]*   **Development codes** EM 379 (Evans Medical Ltd); MC 25 (Murphy Chemical Ltd)

PHYSICAL CHEMISTRY

guazatine

Composition The approved common name guazatine was originally defined as applying to 1,1'-iminodi(octamethylene)diguanidine (BSI used the name *guanoctine* from 1970–1972). It is now known that the material marketed commercially is a reaction mixture. Produced by the amidination of tech. iminodi(octamethylene)= diamine, commercial guazatine contains numerous guanidines, in which the amino and imino groups of the polyamine chain form part, and polyamines; many of these bases are fungicidal. A replacement common name, iminoctadine (*q.v.*) has been established for 1,1'-iminodi(octamethylene)diguanidine.

guazatine acetates

Form Brown solid (tech.).   **M.p.** *c.* 60 °C   **V.p.** <1 × 10^{-2} mPa (50 °C)

K$_{ow}$ logP = –1.2 (pH 3), –0.9 (pH 10).   **S.g./density** 1.09 (20 °C)   **Solubility** In water >3 kg/l (room temperature). In methanol 510, *N*-methylpyrrolidone 1000, dimethyl sulfoxide, dimethylformamide 500, ethanol *c.* 200 (all in g/l, 20 °C). Very sparingly soluble in xylene and other hydrocarbons.   **Stability** No significant hydrolysis at pH 5, 7 and 9 after 1 month at 25 °C.   **pKa** Acetate salt of strong base

COMMERCIALISATION
History Fungicidal properties of the salts of 1,1'-iminodi(octamethylene)=
diguanidine reported by W. S. Catling et al. (*Congr. Plant Pathol., 1st,* 1968, p. 27
(Abstr.)). The triacetate was introduced by Evans Medical Ltd, developed by
Murphy Chemical Ltd (now part of DowElanco) and then by KenoGard VT AB
(now Rhône-Poulenc Agro).   **Patent** GB 1114155 to Murphy Chemical Ltd;
SE 417569 to KenoGard   **Manufacturers** Rhône-Poulenc

APPLICATIONS
guazatine acetates
Uses Guazatine acetates are used as seed treatments for control of *Septoria
nodorum, Tilletia caries, Fusarium* spp., and *Helminthosporium* spp. in cereals and
maize; (also acts as a bird repellent). Control of *Pyricularia oryzae* (rice blast) in
rice, *Cercospora* spp. in peanuts and soya beans, *Septoria* spp. in wheat, and
pineapple disease in sugar cane. Also used as a post-harvest dip for seed potatoes,
citrus fruit, and pineapples, and as a wood protectant.   **Formulation types** LS; SL;
DS.   **Mixtures** *(guazatine acetates +)* imazalil; fenfuram; fenfuram + imazalil;
carboxin; lindane; propiconazole; thiabendazole.   **Selected tradenames** 'Kenopel'
(Rhône-Poulenc); 'Panoctine' (Rhône-Poulenc)

ANALYSIS
Product analysis by potentiometric titration with acid.   **Residues** determined by
hydrolysis, followed by gc-ms determination (H. Kobayashi et al., *Nihon Noyaku
Gakkaishi,* 1977, **2,** 427).

MAMMALIAN TOXICOLOGY
guazatine
ADI (JMPR) 0.03 mg/kg b.w. [1978].

guazatine acetates
Reviews *Pesticide residues in food – 1978.* FAO Plant Production and Protection
Paper 15, 1979. *Pesticide residues in food – 1979 evaluations.* FAO Plant Production
and Protection Paper 15 Sup, 1979.   **Oral** Acute oral LD_{50} for rats 360 mg/kg.
Skin and eye Acute percutaneous LD_{50} for rats >1000, rabbits 1176 mg/kg. May
cause eye irritation.   **Inhalation** LC_{50} for rats 225 mg/m^3.   **NOEL** (2 y) for rats
17.5 mg/kg daily; (1 y) for dogs 0.9 mg/kg daily.   **Other** Non-teratogenic and
non-carcinogenic (rats).   **Toxicity class** WHO (a.i.) II; EPA (formulation) II

ECOTOXICOLOGY
guazatine acetates
Birds LD_{50} for pigeons 82 mg/kg, with emetic effect.   **Fish** LC_{50} (96 h) for
rainbow trout 1.41 mg/l.   **Bees** Not toxic to bees when used as directed; LD_{50}

(contact) >200 µg/bee. **Worms** LC_{50} (14 d) for earthworms >1000 mg/kg.
Daphnia LC_{50} (48 h) 0.15 mg/l.

ENVIRONMENTAL FATE
Animals In rats, following oral administration, rapid and complete degradation occurs, with elimination in the urine and faeces within 2–3 days.
Soil/Environment When applied directly to soil, guazatine is strongly bound temporarily, before undergoing deguanidation and breakdown of the octyl chain, followed by mineralisation to CO_2. It is immobile in soil and is therefore unlikely to exceed 0.1 µg/l in groundwater.

GY-81 *Fungicide, insecticide, nematicide*

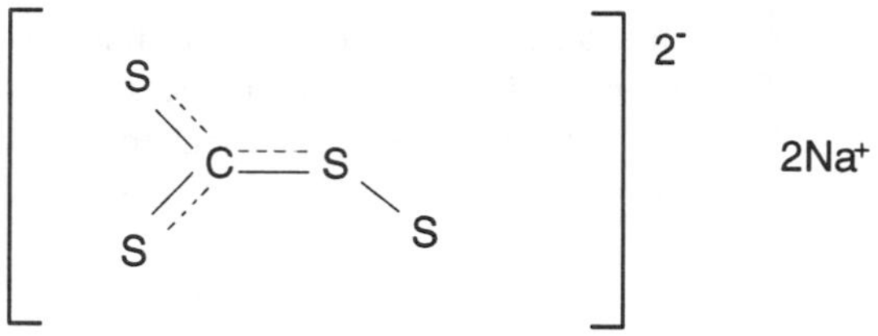

NOMENCLATURE
IUPAC name sodium tetrathio(peroxocarbonate)
Chemical Abstracts name disodium carbono(dithioperoxo)dithioate
CAS RN *[7345–69–9]* **Development codes** GY-81 (Unocal)

PHYSICAL CHEMISTRY
Mol. wt. 186.2 **M.f.** CNa_2S_4 **Form** Hygroscopic orange crystalline solid with an odour resembling rotten eggs. **Solubility** Very soluble in water (>50% at 20 °C). **Stability** Stable at normal and elevated temperatures. Oxidised by metal ions with reduction potentials >0.14 v. Photolysis DT_{50} 175–1013 mins.

COMMERCIALISATION
Patent US 5288753 and earlier; EP 0141844 **Manufacturers** Entek

APPLICATIONS
Mode of action Degrades in the soil to release the broad-spectrum biocide carbon disulfide. **Uses** Water-soluble contact fumigant for the management of grape

phylloxera, plant parasitic nematodes and various soil-borne pathogens causing root rot. **Formulation types** SL. **Selected tradenames** 'Enzone' (Entek)

MAMMALIAN TOXICOLOGY
Oral Acute oral LD_{50} for rats 631 mg/kg. **Skin and eye** Acute percutaneous LD_{50} for rabbits >2.0 g/kg. Non-sensitising to skin of guinea pigs. Moderate eye irritation and severe dermal irritation in rabbits. **Inhalation** LC_{50} (4 h) for male rats 4.73, female rats 3.17 mg/l. **ADI** 6 mg/d (as carbon disulfide). **Other** Non-mutagenic in the following tests: Ames, CHO/HGPRT cell mutation, and unscheduled DNA synthesis (rat hepatocytes). Non-teratogenic in rats and rabbits.

ECOTOXICOLOGY
Birds LD_{50} for bobwhite quail 1180 mg/kg. LC_{50} (5 d) for bobwhite quail and mallard ducks >5620 ppm. **Fish** LC_{50} (96 h) for rainbow trout 6.7, bluegill sunfish 21 mg/l. **Bees** LD_{50} >25 µg/bee. **Daphnia** LC_{50} (48 h) 6.6 mg/l.

ENVIRONMENTAL FATE
Soil/Environment Rapidly degrades in the soil, releasing carbon disulfide gas which in turn is rapidly dissipated, hence minimising the potential for groundwater contamination. Routes of dissipation include evaporation and biological oxidation. The latter produces carbonate and sulfate, both of which are plant nutrients. Dissipation by the combined mechanisms is complete within 4–7 d in most agricultural soils.

387 halfenprox   *Acaricide*

Non-ester pyrethroid

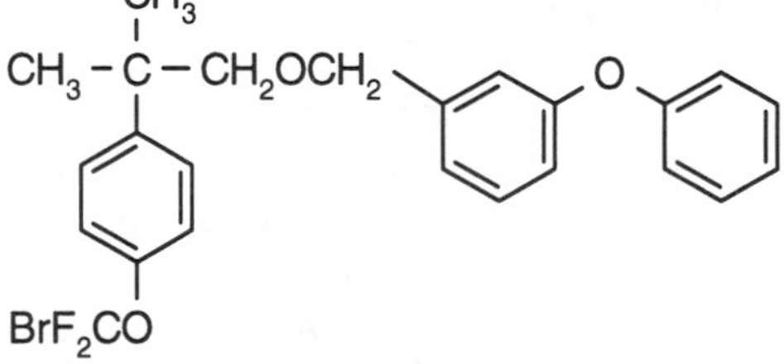

NOMENCLATURE
Common name halfenprox (BSI, pa E-ISO, (m) pa F-ISO)
IUPAC name 2-(4-bromodifluoromethoxyphenyl)-2-methylpropyl 3-phenoxybenzyl ether

Chemical Abstracts name 1-[[2-[4-(bromodifluoromethoxy)phenyl]-2-methyl=
propoxy]methyl]-3-phenoxybenzene
Other names [fubfenprox] (rejected BSI name proposal)
CAS RN *[111872–58–3]* **Development codes** MTI-732 (Mitsui)

PHYSICAL CHEMISTRY
Mol. wt. 477.3 **M.f.** $C_{24}H_{23}BrF_2O_3$ **Form** Colourless clear liquid.
B.p. 291 °C/760 mmHg **V.p.** 7.79×10^{-4} mPa (25 °C) **K$_{ow}$** logP = 4.1
S.g./density 1.318 (20 °C) **Solubility** In water 0.05 µg/l (25 °C).
Stability Degradation is <10% at 55 °C. Stable in buffer solution (pH 4, 7, 9).

COMMERCIALISATION
History Miticide introduced in Japan (1987) by Mitsui Toatsu Chemicals, Inc.
Manufacturers Mitsui Toatsu

APPLICATIONS
Mode of action Broad spectrum acaricide with contact action and moderate
residual activity. Also inhibits the hatching stage of eggs. **Uses** Control of all
stages of red spider mites, fruit tree red mites, two-spotted spider mites, and rust
mites on citrus fruit, vines, top fruit, vegetables, tea, and ornamentals.
Phytotoxicity No phytotoxicity has been observed at field use rates.
Formulation types EC; CS. **Selected tradenames** 'Anniverse' (Japan)
(Mitsui Toatsu); 'Sirbon' (France) (Mitsui Toatsu)

ANALYSIS
By glc with ECD.

MAMMALIAN TOXICOLOGY
Oral Acute oral LD_{50} for male rats 132, female rats 159, male mice 146, female
mice 121 mg/kg. **Skin and eye** Acute percutaneous LD_{50} for rats >2000 mg/kg.
Inhalation LC_{50} (4 h) for male rats 1.38, female rats 0.36 mg/l.

ECOTOXICOLOGY
Birds LD_{50} for bobwhite quail 1884 mg/kg. **Fish** TLm (48 h) for carp
0.0035 mg/l. **Bees** LC_{50} 27 ppm. **Worms** LC_{50} (7 d) 218 ppm. **Daphnia** LC_{50}
0.031 µg/l (48 h).

ENVIRONMENTAL FATE
Animals From rats, the parent compound and 2-(4-bromodifluoromethoxyphenyl)-=
2-methylpropyl 3-(4-hydroxyphenoxy)benzyl ether are recovered. **Plants** In tea,
the parent compound and 2-(4-dibromodifluoromethoxyphenyl)-2-methylpropyl
3-phenoxybenzoate are recovered. **Soil/Environment** DT_{50} in soil c. 10 d.

diacylhydrazine

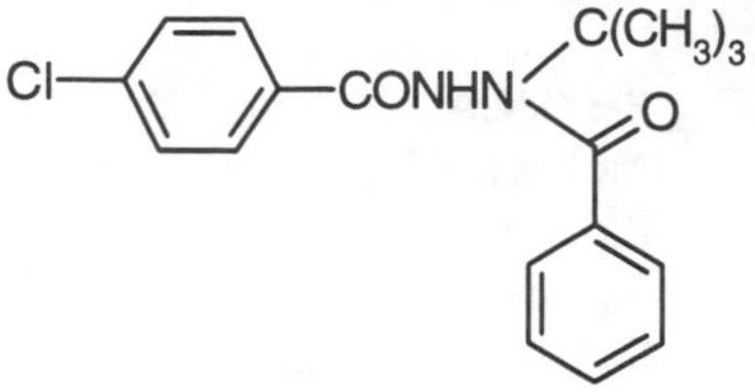

NOMENCLATURE
Common name halofenozide (BSI, pa ISO)
IUPAC name *N*-tert-butyl-*N'*-(4-chlorobenzoyl)benzohydrazide
Chemical Abstracts name 4-chlorobenzoate 2-benzoyl-2-(1,1-dimethylethyl)=
hydrazide
CAS RN *[112226–61–6]* **Development codes** RH-0345

PHYSICAL CHEMISTRY
Mol. wt. 330.8 **M.f.** $C_{18}H_{19}ClN_2O_2$ **Form** White crystalline powder.
M.p. >200 °C **V.p.** <1.3×10^{-2} mPa **K_{ow}** logP = 3.22 **Solubility** In water
12.3 ppm. In isopropanol 3.1%, cyclohexanone 15.4%, aromatic solvents 0.01–1%.
Stability Stable to heat, light and water; hydrolysis DT_{50} 310 d (pH 5), 481 d
(pH 7), 226 d (pH 9).

COMMERCIALISATION
Manufacturers RohMid (j.v. of Rohm & Haas and Cyanamid)

APPLICATIONS
Biochemistry Ecdysone agonist which acts by binding to the receptor site of the
insect moulting hormone, ecdysone. **Mode of action** Systemic, ingested
insecticide, active by root application. Interferes with moulting, primarily affecting
larval stages of insects. Also reduces fecundity in treated adults and has some
ovicidal properties. **Uses** Control of Coleoptera and Lepidoptera in turf and
ornamentals at 0.5–2.0 lb/a. **Formulation types** SC; GR.
Selected tradenames 'Mach 2' (RohMid)

MAMMALIAN TOXICOLOGY
Oral Acute oral LD_{50} for rats 2850, mice 2214 mg/kg. **Skin and eye** Acute
percutaneous LD_{50} for rats and rabbits >2000 mg/kg. Moderately irritating to
eyes, not irritating to skin (rabbits). Skin sensitiser (guinea pigs). **Inhalation** LC_{50}
for rats >2.7 mg/l. **NOEL** (90 d) for dogs 3.8, rats 5.7 mg/kg daily.

ECOTOXICOLOGY
Birds Acute oral LD_{50} for quail >2250 mg/kg. Acute dietary LC_{50} for quail 4522,

mallard ducks >5000 ppm. **Fish** LC_{50} for bluegill sunfish >8.4, trout >8.6, sheepshead minnow >8.8 mg/l. **Bees** LD_{50} (contact) >100 μg/bee. **Worms** LC_{50} >980 mg/kg. **Daphnia** LC_{50} 3.6 mg/l. **Algae** EC_{50} 0.82 mg/l. **Other aquatic spp.** EC_{50} for shrimps 3.7, molluscs 1.2 mg/l.

ENVIRONMENTAL FATE
Soil/Environment Aerobic DT_{50} (lab.) 68–72 d (silt loam), 653–818 d (sandy loam); soil dissipation DT_{50} (field) 46–267 d; turf DT_{50} 3–7 d, DT_{50} for soil photolysis 129 d. K_{oc} 224–279.

389 # halosulfuron-methyl   *Herbicide*

sulfonylurea

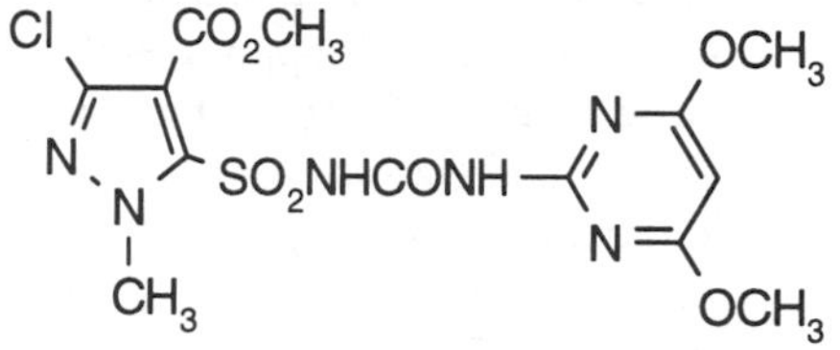

NOMENCLATURE
halosulfuron-methyl
Common name halosulfuron-methyl (BSI, pa E-ISO)
IUPAC name methyl 3-chloro-5-(4,6-dimethoxypyrimidin-2-ylcarbamoyl= sulfamoyl)-1-methylpyrazole-4-carboxylate
CAS RN *[100784–20–1]* **Development codes** A-841101; NC-319 (both Nissan); MON-12000 (Monsanto)

halosulfuron
Common name halosulfuron (BSI, pa E-ISO)
IUPAC name 3-chloro-5-(4,6-dimethoxypyrimidin-2-ylcarbamoylsulfamoyl)-= 1-methylpyrazole-4-carboxylic acid
Chemical Abstracts name 3-chloro-5-[[[[(4,6-dimethoxy-2-pyrimidinyl)amino]= carbonyl]amino]sulfonyl]-1-methyl-1*H*-pyrazole-4-carboxylic acid
CAS RN *[135397–30–7]*

PHYSICAL CHEMISTRY
halosulfuron-methyl
Mol. wt. 434.8 **M.f.** $C_{13}H_{15}ClN_6O_7S$ **Form** White powder.
M.p. 175.5–177.2 °C **V.p.** <0.01 mPa (25 °C) **K_{ow}** logP = −0.0186 (pH 7,

23 ± 2 °C)   **S.g./density** 1.618 g/ml (25 °C)   **Solubility** In water 0.015 (pH 5), 1.65 (pH 7) g/l (20 °C). In methanol 1.62 g/l (20 °C).   **Stability** Stable under normal storage conditions.   **pKa** 3.44 (22 °C)

halosulfuron
Mol. wt. 420.8   **M.f.** $C_{12}H_{13}CIN_6O_7S$

COMMERCIALISATION
History Reported by K. Suzuki *et al.* (*Proc. Br. Crop Prot. Conf. – Weeds*, 1991, **1**, 31). Developed jointly by Monsanto Co. and Nissan Chemical Industries, Ltd. Registered in US in 1994.   **Manufacturers** Nissan

APPLICATIONS
halosulfuron-methyl
Biochemistry Branched chain amino acid synthesis (acetolactate synthase or ALS) inhibitor. Acts by inhibiting biosynthesis of the essential amino acids valine and isoleucine, hence stopping cell division and plant growth. Selectivity derives from rapid metabolism in the crop. Metabolic basis of selectivity in sulfonylureas reviewed (M. K. Koeppe & H. M. Brown, *Agro-Food-Industry*, **6**, 9–14 (1995)).
Mode of action Systemic herbicide, absorbed by the root system and/or leaf surface, and translocated to meristem tissues.   **Uses** Halosulfuron-methyl has demonstrated activity for the control of annual broad-leaved weeds and nutsedge spp. in maize, sugar cane, rice and turf. Efficacy has been observed with both pre-emergence (70–90 g/ha) and post-emergence (18–35 g/ha) applications to weeds in maize; pre-emergence activity can be achieved with preplant incorporated, pre-emergence, early preplant, and reduced tillage systems. Applied in combination with pre-emergence grass compounds, halosulfuron-methyl will provide broad-spectrum weed control in maize; furilazole safener is required for maize tolerance to pre-emergence applications.   **Phytotoxicity** Non-phytotoxic to maize, sugar cane, rice, wheat, barley and turf.   **Formulation types** WP; WG.
Mixtures (*halosulfuron-methyl* +) furilazole.   **Selected tradenames** 'Permit' (Nissan, Monsanto)

ANALYSIS
Details from Nissan.

MAMMALIAN TOXICOLOGY
halosulfuron-methyl
Oral Acute oral LD_{50} for rats 8865 mg/kg.   **Skin and eye** Acute percutaneous LD_{50} for rabbits >2000 mg/kg. Non-irritating to skin; slightly irritating to eyes (rabbits). Non-sensitising to skin (guinea pigs).   **Inhalation** LC_{50} (4 h) for rats >6.0 mg/l.   **NOEL** (104 w) for male rats 2500, female rats 1000 ppm; (18 mo) for male mice 3000, female mice 7000 ppm; (1 y) for male dogs 1, female dogs 10 mg/kg daily.   **ADI** 0.1 mg/kg.   **Toxicity class** EPA (formulation) IV (75 WP)

ECOTOXICOLOGY
halosulfuron-methyl
Birds Acute oral LD_{50} for bobwhite quail >2250 mg/kg. Five-day dietary LC_{50} for bobwhite quail and mallard ducks >5620 ppm.   **Fish** LC_{50} (96 h) for bluegill sunfish >118, rainbow trout >131 mg/l.   **Bees** LD_{50} (dermal) >100 µg/bee.
Worms LC_{50} for earthworms >1000 mg/kg.   **Daphnia** EC_{50} (48 h) >107 mg/l.
Algae EC_{50} (5 d) for green algae 0.0053 mg/l.   **Other aquatic spp.** EC_{50} (96 h, flow through) for oysters ≥94, mysid shrimp (*Mysidopsis bahia*) ≥109 mg/l IC_{30} (11 d) for *Lemna gibba* 0.038 µg/l.

ENVIRONMENTAL FATE
Animals Rapidly eliminated from rats, in urine and faeces. The major metabolite is desmethyl-halosulfuron-methyl.   **Plants** The major metabolite in maize is 3-chloro-1-methyl-5-sulfamoylpyrazole-4-carboxylic acid (cleavage of urea bridge and ester hydrolysis).   **Soil/Environment** Extensively metabolised in soils by cleavage of the sulfonylurea bridge and/or pyrimidine ring opening. DT_{50} <18 d.

390 haloxyfop *Herbicide*

2-(4-aryloxyphenoxy)propionic acid

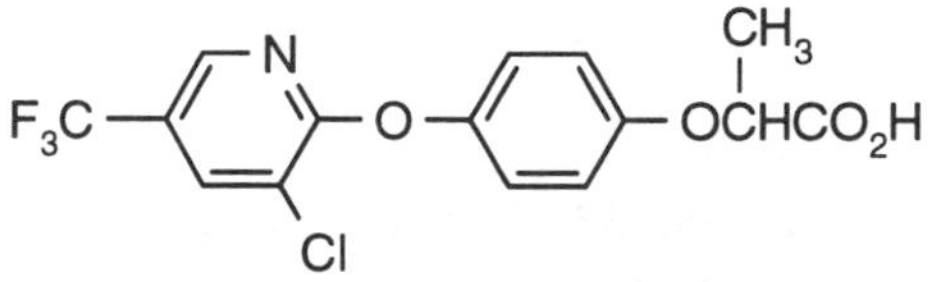

NOMENCLATURE
haloxyfop
Common name haloxyfop (BSI, ANSI, draft E-ISO, (*m*) draft F-ISO, WSSA)
IUPAC name (*RS*)-2-[4-(3-chloro-5-trifluoromethyl-2-pyridyloxy)phenoxy]= propionic acid
Chemical Abstracts name (±)-2-[4-[[3-chloro-5-(trifluoromethyl)-2-pyridinyl]= oxy]phenoxy]propanoic acid
CAS RN [69806–34–4] (unstated stereochemistry)
Development codes Dowco 453

haloxyfop-etotyl
CAS RN [87237–48–7] (unstated stereochemistry)   **EEC no.** 402–560–5
Development codes Dowco 453 EE

haloxyfop-methyl
CAS RN *[69806–40–2]* (unstated stereochemistry)   **Development codes**
Dowco 453 ME

haloxyfop-P-methyl
Other names haloxyfop-R   **CAS RN** *[72619–32–0]*   **Development codes** DE-535

PHYSICAL CHEMISTRY
haloxyfop
Mol. wt. 361.7   **M.f.** $C_{15}H_{11}ClF_3NO_4$ **Form** Colourless crystals.
M.p. 107–108 °C   **V.p.** <1.33×10^{-3} mPa (25 °C) K_{ow} logP = 1.34 (20 °C,
unstated pH)   **S.g./density** 1.64   **Solubility** In water 43.4 mg/l (pH 2.6, 25 °C);
1.590 (pH 5), 6.980 (pH 9) (both in mg/l, 20 °C). In acetone, methanol,
isopropanol >1000, dichloromethane 459, ethyl acetate 518, toluene 118, xylene
74, hexane 0.17 (all in g/l).   **Stability** DT_{50} in water 78 d (pH 5), 73 d (pH 7),
51 d (pH 9).   **pKa** 2.9

haloxyfop-etotyl
Mol. wt. 433.8   **M.f.** $C_{19}H_{19}ClF_3NO_5$ **Form** Colourless crystals.   **M.p.** 58–61 °C
V.p. 1.64×10^{-5} mPa (20 °C, OECD 104) K_{ow} logP = 4.33 (20 °C, OECD 107)
S.g./density 1.34   **Solubility** In water 0.58, 1.91 (pH 5), 1.28 (pH 9.2) (all in mg/l,
20 °C). In dichloromethane 2760, xylene 1250, acetone, ethyl acetate, toluene
>1000, methanol 233, isopropanol 52, hexane 44 (all in g/l, 20 °C).
Stability Hydrolysed to haloxyfop under acidic and alkaline conditions. In water,
DT_{50} (25 °C) 33 d (pH 5), 5 d (pH 7), a few h (pH 9).

haloxyfop-methyl
Mol. wt. 375.7   **M.f.** $C_{16}H_{13}ClF_3NO_4$ **M.p.** 55–57 °C   **V.p.** 0.80 mPa (25 °C)
K_{ow} logP = 4.07   **Solubility** In water 9.3 mg/l (25 °C). In acetonitrile 4.0, acetone
3.5, dichloromethane 3.0, xylene 1.27 (all in kg/kg, 20 °C).

haloxyfop-P-methyl
Composition The (R)- isomer of haloxyfop-methyl.   **Mol. wt.** 375.7
M.f. $C_{16}H_{13}ClF_3NO_4$ **B.p.** >280 °C   **V.p.** 0.328 mPa (25 °C) K_{ow} logP = 4.0479
Solubility In water 8.74 mg/l (25 °C). In acetone, cyclohexanone,
dichloromethane, ethanol, methanol, toluene, xylene >1 kg/l (20 °C).

COMMERCIALISATION
History The methyl and 2-ethoxyethyl (etotyl) esters introduced as herbicides by
Dow Chemical Co. (now DowElanco). The methyl ester of the (R)- isomer
reported by Bayer S.A. (M. C. Botte *et al.*, *Proc. 15th Columa Conf.*, (1992) **1**, 397)
and introduced by DowElanco.   **Manufacturers** DowElanco

APPLICATIONS

Biochemistry Fatty acid synthesis inhibitor. **Mode of action** Haloxyfop-ethoxyethyl and -methyl are selective herbicides, absorbed by the foliage and roots, and hydrolysed to haloxyfop, which is translocated to meristematic tissues, and inhibits their growth. **Uses** Haloxyfop-ethoxyethyl and both forms of haloxyfop-methyl are used pre- and post-emergence for control of annual and perennial grasses in sugar beet, fodder beet, oilseed rape, potatoes, leaf vegetables, onions, flax, sunflowers, soya beans, vines, strawberries, and other crops.
Formulation types EC. **Compatibility** Compatible with many other grass herbicides, and with post-emergence broad-leaved herbicides.

haloxyfop-etotyl
Selected tradenames 'Gallant' (DowElanco)

haloxyfop-methyl
Selected tradenames 'Verdict' (DowElanco)

haloxyfop-P-methyl
Selected tradenames 'Gallant Super' (DowElanco)

ANALYSIS

Product analysis of haloxyfop-etotyl by hplc. **Residues** determined by glc. Details available from DowElanco.

MAMMALIAN TOXICOLOGY

haloxyfop
Oral Acute oral LD_{50} for male rats 337 mg/kg. **Skin and eye** Acute percutaneous LD_{50} for rabbits >5000 mg/kg. **NOEL** (2 y) for rats 0.065 mg/kg daily; no increase in hepatotoxicity. **ADI** (JMPR) 0.0003 mg/kg b.w. [1995].
Toxicity class WHO (a.i.) II

haloxyfop-etotyl
Oral Acute oral LD_{50} for male rats 531, female rats 518 mg/kg.
Skin and eye Acute percutaneous LD_{50} for rats >2000, rabbits >5000 mg/kg; non-irritant to skin; moderate eye irritant (rabbits). No sensitisation (guinea pigs).
NOEL for rats 0.065 mg/kg daily. **EC risk** Xn (R22)

haloxyfop-methyl
Oral Acute oral LD_{50} for male rats 393, female rats 599 mg/kg.
Skin and eye Acute percutaneous LD_{50} for rabbits >5000 mg/kg; non-irritant to skin; moderate eye irritant (rabbits).

haloxyfop-P-methyl
Oral Acute oral LD_{50} for male rats 300, female rats 623 mg/kg.

Skin and eye Acute percutaneous LD_{50} for rats >2000 mg/kg. Non-irritant to skin; slightly irritating to eyes (rabbits).

ECOTOXICOLOGY
haloxyfop
Birds Acute oral LD_{50} for mallard ducks >2150 mg/kg. Eight-day dietary LC_{50} for mallard ducks and bobwhite quail >5620 mg/kg diet. **Fish** LC_{50} (96 h) for trout >800 mg/l. **Daphnia** LC_{50} (48 h) 96.4 mg/l.

haloxyfop-etotyl
Birds Acute oral LD_{50} for mallard ducks >2150 mg/kg. Eight-day dietary LC_{50} for mallard ducks and bobwhite quail >5620 mg/kg diet. **Fish** LC_{50} (96 h) for fathead minnows 0.54, bluegill sunfish 0.28, rainbow trout 1.18 mg/l. **Bees** LD_{50} (48 h, contact and oral) 100 µg/bee. **Daphnia** LC_{50} (48 h) 4.64 mg/l.

haloxyfop-methyl
Birds Eight-day dietary LC_{50} for mallard ducks and bobwhite quail >5620 mg/kg diet. **Fish** LC_{50} (96 h) for rainbow trout 0.38 mg/l. **Bees** LD_{50} (contact, 48 h) >100 µg/bee. **Daphnia** LC_{50} (48 h) 4.64 mg/l.

haloxyfop-P-methyl
Birds Acute oral LD_{50} for bobwhite quail 1159 mg/kg. **Fish** LC_{50} (96 h) for rainbow trout 0.7 mg/l. **Bees** LD_{50} (48 h, oral and contact) >100 µg/bee. **Daphnia** LC_{50} (48 h) 6.12 mg/l.

ENVIRONMENTAL FATE
haloxyfop
Soil/Environment DT_{50} 55–100 d (depending on soil type) (pH 7.0, 20–25 °C); K_{oc} 89 on clay soil (pH 6.9, 3.82% organic carbon).

haloxyfop-etotyl
Animals In mammalian systems, haloxyfop-ethoxyethyl is rapidly hydrolysed to the parent acid haloxyfop. **Soil/Environment** In soil, haloxyfop-ethoxyethyl is converted to haloxyfop. DT_{50} >1 d on silty clay loam (20 °C). Adsorption in silty clay loam soil (pH 7.0, 1.97% organic carbon) K_{oc} 128. In silty clay loam (20 °C) DT_{50} >1 d.

haloxyfop-methyl
Animals In mammalian systems, haloxyfop-methyl is rapidly converted to the parent acid haloxyfop, and subsequently eliminated as the (*R*)- isomer in urine and faeces. **Soil/Environment** In soil, DT_{50} <24 h, forming the parent acid, which is then subject to stereochemical inversion and degradation processes in microbially active soils.

haloxyfop-P-methyl

Animals In mammals, haloxyfop-methyl (R)- isomer is rapidly converted to the parent acid and eliminated. **Soil/Environment** DT_{50} <24 h, forming the parent acid, which is then microbially degraded.

HC-252 *Herbicide*

diphenyl ether

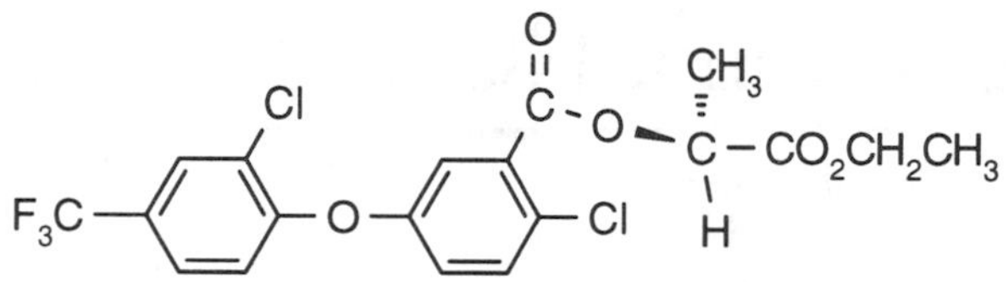

NOMENCLATURE

IUPAC name ethyl *O*-[2-chloro-5-(2-chloro-α,α,α-trifluoro-*p*-tolyloxy)benzoyl]-= L-lactate

Chemical Abstracts name ethyl (*S*)-2-chloro-5-[2-chloro-4-(trifluoromethyl)= phenoxy]benzoate

CAS RN *[131086–42–5]* **Development codes** HC-252 (Budapesti Vegyimüvek)

PHYSICAL CHEMISTRY

Mol. wt. 451.0 **M.f.** $C_{19}H_{15}Cl_2F_3O_5$ (for ethyl ester) **Form** Brown oily viscous liquid (tech.). **Solubility** Very soluble in acetone, methanol and xylene.

COMMERCIALISATION

History Reported by J. Bakos *et al.* (*Proc. Br. Crop Prot. Conf. – Weeds*, 1991, **1**, 83). **Manufacturers** Budapesti Vegyimüvek

APPLICATIONS

Biochemistry Protoporphyrinogen oxidase inhibitor; acts by tetrapyrrole and protoporphyrin accumulation. Tetrapyrroles are photosensitisers which cause photo-oxidation, leading to leaf necrosis and plant death.
Mode of action Contact herbicide. **Uses** Control of broad-leaved weeds in winter wheat, winter barley, peas, soya beans and peanuts.
Formulation types EC.

MAMMALIAN TOXICOLOGY

Oral Acute oral LD_{50} for male rats 843, female rats 963, male mice 1269, female mice 1113 mg/kg. **Skin and eye** Acute percutaneous LD_{50} for rats >5000, rabbits

>2000 mg/kg. Non-irritating to skin; moderate eye irritant (rabbits).
Inhalation LC_{50} (14 d) for male rats 9679, female rats 9344 mg/m^3 air.
Other Non-mutagenic in *Salmonella* micronucleus test, *in vitro* chromosome aberration test, and *in vitro* mouse medulla micronucleus test.

ENVIRONMENTAL FATE
Soil/Environment Since HC-252 is a contact herbicide, there is no soil residual effect, and no soil persistence problems for following crops.

392 gamma-HCH   *Insecticide*

organochlorine

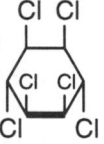

NOMENCLATURE
Common name For mixed isomers: BHC (E-ISO, (*m*) F-ISO, JMAF); HCH (BSI, also alternative E-ISO and F-ISO names); benzene hexachloride (BAN, ESA, EPA, USA); HKhTsH (USSR); [hexachloran] (USSR); hexaklor (Sweden). For the gamma isomer: gamma-HCH (BSI); gamma-BHC or gamma-HCH (E-ISO, (*m*) F-ISO); gamma benzene hexachloride (ESA, EPA, BAN, USA); gamma-HKhTsH (USSR). For material containing ≥99% gamma isomer: lindane (BSI, E-ISO, (*m*) F-ISO, ESA)
IUPAC name 1,2,3,4,5,6-hexachlorocyclohexane (mixed isomers)
Chemical Abstracts name 1,2,3,4,5,6-hexachlorocyclohexane (mixed isomers); stereoisomers: (1α,2α,3β,4α,5β,6β)- alpha; (1α,2β,3α,4β,5α,6β)- beta; (1α,2α,3β,4α,5α,6β)- gamma; (1α,2α,3α,4β,5α,6β)- delta; (1α,2α,3α,4β,5β,6β)- epsilon; (1α,2α,3α,4α,5β,6β)- eta; (1α,2α,3α,4α,5α,6β)- theta; (1α,2α,3α,4α,5α,6α)- zeta
CAS RN *[608–73–1]* (formerly *[39284–22–5]*) mixed isomers; *[319–84–6]* alpha-HCH; *[319–85–7]* beta-HCH; *[58–89–9]* gamma-HCH; *[319–86–8]* delta-HCH; *[6108–10–7]* epsilon-HCH **EEC no.** 210–168–9 (for HCH); 200–401–2 (for lindane) **Official codes** OMS 17 (lindane); ENT 8601 (HCH); ENT 9232 (alpha-HCH); ENT 7796 (gamma-HCH); ENT 9234 (delta-HCH)

PHYSICAL CHEMISTRY
Composition HCH, mixed isomers produced by the chlorination of benzene under u.v. light, has no precise physical properties. Gamma-HCH is isolated by selective

crystallisation of HCH.   **Mol. wt.** 290.8   **M.f.** $C_6H_6Cl_6$ **Form** Colourless crystals.
M.p. 112.5–113.5 °C   **V.p.** 5.6 mPa (20 °C)   **Solubility** In water 7.3 (25 °C), 12
(35 °C) (both in mg/l). In acetone 43.5, methanol 7.4, ethanol 6.4, benzene 28.9,
toluene 27.6, xylene 24.7, diethyl ether 20.8, petroleum ether 2.9, ethyl acetate
35.7, chloroform 24.0, carbon tetrachloride 6.7, cyclohexanone 36.7, dioxane
31.4, acetic acid 12.8 (all in g/l, 20 °C).   **Stability** Extremely stable to light, air,
temperatures up to 180 °C, and to acids. In alkalis, undergoes
dehydrochlorination. DT_{50} (22 °C) 191 d (pH 7), 11 h (pH 9).

COMMERCIALISATION
History Insecticidal properties reported by A. Dupire & M. Racourt (*C. R. Hebd.
Seances Acad. Agric. Fr.*, 1942, **20**, 470) and by R. E. Slade (*Chem. Ind. (London)*,
1945, p. 134). Its insecticidal activity is mainly due to gamma-HCH, which was
introduced by ICI Plant Protection Ltd (now Zeneca Agrochemicals).
Patent US 2502258   **Manufacturers** Hindustan; India Pesticides; Inquinosa

APPLICATIONS
Biochemistry Antagonist of the GABA receptor-chloride channel complex.
Mode of action Insecticide with contact, stomach, and respiratory action.
Uses Control of a broad spectrum of phytophagous and soil-inhabiting insects,
public-health pests, and animal ectoparasites. Used on a wide range of crops
(pests controlled include: Aphididae, larvae of Coleoptera, Curculionidae,
Diplopoda, Diptera, Lepidoptera, Symphyla and Thysanoptera), in stored product
warehouses and storerooms and in public health applications (for control of
Blattodea, Culicidae, Muscidae, Siphonaptera) and in seed treatments (often in
combination with fungicides).   **Phytotoxicity** Non-phytotoxic when used as
recommended, but phytotoxic to cucurbits and hydrangeas when applied to the
foliage.   **Formulation types** SC; EC; FU; LS; WP; GR; DP; UL.   **Mixtures**
(*gamma-HCH +*) carboxin; tecnazene; captan; organomercury compounds;
diazinon; malathion; thiram; dimethoate; methoxychlor; and many other fungicides
and insecticides.   **Compatibility** Incompatible with alkaline materials.
Selected tradenames 'Gamma-Col' (Zeneca); 'Lindamul' (Rhône-Poulenc); 'Lintox'
(Siapa); 'Steward' (Atlas Crop Protection)

ANALYSIS
Product analysis by a cryoscopic method (*CIPAC Handbook*, 1970, **1**, 71; FAO
Specification (CP/34); *Organochlorine Insecticides 1973*; *AOAC Methods*, 1995,
949.05), by glc (*ibid.*, 984.05; *CIPAC Handbook*, 1985, **1C**, 2135), by i.r.
spectrophotometry (*AOAC Methods*, 1995, 947.01) or by potentiometric titration
(*CIPAC Handbook*, 1994, **F**, 190). Hydrolysable chlorine determined (*ibid.*,
183–189).   **Residues** determined by glc or by tlc or paper chromatography (*AOAC
Methods*, 1995, 970.52; *Pestic. Anal. Man.*, 1979, **I**, 201-A, 201-G, 201-I; *Analyst
(London)*, 1979, **104**, 425; *Man. Pestic. Residue Anal.*, 1987, **I**, 5, 6, S8-S10, S12,
S19; *Anal. Methods Residues Pestic.*, 1988, Part I, M1, M12; A. Ambrus *et al.*,

J. Assoc. Off. Anal. Chem., 1981, **64**, 733). In **drinking water,** HCH isomers determined by gc with ECD (*AOAC Methods*, 1995, 990.06).

MAMMALIAN TOXICOLOGY
Reviews *Pesticide residues in food – 1989.* FAO Plant Production and Protection Paper 99, 1989. *Pesticide residues in food – 1989 evaluations. Part II – Toxicology.* FAO Plant Production and Protection Paper 100/2, 1990. See also E. Ullmann, *Lindane,* 1972, Supplements 1974, 1976, 1983. *Pesticide Outlook,* 1990, **1**(4), 10–15. **IARC** 5, 20, 42 **EHC** 124 (for lindane; WHO, 1991), 123 (for α- and β- isomers; WHO, 1992). **Oral** Acute oral LD_{50} values vary with test conditions, especially the carrier: for rats 88–270, mice 59–246 mg/kg. Young animals are especially sensitive. **Skin and eye** Acute percutaneous LD_{50} for rats 900–1000 mg/kg. Skin and eye irritant. **NOEL** (2 y) for rats 25, dogs 50 mg/kg diet. **ADI** (JMPR) 0.008 mg/kg b.w. (for lindane) [1989]; (JMPR) no ADI (for tech., mixtures of isomers) [1973]. **Toxicity class** WHO (a.i.) II; EPA (formulation) II **EC risk** T (R25); Xn (R21); (R40) (for HCH): T (R23/24/25); Xi (R36/38) (for lindane) **PIC** Yes (lindane and gamma-HCH)

ECOTOXICOLOGY
Birds Acute oral LD_{50} for bobwhite quail 120–130 mg/kg. **Fish** LC_{50} (48 h) for guppies 0.16–0.3 mg/l. **Bees** Toxic to bees.

ENVIRONMENTAL FATE
Animals In insects, less-chlorinated, unsaturated metabolites, such as pentachlorocyclohexene (PCCH), are formed. In rats, following oral administration, lindane is found in the milk, body fat, and kidneys, but rapid elimination occurs. In addition to PCCH, 1,2,4-trichlorobenzene and isomeric trichlorophenols are formed as metabolites, and are excreted as the glucuronic acid derivatives. **Soil/Environment** For reviews on the degradation and persistence of lindane in soils, see E. Ulmann (ed.), *Lindane, Monograph of an Insecticide,* 1972, 79–107 and R. Engst *et al., Residue Rev.,* 1977, **68**, 59.

393 *Helicoverpa zea* NPV *Biological agent*

nuclear polyhedrosis virus

NOMENCLATURE
Scientific name *Helicoverpa zea* nuclear polyhedrosis virus

PROPERTIES
Stability Not stable above 32 °C.

COMMERCIALISATION
Manufacturers Thermo Trilogy

APPLICATIONS
Mode of action Viral insecticide, active by ingestion and most effective on early instars. The polyhedral occlusion bodies of the virus (OBs) dissolve in the insect midgut and replicate. Virus progeny enter the hemocoel and infect virtually all cell types. Shortly after death, the integument ruptures, releasing billions of OBs.
Uses Biological insecticide for control of *Helicoverpa zea* (cotton bollworm) and *Heliothis virescens* (tobacco budworm).　**Formulation types** LC.　**Compatibility** Not compatible with strong oxidisers, acids, bases and chlorinated water.
Selected tradenames 'Gemstar' (Thermo Trilogy)

MAMMALIAN TOXICOLOGY
Baculoviruses (a family which includes the NPVs) have been found only in invertebrates; no baculovirus is known to infect vertebrates. The product does not reproduce in mammalian cells.

ECOTOXICOLOGY
Baculoviruses have been found only in invertebrates; no baculovirus is known to infect vertebrates or plants. No adverse effect on fish, wildlife or beneficial organisms has been observed.

ENVIRONMENTAL FATE
The virus is naturally-occurring.

394　heptachlor　　　　　　　　　　　　　　*Insecticide*

cyclodiene organochlorine

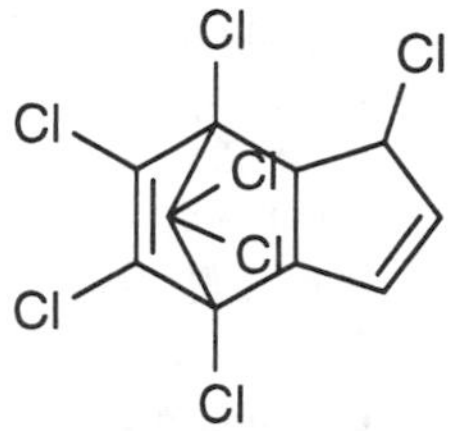

NOMENCLATURE
Common name heptachlor (BSI, E-ISO, ESA, JMAF); heptachlore ((*m*) F-ISO)

IUPAC name 1,4,5,6,7,8,8-heptachloro-3a,4,7,7a-tetrahydro-4,7-methanoindene
Chemical Abstracts name 1,4,5,6,7,8,8-heptachloro-3a,4,7,7a-tetrahydro-=
4,7-methano-1H-indene
CAS RN *[76–44–8]* EEC no. 200–962–3; (heptachlor epoxide, 213–831–0)
Development codes E 3314; Velsicol 104 Official codes OMS 193; ENT 15 152

PHYSICAL CHEMISTRY
Composition Tech. grade is *c.* 72% heptachlor and 28% related compounds.
Mol. wt. 373.3 M.f. $C_{10}H_5Cl_7$ Form Crystalline solid; (tech., waxy solid).
M.p. 95–96 °C (pure); 55–60 °C (tech.) B.p. 135–145 °C V.p. 53 mPa (25 °C,
pure) S.g./density 1.58 (20 °C) (tech.) Solubility In water 0.056 mg/l
(25–29 °C). Soluble in many organic solvents, e.g. in acetone 75, benzene 106,
xylene 102, cyclohexanone 119, carbon tetrachloride 113, ethanol 4.5 (all in
g/100 ml). Tech. in cyclohexanone 1.65 kg/l, ethanol 62.5 g/l, deodorised
kerosene 263 g/l, xylene 1.41 kg/l (20–30 °C). Stability Stable in daylight, air,
moisture, and moderate heat (up to 160 °C). Not readily dehydrochlorinated but
is susceptible to epoxidation.

COMMERCIALISATION
History It was isolated from tech. chlordane, and its insecticidal properties
reported by W. M. Rogoff & R. L. Metcalf (*J. Econ. Entomol.*, 1951, **44**, 910).
Introduced by Velsicol Chemical Corp. Patent US 3437664
Manufacturers Velsicol

APPLICATIONS
Biochemistry Antagonist of the GABA receptor-chloride channel complex.
Mode of action Non-systemic insecticide with contact, stomach, and some
respiratory action. Uses Control of termites, ants, and soil insects in cultivated
and uncultivated soils. Applied as a seed treatment, soil treatment, or directly to
foliage. Also used for control of household insects. Formulation types WP; EC;
DP; GR; Seed treatment. Compatibility Incompatible with alkaline materials.

ANALYSIS
Product analysis by glc (*AOAC Methods*, 1995, 968.04) or by determination of a
labile chlorine substituent (*ibid.*, 962.07; *CIPAC Handbook*, 1970, **1**, 420; FAO
Specification (CP/47); *Organochlorine Insecticides 1973*). Residues determined by
glc or by tlc or paper chromatography (*AOAC Methods*, 1995, 970.52; *Pestic. Anal.
Man.*, 1979, 201-A, 201-G, 201-I; A. Ambrus et *al.*, *J. Assoc. Off. Anal. Chem.*,
1981, **64**, 773; *Analyst (London)*, 1979, **104**, 425; H. K. Suzuki et *al.*, *Anal. Methods
Pestic. Plant Growth Regul.*, 1978, **10**, 73; *Man. Pestic. Residue Anal.*, 1987, **I**, 5, 6,
S8-S10, S12, S19; *Anal. Methods Residues Pestic.*, 1988, Part I, M1, M12). In
drinking water by gc with ECD (*AOAC Methods*, 1995, 990.06).

MAMMALIAN TOXICOLOGY

Reviews *Pesticide residues in food – 1994*, FAO Plant Production and Protection Paper, 127, 1995. *Pesticide residues in food – 1991 evaluations. Part II – Toxicology.* World Health Organisation, WHO/PCS/92.52, 1992. **IARC** 5, 20, 53 **EHC** 38 (WHO, 1984). **Oral** Acute oral LD_{50} for rats 147–220, guinea pigs 116, mice 68 mg/kg. **Skin and eye** Acute percutaneous LD_{50} for rabbits 200–2000, rats 119–250 mg/kg. Irritating to eyes, not a skin irritant (rabbits). **Inhalation** LC_{50} (4 h) for rats exposed to heptachlor in an aerosol >2.0 but <200 mg/l air **NOEL** The target organ is the liver with NOEL in rats at ≤5 mg/kg diet and in dogs at 0.5–1.0 mg/kg diet. A 3-generation feeding study gave NOEL in rats 7 mg/kg diet, and in dogs (2-generation reproduction study) 1 mg/kg diet. Heptachlor epoxide is not teratogenic in rabbits at 5 mg/kg daily. **ADI** (JMPR) 0.0001 mg/kg b.w. (PTDI) [1994]. **Other** Animal experiments have indicated that heptachlor is carcinogenic in some mice and rat strains, perhaps acting as a promoter, rather than an initiator, of carcinogenesis (S. S. Epstein, *Sci. Total Environ.*, 1976, **6**, 103–154). A committee of the United States National Academy of Sciences ruled that heptachlor is a carcinogen to certain mouse strains. *In vitro* and *in vivo* experiments showed that heptachlor is non-mutagenic, and there was also no evidence of teratogenicity in rabbits. **Toxicity class** WHO (a.i.) II; EPA (formulation) II **EC risk** T (R24/25); (R40); (R33) (heptachlor epoxide T (R25); (R40); (R33)) **PIC** Yes

ECOTOXICOLOGY

Birds Acute oral LD_{50} for mallard ducklings >2000 mg/kg. Eight-day dietary LC_{50} for bobwhite quail 450–700, Japanese quail 80–95, pheasants 250–275 mg/kg diet. **Fish** LC_{50} (96 h) for rainbow trout 7, bluegill sunfish 26, fathead minnow 78–130 μg/l.

ENVIRONMENTAL FATE

Animals The principal metabolites in animals are heptachlor epoxide (found in the tissues, faeces, and urine), and 1-exo-hydroxychlordene epoxide, a hydrophilic metabolite excreted in the urine. There is a strong tendency for the epoxide to accumulate in body fat. **Plants** In plants, heptachlor epoxide is a metabolite. **Soil/Environment** In water, heptachlor rapidly undergoes hydrolysis to 1-hydroxy= chlordene, which then undergoes microbial epoxidation to 1-hydroxy-= 2,3-epoxychlordene. Similar degradation pathways occur in moist soil. The half-life in soil is 9–10 months when used at agricultural rates.

organophosphorus

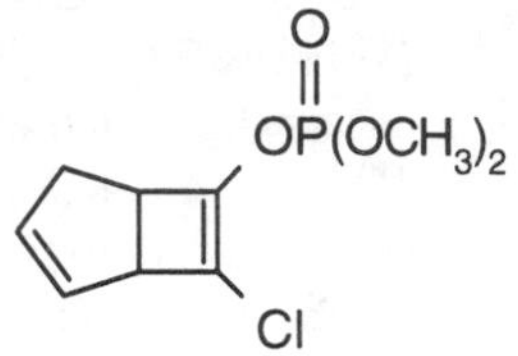

NOMENCLATURE
Common name heptenophos (BSI, E-ISO, (*m*) F-ISO)
IUPAC name 7-chlorobicyclo[3.2.0]hepta-2,6-dien-6-yl dimethyl phosphate
Chemical Abstracts name 7-chlorobicyclo[3.2.0]hepta-2,6-dien-6-yl dimethyl phosphate
CAS RN *[23560–59–0]* **EEC no.** 245-737-0 **Development codes** Hoe 02 982; AE F002982 **Official codes** OMS 1845

PHYSICAL CHEMISTRY
Composition Tech. grade is ≥93% pure. **Mol. wt.** 250.6 **M.f.** $C_9H_{12}ClO_4P$
Form Light brown liquid with odour typical of a phosphate ester.
B.p. 64 °C/0.075 mmHg **V.p.** 65 mPa (15 °C); 170 mPa (25 °C)
K$_{ow}$ logP = 2.32 (OECD 117) **Henry** 5.73×10^{-5} (20 °C), 2.33×10^{-4} (25 °C) (both in Pa m^3 mol^{-1}) **S.g./density** 1.28 (20 °C) **Solubility** In water 2.2 g tech./l (20 °C). Readily miscible with most organic solvents, e.g. in acetone, methanol, xylene >1, hexane 0.13 (all in kg/l, 25 °C). **Stability** Hydrolysed in acidic and alkaline media. **F.p.** 165 °C (Cleveland, open); 152 °C (Pensky-Martens, closed)

COMMERCIALISATION
History First described under an erroneous structure (CAS RN *[34783–40–9]*, *Biul. Inst. Ochr. Rosl.*, 1971, No.50, p. 109); insecticidal and acaricidal properties against crop pests reported by R. T. Hewson (*Proc. Br. Insectic. Fungic. Conf., 8th*, 1975, **2**, 697) and veterinary pests by W. Bonin (*ibid.*, p. 705). Introduced by Hoechst AG. **Patent** DE 1643608; GB 1194603; US 3600474; US 3705240; US 3810919 to Ciba-Geigy **Manufacturers** AgrEvo

APPLICATIONS
Biochemistry Cholinesterase inhibitor. **Mode of action** Systemic insecticide with contact, stomach, and respiratory action. Readily penetrates the plant tissues, and is rapidly translocated to all parts of the plant, primarily acropetally. **Uses** An insecticide with quick initial action and short residual effect. It penetrates plant tissue and is rapidly translocated, controlling sucking insects (especially aphids) and certain Diptera, in field and glasshouse crops. Also effective against ectoparasites (fleas, lice, mites and ticks) of cattle, dogs, pigs, sheep and pets.
Phytotoxicity Non-phytotoxic when used as directed, except to some varieties of

apple, sweet cherry, and rose.   **Formulation types** EC.   **Mixtures** *(heptenophos +)* deltamethrin; permethrin.   **Selected tradenames** 'Hostaquick' (crop protection use) (AgrEvo); 'Ragadan' (veterinary use) (Hoechst)

ANALYSIS
Details available from AgrEvo.

MAMMALIAN TOXICOLOGY
EHC 63 (WHO, 1986; a general review of organophosphorus insecticides).
Oral Acute oral LD_{50} for rats 96–121, dogs 500–1000 mg/kg b.w. daily.
Skin and eye Acute percutaneous LD_{50} for rats >2000 mg/kg. Mild eye irritant.
Inhalation LC_{50} (4 h) for rats 0.95 mg/l air.   **NOEL** (2 y) for dogs 12, rats 15 mg/kg diet.   **ADI** 0.003 mg/kg b.w.   **Toxicity class** WHO (a.i.) Ib
EC risk T (R25)

ECOTOXICOLOGY
Birds Acute oral LD_{50} for Japanese quail 17–55 mg/kg (depending on carrier and sex) for pure a.i.   **Fish** LC_{50} (96 h) for trout 0.056, carp 24 mg/l.   **Bees** Toxic to bees.   **Worms** LC_{50} (14 d) for *Eisenia foetida* 98 mg/kg dry, artificial soil.
Other beneficial spp. Harmless to *Chrysopa* larvae.   **Daphnia** LC_{50} (48 h) 2.2 μg/l.
Algae EC_{50} (72 h) 20 mg/l.

ENVIRONMENTAL FATE
Animals In rats, following oral administration, 90% is excreted in the urine and 6% in the faeces in metabolised form within 6 days.   **Plants** Four days after application to lettuce, the compound is completely metabolised to polar, water-soluble compounds without accumulation of any intermediate.   **Soil/Environment** In soil, rapidly degraded by micro-organisms, DT_{50} (field) 1.4 d, DT_{90} (field) 4.6 d, depending on temperature and soil type; no accumulation. In lab (aerobic), DT_{50} <4 h, DT_{90} 40 h (20 °C). Degradation in water phase of water sediment system is very quick, DT_{50} 27–77 h.

396 *Heterorhabditis bacteriophora* and *H. megidis* *Biological agent*

nematode

NOMENCLATURE
Scientific name *Heterorhabditis bacteriophora*; *Heterorhabditis megidis*
Strain MicroBio Strain UK 211; HW79 (Dutch strain)

PROPERTIES
Form Infective juveniles.   **Stability** Stable for 1 to 3 months at 5 °C.

COMMERCIALISATION
Production Produced by liquid fermentation.   **History** First introduced in the UK in 1990 by AGC Ltd (now MicroBio Ltd).   **Manufacturers** Andermatt; Arbico; Biobest; Koppert; M&R Durango; MicroBio; Neudorff

APPLICATIONS
Mode of action Nematodes actively seek target pest in soil or compost. They penetrate either via body openings or directly through the cuticle. Upon entry, symbiotic bacteria (*Photorhabdus luminescens*) are released and infect the host, resulting in rapid death. Nematodes then reproduce in the cadaver.   **Uses** For control of vine weevil (*Otiorhynchus* sp.) in horticultural and fruit crops. Applied at $0.5 \times 10^6/m^2$ for protected crops and container-grown plants, $1.0 \times 10^6/m^2$ for outdoor, field-grown crops.   **Phytotoxicity** Non-phytotoxic and non-phytopathogenic.   **Formulation types** Mixed with inert carrier.
Compatibility Incompatible with nematicides such as aldicarb and organophosphate insecticides such as fonofos.
Selected tradenames 'Dickmaulrüsslernematoden' (*H. megidis*) (Andermatt); 'Larvanem' (*H. megidis*) (Koppert); 'Nemasys H' (*H. megidis*) (MicroBio, Biobest)

ANALYSIS
Biotest with *Otiorhynchus* sp., or with larvae of *Galleria mellonella* or *Tenebrio molitor*. Alternatively, numbers of nematodes can be counted using a microscope and counting chamber.

MAMMALIAN TOXICOLOGY
Non-toxic.

397 hexachlorobenzene   *Fungicide*

organochlorine

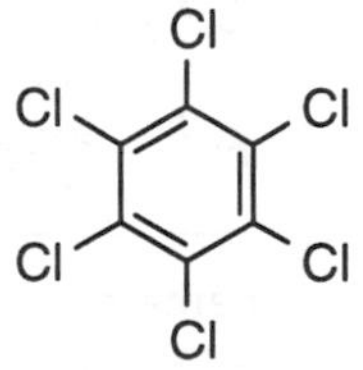

NOMENCLATURE
Common name hexachlorobenzene (BSI, E-ISO, F-ISO, accepted in lieu of a common name)

IUPAC name hexachlorobenzene
Chemical Abstracts name hexachlorobenzene
Other names HCB   **CAS RN** *[118–74–1]*   **EEC no.** 204–273–9

PHYSICAL CHEMISTRY
Mol. wt. 284.8   **M.f.** C_6Cl_6 **Form** Colourless crystals.   **M.p.** 226 °C; tech.
≥220 °C   **B.p.** 323–326 °C   **V.p.** 1.45 mPa (20 °C)   **S.g./density** 2.044 (23 °C)
Solubility Practically insoluble in water. Soluble in hot benzene, chloroform, carbon
disulfide, and diethyl ether. Sparingly soluble in carbon tetrachloride. Practically
insoluble in cold alcohols.   **Stability** Very stable, even to acids and alkalis.

COMMERCIALISATION
History Fungicidal seed treatment reported by H. Yersin *et al.* (*C. R. Seances Acad.
Agric. Fr.*, 1945, **31**, 24).

APPLICATIONS
Mode of action Selective fungicide. Acts by fumigant action on fungal spores.
Uses Seed treatment for control of common bunt and dwarf bunt of wheat.
Formulation types DS; WS.

ANALYSIS
Residues determined by glc (*AOAC Methods*, 1995, 977.19, 980.22; *Pestic. Anal.
Man.*, 1979, **I**, 201-A, 201-G; *Analyst (London)*, 1979, **104**, 425; A. Ambrus *et al.*,
J. Assoc. Off. Anal. Chem., 1981, **64**, 773; *Man. Pestic. Residue Anal.*, 1987, **I**, 1, 5,
6, S9, S10, S12, S19; *Anal. Methods Residues Pestic.*, 1988, Part I, M1, M12). In
drinking water by gc with ECD (*AOAC Methods*, 1995, 990.06).

MAMMALIAN TOXICOLOGY
Reviews *Pesticide residues in food – 1978*. FAO Plant Production and Protection
Paper 15, 1979. *Pesticide residues in food: 1978 evaluations*. FAO Plant Production
and Protection Paper 15 Sup., 1979. See also *WHO Technical Report Series* No. 555
(1974), P. Cooper (*Food Cosmet. Toxicol.*, 1976, **14**, 351–353) and W. Hayes & E.
Hawes (*Handbook of Pesticide Toxicology*, 1991, **3**, 1422–1433).   **IARC** 20
Oral Acute oral LD_{50} for rats 10 000 mg/kg; guinea pigs tolerated doses
>3000 mg/kg.   **Skin and eye** Acute percutaneous LD_{50} not available. May cause
slight skin irritation.   **ADI** (JMPR) Conditional ADI withdrawn [1978].
Other There is evidence that hexachlorobenzene is carcinogenic in mice and
hamsters, and may represent a carcinogenic risk to man. Non-mutagenic in *in vitro*
and *in vivo* studies, but is foetotoxic and produces some teratogenic effects. Has
caused a serious outbreak of porphyria in humans.   **Toxicity class** WHO (a.i.) Ia;
EPA (formulation) IV   **EC risk** (also R45); T (R48/25)   **PIC** Yes

ECOTOXICOLOGY
Fish LC_{50} (96 h) for five freshwater species 0.05–0.2 mg/l.   **Bees** Not toxic to
bees.

hexachlorobenzene 673

MAIN ENTRIES

ENVIRONMENTAL FATE

Animals The principal metabolites in mammals are pentachlorophenol, tetrachlorohydroquinone and pentachlorothiophenol, with lesser amounts of tetrachlorobenzene, pentachlorobenzene, 2,4,5- and 2,4,6-trichlorophenols and 2,3,4,6- and 2,3,5,6-tetrachlorophenols.

398 hexaconazole *Fungicide*

azole

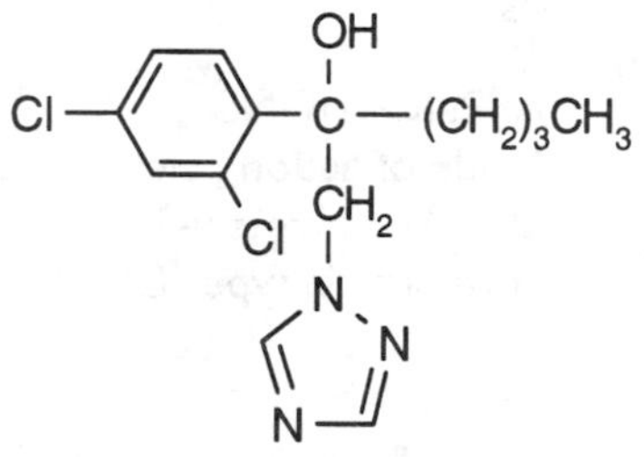

NOMENCLATURE

Common name hexaconazole (BSI, ANSI, draft E-ISO, (*m*) draft F-ISO)
IUPAC name (*RS*)-2-(2,4-dichlorophenyl)-1-(1*H*-1,2,4-triazol-1-yl)hexan-2-ol
Chemical Abstracts name (±)-α-butyl-α-(2,4-dichlorophenyl)-1*H*-1,2,4-triazole-=1-ethanol
CAS RN *[79983–71–4]*   **Development codes** PP523; ICIA0523

PHYSICAL CHEMISTRY

Composition Tech. is >85% pure.   **Mol. wt.** 314.2   **M.f.** $C_{14}H_{17}Cl_2N_3O$
Form White crystalline solid.   **M.p.** 110–112 °C   **V.p.** 0.018 mPa (20 °C)
K_{ow} logP = 3.9 (20 °C)   **S.g./density** 1.29 g/cm^3 (25 °C)   **Solubility** In water 0.017 g/l (20 °C). In dichloromethane 336, methanol 246, acetone 164, ethyl acetate 120, toluene 59, hexane 0.8 (all in g/l, 20 °C).   **Stability** Stable for at least 6 years at ambient temperatures. Stable to hydrolysis and photolysis in water. Formulations stable in sales containers for 6 months at 50 °C and 2 years at ambient temperature.

COMMERCIALISATION

History Fungicide introduced by ICI Agrochemicals (now Zeneca Agrochemicals).
Manufacturers Rallis; Zeneca

APPLICATIONS

Biochemistry Inhibits ergosterol biosynthesis (steroid demethylation inhibitor).

Mode of action Systemic fungicide with protective and curative action.
Uses Control of many fungi, particularly Ascomycetes and Basidiomycetes, e.g. *Podosphaera leucotricha* and *Venturia inaequalis* on apples, *Guignardia bidwellii* and *Uncinula necator* on vines, *Hemileia vastatrix* on coffee, and *Cercospora* spp. on peanuts. Also used on bananas, cucurbits, peppers and other crops.
Phytotoxicity Non-phytotoxic when used as directed. Some injury noted on McIntosh apples. **Formulation types** SC; SG; OL. **Mixtures** *(hexaconazole +)* captan; sulfur; carbendazim; mancozeb. **Selected tradenames** 'Anvil' (fruit) (Zeneca); 'Planete' (cereals) (Zeneca); 'Contaf' (Rallis)

ANALYSIS
Product by gc and FID. **Residues** by gc with either NPD, thermionic or ECD detection. See K. J. Harradine *et al.*, in *Comp. Anal. Profiles*, Chapt. 2.

MAMMALIAN TOXICOLOGY
Reviews *Pesticide residues in food – 1990*. FAO Plant Production and Protection Paper, 102, 1990. *Pesticide residues in food – 1990 evaluations. Toxicology*. World Health Organisation, WHO/PCS/91.47, 1991. **Oral** Acute oral LD_{50} for male rats 2189, female rats 6071 mg/kg. **Skin and eye** Acute percutaneous LD_{50} for rats >2000 mg/kg. Mild irritant to eyes; non-irritating to skin (rabbits). Moderate skin sensitiser (guinea pigs). **Inhalation** LC_{50} (4 h) for rats >5.9 mg/l. **NOEL** for rats 2.5, rabbits 50 mg/kg daily. **ADI** (JMPR) 0.005 mg/kg b.w. [1990].
Other Non-mutagenic. **Toxicity class** WHO (a.i.) III (Table 5); EPA (formulation) IV

ECOTOXICOLOGY
Birds Acute oral LD_{50} for mallard ducks >4000 mg/kg. **Fish** LC_{50} (96 h) for rainbow trout 3.4, mirror carp 5.94 mg/l. **Bees** Acute oral and contact LD_{50} for honeybees >0.1 mg/bee. **Daphnia** LC_{50} (48 h) 2.9 mg/l.

ENVIRONMENTAL FATE
Animals Readily excreted by mammals, with no significant retention in organs or tissues. **Plants** For details of the metabolism of hexaconazole in cereals, see M. W. Skidmore *et al.*, *Br. Crop Prot. Conf. – Pests Dis.*, 1990, **3**, 1035–1040.
Soil/Environment Rapidly degraded in soils in laboratory tests.

hexaflumuron *Insecticide*

benzoylurea

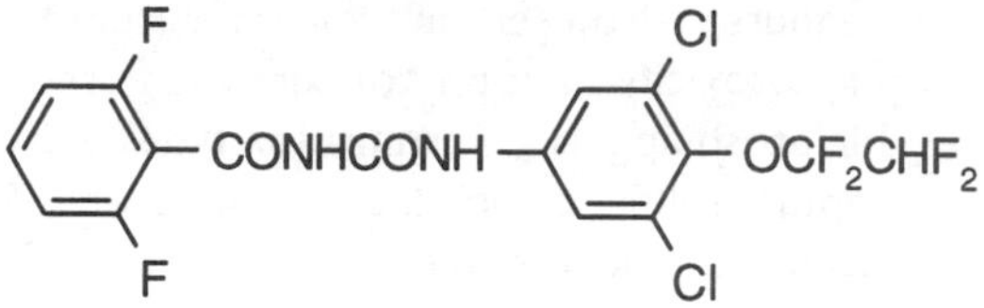

NOMENCLATURE
Common name hexaflumuron (BSI, ANSI, draft E-ISO)
IUPAC name 1-[3,5-dichloro-4-(1,1,2,2-tetrafluoroethoxy)phenyl]-3-=
(2,6-difluorobenzoyl)urea
Chemical Abstracts name N-[[[3,5-dichloro-4-(1,1,2,2-tetrafluoroethoxy)=
phenyl]amino]carbonyl]-2,6-difluorobenzamide
CAS RN *[86479–06–3]* **Development codes** XRD-473; DE-473

PHYSICAL CHEMISTRY
Mol. wt. 461.1 **M.f.** $C_{16}H_8Cl_2F_6N_2O_3$ **Form** Colourless solid. **M.p.** 202–205 °C
V.p. 0.059 mPa (25 °C) **K$_{ow}$** logP = 5.68 **Solubility** In water 0.027 mg/l (18 °C).
In methanol 11.3, xylene 5.2 (both in g/l, 20 °C). **Stability** Undergoes 60%
hydrolysis in 35 d (pH 9). Photolysis DT_{50} 6.3 d (pH 5.0, 25 °C).

COMMERCIALISATION
History Insecticide reported by R. J. Sbragia et al. (*Proc. Int. Congr. Plant Prot.*,
10th, 1983, **1**, 417). Introduced in Latin America in 1987 by Dow (now
DowElanco Ltd). **Manufacturers** DowElanco

APPLICATIONS
Biochemistry Chitin synthesis inhibitor. **Mode of action** Ingested, systemic
insecticide. **Uses** Control of larvae of Lepidoptera, Coleoptera, Homoptera, and
Diptera on top fruit, cotton, and potatoes. Also for control of termites.
Formulation types EC; SC. **Mixtures** (*hexaflumuron +*) chlorpyrifos.
Selected tradenames 'Consult' (DowElanco); 'Consol' (DowElanco)

ANALYSIS
Product and **residue** analysis by glc or hplc. Details available from DowElanco Ltd.

MAMMALIAN TOXICOLOGY
Oral Acute oral LD_{50} for rats >5000 mg/kg. **Skin and eye** Acute percutaneous
LD_{50} for rats >5000 mg/kg. Non-irritating to skin and eyes (rabbits); no skin
sensitisation (guinea pigs). **Inhalation** LC_{50} (4 h) >2.5 mg/l. **NOEL** (2 y) for rats
75 mg/kg daily; (1 y) for dogs 0.5 mg/kg daily. **Toxicity class** WHO (a.i.) III
(Table 5)

ECOTOXICOLOGY
Birds Acute oral LD$_{50}$ for mallard ducks >2000 mg/kg. **Fish** LC$_{50}$ (96 h) for rainbow trout >maximum solubility. **Bees** LD$_{50}$ (oral and contact) >0.1 mg/bee. **Daphnia** LC$_{50}$ (96 h) 0.0001 mg/l. Significant hazard to *Daphnia* spp. only, in field conditions.

ENVIRONMENTAL FATE
Soil/Environment Soil K$_d$ 147–1326, K$_{om}$ 3096–41 170; strongly adsorbed on a wide range of soils.

400 hexazinone   *Herbicide*

1,2,4-triazinone

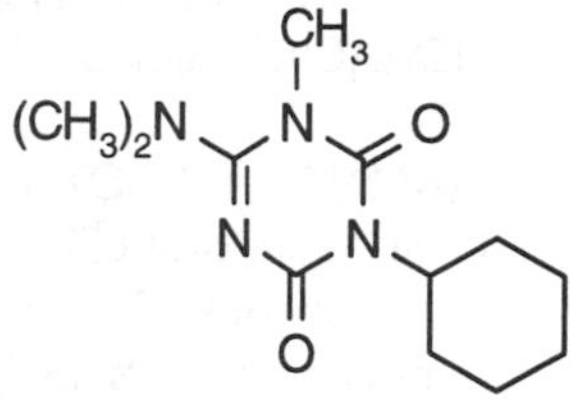

NOMENCLATURE
Common name hexazinone (BSI, E-ISO, (*m*) F-ISO, ANSI, WSSA)
IUPAC name 3-cyclohexyl-6-dimethylamino-1-methyl-1,3,5-triazine-= 2,4(1*H*,3*H*)-dione
Chemical Abstracts name 3-cyclohexyl-6-(dimethylamino)-1-methyl-= 1,3,5-triazine-2,4(1*H*,3*H*)-dione
CAS RN *[51235–04–2]* **Development codes** DPX A3674

PHYSICAL CHEMISTRY
Mol. wt. 252.3 **M.f.** C$_{12}$H$_{20}$N$_4$O$_2$ **Form** Colourless, odourless crystals.
M.p. 115–117 °C **B.p.** Decomposes on distillation. **V.p.** 0.03 mPa (extrapolated) (25 °C); 8.5 mPa (86 °C) **K$_{ow}$** logP = 1.2 (pH 7) **S.g./density** 1.25 **Solubility** In water 33 g/kg (25 °C). In chloroform 3880, methanol 2650, benzene 940, dimethylformamide 836, acetone 792, toluene 386, hexane 3 (all in g/kg, 25 °C).
Stability Stable in aqueous media between pH 5 and pH 9 and below 37 °C. Decomposed by strong acids and strong alkalis. Stable to light.

COMMERCIALISATION
History Herbicide reported by T. J. Hernandez *et al.* (*Proc. North Cent. Weed Control Conf.*, 1974, p. 138). Introduced by E. I. du Pont de Nemours Co.
Patent US 3902887 **Manufacturers** Du Pont

APPLICATIONS

Biochemistry Inhibits photosynthesis. **Mode of action** Non-selective, primarily contact herbicide, absorbed by the leaves and roots, with translocation acropetally. **Uses** A post-emergence contact herbicide effective against many annual and biennial weeds and, except for *Sorghum halepense*, most perennial weeds at 6–12 kg a.i./ha. Used for selective control in alfalfa, pineapple, sugar cane and in plantations of certain coniferous species; also on non-crop areas, but not on sites adjacent to deciduous trees or other desirable plants. **Phytotoxicity** Should not be used for weed control in larch (*Larix* spp.) or in areas planted with deciduous species. **Formulation types** SL; SP; GR. **Mixtures** *(hexazinone +)* diuron; bromacil + diuron. **Selected tradenames** 'Velpar' (Du Pont)

ANALYSIS

Product analysis by hplc or by glc with FID (C. L. McIntosh *et al.*, *Anal. Methods Pestic. Plant Growth Regul.*, 1984, **13**, 267). **Residues** determined by glc (*idem, ibid.*; R. F. Holt, *J. Agric. Food Chem.*, 1981, **29**, 165). In **drinking water**, by gc with FID (*AOAC Methods*, 1995, 991.07).

MAMMALIAN TOXICOLOGY

Oral Acute oral LD_{50} for rats 1690, guinea pigs 860 mg/kg. **Skin and eye** Acute percutaneous LD_{50} for rabbits >5278 mg/kg. Reversible irritant to eyes (rabbits); non-irritating to skin (guinea pigs). **Inhalation** LC_{50} (1 h) for rats >7.48 mg/l. **NOEL** (2 y) for rats 200, mice 200 ppm; (1 y) for dogs 200 ppm. **Toxicity class** WHO (a.i.) III; EPA (formulation) II

ECOTOXICOLOGY

Birds Acute oral LD_{50} for bobwhite quail 2258 mg/kg. Eight-day dietary LC_{50} for bobwhite quail and mallard ducklings >10 000 mg/kg diet. **Fish** LC_{50} (96 h) for rainbow trout 320–420, fathead minnow 274, bluegill sunfish 370–420 mg/l. **Bees** Not toxic to bees; LD_{50} >60 μg/bee. **Daphnia** LC_{50} (48 h) 442 mg/l.

ENVIRONMENTAL FATE

Animals Major urinary metabolites in rats are 3-(4-hydroxycyclohexyl)-6-= (dimethylamino)-1-methyl-1,3,5-triazine-2,4-(1*H*,3*H*)-dione, 3-cyclohexyl-6-= (methylamino)-1-methyl-1,3,5-triazine-2,4-(1*H*,3*H*)-dione, and 3-(4-hydroxy= cyclohexyl)-6-(methylamino)-1-methyl-1,3,5-triazine-2,4-(1*H*,3*H*)-dione. **Soil/Environment** Microbial degradation occurs in soil and natural waters. The triazine ring is broken, with the liberation of CO_2. DT_{50} in soil c. 1–6 mo, depending on climate and soil type.

hexythiazox

Acaricide

NOMENCLATURE
Common name hexythiazox (BSI, draft E-ISO, (*m*) draft F-ISO)
IUPAC name (4*RS*,5*RS*)-5-(4-chlorophenyl)-*N*-cyclohexyl-4-methyl-2-oxo-=
1,3-thiazolidine-3-carboxamide
Chemical Abstracts name *trans*-5-(4-chlorophenyl)-*N*-cyclohexyl-4-methyl-2-oxo-=
3-thiazolidinecarboxamide
CAS RN *[78587–05–0]*   **Development codes** NA-73

PHYSICAL CHEMISTRY
Mol. wt. 352.9   **M.f.** $C_{17}H_{21}CIN_2O_2S$ **Form** Colourless crystals.
M.p. 108.0–108.5 °C   **V.p.** 0.0034 mPa (20 °C) K_{ow} logP = 2.53   **Solubility** In
water 0.5 mg/l (20 °C). In chloroform 1379, xylene 362, methanol 206, acetone
160, acetonitrile 28.6, hexane 4 (all in g/l, 20 °C).   **Stability** Stable to light, air and
heat, and in acidic and alkaline media. Stable to 300 °C. Aqueous solution in
sunlight DT_{50} 16.7 d.

COMMERCIALISATION
History Acaricide reported by T. Yamada (*Proc. Int. Congr. Pestic. Chem., Kyoto, 5th,*
1982), see also *Jpn. Pestic. Inf.,* 1984, No. 44, p. 21. Introduced by Nippon Soda
Co., Ltd.   **Patent** GB 2059961; US 4442116   **Manufacturers** Nippon Soda

APPLICATIONS
Mode of action Non-systemic acaricide with contact and stomach action. Good
translaminar activity. Has ovicidal, larvicidal, and nymphicidal activity. Not active
against adults, but eggs laid by treated females are non-viable.   **Uses** Control of
eggs and larvae of many phytophagous mites (particularly *Panonychus, Tetranychus,*
and *Eotetranychus* spp.) on fruit, vines, vegetables, and cotton.
Formulation types WP; EC; FU.   **Mixtures** (*hexythiazox +*) propargite.
Selected tradenames 'Nissorun' (Nippon Soda); 'Cesar' (AgrEvo); 'Hexygon'
(Gowan); 'Matacar' (Sipcam); 'Ordoval' (BASF); 'Savey' (Gowan); 'Zeldox'
(Zeneca)

ANALYSIS
Product and **residue** analysis by hplc (M. Tokieda *et al.*, *J. Pestic. Sci.*, 1987, **12,** 711).

MAMMALIAN TOXICOLOGY
Reviews *Pesticide residues in food – 1991*. FAO Plant Production and Protection Paper, 111, 1991. *Pesticide residues in food – 1991 evaluations. Part II – Toxicology.* World Health Organisation, WHO/PCS/92.52, 1992. **Oral** Acute oral LD_{50} for rats and mice >5000 mg/kg. **Skin and eye** Acute percutaneous LD_{50} for rats >5000 mg/kg. Mild eye irritant; non-irritating to skin (rabbits). Non-sensitising to skin (guinea pigs). **Inhalation** LC_{50} (4 h) for rats >2 mg/l air. **NOEL** (2 y) for rats 23.1 mg/kg b.w.; (1 y) for dogs 2.87 mg/kg b.w.; (90 d) for rats 70 mg/kg diet. **ADI** (JMPR) 0.03 mg/kg [1991]. **Other** Non-teratogenic. Not mutagenic in the Ames assay. **Toxicity class** WHO (a.i.) III (Table 5); EPA (formulation) IV

ECOTOXICOLOGY
Birds Acute oral LD_{50} for mallard ducks >2510, Japanese quail >5000 mg/kg. Eight-day dietary LC_{50} for mallard ducks and bobwhite quail >5620 mg/kg diet. **Fish** LC_{50} (96 h) for rainbow trout >300, bluegill sunfish 11.6 mg/l; (48 h) for carp 3.7 mg/l. **Bees** Not toxic to bees. LD_{50} by topical application >200 µg/bee. **Daphnia** LC_{50} (48 h) 1.2 mg/l.

ENVIRONMENTAL FATE
Animals The main metabolite in urine and faeces is 5-(4-chlorophenyl)-N-= (*cis*-4-hydroxycyclohexyl)-4-methyl-*trans*-2-oxothiazolidine-3-carboxamide. **Soil/Environment** DT_{50} in clay loam at 15 °C, 8 d. In soil, undergoes oxidation to the corresponding hydroxy and carbonyl compounds. K_{oc} 6200.

402 *Hippodamia convergens*   *Biological agent*

Coleoptera

Ladybird, consumer of aphids.

NOMENCLATURE
Scientific name *Hippodamia convergens* **Other names** ladybird; ladybug

PROPERTIES
Stability Shelf life 2–3 days at 8–10 °C in a dark place.

COMMERCIALISATION
Manufacturers Arbico; Biobest; Koppert; Kunafin; M&R Durango; Nature's
Alternative; Rincon-Vitova

APPLICATIONS
Mode of action The ladybirds and their larvae eat all stages of aphids.   **Uses** For
control of aphids in protected crops.   **Phytotoxicity** Non-phytotoxic and non-
phytopathogenic.   **Formulation types** Shredded wood.
Selected tradenames 'Aphidamia' (Koppert)

MAMMALIAN TOXICOLOGY
Non-toxic.

403 hydramethylnon
Insecticide

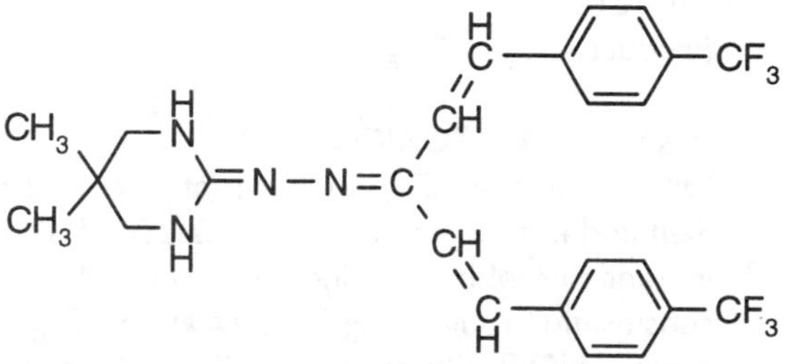

NOMENCLATURE
Common name hydramethylnon (BSI, ANSI, draft E-ISO); hydraméthylnone
((*f*) draft F-ISO)
IUPAC name 5,5-dimethylperhydropyrimidin-2-one 4-trifluoromethyl-=
α-(4-trifluoromethylstyryl)cinnamylidenehydrazone
Chemical Abstracts name tetrahydro-5,5-dimethyl-2(1*H*)-pyrimidinone
[3-[4-(trifluoromethyl)phenyl]-1-[2-[4-(trifluoromethyl)phenyl]ethenyl]-=
2-propenylidene]hydrazone
CAS RN *[67485–29–4]*   **Development codes** AC 217 300; CL 217 300

PHYSICAL CHEMISTRY
Composition Tech. grade is ≥95% pure.   **Mol. wt.** 494.5   **M.f.** $C_{25}H_{24}F_6N_4$
Form Yellow to orange crystals.   **M.p.** 185–190 °C   **V.p.** 0.0027 mPa (25 °C); a
value of 0.0008 mPa (45 °C) has also been reported. K_{ow} logP = 2.31
Solubility In water 0.005–0.007 mg/l (25 °C). In acetone 360, ethanol 72,

1,2-dichloroethane 170, methanol 230, isopropanol 12, xylene 94, chlorobenzene 390 (all in g/l, 20 °C). **Stability** When stored in original unopened container, stable in excess of 24 months at 25 °C, 12 months at 37 °C, and 3 months at 45 °C. Undergoes photolysis in sunlight (DT_{50} c. 1 h). DT_{50} in aqueous suspension (25 °C) 24–33 d (pH 4.9), 10–11 d (pH 7.03), 11–12 d (pH 8.87).

COMMERCIALISATION
History Insecticide reported by J. B. Lovell (*Proc. Br. Crop Prot. Conf., Pests – Dis.,* 1979, **2**, 575). Introduced by American Cyanamid Co. **Manufacturers** Cyanamid

APPLICATIONS
Mode of action Non-systemic insecticide with stomach action. **Uses** Selective control of agricultural and household Formicidae (especially *Camponotus, Iridomyrmex, Monomorium, Solenopsis* and *Pogonomyrmex* spp. and *Pheidole megacephala*) using baits, and Blattellidae (especially *Blatta, Blattella, Periplaneta* and *Supella* spp.). Due to slow action, can be carried into nest by worker ants and kill the queen. **Phytotoxicity** No phytotoxicity observed. **Formulation types** PA; RB. **Selected tradenames** 'Amdro' (Cyanamid); 'Combat' (household use) (Cyanamid); 'Maxforce' (professional use) (Cyanamid)

ANALYSIS
Product analysis by glc.

MAMMALIAN TOXICOLOGY
Oral Acute oral LD_{50} for male rats 1131, female rats 1300 mg/kg.
Skin and eye Acute percutaneous LD_{50} for rabbits >5000 mg/kg; not a skin irritant to rabbits or guinea pigs, reversible irritant to eyes of rabbits. No skin sensitisation (guinea pigs). **Inhalation** LC_{50} (4 h) for rats >5 mg/l air (aerosol or dust). **NOEL** (28 d) for rats 75 mg/kg diet; (90 d) for rats 50 mg/kg diet; (2 y) for rats 50 mg/kg diet; (18 mo) for mice 25 mg/kg diet; (90 d) for beagles 3.0 mg/kg daily; (6 mo) for beagles 3.0 mg/kg daily. **Other** Non-teratogenic and non-embryotoxic in rats and rabbits. Non-mutagenic.
Toxicity class WHO (a.i.) III; EPA (formulation) III

ECOTOXICOLOGY
Birds Acute oral LD_{50} for mallard ducks >2510, bobwhite quail 1828 mg/kg.
Fish Toxic under laboratory conditions using a solvent, but very little risk to fish is expected under normal field conditions because of the compound's low solubility in water and rapid degradation in sunshine. LC_{50} (96 h) for bluegill sunfish 1.70, rainbow trout 0.16, channel catfish 0.10 mg/l; for carp 0.67, 0.39, and 0.34 mg/l (24, 48, and 72 h, respectively). **Bees** Dust is non-toxic topically to honeybees at 0.03 mg/bee. **Daphnia** LC_{50} (48 h) 1.14 mg/l; little hazard would be expected under field conditions because of the very low solubility in water.

ENVIRONMENTAL FATE

Animals In rats, following oral administration, rapidly eliminated in the faeces and urine. No residues were detectable in the milk or tissues of goats (0.2 mg/kg in the daily diet for 8 days). No residues were found in the milk or tissues of cows (0.05 mg/kg for 21 consecutive days).   **Plants** Residues in grass 4 months after treatment were <0.01 ppm. Negligible residues were found in radishes, barley, and french beans planted 3 months after treatment of the soil.

Soil/Environment Rapidly degraded in sunlight by photolysis (DT_{30} c. 1 h). Half life in sandy loam is c. 7 d; half-life when incorporated in sandy loam is c. 28 d. Formulated bait decomposes rapidly in daylight. Not mobile and does not leach. Low bioaccumulation potential.

404 2-hydrazinoethanol *Plant growth regulator*

$$HOCH_2CH_2NHNH_2$$

NOMENCLATURE
IUPAC name 2-hydrazinoethanol; 2-hydroxyethylhydrazine
Chemical Abstracts name 2-hydrazinoethanol
Other names BOH   **CAS RN** *[109–84–2]*

PHYSICAL CHEMISTRY
Mol. wt. 76.1   **M.f.** $C_2H_8N_2O$ **Form** Colourless liquid.   **M.p.** −70 °C
B.p. 110–130 °C/17.5 mmHg   **Solubility** Miscible with water and lower alcohols (room temperature).   **Stability** Stable in the cool; in the dark; dilute solutions easily oxidised.

COMMERCIALISATION
History Introduced as a plant growth regulator by Olin Corp (who no longer manufacture or market it).

APPLICATIONS
Uses Used to promote flowering of Bromeliaceae.   **Formulation types** SL.

405 hydrogen cyanide *Insecticide, rodenticide*

HCN

NOMENCLATURE

hydrogen cyanide

Common name hydrogen cyanide (E-ISO, accepted in lieu of a common name); acide cyanhydrique (F-ISO, accepted in lieu of a common name); hydrocyanic acid (ESA, accepted in lieu of a common name)
IUPAC name hydrogen cyanide
Chemical Abstracts name hydrocyanic acid
Other names prussic acid **CAS RN** *[74–90–8]* **EEC no.** 200–821–6

calcium cyanide

Common name calcium cyanide (E-ISO, ESA); cyanure de calcium (F-ISO)
CAS RN *[592–01–8]* **EEC no.** 209–740–0

sodium cyanide

Common name sodium cyanide (E-ISO); cyanure de sodium (F-ISO); hydrogen cyanide (JMAF)
CAS RN *[143–33–9]*

PHYSICAL CHEMISTRY

hydrogen cyanide

Mol. wt. 27.0 **M.f.** CHN **Form** Colourless, flammable liquid, with an odour of almonds. **M.p.** –15 °C **B.p.** 26.5 °C **V.p.** 961 kPa (25 °C) **S.g./density** 0.699 (20 °C) **Solubility** Completely miscible with water. Soluble in ethanol and diethyl ether. **Stability** Stable, in the pure form. In the liquid form and in aqueous solution, stabilised by the addition of small quantities of mineral acids. **pKa** Weak acid

calcium cyanide

Mol. wt. 92.1 **M.f.** C_2CaN_2

sodium cyanide

Mol. wt. 49.0 **M.f.** CNNa **M.p.** 564 °C **Solubility** Very soluble in water; slightly soluble in ethanol. **Stability** Readily hydrolysed, liberating hydrogen cyanide.

COMMERCIALISATION

History Hydrogen cyanide used as insecticidal fumigant in 1886 by D. W. Coquillett (cited by L. O. Howard, *U.S. Dep. Agric. Yearbook*, 1899, p. 150). Introduced as the liquid acid and as salts from which the acid is liberated *in situ* by the action of moisture. Calcium cyanide was introduced by American Cyanamid

Co. (who no longer manufacture or market it) and by Degesch AG; sodium cyanide introduced by ICI Plant Protection Division (now Zeneca Agrochemicals).

sodium cyanide
Manufacturers Zeneca

APPLICATIONS
Biochemistry Acts by conjugation with ferricytochrome oxidase, thus preventing oxygen transfer from haemoglobin to tissue cells. **Mode of action** Insecticide and rodenticide with respiratory action. **Uses** Hydrogen cyanide is used as an insecticide and rodenticide for fumigating enclosed spaces (stored grain in warehouses, etc.) and, in some countries, glasshouses. Calcium cyanide and sodium cyanide are used to fumigate rabbit burrows, rat runs, etc., by the addition of acid, which releases hydrogen cyanide. **Phytotoxicity** Can have phytotoxic effects, especially on damp foliage. **Formulation types** GA; GE.

hydrogen cyanide
Selected tradenames 'Cyanosil' (Detia Degesch)

sodium cyanide
Selected tradenames 'Cymag' (Zeneca)

ANALYSIS
Product analysis by titration with silver nitrate (*AOAC Methods,* 1984, 6.113, 6.118; 1965, 4.088–4.089, 4.093–4.094). **Residues** determined by recommended methods (*ibid.,* 1984, 26.150–26.151, 19.106; *Pestic. Anal. Man.,* 1979, **II**; *Anal. Methods Residues Pestic.,* 1988, Part II; K. A. Scudamore, *Anal. Methods Pestic. Plant Growth Regul.,* 1988, **16**, 207, 239). Methods reviewed by J. L. Daft in *Comp. Analyt. Profiles,* Chapter 11.

MAMMALIAN TOXICOLOGY
hydrogen cyanide
Reviews *Evaluation of the toxicity of pesticide residues in food.* FAO Meeting Report, No. PL/1965/10; WHO/Food Add./26.65, 1965. *Evaluation of the hazards of consumers resulting from the use of fumigants in the protection of food.* PL/1965/10/2; WHO/Food Add./28.65, 1965. **Oral** Acute oral LD_{50} for rats 10–15 mg/kg. **Skin and eye** Acute percutaneous LD_{50} not available. Rapidly absorbed by skin. Eye irritant. **Inhalation** Exposure for 30 min at 0.36 mg/l air is fatal to man. **ADI** (JMPR) 0.05 mg/kg b.w. [1965]. **Other** Hydrogen cyanide and its salts are very toxic to mammals.
Toxicity class EPA (formulation) I **EC risk** F+ (R12); T+ (R26): (for preparations, concn. ≥7%, T+ (R26/27/28); for 1%≤ concn. <7%, T (R23/24/25); for 0.1%≤ concn. <1%, Xn (R20/21/22)

calcium cyanide
Toxicity class WHO (a.i.) Ia **EC risk** T+ (R28); (R32)

sodium cyanide
Oral Acute oral LD$_{50}$ for rats 6.44 mg/kg. **Skin and eye** Acute percutaneous
LD$_{50}$ for guinea pigs 4.5 mg/kg. **Toxicity class** WHO (a.i.) Ib
EC risk T+ (R26/27/28); (R32)

ECOTOXICOLOGY
hydrogen cyanide
Birds Toxic to birds. **Fish** Toxic to fish. **Bees** Toxic to bees, but not hazardous if
used as directed.

ENVIRONMENTAL FATE
Animals The cyanide ion is metabolised mainly to thiocyanate.

406 hydroprene *Insecticide*

juvenile hormone mimic

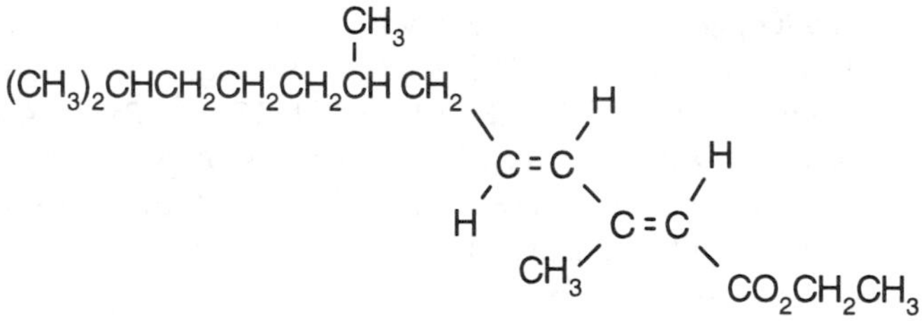

$$(CH_3)_2CHCH_2CH_2CH_2\underset{\underset{\displaystyle CH_3}{|}}{CH}CH_2$$

NOMENCLATURE
Common name hydroprene (BSI, ANSI, ESA, draft E-ISO)
IUPAC name ethyl (E,E)-(RS)-3,7,11-trimethyldodeca-2,4-dienoate
Chemical Abstracts name (E,E)-(±)-ethyl 3,7,11-trimethyl-2,4-dodecadienoate
CAS RN [41205–09–8] (E,E)-(±)- form; [65733–18–8] (E,E)-(S)- isomer;
[65733–19–9] (E,E)-(R)- isomer; [41096–46–2] (E,E)- isomers
Development codes ZR 512 (Sandoz) **Official codes** OMS 1696

PHYSICAL CHEMISTRY
Mol. wt. 266.4 **M.f.** C$_{17}$H$_{30}$O$_2$ **Form** Tech. grade is an amber liquid.
B.p. 174 °C/19 mmHg; 138–140 °C/1.25 mmHg **V.p.** 25–40 mPa (25 °C)
K$_{ow}$ logP = 3.06 **S.g./density** 0.892 (25 °C) **Solubility** In water 2 mg/l (pH 7,
20 °C). Soluble in common organic solvents. **Stability** Stable for more than
3 years under normal storage conditions.

COMMERCIALISATION
History Insect growth regulator reported by C. A. Henrick et al. (*J. Agric. Food Chem.*, 1973, **21**, 354). Introduced by Zoecon Corp. (acquired by Sandoz AG, now Novartis Crop Protection AG).   **Patent** US 4021461   **Manufacturers** Novartis

APPLICATIONS
Mode of action Insect growth regulator (juvenile hormone mimic), preventing metamorphosis to viable adults when applied to larval stages. The (E,E) (S) isomer is more effective against insects than is the (E,E)-(R)- isomer (C. A. Henrick et al., *J. Agric. Food Chem.*, 1978, **26**, 542).   **Uses** Control of Coleoptera, Homoptera, and Lepidoptera. Also control of cockroaches in residential buildings and non-food areas of industrial and commercial premises.   **Formulation types** EC; Aerosol; KN.   **Mixtures** *(hydroprene +)* permethrin.   **Selected tradenames** 'Gencor' (Novartis)

ANALYSIS
Product analysis by gc-ms.   **Residues** determined by glc with FID.

MAMMALIAN TOXICOLOGY
Oral Acute oral LD_{50} for rats >5000, dogs >10 000 mg/kg.   **Skin and eye** Acute percutaneous LD_{50} for rats >5000, rabbits >5100 mg/kg. Non-irritating to skin; mild eye irritant (rabbits).   **Inhalation** LC_{50} (4 h) for rats >5.5 mg/l air.
NOEL (90 d) for rats 50 mg/kg b.w. daily.   **Other** Non-mutagenic, non-teratogenic.   **Toxicity class** WHO (a.i.) III (Table 5); EPA (formulation) IV

ECOTOXICOLOGY
Fish LC_{50} (96 h) for bluegill sunfish and trout >100 mg/l.   **Bees** LD_{50} for adult bees (oral and contact) >1000 µg/bee. Bee larvae affected at 0.1 µg/bee.

ENVIRONMENTAL FATE
Plants In plants, degradation principally involves ester hydrolysis, O-demethylation, and oxidative splitting of the double bond.   **Soil/Environment** In soil, rapidly decomposed, DT_{50} is only a few days.

407 8-hydroxyquinoline sulfate

Fungicide, bactericide

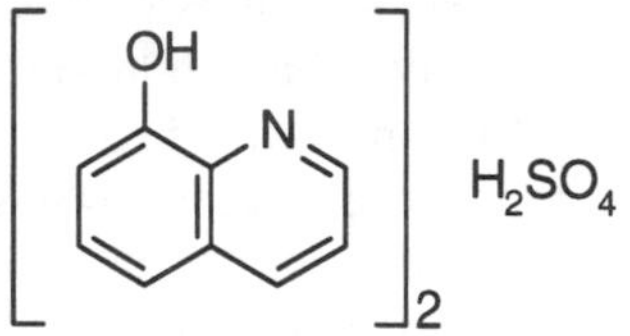

NOMENCLATURE
8-hydroxyquinoline sulfate
Common name 8-hydroxyquinoline sulfate (E-ISO, accepted in lieu of a common name); sulfate d'hydroxy-8 quinoléine (F-ISO)
IUPAC name bis(8-hydroxyquinolinium) sulfate
Chemical Abstracts name 8-quinolinol sulfate (2:1) (salt)
CAS RN *[134–31–6]* for 8-quinolinol sulfate; *[12557–04–9]* for anhydrous mixture **EEC no.** 205–137–1

potassium hydroxyquinoline sulfate
Common name potassium hydroxyquinoline sulfate (British Pharmacopoeia Commission)

PHYSICAL CHEMISTRY
8-hydroxyquinoline sulfate
Mol. wt. 388.4 **M.f.** $C_{18}H_{16}N_2O_6S$ **Form** Yellow crystalline powder.
M.p. 175–178 °C (free base 76 °C) **B.p.** 267 °C (free base) **V.p.** Practically zero.
Solubility In water (20 °C) 300 g/l. Slightly soluble in glycerine. Sparingly soluble in alcohols. Practically insoluble in diethyl ether. Free base: very slightly soluble in water. Readily soluble in warm alcohols, acetone, chloroform, and benzene.
Stability Salt and base are very stable. Forms sparingly-soluble salts with numerous metal ions.

potassium hydroxyquinoline sulfate
Composition The name potassium hydroxyquinoline sulfate is accepted by the British Pharmacopoeia Commission for an equimolecular mixture of bis(8-hydroxyquinolinium) sulfate monohydrate and potassium sulfate.
Form Yellowish-white solid. **M.p.** 172–184 °C **Solubility** Very soluble in water. Hot ethanol dissolves the bis(8-hydroxyquinolinium) sulfate, leaving a residue of potassium sulfate; insoluble in diethyl ether. **Stability** Alkali liberates quinolin-8-ol which precipitates heavy metals.

COMMERCIALISATION

History Use of bis(8-hydroxyquinolinium) sulfate as fungicide on plants reported by G. Fron (*Rev. Pathol. Entomol. Veg.*, 1936, **23**, 131).   **Manufacturers** Probelte

APPLICATIONS

potassium hydroxyquinoline sulfate

Mode of action Systemic fungicide and bactericide.   **Uses** Potassium hydroxyquinoline sulfate is used for control of grey mould (*Botrytis cinerea*) in vine grafting. Also used as a soil sterilant for control of soil-borne diseases such as damping-off in seed beds of vegetables and ornamentals; and as a general disinfectant in horticulture.   **Formulation types** SP; SL.
Selected tradenames 'Cryptonol' (Novartis)

ANALYSIS

Product analysis by reaction with bromine and titration of the excess or by titration with alkali (both for the quinolin-8-ol content) and atomic absorption spectroscopy (for potassium).

MAMMALIAN TOXICOLOGY

8-hydroxyquinoline sulfate

Oral Acute oral LD_{50} for rats 1250, mice 500 mg/kg.   **Skin and eye** Acute percutaneous LD_{50} for rats >4000 mg/kg (67% SP formulation).
Toxicity class WHO (a.i.) III (Table 5); EPA (formulation) III   **EC risk** Xn (R22)

ECOTOXICOLOGY

8-hydroxyquinoline sulfate

Birds Non-toxic.   **Fish** Non-toxic.   **Bees** Not toxic to bees when used as directed.

ENVIRONMENTAL FATE

Animals In mammals, metabolism involves conjugation with glucuronic acid. Following oral administration, *c.* 95% is eliminated within 24–36 hours, principally as metabolites in the urine.

408 hymexazol

Fungicide

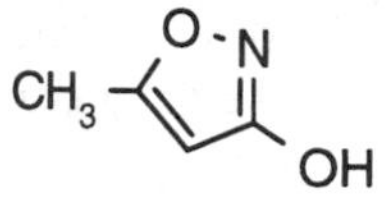

NOMENCLATURE

Common name hymexazol (BSI, E-ISO, (*m*) F-ISO); hydroxyisoxazole (JMAF)

IUPAC name 5-methylisoxazol-3-ol
Chemical Abstracts name 5-methyl-3(2H)-isoxazolone (*C.A.* preferred tautomer)
CAS RN *[10004–44–1]*   **Development codes** F-319; SF-6505 (both Sankyo)

PHYSICAL CHEMISTRY
Composition Tech. is 99% pure.   **Mol. wt.** 99.1   **M.f.** $C_4H_5NO_2$ **Form** Tech.
forms colourless crystals.   **M.p.** 86–87 °C   **V.p.** 182 mPa (25 °C)
K_{ow} logP = 0.480 (unstated pH)   **Solubility** In water 65.1 (pure), 58.2 (pH 3), 67.8
(pH 9) (all in g/l, 20 °C). In acetone 730, dichloromethane 602, ethyl acetate 437,
hexane 12.2, methanol 968, toluene 176 (all in g/l, 20 °C).   **Stability** Stable under
alkaline conditions, and relatively stable under acidic conditions.   Stable to light
and heat.   **pKa** 5.92, weak acid

COMMERCIALISATION
History Fungicidal properties reported by I. Iwai & N. Nakamura (*Chem. Pharm.
Bull.*, 1966, **14**, 1277); chemistry, biological properties and toxicology reviewed by
K. Tomita *et al.* (*Sankyo Kenkyusho Nempo*, 1973, **25**, 1). Introduced by Sankyo
Co., Ltd.   **Patent** JP 518249; JP 532202   **Manufacturers** Sankyo

APPLICATIONS
Mode of action Systemic soil and seed fungicide.   **Uses** Control of soil-borne
diseases caused by *Fusarium*, *Aphanomyces*, *Pythium*, *Corticium* and *Typhula* spp.,
etc. in rice, sugar beet, fodder beet, vegetables, cucurbits, ornamentals,
carnations, and forest tree seedlings. Applied as a soil drench at 30–60 g a.i./hl or
by soil incorporation, and also used as a seed dressing for sugar beet and fodder
beet at 5–90 g/kg seed. Also exhibits some plant growth stimulation activity.
Formulation types DP; SL; WP; WS.   **Mixtures** *(hymexazol +)* metalaxyl.
Selected tradenames 'Tachigaren' (Sankyo)

ANALYSIS
Product analysis by glc (T. Nakamura *et al.*, *Anal. Methods Pestic. Plant Growth
Regul.*, 1978, **10**, 215) or by hplc.   **Residues** determined by glc of a derivative
(*idem, ibid.*). Details available from Sankyo Co. Ltd.

MAMMALIAN TOXICOLOGY
Oral Acute oral LD_{50} for male rats 4678, female rats 3909, male mice 2148,
female mice 1968 mg/kg.   **Skin and eye** Acute percutaneous LD_{50} for rats
>10 000, rabbits >2000 mg/kg. Irritant to eyes and mucous membranes; not
irritant to skin.   **NOEL** (2 y) for male rats 19, female rats 20, dogs 15 mg/kg b.w.
daily.   **Other** Non-mutagenic, non-carcinogenic and non-teratogenic.
Toxicity class WHO (a.i.) III (Table 5); EPA (formulation) III

ECOTOXICOLOGY
Birds Acute oral LD_{50} for male Japanese quail 1698, female Japanese quail 1737,

chickens >1000 mg/kg.   **Fish** LC_{50} (48 h) for carp 165, Japanese killifish >40 mg/l.
Bees Not hazardous to bees. LD_{50} (48 h, oral and contact) >100 µg/bee.

ENVIRONMENTAL FATE
Animals In mammals, following oral administration, hymexazol is metabolised to
glucuronides.   **Plants** In plants, hymexazol undergoes degradation to *O*- and *N*-
glucosides.   **Soil/Environment** In soil, hymexazol is degraded to 5-methyl-2-=
(3*H*)-oxazolone, DT_{50} 2–25 d.

409 ICIA0858 *Fungicide*

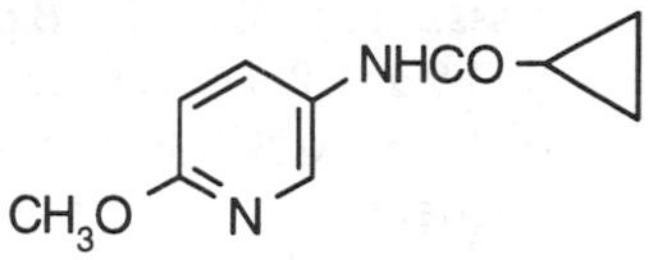

NOMENCLATURE
CAS RN *[112860–04–5]*   **Development codes** ICIA0858; SC-0858

PHYSICAL CHEMISTRY
Mol. wt. 192.2   **M.f.** $C_{10}H_{12}N_2O_2$

APPLICATIONS
Uses Under evaluation as a broad spectrum fungicide.

410 imazalil *Fungicide*

azole

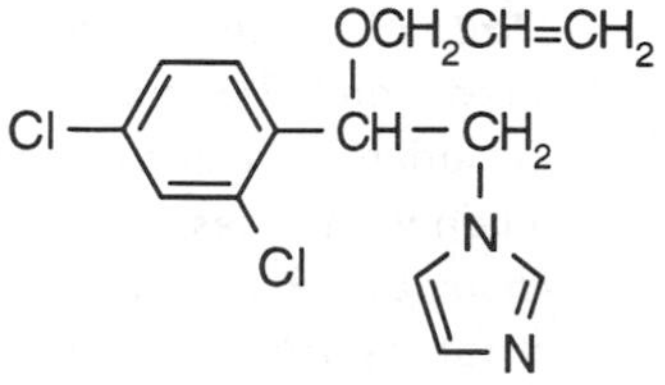

NOMENCLATURE
imazalil
Common name imazalil (BSI, E-ISO, (*m*) F-ISO, ANSI); chloramizol (Republic of

South Africa); enilconazole (BAN, INN)
IUPAC name (±)-1-(β-allyloxy-2,4-dichlorophenylethyl)imidazole;
(±)-allyl 1-(2,4-dichlorophenyl)-2-imidazol-1-ylethyl ether
Chemical Abstracts name (±)-1-[2-(2,4-dichlorophenyl)-2-(2-propenyloxy)ethyl]-=
1*H*-imidazole
CAS RN *[35554–44–0]* (unstated stereochemistry); *[73790–28–0]*
EEC no. 252–615–0   **Development codes** R023979 for base; R018531 for nitrate

imazalil sulfate
CAS RN *[60534–80–7]* (unstated stereochemistry); *[83918–57–4]*
EEC no. 261–351–5   **Development codes** R027180 (Janssen)

PHYSICAL CHEMISTRY
imazalil
Mol. wt. 297.2   **M.f.** $C_{14}H_{14}Cl_2N_2O$ **Form** Slightly yellow to brown crystalline
mass.   **M.p.** 52.7 °C   **B.p.** >340 °C   **V.p.** 0.158 mPa (20 °C) K_{ow} logP = 3.82
(pH 9.2 buffer)   **S.g./density** 1.348 g/ml (26 °C)   **Solubility** In water 0.18 g/l
(pH 7.6, 20 °C). In acetone, dichloromethane, ethanol, methanol, isopropanol,
xylene, toluene, benzene >500, hexane 19 (all in g/l, 20 °C). Also soluble in
heptane and petroleum ether.   **Stability** Very stable to hydrolysis in dilute acids
and alkalis at room temperature, in the absence of light. Stable to temperatures up
to 285 °C. Stable to light under normal storage conditions.   **pKa** 6.53, weak base
F.p. 192 °C

imazalil sulfate
Mol. wt. 395.2   **M.f.** $C_{14}H_{16}Cl_2N_2O_5S$ **Form** Almost white to beige-coloured
powder.   **Solubility** Freely soluble in water, alcohols, and slightly soluble in apolar
organic solvents.

COMMERCIALISATION
History Fungicide reported by E. Laville (*Fruits*, 1973, **28**, 545). Introduced by
Janssen Pharmaceutica.   **Manufacturers** Janssen; Sanachem

APPLICATIONS
Biochemistry Steroid demethylation (ergosterol biosynthesis) inhibitor.
Mode of action Systemic fungicide, with protective and curative action.
Uses Control of a wide range of fungal diseases on fruit, vegetables, and
ornamentals, e.g. powdery mildews on cucurbits and ornamentals; powdery
mildew on roses; storage diseases (particularly *Penicillium*, *Gloeosporium*,
Phomopsis, *Phoma* spp., etc.) of citrus fruit, pome fruit, bananas, and seed
potatoes. Also used as a seed dressing, for control of diseases (particularly
Fusarium and *Helminthosporium* spp.) of cereals. It is particularly active against
benzimidazole-resistant strains of plant-pathogenic fungi. Typical use rates are: for
seed treatment 4–5 g a.i./100 kg seed; for ornamentals and vegetables 5–30 g/hl;

and for post-harvest treatment 2–4 g/t fruit.   **Formulation types** EC; SP; Soluble liquid.   **Mixtures** *(imazalil +)* guazatine acetates; fenfuram; fuberidazole; furmecyclox; carboxin; nuarimol; thiabendazole; thiophanate-methyl; carbendazim; triadimenol; fenfuram + guazatine acetates; fuberidazole + triadimenol; fenfuram + quintozene + thiabendazole; methfuroxam + thiabendazole; fenpiclonil; carboxin + fenpiclonil; iprodione; fuberidazole + triadimenol; carboxin + thiabendazole; anthraquinone + fenfuram + lindane.   **Compatibility** Incompatible with alkaline materials.   **Selected tradenames** 'Fungaflor' (sulfate) (Janssen); 'Fungazil' (sulfate) (Janssen); 'Deccozil' (Elf Atochem); 'Flo-Pro' (Gustafson); 'Florasan' (Siapa); 'Freshgard' (FMC); 'Magnate' (Makhteshim-Agan)

ANALYSIS
Product analysis by glc with FID (*CIPAC Handbook*, 1992, **E**, 95–100).
Residues determined by glc (R. Woestenborghs *et al.*, *Med. Fac. Landbouwwet. Rijksuniv. Gent*, 1988, **53**, 1425) or by hplc (G. R. Cayley *et al.*, *Pestic. Sci.*, 1981, **12**, 103).

MAMMALIAN TOXICOLOGY
imazalil
Reviews *Pesticide residues in food – 1991*. FAO Plant Production and Protection Paper, 111, 1991. *Pesticide residues in food – 1991 evaluations. Part II – Toxicology*. World Health Organisation, WHO/PCS/92.52, 1992.   **Oral** Acute oral LD_{50} for rats 227–343, dogs >640 mg/kg.   **Skin and eye** Acute percutaneous LD_{50} for rats 4200–4880 mg/kg.   **Inhalation** LC_{50} (4 h) for rats 16 mg/l air (with 200 g/l EC). **NOEL** for rats and dogs 2.5 mg/kg b.w. daily.   **ADI** (JMPR) 0.03 mg/kg [1991]. **Toxicity class** WHO (a.i.) II; EPA (formulation) II   **EC risk** Xn (R22); Xi (R36)

imazalil sulfate
EC risk Xn (R22); Xi (R41)

ECOTOXICOLOGY
imazalil
Birds LD_{50} for ring-necked pheasant 2000 mg/kg. LC_{50} (8 d) for mallard ducks >2510 mg/kg b.w.   **Fish** LC_{50} (96 h) for rainbow trout 1.5, bluegill sunfish 4.04 mg/l.   **Bees** Not hazardous to bees when used as directed. LD_{50} (oral) 40 µg/bee.   **Daphnia** LC_{50} (48 h) 3.5 mg/l.

ENVIRONMENTAL FATE
Animals In rats, following oral administration, 90% is eliminated in the metabolised form within 4 days.   **Plants** In plants, imazalil is transformed into α-(2,4-dichlorophenyl)-1*H*-imidazole-1-ethanol.   **Soil/Environment** DT_{50} 30–170 d. Soil adsorption K 182 (clay loam), 209 (sandy loam), 68 (sandy soil).

imidazolinone

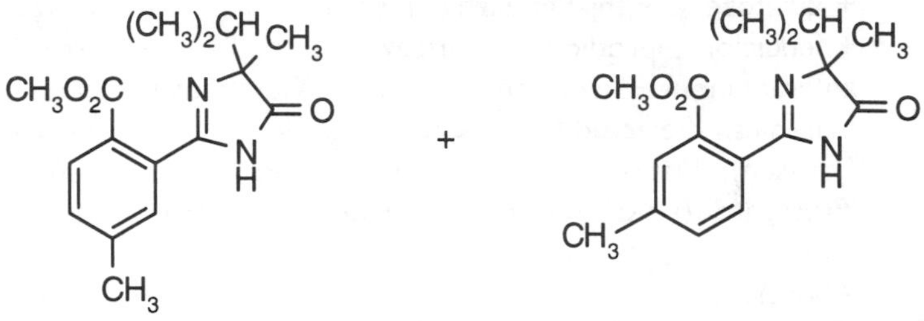

NOMENCLATURE
imazamethabenz-methyl
Common name imazamethabenz-methyl
IUPAC name A reaction product comprising methyl (±)-6-(4-isopropyl-4-methyl-=
5-oxo-2-imidazolin-2-yl)-*m*-toluate and methyl (±)-2-(4-isopropyl-4-methyl-5-oxo-=
2-imidazolin-2-yl)-*p*-toluate
Chemical Abstracts name methyl (±)-2-[4,5-dihydro-4-methyl-4-(1-methylethyl)-=
5-oxo-1*H*-imidazol-2-yl]-4-methylbenzoate with methyl (±)-2-[4,5-dihydro-=
4-methyl-4-(1-methylethyl)-5-oxo-1*H*-imidazol-2-yl]-5-methylbenzoate
CAS RN *[81405–85–8]* **Development codes** AC 222 293; CL 222 293

imazamethabenz
Common name imazamethabenz (BSI, ANSI, draft E-ISO, WSSA)
CAS RN *[100728–84–5]* **Development codes** AC 263 840; CL 263 840

PHYSICAL CHEMISTRY
imazamethabenz-methyl
Composition A mixture of *m*- and *p*- isomers. **Mol. wt.** 288.3
M.f. $C_{16}H_{20}N_2O_3$ **Form** Off-white fine to lumpy powder with slight musty odour.
M.p. 113–153 °C **V.p.** 1.5 × 10^{-3} mPa (25 °C) **K_{ow}** logP = 1.54 (*p*- isomer), 1.82
(*m*- isomer) **S.g./density** 0.3 (20 °C) **Solubility** In distilled water, *m*- isomer
1370, *p*- isomer 857 mg/kg. Mixed isomers in acetone 230, dimethyl sulfoxide
216, isopropanol 183, methanol 309, toluene 45, *n*-heptane 0.6, *n*-hexane 0.4 (all
in g/kg, 25 °C). **Stability** Imazamethabenz-methyl is stable in storage at 25 °C for
24 months, at 37 °C for 12 months, or at 45 °C for 3 months. Hydrolysis is rapid
at pH 9 but slow at pH 5 and pH 7. **pKa** 2.9 (23.5 °C) **F.p.** >93 °C (closed cup)

imazamethabenz
Mol. wt. 274.3 **M.f.** $C_{15}H_{18}N_2O_3$

COMMERCIALISATION
History Herbicidal activity of imazamethabenz-methyl reported by D. L. Shaner et

al. (*Proc. Br. Crop Prot. Conf. – Weeds*, 1982, **1**, 25) and K. Kirkland & N. E. Shafer (*ibid.*, p. 33). Introduced by American Cyanamid Co; US registration 1988.
Patent US 4188487   **Manufacturers** Cyanamid

APPLICATIONS
imazamethabenz-methyl
Biochemistry Inhibits branched-chain amino acid synthesis, hence protein synthesis, and also DNA synthesis (J. B. Pillmoor & J. C. Caseley, *Pestic. Biochem. Physiol.*, 1987, **27**, 340).   **Mode of action** Selective systemic herbicide, absorbed through the roots and leaves, with translocation to the meristematic regions.
Uses Effective post-emergence (at 0.25–0.7 kg a.e./ha) in controlling *Avena* spp., *Alopecurus myosuroides*, *Apera spica-venti* and some dicotyledonous weeds in wheat, barley, rye and sunflowers.   **Phytotoxicity** Cereals and sunflowers are highly tolerant to post-emergence applications. Sugar beet, beetroot, lentils, rape, mustard, tomatoes and broccoli should not be planted for at least 15 months following application.   **Formulation types** SC; EC; SG.
Mixtures (*imazamethabenz-methyl* +) pendimethalin; mecoprop.
Compatibility Formulations not compatible with amine formulations of phenoxy (aryloxyalkanoic) herbicides. Not compatible with concentrated acids or bases; strong oxidising agents may cause vigorous decomposition.
Selected tradenames 'Assert' (Cyanamid); 'Dagger' (Cyanamid)

ANALYSIS
Product by gc.   **Residues** in cereals and sunflower seeds, by gc.   Details from Cyanamid.

MAMMALIAN TOXICOLOGY
imazamethabenz-methyl
Oral Acute oral LD_{50} for male and female rats >5000 mg/kg.   **Skin and eye** Acute percutaneous LD_{50} for male and female rabbits >2000 mg/kg. Non-irritating to skin; reversible eye irritant (rabbits). Not a skin sensitiser in guinea pigs.
Inhalation LC_{50} for male and female rats >5.8 mg/l.   **NOEL** (2 y) for rats 12.5 mg/kg b.w. daily (250 mg/kg diet); (1 y) for dogs 6.25 mg/kg b.w. daily (250 mg/kg diet).   **ADI** 0.0625 mg/kg b.w.   **Other** Not mutagenic in the dominant lethal assay in rats. Not mutagenic in the Ames assay.
Toxicity class WHO (a.i.) III (Table 5); EPA (formulation) III (tech.)

ECOTOXICOLOGY
imazamethabenz-methyl
Birds Acute oral LD_{50} for bobwhite quail and mallard ducks >2150 mg/kg. Eight-day dietary LC_{50} for bobwhite quail and mallard ducks >5000 mg/kg diet.
Fish LC_{50} (96 h) for bluegill sunfish and rainbow trout >100 mg/l.   **Bees** LD_{50} (contact) >100 µg/bee.   **Worms** >123 ppm.   **Daphnia** LC_{50} (48 h) >100 mg/l.

MAIN ENTRIES

ENVIRONMENTAL FATE

Animals Rapidly excreted by animals (rats, goats, hens) after ingestion; low levels of residues found in blood and tissues of rats, in milk and tissues of lactating goats, and in the eggs and tissues of laying hens. **Plants** In plants, the methyl group of the benzene moiety is oxidised to a hydroxymethyl group; glucoside conjugates have been identified as secondary metabolites.

Soil/Environment Imazamethabenz-methyl is slowly degraded to corresponding free acids in sandy loam and clay loam soils under both aerobic and anaerobic conditions; DT_{50} 25–45 d; hydrolysis increases with increasing pH. Subject to photolytic degradation in water and on soil surfaces. Half-life of residues in the soil is c. 30–105 d, with no accumulation of acid metabolites (R. Allen & J. C. Caseley, *Proc. 1987 Br. Crop. Prot. Conf. – Weeds*, 569). Low leaching potential: adsorption Freundlich K ranges from 0.18 (sandy loams) to 8.5 (silt loams) for *p-* isomer; 0.32 to 10.5 for *m-* isomer.

412 imazamox *Herbicide*

imidazolinone

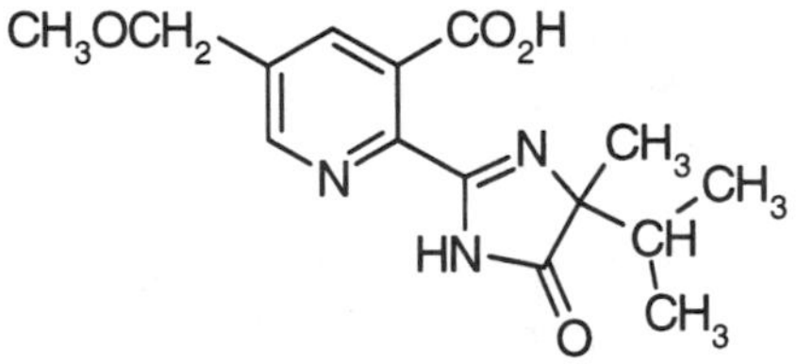

NOMENCLATURE

Common name imazamox (BSI, pa ISO, ANSI)
IUPAC name (*RS*)-2-(4-isopropyl-4-methyl-5-oxo-2-imidazolin-2-yl)-5-methoxy= methylnicotinic acid
Chemical Abstracts name 2-[4,5-dihydro-4-methyl-4-(1-methylethyl)-5-oxo-= 1*H*-imidazol-2-yl]-5-(methoxymethyl)-3-pyridinecarboxylic acid
CAS RN *[114311–32–9]* **Development codes** AC 299,263; CL 299,263

PHYSICAL CHEMISTRY

Composition 93% pure. **Mol. wt.** 305.3 **M.f.** $C_{15}H_{19}N_3O_4$ **Form** Odourless solid. **M.p.** 166.0–166.7 °C (tech.) **V.p.** $<1.3 \times 10^{-2}$ mPa K_{ow} logP = –1.03 (pH 5, uncorrected for dissociation), –2.4 (pH 7, uncorrected), and 0.73 (pH 5 & 6, corrected) **Solubility** In acetone 2.93 g/100 ml.

COMMERCIALISATION

History EUP first granted in 1995. **Manufacturers** Cyanamid

APPLICATIONS
Biochemistry ALS/AHAS inhibitor. Selectivity is due to metabolism in the crop.
Mode of action Post-emergence herbicide translocated to growing points. Plants wilt and turn brown. **Uses** Under development for post-emergence weed control in soya beans and other legumes grown in rotation with sugar beet and other crops where more persistent imidazolinones are not suited. Application rates 0.032–0.04 lb/a. **Formulation types** DG. **Selected tradenames** 'Odyssey' (Canada) (mixture) (Cyanamid); 'Raptor' (USA) (Cyanamid); 'Sweeper' (Latin America) (Cyanamid)

MAMMALIAN TOXICOLOGY
Oral Acute oral LD_{50} for male and female rats >5000 mg/kg. **Skin and eye** Acute percutaneous LD_{50} for male and female rats >4000 mg/kg. **Inhalation** LC_{50} (4 h) for rats >6.3 mg/l. **NOEL** (1 y) for dogs 1165 mg/kg b.w. daily.

ECOTOXICOLOGY
Birds Acute oral LD_{50} (14 d) for bobwhite quail >1846 mg/kg. **Fish** LC_{50} (96 h) for rainbow trout 122 mg/l. **Bees** LD_{50} (contact) >25 µg/bee. **Daphnia** NOEC 122 mg/l.

ENVIRONMENTAL FATE
Soil/Environment Degraded microbially; metabolites show moderate to strong binding potential; unlikely to leach.

413 imazapyr   *Herbicide*

imidazolinone

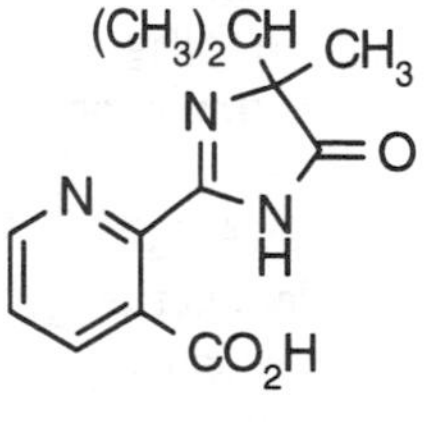

NOMENCLATURE
Common name imazapyr (BSI, ANSI, draft E-ISO, (*m*) draft F-ISO)
IUPAC name 2-(4-isopropyl-4-methyl-5-oxo-2-imidazolin-2-yl)nicotinic acid
Chemical Abstracts name 2-[4,5-dihydro-4-methyl-4-(1-methylethyl)-5-oxo-=1*H*-imidazol-2-yl]-3-pyridinecarboxylic acid
CAS RN *[81334–34–1]* unstated stereochemistry **Development codes** AC 252 925; CL 252 925

PHYSICAL CHEMISTRY
Mol. wt. 261.3 **M.f.** $C_{13}H_{15}N_3O_3$ **Form** White to tan powder with a slight odour of acetic acid. **M.p.** 169–173 °C **V.p.** <0.013 mPa (60 °C) K_{ow} logP = 0.11 (22 °C, unstated pH) **Solubility** In water 9.74 g/l (15 °C), 11.3 g/l (25 °C). In acetone 3.39, dimethyl sulfoxide 47.1, hexane 0.00095, methanol 10.5, methylene chloride 8.72, toluene 0.180 (all in g/100 ml). **Stability** Stable for at least 2 years at 25 °C, 1 year at 37 °C, and 3 months at 45 °C. Stable in aqueous media at pH 5 to pH 9 in the dark; avoid storage >45 °C. In solution, the acid is decomposed in simulated sunlight; DT_{50} 6 d at pH 5 to pH 9. **pKa** pKa_1 1.9, pKa_2 3.6, pKa_3 11

COMMERCIALISATION
History Herbicide reported by P. L. Orwick et al. (*Proc. South. Weed Sci. Soc. Annu. Mtg., 36th,* 1983, p. 291). Introduced by American Cyanamid Co.. **Manufacturers** Cyanamid

APPLICATIONS
Biochemistry Blocks biosynthesis of valine, leucine, and isoleucine through inhibition of acetohydroxy acid synthase, thus causing disruption of protein synthesis, which leads to interference in DNA synthesis and cell growth. This results in chlorosis and tissue necrosis of new leaves. **Mode of action** Non-selective systemic herbicide, absorbed by the foliage and roots, with rapid translocation in the xylem and phloem to the meristematic regions, where it accumulates. **Uses** Imazapyr-isopropylammonium is used for pre- and post-emergence control of annual and perennial grasses, sedges and broad-leaved weeds, as well as many brush and deciduous tree species. Used in non-crop areas, such as industrial sites, railways, roads, drainage channels, etc. at 0.25–1.7 kg a.i./ha; in forestry management at 0.25–1.7 kg/ha; and in plantations of rubber trees and oil palms at 0.125–1.0 kg/ha. **Phytotoxicity** Non-phytotoxic to conifers. Rubber trees and oil palms show good tolerance, though application onto foliage is not selective and care should be taken to avoid contact of the spray with the foliage of oil palms and the leaves and young bark of rubber trees. **Formulation types** SL; GR; EC; SL. **Mixtures** (*imazapyr +*) diuron.

ANALYSIS
Product by gc. **Residues** (1) In raspberries and bilberries, by gc; (2) in pine branches, leaves and forest floors, by hplc. Details from Cyanamid. (3) In soil, by hplc after extraction with ammonium bicarbonate and clean up (C. S. Helling & M. A. Doherty, *Pestic. Sci.,* **45,** 21–26 (1995)).

MAMMALIAN TOXICOLOGY
Oral Acute oral LD_{50} for male and female rats >5000, female mice >2000, male and female rabbits 4800 mg/kg. **Skin and eye** Acute percutaneous LD_{50} for male and female rabbits >2000, male and female rats >2000 mg/kg. Eye irritant; mild

skin irritant (rabbits). No skin sensitisation (guinea pigs).   **Inhalation** LC_{50} for male and female rats >1.3 mg/l air.   **NOEL** (13 w) for rats 10 000 mg/kg b.w. (highest dose tested). No teratogenic or foetotoxic effects were observed at 1000 mg/kg b.w. in rats or 400 mg/kg b.w. in rabbits (highest doses tested).   **Other** Acute i.p. LD_{50} for male rats 2500, female rats 2500–3200 mg/kg. Non-mutagenic, non-carcinogenic.   **Toxicity class** WHO (a.i.) III (Table 5); EPA (formulation) IV

ECOTOXICOLOGY
Birds Acute oral LD_{50} for bobwhite quail and mallard ducks >2150 mg/kg. Eight-day dietary LC_{50} for bobwhite quail and mallard ducks >5000 mg/kg.   **Fish** LC_{50} (96 h) for rainbow trout, bluegill sunfish, and channel catfish >100 mg/l.
Bees Dermal LD_{50} for honeybees >100 μg/bee.   **Daphnia** LC_{50} (48 h) >100 mg/l.

ENVIRONMENTAL FATE
Animals In rats, following oral administration, c. 87% of the dose was excreted in the urine and faeces within 24 hours. In muscle and fat tissues and blood, residual levels were <0.01 mg/kg at both 24 and 192 hours.   **Plants** Following foliar application, residues in plants decline rapidly in the first 24 hours and are exuded from the roots into the soil. The major residue in plants is the parent compound.
Soil/Environment Residual activity in the soil ranges from 6 months to 2 years in temperate climates and 3 to 6 months in tropical climates. The major residue in soil is the parent compound. Bioaccumulation in the environment is highly unlikely.

414   imazaquin   *Herbicide*

imidazolinone

NOMENCLATURE
imazaquin
Common name imazaquin (BSI, ANSI, draft E-ISO, WSSA); imazaquine ((*m*) draft F-ISO)
IUPAC name (*RS*)-2-(4-isopropyl-4-methyl-5-oxo-2-imidazolin-2-yl)quinoline-= 3-carboxylic acid

Chemical Abstracts name (±)-2-[4,5-dihydro-4-methyl-4-(1-methylethyl)-5-oxo-= 1*H*-imidazol-2-yl]-3-quinolinecarboxylic acid
CAS RN *[81335–37–7]* unstated stereochemistry   **Development codes** AC 252 214; CL 252 214

imazaquin-ammonium
CAS RN *[81335–47–9]* unstated stereochemistry

PHYSICAL CHEMISTRY
imazaquin
Composition >95% pure.   **Mol. wt.** 311.3   **M.f.** $C_{17}H_{17}N_3O_3$ **Form** Tan solid with slightly pungent odour.   **M.p.** 219–224 °C (decomp.)   **V.p.** <0.013 mPa (60 °C)   **K_{ow}** logP = 0.34 (22 °C)   **Solubility** In water 60–120 mg/l (25 °C). In toluene 0.4, dimethylformamide 68, dimethyl sulfoxide 159, dichloromethane 14 (all in g/l, 25 °C).   **Stability** Stable for 3 months at 45 °C, 2 y at room temperature when stored in the dark. Rapidly degraded on exposure to u.v. light. **pKa** 3.8

imazaquin-ammonium
Mol. wt. 328.4   **M.f.** $C_{17}H_{20}N_4O_3$ **Solubility** In water 160 g/l (pH 7, 20 °C).

COMMERCIALISATION
History Herbicide reported by P. L. Orwick *et al.* (*Abstr. 1983 Weed Sci. Soc. Annu. Mtg., 36th*, 1983, pp. 90, 91), P. K. Martin (*ibid.*, p. 90, No. 46) and H. M. Hackworth *et al.* (*ibid.*, p. 91). Introduced by American Cyanamid Co.; registered in US in 1986.   **Patent** US 4798619   **Manufacturers** Cyanamid

APPLICATIONS
imazaquin-ammonium
Biochemistry Blocks biosynthesis of branched chain amino acids valine, leucine, and isoleucine through inhibition of acetohydroxy acid synthase, thus causing disruption of protein synthesis, which leads to interference in DNA synthesis and cell growth (D. L. Shaner et al., *Plant Physiol.*, 1984, **76**, 545). Selectivity is due to detoxication in the crop.   **Mode of action** Selective systemic herbicide, absorbed by the roots and foliage, with translocation in the xylem and phloem throughout the plant, and accumulation in the meristematic regions.   **Uses** Pre-planting, pre-emergence or early post-emergence control of broad-leaved weeds in soya beans at 70–140 g a.e./ha. Also controls or reduces competition from grasses and sedges.   **Phytotoxicity** Non-phytotoxic to soya beans.   **Formulation types** DG; SL. **Mixtures** (*imazaquin-ammonium* +) trifluralin; pendimethalin; acifluorfen; imazethapyr; dimethenamid.   **Compatibility** Compatible with soil-applied grass herbicides. If mixed with post-emergence grass herbicides, reduced activity against grasses will result due to herbicide antagonism.   **Selected tradenames** 'Scepter' (Cyanamid)

MAMMALIAN TOXICOLOGY

imazaquin

Oral Acute oral LD_{50} for male and female rats >5000, female mice 2363 mg/kg.
Skin and eye Acute percutaneous LD_{50} for male and female rabbits >2000 mg/kg.
Non-irritating to eyes; mildly irritating to skin (rabbits). No skin sensitisation
(guinea pigs).   **Inhalation** LC_{50} (4 h) for rats >5.7 mg/l air.   **NOEL** (90 d) for rats
10 000 mg/kg diet; (2 y) for rats 5000 mg/kg.   **ADI** 0.25 mg/kg b.w.
Other Non-carcinogenic, non-mutagenic and non teratogenic.
Toxicity class WHO (a.i.) III (Table 5); EPA (formulation) III

ECOTOXICOLOGY

imazaquin

Birds Acute oral LD_{50} for bobwhite quail and mallard ducks >2150 mg/kg. Eight-
day dietary LC_{50} for bobwhite quail and mallard ducks >5000 mg/kg diet.
Fish LC_{50} (96 h) for channel catfish 320, bluegill sunfish 410, rainbow trout
280 mg/l.   **Bees** Contact LD_{50} for honeybees >100 µg/bee.   **Daphnia** LC_{50}
(48 h) 280 mg/l.

ENVIRONMENTAL FATE

Animals Imazaquin is metabolically inert in rats. Following oral administration,
almost all was excreted in the urine as the unchanged compound within 2 days.
No accumulation in blood or tissues.   **Plants** Rapidly metabolised by soya bean
plants to inactive compounds, formed by opening of the imidazolinone ring at the
ring amide (D. L. Shaner & P. A. Robson, *Weed Sci.*, 1985, **33**, 469).
Soil/Environment Steadily degraded in soil by microbial activity and photolysis, but
may remain active in the soil for several weeks to several months, depending on
environmental conditions (G. Basham *et al.*, *Weed Sci.*, 1987, **35**, 576). DT_{50} 60 d;
K_{oc} 20.

415 imazethapyr   *Herbicide*

imidazolinone

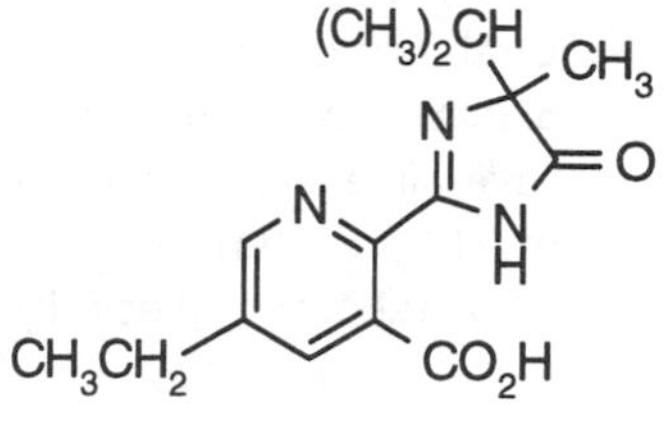

NOMENCLATURE
Common name imazethapyr (BSI, ANSI, draft E-ISO, (*m*) draft F-ISO)

IUPAC name (RS)-5-ethyl-2-(4-isopropyl-4-methyl-5-oxo-2-imidazolin-2-yl)= nicotinic acid
Chemical Abstracts name ($\pm$)-2-[4,5-dihydro-4-methyl-4-(1-methylethyl)-5-oxo-= 1H-imidazol-2-yl]-5-ethyl-3-pyridinecarboxylic acid
CAS RN *[81335–77–5]* **Development codes** AC 263 499; CL 263 499

PHYSICAL CHEMISTRY
Mol. wt. 289.3 **M.f.** $C_{15}H_{19}N_3O_3$ **Form** Colourless crystals. **M.p.** 169–173 °C
B.p. Decomp. 180 °C **V.p.** <0.013 mPa (60 °C) **K_{ow}** logP = 1.04 (pH 5), 1.49
(pH 7), 1.20 (pH 9) (all 25 °C) **S.g./density** 1.10–1.12 (21 °C) **Solubility** In
water 1.4 g/l (25 °C). In acetone 48.2, methanol 105, toluene 5, dichloromethane
185, dimethyl sulfoxide 422, isopropanol 17, heptane 0.9 (all in g/l, 25 °C).
Stability Rapidly degraded in sunlight. DT_{50} c. 3 d. **pKa** pK_1 2.1, pK_2 3.9

COMMERCIALISATION
History Herbicide reported by T. Malefyt *et al.* (*Abstr. 1984 Weed Sci. Soc. Mtg.,
Miami*, p. 18, Abstract 49). Introduced by American Cyanamid Co.
Patent US 4798619 **Manufacturers** Cyanamid

APPLICATIONS
Biochemistry Acts by inhibition of acetohydroxy acid synthase, reducing levels of
branched chain amino acids valine, leucine and isoleucine, and leading to disruption
of protein and DNA synthesis. **Mode of action** Systemic herbicide, absorbed by
the roots and foliage, with translocation in the xylem and phloem, and
accumulation in the meristematic regions. **Uses** Control of many major annual
and perennial grass and broad-leaved weeds in soya beans and other leguminous
crops. Applied pre-plant incorporated, pre-emergence, or post-emergence.
Phytotoxicity Non-phytotoxic to soya beans and other leguminous crops, when
used as directed. **Formulation types** SL. **Mixtures** *(imazethapyr +)* pendimethalin;
trifluralin; atrazine; dicamba. **Compatibility** Not to be tank-mixed with
post-emergence grass herbicides. **Selected tradenames** 'Hammer' (ammonium
salt) (Cyanamid); 'Overtop' (ammonium salt) (Cyanamid); 'Pivot' (ammonium salt)
(Cyanamid); 'Pursuit' (ammonium salt) (Cyanamid); 'Pursuit DG' (acid)
(Cyanamid)

MAMMALIAN TOXICOLOGY
Oral Acute oral LD_{50} for male and female rats, and female mice >5000 mg/kg.
Skin and eye Acute percutaneous LD_{50} for rabbits >2000 mg/kg; mild skin and
reversible eye irritant. **Inhalation** LC_{50} for rats 3.27 mg/l air (analytical), 4.21
mg/l (gravimetric). **NOEL** (2 y) for rats >10 000 mg/kg diet; (1 y) for dogs
>10 000 mg/kg diet (highest dose tested). **Other** Non-mutagenic in the Ames
test. **Toxicity class** WHO (a.i.) III (Table 5); EPA (formulation) III

ECOTOXICOLOGY

Birds Acute oral LD_{50} for bobwhite quail and mallard ducks >2150 mg/kg.
Fish LC_{50} (96 h) for bluegill sunfish 420, rainbow trout 340, channel catfish
240 mg/l. **Bees** Topical LD_{50} for honeybees >0.1 mg/bee. **Worms** I_{50}
>10 000 mg/kg. **Daphnia** LC_{50} (48 h) <1000 mg/l. **Algae** NOEL for *Selenastrum
capricornutum* 50 mg/l. **Other aquatic spp.** I_{50} for *Lemna gibba* 4.38 µg/l.

ENVIRONMENTAL FATE

Animals In rats, following oral administration, 92% was excreted in the urine and
5% in the faeces within 24 hours. Residue levels in blood, liver, kidney, muscle, and
fat tissues were <0.01 ppm after 48 hours. **Plants** Rapidly metabolised in non-
susceptible plants; half-life in soya beans 1.6 days. The primary metabolic route in
maize is oxidative hydroxylation at the α-carbon atom of the ethyl substituent on
the pyridine ring. **Soil/Environment** Half-life in soil 1–3 months.

MAIN ENTRIES

416 imazosulfuron *Herbicide*

sulfonylurea

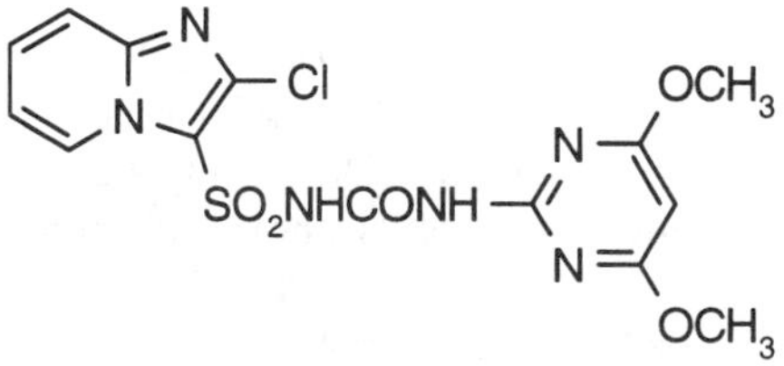

NOMENCLATURE

Common name imazosulfuron (BSI, pa E-ISO)
IUPAC name 1-(2-chloroimidazo[1,2-*a*]pyridin-3-ylsulfonyl)-3-(4,6-dimethoxy=
pyrimidin-2-yl)urea
Chemical Abstracts name 2-chloro-*N*[[(4,6-dimethoxy-2-pyrimidinyl)amino]=
carbonyl]imidazo[1,2-*a*]pyridine-3-sulfonamide
CAS RN *[122548–33–8]* **Development codes** TH-913

PHYSICAL CHEMISTRY

Mol. wt. 412.8 **M.f.** $C_{14}H_{13}ClN_6O_5S$ **Form** Crystalline powder.
M.p. 183–184 °C (decomp.) **V.p.** 4.5×10^{-5} mPa (25 °C) **K_{ow}** logP = 0.049
(pH 7, 25 °C) **S.g./density** 1.574 g/ml (25.5 °C) **Solubility** In water 6.75
(pH 5.1), 67 (pH 6.1), 308 (pH 7.0) (all in mg/l, 25 °C). In acetonitrile 2500,
ethyl acetate 2200, acetone 4800, dichloromethane 12 900, xylene 400 mg/l
(25 °C). **pKa** 4.0

COMMERCIALISATION
Manufacturers Takeda

APPLICATIONS
Biochemistry Branched chain amino acid synthesis (acetolactate synthase or ALS) inhibitor. Acts by inhibiting biosynthesis of the essential amino acids valine and isoleucine, hence stopping cell division and plant growth. Selectivity derives from rapid metabolism in the crop. Metabolic basis of selectivity in sulfonylureas reviewed (M. K. Koeppe & H. M. Brown, *Agro-Food-Industry*, **6**, 9–14 (1995)).
Mode of action Absorbed by plants mainly through the roots and translocated throughout the plant. Inhibits shoot growth and retards root development.
Uses Controls most annual (excluding *Echinochloa oryzicola*) and perennial broad-leaved weeds and sedges in paddy rice (75–95 g a.i./ha) and turf (500–1000 g a.i./ha). **Formulation types** GR; SC. **Mixtures** *(imazosulfuron +)* mefenacet + daimuron; esprocarb + daimuron; pretilachlor + daimuron + dimethametryn; pyributicarb + daimuron. **Selected tradenames** 'Sibatito' (Takeda); 'Takeoff' (Takeda)

ANALYSIS
Product and **residue** analysis by hplc. Methods for sulfonylurea residues in crops, soil and water reviewed (A. C. Barefoot *et al.*, *Proc. Br. Crop Prot. Conf. – Weeds*, 1995, **2**, 707).

MAMMALIAN TOXICOLOGY
Oral Acute oral LD_{50} for rats and mice >5000 mg/kg. **Skin and eye** Acute percutaneous LD_{50} for male and female rats >2000 mg/kg. Not irritant to eyes and skin (rabbits); not a skin sensitiser (guinea pigs). **Inhalation** LC_{50} (4 h) for rats >2.4 mg/l air. **NOEL** (2 y) for male rats 106.1, female rats 132.46 mg/kg b.w. daily; (1 y) for male and female dogs 75 mg/kg b.w. daily. Not oncogenic in rats and mice. Not teratogenic in rats and rabbits. **Other** No mutagenic effects in Ames, DNA-repair and chromosomal aberration tests.

ECOTOXICOLOGY
Birds Acute oral LD_{50} for bobwhite quail and mallards ducks >2250 mg/kg. Five-day dietary LC_{50} for bobwhite quail and mallard ducks >5620 ppm.
Fish LC_{50} (48 h) for carp >10 mg/l. **Bees** LD_{50} (48 h) (oral) 48.2 µg/bee, (contact) 66.5 µg/bee. **Daphnia** LC_{50} (3 h) >40 mg/l.

azole

S — CH₂ —⟨ ⟩— Cl structure:

The molecular structure shows: Cl and Cl attached to a dichlorophenyl ring connected by —N=C with a $S-CH_2$ group linking to a 4-chlorophenyl ring, and a CH_2 group linking to a 1,2,4-triazole ring.

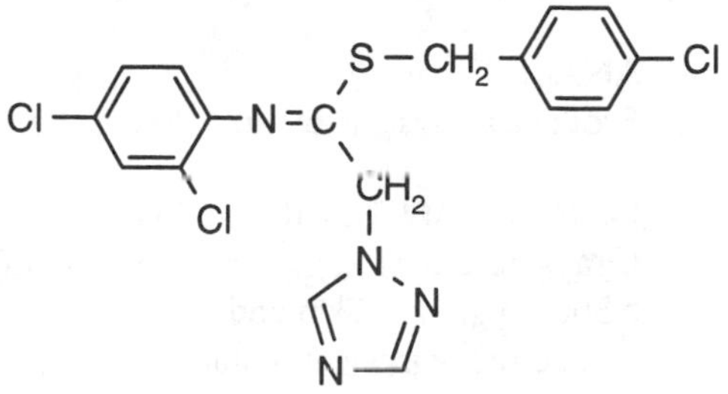

NOMENCLATURE
Common name imibenconazole (BSI, draft E-ISO)
IUPAC name 4-chlorobenzyl N-(2,4-dichlorophenyl)-2-(1H-1,2,4-triazol-1-yl)=
thioacetamidate
Chemical Abstracts name (4-chlorophenyl)methyl N-(2,4-dichlorophenyl)-=
1H-1,2,4-triazole-1-ethanimidothioate
CAS RN *[86598–92–7]* unstated stereochemistry   **Development codes** HF-6305
(tech.); HF-8505 (formulation)

PHYSICAL CHEMISTRY
Mol. wt. 411.7   **M.f.** $C_{17}H_{13}Cl_3N_4S$ **Form** Pale yellowish crystals.
M.p. 89.5–90 °C   **V.p.** 8.5×10^{-5} mPa (25 °C) $\mathbf{K_{ow}}$ logP = 4.94   **Solubility** In
water 1.7 mg/l (20 °C). In acetone 1063, benzene 580, xylene 250, methanol
120 (all in g/l, 25 °C).   **Stability** Stable in weak alkali; unstable in acid and in strong
alkali.

COMMERCIALISATION
History Fungicide reported by H. Ohyama et al. (*Abstr. 9th Annu. Meeting Pestic.
Sci. Soc. Jpn.*, 1984, p. 120; *Proc. 1988 Br. Crop Prot. Conf. – Pests Dis.*, **2**, 519).
Introduced by Hokko Chemical Industry Co., Ltd.   **Patent** JP 85026391;
US 4512989; GB 2109371   **Manufacturers** Hokko

APPLICATIONS
Biochemistry Ergosterol biosynthesis inhibition (steroid demethylation inhibitor); at
higher doses (≥10 ppm), it is physically incorporated into the cell membrane and
causes direct physical destruction of the cell (Y. Ogawa, *Agchem. Japan*, **67**, 20
(1995)).   **Mode of action** Curative and preventative systemic fungicide. Inhibits
growth of both the germ tube and mycelium.   **Uses** Control of scab, powdery
mildew, *Alternaria* leaf spot, sooty blotch, fly speck and rust on apples; scab and
rust on pears; scab on Japanese apricot; powdery mildew and anthracnose on
vines; scab on peaches; scab on citrus; powdery mildew on melon and water
melon; brown leaf spot on peanut; anthracnose, brown round spot and blister

blight on tea; black spot and powdery mildew on roses; and rusts on chrysanthemums. **Formulation types** EC; WP; WG. **Mixtures** (*imibenconazole +*) mancozeb. **Compatibility** Incompatible with pesticides which are strongly acidic. **Selected tradenames** 'Manage' (Hokko)

ANALYSIS
Product analysis is by hplc. **Residue** analysis is by glc with NPD.

MAMMALIAN TOXICOLOGY
Oral Acute oral LD_{50} for male rats 2800, female rats 3000, male and female mice >5000 mg/kg. **Skin and eye** Acute percutaneous LD_{50} for male and female rats >2000 mg/kg. Slight irritant to eyes, non-irritant to skin (rabbits). Slight skin sensitisation (guinea pigs). **Inhalation** LC_{50} (4 h) for rats >1020 mg/m^3. **NOEL** (2 y) for rats 100 mg/kg diet. **Other** Non-mutagenic. **Toxicity class** WHO (a.i.) III (Table 5)

ECOTOXICOLOGY
Birds Acute oral LD_{50} for bobwhite quail and mallard ducks >2250 mg/kg. **Fish** LC_{50} (96 h) for bluegill sunfish 1.0, rainbow trout 0.67, carp 0.84 mg/l. **Bees** LD_{50} (oral) >125 µg/bee, (contact) >200 µg/bee. **Worms** LC_{50} for earthworms >1000 mg/kg soil. **Daphnia** LC_{50} (6 h) >100 mg/l. **Algae** EC_{50} for algae >1000 mg/l.

ENVIRONMENTAL FATE
Animals Imibenconazole orally administered to rats is rapidly metabolised and eliminated. The major metabolite is 2',4'-dichloro-(1*H*-1,2,4-triazol-1-yl)= acetanilide. **Plants** Imibenconazole applied to grapes and apples is degraded or metabolised rapidly. The main metabolite is 2',4'-dichloro-(1*H*-1,2,4-triazol-1-yl)= acetanilide. **Soil/Environment** Rapidly degrades in soil: DT_{50} (lab.) 4–20 d, (field) 1–28 d. K_{oc} 2813–23 391.

418 imidacloprid

Insecticide

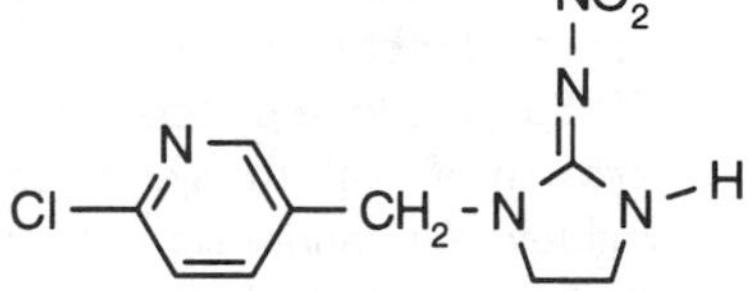

NOMENCLATURE
Common name imidacloprid (BSI, draft E-ISO)

IUPAC name 1-(6-chloro-3-pyridylmethyl)-*N*-nitroimidazolidin-2-ylideneamine
Chemical Abstracts name 1-[(6-chloro-3-pyridinyl)methyl]-*N*-nitro-=
2-imidazolidinimine
CAS RN *[138261–41–3]*   **Development codes** NTN 33 893 (Bayer)

PHYSICAL CHEMISTRY
Mol. wt. 255.7   **M.f.** $C_9H_{10}ClN_5O_2$ **Form** Colourless crystals with a weak
characteristic odour.   **M.p.** 144 °C   **V.p.** 4×10^{-7} mPa (20 °C) K_{ow} logP − 0.57
(22 °C)   **Henry** 2×10^{-10} Pa m^3 mol^{-1} (calc.)   **S.g./density** 1.54 (20 °C)
Solubility In water 0.61 g/l (20 °C). In dichloromethane 55, isopropanol 1.2,
toluene 0.68, *n*-hexane <0.1 (all in g/l, 20 °C).   **Stability** Stable to hydrolysis at
pH 5–11.

COMMERCIALISATION
History Insecticide reported by A. Elbert *et al.* (*Proc. Br. Crop Prot. Conf. – Pests
Dis.*, 1990, **1**, 21). Introduced by Bayer AG and Nihon Tokushu Noyaku Seizo KK.
Patent EP 0192060   **Manufacturers** Bayer

APPLICATIONS
Biochemistry Acts on the central nervous system, causing blockage of postsynaptic
nicotinergic acetylcholine receptors.   **Mode of action** Systemic insecticide with
contact and stomach action. Readily taken up by the plant and further distributed
acropetally, with good root-systemic action.   **Uses** Control of sucking insects,
including ricehoppers, aphids, thrips and whiteflies. Also effective against soil
insects, termites and some species of biting insects, such as rice water weevil and
Colorado beetle. Has no effect on nematodes and spider mites. Used as a seed
dressing, as soil treatment and as foliar treatment in different crops, e.g. rice,
cotton, cereals, maize, sugar beet, potatoes, vegetables, citrus fruit, pome fruit and
stone fruit.   **Formulation types** DP; GR; SC; WG; WP; WS; FS; SL.
Selected tradenames 'Admire' (Bayer, Bayer Corp., Nihon Bayer); 'Confidor'
(Bayer); 'Gaucho' (Bayer, Bayer Corp.)

ANALYSIS
Residues by hplc (F-J. Placke & E. Weber, *Pflanzenschutz-Nachrich. Bayer,* 93/2, **46**,
109–182 (1993)). A method has been developed based on the presence of the
6-chloropyridinylmethylene moiety in all known plant and animal metabolites.

MAMMALIAN TOXICOLOGY
Oral Acute oral LD$_{50}$ for male and female rats *c.* 450, mice *c.* 150 mg/kg.
Skin and eye Acute percutaneous LD$_{50}$ (24 h) for rats >5000 mg/kg. Non-
irritating to eyes and skin (rabbits). Not a skin sensitiser.   **Inhalation** LC$_{50}$ (4 h) for
rats >5323 mg/m^3 dust, 69 mg/m^3 air (aerosol).   **NOEL** (2 y) for male rats 100,
female rats 300, mice 330 mg/kg diet; (52 w) for dogs 500 mg/kg diet.
ADI 0.057 mg/kg b.w.   **Other** Not mutagenic or teratogenic.
Toxicity class WHO (a.i.) II; EPA (formulation) II

ECOTOXICOLOGY

Birds Acute oral LD_{50} for Japanese quail 31, bobwhite quail 152 mg/kg. Dietary LC_{50} (5 d) for bobwhite quail 2225, mallard duck >5000 mg/kg. **Fish** LC_{50} (96 h) for golden orfe 237, rainbow trout 211 mg/l. **Bees** Harmful to honeybees by direct contact, but no problems expected when not sprayed into flowering crop or when used as a seed treatment. **Worms** LC_{50} for *Eisenia foetida* 10.7 mg/kg dry soil. **Daphnia** LC_{50} (48 h) 85 mg/l. **Algae** E_rC_{50} for *Scenedesmus subspicatus* >10 mg/l.

ENVIRONMENTAL FATE

Animals After oral administration of methylene-^{14}C- and 4,5-imidazolidine-^{14}C-labelled imidacloprid to rats, the radioactivity was quickly and almost completely absorbed from the gastro-intestinal tract and quickly eliminated (96% within 48 hours, mainly via the urine). Only c. 15% was eliminated as unchanged parent compound; the most important metabolic steps were hydroxylation at the imidazolidine ring, hydrolysis to 6-chloronicotinic acid, loss of the nitro group with formation of the guanidine and conjugation of the 6-chloronicotinic acid with glycine. All metabolites found in the edible organs and tissues of farm animals contained the 6-chloronicotinic acid moiety. Imidacloprid is also quickly largely eliminated from hens and goats. **Plants** Metabolism was investigated on rice (after soil treatment), maize (seed treatment), potatoes (granule or spray application), aubergines (granules) and tomatoes (spray treatment). In all cases, imidacloprid is metabolised by loss of the nitro group, hydroxylation at the imidazolidine ring, hydrolysis to 6-chloronicotinic acid and formation of conjugates; all metabolites contained the 6-chloropyridinylmethylene moiety. **Soil/Environment** In lab. studies, the most important metabolic steps were oxidation at the imidazolidine ring, reduction or loss of the nitro group, hydrolysis to 6-chloronicotinic acid and mineralisation; these processes were strongly accelerated by vegetation. Imidacloprid shows a medium adsorption to soil. Column leaching tests (with prior ageing) with a.i. and various formulations showed that imidacloprid and soil metabolites are to be classified as immobile; leaching into deeper soil layers is not to be expected if imidacloprid is used as recommended. Stable to hydrolysis under sterile conditions (under exclusion of light). Environmental DT_{50} c. 4 h (calc., based on tests of direct photolysis in aqueous solutions). Besides sunlight, the microbial activity of a water/sediment system is an important factor for the degradation of imidacloprid.

guanidine

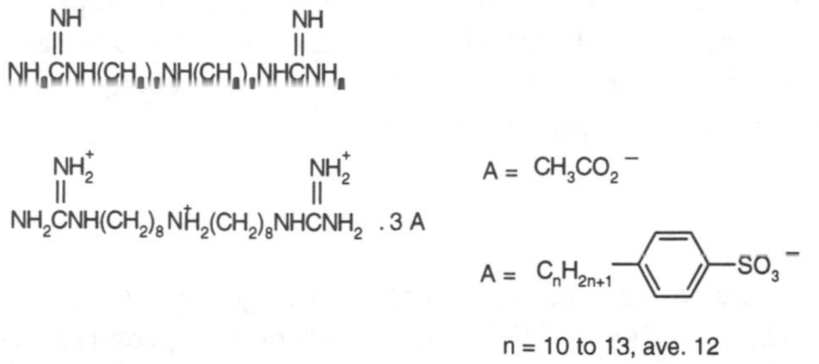

NOMENCLATURE
iminoctadine
Common name iminoctadine (BSI, draft E-ISO, (f) draft F-ISO); [guazatine] (see notes under Composition); [guanoctine] (see notes under Composition)
IUPAC name 1,1'-iminodi(octamethylene)diguanidine; bis(8-guanidino-octyl)amine
Chemical Abstracts name *N,N'''*-(iminodi-8,1-octanediyl)bisguanidine
CAS RN *[13516–27–3]* **EEC no.** 236–855–3

iminoctadine triacetate
IUPAC name 1,1'-iminodi(octamethylene)diguanidinium triacetate
CAS RN *[39202–40–9]* **Development codes** DF 125

iminoctadine tris(albesilate)
IUPAC name 1,1'-iminodi(octamethylene)diguanidinium tris(alkylbenzenesulfonate)
Other names iminoctadine tris(albesilate) **CAS RN** *[99257–43–9]*
Development codes DF-250

PHYSICAL CHEMISTRY
iminoctadine
Composition The name [guazatine] was applied to this structure (BSI, E-ISO, (f) F-ISO) until 1986 (BSI used the name [guanoctine] from 1970–1972), but the name now applies to a reaction mixture (see entry under guazatine).
Mol. wt. 355.6 **M.f.** $C_{18}H_{41}N_7$ **pKa** Strong base

iminoctadine triacetate
Mol. wt. 535.7 **M.f.** $C_{24}H_{53}N_7O_6$ **Form** Colourless crystals.
M.p. 143.0–144.2 °C. **V.p.** <0.4 mPa (23 °C) **K_{ow}** logP = –2.33 (pH 7) (for a.i.)
Solubility In water 764 g/l. In ethanol 117, methanol 777 g/l (25 °C).

iminoctadine tris(albesilate)

Composition The counterion is a mixture of C_{10} to C_{13} alkylbenzenesulfonates; weight average corresponds to C_{12}. **Mol. wt.** 1335 (average)
M.f. $C_{72}H_{131}N_7O_9S_3$ (ave.) **Form** Pale brown waxy solid (tech.).
M.p. 92–96 °C **V.p.** $<1.6 \times 10^{-1}$ mPa (60 °C) **K_{ow}** logP = 2.05 (pH 7)
Solubility In water 6 mg/l (20 °C). In methanol 5660, ethanol 3280, isopropanol 1800, benzene 0.22, acetone 0.55 g/l; insoluble in acetonitrile, dichloromethane, *n*-hexane, xylene, carbon disulfide and ethyl acetate (20 °C). **Stability** Stable in alkaline or acidic aqueous media at room temperature.

COMMERCIALISATION

iminoctadine

History Fungicidal properties of its salts were described originally by W. S. Catling *et al.* (*Congr. Plant Pathol.*, *1st*, 1968, p. 27 (Abstr.)), with details given later (*Jpn. Pestic. Inf.*, 1986, No. 49, p. 7).

iminoctadine triacetate

History Introduced by Evans Medical Ltd, developed by Murphy Chemical Ltd (now part of DowElanco) and introduced in Japan (1983) by Dainippon Ink and Chemicals Inc. **Manufacturers** Dainippon

iminoctadine tris(albesilate)

History Developed by Dainippon Ink & Chemicals, Inc. and reported by E. Adachi and K. Nakajima (*Agrochem. Japan*, **66**, 18 (1995)). First registered in Japan in 1994. **Patent** US 4659739; EP 155509 **Manufacturers** Dainippon

APPLICATIONS

iminoctadine

Biochemistry Affects the functions of the fungal cell membrane and lipid biosynthesis, at a site different from that of the steroid demethylation inhibiting fungicides. **Mode of action** Protective fungicide.

iminoctadine triacetate

Uses Fungicide used on cereals as seed treatment or foliar spray against *Fusarium*, *Septoria*, *Helminthosporium* and *Tilletia* spp.; on citrus pre-harvest against *Alternaria* and *Penicillium* spp.; on apples as a dormant spray against *Valsa ceratosperma*, or foliar spray against *Alternaria*, *Botryosphaeria*, *Gloeodes* or *Zygopiala* spp.
Selected tradenames 'Befran' (Dainippon)

iminoctadine tris(albesilate)

Uses Fungicide used on fruit, vegetables, and field crops against a wide range of fungal pathogens: scab (*Venturia* spp.) on apples and pears; scab (*Cladosporium* spp.) on peaches and Japanese apricot; grey mould on citrus, persimmon, onion, cucumber, tomato and lettuce; powdery mildew on persimmon, water melon, cucumber and strawberry; anthracnose (*Colletotrichum* spp.) on water melon, tea and kidney bean; *Sclerotinia* disease on water melon and lettuce; *Alternaria* blotch, ring rot, blotch, fly speck and sooty blotch on apples; black spot and *Physalospora* canker on pears; brown rot and *Phomopsis* rot on peaches; anthracnose (*Gloeosporium* spp.) and angular leaf spot on persimmon; *Botryosphaeria* disease on kiwi fruit; grey leaf spot on loquat; gummy stem blight on water melon; grey blight and shoot blight (*Pestalotia* spp.) on tea; stem blight and *Stemphylium* disease on asparagus; leaf blight on carrot; leaf mould on tomato; scab (*Fusarium roseum*) on wheat; early blight on potato; *Ramularia* leaf spot on sugar beet. **Formulation types** WP. **Mixtures** (*iminoctadine tris(albesilate)* +) iprodione; thiram; mancozeb. **Selected tradenames** 'Bellkute' (Dainippon)

ANALYSIS
Product analysis by hplc. **Residues** determined by gc after conversion to a pyrimidine derivative or by hplc (Y. Mori *et al.*, *Nihon Noyaku Gakkaishi*, 1982, **7**, 53).

MAMMALIAN TOXICOLOGY
iminoctadine
Toxicity class WHO (a.i.) II **EC risk** Xn (R21/22); Xi (R36/38)

iminoctadine triacetate
Oral Acute oral LD_{50} for rats c. 300, mice 400 mg/kg. **Skin and eye** Acute percutaneous LD_{50} for rats c. 1500 mg/kg; a mild skin and mild eye irritant, but not a skin sensitiser. **Inhalation** LC_{50} (4 h) for rats 0.073 mg/l air (for 25% liquid formulation). **NOEL** 0.356 mg/kg daily. **Other** Not embryotoxic at 8 mg/kg. Not mutagenic in Ames and Rec assays.

iminoctadine tris(albesilate)
Oral Acute oral LD_{50} for male and female rats 1400, male mice 4300, female mice 3200 mg/kg. **Skin and eye** Acute percutaneous LD_{50} for male and female rats >2000 mg/kg. Slight eye and skin irritation (rabbits). Not a skin sensitiser (guinea pigs). **Inhalation** LC_{50} 1.0 mg/l. **NOEL** 0.9 mg/kg daily. **Other** Negative in Ames, Rec and chromosome aberration tests.

ECOTOXICOLOGY
iminoctadine triacetate
Birds Acute oral LD_{50} for mallard ducks 985 mg/kg. **Fish** LC_{50} (96 h) for carp 200, rainbow trout 36 mg/l. **Bees** LD_{50} (oral and contact) >0.1 mg/bee. **Daphnia** LC_{50} (48 h) 2.1 mg/l

iminoctadine tris(albesilate)
Birds Japanese quail 1827 mg/kg. **Fish** LC_{50} (48 h) for rainbow trout 4.5 mg/l, for carp 14.4 ppm. **Worms** LC_{50} for earthworms >1000 ppm. **Daphnia** LC_{50} (3 h) for *D. carinata* >100 ppm.

ENVIRONMENTAL FATE
iminoctadine triacetate
Soil/Environment Soil DT_{50} 90 d (diluvial sandy loam), 122 d (volcanic ash loamy upland soil), 75 d (colluvial clayey loamy upland soil), 28 d (volcanic ash loamy upland soil).

iminoctadine tris(albesilate)
Soil/Environment K_{oc} 214 453.

420 imiprothrin *Insecticide*

pyrethroid

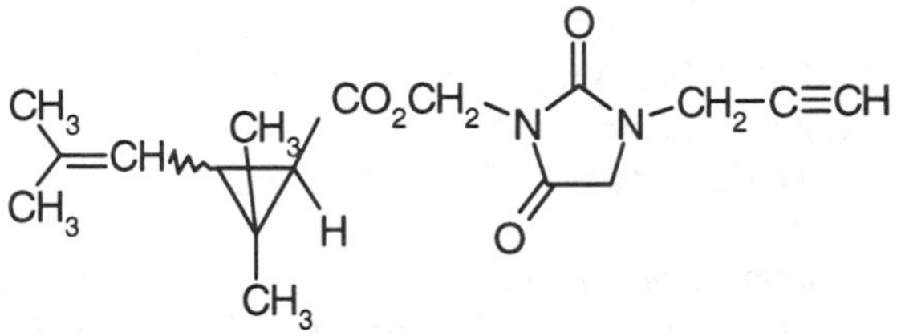

Mixture of isomers

NOMENCLATURE
Common name imiprothrin (BSI, pa E-ISO)
IUPAC name A mixture containing 20% of 2,5-dioxo-3-prop-2-ynylimidazolidin-=
1-ylmethyl (1R,3S)-2,2-dimethyl-3-(2-methylprop-1-enyl)cyclopropanecarboxylate and 80% of 2,5-dioxo-3-prop-2-ynylimidazolidin-1-ylmethyl (1R,3R)-2,2-dimethyl-=
3-(2-methylprop-1-enyl)cyclopropanecarboxylate
Chemical Abstracts name [2,5-dioxo-3-(2-propynyl)-1-imidazolidinyl]methyl 2,2-dimethyl-3-(2-methyl-1-propenyl)cyclopropanecarboxylate
CAS RN *[72963–72–5]* **Development codes** S-41311

PHYSICAL CHEMISTRY
Mol. wt. 318.4 **M.f.** $C_{17}H_{22}N_2O_4$ **Form** Viscous liquid. **V.p.** 1.8×10^{-3} mPa (25 °C) **K_{ow}** logP = 2.9 (25 °C) **S.g./density** 1.1 (20 °C) **Solubility** In water 93.5 mg/l (25 °C).

COMMERCIALISATION
Manufacturers Sumitomo

APPLICATIONS
Uses Household insecticide giving rapid knockdown against cockroaches and other

crawling insects. **Formulation types** AE. **Selected tradenames** 'Pralle'
(Sumitomo)

MAMMALIAN TOXICOLOGY
Oral Acute oral LD_{50} for male rats 1800, female rats 900 mg/kg.
Skin and eye Acute percutaneous LD_{50} for male and female rats >2000 mg/kg.
Inhalation LC_{50} for male and female rats >1200 mg/kg.

421   inabenfide   *Plant growth regulator*

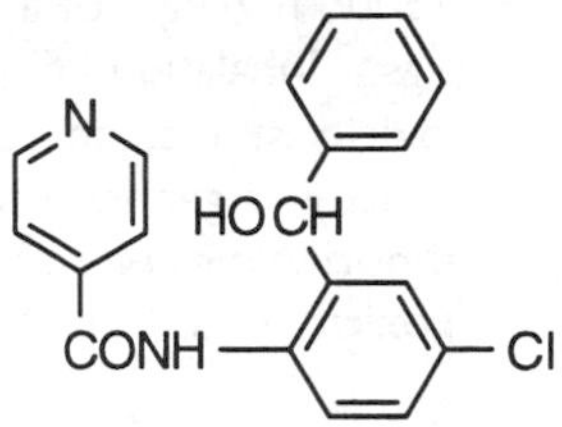

NOMENCLATURE
Common name inabenfide (BSI, draft E-ISO, (*m*) draft F-ISO)
IUPAC name 4'-chloro-2'-(α-hydroxybenzyl)isonicotinanilide
Chemical Abstracts name *N*-[4-chloro-2-(hydroxyphenylmethyl)phenyl]-=
4-pyridinecarboxamide
CAS RN *[82211–24–3]* unstated stereochemistry **Development codes**
CGR-811 (Chugai)

PHYSICAL CHEMISTRY
Mol. wt. 338.8 **M.f.** $C_{19}H_{15}ClN_2O_2$ **Form** Pale yellow-brown or colourless
crystals. **M.p.** 210–212 °C **V.p.** 0.063 mPa (20 °C) K_{ow} logP = 3.13
Solubility In water 1 mg/l (30 °C). In acetone 3.6, ethyl acetate 1.43, xylene 0.58,
methanol 2.35, ethanol 1.61, chloroform 0.59, dimethylformamide 6.72,
acetonitrile 0.58, tetrahydrofuran 1.61, hexane 8×10^{-4} (all in g/l, 30 °C).
Stability Stable to heat and sunlight. Slightly unstable in alkaline media; after
2 weeks at 40 °C, hydrolysed 16.2% (pH 2), 49.5% (pH 5), 83.9% (pH 7), 100%
(pH 11).

COMMERCIALISATION
History Plant growth regulator reported by K. Nakamura (*Jpn. Pestic. Inf.*, 1987,
No. 51, p. 23). Introduced in Japan (1986) by Chugai Pharmaceutical Co. Ltd
(agrochemical interests now Eikou Kasei Co Ltd). **Patent** EP 48998; US 4377407;
JP 6341393 **Manufacturers** Eikou Kasei

APPLICATIONS
Biochemistry Inhibits gibberellin biosynthesis.   **Mode of action** Plant growth
regulator which shortens lower internodes and upper leaf blades.   **Uses** Used to
increase resistance to lodging in rice, being applied to the soil surface under
submerged conditions at 1.5–2.4 kg a.i./ha.   **Phytotoxicity** Non-phytotoxic to
rice.   **Formulation types** GR; WP.   **Mixtures** (inabenfide +) pyroquilon;
probenazole.   **Selected tradenames** 'Seritard' (Eikou Kasei)

ANALYSIS
Product and **residue** analysis by hplc.

MAMMALIAN TOXICOLOGY
Reviews *J. Pestic. Sci.*, **13**, 391–394 (1988).   **Oral** Acute oral LD_{50} for rats and
mice >15 000 mg/kg.   **Skin and eye** Acute percutaneous LD_{50} for rats and mice
>5000 mg/kg; not a skin or eye irritant to rabbits, nor skin sensitiser to guinea
pigs.   **Inhalation** LC_{50} (4 h) for rats >0.46 mg/l air.   **NOEL** In 6 mo and 2 y
toxicity studies on dogs and rats, no ill-effects were observed. In reproductive
toxicity (3 generations) and teratogenicity studies on rats and rabbits, no
abnormalities were observed.   **Other** Not mutagenic in Ames assay.
Toxicity class WHO (a.i.) III (Table 5); EPA (formulation) IV

ECOTOXICOLOGY
Fish LC_{50} (48 h) for carp >30, grey mullet 11 mg/l.   **Daphnia** LC_{50} (3 h)
>30 mg/l.

ENVIRONMENTAL FATE
Animals The major urinary metabolite in the rat is 4-hydroxyinabenfide
(H. Kinoshita et al., *Xenobiotica*, 1987, **17**, 925).   **Plants** Metabolised to inabenfide
ketone.   **Soil/Environment** Half-life under Japanese paddy field conditions,
c. 4 mo.

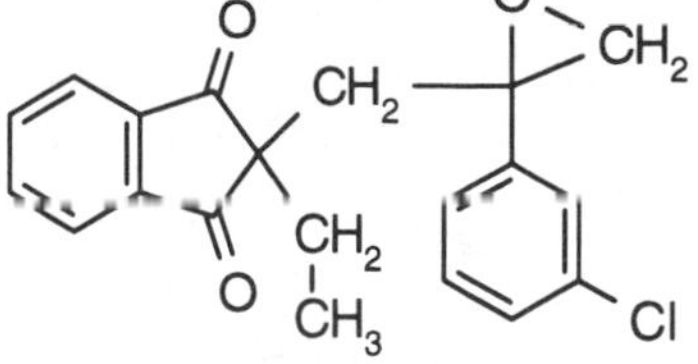

NOMENCLATURE
Common name indanofan (BSI, pa ISO)
IUPAC name (RS)-2-[2-(3-chlorophenyl)-2,3-epoxypropyl]-2-ethylindan-1,3-dione
Chemical Abstracts name 2-[[2-(3-chlorophenyl)oxiranyl]methyl]-2-ethyl-=
1H-indene-1,3(2H)-dione
CAS RN [133220–30–1]   **Development codes** MK-243; MX 70906

PHYSICAL CHEMISTRY
Mol. wt. 340.8   **M.f.** $C_{20}H_{17}ClO_3$ **M.p.** 60.0–62.5 °C   **V.p.** 2.66×10^{-3} mPa
(25 °C)   **Solubility** In water 14 ppm (25 °C).

COMMERCIALISATION
History Reported by O. Ikeda *et al., 8th IUPAC International Congress of Pesticide Chemistry*, Washington, 1994. Under development by Mitsubishi Chemical Corp.
Manufacturers Mitsubishi Chemical

APPLICATIONS
Uses Under development for both pre-emergence and post-emergence weed control in transplanted rice, controlling *Echinochloa crus-galli*, *Monochoria vaginalis*, *Lindernia procumbens*, *Cyperus difformis*, *Scirpus juncoides* and *Eleocharis acicularis*. Also under evaluation for pre-emergence weed control in wheat and barley.
Formulation types EC; SC; WP; GR.

MAMMALIAN TOXICOLOGY
Oral Acute oral LD_{50} for rats >1000 mg/kg.   **Skin and eye** Acute percutaneous LD_{50} for rats >2000 mg/kg. Mild eye irritant, not a skin irritant.   **Other** Negative in Ames test.

ECOTOXICOLOGY
Fish LC_{50} (48 h) for carp 5.0 ppm.

MAIN ENTRIES

indol-3-ylacetic acid *Plant growth regulator*

auxin

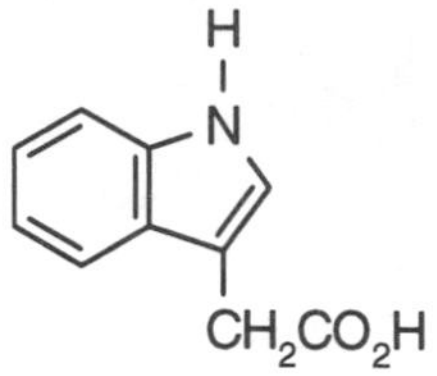

CH₂CO₂H structure shown.

NOMENCLATURE
Common name AIA (France)
IUPAC name indol-3-ylacetic acid; β-indoleacetic acid
Chemical Abstracts name 1*H*-indole-3-acetic acid
Other names IAA   **CAS RN** *[87–51–4]*

PHYSICAL CHEMISTRY
Mol. wt. 175.2   **M.f.** $C_{10}H_9NO_2$ **Form** Pale salmon-coloured crystals.
M.p. 165–169 °C   **V.p.** <0.02 mPa (60 °C)   **Solubility** In water 1.5 g/l (20 °C). In
ethanol 100–1000, acetone 30–100, diethyl ether 30–100, chloroform 10–30 (all
in g/l).   **Stability** Very stable in neutral, acidic and alkaline media. Unstable to light.
pKa 4.75

COMMERCIALISATION
History The main naturally-occurring hormone in higher plants.
Manufacturers ACF Chemiefarma

APPLICATIONS
Mode of action Plant growth regulator which affects cell division and cell
elongation.   **Uses** Used to stimulate rooting of cuttings of herbaceous and woody
ornamentals.   **Formulation types** TB; DP.   **Selected tradenames** 'Rhizopon A'
(Fargro)

ANALYSIS
Product by u.v. spectrophotometry; details from ACF.   **Residues** (1) by hplc with
u.v. detection (V. G. Rivera et al., *J. Chromatogr.*, 1986, **358**(1), 243–252); details
from ACF. (2) By glc with FID (D. R. de Yoe et al., *Can. J. For. Res.*, 1976, **6**(3),
429–435).

MAMMALIAN TOXICOLOGY
Skin and eye Acute percutaneous LD_{50} for mice 1000 mg/kg.

ECOTOXICOLOGY
Bees Not toxic to bees.

ENVIRONMENTAL FATE
Soil/Environment Rapidly degraded in soil.

424　4-indol-3-ylbutyric acid

Plant growth regulator

auxin

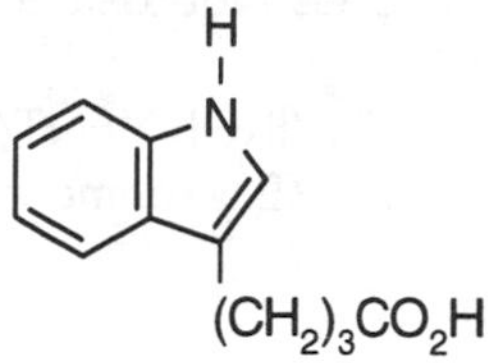

$(CH_2)_3CO_2H$

NOMENCLATURE
Common name 4-indol-3-ylbutyric acid (draft BSI, ISO-E, accepted in lieu of a common name)
IUPAC name 4-(indol-3-yl)butyric acid
Chemical Abstracts name 1*H*-indole-3-butanoic acid
Other names IBA　**CAS RN** *[133–32–4]*

PHYSICAL CHEMISTRY
Mol. wt. 203.2　**M.f.** $C_{12}H_{13}NO_2$　**Form** Colourless to pale yellow crystals.
M.p. 123–125 °C　**V.p.** <0.01 mPa (25 °C)　**Solubility** In water at 20 °C, 250 mg/l. In benzene >1000, acetone, ethanol, and diethyl ether 30–100, chloroform 0.01–0.1 (all in g/l).　**Stability** Very stable in neutral, acidic and alkaline media.　**F.p.** Non-flammable.

COMMERCIALISATION
History The ability of the acid and its simple esters to stimulate root formation in cuttings reported by P. W. Zimmerman & F. Wilcoxon (*Contrib. Boyce Thompson Inst.*, 1935, **7**, 209) and by P. W. Zimmerman & A. E. Hitchcock (*ibid.*, p. 439). Introduced by Union Carbide Corp. and by May & Baker Ltd (both now Rhône-Poulenc Agrochimie).　**Patent** US 3051723　**Manufacturers** Rhône-Poulenc

APPLICATIONS
Mode of action Plant growth regulator which acts on cell division and cell elongation.　**Uses** Used to stimulate rooting of cuttings of herbaceous and woody ornamentals.　**Formulation types** DP.　**Mixtures** (*4-indol-3-ylbutyric acid +*) 2-(1-naphthyl)acetic acid; 2-(1-naphthyl)acetic acid + thiram; captan +

4-indol-3-ylbutyric acid　717

2-(1-naphthyl)acetic acid.   **Selected tradenames** 'Seradix' (Rhône-Poulenc); 'Chryzoplus' (Fargro); 'Synergol' (mixture) (Hortichem)

ANALYSIS
By hplc. Details available from Rhône-Poulenc.

MAMMALIAN TOXICOLOGY
Oral Acute oral LD_{50} for mice 100 mg/kg.   **Other** Acute i.p. LD_{50} for mice 100 mg/kg (H. H. Anderson *et al.*, *Proc. Soc. Exp. Biol. Med.*, 1936, **34**, 138).
Toxicity class EPA (formulation) III

ECOTOXICOLOGY
Bees Not toxic to bees.

ENVIRONMENTAL FATE
Soil/Environment Rapidly degraded in soil.

425 ioxynil   *Herbicide*

hydroxybenzonitrile

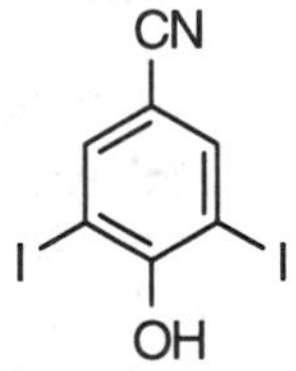

NOMENCLATURE
ioxynil
Common name ioxynil (BSI, E-ISO, (*m*) F-ISO, WSSA)
IUPAC name 4-hydroxy-3,5-di-iodobenzonitrile
Chemical Abstracts name 4-hydroxy-3,5-diiodobenzonitrile
CAS RN *[1689–83–4]*   **EEC no.** 216–881–1   **Development codes** ACP 63–303 (Amchem); M&B 8873 (May & Baker)

ioxynil octanoate
Common name ioxynil octanoate (BSI, E-ISO, (*m*) F-ISO, WSSA); ioxynil (JMAF)
IUPAC name 4-cyano-2,6-di-iodophenyl octanoate
Chemical Abstracts name 4-cyano-2,6-diiodophenyl octanoate
CAS RN *[3861–47–0]*   **EEC no.** 223–375–4   **Development codes** 15830 RP; M&B 11 641

718 ioxynil

ioxynil-sodium
CAS RN *[2961–62–8]*

PHYSICAL CHEMISTRY
ioxynil
Composition Tech. grade is *c.* 95% pure. **Mol. wt.** 370.9 **M.f.** $C_7H_3I_2NO$
Form Colourless solid (tech., cream-coloured with faint phenolic odour).
M.p. 212–213 °C (sublimes at *c.* 110 °C/0.1 mm Hg) **V.p.** <1 mPa (20 °)
K$_{ow}$ logP = 0.89 (unstated pH); logP = 3.43 (25 °C, unionised, K. Chamberlain *et al., Pestic. Sci.*, **47**, 265 (1996)) **Solubility** In water 50 mg/l (20 °C). In acetone
70, ethanol, methanol 20, cyclohexanone 140, tetrahydrofuran 340,
dimethylformamide 740, chloroform 10, carbon tetrachloride <1 (all in g/l,
25 °C). Potassium salt: In water 107 g/l (20–25 °C). In acetone 60, 20% aqueous
acetone 560, tetrahydrofurfuryl alcohol 750, methyl cellosolve 770 (all in g/l,
20–25 °C). **Stability** Stable in storage but readily hydrolysed by alkali.
Decomposed by u.v. light. Acidic in reaction, and forms salts. **pKa** 3.96

ioxynil octanoate
Mol. wt. 497.1 **M.f.** $C_{15}H_{17}I_2NO_2$ **M.p.** 59–60 °C **V.p.** 3.7 mPa (105 °C)
Solubility Practically insoluble in water (20–25 °C). In acetone 100, benzene,
chloroform 650, cyclohexanone, xylene 500, dichloromethane 700, ethanol 150
(all in g/l). **Stability** Stable to storage but readily hydrolysed by alkali.

ioxynil-sodium
Mol. wt. 392.9 **M.f.** $C_7H_2I_2NNaO$ **M.p.** *c.* 360 °C. **Solubility** In water 140 g
a.e./l (20–25 °C). In acetone 120, 20% *m/V* aqueous acetone 670, 2-methoxy=
ethanol 640, tetrahydrofurfuryl alcohol 650 (all in g/l).

COMMERCIALISATION
History Herbicidal properties of ioxynil reported independently by R. L. Wain
(*Nature (London)*, 1963, **200,** 28), by K. Carpenter & B. J. Heywood (*ibid.*, p. 28)
and by R. D. Hart *et al.* (*Proc. Br. Weed Control Conf., 7th*, 1964, p. 3).
Development reviewed by B. J. Heywood (*Chem. Ind. (London)*, 1966, p. 1946).
Introduced by May & Baker Ltd and by Amchem Products Inc. (both now within
Rhône-Poulenc Group). **Patent** GB 1067033 to May & Baker; US 3397054 to
Union Carbide **Manufacturers** CFPI; Makhteshim-Agan; Rhône-Poulenc;
Sanachem

APPLICATIONS
Biochemistry Inhibits photosynthesis, and uncouples oxidative phosphorylation.
Mode of action Selective contact herbicide with some systemic activity. Absorbed
by the foliage, with limited translocation. **Uses** Ioxynil and its salts and esters are
used for post-emergence control of a wide range of annual broad-leaved weeds,
especially young seedlings of the Polygonaceae, Compositae, and Boraginaceae, in

cereals, onions, leeks, garlic, shallots, flax, sugar cane, forage grasses, lawns, and newly-sown turf. Often used in combination with other herbicides, in order to extend the spectrum of control. **Formulation types** EC; SL. **Mixtures** *(ioxynil +)* bromoxynil; mecoprop; linuron; 2,4-D; dichlorprop + MCPA; bromoxynil + isoproturon; 2,4-D + dicamba; bromoxynil + mecoprop; bromoxynil + dichlorprop; clopyralid + mecoprop; isoproturon + mecoprop; bromoxynil + MCPA + mecoprop; and many more. **Compatibility** Incompatible with liquid fertilisers and water-soluble formulations of 2,4-D. **Selected tradenames** 'Actril' (Rhône-Poulenc); 'Totril' (Rhône-Poulenc); 'Iotril' (Makhteshim-Agan); 'Maestro II' (mixture) (CFPI); 'Mextrol' (CFPI); 'Sanoxynil' (octanoate) (Sanachem); 'Trionyl' (mixture) (Agriphar)

ANALYSIS

ioxynil

Product analysis by rp hplc with u.v. detection (*CIPAC Handbook*, 1992, **E**, 101–4), or by titration or by estimation of iodine (H. S. Segal & M. L. Sutherland, *Anal. Methods Pestic., Plant Growth Regul. Food Addit.*, 1967, **5**, 423; *Anal. Methods Pestic. Plant Growth Regul.*, 1972, **6**, 654). **Residues** determined by glc of derivatives (*loc. cit.*), by i.r. spectrometry as ioxynil, or by hplc after alkylation (E.G. Cotterill, *Analyst*, 1982, **107**, 76).

ioxynil octanoate

Product determined by gc (*CIPAC Handbook*, 1995, **G**, 94–97).

MAMMALIAN TOXICOLOGY

ioxynil

Oral Acute oral LD_{50} for rats 110, mice 230 mg/kg. **Skin and eye** Acute percutaneous LD_{50} for rats >2000 mg a.i./kg. **Inhalation** LC_{50} (6 h) for rats >3 mg/l air. **Other** Non-mutagenic in bacterial systems. **Toxicity class** WHO (a.i.) II; EPA (formulation) II   **EC risk** (R63); T (R25); Xn (R21)

ioxynil octanoate

Oral Acute oral LD_{50} for rats 190, mice 240 mg (formulated)/kg.
Skin and eye Acute percutaneous LD_{50} for rats >912, mice 1240 mg/kg.
NOEL (90 d) for rats 4 mg/kg daily. **Toxicity class** WHO (a.i.) II
EC risk (R63); Xn (R22)

ioxynil-sodium

Oral Acute oral LD_{50} for rats 120, mice 190 mg (formulated)/kg.
Skin and eye Acute percutaneous LD_{50} for rats 210 mg (formulated)/kg.
NOEL (90 d) for rats 5.5 mg/kg daily.

ECOTOXICOLOGY

ioxynil

Birds Acute oral LD_{50} for pheasants 75, hens 200 mg/kg. **Fish** LC_{50} (96 h) for

rainbow trout 8.5 mg/l (as phenol).   **Bees** LD$_{50}$ (48 h, oral) 4 μg/bee.
Worms LD$_{50}$ (7 & 14 d) for *Eisenia foetida c.* 35 mg/kg soil.   **Daphnia** EC$_{50}$ (48 h)
3.9 mg/l.   **Algae** EC$_{50}$ (96 h) for *Scenedesmus subspicatus* 24 mg/l.

ioxynil octanoate
Birds Acute oral LD$_{50}$ for pheasants 1000, mallard ducks 1200 mg/kg.   **Fish** LC$_{50}$
(48 h) for harlequin fish 4 mg/l.

ioxynil-sodium
Birds Acute oral LD$_{50}$ for pheasants 35 mg (formulated)/kg.   **Fish** LC$_{50}$ (48 h) for
harlequin fish 3.3 mg/l.   **Bees** Not toxic to bees.

ENVIRONMENTAL FATE
Plants In plants, the ester and nitrile groups are hydrolysed, and deiodination also
occurs.   **Soil/Environment** In soil, DT$_{50}$ c. 10 d. Degraded by hydrolysis and
deiodination to less toxic substances such as hydroxybenzoic acid. Low mobility;
the majority of ioxynil and its metabolites remained in the top 8 cm of all soil
types studied.

426 ipconazole   *Fungicide*

azole

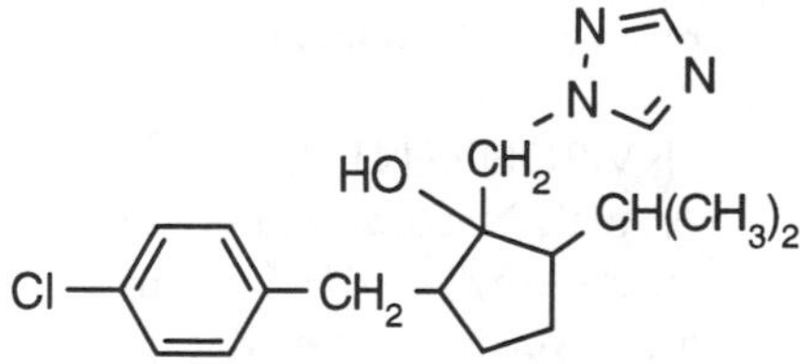

NOMENCLATURE
Common name ipconazole (BSI, pa E-ISO)
IUPAC name (1*RS*,2*SR*,5*RS*;1*RS*,2*SR*,5*SR*)-2-(4-chlorobenzyl)-5-isopropyl-=
1-(1*H*-1,2,4-triazol-1-ylmethyl)cyclopentanol
Chemical Abstracts name 2-[(4-chlorophenyl)methyl]-5-(1-methylethyl)-=
1-(1*H*-1,2,4-triazol-1-ylmethyl)cyclopentanol
CAS RN *[125225–28–7]* (unstated stereochemistry)
Development codes KNF-317

PHYSICAL CHEMISTRY
Composition The common name ipconazole refers to a mixture of the

(1*RS*,2*SR*,5*RS*) isomer pair (I) (hydroxy, benzyl and isopropyl groups on the same side of the cyclopentyl ring) and the (1*RS*,2*SR*,5*SR*) isomer pair (II).
Mol. wt. 333.9 **M.f.** $C_{18}H_{24}ClN_3O$ **Form** Colourless crystals. **M.p.** 88–90 °C
V.p. 3.58×10^{-3} mPa (25 °C) (isomer pair I); 6.99×10^{-3} mPa (25 °C) (isomer pair II) **K$_{ow}$** logP = 4.21 (25 °C) **Solubility** In water 6.93 mg/l (20 °C).
Stability Good thermal and hydrolytic stability.

COMMERCIALISATION
Patent EP 0267778; US 4938792 **Manufacturers** Kureha

APPLICATIONS
Biochemistry Inhibitor of ergosterol biosynthesis. **Uses** Control of a wide range of seed diseases in rice and other crops. It is particularly effective against 'Bakanae' disease, Helminthosporium leaf spot and blast on rice. The component isomer pairs are of equal activity. **Formulation types** WP. **Mixtures** *(ipconazole +)* copper hydroxide. **Selected tradenames** 'Techlead' (Kureha)

ANALYSIS
Product analysis by hplc or gc.

MAMMALIAN TOXICOLOGY
Oral Acute oral LD_{50} for rats 1338 mg/kg. **Skin and eye** Acute percutaneous LD_{50} for rats >2000 mg/kg. Not irritating to skin, slightly irritant to eyes (rabbits). Not a skin sensitiser.

ECOTOXICOLOGY
Fish LC_{50} (48 h) for carp 2.5 mg/l. **Other aquatic spp.** LC_{50} (3 h) for *Moina macrocopa* >40 mg/l.

ENVIRONMENTAL FATE
Animals Metabolites in the rat are 2-(4-chlorobenzyl)-5-(1-hydroxy-1-= methylethyl)-1-(1*H*-1,2,4-triazol-1-ylmethyl)cyclopentanol, 2-(4-chlorobenzyl)-= 5-(2-hydroxy-1-methylethyl)-1-(1*H*-1,2,4-triazol-1-ylmethyl)cyclopentanol and 1,2,4-triazole. **Plants** In rice, metabolites are 2-(4-chlorobenzyl)-5-(1-hydroxy-= 1-methylethyl)-1-(1*H*-1,2,4-triazol-1-ylmethyl)cyclopentanol, 2-(4-chlorobenzyl)-= 5-(2-hydroxy-1-methylethyl)-1-(1*H*-1,2,4-triazol-1-ylmethyl)cyclopentanol and 2-[1-(4-chlorophenyl)hydroxymethyl]-5-isopropyl-1-(1*H*-1,2,4-triazol-1-ylmethyl)= cyclopentanol.

organophosphate ester

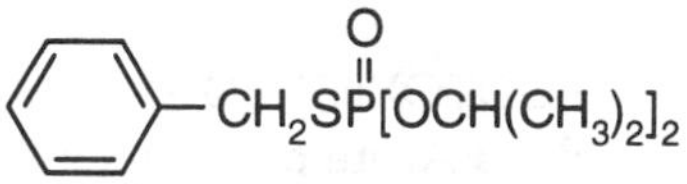

NOMENCLATURE
Common name iprobenfos (BSI, draft E-ISO, (*m*) draft F-ISO); IBP (JMAF)
IUPAC name S-benzyl O,O-di-isopropyl phosphorothioate
Chemical Abstracts name O,O-bis(1-methylethyl) S-(phenylmethyl)
phosphorothioate
CAS RN *[26087–47–8]*   **EEC no.** 247–449–0

PHYSICAL CHEMISTRY
Composition Tech. is *c.* 94% pure.   **Mol. wt.** 288.3   **M.f.** $C_{13}H_{21}O_3PS$
Form Colourless, transparent oil (tech., yellow oil).   **B.p.** 126 °C/0.04 mmHg
V.p. 0.247 mPa (20 °C) K_{ow} logP = 3.21   **S.g./density** 1.103 (20 °C)
Solubility In water 430 mg/l (20 °C). In acetone, acetonitrile, methanol, xylene
>1 (all in kg/l).   **Stability** DT_{50} in water 7230–7793 h (pH 4–9).

COMMERCIALISATION
History Reported by Kumiai Chemical Industry Co., Ltd. (M. Kado *et al.*, *Ann.*
Phytopath. Soc. Jap., **34**(3), 188 (1968)). Reviewed by Y. Uesugi (*Jpn. Pestic. Inf.*,
1970, No. 2, p. 11). Introduced by Kumiai.   **Manufacturers** Hanwha; Ihara/Kumiai

APPLICATIONS
Biochemistry Phospholipid synthesis inhibitor.   **Mode of action** Systemic fungicide
with protective and curative action. Absorbed through the roots and leaves.
Translocation and metabolism in the rice plant have been reported (H. Yamamoto
et al., *Agric. Biol Chem.*, 1973, **37**, 1553).   **Uses** Control of leaf and ear blast
(*Pyricularia oryzae*), stem rot (*Helminthosporium* spp.) and sheath blight (*Rhizoctonia*
solani) in rice.   **Phytotoxicity** Non-phytotoxic to rice. Injury has been noted on
beans, soya beans and aubergines.   **Formulation types** EC; DP; GR.
Mixtures (*iprobenfos +*) mepronil; tricyclazole.   **Compatibility** Combined use with
some insecticides enhances their insecticidal effect against leafhoppers, especially
against insecticide-resistant strains.   **Selected tradenames** 'Kitazin P' (Kumiai)

ANALYSIS
Product analysis by glc with FID (*CIPAC Handbook*, 1988, **D**, 110).
Residues determined by glc with FPD.

MAMMALIAN TOXICOLOGY
Oral Acute oral LD_{50} for male rats 790, female rats 680, male mice 1830, female

MAIN ENTRIES

mice 1760 mg/kg.   **Skin and eye** Acute percutaneous LD_{50} for mice 4000 mg/kg.
Inhalation LC_{50} (4 h) for male rats 1.12, female rats 0.34 mg/l air.   **NOEL** (2 y) for
male rats 0.036, female rats 0.45 mg/kg b.w. daily.   **Toxicity class** WHO (a.i.) III;
EPA (formulation) II   **EC risk** Xn (R22)

ECOTOXICOLOGY
Birds Acute oral LD_{50} for cocks 705 mg/kg.   **Fish** LC_{50} for carp 5.1 mg/l.

ENVIRONMENTAL FATE
Animals Three metabolites were identified in urine and faeces of rats.   **Plants** For
metabolism in plants, see H. Yamamoto et al., (Agric. Biol. Chem., 1973, **37**, 1553).
Soil/Environment DT_{50} 15 d (two soils). K_{oc} 5030.

428 iprodione *Fungicide*

dicarboximide

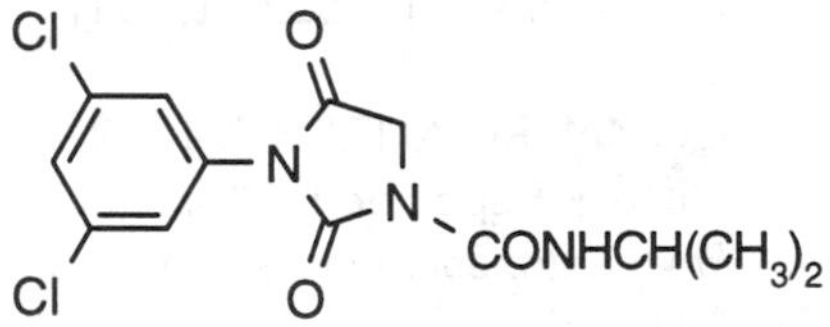

NOMENCLATURE
Common name iprodione (BSI, E-ISO, (m) F-ISO, ANSI)
IUPAC name 3-(3,5-dichlorophenyl)-N-isopropyl-2,4-dioxoimidazolidine-=
1-carboxamide
Chemical Abstracts name 3-(3,5-dichlorophenyl)-N-(1-methylethyl)-2,4-dioxo-=
1-imidazolidinecarboxamide
Other names [glycophene] (rejected common name proposal)
CAS RN [36734–19–7]   **EEC no.** 253–178–9   **Development codes** 26 019 RP

PHYSICAL CHEMISTRY
Composition Tech. is ≥96% pure.   **Mol. wt.** 330.2   **M.f.** $C_{13}H_{13}Cl_2N_3O_3$
Form White, odourless, non-hygroscopic crystals or powder.   **M.p.** 134 °C; (tech.
128–128.5 °C)   **V.p.** 5×10^{-4} mPa (25 °C) K_{ow} logP = 3.0 (pH 3 and 5)
S.g./density 1.00 (20 °C); (tech. 1.434–1.435)   **Solubility** In water 13 mg/l (20
°C). In n-octanol 10, acetonitrile 168, toluene 150, ethyl acetate 225, acetone 342,
dichloromethane 450, hexane 0.59 (all in g/l, 20 °C).   **Stability** Relatively stable in
acid media, but decomposed in alkaline media. DT_{50} 1–7 d (pH 7), <1 h (pH 9).
Aqueous solutions are degraded by u.v. light, but are relatively stable in simulated
sunlight.

COMMERCIALISATION

History Fungicide reported by L. Lacroix et al. (*Phytiatr. Phytopharm.*, 1974, **23**, 165). Introduced by Rhône-Poulenc Agrochimie. **Patent** GB 1312536; US 3755350; FR 2120222 **Manufacturers** Rhône-Poulenc; Sinon

APPLICATIONS

Mode of action Contact fungicide with protective and curative action. Inhibits germination of spores and growth of fungal mycelium. **Uses** Control of *Botrytis*, *Monilia*, *Sclerotinia*, *Alternaria*, *Corticium*, *Fusarium*, *Helminthosporium*, *Phoma*, *Rhizoctonia*, *Typhula* spp., etc. Used mainly on sunflowers, cereals, fruit trees, berry fruit, oilseed rape, rice, cotton, vegetables, and vines as a foliar spray at 0.5–1.0 kg a.i./ha, and on turf at 3–12 kg/ha. Can also be used as a post-harvest dip, as a seed treatment, or as a dip or spray at planting. **Formulation types** WP; SC; WG; FS; SU; EC; DP. **Mixtures** (*iprodione* +) carbendazim; thiophanate-methyl; bromuconazole; thiram; diniconazole; imazalil; chlorothalonil; fosetyl-aluminium. **Selected tradenames** 'Kidan' (Rhône-Poulenc); 'Rovral' (Rhône-Poulenc); 'Verisan' (Rhône-Poulenc)

ANALYSIS

Product analysis by hplc or glc (*CIPAC Handbook*, 1995, **G**, 98–104; L. Lacroix et al., *Anal. Methods Pestic. Plant Growth Regul.*, 1980, **11**, 247). **Residues** determined by glc with ECD (*idem, ibid.*; *Man. Pestic. Residue Anal.*, 1987, **I**, 6, S8, S19; *Anal. Methods Residues Pestic.*, 1988, Part I, M1, M12).

MAMMALIAN TOXICOLOGY

Reviews *Pesticide residues in food – 1995*, FAO Plant Production and Protection Paper 133, 1996. *Pesticide residues in food – 1995 evaluations; Part II – Toxicology & Environment*. World Health Organisation, WHO/PCS/96.48, 1996. **Oral** Acute oral LD_{50} for rats and mice >2000 mg/kg. **Skin and eye** Acute percutaneous LD_{50} for rats and rabbits >2000 mg/kg. Non-irritating to skin and eyes (rabbits). **Inhalation** LC_{50} (4 h) for rats >5.16 mg/l air. **NOEL** (2 y) for rats 150 mg/kg diet; (1 y) for dogs 18 mg/kg b.w. **ADI** (JMPR) 0.06 mg/kg b.w. [1995]. **Toxicity class** WHO (a.i.) III (Table 5); EPA (formulation) IV

ECOTOXICOLOGY

Birds Acute oral LD_{50} for bobwhite quail >2000, mallard ducks >10 400 mg/kg. **Fish** LC_{50} (96 h) for rainbow trout 4.1, bluegill sunfish 3.7 mg/l. **Bees** Contact LD_{50} >0.4 mg/bee. **Worms** LC_{50} for earthworms >1000 mg/kg soil. **Other beneficial spp.** Harmless. **Daphnia** LC_{50} (48 h) 0.25 mg/l. **Algae** EC_{50} (120 h) for *Selenastrum capricornutum* 1.9 mg/l.

ENVIRONMENTAL FATE

Animals In rats, ruminants and birds, iprodione is rapidly eliminated. It also undergoes extensive metabolism, by hydrolysis and rearrangement reactions.

Plants Metabolism studies in cereals, fruit, leafy and oily crops showed that iprodione is the dominant component of the total residue resulting from foliar application. **Soil/Environment** Rapidly metabolised in soil, with formation of CO_2. DT_{50} (lab.) 20–80 d; (field) 20–160 d. K_{oc} 373 to 1551. Rate of degradation increases with successive treatments, hence accumulation does not occur.

429 isazofos

Nematicide, insecticide

organophosphorus

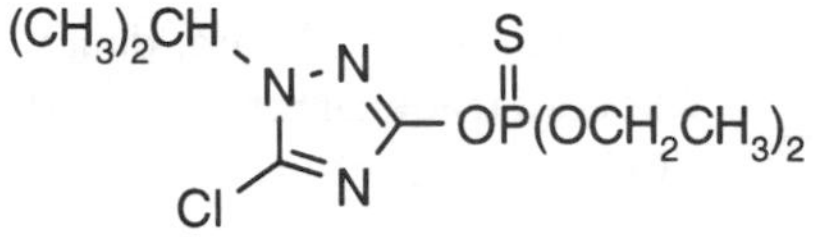

NOMENCLATURE
Common name isazofos (BSI, E-ISO, (*m*) F-ISO, ANSI)
IUPAC name *O*-5-chloro-1-isopropyl-1*H*-1,2,4-triazol-3-yl *O,O*-diethyl phosphorothioate
Chemical Abstracts name *O*-[5-chloro-1-(1-methylethyl)-1*H*-1,2,4-triazol-3-yl] *O,O*-diethyl phosphorothioate
CAS RN *[42509–80–8]* **Development codes** CGA 12 223

PHYSICAL CHEMISTRY
Mol. wt. 313.7 **M.f.** $C_9H_{17}CIN_3O_3PS$ **Form** Yellow liquid. **B.p.** 120 °C (36 Pa)
V.p. 7.45 mPa (20 °C) **K$_{ow}$** logP = 2.99 **S.g./density** 1.23 (20 °C) **Solubility** In water 168 mg/l (20 °C). Miscible with organic solvents, e.g. chloroform, methanol, benzene. **Stability** Hydrolysed more rapidly in alkalis than in acids; on hydrolysis DT_{50} (calc.) (20 °C) 85 d at pH 5, 48 d at pH 7, and 19 d at pH 9. Decomposes above 200 °C.

COMMERCIALISATION
History Insecticide and nematicide reported by D. Dawes *et al.* (*Meded. Fac. Landbouwwet. Rijksuniv. Gent*, 1974, **39**, 727) and F. Bachmann & D. Dawes (*ibid.*, p. 801). Developed by Ciba-Geigy AG (now Novartis Crop Protection AG).
Patent BE 792452; GB 1419131; GB 1419132 **Manufacturers** Novartis

APPLICATIONS
Biochemistry Cholinesterase inhibitor. **Mode of action** Nematicide and systemic insecticide with contact and stomach action. **Uses** Isazofos is a soil-applied nematicide with contact and systemic properties for use on bananas (against

Radopholus similis and *Cosmopolites sordidus*), citrus, cotton, beet, maize, rice, vegetables, sugar cane, and in turf against soil-dwelling insects.
Phytotoxicity Phytotoxic to potatoes and tobacco. **Formulation types** EC; GR.
Selected tradenames 'Miral' (Novartis)

ANALYSIS
Product analysis by glc. **Residues** determined by glc with TID. Details available from Novartis.

MAMMALIAN TOXICOLOGY
EHC 63 (WHO, 1986; a general review of organophosphorus insecticides).
Oral Acute oral LD_{50} for rats 40–60 mg/kg. **Skin and eye** Acute percutaneous LD_{50} for male rats >3100, female rats 118 mg/kg. Mild skin irritant; very slight eye irritant (rabbits). **Inhalation** LC_{50} (4 h) for rats 0.24 mg/l air. **NOEL** (90 d) for rats 2 mg/kg diet (0.2 mg/kg daily), for dogs 2 mg/kg diet (0.05 mg/kg daily).
Toxicity class WHO (a.i.) Ib; EPA (formulation) I **EC risk** R24/25, R26, R43

ECOTOXICOLOGY
Birds Toxic to birds. Acute oral LD_{50} for mallard ducks 61, bobwhite quail 11.1 mg/kg. LC_{50} (8 d) for bobwhite quail 81 ppm. **Fish** LC_{50} (96 h) for bluegill sunfish 0.01, carp 0.22, trout 0.008–0.019 mg/l. **Bees** Toxic to honeybees.
Daphnia LC_{50} (48 h) 0.0014 mg/l.

ENVIRONMENTAL FATE
Animals It is extensively metabolised and rapidly excreted, primarily in the urine. No accumulation takes place in the body. **Soil/Environment** DT_{50} 10 d (lab.).

430 isofenphos   *Insecticide*

organophosphorus

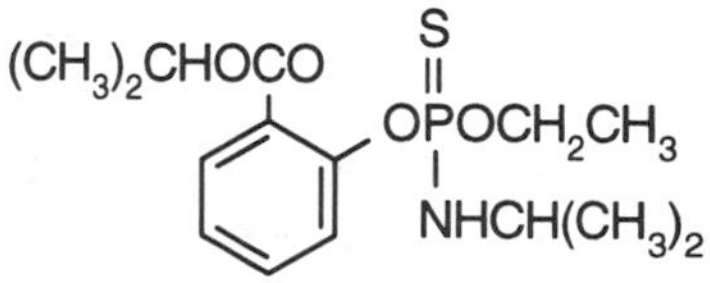

(CH₃)₂CHOCO — S‖OPOCH₂CH₃ / NHCH(CH₃)₂

NOMENCLATURE
Common name isofenphos (BSI, E-ISO); isophenphos ((*m*) F-ISO)
IUPAC name O-ethyl O-2-isopropoxycarbonylphenyl N-isopropylphosphoramido=thioate; isopropyl O-[ethoxy-N-isopropylamino(thiophosphoryl)]salicylate

Chemical Abstracts name 1-methylethyl 2-[[ethoxy[(1-methylethyl)amino]=
phosphinothioyl]oxy]benzoate
CAS RN *[25311–71–1]* **EEC no.** 246-814-1 **Development codes**
BAY SRA 12 869; BAY 92114

PHYSICAL CHEMISTRY
Mol. wt. 345.4 **M.f.** $C_{15}H_{24}NO_4PS$ **Form** Colourless oil; tech. has a
characteristic smell. **V.p.** 0.22 mPa (20 °C), 0.44 mPa (25 °C) **K_{ow}** logP = 4.04
(21 °C) **Henry** 4.2×10^{-3} Pa m^3 mol^{-1} (20 °C) **S.g./density** 1.131 (20 °C)
Solubility In water 18 mg/l (20 °C). In isopropanol, hexane, dichloromethane,
toluene >200 (all in g/l, 20 °C). **Stability** DT_{50} for aqueous hydrolysis 2.8 y
(pH 4), >1 y (pH 7), >1 y (pH 9) (22 °C). Photodegradation on soil surface in the
laboratory is extremely rapid. Under natural light, photodegradation is not so fast.
F.p. >115 °C (tech.)

COMMERCIALISATION
History Insecticide reported by B. Homeyer (*Meded. Fac. Landbouwwet., Rijksuniv.
Gent,* 1974, **39,** 789). Introduced by Bayer AG. **Patent** DE 1668047
Manufacturers Bayer

APPLICATIONS
Biochemistry Cholinesterase inhibitor. **Mode of action** Systemic insecticide with
contact and stomach action. Translocated from roots to a limited extent.
Uses Control of cabbage root flies, carrot flies, onion flies, corn rootworms,
white grubs, wireworms, and other soil insects in vegetables, oilseed rape, maize,
bananas, turf, sugar cane, and other crops; and also controls thrips in citrus and
onions. **Formulation types** GR; EC; WP; DS. **Mixtures** *(isofenphos +)* thiram;
fenamiphos; phoxim; carbofuran; disulfoton. **Selected tradenames** 'Oftanol'
(Bayer)

ANALYSIS
Product analysis by glc with FID or TCD (*CIPAC Handbook,* 1988, **D,** 114; *AOAC
Methods,* 1995, 987.01). **Residues** determined by glc with FID (*Man. Pestic.
Residue Anal.,* 1987, 6; *Anal. Methods Residues Pestic.,* 1988, Part I, M5, M13;
K. Wagner, *Pflanzenschutz-Nachr. (Engl. Ed.),* 1976, **29,** 67). Methods for the
determination of residues are available from Bayer.

MAMMALIAN TOXICOLOGY
Reviews *Pesticide residues in food – 1986.* FAO Plant Production and Protection
Paper 77, 1986. *Pesticide residues in food – 1986 evaluations. Part II – Toxicology.*
FAO Plant Production and Protection Paper 78/2, 1987. **EHC** 63 (WHO, 1986; a
general review of organophosphorus insecticides). **Oral** Acute oral LD_{50} for male
and female rats *c.* 20, mice *c.* 125 mg/kg. **Skin and eye** Acute percutaneous LD_{50}
for male and female rats *c.* 70 mg/kg. Slightly irritating to eyes and skin (rabbits).

Inhalation LC_{50} (4 h) for male rats c. 0.5 mg/l, female rats 0.3 mg/l air (aerosol). **NOEL** (2 y) for rats 1, dogs 2, mice 1 mg/kg diet. **ADI** (JMPR) 0.001 mg/kg b.w. [1986]. **Toxicity class** WHO (a.i.) Ib; EPA (formulation) I **EC risk** T (R24/25)

ECOTOXICOLOGY
Birds Acute oral LD_{50} for bobwhite quail 8.7, mallard ducks 32–36 mg/kg. Five-day LC_{50} for bobwhite quail 145, mallard ducks 4908 mg/kg diet. **Fish** LC_{50} (69 h) for golden orfe 6.19, bluegill sunfish 2.2, rainbow trout 3.3 mg/l. **Bees** Toxic to bees. **Worms** LC_{50} for *Eisenia foetida* 404 mg/kg dry soil. **Daphnia** LC_{50} (48 h) 0.0039–0.0073 mg/l. **Algae** E_rC_{50} for *Scenedesmus subspicatus* 6.8 mg/l.

ENVIRONMENTAL FATE
Animals For a study of metabolism in rats and in rat liver microsomal systems, see M. Ueji & C. Tomizawa, *J. Pestic. Sci.*, 1987, **12**, 245, 269. Elimination is very quick; almost 95% is excreted within 24 h, in urine and faeces. **Plants** The most important metabolites are salicylic acid and dihydroxybenzol acid. **Soil/Environment** Mobility is medium to low. Degradation in different soils is slow.

431 isoprocarb   *Insecticide*

carbamate

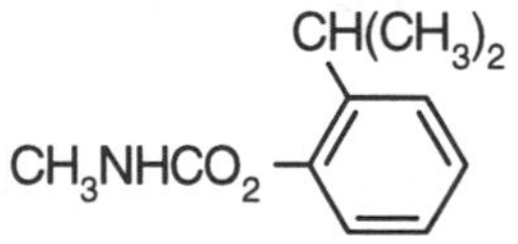

NOMENCLATURE
Common name isoprocarb (BSI, E-ISO); isoprocarbe ((m) F-ISO); MIPC (JMAF) **IUPAC name** o-cumenyl methylcarbamate; 2-isopropylphenyl methylcarbamate **Chemical Abstracts name** 2-(1-methylethyl)phenyl methylcarbamate **CAS RN** [2631–40–5] **EEC no.** 220–114–6 **Development codes** Bayer KHE 0145; BAY 105807 **Official codes** OMS 32; ENT 25 670

PHYSICAL CHEMISTRY
Mol. wt. 193.2 **M.f.** $C_{11}H_{15}NO_2$ **Form** Colourless crystals. **M.p.** 93–96 °C **B.p.** 128–129 °C/20 mmHg **V.p.** 2.8 mPa (20 °C) K_{ow} logP = 2.30 (25 °C) **S.g./density** 0.62 **Solubility** In water 0.265 g/l. In acetone 400, methanol 125 (both in g/l). **Stability** Hydrolysed in alkaline media.

COMMERCIALISATION
History Insecticide reported (*Jpn. Pestic. Inf.*, 1969, No. 1, p. 22). Introduced by Bayer AG and Mitsubishi Chemical Industries Ltd (now Mitsubishi Chemical Corp.). **Manufacturers** Chunhu; Jin Hung; Kuo Ching; Mitsubishi Chemical; Planters Products; Shenzhen Jiangshan; Sinon; Taiwan Tainan Giant

APPLICATIONS
Biochemistry Cholinesterase inhibitor. **Mode of action** Insecticide with contact and stomach action. Fast-acting with moderately long residual activity.
Uses Control of leafhoppers, planthoppers, aphids, capsids, bugs, etc., on rice, cocoa, sugar cane, vegetables and other crops. **Formulation types** WP; DP; EC; GR; HN. **Mixtures** (*isoprocarb* +) cartap hydrochloride; diazinon; dimethylvinphos; fenthion; monocrotophos. **Compatibility** Incompatible with alkaline materials. **Selected tradenames** 'Etrofolan' (Bayer); 'Isso' (Sanonda); 'Mipcin' (Mitsubishi Chemical)

ANALYSIS
Product analysis by hplc (*CIPAC Handbook*, 1988, **D**, 118). **Residues** determined by glc (*Nihon Noyaku Gakkaishi*, 1978, **3**, 119). Methods for the determination of residues are available from Bayer.

MAMMALIAN TOXICOLOGY
EHC 64 (WHO, 1986; a review of carbamate insecticides in general). **Oral** Acute oral LD_{50} for rats *c*. 450 mg/kg. **Skin and eye** Acute percutaneous LD_{50} for rats >500 mg/kg. Slightly irritating to eyes and skin (rabbits). **Inhalation** LC_{50} (4 h) for rats >0.5 mg/l (aerosol). **NOEL** (90 d) for rats 300, dogs 500 mg/kg diet.
Toxicity class WHO (a.i.) II; EPA (formulation) II **EC risk** Xn (R22)

ECOTOXICOLOGY
Fish LC_{50} (96 h) for carp 10–20, golden orfe 20–40 mg/l. **Bees** Harmful to honeybees. **Daphnia** LC_{50} (3 h) 0.30 mg/l.

ENVIRONMENTAL FATE
Animals Metabolites are 2-isopropylphenol and 2-(1-hydroxy-1-methylethyl)-=phenyl *N*-methylcarbamate. Metabolism of carbamate insecticides is reviewed (M. Cool & C. K. Jankowski in "*Insecticides*"). **Plants** As for animals.
Soil/Environment DT_{50} 3–20 d under paddy conditions. Low mobility.

$$\left[\begin{array}{c} S \\ S \end{array}\right] \hspace{-0.5em} = \hspace{-0.5em} \left\langle \begin{array}{c} CO_2CH(CH_3)_2 \\ CO_2CH(CH_3)_2 \end{array} \right.$$

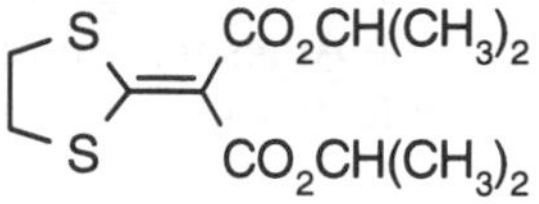

NOMENCLATURE
Common name isoprothiolane (BSI, JMAF, draft E-ISO, (*m*) draft F-ISO)
IUPAC name di-isopropyl 1,3-dithiolan-2-ylidenemalonate
Chemical Abstracts name bis(1-methylethyl)-1,3-dithiolan-2-ylidenepropanedioate
Other names IPT   **CAS RN** *[50512–35–1]*   **Development codes** SS 11 946;
NNF-109 (both Nihon Nohyaku)

PHYSICAL CHEMISTRY
Composition Tech. is ≥96%.   **Mol. wt.** 290.4   **M.f.** $C_{12}H_{18}O_4S_2$ **Form** Colourless
odourless crystals; (tech., yellow solid with irritating smell).   **M.p.** 54–54.5 °C;
(tech. 50–51 °C)   **B.p.** 167–169 °C/0.5 mmHg   **V.p.** 1.9×10^3 mPa (25 °C)
K_{ow} logP = 3.3 (25 °C)   **S.g./density** 1.044   **Solubility** In water 54 mg/l (25 °C).
In methanol 1510, ethanol 760, acetone 4060, chloroform 4130, benzene 2770,
n-hexane 10 (all in g/l, 25 °C).   **Stability** Stable to acid, alkali, light and heat.

COMMERCIALISATION
History Fungicide reported by F. Araki *et al.* (*Proc. Insectic. Fungic. Conf., 8th*,
1975, **2**, 715). Introduced by Nihon Nohyaku Co., Ltd.   **Manufacturers** Nihon
Nohyaku

APPLICATIONS
Biochemistry Inhibits penetration and elongation of infection hyphae, by inhibiting
formation of infection peg or cellulase secretion.   **Mode of action** Systemic
fungicide with protective and curative action, absorbed by the leaves and roots,
with translocation acropetally and basipetally. The mobility in rice plants of various
analogues reported (M. Uchida, *Pestic. Biochem. Physiol.*, 1980, **14**, 249).
Uses Control of rice blast (*Pyricularia oryzae*), rice stem rot, and *Fusarium* leaf spot
on rice. Also reduces the insect population (particularly planthoppers) on rice.
Phytotoxicity Phytotoxic only to cucurbitaceae.   **Formulation types** GR; EC; WP;
DP; driftless dust.   **Selected tradenames** 'Fuji-one' (Nihon Nohyaku)

ANALYSIS
Product analysis by glc with FID (*CIPAC Handbook*, 1992, **E**, 105–9; T. Hattori &
M. Kanauchi, *Anal. Methods Pestic. Plant Growth Regul.*, 1978, **10**, 229).
Residues determined by glc with ECD (*idem, ibid.*).

MAMMALIAN TOXICOLOGY
Oral Acute oral LD$_{50}$ for male rats 1190, female rats 1340, male mice
1340 mg/kg. **Skin and eye** Acute percutaneous LD$_{50}$ for male and female rats
>10 250 mg/kg. Slight eye irritant, not a skin irritant. **Other** Non-mutagenic in
the Ames test. In reproduction and teratogenicity studies on rats, no adverse
effects were observed. **Toxicity class** WHO (a.i.) III; EPA (formulation) III

ECOTOXICOLOGY
Birds Acute oral LD$_{50}$ for male Japanese quail 4710, female Japanese quail
4180 mg/kg. **Fish** LC$_{50}$ (48 h) for rainbow trout 6.8, carp 7.0 mg/l.
Daphnia LC$_{50}$ (3 h) 62 ppm.

433 isoproturon   *Herbicide*

urea

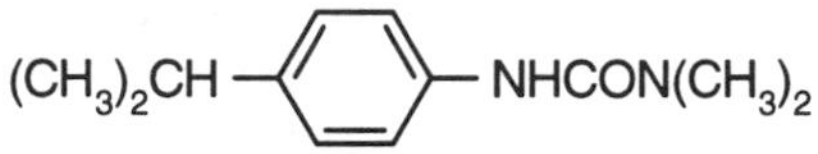

$$(CH_3)_2CH - \!\!\!\!\bigcirc\!\!\!\!- NHCON(CH_3)_2$$

NOMENCLATURE
Common name isoproturon (BSI, E-ISO, (m) F-ISO)
IUPAC name 3-(4-isopropylphenyl)-1,1-dimethylurea; 3-p-cumenyl-
1,1-dimethylurea
Chemical Abstracts name *N,N*-dimethyl-*N'*-[4-(1-methylethyl)phenyl]urea
Other names IPU; [ipuron] **CAS RN** *[34123–59–6]* **EEC no.** 251–835–4
Development codes Hoe 16410; AE F016410; CGA 18 731; 35689 RP;
LS 6912999

PHYSICAL CHEMISTRY
Composition Tech. is ≥98.5% pure. **Mol. wt.** 206.3 **M.f.** $C_{12}H_{18}N_2O$ **Form**
Colourless crystals. **M.p.** 158 °C; tech. 153–156 °C **V.p.** 3.15 × 10^{-3} mPa
(20 °C); 8.1 × 10^{-3} mPa (25 °C) **K$_{ow}$** logP = 2.5 (20 °C) **Henry** 1.46 × 10^{-5} Pa
m^3 mol^{-1} **S.g./density** 1.2 (20 °C) **Solubility** In water 65 mg/l (22 °C). In
methanol 75, dichloromethane 63, acetone 38, benzene 5, xylol 4, *n*-hexane *c.* 0.2
(all in g/l, 20 °C). **Stability** Very stable to light, acids, and alkalis. Hydrolytically
cleaved by strong alkalis on heating.

COMMERCIALISATION
History Herbicide introduced by Hoechst AG (now AgrEvo GmbH), Ciba-Geigy
AG (now Novartis Crop Protection AG, who no longer manufacture or market it)

and Rhône-Poulenc Agrochimie.   **Patent** GB 1407587 to Ciba-Geigy
Manufacturers AgrEvo; Atul; Gharda; Griffin; Jiangsu; Makhteshim-Agan;
Rhône-Poulenc; United Phosphorus Ltd

APPLICATIONS
Biochemistry Photosynthetic electron transport inhibitor.
Mode of action Selective systemic herbicide, absorbed by the roots and leaves,
with translocation.   **Uses** Pre- and post-emergence control of annual grasses
(*Alopecurus myosuroides, Apera spica-venti, Avena fatua* and *Poa annua*) and many
annual broad-leaved weeds in spring and winter wheat (except durum wheat),
spring and winter barley, winter rye, and triticale at 1.0–1.5 kg a.i./ha.
Phytotoxicity Non-phytotoxic to cereals, except durum wheats.
Formulation types SC; WP.   **Mixtures** *(isoproturon +)* neburon; mecoprop;
tri-allate; amidosulfuron; bifenox; fenoxaprop; fluoroglycofen; bromoxynil +
ioxynil; bentazone + dichlorprop; ioxynil + mecoprop; diflufenican; bifenox +
mecoprop; trifluralin; bromoxynil + ioxynil + mecoprop; fluroxypyr; bifenox +
neburon; isoxaben; simazine; and many more.   **Selected tradenames** 'Alon'
(AgrEvo); 'Arelon' (AgrEvo); 'Dhanulon' (Dhanuka); 'Iprofile' (Mirfield); 'Isoguard'
(Gharda); 'Guideline' (Barclay); 'Panron' (Sanonda); 'Pasport' (Searle India);
'Protugan' (Makhteshim-Agan); 'Strong' (Rhône-Poulenc); 'Turonex' (Agriphar)

ANALYSIS
Product analysis by hplc (*CIPAC Handbook*, 1992, **E**, 110–5; *Anal. Methods Pestic.
Plant Growth Regul.,* 1982, **12**) or by titration of the dimethylamine liberated on
hydrolysis.   **Residues** determined by glc (*Man. Pestic. Anal.* Deutsche
Forschungsgemeinschaft, 1978, **1**, 241); Anal. Methods of Pesticide Residues in
Foodstuffs, General Inspectorate for Health Protection, Netherlands, June 1996.

MAMMALIAN TOXICOLOGY
Oral Acute oral LD_{50} for rats 1826–2417, mice 3350 mg/kg.   **Skin and eye** Acute
percutaneous LD_{50} for rats >2000 mg/kg. Non-irritating to skin and eyes
(rabbits).   **Inhalation** LC_{50} (4 h) for rats >1.95 mg/l air.   **NOEL** (90 d) for rats
400, dogs 50 mg/kg diet; (2 y) for rats 80 mg/kg diet.   **ADI** 0.0062 mg/kg b.w.
Toxicity class WHO (a.i.) III   **EC risk** Xn (R40/22)

ECOTOXICOLOGY
Birds Acute oral LD_{50} for Japanese quail 3042–7926, pigeons >5000 mg/kg.
Fish LC_{50} (96 h) for golden orfe 129, bluegill sunfish >100, guppies 90, rainbow
trout 37, carp 193, catfish 9 mg/l.   **Bees** Not toxic to bees; LD_{50} (48 h, oral)
>50->100 µg/bee.   **Worms** LC_{50} (14 d) for *Eisenia foetida* >1000 mg/kg dry
artificial soil.   **Other beneficial spp.** A dose of up to 1.5 kg/ha (as 'Arelon') was
harmless to adult female *Aleochara bilineata*.   **Daphnia** LC_{50} (48 h) 507 mg/l.
Algae LC_{50} (72 h) 0.03 mg/l.

ENVIRONMENTAL FATE

Animals In rats, following oral administration, 50% is eliminated within the first 8 hours, predominantly in the urine. **Plants** In plants, degradation is mainly via hydroxylation of the isopropyl group to 1,1-dimethyl-3-[4-(2'-hydroxy-2'-propyl)= phenyl]urea; N-dealkylation also occurs. **Soil/Environment** Undergoes enzymic and microbial demethylation at the nitrogen, and hydrolysis of the phenylurea to 4-isopropylaniline. DT_{50} in soil 6–28 d; rate of degradation increases 3x between 10 °C and 30 °C (sandy soil) and 10x in an organic soil over the same temperature range.

434 isouron
Herbicide

urea

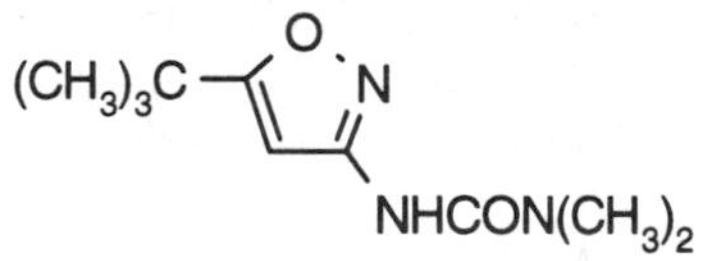

NOMENCLATURE

Common name isouron (BSI, draft E-ISO, (*m*) draft F-ISO, ANSI, WSSA, JMAF); isuron ((*m*) France)
IUPAC name 3-(5-*tert*-butylisoxazol-3-yl)-1,1-dimethylurea
Chemical Abstracts name N'-[5-(1,1-dimethylethyl)-3-isoxazolyl]-N,N-dimethylurea
CAS RN [55861–78–4] **Development codes** SSH 43 (Shionogi)

PHYSICAL CHEMISTRY

Mol. wt. 211.3 **M.f.** $C_{10}H_{17}N_3O_2$ **Form** Colourless crystals. **M.p.** 119–120 °C
V.p. 0.051 mPa (25 °C) **K_{ow}** logP = 1.98 **S.g./density** 1.23 **Solubility** In water 300 mg/l (25 °C). In ethanol 357, acetone 270, xylene 240 (all in g/l, 25 °C).
Stability Stable in sunlight, but decomposed slowly in aqueous solution.
F.p. 156 °C (seta flash tester)

COMMERCIALISATION

History Herbicide reported by H. Yukinaga et al. (*Proc. Asian-Pacific Weed Sci. Soc. Conf., 7th*, 1979, pp. 29, 33) and by H. Yukinaga & K. Katagiri (*Jpn. Pestic. Inf.*, 1980, No. 37, p. 40). Introduced in Japan (1980) by Shionogi & Co., Ltd.
Manufacturers Shionogi

APPLICATIONS

Biochemistry Photosynthetic electron transport inhibitor.

Mode of action Selective systemic herbicide, absorbed principally by the roots, with translocation. **Uses** Pre- and post-emergence control of broad-leaved and grass weeds in sugar cane, pineapples, etc. at 0.5–1.5 kg a.i./ha. Also used for total weed control on non-crop land at 2.5–10 kg/ha. **Formulation types** WP; GR; SC. **Mixtures** *(isouron +)* chlorthiamid + diuron; flupropanate + diuron + dalapon. **Selected tradenames** 'Isoxyl' (Shionogi)

ANALYSIS

Product and **residue** analysis by glc of a derivative.

MAMMALIAN TOXICOLOGY

Oral Acute oral LD_{50} for male rats 630, female rats 760, male mice 520, female mice 530 mg/kg. **Skin and eye** Acute percutaneous LD_{50} for male and female rats >5000 mg/kg. Non-irritating to skin; slight eye irritant (rabbits). **Inhalation** LC_{50} (8 h) for rats >0.415 mg/l. **NOEL** (2 y) for male rats 7.26, female rats 8.77, male mice 3.42, female mice 16.6 mg/kg daily. **ADI** 0.0342 mg/kg. **Other** Acute i.p. LD_{50} for male rats 270, female rats 315 mg/kg. Not mutagenic to *Bacillus subtilis, Escherichia coli* and *Salmonella typhimurium*. **Toxicity class** WHO (a.i.) III

ECOTOXICOLOGY

Birds Acute oral LD_{50} for bobwhite quail >2000 mg/kg. **Fish** LC_{50} (48 h) for carp 75, Japanese killifish 173 mg/l; (96 h) for bluegill sunfish *c.* 140, rainbow trout 110–140 mg/l. **Bees** Acute oral LC_{50} (72 h) for honeybees 1600 mg. a.i./kg (50% WP).

ENVIRONMENTAL FATE

Plants Metabolised in plants by two major degradation pathways, *N*-demethylation or hydroxylation of the *tert*-butyl group. **Soil/Environment** Some microbial degradation occurs in soil. DT_{50} in soil *c.* 22 d.

435 isoxaben   *Herbicide*

amide

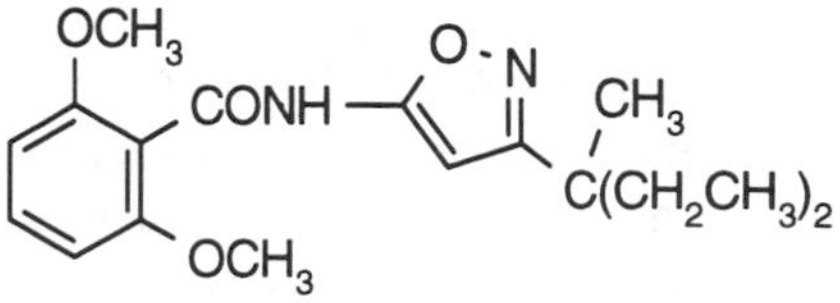

NOMENCLATURE

Common name isoxaben (BSI, ANSI, draft E-ISO, *(m)* draft F-ISO); [benzamizole]

(former use; still used in New Zealand)
IUPAC name N-[3-(1-ethyl-1-methylpropyl)isoxazol-5-yl]-2,6-dimethoxybenzamide
Chemical Abstracts name N-[3-(1-ethyl-1-methylpropyl)-5-isoxazolyl]-=
2,6-dimethoxybenzamide
CAS RN *[82558–50–7]* **Development codes** EL-107

PHYSICAL CHEMISTRY
Composition The product typically contains *c.* 2% of an isomer N-[3-(1,1-=
dimethylbutyl)-5-isoxazolyl]-2,6-dimethoxybenzamide.   **Mol. wt.** 332.4
M.f. $C_{18}H_{24}N_2O_4$ **Form** Colourless crystalline solid.   Forms a monohydrate.
M.p. 176–179 °C   **V.p.** 5.5×10^{-4} mPa (25 °C) K_{ow} logP = 3.94 (pH 5.1, 20 °C)
S.g./density (tech.) 0.58 (22 °C)   **Solubility** In water 1.42 mg/l (pH 7, 20 °C). In
methanol, ethyl acetate, dichloromethane 50–100, acetonitrile 30–50, toluene
4–5, hexane 0.07–0.08 (all in g/l, 25 °C).   **Stability** Stable to hydrolysis at pH 5–9.
Susceptible to photodegradation in aqueous solution.

COMMERCIALISATION
History Herbicide reported by F. Huggenberger *et al.* (*Proc. Br. Crop Prot. Conf.-*
Weeds, 1982, **1,** 47). Introduced in France (1984) by Eli Lilly & Co. (now
DowElanco).   **Patent** GB 2084140   **Manufacturers** DowElanco

APPLICATIONS
Mode of action Selective herbicide, absorbed principally by the roots, with
translocation to stems and leaves. Disrupts root and stem development in
germinating seeds.   **Uses** Pre-emergence control of a broad spectrum of autumn-
and spring-germinating broad-leaved weeds in winter cereals, turf, fruit, vines and
ornamental trees and shrubs. Weeds controlled include *Chamomilla* (*Matricaria*)
spp., *Stellaria media, Bilderdykia, Polygonum, Veronica* and *Viola* spp.
Phytotoxicity For winter barley, wheat, rye and oats, non-phytotoxic at the
recommended application rates. Some rotational crops may be injured if planted
immediately following the treated crop.   **Formulation types** SC.
Mixtures (*isoxaben* +) oryzalin; trifluralin.   **Selected tradenames** 'Flexidor'
(DowElanco)

ANALYSIS
Product and **residue** analysis by hplc with u.v. detection. See B. S. Rutherford in
Comp. Anal. Profiles, Chapt. 8 for details.

MAMMALIAN TOXICOLOGY
Oral Acute oral LD_{50} for rats and mice >10 000, dogs >1000 mg/kg.
Skin and eye Acute percutaneous LD_{50} for rabbits >2000 mg/kg. Slight skin and
eye irritant; not a skin sensitiser (rabbits).   **Inhalation** LC_{50} (1 h) for rats
>1.99 mg/l.   **NOEL** (2 y) for rats 5.6 mg/kg daily. In 3 mo feeding trials, diets
containing 1.25% compound caused only increases in liver and kidney weights, and

elevated levels of hepatic microsomal enzymes. In dogs receiving doses up to 1000 mg/kg daily, the only effects noted were increased levels of hepatic microsomal enzymes. **ADI** 0.056 mg/kg b.w. **Other** Acute i.p. LD_{50} for rats >2000, mice >5000 mg/kg. **Toxicity class** WHO (a.i.) III (Table 5); EPA (formulation) IV

ECOTOXICOLOGY

Birds Acute oral LD_{50} for bobwhite quail >2000 mg/kg. Five-day dietary LC_{50} for mallard ducks and bobwhite quail >5000 mg/kg diet. **Fish** LC_{50} (96 h) for bluegill sunfish and rainbow trout >1.1 mg/l. **Bees** No significant hazard to bees under field conditions. LD_{50} >100 µg/bee. **Daphnia** LC_{50} (48 h) >1.3 mg/l. **Algae** EC_{50} (14 d) for *Selenastrum capricornutum* >1.4 mg/l.

ENVIRONMENTAL FATE

Animals In rats, following oral administration, 90% is excreted in the faeces within 48 hours; *c.* 10% of the absorbed isoxaben undergoes conversion into *c.* 15–20 metabolites, which are excreted in the urine. There is no accumulation of either parent or metabolites in cell tissue. **Plants** In plants, isoxaben is extensively metabolised, primarily through hydroxylation of the alkyl side-chain. For details of absorption, translocation and metabolism in oilseed rape and wheat, see F. Cabanne, *Weed Res.*, 1987, **27**, 135. **Soil/Environment** In soil, relatively low mobility with moderate persistence. Half-life in soil *c.* 3–4 mo; *N*-[3-(1-hydroxy-= 1-methylpropyl)-5-isoxazolyl]-2,6-dimethoxybenzamide is a metabolite. Batch adsorption K_d 6.4–13.0 (4 soils). Lysimeter study confirms lack of mobility.

436 isoxaflutole   *Herbicide*

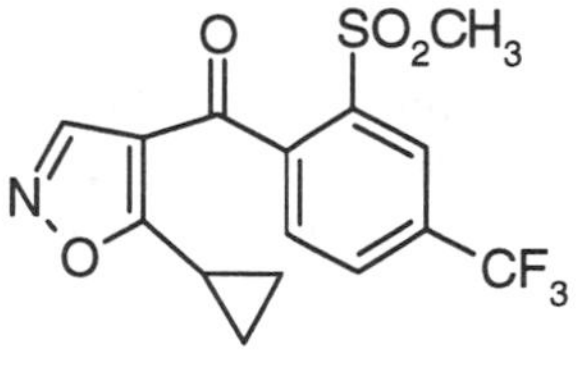

NOMENCLATURE
Common name isoxaflutole (BSI, pa ISO)
IUPAC name 5-cyclopropyl-1,2-oxazol-4-yl α,α,α-trifluoro-2-mesyl-*p*-tolyl ketone
Chemical Abstracts name (5-cyclopropyl-4-isoxazolyl)[2-(methylsulfonyl)-= 4-(trifluoromethyl)phenyl]methanone
CAS RN *[141112–29–0]* **Development codes** RPA 201772

PHYSICAL CHEMISTRY
Composition Tech. is c. 98% pure.   **Mol. wt.** 359.3   **M.f.** $C_{15}H_{12}F_3NO_4S$
Form Off-white or pale yellow solid.   **M.p.** 140 °C   **V.p.** 1×10^{-3} mPa (25 °C)
K_{ow} 2.32   **Henry** 1.87×10^{-5} Pa m^3 mol^{-1} (20 °C)   **S.g./density** 1.590
Solubility In water 6.2 mg/l (pH 5.5, 20 °C).   **Stability** Stable to heat (14 d at
54 °C) and to light. DT_{50} in water 1 d at pH 7.

COMMERCIALISATION
History Reported by B. M. Luscombe et al. (Proc. Br. Crop Prot. Conf. – Weeds,
1995, **1**, 35).   **Patent** EP 0527036   **Manufacturers** Rhône-Poulenc

APPLICATIONS
Biochemistry p-Hydroxyphenyl pyruvate dioxygenase inhibitor. This enzyme
converts p-hydroxyphenyl pyruvate to homogentisate, a key step in plastoquinone
biosynthesis. Inhibition leads to indirect inhibition of carotenoid biosynthesis,
giving rise to chlorosis of new growth.   **Mode of action** Systemic by either root or
foliar uptake.   **Uses** Broad-spectrum grass and broad-leaved weed control in
maize. Applied at 75–140 g/ha pre-emergence or pre-plant; the spectrum can be
enhanced by mixture with other active ingredients.   **Formulation types** WG; SC;
WP.   **Selected tradenames** 'Balance' (Rhône-Poulenc); 'Merlin' (Rhône-Poulenc)

ANALYSIS
By hplc with u.v. detection.

MAMMALIAN TOXICOLOGY
Oral Acute oral LD_{50} for rats >5000 mg/kg.   **Skin and eye** Acute percutaneous
LD_{50} for rabbits >2000 mg/kg. Not a skin irritant; minimal eye irritation (rabbits).
Not a skin sensitiser.   **Inhalation** LC_{50} (4 h) for rats >5.23 mg/l.   **NOEL** for rats
2 mg/kg daily.   **Other** Non-mutagenic, non-neurotoxic.

ECOTOXICOLOGY
Birds Acute oral LD_{50} (14 d) for quail and mallard duck >2150 mg/kg; 8-d dietary
LC_{50} >5000 ppm.   **Fish** Non-toxic at limit of water solubility.   **Bees** LD_{50} (oral
and contact) >100 µg/bee.   **Worms** Non-toxic at 1000 mg/kg.   **Daphnia** Non-
toxic at limit of water solubility.   **Algae** EC_{50} for Selenastrum capricornutum
0.016 mg/l.   **Other aquatic spp.** EC_{50} (96h) for Eastern oyster (Crassostrea
virginica) 3.4 mg/l, mysid shrimp (Mysidopsis bahia) 18 µg/l.

ENVIRONMENTAL FATE
Animals Following oral administration, isoxaflutole is rapidly excreted.
Plants Plant metabolism study demonstrated that residue levels at harvest are very
low, and comprise mainly a non-toxic metabolite.   **Soil/Environment** In laboratory
soil studies, degradation proceeds via hydrolysis and microbial degradation, with
final mineralisation to CO_2. Isoxaflutole and its major metabolites are potentially

mobile in soil under simulated high rainfall; however field studies indicate that residues remain in the surface horizons due to the rate of degradation; after 4 months, virtually no residues remain in the soil.

437   isoxathion   *Insecticide*

organophosphorus

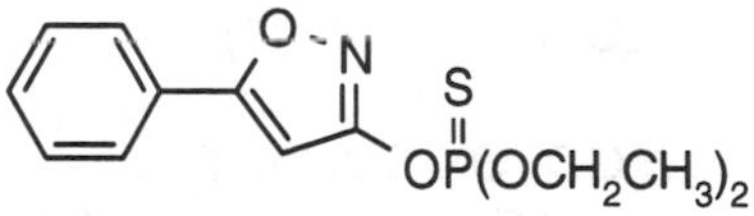

NOMENCLATURE
Common name isoxathion (BSI, JMAF, draft E-ISO, (*m*) draft F-ISO)
IUPAC name *O,O*-diethyl *O*-5-phenylisoxazol-3-yl phosphorothioate
Chemical Abstracts name *O,O*-diethyl *O*-(5-phenyl-3-isoxazolyl) phosphorothioate
CAS RN *[18854–01–8]*   **EEC no.** 242–624–8   **Development codes** E-48; SI-6711 (Sankyo)

PHYSICAL CHEMISTRY
Mol. wt. 313.3   **M.f.** $C_{13}H_{16}NO_4PS$ **Form** Pale yellow liquid.
B.p. 160 °C/0.15 mmHg   **V.p.** <0.133 mPa (25 °C) K_{ow} logP = 3.88 (pH 6.3)
Solubility In water 1.9 mg/l (25 °C). Readily soluble in organic solvents.
Stability Unstable to alkalis. Decomposes at 160 °C.

COMMERCIALISATION
History Insecticide reported by N. Sampei *et al.* (*Sankyo Kenkyusho Nempo*, 1970, **22**, 221). Introduced by the Sankyo Co., Ltd.   **Patent** JP 525850
Manufacturers Sankyo

APPLICATIONS
Biochemistry Cholinesterase inhibitor.   **Mode of action** Insecticide with contact and stomach action.   **Uses** Effective against Aphididae, Coccidae, Diaspididae and Margarodidae (at 500–1000 g a.i./ha) in cabbages, citrus and ornamentals; also against Agromyzidae, Cicadellidae, Delphacidae, Ephyridae and Pyralidae (at 600–900 g/ha) in paddy rice; against Coleoptera (at 1.2 kg/ha) in turf and trees; against Noctuidae and Pieridae (at 500–1000 g/ha) in pome fruit, citrus fruit and vegetables.   **Formulation types** EC; WP; MG; DP.   **Mixtures** *(isoxathion +)* dichlorvos; methomyl.   **Compatibility** Incompatible with alkaline materials.
Selected tradenames 'Karphos' (Sankyo)

ANALYSIS
Product analysis by glc with FID (T. Nakamura & K. Yamaoka, *Anal. Methods Pestic. Plant Growth Regul.*, 1978, **10**, 83). **Residues** determined by glc with FPD or FTD (*idem, ibid.*).

MAMMALIAN TOXICOLOGY
EHC 63 (WHO, 1986; a general review of organophosphorus insecticides).
Oral Acute oral LD_{50} for rats 112, mice 98.4 mg/kg. **Skin and eye** Acute percutaneous LD_{50} for rats >450, mice 193 mg/kg. Not irritant to skin.
NOEL (2 y) for rats 1.2 mg/kg b.w. **Other** Non-mutagenic, non-carcinogenic and non-teratogenic. **Toxicity class** WHO (a.i.) Ib; EPA (formulation) II
EC risk T (R24/25)

ECOTOXICOLOGY
Fish LC_{50} (48 h) for carp 2.13 mg/l.

ENVIRONMENTAL FATE
Soil/Environment DT_{50} in soil 9–40 d.

438 kasugamycin   *Fungicide, bactericide*

aminoglycoside antibiotic

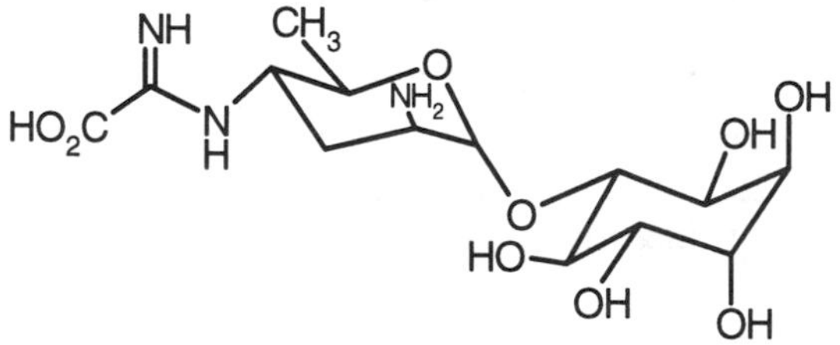

NOMENCLATURE
kasugamycin
Common name kasugamycin (JMAF)
CAS RN *[6980–18–3]*

kasugamycin hydrochloride hydrate
IUPAC name 1L-1,3,4/2,5,6-1-deoxy-2,3,4,5,6-pentahydroxycyclohexyl 2-amino-2,3,4,6-tetradeoxy-4-(α-iminoglycino)-α-D-*arabino*-hexopyranoside hydrochloride hydrate; [5-amino-2-methyl-6-(2,3,4,5,6-pentahydroxycyclo= hexyloxy)tetrahydropyran-3-yl]amino-α-iminoacetic acid hydrochloride hydrate

Chemical Abstracts name 3-O-[2-amino-4-[(carboxyiminomethyl)amino]-2,3,4,6-= tetradeoxy-α-D-*arabino*-hexopyranosyl]-D-*chiro*-inositol hydrochloride hydrate
CAS RN *[19408–46–9]*

PHYSICAL CHEMISTRY

kasugamycin
Mol. wt. 379.4   **M.f.** $C_{14}H_{25}N_3O_9$ **pKa** pKa_1 3.23, pKa_2 7.73, pKa_3 11.0

kasugamycin hydrochloride hydrate
Mol. wt. 433.8   **M.f.** $C_{14}H_{28}ClN_3O_{10}$ **Form** Colourless needle crystals.
M.p. 202–204 °C (decomp.)   **V.p.** $<1.3 \times 10^{-5}$ mPa (25 °C) K_{ow} logP <1
S.g./density 0.43 g/cm^3 (25 °C)   **Solubility** In water 125 g/l (25 °C). In methanol
2.76, acetone, xylene <1 (all in mg/kg, 25 °C).   **Stability** Very stable at room
temperature. Stable in weak acids, but unstable in strong acids and alkalis. DT_{50}
(50 °C) 47 d (pH 5), 14 d (pH 9).   **Specific rotation** $[\alpha]_D^{25}$ +120° (c 1.6 H_2O)

COMMERCIALISATION

Production By fermentation of *Streptomyces kasugaensis*.   **History** Kasugamycin
was discovered by H. Umezawa *et al.* (*J. Antibiot. (Tokyo)*, 1965, **18**, 101),
fungicidal activity reported by T. Ishiyama *et al.* (*ibid.*, p. 115). Introduced by the
Institute of Microbial Chemistry and by Hokko Chemical Industry Co., Ltd.
Patent JP 42006818; BE 657659; GB 1094566   **Manufacturers** Hokko

APPLICATIONS

kasugamycin hydrochloride hydrate
Biochemistry Protein synthesis inhibitor. Inhibits binding of Met-RNA to the
mRNA-30S complex, thereby preventing amino acid incorporation.
Mode of action Systemic fungicide and bactericide with protective and curative
action. Inhibits hyphal growth of *Pyricularia oryzae* on rice, preventing lesion
development; comparatively weak inhibitory action to spore germination,
appressoria formation on the plant surface or penetration into the epidermal cell.
Rapidly taken up into plant tissue and translocated. In contrast, against
Cladosporium fulvum on tomato, inhibition of sporulation is strong, but inhibition of
hyphal growth is weak.   **Uses** Control of rice blast (*Pyricularia oryzae*) and some
other diseases (particularly bacterial grain rot, bacterial seedling blight, and
bacterial brown stripe caused by *Pseudomonas* spp.) in rice. Also used to control
other plant diseases, e.g. leaf mould and bacterial canker in tomatoes, bean halo
blight, scab on apples and pears, *Cercospora* leaf spot on sugar beet and celery,
bacterial soft rot on potatoes and carrots, anthracnose and bacterial spot on
cucumbers, bacterial diseases of citrus fruit, bacterial diseases (particularly
Pseudomonas spp.) of ornamentals, etc.   **Phytotoxicity** Non-phytotoxic to rice,
tomatoes, sugar beet, potatoes and other vegetables, but slight injury has been
noted to peas, beans, soya beans, grapes, citrus, and apples.
Formulation types WP; DP; GR; UL; SL.   **Mixtures** *(kasugamycin +)* Bordeaux

MAIN ENTRIES

mixture; copper oxychloride; phthalide; phthalide + silafluofen; phthalide + silafluofen + tebufenozide; phthalide + validamycin + etofenprox; dichlofenthion + thiram. **Compatibility** Incompatible with pesticides which are strongly alkaline. **Selected tradenames** 'Kasugamin' (Hokko); 'Kasumin' (Hokko)

ANALYSIS
Product analysis by cup assay with *Pseudomonas fluorescens* (NIHJ B-254). **Residues** determined by cup assay with *Pyricularia oryzae* (P2) (*J. Antibiot. (Tokyo)*, 1968, **21**, 49).

MAMMALIAN TOXICOLOGY
kasugamycin hydrochloride hydrate
Oral Acute oral LD_{50} for male rats >5000 mg/kg. **Skin and eye** Acute percutaneous LD_{50} for rabbits >2000 mg/kg. Non-irritating to eyes and skin (rabbits). **Inhalation** LC_{50} (4 h) for rats >2.4 mg/l. **NOEL** (2 y) for rats 300, dogs 800 mg/kg diet. Non-mutagenic and non-teratogenic in rats, and without effect on reproduction. **Toxicity class** WHO (a.i.) III (Table 5); EPA (formulation) IV

ECOTOXICOLOGY
kasugamycin hydrochloride hydrate
Birds Acute oral LD_{50} for male Japanese quail >4000 mg/kg. **Fish** LC_{50} (48 h) for carp and goldfish >40 mg/l. **Bees** LD_{50} (contact) >40 µg/bee. **Daphnia** LC_{50} (6 h) >40 mg/l.

ENVIRONMENTAL FATE
Animals Kasugamycin hydrochloride hydrate orally administered to rabbits was mostly excreted in the urine within 24 h. When injected intravenously to dogs, it was mostly excreted within 8 h. After oral administration to rats at 200 mg/kg, no residues were detected in eleven organs or blood; 96% of administered dose remained in the digestive tract 1 h after administration. **Plants** Degraded to kasugamycinic acid and kasuganobiosamine; finally degraded to ammonia, oxalic acid, CO_2 and water. **Soil/Environment** Degradation proceeds as in plants.

kresoxim-methyl   *Fungicide*

strobilurin analogue

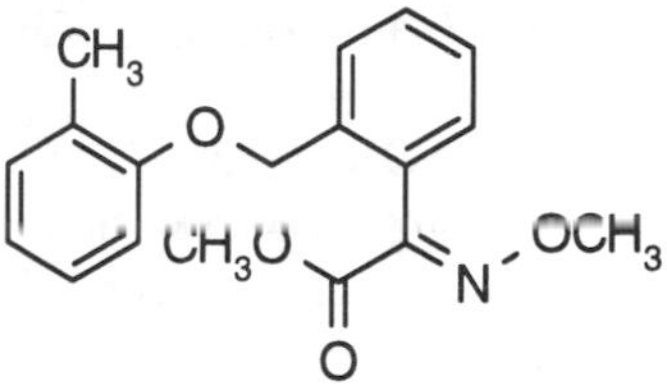

NOMENCLATURE
Common name kresoxim-methyl (BSI, pa ISO)
IUPAC name methyl (*E*)-2-methoxyimino-[2-(*o*-tolyloxymethyl)phenyl]acetate;
methyl (*E*)-methoxyimino[α-(*o*-tolyloxy)-*o*-tolyl]acetate
Chemical Abstracts name (*E*)-α-(methoxyimino)-2-[(2-methylphenoxy)methyl]=
benzeneacetic acid methyl ester
CAS RN *[143390–89–0]*   **Development codes** BAS 490F

PHYSICAL CHEMISTRY
Mol. wt. 313.4   **M.f.** $C_{18}H_{19}NO_4$ **Form** White, mildly aromatic crystals.
M.p. 97.2–101.7 °C   **V.p.** 2.3 × 10^{-3} mPa (20 °C)   **K_{ow}** logP = 3.4 (pH 7, 25 °C)
Henry 3.6 × 10^{-4} Pa m³ mol⁻¹   **S.g./density** 1.258 kg/l (20 °C)   **Solubility** In water
at 20 °C, 2 mg/l.   **Stability** Not hydrolysed within 24 h (pH 7, 20 °C).   **pKa** None
in range 2–12

COMMERCIALISATION
History Discovered by BASF AG in 1983. Reported by E. Ammermann *et al.*,
(*Proc. Br. Crop Prot. Conf. – Pests Dis.*, 1992, **1**, 403). In production since 1995 and
introduced in 1996.   **Patent** EP 253213; US 4829085   **Manufacturers** BASF

APPLICATIONS
Biochemistry Inhibits mitochondrial respiration by inhibiting cyctochrome c
reductase. Selectivity is partly due to enzymic de-esterification within plants.
Mode of action Fungicide with protective, curative, eradicative and long residual
disease control; acts by inhibiting spore germination. Redistribution via the vapour
phase contributes to activity.   **Uses** Control of scab in apples and pears; powdery
mildew on apples, vines, cucurbits and sugar beet; mildew, scald, net blotch and
glume blotch on cereals (in combination with fenpropimorph); blast and sheath
blight on rice; and downy mildew on vines and vegetables.   **Formulation types** SC;
SE; WG.   **Mixtures** (*kresoxim-methyl +*) fenpropimorph; BAS 480F; BAS 480F +
fenpropimorph.   **Selected tradenames** 'Allegro' (mixture) (BASF); 'Mentor'
(mixture) (BASF); 'Stroby' (BASF, Nissan)

MAIN ENTRIES

MAMMALIAN TOXICOLOGY
Oral Acute oral LD_{50} for rats >5000 mg/kg.   **Skin and eye** Acute percutaneous LD_{50} for rats >2000 mg/kg. Non-irritating to skin and eyes (rabbits).
Inhalation LC_{50} (4 h) for rats >5.6 mg/l.   **NOEL** for rats 800 ppm.
ADI 0.4 mg/kg b.w.   **Other** Negative in the Ames test, and non-teratogenic.

ECOTOXICOLOGY
Although it is toxic to aquatic species, exposure tests and ecological studies have shown that there is no danger of permanent damage to aquatic organisms when kresoxim-methyl is used as recommended.   **Birds** LD_{50} (14 d) for quail >2150 mg/kg.   **Fish** LC_{50} (96 h) in range 0.681–1.0 mg/l.   **Bees** LD_{50} (48 h) (contact) >20 µg/bee.   **Worms** LC_{50} >937 mg/kg.   **Daphnia** EC_{50} (48 h) 0.186 mg/l.   **Algae** EC_{50} (0–72 h) 63 µg/l.

ENVIRONMENTAL FATE
Animals In warm-blooded animals such as the rat, kresoxim-methyl is quickly broken down.   **Plants** Residues in cereals and pome fruit at harvest are <0.05 mg/kg, in grapes and vegetables <1 mg/kg.   **Soil/Environment** In soil DT_{90} (lab.) <3 d; the main metabolite is the corresponding acid. K_{oc} 219–372; K_{oc} of the acid 17–24. In water DT_{50} 34 d (pH 7), 7 h (pH 9).

440 KTU 3616 *Fungicide*

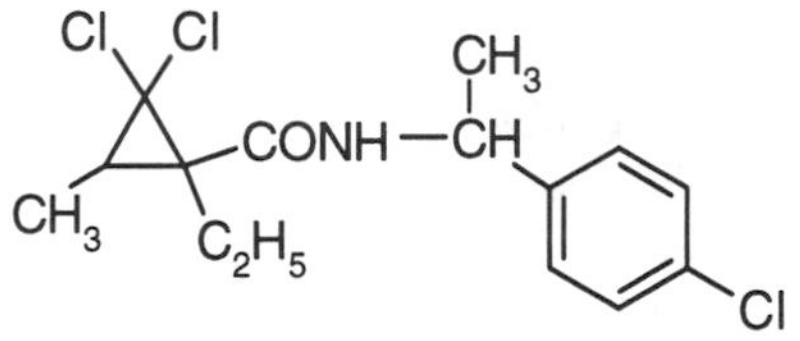

NOMENCLATURE
Common name carpropamid (BSI proposed)
IUPAC name 2,2-dichloro-*N*-[1-(4-chlorophenyl)ethyl]-1-ethyl-3-methylcyclo= propanecarboxamide
Chemical Abstracts name *trans*-2,2-dichloro-*N*-[(1-(4-chlorophenyl)ethyl]-= 1-ethyl-3-methylcyclopropanecarboxamide
CAS RN *[104030–54–8]*; *AR enantiomer [127641–62–7]*; *BR enantiomer [127640–90–8]*   **Development codes** KTU 3616 (Bayer, Nihon Bayer)

PHYSICAL CHEMISTRY

Composition Diastereoisomeric mixture (A : B c. 1 : 1; *R/S* c. 95 : 5); the (*1R,3S,1R*)- enantiomer is designated *AR*, the (*1S,3R,1R*)- enantiomer *BR*. **Mol. wt.** 334.7 **M.f.** $C_{15}H_{18}Cl_3NO$ **Form** Colourless powder; (tech. is a white to yellowish powder). **M.p.** 145–149 °C (tech.) **V.p.** AR: 2×10^{-3} mPa, BR: 3×10^{-3} mPa (both 20°C) **K_{ow}** AR: logP = 4.23, BR: logP = 4.28 (both 22 °C) **Henry** 4×10^{-4} Pa m^3 mol^{-1} (AR, 20 °C) **S.g./density** 1.17 (20 °C) **Solubility** In water 1 7 (AR), 1 9 (BR) (both in mg/l, pH 7, 20 °C).

COMMERCIALISATION

History Discovered and developed jointly by Nihon Bayer Agrochem. K.K. and Bayer AG. Reported by T. Hattori *et al.* (*Proc. Br. Crop Prot. Conf., Pests Dis.*, 1994, **2**, 517). To be launched in 1997, firstly in Korea as a seed dressing product in rice, by Hannong Co. **Manufacturers** Bayer

APPLICATIONS

Biochemistry In *Pyricularia oryzae*, KTU 3616 inhibits melanin biosynthesis, by inhibiting the dehydration reactions from scytalone to 1,3,8-trihydroxynaphthalene and from vermelone to 1,8-dihydroxynaphthalene. **Mode of action** Systemic, protective fungicide. **Uses** Systemic fungicide with specific action against *Pyricularia oryzae*; it does not provide curative properties, so protective treatment is essential. Can be used for seed treatment (300–400 g a.i./dt, seedling box treatment (400 g a.i./ha), seedling box drenching (300 g a.i./ha), spraying (100–150 g a.i./ha) and dusting (150–200 g a.i./ha). **Formulation types** DP; GR; SC; WS; FS. **Mixtures** (*KTU 3616* +) imidacloprid; imidacloprid + fludioxonil. **Selected tradenames** 'Win', 'Win Admire' (Bayer); 'Arcado' (Bayer); 'Seed One' (Hannong); 'Zabara' (Hannong)

MAMMALIAN TOXICOLOGY

Oral Acute oral LD_{50} for male and female rats, and male and female mice >5000 mg/kg. **Skin and eye** Acute percutaneous LD_{50} for male and female rats >2000 mg/kg. No eye or skin irritation (rabbits); not a skin sensitiser (guinea pigs). **Inhalation** LC_{50} for male and female rats >5000 mg/m^3 (dust). **NOEL** (2y) for rats and mice 400 mg/kg diet; (1 y) for dogs 200 mg/kg diet. **ADI** 0.034 mg/kg b.w. (Bayer proposed; Colombia, Korea); 0.014 mg/kg b.w. (Japan). **Other** No mutagenic effects in *in vitro* or *in vivo* tests.

ECOTOXICOLOGY

Fish LC_{50} (96 h) for carp 6.9, trout 10 mg/l. **Daphnia** LC_{50} (3 h) 410 mg/l. **Algae** E_rC_{50} (72 h) >2 mg/l.

ENVIRONMENTAL FATE

Animals After oral administration of radiolabelled KTU 3616 to rats, the radioactivity was readily excreted via faeces and urine. KTU 3616 was metabolised oxidatively, mainly in the liver. **Plants** After treatment of rice plants via soil (nursery box application) or nutrient medium, KTU 3616 was absorbed by the roots and translocated to the shoots. The major residue in rice was KTU 3616. **Soil/Environment** KTU 3616 was metabolised oxidatively under paddy soil conditions; CO_2 was the major metabolite. Field DT_{50} (paddy soil in Korea) 53–89 d. KTU 3616 was hydrolytically stable in sterile aqueous buffer solutions. In natural water, it was photodegraded.

441 KWG 4168 *Fungicide*

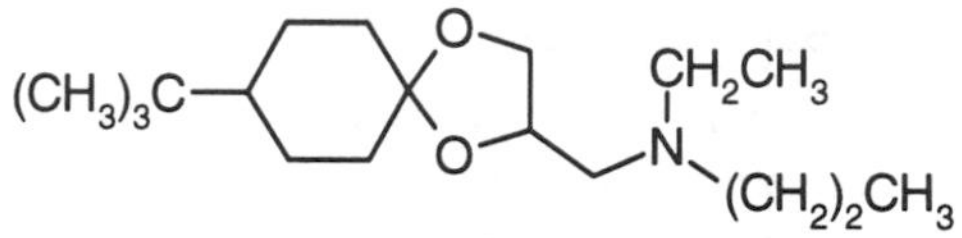

NOMENCLATURE
Common name spiroxamine (BSI proposed)
IUPAC name 8-*tert*-butyl-1,4-dioxaspiro[4.5]decan-2-ylmethyl(ethyl)(propyl)amine
Chemical Abstracts name 8-(1,1-dimethylethyl)-*N*-ethyl-*N*-propyl-1,4-dioxaspiro= [4,5]decane-2-methanamine
CAS RN *[118134–30–8]* **Development codes** KWG 4168

PHYSICAL CHEMISTRY
Composition Comprises 2 diastereoisomers, A and B in the proportions 49–56% and 51–44% respectively. **Mol. wt.** 297.5 **M.f.** $C_{18}H_{35}NO_2$ **Form** Faintly yellowish liquid; tech. is a light brown oily liquid. **V.p.** A: 9.7 mPa (20 °C); B: 17 mPa (25 °C) (both extrapolated) **K$_{ow}$** A: logP = 2.79; B: logP = 2.92 (unstated pH) **Henry** A: 2.5 × 10^{-3}; B: 5.0 × 10^{-3} (both in Pa m^3 mol^{-1}, 20 °C, calc.) **S.g./density** A and B both 0.930 (20 °C) **Solubility** In water, mixture of A and B: >200 × 10^3 (pH 3, mg/l, 20 °C); A: 470 (pH 7), 14 (pH 9); B: 340 (pH 7), 10 (pH 9) (both diastereoisomers in mg/l, 20 °C). **Stability** Stable to hydrolysis and photodegradation; provisional photolytic DT_{50} 50.5 d (25°C). **pKa** 6.9, base **F.p.** 147 °C

COMMERCIALISATION
History Discovered in 1987. Reported by S. Dutzmann *et al.* (*Proc. Br. Crop Prot. Conf. – Pests Dis.*, 1996, **1**, 47). **Manufacturers** Bayer

APPLICATIONS
Biochemistry New sterol biosynthesis inhibitor, acting mainly by inhibition of

Δ^{14}-reductase.   **Mode of action** Protective, curative and eradicative fungicide. Readily penetrates into the leaf tissue, followed by acropetal translocation to the leaf tip. Uniformly distributed within the whole leaf.   **Uses** Under development by Bayer AG as a systemic foliar fungicide in wheat and barley, with rapid initial effect, and prolonged activity, for control of powdery mildew (*Erysiphe graminis* f.sp. *tritici* and f.sp. *hordei*). Also effective against rust diseases (*Puccinia* spp.) of wheat and barley, *Rhynchosporium secalis* and *Pyrenophora teres*.

MAMMALIAN TOXICOLOGY
Oral Acute oral LD_{50} for male rats *c*. 595, female rats 500–560 mg/kg.
Skin and eye Acute percutaneous LD_{50} for male rats >1600, female rats *c*. 1068 mg/kg b.w. Severe skin irritant; not an eye irritant (rabbits).   **Inhalation** LC_{50} (4 h) for male rats *c*. 2772, female rats *c*. 1982 mg/m^3.   **NOEL** (2 y) for rats 70, mice 20 mg/kg diet; (1 y) for dogs 75 mg/kg diet.   **ADI** 0.025 mg/kg b.w.   **Other** Not genotoxic; no specific effects on reproduction.   **Toxicity class** WHO (a.i.) III; EPA (formulation) II   **EC risk** Xn (R21/22, R38, R43)

ECOTOXICOLOGY
Birds LD_{50} 565 mg/kg.   **Fish** LD_{50} (96 h) for rainbow trout 18.5 mg/l.   **Bees** No risk to bees.   **Worms** LC_{50} ≥1000 mg/kg dry wt. substrate.

442 lactofen   *Herbicide*

diphenyl ether

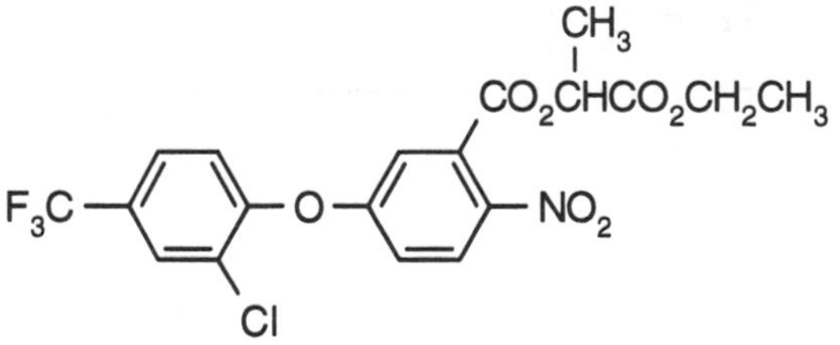

NOMENCLATURE
Common name lactofen (ANSI, WSSA)
IUPAC name ethyl *O*-[5-(2-chloro-α,α,α-trifluoro-*p*-tolyloxy)-2-nitrobenzoyl]-= DL-lactate
Chemical Abstracts name (±)-2-ethoxy-1-ethyl-2-oxoethyl 5-[2-chloro-= 4-(trifluoromethyl)phenoxy]-2-nitrobenzoate
CAS RN *[77501–63–4]* unstated stereochemistry   **Development codes** PPG-844

PHYSICAL CHEMISTRY
Mol. wt. 461.8 **M.f.** $C_{19}H_{15}ClF_3NO_7$ **Form** Dark brown to tan. **Solubility** In water <1 mg/l (20 °C).

COMMERCIALISATION
History Herbicide introduced by PPG Industries.

APPLICATIONS
Biochemistry Protoporphyrinogen oxidase inhibitor. **Uses** Control of broad-leaved weeds in cereals, potatoes, soya beans, rice and peanuts.
Formulation types EC. **Selected tradenames** 'Cobra' (Valent, PPG, Sumitomo)

MAMMALIAN TOXICOLOGY
Oral Acute oral LD_{50} for rats >5000 mg/kg. **Skin and eye** Acute percutaneous LD_{50} for rats 2000 mg/kg. Formulation severely irritating to eyes.
Toxicity class EPA (formulation) IV

lambda-cyhalothrin

See entry 180.

443 lenacil   *Herbicide*

uracil

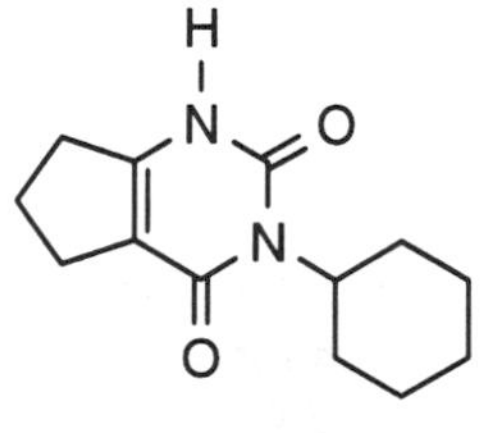

NOMENCLATURE
Common name lenacil (BSI, E-ISO, ANSI, WSSA, JMAF); lenacile ((m) F-ISO)
IUPAC name 3-cyclohexyl-1,5,6,7-tetrahydrocyclopentapyrimidine-2,4(3H)-dione
Chemical Abstracts name 3-cyclohexyl-6,7-dihydro-1H-cyclopentapyrimidine-=2,4(3H,5H)-dione

Other names 3-cyclohexyl-5,6-trimethyleneuracil   **CAS RN** *[2164–08–1]*
Development codes Du Pont B634

PHYSICAL CHEMISTRY
Mol. wt. 234.3   **M.f.** $C_{13}H_{18}N_2O_2$ **Form** White to tan crystals.
M.p. 315.6–316.8 °C   **V.p.** 2×10^{-4} mPa (25 °C)   **K_{ow}** logP = 2.31
S.g./density 1.32   **Solubility** In water 6 mg/l (25 °C). Slightly soluble in most
organic solvents, e.g. diethyl ether 0.3, hexane 0.5, benzene 0.7, acetone,
chloroform 0.9, xylene, methyl isobutyl ketone 2, ethanol 2.4, cyclohexanone 4,
dimethyl sulfoxide 6, dimethylformamide 8 (all in g/kg, 25 °C).   **Stability** Stable up
to the melting point. Stable in water and in aqueous acids. Decomposed by hot
alkalis.   **pKa** 10.3, v. weak acid

COMMERCIALISATION
History Herbicide reported by G. W. Cussans (*Proc. Br. Weed Control Conf., 7th,*
1964, **2**, 671). Introduced by E. I. du Pont de Nemours Co.   **Patent** US 3235360
Manufacturers Du Pont; Gedeon Richter

APPLICATIONS
Biochemistry Photosynthetic electron transport inhibitor.   **Mode of action**
Selective, systemic herbicide, absorbed by the roots.   **Uses** Control of annual
grass and broad-leaved weeds in sugar beet, fodder beet, beetroot, sweet
potatoes, spinach, strawberries, flax, black salsify, and ornamental plants and
shrubs. Applied either pre-plant soil-incorporated or pre-emergence.
Formulation types WP; SC.   **Mixtures** *(lenacil +)* chloridazon; benzthiazuron;
linuron; DCU; isocarbamid; propham; proximpham; cycloate; proximpham +
propham; phenmedipham; neburon.   **Selected tradenames** 'Venzar' (Du Pont);
'Adol' (Gedeon Richter)

ANALYSIS
Product determined by glc.   **Residues** determined by glc (H. L. Pease, *J. Sci. Food
Agric.*, 1966, **17**, 121; H. J. Jarczyk, *Pflanzenschutz-Nachr.* (Eng. Ed.), 1977, **30**, 3)
or by hplc (T. H. Byast, *J. Chromatogr.*, 1977, **134**, 211).

MAMMALIAN TOXICOLOGY
Oral Acute oral LD_{50} for rats >11 000 mg/kg.   **Skin and eye** Acute percutaneous
LD_{50} for rabbits >5000 mg/kg; slight eye irritant (rabbits, WP formulation).
Inhalation LC_{50} >5.2 mg/l.   **NOEL** In 2 y feeding trials with rats, no adverse effect
observed.   **Toxicity class** WHO (a.i.) III (Table 5); EPA (formulation) IV

ECOTOXICOLOGY
Birds Acute oral LD_{50} for Pekin ducks >5620 ppm. Eight-day dietary LC_{50} for
bobwhite quail 2300 mg/kg diet.   **Fish** LC_{50} (96 h) for carp 10, minnow >2.0,
trout >2.0 mg/l. LC_{50} (21 d) for trout >2.3 mg/l.   **Bees** Not toxic to bees;

LD_{50} >25 µg/bee. **Worms** LC_{50} for earthworms >10 000 mg/kg. **Daphnia** LC_{50} (48 h) >8.4 mg/l.

ENVIRONMENTAL FATE
Soil/Environment In soil, complete microbial degradation occurs within 5–6 months. DT_{50} (25 °C) aerobic 82–150 d. K_d 0.35–2.8; K_{oc} 75–254.

444 *Leptomastix dactylopii*   *Biological agent*

Hymenoptera

Wasp parasite of mealybugs.

NOMENCLATURE
Scientific name *Leptomastix dactylopii*

PROPERTIES
Stability Storage is not possible.

COMMERCIALISATION
Manufacturers Arbico; BCP; Biobest; Bugs for Bugs; Koppert; Neudorff; Sautter & Stepper

APPLICATIONS
Mode of action The parasitic wasp lays an egg in the mealybug larva, preferring the third larval stage or even the adult. The egg develops and an adult wasp emerges from the parasitised mealybug.   **Uses** For control of citrus mealybug (*Planococcus citri*) in horticultural and fruit crops.   **Phytotoxicity** Non-phytotoxic and non-phytopathogenic.   **Formulation types** Adult parasitic wasp without carrying material.

MAMMALIAN TOXICOLOGY
Non-toxic.

lindane

See under gamma-HCH (entry 392).

445 linuron
Herbicide

urea

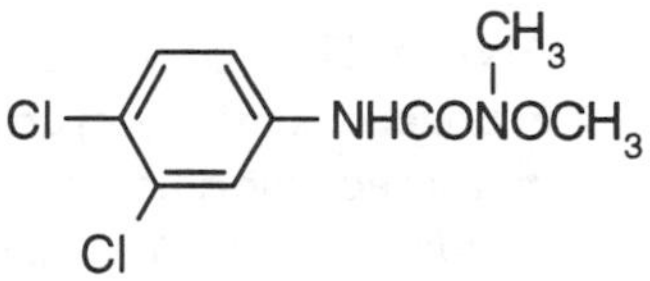

NOMENCLATURE
Common name linuron (BSI, E-ISO, (m) F-ISO, ANSI, WSSA, JMAF)
IUPAC name 3-(3,4-dichlorophenyl)-1-methoxy-1-methylurea
Chemical Abstracts name N'-(3,4-dichlorophenyl)-N-methoxy-N-methylurea
CAS RN [330–55–2] **EEC no.** 206–356–5 **Development codes** DPX-Z326
(Du Pont); AE F002810 (AgrEvo); Hoe 02 810 (Hoechst AG)

PHYSICAL CHEMISTRY
Composition Tech. is ≥94% pure. **Mol. wt.** 249.1 **M.f.** $C_9H_{10}Cl_2N_2O_2$
Form Colourless crystals. **M.p.** 93–95 °C **V.p.** 0.051 mPa (20 °C); 7.1 mPa
(50 °C) (EC method) **K_{ow}** logP = 3.00 **Henry** 2.0×10^{-4} (20 °C, Pa m^3 mol^{-1})
S.g./density 1.49 (20 °C) **Solubility** In water 63.8 mg/l (20 °C, pH 7). In acetone
500, benzene, ethanol 150, xylene 130 (all in g/kg, 25 °C). Readily soluble in
dimethylformamide, chloroform, and diethyl ether. Moderately soluble in aromatic
hydrocarbons. Sparingly soluble in aliphatic hydrocarbons. **Stability** Stable at m.p.
and in aqueous solution at pH 5, 7 & 9; DT_{50} 945 d at all three pH values.

COMMERCIALISATION
History Herbicide reported by K. Härtel (*Meded. Landbouwhogesch. Opzoekingsstn.
Staat Gent,* 1962, **27,** 1275). Introduced by E. I. du Pont de Nemours and Co. and
by Hoechst AG (now AgrEvo GmbH). **Patent** DE 1028986; GB 852422 both to
Hoechst **Manufacturers** AgrEvo; Drexel; Griffin; Makhteshim-Agan

APPLICATIONS
Biochemistry Photosynthetic electron transport inhibitor.
Mode of action Selective systemic herbicide, absorbed principally by the roots but
also by the foliage, with translocation primarily acropetally in the xylem. **Uses**
Pre- and post-emergence control of annual grass and broad-leaved weeds, and

some seedling perennial weeds in asparagus, artichokes, carrots, parsley, fennel, parsnips, herbs and spices, celery, celeriac, onions, leeks, garlic, potatoes, peas, field beans, soya beans, cereals, maize, sorghum, cotton, flax, sunflowers, sugar cane, ornamentals, established vines, bananas, cassava, coffee, tea, rice, peanuts, ornamental trees and shrubs, and other crops. **Formulation types** WP; SC; EC. **Mixtures** *(linuron +)* alachlor; atrazine; chlorpropham; cyanazine; ioxynil; lenacil; metribuzin; monolinuron; propachlor; trietazine; trifluralin; terbutryn; terbacil; amitrole; nitralin; metoxuron; monalide; bromoxynil + ioxynil + mecoprop; bromoxynil + ioxynil + dichlorprop; trietazine + trifluralin; butralin; pendimethalin; simazine. **Selected tradenames** 'Lorox' (Du Pont); 'Afalon' (AgrEvo); 'Linex' (Griffin); 'Linnet' (PBI); 'Linurex' (Makhteshim-Agan); 'Siolcid' (Siapa)

ANALYSIS

Product analysis by hydrolysis to 2,4-dichloroaniline, a derivative of which is measured colorimetrically, or by non-aqueous titration with perchloric acid (*CIPAC Handbook*, 1980, **1A**, 1281) or by hplc (*Anal Methods Pestic. Plant Growth Regul.*, 1982, 12). Impurities by tlc (*CIPAC Handbook*, 1994, **F**, 336).
Residues determined by hplc or glc (H. L. Pease et al., *Anal. Methods Pestic., Plant Growth Regul. Food Addit.*, 1967, **5**, 433; *Anal. Methods Pestic. Plant Growth Regul.*, 1972, **6**, 659; *Man. Pestic. Anal. Deutsche Forschungsgemeinschaft*, 1987, **1**, 241). **In water**, by lc with u.v. detection (*AOAC Methods*, 1995, 992.14, 10.7.01). Details available from AgrEvo.

MAMMALIAN TOXICOLOGY

Oral Acute oral LD_{50} for rats 1500–4000 mg/kg. **Skin and eye** Acute percutaneous LD_{50} for rats >2000 mg/kg. Mild skin irritant (rabbits); not a skin sensitiser (guinea pigs). **Inhalation** LC_{50} (4 h) for rats >6.15 mg/l air. **NOEL** (1 y) for dogs 25 ppm (0.9 mg/kg b.w.). Tumour promoter in rats. **ADI** 0.008 mg/kg b.w. **Toxicity class** WHO (a.i.) III (Table 5); EPA (formulation) III **EC risk** Xn (R40)

ECOTOXICOLOGY

Birds Acute oral LD_{50} for bobwhite quail 940 mg/kg. Eight-day dietary LC_{50} for mallard ducklings 3083 ppm / >5000 mg/kg (values from separate studies), ring-necked pheasants 3438 ppm, Japanese quail >5000 ppm diet. **Fish** LC_{50} (96 h) for rainbow trout 3.15, sheepshead minnow 0.89 mg/l; NOEC 0.49 mg/l. **Bees** LD_{50} (oral) >1600 µg/g bee b.w. **Worms** LC_{50} for *Eisenia foetida* >1000 mg/kg soil. **Other beneficial spp.** Harmless to other beneficial arthropods. **Daphnia** LC_{50} (48 h) 0.75 mg/l, 0.12 mg/l (separate studies). **Other aquatic spp.** LC_{50} (96 h) for mysid shrimp (*Mysidopsis bahia*) 3.4 mg/l; NOEC 2.1 mg/l.

ENVIRONMENTAL FATE

Animals The major metabolites arise from demethylation and demethoxylation.
Plants In plants, metabolism involves demethylation and demethoxylation.

Soil/Environment Microbial degradation is the primary factor in disappearance from soil. Half-life under field conditions is *c.* 2–5 mo (F. Kempson-Jones & R. J. Hance, *Pestic. Sci.*, 1979, **10**, 449). In soil DT_{50} 38–67 d. Soil adsorption K_{oc} 500–600.

---

446 lufenuron · *Insecticide, acaricide*

benzoylurea

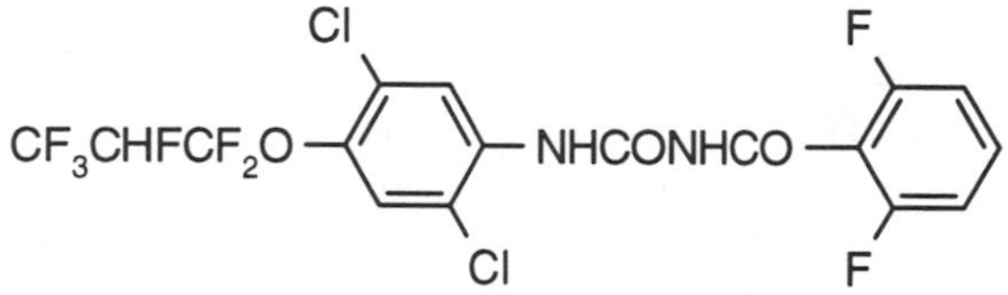

CF₃CHFCF₂O — [2,5-dichlorophenyl] — NHCONHCO — [2,6-difluorophenyl]

NOMENCLATURE
Common name lufenuron (BSI, pa E-ISO, INN)
IUPAC name (RS)-1-[2,5-dichloro-4-(1,1,2,3,3,3-hexafluoropropoxy)phenyl]-= 3-(2,6-difluorobenzoyl)urea
Chemical Abstracts name *N*-[[[2,5-dichloro-4-(1,1,2,3,3,3-hexafluoropropoxy)= phenyl]amino]carbonyl]-2,6-difluorobenzamide
CAS RN [103055–07–8]   **Development codes** CGA 184699

PHYSICAL CHEMISTRY
Mol. wt. 511.2   **M.f.** $C_{17}H_8Cl_2F_8N_2O_3$ **Form** Colourless crystals.
M.p. 164.7–167.7 °C (OECD 102)   **V.p.** $<4 \times 10^{-3}$ mPa (25 °C) (OECD 104)
K_{ow} logP = 5.12 (25 °C) (OECD 117)   **S.g./density** 1.66 at 20 °C (OECD 109)
Solubility In water at 25 °C, <0.06 mg/l. In ethanol 41, acetone 460, toluene 72, *n*-hexane 0.13, *n*-octanol 8.9 g/l (25 °C).   **Stability** Stable in air and light. In water, DT_{50} 32 d (pH 9), 70 d (pH 7), 160 d (pH 5).

COMMERCIALISATION
History Introduced in poster session at RSC/SCI conference 'Chemistry of Insecticides', Oxford, July 1989.   **Manufacturers** Novartis

APPLICATIONS
Biochemistry Inhibits chitin synthesis.   **Mode of action** Acts mostly by ingestion; larvae are unable to moult, and also cease feeding.   **Uses** Insect growth regulator for control of Lepidoptera and Coleoptera larvae on cotton, maize and vegetables; and citrus whitefly and rust mites on citrus fruit. Also for the control of fleas on pets, and of cockroaches and fleas in houses.   **Formulation types** EC.

Compatibility Not compatible with pesticides with alkaline reaction (lime sulfur, copper).   **Selected tradenames** 'Match' (Novartis)

ANALYSIS
Residue analysis by hplc. Details available from Novartis.

MAMMALIAN TOXICOLOGY
Oral Acute oral LD_{50} for rats >2000 mg/kg.   **Skin and eye** Acute percutaneous LD_{50} for rats >2000 mg/kg. Non-irritating to eyes and skin (rabbits). Non-sensitising to skin (guinea pigs).   **Inhalation** LC_{50} (4 h, 20 °C) for rats >2.35 mg/l air.   **NOEL** (2 y) for rats 2.0 mg/kg b.w. daily.   **Toxicity class** WHO (a.i.) III
EC risk R43

ECOTOXICOLOGY
Birds Acute oral LD_{50} for bobwhite quail and mallard ducks >2000 mg/kg. Eight-day dietary LC_{50} for bobwhite quail and mallard ducks >5200 mg/kg.   **Fish** LC_{50} (96 h) for rainbow trout >73, carp >63, bluegill sunfish >29, catfish >45 mg/l.
Bees Oral LC_{50} >38 μg/bee. LD_{50} (topical) >8.0 μg/bee.

ENVIRONMENTAL FATE
Animals Major route of elimination was via faeces, with very little degradation.
Plants No metabolites occurred in significant amounts in the investigated target crops (cotton, tomatoes).   **Soil/Environment** Lufenuron was rapidly degraded in biologically active soils under aerobic conditions. DT_{50} 13–20 d. Lufenuron showed a very strong adsorption onto soil particles: K_{oc} (mean value) 38 mg/g o.c.

447 *Macrolophus caliginosus*   *Biological agent*

predatory bug

NOMENCLATURE
Scientific name *Macrolophus caliginosus*

PROPERTIES
Stability Shelf life 2–3 d at 8–10 °C in a cool, dark place.

COMMERCIALISATION
Manufacturers BCP; Biobest; Koppert; Novartis BCM

Mode of action The predatory bug sucks dry all whitefly stages, with preference for larvae. **Uses** For control of glasshouse whitefly (*Trialeurodes vaporariorum*) and tobacco whitefly (*Bemisia tabaci*) on protected crops. **Phytotoxicity** Non-phytotoxic and non-phytopathogenic. **Formulation types** Mixed with vermiculite. **Selected tradenames** 'Mirical' (Koppert)

MAMMALIAN TOXICOLOGY
Non-toxic.

448 malathion   *Insecticide, acaricide*

organophosphorus

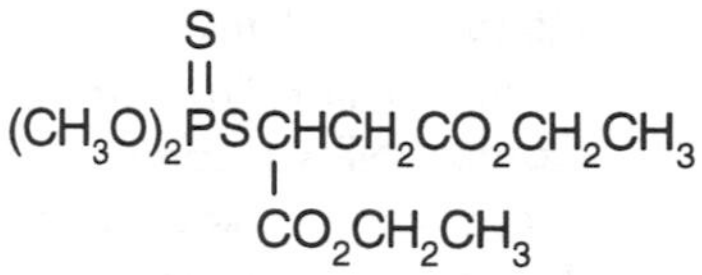

$$(CH_3O)_2\overset{\overset{S}{\|}}{P}SCHCH_2CO_2CH_2CH_3$$
$$\underset{CO_2CH_2CH_3}{|}$$

NOMENCLATURE
Common name malathion (BSI, E-ISO, (*m*) F-ISO, ESA, BAN); maldison (Australia, New Zealand); malathon (JMAF); mercaptothion (South Africa); carbofos (USSR); no name (Germany); mercaptotion (Argentina)
IUPAC name diethyl (dimethoxythiophosphorylthio)succinate; S-1,2-bis(ethoxy= carbonyl)ethyl O,O-dimethyl phosphorodithioate
Chemical Abstracts name diethyl [(dimethoxyphosphinothioyl)thio]butanedioate
CAS RN *[121–75–5]* **EEC no.** 204–497–7 **Development codes** EI 4049 (Cyanamid) **Official codes** OMS 1; ENT 17 034

PHYSICAL CHEMISTRY
Composition Tech. grade is *c.* 95% pure. **Mol. wt.** 330.3 **M.f.** $C_{10}H_{19}O_6PS_2$
Form Tech. grade is a clear amber liquid. **M.p.** 2.85 °C
B.p. 156–157 °C/0.7 mmHg **V.p.** 5.3 mPa (30 °C) **K_{ow}** logP = 2.75
S.g./density 1.23 (25 °C) **Solubility** In water 145 mg/l (25 °C). Miscible with most organic solvents, e.g. alcohols, esters, ketones, ethers, aromatic hydrocarbons. Slightly soluble in petroleum ether and some types of mineral oil.
Stability Relatively stable in neutral, aqueous media. Decomposed by acids and alkalis.

COMMERCIALISATION
History Insecticide reported by G. A. Johnson *et al.* (*J. Econ. Entomol.*, 1952, **45**,

279). Introduced by American Cyanamid Co; rights transferred to Cheminova in 1991. **Patent** US 2578652 **Manufacturers** Aimco; Cheminova; Ficom; Hindustan; Tekchem

APPLICATIONS
Biochemistry Cholinesterase inhibitor; proinsecticide, activated by metabolic oxidative desulfuration to the corresponding oxon. **Mode of action** Non-systemic insecticide and acaricide with contact, stomach, and respiratory action.
Uses Used to control Coleoptera, Diptera, Hemiptera, Hymenoptera and Lepidoptera in a wide range of crops, including cotton, pome, soft and stone fruit, potatoes, rice and vegetables. Used extensively to control major arthropod disease vectors (Culicidae) in public health programmes, ectoparasites (Diptera, Acari, Mallophaga) of cattle, poultry, dogs and cats, human head and body lice (Anoplura), household insects (Diptera, Orthoptera), and for the protection of stored grain. **Phytotoxicity** Non-phytotoxic in general, if used as recommended, but glasshouse cucurbits and beans, certain ornamentals, and some varieties of apple, pear, and grape may be injured. **Formulation types** EC; WP; DP; UL.
Mixtures (malathion +) fenitrothion; parathion; parathion-methyl; dichlorvos; methoxychlor + parathion; piperonyl butoxide + pyrethrins.
Compatibility Incompatible with alkaline materials (residual toxicity may be decreased). **Selected tradenames** 'Cekumal' (Cequisa); 'Fyfanon' (Cheminova); 'Malathane' (Agriphar); 'Malatox' (Pesticides India, Siapa); 'Malixol' (Rhône-Poulenc); 'Maltox' (Aimco); 'MLT' (Sumitomo); 'White Star' (Sanonda)

ANALYSIS
Product analysis by glc (*CIPAC Handbook*, 1983, **1B**, 1852; *AOAC Methods*, 1995, 979.05). **Residues** determined by glc or by tlc or paper chromatography, or by single sweep oscillographic polarography (*ibid.*, 968.24, 970.52, 970.53; *Pestic. Anal. Man.*, 1979, **I**, 201-A, 201-G, 201-H, 201-I; *Analyst (London)*, 1973, **98**, 19; 1977, **102**, 858; 1980, **105**, 515; 1985, **110**, 765; *Man. Pestic. Residue Anal.*, 1987, **I**, 3, 5, 6, S8, S10, S13, S17, S19; *Anal. Methods Residues Pestic.*, 1988, Part I, M2, M5, M10).

MAMMALIAN TOXICOLOGY
Reviews *Pesticide residues in food.* FAO Agricultural Studies, No. 73; WHO Technical Report Series, No. 370, 1967. *Evaluation of some pesticide residues in food.* FAO/PL:CP/15; WHO/ Food Add./67.32, 1967. **IARC** 30 **EHC** 63 (WHO, 1986; a general review of organophosphorus insecticides). **Oral** Acute oral LD$_{50}$ for rats 1375–2800, mice 775–3320 mg/kg. **Skin and eye** Acute percutaneous LD$_{50}$ (24 h) for rabbits 4100 mg/kg. **NOEL** In 21 mo feeding trials, rats receiving 100 mg/kg diet showed normal weight gain. **ADI** (JMPR) 0.02 mg/kg b.w. [1966]. **Toxicity class** WHO (a.i.) III; EPA (formulation) III
EC risk Xn (R22)

ECOTOXICOLOGY
Birds Five-day dietary LC_{50} for bobwhite quail 3500, ringneck pheasant 4320 mg/kg diet.   **Fish** LC_{50} (96 h) for bluegill sunfish 0.1, largemouth bass 0.28 mg/l.
Bees Toxic to bees. LD_{50} (topical) 0.71 μg/bee.

ENVIRONMENTAL FATE
Animals In mammals, following oral administration, the major part of the dose is excreted in the urine and faeces within 24 hours. Degradation is by oxidative desulfuration by liver microsomal enzymes, leading to the formation of malaoxon; malathion and malaoxon are hydrolysed and thus detoxified by carboxylesterases. In insects, metabolism involves hydrolysis of the carboxylate and phosphorodithioate esters, and oxidation to malaoxon.

449 maleic hydrazide   *Plant growth regulator*

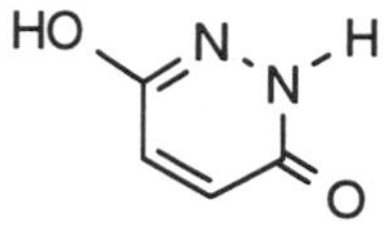

NOMENCLATURE
maleic hydrazide
Common name maleic hydrazide (BSI, E-ISO, JMAF, accepted in lieu of a common name); hydrazide maléique (F-ISO, accepted in lieu of a common name)
IUPAC name 6-hydroxy-2H-pyridazin-3-one; 1,2-dihydropyridazine-3,6-dione
Chemical Abstracts name 6-hydroxy-3(2H)-pyridazinone; 1,2-dihydro-= 3,6-pyridazinedione
CAS RN [10071–13–3] (tautomer drawn); [123–33–1] (dione tautomer)

maleic hydrazide potassium salt
CAS RN [51542–52–0]

PHYSICAL CHEMISTRY
maleic hydrazide
Composition The dry tech. grade is >99% pure.   **Mol. wt.** 112.1   **M.f.** $C_4H_4N_2O_2$
Form The dry tech. grade is a white crystalline solid.   **M.p.** 298–300 °C
V.p. <1 × 10^{-2} mPa (25 °C) K_{ow} logP = –1.96 (pH 7), –0.56 (unionised, 25°C, K. Chamberlain *et al.*, *Pestic. Sci.*, **47**, 265 (1996))   **S.g./density** 1.61 (25 °C)
Solubility In water 4.507; at pH 4.3, 4.417 (both in g/l, 25 °C). In methanol 4.179, hexane, toluene <0.001 (all in g/l, 25 °C).   **Stability** Degraded by light,

DT_{50} (25 °C) 58 d (pH 5, 7), 34 d (pH 9). Stable to hydrolysis, but decomposed by oxidising agents and strong acids. No degradation in the package at 25 °C over 1 y. **pKa** 5.62 (20 °C)

maleic hydrazide potassium salt
Mol. wt. 150.2 **M.f.** $C_4H_3KN_2O_2$ **Solubility** In water 400 g/kg (25 °C).

COMMERCIALISATION
History Plant growth regulating properties reported by D. L. Schoene & O. L. Hoffmann (*Science*, 1949, **109**, 588). Introduced by U.S. Rubber Co. (now Uniroyal Chemical Inc.). **Patent** US 2575954; US 2614916; US 2614917; US 2805926
Manufacturers Drexel; Fair; Uniroyal

APPLICATIONS
Mode of action Plant growth regulator, absorbed by the leaves and roots, with translocation in the xylem and phloem. Inhibits cell division in the meristematic regions, but not cell extension. Also has some herbicidal activity.
Uses Suppression of grass growth on lawns, roadside verges, embankments, and amenity areas; and of growth of shrubs and trees. Inhibition of sprouting in potatoes, onions, beets, swedes and carrots in storage. Prevention of sucker development in tobacco. Induction of dormancy in citrus fruit. Also used as a herbicide in mixtures with 2,4-D.

maleic hydrazide
Formulation types SL; SG. **Compatibility** Incompatible with pesticides which are highly alkaline in reaction. In the presence of heavy-metal and iron, zinc, calcium and magnesium ions, sparingly-soluble salts are formed.
Selected tradenames 'MH' (Uniroyal); 'Regulox' (Rhône-Poulenc)

maleic hydrazide potassium salt
Formulation types SC; SG. **Selected tradenames** 'Royal MH-30' (Uniroyal)

ANALYSIS
Product analysis by hplc, details from Uniroyal. **Residues** determined by hydrolysis to hydrazine which is determined by colorimetry (*AOAC Methods*, 1995, 963.24; *Anal. Methods Pestic., Plant Growth Regul. Food Addit.*, 1964, **4**, 147).

MAMMALIAN TOXICOLOGY
maleic hydrazide
Oral Acute oral LD_{50} for rats >5000 mg/kg. **Skin and eye** Acute percutaneous LD_{50} for rabbits >5000 mg/kg. Slight eye, mild skin irritation (rabbit).
Inhalation LC_{50} (4 h) for rats >3.2 mg/l. **ADI** 0.3 mg/kg b.w.
Toxicity class WHO (a.i.) III (Table 5); EPA (formulation) III **PIC** Yes

maleic hydrazide potassium salt
Reviews *Pesticide residues in food – 1984.* FAO Plant Production and Protection Paper 62, 1985. *Pesticide residues in food – 1984 evaluations.* FAO Plant Production and Protection Paper 67, 1985.   **IARC** 4, 42   **Oral** Acute oral LD_{50} for rats 3900 mg/kg.   **Skin and eye** Acute percutaneous LD_{50} not available. Not a sensitiser (guinea pigs).   **Inhalation** LC_{50} (4 h) for rats >4.03 mg/l.   **NOEL** In chronic feeding trials, NOEL for rats 25, dogs 19 mg/kg b.w. daily. Not carcinogenic in rodents.   **ADI** (JMPR) 5 mg/kg b.w. (for 99.9% pure potassium or sodium salt containing ≤1 mg hydrazine/kg) [1984].

ECOTOXICOLOGY
maleic hydrazide
Birds Acute oral LD_{50} for mallard ducks >4640 mg/kg. Eight-day dietary LC_{50} for mallard ducks and bobwhite quail >10 000 ppm.   **Fish** LC_{50} (96 h) for rainbow trout >1435, bluegill sunfish 1608 mg/l.   **Daphnia** LC_{50} (48 h) 108 ppm.
Algae IC_{50} (96 h) for *Chlorella* >100 mg/l.

maleic hydrazide potassium salt
Birds Acute oral LD_{50} for mallard ducks >2250 mg/kg. Eight-day dietary LC_{50} for mallard ducks >5620 mg/kg.   **Fish** LC_{50} (96 h) for rainbow trout >1000 mg/l.
Bees Not toxic to bees; LD_{50} >100 μg/bee.   **Worms** LC_{50} for earthworms >1000 ppm.   **Daphnia** LC_{50} (48 h) >1000 mg/l.   **Algae** LC_{50} (5 d) for *Selenastrum* >9.84 mg/l.

ENVIRONMENTAL FATE
Animals Rabbits administered a single dose of 100 mg/kg of maleic hydrazide excreted 43–62% of the dose unchanged within 48 hours (J. M. Barnes *et al.*, *Nature (London)*, 1957, **180**, 62–64).   **Plants** Various acids, e.g. succinic, fumaric, and maleic, are found as metabolites in plants.   **Soil/Environment** DT_{50} in soil c. 11 h. Rapid photochemical degradation occurs in water.

450   *Mamestra brassicae* NPV   *Biological agent*

nuclear polyhedrosis virus

NOMENCLATURE
Scientific name *Mamestra brassicae* nuclear polyhedrosis virus
Other names MbNPV

PROPERTIES
Composition 'Mamestrin' contains 2.5×10^{12} PIB/l.   **Stability** 'Mamestrin' has a shelf life of 2 years, when stored at 4 °C.

COMMERCIALISATION
Production MbNPV is multiplied *in vivo* on *M. brassicae* infested at the larval stage, and isolated by centrifugation techniques. 'Mamestrin' is then formulated as a suspension enclosing the entomopathogenic germ and specific additives.
History An isolate of this baculovirus was obtained from a virosed larva of *Mamestra brassicae* (Lepidoptera: Noctuidae) or cabbage armyworm, collected in France by I.N.R.A. researchers. This isolate was developed by Natural Plant Protection (NPP), France.   **Patent** FR 8717748; EP 90401016   **Manufacturers** NPP

APPLICATIONS
Mode of action Acts by ingestion and is active against the larval stage of target insects. First to third stage instar die within 7 d of treatment.   **Uses** 'Mamestrin' is registered in France for the control of *M. brassicae* and may be used against *Heliothis armigera, Phthorimaea operculella, Plutella xylostella*. It should be applied to foliage at a dose of 4 l/ha/application. Has potential for integrated pest management.   **Formulation types** Liquid.   **Selected tradenames** 'Mamestrin' (NPP)

ANALYSIS
Activity is measured by biological titration on *M. brassicae*.

MAMMALIAN TOXICOLOGY
Oral Acute oral LD_{50} for male rats $>10.2 \times 10^9$ PIB/kg b.w., female rats $>10.95 \times 10^9$ PIB/kg b.w.   **Skin and eye** No skin penetration.   **Inhalation** LC_{50} for rats $>7.5 \times 10^{10}$ PIB/kg b.w. No toxicity, infectivity or pathogenicity was detected after an intranasal installation in rats at 7.5×10^{10} PIB/kg b.w.   **NOEL** (99 d) for mice $>1.5 \times 10^9$ PIB/kg b.w. daily.   **Other** There is no evidence of acute or chronic toxicity, eye or skin irritation to mammals; nevertheless, it may induce a slight hypersensitivity. No allergic symptoms or health problems have been observed by research workers, manufacturing staff or users.

ECOTOXICOLOGY
Birds Not toxic and not infective to birds.   **Fish** Not toxic to fish.   **Bees** LD_{50} (oral and contact) $>5 \times 10^4$ PIB/bee.

ENVIRONMENTAL FATE
Soil/Environment MbNPV occurs naturally. After one year, the level of MbNPV remaining in treated soil is equivalent to natural levels.

mancopper

Fungicide

alkylenebis(dithiocarbamate)

NOMENCLATURE
Common name mancopper (BSI, E-ISO, F-ISO)
IUPAC name ethylenebis(dithiocarbamate) mixed metal complex containing
c. 13.7% manganese and *c.* 4% copper
Chemical Abstracts name [[1,2-ethanediylbis[carbamodithioato]](2-)]manganese
mixture with [[1,2-ethanediylbis[carbamodithioato]](2-)]copper
CAS RN *[53988-93-5]*

COMMERCIALISATION
History Fungicide introduced by Rohm & Haas Co.   **Manufacturers** Rohm & Haas

APPLICATIONS
Mode of action Fungicide with protective action.   **Uses** Control of downy mildew
on vines. Also used as a seed treatment for control of *Septoria* and *Fusarium* spp.
on cereals.   **Formulation types** WP; Seed treatment.   **Mixtures** *(mancopper +)*
lindane.   **Selected tradenames** 'Dithane C-90' (Rohm & Haas)

ANALYSIS
Details from Rohm & Haas.

MAMMALIAN TOXICOLOGY
EHC 78 (WHO, 1988; general review of dithiocarbamates).   **Oral** Acute oral
LD_{50} for rats 9600 mg/kg.

mancozeb

Fungicide

alkylenebis(dithiocarbamate)

$$[-SCSNHCH_2CH_2NHCSSMn-]_x(Zn)_y$$

NOMENCLATURE
Common name mancozeb (BSI, E-ISO); mancozèbe ((*m*) F-ISO); manzeb (JMAF)
IUPAC name manganese ethylenebis(dithiocarbamate) (polymeric) complex with
zinc salt
Chemical Abstracts name [[1,2-ethanediylbis[carbamodithioato]](2-)]manganese
mixture with [[1,2-ethanediylbis[carbamodithioato]](2-)]zinc

CAS RN *[8018–01–7]* formerly *[8065–67–6]*; *[8018–01–7]* is also applied to other mixed manganese and zinc ethylenebis(dithiocarbamate) complexes

PHYSICAL CHEMISTRY
Composition The ISO definition is 'a complex of zinc and maneb containing 20% of manganese and 2.55% of zinc, the salt present being stated (for instance mancozeb chloride)'.   **Form** Greyish-yellow powder.   **M.p.** Decomposes, at 192–204 °C   **V.p.** Negligible at 20 °C   **S.g./density** 1.92   **Solubility** In water 6.2 ppm (pH 7.5, 25 °C). Insoluble in most organic solvents; dissolves in solutions of powerful chelating agents but cannot be recovered from them.   **Stability** Stable under normal, dry storage conditions. Slowly decomposed by heat and moisture. On hydrolysis (25 °C) DT_{50} 20 d (pH 5), 17 h (pH 7), 34 h (pH 9). Mancozeb a.i. is unstable and the tech. not isolated; the formulated product is produced in continuous process.   **F.p.** 137.8 °C (Tag open cup)

COMMERCIALISATION
History Fungicide reported in *Fungic. Nematic. Tests,* 1961, **17**. Introduced by Rohm & Haas Co. and by E. I. du Pont de Nemours and Co.   **Patent** GB 996264; US 3379610; US 2974156 all to Rohm & Haas   **Manufacturers** Aimco; Crystal; Desarrollo Quimico; Du Pont; Elf Atochem; Rohm & Haas; Sanachem; Sinon; United Phosphorus

APPLICATIONS
Mode of action Fungicide with protective action.   **Uses** Control of many fungal diseases (e.g. blight, leaf spot, rust, downy mildew, scab, etc.) in field crops, fruit, nuts, vegetables, ornamentals, etc. Particular uses include control of early and late blights of potatoes and tomatoes (at 1.36 kg a.i./ha); *Rhizoctonia solani* and *Streptomyces scabies* on seed potatoes; leaf spot diseases on celery, cucurbits, beet, berries and currants; rusts on cereals, vegetables, roses, carnations, asparagus, beans, apples and plums (at 1.6 kg/ha); downy mildews on hops, vines, onions, leeks, lettuce, cucurbits, ornamentals and tobacco; *Gloeodes pomigena*, *Glomerella cingulata*, *Microthyriella rubi* and *Physalospora obtusa* on apples; scab on apples and pears (at 2.4–3.6 kg/ha); sigatoka disease (*Cercospora musae*) in bananas; shot-hole of stone fruit; anthracnose of beans and cucurbits; damping-off diseases of vegetables; black leg of beet; needle cast in forestry; and many seed-borne diseases of cereals. Used for foliar application or as a seed treatment.
Formulation types WP; WG; DP; SC; DS.   **Mixtures** *(mancozeb +)* metalaxyl; benalaxyl; oxadixyl; cymoxanil; carbendazim; captan; fosetyl-aluminium; zineb; copper oxychloride; sulfur; dinocap; folpet; myclobutanil; fenpropimorph; Bordeaux mixture; pyrifenox; penconazole; and many others.
Selected tradenames 'Dithane M-45' (Rohm & Haas); 'Manzate' (Du Pont); 'Aimcozeb' (Aimco); 'Crittox' (Siapa); 'Manex II' (Griffin); 'Micene' (Sipcam); 'Sancozeb' (Sanachem); 'Saver' (Sanonda); 'Uthane' (United Phosphorus); 'Vondozeb' (Elf Atochem); 'Zebra' (Headland)

ANALYSIS

Product analysis by decomposition with acid and measuring the carbon disulfide liberated, either by glc or by a titrimetric method (*CIPAC Handbook*, 1980, **1A,** 1288). Identified colorimetrically (*CIPAC Handbook*, 1994, **F**, 320) or by u.v. absorbance (*ibid.*, 411). **Residues** determined by reaction with acid to form carbon disulfide which is measured by standard methods (*Analyst (London)*, 1981, **106,** 782; *Pestic. Anal. Man.*, 1979, **II**; *Manu. Pestic. Residue Anal.*, 1987, **I**, S21; *Anal. Methods Residues Pestic.*, 1988, Part II).

MAMMALIAN TOXICOLOGY

Reviews *Pesticide residues in food – 1993*, FAO Plant Production and Protection Paper, 122, 1993. *Pesticide residues in food – 1993 evaluations, Part II – Toxicology.* WHO, WHO/PCS/94.4, 1994. **EHC** 78 (WHO, 1988; general review of dithiocarbamates). **Oral** Acute oral LD_{50} for rats >5000 mg/kg. **Skin and eye** Acute percutaneous LD_{50} for rats >10 000, rabbits >5000 mg/kg. Mild to moderate skin irritant, moderate eye irritant. **Inhalation** LC_{50} (4 h) for rats >5.14 mg/l. **ADI** (JMPR) 0.03 mg/kg b.w.; ethylenethiourea 0.004 mg/kg b.w. [1993]. **Other** At very high levels, mancozeb has caused birth defects in test animals; ethylenethiourea, a trace contaminant and breakdown product of mancozeb, has caused thyroid effects, tumours and birth defects in laboratory animals. **Toxicity class** WHO (a.i.) III (Table 5); EPA (formulation) IV **EC risk** Xi (R37); R43

ECOTOXICOLOGY

Birds In ten-day dietary studies, there were no mortalities in mallard ducks at 6400 mg/kg daily, and in Japanese quail at 3200 mg/kg daily. **Fish** LC_{50} (48 h) for goldfish 9.0, rainbow trout 2.2, catfish 5.2, carp 4.0 mg/l. **Bees** LC_{50} 0.193 mg/bee.

ENVIRONMENTAL FATE

Plants Extensively metabolised in plants, forming ethylenethiourea, ethylenethiuram monosulfide, ethylenethiuram disulfide, and sulfur as transitory intermediates. Terminal metabolites are natural products, especially those derived from glycine. **Soil/Environment** Rapidly degraded in the environment by hydrolysis, oxidation, photolysis, and metabolism. DT_{50} in soil *c.* 6–15 d. K_{oc} >2000.

alkylenebis(dithiocarbamate)

$$[-SCSNHCH_2CH_2NHCS_2Mn-]_x$$

NOMENCLATURE
Common name maneb (BSI, E-ISO, JMAF); manèbe ((*m*) F-ISO)
IUPAC name manganese ethylenebis(dithiocarbamate) (polymeric)
Chemical Abstracts name [1,2-ethanediylbis[carbamodithioato](2-)]manganese
Other names MEB   **CAS RN** *[12427–38–2]*   **EEC no.** 235–654–8
Official codes ENT 14 875

PHYSICAL CHEMISTRY
Mol. wt. 265.3   **M.f.** $C_4H_6MnN_2S_4$ **Form** Yellow crystalline solid.
M.p. Decomposes without melting at 192–204 °C   **V.p.** Negligible at 20 °C
S.g./density 1.92   **Solubility** Practically insoluble in water and in common organic
solvents. Soluble in chelating agents (e.g. sodium salts of ethylenediamine=
tetraacetic acid) with the formation of complexes.   **Stability** Stable to light.
Decomposes on prolonged exposure to air or moisture. On hydrolysis DT_{50}
<24 h (pH 5, 7 or 9). Etem is one of the products formed on contact with
moisture (R. A. Ludwig *et al.*, *Can. J. Bot.*, 1955, **33,** 42; C. W. Pluijgers *et al.*,
Tetrahedron Lett., 1971, p. 1371).

COMMERCIALISATION
History Introduced by E. I. du Pont de Nemours and Co. (who no longer
manufacture or market it) and by Rohm & Haas Co.   **Patent** US 2504404;
US 2710822 both to Du Pont   **Manufacturers** Crystal; Desarrollo Quimico;
Drexel; Elf Atochem

APPLICATIONS
Mode of action Fungicide with protective action.   **Uses** Control of many fungal
diseases (e.g. blight, leaf spot, rust, downy mildew, scab, etc.) in field crops, fruit,
nuts, vegetables, ornamentals, turf, etc. Particular uses include control of early
and late blights of potatoes and tomatoes; *Rhizoctonia solani* and *Streptomyces
scabies* on seed potatoes; leaf spot diseases on celery, beet and currants; rusts on
cereals, roses, carnations, asparagus, beans, apples, plums and currants; downy
mildews on hops, vines, onions, ornamentals and tobacco; *Gloeodes pomigena*,
Glomerella cingulata, *Microthyriella rubi* and *Physalospora obtusa* on apples; scab on
apples and pears; sigatoka disease (*Cercospora musae*) on bananas; anthracnose of
beans; tulip fire; needle cast in forestry; and many seed-borne diseases of cereals.
Used for foliar application or as a seed treatment.   **Phytotoxicity** Morello cherries
and some varieties of apple and cucurbits may be injured.   **Formulation types** WP;
WG; SC; Seed treatment.   **Mixtures** *(maneb +)* zinc oxide; carbendazim;
tridemorph; fentin acetate; zineb; thiophanate-methyl; metalaxyl; vinclozolin;

carbendazim + tridemorph; carbendazim + sulfur; carbendazim + chlorothalonil; carbendazim + fenarimol; and many others.   **Selected tradenames** 'Dithane M-22' (Rohm & Haas); 'Brestan' (mixture) (AgrEvo); 'Manex' (Crystal, Griffin); 'Manzate' (Du Pont); 'Mazin' (Unicrop); 'Multi-W FL' (PBI); 'Policritt' (Siapa); 'Trimangol' (Elf Atochem)

ANALYSIS
Product analysis by decomposition with acid and measurement of the liberated carbon disulfide, either by glc or by colorimetry of a derivative (*AOAC Methods*, 1995, 965.15, 991.33; *CIPAC Handbook*, 1970, **1**, 463; 1980, **1A**, 1293; 1992, **E**, 116–122). Identified colorimetrically (*CIPAC Handbook*, 1994, **F**, 320) or by u.v. absorbance (*ibid.*, 411).   **Residues** determined by reaction to form carbon disulfide which is measured by standard methods (*Analyst (London)*, 1981, **106**, 782; *Pestic. Anal. Man.*, 1979, **II**; *Manu. Pestic. Residue Anal.*, 1987, **I**, S21; *Anal. Methods Residues Pestic.*, 1988, Part II).

MAMMALIAN TOXICOLOGY
Reviews *Pesticide residues in food – 1993*, FAO Plant Production and Protection Paper, 122, 1993. *Pesticide residues in food – 1993 evaluations, Part II – Toxicology.* WHO, WHO/PCS/94.4, 1994.   **IARC** 12   **EHC** 78 (WHO, 1988; general review of dithiocarbamates).   **Oral** Acute oral LD_{50} for rats >5000 mg/kg.
Skin and eye Acute percutaneous LD_{50} for rats and rabbits >5000 mg/kg. Moderate eye irritant; non-irritating to skin (rabbits). May cause irritation to nose and throat.   **Inhalation** LC_{50} (4 h) for rats >3.8 mg/l air.   **NOEL** In 2 y feeding trials, no ill-effect observed in rats receiving 250 mg/kg diet, and at 2500 mg/kg diet showed signs of toxicity; in 1 y trials with dogs, no effect observed at 20 mg/kg daily but toxicity observed at 75 mg/kg daily. At very high levels, maneb has caused birth defects in test animals; ethylenethiourea, a trace contaminant and degradation product of maneb, has caused thyroid effects, tumours and birth defects in laboratory animals.   **ADI** (JMPR) 0.03 mg/kg b.w.; ethylenethiourea 0.004 mg/kg b.w. [1993].   **Toxicity class** WHO (a.i.) III (Table 5); EPA (formulation) IV   **EC risk** Xi (R37); R43

ECOTOXICOLOGY
Birds Eight-day dietary LC_{50} for mallard ducks and bobwhite quail >10 000 mg/kg diet.   **Fish** LC_{50} (48 h) for carp 1.8 mg/l.   **Bees** Not toxic to bees.

ENVIRONMENTAL FATE
Animals In rats, metabolites included ethylenediamine, ethylenebis(thiuram) monosulfide and ethylenebis(thiourea) (H. Seidler *et al.*, *Nahrung*, 1971, **15**, 177–185).   **Plants** In plants, the principal metabolite is ethylenethiourea, which rapidly undergoes further metabolism. Ethylenethiuram monosulfide, ethylenethiuram disulfide, and sulfur are also metabolites.
Soil/Environment Rapidly degraded in the environment by hydrolysis, oxidation,

photolysis, and metabolism. Soil DT_{50} c. 25 d (loamy sand in dark, aerobic conditions).

454 MB-599   *Insecticide synergist*

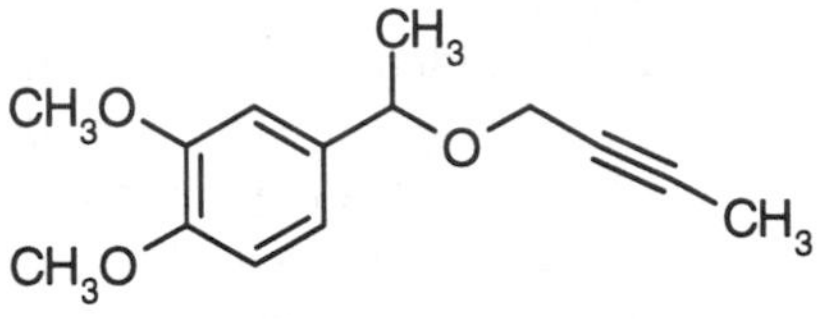

NOMENCLATURE
IUPAC name (±)-(3,4-dimethoxyphenyl)-1-ethylbut-2-ynyl ether
Chemical Abstracts name 4-[1-(2-butynyloxy)ethyl]-1,2-dimethoxybenzene
CAS RN *[185676–84–0]* **Development codes** MB-599; Mi 18

PHYSICAL CHEMISTRY
Composition Tech. is ≥95%. **Mol. wt.** 234.3 **M.f.** $C_{14}H_{18}O_3$ **Form** Colourless or slightly yellow viscous liquid. **B.p.** 120 °C/0.1 mmHg **K_{ow}** logP = c. 2.4
S.g./density 1.079 **Solubility** In water c. 1 mg/l. In hexane 20 g/100 ml; miscible with acetone, dichloromethane and benzene. **Stability** Stable in neutral and basic (boiling 40% NaOH, 3 h) conditions; in HCl at pH 1, DT_{50} 127.3 h (25 °C). Photochemically stable; practically no change after 16 h at 254 and 356 nm using 75W lamp.

COMMERCIALISATION
History Reported by I. Székely et al. (*Proc. Br. Crop Prot. Conf. – Pests Dis.*, 1996, **2**, 473). Under development by Chinoin Agchem Business Unit.
Manufacturers Chinoin

APPLICATIONS
Biochemistry Cytochrome P-450 inhibitor. **Selected tradenames** 'Censor' (Chinoin)

ANALYSIS
Product and **residue** analysis by gc. Details available from Chinoin Agchem Business Unit.

MAMMALIAN TOXICOLOGY
Oral Acute oral LD_{50} for male rats 1290, female rats 697 mg/kg.

Skin and eye Acute percutaneous LD_{50} for rats >2000 mg/kg. Slight eye irritant, not a skin irritant (rabbits).   **Other** Negative in SCE and Ames tests.
Toxicity class WHO (a.i.) III; EPA (formulation) III (proposed)

455 MCPA   *Herbicide*

aryloxyalkanoic acid

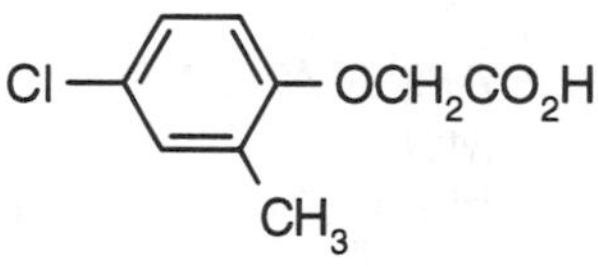

NOMENCLATURE
MCPA
Common name MCPA (BSI, E-ISO, F-ISO, WSSA); 2,4-MCPA ((*m*) France); 2M-4Kh, (USSR); [metaxon] (former name, USSR); MCP (JMAF)
IUPAC name (4-chloro-2-methylphenoxy)acetic acid
Chemical Abstracts name (4-chloro-2-methylphenoxy)acetic acid
CAS RN *[94–74–6]*   **EEC no.** 202–360–6

PHYSICAL CHEMISTRY
MCPA
Composition Tech. grade is 85–99% pure.   **Mol. wt.** 200.6   **M.f.** $C_9H_9ClO_3$
Form Colourless crystals.   **M.p.** 119–120.5 °C; 115–117 °C (tech.)
V.p. 2.3×10^{-2} mPa (20 °C) K_{ow} logP = 2.75 (pH 1), 0.46 (pH 5) (25 °C)
Solubility In water 734 mg/l (25 °C). In ethanol 1530, diethyl ether 770, toluene 26.5, xylene 49, heptane 5 (all in g/l, 25 °C).   **Stability** The acid is chemically very stable. Forms water-soluble alkali-metal and amine salts, although precipitation of calcium or magnesium salts may occur in hard water.   **pKa** 3.07

MCPA-butotyl
Mol. wt. 300.8   **M.f.** $C_{15}H_{21}ClO_4$ **Solubility** Readily soluble in organic solvents.

MCPA-isoctyl
Mol. wt. 312.8   **M.f.** $C_{17}H_{25}ClO_3$

MCPA-potassium
Mol. wt. 238.7   **M.f.** $C_9H_8ClKO_3$

MCPA-sodium
Mol. wt. 222.6   **M.f.** C$_9$H$_8$ClNaO$_3$ **Solubility** In water 270 g/l. In methanol 340, benzene 1 (both in g/l).

MCPA-dimethylammonium
Mol. wt. 245.7   **M.f.** C$_{11}$H$_{16}$ClNO$_3$

COMMERCIALISATION
History Its plant-growth regulating activity reported by R. E. Slade (*Nature (London)*, 1945, **155,** 498). Introduced as a herbicide by ICI Plant Protection Division (now Zeneca Agrochemicals) and, later, numerous other firms.
Manufacturers BASF; Nissan; Nufarm B.V.; Sanachem; United Phosphorus Ltd

APPLICATIONS
MCPA
Mode of action Selective, systemic, hormone-type herbicide, absorbed by the leaves and roots, with translocation. Concentrates in the meristematic regions, where it inhibits growth.   **Uses** Post-emergence control of annual and perennial broad-leaved weeds (including thistles and docks) in cereals (alone or undersown), herbage seed crops, flax, rice, vines, peas, potatoes, asparagus, grassland, turf, under fruit trees, and on roadside verges and embankments, etc. at 0.28–2.25 kg a.i./ha. Control of broad-leaved and woody weeds in forestry. Control of aquatic broad-leaved weeds. Often used in combination with other herbicides.
Phytotoxicity Phytotoxic to vines, vegetables, cotton, and ornamentals.
Formulation types SL; SP; EC; SL.   **Mixtures** (*MCPA* +) 2,4-DB; dichlorprop; dichlorprop-P; dicamba; MCPB; mecoprop; bromoxynil; clopyralid; 2,4-D; flurenol-butyl; and many combinations.   **Selected tradenames** 'Agroxone' (Marks); 'Agricorn' (FCC); 'Agritox' (Nufarm UK); 'Meadowman' (Barclay); 'Dicopur M' (Nufarm GmbH); 'Empal' (Unicrop); 'Sanaphen-M' (Sanachem); 'Selectyl' (Agriphar)

MCPA-isoctyl
Selected tradenames 'Brominal Plus' (Rhône-Poulenc); 'Printormona' (Nufarm UK); 'Rhonox' (Nufarm UK)

MCPA-potassium
Selected tradenames 'Agroxone' (Zeneca); 'Blagal' (Rhône-Poulenc); 'Cydexone 400' (Nufarm UK); 'Erbitox E30' (Siapa); 'Rambasan 400' (Nufarm UK)

MCPA-sodium
Selected tradenames 'Chiptox' (Nufarm UK); 'Erbitox E30' (Siapa)

MCPA-dimethylammonium
Selected tradenames 'Actril M75' (Rhône-Poulenc); 'Agritox' (Nufarm UK);

'Erbitox Combi' (Siapa); 'Kailan' (Rhône-Poulenc); 'Rhomene' (Nufarm UK); 'Spear' (Headland)

ANALYSIS
Product analysis by i.r. spectrometry (*AOAC Methods*, 1995, 971.07; *CIPAC Handbook*, 1970, **1**, 483; 1980, **1A**, 1295), by hplc (*ibid.*, 1985, **1C**, 2137; *AOAC Methods*, 1995, 980.07) or by glc of a derivative (*CIPAC Handbook*, 1985, **1C**, 2257; 1994, **F**, 292–319); FAO Specification 2/TC/S/F (1992). Free phenol impurity determined by glc (*CIPAC Handbook*, 1994, **F**, 197), hplc (*ibid.*, 1994, **F**, 362) or electrochemically (*ibid.*, 1994, **F**, 368). **Residues** determined by glc of a derivative (*J. Lest, Anal. Methods Pestic., Plant Growth Regul. Food Addit.*, 1967, **5**, 439; *Anal. Methods Pestic. Plant Growth Regul.*, 1972, **6**, 663) or by hplc (*M. Meier et al., Fresenius' Z. Anal. Chem.*, 1989, **334** (3), 235).

MAMMALIAN TOXICOLOGY
MCPA
Oral Acute oral LD_{50} for rats 900–1160, mice 550 mg/kg.   **Skin and eye** Acute percutaneous LD_{50} for rats >4000 mg/kg.   **Inhalation** LC_{50} (4 h) for rats >6.36 mg/l.   **NOEL** (24 mo) for rats 20 ppm (c. 1.33 mg/kg daily), for mice 100 ppm (c. 18 mg/kg daily).   **Toxicity class** WHO (a.i.) III; EPA (formulation) III   **EC risk** Xn (R22); Xi (R38, R41), (for salts and esters, Xn (R20/21/22))

ECOTOXICOLOGY
MCPA
Birds Acute oral LD_{50} for bobwhite quail 377 mg/kg.   **Fish** LC_{50} (96 h) for rainbow trout 232 mg/l.   **Bees** LD_{50} 0.104 mg/bee.   **Daphnia** LC_{50} >100 mg/l.

ENVIRONMENTAL FATE
Animals In rats, following oral administration, MCPA is rapidly absorbed and excreted almost exclusively in the urine, with only a small proportion in the faeces. Only moderate metabolism occurs, and there is only a small amount of conjugate formation.   **Plants** In winter wheat, MCPA is hydroxylated at the methyl group with formation of 2-hydroxymethyl-4-chlorophenoxyacetic acid.
Soil/Environment In soil, degraded to 4-chloro-2-methylphenol, followed by ring hydroxylation and ring opening. DT_{50} <7 d after initial 'lag phase'. Duration of residual activity in soil is c. 3–4 mo, following an application rate of 3 kg/ha.

MAIN ENTRIES

aryloxyalkanoic acid

Cl—⟨benzene ring⟩—OCH$_2$COSCH$_2$CH$_3$
                  CH$_3$

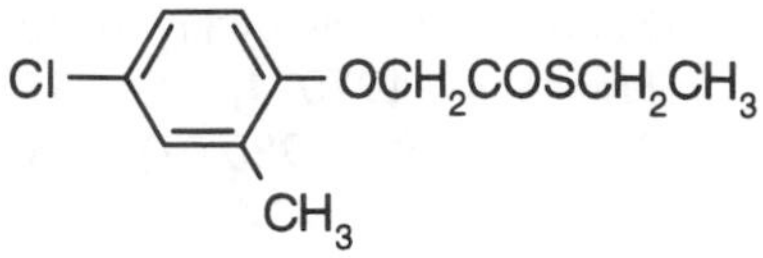

NOMENCLATURE
Common name MCPA-thioethyl (BSI, E-ISO, (*m*) F-ISO); phenothiol (JMAF); no name (USA)
IUPAC name *S*-ethyl 4-chloro-*o*-tolyloxythioacetate
Chemical Abstracts name *S*-ethyl (4-chloro-2-methylphenoxy)ethanethioate
CAS RN *[25319–90–8]* **Development codes** HOK-7501

PHYSICAL CHEMISTRY
Composition Tech. grade is 92% pure. **Mol. wt.** 244.7 **M.f.** $C_{11}H_{13}ClO_2S$
Form White needle crystals (tech., brown crystals). **M.p.** 41–42 °C
B.p. 165 °C/7 mmHg **V.p.** 21 mPa (20 °C) **K$_{ow}$** logP = 4.05 **Solubility** In water 2.3 mg/l (25 °C). In acetone, xylene >1000, hexane 290 (all in g/l, 25 °C).
Stability Stable in acidic media, but comparatively unstable in alkaline media. In water DT$_{50}$ (25 °C) 22 d (pH 7), 2 d (pH 9). Stable below 200 °C.

COMMERCIALISATION
History Herbicide reported by T. Ohi *et al.* (*Zasso Kenkyu*, 1969, **9**, 46). Introduced by Hokko Chemical Industry Co., Ltd. **Patent** US 3708278; GB 1263169 **Manufacturers** Hokko

APPLICATIONS
Mode of action Selective, systemic, hormone-type herbicide, absorbed by the leaves and roots, with translocation. Concentrates in the meristematic regions, where it inhibits growth. **Uses** Post-emergence control of annual and perennial broad-leaved, and cyperaceous, weeds (including *Chenopodium album*, *Convolvulus arvensis*, *Cyperus* spp., *Monochoria vaginalis*, *Polygonum aviculare* and *Sagittaria pygmaea*) in orchards, paddy rice, and cereals. **Formulation types** GR; EC.
Mixtures *(MCPA-thioethyl +)* propanil; simetryn; simetryn + dimepiperate.
Selected tradenames 'Fenobit' (Hokko); 'Herbit' (Hokko); 'Zero One' (Hokko)

ANALYSIS
Product and **residue** analysis by glc.

MAMMALIAN TOXICOLOGY
Oral Acute oral LD$_{50}$ for male rats 790, female rats 877, male mice 811, female

mice 749 mg/kg.   **Skin and eye** Acute percutaneous LD_{50} for male mice >1500 mg/kg. Application (24 h) of tech. to skin and eyes of rabbits produced no irritation.   **Inhalation** LC_{50} (4 h) for rats >44 mg/m^3.   **NOEL** (90 d) for rats and mice 300 mg/kg diet; (2 y) for rats 100, mice 20 mg/kg diet. No reproductive or teratogenic effects in rats. Non-mutagenic.   **Other** Acute i.p. LD_{50} for male rats 530, female rats 570 mg/kg.   **Toxicity class** WHO (a.i.) III

ECOTOXICOLOGY
Birds Acute oral LD_{50} for Japanese quail >3000 mg/kg.   **Fish** LC_{50} (48 h) for carp 2.5 mg/l.   **Bees** LD_{50} (contact) >40 µg/bee.   **Daphnia** LC_{50} (6 h) 4.5 mg/l.

457 MCPB   *Herbicide*

aryloxyalkanoic acid

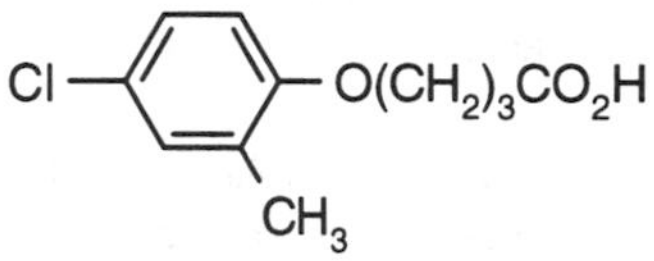

NOMENCLATURE
MCPB
Common name MCPB (BSI, E-ISO, (*m*) F-ISO, WSSA, JMAF); 2,4-MCPB (France); 2M-4Kh-M (USSR)
IUPAC name 4-(4-chloro-*o*-tolyloxy)butyric acid
Chemical Abstracts name 4-(4-chloro-2-methylphenoxy)butanoic acid
CAS RN *[94–81–5]*   **EEC no.** 202–365–3   **Development codes** MB 3046 (Rhône-Poulenc)

MCPB-sodium
CAS RN *[6062–26–6]*

MCPB-ethyl
Common name MCPB-ethyl
IUPAC name ethyl 4-(4-chloro-*o*-tolyloxy)butyrate
Chemical Abstracts name ethyl 4-(4-chloro-2-methylphenoxy)butanoate
CAS RN *[10443–70–6]*

PHYSICAL CHEMISTRY
MCPB
Composition Tech. grade is *c.* 92% pure.   **Mol. wt.** 228.7   **M.f.** $C_{11}H_{13}ClO_3$
Form Colourless crystals; (tech., beige to brown flake).   **M.p.** 100 °C; (tech.,

95–100 °C)   **B.p.** >280 °C   **V.p.** 5.77×10^{-2} mPa (20 °C); 9.83×10^{-2} mPa (25 °C) $\mathbf{K_{ow}}$ logP = 2.79 (unstated pH)   **S.g./density** 1.254 g/cm^3 (22 °C)
Solubility In water 44 mg/l (room temperature). In acetone 313, dichloromethane 160, ethanol 150, hexane 65, toluene 8 (all in g/l, room temperature). The common alkali-metal and amine salts are readily soluble in water, although precipitation of calcium or magnesium salts may occur in hard water, but sparingly soluble in organic solvents.   **Stability** The acid is chemically very stable. Stable to sunlight. Stable to aluminium, tin and iron up to 150 °C.   Forms water-soluble alkali-metal and amine salts, although precipitation of calcium or magnesium salts may occur in hard water.   **pKa** 4.84

MCPB-sodium
Mol. wt. 250.7   **M.f.** $C_{11}H_{12}ClNaO_3$

MCPB-ethyl
Mol. wt. 256.7   **M.f.** $C_{13}H_{17}ClO_3$ **Form** Colourless liquid   **M.p.** −1 °C
B.p. 165–168 °C/0.5 mmHg   **V.p.** 1.9×10^{-1} mPa (20 °C) $\mathbf{K_{ow}}$ logP = 4.26
S.g./density 1.138   **Solubility** In water 10 ppm (20 °C).   **Stability** Unstable in alkali.

COMMERCIALISATION
History Herbicide reported by R. L. Wain & F. Wightman (*Proc. Roy. Soc.,* 1955, **142B**, 525). Introduced by May & Baker Ltd (now Rhône-Poulenc Agrochimie).

MCPB
Patent GB 758980   **Manufacturers** Marks

MCPB-ethyl
Manufacturers Nippon Kayaku

APPLICATIONS
MCPB-sodium
Biochemistry It owes its selectivity to the ability of susceptible plants to translocate it (R. C. Kirkwood *et al., Pestic. Sci.,* 1972, **3**, 307) and to oxidise it to MCPA which is the real toxicant.   **Mode of action** Selective, systemic, hormone-type herbicide, absorbed by the leaves and roots, with translocation.   **Uses** Post-emergence control of annual and perennial broad-leaved weeds (including thistles and docks) in cereals (alone or undersown), clovers, sainfoin, peas, peanuts, and grassland. Control of broad-leaved and woody weeds in forestry.
Phytotoxicity Phytotoxic to vines, oilseed rape, and beet.   **Formulation types** SC; SL.   **Mixtures** *(MCPB-sodium +)* MCPA; bentazone; bentazone + MCPA; simetryn + thiobencarb; quinoclamine + simetryn.   **Selected tradenames** 'Bellmac Straight' (United Phosphorus Ltd); 'Madek' (Agro-Kanesho); 'Tropotox' (Unicrop)

MCPB-ethyl
Uses Used in combination with simetryn, thiobencarb, bentazone, etc. to control broad-leaved weeds in paddy fields.   **Mixtures** *(MCPB-ethyl +)* simetryn; thiobencarb + simetryn; bentazone + simetryn.   **Selected tradenames** 'MCPB' (Nippon Kayaku)

ANALYSIS
MCPB
Product determined by glc (*CIPAC Handbook*, 1994, **F**, 292–319). Free phenol impurity by glc (*ibid.*, 1994, **F**, 197), or by hplc (*ibid.*, 1994, **F**, 362).

MCPB-sodium
Product analysis by titration of extractable acids (*CIPAC Proc.*, 1981, **3**, 277) or by glc (*CIPAC Handbook*, 1985, **1C**, 2257).   **Residues** determined by glc of a derivative with ECD (A. Smith *et al.*, *Weed Res.*, 1981, **21**, 179; A. Guardiglii, *Anal. Methods Pestic. Plant Growth Regul.*, 1976, **8**, 397).

MCPB-ethyl
Product and **residue** analysis by glc.

MAMMALIAN TOXICOLOGY
MCPB
Oral Acute oral LD_{50} for rats 4700 mg/kg.   **Skin and eye** Acute percutaneous LD_{50} for rats >2000 mg/kg. Eye irritant; not a skin irritant. Not a skin sensitiser. **Inhalation** LC_{50} (4 h) for rats >1.14 mg/l air.   **NOEL** (90 d) for rats 100 ppm diet. No organ or histopathological changes at 2500 ppm. Not teratogenic in rats or rabbits.   **Toxicity class** WHO (a.i.) III; EPA (formulation) III   **EC risk** Xn (R22) (also applies to salts and esters)

MCPB-ethyl
Reviews *J. Pestic. Sci.*, **17**, 539–543 (1992).   **Oral** Acute oral LD_{50} for male rats 1780, female rats 1420, male mice 1160, female mice 1550 mg/kg.
Skin and eye Acute percutaneous LD_{50} for male rats >4000 mg/kg.
Inhalation LC_{50} for male rats 3.0–4.5, female rats >5.0 mg/l.   **NOEL** (2 y) for male rats 17.5, female rats 19.4 mg/kg.   **EC risk** Xn (R22)

ECOTOXICOLOGY
MCPB
Birds LC_{50} for birds >20 000 mg/kg.   **Fish** LC_{50} (48 h) for rainbow trout 75, fathead minnows 11 mg/l.   **Bees** Not toxic to bees.

MCPB-sodium
Birds Acute oral LD_{50} for bobwhite quail 282 mg/kg. LC_{50} (8 d) for bobwhite quail and mallard ducklings >5000 ppm. Determinations made on aqueous solution.

MCPB-ethyl
Fish LC_{50} for carp 5.6 ppm.   **Daphnia** LC_{50} 3.3 ppm.

ENVIRONMENTAL FATE
Animals Studies in cattle show that MCPB is excreted in the urine, either unchanged or as MCPA. Levels in milk were at or below the method level of detection.   **Plants** In susceptible plants, undergoes β-oxidation to MCPA, which is subsequently degraded to 4-chloro-2-methylphenol, followed by ring hydroxylation and ring opening.   **Soil/Environment** In soil, metabolism involves degradation of the side-chain to 4-chloro-2-methylphenol, ring hydroxylation, and ring opening. Duration of residual activity in soil is *c.* 6 w.

458 mecarbam   *Insecticide, acaricide*

organophosphorus

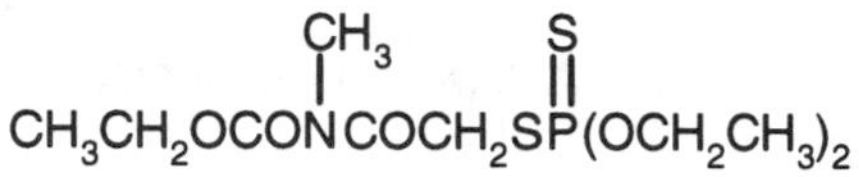

$$CH_3CH_2OCON(CH_3)COCH_2SP(S)(OCH_2CH_3)_2$$

NOMENCLATURE
Common name mecarbam (BSI, E-ISO, JMAF); mécarbame ((*m*) F-ISO); no name (France)
IUPAC name *S*-(*N*-ethoxycarbonyl-*N*-methylcarbamoylmethyl) *O,O*-diethyl phosphorodithioate; ethyl *N*-(diethoxythiophosphorylthio)acetyl-*N*-methyl= carbamate; ethyl (diethoxyphosphinothioylthio)acetyl(methyl)carbamate
Chemical Abstracts name ethyl 6-ethoxy-2-methyl-3-oxo-7-oxa-5-thia-2-aza-= 6-phosphanonanoate 6-sulfide
CAS RN *[2595-54-2]*   **EEC no.** 219-993-9   **Development codes** P 474; MC 474 (both Murphy)

PHYSICAL CHEMISTRY
Composition Tech. grade has purity ≥85%.   **Mol. wt.** 329.4   **M.f.** $C_{10}H_{20}NO_5PS_2$
Form Light brown to pale yellow oil; (tech., pale yellow to brown oil).
B.p. 144 °C/0.02 mmHg   **V.p.** Negligible at room temperature.
S.g./density 1.222 (20 °C)   **Solubility** In water <1 g/l (room temperature). In aliphatic hydrocarbons <50 g/kg (room temperature). Miscible with alcohols, esters, ketones, and aromatic and chlorinated hydrocarbons at room temperature.
Stability Subject to hydrolysis below pH 3.

COMMERCIALISATION
History Insecticide reported by M. Pianka (*Chem. Ind. (London)*, 1961, p. 324).

Introduced by Murphy Chemical Ltd (now part of DowElanco).
Patent GB 867780

APPLICATIONS

Biochemistry Cholinesterase inhibitor. **Mode of action** Insecticide and acaricide with slight systemic properties, and contact and stomach action. Long residual activity. **Uses** Control of aphids, suckers, whitefly, scale insects, mealybugs and red spider mites on citrus, apple, olive, and other fruit trees; leafhoppers, planthoppers and miners on rice; whitefly, leafhoppers and thrips on cotton; and root fly larvae on onions, carrots, celery and brassicas. **Phytotoxicity** Highly phytotoxic to aubergines. **Formulation types** EC. **Mixtures** *(mecarbam +)* carbaryl. **Compatibility** Incompatible with highly alkaline materials. **Selected tradenames** 'Murfotox' (Efthymiadis)

ANALYSIS

Product analysis by glc with FID (V. P. Lynch, *Anal. Methods Pestic. Plant Growth Regul.*, 1976, **8**, 135). **Residues** determined by glc with ECD (*idem, ibid.; Man. Pestic. Residue Anal.*, 1987, **I**, 6, S19; *Anal. Methods Residues Pestic.*, 1988, Part I, M2). Details available from DowElanco.

MAMMALIAN TOXICOLOGY

Reviews *Pesticide residues in food – 1986.* FAO Plant Production and Protection Paper 77, 1986. *Pesticide residues in food – 1986 evaluations. Part II – Toxicology.* FAO Plant Production and Protection Paper 78/2, 1987. **EHC** 63 (WHO, 1986; a general review of organophosphorus insecticides). **Oral** Acute oral LD_{50} for rats 36–53, mice 106 mg/kg. **Skin and eye** Acute percutaneous LD_{50} for rats >1220 mg/kg. **Inhalation** LC_{50} (6 h) for rats 0.7 mg/l air. **NOEL** In 0.5 y feeding trials, rats receiving 1.6 mg/kg daily suffered no ill-effect, but at 4.56 mg/kg daily slight depression of growth rate. **ADI** (JMPR) 0.002 mg/kg b.w. [1986]. **Toxicity class** WHO (a.i.) Ib; EPA (formulation) I **EC risk** T (R24/25)

ECOTOXICOLOGY

Bees Toxic to bees.

ENVIRONMENTAL FATE

Animals Metabolism in the rat was principally by hydrolysis, oxidative desulfuration and degradation of the carbamoyl moiety: O-de-ethylation was a minor pathway. **Soil/Environment** Persists in the soil for 4–6 weeks.

aryloxyalkanoic acid

$$Cl - \text{(ring)} - OCHCO_2H \quad (CH_3)$$

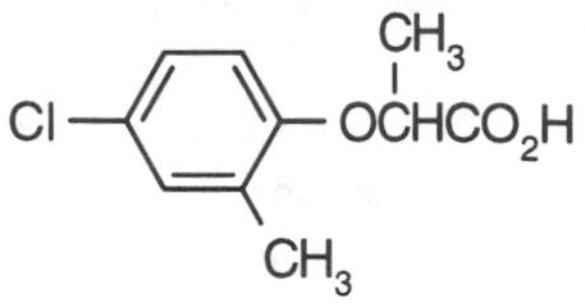

NOMENCLATURE
Common name mecoprop (BSI, E-ISO, (*m*) F-ISO, WSSA); mechlorprop
(Denmark); MCPP (JMAF); mécoprop (France (for a mixture of isomers of
2-(chloro-*o*-tolyloxy)propionic acid))
IUPAC name (*RS*)-2-(4-chloro-*o*-tolyloxy)propionic acid
Chemical Abstracts name (±)-2-(4-chloro-2-methylphenoxy)propanoic acid
Other names CMPP **CAS RN** *[7085–19–0]* racemate; *[93–65–2]* (formerly used
for mecoprop) unstated stereochemistry **EEC no.** 202–264–4
Development codes RD 4593 (AgrEvo)

PHYSICAL CHEMISTRY
Mol. wt. 214.6 **M.f.** $C_{10}H_{11}ClO_3$ **Form** Colourless crystals. **M.p.** 94–95 °C
V.p. 0.31 mPa (20 °C) **K_{ow}** logP = 0.1004 (pH 7), 3.2 (unionised, 25 °C,
K. Chamberlain et al., *Pestic. Sci.*, **47**, 265 (1996)) **Solubility** In water 734 mg/l
(25 °C). In acetone, diethyl ether, ethanol >1000, ethyl acetate 825, chloroform
339 (all in g/kg, 20 °C). Salts in water: potassium 920, sodium 500,
diethanolamine 580, dimethylamine 660 (all in g/l, 20 °C). **Stability** Stable to
heat, and to hydrolysis, reduction, and atmospheric oxidation. Mecoprop is acidic,
and forms salts, many of which are water-soluble. **pKa** 3.78 (20–25 °C)

COMMERCIALISATION
History Plant growth regulating activity described by C. H. Fawcett et al. (*Ann.
Appl. Biol.*, 1953, **40**, 232) and its use as a herbicide by G. B. Lush & E. L. Leafe
(*Proc. Br. Weed Control Conf., 3rd*, 1956, pp. 625, 633). Introduced by The Boots
Co. Ltd Agricultural Division (now AgrEvo GmbH, who do not manufacture or
market it). **Patent** GB 820180; GB 822973; GB 825875 **Manufacturers** Marks;
Nufarm B.V.; United Phosphorus Ltd

APPLICATIONS
Mode of action Selective, systemic, hormone-type herbicide, absorbed by the
leaves, with translocation to the roots. Only the (*R*)-(+) isomer is herbicidally
active. **Uses** Post-emergence control of broad-leaved weeds (especially cleavers,
chickweed, clovers and plantains) in wheat, barley, oats, herbage seed crops
(including undersown), grassland, and under fruit trees and vines; and control of
docks (*Rumex* spp.) in meadows and pastures. Often used in combination with

other herbicides. **Formulation types** EC; SL. **Mixtures** *(mecoprop +)* bentazone; bifenox; clopyralid; cyanazine; dichlorprop; dicamba; MCPA; ioxynil; bromoxynil; 2,4-D; 2,4-DB; isoproturon; and many combinations.
Selected tradenames 'Actril M' (Rhône-Poulenc); 'Atlas CMPP' (Atlas Crop Protection); 'Clenecorn' (FCC); 'Compitox' (Nufarm UK); 'Mega P' (Nufarm B.V.); 'Propal' (Unicrop); 'Propionyl' (Agriphar)

ANALYSIS
Product analysis by hplc (*AOAC Methods*, 1995, 984.07; *CIPAC Handbook*, 1985, **1C**, 2157), by glc of a derivative (*CIPAC Handbook*, 1985, **1C**, 2257; 1994, **G**, 292–319), or by titration of extractable acids (*CIPAC Handbook*, 1980, **1A**, 1297; 1983, **1B**, 1860). Free phenol impurity determined by glc (*ibid.*, 1994, **F**, 197), hplc (*ibid.*, 1994, **F**, 362) or electrochemically (*ibid.*, 1994, **F**, 368).
Residues determined by glc of a derivative (*BBA Kurzfass. Analytik Pflanzenschutzmitteln in Wasser* Teil 1, 2. Auflage 1989, S.225; *Rückstandsanalytik Pflanzenschutzmitteln*, 1976, W4,4 & 1982, XII 6, 6).

MAMMALIAN TOXICOLOGY
Oral Acute oral LD_{50} for rats 930–1166, mice 650 mg/kg. **Skin and eye** Acute percutaneous LD_{50} for rabbits 900, rats >4000 mg/kg. Skin irritant; highly irritating to eyes. **Inhalation** LC_{50} (4 h) for rats >12.5 mg/l air. **NOEL** (21 d) for rats 65 mg/kg daily; (90 d) for rats 4.5–13.5, dogs 4 mg/kg daily; (2 y) for rats 1.1 mg/kg. Rats receiving 100 mg/kg diet for 210 d suffered only a slight enlargement of the kidneys. **Toxicity class** WHO (a.i.) III; EPA (formulation) III
EC risk Xn (R22); Xi (R38, R41), (for salts, Xn (R20/21/22))

ECOTOXICOLOGY
Birds Acute oral LD_{50} for Japanese quail 740 mg/kg. **Fish** LC_{50} (96 h) for trout 150–220, bluegill sunfish >100 mg/l. **Bees** Not toxic to honeybees; LD_{50} (oral) >10 µg/bee, (contact) >100 µg/bee. **Daphnia** LC_{50} (48 h) 420 mg/l.

ENVIRONMENTAL FATE
Animals In mammals, following oral administration, mecoprop is predominantly eliminated unchanged in the urine. **Plants** In plants, mecoprop is hydroxylated at the methyl group with formation of 2-hydroxymethyl-4-chlorophenoxypropionic acid. The hydroxylation of the aromatic ring observed to a slight extent is obviously a minor metabolic pathway in the plant. **Soil/Environment** In soil, degraded predominantly by micro-organisms to 4-chloro-2-methylphenol, followed by ring hydroxylation at the 6-position and ring opening. DT_{50} in soil 7–13 d. Duration of residual activity in soil is c. 2 mo. K_{oc} 12–25.

aryloxyalkanoic acid

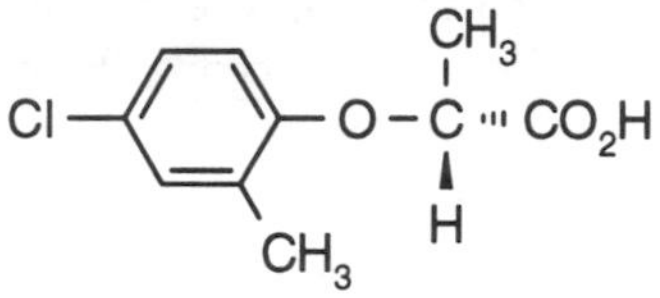

NOMENCLATURE

Common name mecoprop-P (BSI, draft E-ISO, (*m*) draft F-ISO)
IUPAC name (*R*)-2-(4-chloro-*o*-tolyloxy)propionic acid
Chemical Abstracts name (+)-(*R*)-2-(4-chloro-2-methylphenoxy)propanoic acid
CAS RN *[16484–77–8]* previously *[18221–59–5]*, *[94596–45–9]*. Salts: isobutyl
[101012–85–5]; dimethylammonium *[66423–09–4]*; potassium *[66423–05–0]*
EEC no. 240–539–0   **Development codes** BAS 03 729H

PHYSICAL CHEMISTRY

Mol. wt. 214.6   **M.f.** $C_{10}H_{11}ClO_3$ **Form** White crystals with a weak intrinsic
odour.   **M.p.** 94.6–96.2 °C; (tech. 84–91 °C)   **V.p.** 0.4 mPa (20 °C)
K_{ow} logP = –0.23 (pH 7, 25 °C)   **Henry** 1.0×10^{-4} Pa m^3 mol^{-1} **S.g./density**
c. 1.31 (20 °C)   **Solubility** In water 860 mg/l (pH 7, 20 °C). In acetone, diethyl
ether, ethanol >1000, dichloromethane 730, toluene 330, hexane 9 (all in g/kg,
20 °C).   **Stability** Stable to heat, light, and stable in the range pH 3 to pH 9.
Specific rotation $[\alpha]_D$ +35.2° (acetone), –17.1° (benzene), +21° (chloroform),
+28.1° (ethanol) (M. Matell, *Ark. Kemi.*, 1953, **6**, 365). The simple salts are
laevorotatory, $[\alpha]_D$ –14.1° for mecoprop-P-sodium (*idem, ibid.*).   **pKa** 3.68
(20 °C)

COMMERCIALISATION

History Mecoprop-P was shown to be the herbicidal stereoisomer of mecoprop
(the racemate) by M. Matell (*Ark. Kemi.*, 1953, **6**, 365), by B. Aberg
(*Lantbrukshoegksk. Ann.*, 1963, **29**, 3) and by W. O. G. Nuyken *et al.* (*Meded. Fac.
Landbouwwet. Rijksuniv. Gent*, 1987, **52**, 1139). Introduced in Germany (1987) by
BASF AG.   **Manufacturers** BASF; Marks

APPLICATIONS

Mode of action Selective, systemic hormone-type herbicide, absorbed by the
leaves, with translocation to the roots.   **Uses** Post-emergence control of broad-
leaved weeds (especially cleavers, chickweed, clovers and plantains) in wheat,
barley, oats, herbage seed crops, and grassland. Often used in combination with
other herbicides at 1.2–1.5 kg a.i./ha.   **Phytotoxicity** Slightly phytotoxic to winter
rye (though this is only temporary).   **Formulation types** EC; SL.   **Mixtures**
(*mecoprop-P* +) 2,4-D; dichlorprop-P + MCPA; bentazone; dicamba + MCPA; and

many other combinations. **Selected tradenames** 'Duplosan KV' (BASF); 'Optica' (Marks)

ANALYSIS
Product analysis by rp hplc (*CIPAC Handbook,* 1995, **G**, 109–115).
Residues determined by gc-ms.

MAMMALIAN TOXICOLOGY
Oral Acute oral LD_{50} for rats 1050 mg/kg.   **Skin and eye** Acute percutaneous LD_{50} for rats >4000 mg/kg.   **Inhalation** LC_{50} (4 h) for rats >5.6 mg/l.
NOEL (2 y) for rats 1.1 mg/kg b.w. daily.   **Other** Non-carcinogenic and non-oncogenic.   **Toxicity class** WHO (a.i.) III; EPA (formulation) III

ECOTOXICOLOGY
Birds Acute oral LD_{50} for quail *c.* 500 mg/kg. Dietary LC_{50} for bobwhite quail >5600 mg/kg diet (dimethylammonium salt).   **Fish** LC_{50} (96 h) for trout 150–220 mg/l.   **Bees** Not toxic to honeybees; LD_{50} (48 h) >25 µg/bee (dimethylammonium salt).   **Daphnia** EC_{50} (48 h) >100 mg/l.   **Algae** EC_{50} (72 h) for *Pseudokirchneriella subcapitata* 270 mg/l.

ENVIRONMENTAL FATE
Animals In mammals, following oral administration, mecoprop-P is predominantly eliminated unchanged in the urine.   **Plants** In plants, mecoprop-P is metabolised in a similar manner to that in soil.   **Soil/Environment** In soil, degraded predominantly by micro-organisms to 4-chloro-2-methylphenol, followed by ring hydroxylation at the 6-position and ring opening. Duration of residual activity in soil is *c.* 2 mo.

461 mefenacet   *Herbicide*

oxyacetamide

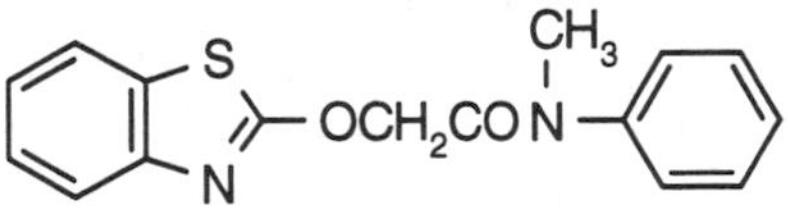

NOMENCLATURE
Common name mefenacet (BSI, draft E-ISO, (*m*) draft F-ISO)
IUPAC name 2-(1,3-benzothiazol-2-yloxy)-*N*-methylacetanilide; 2-benzothiazol-2-yloxy-*N*-methylacetanilide
Chemical Abstracts name 2-(2-benzothiazolyloxy)-*N*-methyl-*N*-phenylacetamide
CAS RN *[73250–68–7]*   **Development codes** FOE 1976 (Bayer); NTN 801 (Nihon Bayer)

PHYSICAL CHEMISTRY
Mol. wt. 298.4   **M.f.** $C_{16}H_{14}N_2O_2S$ **Form** Colourless, odourless crystals.
M.p. 134.8 °C   **V.p.** 6.4×10^{-4} mPa (20 °C); 11 mPa (100 °C) K_{ow} logP = 3.23
Solubility In water 4 mg/l (20 °C). In dichloromethane >200, hexane 0.1–1.0,
toluene 20–50, isopropanol 5–10 (all in g/l, 20 °C).   **Stability** Stable to light. In
storage, 94.8% remains unchanged after 6 months at 30 °C. Stable to hydrolysis
at pH 4–9.

COMMERCIALISATION
History Herbicide reported by R. R. Schmidt et al. (*Meded. Fac. Landbouwwet.
Rijksuniv. Gent,* 1984, **49**, 1075). Introduced in Japan (1987) by Nihon Bayer
Agrochem K.K.   **Patent** DE 2822155; DE 2903966; DE 3038636; DE 3323334;
DE 3344236; DE 3418167; DE 3422861   **Manufacturers** Bayer

APPLICATIONS
Biochemistry Inhibits cell division and growth.   **Mode of action** Selective herbicide.
Uses Used pre-emergence and early post-emergence, mainly in transplanted rice,
at 1.0–1.6 kg/ha to control grass weeds (with a specific action against *Echinochloa
crus-galli*). Preferably used in mixtures with other herbicides.
Formulation types GR; WP; SC.   **Mixtures** *(mefenacet +)* bensulfuron-methyl;
molinate; bromobutide + pyrazolynate; pyrazosulfuron-ethyl; thiobencarb +
bensulfuron-methyl; imazosulfuron; naproanilide; pyriminobac + bensulfuron-
methyl; pyributicarb + imazosulfuron; cyhalofop + pyrazosulfuron.
Selected tradenames 'Act' (mixture) (Nihon Bayer, Nissan, Yashima); 'Batl'
(mixture) (Nihon Bayer, Takeda); 'Hinochloa' (Bayer); 'Rancho' (Bayer);
'Wolf-Ace' (mixture) (Du Pont, Kumiai); 'Zark' (mixture) (Du Pont, Kumiai,
Nihon Bayer)

ANALYSIS
Product analysis by hplc. Details and methods for determination of **residues**
available from Bayer.

MAMMALIAN TOXICOLOGY
Oral Acute oral LD_{50} for rats, mice, and dogs >5000 mg/kg.   **Skin and eye** Acute
percutaneous LD_{50} for rats and mice >5000 mg/kg. Not irritating to skin and eyes
(rabbits).   **Inhalation** LC_{50} (4 h) for rats >0.02 mg/l (dust).   **NOEL** (2 y) for rats
100, mice 300 mg/kg diet.   **ADI** 0.0036 mg/kg b.w.   **Toxicity class** WHO (a.i.)
III (Table 5); EPA (formulation) IV

ECOTOXICOLOGY
Birds Five-day LC_{50} for bobwhite quail >5000 mg/kg diet.   **Fish** LC_{50} (96 h) for
carp 6.0, trout 6.8, golden orfe 11.5 mg/l.   **Worms** LC_{50} (28 d) for *Eisenia foetida*
>1000 mg/kg dry substrate.   **Daphnia** LC_{50} (48 h) 1.81 mg/l.   **Algae** EC_{50} (96 h)
for *Scenedesmus subspicatus* 0.18 mg/l.

ENVIRONMENTAL FATE

Animals Degradation in the rat is to N-methylaniline which is subsequently demethylated, acetylated and hydroxylated to 4-aminophenol and its sulfate- and glucuronide- conjugates.   **Plants** Metabolised to 4-aminophenol via oxidation of an intermediate N-methylaniline. Other metabolites found are benzothiazolone and benzothiazolyloxy acetic acid; both are degraded by hydroxylation.

Soil/Environment In soil, mefenacet is strongly adsorbed and therefore shows little movement. DT$_{50}$ in soil is a few weeks; metabolites formed are benzothiazole and benzothiazolyloxyacetic acid. In sterile aqueous buffer solutions, mefenacet undergoes slow hydrolysis at all pH levels; however in natural water, it is degraded more rapidly.

462 mefenpyr-diethyl   *Herbicide safener*

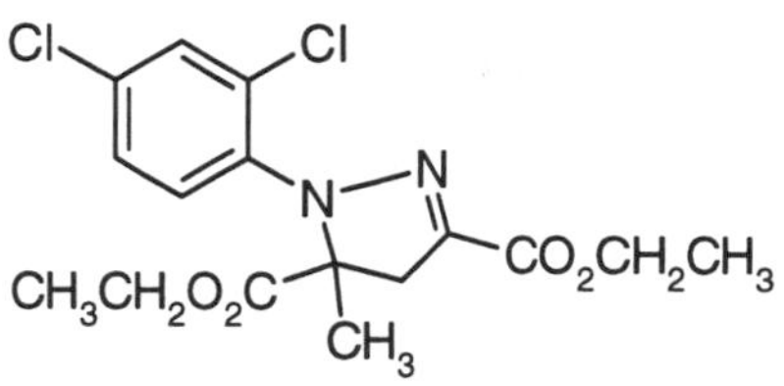

NOMENCLATURE

Common name mefenpyr-diethyl (BSI, pa ISO)

IUPAC name diethyl (RS)-1-(2,4-dichlorophenyl)-5-methyl-2-pyrazoline-= 3,5-dicarboxylate

Chemical Abstracts name diethyl 1-(2,4-dichlorophenyl)-4,5-dihydro-5-methyl-= 1H-pyrazole-3,5-dicarboxylate

CAS RN [135590–91–9]; (diacid [135591–00–3])   **Development codes** Hoe 107892; AE F107892

PHYSICAL CHEMISTRY

Mol. wt. 373.2; (diacid 317.1)   **M.f.** $C_{16}H_{18}Cl_2N_2O_4$; (diacid $C_{12}H_{10}Cl_2N_2O_4$)
Form White to light beige crystals.   **M.p.** 50–52 °C   **V.p.** 6.3×10^{-3} mPa (20 °C)
K$_{ow}$ logP = 3.83   **S.g./density** c. 1.31 (20 °C)   **Solubility** In water 20 mg/kg (pH 6.2, 20 °C). In acetone >500, toluene >400, ethyl acetate >400, methanol >400 (all in g/l, 20 °C).   **Stability** Hydrolysed by acids and alkali.

COMMERCIALISATION

Manufacturers AgrEvo

APPLICATIONS
Uses Used as a herbicide safener in combination with fenoxaprop-P-ethyl, for selective weed control in wheat, rye, triticale and some barley varieties.
Formulation types SC; SE; EW.

MAMMALIAN TOXICOLOGY
Oral Acute oral LD_{50} for rats and mice >5000 mg/kg. **Skin and eye** Acute percutaneous LD_{50} for rats >4000 mg/kg. **Inhalation** LC_{50} (4 h) for rats >1.32 mg/l. **NOEL** (2 y) for rats 48, mice 71 mg/kg b.w. daily.

ECOTOXICOLOGY
Birds Acute oral LD_{50} for Japanese quail >2000 mg/kg. **Fish** LC_{50} (96 h) for carp 2.4, rainbow trout 4.2 mg/l. **Bees** LD_{50} (oral, 48 h) >900 µg/bee; (contact) >700 µg/bee. **Worms** LC_{50} (14 d) for *Eisenia foetida* >1000 mg/kg soil. **Daphnia** LC_{50} (48 h) 53 mg/l. **Algae** E_bC_{50} (72 h) for *Scenedesmus subspicatus* 5.8 mg/l.

463 mefluidide   *Plant growth regulator, herbicide*

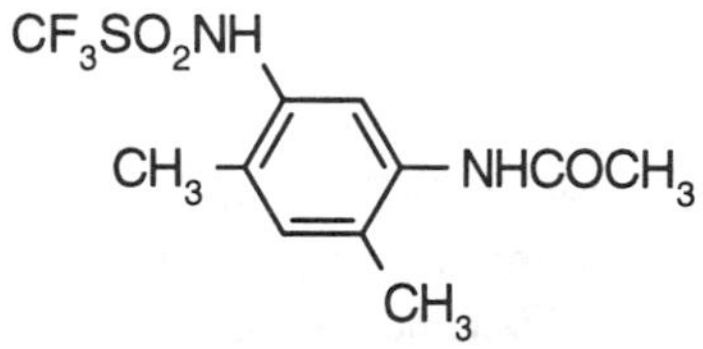

NOMENCLATURE
Common name mefluidide (BSI, E-ISO, (*m*) F-ISO, ANSI, WSSA)
IUPAC name 5'-(1,1,1-trifluoromethanesulfonamido)aceto-2',4'-xylidide
Chemical Abstracts name *N*-[2,4-dimethyl-5-[[(trifluoromethyl)sulfonyl]amino]= phenyl]acetamide
CAS RN *[53780–34–0]* **Development codes** MBR 12 325 (3M)

PHYSICAL CHEMISTRY
Mol. wt. 310.3 **M.f.** $C_{11}H_{13}F_3N_2O_3S$ **Form** Colourless, odourless crystals.
M.p. 183–185 °C **V.p.** <10 mPa (25 °C) **K_{ow}** logP = 2.02 (25 °C, unionised, K. Chamberlain *et al.*, *Pestic. Sci.*, **47**, 265 (1996)) **Solubility** In water 180 mg/l (23 °C). In acetone 350, methanol 310, acetonitrile 64, ethyl acetate 50, *n*-octanol 17, diethyl ether 3.9, dichloromethane 2.1, benzene 0.31, xylene 0.12 (all in g/l, 23 °C). **Stability** Stable at elevated temperatures, but the acetamido moiety is

hydrolysed on refluxing mefluidide in acidic or alkaline solutions; degraded in aqueous solutions exposed to u.v. radiation. **pKa** 4.6

COMMERCIALISATION

History Plant growth regulator reported (*Proc. North Cent. Weed Control Conf.,* 1974). Introduced by the 3M Company. All rights purchased by PBI/Gordon in 1989. **Patent** US 3894078 **Manufacturers** PBI/Gordon

APPLICATIONS

Mode of action Plant growth regulator and herbicide which inhibits growth and development of meristematic regions. **Uses** Inhibition of growth and suppression of seed production of perennial grasses in turf, lawns, grassland, industrial areas, amenity areas, and areas where grass-cutting is difficult (e.g. roadside verges and embankments). Inhibition of growth of ornamental trees and shrubs. Enhancement of the sucrose content of sugar cane. Control of growth and seed production of weeds (particularly *Sorghum halepense* and volunteer cereals) in soya beans and other crops. Application rates range from 0.3 to 1.1 kg a.i./ha.
Formulation types SL. **Compatibility** Compatible with growth-regulator type herbicides. Incompatible with liquid fertilisers which are acidic in nature.
Selected tradenames 'Embark' (PBI/Gordon)

ANALYSIS

Analysis by glc of a derivative.

MAMMALIAN TOXICOLOGY

Oral Acute oral LD_{50} for rats >4000, mice 1920 mg/kg. **Skin and eye** Acute percutaneous LD_{50} for rabbits >4000 mg/kg. Mild eye irritant; non-irritating to skin (rabbits). **NOEL** (90 d) for rats 6000, dogs 1000 mg/kg diet. Non-mutagenic and non-teratogenic. **Other** No mutagenic effect observed in *Salmonella typhimurium*. **Toxicity class** WHO (a.i.) III

ECOTOXICOLOGY

Birds Acute oral LD_{50} for mallard ducks and bobwhite quail >4620 mg/kg. Five-day dietary LC_{50} for mallard ducks and bobwhite quail >10 000 mg/kg diet (observed on the 8th day). **Fish** LC_{50} (96 h) for rainbow trout and bluegill sunfish >100 mg/l. **Bees** Not toxic to bees.

ENVIRONMENTAL FATE

Animals In mammals, following oral administration, mefluidide residues are excreted intact. **Soil/Environment** Rapidly degraded in soil, DT_{50} <1 w. 5-Amino-2,4-dimethyltrifluoromethanesulfonanilide is found as a metabolite.

MAIN ENTRIES

anilinopyrimidine

CH₃C≡C, N, NH—⟨phenyl ring⟩

$CH_3C{\equiv}C$ N NH—phenyl, CH₃ on pyrimidine

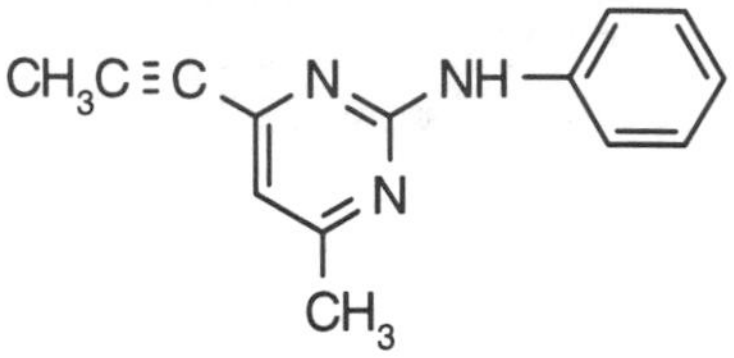

NOMENCLATURE
Common name mepanipyrim (BSI, draft E-ISO)
IUPAC name N-(4-methyl-6-prop-1-ynylpyrimidin-2-yl)aniline
Chemical Abstracts name 4-methyl-N-phenyl-6-(1-propynyl)-2-pyrimidinamine
CAS RN [110235–47–7]   **Development codes** KUF-6201; KIF-3535

PHYSICAL CHEMISTRY
Composition Tech. is >94%.   **Mol. wt.** 223.3   **M.f.** $C_{14}H_{13}N_3$ **Form** White
crystals/powder.   **M.p.** 132.8 °C   **V.p.** 2.32×10^{-2} mPa (20 °C) K_{ow} logP = 3.28
(20 °C)   **S.g./density** 1.2025   **Solubility** In water 3.10 mg/l (20 °C). In acetone
139, methanol 15.4, n-hexane 2.06 g/l (20 °C).   **Stability** Stable in water (DT_{50}
>1 y at pH 4–9). Stable to heat (no change over 14 d at 55°C). Stable to light in
water (DT_{50} 12.9 d).

COMMERCIALISATION
History Fungicide reported by Kumiai Chemical Industry Co., Ltd (S. Maeno *et al.*,
Proc. 1990 Br. Crop Prot. Conf. – Pests Dis., 2, 415). Introduced by Kumiai and Ihara
Chemical Industry Ltd.   **Patent** US 4814338; EP 224339; JP 63208581
Manufacturers Ihara/Kumiai

APPLICATIONS
Biochemistry Inhibits secretion of pathogen proteins such as the cell-wall
degrading enzyme pectinase. Also inhibits uptake by the pathogen of amino acids
and glucose.   **Mode of action** Non-systemic fungicide with preventative action.
Inhibits penetration of pathogen germ tube into host plant by inhibiting elongation
from spore and formation of appressorium. No inhibition of spore formation and
hyphal growth.   **Uses** Control of grey mould on vines, strawberries, tomatoes
and cucumbers; scab on apples and pears; and brown rot on peaches at
0.2–0.75 kg/ha.   **Formulation types** SC; WP.   **Mixtures** *(mepanipyrim +)* thiram;
ziram; vinclozolin.   **Selected tradenames** 'Frupica' (Kumiai)

ANALYSIS
Product analysis by hplc.   **Residue** analysis by glc with NPD. Details from Kumiai.

MAMMALIAN TOXICOLOGY

Oral Acute oral LD_{50} for rats and mice >5000 mg/kg. **Skin and eye** Acute percutaneous LD_{50} for rats >2000 mg/kg. Non-irritating to skin and eyes (rabbits). Non-sensitising to skin (guinea pigs). **Inhalation** LC_{50} (4 h, whole body) for rats >0.59 mg/l. **NOEL** (2 y) for male rats 2.45, female rats 3.07, male mice 56, female mice 68 mg/kg b.w. daily. **ADI** 0.024 mg/kg. **Other** Non-mutagenic, non-teratogenic (rats, rabbits). **Toxicity class** WHO (a.i.) III (Table 5)

ECOTOXICOLOGY

Birds Acute oral LD_{50} for bobwhite quail and mallard ducks >2250 mg/kg. Five-day dietary LC_{50} for bobwhite quail and mallard ducks >5620 mg/kg diet.
Fish LC_{50} (96 h) for bluegill sunfish 3.8, rainbow trout 3.1 mg/l. **Bees** LC_{50} (oral) >1000 mg/l diet, (contact) >100 µg/bee. **Worms** LC_{50} (14 d) for *Eisenia foetida* >1000 mg/kg soil. **Daphnia** LC_{50} (24 h) for *D. carinata* 5.0 mg/l. **Algae** EC_{50} (96 h) for *Selenastrum capricornutum* 1.3 mg/l.

ENVIRONMENTAL FATE

Animals In rats, 96–100% of the dose was excreted via faeces and urine within 96 h. Mepanipyrim was degraded into several metabolites. **Plants** Major metabolites were 2-hydroxypropyl-, 2–3-dihydroxypropyl- and 4-hydroxyphenyl-derivatives of mepanipyrim.

465 mepiquat chloride   *Plant growth regulator*

quaternary ammonium

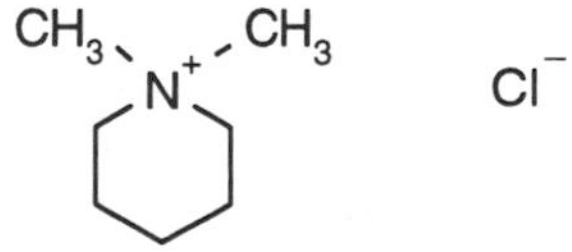

NOMENCLATURE
mepiquat chloride
Common name 1,1-dimethylpiperidinium chloride
CAS RN [24307–26–4]   **Development codes** BAS 083 W

mepiquat
Common name mepiquat (BSI, E-ISO, (m) F-ISO)
IUPAC name 1,1-dimethylpiperidinium
Chemical Abstracts name 1,1-dimethylpiperidinium
CAS RN [15302–91–7]

PHYSICAL CHEMISTRY
mepiquat chloride
Composition ≥99% pure (w/w); tech. concentrate 600 g/l.   **Mol. wt.** 149.7
M.f. $C_7H_{16}ClN$ **Form** Colourless, odourless hygroscopic crystals.   **M.p.** 223 °C
(tech.)   **V.p.** <0.01 mPa (20 °C) K_{ow} logP = −2.82 (pH 7)
S.g./density 1.187 g/cm^3 (tech., 20 °C)   **Solubility** In water at 20 °C, >500 g/kg.
In ethanol 162, chloroform 10.5, acetone, benzene, ethyl acetate, cyclohexane all
<1.0 (all in g/kg at 20 °C).   **Stability** Stable in aqueous media (7 d at pH 1–2 and
pH 12–13, 95 °C). Decomposes at 285 °C. Stable in artificial sunlight.

mepiquat
Mol. wt. 114.2   **M.f.** $C_7H_{16}N$

COMMERCIALISATION
History Plant growth regulating properties of mepiquat chloride reported by
B. Zeeh et al. (*Kem-Kemi*, 1974, **1**, 621). Introduced by BASF AG.
Patent DE 2207575; US 3905798   **Manufacturers** BASF; Gharda

APPLICATIONS
mepiquat chloride
Biochemistry Inhibits the biosynthesis of gibberellic acid.   **Mode of action** Plant
growth regulator, absorbed and translocated throughout the plant.   **Uses** Used on
cotton to reduce vegetative growth and to advance maturation of the bolls; and
to inhibit sprouting in onions, garlic and leeks. Used in combination with ethephon
to prevent lodging (by shortening the stem and strengthening the stem wall) in
cereals, grass seed crops, and flax.   **Formulation types** SL; UL.
Mixtures (*mepiquat chloride* +) ethephon; chlormequat chloride + ethephon.
Selected tradenames 'Pix' (BASF); 'Terpal' (mixture) (BASF); 'Mepex' (Griffin)

ANALYSIS
Product analysis by a gravimetric method or by ion-chromatography (hplc).
Residues determined by ion-chromatography (hplc) of the parent compound or by
glc with FID after conversion to 1-methylpiperidine. Details available from BASF,
method no. CP 160, 1994.

MAMMALIAN TOXICOLOGY
mepiquat chloride
Oral Acute oral LD_{50} for rats 464 mg/kg.   **Skin and eye** Acute percutaneous LD_{50}
for rats >2000 mg/kg. Not irritating to skin and eyes (rabbits). Not a skin
sensitiser.   **Inhalation** LC_{50} (7 h) for rats >3.2 mg/l air.   **NOEL** for rats 3000,
mice 1000 ppm.   **ADI** 1.5 mg/kg b.w.   **Toxicity class** WHO (a.i.) III;
EPA (formulation) II

ECOTOXICOLOGY

mepiquat chloride

Birds Acute oral LD_{50} for bobwhite quail >2000 mg/kg b.w. Dietary LC_{50} for mallard duck and bobwhite quail >10 000 mg/kg diet. **Fish** LC_{50} (96 h) for trout 4300 mg/l. **Bees** Non-toxic to bees. **Worms** LC_{50} (14 d) for *Eisenia foetida* 440 mg/kg soil. **Daphnia** LC_{50} (48 h) 68.5 mg/l. **Algae** EC_{50} (cell volume) (72 h) for *Pseudokirchneriella subcapitata* >1000 mg/l. **Other aquatic spp.** EC_{10} (18 h) for *Pseudomonas putida* 1630 mg/l.

ENVIRONMENTAL FATE

Animals In rats, following oral administration of mepiquat chloride, *c.* 48% is excreted in the urine and *c.* 38% in the faeces, with <1% remaining in the tissues. The unmetabolised material constitutes *c.* 90% in each case.

Soil/Environment DT_{50} of mepiquat chloride between 10 and 97 d at 20±2 °C and 40% of maximum water-holding capacity. K_{oc} 67 – 4685.

466 mepronil *Fungicide*

carboxamide

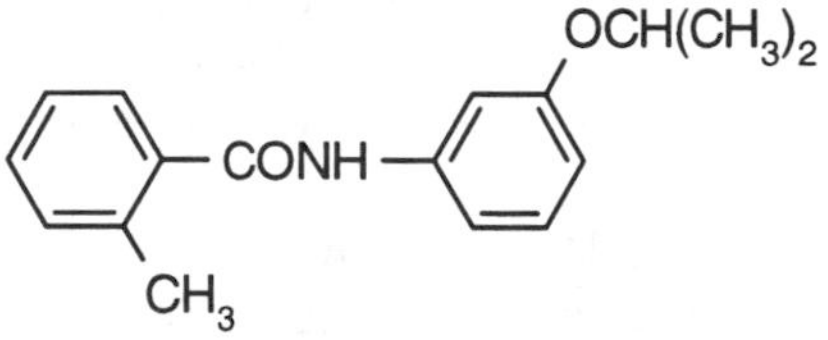

NOMENCLATURE

Common name mepronil (BSI, JMAF, draft E-ISO)
IUPAC name 3'-isopropoxy-*o*-toluanilide
Chemical Abstracts name 2-methyl-*N*-[3-(1-methylethoxy)phenyl]benzamide
CAS RN *[55814–41–0]* **Development codes** B1–2459; KCO-1 (Kumiai)

PHYSICAL CHEMISTRY

Composition Tech. is >94%. **Mol. wt.** 269.3 **M.f.** $C_{17}H_{19}NO_2$ **Form** Colourless crystals. **M.p.** 92–93 °C **V.p.** 0.056 mPa (20 °C) K_{ow} logP = 3.66 **Solubility** In water 12.7 mg/l (20 °C). In acetone >500, methanol >500, acetonitrile 314, benzene 282, hexane 1.1 (all in g/l, 20 °C). **Stability** Stable to light, air, and heat. Stable in neutral, acidic, and weakly alkaline conditions, but hydrolysed in highly alkaline conditions. **F.p.** 225 °C

COMMERCIALISATION

History Reported by Kumiai Chemical Industry Co., Ltd. (I. Chiyomaru *et al.,*

ACS/CSJ Chem. Congr., Div. Pestic. Chem., Hawaii, No. 79 (1979) and S. Kawada *et al., Ann. Phytopath. Soc. Jap.*, **45** (4), 547 (1979)). Reviewed by S. Doi (*Jpn. Pestic. Inf.*, 1981, No. 38, p. 17) and introduced by Kumiai. **Patent** US 3937840; GB 1421112; JP 906789 **Manufacturers** Ihara/Kumiai

APPLICATIONS
Biochemistry Inhibits succinic acid oxidation during metabolic respiration.
Mode of action Systemic fungicide with protective and curative action.
Uses Control of diseases caused by Basidiomycetes (e.g. *Rhizoctonia*, *Puccinia*, *Typhula* spp., etc.) in rice, cereals, potatoes, vegetables, cucumbers, sugar beet, fruit, vines, tobacco, turf grass, ornamentals, etc. Also, when used as a seed or soil treatment, control of damping-off diseases of vegetables and tobacco.
Formulation types DP; SC; WP. **Mixtures** *(mepronil +)* tricyclazole.
Selected tradenames 'Basitac' (Kumiai)

ANALYSIS
Product and **residue** analysis by glc; details available from Kumiai.

MAMMALIAN TOXICOLOGY
Oral Acute oral LD_{50} for rats and mice >10 000 mg/kg. **Skin and eye** Acute percutaneous LD_{50} for rabbits and rats >10 000 mg/kg. Non-irritating to skin and eyes (rabbits). **Inhalation** LC_{50} (6 h) for rats >1.32 mg/l. **NOEL** (2 y) for male rats 5.9, female rats 72.9, male mice 13.7, female mice 17.8 mg/kg b.w. daily.
ADI 0.05 mg/kg. **Other** Non-mutagenic, non-teratogenic in rats and rabbits.
Toxicity class WHO (a.i.) III (Table 5); EPA (formulation) IV

ECOTOXICOLOGY
Birds Acute oral LD_{50} for hens >8000 mg/kg. **Fish** LC_{50} (96 h) for carp 8, rainbow trout 10 mg/l. **Bees** Acute LD_{50} (oral) >0.1 mg/bee, (contact) >1 mg/bee. **Daphnia** LC_{50} for *D. carinata* >10 mg/l.

ENVIRONMENTAL FATE
Animals Almost all of the dose applied to rats was excreted in urine and faeces within 96 h. **Plants** Almost all ^{14}C applied to rice leaf sheath or leaf remained in the part treated, with some translocation upwards. Five metabolites were identified. **Soil/Environment** DT_{50} in flooded soil 46 d (volcanic sandy loam, pH 6.8, T.C. 1.47%), 50.5 d (alluvial, pH 6.5, T.C. 1.18%).

HgCl$_2$

NOMENCLATURE
Common name mercuric chloride (E-ISO, accepted in lieu of a common name); chlorure mercurique (F-ISO)
IUPAC name mercury(II) chloride; mercury dichloride
Chemical Abstracts name mercury chloride (HgCl$_2$)
Other names corrosive sublimate   **CAS RN** *[7487–94–7]*   **EEC no.** 231–299–8

PHYSICAL CHEMISTRY
Mol. wt. 271.5   **M.f.** Cl$_2$Hg   **Form** Colourless crystalline powder.   **M.p.** 277 °C
V.p. 18.6 mPa (35 °C)   **S.g./density** 5.32   **Solubility** In water 69 g/l (20 °C).
Soluble in ethanol, diethyl ether, and pyridine.   **Stability** Readily reduced
chemically or by sunlight to mercury(I) chloride and to metallic mercury. Mercury
chloride oxide is precipitated by alkalis.

COMMERCIALISATION
History Use in crop protection described by H. L. Bolley (*N. D. Agric. Exp. Stn.
Bull.*, 1891, No. 4).

APPLICATIONS
Uses Used in Canada, in combination with mercurous chloride, as a fungicide on
turf.   **Mixtures** *(mercuric chloride +)* mercurous chloride.

ANALYSIS
Product analysis by gravimetery (*AOAC Methods*, 1984, 6.154–6,156) or by
colorimetry (*ibid.*, 6.150–6.153; *CIPAC Handbook*, 1970, **1**, 514). Mercury
impurities determined (*ibid.*, 1994, **F**, 264).   **Residues** determined by atomic
absorption spectrophotometry (*AOAC Methods*, 1984, 25.131–25.135) or by
colorimetry (*ibid.*, 25.138–25.145).

MAMMALIAN TOXICOLOGY
Oral Acute oral LD$_{50}$ for rats 1–5 mg/kg.   **Toxicity class** WHO (a.i.) Ia
EC risk T+ (R28); C (R34); T (R48/24/25)   **PIC** All mercury compounds included
in PIC procedure.

MAIN ENTRIES

468 mercuric oxide *Fungicide, wound protectant*

inorganic

HgO

NOMENCLATURE

Common name mercuric oxide (E-ISO, accepted in lieu of a common name); oxyde mercurique (F-ISO, accepted in lieu of a common name)
IUPAC name mercury(II) oxide; mercury oxide
Chemical Abstracts name mercury oxide (HgO)
Other names yellow oxide of mercury **CAS RN** *[21908–53–2]* formerly *[1344–45–2]*

PHYSICAL CHEMISTRY

Mol. wt. 216.6 **M.f.** HgO **Form** Orange-red to bright red powder (yellow when finely divided). **M.p.** Decomposes at *c.* 500 °C **Solubility** In water 53 mg/l (25 °C). Insoluble in organic solvents. Soluble in dilute acids, forming the corresponding mercury salts. **Stability** Decomposes on exposure to light. Darkens at *c.* 400 °C and decomposes into mercury and oxygen at *c.* 500 °C.

COMMERCIALISATION

History Introduced by Sandoz AG (now Novartis Crop Protection AG, who no longer manufacture or market it). **Manufacturers** United Phosphorus

APPLICATIONS

Uses Wound protectant for pruning cuts and other bark injuries, and control of canker, on fruit trees, rubber trees, vines, and ornamental trees and shrubs.
Formulation types PA.

ANALYSIS

Product analysis by standard methods (titration with potassium iodide after dissolution in acid). Mercury impurities determined (*CIPAC Handbook*, 1994, **F**, 264).

MAMMALIAN TOXICOLOGY

Oral Acute oral LD_{50} for rats 18 mg/kg. Extremely poisonous orally to all animals.
Toxicity class WHO (a.i.) Ib; EPA (formulation) I **EC risk** T+ (R26/27/28); (R33); (for preparations, (% Hg), concn. ≥2%, T+ (R26/27/28, R33), 0.5%≤ concn. <2%, T (R23/24/25, R33), 0.1%≤ concn. <0.5%, Xn (R20/21/22, R33)) (data are for inorganic mercury compounds in general, with indicated exceptions) **PIC** All mercury compounds included in PIC procedure.

ECOTOXICOLOGY

Fish Toxic to fish.

mercurous chloride   *Fungicide, insecticide*

$$Hg_2Cl_2$$

NOMENCLATURE
Common name mercurous chloride (E-ISO, accepted in lieu of a common name); chlorure mercureux (F-ISO, accepted in lieu of a common name)
IUPAC name mercury(I) chloride; dimercury dichloride
Chemical Abstracts name mercury chloride (Hg_2Cl_2)
Other names calomel   **CAS RN** *[7546–30–7]*   **EEC no.** 233–307–5

PHYSICAL CHEMISTRY
Mol. wt. 472.1   **M.f.** Cl_2Hg_2 **Form** White powder.   **M.p.** Sublimes at 400–500 °C.   **S.g./density** 7.15   **Solubility** In water 2 mg/l (25 °C). Soluble in most organic solvents, and also in cold dilute acids.   **Stability** Decomposes slowly in sunlight. Under aqueous conditions, slowly decomposes to mercury and mercuric chloride (more rapid decomposition in the presence of alkali).

COMMERCIALISATION
History Has long been used as an insecticide and, having a lower toxicity to mammals, largely replaced mercuric chloride as recommended by H. Glasgow (*J. Econ. Entomol.*, 1929, **22**, 335).

APPLICATIONS
Uses Soil application for control of *Delia* spp., *Plasmodiophora brassicae* in brassicas; *Sclerotium cepivorum* in onions. Control of dollar spot and *Fusarium* patch on turf.   **Phytotoxicity** Phytotoxic to many crops.   **Formulation types** DP; TC.

ANALYSIS
Product analysis by gravimetery (*AOAC Methods*, 1984, 6.154–6.156) or by colorimetry (*ibid.*, 6.150–6.153; *CIPAC Handbook*, 1970, **1**, 514). Mercury impurities determined (*ibid.*, 1994, **F**, 264).   **Residues** determined by atomic absorption spectrophotometry (*AOAC Methods*, 1984, 25.131–25.135) or by colorimetry (*ibid.*, 25.138–25.145).

MAMMALIAN TOXICOLOGY
Oral Acute oral LD_{50} for rats 210 mg/kg.   **Toxicity class** WHO (a.i.) II; EPA (formulation) II   **EC risk** Xn (R22); Xi (R36/37/38)   **PIC** All mercury compounds included in PIC procedure.

ECOTOXICOLOGY
Fish Toxic to fish.

MAIN ENTRIES

phenylamide (acylalanine type)

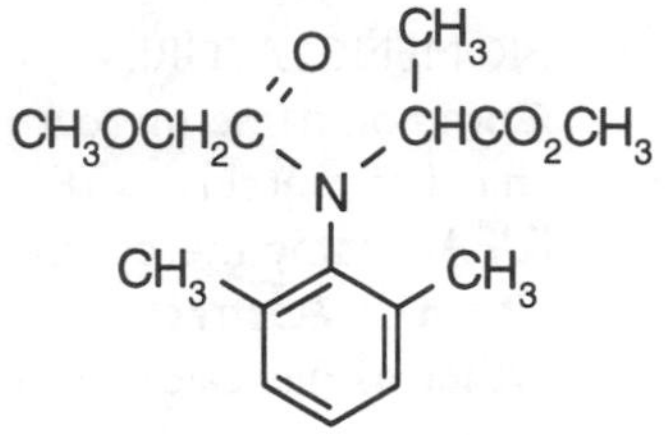

NOMENCLATURE
Common name metalaxyl (BSI, E-ISO, (*m*) F-ISO, ANSI)
IUPAC name methyl *N*-(methoxyacetyl)-*N*-(2,6-xylyl)-DL-alaninate
Chemical Abstracts name methyl *N*-(2,6-dimethylphenyl)-*N*-(methoxyacetyl)-=
DL-alaninate
CAS RN *[57837–19–1]* **EEC no.** 260–979–7 **Development codes** CGA 48 988

PHYSICAL CHEMISTRY
Mol. wt. 279.3 **M.f.** $C_{15}H_{21}NO_4$ **Form** Fine, white powder. **M.p.** Tech.
63.5–72.3 °C **B.p.** 295.9 °C (101 kPa) **V.p.** 0.75 mPa (25 °C) K_{ow} logP = 1.75
(25 °C) **Henry** 1.6×10^{-5} Pa m^3 mol^{-1} (calc.) **S.g./density** 1.20 at 20 °C
Solubility In water 8.4 g/l (22 °C). In ethanol 400, acetone 450, toluene 340,
n-hexane 11, *n*-octanol 68 (all in g/l, 25 °C). **Stability** Stable up to 300 °C. Stable
in neutral and acidic media at room temperature; on hydrolysis DT_{50} (calculated)
(20 °C) >200 d at pH 1, 115 d at pH 9, 12 d at pH 10. **pKa** <<0

COMMERCIALISATION
History Fungicide reported by F. J. Schwinn *et al.* (*Mitt. Biol. Bundesanst. Land-
Fortswirtsch. Berlin-Dahlem*, 1977, **178**, 145) and by P. A. Urech (*Proc. 1977 Br.
Crop. Prot. Conf. – Pests Dis.*, 1977, **2**, 623). Introduced by Ciba-Geigy AG (now
Novartis Crop Protection AG). **Patent** BE 827671; GB 1500581; US 4151299
Manufacturers Jingma; Novartis; Rallis

APPLICATIONS
Biochemistry Inhibits protein synthesis in fungi, by interference with the synthesis
of ribosomal RNA. **Mode of action** Systemic fungicide with protective and
curative action, absorbed through the leaves, stems, and roots. **Uses** Used to
control diseases caused by air- and soil-borne Peronosporales on a wide range of
temperate, subtropical and tropical crops. Foliar sprays with mixtures of metalaxyl
and protectant fungicides are recommended to control air-borne diseases caused
by *Pseudoperonospora humuli* on hops, *Phytophthora infestans* on potatoes,
Peronospora tabacina on tobacco, and *Plasmopara viticola* on vines. Soil applications
of metalaxyl alone are used to control soil-borne pathogens causing root and

lower stem rots on avocado and citrus; also recommended for primary systemic infections of *P. humuli* on hops and in tobacco seed beds. Seed treatments control systemic Peronosporaceae on maize, peas, sorghum and sunflowers, as well as damping-off (*Pythium* spp.) of various crops. **Formulation types** DS; FS; WP; GR. **Mixtures** (*metalaxyl +*) captan; carbendazim; carboxin; copper hydroxide; copper oxinate; copper oxychloride; chlorothalonil; fenpiclonil; folpet; furathiocarb; mancozeb; maneb; thiabendazole; thiram; zineb. **Selected tradenames** 'Ridomil' (Novartis); 'Milor' (Rotam)

ANALYSIS
Product analysis by glc (*CIPAC Handbook*, 1992, **E**, 123–130). **Residues** in plants and soil determined by glc with TID (D. J. Caverley & J. Unwin, *Analyst (London)*, 1980, **106**, 389). Details available from Novartis.

MAMMALIAN TOXICOLOGY
Reviews *Pesticide residues in food – 1982*. FAO Plant Production and Protection Paper 46, 1983. **Oral** Acute oral LD_{50} for rats 633, mice 788, rabbits 697 mg/kg. **Skin and eye** Acute percutaneous LD_{50} for rats >3100 mg/kg; slight irritant to eyes, not irritant to skin of rabbits. No skin sensitisation (guinea pigs). **Inhalation** LC_{50} (4 h) for rats >3600 mg/m^3. **NOEL** for rats 2.5, mice 35.7, dogs 8.0 mg/kg b.w. daily. **ADI** (JMPR) 0.03 mg/kg b.w. [1982]; (Novartis) 0.025 mg/kg b.w. **Other** Not oncogenic, not mutagenic, not teratogenic. **Toxicity class** WHO (a.i.) III **EC risk** R22

ECOTOXICOLOGY
Birds LD_{50} for Japanese quail (7 d) 923, mallard ducks (8 d) 1466 mg/kg. LC_{50} (8 d) for Japanese quail, bobwhite quail and mallard ducks >10 000 mg/kg. **Fish** LC_{50} (96 h) for rainbow trout, carp, and bluegill sunfish >100 mg/l. **Bees** Not toxic to bees; LD_{50} (48 h) (contact) >200, (oral) 269.3 µg/bee. **Worms** LC_{50} (14 d) for *Eisenia foetida* >1000 mg/kg soil. **Other beneficial spp.** Harmless to *Poecilus cupreus* and *Coccinella septempunctata*. **Daphnia** LC_{50} (48 h) >28 mg/l. **Algae** IC_{50} (5 d) for *Scenedesmus subspicatus* 33 mg/l. **Other aquatic spp.** EC_{50} (96 h) for mysid shrimp (*Mysidopsis bahia*) 25, Eastern oyster (*Crassostrea virginica*) 4.6 mg/l.

ENVIRONMENTAL FATE
Animals In mammals, following oral administration, metalaxyl is rapidly absorbed and also rapidly and almost completely eliminated with urine and faeces. Metabolism proceeds via hydrolysis of the ester bond, oxidation of the 2-(6)-methyl group and of the phenyl ring and *N*-dealkylation. Residues in tissues were generally low and there was no evidence for accumulation or retention of metalaxyl or its metabolites. **Plants** Metalaxyl is metabolised by more than 4 types of phase I reaction to form eight metabolites; at phase II, most of the metabolites are sugar conjugated. The types of reaction in phase I are: oxidation

of the phenyl ring, oxidation of the methyl group, cleavage of the methyl ester and N-dealkylation. **Soil/Environment** In soil, DT_{50} 19 d, DT_{90} 90 d. K_{oc} 30–300 ml/g. DT_{50} in water 22–48 d. Photolytically stable in water.

471 metalaxyl-M   *Fungicide*

phenylamide (acylalanine type)

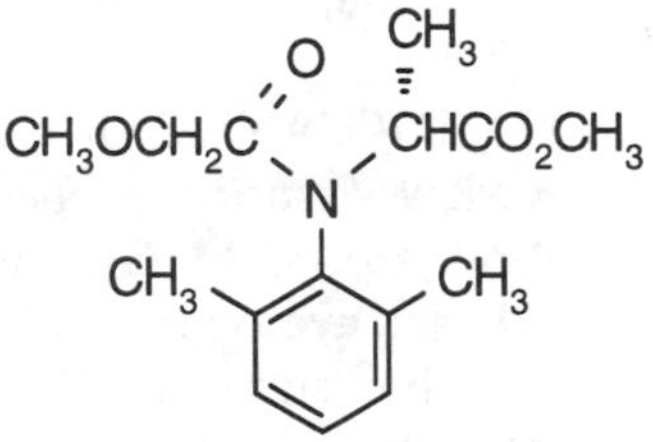

NOMENCLATURE
Common name metalaxyl-M (BSI, E-ISO, (*m*) F-ISO)
IUPAC name methyl *N*-(methoxyacetyl)-*N*-(2,6-xylyl)-D-alaninate; methyl (*R*)-2-{[(2,6-dimethylphenyl)methoxyacetyl]amino}propionate
Chemical Abstracts name methyl *N*-(2,6-dimethylphenyl)-*N*-(methoxyacetyl)-= D-alaninate
Other names R-metalaxyl; CGA 76539; mefenoxam   **CAS RN** *[70630–17–0]*
Development codes CGA 329351

PHYSICAL CHEMISTRY
Composition (*R*)- enantiomer of metalaxyl.   **Mol. wt.** 279.3   **M.f.** $C_{15}H_{21}NO_4$
Form Pale yellow to light brown viscous liquid.   **M.p.** −38.7 °C (glass transition temperature)   **B.p.** Decomp. *c.* 270 °C   **V.p.** 3.3 mPa (25 °C) K_{ow} logP = 1.71
Henry 3.5×10^{-5} Pa m^3 mol^{-1} (calc.)   **S.g./density** 1.125 (20 °C)   **Solubility** In water, 26 g/l (25 °C). In *n*-hexane 59 g/l; miscible with acetone, ethyl acetate, methanol, dichloromethane, toluene, *n*-octanol.   **Stability** Hydrolytically stable under acidic and neutral conditions (DT_{50} >200 d). Under alkaline conditions DT_{50} 116 d (pH 9, 25 °C).   **Specific rotation** Negative.   **F.p.** 179 °C (EEC A.10)

COMMERCIALISATION
History Reported by C. Nuninger *et al.* (*Proc. Br. Crop Prot. Conf. – Pests Dis.*, 1996, **1**, 41). Developed by Ciba-Geigy AG (now Novartis Crop Protection AG) and introduced in the US in 1996.   **Patent** PCT Patent appl. WO96/01559; ZA-P.95/5708   **Manufacturers** Novartis

APPLICATIONS

Biochemistry Inhibits protein synthesis in fungi, by interference with the synthesis of ribosomal RNA. The (R)- (metalaxyl-M) and (S)- isomers have the same mode of action, but differ in effectiveness. **Mode of action** Systemic fungicide with protective and curative action, absorbed through the leaves, stems, and roots. **Uses** Used to control diseases caused by air- and soil-borne Peronosporales on a wide range of temperate, subtropical and tropical crops. Foliar sprays with mixtures of metalaxyl-M and protectant fungicides are recommended to control air-borne diseases caused by *Pseudoperonospora humuli* on hops, *Phytophthora infestans* on potatoes, *Peronospora tabacina* on tobacco, and *Plasmopara viticola* on vines. Soil applications of metalaxyl-M alone are used to control soil-borne pathogens causing root and lower stem rots on avocado and citrus; also recommended for primary systemic infections of *P. humuli* on hops and in tobacco seed beds. Seed treatments control systemic Peronosporaceae on maize, peas, sorghum and sunflowers, as well as damping-off (*Pythium* spp.) of various crops. **Formulation types** DS; EC; FS; GR; WP; WG.

Mixtures *(metalaxyl-M +)* fludioxonil; difenoconazole; mancozeb; folpet; copper compounds; chlorothalonil. **Selected tradenames** 'Apron XL' (Novartis); 'Ridomil Gold' (Novartis)

MAMMALIAN TOXICOLOGY

Oral Acute oral LD_{50} for rats 667 mg/kg. **Skin and eye** Acute percutaneous LD_{50} for rats >2000 mg/kg. Not a skin irritant (rabbit); risk of serious damage to eye (rabbit). Not a skin sensitiser (guinea pigs). **Inhalation** LC_{50} (4 h) for rats >2290 mg/m^3. **NOEL** for rats 2.5, mice 35.7, dogs 8.0 mg/kg b.w. daily. **ADI** 0.025 mg/kg b.w. **Other** Not oncogenic, not mutagenic, not teratogenic. **Toxicity class** WHO (a.i.) II **EC risk** R22, R41

ECOTOXICOLOGY

Birds LD_{50} (14 d) for bobwhite quail 981–1419 mg/kg. LC_{50} (8 d) for bobwhite quail >5620 mg/kg. **Fish** LC_{50} (96 h) for rainbow trout >100 mg/l. **Bees** LD_{50} (48 h, oral) 25 µg/bee. **Worms** LC_{50} (14 d) for *Eisenia foetida* >1000 mg/kg soil. **Other beneficial spp.** Harmless to *Poecilus cupreus* and *Orius insidiosus*. **Daphnia** LC_{50} (48 h) >100 mg/l. **Algae** E_rC_{50} (72 h) for *Scenedesmus subspicatus* 103 mg/l. **Other aquatic spp.** EC_{50} (96 h) for Eastern oyster (*Crassostrea virginica*) 9.7 mg/l.

ENVIRONMENTAL FATE

Soil/Environment DT_{50} in soil 30 d. K_{oc} 120 ml/g. DT_{50} for photolysis (artificial sunlight) 156 h.

MAIN ENTRIES

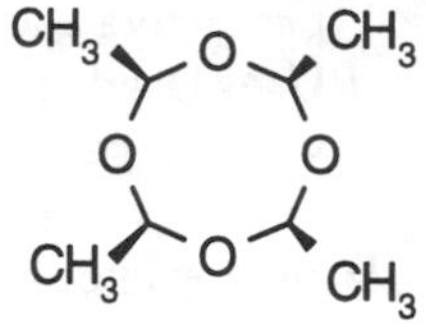

NOMENCLATURE

Common name metaldehyde (BSI, E-ISO, F-ISO, JMAF, accepted in lieu of a common name)

IUPAC name *r*-2,*c*-4,*c*-6,*c*-8-tetramethyl-1,3,5,7-tetroxocane; 2,4,6,8-tetramethyl-1,3,5,7-tetraoxacyclo-octane

Chemical Abstracts name 2,4,6,8-tetramethyl-1,3,5,7-tetraoxacyclooctane; acetaldehyde homopolymer

Other names acetaldehyde tetramer   **CAS RN** *[108–62–3]* for tetramer
EEC no. 203–600–2

PHYSICAL CHEMISTRY

Composition Produced by polymerisation of acetaldehyde; tech. grade contains higher oligomers but mainly the tetramer mentioned above.   **Mol. wt.** 176.2 (tetramer)   **M.f.** $C_8H_{16}O_4$ (tetramer); $(C_2H_4O)_x$ (homopolymer)
Form Crystalline powder.   **M.p.** 246 °C (sealed tube)   **B.p.** Sublimes at 112–115 °C, with partial depolymerisation   **V.p.** Low at room temperature
K_{ow} logP = 0.12   **Henry** 3.5 Pa m^3 mol^{-1} (calc.)   **S.g./density** 1.27 (room temperature)   **Solubility** In water 222 mg/l (20 °C). In toluene 530, methanol 1730 (both mg/l, 20 °C).   **Stability** Depolymerises and sublimes above 112 °C.
F.p. 50–55 °C

COMMERCIALISATION

History Its slug-killing properties reported by G. W. Thomas (*Gard. Chron.*, 1936, **100**, 453).   **Manufacturers** Lonza

APPLICATIONS

Mode of action Molluscicide with contact and stomach action. Metaldehyde-poisoned slugs secrete large quantities of slime, desiccate and die. Their mucus cells are irreversibly destroyed.   **Uses** Control of slugs and snails in agriculture and horticulture.   **Phytotoxicity** Non-phytotoxic when used as recommended. Should not be allowed to contact plant foliage.   **Formulation types** RB; GB; Pellets.   **Selected tradenames** 'Cekumeta' (Cequisa); 'Halizan' (Tamogan); 'Hardy' (Chiltern); 'Meta' (Lonza); 'Metason' (Jewnin-Joffe); 'MifaSlug' (FCC)

ANALYSIS

Product analysis by conversion to acetaldehyde, which is estimated by reaction with sodium hydrogen sulfite and titration with iodine (*CIPAC Handbook*, 1970, **1**, 532) or by reaction with hydroxyammonium chloride and subsequent acid-base titration (*MAFF Ref. Bk.*, 1958, No. 1, p. 58).

MAMMALIAN TOXICOLOGY

Reviews T. Booze & F. Oehme, *Vet. Hum. Toxicol.*, **27**, 11–19 (1985). **Oral** Acute oral LD_{50} for rats 283, mice 425 mg/kg. **Skin and eye** Acute percutaneous LD_{50} for rats >5000 mg/kg. Not a skin irritant (rabbits). Not a skin sensitiser (guinea pigs). **Inhalation** LC_{50} (4 h) for rats >15 mg/l air. **ADI** 0.025 mg/kg b.w. **Toxicity class** WHO (a.i.) III; EPA (formulation) III; II ('Halizan') **EC risk** (R10); Xn (R22)

ECOTOXICOLOGY

Birds Acute oral LD_{50} for quail 170 mg/kg. **Fish** LC_{50} (96 h) for rainbow trout 75 mg/l. **Worms** LC_{50} >50 000 ppm. **Daphnia** EC_{50} (48 h) >90 mg/l. **Algae** EC_{50} (96 h) 73.5 mg/l.

ENVIRONMENTAL FATE

Soil/Environment Aerobic and anaerobic micro-organisms in soil decompose metaldehyde to CO_2 and water.

473 metam *Fungicide, nematicide, herbicide, insecticide*

methyl isothiocyanate precursor

$$CH_3NHCS_2H$$

NOMENCLATURE

metam
Common name metam (since 1990 BSI, E-ISO, F-ISO); metham (WSSA, before 1990 BSI)
IUPAC name methyldithiocarbamic acid
Chemical Abstracts name methylcarbamodithioic acid
CAS RN *[144–54–7]* **Development codes** N-869 (Stauffer)

metam-sodium
Common name metam-sodium (before 1990 E-ISO, F-ISO); karbation (USSR); carbam (JMAF – name also used for ammonium salt)
Other names SMDC **CAS RN** *[137–42–8]*; *[6734–80–1]* (dihydrate)
EEC no. 205–293–0

PHYSICAL CHEMISTRY
metam
Mol. wt. 107.2 **M.f.** $C_2H_5NS_2$

metam-sodium
Mol. wt. 129.2 **M.f.** $C_2H_4NNaS_2$ **Form** Colourless crystals (dihydrate).
M.p. Decomposes without melting. **V.p.** Non-volatile K_{ow} logP <1 (25 °C)
Solubility In water 722 g/l (20 °C). In acetone, ethanol, kerosene, xylene <5 (all in
g/l). Practically insoluble in most other organic solvents. **Stability** Stable in
concentrated aqueous solution, but unstable when diluted. Decomposition is
promoted by acids and heavy-metal salts. Solutions exposed to sunlight DT_{50} 1.6 h
(pH 7, 25 °C); on hydrolysis (25 °C) DT_{50} 23.8 h (pH 5), 180 h (pH 7), 45.6 h
(pH 9).

COMMERCIALISATION
History Fungicidal properties of metam-sodium reported by H. L. Klopping
(Thesis, University of Utrecht, 1951) and by A. J. Overman & D. S. Burgis (*Proc.
Fla. St. Hortic. Soc.*, 1956, **69**, 250). Metam-sodium introduced by Stauffer
Chemical Co. (now Zeneca Agrochemicals) and later by E. I. du Pont de Nemours
and Co. (who no longer manufacturer or market it). **Patent** US 2766554;
US 2791605; GB 789690 all to Stauffer **Manufacturers** Amvac; Aragonesas;
Buckman; Elf Atochem; Lainco; Nufarm Ltd; UCB; Zeneca

APPLICATIONS
metam-sodium
Mode of action Soil fumigant, acting by decomposition to methyl isothiocyanate
(*q.v.*). **Uses** Metam-sodium is a soil sterilant, applied prior to planting out edible
crops, which controls soil fungi, nematodes, weed seeds, and soil insects.
Phytotoxicity Highly phytotoxic. **Formulation types** SC; SL.
Selected tradenames 'Vapam' (Zeneca); 'Arapam' (Aragonesas); 'Trimaton'
(Elf Atochem); 'Unifume' (Unicrop)

ANALYSIS
Product analysis by Fourier transform i.r. or by hydrolysis to form carbon disulfide
which is reacted and titrated with iodine (*CIPAC Handbook*, 1970, **1**, 537; *ibid.*,
1992, **E**, 131–133). **Residues** in crops and soils determined by analysing them for
methyl isothiocyanate by glc. Details available from Zeneca. See also R. A. Gray
(*Anal. Methods Pestic., Plant Growth Regul. Food Addit.*, 1964, **3**, 177; *Anal. Methods
Pestic. Plant Growth Regul.*, 1972, **6**, 717).

MAMMALIAN TOXICOLOGY
metam-sodium
Oral Acute oral LD_{50} for male rats 1800, female rats 1700, mice 285 mg/kg.
Acute oral LD_{50} for rats of the methyl isothiocyanate formed in soil is 97 mg/kg.

Skin and eye Acute percutaneous LD_{50} for rabbits 1300 mg/kg; mild irritant to skin, corrosive to eyes. Any contact with skin or organs should be treated as a burn. **Inhalation** LC_{50} (4 h) for rats >4.7 mg/l air; in 65 exposure-days trial on rats (6 h/d, 5 d/w) NOEL 0.045 mg/l air. **Toxicity class** WHO (a.i.) II; EPA (formulation) II **EC risk** Xn (R21/22); Xi (R41); (R31)

ECOTOXICOLOGY
metam-sodium
Birds Five-day dietary LC_{50} for mallard ducks and Japanese quail >5000 mg/kg diet. **Fish** LC_{50} (96 h) for guppy 4.2, bluegill sunfish 0.39, rainbow trout 0.079 mg/l. **Bees** Non-toxic to bees when used as directed.

ENVIRONMENTAL FATE
Soil/Environment In soil, rapidly decomposes to methyl isothiocyanate, *q.v.*; DT_{50} 23 min. to 4 d.

474 metamitron   *Herbicide*

1,2,4-triazinone

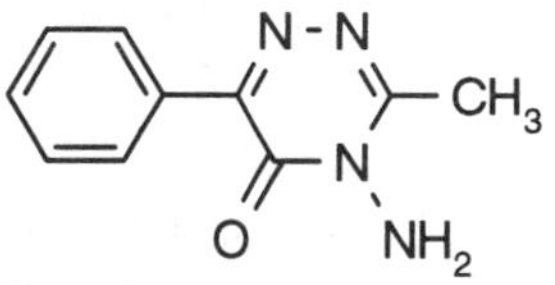

NOMENCLATURE
Common name metamitron (BSI, E-ISO); métamitrone ((f) F-ISO); methiamitron (Belgium)
IUPAC name 4-amino-4,5-dihydro-3-methyl-6-phenyl-1,2,4-triazin-5-one; 4-amino-3-methyl-6-phenyl-1,2,4-triazin-5(4*H*)-one
Chemical Abstracts name 4-amino-3-methyl-6-phenyl-1,2,4-triazin-5(4*H*)-one
CAS RN *[41394–05–2]* **EEC no.** 255–349–3 **Development codes** BAY DRW 1139; BAY 134028

PHYSICAL CHEMISTRY
Mol. wt. 202.2 **M.f.** $C_{10}H_{10}N_4O$ **Form** Yellowish odourless crystals.
M.p. 166.9 °C. **V.p.** 8.6×10^{-4} mPa (20 °C); 2×10^{-3} mPa (25 °C)
K_{ow} logP = 0.83 **Henry** 1×10^{-7} Pa m^3 mol^{-1} (20 °C, calc.)
S.g./density 1.35 g/cm^3 (22.5 °C) **Solubility** In water 1.7 g/l (20 °C). In dichloromethane 30–50, cyclohexanone 10–50, isopropanol 5.7, toluene 2.8, hexane <0.1, methanol 23, ethanol 1.1, chloroform 29 (all in g/l, 20 °C).

Stability Very stable in acidic media; decomposed by strong alkalis (pH>10); DT_{50} (22 °C) 410 d (pH 4), 740 h (pH 7), 230 h (pH 9). Photodecomposition on soil surfaces is very rapid and extremely rapid in water.

COMMERCIALISATION
History Herbicide reported by R. R. Schmidt et al. (*3rd Int. Meeting Selective Weed Control in Beet*, 1975, **1**, 713) and H. Hack (*ibid.*, p. 729) and reviewed by H. Lembrich (*Pflanzenschutz-Nachr. (Engl. Ed.)*, 1978, **31**, 197). Introduced in France (1975) by Bayer AG. **Patent** BE 799854; GB 1368416 **Manufacturers** Bayer; Feinchemie Schwebda; United Phosphorus

APPLICATIONS
Biochemistry Photosynthetic electron transport inhibitor.
Mode of action Selective systemic herbicide, absorbed predominantly by the roots, but also by the leaves, with translocation acropetally. **Uses** Used against grass and broad-leaved weeds in sugar and fodder beets. Applied pre-drilling incorporated, pre- and post-emergence (post-emergence as sequential treatment tank-mixed with oil or other herbicides). Also used in mangold, red beet and certain strawberry varieties. Rates 0.7–3.5 kg a.i./ha for all crops.
Phytotoxicity High selectivity in sugar and fodder beet. **Formulation types** WP; WG. **Mixtures** (*metamitron +)* ethofumesate + phenmedipham; ethofumesate + phenmedipham + desmedipham; lenacil. **Compatibility** Compatible with other beet herbicides. **Selected tradenames** 'Goltix' (Bayer); 'Tornado' (Feinchemie Schwebda)

ANALYSIS
Product analysis by hplc (*CIPAC Handbook*, 1988, **D**, 124) or i.r. spectroscopy, details from Bayer AG. **Residues** in plants determined by glc, details from Bayer AG, and in soil by hplc.

MAMMALIAN TOXICOLOGY
Oral Acute oral LD_{50} for rats c. 1200, mice c. 650, dogs >1000 mg/kg.
Skin and eye Acute percutaneous LD_{50} for rats >4000 mg/kg. No irritation of skin and eyes (rabbit). **Inhalation** LC_{50} (4 h) for rats >0.33 mg/l air (aerosol).
NOEL (2 y) for rats 250, dogs 100 mg/kg diet; (87 w) for mice 56 mg/kg diet.
ADI 0.025 mg/kg b.w. **Toxicity class** WHO (a.i.) III; EPA (formulation) III

ECOTOXICOLOGY
Birds Acute oral LD_{50} for Japanese quail 1875–1930 mg/kg. **Fish** LC_{50} (96 h) for golden orfe 443, rainbow trout 326 mg/l. **Bees** Not toxic to bees. **Worms** LC_{50} for *Eisenia foetida* >1000 mg/kg dry soil. **Daphnia** LC_{50} (48 h) 101.7–206 mg/l. **Algae** E_rC_{50} for *Scenedesmus subspicatus* 0.22 mg/l.

ENVIRONMENTAL FATE
Animals In mammals, following oral administration, elimination occurs within 48 hours, approximately equally in the urine and faeces (c. 98%). **Plants** In sugar beet, the principal metabolite is 3-methyl-6-phenyl-1,2,4-triazin-5(4H)-one. **Soil/Environment** In soil, metamitron is degraded very rapidly. Leaching behaviour can be classified as medium mobile; no leaching into groundwater occurred. Rapid photodecomposition on soil surfaces and in aqueous solution is an important process for the degradation of metamitron in the environment.

---

475 metazachlor   *Herbicide*

chloroacetanilide

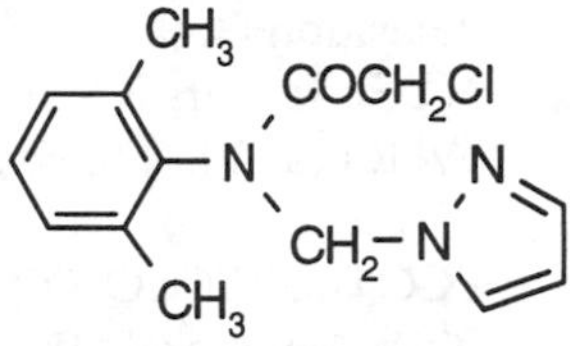

NOMENCLATURE
Common name metazachlor (BSI, E-ISO); métazachlore ((m) F-ISO)
IUPAC name 2-chloro-N-(pyrazol-1-ylmethyl)acet-2',6'-xylidide
Chemical Abstracts name 2-chloro-N-(2,6-dimethylphenyl)-N-(1H-pyrazol-=
1-ylmethyl)acetamide
CAS RN [67129–08–2]   **EEC no.** 266–583–0   **Development codes** BAS 479 00 H

PHYSICAL CHEMISTRY
Composition ≥94% pure.   **Mol. wt.** 277.8   **M.f.** $C_{14}H_{16}ClN_3O$ **Form** Yellowish crystals; (tech., beige solid).   **M.p.** c. 85 °C   **V.p.** 0.093 mPa (20 °C)
K_{ow} logP = 2.13 (pH 7, 22 °C).   **Henry** 5.741×10^{-5} Pa m^3 mol^{-1}
S.g./density c. 1.31 (20 °C)   **Solubility** In water 430 mg/l (20 °C). In acetone, chloroform >1000, ethyl acetate 590, ethanol 200 (all in g/kg, 20 °C).
Stability Stable for at least 2 years at up to 40 °C.

COMMERCIALISATION
History Herbicide introduced by BASF AG.   **Manufacturers** BASF

APPLICATIONS
Mode of action Selective herbicide, absorbed by the hypocotyls and roots. Inhibits germination.   **Uses** Pre-emergence and early post-emergence control of winter and annual grasses (such as *Alopecurus myosuroides, Apera spica-venti, Avena fatua,*

Digitaria sanguinalis, Echinochloa crus-galli, Poa annua and *Setaria* spp.), broad-leaved weeds (*Amaranthus, Anthemis, Matricaria, Polygonum, Sinapis, Solanum, Stellaria, Urtica* and *Veronica* spp.) in artichokes, broccoli, asparagus, Brussels sprouts, cabbages, cauliflower, sweetcorn, garlic, horseradish, kale, leeks, maize, white mustard, onions, peanuts, pome fruits, potatoes, radish, rape, soya beans, stone fruits, strawberries, sugar cane, sunflowers, tobacco and turnips. Applied at 1.0–1.5 kg a.i./ha. **Formulation types** SC. **Selected tradenames** 'Butisan S' (BASF)

ANALYSIS
Product analysis by rp hplc with u.v. detection (*CIPAC Handbook*, 1992, **E**, 134–8). Methods for **residue** analysis (based on 2,6-dimethyl aniline) available from BASF.

MAMMALIAN TOXICOLOGY
Oral Acute oral LD_{50} for rats 2150 mg/kg. **Skin and eye** Acute percutaneous LD_{50} for rats >6810 mg/kg. No irritation of mucous membranes (rabbits). **Inhalation** LC_{50} (4 h) for rats >34.5 mg/l. **NOEL** In long-term feeding trials, **NOEL** for rats 3.6, mice 19 mg/kg b.w. **ADI** 0.036 mg/kg. **Toxicity class** WHO (a.i.) III (Table 5)

ECOTOXICOLOGY
Birds Acute oral LD_{50} for bobwhite quail >2000 mg/kg. LC_{50} for bobwhite quail and mallard ducks >5620 mg/kg b.w. **Fish** LC_{50} (96 h) for rainbow trout 4, carp 15 mg/l. **Bees** Not toxic to bees; highest concentration tested 3.6%. **Worms** LC_{50} (14 d) 440 mg/kg soil. **Daphnia** LC_{50} (48 h) 22 mg/l. **Algae** EC_{50} (96 h) for green algae (*Chlorella fusca*) 1.63 mg/l.

ENVIRONMENTAL FATE
Animals Rats: after oral administration, the a.i. was well resorbed, vigorously metabolised and eliminated mainly and rapidly via the kidneys in the form of polar conjugates (mainly glucuronides). The metabolism of the phase 1 reactions consists mainly of oxidative processes and acts on various sites on the active ingredient molecule: hydroxylation in the pyrazole ring; oxidation of a methyl group in the 2,6-dimethylphenyl ring to the corresponding methylol compound and carboxylic acid; substitution of the aliphatically bonded chlorine in the chloroacetic acid moiety; and a combination of several of these steps.
Plants After pre-emergence application, the ^{14}C-phenyl labelled active ingredient was taken up by oilseed rape plants (0.55 mg/kg 36 days after sowing and 0.43 mg/kg, day 78). In rape straw, the residue increased to 1.25 mg/kg (day 97) as the result of the loss of water in drying. The residues in rape seed were very low: 0.01 mg/kg. Metazachlor was extensively metabolised; the intact a.i. was no longer detectable at the time of harvesting. About 60% of the residue taken up by the plants still contained the unchanged 2,6-dimethylaniline moiety, but it was not possible to identify individual metabolites. **Soil/Environment** Laboratory and field

trials indicate that microbial degradation in aerobic soil is rapid; DT_{50} (lab.) 1–23 d; DT_{50} in soils fresh from the field ≤77 d, soil temperatures down to 10 °C. In field trials DT_{50} 3–9 d, DT_{90} 35–97 d. Metabolism is mainly by conjugation with glutathione and subsequent degradation; the main metabolites (≥10%) were metazachlor oxalic acid and metazachlor sulfonic acid ($COCH_2Cl$ side-chain replaced respy. by $COCO_2H$ and $COCH_2SO_3H$). Lysimeter and outdoor studies indicate that metazachlor is rapidly degraded in the soil, does not accumulate, and that there is no detectable displacement of the a.i. or its metabolites into deeper layers of the soil (a depth of >30 cm). These findings are supported by the results from raw water monitoring programmes.

476 metconazole   *Fungicide*

azole

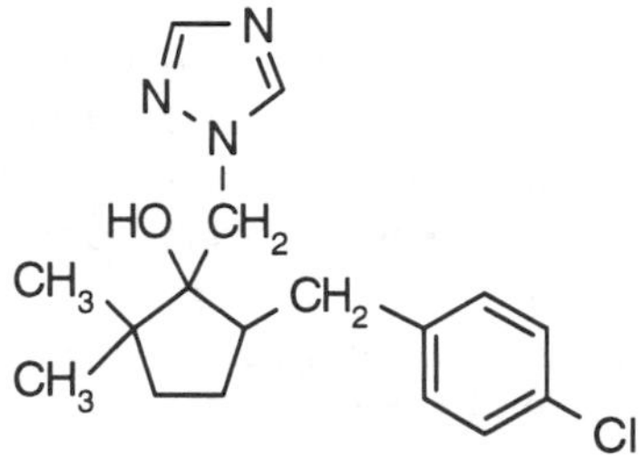

NOMENCLATURE
Common name metconazole (BSI, pa ISO (for *cis:trans* mixture)))
IUPAC name (1*RS*,5*RS*;1*RS*,5*SR*)-5-(4-chlorobenzyl)-2,2-dimethyl-= 1-(1*H*-1,2,4-triazol-1-ylmethyl)cyclopentanol
Chemical Abstracts name 5-[(4-chlorophenyl)methyl]-2,2-dimethyl-= 1-(1*H*-1,2,4-triazol-1-ylmethyl)cyclopentanol
CAS RN [125116–23–6] (unstated stereochemistry)
Development codes WL136184; KNF-S-474; AC 189635; WL 147281; AC 900,768

PHYSICAL CHEMISTRY
Composition The common name metconazole applies to the (1*RS*,5*RS*;1*RS*,5*SR*)- isomers. Tech. material is a mixture of *cis*- and *trans*- isomers, predominantly *cis*-, (1*RS*,5*SR*) (hydroxy and benzyl groups on the same side of the cyclopentyl ring), which is more fungicidally active. **Mol. wt.** 319.8 **M.f.** $C_{17}H_{22}ClN_3O$
Form Off-white odourless crystals. **M.p.** 110–113 °C; (tech. 100.0–108.4 °C)
B.p. *c.* 285 °C **V.p.** 1.23×10^{-2} mPa (20 °C) K_{ow} logP = 3.85 (25 °C)

Henry $<1.29 \times 10^{-4}$ Pa m^3 mol^{-1} **S.g./density** 1.307 kg/m^3 **Solubility** In water, 15 mg/l (20 °C). In methanol 235, acetone 238.9 g/l (20 °C). **Stability** Good thermal and hydrolytic stability. **F.p.** Not autoflammable up to 400 °C

COMMERCIALISATION
History Discovered by Kureha Chemical Industry Co. Ltd. Reported by A. J. Sampson et al. (*Proc. Br. Crop Prot. Conf. – Pests Dis.*, 1992, **1**, 419). Developed jointly with Shell group (later American Cyanamid Co.); introduced by Cyanamid Agro in France in 1994. **Patent** EP 267778 **Manufacturers** Kureha

APPLICATIONS
Biochemistry Ergosterol biosynthesis inhibitor. Interferes with synthesis of fungal cell membrane. **Mode of action** Applied post-emergence, exhibits penetrant, local and acropetal systemicity. **Uses** Control of a wide range of foliar diseases on cereals, at 90 g/ha, and on other crops. It is particularly effective against *Septoria* and rust diseases on cereals. **Phytotoxicity** Exhibits some plant growth regulant activity such as thickening of leaves, stunting and some yellowing. **Formulation types** EC; SL; SC. **Selected tradenames** 'Caramba' (Cyanamid)

MAMMALIAN TOXICOLOGY
Oral Acute oral LD$_{50}$ for rats 661 mg/kg. **Skin and eye** Acute percutaneous LD$_{50}$ for rats >2000 mg/kg. Not irritant to skin; slight irritant to eyes (rabbits). Not a skin sensitiser. **Inhalation** LC$_{50}$ (4 h) for rats >5.6 mg/l. **NOEL** (104 w) for rats 4.8 mg/kg b.w. daily; (52 w) for dogs 11.1 mg/kg b.w. daily; (90 d) for mice 5.5, rats 6.8, dogs 2.5 mg/kg b.w. daily. **ADI** 0.01 mg/kg b.w. **Other** Negative in the Ames test.

ECOTOXICOLOGY
Birds Acute oral LD$_{50}$ for bobwhite quail 790 mg/kg. Acute dietary LC$_{50}$ for mallard duck >5200 mg/kg. **Fish** LC$_{50}$ (96 h) for rainbow trout 2.2–4.0, fathead minnow 3.9, common carp 3.99 mg/l. **Bees** Practically non-toxic to bees; oral LD$_{50}$ (24 h) 97 µg/bee. **Worms** Practically non-toxic to earthworms. **Other beneficial spp.** 60 SL formulation: harmless to moderately harmful for parasitic wasp; slightly harmful for ladybird; harmful for predatory mite; harmless for ground beetle. **Daphnia** LC$_{50}$ (48 h) 3.6–4.4 mg/l. **Algae** EC$_{50}$ (72 h) for *Selenastrum capricornutum* 1.7–2.2 mg/l.

ENVIRONMENTAL FATE
Soil/Environment Log K$_{oc}$ 2.86–3.24.

urea

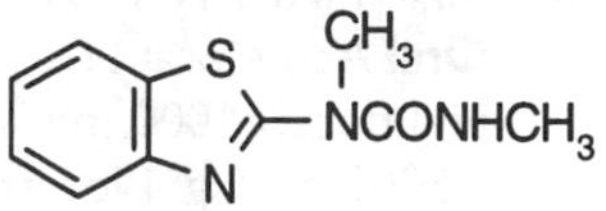

NOMENCLATURE
Common name methabenzthiazuron (BSI, E-ISO, (*m*) F-ISO); methibenzuron (WSSA); no name (Belgium, Canada, USA)
IUPAC name 1-(1,3-benzothiazol-2-yl)-1,3-dimethylurea; 1-benzothiazol-2-yl-= 1,3-dimethylurea
Chemical Abstracts name *N*-2-benzothiazolyl-*N,N'*-dimethylurea
CAS RN *[18691–97–9]* **Development codes** Bayer 74 283; S 25128

PHYSICAL CHEMISTRY
Mol. wt. 221.3 **M.f.** $C_{10}H_{11}N_3OS$ **Form** Colourless odourless crystals.
M.p. 119–121 °C **V.p.** 5.9×10^{-3} mPa (20 °C); 1.5×10^{-2} mPa (25 °C)
K_{ow} logP = 2.64 **Solubility** In water 59 mg/l (20 °C). In acetone 115.9, methanol 65.9, dimethylformamide *c.* 100, dichloromethane >200, isopropanol 20–50, toluene 50–100, hexane 1–2 (all in g/l, 20 °C). **Stability** Unstable in strong acids and alkalis; DT_{50} (22 °C) >1 y (pH 4 to 9). Direct photolysis is very slow (DT_{50} >1 y); the presence of humic substances increases the rate of photodegradation.

COMMERCIALISATION
History Herbicide reported by H. Hack (*Pflanzenschutz-Nachr. (Engl. Ed.)*, 1969, **22**, 341). Introduced in Germany (1968) by Bayer AG (who no longer manufacture or market it). **Patent** GB 1085430

APPLICATIONS
Biochemistry Photosynthetic electron transport inhibitor.
Mode of action Selective herbicide, absorbed primarily through the roots and, to a lesser extent, through the leaves. **Uses** Used at 1.4–2.8 kg a.i./ha to control a broad spectrum of broad-leaved weeds and grasses in cereals, legumes, maize, garlic and onions. Used in combination with other compounds in vineyards and orchards. **Phytotoxicity** 'Tribunil' is well tolerated in crops listed, when the various rates and timings are observed. **Formulation types** WP.
Mixtures (*methabenzthiazuron* +) chlorsulfuron; diuron; isoxaben; amitrole + MCPA. **Compatibility** Should not be used in combination with urea or other liquid fertilisers. **Selected tradenames** ['Tribunil'] (Bayer)

ANALYSIS
Product analysis by hplc with u.v. detection (*CIPAC Handbook*, 1995, **G**, 116–121).
Residues in plants and soil determined by glc with FID (H. J. Jarczyk,

Pflanzenschutz-Nachr. (Engl. Ed.), 1972, **25**, 21) or by hplc. Methods are also available from Bayer.

MAMMALIAN TOXICOLOGY
Oral Acute oral LD_{50} for rats >5000, mice and guinea pigs >2500, rabbits, cats, and dogs >1000 mg/kg. **Skin and eye** Acute percutaneous LD_{50} for rats >5000 mg/kg. Not irritating to skin and eyes (rabbits). **Inhalation** LC_{50} (4 h) for rats 5.12 mg/l air (dust). **NOEL** (2 y) for rats and mice 150, dogs 200 mg/kg diet. **ADI** 0.05 mg/kg b.w. **Toxicity class** WHO (a.i.) III (Table 5); EPA (formulation) IV

ECOTOXICOLOGY
Fish LC_{50} (96 h) for rainbow trout 15.9, golden orfe 29 mg/l. **Bees** Not toxic to bees. **Daphnia** LC_{50} (48 h) 30.6 mg/l.

ENVIRONMENTAL FATE
Animals In rats, ^{14}C-methabenzthiazuron is rapidly metabolised and the radioactivity excreted in the urine, 97% of the dose being recovered within 48 hours. It is metabolised by a combination of side-chain hydrolysis and ring hydroxylation followed by sulfate ester conjugation. The main metabolites are 6-hydroxy-(2-methylamino)-benzothiazole and 6-hydroxy-*N*-benzothiazoyl-= *N*-methyl-*N*'-methylurea and their corresponding sulfate esters. **Plants** In several plant species, similar metabolites were detected. The main metabolites are 1-hydroxymethyl-3-methyl-3-(benzothiazol-2-yl)urea, its glucoside and 3-methyl-= 3-(benzothiazol-2-yl)urea. **Soil/Environment** In soil, methabenzthiazuron is strongly adsorbed. Duration of residual activity is *c.* 3 months.

478 methacrifos   *Insecticide, acaricide*

organophosphorus

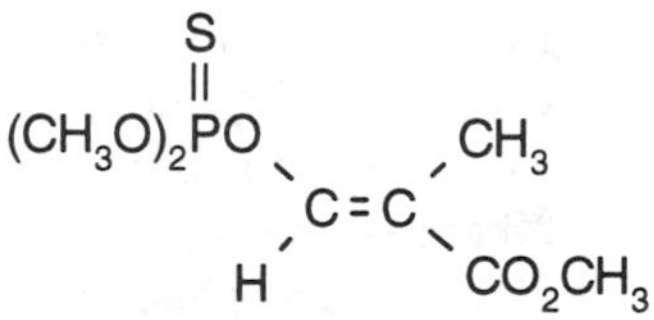

$$
\begin{array}{c}
\text{S} \\
\parallel \\
(CH_3O)_2PO
\end{array}
\diagdown
\begin{array}{c}
CH_3 \\
C=C \\
\end{array}
\diagup
\begin{array}{c}
CH_3 \\
\\
CO_2CH_3
\end{array}
$$

NOMENCLATURE
Common name methacrifos (BSI, E-ISO, (*m*) F-ISO)
IUPAC name methyl (*E*)-3-(dimethoxyphosphinothioyloxy)-2-methacrylate; (*E*)-*O*-2-methoxycarbonylprop-1-enyl *O,O*-dimethyl phosphorothioate

Chemical Abstracts name (E)-methyl 3-[(dimethoxyphosphinothioyl)oxy]-2-=
methyl-2-propenoate
CAS RN [62610–77–9] (E)-isomer; [30864–28–9] unstated stereochemistry
Development codes CGA 20 168   **Official codes** OMS 2005

PHYSICAL CHEMISTRY
Mol. wt. 240.2   **M.f.** $C_7H_{13}O_5PS$ **Form** Colourless liquid.
B.p. 90 °C/0.01 mmHg   **V.p.** 160 mPa (20 °C) K_{ow} logP ≥ 3.0
S.g./density 1.225 (20 °C)   **Solubility** In water 400 mg/l (20 °C). Miscible with
many organic solvents, e.g. methanol, benzene, hexane, and dichloromethane.
Stability Relatively unstable in alkaline conditions; on hydrolysis DT_{50} (calculated)
(20 °C) 66 d (pH 1), 29 d (pH 7), 9.5 d (pH 9). Decomposes at c. 200 °C.
F.p. 69–73 °C

COMMERCIALISATION
History Insecticide reported by R. Wyniger et al. (Proc. 1977 Br. Crop Prot. Conf. –
Pests Dis., 1978, **3**, 1033). Introduced by Ciba-Geigy AG (now Novartis Crop
Protection AG).   **Patent** BE 766000; GB 1342630   **Manufacturers** Novartis

APPLICATIONS
Biochemistry Cholinesterase inhibitor.   **Mode of action** Insecticide and acaricide
with respiratory, contact, and stomach action. Gives rapid knockdown, and has
long residual activity.   **Uses** Insecticide and acaricide with vapour, contact, and
ingested action. Mainly used for control of arthropod pests in stored products by
incorporation or by surface treatment.   **Phytotoxicity** Should not be used on plant
foliage.   **Formulation types** DP; EC.   **Compatibility** Incompatible with strongly
acidic or alkaline materials.   **Selected tradenames** 'Damfin' (Novartis)

ANALYSIS
Product analysis by glc.   **Residues** in grain determined by glc with FPD or TID
(J. Desmarchelier et al., Pestic. Sci., 1977, **8**, 473). Details available from Novartis.

MAMMALIAN TOXICOLOGY
Reviews Pesticide residues in food – 1990. FAO Plant Production and Protection
Paper, 102, 1990. Pesticide residues in food – 1990 evaluations. Toxicology. World
Health Organisation, WHO/PCS/91.47, 1991.   **EHC** 63 (WHO, 1986; a general
review of organophosphorus insecticides).   **Oral** Acute oral LD_{50} for rats 678
mg/kg.   **Skin and eye** Acute percutaneous LD_{50} for rats >3100 mg/kg. Mild skin
irritant; non-irritating to eyes (rabbits).   **Inhalation** LC_{50} (6 h) for rats 2.2 mg/l air.
NOEL (2 y) for rats 0.6 mg/kg b.w. daily.   **ADI** (JMPR) 0.006 mg/kg b.w. [1990].
Toxicity class WHO (a.i.) II

ECOTOXICOLOGY
Birds Acute oral LD_{50} for Japanese quail 116 mg/kg.   **Fish** LC_{50} (96 h) for carp
30.0, rainbow trout 0.4 mg/l.

ENVIRONMENTAL FATE
Animals Rapid excretion; metabolism via hydrolysis of the methyl esters as well as cleavage of the phosphoric and vinyl ester bonds, followed by mineralisation to CO_2.

479 methamidophos *Insecticide, acaricide*

organophosphorus

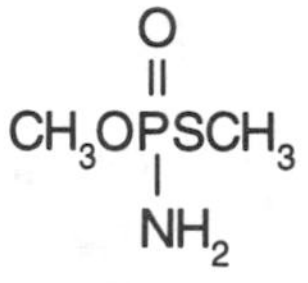

$$CH_3OPSCH_3$$

(structure: $CH_3O\overset{\displaystyle O}{\underset{\displaystyle NH_2}{\|}}PSCH_3$)

NOMENCLATURE
Common name methamidophos (BSI, E-ISO, (*m*) F-ISO, ANSI)
IUPAC name *O,S*-dimethyl phosphoramidothioate
Chemical Abstracts name *O,S*-dimethyl phosphoramidothioate
Other names acephate-met   **CAS RN** *[10265–92–6]*   **EEC no.** 233–606–0
Development codes Ortho 9006 (Chevron); Bayer 71 628; SRA 5172 (Bayer)
Official codes ENT 27 396

PHYSICAL CHEMISTRY
Mol. wt. 141.1   **M.f.** $C_2H_8NO_2PS$ **Form** Colourless crystals with a mercaptan-like odour.   **M.p.** 44.9 °C; (tech. 20–25 °C)   **V.p.** 2.3 mPa (20 °C), 4.7 mPa (25 °C). K_{ow} logP = –0.8 (20 °C)   **Henry** <1.6 × 10^{-6} Pa m^3 mol^{-1} (calc., 20 °C)
S.g./density 1.27 (20 °C)   **Solubility** In water >200 g/l (20 °C). In isopropanol >200, dichloromethane >200, hexane 0.1–1, toluene 2–5 (all in g/l, 20 °C).
Stability Stable at ambient temperature but decomposes on heating without boiling. Stable at pH 3–8. Hydrolysed in acids and alkalis, DT_{50} (22 °C) 1.8 y (pH 4), 120 h (pH 7), 70 h (pH 9). Photodegradation is of minor importance.
F.p. 66°C (EU A.9/ASTN-D56)

COMMERCIALISATION
History Insecticide reported by I. Hammann (*Pflanzenschutz-Nachr. (Engl. Ed.),* 1970, **23**, 133). Introduced by Chevron Chemical Co. and by Bayer AG .
Patent US 3309266; DE 1210835 to Bayer   **Manufacturers** Bayer; Crystal; Jin Hung; Jingma; Pilarquim; Productos OSA; Q.E.A.C.A.; Shenzhen Jiangshan; Sinon; Taiwan Tainan Giant; Tekchem; Tomen; Westrade

APPLICATIONS
Biochemistry Cholinesterase inhibitor.   **Mode of action** Systemic insecticide and

acaricide with contact and stomach action. Absorbed by the roots and leaves.
Uses Control of chewing and sucking insects, and spider mites on ornamentals,
potatoes, pome fruit, stone fruit, citrus fruit, vines, hops, brassicas, beet, cotton,
maize, tobacco, and other crops. At 0.5–1.0 kg/ha, its contact activity persists for
7–21 d. **Formulation types** SL. **Mixtures** (methamidophos +) cyfluthrin;
parathion; triflumuron; beta-cyfluthrin. **Compatibility** Incompatible with alkaline
materials. **Selected tradenames** 'Monitor' (Bayer, Tomen, Valent); 'Tamaron'
(Bayer); 'Cekumidofos' (Cequisa); 'Giant' (Sanonda), 'Jiaanlin' (Shenzhen
Jiangshan); 'Methaphos' (Efthymiadis); 'MTD-600' (Westrade); 'Patrole'
(Productos OSA); 'Pilaron' (Pilarquim)

ANALYSIS
Product analysis by rplc with u.v. detection (*AOAC Methods*, 1995, 992.01; *CIPAC
Handbook*, 1992, **E**, 139–144), by i.r. spectrometry or by glc (J. B. Leary, *Anal.
Methods Pestic. Plant Growth Regul.*, 1973, **7**, 339). **Residues** determined by glc
with FID (*idem, ibid.*; *Manu. Pestic. Residue Anal.*, 1987, **I**, 6, S19; *Anal. Methods
Residues Pestic.*, 1988, Part I, M5; M. A. Luke et al., *J. Ass. Off. Anal. Chem.*, 1981,
64, 1187; A. Ambrus et al., *ibid.*, p. 733; *AOAC Methods*, 1995, 985.22). Details
available from Bayer.

MAMMALIAN TOXICOLOGY
Reviews *Pesticide residues in food – 1990*, FAO Plant Production and Protection
Paper, 103, 1990. *Pesticide residues in food – 1990 evaluations*, Toxicology, WHO,
WHO/PCS/91.47, 1991. **EHC** 63 (WHO, 1986; a general review of
organophosphorus insecticides). **Oral** Acute oral LD_{50} for rats c. 20, guinea pigs
30–50, rabbits, cats and dogs 10–30 mg/kg. **Skin and eye** Acute percutaneous
LD_{50} for rats c. 130 mg/kg; not irritating to skin, slightly irritating to eyes
(rabbits). **Inhalation** LC_{50} (4 h) for rats 0.2 mg/l (aerosol). **NOEL** (2 y) for rats
2, mice 5 mg/kg diet; (12 mo) for dogs 2 mg/kg diet. **ADI** (JMPR) 0.004 mg/kg
b.w. [1990]. **Toxicity class** WHO (a.i.) Ib; EPA (formulation) I
EC risk T+ (R28); T (R24); Xi (R36) **PIC** Yes

ECOTOXICOLOGY
Birds Acute oral LD_{50} for bobwhite quail 10–11, mallard ducks 29.5 mg/kg. Five-
day LC_{50} for mallard ducks 1302, bobwhite quail 42–92 mg/kg diet. **Fish** LC_{50}
(96 h) for rainbow trout 40, golden orfe 47.7 mg/l. **Bees** Toxic to bees.
Worms LC_{50} for *Eisenia foetida* 73 mg/kg dry soil. **Daphnia** LC_{50} (48 h)
0.27 mg/l. **Algae** E_rC_{50} for *Scenedesmus subspicatus* >178 mg/l.

ENVIRONMENTAL FATE
Animals In rats and farm animals, radiolabelled methamidophos was absorbed
rapidly and distributed uniformly among all organs and tissues. More than half of
the radioactivity was rapidly eliminated from the body, mainly via urine and
respiratory air. Radioactivity remaining in the animal was incorporated into

endogenous compounds (carbon-1 pool) and eliminated with the natural turnover of these compounds. Metabolism in the rat was by deamination and demethylation. **Plants** After treatment with methamidophos via the roots, the a.i. was taken up rapidly and translocated into the leaves with the transpiration flow. However, a more prolonged uptake via the roots cannot be expected because of rapid degradation in soil. After foliar treatment, methamidophos was taken up rapidly; however, there was little translocation into untreated parts of the plant. Deaminated methamidophos was a major metabolite.

Soil/Environment Methamidophos was degraded very rapidly in soil, by demethylation and deamination to form CO_2. It was only slightly adsorbed by soil but leaching into deeper soil layers can be ruled out because of rapid degradation. In aqueous buffer solutions, hydrolysis at environmentally relevant pH levels is slow; photolysis may contribute to the degradation. However, in natural water, methamidophos was degraded much more rapidly. Metabolites included deaminated methamidophos, desmethyl-methamidophos and dimethyl disulfide. Low vapour pressure and a short lifetime in air make it unlikely that methamidophos will accumulate in the atmosphere.

480 methasulfocarb

Fungicide, plant growth regulator

$$CH_3NHCOS \longsquare OSO_2CH_3$$

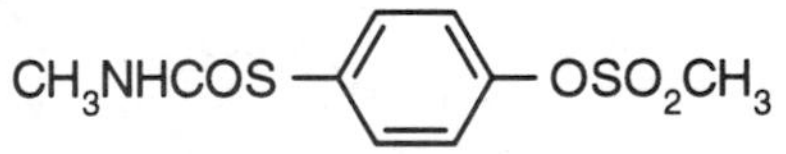

NOMENCLATURE
Common name methasulfocarb (BSI, draft E-ISO, (*m*) draft F-ISO)
IUPAC name S-4-methylsulfonyloxyphenyl methylthiocarbamate
Chemical Abstracts name S-[4-[(methylsulfonyl)oxy]phenyl] methylcarbamothioate
CAS RN *[66952–49–6]* **Development codes** NK-191 (Nippon Kayaku)

PHYSICAL CHEMISTRY
Mol. wt. 261.3 **M.f.** $C_9H_{11}NO_4S_2$ **Form** Colourless crystals.
M.p. 137.5–138.5 °C **Solubility** In water 480 mg/l. Soluble in acetone, alcohols, and benzene. **Stability** Stable to light.

COMMERCIALISATION
History Fungicide and plant growth regulator reported by K. Ohmori (*Jpn. Pestic.*

Inf., 1985, No. 46, p. 17). Introduced in Japan (1984) by Nippon Kayaku Co., Ltd.
Patent US 4126696; DE 2745229; JP 5347527   **Manufacturers** Nippon Kayaku

APPLICATIONS
Uses Soil fungicide for control of *Corticium, Fusarium, Mucor, Pseudomonas, Pythium, Rhizopus, Rhizoctonia,* and *Trichoderma* spp. in rice.
Formulation types DP.   **Selected tradenames** 'Kayabest' (Nippon Kayaku)

ANALYSIS
Product and **residue** analysis by glc.

MAMMALIAN TOXICOLOGY
Oral Acute oral LD_{50} for rats 112–119, male mice 342, female mice 262 mg/kg.
Skin and eye Acute percutaneous LD_{50} for rats and mice >5000 mg/kg.
Inhalation LC_{50} (4 h) for rats >0.44 mg/l air.   **Other** Non-mutagenic to mice.
Non-teratogenic to rats.   **Toxicity class** WHO (a.i.) II

ECOTOXICOLOGY
Fish LC_{50} (48 h) for carp 1.95 mg/l.

481   methidathion   *Insecticide, acaricide*

organophosphorus

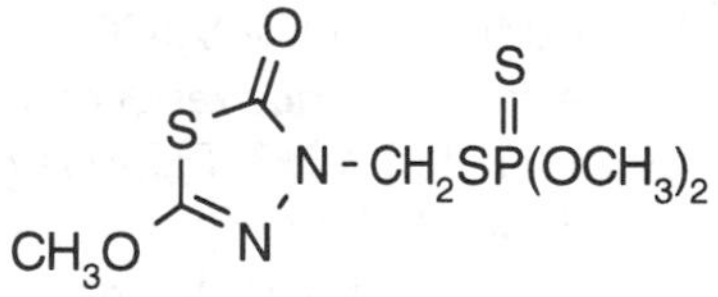

NOMENCLATURE
Common name methidathion (BSI, E-ISO, F-ISO, ANSI, ESA); DMTP (JMAF)
IUPAC name *S*-2,3-dihydro-5-methoxy-2-oxo-1,3,4-thiadiazol-3-ylmethyl
O,O-dimethyl phosphorodithioate; 3-dimethoxyphosphinothioylthiomethyl-=
5-methoxy-1,3,4-thiadiazol-2(3*H*)-one
Chemical Abstracts name *S*-[(5-methoxy-2-oxo-1,3,4-thiadiazol-3(2*H*)-yl)methyl]
O,O-dimethyl phosphorodithioate
CAS RN *[950–37–8]*   **EEC no.** 213–449–4   **Development codes** GS 13 005
(Ciba)   **Official codes** OMS 844; ENT 27 193

PHYSICAL CHEMISTRY
Mol. wt. 302.3   **M.f.** $C_6H_{11}N_2O_4PS_3$ **Form** Colourless crystals.   **M.p.** 39–40 °C

V.p. 2.5×10^{-1} mPa (20 °C)   **K_{ow}** logP = 2.2 (OECD 107)   **S.g./density** 1.51 (20 °C) (OECD 109)   **Solubility** In water 200 mg/l (25 °C). In ethanol 150, acetone 670, toluene 720, hexane 11, *n*-octanol 14 (all in g/l, 20 °C). **Stability** Rapidly hydrolysed in alkaline and strongly acidic media; DT_{50} (25 °C) 30 min at pH 13. Relatively stable to hydrolysis in neutral and slightly acidic media.

COMMERCIALISATION
History Insecticide reported by H. Grob *et al.* (*Proc. Br. Insectic. Fungic. Conf., 3rd,* 1965, p. 451). Introduced by J. R. Geigy S.A. (now Novartis Crop Protection AG). **Patent** BE 623246; GB 1008451   **Manufacturers** Makhteshim-Agan; Novartis

APPLICATIONS
Biochemistry Cholinesterase inhibitor.   **Mode of action** Non-systemic insecticide and acaricide with contact and stomach action.   **Uses** Control of a wide range of sucking and chewing insects (especially scale insects) and spider mites in many crops, e.g. pome fruit, stone fruit, citrus fruit, vines, olives, hops, cotton, potatoes, beet, alfalfa, oilseed rape, maize, sunflowers, safflowers, tobacco, hazels, and some vegetables.   **Formulation types** EC; WP; UL.   **Compatibility** Mixing with alkaline materials may reduce effectiveness.   **Selected tradenames** 'Supracide' (Novartis); 'Suprathion' (Makhteshim-Agan)

ANALYSIS
Product analysis by glc.   **Residues** determined by glc (A. Ambrus *et al., J. Assoc. Off. Anal. Chem.,* 1981, **64**, 773; *Man. Pestic. Residue Anal.,* 1987, **I**, 6, S8, S13, S19; *Anal. Methods Residues Pestic.,* 1988, Part I, M2, M5, M12; D. O. Eberle & R. Suter, *Anal. Methods Pestic. Plant Growth Regul.,* 1976, **8**, 141).

MAMMALIAN TOXICOLOGY
Reviews *Pesticide residues in food – 1992.* FAO Plant Production and Protection Paper, 116, 1993. *Pesticide residues in food – 1992 evaluations. Part II – Toxicology.* World Health Organisation, WHO/PCS/93.34.   **EHC** 63 (WHO, 1986; a general review of organophosphorus insecticides).   **Oral** Acute oral LD_{50} for rats 25–54, mice 25–70, rabbits 63–80, guinea pigs 25 mg/kg.   **Skin and eye** Acute percutaneous LD_{50} for rabbits 200, rats 1546 mg/kg. Mild skin irritant; non-irritating to eyes (rabbits).   **Inhalation** LC_{50} (4 h) for rats 3.6 mg/l air. **NOEL** Human volunteers tolerated daily oral doses of up to 0.11 mg/kg for at least 42 d without reaction. In 2 y feeding trials, NOEL for rats 4 mg/kg diet (0.15 mg/kg daily), for dogs 0.25 mg/kg daily.   **ADI** 0.001 mg/kg b.w. [1992]. **Toxicity class** WHO (a.i.) Ib; EPA (formulation) I   **EC risk** R24, R28

ECOTOXICOLOGY
Birds Acute oral LD_{50} for mallard ducks 23.6–28 mg/kg. LC_{50} (8 d) for bobwhite quail 224 ppm.   **Fish** LC_{50} (96 h) for rainbow trout 0.01, bluegill sunfish 0.002 mg/l.   **Bees** Slightly toxic to bees.

ENVIRONMENTAL FATE
Animals In mammals, methidathion is rapidly metabolised and excreted. **Plants** In plants, rapid metabolism occurs. The overall metabolic pattern indicates hydrolysis of the ester bond, cleavage of the heterocyclic moieties into fragments which are further oxidised to CO_2. **Soil/Environment** Methidathion and its metabolites have a low mobility in soils. The compound is rapidly degraded in soil and water by chemical, photolytic and biological processes. DT_{50} 3–18 d (laboratory and field results).

482 methiocarb

Molluscicide, insecticide, acaricide, bird repellent

carbamate

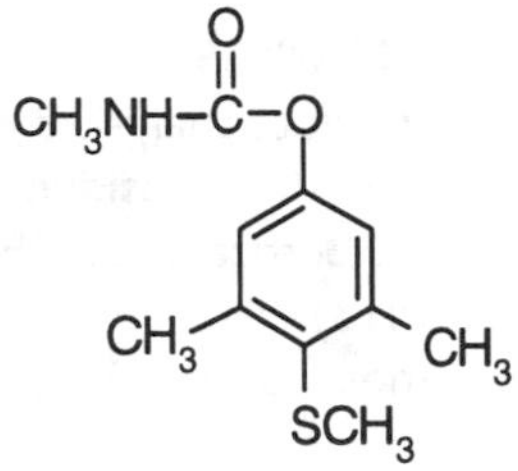

NOMENCLATURE
Common name methiocarb (BSI, Canada, New Zealand, Republic of South Africa, Turkey, ESA, E-ISO); methiocarbe ((*m*) F-ISO); mercaptodimethur (alternative name, E-ISO, (*m*) F-ISO, France, Germany); no name (Eire, USA)
IUPAC name 4-methylthio-3,5-xylyl methylcarbamate
Chemical Abstracts name 3,5-dimethyl-4-(methylthio)phenyl methylcarbamate
CAS RN *[2032–65–7]* **EEC no.** 217–991–2 **Development codes** Bayer 37 344; H 321 (Bayer) **Official codes** OMS 93; ENT 25 726

PHYSICAL CHEMISTRY
Mol. wt. 225.3 **M.f.** $C_{11}H_{15}NO_2S$ **Form** Colourless crystals with a phenol-like odour. **M.p.** 119 °C **V.p.** 0.015 mPa (20 °C); 0.036 mPa (25 °C)
K_{ow} logP = 3.08 (20 °C) **Henry** 1.2×10^{-4} Pa m³ mol⁻¹ (20 °C)
S.g./density 1.236 (20 °C) **Solubility** In water 27 mg/l (20 °C). In dichloromethane >200, isopropanol 53, toluene 33, hexane 1.3 (all in g/l, 20 °C).
Stability Unstable in highly alkaline media. Hydrolysis DT_{50} (22 °C) >1 y (pH 4), <35 d (pH 7), 6 h (pH 9). Photodegradation contributes to the overall elimination of methiocarb from the environment; DT_{50} 6–16 d.

COMMERCIALISATION
History Insecticide reported by G. Unterstenhöfer (*Pflanzenschutz-Nachr. (Engl. Ed.)*, 1962, **15**, 181). Introduced by Bayer AG.   **Patent** FR 1275658; DE 1162352
Manufacturers Bayer

APPLICATIONS
Biochemistry Cholinesterase inhibitor.   **Mode of action** Molluscicide with neurotoxic action. Non-systemic insecticide and acaricide with contact and stomach action.   **Uses** Control of slugs and snails in a wide range of agricultural situations. Broad-range control of Lepidoptera, Coleoptera, Diptera, Thysanoptera and Homoptera (including soil insects), and spider mites in pome fruit, stone fruit, citrus fruit, strawberries, hops, potatoes, beet, maize, oilseed rape, vegetables, and ornamentals. Used as seed treatment for control of frit flies on maize, flea beetles on oilseed rape, and leaf miners on beet; and also acts as a bird repellent.   **Phytotoxicity** May lead to fruit thinning on apple trees, if applied earlier than four weeks after petal fall.   **Formulation types** DP; WP; SC; RB; GB; Seed treatment.   **Compatibility** Incompatible with alkaline materials.
Selected tradenames 'Draza' (Bayer); 'Mesurol' (Bayer)

ANALYSIS
Product analysis by hplc (*CIPAC Handbook*, 1988, **D**, 130; *AOAC Methods*, 1995, 984.10).   **Residues** determined by rplc (*ibid.*, 985.23) or by glc (M. C. Bowman & M. Beroza, *ibid.*, 1969, **52**, 1054; *Anal. Methods Residues Pestic.*, 1988, Part I, M2, M13). Methods for the determination of residues are available from Bayer. For methods in **drinking water**, see *AOAC Methods*, 1995, 991.06.

MAMMALIAN TOXICOLOGY
Reviews *Pesticide residues in food – 1987*. FAO Plant Production and Protection Paper 84, 1987.   **EHC** 64 (WHO, 1986; a review of carbamate insecticides in general).   **Oral** Acute oral LD_{50} for male and female rats c. 20, mice 52–58, guinea pigs c. 40, dogs 25 mg/kg.   **Skin and eye** Acute percutaneous LD_{50} for male and female rats >5000 mg/kg. Non-irritating to skin and eyes (rabbits).
Inhalation LC_{50} (4 h) for rats >0.3 mg/l air (aerosol); c. 0.5 mg/l (dust).   **NOEL** In 2 feeding trials, NOEL for rats and mice 67, dogs 5 mg/kg diet.   **ADI** (JMPR) 0.001 mg/kg b.w. [1987].   **Toxicity class** WHO (a.i.) II; EPA (formulation) I
EC risk T (R25)

ECOTOXICOLOGY
Birds Acute oral LD_{50} for male mallard ducks 7.1–9.4, Japanese quail 5–10 mg/kg. Seven-day LC_{50} for bobwhite quail: no signs of toxicity as birds were repelled by 'Mesurol'.   **Fish** LC_{50} (96 h) for bluegill sunfish 0.754, rainbow trout 0.436–4.7, golden orfe 3.8 mg/l.   **Bees** Not toxic to bees (depending on mode of application).   **Worms** LC_{50} for *Eisenia foetida* >200 mg/kg dry soil.

Daphnia LC_{50} (48 h) 0.019 mg/l.   **Algae** E_rC_{50} for *Scenedesmus subspicatus* 1.15 mg/l.

ENVIRONMENTAL FATE

Animals In dogs and mice, following oral administration, methiocarb is rapidly absorbed and excreted, principally in the urine, with only a small proportion in the faeces. Metabolism involves hydrolysis, oxidation, and hydroxylation, followed by excretion in free or conjugated form. There is a continuous decrease of activity in all organs. Metabolism of carbamate insecticides is reviewed (M. Cool & C. K. Jankowski in "*Insecticides*").   **Plants** In plants, the methylthio group is oxidised to sulfoxide and sulfone, with hydrolysis to the corresponding thiophenol, methylsulfoxide-phenol, and methylsulfonyl-phenol (M. C. Bowman & M. Beroza, *J. Assoc. Off. Anal. Chem.*, 1969, **52**, 1054–1063).   **Soil/Environment** Degradation in soil is rapid. The important metabolites are methylsulfinylphenol and methylsulfonylphenol.

483   methomyl

Insecticide, acaricide

oxime carbamate

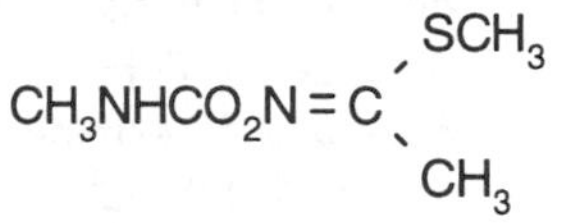

$$CH_3NHCO_2N=C\overset{\displaystyle SCH_3}{\underset{\displaystyle CH_3}{}}$$

NOMENCLATURE

Common name methomyl (BSI, E-ISO, (*m*) F-ISO, ANSI, ESA, JMAF)
IUPAC name *S*-methyl *N*-(methylcarbamoyloxy)thioacetimidate
Chemical Abstracts name methyl *N*-[[(methylamino)carbonyl]oxy]= ethanimidothioate
CAS RN *[16752–77–5]*   **EEC no.** 240–815–0   **Development codes** DPX-X1179
Official codes OMS 1196

PHYSICAL CHEMISTRY

Composition Methomyl is a mixture of (*Z*)- and (*E*)- isomers, the former predominating.   **Mol. wt.** 162.2   **M.f.** $C_5H_{10}N_2O_2S$ **Form** Colourless crystals with a slight sulfurous odour.   **M.p.** 78–79 °C   **V.p.** 0.72 mPa (25 °C)
K_{ow} logP = 0.093   **Henry** 2.13×10^{-6} Pa m^3 mol^{-1} **S.g./density** 1.2946 (25 °C)
Solubility In water 57.9 g/l (25 °C). In methanol 1000, acetone 730, ethanol 420, isopropanol 220, toluene 30 (all in g/kg, 25 °C). Sparingly soluble in hydrocarbons.   **Stability** At room temperature, aqueous solutions undergo slow decomposition. The rate of decomposition increases at higher temperatures, in the presence of sunlight, on exposure to air, and in alkaline media.

COMMERCIALISATION

History Insecticide reported by G. A. Roodhans & N. B. Joy (*Meded. Rijksfac. Landbouwwet. Gent*, 1968, **33**, 833). Introduced by E. I. du Pont de Nemours & Co. **Patent** US 3576834; US 3639633 **Manufacturers** Chunhu; Crystal; Du Pont; Jiangsu; Kuo Ching; Makhteshim-Agan; Rhône-Poulenc; Rotam; Sanachem; Sinon; Shenzhen Jiangshan; Sundat

APPLICATIONS

Biochemistry Cholinesterase inhibitor. **Mode of action** Systemic insecticide and acaricide with contact and stomach action. **Uses** Control of a wide range of insects (particularly Lepidoptera, Hemiptera, Homoptera, Diptera and Coleoptera) and spider mites in fruit, vines, olives, hops, vegetables, ornamentals, field crops, cucurbits, flax, cotton, tobacco, soya beans, etc. Also used for control of flies in animal and poultry houses and dairies. **Phytotoxicity** Non-phytotoxic when used as recommended, except to some varieties of apple.
Formulation types WP; SL; SP. **Mixtures** (*methomyl +*) dicofol; muscalure; tetradifon. **Selected tradenames** 'Lannate' (Du Pont); 'Dunet' (Dhanuka); 'Kuik' (Rotam); 'Methavin' (Rhône-Poulenc); 'Methomex' (Makhteshim-Agan); 'Methosan' (Sanachem); 'Nudrin' (Cyanamid); 'Pilarmate' (Pilarquim); 'Sathomyl' (Sanonda)

ANALYSIS

Product analysis by hplc (J. E. Thean et al., *J. Assoc. Off. Anal. Chem.*, 1978, **61**, 15; R. E. Leitch & H. L. Pease, *Anal. Methods Pestic. Plant Growth Regul.*, 1973, **7**, 331). **Residues** determined by glc with FPD (*idem, ibid.*; R. T. Krause, *J. Assoc. Off. Anal. Chem.*, 1980, **63**, 1114; M. A. Luke et al., *ibid.*, 1981, **64**, 1187; A. Ambrus et al., *ibid.*, p. 733) or by rplc (*AOAC Methods*, 1995, 985.23). Details available from Du Pont. For methods in **drinking water**, see *AOAC Methods*, 1995, 991.06.

MAMMALIAN TOXICOLOGY

Reviews *Pesticide residues in food – 1989*. FAO Plant Production and Protection Paper 99, 1989. *Pesticide residues in food – 1989 evaluations. Part II – Toxicology*. FAO Plant Production and Protection Paper 100/2, 1990. *Pesticide residues in food – 1991*. FAO Plant Production and Protection Paper 111, 1991. **EHC** 178 (WHO, 1996). Methomyl Health & Safety Guide: 97 (WHO 1995). **Oral** Acute oral LD_{50} for male rats 17 mg a.i./kg, for female rats 24 mg/kg.
Skin and eye Acute percutaneous LD_{50} for male rabbits >5000 mg a.i./kg; irritant to eyes of rabbits but not to skin of guinea pigs. **Inhalation** LC_{50} (4 h) for rats 0.3 mg/l air (aerosol). **NOEL** (2 y) for rats 100, mice 50, dogs 100 mg/kg diet. **ADI** (JMPR) 0.03 mg/kg b.w. [1989]. **Toxicity class** WHO (a.i.) Ib; EPA (formulation) I, IV **EC risk** T+ (R28)

ECOTOXICOLOGY

Birds Acute oral LD_{50} for mallard ducks 15.9, pheasants 15.4 mg/kg. Eight-day dietary LC_{50} for Pekin ducks 1890, bobwhite quail 3680 mg/kg diet. **Fish** LC_{50} (96 h) for rainbow trout 3.4, bluegill sunfish 0.9 mg/l. **Bees** Toxic to bees, contact LD_{50} 0.1 µg/bee, but not hazardous when the spray has dried. **Daphnia** LC_{50} (48 h) 28.7 µg/l.

ENVIRONMENTAL FATE

Animals In rats, methomyl was rapidly converted to methomyl methylol, oxime, sulfoxide and sulfoxide oxime; these unstable intermediates were converted to acetonitrile and CO_2, which were eliminated primarily via respiration and in the urine. Metabolism of carbamate insecticides is reviewed (M. Cool & C. K. Jankowski in "*Insecticides*"). **Plants** DT_{50} following leaf application c. 3–5 d. Rapidly degraded to CO_2 and acetonitrile, with incorporation into natural plant components (J. Harvey & R. W. Reiser, *Agric. Food Chem.*, 1973, **21**, 775). **Soil/Environment** Rapidly degraded in soil. DT_{50} in groundwater samples <0.2 d (J. H. Smelt, *Pestic. Sci.*, 1983, **14**, 173–181). K_{oc} 72.

484 methoprene   *Insecticide*

juvenile hormone mimic

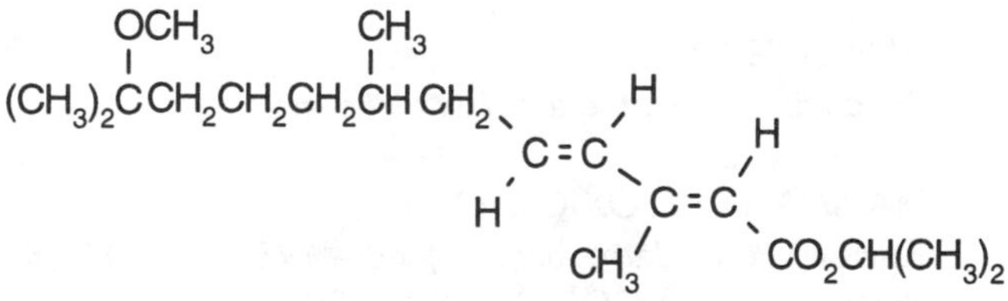

NOMENCLATURE

Common name methoprene (BSI, E-ISO, (*m*) F-ISO, ANSI, ESA)
IUPAC name isopropyl (*E,E*)-(*RS*)-11-methoxy-3,7,11-trimethyldodeca-= 2,4-dienoate
Chemical Abstracts name (*E,E*)-(±)-1-methylethyl 11-methoxy-3,7,11-trimethyl-= 2,4-dodecadienoate
CAS RN [40596–69–8] (*E,E*)- isomers; [41205–06–5] (*E,E*)-(±) isomers; [65733–16–6] (*E,E*)-(*S*)- isomer; [65733–17–7] (*E,E*)-(*R*)- isomer
Development codes ZR 515 (Sandoz) **Official codes** OMS 1697

PHYSICAL CHEMISTRY

Mol. wt. 310.5 **M.f.** $C_{19}H_{34}O_3$ **Form** Tech. grade is an amber liquid.
B.p. 100 °C/0.05 mmHg **V.p.** 3.15 mPa (25 °C) **K$_{ow}$** logP = 5.212

S.g./density 0.921 (25 °C) **Solubility** In water 1.4 mg/l (room temperature). Miscible with all common organic solvents. **Stability** Stable in water, organic solvents, and in the presence of aqueous acids and alkalis. Sensitive to u.v. light. **F.p.** 96 °C (closed cup)

COMMERCIALISATION
History Insect growth regulator reported by C. A. Henrick et al. (*J. Agric. Food Chem.*, 1973, **21**, 354). Introduced by Zoecon Corp. (acquired by Sandoz AG, now Novartis Crop Protection AG). **Patent** US 3904662; US 3912815
Manufacturers Novartis

APPLICATIONS
Mode of action Insect growth regulator (juvenile hormone mimic), preventing metamorphosis to viable adults when applied to larval stages. **Uses** Control of many insect pests (especially Diptera and Pharaoh's ants, but also Coleoptera, Homoptera and Siphonaptera) in public health, stored commodities (including tobacco), food handling, processing and storage establishments, mushroom houses, on animals, and on plants (including glasshouse plants). Particular uses include control of mosquito larvae; sciarid flies in mushroom houses; cigarette beetles and tobacco moths in stored tobacco; Pharaoh's ants; leaf miners on glasshouse chrysanthemums; stored product pests in food and tobacco processing plants and warehouses, etc. **Formulation types** AE; SC; EC; BR; SL; CB.
Mixtures (*methoprene +*) permethrin; pyrethrins. **Selected tradenames** 'Precor' (Novartis)

ANALYSIS
Product and **residue** analysis by gc-ms.

MAMMALIAN TOXICOLOGY
Reviews *Pesticide residues in food – 1987*. FAO Plant Production and Protection Paper 84, 1987. CODEX (1984). EPA FR 59:11570 (11 Mar 94). **Oral** Acute oral LD_{50} for rats >34 600, dogs >5000 mg/kg. **Skin and eye** Acute percutaneous LD_{50} for rabbits 3500 mg/kg. Non-irritating to skin and eyes (rabbits).
Inhalation LC_{50} for rats >210 mg/l air. **NOEL** In 2 y feeding trials, rats receiving 5000 mg/kg diet and mice receiving 2500 mg/kg diet showed no ill-effects. No teratogenic effects on rats at 1000 mg/kg and on rabbits at 500 mg/kg. No mutagenic effects on rats at 2000 mg/kg. No reproductive adverse effects in 3-generation reproduction studies on rats at 2500 mg/kg diet.
ADI (JMPR) 0.1 mg/kg b.w. [1987]. **Toxicity class** WHO (a.i.) III (Table 5); EPA (formulation) IV

ECOTOXICOLOGY
Birds Eight-day dietary LC_{50} for chickens >4640 mg/kg. **Fish** LC_{50} (96 h) for bluegill sunfish 4.6, trout 4.4 mg/l. **Bees** Non-toxic to adult bees; LD_{50} (oral and

topical) >1000 µg/bee. Bee larvae are sensitive at 0.2 µg/bee. **Daphnia** LC$_{50}$ (48 h) 360 µg/l.

ENVIRONMENTAL FATE
Animals In mammals, the secondary metabolite cholesterol has been identified.
Plants In plants, degradation principally involves ester hydrolysis, O-demethylation, and oxidative cleavage of the double bond at the 4-position. In alfalfa and rice, the principal metabolite is 7 methoxycitronellal. **Soil/Environment** In soil, rapidly degraded, DT$_{50}$ c. 10 d.

485 methoxychlor *Insecticide*

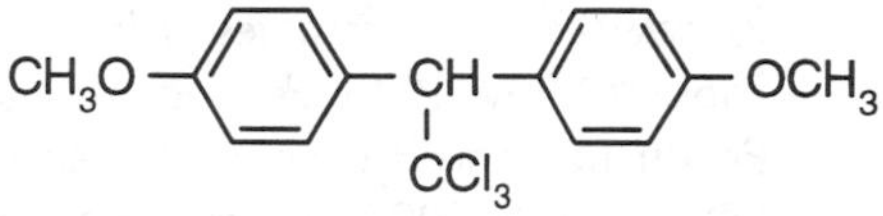

$$CH_3O—\bigcirc—CH—\bigcirc—OCH_3$$
$$\underset{CCl_3}{|}$$

NOMENCLATURE
Common name methoxychlor (BSI, E-ISO, ESA, JMAF); méthoxychlore ((m) F-ISO)
IUPAC name 1,1,1-trichloro-2,2-bis(4-methoxyphenyl)ethane
Chemical Abstracts name 1,1'-(2,2,2-trichloroethylidene)bis[4-methoxybenzene]
Other names DMTD   **CAS RN** *[72–43–5]*   **Official codes** OMS 466; ENT 1716

PHYSICAL CHEMISTRY
Composition Tech. grade contains ≥88% methoxychlor and ≤12% related isomers.
Mol. wt. 345.7   **M.f.** C$_{16}$H$_{15}$Cl$_3$O$_2$ **Form** Colourless crystals; (tech., grey powder).   **M.p.** 89 °C (tech., 77 °C)   **V.p.** Very low   **S.g./density** 1.41 (25 °C)
Solubility In water 0.1 mg/l (25 °C). Readily soluble in aromatic, chlorinated, and ketonic solvents, and vegetable oils. In chloroform and xylene 440, methanol 50 (all in g/kg, 22 °C).   **Stability** Stable to oxidising agents and to u.v. irradiation. Reacts with alkalis, especially in the presence of catalytically-active metals, with loss of hydrogen chloride, but more slowly than DDT. Colour turns to pink or tan on exposure to light.

COMMERCIALISATION
History Insecticide reported by P. Läuger *et al.* (*Helv. Chim. Acta,* 1944, **27**, 892). Introduced by J. R. Geigy AG (now Novartis Crop Protection AG) and by E. I. du Pont de Nemours and Co. (neither company now manufactures or markets it).
Patent CH 226180; GB 547871 to Ciba-Geigy; US 2420928; US 2477655; GB 624561 to Du Pont   **Manufacturers** Kincaid

APPLICATIONS
Mode of action Insecticide with contact and stomach action. **Uses** Control of a wide range of insect pests (particularly chewing insects) in field crops, forage crops, fruit, vines, flowers, vegetables, and in forestry. Used also for control of insect pests in animal houses, dairies, and in household and industrial premises. **Formulation types** AE; WP; EC; DP; GR. **Mixtures** *(methoxychlor +)* malathion + parathion; piperonyl butoxide + pyrethrins. **Compatibility** Incompatible with highly alkaline materials.

ANALYSIS
Product analysis by total chlorine content (H. L. Pease, *J. Assoc. Off. Anal. Chem.*, 1976, **58**, 40). **Residues** determined by glc or by tlc or paper chromatography (*AOAC Methods*, 1995, 970.52; J. Solomon & W. L. Lockhart, *ibid.*, 1978, **60**, 690; *Anal. Methods Pestic. Plant Growth Regul.*, 1972, **6**, 441). In **drinking water** by gc with ECD (*AOAC Methods*, 1995, 990.06).

MAMMALIAN TOXICOLOGY
Reviews *Pesticide residues in food – 1977*. FAO Plant Production and Protection Paper 10 Rev, 1978. *Pesticide residues in food: 1977 evaluations*. FAO Plant Production and Protection Paper 10 Sup, 1978. **IARC** 5, 20 **Oral** Acute oral LD_{50} for rats 6000 mg/kg. **Skin and eye** Acute percutaneous LD_{50} for rabbits >2000 mg/kg; not a skin irritant. **NOEL** In 1 y feeding trials, no toxic effect observed in dogs receiving 300 mg/kg daily; in 2 y trials in rats, no effect at 200 mg/kg diet, but some reduction in growth occurred at 1600 mg/kg diet. **ADI** (JMPR) 0.1 mg/kg b.w. [1977]. **Toxicity class** WHO (a.i.) III (Table 5)

ECOTOXICOLOGY
Birds Acute oral LD_{50} for mallard ducks >2000 mg/kg. Eight-day dietary LC_{50} for bobwhite quail and ring-necked pheasants >5000 mg/kg diet. **Fish** LC_{50} (24 h) for rainbow trout 0.052, bluegill sunfish 0.067 mg/l. **Daphnia** LC_{50} (48 h) 0.00078 mg/l.

ENVIRONMENTAL FATE
Animals Degradation in animals is principally by O-dealkylation to the corresponding phenol and diphenol, and by dehydrochlorination to 4,4'-dihydroxybenzophenone (K. S. Kapoor, *J. Agric. Food Chem.*, 1970, **18**, 1145–1152; *J. Pestic. Sci.*, 1989, **14**, 85; M. D. Corbett et al., *Chem. Res. Toxicol.*, 1990, **3**, 8). **Soil/Environment** DT_{50} in water c. 46 d.

methylarsonic acid   *Herbicide*

organoarsenic

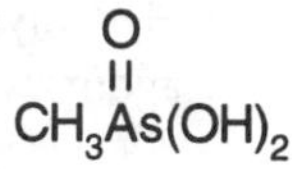

$$\overset{\displaystyle O}{\underset{\displaystyle CH_3As(OH)_2}{\|}}$$

NOMENCLATURE
methylarsonic acid
Common name MAA (WSSA); methylarsonic acid (E-ISO draft proposal in lieu of common name); CMA (WSSA for calcium acid salt); MAMA (WSSA for monoammonium salt)
IUPAC name methylarsonic acid. Salts include calcium bis(hydrogen methylarsonate); ammonium hydrogen methylarsonate
Chemical Abstracts name methylarsonic acid
CAS RN *[124–58–3]*; (*[5902–95–4]* calcium hydrogen methylarsonate)

MSMA
Common name MSMA (WSSA)
IUPAC name sodium hydrogen methylarsonate
Chemical Abstracts name monosodium methylarsonate
CAS RN *[2163–80–6]*

DSMA
Common name DSMA (WSSA, JMAF)
IUPAC name disodium methylarsonate
Chemical Abstracts name disodium methylarsonate
CAS RN *[144–21–8]*

PHYSICAL CHEMISTRY
methylarsonic acid
Mol. wt. 140.0   **M.f.** CH_5AsO_3 **M.p.** 161 °C   **Solubility** Freely soluble in water; soluble in ethanol.   **pKa** Strong dibasic acid.

MSMA
Composition MSMA forms a sesquihydrate. Tech. material is 51% MSMA.
Mol. wt. 162.0   **M.f.** CH_4AsNaO_3 **Form** Colourless crystals (hydrate); clear light yellow liquid (tech.)   **M.p.** 113–116 °C (sesquihydrate)   **B.p.** 110±2 °C
V.p. 1×10^{-2} mPa (25 °C) K_{ow} logP <0   **S.g./density** 1.535 g/ml (25 °C)
Solubility In water at 20 °C, 1.4 kg/kg (anhydrous salt). Soluble in methanol; insoluble in most organic solvents.   **Stability** Stable to hydrolysis. Decomposed by strong oxidising and reducing agents.   **pKa** 9.02

DSMA
Composition DSMA forms a hexahydrate.   **Mol. wt.** 183.9   **M.f.** $CH_3AsNa_2O_3$

Form Colourless crystals (hexahydrate). **M.p.** 132–139 °C **Solubility** In water at 20 °C, 279 g/kg (anhydrous salt). Soluble in methanol, but practically insoluble in most organic solvents. **Stability** Loses water at elevated temperatures. Hydrolysed to MSMA at pH 6–7. Decomposed by strong oxidising and reducing agents.

COMMERCIALISATION

History MSMA and DSMA introduced by Ansul Chemical Co. (which no longer exists), by Diamond Shamrock Co. (now ISK Biosciences Corp.) and by Vineland Chemical Co. **Patent** US 2678265; US 2889347 to Ansul
Manufacturers Ancom; Crystal; Drexel; ISK Biosciences; Luxembourg; Sanachem; Sinon

APPLICATIONS

Mode of action These are selective, contact herbicides with some systemic properties.

methylarsonic acid
Formulation types SC.

MSMA

Uses Used to control grass weeds: in cotton at 2.24 kg/ha; in sugar cane at 3.3 kg/ha; as directed sprays under tree crops; on non-crop areas.
Formulation types SL. **Mixtures** *(MSMA +)* dimethylarsinic acid + sodium dimethylarsinate. **Selected tradenames** 'Bueno' (and others) (ISK Biosciences); 'Puedemas' (Productos OSA); 'Target MSMA' (Luxembourg)

DSMA

Uses Recommended for control of grass weeds in cotton at 2.5 kg/ha, citrus, turf and on uncropped land. **Formulation types** SP. **Selected tradenames** 'Ansar 8100' (ISK Biosciences); 'Dry DSMA' (Luxembourg)

ANALYSIS

Product analysis by acid-base titration; total arsenic determined by wet oxidation (E. A. Dietz & L. O. Moore, *Anal. Methods Pestic. Plant Growth Regul.*, 1978, **10**, 385). **Residues** determined by estimation of total arsenic, reduction to arsine and colorimetry *(idem, ibid.)*.

MAMMALIAN TOXICOLOGY
methylarsonic acid
Oral Acute oral LD_{50} for rats 2833 mg/kg. **Skin and eye** Acute percutaneous LD_{50} not available. Only a mild irritant to skin of rabbits. **Inhalation** LC_{50} (4 h) for rats 2.2 mg/l. **NOEL** (2 y) for male mice 200, female mice 50 ppm; (12 mo) for dogs 2 mg/kg daily. **Toxicity class** WHO (a.i.) III **EC risk** Not

specified, but arsenic compounds in general (with exceptions) are assigned T (R23/25); (preparations, 0.1%≤ concn. <0.2%, Xn (R20/22))

MSMA
Oral Acute oral LD_{50} for young albino rats 900 mg/kg. **Skin and eye** Acute percutaneous LD_{50} not available. Mild irritant to skin and eyes (rabbits). **Inhalation** LC_{50} (4 h) >20 mg/l. **Toxicity class** EPA (formulation) III

DSMA
Oral Acute oral LD_{50} for young albino rats 1800 mg/kg. **Skin and eye** Acute percutaneous LD_{50} for rabbits 10 000 mg/kg. Mild irritant to skin and eyes (rabbits). **Toxicity class** EPA (formulation) III

ECOTOXICOLOGY
methylarsonic acid
Birds LD_{50} for bobwhite quail 425 mg/kg. Dietary LC_{50} (8 d) for bobwhite quail 1667, mallard duck >2866 ppm. **Bees** LD_{50} (oral) 68 μg/bee; topical NOEL 36 μg/bee. **Daphnia** LC_{50} (48 h) 77.5 mg/l.

MSMA
Birds LD_{50} for quail 425.2 g/kg. Five-day dietary LC_{50} for quail 1667, mallard ducks >2866 ppm. **Fish** LC_{50} (96 h) for bluegill sunfish >51, trout >167 mg/l. NOEC for bluegill 93.2, trout 167 mg/l. **Bees** LD_{50} 68 μg/bee; NOEL 36 μg/bee. **Daphnia** LC_{50} (48 h) 77.5 mg/l. **Algae** (for 51% MSMA) EC_{50} (5 d) for *Selenastrum* 7.6 mg/l. **Other aquatic spp.** (for 51% MSMA) EC_{25} (14 d) for *Lemna gibba* 107.3 mg/l; EC_{50} (14 d) 145.9 mg/l.

ENVIRONMENTAL FATE
MSMA
Soil/Environment Terrestrial dissipation, DT_{50} 55 d, DT_{50} for cacodylic acid (primary metabolite) 88 d. MSMA and cacodylic acid are found primarily in upper 6 inches of soil. K_{oc} 250 (sandy soil) to 2850 (silty loam); K_d 0.5 (sandy soil) to 39.4 (sandy loam).

487   methyl bromide   *Soil sterilant, Fumigant*

CH₃Br

$$CH_3Br$$

NOMENCLATURE
Common name methyl bromide (BSI, E-ISO, ESA, JMAF, accepted in lieu of a
common name); bromure de méthyle (F-ISO)
IUPAC name bromomethane
Chemical Abstracts name bromomethane
Other names monobromomethane   **CAS RN** *[74–83–9]*   **EEC no.** 200–813–2

PHYSICAL CHEMISTRY
Mol. wt. 94.9   **M.f.** CH_3Br **Form** Colourless, odourless gas at room temperature.
Chloroform-like odour at high concentrations.   **M.p.** –93 °C   **B.p.** 3.6 °C
V.p. 190 kPa (20 °C)   **S.g./density** 1.732 (0 °C)   **Solubility** In water 17.5 g/l
(20 °C); it forms a crystalline hydrate with ice-water. Readily soluble in most
organic solvents, e.g. lower alcohols, ethers, esters, ketones, aromatic
hydrocarbons, halogenated hydrocarbons, and carbon disulfide.
Stability Hydrolysed very slowly in water, and more rapidly in alkaline media.
F.p. Non-flammable.   **Other properties** Specific heat 0.5 J/g (0 °C).

COMMERCIALISATION
History Insecticidal properties reported by Le Goupil (*Rev. Pathol. Veg. Entomol.
Agric. Fr.*, 1932, **19,** 169). Introduced by Dow Chemical Co. (now DowElanco,
who no longer manufacture or market it).   **Manufacturers** Elf Atochem; Great
Lakes

APPLICATIONS
Mode of action Fumigant insecticide and nematicide.   **Uses** Multi-purpose
fumigant used for insecticidal, acaricidal, and rodenticidal control in mills,
warehouses, grain elevators, ships, etc., and in stored products; soil fumigation for
control of insects, nematodes, soil-borne diseases, and weed seeds; and
glasshouse and mushroom-house fumigation.   **Phytotoxicity** Extremely phytotoxic.
Formulation types GA.   **Mixtures** *(methyl bromide +)* chloropicrin.
Selected tradenames 'Brom-O-Gas' (mixture) (Great Lakes); 'Haltox' (Degesch);
'Meth-O-Gas 100' (Great Lakes)

ANALYSIS
Methyl bromide may be detected by the halide lamp, a non-specific test given by
all volatile halides; by commercially available detector tubes (H. K. Heseltine, *Pest.
Tech.*, 1959, **1,** 253).   Mixtures with chloropicrin are analysed by i.r. spectroscopy.
Commercial thermal conductivity meters used for its estimation in air (H. K.
Heseltine *et al., Chem. Ind. (London)*, 1958, p. 1287); or glc (S. G. Heuser & K. A.
Scudamore, *J. Sci. Food Agric.*, 1969, **20,** 566; *idem, Analyst (London)*, 1968, **93,** 252;

1974, **104**, 425; *Anal. Methods Residues Pestic.*, 1988, Part I, M8; K. A. Scudamore, *Anal. Methods Pestic. Plant Growth Regul.*, 1988, **16**, 207, 241), also glc of a derivative (R. J. Fairall & K. A. Scudamore, *Analyst (London)*, 1980, **105**, 251). It reacts with soil or organic material to form bromide ion. Methods for **residues** in food reviewed, with details of method (J. L. Daft in *Comp. Anal. Profiles*, p. 272); residues of methyl bromide and bromide ion may be distinguished (S. G. Heuser & K. A. Scudamore, *Pestic. Sci.*, 1970, **1**, 244).

MAMMALIAN TOXICOLOGY
Reviews *Pesticide residues in food*. FAO Agricultural Studies, No. 73; WHO Technical Report Series, No. 370, 1967. *Evaluation of some pesticide residues in food*. FAO/PL:CP/15; WHO/Food Add./67.32, 1967. **IARC** 41, 45 **EHC** 166 (WHO, 1995). **Skin and eye** Acute percutaneous LD_{50} not available. Liquid can cause eye and skin burns. **Inhalation** LC_{50} (4 h) for rats 3.03 mg/l air (N. Kato et al., *Ind. Health*, **24**, 87–103 (1986)). Highly toxic to man, with a threshold limit value of 0.019 mg/l air (ACGIH). **ADI** (JMPR) 1.0 mg/kg b.w. as bromide ion [1966]. **Other** In many countries, its use is restricted to trained personnel. **Toxicity class** EPA (formulation) II **EC risk** T (R23); Xi (R36/37/38)

ECOTOXICOLOGY
Birds Acute oral LD_{50} for bobwhite quail 73 mg/kg. **Fish** LC_{50} (96 h) 3.9 mg/l. **Bees** Not hazardous to bees when used as directed. **Daphnia** EC_{50} (48 h) 2.6 mg/l.

ENVIRONMENTAL FATE
Animals Metabolism not totally elucidated; inorganic bromide ion is formed.
Plants Metabolism not totally elucidated; inorganic bromide ion is formed.
Soil/Environment Residues of bromide ion and methyl bromide in soils fumigated with the latter and their effects were reviewed by G. A. Maw & R. J. Kempton (*Soils Fert.*, 1973, **36**, 41).

188 methyldymron   *Herbicide*

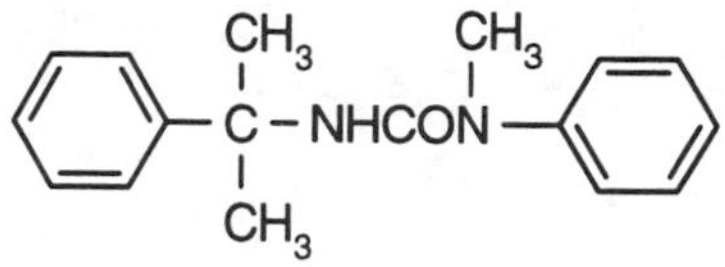

NOMENCLATURE
Common name methyldymron (JMAF)

IUPAC name 1-methyl-3-(1-methyl-1-phenylethyl)-1-phenylurea;
3-(α,α-dimethylbenzyl)-1-methyl-1-phenylurea
Chemical Abstracts name N-methyl-N'-(1-methyl-1-phenylethyl)-N-phenylurea
CAS RN *[42609–73–4]* **Development codes** K-1441; SK-41 (SDS Biotech)

PHYSICAL CHEMISTRY
Mol. wt. 268.4 **M.f.** $C_{17}H_{20}N_2O$ **Form** Colourless, odourless crystals.
M.p. 72 °C **K_{ow}** logP = 3.01 **S.g./density** 1.1–1.2 **Solubility** In water 120 mg/l
(20 °C). In acetone 913, hexane 8.2, methanol 637 (all in g/l at 20 °C).
Stability Stable to heat and light.

COMMERCIALISATION
History Herbicide reported by T. Takematsu *et al.*, *Proc. 5th Asian-Pacific Weed Sci.
Soc. Conf.*, 1975 p. 121. Introduced in Japan (1978) by Showa Denko (now SDS
Biotech.). **Patent** JP 7242493 **Manufacturers** SDS Biotech

APPLICATIONS
Biochemistry Cell division inhibitor. **Mode of action** Selective herbicide, absorbed
principally by the roots, with translocation to the meristem. **Uses** Pre- and early
post-emergence control of cyperaceous and some annual grass weeds in turf at
2–10 kg/ha. **Formulation types** WP. **Mixtures** *(methyldymron +)* 2,4-D.
Selected tradenames 'Stacker' (SDS Biotech)

ANALYSIS
Product and **residue** analysis by glc and hplc; details available from SDS Biotech.

MAMMALIAN TOXICOLOGY
Oral Acute oral LD_{50} for male mice 5000, female mice 5269, male rats 5852,
female rats 3948 mg/kg. **Skin and eye** Acute percutaneous LD_{50} for rats
>2000 mg/kg. **Inhalation** LC_{50} >4.85 mg/l. **NOEL** (90 d) for male rats 26.4,
female rats 25.6 mg/kg diet. **Toxicity class** WHO (a.i.) III (Table 5)

ECOTOXICOLOGY
Fish LC_{50} (48 h) for carp 14 mg/l. **Daphnia** LC_{50} (3 h) for *D. pulex* >40 ppm.

ENVIRONMENTAL FATE
Plants Rapidly metabolised in plants. **Soil/Environment** In soil DT_{50} 9–19 d.

methyl isothiocyanate

Nematicide, fungicide, insecticide, herbicide

CH_3NCS

NOMENCLATURE
Common name methyl isothiocyanate (BSI, E-ISO, JMAF, accepted in lieu of a common name); isothiocyanate de méthyle (F-ISO, accepted in lieu of a common name)
IUPAC name methyl isothiocyanate
Chemical Abstracts name isothiocyanatomethane
Other names MIT; MITC   **CAS RN** *[556–61–6]*   **EEC no.** 209–132–5

PHYSICAL CHEMISTRY
Composition Tech. is ≥94.5% pure.   **Mol. wt.** 73.1   **M.f.** C_2H_3NS
Form Colourless crystals with a horseradish-like odour.   **M.p.** 35–36 °C; (tech. 25.3–27.6 °C)   **B.p.** 118–119 °C   **V.p.** 2.13 kPa (25 °C) K_{ow} logP = 1.37 (calc.)
S.g./density 1.069 (37 °C); tech., 1.0537 (40 °C)   **Solubility** In water 8.2 g/l (20 °C). Readily soluble in common organic solvents, such as ethanol, methanol, acetone, cyclohexanone, dichloromethane, chloroform, carbon tetrachloride, benzene, xylene, petroleum ether, and mineral oils.   **Stability** Unstable and reactive. Rapidly hydrolysed by alkalis, more slowly in acidic and neutral solutions, DT_{50} 85 h (pH 5), 490 h (pH 7), 110 h (pH 9) (25 °C). Sensitive to oxygen and to light. Stable up to 200 °C.   **pKa** 12.3   **F.p.** 26.9 °C

COMMERCIALISATION
History Nematicidal activity reported by E. A. Pieroh *et al.* (*Anz. Schaedlingsskd.*, 1959, **32**, 183). Introduced by Schering AG (now AgrEvo GmbH).
Patent US 3113908   **Manufacturers** AgrEvo

APPLICATIONS
Uses Multi-purpose soil fumigant for control of nematodes, soil fungi, soil insects, and weed seeds.   **Phytotoxicity** Phytotoxic to all green plants.
Formulation types EC.   **Mixtures** *(methyl isothiocyanate +)* 1,3-dichloropropene; chloropicrin + 1,3-dichloropropene.   **Selected tradenames** 'Trapex' (AgrEvo)

ANALYSIS
Product analysis by glc or by hplc.   **Residues** determined by glc with FPD (M. Ottnad *et al.*, *Anal. Methods Pestic. Plant Growth Regul.*, 1978, **10**, 563). Details available from AgrEvo.

MAMMALIAN TOXICOLOGY
Oral Acute oral LD_{50} for rats 72–220, mice 90–104 mg/kg.   **Skin and eye** Acute percutaneous LD_{50} for rats 2780, male mice 1870, rabbits 263 mg/kg. Strongly irritating to eyes and skin (rabbits).   **Inhalation** LC_{50} (1 h) for rats 1.9 mg/l air. **NOEL** (2 y) for rats 10 mg/l drinking water (0.37–0.56 mg/kg b.w. daily); (2 y) for mice 20 mg/l drinking water (3.48 mg/kg b.w. daily); (1 y) for dogs 0.4 mg/kg b.w. daily (by gavage).   **Other** Not carcinogenic.   **Toxicity class** WHO (a.i.) II **EC risk** T (R23/25); C (R34); (R43)

ECOTOXICOLOGY
Birds Acute oral LD_{50} for mallard ducks 136 mg/kg. Five-day dietary LC_{50} for mallard ducks 10 936, pheasants >5000 mg/kg diet.   **Fish** LC_{50} (96 h) for bluegill sunfish 0.14, rainbow trout 0.09, mirror carp 0.37–0.57 mg/l.   **Bees** Not hazardous to bees when used as directed.   **Daphnia** LC_{50} (48 h) 0.055mg/l. **Algae** EC_{50} (96 h) 0.248 mg/l; NOEC (96 h) 0.125 mg/l.

ENVIRONMENTAL FATE
Soil/Environment In damp soil, degradation and evaporation of the bulk of the substance occurs within 3 weeks at 18–20 °C soil temperature, 4 weeks at 6–12 °C, and 8 weeks at 0–6 °C. At the lower temperatures, leaching is more significant and is dependent on soil moisture. The potential for groundwater contamination is small due to low leaching and fast degradation.

490   metiram   *Fungicide*

alkylenebis(dithiocarbamate)

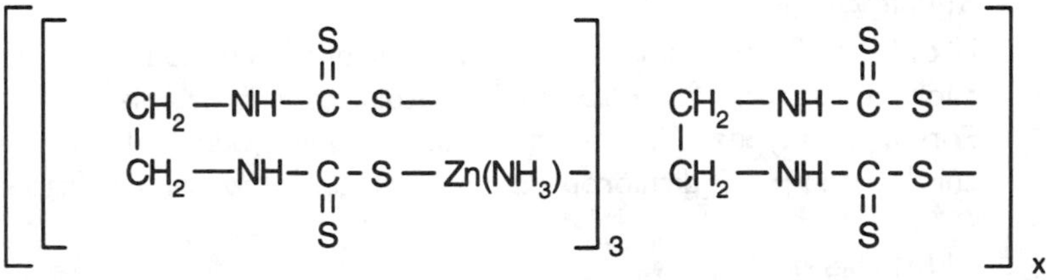

NOMENCLATURE
Common name metiram (New Zealand, JMAF); métirame zinc (France)
IUPAC name zinc ammoniate ethylenebis(dithiocarbamate) - poly(ethylenethiuram disulfide)

Chemical Abstracts name metiram (composition not specified)
CAS RN *[9006–42–2]* **Development codes** FMC 9102 (FMC); BAS 222 F (BASF)

PHYSICAL CHEMISTRY
Composition No common name was accepted by ISO (and metiram was withdrawn by BSI) because the product appears to be a mixture rather than a complex. The dry form of the tech. is 'Metiram TC 85%'. **Mol. wt.** $(1088.7)_x$ **M.f.** $[C_{16}H_{33}N_{11}S_{16}Zn_3]_x$ **Form** Yellow powder (tech.). **M.p.** Decomposes at c. 156 °C **V.p.** <0.010 mPa (20 °C) **K$_{ow}$** logP = 0.3 (pH 7) **Henry** <5.4 × 10^{-3} Pa m^3 mol^{-1} (calc.) **S.g./density** 1.860 (20 °C) **Solubility** Practically insoluble in water. Soluble in pyridine (with decomposition). Practically insoluble in organic solvents (e.g. ethanol, acetone, benzene). **Stability** Stable at 30 °C. Slowly decomposed by light. Non-hygroscopic. Decomposed by strong acids and strong alkalis.

COMMERCIALISATION
History Fungicide introduced in Germany (1958) by BASF AG.
Patent GB 840211; US 3248400; DE 1085709 **Manufacturers** BASF

APPLICATIONS
Biochemistry Inactivates fungal proteins by binding to essential thiol groups.
Mode of action Non-systemic foliar fungicide with protective action.
Uses Control of a wide range of diseases on many crops, e.g. scab on pome fruit; shot-hole on stone fruit; rust on currants and plums; downy mildew, red fire disease, and black rot on vines; *Phytophthora* late blight on potatoes and tomatoes; *Septoria* leaf spot on celery and celeriac; downy mildew on head lettuce; rust on asparagus; anthracnose on beans; downy mildews on hops and tobacco; downy mildews and rusts on ornamentals; diseases of cotton, peanuts, pecans, etc. Application rates in the range 1.5–4.0 kg/ha. Used in forestry to combat needle cast in conifers. Also used as a seed treatment for control of emergence diseases of vegetables and ornamentals. **Formulation types** WG. **Mixtures** *(metiram +)* cymoxanil; nitrothal-isopropyl. **Compatibility** Incompatible with highly alkaline materials. **Selected tradenames** 'Polyram' (BASF)

ANALYSIS
Product analysis by acid hydrolysis to produce carbon disulfide which is estimated by titration. **Residues** determined by hydrolysis to carbon disulfide which is measured by glc or by colorimetry of a derivative (*Analyst (London)*, 1981, **106**, 782; G. E. Keppel, *J. Assoc. Off. Anal. Chem.*, 1971, **54**, 528).

MAMMALIAN TOXICOLOGY
Reviews *Pesticide residues in food – 1993*, FAO Plant Production and Protection Paper, 122, 1993. *Pesticide residues in food – 1993 evaluations, Part II – Toxicology*. WHO, WHO/PCS/94.4, 1994. **EHC** 78 (WHO, 1988; general review of

dithiocarbamates).   **Oral** Acute oral LD_{50} for rats >10 000, mice >5400, guinea pigs 2400–4800 mg/kg.   **Skin and eye** Acute percutaneous LD_{50} for rats >2000 mg/kg. Mild skin and eye irritant.   **Inhalation** LC_{50} (4 h) for rats >5.7 mg/l air.   **NOEL** (2 y) for rats 3.1 mg/kg.   **ADI** (JMPR) 0.03 mg/kg b.w. [1993].   **Toxicity class** WHO (a.i.) III (Table 5); EPA (formulation) IV

ECOTOXICOLOGY
Birds Slightly toxic to birds.   **Fish** LC_{50} (96 h) for carp 85, rainbow trout 1.1 mg/l. LC_{50} (48 h) for harlequin fish 17 mg/l.   **Bees** Oral LD_{50} >40 μg/bee. Contact LD_{50} >16 μg/bee.   **Worms** LC_{50} (14 d) >1000 ppm.   **Daphnia** EC_{50} (48 h) 2.55 mg/l.   **Algae** EC_{50} (96 h) for *Chlorella* 0.3 mg/l.

ENVIRONMENTAL FATE
Soil/Environment As with other dithiocarbamate pesticides, metiram is degraded to derivatives of thiourea, thiuram monosulfide, thiuram disulfide, and sulfur.

491 metobenzuron   *Herbicide*

urea

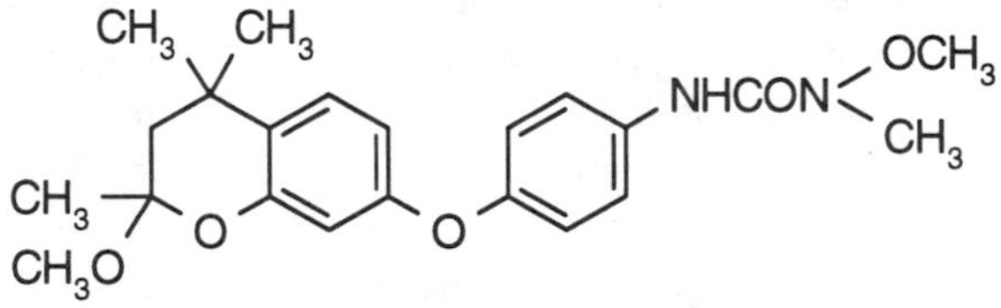

NOMENCLATURE
Common name metobenzuron (BSI, pa E-ISO)
IUPAC name (±)-1-methoxy-3-[4-(2-methoxy-2,4,4-trimethylchroman-7-yloxy)= phenyl]-1-methylurea
Chemical Abstracts name (±)-N'-[4-[(3,4-dihydro-2-methoxy-2,4,4-trimethyl-= 2H-1-benzopyran-7-yl)oxy]phenyl]-N-methoxy-N-methylurea
CAS RN *[111578–32–6]*   **Development codes** UMP-488 (Mitsui Petrochemical)

PHYSICAL CHEMISTRY
Mol. wt. 400.5   **M.f.** $C_{22}H_{28}N_2O_5$ **Form** White powder.   **M.p.** 101.0–102.5 °C   **V.p.** 2.47 mPa (22.3 °C)   **Solubility** In water 0.4 mg/l (25 °C).

COMMERCIALISATION
History Reported by T. Morimoto *et al.* (*Proc. Br. Crop Prot. Conf., Weeds*, 1993, **1**, 67). Introduced by Mitsui Petrochemical Industries Ltd.   **Manufacturers** Mitsui Petrochemical

APPLICATIONS

Biochemistry Inhibitor of photosynthesis. **Uses** Selective post-emergence herbicide for control of a wide range of broad-leaved weeds, including triazine-resistant weeds, in maize. Application rates 125–250 g/ha.
Compatibility Combination with 2,4-D increases efficacy and widens the weed spectrum.

MAMMALIAN TOXICOLOGY

Oral Acute oral LD_{50} for rats and mice >10 000 mg/kg. **Skin and eye** Acute percutaneous LD_{50} for rabbits >2000 mg/kg. Non-irritating to skin and slightly irritating to eyes (rabbits). **Other** Non-mutagenic in the Ames Test.

ENVIRONMENTAL FATE

Soil/Environment DT_{50} in soil 4 h (under irradiation), 8 h (in the dark). Low mobility in sandy loam soil.

492 metobromuron *Herbicide*

urea

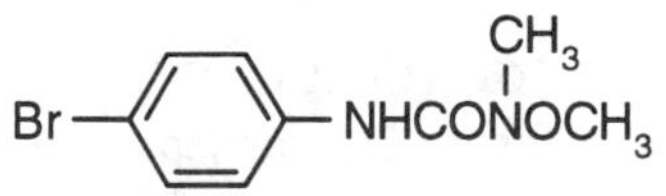

NOMENCLATURE

Common name metobromuron (BSI, E-ISO, (*m*) F-ISO, ANSI, WSSA)
IUPAC name 3-(4-bromophenyl)-1-methoxy-1-methylurea
Chemical Abstracts name *N'*-(4-bromophenyl)-*N*-methoxy-*N*-methylurea
CAS RN *[3060–89–7]* **EEC no.** 221–301–5 **Development codes** C 3126 (Ciba)

PHYSICAL CHEMISTRY

Mol. wt. 259.1 **M.f.** $C_9H_{11}BrN_2O_2$ **Form** Colourless crystals. **M.p.** 95.5–96 °C
V.p. 0.40 mPa (20 °C) **K_{ow}** logP = 2.41 **Henry** 3.1×10^{-4} Pa m^3 mol^{-1} (calc.)
S.g./density 1.60 (20 °C) **Solubility** In water 330 mg/l (20 °C). In acetone 500, dichloromethane 550, methanol 240, toluene 100, octanol 70, chloroform 62.5, hexane 2.6 (all in g/l, 20 °C). **Stability** Very stable in neutral, dilute acid and dilute alkaline media. Hydrolysed by strong acids and alkalis. DT_{50} (20 °C) 150 d (pH 1), >200 d (pH 9), 83 d (pH 13).

COMMERCIALISATION

History Herbicide reported by J. Schuler & L. Ebner (*Proc. Br. Weed Control Conf.,*

7th, 1964, p. 450). Introduced by Ciba AG (now Novartis Crop Protection AG). **Patent** BE 662268; GB 965313 **Manufacturers** Novartis

APPLICATIONS
Biochemistry Photosynthetic electron transport inhibitor. **Mode of action** Selective, systemic herbicide, absorbed by the roots and leaves. **Uses** Pre-emergence control of annual broad-leaved weeds and grasses in beans, potatoes, tomatoes, tobacco, soya beans, maize, cornsalad, artichokes, and sugar beet at 1.5–2.5 kg/ha. **Phytotoxicity** Some bean and tobacco varieties may be injured. **Formulation types** WP; SC. **Mixtures** (metobromuron +) metolachlor; terbutryn. **Selected tradenames** 'Patoran' (Novartis); 'Pattonex' (Makhteshim-Agan)

ANALYSIS
Product analysis by glc. **Residues** determined by alkaline hydrolysis to 4-bromoaniline, derivatives of which are measured by colorimetry or by glc with ECD (G. Voss et al., Anal. Methods Pestic. Plant Growth Regul., 1973, **7**, 569).

MAMMALIAN TOXICOLOGY
Oral Acute oral LD_{50} for rats 2603 mg/kg. **Skin and eye** Acute percutaneous LD_{50} for rats >3000, rabbits >10 200 mg/kg. Not a skin or eye irritant (rabbits). **Inhalation** LC_{50} (4 h) for rats >1.1 mg/l air. **NOEL** (2 y) for rats 250 mg/kg diet (17 mg/kg daily), for dogs 100 mg/kg diet (3 mg/kg daily). **ADI** 0.008 mg/kg b.w. **Toxicity class** WHO (a.i.) III (Table 5); EPA (formulation) III

ECOTOXICOLOGY
Birds LD_{50} for Japanese quail 565, Pekin ducks 6643 mg/kg b.w. LC_{50} (7 d) for Japanese quail >10 000, mallard ducks >24 300, bobwhite quail 18 100 mg/kg diet. **Fish** LC_{50} (96 h) for rainbow trout 36, bluegill sunfish 40, crucian carp 40 mg/l. **Bees** Not toxic to bees. LC_{50} (oral) >325, contact >130 µg/bee. **Worms** LC_{50} (14 d) for earthworms 467 mg/kg (practically non-toxic). **Daphnia** LC_{50} (48 h) 44 mg/l. **Algae** EC_{50} (5 d) 0.26 mg/l.

ENVIRONMENTAL FATE
Animals Metabolism proceeds mainly via N-demethylation, N-dealkoxylation, phenyl ring hydroxylation, cleavage of the ureido group and subsequent conjugation with sulfuric acid. **Plants** In plants, the methyl and methoxy groups are split off, and the urea moiety is further degraded to 4-bromoaniline. **Soil/Environment** DT_{50} in soil 30 d; strongly bound to soil (K_{oc} 184 ml/g), indicating low leaching potential. In water, it is stable to abiotic degradation, but in biologically active systems DT_{50} 2–20 w.

chloroacetanilide

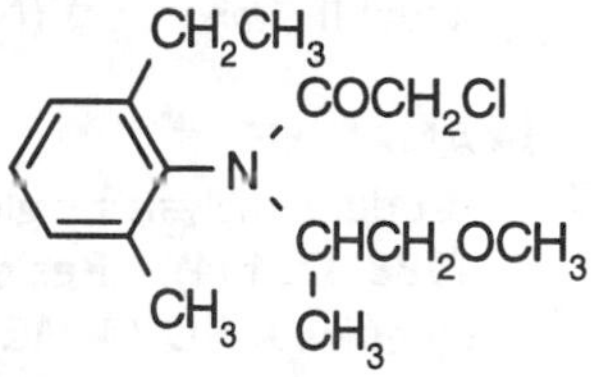

NOMENCLATURE
Common name metolachlor (BSI, E-ISO, ANSI, WSSA); métolachlore ((*m*) F-ISO)
IUPAC name 2-chloro-6'-ethyl-*N*-(2-methoxy-1-methylethyl)aceto-*o*-toluidide
Chemical Abstracts name 2-chloro-*N*-(2-ethyl-6-methylphenyl)-*N*-(2-methoxy-=
1-methylethyl)acetamide
CAS RN *[51218–45–2]* **EEC no.** 257–060–8 **Development codes** CGA 24 705

PHYSICAL CHEMISTRY
Composition Racemic mixture of 1*S*- and 1*R*- isomers. **Mol. wt.** 283.8
M.f. $C_{15}H_{22}ClNO_2$ **Form** Colourless to light tan liquid. **M.p.** −62.1 °C
B.p. 100 °C/0.001 mmHg **V.p.** 4.2 mPa (25 °C) (OECD 104) K_{ow} logP = 2.9
(25 °C) **Henry** 2.4×10^{-3} Pa m^3 mol^{-1} (calc.) **S.g./density** 1.12 (20 °C)
Solubility In water 488 mg/l (25 °C, OECD 105). Miscible with benzene, toluene,
ethanol, acetone, xylene, hexane, dimethylformamide, ethylene dichloride,
cyclohexanone, methanol, octanol, and dichloromethane. Insoluble in ethylene
glycol, propylene glycol, and petroleum ether. **Stability** Stable up to *c.* 275 °C.
Hydrolysed by strong alkalis and strong mineral acids. On hydrolysis in buffer
(20 °C) DT$_{50}$ (calc.) >200 d (2≤ pH ≤10). **F.p.** 190 °C (1013 mbar)

COMMERCIALISATION
History Herbicide reported by H. R. Gerber *et al.* (*Proc. Br. Weed Control Conf.,*
12th, 1974, **2**, 787). Introduced by Ciba-Geigy AG (now Novartis Crop Protection
AG). **Patent** BE 800471; GB 1438311; GB 1438312 **Manufacturers** Novartis

APPLICATIONS
Biochemistry Cell division inhibitor. Maize tolerance is attributed to rapid
detoxification by glutathione transferases. **Mode of action** Selective herbicide,
absorbed predominantly by the hypocotyls and shoots. Inhibits germination.
Uses Control of annual grasses and some broad-leaved weeds in maize, sorghum,
cotton, sugar beet, fodder beet, sugar cane, potatoes, peanuts, soya beans,
safflowers, sunflowers, various vegetables, fruit and nut trees, and woody
ornamentals. Applied pre-emergence, pre-plant-incorporated or early post-
emergence at 1.0–2.5 kg a.i./ha. Often used in combination with broad-leaved
herbicides, to extend the spectrum of activity. **Phytotoxicity** Well tolerated by

most broad-leaved crops, maize, sorghum (safened with fluxofenim or oxabetrinil). **Formulation types** EC; GR; FW; SC. **Mixtures** *(metolachlor +)* ametryn; atrazine; chloridazon; metobromuron; prometryn; fluometuron; terbuthylazine; metribuzin; pendimethalin; flumetsulam. **Selected tradenames** 'Dual' (Novartis); 'Dual S', 'Dual II' (mixtures) (Novartis)

ANALYSIS
Product analysis by glc with FID (*AOAC Methods*, 1995, 985.06; *CIPAC Handbook*, 1988, **D,** 134). **Residues** determined by glc with TID or MCD. In **drinking water**, by gc with FID (*AOAC Methods*, 1995, 991.07). Details available from Novartis.

MAMMALIAN TOXICOLOGY
Reviews IPCS: The WHO recommended classification of pesticides by hazard and guidelines to classification 1994–95. **Oral** Acute oral LD_{50} for rats 2780 mg/kg. **Skin and eye** Acute percutaneous LD_{50} for rats >3170 mg/kg. Mild skin and eye irritant (rabbits). May cause skin sensitisation (guinea pigs). **Inhalation** LC_{50} (4 h) for rats >1.75 mg/l air. **NOEL** (90 d) for rats 300 mg/kg diet (*c.* 15 mg/kg daily), for mice 100 mg/kg diet (*c.* 100 mg/kg daily), for dogs 300 mg/kg diet (*c.* 9.7 mg/kg daily). **ADI** 0.1 mg/kg b.w. **Toxicity class** WHO (a.i.) III; EPA (formulation) III **EC risk** R43

ECOTOXICOLOGY
Birds Acute oral LD_{50} for mallard ducks and bobwhite quail >2150 mg/kg. Eight-day dietary LC_{50} for bobwhite quail and mallard ducks >10 000 mg/kg. **Fish** LC_{50} (96 h) for rainbow trout 3.9, carp 4.9, bluegill sunfish 10 mg/l. **Bees** LD_{50} (oral and contact) >110 µg/bee. **Worms** LC_{50} (14 d) for earthworms 140 mg/kg soil. **Daphnia** LC_{50} (48 h) 25 mg/l. **Algae** EC_{50} for *Scenedesmus subspicatus* 0.1 mg/l.

ENVIRONMENTAL FATE
Animals Rapidly oxidised by rat liver microsomal oxygenases via dechlorination, *O*-demethylation and side-chain oxidation (*J. Agric. Food Chem.*, 1989, **37**, 1088). **Plants** In plants, metabolism involves natural product conjugation of the chloroacetyl group, and hydrolysis and sugar conjugation at the ether group. Final metabolites are polar, water-soluble, and non-volatile. **Soil/Environment** DT_{50} in soil *c.* 20 d (field). K_{oc} 121–309.

carbamate

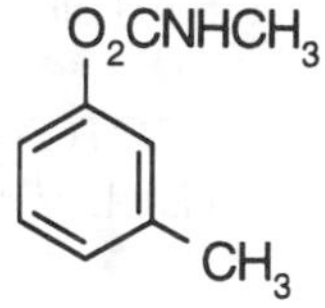

NOMENCLATURE
Common name metolcarb (BSI, E-ISO); métholcarb ((*m*) F-ISO); MTMC (JMAF)
IUPAC name *m*-tolyl methylcarbamate
Chemical Abstracts name 3-methylphenyl methylcarbamate
CAS RN *[1129–41–5]* **EEC no.** 214–446–0 **Development codes** C-3 (Nihon Nohyaku)

PHYSICAL CHEMISTRY
Mol. wt. 165.2 **M.f.** $C_9H_{11}NO_2$ **Form** Tech. grade is a colourless solid.
M.p. 76–77 °C **V.p.** 145 mPa (20 °C) **Solubility** In water 2.6 g/l (30 °C). Soluble in polar organic solvents e.g. in cyclohexanone 790, xylene 100 (both in g/kg, 30 °C), methanol 880 g/kg (room temperature). Sparingly soluble in non-polar solvents.

COMMERCIALISATION
History Insecticide introduced by Nihon Nohyaku Co., Ltd (who no longer manufacture or market it) and by Sumitomo Chemical Co., Ltd.
Manufacturers Chunhu; Shenzhen Jiangshan

APPLICATIONS
Biochemistry Cholinesterase inhibitor. **Mode of action** Systemic insecticide with contact and respiratory action. **Uses** Control of planthoppers, leafhoppers, and other sucking insects on rice. Also used for control of citrus mealybugs, onion thrips, Mediterranean fruit flies, pink bollworms and cotton aphids.
Formulation types DP; EC; MG; WP. **Mixtures** *(metolcarb +)* propaphos.
Compatibility Incompatible with alkaline materials.
Selected tradenames 'Metacrate' (Sumitomo); 'Ofunack M' (mixture) (Mitsui Toatsu)

ANALYSIS
Product analysis by glc or by hplc (S. Sakaue *et al.*, *Nippon Nogei Kagaku Kaishi*, 1981, **55**, 1237). **Residues** determined by glc of a derivative with ECD (J. Miyamoto *et al.*, *Nihon Noyaku Gakkaishi*, 1978, **3**, 119).

MAIN ENTRIES

MAMMALIAN TOXICOLOGY
EHC 64 (WHO, 1986; a review of carbamate insecticides in general). **Oral** Acute oral LD_{50} for male rats 580, female rats 498, mice 109 mg/kg. **Skin and eye** Acute percutaneous LD_{50} for rats >2000 mg/kg. **Inhalation** LC_{50} for rats 0.475 mg/l air. **Toxicity class** WHO (a.i.) II **EC risk** Xn (R22)

ECOTOXICOLOGY
Fish Low toxicity to fish.

ENVIRONMENTAL FATE
Animals Metabolism of carbamate insecticides is reviewed (M. Cool & C. K. Jankowski in "*Insecticides*").

495 metosulam   *Herbicide*

triazolopyrimidine sulfonanilide

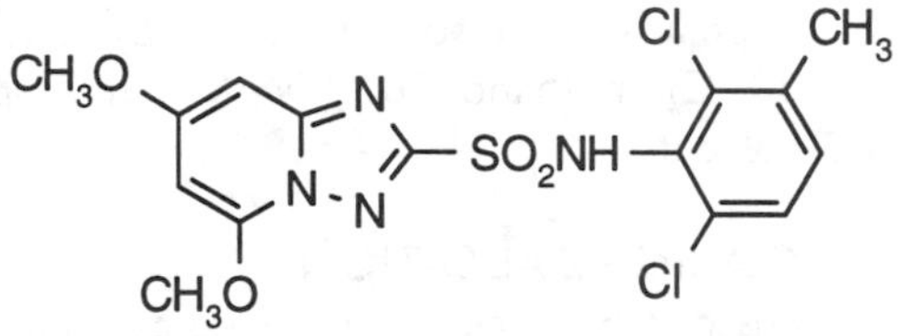

NOMENCLATURE
Common name metosulam (BSI, pa E-ISO, ANSI); métosulame ((m) pa F-ISO)
IUPAC name 2',6'-dichloro-5,7-dimethoxy-3'-methyl[1,2,4]triazolo[1,5-*a*]= pyrimidine-2-sulfonanilide
Chemical Abstracts name *N*-(2,6-dichloro-3-methylphenyl)-5,7-dimethoxy[1,2,4]= triazolo[1,5-*a*]pyrimidine-2-sulfonamide
CAS RN *[139528–85–1]* **Development codes** XDE 511; DE 511; XRD 511

PHYSICAL CHEMISTRY
Mol. wt. 418.3 **M.f.** $C_{14}H_{13}Cl_2N_5O_4S$ **Form** Cream to tan coloured powder.
M.p. 210–211.5 °C **V.p.** 4×10^{-10} mPa (20 °C, Knudsen effusion)
K_{ow} logP = 0.9778 (distilled water), 2.12 (pH 5), 2.46 (pH 7), 3.08 (pH 9) (all at 20 °C) **S.g./density** 1.49 (20 °C) **Solubility** In water 200 (distilled water, pH 7.5), 100 (pH 5.0), 700 (pH 7.0), 5600 (pH 9.0) (all in mg/l, 20 °C). In acetone, acetonitrile, dichloromethane >5.0 g/l; in *n*-octanol, hexane and toluene ≤0.2 g/l. **Stability** Under normal storage conditions, decomposes above m.p. Little photolytic instability (DT_{50} 140 d – Xenon Arc); stable to hydrolysis across environmentally normal range. **pKa** 4.8

COMMERCIALISATION

History Herbicide reported by M. Snel *et al., Med. Fac. Landbouww. Univ. Gent,* 1993, 58/32, p. 845. 'Sansac' (mixture with 2,4-D EHE) launched in Turkey in 1994. **Manufacturers** DowElanco

APPLICATIONS

Biochemistry Acetolactate synthase (ALS) inhibitor. Selectivity in wheat is due to rapid metabolic deactivation. **Mode of action** Herbicide readily taken up by roots and foliage. **Uses** Post-emergence control of many important broad-leaved weeds (including *Galium aparine, Stellaria media,* and all members of the Brassicaceae) in wheat, barley and rye; and pre- or post-emergence control of many important broad-leaved weeds In maize (including *Chenopodium* spp., *Amaranthus retroflexus, Solanum nigrum* and *Polygonum persicaria*).
Phytotoxicity Non-phytotoxic to crops for which use is recommended.
Formulation types SC; SE; WG. **Mixtures** *(metosulam +)* fluroxypyr-meptyl; 2,4-D EHE. **Selected tradenames** 'Pronto' (France) (mixture) (DowElanco); 'Sansac' (Turkey) (DowElanco); 'Sinal' (Middle East/Africa) (DowElanco)

ANALYSIS

An hplc procedure based on extractive methylation is available on request from DowElanco.

MAMMALIAN TOXICOLOGY

Oral Acute oral LD_{50} for rats and mice >5000 mg/kg. **Skin and eye** Acute percutaneous LD_{50} for rabbits >2000 mg/kg. Not a skin sensitiser (guinea pigs). **Inhalation** LC_{50} (4 h) for rats >1.9 mg/l. **NOEL** (2 y) for rats 5 mg/kg b.w. daily; (18 mo) for mice 1000 mg/kg b.w. daily. **ADI** 0.02 mg/kg b.w.
Toxicity class WHO (a.i.) III (Table 5)

ECOTOXICOLOGY

Birds Acute oral LD_{50} for mallard ducks and bobwhite quail >2000 mg/kg.
Fish LC_{50} (96 h) for rainbow trout, bluegill sunfish and fathead minnows >solubility limit of a.i. **Bees** Not toxic to bees; LD_{50} (48 h) (oral) >50, (contact) >100 μg/bee. **Worms** LC_{50} (14 d) for earthworms >1000 ppm. **Daphnia** LC_{50} (48 h) >solubility limit of a.i. **Algae** LC_{50} (72 h) for green algae 75 μg/l.

ENVIRONMENTAL FATE

Animals Metosulam is rapidly absorbed following oral administration (DT_{50} <1 h), extensively metabolised in rodents, much less in dogs, and excreted with metabolites 3-hydroxy (aliphatic oxidation) and 5-hydroxy (O-demethylation) in urine (DT_{50} 54–60 h in rodents, 73 h in dogs). *In vitro* percutaneous absorption is very low in humans and rats (<1% of applied in 24 h). **Plants** Metosulam is poorly absorbed from foliar application to wheat (<5% of applied), so there is little residue accumulation. Metabolised by hydroxylation of the ring methyl, to give a 3-hydroxymethyl- metabolite and its glycoside; these are the only major products found besides the parent molecule. **Soil/Environment** Laboratory aerobic degradation DT_{50} averaged 6 d (4 soils) at 20 °C, and 40% moisture holding capacity. Field DT_{50} in the 0–10 cm horizon has a mean value of 25 d. Metosulam degrades via the 5- and 7- hydroxy analogues to 5-amino-N-= (2,6-dichloro-3-methylphenyl)-1H-1,2,4-triazole-3-sulfonamide and CO_2. Mean K_{oc} (9 soils) <500. Spring treated lysimeters recorded no component >0.1 µg/l in percolate following two successive annual treatments at c. 25 g a.i./ha.

496 metoxuron   *Herbicide*

urea

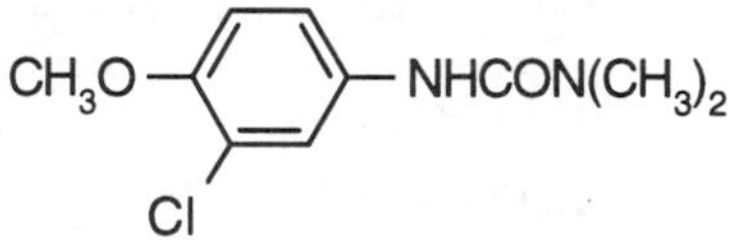

NOMENCLATURE
Common name metoxuron (BSI, E-ISO, (*m*) F-ISO)
IUPAC name 3-(3-chloro-4-methoxyphenyl)-1,1-dimethylurea
Chemical Abstracts name N'-(3-chloro-4-methoxyphenyl)-N,N-dimethylurea
CAS RN *[19937–59–8]* **Development codes** SAN 6915H; SAN 7102H

PHYSICAL CHEMISTRY
Mol. wt. 228.7 **M.f.** $C_{10}H_{13}ClN_2O_2$ **Form** Colourless crystals.
M.p. 126–127 °C **V.p.** 4.3 mPa (20 °C) **K_{ow}** logP = 1.60±0.04 (23 °C)
S.g./density Bulk density 0.8 (20 °C) **Solubility** In water 678 mg/l (24 °C).
Soluble in acetone, cyclohexanone, acetonitrile and hot ethanol. Moderately soluble in diethyl ether, benzene, toluene and cold ethanol. Practically insoluble in petroleum ether. **Stability** Stable during storage under normal conditions; stable 4 w at 54 °C. Hydrolysed by strong acids and alkalis; DT_{50} (50 °C) 18 d (pH 3), 21 d (pH 5), 24 d (pH 7), >30 d (pH 9), 26 d (pH 11). In solution, decomposed by u.v. light.

COMMERCIALISATION
History Herbicide reported by W. Berg (*Z. Pflanzenkr. Pflanzenpathol. Pflanzenschutz.*, 1968, **75**, 233). Introduced by Sandoz AG (now Novartis Crop Protection AG). **Patent** GB 1165160; FR 1497868

APPLICATIONS
Biochemistry Photosynthetic electron transport inhibitor.

Mode of action Selective herbicide, absorbed by the leaves and roots, with translocation. **Uses** Pre- and post-emergence control of some grasses (*Agrostis* spp., *Alopecurus myosuroides*, *Avena fatua* and *ludoviciana*, *Bromus sterilis*, *Lolium* spp.) and annual broad-leaved weeds in winter wheat, winter barley, winter rye, some varieties of spring wheat, and carrots at 2.4–4 kg a.i./ha. Also used for haulm destruction in ware potatoes; and for pre-harvest defoliation in hemp, flax, and tomatoes. **Phytotoxicity** Some damage may occur with some varieties of cereal. **Formulation types** GR; WP; SC; WG. **Mixtures** (*metoxuron +*) simazine; fentin hydroxide. **Selected tradenames** 'Dosanex' (Cyanamid); 'Deftor' (AgrEvo); 'Investt' (Searle India); 'Sulerex' (Siapa)

ANALYSIS
Product analysis by hydrolysis and titration of the liberated dimethylamine, or by tlc (*AOAC Methods*, 1995, 977.06; *CIPAC Handbook*, 1980, **1A**, 1304) or by hplc. **Residues** are determined colorimetrically (M. Wisson et *al.*, *Anal. Methods Pestic. Plant Growth Regul.*, 1976, **8**, 417). Details of methods available from Cyanamid.

MAMMALIAN TOXICOLOGY
Oral Acute oral LD_{50} for rats 3200 mg/kg. **Skin and eye** Acute percutaneous LD_{50} for albino rats >2000 mg/kg. **Inhalation** LC_{50} (2 w) >5 mg/l air. **NOEL** In 90 d feeding trials, rats receiving 1250 mg/kg diet and dogs receiving 2500 mg/kg diet showed no ill-effects. **Toxicity class** WHO (a.i.) III (Table 5)

ECOTOXICOLOGY
Birds In 42-day feeding trials, chicks receiving 1250 mg/kg diet showed no significant abnormalities. **Fish** LC_{50} (96 h) for rainbow trout 18.9 mg/l. **Bees** Not toxic to bees; LD_{50} (oral) 850 ppm. **Worms** LC_{50} for earthworms >1000 mg/kg soil. **Daphnia** LC_{50} (24 h) 215.6 mg/l.

ENVIRONMENTAL FATE
Plants Degradation in plants involves demethylation of the terminal nitrogen and hydrolysis of the urea moiety. **Soil/Environment** In soil, breakdown involves demethylation of the terminal nitrogen, further degradation to 3-chloro-= 4-methoxyaniline, ring hydroxylation, and ring cleavage. DT_{50} in soil *c.* 10–30 d.

Herbicide

1,2,4-triazinone

$(CH_3)_3C$ —⟨N-N⟩— SCH_3 ... O ... N ... NH_2

NOMENCLATURE
Common name metribuzin (BSI, E-ISO, WSSA); métribuzine ((f) F-ISO)
IUPAC name 4-amino-6-*tert*-butyl-4,5-dihydro-3-methylthio-1,2,4-triazin-5-one;
4-amino-6-*tert*-butyl-3-methylthio-1,2,4-triazin-5(4*H*)-one
Chemical Abstracts name 4-amino-6-(1,1-dimethylethyl)-3-(methylthio)-=
1,2,4-triazin-5(4*H*)-one
CAS RN *[21087–64–9]*   **EEC no.** 244-209-7   **Development codes** Bayer 94 337;
DIC 1468 (both Bayer); DPX-G2504 (Du Pont)

PHYSICAL CHEMISTRY
Mol. wt. 214.3   **M.f.** $C_8H_{14}N_4OS$ **Form** White crystals with a weak characteristic
odour.   **M.p.** 126.2 °C   **B.p.** 132 °C/2 Pa   **V.p.** 0.058 mPa (20 °C)
K_{ow} logP = 1.6 (pH 5.6, 20 °C)   **Henry** 1×10^{-5} Pa m^3 mol^{-1} (20 °C, calc.)
S.g./density 1.28 (20 °C)   **Solubility** In water 1.05 g/l (20 °C). In
dimethylformamide 1780, cyclohexanone 1000, chloroform 850, acetone 820,
methanol 450, dichloromethane 340, benzene 220, *n*-butanol 150, ethanol 190,
toluene 87, xylene 90, isopropanol 77, hexane 1.0 (all in g/l, 20 °C).
Stability Relatively stable to u.v. irradiation. At 20 °C, stable to dilute acids and
alkalis; DT_{50} (37 °C) 6.7 h (pH 1.2); DT_{50} (70 °C) 569 h (pH 4), 47 d (pH 7),
191 h (pH 9). Photodecomposition in water is very rapid (DT_{50} <1 d). On soil
surfaces under natural light conditions, DT_{50} 14–25 d.

COMMERCIALISATION
History Herbicide reported by W. Draber et al. (*Naturwissenschaften*, 1968, **55**,
446) and reviewed by L. Eue (*Pflanzenschutz-Nachr. (Engl. Ed.)*, 1972, **25**, 175).
Introduced by Bayer AG and E. I. du Pont de Nemours and Co.
Patent BE 697083, DE 1795784 both to Bayer; US 3905801 to Du Pont
Manufacturers Bayer; Du Pont; Feinchemie Schwebda; Jingma; Rallis

APPLICATIONS
Biochemistry Photosynthetic electron transport inhibitor. Selectivity is due to
metabolism (mostly conjugation) within the plant (C. Fedtke, *Proc. Br. Crop Prot.
Conf.*, (1993) **1**, 221).   **Mode of action** Selective systemic herbicide, absorbed
predominantly by the roots, but also by the leaves, with translocation acropetally
in the xylem.   **Uses** Pre- and post-emergence control of many grasses and broad-
leaved weeds in soya beans, potatoes, tomatoes, sugar cane, alfalfa, asparagus,

maize and cereals at 0.35–0.7 kg a.i./ha.   **Phytotoxicity** Phytotoxic to many crops, including crucifers, cucurbits, lettuce, onions, sugar beet, sunflowers, flax, strawberries, sweet potatoes, and tobacco.   **Formulation types** WP; WG; SC. **Mixtures** *(metribuzin +)* chlorimuron; metolachlor; trifluralin; isoproturon; 2,4-D; mecoprop; BAY FOE 5043.   **Compatibility** Compatible with most other herbicides, except in highly concentrated mixtures.   **Selected tradenames** 'Sencor' (Bayer); 'Lexone' (Du Pont); 'Mistral' (Feinchemie Schwebda)

ANALYSIS
Product analysis by glc with FID (*AOAC Methods*, 1995, 984.11; *CIPAC Handbook*, 1988, **1D**, 124) or by i.r. spectrometry (J. W. Betker *et al.*, *J. Assoc. Off. Anal. Chem.*, 1976, **59**, 278).   **Residues** determined by glc (C. A. Anderson, *Anal. Methods Pestic. Plant Growth Regul.*, 1976, **8**, 453). In **drinking water**, by gc with FID (*AOAC Methods*, 1995, 991.07). Details available from Bayer AG.

MAMMALIAN TOXICOLOGY
Reviews *Toxikologie der Herbizide*, Deutsche Forschungsgemeinschaft [Data collection on toxicology of herbicides], 7. Lieferung, Weinheim, 1981. **Oral** Acute oral LD_{50} for rats *c.* 2000, mice *c.* 700, guinea pigs *c.* 250, cats >500 mg/kg.   **Skin and eye** Acute percutaneous LD_{50} for rats >20 000 mg/kg; not irritating to skin and eyes (rabbits).   **Inhalation** LC_{50} (4 h) for rats >0.65 mg/l air (dust).   **NOEL** (2 y) for rats and dogs 100, mice *c.* 800 mg/kg diet. **ADI** 0.013 mg/kg b.w.   **Toxicity class** WHO (a.i.) III (Table 5); EPA (formulation) III   **EC risk** Xn (R22)

ECOTOXICOLOGY
Birds Acute oral LD_{50} for bobwhite quail 168, mallard ducks 460–680 mg/kg. Five-day dietary LC_{50} for bobwhite quail and mallard ducks >4000 mg/kg diet. **Fish** LC_{50} (96 h) for bluegill sunfish 80, rainbow trout 76, goldfish >10, catfish >10 ppm.   **Bees** Not toxic to bees; LD_{50} 35 µg/bee.   **Worms** LC_{50} for *Eisenia foetida* 331.8 mg/kg dry soil.   **Daphnia** LC_{50} (48 h) 4.5–35 mg/l.   **Algae** E_rC_{50} for *Scenedesmus subspicatus* 0.021 mg/l.

ENVIRONMENTAL FATE
Animals In mammals, following oral administration, 98% elimination occurs within 96 hours, about equally in the urine and the faeces.   **Plants** In plants, metribuzin undergoes oxidative deamination and further degradation to water-soluble conjugates.   **Soil/Environment** Metribuzin is rapidly degraded in soil; microbial breakdown is the major mechanism of loss. Degradation involves deamination, followed by further degradation to water-soluble conjugates. Photodecomposition on soil surfaces and in aqueous solution is an important process for the degradation of metribuzin in the environment.

sulfonylurea

The molecular structure shows a benzene ring with CO₂CH₃ group and SO₂NHCONH group connected to a 1,3,5-triazine ring bearing OCH₃ and CH₃ substituents.

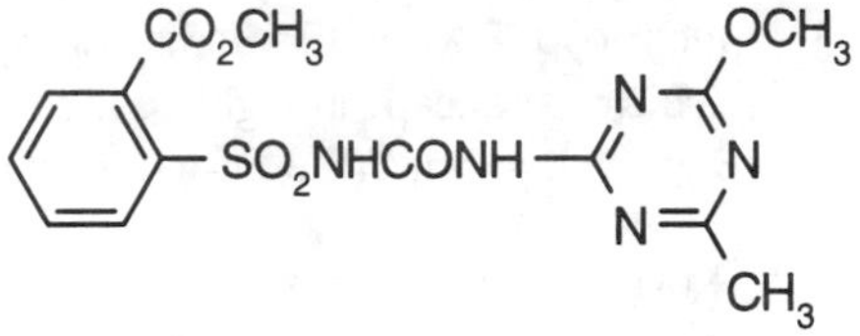

NOMENCLATURE

metsulfuron-methyl

Chemical Abstracts name methyl 2-[[[[(4-methoxy-6-methyl-1,3,5-triazin-2-yl)= amino]carbonyl]amino]sulfonyl]benzoate

CAS RN *[74223–64–6]* **Development codes** DPX-T6376

metsulfuron

Common name metsulfuron (BSI, WSSA, ANSI, draft E-ISO, (*m*) draft F-ISO)

IUPAC name 2-(4-methoxy-6-methyl-1,3,5-triazin-2-ylcarbamoylsulfamoyl)benzoic acid

Chemical Abstracts name 2-[[[[(4-methoxy-6-methyl-1,3,5-triazin-2-yl)amino]= carbonyl]amino]sulfonyl]benzoic acid

CAS RN *[79510–48–8]*

PHYSICAL CHEMISTRY

metsulfuron-methyl

Mol. wt. 381.4 **M.f.** $C_{14}H_{15}N_5O_6S$ **Form** Colourless crystals; (tech., off-white solid with a faint ester-like odour). **M.p.** 158 °C **V.p.** 3.3×10^{-7} mPa (25 °C) K_{ow} logP = −1.74 (pH 7) **Henry** 2.3×10^{-10} Pa m³ mol⁻¹ (calc.)

S.g./density 1.47 **Solubility** In water 0.55 (pH 5), 2.79 (pH 7), 213 (pH 9) (all in g/l, 25 °C). In xylene 0.58, ethanol 2.3, methanol 7.3, acetone 36, dichloromethane 121 (all in g/l, 20 °C). **Stability** Stable in air up to *c.* 140 °C, and in neutral and alkaline solutions at 25 °C. **pKa** 3.3

metsulfuron

Mol. wt. 367.3 **M.f.** $C_{13}H_{13}N_5O_6S$ **Form** Solid.

COMMERCIALISATION

History Herbicidal properties of metsulfuron-methyl reported by R. I. Doig *et al.* (*Proc. Int. Congr. Plant Prot., 10th*, 1983, **1**, 324). Introduced by E. I. du Pont de Nemours and Co. **Patent** US 4370480 **Manufacturers** Du Pont; IPESA

APPLICATIONS

metsulfuron-methyl

Biochemistry Branched chain amino acid synthesis (acetolactate synthase or ALS) inhibitor. Acts by inhibiting biosynthesis of the essential amino acids valine and isoleucine, hence stopping cell division and plant growth. Selectivity derives from rapid metabolism in the crop. Metabolic basis of selectivity in sulfonylureas reviewed (M. K. Koeppe & H. M. Brown, *Agro-Food-Industry*, **6**, 9–14 (1995)). **Mode of action** Selective systemic herbicide absorbed through the roots and foliage, with rapid translocation both acropetally and basipetally. Susceptible plants cease growth almost immediately after post-emergence treatment and are killed in 7–21 days. Surfactants increase the activity of metsulfuron methyl on certain broad-leaved weeds. **Uses** Metsulfuron-methyl controls a wide range of annual and perennial broad-leaved weeds in wheat, barley, rice and oats, by either pre- or post-emergence application, at 4–7.5 g/ha post-emergence. **Formulation types** WG. **Mixtures** *(metsulfuron-methyl +)* chlorsulfuron; thifensulfuron-methyl; chlorimuron-ethyl; 2,4-D-sodium; bensulfuron-methyl. **Selected tradenames** 'Ally' (Du Pont); 'Malban' (IPESA); 'Quit' (Sanonda)

ANALYSIS

Product analysis by hplc (L. W. Hershberger & D. E. Brennan, *Anal. Methods Pestic. Plant Growth Regul.*, 1988, **16**, 83). **Residues** determined by hplc (*idem, ibid.*). Methods for sulfonylurea residues in crops, soil and water reviewed (A. C. Barefoot et al., *Proc. Br. Crop Prot. Conf. – Weeds*, 1995, **2**, 707).

MAMMALIAN TOXICOLOGY

metsulfuron-methyl

Oral Acute oral LD_{50} for male and female rats >5000 mg/kg. **Skin and eye** Acute percutaneous LD_{50} for rabbits >2000 mg/kg; mild skin irritant to guinea pigs, but not a skin sensitiser; moderate but reversible eye irritant. **Inhalation** LC_{50} (4 h) for male and female rats >5 mg/l air. **NOEL** (2 y) for rats 500, male dogs 500, female dogs 5000 mg/kg diet. **ADI** 0.0125 mg/kg (Germany). **Other** Non-mutagenic in the Ames test. Non-teratogenic. **Toxicity class** WHO (a.i.) III (Table 5); EPA (formulation) IV

ECOTOXICOLOGY

metsulfuron-methyl

Birds Acute oral LD_{50} for mallard ducks >2510 mg/kg. Eight-day dietary LC_{50} for mallard ducks and bobwhite quail >5620 mg/kg diet. **Fish** LC_{50} (96 h) for rainbow trout and bluegill sunfish >150 mg/l. **Bees** Non-toxic to bees; LD_{50} >25 µg/bee. **Worms** >1000 mg/kg. **Daphnia** LC_{50} (48 h) >150 mg/l. **Algae** NOEC for green algae 100 mg/l. **Other aquatic spp.** NOEC for *Lemna gibba* 0.16 µg/l.

ENVIRONMENTAL FATE

Animals In mammals, following oral administration, metsulfuron-methyl is excreted predominantly unchanged. The methoxycarbonyl and sulfonylurea groups are only partly degraded, by O-demethylation and hydroxylation. **Plants** In plants, undergoes complete degradation within a few days, by hydrolysis and conjugation. In addition to the hydroxymethyl analogue, other metabolites identified include methyl 2-(aminosulfonyl)benzoate and 2-(aminosulfonyl)benzoic acid. Rapidly metabolised within cereal plants. **Soil/Environment** In soil, metsulfuron-methyl is broken down both by chemical hydrolysis and by microbial degradation. DT_{50} varies from 1 to 5 w, with breakdown being more rapid at lower soil pH, at higher temperatures, and at higher levels of soil moisture. K_{oc} 35 (pH 7).

499 mevinphos   *Insecticide, acaricide*

organophosphorus

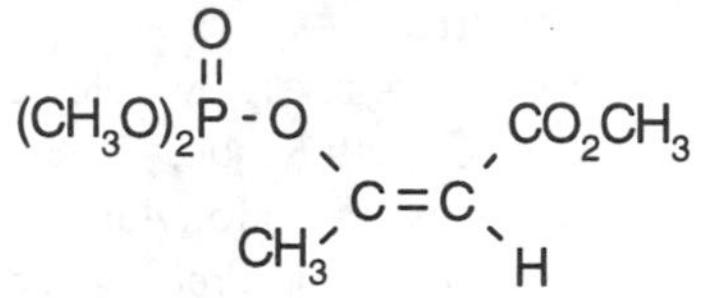

(Z)

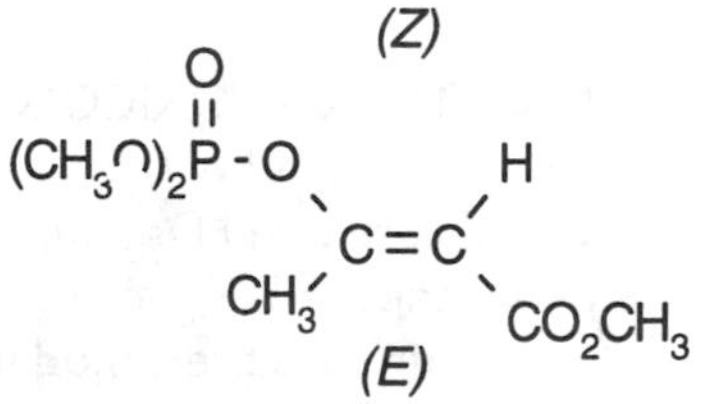

(E)

NOMENCLATURE
Common name mevinphos (BSI, E-ISO, (*m*) F-ISO, ESA). The isomer should be stated, as (*E*)- or (*Z*)- [or *cis* or *trans* (with respect to the carbon chain)]
IUPAC name 2-methoxycarbonyl-1-methylvinyl dimethyl phosphate; methyl 3-(dimethoxyphosphinoyloxy)but-2-enoate
Chemical Abstracts name methyl 3-[(dimethoxyphosphinyl)oxy]-2-butenoate
CAS RN [26718–65–0], formerly [298–01–1] (*E*)- isomer; [338–45–4] (*Z*)- isomer; [7786–34–7] (*Z*)- + (*E*)- isomers   **EEC no.** 232–095–1
Development codes OS-2046 (Cyanamid)   **Official codes** ENT 22 374

PHYSICAL CHEMISTRY
Composition Tech. contains >60% *m*/*m* of the (*E*)- isomer and c. 20% *m*/*m* of the (*Z*)- isomer.   **Mol. wt.** 224.1   **M.f.** $C_7H_{13}O_6P$ **Form** Colourless liquid; (tech., pale yellow liquid).   **M.p.** (*E*)- isomer 21 °C; (*Z*)- isomer 6.9 °C
B.p. 99–103 °C/0.3 mmHg   **V.p.** 17 mPa (20 °C)   **K_{ow}** logP = 0.127
S.g./density 1.24 (20 °C); (*E*)- isomer 1.235; (*Z*)- isomer 1.245
Solubility Completely miscible with water and most organic solvents, e.g. alcohols, ketones, aromatic hydrocarbons, and chlorinated hydrocarbons. Slightly soluble in

aliphatic hydrocarbons, petroleum ether, ligroin, and carbon disulfide.
Stability Stable at ambient temperatures but hydrolysed in aqueous alkaline solution, DT_{50} 120 d (pH 6), 35 d (pH 7), 3 d (pH 9), 1.4 h (pH 11).

COMMERCIALISATION
History Insecticide reported by R. A. Corey et al. (*J. Econ. Entomol.*, 1953, **45**, 386). Introduced by Shell Chemical Co., USA (now American Cyanamid Co.), and later by Amvac Corp. **Patent** US 2685552 **Manufacturers** Amvac, Comlets; Hui Kwang

APPLICATIONS
Biochemistry Cholinesterase inhibitor. **Mode of action** Systemic insecticide and acaricide with contact, stomach, and respiratory action. Short residual activity.
Uses Control of chewing and sucking insects, and spider mites on a wide range of crops, including pome fruit, stone fruit, berry fruit, citrus fruit, strawberries, melons, vines, hops, water melons, vegetables, cucurbits, aubergines, ornamentals, beet, potatoes, cotton, oilseed rape, cereals, clover, lupins, trefoil, okra, sorghum, etc. **Formulation types** EC; SL. **Compatibility** Incompatible with pesticides which are alkaline in reaction. **Selected tradenames** 'Phosdrin' (Cyanamid, Amvac); 'Duraphos' (Amvac); 'Mevindrin' (Hui Kwang)

ANALYSIS
Product analysis by glc with FID (*CIPAC Proc.*, 1981, **3**, 283). **Residues** determined by glc with FPD (*Pestic. Anal. Man.*, 1979, **I**, 201-H, 201-I; *Man. Pestic. Residue Anal.*, 1987, **I**, 3, 6, S8, S13, S17, S19; *Anal. Methods Residues Pestic.*, 1988, Part I, M2, M5, M12; A. Ambrus et al., *J. Assoc. Off. Anal. Chem.*, 1981, **64**, 733). In **drinking water**, by gc with FID (*AOAC Methods*, 1995, 991.07). Details available from Amvac Chemical Corp.

MAMMALIAN TOXICOLOGY
Reviews *Pesticide residues in food*. FAO Agricultural Studies, No. 90; WHO Technical Report Series, No. 525, 1973. *1972 Evaluations of some pesticide residues in food*. AGP:1972/M/9/1; WHO Pesticide Residues Series No. 2, 1973.
EHC 63 (WHO, 1986; a general review of organophosphorus insecticides).
Oral Acute oral LD_{50} for rats 3–12, mice 7–18 mg/kg. **Skin and eye** Acute percutaneous LD_{50} for rats 4–90, rabbits 16–33 mg/kg. Mild irritant to skin and eyes (rabbits). **Inhalation** LC_{50} (1 h) for rats 0.125 mg/l air. **NOEL** In 2 y feeding trials, rats receiving 4 mg/kg diet and dogs receiving 5 mg/kg diet showed no ill-effects. **ADI** (JMPR) 0.0015 mg/kg b.w. [1972]. **Toxicity class** WHO (a.i.) Ia; EPA (formulation) I **EC risk** T+ (R27/28)

ECOTOXICOLOGY
Birds Acute oral LD_{50} for mallard ducks 4.63, chickens 7.52, pheasants 1.37 mg/kg. **Fish** LC_{50} (48 h) for rainbow trout 0.017, bluegill sunfish 0.037 mg/l.
Bees Toxic to bees; LD_{50} 0.027 µg/bee.

ENVIRONMENTAL FATE
Animals In mammals, following oral administration, elimination occurs in 3–4 days in the form of metabolites in the urine and faeces.   **Plants** In plants, hydrolysed rapidly to less toxic products, including phosphoric acid dimethyl ester and phosphoric acid. The (E)- isomer is more rapidly degraded than the (Z)- isomer. Metabolism and degradation have been reviewed (K. I. Beynon et al., Residue Rev., 1973, **47**, 55).

500 milbemectin   *Acaricide, insecticide*

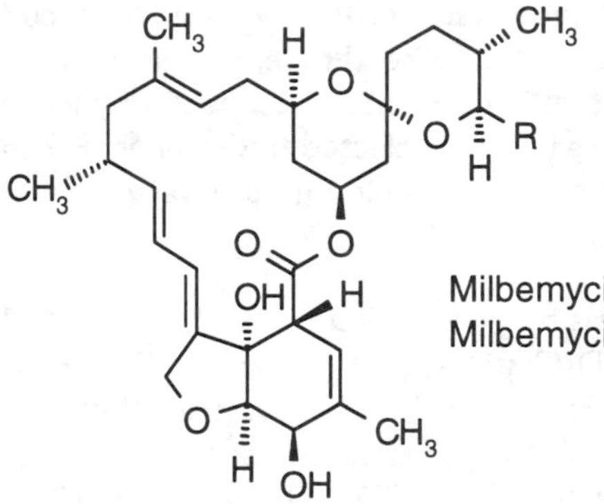

Milbemycin A$_3$: R = CH$_3$
Milbemycin A$_4$: R = CH$_2$CH$_3$

NOMENCLATURE
Common name milbemectin (BSI, pa E-ISO)
IUPAC name A mixture of: (10E,14E,16E,22Z)-(1R,4S,5'S,6R,6'R,8R,13R,20R,21R,= 24S)-21,24-dihydroxy-5',6',11,13,22-pentamethyl-3,7,19-trioxatetracyclo= [15.6.1.1^{4,8}.0^{20,24}]pentacosa-10,14,16,22-tetraene-6-spiro-2'-tetrahydropyran-= 2-one and (10E,14E,16E,22Z)-(1R,4S,5'S,6R,6'R,8R,13R,20R,21R,24S)-6'-ethyl-21,= 24-dihydroxy-5',11,13,22-tetramethyl-3,7,19-trioxatetracyclo[15.6.1.1^{4,8}.0^{20,24}]= pentacosa-10,14,16,22-tetraene-6-spiro-2'-tetrahydropyran-2-one in the ratio 3 to 7
Chemical Abstracts name A$_3$: (6R,25R)-5-O-demethyl-28-deoxy-6,28-epoxy-25-= methylmilbemycin B; A$_4$: (6R,25R)-5-O-demethyl-28-deoxy-6,28-epoxy-25-ethyl= milbemycin B
CAS RN A$_3$: *[51596–10–2]*; A$_4$: *[51596–11–3]*   **Development codes** B-41; E-187; SI-8601 (Sankyo)

PHYSICAL CHEMISTRY
Composition A mixture of the homologues milbemycin A$_3$ (methyl) and

milbemycin A$_4$ (ethyl) in the ratio 3 to 7. **Mol. wt.** A$_3$: 528.7; A$_4$: 542.7
M.f. A$_3$: C$_{31}$H$_{44}$O$_7$; A$_4$: C$_{32}$H$_{46}$O$_7$ **M.p.** A$_3$: 212–215 °C; A$_4$: 212–215 °C
V.p. A$_3$: <1.3 × 10^{-5}, A$_4$: <1.3 × 10^{-5} (both in mPa, 20 °C) **K$_{ow}$** A$_3$: logP = 5.3;
A$_4$: logP = 5.9 **S.g./density** A$_3$: 1.1270, A$_4$: 1.1265 (both 25 °C) **Solubility**
A$_3$: In water 0.88 ppm (20 °C). In methanol 64.8, ethanol 41.9, acetone 66.1,
n-hexane 1.4, benzene 143.1, ethyl acetate 69.5 (all in g/l, 20 °C). A$_4$: In water
7.2 ppm (20 °C). In methanol 458.8, ethanol 234.0, acetone 365.3, *n*-hexane 6.5,
benzene 524.2, ethyl acetate 320.4 (all in g/l, 20 °C).

COMMERCIALISATION
Manufacturers Sankyo

APPLICATIONS
Mode of action Contact and stomach action. **Uses** Control of citrus red mites
and pink citrus rust mites on citrus fruit, Kanzawa spider mites on tea, and spider
mites on aubergines. **Formulation types** EC. **Selected tradenames** 'Milbeknock'
(Sankyo)

MAMMALIAN TOXICOLOGY
Oral Acute oral LD$_{50}$ for male mice 324, female mice 313, male rats 762, female
rats 456 mg/kg. **Skin and eye** Acute percutaneous LD$_{50}$ for male and female rats
>5000 mg/kg. **Other** Not carcinogenic, not teratogenic, not mutagenic.

501 molinate *Herbicide*

thiocarbamate

NOMENCLATURE
Common name molinate (BSI, E-ISO, (*m*) F-ISO, JMAF); no name (Germany)
IUPAC name *S*-ethyl azepane-1-carbothioate; *S*-ethyl perhydroazepin-=
1-carbothioate; *S*-ethyl perhydroazepine-1-thiocarboxylate
Chemical Abstracts name *S*-ethyl hexahydro-1*H*-azepine-1-carbothioate
CAS RN *[2212–67–1]* **EEC no.** 218–661–0 **Development codes** R-4572
(Stauffer) **Official codes** OMS 1373

PHYSICAL CHEMISTRY
Composition Tech. material is 95% pure. **Mol. wt.** 187.3 **M.f.** C$_9$H$_{17}$NOS

molinate 847

Form Clear liquid with an aromatic odour; (tech. is an amber liquid).
B.p. 136.5 °C/10 mmHg   **V.p.** 746 mPa (25 °C) K_{ow} logP = 2.88
S.g./density 1.063 (20 °C)   **Solubility** In water 88 mg/l (20 °C). Miscible with most common organic solvents, e.g. acetone, methanol, ethanol, kerosene, methyl isobutyl ketone, benzene, xylene.   **Stability** Stable for at least 2 years at room temperature and at least 1 month at 120 °C. Relatively stable to hydrolysis by acids and alkalis (pH 5–9) at 40 °C. Unstable to light.   **F.p.** >100 °C

COMMERCIALISATION
History Introduced by Stauffer Chemical Co. (now Zeneca Agrochemicals).
Patent US 3198786; US 3573031   **Manufacturers** Herbex; Nufarm Ltd; Sagrochem; Zeneca

APPLICATIONS
Biochemistry Inhibits lipid metabolism.   **Mode of action** Selective systemic herbicide, rapidly absorbed by the roots, with translocation acropetally to the leaves. Inhibits germination.   **Uses** Control of germinating broad-leaved and grass weeds, particularly *Echinochloa* spp., in rice at 2–4 kg a.i./ha. It is applied either before planting to water-seeded or shallow soil-seeded rice or post-flood, post-emergence in other types of rice culture.   **Phytotoxicity** Non-phytotoxic to rice.
Formulation types EC; GR.   **Mixtures** *(molinate +)* propanil; dimepiperate; thiobencarb.   **Selected tradenames** 'Ordram' (Zeneca); 'Sakkimol' (Sagrochem)

ANALYSIS
Product analysis by glc (*AOAC Methods*, 1995, 974.05; *CIPAC Handbook*, 1983, **1B**, 1866).   **Residues** in crops and soil determined by glc or colorimetry after conversion to a suitable derivative. See also: G. R. Patchett & G. H. Batchelder, *Anal. Methods Pestic., Plant Growth Regul. Food Addit.*, 1967, **5**, 469; G. R. Patchett *et al.*, *Anal. Methods Pestic. Plant Growth Regul.*, 1972, **6**, 668. In **drinking water**, by gc with FID (*AOAC Methods*, 1995, 991.07). Analytical methods available from Zeneca.

MAMMALIAN TOXICOLOGY
EHC 76 (WHO, 1988; general review of thiocarbamates).   **Oral** Acute oral LD_{50} for male rats 369, female rats 450, mice 795 mg/kg.   **Skin and eye** Acute percutaneous LD_{50} for rabbits >4640 mg/kg. Moderate eye irritant; mild skin irritant (rabbits).   Not a skin sensitiser (guinea pigs).   **Inhalation** LC_{50} (4 h) for rats 1.36 mg/l.   **NOEL** (90 d) for rats 8, dogs 20 mg/kg daily.   **Toxicity class** WHO (a.i.) II; EPA (formulation) IV   **EC risk** Xn (R22)

ECOTOXICOLOGY
Birds Five-day dietary LC_{50} for mallard ducklings 13 000 mg/kg diet.   **Fish** LC_{50} (96 h) for rainbow trout 1.3, bluegill sunfish 29, goldfish 30 mg/l. At recommended rates, no detectable effects on fish in ditches draining water from treated rice fields in California.

ENVIRONMENTAL FATE

Animals In rats, following oral administration, molinate is rapidly metabolised. Within 72 hours, c. 50% is eliminated as the metabolite CO_2, 25% excreted in the urine, and 5–20% excreted in the faeces.   **Plants** In plants, molinate is rapidly metabolised to CO_2 and naturally-occurring plant constituents.

Soil/Environment In soil, microbial breakdown involves hydrolysis to ethyl mercaptan, dialkylamine and CO_2. In aerobic soil (pH 4.9–5.9) DT_{50} 8–25 d; in flooded soil 40–160 d. Soil adsorption K. 0.74 2.01 mg/kg In soils with 0.5–2.2% o.m. and pH 5.5–7.8.

502 monocrotophos

Insecticide, acaricide

organophosphorus

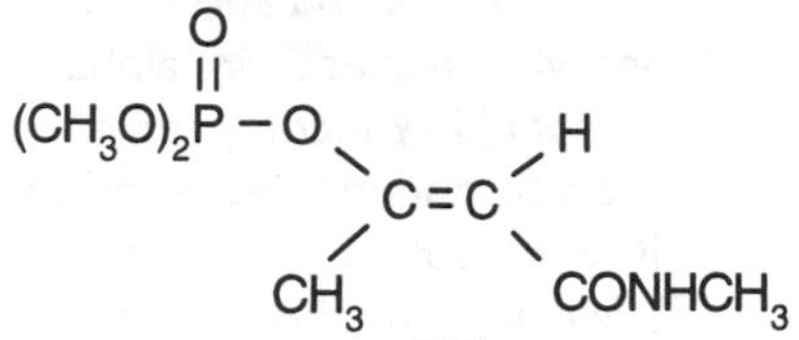

$$(CH_3O)_2P-O$$

with structure showing

(E)

NOMENCLATURE

Common name monocrotophos (BSI, E-ISO, (*m*) F-ISO, ESA, JMAF)

IUPAC name dimethyl (*E*)-1-methyl-2-(methylcarbamoyl)vinyl phosphate; 3-dimethoxyphosphinoyloxy-*N*-methylisocrotonamide

Chemical Abstracts name (*E*)-dimethyl 1-methyl-3-(methylamino)-3-oxo-= 1-propenyl phosphate

CAS RN [6923–22–4] (formerly [919–44–8]) monocrotophos; [919–44–8] (*Z*)- analogue; [2157–98–4] (*E*)- + (*Z*)- compounds   **EEC no.** 230–042–7

Development codes C 1414 (Ciba); SD 9129 (Shell)   **Official codes** OMS 834; ENT 27 129

PHYSICAL CHEMISTRY

Mol. wt. 223.2   **M.f.** $C_7H_{14}NO_5P$ **Form** Colourless, hygroscopic crystals; (tech., ≥74%: dark brown semi-solid).   **M.p.** 54–55 °C   **B.p.** 125 °C/0.0005 mmHg

V.p. 2.9×10^{-1} mPa (20 °C); 9.8×10^{-1} mPa (separate studies) K_{ow} logP = −0.22 (calc.)   **S.g./density** 1.22 kg/l (20 °C)   **Solubility** In water 100% (20 °C). In methanol 100%, acetone 70%, *n*-octanol 25%, toluene 6% (all at 20 °C). Sparingly soluble in kerosene and diesel oil.   **Stability** Decomposes >38 °C, thermal run-away reactions can occur >55 °C; on hydrolysis (20 °C) DT_{50} (calc.) 96 d (pH 5),

66 d (pH 7), 17 d (pH 9); unstable in short-chain alcohols. Decomposes on some inert materials (care should be taken when carrying out chromatography).

COMMERCIALISATION
History Introduced by Ciba AG (now Novartis Crop Protection AG) and by Shell Chemical Co (now American Cyanamid Co.). **Patent** BE 552284; GB 829576 both to Ciba-Geigy **Manufacturers** Bharat; Comlets; Crystal; Cyanamid; Hui Kwang; India Pesticides; Makhteshim-Agan; Nagarjuna; Novartis; Pesticides India; United Phosphorus; Q.E.A.C.A.; Rallis; Sinon; Sundat; Taiwan Tainan Giant; Voltas

APPLICATIONS
Biochemistry Cholinesterase inhibitor. **Mode of action** Systemic insecticide and acaricide with contact and stomach action. Penetrates plant tissue rapidly.
Uses Control of a broad spectrum of pests, including sucking, chewing, and boring insects, and spider mites on cotton, citrus, olives, rice, maize, sorghum, sugar cane, sugar beet, peanuts, potatoes, soya beans, vegetables, ornamentals, and tobacco. **Phytotoxicity** Non-phytotoxic when used as directed, although slight injury may be caused to some varieties of apple, pear, cherry, peach, and sorghum. **Formulation types** SL; UL. **Mixtures** (*monocrotophos +*) parathion-methyl; cypermethrin; alpha-cypermethrin; parathion-methyl + sulfur.
Compatibility Incompatible with pesticides which are alkaline in reaction.
Selected tradenames 'Azodrin' (Cyanamid); 'Nuvacron' (Novartis); 'Apadrin' (Rhône-Poulenc); 'Balwan' (Rallis); 'Croton' (Searle India); 'Crotos' (Siapa); 'Efacron' (Efthymiadis); 'Macabre' (Sanonda); 'Monocron' (Makhteshim-Agan); 'Monodhan' (Dhanuka); 'Monodrin' (Hui Kwang); 'Monostar' (Shaw Wallace); 'Monovol' (Voltas); 'Phoskill' (United Phosphorus); 'Pilardrin' (Pilarquim)

ANALYSIS
Product analysis by rp hplc with u.v. detection (*CIPAC Handbook*, 1992, **E**, 145–150), or by glc, details available from Cyanamid or Novartis.
Residues determined by glc using phosphorus-sensitive detectors (*Pestic. Anal. Man.*, 1979, **I**, 201-H, 201-I; *ibid*, 1979, **II**; A. Ambrus *et al.*, *J. Assoc. Off. Anal. Chem.*, 1981, **64**, 733; *AOAC Methods*, 1995, 985.22; *Man. Pestic. Residue Anal.*, 1987, **I**, S19; *Anal. Methods Residues Pestic.*, 1988, Part I, M2, M5; *Anal. Methods Pestic. Plant Growth Regul.*, 1972, **6**, 287).

MAMMALIAN TOXICOLOGY
Reviews *Pesticide residues in food – 1995*, FAO Plant Production and Protection Paper 133, 1996. *Pesticide residues in food – 1995 evaluations; Part II – Toxicology & Environment*. World Health Organisation, WHO/PCS/96.48, 1996. **EHC** 63 (WHO, 1986; a general review of organophosphorus insecticides). **Oral** Acute oral LD_{50} for male rats 18, female rats 20 mg/kg. **Skin and eye** Acute percutaneous LD_{50} for rabbits 130–250, male rats 126, female rats 112 mg/kg. Non-irritating to skin and eyes (rabbits). **Inhalation** LC_{50} (4 h) for rats

c. 0.08 mg/l air.   **NOEL** (2 y) estimated by WHO (JMPR, 1972) as: for rats 0.5 mg/kg diet (0.025 mg/kg daily); for dogs 0.5 mg/kg diet (0.0125 mg/kg daily). **ADI** (JMPR) 0.0006 mg/kg b.w. [1993].   **Toxicity class** WHO (a.i.) Ib; EPA (formulation) I   **EC risk** R24/28, R40

ECOTOXICOLOGY
Birds Acute oral LD_{50} (14 d) for mallard ducklings 4.8, male Japanese quail 3.7, male bobwhite quail 0.94, chickens 6.7, young pheasants 2.8, partridges 6.5, pigeons 2.8, house sparrows 1.5 mg/kg.   **Fish** LC_{50} (48 h) for rainbow trout 7 mg/l; (24 h) for rainbow trout 12, bluegill sunfish 23 mg/l.   **Bees** Highly toxic to bees. LD_{50} (oral) 0.028–0.033 mg/bee; (topical) 0.025–0.35 mg/bee. **Daphnia** LC_{50} (24 h) 0.24 µg/l.   **Other aquatic spp.** LC_{50} (96 h) for *Gammarus fasciatus* 0.3, *Macrobrachium lamerrii* 1.9, *Crassostrea virginica* >1 mg/l.

ENVIRONMENTAL FATE
Metabolism and breakdown in plant, animals and soil has been reviewed (K. I. Beynon *et al., Residue Rev.,* 1973, **47**, 55).   **Animals** In mammals, following oral administration, 60–65% is excreted within 24 hours, predominantly in the urine. **Soil/Environment** Rapidly degraded in soil; DT_{50} (lab.) 1–5 d.

503   monolinuron   *Herbicide*

urea

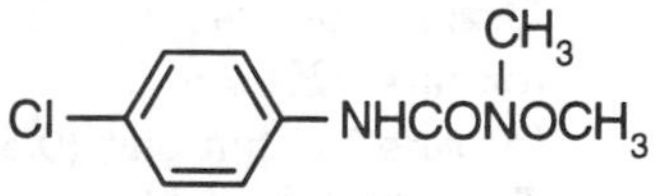

NOMENCLATURE
Common name monolinuron (BSI, E-ISO, (*m*) F-ISO, WSSA)
IUPAC name 3-(4-chlorophenyl)-1-methoxy-1-methylurea
Chemical Abstracts name *N'*-(4-chlorophenyl)-*N*-methoxy-*N*-methylurea
CAS RN *[1746–81–2]*   **EEC no.** 217–129–5   **Development codes** AE F002747; Hoe 002 747

PHYSICAL CHEMISTRY
Mol. wt. 214.6   **M.f.** $C_9H_{11}ClN_2O_2$ **Form** Colourless crystals.   **M.p.** 80–83 °C **V.p.** 1.3 mPa (20 °C); 100 mPa (50 °C)   **K_{ow}** logP = 2.20   **Henry** 5.649×10^{-4} (22 °C), 7.302×10^{-4} (25 °C) Pa m^3 mol^{-1} **S.g./density** 1.3 (20 °C)   **Solubility** In water 735 mg/l (25 °C). Readily soluble in common organic solvents, e.g. alcohols, acetone, dioxane, xylene, chloroform, diethyl ether.   **Stability** Stable in solution,

but slowly decomposed in acidic and alkaline media. Very stable in dry, neutral conditions. Degradation is accelerated by u.v. light. Decomposes at 220 °C.

COMMERCIALISATION

History Herbicide reported by K. Härtel (*Meded. Landbouwhogesch. Opzoekingsstn. Staat Gent*, 1962, **27**, 1275) and history summarised by H. Maier-Bode & K. Härtel (*Residue Rev.*, 1981, **77**, 1). Introduced by Hoechst AG (now AgrEvo GmbH). **Patent** DE 1028986; GB 852422 **Manufacturers** AgrEvo

APPLICATIONS

Biochemistry Photosynthetic electron transport inhibitor.
Mode of action Selective systemic herbicide, absorbed by the roots and leaves, with translocation. **Uses** Control of annual broad-leaved weeds and some annual grasses in asparagus, berry fruit, maize, dwarf french beans, field beans, vines, leeks, onions, potatoes, herbs, alfalfa, flowers, and ornamental shrubs and trees. Applied either pre- or post-emergence. **Formulation types** WP; EC.
Mixtures (*monolinuron* +) linuron; paraquat dichloride. **Selected tradenames** 'Aresin' (AgrEvo); 'Arresin' (AgrEvo)

ANALYSIS

Product analysis method available from AgrEvo. Impurities determined by tlc (*CIPAC Handbook*, 1994, **F**, 336). **Residues** determined by colorimetry (R. L. Dalton & H. L. Pease, *J. Assoc. Off. Agric. Chem.*, 1962, **45**, 377) or by glc (*Man. Pestic. Resid. Anal.*, Deutsche Forschungsgemeinschaft, 1987, **1**, 241). Details available from AgrEvo.

MAMMALIAN TOXICOLOGY

Oral Acute oral LD_{50} for rats 1430–2490. **Skin and eye** Acute percutaneous LD_{50} for rats >2000 mg/kg. **Inhalation** LC_{50} (4 h) for rats >3.39 mg/l air. **NOEL** (2 y) for rats 10 ppm diet (0.5 mg/kg b.w. daily). **ADI** 0.005 mg/kg b.w.
Toxicity class WHO (a.i.) III (Table 5) **EC risk** Xn (R22)

ECOTOXICOLOGY

Birds Acute oral LD_{50} for bobwhite quail 1260, Japanese quail >1690, mallard ducks >500 mg/kg. **Fish** LC_{50} (96 h) for carp 74, rainbow trout 56–75 mg/l.
Bees LD_{50} (oral) >296.3 μg/g b.w. **Worms** LC_{50} (14 d) for *Eisenia foetida* >1000 mg/kg. **Other beneficial spp.** Harmless to beneficials. **Daphnia** LC_{50} (48 h) 32.5 mg/l.

ENVIRONMENTAL FATE

Plants Degradation in plants is similar to that in soil. **Soil/Environment** In soil, breakdown involves cleavage of the methyl and methoxy groups on the terminal nitrogen atom, simultaneous ring hydroxylation and formation of 3-(2-hydroxy-= 4-chlorophenyl)urea and the corresponding 3-hydroxy compound. Further degradation presumably involves formation of the aniline derivative and ring splitting. Half-life in soil is *c.* 45–60 d (Börner, *Z. Pflanzenkr. Pflanzenpathol. Pflanzenschutz,* 1965, **72**, 516). Soil adsorption K_{oc} 250–500.

504 muscalure

Insect pheromone

pheromone

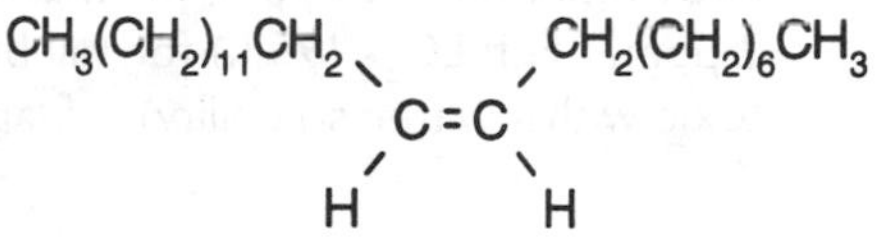

$$CH_3(CH_2)_{11}CH_2 \diagdown \qquad \diagup CH_2(CH_2)_6CH_3$$
$$C = C$$
$$H \diagup \qquad \diagdown H$$

NOMENCLATURE
Common name muscalure (ESA)
IUPAC name (Z)-tricos-9-ene
Chemical Abstracts name (Z)-9-tricosene
CAS RN *[27519–02–4]* (Z)- isomer; *[35857–62–6]* (E)- isomer

PHYSICAL CHEMISTRY
Composition The mating hormone of the housefly (*Musca domestica*). The (E)-isomer is a minor component. **Mol. wt.** 322.6 **M.f.** $C_{23}H_{46}$ **Form** Colourless to amber oil. **M.p.** <0 °C **B.p.** 378 °C; 190 °C/0.5 mmHg **V.p.** 4.7 mPa (27 °C) K_{ow} logP = 4.09 **S.g./density** 0.800 (20 °C) **Solubility** In water 0.3 mg/l (pH 7, 25 °C). Soluble in hydrocarbons, alcohols, ketones, esters. **Stability** Stable to light and temperatures up to 50 °C for 1 year. **F.p.** >113 °C (closed cup)

COMMERCIALISATION
History Sex pheromone of female *Musca domestica* L. isolated by Carlson *et al.* (*Science,* 1971, **174,** 76). Developed as an insect attractant by Zoecon Industries Ltd (acquired by Sandoz AG, now Novartis Crop Protection AG).
Manufacturers Cyclo; Denka

APPLICATIONS
Mode of action Acts as an attractant and by disruption of mating. **Uses** For housefly control. Used mainly in combination with an insecticide to attract and kill.
Formulation types GB. **Mixtures** (*muscalure* +) methomyl.
Selected tradenames 'Muscamone' (Novartis); 'Flybait' (mixture) (Denka)

ANALYSIS
See J. Ko *et al.*, JAOAC Intl., **75** (5), 878–82 (1992).

MAMMALIAN TOXICOLOGY

Oral Acute oral LD_{50} for rats >5000 mg/kg. **Skin and eye** Acute percutaneous LD_{50} for rabbits >2000 mg/kg. Not irritating to eyes or skin (rabbits). Moderate skin sensitiser (guinea pig). **Inhalation** LC_{50} (4 h) >5.71 g/m³. **Other** Not mutagenic in the Ames test, not teratogenic (rats) at >5 g/kg.
Toxicity class WHO (a.i.) III; EPA (formulation) IV ('Flybait')

ECOTOXICOLOGY

Birds Acute oral LD_{50} for mallard ducks >4640 mg/kg. LC_{50} for mallard ducks and bobwhite quail >4640 mg/kg. NOEL in one-generation reproductive study for bobwhite quail >20 ppm, for mallard ducks 0.1 ppm; reproduction hazard at 2 ppm. **Fish** LC_{50} (96 h) for rainbow trout and bluegill sunfish >1000 mg/l (not toxic within water solubility). **Daphnia** LC_{50} (48 h) 265.7 µg/l.

505 myclobutanil *Fungicide*

azole

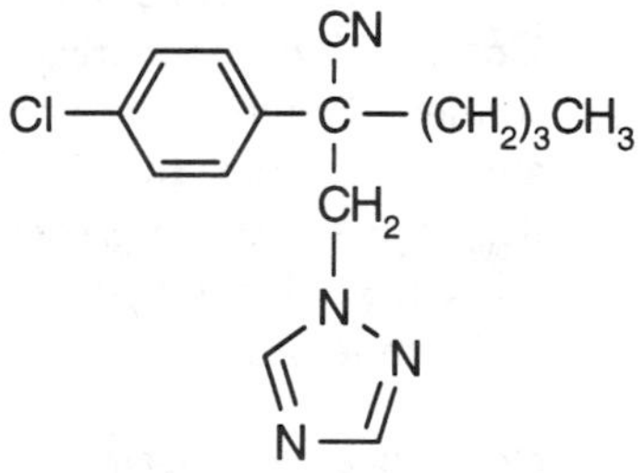

NOMENCLATURE

Common name myclobutanil (BSI, ANSI, draft E-ISO, (m) draft F-ISO)
IUPAC name 2-p-chlorophenyl-2-(1H-1,2,4-triazol-1-ylmethyl)hexanenitrile; 2-(4-chlorophenyl)-2-(1H-1,2,4-triazol-1-ylmethyl)hexanenitrile
Chemical Abstracts name α-butyl-α-(4-chlorophenyl)-1H-1,2,4-triazole-= 1-propanenitrile
CAS RN [88671–89–0] **Development codes** RH-3866

PHYSICAL CHEMISTRY

Mol. wt. 288.8 **M.f.** $C_{15}H_{17}ClN_4$ **Form** Pale yellow solid. **M.p.** 63–68 °C (tech.) **B.p.** 202–208 °C/1 mmHg **V.p.** 0.213 mPa (25 °C) **K$_{ow}$** logP = 2.94 (pH 7–8, 25 °C) **Solubility** In water 142 mg/l (25 °C). Soluble in common organic solvents, e.g. ketones, esters, alcohols and aromatic hydrocarbons, all 50–100 g/l. Insoluble in aliphatic hydrocarbons. **Stability** Stable under normal storage conditions.

Aqueous solutions decompose on exposure to light, DT_{50} 222 d (sterile water), 0.8 d (sensitised sterile water), 25 d (pond water); no hydrolysis (28 °C) in 28 d (pH 5, 7, and 9).

COMMERCIALISATION
History Systemic fungicide reported by C. Orpin et al. (*Proc. 1986 Br. Crop Prot. Conf.- Pests Dis.*, **1b**, 55) and developed by Rohm & Haas Co.
Manufacturers Rohm & Haas

APPLICATIONS
Biochemistry Inhibits ergosterol biosynthesis (steroid demethylation inhibitor).
Mode of action Systemic fungicide with protective and curative action.
Uses Control of Ascomycetes, Fungi Imperfecti, and Basidiomycetes on a wide variety of crops. For example, used as a foliar treatment for control of scab and powdery mildew in apples and pears; powdery mildew, shot-hole, blossom blight, and rust in stone fruit; powdery mildew in vines and cucurbits; powdery mildew and rusts on ornamentals; rusts on perennial grasses grown for seed; and various diseases of wheat; as a seed treatment for control of seed- and soil-borne diseases in barley, maize, cotton, rice and wheat; and as a post-harvest drench or dip. **Formulation types** WP; EC; SC; Seed treatment. **Mixtures** (*myclobutanil +*) captan; dinocap; mancozeb; sulfur; ziram. **Selected tradenames** 'Systhane' (Rohm & Haas)

ANALYSIS
Product by gc. **Residues** by gc. Details from Rohm & Haas.

MAMMALIAN TOXICOLOGY
Reviews *Pesticide residues in food – 1992*. FAO Plant Production and Protection Paper, 116, 1993. *Pesticide residues in food – 1992 evaluations. Part II – Toxicology*. WHO/PCS/93.34. **Oral** Acute oral LD_{50} for male rats 1600, female rats 2290 mg/kg. **Skin and eye** Acute percutaneous LD_{50} for rabbits >5000 mg/kg. Not irritating to their skin but mildly to their eyes; not a skin sensitiser to guinea pigs. **NOEL** (90 d) for rats and female dogs 100, male dogs 10 mg/kg diet. No reproductive effects were observed in rats at 200 mg/kg, but there were some reproductive effects in male rats at 1000 mg/kg. **ADI** (JMPR) 0.03 mg/kg b.w. [1992]. **Other** Non-teratogenic in rats and rabbits, and various mutagenicity tests proved negative. Not mutagenic in the Ames assay. **Toxicity class** WHO (a.i.) III

ECOTOXICOLOGY
Birds Acute oral LD_{50} for bobwhite quail 510, grey partridge 1635 mg/kg.
Fish LC_{50} (96 h) for bluegill sunfish 2.4, rainbow trout 4.2 mg/l. **Bees** Non-toxic to bees. **Daphnia** LC_{50} (48 h) 11 mg/l.

ENVIRONMENTAL FATE
Animals In animals, undergoes oxidation at the butyl group to a ketone and an

alcohol, with partial conjugation to a glucoside. Following oral administration, rapidly excreted in the faeces and urine. **Plants** In plants, metabolism is the same as in animals. **Soil/Environment** In soil, DT_{50} 66 d (silt loam). Decomposition is through highly polar triazole compounds, with further degradation by ring splitting. No degradation under anaerobic conditions.

506 nabam   *Fungicide, algicide*

alkylenebis(dithiocarbamate)

$$NaSCSNHCH_2CH_2NHCSSNa$$

NOMENCLATURE
Common name nabam (BSI, ISO-E); nabame ((*m*) F-ISO)
IUPAC name disodium ethylenebis(dithiocarbamate)
Chemical Abstracts name disodium 1,2-ethanediylbis(carbamodithioate)
CAS RN *[142–59–6]* **EEC no.** 205–547–0

PHYSICAL CHEMISTRY
Mol. wt. 256.3 **M.f.** $C_4H_6N_2Na_2S_4$ **Form** Colourless crystals (hexahydrate).
M.p. Decomposes on heating, without melting. **V.p.** Negligible. **Solubility** In water *c.* 200 g/l (room temperature). Insoluble in common organic solvents.
Stability Decomposed by light, moisture and heat. Stable as an aqueous solution. On aeration, aqueous solutions deposit yellow mixtures of which the main fungicidal components are sulfur and etem (C. W. Pluijgers *et al.*, *Tetrahedron Lett.*, 1971, p. 1371; M. Alvarez *et al.*, *ibid.*, 1973, p. 939; W. R. Benson *et al.*, *J. Assoc. Off. Anal. Chem.*, 1972, **55**, 44).

COMMERCIALISATION
History Fungicide reported by A. E. Dimond *et al.* (*Phytopathology*, 1943, **33**, 1095). Introduced by E. I. du Pont de Nemours and Co. and by Rohm & Haas Co. (though neither company now manufactures it). **Patent** US 2317765

APPLICATIONS
Mode of action Fungicide with protective action, and algicide. **Uses** Control of algae in rice fields. Control of some fungal diseases of cotton, capsicums, and onions, when applied to the soil. Often combined with zinc sulfate (to form zineb) or manganese sulfate (to form maneb). **Phytotoxicity** Phytotoxic to plant foliage.
Formulation types SL. **Mixtures** (*nabam +*) manganese sulfate; zinc sulfate.

ANALYSIS
Product analysis by hydrolysis to carbon disulfide which is converted to dithiocarbonate and this is measured by titration with iodine (*CIPAC Handbook*, 1970, **1**, 539; 1983, **1B**, 1869; 1992, **E**, 151–153; *AOAC Methods*, 1995, 965.15).

MAMMALIAN TOXICOLOGY
Reviews *Pesticide residues in food – 1977*. FAO Plant Production and Protection Paper 10 Rev, 1978, *Pesticide residues in food: 1977 evaluations*. FAO Plant Production and Protection Paper 10 Sup, 1978. **EHC** 78 (WHO, 1988; general review of dithiocarbamates). **Oral** Acute oral LD_{50} for rats 395, mice 580 mg/kg. **Skin and eye** Acute percutaneous LD_{50} not available. Irritating to skin and mucous membranes. **NOEL** A goitrogenic effect noted in rats receiving 1000–2500 mg/kg diet for 10 d (R. B. Smith *et al.*, *J. Pharmacol. Exp. Ther.*, 1953, **109**, 159). **ADI** (JMPR) Temporary ADI withdrawn [1977]. **Other** Acute i.p. LD_{50} for rats 500 mg/kg. **Toxicity class** WHO (a.i.) II; EPA (formulation) II **EC risk** Xn (R22); Xi (R37); R43

ECOTOXICOLOGY
Fish Moderately toxic to fish. **Bees** Not toxic to bees.

ENVIRONMENTAL FATE
Plants In plants, the principal metabolite is ethylenethiourea. Other metabolites include ethylenethiuram monosulfide, ethylenethiuram disulfide, and sulfur.

07 naled

Insecticide, acaricide

organophosphorus

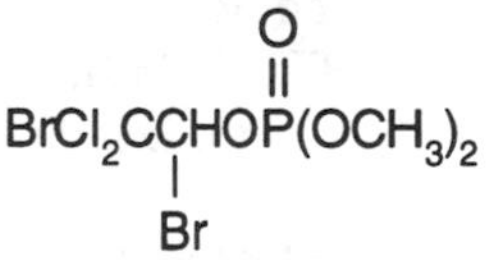

$$BrCl_2CCHOP(OCH_3)_2$$

with O double-bonded above P and Br below the CCH carbon.

NOMENCLATURE
Common name naled (BSI, E-ISO, (*m*) F-ISO, ANSI, ESA); bromchlophos (Republic of South Africa); dibrom (Denmark); BRP (JMAF)
IUPAC name 1,2-dibromo-2,2-dichloroethyl dimethyl phosphate
Chemical Abstracts name 1,2-dibromo-2,2-dichloroethyl dimethyl phosphate
CAS RN *[300–76–5]* **EEC no.** 206–098–3 **Development codes** RE-4355
Official codes OMS 75; ENT 24 988

PHYSICAL CHEMISTRY
Composition Tech. grade is c. 93% pure.   **Mol. wt.** 380.8   **M.f.** $C_4H_7Br_2Cl_2O_4P$
Form Colourless liquid with a slightly pungent odour. Tech. grade is yellow.
M.p. 26–27.5 °C   **B.p.** 110 °C/0.5 mmHg   **V.p.** 266 mPa (20 °C)
S.g./density 1.96 at 20 °C   **Solubility** Practically insoluble in water. Readily soluble
in aromatic and chlorinated solvents. Slightly soluble in aliphatic solvents and
mineral oils.   **Stability** Stable when dry, but rapidly hydrolysed in aqueous media
(over 90% in 48 h at room temperature), and more rapidly in alkaline and acidic
media. Degraded by sunlight. In the presence of metals and reducing agents,
bromine is lost and dichlorvos (q.v.) is formed.

COMMERCIALISATION
History Insecticide reported by J. M. Grayson & B. D. Perkins (*Pest Control*, 1960,
28(6), 9). Introduced by Chevron Chemical Co.   **Patent** GB 855157; US 2971882
Manufacturers Amvac

APPLICATIONS
Biochemistry Cholinesterase inhibitor; activity may be due to *in vivo*
debromination, forming dichlorvos (q.v.).   **Mode of action** Non-systemic
insecticide and acaricide with contact and stomach action, and some respiratory
action. Fast-acting.   **Uses** Control of spider mites, aphids, and other insects on
many crops, including fruit, vegetables, ornamentals, beet, hops, cotton, rice,
alfalfa, soya beans, tobacco, mushrooms, glasshouse crops, and in forestry. Also
used in animal houses and in public health, for control of insects such as flies, ants,
fleas, cockroaches, silverfish, etc.; and for control of mosquitoes.
Phytotoxicity Injury may be caused to apples, pears, melons, cherries, plums,
peaches, nectarines, beans, cotton, and some varieties of ornamentals.
Formulation types EC; DP; UL.   **Compatibility** Incompatible with alkaline
materials.   **Selected tradenames** 'Dibrom' (Valent); 'Bromex' (Makhteshim-Agan)

ANALYSIS
Product analysis by glc (D. E. Pack et al., *Anal. Methods Pestic., Plant Growth Regul.
Food Addit.*, 1964, **2**, 125; J. B. Leary, *Anal. Methods Pestic. Plant Growth Regul.*,
1972, **6**, 350).   **Residues** determined by glc (D. E. Pack et al., *loc. cit.*; J. B. Leary,
loc. cit.). Details of methods available from Chevron Chemical Co.

MAMMALIAN TOXICOLOGY
EHC 63 (WHO, 1986; a general review of organophosphorus insecticides).
Oral Acute oral LD_{50} for rats 430 mg/kg.   **Skin and eye** Acute percutaneous LD_{50}
for rabbits 1100 mg/kg. Irritating to skin; causes eye burns.   **Inhalation** No harm
was suffered by mice exposed for 6 hours to 1.5 mg/l air.   **NOEL** In 2 y feeding
trials, no ill-effect observed in albino rats receiving 100 mg tech. (91% pure)/kg
diet.   **Toxicity class** WHO (a.i.) II; EPA (formulation) I   **EC risk** Xn (R21/22);
Xi (R36/38)

ECOTOXICOLOGY
Fish LC$_{50}$ (24 h) for goldfish 2–4 mg/l; no mortality to mosquito fish when applied at 560 g/ha. **Bees** Toxic to bees. **Other aquatic spp.** LC$_{50}$ for crabs 0.33 mg/l. No mortality to tadpoles when applied at 560 g/ha.

ENVIRONMENTAL FATE
Plants In plants, the bromine atoms are reductively cleaved, with formation of dichlorvos (*q.v.*), which evaporates or undergoes rapid hydrolytic cleavage.

508 naphthenic acid   *Fungicide*

NOMENCLATURE
naphthenic acid
Common name naphthenic acid (BSI, E-ISO, accepted in lieu of a common name)
IUPAC name naphthenic acid, a group of carboxylic acids derived from a fraction of crude petroleum oils and includes alkylcyclopentanecarboxylic acids and alkylcyclohexanecarboxylic acids containing 9 or 10 carbon atoms

copper naphthenate
Common name copper naphthenate (BSI, E-ISO); naphténate de cuivre (F-ISO)
Chemical Abstracts name copper naphthenate
CAS RN *[1338–02–9]* **EEC no.** 215–657–0

zinc naphthenate
Common name naphténate de zinc (F-ISO)
Chemical Abstracts name zinc naphthenate

PHYSICAL CHEMISTRY
naphthenic acid
V.p. <133 mPa (100 °C)

copper naphthenate
Form Viscous green oil with unpleasant odour. **V.p.** <133 mPa (100 °C)
Solubility Practically insoluble in water. Soluble in most organic solvents; in petroleum oil 900 g/kg. **Stability** Stable to natural daylight. Not compatible with concentrated acids and alkalis.

zinc naphthenate
Form Almost colourless viscous oil.   **Solubility** Practically insoluble in water; in petroleum distillate 900 g/kg; soluble in most organic solvents.   **Stability** Stable to natural daylight. Not compatible with concentrated acids and alkalis.

COMMERCIALISATION
History Copper naphthenate and zinc naphthenate have long been used to protect wood from fungal attack. Naphthenic acid has been reviewed (E. R. Littleman & J. R. M. Klotz, *Chem. Rev.*, 1942, **30**, 97).

APPLICATIONS
Mode of action The salts are non-systemic fungicides.   **Uses** Copper and zinc naphthenates are used as fungicides in wood preservation, controlling Basidiomycetes and soft rot fungi responsible for decay of timber. Also as a tree wound dressing to prevent canker; and as an animal repellent to reduce damage to deciduous and coniferous trees.   **Phytotoxicity** Phytotoxic to growing plants.
Formulation types EC.   **Compatibility** Not compatible with concentrated acids or bases.

ANALYSIS
Product (1) by electrolytic estimation (*AOAC Methods*, 1995, 920.24*), atomic absorption spectrophotometry (BS 5666: Part 4: 1979; Part 5: 1986), by titration for copper (*AOAC Methods*, 1995, 964.03*) or colorimetry (for copper). (2) By glc with FID (for acid component).   **Residues** (1) by atomic absorption spectrophotometry or colorimetry (for copper). (2) By glc with FID (for acid component).

MAMMALIAN TOXICOLOGY
copper naphthenate
Oral Acute oral LD_{50} for rats 110, mice 6400–7200 mg/kg.   **Skin and eye** Acute percutaneous LD_{50} for rabbits >2000 mg/kg. Non-irritating to eyes (rabbits).
Inhalation LC_{50} (1 h) for rats >5.5 mg/l.   **EC risk** (R10); Xn (R22)

zinc naphthenate
Oral Acute oral LD_{50} for rats 4920 mg/kg.

2-(1-naphthyl)acetamide

Plant growth regulator

synthetic auxin

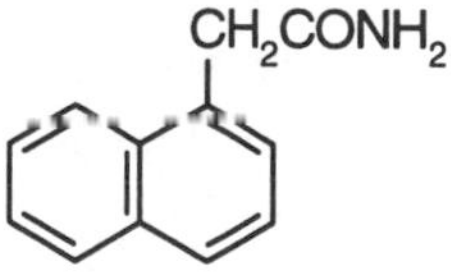

$$CH_2CONH_2$$

NOMENCLATURE
Common name 2-(1-naphthyl)acetamide (draft BSI, E-ISO; used in lieu of a common name)
IUPAC name 2-(1-naphthyl)acetamide
Chemical Abstracts name 1-naphthaleneacetamide
Other names α-naphthaleneacetamide; NAD; NAAm   **CAS RN** *[86–86–2]*

PHYSICAL CHEMISTRY
Mol. wt. 185.2   **M.f.** $C_{12}H_{11}NO$ **Form** Colourless crystals.   **M.p.** 184 °C
V.p. $<1 \times 10^{-2}$ mPa   **Solubility** In water 39 mg/kg (40 °C). Soluble in acetone, ethanol, and isopropanol. Insoluble in kerosene.   **Stability** Stable under normal storage conditions.   **F.p.** Non-flammable.

COMMERCIALISATION
History Introduced as a thinning agent for apples and pears by Amchem Products, Inc. (now Rhône-Poulenc Agrochimie).   **Manufacturers** Amvac

APPLICATIONS
Mode of action Acts by inducing formation of an abscission zone in the peduncle.
Uses Used for thinning many varieties of apples and pears; and to prevent premature fruit fall in apples and cherries.   **Formulation types** WP.
Selected tradenames 'Amid-Thin' (Rhône-Poulenc)

ANALYSIS
Product analysis by glc.   **Residues** determined by glc or hplc. Details available from Rhône-Poulenc, USA.

MAMMALIAN TOXICOLOGY
Oral Acute oral LD_{50} for rats c. 1690 mg/kg.   **Skin and eye** Acute percutaneous LD_{50} for rabbits >2000 mg/kg; minimally irritating to skin, severely to eyes.
Toxicity class WHO (a.i.) III (Table 5); EPA (formulation) III

MAIN ENTRIES

510 2-(1-naphthyl)acetic acid

Plant growth regulator

synthetic auxin

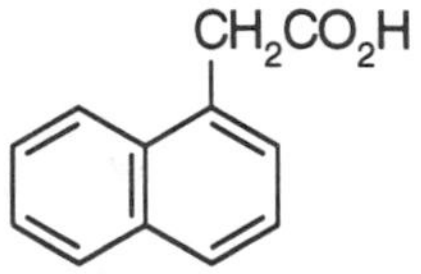

$$CH_2CO_2H$$

NOMENCLATURE
2-(1-naphthyl)acetic acid
Common name 1-naphthylacetic acid (BSI, E-ISO, accepted in lieu of a common name); 1-naphthaleneacetic acid (alternative BSI, E-ISO name, in lieu of a common name); acide naphtylacétique (F-ISO, accepted in lieu of a common name)
IUPAC name 2-(1-naphthyl)acetic acid
Chemical Abstracts name 1-naphthaleneacetic acid
Other names NAA; α-naphthaleneacetic acid   **CAS RN** *[86–87–3]*

ethyl 2-(1-naphthyl)acetate
CAS RN *[2122–70–5]*

PHYSICAL CHEMISTRY
2-(1-naphthyl)acetic acid
Mol. wt. 186.2   **M.f.** $C_{12}H_{10}O_2$ **Form** Colourless crystals.   **M.p.** 134–135 °C
V.p. <0.01 mPa (25 °C)   **Solubility** In water 420 mg/l (20 °C). In xylene 55, carbon tetrachloride 10.6 (both in g/l, 26 °C). Readily soluble in alcohols and acetone. Soluble in diethyl ether and chloroform. Alkali-metal and amine salts are readily soluble in water.   **Stability** Very stable in storage.

ethyl 2-(1-naphthyl)acetate
Mol. wt. 214.3   **M.f.** $C_{14}H_{14}O_2$ **Form** Colourless liquid.   **B.p.** 175 °C
S.g./density 1.106 (25 °C)   **Solubility** Practically insoluble in water (20 °C). Readily soluble in acetone, carbon disulfide, ethanol, isopropanol (20 °C). Slightly soluble in kerosene, diesel oil.   **Stability** Stable.

COMMERCIALISATION
History Plant growth-regulating activity reported by F. E. Gardiner et al. (*Science*, 1939, **90**, 208). Introduced by Amchem Products, Inc. (now Rhône-Poulenc Agrochimie) and ICI Plant Protection Division (now Zeneca Agrochemicals).
Manufacturers Amvac; Rallis

APPLICATIONS

2-(1-naphthyl)acetic acid

Mode of action Promotes elongation and formation of roots, and acts as a growth inhibitor at higher concentrations. **Uses** Used to stimulate root formation in cuttings of woody and herbaceous plants, and vines; to prevent premature flower and fruit drop in apples, pears, grapes, guavas, mangoes, water melons, pawpaws, citrus fruits, aubergines, cucumbers, cotton, soya beans, etc.; to improve setting in fruit; as a fruit thinning agent in apples, pears, olives, citrus, and other fruit; and to induce flowering in pineapples. **Formulation types** SL; SP; WP; EC; DP; TB. **Mixtures** *(2-(1-naphthyl)acetic acid +)* dichlorophen; 2-(1-naphthyl)acetamide; 4-indol-3-ylbutyric acid; 4-indol-3-ylbutyric acid + thiram; captan. **Compatibility** Solutions of the heavy-metal salts of other acids form sparingly-soluble salts of 1-naphthylacetic acid. **Selected tradenames** 'Fruitone-N' (Rhône-Poulenc); 'Drofix' (Productos OSA); 'Fruit Fix' (Amvac); 'Rhizopon B' (Fargro); 'Synergol' (mixture) (Hortichem); 'Tipoff' (Unicrop)

ethyl 2-(1-naphthyl)acetate

Mode of action Plant growth regulator with auxin-like activity. **Uses** Used to control aerial suckers from the roots of apple, pear, plum, cherry, and other fruit trees; to reduce water-shoot production in pruned apple and pear trees; and to inhibit sprouting at pruning points on ornamental trees. **Formulation types** AE; AL; Coating agent.

ANALYSIS

2-(1-naphthyl)acetic acid

Product analysis by glc of a derivative. **Residues** determined by u.v. spectrophotometry (*AOAC Methods,* 1995, 970.54) or by hplc. Details available from Rhône-Poulenc, USA, or Zeneca.

ethyl 2-(1-naphthyl)acetate

Product by glc.

MAMMALIAN TOXICOLOGY

2-(1-naphthyl)acetic acid

Oral Acute oral LD_{50} for rats *c.* 1000–5900 mg/kg (acid), for mice *c.* 700 mg/kg (sodium salt). **Skin and eye** Acute percutaneous LD_{50} for rabbits >5000 mg/kg. Slight to moderate irritation of skin on prolonged contact; severe irritation to eyes (rabbits). **Inhalation** LC_{50} (1 h) >20 000 ppm. **Toxicity class** WHO (a.i.) III (Table 5); EPA (formulation) III

ethyl 2-(1-naphthyl)acetate

Oral Acute oral LD_{50} for rats *c.* 3580 mg/kg. **Skin and eye** Acute percutaneous LD_{50} for rabbits >2000 mg/kg. Slight to moderate irritation to skin on prolonged contact; non-irritating to eyes (rabbits). **Inhalation** LC_{50} for rats >206.5 mg/l.

ECOTOXICOLOGY
2-(1-naphthyl)acetic acid
Birds Eight-day dietary LC_{50} for mallard ducks and bobwhite quail >10 000 mg/kg diet. **Fish** LC_{50} (96 h) for rainbow trout 57, bluegill sunfish 82 mg a.i./l.
Bees Not toxic to bees when used as directed. **Daphnia** LC_{50} (48 h) 360 mg/l.

511 (2-naphthyloxy)acetic acid

Plant growth regulator

---

synthetic auxin

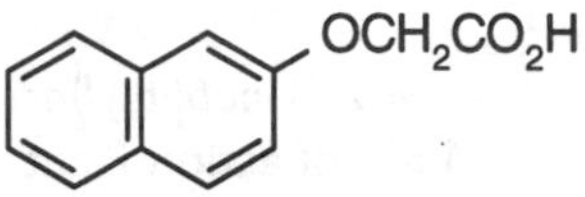

NOMENCLATURE
Common name (2-naphthyloxy)acetic acid (BSI, E-ISO, accepted in lieu of a common name); acide naphtyloxyacétique (F-ISO)
IUPAC name (2-naphthyloxy)acetic acid
Chemical Abstracts name (2-naphthalenyloxy)acetic acid
CAS RN *[120–23–0]*

PHYSICAL CHEMISTRY
Mol. wt. 202.2 **M.f.** $C_{12}H_{10}O_3$ **Form** Colourless crystals; (tech., green crystals).
M.p. 156 °C **Solubility** Sparingly soluble in water. Soluble in ethanol, acetic acid, and diethyl ether. Forms water-soluble alkali-metal and amine salts.

COMMERCIALISATION
History Effect of increasing fruit set reported by S. C. Bausor (*Am. J. Bot.*, 1939, **26**, 415). Introduced by Synchemicals Ltd.

APPLICATIONS
Mode of action Plant growth regulator, absorbed by the leaves and roots.
Uses Used as a fruit setting spray on tomatoes, strawberries, blackberries, capsicums, aubergines, grapes, and pineapples. **Formulation types** SL; EC.
Mixtures ((2-naphthyloxy)acetic acid +) 4-CPA. **Selected tradenames** 'Betapal' (Unicrop); 'Tomatosa' (Productos OSA)

ANALYSIS
Residues determined by hplc (T. E. Archer & J. D. Stokes, *J. Agric. Food Chem.*, 1978, **26**, 452).

MAMMALIAN TOXICOLOGY
Oral Acute oral LD$_{50}$ for rats 1000 mg/kg. **Toxicity class** WHO (a.i.) III

ECOTOXICOLOGY
Bees Not toxic to bees.

ENVIRONMENTAL FATE
Plants In plants, undergoes degradation to 2-naphthol, followed by ring hydroxylation and ring opening.

512 naproanilide   *Herbicide*

alkanamide

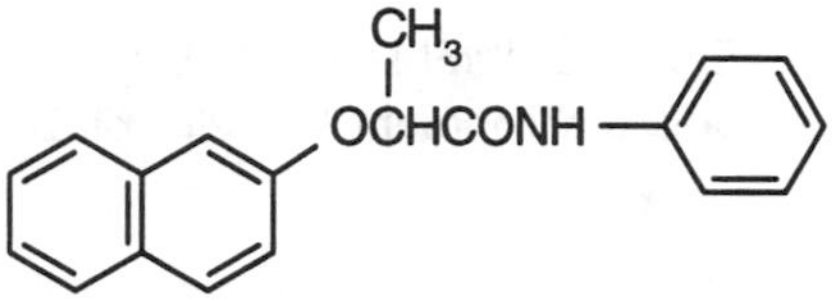

NOMENCLATURE
Common name naproanilide (JMAF)
IUPAC name *N*-phenyl-2-(2-naphthyloxy)propionamide
Chemical Abstracts name 2-(2-naphthalenyloxy)-*N*-phenylpropanamide
CAS RN *[52570–16–8]* unstated stereochemistry **Development codes** MT-101 (Mitsui)

PHYSICAL CHEMISTRY
Mol. wt. 291.3 **M.f.** C$_{19}$H$_{17}$NO$_2$ **Form** Whitish crystals. **M.p.** 128 °C
V.p. 2.8 × 10^{-6} mPa (20 °C) **S.g./density** 1.256 (25 °C) **Solubility** In water at 27 °C, 0.75 mg/l. In acetone 171, toluene 42, ethanol 17, benzene 46 (all in g/l, 27 °C). **Stability** Dilute aqueous solutions slowly degraded by u.v. light; solid stable to light. **F.p.** 110 °C

COMMERCIALISATION
History Herbicide reported by S. Fujisawa (*Jpn. Pestic. Inf.*, 1981, No 39, p. 19). Introduced by Mitsui Toatsu Chemicals, Inc. **Manufacturers** Mitsui Toatsu

APPLICATIONS
Biochemistry The active principle is thought to be the corresponding carboxylic acid. **Mode of action** Selective herbicide, absorbed through stems and roots.

Uses Controls annual and some perennial weeds, but not *Echinochloa crus-galli*, in paddy rice at 2–3 kg a.i./ha. Gives good control of young *Sagittaria pygmaea*. **Phytotoxicity** Phytotoxic to direct seeding rice at high dosages. **Formulation types** GR. **Mixtures** (*naproanilide* +) thiobencarb; pretilachlor; butachlor; mefenacet; bromobutide. **Selected tradenames** 'Uribest' (Mitsui Toatsu)

ANALYSIS
Product and **residue** analysis by hplc (Gotoh, *J. Pestic. Sci.*, 1977, (4), p. 2).

MAMMALIAN TOXICOLOGY
Oral Acute oral LD_{50} for rats >15 000, mice >20 000 mg/kg. **Skin and eye** Acute percutaneous LD_{50} for rats >3000, mice >5000 mg/kg. **Toxicity class** WHO (a.i.) IV (tech); EPA (formulation) IV

ECOTOXICOLOGY
Fish TLm (72 h) >100 ppm. **Daphnia** LC_{50} (6 h) 40 mg/l.

ENVIRONMENTAL FATE
Soil/Environment DT_{50} in soil 2–7 d.

513 napropamide *Herbicide*

alkanamide

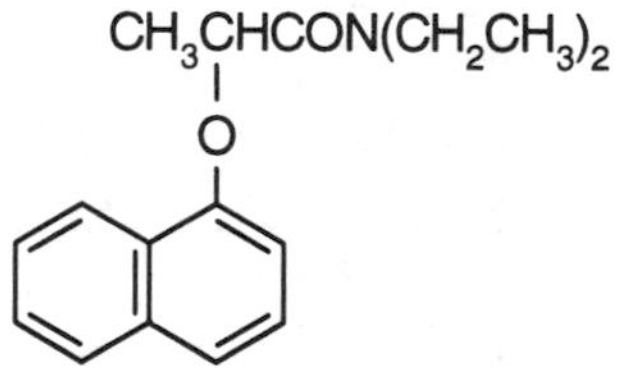

CH₃CHCON(CH₂CH₃)₂

NOMENCLATURE
Common name napropamide (BSI, E-ISO, (*m*) F-ISO, WSSA, JMAF)
IUPAC name (*RS*)-*N,N*-diethyl-2-(1-naphthyloxy)propionamide
Chemical Abstracts name *N,N*-diethyl-2-(1-naphthalenyloxy)propanamide
CAS RN [15299–99–7] unstated stereochemistry; [41643–35–0] (*R*)-(–)- isomer; [41643–36–1] (*S*)-(+)- isomer **Development codes** R-7465 (Zeneca)

PHYSICAL CHEMISTRY
Composition Tech. grade is 92–96% pure. **Mol. wt.** 271.4 **M.f.** $C_{17}H_{21}NO_2$
Form Colourless crystals; (tech., brown solid). **M.p.** 74.8–75.5 °C; (tech.

68–70 °C)   **V.p.** 0.53 mPa (25 °C) K_{ow} logP = 3.3 (25 °C)   **S.g./density** Bulk density 0.584 g/ml   **Solubility** In water 73 mg/l (25 °C). In acetone, ethanol >1000, xylene 505, kerosene 62, hexane 15 (all in g/l, 20 °C); miscible with acetone, ethanol, methyl isobutyl ketone.   **Stability** No decomposition occurs over 16 h at 100 °C. Decomposed by sunlight; DT_{50} 25.7 min. Stable to hydrolysis between pH 4 and 10 at 40 °C.   **F.p.** >104 °C

COMMERCIALISATION
History Herbicide reported by B. J. van den Brink *et al.* (*Symp. New Herbic., 3rd,* 1969, p. 35). Introduced by Stauffer Chemical Co. (now Zeneca Agrochemicals). **Patent** US 3480671; US 3718455   **Manufacturers** Zeneca

APPLICATIONS
Biochemistry Cell division inhibitor.   **Mode of action** Selective systemic herbicide, absorbed by the roots, with translocation acropetally. Inhibits root development and growth.   **Uses** Pre-emergence control of annual grasses and broad-leaved weeds in asparagus, rhubarb, cucurbits, brassicas, oilseed rape, tomatoes, capsicums, potatoes, peas, nuts, fruit trees and bushes (including citrus), vines, strawberries, sunflowers, safflowers, ornamentals, tobacco, olives, figs, mint, turf, and other crops at 2–6 kg/ha. The (R)-(–)- isomer is 8 times as toxic to 3 weed species as the (S)-(+)- isomer (J. H. Chan *et al., J. Agric. Food Chem.,* 1975, **23**, 1008).   **Phytotoxicity** Phytotoxic to wheat and barley.   **Formulation types** EC; WP; GR; SC.   **Mixtures** *(napropamide +)* monolinuron; nitralin; simazine; trifluralin; tefluthrin; tebutam.   **Selected tradenames** 'Devrinol' (Zeneca)

ANALYSIS
Product analysis by glc.   **Residues** in crops and soil determined by glc (G. G. Patchett *et al., Anal. Methods Pestic. Plant Growth Regul.,* 1976, **8**, 347). In **drinking water**, by gc with FID (*AOAC Methods*, 1995, 991.07). Details of analytical methods available from Zeneca.

MAMMALIAN TOXICOLOGY
Oral Acute oral LD_{50} for male rats >5000 mg tech./kg, female rats 4680 mg/kg.
Skin and eye Acute percutaneous LD_{50} for rabbits >4640, guinea pigs >2000 mg/kg. Moderate eye irritant, not a skin irritant (rabbits). Not a skin sensitiser (guinea pigs).   **Inhalation** LC_{50} (4 h) for rats >5 mg/l.   **NOEL** (2 y) for rats 30 mg/kg b.w. daily; (90 d) for dogs 40 mg/kg b.w. daily. Developmental toxicity NOEL for rats and rabbits 1000 mg/kg b.w. daily; multigeneration study (rats) 30 mg/kg b.w. daily.   **ADI** 0.1 mg/kg.   **Other** Negative genotoxicity. Not oncogenic.   **Toxicity class** WHO (a.i.) III (Table 5); EPA (formulation) III

ECOTOXICOLOGY
Birds In 7-day feeding trials, no-effect level for bobwhite quail >5600 mg/kg.
Fish LC_{50} (96 h) for bluegill sunfish 30, rainbow trout 16.6, goldfish >10 mg/l.
Bees LD_{50} 0.121 mg/bee.   **Daphnia** LC_{50} (48 h) 24 mg/l.

ENVIRONMENTAL FATE

Animals In mammals, following oral administration, napropamide is rapidly and extensively metabolised. The majority of the dose is excreted via urine and faeces. Similar effects have been observed in poultry. **Plants** Rapidly metabolised via ring hydroxylation and N-dealkylation, followed by conjugation with sugars to produce water soluble metabolites. **Soil/Environment** K_{oc} 600 (range 470–1200). In aerobic laboratory soil, degradation is slow, DT_{50} 446 d (30 °C); however, in the field, typical DT_{50} is 25 d (range 9–131 d in N. America/Germany). Photodegradation is an important mechanism for loss from soil. Degradates identified in soil are 1-naphthoxypropionic acid, 2-(α-naphthoxy)-N-ethyl-= N-hydroxyethylpropionamide, 2-(α-naphthoxy)-N-ethylpropionamide, 2-(α-naphthoxy)propionic acid, 2-hydroxy-1,4-naphthoquinone, 1,4-naphthoquinone and o-phthalic acid.

514 naptalam   *Herbicide*

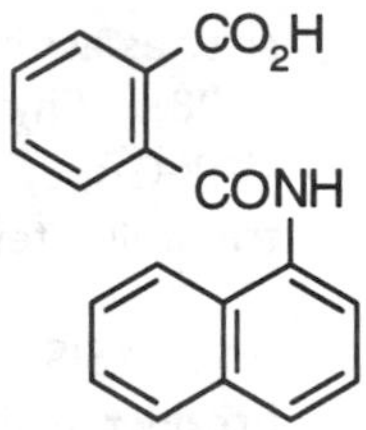

NOMENCLATURE

naptalam

Common name naptalam (BSI, E-ISO, WSSA); naptalame ((*m*) F-ISO); [alanap] (former exception, Turkey)

IUPAC name N-1-naphthylphthalamic acid

Chemical Abstracts name 2-[(1-naphthalenylamino)carbonyl]benzoic acid

Other names NPA   **CAS RN** *[132–66–1]*

naptalam-sodium

Common name NPA (JMAF)

CAS RN *[132–67–2]*

PHYSICAL CHEMISTRY

naptalam

Mol. wt. 291.3   **M.f.** $C_{18}H_{13}NO_3$ **Form** Crystalline solid.   **M.p.** 185 °C

V.p. <133 Pa (20 °C) K_{ow} logP = 0.104 (pH 5), 0.004 (pH 7), −0.036 (pH 9)

S.g./density 1.36 (20 °C)   **Solubility** In water 200 mg/l (20 °C). In acetone 5, dimethylformamide 39, dimethyl sulfoxide 43, methyl ethyl ketone 4, isopropanol 2, carbon tetrachloride 0.1 (all in g/kg). Practically insoluble in benzene, hexane, and xylene. **Stability** Hydrolysed in solutions of pH >9.5; unstable at elevated temperatures, forming N-(1-naphthyl)phthalimide. **pKa** 4.6

naptalam-sodium
Mol. wt. 313.3 **M.f.** $C_{18}H_{12}NNaO_3$ **Solubility** In water 300 g/kg (20 °C). In acetone 17, dimethylformamide 50, methyl ethyl ketone 6, isopropanol 21, benzene 0.5, xylene 0.4 (all in g/kg).

COMMERCIALISATION
History Plant growth-regulating activity of N-arylphthalamic acids reported by O. L. Hoffman & A. E. Smith (*Science*, 1949, **109**, 588). Naptalam-sodium introduced by Uniroyal Chemical Co., Inc. **Patent** US 2556664; US 2556665 **Manufacturers** Uniroyal

APPLICATIONS
Biochemistry Inhibits IAA transport. **Mode of action** Selective herbicide, absorbed predominantly by the roots, but also, to some extent, by the foliage, with accumulation in meristematic tissue. Inhibits seed germination. **Uses** Pre-emergence control of many broad-leaved weeds and some grasses in cucurbits, asparagus, peanuts, soya beans, and established woody ornamentals at 2.0–5.5 kg/ha. **Phytotoxicity** Phytotoxic to beet, tomatoes, spinach, and lettuce.

naptalam
Formulation types GR; SL; WP. **Compatibility** Incompatible with pesticides which may cause amine transfer or replacement. **Selected tradenames** 'Alanap' (Uniroyal)

naptalam-sodium
Selected tradenames 'Alanap-L' (Uniroyal)

ANALYSIS
Product analysis by u.v. spectrometry, details from Uniroyal Inc.
Residues determined by hydrolysis to 1-naphthylamine, a derivative of which is measured by colorimetry (A. E. Smith & G. M. Stone, *Anal. Methods Pestic., Plant Growth Regul. Food Addit.*, 1964, **4**, 1).

MAMMALIAN TOXICOLOGY
naptalam
Oral Acute oral LD_{50} for rats 1770 mg/kg. **Skin and eye** Acute percutaneous LD_{50} for rabbits >5000 mg/kg. Severe eye irritant, mild skin irritant (rabbits).
Inhalation LC_{50} (4 h) for rats >2.07 mg/l air. **NOEL** In chronic feeding trials, NOEL for rats 30, dogs 5 mg/kg b.w. daily. **ADI** 0.05 mg/kg. **Other** Not a bacterial mutagen. **Toxicity class** WHO (a.i.) III (Table 5); EPA (formulation) I

ECOTOXICOLOGY
naptalam
Birds LD_{50} for mallard ducks >4640 mg/kg. Eight-day dietary LC_{50} for mallard ducks and bobwhite quail >10 000 ppm.   **Fish** LC_{50} (96 h) for rainbow trout 76.1, bluegill sunfish 354 ppm.   **Bees** Not toxic to bees.   **Daphnia** LC_{50} (48 h) 118.5 mg/l.

ENVIRONMENTAL FATE
Plants In plants, naptalam rapidly breaks down into phthalic acid and 1-naphthylamine.   **Soil/Environment** Duration of residual activity is *c.* 3–8 w, depending on soil type and moisture content. (E. Smith *et al., J. Agric. Food Chem.*, 1957, **5**, 748). For sandy loam adsorption K_d 14.98, K_{oc} 2152.

515 natamycin   *Fungicide*

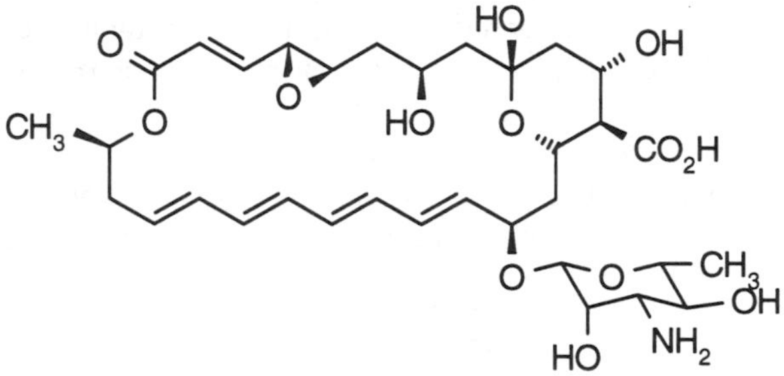

NOMENCLATURE
Common name natamycin (BAN); pimaricin (traditional name); tennecetin (traditional name)
IUPAC name (8E,14E,16E,18E,20E)-(1R,3S,5R,7R,12R,22R,24S,25R,26S)-22-= (3-amino-3,6-dideoxy-β-D-mannopyranosyloxy)-1,3,26-trihydroxy-12-methyl-= 10-oxo-6,11,28-trioxatricyclo[22.3.1.0⁵,⁷]octacosa-8,14,16,18,20-pentaene-= 25-carboxylic acid
Chemical Abstracts name [1R-(1R*,3S*,5R*,7R*,8E,12R*,14E,16E,18E,20E,22R*,= 24S*25R*,26S*)]-22-[(3-amino-3,6-dideoxy-β-D-mannopyranosyl)oxy]-1,3,26-= trihydroxy-12-methyl-10-oxo-6,11,28-trioxatricyclo[22.3.1.0⁵,⁷]octacosa-= 8,14,16,18,20-pentaene-25-carboxylic acid
Other names myprozine   **CAS RN** *[7681–93–8]*   **EEC no.** 231–683–5

PHYSICAL CHEMISTRY
Composition Exists as a trihydrate. Tech. is >87%. **Mol. wt.** 665.7 (anhydrous) **M.f.** $C_{33}H_{47}NO_{13}$ **Form** White crystals. **M.p.** 280–300 °C (decomp.) **Solubility** In water 4.1 g/l (20–22 °C). In dimethyl sulfoxide >200, methanol 97, ethanol 5.5, acetone 0.85, petroleum spirit 0.1, benzene 0.05 (all in g/l, 20–22 °C) (J. R. Marsh & P. J. Weiss, *J. Assoc. Off. Anal. Chem.*, 1967, **50**, 457). **Stability** Stable when dry, but sensitive to light. Forms water-soluble salts with acids or alkalis.

COMMERCIALISATION
Production By fermentation of *Streptomyces natalensis* and *S. chattanoogensis*. **History** Structure established by B. T. Golding *et al.* (*Tetrahedron Lett.*, 1966, p. 3551), W. E. Meyer (*Chem. Commun.*, 1968, p. 470) and G. Gaudiano *et al.* (*Chem. Ind. (Milan)*, 1966, **48**, 1327). Stereochemistry revised, J. M. Lancelin & J. M. Beau, *J. Am. Chem. Soc.*, 1990, **112**, 4060, 6749; A. J. Duplantier & S. Masamune, *ibid.*, p. 7079. Introduced as a fungicide by Gist-Brocades N.V. **Patent** GB 712547; GB 844289; US 3892850 **Manufacturers** Gist-Brocades

APPLICATIONS
Uses Control of diseases of bulbs, especially basal rot (*Fusarium oxysporum*) of daffodils, preferably combined with a hot-water treatment.
Formulation types WP. **Selected tradenames** 'Delvolan' (Gist-Brocades)

ANALYSIS
Product analysis is either by bioassay with a suitable micro-organism, and confirmed by the effect of the enzyme pimaricinase, or by u.v. spectrometry. **Residues** may be determined by bioassay.

MAMMALIAN TOXICOLOGY
Oral Acute oral LD_{50} for rats 2730–4670 mg/kg (G. J. Levinskas, *Toxicol. Appl. Pharmacol.*, 1966, **8**, 97). **Skin and eye** No acute toxicity, even at high doses. Not a skin or eye irritant (rabbits). Not a skin sensitiser. **Toxicity class** WHO (a.i.) III

ECOTOXICOLOGY
Fish Not toxic to fish.

ENVIRONMENTAL FATE
Readily biodegradable.

urea

CH₃ structure: Cl—(ring)—$NHCON(CH_2)_3CH_3$ with CH_3 on N, and Cl at two positions

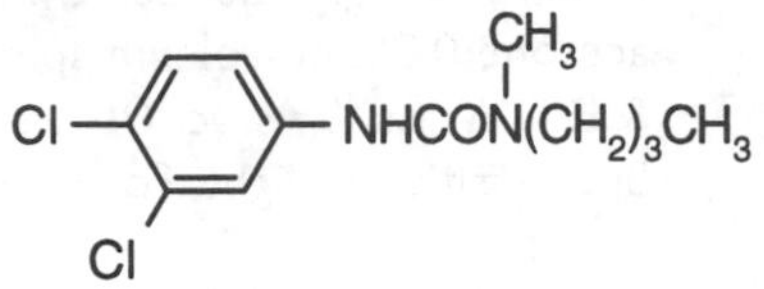

NOMENCLATURE
Common name neburon (BSI, E-ISO, (*m*) F-ISO, ANSI, WSSA); neburea (Republic of South Africa)
IUPAC name 1-butyl-3-(3,4-dichlorophenyl)-1-methylurea
Chemical Abstracts name *N*-butyl-*N'*-(3,4-dichlorophenyl)-*N*-methylurea
CAS RN *[555–37–3]*   **EEC no.** 209–096–0

PHYSICAL CHEMISTRY
Mol. wt. 275.2   **M.f.** $C_{12}H_{16}Cl_2N_2O$ **Form** Colourless crystals.
M.p. 102–103 °C   **Solubility** In water 5 mg/l (25 °C). Sparingly soluble in hydrocarbon solvents.   **Stability** Very stable to moisture and atmospheric oxidation in neutral media. Hydrolysed by acids and alkalis.

COMMERCIALISATION
History Herbicide reported by H. C. Bucha & C. W. Todd (*Science*, 1951, **144**, 493). Introduced by E. I. du Pont de Nemours and Co. (who no longer manufacture or market it).   **Patent** US 2655444; US 2655445
Manufacturers Makhteshim-Agan; Rhône-Poulenc; United Phosphorus Ltd

APPLICATIONS
Biochemistry Photosynthetic electron transport inhibitor.
Mode of action Selective herbicide, absorbed through the roots.   **Uses** Pre-emergence control of annual broad-leaved weeds and grasses in beans, peas, alfalfa, garlic, cereals, beet, strawberries, ornamentals, and in forestry at 2–3 kg a.i./ha.   **Formulation types** WP.   **Mixtures** (*neburon* +*)* butralin; chlomethoxyfen; flurochloridone; isoproturon; lenacil; nitrofen; pendimethalin; terbutryn; bifenox + isoproturon.   **Selected tradenames** 'Granurex' (Rhône-Poulenc); 'Neburex' (Makhteshim-Agan)

ANALYSIS
Residues determined by glc with ECD (*Anal. Methods Pestic., Plant Growth Regul. Food Addit.*, 1964, **4**, 157; *Anal. Methods Pestic. Plant Growth Regul.*, 1972, **6**, 664). **In water**, determined by lc with u.v. detection (*AOAC Methods*, 1995, 992.14, 10.7.01).

MAMMALIAN TOXICOLOGY

Oral Acute oral LD_{50} for rats >11 000 mg/kg.   **Skin and eye** Acute percutaneous LD_{50} not available. A 15% suspension in dimethyl phthalate is a mild skin irritant (guinea pigs).   **Toxicity class** WHO (a.i.) III (Table 5)

ECOTOXICOLOGY

Fish Mortality of four fish species was 90% at 0.6–0.9 mg/l (96 h).   **Bees** Low toxicity to bees.

ENVIRONMENTAL FATE

Soil/Environment As with other phenylurea herbicides, degradation involves dealkylation of the terminal nitrogen atom, ring hydroxylation, degradation to dichlorohydroxyaniline, presumably followed by ring opening. Duration of residual activity in soil is *c.* 3–4 mo (W. Bailey & J. L. White, *Residue Rev.*, 1965, **10**, 97).

517 nickel bis(dimethyldithiocarbamate)

Fungicide, bactericide

dithiocarbamate

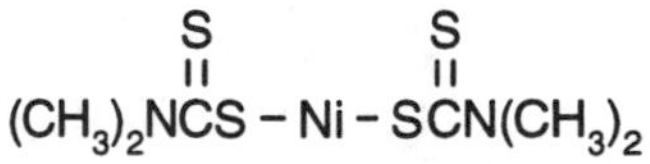

$$(CH_3)_2NC\overset{\overset{\displaystyle S}{\|}}{}S - Ni - S\overset{\overset{\displaystyle S}{\|}}{C}N(CH_3)_2$$

NOMENCLATURE

IUPAC name nickel bis(dimethyldithiocarbamate)
Chemical Abstracts name nickel dimethylcarbamodithioate
Other names organic nickel (JMAF)   **CAS RN** *[15521–65–0]*
Development codes M-1; DDC-Ni

PHYSICAL CHEMISTRY

Composition Tech. is ≥96%.   **Mol. wt.** 299.1   **M.f.** $C_6H_{12}N_2NiS_4$ **Form** Green amorphous powder.   **M.p.** Decomposes above 250 °C   **Solubility** Practically insoluble in water. In chloroform 287, dimethylformamide 560, tetrahydrofuran 475 (all in mg/l, 20 °C).   **Stability** Stable in acidic and alkaline media at room temperature.

COMMERCIALISATION

History Fungicide introduced by Mikasa Chemical Co. (who merged with Yashima Chemical Co. in 1996).   **Manufacturers** Yashima

APPLICATIONS
Mode of action Foliar fungicide and bactericide with protective action. **Uses** Control of bacterial leaf blight and bacterial grain rot in paddy rice, at 180–240 g/10a. **Formulation types** DP; WP. **Selected tradenames** 'Sankel' (Yashima)

ANALYSIS
Residues by decomposition with acid and determination of the liberated carbon disulfide by spectrophotometry (*AOAC Methods*, 1984, 139; *Proc. 2nd Congr. Pestic. Sci. Soc. Japan*, 1977, 127).

MAMMALIAN TOXICOLOGY
EHC 78 (WHO, 1988; general review of dithiocarbamates). **Oral** Acute oral LD_{50} for rats >36 000, mice >30 000 mg/kg. **Skin and eye** Acute percutaneous LD_{50} for rats >5000 mg/kg. **NOEL** (2 y) for rats 200 mg/kg diet.

ECOTOXICOLOGY
Fish LC_{50} (48 h) for carp 360 mg/l.

518 niclosamide   *Molluscicide*

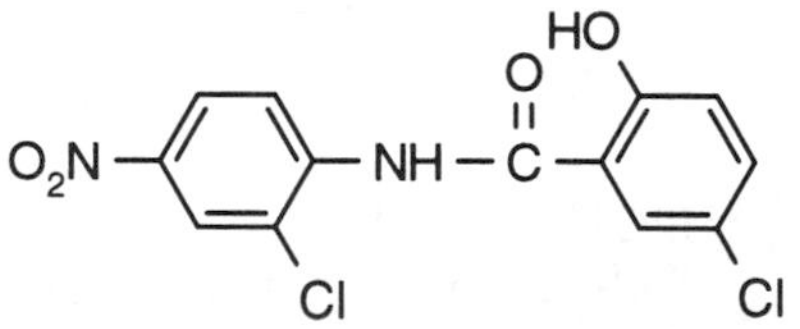

NOMENCLATURE
niclosamide
Common name niclosamide (BSI, E-ISO, (*m*) F-ISO, BAN, Germany (for veterinary use))
IUPAC name 2',5-dichloro-4'-nitrosalicylanilide
Chemical Abstracts name 5-chloro-*N*-(2-chloro-4-nitrophenyl)-2-hydroxy= benzamide
CAS RN *[50–65–7]* **Development codes** Bayer 25 648; Bayer 73; SR 73

niclosamide-olamine
Common name niclosamide-olamine (BSI, E-ISO, (*m*) F-ISO, BAN); clonitralid (Germany, for public health use)
CAS RN *[1420–04–8]* **Development codes** SR73

PHYSICAL CHEMISTRY

niclosamide

Mol. wt. 327.1   **M.f.** $C_{13}H_8Cl_2N_2O_4$ **Form** Almost colourless crystals.
M.p. 230 °C   **V.p.** <1 mPa (20 °C) K_{ow} logP = 1 (pH 9.6)   **Solubility** In water
1.6 (pH 6.4), 110 (pH 9.1) (both in mg/l, 20 °C). Soluble in common organic
solvents such as ethanol and diethyl ether.   **Stability** Hydrolysed in aqueous
solution, DT_{50} (20 °C) c. 7 d (pH 6.9); 18.8 d (pH 13.3). Stable to heat.
Decomposes under u.v. irradiation. Hydrolysed by concentrated acid or alkali.

niclosamide-olamine

Mol. wt. 388.2   **M.f.** $C_{15}H_{15}Cl_2N_3O_5$ **Form** Yellow crystalline solid.
M.p. Decomposes at 208 °C   **V.p.** <<1 mPa (20 °C)   **Solubility** In water 0.1 g/l
(20 °C). In *n*-hexane, toluene <0.1, dichloromethane 0.015, isopropanol 0.25 (all
in g/l, 20 °C).

COMMERCIALISATION

History Molluscicidal properties described by R. Gönnert & E. Schraufstätter (*Proc.
Int. Conf. Trop. Med. Malar.*, 1958, **2**, 5) and development discussed by R. Gönnert
et al. (*Z. Naturforsch., Teil B*, 1961, **16**, 95). The olamine salt introduced as a
molluscicide by Bayer AG.   **Patent** DE 1126374; US 3079297; US 3113067
Manufacturers Bayer

APPLICATIONS

Mode of action Molluscicide with respiratory and stomach action.   **Uses** Control
of golden apple snail in rice. Control of schistosomiasis and fascioliasis in man by
killing fresh-water snails which act as intermediate hosts. Also used for veterinary
control of tapeworm infestations.

niclosamide

Formulation types EC.   **Compatibility** Incompatible with acidic materials.
Selected tradenames 'Bayluscide (EC250)' (Bayer)

niclosamide-olamine

Formulation types WP.   **Selected tradenames** 'Bayluscide (WP70)' (Bayer)

ANALYSIS

Product analysis by redox titration (WHO *Specifications Pestic. Used in Public
Health*, p. 309).   **Residues** in water measured by colorimetry (R. Strufe,
Pflanzenschutz-Nachr. (Engl. Ed.), 1962, **15**, 42) or by titration (details from Bayer
AG).

MAMMALIAN TOXICOLOGY

niclosamide

Oral Acute oral LD_{50} for rats ≥5000 mg/kg.   **Skin and eye** Acute percutaneous
LD_{50} for rats >1000 mg/kg (EC250). Strong eye irritant; skin reacts after repeated

and long-lasting exposure (rabbits). **Inhalation** LC_{50} for rats (1 h)
20 mg/l air. **NOEL** (2 y) for male rats 2000, female rats 8000, mice 200 mg/kg
diet; (1 y) for dogs 100 mg/kg b.w. **ADI** (proposed) 3 mg/kg b.w. **Other** No
relevant mutagenic or embryotoxic effect. **Toxicity class** WHO (a.i.) III (Table 5);
EPA (formulation) II

niclosamide-olamine
Oral Acute oral LD_{50} for rats >5000 mg tech./kg. **Skin and eye** Acute
percutaneous LD_{50} for rats >2000 mg (as 70% WP)/kg. **Inhalation** LC_{50} (4 × 1 h)
3630–8224 mg/m^3 air (as 70% WP).

ECOTOXICOLOGY
niclosamide
Birds LD_{50} for mallard ducks >500 mg/kg. **Fish** LC_{50} (96 h) for golden orfe
0.1 mg/l. **Bees** No significant mortality effects. **Daphnia** LC_{50} (48 h) 0.2 mg/l.

ENVIRONMENTAL FATE
Animals Following oral administration, ^{14}C-niclosamide was absorbed and
metabolised in the rat. The major metabolite in the urine was the reduced
compound 2',5-dichloro-4'-aminosalicylanilide *[10558–45–9]*; several labile
conjugates were also detected. The major constituent in the faeces was unchanged
niclosamide although considerable amounts of 2',5-dichloro-4'-aminosalicylanilide
were also present; parent compound is present not only because of non-
absorption, but also because of release from the biliary conjugate by
β-glucuronidase of the intestinal microflora (L. A. Griffiths & V. Facchini, "The
major metabolite of niclosamide: Identification by mass spectrometry" in *Recent
Developments in Mass Spectrometry in Biochemistry and Medicine*, **2**, 121–126
(1979)). Another study indicates that niclosamide is very poorly absorbed after
dermal application. Radioactivity in the urine and faeces after application of
^{14}C-niclosamide accounted for <2 % and 10 % of the labelled compound applied
to pig and rat skin, respectively; *c.* 20 % was recovered from the area of
application (P. Brennan *et al.,* "Dermal absorption of niclosamide in rats and
minipigs" in *Biopharmaceutics & Drug Disposition,* Vol. 12, 547–556). Studies in fish
with niclosamide and its 2-aminoethanol salt, indicate that niclosamide is rapidly
excreted, as the glucuronide conjugate, and that there is little biomagnification; see
D. P. Schultz & P. D. Harman, *J. Agr. Food Chem.,* **26**, 1226–1230 (1978), J. L. Allen
et al., "Excretion of the Lampricide Bayer 73 by Rainbow Trout", Special Technical
Publication, Philadelphia, 667, 52–61 (1979) and M. S. M. Marzouk,
"Laboruntersuchungen an Karpfen über Rückstände von Bayluscid und dessen
Einwirkung auf das Karpfenblut", PhD Thesis, Univ. Munich, Germany (1981).
Soil/Environment There was a rapid decline in niclosamide residues in paddy
water; degradation followed pseudo-first order kinetics, DT_{50} 0.3 d. At harvest,
niclosamide residues were below the detection limit of 0.03 mg/kg in rice leaves,
stalk and grain, indicating that the use of niclosamide as a molluscicide in rice

production does not lead to persistent residues in the rice paddy ecosystem (S. M. F. Calumpang *et al., Bull. Environm. Contam. Tox.*, **55**, 494–501 (1995)). An aqueous solution of ^{14}C-niclosamide was 95% degraded after 14 d exposure to long-wavelength u.v. light. No degradation occurred within 56 days either in buffered solution (pH 5.0, 6.9 and 8.7) or in pond water (initial pH 7.8) (D. P. Schultz & P. D. Harman, U.S. Fish and Wildlife Service: Investigations in Fish Control, **83**, 1–5 (1978)).

519   nicosulfuron   *Herbicide*

sulfonylurea

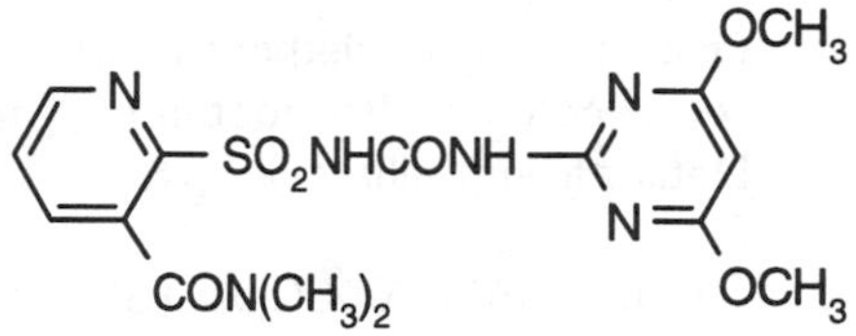

NOMENCLATURE
Common name nicosulfuron (BSI, ANSI, draft E-ISO); no name (Brazil)
IUPAC name 2-(4,6-dimethoxypyrimidin-2-ylcarbamoylsulfamoyl)-*N,N*-dimethyl= nicotinamide; 1-(4,6-dimethoxypyrimidin-2-yl)-3-(3-dimethylcarbamoyl-2-pyridyl= sulfonyl)urea
Chemical Abstracts name 2-[[[[(4,6-dimethoxy-2-pyrimidinyl)amino]carbonyl]= amino]sulfonyl]-*N,N*-dimethyl-3-pyridinecarboxamide
CAS RN *[111991–09–4]*   **Development codes** SL-950; MU-495 (both Ishihara); DPX-V9360 (Du Pont)

PHYSICAL CHEMISTRY
Mol. wt. 410.4   **M.f.** $C_{15}H_{18}N_6O_6S$ **Form** Colourless crystals.   **M.p.** 141–144 °C
V.p. 1.6×10^{-11} mPa   **K$_{ow}$** logP = –0.36 (pH 5), –1.8 (pH 7), –2 (pH 9)
Henry 0.180 Pa m^3 mol^{-1} (20 °C)   **S.g./density** 0.313 (bulk)   **Solubility** In buffered water 3.59 (pH 5), 12.2 (pH 7), 39.2 (pH 9) (all in g/kg, 25 °C). In acetone 18, ethanol 4.5, chloroform, dimethylformamide 64, acetonitrile 23, toluene 0.370, hexane <0.02, dichloromethane 160 (all in g/kg, 25 °C).   **Stability** Hydrolysis DT$_{50}$ 15 d (pH 5); stable at pH 7 & 9.   **pKa** 4.6 (25 °C)   **F.p.** >200 °C (Cleveland open cup)

COMMERCIALISATION
History Herbicide introduced by E. I. du Pont de Nemours and Co. and by Ishihara Sangyo Kaisha, Ltd.   **Patent** US 4789393   **Manufacturers** Du Pont; IPESA; Ishihara Sangyo

APPLICATIONS

Biochemistry Branched chain amino acid synthesis (acetolactate synthase or ALS) inhibitor. Acts by inhibiting biosynthesis of the essential amino acids valine and isoleucine, hence stopping cell division and plant growth. Crop selectivity derives from selective metabolism (pyrimidine-5-hydroxylation, followed by conjugation with glucose).   **Mode of action** Selective systemic herbicide, absorbed by the foliage and roots, with rapid translocation to the meristematic tissues.
Uses Selective post-emergence control in maize of annual grass weeds including *Setaria, Echinochloa, Digitaria, Panicum, Lolium,* and *Avena* spp., broad-leaved weeds including *Amaranthus* spp. and Cruciferae, and perennials such as *Sorghum halepense* and *Agropyron repens.* Applied at 35–70 g/ha.   **Formulation types** SC; WG.   **Selected tradenames** 'Accent' (Du Pont); 'Dasul' (Ishihara Sangyo, Dr Maag); 'Milagro' (Ishihara Sangyo, Zeneca); 'Nostoc' (IPESA)

ANALYSIS

Product by hplc. Methods for sulfonylurea **residues** in crops, soil and water reviewed (A. C. Barefoot *et al., Proc. Br. Crop Prot. Conf. – Weeds,* 1995, **2**, 707). Details from Ishihara Sangyo.

MAMMALIAN TOXICOLOGY

Reviews Material Safety Data sheets (available from Du Pont).   **Oral** Acute oral LD_{50} for rats and mice >5000 mg/kg.   **Skin and eye** Acute percutaneous LD_{50} for rats >2000 mg/kg. Non-sensitising to skin (rabbits). Moderate irritant to eyes of rabbits; not skin sensitiser to guinea pigs.   **Inhalation** LC_{50} for rats (4 h) 5.47 mg/l. **NOEL** In 28 d feeding trials on rats and mice, no adverse effect up to 30 g/kg diet. **Other** Non-mutagenic in the Ames test.   **Toxicity class** WHO (a.i.) III (Table 5); EPA (formulation) IV

ECOTOXICOLOGY

Birds Dietary oral LD_{50} for bobwhite quail >2250 mg/kg. Dietary LC_{50} for mallard ducks and bobwhite quail >5620 ppm.   **Fish** LC_{50} (96 h) for bluegill sunfish and rainbow trout >1000 mg/l.   **Bees** LD_{50} (contact) >20 µg/bee; dietary LC_{50} (48 h) >1000 ppm. NOEC 500 ppm.   **Worms** LC_{50} (14 d) for earthworms >1000 mg/kg.   **Daphnia** LC_{50} (48 h) >1000 mg/l.   **Algae** NOEC (96 h) for green algae 100 mg/l.

ENVIRONMENTAL FATE

Animals In the goat following a dose of 60 ppm, <0.1 ppm was found in tissues and milk; therefore nicosulfuron and its metabolites do not bioaccumulate. Hydrolysis of the sulfonylurea bridge and hydroxylation were the main metabolic pathways.   **Plants** Degraded rapidly in maize, DT_{50} 1.5–4.5 d. Residues <0.02 ppm in all crops.   **Soil/Environment** Soil DT_{50} (aerobic) 26 d (pH 6.1, 5.1% o.m., 25 °C). In four sandy loams, K_d (25 °C) 0.16 (pH 6.6, 1.1% o.m.) to 1.73 (pH 5.4, 4.3% o.m.). Photolysis DT_{50} (soil) 60–67 d; (water) 14–19 d

(pH 5), 200–250 d (pH 7), 180–200 d (pH 9). Values from separate studies
were: Soil DT_{50} 24–43 d (20 °C); DT_{90} 80–143 d (20 °C). K_d 0.05–0.7. In water,
DT_{50} 15 d (pH 5, 20 °C).

520 nicotine

Insecticide

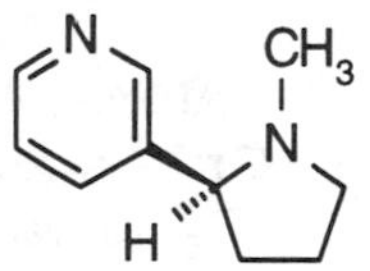

NOMENCLATURE
Common name nicotine (BSI, E-ISO, F-ISO, ESA, in lieu of a common name);
nicotine sulfate (E-ISO (from 1984), JMAF, for sulfate salt); sulfate de nicotine
(F-ISO)
IUPAC name (S)-3-(1-methylpyrrolidin-2-yl)pyridine
Chemical Abstracts name (S)-3-(1-methyl-2-pyrrolidinyl)pyridine
CAS RN [54–11–5] (S)- isomer; [22083–74–5] (RS)- isomers; [75202–10–7]
unstated stereochemistry **EEC no.** 200–193–3

PHYSICAL CHEMISTRY
Composition The predominant component of the crude alkaloid extract is
(S)-(–)- nicotine; small amounts of related alkaloids may be present. Manufactured
nicotine may be the racemic mixture. **Mol. wt.** 162.2 **M.f.** $C_{10}H_{14}N_2$
Form Colourless liquid (darkens rapidly on exposure to light and air).
M.p. −80 °C **B.p.** 246–247 °C **V.p.** 5.65 Pa (25 °C) **K_{ow}** logP = 0.93 (25 °C,
unionised, K. Chamberlain et al., Pestic Sci., **47**, 265 (1996)) **S.g./density** 1.01
(20 °C) **Solubility** Miscible with water below 60 °C, forming a hydrate, and above
210 °C. Miscible with diethyl ether, ethanol; readily soluble in most organic
solvents. **Stability** On exposure to air, darkens and becomes viscous. Forms salts
with acids. **Specific rotation** $[\alpha]_D^{20}$ −161.55° **pKa** pKa_1 3.1, pKa_2 8.2
(K. Chamberlain et al., Pestic Sci., **47**, 265 (1996)) **F.p.** 101 °C

COMMERCIALISATION
History Extracts of tobacco have long been used against sucking insects but have
been replaced by tech. nicotine and nicotine sulfate. **Manufacturers** Nicobrand

APPLICATIONS
Mode of action Non-systemic insecticide with predominantly respiratory action,

nicotine 879

but also slight contact and stomach action. **Uses** Control of aphids, thrips, whitefly, and other insects on glasshouse ornamentals; and aphids and other insects on a range of crops, including fruit, vines, vegetables, and ornamentals. **Formulation types** DP; Fumigant; SL. **Selected tradenames** 'Nico Soap' (United Phosphorus Ltd); 'No-Fid' (Hortichem); 'XL All Nicotine' (Vitax)

ANALYSIS
Product analysis by steam distillation and precipitation as silicotungstate (*AOAC Methods*, 1995, 920.35; *MAFF Ref. Bk.*, No.1, 1958, p. 59; *CIPAC Handbook*, 1970, **1**, 543; 1983, **1A**, 1316). **Residues** determined by glc (R. J. Martin, *J. Assoc. Off. Anal. Chem.*, 1967, **50**, 939) or by colorimetry (*AOAC Methods*, 1995, 964.20).

MAMMALIAN TOXICOLOGY
Oral Acute oral LD_{50} for rats 50–60 mg/kg. **Skin and eye** Acute percutaneous LD_{50} for rabbits 50 mg/kg. Readily absorbed through the skin. Toxic to man by skin contact. **Inhalation** Toxic to man by inhalation. **Other** Lethal oral dose for man is stated to be 40–60 mg. **Toxicity class** WHO (a.i.) Ib; EPA (formulation) I **EC risk** T+ (R27); T (R25); (salts are assigned T+ (R26/27/28))

ECOTOXICOLOGY
Birds Toxic to birds. **Fish** LC_{50} for larval rainbow trout 4 mg/l. **Bees** Toxic to bees, but has a repellent effect. **Daphnia** LC_{50} for *D. Pulex* 0.24 mg/l.

ENVIRONMENTAL FATE
Soil/Environment Nicotine decomposes relatively quickly under the influence of light and air.

521 nitenpyram   *Insecticide*

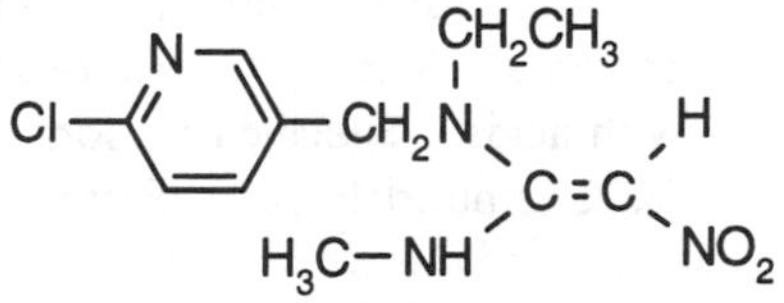

NOMENCLATURE
Common name nitenpyram (BSI, pa E-ISO)
IUPAC name (*E*)-*N*-(6-chloro-3-pyridylmethyl)-*N*-ethyl-*N'*-methyl-2-nitro= vinylidenediamine
Chemical Abstracts name *N*-[(6-chloro-3-pyridinyl)methyl]-*N*-ethyl-*N'*-methyl-= 2-nitro-1,1-ethenediamine

CAS RN *[120738–89–8]; [150824–47–8]* ((E)- isomer)
Development codes TI-304 (Takeda)

PHYSICAL CHEMISTRY
Mol. wt. 270.7   **M.f.** $C_{11}H_{15}ClN_4O_2$ **Form** Pale yellow crystals.   **M.p.** 83–84 °C
V.p. 1.1×10^{-6} mPa (20 °C)   **K_{ow}** logP = –0.64 (25 °C, unstated pH)
S.g./density 1.40 (26 °C)   **Solubility** In water 840 g/l (pH 7.0, 20 °C). In
chloroform 700, acetone 290, xylene 4.5 (all g/l, 20 °C).   **pKa** pKa_1 3.1,
pKa_2 11.5

COMMERCIALISATION
Manufacturers Takeda

APPLICATIONS
Uses Control of aphids, thrips, leafhoppers, whitefly, and other sucking insects on
rice and glasshouse crops.   **Formulation types** SP; GR.
Selected tradenames 'Bestguard' (Takeda)

ANALYSIS
Product and **residue** analysis by hplc. Details from Takeda.

MAMMALIAN TOXICOLOGY
Oral Acute oral LD_{50} for male rats 1680, female rats 1575, male mice 867, female
mice 1281 mg/kg.   **Skin and eye** Acute percutaneous LD_{50} for rats >2000 mg/kg.
Very slightly irritating to eyes, not irritating to skin (rabbits). Not a skin sensitiser
(guinea pigs).   **Inhalation** LC_{50} (4 h) for rats >5.8 g/m^3 air.   **NOEL** (2 y) for male
rats 129, female rats 53.7 mg/kg b.w. daily; (1 y) for male and female dogs
60 mg/kg b.w. daily.   **Other** Not oncogenic in rats and mice. Not teratogenic in
rats and rabbits.

ECOTOXICOLOGY
Birds Acute oral LD_{50} for bobwhite quail >2250, mallard ducks 1124 mg/kg.
Dietary LC_{50} (5 d) for bobwhite quail and mallard ducks >5620 ppm.   **Fish** LC_{50}
(96 h) for carp >1000 mg/l.   **Worms** LC_{50} (14 d) 32.2 mg/kg.   **Daphnia** LC_{50}
(24 h) >10 000 mg/l.   **Algae** NOEC (120 h) 6.25 mg/l.

ENVIRONMENTAL FATE
Soil/Environment DT_{50} in soil 1–15 d, depending on soil type.

522 nithiazine
Insecticide

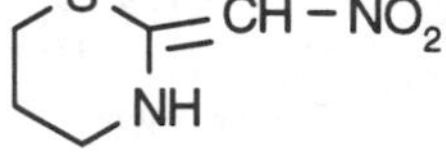

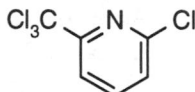

S—CH–NO$_2$
NH

NOMENCLATURE
Common name nithiazine (BSI, pa E-ISO, ANSI)
IUPAC name 2-nitromethylene-1,3-thiazinane
Chemical Abstracts name tetrahydro-2-(nitromethylene)-2H-1,3-thiazine
CAS RN [58842–20–9]   **Development codes** IN-A0159 (Du Pont); SD35651
(Shell/Du Pont)

PHYSICAL CHEMISTRY
Mol. wt. 160.2   **M.f.** C$_5$H$_8$N$_2$O$_2$S

COMMERCIALISATION
History Insecticide reported by S. B. Soloway *et al.* (*Pestic. Venom Toxic.* (*Selected Papers Internat. Congr. Entomol.*, 15th, 1976), 1978, p. 153). Notable as original member of "nitromethylene" group of insecticides. Evaluated originally by Shell Research Ltd, now DuPont Agricultural Products.

523 nitrapyrin
Bactericide, nitrification inhibitor

Cl$_3$C—N—Cl

NOMENCLATURE
Common name nitrapyrin (BSI, E-ISO, ANSI); nitrapyrine ((f) F-ISO)
IUPAC name 2-chloro-6-trichloromethylpyridine
Chemical Abstracts name 2-chloro-6-(trichloromethyl)pyridine
CAS RN [1929–82–4]   **EEC no.** 217–682–2   **Development codes** Dowco 163

PHYSICAL CHEMISTRY
Mol. wt. 230.9   **M.f.** C$_6$H$_3$Cl$_4$N   **Form** Colourless crystals.   **M.p.** 62.5–62.9 °C
B.p. 136–137.5 °C/11 mmHg   **V.p.** 6.4 × 10^2 mPa (25 °C)   **K$_{ow}$** logP = 3.32
Henry 1.4 Pa m^3 mol^{-1} (25 °C, calc.)   **S.g./density** 1.579 (25 °C)   **Solubility** In

water 72 ppm (25 °C). In anhydrous ammonia 540 g/kg (22 °C); in ethanol 300 g/kg (20 °C). In acetone 1.98, dichloromethane 1.85, xylene 1.04 (all in kg/kg, 26 °C).   **Stability** Stable under normal conditions.

COMMERCIALISATION
History Introduced as soil bactericide by Dow Chemical Co. (now DowElanco).
Patent US 3135594; GB 960109   **Manufacturers** DowElanco

APPLICATIONS
Uses Acts as a nitrogen stabiliser because of its highly selective action as a soil bactericide controlling *Nitrosomonas* spp., the bacteria which oxidise ammonium ions in the soil.   **Formulation types** EC.   **Selected tradenames** 'N-Serve' (DowElanco)

ANALYSIS
Product analysis by glc, details available from DowElanco.   **Residues** of 6-chloropyridine-2-carboxylic acid, the main degradation product, determined by glc of the methyl ester (D. J. Jensen, *ibid.*, 1971, **19**, 897; C. T. Redeman, *Bull. Environ. Contam. Toxicol.*, 1967, **2**, 289).

MAMMALIAN TOXICOLOGY
Reviews EPA May 18, 1992 and many earlier.   **Oral** Acute oral LD_{50} for rats 1072–1231 mg/kg.   **Skin and eye** Acute percutaneous LD_{50} for rabbits 2830 mg/kg.   **Inhalation** LC_{50} >2.75 ppm vapour (time-weighted average).
NOEL (94 d) for rats 300, dogs 600 mg/kg diet. In 2 y feeding studies on the metabolite 6-chloropyridine-2-carboxylic acid, NOEL for rats 15, dogs 50 mg/kg daily.   **ADI** 3.0 mg/kg.   **Toxicity class** WHO (a.i.) III   **EC risk** Xn (R22)

ECOTOXICOLOGY
Birds Acute oral LD_{50} for chickens 235 mg/kg. Eight-day dietary LC_{50} for mallard ducks 1466, Japanese quail 820 mg/kg.   **Fish** LC_{50} for channel catfish 5.8 mg/l.
Other beneficial spp. Causes no mortality of Ramshorn snails at 10 mg/l.
Daphnia LC_{50} >10 mg/l.

ENVIRONMENTAL FATE
Animals Metabolism is the same as in soil.   **Plants** Metabolism is the same as in soil.   **Soil/Environment** Hydrolysed in soil to 6-chloropyridine-2-carboxylic acid (6-chloropicolinic acid), which is absorbed by plants and is the principal metabolite (C. T. Redeman et al., *J. Agric. Food Chem.*, 1964, **12**, 207; 1965, **13**, 518). Hydrolysis DT_{50} 2 d (pH 7); water photolysis DT_{50} 12 h. DT_{50} (aerobic soil metabolism) 6.42 d; (anaerobic) *c.* 2.5 h. K_d 0.4–133; K_{oc} 250–9100.

$$CO_2CH(CH_3)_2$$

$$O_2N \qquad CO_2CH(CH_3)_2$$

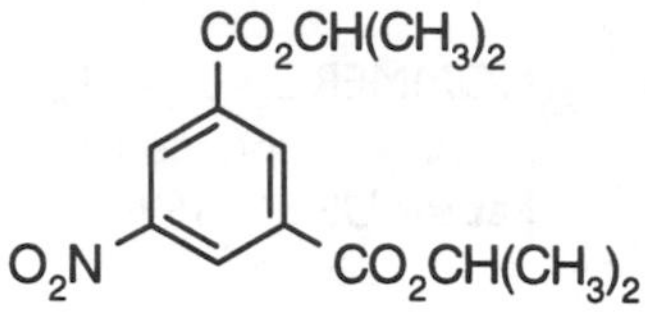

NOMENCLATURE
Common name nitrothal-isopropyl (BSI, E-ISO, (*m*) F-ISO); nitrothale-isopropyl (France)
IUPAC name di-isopropyl 5-nitroisophthalate
Chemical Abstracts name bis(1-methylethyl) 5-nitro-1,3-benzenedicarboxylate
CAS RN *[10552–74–6]* **EEC no.** 234–139–5 **Development codes** BAS 300 00F

PHYSICAL CHEMISTRY
Mol. wt. 295.3 **M.f.** $C_{14}H_{17}NO_6$ **Form** Yellow crystals. **M.p.** 65 °C
V.p. <0.01 mPa (20 °C) **K$_{ow}$** logP = 2.04 (pH 7) **Solubility** In water 2.7 mg/l
(20 °C). In acetone, benzene, chloroform, ethyl acetate >100, diethyl ether 86.5,
ethanol 6.6 (all in g/100 g, 20 °C). **Stability** Stable under normal storage
conditions. Hydrolysed by strong alkalis. **F.p.** 400 °C

COMMERCIALISATION
History Fungicide reported by W. H. Phillips *et al.* (*Proc. Br. Insectic. Fungic. Conf.*,
7th, 1973, **2**, 673). Introduced by BASF AG. **Manufacturers** BASF

APPLICATIONS
Mode of action Non-systemic contact fungicide with protective action.
Uses Used in combination with other fungicides for control of powdery mildews
on apples, vines, hops, vegetables and ornamentals; and scab on apples.
Formulation types WP. **Mixtures** (*nitrothal-isopropyl +*) sulfur; metiram; captan;
captan + metiram. **Selected tradenames** 'Kumulan' (mixture with sulfur) (BASF);
'Pallinal' (mixture with metiram) (BASF)

ANALYSIS
Product analysis by glc with FID (*CIPAC Handbook*, 1992, **E**, 154).
Residues determined by glc with ECD (*Rückstandsanalytik von
Pflanzenschutzmitteln*, 10, Lieferung – 416 (1989), XII, 6, p. 19). Details available
from BASF.

MAMMALIAN TOXICOLOGY
Oral Acute oral LD$_{50}$ for rats >6400 mg tech./kg. **Skin and eye** Acute
percutaneous LD$_{50}$ for rats >2500, rabbits >4000 mg/kg. Slight irritation of eyes

and mucous membranes (rabbits). **Inhalation** LC_{50} (8 h) for rats >2.8 mg/l.
NOEL for rats 4.4 mg/kg b.w. **Toxicity class** WHO (a.i.) III (Table 5);
EPA (formulation) IV

ECOTOXICOLOGY
Fish LC_{50} (96 h) for trout 0.56 mg/l. **Bees** LD_{50} (oral) >100 μg/bee.
Worms LC_{50} (14 d) 1200 mg/kg soil. **Daphnia** LC_{50} (48 h) 2.84 mg/l.
Algae EC_{50} (96 h) 1.72 mg/l.

ENVIRONMENTAL FATE
Animals In rats, following 7 daily oral administrations, c. 85% is eliminated in the
urine and c. 12.5% in the faeces within 6 days of the last dose. **Plants** Metabolites
are 3-carbo-isopropoxy-5-nitrobenzoic acid and 5-nitroisophthalic acid.
Soil/Environment DT_{50} in soil <1–12 d. Adsorption coefficient K_A 8.55–136.4,
depending on soil type.

525   nonanoic acid   *Herbicide, plant growth regulator*

$$CH_3(CH_2)_7CO_2H$$

NOMENCLATURE
IUPAC name nonanoic acid
Chemical Abstracts name nonanoic acid
Other names pelargonic acid; nonoic acid; fatty acids   **CAS RN** *[112–05–0]*
Development codes JT-101

PHYSICAL CHEMISTRY
Composition Tech. is 90–95% C_9 linear acid. **Mol. wt.** 158.2 **M.f.** $C_9H_{18}O_2$
Form White liquid with a fatty acid odour. **M.p.** 12 °C **B.p.** 230–237 °C
V.p. 1×10^5 mPa (20 °C) **K_{ow}** logP = 3.45 **S.g./density** 0.9 (25 °C)
Solubility In water 0.032 g/l (30 °C). **pKa** 4.95 **F.p.** >93 °C (Pensky-Martin
closed cup)

COMMERCIALISATION
Manufacturers Mycogen

APPLICATIONS
Mode of action Post-emergence, contact herbicide and plant growth regulator.
Uses Used for blossom thinning in apples and pears. Also as a herbicide, for
annual weeds in potatoes and peanuts. Total herbicide in non-crop areas.
Formulation types EC. **Selected tradenames** 'Grantrico' (Japan Tobacco); 'Scythe'
(Mycogen)

MAMMALIAN TOXICOLOGY
Oral Acute oral LD_{50} for rats and mice >5000 mg/kg. **Skin and eye** Acute percutaneous LD_{50} for rats >2000 mg/kg. **Inhalation** LC_{50} (4 h) for rats >5.29 mg/l.

ECOTOXICOLOGY
Birds Dietary LC_{50} for mallard duck >5620 ppm. **Fish** LC_{50} (48 h) for carp 59.2 ppm. **Bees** LD_{50} (contact) >25 µg/bee. **Daphnia** LC_{50} (3 h) for *D. similis* >100 ppm.

526 norflurazon   *Herbicide*

pyridazinone (CBI)

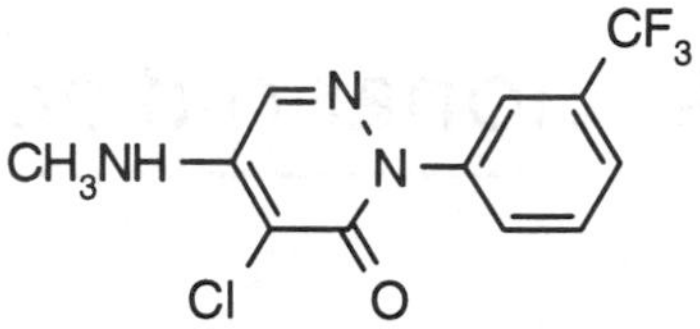

NOMENCLATURE
Common name norflurazon (BSI, E-ISO, ANSI, WSSA); norflurazone ((*m*) F-ISO)
IUPAC name 4-chloro-5-methylamino-2-(α,α,α-trifluoro-*m*-tolyl)pyridazin-=
3(2*H*)-one
Chemical Abstracts name 4-chloro-5-(methylamino)-2-[3-(trifluoromethyl)=
phenyl]-3(2*H*)-pyridazinone
CAS RN *[27314–13–2]*   **Development codes** H 52 143; H 9789 (Sandoz)

PHYSICAL CHEMISTRY
Mol. wt. 303.7   **M.f.** $C_{12}H_9ClF_3N_3O$ **Form** White to greyish brown crystalline powder.   **M.p.** 174–180 °C   **V.p.** 3.857×10^{-3} mPa (25 °C)   **K_{ow}** logP = 2.45 ± 0.02 (pH 6.5, 25 °C)   **Henry** 3.44×10^{-10} Pa m^3 mol^{-1} (calc.)   **Solubility** In water 33.7 mg/l (25 °C). In ethanol 142, acetone 50, xylene 2.5 (all in g/l, 25 °C). Sparingly soluble in hydrocarbons.   **Stability** Stable in aqueous solution (<8% loss after 24 d at 50 °C). Stable under alkaline and acidic conditions. Shelf-life at 20 °C is ≥4 y. Rapidly degraded by sunlight.   **pKa** No dissociation in range pH 1 to 12 (25 °C)

COMMERCIALISATION
History Herbicide introduced in 1968 by Sandoz AG (now Novartis Crop Protection AG).   **Patent** US 3644355   **Manufacturers** Novartis

APPLICATIONS

Biochemistry Blocks carotenoid biosynthesis, by inhibition of phytoene desaturase. Carotenoids dissipate the oxidative energy of singlet oxygen produced during photosynthesis; in their absence in norflurazon-treated plants, singlet oxygen causes peroxidation and destruction of chlorophyll and membrane lipids.
Mode of action Selective herbicide, absorbed by the roots and translocated acropetally in the xylem. Causes interveinal whitening of leaf and stem tissue in emerging susceptible seedlings, leading to necrosis and death. **Uses** Pre-emergence control of grasses and sedges, including *Digitaria* spp. (crabgrasses), *Echinochloa* spp. (barnyardgrass), *Setaria* spp. (foxtails) and *Eleocharis acicularis* (spikerush), as well as broad-leaved weeds, such as *Sida spinosa* (prickly sida), *Portulaca oleracea* (purslane), *Salsola* spp. (Russian thistle) and *Capsella bursa-pastoris* (shepherd's purse). Used at *c.* 0.5–2 kg/ha in cotton, soya beans and peanuts, and at 1.5–4 kg/ha in nuts, citrus, vines, pome fruit, stone fruit, ornamentals, hops, and industrial vegetation management.
Formulation types GR; WG. **Mixtures** *(norflurazon +)* diuron.
Selected tradenames 'Evital' (Novartis); 'Solicam' (Novartis); 'Zorial' (Novartis)

ANALYSIS

Product analysis by glc with FID (S. S. Brady *et al., Anal. Methods Pestic. Plant Growth Regul.*, 1978, **10**, 415), or by tlc, followed by u.v. spectrometry of the eluted compound (details from Novartis). **Residues** determined by glc with ECD (S. S. Brady, *loc. cit.*). In **drinking water**, by gc with FID (*AOAC Methods*, 1995, 991.07).

MAMMALIAN TOXICOLOGY

Oral Acute oral LD_{50} for rats >9000 mg/kg. **Skin and eye** Acute percutaneous LD_{50} for rats >5000, rabbits >20 000 mg/kg. Non-irritating to skin.
Inhalation LC_{50} (1 h) for rats >200 mg/m^3 (for 80 WP formulation, now superseded). **NOEL** (6 mo) for male dogs 1.53, female dogs 1.58 mg/kg b.w. daily; (2 y) for rats 19 mg/kg b.w. daily. **ADI** 0.02 mg/kg b.w. (US). **Other** No teratogenic or mutagenic activity and no adverse effects on reproduction.
Toxicity class WHO (a.i.) III (Table 5); EPA (formulation) IV

ECOTOXICOLOGY

Birds Acute oral LD_{50} for mallard ducks >2510 mg/kg. **Fish** LC_{50} (96 h) for rainbow trout 8.1, bluegill sunfish 16.3 mg/l. **Bees** Non-toxic to bees at 0.235 mg/bee. **Worms** LC_{50} (14 d) >1000 mg/kg soil. **Daphnia** LC_{50} (96 h) >15 mg/l. **Algae** EC_{50} (5 d) for *Selenastrum capricornutum* 0.0176 mg/l.
Other aquatic spp. EC_{50} (14 d) for *Lemna gibba* 0.0875 mg/l.

ENVIRONMENTAL FATE

Animals In rats, norflurazon undergoes demethylation, followed principally by conversion to a sulfoxide (*J. Agric. Food Chem.*, 1989, **37**, 1413). **Plants** In plants, norflurazon undergoes *N*-demethylation to desmethylnorflurazon, and this is further hydrolytically dechlorinated. **Soil/Environment** Dissipated in soil by photodegradation and volatilisation. DT_{50} in soil (US field) *c.* 6–9 mo. K_{oc} (sandy loam) 1055.56, (Mississippi loam) 433.96, (Mississippi sediment) 1055.56, (Keaton sandy loam) 289.47, (Biggs clay) 387.48.

527 novaluron   *Insecticide*

benzoylurea

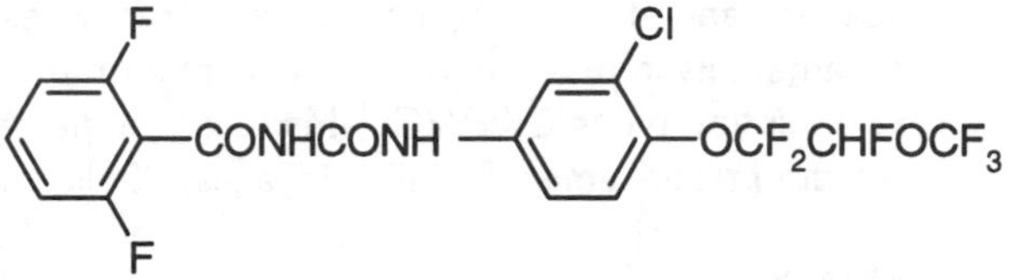

NOMENCLATURE
Common name novaluron (BSI, pa E-ISO)
IUPAC name (±)-1-[3-chloro-4-(1,1,2-trifluoro-2-trifluoromethoxyethoxy)= phenyl]-3-(2,6-difluorobenzoyl)urea
Chemical Abstracts name (±)-*N*-[[[3-chloro-4-[1,1,2-trifluoro-2-(trifluoro= methoxy)ethoxy]phenyl]amino]carbonyl]-2,6-difluorobenzamide
CAS RN *[116714–46–6]* **Development codes** GR 572 (Agrimont); MCW-275 (Makhteshim)

PHYSICAL CHEMISTRY
Composition Tech. is 96%. **Mol. wt.** 492.7 **M.f.** $C_{17}H_9ClF_8N_2O_4$
M.p. 176–179 °C **V.p.** <0.5 mPa (40 °C, OECD 104, EEC A8) K_{ow} logP = 5.27 (OECD 107, EEC A4) **S.g./density** 1.66 (22 °C) **Solubility** In water 53.07 µg/l (25 °C, OECD 105, EEC A6). Soluble in organic solvents. **F.p.** 202 °C (closed cup, EEC A9, ASTM E502–84 and ASTM D 32786–82)

COMMERCIALISATION
History Developed by Isagro and subsequently sold to Makhteshim Chemical Works. **Patent** US 4980376

APPLICATIONS
Biochemistry Chitin synthesis inhibitor, affecting moulting. **Mode of action** Acts by ingestion and contact. Causes abnormal endocuticular deposition and abortive moulting. **Uses** Under development by Makhteshim Chemical Works for control of Lepidoptera, whitefly and agromyzid leaf miners in top fruit, vegetables, cotton

and maize (I. Ishaaya et al., Proc. Br. Crop Prot. Conf. – Pests Dis., 1996, **3**, 1013).
Formulation types EC; SC.   **Selected tradenames** 'Rimon' (Makhteshim)

MAMMALIAN TOXICOLOGY
Oral Acute oral LD_{50} for rats >5000 mg/kg.   **Skin and eye** Acute percutaneous
LD_{50} for rats >2000 mg/kg. Not irritating to eyes and skin (rabbits). Not a skin
sensitiser (guinea pigs).   **Inhalation** LC_{50} (4 h) for rats >5.15 mg/l air.
NOEL (90 d) for rats 100 ppm.

ECOTOXICOLOGY
Birds Acute oral LD_{50} for mallard duck >2000 mg/kg. Dietary LC_{50} (5 d) for
bobwhite quail and mallard duck >5200 ppm.   **Fish** LC_{50} (96 h) for rainbow trout
and bluegill sunfish >1 mg/l.   **Worms** Non-toxic.   **Other beneficial spp.** Non-
toxic.   **Daphnia** LC_{50} (48 h) 58 μg/l.   **Algae** Non-toxic.   **Other aquatic spp.** Non-
toxic.

ENVIRONMENTAL FATE
Animals The major route of elimination was via the faeces.   **Plants** In potatoes
and apples, a.i. remains unchanged.   **Soil/Environment** DT_{50} (aerobic) 68.5–75.5 d
(sandy loam and loamy sand soils).

28   nuarimol   *Fungicide*

pyrimidinyl carbinol

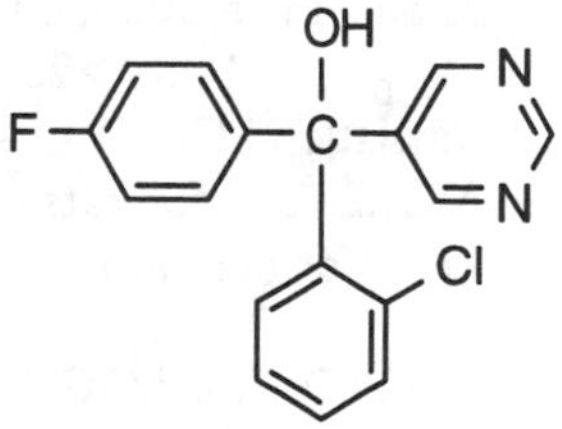

NOMENCLATURE
Common name nuarimol (BSI, E-ISO, (m) F-ISO, ANSI)
IUPAC name (±)-2-chloro-4'-fluoro-α-(pyrimidin-5-yl)benzhydryl alcohol
Chemical Abstracts name (±)-α-(2-chlorophenyl)-α-(4-fluorophenyl)-=
5-pyrimidinemethanol
CAS RN *[63284–71–9]* unstated stereochemistry   **Development codes** EL-228

PHYSICAL CHEMISTRY
Mol. wt. 314.7   **M.f.** $C_{17}H_{12}ClFN_2O$ **Form** Colourless crystals.

M.p. 126–127 °C   **V.p.** <0.0027 mPa (25 °C) K_{ow} logP = 3.18 (pH 7)
Solubility In water 26 mg/l (pH 7, 25 °C). In acetone 170, methanol 55, xylene 20 (all in g/l, 25 °C). Readily soluble in acetonitrile, benzene, and chloroform. Slightly soluble in hexane.   **Stability** Decomposed rapidly by u.v. light. Stable at 52 °C.

COMMERCIALISATION
History Fungicide introduced in Greece (1980) by Eli Lilly & Co. (now DowElanco).   **Patent** GB 1218623   **Manufacturers** DowElanco

APPLICATIONS
Biochemistry Steroid demethylation (ergosterol biosynthesis) inhibitor.
Mode of action Systemic foliar fungicide with curative and protective action. Interferes with the completion of division of sporids.   **Uses** Control of a wide range of pathogenic fungi, such as *Cercosporella* spp., *Septoria* spp., *Ustilago* spp., powdery mildew, leaf spot, etc. in cereals, both as a foliar spray and as a seed treatment; powdery mildews on pome fruit, stone fruit, vines, hops, cucurbits, and other crops; scab on apples; etc.   **Formulation types** EC; SC; SL; WP; Seed treatment.   **Mixtures** *(nuarimol +)* imazalil; captan; chlorothalonil; mancozeb; anthraquinone + maneb; sulfur; tridemorph.   **Selected tradenames** 'Trimidal' (DowElanco)

ANALYSIS
Product analysis by glc with FID (E. W. Day & O. D. Decker, *Anal. Methods Pestic. Plant Growth Regul.*, 1984, **13**, 183).   **Residues** in soil and plant tissue determined by glc with ECD (*idem, ibid.*). Details available from DowElanco.

MAMMALIAN TOXICOLOGY
Oral Acute oral LD_{50} for male rats 1250, female rats 2500, male mice 2500, female mice 3000, beagle dogs 500 mg/kg.   **Skin and eye** Acute percutaneous LD_{50} for rabbits >2000 mg/kg. No skin irritation; slight eye irritant (rabbits).
Inhalation LC_{50} (1 h) – rats essentially unaffected by 0.37 mg tech./l air.
NOEL (2 y) for rats and mice 50 mg/kg diet.   **Toxicity class** WHO (a.i.) III; EPA (formulation) III

ECOTOXICOLOGY
Birds Acute oral LD_{50} for bobwhite quail 200 mg/kg.   **Fish** LC_{50} (96 h) for bluegill sunfish *c.* 12.1 mg/l.   **Bees** Not toxic to bees; LC_{50} (contact) >1 g/l.
Worms NOEC (14 d) for earthworms 100 g/kg soil.   **Daphnia** LC_{50} (48 h) >25 mg/l.   **Algae** EC_{50} (96 h) for *Selenastrum capricornutum* 2.5 mg/l.

ENVIRONMENTAL FATE
Animals In rats, following oral administration, elimination is rapid.
Plants Numerous photodegradation products are formed.   **Soil/Environment** Microbial degradation is accelerated by light; DT_{50} (lab.) 344 d, (field) *c.* 150 d. K_{oc} 2–6 depending on soil type; K_d not determinable due to strong absorption.

octhilinone

Wound protectant, fungicide, bactericide

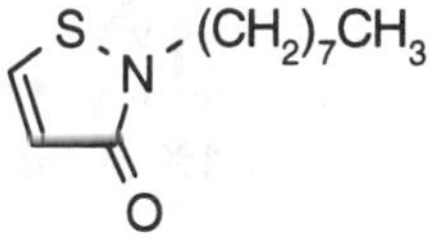

$$S\diagdown_{N}\diagup(CH_2)_7CH_3$$
$$\diagdown O$$

NOMENCLATURE
Common name octhilinone (BSI, E-ISO, (*m*) F-ISO, ANSI)
IUPAC name 2-octylisothiazol-3(2*H*)-one
Chemical Abstracts name 2-octyl-3(2*H*)-isothiazolone
CAS RN *[26530–20–1]* **Development codes** RH 893

PHYSICAL CHEMISTRY
Mol. wt. 213.3 **M.f.** $C_{11}H_{19}NOS$ **Form** A light, golden yellow, clear liquid with a very weak sharp smell. **B.p.** 120 °C/0.01 mmHg **V.p.** 4.9 mPa (25 °C)
K_{ow} logP = 2.45 (24 °C) **Solubility** In distilled water 0.05% (25 °C). In methanol, toluene >800, ethyl acetate >900, hexane 64 (all in g/l). **Stability** Stable to light.

COMMERCIALISATION
History Fungicide introduced by Rohm & Haas Ltd. **Manufacturers** Rohm & Haas

APPLICATIONS
Uses Control of apple and pear canker (*Nectria galligena*) and other fungal and bacterial diseases of top fruit and citrus fruit (e.g. *Phytophthora* and *Ceratocystis* spp.). Also provides a barrier to reinfection, and promotes rapid wound callousing of treated wounds and pruning cuts. **Formulation types** PA; Coating agent.
Selected tradenames 'Pancil-T' (Rohm & Haas)

ANALYSIS
Product analysis by hplc. Details available from Rohm & Haas.

MAMMALIAN TOXICOLOGY
Oral Acute oral LD_{50} for rats 1470 mg/kg. **Skin and eye** Acute percutaneous LD_{50} for rabbits 4.22 ml/kg; not a primary irritant to them. **Inhalation** LC_{50} (4 h) for rats 0.58 mg/l. **NOEL** In 18 mo feeding study, non-carcinogenic to mice at 887 ppm (150 mg/kg daily). **Toxicity class** WHO (a.i.) III

ECOTOXICOLOGY
Birds LD_{50} for bobwhite quail 346, mallard ducks >887 mg/kg. Eight-day dietary LC_{50} for bobwhite quail and mallard ducks >5620 ppm. **Fish** LC_{50} (96 h) for bluegill sunfish 0.196, fathead minnow 0.140, rainbow trout 0.065, channel catfish

0.177 mg/l. **Daphnia** LC$_{50}$ (48 h) 0.180 mg/l.

ENVIRONMENTAL FATE
Soil/Environment River water die away tests at an initial concentration of 1 ppm of a.i. showed that 40–100% of the mildewicide was destroyed within one month. The a.i. was also strongly adsorbed by clays and soil. No hazardous accumulation was observed in fish exposed to water containing the a.i. In activated sludge tests, only 1% of the initial a.i. appeared in the effluent from the sludge.

530 2-(octylthio)ethanol   *Insect repellent*

$$CH_3(CH_2)_7SCH_2CH_2OH$$

NOMENCLATURE
IUPAC name 2-(octylthio)ethanol
Chemical Abstracts name 2-(octylthio)ethanol
Other names 2-hydroxyethyl octyl sulfide   **CAS RN** *[3547–33–9]*
EEC no. 222–598–4   **Development codes** Phillips R-874

PHYSICAL CHEMISTRY
Mol. wt. 190.3   **M.f.** C$_{10}$H$_{22}$OS   **Form** Pale amber liquid, with a mild mercaptan-like odour.   **M.p.** 0 °C   **B.p.** 98 °C/0.1 mmHg   **V.p.** 79 mPa (25 °C)
K$_{ow}$ logP = 0.561   **S.g./density** 0.925–0.935 (24 °C)   **Solubility** Slightly soluble in water; miscible with most organic solvents including refined kerosene, though a co-solvent such as isopropanol is required with the latter to maintain solution at low temperatures.   **Stability** Stable at pH 6–8; and to heat and light.

COMMERCIALISATION
History Insect-repelling properties reported by L. D. Goodhue (*J. Econ. Entomol.*, 1960, **53**, 805). Introduced by Phillips Petroleum Co. and by the McLaughlin Gormley King Co.   **Patent** US 2863799 to Phillips Petroleum Co.

APPLICATIONS
Uses Its chief use is as a repellent to Blattodea, at 1 g a.i./m^2. Also as a repellent using pressurised 'patio' foggers against Formicidae, Muscidae and crawling insects.
Formulation types EC; OL.   **Selected tradenames** 'MGK Repellent 874' (MGK)

ANALYSIS
Details of a glc method available from McLaughlin Gormley King Co.

MAMMALIAN TOXICOLOGY
Oral Acute oral LD_{50} for rats 8530 mg/kg.   **Skin and eye** Acute percutaneous LD_{50} for albino rabbits 13 590 mg/kg. Application of 0.05 mg to the cornea of albino rats produced corneal necrosis in 2 of the 5 eyes tested, but they healed without opacities.   **Inhalation** LC_{50} (4 h) for rats 6.12 mg/l.   **NOEL** In 90 d feeding trials, rats receiving 20 000 mg/kg diet showed no adverse effect. **EC risk** Xi (R41)

ECOTOXICOLOGY
Birds Dietary LC_{50} (8 d) for mallard ducks and bobwhite quail >8000 mg/kg diet. **Fish** LC_{50} (96 h) for rainbow trout 2.9, bluegill sunfish 2.7 mg/l.   **Daphnia** LC_{50} (48 h) 0.38 mg/l.

531 ofurace   *Fungicide*

---

phenylamide (acylamino butyrolactone type)

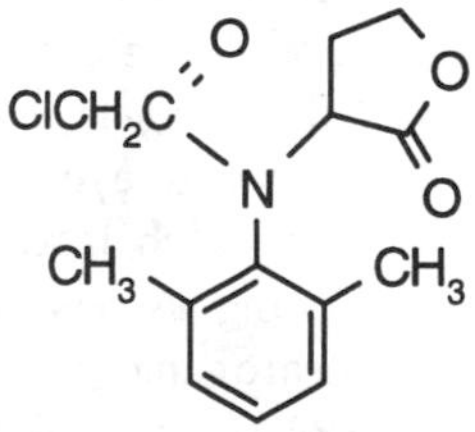

NOMENCLATURE
Common name ofurace (BSI, draft E-ISO, (*m*) draft F-ISO, ANSI)
IUPAC name (±)-α-(2-chloro-*N*-2,6-xylylacetamido)-γ-butyrolactone
Chemical Abstracts name (±)-2-chloro-*N*-(2,6-dimethylphenyl)-*N*-(tetrahydro-= 2-oxo-3-furanyl)acetamide
CAS RN *[58810–48–3]* unstated stereochemistry   **EEC no.** 261–451–9
Development codes RE 20615; AE F057623

PHYSICAL CHEMISTRY
Composition Tech. grade is ≥97% pure.   **Mol. wt.** 281.7   **M.f.** $C_{14}H_{16}ClNO_3$
Form Colourless crystals; tech. is an off-white to light beige crystalline powder.
M.p. 145–146 °C   **V.p.** 2×10^{-2} mPa (20 °C) K_{ow} logP = 1.39 (20 °C)
Henry 3.9×10^{-5} Pa m³ mol⁻¹ (calc.)   **S.g./density** 1.43 (20 °C)   **Solubility** In

water 146 mg/l (20 °C). In acetone 60–75, 1,2-dichloroethane 300–600, ethyl acetate 25–30, methanol 25–30, p-xylene 8.6, heptane 0.0322 (all in g/l). **Stability** Hydrolysis is rapid in alkali, DT_{50} 7 h (pH 9, 35 °C), but stable in acidic media and at elevated temperatures. Photochemical degradation in water, DT_{50} 7 d.

COMMERCIALISATION
History Systemic fungicide introduced by Chevron Chemical Co. Rights transferred to Schering AG (now AgrEvo GmbH) in 1992.

APPLICATIONS
Biochemistry Unknown. **Mode of action** Systemic fungicide with curative and protective action. Absorbed rapidly by the foliage and roots, with translocation both acropetally and basipetally. Acts by inhibiting spore germination or blocking mycelium formation. **Uses** In combination with other fungicides, for control of Phycomycetes, particularly downy mildew of vines, and late blights of potatoes and tomatoes. **Formulation types** SC; WP. **Mixtures** (ofurace +) folpet; mancozeb; metiram; cymoxanil + folpet; cymoxanil + metiram.
Selected tradenames 'Patafol' (mixture) (AgrEvo); 'Vamin' (mixture) (AgrEvo)

ANALYSIS
Product analysis by hplc. **Residues** determined by gc.

MAMMALIAN TOXICOLOGY
Oral Acute oral LD_{50} for male rats 3500, female rats 2600, mice >5000, rabbits >5000 mg/kg. **Skin and eye** Acute percutaneous LD_{50} for rabbits >5000 mg/kg. Moderate eye irritant; mild skin irritant (rabbits). Not a skin sensitiser (guinea pigs). **Inhalation** LC_{50} for rats 2060 mg/m^3. **NOEL** Long-term NOEL for rats 2.5 mg/kg b.w. daily. **ADI** 0.03 mg/kg b.w. **Other** Not mutagenic, not teratogenic, not carcinogenic. **Toxicity class** WHO (a.i.) III (Table 5); EPA (formulation) III

ECOTOXICOLOGY
Birds Acute oral LD_{50} for red-legged partridge >5000 mg/kg. **Fish** LC_{50} (96 h) for rainbow trout 29, golden orfe 57 mg/l. **Bees** Non-toxic to bees; LD_{50} (oral) >58 µg/bee. **Daphnia** EC_{50} (48 h) 46 mg/l.

ENVIRONMENTAL FATE
Animals Ofurace is rapidly excreted, undergoing extensive phase I and phase II biotransformation. **Plants** Similar metabolic pathways have been detected in vines, tomatoes, and potatoes. Ofurace is relatively stable on the plant surface, but, once it has penetrated into plant material, it is metabolised by hydroxylation and conjugation. Ofurace has been defined as the residue to be monitored.
Soil/Environment Degraded in soil, field DT_{50} c. 26 d. Only moderately adsorbed

to soil and is considered to be moderately mobile. Undergoes photochemical degradation in water and is degraded in water sediment systems.

532 oleic acid (fatty acids)

Herbicide, insecticide, fungicide

fatty acid

$$CH_3(CH_2)_7CH=CH(CH_2)_7CO_2M$$

$$M = H, Na \text{ or } K$$

NOMENCLATURE
IUPAC name oleic acid
Chemical Abstracts name (Z)-9-octadecenoic acid
Other names fatty acids   **CAS RN** (Z)- isomer *[112–80–1]*; (E)- isomer *[112–79–8]*; unspecified stereochemistry *[2027–47–6]*
Development codes MYX-6121 (Mycogen); JT-201 (Japan Tobacco, for potassium salt); OK-8905 (Otsuka, for sodium salt)

PHYSICAL CHEMISTRY
Composition Fatty acids (or soaps) comprise oleic acid, usually as potassium or sodium salts; other long-chain fatty acids may also be present. 'M-Pede' is a mixture of oleic and linoleic acids, as their potassium salts. 'Oleate' is the sodium salt of oleic acid. Oleic acid is also a major constituent of neem oil (*q.v.*, under azadirachtin).   **Mol. wt.** 282.5   **M.f.** $C_{18}H_{34}O_2$ **M.p.** 13 °C (oleic acid)

COMMERCIALISATION
History Insecticidal soaps (sodium or potassium salts of a mixture of long-chain fatty acids) have been used as insecticides for many years.
Manufacturers Mycogen; Otsuka

APPLICATIONS
Mode of action Destroys insect cuticle.   **Uses** Potassium and sodium salts used as an insecticide for control of soft-bodied pests (aphids, whitefly, spider mites) on vegetables, fruit and ornamentals, and as a fungicide for control of powdery mildew. The acids may be used for weed control in non-cultivation sites. 'M-Pede' is also used as a surfactant to improve the activity of other insecticides.
Selected tradenames 'M-Pede' (potassium salts, mixture) (Mycogen); 'Savona' (potassium salts) (Koppert); 'Oleate' (sodium salt) (Otsuka)

533 omethoate

Insecticide, acaricide

organophosphorus

$$CH_3NHCOCH_2SP(OCH_3)_2$$
with O double-bonded to P above:

$$\overset{\displaystyle O}{\overset{\displaystyle \|}{CH_3NHCOCH_2SP(OCH_3)_2}}$$

NOMENCLATURE
Common name omethoate (BSI, E-ISO, (*m*) F-ISO); no name (Italy)
IUPAC name *O,O*-dimethyl *S*-methylcarbamoylmethyl phosphorothioate
Chemical Abstracts name *O,O*-dimethyl *S*-[2-(methylamino)-2-oxoethyl] phosphorothioate
CAS RN *[1113–02–6]* **EEC no.** 214–197–8 **Development codes** Bayer 45 432; S 6876 (Bayer)

PHYSICAL CHEMISTRY
Mol. wt. 213.2 **M.f.** $C_5H_{12}NO_4PS$ **Form** Colourless liquid with a mercaptan odour. **M.p.** Solidifies −28 °C (tech.) **B.p.** Decomposes at *c.* 135 °C
V.p. 3.3 mPa (20 °C) **K_{ow}** logP = −0.74 (20 °C) **S.g./density** 1.32 (20 °C)
Solubility Readily soluble in water, alcohols, acetone, and many hydrocarbons. Slightly soluble in diethyl ether. Almost insoluble in petroleum ether.
Stability Hydrolysed in alkaline media; relatively slowly hydrolysed in acidic media: DT_{50} (estimated) 102 d (pH 4), 17 d (pH 7), 28 h (pH 9) (22 °C). **F.p.** 128 °C (tech.)

COMMERCIALISATION
History Insecticide reported by R. Santi & P. de Pietri-Tonelli (*Nature (London)*, 1959, **183**, 398). Introduced by Bayer AG. **Patent** DE 1251304
Manufacturers Bayer

APPLICATIONS
Biochemistry Cholinesterase inhibitor. **Mode of action** Systemic insecticide and acaricide with contact and stomach action. **Uses** Control of spider mites, aphids (including woolly aphids), beetles, caterpillars, scale insects, thrips, suckers, frit flies, etc. on fruit, hops, cereals, rice, potatoes, ornamentals, and other crops.
Phytotoxicity May be phytotoxic to some varieties of peach.
Formulation types AE; EC; SL; UL. **Mixtures** (*omethoate* +) tetradifon; dicofol + tetradifon; cyfluthrin. **Compatibility** Incompatible with alkaline materials.
Selected tradenames 'Folimat' (Bayer)

ANALYSIS
Product analysis by rp hplc (*CIPAC Handbook*, 1992, **E**, 159–162).
Residues determined by glc with FID (*Man. Pestic. Residue Anal.*, 1987, **I**, 6, S13, S17, S19; *Anal. Methods Residues Pestic.*, 1988, Part I, M2, M5, M12; *Analyst*

(London), 1977, **102**, 858; *AOAC Methods*, 1995, 985.22). Details available from Bayer AG.

MAMMALIAN TOXICOLOGY

Reviews *Pesticide residues in food – 1985*. FAO Plant Production and Protection Paper 68, 1986. *Pesticide residues in food – 1985 evaluations. Part II – Toxicology.* FAO Plant Production and Protection Paper 72/2, 1986. **EHC** 63 (WHO, 1986; a general review of organophosphorus insecticides). **Oral** Acute oral LD_{50} for rats c. 25 mg/kg. **Skin and eye** Acute percutaneous LD_{50} for rats 200 mg/kg (24 h). Not irritating to the skin, slightly irritating to the eyes (rabbits). **Inhalation** LC_{50} (4 h) for rats c. 0.3 mg/l (aerosol). **NOEL** NOAEL (2 y) for rats 0.3, mice 10 ppm. NOEL (1 y) for dogs 0.025 mg/kg b.w. **ADI** (JMPR) 0.0003 mg/kg b.w. [1985]. **Toxicity class** WHO (a.i.) Ib; EPA (formulation) I **EC risk** T (R25); Xn (R21)

ECOTOXICOLOGY

Birds Acute oral LD_{50} for male Japanese quail 79.7, female Japanese quail 83.4 mg/kg. **Fish** LC_{50} (96 h) for golden orfe 30, rainbow trout 9.1 mg/l. **Bees** Toxic to bees. **Worms** LC_{50} for *Eisenia foetida* 46 mg/kg dry soil. **Daphnia** LC_{50} (48 h) 0.022 mg/l. **Algae** E_rC_{50} for *Scenedesmus subspicatus* 167.5 mg/l.

ENVIRONMENTAL FATE

Animals Omethoate is not accumulated in animal tissues or fat. The main metabolites in the urine are O-demethylomethoate and N-methyl-2-(methyl= dithio)acetamide. **Plants** Omethoate is rapidly taken up by plants. Demethylation and hydrolysis of P-S bonds are the main metabolic steps. The main metabolites are 3-hydroxy-3-[(2-methylamino-2-oxo-ethyl)thio]propionic acid and its oxidation products. **Soil/Environment** Omethoate has a relatively high mobility in soil but is very rapidly metabolised; DT_{50} only a few days. The main metabolite is CO_2. Aged leaching studies revealed that metabolites have only a low leaching potential.

534 orbencarb

Herbicide

thiocarbamate

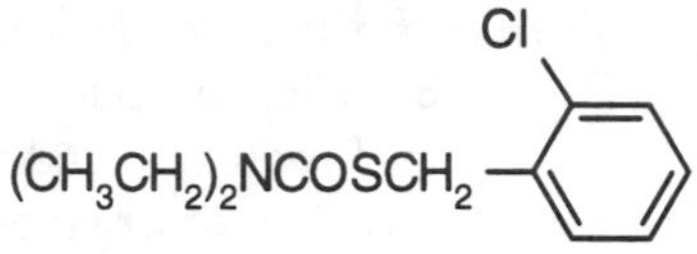

$$(CH_3CH_2)_2NCOSCH_2 -$$

NOMENCLATURE
Common name orbencarb (BSI, draft E-ISO, draft (*m*) F-ISO); orthobencarb (JMAF)
IUPAC name S-2-chlorobenzyl diethylthiocarbamate
Chemical Abstracts name S-[(2-chlorophenyl)methyl] diethylcarbamothioate
CAS RN *[34622–58–7]* **Development codes** B-3356 (Kumiai)

PHYSICAL CHEMISTRY
Composition Tech. is >95%. **Mol. wt.** 257.8 **M.f.** $C_{12}H_{16}CINOS$
Form Colourless liquid; (tech., yellowish liquid). **M.p.** 9.0 °C
B.p. 158 °C/1 mmHg **V.p.** 12.4 mPa (20 °C) **K_{ow}** logP = 3.43
S.g./density 1.176 (20 °C) **Solubility** In water 24 mg/l (20–27 °C). Very soluble in organic solvents, e.g. acetone, xylene, hexane, ethanol and benzene (all >1 kg/l, room temperature). **Stability** Stable to hydrolysis 60 d at pH 5 to pH 9 (20 °C). Slightly decomposed by sunlight in aqueous solution.

COMMERCIALISATION
History Herbicide reported by S. Iori *et al.* (*Shibakusa Kenkyu*, 1975, **4**(2), 63). Introduced in Japan (1970) by Kumiai Chemical Industry Co., Ltd.
Patent US 3816500; JP 1065662; JP 1202209 **Manufacturers** Ihara/Kumiai

APPLICATIONS
Mode of action Systemic herbicide which inhibits formation and elongation of the leaves of germinated weeds and seedlings by absorption through roots, seeds, and mesocotyl, and translocation to the growth points. **Uses** Pre-emergence control of annual grasses (except wild oats) and broad-leaved weeds in barley, wheat, rye, maize, soya beans, cotton and turf at 4–5 kg/ha. **Formulation types** EC.
Mixtures *(orbencarb +)* linuron; atrazine. **Selected tradenames** 'Lanray' (Kumiai)

ANALYSIS
Product and **residue** analysis by glc.

MAMMALIAN TOXICOLOGY
EHC 76 (WHO, 1988; general review of thiocarbamates). **Oral** Acute oral LD_{50} for male rats 800, female rats 820, male mice 935, female mice 1010 mg/kg.

Skin and eye Acute percutaneous LD_{50} for rats >10 000 mg/kg. Not a skin or eye irritant (rabbits).   **Inhalation** LC_{50} (4 h) for male rats 4.32, female rats 2.94 mg/l air.   **NOEL** (90 d) for male rats 1.7, female rats 1.8 mg/kg b.w. daily.
Toxicity class EPA (formulation) III

ECOTOXICOLOGY
Birds Acute oral LD_{50} for bobwhite quail and mallard duck >2000 mg/kg b.w.
Fish LC_{50} (48 h) for carp 3.4 mg/l; (96 h) for rainbow trout 1.88 mg/l.   **Bees** Not toxic to bees. Acute oral LD_{50} 103 µg/bee.   **Daphnia** LC_{50} (24 h) 2.88 mg/l.
Algae LC_{50} (96 h) 5 mg/l.

ENVIRONMENTAL FATE
Animals For metabolism in rats, see T. Unai et al., J. Pestic. Sci. 1986, **11**, 527.
Plants In plants, undergoes relatively rapid degradation. By harvest time, the residue levels are below the limits of detection. For metabolism in soya bean plants, see T. Unai et al., ibid.   **Soil/Environment** For degradation in soil, see T. Unai et al., ibid.

535 *Orius* spp.   *Biological agent*

Hemiptera

Predatory bug.

NOMENCLATURE
Scientific name *Orius laevigatus* (Fieber); *Orius majusculus* (Reuter 1861); *Orius albidepennis* (Reuter); *Orius insidiousus* (Say).

PROPERTIES
Stability Storage transportation conditions are 8 °C and 70–80% r.h. The product should be applied within 18 h of receipt.

COMMERCIALISATION
History *Orius* was first reported a predator of thrips in 1914 and again in the 1930s. Laboratory studies in the USA in the late 1970s and early 1980s led to the development of commercial production in Canada and Europe in the early 1990s.
Manufacturers Arbico; BCP; Koppert; Nature's Alternative; Neudorff; Novartis BCM; Rincon-Vitova; Sautter & Stepper

APPLICATIONS
Mode of action Hemipteran, true bug, species with sucking mouth parts, attacking

a very wide range of small arthropods, with some preference for thrips. Species used in biological control will also develop on a diet of pollen. The life-cycle consists of egg and adult, with five intervening nymphal stages. Eggs are laid partially in or under the cuticle of selected plants. Development depends on temperature, humidity, host and day length (c. 13 to 30+ days). **Uses** *Orius* species are used to control thrips pests, in particular the Western Flower Thrips (WFT), *Frankliniella occidentalis*. Northern European species are subject to diapause periods so that introduction into glasshouses, especially where temperatures are 20 °C or below, may inhibit second generation development. Therefore, a two tier control system is employed, using the predatory mite *Amblyseius cucumeris* (q.v.) in winter-planted crops during the first quarter of the year, supplemented by *Orius* introductions in early spring. *Orius* species are used in all vegetable and many ornamental crops grown under protective cultivation. Such species are not known to cause vegetative damage, however, laboratory trials have shown them to have equal preference for WFT and the beneficial predatory mite *Phytoseiulus persimilis* (q.v.). **Phytotoxicity** Non-phytotoxic and non-phytopathogenic. **Formulation types** Nymphs and adults mixed with vermiculite and buckwheat shells. **Selected tradenames** 'Ori-line' (Novartis BCM); 'Thripor' (Koppert)

MAMMALIAN TOXICOLOGY
There is no evidence of acute or chronic toxicity, eye or skin irritation or hypersensitivity to mammals. No allergic response or health problems have been observed in production staff or horticultural workers.

536 oryzalin   *Herbicide*

2,6-dinitroaniline

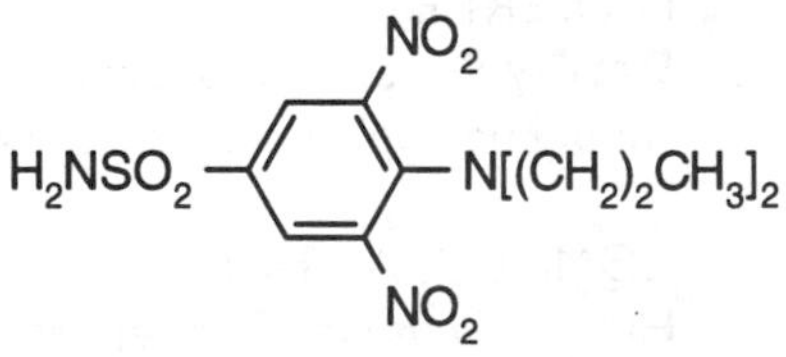

NOMENCLATURE
Common name oryzalin (BSI, E-ISO, (m) F-ISO, ANSI, WSSA)
IUPAC name 3,5-dinitro-N^4,N^4-dipropylsulfanilamide
Chemical Abstracts name 4-(dipropylamino)-3,5-dinitrobenzenesulfonamide
CAS RN [19044–88–3] **Development codes** EL-119

PHYSICAL CHEMISTRY
Mol. wt. 346.4 **M.f.** $C_{12}H_{18}N_4O_6S$ **Form** Tech. grade forms yellow-orange crystals. **M.p.** 141–142 °C; (tech. 138–143 °C) **B.p.** Decomp. at 265 °C **V.p.** <0.0013 mPa (25 °C) K_{ow} logP = 3.73 (pH 7) **Solubility** In water 2.6 mg/l (25 °C). In acetone >500, methyl cellosolve 500, acetonitrile >150, methanol 50, dichloromethane >30, benzene 4, xylene 2 (all in g/l, 25 °C). Insoluble in hexane. **Stability** Stable under normal storage conditions. Does not hydrolyse in aqueous solution at pH 5, 7 or 9. Decomposed by u.v. irradiation; aqueous photolysis DT_{50} 1.4 h in natural sunlight. **pKa** 9.4, v. weak acid

COMMERCIALISATION
History Herbicide reported by J. V. Gramlich et al. (*Abstr. Weed Sci. Soc. Am.*, 1969). Introduced in Bulgaria (1973) by Eli Lilly & Co. (now DowElanco). **Patent** US 3367949 **Manufacturers** DowElanco

APPLICATIONS
Biochemistry Cell division inhibitor. **Mode of action** Selective herbicide. Affects physiological growth processes associated with seed germination. **Uses** Pre-emergence control of many annual grasses and broad-leaved weeds in cotton, fruit trees, nut trees, vines, ornamentals, soya beans, peanuts, oilseed rape, sunflowers, alfalfa, peas, sweet potatoes, mint, and non-crop areas. **Formulation types** WG; WP; SC. **Mixtures** (*oryzalin +*) isoxaben. **Compatibility** Incompatible with alkaline materials. **Selected tradenames** 'Surflan' (DowElanco)

ANALYSIS
Product can be analysed by spectrophotometry (O. D. Decker & W. S. Johnson, *Anal. Methods Pestic. Plant Growth Regul.*, 1976, **8**, 433) or by hplc with u.v. detection (S. D. West in *Comp. Anal. Profiles*, Chapt. 9). **Residues** by gc of a methylated derivative, with ECD; other residue methods briefly reviewed (S. D. West, *loc. cit.*).

MAMMALIAN TOXICOLOGY
Oral Acute oral LD_{50} for rats and gerbils >10 000, cats 1000, dogs >1000 mg/kg. **Skin and eye** Acute percutaneous LD_{50} for rabbits >2000 mg/kg. Slightly irritating to skin; non-irritating to eyes (rabbits). **Inhalation** LC_{50} (4 h) for rats >3.1 mg/l air (nominal 4.8 mg/l). **NOEL** (2 y) for rats 300 mg/kg diet (c. 12–14 mg/kg b.w. daily), for mice 1350 mg/kg diet (c. 100 mg/kg b.w. daily). **ADI** 0.12 mg/kg. **Toxicity class** WHO (a.i.) III (Table 5); EPA (formulation) III

ECOTOXICOLOGY
Birds Acute oral LD_{50} for chickens >1000, bobwhite quail and mallard ducks >500 mg/kg. Five-day dietary LC_{50} >5000 mg/kg. **Fish** LC_{50} for bluegill sunfish 2.88, rainbow trout 3.26 mg/l. LC_{50} (96 h) for goldfish fingerlings >1.4 mg/l. **Bees** LD_{50} (oral) 25 μg/bee, (contact) 11 μg/bee. **Worms** NOEC (14 d) for

earthworms >102.6 mg/kg soil.   **Daphnia** LC_{50} (48 h) 1.4 mg/l.
Other aquatic spp. EC_{50} (14 d) for *Lemna gibba* 0.015 mg/l, NOEC 0.006 mg/l.

ENVIRONMENTAL FATE

Animals In rats, elimination was virtually complete within 72 hours, c. 60% in faeces and 40% in urine. In rabbits, the ratio was reversed. Numerous metabolites have been detected in both urine and faeces. Biliary excretion accounted for a major portion of the radioactivity in the rat faeces. The pattern of metabolites was similar in rat and rabbit.   **Plants** No oryzalin residues were detected in soya bean plants. Trace amounts of radioactivity were associated with plant constituents, but no metabolites were identifiable.   **Soil/Environment** In soil, microbial degradation occurs rapidly, aerobic metabolism being slower than anaerobic (DT_{50} 1.2 mo and 10 d respectively in the same soil). This involves dealkylation of the amine nitrogen, and reduction of the nitro groups. Oxidation, dimerisation and ring formation are also involved in the complex process. K_{oc} 700–1100, K_d 2.1–12.9, in soils with o.m. 0.5–2.0%.

537 oxabetrinil   *Herbicide safener*

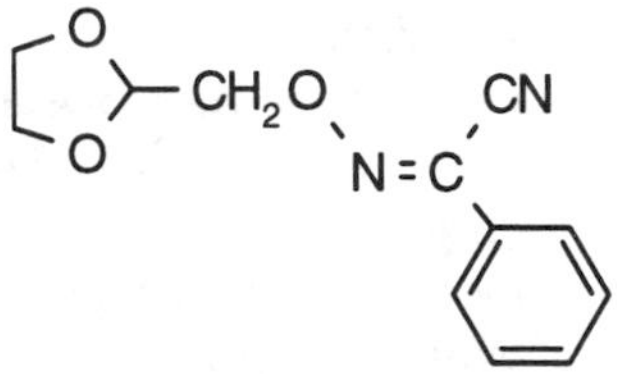

NOMENCLATURE

Common name oxabetrinil (BSI, draft E-ISO, (*m*) draft F-ISO)
IUPAC name (*Z*)-1,3-dioxolan-2-ylmethoxyimino(phenyl)acetonitrile
Chemical Abstracts name α-[(1,3-dioxolan-2-yl)methoxyimino]benzeneacetonitrile
CAS RN *[74782–23–3]* unstated stereochemistry
Development codes CGA 92 194

PHYSICAL CHEMISTRY

Composition Material comprises (*Z*)- and (*E*)- isomers.   **Mol. wt.** 232.2
M.f. $C_{12}H_{12}N_2O_3$ **Form** Colourless crystals.   **M.p.** 77.7 °C   **V.p.** 0.53 mPa
(20 °C) K_{ow} logP = 2.76 (rp-tlc)   **S.g./density** 1.33 (20 °C)   **Solubility** In water
20 mg/l (20 °C). In acetone 250, cyclohexanone 300, toluene 220, methanol 30, hexane 5.6, *n*-octanol 12, xylene 150, dichloromethane 450 (all in g/kg, 20 °C).
Stability Stable up to 240 °C. No significant hydrolysis at pH 5–9 within 30 d.

COMMERCIALISATION
History Herbicide safener reported by T. R. Dill *et al.* (*Abstr. Annu. Meeting Weed Sci. Soc. Am.*, 1982, 20). Introduced by Ciba-Geigy AG (now Novartis Crop Protection AG). **Patent** EP 11047 **Manufacturers** Novartis

APPLICATIONS
Biochemistry Stimulates metolachlor metabolism by inducing glutathione S transferase. **Mode of action** Herbicide safener (seed treatment). **Uses** A herbicide safener which protects sorghum hybrids, as well as various yellow endosperm, sweet sorghum and Sudan grass varieties, from injury by metolachlor. Applied as a seed treatment at 1–2 g a.i./kg seed, allowing the safe use of metolachlor for the control of a wide range of grasses. Sorghum tolerance is maintained when metolachlor is applied in combination with 1,3,5-triazines (atrazine, propazine, terbuthylazine, terbutryn) for additional activity on broad-leaved weeds. **Formulation types** WP. **Compatibility** Can be used in combination with triazines (atrazine, propazine, terbuthylazine, terbutryn) for additional broad-leaved weed control, and with fungicide/insecticide seed treatments. **Selected tradenames** 'Concep II' (Novartis)

ANALYSIS
Residues determined by glc using FID. Details available from Novartis.

MAMMALIAN TOXICOLOGY
Oral Acute oral LD_{50} for rats >5000 mg/kg. **Skin and eye** Acute percutaneous LD_{50} for rats >5000 mg/kg. Minimal skin and eye irritation (rabbits). Not a skin sensitiser. **Inhalation** LC_{50} (4 h) for rats >1.45 mg/l air. **NOEL** 90 d NOAEL for rats 1500 ppm (118 mg/kg b.w.), for dogs 250 ppm (9.4 mg/kg b.w.). **Toxicity class** WHO (a.i.) III (Table 5) **EC risk** R23

ECOTOXICOLOGY
Birds LD_{50} for Japanese quail >2500 mg/kg; (8 d) for bobwhite quail >5000, Pekin ducks >1000 mg/kg. **Fish** LC_{50} (96 h) for trout 7.1, bluegill sunfish 12 mg/l. **Bees** LD_{50} (oral, 24 h) >20 µg/bee, (contact) >1000 ppm. **Daphnia** LC_{50} (48 h) 8.5 mg/l. **Algae** EC_{50} (96 h) for *Selenastrum capricornutum* 10.7 mg/l.

ENVIRONMENTAL FATE
Animals In rats, following ingestion, oxabetrinil is rapidly and fully excreted, predominantly via urine. **Plants** Opening of the dioxolane ring, hydroxylation, glycosylation to water-soluble conjugation products.

(CH₃)₃C—[structure: 1,3,4-oxadiazol-2(3H)-one ring with N-N, C=O, attached to 2,4-dichloro-5-(prop-2-ynyloxy)phenyl]

HC≡C—CH₂—O

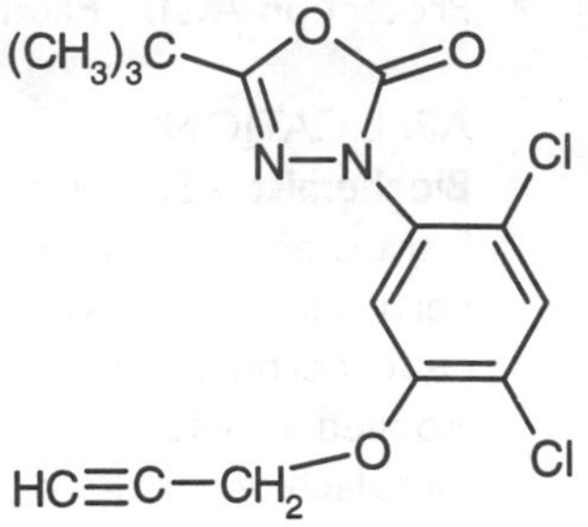

NOMENCLATURE
Common name oxadiargyl (pa ISO)
IUPAC name 5-*tert*-butyl-3-[2,4-dichloro-5-(prop-2-ynyloxy)phenyl]-=
1,3,4-oxadiazol-2(3*H*)-one
Chemical Abstracts name 3-[2,4-dichloro-5-(2-propynyloxy)phenyl]-=
5-(1,1-dimethylethyl)-1,3,4-oxadiazol-2(3*H*)-one
CAS RN *[39807–15–3]*   **EEC no.** 254–637–6   **Development codes** RP 020630

PHYSICAL CHEMISTRY
Mol. wt. 341.2   **M.f.** $C_{15}H_{14}Cl_2N_2O_3$ **Form** White to beige powder with no
characteristic odour.   **M.p.** 131 °C   **V.p.** 2.5×10^{-3} mPa (25 °C).
K$_{ow}$ logP = 3.95   **Henry** 9.1×10^{-4} Pa m^3 mol^{-1} (calc.)   **S.g./density** 1.484 (20 °C)
Solubility In water 0.37 mg/l (20 °C).   **Stability** Stable to heat (15 d, 54 °C), to
light and in water.

COMMERCIALISATION
History Introduced in Latin America in 1996.   **Manufacturers** Rhône-Poulenc

APPLICATIONS
Biochemistry Protoporphyrinogen IX oxidase inhibitor.   **Mode of action** Selective
herbicide active mainly pre-emergence; effects begin at germination. It is not
absorbed by the plant. Efficacy is not dependent on soil texture and type.
Uses Pre-emergence herbicide active on broad-leaved weeds (Amaranthus,
Bidens, Chenopodium, Malva, Monochoria, Polygonum, Portulaca, Potamogeton,
Raphanus, Solanum, Sonchus, Rotala), grasses (Echinochloa, Leptochloa,
Brachiaria, Cenchrus, Digitaria, Eleusine, Panicum and wild rice) and annual sedges
in rice (at 50–150 g/ha), upland crops (sunflower, potato, vegetables and sugar
cane at 300–500 g/ha) and perennial crops (fruit trees and citrus at 500–1500
g/ha).   **Formulation types** SC; WG; WP; EC.   **Selected tradenames** 'Raft'
(Rhône-Poulenc); 'Topstar' (Rhône-Poulenc)

ANALYSIS
By hplc with u.v. detection.

MAMMALIAN TOXICOLOGY
Oral Acute oral LD_{50} for rats >5000 mg/kg. **Skin and eye** Acute percutaneous LD_{50} for rabbits <2000 mg/kg. Not a skin irritant, minimal eye irritation (rabbits). **Inhalation** LC_{50} (4 h) for rats >5.16 mg/l. **Other** Not genotoxic.

ECOTOXICOLOGY
Birds Acute oral LD_{50} (14 d) for quail >2000 mg/kg; dietary 8-d LC_{50} for quail and mallard duck >5200 ppm. **Fish** Not toxic at limit of water solubility. **Bees** LD_{50} (oral and contact) >200 µg/bee. **Worms** Non-toxic at 1000 mg/kg. **Daphnia** Not toxic at limit of water solubility. **Algae** EC_{50} (120 h) for *Selenastrum capricornutum* 1.2 µg/l.

ENVIRONMENTAL FATE
Animals Studies on goat and hen demonstrated that oxadiargyl is rapidly excreted; there is no evidence of accumulation in milk, eggs or edible tissues. **Plants** Studies on lemon, sunflower and rice demonstrated very low levels of residues at harvest, mainly parent compound. **Soil/Environment** Strongly adsorbed to soil (K_{oc} 1000–3000). DT_{50} 18–72 d (20–30 °C). Unlikely to leach or accumulate in soil.

539 oxadiazon *Herbicide*

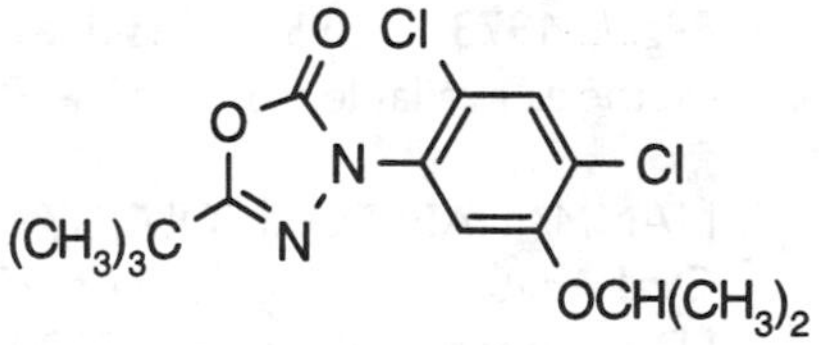

NOMENCLATURE
Common name oxadiazon (BSI, E-ISO, (*m*) F-ISO, ANSI, WSSA, JMAF)
IUPAC name 5-*tert*-butyl-3-(2,4-dichloro-5-isopropoxyphenyl)-1,3,4-oxadiazol-=2(3*H*)-one
Chemical Abstracts name 3-[2,4-dichloro-5-(1-methylethoxy)phenyl]-=5-(1,1-dimethylethyl)-1,3,4-oxadiazol-2(3*H*)-one
CAS RN *[19666–30–9]* **EEC no.** 243–215–7 **Development codes** 17 623 RP

PHYSICAL CHEMISTRY
Composition Tech. is ≥94% pure. **Mol. wt.** 345.2 **M.f.** $C_{15}H_{18}Cl_2N_2O_3$
Form Colourless, odourless crystals. **M.p.** 87 °C **V.p.** 0.1 mPa (25 °C)
K_{ow} logP = 4.91 (20 °C) **Henry** 3.5 × 10^{-2} Pa m^3 mol^{-1} (calc.) **Solubility** In water
1.0 mg/l (20 °C). In methanol, ethanol c. 100, cyclohexane 200, acetone,
isophorone, methyl ethyl ketone, carbon tetrachloride c. 600, toluene, benzene,
chloroform c. 1000 (all in g/l, 20 °C). **Stability** Stable in neutral or acidic medium,
relatively unstable in alkali; DT_{50} 38 d (pH 9, 25 °C).

COMMERCIALISATION
History Herbicide reported by L. Burgaud et al. (*Symp. New Herbic.*, 3rd, 1969,
p. 201). Introduced by Rhône-Poulenc Agrochimie. **Patent** GB 1110500;
US 3385862 **Manufacturers** Rhône-Poulenc

APPLICATIONS
Biochemistry Protoporphyrinogen oxidase inhibitor. **Mode of action** Selective
contact herbicide. **Uses** Pre-emergence control of bindweed, annual broad-
leaved weeds and grasses, and post-emergence control of bindweed and annual
broad-leaved weeds in carnations, gladioli, roses, fruit trees and bushes (including
citrus), vines, ornamental trees and shrubs, hops, cotton, rice, soya beans,
sunflowers, onions, and turf. Effective against mono- and dicotyledonous weeds in
rice at c. 1 kg/ha; in orchards and vineyards at 2 kg/ha post-em. or 4 kg/ha pre-
em. **Phytotoxicity** Carnations are tolerant of over-the-top, post-emergence
application. Not to be used on red fescue, bentgrass turf, dichronda or
centipedegrass. **Formulation types** AL; GR; EC; WP; SC. **Mixtures** (*oxadiazon +*)
aclonifen; butachlor; carbetamide; propanil; 2,4-D; diuron; simazine; diflufenican;
glyphosate. **Selected tradenames** 'Ronstar' (Rhône-Poulenc)

ANALYSIS
Product analysis by glc (J. Desmoras et al., *Anal. Methods Pestic. Plant Growth
Regul.*, 1973, **7**, 595). **Residues** determined by glc (*idem, ibid.*). Other glc and hplc
methods available from Rhône-Poulenc.

MAMMALIAN TOXICOLOGY
Oral Acute oral LD_{50} for rats >5000 mg/kg. **Skin and eye** Acute percutaneous
LD_{50} for rats and rabbits >2000 mg/kg. Slightly irritating to eyes, negligible irritant
to skin (rabbits). **Inhalation** LC_{50} (4 h) for rats >2.77 mg/l. **NOEL** In 2 y feeding
trials, rats and mice receiving 10 mg/kg diet showed no ill-effects.
Toxicity class WHO (a.i.) III (Table 5); EPA (formulation) IV

ECOTOXICOLOGY
Birds Acute oral LD_{50} (24 d) for mallard ducks >1000, bobwhite quail
>2150 mg/kg. **Fish** LC_{50} (96 h) for rainbow trout and bluegill sunfish 1.2 mg/l.
Bees LD_{50} >400 µg/bee with repellent effect. Mortality is negligible by direct

contact at doses up to 27 kg a.i./ha. **Worms** Not toxic at recommended rate.
Daphnia EC_{50} (48 h) >2.4 mg/l. **Algae** EC_{50} 6–3000 µg/l.

ENVIRONMENTAL FATE
Animals In mammals, following oral administration, 93% is eliminated within
72 hours, predominantly in the urine. **Plants** Oxadiazon penetrates plants
primarily via shoot and leaves and is rapidly metabolised. Metabolites do not
accumulate in the plant. **Soil/Environment** Strongly adsorbed by soil colloids and
humus, with very little migration or leaching. Negligible loss due to volatilisation.
Half-life in soil is c. 3–6 mo. See D. Ambrosi et al., J. Agric. Food Chem., 1977, **25**,
868. K_{oc} 1400 (silt loam) to 3200 (sand) at 25 °C.

540 oxadixyl   *Fungicide*

phenylamide (acylamino oxazolidinone type)

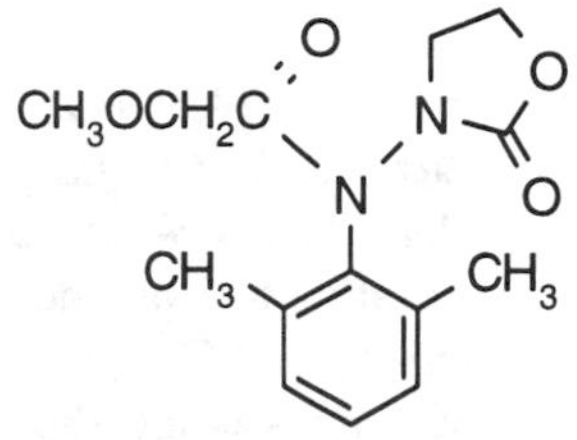

NOMENCLATURE
Common name oxadixyl (BSI, draft E-ISO, draft F-ISO)
IUPAC name 2-methoxy-N-(2-oxo-1,3-oxazolidin-3-yl)aceto-2',6'-xylidide
Chemical Abstracts name N-(2,6-dimethylphenyl)-2-methoxy-N-(2-oxo-=
3-oxazolidinyl)acetamide
CAS RN [77732–09–3] **Development codes** SAN 371F

PHYSICAL CHEMISTRY
Mol. wt. 278.3 **M.f.** $C_{14}H_{18}N_2O_4$ **Form** Colourless, odourless crystals.
M.p. 104–105 °C **V.p.** 0.0033 mPa (20 °C) K_{ow} logP = 0.65–0.8 (22–24 °C)
S.g./density Bulk density 0.5 kg/l **Solubility** In water 3.4 g/kg (25 °C). In acetone
344, dimethyl sulfoxide 390, methanol 112, ethanol 50, xylene 17, diethyl ether
6 (all in g/kg, 25 °C). **Stability** Stable under normal conditions, and when stored
at 70 °C for 2–4 weeks. A 20 ppm solution of the a.i. is stable in aqueous buffer
solutions of pH 5, 7 and 9 at room temperature, DT_{50} c. 4 y.

COMMERCIALISATION
History Systemic fungicide reported by U. Gisi et al. (Meded. Fac. Landbouwwet.

Rijksuniv. Gent, 1983, **48**, 541). Introduced by Sandoz AG (now Novartis Crop Protection AG). **Patent** GB 2058059 **Manufacturers** Novartis

APPLICATIONS
Mode of action Systemic fungicide with curative and protective action. Rapidly absorbed by the leaves and roots, with translocation principally acropetally, but also basipetally and by translaminar movement. Exhibits a synergistic effect with contact fungicides. **Uses** In combination with contact fungicides, for control of Peronosporales, such as downy mildews, late blights, and rusts, in vines, maize, potatoes, tobacco, hops, sunflowers, citrus, fruit and vegetables, and as a seed dressing on cotton, peas, and sunflowers. **Formulation types** FG; WP.
Mixtures *(oxadixyl +)* mancozeb; propineb; cymoxanil + mancozeb; cymoxanil + maneb; captafol + cymoxanil + folpet; copper oxychloride; cymoxanil + folpet; dichlofluanid; zineb; folpet; captafol; cymoxanil; chlorothalonil.
Selected tradenames 'Sandofan' (Novartis); 'Anchor' (Gustafson)

ANALYSIS
By rp hplc with u.v. detection (*CIPAC Handbook*, 1995, **G**, 122–127) or by glc with TSD; details available from Novartis.

MAMMALIAN TOXICOLOGY
Oral Acute oral LD_{50} for male rats 3480, female rats 1860 mg/kg.
Skin and eye Acute percutaneous LD_{50} for rats and rabbits >2000 mg/kg. Non-irritating to skin and eyes (rabbits), nor a skin sensitiser to guinea pigs.
Inhalation LC_{50} (6 h) for male and female rats >5.6 mg/l air. **NOEL** (1 y) for dogs 500 mg/kg diet; (90 d and lifetime) for rats 250 mg/kg diet. Not teratogenic in rabbits (up to 200 mg/kg b.w. daily) or rats (up to 1000 mg/kg b.w. daily) and no significant effect on reproduction in rats (up to 1000 mg/kg diet). **Other** Acute i.p. LD_{50} for male rats 490, female rats 550 mg/kg. Not mutagenic in the Ames, micronucleus and other standard assays. **Toxicity class** WHO (a.i.) III; EPA (formulation) III **EC risk** R22

ECOTOXICOLOGY
Birds Acute oral LD_{50} for mallard ducks >2510 mg/kg. Eight-day dietary LC_{50} for mallard ducks and Japanese quail >5620 mg/kg. **Fish** LC_{50} (96 h) for carp >300, rainbow trout >320, bluegill sunfish 360 mg/l. Not bio-accumulable in fish.
Bees LD_{50} >100 μg/bee by contact and >200 μg/bee by oral administration.
Worms LD_{50} (14 d) for earthworms >1000 ppm (mg a.i./kg dry soil).
Daphnia LC_{50} (48 h) 530 mg/l. **Algae** IC_{50} for *Scenedesmus subspicatus* 46 mg/l.

ENVIRONMENTAL FATE
Animals In rats, following oral administration, absorption is rapid and almost complete, with 81–92% eliminated within 144 hours, in the urine and faeces. Extensive metabolism occurs, mainly by hydrolysis at various points on the

methoxyacetamide moiety, and oxidation of the methyl group on the phenyl ring to the corresponding alcohol. Similar metabolites are found in soil and in plants. **Plants** In plants, the major part (>94%) of the applied dosage remains unchanged. A maximum of 9% penetrates the leaf surface after 42 days, and c. 42% of this is metabolised. **Soil/Environment** In laboratory experiments, DT_{50} was 6–9 mo. However, in the field, DT_{50} values of 2–3 mo were observed. In soil, the main metabolite is oxadixyl acid.

541 oxamyl $\qquad$ *Insecticide, acaricide, nematicide*

oxime carbamate

$$(CH_3)_2NCOC = NOCONHCH_3$$
$$\underset{SCH_3}{|}$$

NOMENCLATURE
Common name oxamyl (BSI, E-ISO, (m) F-ISO, ANSI, ESA); oxamil (JMAF)
IUPAC name N,N-dimethyl-2-methylcarbamoyloxyimino-2-(methylthio)acetamide
Chemical Abstracts name methyl 2-(dimethylamino)-N-[[(methylamino)carbonyl]= oxy]-2-oxoethanimidothioate
CAS RN *[23135–22–0]* **EEC no.** 245–445–3 **Development codes** DPX-D1410

PHYSICAL CHEMISTRY
Mol. wt. 219.3 **M.f.** $C_7H_{13}N_3O_3S$ **Form** Colourless crystals, with a garlic-like odour. **M.p.** At 100–102 °C, changes to a dimorphic form which melts at 108–110 °C. **B.p.** Decomposes on distillation. **V.p.** 0.051 mPa (25 °C) K_{ow} logP = –0.44 (pH 5) **Henry** 3.9×10^{-8} Pa m^3 mol^{-1} **S.g./density** 0.97 at 25 °C **Solubility** In water 280 g/l (25 °C). In methanol 1440, ethanol 330, acetone 670, toluene 10 (all in g/kg, 25 °C). **Stability** Solid and formulations are stable; aqueous solutions decompose slowly; DT_{50} >31 d (pH 5), 8 d (pH 7), 3 h (pH 9); accelerated by aeration and sunlight.

COMMERCIALISATION
History Introduced by E. I. du Pont de Nemours and Co. **Patent** US 3530220; US 3658870 **Manufacturers** Du Pont

APPLICATIONS
Biochemistry Cholinesterase inhibitor. **Mode of action** Contact and systemic insecticide, acaricide, and nematicide. Absorbed by the foliage and roots, with translocation (C. A. Peterson et al., *Pestic. Biochem. Physiol.*, 1978, **8**, 1). **Uses** Control of chewing and sucking insects (including soil insects, but not

wireworms), spider mites, and nematodes in ornamentals, fruit trees, vegetables, cucurbits, beet, bananas, pineapples, peanuts, cotton, soya beans, tobacco, potatoes and other crops. **Phytotoxicity** Non-phytotoxic when used as directed. Some strawberry varieties may be injured. **Formulation types** GR; SL. **Mixtures** *(oxamyl +)* chlorfenvinphos. **Compatibility** Incompatible with alkaline materials. **Selected tradenames** 'Vydate' (Du Pont)

ANALYSIS

Product analysis by hplc (R. F. Holt & R. E. Leitch, *Anal. Methods Pestic. Plant Growth Regul.*, 1978. **10**, 111). **Residues** determined by glc with FPD *(idem, ibid.; Man. Pestic. Residue Anal.*, 1987, **I**, 6; *Anal. Methods Residues Pestic.*, 1988, Part I, M13; A. Ambrus *et al., J. Assoc. Off. Anal. Chem.*, 1981, **64**, 733) or by rplc (*AOAC Methods*, 1995, 985.23). For methods in **drinking water**, see *AOAC Methods*, 1995, 991.06.

MAMMALIAN TOXICOLOGY

Reviews *Pesticide residues in food – 1985.* FAO Plant Production and Protection Paper 68, 1986. *Pesticide residues in food – 1985 evaluations. Part II – Toxicology.* FAO Plant Production and Protection Paper 72/2, 1986. *Pesticide residues in food – 1986.* FAO Plant Production and Protection Paper 78, 1986. **EHC** 64 (WHO, 1986; a review of carbamate insecticides in general). **Oral** Acute oral LD_{50} for rats 5.4 mg/kg. **Skin and eye** Acute percutaneous LD_{50} for male rabbits 5027, female rabbits >2000 mg/kg. Not irritating to skin, not a skin sensitiser on guinea pigs. **Inhalation** LC_{50} (1 h) (atomised spray) for male rats 0.17, female rats 0.12 mg/l air. **NOEL** (2 y) for rats 50 mg a.i./kg diet; for dogs 100 mg/kg diet. **ADI** (JMPR) 0.03 mg/kg b.w. [1985]. **Toxicity class** WHO (a.i.) Ib; EPA (formulation) I **EC risk** T+ (R26/28); Xn (R21)

ECOTOXICOLOGY

Birds Acute oral LD_{50} for quail 4.18 mg a.i./kg. Dietary LC_{50} (8 d) for bobwhite quail 340, mallard duck 766 ppm. **Fish** LC_{50} (96 h) for bluegill sunfish 5.6, goldfish 27.5, rainbow trout 4.2 mg/l. **Bees** Toxic to bees; LD_{50} (oral) 0.078–0.11 μg/bee; (contact) 0.27–0.36 μg/bee. **Daphnia** LC_{50} (48 h) 5.7 mg/l.

ENVIRONMENTAL FATE

Animals In rats, oxamyl was hydrolysed to an oximino metabolite (methyl *N*-hydroxy-*N'*,*N'*-dimethyl-1-thiooxamimidate) or converted enzymically via *N*,*N*-dimethyl-1-cyanoformamide to *N*,*N*-dimethyloxamic acid. Conjugates of the oximino compound, the acid, and their monomethyl derivatives constituted over 70% of the metabolites excreted in the urine and faeces (J. Harvey & J. C-Y. Han, *J. Agric. Food Chem.*, 1978, **26**, 902–910). Metabolism of carbamate insecticides is reviewed (M. Cool & C. K. Jankowski in *"Insecticides"*). **Plants** In plants, oxamyl hydrolyses to the corresponding oximino compound which in turn conjugates with glucose. Total breakdown into natural products has been demonstrated (J. Harvey et al., *J. Agric. Food Chem.*, 1978, **26**, 529–536).   **Soil/Environment** Degraded rapidly in soil, DT_{50} c. 7 d. DT_{50} in groundwater (lab. study) 20 d (anaerobic), 20–400 d (aerobic) (J. H. Smelt *et al.*, *Pestic. Sci.*, 1983, **14**, 173–181). K_{oc} 25.

542 oxasulfuron   *Herbicide*

sulfonylurea

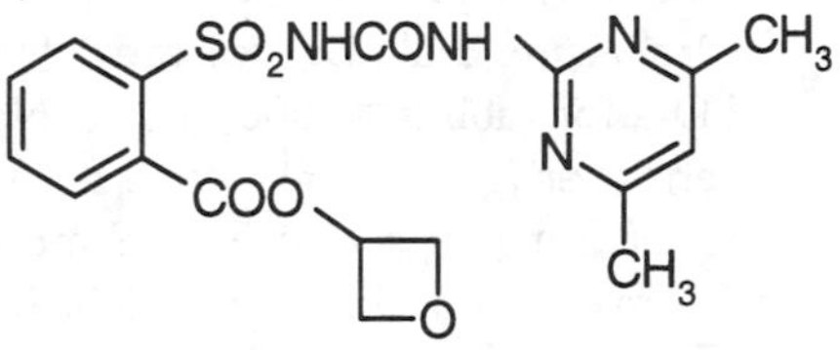

NOMENCLATURE
Common name oxasulfuron (BSI, pa ISO)
IUPAC name oxetan-3-yl 2-[(4,6-dimethylpyrimidin-2-yl)-carbamoylsulfamoyl]= benzoate
Chemical Abstracts name 3-oxetanyl 2-[[[[(4,6-dimethyl-2-pyrimidinyl)amino]= carbonyl]amino]sulfonyl]benzoate
CAS RN *[144651–06–9]*   **Development codes** CGA-277476

PHYSICAL CHEMISTRY
Mol. wt. 406.4   **M.f.** $C_{17}H_{18}N_4O_6S$ **Form** White odourless powder.
M.p. 158 °C (decomp.)   **V.p.** $<2 \times 10^{-3}$ mPa (25 °C) K_{ow} logP = 0.75 (pH 5), –0.81 (pH 7), –2.2 (pH 8.9)   **Henry** $<3.2 \times 10^{-5}$ Pa m^3 mol^{-1} (calc.)
S.g./density 1.41   **Solubility** In water 52 ppm (pH 5.1, 25 °C); in buffer, 63 (pH 5.0), 1700 (pH 6.8), 19 000 (pH 7.8) mg/l (15 °C). In methanol 1500, acetone 9300, toluene 320, *n*-octanol 99, *n*-hexane 2.2, ethyl acetate 2300, dichloromethane 6900 (all in mg/l, 25 °C).   **pKa** 5.1

COMMERCIALISATION
History Reported by R. L. Brooks *et al.* (*Proc. Br. Crop Prot. Conf. – Weeds*, 1995, **1**, 79).   **Patent** US 5209771; EP 0496701   **Manufacturers** Novartis

APPLICATIONS
Biochemistry Branched chain amino acid synthesis (acetolactate synthase or ALS) inhibitor. Acts by inhibiting biosynthesis of the essential amino acids valine and isoleucine, hence stopping cell division and plant growth. Selectivity derives from rapid metabolism in the crop. Metabolic basis of selectivity in sulfonylureas

reviewed (M. K. Koeppe & H. M. Brown, *Agro-Food-Industry,* **6**, 9–14 (1995)).
Mode of action Readily taken up by shoots and roots and translocated to
meristematic tissues. Leaves turn yellow or red, followed by death after 1 to 3
weeks. **Uses** Under development for broad-leaved and grass weed control in
soya beans at 60–90 g/ha. **Formulation types** WG. **Selected tradenames**
'Expert' (Novartis); 'Dynam' (Novartis)

ANALYSIS
Methods for sulfonylurea **residues** in crops, soil and water reviewed (A. C.
Barefoot et al., *Proc. Br. Crop Prot. Conf. – Weeds,* 1995, **2**, 707).

MAMMALIAN TOXICOLOGY
Oral Acute oral LD_{50} for rats >5000 mg/kg. **Skin and eye** Acute percutaneous
LD_{50} for rabbits >2000 mg/kg. No skin or eye irritation (rabbits). Not a skin
sensitiser (guinea pig). **Inhalation** LC_{50} for rats >5.08 mg/l air. **NOEL** (2 y) for
rats 8.3 mg/kg b.w. daily; (18 mo) for mice 1.5 mg/kg b.w. daily; (1 y) for dogs
1.3 mg/kg b.w. daily. **ADI** 0.0026 mg/kg b.w. **Toxicity class** WHO (a.i.) III
(Table 5)

ECOTOXICOLOGY
Birds Acute oral LD_{50} for quail and mallard duck >2250 mg/kg. Dietary LC_{50} for
quail and mallard duck >5620 ppm. **Fish** LC_{50} for bluegill sunfish >111, trout
>116 mg/l. **Bees** LD_{50} >25 µg/bee. **Worms** LC_{50} for earthworms
>1000 mg/kg. **Daphnia** LC_{50} >89.4 mg/l. **Algae** EC_{50} for *Selenastrum
capricornutum* 0.145, for *Navicula* >20 mg/l.

ENVIRONMENTAL FATE
Animals The majority (70–80%) of the applied dose is excreted in urine with no
accumulation in tissues; depletion $t_{1/2}$ c. 7–14 h. The metabolic route involves
hydroxylation of the pyrimidine methyl, hydrolysis of the oxetane ring and
cleavage of the sulfonylurea bridge. **Plants** The major metabolite is saccharine
(0.002 ppm in mature beans); small amounts of oxetane alcohol are also formed.
Metabolism follows a similar route to animals. **Soil/Environment** Soil DT_{50} (lab.)
10 d, (field) <2 w. Breakdown is primarily microbial and hydrolytic. Independent
of soil pH, organic matter content or soil structure. K_{oc} 30.6 (4 soils).

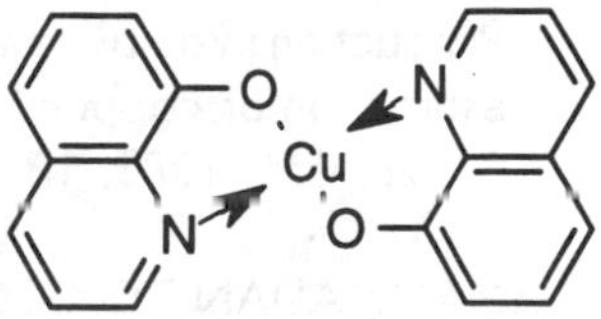

NOMENCLATURE
Common name oxine-copper (BSI, JMAF); oxine copper (E-ISO); oxine-Cu (alternative E-ISO, *(m)* F-ISO); oxine-cuivre (*(m)* F-ISO); oxyquinoléate de cuivre (France); copper 8-quinolinolate (Canada)
IUPAC name bis(quinolin-8-olato-*O,N*)copper; cupric 8-quinolinoxide
Chemical Abstracts name bis(8-quinolinato-N^1,O^8)copper
CAS RN *[10380–28–6]* **EEC no.** 233–84–19
Development codes Ro 17–0099/000 (Maag); CGA 281881

PHYSICAL CHEMISTRY
Mol. wt. 351.9 **M.f.** $C_{18}H_{12}CuN_2O_2$ **Form** Olive-green powder.
M.p. Decomposes above 270 °C **V.p.** 4.6×10^{-5} mPa (25 °C) (EEC A.4)
K_{ow} logP = 2.46 (distilled water, 25 °C). **S.g./density** 1.63 (20 °C) **Solubility** In water 0.07 mg/l (pH 7, 25 °C). In methanol 116, *n*-hexane <0.01 (both in mg/l, 25 °C). **Stability** DT_{50} 60–96 h (pH 7.0). Chemically inert. Stable to u.v. light.

COMMERCIALISATION
History The fungicidal properties of this complex of copper with quinolin-8-ol, the latter known by the trivial name oxine, reported by D. Powell (*Phytopathology*, 1946, **36**, 572). **Manufacturers** La Quinoleine; Tomono

APPLICATIONS
Mode of action Non-systemic, protective fungicide. **Uses** Seed treatment for control of glume blotch, bunt, and snow mould of wheat; *Cercospora*, *Phoma*, and *Pythium* spp. in beet; *Alternaria* and *Botrytis* spp. in flax; *Alternaria* and *Phoma* spp. in oilseed rape; *Sclerotinia sclerotiorum* in sunflowers; and leaf spot on beans and peas; application rates range from 20–100 g/100 kg seed; also for control of leaf spot on celery; and scab and canker on pome fruit. Also used for sealing wounds and pruning cuts on trees, and for treatment of fruit- and potato-handling equipment. **Phytotoxicity** Non-phytotoxic when used as directed. Not to be used on copper-sensitive plants. **Formulation types** WP; SC; DS; PA.
Mixtures *(oxine-copper +)* bitumen; anthraquinone; lindane; anthraquinone + carboxin; endosulfan + lindane; anthraquinone + endosulfan + lindane; carboxin + lindane; petroleum wax; anthraquinone + flutriafol; anthraquinone + prochloraz; anthraquinone + tefluthrin; anthraquinone + ethirimol + flutriafol; fuberidazole.

Selected tradenames 'Okishindo' (Tomono); 'Quinolate' (La Quinoleine); 'Quinondo' (Agro-Kanesho)

ANALYSIS
Product analysis by reaction with sulfuric acid and standard methods for estimation of copper (*AOAC Methods*, 1984, 6.015–6.016; *CIPAC Handbook*, 1970, **1**, 226; *ibid.*, 1983, **1B**, 1915, 1918).

MAMMALIAN TOXICOLOGY
Oral Acute oral LD_{50} for rats 4700, mice 9000 mg/kg.   **Skin and eye** Acute percutaneous LD_{50} for rats >2000 mg/kg. Non-irritating to skin, irritating to eyes (rabbits).   **Inhalation** LC_{50} for rats 150 mg/m³ air.   **NOEL** (2 y) for rats 2, dogs 5 mg/kg diet; (80 w) for mice 400 mg/kg diet.   **ADI** 0.02 mg/kg.
Other Non-carcinogenic, non-mutagenic and non-teratogenic.   **Toxicity class** WHO (a.i.) III (Table 5); EPA (formulation) I (water-base); II, III (petroleum solvent-base)

ECOTOXICOLOGY
Birds LD_{50} (8 d) for bobwhite quail 618, mallard ducks >2000 mg/kg. LC_{50} (8 d) for bobwhite quail 3428, mallard ducks >2000 mg/kg.   **Fish** LC_{50} (96 h) for bluegill sunfish 21.6, rainbow trout 8.94 µg/l.   **Bees** Non-toxic to bees.
Daphnia LC_{50} (48 h) 177 µg/l.   **Algae** EC_{50} (5 d) 2.20–15.4 µg/l.

ENVIRONMENTAL FATE
Animals In rats, between 58–62% is excreted in urine and about 25% eliminated in faeces within 72 h. Metabolites were mainly parent and 8-hydroxyquinoline and a small amount of glucuronide and sulfate conjugates.   **Plants** The product does not penetrate and is not translocated after foliar application after one week in lettuce and up to four weeks in apple. No metabolites were observed in the surface wash-off.   **Soil/Environment** DT_{50} 2 d (20 °C, 63% MWC). No tendency to leach (90% remained in the 0–6 cm top soil layer).

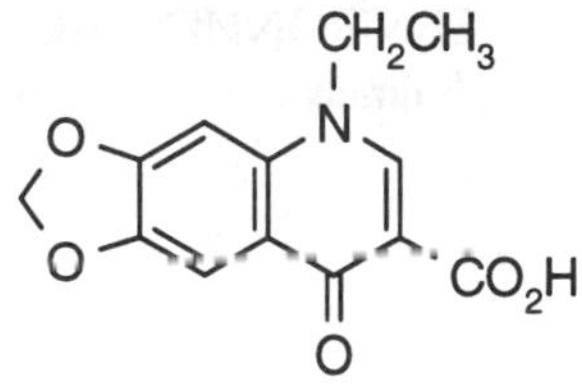

NOMENCLATURE
Common name oxolinic acid (BSI, BAN, INN, draft E-ISO)
IUPAC name 5-ethyl-5,8-dihydro-8-oxo[1,3]dioxolo[4,5-g]quinoline-7-carboxylic acid
Chemical Abstracts name 5-ethyl-5,8-dihydro-8-oxo-1,3-dioxolo[4,5-g]quinoline-=7-carboxylic acid
CAS RN *[14698–29–4]* **Development codes** S-0208 (Sumitomo)

PHYSICAL CHEMISTRY
Mol. wt. 261.2 **M.f.** $C_{13}H_{11}NO_5$ **Form** Colourless crystals; (tech., light brown crystalline solid). **M.p.** >250 °C **V.p.** <0.147 mPa (100 °C)
S.g./density 1.5–1.6 g/cm^3 (23 °C) **Solubility** In water 3.2 mg/l (25 °C). In hexane, xylene, methanol <10 (all in g/kg, 20 °C).

COMMERCIALISATION
History Bactericide reported by D. Kaminksy & R. I. Meltzer (*J. Med. Chem.*, 1968, **11**, 160). Introduced against urinary tract infections by Warner-Lambert. Activity against plant-pathogenic bacteria reported by Y. Hikichi et al. (*Jpn. Pestic. Inf.*, 1989, No. 55, p. 21). Introduced in Japan (1989) by Sumitomo Chemical Co., Ltd as seed treatment. **Patent** US 3287458 to Warner-Lambert
Manufacturers Sumitomo

APPLICATIONS
Mode of action Systemic bactericide. **Uses** Preventative and curative control of Gram-negative bacteria, including *Pseudomonas* and *Erwinia* species, in rice crops.
Formulation types SD; WP; DP. **Selected tradenames** 'Starner' (Sumitomo)

ANALYSIS
Details from Sumitomo.

MAMMALIAN TOXICOLOGY
Oral Acute oral LD$_{50}$ for male rats 630, female rats 570 mg/kg.
Skin and eye Acute percutaneous LD$_{50}$ for male and female rats >2000 mg/kg. Non-irritant to eyes and skin (rabbits). **Inhalation** LC$_{50}$ (4 h) for male rats 2.45, female rats 1.70 mg/l.

Fish LC_{50} (48 h) for carp >10 mg/l.

ENVIRONMENTAL FATE
Animals For details of metabolism, see Crew *et al.*, *Xenobiotica*, 1971, **1**, 193.

545 oxycarboxin   *Fungicide*

carboxamide

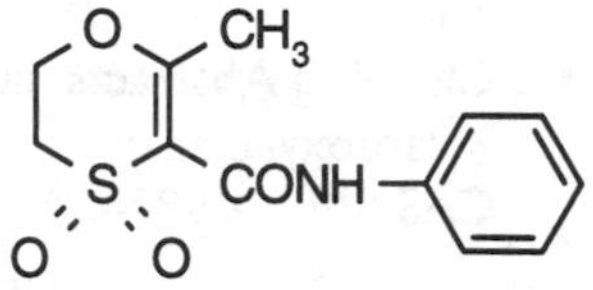

NOMENCLATURE
Common name oxycarboxin (BSI, E-ISO, ANSI, JMAF); oxycarboxine ((*f*) F-ISO)
IUPAC name 5,6-dihydro-2-methyl-1,4-oxathi-ine-3-carboxanilide 4,4-dioxide
Chemical Abstracts name 5,6-dihydro-2-methyl-*N*-phenyl-1,4-oxathiin-=
3-carboxamide 4,4-dioxide
CAS RN *[5259–88–1]* **EEC no.** 226–066–2 **Development codes** F 461
(Uniroyal)

PHYSICAL CHEMISTRY
Composition Tech. grade is ≥97% pure. **Mol. wt.** 267.3 **M.f.** $C_{12}H_{13}NO_4S$
Form White solid; (tech., brownish grey solid). **M.p.** 119.5–121.5 °C
V.p. <5.6 × 10^{-3} mPa (25 °C) **K_{ow}** logP = 0.772 **S.g./density** 1.41 g/cm^3
Solubility In water 1.4 g/l (25 °C). In acetone 83.7 g/l, hexane 8.8 mg/l (25 °C).
Stability Stable at 55 °C for 18 d. Hydrolysis DT_{50} 44 d (pH 6, 25 °C).

COMMERCIALISATION
History Fungicide reported by B. von Schmeling & M. Kulka (*Science*, 1966, **152,**
659). Introduced by Uniroyal Inc. **Patent** US 3399214; US 3402241; US 3454391
Manufacturers Jin Hung; Uniroyal

APPLICATIONS
Mode of action Systemic fungicide with curative action. Application is primarily
foliar, but also absorbed by the roots of the plant. **Uses** Control of rust diseases
on ornamentals (particularly geraniums, chrysanthemums, carnations, and roses),
cereals, and nursery trees; and fairy rings on turf. Applied at 200–1500 g/ha.
Formulation types WP; EC. **Mixtures** *(oxycarboxin +)* captan; mancozeb.

Compatibility Not to be mixed with insecticides and acaricides, as plant injury may occur. **Selected tradenames** 'Plantvax' (Uniroyal); 'Carboject' (Mauget)

ANALYSIS
Product analysis by hplc or by i.r. spectrometry, details available from Uniroyal Chemical Co. **Residue** analysis by hydrolysis and determination of the aniline so formed by glc (H. R. Siskin & J. E. Newell, *J. Agric. Food Chem.*, 1971, **19**, 738).

MAMMALIAN TOXICOLOGY
Oral Acute oral LD_{50} for male rats 5816, female rats 1632 mg/kg.
Skin and eye Acute percutaneous LD_{50} for rabbits >5000 mg/kg. Mild eye irritant; not a skin irritant (rabbits). **Inhalation** LC_{50} (4 h) for rats >5000 mg/l.
NOEL (2 y) for rats 15, dogs 75 mg/kg b.w. daily. **ADI** 0.15 mg/kg.
Toxicity class WHO (a.i.) III (Table 5); EPA (formulation) IV **EC risk** Xn (R22)

ECOTOXICOLOGY
Birds LD_{50} for mallard ducks 1250 mg/kg. Dietary LC_{50} (8 d) for mallard ducks >4640, bobwhite quail >10 000 ppm. **Fish** LC_{50} (96 h) for rainbow trout 19.9, bluegill sunfish 28.1 mg/l. **Bees** LD_{50} (contact) >181 µg/bee. **Daphnia** LC_{50} (48 h) 69.1 mg/l. **Algae** EC_{50} (96 h) for *Chlorella* 19.0 mg/l.

ENVIRONMENTAL FATE
Soil/Environment Aerobic soil metabolism: DT_{50} in sandy loam 2.5–8 w.

546 oxydemeton-methyl *Insecticide*

organophosphorus

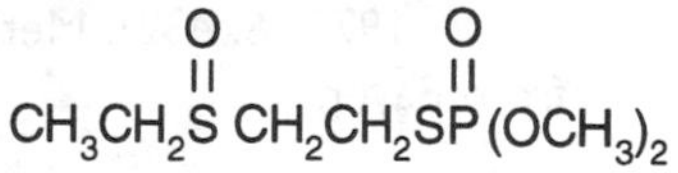

$$CH_3CH_2S\ CH_2CH_2SP(OCH_3)_2$$

NOMENCLATURE
Common name oxydemeton-methyl (BSI, E-ISO, (*m*) F-ISO); oxydemetonmethyl (ESA); oxydemeton-metyl (USSR); [metilmerkaptofosoksid] (former USSR name)
IUPAC name S-2-ethylsulfinylethyl O,O-dimethyl phosphorothioate
Chemical Abstracts name S-[2-(ethylsulfinyl)ethyl] O,O-dimethyl phosphorothioate
CAS RN *[301–12–2]* **EEC no.** 206–110–7 **Development codes** Bayer 21 097; R 2170 (Bayer) **Official codes** ENT 24 964

PHYSICAL CHEMISTRY
Mol. wt. 246.3 **M.f.** $C_6H_{15}O_4PS_2$ **Form** Colourless liquid. **M.p.** <–20 °C

B.p. 106 °C/0.01 mmHg   **V.p.** 3.8 mPa (20 °C) K_{ow} logP = –0.74 (21 °C)
Henry <<1 × 10^{-5} Pa m^3 mol^{-1} (calc.)   **S.g./density** 1.289 (20 °C)   **Solubility**
Completely miscible with water. Soluble in common organic solvents, except in
petroleum ether.   **Stability** Relatively slowly hydrolysed in acidic media, but rapidly
hydrolysed in alkaline media; DT_{50} (estimated) 107 d (pH 4), 46 d (pH 7), 2 d
(pH 9) (22 °C).   **F.p.** 113 °C

COMMERCIALISATION
History Insecticide reported by G. Schrader (*Die Entwicklung neuer insektizider
Phosphorsäure-Ester*). Introduced by Bayer AG.   **Patent** DE 947368; US 2963505
Manufacturers Aimco; Bayer India

APPLICATIONS
Biochemistry Cholinesterase inhibitor.   **Mode of action** Systemic insecticide with
contact and stomach action. Has quick knockdown effect.   **Uses** Control of
aphids, sawflies, suckers, and other sucking insects on fruit, vines, vegetables,
cereals, and ornamentals.   **Phytotoxicity** Some ornamentals may be injured, more
especially in combination with other pesticides.   **Formulation types** EC; SL.
Mixtures (*oxydemeton-methyl +*) parathion; trichlorfon; beta-cyfluthrin; fenvalerate.
Compatibility Incompatible with alkaline materials.
Selected tradenames 'Metasystox R' (Bayer, Gowan); 'Aimcosystox' (Aimco);
'Dhanusystox' (Dhanuka)

ANALYSIS
Product analysis by hplc (*AOAC Methods*, 1995, 991.05; *CIPAC Handbook*, 1992, **E**,
163–5), by oxidation to phosphoric acid which is measured by standard
colorimetric methods (*CIPAC Handbook*, 1983, **1B**, 1871), or by reduction of the
sulfoxide group by titanium(III) sulfate and titration of the excess (details from
Bayer AG).   **Residues** determined, after oxidation to the corresponding sulfone
(demeton-S-methylsulphon), by glc with FID (K. Wagner & J. S. Thornton,
Pflanzenschutz-Nachr. (Engl. Ed.), 1977, **30**, 1; *Anal. Methods Pestic. Plant Growth
Regul.*, 1972, **6**, 432). Methods for the determination of residues are available
from Bayer.

MAMMALIAN TOXICOLOGY
Reviews *Pesticide residues in food – 1989*. FAO Plant Production and Protection
Paper 99, 1989. *Pesticide residues in food – 1989 evaluations. Part II – Toxicology.*
FAO Plant Production and Protection Paper 100/2, 1990.   **EHC** 63 (WHO,
1986; a general review of organophosphorus insecticides).   **Oral** Acute oral
LD_{50} for rats c. 50 mg/kg.   **Skin and eye** Acute percutaneous LD_{50} for rats
c. 130 mg/kg. Mild skin irritant, eye irritant (rabbits; 50% in MIBK).
Inhalation LC_{50} (4 h) for rats 471 mg/m^3 (50% in MIBK).   **NOEL** (2 y) for rats 1,
mice 30 mg/kg diet; (1 y) for dogs 0.25 mg/kg b.w.   **ADI** (JMPR) 0.0003 mg/kg
for sum of demeton-S-methyl, demeton-S-methylsulphon and oxydemeton-methyl

[1989].   **Other** Not a primary embryotoxin, not a mutagen.   **Toxicity class** WHO (a.i.) Ib; EPA (formulation) I   **EC risk** T (R24/25)

ECOTOXICOLOGY

Birds LD_{50} for bobwhite quail 34–37 mg/kg. LC_{50} (5 d) for mallard ducks >5000, bobwhite quail 434 mg/kg diet.   **Fish** LC_{50} (96 h) for rainbow trout 17, golden orfe 447.3, bluegill sunfish 1.9 mg/l.   **Bees** Toxic to bees.   **Worms** LC_{50} for *Eisenia foetida* 115 mg/kg dry soil.   **Daphnia** LC_{50} (48 h) 0.19 mg/l.   **Algae** E_rC_{50} for *Scenedesmus subspicatus* 49 mg/l.

ENVIRONMENTAL FATE

Animals Elimination is very quick; almost 99% is excreted within 48 h in the urine. **Plants** Oxydemeton-methyl is very quickly metabolised in sugar beet. Besides oxidation to biologically-active demeton-S-methylsulphon, the main metabolic reactions are hydrolysis and subsequent dimerisation as well as the formation of conjugates.   **Soil/Environment** Degradation in different soils is extremely rapid. The main metabolic routes involve oxidation of the sulfoxide to a sulfone group, and oxidative and hydrolytic cleavage of the side-chain, with the formation of dimethylphosphoric acid and phosphoric acid.

547   oxyfluorfen   *Herbicide*

diphenyl ether

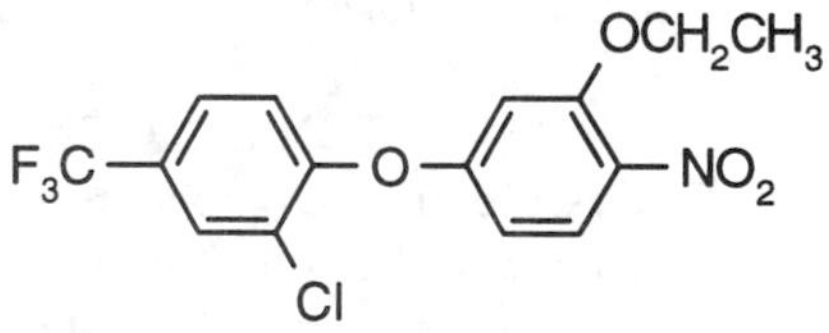

NOMENCLATURE

Common name oxyfluorfen (BSI, E-ISO, ANSI, WSSA); oxyfluorfène ((m) F-ISO)
IUPAC name 2-chloro-α,α,α-trifluoro-*p*-tolyl 3-ethoxy-4-nitrophenyl ether
Chemical Abstracts name 2-chloro-1-(3-ethoxy-4-nitrophenoxy)-4-(trifluoro= methyl)benzene
CAS RN *[42874–03–3]*   **EEC no.** 255–983–0   **Development codes** RH-2915

PHYSICAL CHEMISTRY

Mol. wt. 361.7   **M.f.** $C_{15}H_{11}ClF_3NO_4$ **Form** Orange crystalline solid.
M.p. 85–90 °C; (tech., 65–84 °C)   **B.p.** 358.2 °C (decomp.)   **V.p.** (pure a.i.) 0.0267 mPa (25 °C) K_{ow} logP = 4.47   **S.g./density** 1.35 (73 °C)   **Solubility** In

water 0.116 mg/l (25 °C). Readily soluble in most organic solvents, e.g. acetone 72.5, cyclohexanone, isophorone 61.5, dimethylformamide >50, chloroform 50–55, mesityl oxide 40–50 (all in g/100 g, 25 °C).   **Stability** No significant hydrolysis in 28 d at pH 5–9 (25 °C). Decomposed rapidly by u.v. irradiation; DT_{50} 3 d (room temperature). Stable up to 50 °C.

COMMERCIALISATION
History Herbicide reported by R. Y. Yih & C. Swithenbank (*J. Agric. Food Chem.*, 1975, **23**, 592). Introduced by Rohm & Haas Co.   **Patent** US 3798276
Manufacturers Makhteshim-Agan; Rohm & Haas

APPLICATIONS
Biochemistry Protoporphyrinogen oxidase inhibitor.   **Mode of action** Selective contact herbicide, absorbed more readily by the foliage (and especially the shoots) than by the roots, with very little translocation.   **Uses** Control of annual broad-leaved weeds and grasses in a variety of tropical and subtropical crops, by pre- or post-emergence application at rates in the range 0.25 – 2.0 kg a.i./ha. Particular crops include tree fruit (including citrus), vines, nuts, cereals, maize, soya beans, peanuts, rice, cotton, bananas, peppermint, onions, garlic, ornamental trees and shrubs, and conifer seedbeds.   **Phytotoxicity** Soya beans and cotton may be injured by contact with oxyfluorfen.   **Formulation types** EC; GR.
Mixtures (*oxyfluorfen +*) oryzalin; pendimethalin; propyzamide; dalapon-sodium; propyzamide + terbuthylazine.   **Selected tradenames** 'Goal' (Rohm & Haas); 'Galigan' (Makhteshim-Agan)

ANALYSIS
Product and **residues** determined by glc (I. Adler & C. K. Hoffman, *Anal. Methods. Pestic. Plant Growth Regul.*, **11**, 331–341).

MAMMALIAN TOXICOLOGY
Oral Acute oral LD_{50} for rats and dogs >5000 mg/kg.   **Skin and eye** Acute percutaneous LD_{50} for rabbits >10 000 mg/kg. Mild to moderate eye irritant; mild skin irritant (rabbits).   **Inhalation** LC_{50} >5.4 mg/l.   **NOEL** In chronic dietary trials, NOEL for rats 40, dogs 100, mice 2 mg/kg diet.   **ADI** 0.003 mg/kg.
Toxicity class WHO (a.i.) III (Table 5); EPA (formulation) IV

ECOTOXICOLOGY
Birds Acute LD_{50} for bobwhite quail >2150 mg/kg. Eight-day dietary LC_{50} for mallard ducks and bobwhite quail >5000 ppm.   **Fish** LC_{50} (96 h) for bluegill sunfish 0.2, trout 0.41, channel catfish 0.4 mg/l.   **Bees** Not toxic to honeybees at 0.025 mg a.i./bee.   **Daphnia** LC_{50} (48 h) 1.5 mg a.i./l.

ENVIRONMENTAL FATE
Animals For details of metabolism, see I. L. Adler *et al.*, *J. Agric. Food Chem.*, 1977,

25, 1339.   **Plants** Not readily metabolised in plants.   **Soil/Environment** Strongly adsorbed on soil, not readily desorbed, and shows negligible leaching. K_{oc} from 2891 (sand) to 32 381 (silty clay loam). Photodecomposition in water is rapid, and on soil is slow. Microbial degradation is not a major factor. Field dissipation DT_{50} 5–55 d; soil DT_{50} (in dark) (aerobic) 292 d, (anaerobic) c. 580 d.

548 paclobutrazol   *Plant growth regulator*

azole

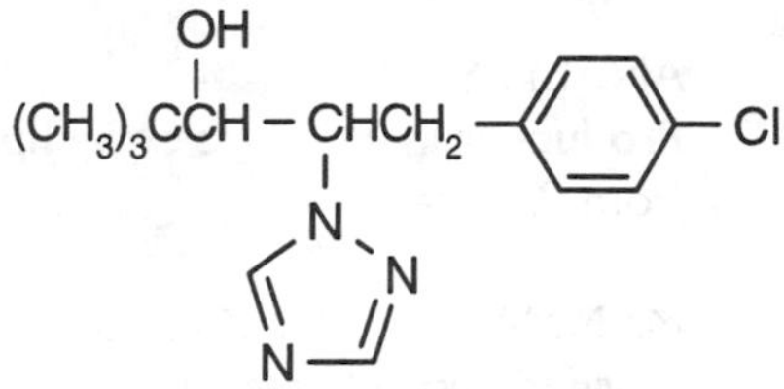

NOMENCLATURE
Common name paclobutrazol (BSI, draft E-ISO, (m) draft F-ISO, ANSI)
IUPAC name (2RS,3RS)-1-(4-chlorophenyl)-4,4-dimethyl-2-(1H-1,2,4-triazol-1-yl)= pentan-3-ol
Chemical Abstracts name (R*,R*)-(±)-β-[(4-chlorophenyl)methyl]-= α-(1,1-dimethylethyl)-1H-1,2,4-triazole-1-ethanol
CAS RN *[76738–62–0]* stated stereoisomers   **Development codes** PP333

PHYSICAL CHEMISTRY
Composition Tech. material is 90% pure.   **Mol. wt.** 293.8   **M.f.** $C_{15}H_{20}ClN_3O$
Form White crystalline solid.   **M.p.** 165–166 °C   **V.p.** 0.001 mPa (20 °C)
K_{ow} logP = 3.2   **S.g./density** 1.22 g/ml   **Solubility** In water 26 mg/l (20 °C). In acetone 110, cyclohexanone 180, dichloromethane 100, hexane 10, xylene 60, methanol 150, propylene glycol 50 (all in g/l, 20 °C).   **Stability** Stable for more than 2 years at 20 °C, and more than 6 months at 50 °C. Stable to hydrolysis (pH 4–9), and not degraded by u.v. light (pH 7, 10 days).

COMMERCIALISATION
History Plant growth regulator reported by B. G. Lever *et al.* (*Proc. Br. Crop Prot. Conf. – Weeds*, 1982, **1**, 3) and introduced by ICI Agrochemicals (now Zeneca Agrochemicals).   **Patent** GB 1595696; US 1595697   **Manufacturers** Zeneca

APPLICATIONS
Biochemistry Inhibits gibberellin and sterol biosynthesis and hence the rate of cell

division.   **Mode of action** Plant growth regulator taken up into the xylem through the leaves, stems, or roots, and translocated to growing sub-apical meristems. Produces more compact plants and enhances flowering and fruiting.   **Uses** Used on fruit trees to inhibit vegetative growth and to improve fruit set; on pot-grown ornamentals and flower crops (e.g. chrysanthemums, begonias, freesias, poinsettias and bulbs) to inhibit growth; on rice to increase tillering, reduce lodging, and increase yield; on turf to retard growth; and on grass seed crops to reduce height and prevent lodging. To be applied as a foliar spray, as a soil drench, or by trunk injection. Has some fungicidal activity against mildew and rusts. **Phytotoxicity** Non-phytotoxic, though it intensifies greening. Some spotting has been noted on periwinkle foliage at higher temperatures.   **Formulation types** SC; WP.   **Selected tradenames** 'Bonzi' (Zeneca); 'Clipper' (Zeneca); 'Cultar' (Zeneca); 'Parlay' (Zeneca); 'Multeffect' (Sanonda)

ANALYSIS
Product analysis by glc or by hplc.   **Residues** determined by tlc. Details available from Zeneca.

MAMMALIAN TOXICOLOGY
Reviews *Pesticide residues in food – 1988. FAO Plant Production and Protection Paper 92, 1988. Pesticide residues in food – 1988 evaluations. Part II – Toxicology. FAO Plant Production and Protection Paper 93/2, 1989.*   **Oral** Acute oral LD_{50} for male rats 2000, female rats 1300, male mice 490, female mice 1200, guinea pigs 400–600, male rabbits 840, female rabbits 940 mg/kg.   **Skin and eye** Acute percutaneous LD_{50} for rats and rabbits >1000 mg/kg. Mild skin irritant, moderate eye irritant (rabbits). Not a skin sensitiser (guinea pigs).   **Inhalation** LC_{50} (4 h) for male rats 4.79, female rats 3.13 mg/l air.   **NOEL** (1 y) for dogs 75 mg/kg daily; (2 y) for rats 250 mg/kg diet.   **ADI** (JMPR) 0.1 mg/kg b.w. [1988].   **Other** Not mutagenic.   **Toxicity class** WHO (a.i.) III; EPA (formulation) IV

ECOTOXICOLOGY
Birds Acute oral LD_{50} for mallard ducks >7900 mg/kg.   **Fish** LC_{50} (96 h) for rainbow trout 27.8 mg/l. NOEC 3.3 mg/l.   **Bees** Acute oral NOEL >0.002 mg/bee; acute percutaneous NOEL >0.040 mg/bee.   **Daphnia** LC_{50} (48 h) 33.2 mg/l.   **Algae** EC_{50} (cell volume) for *Chlorella fusca* 180 μmol/l (*Pestic. Sci.*, **47**, 337 (1996)).

ENVIRONMENTAL FATE
Soil/Environment Soil DT_{50} 0.5–1.0 y in general; in calcareous clay loam (pH 8.8, 14% o.m.) DT_{50} <42 d, in coarse sandy loam (pH 6.8, 4% o.m.) DT_{50} >140 d.

549

Paecilomyces fumosoroseus   *Insecticide*

entomopathogenic fungus

NOMENCLATURE
Scientific name *Paecilomyces fumosoroseus*   **Strain** Apopka Strain 97

COMMERCIALISATION
History Reported by G. Sterk *et al.* (*Proc. Br. Crop Prot. Conf. – Pests Dis.,* 1996, **2**, 461).

APPLICATIONS
Uses For control of whitefly (*Trialeurodes vaporariorum* and *Bemisia tabaci*) in glasshouse crops.   **Formulation types** WG.   **Selected tradenames** 'PFR 97' (Thermo Trilogy)

MAIN ENTRIES

---

paraquat dichloride   *Herbicide*

bipyridylium

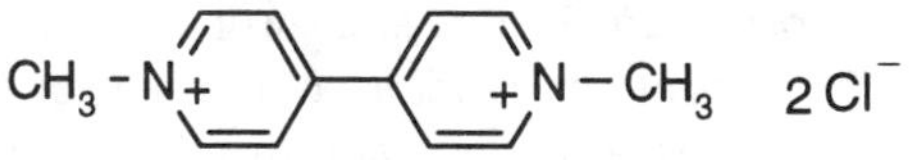

$$CH_3-N^+\!\!\!\!\!\!\diagup\!\!\!\!\!\diagdown\!\!\!\!\!\!\diagup\!\!\!\!\!\diagdown\!\!\!\!\!\!^+N-CH_3 \quad 2\,Cl^-$$

NOMENCLATURE
paraquat dichloride
Other names methyl viologen   **CAS RN** *[1910–42–5]*
Development codes PP148; PP910 for bis(methyl sulfate)

paraquat
Common name paraquat (BSI, E-ISO, (*m*) F-ISO, ANSI, WSSA, JMAF); no name (Germany)
IUPAC name 1,1'-dimethyl-4,4'-bipyridinium
Chemical Abstracts name 1,1'-dimethyl-4,4'-bipyridinium
CAS RN *[4685–14–7]*; *[2074–50–2]* for bis(methyl sulfate)   **EEC no.** 225–141–7

PHYSICAL CHEMISTRY
paraquat dichloride
Composition Not normally isolated from the tech. products, which are >95% pure.   **Mol. wt.** 257.2   **M.f.** $C_{12}H_{14}Cl_2N_2$ **Form** Colourless, hygroscopic crystals.

M.p. Decomposes at *c.* 300 °C **V.p.** <0.1 mPa **S.g./density** 1.24–1.26 (20 °C)
Solubility In water *c.* 700 g/l (20 °C). Practically insoluble in most other organic
solvents. **Stability** Stable in neutral and acidic media, but readily hydrolysed in
alkaline media. Photochemically decomposed by u.v. irradiation in aqueous
solution.

paraquat
Mol. wt. 186.3 **M.f.** $C_{12}H_{14}N_2$

COMMERCIALISATION
History Herbicidal properties of the dichloride and bis(methyl sulfate) described
by R. C. Brian (*Nature (London)*, 1958, **181**, 446) and their properties reviewed by
A. Calderbank (*Adv. Pest Control Res.*, 1968, **8**, 127). Both salts (though only the
former is still sold) were introduced by ICI Plant Protection Division (now Zeneca
Agrochemicals). **Patent** GB 813531 **Manufacturers** Comlets; Crystal; Kuo Ching;
Pilarquim; Productos OSA; Sanex; Sinon; Zeneca

APPLICATIONS
paraquat dichloride
Biochemistry During photosynthesis, superoxide is generated, which damages cell
membranes and cytoplasm. **Mode of action** Non-selective contact herbicide,
absorbed by the foliage, with some translocation in the xylem. **Uses** Broad-
spectrum control of broad-leaved weeds and grasses in fruit orchards (including
citrus), plantation crops (bananas, coffee, cocoa palms, coconut palms, oil palms,
rubber, etc.), vines, olives, tea, alfalfa, onions, leeks, sugar beet, asparagus,
ornamental trees and shrubs, in forestry, etc. Also used for general weed control
on non-crop land; as a defoliant for cotton and hops; for destruction of potato
haulms; as a desiccant for pineapples, sugar cane, soya beans, and sunflowers; for
strawberry runner control; in pasture renovation; and for control of aquatic
weeds. **Formulation types** SL. **Mixtures** (*paraquat dichloride* +) diuron; MCPA;
diquat dibromide; simazine; monolinuron; diquat dibromide + simazine; amitrole +
ammonium thiocyanate; amitrole + diquat dibromide + simazine; linuron +
metolachlor; asulam. **Compatibility** Incompatible with alkaline materials, anionic
surfactants, and clay-containing inert materials. **Selected tradenames**
'Gramoxone' (Zeneca); 'Efoxon' (Efthymiadis); 'Herbaxon' (Westrade); 'Osaquat'
(Productos OSA); 'Pilarxone' (Pilarquim); 'Total' (Barclay); 'Weedless' (Sanonda)

ANALYSIS
Product analysis by colorimetry (*AOAC Methods*, 1995, 969.09; *CIPAC Handbook*,
1970, **1**, 547; 1992, **E**, 166–168; 1995, **G**, 128); impurities determined by glc (*ibid.*,
1980, **1A**, 1317; FAO Specification (CP/50), *Herbicides 1977*, pp. 52, 54); in
mixture with diquat, by colorimetry (*CIPAC Handbook*, 1992, **E**, 73–78; *ibid.*, 1995,
G, 47–49). **Residues** determined by colorimetry after reduction (*Pestic. Anal.
Man.*, 1979, **II**; A. Calderbank & S. H. Yuen, *Analyst (London)*, 1965, **90**, 99;

P. F. Lott et al., J. Chromatogr. Sci., 1978, **16**, 390; J. B. Leary, Anal. Methods Pestic. Plant Growth Regul., 1978, **10**, 321). Residues in potatoes by rplc with dual channel u.v. detection (AOAC Methods, 1995, 992.17). Details of methods available from Zeneca.

MAMMALIAN TOXICOLOGY
paraquat dichloride
Reviews Pesticide residues in food – 1986. FAO Plant Production and Protection Paper **77**, 1986. Pesticide residues in food – 1986 evaluations. Part II – Toxicology. FAO Plant Production and Protection Paper 78/2, 1987. **EHC** 39 (WHO, 1984).
Oral Acute oral LD_{50} for rats 157, mice 104, guinea pigs 22–42, dogs 25–50, cats 40–50, cows 50–75, sheep 50–75 mg/kg. Lethal dose for man is c. 30 mg/kg.
Skin and eye Acute percutaneous LD_{50} for rabbits 236–500 mg paraquat ion/kg. Irritating to skin and eyes (rabbits). Absorption through intact human skin is minimal; exposures can cause irritation and a delay in the healing of cuts and wounds; can cause temporary damage to nails. Not a skin sensitiser (guinea pigs). **Inhalation** No vapour toxicity. If inhaled, may cause nose bleeding.
NOEL (2 y) for dogs 34, rats 170 mg/kg diet. **ADI** (JMPR) 0.004 mg/kg b.w. (as paraquat ion) [1986]. **Other** Has serious delayed effects if absorbed.
Toxicity class WHO (a.i.) II; EPA (formulation) II **EC risk** T (R24/25); Xi (R36/37/38) (applies to all salts) **PIC** Yes

ECOTOXICOLOGY
paraquat dichloride
Birds Acute oral LD_{50} for hens 262–380 mg/kg. LC_{50} (5 d) for bobwhite quail 981, Japanese quail 970, mallard ducks 4048 mg/kg. **Fish** Depends on the formulation and wetter used, LC_{50} (96 h) for rainbow trout 32, brown trout 2.5–13 mg/l.
Bees Non-toxic to bees.

ENVIRONMENTAL FATE
Animals In rats, following oral administration, 76–90% of the dose was excreted in the faeces, and 11–20% in the urine. **Plants** On plant surfaces, photochemical degradation occurs. Degradation products which have been isolated include 1-methyl-4-carboxypyridinium chloride and methylamine hydrochloride.
Soil/Environment Clays and o.m. rapidly and strongly adsorb paraquat, resulting in complete deactivation; typical Strong Adsorption Capacities vary from 20–3000 mg/kg soil, depending on clay/o.m. content. Desorption requires digestion with 12N H_2SO_4 for several hours.

551 parathion

Insecticide, acaricide

organophosphorus

$$O_2N \underset{}{\longleftarrow} \bigcirc \underset{}{\longrightarrow} O\overset{\overset{S}{\parallel}}{P}(OCH_2CH_3)_2$$

NOMENCLATURE
Common name parathion (BSI, E-ISO, F-ISO, ESA, JMAF)
IUPAC name *O,O*-diethyl *O*-4-nitrophenyl phosphorothioate
Chemical Abstracts name *O,O*-diethyl *O*-(4-nitrophenyl) phosphorothioate
Other names parathion-ethyl; thiophos; ethyl parathion; S.N.P.
CAS RN *[56–38–2]* **EEC no.** 200–271–7 **Development codes** ACC 3422
(Cyanamid); BAY 9491 (Bayer) **Official codes** OMS 19; ENT 15 108

PHYSICAL CHEMISTRY
Composition Tech. grade is 96–98%. **Mol. wt.** 291.3 **M.f.** $C_{10}H_{14}NO_5PS$
Form Pale yellow liquid with a phenol-like odour. **M.p.** 6.1 °C
B.p. 150 °C/80 Pa **V.p.** 0.89 mPa (20 °C) **K_{ow}** logP = 3.83 **Henry** 0.0302 Pa
m^3 mol^{-1} **S.g./density** 1.2694 **Solubility** In water 11 mg/l (20 °C). Completely
miscible with most organic solvents, e.g. dichloromethane >200, isopropanol,
toluene, hexane 50–100 (all in g/l, 20 °C). **Stability** Hydrolysed very slowly in
acidic media (pH 1–6), more rapidly in alkaline media; DT_{50} (22 °C) 272 d (pH 4),
260 d (pH 7), 130 d (pH 9). Isomerises on heating above 130 °C, to the
O,S-diethyl isomer. **F.p.** 174 °C (tech.)

COMMERCIALISATION
History Discovered by G. Schrader (cited by H. Martin & H. Shaw, *BIOS, Final
Report*, 1946, No. 1095). Introduced by American Cyanamid Co., ICI Plant
Protection Ltd (now Zeneca Agrochemicals), Monsanto Chemical Co. (none of
which now manufacture or market it), and Bayer AG. **Patent** DE 814152;
US 1893018; US 2842063 **Manufacturers** Cheminova

APPLICATIONS
Biochemistry Cholinesterase inhibitor; proinsecticide, activated by metabolic
oxidative desulfuration to the corresponding oxon. **Mode of action** Non-systemic
insecticide and acaricide with contact, stomach, and some respiratory action.
Uses Control of sucking and chewing insects (including soil insects) and mites in a
very wide range of crops, such as cereals, fruit (including citrus), vines, hops,
vegetables, ornamentals, cotton, and field crops. **Phytotoxicity** Non-phytotoxic,
except to some ornamentals, cucurbits, sorghum, and some varieties of apple,
pear, and tomato. **Formulation types** WP; EC; GR; DP; CS; Aerosol.
Mixtures *(parathion +)* endosulfan; ethion; lindane; oxydemeton-methyl; petroleum
oil; tetradifon; malathion; malathion + methoxychlor; parathion-methyl;

thiometon; methamidophos.   **Compatibility** Incompatible with alkaline materials.
Selected tradenames 'E605' (Bayer); 'Chimac Par H' (Agriphar); 'Fighter'
(Sanonda); 'Fostox E' (Siapa)

ANALYSIS
Product analysis by glc (*AOAC Methods*, 1995, 978.06, 980.11; *CIPAC Handbook*,
1985, **1C**, 2169) or hplc (*ibid.*, 1983, **1B**, 1875; *AOAC Methods*, 1995, 978.07*).
Residues determined by glc, tlc or paper chromatography, or by single sweep
oscillographic polarography (A. Ambrus et al., *J. Assoc. Off. Anal. Chem.*, 1981, **64**,
733; *AOAC Methods*, 1995, 968.24, 970.52, 970.53, 974.22; *Analyst (London)*, 1977,
102, 858; *Man. Pestic. Residue Anal.*, 1987, **I**, 3–6, S8, S10, S13, S17, S19; *Anal.
Methods Residues Pestic.*, 1988, Part I, M2, M5, M12). Methods for the
determination of residues are available from Bayer.

MAMMALIAN TOXICOLOGY
Reviews *Pesticide residues in food – 1995*, FAO Plant Production and Protection
Paper 133, 1996. *Pesticide residues in food – 1995 evaluations; Part II – Toxicology &
Environment*. World Health Organisation, WHO/PCS/96.48, 1996.   **IARC** 30
EHC 63 (WHO, 1986; a general review of organophosphorus insecticides).
Oral Acute oral LD_{50} for rats c. 2, mice c. 12, guinea pigs c. 10 mg/kg.
Skin and eye Acute percutaneous LD_{50} for male rats 71, female rats 76 mg/kg.
Not irritating to skin and eyes (rabbits). Not a skin sensitiser.   **Inhalation** LC_{50}
(4 h) for male and female rats 0.03 mg/l (aerosol).   **NOEL** (2 y) for rats 2 mg/kg
diet; (18 mo) for mice <60 mg/kg diet; (12 mo) for dogs <0.01 mg/kg b.w. daily.
ADI (JMPR) 0.004 mg/kg b.w. [1995].   **Toxicity class** WHO (a.i.) Ia;
EPA (formulation) I   **EC risk** T+ (R27/28)   **PIC** Yes

ECOTOXICOLOGY
Fish LC_{50} (96 h) for rainbow trout 1.5, golden orfe 0.58 mg/l.   **Bees** Toxic to
bees.   **Worms** LC_{50} for *Eisenia foetida* 267 mg/kg dry soil.   **Daphnia** LC_{50} (48 h)
0.0025 mg/l.   **Algae** E_rC_{50} for *Scenedesmus subspicatus* 0.5 mg/l.

ENVIRONMENTAL FATE
Animals In rats, following oral administration, 4-nitrophenol and paraoxon were
the principal metabolites excreted (M. P. Carver, *Pestic. Biochem. Physiol.*, 1990, **38**,
254; L. G. Sultatos & C. L. Gagliardi, *Biochem. Pharmacol.*, 1990, **39**, 799; R. G.
Hall, *Bull. Environ. Contam. Toxicol.*, 1990, **44**, 629).   **Plants** The major metabolites
are paraoxon, diethylphosphate, 4-nitrophenol and photometabolites S-ethyl
parathion and S-phenyl parathion.   **Soil/Environment** Based on K_{oc} values and
leaching studies, parathion can be classified as a compound with low mobility. In
biologically-active soils, parathion is rapidly degraded under laboratory conditions
as well as in the field. The degradation results in CO_2 via very short-lived
intermediate products such as paraoxon, aminoparathion and 4-nitrophenol.

organophosphorus

$O_2N \diagdown\!\!\!\diagup\!\!\!-\!\!\!\diagup\!\!\!\diagdown\!\!\!- OP(OCH_3)_2$ with S double-bonded above the P

NOMENCLATURE
Common name parathion-methyl (BSI, E-ISO, (*m*) F-ISO); methyl parathion (ESA, JMAF); metaphos (USSR)
IUPAC name *O,O*-dimethyl *O*-4-nitrophenyl phosphorothioate
Chemical Abstracts name *O,O*-dimethyl *O*-(4-nitrophenyl) phosphorothioate
CAS RN *[298–00–0]* **EEC no.** 206–050–1 **Development codes** E-120;
BAY 11405 **Official codes** OMS 213; ENT 17 292

PHYSICAL CHEMISTRY
Mol. wt. 263.2 **M.f.** $C_8H_{10}NO_5PS$ **Form** Colourless, odourless crystals; (tech., light to dark tan-coloured liquid). **M.p.** 35–36 °C; (tech. *c.* 29 °C)
B.p. 154 °C/136 Pa **V.p.** 0.2 mPa (20 °C); 0.41 mPa (25 °C) **K_{ow}** logP = 3.0
Henry 8.57×10^{-3} Pa m^3 mol^{-1} **S.g./density** 1.358 (20 °C); (tech., 1.20–1.22)
Solubility In water 55 mg/l (20 °C). Readily soluble in common organic solvents, e.g. dichloromethane, toluene >200, hexane 10–20 (all in g/l, 20 °C). Sparingly soluble in petroleum ether and some types of mineral oil. **Stability** Hydrolysed in alkaline and acidic media (*c.* 5× more rapidly than parathion); DT_{50} (25 °C) 68 d (pH 5), 40 d (pH 7), 33 d (pH 9). Isomerises on heating, to the *O,S*-dimethyl analogue. Photodegrades in water. **F.p.** >150 °C (tech.)

COMMERCIALISATION
History Insecticidal properties reported by G. Schrader (*Angew. Chem. Monograph* No. 52 (2nd Ed.), 1952). Introduced by Bayer AG. **Patent** DE 814142
Manufacturers Bayer; Cheminova; Rallis; Rotam; Sanonda; Shenzhen Jiangshan; Sinon; Tekchem

APPLICATIONS
Biochemistry Cholinesterase inhibitor. **Mode of action** Non-systemic insecticide and acaricide with contact, stomach, and some respiratory action. **Uses** Control of chewing and sucking insects in a very wide range of crops, such as cereals, fruit (including citrus), vines, vegetables, ornamentals, cotton, and field crops.
Phytotoxicity Non-phytotoxic when used as recommended. Alfalfa, sorghum, certain varieties of apple, cucurbit leaves (when wet), peach leaves, roses and some other ornamental plants may suffer slight injury. **Formulation types** WP; EC; DP; UL; CS. **Mixtures** *(parathion-methyl +)* parathion; endosulfan; monocrotophos; dicofol; dicofol + sulfur; lindane; carbophenothion; carbophenothion + sulfur; fenson; phosalone; tetradifon; omethoate; dicofol +

ethion; ethion; cypermethrin; propargite; sulfur; petroleum oils.
Compatibility Not compatible with alkaline materials. **Selected tradenames**
'Folidol-M' (Bayer); 'Metacide' (Bayer); 'Cekumethion' (Cequisa); 'Dhanuman'
(Dhanuka); 'Fostox metil' (Siapa); 'Jiajiduiliulin' (Shenzhen Jiangshan);
'Morfos Methyl' (Efthymiadis); 'Parataf' (Rallis); 'Paratox' (Aimco); 'Penncap-M'
(Elf Atochem); 'Sweeper' (Sanonda); 'Thionyl' (Agriphar)

ANALYSIS
Product analysis by glc (*AOAC Methods*, 1995, 977.04, 980.11; *CIPAC Handbook*,
1985, **1C**, 2169) or hplc (*ibid.*, 1983, **1B**, 1879; *AOAC Methods*, 1995, 977.05*) or
by hydrolysis to 4-nitrophenol which is determined colorimetrically (*CIPAC
Handbook*, 1970, **1**, 568; 1980, **1A**, 568); with S-methyl isomer, by rplc with u.v.
detection (*CIPAC Handbook*, 1992, **E**, 169–172). **Residues** determined by glc, tlc
or paper chromatography, or by single sweep oscillographic polarography (*Man,
Pestic. Residue Anal.*, 1987, **I**, 3, 5, 6, S8, S13, S17, S19; *Anal. Methods Residues
Pestic.*, 1988, Part I, M2, M5, M12; *AOAC Methods*, 1995, 968.24, 970.52, 970.53).
In **drinking water**, by gc with FID (*AOAC Methods*, 1995, 991.07). Methods for the
determination of residues are available from Bayer.

MAMMALIAN TOXICOLOGY
Reviews *Pesticide residues in food – 1995*, FAO Plant Production and Protection
Paper 133, 1996. *Pesticide residues in food – 1995 evaluations; Part II – Toxicology &
Environment*. World Health Organisation, WHO/PCS/96.48, 1996. **IARC** 30
EHC 145 (WHO, 1993), 63 (WHO, 1986; a general review of organophosphorus
insecticides). **Oral** Acute oral LD_{50} for rats *c*. 3, male mice *c*. 30, male and
female rabbits 19 mg/kg. **Skin and eye** Acute percutaneous LD_{50} for male and
female rats *c*. 45 mg/kg (24 h); not irritating to skin and eyes (rabbits). Not a skin
sensitiser. **Inhalation** LC_{50} (4 h) for rats *c*. 0.17 mg/l air (aerosol). **NOEL** (2 y)
for rats 2, mice 1 mg/kg diet; (12 mo) for dogs 0.3 mg/kg b.w. daily. **ADI** (JMPR)
0.003 mg/kg b.w. [1995]. **Toxicity class** WHO (a.i.) Ia; EPA (formulation) I
EC risk T+ (R28); T (R24) **PIC** Yes

ECOTOXICOLOGY
Birds Five-day LC_{50} for mallard ducks 1044 mg (as EC 480)/kg. **Fish** LC_{50} (96 h)
for rainbow trout 2.7, golden orfe 6.9 mg/l. **Bees** Toxic to bees. **Worms** LC_{50}
for *Eisenia foetida* 40 mg/kg dry soil. **Daphnia** LC_{50} (48 h) 0.0073 mg/l.
Algae E_rC_{50} for *Scenedesmus subspicatus* 3 mg/l.

ENVIRONMENTAL FATE
EHC 145 concludes that, because of toxicity to bees and other beneficial
organisms, spraying should be timed with great care. **Animals** In mammals,
following oral administration, parathion-methyl is almost completely excreted in
the urine within 24 hours. The major metabolites in man are 4-nitrophenol and
dimethyl phosphate. **Plants** The major metabolites formed in plants are

4-nitrophenol, 4-nitrophenyl glucopyranoside and *P-S*-demethyl parathion-methyl. **Soil/Environment** Based on the K_{oc} values and leaching studies, parathion-methyl can be classified as a compound with low/medium mobility. In biologically-active soils, parathion-methyl is rapidly degraded. As for phosphorothioates in general, metabolism is by oxidation to the phosphate, demethylation of the ester groups, and hydrolysis to phosphorothioic acid, phosphoric acid, and 4-nitrophenol.

553 pebulate   *Herbicide*

thiocarbamate

$$CH_3(CH_2)_3NCOS(CH_2)_2CH_3$$
$$| $$
$$CH_2CH_3$$

NOMENCLATURE
Common name pebulate (BSI, E-ISO, (m) F-ISO, WSSA, JMAF)
IUPAC name *S*-propyl butyl(ethyl)thiocarbamate
Chemical Abstracts name *S*-propyl butylethylcarbamothioate
CAS RN [1114–71–2] **EEC no.** 214-215-4 **Development codes** R-2061
(Zeneca)

PHYSICAL CHEMISTRY
Composition Tech. material is 95% pure. **Mol. wt.** 203.3 **M.f.** $C_{10}H_{21}NOS$
Form Colourless or yellow liquid with an aromatic odour. **B.p.** 142 °C/21 mmHg
V.p. 9 Pa (30 °C); 4.7 Pa (25 °C) **K_{ow} logP** = 3.83 **S.g./density** 0.956 (20 °C)
Solubility In water 60 mg/l (20 °C). Miscible with most organic solvents, e.g. acetone, benzene, toluene, xylene, methanol, isopropanol, kerosene.
Stability Stable up to 200 °C. In water, DT_{50} (40 °C) 11 d (pH 4 and pH 10), 12 d (pH 7). **F.p.** 124 °C

COMMERCIALISATION
History Herbicide reported by E. O. Burt *(Proc. South. Weed Conf., 12th, 1959, p. 19)*. Introduced by Stauffer Chemical Co. (now Zeneca Agrochemicals).
Patent US 3175897 **Manufacturers** Zeneca

APPLICATIONS
Biochemistry Inhibitor of very long-chain fatty acid synthesis. The sulfoxide may be the active form (P. B. Barrett & J. L. Harwood, *Proc. Br. Crop. Prot. Conf., 1993.* **1**, 183). **Mode of action** Selective systemic herbicide, absorbed by the roots, with translocation throughout the plant. Acts by inhibition of germination.
Uses Control of annual grasses (including nutgrasses) and some broad-leaved

weeds in sugar beet, tomatoes, and tobacco. Applied pre-emergence at 4–6 kg a.i./ha with soil incorporation. **Formulation types** EC; GR.
Selected tradenames 'Tillam' (Zeneca)

ANALYSIS
Product analysis by glc (*AOAC Methods*, 1995, 974.05; *CIPAC Handbook*, 1983, **1B**, 1887). **Residues** in crops and soils determined by glc (G. G. Patchett *et al.*, *Anal. Methods Pestic., Plant Growth Regul. Food Addit.*, 1964, **4**, 343; W. Y. Ja *et al.*, *Anal. Methods Pestic. Plant Growth Regul.*, 1972, **6**, 698). In **drinking water**, by gc with FID (*AOAC Methods*, 1995, 991.07). Analytical methods available from Zeneca.

MAMMALIAN TOXICOLOGY
EHC 76 (WHO, 1988; general review of thiocarbamates). **Oral** Acute oral LD_{50} for rats 1120, mice 1652 mg/kg. **Skin and eye** Acute percutaneous LD_{50} for rabbits 4640 mg/kg. Slight irritation of skin, eyes and mucous membranes (rabbits). Not a skin sensitiser (guinea pigs). **Inhalation** LC_{50} (4 h) for female rats >3.5 mg/l. **NOEL** (90 d) for rats 16, dogs 20 mg/kg daily; (1 y) for dogs 5 mg/kg daily. Not carcinogenic or teratogenic. **Toxicity class** WHO (a.i.) II; EPA (formulation) III **EC risk** Xn (R22)

ECOTOXICOLOGY
Birds Seven-day dietary LC_{50} for bobwhite quail 8400 mg/kg diet. **Fish** LC_{50} (96 h) for rainbow trout and bluegill sunfish *c.* 7.4 mg/l; (48 h) for silver mullet 6.25, killifish 7.78 mg/l. **Bees** Non-toxic to bees at 0.011 mg/bee.
Daphnia LC_{50} (48 h) 5.9 mg/l.

ENVIRONMENTAL FATE
Animals In animals, pebulate is rapidly metabolised. In rats, following oral administration, *c.* 50% was expired as CO_2, 25% excreted in the urine, and 5% excreted in the faeces within 3 days. **Plants** In plants, pebulate is rapidly metabolised to CO_2 and naturally-occurring plant constituents.
Soil/Environment In soil, pebulate disappears mainly by microbial degradation to the mercaptan, ethylbutylamine, and CO_2. DT_{50} in soil 2–3 w.

azole

$$CH_3CH_2CHCO_2(CH_2)_3CH=CH_2$$

$$CH_2-N-C=O$$

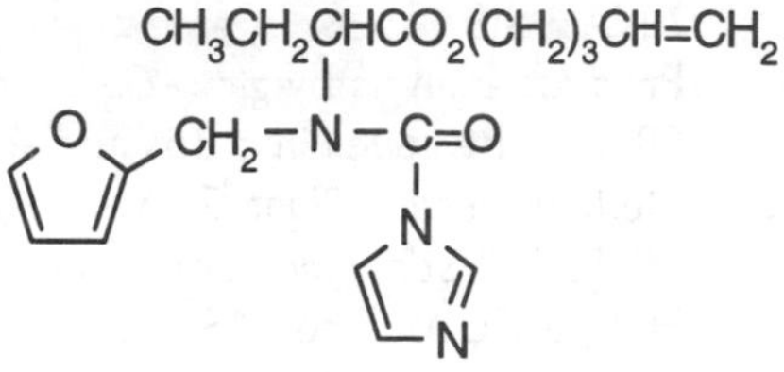

NOMENCLATURE
Common name pefurazoate (BSI, draft E-ISO)
IUPAC name pent-4-enyl N-furfuryl-N-imidazol-1-ylcarbonyl-DL-homoalaninate
Chemical Abstracts name 4-pentenyl 2-[(2-furanylmethyl)(1H-imidazol-=
1-ylcarbonyl)amino]butanoate
CAS RN *[101903–30–4]* **Development codes** UR-0003; UHF-8615 (Hokko,
Ube)

PHYSICAL CHEMISTRY
Mol. wt. 345.4 **M.f.** $C_{18}H_{23}N_3O_4$ **Form** Pale brownish liquid. **B.p.** Decomposes
at 235 °C **V.p.** 0.648 mPa (23 °C) **K_{ow}** logP = 3 **S.g./density** 1.152 (20 °C)
Solubility In water 443 mg/l (25 °C). In hexane 12.0, cyclohexane 36.9, dimethyl
sulfoxide, ethanol, acetone, acetonitrile, chloroform, ethyl acetate, toluene >1000
(all in g/l, 25 °C). **Stability** Stable in acidic media, slightly unstable in alkaline
media. Slightly unstable to sunlight. Decomposition *c.* 1% after 90 days at 40 °C.

COMMERCIALISATION
History Fungicide reported by M. Takenaka & I. Yamane (*Jpn. Pestic. Inf.,* 1990, No.
57, 7) and developed by Hokko Chemical Industry Co., Ltd and Ube Industries
Ltd. **Manufacturers** Ube

APPLICATIONS
Biochemistry Ergosterol biosynthesis inhibitor. **Mode of action** Inhibits the growth
of germ tubes and hyphae. **Uses** Control of seed-borne fungal diseases,
particularly *Fusarium moniliforme* (bakanae disease), *Pyricularia oryzae* (rice blast),
and *Cochliobolus miyabeanus* (brown spot) on rice plants. Also controls rice
seedling blight caused by soil-borne pathogens (*Trichoderma viride* and *Fusarium*
spp.) in nursery box cultivation, and other fungal pathogens on seeds of wheat and
barley, and bulbs of flowers. Moderate activity against the gram-positive bacterium
causing tomato canker (*Corynebacterium michiganense*). **Formulation types** WP;
EC; SC. **Selected tradenames** 'Healthied' (Hokko, Ube)

ANALYSIS
Product analysis is by hplc.

MAMMALIAN TOXICOLOGY
Reviews *Japan Pesticide Information*, **57**, 33 (1990).   **Oral** Acute oral LD_{50} for male rats 981, female rats 1051, male mice 1299, female mice 946 mg/kg.
Skin and eye Acute percutaneous LD_{50} for rats >2000 mg/kg. Slightly irritating to eyes; non-irritating to skin (rabbits). Non-sensitising to skin (guinea pigs).
Inhalation LC_{50} for rats >3450 mg/m^3.   **NOEL** (90 d) for rats 50 mg/kg diet.
Other Non-teratogenic in rats and rabbits.

ECOTOXICOLOGY
Birds Acute oral LD_{50} for adult Japanese quail 2380, chicks 4220 mg/kg.
Fish LC_{50} (48 h) for rainbow trout 4.0, bluegill sunfish 12.0, carp 16.9, killifish 12.0, goldfish 20.0, loach 15.0 mg/l.   **Bees** LD_{50} (topical) for honeybees 100 µg/bee.   **Daphnia** LC_{50} (6 h) >100 mg/l.

ENVIRONMENTAL FATE
Animals Pefurazoate orally administered to rats is rapidly metabolised and most of the metabolites are excreted in the urine and faeces within 24 hours. Pefurazoate is not detected in any organs, tissues, and excretions 1 hour after oral administration.   **Plants** Neither pefurazoate nor its metabolites were detected in the grain and straw of rice plants derived from treated seeds. Pefurazoate absorbed into rice seeds is also rapidly metabolised and not detected in the roots and shoots of young seedlings (*Japan Pesticide Information*, **57**, 33 (1990)).
Soil/Environment Pefurazoate incubated at 28 °C in soil under flooded conditions, DT_{50} 7–16 d. In paddy soil under field conditions, pefurazoate is decomposed more rapidly, DT_{50} <2 d.

555   penconazole   *Fungicide*

azole

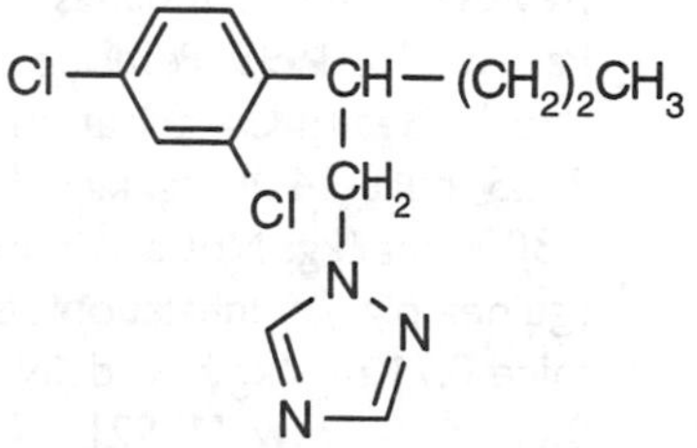

NOMENCLATURE
Common name penconazole (BSI, draft E-ISO, (*m*) draft F-ISO)
IUPAC name 1-(2,4-dichloro-β-propylphenethyl)-1*H*-1,2,4-triazole

Chemical Abstracts name 1-[2-(2,4-dichlorophenyl)pentyl]-1H-1,2,4-triazole
CAS RN *[66246–88–6]* **EEC no.** 266–275–6 **Development codes** CGA 71 818

PHYSICAL CHEMISTRY
Mol. wt. 284.2 **M.f.** $C_{13}H_{15}Cl_2N_3$ **Form** Fine white powder. **M.p.** 57.6–60.3 °C
V.p. 0.37 mPa (25 °C) K_{ow} logP = 3.72 (pH 5.7, 25 °C). **Henry** 6.6×10^{-4} Pa
m^3 mol^{-1} (calc.) **S.g./density** 1.30 (20 °C) **Solubility** In water 73 mg/l (25 °C). In
ethanol 730, acetone 770, toluene 610, *n*-hexane 22, *n*-octanol 400 (all in g/l,
25 °C). **Stability** Stable to hydrolysis (pH 1–13), and to temperatures up to
350 °C. **pKa** 1.51, v. weak base

COMMERCIALISATION
History Fungicide reported by J. Eberle *et al.* (*Proc. Int. Congr. Plant Prot., 10th,*
1983, 1, 376). Introduced by Ciba-Geigy AG (now Novartis Crop Protection AG)
as an agricultural fungicide, having been invented by Janssen Pharmaceutical NV.
Patent GB 1589852; BE 857570 (to Janssen Pharmaceutical)
Manufacturers Novartis

APPLICATIONS
Biochemistry Sterol demethylation inhibitor; inhibits cell membrane ergosterol
biosynthesis, stopping development of the fungi. **Mode of action** Systemic
fungicide with protective and curative action. Absorbed by the leaves, with
translocation acropetally. **Uses** Control of powdery mildew, pome fruit scab and
other pathogenic Ascomycetes, Basidiomycetes and Deuteromycetes on vines,
cucurbits, pome fruit, stone fruit, ornamentals, hops, tobacco and vegetables.
Formulation types WP; EC; EW. **Mixtures** *(penconazole +)* captan; sulfur.
Selected tradenames 'Topas' (Novartis)

ANALYSIS
Product analysis by glc. **Residues** determined by glc. Details available from
Novartis.

MAMMALIAN TOXICOLOGY
Reviews *Pesticide residues in food – 1992.* FAO Plant Production and Protection
Paper, 116, 1993. *Pesticide residues in food – 1992 evaluations. Part II – Toxicology.*
World Health Organisation, WHO/PCS/93.34. **Oral** Acute oral LD_{50} for rats
2125, mice 2444 mg/kg. **Skin and eye** Acute percutaneous LD_{50} for rats
>3000 mg/kg. Not a skin irritant; irritating to eyes (rabbits). Not a skin sensitiser
(guinea pigs). **Inhalation** LC_{50} (4 h) >4000 mg/m^3. **NOEL** (2 y) for rats 3.8,
mice 0.71 mg/kg b.w. daily; (1 y) for dogs 3.3 mg/kg b.w. daily. **ADI** (JMPR)
0.03 mg/kg b.w. [1992]. **Other** Not mutagenic, not teratogenic, not oncogenic.
Toxicity class WHO (a.i.) III (Table 5)

ECOTOXICOLOGY
Birds Acute oral LD_{50} (8 d) for Japanese quail 2424, Pekin ducks >3000, mallard

ducks >1590 mg/kg. LC_{50} (8 d) for bobwhite quail and mallard ducks >5620 ppm. **Fish** LC_{50} (96 h) for rainbow trout 1.7–4.3, carp 3.8–4.6, bluegill sunfish 2.1–2.8 mg/l. **Bees** Non-toxic to bees; LD_{50} (oral and topical) >5 μg/bee. **Worms** LC_{50} (14 d) for earthworms >1000 mg/kg. **Daphnia** IC_{50} (48 h) 7–11 mg/l. **Algae** IC_{50} (5 d) for *Scenedesmus subspicatus* 3.0 mg/l; EC_{50} (5 d) for *Selenastrum capricornutum* 0.83 mg/l.

ENVIRONMENTAL FATE
Animals After oral administration, penconazole is rapidly eliminated practically to entirety with urine and faeces. Residues in tissues were not significant and there was no evidence for accumulation. **Plants** Metabolic pathways are hydroxylation of the propyl side chain, conjugation to glucosides or metabolism to triazolylalanine and triazolylacetic acid. **Soil/Environment** DT_{50} in soil is 133–343 d, depending on soil type. DT_{50} for photolysis is 4 d (natural sunlight).

556 pencycuron   *Fungicide*

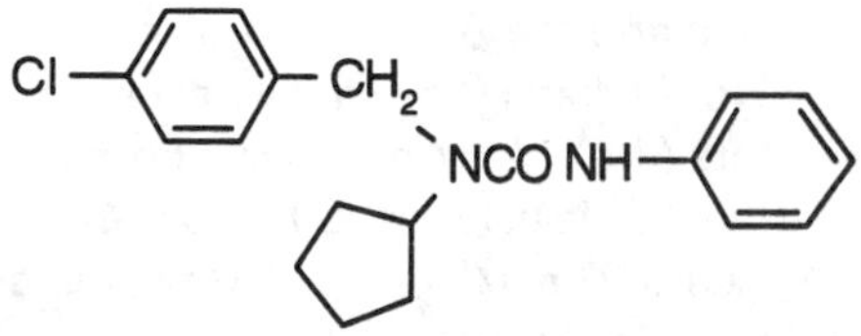

NOMENCLATURE
Common name pencycuron (BSI, draft E-ISO, (*m*) draft F-ISO)
IUPAC name 1-(4-chlorobenzyl)-1-cyclopentyl-3-phenylurea
Chemical Abstracts name *N*-[(4-chlorophenyl)methyl]-*N*-cyclopentyl-*N'*-phenyl=
urea
CAS RN *[66063–05–6]*   **Development codes** NTN 19 701 (Bayer)

PHYSICAL CHEMISTRY
Mol. wt. 328.8   **M.f.** $C_{19}H_{21}ClN_2O$ **Form** Colourless, odourless crystals.
M.p. 128 °C (modification A); 132 °C (modification B)   **V.p.** 5×10^{-7} mPa
(20 °C, extrapolated) K_{ow} logP = 4.68   **Henry** 5×10^{-7} Pa m^3 mol^{-1} (20 °C)
S.g./density 1.22 (20 °C)   **Solubility** In water 0.3 mg/l (20 °C). In dichloromethane 200–500, toluene 20–50, isopropanol 2–5, hexane <1 (all in g/l, 20 °C).   **Stability** On hydrolysis (25 °C) DT_{50} 280 d (pH 4), 22 y (pH 7), 17 y (pH 9). Photodegrades in water and on soil surfaces.

COMMERCIALISATION
History Fungicide reported by P.-E. Frohberger & F. K. Grossman (*Mitt. Biol. Bundesanst. Land-Forstwirtsch., Berlin-Dahlem,* 1981, **203,** 230). Introduced by Bayer AG. **Patent** BE 856922; DE 2732257 **Manufacturers** Bayer

APPLICATIONS
Mode of action Non-systemic fungicide with protective action. **Uses** Control of diseases caused by *Rhizoctonia solani* and *Pellicularia* spp. in potatoes, rice, cotton, sugar beet, vegetables, ornamentals and turf. In particular, control of black scurf of potatoes, sheath blight of rice, and damping-off of ornamentals. Can be used as a foliar spray and dust application, as a seed treatment, or by soil incorporation. **Formulation types** WP; DP; DS; SC; FS. **Mixtures** *(pencycuron +)* captan; tolylfluanid; edifenphos; imazalil; captan. **Compatibility** Compatible with a number of other fungicides and insecticides, e.g. edifenphos, captan, thiram, fenthion. **Selected tradenames** 'Monceren' (Bayer)

ANALYSIS
Product analysis by lc (*CIPAC Handbook,* 1992, **E,** 173–8). **Residues** determined by glc, details available from Bayer AG.

MAMMALIAN TOXICOLOGY
Oral Acute oral LD_{50} for rats, mice and dogs >5000, cats >1000 mg/kg.
Skin and eye Acute percutaneous LD_{50} (24 h) for rats and mice >2000 mg/kg. Non-irritating to skin and eyes (rabbits). Not a skin sensitiser. **Inhalation** LC_{50} for rats (4 h) >0.27 mg/l air (aerosol); >5.13 mg/l air (dust). **NOEL** (2 y) for male rats 50, female rats 500, dogs 100, male and female mice 500 mg/kg diet. **ADI** 0.02 mg/kg b.w. **Toxicity class** WHO (a.i.) III (Table 5); EPA (formulation) IV

ECOTOXICOLOGY
Birds LD_{50} for bobwhite quail >2000 mg/kg. **Fish** LC_{50} (96 h) for carp 8.8, rainbow trout >690 mg/l. **Bees** Not hazardous to bees; LD_{50} (oral and contact) >100 μg/bee. **Worms** LC_{50} (14 d) >1000 mg/kg. **Other beneficial spp.** Not toxic for carabid beetles. **Daphnia** LC_{50} (48 h) 0.27–0.67 mg/l. **Algae** EC_{50} (96 h) for *Scenedesmus* 0.56 mg/l.

ENVIRONMENTAL FATE
Animals In rats, following oral administration, up to 74% is eliminated within 3 days in the urine and faeces, either as the unchanged material or as metabolites. Eleven metabolites have been identified (I. Ueyama *et al., J. Agric. Food Chem.,* 1982, **30**(6), 1061–1067). The main metabolic pathway comprises hydroxylation of the phenyl group to various diol and triol compounds which are frequently eliminated as sulfate and glucuronic acid conjugates. **Plants** Degradation only to a low degree. The hydroxylated metabolites (partly as conjugates) can be

considered to be the main metabolites.   **Soil/Environment** Leaching and
adsorption behaviour in soil characterises the compound as slightly mobile.
Degraded in soil very effectively in laboratory studies. Based on determined DT_{50}
values, the compound can be classified as moderately stable to stable. The main
metabolites are derivatives of *p*-chlorobenzylamine and *p*-chlorobenzylformamide.

557   pendimethalin   *Herbicide*

2,6-dinitroaniline

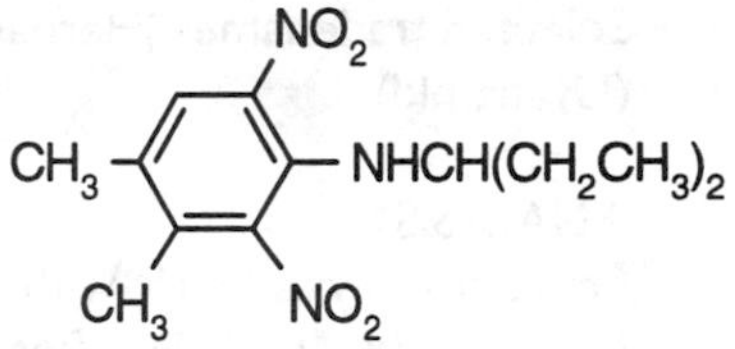

NOMENCLATURE
Common name pendimethalin (BSI, E-ISO, ANSI, WSSA); pendiméthaline
((*f*) F-ISO); [penoxalin] (former WSSA name)
IUPAC name *N*-(1-ethylpropyl)-2,6-dinitro-3,4-xylidine
Chemical Abstracts name *N*-(1-ethylpropyl)-3,4-dimethyl-2,6-dinitrobenzenamine
CAS RN *[40487–42–1]*   **EEC no.** 254–938–2   **Development codes** AC 92 553

PHYSICAL CHEMISTRY
Composition Tech. grade is 90% pure.   **Mol. wt.** 281.3   **M.f.** $C_{13}H_{19}N_3O_4$
Form Orange-yellow crystals.   **M.p.** 54–58 °C   **B.p.** Decomposes on distillation.
V.p. 4.0 mPa (25 °C) K_{ow} logP = 5.18   **S.g./density** 1.19 at 25 °C   **Solubility** In
water 0.3 mg/l (20 °C). In acetone 700, xylene 628, corn oil 148, heptane 138,
isopropanol 77 (all in g/l, 26 °C). Readily soluble in benzene, toluene,
chloroform, and dichloromethane. Slightly soluble in petroleum ether and petrol.
Stability Very stable in storage; store above 5 °C and below 130 °C. Stable to
acids and alkalis. Slowly decomposed by light. DT_{50} in water <21 d.

COMMERCIALISATION
History Herbicide reported by P. L. Sprankle (*Proc. Br. Weed Control Conf., 12th,*
1974, **2,** *825*). Introduced by American Cyanamid Co.   **Patent** BE 816837;
US 4199669   **Manufacturers** Cyanamid; Feinchemie Schwebda; Rallis

APPLICATIONS
Biochemistry Inhibits cell division and cell elongation.   **Mode of action** Selective
herbicide, absorbed by the roots and leaves. Affected plants die shortly after

germination or following emergence from the soil. **Uses** Control of most annual grasses and many annual broad-leaved weeds at 0.4–2.0 kg a.i./ha in cereals, onions, leeks, garlic, fennel, maize, sorghum, rice, soya beans, peanuts, brassicas, carrots, celery, black salsify, peas, field beans, lupins, evening primrose, tulips, potatoes, cotton, hops, pome fruit, stone fruit, berry fruit (including strawberries), citrus fruit, lettuce, aubergines, capsicums, established turf, and in transplanted tomatoes, sunflowers, and tobacco. Applied pre-plant incorporated, pre-emergence, pre-transplanting, or early post-emergence. Also used for control of suckers in tobacco. **Phytotoxicity** Injury to maize may occur if used as a pre-plant, soil-incorporated treatment. **Formulation types** EC; GR; WG; WP. **Mixtures** *(pendimethalin +)* neburon; atrazine; linuron; chlorotoluron; imazamethabenz; isoproturon; imazaquin; propachlor; prometryn; metolachlor; metolachlor + terbuthylazine; metobromuron; atrazine + metolachlor. **Selected tradenames** 'Herbadox' (Cyanamid); 'Prowl' (Cyanamid); 'Stomp' (Cyanamid)

ANALYSIS
Product analysis by glc with FID (J. C. Wyckoff, *Anal. Methods Pestic. Plant Growth Regul.*, 1978, **10**, 461). **Residues** of the 4-hydroxymethyl analogue (after formation of a derivative) and of pendimethalin determined by glc with ECD (*idem, ibid.*).

MAMMALIAN TOXICOLOGY
Oral Acute oral LD_{50} for male rats 1250, female rats 1050, male mice 1620, female mice 1340, rabbits >5000, beagle dogs >5000 mg/kg. **Skin and eye** Acute percutaneous LD_{50} for rabbits >5000 mg/kg. Non-irritating to skin and eyes (rabbits). **NOEL** In 2 y feeding trials, rats receiving 100 mg/kg diet showed no ill-effects. **Toxicity class** WHO (a.i.) III; EPA (formulation) III **EC risk** Xn (R22)

ECOTOXICOLOGY
Birds Eight-day dietary LC_{50} for bobwhite quail 4187, mallard ducks 10 388 mg/kg diet. **Fish** LC_{50} (96 h) for rainbow trout 0.14, bluegill sunfish 0.2, channel catfish 0.42 mg/l. **Bees** LD_{50} (topical) >50 µg/bee.

ENVIRONMENTAL FATE
Animals In rats, the major metabolic routes for pendimethalin involve hydroxylation of the 4-methyl and N-1-ethyl groups, oxidation of these alkyl groups to carboxylic acids, nitro-reduction, cyclisation and conjugation (J. Zulian *J. Agric. Food Chem.*, 1990, **38**, 1743). **Plants** In plants, the 4-methyl group on the benzene ring is oxidised to the carboxylic acid via the alcohol. The amino nitrogen is also oxidised. At harvest time, residues in crops are below the validated sensitivity of the analytical method (0.05 ppm). **Soil/Environment** In soil, the 4-methyl group on the benzene ring is oxidised to the carboxylic acid via the alcohol; the amino nitrogen is also oxidised. DT_{50} in soil is 3–4 mo (A. Walker

& W. Bond, *Pestic. Sci.*, 1977, **8**, 359). K_d ranges from 2.23 (0.01% o.m., pH 6.6) to 1638 (16.9% o.m., pH 6.8) (H. J. Pedersen *et al.*, *Pestic. Sci.*, 1995, **44**, 131).

558 pentachlorophenol

Insecticide, fungicide, herbicide

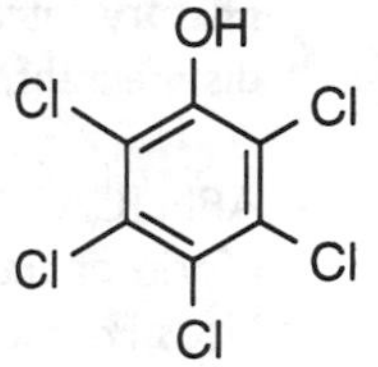

NOMENCLATURE
pentachlorophenol
Common name PCP (WSSA, JMAF); pentachlorophenol (BSI, E-ISO, F-ISO, accepted in lieu of a common name)
IUPAC name pentachlorophenol
Chemical Abstracts name pentachlorophenol
CAS RN *[87–86–5]* **EEC no.** 201–778–6

sodium pentachlorophenoxide
Common name sodium pentachlorophenoxide (BSI, E-ISO); pentachlorophénate de sodium (F-ISO)
IUPAC name sodium pentachlorophenoxide; sodium pentachlorophenate

pentachlorophenyl laurate
IUPAC name pentachlorophenyl laurate
Chemical Abstracts name pentachlorophenyl dodecanoate
CAS RN *[3772–94–9]*

PHYSICAL CHEMISTRY
pentachlorophenol
Mol. wt. 266.3 **M.f.** C_6HCl_5O **Form** Colourless crystals, with a phenolic odour; (tech., dark grey). **M.p.** 191 °C; (tech., 187–189 °C) **B.p.** 309–310 °C (decomp.) **V.p.** 16 Pa (100 °C). **K_{ow}** logP = 5.1 (25 °C, unionised, K. Chamberlain *et al.*, *Pestic Sci.*, **47**, 265 (1996)) **S.g./density** 1.98 (22 °C) **Solubility** In water 80 mg/l (30 °C). Soluble in most organic solvents, e.g. acetone

215 g/l (20 °C). Slightly soluble in carbon tetrachloride and paraffins. The sodium, calcium and magnesium salts are soluble in water. **Stability** Relatively stable and non-hygroscopic. **pKa** 4.71 **F.p.** Not flammable.

sodium pentachlorophenoxide
Mol. wt. 288.3 **M.f.** C_6Cl_5NaO **Solubility** Crystallises from water as a monohydrate, solubility in water 330 g/l (25 °C). Insoluble in petroleum oils.

pentachlorophenyl laurate
Mol. wt. 448.6 **M.f.** $C_{18}H_{23}Cl_5O_2$

COMMERCIALISATION
History Introduced c. 1936 as a timber preservative and later used as a general disinfectant. **Manufacturers** Excel; KMG-Bernuth; Vulcan

APPLICATIONS
Mode of action Insecticide, fungicide, and non-selective contact herbicide.
Uses Pentachlorophenol is used to control termites and, frequently, as an ester (such as pentachlorophenyl laurate) to protect wood from fungal rots and wood-boring insects, and as a general herbicide. The sodium salt is used as a general disinfectant, e.g. for trays in mushroom houses. **Formulation types** GR; WP; OL.
Selected tradenames 'Biocel Sg 85' (Excel)

ANALYSIS
Product analysis by titration with alkali (*MAFF Ref. Bk.*, 1958, No. 1, p. 64).
Residues determined by colorimetry of derivatives (W. W. Kilgore & K. W. Cheng, *Anal. Methods Pestic., Plant Growth Regul. Food Addit.*, 1967, **5**, 313; *Anal. Methods Pestic. Plant Growth Regul.*, 1972, **6**, 581) or by glc (*AOAC Methods*, 1995, 985.24). In **drinking water**, by conversion to methyl ester with diazomethane, then gc with ECD (*AOAC Methods*, 1995, 992.32, 10.7.03).

MAMMALIAN TOXICOLOGY
pentachlorophenol
EHC 71 (WHO, 1987). **Oral** Acute oral LD_{50} for rats 210 mg/kg.
Skin and eye Acute percutaneous LD_{50} not available. Irritating to skin (the solid and aqueous solutions >10 g/l), eyes, and mucous membranes. **NOEL** No deaths occurred among dogs and rats receiving 3.9–10 mg daily for 70–190 d.
Toxicity class WHO (a.i.) Ib; EPA (formulation) II **EC risk** (R40); T+ (R26); T (R24/25); Xi (R36/37/38) **PIC** Yes

sodium pentachlorophenoxide
EC risk As for the phenol.

ECOTOXICOLOGY
sodium pentachlorophenoxide
Fish LC$_{50}$ (48 h) for rainbow and brown trout 0.17 mg/l (J. S. Alabaster, *Proc. Br. Weed Control Conf., 4th*, 1958, p. 84).

ENVIRONMENTAL FATE
Animals The principal metabolites in female rats were tetrachloromonophenols, diphenols and hydroquinones (G. Renner & C. Hopper, *Xenobiotica*, 1990, **20**, 573). **Soil/Environment** Very persistent in the environment.

pentanochlor   *Herbicide*

anilide

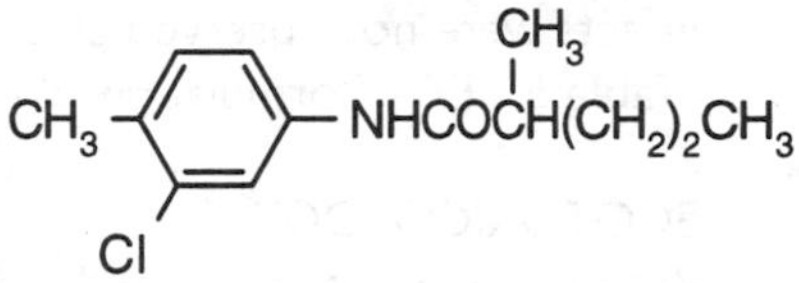

CH$_3$—⟨benzene ring⟩—NHCOCH(CH$_2$)$_2$CH$_3$ with CH$_3$ substituent; Cl on ring

NOMENCLATURE
Common name pentanochlor (BSI, E-ISO); pentanochlore ((*f*) F-ISO); solan (WSSA, Canada, ex ANSI); CMMP (JMAF)
IUPAC name 3'-chloro-2-methylvalero-*p*-toluidide
Chemical Abstracts name *N*-(3-chloro-4-methylphenyl)-2-methylpentanamide
CAS RN *[2307–68–8]*   **Development codes** FMC 4512

PHYSICAL CHEMISTRY
Mol. wt. 239.7   **M.f.** C$_{13}$H$_{18}$ClNO   **Form** Colourless solid; (tech. grade, colourless to pale cream powder).   **M.p.** 85–86 °C (pure); 82–86 °C (tech.)
V.p. Very low at room temperature.   **S.g./density** 1.106 (20 °C)   **Solubility** In water 8–9 mg/l (20 °C). In di-isobutyl ketone 460, isophorone 550, methyl isobutyl ketone 520, xylene 200–300, pine oil 410 (all in g/kg, 20 °C).
Stability Stable to hydrolysis in neutral media at room temperature.

COMMERCIALISATION
History Herbicide reported by D. H. Moore (*Proc. Northeast. Weed Control Conf.*, 1960, p. 86). Introduced by FMC Corp. (who no longer manufacture or market it).
Patent GB 869169; US 3020142

APPLICATIONS
Biochemistry Inhibits photosynthesis.   **Mode of action** Selective, contact herbicide

with absorption via the leaves.   **Uses** Selective pre- and post-emergence herbicidal control in carrot, celeriac, celery, fennel, parsley, and parsnip at <4 kg a.i./ha. Used pre-sowing on tomatoes and some flower crops, and as a contact spray in carnations, chrysanthemums, fruit trees, ornamental trees and shrubs, roses, sweet peas, and tomatoes. With chlorpropham, used as above, though also on narcissi and tulips.   **Formulation types** EC.   **Mixtures** *(pentanochlor +)* chlorpropham.   **Selected tradenames** 'Croptex Bronze' (Hortichem)

ANALYSIS
Residues determined by hydrolysis to 3-chloro-*p*-toluidine which is estimated by glc.

MAMMALIAN TOXICOLOGY
Oral Acute oral LD_{50} for rats >10 000 mg/kg.   **Skin and eye** Acute percutaneous LD_{50} not available. Slight irritation of skin and mucous membranes.   **NOEL** In 140 d feeding trials, rats receiving 20 000 mg/kg diet suffered no effect on body weight or survival, but histopathological changes were found in the liver; these effects were not observed at 2000 mg/kg diet.   **Toxicity class** WHO (a.i.) III (Table 5); EPA (formulation) IV

ECOTOXICOLOGY
Bees Not very toxic to bees.

ENVIRONMENTAL FATE
Soil/Environment Similar to 3,4-dichloroaniline-based acylanilide herbicides; converted to 3-chloro-4-methylaniline. This does not accumulate but is converted to 3,3'-dichloro-4,4'-dimethylazobenzene.

560 pentoxazone   *Herbicide*

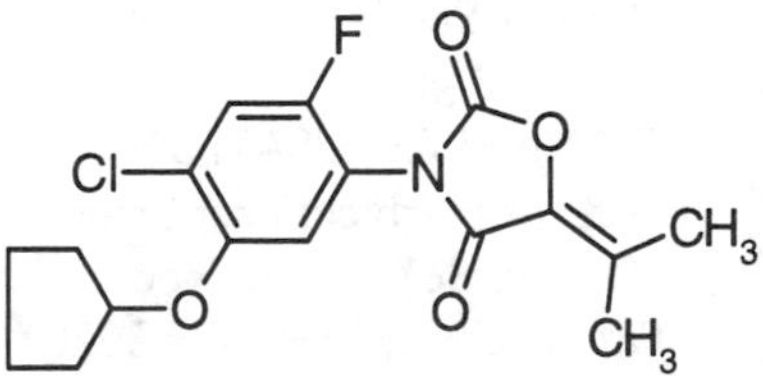

NOMENCLATURE
Common name pentoxazone (BSI, pa ISO)

IUPAC name 3-(4-chloro-5-cyclopentyloxy-2-fluorophenyl)-5-isopropylidene-=
1,3-oxazolidine-2,4-dione
Chemical Abstracts name 3-[4-chloro-5-(cyclopentyloxy)-2-fluorophenyl]-=
5-(1-methylethylidene)-2,4-oxazolidinedione
CAS RN *[110956–75–7]* **Development codes** KPP-314

PHYSICAL CHEMISTRY
Mol. wt. 353.8 **M.f.** $C_{17}H_{17}ClFNO_4$ **Form** Colourless, odourless crystalline
powder. **M.p.** 104 °C **V.p.** $<1.11 \times 10^{-2}$ mPa (25 °C) **K_{ow}** logP = 0.67
S.g./density 1.418 (25 °C) **Solubility** In water 0.216 mg/l (25 °C).
Stability Stable to heat, light and acid; unstable in alkali.

COMMERCIALISATION
Manufacturers Kaken

APPLICATIONS
Uses Pre-emergence control of *Echinochloa* in rice at 200–450 g/ha.
Formulation types SC; GR. **Mixtures** *(pentoxazone +)* imazosulfuron;
pyrazosulfuron-ethyl; cyclosulfamuron; bromobutide; daimuron; cumyluron.
Selected tradenames 'Wechser' (Kaken)

ANALYSIS
By hplc.

MAMMALIAN TOXICOLOGY
Oral Acute oral LD_{50} for male and female rats and male and female mice
>5000 mg/kg. **Skin and eye** Acute percutaneous LD_{50} for male and female rats
>2000 mg/kg. **Inhalation** LC_{50} (4 h) for male and female rats >5100 mg/m³.

ECOTOXICOLOGY
Birds Acute oral LD_{50} for male and female bobwhite quail >2250 mg/kg.
Fish LC_{50} (96 h) for carp 21.4 ppm. **Bees** LC_{50} (oral) >458.5 ppm, (contact)
98.7 µg/bee. **Worms** LC_{50} (14 d) >851 ppm. **Daphnia** LC_{50} (24 h) >38.8 ppm.
Algae EC_{50} (72 h) for *Selenastrum capricornutum* 1.31 ppb.

ENVIRONMENTAL FATE
Animals Male and female rats excrete >95% of administered dose, mainly via the
faeces, within 48 h. **Soil/Environment** In soil, DT_{50} <29 d (two types of flooded
soil, 28 °C). In water DT_{50} 1.4 d (pH 8.0, 20 °C).

pyrethroid

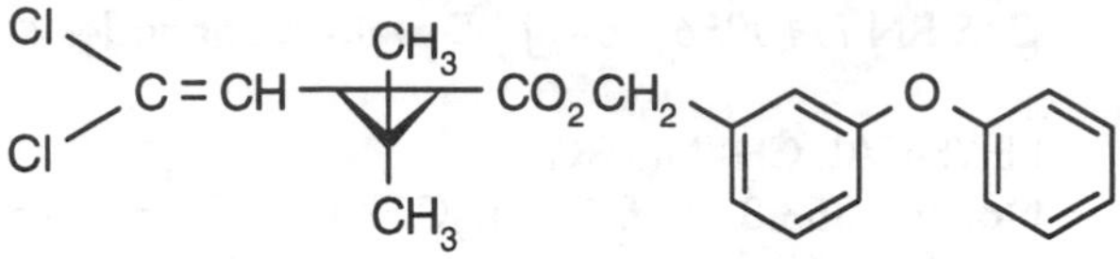

NOMENCLATURE
Common name permethrin (BSI, E-ISO, ANSI, ESA, BAN); perméthrine (F-ISO); no name (Eire, Bangladesh)
IUPAC name 3-phenoxybenzyl (1*RS*,3*RS*;1*RS*,3*SR*)-3-(2,2-dichlorovinyl)-= 2,2-dimethylcyclopropanecarboxylate *Roth*: 3-phenoxybenzyl (1*RS*)-*cis-trans*-3-(2,2-dichlorovinyl)-2,2-dimethylcyclopropanecarboxylate
Chemical Abstracts name (3-phenoxyphenyl)methyl 3-(2,2-dichloroethenyl)-= 2,2-dimethylcyclopropanecarboxylate
CAS RN *[52645–53–1]*, formerly *[57608–04–5]* and *[63364–00–1]*
EEC no. 258–067–9 **Development codes** NRDC 143; FMC 33 297; PP557 (Zeneca); WL 43 479 (Shell); LE 79–519 (Rhône-Poulenc)
Official codes OMS 1821

PHYSICAL CHEMISTRY
Mol. wt. 391.3 **M.f.** $C_{21}H_{20}Cl_2O_3$ **Form** Tech. grade is a yellow-brown to brown liquid, which sometimes tends to crystallise partly at room temperature.
M.p. 34–35 °C; *cis*- isomers 63–65 °C; *trans*- 44–47 °C **B.p.** 200 °C/0.1 mmHg; >290 °C/760 mmHg **V.p.** 0.07 mPa (20 °C); *cis*- 0.0025 mPa; *trans*- 0.0015 mPa (both 20 °C) (D. Wells *et al.*, *Pestic. Sci.*, 1986, **17**, 473). K_{ow} logP = 6.1 (20 °C)
S.g./density 1.19–1.27 at 20 °C **Solubility** In water *c.* 0.2 mg/l (30 °C). In xylene, hexane >1000, methanol 258 (all in g/kg, 25 °C). **Stability** Stable to heat (≥2 y at 50 °C), more stable in acid than alkaline media with optimum stability *c.* pH 4; some photochemical degradation observed in laboratory studies but field data indicate this does not adversely affect biological performance. **F.p.** 'Perigen' >100 °C

COMMERCIALISATION
History Insecticide reported by M. Elliott *et al.* (*Proc. Br. Insectic. Fungic. Conf.*, 7th, 1973, **2**, 721; *Nature (London)*, 1973, **246**, 169). Developed by FMC Corp., ICI Agrochemicals (now Zeneca Agrochemicals), Mitchell Cotts Chemicals, Penick Corp., Shell International Chemical Co., Ltd (now American Cyanamid Co.), Sumitomo Chemical Co., Ltd and the Wellcome Foundation. **Patent** GB 1413491 to NRDC **Manufacturers** Cyanamid; FMC; Mitchell Cotts; Mitsu; Sanachem; Sumitomo; United Phosphorus; Zeneca

APPLICATIONS

Mode of action Non-systemic insecticide with contact and stomach action, having a slight repellent effect. **Uses** A contact insecticide effective against a broad range of pests. It controls leaf- and fruit-eating Lepidoptera and Coleoptera in cotton at 100–150 g a.i./ha, in fruit at 25–50 g/ha, in tobacco, vines and other crops at 50–200 g/ha, in vegetables at 40–70 g/ha. It has good residual activity on treated plants. It is effective against a wide range of animal ectoparasites, provides >60 d residual control of biting flies in animal housing at 200 mg a.i. (as EC)/m^2 wall or 30 mg a.i. (as WP)/m^2 wall, and is effective as a wool preservative at 200 mg/kg wool. It provides >120 d control of Blattodea, Diptera, Hymenoptera and other crawling insects at 100 mg a.i. (as WP)/m^2, also flying insects.

Phytotoxicity Non-phytotoxic when used as directed (except that some ornamentals may be injured). **Formulation types** EC; WP; UL; Fumigant; Aerosol; DP; WG. **Mixtures** *(permethrin +)* dimethoate; pyrethrins; malathion; tetramethrin; heptenophos; bioallethrin + piperonyl butoxide; pirimiphos-methyl; chlorpyrifos-methyl + pyrethrins; bioallethrin S-cyclopentenyl isomer; bioallethrin S-cyclopentenyl isomer + piperonyl butoxide. **Compatibility** Mixing with calcium nitrate is not recommended. **Selected tradenames** 'Ambush' (crop protection) (Zeneca); 'Assithrin' (Delicia); 'Cliper' (Cequisa); 'Coopex' (mixture) (AgrEvo); 'Corsair' (Rhône-Poulenc); 'Dragnet' (termiticide) (FMC); 'Dragon' (public health) (Zeneca); 'Eksmin' (Sumitomo); 'Kafil' (crop protection) (Zeneca); 'Outflank' (veterinary) (Cyanamid); 'Perkill' (United Phosphorus); 'Permetiol' (Agriphar); 'Permit' (Sanonda); 'Pounce' (FMC); 'Pramex' (Roussel Uclaf Corp); 'Sanathrin' (Sanachem); 'Talcord' (agronomic use) (Cyanamid)

ANALYSIS

Product analysis of pyrethroids reviewed by E. Papadopoulou-Mourkidou in *Comp. Analyt. Profiles*. By glc with FID (*CIPAC Handbook*, 1985, **1C**, 2172; *AOAC Methods*, 1995, 986.03; H. Swaine & M. J. Tandy, *Anal. Methods Pestic. Plant Growth Regul.*, 1984, **13**, 103). Identity also by glc or hplc (*CIPAC Handbook*, 1994, **F**, 404). **Residues** determined by glc with ECD (*idem, ibid.*; A. Ambrus *et al.*, *J. Assoc. Off. Anal. Chem.*, 1981, **64**, 733; *Man. Pesticide Residue Anal.*, 1987, **I**, 6, S19; *Anal. Methods Residues Pestic.*, 1988, Part I, M11). Permethrin isomers in **drinking water** by gc with ECD (*AOAC Methods*, 1995, 990.06).

MAMMALIAN TOXICOLOGY

Reviews *Pesticide residues in food – 1987*. FAO Plant Production and Protection Paper 84, 1987. *Pesticide residues in food – 1987 evaluations. Part II – Toxicology*. FAO Plant Production and Protection Paper 86/2, 1988. See also A. J. Gray & D. M. Soderlund, Chapt. 5 in *"Insecticides"*. **IARC** 53 **EHC** 94 (WHO, 1990). **Oral** Oral LD_{50} values of permethrin depend on such factors as: carrier, *cis/trans* ratio of the sample, the test species, its sex, age and degree of fasting; values reported sometimes differ markedly. Values for a *cis/trans* ratio of *c.* 40:60 are: for rats 430–4000, mice 540–2690 mg/kg; with a 20:80 ratio, the LD_{50} is

c. 6000 mg/kg.   **Skin and eye** Acute percutaneous LD_{50} for rats >2500, rabbits >2000 mg/kg. Mild eye and skin irritant (rabbits). Moderate skin sensitiser. **Inhalation** LC_{50} (3 h) for mice and rats >685 mg/m^3 air; (separate study gives >13 800 mg/m^3).   **NOEL** In 2 y feeding trials, rats receiving 100 mg/kg diet showed no ill-effects.   **ADI** (JMPR) 0.05 mg/kg b.w. (for nominal *cis/trans* 40:60 and 25:75 isomers) [1987].   **Other** No mutagenic, teratogenic, or carcinogenic activity.   **Toxicity class** WHO (a.i.) II; EPA II ('Ambush'); III ('Outflank') **EC risk** Xn (R22)

ECOTOXICOLOGY
Birds Typical oral LD_{50} values for a *cis/trans* ratio of c. 40:60 are: for chickens >3000, Japanese quail >13 500 mg/kg.   **Fish** LC_{50} (96 h) for rainbow trout 2.5 µg/l; (48 h) for rainbow trout 5.4, bluegill sunfish 1.8 µg/l.   **Bees** Toxic to bees; LD_{50} (24 h) oral 0.098, topical 0.029 µg/bee.   **Daphnia** LC_{50} (48 h) 0.6 µg/l.

ENVIRONMENTAL FATE
EHC 94 concludes that permethrin is not likely to be a hazard to the environment when used as recommended.   **Animals** In mammals, there is hydrolysis of the ester bond, hydroxylation, and elimination as the glucoside conjugate (M. Elliott et al., J. Agric. Food Chem., 1976, **24**, 270; L. C. Gaughan et al., ibid., 1978, **26**, 613). **Soil/Environment** In soil and water, degradation is rapid. DT_{50} in soil <38 d (pH 4.2–7.7, o.m. 1.3–51.3%) (R. L. Holmstead et. al., J. Agric. Food Chem., 1978, **26**, 590).

562 petroleum oils

Acaricide, insecticide, herbicide, adjuvant

NOMENCLATURE
Names petroleum oils; mineral oils; white oils (refined grades); paraffin oils; adjuvant oils; spray oils

PHYSICAL CHEMISTRY
Composition They consist largely of aliphatic hydrocarbons, both saturated and unsaturated, the content of the latter being reduced by refinement. Produced by the distillation and refinement of crude mineral oils, those used as pesticides generally distil >310 °C. They may be classified by the proportion distilling at

335 °C, namely: 'light' (67–79%), 'medium' (40–49%) and 'heavy' (10–25%). Adjuvant oils are highly refined self-emulsifying oils which form a quick-breaking emulsion, spread quickly and help penetration of the active ingredient into the plant and pest. **B.p.** Fraction distilling above 310 °C **S.g./density** 0.65–1.06 (crude oil); 0.78–0.80 (kerosenes); 0.82–0.92 (spray oils) **F.p.** 'Actipron' ≥204 °C **Other properties** 'Actipron' dynamic viscosity c. 30 cSt (40 °C).

COMMERCIALISATION
History Oils of higher distillation range came into use c. 1922. More recently, highly refined oils have been used as adjuvants to increase the effectiveness of some other pesticides.

APPLICATIONS
Mode of action Acaricide and insecticide with ovicidal activity; contact herbicide; adjuvant for herbicides. **Uses** Control of overwintering forms (particularly winter eggs) of spider mites, aphids, and scale insects on fruit trees (pome fruit, stone fruit, citrus fruit), vines, and some ornamentals; and scale insects and summer eggs of spider mites and other insects on fruit trees (pome fruit, stone fruit, citrus fruit), vines, olives, bananas, some ornamentals, and glasshouse crops. Control of grass and broad-leaved weeds in umbelliferous crops (including carrots, celery, caraway, chervil, coriander, dill, lovage, and parsley), and in tree nurseries. Also used as a surfactant and as an adjuvant to enhance the activity of herbicides. **Phytotoxicity** Some petroleum oils are phytotoxic to plants, and must therefore be used only during the dormant period. **Formulation types** Oil; EC. **Compatibility** Incompatible with sulfur-containing materials. **Selected tradenames** 'Actipron' (BP); 'Spraying oil' (Hortichem)

ANALYSIS
Formulation analysis is by standard methods (*MAFF Ref. Bk.*, 1958, No. 1, p. 66; *CIPAC Handbook*, 1970, **1**, 582).

MAMMALIAN TOXICOLOGY
Oral Acute oral LD_{50} for rats and mice >4300 mg/kg for adjuvant oils, ('Actipron'). **Other** No toxicological problem due to petroleum oils has been reported in practice. **EC risk** Petroleum extracts are classified as (R45)

ENVIRONMENTAL FATE
Soil/Environment No risk of polynuclear aromatic compounds entering the food chain.

MAIN ENTRIES

563 phenmedipham   *Herbicide*

bis-carbamate

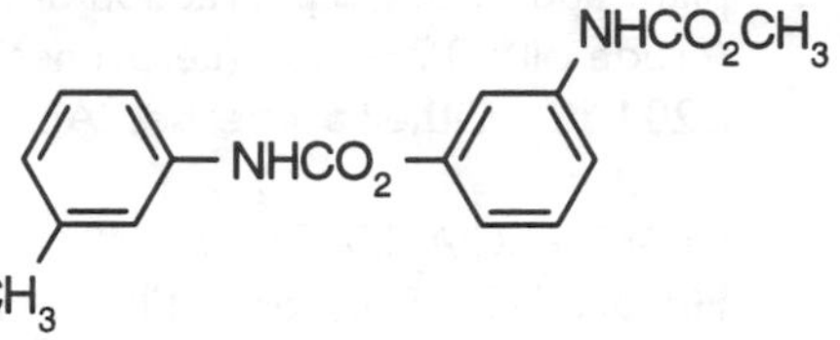

NOMENCLATURE
Common name phenmedipham (BSI, E-ISO, ANSI, WSSA, JMAF); phenmédiphame ((*m*) F-ISO)
IUPAC name methyl 3-(3-methylcarbaniloyloxy)carbanilate;
3-methoxycarbonylaminophenyl 3'-methylcarbanilate
Chemical Abstracts name 3-[(methoxycarbonyl)amino]phenyl (3-methylphenyl)= carbamate
Other names PMP   **CAS RN** *[13684–63–4]*   **Development codes** SN 38 584; ZK 15320; EP-452

PHYSICAL CHEMISTRY
Composition Tech. grade is >97% pure.   **Mol. wt.** 300.3   **M.f.** $C_{16}H_{16}N_2O_4$
Form Colourless crystals.   **M.p.** 143–144 °C; 140–144 °C (tech.)
V.p. 1.33×10^{-6} mPa (25 °C)   **K_{ow}** logP = 3.59 (pH 3.9)   **Henry** 5×10^{-8} Pa m^3 mol^{-1} (calc.)   **S.g./density** 0.34–0.54 (20 °C)   **Solubility** In water 4.7 mg/l (room temperature). Soluble in polar organic solvents. In acetone, cyclohexanone *c.* 200, methanol *c.* 50, chloroform 20, benzene 2.5, hexane *c.* 0.5, dichloromethane 16.7, ethyl acetate 56.3, toluene *c.* 0.97, 2,2,4-trimethylpentane 1.16 (all in g/l, 20 °C).
Stability Stable up to 200 °C. Very stable in acidic media, but hydrolysed in neutral and alkaline media; DT_{50} (22 °C) 50 d (pH 5), 14.5 h (pH 7), 10 min (pH 9). In solution (pH 3.8) irradiated at 280nm, DT_{50} 9.7 d.   **pKa** <0.1

COMMERCIALISATION
History Herbicide reported by F. Arndt & C. Kötter (*Abstr. Int. Congr. Plant Prot., 6th, Vienna*, 1967, p. 433). Introduced by Schering AG (now AgrEvo GmbH).
Patent GB 1127050   **Manufacturers** AgrEvo; Griffin; United Phosphorus Ltd

APPLICATIONS
Biochemistry Photosynthetic electron transport inhibitor.
Mode of action Selective systemic herbicide, absorbed through the leaves, with translocation primarily in the apoplast.   **Uses** Used post-emergence at 1 kg a.i./ha in beet crops, especially sugar beet, after the emergence of most broad-leaved weeds and before they develop more than 2–4 true leaves; also used on strawberries, spinach, peas, mangold, and red beet.   **Phytotoxicity** Non-phytotoxic to beet crops.   **Formulation types** EC; SC; oil-SC; SL; WG.

Mixtures *(phenmedipham +)* desmedipham; ethofumesate; ethofumesate + desmedipham; ethofumesate + metamitron; chloridazon. **Compatibility** Incompatible with alkaline substances. **Selected tradenames** 'Betanal' (AgrEvo); 'Alegro' (Rhône-Poulenc); 'Beetup' (United Phosphorus Ltd); 'Beta' (Stefes); 'Betapost' (Agriphar); 'Dephend' (Headland); 'Fender' (ISK Biosciences); 'Herbasan' (Mirfield); 'Kemifam' (AgrEvo); 'Kontakt' (Feinchemie Schwebda); 'Protrum K' (Atlas Crop Protection); 'Punter' (Barclay); 'Spin-aid' (NOR-AM); 'Vangard' (FCC)

ANALYSIS
Product analysis by titration (*CIPAC Proc.*, 1980, **2**, 215; *CIPAC Handbook*, 1985, **1C**, 2181) or by hplc (*ibid.*, p. 2181); method available from AgrEvo. **Residues** by glc (K. Kossmann & N. A. Jenny, *Anal. Methods Pestic. Plant Growth Regul.*, 1973, **7**, 611); in soil by hplc or by hydrolysis to *m*-toluidine, derivatives of which are determined by glc with ECD or by colorimetry (K. Kossmann, *Weed Res.*, 1970, **10**, 340). Details available from AgrEvo.

MAMMALIAN TOXICOLOGY
Oral Acute oral LD_{50} for rats and mice >8000, guinea pigs and dogs >4000 mg/kg. **Skin and eye** Acute percutaneous LD_{50} for rabbits 1000, rats 2500 mg/kg. Not a skin sensitiser. **Inhalation** LC_{50} (4 h) for rats >7.0 mg/l. **NOEL** (2 y) for rats 100, dogs 1000 mg/kg; (90 d) for rats and dogs 200 mg/kg diet. **ADI** 0.03 mg/kg. **Other** Acute i.p. LD_{50} for rats >5000 mg/kg. **Toxicity class** WHO (a.i.) III (Table 5); EPA (formulation) IV

ECOTOXICOLOGY
Birds Acute oral LD_{50} for chickens >2500, mallard ducks >2100 mg/kg. Dietary LC_{50} (8 d) for mallard ducks and bobwhite quail >6000 mg/kg diet. **Fish** LC_{50} (96 h) for rainbow trout 1.4 – 3.0, bluegill sunfish 3.98 mg/l. LC_{50} (96 h) for harlequin fish 16.5 mg/l (15.9% EC formulation). **Bees** Not toxic to bees; LD_{50} (oral) >23, (contact) 50 μg/bee. **Worms** EC_{50} (14 d) >156 mg/kg soil. **Daphnia** LC_{50} (72 h) 3.8 mg/l. **Algae** IC_{50} (96 h) 0.13 mg/l.

ENVIRONMENTAL FATE
Animals In mammals, following oral administration, 99% is excreted within 72 hours, mainly in the urine. Hydrolysis to methyl *N*-(3-hydroxyphenyl)carbamate and conjugation to glucuronides and ethereal sulfates are the major steps in metabolism. **Plants** Methyl *N*-(3-hydroxyphenyl)carbamate is the major metabolite in plants. **Soil/Environment** DT_{50} in soil *c.* 25 d, DT_{90} *c.* 108 d. Metabolites include methyl *N*-(3-hydroxyphenyl)carbamate and *m*-aminophenol, and subsequently complexes with soil components (R. Senawana *et al.*, *Bull. Environ. Contam. Toxicol.*, 1971, **6**, 322–327). Phenmedipham does not accumulate in soil, nor is there any relevant uptake by following crops. Due to its favourable physico-chemical parameters, no risk of groundwater contamination is to be expected. K_{oc} 2400.

564 phenothrin [(1R)-*trans*- isomer]

pyrethroid

(1R)-*cis*-

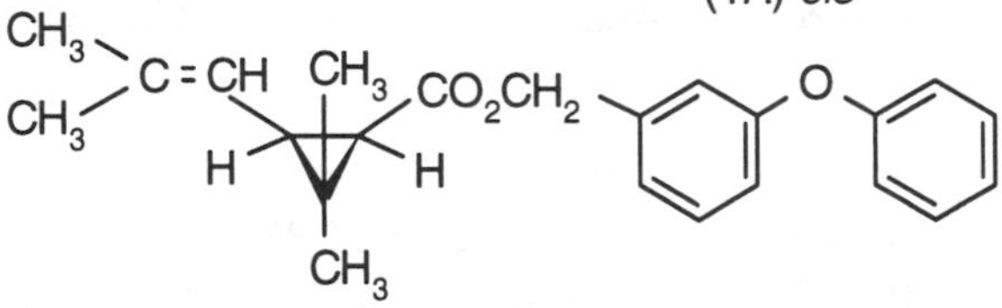

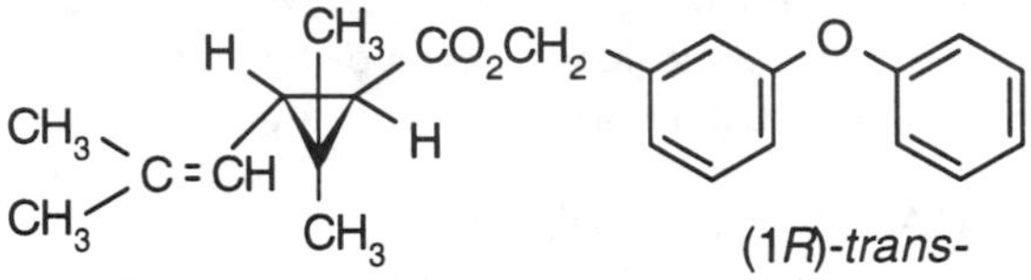

(1R)-*trans*-

NOMENCLATURE
Common name phenothrin (see composition) (BSI, E-ISO, BAN for (1RS)-
cis-trans- isomers); phénothrine ((*f*) F-ISO)
IUPAC name (for phenothrin): 3-phenoxybenzyl (1RS,3RS;1RS,3SR)-2,2-=
dimethyl-3-(2-methylprop-1-enyl)cyclopropanecarboxylate (see Composition)
Alt: 3-phenoxybenzyl (±)-*cis-trans*-chrysanthemate *Roth*: 3-phenoxybenzyl
(1RS)-*cis-trans*-2,2-dimethyl-3-(2-methylprop-1-enyl)cyclopropanecarboxylate
Chemical Abstracts name (3-phenoxyphenyl)methyl 2,2-dimethyl-3-(2-methyl-=
1-propenyl)cyclopropanecarboxylate
Other names d-phenothrin for (1R)-rich grade **CAS RN** [26002–80–2] (1RS)-=
cis-trans- isomers; [51186–88–0] (1R)-*cis*- isomer; [26046–85–5] (1R)-*trans*-
isomer **Development codes** S-2539 (Sumitomo) **Official codes** OMS 1809 for
phenothrin; OMS 1810 for (1R)-*cis-trans*- isomers; ENT 27 972 for (1R)-*cis-trans*-
isomers

PHYSICAL CHEMISTRY
Composition The common name phenothrin refers to a mixture of (1RS)-*cis*-=
trans- isomers. *d*-Phenothrin contains ≥95% (1R)- isomers, ≥75% *trans*- isomer.
Mol. wt. 350.5 **M.f.** $C_{23}H_{26}O_3$ **Form** Pale yellow to yellow-brown clear liquid
with a faint characteristic odour. **B.p.** >290 °C/760 mmHg **V.p.** 1.9×10^{-2} mPa
(21.4 °C) **S.g./density** 1.06 (20 °C) **Solubility** In water <9.7 µg/l (25 °C). In
methanol >5.0, hexane >4.96 g/ml (25 °C). **Stability** Stable under normal
storage conditions. Hydrolysed by alkalis. **F.p.** 107 °C

COMMERCIALISATION
History Insecticidal activity of the (1R)-*cis-trans*- isomers reported by K. Fujimoto

et al. (*Agric. Biol. Chem.*, 1973, **37**, 2681). Introduced by Sumitomo Chemical Co., Ltd. **Patent** JP 1027088; US 3934028 **Manufacturers** Sumitomo

APPLICATIONS
Mode of action Non-systemic insecticide with contact and stomach action. Gives rapid knockdown. **Uses** Used to control injurious and nuisance insects of public health (Acari, *Cimex lectularius*, Culicidae, Muscidae, Siphonaptera and *Pediculus humanus*) Also used to protect stored grain. **Formulation types** Aerosol; EC; OL. **Mixtures** *(d-phenothrin +)* allethrin; piperonyl butoxide; tetramethrin; allethrin + piperonyl butoxide; piperonyl butoxide + tetramethrin. **Compatibility** Incompatible with alkaline materials. **Selected tradenames** 'Sumithrin' (*d*-phenothrin) (Sumitomo)

ANALYSIS
Product analysis of pyrethroids reviewed by E. Papadopoulou-Mourkidou in *Comp. Analyt. Profiles*. By hplc (S. Sakaue et al., *ibid.*, 1981, **45**, 1135) or by glc with FID (*idem, ibid.*; Y. Takimoto et al., *Anal. Methods Pestic. Plant Growth Regul.*, 1984, **13**, 133). **Residues** determined by glc with ECD of a derivative (*idem, ibid.*). Details available from Sumitomo Chemical Co.

MAMMALIAN TOXICOLOGY
Reviews *Pesticide residues in food – 1988*. FAO Plant Production and Protection Paper 92, 1988. *Pesticide residues in food – 1988 evaluations. Part II – Toxicology*. FAO Plant Production and Protection Paper 93/2, 1989. See also A. J. Gray & D. M. Soderlund, Chapt. 5 in *"Insecticides"*. **EHC** 96 (WHO, 1990). **Oral** Acute oral LD_{50} for rats >5000 mg/kg. **Skin and eye** Acute percutaneous LD_{50} for rats >2000 mg/kg. **Inhalation** LC_{50} (4 h) for rats >3760 mg/m^3. **ADI** (JMPR) 0.07 mg/kg b.w. (for *d*-phenothrin) [1988]. **Toxicity class** WHO (a.i.) III (Table 5); EPA (formulation) IV

ECOTOXICOLOGY
Birds Acute oral LD_{50} for bobwhite quail >2500 mg/kg. **Fish** LC_{50} (96 h) for rainbow trout 2.7, bluegill sunfish 16 µg/l. **Bees** Toxic to bees.

ENVIRONMENTAL FATE
Soil/Environment Undergoes degradation by photo-oxidation. For details, see L. O. Ruzo et al., *J. Agric. Food Chem.*, 1982, **30**, 110.

565 phenthoate

Insecticide, acaricide

organophosphorus

$$\text{C}_6\text{H}_5-\underset{\underset{\text{CO}_2\text{CH}_2\text{CH}_3}{|}}{\text{CHSP(OCH}_3)_2}\overset{\overset{\text{S}}{||}}{}$$

NOMENCLATURE
Common name phenthoate (BSI, E-ISO, (*m*) F-ISO); PAP (JMAF)
IUPAC name S-α-ethoxycarbonylbenzyl O,O-dimethyl phosphorodithioate
Chemical Abstracts name ethyl α-[(dimethoxyphosphinothioyl)thio]=
benzeneacetate
Other names dimephenthoate **CAS RN** *[2597–03–7]* **EEC no.** 219–997–0
Development codes L 561 (Agrimont); S-2940 (Sumitomo)
Official codes OMS 1075; ENT 27 386

PHYSICAL CHEMISTRY
Mol. wt. 320.4 **M.f.** $C_{12}H_{17}O_4PS_2$ **Form** Colourless crystals; (tech., reddish-
yellow oil). **M.p.** 17–18 °C **B.p.** 186–187 °C/5 mmHg **V.p.** 5.3 mPa (40 °C)
K$_{ow}$ logP = 3.69 **S.g./density** 1.226 (20 °C) **Solubility** In water 10 mg/l (25 °C).
Readily soluble in methanol, ethanol, acetone, hexane, xylene, benzene, carbon
disulfide, chloroform, methylene dichloride, acetonitrile, tetrahydrofuran; in
n-hexane 116, kerosene 340 (both in g/l, 25 °C) **Stability** Decomposes 180 °C.
In water, stable under neutral and acidic conditions; degrades under alkaline
conditions. **F.p.** 165–170 °C

COMMERCIALISATION
History Introduced by Montecatini S.p.A. (now Isagro Srl). **Patent** GB 834814;
US 2947662 **Manufacturers** Aimco; Hanwha; Nissan

APPLICATIONS
Biochemistry Cholinesterase inhibitor. **Mode of action** Non-systemic insecticide
and acaricide with contact and stomach action. **Uses** Control of aphids, scale
insects, jassids, lepidopterous larvae, bollworms, borers, leafhoppers,
planthoppers, mealybugs, lace bugs, suckers, thrips, whitefly, spider mites, etc. in
citrus fruit, pome fruit, olives, Japanese persimmons, chestnuts, mulberries, cotton,
cereals, maize, rice, coffee, tea, sunflowers, sugar cane, tobacco, vegetables,
cucurbits, and ornamentals. Also used for mosquito control (adults and larvae).
Phytotoxicity Phytotoxic to some varieties of grape, fig, and peach, and red-
skinned varieties of apple. **Formulation types** EC; DP. **Compatibility** Not
compatible with pesticides which are alkaline in reaction. **Selected tradenames**
'Elsan' (Nissan); 'Cidial' (Isagro); 'Aimsan' (Aimco)

ANALYSIS

Product analysis by glc (B. Bazzi *et al.*, *Anal. Methods Pestic. Plant Growth Regul.*, 1976, **8**, 159). **Residues** determined by glc (B. Bazzi *et al.*, *loc. cit.*; *Man. Pestic. Residue Anal.*, 1987, **I**, 6, S19; *Anal. Methods Residues Pestic.*, 1988, Part I, M11; A. Ambrus *et al.*, *J. Assoc. Off. Anal. Chem.*, 1981, **64**, 733). Details available from Isagro.

MAMMALIAN TOXICOLOGY

Reviews *Pesticide residues in food – 1984*. FAO Plant Production and Protection Paper 62, 1985. *Pesticide residues in food – 1984 evaluations*. FAO Plant Production and Protection Paper 67, 1985. **EHC** 63 (WHO, 1986; a general review of organophosphorus insecticides). **Oral** Acute oral LD_{50} for male rats 410, mice 249–270, dogs >500, guinea pigs 377, rabbits 72 mg/kg. **Skin and eye** Acute percutaneous LD_{50} for rats >5000, mice 2620 mg/kg. Non-irritating to skin (rabbits). **Inhalation** LC_{50} (4 h) for rats 3.17 mg/l air. **NOEL** (104 w) for dogs 0.29 mg/kg daily. **ADI** (JMPR) 0.003 mg/kg b.w. [1984]. **Toxicity class** WHO (a.i.) II; EPA (formulation) II **EC risk** Xn (R21/22)

ECOTOXICOLOGY

Birds Acute oral LD_{50} for pheasants 218, quail 300 mg/kg. **Fish** TLm (48 h) for carp 2.5, goldfish 2.4 ppm. **Bees** Toxic to bees. LD_{50} 0.306 μg/bee.

ENVIRONMENTAL FATE

Animals In mammals, phenthoate is degraded with almost equal facility by hydrolysis of the carboethoxy moiety, cleavage of the P-S or C-S bond and removal of the methoxy group by either direct demethylation or hydrolytic cleavage of the P-O bond. The following metabolites were identified in either urine or faeces: demethyl phenthoate, demethyl phenthoate acid, demethyl phenthoate oxon acid, *O,O*-dimethyl phosphorodithioic and phosphorothioic acids. **Plants** In plants, there is oxidation to the phosphorothioate, followed by hydrolysis. Phosphoric acid, dimethyl and monomethyl phosphate have been identified as metabolites. **Soil/Environment** DT_{50} ≤1 d in both upland and submerged soil. The degradation product is phenthoate acid.

566 phenylmercury acetate *Fungicide*

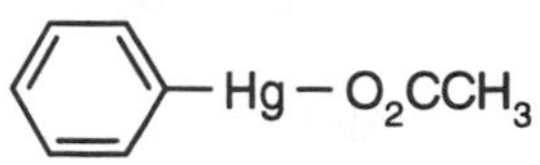

NOMENCLATURE

Common name PMA (JMAF); phenylmercury acetate (BSI, E-ISO, accepted in lieu of a common name); acétate de phénylmercure (F-ISO, accepted in lieu of a common name)

IUPAC name phenylmercury acetate
Chemical Abstracts name (acetato-O)phenylmercury
Other names PMA   **CAS RN** *[62–38–4]*   **EEC no.** 200–532–5

PHYSICAL CHEMISTRY
Mol. wt. 336.7   **M.f.** $C_8H_8HgO_2$ **Form** Colourless crystals.   **M.p.** 149–153 °C (decomp.)   **V.p.** 1.2 mPa (35 °C)   **Solubility** In water 4.37 g/l (15 °C). In acetone 48, methanol 34, 95% ethanol 17, benzene 15 (all in g/l, 15 °C).   **Stability** Very stable to dilute acids. In the presence of alkalis, phenylmercury hydroxide is formed.

COMMERCIALISATION
History Fungicidal properties of the corresponding chloride described by E. Riehm (*Zentralbl. Bakteriol. Parasitenkd. Infektionskr. Hyg. Abt. 2*, 1914, **40**, 424); the acetate was marketed as a seed treatment ('Ceresan') by I. G. Farbenindustrie (now Bayer AG) in 1932 and its toxicity to crabgrass reported by J. A. De France (*Greenkeepers Rep.*, 1947, **15**(1), 30).   **Manufacturers** United Phosphorus

APPLICATIONS
Mode of action Fungicide with eradicant action. Has some herbicidal activity.
Uses Mainly used as a seed treatment for control of seed-borne diseases in cereals (e.g. wheat bunt, snow mould and leaf stripe in barley, loose smut in oats), beet, cotton, flax, rice, sorghum, turf, and ornamentals. Also for control of *Fusarium* spp. in tulips (by bulb dipping). Has been used as a selective herbicide to control crabgrass in lawns.   **Phytotoxicity** Defoliation occurs if applied to crab apples.   **Formulation types** DS; LS; WS.   **Mixtures** *(phenylmercury acetate +)* carboxin; lindane; methfuroxam.   **Compatibility** Incompatible with alkaline substances, oils, and many other fungicides.   **Selected tradenames** 'Unisan' (United Phosphorus)

ANALYSIS
Product analysis by digestion with concentrated sulfuric and nitric acids and titration of the mercury with a thiocyanate (*AOAC Methods*, 1984, 6.150–6.153) or by gravimetry (*ibid.*, 6.154–6.156). Mercury impurities determined (*CIPAC Handbook*, 1994, **F**, 264).   **Residues** determined by a similar oxidative digestion with colorimetry of a complex with dithizone (*AOAC Methods*, 1984, 25.138–25.145; *EPPO Recommended Methods, Series A, No.* 37) or by atomic absorption spectroscopy (*AOAC Methods*, 1984, 25.131–25.137).

MAMMALIAN TOXICOLOGY
Reviews *Pesticide residues*, FAO Rept. No. PL:1967/M/11; WHO Tech. Rept. Series, No. 391, 1968. *1967 Evaluations of some pesticide residues in food,*

FAO/PL:1967/M 11/1; WHO/Food Add./68.30, 1968. **EHC** 1 (WHO, 1976), 86 (WHO, 1989). **Oral** Acute oral LD_{50} for rats 24, mice 70 mg/kg.
Skin and eye Acute percutaneous LD_{50} not available. Absorbed through the skin; may cause dermatitis and allergy. **Inhalation** There is some vapour hazard, so conditions under which seed is treated are controlled. **ADI** (JMPR) No ADI [1967]. **Other** Teratogenic to rats. **Toxicity class** WHO (a.i.) Ia; EPA (formulation) I **EC risk** T (R25, R48/24/25); C (R34) **PIC** All mercury compounds included in PIC procedure.

ECOTOXICOLOGY
Birds Acute oral LD_{50} for chickens 60 mg/kg. **Bees** Not hazardous to bees when used as prescribed.

567 2-phenylphenol *Fungicide*

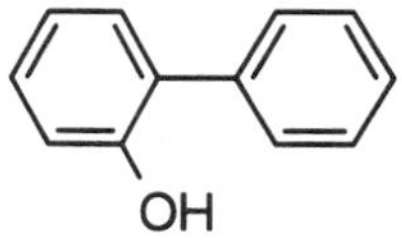

NOMENCLATURE
2-phenylphenol
Common name 2-phenylphenol (BSI, E-ISO, accepted in lieu of a common name); phényl-2 phenol (F-ISO)
IUPAC name biphenyl-2-ol
Chemical Abstracts name [1,1'-biphenyl]-2-ol
Other names 2-hydroxybiphenyl; orthophenylphenol **CAS RN** *[90–43–7]*
EEC no. 201–993–5

sodium 2-phenylphenoxide
CAS RN *[132–27–4]* **EEC no.** 205–055–6

PHYSICAL CHEMISTRY
2-phenylphenol
Mol. wt. 170.2 **M.f.** $C_{12}H_{10}O$ **Form** Colourless to pinkish crystals. **M.p.** 57 °C
B.p. 286 °C **V.p.** 0.9 kPa (140 °C) **S.g./density** 1.217 (25 °C) **Solubility** In water 0.7 g/l (25 °C). Soluble in most organic solvents, including ethanol, ethylene glycol, isopropanol, glycol ethers, and polyglycols.

sodium 2-phenylphenoxide
Mol. wt. 192.2 **M.f.** $C_{12}H_9NaO$ **Solubility** *c.* 1.1 kg/kg water (35 °C), giving a solution of pH 12.0–13.5.

COMMERCIALISATION
History Fungicidal properties reported by R. G. Tomkins (*Rep. Food Investigation Board 1936*, p. 149).

APPLICATIONS
Mode of action Disinfectant and fungicide with protective action. **Uses** Post-harvest control of storage diseases of apples, citrus fruit, stone fruit, tomatoes, cucumbers, capsicums, gladioli, etc. (by impregnation of fruit wrappers, crates, etc., or by application as a wax directly to the fruit); and disinfection of seed boxes. Also used to control apple canker (by application during the dormant period). **Phytotoxicity** Should not be used on growing plants.
Formulation types Coating agent.

ANALYSIS
Residues in fruit wrappers or citrus peel determined by colorimetry of derivatives (R. Mestres *et al., Trav. Soc. Pharm. Montpellier*, 1975, **35**, 81), by hplc (J. E. Farrow *et al., Analyst (London)*, 1977, **102**, 752) or by glc (Y. Kitada *et al., J. Assoc. Off. Anal. Chem.*, 1982, **65**, 1302).

MAMMALIAN TOXICOLOGY
2-phenylphenol
Reviews *Pesticide residues in food – 1990.* FAO Plant Production and Protection Paper, 102, 1990. *Pesticide residues in food – 1990 evaluations. Toxicology.* World Health Organisation, WHO/PCS/91.47, 1991. **IARC** 30 **Oral** Acute oral LD_{50} for male rats 2700, mice 2000 mg/kg. **Skin and eye** Acute percutaneous LD_{50} not available. Mild skin irritant. **NOEL** In 2 y feeding trials, rats receiving 2 g/kg diet showed no ill-effect (H. C. Hodge *et al., J. Pharmacol. Exp. Ther.*, 1952, **104**, 202). **ADI** (JMPR) 0.02 mg/kg b.w. [1990]. **Toxicity class** WHO (a.i.) III (Table 5) **EC risk** Xi (R36/38)

sodium 2-phenylphenoxide
Reviews As for the phenol. **ADI** See the phenol. **EC risk** Xn (R22); Xi (R38, R41)

ECOTOXICOLOGY
2-phenylphenol
Fish Toxic to fish.

ENVIRONMENTAL FATE
Animals Excreted in mammals principally as the parent compound and as the glucuronide and sulfate conjugates.

N-phenylphthalamic acid

Plant growth regulator

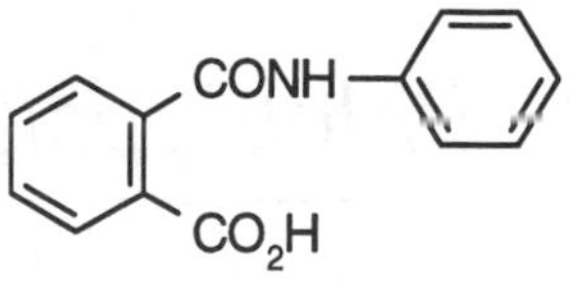

NOMENCLATURE
Other names phthalanilic acid   **CAS RN** *[4727–29–1]*

PHYSICAL CHEMISTRY
Composition Tech. is ≥97%.   **Mol. wt.** 241.2   **M.f.** $C_{14}H_{11}NO_3$ **Form** Greyish white powder.   **M.p.** 169 °C (decomp.)   **S.g./density** Bulk density 390±50 g/dm^3
Solubility In water 20 mg/l (20 °C). Readily soluble in methanol, ethanol, acetone, acetonitrile.   **Stability** Stable in neutral media. Hydrolysed in strong acid or base. Decomposes slowly above 100 °C, or in u.v. light, to give *N*-phenylphthalimide.
pKa 2–3

COMMERCIALISATION
History Developed by Neviki Research Institute for the Heavy Chemical Industry of Hungary. Introduced in 1982. Marketed by Chemol.   **Manufacturers** A. G. Nagy

APPLICATIONS
Mode of action Increases working life of stigma, promotes pollination.
Uses Used to encourage fruit set and promote yield in tomatoes, peppers, beans, peas and other vegetables, soya beans, alfalfa, rape, lupin, sunflower, grapes, pome and stone fruit, and arable crops.   **Formulation types** WP.
Selected tradenames 'Nevirol' (Chemol)

ANALYSIS
By hplc and i.r. spectrophotometry.

MAMMALIAN TOXICOLOGY
Oral LD_{50} for male and female rats, and male and female mice >5000 mg/kg.
Skin and eye Acute percutaneous LD_{50} for rats and rabbits >2000 mg/kg. Mildly irritant to eyes, not irritant to skin (rabbits). Not a skin sensitiser (guinea pigs).
Inhalation LC_{50} (4 h) for rats 5300 mg/m^3.   **Other** Acute i.p. LD_{50} for male rats 1821.0, female rats 1993.7 mg/kg.   **Toxicity class** WHO (a.i.) III (Table 5)

ECOTOXICOLOGY
Birds LD_{50} for male Japanese quail and pheasants >10 700 mg/kg.   **Fish** LC_{50}

MAIN ENTRIES

(96 h) for carp 650, goldfish 1000, pike 360 mg/l. **Bees** LD_{50} (oral) >1000 μg/10 bees. **Daphnia** EC_{50} (96 h) 42 mg/l. **Algae** EC_{50} (96 h) 74 mg/l.

569 *Phlebiopsis gigantea*   *Biological agent*

fungus

NOMENCLATURE
Scientific name *Phlebiopsis gigantea* (Fr.) Jul. **Other names** *Phlebia gigantea*; *Peniophora gigantea*

PROPERTIES
Composition Product contains 10^6-10^7 cfu/g. **Stability** The product remains active for 1–2 weeks at room temperature, and for 12 months below 8 °C, when stored in an unopened package.

COMMERCIALISATION
Production By fermentation. **History** It has long been known that *Phlebiopsis gigantea* is effective in preventing the aerial infection of *Heterobasidion annosum* in pine and spruce stumps. The strain of *P. gigantea* in 'Rotstop' was originally isolated in 1987 by the Finnish Forest Research Institute from a Norway spruce log left in the forest. It was used in a stump treatment on spruce and pine in 1988 and re-isolated in 1989 from a spruce stump. In 1991, it was formulated into a preparation by Kemira Oy (K. Korhonen et al., *Control of Heterobasidion annosum by stump treatment with 'Rotstop', a new commercial formulation of Phlebiopsis gigantea* in Proc. of 8th Int. Conf. on Root and Butt Rots. Wik, Sweden and Haikko, Finland, August 9–16, 1993. Eds. M. Johanson & J. Stenlid. IUFRO, Uppsala, Sweden, pp. 675–685). **Manufacturers** Kemira Agro

APPLICATIONS
Mode of action Competes with the pathogen for living space. **Uses** Biological control agent against root and butt rot (*Heterobasidion annosum*, also known as *Fomes annosus*) in pine and spruce. Recommended application time is during the period that *H. annosum* is capable of spreading and growing on stumps (during the vegetative period whenever the temperature rises above +8 °C). The product is mixed with water, and the suspension is sprayed on the stump surface with a spraying device on the harvester head at tree felling, or manually after felling.
Formulation types WP. **Compatibility** 'Rotstop' is incompatible with chemical fungicides. Authorisation for use together with chemicals has not been applied for.
Selected tradenames 'Rotstop' (Kemira Agro)

ANALYSIS
The viability and purity of the product is tested by determining the number of colony-forming units by plate count methods. The growth rate and lignolytic activity of *P. gigantea* on a substrate containing milled wood as sole energy source is used to further check the vitality of the fungus. In the presence of *Heterobasidion annosum*, it can be used as an indicative efficacy test. The efficacy can also be tested by treating the upper surface of freshly cut tree stem pieces with the product and spores of *H. annosum*

MAMMALIAN TOXICOLOGY
Application of the product does not cause additional exposure to the fungus more dangerous than the natural exposure level. No allergic responses or health problems have been observed by research workers, manufacturing staff or users. Inhalation of the fine powder and skin contact should be avoided by using normal protective equipment.

ENVIRONMENTAL FATE
The fungus belongs to the natural microflora of coniferous forests.
Soil/Environment The strain is a very common naturally-occurring saprophytic, wood-decomposing, white-rot fungus. The application of 'Rotstop' increases the population of *P. gigantea* locally on and in the stump. Long-term use of the product might increase the occurrence of the fungus and thus increase its natural control effect on *H. annosum*.

570 phorate   *Insecticide, acaricide, nematicide*

organophosphorus

$$CH_3CH_2SCH_2S\overset{\displaystyle S}{\overset{\|}{P}}(OCH_2CH_3)_2$$

NOMENCLATURE
Common name phorate (BSI, E-ISO, (*m*) F-ISO, ANSI, ESA); [timet] (former exception, USSR)
IUPAC name *O,O*-diethyl *S*-ethylthiomethyl phosphorodithioate
Chemical Abstracts name *O,O*-diethyl *S*-[(ethylthio)methyl] phosphorodithioate
CAS RN *[298-02-2]*   **EEC no.** 206-052-2   **Development codes** EI 3911; AC 3911 (Cyanamid)   **Official codes** ENT 24 042

PHYSICAL CHEMISTRY
Composition Tech. phorate is >90% pure.   **Mol. wt.** 260.4   **M.f.** $C_7H_{17}O_2PS_3$

Form Colourless liquid (tech.). **M.p.** <−15 °C (tech.)
B.p. 118–120 °C/0.8 mmHg (tech.) **V.p.** 85 mPa (25 °C) **K$_{ow}$** logP = 3.92
Henry 5.9 × 10^{-1} Pa m^3 mol^{-1} (calc.) **S.g./density** 1.167 (tech., 25 °C)
Solubility In water 50 mg/l (25 °C). Miscible with alcohols, ketones, ethers, esters, aromatic, aliphatic and chlorinated hydrocarbons, dioxane, vegetable oils, and other organic solvents. **Stability** Stable under normal storage conditions for at least 2 years. Aqueous solutions degraded by light (DT$_{50}$ 1.1 d); stability to hydrolysis optimum at pH 5–7, DT$_{50}$ 3.2 d (pH 7), 3.9 d (pH 9). **F.p.** >110 °C (Setaflash closed cup)

COMMERCIALISATION
History Introduced by American Cyanamid Co. **Patent** US 2586655; US 2596076; US 2970080 (Cyanamid); US 2759010 (Bayer) **Manufacturers** Cyanamid; Pesticides India; United Phosphorus Ltd; Voltas

APPLICATIONS
Biochemistry Cholinesterase inhibitor; activity derives from metabolic conversion to phorate oxon, the phosphorothioate. **Mode of action** Systemic insecticide and acaricide with contact and stomach action. **Uses** Control of Agromyzidae, Aleyrodidae, Aphididae, Chrysomelidae, Noctuidae, Pyralidae, Tetranychidae and certain nematodes (*Meloidogyne* spp.) in brassicas, beetroot, sugar beet, fodder beet, carrots, field beans, broad beans, celery, maize, sorghum, wheat, potatoes, tomatoes, hops, soya beans, sunflowers, sugar cane, alfalfa, cotton, coffee, rice, peanuts, and some ornamentals. **Phytotoxicity** Phytotoxic to apples and tobacco. Phytotoxicity is possible in beet, carrots, beans, maize, and tomatoes if granules come into direct contact with seeds in the furrow. **Formulation types** GR.
Mixtures *(phorate +)* flucythrinate; terbufos. **Compatibility** Incompatible with alkaline compounds and with water-containing preparations.
Selected tradenames 'Thimet' (Cyanamid); 'Cekuforatox' (Cequisa); 'Dhan' (Dhanuka); 'Kurunai' (Ramcides); 'Umet' (United Phosphorus); 'Volphor' (Voltas); 'Warrant' (Searle India)

ANALYSIS
Product analysis by i.r. spectrometry (*CIPAC Handbook*, 1983, **1B**, 1890; *AOAC Methods*, 1995, 964.05). **Residues** of phorate and its oxidation products determined by glc (*Pestic. Anal. Man.*, 1979, **I**, 201-A, 201-G, 201-H, 201-I; *Man. Pestic. Residue Anal.*, 1987, **I**, 3, 6, S8, S13, S16, S17, S19; *Anal. Methods Residues Pestic.*, 1988, Part I, M2, M5; *Analyst (London)*, 1980, **105**, 515; A. Ambrus *et al.*, *J. Assoc. Off. Anal. Chem.*, 1981, **64**, 733). Details available from American Cyanamid Co.

MAMMALIAN TOXICOLOGY
Reviews *Pesticide residues in food – 1994*, FAO Plant Production and Protection Paper, 127, 1995. *Pesticide residues in food – 1994 evaluations, Pt. II – Toxicology,*

WHO, WHO/PCS/95.2, 1995. **EHC** 63 (WHO, 1986; a general review of organophosphorus insecticides). **Oral** Acute oral LD_{50} for male rats 3.7, female rats 1.6, mice c. 6 mg/kg. **Skin and eye** Acute percutaneous LD_{50} for male rats 6.2, female rats 2.5, guinea pigs 20–30, male rabbits 5.6, female rabbits 2.9, guinea pigs 30.0 mg/kg. Values for GR depend on a.i. content, carrier, test method and animal species – typical values include: male rats 98–137 mg a.i. (as GR)/kg, male rabbits 93–245 mg a.i. (as GR)/kg. **Inhalation** LC_{50} (1 h) for male rats 0.06, female rats 0.011 mg/l. **NOEL** In 90 d feeding trials, rats receiving 6 mg/kg diet showed no ill-effects other than depression of cholinesterase levels. **ADI** (JMPR) 0.0005 mg/kg b.w. [1994]. **Other** Not mutagenic, not teratogenic, not carcinogenic. **Toxicity class** WHO (a.i.) Ia; EPA (formulation) I **EC risk** T+ (R27/28)

ECOTOXICOLOGY
Birds Acute oral LD_{50} for mallard ducks 0.62, ring-necked pheasants 7.1 mg/kg. **Fish** LC_{50} (96 h) for rainbow trout 0.013, channel catfish 0.28 mg/l. **Bees** Toxic to bees; LD_{50} (topical) 10 μg/bee.

ENVIRONMENTAL FATE
Animals In animals, phorate is metabolically oxidised to the sulfoxide and sulfone, and their phosphorothioate analogues, followed by hydrolysis to dithio-, thio-, and orthophosphoric acids. **Plants** Degradation is similar to that in animals.
Soil/Environment In soil, metabolic oxidation gives the sulfoxide and sulfone, and their phosphorothioate analogues, and these then undergo hydrolysis, although the sulfone can persist under certain conditions (D. L. Suett, *Pestic. Sci.*, 1975, **6**, 385). Half-life in soil 2–14 d. Soil K_{oc} 543.

571 phosalone   *Insecticide, acaricide*

organophosphorus

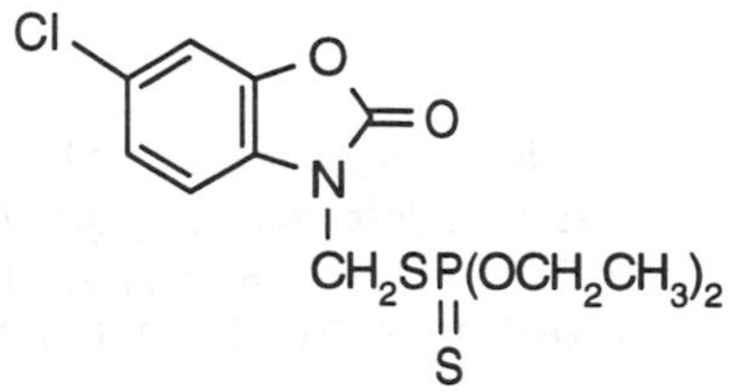

NOMENCLATURE
Common name phosalone (BSI, E-ISO, (f) F-ISO, ANSI, ESA, JMAF); [benzofos] (former exception, USSR)

IUPAC name S-6-chloro-2,3-dihydro-2-oxobenzoxazol-3-ylmethyl O,O-diethyl phosphorodithioate
Chemical Abstracts name S-[(6-chloro-2-oxo-3(2H)-benzoxazolyl)methyl] O,O-diethyl phosphorodithioate
Other names benzphos   **CAS RN** [2310–17–0]   **EEC no.** 218–996–2
Development codes 11 974 RP; NPH 1090   **Official codes** ENT 27 163

PHYSICAL CHEMISTRY
Mol. wt. 367.8   **M.f.** $C_{12}H_{15}ClNO_4PS_2$ **Form** Colourless crystals, with an odour of garlic.   **M.p.** Tech. 42–48 °C   **V.p.** <0.06 mPa (25 °C) K_{ow} logP = 4.01 (20 °C)   **Henry** 7.4×10^{-3} Pa m^3 mol^{-1} (calc.)   **S.g./density** 1.338 g/ml (20 °C)
Solubility In water 3.05 mg/l (25 °C). In ethyl acetate, acetone, acetonitrile, benzene, chloroform, dichloromethane, dioxane, methyl ethyl ketone, toluene, xylene c. 1000, hexane 11 (all in g/l, 20 °C).   **Stability** Hydrolysed by strong alkalis and acids; DT_{50} 9 d (pH 9).

COMMERCIALISATION
History Insecticide reported by J. Desmoras et al. (Phytiatr. Phytopharm., 1963, **12**, 199). Introduced by Rhône-Poulenc Agrochimie.   **Patent** GB 1005372; BE 609209; FR 1482025   **Manufacturers** Rhône-Poulenc

APPLICATIONS
Biochemistry Cholinesterase inhibitor.   **Mode of action** Non-systemic insecticide and acaricide showing localised penetration of plant cuticle; with contact and stomach action.   **Uses** A non-systemic acaricide and insecticide used primarily in pome and stone fruit trees. Effective against Coleoptera, Homoptera (Aphididae), Lepidoptera (Cydia pomonella) and Thysanoptera on fruit trees. It is selective of most beneficial insects and widely used in integrated pest management programmes. Also used in grapes, oilseed rape, ornamentals, potatoes and vegetables.   **Phytotoxicity** Non-phytotoxic when used at the recommended application rates. At higher application rates, Golden Delicious and other yellow apple varieties have been injured.   **Formulation types** EC; WP; SC.
Mixtures (phosalone +) parathion; cypermethrin; dimethoate.
Selected tradenames 'Zolone' (Rhône-Poulenc)

ANALYSIS
Product analysis by glc or hplc (CIPAC Handbook, 1988, **D**, 141).
Residues determined by glc (A. Ambrus et al., J. Assoc. Off. Anal. Chem., 1981, **64**, 733; Man. Pestic. Residue Anal., 1987, **I**, 5, 6, S8, S19; Anal. Methods Residues Pestic., 1988, Part I, M2, M5, M12).

MAMMALIAN TOXICOLOGY
Reviews Pesticide residues in food – 1993 evaluations, Part II – Toxicology. WHO, WHO/PCS/94.4, 1994.   **EHC** 63 (WHO, 1986; a general review of

organophosphorus insecticides).   **Oral** Acute oral LD_{50} for rats 120 mg/kg.
Skin and eye Acute percutaneous LD_{50} for rats 1500 mg/kg.   **Inhalation** LC_{50} (4 h)
for female rats 0.7 mg/l.   **NOEL** (2 y) for rats 2.5 mg/kg b.w.   **ADI** (JMPR)
0.001 mg/kg b.w. [1993].   **Toxicity class** WHO (a.i.) II   **EC risk** T (R25);
Xn (R21)

ECOTOXICOLOGY
Birds Acute oral LD_{50} for mallard duck >2150 mg/kg. Dietary LC_{50} (8 d) for
bobwhite quail 2033, mallard duck 1659 ppm diet.   **Fish** LC_{50} (96 h) for rainbow
trout 0.63, carp 2.1 mg/l.   **Worms** Moderately toxic.   **Daphnia** EC_{50} (48 h)
0.74 µg/l.   **Algae** Low toxicity.

ENVIRONMENTAL FATE
Animals Rapidly eliminated, primarily via urine.   **Plants** Rapidly degraded via
oxidation, cleavage, hydrolysis and dechlorination.   **Soil/Environment** Low
mobility in soils; strongly adsorbed and rapidly degraded; DT_{50} c. 1–4 d.

572 phosmet   *Insecticide, acaricide*

organophosphorus

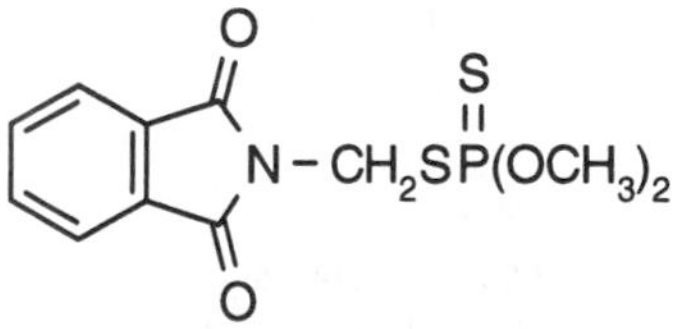

NOMENCLATURE
Common name phosmet (BSI, E-ISO, (*m*) F-ISO, ESA); phtalofos (USSR);
PMP (JMAF)
IUPAC name *O,O*-dimethyl *S*-phthalimidomethyl phosphorodithioate;
N-(dimethoxyphosphinothioylthiomethyl)phthalimide
Chemical Abstracts name *S*-[(1,3-dihydro-1,3-dioxo-2*H*-isoindol-2-yl)methyl]
O,O-dimethyl phosphorodithioate
CAS RN *[732–11–6]*   **EEC no.** 211–987–4   **Development codes** R-1504
(Zeneca)   **Official codes** OMS 232; ENT 25 705

PHYSICAL CHEMISTRY
Composition Tech. material is 92% pure.   **Mol. wt.** 317.3   **M.f.** $C_{11}H_{12}NO_4PS_2$
Form Colourless crystals; (tech., off-white or pink waxy solid).
M.p. 72.0–72.7 °C; (tech., 66–69 °C)   **V.p.** 0.065 mPa (25 °C) K_{ow} logP = 2.95

Solubility In water 25 mg/l (25 °C). In acetone 650, benzene 600, toluene, methyl isobutyl ketone 300, xylene 250, methanol 50, kerosene 5 (all in g/l, 25 °C).
Stability Rapidly hydrolysed in alkaline media; relatively stable in acidic conditions; DT_{50} (20 °C) 13 d (pH 4.5), <12 h (pH 7), <4 h (pH 8.3). Decomposes rapidly above 100 °C. Decomposes in sunlight in aqueous solution or on glass plates.
F.p. >106 °C

COMMERCIALISATION
History Insecticide reported by B. A. Butt & J. C. Keller (*J. Econ. Entomol.*, 1961, **54,** 813). Introduced by Stauffer Chemical Co. (now Zeneca Agrochemicals) and now marketed by Gowan Company and others.   **Patent** US 2767194
Manufacturers General Quimica; Inquinosa; Productos OSA; Zeneca

APPLICATIONS
Biochemistry Cholinesterase inhibitor.   **Mode of action** Non-systemic insecticide and acaricide with predominantly contact action.   **Uses** Control of lepidopterous larvae, aphids, suckers, fruit flies, and spider mites on pome fruit, stone fruit, citrus fruit, ornamentals and vines; Colorado beetles on potatoes; boll weevils on cotton; olive moths and olive thrips on olives; blossom beetles on oilseed rape; leaf beetles and weevils on alfalfa; European corn borers on maize and sorghum; sweet potato weevils on sweet potatoes in storage; etc. at 0.5–1.0 kg a.i./ha. Also used as an animal ectoparasiticide.   **Formulation types** WP; EC; DP; SL.
Mixtures *(phosmet +)* carbophenothion.   **Compatibility** Incompatible with alkaline materials.   **Selected tradenames** 'Cekumet' (Cequisa); 'Fosdan' (General Quimica); 'Imidan' (Gowan); 'Inovat' (Productos OSA); 'Inovitan' (Efthymiadis); 'Prolate' (Gowan)

ANALYSIS
Product analysis by capillary gc.   **Residues** in crops determined by capillary gc (M. A. Luke *et al.*, *J. Assoc. Off. Anal. Chem.*, 1981, **64,** 1187; G. H. Batchelder *et al.*, *Anal. Methods Pestic., Plant Growth Regul. Food. Addit.*, 1967, **5,** 257; J. E. Barney *et al.*, *Anal. Methods Pestic. Plant Growth Regul.*, 1972, **6,** 408). Details available from Zeneca.

MAMMALIAN TOXICOLOGY
Reviews *Pesticide residues in food – 1994*, FAO Plant Production and Protection Paper, 127, 1995. *Pesticide residues in food – 1994 evaluations, Part II – Toxicology*, WHO, WHO/PCS/95.2, 1995.   EHC 63 (WHO, 1986; a general review of organophosphorus insecticides).   **Oral** Acute oral LD_{50} for male rats 113, female rats 160 mg/kg.   **Skin and eye** Acute percutaneous LD_{50} for albino rabbits >5000 mg/kg; mild skin and eye irritant (rabbits). Not a skin sensitiser (guinea pigs).   **Inhalation** LC_{50} (1 h) for rats 2.76 mg/l air.   **NOEL** (2 y) for rats and dogs 40 mg/kg diet. Not carcinogenic or teratogenic.   **ADI** (JMPR) 0.01 mg/kg b.w. [1994].   **Toxicity class** WHO (a.i.) II; EPA (formulation) II   **EC risk** Xn (R21/22)

ECOTOXICOLOGY

Birds LC_{50} (5 d) for bobwhite quail 507, mallard ducks >5000 mg/kg diet.
Fish LC_{50} (96 h) for bluegill sunfish 0.07, rainbow trout 0.23 mg/l. **Bees** LD_{50}
0.001 mg/bee. **Daphnia** LC_{50} (48 h) 8.5 µg/l.

ENVIRONMENTAL FATE

Animals In animals, rapid metabolism occurs to phthalamic acid, phthalic acid, and
phthalic acid derivatives, which have been isolated from urine as phosmet
metabolites (J. B. McBain et al., J. Agric. Food Chem., 1968, **16**, 813–820).
Plants In plants, phosmet is rapidly broken down to non-toxic metabolites.
Soil/Environment Rapidly broken down in the soil.

MAIN ENTRIES

573 phosphamidon *Insecticide, acaricide*

organophosphorus

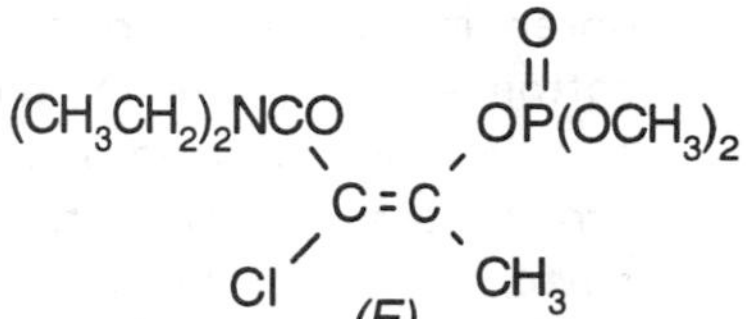

NOMENCLATURE

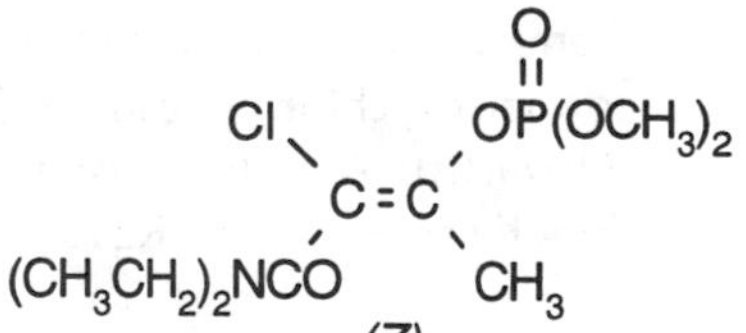

Common name phosphamidon (BSI, E-ISO, (m) F-ISO, ANSI, ESA)
IUPAC name 2-chloro-2-diethylcarbamoyl-1-methylvinyl dimethyl phosphate;
2-chloro-3-dimethoxyphosphinoyloxy-N,N-diethylbut-2-enamide
Chemical Abstracts name 2-chloro-3-(diethylamino)-1-methyl-3-oxo-1-propenyl
dimethyl phosphate
CAS RN [13171–21–6] (E)- + (Z)- isomers; [23783–98–4] (Z)- isomer;
[297–99–4] (E)- isomer **EEC no.** 236–116–5 **Development codes** C 570 (Ciba)
Official codes OMS 1325; ENT 25 515

PHYSICAL CHEMISTRY

Composition The commercial compound contains 70% *m/m* (Z)- isomer
(β- isomer) (which has the greater insecticidal activity) and 30% (E)- isomer

phosphamidon 965

(α- isomer).   **Mol. wt.** 299.7   **M.f.** $C_{10}H_{19}ClNO_5P$ **Form** Pale yellow liquid.
B.p. 162 °C/1.5 mmHg; 94 °C/0.04 mmHg   **V.p.** 2.2 mPa (25 °C)
K_{ow} logP = 0.79   **S.g./density** 1.21 (25 °C)   **Solubility** Miscible with water,
acetone, dichloromethane, toluene and other common organic solvents, with the
exception of aliphatic hydrocarbons, e.g. solubility in hexane 32 g/l (25 °C).
Stability Rapidly hydrolysed by alkalis: DT_{50} (calc.) (20 °C) 60 d (pH 5), 54 d
(pH 7), 12 d (pH 9).

COMMERCIALISATION
History Insecticide reported by F. Bachmann & J. Meierhans (*Bull. Cent. Int.
Antiparasit.*, 1956, Nov., p. 18). Introduced by Ciba AG (now Novartis Crop
Protection AG).   **Patent** BE 552284; GB 829576   **Manufacturers** Aimco;
Hui Kwang; Novartis; Pilarquim; Rallis; United Phosphorus

APPLICATIONS
Biochemistry Cholinesterase inhibitor.   **Mode of action** Systemic insecticide and
acaricide with stomach and slight contact action. Absorbed by the leaves and
roots.   **Uses** Control of sucking, chewing, and boring insects, and spider mites on
a very wide range of crops. Specific uses include control of leaf beetles and stem
borers in rice, stem borers in sugar cane, Colorado beetles in potatoes, thrips in
cotton, etc. Also used to control aphids, sawflies, suckers, fruit flies, leaf miners,
moth and beetle larvae, and many other insects in fruit, vines, olives, vegetables,
ornamentals, cereals, beet, maize, alfalfa, many other crops, and in forestry.
Phytotoxicity Non-phytotoxic, except to some varieties of cherry, plum, peach,
maple and sorghum.   **Formulation types** SL; SC; EC; UL.
Compatibility Incompatible with alkaline materials. Should not be mixed with
copper oxychloride, captan, folpet, or sulfur.   **Selected tradenames** 'Dimecron'
(Novartis); 'Aimphon' (Aimco); 'Kinadon' (United Phosphorus); 'Phosron'
(Hui Kwang); 'Rilan' (Rallis); 'Rimdon' (Ramcides)

ANALYSIS
Product analysis by rplc with u.v. detection (*AOAC Methods*, 1995, 993.01; *CIPAC
Handbook*, 1992, **E**, 179–183), or by glc (A. A. Carlstrom, *J. Assoc. Off. Anal.
Chem.*, 1972, **55**, 1331).   **Residues** determined by glc with FPD (A. Ambrus, *ibid.*,
1981, **64,** 733; *Man. Pestic. Residue Anal.*, 1987, **I**, 6, S13, S19; *Anal. Methods
Residues Pestic.*, 1988, Part I, M5, M12; *Pestic. Anal. Man.*, 1979, **I**, 201-H, 201-I).

MAMMALIAN TOXICOLOGY
Reviews *Pesticide residues in food – 1986*. FAO Plant Production and Protection
Paper 77, 1986. *Pesticide residues in food – 1986 evaluations. Part II – Toxicology*.
FAO Plant Production and Protection Paper 78/2, 1987. Residue Reviews, Volume
37.   **EHC** 63 (WHO, 1986; a general review of organophosphorus insecticides).
Oral Acute oral LD_{50} for rats 17.9–30 mg/kg.   **Skin and eye** Acute percutaneous
LD_{50} for rats 374–530, rabbits 267 mg/kg. Slight skin irritation; moderate eye

irritation (rabbits).   **Inhalation** LC_{50} (4 h) for rats *c.* 0.18, mice 0.033 mg/l air.
NOEL (2 y) for rats 1.25, dogs 0.1 mg/kg b.w. daily.   **ADI** (JMPR) 0.0005 mg/kg
b.w. [1986].   **Toxicity class** WHO (a.i.) Ia; EPA (formulation) I
EC risk R24/28, R40

ECOTOXICOLOGY
Birds Acute oral LD_{50} for Japanese quail 3.6–7.5, mallard ducks 3.8 mg/kg.
LC_{50} (8 d) for Japanese quail 90–250 ppm.   **Fish** LC_{50} (96 h) for rainbow trout
7.8, fathead minnow 100 mg/l.   **Bees** Highly toxic to bees.   **Daphnia** LC_{50} (48 h)
0.01–0.22 mg/l.   **Other aquatic spp.** Highly toxic to crustaceans.

ENVIRONMENTAL FATE
Animals In mammals, following oral administration, 85–90% of the dose is
excreted within 24 hours, almost entirely in the urine. Complete metabolism
occurs during the passage, by oxidative dealkylation of the amide group and
hydrolysis of the phosphorus-ester bond. Metabolism and breakdown in plant,
animals and soil has been reviewed (K. I. Beynon *et al.*, *Residue Rev.*, 1973, **47**, 55).
Plants In plants, an ethyl group is split off from the amide group, and,
simultaneously or subsequently, the ester bond between the side-chain and the
phosphorus atom is hydrolytically cleaved. Dechlorination also occurs, as does
further degradation to small fragments.

MAIN ENTRIES

574 phosphine   *Insecticide, rodenticide*

$$PH_3$$

NOMENCLATURE
phosphine
IUPAC name phosphine
Chemical Abstracts name phosphine
Other names hydrogen phosphide   **CAS RN** *[7803–51–2]*   **EEC no.** 232–260–8

aluminium phosphide
Common name aluminium phosphide (E-ISO, accepted in lieu of a common
name); aluminum phosphide (JMAF, accepted in lieu of a common name)
IUPAC name aluminium phosphide
Chemical Abstracts name aluminum phosphide
CAS RN *[20859–73–8]*   **EEC no.** 244–088–0

magnesium phosphide
IUPAC name magnesium phosphide
Chemical Abstracts name magnesium phosphide
CAS RN [12057–74–8]   **EEC no.** 235–023–7

zinc phosphide
Common name zinc phosphide (E-ISO, JMAF, accepted in lieu of a common name); phosphure de zinc (F-ISO, accepted in lieu of a common name)
IUPAC name trizinc diphosphide
Chemical Abstracts name zinc phosphide
CAS RN [1314–84–7]   **EEC no.** 215–244–5

PHYSICAL CHEMISTRY
phosphine
Mol. wt. 34.0   **M.f.** H_3P **Form** Colourless, odourless, flammable gas; (tech., garlic or rotting fish odour).   **M.p.** –132.5 °C   **B.p.** –87.4 °C   **V.p.** High
Henry 33 269 Pa m^3 mol^{-1} **S.g./density** Gaseous 1.18 (air = 1)   **Solubility** In water 26 cm^3/100 ml (17 °C). In ethanol 0.5, ether 2, oil of turpentine 3.25 (all in vol. phosphine per vol. solvent, 18 °C). In cyclohexanol 285.6 cm^3/100ml (26 °C).
Stability Oxidised to phosphoric acid by oxidising agents and atmospheric oxygen.
F.p. It is spontaneously flammable in air (due to the presence of traces of other hydrides of phosphorus) with an explosion limit of 26.1–27.1 mg/l.

aluminium phosphide
Mol. wt. 58.0   **M.f.** AlP   **Form** Dark grey or yellowish crystals.   **M.p.** >1000 °C
V.p. Very low, even at 1000 °C   **S.g./density** 2.85 (25 °C)   **Stability** Though stable when dry, it reacts with moist air, violently with acids, producing phosphine.

magnesium phosphide
Mol. wt. 134.9   **M.f.** Mg_3P_2 **Form** Yellow-green crystals.   **M.p.** >750 °C
S.g./density 2.055   **Stability** Stable when dry, but reacts with atmospheric moisture, and violently with acids, producing phosphine; used to generate this fumigant, reacting more rapidly than aluminium phosphide.

zinc phosphide
Composition Tech. grade is 80–95% pure.   **Mol. wt.** 258.1   **M.f.** P_2Zn_3
Form Amorphous grey-black powder with a garlic-like odour.   **M.p.** 420 °C (when heated in the absence of oxygen)   **V.p.** Negligible in the dry state.
S.g./density 4.55   **Solubility** Practically insoluble in water (decomposes slowly). Slightly soluble in carbon disulfide and benzene. Practically insoluble in alcohols.
Stability Stable when dry, but decomposes slowly in moist air; it is decomposed violently by acids to produce phosphine, which is a potent mammalian poison, and impurities which render the gas spontaneously flammable.

COMMERCIALISATION
aluminium phosphide
History Introduced as a source of fumigant insecticide by Dr. Werner Freyberg Chemische Fabrik (now Detia Freyberg).   **Patent** GB 461997; US 2117158
Manufacturers Ag Pesticides; Aimco; Detia Freyberg; Excel; Pestcon; Shenzhen Jiangshan; United Phosphorus; Young IL

magnesium phosphide
History Introduced by Degesch AG to generate phosphine and so fumigate stored foodstuffs.   **Patent** DE 923999 to Edmund   **Manufacturers** Aimco; Detia Freyberg; United Phosphorus

zinc phosphide
History It has long been used as a poison against rodents.   **Manufacturers** Ag Pesticides; Aimco; Excel; Hacco; Lipha; Motomco; United Phosphorus

APPLICATIONS
aluminium phosphide
Mode of action Insecticide and rodenticide which is a respiratory, metabolic, and nerve poison. Evolves a non-flammable mixture of phosphine (the toxicant), ammonia and carbon dioxide.   **Uses** Fumigation control of insect and rodent pests in stored grains (wheat, rye, barley, rice, sorghum, maize, etc.), seed grains, grain products (flour, noodles, semolina, etc.), pulses (peas, beans, lentils, etc.), tobacco, tapioca (roots and flour), oil seeds, expeller cake, nuts, nut kernels, dried fruit, coffee beans, cocoa beans, tea, etc.; and in empty warehouses, silos, packing materials, transport containers, etc.   **Phytotoxicity** Living plants, fresh vegetables and fruits, with few exceptions, should not be fumigated.   **Formulation types** GE; Fumigant.   **Selected tradenames** 'Agtoxin' (Ag Pesticides); 'Al-Phos' (Aimco); 'Celphide' (Excel); 'Celphos' (Excel); 'Phostek' (Killgerm); 'Phostoxin' (Detia Degesch); 'Quickphos' (United Phosphorus); 'Shaphos' (Sanonda)

magnesium phosphide
Mode of action Insecticide and rodenticide which is a respiratory, metabolic, and nerve poison. Liberates phosphine, which is the toxicant.   **Uses** As for aluminium phosphide. Also used for control of moles, voles, rats, hamsters, and rabbits by fumigation of burrows.   **Phytotoxicity** Living plants, fresh vegetables and fruits, with few exceptions, should not be fumigated.   **Formulation types** GE; Fumigant.   **Selected tradenames** 'Magtoxin' (Detia Degesch); 'Magnaphos' (United Phosphorus)

zinc phosphide
Mode of action Rodenticide (for single ingestion). Reacts with stomach acids to liberate poisonous phosphine, which enters the bloodstream, and results in damage to the liver, kidneys and heart.   **Uses** Bait rodenticide for control of rats,

mice, voles, ground squirrels, and gophers. Also used in tracking powder form for house mouse control. **Formulation types** AB; RB; SB; TP; PA.
Selected tradenames 'Agzinphos' (Ag Pesticides); 'Commando' (Excel); 'Denkarin Grains' (Denka); 'Ratol' (United Phosphorus); 'Rattekal-Plus' (Delicia); 'Ridall-Zinc' (Lipha); 'Zawa' (Sanonda); 'Zinc-Tox' (Aimco)

ANALYSIS
phosphine
Phosphine present during fumigation can be determined using commercially available detector tubes, by glc (B. Chakrabarti & H. E. Wainman, *Chem. Ind. (London)*, 1972, p. 300) or by aspiration through aqueous mercuric chloride and measuring the change in electrical conductivity (A. H. Harris, *Proc. GASGA Tech. Seminar*, TDRI, Slough, 1986). Methods for **residues** in foods reviewed, with details by J. L. Daft in *Comp. Anal. Profiles*, p. 274.

aluminium phosphide
Product and **residue** analysis depend upon determining the phosphine liberated by acid treatment. Measurement is by glc (B. Berck et al., *J. Agric. Food Chem.*, 1970, **18**, 143; T. Dumas, *J. Assoc. Off. Anal. Chem.*, 1978, **61**, 51; *Anal. Methods Residues Pestic.*, 1988, Part II, M8; K. A. Scudamore, *Anal. Methods Pestic. Plant Growth Regul.*, 1988, **16**, 251; K. A. Scudamore & G. Goodship, *Pestic. Sci.*, 1986, **37**, 385).

magnesium phosphide
Product analysis by determining the phosphine, liberated on treatment with acid, by glc (T. Dumas, *J. Assoc. Off. Anal. Chem.*, 1978, **61**, 51) or phosphate produced after reaction with bromine water (R. B. Bruce et al., *J. Agric. Food Chem.*, 1962, **10**, 18).

zinc phosphide
Product analysis by reaction with acid, the phosphine produced is estimated by titration (*CIPAC Handbook*, 1970, **1**, 703), or oxidised to phosphoric acid which is estimated by standard methods (J. W. Elmore & F. R. Roth, *J. Assoc. Off. Agric. Chem.*, 1943, **26**, 559; 1947, **30**, 213; B. L. Griswold et al., *Anal. Chem.*, 1951, **23**, 192).

MAMMALIAN TOXICOLOGY
phosphine
Reviews *Pesticide residues in food.* FAO Agricultural Studies, No. 73; WHO Technical Report Series, No. 370, 1967. *Evaluation of some pesticide residues in food.* FAO/PL:CP/15; WHO/Food Add./67.32, 1967. **EHC** 73 (WHO, 1988; a review of phosphine and metal phosphides). **Skin and eye** Acute percutaneous LD_{50} not available; no absorption through the skin. **Inhalation** Powerful respiratory poison. LC_{50} (4 h) for rats 11 ppm (0.015 mg/l) (R. S. Waritz & R. M. Brown, *Am. Ind. Assoc. J.*, **36**, 452–458 (1975)). Inhalation at 10 mg/m^3 can cause

death within 6 hours, and at 300 ml gas/m³ for one hour there is danger to life. No symptoms of chronic poisoning are observed. **ADI** Not necessary on basis of no residue in food [1966]. **Other** Phosphine is a potent, acute mammalian poison, but feeding trials with fumigated foodstuffs have shown no chronic effects on rats (U. Hackenberg, *Toxicol. Appl. Pharmacol.*, 1972, **23**, 147). **Toxicity class** EPA (formulation) I

aluminium phosphide
Oral Acute oral LD_{50} for rats 8.7 mg/kg. **EC risk** F (R15/29); T+ (R28); (R32)

magnesium phosphide
Oral Acute oral LD_{50} for rats 11.2 mg/kg. **EC risk** F (R15/29); T+ (R28)

zinc phosphide
Oral Acute oral LD_{50} for rats 45.7, sheep 60–70 mg/kg (M. A. Nekrasova, *Sb. Rab., Leningr. Vet. Inst.*, 1964, No. 25, 372). **Skin and eye** Acute percutaneous LD_{50} for rabbits 2000–5000 mg/kg. Non-irritating to skin and eyes. **Toxicity class** WHO (a.i.) Ib; EPA (formulation) I **EC risk** F (R15/29); T+ (R28); (R32)

ECOTOXICOLOGY
phosphine
Fish LC_{50} (96 h) for rainbow trout 9.7×10^{-3} ppm. **Daphnia** EC_{50} (24 h) 0.2 mg/l.

zinc phosphide
Birds Acute oral LD_{50} for mallard ducks 37.5, bobwhite quail 13.5, pheasants 9 mg/kg (D. W Hayne, *Mich. Agric. Exp. Stn., Q. Bull.*, 1951, No. 33, 412); for fowls, the lethal dose is 7–17 mg/kg (G. D. Shearer, *J. Comp. Pathol. Therap.*, 1945, **55**, 301). **Fish** Acute LC_{50} for bluegill sunfish 0.8, rainbow trout 0.5 mg/l.

ENVIRONMENTAL FATE
Animals In mammals, phosphine is probably metabolised to non-toxic phosphates. **Plants** In stored products, phosphine undergoes oxidation to phosphoric acid.

575 phoxim

Insecticide

organophosphorus

$$\text{C}_6\text{H}_5 - \overset{\underset{|}{\text{CN}}}{\text{C}} = \text{NO}\overset{\underset{\|}{\text{S}}}{\text{P}}(\text{OCH}_2\text{CH}_3)_2$$

NOMENCLATURE

Common name phoxim (BSI, E-ISO, ESA); phoxime ((*f*) F-ISO)
IUPAC name *O,O*-diethyl α-cyanobenzylideneamino-oxyphosphonothioate;
2-(diethoxyphosphinothioyloxyimino)-2-phenylacetonitrile
Chemical Abstracts name α-[[(diethoxyphosphinothioyl)oxy]imino]benzene-=
acetonitrile
CAS RN *[14816–18–3]* **EEC no.** 238–887–3 **Development codes** BAY 5621;
Bayer 77 488; BAY SRA 7502 **Official codes** OMS 1170

PHYSICAL CHEMISTRY

Mol. wt. 298.3 **M.f.** $C_{12}H_{15}N_2O_3PS$ **Form** Yellow liquid; (tech., reddish-brown
oil). **M.p.** 6.1 °C **B.p.** Decomposes on distillation. **V.p.** 2.1 mPa (20 °C)
K_{ow} logP = 3.38 **S.g./density** 1.178 (20 °C) **Solubility** In water 1.5 mg/l (20 °C).
In toluene, *n*-hexane, dichloromethane, isopropanol >200 g/l. Slightly soluble in
aliphatic hydrocarbons, vegetable and mineral oils. **Stability** Relatively slowly
hydrolysed; DT_{50} (estimated) 26.7 d (pH 4), 7.2 d (pH 7), 3.1 d (pH 9) (22 °C).
Gradually decomposed under u.v. irradiation.

COMMERCIALISATION

History Insecticide reported by A. Wybou & I. Hammann (*Meded. Rijksfac.
Landbouwwet. Gent*, 1968, **33**, 817). Introduced by Bayer AG. **Patent** BE 678139;
DE 1238902 **Manufacturers** Bayer

APPLICATIONS

Biochemistry Cholinesterase inhibitor. **Mode of action** Non-systemic insecticide
with contact and stomach action. Short duration of activity. **Uses** Control of
stored-product insects in granaries, mills, silos, ships, etc.; ants and other insects in
households and in public health; caterpillars (mainly *Spodoptera* spp.) and soil
insects in maize, vegetables, potatoes, beet and cereals at 5 kg/ha; also migratory
locusts. **Phytotoxicity** May exhibit some phytotoxicity to cotton.
Formulation types EC; KN; GR; DP; Seed treatment; UL; WP. **Mixtures** *(phoxim +)*
cyfluthrin; isofenphos; propoxur; carbofuran. **Compatibility** Incompatible with
alkaline materials. **Selected tradenames** 'Baythion' (public health) (Bayer);
'Volaton' (agricultural use) (Bayer)

ANALYSIS

Product analysis by hplc (*CIPAC Handbook*, 1985, **1C**, 2187), details available from

Bayer AG. **Residues** determined by glc (*Man. Pestic. Residue Anal.*, 1987, **I**, 6, S19; *Anal. Methods Residues Pestic.*, 1988, Part I, M2, M12; A. Ambrus et. al., *J. Assoc. Off. Anal. Chem.*, 1981, **64**, 733). Methods for the determination of residues are available from Bayer.

MAMMALIAN TOXICOLOGY
Reviews *Pesticide residues in food – 1984*. FAO Plant Production and Protection Paper 62, 1985. *Pesticide residues in food – 1984 evaluations*. FAO Plant Production and Protection Paper 67, 1985. **Oral** Acute oral LD_{50} for rats >2000 mg/kg. **Skin and eye** Acute percutaneous LD_{50} for rats >5000 μl/kg. Not irritating to eyes or skin (rabbits). **Inhalation** LC_{50} (4 h) for rats >4.0 mg/l air (aerosol). **NOEL** (2 y) for rats 15, mice 1 mg/kg diet; (1 y) for male dogs 0.3, female dogs 0.1 mg/kg diet. **ADI** (JMPR) 0.001 mg/kg b.w. [1984]. **Toxicity class** WHO (a.i.) II; EPA (formulation) III **EC risk** Xn (R22)

ECOTOXICOLOGY
Birds LD_{50} for hens 40 mg/kg. **Fish** LC_{50} (96 h) for rainbow trout 0.53, bluegill sunfish 0.22 mg/l. **Bees** Toxic to bees by contact and respiratory action. **Daphnia** LC_{50} (48 h) 0.00081 mg/l (80% premix).

ENVIRONMENTAL FATE
Animals Rapidly metabolised in the mouse to diethylphosphoric acid and desethyl phoxim. There is also unusually rapid metabolism of the oxon, and the nitrile group is also metabolised to phoxim carboxylic acid. Elimination is very quick; almost 97% is excreted within 24 h in the urine and faeces. **Plants** In cotton, photochemical degradation and metabolism involve isomerisation to *O,O*-diethyl *S*-α-cyanobenzylideneaminothiophosphonoate and tetraethyl diphosphate. **Soil/Environment** In soil, photochemical isomerisation occurs to diethoxyphosphorylthioiminophenylacetonitrile. Tetraethyl diphosphate and tetraethyl phosphorodithioate are further metabolites. Degradation in soil is very rapid. For further details, see G. Draeger (*Pflanzenschutz-Nachr. Bayer*, 1977, **30**, 28).

NOMENCLATURE
Common name phthalide (JMAF); fthalide (JMAF, alternative spelling)
IUPAC name 4,5,6,7-tetrachlorophthalide
Chemical Abstracts name 4,5,6,7-tetrachloro-1(3*H*)-isobenzofuranone
Other names TCP   **CAS RN** *[27355–22–2]*   **Development codes** KF-32 (Kureha);
Bayer 96 610

PHYSICAL CHEMISTRY
Mol. wt. 271.9   **M.f.** $C_8H_2Cl_4O_2$ **Form** Colourless crystals.   **M.p.** 209–210 °C
V.p. 3×10^{-3} mPa (23 °C)   **K_{ow}** logP = 3.01   **Solubility** In water at 25 °C,
2.5 mg/l. In acetone 8.3, benzene 16.8, dioxane 14.1, ethyl alcohol 1.1,
tetrahydrofuran 19.3 (all in g/l, 25 °C).   **Stability** Stable for 12 h at pH 2 (2.5 ppm
aq. solution); in weak alkali DT_{50} c. 10 d (pH 6.8, 5–10 °C, 2.0 ppm aq. solution);
15% ring opening in 12 h (pH 10, 25 °C, 2.5 ppm aq. solution). Stable to heat and
light.

COMMERCIALISATION
History Fungicide reported by K. Nambu (*Jpn. Pestic. Inf.*, 1972, No. 10, p. 73) and
by K. Wagner & H. Scheinflug (*Pflanzenchutz-Nachr. (Engl. Ed.)*, 1975, **28**, 210).
Introduced by Kureha Chemical Co., Ltd and manufactured by a process licensed
from Bayer AG.   **Patent** JP 575584 to Kureha; DE 1643347 to Bayer
Manufacturers Kureha

APPLICATIONS
Biochemistry Anti-penetrant action with melanin biosynthesis inhibition (reduction
of 1,3,8-trihydroxynaphthalene).   **Mode of action** Foliar fungicide with protective
action.   **Uses** Control of rice blast (*Pyricularia oryzae*).   **Formulation types** DP;
WP; SC.   **Mixtures** *(phthalide +)* ferimzone.   **Compatibility** Incompatible with
pesticides which are strongly alkaline.   **Selected tradenames** 'Rabcide' (Kureha);
'Blasin' (mixture) (Takeda); 'Kasurabcide' (Hokko)

ANALYSIS
Product analysis is by glc with TCD.   **Residues** may be determined by glc with
ECD (H. Nagayoshi *et al.*, *Bull. Agric. Chem. Inspect. Stn.*, 1973, **13**, 27). Details
available from Kureha Chemical Industry Co., Ltd.

MAMMALIAN TOXICOLOGY

Reviews *J. Pestic. Sci.*, **15** (2), 311–314 (1990).   **Oral** Acute oral LD_{50} for rats and mice >10 000 mg/kg.   **Skin and eye** Acute percutaneous LD_{50} for rats and mice >10 000 mg/kg; non-irritating to eyes and shaved skin of rabbits.
Inhalation LC_{50} (4 h) for rats >4.1 g/m^3.   **NOEL** (2 y) for rats 2000, mice 100 mg/kg diet (M. Ishida & K. Nambu, *J. Pestic. Sci.*, **3**, 10–26).   **Other** Acute i.p. LD_{50} for male rats 9780, female rats 15 000, mice 10 000 mg/kg.
Toxicity class WHO (a.i.) III (Table 5); EPA (formulation) IV

ECOTOXICOLOGY

Birds No effect on hens fed 1.5 mg/kg for 7 days, and for another 3 days at 15 mg/kg.   **Fish** LC_{50} (48 h) for young carp >320 mg a.i. (as tech. or DP)/l, 135 mg a.i. (as 50 % WP)/l.   **Bees** Non-toxic to bees; LD_{50} (contact) >0.4 mg/bee.   **Daphnia** LC_{50} (3 h) >40 ppm.

ENVIRONMENTAL FATE

Animals In rats, principal metabolites are 2-hydroxymethyl-3,4,5,6-tetrachloro= benzoic acid and its oxidation products.   **Plants** 4,7-Dichlorophthalide and 4,6,7-trichlorophthalide are formed in rice.   **Soil/Environment** Principal metabolites in soil are 2-hydroxymethyl-3,4,5,6-tetrachlorobenzoic acid and its oxidation products (K. Aoki et al., *Jpn. Pestic. Inf.*, 1979, **36**, 32).

MAIN ENTRIES

577 *Phytoseiulus persimilis*　　　　　*Biological agent*

Acari

Spider mite-consuming mite.

NOMENCLATURE

Scientific name *Phytoseiulus persimilis* Athias-Henriot   **Other names** *Phytoseiulus riegeli* Dosse (Chant 1959); *Amblyseius tardi* Lambardini (Kennett and Caltagirone 1968)

PROPERTIES

Stability The product should be used within 18 h of receipt.

COMMERCIALISATION

Production It is cultured on *Tetranychus urticae* feeding on bean plants (*Phaseolus vulgaris*).   **History** P. persimilis was identified as an important predator of spider mites by Dosse in 1960. GCRI in the UK prepared a critical series of experiments which demonstrated the efficacy of this predator on a commercial crop. The mite is now widely used on crops in many countries. **Manufacturers** Applied Bio-nomics; Arbico; BCP; Bio Protection; Biobest; English Woodlands; Hawkesbury; Koppert; Nature's Alternative; Neudorff; Novartis BCM; Rincon-Vitova; Sautter & Stepper

APPLICATIONS
Mode of action All active stages of *P. persimilis*, except the larva, feed, and all stages of the host, including the egg, are attacked. Speed of development depends on the temperature and humidity of its environment. Development time, egg to egg, ranges from 25 days at 15 °C to 5 days at 30 °C. The optimum range for reproduction is 17 to 28 °C. The range at which predation occurs is between 17 to 30 °C, with an optimum consumption at 26 °C. The critical relative humidity levels are 50 and 85 per cent. **Uses** *P. persimilis* is used to control *Tetranychus* species in most protected vegetable crops and in protected ornamentals, on a global scale. In field crops, it is mainly used on strawberries, raspberries and beans, with some applications on vines and melons. Because it has an intrinsic rate of increase about twice that of its host, it is capable of controlling populations well in excess of its own. However, general recommendations are to apply from one up to four per plant when the spider mite population is still at a low level. Since it is specific to spider mites, it can be employed with complete safety in multiple agent biological control programmes. **Phytotoxicity** Non-phytotoxic and non-phytopathogenic. **Formulation types** Mixed in vermiculite or on leaf pieces. **Selected tradenames** 'Phyto-line p' (Novartis BCM); 'Spidex', 'Spidex-T', 'Spidex-CPR' (Koppert)

MAMMALIAN TOXICOLOGY
Allergic reactions, in the form of rhinitis and skin irritation, can occur in production staff, if not protected. Cases of allergy in workers applying the predator to the crop have not been reported. There is no evidence of any form of hypersensitivity in other mammals.

picloram *Herbicide*

pyridinecarboxylic acid

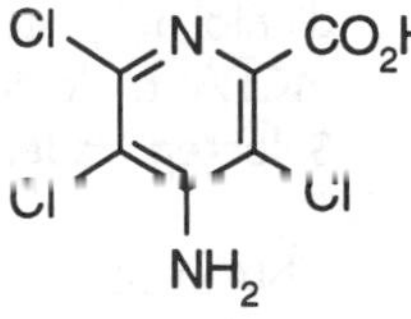

NOMENCLATURE
picloram
Common name picloram (BSI, E-ISO, ANSI, WSSA, JMAF); piclorame ((*m*) F-ISO)
IUPAC name 4-amino-3,5,6-trichloropyridine-2-carboxylic acid;
4-amino-3,5,6-trichloropicolinic acid
Chemical Abstracts name 4-amino-3,5,6-trichloro-2-pyridinecarboxylic acid
CAS RN *[1918–02–1]*

picloram-potassium
CAS RN *[2545–60–0]* potassium salt (1:1); *[11562–68–2]* potassium salt
(unstated composition)

PHYSICAL CHEMISTRY
picloram
Mol. wt. 241.5 **M.f.** $C_6H_3Cl_3N_2O_2$ **Form** Colourless powder with a chlorine-like
odour. **M.p.** Decomposes at *c.* 215 °C without melting **V.p.** 0.082 mPa (35 °C)
Solubility In water 430 mg/l (25 °C). In acetone 19.8, ethanol 10.5, isopropanol
5.5, acetonitrile 1.6, diethyl ether 1.2, dichloromethane 0.6, benzene 0.2, carbon
disulfide <0.05 (all in g/l, 25 °C). **Stability** Very stable to acids and alkalis, but
decomposed by hot concentrated alkalis. Readily forms water-soluble alkali-metal
and amine salts. In aqueous solution, decomposed by u.v. irradiation, DT_{50} 2.6 d
(25 °C). **pKa** 2.3 (22 °C)

picloram-potassium
Mol. wt. 279.6 **M.f.** $C_6H_2Cl_3KN_2O_2$ **Solubility** In water 400 g/l (25 °C).

COMMERCIALISATION
History Herbicide reported by E. R. Laning (*Down Earth,* 1963, **19,** 3). Introduced
by Dow Chemical Co. (now DowElanco). **Patent** US 3285925
Manufacturers DowElanco

APPLICATIONS
Mode of action Selective systemic herbicide, absorbed rapidly by the roots and
leaves, and translocated both acropetally and basipetally, accumulating in new
growth. **Uses** Control of most annual and perennial broad-leaved weeds (except

MAIN ENTRIES

crucifers), including woody weeds, bracken, ferns, docks, etc. on grassland and non-crop areas. **Phytotoxicity** Phytotoxic to most broad-leaved crops (except crucifers). Non-phytotoxic to established grasses, but seedling grasses may be susceptible. **Formulation types** SL. **Mixtures** *(picloram +)* bromacil; 2,4-D; dichlorprop; atrazine + diuron; bromacil + diuron; 2,4-D + dichlorprop; 2,4-D + MCPA; triclopyr; amitrole + atrazine + simazine; 2,4-D + MCPA + mecoprop. **Selected tradenames** 'Tordon' (potassium salt) (DowElanco)

ANALYSIS
Product analysis by hplc (*CIPAC Handbook*, 1983, **1B,** 1893; *AOAC Methods*, 1995, 976.03). **Residues** determined by glc of derivatives (J. R. Ramsey, *Anal. Methods Pestic., Plant Growth Regul. Food Addit.*, 1967, **5,** 507; *Anal. Methods Pestic. Plant Growth Regul.*, 1972, **6,** 700) or by pulse polarography (J. L. Whitaker & J. Osteryoung, *J. Agric. Food Chem.*, 1980, **28,** 89). In **drinking water**, by conversion to methyl ester with diazomethane, then gc with ECD (*AOAC Methods*, 1995, 992.32, 10.7.03). Details available from DowElanco.

MAMMALIAN TOXICOLOGY
picloram
Oral Acute oral LD_{50} for male rats >5000, mice 2000–4000, rabbits c. 2000, guinea pigs c. 3000, sheep >1000, cattle >750 mg/kg. **Skin and eye** Acute percutaneous LD_{50} for rabbits >2000 mg/kg. Moderate eye irritant; mild skin irritant (rabbits). No skin sensitisation. **NOEL** (2 y) for rats 20 mg/kg daily. **ADI** 0.2 mg/kg. **Toxicity class** WHO (a.i.) III (Table 5); EPA (formulation) IV

picloram-potassium
Oral Acute oral LD_{50} for male rats >5000 mg/kg. **Skin and eye** Acute percutaneous LD_{50} for rabbits >2000 mg/kg. Moderate eye irritant, not a skin irritant (rabbits). Positive skin sensitiser. **Inhalation** LC_{50} for rats >1.63 mg/l.

ECOTOXICOLOGY
picloram
Birds Acute oral LD_{50} for chicks c. 6000 mg/kg. LC_{50} for mallard ducks, bobwhite quail >10 000 mg/kg diet. **Fish** LC_{50} (96 h) for bluegill sunfish 19.4, fathead minnow 55.3 mg/l. **Bees** Not toxic to bees; LC_{50} >1000 mg/kg. **Daphnia** LC_{50} (48 h) 50.7 mg/l. **Algae** EC_{50} for *Selenastrum* 36.9 mg/l.

picloram-potassium
Birds Acute oral LD_{50} for mallard ducks and bobwhite quail >10 000 mg/kg diet. **Fish** LC_{50} (96 h) for bluegill sunfish 24 mg/l. **Bees** LD_{50} >100 µg/bee. **Daphnia** LC_{50} 68.3 mg/l. **Algae** EC_{25} for *Selenastrum* 52.6 mg/l.

ENVIRONMENTAL FATE
Animals In mammals, following oral administration, picloram is rapidly excreted in

an unchanged form.   **Plants** On plant surfaces, photodecomposition occurs, possibly with cleavage of the pyridine ring.   **Soil/Environment** Degraded by light, more rapidly on the soil surface or in clear, moving water. Degraded slowly by soil micro-organisms, DT_{50} 30–90 d.

pindone   *Rodenticide*

indandione anticoagulant

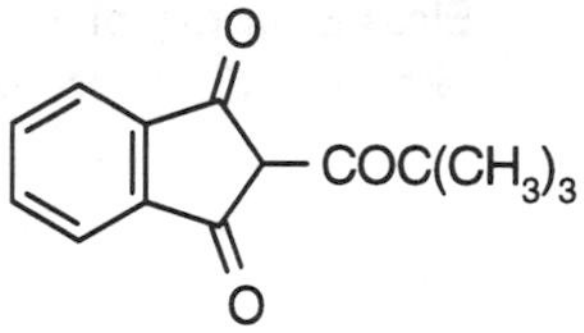

NOMENCLATURE
Common name pindone (BSI, E-ISO, (*f*) F-ISO); pivaldione (France); [pival] (former exception, Turkey); no name (Portugal)
IUPAC name 2-pivaloylindan-1,3-dione
Chemical Abstracts name 2-(2,2-dimethyl-1-oxopropyl)-1*H*-indene-1,3(2*H*)-dione
CAS RN *[83–26–1]*   **EEC no.** 201–462–8

PHYSICAL CHEMISTRY
Mol. wt. 230.3   **M.f.** $C_{14}H_{14}O_3$ **Form** Yellow crystals.   **M.p.** 108.5–110.5 °C
V.p. Very low   **Solubility** In water 18 mg/l (25 °C). Soluble in most organic solvents. Readily soluble in aqueous alkalis or ammonia to give bright yellow salts.
Stability Very stable.

COMMERCIALISATION
History Insecticidal properties reported by L. B. Kilgore *et al.* (*Ind. Eng. Chem.*, 1942, **34**, 494). Introduced by Kilgore Chemical Co.   **Patent** US 2310949
Manufacturers Motomco

APPLICATIONS
Biochemistry Anticoagulant; inhibits blood coagulation by blocking prothrombin formation.   **Uses** Control of rats and mice in baits containing 250 mg/kg. Bait shyness does not occur after feeding. Renders cereal bait resistant to insect infestation and fungal infection.   **Formulation types** CB; AB; TP; RB.
Compatibility Compatible with other rodenticides.   **Selected tradenames** 'Pival' (Motomco, Kilgore); 'Pivalyn' (sodium salt) (Motomco, Kilgore)

ANALYSIS
Analysis by colorimetry (J. B. La Clair, *J. Assoc. Off. Agric. Chem.*, 1955, **38**, 299).

MAMMALIAN TOXICOLOGY
Oral Acute oral LD_{50} for rats 280, rabbits 150–170, dogs 75–100 mg/kg. Chronic oral LD_{50} for rabbits 0.52, dogs 2.5, sheep >12 mg/kg/day. **Other** Acute LD_{50} by injection for rats is *c.* 50 mg/kg, but it is more toxic when given in small daily doses of 15–35 mg/kg. Dogs are killed by daily doses of 2.5 mg/kg (J. R. Beauregard *et al.*, *J. Agric. Food Chem.*, 1955, **3**, 124; J. P. Saunders *et al.*, *ibid.*, p. 762). **Toxicity class** WHO (a.i.) II **EC risk** T (R25, R48/25)

ECOTOXICOLOGY
Birds Eight-day dietary LC_{50} for mallard ducks 250, bobwhite quail 1560 mg/kg. **Fish** LC_{50} (96 h) for bluegill sunfish 1.6, rainbow trout 0.21 mg/l.

580 piperalin   *Fungicide*

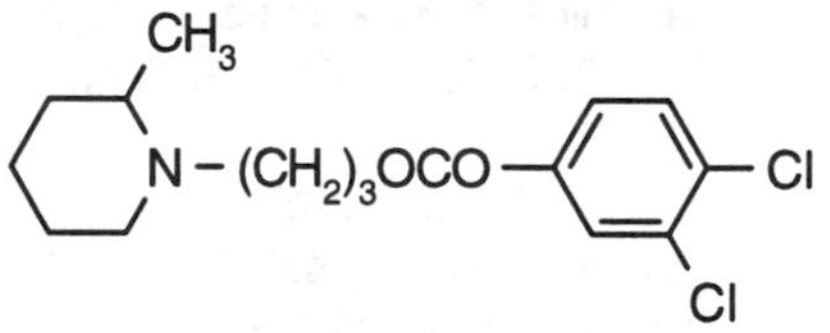

NOMENCLATURE
Chemical Abstracts name 3-(2-methylpiperidino)propyl 3,4-dichlorobenzoate
Other names piperalin **CAS RN** *[3478–94–2]*

PHYSICAL CHEMISTRY
Mol. wt. 325.2 **M.f.** $C_{16}H_{16}Cl_2NO_2$ **Form** Viscous amber liquid.
B.p. 156–157 °C/20 mmHg **V.p.** 3.99×10^5 mPa (25 °C) K_{ow} logP = 4.31
(pH 5, 7, 9; 21 °C) **Solubility** Emulsifies in water. **F.p.** 71.1 °C

COMMERCIALISATION
Manufacturers SePRO

APPLICATIONS
Uses Control of powdery mildew on roses, lilacs, dahlias, zinnia, chrysanthemums, phlox, and certain other ornamentals in commercial greenhouses.
Phytotoxicity Used as directed, piperalin will not harm foliage or floral tissue.

Formulation types SC. **Compatibility** Compatible with the surfactants 'Triton B-1956', 'No-Foam B', and 'Stepanol WA Paste'. **Selected tradenames** 'Pipron' (SePRO)

MAMMALIAN TOXICOLOGY
Oral Acute oral LD_{50} for rats 2500 mg/kg. **Skin and eye** Acute percutaneous LD_{50} not available. Irritating to eyes and skin (rabbits). Non-sensitising to skin (guinea pigs). **NOEL** No adverse effects; not carcinogenic. **Other** Not mutagenic or teratogenic.

581 piperonyl butoxide *Insecticide synergist*

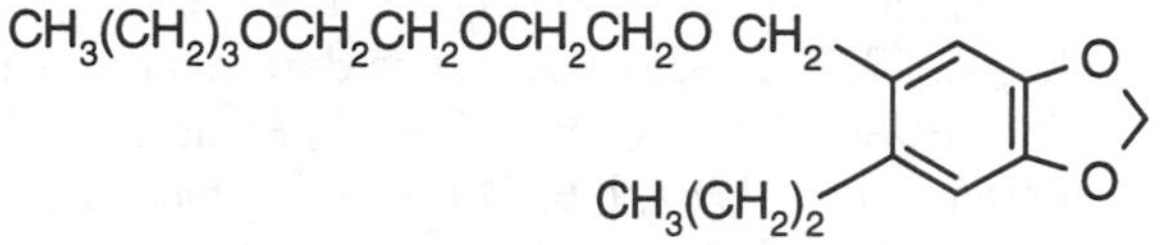

$$CH_3(CH_2)_3OCH_2CH_2OCH_2CH_2O\ CH_2$$
$$CH_3(CH_2)_2$$

NOMENCLATURE
Common name piperonyl butoxide (BAN; accepted in lieu of a common name by BSI, E-ISO, ESA); piperonyl butoxyde (F-ISO)
IUPAC name 5-[2-(2-butoxyethoxy)ethoxymethyl]-6-propyl-1,3-benzodioxole; 2-(2-butoxyethoxy)ethyl 6-propylpiperonyl ether
Chemical Abstracts name 5-[[2-(2-butoxyethoxy)ethoxy]methyl]-6-propyl-= 1,3-benzodioxole
Other names PBO **CAS RN** *[51–03–6]* **Official codes** ENT 14 250

PHYSICAL CHEMISTRY
Composition 90% min. piperonyl butoxide. **Mol. wt.** 338.4 **M.f.** $C_{19}H_{30}O_5$
Form Colourless liquid; (tech. is a yellow oil). **B.p.** 180 °C/1 mmHg (tech.)
V.p. 2.0×10^{-2} mPa (60 °C, gas saturation method) K_{ow} logP = 4.75
Henry <2.3×10^{-6} Pa m^3 mol^{-1} (calc.) **S.g./density** 1.060 (20 °C, ENDQC-4 method) **Solubility** In water 14.3 mg/l (25 °C). Soluble in all common organic solvents including mineral oils and fluorinated aliphatic hydrocarbons (aerosol propellants). **Stability** Essentially stable to hydrolysis at pH 5, 7 and 9 in sterile buffers in the dark at 25 °C. Rapidly degraded in aqueous solution (pH 7) in sunlight (DT_{50} 8.4 h). **F.p.** 140 °C (ASTM D93) **Other properties** Viscosity 40 cP (25 °C)

COMMERCIALISATION
History Activity as a synergist for pyrethrins described by H. Wachs (*Science*, 1947, **105**, 530). **Patent** US 2485681; US 2550737 **Manufacturers** Endura; Prentiss

APPLICATIONS
Biochemistry Inhibits mixed function oxidases (MFO) of the insect so that its natural detoxification system is blocked and the efficacy of applied insecticides is increased. **Uses** A synergist for the pyrethrins and related insecticides. **Formulation types** Aerosol; Emulsion; Oil. **Mixtures** *(piperonyl butoxide +)* pyrethrins; resmethrin; tetramethrin; allethrin; pyrethrins + resmethrin; dichlorvos + pyrethrins; pirimiphos-methyl + pyrethrins; phenothrin; bioallethrin + permethrin; bendiocarb + pyrethrins; deltamethrin; fenitrothion + tetramethrin. **Selected tradenames** 'Butacide' (AgrEvo); 'Prentox' (Prentiss)

ANALYSIS
Product analysis by glc (*CIPAC Handbook*, 1988, **1C**, 2190, 2209; *Br. Pharmacopoeia Vet.*, 1977, p. 64; *AOAC Methods*, 1995, 982.02) or by colorimetry of a derivative (*ibid.*, 960.11*; *CIPAC Handbook*, 1980, **1A**, 1325; *Anal. Methods Pestic. Plant Growth Regul.*, 1972, **6**, 458). **Residues** determined by colorimetry of a derivative (*idem, ibid.*; *AOAC Methods*, 1995, 960.43; *Pestic. Anal. Man.*, 1979, **II**; *Man. Pestic. Residue Anal.*, 1987, **I**, 6, S19; *Anal. Methods Residues Pestic.*, 1988, Part II; R. T. Krause & E. M. August, *J. Assoc. Off. Anal. Chem.*, 1983, **66**, 234, 1018).

MAMMALIAN TOXICOLOGY
Reviews *Pesticide residues in food – 1995*, FAO Plant Production and Protection Paper 133, 1996. *Pesticide residues in food – 1995 evaluations; Part II – Toxicology & Environment*. World Health Organisation, WHO/PCS/96.48, 1996. **IARC** 30 **Oral** Acute oral LD_{50} for rats and rabbits c. 7500 mg/kg. **Skin and eye** Acute percutaneous LD_{50} for rats >7950, rabbits 1880 mg/kg. Not irritant to eyes or skin; not a skin sensitiser. **Inhalation** LC_{50} for rats >5.9 mg/l. **NOEL** (2 y) for mice and rats 30; (1 y) for dogs 16 mg/kg b.w. daily. **ADI** (JMPR) 0.2 mg/kg b.w. [1995]. **Other** Not teratogenic, mutagenic or carcinogenic. **Toxicity class** WHO (a.i.) III (Table 5); EPA (formulation) IV

ECOTOXICOLOGY
Birds Acute oral LD_{50} for bobwhite quail >2250 mg/kg. **Fish** LC_{50} (24 h) for carp 5.3 mg/l. **Bees** LD_{50} >25 µg/bee. **Daphnia** LC_{50} (24 h) 2.95 mg/l. **Algae** EC_{50} (cell volume) for *Chlorella fusca* 44 µmol/l (*Pestic. Sci.*, **47**, 337 (1996)).

ENVIRONMENTAL FATE
Animals In mammals (and also in insects), oxidative attack on the carbon atom of the methylenedioxy group leads to the formation of the dihydroxyphenyl compound. Oxidative degradation of the side-chain also occurs. Elimination is as

the glucoside or amino acid derivative. **Soil/Environment** DT_{50} for aerobic soil metabolism c. 14 d. K_{oc} 399–830. Although mobile in sandy soil, it is not expected to leach under outdoor conditions, where rapid degradation occurs. Degradation in soil or water is mainly via oxidation of the butyl side-chain to form methylenedioxypropylbenzyl alcohol followed by the corresponding aldehyde, ultimately with mineralisation to CO_2; there is no accumulation of metabolites.

582 piperophos *Herbicide*

organophosphorus herbicide

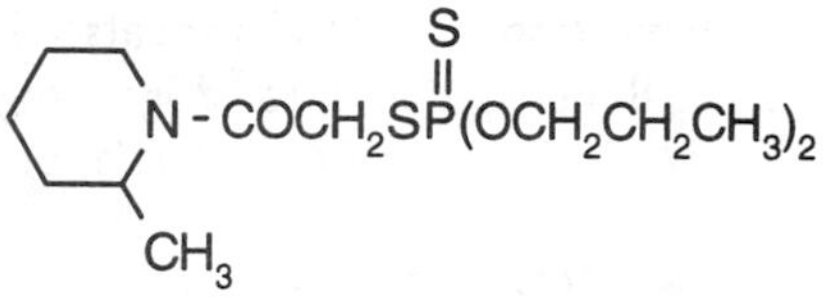

NOMENCLATURE
Common name piperophos (BSI, E-ISO, (m) F-ISO)
IUPAC name S-2-methylpiperidinocarbonylmethyl O,O-dipropyl phosphorodithioate
Chemical Abstracts name S-[2-(2-methyl-1-piperidinyl)-2-oxoethyl] O,O-dipropyl phosphorodithioate
CAS RN [24151–93–7] **Development codes** C 19 490 (Ciba)

PHYSICAL CHEMISTRY
Mol. wt. 353.5 **M.f.** $C_{14}H_{28}NO_3PS_2$ **Form** Pale yellow, slightly viscous clear liquid with a somewhat sweet odour. **B.p.** >250 °C; thermal decomposition begins before c. 190 °C **V.p.** 0.032 mPa (20 °C) K_{ow} logP = 4.3 **Henry** 4.5×10^{-4} Pa m^3 mol^{-1} (calc.) **S.g./density** 1.13 (20 °C) **Solubility** In water 25 mg/l (20 °C). Miscible with benzene, hexane, acetone, dichloromethane and octanol. **Stability** Stable under normal storage conditions. Slowly hydrolysed at pH 9; DT_{50} (20 °C) (calc.) >200 d (5 ≤ pH ≤7), 178 d (pH 9).

COMMERCIALISATION
History Herbicide reported by D. H. Green & L. Ebner (*Proc. Br. Weed Control Conf., 11th*, 1972, **2**, 822). Introduced by Ciba-Geigy AG (now Novartis Crop Protection AG). **Patent** BE 725992; GB 1255946 **Manufacturers** Novartis

APPLICATIONS
Mode of action Selective, systemic herbicide, absorbed by the roots, coleoptiles,

and leaves of young plants. **Uses** A selective herbicide active against annual grasses and sedges in direct seeded or transplanted rice. Used, in combination with dimethametryn, for the control of both grass and broad-leaved weeds. In tropical regions, piperophos is applied in combination with 2,4-D or cinosulfuron to widen the spectrum against broad-leaved weeds. **Formulation types** EC; WP. **Mixtures** *(piperophos +)* dimethametryn; 2,4-D; cinosulfuron. **Selected tradenames** 'Rilof' (Novartis)

ANALYSIS
Product analysis by glc. **Residues** determined by glc with TID. Details of methods available from Novartis.

MAMMALIAN TOXICOLOGY
Oral Acute oral LD_{50} for rats 324 mg/kg. **Skin and eye** Acute percutaneous LD_{50} for rats >2150 mg/kg. Non-irritant to skin; slight irritant to eyes (rabbits). **Inhalation** LC_{50} (1 h) for rats >1.96 mg/l air. **NOEL** (90 d) for rats 10 mg/kg diet (0.8 mg/kg daily); for dogs 5 mg/kg diet (0.15 mg/kg daily). **Toxicity class** WHO (a.i.) II; EPA (formulation) II **EC risk** Xn (R22, R43)

ECOTOXICOLOGY
Birds Eight-day dietary LC_{50} for Japanese quail 11 629 ppm. **Fish** LC_{50} (96 h) for rainbow trout 6, crucian carp 5 mg/l. **Bees** LD_{50} (oral) >22 µg/bee, (contact) 30 µg/bee. **Worms** LC_{50} (14 d) for earthworms 180 mg/kg soil. **Daphnia** LC_{50} (48 h) 0.0033 mg/l. **Algae** EC_{50} (5 d) for *Scenedesmus subspicatus* 0.059 mg/l.

ENVIRONMENTAL FATE
Animals In urine, no unchanged piperophos was present, indicating extensive degradation of the compound. Degradation proceeds via hydrolysis of the thiolo phosphate followed by methylation of the sulfur or hydroxylation of the piperidine moiety. Hydroxylation at the γ-carbon leads via ring opening to carboxylic acids. The derivatives are conjugated with glucuronic acid. **Plants** Degradation of piperophos takes place rapidly in the plant, forming 2-(2'-methyl-1'-piperidinyl)-= 2-oxoethane sulfonic acid, 2-(2'-methyl-1'-piperidinyl)-2-oxoethanoic acid and a fraction of unknown polar substances. The proposed pathway involves hydrolysis to the corresponding sulfhydryl and hydroxyl derivatives followed by further oxidation to the sulfonic acid and oxalic acid derivatives. Injection experiments in plants confirmed the capacity of the rice plant to oxidise piperophos derivatives rapidly. **Soil/Environment** DT_{50} (field) <30 d; rapid dissipation by biodegradation from soil and paddy systems. Relatively stable to hydrolysis and photolysis.

pirimicarb *Insecticide*

carbamate

CH₃, N, N(CH₃)₂ (structure diagram)

$$\text{CH}_3 \quad \text{N} \quad \text{N(CH}_3)_2$$

$$\text{CH}_3 \quad \text{N}$$

$$\text{OCON(CH}_3)_2$$

MAIN ENTRIES

NOMENCLATURE
Common name pirimicarb (BSI, E-ISO, ANSI, JMAF); pirimicarbe ((m) F-ISO); pyrimicarbe (France)
IUPAC name 2-dimethylamino-5,6-dimethylpyrimidin-4-yl dimethylcarbamate
Chemical Abstracts name 2-(dimethylamino)-5,6-dimethyl-4-pyrimidinyl dimethylcarbamate
CAS RN *[23103–98–2]* **EEC no.** 245–430–1 **Development codes** PP062
Official codes ENT 27 766; OMS 1330

PHYSICAL CHEMISTRY
Composition Tech. material is 95% pure. **Mol. wt.** 238.3 **M.f.** $C_{11}H_{18}N_4O_2$
Form Colourless solid. **M.p.** 90.5 °C; (tech. 87.3–90.7 °C) **V.p.** 0.97 mPa
(25 °C) **K_{ow}** logP = 1.7 (unionised) **S.g./density** 1.21 g/ml (25 °C) **Solubility** In water 3.0 g/l (pH 7.4, 20 °C). In acetone 4.0, ethanol 2.5, xylene 2.9, chloroform 3.3 (all in g/l, 25 °C). **Stability** Stable under normal storage conditions for more than 2 years. Hydrolysed by boiling with strong acids and alkalis. Aqueous solutions are unstable to u.v. light; DT_{50} <1 d (pH 5, 7 or 9). **pKa** 4.54, weak base (K. Chamberlain et al., *Pestic Sci.*, **47**, 265 (1996))

COMMERCIALISATION
History Aphicide reported by F. L. C. Baranyovits & R. Ghosh (*Chem. Ind. (London)*, 1969, p. 1018). Introduced by ICI Plant Protection Division (now Zeneca Agrochemicals). **Patent** GB 1181657 **Manufacturers** Jiangsu; Zeneca

APPLICATIONS
Biochemistry Cholinesterase inhibitor. **Mode of action** Selective systemic insecticide with contact, stomach, and respiratory action. Absorbed by the roots, and translocated through the xylem. Penetrates the leaves, but is not translocated extensively. **Uses** A selective aphicide used in cereals, fruit, ornamentals, strawberries, potatoes, sugar beet, fodder beet, cotton, oilseed rape, tobacco, and glasshouse crops, effective against organophosphorus-resistant *Myzus persicae*.
Formulation types AE; WP; WG; FU; EC; DP. **Mixtures** (pirimicarb +) bupirimate + triforine; chlorpyrifos; lambda-cyhalothrin; endosulfan; phosalone; deltamethrin. **Selected tradenames** 'Aphox' (Zeneca); 'Pirimor' (Zeneca); 'Pilly' (Sanonda); 'Pirimisect' (Barclay)

ANALYSIS

Product analysis by glc (J. E. Bagness & W. G. Sharples, *Analyst (London)*, 1974, **99**, 225; *AOAC Methods*, 1995, 982.08; *CIPAC Proc.*, 1981, **3**, 308). Identity also by glc, tlc, i.r. or nmr (*CIPAC Handbook*, 1994, **F**, 406). **Residues** determined by glc with FID or by colorimetry (*Man. Pestic. Residue Anal.*, 1987, **I**; D. J. W. Bullock, *Anal. Methods Pestic. Plant Growth Regul.*, 1973, **7**, 399). Details available from Zeneca.

MAMMALIAN TOXICOLOGY

Reviews *Pesticide residues in food – 1982*. FAO Plant Production and Protection Paper 46, 1983. *Pesticide residues in food: 1982 evaluations*. FAO Plant Production and Protection Paper 49, 1983. **EHC** 64 (WHO, 1986; a review of carbamate insecticides in general). **Oral** Acute oral LD_{50} for female rats 147, mice 107, dogs 100–200 mg/kg. **Skin and eye** Acute percutaneous LD_{50} for rats and rabbits >500 mg/kg. Non-irritating to skin, mild irritant to eyes (rabbits). Not a skin sensitiser (guinea pigs). **Inhalation** LC_{50} (6 h) for rats 0.3 mg/l. **NOEL** (2 y) for dogs 1.8 mg/kg daily; for rats 250 mg/kg diet (12.5 mg/kg daily). Not carcinogenic; no adverse reproductive effects. **ADI** (JMPR) 0.02 mg/kg b.w. [1982]. **Toxicity class** WHO (a.i.) II; EPA (formulation) II **EC risk** T (R25)

ECOTOXICOLOGY

Birds Acute oral LD_{50} for poultry 25–50, mallard ducks 17.2, bobwhite quail 8.2 mg/kg. **Fish** LC_{50} (96 h) for rainbow trout 29, bluegill sunfish 55 mg/l. **Bees** Not toxic to bees; LD_{50} (24 h) (oral) 3.5, (contact) 51 μg/bee (tech.). **Worms** LC_{50} >60 mg/kg. **Other beneficial spp.** Not toxic to Collembola (J. A. Wiles & G. K. Frampton, *Pestic. Sci.*, **47**, 273 (1996)). **Daphnia** LC_{50} (24 h) 0.08 ppm. **Algae** NOEC (96 h) for green algae 50 mg/l.

ENVIRONMENTAL FATE

Animals In mammals, the major metabolites are 2-dimethylamino-5,6-dimethyl-=4-hydroxypyrimidine, 2-methylamino-5,6-dimethyl-4-hydroxypyrimidine, 2-amino-5,6-dimethyl-4-hydroxypyrimidine and 2-dimethylamino-6-hydroxy=methyl-5-methyl-4-hydroxypyrimidine. **Soil/Environment** For details of degradation in soil, see I. R. Hill, *ACS Symp. Series*, 1976, **29**, 358. DT_{50} in soil 7–234 d, according to soil type (range o.m. 1.7–51.9%, pH 5.5–8.1).

organophosphorus

$$\text{(CH}_3\text{CH}_2\text{)}_2\text{N} \overset{\displaystyle \text{CH}_3}{\underset{\displaystyle \text{N}}{\overset{\displaystyle}{\big|}}} \quad \overset{\displaystyle S}{\underset{\displaystyle}{\overset{\displaystyle \parallel}{\text{—OP(OCH}_2\text{CH}_3\text{)}_2}}}$$

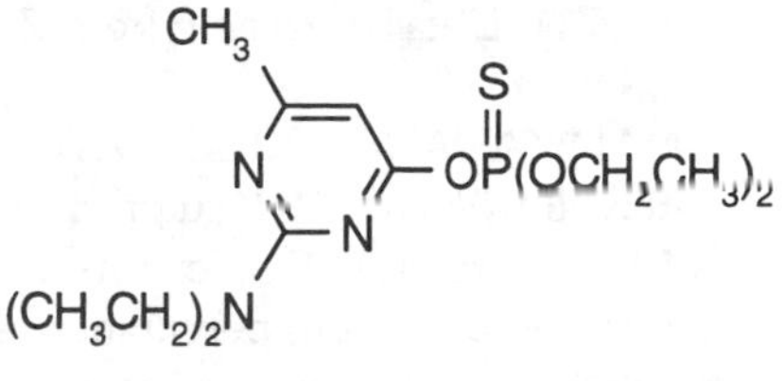

NOMENCLATURE
Common name pirimiphos-ethyl (BSI, E-ISO, ANSI, BAN, ESA); pyrimiphos-éthyl ((m) F-ISO)
IUPAC name O,O-diethyl O-2-diethylamino-6-methylpyrimidin-4-yl phosphorothioate
Chemical Abstracts name O-[2-(diethylamino)-6-methyl-4-pyrimidinyl] O,O-diethyl phosphorothioate
CAS RN [23505–41–1]   **EEC no.** 245–704–0   **Development codes** PP211

PHYSICAL CHEMISTRY
Composition Tech. material is 95% pure.   **Mol. wt.** 333.4   **M.f.** $C_{13}H_{24}N_3O_3PS$
Form Straw-coloured liquid; (tech. is a clear red-brown liquid with a mercaptan-like odour).   **M.p.** 15–18 °C (tech.)   **B.p.** Decomposes above c. 194 °C
V.p. 0.68 mPa (20 °C), 39 mPa (25 °C)   **K_{ow}** logP = 5.0 (unionised)
S.g./density 1.14 (20 °C)   **Solubility** In water 2.3 mg/l (pH 7.2). Miscible with most organic solvents.   **Stability** Stable for at least 1 year at room temperature, and for at least 5 days at 80 °C. Hydrolysed by acid and base; DT_{50} (25 °C) 52–1200 d (pH range 5.5–8.5), most stable pH 7–8.   **F.p.** >60 °C

COMMERCIALISATION
History Insecticide introduced by ICI Plant Protection Division (now Zeneca Agrochemicals).   **Patent** GB 1019227; GB 1205000   **Manufacturers** Zeneca

APPLICATIONS
Biochemistry Cholinesterase inhibitor.   **Mode of action** Insecticide with contact and respiratory action.   **Uses** Broad-spectrum control of Diptera and Coleoptera soil pests and some foliar pests on bananas and other fruit, field crops, ornamentals, turf, vegetables, etc. Also used as a seed treatment, and in compost to control sciarid and phorid flies on mushrooms.   **Phytotoxicity** Relatively non-phytotoxic, except at high rates of seed dressing.   **Formulation types** DS; EC; GR; CG.   **Compatibility** Can be used as a seed treatment in combination with fungicides.   **Selected tradenames** 'Primicid' (Zeneca)

MAIN ENTRIES

ANALYSIS
Product analysis by glc with FID (*CIPAC Handbook*, 1988, **D**, 146). Identity also by glc, tlc, i.r. or nmr (*CIPAC Handbook*, 1994, **F**, 406). **Residues** determined by glc with FPD or FTD (D. J. W. Bullock, *Anal. Methods Pestic. Plant Growth Regul.*, 1976, **8**, 171). Details available from Zeneca.

MAMMALIAN TOXICOLOGY
EHC 63 (WHO, 1986; a general review of organophosphorus insecticides).
Oral Acute oral LD_{50} for rats 140–200, guinea pigs 50–100, cats 25–50 mg/kg.
Skin and eye Acute percutaneous LD_{50} for male rats 1000–2000 mg/kg. Non-irritating to skin; may cause eye irritation (rabbits). Not a skin sensitiser (guinea pigs). **Inhalation** LC_{50} (6 h) for rats >5 ppm (no toxic effects over 3 w).
NOEL (90 d) for rats 1.6 mg/kg diet (0.08 mg/kg daily); for dogs 0.2 mg/kg daily. For rats receiving 27 mg/kg diet (1.6 mg/kg daily) and dogs 2 mg/kg daily, the only effect was on the cholinesterase levels. **Toxicity class** WHO (a.i.) Ib; EPA (formulation) II **EC risk** T (R25); Xn (R21)

ECOTOXICOLOGY
Birds Acute oral LD_{50} for mallard ducks 2.5, bobwhite quail 10–20 mg/kg.
Fish LC_{50} (96 h) for common carp 0.22, brown trout 0.02 mg/l. **Bees** Toxic to bees. **Daphnia** LC_{50} (48 h) 0.3 µg/l.

ENVIRONMENTAL FATE
Soil/Environment Half-life in soil 21–70 d (non-sterile, aerobic, o.m. 1.8–6.3%, pH 6.0–7.5). For details of persistence in soil, see D. L. Suett, *Pestic. Sci.*, 1975, **6**, 385.

585 pirimiphos-methyl *Insecticide, acaricide*

organophosphorus

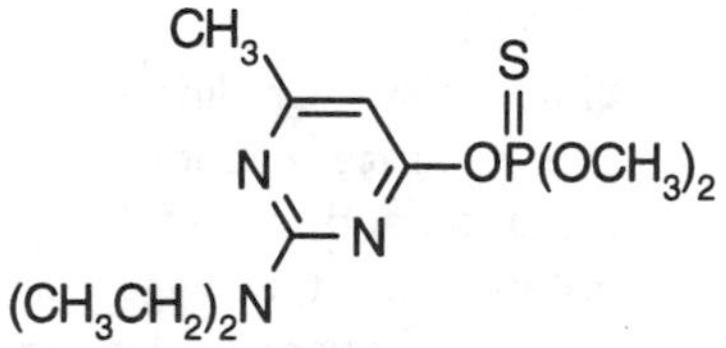

$(CH_3CH_2)_2N$

NOMENCLATURE
Common name pirimiphos-methyl (BSI, E-ISO, ANSI, ESA); pyrimiphos-méthyl ((*m*) F-ISO)
IUPAC name *O,O*-dimethyl *O*-2-diethylamino-6-methylpyrimidin-4-yl phosphorothioate

Chemical Abstracts name *O*-[2-(diethylamino)-6-methyl-4-pyrimidinyl]
O,O-dimethyl phosphorothioate
CAS RN *[29232-93-7]* **EEC no.** 249-528-5 **Development codes** PP511
Official codes OMS 1424

PHYSICAL CHEMISTRY
Composition Tech. material is 88% pure. **Mol. wt.** 305.3 **M.f.** $C_{11}H_{20}N_3O_3PS$
Form Straw-coloured liquid. **M.p.** 15–18 °C (tech.) **B.p** Decomposes on
distillation. **V p.** 2 mPa (20 °C), 15 mPa (30 °C), 23 mPa (40 °C) **K_{ow}** logP = 4.2
(20 °C, unionised) **Henry** 7.2×10^{-2} Pa m^3 mol^{-1} **S.g./density** 1.17 (20 °C);
1.157 (30 °C) **Solubility** In water 9.9 (pH 5.2), 8.6 (pH 7.3), 9.3 (pH 9.3) (all in
mg/l, 30 °C). Miscible with most organic solvents, e.g. alcohols, ketones,
halogenated hydrocarbons. **Stability** Hydrolysed by concentrated acids and alkalis;
DT_{50} 7.5–35 d (pH range 5.8–8.5, most stable at pH 7). In sunlight, aqueous
solution had DT_{50} 1 d. **pKa** 3.71, weak base (K. Chamberlain *et al.*, *Pestic. Sci.*,
47, 265 (1996)) **F.p.** >46 °C

COMMERCIALISATION
History Insecticide introduced by ICI Plant Protection Division (now Zeneca
Agrochemicals). **Patent** GB 1019227; GB 1204552 **Manufacturers** Zeneca

APPLICATIONS
Biochemistry Cholinesterase inhibitor. **Mode of action** Broad-spectrum insecticide
and acaricide with contact and respiratory action. Penetrates the leaf tissue and
exhibits translaminar action. **Uses** Control of a wide range of insects and mites in
warehouses, stored grain, animal houses, domestic and industrial premises;
chewing insects, sucking insects, boring insects, and mites on vegetables,
ornamentals, bulb flowers, sugar cane, maize, sorghum, rice, citrus and other fruit,
olives, vines, alfalfa, cereals, etc.; and glasshouse pests (especially whitefly, thrips,
mealybugs, aphids, and mites) on tomatoes, cucumbers, capsicums, aubergines,
and other glasshouse crops. **Formulation types** AE; DP; EC; FU; UL; HN; LS; SG;
KN. **Mixtures** *(pirimiphos-methyl +)* synergized pyrethrins; permethrin;
cypermethrin; dichlorvos; petroleum oils; buprofezin; piperonyl butoxide.
Compatibility Miscible with common insecticides and fungicides. Mixing with
strongly alkaline or acidic substances should be avoided.
Selected tradenames 'Actellic' (Zeneca); 'Actellifog' (Hortichem)

ANALYSIS
Product analysis by glc with FID (*AOAC Methods*, 1995, 991.34, 7.7.24; *CIPAC
Handbook*, 1985, **1C**, 2192). Identity also by glc, tlc, i.r. or nmr (ibid., 1994, **F**,
406). **Residues** determined by glc with FPD or FTD (*Analyst* (London), 1980, **105**,
515; 1985, **110**, 765; A. Ambrus *et al.*, *J. Assoc. Off. Anal. Chem.*, 1981, **64**, 733;
Manu. Pestic. Residue Anal., 1987, **I**, S8, S19; *Anal. Methods Residues Pestic.*, 1988,
Part I, M9; Part II). Details available from Zeneca.

ENTRIES

MAMMALIAN TOXICOLOGY
Reviews *Pesticide residues in food – 1992.* FAO Plant Production and Protection Paper, 116, 1993. *Pesticide residues in food – 1992 evaluations. Part II – Toxicology.* World Health Organisation, WHO/PCS/93.34. **EHC** 63 (WHO, 1986; a general review of organophosphorus insecticides). **Oral** Acute oral LD_{50} for female rats 2050, mice 1180 mg/kg. **Skin and eye** Acute percutaneous LD_{50} for female rats >4592 mg/kg. Slight eye and skin irritation (rabbits). Not a skin sensitiser (guinea pigs). **Inhalation** LC_{50} (4 h) for rats >5.04 mg/l. **NOEL** (2 y) for rats 10 mg/kg diet. No teratogenic effects, and no concentration in adipose tissue. NOEL (90 d) for dogs 20, rats 8 mg/kg diet. **ADI** (JMPR) 0.03 mg/kg b.w. [1992].
Toxicity class WHO (a.i.) III; EPA (formulation) III **EC risk** Xn (R22)

ECOTOXICOLOGY
Birds Acute oral LD_{50} for hens 30–60, bobwhite quail 140 mg/kg. **Fish** LC_{50} (96 h) for rainbow trout 0.64 mg/l; (48 h) for mirror carp 1.4 mg/l. **Bees** Toxic to bees. **Daphnia** EC_{50} (48 h) 1.4 mg/l; EC_{50} (21 d) 0.08 µg/l.

ENVIRONMENTAL FATE
Animals In mammals, the P-O bond is cleaved extensively and *N*-dealkylation and/or conjugation is a further step in the metabolism of the pyrimidine leaving group. **Plants** Rapidly evaporates. After 2–3 days, <10% remains on plants, including the degradation product *O*-2-ethylamino-6-methylpyrimidin-4-yl *O,O*-dimethyl phosphorothioate. On stored cereals, DT_{50} >2 mo.
Soil/Environment DT_{50} in soil <30 d.

antibiotic

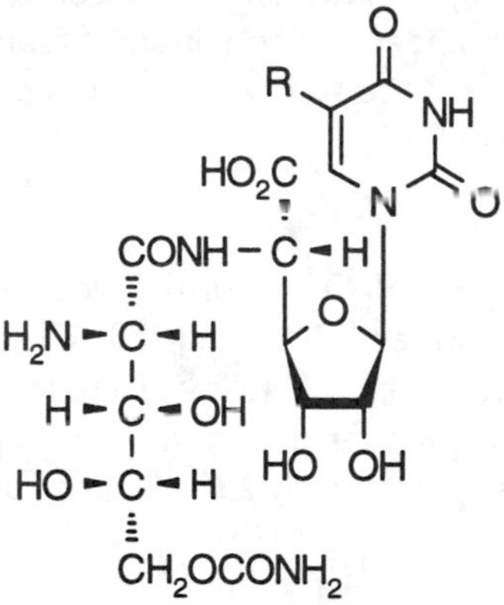

polyoxin B: R = − CH$_2$OH

polyoxorim: R = − CO$_2$H

NOMENCLATURE
polyoxins
Common name polyoxins (JMAF)
CAS RN *[11113–80–7]*

polyoxin B
IUPAC name 5-(2-amino-5-*O*-carbamoyl-2-deoxy-L-xylonamido)-1,5-dideoxy-1-=
(1,2,3,4-tetrahydro-5-hydroxymethyl-2,4-dioxopyrimidin-1-yl)-β-D-allofuranuronic
acid
Chemical Abstracts name 5-[[2-amino-5-*O*-(aminocarbonyl)-2-deoxy-L-xylonoyl]=
amino]-1,5-dideoxy-1-[3,4-dihydro-5-(hydroxymethyl)-2,4-dioxo-1(2*H*)-=
pyrimidinyl]-β-D-allofuranuronic acid
CAS RN *[19396–06–6]*

polyoxorim
Common name polyoxorim (BSI, pa ISO); polyoxin D (JMAF)
IUPAC name 5-(2-amino-5-*O*-carbamoyl-2-deoxy-L-xylonamido)-1-(5-carboxy-=
1,2,3,4-tetrahydro-2,4-dioxopyrimidin-1-yl)-1,5-dideoxy-β-D-allofuranuronic acid
Chemical Abstracts name 5-[[2-amino-5-*O*-(aminocarbonyl)-2-deoxy-L-xylonoyl]=
amino]-1-(5-carboxy-3,4-dihydro-2,4-dioxo-1-(2*H*)-pyrimidinyl)-1,5-dideoxy-β-=
D-allofuranuronic acid
CAS RN *[22976–86–9]*; zinc salt *[146659–78–1]*

PHYSICAL CHEMISTRY
polyoxins
Composition Polyoxin complex consists of polyoxin B and several other polyoxins
of lower potency.

polyoxin B
Mol. wt. 507.4   **M.f.** $C_{17}H_{25}N_5O_{13}$ **Form** Amorphous powder.   **M.p.** >160 °C (decomp.)   **Solubility** In water 1 kg/l (20 °C). In acetone, methanol and common organic solvents <100 mg/l.   **Stability** Hygroscopic, and thus should be stored in tightly closed containers under dry conditions. Stable between pH 1 and pH 8.
Specific rotation $[\alpha]_D^{20}$ +34° (c = 1, water)   **pKa** pKa_1 (carboxyl) 3.0, pKa_2 (amino) 6.9, pKa_3 (uracil) 9.4

polyoxorim
Mol. wt. 521.4   **M.f.** $C_{17}H_{23}N_5O_{14}$ **Form** Colourless crystals.   **M.p.** >190 °C (decomp.)   **Solubility** In water <200 mg/l (20 °C) (zinc salt). In acetone and methanol <200 mg/l (zinc salt).   **Stability** Hygroscopic, and thus should be stored in tightly closed containers under dry conditions.   **Specific rotation** $[\alpha]_D^{20}$ +30° (c = 1, water)   **pKa** pKa_1 (carboxyl) 2.6, pKa_2 (carboxyl) 3.7, pKa_3 (amino) 7.3, pKa_4 (uracil) 9.4

COMMERCIALISATION
Production These polyoxins are produced by fermentation of *Streptomyces cacaoi* var. *asoensis*.   **History** Polyoxin B isolated and its structure elucidated by K. Isono *et al.* (*Agric. Biol. Chem.*, 1965, **29**, 848; *Tetrahedron Lett.*, 1968, 1133). Polyoxorim (polyoxin D) was isolated by S. Suzuki *et al.* (*Agric. Biol. Chem.*, 1966, **30**, 817) and its structure established by K. Isono *et al.* (*ibid.*, 1968, **32**, 1193). Their fungicidal activity reported by J. Eguchi *et al.* (*Nippon Shokubutsu Byori Gakkaiho*, 1968, **34**, 280) and by S. Sasaki *et al.* (*ibid.*, p. 272) respectively. Polyoxin B was introduced by Hokko Chemical Ind. Co. Ltd, Kaken Pharmaceutical Co. Ltd, Kumiai Chemical Industry Co. Ltd and by Nihon Nohyaku Co. Ltd; the zinc salt of polyoxorim was introduced as a fungicide by Kaken, Kumiai and Nihon Nohyaku.

polyoxin B
Patent JP 493008   **Manufacturers** Hokko; Kaken

polyoxorim
Patent JP 577960   **Manufacturers** Kaken

APPLICATIONS
Biochemistry Act by inhibition of cell wall chitin synthesis.
Mode of action Systemic fungicides with protective action.

polyoxin B
Uses Polyoxin B is used for control of *Alternaria* spp. and powdery mildews in apples and pears; *Botrytis cinerea* in vines and aubergines; powdery mildews in roses, chrysanthemums, capsicums, and melons; blight of carnations; powdery mildew, brown spot, and grey mould in tobacco; powdery mildew and grey mould in strawberries; leaf mould, early blight, and grey mould in tomatoes; powdery

mildew, grey mould, gummosis, *Sclerotinia* rot, and *Corynespora melonis* in cucumbers; *Alternaria* blight in carrots; purple blotch in leeks; etc.
Formulation types WP; EC; SG.   **Mixtures** *(polyoxin B +)* captan; oxine-copper; iminoctadine triacetate.   **Compatibility** Incompatible with alkaline materials.
Selected tradenames 'Polyoxin AL' (Kaken, Kumiai, Nihon Nohyaku, Hokko)

polyoxorim
Uses Polyoxorim (polyoxin D) zinc salt is used for control of sheath blight (*Rhizoctonia solani*) in rice; canker in apples and pears; *Rhizoctonia solani*, Drechslera, *Bipolaris*, *Curvularia*, and *Helminthosporium* spp. in lawn turf; etc.
Formulation types WP; PA.   **Mixtures** *(polyoxorim zinc salt +)* thiram; propiconazole.   **Compatibility** Incompatible with alkaline materials.
Selected tradenames 'Kakengel' (zinc salt) (Kaken); 'Polyoxin Z' (zinc salt) (Kaken); 'Stopit' (Kaken)

ANALYSIS
Product analysis by bioassay using *Alternaria mali* ACI-1157 (polyoxin B) and *Rhizoctonia solani* (=*Pellicularia sasakii*) ACI-1134 (polyoxorim).
Residues determined by bioassay with *A. mali* AKI-3 (polyoxin B) and *R. solani* ACI-1134 (polyoxorim).

MAMMALIAN TOXICOLOGY
polyoxin B
Oral Acute oral LD_{50} for male rats 21 000, female rats 21 200, male mice 27 300, female mice 22 500 mg/kg.   **Skin and eye** Acute percutaneous LD_{50} for rats >2000 mg/kg. Non-irritant to mucous membranes and skin (rats).
Inhalation LC_{50} (6 h) for rats 10 mg/l air.   **NOEL** (2 y) for rats and mice >48 000 mg/kg diet.   **Toxicity class** EPA (formulation) IV

polyoxorim
Oral Acute oral LD_{50} for male and female rats >9600 mg/kg.   **Skin and eye** Acute percutaneous LD_{50} for rats >750 mg/kg.   **Inhalation** LC_{50} (4 h) for male rats 2.44, female rats 2.17 mg/l air.   **NOEL** (2 y) for rats >50 000, mice >40 000 mg/kg diet.   **Toxicity class** EPA (formulation) III (WP)

ECOTOXICOLOGY
polyoxin B
Fish LC_{50} (48 h) for carp >40 mg/l. Japanese killifish unaffected by 100 mg/l for 72 h.   **Daphnia** LC_{50} (3 h) for *D. pulex* >40 mg/l.   **Other aquatic spp.** LC_{50} (3 h) for *Moina macrocopa* >40 mg/l.

polyoxorim
Fish LC_{50} (48 h) for carp >40 mg/l.   **Daphnia** LC_{50} (3 h) for *D. pulex* >40 mg/l.
Other aquatic spp. LC_{50} (3 h) for *Moina macrocopa* >40 mg/l.

ENVIRONMENTAL FATE
polyoxin B
Soil/Environment In upland conditions at 25 °C, DT_{50} <2 d (two soils, o.c. 6.2%, pH 6.3, moisture 23.3% and o.c. 1.1%, pH 6.8, moisture 63.6% respy.).

polyoxorim
Soil/Environment In flooded soil at 25 °C, DT_{50} <10 d (two soil types, o.c. 2.5%, pH 6.0 and o.c. 9.6%, pH 6.0 respy). In upland conditions at 25 °C, DT_{50} <7 d (two soil types, o.c. 0.6%, pH 6.4, moisture 10.7% and o.c. 6.2%, pH 6.3, moisture 61.9% respy). In water DT_{50} 4 h (pH 5.5, 24 °C), 8 h (pH 5.8, 26.5 °C).

587 prallethrin   *Insecticide*

pyrethroid

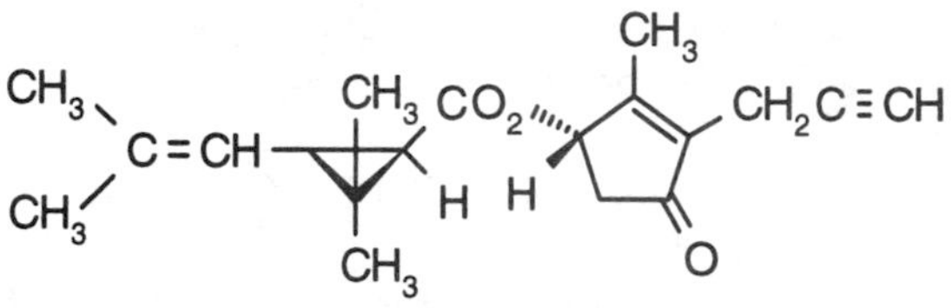

NOMENCLATURE
Common name prallethrin (BSI, draft E-ISO (stereochemistry unspecified)); pralléthrine ((f) draft F-ISO); no name (France)
IUPAC name *Roth*: (S)-2-methyl-4-oxo-3-prop-2-ynylcyclopent-2-enyl (1R)-*cis-trans*-2,2-dimethyl-3-(2-methylprop-1-enyl)cyclopropanecarboxylate
Chemical Abstracts name (S)-2-methyl-4-oxo-3-(2-propynyl)-2-cyclopenten-1-yl 2,2-dimethyl-3-(2-methyl-1-propenyl)cyclopropanecarboxylate
CAS RN *[23031–36–9]* **Development codes** S-4068SF (Sumitomo)
Official codes OMS 3033

PHYSICAL CHEMISTRY
Mol. wt. 300.4 **M.f.** $C_{19}H_{24}O_3$ **Form** Yellow to yellow-brown liquid.
B.p. >313.5 °C/760 mmHg **V.p.** <0.013 mPa (23.1 °C) **K_{ow}** logP = 4.49 (25 °C)
S.g./density 1.03 (20 °C) **Solubility** In water 8 mg/l (25 °C). In hexane, methanol, xylene >500 (all in g/kg, 20–25 °C). **Stability** Under normal storage conditions, stable for at least 2 years.

COMMERCIALISATION
History Insecticide reported by T. Mutsunaga *et al.* (*Eisei Dobutsu*, 1987, **38**, 219). Introduced by Sumitomo Chemical Co., Ltd. **Manufacturers** Sumitomo

APPLICATIONS
Mode of action Insecticide with contact action. **Uses** Used to control public health insects (Blattodea, Culicidae, Muscidae). **Formulation types** AE; EC; EW; KN; Oil; Mat. **Compatibility** Incompatible with alkaline materials.
Selected tradenames 'Etoc' (Sumitomo)

ANALYSIS
Product by gc. Details from Sumitomo.

MAMMALIAN TOXICOLOGY
Oral Acute oral LD_{50} for male rats 640, female rats 460 mg/kg. **Skin and eye** Acute percutaneous LD_{50} for rats >5000 mg/kg. Non-irritating to the eyes and skin (rabbits). Not a skin sensitiser (guinea pigs). **Inhalation** LC_{50} (4 h) for rats 288–333 mg/m^3. **Toxicity class** WHO (a.i.) II; EPA (formulation) III

588 pretilachlor   *Herbicide*

chloroacetanilide

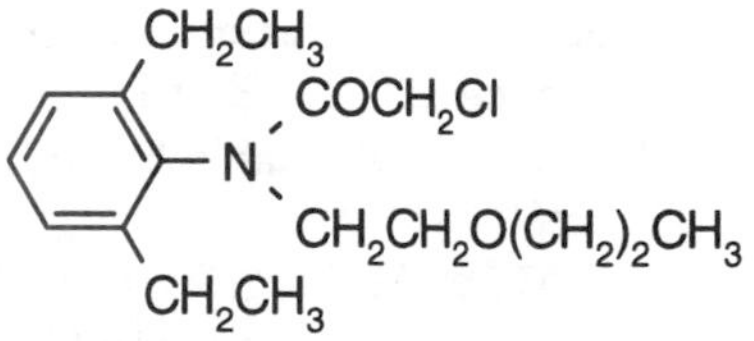

NOMENCLATURE
Common name pretilachlor (BSI, E-ISO); prétilachlore ((m) F-ISO)
IUPAC name 2-chloro-2',6'-diethyl-N-(2-propoxyethyl)acetanilide
Chemical Abstracts name 2-chloro-N-(2,6-diethylphenyl)-N-(2-propoxyethyl)= acetamide
CAS RN [51218–49–6] **Development codes** CGA 26 423

PHYSICAL CHEMISTRY
Mol. wt. 311.9 **M.f.** $C_{17}H_{26}ClNO_2$ **Form** Colourless liquid.
B.p. 135 °C/0.001 mmHg **V.p.** 0.133 mPa (20 °C) **K_{ow}** logP = 4.08 (shake flask) **Henry** 8.1×10^{-4} Pa m^3 mol^{-1} **S.g./density** 1.076 (20 °C) **Solubility** In water 50 mg/l (20 °C). Very soluble in benzene, hexane, methanol and dichloromethane. **Stability** Relatively stable to hydrolysis; DT_{50} (calc.) (20 °C) >200 d (pH 1–pH 9), 14 d (pH 13). **F.p.** 129 °C (EEC A.9)

COMMERCIALISATION
History Introduced by Ciba-Geigy AG (now Novartis Crop Protection AG) in 1988. **Patent** BE 800471; GB 1438311; GB 1438312 **Manufacturers** Novartis

APPLICATIONS
Biochemistry Cell division inhibitor. **Mode of action** Selective herbicide. It is taken up readily by the hypocotyls, mesocotyls and coleoptiles, and, to a lesser extent, by the roots of germinating weeds. **Uses** Herbicide effective against main annual grasses, broad-leaved weeds and sedges in transplanted and seeded rice.
Phytotoxicity Applied alone, pretilachlor will cause injury to direct seeded rice.
Formulation types EC; GR. **Mixtures** *(pretilachlor +)* fenclorim; clomeprop; dimethametryn + pyrazolynate; pyrazosulfuron-ethyl + esprocarb + dimethametryn. **Selected tradenames** 'Rifit' (transplanted rice) (Novartis); 'Sofit' (direct sown rice) (mixture) (Novartis); 'Solnet' (transplanted rice) (Novartis); 'Gorbo' (Takeda); 'Sparkstar' (mixture) (Nissan)

ANALYSIS
Product analysis by glc. **Residues** determined by glc with MCD or TID. Details available from Novartis.

MAMMALIAN TOXICOLOGY
Oral Acute oral LD_{50} for rats 6099 mg/kg. **Skin and eye** Acute percutaneous LD_{50} for rats >3100 mg/kg. Moderate irritant to skin; non-irritant to eyes (rabbits). **Inhalation** LC_{50} (4 h) for rats >2.8 mg/l air. **NOEL** (2 y) for rats 30, mice 300 ppm; (0.5 y) for dogs 300 ppm. **ADI** 0.018 mg/kg. **Toxicity class** WHO (a.i.) III (Table 5) **EC risk** R38, R43

ECOTOXICOLOGY
Birds Non-toxic to birds; LD_{50} for Japanese quail >10 000 mg/kg. LC_{50} for Japanese quail >1000 ppm. **Fish** LC_{50} (96 h) for rainbow trout 0.9, catfish 2.7, crucian carp 2.3 mg/l. **Bees** LD_{50} (contact) 93 µg/bee. **Daphnia** LC_{50} 13 mg/l. **Algae** EC_{50} for *Selenastrum capricornutum* 0.002 mg/l.

ENVIRONMENTAL FATE
Animals Substitution of the chlorine atom for glutathione to form a conjugate. Cleavage of the ether bond to yield an ethyl alcohol derivative. Both metabolites are susceptible to further degradation. **Plants** Substitution of the chlorine atom to form a conjugate. Cleavage of the ether bond to an ethyl alcohol derivative. Hydrolytic and reductive removal of the chlorine atom. **Soil/Environment** Applied to paddy water, pretilachlor disappeared from the water by adsorption to the soil, where it is rapidly degraded under practical conditions, median DT_{50} (lab.) 30 d. Due to strong soil adsorption, unlikely to leach.

sulfonylurea

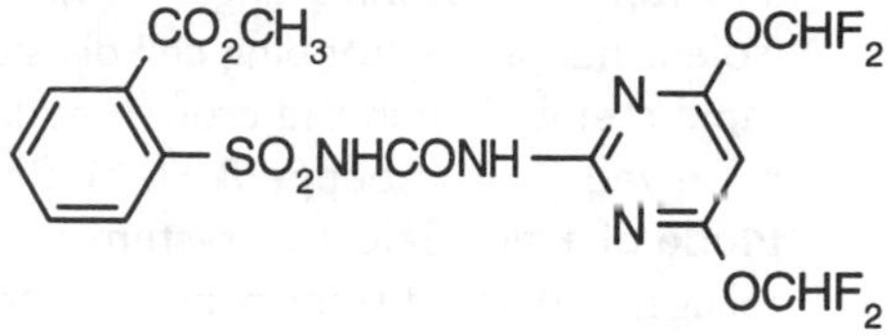

NOMENCLATURE

primisulfuron-methyl
Common name primisulfuron-methyl (BSI, draft E-ISO)
CAS RN *[86209–51–0]* **Development codes** CGA 136872

primisulfuron
Common name primisulfuron (BSI, draft E-ISO)
IUPAC name 2-[4,6-bis(difluoromethoxy)pyrimidin-2-ylcarbamoylsulfamoyl]=
benzoic acid
Chemical Abstracts name 2-[[[[[4,6-bis(difluoromethoxy)-2-pyrimidinyl]amino]=
carbonyl]amino]sulfonyl]benzoic acid
CAS RN *[113036–87–6]*

PHYSICAL CHEMISTRY

primisulfuron-methyl
Mol. wt. 468.3 **M.f.** $C_{15}H_{12}F_4N_4O_7S$ **Form** Fine white powder.
M.p. 194.8–197.4 °C (decomp.) **V.p.** $<5 \times 10^{-3}$ mPa (25 °C) (OECD 104)
K_{ow} logP = 0.06 (25 °C, unstated pH) **Henry** <0.04 Pa m^3 mol^{-1} (calc.)
S.g./density 1.61 (20 °C) **Solubility** In water 3.3 (pH 5), 243 (pH 7), 5280
(pH 9) (all in mg/l). In acetone 35 000, ethanol 1000, toluene 570, *n*-octanol 7.7,
n-hexane 1.5 (all in mg/l, 25 °C). **Stability** Stable for at least 3 years at room
temperature. On hydrolysis (50 °C) DT_{50} 10 h (pH 3 & pH 5), >300 h (pH 7 &
pH 9), 120 h (pH 10). Stable up to 100 °C. **pKa** 3.47

primisulfuron
Mol. wt. 454.3 **M.f.** $C_{14}H_{10}F_4N_4O_7S$

COMMERCIALISATION

History Herbicidal activity of primisulfuron-methyl reported by W. Maurer et al.
(*Proc. 1987 Br. Crop Prot. Conf. – Weeds*, **1**, 41). Introduced by Ciba-Geigy AG
(now Novartis Crop Protection AG). **Patent** EP 84020; US 4478635
Manufacturers Novartis

APPLICATIONS

primisulfuron-methyl

Biochemistry Branched chain amino acid synthesis (acetolactate synthase or ALS) inhibitor. Acts by inhibiting biosynthesis of the essential amino acids valine and isoleucine, hence stopping cell division and plant growth. Selectivity derives from rapid metabolism in the crop. Metabolic basis of selectivity in sulfonylureas reviewed (M. K. Koeppe & H. M. Brown, *Agro-Food-Industry*, **6**, 9–14 (1995)). **Mode of action** Selective systemic herbicide, absorbed through the roots and foliage, with rapid translocation both acropetally and basipetally. **Uses** Post-emergence control of problem grass weeds, e.g. *Sorghum bicolor* (shattercane), *Sorghum almum*, *Sorghum halepense* (johnsongrass) and *Agropyron repens*, and many broad-leaved weeds in maize, at 20–40 g/ha. **Phytotoxicity** Maize hybrids differ in their sensitivity to primisulfuron-methyl. **Formulation types** WP; WG. **Compatibility** Compatible in tank mixtures with atrazine, bromoxynil, cyanazine, dicamba and 2,4-D. **Selected tradenames** 'Beacon' (Novartis); 'Tell' (Novartis)

ANALYSIS

Residues determined by hplc. Methods for sulfonylurea residues in crops, soil and water reviewed (A. C. Barefoot *et al.*, *Proc. Br. Crop Prot. Conf. – Weeds*, 1995, **2**, 707).

MAMMALIAN TOXICOLOGY

primisulfuron-methyl

Oral Acute oral LD_{50} for rats >5050 mg/kg. **Skin and eye** Acute percutaneous LD_{50} for rats >2010, mice >2100 mg/kg. Slightly irritating to eyes (rabbits); non-irritating to skin (rabbits). No skin sensitisation (guinea pigs). **Inhalation** LC_{50} (4 h) for rats >4.8 mg/l air. **NOEL** (2 y) for rats 13 mg/kg b.w. daily; (19 mo) for mice 45 mg/kg b.w. daily; (1 y) for dogs 25 mg/kg b.w. daily. **ADI** 0.13 mg/kg b.w. **Other** Non-mutagenic in several tests (e.g., DNA repair and micronucleus). **Toxicity class** WHO (a.i.) III (Table 5); EPA (formulation) IV

ECOTOXICOLOGY

primisulfuron-methyl

Birds Oral LD_{50} for bobwhite quail and mallard duck >2150 mg/kg. **Fish** LC_{50} (96 h) for rainbow trout 70 mg/l. **Bees** Non-toxic to honeybees; LC_{50} (48 h, contact) >100 µg/bee. **Worms** LD_{50} (14 d) >100 mg/kg soil. **Daphnia** LC_{50} (48 h) 260–480 mg/l. **Algae** EC_{50} (7 d) for *Selenastrum* 24, *Anabaena* 176, *Navicula* >227, *Skeletonema* >225 µg/l.

ENVIRONMENTAL FATE

Animals The major pathway in the metabolism of primisulfuron-methyl in the rat and other large animals involves hydroxylation of the pyrimidine ring, followed by hydrolysis and/or glucuronic acid conjugation. **Plants** In maize, a major metabolite is the *O*-glucoside of 5-hydroxy-primisulfuron. At harvest time, no

residues >0.05 mg/kg are detectable in the grain and fodder of maize.
Soil/Environment Weakly absorbed by soil components, but field studies and lysimeter results indicate very low leaching of primisulfuron-methyl. Degradation in the soil is primarily by chemical means, with limited degradation by microbes; DT_{50} (field) 4–29 d.

590 probenazole *Fungicide, bactericide*

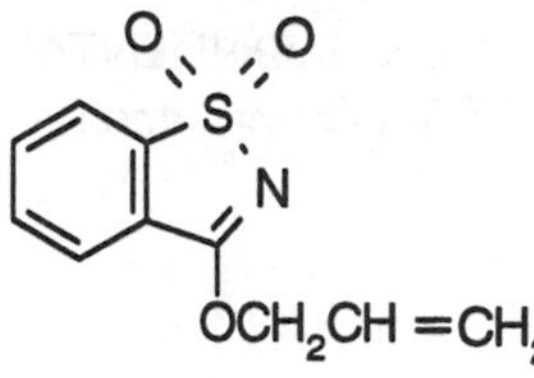

NOMENCLATURE
Common name probenazole (JMAF)
IUPAC name 3-allyloxy-1,2-benz[*d*]isothiazole 1,1-dioxide
Chemical Abstracts name 3-(2-propenyloxy)-1,2-benzisothiazole 1,1-dioxide
CAS RN *[27605–76–1]*

PHYSICAL CHEMISTRY
Mol. wt. 223.2 **M.f.** $C_{10}H_9NO_3S$ **Form** Colourless crystals. **M.p.** 138–139 °C
Solubility Slightly soluble in water (c. 150 mg/l). Readily soluble in acetone, dimethylformamide and chloroform. Slightly soluble in methanol, ethanol, diethyl ether and benzene. Very slightly soluble in hexane and ligroin.

COMMERCIALISATION
History Fungicide reported by M. Uchiyama *et al.* (*Agric. Biol. Chem.*, 1973, **37,** 737) and reviewed by T. Ohashi (*Jpn. Pestic. Inf.*, 1980, No. 37, p. 37). Introduced by Meiji Seika Kaisha Ltd in collaboration with Hokko Chemical Industry Co., Ltd.
Manufacturers Meiji Seika

APPLICATIONS
Mode of action Systemic fungicide and bactericide, absorbed by the roots and translocated acropetally. **Uses** Control of rice blast, bacterial leaf blight, and bacterial grain rot in paddy rice at 2.4–3.2 kg a.i./ha; and angular leaf spot in cucumbers. **Formulation types** GR. **Selected tradenames** 'Oryzemate' (Meiji Seika, Hokko)

ANALYSIS
Product by glc; details from Meiji Seika.

MAMMALIAN TOXICOLOGY
Oral Acute oral LD_{50} for rats 2030, mice 2750–3000 mg/kg. **Skin and eye** Acute percutaneous LD_{50} for rats >5000 mg/kg. **NOEL** In chronic toxicity studies on rats, NOEL was 110 mg/kg. Non-mutagenic. Not teratogenic in rats (600 mg/kg diet). **Other** Not mutagenic in standard tests. **Toxicity class** WHO (a.i.) III (Table 5)

ECOTOXICOLOGY
Fish LC_{50} (48 h) for carp 6.3, Japanese killifish >6.0 mg/l.

ENVIRONMENTAL FATE
Soil/Environment Half-life in soil <24 h (alluvial or volcanic soil).

591 prochloraz   *Fungicide*

azole

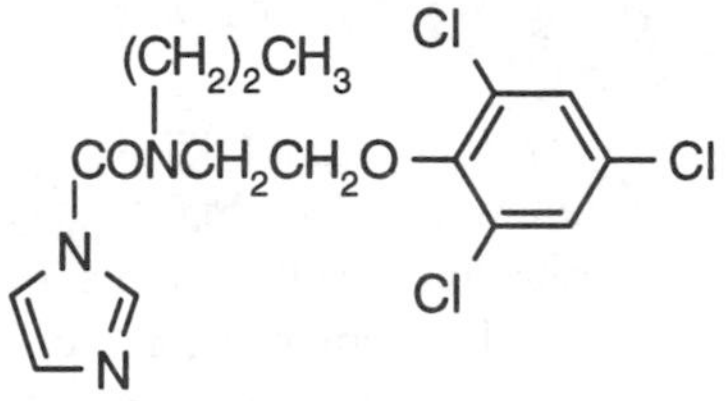

NOMENCLATURE
Common name prochloraz (BSI, E-ISO, (*m*) F-ISO, ANSI)
IUPAC name *N*-propyl-*N*-[2-(2,4,6-trichlorophenoxy)ethyl]imidazole-1-= carboxamide; 1-{*N*-propyl-*N*-[2-(2,4,6-trichlorophenoxy)ethyl]}carbamoylimidazole
Chemical Abstracts name *N*-propyl-*N*-[2-(2,4,6-trichlorophenoxy)ethyl]-= 1*H*-imidazole-1-carboxamide
CAS RN *[67747–09–5]*   **Development codes** BTS 40 542 (AgrEvo)

PHYSICAL CHEMISTRY
Composition Tech. grade is *c.* 97% pure. **Mol. wt.** 376.7 **M.f.** $C_{15}H_{16}Cl_3N_3O_2$
Form Odourless, colourless crystals; tech. grade is a mildly aromatic, golden brown liquid that tends to solidify on cooling. **M.p.** 46.5–49.3 °C (>99% pure)
B.p. 208–210 °C/0.2 mmHg (decomp.) **V.p.** 1.5×10^{-1} mPa (25 °C), 9.0×10^{-2} mPa (20 °C) **K_{ow}** logP = 4.12 (unionised) **Henry** 1.64×10^{-3} Pa m^3

mol^{-1} (calc.)   **S.g./density** 1.42 (20 °C)   **Solubility** In water 34.4 mg/l (25 °C).
Readily soluble in a wide range of organic solvents, e.g. chloroform, diethyl ether,
toluene, xylene 2.5, acetone 3.5, hexane c. 7.5 × 10^{-3} (all in kg/l, 25 °C).
Stability Stable in water at pH 7 and 20 °C. Decomposes in concentrated acids
and alkalis, in the presence of sunlight, and on prolonged heating at high
temperatures (200 °C).   **pKa** 3.8, weak base   **F.p.** 160 °C (closed cup)

COMMERCIALISATION
History Fungicide reported by R. J. Birchmore et al. (*Proc. Br. Crop. Prot. Conf. –
Pests Dis., 1977, 2, 593*). Introduced by The Boots Co. Ltd (now AgrEvo GmbH).
Patent GB 1469772; US 3991071; US 4080462   **Manufacturers** AgrEvo; Griffin

APPLICATIONS
Biochemistry Steroid demethylation (ergosterol biosynthesis) inhibitor.
Mode of action Fungicide with protective and eradicant action.   **Uses** A
protectant and eradicant fungicide effective against a wide range of diseases
affecting field crops, fruit, turf, and vegetables. An EC is recommended for use in
cereals (400–600 g a.i./ha) against *Pseudocercosporella, Pyrenophora,
Rhynchosporium,* and *Septoria* spp., with useful activity against *Erysiphe* spp.; in
oilseed rape (500 g/ha) against *Alternaria, Botrytis, Pyrenopeziza* and *Sclerotinia* spp.
Useful activity is also shown against *Ascochyta* and *Botrytis* spp. in field legumes;
and *Cercospora* and *Erysiphe* spp. in beet. Good activity against storage or transit
diseases of citrus and tropical fruit when applied as a dip treatment
(0.5–0.7 g a.i./l). A WP is recommended in mushrooms against *Verticillium
fungicola* and *Mycogone perniciosa,* and in rice against *Pyricularia.* A seed treatment
(0.2–0.5 g/kg) will control several cereal diseases caused by *Cochliobolus, Fusarium,
Pyrenophora* and *Septoria* spp., and, in flax, *Alternaria.*   **Formulation types** EC; EW;
FS; LS; WP.   **Mixtures** *(prochloraz +)* carbendazim; mancozeb; fenpropimorph;
cyproconazole; fenpropidin; carboxin; fenpropidin + fenpropimorph; triadimefon;
fenbuconazole; fenbuconazole + carbendazim; chlorothalonil; bromuconazole;
fluquinconazole.   **Compatibility** Forms a complex with some metal ions, e.g.
prochloraz-manganese used for WP formulations.   **Selected tradenames** 'Sportak'
(AgrEvo); 'Eyetak' (Barclay); 'Mirage' (Makhteshim-Agan)

ANALYSIS
Product analysis by hplc and glc.   **Residue** analysis by glc. Details available from
AgrEvo.

MAMMALIAN TOXICOLOGY
Reviews *Pesticide residues in food – 1983.* FAO Plant Production and Protection
Paper 56, 1984. *Pesticide residues in food: 1983 evaluations.* FAO Plant Production
and Protection Paper 61, 1985.   **Oral** Acute oral LD$_{50}$ for rats 1600–2400, mice
2400 mg/kg.   **Skin and eye** Acute percutaneous LD$_{50}$ for rats >2100, rabbits
>3000 mg/kg. Not irritating to skin, slight eye irritation (rabbits).   **Inhalation** LC$_{50}$

(4 h) for rats >2.16 mg/l air. **NOEL** (2 y) for dogs 30 mg/kg diet (0.92 mg/kg daily). **ADI** (JMPR) 0.01 mg/kg b.w. [1983]. **Toxicity class** WHO (a.i.) III **EC risk** Xn (R22)

ECOTOXICOLOGY
Birds Acute oral LD_{50} for bobwhite quail 662, mallard ducks >1954 mg/kg. Dietary LC_{50} (5 d) for bobwhite quail and mallard ducks >5200 mg/kg. **Fish** LC_{50} (96 h) for rainbow trout 1.5, bluegill sunfish 2.2 mg/l. **Bees** Low toxicity to bees. LD_{50} (topical) 50 μg/bee; (oral) 60 μg/bee. **Worms** LC_{50} for earthworms (*Eisenia foetida*) 207 mg/kg soil. **Other beneficial spp.** Low toxicity to a range of beneficial arthropods. **Daphnia** LC_{50} (48 h) 4.3 mg/l. **Algae** E_bC_{50} (72 h) for *Selenastrum capricornutum* 0.1 mg/l, E_rC_{50} 1.54 mg/l. **Other aquatic spp.** EC_{50} (96 h) for Eastern oyster (*Crassostrea virginica*) 0.95, mysid shrimp (*Mysidopsis bahia*) 0.77 mg/l.

ENVIRONMENTAL FATE
Animals In all species examined, prochloraz is rapidly metabolised initially by cleavage of the imidazole ring and quantitatively eliminated from the body following oral administration. Whilst absorption following dermal exposure is low, residues in plasma and tissues are again rapidly eliminated from the body.
Plants The primary plant metabolite, N-formyl-N'-1-propyl-N-(2-(2,4,6-trichloro= phenoxy)ethyl) urea is formed from cleavage of the imidazole ring. This is degraded to N-propyl-N-(2-(2,4,6-trichlorophenoxy)ethyl)urea which occurs in both free and conjugated forms. Other metabolites include: 2-(2,4,6-trichloro= phenoxy)ethanol, 2-(2,4,6-trichlorophenoxy)acetic acid, traces of 2,4,6-trichloro= phenol and conjugates of the above. Little unchanged prochloraz is present.
Soil/Environment Degrades in the soil to a range of mainly volatile metabolites (degradation is not pH-dependent). Prochloraz is well adsorbed onto soil particles, and is not readily leached; K_d 152 (sandy loam), 256 (silty clay loam). In a further study, mean K_{oc} 1463. Possesses low toxicity to a wide range of soil microflora and microfauna, but has inhibitory effects on soil fungi. DT_{50} under field conditions is 5–37 d.

dicarboximide

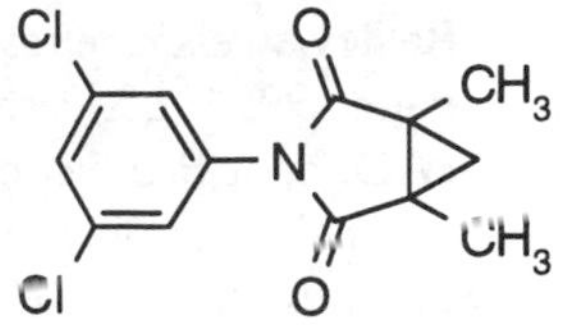

NOMENCLATURE
Common name procymidone (BSI, E-ISO, (f) F-ISO, JMAF)
IUPAC name N-(3,5-dichlorophenyl)-1,2-dimethylcyclopropane-1,2-dicarboximide
Chemical Abstracts name 3-(3,5-dichlorophenyl)-1,5-dimethyl-3-azabicyclo=
[3.1.0]hexane-2,4-dione
CAS RN *[32809–16–8]*   **Development codes** S-7131 (Sumitomo)

PHYSICAL CHEMISTRY
Mol. wt. 284.1   **M.f.** $C_{13}H_{11}Cl_2NO_2$ **Form** Colourless crystals; (tech., light brown solid).   **M.p.** 166–166.5 °C; (tech., 164–166 °C)   **V.p.** 18 mPa (25 °C); 10.5 mPa (20 °C)   **K_{ow}** logP = 3.14 (26 °C)   **S.g./density** 1.452 (25 °C)   **Solubility** In water 4.5 mg/l (25 °C). Slightly soluble in alcohols. In acetone 180, xylene 43, chloroform 210, dimethylformamide 230, methanol 16 (all in g/l, 25 °C).
Stability Stable under normal storage conditions. Stable to light, heat and moisture.

COMMERCIALISATION
History Fungicide reported by Y. Hisada *et al.* (*J. Pestic. Sci.*, 1976, **1**, 145). Introduced by Sumitomo Chemical Co., Ltd.   **Patent** GB 1298261; US 3903090
Manufacturers Sumitomo

APPLICATIONS
Biochemistry Inhibitor of triglyceride synthesis in fungi.   **Mode of action** Systemic fungicide with protective and curative properties. Absorbed through the roots, with translocation to leaves and flowers.   **Uses** Control of *Botrytis*, *Sclerotinia*, *Monilia*, and *Helminthosporium* spp. on fruit (including top fruit, strawberries and raspberries), vines, vegetables (including tomatoes, peas and beans), ornamentals, cereals, sunflowers, oilseed rape, soya beans, peanuts, tobacco, etc. Usually applied at 0.5–1.0 kg a.i./ha.   **Phytotoxicity** Non-phytotoxic when used as directed. Not to be applied to cyclamen after the bud sprouting stage.
Formulation types HN; SP; WP; SC; DP; WG.   **Mixtures** (*procymidone +*) chlorothalonil; maneb + zineb; thiram.   **Compatibility** Incompatible with alkaline materials.   **Selected tradenames** 'Sumisclex' (Sumitomo); 'Sumilex' (Sumitomo)

ANALYSIS
Product analysis by glc (M. Horiba, *Agric. Biol. Chem.*, 1982, **46**, 1095).

MAIN ENTRIES

Residues determined by glc (*Man. Pestic. Residue Anal.*, 1987, **I**, 6, S18, S19). Details available from Sumitomo Chemical Co.

MAMMALIAN TOXICOLOGY
Reviews *Pesticide residues in food – 1989.* FAO Plant Production and Protection Paper 99, 1989. *Pesticide residues in food – 1989 evaluations. Part II – Toxicology.* FAO Plant Production and Protection Paper 100/2, 1990. **Oral** Acute oral LD_{50} for male rats 6800, female rats 7700 mg/kg. **Skin and eye** Acute percutaneous LD_{50} for rats >2500 mg/kg. Non-irritating to skin and eyes (rabbits).
Inhalation LC_{50} (4 h) for rats >1500 mg/m^3. **NOEL** (90 d) for dogs 3000 mg/kg; (2 y) for male rats 1000, female rats 300 mg/kg. No mutagenic or carcinogenic effects. **ADI** (JMPR) 0.1 mg/kg b.w. [1989]. **Toxicity class** WHO (a.i.) III (Table 5)

ECOTOXICOLOGY
Fish LC_{50} (96 h) for bluegill sunfish 10.3, rainbow trout 7.2 mg/l. **Bees** Not toxic to bees.

ENVIRONMENTAL FATE
Animals In animals, rapidly and completely eliminated in the faeces and urine.
Soil/Environment Persists in soil for *c.* 4–12 weeks, depending on humus content.

593 prodiamine   *Herbicide*

2,6-dinitroaniline

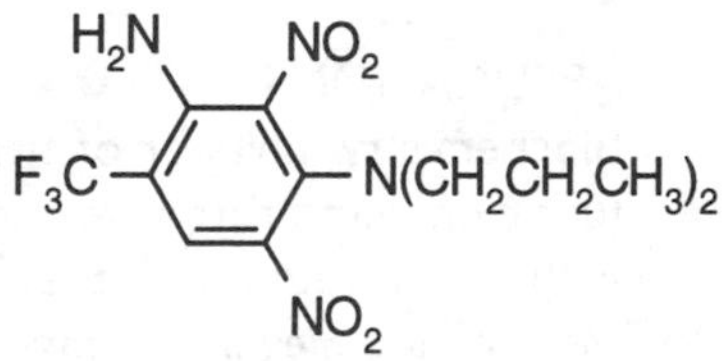

NOMENCLATURE
Common name prodiamine (BSI, E-ISO, (*f*) F-ISO, ANSI, WSSA)
IUPAC name 5-dipropylamino-α,α,α-trifluoro-4,6-dinitro-*o*-toluidine; 2,6-dinitro-N^1,N^1-dipropyl-4-trifluoromethyl-*m*-phenylenediamine
Chemical Abstracts name 2,4-dinitro-N^3,N^3-dipropyl-6-(trifluoromethyl)-= 1,3-benzenediamine
CAS RN [29091–21–2] **Development codes** USB-3153 (Borax); CN-11–2936 (Velsicol); SAN 745H (Sandoz)

PHYSICAL CHEMISTRY
Mol. wt. 350.3 **M.f.** $C_{13}H_{17}F_3N_4O_4$ **Form** Yellow crystals. **M.p.** 122.5–124 °C
V.p. 0.0033 mPa (25 °C) **K_{ow}** logP = 4.10±0.07 (25 °C, unstated pH)
S.g./density 1.47 (25 °C) **Solubility** In water 0.013 mg/l (25 °C). In acetone 226,
acetonitrile 45, benzene 74, chloroform 93, ethanol 7, hexane 20, xylene 35.4 (all
in g/l, 20 °C). **Stability** Moderately stable to light. Decomposes at 240 °C.
pKa 11.5, strong base **F.p.** 196 °C

COMMERCIALISATION
History Herbicide discovered by US Borax, evaluated by Velsicol Chemical Corp.
and marketed by Sandoz AG (now Novartis Crop Protection AG).
Manufacturers Novartis

APPLICATIONS
Biochemistry Cell division inhibitor. **Mode of action** Selective herbicide which
inhibits spindle fibre formation and hence cell division, and root and shoot growth.
Uses Pre-plant or pre-emergence control of annual grasses and broad-leaved
weeds in cotton, alfalfa, nonbearing fruit trees, vines, nuts, ornamentals, soya
beans, turf, etc. at rates of 0.375–2.0 kg a.i./ha. **Formulation types** WG.
Selected tradenames 'Barricade' (USA) (Novartis); 'Kusablock' (Japan) (Novartis)

ANALYSIS
Product analysis by gc-FID, and **residue** analysis by gc-ECD.

MAMMALIAN TOXICOLOGY
Oral Acute oral LD_{50} for rats >5000 mg/kg. **Skin and eye** Acute percutaneous
LD_{50} for rats >2000 mg/kg. Irritating to eyes; non-irritating to skin (rabbits). Not
a skin sensitiser. **Inhalation** LC_{50} (4 h) >0.256 mg/m³ (maximum achievable).
NOEL (1 y) for dogs 20.0 mg/kg daily; (1.5 y) for mice 60 mg/kg daily; (2 y) for
rats 7.2 mg/kg daily. **Toxicity class** WHO (a.i.) III (Table 5)

ECOTOXICOLOGY
Birds Acute oral LD_{50} for bobwhite quail >2250 mg/kg. Eight-day dietary LC_{50} for
bobwhite quail >10 000 mg/kg daily. **Fish** LC_{50} (96 h) for rainbow trout >829,
bluegill sunfish >552 µg/l. **Bees** LD_{50} (topical) >100 µg/bee. **Daphnia** LC_{50}
(48 h) >658 µg/l.

ENVIRONMENTAL FATE
Animals In rats, following oral administration, 58.4% is excreted in the faeces and
28.6% in the urine within 48 hours. **Soil/Environment** Prodiamine is subject to
photodegradation. Metabolism involves reduction of the nitro groups. DT_{50} in
sandy loam (field) 69 d. Strongly adsorbed on soil; K_{oc} and K_d for sand texture
19 540, 19.54, for sandy loam texture 12 860, 398.5, for Kenya loam texture 5440,
120.

organophosphorus

NOMENCLATURE
Common name profenofos (BSI, E-ISO, (*m*) F-ISO, ANSI, ESA)
IUPAC name O-4-bromo-2-chlorophenyl O-ethyl S-propyl phosphorothioate
Chemical Abstracts name O-(4-bromo-2-chlorophenyl) O-ethyl S-propyl
phosphorothioate
CAS RN *[41198–08–7]* **EEC no.** 255–255–2 **Development codes** CGA 15 324
Official codes OMS 2004

PHYSICAL CHEMISTRY
Mol. wt. 373.6 **M.f.** $C_{11}H_{15}BrClO_3PS$ **Form** Pale yellow liquid with garlic-like
odour. **B.p.** 100 °C/1.80 Pa **V.p.** 1.24×10^{-1} mPa (25 °C) (OECD 104)
K_{ow} logP = 4.44 (OECD 107) **S.g./density** 1.455 (20 °C) **Solubility** In water
28 mg/l (25 °C). Readily miscible with most organic solvents. **Stability** Relatively
stable under neutral and slightly acidic conditions. Unstable under alkaline
conditions; on hydrolysis DT_{50} (calc.) (20 °C) 93 d (pH 5), 14.6 d (pH 7), 5.7 h
(pH 9).

COMMERCIALISATION
History Insecticide reported by F. Buholzer (*Proc. Br. Insectic. Fungic. Conf., 8th,*
1975, **2**, 659). Introduced by Ciba-Geigy AG (now Novartis Crop Protection AG).
Patent BE 789937; GB 1417116 **Manufacturers** Novartis; Pesticides India

APPLICATIONS
Biochemistry Cholinesterase inhibitor. The separate optical isomers, due to the
chiral phosphorus atom, show different types of insecticidal activity and ability to
inhibit acetylcholinesterase (H. Leader & J. E. Casida, *J. Agric. Food Chem.*, 1982,
30, 546). **Mode of action** Non-systemic insecticide and acaricide with contact and
stomach action. Exhibits a translaminar effect. Has ovicidal properties.
Uses Control of insects (particularly Lepidoptera) and mites on cotton, maize,
sugar beet, soya beans, potatoes, vegetables, tobacco, and other crops.
Phytotoxicity Slight reddening of cotton may occur. **Formulation types** EC; UL.
Mixtures *(profenofos +)* cypermethrin; deltamethrin.
Selected tradenames 'Curacron' (Novartis); 'Sanofos' (Sanonda)

ANALYSIS

Product analysis by glc.   **Residues** determined by glc with TID.   Details available from Novartis.

MAMMALIAN TOXICOLOGY

Reviews *Pesticide residues in food – 1990*. FAO Plant Production and Protection Paper, 102, 1990. *Pesticide residues in food – 1990 evaluations. Part II – Toxicology.* World Health Organisation, WHO/PCS/91.47, 1991.   **EHC** 63 (WHO, 1986; a general review of organophosphorus insecticides).   **Oral** Acute oral LD_{50} for rats 358, rabbits 700 mg/kg.   **Skin and eye** Acute percutaneous LD_{50} for rats c. 3300, rabbits 172 mg/kg. Moderate eye irritant; mild skin irritant (rabbits).
Inhalation LC_{50} (4 h) for rats c. 3 mg/l air.   **NOEL** (using EC formulation 380 g a.i./l) for rats (2 y) 0.3 mg a.i./kg diet; for lifetime study 1.0 mg a.i./kg diet; for mice 0.08 mg/kg diet.   **ADI** (JMPR) 0.01 mg/kg b.w. [1990].
Toxicity class WHO (a.i.) II; EPA (formulation) II   **EC risk** Xn (R20/21/22)

ECOTOXICOLOGY

Birds LC_{50} (8 d) for bobwhite quail 70–200, Japanese quail >1000, mallard ducks 150–612 ppm.   **Fish** LC_{50} (96 h) for rainbow trout 0.08, crucian carp 0.09, bluegill sunfish 0.3 mg/l.   **Bees** Toxic to bees.   **Other aquatic spp.** Highly toxic to crustaceans.

ENVIRONMENTAL FATE

Animals Rats rapidly excrete ^{14}C-profenofos after oral administration. The predominant metabolic pathway involves stepwise dealkylation and hydrolysis, followed by conjugation.   **Plants** In cotton, Brussels sprouts and lettuce, the compound is rapidly taken up and metabolised. The overall metabolic pattern indicates degradation to polar metabolites.   **Soil/Environment** Mean half-life in soil (lab. and field) is c. 1 w.

595 prohexadione-calcium

Plant growth regulator

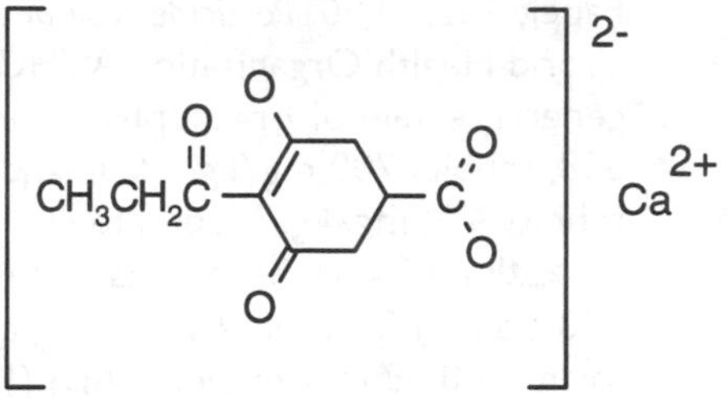

NOMENCLATURE

prohexadione-calcium
IUPAC name calcium 3-oxido-5-oxo-4-propionylcyclohex-3-enecarboxylate
Chemical Abstracts name calcium 3,5-dioxo-4-(1-oxopropyl)cyclohexane=
carboxylate
CAS RN *[127277–53–6]*   **Development codes** KUH 833; KIM-112 (Kumiai);
BX-112 (Ihara)

prohexadione
Common name prohexadione (BSI, draft E-ISO)
IUPAC name 3,5-dioxo-4-propionylcyclohexanecarboxylic acid
Chemical Abstracts name 3,5-dioxo-4-(1-oxopropyl)cyclohexanecarboxylic acid
CAS RN *[88805–35–0]*

PHYSICAL CHEMISTRY

prohexadione-calcium
Composition Tech. is >89%.   **Mol. wt.** 250.3   **M.f.** $C_{10}H_{10}CaO_5$
Form Odourless, fine white powder.   **M.p.** >360 °C   **V.p.** 1.33×10^{-2} mPa
(20 °C)   **K_{ow}** logP = –2.90   **Henry** 1.92×10^{-5} Pa m^3 mol^{-1} **S.g./density** 1.460
Solubility In water 174 mg/l (20 °C). In methanol 1.11, acetone 0.038 mg/l
(20 °C).   **Stability** Stable in water; DT_{50} (20 °C) 5 d (pH 5), 83 d (pH 9). Stable
to heat (200 °C). Stable in sunlight in water, DT_{50} 4 d.

prohexadione
Mol. wt. 212.2   **M.f.** $C_{10}H_{12}O_5$

COMMERCIALISATION

History Reported by Kumiai Chemical Industry Co., Ltd (Y. Toyokawa *et al., Proc.*
Ann. Meeting Soc. Chem. Regul. Pl. (Jap.), 68 (1987) and T. Miyazawa *et al., Proc. Br.*
Crop Prot. Conf. Weeds, **3**, 967 (1991)). The calcium salt (1:1) developed by Kumiai

Chemical Industry Co., Ltd and Ihara Chemical Industry Co., Ltd. Introduced in Japan in 1994. **Patent** EP 123001; US 4678496 **Manufacturers** Ihara/Kumiai

APPLICATIONS
Biochemistry Inhibits 3β-hydroxylation of GA_{20} to GA_1 in gibberellin biosynthesis. The reduced level of gibberellins leads to growth retardation of plants.
Mode of action Plant growth regulator and retardant. Foliar applied and absorbed via green tissue; translocated basipetally as well as acropetally within plants.
Uses As an anti-lodging agent in small grain cereals. Also could be used as a growth retardant in turf, peanuts, flowers and to inhibit new twig elongation of fruit trees. **Formulation types** SC; WP. **Mixtures** (prohexadione-calcium +) mepiquat chloride. **Selected tradenames** 'Viviful' (Kumiai)

ANALYSIS
Product and **residues** by hplc. Details available from Kumiai.

MAMMALIAN TOXICOLOGY
prohexadione-calcium
Oral Acute oral LD_{50} for rats and mice >5000 mg/kg. **Skin and eye** Acute percutaneous LD_{50} for rats >2000 mg/kg. Slightly irritating to eyes, not irritating to skin (rabbits). **Inhalation** LC_{50} (4 h, whole body) for rats >4.21 mg/l air.
NOEL (2 y) for male rats 93.9, female rats 114, male mice 279, female mice 351 mg/kg b.w. daily; (1 y) for male and female dogs 20 mg/kg b.w. daily.
Other Non-mutagenic, non-teratogenic (rats and rabbits).

ECOTOXICOLOGY
prohexadione-calcium
Birds Acute oral LD_{50} for bobwhite quail and mallard duck >2000 mg/kg b.w. Dietary LC_{50} (5 d) for bobwhite quail and mallard duck >5200 mg/kg diet.
Fish LC_{50} (96 h) for carp >150 ppm, rainbow trout and bluegill sunfish >100 mg/l. **Bees** LD_{50} (oral and contact) >100 μg/bee. **Worms** LC_{50} (14 d) for *Eisenia foetida* >1000 mg/kg soil. **Daphnia** LC_{50} (48 h) for *D. carinata* >150 mg/l. **Algae** EC_{50} (120 h) for *Selenastrum capricornutum* >100 mg/l.

ENVIRONMENTAL FATE
Animals More than 90% of applied ^{14}C was excreted in urine and faeces.
Plants Prohexadione-calcium applied to plants is ultimately degraded to natural substances. **Soil/Environment** Prohexadione-calcium applied to soil was degraded within a few hours. K_{oc} 4330.

1,3,5-triazine

$$CH_3O-\!\!\!<\!\!\!\begin{array}{c} N \\ \\ N \end{array}\!\!\!>\!\!\!-NHCH(CH_3)_2$$

(structure with NHCH(CH$_3$)$_2$ below)

NOMENCLATURE
Common name prometon (BSI, E-ISO, ANSI, WSSA); prométone ((*m*) F-ISO); no name (Germany)
IUPAC name N^2,N^4-di-isopropyl-6-methoxy-1,3,5-triazine-2,4-diamine
Chemical Abstracts name 6-methoxy-N,N'-bis(1-methylethyl)-1,3,5-triazine-= 2,4-diamine
CAS RN *[1610–18–0]*   **Development codes** G 31 435 (Ciba)

PHYSICAL CHEMISTRY
Mol. wt. 225.3   **M.f.** $C_{10}H_{19}N_5O$ **Form** Colourless powder.   **M.p.** 91–92 °C
V.p. 0.306 mPa (20 °C)   **S.g./density** 1.088 (20 °C)   **Solubility** In water 750 mg/l (20 °C). In benzene >250, methanol, acetone >500, dichloromethane 350, toluene 250 (all in g/l, 20 °C).   **Stability** Stable to hydrolysis at 20 °C in neutral, slightly acidic or slightly alkaline media. Hydrolysed by acids or alkalis on warming. Decomposed by u.v. irradiation.   **pKa** 4.3 (21 °C), weak base

COMMERCIALISATION
History Herbicide reported by H. Gysin & E. Knüsli (*Adv. Pest Control Res.*, 1960, **3**, 289). Introduced by J. R. Geigy S.A. (now Novartis Crop Protection AG).
Patent CH 337019; GB 814948   **Manufacturers** Novartis

APPLICATIONS
Biochemistry Inhibits photosynthesis.   **Mode of action** Non-selective systemic herbicide, absorbed by the foliage and roots, with translocation acropetally.
Uses Control of most annual and many perennial broad-leaved weeds, grasses, and brush weeds on non-crop areas at 10–20 kg a.i./ha. Gives control for a full season or longer.   **Phytotoxicity** Not to be used near desired plants.
Formulation types WP; EC.   **Mixtures** *(prometon +)* atrazine; simazine + sodium chlorate + sodium metaborate.   **Selected tradenames** 'Pramitol' (USA) (Novartis)

ANALYSIS
Product analysis by glc with FID (*AOAC Methods*, 1995, 971.08; K. Hofberg *et al.*, *J. Assoc. Off. Anal. Chem.*, 1973, **56**, 586; B. G. Tweedy & R. A. Kahrs, *Anal. Methods Pestic. Plant Growth Regul.*, 1978, **10**, 493; *CIPAC Handbook*, 1985, **1C**, 2202) or by acidimetric titration (*ibid.*, 2202).   **Residues** determined by glc with ECD (*idem, ibid.*; E. Knüsli, *ibid.*, 1972, **6**, 679) or FID (K. Ramsteiner *et al.*,

J. Assoc. Off. Anal. Chem., 1974, **57**, 192). In **drinking water**, by gc with FID (*AOAC Methods*, 1995, 991.07).

MAMMALIAN TOXICOLOGY

Oral Acute oral LD_{50} for rats 2980, mice 2160 mg/kg. **Skin and eye** Acute percutaneous LD_{50} for rabbits >2000 mg/kg. Mild skin irritant; irritating to eyes (rabbits). **Inhalation** LC_{50} (4 h) for rats >3.26 mg/l air. **NOEL** (90 d) for rats 5.4 mg/kg b.w. daily. **ADI** Not established, not used in crops.
Toxicity class WHO (a.i.) III (Table 5); EPA (formulation) III

ECOTOXICOLOGY

Birds Slightly toxic to birds. **Fish** LC_{50} (96 h) for rainbow trout 12, crucian carp 70, bluegill sunfish 40 mg/l. **Bees** Non-toxic to bees.

ENVIRONMENTAL FATE

Animals For metabolism in rats, see J. E. Bakke *et al.*, *J. Agric. Food Chem.*, 1967, **15**, 628. **Soil/Environment** In soil, microbial degradation is the most important route, involving hydrolytic cleavage of the methoxy group to give hydroxy metabolites, and dealkylation of the side-chains. Persistence in soil can be as much as one year, depending on soil type, moisture, and the application rate used.

597 prometryn   *Herbicide*

1,3,5-triazine

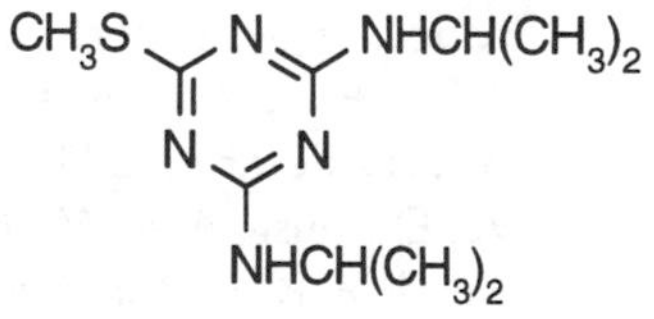

NOMENCLATURE

Common name prometryn (BSI from 1984, E-ISO, ANSI, WSSA); prometryne ((*f*) F-ISO, JMAF and, until 1984, BSI)
IUPAC name N^2,N^4-di-isopropyl-6-methylthio-1,3,5-triazine-2,4-diamine
Chemical Abstracts name *N,N*'-bis(1-methylethyl)-6-(methylthio)-1,3,5-triazine-= 2,4-diamine
CAS RN *[7287–19–6]* **EEC no.** 230–711–3 **Development codes** G 34 161 (Ciba)

PHYSICAL CHEMISTRY

Composition 97% pure. **Mol. wt.** 241.4 **M.f.** $C_{10}H_{19}N_5S$ **Form** White powder.

M.p. 118–120 °C   **B.p.** >300 °C/100 kPa   **V.p.** 0.165 mPa (25 °C) (OECD 104)
K$_{ow}$ logP = 3.1 (25 °C, unionised)   **Henry** 1.2 × 10^{-3} Pa m^3 mol^{-1} (calc.)
S.g./density 1.15 (20 °C)   **Solubility** In water 33 mg/l (25 °C). In acetone 300,
ethanol 140, hexane 6.3, toluene 200, *n*-octanol 110 (all in g/l, 25 °C).
Stability Stable to hydrolysis at 20 °C in neutral, slightly acidic or slightly alkaline
media. Hydrolysed by warm acids and alkalis. Decomposed by u.v. irradiation.
pKa 4.1, weak base

COMMERCIALISATION
History Herbicide reported by H. Gysin (*Chem. Ind. (London)*, 1962, p. 1393).
Introduced by J. R. Geigy S.A. (now Novartis Crop Protection AG).
Patent CH 337019; GB 814948   **Manufacturers** Griffin; Makhteshim-Agan;
Novartis

APPLICATIONS
Biochemistry Photosynthetic electron transport inhibitor. Also inhibits oxidative
phosphorylation.   **Mode of action** Selective systemic herbicide, absorbed by the
leaves and roots, with translocation acropetally through the xylem from the roots
and foliage, and accumulation in the apical meristems.   **Uses** Used pre-emergence
at 0.8–2.5 kg/ha in cotton, sunflower, peanuts, potatoes, carrots, peas and beans;
post-emergence at 0.8–1.5 kg/ha in cotton, potatoes, carrots, celery and leeks.
Phytotoxicity Used as a directed spray on cotton while young to avoid foliar injury.
Formulation types FW; WP; SC.   **Mixtures** (*prometryn* +) metolachlor;
fluometuron; ametryn; terbutryn; simazine.   **Selected tradenames** 'Caparol' (USA)
(Novartis); 'Gesagard' (Novartis); 'Cotton-Pro' (Griffin); 'Efmetryn' (Efthymiadis);
'Prometrex' (Makhteshim-Agan)

ANALYSIS
Product analysis by glc (*CIPAC Handbook*, 1980, **1A**, 1328; *AOAC Methods*, 1995,
971.08; R. T. Murphy *et al.*, *J. Assoc. Off. Anal. Chem.*, 1971, **54**, 703).
Residues determined by glc with ECD or FID (K. Ramsteiner *et al.*, *ibid.*, 1974, **57**,
192; E. Knüsli, *Anal. Methods Pestic. Plant Growth Regul.*, 1972, **6**, 680; B. G.
Tweedy & R. A. Kahrs, *ibid.*, 1978, **10**, 493).

MAMMALIAN TOXICOLOGY
Oral Acute oral LD$_{50}$ for rats 5233 mg/kg.   **Skin and eye** Acute percutaneous
LD$_{50}$ for rats >3100, rabbits >2020 mg/kg; not irritant to skin and slightly irritant
to eyes of rabbits. Not a skin sensitiser (guinea pigs).   **Inhalation** LC$_{50}$ (4 h) for
rats >5170 mg/m^3.   **NOEL** (2 y) for rats 750, dogs 150 ppm; (21 mo) for mice
10 ppm.   **ADI** 0.01 mg/kg.   **Toxicity class** WHO (a.i.) III (Table 5);
EPA (formulation) III (80 WP), II (4 L)

ECOTOXICOLOGY
Birds Eight-day dietary LC$_{50}$ for bobwhite quail >5000, mallard ducks >500 ppm.

Fish LC_{50} (96 h) for rainbow trout 5.5, bluegill sunfish 7.9 mg/l. **Bees** Not toxic to bees; LD_{50} (oral) >99 μg/bee, (contact) >130 μg/bee. **Worms** LC_{50} (14 d) for earthworms 153 mg/kg soil. **Daphnia** LC_{50} (48 h) 12.66 mg/l. **Algae** EC_{50} (5 d) for *Selenastrum capricornutum* 0.023 mg/l. **Other aquatic spp.** EC_{50} for mysid shrimp (*Mysidopsis bahia*) 1.7 mg/l.

ENVIRONMENTAL FATE

Animals For metabolism in rats and rabbits, see C. Boehme & F. Baer, *Food Cosmet. Toxicol.*, 1967, **5**, 23. **Plants** Metabolised by tolerant plants, and, to a lesser extent, by sensitive plants, by oxidation of the methylthio group to hydroxy metabolites, and by dealkylation of the side-chains. Degradation in plants is generally slow. **Soil/Environment** In soil, microbial degradation takes place, with oxidation of the methylthio group to hydroxy metabolites, and dealkylation of the side-chains. Median DT_{50} in soil 50 d (14–158 d). K_{oc} 400, indicating low mobility in soil. Degradation in aquatic systems is caused by microbial processes; further dissipation results from adsorption to the suspended and bottom sediments.

598 propachlor

Herbicide

chloroacetanilide

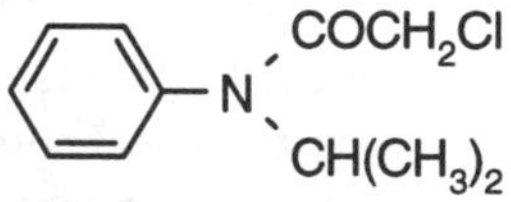

NOMENCLATURE

Common name propachlor (BSI, E-ISO, WSSA); propachlore ((*m*) F-ISO)
IUPAC name 2-chloro-*N*-isopropylacetanilide; α-chloro-*N*-isopropylacetanilide
Chemical Abstracts name 2-chloro-*N*-(1-methylethyl)-*N*-phenylacetamide
CAS RN *[1918–16–7]* **EEC no.** 217–638–2 **Development codes** CP 31 393 (Monsanto)

PHYSICAL CHEMISTRY

Composition 96.5% pure. **Mol. wt.** 211.7 **M.f.** $C_{11}H_{14}ClNO$ **Form** Light tan solid. **M.p.** 77 °C (pure); 67–76 °C (tech.) **B.p.** 110 °C/0.03 mmHg
V.p. 10.5 mPa (25 °C) K_{ow} logP = 1.5–2.3 **S.g./density** 1.134 (25 °C)
Solubility In water 580 mg/l (25 °C). In acetone 448, benzene 737, toluene 342, ethanol 408, xylene 239, chloroform 602, carbon tetrachloride 174, diethyl ether 219 (all in g/kg, 25 °C). Slightly soluble in aliphatic hydrocarbons.
Stability Decomposed in alkaline and strongly acidic media. Decomposes at 170 °C. Stable to u.v. light. **F.p.** 173.8 °C (open cup ASTM)

COMMERCIALISATION

History Herbicide reported by D. D. Baird et al. (*Proc. North Cent. Weed Control Conf.*, 1964). Introduced by Monsanto Chemical Co. **Patent** US 2863752 **Manufacturers** Makhteshim-Agan; Monsanto; Sagrochem

APPLICATIONS

Biochemistry Acts by inhibition of cell elongation and protein synthesis, and possibly also nucleic acid synthesis. **Mode of action** Selective herbicide, absorbed mainly by germinating seedling shoots, and secondarily by roots, with translocation throughout the plant, and higher concentrations in vegetative parts than in reproductive parts. **Uses** Used pre-emergence, pre-planting incorporated or early post-emergence to control annual grasses and some broad-leaved weeds in beans, brassicas, cotton, peanuts, leeks, maize, onions, peas, roses, ornamental trees and shrubs, soya beans and sugar cane at 3.36–6.72 kg a.i./ha. **Formulation types** WP; SC; GR. **Mixtures** *(propachlor +)* atrazine; chlorthal-dimethyl; chloridazon. **Selected tradenames** 'Ramrod' (Monsanto); 'Albrass' (Zeneca); 'Atlas Propachlor' (Atlas Crop Protection); 'Prolex' (Makhteshim-Agan); 'Satecid' (Sagrochem)

ANALYSIS

Product analysis by glc with FID (*AOAC Methods*, 1995, 986.05; *CIPAC Handbook*, 1988, **D**, 151). **Residues** determined by glc with ECD. In **drinking water** by gc with ECD (*AOAC Methods*, 1995, 990.06). Details available from Monsanto Co.

MAMMALIAN TOXICOLOGY

EHC 147 (WHO, 1993). **Oral** Acute oral LD_{50} for rats 550–1700 mg tech./kg. **Skin and eye** Acute percutaneous LD_{50} for rabbits >20 000 mg/kg. Slight skin irritation; moderate eye irritation (rabbits). Contact dermal sensitisation observed in guinea pigs. Formulations may produce skin sensitisation in susceptible individuals. **Inhalation** LC_{50} (4 h) for rats >1.2 mg/l **NOEL** (2 y) for rats 2.6 mg/kg b.w.; (18 mo) for mice 1.6 mg/kg b.w.; (1 y) for dogs 9 mg/kg b.w. No evidence of carcinogenic, teratogenic, or reproductive effects. **Toxicity class** WHO (a.i.) III; EPA (formulation) III **EC risk** Xn (R22); Xi (R36); (R43)

ECOTOXICOLOGY

Birds Acute oral LD_{50} for bobwhite quail 91 mg/kg. Eight-day dietary LC_{50} for bobwhite quail and mallard ducks >5620 mg/kg diet. **Fish** LC_{50} (96 h) for bluegill sunfish >1.4, rainbow trout 0.17 mg/l. **Bees** LC_{50} (48 h, oral) >1000 ppm; LD_{50} (48 h, contact) >25 µg/bee. **Worms** EC_{50} (14 d) for earthworms 217.9 mg/kg soil. **Daphnia** LC_{50} (48 h) 7.8 mg/l. **Algae** TL_{50} (72 h) for *Selenastrum capricornutum* 0.029 mg/l.

ENVIRONMENTAL FATE
EHC 147 notes that propachlor is rapidly degraded by micro-organisms in soil and water, and concludes that propachlor does not bioconcentrate or biomagnify. High toxicity to some aquatic organisms suggests that direct contamination of water should be avoided. Rapidly eliminated from mammals. **Animals** In mammals, following oral administration, excreted in the urine within 72 hours. Metabolised in the liver and possibly in the kidneys. Glutathione conjugation is the main metabolic step (J. E. Bakke et al., Science, 1980, **210**, 433). **Plants** Rapidly metabolised in plants to polar compounds and to the 2-hydroxy analogue; glutathione conjugation is involved (G. L. Lamoureu et al., J. Agric. Food Chem., 1971, **19**, 346–351). **Soil/Environment** In soil, microbial degradation involving dechlorination occurs (N. J. Novick et al., J. Agric. Food Chem., 1986, **34**, 721–725). DT_{50} 4 d (3 soil types). Persists in soil for 4–6 weeks. Loss by photodecomposition or volatilisation is negligible.

99 propamocarb hydrochloride   *Fungicide*

carbamate

$$(CH_3)_2N(CH_2)_3NHCO_2(CH_2)_2CH_3 \quad .HCl$$

NOMENCLATURE
propamocarb hydrochloride
IUPAC name propyl 3-(dimethylamino)propylcarbamate hydrochloride
Chemical Abstracts name propyl [3-(dimethylamino)propyl]carbamate hydrochloride
CAS RN *[25606–41–1]* **EEC no.** 247–125–9 **Development codes** AE B066752; SN 66 752; Zk 66752

propamocarb
Common name propamocarb (BSI, E-ISO, ANSI); propamocarbe ((m) F-ISO)
IUPAC name propyl 3-(dimethylamino)propylcarbamate
Chemical Abstracts name propyl [3-(dimethylamino)propyl]carbamate
CAS RN *[24579–73–5]* **Development codes** SN 39744

PHYSICAL CHEMISTRY
propamocarb hydrochloride
Composition The aqueous concentrate contains 780 g/l propamocarb hydrochloride. **Mol. wt.** 224.7 **M.f.** $C_9H_{21}ClN_2O_2$ **Form** Colourless, faintly aromatic, hygroscopic crystals. **M.p.** 45–55 °C **V.p.** 3.85×10^{-2} mPa (20 °C) **K_{ow}** logP = –2.6 **Henry** 8.65×10^{-9} Pa m³ mol⁻¹ (20 °C, calc.)

S.g./density 1.085 g/ml (20 °C, aqueous concentrate) **Solubility** In water 1005 g/l (20 °C). In methanol 656, dichloromethane >626, acetone 560, ethyl acetate 4.34, toluene 0.14, hexane <0.01 (all in g/l, 20 °C). **Stability** Stable to hydrolysis, to temperatures up to 400 °C, and to photolysis.

propamocarb
Mol. wt. 188.3 **M.f.** $C_9H_{20}N_2O_2$ **V.p.** 730 mPa (25 °C) **K$_{ow}$** logP = 0.84 (20 °C) **Henry** 1.5×10^{-4} Pa m^3 mol^{-1} (25 °C) **Solubility** In water >900 g/l (pH 7.0, 20 °C). In hexane >883, methanol >933, dichloromethane >937, toluene >852, acetone >921, ethyl acetate >856 (all in g/l, 20 °C). **pKa** 9.5, strong base (K. Chamberlain et al., *Pestic Sci.*, **47**, 265 (1996))

COMMERCIALISATION
History Fungicide reported by E. A. Pieroh et al. (*Meded. Fac. Landbouwwet. Rijksuniv. Gent*, 1978, **43**, 933). The hydrochloride introduced by Schering AG (now AgrEvo GmbH). **Patent** DE 1567169; DE 1643040 **Manufacturers** AgrEvo; Agriphar

APPLICATIONS
Biochemistry Multi-site inhibitor. **Mode of action** Systemic fungicide with protective action. Absorbed by the roots and leaves, and transported acropetally. **Uses** Specific control of phycomycetous diseases (*Pythium*, *Phytophthora*, *Aphanomyces*, *Bremia*, *Peronospora*, and *Pseudoperonospora* spp.). In particular, control of *Pythium* and *Phytophthora* spp. in vegetables, ornamentals, glasshouse tomatoes, glasshouse cucumbers, tulips, tobacco, and in forestry seedbeds; *Pythium* blight on turf; downy mildew on lettuce, cucurbits and cabbages; *Phytophthora infestans* in potatoes and tomatoes; *Phytophthora cactorum* on strawberries; etc. Applied to the soil, or used as a dip treatment (for bulbs and tubers) or seed treatment, and as a foliar spray in late-season applications. **Formulation types** SC. **Selected tradenames** 'Banol' (NOR-AM); 'Previcur' (AgrEvo); 'Proplant' (Agriphar)

ANALYSIS
Product analysis by lc with u.v. detection (*CIPAC Handbook*, 1992, **E**, 184–6). **Residues** determined by glc with FID (I. A. Gentile & E. Passera, *J. Chromatog.*, 1982, **236**, 254). Details available from AgrEvo.

MAMMALIAN TOXICOLOGY
propamocarb hydrochloride
Reviews *Pesticide residues in food – 1986*. FAO Plant Production and Protection Paper 77, 1986. *Pesticide residues in food – 1986 evaluations. Part II – Toxicology*. FAO Plant Production and Protection Paper 78/2, 1987. **Oral** Acute oral LD$_{50}$ for rats 2000–2900, mice 2650–2800, dogs c. 1450 mg/kg. **Skin and eye** Acute percutaneous LD$_{50}$ for rats and mice >3000 mg/kg. Not a skin or eye irritant (rabbits). Not a skin sensitiser. **Inhalation** LC$_{50}$ (4 h) for rats >0.0057 mg/l air.

NOEL (2 y) for rats 1000, dogs 3000 ppm diet (41 and 70 mg/kg b.w. respy.).
ADI (JMPR) 0.1 mg/kg b.w. [1986].   Other Acute i.p. LD$_{50}$ for rats
437–460 mg/kg. Not mutagenic.

propamocarb
Inhalation LC$_{50}$ (4 h) for rats >3.96 mg/l air.   Toxicity class WHO (a.i.) III
(Table 5); EPA (formulation) IV

ECOTOXICOLOGY
propamocarb hydrochloride
Birds Acute oral LD$_{50}$ for pheasants 3050, mallard ducks >6290 mg/kg. LC$_{50}$ (5 d)
tor pheasants >52 000, Japanese quail >25 000, mallard ducks 12 915 ppm diet.
Fish LC$_{50}$ (96 h) for sheepshead minnow >100, carp 155, bluegill sunfish 275,
trout 275 mg/l.   Bees LD$_{50}$ >0.1 mg/bee ('Previcur N').   Worms LC$_{50}$ (14 d)
>1000 mg/kg soil.   Other beneficial spp. Not toxic to Argsope argentata (web-
building spider), Aleochara sp. (staphylinid beetle); not harmful to Poecilus cupreus
(carabid beetle); slightly harmful to C. septemfunctata (ladybird).   Daphnia LC$_{50}$
(48 h) 280 mg/l.   Algae E$_r$C$_{50}$ (96 h) for Scenedesmus quadricauda 560 mg/l,
E$_b$C$_{50}$ 360 mg/l.   Other aquatic spp. EC$_{50}$ (96 h) for Eastern oyster (Crassostrea
virginica) 43.9 mg/l; LC$_{50}$ (96 h) for mysid shrimp (Mysidopsis bahia) >100 mg/l.

ENVIRONMENTAL FATE
Animals Rapidly absorbed and almost totally excreted (>90% in 24 h), mainly via
urine. Mineralisation occurs via oxidation and hydrolytic decomposition.
Plants Mainly unchanged in plants.   Soil/Environment Rapidly degraded in soil by
microbial processes, following a brief lag phase, DT$_{50}$ <30 d, DT$_{90}$ <70 d.
Propamocarb hydrochloride is retained in the upper soil layer (4–20 cm) and little
is found in leachate. Stable in aqueous medium, but rapidly degraded by aquatic
micro-organisms (up to 97% in 35 d). It is adsorbed onto sediment, but with
limited desorption.

00 propanil

Herbicide

anilide

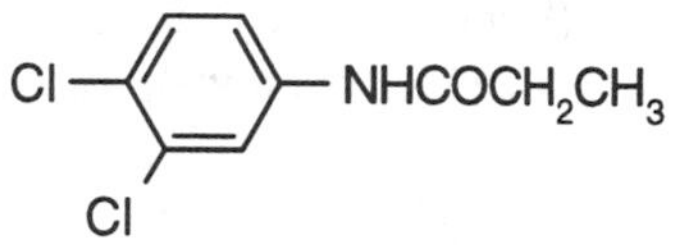

Cl—⟨benzene ring⟩— NHCOCH$_2$CH$_3$
   |
   Cl

NOMENCLATURE
Common name propanil (BSI, E-ISO, (m) F-ISO, WSSA); DCPA (JMAF); no name
(Germany)

IUPAC name 3',4'-dichloropropionanilide
Chemical Abstracts name N-(3,4-dichlorophenyl)propanamide
Other names 3,4-DCPA   **CAS RN** *[709–98–8]*   **EEC no.** 211–914–6
Development codes FW-734 (Rohm & Haas); Bayer 30 130; S 10145 (Bayer)

PHYSICAL CHEMISTRY
Mol. wt. 218.1   **M.f.** $C_9H_9Cl_2NO$ **Form** Colourless odourless crystals; (tech.,
brown crystals).   **M.p.** 91.5 °C   **V.p.** 0.02 mPa (20 °C); 0.05 mPa (25 °C)
K_{ow} logP = 3.3 (20 °C)   **S.g./density** 1.41 g/cm^3 (22 °C)   **Solubility** In water
130 mg/l (20 °C). In isopropanol, dichloromethane >200, toluene 50–100,
hexane <1 (all in g/l, 20 °C). In benzene 7×10^4, acetone 1.7×10^6, ethanol
1.1×10^6 (all in ppm, 25 °C).   **Stability** Hydrolysed in strongly acidic and alkaline
media to 3,4-dichloroaniline and propionic acid; stable at normal pH range: DT_{50}
(22 °C) >>1 y (pH 4, 7, 9). Rapidly degraded in water by sunlight; photolysis
DT_{50} 12–13 h.

COMMERCIALISATION
History Herbicide (*Proc. South. Weed Control Conf.*, 1960, p. 20). Introduced by
Rohm & Haas Co. in 1961, later by Bayer AG (1965) and by Monsanto Chemical
Co.   **Patent** DE 1039779; GB 903766 to Bayer   **Manufacturers** Bayer; Cedar;
Defensa; Griffin; Hodogaya; Inquiport; Nufarm Ltd; Proficol; Rohm & Haas; Tifa;
Westrade

APPLICATIONS
Biochemistry Photosynthetic electron transport inhibitor.
Mode of action Selective contact herbicide with a short duration of activity.
Uses Contact herbicide used post-emergence in rice at 2.5–5.0 kg/ha to control
broad-leaved and grass weeds including *Amaranthus retroflexus*, *Digitaria* spp.,
Echinochloa spp., *Panicum* spp., and *Setaria* spp. Also used, in mixture with MCPA,
in wheat. A mixture with carbaryl is used in citrus crops grown in sod culture.
Phytotoxicity Phytotoxic to many broad-leaved crops. Normally phytotoxic to
crops which have been treated with organophosphate or carbamate insecticides.
Formulation types EC; SC; UL.   **Mixtures** *(propanil +)* bentazone; molinate;
MCPA; carbaryl; fenoprop; bifenox; MCPA + pyrazoxyfen; MCPA + mecoprop;
MCPA-thioethyl; MCPA-thioethyl + pyridate.   **Compatibility** Incompatible with a
number of pesticides, particularly carbamates and organophosphates. Incompatible
with liquid fertilisers.   **Selected tradenames** 'Stam' (Rohm & Haas); 'Stampede'
(Rohm & Haas); 'Surcopur' (Bayer); 'Nox' (Crystal); 'Propasint' (Westrade);
'RiceNil' (Gilmore); 'Riselect' (Isagro)

ANALYSIS
Product analysis by glc (I. L. Adler & W. J. Zogorski, *Anal. Methods Pestic. Plant
Growth Regul.*, 1984, **13**, 281).   **Residues** determined by glc (*idem, ibid.; ibid.*,
1972, **6**, 692) or colorimetry of the 3,4-dichloroaniline formed on hydrolysis

(C. F. Gordon et al., Anal. Methods Pestic., Plant Growth Regul. Food Addit., 1964, **4**, 235). **In water**, by lc with u.v. detection (AOAC Methods, 1995, 992.14, 10.7.01). Methods for the determination of residues also available from Bayer.

MAMMALIAN TOXICOLOGY
Oral Acute oral LD_{50} for rats >2500, mice c. 1800 mg/kg.   **Skin and eye** Acute percutaneous LD_{50} for rats (24 h) >5000 mg/kg; not irritating to skin and eyes (rabbits). Not a skin sensitiser (guinea pigs).   **Inhalation** LC_{50} (4 h) for rats >1.25 mg/l air (dust).   **NOEL** (2 y) for rats 400, dogs 600 mg/kg diet. **ADI** 0.005 mg/kg b.w. (US).   **Other** Not mutagenic, not carcinogenic. **Toxicity class** WHO (a.i.) III (Table 5); EPA III ('Propanex'), II ('Supernox') **EC risk** Xn (R22)

ECOTOXICOLOGY
Birds Acute oral LD_{50} for mallard ducks 375, bobwhite quail 196 mg/kg. Dietary LC_{50} (5 d) for mallard ducks 5627, bobwhite quail 2861 ppm   **Fish** LC_{50} (48 h) for carp 8–11 mg/l.   **Daphnia** LC_{50} (48 h) 4.8 mg/l.

ENVIRONMENTAL FATE
Animals The major metabolism pathway for propanil in microsomal incubations was acylamidase hydrolysis to 3,4-dichloroaniline (D. C. McMillan et al., Tox. Appl. Toxicol., 1990, **103**, 90).   **Plants** In rice, propanil is hydrolysed by an aryl acylamidase to 3,4-dichloroaniline and propionic acid as metabolic intermediates. **Soil/Environment** In soil, rapid microbial degradation to the aniline derivative occurs. Duration of activity in warm, moist conditions is only a few days. Degradation products are propionate which is rapidly metabolised to CO_2, and 3,4-dichloroaniline which is bound (80% in 27 h) to the soil. K_{oc} 239–800.

601 propaphos   *Insecticide*

organophosphorus

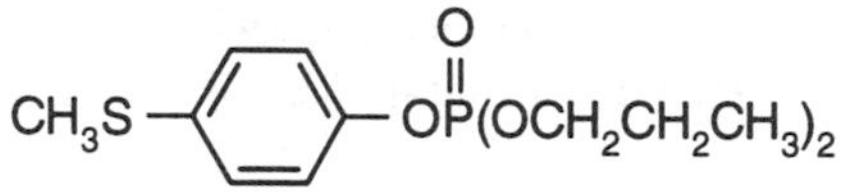

CH$_3$S —⟨benzene ring⟩— O—P(OCH$_2$CH$_2$CH$_3$)$_2$ with =O

NOMENCLATURE
Common name propaphos (BSI, JMAF, draft E-ISO); propafos (draft (m) F-ISO)
IUPAC name 4-(methylthio)phenyl dipropyl phosphate

Chemical Abstracts name 4-(methylthio)phenyl dipropyl phosphate
CAS RN *[7292–16–2]* **Development codes** NK-1158 (Nippon Kayaku)

PHYSICAL CHEMISTRY
Mol. wt. 304.3 **M.f.** $C_{13}H_{21}O_4PS$ **Form** Colourless liquid.
B.p. 175–177 °C/0.85 mmHg **V.p.** 0.12 mPa (25 °C) **K_{ow}** logP = 3.67
S.g./density 1.1504 (20 °C) **Solubility** In water 125 mg/l (25 °C). Soluble in most
organic solvents. **Stability** Stable <230 °C; in neutral and acidic media, but is
slowly decomposed in alkaline media.

COMMERCIALISATION
History Insecticide reported (*Jpn. Pestic. Inf.*, 1970, No. 4, p. 7). Introduced by
Nippon Kayaku Co., Ltd. **Patent** JP 482500; JP 462729 **Manufacturers** Nippon
Kayaku

APPLICATIONS
Biochemistry Cholinesterase inhibitor. **Mode of action** Systemic insecticide with
contact and stomach action. **Uses** Control of *Chilo suppressalis*, *Laodelphax
striatella*, *Nephotettix cincticeps*, *Oulema oryzae* (including strains resistant to other
organophosphorus and carbamate insecticides) in paddy rice.
Formulation types GR; DP. **Mixtures** (*propaphos* +) metolcarb; benfuracarb;
carbaryl. **Selected tradenames** 'Kayaphos' (Nippon Kayaku)

ANALYSIS
Product and **residue** analysis by glc (S. Asaka *et al.*, *Noyaku Kagayu*, 1975, **3**, 36).

MAMMALIAN TOXICOLOGY
Reviews *J. Pestic. Sci.*, **14**, 511–515 (1989). **EHC** 63 (WHO, 1986; a general
review of organophosphorus insecticides). **Oral** Acute oral LD_{50} for rats 70,
mice 90, rabbits 82.5 mg/kg. **Skin and eye** Acute percutaneous LD_{50} for rats 88.5
mg/kg. **Inhalation** LC_{50} for rats 39.2 mg/m^3. **NOEL** (2 y) for rats 0.08, mice
0.05 mg/kg b.w. **Toxicity class** WHO (a.i.) Ib

ECOTOXICOLOGY
Birds LD_{50} for chickens 2.5–5.0 mg/kg. **Fish** LC_{50} (48 h) for carp 4.8 mg/l.
Bees Toxic to bees. **Daphnia** Toxic.

ENVIRONMENTAL FATE
Plants For metabolism in plants, see Y. Fujii *et al.*, *J. Pestic. Sci.*, **5** (1), 55–62
(1980).

2-(4-aryloxyphenoxy)propionic acid

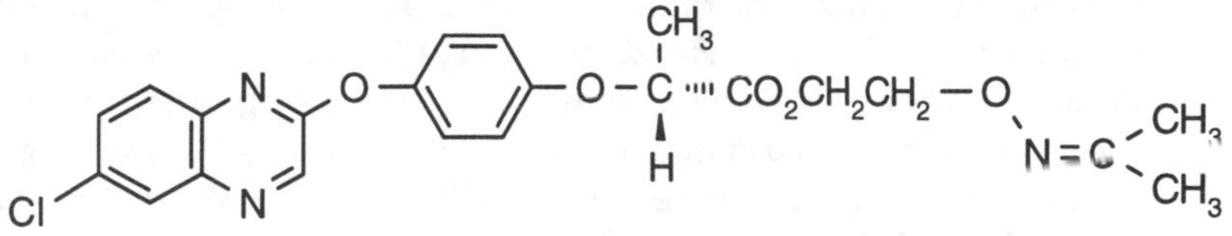

NOMENCLATURE
Common name propaquizafop (BSI, draft E-ISO, (*m*) draft F-ISO)
IUPAC name 2-isopropylideneamino-oxyethyl (*R*)-2-[4-(6-chloroquinoxalin-=
2-yloxy)phenoxy]propionate
Chemical Abstracts name (*R*)-2-[[(1-methylethylidene)amino]oxy]ethyl
2-[4-[(6-chloro-2-quinoxalinyl)oxy]phenoxy]propanoate
CAS RN *[111479–05–1]* **Development codes** Ro 17–3664/000 (Maag);
CGA 233 380 (Ciba)

PHYSICAL CHEMISTRY
Composition (*R*)- isomer. **Mol. wt.** 443.9 **M.f.** $C_{22}H_{22}ClN_3O_5$ **Form** Colourless
crystals. **M.p.** 66.3 °C **V.p.** 4.4×10^{-7} mPa (25 °C) (OECD 104)
K_{ow} logP = 4.78 (25 °C) **Henry** 9.2×10^{-8} Pa m^3 mol^{-1} (calc.) **S.g./density** 1.30
(20 °C) **Solubility** In water 0.63 mg/l (25 °C). In ethanol 59, acetone 730,
toluene 630, *n*-hexane 37, *n*-octanol 16 (all in g/l, 25 °C). **Stability** Stable ≥2 y in
closed container at room temperature. Moderately stable at acid and neutral pH.
Rapidly hydrolysed under alkaline conditions. Stable to u.v. light.

COMMERCIALISATION
History Herbicide reported by P. F. Bocion *et al.* (*Proc. 1987 Br. Crop Prot. Conf. –
Weeds*, **1**, 55). Introduced by Dr R. Maag Ltd. **Patent** EP 52798, US 4545807
Manufacturers Novartis

APPLICATIONS
Biochemistry Fatty acid synthesis inhibitor (ACCase). **Mode of action** Systemic
post-emergence herbicide. Absorbed by foliage and roots and translocated
throughout the plant. Treated grasses cease growth within 3–4 days, show
chlorosis of younger plant tissues, followed by a progressive collapse of the entire
plant 10–20 days later. **Uses** Control of a wide range of annual and perennial
grasses (including *Sorghum halepense*, *Agropyron repens* and *Cynodon dactylon*) in
soya beans, cotton, sugar beet, potatoes, peanuts, peas, oilseed rape and
vegetables. **Phytotoxicity** Occasionally soya beans and peas may show some
chlorotic and necrotic spots on treated leaves at higher dosage rates; this does not
affect further vegetative growth or yield. Not to be used on cucurbits.
Formulation types EC. **Selected tradenames** 'Agil' (Novartis)

MAIN ENTRIES

ANALYSIS
Product analysis by hplc. **Residues** determined by hplc or by glc with ECD.

MAMMALIAN TOXICOLOGY
Oral Acute oral LD_{50} for rats >5000, mice 3009 mg/kg. **Skin and eye** Acute percutaneous LD_{50} for rats >2000 mg/kg. Non-irritant to skin; moderate eye irritation (rabbits). No allergic sensitisation in open epicutaneous test (guinea pigs); sensitising in Maximisation test. **Inhalation** LC_{50} (4 h) 2.5 mg/l air. **NOEL** (2 y) for rats and mice 1.5 mg/kg b.w. daily; (1 y) for dogs 20 mg/kg b.w. daily. **ADI** 0.015 mg/kg b.w. **Toxicity class** WHO (a.i.) III (Table 5) **EC risk** R43

ECOTOXICOLOGY
Birds LC_{50} (10 d) for mallard ducks and (14 d) for bobwhite quail >6593 mg/kg feed. **Fish** LC_{50} (96 h) for rainbow trout 1.2, carp 0.19, bluegill sunfish 0.34 mg/l. **Bees** LD_{50} (48 h) (oral) >20 µg/bee, (contact) >200 µg/bee. **Worms** LC_{50} (14 d) >1000 mg/kg soil. **Daphnia** EC_{50} (acute) >>2 mg/l. **Algae** EC_{50} (96 h) for *Selenastrum capricornutum* >2.1 mg/l.

ENVIRONMENTAL FATE
Animals Rapidly eliminated in the rat via urine and faeces following oral administration. **Plants** Rapidly broken down in soya bean, sugar beet and cotton foliage to the free acid, followed by further metabolism to the quinoxaline oxyphenol. **Soil/Environment** Fast degradation to the free acid (DT_{50} c. 3 d) and to further metabolites. DT_{50} (field) for parent and metabolites 15–26 d after spring application. DT_{50} for aquatic degradation and dissipation, for parent and metabolites c. 20 d. No tendency for bioaccumulation.

603 propargite   *Acaricide*

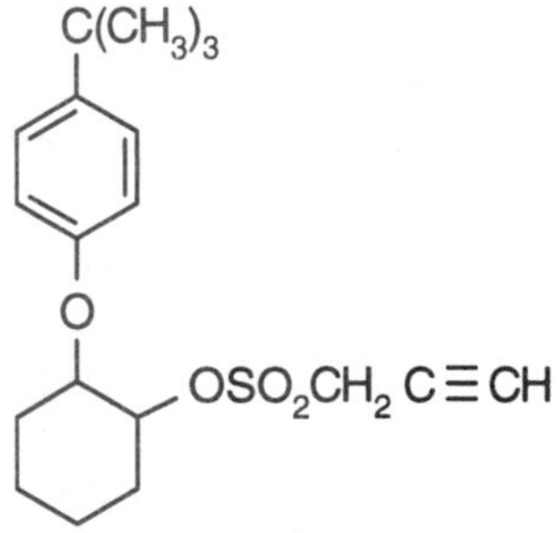

NOMENCLATURE
Common name propargite (BSI, E-ISO, (m) F-ISO, ANSI, ESA); BPPS (JMAF)

IUPAC name 2-(4-*tert*-butylphenoxy)cyclohexyl prop-2-ynyl sulfite
Chemical Abstracts name 2-[4-(1,1-dimethylethyl)phenoxy]cyclohexyl 2-propynyl sulfite
CAS RN *[2312–35–8]* **EEC no.** 219–006–1 **Development codes** DO 14
(Uniroyal) **Official codes** ENT 27 226

PHYSICAL CHEMISTRY
Composition Tech. grade is >85% *m*/*m* pure. **Mol. wt.** 350.5 **M.f.** $C_{19}H_{26}O_4S$
Form Dark reddish-brown viscous liquid (tech,). **V.p.** 0.006 mPa (25 °C)
K$_{ow}$ logP = 3.73 **S.g./density** 1.1130 (20 °C) **Solubility** In water 632 mg/l
(25 °C). Miscible with most organic solvents, such as acetone, benzene, ethanol,
hexane, heptane, and methanol. **Stability** DT$_{50}$ 80 d (pH 7); decomposed by
strong acids and alkalis (pH >10). No degradation in packaging at 20 °C for 1 y
and 50% r.h. About 1% decomposition after 14 d at 55 °C in water-saturated air.
About 3% decomposition after continuous exposure to simulated sunlight for 7 d
at 20 °C. **pKa** >12 **F.p.** 71.4 °C (Pensky-Marten closed cup)

COMMERCIALISATION
History Acaricide introduced by Uniroyal. **Patent** US 3272854; US 3463859
Manufacturers Jin Hung; Uniroyal

APPLICATIONS
Mode of action Non-systemic acaricide with predominantly contact action, and
long residual activity. **Uses** Control of many species of phytophagous mites
(particularly motile stages) on a variety of crops including vines, fruit trees
(including top fruit, stone fruit, citrus fruit), hops, nuts, tomatoes, vegetables,
ornamentals, cotton, maize, peanuts, and sorghum. Used at rates from
0.75–1.8 kg/ha on row crops and foliar sprays of 0.85–3.0 kg/ha on perennial
fruit and nut crops. **Phytotoxicity** Phytotoxic to pears, strawberries, roses, and
cotton under 10 inches in height. Citrus fruit and beans may also show some
injury. **Formulation types** WP; EC; EW. **Mixtures** *(propargite +)* tetradifon;
parathion-methyl; hexythiazox; clofentezine. **Compatibility** Incompatible with
alkaline materials, oil sprays, and pesticides containing a large amount of
petroleum solvents. **Selected tradenames** 'Omite' (Uniroyal); 'Retador'
(Productos OSA)

ANALYSIS
Product analysis by i.r. spectroscopy, details available from Uniroyal Chemical Co.
Residues determined by glc (*Man. Pestic. Residue Anal.*, 1987, **I**, 6; *Anal. Methods
Residues Pestic.*, 1988, Part I, M1; A. Ambrus *et al.*, *J. Assoc. Off. Anal. Chem.*, 1981,
64, 733; G. M. Stone, *Anal. Methods Pestic. Plant Growth Regul.*, 1973, **7**, 355).

MAMMALIAN TOXICOLOGY
Reviews *Pesticide residues in food – 1982.* FAO Plant Production and Protection

Paper 46, 1983. *Pesticide residues in food: 1982 evaluations.* FAO Plant Production and Protection Paper 49, 1983. **Oral** Acute oral LD_{50} for rats 2800, rabbits 311 mg/kg. **Skin and eye** Acute percutaneous LD_{50} for rabbits 4000 mg/kg; severe eye and skin irritant to rabbits, not a skin sensitiser to guinea pigs. **Inhalation** LC_{50} (4 h) for rats 0.89 mg/l. **NOEL** In chronic feeding studies in rats and dogs, NOEL 4 mg/kg b.w. daily. Carcinogenic in rats. **ADI** (JMPR) 0.15 mg/kg b.w. [1982]. **Other** Acute i.p. LC_{50} for male rats 260, female rats 172 mg/kg. **Toxicity class** WHO (a.i.) III; EPA (formulation) I **EC risk** Xn (R22); Xi (R36)

ECOTOXICOLOGY
Birds Acute oral LD_{50} for mallard ducks >4640 mg/kg. Five-day dietary LC_{50} for mallard ducks >4640, bobwhite quail 3401 mg/kg diet. **Fish** LC_{50} (96 h) for rainbow trout 0.118, bluegill sunfish 0.168, catfish 0.04, sheepshead minnow 0.06 mg/l. **Bees** LD_{50} (48 h, contact) 15 µg/bee. **Daphnia** LC_{50} (48 h) 0.092 mg/l. **Other aquatic spp.** LC_{50} (96 h) for grass shrimp 0.101 mg/l; for Quahog clam embryo larvae 0.110 mg/l.

ENVIRONMENTAL FATE
Animals In mammals, propargite is hydrolysed at the sulfite ester linkage to 1-[4-(1,1-dimethylethyl)phenoxy]-2-cyclohexanol and subsequent hydroxylation of the *tert*-butyl side-chain. Additional metabolites are formed by further oxidation or sulfation of the *tert*-butyl group and oxidation of the cyclohexyl moiety.
Plants Although propargite is considered a non-systemic pesticide, a small portion of the applied dose penetrates the outside layer of the foliage and undergoes the same metabolism as is observed in animals. In most fruits (applies, citrus), propargite stays mainly on the surface and, to a lesser extent, in the peel; only trace residues were detected in the pulp. **Soil/Environment** DT_{50} in most soils 7–14 w. K_{oc} 23 000–90 000, relatively immobile in soil.

604 propazine   *Herbicide*

1,3,5-triazine

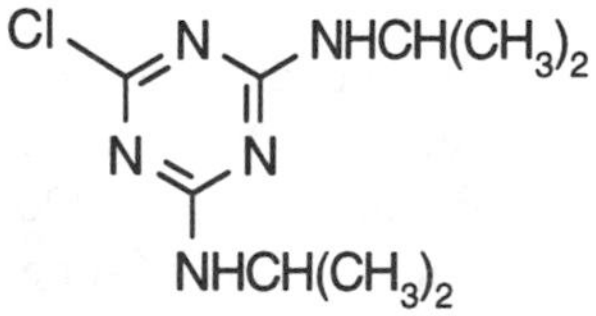

NOMENCLATURE
Common name propazine (BSI, E-ISO, (*f*) F-ISO, ANSI, WSSA, JMAF); no name (Germany)

IUPAC name 6-chloro-N^2,N^4-di-isopropyl-1,3,5-triazine-2,4-diamine
Chemical Abstracts name 6-chloro-N,N'-bis(1-methylethyl)-1,3,5-triazine-=
2,4-diamine
CAS RN *[139–40–2]* **EEC no.** 205–359–9 **Development codes** G 30 028 (Ciba)

PHYSICAL CHEMISTRY
Mol. wt. 229.7 **M.f.** $C_9H_{16}CIN_5$ **Form** Colourless powder. **M.p.** 212–214 °C
V.p. 0.0039 mPa (20 °C) **S.g./density** 1.162 (20 °C) **Solubility** In water 5.0 mg/l
(20 °C). In benzene, toluene 6.2, diethyl ether 5.0, carbon tetrachloride 2.5 (all in
g/kg, 20 °C). **Stability** Stable in neutral, slightly acidic, and slightly alkaline media.
Hydrolysed by acids and alkalis on heating, with the formation of hydroxy=
propazine. **pKa** 1.7 (21 °C), v. weak base

COMMERCIALISATION
History Herbicide reported by H. Gysin & E. Knüsli (*Proc. Int. Congr. Crop Protect.*,
4th, 1957, **1**, 549). Introduced by J. R. Geigy S.A. (now Novartis Crop Protection
AG, who no longer manufacture or market it). **Patent** BE 540947; GB 814947
Manufacturers Makhteshim-Agan

APPLICATIONS
Biochemistry Photosynthetic electron transport inhibitor. It is metabolised in
tolerant plants to the corresponding 6-hydroxy compound.
Mode of action Selective systemic herbicide, absorbed by the roots, with
translocation acropetally in the xylem, and accumulation in the apical meristems
and leaves. **Uses** Pre-planting or pre-emergence control of grasses and broad-
leaved weeds in sorghum and umbelliferous crops (carrots, chervil, parsley, etc.) at
0.5–3.0 kg a.i./ha. **Phytotoxicity** Sugar beet and many vegetables are sensitive.
Formulation types WP; WG; Liquid. **Mixtures** *(propazine +)* metolachlor;
chlorpropham; prometryn; propachlor. **Selected tradenames** ['Gesamil']
(Novartis); ['Milogard'] (Novartis); 'Milo-Pro' (Griffin); 'Prozinex' (Makhteshim-
Agan)

ANALYSIS
Product analysis by glc with FID (*CIPAC Handbook*, 1980, **1A**, 1333; *AOAC Methods*,
1995, 971.08; B. G. Tweedy & R. A. Kahrs, *Anal. Methods Pestic.*, *Plant Growth
Regul.*, 1978, **10**, 493). **Residues** determined by glc with ECD or FID (E. Knüsli,
ibid., 1972, **6**, 234, 679; K. Ramsteiner *et al.*, *J. Assoc. Off. Anal. Chem.*, 1974, **57**,
192). In **drinking water**, by gc with FID (*AOAC Methods*, 1995, 991.07).

MAMMALIAN TOXICOLOGY
Oral Acute oral LD_{50} for rats >7000 mg/kg. **Skin and eye** Acute percutaneous
LD_{50} for rats >3100, rabbits >10 200 mg/kg. Mild eye and skin irritant (rabbits).
Inhalation LC_{50} (4 h) for rabbits >2.04 mg/l air. **NOEL** In 130 d feeding trials,
male and female rats receiving 250 mg/kg diet showed no ill-effects. In 90 d

feeding trials with a WP (800 g a.i./kg), NOEL for rats 200 mg a.i./kg diet (13 mg/kg daily); for dogs 200 mg a.i./kg diet (7 mg/kg daily). **Toxicity class** WHO (a.i.) III (Table 5); EPA (formulation) IV   **EC risk** (R40)

ECOTOXICOLOGY
Birds Eight-day dietary LC_{50} for bobwhite quail and mallard ducks >10 000 mg/kg. **Fish** LC_{50} (96 h) for rainbow trout 17.5, bluegill sunfish >100, goldfish >32.0 mg/l. **Bees** Not toxic to bees.

ENVIRONMENTAL FATE
Animals In mammals, following oral administration, 42–46% is eliminated in metabolised form within 24 hours. For details of metabolites, see J. E. Bakke et al., J. Agric. Food Chem., 1967, **15**, 628–631.   **Plants** In tolerant plants, the chlorine atom is hydrolysed, to give 6-hydroxypropazine. Both substituted amino groups are dealkylated, presumably followed by ring opening and decomposition. In susceptible plants, propazine is not readily metabolised to non-phytotoxic compounds, but instead accumulates and causes death.   **Soil/Environment** In soil, microbial degradation occurs, with hydrolysis of the chlorine atom to give hydroxypropazine, dealkylation of both substituted amino groups, presumably followed by ring opening and decomposition. DT_{50} 80–100 d.

605 propetamphos   *Insecticide, acaricide*

organophosphorus

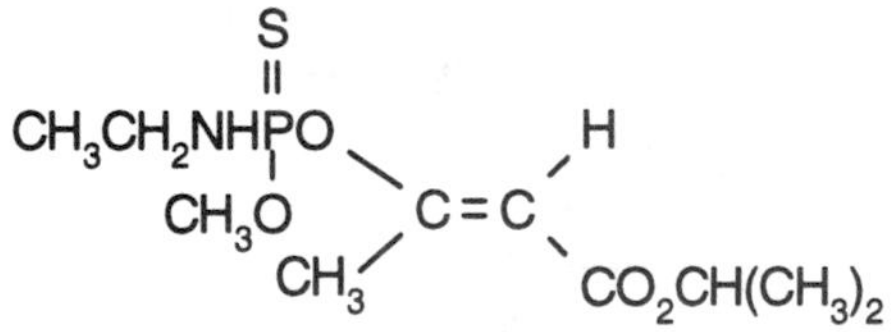

NOMENCLATURE
Common name propetamphos (BSI, E-ISO, (m) F-ISO, ANSI, ESA); propetamfos (Austria)
IUPAC name (E)-O-2-isopropoxycarbonyl-1-methylvinyl O-methyl ethylphosphoramidothioate
Chemical Abstracts name 1-methylethyl (E)-3-[[(ethylamino)methoxyphosphino= thioyl]oxy]-2-butenoate
CAS RN [31218–83–4]   **EEC no.** 250–517–2   **Development codes** SAN 52 1391
Official codes OMS 1502

PHYSICAL CHEMISTRY

Mol. wt. 281.3 **M.f.** $C_{10}H_{20}NO_4PS$ **Form** Yellowish oily liquid (tech.).
B.p. 87–89 °C/0.005 mmHg **V.p.** 1.9 mPa (20 °C) K_{ow} logP = 3.82
S.g./density 1.1294 (20 °C) **Solubility** In water 110 mg/l (24 °C). Readily miscible with acetone, ethanol, methanol, hexane, diethyl ether, dimethyl sulfoxide, chloroform and xylene. **Stability** Stable during storage (shelf life $\geq$2 y at 20 °C) and to light. On hydrolysis (25 °C) DT_{50} 11 d (pH 3), 1 y (pH 6), 41 d (pH 9). Stable in aqueous solution (5 ppm level) in sunlight; no degradation during 70 h.
pKa 13.67 (23 °C)

COMMERCIALISATION

History Insecticide reported by J. P. Leber (*Proc. Int. IUPAC Congr., 2nd,* 1974, **1,** 381). Introduced by Sandoz AG (now Novartis Crop Protection AG).
Manufacturers Novartis

APPLICATIONS

Biochemistry Cholinesterase inhibitor. **Mode of action** Insecticide with contact and stomach action. Long residual activity. **Uses** Control of household and public health pests, especially cockroaches, flies, fleas, mosquitoes, clothes moths, ants and animal ectoparasites (ticks, lice, mites). **Formulation types** WP; EC; DP; LA; Aerosol; EW. **Mixtures** (*propetamphos +*) dichlorvos; hydroprene.
Selected tradenames 'Safrotin' (Novartis)

ANALYSIS

Product analysis by glc with FID (*CIPAC Handbook,* 1992, **E,** 187–192) or by paper chromatography with subsequent phosphorus determination by standard colorimetric methods. **Residues** determined by glc. Details available from Novartis.

MAMMALIAN TOXICOLOGY

EHC 63 (WHO, 1986; a general review of organophosphorus insecticides).
Oral Acute oral LD_{50} for male rats 119, female rats 59.5 mg/kg.
Skin and eye Acute percutaneous LD_{50} for male rats 2825, female rats >2260 mg/kg. **Inhalation** LC_{50} (4 h) for male rats >1.5, female rats 0.69 mg/l air.
NOEL (2 y) for rats 6 mg/kg diet. **Toxicity class** WHO (a.i.) Ib; EPA (formulation) II

ECOTOXICOLOGY

Birds Acute oral LD_{50} for mallard ducks 197 mg/kg. **Fish** LC_{50} (96 h) for carp 7.0 mg/l; for rainbow trout 4.6 mg/kg diet. **Daphnia** LC_{50} (48 h) 14.5 µg/l.
Algae LC_{50} (96 h) for green algae 2.9 mg/l.

ENVIRONMENTAL FATE

Animals In rats, propetamphos is completely metabolised and rapidly excreted

mainly via urine and exhaled air. The oxidative metabolic process leading to CO_2 formation predominates at lower doses. Propetamphos is detoxified through hydrolytic reactions involving the phosphorus and carboxylic ester bonds followed by conjugation, and through oxidation processes leading ultimately to CO_2.
Soil/Environment Residual activity indoors can be as much as 2–3 months.

606 propham *Herbicide, plant growth regulator*

carbamate

NOMENCLATURE
Common name propham (BSI, E-ISO, WSSA); prophame (F-ISO); IFK (USSR)
IUPAC name isopropyl phenylcarbamate; isopropyl carbanilate
Chemical Abstracts name 1-methylethyl phenylcarbamate
Other names IPC   **CAS RN** *[122–42–9]*

PHYSICAL CHEMISTRY
Composition Tech. grade is 99% pure.   **Mol. wt.** 179.2   **M.f.** $C_{10}H_{13}NO_2$
Form Colourless crystals.   **M.p.** 87.0–87.6 °C; (tech. 86.5–87.5 °C)
B.p. Sublimes on heating.   **V.p.** Considerable at 85 °C (sublimes slowly at room temperature).   **S.g./density** 1.09 (20 °C)   **Solubility** In water 250 mg/l (20 °C). Soluble in esters, alcohols, acetone, benzene, cyclohexane and xylene.
Stability Stable up to 100 °C. Not sensitive to light. Hydrolysed slowly in acidic and alkaline media.

COMMERCIALISATION
History Plant growth regulating properties reported by W. G. Templeman & W. A. Sexton (*Nature (London)*, 1945, **156**, 630). Introduced as a herbicide by ICI Plant Protection Division (now Zeneca Agrochemicals).   **Patent** GB 574995

APPLICATIONS
Biochemistry Inhibits root and epicotyl growth, normal cell division, protein and amylase synthesis, photolytic activity of isolated chloroplasts, and affects activity of messenger RNA.   **Mode of action** Selective systemic herbicide and growth regulator, absorbed by the roots and coleoptiles, and readily translocated acropetally.   **Uses** Control of many annual grasses and some broad-leaved weeds in alfalfa, clover, sugar beet, spinach, lettuce, peas, beans, flax, safflowers, lentils, and perennial grass-seed crops at 2.3–5.0 kg a.i./ha. Also in combination with

other herbicides for weed control in beetroot, fodder beet, lettuce and mangels. Applied pre-planting, pre-emergence, or post-emergence. Also used as a sprouting inhibitor for ware potatoes, often in combination with chlorpropham.
Formulation types WP; EC; GR; SC; DP. **Mixtures** *(propham +)* chlorpropham; lenacil; fenuron; chloridazon + fenuron; diuron + isoprocarb; chlorpropham + fenuron; chlorpropham + ethofumesate; proximpham + lenacil; chlorpropham + proximpham; chloridazon + chlorpropham + fenuron; chlorpropham + ethofumesate + fenuron; chlorpropham + diuron.

ANALYSIS
Product analysis by hydrolysis, the liberated carbon dioxide being determined titrimetrically *(CIPAC Handbook,* 1970, **1,** 593) or by glc *(ibid.,* 1985, **1C,** 2025). **Residues** determined by hydrolysis to aniline, a derivative of which is estimated by glc (L. N. Gard & C. E. Ferguson, *Anal. Methods Pestic., Plant Growth Regul. Food Addit.,* 1964, **4,** 139; *Anal. Methods Pestic. Plant Growth Regul.,* 1972, **6,** 657). In **water**, by lc with u.v. detection *(AOAC Methods,* 1995, 992.14, 10.7.01).

MAMMALIAN TOXICOLOGY
Reviews *Pesticide residues in food – 1992.* FAO Plant Production and Protection Paper, 116, 1993. *Pesticide residues in food – 1992 evaluations. Part II – Toxicology.* World Health Organisation, WHO/PCS/93.34. **IARC** 12 **Oral** Acute oral LD_{50} for rats 5000, mice 3000 mg/kg. **NOEL** In 30 d feeding trials, rats receiving 10 000 mg/kg diet showed no ill-effect and there was no evidence of carcinogenic activity (W. C. Heuper, *Ind. Med.,* 1952, **21,** 71). **ADI** (JMPR) No ADI [1992]. **Other** i.p. LD_{50} for rats 600, mice 1000 mg/kg. **Toxicity class** WHO (a.i.) III (Table 5); EPA (formulation) IV

ECOTOXICOLOGY
Birds Acute oral LD_{50} for mallard ducks >2000 mg/kg. **Fish** LC_{50} (48 h) for bluegill sunfish 32, guppies 35 mg/l. **Bees** Not hazardous to bees when used as recommended. **Algae** EC_{50} (cell volume) for *Chlorella fusca* 111 μmol/l (*Pestic. Sci.,* **47,** 337 (1996)).

ENVIRONMENTAL FATE
Animals In mammals, the major metabolic route is *N*-oxidation to hydroxy-IPC. **Plants** In soya beans, the only major metabolite identified is isopropyl *N*-2-hydroxycarbanilate, which is found as water-soluble conjugates of glucose or other plant components. **Soil/Environment** In soil, readily degraded by micro-organisms, with enzymic hydrolysis of the ester bond and degradation of the unstable *N*-phenylcarbamic acid to aniline and CO_2, with further microbial degradation of aniline. DT_{50} in soil *c.* 15 d (16 °C), 5 d (29 °C), but can vary greatly with microbial activity and moisture content of the soil.

azole

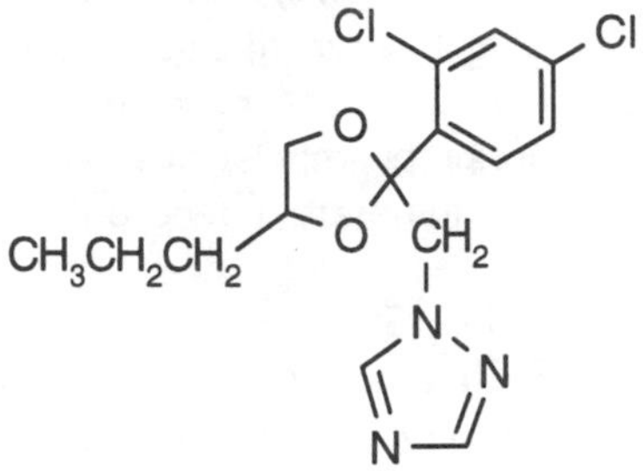

NOMENCLATURE
Common name propiconazole (BSI, draft E-ISO, (*m*) draft F-ISO)
IUPAC name (±)-1-[2-(2,4-dichlorophenyl)-4-propyl-1,3-dioxolan-2-ylmethyl]-=
1*H*-1,2,4-triazole
Chemical Abstracts name 1-[[2-(2,4-dichlorophenyl)-4-propyl-1,3-dioxolan-2-yl]=
methyl]-1*H*-1,2,4-triazole
CAS RN [60207–90–1] unstated stereochemistry   **EEC no.** 262–104–4
Development codes CGA 64 250

PHYSICAL CHEMISTRY
Mol. wt. 342.2   **M.f.** $C_{15}H_{17}Cl_2N_3O_2$ **Form** Yellowish, odourless, viscous liquid
(tech.).   **B.p.** 120 °C (1.9 Pa); >250 °C (101 kPa)   **V.p.** 5.6×10^{-2} mPa (25 °C)
K_{ow} logP = 3.72 (pH 6.6, 25 °C)   **Henry** 9.2×10^{-5} Pa m^3 mol^{-1} (calc.)
S.g./density 1.29 (20 °C)   **Solubility** In water 100 mg/l (20 °C). In *n*-hexane
47 g/l. Completely miscible with ethanol, acetone, toluene and *n*-octanol (25 °C).
Stability Stable up to 320 °C; no significant hydrolysis.   **pKa** 1.09, v. weak base

COMMERCIALISATION
History Fungicide described by Janssen Pharmaceutica. Developed for agricultural
uses by Ciba-Geigy AG (now Novartis Crop Protection AG) and reported by P. A.
Urech et al. (*Proc. Br. Crop Prot. Conf.*, 1979, **2**, 508).   **Patent** GB 1522657;
BE 835579 to Janssen   **Manufacturers** Defensa; Novartis

APPLICATIONS
Biochemistry Steroid demethylation (ergosterol biosynthesis) inhibitor.
Mode of action Systemic foliar fungicide with protective and curative action, with
translocation acropetally in the xylem.   **Uses** Systemic foliar fungicide with a broad
range of activity. On cereals, it controls diseases caused by *Cochliobolus sativus*,
Erysiphe graminis, Leptosphaeria nodorum, Puccinia spp., *Pyrenophora teres,*
Pyrenophora tritici-repentis, Rhynchosporium secalis, and *Septoria* spp. In banana,
control of *Mycosphaerella musicola* and *Mycosphaerella fijiensis* var. *difformis*. Other
uses are in turf against *Sclerotinia homeocarpa, Rhizoctonia solani, Puccinia* spp. and

Erysiphe graminis; in rice against *Rhizoctonia solani, Helminthosporium oryzae,* and dirty panicle complex; in coffee against *Hemileia vastatrix*; in peanuts against *Cercospora* spp.; in stone fruit against *Monilinia* spp., *Podosphaera* spp., *Sphaerotheca* spp. and *Traneschelia* spp.; in maize against *Helminthosporium* spp. **Formulation types** EC; SC; Emulsifiable gel. **Mixtures** *(propiconazole +)* carbendazim; chlorothalonil; cyprodinil; difenoconazole; fenbuconazole; fenpropidin; fenpropimorph; tridemorph; pyrazophos; pyroquilon; tebuconazole; carbendazim + chlorothalonil; fenpropidin + fenpropimorph; fenpropidin + tebuconazole. **Selected tradenames** 'Tilt' (Novartis); 'Radar' (Zeneca); 'Bumper' (Makhteshim-Agan); 'Juno' (Defensa)

ANALYSIS
Product analysis by glc with FID (*CIPAC Handbook*, 1995, **G**, 129–136). **Residue** analysis by glc with FID (*Methodensammlung Rückstandanal. Pflanzenschutzmitteln*, 1987, S19, 624). Details available from Novartis.

MAMMALIAN TOXICOLOGY
Reviews *Pesticide residues in food – 1987. FAO Plant Production and Protection Paper 84, 1987. Pesticide residues in food – 1987 evaluations. Part II – Toxicology. FAO Plant Production and Protection Paper 86/2, 1988.* **Oral** Acute oral LD_{50} for rats 1517, mice 1490 mg/kg. **Skin and eye** Acute percutaneous LD_{50} for rats >4000, rabbits >6000 mg/kg; non-irritating to skin and eyes (rabbits). No sensitisation (guinea pigs). **Inhalation** LC_{50} (4 h) for rats >5800 mg/m^3. **NOEL** (2 y) for rats 3.6, mice 10 mg/kg b.w. daily; (1 y) for dogs 1.9 mg/kg b.w. daily. **ADI** 0.02 mg/kg b.w. **Other** Not mutagenic, not teratogenic. No carcinogenic potential of relevance for human exposure. **Toxicity class** WHO (a.i.) II **EC risk** R22

ECOTOXICOLOGY
Birds Acute oral LD_{50} for Japanese quail 2223, bobwhite quail 2825, mallard ducks >2510, Pekin ducks >6000 mg/kg. LC_{50} (8 d) for Japanese quail >1000, bobwhite quail >5620, mallard ducks >5620, Pekin ducks >1000 ppm. **Fish** LC_{50} (96 h) for carp 6.8, rainbow trout 5.3, golden orfe 5.1, spot 2.6 mg/l. **Bees** Not toxic to bees; LD_{50} (contact and oral) >100 µg/bee. **Worms** No toxic effects against *Lumbricus rebellus.* **Other beneficial spp.** Under field conditions, not expected to have any significant impact. **Daphnia** EC_{50} 4.8 mg/l. **Algae** EC_{50} 0.02–13.6 mg/l for three freshwater algae and two diatom species. **Other aquatic spp.** LC_{50} (96 h) for crayfish 42 mg/l. EC_{50} (96 h) for mysid shrimp (*Mysidopsis bahia*) 0.5 mg/l.

ENVIRONMENTAL FATE
Animals After oral adminstration to the rat, propiconazole is rapidly absorbed and also rapidly and almost completely eliminated with urine and faeces. Residues in tissues were generally low and there was no evidence for accumulation or

retention of propiconazole or its metabolites. The major sites of enzymic attack are the propyl side-chain and the cleavage of the dioxolane ring, together with some attack at the 2,4-dichlorophenyl and 1,2,4-triazole rings. In the mouse, the major metabolic pathway is via cleavage of the dioxolane ring (R. Bissig & W. Muecke, *Br. Crop Prot. Conf. – Pests Dis.*, 1988, **2**, 675–680).
Plants Degradation proceeds through hydroxylation of the *n*-propyl side-chain and deketalisation of the dioxolan ring. After cleavage of triazole, triazole-alanine is formed as the main metabolite. Metabolites are conjugated mostly as glucosides. For details of metabolites of propiconazole in wheat, rice and vines, see B. Donzel et al., *IUPAC 7th Int. Congr. Pestic. Chem.*, 1990, **2**, 160. **Soil/Environment** DT_{50} in aerobic soils (25 °C) 40–70 d. The main degradation pathways are hydroxylation of the propyl side-chain and the dioxolane ring, and finally formation of 1,2,4-triazole. $K_{oc\ (ads)}$ 950 ml/g, immobile in soil.

608 propineb   *Fungicide*

alkylenebis(dithiocarbamate)

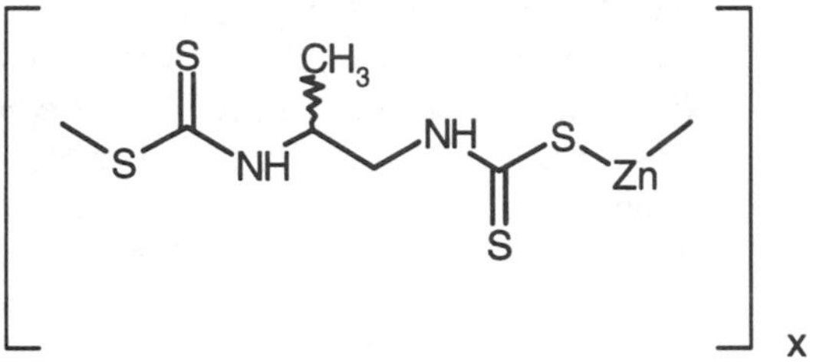

NOMENCLATURE
Common name propineb (BSI, E-ISO, JMAF); propinèbe ((m) F-ISO)
IUPAC name polymeric zinc propylenebis(dithiocarbamate)
Chemical Abstracts name [[(1-methyl-1,2-ethanediyl)bis[carbamodithioato]](2–)]= zinc homopolymer
CAS RN *[12071-83-9]* (monomer); *[9016-72-2]* formerly *[31530-30-0]* (homopolymer) **EEC no.** 235-134-0 **Development codes** Bayer 46 131; LH 30/Z (Bayer)

PHYSICAL CHEMISTRY
Mol. wt. 289.8 (theoretical monomer) **M.f.** $(C_5H_8N_2S_4Zn)_x$ **Form** White powder with a slight, characteristic smell. **M.p.** Decomposes above 150 °C **V.p.** <1 mPa (20 °C) **K_{ow}** logP = –0.26 (PTU, main metabolite) (20 °C) **S.g./density** 1.813 g/ml (23 °C) **Solubility** In water 0.01 g/l (20 °C). In toluene, hexane, dichloromethane <0.1 g/l.; in DMF + DMSO > 200 g/l. **Stability** Stable

when dry. Decomposed by moisture, and in acidic and alkaline media; DT_{50} (22 °C) (estimated) 1 d (pH 4), c. 1 d (pH 7), >2 d (pH 9).

COMMERCIALISATION
History Fungicide reported by H. Goeldner (*Pflanzenschutz-Nachr. (Engl. Ed.)*, 1963, **16**, 49) and reviewed by F. Grewe (*ibid.*, 1967, **20**, 581). Introduced by Bayer AG. **Patent** BE 611960 **Manufacturers** Bayer

APPLICATIONS
Biochemistry Multisite activity, characteristic for dithiocarbamate fungicides.
Mode of action Basic foliar fungicide with protective action. Conidia or germinating conidia are killed by contact. **Uses** Control of downy mildew, black rots, red fire disease, and grey mould on vines; scab and brown rot on apples and pears; leaf spot diseases on stone fruit; *Alternaria* and *Phytophthora* blights, downy mildew, *Septoria* leaf spot, and leaf mould in tomatoes; *Phytophthora* and *Alternaria* blight of potatoes; downy mildew (blue mould) on tobacco; rusts and leaf spot diseases on ornamentals; rusts, leaf spot diseases and downy mildews on vegetables. Also used on citrus fruit and berry fruit, rice and tea.
Phytotoxicity Non-phytotoxic when used as recommended, also in young sensitive growing stages. **Formulation types** WP; DP; WG. **Mixtures** (propineb +) copper oxychloride; triadimefon; triadimenol; chinomethionat; cymoxanil; oxadixyl; cymoxanil + triadimefon; copper oxychloride + oxadixyl; bitertanol; cymoxanil + copper oxychloride; cymoxanil + oxadixyl. **Compatibility** Compatible with most other powder-formulated pesticides. Not stable when combined with alkaline materials. **Selected tradenames** 'Antracol' (Bayer)

ANALYSIS
Product analysis by titration with iodine, after hydrolysis and conversion of the liberated carbon disulfide to dithiocarbonate (*CIPAC Proc.*, 1981, **2**, 313). Identified colorimetrically (*CIPAC Handbook*, 1994, **F**, 320). **Residues** determined by colorimetry of a complex formed with the carbon disulfide produced on hydrolysis (*Analyst (London)*, 1980, **105**, 515; K. Vogeler, *Pflanzenschutz-Nachr. (Engl. Ed.)*, 1967, **20**, 525) or by polarography (*idem, ibid.*). Methods for the determination of residues are available from Bayer.

MAMMALIAN TOXICOLOGY
Reviews *Pesticide residues in food – 1993*. FAO Plant Production and Protection Paper, 122, 1993. *Pesticide residues in food – 1993 evaluations. Part II – Toxicology*. World Health Organisation, WHO/PCS/94.4, 1994. **EHC** 78 (WHO, 1988; general review of dithiocarbamates). **Oral** Acute oral LD_{50} for rats >5000 mg/kg. **Skin and eye** Acute percutaneous LD_{50} for rats >5000 mg/kg. Not irritating to eyes and skin (rabbits). **Inhalation** LC_{50} (4 h) for rats >0.7 mg/l air (aerosol). **NOEL** (2 y) for rats 50, mice 800, dogs 1000 mg/kg diet.

ADI (JMPR) 0.007 mg/kg b.w. [1993].   **Toxicity class** WHO (a.i.) III (Table 5);
EPA (formulation) IV

ECOTOXICOLOGY
Birds LD_{50} for Japanese quail >5000 mg/kg.   **Fish** LC_{50} (96 h) for rainbow trout
1.9, golden orfe 133 mg/l.   **Bees** Not toxic to bees; LD_{50} (oral; 70 WP, 70 WG)
>100 µg/bee.   **Worms** LD_{50} (14 d) for 70 WP and 70 WG >1000 µg/kg dry
substrate.   **Other beneficial spp.** Effects on non-target insects are unlikely; only
predatory mites are sensitive. Under field conditions, only moderate effects were
observed.   **Daphnia** LC_{50} (48 h) 4.7 mg/l.   **Algae** E_rC_{50} (96 h) 2.7 mg/l.

ENVIRONMENTAL FATE
Animals Elimination is quick; almost 91% is excreted within 48 h in the urine and
faeces, and 7% with exhaled air.   **Plants** Residues of the applied compound
including its metabolite propylenethiourea (PTU) were found mainly on the plant
surface. Only the metabolites propyleneurea and 4-methyl-imidazoline were taken
up into the plant, in very small amounts. PTU has to be considered as a relevant
residue with regard to its toxicological properties.   **Soil/Environment** For
behaviour in soil, see W. Mittelstaedt & F. Fuehr, *Landwirtsch. Forsch.*, 1977, **30**,
221. Degradation is very rapid. Propineb can be classified as not mobile in soils.

609 propisochlor   *Herbicide*

chloroacetanilide

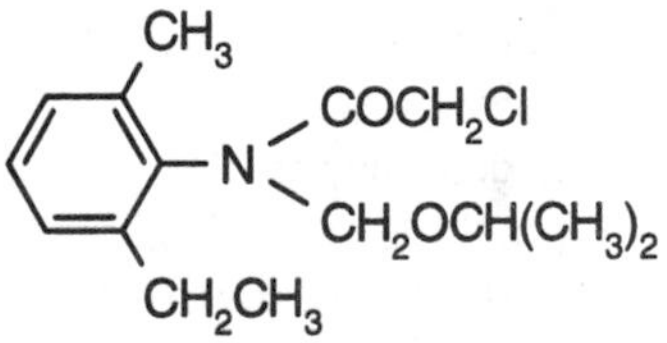

NOMENCLATURE
Common name propisochlor (BSI, pa E-ISO); propisochlore ((m) pa F-ISO)
IUPAC name 2-chloro-6'-ethyl-N-isopropoxymethylaceto-o-toluidide
Chemical Abstracts name 2-chloro-N-(2-ethyl-6-methylphenyl)-N-[(1-methyl=
ethoxy)methyl]acetamide
CAS RN *[86763–47–5]*

PHYSICAL CHEMISTRY
Composition ≥95% pure.   **Mol. wt.** 283.8   **M.f.** $C_{15}H_{22}ClNO_2$ **Form** Light brown

to purple aromatic oil. **M.p.** 21.6 °C **B.p.** Decomposes above 243 °C
V.p. 1 mPa (20 °C) K_{ow} logP = 3.50 (20 °C) **S.g./density** 1.097 g/cm^3 (20 °C)
Solubility In water 184 mg/l (20 °C). Soluble in most organic solvents.
Stability Hydrolytically stable; after 5 d at 50°C (pH 4, 7, 9), <10% decomposition.
Decomposes above 243 °C. **F.p.** 175 °C (Marcusson open cup), 110 °C
(closed cup)

COMMERCIALISATION
Patent HU 208224 **Manufacturers** Nitrokémia

APPLICATIONS
Biochemistry Protein synthesis inhibitor. **Mode of action** Selective herbicide,
absorbed by the shoots of germinating plants. **Uses** Pre-emergence or pre-plant
incorporated for control of annual grasses and some broad-leaved weeds in
maize, sunflowers, soya beans, potatoes and peas (1.1–1.8 kg/ha). It can be used
in beans and sweet lupins (1.1–1.8 kg/ha) and onions (1.4–1.8 kg/ha).
Formulation types EC. **Selected tradenames** 'Proponit' (Nitrokémia)

ANALYSIS
By glc and hplc.

MAMMALIAN TOXICOLOGY
Oral Acute oral LD_{50} for male rats 3433, female rats 2088 mg/kg.
Skin and eye Acute percutaneous LD_{50} for male and female rats >2000 mg/kg.
Inhalation LC_{50} for male and female rats >5000 mg/m^3. **NOEL** (90 d) for rats
250 mg/kg daily (25 mg/kg b.w.). **ADI** 2.5 mg/kg. **Toxicity class** WHO (a.i.) III

ECOTOXICOLOGY
Birds Acute oral LD_{50} for mallard ducks 2000, Japanese quail 688 mg/kg.
Eight-day LC_{50} for quail and ducks 5000 mg/kg. **Fish** LC_{50} (96 h) for carp 7.94,
rainbow trout 0.25 mg/l. **Bees** Non-toxic; LD_{50} (oral and contact) 100 µg/bee.
Worms Not dangerous. **Other beneficial spp.** Not dangerous to soil micro-
organisms. **Daphnia** LC_{50} (96 h) 6.19 mg/l. **Algae** EC_{50} for *Selenastrum
capricornutum* 2.8 µg/l.

ENVIRONMENTAL FATE
Animals In rats, most of applied radioactivity is excreted rapidly, mainly within
24 h. **Soil/Environment** Decomposed by microbial action; typical soil DT_{50}
10–15 d. K_{oc} 333.3 (acid, sandy soil), 364.4 (acid loamy soil), 493.5 (alkaline loamy
soil).

carbamate

$$\text{OCONHCH}_3$$
$$\text{OCH(CH}_3)_2$$

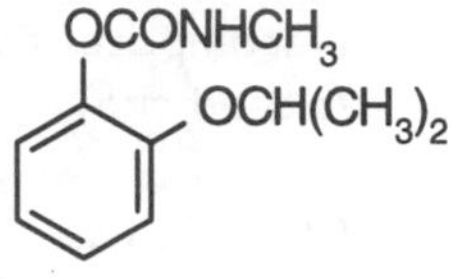

NOMENCLATURE
Common name propoxur (BSI, E-ISO, F-ISO, ESA); PHC (JMAF); [arprocarb] (former BSI name)
IUPAC name 2-isopropoxyphenyl methylcarbamate
Chemical Abstracts name 2-(1-methylethoxy)phenyl methylcarbamate
CAS RN *[114–26–1]* **EEC no.** 204–043–8 **Development codes** Bayer 39 007; BOQ 5812315 (Bayer) **Official codes** OMS 33; ENT 25 671

PHYSICAL CHEMISTRY
Mol. wt. 209.2 **M.f.** $C_{11}H_{15}NO_3$ **Form** Colourless crystals; (tech., white to cream-coloured crystals). **M.p.** 90 °C (crystal form I), 87.5 °C (crystal form II, unstable) **B.p.** Decomposes on distillation. **V.p.** 1.3 mPa (20 °C); 2.8 mPa (25 °C) K_{ow} logP = 1.56 **Henry** 1.5×10^{-4} Pa m^3 mol^{-1} (20 °C) **S.g./density** 1.18 (20 °C) **Solubility** In water 1.9 g/l (20 °C). Soluble in most organic solvents, e.g. isopropanol >200, toluene 50–100, hexane 1–2 (all in g/l, 20 °C). **Stability** Stable in water at pH 7. Hydrolysed by strong alkali; DT_{50} (22 °C) 1 y (pH 4), 93 d (pH 7), 30 h (pH 9); DT_{50} (20 °C) 40 min (pH 10). Direct photodegradation is not a major contributor to the overall elimination of propoxur from the environment (DT_{50} 5–10 d); indirect photodecomposition (addition of humic acid) is more rapid (DT_{50} 88 h).

COMMERCIALISATION
History Insecticide reported by G. Unterstenhöfer (*Meded. Landbouwhogesch. Opzoekingsstn. Staat Gent,* 1963, **28**, 758). Introduced by Bayer AG.
Patent US 3111539; DE 1108202 **Manufacturers** Bayer; Crystal; Kuo Ching; Pilarquim; Sanachem; Sanex; Taiwan Tainan Giant

APPLICATIONS
Biochemistry Cholinesterase inhibitor. **Mode of action** Non-systemic insecticide with contact and stomach action. Gives rapid knockdown, and has long residual activity. **Uses** Control of cockroaches, flies, fleas, mosquitoes, bugs, ants, millipedes and other insect pests in food storage areas, houses, animal houses, etc. Control of sucking and chewing insects (including aphids) in fruit, vegetables, ornamentals, vines, maize, alfalfa, soya beans, cotton, sugar cane, rice, cocoa, forestry, etc.; also against migratory locusts and grasshoppers.
Phytotoxicity Chrysanthemums, carnations, and hydrangeas may be injured at

higher dose rates. Blossom thinning may occur on fruit trees.

Formulation types AE; WP; EC; DP; FU; RB; UL; GR; SL; Oilspray.

Mixtures *(propoxur +)* dichlorvos; methiocarb; pyrethrins; cyfluthrin + dichlorvos; azinphos-methyl; phoxim; dichlorvos + methoprene.   **Compatibility** Mixtures with alkaline substances are not stable.   **Selected tradenames** 'Baygon' (public health) (Bayer); 'Unden' (for agricultural use) (Bayer); 'Mitoxur' (Sanachem); 'Proper' (Sanonda)

ANALYSIS

Product analysis by hplc (*CIPAC Handbook*, 1988, **D**, 155; *AOAC Methods*, 1995, 984.12) or by i.r. spectrometry (C. A. Anderson, *Anal. Methods Pestic. Plant Growth Regul.*, 1973, **7**, 163).   **Residues** determined by glc (*idem., ibid.; AOAC Methods*, 1995, 975.40; M. A. Luke *et al.*, *J. Assoc. Off. Anal. Chem.*, 1981, **64**, 1187; A. Ambrus *et al.*, *ibid.*, p. 733; *Man. Pestic. Residue Anal.*, 1987, **I**, 6, S19; *Anal. Methods Residues Pestic.*, 1988, Part I, M5). Methods for the determination of residues are available from Bayer. For methods in **drinking water**, see *AOAC Methods*, 1995, 991.06.

MAMMALIAN TOXICOLOGY

Reviews *Pesticide residues in food – 1989*. FAO Plant Production and Protection Paper 99, 1989. *Pesticide residues in food – 1989 evaluations. Part II – Toxicology.* FAO Plant Production and Protection Paper 100/2, 1990.   **EHC** 64 (WHO, 1986; a review of carbamate insecticides in general).   **Oral** Acute oral LD_{50} for male and female rats *c.* 50 mg/kg.   **Skin and eye** Acute percutaneous LD_{50} (24 h) for male and female rats >5000 mg/kg. Non-irritating to skin, slightly irritating to eyes (rabbits).   **Inhalation** LC_{50} for rats (4 h) >0.5 mg/l (aerosol); 0.654 mg/l air (dust).   **NOEL** (2 y) for rats 200, mice 500 mg/kg diet; (12 mo) for dogs 200 mg/kg diet.   **ADI** (JMPR) 0.02 mg/kg b.w. [1989].   **Toxicity class** WHO (a.i.) II; EPA (formulation) II   **EC risk** T (R25)

ECOTOXICOLOGY

Birds Five-day dietary LC_{50} for bobwhite quail 2828, mallard ducks >5000 mg/kg diet.   **Fish** LC_{50} (96 h) for bluegill sunfish 6.2–6.6, rainbow trout 3.7–13.6, golden orfe 12.4 mg/l.   **Bees** Highly toxic to bees.   **Daphnia** LC_{50} (48 h) 0.15 mg/l.

ENVIRONMENTAL FATE

Animals In rats, the principal metabolites are 2-hydroxyphenyl-*N*-methylcarbamate and 2-isopropoxyphenol; minor metabolites include 5-hydroxy propoxur and *N*-hydroxymethyl propoxur. Elimination is very rapid, 96% excreted in the urine. Metabolism of carbamate insecticides is reviewed (M. Cool & C. K. Jankowski in "*Insecticides*").   **Plants** The primary metabolite is demethylpropoxur (max. amount 2.7–3.6%).   **Soil/Environment** The mobility of propoxur in the soil is relatively high. The compound rapidly degrades in different soils.

611 propyzamide *Herbicide*

amide

NOMENCLATURE
Common name propyzamide (BSI, E-ISO, (*m*) F-ISO, JMAF); pronamide (WSSA)
IUPAC name 3,5-dichloro-*N*-(1,1-dimethylpropynyl)benzamide
Chemical Abstracts name 3,5-dichloro-*N*-(1,1-dimethyl-2-propynyl)benzamide
CAS RN *[23950–58–5]* **Development codes** RH 315

PHYSICAL CHEMISTRY
Mol. wt. 256.1 **M.f.** $C_{12}H_{11}Cl_2NO$ **Form** Colourless, odourless powder.
M.p. 155–156 °C **V.p.** 0.058 mPa (25 °C) **K_{ow}** logP = *c*. 3.1–3.2 **Solubility** In
water 15 mg/l (25 °C). In methanol, isopropanol 150, cyclohexanone, methyl
ethyl ketone 200, dimethyl sulfoxide 330 (all in g/l). Moderately soluble in
benzene, xylene, carbon tetrachloride. Slightly soluble in petroleum ether.
Stability Decomposes above melting point. Degraded photolytically on soil thin
films, DT_{50} 13–57 d in artificial sunlight. In solution for 28 d (pH 5–9, 20 °C)
<10% loss.

COMMERCIALISATION
History Herbicide introduced by Rohm & Haas Co. **Patent** GB 1209068;
US 3534098; US 3640699 **Manufacturers** Rohm & Haas; United Phosphorus Ltd

APPLICATIONS
Biochemistry Inhibits photosynthesis. Cell division inhibitor.
Mode of action Selective systemic herbicide, absorbed by the roots, with
translocation. **Uses** Selective control of many annual and perennial grasses, and
some broad-leaved weeds in fruit, vines, lettuce, endive, chicory, brassicas, oilseed
rape, legumes, alfalfa, clover, trefoil, sainfoin, artichokes, sugar beet, roses,
ornamental trees and shrubs, on fallow land, and in forestry. Also control of *Poa
annua* in Bermuda grass, *Zoysia*, and certain other turf species. Used either pre-
emergence or early post-emergence. Rates vary from 0.56–2.2 kg a.i./ha.
Phytotoxicity Certain varieties of lettuce may be injured at higher dose rates.
Formulation types WP; GR; SC. **Mixtures** *(propyzamide +)* clopyralid; diuron;
simazine; oxyfluorfen; chlorpropham; terbuthylazine; oxyfluorfen + terbuthylazine.
Compatibility Compatible with many other herbicides to extend weed spectrum.
Selected tradenames 'Kerb' (Rohm & Haas); 'Judo' (Headland); 'Piza' (Barclay);
'Rapier' (United Phosphorus Ltd)

ANALYSIS

Product analysis by glc (I. L. Adler *et al.*, *J. Assoc. Off. Anal. Chem.*, 1972, **55**, 802; *idem, Anal. Methods Pestic. Plant Growth Regul.*, 1976, **8**, 443).
Residues determined by glc of a derivative (*idem, loc. cit.*) or by glc with ECD. In **drinking water**, by gc with FID (*AOAC Methods*, 1995, 991.07).

MAMMALIAN TOXICOLOGY

Reviews *Pronamide: Reregistration Eligibility Decision*, US EPA, May 1994.
Oral Acute oral LD_{50} for male rats 8350, female rats 5620, dogs >10 000 mg/kg.
Skin and eye Acute percutaneous LD_{50} for rabbits >3160 mg/kg. Slightly irritating to eyes and skin. **Inhalation** LC_{50} for rats >5.0 mg/l. **NOEL** Chronic studies indicate NOEL for dogs 300, rats 200, mice 13 mg/kg diet. **ADI** 0.08 mg/kg.
Toxicity class WHO (a.i.) III (Table 5); EPA (formulation) IV

ECOTOXICOLOGY

Birds Acute oral LD_{50} for Japanese quail 8770, mallard ducks >14 mg/kg. Eight-day dietary LC_{50} for bobwhite quail and mallard ducks >10 000 ppm. **Fish** LC_{50} (96 h) for rainbow trout >4.7, carp >5.1 mg a.i./l. **Bees** Not hazardous to bees; LD_{50} >100 µg a.i./bee. **Worms** LC_{50} for earthworms >346 ppm. **Daphnia** LC_{50} (48 h) >5.6 mg a.i./l

ENVIRONMENTAL FATE

Animals For metabolism in animals and plants, see R. Y. Yih *et al.*, *J. Agric. Food Chem.*, 1971, **19**, 314–324; J. D. Fisher, *ibid.*, 1974, **22**, 606–608; J. M. Cantier *et al.*, *Pestic. Sci.*, 1986, **17**, 235. **Plants** See under Animals.
Soil/Environment DT_{50} in soil (25 °C) *c.* 30 d. For further metabolism in soil, see refs. given under Animals. Duration of residual activity in soil, following an application rate of 1–4 kg/ha, is *c.* 2–6 months. K_{oc} 800. K_d ranges from 0.04 (0.01% o.m., pH 6.6) to 72.2 (16.9% o.m., pH 6.8) (H. J. Pedersen *et al.*, *Pestic. Sci.*, 1995, **44**, 131).

612 prosulfocarb   *Herbicide*

thiocarbamate

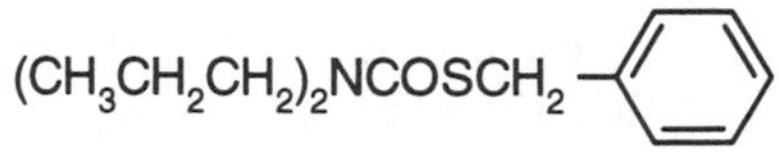

$$(CH_3CH_2CH_2)_2NCOSCH_2-\!\!\bigcirc$$

NOMENCLATURE

Common name prosulfocarb (BSI, draft E-ISO)
IUPAC name *S*-benzyl dipropylthiocarbamate

Chemical Abstracts name S-(phenylmethyl) dipropylcarbamothioate
CAS RN *[52888–80–9]* EEC no. 401–730–6 Development codes SC-0574;
ICIA0574

PHYSICAL CHEMISTRY
Composition Tech. material is 95% pure. Mol. wt. 251.4 M.f. $C_{14}H_{21}NOS$
Form Colourless liquid; (tech. is a light yellow liquid with a slightly sweet odour).
B.p. 129 °C/33 Pa V.p. 0.069 mPa (25 °C) K_{ow} logP = 4.65 (25 °C)
S.g./density 1.042 Solubility In water 13.2 mg/l (20 °C). Miscible with acetone,
chlorobenzene, ethanol, xylene, ethyl acetate and kerosene. Stability Stable for at
least 2 months at 52 °C. Hydrolysed in aqueous solution (pH 7 and 25 °C), DT_{50}
25 d. F.p. >100 °C

COMMERCIALISATION
History Herbicide reported by J. L. Glasgow et al. (*Proc. 1987 Br. Crop. Prot. Conf.
– Weeds*, **1**, 27). Discovered by Stauffer Chemical Co. (now Zeneca
Agrochemicals) and introduced in Belgium (1988) by ICI Agrochemicals (also now
Zeneca Agrochemicals). Manufacturers Zeneca

APPLICATIONS
Biochemistry Inhibits lipid metabolism. Mode of action Selective herbicide,
absorbed by the leaves and roots, which acts by inhibiting growth in the
meristematic region. It causes dark greening, twisting, inhibition of shoots and
roots, and a failure of leaf emergence from coleoptiles. Uses Pre-emergence and
early post-emergence control of a wide range of grass and broad-leaved weeds in
winter wheat, winter barley and rye at 3–4 kg a.i./ha. Controls especially *Galium
aparine*, but also provides some control of *Poa annua*, *Alopecurus myosuroides*,
Stellaria media, *Sinapis arvensis*, *Veronica* spp., and *Lolium multiflorum*.
Phytotoxicity Post-emergence applications may cause some injury to winter barley.
Formulation types EC. Selected tradenames 'Boxer' (Zeneca); 'Defi' (Zeneca)

ANALYSIS
Product and residue analysis by glc. Details available from Zeneca.

MAMMALIAN TOXICOLOGY
EHC 76 (WHO, 1988; general review of thiocarbamates). Oral Acute oral LD_{50}
for male rats 1820, female rats 1958 mg/kg. Skin and eye Acute percutaneous
LD_{50} for rabbits >2000 mg/kg. Slight/mild irritation to skin and eyes (rabbits).
Not a skin sensitiser (guinea pigs). Inhalation LC_{50} (4 h) for rats >4.7 mg/l.
NOEL (2 y) for rats 0.5 mg/kg daily; (18 mo) for mice >65 mg/kg daily.
NOEL for sub-chronic toxicity in rats and dogs 1–10 mg/kg daily. Non-teratogenic
in rats and rabbits. Other Non-mutagenic in the Ames test.
Toxicity class WHO (a.i.) II EC risk Xn (R22, R48/22)

ECOTOXICOLOGY
Birds No-effect level (5 d) for mallard ducks 3160, bobwhite quail 1780 mg/kg.
Fish LC_{50} (96 h) for bluegill sunfish 4.2, rainbow trout 1.7 mg/l.   **Daphnia** LC_{50}
(48 h) 1.3 mg/l.

ENVIRONMENTAL FATE
Plants In plants, hydrolytic decomposition occurs and the thiol group is split off,
giving mercaptan, dipropylamine and CO_2. **Soil/Environment** In soil, undergoes
hydrolytic decomposition, DT_{50} 10–35 d.

613 prosulfuron   *Herbicide*

sulfonylurea

NOMENCLATURE
Common name prosulfuron (BSI, pa E-ISO)
IUPAC name 1-(4-methoxy-6-methyl-1,3,5-triazin-2-yl)-3-[2-(3,3,3-trifluoro=
propyl)phenylsulfonyl]urea
Chemical Abstracts name N-[[3-(4-methoxy-6-methyl-1,3,5-triazin-2-yl)-amino]=
carbonyl]-2-(3,3,3-trifluoropropyl)benzene sulfonamide
CAS RN *[94125–34–5]*   **Development codes** CGA 152005

PHYSICAL CHEMISTRY
Mol. wt. 419.4   **M.f.** $C_{15}H_{16}F_3N_5O_4S$ **Form** Colourless, odourless crystals.
M.p. 155 °C (decomp.)   **V.p.** $<3.5 \times 10^{-3}$ mPa (25 °C) (OECD 104)
K_{ow} log P = 1.5 (pH 5.0), –0.21 (pH 6.9), –0.76 (pH 9.0) (25 °C)
Henry $<3 \times 10^{-4}$ Pa m^3 mol^{-1} **S.g./density** 1.45 g/cm^3 (20 °C)   **Solubility** In
distilled water 29 mg/l (pH 4.5, 25 °C); in buffered water 87 (pH 5.0), 4000
(pH 6.8), 43 000 (pH 7.7) (all in mg/l, 25 °C). In ethanol 8.4, acetone 160, toluene
6.1, n-hexane 0.0064, n-octanol 1.4, ethyl acetate 56, dichloromethane 180
(all in g/l, 25 °C).   **Stability** Hydrolysed rapidly at pH 5, DT_{50} 5–10 d (20 °C),
very slowly at pH 7 and 9. Not degraded photolytically.   **pKa** 3.76

COMMERCIALISATION
History Reported by M. Schulte *et al.* (*Proc. Br. Crop Prot. Conf.* 1993, **1**, 53).

Introduced by Ciba-Geigy AG (now Novartis Crop Protection AG).
Manufacturers Novartis

APPLICATIONS
Biochemistry Branched chain amino acid synthesis (acetolactate synthase or ALS)
inhibitor. Acts by inhibiting biosynthesis of the essential amino acids valine and
isoleucine, hence stopping cell division and plant growth. Selectivity derives from
rapid metabolism in the crop. Metabolic basis of selectivity in sulfonylureas
reviewed (M. K. Koeppe & H. M. Brown, *Agro-Food-Industry*, **6**, 9–14 (1995)).
Mode of action Absorbed via leaves and roots, and translocated within xylem as
well as phloem to the site of action. Plant death occurs 1–3 weeks after
application. **Uses** Post-emergence application in maize, sorghum, cereals, pasture
and turf for broad-spectrum control of *Amaranthus, Abutilon, Chenopodium,
Polygonum, Rumex, Stellaria* and other annual broad-leaved weeds at 12–40 g
a.i./ha. **Formulation types** WG. **Mixtures** *(prosulfuron +)* bromoxynil;
primisulfuron-methyl; terbuthylazine. **Compatibility** Not compatible with
organophosphate insecticides. **Selected tradenames** 'Exceed' (USA) (mixture)
(Novartis); 'Peak' (Novartis)

ANALYSIS
Product by l.c. **Residues** by hplc and l.c. Methods for sulfonylurea residues
in crops, soil and water reviewed (A. C. Barefoot *et al., Proc. Br. Crop Prot.
Conf. – Weeds,* 1995, **2**, 707).

MAMMALIAN TOXICOLOGY
Oral Acute oral LD_{50} for rats 986, mice 1247 mg/kg. **Skin and eye** Acute
percutaneous LD_{50} for rabbits >2000 mg/kg. Non-irritating to skin and eyes
(rabbits). Non-sensitising to skin (guinea pigs). **Inhalation** LC_{50} (4 h) for rats
>5000 mg/m³. **NOEL** (2 y) for rats 200 mg/kg diet (8.6 mg/kg daily); (1 y) for
dogs 60 ppm (1.9 mg/kg daily). **ADI** 0.019 mg/kg b.w. **Other** Non-mutagenic;
non-teratogenic (rat and rabbit). **Toxicity class** WHO (a.i.) III; EPA
(formulation) III **EC risk** Xn (R22)

ECOTOXICOLOGY
Birds LD_{50} for mallard ducks 1300, bobwhite quail >2150 mg/kg. Eight-day dietary
LC_{50} for mallard ducks and bobwhite quail >5000 ppm. **Fish** LC_{50} (96 h) for
catfish, rainbow trout and carp >100, bluegill sunfish and sheepshead >155 mg/l.
Bees LD_{50} (48 h) (oral and contact) >100 μg/bee. **Worms** LC_{50} (14 d) for
earthworms >1000 mg/kg. **Other beneficial spp.** No effect on rove beetle,
ground beetle, aphid predator or ladybird, up to an application rate of 30 g/ha.
No effect on respiration/nitrification. **Daphnia** LC_{50} (48 h) >120 mg/l.
Algae EC_{50} for *Selenestrum capricornutum* 0.011, *Anabaena flos-aquae* >0.027,
Navicula pelliculosa >0.084, *Skeletonema costatum* >0.029 mg/l.
Other aquatic spp. EC_{50} for mysid shrimp (*Mysidopsis bahia*) >150, Eastern oyster
(*Crassostrea virginica*) >125 mg/l.

ENVIRONMENTAL FATE

Animals 90–95% is excreted within 48 h. Main metabolic pathway is
O-demethylation and side-chain hydroxylation. **Plants** Main metabolic pathway is
hydroxylation and cleavage of the phenyl and triazine ring systems.
Soil/Environment DT$_{50}$ 5–23 d, depending on temperature, soil moisture and
% o.c.; DT$_{90}$ 30 to <100 d (typical value 60 d). K$_{oc}$ 4–41, depending on % o.c.
Under practical conditions, the high mobility is outweighed by fast degradation.
No prosulfuron is leached deeper than 50 cm into the soil.

514 prothiofos *Insecticide*

organophosphorus

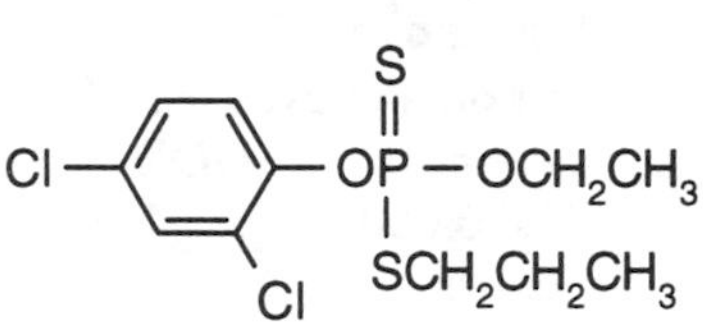

NOMENCLATURE

Common name prothiofos (BSI, E-ISO, (m) F-ISO, JMAF); no name (Eire)
IUPAC name O-2,4-dichlorophenyl O-ethyl S-propyl phosphorodithioate
Chemical Abstracts name O-(2,4-dichlorophenyl) O-ethyl S-propyl
phosphorodithioate
CAS RN [34643–46–4] **Development codes** NTN 8629 (Bayer)
Official codes OMS 2006

PHYSICAL CHEMISTRY

Mol. wt. 345.2 **M.f.** C$_{11}$H$_{15}$Cl$_2$O$_2$PS$_2$ **Form** Colourless liquid with a weak
characteristic odour. **B.p.** 125–128 °C/13 Pa **V.p.** 0.3 mPa (20 °C); 0.6 mPa
(25 °C) **K$_{ow}$** logP = 5.67 (20 °C) **S.g./density** 1.31 (20 °C) **Solubility** In water
0.07 mg/l (20 °C). In dichloromethane, isopropanol, toluene >200 (all in g/l,
20 °C). **Stability** Hydrolysis DT$_{50}$ in buffer (22 °C) 120 d (pH 4), 280 d (pH 7),
12 d (pH 9). Photodegradation DT$_{50}$ 13 h. **F.p.** >110 °C

COMMERCIALISATION

History Insecticide reported by A. Kudamatsu (*Jpn. Pestic. Inf.*, 1976, No. 26,
p. 14). Developed by Nihon Tokushu Noyaku Seizo K.K. (NITOKUNO, now
Nihon Bayer Agrochem K.K.) and by Bayer AG. **Patent** DE 2111414 to Bayer
Manufacturers Bayer

APPLICATIONS

Biochemistry Cholinesterase inhibitor. **Mode of action** Non-systemic insecticide

with contact and stomach action.   **Uses** Control of leaf-eating caterpillars, *Pseudococcus* spp., thrips, cockchafer larvae, cutworms, etc. in a range of crops, including vegetables, fruit, maize, sugar cane, sugar beet, tea, tobacco and ornamentals.   **Phytotoxicity** Russetting is possible with Golden Delicious apples.
Formulation types WP; EC.   **Compatibility** Not compatible with alkaline materials.
Selected tradenames 'Tokuthion' (agricultural use) (Bayer)

ANALYSIS
Product analysis by glc.   **Residues** determined by glc (E. Möllhoff, *Pflanzenschutz-Nachr. (Engl. Ed.)*, 1975, **28**, 382). Methods for the determination of residues are available from Bayer.

MAMMALIAN TOXICOLOGY
EHC 63 (WHO, 1986; a general review of organophosphorus insecticides).
Oral Acute oral LD_{50} for male rats 1569, female rats 1390, mice *c.* 2200 mg/kg.
Skin and eye Acute percutaneous LD_{50} (24 h) for rats >5000 mg/kg; not irritating to skin and eyes (rabbit). Skin sensitiser.   **Inhalation** LC_{50} (4 h) for rats >2.7 mg/l air (aerosol).   **NOEL** (2 y) for rats 5, mice 1, dogs 0.4 mg/kg diet.
ADI 0.0001 mg/kg b.w.   **Toxicity class** WHO (a.i.) II; EPA (formulation) III
EC risk Xn (R22); Xi (R43)

ECOTOXICOLOGY
Birds Acute oral LD_{50} for Japanese quail 100–200 mg/kg.   **Fish** LC_{50} (96 h) for golden orfe 4–8, rainbow trout 0.5–1 mg/l (500 g/l EC formulation).   **Bees** Not hazardous to bees if used as recommended.   **Daphnia** LC_{50} (48 h) 0.014 mg/l.
Algae E_rC_{50} for *Scenedesmus subspicatus* 2.3 mg/l.

ENVIRONMENTAL FATE
Animals In rats, prothiofos is rapidly absorbed and metabolised. At 72 h after administration, 98% of the dose was excreted. Metabolic routes include oxidation to prothiofos-oxon, hydrolysis to 2,4-dichlorophenol which is further conjugated, and cleavage of the propyl group of prothiofos and prothiofos-oxon, with formation of 2,4-dichlorophenylethyl hydrogen phosphorothioate and 2,4-dichlorophenylethyl hydrogen phosphate.   **Plants** In plants, prothiofos is hydrolysed to 2,4-dichlorophenol which is further conjugated. Formation of prothiofos-oxon and cleavage of the propyl group also occurs.
Soil/Environment Prothiofos is very strongly adsorbed in soil. DT_{50} under field conditions is 1–2 mo. In soil, dechlorination to 4-chloroprothiofos and oxidation to prothiofos-oxon occurs, as well as hydrolysis to 2,4-dichlorophenol which is finally metabolised to CO_2.

pymetrozine

azomethine

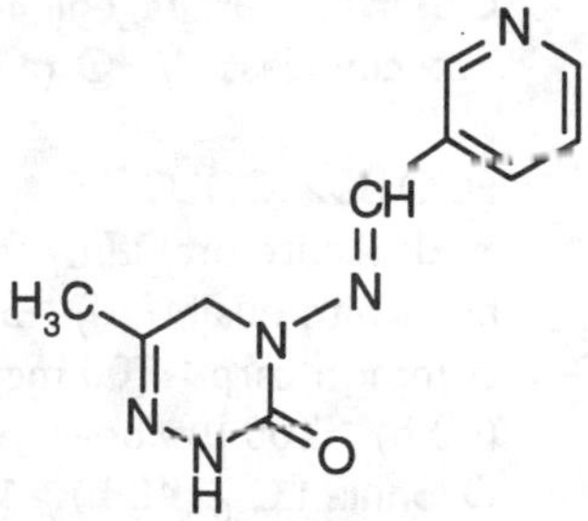

NOMENCLATURE

Common name pymetrozine (BSI, pa E-ISO)

IUPAC name (*E*)-4,5-dihydro-6-methyl-4-(3-pyridylmethyleneamino)-= 1,2,4-triazin-3(2*H*)-one

Chemical Abstracts name (*E*)-4,5-dihydro-6-methyl-4-[(3-pyridinylmethylene)= amino]-1,2,4-triazin-3(2*H*)-one

CAS RN *[123312–89–0]* **Development codes** CGA 215944

PHYSICAL CHEMISTRY

Mol. wt. 217.2 **M.f.** $C_{10}H_{11}N_5O$ **Form** Colourless crystals. **M.p.** 217 °C
V.p. <4 × 10^{-3} mPa (25 °C) (OECD 104) **K_{ow}** logP = –0.18 (OECD 107)
S.g./density 1.36 (20 °C) **Solubility** In water at 25 °C, 0.29 g/l. In ethanol 2.25,
hexane <0.001 (both in g/l, 20 °C). **Stability** Stable in air (OECD 113/DTA).
Hydrolysis DT_{50} 4.3 h (pH 1), 25 d (pH 5).

COMMERCIALISATION

History Reported by C. R. Flückiger *et al.* (*Proc. Br. Crop Prot. Conf. – Pests Dis.*,
1992, **1**, 43). Introduced by Ciba-Geigy (now Novartis Crop Protection AG).
Manufacturers Novartis

APPLICATIONS

Biochemistry Novel, unidentified biochemistry. **Mode of action** Insecticide
selective against Homoptera, causing them to stop feeding. **Uses** Control of
aphids and whitefly in vegetables, ornamentals, cotton, field crops, deciduous fruit
and citrus fruit; both juvenile and adult stages are susceptible. Also control of
planthoppers in rice. Its selective action makes it especially useful in IPM
programmes. **Formulation types** WP. **Selected tradenames** 'Chess' (Novartis)

ANALYSIS

Residue determination by hplc and u.v. detection. Details available from Novartis.

MAIN ENTRIES

MAMMALIAN TOXICOLOGY

Oral Acute oral LD_{50} for rats 5820 mg/kg.   **Skin and eye** Acute percutaneous LD_{50} for rats >2000 mg/kg. Non-irritating to the skin and eyes (rabbits). Not a skin sensitiser to guinea pigs.   **Inhalation** LC_{50} (4 h) >1800 mg/m³ air.
Other Non-mutagenic in 5 assay tests including the Ames test.
Toxicity class WHO (a.i.) III

ECOTOXICOLOGY

Birds Acute oral LD_{50} for bobwhite quail, mallard duck >2000 mg/kg. LC_{50} for bobwhite quail (8 d) >5200 ppm.   **Fish** LC_{50} (96 h) for rainbow trout and common carp >100 mg/l.   **Bees** Oral LD_{50} (48 h) >117 µg/bee; contact LD_{50} (48 h) >200 µg/bee.   **Worms** LC_{50} (14 d) for *Eisenia foetida* 1098 mg/kg.
Daphnia LC_{50} (48 h) >100 mg/l.   **Algae** LC_{50} (72 h) for *Scenedesmus* sp. 47.1, *Selenastrum* sp. 58 mg/l.

ENVIRONMENTAL FATE

Soil/Environment Slightly mobile in soils, and rapidly degraded (DT_{50} c. 8–38 d). K_{om} = 292.

616   pyraclofos   *Insecticide*

organophosphorus

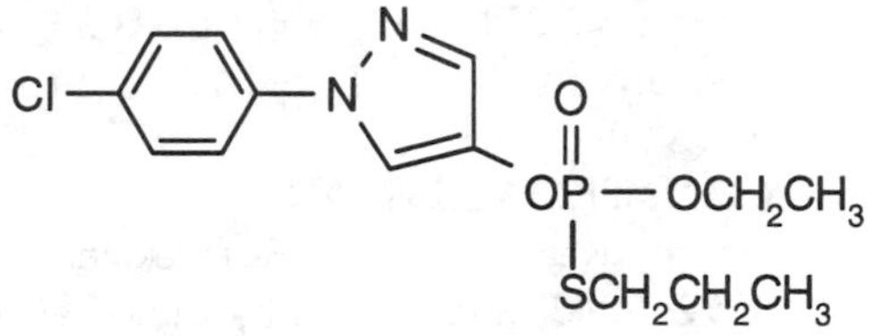

NOMENCLATURE

Common name pyraclofos (BSI, draft E-ISO, (m) draft F-ISO)
IUPAC name (RS)-[O-1-(4-chlorophenyl)pyrazol-4-yl O-ethyl S-propyl phosphorothioate]
Chemical Abstracts name (±)-O-[1-(4-chlorophenyl)-1H-pyrazol-4-yl] O-ethyl S-propyl phosphorothioate
CAS RN *[77458–01–6]* unstated stereochemistry   **Development codes** TIA-230 (Takeda)   **Official codes** OMS 3040

PHYSICAL CHEMISTRY

Mol. wt. 360.8   **M.f.** $C_{14}H_{18}ClN_2O_3PS$ **Form** Pale yellow oil.
B.p. 164 °C/0.01 mmHg   **V.p.** 1.6×10^{-3} mPa (20 °C) K_{ow} logP = 3.77 (20 °C)

S.g./density 1.271 (28 °C)   **Solubility** In water 33 mg/l (20 °C). Miscible with most organic solvents.   **Stability** On hydrolysis DT_{50} (25 °C, pH 7) 29 d.

COMMERCIALISATION

History Insecticide reported by Y. Kono et al. (*Pestic. Biochem. Physiol.*, 1983, **20**, 225). Introduced in Japan (1989) by Takeda Chemical Industries Ltd.
Patent US 4474775   **Manufacturers** Takeda

APPLICATIONS

Biochemistry Cholinesterase inhibitor.   **Mode of action** Respiratory, contact and stomach action.   **Uses** Control of Lepidoptera (*Spodoptera* and *Heliothis* spp.), Coleoptera, Acarina, and nematodes in fruit, vegetables, field crops, ornamentals, and in forestry at 0.25–1.5 kg a.i./ha. Also used in public health.
Phytotoxicity Pyraclofos is slightly phytotoxic to fruit trees such as apple, Japanese pear, peach, and citrus, depending on varieties, stages and climatic conditions.
Formulation types EC; WP; GR.   **Selected tradenames** 'Boltage' (Takeda); 'Voltage' (Takeda)

ANALYSIS

Product and **residue** analysis is by hplc.

MAMMALIAN TOXICOLOGY

EHC 63 (WHO, 1986; a general review of organophosphorus insecticides).
Oral Acute oral LD_{50} for male and female rats 237, male mice 575, female mice 420 mg/kg.   **Skin and eye** Acute percutaneous LD_{50} for rats >2000 mg/kg. Non-irritating to eyes and skin (rabbits). Not a skin sensitiser (guinea pigs).
Inhalation LC_{50} for male rats 1.69, female rats 1.46 mg/l.   **NOEL** (2 y) for male rats 0.101, female rats 0.120, male mice 1.03, female mice 1.28 mg/kg b.w. daily.
Other Not oncogenic in rats and mice. Not teratogenic in rats and rabbits.
Toxicity class WHO (a.i.) II

ECOTOXICOLOGY

Birds LC_{50} for bobwhite quail 164, mallard ducks 348 mg/kg diet.   **Fish** LC_{50} (72 h) for carp 0.028, Japanese killifish 1.9 mg/l.   **Bees** LD_{50} (contact) for honeybees 0.953 µg/bee.   **Other aquatic spp.** LC_{50} (3 h) for *Moina macrocopa* 0.052 mg/l.

ENVIRONMENTAL FATE

Animals In rats, following oral administration, >90 % of the dose is eliminated within 24 hours, principally in the urine.   **Soil/Environment** DT_{50} in soil 3–38 d, depending on soil type.

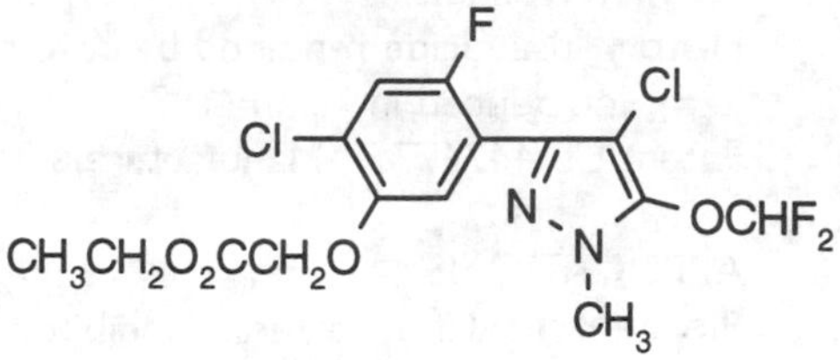

NOMENCLATURE
Common name pyraflufen-ethyl (BSI, pa ISO)
IUPAC name ethyl 2-chloro-5-(4-chloro-5-difluoromethoxy-1-methylpyrazol-=
3-yl)-4-fluorophenoxyacetate
Chemical Abstracts name ethyl 2-chloro-5-[4-chloro-5-(difluoromethoxy)-=
1-methyl-1*H*-pyrazol-3-yl]-4-fluorophenoxyacetate
CAS RN *[129630–17–7]*   **Development codes** ET-751

PHYSICAL CHEMISTRY
Mol. wt. 413.2   **M.f.** $C_{15}H_{13}Cl_2F_3N_2O_4$; (acid $C_{13}H_9Cl_2F_3N_2O_4$)   **Form** Pale
brown crystals (tech.).   **M.p.** 126–127 °C   **V.p.** 4.79 mPa   **Solubility** In water
<1 ppm (25 °C). In xylene 2.9, acetone 25.0, ethanol 1.0, ethyl acetate 14.5 (all in
% wt./wt., 25 °C).

COMMERCIALISATION
History Reported by Nihon Nohyaku Co., Ltd. (Y. Miura *et al., Proc. Br. Crop Prot.
Conf., Weeds,* 1993, **1**, 35).   **Manufacturers** Nihon Nohyaku

APPLICATIONS
Biochemistry Acts by inhibition of protoporphyrinogen-IX oxidase. Selectivity
between wheat and *Galium aparine* is due to differences in deposition, absorption
and metabolism (S. Murata *et al., Proc. Br. Crop Prot. Conf. – Weeds,* 1995, **1**, 243).
Mode of action Contact herbicide; when applied to foliage, it is readily absorbed
into plant tissues, and rapid necrosis or desiccation of stems and leaves is induced
in the presence of light.   **Uses** Selective post-emergence control of broad-leaved
weeds, especially *Galium aparine, Matricaria inodora, Lamium purpureum* and
Stellaria media, in cereals.   **Formulation types** EC.

MAMMALIAN TOXICOLOGY
Oral Acute oral LD_{50} for rats >5000 mg/kg.   **Skin and eye** Acute percutaneous
LD_{50} not available. Non-irritant to skin and slightly irritant to eyes (rabbits).
Other Non-mutagenic in the Ames test.

ECOTOXICOLOGY
Fish LC_{50} (48 h) for carp and rainbow trout >10 ppm.

ENVIRONMENTAL FATE
Plants Metabolism proceeds via de-esterification followed by *N*-demethylation.

₅18 pyrazolynate   *Herbicide*

pyrazole (herbicide)

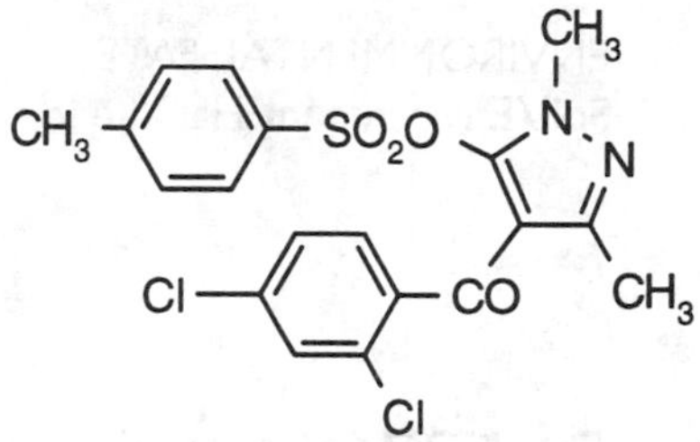

NOMENCLATURE
Common name pyrazolynate (BSI, draft E-ISO); pyrazolate (JMAF)
IUPAC name 4-(2,4-dichlorobenzoyl)-1,3-dimethylpyrazol-5-yl toluene-4-sulfonate
Chemical Abstracts name (2,4-dichlorophenyl)[1,3-dimethyl-5-[[(4-methyl=
phenyl)sulfonyl]oxy]-1*H*-pyrazol-4-yl]methanone
CAS RN *[58011–68–0]* **Development codes** A-544; H-468T; SW-751 (all
Sankyo)

PHYSICAL CHEMISTRY
Mol. wt. 439.3 **M.f.** $C_{19}H_{16}Cl_2N_2O_4S$ **Form** Colourless, rod-shaped crystals.
M.p. 117.5–118.5 °C **V.p.** <1.3 × 10⁻² mPa (20 °C) **Solubility** In water
0.056 mg/l (25 °C). In ethanol 14, ethyl acetate 118, 1,4-dioxane 256, hexane
0.6 (all in g/l). **Stability** Aqueous solutions hydrolyse readily.

COMMERCIALISATION
History Herbicide reported by M. Ishida *et al.*, *Sankyo Kenkyusho Nenpo*, 1984,
36, 44. Introduced in Japan (1980) by Sankyo Co., Ltd. **Patent** JP 1001829;
GB 1463473 **Manufacturers** Sankyo

APPLICATIONS
Biochemistry Chlorophyll biosynthesis inhibitor. **Uses** Control of grasses, sedges,
Potamogeton distinctus, *Sagittaria trifolium*, and *Alisma canaliculatum* in paddy rice.
Formulation types GR. **Mixtures** *(pyrazolynate +)* bromobutide; dimethametryn +
pretilachlor. **Selected tradenames** 'Sanbird' (Sankyo)

ANALYSIS

Product analysis by hplc.   **Residues** are determined by glc (M. Ishida et al., Sankyo Kenkyusho Nenpo, 1984, **36**, 44).

MAMMALIAN TOXICOLOGY

Oral Acute oral LD_{50} for male rats 9550, female rats 10 233, male mice 10 070, female mice 11 092 mg/kg.   **Skin and eye** Acute percutaneous LD_{50} for rats >5000 mg/kg. Non-irritating to skin.   **Other** Non-mutagenic.
Toxicity class WHO (a.i.) III (Table 5)

ECOTOXICOLOGY

Fish LC_{50} for carp 92 mg/l.

ENVIRONMENTAL FATE

Soil/Environment Half-life in soil 10–20 d.

619 pyrazophos   *Fungicide*

organophosphate ester

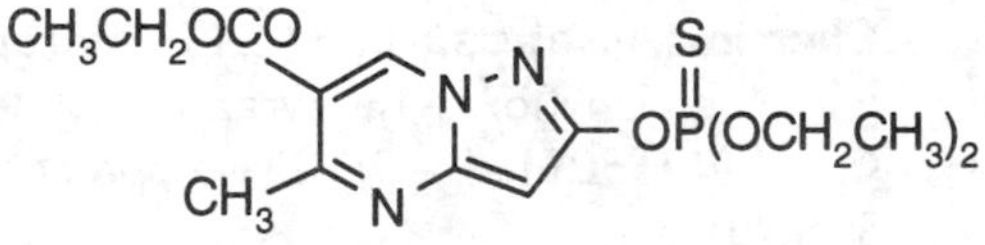

NOMENCLATURE

Common name pyrazophos (BSI, E-ISO, (m) F-ISO)
IUPAC name ethyl 2-diethoxyphosphinothioyloxy-5-methylpyrazolo[1,5-a]= pyrimidine-6-carboxylate; O,O-diethyl O-6-ethoxycarbonyl-5-methylpyrazolo= [1,5-a]pyrimidin-2-yl phosphorothioate
Chemical Abstracts name ethyl 2-[(diethoxyphosphinothioyl)oxy]-5-methyl= pyrazolo[1,5-a]pyrimidine-6-carboxylate
CAS RN [13457–18–6]   **EEC no.** 236–656–1   **Development codes** Hoe 02 873

PHYSICAL CHEMISTRY

Composition Tech. grade is ≥94% pure.   **Mol. wt.** 373.4   **M.f.** $C_{14}H_{20}N_3O_5PS$
Form Colourless crystals.   **M.p.** 51–52 °C   **B.p.** Decomposition begins at 160 °C
V.p. 0.22 mPa (50 °C) (Antoine) K_{ow} logP = 3.8   **Henry** 2.578×10^{-4} Pa m^3 mol^{-1} (calc.)   **S.g./density** 1.348 at 25 °C   **Solubility** In water 4.2 mg/l (25 °C). Readily soluble in most organic solvents, e.g. xylene, benzene, carbon tetrachloride, dichloromethane, trichloroethylene. In acetone, toluene, ethyl

acetate >400, hexane 16.6 (all in g/l, 20 °C).   **Stability** Hydrolysed by acids and alkalis. Unstable in undiluted form.   **F.p.** 34±2 °C (Abel-Pensky closed method)

COMMERCIALISATION
History Fungicide reported by F. M. Smit (*Meded. Rijksfac. Landbouwwet. Gent,* 1969, **34,** 763). Introduced by Hoechst AG (now AgrEvo GmbH).
Patent DE 1545790; GB 1145306   **Manufacturers** AgrEvo

APPLICATIONS
Biochemistry Prevents development of appressoria by fungal conidia. Inhibits melanin biosynthesis in some species.   **Mode of action** Systemic fungicide with protective and curative action. Absorbed by the leaves and shoots, and translocated acropetally within the plant.   **Uses** Control of *Erysiphe,* *Helminthosporium* and *Rhynchosporium* in cereals, *Podosphaera* in pome fruit, *Erysiphe* and *Sphaerotheca* in cucurbits, *Erysiphe* in tomatoes, *Oidium* and *Podosphaera* in stone fruit, and *Uncinula* in grapes. Application rates vary from 100–600 g/ha, depending on crop.   **Phytotoxicity** Non-phytotoxic when used as recommended (except to some varieties of grape).   **Formulation types** EC; WP.
Mixtures *(pyrazophos +)* captafol; carbendazim; propiconazole; carbendazim + cyproconazole; carbendazim + flutriafol; carbendazim + flusilazole; carbendazim + maneb.   **Selected tradenames** 'Afugan' (AgrEvo, Procida)

ANALYSIS
Product analysis by hplc (*CIPAC Handbook,* 1985, **1C,** 2206) or by glc.
Residues determined by glc (J. Asshauer *et al., Anal. Methods Pestic. Plant Growth Regul.,* 1978, **10,** 237; *Man. Pestic. Residue Anal.,* 1987, **I,** 6, S19; *Anal. Methods Residues Pestic.,* 1988, Part I, M2, M5, M12; A. Ambrus *et al., J. Assoc. Off. Anal. Chem.,* 1981, **64,** 773). Details available from AgrEvo.

MAMMALIAN TOXICOLOGY
Reviews *Pesticide residues in food – 1992.* FAO Plant Production and Protection Paper, 116, 1993. *Pesticide residues in food – 1992 evaluations. Part II – Toxicology.* World Health Organisation, WHO/PCS/93.34.   **Oral** Acute oral LD_{50} for rats 151–778 mg/kg (depending on sex and carrier).   **Skin and eye** Acute percutaneous LD_{50} for rats >2000 mg/kg. Non-irritating to skin; slightly irritating to eyes (rabbits).   **Inhalation** LC_{50} (4 h) for rats 1220 mg/m^3 air.   **NOEL** (2 y) for rats 5 mg/kg diet. A 3-generation test on rats showed no effect at 50 mg/kg diet.
ADI (JMPR) 0.004 mg/kg b.w. [1992].   **Toxicity class** WHO (a.i.) II; EPA (formulation) II   **EC risk** Xn (R22)

ECOTOXICOLOGY
Birds Acute oral LD_{50} for quail (depending on carrier and sex) 118–480 mg tech./kg. Dietary LC_{50} (14 d) for mallard duck *c.* 340, bobwhite quail *c.* 300 mg/kg.   **Fish** LC_{50} (96 h) for carp 2.8–6.1, rainbow trout 0.48–1.14, bluegill

sunfish 0.28 mg/l.   **Bees** LD_{50} (24 h, contact) 0.25 µg/bee.   **Worms** LC_{50} (14 d) for *Eisena foetida* >1000 mg/kg soil.   **Other beneficial spp.** 'Afugan 30 EC' rated harmful to *Poecilus cupreus*, *Syrphus corollae*, *Chrysoperla carnea* and *Aleochara bilineata*; harmless to *Pardosa amentata* (IOBC scheme).   **Daphnia** LC_{50} (48 h) 0.36 µg/l (soft water); 0.63 µg/l (hard water). NOEL 0.18 µg/l in hard and soft water.   **Algae** LC_{50} (72 h) for *Scenedesmus subspicatus* 65.5 mg/l.

ENVIRONMENTAL FATE

Animals Extensively metabolised and rapidly eliminated in rats, DT_{50} 4–5 h. Excretion mainly in urine; the most important metabolite is ethyl 2-hydroxy-= 5-methyl-6-pyrazolo[1,5-*a*]pyrimidine-carboxylate, partly excreted as a sulfate conjugate.   **Plants** Half-life in wheat leaves *c.* 19 days. Following hydrolysis of the phosphate bond, a β-glucoside of the pyrazolopyrimidine is formed.
Soil/Environment Soil degradation occurs by cleavage of the phosphoric acid group, saponification of the carboxylate fraction and further degradation of the heterocyclic ring system, ultimately to CO_2. Degradation rates vary with soil type and properties, with no obvious correlation with any single soil characteristic. DT_{50} 10–21 d, DT_{90} 111–235 d (field). Strongly absorbed to soils (K_{oc} 1332–2670, calc.). Column studies and leaching models indicate pyrazophos will not leach.

620   pyrazosulfuron-ethyl   *Herbicide*

sulfonylurea

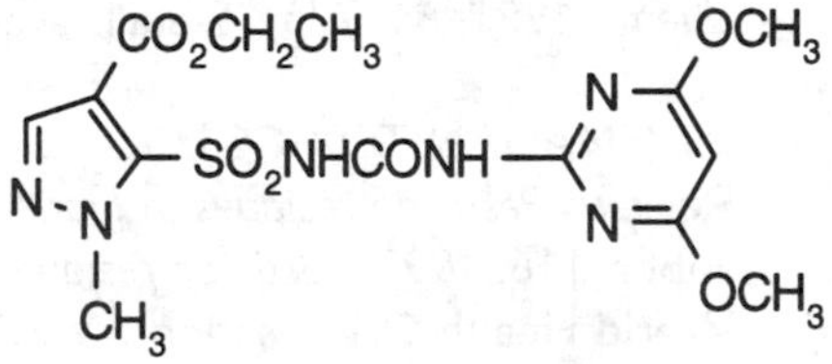

NOMENCLATURE
pyrazosulfuron-ethyl
Common name pyrazosulfuron-ethyl
IUPAC name ethyl 5-(4,6-dimethoxypyrimidin-2-ylcarbamoylsulfamoyl)-1-methyl= pyrazole-4-carboxylate
Chemical Abstracts name ethyl 5-[[[[(4,6-dimethoxy-2-pyrimidinyl)amino]= carbonyl]amino]sulfonyl]-1-methyl-1*H*-pyrazole-4-carboxylate
CAS RN *[93697–74–6]*   **Development codes** NC-311

pyrazosulfuron
Common name pyrazosulfuron (BSI, draft E-ISO, (*m*) draft F-ISO)

IUPAC name 5-(4,6-dimethoxypyrimidin-2-ylcarbamoylsulfamoyl)-1-methyl=
pyrazole-4-carboxylic acid
Chemical Abstracts name 5-[[[[(4,6-dimethoxy-2-pyrimidinyl)amino]carbonyl]=
amino]sulfonyl]-1-methyl-1H-pyrazole-4-carboxylic acid
CAS RN *[98389–04–9]*

PHYSICAL CHEMISTRY
pyrazosulfuron-ethyl
Mol. wt. 414,4   **M.f.** $C_{14}H_{18}N_6O_7S$ **Form** Colourless crystals.   **M.p.** 181–182 °C
V.p. 0.0147 mPa (20 °C)   **K$_{ow}$** logP = 1.3 (unstated pH)   **S.g./density** 1.44
(20 °C)   **Solubility** In water 0.0145 g/l (20 °C). In methanol 0.7, hexane 0.2,
benzene 15.6, chloroform 234.4, acetone 31.7 (all in g/l, 20 °C).   **Stability** Stable
at 50 °C for 6 months. Relatively stable at pH 7. Unstable in acidic or alkaline
media.

pyrazosulfuron
Mol. wt. 386.3   **M.f.** $C_{12}H_{14}N_6O_7S$

COMMERCIALISATION
History Pyrazosulfuron-ethyl reported as a herbicide by S. Kobayashi (*Jpn. Pestic.
Inf.*, 1989, No. 55, p. 17). Introduced by Nissan Chemical Industries, Ltd in 1990.
Manufacturers Nissan

APPLICATIONS
pyrazosulfuron-ethyl
Biochemistry Branched chain amino acid synthesis (acetolactate synthase or ALS)
inhibitor. Acts by inhibiting biosynthesis of the essential amino acids valine and
isoleucine, hence stopping cell division and plant growth. Selectivity derives from
rapid metabolism (demethylation of methoxy group) in the crop. Metabolic basis
of selectivity in sulfonylureas reviewed (M. K. Koeppe & H. M. Brown, *Agro-Food-
Industry*, **6**, 9–14 (1995)).   **Mode of action** Systemic herbicide, absorbed by roots
and/or leaves and translocated to the meristem.   **Uses** Control of annual and
perennial broad-leaved weeds and sedges, pre- or post-emergence in wet-sown
and transplanted rice crops, at 15–30 g/ha.   **Formulation types** GR; SC; WP; WG.
Mixtures (*pyrazosulfuron-ethyl* +) esprocarb + pretilachlor + dimethametryn;
mefenacet; esprocarb; pretilachlor; cafenstrole; pretilachlor + cyhalofop-methyl;
cafenstrole + cyhalofop-methyl.   **Selected tradenames** 'Sirius' (Nissan); 'Act'
(mixture) (Nihon Bayer, Nissan); 'Agreen' (Nissan); 'Sparkstar' (mixture) (Nissan,
Novartis, Zeneca); 'Billy' (Sanonda)

ANALYSIS
Product by hplc. Methods for sulfonylurea **residues** in crops, soil and water
reviewed (A. C. Barefoot *et al.*, *Proc. Br. Crop Prot. Conf. – Weeds*, 1995, **2**, 707).
Details from Nissan.

MAMMALIAN TOXICOLOGY
pyrazosulfuron-ethyl
Oral Acute oral LD_{50} for rats and mice >5000 mg/kg. **Skin and eye** Acute percutaneous LD_{50} for rats >2000 mg/kg. Non-irritating to skin and eyes (rabbits). Non-sensitising to skin (guinea pigs). **Inhalation** LC_{50} for rats >3.9 mg/l air. **NOEL** (78 w) for mice 4.3 mg/kg daily. **ADI** 0.043 mg/kg.
Other Non-mutagenic in the Ames test. Non-teratogenic in rats or rabbits.
Toxicity class WHO (a.i.) III (Table 5)

ECOTOXICOLOGY
pyrazosulfuron-ethyl
Birds Acute oral LD_{50} for bobwhite quail >2250 mg/kg. **Fish** LC_{50} (96 h) for rainbow trout and bluegill sunfish >180 mg/l; (48 h) for carp >30 mg/l. **Bees** LD_{50} (contact) >100 µg/bee. **Daphnia** TLm (3 h) >40 ppm. **Algae** I_{50} for green alga (*Scenedesmus acutus*) 1 ppm.

ENVIRONMENTAL FATE
Animals In rats, after 48 h, 80% of applied pyrazosulfuron-ethyl is excreted in urine and faeces. The major metabolic reaction is demethylation of the methoxy group. **Soil/Environment** In soil DT_{50} <15 d. In water DT_{50} in buffer solution (pH 7), paddy fields or river water are *c*. 28 d.

621 pyrazoxyfen   *Herbicide*

pyrazole (herbicide)

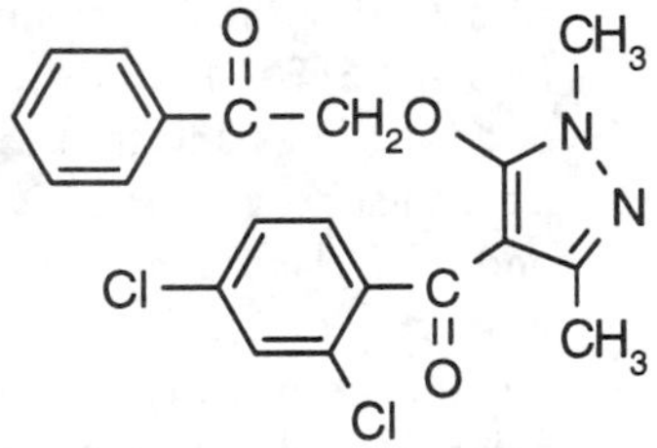

NOMENCLATURE
Common name pyrazoxyfen (BSI, draft E-ISO); pyrazoxyfène ((*m*) draft F-ISO)
IUPAC name 2-[4-(2,4-dichlorobenzoyl)-1,3-dimethylpyrazol-5-yloxy]= acetophenone
Chemical Abstracts name 2-[[4-(2,4-dichlorobenzoyl)-1,3-dimethyl-1*H*-pyrazol-= 5-yl]oxy]-1-phenylethanone
CAS RN *[71561–11–0]* **Development codes** SL-49 (Ishihara)

PHYSICAL CHEMISTRY
Mol. wt. 403.3 **M.f.** $C_{20}H_{16}Cl_2N_2O_3$ **Form** Colourless crystals.
M.p. 111–112 °C **V.p.** 0.048 mPa (25 °C) **K$_{ow}$** logP = 3.69 **S.g./density** 1.37
(20 °C) **Solubility** In water 0.9 g/l (20 °C). In acetone 223, benzene 325, ethanol
14, chloroform 1068, hexane 900, xylene 116, toluene 200 (all in g/l, 20 °C).
Stability Stable to acid, alkali, light and heat.

COMMERCIALISATION
History Herbicide reported by F. Kimura (*Jpn. Pestic. Inf.*, 1984, No. 45, 24).
Introduced by Ishihara Sangyo Kaisha, Ltd. **Manufacturers** Ishihara Sangyo

APPLICATIONS
Biochemistry Chlorophyll synthesis inhibitor. **Mode of action** Selective systemic
herbicide which is absorbed through the young stems and roots of weeds, with
translocation into the whole plant. **Uses** Applied pre- or post-weed emergence
at 3 kg a.i./ha after transplanting paddy rice, to control annual and perennial
weeds. Cannot be used successfully in upland crops. Can be used in direct-seeded
rice at temperatures <35 °C. **Phytotoxicity** Non-phytotoxic to rice. Can be used
on direct-seeded rice but, at temperatures >35 °C, temporary crop damage may
occur. **Formulation types** GR; SC. **Mixtures** (*pyrazoxyfen +*) bromobutide;
pretilachlor; thenylchlor; thenylchlor + bromobutide.
Selected tradenames 'Paicer' (Ishihara Sangyo); 'Mondaris' (Zeneca)

ANALYSIS
Product by hplc. Details from Ishihara Sangyo.

MAMMALIAN TOXICOLOGY
Oral Acute oral LD$_{50}$ for male rats 1690, female rats 1644, mice 8450 mg/kg.
Skin and eye Acute percutaneous LD$_{50}$ for rats >5000 mg/kg. **Inhalation** LC$_{50}$ for
rats >0.28 mg/kg. **Toxicity class** WHO (a.i.) III; EPA (formulation) III

ECOTOXICOLOGY
Fish LC$_{50}$ (48 h) for carp 2.5, rainbow trout 0.79, killifish 2.7 mg/l. **Daphnia** LC$_{50}$
(3 h) 127 mg/l.

ENVIRONMENTAL FATE
Soil/Environment In soil, DT$_{50}$ 4–15 d. K$_d$ 109–439.

MAIN ENTRIES

natural pyrethrin

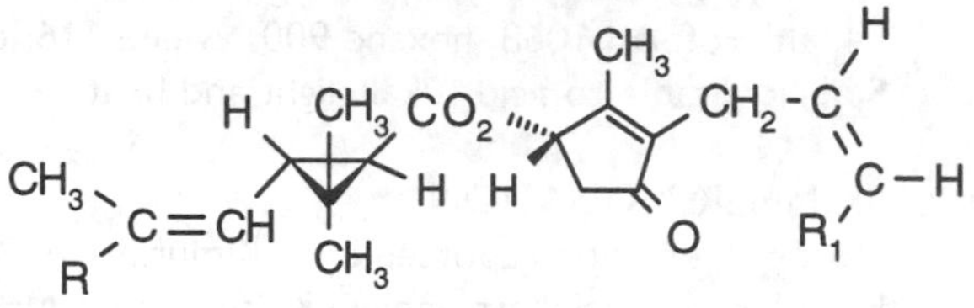

R = - CH$_3$ or - CO$_2$CH$_3$
R$_1$ = - CH=CH$_2$ or - CH$_3$ or - CH$_2$CH$_3$

NOMENCLATURE
pyrethrins (pyrethrum)
Common name pyrethrins (BSI, E-ISO, ESA, JMAF); pyrèthres (F-ISO)
CAS RN *[8003–34–7]*

pyrethrins (chrysanthemates)
IUPAC name for pyrethrin I, *Roth*: (Z)-(S)-2-methyl-4-oxo-3-(penta-2,4-dienyl)=
cyclopent-2-enyl (1R)-*trans*-2,2-dimethyl-3-(2-methylprop-1-enyl)cyclopropane=
carboxylate; or (Z)-(S)-2-methyl-4-oxo-3-(penta-2,4-dienyl)cyclopent-2-enyl
(+)-*trans*-chrysanthemate; for cinerin I, *Roth*: (Z)-(S)-3-(but-2-enyl)-2-methyl-=
4-oxocyclopent-2-enyl (1R)-*trans*-2,2-dimethyl-3-(2-methylprop-1-enyl)cyclo=
propanecarboxylate; or (Z)-(S)-3-(but-2-enyl)-2-methyl-4-oxocyclopent-2-enyl
(+)-*trans*-chrysanthemate; for jasmolin I, *Roth*: (Z)-(S)-2-methyl-4-oxo-3-(pent-=
2-enyl)cyclopent-2-enyl (1R)-*trans*-2,2-dimethyl-3-(2-methylprop-1-enyl)cyclo=
propanecarboxylate; or (Z)-(S)-2-methyl-4-oxo-3-(pent-2-enyl)cyclopent-2-enyl
(+)-*trans*-chrysanthemate
Chemical Abstracts name for pyrethrin I, [1R-[1α[S*(Z)],3β]]-2-methyl-4-oxo-=
3-(2,4-pentadienyl)cyclopenten-1-yl 2,2-dimethyl-3-(2-methyl-1-propenyl)cyclo=
propanecarboxylate; for cinerin I, [1R-[1α[S*(Z)],3β]]-3-(2-butenyl)-2-methyl-=
4-oxo-2-cyclopenten-1-yl 2,2-dimethyl-3-(2-methyl-1-propenyl)cyclopropane=
carboxylate; for jasmolin I, [1R-[1α[S*(Z)],3β]]-2-methyl-4-oxo-3-(2-pentenyl)-=
2-cyclopenten-1-yl 2,2-dimethyl-3-(2-methyl-1-propenyl)cyclopropanecarboxylate
CAS RN *[121–21–1]* pyrethrin I; *[2540–06–6]* cinerin I; *[4466–14–2]* jasmolin I
EEC no. 204–455–8 (pyrethrin I); 246–948–0 (cinerin I)

pyrethrins (pyrethrates)
IUPAC name for pyrethrin II, *Roth*: (Z)-(S)-2-methyl-4-oxo-3-(penta-2,4-dienyl)=
cyclopent-2-enyl (E)-(1R)-*trans*-3-(2-methoxycarbonylprop-1-enyl)-2,2-dimethyl=
cyclopropanecarboxylate; or (Z)-(S)-2-methyl-4-oxo-3-(penta-2,4-dienyl)cyclo=
pent-2-enyl pyrethrate; for cinerin II, *Roth*: (Z)-(S)-3-(but-2-enyl)-2-methyl-4-oxo=
cyclopent-2-enyl (E)-(1R)-*trans*-3-(2-methoxycarbonylprop-1-enyl)-2,2-dimethyl=
cyclopropanecarboxylate; or (Z)-(S)-3-(but-2-enyl)-2-methyl-4-oxocyclopent-=

2-enyl pyrethrate; for jasmolin II, *Roth*: (Z)-(S)-2-methyl-4-oxo-3-(pent-2-enyl)=
cyclopent-2-enyl (E)-(1R)-*trans*-3-(2-methoxycarbonylprop-1-enyl)-2,2-dimethyl=
cyclopropanecarboxylate; or (Z)-(S)-2-methyl-4-oxo-3-(pent-2-enyl)cyclopent-=
2-enyl pyrethrate.
Chemical Abstracts name for pyrethrin II, [1R-[1α[S*(Z)],3β(E)]]-2-methyl-=
4-oxo-3-(2,4-pentadienyl)-2-cyclopenten-1-yl 3-(3-methoxy-2-methyl-3-oxo-=
1-propenyl)-2,2-dimethylcyclopropanecarboxylate; for cinerin II, [1R-[1α[S*(Z)],=
3β(E)]]-3-(2-butenyl)-2-methyl-4-oxo-2-cyclopenten-1-yl 3-(3-methoxy-2-methyl-
3-oxo-1-propenyl)-2,2-dimethylcyclopropanecarboxylate; for jasmolin II, [1R-[1α=
[S*(Z)],3β(E)]]-2-methyl-4-oxo-3-(2-pentenyl)-2-cyclopent-1-enyl 3-(3-methoxy-=
2-methyl-3-oxo-1-propenyl)-2,2-dimethylcyclopropanecarboxylate
CAS RN *[121–29–9]* pyrethrin II; *[1172–63–0]* cinerin II; *[121–20–0]* jasmolin II
EEC no. 204–462–6 (pyrethrin II); 204–454–2 (cinerin II)

PHYSICAL CHEMISTRY
pyrethrins (pyrethrum)
Composition Pyrethrum extract is defined as a mixture of three naturally-
occurring, closely related insecticidal esters of chrysanthemic acid, Pyrethrins I,
and the three corresponding esters of pyrethrin acid, Pyrethrins II. In the USA, it is
standardised as 45–55% w/w total pyrethrins, but samples may be 20%; the
proportion of Pyrethrins I to II is typically in the range 0.8–2.8; the ratio of
individual esters (pyrethrins: cinerins: jasmolins) is 71:21:7. In Europe, pyrethrum
extract is 25±0.5% pyrethrins. The three components of Pyrethrins I are pyrethrin
I (R = CH_3,R_1 = CH:CH_2); jasmolin I (R = CH_3, R_1 = CH_2CH_3); cinerin I
(R = CH_3,R_1 = CH_3); the components of Pyrethrins II correspond (R = CO_2CH_3).
Form Refined extract is a pale yellow mobile oil with a faint flowery odour;
unrefined extract is a dark greenish brown viscous liquid. Powder (ground flowers)
is tan colour. **S.g./density** 0.84–0.86 (25% pale extract), c. 0.9 (oleoresin crude)
Solubility Pyrethrins are soluble in organic solvents, e.g. alcohols, hydrocarbons,
aromatics, esters, etc. **Stability** Stable >10 y in absence of light, at ambient
temperature. In light, rapid oxidation and inactivation occurs; DT_{50} in sunlight
10–12 min. Also destroyed by alkali, and by clay. Heat >200 °C leads to formation
of less-active iso-pyrethrins. **F.p.** 76 °C (Abel)

pyrethrins (chrysanthemates)
Mol. wt. pyrethrin I: 328.4; cinerin I: 316.4; jasmolin I: 330.4 **M.f.** pyrethrin I:
$C_{21}H_{28}O_3$; cinerin I: $C_{20}H_{28}O_3$; jasmolin I: $C_{21}H_{30}O_3$ **B.p.** 170 °C/0.1 mmHg
(pyrethrin I) **V.p.** 2.7 mPa (pyrethrin I) **K$_{ow}$** logP = 5.9 (pyrethrin I)
Solubility In water 0.2 ppm (pyrethrin I). **Specific rotation** $[\alpha]_D^{20}$ −14° (iso-octane)
(pyrethrin I)

pyrethrins (pyrethrates)
Mol. wt. pyrethrin II 372.4; cinerin II 360.4; jasmolin II 374.5 **M.f.** pyrethrin II:
$C_{22}H_{28}O_5$; cinerin II: $C_{21}H_{28}O_5$; jasmolin II: $C_{22}H_{30}O_5$ **B.p.** 200 °C/0.1 mmHg

(pyrethrin II)   **V.p.** 5.3 × 10^{-2} mPa (pyrethrin II)   **K$_{ow}$** logP = 4.3 (pyrethrin II)
Solubility In water 9.0 ppm (pyrethrin II).   **Specific rotation** [α]$_D^{20}$ +14.7°
(iso-octane) (pyrethrin II)

COMMERCIALISATION
Production Pyrethrum is extracted from the flower *Tanacetum* (= *Chrysanthemum*
= *Pyrethrum*) *cinerariaefolium*. The extract is refined using methanol (Pyrethrum
Board of Kenya and MGK) or supercritical carbon dioxide (Agropharm).
History Pyrethrum was identified in antiquity in China. It spread West to Persia
(Iran) probably via the Silk Roads during the Middle Ages. Dried powdered flower
heads were known as "Persian Insect Powder". Records of use date from the early
19th century when it was introduced to Dalmatia, France, United States and Japan.
Current production is from East Africa (1930), Ecuador and Papua New Guinea
(1950) and Australia (1980).   **Manufacturers** Agropharm; MGK; Office du Pirètre
au Rwanda (OPYRWA); Pyrethrum Board of Kenya; Tanganyika Extract Co.
(TECO)

APPLICATIONS
Mode of action Non-systemic insecticide with contact action. Causes paralysis
initially, with death occurring later. Has some acaricidal activity.   **Uses** Control of
a wide range of insects and mites in public health, stored products, animal houses,
and on domestic and farm animals. Control of chewing and sucking insects and
spider mites on fruit, vegetables, field crops, ornamentals, glasshouse crops, and
house plants. Normally combined with synergists, e.g. piperonyl butoxide, which
inhibit detoxification.   **Formulation types** AE; DP; EC; Fogging concentrate;
Pressurised liquid CO$_2$; WP; UL.   **Mixtures** *(pyrethrins +)* lindane; resmethrin;
pirimiphos-methyl; piperonyl butoxide; bendiocarb; dichlorvos; tetramethrin;
diazinon; resmethrin + piperonyl butoxide; chlordane + piperonyl butoxide;
piperonyl butoxide + rotenone; lindane + dichlorvos; chlorpyrifos-methyl +
permethrin; bendiocarb + piperonyl butoxide; pirimiphos-methyl + piperonyl
butoxide.   **Compatibility** Incompatible with alkaline substances.
Selected tradenames 'Alfadex' (Novartis); 'Evergreen' (MGK); 'ExciteR' (Prentiss);
'Milon' (Delicia); 'Pycon' (for concentrated mixture with piperonyl butoxide)
(Agropharm); 'Pyrocide' (MGK); 'Pyronyl' (mixture) (Prentiss)

ANALYSIS
Products and extracts of flowers analysed by glc (*AOAC Methods,* 1995, 982.02; D.
B. McClellan, *Anal. Methods Pestic. Plant Growth Regul.,* 1972, **6**, 461; J. Sherma,
ibid., 1976, **8**, 225); or by a titrimetric method (*AOAC Methods,* 1995, 936.05),
though, due to interference by certain adjuvants, only approximate results on
formulations can be obtained, an empirical factor and the determined pyrethrin I
content being used (*CIPAC Handbook,* 1970, **1**, 598; D. B. McClellan, *Anal. Methods
Pestic., Plant Growth Regul. Food Addit.,* 1964, **2**, 399).   **Residues** determined by
colorimetry or by glc (J. Sherma, *loc. cit.*; R. Mestres *et al., Ann. Falsif. Expert.*

Chim., 1979, **72**, 577; W. Specht & M. Tillkes, *Fresenius' Z. Anal. Chem.*, 1980, **301**, 300).

MAMMALIAN TOXICOLOGY
pyrethrins (pyrethrum)
Reviews *Pesticide residues in food.* FAO Agricultural Studies, No. 90; WHO Technical Report Series, No. 525, 1973. *1972 Evaluations of some pesticide residues in food.* AGP:1972/M/9/1; WHO Pesticide Residue Series No. 2, 1973. **Oral** Acute oral LD_{50} for male rats 2370, female rats 1030, mice 273–796 mg/kg. **Skin and eye** Acute percutaneous LD_{50} for rats >1500, rabbits 5000 mg/kg. Slightly irritating to skin and eyes. Constituents of the flowers may cause dermatitis to sensitised individuals, but are removed during the preparation of refined extracts. **Inhalation** LC_{50} (4 h) for rats 3.4 mg/l. **NOEL** for rats 100 ppm. **ADI** (JMPR) 0.04 mg/kg b.w. [1972]. **Other** There is no evidence that synergists increase toxicity of the pyrethrins to mammals. **Toxicity class** WHO (a.i.) II; EPA (formulation) III **EC risk** Xn (R20/21/22)

pyrethrins (chrysanthemates)
EC risk Xn (R20/21/22) (pyrethrin I); Xn (R22) (cinerin I)

pyrethrins (pyrethrates)
EC risk Xn (R20/21/22) (pyrethrin II); Xn (R22) (cinerin II)

ECOTOXICOLOGY
pyrethrins (pyrethrum)
Birds Acute oral LD_{50} for mallard ducks >10 000 mg/kg. **Fish** Highly toxic to fish. LC_{50} (96 h) (static tests) for coho salmon 39, channel catfish 114 mg/l. LC_{50} for bluegill sunfish 10, rainbow trout 5.2 µg/l. **Bees** Toxic to bees, but exhibits a repellent effect. LD_{50} (oral) 22 ng/bee, (contact) 130–290 ng/bee. **Daphnia** LC_{50} 12 µg/l.

ENVIRONMENTAL FATE
Animals In mammals, rapidly degraded in the stomach by hydrolysis of the ester bond to harmless metabolites. For details, see T. J. Class *et al., J. Agric. Food Chem.*, 1990, **38**, 529. **Soil/Environment** In the environment, degradation, promoted by sunlight and u.v. light, begins at the alcohol group and involves the formation of numerous unknown cleavage products.

thiocarbamate

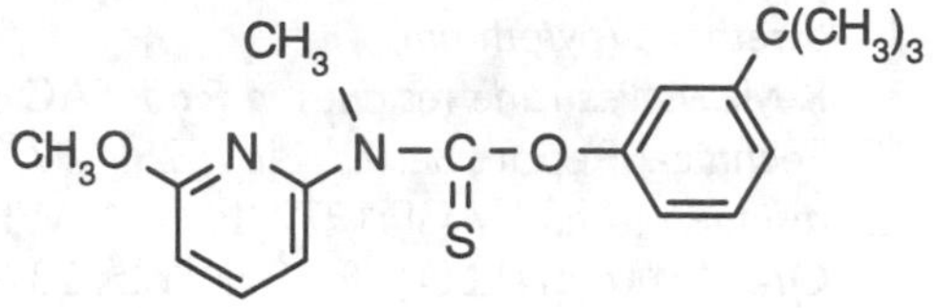

NOMENCLATURE
Common name pyributicarb (BSI, draft E-ISO)
IUPAC name O-3-*tert*-butylphenyl 6-methoxy-2-pyridyl(methyl)thiocarbamate
Chemical Abstracts name O-[3-(1,1-dimethylethyl)phenyl] (6-methoxy-=
2-pyridinyl)methylcarbamothioate
CAS RN [88678–67–5] **Development codes** TSH-888 (Tosoh)

PHYSICAL CHEMISTRY
Mol. wt. 330.4 **M.f.** $C_{18}H_{22}N_2O_2S$ **Form** White crystals. **M.p.** 85.7–86.2 °C
V.p. 0.269 mPa (40 °C) **Solubility** In water 0.32 mg/l (20 °C). In methanol 28,
acetone 780, ethanol 33, xylene 580, chloroform 390, ethyl acetate 560 (all in g/l,
20 °C).

COMMERCIALISATION
History Herbicide introduced by Toyo Soda Mfg. Co. Ltd., (now Tosoh). Acquired
by Dainippon Ink and Chemicals Inc. in 1993. **Manufacturers** Dainippon

APPLICATIONS
Biochemistry Primary target site appears to be squalene epoxidase, an enzyme of
sterol and triterpenoid biosynthesis catalysing the epoxidation of squalene to
squalene epoxide. **Mode of action** Systemic herbicide, absorbed by the roots,
leaves and stem, and translocated to active growth sites where it inhibits
elongation of the roots and aerial parts. **Uses** As a herbicide, pre- or early post-
emergence control of annual and perennial grass weeds, especially *Echinochloa
oryzicola, Cyperus difformis*, and *Monochoria vaginalis*, in paddy rice; and pre-
emergence control of *Digitaria ciliaris, Setaria viridis* and *Poa annua* in turf. As a
fungicide, pre- or early post-emergence control of *Curvularia, Rhizoctonia, Typhula
incarnata* and *Sclerotinia* diseases of turf. **Formulation types** WP; SC.
Mixtures *(pyributicarb +)* bromobutide; bromobutide + benzofenap; bensulfuron-
methyl; imazosulfuron + daimuron; bromobutide + bifenox.
Selected tradenames 'Eigen' (Dainippon); 'Isshintasuke' (Dainippon)

ANALYSIS
Details from Tosoh.

MAMMALIAN TOXICOLOGY
Reviews Tosoh, *Nihon Noyaku Gakkaishi*, **15**, 503 (1990) (in Japanese). **EHC** 76 (WHO, 1988; general review of thiocarbamates). **Oral** Acute oral LD_{50} for rats and mice >5000 mg/kg. **Skin and eye** Acute percutaneous LD_{50} for rats >5000 mg/kg. Non-irritating to eyes; slightly irritating to skin (rabbits). Non-sensitising to skin (guinea pigs). **Inhalation** LC_{50} (4 h) for rats >6520 mg/m^3. **NOEL** (2 y) for rats 0.753 mg/kg.

ECOTOXICOLOGY
Fish LC_{50} (48 h) for carp 11 mg/l. **Daphnia** LC_{50} (3 h) >15 mg/l.

ENVIRONMENTAL FATE
Plants Residues in the unpolished grain of rice plants harvested 113 and 119 days after treatment with pyributicarb (40 kg/ha) were found to be less than the detectable limit (0.005 mg/kg). **Soil/Environment** Half-life in paddy soil is 13–18 d. K_{oc} 1885.

624 pyridaben   *Insecticide, acaricide*

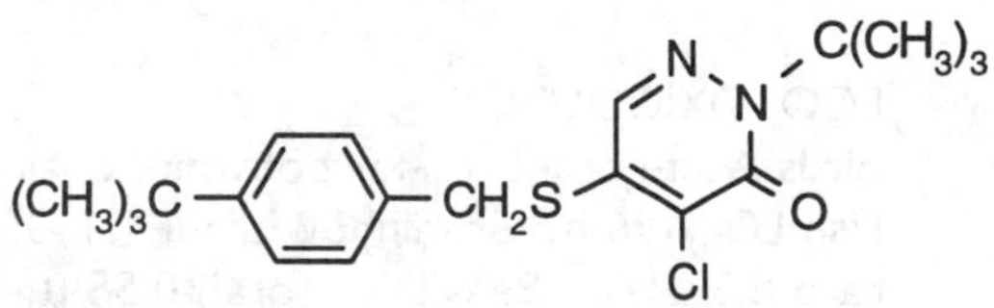

NOMENCLATURE
Common name pyridaben (BSI, draft E-ISO)
IUPAC name 2-*tert*-butyl-5-(4-*tert*-butylbenzylthio)-4-chloropyridazin-3(2*H*)-one
Chemical Abstracts name 4-chloro-2-(1,1-dimethylethyl)-5-[[[4-(1,1-dimethyl= ethyl)phenyl]methyl]thio]-3(2*H*)-pyridazinone
CAS RN *[96489–71–3]* **Development codes** NC-129; NCI-129 (Nissan); BAS-300I (Bayer)

PHYSICAL CHEMISTRY
Mol. wt. 364.9 **M.f.** $C_{19}H_{25}ClN_2OS$ **Form** Colourless crystals.
M.p. 111–112 °C **V.p.** 0.25 mPa (20 °C) **K$_{ow}$** logP = 6.37 (23±1 °C, distilled water) **S.g./density** 1.2 (20 °C) **Solubility** In water 0.012 mg/l (24 °C). In acetone 460, ethanol 57, hexane 10, benzene 110, xylene 390, cyclohexane 320, *n*-octanol 63 (all in g/l, 20 °C). **Stability** Stable at 50 °C for 90 d; unstable to light. Stable to hydrolysis for 30 d in the dark (pH 5, 7 and 9, 25 °C).

COMMERCIALISATION
History Insecticide and acaricide reported by K. Hirata *et al.* (*Proc. 1988 Br. Crop Prot. Conf – Pests Dis.,* **1,** 41). Introduced by Nissan Chemical Industries, Ltd. **Manufacturers** Nissan

APPLICATIONS
Mode of action Non-systemic insecticide and acaricide. Rapid knock-down and long residual activity. Active against all developing stages, especially against the larval and nymph stages. **Uses** Control of Acari, Aleyrodidae, Aphididae, Cicadellidae and Thysanoptera on field crops, fruit trees, ornamentals and vegetables at 5–20 g/hl or 100–300 kg/ha. **Formulation types** EC; WP; SC. **Selected tradenames** 'Sanmite' (Nissan)

ANALYSIS
Details from Nissan.

MAMMALIAN TOXICOLOGY
Oral Acute oral LD_{50} for male rats 1350, female rats 820, male mice 424, female mice 383 mg/kg. **Skin and eye** Acute percutaneous LD_{50} for rats and rabbits >2000 mg/kg. Non-irritant to eyes and skin (rabbits). Non-sensitising to skin (guinea pigs). **Inhalation** LC_{50} for male rats 0.66, female rats 0.62 mg/l air. **NOEL** (78 w) for mice 0.81 mg/kg daily. **Other** Non-mutagenic in Ames, DNA repair, *in vitro* chromosomal (Chinese hamster), and mouse micronucleus tests. **Toxicity class** WHO (a.i.) III; EPA (formulation) I (75 WP)

ECOTOXICOLOGY
Birds Acute oral LD_{50} for bobwhite quail >2250, mallard ducks >2500 mg/kg. **Fish** LC_{50} (96 h) for rainbow trout 1.1–3.1, bluegill sunfish 1.8–3.3 µg/l; (48 h) for carp 8.3 µg/l. **Bees** LD_{50} (oral) 0.55 µg/bee. **Worms** LC_{50} (14 d) 38 mg/kg soil. **Daphnia** EC_{50} (48 h) 0.59 µg/l (nominal concentration). **Algae** Does not significantly affect the average specific growth rate of *Selenastrum capricornutum*.

ENVIRONMENTAL FATE
Animals In the rat, an orally administered dose is excreted mainly in the faeces. Elimination is rapid and almost complete within 96 h. **Plants** After application to citrus and apple, pyridaben degrades gradually photochemically and is not translocated into the pulp. **Soil/Environment** Readily degrades microbiologically in aerobic soils, DT_{50} <21 d. In natural water, DT_{50} 10 d (25 °C, in dark). DT_{50} for aqueous photolysis c. 30 min (pH 7). See also Stability.

pyridaphenthion *Insecticide, acaricide*

organophosphorus

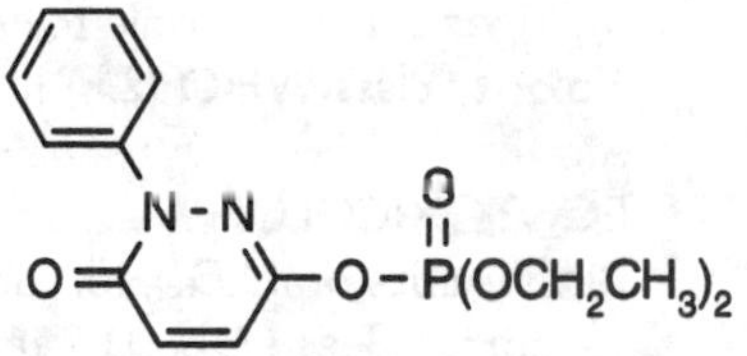

NOMENCLATURE
Common name pyridaphenthion (JMAF)
IUPAC name *O*-(1,6-dihydro-6-oxo-1-phenylpyridazin-3-yl) *O,O*-diethyl phosphorothioate
Chemical Abstracts name *O*-(1,6-dihydro-6-oxo-1-phenyl-3-pyridazinyl) *O,O*-diethyl phosphorothioate
CAS RN *[119–12–0]* **Development codes** NC-250; CL 12503 (Cyanamid)

PHYSICAL CHEMISTRY
Mol. wt. 340.3 **M.f.** $C_{14}H_{17}N_2O_4PS$ **Form** Pale yellow solid. **M.p.** 54.5–56.0 °C
V.p. 0.00147 mPa (20 °C) **K_{ow}** logP = 3.2 **S.g./density** 1.325 (20 °C)
Solubility In water 100 ppm (20 °C). In acetone 3.77, methanol 2.66 (in kg/kg, 21 °C); in diethyl ether 1.01 kg/kg (25 °C).

COMMERCIALISATION
History Insecticide introduced in Japan (1974) by Mitsui Toatsu Chemicals, Inc.
Manufacturers Mitsui Toatsu

APPLICATIONS
Biochemistry Cholinesterase inhibitor. **Mode of action** Insecticide and acaricide with contact and stomach action. **Uses** Control of sucking and chewing insects (e.g. stem borers, leaf rollers, leafhoppers, planthoppers, leaf beetles, skippers, grasshoppers, thrips, aphids, sawflies, lace bugs, caterpillars, onion flies, bean seed flies, etc.) and spider mites on paddy rice, vegetables, fruit, and ornamentals. Also effective against Isoptera in public health. **Formulation types** EC; WP; UL; DP.
Compatibility Incompatible with alkaline products.
Selected tradenames 'Ofunack' (mixture) (Mitsui Toatsu); 'Oreste' (Sipcam Phyteurop)

ANALYSIS
Product and **residue** analysis by glc.

MAMMALIAN TOXICOLOGY
EHC 63 (WHO, 1986; a general review of organophosphorus insecticides).

Oral Acute oral LD$_{50}$ for male rats 769, female rats 850, mice 459, dogs >12 000 mg/kg. **Skin and eye** Acute percutaneous LD$_{50}$ for male rats 2300, female rats 2100 mg/kg. Non-irritating to skin (rabbits). **Inhalation** LC$_{50}$ (4 h) for rats >1133.3 mg/m^3. **Other** In teratogenicity, mutagenicity, carcinogenicity, and multigeneration chronic toxicity studies on rats, no adverse effects were observed. **Toxicity class** WHO (a.i.) III

ECOTOXICOLOGY
Birds Acute oral LD$_{50}$ for Japanese quail 68 mg/kg. **Fish** TLm (48 h) for carp 12 ppm. **Bees** LD$_{50}$ 0.08 µg/bee. **Daphnia** TLm (3 h) 0.02 mg/l.

ENVIRONMENTAL FATE
Animals In mice and rats, the parent compound, O-ethyl-O-(3-oxo-2-phenyl-2H-=pyridazine-6-yl) phosphorothioate and the corresponding phosphate are found. **Plants** In rice, phenyl maleic hydrazide, O,O-diethyl thiophosphoric acid, and PMH glycoside are formed. **Soil/Environment** Soil DT$_{50}$ 11–24 d.

626 pyridate $\qquad$ *Herbicide*

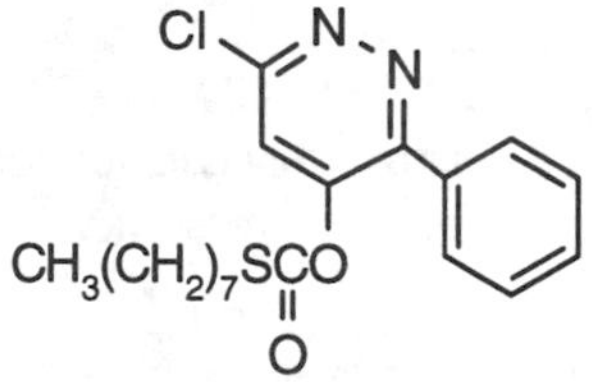

NOMENCLATURE
Common name pyridate (BSI, E-ISO, (*m*) F-ISO, WSSA)
IUPAC name 6-chloro-3-phenylpyridazin-4-yl S-octyl thiocarbonate
Chemical Abstracts name O-(6-chloro-3-phenyl-4-pyridazinyl) S-octyl carbonothioate
CAS RN *[55512–33–9]* **EEC no.** 259–686–7 **Development codes** CL 11 344 (Agrolinz)

PHYSICAL CHEMISTRY
Mol. wt. 378.9 **M.f.** C$_{19}$H$_{23}$ClN$_2$O$_2$S **Form** Colourless crystals; (tech., brown, oily liquid). **M.p.** 27 °C; (tech., 20–25 °C) **B.p.** 220 °C/0.1 mmHg (tech.) **V.p.** 1.3 × 10^{-4} mPa (20 °C) (tech.) **K$_{ow}$** logP >3 **S.g./density** 1.16 (20 °C) (tech.) **Solubility** In water c. 1.5 mg/l (20 °C). Readily soluble in most organic solvents. **Stability** Stable in neutral media. Hydrolysed by strong acids and alkalis.

COMMERCIALISATION
History Herbicide reported by A. Diskus *et al.* (*Proc. Br. Crop Prot. Conf. – Weeds,* 1976, **2**, 717). Introduced by Chemie Linz AG (now DSM Chemie Linz), who (as Agrolinz Melamin GmbH) sold it to Sandoz AG (now Novartis Crop Protection AG) in 1994. **Patent** AT 326409 **Manufacturers** DSM Chemie Linz

APPLICATIONS
Biochemistry Inhibits photosystem II electron transport in photosynthesis.
Mode of action Selective contact herbicide, absorbed predominantly by the leaves.
Uses Post-emergence control of annual broad-leaved weeds, especially *Solanum* spp., *Chenopodium* spp., *Galium aparine* and *Amaranthus retroflexus* (atrazine-resistant biotypes), and some grass weeds in maize, sweet corn, cereals, rice, peanuts and vegetables at 0.9 kg a.i./ha. Often used in combination with atrazine or other products, to extend the spectrum of activity. **Phytotoxicity** Should not be applied in mixture with, or within 14 days of, any other product which may result in de-waxing of crop foliage. **Formulation types** WP; EC.
Mixtures (*pyridate +*) atrazine; bromoxynil; clopyralid; dichlorprop; terbuthylazine; bromoxynil + ioxynil. **Selected tradenames** 'Lentagran' (Novartis); 'Pirate' (Barclay)

ANALYSIS
Product analysis by hplc with u.v. detection (*CIPAC Handbook*, 1995, **G**, 137–142).
Residues determined by hplc. Details available from Novartis.

MAMMALIAN TOXICOLOGY
Oral Acute oral LD_{50} for male and female rats >2000, male mice *c.* 10 000, female mice >10 000 mg/kg. **Skin and eye** Acute percutaneous LD_{50} for rabbits ≥2000 mg/kg. Moderately irritating to skin; non-irritating to eyes (rabbits). Evidence of sensitivity in guinea pigs, but no symptoms observed in man (applicators, technicians, workers). **Inhalation** LC_{50} (4 h) for rats >4.37 mg/l air.
NOEL (28 mo) for rats *c.* 18 mg/kg b.w. daily; (12 mo) for dogs 30 mg/kg b.w. daily (oral gavage). **ADI** 0.18 mg/kg (EU proposed); 0.35 mg/kg (WHO, 1992).
Other Tests for carcinogencity, mutagenicity, and teratogenicity were negative.
Toxicity class WHO (a.i.) III; EPA (formulation) III **EC risk** R43

ECOTOXICOLOGY
Birds Acute oral LD_{50} for bobwhite quail 1502, 5-d-old Pekin ducks and 10-d-old pheasants >10 000 mg/kg. Eight-day dietary LC_{50} for Japanese quail >10 000, bobwhite quail and mallard ducks >5000 mg/kg b.w. **Fish** LC_{50} (96 h) for catfish 48, rainbow trout >1.4–81, bluegill sunfish and carp >100 mg/l. **Bees** Non-toxic to honeybees. Oral LD_{50} >100 µg/bee; contact LD_{50} >160 µg/bee.
Worms LC_{50} for earthworms 799 mg/kg soil. **Other beneficial spp.** Harmless to beneficial arthropods such as *Poecilus* and *Aleochara*. No effect on soil respiration, ammonification and nitrification. **Daphnia** LC_{50} 0.83 and 1 mg/l; in simulated field, 3.5–7.1 mg/l. **Algae** IC_{50} for *Scenedesmus subspicatus* 82.1 mg/l.

MAIN ENTRIES

ENVIRONMENTAL FATE

Animals In rats, following oral administration, pyridate is rapidly and completely hydrolysed to its principal metabolite 3-phenyl-4-hydroxy-6-chloropyridazine, which is detoxified by formation of O- and N-glucuronides. The principal metabolite and the conjugates are rapidly and completely excreted. No cumulative effect after repeated administration. **Plants** In plants, pyridate is hydrolysed into its principal metabolite 3-phenyl-4-hydroxy-6-chloropyridazine, with a half-life ranging from a few minutes to days. This metabolite itself forms conjugates as O- and N-glycosides, which do not possess herbicidal activity.

Soil/Environment Rapidly degraded in soil, DT_{50} <2.8 d. The principal metabolite is 3-phenyl-4-hydroxy-6-chloropyridazine, which has DT_{50} 7–21 d (field), DT_{50} 15–34 d (lab.), DT_{90} 47–112 d (lab.). In biologically-active water, pyridate is rapidly transformed into the same metabolite as in soil, decomposition of which is accelerated by photodegradation, DT_{50} 16 d. Pyridate is not leached from soil; although its metabolite is less strongly bound to soil (K_d 0.3 to 26), lysimeter, water monitoring and field dissipation studies indicate it is unlikely to leach in practice.

627 pyrifenox

Fungicide

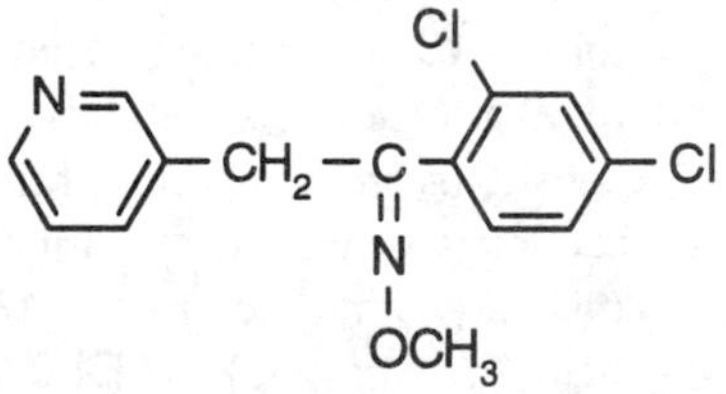

NOMENCLATURE

Common name pyrifenox (BSI, ANSI, draft E-ISO, (*m*) draft F-ISO)
IUPAC name 2',4'-dichloro-2-(3-pyridyl)acetophenone (*E*,*Z*)-*O*-methyloxime
Chemical Abstracts name 1-(2,4-dichlorophenyl)-2-(3-pyridinyl)ethanone *O*-methyloxime
CAS RN *[88283–41–4]* **Development codes** Ro 15–1297/000 (Maag); CGA 179945 (Ciba); NRK-297 (Nippon Kayaku)

PHYSICAL CHEMISTRY

Composition A mixture of *E*- and *Z*- isomers. **Mol. wt.** 295.2
M.f. $C_{14}H_{12}Cl_2N_2O$ **Form** Pale yellow viscous liquid with a slightly sweet odour.
B.p. 212.1 °C **V.p.** 1.7 mPa (25 °C) **K_{ow}** logP = 3.4 (pH 5.0), 3.7 (pH 7.0), 3.7

(pH 9.0) (all 25 °C)   **Henry** 5.8×10^{-3} Pa m^3 mol^{-1} (calc.)   **S.g./density** 1.28 (20 °C)   **Solubility** In water 300 (pH 5.0), 150 (pH 6.7), 130 (pH 9.0) (all in mg/l, 25 °C). In n-hexane 210 g/l (25 °C). Completely miscible in ethanol, acetone, toluene and n-octanol.   **Stability** Stable for >3 years in a closed container at room temperature. Stable to u.v. light and to hydrolysis in water (pH 3, 7, 9; 50 °C). **pKa** 4.61, weak base   **F.p.** 106 °C (1013 mbar)

COMMERCIALISATION
History Fungicide reported by P. Zobrist et al. (*Proc. Br. Crop Prot. Conf. – Pests Dis.*, 1986, **1**, 47). Introduced by Dr R. Maag Ltd.   **Patent** EP 49854 **Manufacturers** Novartis

APPLICATIONS
Biochemistry Ergosterol biosynthesis inhibitor.   **Mode of action** Systemic fungicide with protective and curative action. Absorbed by the leaves and roots, and translocated acropetally.   **Uses** Systemic fungicide used at 40–150 g a.i./ha for the control of powdery mildew, scab and other pathogenic Ascomycetes, Basidiomycetes and Deuteromycetes on grapes, cucurbits, pome fruit, stone fruit, peanuts, sugar beet, ornamentals and vegetables.   **Formulation types** WG; EC. **Mixtures** (*pyrifenox +*) captan.   **Selected tradenames** 'Dorado' (Novartis); 'Podigrol' (Novartis, Nippon Kayaku); 'Rondo' (Novartis)

ANALYSIS
Product analysis by tlc, gc, or hplc; gc with FID is recommended. **Residues** determined by lc, gc or hplc. See R. P. Hanni & A. J. Schuler in *Comp. Anal. Profiles*, Chapt. 3. Details also available from Novartis.

MAMMALIAN TOXICOLOGY
Reviews *J. Pestic. Sci.*, **16**, 355–359 (1991).   **Oral** Acute oral LD$_{50}$ for rats 2912, mice >2000 mg/kg.   **Skin and eye** Acute percutaneous LD$_{50}$ for rats >5000 mg/kg. Weak irritant to skin (humans), non-irritant to eyes (rabbits). Not a skin sensitiser (guinea pigs).   **Inhalation** LC$_{50}$ (4 h) for rats >2048 mg/m^3 air. **NOEL** (2 y) for rats 15; (1.5 y) for mice 45; (1 y) for dogs 10 mg/kg b.w. daily. **ADI** 0.1 mg/kg.   **Other** Not mutagenic, not teratogenic, and not oncogenic. **Toxicity class** WHO (a.i.) III

ECOTOXICOLOGY
Birds Acute oral LD$_{50}$ (14 d) for mallard ducks and bobwhite quail >2000 mg/kg. **Fish** LC$_{50}$ (96 h) for rainbow trout 7.1, mirror carp 12.2, bluegill sunfish 6.6 mg/l. **Bees** LD$_{50}$ (48 h) (oral) 59 μg/bee, (contact) 70 μg/bee.   **Worms** LC$_{50}$ (14 d) for earthworms 733 mg/kg.   **Other beneficial spp.** Harmless or slightly toxic. **Daphnia** EC$_{50}$ (48 h) 3.6 mg/l.   **Algae** EC$_{50}$ (96 h) for *Scenedesmus subspicatus* 0.095 mg/l.

ENVIRONMENTAL FATE

Animals In rats, following oral administration, pyrifenox is rapidly absorbed, metabolised and excreted in the urine and faeces. There are no indications of bioretention in tissue or organs.   **Plants** Relatively rapid degradation in plants. DT_{50} in peanut leaves 4 d, apple leaves 3 d, apple fruit 9 d. The main degradation pathway is by hydrolysis and elimination of the oxime group.

Soil/Environment Moderate soil mobility, no tendency toward bioaccumulation, little potential for environmental persistence, and fairly rapid dissipation in plants, soil, water and animals. R. P. Hanni & A. J. Schuler in *Comp. Anal. Profiles*, Chapt. 3. DT_{50} in soil 50–120 d, K_{oc} 980 ml/g soil.

628   pyrimethanil   *Fungicide*

anilinopyrimidine

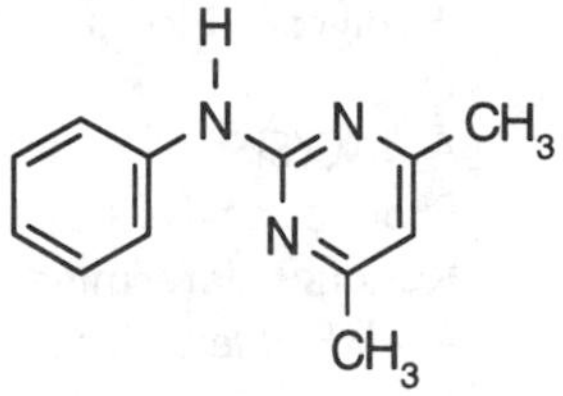

NOMENCLATURE

Common name pyrimethanil (BSI, draft ISO)
IUPAC name *N*-(4,6-dimethylpyrimidin-2-yl)aniline
Chemical Abstracts name 4,6-dimethyl-*N*-phenyl-2-pyrimidinamine
CAS RN *[53112–28–0]*   **Development codes** SN 100309; ZK 100309

PHYSICAL CHEMISTRY

Mol. wt. 199.3   **M.f.** $C_{12}H_{13}N_3$ **Form** Colourless crystals.   **M.p.** 96.3 °C
V.p. 2.2 mPa (25 °C, OECD 104)   **K**$_{ow}$ logP = 2.84 (pH 6.1, 25 °C)
Henry 3.6×10^{-3} Pa m^3 mol^{-1} (calc.)   **S.g./density** 1.15 (20 °C)   **Solubility** In water 0.121 g/l (pH 6.1, 25 °C). In acetone 389, ethyl acetate 617, methanol 176, methylene chloride 1000, *n*-hexane 23.7, toluene 412 (in g/l, 20 °C).
Stability Stable in water within the relevant pH range. Stable for 14 d at 54 °C.
pKa 3.52, weak base (20 °C) (OECD 112)

COMMERCIALISATION

History Reported by G. L. Neumann *et al.* (*Proc. Br. Crop Prot. Conf. – Pests Dis.*,

1992, **1**, 395). Introduced by Schering (now AgrEvo GmbH).
Manufacturers AgrEvo

APPLICATIONS
Biochemistry Inhibition of the secretion enzymes necessary for infection.
Mode of action Protectant (in *Botrytis*) and both protective and curative action
(in *Venturia*). **Uses** Control of grey mould (*Botrytis cinerea*) on vines, fruit,
vegetables and ornamentals. Control of leaf scab (*Venturia inaequalis* or *pirina*)
on pome fruit. **Phytotoxicity** May be phytotoxic in closed systems at ≥80% r.h.
on certain species. **Formulation types** SC. **Mixtures** *(pyrimethanil +)*
chlorothalonil. **Selected tradenames** 'Mythos' (AgrEvo); 'Scala' (AgrEvo)

ANALYSIS
Product and **residue** analysis by hplc. Details available from AgrEvo.

MAMMALIAN TOXICOLOGY
Oral Acute oral LD_{50} for rats 4150–5971, mice 4665–5359 mg/kg.
Skin and eye Acute percutaneous LD_{50} for rats >5000 mg/kg. Non-irritating to
skin and eyes (rabbits) and not a skin sensitiser (guinea pigs). **Inhalation** LC_{50}
(4 h) for rats >>1.98 mg/l. **NOEL** (2 y) for rats 20 mg/kg b.w. daily.
ADI 0.17–0.2 mg/kg. **Other** Negative in mutagenicity tests and non-teratogenic
in rats and rabbits. **Toxicity class** WHO (a.i.) III (Table 5); EPA (formulation) IV
('Scala' and 'Mythos')

ECOTOXICOLOGY
Birds Acute oral LD_{50} for mallard ducks and bobwhite quail >2000 mg/kg.
LC_{50} (5 d) for mallard ducks and bobwhite quail >5200 mg/kg diet. **Fish** LC_{50}
(96 h) for mirror carp 35.4, rainbow trout 10.6 mg/l. **Bees** LD_{50} (oral and
contact) >100 μg/bee. **Worms** LC_{50} (14 d) for earthworms 625 mg/kg dry
soil. **Other beneficial spp.** Classified as harmless to a range of beneficials.
Daphnia LC_{50} (48 h) 2.9 mg/l. **Algae** E_bC_{50} (96 h) 1.2 mg/l.

ENVIRONMENTAL FATE
Animals Rapidly absorbed, extensively metabolised and rapidly excreted in all
species examined. No evidence of accumulation, even on repeated dosing.
Metabolism proceeds by oxidation to phenolic derivatives which are excreted as
glucuronide or sulfate conjugates. **Plants** Little metabolism occurs in fruit;
residues at maturity consist essentially of unchanged parent compound only. For
this reason, a crop residue monitoring method has been developed for the direct
determination of pyrimethanil itself. **Soil/Environment** DT_{50} in laboratory studies
27–82 d; field studies indicate rapid degradation, DT_{50} 7–54 d. K_{oc} 265–751. Low
potential for leaching to groundwater; field studies show minimal movement of
pyrimethanil into deeper soil layers. Pyrimethanil disappears rapidly from surface
water and moderately adsorbs to the sediment, from which it is further degraded.

CH₃ and CH₃ structure with pyrimidine ring: N, N, NH(CH₂)₂O—phenoxy—(CH₂)₂OCH₂CH₃, CH₃CH₂, Cl

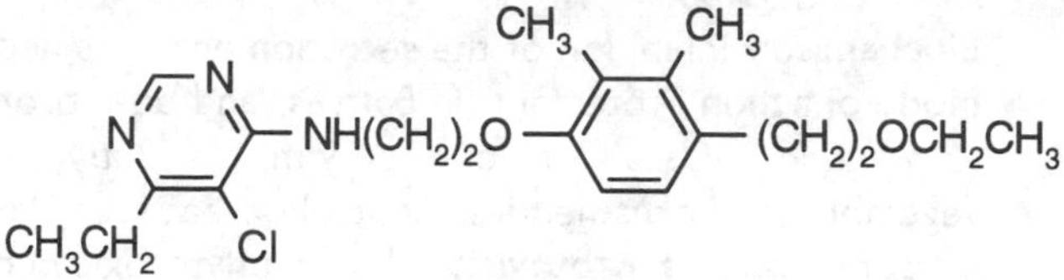

NOMENCLATURE
Common name pyrimidifen (BSI, pa E-ISO)
IUPAC name 5-chloro-N-{2-[4-(2-ethoxyethyl)-2,3-dimethylphenoxy]ethyl}-=
6-ethylpyrimidin-4-amine
Chemical Abstracts name 5-chloro-N-[2-[4-(2-ethoxyethyl)-2,3-dimethyl=
phenoxy]ethyl]-6-ethyl-4-pyrimidinamine
CAS RN *[105779–78–0]* **Development codes** E-787 (Sankyo, Ube); SU-8801,
SU-9118 (Sankyo)

PHYSICAL CHEMISTRY
Mol. wt. 377.9 **M.f.** $C_{20}H_{28}ClN_3O_2$ **Form** Colourless crystals.
M.p. 69.4–70.9 °C **V.p.** 1.2×10^{-9} mPa (25 °C) **S.g./density** 1.22 (20 °C)
Solubility In water 2.17 mg/l (25 °C).

COMMERCIALISATION
History Developed jointly by Sankyo Co., Ltd and Ube Industries.
Manufacturers Sankyo; Ube

APPLICATIONS
Uses Control of all stages of spider mites on apples, pears, vegetables, and tea;
spider mites and rust mites on citrus fruit; and diamond-back moth on vegetables.
Formulation types WP; SC. **Selected tradenames** 'Miteclean' (Sankyo, Ube);
SU-9118 (Sankyo, Ube)

ANALYSIS
Details from Sankyo.

MAMMALIAN TOXICOLOGY
Oral Acute oral LD_{50} for male rats 148, female rats 115, male mice 245, female
mice 229 mg/kg. **Skin and eye** Acute percutaneous LD_{50} for male and female rats
>2000 mg/kg.

ECOTOXICOLOGY
Birds LD_{50} for mallard ducks 445 mg/kg; LC_{50} for mallard ducks >5200 ppm.

Fish LC_{50} (48 h) for carp 0.093 ppm (SC). Bees LD_{50} (oral) 0.638 μg/bee, (contact) 0.660 μg/bee.

630 pyriminobac-methyl *Herbicide*

pyrimidinyloxybenzoic

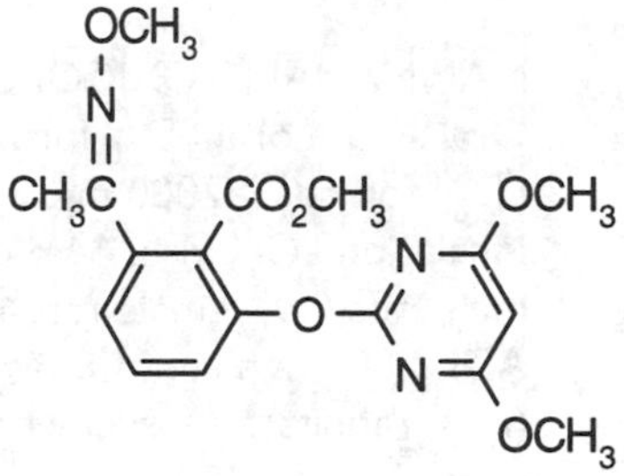

NOMENCLATURE
Common name pyriminobac-methyl (BSI, pa ISO)
IUPAC name methyl 2-(4,6-dimethoxy-2-pyrimidinyloxy)-6-(1-methoxy= iminoethyl)benzoate
Chemical Abstracts name methyl 2-[(4,6-dimethoxypyrimidin-2-yl)oxy]-= 6-[1-(methoxyimino)ethyl]benzoate
CAS RN *[136191–56–5]* (acid); *[136191–64–5]* (methyl ester)
Development codes KIH-6127; KUH-920 (Japan)

PHYSICAL CHEMISTRY
Composition Tech. is >93%: (E)- isomer 75–78%, (Z)- isomer 20–11%.
Mol. wt. 361.4   **M.f.** $C_{17}H_{19}N_3O_6$ **Form** White powder; (tech., pale yellow grains).   **M.p.** Tech. 105 °C; pure (E)- isomer 107–109 °C; pure (Z)- isomer 70 °C
V.p. (E)- isomer 3.5×10^{-2} mPa (25 °C); (Z)- isomer 2.681×10^{-2} mPa (25 °C)
K_{ow} (E)- isomer, logP = 2.98 (21.5 °C); (Z)- isomer, logP = 2.70 (20.6 °C)
S.g./density (E)- isomer 1.3868; (Z)- isomer 1.2734 (both 20 °C)
Solubility (E)- isomer: in water 0.00925, methanol 14.6 (both g/l, 20 °C).
(Z)- isomer: in water 0.175, methanol 14.0 (both g/l, 20 °C).   **Stability** Stable in water (>1 y, pH 4–9), to light, and to heat (not decomposed after 14 d at 55 °C)

COMMERCIALISATION
History Reported by Kumiai Chemical Industry Co., Ltd (R. Hanai *et al.*, *Proc. Br. Crop Prot. Conf., Weeds*, 1993, **1**, 47).   **Patent** US 5118339
Manufacturers Ihara/Kumiai

APPLICATIONS

Biochemistry Acts by inhibition of acetolactate synthase, blocking branched-chain amino acid biosynthesis. **Mode of action** Selective, systemic herbicide, absorbed by foliage. **Uses** Selective control of barnyard grass (*Echinochloa* spp.) in paddy rice, applied early post-emergence at 30–60 g/ha. **Formulation types** WP; GR. **Mixtures** (*pyriminobac-methyl +*) bensulfuron-methyl; cafenstrole; mefenacet; pretilachlor. **Selected tradenames** 'Prosper' (Kumiai)

ANALYSIS

Product and **residue** analysis by glc. Details from Kumiai.

MAMMALIAN TOXICOLOGY

Oral Acute oral LD_{50} for rats >5000 mg/kg. **Skin and eye** Acute percutaneous LD_{50} for rats >2000 mg/kg. Slightly irritant to skin and eyes (rabbits). **Inhalation** LC_{50} (4 h, 14 d observation) for rats >5.5 mg/l air. **NOEL** (2 y) for male rats 0.9, female rats 1.2, male mice 8.1, female mice 9.3 mg/kg b.w. daily. **ADI** 0.009 mg/kg. **Other** Non-mutagenic in the Ames Test. Non-teratogenic (rats, rabbits). **Toxicity class** WHO (a.i.) III

ECOTOXICOLOGY

Birds Acute oral LD_{50} for bobwhite quail >2000 mg/kg. Dietary LC_{50} (5 d) for bobwhite quail and mallard duck >5200 mg/kg diet. **Fish** LC_{50} (96 h) for carp 30.9, rainbow trout 21.2 mg/l. **Bees** LD_{50} (24 h, oral and contact) >200 μg/bee. **Worms** NOEL (14 d) for *Eisenia foetida* >1000 mg/kg soil. **Daphnia** LC_{50} (24 h) for *D. carinata* >20 mg/l.

ENVIRONMENTAL FATE

Animals Almost all of dosed ^{14}C was excreted in urine and faeces. Many metabolites were detected. **Plants** c. 10% of soil-applied ^{14}C-pyriminobac-methyl (4-leaf stage, rice) was distributed throughout the plant; at harvest, residues were mostly in the straw. **Soil/Environment** K_{oc} ((*E*)- isomer) 972, ((*Z*)- isomer) 499.

631 pyriproxyfen   *Insecticide*

juvenile hormone mimic

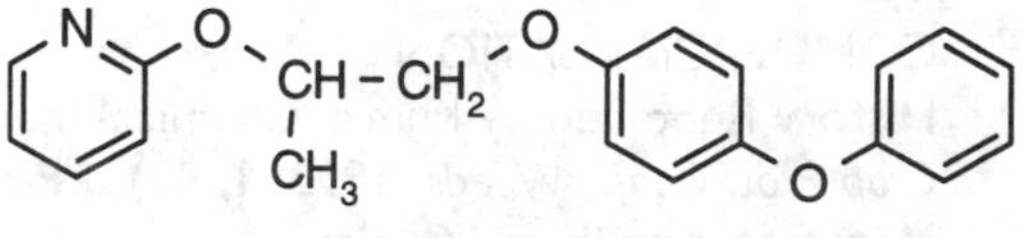

NOMENCLATURE

Common name pyriproxyfen (BSI, draft E-ISO); no name (Sweden)

IUPAC name 4-phenoxyphenyl (*RS*)-2-(2-pyridyloxy)propyl ether
Chemical Abstracts name 2-[1-methyl-2-(4-phenoxyphenoxy)ethoxy]pyridine
CAS RN *[95737–68–1]* unstated stereochemistry **Development codes** S-9318;
S-31183

PHYSICAL CHEMISTRY
Mol. wt. 321.5 **M.f.** $C_{20}H_{19}NO_3$ **Form** Colourless crystals. **M.p.** 45–47 °C
V.p. 0.29 mPa (20 °C) **S.g./density** 1.23 (20 °C) **Solubility** In hexane 400,
methanol 200, xylene 500 (all in g/kg, 20–25 °C).

COMMERCIALISATION
History Insecticide introduced by Sumitomo Chemical Co., Ltd.
Manufacturers Sumitomo

APPLICATIONS
Mode of action Suppressor of embryogenesis and adult formation. **Uses** Control
of public health insect pests (flies, beetles, midges, mosquitoes). Applied to
breeding sites (swamps, livestock houses, etc.). **Formulation types** EC; GR; WG.
Selected tradenames 'Sumilarv' (Sumitomo)

ANALYSIS
Details from Sumitomo.

MAMMALIAN TOXICOLOGY
Oral Acute oral LD_{50} for rats >5000 mg/kg. **Skin and eye** Acute percutaneous
LD_{50} for rats >2000 mg/kg; not an irritant to skin and eyes (rabbits). Not a skin
sensitiser (guinea pigs). **Inhalation** LC_{50} (4 h) for rats >1300 mg/m^3.
Toxicity class WHO (a.i.) III (Table 5)

632 pyrithiobac-sodium   *Herbicide*

pyrimidinyloxybenzoic analogue

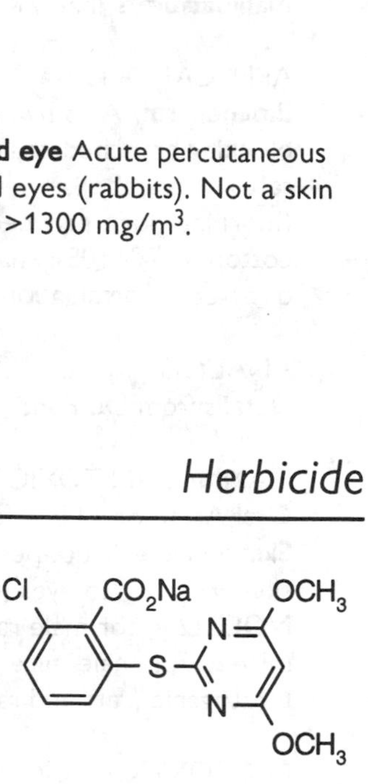

NOMENCLATURE
Common name pyrithiobac-sodium (BSI, pa E-ISO, ANSI)

IUPAC name sodium 2-chloro-6-(4,6-dimethoxypyrimidin-2-ylthio)benzoate
Chemical Abstracts name sodium 2-chloro-6-(4,6-dimethoxypyrimidin-2-ylthio)=
benzoate
CAS RN *[123343–16–8]* (for sodium salt); *[123342–93–8]* (for parent acid)
Development codes KIH-2031 (Kumiai); DPX-PE 350 (Du Pont)

PHYSICAL CHEMISTRY
Composition Tech. is >93%.   **Mol. wt.** 348.7   **M.f.** $C_{13}H_{10}ClN_2NaO_4S$
Form White solid.   **M.p.** 233.8–234.2 °C (decomp.)   **V.p.** 4.80×10^{-6} mPa
(25 °C) K_{ow} logP = 0.6 (pH 5), –0.84 (pH 7)   **S.g./density** 1.609   **Solubility** In
water 264 (pH 5), 705 (pH 7), 690 (pH 9), 728 (unbuffered) (all in g/l, 20 °C). In
acetone 812, methanol 270×10^3, *n*-hexane 10, dichloromethane 8.38 (all in mg/l,
20 °C).   **Stability** Stable in water (32 d, pH 5–9, 27 °C) and to heat (15 d at
54 °C).   **pKa** 2.34

COMMERCIALISATION
History Reported by Kumiai Chemical Industry Co., Ltd (S. Takahashi et *al.*, *Proc.
Br. Crop Prot. Conf. – Weeds*, 1991, **1**, 57–62). Developed jointly by Kumiai, Ihara
Chemical Industry Co., Ltd and E. I. du Pont de Nemours Co. Introduced by Du
Pont in USA in 1995.   **Patent** US 4932999; EP 315889
Manufacturers Ihara/Kumiai

APPLICATIONS
Biochemistry Acts by inhibiting the plant enzyme acetolactate synthase, and thus
blocking branched-chain amino acid biosynthesis. Selectivity is based on differing
rates of metabolism.   **Uses** Control of a wide range of broad-leaved weeds
(morning glory, pigweed, cocklebur) and grasses pre- and post-emergence in
cotton, at 70–105 g/ha.   **Phytotoxicity** Non-phytotoxic to cotton when applied as
directed.   **Formulation types** SP.   **Selected tradenames** 'Staple' (Du Pont, Kumiai)

ANALYSIS
Details from Du Pont.

MAMMALIAN TOXICOLOGY
Oral Acute oral LD_{50} for male rats 3300, female rats 3200 mg/kg.
Skin and eye Acute percutaneous LD_{50} for rabbits >2000 mg/kg. Non-irritating to
skin; irritating to eyes (rabbits).   **Inhalation** LC_{50} (4 h) for rats >6.9 mg/l.
NOEL (2 y) for male rats 58.7, female rats 278 mg/kg b.w. daily; (78 w) for male
mice 217, female mice 319 mg/kg b.w. daily.   **Other** Non-mutagenic. Non-
teratogenic (rats and rabbits).   **Toxicity class** WHO (a.i.) III; EPA (formulation) III

ECOTOXICOLOGY
Birds Acute oral LD_{50} for bobwhite quail >2250 mg/kg b.w. Dietary LC_{50} (5 d) for
mallard duck and bobwhite quail >5620 mg/kg diet.   **Fish** LC_{50} (96 h) for bluegill

sunfish >930, rainbow trout >1000, sheepshead minnow >145 mg/l.   **Bees** LD_{50}
(48 h, contact) >25 µg/bee.   **Daphnia** LC_{50} (48 h) >1100 mg/l.   **Algae** EC_{50}
(5 d) for *Selenestrum capricornutum* 107 µg/l; NOEC for *S. capricornutum*
22.8 µg/l.

ENVIRONMENTAL FATE
Animals More than 90% of dosed ^{14}C was excreted in urine and faeces of rats
Four metabolites were identified.   **Plants** Amounts of pyrithiobac-sodium in
cotton leaves rapidly decreased, following foliar application.
Soil/Environment Microbial and photochemical degradation play a major role in
degradation in the environment; DT_{50} in silty soil 60 d. K_d 0.32 (sandy loam), 0.60,
0.38, 0.75 (3 silty loam soils).

633 pyroquilon   *Fungicide*

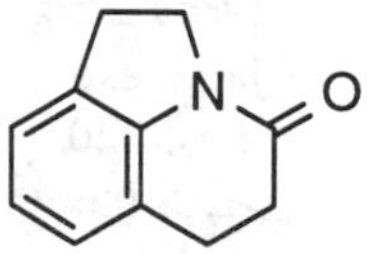

NOMENCLATURE
Common name pyroquilon (BSI, draft E-ISO); pyroquilone ((f) draft F-ISO)
IUPAC name 1,2,5,6-tetrahydropyrrolo[3,2,1-*ij*]quinolin-4-one
Chemical Abstracts name 1,2,5,6-tetrahydro-4*H*-pyrrolo[3,2,1-*ij*]quinolin-4-one
Other names [4-lilolidone]   **CAS RN** *[57369–32–1]*   **Development codes**
CGA 49 104

PHYSICAL CHEMISTRY
Mol. wt. 173.2   **M.f.** $C_{11}H_{11}NO$ **Form** White crystals.   **M.p.** 112 °C   **V.p.** 0.16
mPa (20 °C) K_{ow} logP = 1.57   **S.g./density** 1.29 (20 °C)   **Solubility** In water
4 g/l (20 °C). In acetone 125, benzene 200, dichloromethane 580, isopropanol
85, methanol 240 (all in g/l, 20 °C).   **Stability** Stable to hydrolysis, and to
temperatures up to 320 °C.

COMMERCIALISATION
History Fungicide reported by F. J. Schwinn *et al.* (*Int. Congr. Plant Prot., 9th, Annu.
Mtg. Am. Phytopathol. Soc., 71st*, 1979, Abstract 479). Introduced as an
agricultural fungicide by Ciba-Geigy AG (now Novartis Crop Protection AG),
having been discovered by Pfizer Ltd.   **Patent** GB 1394373 to Pfizer.

APPLICATIONS

Biochemistry Melanin biosynthesis inhibitor (reduction of 1,3,8-trihydroxy= naphthalene).   **Mode of action** Systemic fungicide.   **Uses** Systemic fungicide giving effective preventative control of *Pyricularia oryzae* in rice as foliar spray or seed treatment.   **Formulation types** WP; GR.   **Mixtures** *(pyroquilon +)* inabenfide. **Selected tradenames** 'Coratop' (Novartis); 'Fongarene' (Novartis)

ANALYSIS

Product analysis by glc.   **Residues** determined by hplc. Details available from Novartis.

MAMMALIAN TOXICOLOGY

Oral Acute oral LD_{50} for rats 321, mice 581 mg/kg.   **Skin and eye** Acute percutaneous LD_{50} for rats >3100 mg/kg; not a skin irritant, minimal eye irritant (rabbits). Not a skin sensitiser (guinea pigs).   **Inhalation** LC_{50} (4 h) for rats >5100 mg/m^3.   **NOEL** (2 y) for rats 22.5, mice 1.5 mg/kg b.w. daily; (1 y) for dogs 60.5 mg/kg b.w. daily.   **ADI** 0.015 mg/kg b.w.   **Other** Not teratogenic, not mutagenic, not oncogenic.   **Toxicity class** WHO (a.i.) II   **EC risk** R22

ECOTOXICOLOGY

Birds LC_{50} (8 d) for Japanese quail 794, chickens 431 mg/kg. LC_{50} (8 d) for Japanese quail >10 000 mg/kg.   **Fish** LC_{50} (96 h) for catfish 21, rainbow trout 13, perch 20, guppy 30 mg/l.   **Bees** Practically non-toxic to honeybees; LD_{50} (oral) >20, (contact) >1000 µg/bee.   **Daphnia** LC_{50} (48 h) 60 mg/l.   **Algae** No effect on *Scenedesmus acutus*.

ENVIRONMENTAL FATE

Animals After oral administration, pyroquilon is rapidly metabolised and eliminated via urine and faeces. Residues in tissues were generally low and there was no evidence for accumulation or retention of pyroquilon or its metabolites.
Plants Major metabolites in rice grain were 3,4-dihydro-4-hydroxy-2-oxo= quinoline-8-acetic acid and two other acetic acid derivatives.
Soil/Environment DT_{50} (silty soil) 2, (sandy loam) 18 w. K_d 1.3–42 µg/g soil, little to moderately mobile. Photolysis in water, DT_{50} 10 d.

quinalphos   *Insecticide, acaricide*

organophosphorus

S
||
N — OP(OCH₂CH₃)₂

$$\text{S=P(OCH}_2\text{CH}_3)_2$$

NOMENCLATURE
Common name quinalphos (BSI, E-ISO, (*m*) F-ISO); chinalphos ((*m*) France)
IUPAC name *O,O*-diethyl *O*-quinoxalin-2-yl phosphorothioate
Chemical Abstracts name *O,O*-diethyl *O*-2-quinoxalinyl phosphorothioate
CAS RN *[13593–03–8]* **EEC no.** 237–031–6 **Development codes** Bay 77 049;
SAN 6538; SAN 6626 **Official codes** ENT 27 394

PHYSICAL CHEMISTRY
Mol. wt. 298.3 **M.f.** $C_{12}H_{15}N_2O_3PS$ **Form** Colourless crystals. **M.p.** 31–32 °C
B.p. 142 °C/0.0003 mmHg (decomp.) **V.p.** 0.346 mPa (20 °C) K_{ow} logP = 4.44
(23 °C, 10–100 ppm level) **S.g./density** 1.235 (20 °C) **Solubility** In water
17.8 mg/l (22–23 °C). In hexane, 250 g/l (23 °C). Readily soluble in toluene,
xylene, diethyl ether, ethyl acetate, acetone, acetonitrile, methanol, ethanol.
Slightly soluble in petroleum ether (23 °C). **Stability** A.i. stable 14 d at room
temperature; liquid tech. grade less stable but stable under ambient storage
conditions, when diluted in non-polar organic solvents and in the presence of
stabilising agents. Formulations are stable (shelf life at ave. ann. temp. ≤25 °C
c. 2 y). Susceptible to hydrolysis; DT_{50} (25 °C, 17 ppm and 2.5 ppm) 23 d (pH 3),
39 d (pH 6), 26 d (pH 9).

COMMERCIALISATION
History Insecticide reported by K-J. Schmidt & L. Hammann (*Pflanzenschutz-Nachr.
(Engl. Ed.)*, 1969, **22**, 314). Introduced by Bayer AG (who no longer manufacture
or market it) and by Sandoz AG (now Novartis Crop Protection AG). **Patent** BE
681443; DE 1545817 to Bayer **Manufacturers** Aimco; Ficom; New Chemi;
Novartis; United Phosphorus

APPLICATIONS
Biochemistry Cholinesterase inhibitor. **Mode of action** Insecticide and acaricide
with contact and stomach action. By penetrating the plant tissues through
translaminar action, exhibits a systemic effect. **Uses** Control of many insect pests
of the orders Lepidoptera, Coleoptera, Diptera, Hemiptera, etc. For example,
used against caterpillars on fruit trees, cotton, vegetables, and peanuts; scales on
fruit trees; and the pest complex on rice. Also controls aphids, bollworms, borers,
leafhoppers, mealybugs, mites, planthoppers, thrips, etc. on beet, vines,
ornamentals, potatoes, soya beans, tea, coffee, cocoa, and other crops. Dose

quinalphos 1077

rates 250–500 g a.i./ha of EC and 0.75–1.0 kg a.i./ha of GR.
Phytotoxicity Slightly phytotoxic to certain fruit trees.   **Formulation types** EC; GR;
DP; UL; EW.   **Mixtures** *(quinalphos +)* disulfoton; thiometon; alpha-cypermethrin.
Compatibility Not compatible with alkaline substances.
Selected tradenames 'Ekalux' (Novartis); 'Danulux' (Dhanuka); 'Hubelux'
(Sanonda); 'Quinaal' (Ramcides); 'Quinatox' (Aimco); 'Smash' (Searle India);
'Starlux' (Shaw Wallace)

ANALYSIS
Product analysis by glc or by tlc and subsequent u.v. spectrometric determination
of the eluted compound.   **Residues** determined by glc (M. Wisson *et al., Anal.
Methods Pestic. Plant Growth Regul.,* 1980, **6**, 147).

MAMMALIAN TOXICOLOGY
EHC 63 (WHO, 1986; a general review of organophosphorus insecticides).
Oral Acute oral LD_{50} for male rats 71 mg/kg.   **Skin and eye** Acute percutaneous
LD_{50} for male rats 1750 mg/kg. Non-irritating to skin and eyes (rabbits).
Inhalation LC_{50} (4 h) for rats 0.45 mg/l air.   **NOEL** (2 y) (based on cholinesterase
inhibition) for rats 3 mg/kg diet.   **Other** No teratogenic effects (rats and rabbits);
no mutagenic potential. Cholinesterase inhibitor in rats, mice, and dogs.
Toxicity class WHO (a.i.) II; EPA (formulation) II

ECOTOXICOLOGY
Birds Acute LD_{50} (14 d) for Japanese quail 4.3, mallard ducks 37 mg/kg. Eight-day
dietary LC_{50} for quail 66, mallard ducks 220 mg/kg.   **Fish** LC_{50} (96 h) for carp
3.63, rainbow trout 0.005 mg/l.   **Bees** Very toxic to bees; LD_{50} (oral)
0.07 μg/bee, (topical) 0.17 μg/bee.   **Worms** LC_{50} (7 d) 188 mg/kg soil;
(14 d) 118.4 mg/kg soil.   **Daphnia** LC_{50} (48 h) 0.66 μg/l.

ENVIRONMENTAL FATE
Animals In rats, following oral administration, rapidly absorbed and metabolised to
2-hydroxyquinoxaline (free and as its conjugates), which are excreted in the urine
(*c.* 87%) and bile (*c.* 13%) within a short time.   **Plants** In plants, one-third is
absorbed by the leaf surface and penetrates into the plant, whilst two-thirds
disappears by evaporation within 14 days. The principal metabolite is 2-hydroxy=
quinoxaline (free and as its conjugates).   **Soil/Environment** In soil, rapidly
degraded under aerobic conditions, DT_{50} *c.* 3 w. The hydrolysis product
2-hydroxyquinoxaline does not accumulate in the soil, but is further broken down
to polar metabolites and CO_2. Freundlich K 25–320 mg/kg (o.m. 1.1–35.5%);
DT_{50} 21 d (o.m. 2.6%, pH 6.8, 18–22 °C).

quinolinecarboxylic acid

NOMENCLATURE
Common name quinclorac (BSI, draft E-ISO, (*m*) draft F-ISO)
IUPAC name 3,7-dichloroquinoline-8-carboxylic acid
Chemical Abstracts name 3,7-dichloro-8-quinolinecarboxylic acid
CAS RN *[84087–01–4]* **EEC no.** 402–780–1 **Development codes** BAS 514 H

PHYSICAL CHEMISTRY
Mol. wt. 242.1 **M.f.** $C_{10}H_5Cl_2NO_2$ **Form** Colourless crystals. **M.p.** 274 °C
V.p. <0.01 mPa (20 °C) **K_{ow} logP** = –1.15 (pH 7) **S.g./density** 1.75 **Solubility** In
water 0.065 mg/kg (pH 7, 20 °C). In ethanol, acetone 2 (both in g/kg, 20 °C).
Practically insoluble in other organic solvents. **Stability** Stable to heat and light
and between pH 3 to 9. **pKa** 4.34 (20 °C)

COMMERCIALISATION
History Herbicide reported by E. Haden *et al.* (*Proc. Br. Crop Prot. Conf. – Weeds*,
1985, **1**, 77). Introduced in Spain and Korea (1989) by BASF AG.
Patent EP 60429; US 4497651; US 4632696; DE 3108873 **Manufacturers** BASF

APPLICATIONS
Mode of action Rapidly absorbed through the foliage. Weak auxin activity as
determined in wheat coleoptile elongation test, cucumber root elongation test,
cucumber curvature test and ethylene biosynthesis test in soya beans. No
influence on Hill reaction. Plant response is similar to IAA or auxin-type herbicides
of the class of benzoic acids and pyridine compounds. **Uses** Pre- and post-
emergence control of grass weeds (*Echinochloa* spp., *Aeschynomene* spp., *Sesbania*
spp.) and other weeds in direct-seeded and transplanted rice at 0.25–0.75 kg
a.i./ha. **Phytotoxicity** Non-phytotoxic to transplanted and direct-seeded rice.
Under continuous irrigation conditions, injury may occur in adjacent umbelliferous
crops if connected to the same waterway. **Formulation types** SC; WP; GR.
Mixtures (*quinclorac +*) bentazone; propanil; bensulfuron-methyl; pyrazosulfuron-
ethyl; thiobencarb; pretilachlor; triclopyr; 2,4-D. **Compatibility** Good
combination partner for all rice herbicides insufficiently effective against
Echinochloa. **Selected tradenames** 'Facet' (BASF); 'Queen' (Sanonda)

ANALYSIS
Product analysis by hplc. **Residues** in plant and animal matrices by glc, in soil by
hplc.

MAMMALIAN TOXICOLOGY
Oral Acute oral LD_{50} for rats 2680, mice >5000 mg/kg.   **Skin and eye** Acute percutaneous LD_{50} for rats >2000 mg/kg. Non-irritating to eyes and skin (rabbits).   **Inhalation** LC_{50} (4 h) for rats >5.2 mg/l.   **NOEL** (2 y) for rats 533 mg/kg b.w.   **Other** Not carcinogenic.   **Toxicity class** WHO (a.i.) III (Table 5); EPA (formulation) III   **EC risk** Xi (R43)

ECOTOXICOLOGY
Birds Acute oral LD_{50} for mallard ducks and quail >2000 mg/kg. In 8-day feeding trials, LD_{50} for mallard ducks was >5000 mg/kg.   **Fish** LC_{50} (96 h) for rainbow trout, bluegill sunfish, carp, and minnow were all >100 mg/l.   **Bees** Non-toxic by contact or ingestion.   **Daphnia** LC_{50} (48 h) 113 mg/l.   **Other aquatic spp.** LC_{50} (96 h) for mysid shrimp (*Mysidopsis bahia*) 67, blue crab >100 mg/l. LC_{50} (48 h) for quahog clam >100 mg/l.

ENVIRONMENTAL FATE
Animals More than 90% of radiolabelled quinclorac administered orally to rats is excreted in the urine within 5 days.   **Plants** In plants, systematically translocated to the roots and to the leaves.   **Soil/Environment** Only slightly absorbed by the soil. Depending on soil type and organic matter content, the chemical is relatively mobile, this mobility increasing with higher percolation rates in fields. Quinclorac is degraded by micro-organisms, 3-chloro-8-quinolinecarboxylic acid being a major metabolite. Water regimes causing changes in moisture content in rice soils enhance the microbial degradation. Photolytic decomposition in active paddy water occurs in the presence of sunlight and dissolved humic acids.

636   quinmerac   *Herbicide*

quinolinecarboxylic acid

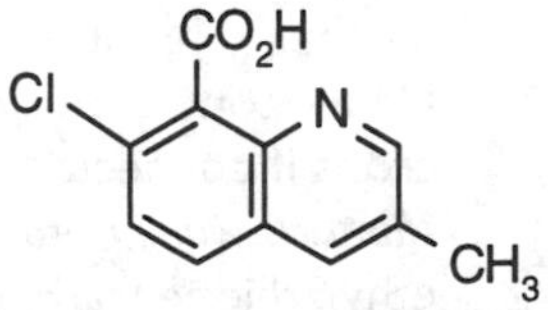

NOMENCLATURE
Common name quinmerac (BSI, draft E-ISO)
IUPAC name 7-chloro-3-methylquinoline-8-carboxylic acid
Chemical Abstracts name 7-chloro-3-methyl-8-quinolinecarboxylic acid
CAS RN *[90717–03–6]*   **Development codes** BAS 518H

PHYSICAL CHEMISTRY
Mol. wt. 221.6 **M.f.** $C_{11}H_8ClNO_2$ **Form** Colourless, odourless solid.
M.p. 244 °C **V.p.** <0.01 mPa (20 °C) K_{ow} logP = –1.11 (pH 7)
S.g./density 1.49 **Solubility** In water (deionised) 223 mg/l, (pH 9) 240 g/l (both
20 °C). In acetone, dichloromethane 2, ethanol 1, n-hexane, toluene, ethyl acetate
<1 (all in g/kg, 20 °C). **Stability** Stable to heat and light, and at pH 3–9.
pKa 4.31 (20 °C)

COMMERCIALISATION
History Herbicide reported by B. Wuerzer et al. (Proc. Br. Crop Prot. Conf. –
Weeds, 1985, **1**, 63) and W. Nuyken et al. (ibid., p. 71). Introduced by BASF AG.
Patent DE 3233089 **Manufacturers** BASF

APPLICATIONS
Biochemistry Auxin; induces formation of 1-aminocyclopropane-1-carboxylic acid,
leading to ethylene formation which induces the formation of abscisic acid
(K. Grossmann & F. Scheltrup, Proc. Br. Crop Prot. Conf. – Weeds, 1995, **1**, 393).
Mode of action Taken up primarily via the roots, in post-emergence applications;
in part also by the leaves. Moist conditions promote the uptake of the a.i., as well
as the rapid onset of activity. Ethylene and abscisic acid production lead to
epinasty, altered uptake and water relations and other effects. **Uses** Control of
Galium aparine, Veronica spp. and other broad-leaved weeds in cereals, oilseed
rape and sugar beet at 0.25–1.0 kg a.i./ha.
Formulation types WP; SC. **Mixtures** (quinmerac +) chloridazon; metazachlor.
Selected tradenames 'Butisan Star' (oilseed rape) (mixture) (BASF); 'Fiesta' (sugar
beet) (mixture) (BASF); 'Gavelan' (BASF)

ANALYSIS
Product Analysis by hplc. Details available from BASF.

MAMMALIAN TOXICOLOGY
Oral Acute oral LD_{50} for rats >5000 mg/kg. **Skin and eye** Acute percutaneous
LD_{50} for rats >2000 mg/kg. Non-irritating to skin or eyes (rabbits).
Inhalation LC_{50} (4 h) for rats > 5.4 mg/l. **NOEL** In feeding trials, NOAEL:
(12 mo) for rats 404, dogs 8 mg/kg b.w.; (78 w) for mice 38 mg/kg b.w.
ADI 0.08 mg/kg b.w. **Other** Non-mutagenic, non-teratogenic.
Toxicity class WHO (a.i.) III (Table 5)

ECOTOXICOLOGY
Birds Acute LD_{50} for bobwhite quail >2000 mg/kg. **Fish** LC_{50} (96 h) for trout
86.8, carp >100 mg/l. **Bees** Not hazardous to bees; LD_{50} (oral and contact)
>200 µg/bee. **Daphnia** LC_{50} (48 h) 148.7 mg/l. **Algae** EC_{50} (72 h) for green
algae (Chlorella fusca) 48.5 mg/l.

ENVIRONMENTAL FATE

Animals Only minor metabolites formed (rat). **Plants** For details of metabolism in rape, wheat and sugar beet, see E. Keller, *IUPAC 7th Int. Congr. Pestic. Chem.*, 1990, **2**, 154. **Soil/Environment** DT_{50} (lab.) 28–85 d (20 °C), (field) 3–33 d. K_{oc} 19–185.

637 quinoclamine   *Herbicide, algicide*

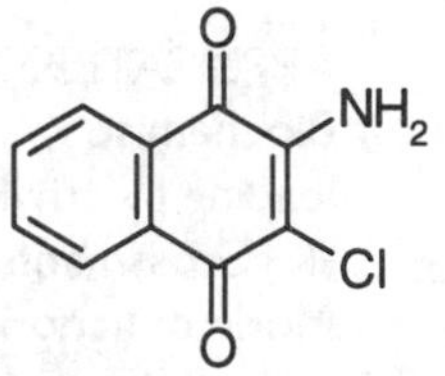

NOMENCLATURE

Common name quinoclamine (BSI, draft E-ISO, (f) draft F-ISO); ACN (JMAF)
IUPAC name 2-amino-3-chloro-1,4-naphthoquinone
Chemical Abstracts name 2-amino-3-chloro-1,4-naphthalenedione
Other names ACNQ   **CAS RN** *[2797–51–5]*   **Development codes** 06K
(Uniroyal)

PHYSICAL CHEMISTRY

Mol. wt. 207.6   **M.f.** $C_{10}H_6ClNO_2$ **Form** Yellow crystals.   **M.p.** 198–200 °C
V.p. 0.06 mPa (25 °C)   **K_{ow}** logP = 1.5   **S.g./density** 1.66   **Solubility** In acetic acid 16, acetone 13, chlorobenzene 5, nitrobenzene 37 (all in g/l, 20 °C).
Stability Stable up to 155 °C, in solution in the dark.

COMMERCIALISATION

History Algicide, fungicide and herbicide originally developed by Uniroyal Inc. (who no longer manufacture or market it). Introduced in Japan (1972) by Agro-Kanesho Co., Ltd.   **Manufacturers** Agro-Kanesho

APPLICATIONS

Uses Control of algae and weeds in paddy rice crops.   **Formulation types** WP; GR.   **Mixtures** *(quinoclamine +)* MCPB + simetryn.
Selected tradenames 'Mogeton' (Agro-Kanesho)

ANALYSIS

Product analysis by glc or hplc.   **Residues** determined by glc with ECD.

MAMMALIAN TOXICOLOGY
Oral Acute oral LD$_{50}$ for male rats 1360, female rats 1600, male mice 1350, female mice 1260 mg/kg.   **Skin and eye** Acute percutaneous LD$_{50}$ for rats >500 mg/kg.   **Inhalation** LC$_{50}$ (4 h) for rats >0.79 mg/l air.   **NOEL** (2 y) for male rats 5.7 mg/kg b.w. daily.   **Toxicity class** WHO (a.i.) III

ECOTOXICOLOGY
Fish LC$_{50}$ (48 h) for carp 0.7 mg/l.   **Daphnia** LC$_{50}$ (3 h) >10 mg/l.

538   quinoxyfen   *Fungicide*

phenoxyquinoline

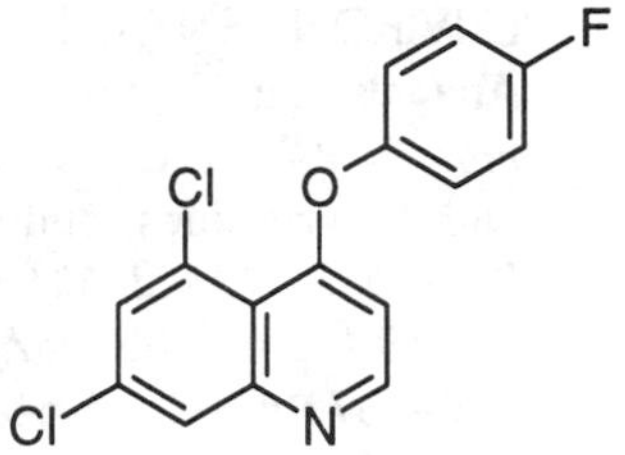

NOMENCLATURE
Common name quinoxyfen (ANSI, BSI, pa ISO)
IUPAC name 5,7-dichloro-4-quinolyl 4-fluorophenyl ether
Chemical Abstracts name 5,7-dichloro-4-(4-fluorophenoxy)quinoline
CAS RN *[124495–18–7]*   **Development codes** DE-795; LY214352

PHYSICAL CHEMISTRY
Mol. wt. 308.1   **M.f.** C$_{15}$H$_8$Cl$_2$FNO   **Stability** In dark, at 25 °C, stable to hydrolysis at pH 7 and 9; DT$_{50}$ 75 d (pH 4). Degraded more rapidly in light. Water/sediment studies also reported (G. L. Reeves *et al., Proc. Br. Crop Prot. Conf. – Pests Dis.,* 1996, **3**, 1169).

COMMERCIALISATION
History Reported by C. Longhurst *et al.* (*Proc. Br. Crop Prot. Conf. – Pests Dis.,* 1996, **1**, 27). Under development by DowElanco.

APPLICATIONS
Biochemistry Not known, but not a sterol demethylation or a mitochondrial electron transport inhibitor.   **Mode of action** Mobile, protectant fungicide acting through inhibition of appressorial development; not effective as an eradicant.

Active through systemic acropetal and basipetal movement, and by vapour transfer. **Uses** Under development for control of cereal powdery mildew; offers long-term (up to 70 d) protection. No cross-resistance with current mildewicides. **Formulation types** SC.

MAMMALIAN TOXICOLOGY
Oral Acute oral LD_{50} for rats >5000 mg/kg. **Skin and eye** Acute percutaneous LD_{50} for rabbits >2000 mg/kg. Mild eye irritant, not a skin irritant (rabbits). Skin sensitisation (guinea pigs) depends on the test. **Inhalation** LC_{50} for rats >3.38 mg/l. **ADI** 0.2 mg/kg. **Other** Not mutagenic, teratogenic or oncogenic.

ECOTOXICOLOGY
Birds LD_{50} >5620 mg/kg diet (2 species). **Fish** LC_{50} >0.2 mg/l (2 species). **Bees** LC_{50} (oral) >1000 mg/kg honey. **Other beneficial spp.** Harmless to 4 beneficial insect species. No effect on soil micro-organisms at 4000 g/ha.

ENVIRONMENTAL FATE
Animals Metabolism in goat and hen studied (G. L. Reeves et al., Proc. Br. Crop Prot. Conf. – Pests Dis., 1996, **3**, 1169). **Plants** Only slightly metabolised in wheat, with low residues found in the crop (G. L. Reeves et al., Proc. Br. Crop Prot. Conf. – Pests Dis., 1996, **3**, 1169). **Soil/Environment** DT_{50} 123–454 d; K_{oc} 15 415–34 985 mg/g (non-leaching). DT_{50} (aerobic) 224–508 d (5 soils), DT_{90} 730–1673 d. DT_{50} (anaerobic) 289 d, DT_{90} 959 d. The main metabolite in the soil (also classed as non-leaching) was formed by hydroxylation at the 3-position of the quinoline ring; a minor metabolite formed by cleavage of the ether bridge was observed, especially in acidic soil. There was minimal photolysis on the soil surface (DT_{50} >1 y, Southern England). Photolysis in water, DT_{50} 1.5 h (June), 22.8 h (December).

639 quintozene   *Fungicide*

aromatic hydrocarbon derivative

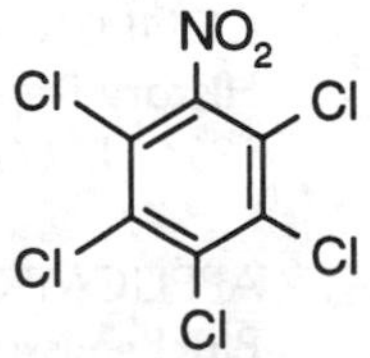

NOMENCLATURE
Common name quintozene (BSI, E-ISO, (*m*) F-ISO); PKhNB (USSR); PCNB (JMAF)

IUPAC name pentachloronitrobenzene
Chemical Abstracts name pentachloronitrobenzene
CAS RN *[82–68–8]* **EEC no.** 201–435–0

PHYSICAL CHEMISTRY
Composition Tech. grade is 99% pure. **Mol. wt.** 295.3 **M.f.** $C_6Cl_5NO_2$
Form Colourless needles; (tech., pale yellow crystals). **M.p.** 143–144 °C;
142–145 °C (tech.) **B.p.** 328 °C (with slight decomposition) **V.p.** 12.7 mPa
(25 °C) **K_{ow}** logP = 5.1 **S.g./density** 1907 kg/m³ (21 °C) **Solubility** In water
0.1 mg/l (20 °C). In toluene 1140, methanol 20, heptane 30 (all in g/l).
Stability Stable to heat. Stable in acidic media, but hydrolysed by alkalis. Some
surface colouring after 10 h exposure to sunlight.

COMMERCIALISATION
History Introduced as a fungicide by I. G. Farbenindustrie AG (now Bayer AG,
who no longer manufacture or market it). **Patent** DE 682048
Manufacturers Amvac; Mitsui Toatsu; Rhône-Poulenc; Uniroyal

APPLICATIONS
Mode of action Seed and soil, contact fungicide. **Uses** Control of damping-off
diseases in brassicas, lettuce, cotton, flower crops, tomatoes, etc.; *Rhizoctonia* spp.
in brassicas, lettuce, tomatoes, and flower crops; bulb rot and tulip fire in tulips;
dollar spot, red thread, and snow mould in turf; *Rhizoctonia solani* and scab in
potatoes; bunt and dwarf bunt of wheat; smut and white rot in onions; white rot
in garlic; club root of brassicas; and soil-borne *Sclerotium, Sclerotinia,* and *Botrytis*
spp. in many glasshouse crops. Also used on peanuts, bananas, beans, peas, rice,
maize, safflowers, sorghum, soya beans, etc. Applied at 1–1.5 kg/ha.
Formulation types DP; WP; EC; SC; GR; Seed treatment. **Mixtures** *(quintozene +)*
carbaryl; etridiazole; fuberidazole; fuberidazole + triadimenol; carbendazim +
imazalil. **Compatibility** Incompatible with alkaline materials.
Selected tradenames 'Brassicol' (AgrEvo); 'Folosan' (Uniroyal); 'Kobutol' (Hokko);
'Terraclor' (Uniroyal); 'RTU' (Gustafson)

ANALYSIS
Product analysis by glc with FID (*CIPAC Handbook*, 1985, **1C**, 2213; *AOAC Methods*,
1995, 982.04). **Residues** determined by glc (*Pestic. Anal. Man.*, 1979, **I**, 201A,
201G, 201I; *Man. Pestic. Residue Anal.*, 1987, **I**, 4–6, S8, S9, S12, S19; *Anal.
Methods Residues Pestic.*, 1988, Part I, M1, M12).

MAMMALIAN TOXICOLOGY
Reviews *Pesticide residues in food – 1995*, FAO Plant Production and Protection
Paper 133, 1996. *Pesticide residues in food – 1995 evaluations; Part II – Toxicology &
Environment.* World Health Organisation, WHO/PCS/96.48, 1996. Monograph for
the placement of quintozene onto annex 1 of European directive 91/414 (1997).

IARC 5 EHC 41 (WHO, 1984). **Oral** Acute oral LD_{50} for rats >5000 mg/kg.
Skin and eye Acute percutaneous LD_{50} for rabbits >5000 mg/kg. Not irritating to skin, slightly irritating to eyes (rabbits). **Inhalation** LC_{50} (4 h) for rats >1.7 mg/l.
NOEL (2 y, oncogenic) for rats 1 mg/kg b.w. daily, (1 y, feeding) for dogs 3.75 mg/kg b.w. daily. **ADI** (JMPR) 0.01 mg/kg b.w. [1995].
Toxicity class WHO (a.i.) III (Table 5); EPA (formulation) III **EC risk** Xi (R43)

ECOTOXICOLOGY
Birds LD_{50} for mallard ducks 2000 mg/kg. Dietary LC_{50} (8 d) for mallard ducks and bobwhite quail >5000 ppm. **Fish** LC_{50} (96 h) for rainbow trout 0.55, bluegill sunfish 0.1 ppm. **Bees** LD_{50} (contact) >100 μg/bee. **Daphnia** LC_{50} (48 h) 0.77 mg/l. **Other aquatic spp.** LC_{50} (96 h) for shrimps 0.012, oyster 0.029 ppm.

ENVIRONMENTAL FATE
Animals In mammals, the major routes of elimination are as unchanged material in the faeces or as metabolites in the urine. In rats, sheep and monkeys, the principal metabolite is pentachloroaniline (formed by reduction of the nitro group). Other metabolites include pentachlorophenol, pentachlorothioanisole, pentachloro= benzene, bis-methyl-tetrachlorobenzene, methyl pentachlorophenyl sulfide and N-acetyl-S-pentachlorophenylcysteine (E. J. Kuchar et al., J. Agric. Food Chem., 1969, **17**, 1237–1240; W. Koegel et al., Chemosphere, 1979, 89–105; P. W. Aschbacher & V. J. Feil, J. Agric. Food Chem., 1983, **31**, 1150).
Plants In plants, quintozene undergoes conversion to pentachloroaniline, methylthiopentachlorobenzene and a variety of chlorophenyl methyl sulfoxides and sulfones. **Soil/Environment** Persists in soil, with a half-life of c. 4–10 months. Part is lost from the soil by volatilisation. Biodegradation occurs, mainly to pentachloroaniline and also methylthiopentachlorobenzene. For details of effects on soil organisms, see E. R. Ingham, Crop Prot., 1985, **4**, 3–32. K_{oc} for adsorption 6030 (silt loam), 2966 (sand); for desorption 9584 (silt loam), 3285 (sand).

2-(4-aryloxyphenoxy)propionic acid

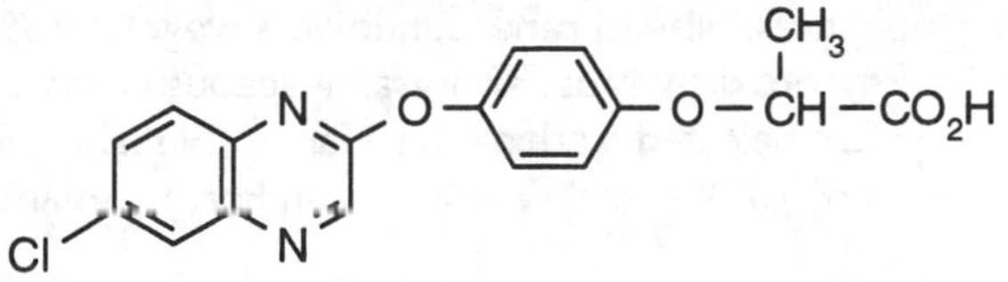

NOMENCLATURE
quizalofop
Common name quizalofop (BSI, ANSI, draft E-ISO, (*m*) draft F-ISO)
IUPAC name (*RS*)-2-[4-(6-chloroquinoxalin-2-yloxy)phenoxy]propionic acid
Chemical Abstracts name (±)-2-[4-[(6-chloro-2-quinoxalinyl)oxy]phenoxy]=
propanoic acid
CAS RN [76578–12–6]

quizalofop-ethyl
IUPAC name ethyl (*RS*)-2-[4-(6-chloroquinoxalin-2-yloxy)phenoxy]propionate
CAS RN [100760–10–9]   **Development codes** NCI-96683; NC-302 (both
Nissan); FBC-32 197 (FBC Limited); DPX-Y6202 (Du Pont); EXP 3864
(Rhône-Poulenc)

PHYSICAL CHEMISTRY
quizalofop
Mol. wt. 344.8   **M.f.** $C_{17}H_{13}CIN_2O_4$

quizalofop-ethyl
Mol. wt. 372.8   **M.f.** $C_{19}H_{17}CIN_2O_4$ **Form** Colourless crystals.
M.p. 91.7–92.1 °C   **B.p.** 220 °C/0.2 mmHg   **V.p.** 8.65×10^{-4} mPa (20 °C)
K_{ow} logP = 4.28 (23±1 °C, distilled water)   **S.g./density** 1.35 (20 °C)
Solubility In water 0.3 mg/l (20 °C). In benzene 290, xylene 120, acetone 111,
ethanol 9, hexane 2.6 (all in g/l, 20 °C).   **Stability** Stable at 50 °C for 90 d.
Stable in organic solvents at 40 °C for 90 d. Unstable to light (DT_{50} 10–30 d).
Stable at pH 3–7.

COMMERCIALISATION
History Herbicidal activity of quizalofop-ethyl (the ethyl ester) reported by G.
Sakata *et al.* (*Proc. 10th Int. Congr. Plant Prot.*, 1983, **1**, 315). Introduced by Nissan
Chemical Industries, Ltd.   **Patent** GB 2042539   **Manufacturers** Nissan; IPESA

APPLICATIONS
quizalofop-ethyl
Biochemistry Acetyl CoA carboxylase inhibitor; inhibition of fatty acid

biosynthesis. **Mode of action** Systemic herbicide, absorbed from the leaf surface, with translocation throughout the plant, moving in both the xylem and phloem, and accumulating in the meristematic tissue. **Uses** Selective post-emergence control of annual and perennial grass weeds in potatoes, soya beans, sugar beet, peanuts, oilseed rape, sunflowers, vegetables, cotton, and flax.
Phytotoxicity Most non-graminaceous crops are tolerant. **Formulation types** EC; SC. **Selected tradenames** 'Targa' (Nissan, Rhône-Poulenc); 'Pilot' (Nissan, AgrEvo); 'Assure' (Canada) (Du Pont); 'Tolan' (IPESA)

ANALYSIS
Product by hplc. **Residues** by gc. Details from Nissan.

MAMMALIAN TOXICOLOGY
quizalofop-ethyl
Oral Acute oral LD_{50} for male rats 1670, female rats 1480, male mice 2350, female mice 2360 mg/kg. **Skin and eye** Acute percutaneous LD_{50} for rats and mice >5000 mg/kg; not a skin irritant to rabbits nor a sensitiser to guinea pigs. **Inhalation** LC_{50} (4 h) for rats 5.8 mg/l. **NOEL** (104 w) for rats 0.9 mg/kg daily; (78 w) for mice 1.55 mg/kg daily; (52 w) for dogs 13.4 mg/kg daily. **ADI** 0.009 mg/kg. **Other** Non-mutagenic in the Ames and REC tests. **Toxicity class** WHO (a.i.) III; EPA (formulation) III

ECOTOXICOLOGY
quizalofop-ethyl
Birds Acute oral LD_{50} for mallard ducks and bobwhite quail >2000 mg/kg. **Fish** LC_{50} (96 h) for rainbow trout 10.7, bluegill sunfish 2.8 mg/l. **Bees** LD_{50} (contact) >50 μg/bee. **Daphnia** LC_{50} (96 h) 2.1 mg/l. **Algae** EC_{50} (96 h) for green algae (*Scenedesmus panonicus*) >3.2 mg/l.

ENVIRONMENTAL FATE
Animals In mammals, following oral administration, the parent is rapidly metabolised; almost all of the dose is eliminated mainly in the faeces within 3 d. **Plants** In broad-leaved plants, absorption and translocation is very limited, and most parent applied stays on and/or in the treated leaves. Unchanged parent is the main component in/on treated leaves. **Soil/Environment** In soil, degrades rapidly to quizalofop; DT_{50} <1 d.

quizalofop-P *Herbicide*

2-(4-aryloxyphenoxy)propionic acid

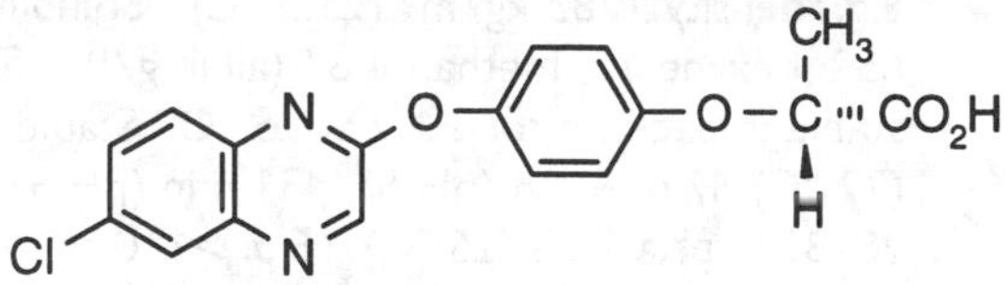

NOMENCLATURE
quizalofop-P
Common name quizalofop-P (BSI, draft E-ISO)
IUPAC name (R)-2-[4-(6-chloroquinoxalin-2-yloxy)phenoxy]propionic acid
Chemical Abstracts name (R)-2-[4-[(6-chloro-2-quinoxalinyl)oxy]phenoxy]=
propanoic acid
CAS RN *[94051–08–8]*

quizalofop-P-ethyl
Common name quizalofop-P-ethyl (BSI, draft E-ISO)
IUPAC name ethyl (R)-2-[4-(6-chloroquinoxalin-2-yloxy)phenoxy]propionate
CAS RN *[100646–51–3]*   **Development codes** D(+) NC-302 (Nissan);
DPX-79376 (Du Pont)

quizalofop-P-tefuryl
Common name quizalofop-P-tefuryl
IUPAC name (±)-tetrahydrofurfuryl-(R)-2-[4-(6-chloroquinoxalin-2-yloxy)=
phenoxy]propionate
Chemical Abstracts name (RS)-2-tetrahydrofuranylmethyl (R)-2-[4-(6-chloro-=
2-quinoxalinyloxy)phenoxy]propanoate
CAS RN *[119738–06–6]*   **Development codes** UBI C4874 (Uniroyal)

PHYSICAL CHEMISTRY
quizalofop-P
Mol. wt. 344.8   **M.f.** $C_{17}H_{13}ClN_2O_4$

quizalofop-P-ethyl
Mol. wt. 372.8   **M.f.** $C_{19}H_{17}ClN_2O_4$ **Form** White crystalline solid.
M.p. 76–77 °C   **B.p.** 220 °C/26.6 Pa   **V.p.** 1.1×10^{-4} mPa (20 °C)
K_{ow} logP = 4.66 (23±1 °C)   **S.g./density** 1.35 g/cm^3 **Solubility** In water 0.4 mg/l
(20 °C). In acetone 650, ethanol 22, hexane 5, xylene 360 (all in g/l, 20 °C).
Stability Stable in neutral and acidic media, but unstable in alkaline media; DT_{50}
19 h (pH 9). Stable at high temperatures and in organic solvents.
Specific rotation $[\alpha]_D^{20}$ +35.9° (20 °C)

quizalofop-P-tefuryl
Mol. wt. 428.9 **M.f.** $C_{22}H_{21}CIN_2O_5$ **Form** Thick yellow liquid, which can crystallise upon standing at room temperature. **M.p.** 59–68 °C; (tech. is a liquid)
V.p. 7.9×10^{-3} mPa (25 °C, gas saturation method) K_{ow} logP = 4.32 (25 °C)
S.g./density 1283 kg/m^3 (21.5 °C) **Solubility** In water 4 mg/l (25 °C). In toluene 652, hexane 12, methanol 64 (all in g/l). **Stability** Stable for $\geq$ 14 d at 55 °C; stable in package for $\geq$1 y at 25 °C. Stable $\geq$7 d in sunlight. In water, DT$_{50}$ (22 °C) 82 d, 8.1 d (pH 5), 431 min (pH 9). **Specific rotation** $[\alpha]_D$ 31.9°; tech. 28–32° **pKa** 1.25 (25 °C) **F.p.** >110 °C (Pensky Martens)

COMMERCIALISATION
History Herbicidal activity of quizalofop-ethyl (ethyl ester of the racemate) reported by G. Sakata *et al.* (*Proc. 10th Int. Congr. Plant Prot.*, 1983, **1**, 315). The herbicidal enantiomers (quizalofop-P-ethyl) introduced by Nissan Chemical Industries, Ltd and quizalofop-P-tefuryl ((±)-tetrahydrofurfuryl ester) (A. R. Bell & A. S. Peddie, *Proc. 1989 Br. Crop Prot. Conf. – Weeds*, **1**, 65) by Uniroyal Chemical Co., Inc.

quizalofop-P
Manufacturers Nissan; IPESA

quizalofop-P-tefuryl
Manufacturers IPESA; Uniroyal

APPLICATIONS
quizalofop-P-ethyl
Biochemistry Acetyl CoA carboxylase inhibitor; inhibition of fatty acid biosynthesis. **Mode of action** Systemic herbicide, absorbed from the leaf surface, with translocation throughout the plant, moving in both the xylem and phloem, and accumulating in the meristematic tissue. **Uses** Selective post-emergence control of annual and perennial grass weeds in potatoes, soya beans, sugar beet, peanuts, oilseed rape, sunflowers, vegetables, cotton, and flax.
Phytotoxicity Most non-graminaceous crops are tolerant. **Formulation types** EC; SC. **Compatibility** Can be used in combination with post-emergence broad-leaved herbicides. **Selected tradenames** 'Targa D+' (Nissan); 'Targa Super' (Nissan); 'Assure II' (USA) (Du Pont); 'CoPilot' (Nissan); 'Omega' (Argentina) (Du Pont); 'Pilot D' (Nissan); 'Pilot Super' (Nissan); 'Sheriff' (Du Pont); 'Herban LPU' (IPESA); 'Mostar' (IPESA); 'Targa Prestige' (Rhône-Poulenc)

quizalofop-P-tefuryl
Mode of action Systemic herbicide, absorbed from the leaf surface, with translocation throughout the plant, moving in both the xylem and phloem, and accumulating in the meristematic tissue. **Uses** Control of annual grasses such as *Avena fatua* and *Alopecurus myosuroides*, and perennial grasses such as *Sorghum*

halepense and *Elymus repens,* in potatoes, flax, oilseed rape, peas, sugar beet, cotton, and soya beans. **Formulation types** EC. **Compatibility** Certain post-emergence broad-leaved herbicides are incompatible with quizalofop-P-tefuryl. **Selected tradenames** 'Pantera' (Uniroyal); 'Logico' (IPESA)

ANALYSIS
quizalofop-P-ethyl
Products by hplc. **Residues** by glc. Details from Nissan.

quizalofop-P-tefuryl
Details from Uniroyal.

MAMMALIAN TOXICOLOGY
quizalofop-P-ethyl
Oral Acute oral LD_{50} for male rats 1210, female rats 1182, male mice 1753, female mice 1805 mg/kg. **NOEL** (90 d) for rats 7.7 mg/kg daily. **Toxicity class** EPA (formulation) III

quizalofop-P-tefuryl
Oral Acute oral LD_{50} for rats 1012 mg/kg. **Skin and eye** Acute percutaneous LD_{50} for rabbits >2000 mg/kg. Mild eye irritation, non-irritating to skin (rabbits). Non-sensitising to skin (guinea pigs). **Inhalation** LC_{50} (4 h) for rats >4.8 mg/l (12% EC). **NOEL** In chronic oncogenic feeding studies, NOEL for rats 1.25, dogs 19 mg/kg b.w. daily. **ADI** 0.01 mg/kg. **Toxicity class** EPA (formulation) III

ECOTOXICOLOGY
quizalofop-P-ethyl
Birds Acute oral LD_{50} for mallard ducks and bobwhite quail >2000 mg/kg. **Fish** LC_{50} (96 h) for rainbow trout >0.5 mg/l. **Bees** Assumed similar to racemate. **Worms** LC_{50} >1000 mg/kg. **Daphnia** LC_{50} (48 h) 0.29 mg/l. **Algae** Assumed similar to racemate.

quizalofop-P-tefuryl
Birds Acute oral LD_{50} for bobwhite quail and mallard ducks >2150 mg/kg. LC_{50} (8 d) for bobwhite quail and mallard ducks >5000 ppm. **Fish** LC_{50} (96 h) for trout 0.51, sunfish 0.23 mg/l. **Bees** LD_{50} >100 µg/bee. **Daphnia** LC_{50} (48 h) >1.5 mg/l. **Algae** LC_{50} (5 d) for *Selenastrum* >1.9 mg/l.

ENVIRONMENTAL FATE
quizalofop-P-ethyl
Animals The degradation pattern is the same as that for quizalofop-ethyl.
Plants The degradation pattern is the same as that for quizalofop-ethyl.
Soil/Environment In soil, degrades rapidly to quizalofop-P; DT_{50} ≤1 d.

Animals Main metabolite is quizalofop acid. **Plants** Main metabolite is quizalofop acid. **Soil/Environment** Aerobic DT_{50} 4.7 h (sandy loam). Adsorption K_d 8.69, K_{oc} 477 (both in sandy loam).

642   resmethrin   *Insecticide*

pyrethroid

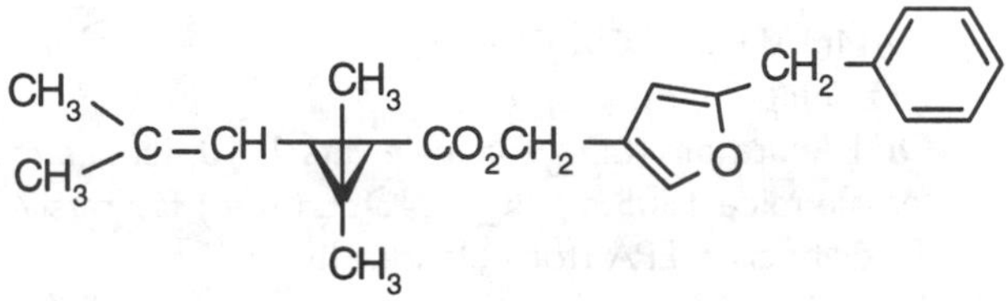

NOMENCLATURE
Common name resmethrin ((the ratio of *cis-trans* isomers should be stated) BSI, E-ISO, ANSI, ESA, JMAF); resméthrine ((f) F-ISO)
IUPAC name 5-benzyl-3-furylmethyl (1*RS*,3*RS*;1*RS*,3*SR*)-2,2-dimethyl-3-(2-methyl= prop-1-enyl)cyclopropanecarboxylate *Alt*: 5-benzyl-3-furylmethyl (±)-*cis-trans*-= chrysanthemate *Roth*: 5-benzyl-3-furylmethyl (1*RS*)-*cis-trans*-2,2-dimethyl-= 3-(2-methylprop-1-enyl)cyclopropanecarboxylate
Chemical Abstracts name [5-(phenylmethyl)-3-furanyl]methyl 2,2-dimethyl-= 3-(2-methyl-1-propenyl)cyclopropanecarboxylate
CAS RN *[10453–86–8]* unstated stereochemistry   **EEC no.** 233–940–7
Development codes NRDC 104; FMC 17 370; SBP 1382 (Penick Corp./RUC)
Official codes OMS 1206

PHYSICAL CHEMISTRY
Composition It contains 20–30% (1*RS*)-*cis*- and 80–70% (1*RS*)-*trans*- isomers. Tech. grade contains ≥84.5 % total isomers.   **Mol. wt.** 338.4   **M.f.** $C_{22}H_{26}O_3$
Form Colourless crystals; (tech., yellow to brown waxy solid).   **M.p.** 56.5 °C
(pure (1*RS*)-*trans*- isomer)   **B.p.** decomp. >180 °C   **V.p.** <0.01 mPa (25 °C)
K_{ow} logP = 5.43 (25 °C)   **S.g./density** 0.958–0.968 (20 °C), 1.035 (30 °C)
Solubility In water 37.9 µg/l (25 °C). In acetone *c.* 30%, chloroform, dichloromethane, ethyl acetate, toluene >50%, xylene >40%, ethanol, *n*-octanol *c.* 6%, *n*-hexane *c.* 10%, isopropyl ether *c.* 25%, methanol *c.* 3% (all *m/v*, 20 °C).
Stability Stable to heat and to oxidation. Decomposes rapidly on exposure to air and light (more slowly than pyrethrins) (J. H. Fales *et al.*, CSMA *Proc. 55th Annu. Meeting*, December 1968; A. B. Hadaway *et al.*, *Bull. W.H.O.*, 1970, **42**, 387). Unstable in alkaline media.   **Specific rotation** $[\alpha]_D$ –1° to +1°   **F.p.** 129 °C

COMMERCIALISATION
History Insecticide reported by M. Elliott et al. (*Nature (London)*, 1967, **213**, 493). Selective isomers are also available, see bioresmethrin. Resmethrin was introduced under licence by FMC Corp., Mitchell Cotts Chemicals, Penick Corp., Sumitomo Chemical Co., Ltd. **Patent** GB 1168797; GB 1168798; GB 1168799 all to NRDC **Manufacturers** AgrEvo EH; Mitchell Cotts; Sumitomo

APPLICATIONS
Mode of action Non-systemic insecticide with contact action. **Uses** A potent contact insecticide effective against a wide range of insects (I. C. Brooks et al., *Soap. Chem. Spec.*, 1969, **45**(3), 62). The toxicity to normal *Musca domestica* is 20× that of natural pyrethrum (M. Elliott et al., loc. cit.) but it is not synergised by pyrethrum synergists. Used to control agricultural, horticultural, household and public health insect pests, often in combination with more persistent insecticides. **Formulation types** AE; EC; UL; WP. **Mixtures** (*resmethrin +*) tetramethrin; bioallethrin; malathion; pyrethrins; piperonyl butoxide; piperonyl butoxide + pyrethrins; piperonyl butoxide + tetramethrin; lindane + tetramethrin; fenitrothion + permethrin. **Compatibility** Compatible with other neutral insecticides and fungicides. **Selected tradenames** 'Chryson' (Sumitomo); 'Synthrin' (AgrEvo USA)

ANALYSIS
Product analysis of pyrethroids reviewed by E. Papadopoulou-Mourkidou in *Comp. Analyt. Profiles*. By glc with FID (B. B. Brown, *Anal. Methods Pestic. Plant Growth Regul.*, 1973, **7**, 441). The *cis-* and *trans-* isomers can be separated and estimated by hplc (J. H. Zehner & R. A. Simonaitis, *J. Assoc. Off. Anal. Chem.*, 1976, **59**, 1101) or by glc (A. Murano et al., *Agric. Biol. Chem.*, 1971, **35**, 1200). **Residues** determined by glc (*Pestic. Anal. Man.*, 1979, **II**; R. Mestres et al., *Annal. Falsif. Expert. Chim.*, 1979, **72**, 577).

MAMMALIAN TOXICOLOGY
Reviews A. J. Gray & D. M. Soderlund, Chapt. 5 in "*Insecticides*". **EHC** 92 (WHO, 1989). **Oral** Acute oral LD_{50} for rats >2500 mg/kg. **Skin and eye** Acute percutaneous LD_{50} for rats >3000 mg/kg. Non-irritating to skin and eyes. No skin sensitisation (guinea pigs). **Inhalation** LC_{50} (4 h) for rats >9.49 g/m^3 air. **NOEL** (90 d) for rats >3000 mg/kg. No teratogenic effects observed in rabbits receiving 100 mg/kg daily, in mice at 50 mg/kg daily, or rats at 80 mg/kg daily. No oncogenic effects noted in a 112 w trial on rats up to 5000 ppm; NOEL 500 ppm in diet. In an 85 w feeding trial with mice, no oncogenic effects found up to 1000 ppm in diet. **ADI** (EPA) 0.125 mg/kg. **Other** No carcinogenic, mutagenic or teratogenic effects have been demonstrated. **Toxicity class** WHO (a.i.) III; EPA (formulation) III **EC risk** Xn (R22)

ECOTOXICOLOGY
Birds Acute oral LD_{50} for California quail >2000 mg/kg. **Fish** Toxic to fish; LC_{50}

(96 h) for yellow perch 2.36, sheepshead minnow 11, bluegill sunfish 17 µg/l.
Bees Toxic to bees; LD_{50} (oral) 0.069, (contact) 0.015 µg/bee. **Daphnia** LC_{50}
(48 h) 3.7 µg/l. **Other aquatic spp.** LC_{50} (96 h) for pink shrimp 1.3 µg/l.

ENVIRONMENTAL FATE
Animals Metabolism in laying hens was principally by ester hydrolysis and
oxidation, followed by conjugation (*J. Agric. Food Chem.*, 1989, **37**, 800).
Plants Metabolism of ^{14}C-resmethrin on glasshouse grown tomatoes, glasshouse
grown lettuce and field grown wheat demonstrated very rapid degradation
resulting in no parent residues 5 days after application. Numerous degradation
products were observed in very low amounts.

643 RH-2485   *Insecticide*

diacylhydrazine

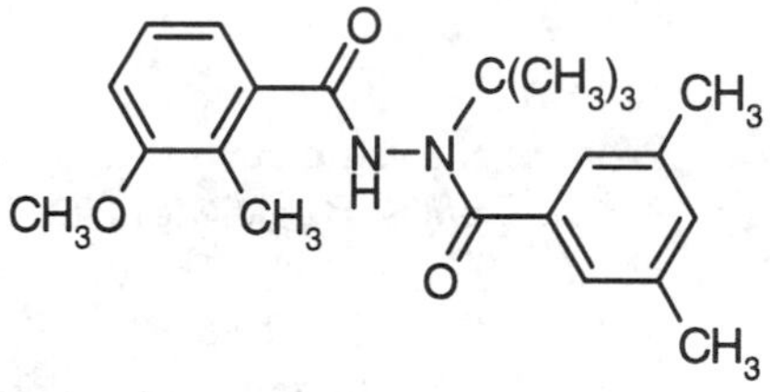

NOMENCLATURE
Common name methoxyfenozide (BSI proposed)
IUPAC name *N-tert*-butyl-*N*'-(3-methoxy-*o*-toluoyl)-3,5-xylohydrazide
CAS RN *[161050–58–4]* **Development codes** RH-2485; RH-112,485

PHYSICAL CHEMISTRY
Mol. wt. 368.5 **M.f.** $C_{22}H_{28}N_2O_3$ **Form** White powder. **M.p.** 204–205 °C
V.p. <5.3 × 10^{-3} mPa (25 °C) **K_{ow}** logP = 3.7 (shake flask) **Solubility** In water
<1 mg/l. In DMSO 11, cyclohexanone 9.9, acetone 9 (all in g/100 g). **Stability**
Stable at 25 °C and to hydrolysis at pH 5, 7 and 9.

COMMERCIALISATION
History Reported by D. P. Le *et al.* (*Proc. Br. Crop Prot. Conf. – Pests Dis.*, 1996, **2**,
481).

APPLICATIONS
Biochemistry Second generation ecdysone agonist. Causes cessation of feeding
and premature lethal moult. **Mode of action** Active primarily by ingestion, also

with contact, ovicidal and root-systemic activity. Does not have translaminar or phloem-systemic properties. **Uses** Under development by Rohm & Haas for control of lepidopterous larvae at 20 to 300 g/ha. **Formulation types** WP.

MAMMALIAN TOXICOLOGY
Oral Acute oral LD_{50} for rats and mice >5000 mg/kg. **Skin and eye** Acute percutaneous LD_{50} for rats >2000 mg/kg. Not irritating to eyes, slightly irritating to skin (rabbits). Not a skin sensitiser (guinea pigs). **Inhalation** LC_{50} for rats >4.3 mg/l. **Other** Negative in Ames test.

ECOTOXICOLOGY
Birds LC_{50} (8 d dietary) for mallard duck and bobwhite quail >5620 mg/kg diet. **Fish** LC_{50} (96 h) for bluegill sunfish >4.3 mg/l. **Bees** Non-toxic (contact) at 100 μg/bee. **Worms** LC_{50} for earthworms (14 d) >1213 mg/kg soil. **Other beneficial spp.** Not toxic to a range of insect species. **Daphnia** LC_{50} (48 h) 3.7 mg/l.

644 rimsulfuron *Herbicide*

sulfonylurea

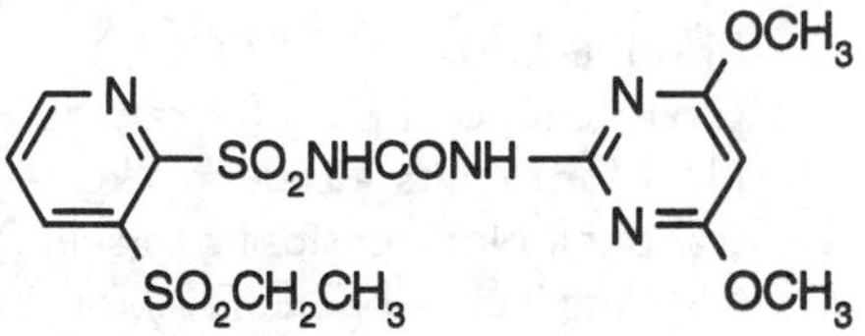

NOMENCLATURE
Common name rimsulfuron (BSI, pa E-ISO)
IUPAC name 1-(4,6-dimethoxypyrimidin-2-yl)-3-(3-ethylsulfonyl-2-pyridyl= sulfonyl)urea
Chemical Abstracts name N-[[(4,6-dimethoxy-2-pyrimidinyl)amino]carbonyl]-= 3-(ethylsulfonyl)-2-pyridinesulfonamide
CAS RN [122931–48–0] **Development codes** DPX-E9636

PHYSICAL CHEMISTRY
Composition Tech. is 99%. **Mol. wt.** 431.4 **M.f.** $C_{14}H_{17}N_5O_7S_2$
Form Colourless crystals. **M.p.** 176–178°C **V.p.** 1.5×10^{-3} mPa (25 °C)
K_{ow} logP = 0.288 (pH 5), −1.47 (pH 7) (25 °C) **S.g./density** 0.784 (25 °C)
Solubility In water (25 °C) <10 mg/l (unbuffered); 7.3 g/l (buffered, pH 7).
Stability On hydrolysis (25 °C) DT_{50} 4.6 d (pH 5), 7.2 d (pH 7), 0.3 d (pH 9). pKa 4.0

COMMERCIALISATION
History Herbicide reported by H. L. Palm *et al.* (*Proc. 1989 Br. Crop Prot. Conf. – Weeds*, **1**, 23). Introduced by E. I. du Pont de Nemours and Co.
Manufacturers Du Pont

APPLICATIONS
Biochemistry Branched chain amino acid synthesis (acetolactate synthase or ALS) inhibitor. Acts by inhibiting biosynthesis of the essential amino acids valine and isoleucine, hence stopping cell division and plant growth. Crop selectivity derives from rapid, selective metabolism (by contraction of the sulfonylurea group and ring migration); see Environmental Fate. **Mode of action** Selective systemic herbicide, absorbed by the foliage and roots, with rapid translocation to the meristematic tissues. **Uses** Rimsulfuron is a post-emergence sulfonylurea herbicide that effectively controls most annual and perennial grasses and several broad-leaved weeds in maize. Also used in tomatoes and potatoes. The target rate for most situations is 15 g a.i./ha. Rimsulfuron has a wide crop-safety margin under most conditions. **Formulation types** WG. **Selected tradenames** 'Titus' (Europe, Africa, Mediterranean) (Du Pont)

ANALYSIS
Product by glc. **Residues** by hplc. Details from Du Pont. Methods for sulfonylurea residues in crops, soil and water reviewed (A. C. Barefoot *et al.*, *Proc. Br. Crop Prot. Conf. – Weeds*, 1995, **2**, 707).

MAMMALIAN TOXICOLOGY
Oral Acute oral LD_{50} for rats >5000 mg/kg. **Skin and eye** Acute percutaneous LD_{50} for rabbits >2000 mg/kg. Non-irritating to skin; moderately irritating to eyes (rabbits). Non-sensitising to skin (guinea pigs). **Inhalation** LC_{50} (4 h) for rats >5.4 mg/l air. **NOEL** (2 y) for male rats 300, female rats 3000 ppm; (18 mo) for mice 2500 ppm; (1 y) for dogs 50 ppm. NOEL in 2-generation rat reproduction study 3000 ppm. Not teratogenic or oncogenic. **Other** Non-mutagenic in the Ames test. **Toxicity class** WHO (a.i.) III (Table 5)

ECOTOXICOLOGY
Birds Acute oral LD_{50} for bobwhite quail >2250, mallard ducks >2000 mg/kg. Dietary LC_{50} for bobwhite quail and mallard ducks >5620 ppm. **Fish** LC_{50} (96 h) for bluegill sunfish and rainbow trout >390, carp >900, sheepshead minnow 110 mg/l. **Bees** LD_{50} (contact) >100 µg/bee, (dietary) >1000 ppm.
Worms LC_{50} (14 d) for earthworms (*Eisenia foetida*) >1000 mg/kg.
Daphnia LC_{50} (48 h) >360 mg/l.

ENVIRONMENTAL FATE
Animals Rapidly metabolised and excreted in the urine and faeces. **Plants** Half-life in maize 6 h, blackgrass 46 h, Johnsongrass 25 d and *Sorghum bicolor* 52 d. See also

Mode of action and Stability.   **Soil/Environment** Degraded rapidly in soil, predominantly via chemical pathways (microbial degradation plays a minor role). Major metabolite is [1-(3-ethylsulfonyl)-2-pyridinyl]-4,6-dimethoxy-2-pyrimidine= amine. Rates of degradation are influenced by pH, and the compound is most stable in neutral pH soil and degrades more rapidly in alkaline and acidic soils. DT_{50} in soil 10–20 d (25 °C, lab. study).

645   rotenone   *Insecticide, acaricide*

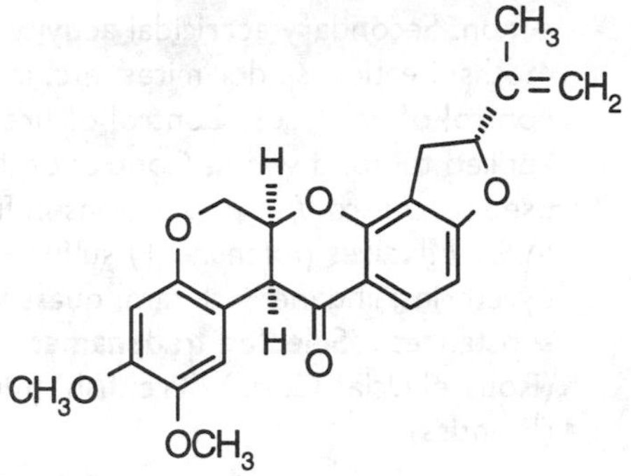

NOMENCLATURE
Common name rotenone (BSI, E-ISO, F-ISO, ESA, accepted in lieu of a common name); derris (JMAF)
IUPAC name (2R,6aS,12aS)-1,2,6,6a,12,12a-hexahydro-2-isopropenyl-= 8,9-dimethoxychromeno[3,4-b]furo[2,3-h]chromen-6-one
Chemical Abstracts name [2R-(2α,6aα,12aα)]-1,2,12,12a-tetrahydro-= 8,9-dimethoxy-2-(1-methylethenyl)[1]benzopyrano[3,4-b]furo[2,3-h][1]benzo= pyran-6(6aH)-one
Other names (for the plant extract) derris root; tuba-root; aker-tuba; (for the plants) barbasco; cubé; haiari; nekoe; timbo   **CAS RN** *[83–79–4]*
EEC no. 201–501–9   **Official codes** ENT 133

PHYSICAL CHEMISTRY
Mol. wt. 394.4   **M.f.** $C_{23}H_{22}O_6$ **Form** Colourless crystals.   **M.p.** 163 °C; 181 °C (dimorphic).   **V.p.** <1 mPa (20 °C)   **Solubility** Very slightly soluble in water (15 mg/l, 100 °C). Readily soluble in acetone, carbon disulfide, ethyl acetate and chloroform. Less readily soluble in diethyl ether, alcohols, petroleum ether and carbon tetrachloride. Crystallises from some solvents with the formation of solvates (H. A. Jones, *J. Am. Chem. Soc.*, 1931, **33**, 2738).   **Stability** Decomposes

on exposure to light and air. Racemised by alkalis to less insecticidal compounds, more rapidly in certain solvents. Extracts from derris roots can be stabilised with phosphoric acid. **Specific rotation** $[\alpha]_D^{20}$ −231° (benzene)

COMMERCIALISATION
History Derris root has long been used as a fish poison and its insecticidal properties were known to the Chinese well before it was isolated by E. Geoffrey (*Ann. Inst. Colon. (Marseilles)*, 1895, **2,** 1); its structure was established by E. B. LaForge et al. (*Chem. Rev.*, 1933, **12,** 181). **Manufacturers** AgrEvo EH; Prentiss; Tifa

APPLICATIONS
Biochemistry Probably acts on the electron-transport chain.
Mode of action Selective non-systemic insecticide with contact and stomach action. Secondary acaricidal activity. **Uses** Control of aphids, thrips, suckers, moths, beetles, spider mites, etc. in fruit and vegetable cultivation. Insecticidal control of premises. Control of fire ants. Control of mosquito larvae when applied to pond water. Control of lice, ticks, and warble flies on animals. Also used to control fish populations in fish management. **Formulation types** DP; EC; WP. **Mixtures** (*rotenone* +) sulfur; carbaryl; piperonyl butoxide + pyrethrins; pyrethrins; lindane + thiram; quassia. **Compatibility** Not compatible with alkaline substances. **Selected tradenames** 'Chem Sect' (Tifa); 'Cube root' (Tifa); 'Noxfire' (Roussel Uclaf Corp.); 'Prenfish' (mixture) (Prentiss); 'Synpren fish' (mixture) (Prentiss)

ANALYSIS
Product analysis by i.r. spectrometry (*AOAC Methods*, 1995, 961.03*), by rplc (*ibid.*, 983.06; *CIPAC Handbook*, 1985, **1C,** 2217), or by solvent extraction and crystallisation (*AOAC Methods*, 1995, 938.01); the purity of a benzene solution of rotenone may be checked by polarimetry (*CIPAC Handbook*, 1970, **1,** 610). **Residues** determined by hplc (M. C. Bowman et al., *J. Assoc. Off. Anal. Chem.*, 1978, **61,** 1445).

MAMMALIAN TOXICOLOGY
Oral Acute oral LD_{50} for white rats 132–1500, white mice 350 mg/kg.
Other Estimated lethal dose for humans 300–500 mg/kg; more toxic when inhaled than when ingested. Very toxic to pigs. **Toxicity class** WHO (a.i.) II; EPA III, (EC formulation I) **EC risk** T (R25); Xi (R36/37/38)

ECOTOXICOLOGY
Fish LC_{50} (96 h) for rainbow trout 31, bluegill sunfish 23 μg/l. **Bees** Not toxic alone to bees, but toxic in combination with pyrethrum.

ENVIRONMENTAL FATE
Animals In rat liver and in insects, the furan ring is enzymically opened and

cleaved, leaving behind a methoxy group. The principal metabolite is rotenonone. An alcohol has been found as a further metabolite, this being formed via oxidation of a methyl group of the isopropenyl residue (I. Yamamoto, *Residue Rev.*, 1969, **25**, 161).

646   RU 15525   *Insecticide*

pyrethroid

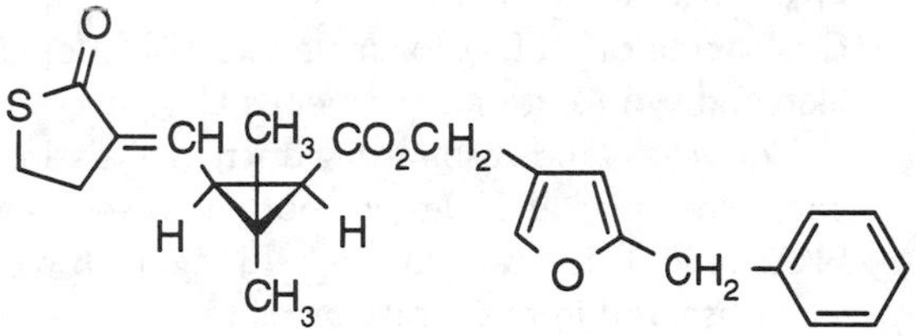

NOMENCLATURE
IUPAC name 5-benzyl-3-furylmethyl (*E*)-(1*R*,3*S*)-2,2-dimethyl-3-(2-oxothiolan-= 3-ylidenemethyl)cyclopropanecarboxylate *Roth*: 5-benzyl-3-furylmethyl (*E*)-(1*R*)-*cis*-2,2-dimethyl-3-(2-oxothiolan-3-ylidenemethyl)cyclopropane= carboxylate
Chemical Abstracts name [1*R*-[1α,3α(*E*)]]-[5-(phenylmethyl)-3-furanyl]methyl 3-[(dihydro-2-oxo-3(2*H*)-thienylidene)methyl]-2,2-dimethylcyclopropane= carboxylate
Other names kadethrin; kadethrine   **CAS RN** *[58769–20–3]*
EEC no. 261–433–0   **Development codes** RU 15 525   **Official codes** ENT 29 117; AI3–29 117

PHYSICAL CHEMISTRY
Composition Tech. grade is ≥93% pure *m/m* of stated stereoisomer.
Mol. wt. 396.5   **M.f.** $C_{23}H_{24}O_4S$ **Form** Yellow-brown, viscous oil.   **M.p.** 31 °C
V.p. <0.1 mPa at 20 °C   **Solubility** Practically insoluble in water. Soluble in dichloromethane, ethanol (10 g/kg), benzene, toluene, xylene, acetone, piperonyl butoxide. Slightly soluble in kerosene.   **Stability** Hydrolysed by aqueous alkalis. Rapidly decomposed by light (more slowly in mineral oils). Unstable to heat.
Specific rotation $[\alpha]_D^{20}$ +16 to +19° (c. 50 g tech./l toluene)   **F.p.** >100 °C

COMMERCIALISATION
History Insecticide reported by J. Martel & J. Buendia (*Int. Congr. Pestic. Chem. IUPAC, 3rd,* 1974) and by J. Lhoste & F. Rauch (*Pestic. Sci.,* 1976, **7**, 247). Introduced by Roussel Uclaf (now AgrEvo GmbH).   **Patent** FR 2097244; US 3842177   **Manufacturers** AgrEvo

APPLICATIONS

Mode of action Insecticide with contact action. Gives rapid knockdown.
Uses Control of household insect pests, particularly houseflies, mosquitoes, and cockroaches. Normally used in combination with other pyrethroid insecticides and synergists. **Formulation types** TC; Aerosol; OL. **Mixtures** *(RU 15525 +)* bioresmethrin + piperonyl butoxide; permethrin. **Compatibility** Incompatible with alkaline materials. **Selected tradenames** 'Kadethrin'/'Kadethrine' (AgrEvo)

ANALYSIS

Product analysis by glc or hplc, details available from AgrEvo.

MAMMALIAN TOXICOLOGY

Oral Acute oral LD_{50} for male rats 1324, female rats 650, dogs >1000 mg/kg.
Skin and eye Acute percutaneous LD_{50} for female rats >3200 mg/kg. Slight irritant to skin, eyes, and respiratory tract. **Inhalation** Inhalation trials (10 d) with rats and guinea pigs at 200 times normal dose in aerosols tolerated perfectly.
NOEL (90 d) for rats 25, dogs 15 mg/kg b.w. daily. **Other** No teratogenic effect was observed in mice, rats or rabbits.

ECOTOXICOLOGY

Fish Toxic to fish; LC_{50} (96 h) for rainbow trout 0.13 µg/l. **Bees** Toxic to bees.

ENVIRONMENTAL FATE

Animals In rats, following oral administration, there is rapid metabolism (by opening of the thiolactone ring, cleavage of the ester, and ring hydroxylation), and rapid excretion. See K. Ohsawa & J. E. Casida (*J. Agric. Food Chem.*, 1980, **28**, 250–255).

647 S421

Insecticide synergist

$$(CCl_3CHClCH_2)_2O$$

NOMENCLATURE

IUPAC name bis(2,3,3,3-tetrachloropropyl) ether
Chemical Abstracts name 1,1'-oxybis[2,3,3,3-tetrachloropropane]
CAS RN *[127–90–2]* **Development codes** S421 (Sankyo, BASF)

PHYSICAL CHEMISTRY

Mol. wt. 377.7 **M.f.** $C_6H_6Cl_8O$ **Form** Yellow liquid. **M.p.** Still viscous at −50 °C
B.p. 144–150 °C/1 mmHg **V.p.** High. **S.g./density** 1.64–1.66 (20 °C)

Solubility Practically insoluble in water. Miscible with common organic solvents, e.g. ether, benzene, chloroform, trichloroethylene, dichloromethane, dioxane, methyl ethyl ketone, petroleum, diesel oil, etc. **Stability** Unstable to alkalis. To prevent loss of hydrogen chloride during storage, 1% stabiliser (epichlorohydrin) is added. **F.p.** 177 °C

COMMERCIALISATION
Manufacturers Sankyo

APPLICATIONS
Uses Acts as a synergist to increase the insecticidal activity of pyrethrins, allethrin, and carbaryl. **Mixtures** (*S421* +) pyrethrins; piperonyl butoxide + pyrethrins.

ANALYSIS
Product by determination of total chlorine by one of the usual methods. Because tech. S421 contains volatile chlorine-containing by-products, heating the sample at about 80 °C under vacuum to constant weight is recommended.

MAMMALIAN TOXICOLOGY
Oral Acute oral LD_{50} for rats 8 ml/kg (24 h) and 2.5 ml/kg (8 d), for mice 12 and 3.5 ml/kg, respy. **Skin and eye** Acute percutaneous LD_{50} not available. Slight irritation of skin and mucous membranes. **NOEL** Repeated oral feeding to rabbits of 0.2 ml causes liver damage.

ECOTOXICOLOGY
Fish LC_{50} (96 h) for guppies 1.7 mg/l.

648 sethoxydim   *Herbicide*

cyclohexanedione oxime

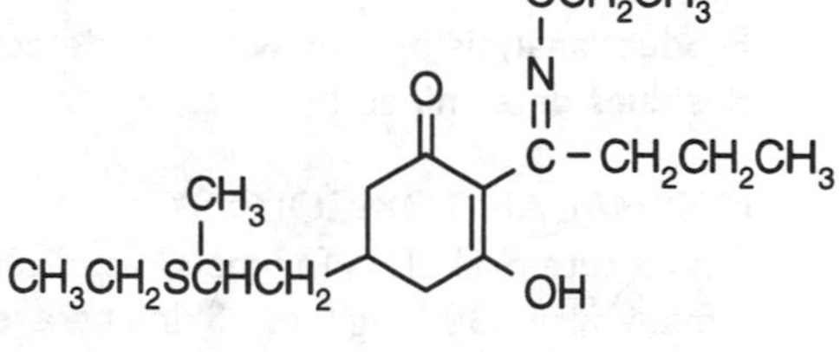

NOMENCLATURE
Common name sethoxydim (BSI, draft E-ISO); séthoxydime ((*f*) draft F-ISO)
IUPAC name (±)-(*EZ*)-2-(1-ethoxyiminobutyl)-5-[2-(ethylthio)propyl]-3-hydroxy=cyclohex-2-enone

Chemical Abstracts name (±)-2-[1-(ethoxyimino)butyl]-5-[2-(ethylthio)propyl]-3-=
hydroxy-2-cyclohexen-1-one (i); first publication on the compound presented the
structure as a tautomer, 2-[1-(ethoxyamino)butylidene]-5-[2-(ethylthio)propyl]-=
1,3-cyclohexanedione (ii)
CAS RN *[74051–80–2]* (i); *[71441–80–0]* (ii)   **Development codes** NP-55
(Nippon Soda); BAS 90 520H (BASF AG); SN 81 742 (AgrEvo)

PHYSICAL CHEMISTRY
Mol. wt. 327.5   **M.f.** $C_{17}H_{29}NO_3S$ **Form** Oily, odourless liquid.
B.p. >90 °C/3 × 10^{-5} mmHg   **V.p.** <0.013 mPa K_{ow} logP = 4.51 (pH 5), 1.65
(pH 7)   **S.g./density** 1.043 at 25 °C   **Solubility** In water 25 (pH 4), 4700 (pH 7)
(both in mg/l, 20 °C). Soluble in most common organic solvents e.g. acetone,
benzene, ethyl acetate, hexane, methanol all >1 kg/kg (25 °C).   **Stability** The
commercial product is stable for at least 2 y under normal storage conditions. At
10 mg/l, 12 h/d illumination with xenon lamp, DT_{50} is 5.5 d (pH 8.7, 25 °C).

COMMERCIALISATION
History Introduced in USA (1983) by Nippon Soda Co., Ltd.   **Patent** JP 52112945;
US 4249937; UK 1589003; DE 2822304.   **Manufacturers** Nippon Soda

APPLICATIONS
Biochemistry Acetyl CoA carboxylase (fatty acid synthesis) inhibitor. Inhibits
mitosis.   **Mode of action** Selective systemic herbicide, absorbed predominantly by
the foliage, and, to a lesser extent, by the roots. Translocated rapidly both
acropetally and basipetally.   **Uses** Control of annual (at 0.20–0.25 kg a.i./ha) and
perennial (0.2–0.5 kg/ha) grasses (except *Poa* spp.) in broad-leaved crops,
including cotton, oilseed rape, soya beans, sugar beet, fodder beet, sunflowers,
spinach, potatoes, tobacco, peanuts, strawberries, alfalfa, flax, and vegetables.
Phytotoxicity Non-phytotoxic to broad-leaved crops, but phytotoxic to most
monocotyledonous crops (except onions, garlic, and asparagus).
Formulation types EC.   **Compatibility** Incompatible with organic and inorganic
copper compounds.   **Selected tradenames** 'Nabu' (Nippon Soda); 'Fervinal'
(AgrEvo); 'Poast' (BASF)

ANALYSIS
Product analysis by hplc with u.v. detection (*CIPAC Handbook*, 1992, **E**, 193–6).
Residues determined by hplc.

MAMMALIAN TOXICOLOGY
Oral Acute oral LD_{50} for male rats 3200, female rats 2676, male mice 5600,
female mice 6300 mg/kg.   **Skin and eye** Acute percutaneous LD_{50} for rats and
mice >5000 mg/kg. Non-irritating to skin and eyes (rabbits); no skin sensitisation.
Inhalation LC_{50} (4 h) for rats >6.28 mg/l air.   **NOEL** (2 y) for rats 17.2, mice
13.7 mg/kg b.w. daily.   **ADI** 0.14 mg/kg (Japan).   **Toxicity class** WHO (a.i.) III;
EPA (formulation) III

ECOTOXICOLOGY

Birds Acute oral LD$_{50}$ for Japanese quail >5000 mg/kg.   **Fish** LC$_{50}$ (48 h) for carp 153, trout 38 mg tech./l.   **Bees** No significant hazard to bees.   **Daphnia** LC$_{50}$ (3 h) 1.5 mg/l.

ENVIRONMENTAL FATE

Animals In rats, following oral administration, 78.5% is eliminated in the urine and 20.1% in the faeces within 48 h.   **Plants** In soya beans, the parent molecule is oxidised, structurally rearranged, and conjugated. Transformation to metabolites is very rapid.   **Soil/Environment** DT$_{50}$ in soil <l d at 15 °C. Metabolism involves molecular rearrangement, oxidation and conjugation processes.

649 siduron   *Herbicide*

urea

NOMENCLATURE

Common name siduron (BSI, E-ISO, (*m*) F-ISO, ANSI, WSSA, JMAF)
IUPAC name 1-(2-methylcyclohexyl)-3-phenylurea
Chemical Abstracts name *N*-(2-methylcyclohexyl)-*N*'-phenylurea
CAS RN *[1982–49–6]*   **Development codes** Du Pont 1318

PHYSICAL CHEMISTRY

Mol. wt. 232.3   **M.f.** C$_{14}$H$_{20}$N$_2$O   **Form** Colourless crystals.   **M.p.** 133–138 °C
V.p. 5.3 × 10^{-4} mPa (25 °C)   **K$_{ow}$** logP = 3.8   **Henry** 6.8 × 10^{-6} Pa m^3 mol^{-1} (calc.)   **S.g./density** 1.08 (25 °C)   **Solubility** In water 18 mg/l (25 °C). In dimethylacetamide 367, dimethylformamide 260, ethanol 160, isophorone, dichloromethane 118 (all in g/kg, 25 °C).   **Stability** Stable up to its melting point and in neutral aqueous solutions. Decomposes slowly in acid and alkaline media.

COMMERCIALISATION

History Herbicide reported by R. W. Varner *et al.* (*Proc. Br. Weed Control Conf., 7th*, 1964, p. 38). Introduced by E. I. du Pont de Nemours and Co (who no longer market it). Marketed by Gowan since 1994.   **Patent** US 3309192
Manufacturers Raschig

APPLICATIONS
Biochemistry Photosynthetic electron transport inhibitor.
Mode of action Selective herbicide, absorbed by the roots, with translocation in the xylem.   **Uses** Pre-emergence control of *Digitaria* spp. and annual grass weeds in newly-seeded grass (at 2–6 kg a.i./ha) or established turf (at 8–12 kg/ha).
Phytotoxicity Bermuda grass and certain bent grasses may be injured when applied pre-emergence.   **Formulation types** WP.   **Selected tradenames** 'Tupersan' (Gowan)

ANALYSIS
Product analysis by hplc.   **Residues** determined by colorimetry (R. L. Dalton & H. L. Pease, *J. Assoc. Off. Agric. Chem.*, 1962, **45**, 377).

MAMMALIAN TOXICOLOGY
Oral Acute oral LD_{50} for rats >7500 mg/kg.   **Skin and eye** Acute percutaneous LD_{50} for rabbits >5500 mg/kg (maximum feasible dose).   **Inhalation** LC_{50} (4 h) for rats >5.8 mg/l.   **NOEL** (2 y) for rats 500, dogs 2500 mg/kg diet.
Toxicity class WHO (a.i.) III (Table 5); EPA (WP formulation) III

ECOTOXICOLOGY
Birds Eight-day dietary LC_{50} for mallard ducklings and bobwhite quail >10 000 mg/kg diet.   **Fish** LC_{50} (96 h) for rainbow trout 14, bluegill sunfish 16 mg/l.   **Daphnia** EC_{50} (48 h) 18 mg/l.   **Algae** EC_{50} (120 h) for *Selenastrum capricornutum* 250 µg/l.

ENVIRONMENTAL FATE
Plants In studies with ^{14}C-labelled siduron, no metabolites were detected in barley plants after an 8-day absorption period.   **Soil/Environment** In soil, resists leaching. Microbial degradation occurs, DT_{50} *c*. 120–150 d; metabolites include methylcyclohexylurea, methylcyclohexylamine, phenylurea and aniline. In water, photolysis DT_{50} 290 d (natural sunlight).

non-ester pyrethroid

CH$_3$CH$_2$O— ⬡ —Si—(CH$_2$)$_3$— ⬡ —F, with Si bearing CH$_3$ (top) and CH$_3$ (bottom), and the right ring bearing O—⬡ (phenoxy)

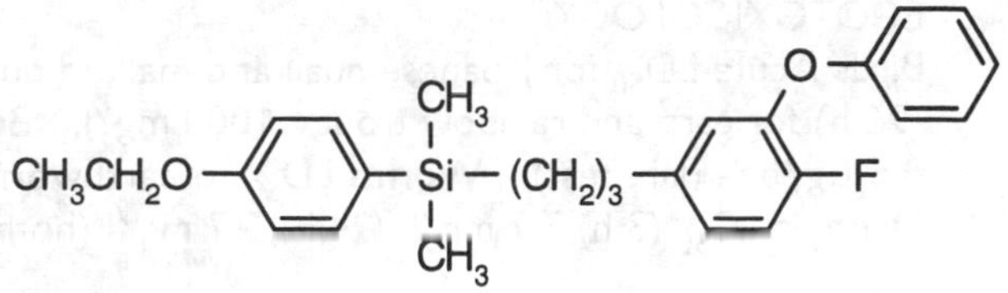

NOMENCLATURE

Common name silafluofen (BSI, pa E-ISO)
IUPAC name (4-ethoxyphenyl)[3-(4-fluoro-3-phenoxyphenyl)propyl](dimethyl)=
silane
Chemical Abstracts name (4-ethoxyphenyl)[3-(4-fluoro-3-phenoxyphenyl)propyl]=
dimethylsilane
Other names silaneophan **CAS RN** *[105024–66–6]*
Development codes Hoe 084498; Hoe 498

PHYSICAL CHEMISTRY

Mol. wt. 408.6 **M.f.** C$_{25}$H$_{29}$FO$_2$Si **Form** Liquid. **B.p.** Decomposes without boiling above 170 °C **V.p.** 2.5 × 10^{-3} mPa (20 °C) (vapour pressure balance) **K$_{ow}$** logP = 8.2 **S.g./density** 1.08 (20°C) **Solubility** In water at 20 °C, 0.001 mg/l. Soluble in most organic solvents. **Stability** Stable over a period of 2 years, when stored in original unopened containers at 20 °C. **F.p.** >100 °C (closed cup)

COMMERCIALISATION

History Patented by Katsuda in Japan in 1986, and by Hoechst (now AgrEvo GmbH) in Europe in 1987. Introduced by Dainihon Jochugiku in Japan as a termiticide. **Manufacturers** AgrEvo

APPLICATIONS

Mode of action A broad-spectrum insecticide which acts mainly as a stomach poison, but also by contact. **Uses** Control of Coleoptera, Diptera, Heteroptera, Homoptera, Isoptera, Lepidoptera, Orthoptera, and Thysanoptera.
Formulation types EO; EC; DP; WP. **Compatibility** Good compatibility when mixed with Bordeaux mixture or DCPA. **Selected tradenames** 'Joker' (AgrEvo); 'Silatop' (AgrEvo)

ANALYSIS

Details available from AgrEvo.

MAMMALIAN TOXICOLOGY

Oral Acute oral LD$_{50}$ for male and female rats >5000 mg/kg. **Skin and eye** Acute percutaneous LD_{50} for male and female rats >5000 mg/kg. **Inhalation** LC_{50} (4 h) for rats >6.61 mg/l air. **Other** Non-teratogenic and non-mutagenic.

ECOTOXICOLOGY
Birds Acute LD_{50} for Japanese quail and mallard ducks >2000 mg/kg. **Fish** LC_{50} (96 h) for carp and rainbow trout >1000 mg/l. **Bees** Oral LD_{50} (24 h) 0.5 μg/bee (lab. test). **Worms** LD_{50} to earthworms >1000 mg/kg. **Daphnia** LC_{50} (3 h) 7.7 mg/l; (24 h) 1.7 mg/l (both Japanese guidelines).

651 simazine   *Herbicide*

1,3,5-triazine

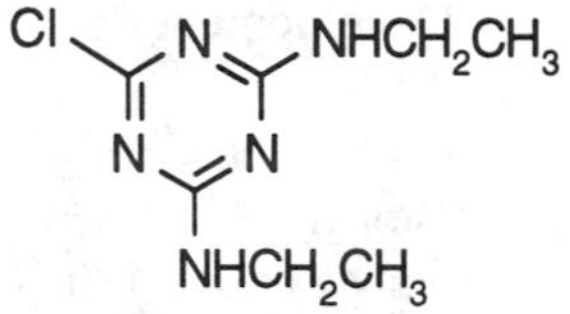

NOMENCLATURE
Common name simazine (BSI, E-ISO, (f) F-ISO, ANSI, WSSA); CAT (JMAF)
IUPAC name 6-chloro-N^2,N^4-diethyl-1,3,5-triazine-2,4-diamine
Chemical Abstracts name 6-chloro-N,N'-diethyl-1,3,5-triazine-2,4-diamine
CAS RN *[122–34–9]* **EEC no.** 204–535–2 **Development codes** G 27 692 (Ciba)

PHYSICAL CHEMISTRY
Composition Tech. grade is ≥97% pure. **Mol. wt.** 201.7 **M.f.** $C_7H_{12}ClN_5$
Form Colourless powder. **M.p.** 225–227 °C (decomp.) **V.p.** 2.94×10^{-3} mPa
(25 °C) (OECD 104) **K_{ow}** logP = 2.1 (25 °C, unionised) **Henry** 5.6×10^{-3} Pa m^3
mol^{-1} (calc.) **S.g./density** 1.33 (22 °C) **Solubility** In water 6.2 mg/l (pH 7,
20 °C). In ethanol 570, acetone 1500, toluene 130, n-octanol 390, n-hexane 3.1
(all in mg/l, 25 °C). **Stability** Relatively stable in neutral, weakly acidic and weakly
alkaline media. Rapidly hydrolysed by stronger acids and bases; DT_{50} (calc.) 8.8 d
(pH 1), 3.7 d (pH 13) (20 °C). Decomposed by u.v. irradiation (c. 90% in 96 h).
pKa 1.62 (20 °C), v. weak base

COMMERCIALISATION
History Herbicide reported by A. Gast et al. (*Experientia*, 1956, **12**, 146).
Introduced by J. R. Geigy S.A. (now Novartis Crop Protection AG).
Patent BE 540590; GB 894947 **Manufacturers** Drexel; Makhteshim-Agan;
Novartis; Oxon Italia; Rallis; Sanachem

APPLICATIONS

Biochemistry Photosynthetic electron transport inhibitor.
Mode of action Selective systemic herbicide, absorbed principally through the roots, but also through the foliage, with translocation acropetally in the xylem, accumulating in the apical meristems and leaves. **Uses** Control of most germinating annual grasses and broad-leaved weeds in pome fruit, stone fruit, bush and cane fruit, citrus fruit, vines, strawberries, nuts, olives, pineapples, field beans, french beans, peas, maize, sweet corn, asparagus, hops, alfalfa, lupins, oilseed rape, artichokes, sugar cane, cocoa, coffee, rubber, oil palms, tea, turf and ornamentals. Applied at rates up to 1.5 kg/ha within the EU and up to 2–3 kg/ha in perennial crops in the tropics and subtropics. **Phytotoxicity** Phytotoxic to a number of crops, including sugar beet, tobacco, tomatoes, cucurbits, clover, rice, soya beans, lettuce, oats, and many vegetables (e.g. spinach, onions, carrots, crucifers, etc.). **Formulation types** GR; SC; WP; WG. **Mixtures** *(simazine +)* ametryn; 2,4-DES; metoxuron; trietazine; dichlobenil; propyzamide; and many others. **Selected tradenames** 'Caliber' (USA) (Novartis); 'Gesatop' (Novartis); 'Princep' (USA) (Novartis); 'Princep Caliber' (Novartis); 'Amizina' (Sipcam); 'Luserb' (Siapa); 'Sanasim' (Sanachem); 'Simanex' (Makhteshim-Agan); 'Simatylone LA' (Agriphar); 'Tafazine' (Rallis); 'Trinovin' (mixture) (Efthymiadis)

ANALYSIS

Product analysis by glc with FID (*CIPAC Handbook*, 1980, **1A**, 1343; B. G. Tweedy & R. A. Kahrs, *Anal. Methods Pestic. Plant Growth Regul.*, 1979, **10**, 493; *AOAC Methods*, 1995, 971.08). **Residues** determined by glc with DMC or FPD (K. Ramsteiner *et al.*, *J. Assoc. Off. Anal. Chem.*, 1974, **57**, 192; B. G. Tweedy & R. A. Kahrs, *loc. cit.*) or by hplc. In **drinking water**, by gc with FID (*AOAC Methods*, 1995, 991.07).

MAMMALIAN TOXICOLOGY

Oral Acute oral LD_{50} for rats, mice, and rabbits >5000 mg/kg.
Skin and eye Acute percutaneous LD_{50} for rats >2000. Non-irritating to skin and eyes (rabbits). **Inhalation** LC_{50} (4 h) for rats >5.5 mg/l. **NOEL** (2 y) for female rats 0.5 mg/kg b.w. daily; (1 y) for female dogs 0.8 mg/kg b.w. daily; (95 w) for mice 5.7 mg/kg b.w. daily. **ADI** 0.005 mg/kg b.w. **Toxicity class** WHO (a.i.) III (Table 5); EPA (formulation) IV

ECOTOXICOLOGY

Birds Acute oral LD_{50} for mallard ducks >2000 mg/kg. Eight-day dietary LC_{50} for mallard ducks 10 000 mg/kg, Japanese quail >5000 mg/kg. **Fish** LC_{50} (96 h) for bluegill sunfish 90, rainbow trout >100, crucian carp >100, guppies 49 mg/l.
Bees LD_{50} (48 h oral and topical) >99 µg/bee. **Worms** LC_{50} (14 d) for earthworms >1000 mg/kg. **Daphnia** LC_{50} (48 h) >100 mg/l; (21 d) 0.29 mg/l.
Algae EC_{50} (72 h) for *Scenedesmus subspicatus* 0.042 mg/l; (5 d) for *Selenastrum capricornutum* 0.26 mg/l.

ENVIRONMENTAL FATE

Animals In mammals, following oral administration, 65–97% is eliminated within 24 h as the de-ethylated metabolite (Y. Deng et al., J. Agric. Food Chem., 1990, **38**, 1411). Elimination of low rates is primarily in the urine of rats, with a shift to faecal elimination at high doses. Excretion is rapid (c. 90% in 48 h). Degradation to desethylsimazine and bis-desethylsimazine (diaminochlorotriazine) is the primary metabolic pathway. **Plants** Readily metabolised by tolerant plants to the herbicidally-inactive 6-hydroxy analogue and amino acid conjugates. The hydroxysimazine is further degraded by dealkylation of the side-chains and by hydrolysis of the resulting amino groups on the ring, with evolution of CO_2. In sensitive plants, unaltered simazine leads to chlorosis and death.

Soil/Environment Major metabolites under all conditions are desethylsimazine and hydroxysimazine. Microbial breakdown in soil results in degradation of simazine at very variable rates; DT_{50} 27–102 d, median 49 d; temperature and soil moisture are the main factors affecting rates. K_{oc} 103–277, median 160; K_d 0.37–4.66 (12 soils). Under field conditions, simazine has a low leaching potential. Loss by direct photodecomposition is insignificant. Indirect photodecomposition in the presence of photosensitisers such as humic acids is however likely.

652 simetryn   *Herbicide*

1,3,5-triazine

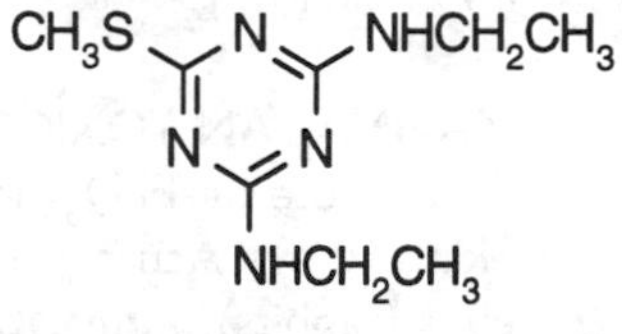

NOMENCLATURE

Common name simetryn (BSI (since 1984), E-ISO, WSSA); simetryne ((*f*) F-ISO, Canada, JMAF, BSI (before 1984))

IUPAC name N^2,N^4-diethyl-6-methylthio-1,3,5-triazine-2,4-diamine; 2,4-bis(ethylamino)-6-methylthio-1,3,5-triazine

Chemical Abstracts name N,N'-diethyl-6-(methylthio)-1,3,5-triazine-2,4-diamine

CAS RN *[1014–70–6]* **EEC no.** 213–801–7 **Development codes** G 32 911 (Ciba-Geigy AG)

PHYSICAL CHEMISTRY

Mol. wt. 213.3 **M.f.** $C_8H_{15}N_5S$ **Form** Whitish crystals. **M.p.** 82–83 °C
V.p. 9.5×10^{-2} mPa (20 °C) **K_{ow}** logP = 2.6 (calc., unionised) **S.g./density** 1.02

Solubility In water 400 mg/l (20 °C). In methanol 380, acetone 400, toluene 300, *n*-hexane 4, *n*-octanol 160 (all in g/l, 20 °C). **pKa** 4.0, weak base

COMMERCIALISATION
History Herbicide reported by J. R. Geigy S.A. (now Novartis Crop Protection AG). Introduced in Japan by Nihon Nohyaku Co., Ltd, Nippon Kayaku Co., Ltd (and formerly Sankyo Co., Ltd and Hokko Chemical Industry Co., Ltd).
Patent CH 337019 to Ciba-Geigy **Manufacturers** Nippon Kayaku

APPLICATIONS
Biochemistry Photosynthetic electron transport inhibitor.
Mode of action Selective herbicide. **Uses** Used in combination with thiobencarb to control broad-leaved weeds in rice. **Formulation types** GR.
Mixtures *(simetryn +)* thiobencarb; thiobencarb + MCPB-ethyl; quinoclamine + MCPB. **Selected tradenames** 'Gy-bon' (Nihon Nohyaku, Novartis)

ANALYSIS
Product analysis by titration or by glc (*CIPAC Handbook*, 1983, **1B**, 1984). In **drinking water**, by gc with FID (*AOAC Methods*, 1995, 991.07).

MAMMALIAN TOXICOLOGY
Oral Acute oral LD_{50} for rats 750–1195 mg/kg. **Skin and eye** Acute percutaneous LD_{50} for rats >3200 mg/kg. Not a skin or eye irritant (rabbit).
NOEL (2 y) for rats 1.2 mg/kg daily (2.5 mg/kg b.w.), mice 56 mg/kg daily, dogs 10.5 mg/kg daily. **ADI** 0.025 mg/kg b.w. (provisional value, Japan).
Toxicity class WHO (a.i.) III; EPA (formulation) III **EC risk** Xn (R22)

ECOTOXICOLOGY
Fish LC_{50} (96 h) for trout 7, guppy 5.2 ppm. **Bees** Non-toxic. **Algae** Toxic.

ENVIRONMENTAL FATE
Animals DT_{50} in rats *c.* 10 h. Excreted in urine and faeces; two oxidative metabolic pathways. **Plants** Selectivity and metabolism of simetryn reviewed (H. Matsumoto & K. Ishizuka, *Weed. Res. (Japan)*, **26**, 1981)

653 sodium chlorate *Herbicide*

$NaClO_3$

NOMENCLATURE
Common name sodium chlorate (E-ISO, used in lieu of a common name); chlorate

de sodium (F-ISO, used in lieu of a common name)
IUPAC name sodium chlorate
Chemical Abstracts name sodium chlorate
CAS RN *[7775–09–9]* **EEC no.** 231–887–4

PHYSICAL CHEMISTRY
Mol. wt. 106.4 **M.f.** $ClNaO_3$ **Form** Colourless powder. **M.p.** 248 °C
B.p. Decomposes at *c.* 300 °C **V.p.** Negligible at room temperature.
S.g./density 2.490 (15 °C) **Solubility** In water 790 g/l (0 °C), 2300 g/l (100 °C).
In 90% alcohol 16 g/kg. Soluble in glycerol. **Stability** Stable at normal
temperatures if no acidic or oxidisable substances or ammonium salts are present.
A potent oxidising agent, reacting with organic materials, so creating a serious fire
hazard, for example with splashed clothing. Marketed formulations contain a fire
depressant.

COMMERCIALISATION
History Used as a herbicide since *c.* 1910. **Manufacturers** Caffaro; Drexel;
Kerr-McGee

APPLICATIONS
Mode of action Non-selective herbicide, absorbed through the roots and foliage,
with translocation basipetally in the xylem, as the phloem tissue is killed. The rate
of respiration is increased, catalase activity decreased, and the plant's food
reserves are depleted. **Uses** Total weed control on non-crop land, applied at up
to 600 kg/ha. Also used as a defoliant and desiccant in cotton, safflowers,
sunflowers, lupins, alfalfa, clover, field beans, soya beans, flax, rice, etc. Has a soil-
sterilant effect. **Phytotoxicity** Phytotoxic to all crops. **Formulation types** SP; SC;
SL. **Mixtures** *(sodium chlorate +)* atrazine; atrazine + 2,4-D; atrazine + sodium
metaborate; bromacil + sodium metaborate; diuron + sodium metaborate.
Compatibility Not compatible with herbicides which are susceptible to oxidation.
Selected tradenames 'Dervan' (Caffaro); 'Kusatol' (Hodogaya); 'Sochlor' (Simplot)

ANALYSIS
Product analysis is by titration (*CIPAC Handbook*, 1970, **1**, 626; *ibid.*, 1992, **E**,
197–201; *Herbicides 1977*, p. 41). **Residues** in soil may be determined by
colorimetry of a derivative.

MAMMALIAN TOXICOLOGY
Oral Acute oral LD_{50} for rats 1200 mg/kg (E. F. Edson, *Pharm. J.*, 1960, **185**, 361).
Skin and eye Acute percutaneous LD_{50} not available. Irritating to skin and mucous
membranes. **Other** A dose of 5–10 g can prove fatal in adults, as can a dose of
2 g in small children. Cases of chronic toxicity are unknown. **Toxicity class**
WHO (a.i.) III; EPA (formulation) III **EC risk** O (R9); Xn (R22)

ECOTOXICOLOGY
Fish Toxic to fish; LC_{50} (24 h) for rainbow trout 4200, harlequin fish 8600 mg/l.
Bees Not toxic to bees. Not applied when plants are in flower. **Worms** Not toxic to worms. **Daphnia** Not toxic. **Algae** Toxic to algae.

ENVIRONMENTAL FATE
Soil/Environment Remains in soil for 0.5 to 5 years, depending upon rate of application, soil type, fertility, organic matter content, moisture, and weather conditions.

54 sodium fluoride   *Insecticide*

NaF

NOMENCLATURE
Common name sodium fluoride (E-ISO, used in lieu of a common name); fluorure de sodium (F-ISO, used in lieu of a common name)
IUPAC name sodium fluoride
Chemical Abstracts name sodium fluoride
CAS RN *[7681–49–4]*

PHYSICAL CHEMISTRY
Mol. wt. 42.0 **M.f.** FNa **Form** Colourless powder. **M.p.** 993 °C
V.p. Negligible at room temperature **S.g./density** 2.8 **Solubility** In water 42.2 g/l (18 °C). Slightly soluble in ethanol.

COMMERCIALISATION
History Introduced in baits against insects in stores.

APPLICATIONS
Mode of action Insecticide with stomach and contact action. **Uses** Used in insect baits and as a timber preservative. **Phytotoxicity** Highly phytotoxic.

ANALYSIS
Product analysis is by determination of the fluorine content by titrimetric methods (*AOAC Methods,* 1995, 921.04, 929.04, 933.03).

MAMMALIAN TOXICOLOGY
Oral Highly toxic, the lowest oral lethal dose being 28–100 mg/kg for a range of vertebrates. Acute oral LD_{50} for rats 180 mg/kg (*Am. Ind. Hyg. Assoc. J.,* 1969, **30**, 470). **Toxicity class** WHO (a.i.) II

MAIN ENTRIES

655 sodium fluoroacetate   *Rodenticide*

FCH_2CO_2Na

NOMENCLATURE
Common name sodium fluoroacetate (E-ISO, accepted in lieu of a common name); fluoroacétate de sodium (F-ISO, accepted in lieu of a common name); fluoroacetic acid (BSI, for parent acid)
IUPAC name sodium fluoroacetate
Chemical Abstracts name sodium fluoroacetate
Other names Compound 1080 **CAS RN** *[62–74–8]* **EEC no.** 200–548–2

PHYSICAL CHEMISTRY
Mol. wt. 100.0 **M.f.** $C_2H_2FNaO_2$ **Form** Colourless hygroscopic powder.
M.p. Decomposes *c.* 200 °C **V.p.** Non-volatile. **Solubility** Very soluble in water. Almost insoluble in ethanol, acetone, and petroleum oils.

COMMERCIALISATION
History Rodenticidal properties reported by E. R. Kalmbeck (*Science*, 1945, **102**, 232).

APPLICATIONS
Biochemistry Tricarboxylic acid cycle inhibitor. **Uses** Used in baits for control of rodents. In Australia, also used for control of wild rabbits, wild dogs and wild pigs.
Formulation types CB.

ANALYSIS
Product analysis is by conversion to lead chloride fluoride (*AOAC Methods*, 1984, 6.019–6.062); sodium fluoride is a usual contaminant (*WHO Manual* (2nd Ed.), SRT/5). For **residues**, see *AOAC Methods* 1995, 949.09, 949.10.

MAMMALIAN TOXICOLOGY
Oral Acute oral LD_{50} for *Rattus norvegicus* 0.22 mg/kg (S. H. Dieke & C. P. Richter, *U.S. Public Health Rep.*, 1946, **61**, 672). **Toxicity class** WHO (a.i.) Ia
EC risk T+ (R26/27/28)

56 sodium hexafluorosilicate   *Insecticide*

inorganic fluoride

Na_2SiF_6

NOMENCLATURE
Common name sodium hexafluorosilicate (E-ISO, accepted in lieu of a common name); fluorosilicate de sodium (F-ISO, accepted in lieu of a common name); sodium fluorosilicate (JMAF, accepted in lieu of a common name)
IUPAC name disodium hexafluorosilicate; sodium fluorosilicate
Chemical Abstracts name disodium hexafluorosilicate
Other names sodium fluosilicate; sodium silicofluoride   **CAS RN** *[16893–85–9]*

PHYSICAL CHEMISTRY
Mol. wt. 188.1   **M.f.** F_6Na_2Si **Form** Colourless crystals.   **M.p.** Decomposes on melting.   **S.g./density** 2.68   **Solubility** In water 6.5 g/l (17.5 °C). Very slightly soluble in ethanol.

COMMERCIALISATION
History Introduced as an insecticide by Panorama Chemicals (Pty) Ltd.

APPLICATIONS
Uses Pre- or early post-emergence control of soil insect pests, especially larvae of *Noctua pronuba*, *Agrotis* and *Euxoa* spp. in agricultural and horticultural crops. Applied at 1 kg a.i./ha for overall application, or ≥0.5 kg/ha for row application.
Formulation types RB.   **Selected tradenames** 'Prodan' (Tamogan)

ANALYSIS
Product analysis by titration (*AOAC Methods*, 1995, 921.04, 945.05).

MAMMALIAN TOXICOLOGY
Oral Acute oral LD_{50} for rats 125 mg/kg (*Handbook of Toxicology*, 1956, **1**, 278).
Toxicity class WHO (a.i.) II

MAIN ENTRIES

57 *Spodoptera exigua* NPV   *Biological agent*

nuclear polyhedrosis virus

NOMENCLATURE
Scientific name *Spodoptera exigua* multicapsid nuclear polyhedrosis virus

PROPERTIES
Composition Rod-shaped, elongated particles enclosed in a protein crystalline matrix (occlusion body). **Form** Tan to brown powder. **Solubility** Insoluble in water.

COMMERCIALISATION
Manufacturers Thermo Trilogy

APPLICATIONS
Mode of action Infectious virus active by ingestion. Caterpillars cease feeding after c. 4 d and die after 5–10 d. **Uses** For control of beet armyworm in various crops. **Formulation types** WP; liquid concentrate. **Selected tradenames** 'Spod-X' (Thermo Trilogy, Brinkmann)

658 SSF-126 *Fungicide*

strobilurin analogue

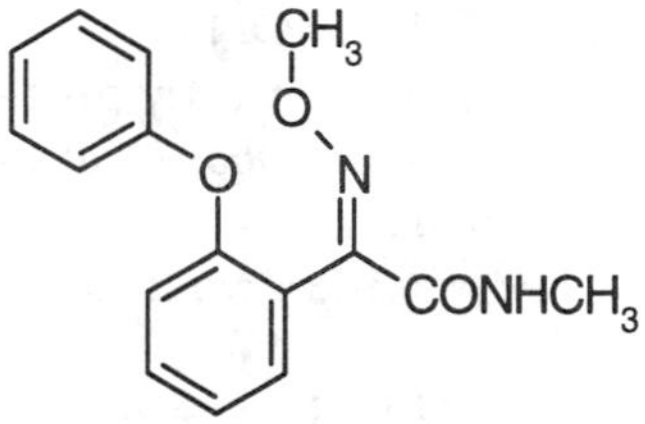

NOMENCLATURE
Common name metominostrobin (BSI proposed); fenominostrobin (BSI proposed)
IUPAC name (E)-2-methoxyimino-N-methyl-2-(2-phenoxyphenyl)acetamide
Chemical Abstracts name (E)-α-methoxyimino-N-methyl-2-phenoxybenzene= acetamide
CAS RN [133408–50–1] **Development codes** SSF-126

PHYSICAL CHEMISTRY
Composition >97% pure. **Mol. wt.** 284.3 **M.f.** $C_{16}H_{16}N_2O_3$ **Form** White crystals. **M.p.** 87–89 °C **V.p.** 0.018 mPa (25 °C, gas saturation method) **K$_{ow}$** logP = 2.32 (20 °C) **S.g./density** 1.27–1.30 (20 °C) **Solubility** In water 0.128 g/l (20 °C). In dichloromethane 1380, chloroform 1280, dimethyl sulfoxide 940 (all g/l, 25 °C). **Stability** Stable to heat and to acidic and alkaline media. Slightly unstable to light. **F.p.** 226 °C (Seta flash tester)

COMMERCIALISATION
Manufacturers Shionogi

APPLICATIONS
Biochemistry Respiratory electron transport inhibitor. Acts by inhibiting cytochrome pathway, between cytochrome b and cytochrome c_1 (Complex III). **Mode of action** Systemic fungicide with protective and curative action. **Uses** For control of *Pyricularia oryzae* on rice. at 1.5–1.8 kg/ha; applied before or at outbreak of infection symptoms. **Formulation types** GR; WP. **Selected tradenames** 'Oribright' (Shionogi)

MAMMALIAN TOXICOLOGY
Oral Acute oral LD_{50} for male rats 776, female rats 708 mg/kg. **Skin and eye** Acute percutaneous LD_{50} for male and female rats >2000 mg/kg. Not a skin irritant. **Inhalation** LC_{50} (4 h) for male and female rats >1880 mg/m^3.

ECOTOXICOLOGY
Fish LC_{50} (96 h) for carp 17.5, killifish 22.5 mg/l. **Bees** LD_{50} (oral) >50 000, (contact) >100 000 ppm. **Worms** LC_{50} (14 d) for *Eisenia foetida* 114 mg/kg; NOEC (14 d) 56.2 mg/kg. **Other beneficial spp.** LC_{50} (5 d) for silkworm 250 ppm. **Daphnia** EC_{50} (48 h) for *D. pulex* 14.0, (24 h) 22.3 (both mg/l). **Algae** EC_{50} (72 h) for *Chlorella vulgaris* 51.0 mg/l; NOEC (72 h) 10 mg/l.

ENVIRONMENTAL FATE
Soil/Environment Soil DT_{50} 339–346 d.

659 *Steinernema* spp. *Biological agent*

nematode

NOMENCLATURE
Steinernema carpocapsae
Other names *Neoaplectana carpocapsae*

Steinernema feltiae
Strain MicroBio Strain UK 76
Other names *Neoaplectana feltiae* (Filipjev); *Neoaplectana bibionis*

Steinernema glaseri
Development codes B-326

Steinernema scapterisci
Development codes B-319 (Biosys)

PROPERTIES
Steinernema feltiae
Form Infective juveniles with inert carrier. **Stability** Stable for 4 weeks to 3 months at 5 °C.

Steinernema scapterisci
Composition 'Otinem S' is a clay formulation containing 60 million infective-stage nematodes per pack. **Stability** Shelf-life of product is 2 months. Should be stored at 10 °C.

COMMERCIALISATION
Steinernema carpocapsae
Manufacturers Arbico; M&R Durango

Steinernema feltiae
History First sold in UK in 1989. **Manufacturers** Andermatt; Koppert; MicroBio

Steinernema glaseri
History Reported by R. Georgis et al. (*Proc. Br. Crop Prot. Conf. – Pests Dis.*, 1992, **1**, 73). Introduced by Biosys (now Thermo Trilogy).

Steinernema scapterisci
Production *S. scapterisci* is cultured on a sterile undefined medium and the infective juveniles harvested by concentration. **History** Reported by R Georgis et al. (*Proc. Br. Crop Prot. Conf. – Pests Dis.*, 1992, **1**, 73). Introduced by Biosys (now Thermo Trilogy, who no longer market it) and licensed to Ecogen. *S. scapterisci* is a wild-type entomopathogenic nematode originally isolated in South America by University of Florida researchers from populations of mole crickets during a natural epizootic. **Manufacturers** Ecogen

APPLICATIONS
Steinernema spp.
Mode of action Parasitic nematode which searches for, enters and kills the target pests.

Steinernema carpocapsae
Uses Control of black vine weevil (*Otiorhynchus sulcatus*) in blackcurrants, strawberries and ornamentals; and cutworms (*Agrotis* spp.) in vegetables and turf.

Formulation types WG.   **Selected tradenames** 'BioSafe-N' (Thermo Trilogy); 'BioVector' (Thermo Trilogy); 'Vector' (Thermo Trilogy)

Steinernema feltiae
Mode of action Nematodes actively seek target pest in soil or compost and penetrate via body openings. Upon entry, symbiotic bacteria (*Xenorhabdus bovienii*) are released and infect the host, resulting in rapid death. Nematodes then reproduce in the cadaver.   **Uses** For control of sciarid flies (fungus gnats, *Bradysia paupera, Lycoriella auripila*) in protected crops.   **Formulation types** WG.
Selected tradenames 'Entonem' (Koppert); 'Nemasys' (for horticultural crops) (MicroBio, Biobest); 'Nemasys M' (for mushrooms) (MicroBio, Biobest); 'Sciarid' (for mushrooms) (Koppert); 'Traunem' (Andermatt); 'Magnet' (for mushrooms) (Thermo Trilogy); 'X-Gnat' (for ornamentals) (Thermo Trilogy)

Steinernema glaseri
Uses Control of white grubs (Scarabaeidae) in turf.

Steinernema riobravis
Uses Control of large nymph and adult mole crickets (*Scapteriscus* spp.) in turf; and citrus weevils (*Pachnaeus litus*), sugar cane rootstalk borer (*Diaprepes abbreviatus*) and other pests in citrus.   **Formulation types** WG.
Selected tradenames 'Biovector 335' (Thermo Trilogy); 'Devour' (Thermo Trilogy); 'Vector MC' (Thermo Trilogy)

Steinernema scapterisci
Mode of action Infective juveniles actively enter mole crickets though natural body openings (i.e. mouth, anus, spiracles) and penetrate into the blood cavity where they release a symbiotically associated bacterium. The bacterium quickly multiplies, releasing toxins that paralyse and kill the insect, usually within 36–48 hours.   **Uses** *S. scapterisci* has activity primarily against mole crickets infesting turf grass, and sees most use on golf courses.   **Phytotoxicity** Not phytotoxic.   **Formulation types** On clay.   **Compatibility** *S. scapterisci* is compatible with a wide range of pesticides, but is susceptible to certain nematicides.
Selected tradenames 'Otinem S' (Ecogen)

ANALYSIS
Steinernema feltiae
Biotest with Sciarids or larvae of *Tenebrio molitor*.

Steinernema scapterisci
The activity of *S. scapterisci* is measured in terms of infective juvenile count and by insect bioassay.

MAMMALIAN TOXICOLOGY

Steinernema scapterisci

Neither *S. scapterisci* nor its bacterial symbiote have shown evidence of toxicity, infectivity, irritation, or hypersensitivity to mammals. No allergic responses or health problems have been observed by research workers, manufacturing staff, or users.

660 *Streptomyces griseoviridis*   *Biological agent*

bacterium

NOMENCLATURE

Scientific name *Streptomyces griseoviridis* Anderson et al.   **Strain** K61

PROPERTIES

Stability The formulated material is stable for 12 months when stored in an unopened package below 8 °C.

COMMERCIALISATION

Production 'Mycostop' is produced by fermentation and freeze-drying of the organism, and contains the minimum of 10^8 colony forming units per gram of the product.   **History** The development of a commercial *Streptomyces* biofungicide originated from the disease-suppressive properties of Finnish light coloured *Sphagnum* peat. In the preliminary studies carried out by the Department of Plant Pathology at the University of Helsinki, streptomycetes were found to be common microbes in horticultural peat. In the screening tests, a number of *Streptomyces* strains proved to be effective antagonists to soil-borne and seed-borne fungal pathogens. One of these promising strains, K61, was chosen for further development. The commercial formulation was developed by Kemira Oy (O. Mohammadi, *Proc. of the 3rd Int. Workshop on Plant Growth-Promoting Rhizobacteria*, 1994, 282–284).   **Manufacturers** Kemira Agro

APPLICATIONS

Mode of action The efficacy of 'Mycostop' is based on several modes of action: competition for living space and nutrients, lysis of the cell wall of fungal pathogens by extracellular enzymes, and production of antifungal metabolites. 'Mycostop' has also shown growth stimulation in healthy crops.   **Uses** Primary targets of 'Mycostop' are *Fusarium* fungi which cause wilt, root and basal rot diseases in vegetables, ornamentals and herbs. It also controls some other soil-borne and seed-borne diseases such as *Alternaria* and *Phomopsis*. 'Mycostop' can be used in glasshouse cultivation as a dry seed treatment or as an aqueous suspension for the spraying or drenching of growth substrate, or via drip irrigation. No safety period is required between treatment and harvesting. **Formulation types** WP. **Compatibility** 'Mycostop' is compatible with a range of pesticides; most products can be used on the same day. However tank mixtures of 'Mycostop' and other pesticides or concentrated fertiliser solutions are not recommended. **Selected tradenames** 'Mycostop' (Kemira Agro)

ANALYSIS
The viability and purity of the product is tested by determining the number of colony-forming units by plate count methods. The biological efficacy is tested with a bioassay using artificially infected cauliflower seeds.

MAMMALIAN TOXICOLOGY
Not hazardous to animals according to animal studies. No allergic responses or health problems have been observed by research workers, manufacturing staff or users, when the product is used in accordance with the directions for use. Inhalation of the fine powder and skin contact should be avoided by using normal protective equipment.

ECOTOXICOLOGY
Birds NOEL for bobwhite quail or mallard ducks 2.45×10^9 cfu/kg. **Fish** Not toxic; NOEL for rainbow trout 5000 cfu/ml. **Bees** NOEL 9.8×10^8 cfu/kg. **Daphnia** NOEL 10^4 cfu/ml.

ENVIRONMENTAL FATE
Soil/Environment The strain belongs to the natural microflora of soil.

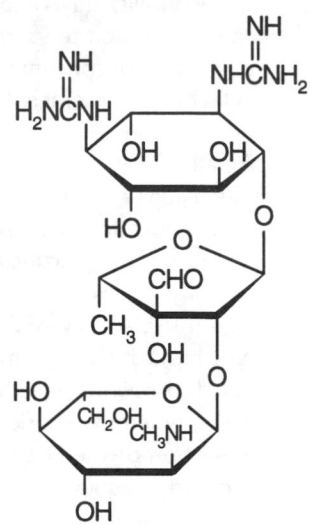

NOMENCLATURE
streptomycin
Common name streptomycin (BSI, E-ISO, BAN, JMAF); streptomycine ((*f*) F-ISO);
no name (Denmark)
IUPAC name O-2-deoxy-2-methylamino-α-L-glucopyranosyl-(1→2)-O-5-deoxy-=
3-C-formyl-α-L-lyxofuranosyl-(1→4)-N^3,N^3-diamidino-D-streptamine; 1,1'-[1-L-=
(1,3,5/2,4,6)-4-[5-deoxy-2-O-(2-deoxy-2-methylamino-α-L-glucopyranosyl)-3-C-=
formyl-α-L-lyxofuranosyloxy]-2,5,6-trihydroxycyclohex-1,3-ylene]diguanidine
Chemical Abstracts name O-2-deoxy-2-(methylamino)-α-L-glucopyranosyl-=
(1→2)-O-5-deoxy-3-C-formyl-α-L-lyxofuranosyl-(1→4)-N,N-bis(aminoimino=
methyl)-D-streptamine
CAS RN *[57–92–1]*

streptomycin sesquisulfate
CAS RN *[3810–74–0]*

PHYSICAL CHEMISTRY
streptomycin
Mol. wt. 581.6 **M.f.** $C_{21}H_{39}N_7O_{12}$ **Stability** Unstable in strong acids and alkalis.

streptomycin sesquisulfate
Mol. wt. 1457.3 **M.f.** $C_{42}H_{84}N_{14}O_{36}S_3$ **Form** Off-white powder. **Solubility** In
water >20 g/l (pH 7, 28 °C). In ethanol 0.9, methanol >20, petroleum ether 0.02
(all in g/l). **Stability** Stable; hygroscopic. **Specific rotation** $[\alpha]_D^{25}$ 84°

COMMERCIALISATION
Production Streptomycin is obtained by fermentatation of *Streptomyces griseus*, isolated as sesquisulfate. **History** Antibacterial antibiotic reported by Schatz *et al.* (*Proc. Soc. Exp. Biol. Med.*, 1944, **55**, 66). Structure (Brink & Folkers, *J. Am. Chem. Soc.*, 1947, **69**, 1234; Wolfrom *et al.*, ibid., 1950, **76**, 3675). **Manufacturers** Merck

APPLICATIONS
Mode of action Bactericide with systemic action. **Uses** Control of bacterial shot-hole, bacterial rots, bacterial canker, bacterial wilts, fire blight, and other bacterial diseases (especially those caused by gram-positive species of bacteria) in pome fruit, stone fruit, citrus fruit, olives, vegetables, potatoes, tobacco, cotton, and ornamentals. **Phytotoxicity** Chlorosis may occur on grapes, pears, peaches, and some ornamentals. **Formulation types** WP; Liquid. **Compatibility** Incompatible with pyrethrins and alkaline materials. **Selected tradenames** 'Agrimycin 17' (sesquisulfate) (Merck); 'AS-50' (sesquisulfate) (Merck)

ANALYSIS
Product analysis by bioassay with suitable bacteria.

MAMMALIAN TOXICOLOGY
streptomycin
Oral Acute oral LD_{50} for mice >10 000 mg/kg. **Skin and eye** Acute percutaneous LD_{50} for male mice 400, female mice 325 mg/kg. May cause allergic skin reaction. **NOEL** In chronic toxicity studies on rats, NOEL was 125 mg/kg. **Other** Acute i.p. LD_{50} for male mice 340, female mice 305 mg/kg.

streptomycin sesquisulfate
Oral Acute oral LD_{50} for rats 9000, mice 9000, hamsters 400 mg/kg.

662 strychnine   *Rodenticide*

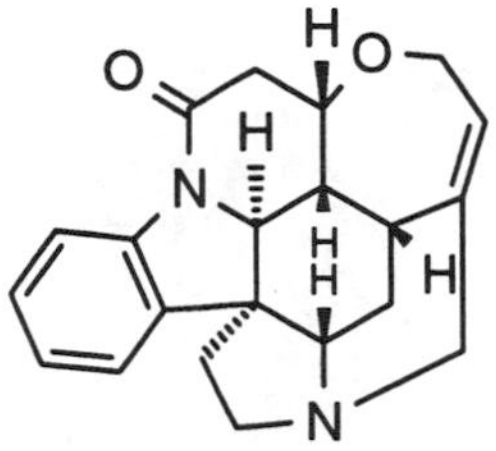

NOMENCLATURE
Common name strychnine (BSI, E-ISO, F-ISO, accepted in lieu of a common name)

Chemical Abstracts name strychnidin-10-one
CAS RN *[57–24–9]* EEC no. 200–319–7

PHYSICAL CHEMISTRY
Mol. wt. 334.4 M.f. $C_{21}H_{22}N_2O_2$ Form Colourless crystals. M.p. 270–280 °C
(decomp.); >199 °C (sulfate) Solubility In water 143 mg/l. In benzene 5.6,
ethanol 6.7, chloroform 200 (all in g/l). Sparingly soluble in diethyl ether and
petroleum spirit. Sulfate solubility in water 30 g/l (15 °C); soluble in ethanol.
Stability Strychnine is a base, and forms water-soluble salts with acids: the
hydrochloride, colourless prisms with 1.5–2.0 mol of water of crystallisation
which are lost at 110 °C; the sulfate, colourless crystals (as pentahydrate)
becoming anhydrous at 100 °C. Specific rotation $[\alpha]_D^{18}$ –139° (chloroform)

COMMERCIALISATION
Production By extracting seeds of *Strychnos* or Loganiaceae spp. History The
physiological properties of the alkaloids in extract of *Strychnos nux-vomica* seeds
have long been known, strychnine being the most important.

APPLICATIONS
Biochemistry Antagonist to the neurotransmitter glycine, acting principally on the
spinal cord. Mode of action Rodenticide absorbed mainly by the intestine.
Uses Used (as strychnine sulfate) for control of moles, squirrels, rabbits, gophers,
mice and other small mammals, and also for control of sparrows, pigeons and
other bird pests. Rats are somewhat resistant, as they normally avoid the bait.
Formulation types RB; CB.

ANALYSIS
Product analysis by acid-base titration (*AOAC Methods*, 1995, 920.211) or by
spectrophotometry (*ibid*, 962.22).

MAMMALIAN TOXICOLOGY
Oral Acute oral LD_{50} for rats 16 mg/kg. Other Lethal dose: for rats 1–30 mg/kg;
for man 30–60 mg/kg. Toxicity class WHO (a.i.) Ib EC risk T+ (R27/28) (base);
T+ (R26/27) (salts)

663 sulcofuron-sodium *Insecticide*

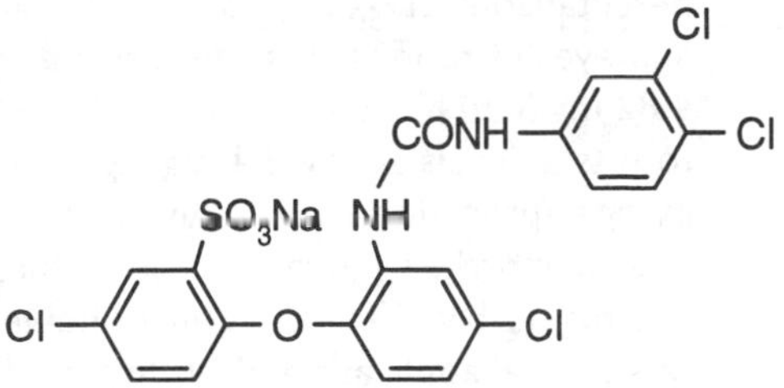

NOMENCLATURE

sulcofuron-sodium

IUPAC name sodium 5-chloro-2-[4-chloro-2-[3-(3,4-dichlorophenyl)ureido]=phenoxy]benzenesulfonate

Chemical Abstracts name sodium 5-chloro-2-[4-chloro-2-[[[(3,4-dichlorophenyl)=amino]carbonyl]amino]phenoxy]benzenesulfonate

CAS RN *[3567–25–7]*

sulcofuron

Common name sulcofuron (BSI, draft E-ISO, draft F-ISO)

CAS RN *[24019–05–4]*

PHYSICAL CHEMISTRY

sulcofuron-sodium

Mol. wt. 544.2 **M.f.** $C_{19}H_{11}Cl_4N_2NaO_5S$ **Form** White odourless powder. **M.p.** 216–231 °C (OECD 102) **V.p.** 1.9×10^{-6} mPa (25 °C) K_{ow} logP = 1.89 (unstated pH) **S.g./density** 1.69 (20 °C) **Solubility** In water 1.24 g/l (pH 6.9, 20 °C). **Stability** Hydrolytically stable at 25 °C (DT_{50} >31 d, pH 5, 7 and 9). **F.p.** >150 °C

sulcofuron

Mol. wt. 522.2 **M.f.** $C_{19}H_{12}Cl_4N_2O_5S$

COMMERCIALISATION

History Moth-proofing agent introduced by Ciba-Geigy AG (now Novartis Crop Protection AG). **Manufacturers** Ciba Specialty Chemicals

APPLICATIONS

Mode of action Stomach poison to wool-feeding insect larvae; inhibits digestion. **Uses** Moth-proofing agent used to control Tineidae and Dermestidae larvae which attack wool and wool blend fabrics. **Formulation types** High concentrate; Liquid. **Selected tradenames** 'Mitin' (Ciba Specialty Chemicals)

MAIN ENTRIES

MAMMALIAN TOXICOLOGY
sulcofuron-sodium
Oral Acute oral LD_{50} for male and female rats 645 mg/kg.   **Skin and eye** Acute percutaneous LD_{50} for male and female rats >2000 mg/kg. Not irritant to skin and eyes of rabbits; not sensitising to guinea pigs.   **Inhalation** LC_{50} (4 h) for rats 4.82 mg/l.   **NOEL** Dietary NOEL 3.1 mg/kg b.w. Dietary application for 90 days to rats at levels above 3.1 mg/kg led only to slight hemolytic anaemia, while gavage application at dose levels of 50 mg/kg and above induced severe ulcerative and haemorrhagic inflammation of the gastrointestinal tract. Dermal NOEL (90 d) 100 mg/kg b.w. The substance showed no teratogenic or embryotoxic effects in rats by oral application of up to 80 mg/kg b.w. (OECD 414). No genotoxic or mutagenic effects were seen in bacteria in several Ames tests as well as in cell cultures in tests for point mutation, chromosome aberration or unscheduled DNA synthesis. *In vivo*, no clastogenic or DNA damaging effects occurred in a micronucleus test and a test for SCE induction. The substance can be regarded as essentially non-mutagenic.   **ADI** 3.1 µg/kg b.w. by oral intake; 100 µg/kg b.w. by dermal intake.

ECOTOXICOLOGY
sulcofuron-sodium
Birds Acute oral LD_{50} for mallard duck (14 d) >2150 mg/kg, bobwhite quail (21 d) 966 mg/kg. Eight-day dietary LC_{50} for bobwhite quail 5076 ppm.   **Fish** LC_{50} (96 h) for zebra fish 14.5, rainbow trout 6.8 mg/l.   **Other beneficial spp.** EC_{50} for *Pseudomonas putida* (cell multiplication test) 160 mg/l. IC_{50} (3 h) for effluent treatment bacteria (respiration inhibition test) >100 mg/l. IC_{50} (4 h) for nitrifying bacteria 43 mg/l.   **Daphnia** LC_{50} (48 h) 9.3 mg/l.   **Algae** EC_{50} (72 h) for *Scenedesmus subspicatus* 2.8 mg/l.

664   sulcotrione   *Herbicide*

triketone

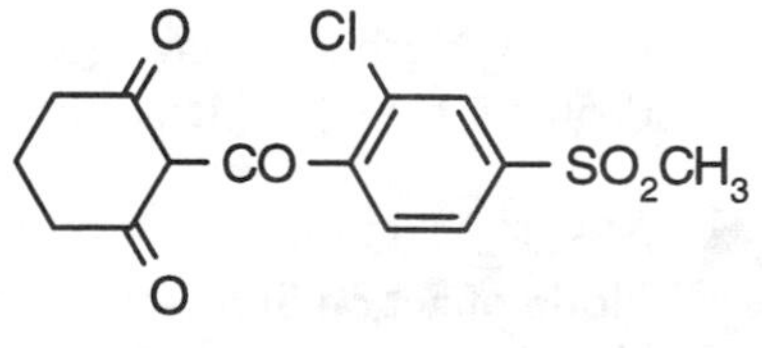

NOMENCLATURE
Common name sulcotrione (BSI, pa E-ISO)
IUPAC name 2-(2-chloro-4-mesylbenzoyl)cyclohexane-1,3-dione

Chemical Abstracts name 2-[2-chloro-4-(methylsulfonyl)benzoyl]-1,3-cyclo=hexanedione
CAS RN *[99105–77–8]*   **Development codes** ICIA0051; SC0051

PHYSICAL CHEMISTRY
Composition Tech. material is 90% pure.   **Mol. wt.** 328.8   **M.f.** $C_{14}H_{13}ClO_5S$
Form White solid; tech. is a light tan solid.   **M.p.** 139 °C; (tech. 131–139 °C)
V.p. 5×10^{-3} mPa (25 °C)   **Solubility** In water at 25 °C, 165 mg/l. Soluble in acetone and chlorobenzene.   **Stability** Stable in water, with or without exposure to sunlight. Thermostable up to 80 °C.   **pKa** 3.13 (23 °C)

COMMERCIALISATION
History Reported by J. M. Beraud *et al.* (*Proc. Br. Crop Prot. Conf. – Weeds*, 1991, **1**, 51). Introduced by ICI (now Zeneca Agrochemicals).   **Manufacturers** Zeneca

APPLICATIONS
Biochemistry Probably acts by directly affecting chlorophyll synthesis.
Mode of action Herbicide, absorbed predominantly by the leaves, but also by the roots.   **Uses** Control of broad-leaved weeds and grasses post-emergence in maize and sugar cane at 200–300 g/ha.   **Formulation types** SC.
Mixtures (*sulcotrione +)* atrazine.   **Selected tradenames** 'Galleon' (Zeneca); 'Mikado' (Zeneca)

MAMMALIAN TOXICOLOGY
Oral Acute oral LD_{50} for rats >5000 mg/kg.   **Skin and eye** Acute percutaneous LD_{50} for rabbits >4000 mg/kg. Low rate of absorption through the skin; non-irritating to skin; mild eye irritant (rabbits). Strong skin sensitiser (guinea pigs).
Inhalation LC_{50} (4 h) >1.6 mg/l.   **Other** Non-teratogenic in rats and rabbits. Non-genotoxic as demonstrated *in vivo*.

ECOTOXICOLOGY
Birds Low toxicity to birds.   **Fish** LC_{50} (96 h) for rainbow trout 227 mg/l.
Bees Low toxicity to bees, both by topical and oral application.   **Daphnia** LC_{50} (48 h) >100 mg/l.

ENVIRONMENTAL FATE
Soil/Environment Rapidly degraded in soil, DT_{50} 15–72 d. No adverse effects on soil micro-organisms.

$$CH_3SO_2NH$$

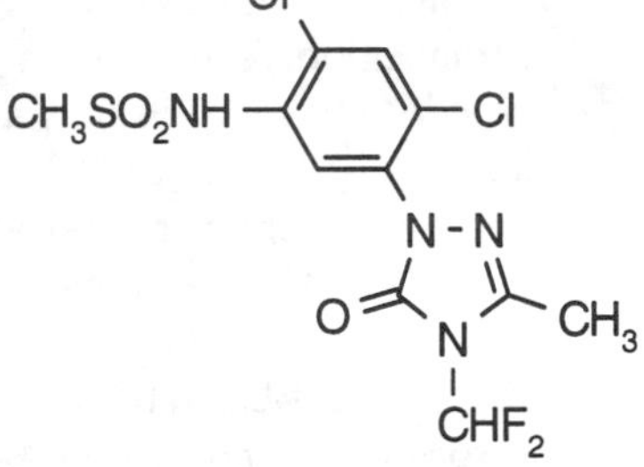

NOMENCLATURE
Common name sulfentrazone (BSI, pa E-ISO, ANSI)
IUPAC name 2',4'-dichloro-5'-(4-difluoromethyl-4,5-dihydro-3-methyl-5-oxo-=
1H-1,2,4-triazol-1-yl)methanesulfonanilide
Chemical Abstracts name N-[2,4-dichloro-5-[4-(difluoromethyl)-4,5-dihydro-=
3-methyl-5-oxo-1H-1,2,4-triazol-1-yl]phenyl]methanesulfonamide
CAS RN *[122836–35–5]* **Development codes** F6285; FMC 97285

PHYSICAL CHEMISTRY
Mol. wt. 387.2 **M.f.** $C_{11}H_{10}Cl_2F_2N_4O_3S$ **Form** Tan solid. **M.p.** 121–123 °C
V.p. 1.3×10^{-4} mPa (25 °C) **S.g./density** 1.21 g/ml (20 °C) **Solubility** In water
0.11 (pH 6), 0.78 (pH 7), 16 (pH 7.5) (all in mg/g, 25 °C). Soluble to some
extent in acetone and other polar organic solvents.

COMMERCIALISATION
History Reported by W. A. van Saun *et al.* (*Proc. Br. Crop Prot. Conf. – Weeds*,
1991, **1**, 77). Introduced by FMC Corp. **Manufacturers** FMC

APPLICATIONS
Biochemistry Protoporphyrinogen oxidase inhibitor (chlorophyll biosynthesis
pathway). **Mode of action** Herbicide absorbed by the roots and foliage, with
translocation primarily in the apoplasm, and limited movement in the phloem.
Uses Control of annual broad-leaved weeds, some grasses and *Cyperus* spp. in
soya beans. Applied pre-emergence or pre-plant incorporated.
Formulation types SC; WG. **Selected tradenames** 'Authority' (USA) (FMC);
'Boral' (Brazil) (FMC); 'Capaz' (Latin America) (FMC)

MAMMALIAN TOXICOLOGY
Oral Acute oral LD_{50} for rats 2855 mg/kg. **Skin and eye** Acute percutaneous
LD_{50} for rabbits >2000 mg/kg. Non-irritating to skin; mild eye irritant (rabbits).
Non-sensitising to skin (guinea pigs). **Inhalation** LC_{50} (4 h) for rats >4.14 mg/l.

NOEL 10 mg/kg daily (rat teratology study).   **Other** Non-mutagenic in the Ames test, mouse lymphoma and *in vivo* mouse micronucleus assay.

ECOTOXICOLOGY
Birds Acute oral LD_{50} for mallard ducks >2250 mg/kg. Eight-day dietary LC_{50} for ducks and quail >5620 mg/kg.   **Fish** LC_{50} (96 h) for bluegill sunfish 93.8, rainbow trout >130 mg/l.   **Daphnia** LC_{50} (48 h) 60.4 mg/l.

ENVIRONMENTAL FATE
Plants In soya beans, over 95% of the parent sulfentrazone is metabolised to the non-polar, hydroxymethyl analogue within 12 hours. This analogue is also rapidly converted, over the same time period, to three polar metabolites, two of which are glycosidic derivatives and one a non-glycoside metabolite.
Soil/Environment Moderately mobile in the soil. Degradation is primarily by microbial means. Not susceptible to photodecomposition. Fairly persistent in the soil. In the field, DT_{50} 110–280 d.

666   sulfluramid

Insecticide

$$CF_3(CF_2)_7SO_2NHCH_2CH_3$$

NOMENCLATURE
Common name sulfluramid (BSI, ANSI, draft E-ISO)
IUPAC name *N*-ethylperfluoro-octane-1-sulfonamide
Chemical Abstracts name *N*-ethyl-1,1,2,2,3,3,4,4,5,5,6,6,7,7,8,8,8-heptadecafluoro-1-octanesulfonamide
CAS RN *[4151–50–2]*   **Development codes** GX 071

PHYSICAL CHEMISTRY
Composition Tech. is 98.0%.   **Mol. wt.** 527.2   **M.f.** $C_{10}H_6F_{17}NO_2S$ **Form** Colourless crystals.   **M.p.** 96 °C; tech. 87–93 °C   **B.p.** 196 °C   **V.p.** 0.057 mPa (25 °C) K_{ow} logP = >6.8 (unionised)   **Solubility** Insoluble in water at 25 °C. In dichloromethane 18.6, hexane 1.4, methanol 833 (all in g/l).   **Stability** Stable >90 d at 50 °C; to light >90 d in closed jar.   **pKa** 9.5, v. weak acid   **F.p.** >93 °C (Seta flash)

COMMERCIALISATION
History Insecticide introduced in USA (1989) by Griffin Corp.   **Patent** US 4921696
Manufacturers Griffin

APPLICATIONS
Biochemistry Uncouples oxidative phosphorylation. **Uses** Used for household control of Formicidae and Blattellidae. **Formulation types** RB.
Selected tradenames 'Finitron' (Griffin); 'Volcano' (Griffin)

ANALYSIS
Product analysis by glc.

MAMMALIAN TOXICOLOGY
Oral Acute oral LD_{50} for rats >5000 mg/kg. **Skin and eye** Acute percutaneous LD_{50} for rabbits >2000 mg/kg. Mild skin irritant, practically non-irritating to eyes (rabbits). **Inhalation** LC_{50} (4 h) for rats >4.4 mg/l. **NOEL** (90 d) for male dogs 33, female dogs 100, rats 10 ppm.

ECOTOXICOLOGY
Birds Acute oral LD_{50} for bobwhite quail 45 mg/kg. LC_{50} (8 d) for bobwhite quail 300, mallard ducks 165 ppm diet. **Fish** LC_{50} (96 h) for fathead minnow >9.9, rainbow trout >7.99 mg/l. **Daphnia** LC_{50} (48 h) 0.39 mg/l.

667 sulfometuron-methyl *Herbicide*

sulfonylurea

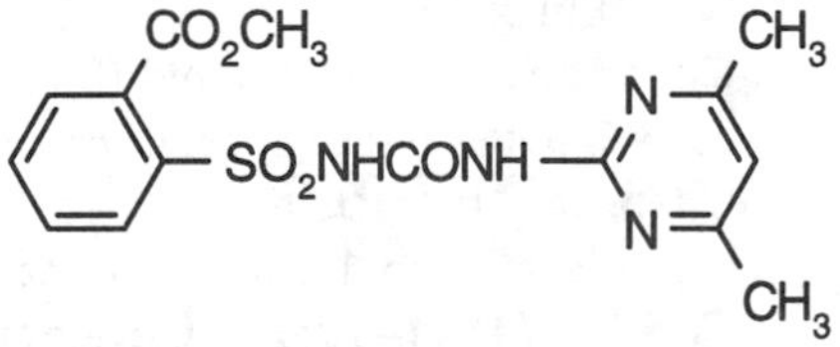

NOMENCLATURE
sulfometuron-methyl
CAS RN *[74222–97–2]* **Development codes** DPX T5648

sulfometuron
Common name sulfometuron (BSI, E-ISO, (*m*) F-ISO, ANSI, WSSA)
IUPAC name 2-(4,6-dimethylpyrimidin-2-ylcarbamoylsulfamoyl)benzoic acid; 2-[3-(4,6-dimethylpyrimidin-2-yl)ureidosulfonyl]benzoic acid
Chemical Abstracts name 2-[[[[(4,6-dimethyl-2-pyrimidinyl)amino]carbonyl]= amino]sulfonyl]benzoic acid
CAS RN *[74223–56–6]* **Development codes** DPX 5648

PHYSICAL CHEMISTRY
sulfometuron-methyl
Composition Tech. is >93%.   **Mol. wt.** 364.4   **M.f.** $C_{15}H_{16}N_4O_5S$
Form Colourless solid (tech.).   **M.p.** 203–205 °C   **V.p.** 7.3×10^{-11} mPa (25 °C)
K_{ow} logP = 1.18 (pH 5), –0.51 (pH 7)   **S.g./density** 1.48   **Solubility** In water
244 mg/l (pH 7, 25 °C). In acetone 3300, acetonitrile 1800, ethyl acetate 650,
diethyl ether 60, hexane <1, methanol 550, methylene chloride 15 000, dimethyl
sulfoxide 32 000, octanol 140, toluene 240 (all in mg/kg, 25 °C).
Stability Aqueous suspensions are stable to hydrolysis at pH 7 to pH 9, DT_{50}
c. 18 d (pH 5).   **pKa** 5.2

sulfometuron
Mol. wt. 350.3   **M.f.** $C_{14}H_{14}N_4O_5S$

COMMERCIALISATION
History The methyl ester introduced as a herbicide by E. I. du Pont de Nemours
and Co.   **Manufacturers** Du Pont

APPLICATIONS
sulfometuron-methyl
Biochemistry Branched chain amino acid synthesis (acetolactate synthase or ALS)
inhibitor. Acts by inhibiting biosynthesis of the essential amino acids valine and
isoleucine, hence stopping cell division and plant growth.   **Mode of action** Broad-
spectrum herbicide, absorbed rapidly by the leaves and roots, with translocation
throughout the plant.   **Uses** Control of annual and perennial grasses (especially
Sorghum halepense) and broad-leaved weeds in non-crop land. Also used for
selective weed control in Bermuda grass and certain other turf grasses; and in
forestry to control woody tree species in pine trees. Applied either pre- or post-
emergence.   **Phytotoxicity** Not to be applied near desirable trees and plants.
Formulation types WG.   **Compatibility** Incompatible with alkaline materials.
Selected tradenames 'Oust' (Du Pont)

ANALYSIS
Product and **residue** analysis by hplc (E. W. Zahnow & T. J. Waeghe, *Anal. Methods
Pestic. Plant Growth Regul.*, 1988, **16**, 105). Details available from Du Pont.

MAMMALIAN TOXICOLOGY
sulfometuron-methyl
Oral Acute oral LD_{50} for male rats >5000 mg/kg.   **Skin and eye** Acute
percutaneous LD_{50} for rabbits >2000 mg/kg; slight skin and eye irritant (rabbits).
No skin sensitisation (guinea pigs).   **Inhalation** LC_{50} (4 h) in rats >11 mg/l air.
NOEL (2 y) for rats 50 mg/kg diet. In 2-generation reproduction studies, NOEL
for rats 500 mg/kg diet. Teratogenicity effects in rats at 1000 mg/kg diet, and
in rabbits at 300 mg/kg diet.   **Toxicity class** WHO (a.i.) III (Table 5);
EPA (formulation) IV

ECOTOXICOLOGY
sulfometuron-methyl
Birds Acute oral LD_{50} for mallard ducks >5000, bobwhite quail >5620 mg/kg.
Fish LC_{50} (96 h) for rainbow trout and bluegill sunfish >12.5 mg/l. **Bees** Contact
LD_{50} >100 µg/bee. **Daphnia** LC_{50} >12.5 ppm.

ENVIRONMENTAL FATE
Animals Metabolised to hydroxylated sulfometuron-methyl.
Soil/Environment Degraded by microbial action and by hydrolysis. DT_{50} in soil
c. 4 w. K_{oc} 85.

668 sulfosulfuron *Herbicide*

sulfonylurea

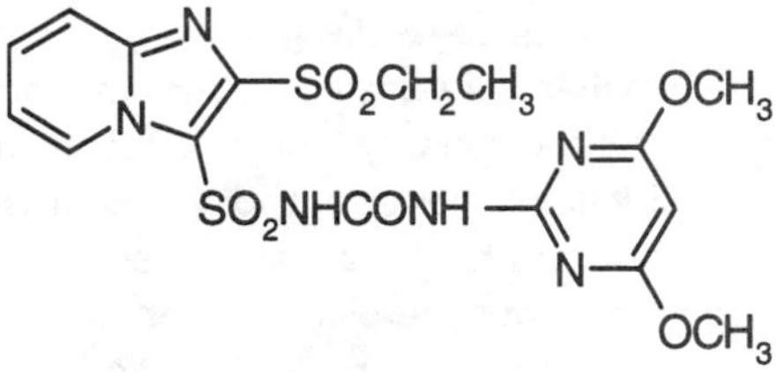

NOMENCLATURE
Common name sulfosulfuron (BSI, pa ISO)
IUPAC name 1-(4,6-dimethoxypyrimidin-2-yl)-3-(2-ethylsulfonylimidazo[1,2-a]=
pyridin-3-yl)sulfonylurea
Chemical Abstracts name *N*-[[(4,6-dimethoxy-2-pyrimidinyl)amino]carbonyl]-=
2-(ethylsulfonyl)imidazo[1,2-*a*]pyridine-3-sulfonamide
CAS RN *[141776–32–1]* **Development codes** MON 37500; TKM 19;
MON 37588 (for WG formulation)

PHYSICAL CHEMISTRY
Mol. wt. 470.48 **M.f.** $C_{16}H_{18}N_6O_7S_2$ **Form** White, odourless solid.
M.p. 201.1–201.7 °C **V.p.** <1 × 10^{-3} mPa **K$_{ow}$** logP <1 (pH 5 to 9 buffer)
Solubility In water 18 (pH 5), 1627 (pH 7), 482 (pH 9) ppm.

COMMERCIALISATION
History Reported by S. K. Parrish *et al.* (*Proc. Br. Crop Prot. Conf. – Weeds*, 1995,
1, 57). Under development by Monsanto Co. and Takeda Chemical Industries, Ltd.
Manufacturers Monsanto; Takeda

APPLICATIONS

Biochemistry By analogy with other sulfonylureas, assumed to be a branched chain amino acid synthesis (acetolactate synthase or ALS) inhibitor, acting by inhibiting biosynthesis of the essential amino acids valine and isoleucine, hence stopping cell division and plant growth. Selectivity derives from rapid metabolism in the crop. Metabolic basis of selectivity in sulfonylureas reviewed (M. K. Koeppe & H. M. Brown, *Agro-Food-Industry*, **6**, 9–14 (1995)). **Uses** Under development for control of grass and broad-leaved weeds in cereals (wheat) at between 10 and 35 g/ha. **Phytotoxicity** Barley and oats are sensitive. Tolerance of durum wheat is variety-specific. **Formulation types** WG.

ANALYSIS

Methods for sulfonylurea residues in crops, soil and water reviewed (A. C. Barefoot *et al.*, *Proc. Br. Crop Prot. Conf. – Weeds*, 1995, **2**, 707).

MAMMALIAN TOXICOLOGY

Oral Acute oral LD_{50} for rats >5000 mg/kg. **Skin and eye** Acute percutaneous LD_{50} for rats >5000 mg/kg. Not a skin irritant; moderate eye irritant (rabbits). Not a skin sensitiser (guinea pigs). **Inhalation** Practically non-toxic. **Toxicity class** EPA (formulation) III

ECOTOXICOLOGY

Birds 5-Day dietary LC_{50} for mallard ducks >5620 ppm. **Fish** LC_{50} (96 h) for rainbow trout >95, carp >91 mg/l. **Bees** LD_{50} (oral) >30, (dermal) >25 µg/bee. **Daphnia** LC_{50} (48 h) >96 mg/l.

ENVIRONMENTAL FATE

Soil/Environment DT_{50} 20–60 d. Despite rapid degradation, rotational injury to sensitive crops can be expected. See S. K. Parrish *et al.*, *Proc. Br. Crop Prot. Conf. – Weeds*, 1995, **1**, 667.

669   sulfotep   *Insecticide, acaricide*

organophosphorus

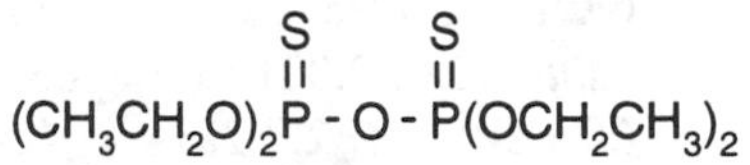

$$(CH_3CH_2O)_2\overset{\displaystyle S}{\overset{\|}{P}} - O - \overset{\displaystyle S}{\overset{\|}{P}}(OCH_2CH_3)_2$$

NOMENCLATURE

Common name sulfotep (BSI, E-ISO, (*m*) F-ISO); sulfotepp (ESA)
IUPAC name *O,O,O',O'*-tetraethyl dithiopyrophosphate

Chemical Abstracts name tetraethyl thiodiphosphate
Other names dithio; dithione; thiotep   **CAS RN** *[3689–24–5]*
EEC no. 222–995–2   **Development codes** Bayer E 393; ASP-47 (Victor Chemical Works)   **Official codes** ENT 16 273

PHYSICAL CHEMISTRY
Mol. wt. 322.3   **M.f.** $C_8H_{20}O_5P_2S_2$ **Form** Pale yellow liquid.
B.p. 136–139 °C/2 mmHg; 92 °C/0.1 mmHg   **V.p.** 14 mPa (20 °C)
K_{ow} logP = 3.99 (20 °C)   **Henry** 4.5 × 10^{-1} Pa m^3 mol^{-1} (20 °C)
S.g./density 1.196 (20 °C)   **Solubility** In water 10 mg/l (20 °C). Miscible with most organic solvents. Sparingly soluble in ligroin and petroleum ether.
Stability Relatively slowly hydrolysed; DT_{50} (estimated) 10.7 d (pH 4), 8.2 d (pH 7), 9.1 d (pH 9) (22 °C).   **F.p.** 102 °C

COMMERCIALISATION
History Insecticide discovered in 1944 by G. Schrader & H. Kükenthal (cited by G. Schrader, *Die Entwicklung insektizider Phosphorsaure-Ester*, 3rd Ed.). Introduced by Bayer AG.   **Patent** DE 848812   **Manufacturers** Bayer

APPLICATIONS
Biochemistry Cholinesterase inhibitor.   **Mode of action** Non-systemic insecticide and acaricide with contact and respiratory action.   **Uses** Fumigation control of aphids, thrips, spider mites, and whitefly on glasshouse crops; and sciarid flies and phorid flies on mushrooms.   **Phytotoxicity** Phytotoxic to some ornamentals, e.g. chrysanthemums, orchids, and some rose varieties.   **Formulation types** Fumigant; VP.   **Selected tradenames** 'Bladafum' (Bayer)

ANALYSIS
Product analysis by i.r. spectroscopy (details from Bayer AG, *CIPAC Proc.*, 1980, **2**, 221).   **Residues** determined by glc (G. Dräger, *Pflanzenschutz-Nachr. (Engl. Ed.)*, 1968, **21**, 359; *Anal. Methods Pestic. Plant Growth Regul.*, 1972, **6**, 483).   Methods for the determination of residues are available from Bayer.

MAMMALIAN TOXICOLOGY
EHC 63 (WHO, 1986; a general review of organophosphorus insecticides).
Oral Acute oral LD_{50} for rats c. 10 mg/kg.   **Skin and eye** Acute percutaneous LD_{50} (7 d) for rats 65 mg/kg; (4 h) for rats 262 mg/kg. Not irritating to eyes and skin (rabbits).   **Inhalation** LC_{50} (4 h) for rats c. 0.05 mg/l air (aerosol).
NOEL (2 y) for rats 10, mice 50 mg/kg diet; (13 w) for dogs 0.5 mg/kg diet.
ADI (BGA) 0.001 mg/kg b.w.   **Toxicity class** WHO (a.i.) Ia; EPA (formulation) I
EC risk T+ (R27/28)

ECOTOXICOLOGY
Birds No data; only used as a glasshouse fumigant.   **Fish** LC_{50} (96 h) for golden

orfe 0.071, rainbow trout 0.00361 mg/l. **Bees** No data; only used as a glasshouse fumigant. **Daphnia** LC_{50} (48 h) 0.002 mg/l. **Algae** E_rC_{50} for *Scenedesmus subspicatus* 7.2 mg/l.

ENVIRONMENTAL FATE
Animals Orally administered sulfotep was absorbed relatively quickly by rats. Elimination took place quickly via the kidney. The main elimination product is the diethylthiophosphate **Plants** Major metabolites are monothiotep, tetraethylpyrophosphate, diethylthiophosphoric acid and diethylphosphoric acid. **Soil/Environment** K_{oc} value and leaching studies indicate that sulfotep can be classified as immobile. Degraded quickly in different soils. Monothiotep and tetraethylpyrophosphate were very short-lived intermediate products which were degraded to simple thiophosphates or phosphates. In aqueous buffer solutions, sulfotep underwent hydrolysis at environmentally relevant pH levels and temperatures. A direct (primary) photochemical degradation in water could not be ascertained. Low tendency to volatilise.

570 sulfur

Fungicide, acaricide

NOMENCLATURE
Common name sulfur (E-ISO, JMAF, ESA, accepted in lieu of a common name); soufre (F-ISO, accepted in lieu of a common name)
IUPAC name sulfur
Chemical Abstracts name sulfur
CAS RN *[7704–34–9]* **EEC no.** 231–722–6

PHYSICAL CHEMISTRY
Mol. wt. 32.1 **M.f.** S_x **Form** Yellow powder, which can exist as various allotropic forms. **M.p.** 114.5 °C (rhombic 112.8 °C, monoclinic 119 °C) **B.p.** 444.6 °C **V.p.** 0.527 mPa (30.4 °C) (rhombic); 8.6 mPa (59.4 °C) **S.g./density** 2.07 (rhombic) **Solubility** Practically insoluble in water. Crystalline forms are soluble in carbon disulfide, but amorphous forms are not. Very slightly soluble in ether and petroleum ether; more readily soluble in hot benzene and acetone. **Stability** Very stable. In the presence of strong alkalis, sulfides can be formed.

COMMERCIALISATION
History Has been used as a pesticide for many years. **Manufacturers** Aimco; BASF; Crystal; Drexel; Elf Atochem; Excel; FMC; Novartis

APPLICATIONS
Mode of action Non-systemic fungicide with protective action. Secondary

acaricidal activity.   **Uses** Control of scab on apples, pears, and peaches; powdery mildews on a range of crops, including fruit, vines, hops, beet, cereals, ornamentals, cucumbers, vegetables, and in forestry; shot-hole of stone fruit; and acarinosis of vines. Also controls mites (particularly eriophyid mites) on a range of crops.   **Phytotoxicity** Phytotoxic, to some extent, to a number of crops, including cucurbits, apricots, raspberries, and certain other 'sulfur-shy' varieties. **Formulation types** DP; WP; SC; MG; WG.   **Mixtures** *(sulfur +)* nitrothal-isopropyl; rotenone; mancozeb; pyrifenox; thiram; triadimefon; triadimenol; vinclozolin; zineb; ziram; carbendazim + maneb; copper oxychloride + cymoxanil; carbendazim; folpet; copper oxychloride + zineb; parathion-methyl; dicofol + parathion-methyl; and many others.   **Compatibility** For reasons of phytotoxicity, mixing with oils should be avoided.   **Selected tradenames** 'Cosan' (AgrEvo); 'Kumulus DF' (BASF); 'Microthiol Special' (Elf Atochem); 'Rasulf' (Ramcides); 'Sulfex' (Excel); 'Sulphotox' (Aimco); 'That' (Stoller); 'Thiovit' (Novartis)

ANALYSIS
Product analysis is by conversion to sodium thiosulfate which is determined by titration (*CIPAC Handbook*, 1970, **1**, 632; *ibid.*, 1992, **E**, 202–210; *FAO Plant Prot. Bull.*, 1961, **9**(5), 80).   **Residues** may be determined by colorimetry (H. A. Ory et al., *Analyst (London)*, 1957, **82**, 189).

MAMMALIAN TOXICOLOGY
Skin and eye Acute percutaneous LD_{50} not available. Irritating to skin, eyes, and mucous membranes.   **Other** Practically non-toxic to humans and animals. **Toxicity class** WHO (a.i.) III (Table 5); EPA (formulation) IV

ECOTOXICOLOGY
Fish Non-toxic to fish.   **Bees** Non-toxic to bees.

ENVIRONMENTAL FATE
Plants Degradation proceeds primarily by microbial reduction in and on plants. **Soil/Environment** In the environment, slight oxidation to the volatile oxides.

671 sulfuric acid   *Herbicide, desiccant*

NOMENCLATURE
Common name sulfuric acid (E-ISO, accepted in lieu of a common name); acide sulfurique (F-ISO)
IUPAC name sulfuric acid
Chemical Abstracts name sulfuric acid

Other names [brown oil of vitriol]; [BOV] (for commercial grade)
CAS RN *[7664–93–9]*

PHYSICAL CHEMISTRY
Mol. wt. 98.08   **M.f.** H_2O_4S **Form** Pure sulfuric acid is a colourless, viscous liquid;
commercial acid is a brown, viscous liquid.   **S.g./density** 1.834 (pure);
1.675–1.710 (15 °C) (commercial acid)

COMMERCIALISATION
History Herbicidal activity reported by C. D. Woods & J. M. Bartlett (*Bull. Me.
Agric. Exp. Stn.,* 1909, No. 167).

APPLICATIONS
Uses Used by contractors for haulm destruction in potatoes and as a pre-harvest
desiccant for linseed and bulbs. Formerly used as a selective herbicide for onions
and cereals, pre-em. for weed control in horticultural crops and for pre-harvest
desiccation of leguminous seed crops.

ANALYSIS
Product analysis usually by determination of the density and reference to standard
tables; or, after cautious dilution with water, by titration with alkali.

672 sulfuryl fluoride   *Insecticide*

SO_2F_2

NOMENCLATURE
Common name sulfuryl fluoride (E-ISO, accepted in lieu of a common name)
IUPAC name sulfuryl fluoride
Chemical Abstracts name sulfuryl fluoride
CAS RN *[2699–79–8]*

PHYSICAL CHEMISTRY
Mol. wt. 102.1   **M.f.** F_2O_2S **Form** Colourless, odourless gas.   **M.p.** −136.7 °C
B.p. −55.2 °C/760 mmHg   **V.p.** 1.7×10^3 kPa (21.1 °C)   **S.g./density** 1.36
(20 °C)   **Solubility** In water 750 mg/kg (25 °C, 1 atmosphere). In ethanol
0.24–0.27, toluene 2.1–2.2, carbon tetrachloride 1.36–1.38 (all in l/l).
Stability Stable to light. Stable up to c. 500 °C when dry. Rapidly hydrolysed by
aqueous alkali, but not by water.

COMMERCIALISATION

History Insecticidal properties reported by E. E. Kenaga (*J. Econ. Entomol.*, 1957, **40**, 1). Introduced by Dow Chemical Co. (now DowElanco). **Patent** US 2875127; US 3092458 **Manufacturers** DowElanco

APPLICATIONS

Mode of action Fumigant insecticide. **Uses** Used as a fumigant for control of Blattodea, Coleoptera, Isoptera, Lepidoptera, and rodents. **Phytotoxicity** Phytotoxic, but with little effect on the germination of weed and crop seeds. **Formulation types** Fumigant. **Selected tradenames** 'Vikane' (DowElanco)

ANALYSIS

Product analysis by glc, by i.r. spectrometry or by an instrument that pyrolyses sulfuryl fluoride (measuring the resultant sulfur dioxide). Atmospheres analysed by trapping in aqueous alkali and titration of the fluorosulfate produced; or a thermal conductivity meter (S. G. Heuser, *Anal. Chem.*, 1963, **35**, 1476). Details of methods for total fluoride residues available from DowElanco. Methods reviewed by J. L. Daft in *Comp. Analyt. Profiles*, Chapter 11.

MAMMALIAN TOXICOLOGY

Oral Acute oral LD_{50} for rats 100 mg/kg. **Inhalation** LC_{50} (4 h) for male rats 1122, female rats 991 ppm; (1 h) for male rats 3730, female rats 3021 ppm. **NOEL** In 90 d inhalation trials, NOEL for rats and rabbits 30 ppm for exposure 6 h/d and 5 d/w. **Other** Humans are not exposed to significant levels when proper fumigation methods are used.

673 sulprofos   *Insecticide*

organophosphorus

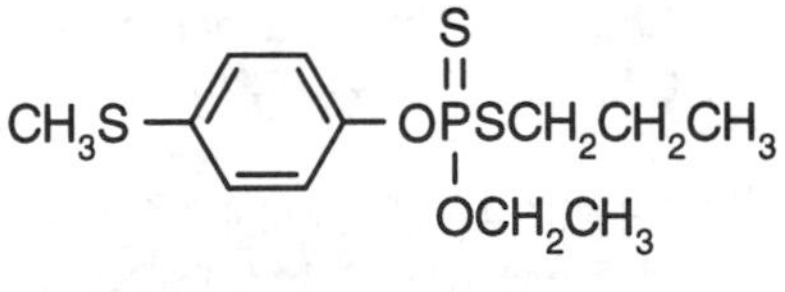

$$CH_3S-\langle\ \rangle-OPSCH_2CH_2CH_3$$

NOMENCLATURE

Common name sulprofos (BSI, E-ISO, (*m*) F-ISO, ESA)
IUPAC name *O*-ethyl *O*-4-(methylthio)phenyl *S*-propyl phosphorodithioate
Chemical Abstracts name *O*-ethyl *O*-[4-(methylthio)phenyl] *S*-propyl phosphorodithioate
CAS RN [35400–43–2] **Development codes** NTN 9306 (Bayer); BAY 123234

PHYSICAL CHEMISTRY

Mol. wt. 322.4   **M.f.** $C_{12}H_{19}O_2PS_3$ **Form** Colourless oil with a mercaptan-like odour; (tech., tan liquid).   **M.p.** −15 °C (tech.)   **B.p.** 125 °C/1 Pa   **V.p.** 0.084 mPa (20 °C); 0.16 mPa (25 °C)   **K_{ow}** logP = 5.48   **S.g./density** 1.20 (20 °C)   **Solubility** In water 0.31 mg/l (20 °C). In isopropanol 400–600, dichloromethane, hexane, toluene >1200 (all in g/l, 20 °C).   **Stability** Hydrolysis in buffer systems, DT_{50} (22 °C) 26 d (pH 4), 151 d (pH 7), 51 d (pH 9). Photodegraded in water and on soil surfaces. For films in sunlight, 50% decomposition occurs in <2 days.   **F.p.** 64 °C

COMMERCIALISATION

History Insecticide reported by G. Zoebelein (*Proc. Conf. Pest Control, Cairo, 4th,* 1978, p. 456). Introduced by Bayer AG.   **Patent** NL 6508899 (1965) to Bayer.   **Manufacturers** Bayer Corp.

APPLICATIONS

Biochemistry Cholinesterase inhibitor.   **Mode of action** Non-systemic insecticide with contact and stomach action.   **Uses** Control of Lepidoptera, Thysanoptera, *Heliothis* spp., and other insects in cotton, soya beans, tobacco, vegetables, and tomatoes.   **Formulation types** EC; UL.   **Selected tradenames** 'Bolstar' (Bayer)

ANALYSIS

Product analysis by glc (*AOAC Methods,* 1995, 980.12). Details of methods for determination of **product** and **residues** available from Bayer AG.

MAMMALIAN TOXICOLOGY

EHC 63 (WHO, 1986; a general review of organophosphorus insecticides).   **Oral** Acute oral LD_{50} for male rats 304, female rats 176, male and female mice c. 1700 mg/kg.   **Skin and eye** Acute percutaneous LD_{50} for male rats 5491, female rats 1064 mg/kg. Not irritating to skin, mild irritant to eyes (rabbits). Not a skin sensitiser.   **Inhalation** LC_{50} (4 h) for male and female rats >4130 μg/l air.   **NOEL** (2 y) for rats 6, mice 2.5, dogs 10 mg/kg diet.   **ADI** 0.003 mg/kg b.w.   **Toxicity class** WHO (a.i.) II; EPA (formulation) II   **EC risk** T (R25); Xn (R21)

ECOTOXICOLOGY

Birds Acute oral LD_{50} for bobwhite quail 47 mg/kg. Five-day dietary LC_{50} for bobwhite quail 99 mg/kg diet.   **Fish** LC_{50} (96 h) for bluegill sunfish 11–14, rainbow trout 23–38 mg/l.   **Daphnia** LC_{50} (48 h) 0.83–1 μg/l.   **Algae** E_rC_{50} for *Selenastrum capricornutum* 64 mg/l.

ENVIRONMENTAL FATE

Animals In rats, following oral administration, c. 92% is eliminated within 24 hours. The main degradation products are the oxygen-analogue, phenol, sulprofos sulfoxide and sulprofos sulfone, and the corresponding oxygen-analogues and phenols. **Plants** The major metabolites are sulprofos sulfoxide and sulprofos sulfone. The corresponding phenols and oxygen-analogues are minor degradation products. **Soil/Environment** The adsorption of sulprofos in various soil types occurred rapidly. The compound is highly adsorbed. The half-life of sulprofos is in the range of some days to several weeks, depending on the soil types. The major metabolites are sulprofos sulfoxide and sulprofos sulfone.

674 SZI-121   *Acaricide*

tetrazine

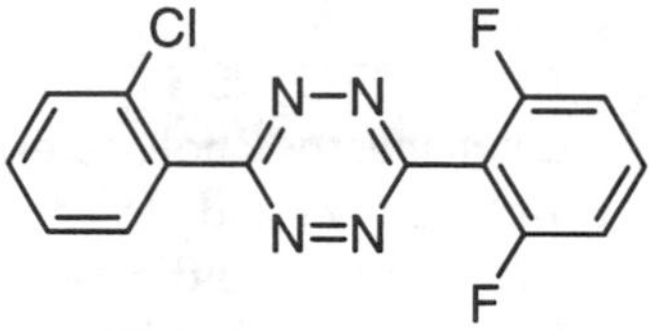

NOMENCLATURE
IUPAC name 3-(2-chlorophenyl)-6-(2,6-difluorophenyl)-1,2,4,5-tetrazine
Chemical Abstracts name 3-(2-chlorophenyl)-6-(2,6-difluorophenyl)-=
1,2,4,5-tetrazine
CAS RN *[162320–67–4]* **Development codes** SZI-121

PHYSICAL CHEMISTRY
Mol. wt. 304.7 **M.f.** $C_{14}H_7ClF_2N_4$ **Form** Magenta crystals. **M.p.** 187–189 °C
V.p. <10^{-2} mPa **K_{ow} logP** = 3.3 (25 °C) **S.g./density** 0.27 **Solubility** In water
0.23 mg/l (pH 7, 25 °C). In acetone 14.4, methanol 3.7, hexane 0.3,
dichloromethane 7.2, benzene 3.2, acetonitrile 13.7 (all in g/l, 25 °C).
Stability Stable to light and air; decomposes above melting point. Stable under
acidic conditions, but hydrolysed at pH >7; DT_{50} 60 h (pH 9, 25 °C, 40%
acetonitrile). **F.p.** 425 °C (closed cup)

COMMERCIALISATION
History Reported by L. Pap *et al.* (*Proc. Br. Crop Prot. Conf., Pests Dis.*, 1994, **1**,
75). **Patent** US 5455237; EP 635499; HU 212613 **Manufacturers** Chinoin

APPLICATIONS
Mode of action Contact ovicide also with translaminar activity, leading to ingestion
by adult mite; also prevents development of mites at chrysalis stages. **Uses** For
control of phytophagous mites, including *Panonychus, Tetranychus, Eriophyes, Aculus,
Vasates (Phyllocoptes)* and *Calepitrimerus* spp., in pome and stone fruits and vines,

at 60–100 g/ha.   **Formulation types** SC.   **Compatibility** Not compatible with alkaline pesticides.   **Selected tradenames** 'Flumite' (Chinoin)

ANALYSIS
Product and **residue** analysis by hplc. Details available from Chinoin Agchem Business Unit.

MAMMALIAN TOXICOLOGY
Oral Acute oral LD_{50} for male rats 979, female rats 594 mg/kg.
Skin and eye Acute percutaneous LD_{50} for male and female rats >2000 mg/kg. Eye irritation index (rabbits) 15, skin irritation index (rabbits) 0; not a skin sensitiser (guinea pigs).   **Inhalation** LC_{50} (4 h) for rats >5000 mg/m^3.
NOEL (3 mo) for rats and dogs 10 mg/kg b.w. daily; (28 d) for rats (oral) 50 mg/kg b.w. daily, (dermal) >500 mg/kg b.w. daily.   **Other** Negative in micronucleus and Ames tests.

ECOTOXICOLOGY
Birds Acute oral LD_{50} for Japanese quail >2000 mg/kg b.w.   **Fish** LC_{50} for rainbow trout >400 mg/l.   **Bees** LD_{50} (oral and contact) >25 µg/bee.
Worms LC_{50} for earthworms >1000 mg/kg dry soil.   **Other beneficial spp.** Not harmful to *Encarsia formosa* and *Phytoseiulus persimilis*.   **Daphnia** LC_{50} (48 h) 0.14 mg/l.   **Algae** Not toxic to *Selenastrum capricornutum*.

ENVIRONMENTAL FATE
Soil/Environment DT_{50} 44 d (acid sandy soil), 30 d (brown forest soil), 38 d (limecoated black soil).

675   tar oils   *Insecticide, herbicide, fungicide*

PHYSICAL CHEMISTRY
Composition Tar oils consist mainly of aromatic hydrocarbons, but contain components soluble in aqueous alkali: the 'phenols' or 'tar acids'; and nitrogenous bases soluble in dilute mineral acids: 'tar bases'. They are produced by the distillation of tars resulting from the high-temperature carbonisation of coal and of coke oven and blast furnace tars.   **Form** Brown to black liquids.   **S.g./density** 1.05–1.11 (15 °C)   **Solubility** Insoluble in water. Soluble in most organic solvents and in dimethyl sulfate.

COMMERCIALISATION
History Although they have been used for wood preservation since 1890, the

introduction of the formulated products known as tar oil washes for crop protection, using the heavy creosote and anthracene oil ranges, dates from c. 1920.

APPLICATIONS
Uses Control of overwintering stages (particularly eggs) of many insects, e.g. aphids, suckers, moths, mealybugs, and scale insects, and overwintering diseases on dormant pome, stone, cane, and bush fruit, and grape vines. Moss and lichen are also controlled. Also used as a hop defoliant, and as a game repellent in forestry.   **Phytotoxicity** Extremely phytotoxic, and so must only be used during the dormant season.   **Formulation types** EC; Emulsion.   **Mixtures** *(tar oil +)* phenols; diuron.

ANALYSIS
Product analysis is by standard methods *(MAFF Ref. Bk., 1958, No. 1, p. 81).*

MAMMALIAN TOXICOLOGY
Skin and eye Acute percutaneous LD_{50} not available. Irritating to skin, eyes and respiratory system. May cause dermatitis, especially in sunlight.

ECOTOXICOLOGY
Fish Toxic to fish.

tau-fluvalinate

See entry 362.

676   2,3,6-TBA *Herbicide*

benzoic acid (auxin)

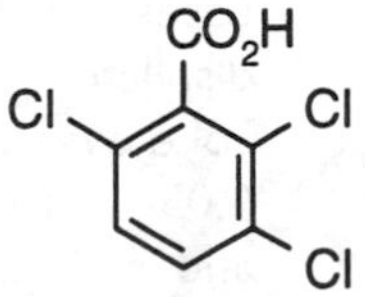

NOMENCLATURE
Common name 2,3,6-TBA (BSI, E-ISO, *(m)* F-ISO, WSSA); acide

trichlorobenzoique (optional alternative in France); TCBA (JMAF)
IUPAC name 2,3,6-trichlorobenzoic acid
Chemical Abstracts name 2,3,6-trichlorobenzoic acid
CAS RN *[50–31–7]*; (2,3,6-TBA-sodium *[2078–42–4]*)   **EEC no.** 200–026–4
Development codes HC-1281 (Heyden)

PHYSICAL CHEMISTRY
Composition Tech. grade is *c.* 60% 2,3,6-TBA. **Mol. wt.** 225.5 **M.f.** $C_7H_3Cl_3O_2$
Form Colourless crystals (tech., colourless to buff crystalline powder).
M.p. 125–126 °C (pure acid); *c.* 80–100 °C (tech.) **B.p.** Decomposes on
distillation.   **V.p.** Very low at room temperature; 3.2 Pa (100 °C) **Solubility** In
water 7.7 g/l (22 °C). Readily soluble in common organic solvents, e.g. acetone
60.7, benzene 23.8, chloroform 23.7, ethanol 63.7, methanol 71.7, xylene 21.0 (all
in g/100 ml).   **Stability** Stable to light, and to temperatures up to 60 °C.

COMMERCIALISATION
History Herbicide reported by H. J. Miller (*Weeds*, 1952, **1**, 185). Introduced by
the Heyden Chemical Corp. and later by E. I. du Pont de Nemours and Co. (who
now neither manufacture nor market it), and currently available in mixtures with
other herbicides.   **Patent** US 2848470; US 3081162

APPLICATIONS
Biochemistry Inhibits oxidative phosphorylation. Auxin type.
Mode of action Systemic growth-regulator herbicide with hormone-like action,
absorbed by the leaves and roots.   **Uses** Used post-emergence, in combinations
with other growth regulator herbicides, in cereals and grass seed crops to control
broad-leaved annual and perennial weeds including *Bilderdykia convolvulus*, *Galium
aparine*, *Polygonum aviculare*, *P. persicaria*, *Chamomilla* and *Matricaria* spp.
Mixtures *(2,3,6-TBA +)* MCPA; dichlorprop + MCPA; dichlorprop + mecoprop;
dicamba + MCPA + mecoprop; MCPA + mecoprop.   **Compatibility** Mixtures with
fertilisers are not recommended.

ANALYSIS
Product analysis by glc (*CIPAC Handbook*, 1985, **1C**, 2257; 1994, **F**, 292–319).
Residues determined by glc of a suitable derivative (J. J. Kirkland & H. L. Pease,
J. Agric. Food Chem., 1964, **12**, 468).

MAMMALIAN TOXICOLOGY
Oral Acute oral LD_{50} for rats 1500, mice 1000, guinea pigs >1500, rabbits
600 mg/kg.   **Skin and eye** Acute percutaneous LD_{50} for rats >1000 mg/kg.
NOEL In 64–69 day feeding trials, rats receiving 10 000 mg/kg diet suffered a
minor disturbance of water metabolism, no trace of which was apparent at
1000 mg/kg diet. It is largely excreted unchanged in the urine.
Toxicity class WHO (a.i.) III; EPA (formulation) III   **EC risk** Xn (R22)

ECOTOXICOLOGY
Birds Acute oral LD$_{50}$ for hens >1500 mg/kg. **Fish** Threshold value for toxic activity 100–150 mg/l for perch and 300 mg/l for roach. One of several determining factors is the content of the second, more toxic, 2,4,5-isomer in the active ingredient. **Bees** Not toxic to bees.

ENVIRONMENTAL FATE
Animals In mammals, excretion is mainly as the unchanged form in the urine.
Soil/Environment Very slow degradation in soil. A stable protein complex may possibly be formed.

677 TCA-sodium   *Herbicide*

halogenated alkanoic acid

$$Cl_3CCO_2Na$$

NOMENCLATURE
Common name TCA-sodium (for sodium trichloroacetate: BSI, E-ISO, (m) F-ISO (all since 1990); also Australia, Canada, New Zealand, USA, JMAF); TCA (for sodium trichloroacetate: BSI, E-ISO, (m) F-ISO (all until 1990, except for countries listed above)); trichloroacetic acid (for the acid: BSI, E-ISO, (m) F-ISO (all until 1990, except for countries listed following)); TCA (for the acid: BSI, E-ISO, (m) F-ISO (all since 1990); also Australia, Canada, New Zealand, USA, JMAF); trichloroacétate de sodium (for sodium trichloroacetate: France – as optional alternative to common name)

TCA-sodium
IUPAC name sodium trichloroacetate
Chemical Abstracts name sodium trichloroacetate
CAS RN *[650–51–1]* **EEC no.** 211–479–2

trichloroacetic acid
IUPAC name trichloroacetic acid
Chemical Abstracts name trichloroacetic acid
CAS RN *[76–03–9]* **EEC no.** 200–927–2

PHYSICAL CHEMISTRY
Composition The name TCA has been used variously to refer either to the acid or to the sodium salt (see Common name). The names used here are trichloroacetic acid and TCA-sodium.

TCA-sodium
Mol. wt. 185.4 **M.f.** $C_2Cl_3NaO_2$ **Form** Yellowish deliquescent powder. **M.p.** Decomposes at 165–200 °C **V.p.** <0.1 mPa (70 °C) **Solubility** In water 1.2 kg/l (25 °C). In methanol 232, acetone 7.6, diethyl ether 0.2, benzene 0.07, carbon tetrachloride 0.04, heptane 0.02 (all in g/l, 25 °C). **Stability** Stable under dry conditions, but decomposes in aqueous, strongly alkaline media.

trichloroacetic acid
Mol. wt. 163.4 **M.f.** $C_2HCl_3O_2$ **Form** Colourless hygroscopic crystals. **M.p.** 55–58 °C **B.p.** 196–197 °C **S.g./density** 1.62 **Solubility** In water 10 kg/l (25 °C). Soluble in diethyl ether, ethanol. **Stability** Stable in the absence of moisture; decomposes to chloroform under alkaline conditions.

COMMERCIALISATION
History Herbicidal properties of TCA-sodium described by K. C. Barrons & A. J. Watson (*North Cent. Weed Control Conf., Res. Rep.*, 1947, pp. 43, 284). Introduced by E. I. du Pont de Nemours and Co. and by Dow Chemical Co. (now DowElanco) (neither company now manufactures or markets it) and currently available from several suppliers.

APPLICATIONS
TCA-sodium
Mode of action Selective systemic herbicide, absorbed mainly by the roots, but also, to a lesser extent, by the foliage. Translocated readily, with accumulation in meristematic tissues. Causes non-selective leaf chlorosis. **Uses** TCA-sodium is a pre-emergence herbicide for control of many annual and perennial grasses (especially couch grass, wild oats, and volunteer cereals) in oilseed rape, sugar beet, fodder beet, asparagus, rhubarb, brassicas, carrots, potatoes, tomatoes, peas, field beans, lettuce, alfalfa, sunflowers, sugar cane, cotton, flax, rice, gladioli, and on non-crop areas, at up to 30 kg a.e./ha. **Formulation types** GR; SG; WG; SP. **Mixtures** (*TCA-sodium* +) atrazine; amitrole + atrazine; diquat dibromide + simazine; amitrole + 2,4-D + MCPA; amitrole + 2,4-D + sodium chlorate. **Selected tradenames** 'Erbitox T95G' (Siapa)

ANALYSIS
Product analysis by decarboxylation with sulfuric acid followed by titration of the excess of acid (*CIPAC Handbook*, 1970, **1**, 691; *AOAC Methods*, 1995, 962.08). **Residues** determined by glc of a derivative with ECD.

MAMMALIAN TOXICOLOGY
TCA-sodium
Oral Acute oral LD_{50} for rats 3200–5000, mice 3600–5600 mg/kg.
Skin and eye Acute percutaneous LD_{50} for rats >2000 mg/kg. Irritant to skin, eyes, and mucous membranes. Corrosive to skin on prolonged contact.

Inhalation LC_{50} (4 h) for rats >365 mg/l air.   **Toxicity class** WHO (a.i.) III
(Table 5); EPA (formulation) III   **EC risk** Xn (R22)

trichloroacetic acid
Oral Acute oral LD_{50} for rats 400 mg/kg.   **Skin and eye** Acute percutaneous LD_{50}
not available. Extremely corrosive to skin.   **Toxicity class** WHO (a.i.) II
EC risk C (R35) (also applies to preparations with concn. ≥10%; for
5%≤ concn. <10%, C (R34); for 1%≤ concn. <5%, Xi (R36/38))

ECOTOXICOLOGY
TCA-sodium
Birds Acute oral LD_{50} for chickens 4280 mg/kg.   **Fish** Not toxic to fish. Threshold
value for toxic activity *c.* 11 000 mg/l for perch.   **Bees** Not toxic to bees.

ENVIRONMENTAL FATE
Soil/Environment In soil, microbial degradation involves chloride formation, but is
slower than for dalapon-sodium. TCA-sodium is not tightly adsorbed, and may be
leached by heavy rainfall. Persists in soil for *c.* 21 to 90 days, depending upon rate
applied, soil moisture, and temperature.

678 tebuconazole   *Fungicide*

azole

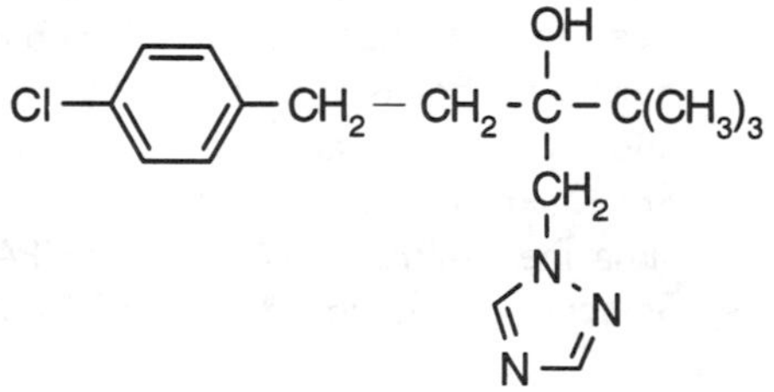

NOMENCLATURE
Common name tebuconazole (BSI, draft E-ISO)
IUPAC name (*RS*)-1-*p*-chlorophenyl-4,4-dimethyl-3-(1*H*-1,2,4-triazol-1-ylmethyl)=
pentan-3-ol
Chemical Abstracts name (±)-α-[2-(4-chlorophenyl)ethyl]-α-(1,1-dimethylethyl)-=
1*H*-1,2,4-triazole-1-ethanol
Other names [fenetrazole]; [terbuconazole]; [terbutrazole]; [ethyltrianol]
CAS RN *[107534–96–3]*   **EEC no.** ELINCS: 403–640–2
Development codes HWG 1608 (Bayer)

PHYSICAL CHEMISTRY

Composition Racemate. **Mol. wt.** 307.8 **M.f.** $C_{16}H_{22}ClN_3O$ **Form** Colourless crystals (tech., a colourless to light brown powder). **M.p.** 105 °C
V.p. 1.7×10^{-3} mPa (20 °C, OECD 104)) **K$_{ow}$** logP = 3.7 (20 °C)
Henry 1×10^{-5} Pa m^3 mol^{-1} (20 °C) **S.g./density** 1.25 (26 °C) **Solubility** In water 36 mg/l (pH 5–9, 20 °C). In dichloromethane >200, isopropanol, toluene 50–100, hexane <0.1 (all in g/l, 20 °C). **Stability** Stable to elevated temperatures, and to photolysis and hydrolysis in pure water, under sterile conditions; hydrolysis DT$_{50}$ >1 y (pH 4–9, 22 °C). See also Environmental Fate.

COMMERCIALISATION

History Fungicide reported by Kuck & Berg (*Mitt. Biol. Bundesanstalt. Land-u-Forstwirtsch. Berlin-Dahlem*, 1986, **232**, 196). Introduced by Bayer AG. **Patent** EP 40345; US 4723984 **Manufacturers** Bayer Corp.

APPLICATIONS

Biochemistry Steroid demethylation (ergosterol biosynthesis) inhibitor.
Mode of action Systemic fungicide with protective, curative, and eradicant action. Rapidly absorbed into the vegetative parts of the plant, with translocation principally acropetally. **Uses** As a **seed dressing**, tebuconazole is effective against various smut and bunt diseases of cereals such as *Tilletia* spp., *Ustilago* spp., and *Urocystis* spp., also against *Septoria nodorum* (seed-borne), and *Sphacelotheca reiliana* in maize. As a **spray**, tebuconazole controls numerous pathogens in various crops: rust species (*Puccinia* spp.), powdery mildew (*Erysiphe graminis*), *Rhynchosporium secalis*, *Septoria* spp., *Pyrenophora* spp., *Cochliobolus sativus* and *Fusarium* spp. in cereals; *Mycosphaerella* spp., *Puccinia* spp. and *Sclerotium rolfsii* in peanuts; *Mycosphaerella* spp. in bananas; *Sclerotinia sclerotiorum* and various pathogens of leaf and stem diseases in oilseed rape; *Exobasidium vexans* in tea; *Phakopsora pachyrhizi* in soya beans; *Monilinia* spp., rust species, powdery mildew and scab in pome and stone fruit; *Botrytis* spp., rust species, powdery mildew fungi, and (with dipping or spraying) *Sclerotium cepivorum* in grapes and some vegetable crops. **Phytotoxicity** Good plant compatibility in most crops with any formulation and achieved in more sensitive crops by appropriate formulations, e.g. WP, WG or SC. **Formulation types** EC; EW; WP; WG; FS; DS; WS; ES; SC.
Mixtures Sprays: *(tebuconazole +)* triadimenol; triadimefon; carbendazim; dichlofluanid; tolylfluanid; fenpropidin; prochloraz; propiconazole; tridemorph. Seed dressings *(tebuconazole +)* triazoxide; imidacloprid; imazalil; guazatine; thiram; fludioxonil; captan; anthraquinone; cypermethrin; triflumuron; cyprodinil; difenoconazole. **Selected tradenames** 'Elite' (spray) (Bayer Corp.); 'Folicur' (spray) (Bayer); 'Horizon' (spray) (Bayer); 'Lynx' (spray) (Bayer Corp.); 'Matador' (spray) (mixture) (Bayer); 'Raxil' (seed treatment) (Bayer); 'Silvacur' (spray) (Bayer, Nihon Bayer)

ANALYSIS
Residues in soil, water and plant material determined by glc (W. Maasfeld, *Pflanzenschutz-Nachr. Bayer (Engl. Edn.),* 1987, **40,** 29; H. Allmendinger, *ibid.,* 1991, **44**, 5). Details available from Bayer.

MAMMALIAN TOXICOLOGY
Reviews *Pesticide residues in food – 1994,* FAO Plant Production and Protection Paper, 127, 1994. *Pesticide residues in food – 1994 evaluations, Pt. II – Toxicology,* WHO, WHO/PCS/95.2, 1995. **Oral** Acute oral LD_{50} for male rats 4000, female rats 1700, mice *c.* 3000 mg/kg. **Skin and eye** Acute percutaneous LD_{50} for rats >5000 mg/kg. Non-irritating to skin, mild irritant to eyes (rabbits). **Inhalation** LC_{50} (4 h) for rats 0.37 mg/l air (aerosol), >5.1 mg/l (dust). **NOEL** (2 y) for rats 300, dogs 100, mice 20 mg/kg diet. **ADI** (JMPR) 0.03 mg/kg b.w. [1994]. **Toxicity class** WHO (a.i.) III; EPA (formulation) II ('Elite' 45DF); III ('Folicur' 3.6F) **EC risk** Xn (R22)

ECOTOXICOLOGY
Birds Acute oral LD_{50} for male Japanese quail 4438, female Japanese quail 2912, bobwhite quail 1988 mg/kg. **Fish** LC_{50} (96 h) for rainbow trout 6.4, golden orfe 8.7 mg/l. **Bees** Non-toxic to bees. **Worms** Acute (14 d) for *Eisenia foetida* 1.381 mg/kg dry soil. **Other beneficial spp.** According to the available data, no harmful effects on beneficials are to be expected under field conditions. **Daphnia** LC_{50} (48 h) 11.5 mg/l. **Algae** E_rC_{50} (96 h) for *Scenedesmus subspicatus* 4.01 mg/l.

ENVIRONMENTAL FATE
Animals After three days, elimination was almost complete (>99%). Tebuconazole was excreted with the urine and the faeces. **Plants** Metabolism studies on wheat, grape and peanut show that tebuconazole is the major terminal residue. The metabolites detected were mainly triazole-containing compounds of no toxicological relevance. In plant tissue, a mean half-life of 12 days could be derived (cereals). **Soil/Environment** The degradation of tebuconazole in soil was slow in laboratory studies. Under field conditions, the compound degraded much more rapidly, and did not accumulate in long-term studies (3–5 y). Since no residues could be detected in deeper soil layers of these and other studies, and adsorption/desorption studies indicated a low mobility in the soil, groundwater contamination through leaching can be excluded. In natural waters, hydrolysis and indirect photolysis occur; in a pond study, the compound dissipated from the water body with DT_{50} 1–3 w. Low vapour pressure and strong adsorption result in low volatilisation into the air.

diacylhydrazine

CH_3CH_2—⟨benzene⟩—$CONHNCO$—⟨benzene with CH_3, CH_3⟩
$C(CH_3)_3$

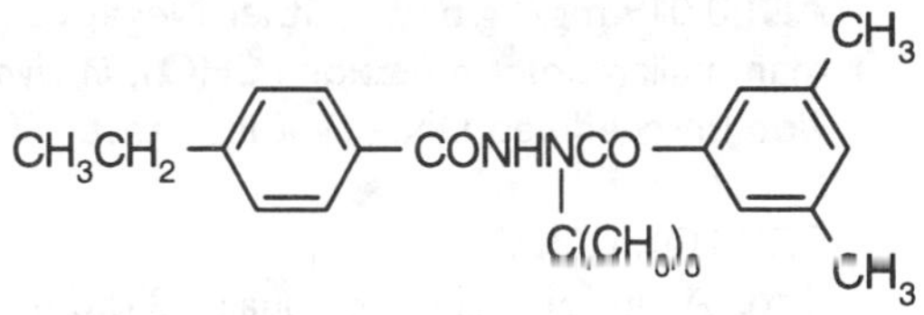

NOMENCLATURE
Common name tebufenozide (BSI, pa E-ISO, ANSI)
IUPAC name *N* tert butyl *N'* (4 ethylbenzoyl) 3,5 dimethylbenzohydrazide
Chemical Abstracts name 3,5-dimethylbenzoic acid 1-(1,1-dimethylethyl)-=
2-(4-ethylbenzoyl)hydrazide
CAS RN *[112410–23–8]*   **Development codes** RH-5992; RH-75922

PHYSICAL CHEMISTRY
Mol. wt. 352.5   **M.f.** $C_{22}H_{28}N_2O_2$ **Form** Off-white powder.   **M.p.** 191 °C
V.p. 3.0×10^{-3} mPa (25 °C, gas saturation method) K_{ow} logP = 4.25 (pH 7)
S.g./density 1.03 (20 °C, pycrometer method)   **Solubility** In water at 25 °C,
0.83 ppm. Slightly soluble in organic solvents.   **Stability** Stable at 94 °C for 7 d.
Stable to light in pH 7 aqueous solution (25 °C). Stable in dark, sterile water 30 d,
(25 °C). DT_{50} in natural pond water 67 d, in light, 30 d (25 °C).

COMMERCIALISATION
History Reported by J. J. Heller *et al.* (*Proc. Br. Crop Prot. Conf. – Pests Dis.*, 1992,
1, 59). Being developed in USA by Rohm & Haas Co., under licence in Europe by
AgrEvo GmbH and in Japan by Hokko Chemical Industry Co., Ltd and Nihon
Nohyaku Co., Ltd.   **Manufacturers** Rohm & Haas

APPLICATIONS
Biochemistry Ecdysone agonist which acts by binding to the receptor site of the
insect moulting hormone, ecdysone.   **Mode of action** Lethally accelerates
moulting process.   **Uses** Control of lepidopteran larvae on rice, fruit, row crops,
nut crops, vegetables, vines, and forestry.   **Formulation types** WP; SC; SU; DP;
GR.   **Selected tradenames** 'Confirm' (Rohm & Haas); 'Mimic' (forestry) (Rohm &
Haas)

ANALYSIS
Details from Rohm & Haas.

MAMMALIAN TOXICOLOGY
Oral Acute oral LD_{50} for rats and mice >5000 mg/kg.   **Skin and eye** Acute
percutaneous LD_{50} for rats >5000 mg/kg. Non-irritating to eyes and skin

(rabbits). Not a skin sensitiser (guinea pigs).   **Inhalation** LC_{50} (4 h) for male rats >4.3, female rats >4.5 mg/l.   **NOEL** (24 mo) for rats 5.5 mg/kg b.w. daily; (18 mo) for mice 8.1 mg/kg b.w. daily; (12 mo) for dogs 1.9 mg/kg b.w. daily. **ADI** 0.019 mg/kg b.w.   **Other** Negative in the Ames test, reverse mutation assay, mammalian point mutation (CHO), *in vivo* and *in vitro* cytogenetic assay, and *in vitro* unscheduled DNA synthesis test.   **Toxicity class** EPA (formulation) III (2 SC)

ECOTOXICOLOGY
Birds Acute oral LD_{50} for quail >2150 mg/kg. Eight-day dietary LC_{50} for mallard ducks and quail >5000 mg/kg.   **Fish** LC_{50} (96 h) for rainbow trout 5.7, bluegill sunfish 3.0 mg/l.   **Bees** LD_{50} (96 h, contact) for honeybees >234 µg/bee. **Worms** LC_{50} for earthworms >1000 mg/kg.   **Other beneficial spp.** Safe to predatory mites, wasps and other beneficial species.   **Daphnia** LC_{50} (48 h) 3.8 mg/l.   **Algae** EC_{50} (120 h) for *Selenastrum* >0.64 mg/l; (96 h) for *Scenedesmus* 0.23 mg/l.   **Other aquatic spp.** LC_{50} (96 h) for mysid shrimp (*Mysidopsis bahia*) 1.4, Eastern oyster (*Crassostrea virginica*) 0.64 mg/l.

ENVIRONMENTAL FATE
Animals In the rat, 16 whole-molecule metabolites are formed as a result of oxidation of the alkyl substituents of the aromatic rings, primarily at the benzylic positions.   **Plants** In apples, grapes, rice and sugar beet, the major component is unchanged tebufenozide. Metabolites which are detected in small amounts result from oxidation of the alkyl substituents of the aromatic ring, primarily at the benzylic position.   **Soil/Environment** Metabolic DT_{50} in soil 66 d; for aerobic, aquatic soil 100 d (25 °C, 3 soil types); for anaerobic aquatic metabolism 179 d (25 °C, silt loam). DT_{50} for field dissipation 4–53 d (4 sites). K_{oc} 351–894. Field dissipation studies indicate no mobility below 30 cm.

680   tebufenpyrad   *Acaricide*

pyrazole (acaricide)

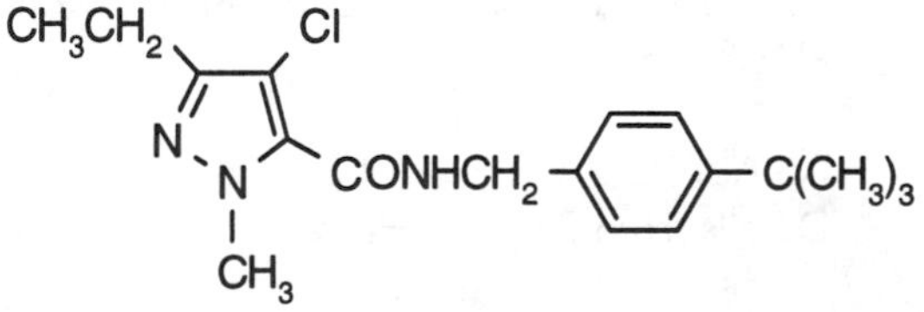

NOMENCLATURE
Common name tebufenpyrad (BSI, draft ISO)

IUPAC name *N*-(4-*tert*-butylbenzyl)-4-chloro-3-ethyl-1-methylpyrazole-=
5-carboxamide
Chemical Abstracts name 4-chloro-*N*[[4-(1,1-dimethylethyl)phenyl]methyl]-=
3-ethyl-1-methyl-1*H*-pyrazole-5-carboxamide
CAS RN *[119168–77–3]*   **Development codes** AC 801757 (Cyanamid);
MK-239 (Mitsubishi)

PHYSICAL CHEMISTRY
Mol. wt. 333.8   **M.f.** $C_{18}H_{24}ClN_3O$ **Form** Colourless crystals.   **M.p.** 61–62 °C
V.p. $<1 \times 10^{-2}$ mPa (25 °C) K_{ow} logP = 4.61 (25 °C)   **S.g./density** 1.0214
Solubility In water at 25 °C, 2.6 mg/l. Soluble in acetone, methanol, chloroform,
acetonitrile, hexane, and benzene.   **Stability** Stable to hydrolysis, DT_{50} >28 d
(pH 5, 7 and 9).

COMMERCIALISATION
History Discovered by Mitsubishi Kasei (now Mitsubishi Chemical Corp.) and
developed in collaboration with American Cyanamid Co.   **Patent** US 4950668;
EP 289879.   **Manufacturers** Mitsubishi Chemical

APPLICATIONS
Biochemistry Mitochondrial respiration inhibitor. Acts as an inhibitor of the
electron transport chain at Site I. Does not act as a specific muscle or nerve
poison.   **Mode of action** Non-systemic acaricide active by contact and ingestion.
Exhibits translaminar movement following application to leaves, and thus inhibits
the development of mite eggs oviposited on the undersides of leaves.
Uses Control of all stages of *Tetranychus*, *Panonychus*, *Origonychus*, and
Eotetranychus spp. on top fruit, vines, citrus, vegetables, hops, ornamentals,
melons, and cotton. Applied at 3.3–10 g a.i./100 l.   **Formulation types** EC; WP;
EW; WG.   **Mixtures** (*tebufenpyrad* +) bifenthrin.
Selected tradenames 'Comanché' (Cyanamid); 'Masai' (Cyanamid); 'Oscar'
(Cyanamid); 'Pyranica' (Mitsubishi)

ANALYSIS
Details from Mitsubishi.

MAMMALIAN TOXICOLOGY
Oral Acute oral LD_{50} for male rats 595, female rats 997, male mice 224, female
mice 210 mg/kg.   **Skin and eye** Acute percutaneous LD_{50} for rats >2000 mg/kg.
Non-irritating to skin, slightly irritating to eyes (rabbits). Not a skin sensitiser
(guinea-pig).   **Inhalation** LC_{50} for male rats 2660, female rats >3090 mg/m^3.
Other No mutagenic effects were observed in the following tests; Ames,
mammalian micronucleus, *Drosophila* wing spot, *in vitro* cultured human
lymphocytes, *in vivo* bone marrow erythrocytes, unscheduled DNA synthesis, and
CHO/HGPRT.   **EC risk** Xn (R20/22)

ECOTOXICOLOGY

Birds Acute oral LD_{50} for mallard ducks >2000 mg/kg. Eight-day dietary LC_{50} for mallard ducks and bobwhite quail >5000 mg/kg diet.   **Fish** LC_{50} (48 h) for carp 0.073 mg/l.   **Bees** Low toxicity to honeybees.   **Daphnia** LC_{50} (3 h) 1.2 mg/l.

ENVIRONMENTAL FATE

Animals Metabolite is N-[4-(1-hydroxymethyl-1-methylethyl)benzyl]-4-chloro-= 3-(1-hydroxyethyl)-1-methylpyrazole-5-carboxamide.   **Plants** As for animals.
Soil/Environment Aerobic degradation occurs in soil, DT_{50} 20–30 d.
K_{oc} 1380–4930.

681   tebupirimfos   *Insecticide*

organophosphorus

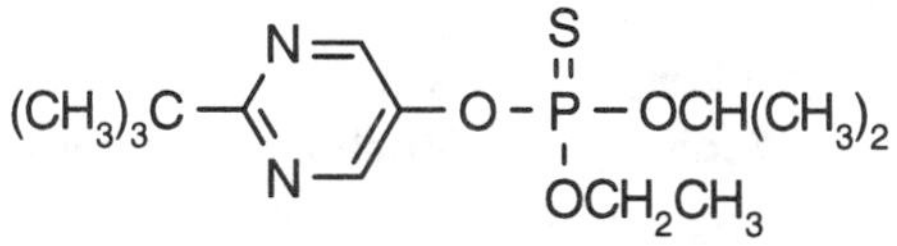

NOMENCLATURE

Common name tebupirimfos (BSI, pa-ISO)
IUPAC name O-(2-*tert*-butylpyrimidin-5-yl) O-ethyl O-isopropyl phosphorothioate
Chemical Abstracts name O-[2-(1,1-dimethylethyl)-5-pyrimidinyl] O-ethyl
O-(1-methylethyl) phosphorothioate
CAS RN *[96182–53–5]*   **Development codes** BAY MAT 7484

PHYSICAL CHEMISTRY

Mol. wt. 318.4   **M.f.** $C_{13}H_{23}N_2O_3PS$ **Form** Colourless to amber crystals.
B.p. 135 °C/1.5 mmHg; 152 °C/760 mmHg   **V.p.** 5 mPa (20 °C)   **Solubility** In water 5.5 mg/l (20 °C, pH 7). Soluble in most organic solvents, e.g. alcohols, ketones and toluene.   **Stability** Hydrolysed under alkaline conditions.

COMMERCIALISATION

History Reported by J. Hartwig *et al.* (*Proc. Br. Crop Prot. Conf. – Pests Dis.*, 1992, **1**, 35). Introduced by Bayer Corp.   **Manufacturers** Bayer

APPLICATIONS

Biochemistry Cholinesterase inhibitor.   **Mode of action** Insecticide with contact action and good residual activity.   **Uses** Control of soil insects, especially *Diabrotica* species, wireworms and Diptera maggots.   **Formulation types** GR.
Mixtures *(tebupirimfos +)* cyfluthrin.   **Selected tradenames** 'Aztec' (Bayer Corp.)

MAMMALIAN TOXICOLOGY
EHC 63 (WHO, 1986; a general review of organophosphorus insecticides).
Oral Acute oral LD_{50} for male rats 2.9–3.6, female rats 1.3–1.8, male mice 14.0,
female mice 9.3 mg/kg. **Skin and eye** Acute percutaneous LD_{50} for male rats
31.0, female rats 9.4 mg/kg. **Inhalation** LC_{50} (4 h) for male rats c. 82, female rats
c. 36 mg/m^3 air (aerosol). **NOEL** (2 y) for rats 1, mice 0.3, dogs 0.7 mg/kg diet.
ADI 0.0002 mg/kg b.w. **Other** Non-mutagenic and non-teratogenic; no evidence
of carcinogenic potential. **Toxicity class** WHO (a.i.) Ia

ECOTOXICOLOGY
Birds LD_{50} for bobwhite quail 20.3 mg/kg. LC_{50} (5 d) for mallard ducks 577,
bobwhite quail 191 mg/kg. **Fish** LC_{50} (96 h) for rainbow trout 2250, golden orfe
2550 mg/kg. **Daphnia** LC_{50} (96 h) 0.078 µg/l. **Algae** E_rC_{50} (96 h, 23 °C) for
Scenedesmus subspicatus 1.8 mg/l.

682 tebutam *Herbicide*

amide

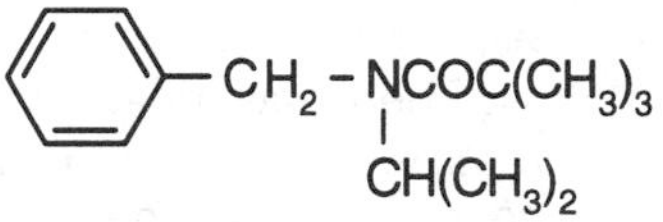

CH$_2$ – NCOC(CH$_3$)$_3$
|
CH(CH$_3$)$_2$

NOMENCLATURE
Common name tebutam (BSI, draft E-ISO); tébutame ((m) draft F-ISO); butam
(ANSI, WSSA); no name (Japan)
IUPAC name *N*-benzyl-*N*-isopropylpivalamide
Chemical Abstracts name 2,2-dimethyl-*N*-(1-methylethyl)-*N*-(phenylmethyl)=
propanamide
CAS RN *[35256–85–0]* **EEC no.** 252–470–3 **Development codes** GPC-5544
(Gulf); Ro 14–9480/000 (Roche); CGA 39625 (Ciba-Geigy)

PHYSICAL CHEMISTRY
Mol. wt. 233.4 **M.f.** $C_{15}H_{23}NO$ **Form** Light brown liquid.
B.p. 95–97 °C/0.1 mmHg **V.p.** c. 89 mPa (25 °C) **K_{ow}** logP = 3
Henry 1.5×10^{-2} Pa m^3 mol^{-1} (calc.) **S.g./density** 0.975 (20 °C) **Solubility** In
water 0.79 g/l (pH 7, 20 °C). Readily soluble in organic solvents, e.g. acetone,
hexane, methanol, toluene, chloroform >500 (all in g/l, 25 °C).
Stability Chemically stable under normal storage conditions. Thermally stable at
temperatures below the boiling point. Stable to light. Stable to hydrolysis at pH 5,
pH 7 and pH 9 (all at 25 °C). **F.p.** >80 °C (with 2% toluene)

COMMERCIALISATION
History Herbicide reported by R. A. Schwartzbeck (*Proc. Br. Crop Prot. Conf. – Weeds*, 1976, **2**, 739). Introduced by Gulf Oil Chemicals Co. and later acquired by Dr R. Maag Ltd (now Novartis Crop Protection AG). **Patent** US 3974218; US 3707366 to Gulf **Manufacturers** Novartis

APPLICATIONS
Biochemistry Cell division inhibitor. **Mode of action** Selective herbicide, acting by inhibition of weed germination and also through root uptake. **Uses** Pre-emergence control of annual grasses, volunteer cereals and broad-leaved weeds in oilseed rape and brassicas. **Phytotoxicity** A large number of crops are susceptible, including cereals, maize, carrots, flax, onions, garlic, lettuce, melons, sorghum, spinach, and sugar beet. **Formulation types** EC. **Mixtures** (*tebutam* +) clomazone. **Selected tradenames** 'Colzor' (mixture) (Novartis); 'Comodor' (Novartis)

ANALYSIS
Product analysis by hplc. **Residues** determined by glc with FID or by hplc.

MAMMALIAN TOXICOLOGY
Oral Acute oral LD_{50} for albino rats 6210, guinea pigs 2025 mg/kg.
Skin and eye Acute percutaneous LD_{50} for rats and albino rabbits >2000 mg/kg; a slight irritant to the eyes and skin of rabbits. **Inhalation** LC_{50} for rats >2.18 mg/l air. There was no adverse effect when albino rats were exposed to an aerosol (66% vol. mixture of tech. product and acetone). **NOEL** (2 y) for rats 600, mice 3000 ppm. **ADI** 0.15 mg/kg b.w. **Toxicity class** WHO (a.i.) III (Table 5)

ECOTOXICOLOGY
Birds Eight-day dietary LC_{50} for mallard ducks and bobwhite quail >5000 mg/kg.
Fish LC_{50} (96 h) for rainbow trout 23, bluegill sunfish 19, moonfish 18.7 mg/l.
Bees Acute oral LD_{50} 100 μg/bee. **Algae** LC_{50} (96 h) for *Scenedesmus* 82 mg/l.

ENVIRONMENTAL FATE
Animals In rats, following oral administration, rapidly excreted in the urine and faeces. **Plants** In plants, tebutam is rapidly metabolised.
Soil/Environment Bound by the organic matter of soil; DT_{50} c. 2 mo.

urea

$(CH_3)_3C$ ⟨S⟩ $N-N$ with CH_3 / $NCONHCH_3$

NOMENCLATURE
Common name tebuthiuron (BSI, E-ISO, (m) F-ISO, ANSI, WSSA)
IUPAC name 1-(5-*tert*-butyl-1,3,4-thiadiazol-2-yl)-1,3-dimethylurea
Chemical Abstracts name N-[5-(1,1-dimethylethyl)-1,3,4-thiadiazol-2-yl]-=
N,N'-dimethylurea
CAS RN *[34014–18–1]* **EEC no.** 251–793–7 **Development codes** EL-103

PHYSICAL CHEMISTRY
Mol. wt. 228.3 **M.f.** $C_9H_{16}N_4OS$ **Form** Colourless, odourless solid.
M.p. 161.5–164 °C (decomp.) **V.p.** 0.27 mPa (25 °C) K_{ow} logP = 1.79 (pH 7,
25 °C) **Solubility** In water 2.5 g/l (25 °C). In benzene 3.7, hexane 6.1,
2-methoxyethanol 60, acetonitrile 60, acetone 70, methanol 170, chloroform 250
(all in g/l, 25 °C). **Stability** Stable at 52 °C (highest storage temperature tested).
Stable in aqueous media between pH 5 and 9. On hydrolysis, DT_{50} (25 °C) >64 d
(pH 3, 6 and 9).

COMMERCIALISATION
History Herbicide reported by J. F. Schwer (*Proc. Br. Weed Control Conf., 12th,*
1974, **2**, *847*). Introduced in Brazil (1974) by Eli Lilly & Co. (now DowElanco).
Patent GB 1266172 **Manufacturers** DowElanco; Sanachem

APPLICATIONS
Biochemistry Photosynthetic electron transport inhibitor.
Mode of action Systemic soil-herbicide with low selectivity, absorbed mainly by the
roots, with ready translocation. **Uses** Broad-spectrum herbicide for control of
herbaceous and woody plants (0.6–4.5 kg a.i./ha), annual weeds (1.3–4.5 kg/ha),
many perennial grass and broad-leaved weeds (2.2–6.8 kg/ha). Uses include
control of total vegetation in non-crop areas, undesirable woody plants in
pastures and rangeland, and grass and broad-leaved weeds in sugar cane.
Phytotoxicity Not to be applied near desirable trees or plants.
Formulation types WP; GR; WG; Pellets. **Mixtures** (*tebuthiuron +*) amitrole +
atrazine; 2,4-D + picloram. **Selected tradenames** 'Spike' (DowElanco);
'Bushwacker' (Rhône-Poulenc); 'Tebusan' (Sanachem)

ANALYSIS
Product analysis by hplc or by glc with FID (A. Loh *et al., Anal. Methods Pestic.*

Plant Growth Regul., 1980, **11**, 351).   **Residues** determined by glc with FPD (*idem, ibid.*). In **drinking water**, by gc with FID (*AOAC Methods*, 1995, 991.07). Details available from DowElanco.

MAMMALIAN TOXICOLOGY
Oral Acute oral LD_{50} for mice 579, rats 644, rabbits 286, dogs >500, cats >200 mg/kg.   **Skin and eye** Acute percutaneous LD_{50} not available. No skin irritation in rabbits when applied at 200 mg/kg; no eye irritation when applied at 71 mg/kg.   **NOEL** In 2 y feeding trials, rats receiving 800 mg/kg diet showed no ill-effects.   **ADI** 0.07 mg/kg daily (based on 2-generation rat reproduction study). **Other** No evidence of teratogenicity (rats and rabbits).   **Toxicity class** WHO (a.i.) III; EPA (formulation) III   **EC risk** Xn (R22)

ECOTOXICOLOGY
Birds Acute oral LD_{50} for chickens, bobwhite quail, and mallard ducks >500 mg/kg. In 1-month feeding trials, chickens receiving 1000 mg/kg showed no ill-effects.   **Fish** LC_{50} (96 h) for rainbow trout 144, goldfish and fathead minnow >160, bluegill sunfish 112 mg/l.   **Bees** LD_{50} >100 μg/bee.   **Daphnia** LC_{50} 297 ppm.   **Algae** EC_{50} for *Anabaena* 4.06, *Navicula* 0.081, *Selenastrum* 0.05, *Skeletonema* 0.05 ppm.   **Other aquatic spp.** EC_{50} for *Lemna* 0.135 ppm.

ENVIRONMENTAL FATE
Animals The major metabolites in mammals were formed by *N*-demethylation of the substituted urea side-chain (D. M. Morton & D. G. Hoffman, *J. Toxicol. Environ. Health*, 1976, **1**, 757–768).   **Plants** In plants, the principal metabolic pathways involve *N*-demethylation and hydroxylation of the *tert*-butyl side-chain.
Soil/Environment Some microbial breakdown occurs in soil, but this is not the predominant mode of degradation. Loss due to photodecomposition and volatilisation is negligible. Half-life in soil is considerably greater in soils with low moisture content, and in high organic soils. Adsorption Freundlich K values range from 0.11 in sand (pH 7.7, o.m. 0.5%) to 1.82 in clay loam (pH 6.9, o.m. 2.0%).

684 tecloftalam *Bactericide*

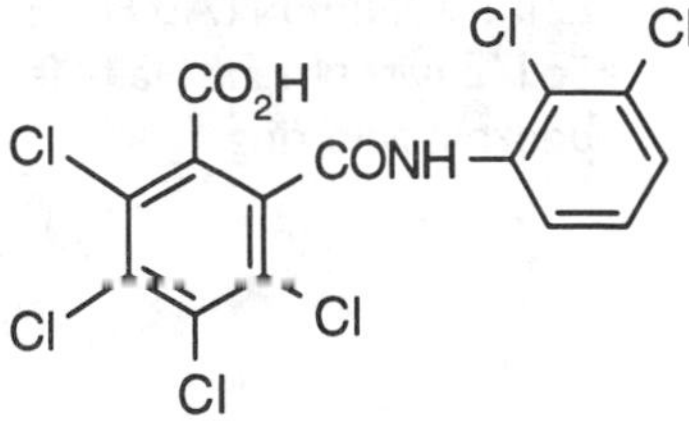

NOMENCLATURE
Common name tecloftalam (BSI, draft E-ISO); técloftalame ((*m*) draft F-ISO)
IUPAC name 3,4,5,6-tetrachloro-*N*-(2,3-dichlorophenyl)phthalamic acid
Chemical Abstracts name 2,3,4,5-tetrachloro-6-[[(2,3-dichlorophenyl)amino]=carbonyl]benzoic acid
CAS RN *[76280–91–6]* **Development codes** F-370; SF-7306; SF 7402

PHYSICAL CHEMISTRY
Mol. wt. 447.9 **M.f.** $C_{14}H_5Cl_6NO_3$ **Form** Colourless solid. **M.p.** 198–199 °C **V.p.** <0.013 mPa (60 °C) **Solubility** In water 14 mg/l (26 °C). In acetone 25.6, benzene 0.95, dimethylformamide 162, dioxane 64.8, ethanol 19.2, ethyl acetate 8.7, methanol 5.4, xylene 0.16 (all in g/l). **Stability** Degraded by sunlight or u.v. light. Hydrolysed in strongly acidic media. Stable under alkaline or neutral conditions.

COMMERCIALISATION
History Bacteriostat reported by Y. Takahi (*Jpn. Pestic. Inf.*, 1985, No. 46, p. 25). Introduced in Japan (1987) by Sankyo Co., Ltd. **Manufacturers** Sankyo

APPLICATIONS
Mode of action Bacteriostat. **Uses** Control of bacterial leaf blight (caused by *Xanthomonas campestris* pv. *oryzae*) in paddy rice. **Formulation types** DP; WP.
Selected tradenames 'Shirahagen-S' (Sankyo)

ANALYSIS
Product analysis by hplc. **Residues** determined by glc with ECD after conversion to the corresponding imide. Details available from Sankyo Co., Ltd.

MAMMALIAN TOXICOLOGY
Oral Acute oral LD_{50} for male rats 2340, female rats 2400, male mice 2010, female mice 2220 mg/kg. **Skin and eye** Acute percutaneous LD_{50} for rats >1500, mice >1000 mg/kg. **Other** Non-teratogenic and no adverse effects on reproduction.

MAIN ENTRIES

ECOTOXICOLOGY
Fish LC_{50} (48 h) for carp 30 mg/l. **Daphnia** LC_{50} (3 h) 300 mg/l.

ENVIRONMENTAL FATE
Soil/Environment Half-life in soil 4–10 d, undergoing loss of chlorine from the benzoic acid ring.

685 tecnazene   *Fungicide, plant growth regulator*

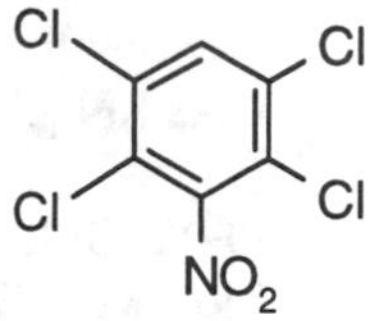

NOMENCLATURE
Common name tecnazene (BSI, E-ISO, (*m*) F-ISO)
IUPAC name 1,2,4,5-tetrachloro-3-nitrobenzene
Chemical Abstracts name 1,2,4,5-tetrachloro-3-nitrobenzene
Other names TCNB   **CAS RN** *[117–18–0]*   **EEC no.** 204–178–2

PHYSICAL CHEMISTRY
Mol. wt. 260.9   **M.f.** $C_6HCl_4NO_2$ **Form** Colourless crystals.   **M.p.** 99 °C
B.p. 304 °C (decomp.)   **V.p.** 240 mPa (15 °C)   **S.g./density** 1.744 (25 °C)
Solubility In water 0.44 mg/l (20 °C). In ethanol *c.* 40 g/l (25 °C).   **Stability** Very stable to acids and bases. Stable to heat up to almost 300 °C. In solution, decomposes slowly when exposed to u.v. radiation.

COMMERCIALISATION
History Fungicide introduced *c.* 1946 by Bayer AG (who no longer manufacture or market it).   **Patent** US 2615801

APPLICATIONS
Mode of action Fungicide with protective and curative action. Plant growth regulator which inhibits sprouting.   **Uses** Control of dry rot in seed potatoes and ware potatoes, and inhibition of sprouting in ware potatoes. Smoke formulations for control of *Botrytis* on various glasshouse crops including chrysanthemums, lettuce, and tomatoes.   **Phytotoxicity** Some varieties of rose are susceptible.
Formulation types DP; GR; Fumigant.   **Mixtures** (*tecnazene* +) lindane;

thiabendazole; organoiodine; carbendazim. **Selected tradenames** 'Fusarex' (Zeneca)

ANALYSIS
Product analysis by glc or by polarography (*CIPAC Handbook,* 1970, **1,** 663).
Residues determined by glc. Details available from Zeneca.

MAMMALIAN TOXICOLOGY
Reviews *Pesticide residues in food – 1994,* FAO Plant Production and Protection Paper, 127, 1995. *Pesticide residues in food – 1994 evaluations, Pt. II – Toxicology,* WHO, WHO/PCS/95.2, 1995. **EHC** 42 (WHO, 1984). **Oral** Acute oral LD_{50} for male rats 2047, female rats 1256 mg/kg. **Skin and eye** Acute percutaneous LD_{50} for rats >2000 mg/kg. Not a skin irritant (rats). Slight/mild eye irritant (rabbits). **Inhalation** LC_{50} (4 h) for rats >2.74 mg/l. **NOEL** In long-term feeding trials, NOEL for rats 50 mg/kg daily; for rabbits 15 mg/kg daily. No evidence of teratogenicity, embryotoxicity, or carcinogenicity. In 2 y feeding trials, no oncogenic response observed in rats receiving 150 mg/kg diet. In 560 d feeding trials, no oncogenic response observed in mice receiving 1500 mg/kg diet.
ADI (JMPR) 0.02 mg/kg b.w. [1994]. **Toxicity class** WHO (a.i.) III (Table 5)
EC risk Xi (R43)

ECOTOXICOLOGY
Fish LC_{50} (96 h) for rainbow trout (flow through) 0.37 mg/l. **Bees** Non-toxic to bees. **Daphnia** LC_{50} (48 h) >0.58 mg/l.

ENVIRONMENTAL FATE
Animals In animals, following oral administration, tecnazene is rapidly absorbed and metabolised. High single oral doses are predominantly excreted unchanged in the faeces. Several metabolites are presumed in the urine, the most important being the mercapturic acid conjugate. **Plants** In plants, substitution of the chlorine atoms by sulfhydryl groups has been observed. **Soil/Environment** In soil, tecnazene is rapidly lost, probably mainly through evaporation.

benzoylurea

F

F Cl

CONHCONH

F

F

F

Cl

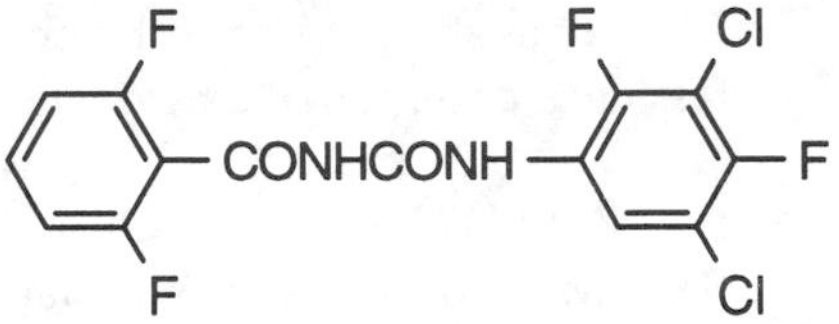

NOMENCLATURE
Common name teflubenzuron (BSI, draft E-ISO, (*m*) draft F-ISO); no name (the Netherlands)
IUPAC name 1-(3,5-dichloro-2,4-difluorophenyl)-3-(2,6-difluorobenzoyl)urea
Chemical Abstracts name *N*-[[(3,5-dichloro-2,4-difluorophenyl)amino]carbonyl]-= 2,6-difluorobenzamide
CAS RN *[83121–18–0]*   **Development codes** CME 134 (Cyanamid)

PHYSICAL CHEMISTRY
Mol. wt. 381.1   **M.f.** $C_{14}H_6Cl_2F_4N_2O_2$ **Form** White to yellowish crystals.
M.p. 222.5 °C   **V.p.** 8×10^{-7} mPa (20 °C)   **K_{ow}** logP = 4.3 (20 °C)   **Solubility** In water 0.019 mg/l (23 °C). In acetone 10, ethanol 1.4, dimethyl sulfoxide 66, dichloromethane 1.8, cyclohexanone 20, hexane 0.05, toluene 0.85 (all in g/l, 20 °C).   **Stability** Stable for more than 2 years at room temperature. On hydrolysis DT_{50} (50 °C) 5 d (pH 7), 4 h (pH 9).

COMMERCIALISATION
History Insecticide reported by H-M. Becher *et al.* (*Proc. Int. Congr. Plant Prot., 10th,* 1983, **1**, 408). Introduced in Thailand (1984) by Celamerck GmbH & Co., (now American Cyanamid Co.).   **Patent** EP 52833   **Manufacturers** Cyanamid

APPLICATIONS
Biochemistry Chitin synthesis inhibitor affecting moulting.   **Mode of action** Non-systemic insect growth regulator with stomach action. May affect fertility of female insects after contact or ingestion.   **Uses** Control of Lepidoptera, Coleoptera, Diptera, Aleyrodidae, Hymenoptera, Psyllidae, and Hemiptera larvae on vines, pome fruit, stone fruit, citrus fruit, cabbages, potatoes, vegetables, soya beans, trees, sorghum, tobacco, and cotton.   Also controls fly and mosquito larvae, and immature stages of major locust species.   **Formulation types** SC; UL.
Mixtures (*teflubenzuron +*) phosalone.   **Selected tradenames** 'Nemolt' (Cyanamid); 'Nomolt' (Cyanamid)

ANALYSIS
Details of methods are available from Cyanamid.

MAMMALIAN TOXICOLOGY

Reviews *Pesticide residues in food – 1994*, FAO Plant Production and Protection Paper, 127, 1995. *Pesticide residues in food – 1994 evaluations, Pt. II – Toxicology*, WHO, WHO/PCS/95.2, 1995. **Oral** Acute oral LD_{50} for rats and mice >5000 mg/kg. **Skin and eye** Acute percutaneous LD_{50} for rats >2000 mg/kg; not irritant to skin or eyes of rabbits. Non-sensitising to skin. **Inhalation** LC_{50} (4 h) for rats >5058 mg dust/m^3. **NOEL** (90 d) for rats 8 mg/kg b.w. daily, for dogs 4.1 mg/kg b.w. daily. **ADI** (JMPR) 0.01 mg/kg b.w. [1994]. **Other** Non-teratogenic. Non-mutagenic in Ames, mammalian cell, and micronucleus assays. **Toxicity class** WHO (a.i.) III (Table 5); EPA (formulation) IV

ECOTOXICOLOGY

Birds Acute oral LD_{50} for quail >2250 mg/kg. Dietary LC_{50} for quail and ducks >5000 mg/kg. **Fish** LC_{50} (96 h) for trout and carp >500 mg/l. **Bees** Non-toxic to bees when used at recommended rates; LD_{50} (topical) >1000 µg/bee. **Other beneficial spp.** Low toxicity to predatory arthropods.

ENVIRONMENTAL FATE

Animals In rats, following oral administration, teflubenzuron and its metabolites are rapidly excreted in the faeces and urine. Three metabolites have been identified in the urine (D. Eichler *et al.*, *6th Int. Congr. Pestic. Chem. IUPAC*, 1986, Abstr. 7A-10). **Plants** Almost no uptake and no metabolism in plants. **Soil/Environment** DT_{50} in soil 2–12 w. Microbial degradation occurs in soil, with rapid metabolism to 3,5-dichloro-2,4-difluorophenylurea as the major metabolite.

pyrethroid

(Z)-(1S)-cis -

(Z)-(1R)-cis -

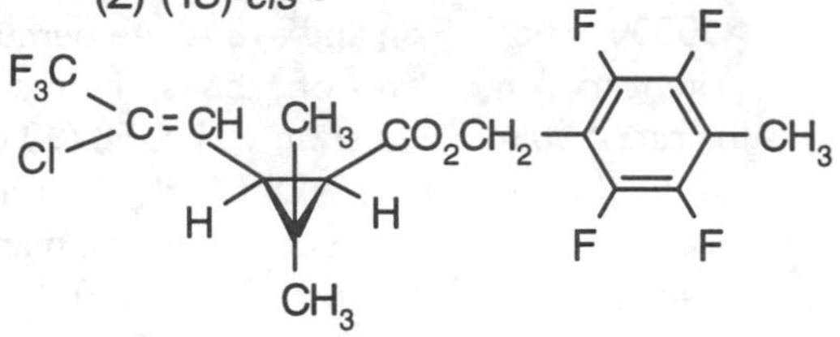

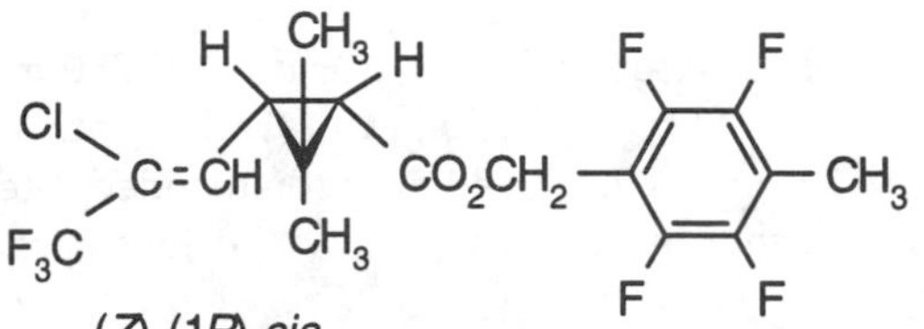

NOMENCLATURE
Common name tefluthrin (BSI, ANSI, draft E-ISO); téfluthrine ((f) draft F-ISO)
IUPAC name 2,3,5,6-tetrafluoro-4-methylbenzyl (Z)-(1RS,3RS)-3-(2-chloro-=
3,3,3-trifluoroprop-1-enyl)-2,2-dimethylcyclopropanecarboxylate
Roth: 2,3,5,6-tetrafluoro-4-methylbenzyl (Z)-(1RS)-cis-3-(2-chloro-=
3,3,3-trifluoroprop-1-enyl)-2,2-dimethylcyclopropanecarboxylate
Chemical Abstracts name [1α,3α(Z)]-(±)-(2,3,5,6-tetrafluoro-4-methylphenyl)=
methyl 3-(2-chloro-3,3,3-trifluoro-1-propenyl)-2,2-dimethylcyclopropane=
carboxylate
CAS RN _[79538–32–2]_   **Development codes** PP993

PHYSICAL CHEMISTRY
Composition Tech. material is _c._ 92% pure.   **Mol. wt.** 418.7   **M.f.** $C_{17}H_{14}ClF_7O_2$
Form Colourless solid; (tech., off-white).   **M.p.** 44.6 °C; (tech. 39–43 °C)
B.p. 153 °C/1 mmHg   **V.p.** 8 mPa (20 °C), 50 mPa (40 °C), 2 Pa (80 °C, by
extrapolation) K_{ow} logP = 6.5 (20 °C)   **S.g./density** 1.48 g/ml (25 °C)
Solubility In water 0.02 mg/l (purified and buffered, pH 5 and 9, 20 °C). In
acetone, hexane, toluene, dichloromethane, ethyl acetate >500, methanol 263 (all
in g/l, 21 °C).   **Stability** Stable for at least 9 months at 15–25 °C. Stable for
>84 days at 50 °C. Stable to hydrolysis at pH 5–7 for >30 days. At pH 9, 7%
hydrolysis in 30 days. At pH 7, 27–30% loss in aqueous solution exposed to
sunlight for 31 days.   **F.p.** 124 °C

COMMERCIALISATION
History Insecticide reported by A. R. Jutsum _et al._ (_Proc. 1986 Br. Crop Prot. Conf._
– Pests Dis., 1, 97). Introduced in Belgium (1986) by ICI Agrochemicals (now
Zeneca Agrochemicals).   **Patent** EP 31199; US 4405640   **Manufacturers** Zeneca

APPLICATIONS
Uses Control of a wide range of soil insect pests, particularly those of the orders Coleoptera, Lepidoptera, and Diptera, in maize, sugar beet, and other crops. **Formulation types** GR; EC; CS. **Mixtures** *(tefluthrin +)* oxine-copper; anthraquinone. **Selected tradenames** 'Force' (Zeneca); 'Fireban' (Uniroyal)

ANALYSIS
Product and **residue** analysis by glc. Details available from Zeneca.

MAMMALIAN TOXICOLOGY
Oral Oral and percutaneous LD_{50} values of tefluthrin depend on such factors as carrier, the test species, its sex, age and degree of fasting; values reported sometimes differ markedly. Acute oral LD_{50} for male rats 22, female rats 35 mg/kg (corn oil carrier), mice 45–46 mg/kg. **Skin and eye** Acute percutaneous LD_{50} for male rats 148–1480 mg/kg, female rats 262 mg/kg. Slight eye and skin irritation (rabbits). No skin sensitisation (guinea pigs). **Inhalation** LC_{50} (4 h) for rats 0.0427 mg/l. **NOEL** (2 y) for rats 25 mg/kg diet; (1 y) for dogs 0.5 mg/kg daily. **Toxicity class** WHO (a.i.) Ib

ECOTOXICOLOGY
Birds Acute oral LD_{50} for mallard ducks 4190, bobwhite quail 730 mg/kg. Sub-acute dietary LC_{50} for mallard ducks 2317, bobwhite quail 15 000 mg/kg. **Fish** LC_{50} (96 h) for rainbow trout 60, bluegill sunfish 130 ng/l. Under field conditions, adsorption of tefluthrin on bottom and suspended sediments should prevent any hazard. **Bees** LD_{50} (contact) 280 ng/bee, (oral) 1880 ng/bee. **Daphnia** LC_{50} (48 h) 70 ng/l.

ENVIRONMENTAL FATE
Animals For a study of the metabolism of tefluthrin in the goat, see *Pestic. Sci.*, 1989, **25**, 375. **Plants** No residues have been found on major crops treated at recommended rates (limit of detection 0.01 mg/kg). **Soil/Environment** DT_{50} in soil 150 d (5 °C), 24 d (20 °C), 17 d (30 °C), partly due to loss by volatilisation. At normal application rates, there was no effect on soil microflora.

organophosphorus

$$(CH_3O)_2\overset{\overset{S}{\|}}{P}O-\!\!\!\left\langle\bigcirc\right\rangle\!\!\!-S-\!\!\!\left\langle\bigcirc\right\rangle\!\!\!-O\overset{\overset{S}{\|}}{P}(OCH_3)_2$$

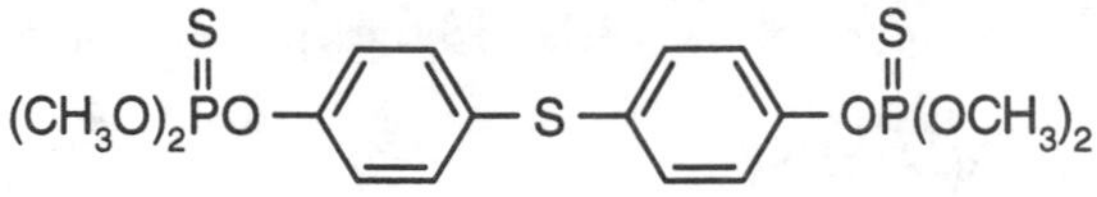

NOMENCLATURE
Common name temephos (BSI, E-ISO, (*m*) F-ISO, ANSI, ESA)
IUPAC name *O,O,O',O'*-tetramethyl *O,O'*-thiodi-*p*-phenylene
bis(phosphorothioate); *O,O,O',O'*-tetramethyl *O,O'*-thiodi-*p*-phenylene
diphosphorothioate
Chemical Abstracts name *O,O'*-(thiodi-4,1-phenylene) bis(*O,O*-dimethyl
phosphorothioate)
CAS RN *[3383–96–8]*   **Development codes** AC 52 160   **Official codes** OMS 786;
ENT 27 165

PHYSICAL CHEMISTRY
Composition Tech. grade is >90% pure.   **Mol. wt.** 466.5   **M.f.** $C_{16}H_{20}O_6P_2S_3$
Form Colourless crystals; (tech., brown, viscous liquid).   **M.p.** 30.0–30.5 °C
K_{ow} logP = 4.91   **S.g./density** 1.32 (tech.)   **Solubility** In water 0.03 mg/l (25 °C).
Soluble in common organic solvents such as diethyl ether, aromatic and
chlorinated hydrocarbons. In hexane 9.6 g/l.   **Stability** Hydrolysed by strong acids
and alkalis (optimum stability is at pH 5 to pH 7).   Do not store at temperatures
above 49 °C.

COMMERCIALISATION
History Introduced by American Cyanamid Co.   **Patent** BE 648531; GB 1039238;
US 3317636   **Manufacturers** Cyanamid; Gharda

APPLICATIONS
Biochemistry Cholinesterase inhibitor.   **Mode of action** Non-systemic insecticide.
Uses Control of larvae of *Anopheles*, *Aedes*, *Culex*, *Simulium* spp., etc., in public
health and agricultural situations. Also used for controlling fleas on dogs and cats,
and lice on humans.   **Formulation types** EC; MG; GR; DP; Fumigant; SG; RB; KN;
Fine grains.   **Mixtures** (*temephos* +) trichlorfon.   **Compatibility** Incompatible with
alkaline substances.   **Selected tradenames** 'Abate' (Cyanamid); 'Temeguard'
(Gharda)

ANALYSIS
Product analysis by hplc (*CIPAC Handbook*, 1985, **1C**, 2230; *AOAC Methods*, 1995,
982.07) or by glc (*J. Assoc. Off. Anal. Chem.*, 1982, **65**, 454).
Residues determined by glc (N. R. Pasarela & E. J. Orloski, *Anal. Methods Pestic.
Plant Growth Regul.*, 1973, **7**, 119).

MAMMALIAN TOXICOLOGY
EHC 63 (WHO, 1986; a general review of organophosphorus insecticides).
Oral Acute oral LD_{50} for male rats 4204, female rats >10 000 mg/kg.
Skin and eye Acute percutaneous LD_{50} (24 h) for rabbits 2181, rats
>4000 mg/kg. Non-irritating to eyes and skin.   **NOEL** (2 y) for rats 300 mg/kg
diet.   **Other** No toxic symptom felt by humans receiving 256 mg/man for 5 d, or
64 mg/man for 28 d (R. L. Laws *et al., Arch. Environ. Health,* 1967, **14**, 289).
Toxicity class WHO (a.i.) III (Table 5); EPA (formulation) III

ECOTOXICOLOGY
Birds Five-day dietary LC_{50} for mallard ducks 1200, ring-necked pheasants
170 mg/kg diet.   **Fish** LC_{50} for rainbow trout 31.8 mg/l.   **Bees** Highly toxic by
direct contact; LD_{50} (topical) 1.55 μg/bee.

ENVIRONMENTAL FATE
Animals In mammals, elimination is mainly of unchanged temephos in the faeces
and urine. Other principal urinary metabolites are sulfate ester conjugates of
4,4'-thiodiphenol, 4,4'-sulfinyldiphenol, and 4,4'-sulfonyldiphenol (R. C. Blinn,
J. *Agric. Food Chem.,* **17**, 118–122).   **Plants** In plants, oxidation to the sulfoxide,
and, to a lesser extent, to the sulfone and the mono- and di-orthophosphates.
Further degradation proceeds very slowly.   **Soil/Environment** Soil adsorption
Freundlich K 73 (loamy sand), 130 (sandy loam), 244 (silt loam), 541 (loam).

689   terbacil   *Herbicide*

uracil

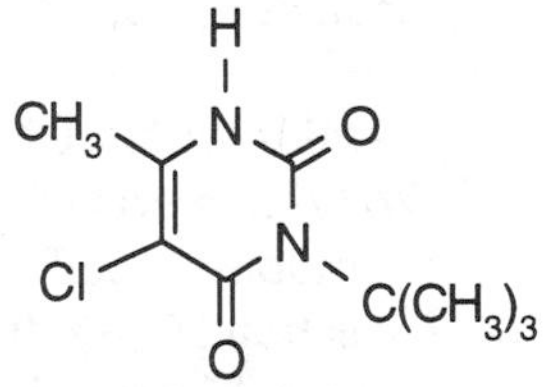

NOMENCLATURE
Common name terbacil (BSI, E-ISO, (*m*) F-ISO, ANSI, WSSA, JMAF)
IUPAC name 3-*tert*-butyl-5-chloro-6-methyluracil
Chemical Abstracts name 5-chloro-3-(1,1-dimethylethyl)-6-methyl-=
2,4(1*H*,3*H*)-pyrimidinedione
CAS RN *[5902–51–2]*   **Development codes** Du Pont Herbicide 732

PHYSICAL CHEMISTRY
Composition Tech. is 97%. **Mol. wt.** 216.7 **M.f.** $C_9H_{13}CIN_2O_2$
Form Colourless crystals. **M.p.** 175–177 °C **B.p.** Sublimation begins below the
m.p. **V.p.** 0.0625 mPa (29.5 °C) **K_{ow}** logP = 1.91 **S.g./density** 1.34 (25 °C)
Solubility In water 710 mg/l (25 °C). In dimethylformamide 337, cyclohexanone
220, methyl isobutyl ketone 121, butyl acetate 88, xylene 65 (all in g/kg, 25 °C).
Sparingly soluble in mineral oils and aliphatic hydrocarbons. Readily soluble in
strong aqueous alkalis. **Stability** Very stable, even at temperatures up to the m.p.
Stable in aqueous alkaline media at room temperature.

COMMERCIALISATION
History Herbicide reported by H. C. Bucha et al. (*Science*, 1962, **137**, 537).
Introduced by E. I. du Pont de Nemours and Co. **Patent** US 3235357; BE 625897
Manufacturers Du Pont

APPLICATIONS
Biochemistry Photosynthetic electron transport inhibitor.
Mode of action Selective herbicide, absorbed primarily through the roots, with
lesser absorption through the leaves and stems. **Uses** Selective control of many
annual and some perennial weeds in apples, citrus, alfalfa, peaches, and sugar cane
at 0.5–4 kg/ha (as area actually treated); also *Cynodon dactylon* and *Sorghum
halepense* in citrus at 4–8 kg/ha. **Formulation types** WP. **Mixtures** *(terbacil +)*
linuron; diuron + linuron; linuron + monolinuron. **Selected tradenames** 'Sinbar'
(Du Pont); 'Geonter' (Chemol)

ANALYSIS
Product analysis by hplc (H. L. Pease et al., *Anal. Methods Pestic. Plant Growth
Regul.*, 1978, **10**, 483) or by i.r. spectrometry, details from Du Pont.
Residues determined by glc (H. L. Pease et al., *loc. cit.*; R. F. Holt & H. L. Pease,
J. Agric. Food Chem., 1977, **25**, 373). In **drinking water**, by gc with FID (*AOAC
Methods*, 1995, 991.07).

MAMMALIAN TOXICOLOGY
Oral Acute oral LD_{50} for rats 934 mg/kg. **Skin and eye** Acute percutaneous LD_{50}
for rabbits >2000 mg/kg. Mild eye irritant, slight skin irritant. Not a skin sensitiser.
Inhalation LC_{50} (4 h) for rats >4.4 mg/l. **NOEL** (2 y) for rats and dogs
250 mg/kg diet. **Toxicity class** WHO (a.i.) III (Table 5); EPA (formulation) IV

ECOTOXICOLOGY
Birds Eight-day dietary LC_{50} for Pekin ducklings >56 000, pheasant chicks
>31 450 mg/kg diet. **Fish** LC_{50} (96 h) for rainbow trout 46.2 mg/l. **Bees** Non-
toxic to bees. **Daphnia** LC_{50} (48 h) 68 ppm. **Other aquatic spp.** LC_{50} (48 h) for
fiddler crab >1000 mg/l.

ENVIRONMENTAL FATE

Animals The principal biotransformation pathways are hydroxylation of the 6-methyl group, and replacement of the 5-chloro group with a hydroxy group (B. C. Mayo et al., Proc. Br. Crop Prot. Conf. – Pests Dis., 1988, **2**, 681).   **Plants** In alfalfa, 12% of terbacil plus its metabolites are still found 6–8 months after application (R. F. Holt & H. L. Pease, J. Agric. Food Chem., 1972, **25**, 373). **Soil/Environment** Undergoes microbial decomposition in moist soil. In top soil, 50% still remains 5–7 months after applying 4.5 kg/ha (P. B. Marriage, Weed Res. 1977, **17**, 219–225).

---

590   terbufos   _Insecticide, nematicide_

organophosphorus

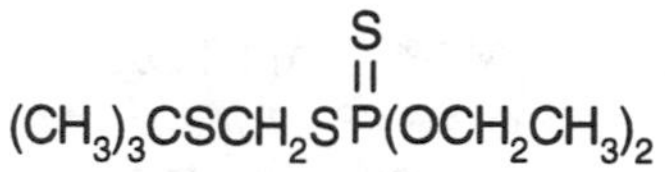

$$\begin{array}{c} S \\ \parallel \\ (CH_3)_3CSCH_2S\,P(OCH_2CH_3)_2 \end{array}$$

NOMENCLATURE
Common name terbufos (BSI, E-ISO, (m) F-ISO, ANSI, ESA)
IUPAC name S-tert-butylthiomethyl O,O-diethyl phosphorodithioate
Chemical Abstracts name S-[[(1,1-dimethylethyl)thio]methyl] O,O-diethyl phosphorodithioate
CAS RN [13071–79–9]   **EEC no.** 235–963–8   **Development codes** AC 92 100
Official codes ENT 27 920

PHYSICAL CHEMISTRY
Composition Tech. grade is ≥85% pure.   **Mol. wt.** 288.4   **M.f.** $C_9H_{21}O_2PS_3$
Form Slightly yellow liquid with a mercaptan-like odour.   **M.p.** −29.2 °C
B.p. 69 °C/0.01 mmHg   **V.p.** 34.6 mPa (25 °C) K_{ow} logP = 2.77
S.g./density 1.105 (24 °C)   **Solubility** In water 4.5 mg/l (27 °C). Readily soluble in most organic solvents, e.g. aromatic hydrocarbons, chlorinated hydrocarbons, alcohols, acetone c. 300 (all in g/l).   **Stability** Stable for more than 2 years at room temperature. Decomposes on prolonged heating above 120 °C. Hydrolysed by strong alkalis (pH >9) and acids (pH <2).   **F.p.** 88 °C (Tag open cup)

COMMERCIALISATION
History Insecticide reported by E. B. Fagan (Proc. Br. Insectic. Fungic. Conf. 7th, 1973, **2**, 695). Introduced by American Cyanamid Co.   **Manufacturers** Cyanamid; Rotam; United Phosphorus Ltd

APPLICATIONS
Biochemistry Cholinesterase inhibitor.   **Mode of action** Soil insecticide and

nematicide with stomach and contact action.   **Uses** Has effective initial and residual activity against soil-dwelling arthropods. Soil application of granules controls *Diabrotica* spp. larvae in maize, *Tetanops myopaeformis* in sugar beet, *Delia* spp. larvae in cabbages and onions, Diplopoda, Symphyla, Elateridae, and other soil-dwelling arthropods. Various above-ground pests are also controlled on plants grown in terbufos-treated soil. It has nematicidal activity, controlling *Radopholus similis, Meloidogyne incognita* and *Helicotylenchus* spp. in bananas, and *Ditylenchus dipsaci* in sugar beet. Rates 0.25–2.0 kg a.i./ha.   **Formulation types** GR.   **Mixtures** *(terbufos +)* phorate.   **Selected tradenames** 'Contraven' (Cyanamid); 'Counter' (Cyanamid, BASF, DowElanco); 'Cyanater' (Siapa); 'Hunter' (United Phosphorus); 'Pilarfox' (Pilarquim); 'Terborox' (Rotam); 'Tertin' (Sanonda)

ANALYSIS
Formulations and **residues** determined by glc (E. J. Orloski, *Anal. Methods Pestic. Plant Growth Regul.,* 1980, **11,** 165). In **drinking water**, by gc with FID (*AOAC Methods*, 1995, 991.07).

MAMMALIAN TOXICOLOGY
Reviews *Pesticide residues in food – 1989.* FAO Plant Production and Protection Paper 99, 1989. *Pesticide residues in food – 1990 evaluations. Toxicology,* World Health Organisation, WHO/PCS/91.47, 1991.   **EHC** 63 (WHO, 1986; a general review of organophosphorus insecticides).   **Oral** Acute oral LD_{50} for male albino rats 1.6 mg tech./kg; for female albino mice 5.4 mg/kg.   **Skin and eye** Acute percutaneous LD_{50} for rats 9.8, rabbits 1.0 mg/kg.   Skin and eye irritant. **Inhalation** LC_{50} (4 h) for male rats 0.0061 mg a.i./l air, for females 0.0012 mg/l air. **NOEL** In 2 y feeding trials, rats receiving 1 mg/kg diet showed no adverse effects other than cholinesterase depression and associated syndrome.   **ADI** (JMPR) 0.0002 mg/kg b.w. [1989].   **Toxicity class** WHO (a.i.) Ia; EPA (formulation) I **EC risk** T+ (R27/28)

ECOTOXICOLOGY
Birds Eight-day dietary LC_{50} for mallard ducks 185, ring-necked pheasants 145 mg/kg diet.   **Fish** LC_{50} (96 h) for rainbow trout 0.01, bluegill sunfish 0.004 mg/l.   **Bees** Not toxic to bees when used as directed. LD_{50} for honeybees by topical application 4.1 µg/bee.

ENVIRONMENTAL FATE
Animals Degradation in animals is the same as in soil.   **Plants** Degradation in plants is the same as in soil.   **Soil/Environment** Oxidative and hydrolytic degradation occurs in soil. No accumulation. Half-life in soil is 9–27 d.

1,3,5-triazine

$$CH_3O \diagdown \quad N \diagdown \quad NHC(CH_3)_3$$

NHCH$_2$CH$_3$

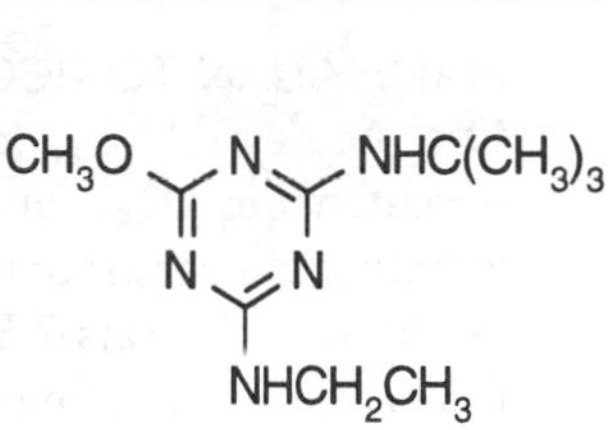

NOMENCLATURE
Common name terbumeton (BSI, E-ISO, (*m*) F-ISO)
IUPAC name N^2-*tert*-butyl-N^4-ethyl-6-methoxy-1,3,5-triazine-2,4-diamine
Chemical Abstracts name *N*-(1,1-dimethylethyl)-*N'*-ethyl-6-methoxy-=
1,3,5-triazine-2,4-diamine
CAS RN *[33693–04–8]*   **EEC no.** 251–637–9   **Development codes** GS 14 259
(Ciba)

PHYSICAL CHEMISTRY
Mol. wt. 225.3   **M.f.** $C_{10}H_{19}N_5O$ **Form** Colourless crystals.   **M.p.** 123–124 °C
V.p. 0.27 mPa (20 °C)   **K$_{ow}$** logP = 3.04   **S.g./density** 1.08 (20 °C)   **Solubility** In
water 130 mg/l (20 °C). In acetone 130, toluene 110, methanol 220,
dichloromethane 360, *n*-octanol 90 (all in g/l, 20 °C).   **Stability** Stable in neutral,
weakly acidic and weakly alkaline media. Hydrolysed by strong acids and alkalis,
especially at higher temperatures, with the formation of hydroxytriazine; DT$_{50}$
(20 °C) (calc.) 29 d (pH 1), 1.6 y (pH 13).

COMMERCIALISATION
History Herbicide reported by A. Gast & E. Fankhauser (*Proc. Br. Weed Control
Conf., 8th*, 1966, **2**, 485). Introduced by J. R. Geigy S.A. (now Novartis Crop
Protection AG).   **Patent** CH 337019; GB 814948   **Manufacturers** Novartis

APPLICATIONS
Biochemistry Photosynthetic electron transport inhibitor.   **Mode of action**
Selective herbicide, absorbed by the leaves and roots.   **Uses** Selective control of
annual and perennial grasses and broad-leaved weeds in citrus orchards. Used in
combination with terbuthylazine for herbicidal control in vineyards, apple and
citrus orchards, and in forestry at 3–6 kg total a.i./ha.   **Phytotoxicity** Should not
be applied to the vine variety Airen. May be phytotoxic to most annual plants, by
drift, etc.   **Formulation types** SC; WP.   **Mixtures** (*terbumeton +*) terbuthylazine.
Selected tradenames 'Caragard' (Novartis)

ANALYSIS
Product analysis by glc or by acidimetric titration; details available from Novartis.
Residues determined by glc with MCD (K. Ramsteiner *et al., J. Assoc. Off. Anal.*

MAIN ENTRIES

Chem., 1974, **57**, 192; E. Knüsli, *Anal. Methods Pestic., Plant Growth Regul. Food Addit.*, 1964, **4**, 13).

MAMMALIAN TOXICOLOGY
Oral Acute oral LD_{50} for rats 651, mice 2343 mg/kg. **Skin and eye** Acute percutaneous LD_{50} for rats >3170 mg/kg. Non-irritating to skin (rats); slightly irritating to eyes (rabbits). **Inhalation** LC_{50} (4 h) for rats >10 mg/l.
NOEL (2 y) for rats 7.5 mg/kg b.w. daily; (18 mo) for mice 25 mg/kg b.w. daily; (13 w) for dogs 25 mg/kg b.w. daily. **ADI** 0.075 mg/kg b.w.
Toxicity class WHO (a.i.) II; EPA (formulation) II **EC risk** Xn (R22)

ECOTOXICOLOGY
Fish LC_{50} (96 h) for rainbow trout 14, channel catfish 10, bluegill sunfish 30, crucian carp 30 mg/l. **Bees** Non-toxic to bees. **Daphnia** LC_{50} (48 h) 40 mg/l.
Algae LC_{50} 0.009 mg/l.

ENVIRONMENTAL FATE
Animals In mammals, following oral administration, terbumeton is rapidly absorbed. More than 95% is eliminated within 96 hours, predominantly in the urine, with c. 25% in the faeces. **Plants** Demethoxylation and dealkylation (primarily de-ethylation) and conjugation are the main metabolic steps in crop plants. **Soil/Environment** In soil, terbumeton undergoes microbial demethylation to the hydroxytriazine. There is uncertainty about subsequent degradation. DT_{50} in soil c. 300 d. Moderately mobile under lab. conditions, but only slightly mobile under typical conditions of use; K_{oc} 37.5–158, K_d 0.6–8.9.

692 terbuthylazine   *Herbicide*

1,3,5-triazine

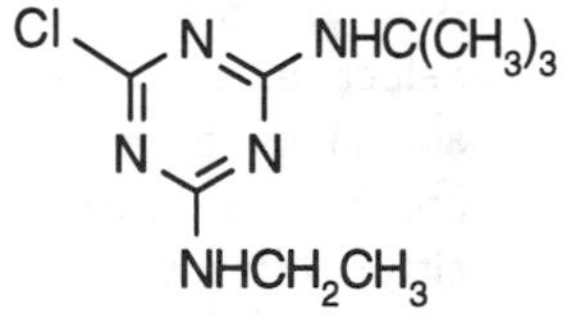

NOMENCLATURE
Common name terbuthylazine (BSI, E-ISO, (f) F-ISO, ANSI, WSSA)
IUPAC name N^2-*tert*-butyl-6-chloro-N^4-ethyl-1,3,5-triazine-2,4-diamine
Chemical Abstracts name 6-chloro-N-(1,1-dimethylethyl)-N'-ethyl-1,3,5-triazine-= 2,4-diamine
CAS RN *[5915–41–3]* **EEC no.** 227–637–9 **Development codes** GS 13 529

PHYSICAL CHEMISTRY

Composition Tech. is ≥96 % pure.   **Mol. wt.** 229.7   **M.f.** $C_9H_{16}ClN_5$
Form Colourless powder.   **M.p.** 177–179 °C   **V.p.** 0.15 mPa (25 °C)
K_{ow} logP = 3.21 (unionised)   **Henry** 4.1×10^{-4} Pa m^3 mol^{-1} (calc.)
S.g./density 1.188 (20 °C)   **Solubility** In water 8.5 mg/l (pH 7, 20 °C). In acetone
41, ethanol 14, n-octanol 12, n-hexane 0.36 (all in g/l, 25 °C).   **Stability** Stable in
neutral, weakly acidic and weakly alkaline media; hydrolysed in acidic or alkaline
media; DT_{50} (calc., 20 °C) 8 d (pH 1), 12 d (pH 13). In natural sunlight DT_{50}
>40 d.   **pKa** 2.0, v. weak base   **F.p.** >150 °C

COMMERCIALISATION

History Herbicide reported by A. Gast & E. Fankhauser (*Proc. Br. Weed Control
Conf., 8th,* 1966, p. 485). Introduced by J. R. Geigy S.A. (now Novartis Crop
Protection AG).   **Patent** BE 540590; GB 814947   **Manufacturers** Makhteshim-
Agan; Novartis; Sanachem

APPLICATIONS

Biochemistry Inhibits photosynthetic electron transport.
Mode of action Herbicide, absorbed mainly by the roots.   **Uses** Broad-spectrum
pre- or post-emergence weed control in maize, sorghum, vines, fruit trees, citrus,
coffee, oil palm, cocoa, olives, potatoes, peas, beans, sugar cane, rubber, and in
forestry in tree nurseries and new plantings. It remains largely in the topsoil,
controlling a wide range of weeds at rates of 0.6–3 kg a.i./ha; high rates are only
recommended as band applications.   **Phytotoxicity** Phytotoxic to many annual
plants and to aquatic plants.   **Formulation types** SC; WG.
Mixtures (*terbuthylazine +*) terbutryn; terbumeton; glyphosate; metolachlor;
metolachlor + pendimethalin.   **Selected tradenames** 'Gardoprim' (Novartis);
'Folar' (mixture) (Novartis); 'Topogard' (mixture) (Novartis); 'Tyllanex'
(Makhteshim-Agan)

ANALYSIS

Product analysis by glc (*CIPAC Handbook,* 1983, **1B,** 1897; *AOAC Methods,* 1995,
981.04, 971.08; FAO Specification CP/61; *J. Assoc. Off. Anal. Chem.,* 1981, **64,**
504); particulars available from Novartis.   **Residues** determined by glc with DMC
or FPD (K. Ramsteiner *et al., ibid.,* 1974, **57,** 192; E. Knüsli, *Anal. Methods Pestic.,
Plant Growth Regul. Food Addit.,* 1964, **4,** 13).

MAMMALIAN TOXICOLOGY

Oral Acute oral LD_{50} for rats 1590–>2000 mg/kg.   **Skin and eye** Acute
percutaneous LD_{50} for rats >2000 mg/kg. No skin or eye irritation. Not a skin
sensitiser.   **Inhalation** LC_{50} (4 h) for rats >5.3 mg/l air.   **NOEL** (1 y) for dogs
0.4 mg/kg b.w. daily; (lifetime) for rats 0.22 mg/kg b.w. daily; (2 y) for mice
15.4 mg/kg b.w. daily.   **ADI** 0.0022 mg/kg.   **Toxicity class** WHO (a.i.) III
(Table 5); EPA (formulation) III   **EC risk** R22

ECOTOXICOLOGY
Birds Acute oral LD_{50} for duck and quail >1000 mg/kg. Eight-day dietary LC_{50} for duck and quail >5620 ppm. **Fish** LC_{50} (96 h) for rainbow trout 3.8–4.6, bluegill sunfish 7.5, carp and catfish 7.0 mg/l. **Bees** LD_{50} (oral and contact) >100 µg/bee. **Worms** LC_{50} (7 d) for earthworms >200 mg/kg soil. **Other beneficial spp.** No effects on bacterial respiration and nitrification in range 10.9–109 mg/kg soil. **Daphnia** LC_{50} (48 h) 21–50.9 mg/l. **Algae** EC_{50} (72 h) for *Scenedesmus subspicatus* 0.016–0.024 mg/l.

ENVIRONMENTAL FATE
Animals In mammals, following oral administration, 72–84% is eliminated in the urine and faeces within 24 h, and almost all within 48 h. A de-ethyl metabolite forms rapidly, followed by conjugates of products formed by oxidation of one methyl group of the *tert*-butyl moiety. All are rapidly excreted. **Plants** Triazine-tolerant plants (e.g. maize) rapidly de-chlorinate terbuthylazine to hydroxy-terbuthylazine. Various amounts of de-ethylated and hydroxy de-ethylated metabolites are produced, depending on the plant species. **Soil/Environment** Adsorption on soils is strong: K_{oc} 162–278, K_d 2.2–25 are typical values for light agricultural soils. Terbuthylazine is only slightly mobile. Microbial degradation proceeds mainly by de-ethylation and hydroxylation, with eventual ring cleavage. DT_{50} 30–60 d in biologically active soil.

693 terbutryn
Herbicide

1,3,5-triazine

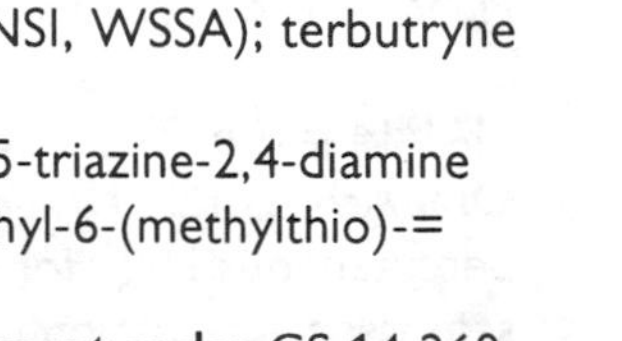

NOMENCLATURE
Common name terbutryn (BSI (from 1984), E-ISO, ANSI, WSSA); terbutryne (BSI (before 1984), (*m*) F-ISO)
IUPAC name N^2-*tert*-butyl-N^4-ethyl-6-methylthio-1,3,5-triazine-2,4-diamine
Chemical Abstracts name *N*-(1,1-dimethylethyl)-*N*'-ethyl-6-(methylthio)-=1,3,5-triazine-2,4-diamine
CAS RN *[886–50–0]* **EEC no.** 212–950–5 **Development codes** GS 14 260

PHYSICAL CHEMISTRY
Composition 96% pure. **Mol. wt.** 241.4 **M.f.** $C_{10}H_{19}N_5S$ **Form** White powder.

M.p. 104–105 °C   **B.p.** 274 °C/101 kPa.   **V.p.** 0.225 mPa (25 °C) (OECD 104)
K$_{ow}$ logP = 3.65 (25 °C, unionised)   **Henry** 1.5 × 10^{-3} Pa m^3 mol^{-1} (calc.)
S.g./density 1.12 (20 °C)   **Solubility** In water 22 mg/l (22 °C). In acetone 220,
hexane 9, *n*-octanol 130, methanol 220, toluene 45 (all in g/l, 20 °C). Also readily
soluble in dioxane, diethyl ether, xylene, chloroform, carbon tetrachloride,
dimethylformamide. Slightly soluble in petroleum ether.   **Stability** Stable under
normal conditions. The methylthio group is hydrolysed in the presence of strong
acids or alkalis. At 70 °C, no significant hydrolysis occurs at pH 5, pH 7, or pH 9
pKa 4.3, weak base

COMMERCIALISATION
History Herbicide reported by A. Gast *et al.* (*Proc. Symp. New Herbic., 2nd,* 1965,
p. 305). Introduced by J. R. Geigy (now Novartis Crop Protection AG).
Patent CH 337019; GB 814948   **Manufacturers** Makhteshim-Agan; Novartis

APPLICATIONS
Biochemistry Photosynthetic electron transport inhibitor.
Mode of action Selective herbicide, absorbed by the roots and foliage, with
translocation acropetally through the xylem, and accumulation in the apical
meristems.   **Uses** Used pre-emergence in winter cereals at 1–2 kg a.i./ha to
control blackgrass and annual meadow grass. Among the autumn-germinating
broad-leaved weeds controlled are chickweed, mayweed, poppies and speedwell,
but cleavers are rather resistant. Other pre-emergence uses are on sugar cane and
sunflowers; and, in mixture with terbuthylazine, on beans, peas and potatoes. In
mixture with metolachlor, used in cotton and peanuts. Also used post-emergence.
(0.2–0.4 kg/ha) in cereals, (1–3 kg/ha) in sugar cane, and as a directed spray in
maize. As 'Clarosan', it is used to control algae and submerged vascular plants in
waterways, reservoirs and fish ponds.   **Phytotoxicity** Not safe for post-emergence
use in cereals which are under stress.   **Formulation types** WP; SC; GR; MG; FW.
Mixtures (*terbutryn* +) terbuthylazine; triasulfuron; metolachlor; atrazine;
metobromuron.   **Selected tradenames** 'Igran' (Novartis); 'Terbutrex'
(Makhteshim-Agan)

ANALYSIS
Product analysis by glc with FID (*CIPAC Handbook,* 1980, **1A,** 1351; *AOAC Methods,*
1995, 971.08; FAO Specification (CP/61); A. H. Hofberg *et al., J. Assoc. Off. Anal.
Chem.,* 1973, **56,** 586).   **Residues** determined by glc with DMC or FPD
(K. Ramsteiner *et al., ibid.,* 1974, **57,** 192; B. G. Tweedy & R. A. Kahrs, *Anal.
Methods Pestic. Plant Growth Regul.,* 1978, **10,** 493; E. Knüsli, *Anal. Methods Pestic.,
Plant Growth Regul. Food Addit.,* 1964, **4,** 13), or by hplc. In **drinking water,** by gc
with FID (*AOAC Methods,* 1995, 991.07).

MAMMALIAN TOXICOLOGY
Oral Acute oral LD$_{50}$ for rats 2500, mice 500 mg/kg.   **Skin and eye** Acute

percutaneous LD_{50} for rats >2000, rabbits >20 000 mg/kg. Not a skin or eye irritant (rabbits). Not a skin sensitiser (guinea pigs). **Inhalation** LC_{50} (4 h) for rats >2200 mg/m³ air. **NOEL** (2 y) for rats 300 ppm, for mice 3000 ppm; (1 y) for dogs 100 ppm. **ADI** 0.027 mg/kg. **Toxicity class** WHO (a.i.) III (Table 5); EPA (formulation) III

ECOTOXICOLOGY
Birds Ten-day dietary LC_{50} for bobwhite quail 5000, mallard ducks >4640 ppm. **Fish** LC_{50} (96 h) for rainbow trout 3, bluegill sunfish 4, carp 4.7 mg/l. **Bees** Not toxic to bees; LD_{50} (oral) >225 µg/bee; (contact) >101 µg/bee. **Worms** LC_{50} for *Eisenia foetida* 170 mg/kg. **Daphnia** LC_{50} (48 h) 2.66 mg/l. **Algae** EC_{50} (7 d) for *Selenastrum capricornutum* 0.013 mg/l. **Other aquatic spp.** EC_{50} (48 h) for Quahog clam 5.6 mg/l.

ENVIRONMENTAL FATE
Animals In mammals, following oral administration, 73–85% is eliminated as the dealkylated hydroxy metabolite in the faeces within 24 hours (Y. Deng *et al.*, *J. Agric. Food Chem.*, 1990, **38**, 1411). **Plants** In plants, terbutryn is degraded in a manner similar to other methylthio-s-triazines, *viz.* by oxidation of the methylthio group to hydroxy metabolites, and by dealkylation of the side-chains. Conjugates are also formed. **Soil/Environment** Soil micro-organisms play an important role in the degradation of terbutryn. Residual activity in soil is 3–10 weeks, depending upon rate of application, soil type, and weather. DT_{50} in soil 14–50 d. K_{oc} 2000, indicating a low leaching potential. Degradation in aquatic systems is caused by microbial processes; photolysis also contributes. Considerable amounts of terbutryn are removed from the water by adsorption to the sediment.

694 tetrachlorvinphos *Insecticide, acaricide*

organophosphorus

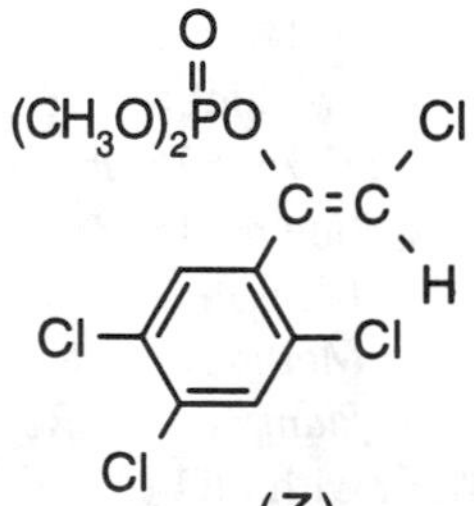

NOMENCLATURE
Common name tetrachlorvinphos (BSI, E-ISO, (*m*) F-ISO); stirofos (ESA); CVMP (JMAF)

IUPAC name (Z)-2-chloro-1-(2,4,5-trichlorophenyl)vinyl dimethyl phosphate
Chemical Abstracts name (Z)-2-chloro-1-(2,4,5-trichlorophenyl)ethenyl dimethyl phosphate
CAS RN [22248–79–9] tetrachlorvinphos; [22350–76–1] the analogous (E)-isomer; [961–11–5] mixed (Z)- + (E)- isomers   **Development codes** SD 8447
Official codes OMS 595; ENT 25 841

PHYSICAL CHEMISTRY
Composition Tech. grade is typically 98% pure.   **Mol. wt.** 366.0
M.f. $C_{10}H_9Cl_4O_4P$ **Form** Tech. grade is an off-white crystalline solid.
M.p. 94–97 °C   **V.p.** 0.0056 mPa (20 °C)   **Solubility** In water 11 mg/l (20 °C). In acetone <200, chloroform, dichloromethane 400, xylene <150 (all in g/kg, 20 °C).
Stability Stable <100 °C; slowly hydrolysed (50 °C), DT_{50} 54 d (pH 3), 44 d (pH 7), 80 h (pH 10.5).

COMMERCIALISATION
History Insecticide reported by R. R. Whetsone et al. (J. Agric. Food Chem., 1966, **14,** 352). Introduced by Shell Chemical Company, USA (now DuPont Agricultural Products, who have transferred rights to Hartz Mountain and Fermenta Animal Health Co.).   **Patent** US 3102842   **Manufacturers** Cyanamid; SDS Biotech

APPLICATIONS
Biochemistry Cholinesterase inhibitor.   **Mode of action** Non-systemic insecticide and acaricide with contact and stomach action.   **Uses** A selective insecticide controlling: Lepidoptera and Diptera in fruit, and Lepidoptera in cotton, maize, rice, tobacco and vegetables. With certain exceptions, not highly active against Hemiptera and other sucking pests and, because of rapid breakdown, is not effective against soil-dwelling pests. It is used against Muscidae in dairies and in livestock barns, against certain ectoparasites on poultry and against pests of stored products. Also used against insects that damage forests and pasture.
Phytotoxicity Russetting may occur on Golden Delicious apples under certain conditions.   **Formulation types** WP; EC; DP; SC.   **Mixtures** (tetrachlorvinphos +) fenbucarb.   **Compatibility** Incompatible with dodine and with alkaline materials.
Selected tradenames 'Debantic' (ISK Biosciences); 'Gardona' (crop protection use) (Cyanamid)

ANALYSIS
Product analysis by glc with FID, details available from Cyanamid.   **Residues** of tetrachlorvinphos and its major metabolites in plants determined by glc (Anal. Methods Pestic. Plant Growth Regul., 1973, **7,** 297; K. I. Beynon & A. N. Wright, J. Sci. Food Agric., 1969, **20,** 250; K. I. Beynon et al., Pestic. Sci., 1970, **1,** 250, 254, 259). In **drinking water**, by gc with FID (AOAC Methods, 1995, 991.07).

MAMMALIAN TOXICOLOGY
EHC 63 (WHO, 1986; a general review of organophosphorus insecticides).

Oral Acute oral LD$_{50}$ for rats 4000–5000, mice 2500–5000 mg/kg.
Skin and eye Acute percutaneous LD$_{50}$ for rabbits >2500 mg/kg. **NOEL** (2 y) for rats 125 mg/kg diet, for dogs 200 mg/kg diet. Reproduction studies on rats receiving 1000 mg/kg diet showed no adverse effects. **Toxicity class** WHO (a.i.) III (Table 5); EPA (formulation) III

ECOTOXICOLOGY
Birds Acute oral LD$_{50}$ for Chukar partridges and mallard ducks >2000 mg/kg; for various other birds 1500–2600 mg/kg. **Fish** LC$_{50}$ (24 h) for five different species 0.3–6.0 mg/l. **Bees** Toxic to bees.

ENVIRONMENTAL FATE
Animals In mammals, rapid metabolism occurs, with elimination of the metabolites within a few days. Metabolites found in the urine of dogs and rats include 2,4,5-trichlorophenylethanediol glucuronide, further glucuronic acid derivatives, 2,4,5-trichloromandelic acid, and 2-chloro-1-(2,4,5-trichlorophenyl)vinylmethyl hydrogen phosphate. No significant residues of tetrachlorvinphos or its metabolites are found in milk or tissues of exposed animals.
Soil/Environment For a review on metabolism and breakdown in plants, animals and soil, see K. I. Beynon *et al.*, *Residue Rev.*, 1973, **47**, 55.

695 tetraconazole   *Fungicide*

azole

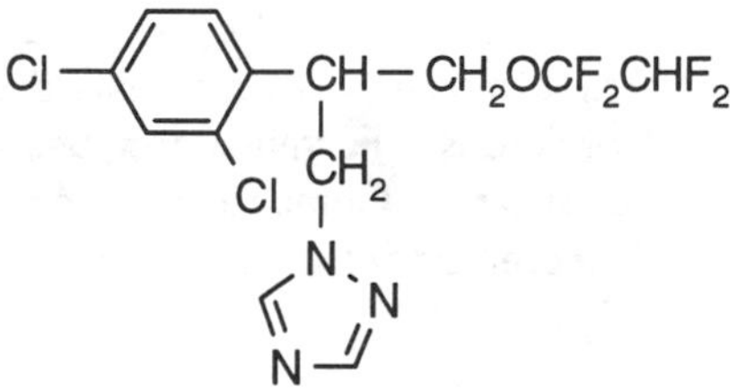

NOMENCLATURE
Common name tetraconazole (BSI, draft E-ISO)
IUPAC name (*RS*)-2-(2,4-dichlorophenyl)-3-(1*H*-1,2,4-triazol-1-yl)propyl 1,1,2,2-tetrafluoroethyl ether
Chemical Abstracts name (±)-1-[2-(2,4-dichlorophenyl)-3-(1,1,2,2-tetrafluoro= ethoxy)propyl]-1*H*-1,2,4-triazole
CAS RN [112281–77–3] unstated stereochemistry **Development codes** M 14360

PHYSICAL CHEMISTRY
Mol. wt. 372.1 **M.f.** $C_{13}H_{11}Cl_2F_4N_3O$ **Form** Colourless, viscous liquid; tech., yellow to yellowish-brown liquid. **M.p.** Pouring point 6 °C **B.p.** Decomp. 240 °C without boiling **V.p.** 0.18 mPa (gas saturation method) **K_{ow}** logP = 3.56 (20 °C) **Henry** 3.6×10^{-4} Pa m^3 mol^{-1} (20 °C, calc.) **S.g./density** 1.432 (20 °C) **Solubility** In water 156 mg/l (pH 7, 20 °C). Readily soluble in 1,2-dichloroethane, acetone, and methanol. **Stability** Stable in dilute aqueous solutions at pH 5 to 9. Stable in water to sunlight.

COMMERCIALISATION
History Fungicide reported by C. Garavaglia et al. (*Proc. 1988 Br. Crop Prot. Conf. – Pests Dis.*, **1**, 49). Introduced by Agrimont S.p.A. (now Isagro S.p.A.). **Patent** EP 234242 **Manufacturers** Isagro

APPLICATIONS
Biochemistry Sterol C^{14}-demethylase inhibitor. **Mode of action** Broad-spectrum systemic fungicide with protective, curative and eradicant properties. Absorbed by the roots, stem and leaves, with translocation acropetally to all parts of the plant, including subsequent growth. **Uses** Control of powdery mildew, brown rust, *Septoria* and *Rhynchosporium* on cereals; powdery mildew and scab on apples; powdery mildew on vines and cucumbers; powdery mildew and beet leaf spot on sugar beet; and powdery mildew and rust on vegetables. **Formulation types** EC; SL; LS. **Selected tradenames** 'Arpege' (Sipcam Phyteurop); 'Domark' (Isagro); 'Arbitre' (Rhône-Poulenc)

ANALYSIS
Product and **residue** analysis by glc or hplc. Details available from Isagro.

MAMMALIAN TOXICOLOGY
Oral Acute oral LD_{50} for male rats 1248, female rats 1031 mg/kg.
Skin and eye Acute percutaneous LD_{50} for rats >2000 mg/kg. Not irritating to skin, slightly irritating to eyes (rabbits). Not a skin sensitiser (guinea pigs). **Inhalation** LC_{50} (4 h) for rat >3.66 mg/l. **NOEL** In 2 y study, NOAEL for rat 80 mg/kg diet. **Other** Non-mutagenic in Ames test. **Toxicity class** WHO (a.i.) II

ECOTOXICOLOGY
Birds Dietary LC_{50} (5 d) for bobwhite quail 650, mallard ducks 422 mg/kg.
Fish LC_{50} (96 h) for rainbow trout 4.8, bluegill sunfish 4.3 mg/l. **Bees** LD_{50} (contact) >100 µg/bee. **Daphnia** LC_{50} (48 h) 3.0 mg/l.

ENVIRONMENTAL FATE
Animals Tetraconazole is readily adsorbed, metabolised and excreted with no significant retention in the tissues. The main metabolite identified in urine of rats is 1,2,4-triazole. **Plants** Extensive metabolism occurs in plants. The identified

metabolites are tetraconazole acid, tetraconazole alcohol, triazolylalanine and triazolylacetic acid. **Soil/Environment** No accumulation in the field. No leaching occurs in standard soils. K_{oc} 531–1922 in 4 soil types.

696 tetradec-11-en-1-yl acetate

Insect pheromone

pheromone

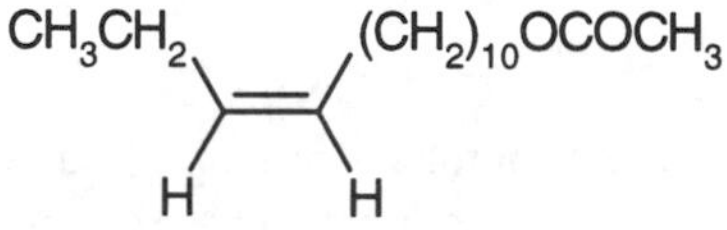

$$CH_3CH_2 \diagup \diagdown (CH_2)_{10}OCOCH_3$$

H   H

(Z)- isomer

NOMENCLATURE
IUPAC name tetradec-11-en-1-yl acetate
Chemical Abstracts name 11-tetradecen-1-yl acetate
Other names tea tortrix pheromone   **CAS RN** *[20711–10–8]* ((Z)- isomer); *[33189–72–9]* ((E)- isomer)

PHYSICAL CHEMISTRY
Composition Composition varies according to application and may comprise either (Z)- or (E)- isomer, or both, possibly in mixture with other pheromones. Physico-chemical data is for (Z)- isomer.   **Mol. wt.** 254.4   **M.f.** $C_{16}H_{30}O_2$
Form Colourless liquid.   **B.p.** 117 °C/1 mmHg K_{ow} logP >4   **S.g./density** 0.875 (20 °C)   **Solubility** Soluble in common organic solvents.   **Stability** Unstable in alkali.

COMMERCIALISATION
History A mating disruption device including the pheromone was registered in Japan in 1983. In use since 1985.   **Manufacturers** Shin-Etsu

APPLICATIONS
Uses For control of tea tortrix (*Homona magnanima*), smaller tea tortrix (*Adoxophyes* sp.) and leaf rollers such as *Adoxophyes orana* and *Platynota stultana*.
Formulation types Tubes.   **Mixtures** *(tetradec-11-en-1-yl acetate +)* codlemone.
Selected tradenames 'Hamaki-con' ((Z)- isomer) (Shin-Etsu); 'Isomate-C Special' ((Z)- isomer, mixture with codlemone) (Shin-Etsu); 'NoMate OLR' (mainly (E)- isomer) (Ecogen)

tetradifon *Acaricide*

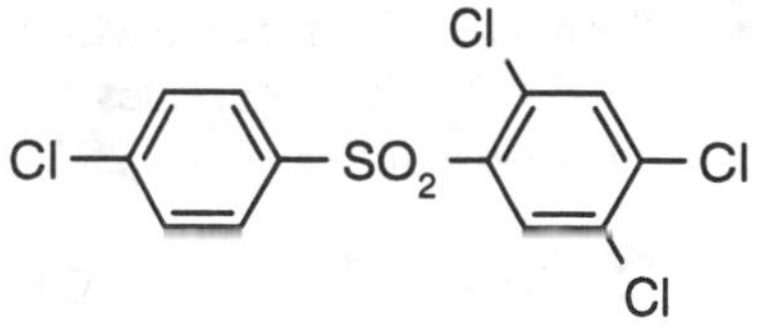

NOMENCLATURE
Common name tetradifon (BSI, E-ISO, (*m*) F-ISO, ANSI, ESA, JMAF); [tedion] (former name, Turkey and USSR); no name (Portugal)
IUPAC name 4-chlorophenyl 2,4,5-trichlorophenyl sulfone
Chemical Abstracts name 1,2,4-trichloro-5-[(4-chlorophenyl)sulfonyl]benzene
CAS RN *[116–29–0]* **Development codes** V-18 (Solvay Duphar)
Official codes ENT 23 737

PHYSICAL CHEMISTRY
Composition Tech. grade tetradifon is ≥95% pure. **Mol. wt.** 356.0
M.f. $C_{12}H_6Cl_4O_2S$ **Form** Colourless crystals; (tech., slightly yellow).
M.p. 148–149 °C (pure); ≥144 °C (tech.) **V.p.** 3.2×10^{-5} mPa (20 °C)
K_{ow} logP = 4.61 **Henry** 1.46×10^{-4} Pa m^3 mol^{-1} **S.g./density** 1.515 (20 °C)
Solubility In water 0.078 mg/l (20 °C). In acetone 82, benzene 148, chloroform 255, cyclohexanone 200, dioxane 223, kerosene 10, methanol 10, toluene 135, xylene 115 (all in g/l, 10 °C). **Stability** Extremely stable, even in strong acids and alkalis. Stable to heat and to sunlight. Resistant to strong oxidising agents.

COMMERCIALISATION
History Acaricide reported by H. O. Huisman *et al.* (*Nature (London)*, 1955, **176,** 515). Introduced by N. V. Philips-Roxane (now Uniroyal Chemical Co., Inc).
Patent NL 81359; US 2812281 **Manufacturers** Uniroyal

APPLICATIONS
Mode of action Non-systemic acaricide. Long-acting, penetrating through the plant tissue. Exhibits contact action on eggs and all non-adult stages, and also acts indirectly by sterilisation of females, leading to the development of non-viable eggs. **Uses** Larvicidal and ovicidal activity on a large range of phytophagous mites. Used on many fruit crops (including citrus), vegetables, cotton, hops, tea and ornamentals. For most crops, the recommended concentration is 0.0125–0.015% a.i.; in cotton, a dosage of 150–300 g a.i./ha is used. **Phytotoxicity** Except for some species of ornamentals (dahlia, ficus, cissus, primula, kalanchoe and some rose cultivars), non-phytotoxic when used as recommended. **Formulation types** EC. **Mixtures** (*tetradifon +)* dicofol; binapacryl; fenitrothion; carbaryl; diazinon + methoxychlor; dicofol + dinocap; cyhexatin; omethoate; dioxathion; parathion;

propargite. **Selected tradenames** 'Tedion V-18' (Uniroyal); 'Duracide' (mixture) (Endura); 'Tedone' (Siapa)

ANALYSIS
Product analysis by glc with FID (*CIPAC Handbook*, 1983, **1B**, 1901; *AOAC Methods*, 1995, 981.02). **Residues** determined by glc (*ibid.*, 976.23; *Anal. Methods Pestic. Plant Growth Regul.*, 1972, **6**, 488; A. van Rossum et al., *ibid.*, 1978, **10**, 119).

MAMMALIAN TOXICOLOGY
EHC 67 (WHO, 1986). **Oral** Acute oral LD_{50} for male rats >14 700 mg/kg. **Skin and eye** Acute percutaneous LD_{50} for rabbits >10 000 mg/kg. Non-irritating to skin; slight eye irritant (rabbits). **Inhalation** LC_{50} (4 h) for rats >3 mg/l air. **NOEL** In 2 y feeding study, NOAEL for rat 300 mg/kg diet. 2-Generation reproduction NOEL for rat 200 mg/kg diet. No teratogenic effects in rats and rabbits. Not mutagenic. **Other** Acute i.p. LD_{50} for rats >2500, mice >500 mg/kg. **Toxicity class** WHO (a.i.) III (Table 5); EPA (formulation) III

ECOTOXICOLOGY
Birds Eight-day dietary LC_{50} for bobwhite quail, Japanese quail, pheasants, mallard ducks >5000 mg/kg diet. **Fish** LC_{50} (96 h) for bluegill sunfish 880, channel catfish 2100, rainbow trout 1200 mg/l. **Bees** Not hazardous to bees when used as recommended; LD_{50} (contact) >1250 µg/bee. **Worms** LD_{50} >5000 mg/kg substrate. **Other beneficial spp.** Harmless at normal rates to the natural enemies of spider mites. **Daphnia** LC_{50} (48 h) >2 ppm. **Algae** EC_{50} (96 h) for *Selenastrum capricornutum* >100 mg/l.

ENVIRONMENTAL FATE
EHC report concludes there is no evidence of risk to the environment when used as recommended. **Animals** In rats, following oral administration of a single dose, 70% was excreted via the bile in the faeces within 48 hours. **Plants** Non-systemic; tetradifon is not metabolised on plants. **Soil/Environment** Strongly and irreversibly bound to soil/humic acid complex; virtually immobile in soil.

pyrethroid

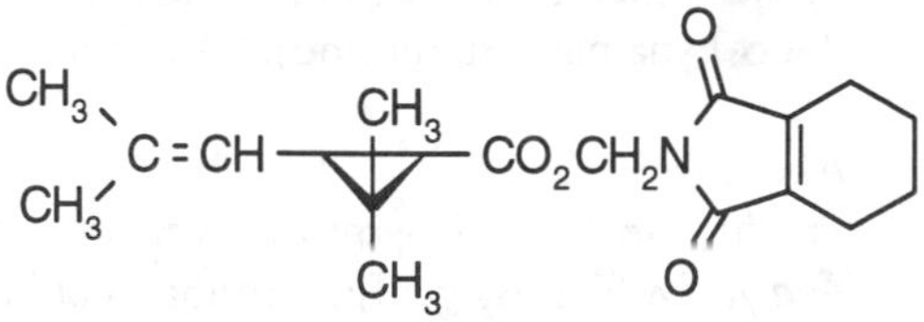

NOMENCLATURE

Common name tetramethrin (BSI, E-ISO, ANSI); tétraméthrine ((*f*) F-ISO); phthalthrin (JMAF)

IUPAC name cyclohex-1-ene-1,2-dicarboximidomethyl (1*RS*,3*RS*;1*RS*,3*SR*)-= 2,2-dimethyl-3-(2-methylprop-1-enyl)cyclopropanecarboxylate
Alt: cyclohex-1-ene-1,2-dicarboximidomethyl (±)-*cis*-*trans*-chrysanthemate
Roth: cyclohex-1-ene-1,2-dicarboximidomethyl (1*RS*)-*cis*-*trans*-2,2-dimethyl-= 3-(2-methylprop-1-enyl)cyclopropanecarboxylate; 3,4,5,6-tetrahydrophthalimido= methyl (±)-*cis*-*trans*-chrysanthemate

Chemical Abstracts name (1,3,4,5,6,7-hexahydro-1,3-dioxo-2*H*-isoindol-2-yl)= methyl 2,2-dimethyl-3-(2-methyl-1-propenyl)cyclopropanecarboxylate

CAS RN *[7696–12–0]*   **Development codes** FMC 9260; SP 1103 (Sumitomo)
Official codes OMS 1011

PHYSICAL CHEMISTRY

Composition Tech. is *c*. 92% pure.   **Mol. wt.** 331.4   **M.f.** $C_{19}H_{25}NO_4$
Form Colourless crystals, with slight pyrethrum-like odour. Tech. tetramethrin is a colourless to light yellow-brown solid.   **M.p.** 68–70 °C; tech. 60–80 °C
V.p. 0.944 mPa (30 °C)   **K_{ow}** logP = 4.6 (25 °C)   **S.g./density** 1.1 (20 °C)
Solubility In water 1.83 mg/l (25 °C). In acetone, ethanol, methanol, hexane, *n*-octanol all >2 g/100 ml.   **Stability** Sensitive to alkalis and strong acids. Stable on storage up to *c*. 50 °C. Stable in ketones, chloroform, xylene, common aerosol propellants, etc. Variable stability with inorganic carriers.

COMMERCIALISATION

History Introduced by Sumitomo Chemical Co., Ltd and later by FMC Corp. (who no longer manufacture or market it).   **Patent** JP 453929; JP 462108; US 3268398 all to Sumitomo   **Manufacturers** Endura; Sumitomo

APPLICATIONS

Mode of action Non-systemic insecticide with contact action. Gives rapid knockdown.   **Uses** Normally used in combination with synergists (e.g. piperonyl butoxide) and other insecticides for control of flies, cockroaches, mosquitoes, wasps, and other insect pests in public health and home and garden use.
Formulation types AE; EC; DP; OE; UL; Oil; EW.   **Mixtures** *(tetramethrin +)*

MAIN ENTRIES

resmethrin; fenitrothion; piperonyl butoxide + pyrethrins; piperonyl butoxide + resmethrin; phenothrin; permethrin; piperonyl butoxide; fenitrothion + lindane; permethrin + piperonyl butoxide. **Compatibility** Incompatible with mineral carriers such as kieselguhr, acidic clays and kaolin. **Selected tradenames** 'Neo-Pynamin' (Sumitomo); 'Duracide' (mixture) (Endura)

ANALYSIS

Product analysis of pyrethroids reviewed by E. Papadopoulou-Mourkidou in *Comp. Analyt. Profiles.* By glc (M. Horiba et al., *Botyu-Kagaku*, 1975, **40**, 123), or by u.v. spectrometry (J. Miyamoto, *Anal. Methods Pestic. Plant Growth Regul.*, 1973, **7**, 345).

MAMMALIAN TOXICOLOGY

Reviews See A. J. Gray & D. M. Soderlund, Chapt. 5 in *"Insecticides"*. **EHC** 98 (WHO, 1990). **Oral** Acute oral LD_{50} for rats >5000 mg/kg. **Skin and eye** Acute percutaneous LD_{50} for rabbits >2000 mg/kg. Non-irritant to skin. **Inhalation** LC_{50} (4 h) for rats >2.74 mg/l air. **NOEL** In 13 w feeding trials, dogs receiving 5000 mg/kg diet showed no ill-effects. In 6 mo feeding trials, no-effect level for rats was 1500 mg/kg diet. **Other** No evidence of oncogenicity. **Toxicity class** WHO (a.i.) III (Table 5); EPA (formulation) IV

ECOTOXICOLOGY

Birds Acute oral LD_{50} for bobwhite quail >1000 mg/kg. **Fish** LC_{50} (96 h) for bluegill sunfish 16 μg/l. **Bees** Toxic to bees.

ENVIRONMENTAL FATE

EHC 98 notes rapid abiotic degradation in air and water, rapid metabolism and excretion in mammals, and no tendency to accumulate in tissues. **Animals** In rats, following oral administration, around 95% of tetramethrin (metabolised) is eliminated in the urine and faeces within 5 days. The principal metabolite is 3-hydroxycyclohexan-1,2-dicarboximide (J. Miyamoto et al., *J. Agric. Biol. Chem.*, 1968, **32**, 628). **Soil/Environment** Degradation involves cleavage of the ester bond, leading to chrysanthemic acid derivatives and phenoxybenzoic acid. These are further metabolised by hydroxylation and conjugation.

tetramethrin [(1R)- isomers]   *Insecticide*

pyrethroid

NOMENCLATURE
IUPAC name cyclohex-1-ene-1,2-dicarboximidomethyl (1R,3R;1R,3S)-=
2,2-dimethyl-3-(2-methylprop-1-enyl)cyclopropanecarboxylate
Alt: cyclohex-1-ene-1,2-dicarboximidomethyl (+)-*cis-trans*-chrysanthemate
Roth: cyclohex-1-ene-1,2-dicarboximidomethyl (1R)-*cis-trans*-2,2-dimethyl-=
3-(2-methylprop-1-enyl)cyclopropanecarboxylate; 3,4,5,6-tetrahydro=
phthalimidomethyl (+)-*cis-trans*-chrysanthemate
Chemical Abstracts name (1,3,4,5,6,7-hexahydro-1,3-dioxo-2H-isoindol-2-yl)=
methyl (1R)-*cis-trans*-2,2-dimethyl-3-(2-methyl-1-propenyl)cyclopropane=
carboxylate
Other names d-tetramethrin   **CAS RN** [7696–12–0] for mixed stereoisomers;
[51348–90–4] for (1R)-*cis*- isomer; [1166–46–7] for (1R)-*trans*- isomer
Official codes OMS 3035

PHYSICAL CHEMISTRY
Composition The name d-tetramethrin is used for a (1R)- rich product containing
cis-trans- isomers in the ratio 20:80.   **Mol. wt.** 331.4   **M.f.** $C_{19}H_{25}NO_4$
Form d-Tetramethrin is a viscous, yellow or brown solid.   **B.p.** 142 °C/0.1 mmHg
V.p. 0.32 mPa (20 °C)   **S.g./density** 1.11 (25 °C)   **Solubility** In water 2–4 mg/l
(23 °C). In hexane, methanol, xylene >500 g/kg.

COMMERCIALISATION
History Insecticide reported by G. Shinjo *et al.* (*Eisei Dobutsu*, 1981, **32**, 221).
Introduced by Sumitomo Chemical Co., Ltd.   **Patent** JP 453929; JP 462108;
JP 1023548; US 3268398; US 3634023 to Sumitomo for d-tetramethrin
Manufacturers Sumitomo

APPLICATIONS
Mode of action Contact insecticide with knockdown action.   **Uses** Used on
Blattodea, Culicidae, Muscidae and other pests of public health. Often mixed with
other insecticides and synergists.   **Formulation types** AE; EC; EO; Concentrates
for the preparation of oil-based and water-based pressurised formulations.
Mixtures (tetramethrin [(1R)- isomers] +) cyphenothrin (1R)- isomers.
Selected tradenames 'Neo-Pynamin Forte' (Sumitomo)

ANALYSIS

Product analysis of pyrethroids reviewed by E. Papadopoulou-Mourkidou in *Comp. Analyt. Profiles*. By glc (M. Horiba et al., *Botyu-Kagaku*, 1975, **40**, 123), or by u.v. spectrometry (J. Miyamoto, *Anal. Methods Pestic. Plant Growth Regul.*, 1973, **7**, 345).

MAMMALIAN TOXICOLOGY

Oral Acute oral LD_{50} for rats >5000 mg/kg.   **Skin and eye** Acute percutaneous LD_{50} for rats >5000 mg/kg.   **Inhalation** LC_{50} (3 h) for rats >1180 mg/m^3.

700   thenylchlor   *Herbicide*

chloroacetanilide

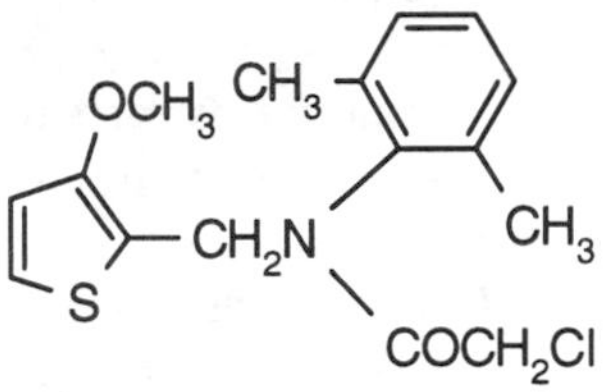

NOMENCLATURE

Common name thenylchlor (BSI, pa E-ISO)
IUPAC name 2-chloro-N-(3-methoxy-2-thenyl)-2',6'-dimethylacetanilide
Chemical Abstracts name 2-chloro-N-(2,6-dimethylphenyl)-N-[(3-methoxy-=
2-thienyl)methyl]acetamide
CAS RN *[96491–05–3]*   **Development codes** NSK-850 (Tokuyama Corp.)

PHYSICAL CHEMISTRY

Composition ≥95% pure.   **Mol. wt.** 323.8   **M.f.** $C_{16}H_{18}ClNO_2S$ **Form** White solid with a slight sulfurous odour.   **M.p.** 72–74 °C   **B.p.** 173–175 °C/0.5 mmHg
V.p. 2.8×10^{-2} mPa (25 °C)   **K$_{ow}$** logP = 3.53 (25 °C)   **S.g./density** 1.19 (25 °C)
Solubility In water 11 ppm (20 °C).   **Stability** Decomposes at 260 °C.
Decomposed by u.v. irradiation (400 nm, 8 h). Stable to acid and base in range pH 3–8.   **F.p.** 224 °C

COMMERCIALISATION

Patent US 4802907   **Manufacturers** Tokuyama

APPLICATIONS

Uses Pre-emergence control of annual grass and broad-leaved weeds in paddy rice. Especially for control of *Echinochloa* spp. up to 2-leaf stage, at 270 g a.i./ha.

Phytotoxicity No crop (rice) injury after post-planting treatment.
Formulation types GR; Flowable wettable powder; EC. **Mixtures** *(thenylchlor +)* bensulfuron-methyl; bromobutide + pyrazoxyfen. **Selected tradenames** 'Kusamets' (Tokuyama); 'Onebest' (Tokuyama)

ANALYSIS
By glc or hplc.

MAMMALIAN TOXICOLOGY
Oral Acute oral LD_{50} for rats and mice >5000 mg/kg. **Skin and eye** Acute percutaneous LD_{50} for rats >2000 mg/kg. **Inhalation** LC_{50} (4 h) for rats >5.67 mg/l. **NOEL** for rats 6.84 mg/kg daily.

ECOTOXICOLOGY
Birds LD_{50} for bobwhite quail >2000 mg/kg. **Fish** TLm (48 h) for carp 0.76 ppm. **Bees** LD_{50} (96 h) >100 μg/bee. **Worms** LD_{50} (14 d) for earthworm >1000 ppm. **Daphnia** LC_{50} (3 h) >100 ppm.

ENVIRONMENTAL FATE
Soil/Environment K_{oc} 480–2846.

theta-cypermethrin

See entry 186.

701 thiabendazole *Fungicide*

benzimidazole

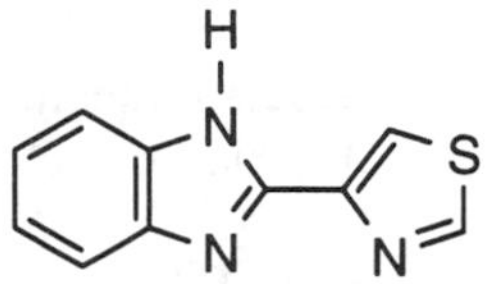

NOMENCLATURE
Common name thiabendazole (BSI, E-ISO, (*m*) F-ISO, BAN, JMAF)
IUPAC name 2-(thiazol-4-yl)benzimidazole; 2-(1,3-thiazol-4-yl)benzimidazole

Chemical Abstracts name 2-(4-thiazolyl)-1*H*-benzimidazole
Other names TBZ   **CAS RN** *[148–79–8]*   **Development codes** MK-360

PHYSICAL CHEMISTRY
Mol. wt. 201.2   **M.f.** $C_{10}H_7N_3S$ **Form** Off-white odourless powder.
M.p. 304–305 °C   **B.p.** Sublimes above 310 °C   **V.p.** Negligible at room
temperature.   **Solubility** In water 10 (pH 2), <0.05 (pH 5–12), >0.05 (pH 12) (all
in g/l, 25 °C). In dimethylformamide 39, benzene 0.23, chloroform 0.08, dimethyl
sulfoxide 80, methanol 9.3 (all in g/l, room temperature).   **Stability** Very stable in
aqueous suspension and in acidic media. Stable to heat and light.

COMMERCIALISATION
History Fungicidal properties reported by H. J. Robinson *et al.* (*J. Invest. Dermatol.*,
1964, **42**, 479) and T. Staron & C. Allard (*Phytriatr.-Phytopharm.*, 1964, **13**, 163)
and its systemic properties in plants by D. C. Erwin *et al.* (*Phytopathology*, 1968,
58, 860). It was originally introduced by Merck & Co. as an anthelmintic and later
as an agricultural fungicide.   **Patent** US 3017415   **Manufacturers** Merck

APPLICATIONS
Mode of action Systemic fungicide with protective and curative action. Absorbed
by the leaves and roots.   **Uses** Control of *Aspergillus, Botrytis, Ceratocystis,
Cercospora, Colletotrichum, Corticium, Diaporthe, Diplodia, Fusarium, Gibberella,
Gloeosporium, Oospora, Penicillium, Phoma, Rhizoctonia, Sclerotinia, Septoria,
Thielaviopsis, Verticillium* spp., etc. in asparagus, avocados, bananas, barley, beans,
cabbage, celery, chicory, cherries, citrus fruit, cotton, some cucurbits, flax,
mangoes, mushrooms, oats, onions, ornamentals, pawpaws, pome fruit, potatoes,
rice, soya beans, strawberries, sugar beet, sweet potatoes, tobacco, tomatoes,
turf, vines, and wheat. Also used for control of storage diseases of fruit and
vegetables; and for control of Dutch elm disease. It is also used as an anthelmintic
in human and veterinary medicine.   **Formulation types** WP; SC; FT; Slurry for
seed treatment; FS; Liquid seed treatment.   **Mixtures** *(thiabendazole +)* tecnazene;
thiram; organoiodine; imazalil; organoiodine + phorate; fenfuram + imazalil +
quintozene; 8-hydroxyquinoline sulfate; carboxin; carboxin + imazalil;
chlorpropham + propham; imazalil + methfuroxam; maneb; maneb + quintozene;
maneb + sulfur; methfuroxam; captan + fosetyl; metalaxyl; ethirimol + flutriafol;
bendiocarb + captan + fosetyl.   **Compatibility** Incompatible with a number of
pesticides, including copper-containing fungicides and highly alkaline materials.
Selected tradenames 'Mertect' (Merck); 'Storite' (Merck); 'Tecto' (Merck,
DowElanco); 'Decco' (Elf Atochem); 'LSP' (Gustafson)

ANALYSIS
Product analysis is by rplc (*CIPAC Handbook*, 1992, **E**, 211–7).   **Residues** in food
crops determined fluorimetrically (J. S. Wood, *Anal. Methods Pestic. Plant Growth
Regul.*, 1976, **8**, 299; *Pestic. Anal. Man.*, 1979, **I**, 201-I; R. Mestres *et al.*, *Ann. Falsif.
Exp. Chim.*, 1974, **67**, 585; 1976, **69**, 369).

MAMMALIAN TOXICOLOGY

Reviews *Evaluation of certain veterinary drug residues in food,* Fortieth report of the Joint FAO/WHO Expert Committee on Food Additives, Technical Report Series No. 832, 1993 (under the name tiabendazole). *Toxicological evaluation of certain veterinary drug residues in food,* WHO Food Additives Series, No 31, 1993. *Residues of some veterinary drugs in animals and foods,* FAO Food and Nutrition Paper 41/5, 1993. **Oral** Acute oral LD_{50} for mice 3600, rats 3100, rabbits 3850 mg/kg **Skin and eye** Acute percutaneous LD_{50} not available. Non-irritating to eyes and skin (rabbits). Not a skin sensitiser (guinea pigs). **Inhalation** Chronic inhalation of 0.07 mg/l air induced no observable clinical effects. **NOEL** In 2 y feeding trials, rats receiving 40 mg/kg daily showed no ill-effects. **ADI** (JMPR) 0.1 mg/kg b.w. [1992]. **Toxicity class** WHO (a.i.) III (Table 5); EPA (formulation) III

ECOTOXICOLOGY

Fish Low toxicity to fish. **Bees** Not toxic to bees. **Daphnia** LC_{50} 0.45 ppm.

ENVIRONMENTAL FATE

Animals In mammals, thiabendazole undergoes hydroxylation at the 5-position. Following oral administration, 87% is eliminated in the urine within 24 hours.

702 thiazopyr *Herbicide*

pyridine

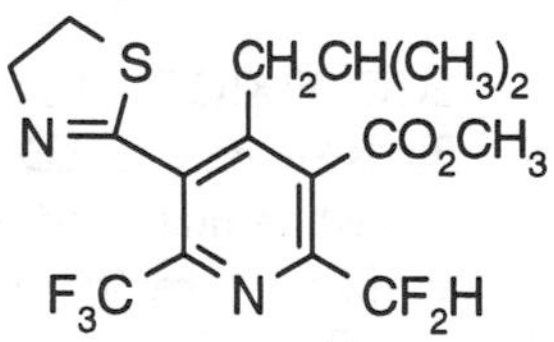

NOMENCLATURE

Common name thiazopyr (BSI, pa E-ISO, ANSI)
IUPAC name methyl 2-difluoromethyl-5-(4,5-dihydro-1,3-thiazol-2-yl)-4-isobutyl-=
6-trifluoromethylnicotinate
Chemical Abstracts name methyl 2-(difluoromethyl)-5-(4,5-dihydro-2-thiazolyl)-=
4-(2-methylpropyl)-6-(trifluoromethyl)-3-pyridinecarboxylate
CAS RN *[117718–60–2]* **Development codes** MON 13200; RH-123652

PHYSICAL CHEMISTRY

Composition Tech. is 93%. **Mol. wt.** 396.4 **M.f.** $C_{16}H_{17}F_5N_2O_2S$ **Form** Light tan

crystalline solid with a sulfur odour.   **M.p.** 77.3–79.1 °C   **V.p.** 0.3 mPa (25 °C)
K$_{ow}$ logP = 3.89 (21 °C)   **S.g./density** 1.373 (25 °C)   **Solubility** In water 2.5 mg/l
(20 °C). In methanol 28.7, hexane 3.06 (both in g/100 ml, 20 °C).
Stability Photostable in dry soil. Aqueous photolysis DT$_{50}$ 15 d.

COMMERCIALISATION
History Introduced by Monsanto Co. and subsequently sold to Rohm & Haas Co.
in 1994.   **Patent** US 4988384   **Manufacturers** Rohm & Haas

APPLICATIONS
Biochemistry Inhibits cell division by disrupting spindle microtubule formation.
Mode of action Symptoms include root growth inhibition and swelling in
meristematic regions; may also show swollen hypocotyls or internodes. Seed
germination is not affected.   **Uses** Pre-emergence control of annual grass and
some broad-leaved weeds in tree fruit, vines, citrus, sugar cane, pineapples, alfalfa,
and forestry. Generally applied at 0.1–0.56 kg/ha.   **Formulation types** EC; GR;
WP.   **Mixtures** Compatible with many products in tank mix.
Selected tradenames 'Visor' (Rohm & Haas)

ANALYSIS
Details from Rohm & Haas.

MAMMALIAN TOXICOLOGY
Oral Acute oral LD$_{50}$ for rats >5000 mg/kg.   **Skin and eye** Acute percutaneous
LD$_{50}$ for rabbits >5000 mg/kg. Slightly irritating to eyes, practically non-irritating
to skin (rabbits). Non-sensitising to skin (guinea pigs).   **Inhalation** LC$_{50}$ (4 h) for
rats >1.2 mg/l air.   **NOEL** (2 y) for rats 0.36 mg/kg b.w. daily; (1 y) for dogs
0.5 mg/kg b.w. daily.

ECOTOXICOLOGY
Birds Acute oral LD$_{50}$ for bobwhite quail 1913 mg/kg. Dietary LC$_{50}$ (5 d) for
bobwhite quail and mallard ducks >5620 mg/kg.   **Fish** LC$_{50}$ (96 h) for bluegill
sunfish 3.4, rainbow trout 3.2 mg/l.   **Bees** LD$_{50}$ >100 µg/bee.   **Worms** LC$_{50}$
(14 d) >1000 mg/kg soil.   **Daphnia** LC$_{50}$ (48 h) 6.1 mg/l.   **Algae** EC$_{50}$ 0.05 mg/l.

ENVIRONMENTAL FATE
Animals Rapidly oxidised by rat liver microsomes via sulfur and carbon oxidations
and via oxidative de-esterification.   **Plants** Studies in several species indicate
thiazopyr is initially metabolised in the dihydrothiazole ring by plant oxygenases to
the sulfoxide, sulfone, hydroxy derivative and thiazole, and is also de-esterified to
the carboxylic acid; P. C. C. Feng et al., Pestic. Sci., **45** (3), 203 (1995).
Soil/Environment In soil, degraded by both soil micro-organisms and hydrolysis.
Soil dissipation studies across multiple locations in the US indicate average DT$_{50}$
64 d (8–150 d). Vertical mobility was found to be minimal, with few detections

below 18 inches. The monoacid metabolite also has limited vertical mobility under normal use. Not significantly photolysed in soil, but DT_{50} 15 d in aqueous solution, indicating limited potential for surface water contamination.

703   thidiazuron   *Plant growth regulator*

phenylurea

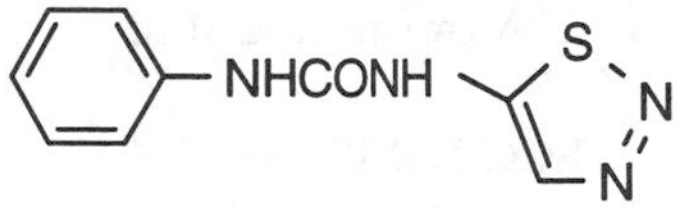

NOMENCLATURE
Common name thidiazuron (BSI, E-ISO, (*m*) F-ISO, ANSI)
IUPAC name 1-phenyl-3-(1,2,3-thiadiazol-5-yl)urea
Chemical Abstracts name *N*-phenyl-*N*'-1,2,3-thiadiazol-5-ylurea
CAS RN [51707–55–2]   **Development codes** SN 49 537

PHYSICAL CHEMISTRY
Mol. wt. 220.2   **M.f.** $C_9H_8N_4OS$ **Form** Colourless, odourless crystals.
M.p. 210.5–212.5 °C (decomp.)   **V.p.** 4×10^{-6} mPa (25 °C) (Langmuir method)
K_{ow} logP = 1.77 (pH 7.3)   **Solubility** In water 31 mg/l (pH 7, 25 °C). In hexane 0.002, methanol 4.20, dichloromethane 0.003, toluene 0.400, acetone 6.67, ethyl acetate 1.1 (all in g/l, 20 °C).   **Stability** Rapidly converted to photoisomer, 1-phenyl-3-(1,2,5-thiadiazol-3-yl)urea in presence of light (λ >290 nm). Hydrolytically stable at room temperature from pH 5–9. No decomposition in accelerated storage stability study (14 d, 54 °C).   **pKa** 8.86

COMMERCIALISATION
History Plant growth regulator reported by F. Arndt *et al.* (*Plant Physiol.*, 1976, **57**, Supplement p. 99). Introduced by Schering AG (now AgrEvo GmbH).
Patent DE 2506690; DE 2214632   **Manufacturers** AgrEvo

APPLICATIONS
Biochemistry Cytokinin activity.   **Mode of action** Plant growth regulator, absorbed by the leaves, which stimulates formation of an abscission layer between the plant stem and the leaf petioles, causing the dropping of entire green leaves.
Uses Used for defoliation of cotton, in order to facilitate harvesting.
Formulation types WP; EC; Oily flowable.   **Mixtures** (*thidiazuron* +) diuron.
Selected tradenames 'Dropp' (AgrEvo)

ANALYSIS

Product analysis by hplc. **Residues** determined by hplc or by glc with ECD of a derivative. Details available from AgrEvo.

MAMMALIAN TOXICOLOGY

Oral Acute oral LD_{50} for mice >5000, rats >4000 mg/kg. **Skin and eye** Acute percutaneous LD_{50} for rats >1000, rabbits >4000 mg/kg. Mild eye irritant; non-irritating to skin (rabbits). No skin sensitisation (guinea pigs). **Inhalation** LC_{50} (4 h) for rats >2.3 mg/l air. **NOEL** (90 d) for rats 200 mg/kg diet; (1 y) for dogs 100 mg/kg diet. No significant effects observed in a 2 y carcinogenicity study in mice and a 3-generation reproduction study in rats. **Other** Acute i.p. LD_{50} for rats 4200 mg/kg. Non-mutagenic. **Toxicity class** WHO (a.i.) III (Table 5); EPA (formulation) III

ECOTOXICOLOGY

Birds Acute oral LD_{50} for Japanese quail >3160 mg/kg. Eight-day dietary LC_{50} for bobwhite quail and mallard ducks >5000 mg/kg diet. **Fish** LC_{50} (96 h) for rainbow trout, bluegill sunfish, and catfish >1000 mg/l. **Bees** Non-toxic to honeybees. **Worms** LC_{50} (14 d) for earthworms >1400 mg/kg. **Daphnia** LC_{50} (48 h) >10 mg/l.

ENVIRONMENTAL FATE

Animals In rats and goats, metabolism involves hydroxylation of the phenyl group, followed by formation of water-soluble conjugates. Following oral administration, the compound is excreted in the urine and faeces within 96 hours. **Plants** Only small amounts of residue (normally <0.1 mg/kg) are likely in cottonseed. **Soil/Environment** Strongly adsorbed by soil. DT_{50} in soil c. 26–144 d (aerobic), 28 d (anaerobic). Essential soil microbial processes are only temporarily influenced, if at all.

704 thifensulfuron-methyl   *Herbicide*

sulfonylurea

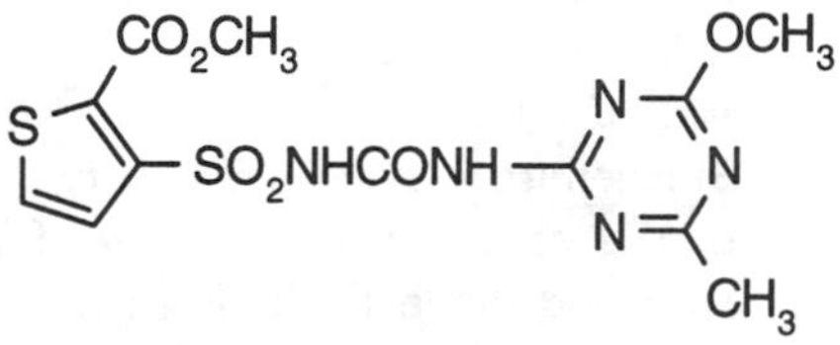

NOMENCLATURE

thifensulfuron-methyl
Chemical Abstracts name methyl 3-[[[[(4-methoxy-6-methyl-1,3,5-triazin-2-yl)=

amino]carbonyl]amino]sulfonyl]-2-thiophenecarboxylate
CAS RN *[79277–27–3]* **Development codes** DPX- M6316

thifensulfuron
Common name thifensulfuron (BSI, ANSI, draft E-ISO); [thiameturon] (WSSA former name)
IUPAC name 3-(4-methoxy-6-methyl-1,3,5-triazin-2-ylcarbamoylsulfamoyl)= thiophen-2-carboxylic acid
Chemical Abstracts name 3-[[[[(4-methoxy-6-methyl-1,3,5-triazin-2-yl)amino]= carbonyl]amino]sulfonyl]-2-thiophenecarboxylic acid
CAS RN *[79277–67–1]*

PHYSICAL CHEMISTRY
thifensulfuron-methyl
Mol. wt. 387.4   **M.f.** $C_{12}H_{13}N_5O_6S_2$ **Form** Colourless, odourless crystals.
M.p. 176 °C   **V.p.** 1.7×10^{-5} mPa (25 °C, Knudsen method) K_{ow} logP = 0.20
(pH 5), 0.02 (pH 7)   **Henry** 4.4×10^{-10} Pa m^3 mol^{-1} **S.g./density** 1.49
Solubility In water 230 (pH 5), 6270 (pH 7) (both in mg/l, 25 °C). In hexane <0.1,
xylene 0.2, ethanol 0.9, methanol, ethyl acetate 2.6, acetonitrile 7.3, acetone 11.9,
dichloromethane 27.5 (all in g/l, 25 °C).   **Stability** Stable at 55 °C. On hydrolysis,
DT_{50} (45 °C) 4.7 h (pH 3), 38 h (pH 5), 250 h (pH 7), 11 h (pH 9).   **pKa** 4.0
(25 °C)

thifensulfuron
Mol. wt. 373.4   **M.f.** $C_{11}H_{11}N_5O_6S_2$

COMMERCIALISATION
History Herbicidal activity of thifensulfuron-methyl reported by R. M. Ambach *et al.* (*Proc. N. Cent. Weed Control Conf.*, 1985, **39**, 1220). Introduced by E. I. du Pont de Nemours and Co.   **Manufacturers** Du Pont

APPLICATIONS
thifensulfuron-methyl
Biochemistry Branched chain amino acid synthesis (acetolactate synthase or ALS) inhibitor. Acts by inhibiting biosynthesis of the essential amino acids valine and isoleucine, hence stopping cell division and plant growth. Crop selectivity derives from metabolism (de-esterification) (M. K. Koeppe & H. M. Brown, *Agro-Food-Industry*, **6**, 9–14 (1995)).   **Mode of action** Selective systemic herbicide, absorbed by the leaves and roots, with translocation acropetally and basipetally throughout the plant.   **Uses** Uses in winter and spring wheat, winter and spring barley, rye, triticale, oats, soya beans, pasture and maize (grain and fodder). Post-emergence control of broad-leaved weeds and also loose silky-bent (*Apera spica-venti*) in winter and spring cereals. Application rates 9–60 g/ha. Normally used in combination with metsulfuron-methyl.   **Phytotoxicity** Treated cereals may

experience a temporary inhibition of growth and colour changes in the leaves, but these have no ultimate effect on the yield.   **Formulation types** TB; WG. **Mixtures** *(thifensulfuron-methyl +)* metsulfuron-methyl; pyridate; flupyrsulfuron-methyl; tribenuron-methyl; rimsulfuron; chlorimuron. **Selected tradenames** 'Harmony' (Du Pont); 'Pinnacle' (Du Pont)

ANALYSIS
For **tech./formulations**, by lc using u.v. For **residues**, by lc using column switching technique.

MAMMALIAN TOXICOLOGY
thifensulfuron-methyl
Oral Acute oral LD_{50} for rats >5000 mg/kg.   **Skin and eye** Acute percutaneous LD_{50} for rabbits >2000 mg/kg. Non-irritating to skin; mild but reversible irritant to eyes. Not a skin sensitiser.   **Inhalation** LC_{50} (4 h) for rats >7.9 mg/l air. **NOEL** (90 d) for rats 100 mg/kg diet, for mice 7500 mg/kg diet, for dogs 1500 mg/kg diet; (2 y) for rats 25 mg/kg diet. NOEL in: reproduction (2-generation) in rats 2500 mg/kg diet; teratogenicity in rats 200 mg/kg daily, in rabbits >650 mg/kg daily.   **ADI** 0.026 mg/kg.   **Other** Non-mutagenic in the Ames and three other mutagenicity tests.   **Toxicity class** WHO (a.i.) III (Table 5); EPA (formulation) IV (TB, WG)

ECOTOXICOLOGY
thifensulfuron-methyl
Birds Acute oral LD_{50} for mallard ducks >2510 mg/kg. Eight-day dietary LC_{50} for mallard ducks and Japanese quail >5620 mg/kg diet.   **Fish** LD_{50} (96 h) for rainbow trout and bluegill sunfish >100, catfish 360 mg/l.   **Bees** Non-toxic to bees. LD_{50} (topical) >12.5 µg/bee.   **Worms** >2000 mg/kg.   **Daphnia** LC_{50} (48 h) 970 mg/l.

ENVIRONMENTAL FATE
Animals In mammals, following oral administration of thifensulfuron-methyl, 70–75% of the unchanged material is excreted in the urine and faeces. The primary degradation mechanism involves hydrolysis of the methoxycarbonyl group, O-demethylation of the heterocyclic ring, and hydrolysis of the sulfonylurea group.   **Plants** In plants, complete degradation of thifensulfuron-methyl occurs within a few days. In tolerant crops such as cereals, the a.i. is almost completely hydrolysed within 24 h. DT_{50} in wheat *c.* 3–4 h, in soya beans *c.* 5–6 h. See also Mode of action.   **Soil/Environment** Thifensulfuron-methyl is broken down in soil to non-active metabolites by microbial degradation and chemical hydrolysis. Under the same soil conditions, it decomposes 20–50 times more rapidly than metsulfuron-methyl. DT_{50} in soil *c.* 6–12 d under natural sunlight (*c.* 14 d in the absence of sunlight); DT_{50} (30 °C, pH 8) only a few hours. K_d (silt loam) 0.6–8.6.

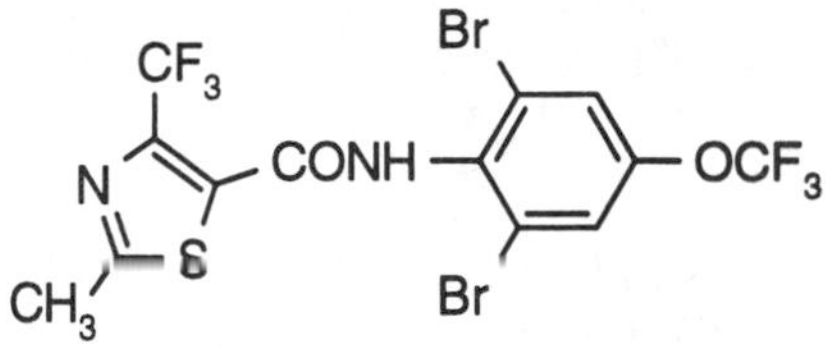

NOMENCLATURE
Common name thifluzamide (BSI, pa E-ISO, ANSI)
IUPAC name 2',6'-dibromo-2-methyl-4'-trifluoromethoxy-4-trifluoromethyl-=
1,3-thiazole-5-carboxanilide
Chemical Abstracts name N-[2,6-dibromo-4-(trifluoromethoxy)phenyl]-2-methyl-=
4-(trifluoromethyl)-5-thiazolecarboxamide
CAS RN *[130000–40–7]* **Development codes** MON 24000; RH-130 753

PHYSICAL CHEMISTRY
Mol. wt. 528.1 **M.f.** $C_{13}H_6Br_2F_6N_2O_2S$ **Form** White to light brown powder.
M.p. 177.9–178.6 °C **Solubility** In water 1.6 mg/l (20 °C). **Stability** Stable at
pH 5.0–9.0.

COMMERCIALISATION
History Reported by P. O'Reilly et al. (*Proc. Br. Crop Prot. Conf. – Pests Dis.*, 1992,
1, 427). Initially developed by Monsanto Co. and subsequently sold to Rohm &
Haas Co. in 1994. **Manufacturers** Rohm & Haas

APPLICATIONS
Biochemistry Inhibition of succinate dehydrogenase in the tricarboxylic acid cycle.
Uses Control of a wide range of Basidiomycete fungi on rice, cereals, field crops
and turf, by foliar application and as a seed treatment. It is particularly effective for
foliar treatment of *Rhizoctonia*, *Puccinia* and *Corticium*, and for seed treatment of
Ustilago, *Tilletia* and *Pyrenophora*. **Formulation types** WP; SC; FS; WG.

MAMMALIAN TOXICOLOGY
Oral Acute oral LD_{50} for rats >5000 mg/kg. **Skin and eye** Acute percutaneous
LD_{50} for rabbits >5000 mg/kg. Moderately irritating to eyes (rabbits); slightly
irritating to skin (rabbits). **Other** Negative in the Ames test.

ECOTOXICOLOGY
Birds LC_{50} for bobwhite quail and mallard ducks >5620 ppm. **Fish** LC_{50} (96 h) for
bluegill sunfish 1.2, rainbow trout 1.3, carp 2.9 mg/l. **Daphnia** LC_{50} (48 h)
2.9 mg/l.

MAIN ENTRIES

thiocarbamate

$$(CH_3CH_2)_2NCOSCH_2 \text{—} \langle \text{benzene ring} \rangle \text{—} Cl$$

NOMENCLATURE
Common name thiobencarb (BSI, E-ISO, ANSI, WSSA); thiobencarbe ((*m*) F-ISO); benthiocarb (JMAF)
IUPAC name S-4-chlorobenzyl diethylthiocarbamate; S-4-chlorobenzyl diethyl(thiocarbamate)
Chemical Abstracts name S-[(4-chlorophenyl)methyl] diethylcarbamothioate
CAS RN *[28249–77–6]* **EEC no.** 248–924–5 **Development codes** B-3015 (Kumiai); IMC3950

PHYSICAL CHEMISTRY
Composition Tech. is *c.* 93% pure. **Mol. wt.** 257.8 **M.f.** $C_{12}H_{16}ClNOS$
Form Pale yellow to brownish-yellow liquid. **M.p.** 3.3 °C
B.p. 126–129 °C/0.008 mmHg **V.p.** 2.93 mPa (23 °C) K_{ow} logP = 3.42
S.g./density 1.145–1.180 (20 °C) **Solubility** In water 30 mg/l (20 °C). Readily soluble in acetone, ethanol, xylene, methanol, benzene, *n*-hexane, and acetonitrile.
Stability Stable in water at pH 5–9 for 30 d (21 °C). Stable to light and heat.

COMMERCIALISATION
History Herbicide reported (*Jpn. Pestic. Inf.*, 1970, No. 2, p. 29). Introduced in Japan (1969) by Kumiai Chemical Industry Co., Ltd. and in USA by Chevron Chemical Co. **Patent** BP 1259471; JP 65740; US 3582314
Manufacturers Ihara/Kumiai

APPLICATIONS
Biochemistry Protein synthesis inhibitor. **Mode of action** Selective herbicide, absorbed by coleoptile, mesocotyl, roots and leaves. Inhibits shoot growth of emerging seedlings. **Uses** Pre-emergence to early post-emergence control of *Echinochloa*, *Leptochloa*, and *Cyperus* spp., and other monocotyledonous and annual broad-leaved weeds in direct-seeded and transplanted rice at 3–6 kg a.i./ha.
Phytotoxicity Non-phytotoxic to rice. **Formulation types** EC; GR.
Mixtures (*thiobencarb +*) bensulfuron-methyl + mefenacet; molinate; propanil; simetryn; linuron; pendimethalin. **Selected tradenames** 'Saturn' (Kumiai); 'Bigturn' (Sanonda); 'Bolero' (Valent); 'Siacarb' (Siapa)

ANALYSIS
Product and **residue** analysis by glc with FID (*CIPAC Handbook*, 1988, **1D**, 159; K. Kojima *et al.*, *USA-Japan Seminar Envir. Toxicol. Pestic.*, 1971; K. Ishikawa *et al.*,

Agric. Biol. Chem., 1971, **35**, 1161; S. K. De Datta *et al.*, *Weed Res.*, 1971, **11**, 41; Y. Ishii, *Jpn. Pestic. Inf.*, 1974, No. 19, p. 21).

MAMMALIAN TOXICOLOGY

EHC 76 (WHO, 1988; general review of thiocarbamates). **Oral** Acute oral LD_{50} for male rats 1033, female rats 1130, male mice 1102, female mice 1402 mg/kg. **Skin and eye** Acute percutaneous LD_{50} for rabbits and rats >2000 mg/kg. Irritating to skin and eyes. **Inhalation** LC_{50} (1 h) for rats 13 mg/l. **NOEL** (2 y) for male rats 0.9 mg/kg b.w. daily, for female rats 1.0 mg/kg b.w. daily; (1 y) for dogs 1.0 mg/kg daily. **ADI** 0.009 mg/kg. **Other** Non-mutagenic, non-teratogenic and non-oncogenic. **Toxicity class** WHO (a.i.) Table II; EPA (formulation) III **EC risk** Xn (R22)

ECOTOXICOLOGY

Birds Acute oral LD_{50} for hens 2629, bobwhite quail >7800, mallard ducks >10 000 mg/kg. Eight-day dietary LC_{50} for bobwhite quail and mallard ducks >5000 mg/kg diet. **Fish** LC_{50} (48 h) for carp 3.6, bluegill sunfish 2.4 mg/l. **Bees** LC_{50} (oral) >100 µg/bee. **Daphnia** LC_{50} (48 h) 0.1 mg/l. **Algae** EC_{50} (120 h) for *Anabaena flos-aquae* >3.1, *Navicula pelliculosa* 0.38, *Selenastrum capricornutum* 0.017, *Skeletonema costatum* 0.073 mg/l. **Other aquatic spp.** EC_{50} (14 d) for *Lemna gibba* 0.99 mg/l.

ENVIRONMENTAL FATE

Animals Thiobencarb was metabolised principally by *S*-oxygenation to the corresponding sulfoxide by liver microsomes from the striped bass (J. R. Cashman *et al.*, *Chem. Res. Toxicol.*, 1990, **3**, 433). **Soil/Environment** Rapidly adsorbed by soil, and not readily leached. Degradation is primarily by microbial breakdown, with little loss from volatilisation and photodegradation. Half-life in soil varies from 2–3 w under aerobic conditions to 6–8 mo under anaerobic conditions. K_{oc} 3170.

707 thiocyclam   *Insecticide*

2-dimethylaminopropane-1,3-dithiol analogue

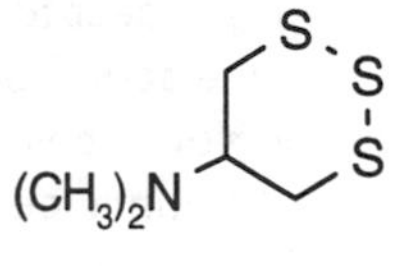

NOMENCLATURE
thiocyclam
Common name thiocyclam (BSI, E-ISO, JMAF); thiocyclame ((*m*) F-ISO)

IUPAC name *N,N*-dimethyl-1,2,3-trithian-5-ylamine
Chemical Abstracts name *N,N*-dimethyl-1,2,3-trithian-5-amine
CAS RN *[31895–21–3]*   Development codes SAN 155I

thiocyclam hydrogen oxalate
IUPAC name *N,N*-dimethyl-1,2,3-trithian-5-ylamine hydrogen oxalate
Chemical Abstracts name *N,N*-dimethyl-1,2,3-trithian-5-amine hydrogen oxalate
CAS RN *[31895–22–4]*   EEC no. 250–859–2   Development codes SAN 155I

PHYSICAL CHEMISTRY
thiocyclam
Mol. wt. 181.3   M.f. $C_5H_{11}NS_3$

thiocyclam hydrogen oxalate
Mol. wt. 271.4   M.f. $C_7H_{13}NO_4S_3$ Form Colourless, odourless solid.
M.p. 125–128 °C (decomp.)   V.p. 0.545 mPa (20 °C) K_{ow} logP = –0.07
(unstated pH)   Henry 1.8×10^{-6} Pa m^3 mol^{-1} (calc.)   S.g./density 0.6
Solubility In water 84 (pH <3.3, 23 °C), 44.1 (pH 3.6, 20 °C), 16.3 (pH 6.8,
20 °C) (all in g/l). In dimethyl sulfoxide 92, methanol 17, ethanol 1.9, acetonitrile
1.2, acetone 0.5, ethyl acetate, chloroform <1, toluene, hexane <0.01 (all in g/l,
23 °C).   Stability Stable during storage, shelf life (20 °C) ≥2 y; degraded by
sunlight, DT_{50} 2–3 d in surface waters; hydrolysis DT_{50} (25 °C) 0.5 y (pH 5),
5–7 d (pH 7–9).   pKa pKa_1 3.95, pKa_2 7.00

COMMERCIALISATION
History Insecticide reported by W. Berg & H. J. Knutti (*Proc. Br. Insectic. Fungic.
Conf. 8th*, 1975, **2**, 683). Its hydrogen oxalate was introduced by Sandoz Agro AG
(now Novartis Crop Protection AG).   Patent DE 2039555
Manufacturers Novartis; SDS Biotech

APPLICATIONS
Biochemistry Analogue or propesticide of the natural toxin nereistoxin.
Mode of action Selective insecticide with contact and stomach action. Limited
systemic activity, with translocation acropetally.   Uses Control of Lepidoptera,
Coleoptera, some Diptera and Thysanoptera. In potatoes for Colorado potato
beetle, in rape for Coleoptera and Lepidoptera pest complexes, in irrigated rice
for stem borers and some other pests, in maize for corn borer and *Tanymecus*, in
sugar beet for sugar beet weevil and other Coleoptera, in sugar cane for sugar
cane stem borer, in fruit trees for Lepidoptera, in vegetables for leaf miners, and
various Lepidoptera and Coleoptera.   Phytotoxicity Non-phytotoxic at
recommended dose to most crops, except for Stark varieties of apple.

thiocyclam hydrogen oxalate
Formulation types WP; SP; GR; Dust.   Mixtures *(thiocyclam hydrogen oxalate +)*

disulfoton; fenobucarb; etofenprox; diclomezine. **Compatibility** Incompatible with copper-containing fungicides. **Selected tradenames** 'Evisect' (Novartis)

ANALYSIS
Product analysis by hplc (M. Wisson *et al.*, *Anal. Methods Pestic. Plant Growth Regul.*, 1980, **11**, 185). **Residues** determined by glc with FPD or ECD (*idem, ibid.*, p. 190).

MAMMALIAN TOXICOLOGY
thiocyclam hydrogen oxalate
Oral Acute oral LD_{50} for male rats 399, female rats 370, male mice 273 mg/kg.
Skin and eye Acute percutaneous LD_{50} for male rats 1000, female rats 880 mg/kg. Non-irritant to skin and eyes. **Inhalation** LC_{50} (1 h) for rats >4.5 mg/l air.
NOEL (2 y) for rats 100 mg/kg diet, for dogs 75 mg/kg diet.
Toxicity class WHO (a.i.) II **EC risk** Xn (R21/22)

ECOTOXICOLOGY
thiocyclam hydrogen oxalate
Birds Acute oral LD_{50} for quail 3.45 mg/kg. Eight-day dietary LC_{50} for quail 340 mg/kg diet. **Fish** LC_{50} (96 h) for carp 1.01, trout 0.04 mg/l.
Bees Moderately toxic to bees; LD_{50} (24 h, oral) 11.9 μg a.i./bee ('Evisect S').
Daphnia LC_{50} (48 h) 2.01 mg/l. **Algae** EC_{50} (96 h) for green algae 3.3 mg/l.

ENVIRONMENTAL FATE
Plants Degradation in plants is the same as in soil. **Soil/Environment** Thiocyclam hydrogen oxalate, via nereistoxin and its oxide, is broken down to smaller fragments. Degradation is accelerated by light. DT_{50} in soil 1 d (o.c. 2.8%, pH 6.8, 22 °C).

708 thiodicarb   *Insecticide, molluscicide*

oxime carbamate

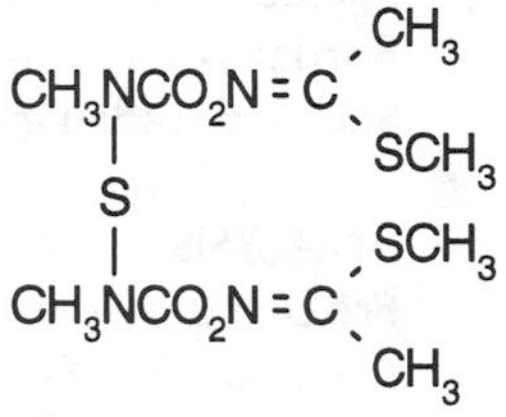

$$CH_3NCO_2N=C\overset{\diagup CH_3}{\diagdown SCH_3}$$

(S linked between the two carbamate groups)

$$CH_3NCO_2N=C\overset{\diagup SCH_3}{\diagdown CH_3}$$

NOMENCLATURE
Common name thiodicarb (BSI, draft E-ISO, (*m*) draft F-ISO, ANSI)

IUPAC name 3,7,9,13-tetramethyl-5,11-dioxa-2,8,14-trithia-4,7,9,12-tetra-=
azapentadeca-3,12-diene-6,10-dione
Chemical Abstracts name dimethyl *N,N'*-[thiobis[(methylimino)carbonyloxy]]=
bis(ethanimidothioate)
CAS RN *[59669–26–0]* **Development codes** CGA 45156; UC 80 502;
UC 51762; RPA 80600 M (Rhône-Poulenc) **Official codes** AI3–29311; OMS 3026

PHYSICAL CHEMISTRY
Composition Tech. grade is 96% pure. **Mol. wt.** 354.5 **M.f.** $C_{10}H_{18}N_4O_4S_3$
Form Colourless crystals; (tech., pale tan crystals). **M.p.** 173–174 °C
V.p. 5.7 mPa (20 °C) **S.g./density** 1.44 g/ml (20 °C) **Solubility** In water 35 mg/l
(25 °C). In dichloromethane 150, acetone 8, methanol 5, xylene 3 (all in g/kg,
25 °C). **Stability** Stable at pH 6, rapidly hydrolysed at pH 9 and slowly at pH 3
(DT_{50} c. 9 d). Aqueous suspensions are decomposed by sunlight. Stable up to
60 °C.

COMMERCIALISATION
History Insecticide reported by A. A. Sousa *et al.* (*J. Econ. Entomol.*, 1977, **70**, 803)
and reviewed by H. S. Yang & D. E. Thurman (*Proc. Br. Crop Prot. Conf. – Pests Dis.*,
1981, **3**, 687). Discovered almost simultaneously by Union Carbide Agricultural
Products Co., Inc. and by Ciba-Geigy AG (now Novartis Crop Protection AG).
Introduced by Union Carbide (now Rhône-Poulenc Agrochimie).
Patent US 4382957 **Manufacturers** Rhône-Poulenc

APPLICATIONS
Biochemistry Cholinesterase inhibitor. **Mode of action** Insecticide with
predominantly stomach action, but also limited contact action. As a seed
treatment, rapidly translocated systemically through the plant. Molluscicide which
provokes paralysis and death. **Uses** Control of all stages of major Lepidoptera
and Coleoptera pests and some Hemiptera and Diptera on cotton, soya beans,
maize, vines, fruit, vegetables, and many other crops at 200–1000 g/ha; seed
treatment rates are 2500–10 000 g/tonne. Also used as a molluscicide for control
of slugs in cereals and oilseed rape. **Formulation types** SC; WP; SG; DP; RB; GB;
Aqueous flowable; WG; Dust. **Compatibility** Incompatible with acidic and alkaline
substances, certain heavy-metal oxides, and salts of certain fungicides such as
maneb, mancozeb (except WP formulations), cuprammonium carbonate, or
Bordeaux mixtures. Not miscible with vegetable oil diluents.
Selected tradenames 'Larvin' (Rhône-Poulenc); 'Skipper' (Pepro, Rhône-Poulenc)

ANALYSIS
Product analysis by hplc. **Residues** determined by glc of a derivative (*Pestic. Anal.
Man.*, **II**, 201J; *Pestic. Reg. Sec.*, 180.407) or by hplc. Details available from Rhône-
Poulenc, USA.

MAMMALIAN TOXICOLOGY

Reviews *Pesticide residues in food – 1986*. FAO Plant Production and Protection Paper 77, 1986. *Pesticide residues in food – 1986 evaluations. Part II – Toxicology.* FAO Plant Production and Protection Paper 78/2, 1987. **EHC** 64 (WHO, 1986; a review of carbamate insecticides in general). **Oral** Acute oral LD_{50} for rats 66 (in water), 120 (in corn oil), dogs >800, monkeys >467 mg/kg. **Skin and eye** Acute percutaneous LD_{50} for rabbits >2000 mg/kg; slightly irritating to their eyes and skin. **Inhalation** LC_{50} (4 h) for rats 0.32 mg/l air. **NOEL** (2 y) for rats 3.75, mice 5.0 mg/kg daily. **ADI** (JMPR) 0.03 mg/kg b.w. [1986]. **Toxicity class** WHO (a.i.) II; EPA (formulation) II

ECOTOXICOLOGY

Birds Acute oral LD_{50} for Japanese quail 2023 mg/kg. Dietary LC_{50} for mallard duck 5620 mg/kg diet. **Fish** LC_{50} (96 h) for bluegill sunfish 1.21, rainbow trout 2.55 mg/l. **Bees** Moderately toxic to bees exposed to direct spray. No hazard to honeybees after spray residues have dried. **Daphnia** LC_{50} (48 h) 0.053 mg/l.

ENVIRONMENTAL FATE

Animals In rats, rapidly degraded to methomyl, which was rapidly converted to methomyl methylol, oxime, sulfoxide and sulfoxide oxime. These unstable intermediates were converted to acetonitrile and CO_2, which were eliminated primarily by respiration and in the urine; a small fraction of acetonitrile was further degraded to acetamide, acetic acid and CO_2 (K. Huhtanen & H. W. Dorough, *Pestic. Biochem. Physiol.*, 1976, **6**, 571–583). **Plants** Major metabolites include thiodicarb methomyl, acetonitrile and CO_2. **Soil/Environment** Rapidly degraded in soils of various types, under both aerobic and anaerobic conditions, by hydrolysis and photolysis. The primary degradation products are methomyl and methomyl oxime. DT_{50} of thiodicarb in soil is 3–8 d, depending upon the soil type.

709 thiofanox

Insecticide, acaricide

oxime carbamate

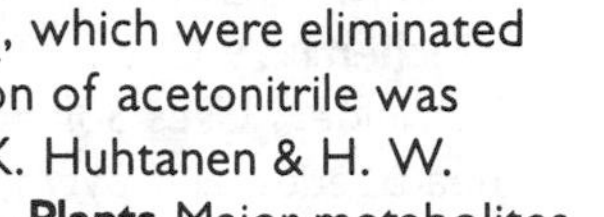

$$CH_3NHCO_2N = \overset{\overset{\textstyle C(CH_3)_3}{|}}{C}CH_2SCH_3$$

NOMENCLATURE

Common name thiofanox (BSI, E-ISO, (*m*) F-ISO, ANSI, ESA); thiofanocarb (Republic of South Africa)
IUPAC name 3,3-dimethyl-1-methylthiobutanone *O*-methylcarbamoyloxime;

1-(2,2-dimethyl-1-methylthiomethylpropylideneamino-oxy)-*N*-methylformamide
Chemical Abstracts name 3,3-dimethyl-1-(methylthio)-2-butanone O-[(methyl=
amino)carbonyl]oxime
CAS RN *[39196–18–4]* **EEC no.** 254–346–4 **Development codes** DS 15 647

PHYSICAL CHEMISTRY
Mol. wt. 218.3 **M.f.** $C_9H_{18}N_2O_2S$ **Form** Colourless solid with a pungent odour.
M.p. 56.5–57.5 °C **V.p.** 22.6 mPa (25 °C) **Solubility** In water 5.2 g/l (22 °C).
Readily soluble in chlorinated and aromatic hydrocarbons, ketones and non-polar
solvents. Slightly soluble in aliphatic hydrocarbons. **Stability** Stable to heat under
normal storage conditions. Relatively stable to hydrolysis at pH 5–9 (under
30 °C). Decomposed by strong acids and alkalis.

COMMERCIALISATION
History Insecticide reported by R. L. Schauer (*Proc. Br. Insectic. Fungic. Conf., 7th,*
1973, **2**, 713). Relationship between chemical structure and biological activity of
analogues reported by T. A. Magee & L. E. Limpel (*J. Agric. Food Chem.*, 1977, **25**,
1376). Introduced by Diamond Shamrock Chemical Co. Later sold to Rhône-
Poulenc Agrochimie. **Manufacturers** Rhône-Poulenc; SDS Biotech

APPLICATIONS
Biochemistry Cholinesterase inhibitor. **Mode of action** Systemic insecticide and
acaricide. **Uses** Soil application for control of black bean aphids, beet leaf miners,
flea beetles, and pygmy mangold beetles on beet; flea beetles on rape; aphids,
capsids, flea beetles, Colorado beetles, and leafhoppers on potatoes; aphids,
midges, flea beetles, soil-inhabiting larvae, and spider mites on cotton; aphids and
spider mites in floriculture, etc. Applied at rates in the range 0.4–3 kg/ha.
Formulation types GR. **Selected tradenames** 'Dacamox' (Rhône-Poulenc,
Embetec)

ANALYSIS
Residues, which are likely to include thiofanox, its sulfoxide and sulfone, are
oxidised to the sulfone, which is determined by glc (W. T. Chin, *J. Agric. Food
Chem.*, 1975, **23**, 963; M. B. Szalkowski *et al.*, *J. Chromatogr.*, 1976, **128**, 426).

MAMMALIAN TOXICOLOGY
EHC 64 (WHO, 1986; a review of carbamate insecticides in general). **Oral** Acute
oral LD_{50} for albino rats 8.5 mg/kg. **Skin and eye** Acute percutaneous LD_{50} for
albino rabbits 39 mg/kg. **NOEL** (90 d) for rats and beagle dogs 1 mg/kg daily.
Clinical symptoms of reversible cholinesterase inhibition at 4.0 mg/kg daily, lasting
for 3–4 hours. Weight gain of rats receiving 100 mg/kg diet was not affected.
Toxicity class WHO (a.i.) Ib; EPA (formulation) I **EC risk** T+ (R27/28)

ECOTOXICOLOGY
Birds Acute oral LD_{50} for mallard ducks 109, bobwhite quail 43 mg/kg. **Fish** LC_{50}

(96 h) for rainbow trout 0.13, bluegill sunfish 0.33 mg/l.   **Bees** Not toxic to bees when used as directed.

ENVIRONMENTAL FATE
Animals Metabolism of carbamate insecticides is reviewed (M. Cool & C. K. Jankowski in "*Insecticides*").   **Plants** Metabolism in plants is the same as in soil. The duration of activity following absorption through the roots is at least 8 weeks. **Soil/Environment** In soil, the methylthio group is rapidly oxidised to the sulfoxide and further, to the sulfone. Further degradation occurs to water-soluble metabolites.

710 thiometon   *Insecticide, acaricide*

organophosphorus

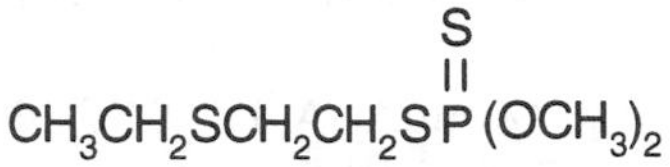

$$CH_3CH_2SCH_2CH_2S\overset{\displaystyle S}{\overset{\displaystyle \|}{P}}(OCH_3)_2$$

NOMENCLATURE
Common name thiometon (BSI, E-ISO, (*m*) F-ISO, JMAF); dithiométon ((*m*) France); [M-81] (former exception, USSR); no name (Germany, Portugal)
IUPAC name S-2-ethylthioethyl O,O-dimethyl phosphorodithioate
Chemical Abstracts name S-[2-(ethylthio)ethyl] O,O-dimethyl phosphorodithioate
CAS RN *[640–15–3]*   **EEC no.** 211–362–6   **Development codes** Bayer 23 129; SAN 1831

PHYSICAL CHEMISTRY
Mol. wt. 246.3   **M.f.** $C_6H_{15}O_2PS_3$ **Form** Colourless liquid, with characteristic odour of sulfur-containing organophosphate esters.   **B.p.** 110 °C/0.1 mmHg
V.p. 39.9 mPa (20 °C)   **K_{ow}** logP = 3.15 (mean, 20 °C)   **Henry** 28.40×10^{-3} Pa m^3 mol^{-1} (calc.)   **S.g./density** 1.209 (20 °C)   **Solubility** In water 200 mg/l (25 °C). Readily soluble in common organic solvents.   Slightly soluble in petroleum ether and mineral oils.   **Stability** Good stability when in non-polar solvents, more unstable in the pure state.   Hydrolysed more rapidly in alkaline media than in acidic media; DT_{50} 90 d (pH 3), 83 d (pH 6), 43 d (pH 9) (5 °C); 25 d (pH 3), 27 d (pH 6), 17 d (pH 9) (25 °C). Shelf-life (of 'Ekatin 50ZP') at 20 °C is *c.* 2 y.

COMMERCIALISATION
History Introduced by Bayer AG (who no longer manufacture or market it) and independently by Sandoz AG (now Novartis Crop Protection AG).
Patent DE 917668 to Bayer; CH 319579 to Sandoz   **Manufacturers** Novartis

APPLICATIONS

Biochemistry Cholinesterase inhibitor. **Mode of action** Systemic insecticide and acaricide with contact and stomach action. **Uses** Control of sucking insects (aphids, woolly aphids, suckers, sawfly larvae, thrips, etc.), bryobia mites, tetranychid mites, and fruit-tree red spider mites (non-OP resistant) on a wide range of crops, including ornamentals, strawberries, citrus and other fruit, swedes, turnips, brassicas, other vegetables, olives, vines, beet, tobacco, cotton, and cereals. **Phytotoxicity** Non-phytotoxic, except for some ornamentals. **Formulation types** EC; UL. **Mixtures** *(thiometon +)* endosulfan; tau-fluvalinate; quinalphos. **Compatibility** Incompatible with alkaline materials. **Selected tradenames** 'Ekatin' (Novartis); 'Mavrik' (mixture) (Novartis)

ANALYSIS

Product analysis by glc (*CIPAC Handbook*, 1992, **E**, 218–221; M. Wisson et al., *Anal. Methods Pestic. Plant Growth Regul.*, 1976, **8**, 239), by i.r. spectroscopy, or by paper chromatography, followed by combustion and subsequent determination of phosphate by standard methods (*CIPAC Handbook*, 1980, **1A**, 1355). **Residues** determined by glc (M. Wisson, *loc. cit.*, p. 244; *Pestic. Anal. Man.*, 1979, **I**, 201-H, 201-I; D. C. Abbott et al., *Pestic. Sci.*, 1970, **1**, 10).

MAMMALIAN TOXICOLOGY

Reviews *Pesticide residues in food – 1979.* FAO Plant Production and Protection Paper 20, 1980. *Pesticide residues in food: 1979 evaluations.* FAO Plant Production and Protection Paper 20 Sup, 1980. **EHC** 63 (WHO, 1986; a general review of organophosphorus insecticides). **Oral** Acute oral LD_{50} for male rats 73, female rats 136 mg/kg. **Skin and eye** Acute percutaneous LD_{50} for male rats 1429, female rats 1997 mg/kg. Not a skin irritant. **Inhalation** LC_{50} (4 h) in rats 1.93 mg/l air (actual value for 'Ekatin 50ZP'). **NOEL** (2 y) for dogs 6 mg/kg diet, for rats 2.5 mg/kg diet. **ADI** (JMPR) 0.003 mg/kg b.w. [1979]. **Toxicity class** WHO (a.i.) Ib; EPA (formulation) II **EC risk** T (R25); Xn (R21)

ECOTOXICOLOGY

Birds Acute oral LD_{50} (14 d) for male mallard ducks 95, female mallard ducks 53 mg/kg. Acute LD_{50} (14 d) for male Japanese quail 46, female Japanese quail 60 mg/kg. **Fish** LC_{50} (96 h) for carp 13.2, rainbow trout 8.0 mg/l (both under static conditions). **Bees** Toxic to bees; LD_{50} (oral) 0.56 µg/bee. **Worms** LC_{50} (7 d) 43.94 mg/kg soil; (14 d) 19.92 mg/kg soil. **Daphnia** LC_{50} (24 h) 8.2 mg/l. **Algae** EC_{50} (96 h) for green algae 12.8 mg/l.

ENVIRONMENTAL FATE

Animals Metabolism in animals is the same as in soil. Metabolism in rat showed that thiometon is rapidly and almost completely metabolised and excreted, mainly via urine. Metabolism proceeds via the conversion of P=S to P=O, oxidation of the sulfide to sulfoxide and sulfone, and hydrolysis to the dimethyl thiolophosphoric acid.   **Plants** Oxidation and hydrolytic reactions are the most important factors in degradation. The main metabolites are thiometon sulfoxide and thiometon sulfone. The corresponding P=O analogues occur only in very small quantities.   **Soil/Environment** In soil, thiometon is metabolised to thiometon sulfoxide and thiometon sulfone. Thiometon: K_{oc} 579 ml/g (low mobility), DT_{50} <1 d; thiometon sulfone: K_{oc} 52 ml/g (highly mobile), DT_{50} <2 d. Due to short soil persistence and high soil adsorption, thiometon is judged to have no risk of leaching into groundwater under field conditions. Experiments confirm that thiometon sulfone does not pose a risk of leaching into groundwater. Neither thiometon nor its main metabolites are likely to persist or accumulate in water/sediment systems under aerobic conditions.

711   thiophanate-methyl

Fungicide, wound protectant

benzimidazole precursor

NOMENCLATURE
Common name thiophanate-methyl (BSI, E-ISO, (*m*) F-ISO, ANSI, JMAF)
IUPAC name dimethyl 4,4'-(*o*-phenylene)bis(3-thioallophanate)
Chemical Abstracts name dimethyl [1,2-phenylenebis(iminocarbonothioyl)]= bis[carbamate]
CAS RN *[23564–05–8]*   **EEC no.** 245–740–7   **Development codes** NF 44 (Nippon Soda)

PHYSICAL CHEMISTRY
Mol. wt. 342.4   **M.f.** $C_{12}H_{14}N_4O_4S_2$ **Form** Colourless crystals.   **M.p.** 172 °C (decomp.)   **V.p.** 0.0095 mPa (25 °C) K_{ow} logP = 1.50   **Solubility** Practically insoluble in water (23 °C). In acetone 58.1, cyclohexanone 43, methanol 29.2, chloroform 26.2, acetonitrile 24.4, ethyl acetate 11.9 (all in g/kg, 23 °C). Slightly soluble in hexane.   **Stability** Stable in neutral, aqueous solution at room temperature. Stable to air and sunlight. Quite stable in acidic solution at room temperature; unstable in alkaline solution; DT_{50} 24.5 h (pH 9, 22 °C). Formulated product is stable ≥2 y below 50 °C.   **pKa** 7.28

COMMERCIALISATION
History Fungicide reported by K. Ishii (*Abstr. Int. Congr. Plant Prot., 7th, Paris,* 1970, p. 200). Introduced by the Nippon Soda Co., Ltd. **Patent** DE 1930540
Manufacturers Aimco; Elf Atochem; Nippon Soda; Rallis

APPLICATIONS
Biochemistry Carbendazim precursor. **Mode of action** Systemic fungicide with protective and curative action. Absorbed by the leaves and roots. **Uses** A fungicide used at 30–50 g a.i./ha and effective against a wide range of fungal pathogens including: eyespot and other diseases of cereals; scab on apples and pears; *Monilia* disease and *Gloeosporium* rot on apples; *Monilia* spp. on stone fruit; canker on fruit trees; powdery mildews on pome fruit, stone fruit, vegetables, cucurbits, strawberries, vines, roses, etc.; *Botrytis* and *Sclerotinia* spp. on various crops; leaf spot diseases on beet, oilseed rape, celery, celeriac, etc.; club root on brassicas; dollar spot, *Corticium*, and *Fusarium* spp. on turf; grey mould in vines; *Pyricularia oryzae* in rice; sigatoka disease in bananas; and many diseases in floriculture. Also used on almonds, pecans, tea, coffee, peanuts, soya beans, tobacco, chestnuts, sugar cane, citrus fruit, figs, hops, mulberries, and many other crops. Used additionally as a wound protectant for pruning cuts on trees.
Formulation types SC; WP; DP; PA. **Mixtures** *(thiophanate-methyl +)* etridiazole; lindane; iprodione; vinclozolin; maneb; thiram; mancozeb; captan; folpet; zineb; gamma-HCH. **Compatibility** Incompatible with alkaline and copper-containing compounds. **Selected tradenames** 'Topsin M' (Nippon Soda, Elf Atochem); 'Aimthyl' (Aimco); 'Cekufanato' (Cequisa); 'Cycosin' (Cyanamid); 'Do' (Sanonda); 'Mildothane' (Rhône-Poulenc); 'Roko' (Wockhardt)

ANALYSIS
Product analysis by hplc (*CIPAC Handbook,* 1988, **D,** 162). **Residues** determined by colorimetry (*Pestic. Anal. Man.,* 1979, **II**; V. L. Miller *et al., J. Assoc. Off. Anal. Chem.,* 1977, **60,** 1154), by glc (A. Ambrus *et al., ibid.,* 1981, **64,** 733), by hplc (S. Ono, *Nihon Noyaku Gakkaishi,* 1982, **7,** 363; N. Shiga *et al., ibid.,* 1977, **2,** 27; *Anal. Methods Residues Pestic.,* 1988, Part I, M3).

MAMMALIAN TOXICOLOGY
Reviews *Pesticide residues in food – 1995,* FAO Plant Production and Protection Paper 133, 1996. *Pesticide residues in food – 1995 evaluations; Part II – Toxicology & Environment.* World Health Organisation, WHO/PCS/96.48, 1996. **Oral** Acute oral LD_{50} for male rats 7500, female rats 6640, male mice 3510, male rabbits 2270 mg/kg. **Skin and eye** Acute percutaneous LD_{50} for male and female rats >10 000 mg/kg. Mild skin and eye irritant. **Inhalation** LC_{50} (4 h) for rats 1.7 mg/l air. **NOEL** (2 y) for rats and mice 160 mg/kg diet, for dogs 50 mg/kg diet. **ADI** (JMPR) 0.02 mg/kg b.w. [1995]. **Toxicity class** WHO (a.i.) III (Table 5); EPA (formulation) IV **EC risk** (R40)

ECOTOXICOLOGY
Birds Acute oral and percutaneous LD_{50} for Japanese quail >5000 mg/kg.
Fish LC_{50} (48 h) for rainbow trout 7.8, carp 11 mg/l.   **Bees** Not toxic to bees;
LD_{50} (topical) >100 μg/bee.   **Daphnia** LC_{50} (48 h) 20.2 mg/l.   **Algae** EC_{50} (96 h)
for *Chlorella* 0.8 mg/l.

ENVIRONMENTAL FATE
Animals In rats, following oral administration, 61% is excreted in the urine and 35%
in the faeces within 90 minutes after the last dose. Metabolism involves cyclisation
to carbendazim (q.v.). The principal metabolite in rats is methyl 5-hydroxy=
benzimidazol-2-carbamate.   **Plants** In plants, cyclisation occurs, leading to the
formation of carbendazim (q.v.).   **Soil/Environment** Soil persistence is *c.* 3–4 w. In
soil, in aqueous solution, and under the influence of u.v. light, cyclisation occurs,
leading to the formation of carbendazim (q.v.). This then undergoes degradation
to 2-aminobenzimidazole and 5-hydroxy-2-aminobenzimidazole. Soil adsorption
K_d 1.2.

712   thiram   *Fungicide*

dimethyldithiocarbamate

NOMENCLATURE
Common name thiram (BSI, E-ISO); thirame ((*m*) F-ISO); TMTD (USSR); thiuram
(JMAF)
IUPAC name tetramethylthiuram disulfide; bis(dimethylthiocarbamoyl) disulfide
Chemical Abstracts name tetramethylthioperoxydicarbonic diamide
CAS RN [137–26–8]   **EEC no.** 205–286–2   **Official codes** ENT 987

PHYSICAL CHEMISTRY
Mol. wt. 240.4   **M.f.** $C_6H_{12}N_2S_4$ **Form** Colourless crystals.   **M.p.** 155–156 °C
V.p. 2.3 mPa (25 °C)   **K_{ow}** logP = 1.73   **S.g./density** 1.29 (20 °C)   **Solubility** In
water 18 mg/l (room temperature). In ethanol <10, acetone 80, chloroform 230
(all in g/l, room temperature). In hexane 0.04, dichloromethane 170, toluene 18,
isopropanol 0.7 (all in g/l, 20 °C).   **Stability** Decomposed in acidic media. Some
deterioration on prolonged exposure to heat, air or moisture. DT_{50} (est.) (22 °C)
128 d (pH 4), 18 d (pH 7), 9 h (pH 9).

COMMERCIALISATION

History Fungicidal properties described by W. H. Tisdale & A. L. Flenner (*Ind. Eng. Chem.*, 1942, **34,** 501). Introduced by E. I. du Pont de Nemours and Co. (who no longer manufacture or market it), by Bayer AG and later by other companies. **Patent** US 1972961 to Du Pont; DE 642532 to I. G. Farbenindustrie **Manufacturers** General Quimica; India Pesticides; UCB; Uniroyal

APPLICATIONS

Mode of action Basic contact fungicide with protective action. **Uses** Protective fungicide applied to foliage to control: *Botrytis* spp. on grapes, soft fruit, lettuce, vegetables and ornamentals; rust on ornamentals; scab and storage diseases on apple and pear; leaf curl and *Monilia* on stone fruit. Used in seed treatments alone or in combination with added insecticides or fungicides to control damping-off diseases (e.g. *Pythium* spp.), and other diseases like *Fusarium* spp. of maize, cotton, cereals, legumes, vegetables and ornamentals. **Formulation types** WP; DP; WS; SC; WG; Liquid seed treatment; FS. **Mixtures** *(thiram +)* benomyl; carboxin; chlozolinate; dichlofenthion; iprodione; isofenphos; tebuconazole; pencycuron; triadimenol; thiophanate-methyl; dicloran; 2-(1-naphthyl)acetic acid + 4-indol-3-ylbutyric acid; ziram; pencycuron; permethrin + petroleum oil; lindane + rotenone; bendiocarb; tebuconazole; triadimenol; thiabendazole; vinclozolin; zineb. **Selected tradenames** 'Pomarsol' (Bayer); 'Aatiram' (AgrEvo); 'Ceku TMTD' (Cequisa); 'Rhodiason' (Rhône-Poulenc); 'Thianosan' (UCB); 'Thiram' (Uniroyal); 'Thiratox' (Efthymiadis); 'Tiurante' (General Quimica)

ANALYSIS

Product analysis by hplc (*CIPAC Handbook*, 1988, **D,** 169) or by hydrolysis to dimethylamine, estimated by titration (*ibid.*, 1970, **1,** 672; 1980, **1A,** 1360; *AOAC Methods*, 1995, 966.08). **Residues** determined by conversion to carbon disulfide, estimated by glc or colorimetry of a derivative (*ibid.*, 972.29; *Analyst (London)*, 1981, **106,** 782). *Selective determination of thiram in crops*, L. Roland in *Med. Fac. Landbouw. Univ. Gent*, 1992, **57,** 1255–1260. Details available from UCB Chemicals.

MAMMALIAN TOXICOLOGY

Reviews *Pesticide residues in food – 1992.* FAO Plant Production and Protection Paper, 116, 1993. *Pesticide residues in food – 1992 evaluations. Part II – Toxicology.* World Health Organisation, WHO/PCS/93.34. **IARC** 12, 53 **EHC** 78 (WHO, 1988). **Oral** Acute oral LD_{50} for rats 2600, mice 1500–2000, rabbits 210 mg/kg. **Skin and eye** Acute percutaneous LD_{50} for rabbits >2000 mg/kg. Moderate eye irritant, slight skin irritant. In percutaneous toxicity test in humans, application of the dry powder to the skin produced very slight erythema in 9% of cases. Skin sensitiser (guinea pigs). **Inhalation** LC_{50} (4 h) for rats 4.42 mg/l air. **NOEL** (2 y) for rats 1.5 mg/kg b.w. daily; (1 y) for dogs 0.75 mg/kg b.w. daily. **ADI** (JMPR) 0.01 mg/kg b.w. [1992]. **Toxicity class** WHO (a.i.) III; EPA (formulation) III **EC risk** Xn (R20/22, R40, R43); Xi (R36/37)

ECOTOXICOLOGY

Birds Acute oral LD_{50} for male ring-necked pheasant 673, mallard duck >2800, starling >100, redwing blackbird >100 mg/kg. LC_{50} (8 d) for ring-necked pheasant >5000, mallard duck >5000, bobwhite quail >3950, Japanese quail >5000 ppm. **Fish** LC_{50} (96 h) for bluegill sunfish 0.0445, rainbow trout 0.128 mg/l. **Bees** LD_{50} (contact) 73.7 μg/bee (75% formulation). **Daphnia** LC_{50} (48 h) 0.21 mg/l.

ENVIRONMENTAL FATE

Animals The main metabolites formed are COS and CS_2. **Plants** The major metabolite in plants is dimethylamine salt of dimethyldithiocarbamic acid; tetramethylthiuram disulfide, tetramethylthiuram monosulfide, tetramethylthiourea, carbon disulfide and sulfur can also be formed. Dimethyldithiocarbamic acid can be present as the free acid or as the metabolic conversion products DDC-β-glucoside, DDC-α-aminobutyric acid and DDC-α-alanine. **Soil/Environment** DT_{50} 0.5 d (sandy soil, pH 6.7).

713 tiocarbazil   *Herbicide*

thiocarbamate

NOMENCLATURE

Common name tiocarbazil (BSI, draft E-ISO, (m) draft F-ISO)
IUPAC name S-benzyl di-sec-butylthiocarbamate
Chemical Abstracts name S-(phenylmethyl) bis(1-methylpropyl)carbamothioate
CAS RN *[36756–79–3]* **EEC no.** 253–190–4 **Development codes** M 3432

PHYSICAL CHEMISTRY

Mol. wt. 279.4 **M.f.** $C_{16}H_{25}NOS$ **Form** Colourless liquid with aromatic odour.
B.p. 130–132 °C/0.1 mmHg **V.p.** 93 mPa (50 °C) **K_{ow}** logP = 4.4
S.g./density 1.023 (20 °C) **Solubility** In water 2.5 mg/l (30 °C). Miscible with polar and non-polar organic solvents. **Stability** Stable to hydrolysis at pH 5.6–8.4. Slightly decomposed after 30 d at 40 °C in aqueous ethanol at pH 1.5. Stable to storage for 60 d at 40 °C, and for 100 h in aqueous solution exposed to sunlight.

COMMERCIALISATION

History Herbicide reported by N. Caracalli *et al. (Proc. Congr. Risicultura, 8th,*

1973, p. 446). Introduced by Montedison S.p.A. (now Isagro S.p.A.) in Italy (1974).
Patent IT 907710; DE 2144700   **Manufacturers** Isagro

APPLICATIONS
Biochemistry Cell elongation inhibitor.   **Mode of action** Selective herbicide,
absorbed by the roots and coleoptiles.   **Uses** Control of *Echinochloa* spp., *Lolium*
perenne, Cyperus spp., and other monocotyledonous weeds in rice paddy fields.
Applied pre- or post-emergence at 4 kg a.i./kg, or as a seed dressing.
Phytotoxicity Non-phytotoxic to rice.   **Formulation types** SL; Seed treatment; GR;
EC.   **Selected tradenames** 'Drepamon' (Isagro)

ANALYSIS
Product and **residue** analysis by tlc (R. Fabbrini & G. Galluzzi, *Anal. Methods Pestic.*
Plant Growth Regul., 1980, **11**, 307). **Product** analysis also by glc method available
from Isagro.

MAMMALIAN TOXICOLOGY
EHC 76 (WHO, 1988; general review of thiocarbamates).   **Oral** Acute oral LD_{50}
for rats, rabbits, and guinea pigs >10 000 mg tech./kg, for mice 8000 mg/kg.
Skin and eye Acute percutaneous LD_{50} for rats and rabbits >1200 mg/kg.
Inhalation LC_{50} for rats >0.18 mg/l air.   **NOEL** In 2 y feeding trials, albino rats and
beagle dogs receiving 1000 mg tech./kg diet suffered no ill-effect except for a
slight weight loss in male dogs. No effect on reproduction in albino rats receiving
300 mg/kg diet for 3 generations.   **Toxicity class** WHO (a.i.) III (Table 5)

ECOTOXICOLOGY
Birds Acute oral LD_{50} for chickens, pheasants, quail >10 000 mg tech./kg.
Fish LC_{50} in several fish species ≥8 mg/l.   **Bees** Not hazardous for honeybees.
Other aquatic spp. LC_{50} for the mollusc *Australorbis glabratus* >60 mg/l.

ENVIRONMENTAL FATE
Animals Tiocarbazil, when administered orally at the dose of 1 g/kg, is almost
totally eliminated 7 days after treatment, either as such in the faeces or as
products deriving from its metabolism, through the urine. In the blood and in the
main organs, minimum concentrations of the product and of its metabolites have
been found 48 hours after treatment.   **Plants** Tiocarbazil undergoes extensive
metabolism in rice: 3 compounds have been identified in plants (except
caryopsides): N,N-di-sec-butylcarbamoylthiolglycolic acid, N,N-di-sec-butyl=
carbamoyl benzylsulfoxide, N,N-di-sec-butylcarbamoyl benzylsulfone. The first by
oxidation of the benzene ring, the other two by oxidation of the sulfur atom.
Soil/Environment Tiocarbazil is strongly adsorbed by the soil (where it largely
localises in the uppermost layer), K_{om} 1711, and undergoes a rapid degradation
due to the attack of soil micro-organisms; 50% of tiocarbazil applied is no longer
present in the soil/water system of a rice field 8–15 days after treatment.

organophosphate ester

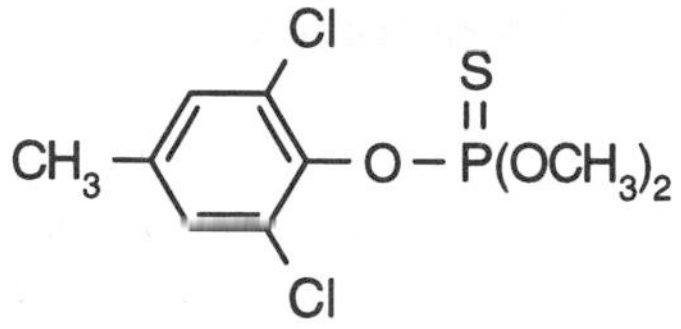

NOMENCLATURE
Common name tolclofos-methyl (BSI, E-ISO, (*m*) F-ISO)
IUPAC name *O*-2,6-dichloro-*p*-tolyl *O,O*-dimethyl phosphorothioate
Chemical Abstracts name *O*-(2,6-dichloro-4-methylphenyl) *O,O*-dimethyl phosphorothioate
CAS RN *[57018–04–9]* **Development codes** S-3349 (Sumitomo)

PHYSICAL CHEMISTRY
Mol. wt. 301.1 **M.f.** $C_9H_{11}Cl_2O_3PS$ **Form** Colourless crystals; (tech., colourless to light brown solid). **M.p.** 78–80 °C **V.p.** 57 mPa (20 °C) **K$_{ow}$** logP = 4.56 (25 °C) **Solubility** In water 1.10 ppm (25 °C). In hexane 3.8%, xylene 36.0%, methanol 5.9%. **Stability** Stable to light, heat and moisture. Decomposed in alkaline and acidic media. **F.p.** 210 °C (Cleveland open cup)

COMMERCIALISATION
History Introduced by Sumitomo Chemical Co., Ltd. **Patent** GB 1467561; US 4039635 **Manufacturers** Sumitomo

APPLICATIONS
Biochemistry Acts by inhibition of phospholipid biosynthesis, leading to inhibition of germination of spores and growth of fungal mycelium. **Mode of action** Non-systemic contact fungicide with protective and curative action. **Uses** Control of soil-borne diseases caused by *Rhizoctonia*, *Corticium*, *Sclerotium* and *Typhula* spp. on potatoes, sugar beet, cotton, peanuts, vegetables, cereals, ornamentals, bulb flowers, lawn turf, etc. Used as a seed, bulb or tuber treatment, soil drench, foliar spray, or by soil incorporation. **Formulation types** WP; DP; EC; SC; Seed treatment. **Mixtures** (*tolclofos-methyl* +) thiram. **Compatibility** Incompatible with alkaline compounds. **Selected tradenames** 'Rizolex' (Sumitomo, Du Pont)

ANALYSIS
Details of **product** and **residue** analysis available from Sumitomo Chemical Co.

MAMMALIAN TOXICOLOGY
Reviews *Pesticide residues in food – 1994*, FAO Plant Production and Protection Paper, 127, 1995. *Pesticide residues in food – 1994 evaluations, Part II – Toxicology,*

MAIN ENTRIES

WHO, WHO/PCS/95.2, 1995.   **Oral** Acute oral LD$_{50}$ for rats 5000 mg/kg.
Skin and eye Acute percutaneous LD$_{50}$ for rats >5000 mg/kg. Non-irritating to
skin and eyes (rabbits).   **Inhalation** LC$_{50}$ for rats >3320 mg/m^3.   **NOEL** In 6 mo
feeding trials, rats showed no ill-effects.   **ADI** (JMPR) 0.07 mg/kg b.w. [1994].
Toxicity class WHO (a.i.) III (Table 5); EPA (formulation) III

ECOTOXICOLOGY
Birds Acute oral LD$_{50}$ for mallard ducks and bobwhite quail >5000 mg/kg.
Fish LC$_{50}$ (96 h) for bluegill sunfish >720 µg/l.

ENVIRONMENTAL FATE
Animals In mammals, tolclofos-methyl is rapidly metabolised mainly by oxidative
desulfuration of the P=S group to a P=O group, oxidation at the 4-methyl group,
and cleavage of the P-O-aryl and P-O-methyl linkages. Excretion is rapid, and is
substantially complete within a few days.   **Soil/Environment** Persistent in the soil,
based on a balance between rapid biodegradability and high physicochemical
stability. Photolytically degraded (8 h sunlight/day), DT$_{50}$ 44 d in water, 15–28 d in
lake and river water, <2 d on the soil surface (including loss by evaporation).
Degradation proceeds by demethylation and hydrolysis to 2,5-dichlorocresol.

715 tolylfluanid   *Fungicide*

N-trihalomethylthio

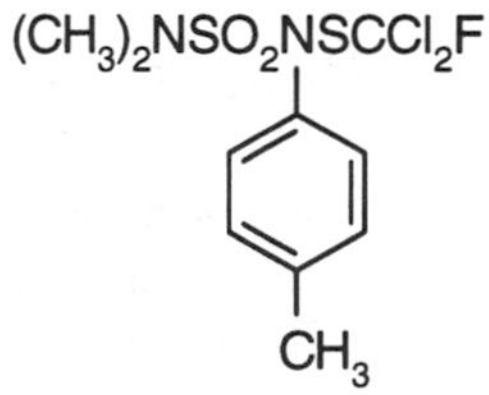

$(CH_3)_2NSO_2NSCCl_2F$

CH$_3$

NOMENCLATURE
Common name tolylfluanid (BSI, E-ISO); tolylfluanide ((*m*) F-ISO)
IUPAC name *N*-dichlorofluoromethylthio-*N'*,*N'*-dimethyl-*N*-*p*-tolylsulfamide
Chemical Abstracts name 1,1-dichloro-*N*-[(dimethylamino)sulfonyl]-1-fluoro-=
N-(4-methylphenyl)methanesulfenamide
CAS RN *[731–27–1]*   **EEC no.** 211–986–9   **Development codes** BAY 49 854;
KUE 13183B (Bayer)

PHYSICAL CHEMISTRY
Mol. wt. 347.2   **M.f.** C$_{10}$H$_{13}$Cl$_2$FN$_2$O$_2$S$_2$ **Form** Colourless to pale yellow,
odourless crystalline powder.   **M.p.** 93 °C   **B.p.** Decomposes on distillation.

V.p. 0.2 mPa (20 °C)   **K$_{ow}$** logP = 3.90 (20 °C)   **S.g./density** 1.52 g/cm^3 (20 °C)
Solubility In water 0.9 mg/l (room temperature). In hexane 5–10,
dichloromethane >250, isopropanol 20–50, toluene >200 (all in g/l, room
temperature).   **Stability** Hydrolysis DT$_{50}$ (22 °C) 12 d (pH 4), 29 h (pH 7),
<10 min (pH 9). Under environmental conditions, hydrolysis occurs much more
rapidly than photolysis.

COMMERCIALISATION
History Fungicide reported by H. Kaspers & F. Grewe (*Abstr. Int. Congr. Crop Prot.,
6th Vienna, 1967*, p. 345). Introduced by Bayer AG.   **Patent** DE 1193498 to Bayer
Manufacturers Bayer

APPLICATIONS
Biochemistry Multi-site action.   **Mode of action** Foliar basic fungicide with
protective action.   **Uses** Used in deciduous fruit crops; when used regularly for
control of *Venturia* spp., there is usually no need for additional steps to control
Podosphaera leucotricha on apples, or red spider mites. Very good activity against
storage diseases when used for final treatments. Recommended for use in IPM
programmes. In grapes, used in rotation with specifically acting fungicides against
Uncinula necator, Plasmopara viticola and *Botrytis cinerea*. In vegetables, against *B.
cinerea, Peronospora* spp. and powdery mildew. In ornamentals, against powdery
mildew, rust and leaf spot diseases.   **Formulation types** WP; WG; WS.
Mixtures (*tolylfluanid +*) tebuconazole.   **Compatibility** Incompatible with liquid
insecticides and with alkaline substances (e.g. Bordeaux mixture, lime sulfur).
Selected tradenames 'Elvaron M' (Bayer); 'Euparen M' (Bayer)

ANALYSIS
Product analysis by hplc (*CIPAC Handbook*, 1985, **1C**, 2083; *ibid.*, 1992, **E**, 222).
Residues determined by glc (R. Brennecke, *Pflanzenschutz-Nachr. Bayer (Engl. Ed.)*,
1988, **41**, 137; 1989, **42**, 223). Details available from Bayer.

MAMMALIAN TOXICOLOGY
Reviews *Pesticide residues in food – 1988*. FAO Plant Production and Protection
Paper 92, 1988. *Pesticide residues in food – 1988 evaluations. Part II – Toxicology*.
FAO Plant Production and Protection Paper 93/2, 1989.   **Oral** Acute oral LD$_{50}$
for rats >5000, mice >1000, guinea pigs 250–500 mg/kg.   **Skin and eye** Acute
percutaneous LD$_{50}$ for rats >5000 mg/kg; severely irritating to skin and
moderately irritating to eyes (rabbit). Skin sensitiser.   **Inhalation** LC$_{50}$ (4 h) for rats
c. 0.16 mg/l air (aerosol).   **NOEL** (2 y) for rats and mice 300 mg/kg diet, for
dogs 12.5 mg/kg b.w.   **ADI** (JMPR) 0.1 mg/kg b.w. [1988].   **Toxicity class**
WHO (a.i.) III (Table 5); EPA (formulation) II   **EC risk** T+ (R26, R36/38, R43)

ECOTOXICOLOGY
Birds LD$_{50}$ for Japanese quail >5000 mg/kg. Five-day dietary LC$_{50}$ for bobwhite
quail >5000 mg/kg.   **Fish** LC$_{50}$ (96 h) for golden orfe 0.06, rainbow trout

MAIN ENTRIES

0.05 mg/l. **Bees** Non-toxic to bees. **Worms** LC_{50} for *Eisenia foetida* >1000 mg/kg dry soil. **Daphnia** LC_{50} (48 h) 0.57 mg/l. **Algae** E_rC_{50} for *Scenedesmus subspicatus* >1.0 mg/l.

ENVIRONMENTAL FATE
Animals In animals, ^{14}C-tolylfluanid is rapidly absorbed and the radioactivity excreted. Tolylfluanid is rapidly hydrolysed to DMST (dimethylamino sulfotoluidide) and then transformed to the main metabolite 4-(dimethyl= aminosulfonylamino)benzoic acid which can be further demethylated. **Plants** In plants, tolylfluanid is rapidly hydrolysed to DMST (see above) which is further hydroxylated and conjugated. **Soil/Environment** In soil, tolylfluanid is rapidly hydrolysed to DMST (see above) with a DT_{50} of a few days. DMST is further degraded to methylaminosulfotoluidide, 4-(dimethylaminosulfonylamino)= benzoic acid, 4-(methylaminosulfonylamino)benzoic acid and finally CO_2. Due to the rapid hydrolysis, leaching into deeper soil layers is very unlikely.

716 *N-m-tolylphthalamic acid*

Plant growth regulator

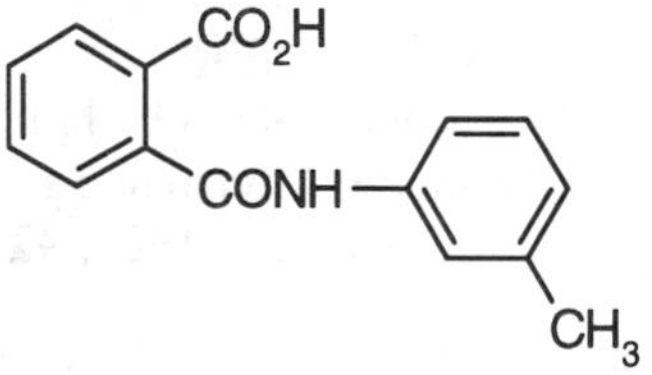

NOMENCLATURE
IUPAC name *N-m*-tolylphthalamic acid
Other names N-m-t   **CAS RN** *[85–72–3]*   **EEC no.** 201–626–9

PHYSICAL CHEMISTRY
Mol. wt. 255.3   **M.f.** $C_{15}H_{13}NO_3$ **Form** White powder.   **M.p.** 152 °C
Solubility In water 0.1 g/l.

COMMERCIALISATION
Manufacturers Makhteshim-Agan

APPLICATIONS
Mode of action Systemic plant growth regulator which prevents abscission of

flowers and young fruit.   **Uses** Increases fruit set of tomatoes, aubergines and lima beans under adverse conditions.   **Formulation types** WP.
Selected tradenames 'Tomaset' (Makhteshim-Agan)

MAMMALIAN TOXICOLOGY
Oral Acute oral LD_{50} for rats >5000 mg/kg.   **Skin and eye** Acute percutaneous LD_{50} for rabbits >20 000 mg/kg.   **Inhalation** LC_{50} for rats >5 mg/l.

'17 tralkoxydim   *Herbicide*

cyclohexanedione oxime

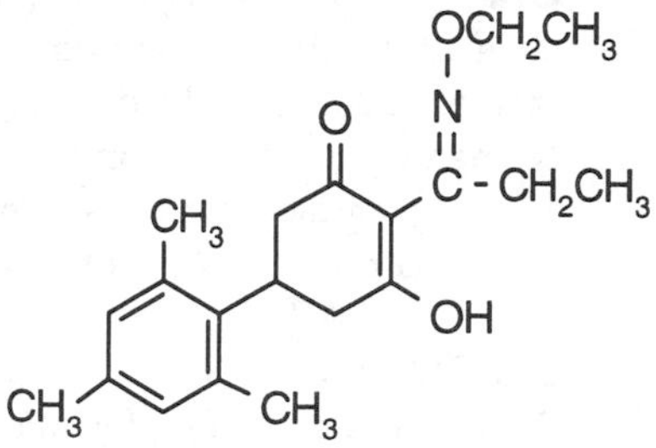

NOMENCLATURE
Common name tralkoxydim (BSI, draft E-ISO); tralkoxydime ((f) draft F-ISO)
IUPAC name 2-[1-(ethoxyimino)propyl]-3-hydroxy-5-mesitylcyclohex-2-enone
Chemical Abstracts name 2-[1-(ethoxyimino)propyl]-3-hydroxy-=
5-(2,4,6-trimethylphenyl)-2-cyclohexen-1-one
CAS RN *[87820–88–0]*   **Development codes** ICIA0604; PP604

PHYSICAL CHEMISTRY
Composition Tech. material is 95% pure.   **Mol. wt.** 329.4   **M.f.** $C_{20}H_{27}NO_3$
Form Colourless, odourless solid.   **M.p.** 106 °C; (tech. 99–104 °C)
V.p. 3.7×10^{-4} mPa (20 °C) K_{ow} logP = 2.1 (20 °C, unstated pH)
S.g./density 1.16 (25 °C)   **Solubility** In water 5 (pH 5), 6.7 (pH 6.5), 9800 (pH 9) (all in mg/l, 20 °C). In hexane 18, methanol 25, acetone 89, ethyl acetate 100, toluene 213, dichloromethane >500 (all in g/l, 24 °C).   **Stability** Stable >1.5 y (15–25 °C). DT_{50} (25 °C) 6 d (pH 5), 114 d (pH 7); after 28 d, 87% unchanged (pH 9).   **pKa** (25 °C) 4.3

COMMERCIALISATION
History Herbicide reported by R. B. Warner et al. (*Proc. 1987 Br. Crop Prot. Conf. – Weeds*, **1**, 19). Introduced by ICI Agrochemicals (now Zeneca Agrochemicals).
Manufacturers Zeneca

APPLICATIONS

Biochemistry Cell division inhibitor. Acetyl CoA carboxylase (fatty acid synthesis) inhibitor. **Mode of action** Selective systemic herbicide, absorbed by the leaves and translocated acropetally in the phloem to the growing points. **Uses** Control of *Avena* spp. and other grass weeds (including *Lolium* spp., *Setaria viridis*, *Phalaris* spp., *Alopecurus myosuroides* and *Apera spica-venti*) in wheat and barley. **Formulation types** EC; SC; WG. **Compatibility** Compatible with a range of broad-leaved weed herbicides. **Selected tradenames** 'Achieve' (Zeneca); 'Grasp' (Zeneca)

ANALYSIS

Product and **residue** analysis by hplc. Details available from Zeneca.

MAMMALIAN TOXICOLOGY

Oral Acute oral LD_{50} for male rats 1324, female rats 934, male mice 1231, female mice 1100, male rabbits >519 mg/kg. **Skin and eye** Acute percutaneous LD_{50} for rats >2000 mg/kg. Mild skin and eye irritant (rabbits). Not a skin sensitiser (guinea pigs). **Inhalation** LC_{50} (4 h) for rats >3.5 mg/l air. **NOEL** (90 d) for rats 12.5 mg/kg diet, for dogs 5 mg/kg diet. Not carcinogenic. **Other** Non-mutagenic in standard tests. Non-teratogenic in rabbits. **Toxicity class** WHO (a.i.) III

ECOTOXICOLOGY

Birds Acute oral LD_{50} for mallard ducks >3020, partridges 4430 mg/kg. Dietary LC_{50} for mallard ducks >7400, quail 6237 mg/kg diet. **Fish** LC_{50} (96 h) for mirror carp >8.2, bluegill sunfish >6.1, rainbow trout >7.2 mg/l. **Bees** LD_{50} (contact) >0.1 mg/bee, (oral) 0.054 mg/bee.

ENVIRONMENTAL FATE

Plants Degrades rapidly in crops. At a limit of detection of 0.02 mg/kg, no residues of tralkoxydim or its metabolites have been found in wheat or barley at harvest, after application at up to double the recommended use rates.
Soil/Environment Degrades rapidly in soil, DT_{50} (aerobic) *c.* 3 d, (20 °C), (flooded soil) *c.* 25 d.

tralomethrin   *Insecticide*

pyrethroid

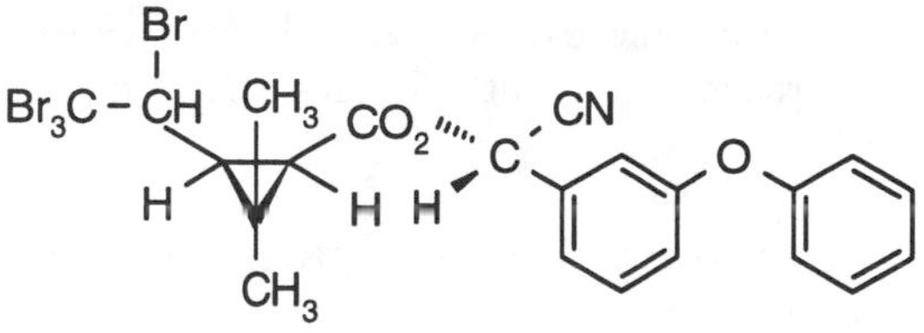

NOMENCLATURE
Common name tralomethrin (BSI, ANSI, draft E-ISO); tralométhrine (draft F-ISO)
IUPAC name (S)-α-cyano-3-phenoxybenzyl (1R,3S)-2,2-dimethyl-3-[(RS)-1,2,2,2-=
tetrabromoethyl]cyclopropanecarboxylate *Roth*: (S)-α-cyano-3-phenoxybenzyl
(1R)-*cis*-2,2-dimethyl-3-[(RS)-1,2,2,2-tetrabromoethyl]cyclopropanecarboxylate
Chemical Abstracts name cyano(3-phenoxyphenyl)methyl 2,2-dimethyl-=
3-(1,2,2,2-tetrabromoethyl)cyclopropanecarboxylate
CAS RN *[66841–25–6]* (unstated stereochemistry)   **Development codes** NU 831;
RU 25474; HAG 107   **Official codes** OMS 3048

PHYSICAL CHEMISTRY
Composition Laboratory grade tralomethrin (>93% a.i.) is a 60:40 mixture of two
active diastereoisomers.   **Mol. wt.** 665.0   **M.f.** $C_{22}H_{19}Br_4NO_3$ **Form** Orange-to-
yellow resinous solid (tech.).   **M.p.** 138–148 °C   **V.p.** 4.8×10^{-6} mPa (25 °C)
K_{ow} logP = *c*. 5   **Henry** 3.95×10^{-10} Pa m^3 mol^{-1} **S.g./density** 1.70 (20 °C)
Solubility In water 80 µg/l. In acetone, dichloromethane, toluene, xylene >1000,
dimethyl sulfoxide >500, ethanol >180 (all in g/l).   **Stability** Stable for 6 months at
50 °C. Acidic media reduce hydrolysis and epimerisation.   **Specific rotation** $[\alpha]_D$
+21–27° (50 g/l toluene)   **F.p.** 26 °C

COMMERCIALISATION
History Discovered and introduced by Roussel Uclaf.   **Patent** FR 2364884
Manufacturers AgrEvo EH

APPLICATIONS
Biochemistry Activity may be due to debromination to form deltamethrin (*q.v.*).
Mode of action Non-systemic insecticide with contact and stomach action.
Uses Insecticide with ingested and contact action. Controls a range of agronomic
pests (Coleoptera, Homoptera, Orthoptera and particularly Lepidoptera in
cereals, coffee, cotton, fruit, maize, oilseed rape, rice, tobacco and vegetables) at
7.5–20 g a.i./ha. If applied before infestations are well established, protects most
crops against plant-sucking Hemiptera. Soil-surface sprays (5–10 g/ha) control
cutworms (*Agrotis* and *Euxoa* spp.). Effective for wood protection, insect
household pests, in public health, and in stored grains (Blattodea, Culicidae,

Muscidae, grain borers and wood-attacking insects). Also very effective on Muscidae, Tabanidae, Ixodidae and other Acari on cattle, sheep and pigs, by application as dip, spray or pour-on. **Formulation types** EC; SC; WP.
Selected tradenames 'Saga' (environmental health uses) (AgrEvo); 'Scout' (crop protection uses) (AgrEvo); 'Tralox' (veterinary uses) (Hoechst Roussel Vet.); 'Tracker' (Du Pont); 'Tralate' (Du Pont)

ANALYSIS
Product analysis by hplc.   **Residues** determined by glc with ECD.

MAMMALIAN TOXICOLOGY
Oral Acute oral LD_{50} for rats 99–3000 mg a.i./kg depending on the carrier used; for dogs >500 mg (in capsules)/kg.   **Skin and eye** Acute percutaneous LD_{50} for rabbits >2000 mg/kg. Moderate skin irritant; mild eye irritant (rabbits).
Inhalation LC_{50} (4 h) for rats >0.286 mg/l air.   **NOEL** (2 y) for rats 0.75 mg/kg b.w. daily; for mice 3 mg/kg b.w. daily; for dogs 1 mg/kg b.w. daily.
ADI 0.0075 mg/kg b.w.   **Other** Tech. not mutagenic or teratogenic in rats or rabbits.   **Toxicity class** WHO (a.i.) II; EPA (formulation) III

ECOTOXICOLOGY
Birds Acute oral LD_{50} for quail >2510 mg/kg. Eight-day dietary LC_{50} for mallard ducks 7716, quail 4300 mg/kg diet.   **Fish** LC_{50} (96 h) for rainbow trout 0.0016, bluegill sunfish 0.0043 mg/l.   **Bees** Non-toxic to honeybees. LD_{50} (contact) 0.12 µg/bee. Low LC_{50} and LD_{50} values under laboratory conditions do not represent significant hazard to bees in normal field use.   **Daphnia** LC_{50} (48 h) 38 ng/l. Low LC_{50} and LD_{50} values under laboratory conditions do not represent significant hazard to aquatic fauna in normal field use.

ENVIRONMENTAL FATE
Animals Tralomethrin is transformed into deltamethrin and further degraded.
Plants Tralomethrin is transformed into deltamethrin.   **Soil/Environment** Strongly adsorbed in soil; DT_{50} 64–84 d. K_d 197–8784; K_{oc} 43 796–675 667; highly immobile in various soils (sandy to clay loam).

pyrethroid

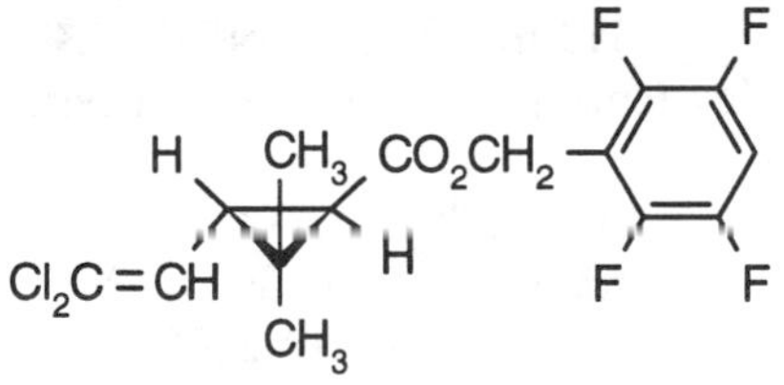

NOMENCLATURE
Common name transfluthrin (BSI, pa E-ISO)
IUPAC name 2,3,5,6-tetrafluorobenzyl (1R,3S)-3-(2,2-dichlorovinyl)-2,2-dimethyl= cyclopropanecarboxylate *Roth*: 2,3,5,6-tetrafluorobenzyl (1R)-*trans*-3-(2,2-= dichlorovinyl)-2,2-dimethylcyclopropanecarboxylate
Chemical Abstracts name (1R-*trans*)-(2,3,5,6-tetrafluorophenyl)methyl 3-(2,2-dichloroethenyl)-2,2-dimethylcyclopropanecarboxylate
Other names benfluthrin (rejected BSI name proposal)   **CAS RN** [118712–89–3]
EEC no. 405–060–5   **Development codes** NAK 4455

PHYSICAL CHEMISTRY
Composition Tech. is ≥92%.   **Mol. wt.** 371.2   **M.f.** $C_{15}H_{12}Cl_2F_4O_2$
Form Colourless crystals.   **M.p.** 32 °C   **B.p.** 135 °C/0.1 mbar   **V.p.** 4.0×10^{-1} mPa (20 °C)   **K_{ow}** logP = 5.46 (20 °C)   **S.g./density** 1.5072 g/cm^3 (23 °C)
Solubility In water 5.7×10^{-5} g/l (20 °C). In organic solvents >200 g/l.
Stability No decomposition after 5 h at 200 °C. DT_{50} in pure water (25 °C) >1 y (pH 5), >1 y (pH 7), 14 d (pH 9).   **Specific rotation** $[\alpha]_D^{29}$ +15.3° (c = 0.5 CHCl$_3$)

COMMERCIALISATION
Patent DE 2714042; DE 3705224   **Manufacturers** Bayer

APPLICATIONS
Biochemistry Effector on presynaptic voltage gate sodium channels in nerve membranes, causing knock-down effects in insects.   **Mode of action** Insecticide active by inhalation and contact; also repellent.   **Uses** Fast-acting insecticide against mosquitoes, flies, cockroaches and whitefly.   **Formulation types** Spraying cans; Vapouriser liquid; Mosquito coils; VP;AL; FU; XX.   **Mixtures** (*transfluthrin* +) propoxur; cyfluthrin; dichlorvos.   **Selected tradenames** 'Baygon' (Bayer); 'Bayothrin' (Bayer)

ANALYSIS
Methods for the determination of residues are available from Bayer.

MAIN ENTRIES

MAMMALIAN TOXICOLOGY

Oral Acute oral LD_{50} for male and female rats >5000, male mice 583, female mice 688, hens >5000 mg/kg.   **Skin and eye** Acute percutaneous LD_{50} (24 h) for male and female rats >5000 mg/kg.   **Inhalation** LC_{50} (4 h) for male and female rats >513 mg/m^3 air.   **NOEL** (24 mo) for male and female rats 20 ppm, for male and female mice 100 ppm.   **Toxicity class** WHO (a.i.) III (Table 5)   **EC risk** Xi (R38)

ECOTOXICOLOGY

Birds Acute oral LD_{50} for Virginian tree quail (*Colinus virginianus*) and canary bird (*Serinus canarius*) >2000 mg/kg.   **Fish** LC_{50} (96 h) for golden orfe 1.25, rainbow trout 0.7 µg/l.   **Daphnia** LC_{50} (48 h) 0.0017 mg/l.   **Algae** EC_{50} (96 h) for *Scenedesmus subspicatus* >0.1 mg/l.

ENVIRONMENTAL FATE

Animals Tetrafluorobenzoic acid and the glucuronate of tetrafluorobenzyl alcohol are formed.   **Soil/Environment** In aquatic ecosystems, tetrafluorobenzyl alcohol and tetrafluorobenzoic acid are formed. DT_{50} in water 2 d (in light), 8 d (in dark).

720   triadimefon   *Fungicide*

azole

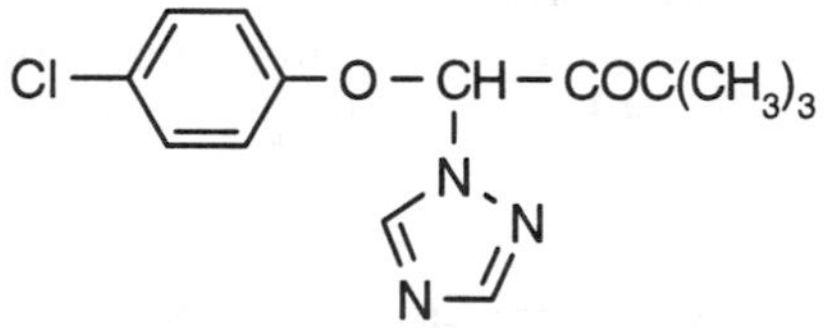

NOMENCLATURE

Common name triadimefon (BSI, E-ISO); triadiméfone ((m) F-ISO)
IUPAC name 1-(4-chlorophenoxy)-3,3-dimethyl-1-(1*H*-1,2,4-triazol-1-yl)= butan-2-one
Chemical Abstracts name 1-(4-chlorophenoxy)-3,3-dimethyl-1-(1*H*-1,2,4-triazol-= 1-yl)-2-butanone
CAS RN *[43121–43–3]* (unstated stereochemistry)   **EEC no.** 256–103–8
Development codes BAY MEB 6447; BAY 129128

PHYSICAL CHEMISTRY

Composition Racemate (i.e. 1:1 mixture of (1*R*)- and (1*S*)- enantiomers).
Mol. wt. 293.8   **M.f.** $C_{14}H_{16}ClN_3O_2$ **Form** Colourless crystals with a weak characteristic odour.   **M.p.** Modification 1: 78 °C; modification 2: 82 °C

V.p. 0.02 mPa (20 °C); 0.06 mPa (25 °C)   **K$_{ow}$** logP = 3.11   **Henry** 9 × 10^{-5} Pa m^3 mol^{-1} (20 °C)   **S.g./density** 1.283 (21.5 °C)   **Solubility** In water 64 mg/l (20 °C). Moderately soluble in most organic solvents except aliphatics. In dichloromethane, toluene >200, isopropanol 99, hexane 6.3 (all in g/l, 20 °C).   **Stability** Stable to hydrolysis; DT$_{50}$ (22 °C) >1 y (pH 3, 6, and 9).

COMMERCIALISATION

History Fungicide reported by P. E. Frohberger (*Mitt. Biol. Bundesanst. Land-Forstwirtsch. Berlin-Dahlem*, 1973, **151**, 61) and by F. Grewe & K. H. Büchel (*ibid.*, p. 208). Introduced by Bayer AG.   **Patent** BE 793867; US 3912752   **Manufacturers** Bayer

APPLICATIONS

Biochemistry Steroid demethylation (ergosterol biosynthesis) inhibitor. **Mode of action** Systemic fungicide with protective, curative and eradicant action. Absorbed by the roots and leaves, with ready translocation in young growing tissues, but less ready translocation in older, woody tissues.   **Uses** Control of powdery mildews in cereals, pome fruit, stone fruit, berry fruit, vines, hops, cucurbits, tomatoes, vegetables, sugar beet, mangoes, ornamentals, turf, flowers, shrubs and trees; rusts in cereals, pines, coffee, seed grasses, turf, flowers, shrubs and trees; *Monilinia* spp. in stone fruit; black rot of grapes; leaf blotch, leaf spot, and snow mould in cereals; pineapple disease butt rot in pineapples and sugar cane; leaf spots and flower blights in flowers, shrubs and trees; and many other diseases of turf.   **Phytotoxicity** Ornamentals may be damaged if used at excessive rates.   **Formulation types** WP; EC; DP; WG; PA.   **Mixtures** *(triadimefon +)* captan; carbendazim; propineb; cymoxanil + propineb; tebuconazole; sulfur. **Compatibility** Compatible with WP formulations of other pesticides. **Selected tradenames** 'Bayleton' (Bayer)

ANALYSIS

Product analysis by hplc (*CIPAC Handbook*, 1985, **1C**, 2236; *AOAC Methods*, 1995, 985.08) or by i.r. spectrometry.   **Residues** of triadimefon and the corresponding alcohol (triadimenol) determined by glc (R. Brennecke, *Pflanzenschutz. Nachr. Bayer (Engl. Ed.)*, 1984, **37**, 68; W. Specht, *ibid.*, 1977, **30**, 55; W. Specht & M. Tillkes, *ibid.*, 1980, **33**, 61). In **drinking water**, by gc with FID (*AOAC Methods*, 1995, 991.07). Methods for the determination of residues are available from Bayer.

MAMMALIAN TOXICOLOGY

Reviews *Pesticide residues in food – 1985*. FAO Plant Production and Protection Paper 68, 1986. *Pesticide residues in food – 1985 evaluations. Part II – Toxicology*. FAO Plant Production and Protection Paper 72/2, 1986.   **Oral** Acute oral LD$_{50}$ for rats and mice *c.* 1000, rabbits 250–500, dogs >500 mg/kg.   **Skin and eye** Acute percutaneous LD$_{50}$ for rats >5000 mg/kg. Mildly irritating to eyes and skin

(rabbits).   **Inhalation** LC_{50} (4 h) for rats 3.27 mg/l air (dust), >0.5 mg/l (air).
NOEL (2 y) for rats 300, mice 50, dogs 330 mg/kg diet.   **ADI** (JMPR) 0.03 mg/kg
b.w. [1985].   **Toxicity class** WHO (a.i.) III; EPA (formulation) III   **EC risk** Xn (R22)

ECOTOXICOLOGY
Birds Acute oral LD_{50} for mallard ducks >4000 mg/kg. Dietary LC_{50} (5 d) for
mallard ducks >10 000, bobwhite quail >4640 mg/kg diet.   **Fish** LC_{50} (96 h) for
bluegill sunfish 11, orfe 13.8, rainbow trout 17.4 mg/l.   **Bees** LD_{50} (contact)
>100 μ/bee.   **Daphnia** LC_{50} (48 h) 11.3 mg/l.   **Algae** E_rC_{50} for *Scenedesmus*
subspicatus 1.71 mg/l.

ENVIRONMENTAL FATE
Animals In mammals, following oral administration, 83–96% is excreted unchanged
in the urine and faeces within 2 to 3 d. However, metabolism occurs in the liver,
mostly to triadimenol (*q.v.*) and its glucuronic acid conjugates. Half-life in blood
plasma is *c.* 2.5 h.   **Plants** In plants, the carbonyl group is reduced to a hydroxyl
group, with the formation of triadimenol (*q.v.*).   **Soil/Environment** In soil, the
carbonyl group is reduced to a hydroxyl group, with the formation of triadimenol
(*q.v.*). DT_{50} of triadimefon in sandy loam *c.* 18 d, in loam *c.* 6 d. K_{oc} 300.

721 triadimenol   *Fungicide*

azole

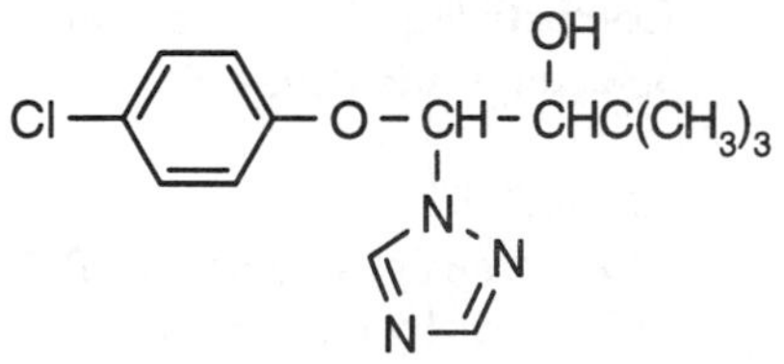

NOMENCLATURE
Common name triadimenol (BSI, E-ISO, (*m*) F-ISO)
IUPAC name (1*RS*,2*RS*;1*RS*,2*SR*)-1-(4-chlorophenoxy)-3,3-dimethyl-=
1-(1*H*-1,2,4-triazol-1-yl)butan-2-ol
Chemical Abstracts name β-(4-chlorophenoxy)-α-(1,1-dimethylethyl)-=
1*H*-1,2,4-triazole-1-ethanol
CAS RN [55219–65–3] unstated stereochemistry; [89482–17–7] diastereoisomer
A; [82200–72–4] diastereoisomer B   **EEC no.** 259–537–6
Development codes BAY KWG 0519

PHYSICAL CHEMISTRY
Composition (1*RS*,2*SR*) is referred to here as diastereoisomer A; (1*RS*,2*RS*) is

referred to here as diastereoisomer B. The ratio of A:B is *c.* 7:3.   **Mol. wt.** 295.8
M.f. $C_{14}H_{18}ClN_3O_2$ **Form** Colourless crystals with a weak characteristic odour.
M.p. A: 138.2 °C; B: 133.5 °C; eutectic A+B: 110 °C; tech. 103–120 °C
V.p. A: 6×10^{-4} mPa; B: 4×10^{-4} mPa (20 °C) K_{ow} A: logP = 3.08;
B: logP = 3.28 (25 °C)   **Henry** A: 3×10^{-6}; B: 4×10^{-6} (Pa m^3 mol^{-1}, 20 °C)
S.g./density A: 1.237; B: 1.299 (22 °C)   **Solubility** In water A: 62; B: 33 mg/l
(20 °C). In dichloromethane 200–500, isopropanol 50–100, hexane 0.1–1.0,
toluene 20 50 (all in g/l, 20 °C).   **Stability** Both diastereoisomers are stable to
hydrolysis; DT_{50} (20 °C) >1 y (pH 4, 7 or 9).

COMMERCIALISATION
History Fungicide reported by P. E. Frohberger (*Pflanzenschutz-Nachr. (Engl. Ed.)*,
1978, **31**, 11). Introduced by Bayer AG.   **Patent** DE 2324010   **Manufacturers**
Bayer

APPLICATIONS
Biochemistry Inhibits steroid demethylation (ergosterol biosynthesis) and other
enzymic processes in phytopathogenic fungi.   **Mode of action** Systemic fungicide
with protective, curative and eradicant action. Absorbed by the roots and leaves,
with ready translocation in young growing tissues, but less ready translocation in
older, woody tissues.   **Uses** Control of powdery mildews, rusts and
Rhynchosporium in cereals, and, when applied as a seed treatment, control of bunt,
smuts, *Typhula* spp., seedling blight, leaf stripe, net blotch and other cereal
diseases. Also used on vegetables, ornamentals, coffee, hops, vines, fruit, tobacco,
sugar cane, bananas and other crops, mainly against powdery mildews, rusts and
various leaf spot diseases.   **Formulation types** DS; WS; DP; SC; EC; WP; EW;
GR; WG; FS.   **Mixtures** (*triadimenol +*) anthraquinone; fuberidazole; imazalil;
carbendazim; cyfluthrin; disulfoton; disulfoton + fenamiphos; tebuconazole;
bitertanol; bitertanol + captan; fuberidazole + imazalil; tridemorph;
chinomethionat; anthraquinone + triazoxide; sulfur.   **Selected tradenames** 'Baytan'
(seed treatment) (Bayer); 'Bayfidan' (spray application) (Bayer); 'Silvacur'
(mixture) (Bayer)

ANALYSIS
Product analysis by gc with FID (*CIPAC Handbook*, 1992, **E**, 224–9).
Residues determined by glc (W. Specht & M. Tillkes, *Pflanzenschutz-Nachr. (Engl.
Ed.)*, 1980, **33**, 61; R. Brennecke, *ibid.*, 1984, **37**, 68; A. Allmendinger, *ibid.*, 1991,
44, 5). Methods for the determination of residues are available from Bayer.

MAMMALIAN TOXICOLOGY
Reviews *Pesticide residues in food – 1989.* FAO Plant Production and Protection
Paper 99, 1989. *Pesticide residues in food – 1989 evaluations. Part II – Toxicology.*
FAO Plant Production and Protection Paper 100/2, 1990.   **Oral** Acute oral LD_{50}
for rats *c.* 700, mice *c.* 1300 mg/kg.   **Skin and eye** Acute percutaneous LD_{50} for

rats >5000 mg/kg. Non-irritating to eyes and skin (rabbits).   **Inhalation** LC_{50} (4 h) for rats >0.9 mg/l air (aerosol).   **NOEL** (2 y) for rats and mice 125, for dogs 600 mg/kg diet.   **ADI** (JMPR) 0.05 mg/kg b.w. [1989].   **Toxicity class** WHO (a.i.) III; EPA (formulation) III   **EC risk** Xn (R22)

ECOTOXICOLOGY
Birds Acute oral LD_{50} for bobwhite quail >2000 mg/kg.   **Fish** LC_{50} (96 h) for golden orfe 17.4–27.3, rainbow trout 14–23.5, bluegill sunfish 15 mg/l.
Bees Non-toxic to honeybees.   **Worms** LC_{50} for *Eisenia foetida* 772 mg/kg dry soil.   **Daphnia** LC_{50} (48 h) 51 mg/l.   **Algae** E_rC_{50} for *Scenedesmus subspicatus* 3.7 mg/l.

ENVIRONMENTAL FATE
Animals In rats, triadimenol was metabolised mainly by oxidation of the *tert*-butyl moiety to the corresponding alcohol and then to carboxylic acid. A small fraction of these compounds was conjugated. Oxidation at the hydroxyl group to the corresponding keto compound (triadimefon, *q.v.*) and its subsequent oxidation at the *tert*-butyl moiety was of secondary importance.   **Plants** The most important breakdown reactions of triadimenol in plants after seed treatment and spray application are conjugation of the a.i. with various sugar compounds (especially hexose) and oxidation at the *tert*-butyl moiety. The resulting primary alcohol is likewise partly conjugated. After seed treatment, the 1,2,4-triazole formed in the soil by hydrolysis was taken up into the plant via the roots and conjugated with various endogenous plant substances.   **Soil/Environment** In soil, triadimenol is a degradation product of triadimefon (*q.v.*). Degradation involving hydrolytic cleavage leads to the formation of 4-chlorophenol. Metabolism of the individual triadimenol enantiomers proceeds at different rates (T. Clark *et al.*, *Proc. Br. Crop Prot. Conf. – Pests Dis.*, 1986, 475). DT_{50} in sandy loam 110–375 d; in loam 240–270 d.

722   tri-allate   *Herbicide*

thiocarbamate

$$[(CH_3)_2CH]_2NCOSCH_2CCl=CCl_2$$

NOMENCLATURE
Common name tri-allate (BSI, E-ISO); triallate ((*m*) F-ISO, WSSA)
IUPAC name S-2,3,3-trichloroallyl di-isopropyl(thiocarbamate);
S-2,3,3-trichloroallyl di-isopropylthiocarbamate
Chemical Abstracts name S-(2,3,3-trichloro-2-propenyl) bis(1-methylethyl)= carbamothioate

CAS RN *[2303–17–5]* **EEC no.** 218–962–7 **Development codes** CP 23 426
(Monsanto)

PHYSICAL CHEMISTRY
Composition Tech. is *c.* 96% pure. **Mol. wt.** 304.7 **M.f.** $C_{10}H_{16}Cl_3NOS$
Form Dark yellow to brown solid; (>30 °C clear brown to dark brown liquid).
M.p. 29–30 °C **B.p.** 117 °C/40 mPa **V.p.** 16 mPa (25 °C) **K$_{ow}$** logP = 4.6
S g /density 1.273 (25 °C) **Solubility** In water 4 mg/l (25 °C) Readily soluble in
common organic solvents such as acetone, diethyl ether, ethyl acetate, ethanol,
benzene, heptane. **Stability** Stable under normal storage conditions. Hydrolysed
by strong acids and alkalis. Stable to light. Decomposition temperature >200 °C.
F.p. >150 °C (closed cup)

COMMERCIALISATION
History Herbicide reported by G. Friesen (*Res. Proc. Nat. Weed Comm., Can.,*
1960). Introduced by Monsanto Co. **Patent** US 3330821; US 3330642
Manufacturers Monsanto

APPLICATIONS
Biochemistry Inhibits fatty acid and lipid biosynthesis; inhibits isoprenoid
biosynthesis (i.e. gibberellic acid); inhibits flavenoid and kaurene biosynthesis (i.e.
anthocyanins). **Mode of action** Selective herbicide, absorbed principally by the
coleoptiles. **Uses** Control of wild oats and some annual grasses in wheat, barley,
rye, field beans, peas, lentils, beet, oilseed rape, maize, flax, alfalfa, clover, vetches,
sainfoin, safflowers, sunflowers, and vegetables at 1.12–1.68 kg a.i./ha. Applied
pre- or post-planting, with soil incorporation. **Phytotoxicity** Oats are susceptible.
Formulation types EC; GR. **Mixtures** *(tri-allate +)* chloridazon; isoproturon;
metoxuron. **Selected tradenames** 'Avadex BW' (Monsanto, Du Pont); 'Far-Go'
(Monsanto)

ANALYSIS
Product analysis by glc. **Residues** in soil, grain or straw determined by glc with
ECD.

MAMMALIAN TOXICOLOGY
EHC 76 (WHO, 1988; general review of thiocarbamates). **Oral** Acute oral LD$_{50}$
for rats 1100 mg/kg. **Skin and eye** Acute percutaneous LD$_{50}$ for rabbits
8200 mg/kg. Slightly irritating to skin and eyes (rabbits). Not a skin sensitiser.
Inhalation Inhalation of saturated air (5.3 mg/l) for 12 hours had no harmful
effect on rats. **NOEL** (2 y) for mice 20 mg/kg diet (*c.* 3.9 mg/kg b.w.), for rats
50 mg/kg diet (*c.* 2.5 mg/kg b.w.); (1 y) for dogs 2.5 mg/kg b.w.
Toxicity class WHO (a.i.) III; EPA (formulation) III **EC risk** Xn (R22)

ECOTOXICOLOGY
Birds Acute oral LD$_{50}$ for bobwhite quail >2251 mg/kg. Eight-day dietary LC$_{50}$ for

MAIN ENTRIES

mallard duck and bobwhite quail >5620 mg/kg diet. **Fish** LC$_{50}$ (96 h) for rainbow trout 1.2, bluegill sunfish 1.3 mg/l. **Bees** Non-toxic to bees.
Other beneficial spp. No adverse effect on soil microflora. **Daphnia** LC$_{50}$ (48 h) 0.43 mg/l. **Algae** EC$_{50}$ (96 h) for *Selenastrum capricornutum* 0.12 mg/l.

ENVIRONMENTAL FATE
Animals Tri-allate metabolism in rats proceeds via three pathways: S-oxidation to sulfur acids, S-oxidation/reduction to thiol derivatives, and C-oxidation of the 2,3,3-trichloropropenethiol moiety. **Plants** 2,3,3-Trichloropropenesulfonic acid is the major detectable crop metabolite. **Soil/Environment** Main loss from soil by microbial action (or by volatilisation, if not incorporated in soil). Metabolism proceeds via hydrolytic cleavage with the formation of dialkylamine, CO$_2$ and mercaptan moieties. The latter is transformed via sulfhydryl group exchange into the corresponding alcohol. DT$_{50}$ (soil) 8–11 w, (water) 3–15 d. K$_{oc}$ 2400.

723 triasulfuron

Herbicide

sulfonylurea

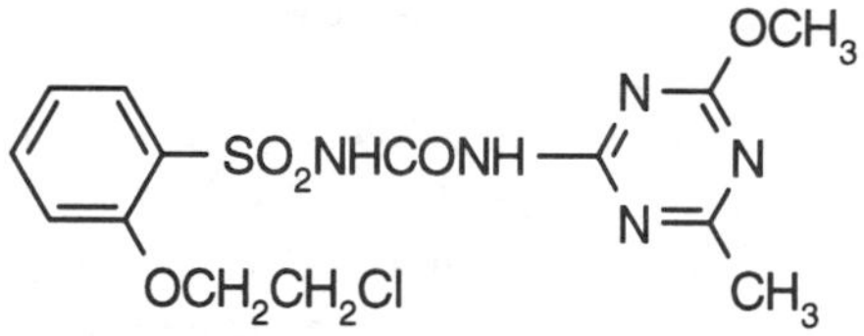

NOMENCLATURE
Common name triasulfuron (BSI, draft E-ISO, (*m*) draft F-ISO)
IUPAC name 1-[2-(2-chloroethoxy)phenylsulfonyl]-3-(4-methoxy-6-methyl-= 1,3,5-triazin-2-yl)urea
Chemical Abstracts name 2-(2-chloroethoxy)-N-[[(4-methoxy-6-methyl-= 1,3,5-triazin-2-yl)amino]carbonyl]benzenesulfonamide
CAS RN *[82097–50–5]* **Development codes** CGA 131 036

PHYSICAL CHEMISTRY
Mol. wt. 401.8 **M.f.** C$_{14}$H$_{16}$ClN$_5$O$_5$S **Form** Fine white powder. **M.p.** 178.1 °C (decomp.) **V.p.** <2 × 10^{-3} mPa (25 °C) (OECD 104) **K$_{ow}$** logP = 1.1 (pH 5.0), −0.59 (pH 6.9), −1.8 (pH 9.0) (25 °C) **Henry** <8 × 10^{-5} Pa m^3 mol^{-1} (calc.)
S.g./density 1.5 g/cm^3 **Solubility** In water 32 (pH 5), 815 (pH 7), 13 500 (pH 8.4) (all in mg/l, 25 °C). In acetone 14, dichloromethane 36, ethyl acetate 4.3 (all in g/l, 25 °C). In ethanol 420, *n*-octanol 130, *n*-hexane 0.04, toluene 300 (all in mg/l, 25 °C). **Stability** Stable for more than 2 years under normal storage

conditions. Partial decomposition below the melting point. On hydrolysis, DT_{50} 8.2 h (pH 1), 3.1 y (pH 7), 4.7 h (pH 10).   **pKa** 4.64 (20 °C)

COMMERCIALISATION
History Herbicide reported by J. Amrein & H. R. Gerber (*1985 Br. Crop Prot. Conf. – Weeds*, **1**, 55). Introduced by Ciba-Geigy AG (now Novartis Crop Protection AG).   **Patent** US 4514212; EP 44808   **Manufacturers** Novartis

APPLICATIONS
Biochemistry Branched chain amino acid synthesis (acetolactate synthase or ALS) inhibitor. Acts by inhibiting biosynthesis of the essential amino acids valine and isoleucine, hence stopping cell division and plant growth. Selectivity derives from rapid metabolism in the crop. Metabolic basis of selectivity in sulfonylureas reviewed (M. K. Koeppe & H. M. Brown, *Agro-Food-Industry*, **6**, 9–14 (1995)).
Mode of action Selective herbicide, absorbed by the leaves and roots, and rapidly translocated to meristems.   **Uses** Control of broad-leaved weeds pre- and post-emergence in wheat, barley and triticale at 5–10 g/ha.   **Formulation types** WG.
Mixtures *(triasulfuron +)* isoproturon; chlorotoluron; methabenzthiazuron; terbutryn; prosulfuron; fluoroglycofen-ethyl; bromoxynil + ioxynil + mecoprop.
Selected tradenames 'Logran' (several products, some mixtures) (Novartis); 'Satis' (mixture) (Novartis)

ANALYSIS
Analysis by glc or by hplc with u.v. detection. Methods for sulfonylurea residues in crops, soil and water reviewed (A. C. Barefoot et al., *Proc. Br. Crop Prot. Conf. – Weeds*, 1995, **2**, 707). Details from Novartis.

MAMMALIAN TOXICOLOGY
Oral Acute oral LD_{50} for rats and mice >5000 mg/kg.   **Skin and eye** Acute percutaneous LD_{50} for rats >2000 mg/kg. Mild skin irritant; non-irritating to eyes (rabbits). Non-sensitising to skin (guinea pigs).   **Inhalation** LC_{50} (4 h) for rats >5.18 mg/l air.   **NOEL** (2 y) for rats 32.1 mg/kg b.w. daily, mice 1.2 mg/kg b.w. daily; (1 y) for dogs 33 mg/kg b.w. daily.   **ADI** 0.012 mg/kg b.w.
Toxicity class WHO (a.i.) III (Table 5); EPA (formulation) IV

ECOTOXICOLOGY
Birds Acute oral LD_{50} for quail and ducks >2150 mg/kg.   **Fish** LC_{50} (96 h) for rainbow trout, carp, catfish, sheepshead minnow and bluegill sunfish >100 mg/l.
Bees Non-toxic to honeybees. LD_{50} (acute and contact) >100 µg/bee.
Worms LC_{50} (14 d) for earthworms >1000 mg/kg soil.   **Daphnia** LC_{50} (96 h) >100 mg/l.   **Algae** EC_{50} (5–14 d) for *Selenastrum* 0.035, *Scenedesmus* 0.77, *Anabaena* 1.7, *Navicula* >100 mg/l.   **Other aquatic spp.** EC_{50} (48 h) for Quahog clam 56 mg/l.

ENVIRONMENTAL FATE

Animals Mainly excreted in the urine in unchanged form. **Plants** In wheat, metabolism is by hydroxylation (*para* to the sulfonyl urea bridge), followed by conjugation of various hydroxy metabolites with glucose. DT_{50} in forage *c.* 3 d. In straw and grain, no residues were detectable at harvest time.

Soil/Environment The degradation behaviour in soil is determined by the soil type, pH, and especially temperature and moisture content. Field studies with silty loam, clay loam and sandy loam showed a median DT_{50} 19 d, depending on soil type.

724 triazamate

Insecticide

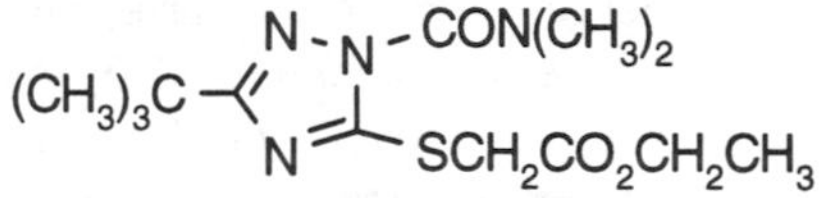

$(CH_3)_3C$ — triazole ring with $N-N-CON(CH_3)_2$ and $SCH_2CO_2CH_2CH_3$

NOMENCLATURE

Common name triazamate (BSI, pa E-ISO)

IUPAC name ethyl (3-*tert*-butyl-1-dimethylcarbamoyl-1*H*-1,2,4-triazol-5-ylthio)= acetate

Chemical Abstracts name ethyl [[1-[(dimethylamino)carbonyl]-3-(1,1-dimethyl= ethyl)-1*H*-1,2,4-triazol-5-yl]thio]acetate

CAS RN *[112143–82–5]*   **Development codes** RH-7988; RH-57988; WL 145158; CL 900050; AC 900,050

PHYSICAL CHEMISTRY

Mol. wt. 314.4   **M.f.** $C_{13}H_{22}N_4O_3S$ **Form** White to light tan crystalline solid with a slight sulfur odour (tech.).   **M.p.** 53 °C   **B.p.** >280 °C   **V.p.** 0.16 mPa (25 °C) K_{ow} logP = 2.69 (pH 7)   **Henry** 4.5×10^{-5} Pa m^3 mol^{-1} (20 °C, calc.) **S.g./density** 1.222 (20.5 °C)   **Solubility** In water, 433 ppm (pH 7, 25 °C). Soluble in dichloromethane and ethyl acetate (tech.).   **Stability** Stable under normal storage conditions and at pH ≤7.0; DT_{50} 220 d (pH 5), 49 h (pH 7), 1 h (pH 9). **pKa** Non-ionising pH 4–9   **F.p.** 189 °C (EEC A9)   **Other properties** Surface tension (90% aqueous saturation) 46.5 mN/m (20 °C, EEC A5).

COMMERCIALISATION

History Aphicide reported by A. Murray *et al.* (*Proc. 1988 Br. Crop Prot. Conf. – Pests Dis.*, **1**, 73). Introduced by Rohm & Haas Co., who entered into an agreement with Shell (now American Cyanamid Co.) in 1991 for development outside US.   **Manufacturers** Great Lakes

APPLICATIONS

Biochemistry Cholinesterase inhibitor. **Mode of action** Acts by contact and ingestion, exhibiting systemic and translaminar movement. **Uses** Control of aphids (including those resistant to carbamate and organophosphorus insecticides) by foliar application on a wide variety of crops at 35–280 g/ha. Suitable for inclusion in integrated pest management programmes because it is minimally toxic to beneficial insects. **Formulation types** EW; WP.
Selected tradenames 'Aphistar' (Rohm & Haas); 'Aztec' (Cyanamid)

MAMMALIAN TOXICOLOGY

Oral Acute oral LD_{50} (tech.) for male rats 100–200, female rats 50–100, mice 54 mg/kg. **Skin and eye** Acute percutaneous LD_{50} for rats >5000 mg/kg. Practically non-irritating to the skin and non-irritating to the eyes (rabbits). Not a skin sensitiser (guinea pigs). **Inhalation** LC_{50} for rats 0.47 mg/l air. **NOEL** for male dogs 0.023, female dogs 0.025, male rats 0.45, female rats 0.58 mg/kg daily. **ADI** 0.004 mg/kg (proposed). **Other** Not mutagenic, not genotoxic, not teratogenic, not oncogenic, not a reproductive toxicant. **EC risk** Xn (R23/25)

ECOTOXICOLOGY

Birds Acute oral LD_{50} (single dose) for quail 8 mg/kg. Dietary LC_{50} (8 d) for mallard ducks 292, quail 411 ppm. **Fish** LC_{50} (96 h) for bluegill sunfish 0.74, rainbow trout 0.53, sheepshead minnow 5.9 mg/l. **Bees** Non-toxic; LD_{50} (96 h, oral) 41 μg/bee; (24 h, contact) >160 μg/bee. **Worms** LC_{50} (14 d) for earthworms 340 mg/kg; NOEC <95 mg/kg. **Other beneficial spp.** *Poecilus cupreus* and *Aleochara bilineata*: no toxic effects at 140 g/ha (lab.). *Coccinella septumpunctata*: 87% effects at 140 g/ha (lab.); no toxic effects in semi-field tests. *Aphidius rhopalosiphi "mummies"*: 20% mortality at 140 g/ha (lab.). *Chrysoperla carnea*: no effect on third instar, 60% effect on second instar at 140 g/ha (lab.); no effects on first instar in extended lab. tests. **Daphnia** LC_{50} (48 h) 0.014 mg/l. **Algae** LC_{50} (72 h) for *Selenastrum capricornutum* 29 mg/l; NOEC (72 h) 9.7 mg/l. **Other aquatic spp.** LC_{50} (96 h) for mysid shrimp (*Mysidopsis bahia*) 160 μg/l.

ENVIRONMENTAL FATE

Very rapidly metabolised by enzyme-catalysed hydrolysis and oxidation in all biological systems studied. Transient metabolites are either degraded further (soils and plants), or are eliminated (vertebrates). **Animals** Hydrolysis, followed by decarbamoylation. **Plants** Hydrolysis, followed by decarbamoylation.
Soil/Environment DT_{50} 1–5 h. K_{oc} 160–378 (5 soils, 25±1 °C). Aqueous photolysis DT_{50} 301 d (pH *c.* 7). Metabolites found in leachates are not acutely toxic to *Daphnia magna*. Neither parent triazamate nor a dimethylcarbamoyl-containing metabolite has any potential for bioaccumulation or for persistence in the environment. Short aerobic half-life coupled with moderate soil adsorption make it unlikely that triazamate will leach.

725 triaziflam   *Herbicide*

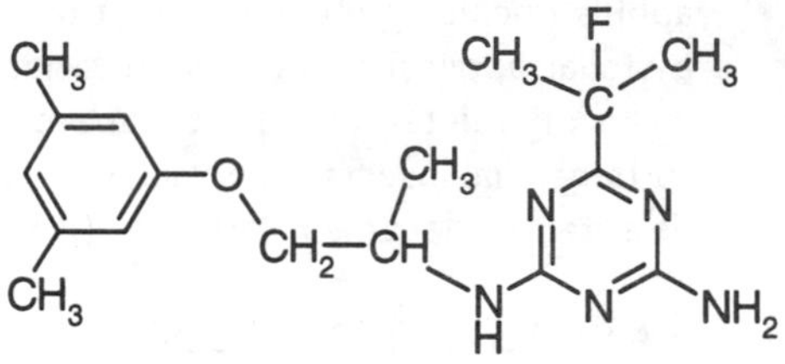

NOMENCLATURE
Common name triaziflam (BSI, pa ISO)
IUPAC name (RS)-N-[2-(3,5-dimethylphenoxy)-1-methylethyl]-6-(1-fluoro-=
1-methylethyl)-1,3,5-triazine-2,4-diamine
Chemical Abstracts name N-[2-(3,5-dimethylphenoxy)-1-methylethyl]-=
6-(1-fluoro-1-methylethyl)-1,3,5-triazine-2,4-diamine
CAS RN [131475–57–5]   **Development codes** IDH-1105

PHYSICAL CHEMISTRY
Mol. wt. 333.4   **M.f.** $C_{17}H_{24}FN_5O$

COMMERCIALISATION
Manufacturers Idemitsu Kosan

APPLICATIONS
Uses Pre- and post-emergence herbicide active on broad-leaved and grass weeds
in rice.

726 triazophos   *Insecticide, acaricide, nematicide*

organophosphorus

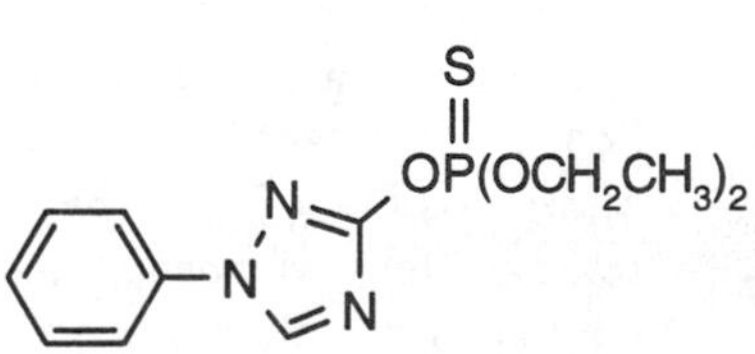

NOMENCLATURE
Common name triazophos (BSI, E-ISO, (m) F-ISO)
IUPAC name O,O-diethyl O-1-phenyl-1H-1,2,4-triazol-3-yl phosphorothioate

Chemical Abstracts name O,O-diethyl O-(1-phenyl-1H-1,2,4-triazol-3-yl) phosphorothioate
CAS RN *[24017–47–8]* **EEC no.** 245–986–5 **Development codes** Hoe 002960; AE F002960

PHYSICAL CHEMISTRY
Composition Tech. grade is ≥92% pure. **Mol. wt.** 313.3 **M.f.** $C_{12}H_{16}N_3O_3PS$
Form Pale yellow oil. **M.p.** 2–5 °C **B.p.** Exothermic decomposition above 140 °C **V.p.** 0.39 mPa (30 °C); 13 mPa (55 °C) **K_{ow}** logP = 3.34
S.g./density 1.24 (20 °C) **Solubility** In water 39 mg/l (pH 7, 20 °C). In acetone, dichloromethane, methanol, isopropanol, ethyl acetate and polyethyleneglycols >500, n-hexane 11.1 (all g/l, 20 °C). **Stability** Stable to light. Hydrolysed by aqueous acids and alkalis.

COMMERCIALISATION
History Insecticide reported by M. Vulic *et al.* (*Abstr. Int. Congr. Plant Prot., 7th, Paris,* 1970, p. 123). Introduced by Hoechst AG (now AgrEvo GmbH).
Patent DE 1670876; DE 1299924; US 3686200 **Manufacturers** AgrEvo

APPLICATIONS
Biochemistry Cholinesterase inhibitor. **Mode of action** Broad-spectrum insecticide and acaricide with contact and stomach action. Non-systemic, but penetrates deeply into plant tissues. **Uses** Control of aphids, thrips, midges, beetles, lepidopterous larvae, cutworms and other soil insects, spider mites and other species of mite, etc. in ornamentals, fruit trees (including citrus), vines, bananas, strawberries, vegetables, oilseed rape, cereals, sugar beet, sugar cane, maize, soya beans, peanuts, guavas, mangoes, oil palms, olives, cotton, coffee, rice, grassland, and in forestry. Also controls some free-living nematodes, particularly in ornamentals and strawberries, and as a bulb dip for tulips and garlic.
Formulation types EC; UL. **Mixtures** (*triazophos* +) deltamethrin.
Selected tradenames 'Hostathion' (AgrEvo); 'Trelka' (AgrEvo); 'Spark' (AgrEvo); 'Try' (Sanonda)

ANALYSIS
Product analysis by glc. **Residues** determined by glc with FPD (W. G. Thier *et al., Anal. Methods Pestic. Plant Growth Regul.,* 1978, **10**, 127).

MAMMALIAN TOXICOLOGY
Reviews *Pesticide residues in food – 1993,* FAO Plant Production and Protection Paper, 122, 1993. *Pesticide residues in food – 1993 evaluations, Part II – Toxicology.* WHO, WHO/PCS/94.4, 1994. **EHC** 63 (WHO, 1986; a general review of organophosphorus insecticides). **Oral** Acute oral LD_{50} for rats 57–59, dogs >320–>500 mg/kg b.w. daily. **Skin and eye** Acute percutaneous LD_{50} for rats >2000 mg/kg. No skin or eye irritation. **Inhalation** LC_{50} (4 h) for rats 0.531 mg/l air. **NOEL** In 2 y feeding trials, rats receiving 1 mg/kg diet and dogs receiving

0.3 mg/kg diet were unaffected, except for inhibition of blood serum cholinesterase. **ADI** (JMPR) 0.001 mg/kg b.w. [1993]. **Toxicity class** WHO (a.i.) Ib **EC risk** T (R24/25)

ECOTOXICOLOGY
Birds Acute oral LD_{50} for Japanese quail 4.2–27.1 mg/kg, depending on carrier and sex. Eight-day LC_{50} for mallard 325 mg/kg diet. **Fish** LC_{50} (96 h) for carp 5.6, golden orfe 7.5–18 mg/l. LC_{50} (21 d) for trout 0.01 mg/l. **Bees** Toxic to honeybees. **Worms** LC_{50} (14 d) for *Eisenia foetida* 187 mg/kg dry artificial soil. **Daphnia** LC_{50} (48 h) 0.003 mg/l. **Algae** LC_{50} (96 h) 1.15 mg/l.

ENVIRONMENTAL FATE
Animals Mainly eliminated via urine (75–94% of applied radioactivity). Excretion DT_{50} <1 d. **Plants** In cotton plants, 1-phenyl-3-hydroxy-1,2,4-triazole is found as a metabolite, and the occurrence of a desethyl derivative of triazophos is presumed (W. G. Thier *et al.*, *Fresenius' Z. Anal. Chem.*, 1973, **267**, 181–186). **Soil/Environment** DT_{50} (aerobic, field) 6–12 d, DT_{90} 39–114 d. DT_{50} (lab.) 7–46 d, DT_{90} 109–181 d. Rapidly degraded in aquatic systems: DT_{50} (elimination from water) <3 d, (elimination from water/sediment system) <11 d; DT_{90} (elimination from the system) <47 d.

727 triazoxide   *Fungicide*

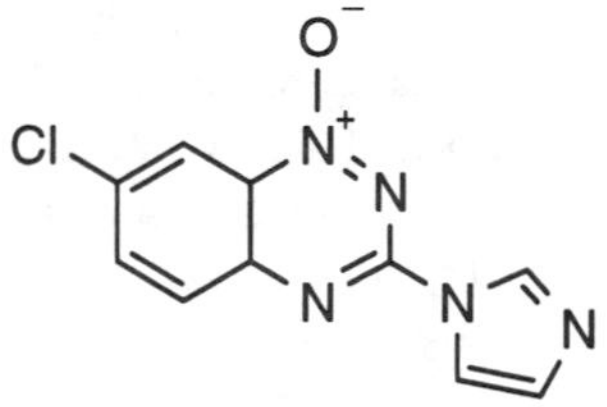

NOMENCLATURE
Common name triazoxide (BSI, draft E-ISO, (*m*) draft F-ISO)
IUPAC name 7-chloro-3-imidazol-1-yl-1,2,4-benzotriazine 1-oxide
Chemical Abstracts name 7-chloro-3-(1*H*-imidazol-1-yl)-1,2,4-benzotriazine 1-oxide
CAS RN *[72459–58–6]* **EEC no.** 276–668–4 **Development codes** SAS 9244 (Bayer)

PHYSICAL CHEMISTRY
Mol. wt. 247.7 **M.f.** $C_{10}H_6ClN_5O$ **Form** Light yellow crystals. **M.p.** 182 °C

V.p. 1×10^{-4} mPa (20 °C); 3×10^{-4} mPa (25 °C)   **K$_{ow}$** logP = 2.04 (23 °C)
Henry 7.3×10^{-7} Pa m^3 mol^{-1} (20 °C)   **S.g./density** 1.577 g/cm^3 (22.5 °C)
Solubility In water 34 (pH 7 & 9), 58 (pH 4) (both mg/l, 20 °C). In hexane 0.05,
dichloromethane 32, isopropanol 1.8, toluene 6.9 (all in g/l, 20 °C).
Stability Photolytically decomposed; DT$_{50}$ (solid state) 1 d, (dissolved in water)
2–5 d. Hydrolytic stability depends strongly on pH; DT$_{50}$ (extrapolated) >>1 y
(pH 4), 3.6 y (pH 7), 22.6 d (pH 9) (22 °C).

COMMERCIALISATION
History Fungicide evaluated by Bayer AG.   **Manufacturers** Bayer

APPLICATIONS
Mode of action Specific non-systemic fungicide.   **Uses** Exclusively as a seed
dressing for control of seed-borne *Pyrenophora graminea* and *Pyrenophora teres* in
barley. Only used in combination with other fungicides.   **Phytotoxicity** Shows
good seed tolerance at recommended doses.   **Formulation types** WS; FS.
Mixtures *(triazoxide +)* anthraquinone + triadimenol; tebuconazole + imidacloprid.
Selected tradenames 'Brio' (mixture) (France) (Bayer); 'Gaucho Orge' (mixture)
(France) (Bayer); 'Raxil S' (mixture) (Bayer)

ANALYSIS
Standard method for **residues** in soil, water, and plants is Bayer method No. 00055
(RA-662/253B), by R. Brennecke: HPLC method for the determination of SAS
9244 residues in soil and water, issued July 24, 1984, and adapted to plant
materials by supplement No. E001 (MR-408/95), issued June 1, 1995. Details
available from Bayer.

MAMMALIAN TOXICOLOGY
Oral Acute oral LD$_{50}$ for rats c. 150 mg/kg.   **Skin and eye** Acute percutaneous
LD$_{50}$ for rats >5000 mg/kg. Non-irritant to skin and eyes (rabbits).
Inhalation LC$_{50}$ (4 h) for rats c. 0.75 mg/l (dust).   **NOEL** (2 y) for rats
1 mg/kg diet, for mice 5 mg/kg diet; (1 y) for dogs 10 mg/kg diet.
ADI 0.0005 mg/kg b.w.   **Toxicity class** WHO (a.i.) II

ECOTOXICOLOGY
Birds Acute oral LD$_{50}$ for Japanese quail 106–111 mg/kg. LC$_{50}$ (5 d) for Japanese
quail 3762 mg/kg diet.   **Fish** LC$_{50}$ for golden orfe 7.4, rainbow trout 0.63 mg/l.
Worms LC$_{50}$ for *Eisenia foetida* >1000 mg/kg dry soil.   **Daphnia** LC$_{50}$ 4.8 mg/l.
Algae E$_r$C$_{50}$ for *Scenedesmus subspicatus* 0.16 mg/l.

ENVIRONMENTAL FATE
Animals A complete absorption of triazoxide is followed by fast distribution in the
body and rapid elimination; almost 100% of the dose is excreted within 48 h in the
urine and faeces. Extensive metabolism of triazoxide was observed in the animal

body (rat).   **Plants** Because of its non-systemic properties, the amount of triazoxide taken up by plants after seed treatment is very low. Residues in the upper plant parts were never found.   **Soil/Environment** Field studies in Germany revealed a medium fast dissipation rate from soil (average DT_{50} c. 2 mo). Low application rate (only 2–3 g a.i./dt seed) and low mobility in soil excludes groundwater contamination through leaching. Triazoxide was found to degrade rapidly in natural water/sediment systems, to which aqueous photolysis greatly contributes (environmental DT_{50} 2.2, 1.7, 4.7, and 9.7 d (calc.) for spring, summer, autumn and winter, respy.). The use of triazoxide exclusively for seed dressing also minimises exposure in the environment.

728 tribenuron-methyl   *Herbicide*

sulfonylurea

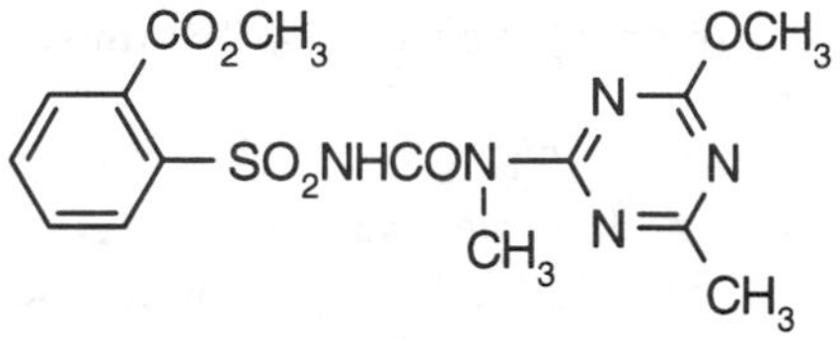

NOMENCLATURE
tribenuron-methyl
Chemical Abstracts name methyl 2-[[[[(4-methoxy-6-methyl-1,3,5-triazin-2-yl)= methylamino]carbonyl]amino]sulfonyl]benzoate
CAS RN *[101200–48–0]*   **EEC no.** 401–190–1   **Development codes** DPX-L5300; L5300

tribenuron
Common name tribenuron (BSI, ANSI, draft E-ISO)
IUPAC name 2-[4-methoxy-6-methyl-1,3,5-triazin-2-yl(methyl)carbamoyl= sulfamoyl]benzoic acid
Chemical Abstracts name 2-[[[[(4-methoxy-6-methyl-1,3,5-triazin-2-yl)methyl= amino]carbonyl]amino]sulfonyl]benzoic acid
CAS RN *[106040–48–6]*

PHYSICAL CHEMISTRY
tribenuron-methyl
Composition Tech. is >95%.   **Mol. wt.** 395.4   **M.f.** $C_{15}H_{17}N_5O_6S$ **Form** Light brown, odourless solid.   **M.p.** 141 °C   **V.p.** 5.2×10^{-5} mPa (25 °C) K_{ow} logP = −0.44 (pH 7)   **Henry** 1.03×10^{-8} Pa m³ mol⁻¹   **S.g./density** 1.5

(25 °C)   **Solubility** In water 0.05 (pH 5), 2.04 (pH 7) (both in g/l, 20 °C). In acetone 43.8, acetonitrile 54.2, carbon tetrachloride 3.12, ethyl acetate 17.5, hexane 0.028, and methanol 3.39 (all in mg/l, 25 °C).   **Stability** Stable at 45 °C. On hydrolysis (45 °C), stable at pH 8–10 but rapid loss at pH <7 or pH >12. Relatively unstable in most organic solvents.   **pKa** 5

tribenuron
Mol. wt. 381.4   **M.f.** $C_{14}H_{15}N_5O_6S$

COMMERCIALISATION
History Herbicide reported by D. T. Ferguson et al. (*Proc. Br. Crop Prot. Conf. – Weeds*, 1985, **1**, 43). The methyl ester introduced in Spain (1986) by E. I. du Pont de Nemours and Co.   **Manufacturers** Du Pont

APPLICATIONS
tribenuron-methyl
Biochemistry Branched chain amino acid synthesis (acetolactate synthase or ALS) inhibitor. Acts by inhibiting biosynthesis of the essential amino acids valine and isoleucine, hence stopping cell division and plant growth. Selectivity derives from rapid metabolism in the crop. Metabolic basis of selectivity in sulfonylureas reviewed (M. K. Koeppe & H. M. Brown, *Agro-Food-Industry*, **6**, 9–14 (1995)).
Mode of action Rapidly absorbed by foliage and roots and translocated throughout the plant. Susceptible plants cease growth almost immediately after post-emergence treatment and are killed in 7–21 days. Surfactants increase the activity of tribenuron-methyl on certain broad-leaved weeds.   **Uses** Post-emergence control of broad-leaved weeds in spring and winter cereals, at 9–30 g/ha.
Formulation types WG; TB.   **Mixtures** *(tribenuron-methyl +)* thifensulfuron-methyl; metribuzin.   **Selected tradenames** 'Express' (USA) (Du Pont); 'Granstar' (Europe) (Du Pont); 'Pointer' (Germany) (Du Pont)

ANALYSIS
Product and **residue** analysis by hplc. Methods for sulfonylurea residues in crops, soil and water reviewed (A. C. Barefoot et al., *Proc. Br. Crop Prot. Conf. – Weeds*, 1995, **2**, 707).

MAMMALIAN TOXICOLOGY
tribenuron-methyl
Oral Acute oral LD_{50} for rats >5000 mg/kg.   **Skin and eye** Acute percutaneous LD_{50} for rabbits >2000 mg/kg. Non-irritating to skin (rabbits); mild (reversible) irritant to eyes. Mildly sensitising to skin (guinea pig maximisation test).
Inhalation LC_{50} (4 h) for rats >5.0 mg/l air.   **NOEL** (2 y) for rats 25 ppm diet; (18 mo) for mice 200 ppm diet (30 mg/kg b.w. daily); (1 y) for dogs 250 ppm diet (8.2 mg/kg b.w. daily); (90 d) for rats 100, for mice 500, for dogs 500 mg/kg diet. Non-teratogenic in rats at 20 mg/kg daily.   **ADI** 0.011 mg/kg.   **Other** Non-

mutagenic in the Ames test and negative in CHO, unscheduled DNA, *in vivo* cytogenetic, *in vivo* mouse micronucleus, and *in vitro* human lymphocyte assays. **Toxicity class** WHO (a.i.) III (Table 5); EPA (formulation) III   **EC risk** Xi (R43)

ECOTOXICOLOGY
tribenuron-methyl
Birds Acute oral LD_{50} for bobwhite quail >2250 mg/kg. Eight-day dietary LC_{50} for bobwhite quail and mallard ducks >5620 mg/kg.   **Fish** LC_{50} (96 h) for bluegill sunfish and rainbow trout >1000 mg/l.   **Bees** LD_{50} for honeybees >100 μg/bee. **Worms** LD_{50} for earthworms >2000 ppm.   **Daphnia** LC_{50} (48 h) 720 mg/l (unfed), 1000 mg/l (fed).   **Algae** EC_{50} (120 h) for green algae and bluegreen algae >11.5 μg/l.

ENVIRONMENTAL FATE
Soil/Environment Half-life of tribenuron-methyl in soil 1–7 days. No significant photodecomposition under field conditions. Degradation in the soil occurs by hydrolysis and by direct microbial degradation. Hydrolysis is affected by soil pH, being faster in acidic than alkaline soils. Losses due to volatilisation are not significant.

729   tribufos

Plant growth regulator

$$[CH_3(CH_2)_3S]_3PO$$

NOMENCLATURE
Common name tribufos (pa ISO)
IUPAC name *S,S,S*-tributyl phosphorotrithioate
Chemical Abstracts name *S,S,S*-tributyl phosphorotrithioate
CAS RN *[78–48–8]*   **Development codes** Chemagro B-1776

PHYSICAL CHEMISTRY
Mol. wt. 314.5   **M.f.** $C_{12}H_{27}OPS_3$ **Form** Colourless to pale yellow liquid, with a mercaptan-like odour.   **M.p.** <–25 °C   **B.p.** 210 °C/750 mmHg   **V.p.** 0.35 mPa (20 °C), 0.71 mPa (25 °C) K_{ow} logP = 3.23   **S.g./density** 1.057 (20 °C)
Solubility Practically insoluble in water (2.3 mg/l at 20 °C). Soluble in aliphatic, aromatic, and chlorinated hydrocarbons, and alcohols.   **Stability** Relatively stable to acids and heat. Hydrolysed slowly under alkaline conditions; DT_{50} (35 °C) 14 d (pH 4.7 to 9). Photolysed slowly.

COMMERCIALISATION
History Defoliant tested by Ethyl Corp. and introduced by Chemagro Corp. in 1965.   **Patent** US 2943107; US 2965467   **Manufacturers** Bayer Corp.

APPLICATIONS
Mode of action Plant growth regulator, absorbed by the leaves. Stimulates formation of an abscission layer between the plant stem and the leaf petioles, causing the dropping of entire green leaves.   **Uses** Used for defoliation of cotton, at 1–2 kg a.i./ha, to facilitate harvesting.   **Phytotoxicity** Very phytotoxic.
Formulation types EC.   **Selected tradenames** 'DEF 6' (Bayer Corp.)

ANALYSIS
Product analysis by i.r. spectroscopy.   **Residues** in cotton seed determined by glc (R. F. Thomas & T. H. Harris, *J. Agric. Food Chem.*, 1965, **13**, 505; D. MacDougall, *Anal. Methods Pestic., Plant Growth Regul. Food Addit.*, 1964, **4**, 89; *Anal. Methods Pestic. Plant Growth Regul.*, 1972, **6**, 627). Details available from Bayer AG.

MAMMALIAN TOXICOLOGY
Oral Acute oral LD_{50} for male rats 435, female rats 234 mg/kg.
Skin and eye Acute percutaneous LD_{50} for rats 850, rabbits *c.* 1000 mg/kg; moderate primary dermal irritant, minimal eye irritant (rabbit). Not a skin sensitiser.   **Inhalation** LC_{50} (4 h) for male rats 4.65, female rats 2.46 mg/l (aerosol).   **NOEL** (2 y) for rats 4 mg/kg diet; (12 mo) for dogs 4 mg/kg diet; (90 w) for mice 10 mg/kg diet.   **ADI** 0.001 mg/kg b.w.   **Toxicity class** WHO (a.i.) II; EPA (formulation) II   **EC risk** Xn (R22, R38)

ECOTOXICOLOGY
Birds Acute oral LD_{50} for bobwhite quail 142–163, mallard duck 500–707 mg/kg. Five-day LC_{50} for bobwhite quail 1643, mallard duck >5000 mg/kg diet.
Fish LC_{50} (96 h) for bluegill sunfish 0.72–0.84, rainbow trout 1.07–1.52 mg/l.
Bees Relatively non-toxic to honeybees.   **Daphnia** LC_{50} (48 h) 0.12 mg/l.

ENVIRONMENTAL FATE
Animals Tribufos is rapidly absorbed and metabolised; 96% of the administered radioactivity was excreted within 72 hours. Metabolism proceeds by hydrolysis followed by methylation and successive oxidation of butylmercaptan, yielding the main metabolite (3-hydroxy)-butylmethylsulfone.   **Plants** Unmetabolised tribufos is the primary residue in treated cotton.   **Soil/Environment** In soil, tribufos is very strongly absorbed, leaching is extremely unlikely. The half-life under field conditions is 2–7 weeks. The main metabolite is 1-butane sulfonic acid.

organophosphorus

$$\underset{\underset{OH}{|}}{Cl_3C\,CHP(OCH_3)_2}$$

$$\overset{\displaystyle O}{\overset{\displaystyle \|}{}}$$

NOMENCLATURE
Common name trichlorfon (BSI – since 1984, E-ISO, (m) F-ISO); chlorophos (USSR); metriphonate (BAN); DEP (JMAF); [trichlorphon] (BSI – before 1984); [dipterex] (former exception, Turkey)
IUPAC name dimethyl 2,2,2-trichloro-1-hydroxyethylphosphonate
Chemical Abstracts name dimethyl (2,2,2-trichloro-1-hydroxyethyl)phosphonate
CAS RN *[52–68–6]* **EEC no.** 200–149–3 **Development codes** Bayer 15 922; Bayer L13/59 **Official codes** OMS 800; ENT 19 763

PHYSICAL CHEMISTRY
Composition Racemate, i.e. 1:1 mixture of the (1R)- and (1S)- enantiomers.
Mol. wt. 257.4 **M.f.** $C_4H_8Cl_3O_4P$ **Form** Colourless crystals with a weak characteristic odour. **M.p.** 78.5 °C; delayed melting to 84 °C **V.p.** 0.21 mPa (20 °C); 0.5 mPa (25 °C) **K_{ow}** logP = 0.43 (20 °C) **Henry** 4.4×10^{-7} Pa m^3 mol^{-1} (20 °C) **S.g./density** 1.73 (20 °C) **Solubility** In water, 120 g/l (20 °C). Readily soluble in common organic solvents (with the exception of aliphatic hydrocarbons and petroleum oils), hexane 0.1–1, dichloromethane, isopropanol >200, toluene 20–50 (all in g/l, 20 °C). **Stability** Subject to hydrolysis and dehydrochlorination. Decomposition proceeds more rapidly with heating, and above pH 6. Rapidly converted by alkalis to dichlorvos (*q.v.*), which is then hydrolysed: DT_{50} (22 °C) 510 d (pH 4), 46 h (pH 7), <30 min (pH 9). Photolysis is slow.

COMMERCIALISATION
History Insecticide reported by G. Unterstenhöfer (*Anz. Schaedlingskd.*, 1957, **30**, 7). First prepared by W. Lorenz, and introduced by Bayer AG.
Patent US 2701225 **Manufacturers** Bayer; Cequisa; Denka; Jin Hung; Makhteshim-Agan; Sinon

APPLICATIONS
Biochemistry Cholinesterase inhibitor; activity may be due to *in vivo* conversion to dichlorvos (*q.v.*). **Mode of action** Non-systemic insecticide with contact and stomach action. **Uses** Insecticidal control in agriculture, horticulture, forestry, food storage, gardening, households, and animal husbandry. In particular, control of Diptera, Lepidoptera, Hymenoptera, Hemiptera, and Coleoptera on many crops. Also used to control household pests such as flies, cockroaches, fleas, bed

bugs, silverfish, ants, etc.; as a fly bait in farm buildings and animal houses; and for control of ectoparasites on domestic animals. **Formulation types** SL; WP; SP; DP; GR; UL; Coating agent; PO; SC; GB. **Compatibility** Incompatible with alkaline materials and oils. **Selected tradenames** 'Dipterex' (Bayer); 'Cekufon' (Cequisa); 'Denkaphon' (Denka); 'Danex' (Makhteshim-Agan); 'Saprofon' (Sanonda)

ANALYSIS
Product analysis by polarography (P. A. Giang & R. I. Caswell, *J. Agric. Food Chem.*, 1957, **5**, 753) or by titration of the chloride ion obtained by hydrolysis (*CIPAC Handbook*, 1970, **1**, 684; FAO Specification (CP/51)). **Residues** determined by glc (*Man. Pestic. Residue Anal.*, 1987, **I**, 6, S13, S19; *Anal. Methods Residues Pestic.*, 1988, Part I, M2, M5, M12; A. Ambrus *et al.*, *J. Assoc. Off. Anal. Chem.*, 1981, **64**, 733). Details available from Bayer AG.

MAMMALIAN TOXICOLOGY
Reviews *Pesticide residues in food – 1978*. FAO Plant Production and Protection Paper 15, 1979. *Pesticide residues in food: 1978 evaluations*. FAO Plant Production and Protection Paper 15 Sup, 1979. **IARC** 30 **EHC** 63 (WHO, 1986; a general review of organophosphorus insecticides), 132 (WHO, 1992). **Oral** Acute oral LD_{50} for male and female rats c. 250 mg/kg. **Skin and eye** Acute percutaneous LD_{50} for male and female rats >5000 mg/kg (24 h); not irritating to skin and eye (rabbit). **Inhalation** LC_{50} (4 h) for male and female rats >0.5 mg/l air (aerosol). **NOEL** (2 y) for rats 100, for mice 300 mg/kg diet; (4 y) for dogs 50 mg/kg diet. **ADI** (JMPR) 0.01 mg/kg b.w. [1978]. **Toxicity class** WHO (a.i.) II; EPA (formulation) II **EC risk** Xn (R22); (R43)

ECOTOXICOLOGY
Fish LC_{50} (96 h) for rainbow trout 0.7, golden orfe 0.52 mg/l. **Bees** Low toxicity to bees and other beneficial insects. **Daphnia** LC_{50} (48 h) 0.00096 mg/l. **Algae** E_rC_{50} for *Scenedesmus subspicatus* >10 mg/l.

ENVIRONMENTAL FATE
EHC 132 notes that trichlorfon is moderately toxic to fish and birds, and moderately to highly toxic to aquatic arthropods, and concludes it should not be sprayed over areas of water. **Animals** In animals, trichlorfon is rapidly absorbed and metabolised. Excretion of the radioactivity in the urine is more or less complete within 6 hours. The major metabolites are dimethylphosphoric acid, monomethylphosphoric acid and conjugates of dichloroacetic acid. **Plants** In plants, trichlorfon is rapidly hydrolysed. The main metabolites are dimethyl- and monomethyl- phosphoric acid and conjugates of dichloroacetic acid and dichloroethanol. **Soil/Environment** Trichlorfon has a relatively high mobility in soil, but it is rapidly metabolised to CO_2. Intermediates are dichloroethanol, dichloroacetic acid and trichloroacetic acid. K_{oc} 20 (±10).

731 *Trichoderma harzianum*   *Fungicide*

fungus

NOMENCLATURE
Scientific name *Trichoderma harzianum*   **Var.** TH11 ('Harzan')   **Strain** T-39
Development codes ABG-8007 (Makhteshim)

PROPERTIES
Form Green-grey solid.   **S.g./density** 0.369   **Solubility** Not soluble in water.
Stability Stable for 1 year, vacuum packed at 20 °C.

COMMERCIALISATION
History Introduced in Israel in 1994.   **Manufacturers** Makhteshim-Agan

APPLICATIONS
Mode of action Competes with pathogenic fungi.   **Uses** 'Trichodex' is for control of *Botrytis* and *Sclerotinia* on vines and vegetables in soil.   **Formulation types** MG.
Selected tradenames 'Harzan' (NPP); 'Trichodex' (Makhteshim)

MAMMALIAN TOXICOLOGY
Oral Acute oral LD_{50} for rats >500 mg/kg.   **Skin and eye** Acute percutaneous LD_{50} not available. Eye irritant, not a skin irritant. Possible skin sensitiser.
Inhalation LC_{50} >0.89 mg/l.   **Other** Non-infectious, non-pathogenic.
Toxicity class EPA (formulation) III-IV

ECOTOXICOLOGY
Birds Acute oral LD_{50} for mallard duck and bobwhite quail >2000 mg/kg.
Fish LC_{50} (96 h) for zebra fish 1.23×10^5 cfu/ml.   **Bees** (oral) non toxic at 1000 ppm.   **Daphnia** LC_{50} (10 d) 1.6×10^4 cfu/ml.

732 *Trichogramma* spp.   *Biological agent*

Hymenoptera

Wasp parasite of Lepidoptera.

NOMENCLATURE
Scientific name *Trichogramma evanescens; T. maidis; T. minutum; T. pretiosum; T. platneri*   **Other names** *T. brassicae*

1236   Trichogramma spp.

PROPERTIES
Form Parasitised eggs of *Ephestia kühniella*. **Stability** Shelf life 2–3 d at 10–15 °C in a dark place.

COMMERCIALISATION
Manufacturers Arbico; Beneficial; Bio Protection; Biobest; Koppert; Kunafin; M&R Durango; Rincon-Vitova; Nature's Alternative

APPLICATIONS
Mode of action The parasitic wasp lays an egg inside the butterfly egg, from which an adult wasp emerges. **Uses** For control of Lepidoptera in protected crops. Various species of *Trichogramma* are used in different situations.
Phytotoxicity Non-phytotoxic and non-phytopathogenic. **Formulation types** Eggs on cards. **Selected tradenames** 'Tricho-strip' (*evanescens*) (Koppert); 'Trichocap' (*maidis*) (BASF); 'Pyratyp' (*maidis*) (BASF)

MAMMALIAN TOXICOLOGY
Non-toxic.

733 triclopyr   *Herbicide*

aryloxyalkanoic acid

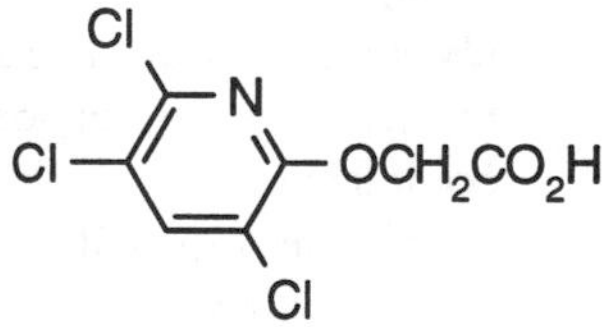

NOMENCLATURE
triclopyr
Common name triclopyr (BSI, E-ISO, F-ISO, ANSI, WSSA); no name (Eire)
IUPAC name 3,5,6-trichloro-2-pyridyloxyacetic acid
Chemical Abstracts name [(3,5,6-trichloro-2-pyridinyl)oxy]acetic acid
CAS RN *[55335–06–3]* **Development codes** Dowco 233

triclopyr-butotyl
CAS RN *[64470–88–8]*

PHYSICAL CHEMISTRY
triclopyr
Mol. wt. 256.5 **M.f.** $C_7H_4Cl_3NO_3$ **Form** Fluffy, colourless solid. **M.p.** 150.5 °C

B.p. Decomposes at 208 °C   **V.p.** 0.2 mPa (25 °C, vapour pressure balance)
K_{ow} logP = 0.42 (pH 5), –0.45 (pH 7), –0.96 (pH 9)   **S.g./density** 1.85 (21 °C)
Solubility In water 0.408 (purified), 7.69 (pH 5), 8.10 (pH 7), 8.22 (pH 9) (all in
g/l, 20 °C). In acetone 581, acetonitrile 92.1, hexane 0.09, toluene 19.2,
dichloromethane 24.9, methanol 665, ethyl acetate 271 (all in g/l).
Stability Stable under normal storage conditions and to hydrolysis, but subject to
photodecomposition, DT_{50} <12 h.   **pKa** 3.97

triclopyr-butotyl
Mol. wt. 356.6   **M.f.** $C_{13}H_{16}Cl_3NO_4$

triclopyr-triethylammonium
Mol. wt. 357.7   **M.f.** $C_{13}H_{19}Cl_3N_2O_3$

COMMERCIALISATION
History Herbicide reported by B. C. Byrd *et al.* (*Proc. West. Soc. Weed Sci.*, 1975,
28, 44). Introduced by Dow Chemical Co. (now DowElanco).
Manufacturers Agriphar; DowElanco

APPLICATIONS
Mode of action Selective systemic herbicide, rapidly absorbed by the foliage and
roots, with translocation throughout the plant, accumulating in meristematic tissue.
Induces auxin-type responses in susceptible species (mainly broad-leaved weeds,
grass weeds being unaffected at normal application rates).

triclopyr
Uses Control of woody plants and many broad-leaved weeds (e.g. nettles, docks,
brambles, gorse, broom) in grassland, uncultivated land, industrial areas,
coniferous forests, plantation crops, and rice fields. In plantation crops (oil palm,
rubber) at 125–250 g a.e./ha for covercrop maintenance, and at 0.72–1.0 kg
a.e./ha to control *Eupatorium odoratum* and other problem plants; in pastures at
1–2 kg/ha to control annual and perennial herbaceous weeds, and at 2–4 kg/ha
against *Rubus* spp. and other woody plants; in forestry at 4–8 kg/ha for site
preparation, and at 1.5–2.0 kg/ha for conifer release; for industrial sites at 2–8
kg/ha; in rangeland at 0.24–1.0 kg/ha.   **Phytotoxicity** Clovers are sensitive, as also
are larch, lodgepole pine, ornamentals, and edible crops.
Selected tradenames 'Trident' (Agriphar)

triclopyr-butotyl
Formulation types EC.   **Mixtures** *(triclopyr-butotyl +)* clopyralid; fluroxypyr;
propanil.   **Selected tradenames** 'Garlon' (DowElanco)

triclopyr-triethylammonium
Selected tradenames 'Garlon 3A' (DowElanco)

ANALYSIS

Product analysis by hplc.   **Residues** determined by glc. Details available from DowElanco.

MAMMALIAN TOXICOLOGY

triclopyr

Oral Acute oral LD_{50} for male rats 692, female rats 577 mg/kg.
Skin and eye Acute percutaneous LD_{50} for rabbits >2000 mg/kg. Mild eye irritant; non-irritating to skin (rabbits).   **Inhalation** LC_{50} (4 h) for rats >256 ppm.
NOEL (2 y) for rats 3.0 mg/kg b.w. daily, for mice 35.7 mg/kg b.w. daily.
ADI 0.005 mg/kg b.w.   **Toxicity class** WHO (a.i.) III; EPA (formulation) III

ECOTOXICOLOGY

triclopyr

Birds Acute oral LD_{50} for mallard ducks 1698 mg/kg. Eight-day dietary LC_{50} for mallard ducks >5000, Japanese quail 3278, bobwhite quail 2935 mg/kg.   **Fish** LC_{50} (96 h) for rainbow trout 117, bluegill sunfish 148 mg/l.   **Bees** Non-toxic to bees; contact LD_{50} >100 μg/bee.   **Daphnia** LC_{50} (48 h) 133 mg/l.   **Algae** EC_{50} (5 d) for *Selenastrum capricornutum* 45 mg/l.

ENVIRONMENTAL FATE

Animals In mammals, following oral administration, excretion is primarily via the urine as the unchanged compound. For details of minor urinary metabolites, see C. Timchalk *et al.*, *Toxicology*, 1990, **62**, 71.   **Plants** In plants, DT_{50} *c.* 3–10 d. The main metabolite is 3,5,6-trichloro-2-methoxypyridine.   **Soil/Environment** In soil, fairly rapid degradation by microbial activity, with an average half-life of 46 d, depending on soil and climatic conditions. The major degradation product is 3,5,6-trichloro-2-pyridinol (which has a soil half-life of 30–90 d), with a smaller amount of 3,5,6-trichloro-2-methoxypyridine. K_{oc} *c.* 59 ml/g; K_d *c.* 87 (unaged samples), *c.* 225 (aged) ml/g.

734 tricyclazole   *Fungicide*

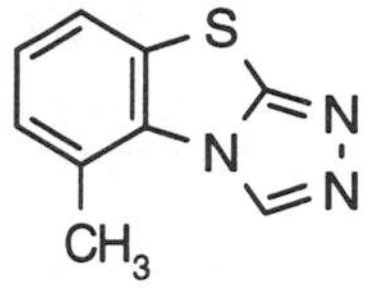

NOMENCLATURE

Common name tricyclazole (BSI, E-ISO, (*m*) F-ISO, ANSI)

IUPAC name 5-methyl-1,2,4-triazolo[3,4-b][1,3]benzothiazole
Chemical Abstracts name 5-methyl-1,2,4-triazolo[3,4-b]benzothiazole
CAS RN *[41814–78–2]*   **EEC no.** 255–559–5   **Development codes** EL-291

PHYSICAL CHEMISTRY
Mol. wt. 189.2   **M.f.** $C_9H_7N_3S$ **Form** Crystalline solid.   **M.p.** 187–188 °C
B.p. 275 °C   **V.p.** 0.027 mPa (25 °C) K_{ow} logP = 1.4   **Solubility** In water at
25 °C, 1.6 g/l. In acetone 10.4, methanol 25, xylene 2.1 (all in g/l, 25 °C).
Stability Stable at 52 °C (highest storage temperature tested). Relatively stable to
u.v. light.

COMMERCIALISATION
History Fungicide reported by J. D. Froyd *et al.* (*Phytopathology*, 1976, **66**, 1135).
Introduced in Philippines (1976) by Eli Lilly & Co. (now DowElanco).
Patent GB 1419121   **Manufacturers** DowElanco

APPLICATIONS
Biochemistry Melanin biosynthesis inhibitor (reduction of 1,3,8-trihydroxy=
naphthalene and vermelon).   **Mode of action** Systemic fungicide, absorbed rapidly
by the roots, with translocation through the plant.   **Uses** Control of rice blast
(*Pyricularia oryzae*) in transplanted and direct-seeded rice. Can be applied as a flat
drench, transplant root soak, or foliar application. One or 2 applications by one or
more of these methods give a season-long control of the disease.
Formulation types WP; DP; GR; WG; SC.   **Mixtures** *(tricyclazole +)* iprobenfos;
mepronil; ferimzone; carbosulfan; imidacloprid; etofenprox.
Selected tradenames 'Beam' (DowElanco); 'Sazole' (Sanonda)

ANALYSIS
Product analysis by glc with FID or by hplc (E. W. Day *et al.*, *Anal. Methods Pestic.
Plant Growth Regul.*, 1980, **11**, 263).   **Residues** in plant tissue determined by glc
with FPD, the main metabolite (hydroxymethyl analogue) first being converted to
a derivative (*idem, ibid.*). In **drinking water**, by gc with FID (*AOAC Methods*, 1995,
991.07). Details available from DowElanco.

MAMMALIAN TOXICOLOGY
Oral Acute oral LD_{50} for rats 314, mice 245, dogs >50 mg/kg.
Skin and eye Acute percutaneous LD_{50} for rabbits >2000 mg/kg. Slight eye
irritant; non-irritating to skin (rabbits).   **Inhalation** LC_{50} (1 h) for rats 0.146 mg/l
air.   **NOEL** (2 y) for rats 9.6 mg/kg b.w.; for mice 6.7 mg/kg b.w.; (1 y) for dogs
5 mg/kg b.w.; 3-generation reproduction for rat 3 mg/kg b.w.   **ADI** 0.03 mg/kg.
Toxicity class WHO (a.i.) II; EPA (formulation) II   **EC risk** Xn (R22)

ECOTOXICOLOGY
Birds Acute oral LD_{50} for mallard ducks and bobwhite quail >100 mg/kg.

Fish LC$_{50}$ (96 h) for bluegill sunfish 16.0, rainbow trout 7.3, goldfish fingerlings 13.5 mg/l.   **Daphnia** LC$_{50}$ (48 h) >20 mg/l; NOEC (21 d) 0.96 mg/l.

ENVIRONMENTAL FATE
Animals Rapid and extensive metabolism.   **Plants** The principal metabolite in plants is the hydroxymethyl analogue.   **Soil/Environment** K$_d$ 4 (loamy sand, pH 6.5, 1.5% o.m.), 45 (loam, pH 5.7, 3.1% o.m.), 21 (clay loam, pH 7.4, 1.9% o.m.), 22 (silty clay loam, pH 5.7, 4.1% o.m.).

35   tridec-4-en-1-yl acetate   *Insect pheromone*

pheromone

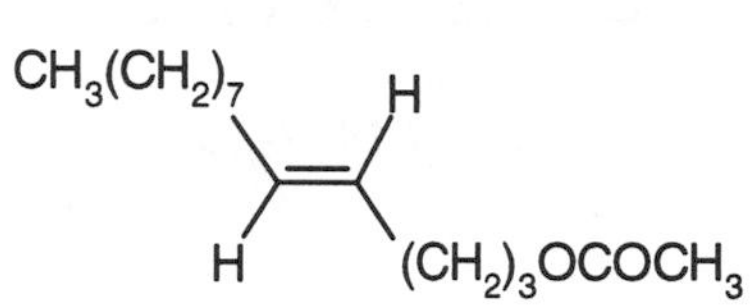

NOMENCLATURE
IUPAC name (E)-tridec-4-en-1-yl acetate
Chemical Abstracts name (E)-4-tridecen-1-ol acetate
Other names tomato pinworm pheromone; *Keiferia lycopersicella* pheromone
CAS RN *[72269–48–8]* ((E)- isomer); *[65954–19–0]* ((Z)- isomer)

PHYSICAL CHEMISTRY
Composition Major component of the pheromone of the tomato pinworm, *Keiferia lycopersicella*, is the (E)- isomer; the (Z)- isomer is a minor component. **Mol. wt.** 240.4   **M.f.** C$_{15}$H$_{28}$O$_2$ **Form** Off-white to light brown, viscous liquid with a sweet odour.

APPLICATIONS
Mode of action Acts by mating disruption.   **Uses** For control of the tomato pinworm (*Keiferia lycopersicella*).   **Formulation types** Microencapsulated; polymer matrix.   **Selected tradenames** 'NoMate TPW' (Ecogen)

morpholine

$$CH_3 - (C_nH_{2n}) - N \diagdown O$$

with CH₃ groups:

$CH_3 - (C_nH_{2n}) - N$ [morpholine ring with two CH_3 substituents]

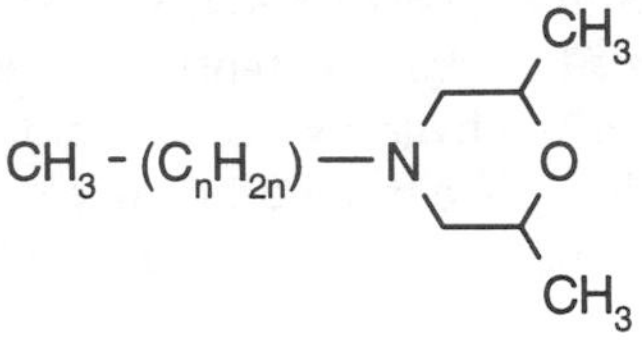

n = 10, 11, 12 (60 − 70%) or 13

NOMENCLATURE
Common name tridemorph (BSI, E-ISO); tridémorphe ((*m*) F-ISO) for a reaction mixture (see below)
IUPAC name 4-*alkyl*-2,6-dimethylmorpholine; for original definition, 2,6-dimethyl-= 4-tridecylmorpholine
Chemical Abstracts name for original structure/main component, 2,6-dimethyl-= 4-tridecylmorpholine
CAS RN *[81412–43–3]* for new definition; *[24602–86–6]* for 4-tridecyl-component **EEC no.** 246–347–3 **Development codes** BASF 220 F

PHYSICAL CHEMISTRY
Composition Originally thought to consist only of tridecyl (C_{13}) isomers, the reaction mixture has now been shown to comprise C_{11} to C_{14} homologues containing 60–70% of 4-tridecyl isomers, 0.2% C_9 and C_{15} homologues and 5% of 2,5-dimethyl isomers. **Mol. wt.** 297.5 (approx.) **M.f.** $C_{19}H_{39}NO$ (approx.)
Form Yellow oily liquid with a slight amine-like odour. **B.p.** 134 °C/0.5 mmHg (tech.) **V.p.** 12 mPa (20 °C) K_{ow} logP = 4.20 (pH 7, 22 °C) **Henry** 3.2 Pa m^3 mol^{-1} (calc.) **S.g./density** 0.86 (tech.) (20 °C) **Solubility** In water 1.1 mg/l (pH 7, 20 °C). Miscible with ethanol, acetone, ethyl acetate, cyclohexane, diethyl ether, benzene, chloroform, olive oil. **Stability** Stable ≤50 °C. On u.v. irradiation of an aqueous solution containing 20 mg/kg, 50% hydrolysis occurs in 16.5 h.
pKa 6.50 (20 °C), base **F.p.** 142 °C (Pensky-Martens)

COMMERCIALISATION
History Fungicide reported by J. Kradel et al. (*Proc. Br. Insectic. Fungic. Conf., 5th,* 1969, **1**, 16) and by E. H. Pommer et al. (*ibid.*, 1969, **2**, 347). Introduced in Germany (1969) by BASF AG. **Patent** DE 1164152 **Manufacturers** BASF

APPLICATIONS
Biochemistry Steroid reduction (ergosterol biosynthesis) inhibitor.
Mode of action Systemic fungicide with eradicant action. Absorbed by the leaves and roots, giving some protective action. **Uses** Control of *Erysiphe graminis* in cereals, *Mycosphaerella* spp. in bananas, *Caticum solmonicolor* in tea, *Oidium hevea* in hevea, and *Exobasidium vexans*. Mixed with carbendazim to extend spectrum of

cereal diseases controlled.   **Phytotoxicity** Scorch may occur on some winter wheat varieties under certain climatic conditions.   **Formulation types** EC.   **Mixtures** *(tridemorph +)* carbendazim + maneb; propiconazole; triadimenol; BAS 480F; fenpropimorph.   **Selected tradenames** 'Calixin' (BASF)

ANALYSIS
Product analysis by acid-base titration.   **Residues** in cereal straw and soil determined by colorimetry of a derivative or by gc-ms. Details available from BASF.

MAMMALIAN TOXICOLOGY
Oral Acute oral LD_{50} for rats 480 mg/kg.   **Skin and eye** Acute percutaneous LD_{50} for rats >4000 mg/kg. Skin irritant, not an eye irritant (rabbits).   **Inhalation** LC_{50} (4 h) for rats 4.5 mg/l air.   **NOEL** (2 y) for rats 30 mg/kg diet (*c.* 1.8 mg/kg daily), for dogs 50 ppm diet.   **ADI** 0.016 mg/kg b.w.   **Toxicity class** WHO (a.i.) II; EPA (formulation) III   **EC risk** Xn (R21/22)

ECOTOXICOLOGY
Birds LD_{50} for quail 1388, duck >2000 mg/kg.   **Fish** LC_{50} (96 h) for trout 3.4 mg/l.   **Bees** LD_{50} (24 h) >200 μg/bee.   **Worms** LC_{50} (14 d) for *Eisenia foetida* 880 mg/kg.   **Daphnia** LC_{50} (48 h) 1.3 mg/l.   **Algae** (96 h) 0.28 mg/l.

ENVIRONMENTAL FATE
Animals In rats, following oral administration, tridemorph is rapidly absorbed, and is almost completely eliminated within 2 days.   **Plants** Residues in cereal grains at harvest are <0.05 mg/kg, mostly polar material. Metabolism proceeds via oxidation of the 4-alkyl side-chain and/or by opening of the morpholine ring.   **Soil/Environment** DT_{50} (lab.) 20–50 d, (field) 14–34 d. K_{oc} 2500–10 000.

737   trietazine   *Herbicide*

1,3,5-triazine

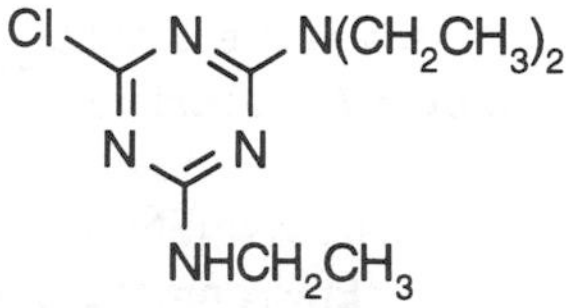

NOMENCLATURE
Common name trietazine (BSI, E-ISO, (*f*) F-ISO, ANSI, JMAF)
IUPAC name 6-chloro-N^2,N^2,N^4-triethyl-1,3,5-triazine-2,4-diamine

Chemical Abstracts name 6-chloro-*N,N,N'*-triethyl-1,3,5-triazine-2,4-diamine
CAS RN *[1912–26–1]* **Development codes** G 27 901 (Ciba-Geigy); NC 1667
(AgrEvo)

PHYSICAL CHEMISTRY
Mol. wt. 229.7 **M.f.** $C_9H_{16}ClN_5$ **Form** Colourless crystals. **M.p.** 102–103 °C
Solubility In water at 25 °C, 20 mg/l. In acetone 170, benzene 200, chloroform
>500, dioxane 100, ethanol 30 (all in g/l, 25 °C). **Stability** Stable in neutral
media. Hydrolysed on heating with strong acids and alkalis, with loss of chlorine.

COMMERCIALISATION
History Herbicide reported by H. Gysin & E. Knüsli (*Proc. Br. Weed Control Conf.,*
4th, 1958, p. 225). Discovered by J. R. Geigy S.A. (now Novartis Crop Protection
AG) and introduced commercially by Fisons Ltd (now AgrEvo GmbH) in mixtures
with other herbicides. **Patent** CH 329277; GB 814947; US 2819855 to Ciba-
Geigy **Manufacturers** AgrEvo

APPLICATIONS
Biochemistry Photosynthetic electron transport inhibitor.
Mode of action Selective herbicide, absorbed by the roots and leaves. **Uses** Used
with linuron for weed control in potatoes, with simazine in peas, and with
terbutryn in potatoes, peas and field beans at 1.6–4.5 kg/ha, according to crop
and soil type. **Formulation types** SC; WP. **Mixtures** (*trietazine* +) terbutryn;
simazine. **Selected tradenames** 'Remtal' (mixture) (AgrEvo); 'Aventox'
(DowElanco)

ANALYSIS
Product and **residues** determined by gc-ms, details available from AgrEvo.

MAMMALIAN TOXICOLOGY
Oral Acute oral LD_{50} for rats 494–841 mg/kg. **Skin and eye** Acute percutaneous
LD_{50} for rats >>600 mg/kg. Non-irritating to skin (rats). **NOEL** In 90 d feeding
trials, rats receiving 16 mg/kg diet showed no ill-effect. **Toxicity class**
WHO (a.i.) III (Table 5) **EC risk** Xn (R22)

ECOTOXICOLOGY
Birds Acute oral LD_{50} for quail 800 mg/kg. **Fish** LC_{50} (24 h) for guppies 5.5 mg/l.
Bees Non-toxic to honeybees.

ENVIRONMENTAL FATE
Soil/Environment General degradation scheme for chlorotriazines: chlorine atom
cleavage to yield hydroxy compounds which are conjugated with amino acids.
Dealkylation of the side-chain, ring cleavage, and evolution of CO_2.

azole

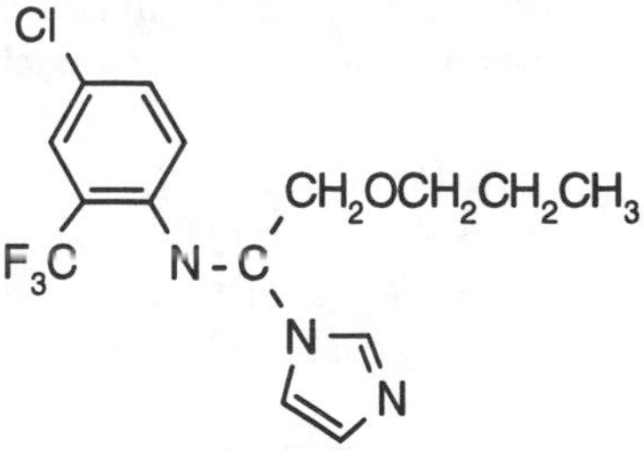

NOMENCLATURE
Common name triflumizole (BSI, JMAF, draft E-ISO, (*f*) draft F-ISO)
IUPAC name (*E*)-4-chloro-α,α,α-trifluoro-*N*-(1-imidazol-1-yl-2-propoxy=
ethylidene)-*o*-toluidine
Chemical Abstracts name (*E*)-1-[1-[[4-chloro-2-(trifluoromethyl)phenyl]imino]-=
2-propoxyethyl]-1*H*-imidazole
CAS RN *[99387–89–0]* ((*E*)- isomer); *[68694–11–1]* (unstated stereochemistry)
Development codes NF-114 (Nippon Soda)

PHYSICAL CHEMISTRY
Mol. wt. 345.7   **M.f.** $C_{15}H_{15}ClF_3N_3O$ **Form** Colourless crystals.   **M.p.** 63.5 °C
V.p. 0.186 mPa (25 °C)   **Solubility** In water 12.5 g/l (20 °C). In chloroform 2220,
hexane 17.6, xylene 639, acetone 1440, methanol 496 (all in g/l, 20 °C).
Stability Unstable in highly alkaline and acidic media. Aqueous solutions degraded
by sunlight, DT_{50} 29 h.   **pKa** 3.7 (25 °C), weak base

COMMERCIALISATION
History Fungicide reported by A. Nakata (*Proc. Int. Congr. Pestic. Chem., Kyoto, 5th*,
1982). Introduced by Nippon Soda Co., Ltd.   **Patent** JP 79119462; US 4208411;
UK 1591212; DE 2814041.   **Manufacturers** Nippon Soda

APPLICATIONS
Biochemistry Steroid demethylation (ergosterol biosynthesis) inhibitor.
Mode of action Systemic fungicide with protective and curative action.
Uses Control of *Gymnosporangium* and *Venturia* spp. in pome fruit, against
powdery Erysiphaceae in fruit and vegetables, and against *Fusarium, Fulvia* and
Monilinia spp. It is also used as a seed treatment against *Helminthosporium, Tilletia*
and *Ustilago* spp. in cereals.   **Formulation types** WP; EC; FU.
Selected tradenames 'Trifmine' (Nippon Soda); 'Procure' (Uniroyal)

ANALYSIS
Product analysis by hplc.   **Residues** determined by hplc.

MAMMALIAN TOXICOLOGY
Oral Acute oral LD_{50} for male rats 715, female rats 695 mg/kg.
Skin and eye Acute percutaneous LD_{50} for rats >5000 mg/kg. Mild eye irritant, not a skin irritant. **Inhalation** LC_{50} (4 h) for rats >3.2 mg/l air. **NOEL** (2 y) for rats 3.7 mg/kg diet. **Toxicity class** WHO (a.i.) III; EPA (formulation) III

ECOTOXICOLOGY
Birds Acute oral LD_{50} for male Japanese quail 2467, female Japanese quail 4308 mg/kg. **Fish** LC_{50} (48 h) for carp 1.26 mg/l. **Bees** LD_{50} for honeybees 0.14 mg/bee. **Daphnia** LC_{50} (3 h) 9.7 mg/l.

ENVIRONMENTAL FATE
Animals For details of metabolism in rats, see T. Tanoue *et al.*, *IUPAC 7th Int. Congr. Pestic. Chem.*, 1990, **2**, 177. **Soil/Environment** In soil, DT_{50} 14 d (on clay). Photolytic degradation leads to the metabolite (*E*)-4-chloro-α,α,α-trifluoro-=N-(1-amino-1-yl-2-propoxyethylidene)-*o*-toluidine. K_{oc} 1083–1663.

739 triflumuron   *Insecticide*

benzoylurea

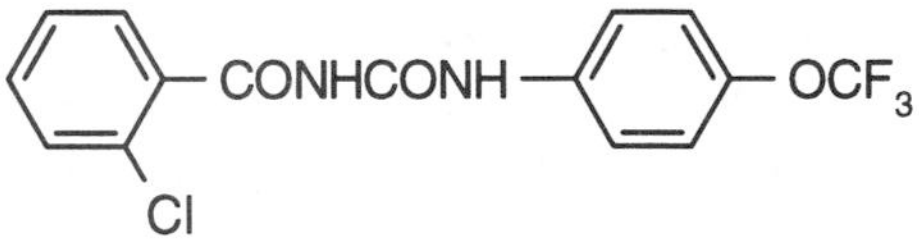

NOMENCLATURE
Common name triflumuron (BSI, draft E-ISO, (*m*) draft F-ISO)
IUPAC name 1-(2-chlorobenzoyl)-3-(4-trifluoromethoxyphenyl)urea
Chemical Abstracts name 2-chloro-*N*-[[[4-(trifluoromethoxy)phenyl]amino]=carbonyl]benzamide
CAS RN *[64628–44–0]* **Development codes** SIR 8514 (Bayer)
Official codes OMS 2015

PHYSICAL CHEMISTRY
Mol. wt. 358.7 **M.f.** $C_{15}H_{10}ClF_3N_2O_3$ **Form** Colourless odourless powder.
M.p. 195 °C **V.p.** 4×10^{-5} mPa (20 °C) **K_{ow}** logP = 4.91 (20 °C)
Henry 3×10^{-3} Pa m^3 mol^{-1} **S.g./density** 1.445 (20 °C) **Solubility** In water at 20 °C, 0.025 mg/l. In dichloromethane 20–50, isopropanol 1–2, toluene 2–5, hexane <0.1 g/l (20 °C). **Stability** Stable to hydrolysis in neutral media and in acids; hydrolysed by alkali: DT_{50} (22 °C) 960 d (pH 4), 580 d (pH 7), 11 d (pH 9).

COMMERCIALISATION
History Insecticide reported by G. Zoebelein et al. (*Int. Congr. Plant Prot., 9th, Annu. Mtg. Am. Phytopathol. Soc., 71st, 1979,* Abstract 309) and by I. Hammann & W. Sirrenberg (*Pflanzenschutz-Nachr. (Engl. Ed.), 1980,* **33,** 1). Introduced by Bayer AG. **Patent** DE 2601780 **Manufacturers** Bayer

APPLICATIONS
Biochemistry Chitin synthesis inhibitor. **Mode of action** Ingested insecticide, acting by inhibition of moulting. **Uses** Control of Lepidoptera, Psyllidae, Diptera, and Coleoptera on fruit, soya beans, vegetables, forest trees, and cotton. Also used against larvae of flies, fleas and cockroaches in public and animal health.
Formulation types SC; WP; EC; OF; UL. **Mixtures** *(triflumuron +)* methamidophos; sulprofos. **Selected tradenames** 'Alsystin' (crop protection) (Bayer); 'Baycidal' (public health) (Bayer); 'Starycide' (animal health) (Bayer)

ANALYSIS
Methods for determination of **residues** available from Bayer.

MAMMALIAN TOXICOLOGY
Oral Acute oral LD_{50} for male and female rats and mice >5000, dogs >1000 mg/kg. **Skin and eye** Acute percutaneous LD_{50} for male and female rats >5000 mg/kg; not irritating to skin and eye (rabbits). Not a skin sensitiser.
Inhalation LC_{50} (4 h) for male and female rats >0.12 mg/l air (aerosol), >1.6 mg/l air (dust). **NOEL** (2 y) for rats 20, mice 20 mg/kg diet; (1 y) for dogs 20 mg/kg diet. **ADI** 0.0072 mg/kg b.w. **Toxicity class** WHO (a.i.) III (Table 5); EPA (formulation) IV

ECOTOXICOLOGY
Birds Acute oral LD_{50} for bobwhite quail 561 mg/kg. **Fish** LC_{50} (96 h) for rainbow trout >320, golden orfe >100 mg/l. **Bees** Toxic to bees. **Worms** LC_{50} (14 d) >1000 mg/kg. **Other beneficial spp.** Adult stages are not affected, slight effects on juvenile stages possible; safe to predatory mites. **Daphnia** LC_{50} (48 h) 0.225 mg/l. **Algae** E_rC_{50} (96 h) for *Scenedesmus* >25 mg/l.

ENVIRONMENTAL FATE
Animals Triflumuron labelled in the 2-chlorobenzoyl moiety was metabolised in rats by hydrolytic cleavage forming metabolites which contained only the 2-chlorophenyl ring and were partly hydroxylated and conjugated. Correspondingly, in experiments with labelling in the 4-trifluoromethoxyphenyl group, metabolites were found which contained only the 4-trifluoromethoxyphenyl ring, partly in hydroxylated form. **Plants** Following spray application to apples, soya beans and potatoes, triflumuron is only slightly metabolised; metabolites were the same as those formed in animals. For residue analyses, it is sufficient to determine the parent compound in the harvested crops.

Soil/Environment Degradation: In laboratory tests, triflumuron was moderately quickly degraded in the soil; degradation in the field was more rapid by a factor of 3–5. Repeated applications over 3 years to soil without vegetation did not result in any accumulation in the soil. In practice-relevant applications in forests, the concentrations of residues found in the soil were very low at all times and declined below the limit of detection after a few months. **Metabolism:** In soil, 50% of applied triflumuron labelled in the 2-chlorobenzoyl moiety was degraded to CO_2 within 112 days and c. 20% of the radioactivity was bound to the soil. When using triflumuron labelled in the 4-trifluoromethoxyphenyl moiety, the compound was mineralised more slowly, while the percentage of bound residues was markedly increased. Metabolism was mainly induced by microbes, and resulted in metabolites which contained just one of the two rings in each case.

740 trifluralin   *Herbicide*

2,6-dinitroaniline

NOMENCLATURE
Common name trifluralin (BSI, E-ISO, ANSI, WSSA, JMAF); trifluraline ((*f*) F-ISO)
IUPAC name α,α,α-trifluoro-2,6-dinitro-N,N-dipropyl-*p*-toluidine
Chemical Abstracts name 2,6-dinitro-N,N-dipropyl-4-(trifluoromethyl)= benzenamine
CAS RN *[1582–09–8]* **EEC no.** 216–428–8 **Development codes** L-36 352 (DowElanco); EL-152

PHYSICAL CHEMISTRY
Mol. wt. 335.3 **M.f.** $C_{13}H_{16}F_3N_3O_4$ **Form** Yellow-orange crystals.
M.p. 48.5–49 °C; (tech., 43–47.5 °C) **B.p.** 96–97 °C/24 Pa **V.p.** 6.1 mPa
(25 °C, EEC A4) **K$_{ow}$** logP = 4.83 (20 °C, EEC A8) **Henry** 15 Pa m^3 mol^{-1} (calc.)
S.g./density 1.36 (22 °C, EEC A3) **Solubility** In water 0.184 (pH 5), 0.221 (pH 7), 0.189 (pH 9) (all in mg/l, EEC A6); tech. 0.343 (pH 5), 0.395 (pH 7), 0.383 (pH 9) (all in mg/l, EEC A6). In acetone, chloroform, acetonitrile, toluene, ethyl acetate >1000, methanol 33–40, hexane 50–67 (all in g/l, 25 °C).
Stability Stable at 52 °C (highest storage temperature tested). Stable to hydrolysis at pH values of 3, 6 and 9 (52 °C). Decomposed by u.v. irradiation (E. Leitis &

D. G. Crosby, *J. Agric. Food Chem.*, 1974, **22**, 842).   **F.p.** 151 °C (closed cup); tech. 153 °C (open cup) (both Pensky-Martens)

COMMERCIALISATION
History Herbicide reported by E. F. Alder *et al.* (*Proc. North Cent. Weed Control Conf.*, 1960, p. 23). Introduced in USA (1961) by Eli Lilly & Co. (now DowElanco). **Patent** US 3257190   **Manufacturers** Budapesti Vegyimüvek; Defensa; DowElanco; Drexel; Makhteshim-Agan; Nortox; Nufarm Ltd; Q.E.A.C.A.; Sanachem; Westrade

APPLICATIONS
Biochemistry Cell division inhibitor.   **Mode of action** Selective soil-herbicide, which acts by entering the seedling in the hypocotyl region. Also inhibits root development.   **Uses** Pre-emergence control of many annual grasses and broad-leaved weeds in brassicas, beans, peas, carrots, parsnips, lettuce, capsicums, tomatoes, artichokes, onions, garlic, vines, strawberries, raspberries, citrus fruit, oilseed rape, peanuts, soya beans, sunflowers, safflowers, ornamentals, cotton, sugar beet, sugar cane, and in forestry. Used with linuron or isoproturon for control of annual grasses and broad-leaved weeds in winter cereals. Normally applied pre-planting with soil incorporation at 0.5–1.0 kg a.i./ha, but post-planting application is also possible for some crops.   **Formulation types** EC; GR. **Mixtures** (*trifluralin +)* linuron; napropamide; metribuzin; clomazone; tebutam; bromoxynil + ioxynil; isoproturon; terbutryn; linuron + trietazine; linuron + neburon; linuron + clomazone; isoxaben; benfluralin.
Selected tradenames 'Treflan' (DowElanco); 'Commerce' (mixture) (FMC); 'Eflurin' (Efthymiadis); 'Ipersan' (Q.E.A.C.A.); 'Olitref' (Budapesti Vegyimüvek); 'Premerlin' (Defensa); 'Sinfluran' (Westrade); 'Tri-4' (Cyanamid); 'Trifluran' (Cequisa); 'Triflurex' (Makhteshim-Agan); 'Trifsan' (Sanachem); 'Trigard' (FCC); 'Trilin' (Griffin); 'Triplen' (Sipcam); 'Tristar' (PBI); 'Zeltoxone' (Zeneca)

ANALYSIS
Product analysis by glc with FID (*CIPAC Handbook*, 1980, **1A**, 1362; *AOAC Methods*, 1995, 973.14) or by u.v. spectrometry (*ibid.*, 973.13).   **Residues** determined by glc with ECD (J. B. Tepe & R. E. Scroggs, *Anal. Methods Pestic., Plant Growth Regul. Food Addit.*, 1967, **5**, 527; *Anal. Methods Pestic. Plant Growth Regul.*, 1972, **6**, 703). In **drinking water** by gc with ECD (*AOAC Methods*, 1995, 990.06). Details from DowElanco.

MAMMALIAN TOXICOLOGY
Oral Acute oral LD_{50} for rats >5000 mg/kg.   **Skin and eye** Acute percutaneous LD_{50} >5000 mg/kg (rabbits). Non-irritating to skin, slightly irritating to eyes (rabbits).   **Inhalation** In rats, 1-hour exposure to 4.83 mg of a 20% milled concentrate/litre of air resulted in 2/10 deaths.   **NOEL** In 2 y feeding trials in rats, the only effect at the low dose of 813 mg/kg in diet was the formation of renal calculi. This has been shown to be reversible in a 90 d study in dogs and

a NOEL established at 2.4 mg/kg daily. NOEL in mice was 73 mg/kg daily.
ADI 0.024 mg/kg. **Toxicity class** WHO (a.i.) III (Table 5); EPA (formulation) III,
IV **EC risk** Material containing <0.5 ppm NPDA is Xi (R36); (R43)

ECOTOXICOLOGY
Birds Acute oral LD_{50} for bobwhite quail >2000 mg/kg. Five-day dietary LC_{50} for
bobwhite quail and mallard ducks >5000 mg/kg. **Fish** LC_{50} (96 h) for young
rainbow trout 0.01–0.04, young bluegill sunfish 0.02–0.09 mg/l. **Bees** LD_{50} (oral)
>50, (contact) >100 μg/bee. **Daphnia** LC_{50} (48 h) 0.56 mg/l; NOEC (21 d)
0.051 mg/l. **Other aquatic spp.** LD_{50} (96 h) for grass shrimp (*Palaemonetes* sp.)
0.64 mg/l.

ENVIRONMENTAL FATE
Animals Degradation in animals is as for soil (J. L. Emmerson & R. C. Anderson,
Toxicol. Appl. Pharmacol., 1966, **9**, 84–97). Following oral administration, *c.* 70% is
eliminated in the urine and 15% in the faeces within 72 hours. **Plants** Degradation
in plants is as for soil. **Soil/Environment** Absorbed by the soil, and is extremely
resistant to leaching. Little lateral movement in the soil. Metabolism involves
dealkylation of the amino group, reduction of the nitro group to an amino group,
partial oxidation of the trifluoromethyl group to a carboxy group, and subsequent
degradation to smaller fragments (T. Golab *et al.*, *J. Agric. Food Chem.*, 1979, **27**,
163); DT_{50} 57–126 d. Duration of residual activity in soil is 6–8 mo. In laboratory
studies, degradation was more rapid under anaerobic conditions, e.g. for loam soil,
DT_{50} (anaerobic) 25 d, DT_{50} (aerobic) 116 d. K_{oc} 6400–13 400: K_d ranges from
3.75 (0.01% o.m., pH 6.6) to 639 (16.9% o.m., pH 6.8) (H. J. Pedersen *et al.*,
Pestic. Sci., **44**, 131 (1995)).

741 triflusulfuron-methyl   *Herbicide*

sulfonylurea

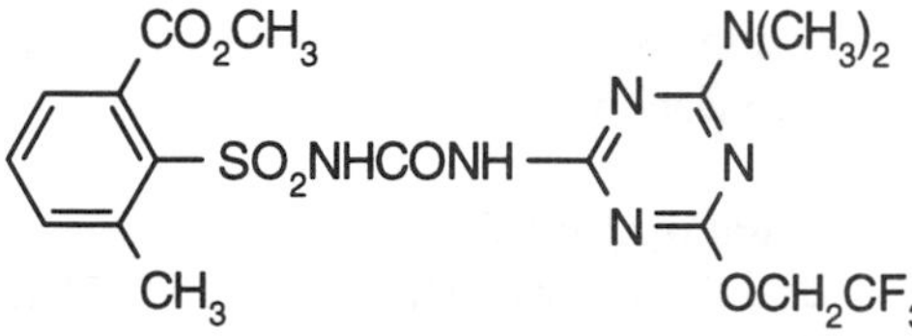

NOMENCLATURE
triflusulfuron-methyl
Common name triflusulfuron-methyl
Other names JT478 **CAS RN** *[126535–15–7]* **Development codes** DPX-66037;
IN 66037

triflusulfuron
Common name triflusulfuron (BSI, pa E-ISO, ANSI)
IUPAC name 2-[4-dimethylamino-6-(2,2,2-trifluoroethoxy)-1,3,5-triazin-=
2-ylcarbamoylsulfamoyl]-*m*-toluic acid
Chemical Abstracts name 2-[[[[[4-(dimethylamino)-6-(2,2,2-trifluoroethoxy)-=
1,3,5-triazin-2-yl]amino]carbonyl]amino]sulfonyl]-3-methylbenzoic acid
CAS RN *[135990–29–3]*

PHYSICAL CHEMISTRY
triflusulfuron-methyl
Composition Tech. is >96%. **Mol. wt.** 492.4 **M.f.** $C_{17}H_{19}F_3N_6O_6S$ **Form** White
crystalline solid. **M.p.** 160–163 °C; tech. 155–158 °C **V.p.** <1 × 10^{-2} mPa
(25 °C) **K_{ow}** logP = 0.96 (pH 7) **Henry** <5.9 × 10^{-5} Pa m^3 mol^{-1}
S.g./density 1.45 **Solubility** In water 1 (pH 3), 3 (pH 5), 110 (pH 7), 11 000
(pH 9) (all in mg/l, 25 °C). In methylene chloride 580, acetone 120, methanol 7,
toluene 2, acetonitrile 80 (all in mg/ml, 25 °C). **Stability** Rapidly hydrolysed in
water, DT_{50} 3.7 d (pH 5), 32 d (pH 7), 36 d (pH 9) (all 25 °C). **pKa** 4.4

triflusulfuron
Mol. wt. 478.4 **M.f.** $C_{16}H_{17}F_3N_6O_6S$

COMMERCIALISATION
History Reported by L. A. Peeples *et al.* (*Proc. Br. Crop Prot. Conf. – Weeds*, 1991,
1, 25). Introduced by E. I. du Pont de Nemours and Co. **Manufacturers** Du Pont

APPLICATIONS
triflusulfuron-methyl
Biochemistry Branched chain amino acid synthesis (acetolactate synthase or ALS)
inhibitor. Acts by inhibiting biosynthesis of the essential amino acids valine and
isoleucine, hence stopping cell division and plant growth. Selectivity is due to rapid
metabolism in sugar beet (M. K. Koeppe *et al.*, *Proc. Br. Crop Prot. Conf.*, 1993, **1**,
177). Metabolic basis of selectivity in sulfonylureas reviewed (M. K. Koeppe & H.
M. Brown, *Agro-Food-Industry*, **6**, 9–14 (1995)). **Mode of action** Post-emergence
selective herbicide. Symptoms occur first in meristematic tissue. **Uses** Post-
emergence control of many annual and perennial broad-leaved weeds in sugar
beet at 10–20 g/ha. **Formulation types** WG. **Mixtures** (*triflusulfuron-methyl* +)
lenacil. **Compatibility** Compatible with other sugar beet herbicides. **Selected
tradenames** 'Debut' (W. Europe) (Du Pont); 'Safari' (E. Europe & Scandinavia)
(Du Pont); 'Upbeet' (USA) (Du Pont)

ANALYSIS
By lc and u.v. detection.

MAMMALIAN TOXICOLOGY
triflusulfuron-methyl
Oral Acute oral LD_{50} for rats >5000 mg/kg. **Skin and eye** Acute percutaneous LD_{50} for rabbits >2000 mg/kg. Non-irritating to skin and eyes (rabbits). Non-sensitising to skin (guinea pigs). **Inhalation** LC_{50} (4 h) for rats >5.1 mg/l.
NOEL (1 y) for dogs 875 ppm; (18 mo) for mice 150 ppm; (2 y) for male rats 100, female rats 750 ppm. **ADI** 0.05 mg/kg (UK). **Other** Non-mutagenic in the Ames test. **Toxicity class** WHO (a.i.) III (Table 5)

ECOTOXICOLOGY
triflusulfuron-methyl
Birds LD_{50} for mallard duck and bobwhite quail 2250 mg/kg. Dietary LC_{50} for mallard duck and bobwhite quail >5620 ppm. **Fish** LC_{50} (96 h) for bluegill sunfish 760, trout 730 mg/l. **Bees** Oral LD_{50} (48 h) >1000 ppm. **Worms** LD_{50} for earthworm >1000 mg/kg. **Daphnia** LC_{50} (48 h) >960 mg/l. **Algae** EC_{50} (120 h) for green algae 0.62 mg/l. **Other aquatic spp.** LC_{50} (14 d) for *Lemna gibba* 9.0 µg/l.

ENVIRONMENTAL FATE
The compound rapidly degrades in water, soil, plants, and animals. The primary metabolic pathways in all the systems are the cleavage of the sulfonylurea bridge yielding methyl saccharin and triazine amine followed by N-demethylation to N-desmethyl triazine amine, and N,N-bis-desmethyl triazine amine. The metabolic pathway is consistent in aquatic, soil, and biological systems.
Soil/Environment Degrades rapidly in soil by chemical and microbial mechanisms. Microbial degradation is important in alkaline conditions, but plays only a minor role in neutral and acidic conditions because chemical hydrolysis is very rapid. Half-life in soil 3 d. Bioaccumulation is unlikely to occur.

742 triforine   *Fungicide*

azole analogue

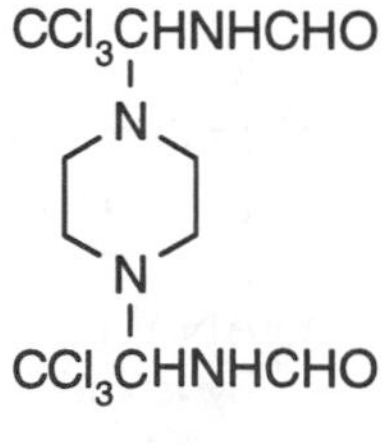

$CCl_3CHNHCHO$

(piperazine ring structure)

$CCl_3CHNHCHO$

NOMENCLATURE
Common name triforine (BSI, E-ISO, (f) F-ISO, ANSI, JMAF)

IUPAC name N,N'-[piperazine-1,4-diylbis[(trichloromethyl)methylene]]=
diformamide; 1,1'-piperazine-1,4-diyldi-[N-(2,2,2-trichloroethyl)formamide]
Chemical Abstracts name N,N'-[1,4-piperazinediylbis(2,2,2-trichloroethylidene)]=
bisformamide
CAS RN *[26644–46–2]*   **Development codes** Cela W524

PHYSICAL CHEMISTRY

Mol. wt. 435.0   **M.f.** $C_{10}H_{14}Cl_6N_4O_2$ **Form** White to light brown crystals.
M.p. 155 °C (decomp.)   **V.p.** 8×10^1 mPa (25 °C)   **K_{ow}** logP = 2.2 (20 °C,
unstated pH)   **Henry** 2.5 Pa m^3 mol^{-1} **S.g./density** 1554 kg/m^3 (20 °C)
Solubility In water 9 mg/l (20 °C). In dimethylformamide 330, dimethyl sulfoxide
476, N-methylpyrrolidone 476, acetone 11, dichloromethane 1, methanol 10 (all in
g/l). Also soluble in tetrahydrofuran; slightly soluble in dioxane and
cyclohexanone; insoluble in benzene, petroleum ether and cyclohexane.
Stability Stable ≤180 °C. Decomposed in aqueous solution exposed to u.v. or
daylight. Decomposed in strongly acidic media (to trichloroacetaldehyde and
piperazine salts) and in strongly alkaline media (to chloroform and piperazine);
DT$_{50}$ (pH 5 to 7, 25 °C) 3.5 d.   **pKa** 10.6, strong base

COMMERCIALISATION

History Fungicide reported by P. Schicks & K. H. Veen (*Proc. Br. Insectic. Fungic.
Conf., 5th*, 1969, **2**, 569). Introduced by Cela GmbH (now American Cyanamid
Co.).   **Patent** DE 1901421 to Boehringer   **Manufacturers** Cyanamid

APPLICATIONS

Biochemistry Ergosterol biosynthesis inhibitor.   **Mode of action** Systemic fungicide
with protective and curative action. Absorbed by the leaves and roots, with
translocation acropetally.   **Uses** Control of powdery mildews on cereals, fruit,
vines, hops, ornamentals, cucurbits, and some vegetables; rusts on cereals, fruit,
ornamentals, and beans; *Monilia* spp. on stone fruit; *Ascochyta* blight on
chrysanthemums; black spot on roses; scab on apples; and fairy rings on turf. Also
suppresses tetranychid mite activity.   **Phytotoxicity** Phytotoxic to certain varieties
of pear.   **Formulation types** EC; WP.   **Mixtures** *(triforine +)* carbendazim.
Compatibility Should not be mixed with wetting agents, spreader-stickers, or other
adjuvants.   **Selected tradenames** 'Saprol' (Cyanamid)

ANALYSIS

Product analysis by tlc.   **Residues** determined by glc (R. Darskus & D. Eichler,
Anal. Methods Pestic. Plant Growth Regul., 1978, **10**, 243). Details available from
Cyanamid.

MAMMALIAN TOXICOLOGY

Reviews *Pesticide residues in food – 1978.* FAO Plant Production and Protection
Paper 15, 1979. *Pesticide residues in food: 1978 evaluations.* FAO Plant Production

and Protection Paper 15 Sup, 1979.   **Oral** Acute oral LD$_{50}$ for rats >16 000, mice >6000, dogs >2000 mg/kg.   **Skin and eye** Acute percutaneous LD$_{50}$ for rabbits and rats >10 000 mg/kg.   **Inhalation** LC$_{50}$ (1 h) for rats >4.5 mg/l.   **NOEL** (2 y) for rats 200 mg/kg diet, for dogs 100 mg/kg diet.   **ADI** (JMPR) 0.02 mg/kg b.w. [1978].   **Other** Acute i.p. LD$_{50}$ for rats >4000 mg/kg.   **Toxicity class** WHO (a.i.) III (Table 5); EPA (formulation) IV

ECOTOXICOLOGY
Birds Acute oral LD$_{50}$ for bobwhite quail >5000 mg/kg.   **Fish** LC$_{50}$ (96 h) for bluegill sunfish and rainbow trout >1000 mg/l.   **Bees** 60 g a.i./hl does not pose a hazard.   **Worms** Low toxicity to earthworms; LD$_{50}$ >1000 mg/kg.
Other beneficial spp. Harmless to beneficial arthropods.   **Daphnia** LC$_{50}$ (48 h) 117 mg/l.   **Algae** EC$_{50}$ for *Scenedesmus capricornutum* 380 mg/l.

ENVIRONMENTAL FATE
Soil/Environment In soil, a range of non-fungitoxic metabolic end-products are formed, presumably including piperazine (J. P. Rouchard *et al.*, *Pestic. Sci.*, 1978, **9**, 74, 139, 587; A. Fuchs & W. Ost, *Arch. Environ. Contam. Toxicol.*, 1976, **4**, 30–43). DT$_{50}$ in soil c. 3 w. Soil adsorption Freundlich K 2.75–6.28 in loamy sand and silty loam (o.c. 2.4–6.3%, pH 5.2–7.9). Does not accumulate in the environment.

743   trimethacarb   *Insecticide, molluscicide*

carbamate

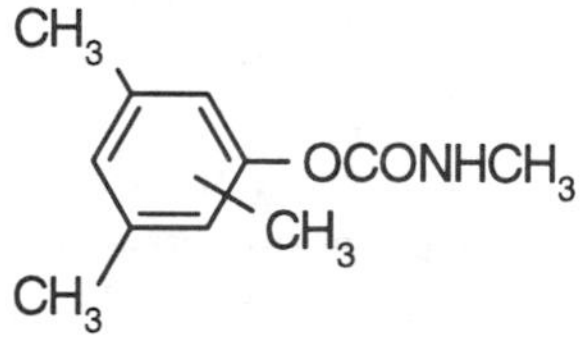

NOMENCLATURE
Common name trimethacarb (BSI, ANSI, draft E-ISO, (m) draft F-ISO)
IUPAC name Reaction product comprising 3,4,5-trimethylphenyl methylcarbamate (I) and 2,3,5-trimethylphenyl methylcarbamate (II) in a ratio between 3.5:1 and 5.0:1 *m/m*
Chemical Abstracts name 2,3,5(or 3,4,5)-trimethylphenol methylcarbamate
CAS RN *[12407–86–2]* trimethacarb; *[2686–99–9]* (I); *[2655–15–4]* (II)
Development codes UC 27 867   **Official codes** OMS 597

PHYSICAL CHEMISTRY
Mol. wt. 193.2   **M.f.** C$_{11}$H$_{15}$NO$_2$ **Form** Buff to brown crystalline solid.

M.p. 105–114 °C   **V.p.** 6.8 mPa (25 °C)   **Solubility** In water >58 mg/kg (23 °C). Not readily soluble in organic solvents.   **Stability** Decomposed by strong acids and alkalis. Stable to light.

COMMERCIALISATION
History Insecticide originally introduced by Shell Development Co. and later by Union Carbide Agrochemicals (now Rhône-Poulenc Agrochimie, who no longer manufacture or market it).   **Manufacturers** Drexel

APPLICATIONS
Biochemistry Cholinesterase inhibitor   **Mode of action** Insecticide with predominantly stomach action, but also some contact action. Long residual activity.   **Uses** Control of corn rootworm larvae in maize. Also controls a wide range of insect and mollusc pests, and acts as a mammal and bird repellent. **Phytotoxicity** Non-phytotoxic, when used as directed. Phytotoxic to the seeds of some crops, including maize, sorghum, wheat, and rice.   **Formulation types** GR; WP.

ANALYSIS
Residues by glc with ECD (S. C. Lau & R. L. Marxmiller, *J. Agric. Food Chem.*, 1970, 413–415).

MAMMALIAN TOXICOLOGY
EHC 64 (WHO, 1986; a review of carbamate insecticides in general).   **Oral** Acute oral LD_{50} for rats 130 mg/kg.   **Skin and eye** Acute percutaneous LD_{50} for rats >2000 mg/kg.

ECOTOXICOLOGY
Fish Toxic to fish.   **Bees** Toxic to bees, in spray formulations.

ENVIRONMENTAL FATE
Animals Metabolism of carbamate insecticides is reviewed (M. Cool & C. K. Jankowski in "*Insecticides*").   **Plants** In plants, metabolism occurs via hydroxylation of the *N*-methyl group and at the 3- and 4-methyl positions. All metabolites may be conjugated as glucosides.   **Soil/Environment** Soil DT_{50} *c.* 60 d in sterile soil.

744 trinexapac-ethyl *Plant growth regulator*

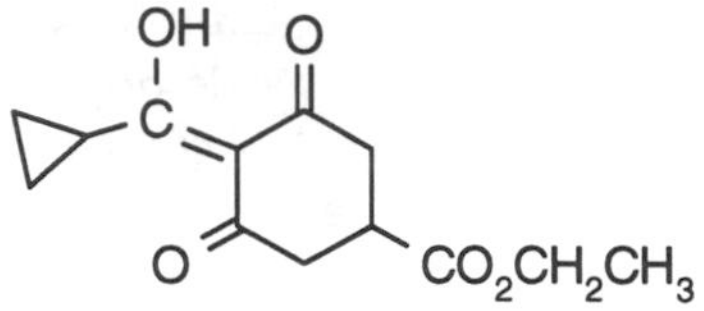

NOMENCLATURE

trinexapac-ethyl

Common name trinexapac-ethyl

IUPAC name ethyl 4-cyclopropyl(hydroxy)methylene-3,5-dioxocyclohexane=carboxylate

Chemical Abstracts name ethyl 4-(cyclopropylhydroxymethylene)-3,5-dioxo=cyclohexanecarboxylate

Other names cimectacarb-ethyl (rejected name) **CAS RN** *[95266–40–3]*

Development codes CGA 163 935

trinexapac

Common name trinexapac (BSI, pa E-ISO)

IUPAC name 4-cyclopropyl(hydroxy)methylene-3,5-dioxocyclohexanecarboxylic acid

CAS RN *[104273–73–6]* **Development codes** CGA 179500

PHYSICAL CHEMISTRY

trinexapac-ethyl

Composition Tech. is ≥92% pure. **Mol. wt.** 252.3 **M.f.** $C_{13}H_{16}O_5$ **Form** White odourless solid; tech. is a yellowish brown liquid (30 °C) and a solid-liquid melt (20 °C) with a slight sweet odour. **M.p.** 36 °C **B.p.** >270 °C **V.p.** 1.6 mPa (20 °C), 2.16 mPa (25 °C) (OECD 104) **K_{ow}** logP = 1.60 (pH 5.3, 25 °C) **S.g./density** 1.215 g/cm^3 (20 °C) **Solubility** In water 2.8 (pH 4.9), 10.2 (pH 5.5), 21.1 (pH 8.2) g/l (25 °C). In ethanol, acetone, toluene, *n*-octanol 100%, *n*-hexane 5% (25 °C). **Stability** Heat stable up to boiling point. Photolytically and hydrolytically stable under normal environmental conditions (pH 6–7, 25 °C). Less stable in alkali. **pKa** 4.57 **F.p.** 133 °C (1013 mbar) (84/449/EEC A9)

trinexapac

Mol. wt. 224.2 **M.f.** $C_{11}H_{12}O_5$ **M.p.** 144.4 °C **B.p.** *c.* 220 °C **V.p.** 2.3×10^{-3} mPa (25 °C, OECD 104, EEC A4) **K_{ow}** logP = 1.8 (pH 2) **S.g./density** 1.41 (20 °C) **Solubility** In water 13 (pH 5.0), 200 (pH 6.8), 260 (pH 8.4) g/l. In acetone 95, ethyl acetate 37, methanol 84, octanol 17 g/l. **pKa** pKa_1 5.32, pKa_2 3.93

COMMERCIALISATION

History First reported by E. Kerber *et al., Proc. Br. Crop Protection Conf.,* 1989, **I**, 83. Introduced by Ciba-Geigy AG (now Novartis Crop Protection AG). **Patent** EP 126713; US 4693745 **Manufacturers** Novartis

APPLICATIONS
trinexapac-ethyl

Biochemistry Inhibits a key enzyme in the biosynthesis of gibberellic acid (GA_1) **Mode of action** Plant growth regulator which reduces stem growth by inhibition of internode elongation. Absorbed by the foliage, with translocation to the growing shoot. **Uses** Used for the prevention of lodging in cereals and in winter oilseed rape at 0.1–0.3 kg/ha. Also used in turf at 0.15–0.5 kg/ha to reduce mowing, and as a maturation promoter in sugar cane at 0.1–0.25 kg/ha. **Phytotoxicity** Could cause inhibition or growth stoppage of some plants, e.g. grasses, aquatic plants and algae. **Formulation types** EC; ME; WP. **Mixtures** *(trinexapac-ethyl +)* ethephon. **Compatibility** Compatibility of little importance because of late timing of application; compatible with certain late-season fungicides.
Selected tradenames 'Metro' (Novartis, La Quinoleine); 'Moddus' (Novartis); 'Primo' (Novartis); 'Sonis' (Novartis, La Quinoleine)

ANALYSIS
All methods use hplc, for both ester and acid metabolite.

MAMMALIAN TOXICOLOGY
trinexapac-ethyl

Oral Acute oral LD_{50} for rats 4460 mg/kg. **Skin and eye** Acute percutaneous LD_{50} for rats >4000 mg/kg; not irritant to skin or eyes of rabbits. Non-sensitising to skin (guinea pigs). **Inhalation** LC_{50} (48 h) for rats >5.3 mg/l. **NOEL** (2 y) for rats 115 mg/kg b.w. daily; (18 mo) for mice 451 mg/kg b.w. daily; (1 y) for dogs 31.6 mg/kg b.w. daily. **ADI** 0.316 mg/kg b.w. **Toxicity class** WHO (a.i.) III (Table 5)

ECOTOXICOLOGY
trinexapac-ethyl

Birds LD_{50} for duck and quail >2000 mg/kg. Eight-day LC_{50} for duck and quail >5000 ppm. **Fish** LC_{50} (96 h) for trout, carp, bluegill sunfish, catfish, fathead minnow 35–180 mg/l. **Bees** Non-toxic; LD_{50} (oral) >293 µg/bee, (contact) >115 µg/bee. **Worms** Low toxicity to earthworms; LC_{50} >93 mg/kg. **Daphnia** LC_{50} (96 h) 142 mg/l.

ENVIRONMENTAL FATE
Animals In rats, goats and hens, 90% excretion occurs within 24 h, all as the acid metabolite. **Plants** Rapid metabolism to the acid, which remains by far the predominant metabolite. **Soil/Environment** In soil, the ester undergoes rapid

degradation, DT_{50} <1 d. The acid has K_d values 1.5–16, K_{oc} 140–600. Further metabolism is rapid, DT_{50} typically 3–5 w. Within 7 w, 50% is mineralised to CO_2.

745 triticonazole   *Fungicide*

azole

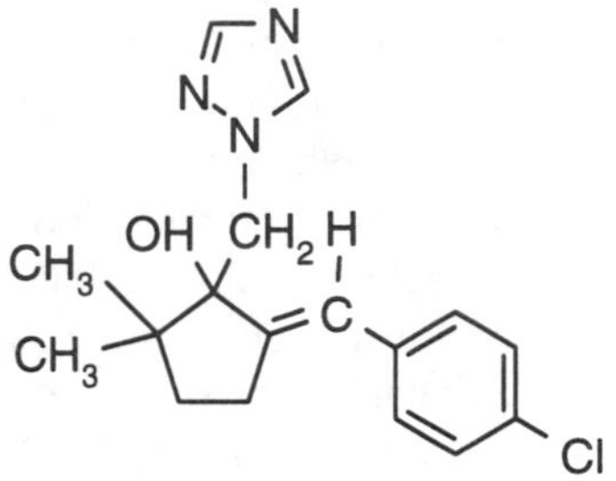

NOMENCLATURE
Common name triticonazole (BSI, pa E-ISO)
IUPAC name (±)-(E)-5-(4-chlorobenzylidene)-2,2-dimethyl-1-(1H-1,2,4-triazol-=
1-ylmethyl)cyclopentanol
Chemical Abstracts name 5-[(4-chlorophenyl)methylene]-2,2-dimethyl-=
1-(1H-1,2,4-triazol-1-ylmethyl)cyclopentanol
CAS RN *[131983–72–7]*   **Development codes** RPA 400727

PHYSICAL CHEMISTRY
Composition Material is a racemic mixture; 95% pure.   **Mol. wt.** 317.8
M.f. $C_{17}H_{20}ClN_3O$ **Form** White powder, odourless at 22 °C.
M.p. 139–140.5 °C   **V.p.** <1 × 10^{-5} mPa (50 °C) K_{ow} logP = 3.29 (20 °C)
Henry <3.9 × 10^{-5} Pa m³ mol⁻¹ (calc.)   **S.g./density** 1.326–1.369 (20 °C)
Solubility In water 7 mg/l (10 °C), independent of pH.   **Stability** Slight decomposition at 180 °C.

COMMERCIALISATION
History Fungicide developed by Rhône-Poulenc Agrochimie, first patented in 1988.
Patent FR 2641277   **Manufacturers** Rhône-Poulenc

APPLICATIONS
Biochemistry Inhibition of sterol demethylation.   **Uses** Seed disinfectant against seed-borne diseases and a preventive treatment against a number of foliar pathogens such as rusts, powdery mildew, leaf spot, eye spot, leaf and net blotch of cereals and head smut of maize.   **Formulation types** FS.

Mixtures *(triticonazole +)* anthraquinone; guazatine; iprodione.
Selected tradenames 'Alios' (Rhône-Poulenc); 'Premis' (Rhône-Poulenc); 'Real' (Rhône-Poulenc)

MAMMALIAN TOXICOLOGY
Oral Acute oral LD_{50} for rats >2000 mg/kg. **Skin and eye** Acute percutaneous LD_{50} for rats >2000 mg/kg. Not a skin or eye irritant. **Inhalation** LC_{50} >1.4 mg/l air. **NOEL** Chronic NOEL for rats 750 ppm (29.4 and 38.3 mg/kg b.w. daily for males and females respectively); for dogs 2.5 mg/kg b.w. **ADI** 0.0025 mg/kg (provisional, French). **Toxicity class** WHO (a.i.) III (Table 5)

ECOTOXICOLOGY
Birds Acute oral LD_{50} for bobwhite quail >2000 mg/kg. **Fish** Low acute toxicity to rainbow trout; LC_{50} >10 mg/l. **Worms** Non-toxic. **Daphnia** LC_{50} (48 h) >9.3 mg/l. **Algae** EC_{50} (96 h) >1.0 mg/l.

ENVIRONMENTAL FATE
Animals 90% Elimination via faeces within 7 days in the rat. **Plants** Metabolised to a dihydroxy metabolite and others. **Soil/Environment** DT_{50} 224–360 d (10 °C).

746 uniconazole *Plant growth regulator*

azole

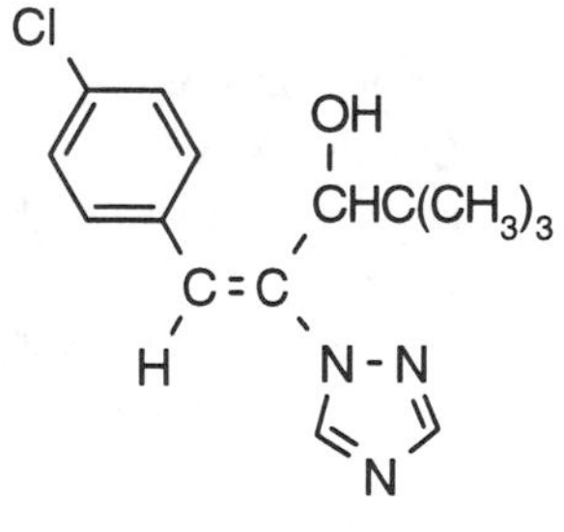

NOMENCLATURE
uniconazole
Common name uniconazole (BSI, ANSI, draft E-ISO, (*m*) draft F-ISO)
IUPAC name (*E*)-(*RS*)-1-(4-chlorophenyl)-4,4-dimethyl-2-(1*H*-1,2,4-triazol-1-yl)= pent-1-en-3-ol
Chemical Abstracts name (*E*)-(±)-β-[(4-chlorophenyl)methylene]-= α-(1,1-dimethylethyl)-1*H*-1,2,4-triazole-1-ethanol
CAS RN *[83657–22–1]* **Development codes** S-07; S-327 (both Sumitomo); XE-1019 (Chevron)

uniconazole-P

Common name uniconazole-P (BSI, pa E-ISO)
IUPAC name (E)-(S)-1-(4-chlorophenyl)-4,4-dimethyl-2-(1H-1,2,4-triazol-1-yl)=
pent-1-en-3-ol
CAS RN *[83657–17–4]* (E)-(S)-(+)- isomer; *[83657–16–3]* (E)-(R)-(–)- isomer;
[76714–83–5] (E)- isomers   **Development codes** S-3307 D

PHYSICAL CHEMISTRY
uniconazole
Mol. wt. 291.8   **M.f.** $C_{15}H_{18}ClN_3O$ **Form** White crystalline solid.
M.p. 147–164 °C   **V.p.** 8.9 mPa (20 °C) K_{ow} logP = 3.67 (25 °C)
S.g./density 1.28 (21.5 °C)   **Solubility** In water 8.41 mg/l (25 °C). In methanol 88,
hexane 0.3, xylene 7 (all in g/kg, 25 °C). Soluble in acetone, ethyl acetate,
chloroform, and dimethylformamide.   **Stability** Good stability under normal
storage conditions.

uniconazole-P
Mol. wt. 291.8   **M.f.** $C_{15}H_{18}ClN_3O$ **Form** White crystalline solid with a faint
characteristic odour.   **M.p.** 152.1–155.0 °C   **V.p.** 5.3 mPa (20 °C)
S.g./density 1.28 (25 °C).   **Solubility** In water 8.41 mg/l (25 °C). In methanol 72,
hexane 0.2 g/l (20 °C).   **Stability** Stable under normal storage conditions.
F.p. 195 °C

COMMERCIALISATION
History Plant growth regulator reported by K. Izumi et al. (*Plant Cell Physiol.*, 1984,
25, 611); uniconazole-P is more potent than the (R)-(–)- isomer. Introduced by
Sumitomo Chemical Co., Ltd and by Valent.   **Manufacturers** Sumitomo

APPLICATIONS
Biochemistry Inhibits gibberellin biosynthesis.   **Mode of action** Plant growth
regulator, absorbed by the stems and roots, with translocation in the xylem to
growing points.   **Uses** Used to reduce lodging in rice; to reduce vegetative growth
and increase flowering of ornamentals; and to reduce vegetative growth and the
need for pruning in trees.   **Formulation types** WP; GR; SL.

uniconazole
Selected tradenames 'Lomica' (Japan) (Sumitomo); 'Sumiseven' (Japan)
(Sumitomo)

uniconazole-P
Selected tradenames 'Sumagic' (Sumitomo, Valent)

ANALYSIS
Product analysis by glc or hplc, details available from Sumitomo.
Residues determined by glc or hplc.

MAMMALIAN TOXICOLOGY
uniconazole
Toxicity class WHO (a.i.) III

uniconazole-P
Oral Acute oral LD_{50} for male rats 2020, female rats 1790 mg/kg.
Skin and eye Acute percutaneous LD_{50} for rats >2000 mg/kg. Not irritating to
skin; minimal eye irritation (rabbits). **Inhalation** LC_{50} (4 h) for rats >2750 mg/m^3
Other Non-mutagenic in Ames assay. **Toxicity class** EPA (formulation) III

ECOTOXICOLOGY
uniconazole-P
Fish LC_{50} (96 h) for rainbow trout 14.8, carp 7.64 mg/l. **Bees** Acute contact
LD_{50} for bees >20 µg/bee.

747 validamycin   *Fungicide*

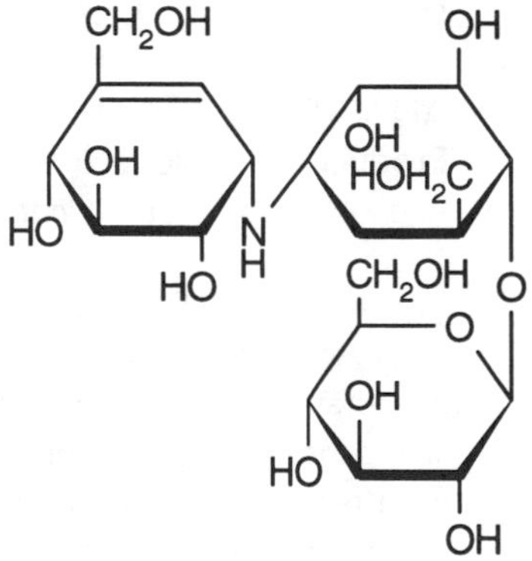

NOMENCLATURE
Common name validamycin (JMAF); validamycin A (Japanese Antibiotics Research
Association)
IUPAC name 1L-(1,3,4/2,6)-2,3-dihydroxy-6-hydroxymethyl-4-[(1S,4R,5S,6S)-=
4,5,6-trihydroxy-3-hydroxymethylcyclohex-2-enylamino]cyclohexyl β-D-gluco=
pyranoside
Chemical Abstracts name [1S-(1α,4α,5β,6α)]-1,5,6-trideoxy-4-O-β-D-gluco=
pyranosyl-5-(hydroxymethyl)-1-[[4,5,6-trihydroxy-3-(hydroxymethyl)-2-cyclo=
hexen-1-yl]amino]-D-*chiro*-inositol
CAS RN [37248–47–8]

PHYSICAL CHEMISTRY
Mol. wt. 497.5 **M.f.** $C_{20}H_{35}NO_{13}$ **Form** Colourless, odourless, hygroscopic powder. **M.p.** 130–135 °C (decomp.) **V.p.** Negligible at room temperature. **Solubility** Readily soluble in water. Soluble in methanol, dimethylformamide and dimethyl sulfoxide. Slightly soluble in ethanol and acetone. Sparingly soluble in diethyl ether and ethyl acetate. **Stability** Stable at room temperature in neutral or alkaline media, but slightly unstable in acidic media. **Specific rotation** $[\alpha]_D^{24}$ +110° (water)

COMMERCIALISATION
Production By the fermentation of *Streptomyces hygroscopicus* var. *limoneus* nov. var. **History** Antibiotic described by T. Iwasa *et al.* (*J. Antibiot.*, 1970, **23**, 595). Correct structure reported by T. Suami *et al.* (*ibid.*, 1980, **33**, 98). Introduced by Takeda Chemical Industries Ltd. **Manufacturers** Takeda

APPLICATIONS
Mode of action Non-systemic antibiotic with fungistatic action. **Uses** Control of *Rhizoctonia solani* in rice, potatoes, vegetables, strawberries, tobacco, ginger and other crops; damping-off diseases of cotton, rice and sugar beet, etc. Applied as a foliar spray, soil drench, seed dressing, or by soil incorporation.
Formulation types DP; SL; DS; Liquid. **Mixtures** *(validamycin +)* fenobucarb; phthalide. **Selected tradenames** 'Validacin' (Takeda); 'Mycin' (Sanonda); 'Solacol' (AgrEvo)

ANALYSIS
Product and **residue** analysis by glc of a derivative with FID (K. Nishi & K. Konishi, *Anal. Methods Pestic. Plant Growth Regul.*, 1976, **8**, 309).

MAMMALIAN TOXICOLOGY
Oral Acute oral LD_{50} for rats and mice >20 000 mg/kg. **Skin and eye** Acute percutaneous LD_{50} for rats >5000 mg/kg. Non-irritating to skin (rabbits). Not a skin sensitiser (guinea pigs). **Inhalation** LC_{50} (4 h) for rats >5 mg/l air. **NOEL** In 90 d feeding trials, rats receiving 1000 mg/kg of diet and mice receiving 2000 mg/kg of diet showed no ill-effects. In 2 y feeding trials, NOEL for rats was 40.4 mg/kg b.w. daily. **Toxicity class** WHO (a.i.) III (Table 5); EPA (formulation) IV

ECOTOXICOLOGY
Fish LC_{50} (72 h) for carp >40 mg/l. **Daphnia** LC_{50} (24 h) for *D. pulex* >40 mg/l.

ENVIRONMENTAL FATE
Animals In animals, cleavage to glucose and an amine residue occurs.
Soil/Environment Rapid microbial degradation in soil; DT_{50} ≤5 h.

organophosphorus

$$\begin{array}{cc}
CH_3 & O \\
| & \| \\
\end{array}$$
$$CH_3NHCOCHSCH_2CH_2SP(OCH_3)_2$$

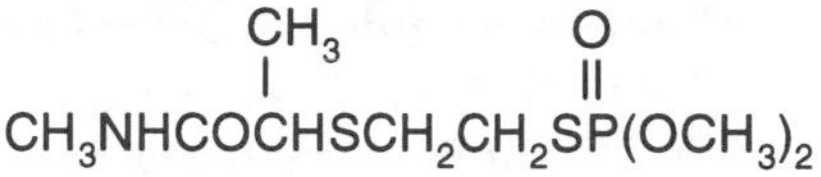

NOMENCLATURE

Common name vamidothion (BSI, E-ISO, (m) F-ISO, JMAF)
IUPAC name *O,O*-dimethyl *S*-2-(1-methylcarbamoylethylthio)ethyl phosphorothioate; 2-(2-dimethoxyphosphinoylthioethylthio)-*N*-methyl= propionamide
Chemical Abstracts name *O,O*-dimethyl *S*-[2-[[1-methyl-2-(methylamino)-2-oxo= ethyl]thio]ethyl] phosphorothioate
CAS RN *[2275–23–2]* **EEC no.** 218–894–8 **Development codes** 10 465 RP; NPH 83 **Official codes** ENT 26 613

PHYSICAL CHEMISTRY

Mol. wt. 287.3 **M.f.** $C_8H_{18}NO_4PS_2$ **Form** Colourless needles; (tech., white waxy solid). **M.p.** *c.* 43 °C; (tech., 40 °C) **V.p.** Negligible (20 °C) **Solubility** Readily soluble in water (4 kg/l), benzene, toluene, methyl ethyl ketone, ethyl acetate, acetonitrile, dichloromethane, cyclohexanone, chloroform (all *c.* 1 kg/l). Almost insoluble in cyclohexane and petroleum ether. **Stability** Undergoes slight decomposition at room temperature, but solutions in organic solvents (methyl ethyl ketone, cyclohexanone) are stable. Decomposed in strong acidic or alkaline media.

COMMERCIALISATION

History Insecticide reported by J. Desmoras *et al.* (*Phytiatr.-Phytopharm.*, 1962, **11**, 107). Introduced by Rhône-Poulenc Agrochimie. **Patent** GB 872823; BE 575106 **Manufacturers** Rhône-Poulenc

APPLICATIONS

Biochemistry Cholinesterase inhibitor. **Mode of action** Systemic insecticide and acaricide. Metabolised in plants to the corresponding sulfoxide which is of similar activity to vamidothion but of greater persistence. **Uses** Systemic insecticide giving persistent control of *Eriosoma lanigerum* and other piercing and sucking Homoptera in cotton, hops, pome and stone fruit, and rice at 37–50 g a.i./hl. **Formulation types** EC. **Selected tradenames** 'Kilval' (Rhône-Poulenc)

ANALYSIS

Product analysis by hplc or glc (J. Desmoras *et al.*, *Anal. Methods Pestic. Plant Growth Regul.*, 1973, **7**, 479). **Residues** determined by glc (*idem, ibid.*; *Man. Pestic. Residue Anal.*, 1987, **I**, 3, 6, S13; *Anal. Methods Residues Pestic.*, 1988,

Part I, M2, M5, M10; A. Ambrus *et al.*, *J. Assoc. Off. Anal. Chem.*, 1981, **64**, 733; A. R. C. Hill, *Analyst (London)*, 1984, **109**, 483).

MAMMALIAN TOXICOLOGY
Reviews *Pesticide residues in food – 1988*. FAO Plant Production and Protection Paper 92, 1988. *Pesticide residues in food – 1988 evaluations. Part II – Toxicology.* FAO Plant Production and Protection Paper 93/2, 1989. **EHC** 63 (WHO, 1986; a general review of organophosphorus insecticides). **Oral** Acute oral LD_{50} for male rats 100–105, female rats 64–67, mice 34–37 mg/kg. For the sulfoxide, acute oral LD_{50} for male rats 160, mice 80 mg/kg. **Skin and eye** Acute percutaneous LD_{50} for mice 1460, rabbits 1160 mg/kg. **Inhalation** LC_{50} (4 h) for rats 1.73 mg/l air. **NOEL** In 90 d feeding trials, the growth rate of rats receiving 50 mg vamidothion/kg diet or 100 mg of its sulfoxide/kg diet was unaffected. **ADI** (JMPR) 0.008 mg/kg b.w. [1988]. **Toxicity class** WHO (a.i.) Ib
EC risk T (R25); Xn (R21)

ECOTOXICOLOGY
Birds Acute oral LD_{50} for pheasants 35 mg/kg. **Fish** LC_{50} (96 h) for zebra fish 590 mg/l. At 10 mg/l, harmless to goldfish (14 d). **Bees** Toxic to bees.
Daphnia EC_{50} (48 h) 0.19 mg/l.

ENVIRONMENTAL FATE
Animals Oxidised to the sulfoxide and sulfone, followed by cleavage of the P-S and S-C bonds, to give water-soluble metabolites. **Plants** Metabolised in plants to the corresponding sulfoxide, which appears within hours after treatment. Also demethylation and hydrolysis to phosphoric acid. After 20 d, all the toxic residues in the plant are in the form of the sulfoxide. **Soil/Environment** In soil, DT_{50} 1.0–1.5 d (aerobic, 22 °C).

749 vernolate   *Herbicide*

thiocarbamate

$$(CH_3CH_2CH_2)_2NCOSCH_2CH_2CH_3$$

NOMENCLATURE
Common name vernolate (BSI, E-ISO, (m) F-ISO, WSSA, JMAF)
IUPAC name S-propyl dipropylthiocarbamate
Chemical Abstracts name S-propyl dipropylcarbamothioate
CAS RN [1929–77–7] **EEC no.** 217–681–7 **Development codes** R-1607 (Zeneca)

PHYSICAL CHEMISTRY
Composition Tech. material is 95% pure. **Mol. wt.** 203.3 **M.f.** $C_{10}H_{21}NOS$
Form Clear liquid, with slight aromatic odour; (tech. is a clear yellow liquid).
B.p. 150 °C/30 mmHg **V.p.** 1.39 Pa (25 °C) **K**$_{ow}$ logP = 3.84 (20 °C)
S.g./density 0.952 g/ml (20 °C) **Solubility** In water at 20 °C, 90 mg/l. Miscible
with common organic solvents, e.g. xylene, methyl isobutyl ketone, kerosene,
acetone, ethanol. **Stability** Stable in neutral media, and relatively stable in acidic
and alkaline media; at pH 7, 50% loss occurs in 13 d at 40 °C. Stable up to 200 °C.
Decomposed by sunlight. **F.p.** 121 °C

COMMERCIALISATION
History Herbicide introduced by Stauffer Chemical Co. (now **Zeneca**
Agrochemicals). **Patent** US 2913327

APPLICATIONS
Biochemistry Inhibits lipid metabolism. **Mode of action** Selective herbicide,
absorbed by the roots, with translocation to the stems and leaves. Germination
inhibitor. **Uses** Control of germinating broad-leaved and grass weeds in peanuts,
soya beans, maize, tobacco and sweet potatoes at 1.5–3.0 kg a.i./ha. Soil
incorporation is necessary, either pre-planting or pre-emergence.
Formulation types EC; GR. **Mixtures** (*vernolate +*) dichlormid; EPTC.
Selected tradenames 'Vernam' (Zeneca)

ANALYSIS
Product analysis by glc (*CIPAC Handbook*, 1983, **1B**, 1905; *AOAC Methods*, 1995,
974.05). **Residues** in crops and soils determined by glc or by colorimetry after
conversion to a derivative (G. G. Patchett & G. H. Batchelder, *Anal. Methods
Pestic., Plant Growth Regul. Food Addit.*, 1967, **5**, 337; W. J. Ja, *Anal. Methods Pestic.
Plant Growth Regul.*, 1972, **6**, 708). In **drinking water**, by gc with FID (*AOAC
Methods*, 1995, 991.07). Details available from Zeneca.

MAMMALIAN TOXICOLOGY
Oral Acute oral LD$_{50}$ for male rats 1500, female rats 1550 mg/kg.
Skin and eye Acute percutaneous LD$_{50}$ for rabbits >5000 mg/kg. Non-irritating to
eyes and skin (rabbits). Not a skin sensitiser (guinea pigs). **Inhalation** LC$_{50}$ (4 h)
for rats >5 mg/l. **NOEL** (51 w) for rats 5 mg/kg daily; (90 d) for rats 32 mg/kg,
for dogs 38 mg/kg daily. **Toxicity class** WHO (a.i.) II; EPA (formulation) III
EC risk Xn (R22)

ECOTOXICOLOGY
Birds Seven-day dietary LC$_{50}$ for bobwhite quail 12 000 mg/kg. **Fish** LC$_{50}$ (96 h)
for rainbow trout 4.6, bluegill sunfish 8.4 mg/l. **Bees** Non-toxic to bees at
0.011 mg/bee.

ENVIRONMENTAL FATE

Plants Readily metabolised by plants to CO_2 and naturally-occurring plant constituents.   **Soil/Environment** In soil, microbial decomposition to mercaptan, amine, isopropanol and CO_2. DT_{50} (27 °C) 8–16 d; (4 °C) >64 d.

750   *Verticillium lecanii*   *Biological agent*

entomopathogenic fungus

NOMENCLATURE

Scientific name *Verticillium lecanii* (Zimm.) Viégas   **Strain** whitefly strain; aphid strain   **Other names** white halo fungus; [*Cephalosporium lecanii*]

PROPERTIES

Form Fungal spores.   **Solubility** Insoluble in water and organic solvents. **Stability** Unstable at temperatures above 35 °C, even for short periods. Refrigeration (4 °C) is necessary for prolonged storage (shelf life 6 months).

COMMERCIALISATION

Production *V. lecanii* isolates are cultured on a sterile undefined medium and the spores harvested by concentration and drying.   **History** An isolate of the Deuteromycete (Moniliales) fungus *V. lecanii* was obtained from the aphid *Macrosiphoniella sanborni* by R. A. Hall (*J. Inverteb. Pathol.*, 1976, **27**, 41) during a natural epidemic. Later he obtained a second isolate, with a different host range, from the glasshouse whitefly *Trialeurodes vaporariorum* (*idem, Ann. Appl. Biol.*, 1979, **101**, 1). These isolates were developed by Tate & Lyle Ltd (who no longer produce or market them).   **Manufacturers** Koppert

APPLICATIONS

Mode of action Insecticide, acting through degradation of cuticle and subsequent fungal growth in haemolymph and tissues of insects. Re-sporulation from dead insects leads to infection of epidemic proportions.   **Uses** 'Mycotal' is used for control of whitefly, with a side-effect on thrips. 'Vertalec' is used to control aphids on protected glasshouse crops. A mutant form is under development by the USDA for control of cyst nematode in soya beans.   **Phytotoxicity** Non-phytotoxic and non-phytopathogenic.   **Formulation types** WP.
Compatibility Susceptible to some fungicides, especially dithiocarbamates.
Selected tradenames 'Mycotal' (whitefly strain) (Koppert); 'Vertalec' (aphid strain) (Koppert)

ANALYSIS

Activity of *V. lecanii* is measured in terms of spore count and efficacy against insects. Spore count measured as viable colony-forming unit count can be assayed by conventional techniques. Efficacy assays are made against *Aphis fabae* for 'Vertalec', and pupae of *Trialeurodes vaporariorum* for 'Mycotal'.

MAMMALIAN TOXICOLOGY

No skin or eye irritation observed. There is no evidence of acute or chronic toxicity, infectivity or hypersensitivity to mammals. No allergic responses or health problems have been observed by research workers, manufacturing staff or users.

'51 vinclozolin   *Fungicide*

dicarboximide

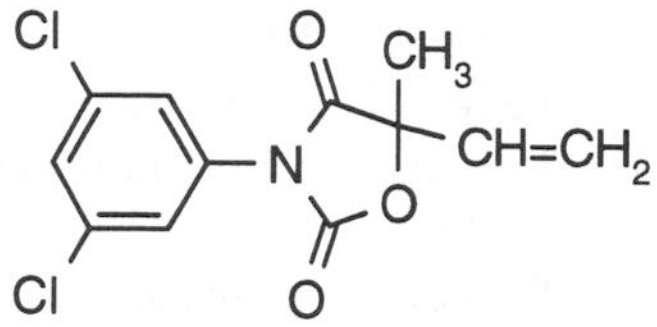

NOMENCLATURE

Common name vinclozolin (BSI, E-ISO, JMAF); vinclozoline ((*m*) F-ISO)
IUPAC name (*RS*)-3-(3,5-dichlorophenyl)-5-methyl-5-vinyl-1,3-oxazolidine-=2,4-dione
Chemical Abstracts name (±)-3-(3,5-dichlorophenyl)-5-ethenyl-5-methyl-=2,4-oxazolidinedione
CAS RN *[50471–44–8]* unstated stereochemistry   **Development codes** BAS 352F

PHYSICAL CHEMISTRY

Composition Tech. is ≥96% *m/m* pure.   **Mol. wt.** 286.1   **M.f.** $C_{12}H_9Cl_2NO_3$
Form Colourless crystals with a slight aromatic odour.   **M.p.** 108 °C (tech.)
B.p. 131 °C/0.05 mm Hg   **V.p.** 0.13 mPa (20 °C) K_{ow} logP = 3 (pH 7)
S.g./density 1.51   **Solubility** In water at 20 °C, 2.6 mg/l. In methanol 1.54, acetone 33.4, ethyl acetate 23.3, *n*-heptane 0.45, toluene 10.9, dichloromethane 47.5 (all in g/100 ml soln., 20 °C).   **Stability** Stable up to 50 °C. Stable for 24 h in acidic media. In 0.1N sodium hydroxide, 50% hydrolysis occurs in 3.8 h.

COMMERCIALISATION

History Fungicide reported by E.-H. Pommer & D. Mangold (*Meded. Fac. Landbouwwet. Rijksuniv. Gent*, 1975, **40**, 713). Introduced in Germany (1976) by BASF AG.   **Patent** DE 2207576   **Manufacturers** BASF

APPLICATIONS
Mode of action Non-systemic fungicide with protective action. Prevents spore germination. **Uses** Control of *Botrytis* and *Sclerotinia* spp. in vines, strawberries, oilseed rape, vegetables, fruit, and ornamentals; *Monilia* spp. in pome fruit and stone fruit; *Sclerotinia*, *Helminthosporium*, and *Corticium* spp. in turf; etc. **Formulation types** WP; SC; WG; FD. **Mixtures** *(vinclozolin +)* carbendazim; chlorothalonil; maneb; sulfur; thiophanate-methyl; thiram. **Selected tradenames** 'Ronilan' (BASF); 'Flotilla' (Barclay)

ANALYSIS
Product analysis by glc with FID (*CIPAC Handbook*, 1988, **D**, 173). **Residues** determined by hydrolysis to 3,5-dichloroaniline, a derivative of which is measured by glc with ECD (*Methodensammlung Rückstandsanal. Pflanzenschutzmitteln*, 1987, **XII**, 6, S8, S19; *Anal. Methods Residues Pestic.*, 1988, Part I, M1, M12; A. Ambrus et *al.*, *J. Assoc. Off. Anal. Chem.*, 1981, **64**, 733). Details available from BASF.

MAMMALIAN TOXICOLOGY
Reviews *Pesticide residues in food – 1995*, FAO Plant Production and Protection Paper 133, 1996. *Pesticide residues in food – 1995 evaluations; Part II – Toxicology & Environment*. World Health Organisation, WHO/PCS/96.48, 1996. **Oral** Acute oral LD_{50} for rats and mice >15 000, guinea pigs *c.* 8000 mg/kg. **Skin and eye** Acute percutaneous LD_{50} for rats >5000 mg/kg. **Inhalation** LC_{50} (4 h) for rats >29.1 mg/l air. **NOEL** (2 y) for rats 1.4 mg/kg b.w.; (1 y) for dogs 2.4 mg/kg b.w. **ADI** (JMPR) 0.01 mg/kg b.w. [1995]. **Other** Vinclozolin has shown antiandrogenic properties in laboratory animals. **Toxicity class** WHO (a.i.) III (Table 5); EPA (formulation) IV

ECOTOXICOLOGY
Birds Acute oral LD_{50} for quail >2510 mg/kg. LC_{50} for quail >5620 mg/kg. **Fish** LC_{50} (96 h) for trout 22–32, guppies 32.5, bluegill sunfish 50 mg/l. **Bees** Not toxic to bees. LD_{50} >200 mg/bee. **Worms** Not toxic to earthworms. **Daphnia** LC_{50} (48 h) 4.0 mg/l.

ENVIRONMENTAL FATE
Animals In the hen, the major metabolic routes are epoxidation of the vinyl group, followed by hydration of the intermediate epoxide, and by hydrolytic cleavage of the heterocyclic ring (G. M. Dean et *al.*, *Proc. Br. Crop Prot. Conf. – Pests Dis.*, 1988, **2**, 693–8). In rats, following oral administration, eliminated in approximately equal proportions in the urine and faeces, with the principal metabolite being N-(3,5-dichlorophenyl)-2-methyl-2,3,4-trihydroxybutanamide. **Plants** In plants, the primary metabolites are (1-carboxy-1-methyl)allyl 3,5-dichlorophenyl= carbamate and N-(3,5-dichlorophenyl)-2-hydroxy-2-methyl-3-butenamide. Alkaline hydrolysis leads to loss of 3,5-dichloroaniline from vinclozolin and its metabolites.

The metabolites exist as conjugates. **Soil/Environment** Metabolism occurs by loss of the vinyl group, cleavage of the 5-membered ring and eventual formation of 3,5-dichloroaniline. K_{oc} 100–735. Soil degradation of vinclozolin takes place with half-lives of several weeks and mainly leads to the formation of bound residues.

'752 vitamin D₃

Rodenticide

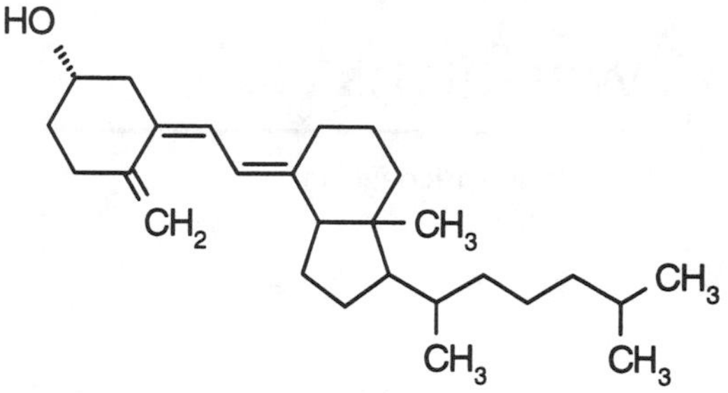

NOMENCLATURE
IUPAC name (3β,5Z,7E)-9,10-secocholesta-5,7,10(19)-trien-3-ol
Chemical Abstracts name (3β,5Z,7E)-9,10-secocholesta-5,7,10(19)-trien-3-ol
Other names cholecalciferol   **CAS RN** *[67–97–0]*

PHYSICAL CHEMISTRY
Mol. wt. 384.6   **M.f.** $C_{27}H_{44}O$ **Form** Light brown resin.   **M.p.** 84–85 °C
Solubility Insoluble in water. Soluble in acetone, chloroform and fatty oils.
Stability Oxidised and inactivated by moist air within a few days.

COMMERCIALISATION
Manufacturers Uniroyal

APPLICATIONS
Biochemistry Hypercalcification in vessels and organs.   **Uses** Control of rats and mice (by single or multiple feeding).   **Formulation types** BB; AB; RB.
Mixtures *(vitamin D₃ +)* coumatetralyl.   **Selected tradenames** 'Racumin D' (Bayer)

MAMMALIAN TOXICOLOGY
Oral Acute oral LD_{50} for rats 43.6 mg/kg.   **Skin and eye** Acute percutaneous LD_{50} for male rats 61, female rats 185, rabbits >2000 mg/kg.   **Inhalation** LC_{50} (4 h) 130–380 mg/m³ air.   **NOEL** A dose of 10–25 μg vitamin D₃ daily in humans produces no toxic effects.   **Toxicity class** EPA (formulation) III

ECOTOXICOLOGY
Birds Oral LD_{50} for mallard duck >2000 mg/kg.

ENVIRONMENTAL FATE
Animals Metabolised in the liver to 25-hydroxycholecalciferol by a NADPH-dependent reaction. This metabolite is then transferred to the kidney and converted to 24,25-, or 1,25-dihydrocholecalciferol by mitochondrial mixed-function oxidases.

753 warfarin   *Rodenticide*

coumarin anticoagulant

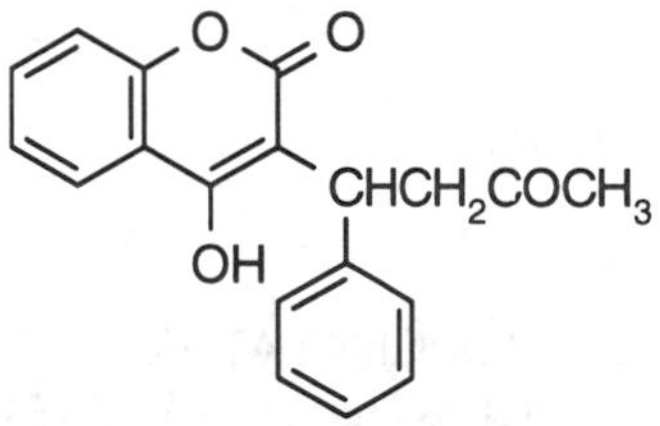

NOMENCLATURE
Common name warfarin (BSI, E-ISO, BAN); warfarine ((*m*) F-ISO); coumafène (France); zoocoumarin (USSR); coumarins (JMAF, also applied to coumatetralyl (*q.v.*)); no name (The Netherlands)
IUPAC name (*RS*)-4-hydroxy-3-(3-oxo-1-phenylbutyl)coumarin; 3-(α-acetonyl=benzyl)-4-hydroxycoumarin
Chemical Abstracts name 4-hydroxy-3-(3-oxo-1-phenylbutyl)-2*H*-1-benzopyran-=2-one
CAS RN *[81–81–2]* unstated stereochemistry; *[5543–58–8]* (R)-(+)- isomer
EEC no. 201–377–6

PHYSICAL CHEMISTRY
Mol. wt. 308.3 **M.f.** $C_{19}H_{16}O_4$ **Form** The racemate forms colourless crystals.
M.p. 161–162 °C **V.p.** 1.5×10^{-3} mPa **Solubility** In water 17 mg/l (20 °C). Very slightly soluble in benzene, diethyl ether and cyclohexane. Moderately soluble in methanol, ethanol, and isopropanol. In acetone 65, chloroform 56, dioxane 100 (all in g/l, 20 °C). Dissolves in aqueous alkalis with the formation of water-soluble salts. Sodium salt: in water, up to 400 g/l; insoluble in organic solvents.
Stability Very stable, even to strong acids. **pKa** It is acidic.

COMMERCIALISATION
History Its anticoagulant properties reported by K. P. Link *et al.* (*J. Biol. Chem.*, 1944, **153,** 5) at the Wisconsin Alumni Research Foundation. **Patent** US 2427578
Manufacturers Aimco

APPLICATIONS
Biochemistry General internal bleeding is induced by reduction of the prothrombin content of the blood. **Mode of action** Anticoagulant rodenticide. Repeated ingestion is necessary to produce toxic symptoms. The (*S*)-(−)- isomer has 7-fold greater rodenticidal activity than the (*R*)-(+)- isomer (B. D. West *et al., J. Am. Chem. Soc.*, 1961, **83,** 2676). **Uses** Control of rats and mice. There is no tendency to bait-shyness. **Formulation types** RB; GB; CB; AB; TP; Gel.
Mixtures *(warfarin +)* pindone; ergocalciferol; sulfaquinoxaline.
Compatibility Compatible with other rodenticides. **Selected tradenames** 'Sakarat' (Killgerm)

ANALYSIS
Product analysis by u.v. spectrometry (*AOAC Methods*, 1995, 960.15; *CIPAC Handbook*, 1970, **1,** 696) or by hplc (K. Hunter, *Anal. Methods Pestic. Plant Growth Regul.*, 1988, **16,** 154). **Residues** determined by hplc of a derivative (*idem, ibid.*).

MAMMALIAN TOXICOLOGY
Oral Acute oral LD_{50} for rats 186, mice 374 mg/kg. Oral LD_{50} for rats 1, pigs 1, cats 3, dogs 3, cattle 200 (all mg/kg daily for 5 days). **Other** Organ damage is observed, as well as inhibition of blood coagulation. Only slightly dangerous to humans and domestic animals when used as directed, but care must be taken with young pigs, which are especially susceptible. **Toxicity class** WHO (a.i.) Ib; EPA (formulation) I **EC risk** (R61); T (also R48/25)

ECOTOXICOLOGY
Birds Poultry are relatively resistant.

ENVIRONMENTAL FATE
Animals Metabolites in mammals include 4-, 6-, 7-, and 8-hydroxycoumarin (R. F. Lawrence *et al., Chirality*, 1990, **2,** 96; F. A. Sutcliffe *et al., Chem. Biol. Interact.*, 1990, **75,** 171).

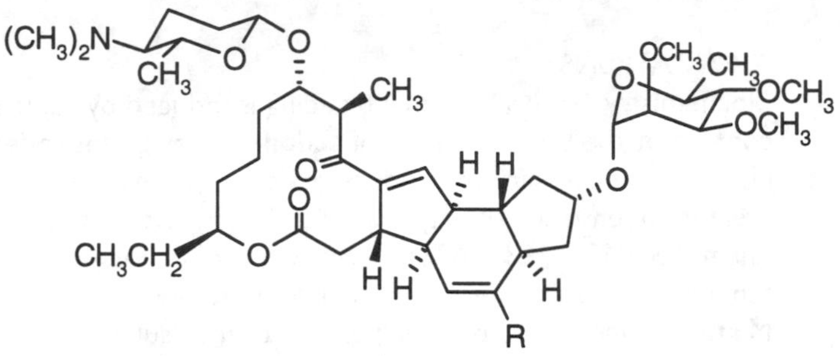

spinosyn A, R = H-

spinosyn D, R = CH₃-

NOMENCLATURE

Common name spinosad (BSI proposed)

IUPAC name A mixture of (2R,3aS,5aR,5bS,9S,13S,14R,16aS,16bR)-2-(6-deoxy-=
2,3,4-tri-O-methyl-α-L-mannopyranosyloxy)-13-(4-dimethylamino-=
2,3,4,6-tetradeoxy-β-D-erythopyranosyloxy)-9-ethyl-=
2,3,3a,5a,6,7,9,10,11,12,13,14,15,16a,16b-hexadecahydro-14-methyl-=
1H-8-oxacyclododeca[b]as-indacene-7,15-dione and
(2R,3aR,5aS,5bS,9S,13S,14R,16aS,16bR)-2-(6-deoxy-2,3,4-tri-O-methyl-=
α-L-mannopyranosyloxy)-13-(4-dimethylamino-2,3,4,6-tetradeoxy-β-D-erytho=
pyranosyloxy)-9-ethyl-2,3,3a,5a,6,7,9,10,11,12,13,14,15,16a,16b-hexadecahydro-=
4,14-dimethyl-1H-8-oxacyclododeca[b]as-indacene-7,15-dione in the proportion
65–95% to 35–5%

Chemical Abstracts name 2-((6-deoxy-2,3,4-tri-O-methyl-α-L-mannopyranosyl)=
oxy)-13-(((5-dimethylamino)tetrahydro-6-methyl-2H-pyran-2-yl)oxy)-9-ethyl-=
2,3,3a,5a,5b,6,9,10,11,12,13,14,16a,16b-tetradecahydro-14-methyl-=
1H-as-indaceno(3,2-d)oxacyclododecin-7,15-dione (spinosyn A), mixture with
2-((6-deoxy-2,3,4-tri-O-methyl-α-L-mannopyranosyl)oxy)-13-(((5-dimethyl=
amino)tetrahydro-6-methyl-2H-pyran-2-yl)oxy)-9-ethyl-=
2,3,3a,5a,5b,6,9,10,11,12,13,14,16a,16b-tetradecahydro-4,14-dimethyl-=
1H-as-indaceno(3,2-d)oxacyclododecin-7,15-dione (spinosyn D)

CAS RN [131929–60–7] (spinosyn A); [131929–63–0] (spinosyn D)

Development codes XDE-105; DE-105

PHYSICAL CHEMISTRY
Mol. wt. 732.0 (spinosyn A); 746.0 (spinosyn D) **M.f.** $C_{41}H_{65}NO_{10}$ (spinosyn A); $C_{42}H_{67}NO_{10}$ (spinosyn D) **Form** Light grey to white crystals (tech.).

COMMERCIALISATION
Production Derived from the fungus *Saccharopolyspora spinosa*.
Manufacturers DowElanco

APPLICATIONS
Mode of action Causes paralysis. **Uses** Under development for use on nematodes, beetles, flies and hoppers, at 50–100 g/ha. **Formulation types** SC, WG.

55 XMC   *Insecticide*

carbamate

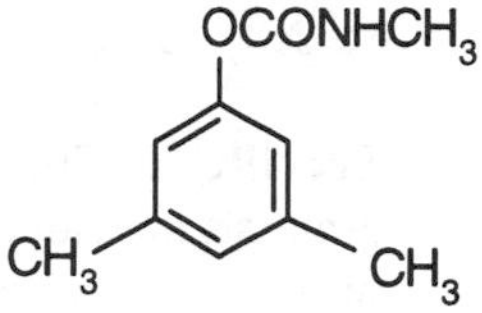

OCONHCH$_3$

CH$_3$ CH$_3$

NOMENCLATURE
Common name XMC (JMAF)
IUPAC name 3,5-xylyl methylcarbamate
Chemical Abstracts name 3,5-dimethylphenyl methylcarbamate
CAS RN *[2655–14–3]* **Development codes** H-69 (Hodogaya)

PHYSICAL CHEMISTRY
Composition Tech. grade is 97% pure. **Mol. wt.** 179.2 **M.f.** $C_{10}H_{13}NO_2$
Form Colourless crystals. **M.p.** 99 °C (tech.) **K$_{ow}$** logP = 2.3 (25 °C)
S.g./density 0.54 **Solubility** In water 0.47 g/l (20 °C). Soluble in most organic solvents; acetone 5.74, benzene 2.04, ethanol 3.52, ethyl acetate 2.77 (all in g/l, 20 °C); also soluble in cyclohexanone and 3,5,5-trimethylcyclohex-2-enone.
Stability Rapidly hydrolysed in alkaline media. Relatively stable to neutral and weakly acidic aqueous solutions. Stable to light and to temperatures up to 90 °C.

COMMERCIALISATION
History Introduced as an insecticide by Hokko Chemical Industry Co., Ltd and Hodogaya Chemical Co., Ltd. **Manufacturers** Hodogaya; Jin Hung

APPLICATIONS
Biochemistry Cholinesterase inhibitor. **Mode of action** Insecticide with predominantly contact action. **Uses** Control of leafhoppers and planthoppers on rice, and tea green leafhoppers on tea. **Formulation types** DP; WP; EC; MG. **Mixtures** *(XMC +)* carbaryl; malathion; fenthion; edifenphos; diazinon; kasugamycin. **Selected tradenames** 'Macbal' (Hodogaya)

ANALYSIS
Product analysis by hydrolysis to 3,5-xylenol, which is measured by u.v. spectroscopy. **Residues** determined by hydrolysis to 3,5-xylenol, a derivative of which is measured by glc.

MAMMALIAN TOXICOLOGY
EHC 64 (WHO, 1986; a review of carbamate insecticides in general). **Oral** Acute oral LD_{50} for rats 542, rabbits 445, mice 245 mg/kg. **NOEL** (90 d) for rats and mice 230 mg/kg b.w. daily. **ADI** 0.0034 mg/kg b.w. (provisional). **Toxicity class** WHO (a.i.) III; EPA (formulation) III **EC risk** Xn (R22)

ECOTOXICOLOGY
Birds Low toxicity to birds. **Fish** LC_{50} (48 h) for carp >40 mg/l. **Bees** Low toxicity to bees.

ENVIRONMENTAL FATE
Animals In insects, metabolism mainly involves hydroxylation of the benzene ring and the ring methyl substituents. **Soil/Environment** In soil, hydrolysed to 3,5-xylenol and *N*-methylcarbamic acid.

756 xylylcarb   *Insecticide*

carbamate

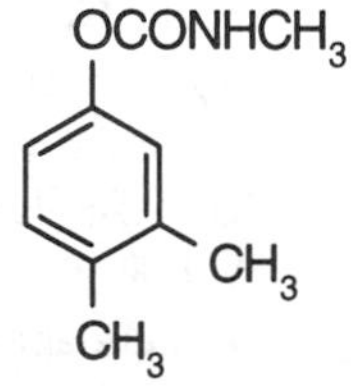

NOMENCLATURE
Common name xylylcarb (BSI, draft E-ISO, *(m)* draft F-ISO); MPMC (JMAF)
IUPAC name 3,4-xylyl methylcarbamate
Chemical Abstracts name 3,4-dimethylphenyl methylcarbamate

CAS RN *[2425–10–7]* **EEC no.** 219–364–9 **Development codes** S-1046
(Sumitomo)

PHYSICAL CHEMISTRY
Mol. wt. 179.2 **M.f.** $C_{10}H_{13}NO_2$ **Form** Colourless solid. **M.p.** 79–80 °C; (tech.,
71.5–76 °C) **V.p.** 121 mPa (25 °C) **Solubility** In water 580 mg/l (20 °C). In
acetonitrile 48.3%, cyclohexanone 43.5%, xylene 11.8% (all at room temperature).
Stability Hydrolysed in alkaline media.

COMMERCIALISATION
History Insecticide reported by R. L. Metcalf *et al.* (*J. Econ. Entomol.*, 1963, **56,**
862). Introduced by Sumitomo Chemical Co., Ltd. **Manufacturers** Sumitomo

APPLICATIONS
Biochemistry Cholinesterase inhibitor. **Mode of action** Non-systemic insecticide.
Uses Control of hoppers and other sucking insects on rice; and leafhoppers,
planthoppers, and scale insects on fruit. **Formulation types** EC; WP; DP; MG.
Selected tradenames 'Meobal' (Sumitomo)

ANALYSIS
Product analysis by hplc (S. Sakaue *et al.*, *Nippon Nogei Kagaku Kaishi*, 1981, **55,**
1237) or by u.v. spectrometry. **Residues** determined by glc of a derivative with
ECD (J. Miyamoto *et al.*, *Nihon Hoyaku Gakkaishi*, 1978, **3,** 119).

MAMMALIAN TOXICOLOGY
EHC 64 (WHO, 1986; a review of carbamate insecticides in general). **Oral** Acute
oral LD_{50} for male rats 375, female rats 325 mg/kg. **Skin and eye** Acute
percutaneous LD_{50} for rats >1000 mg/kg. **Toxicity class** WHO (a.i.) II
EC risk Xn (R22)

ENVIRONMENTAL FATE
Animals Metabolism of carbamate insecticides is reviewed (M. Cool & C. K.
Jankowski in "*Insecticides*").

MAIN ENTRIES

zeta-cypermethrin

See entry 187.

alkylenebis(dithiocarbamate)

$$[-SCSNHCH_2CH_2NHCSSZn-]_x$$

NOMENCLATURE
Common name zineb (BSI, E-ISO, JMAF); zinèbe ((*m*) F-ISO); no name (Germany)
IUPAC name zinc ethylenebis(dithiocarbamate) (polymeric)
Chemical Abstracts name [[1,2-ethanediylbis[carbamodithioato]](2–)]zinc
CAS RN *[12122–67–7]* **EEC no.** 235–180–1 **Official codes** ENT 14 874

PHYSICAL CHEMISTRY
Mol. wt. 275.8 **M.f.** $C_4H_6N_2S_4Zn$ **Form** Pale yellow powder. **M.p.** Decomposes
at 157 °C without melting **V.p.** <0.01 mPa (20 °C) **K$_{ow}$** logP ≤1.3 (20 °C)
Solubility In water at room temperature, *c.* 10 mg/l. Practically insoluble in
common organic solvents. Dissolves in some chelating agents, for example salts of
ethylenediaminetetra-acetic acid, from which it cannot be recovered.
Stability Unstable to light, moisture and heat on prolonged storage (decomposition
is reduced by stabilisers). When precipitated from concentrated solution, a
polymer is formed which is less fungicidal. **F.p.** 90 °C; autoignition temperature
149 °C

COMMERCIALISATION
History J. M. Heuberger & T. F. Manns (*Phytopathology*, 1943, **33**, 113) reported
the addition of zinc sulfate improved the field performance of nabam as a
fungicide. This led to the introduction of zineb by Rohm & Haas Co. (and by E. I.
du Pont de Nemours and Co. who no longer manufacture or market it).
Patent US 2457674; US 3050439 **Manufacturers** Elf Atochem; Rhône-Poulenc

APPLICATIONS
Mode of action Foliar fungicide with protective action. **Uses** Control of downy
mildews in vines, hops, lettuce, onions, spinach, brassicas, oilseed rape, tobacco,
and ornamentals; rusts in currants, berry fruit, plums, vegetables, and ornamentals
(especially carnations and chrysanthemums); red fire disease on vines; potato and
tomato blights; leaf spot diseases on currants, berry fruit, olives, and celery;
anthracnose on beans, vines, and citrus fruit; scab on apples and pears; shot-hole
on stone fruit; *Cercospora* on bananas; tulip fire; *Botrytis*; needle cast in forestry;
etc. **Phytotoxicity** Non-phytotoxic, except for zinc-sensitive plants such as
tobacco and cucurbits. **Formulation types** WP; DP. **Mixtures** *(zineb +)* maneb;
captan; captan + dinocap; copper oxychloride; copper oxychloride + sulfur; sulfur;
ziram; sulfur + ziram; mancozeb. **Compatibility** Incompatible with alkaline or
mercury-containing compounds. **Selected tradenames** 'Dithane Z-78' (Rohm &
Haas); 'AAphytora' (AgrEvo); 'Amitan' (Sipcam); 'Peran' (Efthymiadis); 'Permilan'
(Nufarm GmbH); 'Phytox' (Staehler); 'Tritoftorol' (Elf Atochem); 'Vitex' (Siapa)

ANALYSIS

Product analysis by acid hydrolysis, converting the carbon disulfide evolved into dithiocarbonate which is estimated by titration with iodine (*CIPAC Handbook*, 1970, **1**, 706; 1980, **1A**, 1293; 1992, **E**, 230–5; *AOAC Methods*, 1995, 965.15). Identified colorimetrically (*CIPAC Handbook*, 1994, **F**, 320) or by u.v. absorbance (*ibid.*, 411). **Residues** determined by acid hydrolysis, measuring the carbon disulfide evolved using glc (*Analyst (London)*, 1981, **106**, 782) or by conversion to a copper/amine complex which is estimated by colorimetry (*Pestic. Anal. Man.*, 1979, **II**; *Manu. Pestic. Residue Anal.*, 1987, **I**, S21; *Anal. Methods Residues Pestic.*, 1988, Part II; W. K. Lowen & H. L. Pease, *Anal. Methods Pestic., Plant Growth Regul. Food Addit.*, 1964, **3**, 69).

MAMMALIAN TOXICOLOGY

Reviews *Pesticide residues in food – 1993*, FAO Plant Production and Protection Paper, 122, 1993. *Pesticide residues in food – 1993 evaluations, Part II – Toxicology*. WHO, WHO/PCS/94.4, 1994. **IARC** 12 **EHC** 78 (WHO, 1988; general review of dithiocarbamates). **Oral** Acute oral LD_{50} for rats >5200 mg/kg. **Skin and eye** Acute percutaneous LD_{50} for rats >6000 mg/kg. Slight irritation of skin and mucous membranes. **NOEL** In feeding trials, growth was inhibited in rats within 74 w at 10 000 mg/kg diet. **ADI** (JMPR) 0.03 mg/kg b.w.; ethylenethiourea 0.004 mg/kg b.w. [1993]. **Other** At very high doses, ethylenethiourea, a trace contaminant and breakdown product of zineb, has caused thyroid effects, tumours and birth defects in laboratory animals. **Toxicity class** WHO (a.i.) III (Table 5); EPA (formulation) IV **EC risk** Xi (R37); R43

ECOTOXICOLOGY

Fish LC_{50} for perch 2, roach 6–8 mg/l. **Bees** Not toxic to bees.

ENVIRONMENTAL FATE

Plants Ethylenethiourea is the major metabolite in plants. Ethylenethiuram monosulfide and presumably ethylenethiuram disulfide and sulfur are also formed.

758 ziram *Fungicide, bird & rodent repellent*

dimethyldithiocarbamate

$$[(CH_3)_2NCS_2]_2Zn$$

NOMENCLATURE

Common name ziram (BSI, E-ISO, JMAF); zirame ((*m*) F-ISO)
IUPAC name zinc bis(dimethyldithiocarbamate)

Chemical Abstracts name (T-4)-bis(dimethyldithiocarbamato-S,S')zinc
CAS RN *[137–30–4]*   **EEC no.** 205–288–3

PHYSICAL CHEMISTRY
Mol. wt. 305.8   **M.f.** $C_6H_{12}N_2S_4Zn$ **Form** Colourless powder.   **M.p.** 246 °C;
(tech., 240–244 °C)   **V.p.** $<1 \times 10^{-3}$ mPa (extrapolated) K_{ow} logP = 1.23
(20 °C)   **Henry** <1.9 Pa m^3 mol^{-1} **S.g./density** 1.66 (25 °C)   **Solubility** In water
1.58–18.3 mg/l (20 °C). In acetone 2.88, methanol 0.22, toluene 2.33,
n-hexane 0.07 (all in g/l, 20 °C).   **Stability** Hydrolysis DT_{50} 18 h (pH 7).

COMMERCIALISATION
History Fungicide introduced by E. I. du Pont de Nemours and Co. (who no longer
manufacture or market it).   **Manufacturers** Elf Atochem; FMC; India Pesticides;
UCB

APPLICATIONS
Biochemistry Inhibitor of enzymes containing copper ions or sulfhydryl groups.
Mode of action Basic contact, foliar fungicide with protective action. Repellent to
birds and rodents.   **Uses** Fungicidal control in pome fruit, stone fruit, nuts, vines,
vegetables, and ornamentals. In particular, control of scab in apples and pears,
Monilia, *Alternaria*, *Septoria*, peach leaf curl, shot-hole, rusts, black rot and
anthracnose. Also used as a wildlife repellent, when smeared as a paste onto tree
trunks or sprayed onto ornamentals, dormant fruit trees, and other crops.
Phytotoxicity Non-phytotoxic, except to zinc-sensitive crops such as tobacco and
cucurbits.   **Formulation types** WP; Liquid; PA; WG.   **Mixtures** *(ziram +)*
bitertanol; dodine; myclobutanil; sodium arsenite; sulfur; thiram; zineb; sulfur +
zineb.   **Compatibility** Incompatible with compounds containing iron, copper,
mercury, TEPP, lime, or calcium arsenate.   **Selected tradenames** 'AAvolex'
(Staehler); 'Cekuziram' (Cequisa); 'Crittam' (Siapa); 'Cuman' (Novartis); 'Fuclasin'
(AgrEvo); 'Mezene' (Isagro); 'Pomarsol Z' (Bayer); 'Thionic' (UCB); 'Triscabol'
(Elf Atochem)

ANALYSIS
Product analysis by acid hydrolysis, the carbon disulfide liberated being converted
to dithiocarbonate which is estimated by titration with iodine (*CIPAC Handbook,*
1970, **1**, 716; *ibid.*, 1992, **E**, 236–8; *AOAC Methods*, 1995, 965.15).
Residues determined by acid hydrolysis followed by colorimetry of the liberated
carbon disulfide (W. K. Lowen & H. L. Pease, *Anal. Methods Pestic., Plant Growth
Regul. Food Addit.*, 1964, **3**, 69; G. E. Keppel, *J. Assoc. Off. Anal. Chem.*, 1971, **54**,
528; *Analyst (London)*, 1981, **106**, 782).

MAMMALIAN TOXICOLOGY
Reviews *Pesticide residues in food – 1980*. FAO Plant Production and Protection
Paper 26, 1981. *Pesticide residues in food: 1980 evaluations*. FAO Plant Production

and Protection Paper 26 Sup, 1981. **IARC** 12, 53 **EHC** 78 (WHO, 1988).
Oral Acute oral LD_{50} for rats 320, guinea pigs 100–150, rabbits 100–300 mg/kg.
Skin and eye Acute percutaneous LD_{50} for rabbits >2000 mg/kg. Irritating to
mucous membranes; highly irritating to eyes; not irritating to skin. Skin sensitiser
(guinea pigs). **Inhalation** LC_{50} (4 h) for rats 0.07 mg/l. **NOEL** In 1 y feeding
trials, rats receiving 5 mg a.i./kg daily showed no effect, neither did weanling rats
receiving 100 mg/kg diet for 30 d, nor beagle dogs receiving 100 ppm in the diet
for 13 w. **ADI** (JMPR) 0.02 mg/kg b.w. [1980]. **Other** I.p. LD_{50} for rats, guinea
pigs, and rabbits 5–73, mice 17 mg/kg. **Toxicity class** WHO (a.i.) III;
EPA (formulation) III **EC risk** (R40); Xn (R22); Xi (R36/37/38)

ECOTOXICOLOGY
Birds LD_{50} for bobwhite quail 97 mg/kg. **Fish** LC_{50} (96 h) for rainbow trout
1.9 mg/l. **Bees** Not toxic to bees; LD_{50} >100 µg/bee. **Worms** LC_{50} (7 d)
190 mg/kg soil. **Daphnia** EC_{50} (48 h) 0.048 mg/l.

ENVIRONMENTAL FATE
Animals Ziram, orally administered to rats, was mostly eliminated within 1–2 d,
leaving 1–2% of the dose in the tissues and carcass after 7 d. **Plants** The major
metabolite in plants is dimethylamine salt of dimethyldithiocarbamic acid;
tetramethylthiourea, carbon disulfide and sulfur can also be formed.
Dimethyldithiocarbamic acid can be present as the free acid or as the metabolic
conversion products, DDC-β-glucoside, DDC-α-aminobutyric acid and
DDC-α-alanine. **Soil/Environment** In soil, aerobic DT_{50} 42 h. Unlikely to leach.

759 ZXI 8901 *Insecticide*

pyrethroid

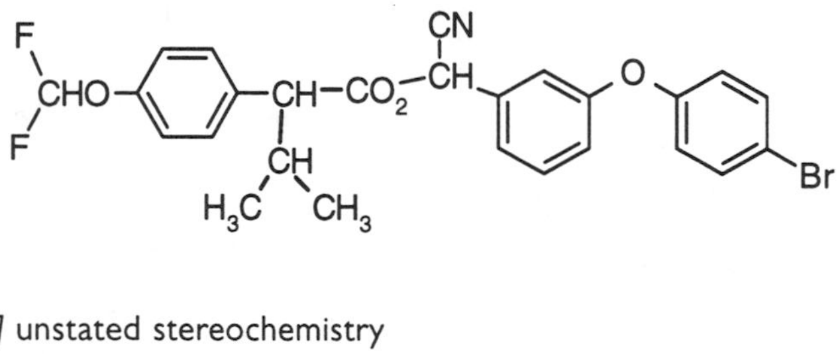

NOMENCLATURE
CAS RN *[160791–64–0]* unstated stereochemistry

PHYSICAL CHEMISTRY
Mol. wt. 450.5 **M.f.** $C_{26}H_{22}BrF_2NO_4$ **V.p.** 1.6×10^{-4} mPa (20 °C)
K_{ow} logP = 4.35 **Solubility** In water 3.56 µg/l (25 °C). Soluble in benzene,

ethanol, cyclohexane.   **Stability** Hydrolysis DT_{50} 15.6 d (pH 5), 8.3 d (pH 7), 4.2 d (pH 9). Aqueous photolysis DT_{50} c. 10 min.

COMMERCIALISATION
History Reported by W-G. Jin *et al.* (*Proc. Br. Crop Prot. Conf. – Pests Dis.*, 1996, **2**, 455).   **Manufacturers** Zhong-Xi

APPLICATIONS
Uses Under development for control of spider mites and Lepidoptera in cereals, cotton, vegetables, fruit, tea and soya beans.

MAMMALIAN TOXICOLOGY
Oral Acute oral LD_{50} for male and female rats >10 000, male and female mice 12 600 mg/kg.   **Skin and eye** Acute percutaneous LD_{50} for male and female rats >20 000 mg/kg. Not a skin or eye irritant (rabbits).   **NOEL** (90 d) for male rats 19.6, female rats 27.4 mg/kg daily.

ECOTOXICOLOGY
Birds Acute LD_{50} for Japanese quail 337 mg/kg.   **Fish** LC_{50} (96 h) for carp 0.48 mg/l.   **Bees** LC_{50} 957.7 mg/l.   **Worms** LC_{50} for *Eisenia foetida* c. 100 mg/kg soil.   **Daphnia** LC_{50} (48 h) 14.3 µg/l.   **Algae** EC_{50} (96 h) for *Scenedesmus obliquus* 0.23 mg/l.

ENVIRONMENTAL FATE
Soil/Environment DT_{50} in soil 4.8–8.8 d.

SUPERSEDED ENTRIES

Materials believed to be no longer manufactured, or marketed for crop protection use.

These entries are short descriptions. For chemical materials, they may include the following information:

Sequential entry number.
A header name, with an indication of the type of name (such as common name, chemical name, etc.).
Chemical Abstracts Service Registry Number (**CAS RN**).
A sentence describing the use, an early scientific reference, and name of company inventing or developing the product.
Approved common name (if different from the name in the header, or if other common names have been approved by national bodies).
IUPAC chemical name.
Chemical Abstracts name (**CA name**).
Molecular formula (**M.f.**).
Other names.
Code numbers (development codes and official codes).
Tradenames. Note that, unlike in the Main entries, obsolete tradenames are not enclosed by brackets.

The last item of information, prefaced **Details**, gives the last Edition of the Pesticide Manual to contain full details of the material, with the entry or page number in that Edition.

For an explanation of this information, if needed, the Guide to use of the Main Entries, page x should be consulted.

The entries in this Section are readily identified in the indexes by an 'S' before the entry number (e.g. S798).

It is difficult, in some cases, to be sure whether or not all commercial activity in a substance has ceased; some of these 'superseded' materials are known to be still in use for non-agricultural purposes. The Editor will be grateful for details of any materials in this Section which are still in commercial, agricultural use.

760   ACD 10614, ACD 10435 (development code).   **CAS RN** *[756–91–2] (i);* *[16202–91–8]* (ii); *[19493–94–8]* monosodium salts. Herbicide evaluated by Allied Chemical Corp. Agrochemical Division (later Hopkins Agricultural Chemical Co.). **IUPAC name** 1,1,1,7,7,7-hexafluoro-4-methyl-2,6-bis(trifluoromethyl)hept-3-ene-= 2,6-diol (i) and tautomer 1,1,1,7,7,7-hexafluoro-4-methylene-2,6-bis(trifluoromethyl)= heptane-2,6-diol (ii); forming monosodium 1,1,1,7,7,7-hexafluoro-4-methyl-2,6-bis= (trifluoromethyl)hept-3-ene-2,6-diolate and monosodium 1,1,1,7,7,7-hexafluoro-4-= methylene-2,6-bis(trifluoromethyl)heptane-2,6-diolate   **CA name** 1,1,1,7,7,7-hexa= fluoro-4-methyl-2,6-bis(trifluoromethyl)-3-heptene-2,6-diol (i) and 1,1,1,7,7,7-hexa= fluoro-4-methylene-2,6-bis(trifluoromethyl)-2,6-heptanediol (ii).   **M.f.** $C_{10}H_8F_{12}O_2$ **Code nos.** ACD 10 614 (i + ii); ACD 10 435 (monosodium salts).   **Details** *PM4*, p. 296.

761   acrylonitrile (IUPAC name).   **CAS RN** *[107–13–1]*. Developed as a fumigant insecticide (*The Chemistry of Acrylonitrile*) by American Cyanamid Co.   **CA name** 2-propenenitrile.   **M.f.** C_3H_3N **Other names** vinyl cyanide   **Code nos.** ENT 54. **Selected tradenames** 'Ventox'.   **Details** *PM5*, p. 3.

762   aldoxycarb (common name).   **CAS RN** *[1646–88–4]*. Insecticide and nematicide reported by M. H. J. Weiden *et al.* (*J. Econ. Entomol.*, 1965, **58**, 154). Introduced by Union Carbide Corp. (later Rhône-Poulenc Agrochimie).   **Common names** aldoxycarb, aldoxycarbe   **IUPAC name** 2-mesyl-2-methylpropionaldehyde *O*-methylcarbamoyl= oxime; 2-methyl-2-methylsulfonylpropionaldehyde *O*-methylcarbamoyloxime **CA name** 2-methyl-2-(methylsulfonyl)propanal *O*-[(methylamino)carbonyl]oxime. **M.f.** $C_7H_{14}N_2O_4S$ **Code nos.** UC 21 865; ENT 29 261; AI3-29 261; AN4-9. **Selected tradenames** 'Standak'.   **Details** *PM10*, entry 18.

763   aldrin (common name).   **CAS RN** *[309–00–2]*. Insecticide reported by C. W. Kearns *et al.* (*J. Econ. Entomol.*, 1949, **42**, 27). Introduced by J. Hyman & Co. and by Shell International Chemical Co. Ltd.   **Common names** HHDN (for pure material), aldrin (for material containing 95% HHDN), aldrine (for material containing 95% HHDN)   **IUPAC name** (1*R*,4*S*,4a*S*,5*S*,8*R*,8a*R*)-1,2,3,4,10,10-hexachloro-1,4,4a,5,8,8a-hexahydro-= 1,4:5,8-dimethanonaphthalene; 1,2,3,4,10,10-hexachloro-1,4,4a,5,8,8a-hexahydro-= exo-1,4-endo-5,8-dimethanonaphthalene   **CA name** (1α,4α,4aβ,5α,8α,8aβ)-= 1,2,3,4,10,10-hexachloro-1,4,4a,5,8,8a-hexahydro-1,4:5,8-dimethanonaphthalene. **M.f.** $C_{12}H_8Cl_6$ **Code nos.** Compound 118 (to Hyman); OMS 194; ENT 15949   **Selected tradenames** 'Octalene'.   **Details** *PM8*, entry 00120.

764   allidochlor (common name).   **CAS RN** *[93–71–0]*. Herbicide reported by P. C. Hamm & A. J. Speziale (*J. Agric. Food Chem.*, 1956, **4**, 518). Introduced by Monsanto Co. **Common names** allidochlor, alidochlore, CDAA   **IUPAC name** *N,N*-diallyl-2-chloro= acetamide   **CA name** 2-chloro-*N,N*-di-2-propenylacetamide.   **M.f.** $C_8H_{12}ClNO$ **Code nos.** CP 6343   **Selected tradenames** 'Randox'; 'Randox T'.   **Details** *PM7*, entry 00140.

765   allyxycarb (common name).   **CAS RN** *[6392–46–7]*. Insecticide introduced by Bayer AG.   **Common names** allyxycarb, allyxycarbe, APC   **IUPAC name** 4-diallyl= amino-3,5-xylyl methylcarbamate   **CA name** 4-(di-2-propenylamino)-3,5-dimethyl=

phenyl methylcarbamate. **M.f.** $C_{16}H_{22}N_2O_2$ **Code nos.** Bayer 50 282; A 546; OMS 773. **Selected tradenames** 'Hydrol'. **Details** *PM4*, p. 10.

766 alorac (common name). **CAS RN** *[19360–02–2]*. Plant growth regulator. **IUPAC name** (Z)-2,3,5,5,5-pentachloro-4-oxopent-2-enoic acid **CA name** (Z)-2,3,5,5,5-pentachloro-4-oxo-2-pentenoic acid. **M.f.** $C_5HCl_5O_3$.

767 ametridione (common name). **CAS RN** *[78168–93–1]*. Herbicide evaluated by Bayer AG. **IUPAC name** 1-amino-6-ethylthio-3-neopentyl-1,3,5-triazine-= 2,4(1*H*,3*H*)-dione **CA name** 1-amino-3-(2,2-dimethylpropyl)-6-(ethylthio)-= 1,3,5-triazine-2,4(1*H*,3*H*)-dione. **M.f.** $C_{10}H_{18}N_4O_2S$ **Code nos.** BAY SSH 0860.

768 amibuzin (common name). **CAS RN** *[76636–10–7]*. Herbicide developed by Bayer AG. **Common names** amibuzin, amibuzine **IUPAC name** 6-*tert*-butyl-= 3-dimethylamino-4-methyl-1,2,4-triazin-5(4*H*)-one **CA name** 3-(dimethylamino)-= 6-(1,1-dimethylethyl)-4-methyl-1,2,4-triazin-5(4*H*)-one. **M.f.** $C_{10}H_{18}N_4O$ **Code nos.** DIC 3202.

769 amidithion (common name). **CAS RN** *[919–76–6]*. Acaricide and insecticide (V. Dittrich & F. Bachman, *Proc. Br. Insectic. Fungic. Conf.*, *2nd*, 1963, p. 421). Introduced by Ciba AG (later Ciba-Geigy AG). **Common names** amidithion, amidiphos **IUPAC name** S-2-methoxyethylcarbamoylmethyl *O*,*O*-dimethyl phosphorodithioate; 2-dimethoxy-= phosphinothioylthio-*N*-(2-methoxyethyl)acetamide **CA name** S-[2-[(2-methoxy= ethyl)amino]-2-oxoethyl] *O*,*O*-dimethyl phosphorodithioate. **M.f.** $C_7H_{16}NO_4PS_2$ **Code nos.** Ciba 2446, ENT 27 160. **Selected tradenames** 'Thiocron'. **Details** *PM2*, p. 16.

770 amidothioate (common name). **CAS RN** *[54381–26–9]*. Acaricide introduced by Nippon Kayaku Co., Ltd. **IUPAC name** *O*-2-chloro-4-methylthiophenyl *O*-methyl ethylphosphoramidothioate **CA name** *O*-[2-chloro-4-(methylthio)phenyl] *O*-methyl ethylphosphoramidothioate. **M.f.** $C_{10}H_{15}ClNO_2PS_2$ **Selected tradenames** 'Mitemate'.

771 aminocarb (common name). **CAS RN** *[2032–59–9]*. Insecticide reported by G. Unterstenhöfer (*Meded. Landbouwhogesch. Opzoekingsstn. Staat Gent*, 1963, **28**, 758). Introduced by Bayer AG. **Common names** aminocarb, aminocarbe **IUPAC name** 4-dimethylamino-*m*-tolyl methylcarbamate **CA name** 4-(dimethylamino)-3-methyl= phenyl methylcarbamate. **M.f.** $C_{11}H_{16}N_2O_2$ **Code nos.** Bayer 44 646; A 363; ENT 25 784; OMS 170. **Selected tradenames** 'Matacil'. **Details** *PM8*, entry 00270.

772 amiprofos-methyl (common name). **CAS RN** *[36001–88–4]*. Herbicide reported by M. Aya *et al.* (*Zasso Kenkyu*, 1973, (15), p. 20). Evaluated by Bayer AG. **IUPAC name** *O*-methyl *O*-2-nitro-*p*-tolyl isopropylphosphoramidothioate **CA name** *O*-methyl *O*-= (4-methyl-2-nitrophenyl) (1-methylethyl)phosphoramidothioate. **M.f.** $C_{11}H_{17}N_2O_4PS$ **Code nos.** BAY NTN 6867; NTN 2925 **Selected tradenames** 'Tokunol M'.

773 amiton; amiton hydrogen oxalate (common name). **CAS RN** *[78-53-5]* amiton; *[3734-97-2]* amiton hydrogen oxalate. Insecticide and acaricide reported by R. Ghosh &

J. F. Newman (*Chem. Ind. (London)*, 1955, p. 118) and by G. L. Baldit (*J. Sci. Food Agric.*, 1958, **9**, 516). Evaluated by Plant Protection Ltd (later ICI Agrochemicals). **IUPAC name** S-2-diethylaminoethyl O,O-diethyl phosphorothioate **CA name** S-[2-(diethyl= amino)ethyl] O,O-diethyl phosphorothioate. **M.f.** $C_{10}H_{24}NO_3PS$; (oxalate $C_{12}H_{26}NO_7PS$) **Code nos.** R 5158 (amiton); R 6199 (amiton hydrogen oxalate); ENT 24 980-X (amiton); ENT 20 993 (amiton hydrogen oxalate). **Selected tradenames** 'Tetram'.

774 **ampropylfos** (common name). **CAS RN** [16606–64–7]. Fungicide discovered by KenoGard AB (later Rhône-Poulenc Agrochimie). **IUPAC name** (RS)-1-aminopropyl= phosphonic acid **CA name** (±)-(1-aminopropyl)phosphonic acid. **M.f.** $C_3H_{10}NO_3P$ **Selected tradenames** 'Appa'. **Details** *PM9,* entry 0375.

775 **anabasine** (common name). **CAS RN** [494–52–0]. Insecticide isolated from *Anabasis aphylla.* **IUPAC name** (S)-3-(piperidin-2-yl)pyridine **CA name** (S)-3-(2-piperidinyl)pyridine. **M.f.** $C_{10}H_{14}N_2$ **Other names** 2-(3-pyridyl)piperidine **Selected tradenames** 'Neonicotine'.

776 **anilazine** (common name). **CAS RN** [101–05–3]. Fungicide reported by C. N. Wolf *et al.* (*Science,* 1955, **121**, 61). Introduced by the Ethyl Corp. and later by Nippon Soda Co., Ltd and by Bayer AG. **Common names** anilazine, triazine **IUPAC name** 4,6-dichloro-N-(2-chlorophenyl)-1,3,5-triazin-2-amine; 2-chloro-N-(4,6-dichloro-= 1,3,5-triazin-2-yl)aniline **CA name** 4,6-dichloro-N-(2-chlorophenyl)-1,3,5-triazin-= 2-amine. **M.f.** $C_9H_5Cl_3N_4$ **Code nos.** B-622 (Ethyl Corp.), ENT 26 058. **Selected tradenames** 'Dyrene'. **Details** *PM10,* entry 29.

777 **anisuron** (common name). **CAS RN** [2689–43–2]. Herbicide. **IUPAC name** 1-(3,4-dichlorophenyl)-1-(4-methoxybenzoyl)-3,3-dimethylurea **CA name** N-(3,4-dichlorophenyl)-N-[(dimethylamino)carbonyl]-4-methoxybenzamide. **M.f.** $C_{17}H_{16}Cl_2N_2O_3$.

778 **antu** (common name). **CAS RN** [86–88–4]. Toxicity to rodents reported by C. F. Richter (*J. Am. Med. Assoc.,* 1945, **129**, 927; *Proc. Soc. Exp. Biol. Med.,* 1946, **63**, 364). **IUPAC name** 1-(1-naphthyl)-2-thiourea **CA name** 1-naphthalenylthiourea. **M.f.** $C_{11}H_{10}N_2S$ **Other names** α-naphthylthiourea. **Details** *PM8,* entry 00450.

779 **aramite** (common name). **CAS RN** [140–57–8]. Acaricide reported by W. D. Harris & J. W. Zukel (*J. Agric. Food Chem.,* 1954, **2**, 140). Introduced by Uniroyal Inc. **IUPAC name** 2-(4-*tert*-butylphenoxy)-1-methylethyl 2-chloroethyl sulfite **CA name** 2-chloroethyl 2-[4-(1,1-dimethylethyl)phenoxy]-1-methylethyl sulfite. **M.f.** $C_{15}H_{23}ClO_4S$ **Code nos.** 88-R, ENT 16 519. **Selected tradenames** 'Aramite'. **Details** *PM3,* p. 73.

780 **arsenous oxide** (chemical name). **CAS RN** [1327–53–3]. Used as a rodenticide since the 16th century. **Common names** arsenious acid, arsenous oxide, oxyde arsenieux **IUPAC name** diarsenic trioxide; arsenic trioxide **CA name** arsenic oxide (As_2O_3). **M.f.** As_2O_3 **Other names** arsenious oxide; white arsenic. **Details** *PM9,* entry 460.

781   athidathion (common name).   CAS RN *[19691–80–6]*. Insecticide reported by K. Rüfenacht *(Helv. Chim. Acta*, 1968, **51**, 518). Introduced by J. R. Geigy S.A. (later Ciba-Geigy AG).   **IUPAC name** *O,O*-diethyl *S*-2,3-dihydro-5-methoxy-2-oxo-= 1,3,4-thiadiazol-3-ylmethyl phosphorodithioate; 2-diethoxyphosphinothioylthio= methyl-5-methoxy-1,3,4-thiadiazol-2(3*H*)-one   **CA name** *O,O*-diethyl *S*-[(5-methoxy-= 2-oxo-1,3,4-thiadiazol-3(2*H*)-yl)methyl] phosphorodithioate.   **M.f.** $C_8H_{15}N_2O_4PS_3$ **Code nos.** G 13 006.

782   atraton (common name).   CAS RN *[1610–17–9]*. Herbicide reported by E. Knüsli (*Phytiatr.-Phytopharm.*, 1958, **7**, 81). Introduced by J. R. Geigy S.A. (later Ciba-Geigy AG).   **Common names** atraton, atratone   **IUPAC name** N^2-ethyl-N^4-isopropyl-= 6-methoxy-1,3,5-triazine-2,4-diamine; 2-ethylamino-4-isopropylamino-6-methoxy-= 1,3,5-triazine   **CA name** *N*-ethyl-6-methoxy-*N'*-(1-methylethyl)-1,3,5-triazine-= 2,4-diamine.   **M.f.** $C_9H_{17}N_5O$ **Code nos.** G 32 293   **Selected tradenames** 'Gestatamin'.

783   aziprotryne (common name).   CAS RN *[4658–28–0]*. Herbicide reported by D. H. Green *et al.* (*C. R. Journ. Etud. Herbic. Conf. COLUMA, 4th*, 1967, **1**, 1). Introduced by Ciba AG (later Ciba-Geigy AG).   **Common names** aziprotryne, aziprotryn   **IUPAC name** 4-azido-*N*-isopropyl-6-methylthio-1,3,5-triazin-2-ylamine   **CA name** 4-azido-= *N*-(1-methylethyl)-6-(methylthio)-1,3,5-triazin-2-amine.   **M.f.** $C_7H_{11}N_7S$ **Code nos.** C 7019   **Selected tradenames** 'Mesoranil'.   **Details** *PM9*, entry 0550.

784   azithiram (common name).   CAS RN *[5834–94–6]*. Fungicide evaluated by ICI Plant Protection Division (later ICI Agrochemicals).   **Common names** azithiram, azithirame   **IUPAC name** bis(3,3-dimethylthiocarbazoyl) disulfide; *N,N'*-bis(dimethyl= amino)thiuram disulfide   **CA name** 2,2,2',2'-tetramethylthioperoxydicarbonic dihydrazide ([[[(CH$_3$)$_2$NNH]C(S)]$_2$S$_2$).   **M.f.** $C_6H_{14}N_4S_4$ **Code nos.** PP447.

785   azobenzene (IUPAC name).   CAS RN *[103–33–3]*. Acaricidal activity reported by W. E. Blauvelt (*N. Y. State Flower Grow. Bull.*, 1945, No. 2). Was mainly used in glasshouses as a smoke.   **CA name** diphenyldiazene.   **M.f.** $C_{12}H_{10}N_2$ **Code nos.** ENT 14 611.   **Details** *PM4*, p. 27.

786   azothoate (common name).   CAS RN *[5834–96–8]*. Insecticide and acaricide introduced by Montecatini S.p.A. (later Agrimont S.p.A.).   **IUPAC name** *O*-4-(4-chlorophenylazo)phenyl *O,O*-dimethyl phosphorothioate   **CA name** *O*-[4-[(4-chlorophenyl)azo]phenyl] *O,O*-dimethyl phosphorothioate.   **M.f.** $C_{14}H_{14}ClN_2O_3PS$ **Code nos.** L 1058, OMS 1089.   **Selected tradenames** 'Slam C'.

787   barban (common name).   CAS RN *[101–27–9]*. Herbicide reported by A. D. Brown (*Proc. North Cent. Weed Control Conf.*, 1958, **15**, 98). Introduced by the Spencer Chemical Co. and later Velsicol Chemical Corp.   **Common names** barban, barbane, barbanate, CBN, [chlorinat]   **IUPAC name** 4-chlorobut-2-ynyl 3-chlorocarbanilate **CA name** 4-chloro-2-butynyl (3-chlorophenyl)carbamate.   **M.f.** $C_{11}H_9Cl_2NO_2$ **Code nos.** CS-847 (Spencer Chemicals)   **Selected tradenames** 'Carbyne', 'B 25'. **Details** *PM8*, entry 00620.

788 **barium carbonate** (IUPAC/Chemical Abstracts name). **CAS RN** *[513–77–9]*. Used as a rodenticide for many years. **Common names** barium carbonate **M.f.** $CBaO_3$. **Details** *PM5*, p. 28.

789 **barium polysulfide** (IUPAC/Chemical Abstracts name). **CAS RN** *[50864–67–0]*. Insecticide and fungicide introduced by Bayer AG. **Common names** barium polysulfide, polysulfure de baryum **M.f.** BaS_x **Selected tradenames** 'Solbar'.

790 **Bayer 22/190** (development code). **CAS RN** *[500–28–7]*. Insecticide introduced by Bayer AG. **IUPAC name** *O*-3-chloro-4-nitrophenyl *O,O*-dimethyl phosphorothioate **CA name** *O*-(3-chloro-4-nitrophenyl) *O,O*-dimethyl phosphorothioate. **M.f.** $C_8H_9ClNO_5PS$ **Code nos.** Bayer 22/190, OMS 217. **Selected tradenames** 'Chlorthion'.

791 **Bayer 22408** (development code). **CAS RN** *[2668–92–0]*. Insecticide, discovered by W. Lorenz, evaluated by Bayer AG. **IUPAC name** *O,O*-diethyl naphthalene-1,8-dicarboximido-oxyphosphonothioate; *O,O*-diethyl naphthalimido-= oxyphosphonothioate; *N*-diethoxyphosphinothioyloxynaphthalene-1,8-dicarboximide **CA name** 2-[(diethoxyphosphinothioyl)oxy]-1*H*-benz[*de*]isoquinoline-1,3(2*H*)-dione. **M.f.** $C_{16}H_{16}NO_5PS$ **Code nos.** Bayer 22408; S 125; ENT 24 970. **Details** *PM1*, p. 159.

792 **Bayer 32394** (development code). Fungicide introduced by Bayer AG. **IUPAC name** tris(1-dodecyl-3-methyl-2-phenylbenzimidazolium) hexacyanoferrate; tris(1-dodecyl-3-methyl-2-phenylbenzimidazolium) ferricyanide **CA name** tris[1-dodecyl-3-methyl-2-phenyl-1*H*-benzimidazolium] hexakis(cyano-*C*)ferrate. **M.f.** $C_{84}H_{111}FeN_{12}$ **Code nos.** Bayer 32394; B 169 Ferricyanide **Selected tradenames** 'Fungilon'. **Details** *PM2*, p. 213.

793 **benodanil** (common name). **CAS RN** *[15310–01–7]*. Fungicide reported by E. H. Pommer *et al.*, at the 39th Deutsche Pflanzenschutztagung, Stuttgart, 1973 and by F. Löcher *et al.* (*Meded. Fac. Landbouwwet. Rijksuniv. Gent*, 1974, **39**, 1079). Introduced by BASF AG. **IUPAC name** 2-iodobenzanilide **CA name** 2-iodo-*N*-phenylbenzamide. **M.f.** $C_{13}H_{10}INO$ **Code nos.** BAS 3170F **Selected tradenames** 'Calirus'. **Details** *PM8*, entry 00710.

794 **benzofluor** (common name). **CAS RN** *[68672–17–3]*. Herbicide and plant growth regulator introduced by 3M Company. **Common names** benzofluor **IUPAC name** 4'-ethylthio-2'-(trifluoromethyl)methanesulfonanilide; 4'-ethylthio-α',α',α'-trifluoro= methanesulfon-*o*-toluidide **CA name** *N*-[4-(ethylthio)-2-(trifluoromethyl)phenyl]= methanesulfonamide. **M.f.** $C_{10}H_{12}F_3NO_2S_2$ **Code nos.** MBR 18 337.

795 **benquinox** (common name). **CAS RN** *[495–73–8]*. Fungicide reported by P. E. Frohberger (*Phytopath. Z.*, 1956, **37**, 427). Introduced by Bayer AG. **Common names** benquinox, tserenox **IUPAC name** 2'-(4-hydroxyiminocyclohexa-2,5-dienylidene)= benzohydrazide; 1,4-benzoquinone 1-benzoylhydrazone 4-oxime **CA name** benzoic acid [4-(hydroxyimino)-2,5-cyclohexadien-1-ylidene]hydrazide. **M.f.** $C_{13}H_{11}N_3O_2$ **Other names** COBH **Code nos.** Bayer 15 080 **Selected tradenames** 'Ceredon',

'Ceredon special', 'Cereline', 'Tillantox', 'Ceredon T', 'Rhizoctol slurry', 'Rhizoctol combi'. **Details** *PM4*, p. 35.

796   bentaluron (common name).   **CAS RN** *[28956–64–1]*. Fungicide evaluated by Ciba AG (later Ciba-Geigy AG).   **IUPAC name** 1-(1,3-benzothiazol-2-yl)-= 3-isopropylurea; 1-benzothiazol-2-yl-3-isopropylurea   **CA name** N-2-benzothiazolyl-= N'-(1-methylethyl)urea.   **M.f.** $C_{11}H_{13}N_3OS$ **Code nos.** CGA 18 734.

797   benzadox; benzadox-ammonium (common name).   **CAS RN** *[5251–93–4]* (benzadox). Herbicidal activity reported *(Farm Chem.*, 1967, **130**, 86). Introduced by Gulf Oil Corp. and by Murphy Chemical Co. (later DowElanco).   **IUPAC name** benzamido-oxyacetic acid   **CA name** [(benzoylamino)oxy]acetic acid.   **M.f.** $C_9H_9NO_4$; (ammonium salt $C_9H_{12}N_2O_4$)   **Code nos.** MC 0035 (Murphy); S 6173 (Gulf)   **Selected tradenames** 'Topcide'.   **Details** *PM4*, p. 38.

798   benzamacril; benzamacril-isobutyl (common name).   **CAS RN** *[88107–27–1]* benzamacril-isobutyl. Fungicide evaluated by FBC Limited (later Schering Agriculture). **IUPAC name** 2-cyano-3-(N-methylbenzylamino)acrylic acid; 3-[benzyl(methyl)amino]-= 2-cyanoacrylic acid   **CA name** 2-cyano-3-[methyl(phenyl)amino]-2-propenoic acid. **M.f.** $C_{12}H_{12}N_2O_2$; (isobutyl ester $C_{16}H_{20}N_2O_2$).

799   benzamorf (common name).   **CAS RN** *[12068–08–5]*. Fungicide introduced by BASF AG.   **Common names** benzamorf, benzamorphe   **IUPAC name** morpholinium 4-dodecylbenzenesulfonate   **CA name** 4-dodecylbenzenesulfonic acid compound with morpholine (1:1).   **M.f.** $C_{22}H_{39}NO_4S$ **Code nos.** BAS 276F.

800   benzipram (common name).   **CAS RN** *[35256–86–1]*. Herbicide introduced by Gulf Oil Corp.   **Common names** benzipram, benziprame   **IUPAC name** N-benzyl-= N-isopropyl-3,5-dimethylbenzamide   **CA name** 3,5-dimethyl-N-(1-methylethyl)-= N-(phenylmethyl)benzamide.   **M.f.** $C_{19}H_{23}NO$ **Code nos.** S-18 510.

801   5-(1,3-benzodioxol-5-yl)-3-hexylcyclohex-2-enone (IUPAC name).   **CAS RN** *[8066–12–4]*. Synergist for insecticides.   **CA name** 5-(5-benzo-1,3-dioxolyl)-3-hexyl-= 2-cyclohexen-1-one.   **M.f.** $C_{19}H_{24}O_3$ **Other names** piperonyl cyclonene.

802   benzoylprop; benzoylprop-ethyl (common name).   **CAS RN** *[22212–55–1]* benzoylprop-ethyl; *[22212-56-2]* benzoylprop. Benzoylprop-ethyl is herbicide reported by T. Chapman et al. (*Symp. New. Herbic.*, 3rd, 1969, p. 40). Introduced by Shell Research Ltd.   **IUPAC name** N-benzoyl-N-(3,4-dichlorophenyl)-DL-alanine (I) **CA name** (I).   **M.f.** $C_{16}H_{13}Cl_2NO_3$; (ethyl ester $C_{18}H_{17}Cl_2NO_3$)   **Code nos.** WL 17 731 **Selected tradenames** 'Suffix'.   **Details** *PM8*, entry 00850.

803   benzthiazuron (common name).   **CAS RN** *[1929–88–0]*. Herbicide reported by H. Hack (International Meeting on Selective Weed Control in Sugar Beet Crops, Marly-le-Roi, 9-10. March 1967). Introduced in The Netherlands (1967) by Bayer AG.   **IUPAC name** 1-(1,3-benzothiazol-2-yl)-3-methylurea; 1-benzothiazol-2-yl-3-methylurea

CA name N-2-benzothiazolyl-N'-methylurea.   **M.f.** $C_9H_9N_3OS$ **Code nos.** Bayer 60 618   **Selected tradenames** 'Gatnon'.   **Details** PM9, entry 0860.

804   binapacryl (common name).   **CAS RN** [485–31–4]. Acaricide and fungicide reported by L. Emmel & M. Czech (Anz. Schaedlingskd., 1960, **33**, 145). Introduced by Hoechst AG.   **IUPAC name** 2-sec-butyl-4,6-dinitrophenyl 3-methylcrotonate; 2-sec-butyl-4,6-dinitrophenyl 3-methylbut-2-enoate   **CA name** 2-(1-methylpropyl)-= 4,6-dinitrophenyl 3-methyl-2-butenoate.   **M.f.** $C_{15}H_{18}N_2O_6$ **Code nos.** Hoe 02 784, ENT 25 793, OMS 571.   **Selected tradenames** 'Acricid', 'Endosan', 'Morocide', 'Morrocid'.   **Details** PM8, entry 00960.

805   biopermethrin (common name).   **CAS RN** [51877–74–8]. Insecticide discovered by M. Elliott et al.   **Common names** biopermethrin, bioperméthrine   **IUPAC name** 3-phenoxybenzyl (1R)-trans-3-(2,2-dichlorovinyl)-2,2-dimethylcyclopropanecarboxylate; 3-phenoxybenzyl (1R,3S)-3-(2,2-dichlorovinyl)-2,2-dimethyl-cyclopropanecarboxylate   **CA name** (1R-trans)-(3-phenoxyphenyl)methyl 3-(2,2-dichloroethenyl)-2,2-dimethyl= cyclopropanecarboxylate.   **M.f.** $C_{21}H_{20}Cl_2O_3$ **Code nos.** NRDC 147, OMS 1823.

806   bis(2-chloroethyl) ether (IUPAC name).   **CAS RN** [111–44–4]. Insecticidal activity reported by R. C. Roark & R. T. Cotton (USDA Tech. Bull., 1929, No. 162); used to fumigate soil.   **Common names** dichloroethyl ether   **IUPAC name** bis(2-chloroethyl) ether; di-(2-chloroethyl) ether   **CA name** 1,1'-oxybis[2-chloroethane].   **M.f.** $C_4H_8Cl_2O$ **Code nos.** ENT 4505.   **Details** PM4, p. 177.

807   1,1-bis(4-chlorophenyl)-2-ethoxyethanol (IUPAC name).   **CAS RN** [6012–83–5]. Acaricide introduced by J. R. Geigy S.A. (later Ciba-Geigy AG).   **CA name** 4-chloro-= α-(4-chlorophenyl)-α-(ethoxymethyl)benzenemethanol.   **M.f.** $C_{16}H_{16}Cl_2O_2$ **Code nos.** Geigy 337, G 23 645   **Selected tradenames** 'Etoxinol'.

808   bis(methylmercury) sulfate (IUPAC name).   **CAS RN** [3810–81–9]. Fungicide introduced by Bayer AG.   **CA name** dimethyl–sulfatodimercury.   **M.f.** $C_2H_6Hg_2O_4S$ **Selected tradenames** 'Cerewet', 'Aretan-nieuw'.   **Details** PM2, p. 47.

809   bisthiosemi (common name).   **CAS RN** [39603–48–0]. Rodenticide introduced by Nippon Kayaku Co., Ltd   **IUPAC name** 1,1'-methylenedi(thiosemicarbazide)   **CA name** 2,2'-methylenebis(hydrazinecarbothioamide).   **M.f.** $C_3H_{10}N_6S_2$ **Code nos.** NK-15 561   **Selected tradenames** 'Kayanex'.   **Details** PM9, entry 8380.

810   bis(tributyltin) oxide (IUPAC name).   **CAS RN** [56–35–9]. Fungicide reported by J. G. A. Luijten & G. J. M. van der Kerk (Investigations in the Field of Organotin Chemistry, 1955). Introduced (1966) by M & T Chemicals Ltd.   **CA name** hexabutyldistannoxane. **M.f.** $C_{24}H_{54}OSn_2$ **Code nos.** ENT 24 979.   **Selected tradenames** 'TBTO'.   **Details** PM9, entry 1180.

811   bromfenvinfos (common name).   **CAS RN** [33399–00–7] unstated stereochemistry; [58580-14-6] (E)- isomer; [58580-13-5] (Z)- isomer. Insecticide. **IUPAC name** 2-bromo-1-(2,4-dichlorophenyl)vinyl diethyl phosphate [(E)- to (Z)- ratio *c.* 1:18]   **CA name** 2-bromo-1-(2,4-dichlorophenyl)ethenyl diethyl phosphate.   **M.f.** $C_{12}H_{14}BrCl_2O_4P$ **Code nos.** IPO-62.

812   bromobonil (common name).   **CAS RN** *[25671–46–9]* unstated stereochemistry. Herbicide evaluated by Boehringer & Sohn (became Shell Agrar).   **IUPAC name** 2,6-dibromo-4-cyanophenyl tetrahydrofurfuryl carbonate   **CA name** 2,6-dibromo-= 4-cyanophenyl (tetrahydro-2-furanyl)methyl carbonate.   **M.f.** $C_{13}H_{11}Br_2NO_4$.

813   1-bromo-2-chloroethane (IUPAC/Chemical Abstracts name).   **CAS RN** *[107–04–0]*. Insecticide used to fumigate soil reported by M. W. Stone & F. B. Foley (*J. Econ. Entomol.*, 1951, **44**, 711).   **M.f.** C_2H_4BrCl **Other names** ethylene chlorobromide; *sym*-chlorobromoethane.   **Details** *PM4*, p. 255.

814   3-bromo-1-chloroprop-1-ene (IUPAC name).   **CAS RN** *[3737–00–6]*. Fumigant insecticide for soil introduced by Shell Development Co.   **CA name** 3-bromo-= 1-chloro-1-propene.   **M.f.** C_3H_4BrCl **Code nos.** CBP-55.

815   bromocyclen (common name).   **CAS RN** *[1715–40–8]*. Veterinary insecticide and acaricide introduced by Hoechst AG.   **Common names** bromocyclen, bromocyclène   **IUPAC name** 5-bromomethyl-1,2,3,4,7,7-hexachlorobicyclo[2.2.1]hept-2-ene; 5-bromomethyl-1,2,3,4,7,7-hexachloro-8,9,10-trinorborn-2-ene   **CA name** 5-(bromomethyl)-1,2,3,4,7,7-hexachlorobicyclo[2.2.1]hept-2-ene.   **M.f.** $C_8H_5BrCl_6$ **Code nos.** OMS 202.   **Selected tradenames** 'Alugan', 'Bromodan'.   **Details** *PM8*, entry 01330.

816   bromophos (common name).   **CAS RN** *[2104–96–3]*. Insecticide reported by R. Immel & G. Geisthardt (*Meded. Landbouwhogesch. Opzoekingsstn. Staat Gent*, 1964, **29**, 1242), and reviewed by D. Eichler (*Residue Rev.*, 1972, **41**, 65). Introduced by C. H. Boehringer Sohn/Cela GmbH (later Shell Agrar GmbH).   **IUPAC name** O-4-bromo-= 2,5-dichlorophenyl O,O-dimethyl phosphorothioate   **CA name** O-(4-bromo-= 2,5-dichlorophenyl) O,O-dimethyl phosphorothioate.   **M.f.** $C_8H_8BrCl_2O_3PS$ **Code nos.** S-1942; OMS 658; ENT 27162.   **Details** *PM8*, entry 01380.

817   bromophos-ethyl (common name).   **CAS RN** *[4824–78–6]*. Insecticide reported by M. S. Mulla *et al.* (*Mosq. News*, 1964, **24**, 312), and reviewed by D. Eichler (*Residue Rev.*, 1972, **41**, 65). Introduced by C. H. Boehringer Sohn/Cela GmbH (later Shell Agrar GmbH).   **IUPAC name** O-4-bromo-2,5-dichlorophenyl O,O-diethyl phosphorothioate   **CA name** O-(4-bromo-2,5-dichlorophenyl) O,O-diethyl phosphorothioate.   **M.f.** $C_{10}H_{12}BrCl_2O_3PS$ **Code nos.** S-2225 (Cyanamid); OMS 659; ENT 27258   **Selected tradenames** 'Nexagan'.   **Details** *PM8*, entry 01390.

818   brompyrazon (common name).   **CAS RN** *[3042–84–0]*. Herbicide reported by A. Fischer (*Weed Res.*, 1962, **2**, 177). Introduced by BASF AG.   **Common names** brompyrazon, brompyrazone   **IUPAC name** 5-amino-4-bromo-2-phenylpyridazin-= 3(2*H*)-one   **CA name** 5-amino-4-bromo-2-phenyl-3(2*H*)-pyridazinone. **M.f.** $C_{10}H_8BrN_3O$ **Code nos.** BAS 2430H   **Selected tradenames** 'Basanor'.   **Details** *PM5*, p. 55.

819 BTS **44584** (development code). **CAS RN** *[68721–60–8]*. Plant growth regulator reported by J. F. Garrod *et al.* (*Monograph Brit. Plant Growth Regulator Group*, 1980, No. 4, p. 67). Evaluated by Boots Co. Ltd (later Schering Agriculture). **IUPAC name** dimethyl(4-piperidinocarbonyloxy-2,5-xylyl)sulfonium toluene-4-sulfonate   **CA name** [2,5-dimethyl-4-[(4-piperidinylcarbonyl)oxy]phenyl]dimethylsulfonium 4-methyl= benzenesulfonate. **M.f.** $C_{23}H_{31}NO_5S_2$ **Code nos.** BTS 44584.

820 **bufencarb** (common name). **CAS RN** *[8065–36–9]* bufencarb; *[2282–34–0]* (i); *[6172–04–8]* (ii). Insecticide introduced by Chevron Chemical Co. **Common names** (for a reaction product) bufencarb, bufencarbe   **IUPAC name** of the main components are: (i) 3-(1-methylbutyl)phenyl methylcarbamate (I); and (ii) 3-(1-ethylpropyl)phenyl methylcarbamate (II)   **CA name** (I) and (II). **M.f.** $C_{13}H_{19}NO_2$ **Code nos.** Ortho 5353; OMS 227; ENT 27 127. **Selected tradenames** 'Bux'. **Details** *PM7*, entry 01450.

821 **buminafos** (common name). **CAS RN** *[51249–05–9]*. Plant growth regulator evaluated by VEB Chemiekombinat Bitterfeld. **IUPAC name** dibutyl 1-butylamino= cyclohexylphosphonate   **CA name** dibutyl 1-(butylamino)cyclohexylphosphonate. **M.f.** $C_{18}H_{38}NO_3P$.

822 **butacarb** (common name). **CAS RN** *[2655–19–8]*. Acaricide reported by J. Fraser *et al.* (*J. Sci. Food Agric.*, 1967, **18**, 372), used as mixture with gamma-HCH against animal ectoparasites. Introduced by Boots Company Ltd (later Schering Agrochemicals). **Common names** butacarb, butacarbe   **IUPAC name** 3,5-di-*tert*-butylphenyl methylcarbamate   **CA name** 3,5-bis(1,1-dimethylethyl)phenyl= methylcarbamate. **M.f.** $C_{16}H_{25}NO_2$ **Code nos.** RD 14 639; BTS 14 639. **Details** *PM5*, p. 59.

823 **butathiofos** (common name). **CAS RN** *[90338–20–8]*. Insecticide under development by DowElanco. **IUPAC name** O-2-*tert*-butylpyrimidin-5-yl O,O-diethyl phosphorothioate   **CA name** O-[2-(1,1-dimethylethyl)-5-pyrimidinyl] O,O-diethyl phosphorothioate. **M.f.** $C_{12}H_{21}N_2O_3PS$ **Code nos.** Dowco 429; XRD 429.

824 **butenachlor** (common name). **CAS RN** *[87310–56–3]*. Herbicide introduced by Agro-Kanesho Co. Ltd. **IUPAC name** (Z)-N-but-2-enyloxymethyl-2-chloro-= 2',6'-diethylacetanilide   **CA name** (Z)-2-chloro-N-[(2-butenyloxy)methyl]-N-= (2,6-diethylphenyl)acetamide. **M.f.** $C_{17}H_{24}ClNO_2$ **Code nos.** KH-218. **Details** *PM10*, entry 90.

825 **buthidazole** (common name). **CAS RN** *[55511–98–3]*. Herbicide reported by R. F. Anderson (*Proc. Int. Velsicol Symp., 8th,* 1974). Introduced by Velsicol Chemical Corp. (later Sandoz AG). **IUPAC name** 3-(5-*tert*-butyl-1,3,4-thiadiazol-2-yl)-4-hydroxy-= 1-methyl-2-imidazolidone   **CA name** 3-[5-(1,1-dimethylethyl)-1,3,4-thiadiazol-2-yl]-= 4-hydroxy-1-methyl-2-imidazolidinone. **M.f.** $C_{10}H_{16}N_4O_2S$ **Code nos.** Vel-5026   **Selected tradenames** 'Ravage'. **Details** *PM7*, entry 01530.

826 **buthiobate** (common name). **CAS RN** *[51308–54–4]*. Fungicide reported by T. Kato *et al.* (*Agric. Biol. Chem.*, 1975, **39**, 169). Introduced by Sumitomo Chemical Co.,

Ltd. **IUPAC name** butyl 4-*tert*-butylbenzyl *N*-(3-pyridyl)dithiocarbonimidate **CA name** butyl [4-(1,1-dimethylethyl)phenyl]methyl 3-pyridinylcarbonimidodithioate. **M.f.** $C_{21}H_{28}N_2S_2$ **Code nos.** S-1358 **Selected tradenames** 'Denmert'. **Details** *PM8*, entry 01540.

827 buthiuron (common name). **CAS RN** *[30043–55–1]*. Herbicide evaluated by Bayer AG. **IUPAC name** 1-(5-butylsulfonyl-1,3,4-thiadiazol-2-yl)-1,3-dimethylurea **CA name** *N*-[5-(butylsulfonyl)-1,3,4-thiadiazol-2-yl]-*N*,*N*'-dimethylurea. **M.f.** $C_9H_{16}N_4O_3S$ **Code nos.** MET 1489.

828 butonate (common name). **CAS RN** *[126–22–7]*. Insecticide reported by B. W. Arthur & J. E. Casida (*J. Agric. Food Chem.*, 1958, **6**, 360). Introduced by Prentiss Drug & Chemical Co. **Common names** butonate, [butilchlorofos] **IUPAC name** 2,2,2-trichloro-1-(dimethoxyphosphinoyl)ethyl butyrate; dimethyl 1-butyryloxy-= 2,2,2-trichloroethylphosphonate **CA name** 2,2,2-trichloro-1-(dimethoxyphosphinyl)= ethyl butanoate. **M.f.** $C_8H_{14}Cl_3O_5P$ **Code nos.** ENT 20 852. **Details** *PM4*, p. 61.

829 butopyronoxyl (common name). **CAS RN** *[532–34–3]*. Insect repellent introduced by US Industrial Chemicals Inc. and by Kilgore Chemicals. **IUPAC name** butyl 3,4-dihydro-2,2-dimethyl-4-oxo-2*H*-pyran-6-carboxylate (I); butyl 5,6-dihydro-= 6,6-dimethyl-4-oxo-4*H*-pyran-2-carboxylate; butyl dihydro-6,6-dimethyl-4-oxopyran-= 2-carboxylate **CA name** (I). **M.f.** $C_{12}H_{18}O_4$ **Other names** dihydropyrone; butyl mesityloxide oxalate **Code nos.** ENT 9. **Selected tradenames** 'Indalone'. **Details** *PM6*, p. 66.

830 2-(2-butoxyethoxy)ethyl piperonylate (IUPAC name). **CAS RN** *[136–63–0]*. Synergist for pyrethrum introduced by Bush Boake & Allen. **IUPAC name** 2-(2-butoxyethoxy)ethyl piperonylate **CA name** 2-(2-butoxyethoxy)ethyl 1,3-benzodioxole-5-carboxylate. **M.f.** $C_{16}H_{22}O_6$ **Other names** BCP **Selected tradenames** 'Bucarpolate'.

831 2-(2-butoxyethoxy)ethyl thiocyanate (IUPAC/Chemical Abstracts name). **CAS RN** *[112–56–1]*. Insecticide reported by D. F. Murphy & C. H. Peet (*J. Econ. Entomol.*, 1932, **25**, 123). Introduced by Rohm & Haas Co. **M.f.** $C_9H_{17}NO_2S$ **Other names** butyl 'Carbitol' thiocyanate; butyl 'Carbitol' rhodanate; beta-butoxy-beta'-thiocyanodiethyl ether **Code nos.** ENT 6. **Selected tradenames** 'Lethane 384'. **Details** *PM5*, p. 62.

832 butoxy(polypropylene glycol) (ESA name). **CAS RN** *[9003–13–8]*. Fly repellent. Introduced by Union Carbide Corp. **IUPAC name** α-hydroxypropyl-*o*-butoxypoly= [oxy(1-methylethylene)] **CA name** α-butyl-*o*-hydroxypoly[oxy(methyl-1,2-ethane= diyl)]. **Code nos.** ENT 8286. **Selected tradenames** 'Crag Fly Repellent'.

833 buturon (common name). **CAS RN** *[3766–60–7]*. Herbicide reported by A. Fischer (*Meded. Landbouwhogesch, Opzoekingsstn. Staat Gent*, 1964, **29**, 719). Introduced by BASF AG. **IUPAC name** 3-(4-chlorophenyl)-1-methyl-1-(1-methylprop-2-ynyl)urea **CA name** *N*'-(4-chlorophenyl)-*N*-methyl-*N*-(1-methyl-2-propynyl)urea. **M.f.**

$C_{12}H_{13}CIN_2O$ **Code nos.** H 95   **Selected tradenames** 'Eptapur'.   **Details** *PM8*, entry 01650.

834   cadmium calcium copper zinc chromate sulfate (IUPAC/Chemical Abstracts name).   **CAS RN** *[12001–20–6]*. Fungicide introduced by Union Carbide Corp. (later Rhône-Poulenc Ag.).   **Code nos.** Fungicide 531   **Selected tradenames** 'Crab Turf Fungicide'.

835   calcium cyanamide (IUPAC/Chemical Abstracts name).   **CAS RN** *[156–62–7]*. Used as fertiliser since 1905 and later introduced as herbicide and defoliant by American Cyanamid Co.   **M.f.** $CCaN_2$ **Selected tradenames** 'Cyanamid'.   **Details** *PM5*, p. 72.

836   cambendichlor (common name).   **CAS RN** *[56141–00–5]*. Herbicide introduced by Velsicol Chemical Corp. (later Sandoz AG).   **Common names** cambendichlor, cambendichlore   **IUPAC name** 2,2′-(phenylimino)diethylene bis(3,6-dichloro-o-anisate)   **CA name** (phenylimino)di-2,1-ethanediyl bis(3,6-dichloro-2-methoxybenzoate).   **M.f.** $C_{26}H_{23}Cl_4NO_6$ **Code nos.** Vel 4207.

837   camphechlor (common name).   **CAS RN** *[8001–35–2]*. Insecticide reported by W. LeRoy Parker & J. R. Beacher (*Del. Univ. Agric. Exp. Stn. Bull.*, 1947, No. 264). Composition examined by J. E. Casida *et al.* (*J. Agric. Food Chem.*, 1974, **22**, 653; 1975, **23**, 991); 26 components account for about 40% of the product; a heptachlorobornane and a mixture of 2 octachlorobornanes. Introduced by Hercules Inc.   **Common names** camphechlor, camphéchlore, toxaphene, polychlorcamphene   **IUPAC name** a reaction mixture of chlorinated camphenes containing 67-69% chlorine   **CA name** toxaphene. **M.f.** $C_{10}H_{10}Cl_8$ (approx.)   **Code nos.** Hercules 3956, ENT 9735.   **Selected tradenames** 'Toxaphene' (except in countries where used as common name).   **Details** *PM8*, entry 01920.

838   carbamorph (common name).   **CAS RN** *[31848–11–0]*. Fungicide introduced by Murphy Chemical Co. (later DowElanco).   **Common names** carbamorph, carbamorphe **IUPAC name** morpholinomethyl dimethyldithiocarbamate   **CA name** 4-morpholinyl= methyl dimethylcarbamodithioate.   **M.f.** $C_8H_{16}N_2OS_2$ **Code nos.** MC 833.

839   carbanolate (common name).   **CAS RN** *[671–04–5]*. Veterinary insecticide and ixodicide introduced by Upjohn Co. (later Nor-Am).   **IUPAC name** 6-chloro-3,4-xylyl methylcarbamate   **CA name** 2-chloro-4,5-dimethylphenyl methylcarbamate.   **M.f.** $C_{10}H_{12}CINO_2$ **Code nos.** U 12 927, OMS 174.   **Selected tradenames** 'Banol'.

840   carbasulam (common name).   **CAS RN** *[1773–37–1]*. Herbicide evaluated by May & Baker Ltd (later Rhône-Poulenc Agrochimie).   **Common names** carbasulam, carbasulame   **IUPAC name** methyl 4-(methoxycarbonylsulfamoyl)carbanilate   **CA name** methyl [[4-[(methoxycarbonyl)amino]phenyl]sulfonyl]carbamate.   **M.f.** $C_{10}H_{12}N_2O_6S$ **Code nos.** M&B 9555.

841   carbon disulfide (IUPAC/Chemical Abstracts name).   **CAS RN** *[75–15–0]*. Used as insecticide in 1854 by Garreau (see *Science*, 1926, **64**, 326).   **Common names** carbon disulfide, carbon disulphide, sulfure de carbone. **M.f.** CS$_2$ **Other names** carbon bisulphide. **Details** *PM8*, entry 02030.

842 carbon tetrachloride (IUPAC name). **CAS RN** *[56–23–5]*. Use as insecticidal fumigant described by W. E. Britton (*Conn. Agric. Exp. Stn. Rep.*, 1908, No. 31). **Common names** carbon tetrachloride, tétrachlorure de carbone **IUPAC name** carbon tetrachloride; tetrachloromethane **CA name** tetrachloromethane. **M.f.** CCl$_4$ **Code nos.** ENT 27 164. **Details** *PM9*, entry 2040.

843 carbophenothion (common name). **CAS RN** *[786–19–6]*. Insecticide and acaricide (*Agric. Chem.*, 1956, **11**(11), 91). Introduced by Stauffer Chemical Co. (later ICI Americas). **IUPAC name** S-4-chlorophenylthiomethyl *O,O*-diethyl phosphoro= dithioate **CA name** S-[[(4-chlorophenyl)thio]methyl] *O,O*-diethyl phosphorodithioate. **M.f.** C$_{11}$H$_{16}$ClO$_2$PS$_3$ **Code nos.** R-1303 (Zeneca); OMS 244; ENT 23708 **Selected tradenames** 'Trithion'. **Details** *PM8*, entry 02050.

844 carboxazole (common name). **CAS RN** *[55808–13–4]*. Herbicide evaluated by Shionogi Co. Ltd. **IUPAC name** methyl 5-*tert*-butyl-1,2-oxazol-3-ylcarbamate; methyl 5-*tert*-butylisoxazol-3-ylcarbamate **CA name** methyl [5-(1,1-dimethylethyl)-= 3-isoxazolyl]carbamate. **M.f.** C$_9$H$_{14}$N$_2$O$_3$ **Code nos.** SSH-42.

845 CECA (common name). **CAS RN** *[17756–81–9]*. Fungicide introduced by Nippon Soda Co., Ltd. **IUPAC name** 2-chloro-*N*-(2-cyanoethyl)acetamide (I) **CA name** (I). **M.f.** C$_5$H$_7$ClN$_2$O **Code nos.** NF 21 **Selected tradenames** 'Udonkor'.

846 CGA 80 000 (development code). **CAS RN** *[79555–80–9]* racemate; *[67932–85–8]* unstated stereochemistry. Fungicide reported by P. Margot *et al.* (*Proc. 1988 Br. Crop Prot. Conf. - Pests Dis.*, **2**, 527). Evaluated by Ciba-Geigy AG. **IUPAC name** (*RS*)-α-[*N*-(3-chloro-2,6-xylyl)-2-methoxyacetamido]-γ-butyrolactone **CA name** (±)-*N*-(3-chloro-2,6-dimethylphenyl)-2-methoxy-*N*-(tetrahydr-2-oxo-3-furanyl)= acetamide. **M.f.** C$_{15}$H$_{18}$ClNO$_4$ **Code nos.** CGA 80 000 **Selected tradenames** 'Vangard'.

847 chlobenthiazone (common name). **CAS RN** *[63755–05–5]*. Fungicide reported by S. Inoue *et al.* (*Proc. 1981 Br. Crop. Prot. Conf. - Pests Dis.*, 1981, **1**, 19). Evaluated by Sumitomo Chemical Co., Ltd. **IUPAC name** 4-chloro-3-methylbenzothiazol-2(3*H*)-one; 4-chloro-3-methyl-1,3-benzothiazol-2(3*H*)-one **CA name** 4-chloro-3-methyl-= 2(3*H*)-benzothiazolone. **M.f.** C$_8$H$_6$ClNOS **Code nos.** S-1901.

848 chloraniformethan (common name). **CAS RN** *[20856–57–9]*. Fungicide reported by A. O. Paulus *et al.* (*Calif. Agric.*, 1968, **22**(3), 10). Introduced by Bayer AG. **Common names** chloraniformethan, chloraniforméthane **IUPAC name** *N*-[2,2,2-trichloro-= 1-(3,4-dichloroanilino)ethyl]formamide **CA name** *N*-[2,2,2-trichloro-= 1-[(3,4-dichlorophenyl)amino]ethyl]formamide. **M.f.** C$_9$H$_7$Cl$_5$N$_2$O **Code nos.** Bayer 79 770 **Selected tradenames** 'Imugan', 'Milfaron'. **Details** *PM4*, p. 91.

849   chloranil (common name).   **CAS RN** *[118–75–2]*. Fungicidal activity of chloranil, long known as a chemical reagent, reported by H. S. Cunningham & E. G. Shavelle (*Phytopathology*, 1940, **30**, 4). Introduced by Uniroyal Inc.   **Common names** chloranil, chloranile   **IUPAC name** tetrachloro-*p*-benzoquinone; 2,3,5,6-tetrachloro-1,4-benzo= quinone   **CA name** 2,3,5,6-tetrachloro-2,5-cyclohexadiene-1,4-dione.   **M.f.** $C_6Cl_4O_2$ **Code nos.** ENT 3797.   **Selected tradenames** 'Spergon'.   **Details** *PM5*, p. 90.

850   chloranocryl (common name).   **CAS RN** *[2164–09–2]*. Herbicide reported by S. W. Bingham & W. K. Porter (*Weeds*, 1961, **9**, 282). Introduced by FMC Corp.   **Common names** chloranocryl, dicryl   **IUPAC name** 3',4'-dichloro-2-methylacrylanilide   **CA name** *N*-(3,4-dichlorophenyl)-2-methyl-2-propenamide.   **M.f.** $C_{10}H_9Cl_2NO$ **Code nos.** FMC 4556   **Selected tradenames** 'Dicryl'.   **Details** *PM2*, p. 85.

851   chlorazifop; chlorazifop-propargyl (common name).   **CAS RN** *[60074–25–1]* chlorazifop, unstated stereochemistry; *[74310–70–6]* chlorazifop, racemate; *[72492–54–7]* chlorazifop, (*R*)- isomer; *[72492–55–8]* chlorazifop, (*S*)- isomer; *[72280–52–5]* formerly *[77107–49–4]* chlorazifop-propargyl unstated stereochemistry; *[74267–69–9]* chlorazifop-propargyl (*R*)- isomer. Chlorazifop-propargyl evaluated as a herbicide by Ciba-Geigy AG (chlorazifop also formed by hydrolysis of isoxapyrifop *in vivo*).   **IUPAC name** (±)-2-[4-(3,5-dichloro-2-pyridyloxy)phenoxy]propionic acid **CA name** (±)-2-[4-[(3,5-dichloro-2-pyridinyl)oxy]phenoxy]propanoic acid. **M.f.** $C_{14}H_{11}Cl_2NO_4$; (propargyl ester $C_{17}H_{13}Cl_2NO_4$)   **Code nos.** CGA 82 725 (chlorazifop-propargyl).

852   chlorazine (common name).   **CAS RN** *[580–48–3]*. Herbicide introduced by J. R. Geigy S.A. (later Ciba-Geigy AG).   **IUPAC name** 6-chloro-N^2,N^2,N^4,N^4-tetraethyl-= 1,3,5-triazine-2,4-diamine; 2-chloro-4,6-bis(diethylamino)-1,3,5-triazine   **CA name** 6-chloro-*N,N,N',N'*-tetraethyl-1,3,5-triazine-2,4-diamine.   **M.f.** $C_{11}H_{20}ClN_5$ **Code nos.** G 25 804.

853   chlorbenside (common name).   **CAS RN** *[103–17–3]*. Acaricide reported by J. E. Cranham *et al.* (*Chem. Ind. (London)*, 1953, p. 1206) and by D. J. Higgons & D. W. Kilbey (*ibid.*, p. 1359). Introduced by The Boots Company Ltd (later Schering Agrochemicals). **IUPAC name** 4-chlorobenzyl 4-chlorophenyl sulfide   **CA name** 1-chloro-= 4-[[(4-chlorophenyl)methyl]thio]benzene.   **M.f.** $C_{13}H_{10}Cl_2S$ **Code nos.** HRS 860; RD 2195; ENT 20 696.   **Selected tradenames** 'Chlorparacide', 'Chlorsulphacide'. **Details** *PM3*, p. 97.

854   chlorbicyclen (common name).   **CAS RN** *[2550–75–6]*. Insecticide introduced by Hercules Inc. Agrochemicals (and later Nor-Am).   **Common names** chlorbicyclen, chlorbicyclène   **IUPAC name** 1,2,3,4,7,7-hexachloro-5,6-bis(chloromethyl)-= 8,9,10-trinorborn-2-ene; 1,2,3,4,7,7-hexachloro-5,6-bis(chloromethyl)bicyclo[2.2.1]= hept-2-ene (I)   **CA name** (I).   **M.f.** $C_9H_6Cl_8$ **Code nos.** Hercules 426; OMS 211; OMS 785.   **Selected tradenames** 'Alodan'.

855   chlorbufam (common name).   **CAS RN** *[1967–16–4]*. Herbicide reported by A. Fischer (*Z. Pflanzenkr. Pflanzenpathol. Pflanzenschutz*, 1960, **67**, 577). Introduced by

BASF AG. **Common names** chlorbufam, chlorbufame, BIPC **IUPAC name**
1-methylprop-2-ynyl 3-chlorocarbanilate **CA name** 1-methyl-2-propynyl (3-chloro=
phenyl)carbamate. **M.f.** $C_{11}H_{10}ClNO_2$ **Selected tradenames** 'BiPC'. **Details** *PM10*,
entry 116.

856 chlordecone (common name). **CAS RN** *[143–50–0]*. Insecticide introduced by
Allied Chemical Corp. Agricultural Div. (later Hopkins Ltd). **IUPAC name**
perchloropentacyclo[5.3.0.0^{2,6}.0^{3,9}.0^{4,8}]decan-5-one; decachloropentacyclo=
[5.2.1.0^{2,6}.0^{3,9}.0^{5,8}]decan-4-one **CA name** 1,1a,3,3a,4,5,5,5a,5b,6-decachloro=
octahydro-1,3,4-metheno-2*H*-cyclobuta[*cd*]pentalen-2-one. **M.f.** $C_{10}Cl_{10}O$
Code nos. GC-1189, ENT 16 391. **Selected tradenames** 'Kepone'. **Details** *PM5*, p. 94.

857 chlordimeform; chlordimeform hydrochloride (common name). **CAS RN**
[6164–98–3] chlordimeform; *[19750–95–9]* chlordimeform hydrochloride. Acaricide
reported by V. Dittrich (*J. Econ. Entomol.*, 1966, **59**, 889). Introduced by by Ciba AG
(later Ciba-Geigy AG) and Schering AG. **Common names** chlordimeform,
chlordiméforme, chlorophenamidine, chlorodimeform **IUPAC name** N^2-(4-chloro-=
o-tolyl)-N^1,N^1-dimethylformamidine **CA name** *N'*-(4-chloro-2-methylphenyl)-=
N,N-dimethylmethanimidamide. **M.f.** $C_{10}H_{13}ClN_2$; (hydrochloride $C_{10}H_{14}Cl_2N_2$)
Code nos. Schering 36 268; C 8514 (Ciba-Geigy); OMS 1209 (chlordimeform);
ENT 27 567 (chlordimeform hydrochloride). **Selected tradenames** 'Fundal',
'Spanone', 'Galecron'. **Details** *PM8*, entry 02270.

858 chloreturon (common name). **CAS RN** *[20782–58–5]*. Herbicide reported by A.
Aamisepp (*Proc. Swed. Weed Conf.*, 1977, **18**, F9-F13). Evaluated by Hoechst AG.
IUPAC name 3-(3-chloro-4-ethoxyphenyl)-1,1-dimethylurea **CA name** *N'*-(3-chloro-=
4-ethoxyphenyl)-*N,N*-dimethylurea. **M.f.** $C_{11}H_{15}ClN_2O_2$.

859 chlorfenac; chlorfenac-sodium (common name). **CAS RN** *[85–34–7]* chlorfenac;
[2439-00-1] chlorfenac-sodium. Herbicide introduced by Amchem Products Inc. (later
Rhône-Poulenc Agrochimie) and Hooker Chemical Corp. **Common names** chlorfenac,
fenac **IUPAC name** (2,3,6-trichlorophenyl)acetic acid **CA name** 2,3,6-trichloro=
benzeneacetic acid. **M.f.** $C_8H_5Cl_3O_2$; (sodium salt $C_8H_4Cl_3NaO_2$) **Selected**
tradenames 'Fenac', 'Fenatrol' (sodium salt). **Details** *PM8*, entry 02300.

860 chlorfenazole (common name). **CAS RN** *[3574–96–7]*. Fungicide introduced by
Celamerck GmbH & Co. KG (later Shell Agrar GmbH). **IUPAC name** 2-(2-chloro=
phenyl)benzimidazole **CA name** 2-(2-chlorophenyl)-1*H*-benzimidazole.
M.f. $C_{13}H_9ClN_2$ **Code nos.** CUR 616.

861 chlorfenethol (common name). **CAS RN** *[80–06–8]*. Acaricide reported by O.
Grummitt (*Science*, 1950, **111**, 361). Introduced by Sherwin-Williams & Co. and later by
Nippon Soda Chemical Co., Ltd. **Common names** chlorfenethol, BCPE **IUPAC name**
1,1-bis(4-chlorophenyl)ethanol **CA name** 4-chloro-α-(4-chlorophenyl)α-methyl=
benzenemethanol. **M.f.** $C_{14}H_{12}Cl_2O$ **Other names** DMC; DCPC **Code nos.**
ENT 9624. **Selected tradenames** 'Dimite', 'Qikron'. **Details** *PM6*, p. 97.

862   chlorfenprop; chlorfenprop-methyl (common name).   **CAS RN** *[59604–11–4]* chlorfenprop (racemate); *[59404–06–7]* (–)- enantiomer; *[59604–10–3]* chlorfenprop-methyl (racemate); *[14437–17–3]* (unstated stereochemistry). Herbicide reported by L. Eue (*Z. Pflanzenkr. Pflanzenpathol. Pflanzenschutz*, 1968, Sonderheft **IV,** 211). Only the (–)- enantiomer is herbicidal (T. Schmidt *et al., Z. Naturforsch. C. Biosci.,* 1976, **31C,** 252). Chlorfenprop-methyl introduced by Bayer AG.   **IUPAC name** (±)-2-chloro-= 3-(4-chlorophenyl)propionic acid   **CA name** (±)-α,4-dichlorobenzenepropanoic acid.   **M.f.** $C_9H_8Cl_2O_2$; (methyl ester $C_{10}H_{10}Cl_2O_2$)   **Code nos.** Bayer 70 533 Selected tradenames 'Bidisin'.   Details *PM7,* entry 02330.

863   chlorfenson (common name).   **CAS RN** *[80–33–1].* Acaricide reported by E. E. Kenaga & R. W. Hummer (*J. Econ. Entomol.,* 1949, **42,** 996). Introduced by Dow Chemical Co. (later DowElanco).   **Common names** chlorfenson, ovatran, chlorofénizon, ovex, CPCBS, [ephirsulphonate]   **IUPAC name** 4-chlorophenyl 4-chlorobenzenesulfonate   **CA name** 4-chlorophenyl 4-chlorobenzenesulfonate.   **M.f.** $C_{12}H_8Cl_2O_3S$ **Code nos.** K 6451, ENT 16 358.   **Selected tradenames** 'Ovotran', 'Ovitox', 'Sappiron', 'Trichlorfenson', 'Mitran'.   **Details** *PM8,* entry 02340.

864   chlorfensulphide (common name).   **CAS RN** *[2274–74–0].* Introduced as a component of an acaricide by Nippon Soda Co., Ltd.   **Common names** chlorfensulphide, chlorfensulfide, CPAS   **IUPAC name** 4-chlorophenyl 2,4,5-trichloro= benzenediazosulfide; 4-chlorophenyl 2,4,5-trichlorophenylazosulfide   **CA name** [(4-chlorophenyl)thio](2,4,5-trichlorophenyl)diazene.   **M.f.** $C_{12}H_6Cl_4N_2S$ **Selected tradenames** 'Milbex'.   **Details** *PM4,* p. 101.

865   chlorflurazole (common name).   **CAS RN** *[3615–21–2].* Herbicide evaluated by Fisons PLC (later Schering Agriculture).   **IUPAC name** 4,5-dichloro-2-trifluoro= methylbenzimidazole   **CA name** 4,5-dichloro-2-(trifluoromethyl)-1*H*-benzimidazole.   **M.f.** $C_8H_3Cl_2F_3N_2$ **Code nos.** NC 3363.

866   chlorfluren; chlorfluren-methyl (common name).   **CAS RN** *[24539–66–0]* chlorfluren; *[22909–50–8]* chlorfluren-methyl. Plant growth regulator introduced by E. Merck (later Shell Agrar GmbH).   **Common names** chlorfluren, chlorflurène   **IUPAC name** 2-chlorofluorene-9-carboxylic acid   **CA name** 2-chloro-9*H*-fluorene-= 9-carboxylic acid.   **M.f.** $C_{14}H_9ClO_2$; (methyl ester $C_{15}H_{11}ClO_2$)   **Code nos.** IT-5732 (chlorfluren-methyl).

867   chlorflurenol-methyl; chlorflurenol (common name).   **CAS RN** *[2536–31–4];* (acid *[2464–37–1]*). Plant growth regulator and herbicide. Effects of fluorene-= 9-carboxylic acids on plant growth reported by G. Schneider (*Naturwissenschaften,* 1964, **51,** 416) who proposed they be called morphactins (G. Schneider *et al., Nature (London),* 1965, **208,** 1013). The methyl ester of chlorflurenol, chlorflurenol-methyl, introduced by E. Merck (later American Cyanamid Co.).   **Common names** (for acid) chlorflurenol, chloroflurénol, chlorflurecol   **IUPAC name** methyl 2-chloro-9-hydroxyfluorene-= 9-carboxylate   **CA name** methyl 2-chloro-9-hydroxy-9*H*-fluorene-9-carboxylate.   **M.f.** $C_{15}H_{11}ClO_3$; (acid $C_{14}H_9ClO_3$)   **Code nos.** IT 3456 (for acid).   **Details** *PM10,* entry 121.

SUPERSEDED ENTRIES

868 chlornitrofen (common name). **CAS RN** *[1836–77–7]*. Herbicide introduced in Japan (1966) by Mitsui Toatsu Chemicals, Inc. **Common names** chlornitrofen, chlornitrofène, CNP **IUPAC name** 4-nitrophenyl 2,4,6-trichlorophenyl ether **CA name** 1,3,5-trichloro-2-(4-nitrophenoxy)benzene. **M.f.** $C_{12}H_6Cl_3NO_3$ **Code nos.** MO-338 **Selected tradenames** 'Mo'. **Details** *PM10*, entry 126.

869 chlorobenzilate (common name). **CAS RN** *[510–15–6]*. Acaricide reported by R. Gasser (*Experientia*, 1952, **8**, 65). Introduced by J. R. Geigy S.A. (later Ciba-Geigy AG). **IUPAC name** ethyl 4,4'-dichlorobenzilate **CA name** ethyl 4-chloro-α-(4-chloro= phenyl)-α-hydroxybenzeneacetate. **M.f.** $C_{16}H_{14}Cl_2O_3$ **Code nos.** G 23 992, ENT 18 596. **Selected tradenames** 'Akar', 'Acaraben' (USA). **Details** *PM10*, entry 128.

870 1-chloro-2,4-dinitronaphthalene (IUPAC/Chemical Abstracts name). **CAS RN** *[2401–85–6]*. Fungicide. **M.f.** $C_{10}H_5ClN_2O_4$.

871 chloromebuform (common name). **CAS RN** *[37407–77–5]*. Acaricide evaluated by Ciba-Geigy AG. **Common names** chloromebuform, chlor-mébuforme **IUPAC name** N^1-butyl-N^2-(4-chloro-o-tolyl)-N^1-methylformamidine **CA name** N-butyl-N'-= (4-chloro-2-methylphenyl)-N-methylmethanimidamide; N-butyl-N'-(4-chloro-= o-tolyl)-N'-methylformamidine. **M.f.** $C_{13}H_{19}ClN_2$ **Code nos.** CGA 22 598 **Selected tradenames** 'Ektomin'.

872 chloromethiuron (common name). **CAS RN** *[28217–97–2]*. Ixodicide reported by M. von Orelli et al. (*Proc. World Vet. Congr., 20th*, 1975, p. 659). Introduced by Ciba-Geigy AG. **IUPAC name** 3-(4-chloro-o-tolyl)-1,1-dimethyl(thiourea) **CA name** N'-(4-chloro-2-methylphenyl)-N,N-dimethylthiourea. **M.f.** $C_{10}H_{13}ClN_2S$ **Code nos.** CGA 13 444 **Selected tradenames** 'Dipofene'. **Details** *PM9*, entry 2640.

873 chloroneb (common name). **CAS RN** *[2675–77–6]*. Fungicide reported by M. J. Fielding & R. C. Rhodes (*Proc. Cotton Dis. Counc.*, 1967, **27**, 56). Introduced by E. I. du Pont de Nemours and Co. **Common names** chloroneb, chloronèbe **IUPAC name** 1,4-dichloro-2,5-dimethoxybenzene **CA name** 1,4-dichloro-2,5-dimethoxybenzene. **M.f.** $C_8H_8Cl_2O_2$ **Code nos.** Soil Fungicide 1823 **Selected tradenames** 'Demosan', 'Teremec'. **Details** *PM10*, entry 129.

874 1-chloro-2-nitropropane (IUPAC/Chemical Abstracts name). **CAS RN** *[2425–66–3]*. Fungicide reported by Bushong (*Seed Soil Treatment Newslett.*, 1962, **4**, 39). Introduced by FMC Corp. **M.f.** $C_3H_6ClNO_2$ **Code nos.** FMC 5916 **Selected tradenames** 'Lanstan'. **Details** *PM2*, p. 104.

875 3-(4-chlorophenyl)-5-methylrhodanine (IUPAC name). **CAS RN** *[6012–92–6]*. Fungicide and nematicide developed by Stauffer Chemical Co. **IUPAC name** 3-(4-chlorophenyl)-5-methylrhodanine; 3-(4-chlorophenyl)-5-methyl-2-thioxo-= 1,3-thiazolidin-4-one **CA name** 3-(4-chlorophenyl)-5-methyl-2-thioxo-= 4-thiazolidinone. **M.f.** $C_{10}H_8ClNOS_2$ **Code nos.** N 244.

1298   *Superseded entries – 875*

876   4-chlorophenyl phenyl sulfone (IUPAC name).   **CAS RN** *[80–00–2]*. Acaricide introduced by Stauffer Chemical Co. (later ICI Americas).   **CA name** 1-chloro-= 4-(phenylsulfonyl)benzene.   **M.f.** $C_{12}H_9ClO_2S$ **Code nos.** R-242   **Selected tradenames** 'Sulphenone'.

877   chloropon (common name).   Herbicide.   **IUPAC name** 2,2,3-trichlorpropionic acid   **CA name** 2,2,3-trichloropropanoic acid.   **M.f.** $C_3H_3Cl_3O_2$.

878   chloropropylate (common name).   **CAS RN** *[5836–10–2]*. Acaricide reported by F. Chabousson (*Phytiatr.-Phytopharm.*, 1956, **5**, 203). Introduced by J. R. Geigy S.A. (later Ciba-Geigy AG).   **IUPAC name** isopropyl 4,4′-dichlorobenzilate   **CA name** 1-methylethyl 4-chloro-α-(4-chlorophenyl)-α-hydroxybenzeneacetate.   **M.f.** $C_{17}H_{16}Cl_2O_3$ **Code nos.** G 24 163, ENT 26 999.   **Selected tradenames** 'Rospin'. **Details** *PM8*, entry 02900.

879   2-chlorovinyl diethyl phosphate (IUPAC name).   **CAS RN** *[311–47–7]*. Insecticide developed by Shell Chemical Co.   **CA name** 2-chloroethenyl diethyl phosphate.   **M.f.** $C_6H_{12}ClO_4P$ **Code nos.** SD 1836, OS 1836.

880   chloroxuron (common name).   **CAS RN** *[1982–47–4]*. Herbicide reported (*Symp. New Herbic.*, *3rd*, 1961, p. 88). Introduced by Ciba AG (later Ciba-Geigy AG). **Common names** chloroxuron, [chloroxifenidim]   **IUPAC name** 3-[4-(4-chloro= phenoxy)phenyl]-=1,1-dimethylurea   **CA name** *N*′-[4-(4-chlorophenoxy)phenyl]-= *N,N*-dimethylurea.   **M.f.** $C_{15}H_{15}ClN_2O_2$ **Code nos.** C 1983   **Selected tradenames** 'Tenoran'.   **Details** *PM9*, entry 2970.

881   2-(4-chloro-3,5-xylyloxy)ethanol (IUPAC name).   **CAS RN** *[5825–79–6]*. Insecticide introduced by Union Carbide Corp. (later Rhône-Poulenc Ag.). **CA name** 2-(4-chloro-3,5-dimethylphenoxy)ethanol.   **M.f.** $C_{10}H_{13}ClO_2$ **Code nos.** Experimental Chemotherapeutant 1182.

882   chloroxynil (common name).   **CAS RN** *[1891–95–8]*. Herbicide evaluated by May & Baker Ltd (later Rhône-Poulenc Agrochimie).   **IUPAC name** 3,5-dichloro-= 4-hydroxybenzonitrile (I)   **CA name** (I).   **M.f.** $C_7H_3Cl_2NO$.

883   chlorphoxim (common name).   **CAS RN** *[14816–20–7]*. Insecticide reported by J. E. Hudson & W. O. Obudho (*Mosq. News*, 1972, **32**, 37). Introduced by Bayer AG. **Common names** chlorphoxim, chlorphoxime   **IUPAC name** 2-(2-chlorophenyl)-= 2-(diethoxyphosphinothioyloxyimino)acetonitrile; *0,0*-diethyl 2-chloro-α-cyano= benzylideneamino-oxyphosphonothioate   **CA name** 7-(2-chlorophenyl)-4-ethoxy-= 3,5-dioxa-6-aza-4-phosphaoct-6-ene-8-nitrile 4-sulfide; formerly 2-chloro-= α-(diethoxyphosphinothioyloxy)imino]benzeneacetonitrile.   **M.f.** $C_{12}H_{14}ClN_2O_3PS$ **Code nos.** BAY SRA 7747, OMS 1197.   **Selected tradenames** 'Baythion C'.   **Details** *PM8*, entry 03010.

884   chlorprazophos (common name).   **CAS RN** *[36145–08–1]*. Insecticide evaluated by Bayer AG.   **IUPAC name** *0*-(3-chloro-7-methylpyrazolo[1,5-*a*]pyrimidin-2-yl)

O,O-diethyl phosphorothioate (I)   **CA name** (I).   **M.f.** $C_{11}H_{15}ClN_3O_3PS$ **Code nos.** HOX 2709.

885   chlorprocarb (common name).   **CAS RN** *[23121–99–5]*. Herbicide introduced by BASF AG.   **Common names** chlorprocarb, chlorprocarbe   **IUPAC name** methyl 3-[1-(chloromethyl)propylcarbamoyloxy]carbanilate   **CA name** 3-[(methoxycarbonyl)= amino]phenyl [1-(chloromethyl)propyl]carbamate; methyl *m*-hydroxycarbanilate [1-(chloromethyl)propyl]carbamate ester. **M.f.** $C_{13}H_{17}ClN_2O_4$ **Code nos.** BAS 379H.

886   chlorquinox (common name).   **CAS RN** *[3495–42–9]*. Fungicide introduced by Fisons Ltd (later Schering Agrochemicals Ltd).   **IUPAC name** 5,6,7,8-tetrachloro= quinoxaline (I)   **CA name** (I).   **M.f.** $C_8H_2Cl_4N_2$ **Selected tradenames** 'Lucel'. **Details** *PM5*, p. 119.

887   chlorthiophos (common name).   **CAS RN** *[60238–56–4]* for mixed isomers; *[21923–23–9]* for isomer (i). Insecticide reported by H. Holtmann & E. Raddatz (*Proc. Br. Insectic. Fungic. Conf.*, 6th, 1971, **2**, 485). Introduced by C. H. Boehringer Sohn/Cela GmbH (later Shell Agrar GmbH).   **IUPAC name** O-2,5-dichloro-4-methylthiophenyl O,O-diethyl phosphorothioate (i, main component); O-4,5-dichloro-2-methylthiophenyl O,O-diethyl phosphorothioate (ii, small quantities); O-2,4-dichloro-5-methylthiophenyl O,O-diethyl phosphorothioate (iii, minor component)   **CA name** O-[2,5-dichloro-= 4-(methylthio)phenyl] O,O-diethyl phosphorothioate (i); O-[4,5-dichloro-2-(methylthio)= phenyl] O,O-diethyl phosphorothioate (ii); O-[2,4-dichloro-5-(methylthio)phenyl] O,O-diethyl phosphorothioate (iii).   **M.f.** $C_{11}H_{15}Cl_2O_3PS_2$ **Code nos.** S 2957; OMS 1342; ENT 27 635.   **Selected tradenames** 'Celathion'.   **Details** *PM7*, entry 03110.

888   ciobutide (common name).   Plant growth regulator discovered by EGYT Pharmacochemical Works.   **IUPAC name** (*RS*)-2-cyano-2-phenylbutyramide   **CA name** (±)-α-cyano-α-ethylbenzeneacetamide.   **M.f.** $C_{11}H_{12}N_2O$.

889   CL 304,415 (development code).   Herbicide safener, reported by G. E. Cary *et al.*, *Proc. 8th IUPAC International Congress of Pesticide Chemistry*, Washington, 1984. Developed by American Cyanamid Co.   **M.f.** $C_{12}H_{12}O_5$ **Code nos.** CL 304,415. **Details** *PM10*, entry 146.

890   climbazole (common name).   **CAS RN** *[38083–17–9]*. Fungicide introduced by Bayer AG.   **IUPAC name** 1-(4-chlorophenoxy)-1-(imidazol-1-yl)-3,3-dimethylbutanone   **CA name** 1-(4-chlorophenoxy)-1-(1*H*-imidazol-1-yl)-3,3-dimethyl-2-butanone.   **M.f.** $C_{15}H_{17}ClN_2O_2$ **Code nos.** BAY MEB 6401   **Selected tradenames** 'Baysan'.   **Details** *PM7*, entry 02760.

891   cliodinate (common name).   **CAS RN** *[69148–12–5]*. Herbicide evaluated by Celamerck GmbH KG (later Shell Agrar GmbH).   **IUPAC name** 2-chloro-3,5-di-iodo-= 4-pyridyl acetate   **CA name** 2-chloro-3,5-di-iodo-4-pyridinyl acetate.   **M.f.** $C_7H_4ClI_2NO_2$ **Code nos.** ASD 2288.

892 cloethocarb (common name). **CAS RN** *[51487–69–5]*. Insecticide reported by V. Harries *et al.* (*Meded. Fac. Landbouwwet. Rijksuniv. Gent,* 1978, **45**, 739). Introduced by BASF AG. **IUPAC name** 2-(2-chloro-1-methoxyethoxy)phenyl methylcarbamate **CA name** 2-(2-chloro-1-methoxyethoxy)phenyl methylcarbamate. **M.f.** $C_{11}H_{14}ClNO_4$ **Code nos.** BAS 2631 **Selected tradenames** 'Lance'. **Details** *PM10*, entry 149.

893 clofop; clofop-isobutyl (common name). **CAS RN** *[59621–49–7]* racemic clofop; *[26129–32–8]* clofop (unstated stereochemistry); *[51337–71–4]* clofop-isobutyl (unstated stereochemistry). Herbicide described by F. Schwerdtle *et al.* (*Mitt. Biol., Bundesanst. Land-Forstwirtsch. Berlin-Dahlem,* 1975, **165**, 171). Clofop-isobutyl introduced by Hoechst AG. **IUPAC name** ($\pm$)-2-[4-(4-chlorophenoxy)phenoxy]= propionic acid **CA name** ($\pm$)-2-[4-(4-chlorophenoxy)phenoxy]propanoic acid. **M.f.** $C_{15}H_{13}ClO_4$; (isobutyl ester $C_{19}H_{21}ClO_4$) **Code nos.** Hoe 22 870. **Details** *PM6*, p. 125.

894 cloproxydim (common name). **CAS RN** *[95480–33–4]*. Herbicide evaluated by Chevron Chemical Co. **Common names** cloproxydim, cloproxydime, [clopropoxydim] **IUPAC name** ($\pm$)-2-[1-(3-chloroallyloxy)iminobutyl]-5-(2-ethylthiopropyl)-3-hydroxy= cyclohex-2-enone **CA name** 2-[1-[[(3-chloro-2-propenyl)oxy]imino]butyl]-= 5-[2-(ethylthio)propyl]-3-hydroxy-2-cyclohexen-1-one. **M.f.** $C_{18}H_{28}ClNO_3S$ **Code nos.** RE-36 290 **Selected tradenames** 'Selectone'.

895 copper bis(3-phenylsalicylate) (IUPAC name). **CAS RN** *[5328–04–1]*. Fungicide. **CA name** bis(2-hydroxy[1,1'-biphenyl]-3-carboxylato-O^2,O^0)copper. **M.f.** $C_{26}H_{18}CuO_6$.

896 copper zinc chromate (IUPAC name). Fungicide evaluated by Union Carbide Corp. (later Rhône-Poulenc Agrochimie). **IUPAC name** copper zinc chromate (indefinite composition). **Code nos.** Crag Fungicide 658; Experimental Fungicide 658.

897 coumachlor (common name). **CAS RN** *[81–82–3]*. Rodenticide reported by M. Reiff & R. Wiesmann (*Acta Trop.,* 1951, **8**, 97). Introduced by J. R. Geigy S.A. (later Ciba-Geigy AG) **Common names** coumachlor, coumachlore **IUPAC name** 3-[1-(4-chlorophenyl)-3-oxobutyl]-4-hydroxycoumarin **CA name** 3-[1-(4-chloro= phenyl)-3-oxobutyl]-4-hydroxy-2*H*-1-benzopyran-2-one. **M.f.** $C_{19}H_{15}ClO_4$ **Code nos.** G 23 133 **Selected tradenames** 'Ratilan', 'Tomorin'. **Details** *PM9*, entry 3320.

898 coumafuryl (common name). **CAS RN** *[117–52–2]* unstated stereochemistry. Rodenticide developed by Amchem Chemical Co. (later Rhône-Poulenc Agrochimie). **Common names** coumafuryl, fumarin, [tomarin] **IUPAC name** 3-[1-(2-furyl)-= 3-oxobutyl]-4-hydroxycoumarin **CA name** 3-[1-(2-furanyl)-3-oxobutyl]-4-hydroxy-= 2*H*-1-benzopyran-2-one. **M.f.** $C_{17}H_{14}O_5$.

899 coumithoate (common name). **CAS RN** *[572–48–5]*. Insecticide introduced by Montecatini S.p.A. (later Agrimont S.p.A.). **IUPAC name** *O,O*-diethyl *O*-(7,8,9,10-tetrahydro-6-oxo-6*H*-benzo[*c*]chromen-3-yl) phosphorothioate; 3-diethoxyphosphinothioyloxy-7,8,9,10-tetrahydrobenzo[*c*]chromen-6-one **CA name**

O,O-diethyl O-(7,8,9,10-tetrahydro-6-oxo-6H-dibenzo[b,d]pyran-3-yl) phosphorothioate.   M.f. $C_{17}H_{21}O_5PS$ Selected tradenames 'Dithion', 'Dition'.

900   CP 17029 (development code).   **CAS RN** *[845–52–3]*. Herbicide introduced by Monsanto Co.   **IUPAC name** N^2,N^4-bis(3-methoxypropyl)-6-methylthio-1,3,5-= triazine-2,4-diamine; 2,4-bis(3-methoxypropylamino)-6-(methylthio)-1,3,5-triazine **CA name** N,N'-bis(3-methoxypropyl)-6-(methylthio)-1,3,5-triazine-2,4-diamine. **M.f.** $C_{12}H_{23}N_5O_2S$ **Other names** MPMT   **Code nos.** CP 17 029   **Selected tradenames** 'Lambast'.

901   credazine (common name).   **CAS RN** *[14491–59–9]*. Herbicide reported by T. Jojima et al. (*Agric. Biol. Chem.*, 1968, **32**, 1376; 1969, **33**, 96). Introduced by Sankyo Chemical Ltd.   **IUPAC name** pyridazin-3-yl o-tolyl ether   **CA name** 3-(2-methyl= phenoxy)pyridazine.   **M.f.** $C_{11}H_{10}N_2O$ **Code nos.** H 722; SW-6701; SW-6721 **Selected tradenames** 'Kusakira'.   **Details** *PM6*, p. 522.

902   crimidine (common name).   **CAS RN** *[535–89–7]*. Rodenticide introduced in the early 1940s by Bayer AG.   **IUPAC name** 2-chloro-N,N,6-trimethylpyrimidin-4-amine; 2-chloro-N,N,6-trimethylpyrimidin-4-ylamine   **CA name** 2-chloro-N,N,6-trimethyl-= 4-pyrimidinamine.   **M.f.** $C_7H_{10}ClN_3$ **Selected tradenames** 'Castrix'.   **Details** *PM7*, entry 03380.

903   crotoxyphos (common name).   **CAS RN** *[7700–17–6]*. Insecticide reported by C. P. Weidenback & R. L. Younger (*J. Econ. Entomol.*, 1962, **55**, 793). Introduced by the Shell Development Co.   **Common names** crotoxyphos, crotoxyfos   **IUPAC name** 1-phenylethyl 3-(dimethoxyphosphinoyloxy)isocrotonate; 1-phenylethyl 3-(dimethoxy= phosphinyloxy)isocrotonate; dimethyl (E)-1-methyl-2-(1-phenylethoxy-carbonyl)vinyl phosphate   **CA name** (E)-1-phenylethyl 3-[(dimethoxyphosphinyl)oxy]-2-butenoate. **M.f.** $C_{14}H_{19}O_6P$ **Code nos.** SD 4294; OMS 239; ENT 24 717.   **Selected tradenames** 'Ciodrin'.   **Details** *PM8*, entry 03390.

904   crufomate (common name).   **CAS RN** *[299–86–5]*. Veterinary insecticide and anthelmintic reported by J. F. Landram & R. J. Shaver (*J. Parasitol.*, 1959, **45**, 55). Introduced by Dow Chemical Co. (later DowElanco).   **IUPAC name** 4-tert-butyl-= 2-chlorophenyl methyl methylphosphoramidate   **CA name** 2-chloro-4-(1,1-dimethyl= ethyl)phenyl methyl methylphosphoramidate.   **M.f.** $C_{12}H_{19}ClNO_3P$ **Code nos.** Dowco 132, ENT 26 602-X.   **Selected tradenames** 'Ruelene'.   **Details** *PM6*, p. 131.

905   CS 708 (development code).   **CAS RN** *[117–26–0]* (i); *[117–27–1]* (ii); *[8027–00–7]* formerly *[8002–82–2]* (i + ii). Insecticide introduced by Commercial Solvents Corp.   **IUPAC name** 1,1-bis(4-chlorophenyl)-2-nitrobutane (i) + 1,1-bis(4-chlorophenyl)-2-nitropropane (ii)   **CA name** 1,1'-(2-nitrobutylidene)= bis[4-chlorobenzene] (i) + 1,1'-(2-nitropropylidene)bis[4-chlorobenzene] (ii).   **M.f.** $C_{16}H_{15}Cl_2NO_2$ (i), $C_{15}H_{13}Cl_2NO_2$ (ii)   **Code nos.** CS 674A (i); CS 645A (ii); CS 708 (i + ii); ENT 18 066 [2:1 mixture of (i) + (ii)]; OMS 467 (ii).   **Selected tradenames** 'Bulan', 'Prolan', 'Dilan'.   **Details** *PM5*, p. 176.

906 cufraneb (common name). **CAS RN** *[11096–18–7]*. Fungicide introduced by Universal Crop Protection. **Common names** cufraneb, cufranèbe **IUPAC name** ethylenebis(dithiocarbamate) mixed metal complex containing not less than 8.15% (*m/m*) of zinc, 8.05% (*m/m*) of manganese, 5.5% (*m/m*) of copper and 1.0% (*m/m*) of iron **CA name** cufraneb (not defined in 9CI; see 8CI). **Details** *PM9, entry 3420.*

907 *m*-cumenyl methylcarbamate (IUPAC name). **CAS RN** *[64–00–6]*. Insecticide introduced by Hercules Inc. (later Nor-Am) and Union Carbide Corp. (later Rhône-Poulenc Agrochimie). **IUPAC name** *m*-cumenyl methylcarbamate; 3-isopropylphenyl methylcarbamate **CA name** 3-(1-methylethyl)phenyl methylcarbamate. **M.f.** $C_{11}H_{15}NO_2$ **Code nos.** AC 5727 (Hercules); UC 10 854 (Union Carbide); ENT 25 500. **Details** *PM1, p. 263.*

908 cupric hydrazinium sulfate (traditional name). **CAS RN** *[33271–65–7]*. Fungicide introduced by Olin Mathieson Chemical Corp. **IUPAC name** copper(II) dihydrazinium disulfate **CA name** bis(hydrazine)bis(hydrogen sulfato)copper. **M.f.** $CuH_{10}N_4O_8S_2$ **Other names** cupric hydrazinium sulfate **Code nos.** Mathieson 466 **Selected tradenames** 'Omazene'.

909 cuprobam (common name). **CAS RN** *[7076–63–3]*. Fungicide. **Common names** cuprobam, cuprobame **IUPAC name** tricopper dichloride dimethyldithiocarbamate **CA name** dichloro(dimethylcarbamodithioato)tricopper. **M.f.** $C_3H_6Cl_2Cu_3NS_2$.

910 cyanatryn (common name). **CAS RN** *[21689–84–9]*. Herbicide reported by V. V. Dovlatyan & F. V. Avetisyan (*Arm. Khim. Zh.*, 1972, **25**, 880) and evaluated by Shell Research Ltd. **Common names** cyanatryn, cyanatryne **IUPAC name** 2-(4-ethyl= amino-6-methylthio-1,3,5-triazin-2-ylamino)-2-methylpropionitrile **CA name** 2-[[4-(ethylamino)-6-(methylthio)-1,3,5-triazin-2-yl]amino]-2-methylpropanenitrile. **M.f.** $C_{10}H_{16}N_6S$ **Code nos.** WL 63 611.

911 2-cyano-3-(2,4-dichlorophenyl)acrylic acid (IUPAC name). **CAS RN** *[6013–05–4]*. Plant growth regulator developed by Ethyl Corp. Agrochemicals (later Nor-Am). **CA name** 2-cyano-3-(2,4-dichlorophenyl)-2-propenoic acid. **M.f.** $C_{10}H_5Cl_2NO_2$ **Code nos.** Ethyl 214.

912 cyanofenphos (common name). **CAS RN** *[13067–93–1]*. Insecticide reported by Y. Nishizawa (*Bull. Agric. Chem. Soc. Japan*, 1960, **24**, 744). Introduced by Sumitomo Chemical Co., Ltd. **Common names** cyanofenphos, cyanophenphos, CYP **IUPAC name** *O*-4-cyanophenyl *O*-ethyl phenylphosphonothioate **CA name** *O*-(4-cyanophenyl) *O*-ethyl phenylphosphonothioate. **M.f.** $C_{15}H_{14}NO_2PS$ **Code nos.** S-4087, ENT 25 832a. **Selected tradenames** 'Surecide'. **Details** *PM7, entry 03490.*

913 cyanthoate (common name). **CAS RN** *[3734–95–0]*. Insecticide and acaricide reported by F. Galbaiti (*Proc. Br. Insectic. Fungic. Conf., 1st*, 1961, **2**, 507). Introduced by Montecatini S.p.A. (later Montedison S.p.A.). **IUPAC name** *S*-[*N*-(1-cyano-1-methyl= ethyl)carbamoylmethyl] *O,O*-diethyl phosphorothioate; *N*-(1-cyano-1-methylethyl)-= 2-(diethoxyphosphinoylthio)acetamide **CA name** *S*-[2-[[(1-cyano-1-methylethyl)=

amino]-2-oxoethyl] *O,O*-diethyl phosphorothioate. **M.f.** $C_{10}H_{19}N_2O_4PS$ **Code nos.** M 1568 **Selected tradenames** 'Tartan'. **Details** *PM5*, p. 138.

914 cyclafuramid (common name). **CAS RN** *[34849–42–8]*. Fungicide introduced by BASF AG. **Common names** cyclafuramid, cyclafuramide **IUPAC name** N-cyclohexyl-= 2,5-dimethyl-3-furamide **CA name** N-cyclohexyl-2,5-dimethyl-3-furancarboxamide. **M.f.** $C_{13}H_{19}NO_2$ **Code nos.** BAS 327F.

915 cycloheximide (common name). **CAS RN** *[66–81–9]*. Antifungal antibiotic isolated by A. Whiffen *et al.* (*J. Bacteriol.*, 1946, **52**, 610), agricultural uses reviewed (J. H. Ford et *al.*, *Plant Dis. Rep.*, 1958, **42**, 680). Introduced for crop protection by the Upjohn Co. (later Nor-Am.). **IUPAC name** 4-[(2R)-2-[(1S,3S,5S)-(3,5-dimethyl-= 2-oxocyclohexyl)]-2-hydroxyethyl]piperidine-2,6-dione; 3-[(2R)-2-[(1S,3S,5S)-= 3,5-dimethyl-2-oxocyclohexyl]-2-hydroxyethyl]glutarimide **CA name** [1S-[1α(S*),3α,5β]]-4-[2-(3,5-dimethyl-2-oxocyclohexyl)-2-hydroxyethyl]-= 2,6-piperidinedione. **M.f.** $C_{15}H_{23}NO_4$ **Selected tradenames** 'Acti-dione', 'Acti-Aid'. **Details** *PM8*, entry 03600.

916 cycluron (common name). **CAS RN** *[2163–69–1]*. Herbicide reported by A. Fischer (*Z. Pflanzenkr. Pflanzenpathol. Pflanzenschutz*, 1960, **67**, 577). Introduced by BASF AG. **Common names** cycluron, COMU **IUPAC name** 3-cyclo-octyl-= 1,1-dimethylurea **CA name** N'-cyclooctyl-N,N-dimethylurea. **M.f.** $C_{11}H_{22}N_2O$ **Other names** OMU **Selected tradenames** 'Alipur'. **Details** *PM6*, p. 141.

917 cyometrinil (common name). **CAS RN** *[78370–21–5]* (Z)- isomer; *[63278–33–1]* unstated stereochemistry. Herbicide safener for 2-chloroacetanilides, introduced by Ciba-Geigy AG. **IUPAC name** (Z)-cyanomethoxyimino(phenyl)= acetonitrile **CA name** (Z)-α-[cyanomethoxyimino]benzeneacetonitrile. **M.f.** $C_{10}H_7N_3O$ **Code nos.** CGA 43 089 **Selected tradenames** 'Concep'. **Details** *PM7*, entry 03670.

918 cypendazole (common name). **CAS RN** *[28559–00–4]*. Systemic fungicide introduced by Bayer AG. **IUPAC name** methyl 1-(5-cyanopentylcarbamoyl)= benzimidazol-2-ylcarbamate **CA name** methyl [1-[[(5-cyanopentyl)amino]carbonyl]-= 1H-benzimidazol-2-yl]carbamate. **M.f.** $C_{16}H_{19}N_5O_3$ **Code nos.** DAM 18 654 **Selected tradenames** 'Folicidin'. **Details** *PM4*, p. 145.

919 cyperquat; cyperquat chloride (common name). **CAS RN** *[48134–75–4]* cyperquat; *[39794–99–5]* cyperquat chloride. Cyperquat chloride introduced as a herbicide by Gulf Oil Chemicals. **IUPAC name** 1-methyl-4-phenylpyridinium (I) **CA name** (I). **M.f.** $C_{12}H_{12}N$; (cyperquat chloride $C_{12}H_{12}ClN$) **Code nos.** S 21 634.

920 cyprazine (common name). **CAS RN** *[22936–86–3]*. Herbicide reported by O. C. Burnside et *al.* (*Proc. North Cent. Weed Control Conf.*, 1969, p. 21). Introduced by Gulf Oil Corp. **IUPAC name** 6-chloro-N^2-cyclopropyl-N^4-isopropyl-1,3,5-triazine-= 2,4-diamine; 2-chloro-4-cyclopropylamino-6-isopropylamino-1,3,5-triazine **CA name**

6-chloro-N-cyclopropyl-N'-(1-methylethyl)-1,3,5-triazine-2,4-diamine. **M.f.** $C_9H_{14}ClN_5$ **Code nos.** S 6115 **Selected tradenames** 'Outfox'. **Details** PM4, p. 146.

921 cyprazole (common name). **CAS RN** [42089–03–2]. Herbicide introduced by Gulf Oil Chemicals. **IUPAC name** N-[5-(2-chloro-1,1-dimethylethyl)-1,3,4-thiadiazol-=2-yl]cyclopropanecarboxamide (I) **CA name** (I). **M.f.** $C_{10}H_{14}ClN_3OS$ **Code nos.** S 19 073.

922 cyprofuram (common name). **CAS RN** [69581–33–5] unstated stereochemistry. Fungicide reported by D. Baumert & H. Buschhaus (Meded. Fac. Landbouwwet. Rijksuniv. Gent, 1982, **47**, 979). Introduced by Schering AG. **IUPAC name** ($\pm$)-α-[N-(3-chloro=phenyl)cyclopropanecarboxamido]-γ-butyrolactone **CA name** N-(3-chlorophenyl)-N-=(tetrahydro-2-oxo-3-furanyl)cyclopropanecarboxamide. **M.f.** $C_{14}H_{14}ClNO_3$ **Code nos.** SN 78 314 **Selected tradenames** 'Vinicur'. **Details** PM8, entry 03740.

923 cypromid (common name). **CAS RN** [2759–71–9]. Herbicide reported by T. R. Hopkins et al. (Proc. Symp. New Herbic., 2nd, Paris, 1965, p. 187). Introduced by Gulf Oil Corp. **Common names** cypromid, cypromide **IUPAC name** 3',4'-dichlorocyclo=propanecarboxanilide **CA name** N-(3,4-dichlorophenyl)cyclopropanecarboxamide. **M.f.** $C_{10}H_9Cl_2NO$ **Code nos.** S 6000 **Selected tradenames** 'Clobber'. **Details** PM3, p. 148.

924 DAEP (common name). **CAS RN** [13265–60–6]. Insecticide developed by Nippon Soda Co., Ltd. **IUPAC name** N-(2-dimethoxyphosphinothioylthioethyl)=acetamide; S-2-acetamidoethyl O,O-dimethyl phosphorodithioate **CA name** S-[2-(acetylamino)ethyl] O,O-dimethyl phosphorodithioate. **M.f.** $C_6H_{14}NO_3PS_2$ **Selected tradenames** 'Amiphos'.

925 DCPM (common name). **CAS RN** [555–89–5]. Acaricide reported by L. R. Jeppson (J. Econ. Entomol., 1946, **39**, 813). Introduced by Dow Chemical Co. (later DowElanco). **IUPAC name** bis(4-chlorophenoxy)methane **CA name** 1,1'-[methylenebis(oxy)]bis[4-chlorobenzene]. **M.f.** $C_{13}H_{10}Cl_2O_2$ **Other names** oxythane **Code nos.** K-1875 **Selected tradenames** 'Neotran'. **Details** PM1, p. 39.

926 decafentin (common name). **CAS RN** [15652–38–7]. Fungicide introduced by Celamerck GmbH & Co. (later Shell Agrar). **IUPAC name** decyltriphenylphosphonium bromochlorotriphenylstannate(IV) **CA name** (TB-5-12)-decyltriphenylphosphonium bromochlorotriphenylstannate(1-). **M.f.** $C_{46}H_{51}BrClPSn$ **Code nos.** A 36 **Selected tradenames** 'Stanoram'.

927 decarbofuran (common name). **CAS RN** [1563–67–3]. Insecticide evaluated by Bayer AG. **IUPAC name** 2,3-dihydro-2-methylbenzofuran-7-yl methylcarbamate **CA name** 2,3-dihydro-2-methyl-7-benzofuranyl methylcarbamate. **M.f.** $C_{11}H_{13}NO_3$ **Code nos.** BAY 62 863.

928 dehydroacetic acid (traditional name). **CAS RN** [520–45–6] (i), [771–03–9] (ii), [16807–48–0] (iii). Long-known as a chemical, fungicidal activity reported by P. A. Wolf

(*Food Technol. (Chicago)*, 1950, **4**, 294). Introduced by Dow Chemical Co. **IUPAC name** 2-acetyl-5-methyl-3-oxopent-4-en-5-olide (i); 3-acetyl-6-methylpyran-2,4-dione (i) **CA name** 3-acetyl-6-methyl-2*H*-pyran-2,4(3*H*)-dione (i); 3-acetyl-4-hydroxy-6-methyl-= 2*H*-pyran-2-one (ii); 3-acetyl-2-hydroxy-6-methyl-4*H*-pyran-4-one (iii). **M.f.** $C_8H_8O_4$ **Other names** dehydroacetic acid; DHA. **Details** *PM6*, p. 152.

929   delachlor (common name).   **CAS RN** *[24353–58–0]*. Herbicide reported by R. F. Husted *et al.* (*Proc. North Cent. Weed Control Conf.*, 1967). Introduced by Monsanto Co.   **Common names** delachlor, délachlore   **IUPAC name** 2-chloro-*N*-(isobutoxy= methyl)acet-2′,6′-xylidide   **CA name** 2-chloro-*N*-(2,6-dimethylphenyl)-*N*-[(2-methyl= propoxy)methyl]acetamide.   **M.f.** $C_{15}H_{22}ClNO_2$ **Code nos.** CP 52 223.   **Details** *PM2*, p. 100.

930   demephion; demephion-O; demephion-S (common name).   **CAS RN** *[682–80–4]* (i); *[2587–90–8]* (ii); *[8065–62–1]* (i + ii). Systemic acaricide and insecticide reported by H. Rueppold (*Wiss. Z. Martin-Luther Univ. Halle-Wittenberg Math.-Naturwiss. Reihe*, 1955, **5**, 219). Introduced initially by VEB Farbenfabrik Wolfen.   **Common names** demephion (for a reaction product comprising demephion-O and demephion-S), demephion-O (i) and demephion-S (ii) (for thiono and thiolo isomers respy.)   **IUPAC name** (i) *O,O*-dimethyl *O*-2-methylthioethyl phosphorothioate; (ii) *O,O*-dimethyl *S*-2-methylthioethyl phosphorothioate   **CA name** (i) *O,O*-dimethyl *O*-[2-(methylthio)= ethyl] phosphorothioate; (ii) *O,O*-dimethyl *S*-[2-(methylthio)ethyl] phosphorothioate. **M.f.** $C_5H_{13}O_3PS_2$ **Selected tradenames** 'Tinox' (Eastern Europe), 'Cymetox', 'Pyracide', 'Atlasetox'.   **Details** *PM6*, p. 153.

931   demeton; demeton-O; demeton-S (common name).   **CAS RN** *[298–03–3]* demeton-O; *[126–75–0]* demeton-S; *[8065–48–3]* (formerly *[8000–97–3]*) demeton. Insecticide reported by G. Unterstenhöfer (*Meded. Landbouwhogesch. Opzoekingsstn. Staat Gent*, 1952, **17**, 75). Introduced by Farbenfabriken Bayer AG (later Bayer AG). **Common names** demeton (for a reaction product comprising demeton-O and demeton-S), demeton-O (i), demeton-S (ii) (for thiono and thiolo isomers respy.), mercaptofostion (i), mercaptostiol (ii), [mercaptofos] (i), [mercaptofos teolevy] (ii) **IUPAC name** (i) *O,O*-diethyl *O*-2-ethylthioethyl phosphorothioate; (ii) *O,O*-diethyl *S*-2-ethylthioethyl phosphorothioate   **CA name** (i) *O,O*-diethyl *O*-[2-(ethylthio)ethyl] phosphorothioate; (ii) *O,O*-diethyl *S*-[2-(ethylthio)ethyl] phosphorothioate.   **M.f.** $C_8H_{19}O_3PS_2$ **Other names** diethyl 2-ethylthioethyl phosphorothionate (i) and diethyl 2-ethylthioethyl phosphorothiolate (ii)   **Code nos.** Bayer 10 756; E-1059; ENT 17 295 (demeton).   **Selected tradenames** 'Systox'.   **Details** *PM8*, entry 03910.

932   demeton-S-methylsulphon (common name).   **CAS RN** *[17040–19–6]*. Insecticide introduced by Bayer AG.   **Common names** demeton-S-methylsulphon, déméton-S-= méthylsulfone, demeton-S-methyl sulphone   **IUPAC name** *S*-2-ethylsulfonylethyl *O,O*-dimethyl phosphorothioate   **CA name** *S*-[2-(ethylsulfonyl)ethyl] *O,O*-dimethyl phosphorothioate.   **M.f.** $C_6H_{15}O_5PS_2$ **Code nos.** Bayer 20 315; E 158; M3/158 **Selected tradenames** 'Metaisosystoxsulfon'.   **Details** *PM9*, entry 3930.

933 **2,4-DEP** (common name). **CAS RN** *[39420–34–3]* (i + ii); *[94–84–8]* (i). Herbicide introduced by Uniroyal Inc. **IUPAC name** mixture of (i) tris[2-(2,4-= dichlorophenoxy)ethyl] phosphite (I) and (ii) bis[2-(2,4-dichlorophenoxy)ethyl] phosphonate (II) **CA name** (I) + (II). **M.f.** $C_{24}H_{21}Cl_6O_6P$ (i), $C_{16}H_{15}Cl_4O_5P$ (ii) **Code nos.** 3Y9 **Selected tradenames** 'Falone'. **Details** *PM2*, p. 143.

934 **dialifos** (common name). **CAS RN** *[10311–84–9]* . Insecticide reported by W. R. Cothran et al. (*J. Econ. Entomol.*, 1967, **60**, 1151). Introduced by Hercules Inc. (later Nor-Am Chemical Co.). **Common names** dialifos, dialiphos, dialifor **IUPAC name** S-2-chloro-1-phthalimidoethyl O,O-diethyl phosphorodithioate; N-[2-chloro-= 1-(diethoxyphosphinothioylthio)ethyl]phthalimide **CA name** S-[2-chloro-= 1-(1,3-dihydro-1,3-dioxo-2H-isoindol-2-yl)ethyl] O,O-diethyl phosphorodithioate. **M.f.** $C_{14}H_{17}ClNO_4PS_2$ **Code nos.** Hercules 14503, ENT 27320 **Selected tradenames** 'Torak'. **Details** *PM8*, entry 03980.

935 **di-allate** (common name). **CAS RN** *[2303–16–4]*. Herbicide reported by L. H. Hannah (*Proc. North Cent. Weed Control Conf.*, 1959, p. 50). Introduced by Monsanto Co. **Common names** di-allate, diallate **IUPAC name** S-2,3-dichloroallyl di-isopropyl= (thiocarbamate) **CA name** S-(2,3-dichloro-2-propenyl) bis(1-methylethyl)= carbamothioate. **M.f.** $C_{10}H_{17}Cl_2NOS$ **Code nos.** CP 15 336 **Selected tradenames** 'Avadex'. **Details** *PM8*, entry 03990.

936 **diamidafos** (common name). **CAS RN** *[1754–58–1]*. Nematicide reported by C. R. Youngson & C. A. I. Goring (*Down Earth*, 1963, **18**(4), 3). Introduced by Dow Chemical Co. (later DowElanco). **IUPAC name** phenyl N,N'-dimethylphosphoro= diamidate (I) **CA name** (I). **M.f.** $C_8H_{13}N_2O_2P$ **Code nos.** Dowco 169 **Selected tradenames** 'Nellite'. **Details** *PM5*, p. 413.

937 **1,2-dibromo-3-chloropropane** (IUPAC/Chemical Abstracts name). **CAS RN** *[96–12–8]*. Nematicidal activity reported by C. W. McBeth & G. B. Bergeson (*Plant Dis. Rep.*, 1955, **39**, 223). Introduced by Dow Chemical Co. (later DowElanco) and Shell Development Co. **Common names** DBCP **M.f.** $C_3H_5Br_2Cl$ **Code nos.** OS1897 (Shell) **Selected tradenames** 'Nemagon', 'Fumazone'. **Details** *PM6*, p. 164.

938 **dibutyl adipate** (IUPAC name). **CAS RN** *[105–99–7]*. Insect repellent introduced by Union Carbide Corp. (later Rhône-Poulenc Agrochimie). **CA name** dibutyl hexanedioate. **M.f.** $C_{14}H_{26}O_4$ **Code nos.** Experimental Tick Repellent 3.

939 **dibutyl phthalate** (IUPAC name). **CAS RN** *[84–74–2]*. Activity as insect repellent reported by F. M. Snyder & F. A. Morton (*J. Econ. Entomol.*, 1947, **40**, 586). **Common names** dibutyl phthalate, phtalate de butyle **CA name** dibutyl 1,2-benzene= dicarboxylate. **M.f.** $C_{16}H_{22}O_4$ **Other names** DBP. **Details** *PM5*, p. 163.

940 **dibutyl succinate** (IUPAC name). **CAS RN** *[141–03–7]*. Insect repellent introduced by Glenn Chemical Co. **CA name** dibutyl butanedioate. **M.f.** $C_{12}H_{22}O_4$ **Selected tradenames** 'Tabatrex', 'Tabutrex'. **Details** *PM5*, p. 164.

941   dicamba-methyl (common name). **CAS RN** *[6597–78–0]*. Plant growth regulator introduced by Velsicol Chemical Co. (later Sandoz AG). **Common names** dicamba-methyl, disugran **IUPAC name** methyl 3,6-dichloro-o-anisate **CA name** methyl 3,6-dichloro-2-methoxybenzoate. **M.f.** $C_9H_8Cl_2O_3$ **Selected tradenames** 'Racuza'.

942   dicapthon (common name). **CAS RN** *[2463–84–5]*. Insecticide reported by T. B. Davich & J. W. Apple (*J. Econ. Entomol.*, 1951, **44**, 528) and by J. C. Gaines (*ibid.*, p. 750). Introduced by American Cyanamid Co. **IUPAC name** O-2-chloro-4-nitrophenyl O,O-dimethyl phosphorothioate **CA name** O-(2-chloro-4-nitrophenyl) O,O-dimethyl phosphorothioate. **M.f.** $C_8H_9ClNO_5PS$ **Code nos.** Experimental Insecticide 4124; OMS 214; ENT 17 035. **Selected tradenames** 'Dicapton'. **Details** *PM3*, p. 169.

943   dichlofenthion (common name). **CAS RN** *[97–17–6]*. Nematicide reported by M. A. Manzelli (*Plant Dis. Rep.*, 1955, **39**, 400). Introduced against soil-dwelling insects and nematodes by Virginia-Carolina Chemical Corp. **Common names** dichlofenthion, ECP **IUPAC name** O-2,4-dichlorophenyl O,O-diethyl phosphorothioate **CA name** O-(2,4-dichlorophenyl) O,O-diethyl phosphorothioate. **M.f.** $C_{10}H_{13}Cl_2O_3PS$ **Code nos.** V-C 13 Nemacide, ENT 17 470. **Selected tradenames** 'Mobilawn'. **Details** *PM6*, p. 167.

944   dichlone (common name). **CAS RN** *[117–80–6]*. Fungicidal properties of this established chemical reported by W. P. ter Horst & E. L. Felix (*Ind. Eng. Chem.*, 1943, **35**, 1255). Introduced by Uniroyal Inc. and by FMC Corp. **IUPAC name** 2,3-dichloro-= 1,4-naphthoquinone **CA name** 2,3-dichloro-1,4-naphthalenedione. **M.f.** $C_{10}H_4Cl_2O_2$ **Code nos.** USR 604 (to Uniroyal), ENT 3776. **Selected tradenames** 'Kolo'. **Details** *PM10*, entry 205.

945   dichloralurea (common name). **CAS RN** *[116–52–9]*. Herbicide introduced by Union Carbide Corp. (later Rhône-Poulenc Ag.). **Common names** DCU, dichloralurea, dichloralurée **IUPAC name** 1,3-bis(2,2,2-trichloro-1-hydroxyethyl)urea **CA name** N,N'-bis(2,2,2-trichloro-1-hydroxyethyl)urea. **M.f.** $C_5H_6Cl_6N_2O_3$ **Code nos.** Crag Herbicide 2 **Selected tradenames** 'Crag Herbicide 2'. **Details** *PM1*, p. 139.

946   dichlorflurenol; dichlorflurenol-methyl (common name). **CAS RN** *[21634–96–8]* dichlorflurenol-methyl. Dichlorflurenol-methyl introduced as a plant growth regulator by E. Merck (later Shell Agrar GmbH). **Common names** dichlorflurenol, dichloroflurénol, [dichlorflurecol] **IUPAC name** 2,7-dichloro-9-hydroxyfluorene-= 9-carboxylic acid **CA name** 2,7-dichloro-9-hydroxy-9H-fluorene-9-carboxylic acid. **M.f.** $C_{14}H_8Cl_2O_3$; (methyl ester $C_{15}H_{10}Cl_2O_3$) **Code nos.** IT-5733.

947   dichlormate (common name). **CAS RN** *[1966–58–1]* dichlormate; *[62046–37–1]* tech. grade containing 2,3-dichloro- analogue. Herbicide reported by R. A. Herrett & R. V. Berthold (*Science*, 1965, **149**, 191). Introduced by Union Carbide Corp. (later Rhône-Poulenc Ag.). **IUPAC name** 3,4-dichlorobenzyl methyl= carbamate **CA name** (3,4-dichlorophenyl)methyl methylcarbamate. **M.f.** $C_9H_9Cl_2NO_2$ **Code nos.** UC 22 463A (pure dichlormate); UC 22 463 (TC grade) **Selected tradenames** 'Rowmate'. **Details** *PM1*, p. 141.

948 1,1-dichloro-2,2-bis(4-ethylphenyl)ethane (IUPAC name). **CAS RN** *[72–56–0].*
Insecticide introduced by Rohm & Haas Co. **CA name** 1,1′-(2,2-dichloroethylidene)=
bis[4-ethylbenzene]. **M.f.** $C_{18}H_{20}Cl_2$ **Other names** ethylan **Code nos.** Q-137.
Details *PM6*, p. 170.

949 O-2,5-dichloro-4-iodophenyl O-ethyl ethylphosphonothioate (IUPAC name).
CAS RN *[25177–27–9].* Insecticide introduced by Ciba AG (later Ciba-Geigy AG). **CA
name** O-(2,5-dichloro-4-iodophenyl) O-ethyl ethylphosphonothioate. **M.f.**
$C_{10}H_{12}Cl_2IO_2PS$ **Code nos.** C 18 244. **Details** *PM2*, p. 231.

950 1,1-dichloro-1-nitroethane (IUPAC/Chemical Abstracts name). **CAS RN**
[594–72–9]. Insecticidal activity reported by W. C. O'Kane & H. W. Smith (*J. Econ.
Entomol.*, 1941, **34**, 438). Introduced by Commercial Solvents Corp. **M.f.** $C_2H_3Cl_2NO_2$
Selected tradenames 'Ethide'. **Details** *PM3*, p. 177.

951 2,4-dichlorophenyl benzenesulfonate (IUPAC/Chemical Abstracts name). **CAS
RN** *[97–16–5].* Acaricide introduced by Allied Chemical Corp. Agrochemical Division
(later Hopkins Agricultural Chemical Co.). **M.f.** $C_{12}H_8Cl_2O_3S$ **Code nos.** EM 293
Selected tradenames 'Genite', 'Genitol'. **Details** *PM4*, p. 179.

952 (RS)-N-(3,5-dichlorophenyl)-2-(methoxymethyl)succinimide (IUPAC name).
CAS RN *[81949–88–4].* Fungicide evaluated by Wacker-Chemie GmbH. **CA name**
(±)-1-(3,5-dichlorophenyl)-3-(methoxymethyl)-2,5-pyrrolidinedione. **M.f.**
$C_{12}H_{11}Cl_2NO_3$ **Other names** [metomeclan] (unadopted proposed common name)
Code nos. Co 6054 **Selected tradenames** 'Drawifol'.

953 N-3,5-dichlorophenylsuccinimide (IUPAC name). **CAS RN** *[24096–53–5].*
Fungicide introduced by Sumitomo Chemical Co., Ltd. **CA name** 1-(3,5-dichloro=
phenyl)-2,5-pyrrolidinedione. **M.f.** $C_{10}H_7Cl_2NO_2$ **Code nos.** S-47127 **Selected
tradenames** 'Ohric'.

954 1,2-dichloropropane (IUPAC/Chemical Abstracts name). **CAS RN** *[78–87–5].*
Its properties as an insecticidal fungicide were described by I. E. Neifert *et al.* (*U.S. Dep.
Agric. Bull.*, 1925, No. 1313). **M.f.** $C_3H_6Cl_2$ **Other names** propylene dichloride
Code nos. ENT 15 406. **Details** *PM8*, entry 04400.

955 1,2-dichloropropane with 1,3-dichloropropene (IUPAC name). **CAS RN**
[8003–19–8]. Properties as a soil fumigant described by W. Carter (*Science*, 1943, **97**,
383). Introduced by Shell Chemical Co. and by Dow Chemical Co., now replaced by
1,3-dichloropropene. **IUPAC name** 1,2-dichloropropane + 1,3-dichloropropene
CA name 1,2-dichloropropane + 1,3-dichloro-1-propene. **Code nos.** ENT 8420.
Selected tradenames 'D-D', 'Vidden D'. **Details** *PM8*, entry 04410.

956 1,3-dichloro-1,1,3,3-tetrafluoropropane-2,2-diol (IUPAC name). **CAS RN**
[993–57–8] (i); *[127–21–9]* (ii). Fungicide reported by D. W. George (Plant Dis. Rep.,
1964, **48**, 162). Introduced by Allied Chemical Corp. **IUPAC name** 1,3-dichloro-=
1,1,3,3-tetrafluoropropane-2,2-diol (i); 1,3-dichloro-1,1,3,3-tetrafluoroacetone

hydrate (ii)   **CA name** 1,3-dichloro-1,1,3,3-tetrafluoro-2,2-propanediol (i); 1,3-dichloro-=
1,1,3,3-tetrafluoro-2-propanone hydrate (ii).   **M.f.** $C_3H_2Cl_2F_4O_2$ **Code nos.**
GC 9832.   **Details** *PM4*, p. 185.

957   3,4-dichlorotetrahydrothiophene 1,1-dioxide (IUPAC/Chemical Abstracts name).
CAS RN *[3001–57–8]*. Nematicide introduced by Diamond Shamrock Chemical Co.
(later Fermenta).   **M.f.** $C_4H_6Cl_2O_2S$ **Other names** dichlorothiolane dioxide   **Selected
tradenames** 'PRD Experimental Nematicide'.

958   2,2-dichlorovinyl 2-ethylsulfinylethyl methyl phosphate (IUPAC name).   **CAS RN**
[7076–53–1]. Insecticide introduced by Celamerck GmbH & Co. (later Shell Agrar
GmbH).   **CA name** 2,2-dichloroethenyl 2-(ethylsulfinyl)ethyl methyl phosphate.   **M.f.**
$C_7H_{13}Cl_2O_5PS$ **Code nos.** Nexion 1378.

959   dichlozoline (common name).   **CAS RN** *[24201–58–9]*. Fungicide evaluated by
Chevron Chemical Co.   **IUPAC name** 3-(3,5-dichlorophenyl)-5,5-dimethyl-=
1,3-oxazolidine-2,4-dione   **CA name** 3-(3,5-dichlorophenyl)-5,5-dimethyl-=
2,4-oxazolidinedione.   **M.f.** $C_{11}H_9Cl_2NO_3$ **Other names** DDOD   **Code nos.**
CS 8890; Ortho 8890   **Selected tradenames** 'Sclex'.

960   diclobutrazol (common name).   **CAS RN** *[75736–33–3]* diclobutrazol;
[66345–62–8] unstated stereochemistry. Fungicide reported by K. J. Bent & A. M.
Skidmore (*Proc. 1979 Br. Crop Prot. Conf. - Pests Dis.*, 1977, **2**, 477). Introduced by
ICI Agrochemicals.   **IUPAC name** (2*RS*,3*RS*)-1-(2,4-dichlorophenyl)-4,4-dimethyl-=
2-(1*H*-1,2,4-triazol-1-yl)pentan-3-ol   **CA name** (*R**,*R**)-(±)-β-[(2,4-dichlorophenyl)=
methyl]-α-(1,1-dimethylethyl)-1*H*-1,2,4-triazole-1-ethanol.   **M.f.** $C_{15}H_{19}Cl_2N_3O$
Code nos. PP296   **Selected tradenames** 'Vigil'.   **Details** *PM9*, entry 4510.

961   dicyclopentadiene (chemical name).   **CAS RN** *[77–73–6]* . Animal repellent
introduced by Staehler.   **CA name** 3a,4,7,7a-tetrahydro-4,7-methano-1*H*-indene.   **M.f.**
$C_{10}H_{12}$.

962   dieldrin (common name).   **CAS RN** *[60–57–1]*. Insecticide reported by C. W.
Kearns et al. (*J. Econ. Entomol.*, 1949, **42**, 127). Introduced by J. Hyman & Co. and later
by the Shell International Chemical Co. Ltd.   **Common names** HEOD (for pure
compound), dieldrin, dieldrin (material containing >85% HEOD), dieldrine   **IUPAC
name** (1*R*,4*S*,4a*S*,5*R*,6*R*,7*S*,8*S*,8a*R*)-1,2,3,4,10,10-hexachloro-1,4,4a,5,6,7,8,8a-octa=
hydro-6,7-epoxy-1,4:5,8-dimethanonaphthalene; 1,2,3,4,10,10-hexachloro-6,7-epoxy-=
1,4,4a,5,6,7,8,8a-octahydro-*endo*-1,4,-*exo*-5,8-dimethanonaphthalene   **CA name**
(1α,2β,2aα,3β,6β,6aα,7β,7aα)-3,4,5,6,9,9-hexachloro-1a,2,2a,3,6,6a,7,7a-octa=
hydro-2,7:3,6-dimethanonaphth[2,3-*b*]oxirene.   **M.f.** $C_{12}H_8Cl_6O$ **Code nos.**
Compound 497 (to Hyman); OMS 18; ENT 16 225.   **Selected tradenames** 'Octalox'.

963   diethamquat; diethamquat dichloride (common name). Dichloride evaluated as
herbicide by ICI Plant Protection Division (later ICI Agrochemicals).   **IUPAC name**
1,1′-bis(diethylcarbamoylmethyl)-4,4′-bipyridinium; 1,1′-bis(diethylcarbamoylmethyl)-=
4,4′-bipyridyl-1,1′-diylium   **CA name** 1,1′-bis[2-(diethylamino)-2-oxoethyl]-=

4,4'-bipyridinium.   **M.f.** $C_{22}H_{32}N_4O_2$; (diethamquat dichloride $C_{22}H_{32}Cl_2N_4O_2$)   **Code nos.** PP831.

964   diethatyl-ethyl; diethatyl (common name).   **CAS RN** *[38727–55–8]*; (acid *[38725–95–0]*). Herbicidal activity of the ethyl ester reported by S. K. Lehman (*Proc. North Cent. Weed Control Conf., 29th,* 1972); introduced by Hercules Inc. Agrochemicals (later Nor-Am Chemical Co.).   **Common names** (for the acid) diethatyl   **IUPAC name** (acid) *N*-chloroacetyl-*N*-(2,6-diethylphenyl)glycine   **CA name** (acid) *N*-(chloroacetyl)-= *N*-(2,6-diethylphenyl)glycine.   **M.f.** $C_{16}H_{22}ClNO_3$; (acid $C_{14}H_{18}ClNO_3$)   **Code nos.** (acid) Hercules 22 234   **Selected tradenames** 'Antor'.   **Details** *PM 9*, entry 4610.

965   *O,O*-diethyl *O*-4-methyl-2-oxo-2*H*-chromen-7-yl phosphorothioate (IUPAC name).   **CAS RN** *[299–45–6]*. Insecticide reported by G. Schrader (*Angew. Chem.*, 1952, Monogr. No. 62, 2nd Ed.). Introduced by Bayer AG.   **IUPAC name** *O,O*-diethyl *O*-4-methyl-2-oxo-2*H*-chromen-7-yl phosphorothioate; *O,O*-diethyl *O*-4-methyl= coumarin-7-yl phosporothioate; 7-diethoxyphosphinothioyloxy-4-methylcoumarin   **CA name** *O,O*-diethyl *O*-(4-methyl-2-oxo-2*H*-1-benzopyran-7-yl) phosphorothioate.   **M.f.** $C_{14}H_{17}O_5PS$ **Code nos.** E 838   **Selected tradenames** 'Potasan'; 'Potasan G'.   **Details** *PM1*, p. 158.

966   *O,O*-diethyl *O*-6-methyl-2-propylpyrimidin-4-yl phosphorothioate (IUPAC name).   **CAS RN** *[5826–91–5]*. Insecticide introduced by J. R. Geigy S.A. (later Ciba-Geigy AG).   **CA name** *O,O*-diethyl *O*-(6-methyl-2-propyl-4-pyrimidinyl) phosphorothioate.   **M.f.** $C_{12}H_{21}N_2O_3PS$ **Code nos.** G 24 622   **Selected tradenames** 'Pyrazinon'.

967   diethyl 5-methylpyrazol-3-yl phosphate (IUPAC name).   **CAS RN** *[108–34–9]*. Insecticide introduced by J. R. Geigy S.A. (later Ciba-Geigy AG).   **CA name** diethyl 5-methyl-1*H*-pyrazol-3-yl phosphate.   **M.f.** $C_8H_{15}N_2O_4P$ **Code nos.** G 24 483, ENT 24 723.   **Selected tradenames** 'Pyrazoxon'.

968   difenopenten; difenopenten-ethyl (common name).   **CAS RN** *[81416–44–6]* difenopenten; *[71101–05–8]* difenopenten-ethyl. Herbicide evaluated by Chevron Chemical Co.   **Common names** difenopenten, difénopentène   **IUPAC name** (*E*)-(±)-4-[4-(α,α,α-trifluoro-*p*-tolyloxy)phenoxy]pent-2-enoic acid   **CA name** (*E*)-(±)-4-[4-[4-(trifluoromethyl)phenoxy]phenoxy]-2-pentenoic acid.   **M.f.** $C_{18}H_{15}F_3O_4$; (ethyl ester $C_{20}H_{19}F_3O_4$)   **Code nos.** XE-773; KK-80 (both for difenopenten-ethyl).

969   difenoxuron (common name).   **CAS RN** *[14214–32–5]*. Herbicide reported by L. Ebner & J. Schuler (*Proc. Br. Weed Control Conf., 7th,* 1964, **2**, 711). Introduced by Ciba-Geigy AG.   **IUPAC name** 3-[4-(4-methoxyphenoxy)phenyl]-1,1-dimethylurea   **CA name** *N'*-[4-(4-methoxyphenoxy)phenyl]-*N,N*-dimethylurea.   **M.f.** $C_{16}H_{18}N_2O_3$ **Code nos.** C 3470   **Selected tradenames** 'Lironion'.   **Details** *PM9*, entry 4700.

970   2,3-dihydro-5,6-diphenyl-1,4-oxathi-ine (IUPAC name).   **CAS RN** *[58041–19–3]*. Plant growth regulator evaluated by Uniroyal Inc.   **CA name** 2,3-dihydro-5,6-diphenyl-= 1,4-oxathiin.   **M.f.** $C_{16}H_{14}OS$ **Code nos.** P 293.

971 **2,3-dihydro-5-phenyl-1,4-dithi-ine 1,1,4,4-tetraoxide** (IUPAC name). **CAS RN** *[34407–87–9]*. Fungicide evaluated by Uniroyal Inc. **CA name** 2,3-dihydro-5-phenyl-= 1,4-dithiin 1,1,4,4-tetraoxide. **M.f.** $C_{10}H_{10}O_4S_2$ **Code nos.** P 368.

972 **dimefox** (common name). **CAS RN** *[115–26–4]*. Insecticide and acaricide reported by H. Kükenthal & G. Schrader (*B.I.O.S. Final Report*, 1946, 1095). Introduced by Fisons Pest Control Ltd (later Schering Agrochemicals). **IUPAC name** tetramethylphosphorodiamidic fluoride (I); bis(dimethylamino)fluorophosphine oxide **CA name** (I). **M.f.** $C_4H_{12}FN_2OP$ **Code nos.** ENT 19 109. **Selected tradenames** 'Pestox XIV'. **Details** *PM6*, p. 196.

973 **dimethrin** (common name). **CAS RN** *[70–38–2]* unstated stereochemistry. Insecticide synthesised by W. F. Barthel, activity reported by P. G. Piquett & W. A. Gersdorff (*J. Econ. Entomol.*, 1958, **51**, 791). **Common names** dimethrin, diméthrine **IUPAC name** 2,4-dimethylbenzyl (1*RS*)-*cis,trans*-2,2-dimethyl-3-(2-methylprop-1-enyl)= cyclopropanecarboxylate; 2,4-dimethylbenzyl (1*RS*,3*RS*;1*RS*,3*SR*)-2,2-dimethyl-= 3-(2-methylprop-1-enyl)cyclopropanecarboxylate; 2,4-dimethylbenzyl (±)-*cis-trans*-= chrysanthemate **CA name** (2,4-dimethylphenyl)methyl 2,2-dimethyl-3-(2-methyl-= 1-propenyl)cyclopropanecarboxylic acid. **M.f.** $C_{19}H_{26}O_2$ **Code nos.** OMS 187.

974 **2-(4,5-dimethyl-1,3-dioxolan-2-yl)phenyl methylcarbamate** (IUPAC/Chemical Abstracts name). **CAS RN** *[7122–04–5]*. Insecticide reported by F. Bachmann & J. B. Legge (*J. Sci. Food Agric.*, 1968 Suppl., p. 39). Introduced by Ciba AG (later Ciba-Geigy AG). **M.f.** $C_{13}H_{17}NO_4$ **Code nos.** C 10 015, ENT 27 410. **Selected tradenames** 'Fondaren'. **Details** *PM3*, p. 198.

975 **5,5-dimethyl-3-oxocyclohex-1-enyl dimethylcarbamate** (IUPAC name). **CAS RN** *[122–15–6]*. Insecticide introduced by J. R. Geigy S.A. (later Ciba-Geigy AG). **CA name** 5,5-dimethyl-3-oxo-1-cyclohexen-1-yl dimethylcarbamate. **M.f.** $C_{11}H_{17}NO_3$ **Code nos.** G 19 258, ENT 24 728. **Selected tradenames** 'Dimetan'.

976 **dimetilan** (common name). **CAS RN** *[644–64–4]*. Insecticide reported by H. Gysin (*Chimia*, 1954, **8**, 205, 221). Introduced by J. R. Geigy S.A. (later Ciba-Geigy AG). **IUPAC name** 1-dimethylcarbamoyl-5-methylpyrazol-3-yl dimethylcarbamate **CA name** 1-[(dimethylamino)carbonyl]-5-methyl-1*H*-pyrazol-3-yl dimethylcarbamate. **M.f.** $C_{10}H_{16}N_4O_3$ **Code nos.** G 22 870; GS 13 332; OMS 479; ENT 25 922. **Selected tradenames** 'Snip'. **Details** *PM6*, p. 205.

977 **dimexano** (common name). **CAS RN** *[1468–37–7]*. Herbicide introduced by Vondelingenplaat N.V. **Common names** dimexano, [dimexan] **IUPAC name** *O,O*-dimethyl dithiobis(thioformate) **CA name** dimethyl thioperoxydicarbonate. **M.f.** $C_4H_6O_2S_4$ **Other names** dimethylxanthic disulfide **Selected tradenames** 'Tri-PE'. **Details** *PM5*, p. 206.

978 **dimidazon** (common name). **CAS RN** *[3295–78–1]*. Herbicide introduced by BASF AG. **Common names** dimidazon, dimidazone, [dimethazone] **IUPAC name**

4,5-dimethoxy-2-phenylpyridazin-3(2*H*)-one   **CA name** 4,5-dimethoxy-2-phenyl-=
3(2*H*)-pyridazinone.   **M.f.** $C_{12}H_{12}N_2O_3$ **Code nos.** BAS 255H.

979   dinex; dinex-diclexine (common name).   **CAS RN** *[131–89–5]* dinex; *[317–83–9]*
dinex-diclexine. Dinex was introduced as an acaricide and insecticide by Dow Chemical
Co. Ltd (later DowElanco), and its dicyclohexylammonium salt (dinex-diclexine) by
Fisons Ltd (later Schering Agriculture).   **Common names** dinex, pédinex, DN
IUPAC name 2-cyclohexyl-4,6-dinitrophenol (I)   **CA name** (I).   **M.f.** $C_{12}H_{14}N_2O_5$;
(dinex-diclexine $C_{24}H_{37}N_3O_5$)   **Other names** DINOCHP (dinex)   **Code nos.**
DN1 (dinex); DN111 (dinex-diclexine); ENT 157 (dinex); ENT 30 828
(dinex-diclexine).   **Selected tradenames** 'Dynone II'.

980   dinocton (common name).   **CAS RN** *[32534–96–6]* (i + II); *[19000–58–9]* (i);
[19000–52–3] (ii); *[32535–08–3]* (iii + iv); *[6465–51–6]* (iii); *[6465–60–7]* (iv).
Acaricide and fungicide reported by M. Pianka (*Proc. Crop Prot. Symp., 2nd., Magdeburg*,
1966). Evaluated by Murphy Chemical Co. (later DowElanco).   **Common names**
dinocton (for a reaction product comprising isomeric dinitro(octyl)phenyl methyl
carbonates)   **IUPAC name** for main components: (i) 2,4-dinitro-6-(1-propylpentyl)=
phenyl methyl carbonate (I); (ii) 2-(1-ethylhexyl)-4,6-dinitrophenyl methyl carbonate
(II); (iii) 2,6-dinitro-4-(1-propylpentyl)phenyl methyl carbonate (III); (iv) 4-(1-ethyl=
hexyl)-2,6-dinitrophenyl methyl carbonate (IV)   **CA name** (I), (II), (III), (IV).   **M.f.**
$C_{16}H_{22}N_2O_7$ **Other names** dinocton-6 (i + ii); dinocton-4 (iii + iv)   **Code nos.**
MC 1945 (i + ii); MC 1947 (iii + iv).   **Details** *PM4*, p. 210 (i + ii), p. 209 (iii + iv).

981   dinofenate (common name).   **CAS RN** *[61614–62–8]*. Herbicide evaluated by
Pennwalt.   **IUPAC name** 2-*sec*-butyl-4,6-dinitrophenyl 2,4-dinitrophenyl carbonate
CA name 2,4-dinitrophenyl 2-(1-methylpropyl)-4,6-dinitrophenyl carbonate.   **M.f.**
$C_{17}H_{14}N_4O_{11}$.

982   dinopenton (common name).   **CAS RN** *[5386–57–2]*. Acaricide reported at 5th
Intern. Pestic. Congr., London, 1983 (M. Pianka, *J. Sci. Food Agric.*, 1966, **17**, 47).
Evaluated by Murphy Chemical Ltd (later DowElanco).   **IUPAC name** isopropyl
2-(1-methylbutyl)-4,6-dinitrophenyl carbonate   **CA name** 2-(1-methylbutyl)-=
4,6-dinitrophenyl 1-methylethyl carbonate.   **M.f.** $C_{15}H_{20}N_2O_7$.

983   dinoprop (common name).   **CAS RN** *[7257–41–2]*. Herbicide and insecticide.
IUPAC name 4,6-dinitro-*o*-cymen-3-ol   **CA name** 3-methyl-2-(1-methylethyl)-=
4,6-dinitrophenol.   **M.f.** $C_{10}H_{12}N_2O_5$.

984   dinosam (common name).   **CAS RN** *[4097–36–3]*. Herbicide reported by A. S.
Crafts (*Plant Physiol.*, 1946, **21**, 345). Introduced by Standard Agricultural Chemicals Inc.
Common names dinosam, dinosame   **IUPAC name** 2-(1-methylbutyl)-4,6-dinitrophenol
(I)   **CA name** (I).   **M.f.** $C_{11}H_{14}N_2O_5$ **Other names** DNAP   **Selected tradenames** 'Sinox
General'.

985   dinoseb (common name).   **CAS RN** *[88–85–7]*. Herbicidal activity of dinoseb
reported by A. S. Crafts (*Science*, 1945, **101**, 417).   Introduced by Dow Chemical Co.

(later DowElanco). **Common names** dinoseb, dinosèbe, DNBP **IUPAC name**
2-sec-butyl-4,6-dinitrophenol **CA name** 2-(1-methylpropyl)-4,6-dinitrophenol. **M.f.**
$C_{10}H_{12}N_2O_5$ **Code nos.** DN 289 (Dow); Hoe 26 150; ENT 1122 **Selected tradenames**
'Premerge' (dinoseb), 'Premerge' (olamine and 2-hydroxypropylammonium salts),
'Premerge 3' (olamine and diolamine salts), 'Aretit', 'Ivosit'. **Details** *PM9*, entry 5190.

986 dinoseb acetate (common name). **CAS RN** *[2813–95–8]* Herbicidal activity
reported by H. Härtel (*Meded. Landbouwhogesch. Opzoekinsstn. Staat Gent,* 1960, **25,**
1422). Introduced by Hoechst AG (later AgrEvo GmbH). **Common names** dinoseb
acetate, dinosèb-acétate, DNBPA **IUPAC name** 2-sec-butyl-4,6-dinitrophenyl acetate
CA name 2-(1-methylpropyl)-4,6-dinitrophenyl acetate. **M.f.** $C_{12}H_{14}N_2O_6$ **Code nos.**
Hoe 02904 **Selected tradenames** 'Phenotan', 'Ivocit'. **Details** *PM9*, entry 5190.

987 dinosulfon (common name). **CAS RN** *[5386–77–6]* unstated stereochemistry.
Acaricide and fungicide reported by M. Pianka & J. D. Edwards (*J. Sci. Food Agric.,* 1968,
19, 60) and M. Pianka & P. J. J. Sweet (*ibid.,* pp. 672, 676). Evaluated by Murphy Chemical
Co. (became DowElanco). **IUPAC name** S-methyl O-2-(1-methylheptyl)-4,6-dinitro=
phenyl thiocarbonate **CA name** S-methyl O-[2-(1-methylheptyl)-4,6-dinitrophenyl]
carbonothioate. **M.f.** $C_{16}H_{22}N_2O_6S$.

988 dinoterbon (common name). **CAS RN** *[6073–72–9]*. Acaricide and fungicide
reported at 5th Intern. Pestic. Congr., London, 1963 (M. Pianka, *J. Sci. Food Agric.,* 1966,
17, 47). Evaluated by Murphy Chemical Co. (became DowElanco). **IUPAC name**
2-tert-butyl-4,6-dinitrophenyl ethyl carbonate **CA name** 2-(1,1-dimethylethyl)-=
4,6-dinitrophenyl ethyl carbonate. **M.f.** $C_{13}H_{16}N_2O_7$ **Code nos.** OMS 1057.

989 dioxabenzofos (common name). **CAS RN** *[3811–49–2]* unstated
stereochemistry. Insecticide reported by M. Eto & Y. Oshima (*Agric. Biol. Chem.,* 1962,
26, 452). Introduced by Sumitomo Chemical Co., Ltd. **Common names** dioxabenzofos,
salithion **IUPAC name** (*RS*)-2-methoxy-4*H*-1,3,2λ^5-benzodioxaphosphinine 2-sulfide;
(*RS*)-2-methoxy-4*H*-1,3,2λ^5-benzodioxaphosphorine 2-sulfide **CA name**
($\pm$)-2-methoxy-4*H*-1,3,2-benzodioxaphosphorin 2-sulfide. **M.f.** $C_8H_9O_3PS$
Selected tradenames 'Salithion'. **Details** *PM10*, entry 245.

990 dioxacarb (common name). **CAS RN** *[6988–21–2]*. Insecticide reported by F.
Bachmann & J. B. Legge (*J. Sci. Food Agric., Suppl.,* 1968, p. 39). Introduced by Ciba AG
(later Ciba-Geigy AG). **Common names** dioxacarb, dioxacarbe **IUPAC name**
2-(1,3-dioxolan-2-yl)phenyl methylcarbamate **CA name** 2-(1,3-dioxolan-2-yl)phenyl
methylcarbamate. **M.f.** $C_{11}H_{13}NO_4$ **Code nos.** C 8353; OMS 1102; ENT 27 389.
Selected tradenames 'Elocron', 'Famid'. **Details** *PM9*, entry 5250.

991 dioxathion (common name). **CAS RN** *[78–34–2]*. Insecticide reported by W. R.
Diveley *et al.* (*J. Am. Chem. Soc.,* 1959, **81,** 139). The cis- is more toxic than the trans-
isomer to flies and rats (B. W. Arthur & J. E. Casida, *J. Econ. Entomol.,* 1959, **52,** 20).
Introduced by Hercules Inc. (later by Nor-Am Chemical Co.). **Common names**
dioxathion, [delnav], dioxane phosphate **IUPAC name** S,S'-(1,4-dioxane-2,3-diyl)
O,O,O',O'-tetraethyl bis(phosphorodithioate) **CA name** S,S'-1,4-dioxane-2,3-diyl

bis(O,O-diethyl phosphorodithioate).   **M.f.** $C_{12}H_{26}O_6P_2S_4$ **Code nos.** Hercules AC258, ENT 22 879.   **Selected tradenames** 'Delnav', 'Deltic'.   **Details** *PM8,* entry 05260.

992   diphenyl sulfone (IUPAC name).   **CAS RN** *[127–63–9].* Acaricide reported by J. K. Eaton & R. G. Davies (*Ann. Appl. Biol.,* 1950, **37**, 471).   **Common names** diphenyl sulfone, diphénylsulfone   **CA name** 1,1'-sulfonylbis[benzene].   **M.f.** $C_{12}H_{10}O_2S$ **Other names** DPS.   **Details** *PM4,* p. 220.

993   dipropetryn (common name).   **CAS RN** *[4147–51–7].* Herbicide reported by G. A. Buchanan & D. L. Thurlow (*Ala. Agric. Exp. Stn. Highlights,* 1968, **15**(2), 7). Introduced by J. R. Geigy S.A. (later Ciba-Geigy AG).   **Common names** dipropetryn, dipropétryne   **IUPAC name** 6-ethylthio-N^2,N^4-di-isopropyl-1,3,5-triazine-2,4-diamine   **CA name** 6-(ethylthio)-*N,N'*-bis(1-methylethyl)-1,3,5-triazine-2,4-diamine.   **M.f.** $C_{11}H_{21}N_5S$ **Code nos.** GS 16 068   **Selected tradenames** 'Cotofor', 'Sancap'. **Details** *PM9,* entry 5310.

994   dipyrithione (common name).   **CAS RN** *[3696–28–4].* Fungicide and bactericide discovered by Olin Chemicals and developed by Yashima Chemical Industry and Sankyo Co., Ltd.   **IUPAC name** di-2-pyridyl disulfide 1,1'-dioxide   **CA name** 2,2'-dithiobis= [pyridine] 1,1'-dioxide.   **M.f.** $C_{10}H_8N_2O_2S_2$ **Code nos.** OSY-20   **Selected tradenames** 'Omadine disulfide', 'Omadine-DS'.   **Details** *PM9,* entry 5325.

995   disul; disul-sodium (common name).   **CAS RN** *[149–26–8]* disul; *[136–78–7]* disul-sodium. Herbicide reported by L. J. King *et al.* (*Contrib. Boyce Thompson Inst.,* 1950, **16**, 191). Introduced by Union Carbide Corp. (later Rhône-Poulenc Ag.). **Common names** disul, 2,4-DES, sesone   **IUPAC name** 2-(2,4-dichlorophenoxy)ethyl hydrogen sulfate (I)   **CA name** (I).   **M.f.** $C_8H_8Cl_2O_5S$; (sodium salt $C_8H_7Cl_2NaO_5S$) **Other names** SES   **Selected tradenames** 'Crag Herbicide I', 'Crag Sesone', 'Herbon' (DES-sodium).   **Details** *PM8,* entry 03950.

996   ditalimfos (common name).   **CAS RN** *[5131–24–8].* Fungicide reported by H. Tolkmith (*Nature (London),* 1966, **211**, 522). Introduced by Dow Chemical Co. (later DowElanco).   **Common names** ditalimfos, ditalimphos   **IUPAC name** O,O-diethyl phthalimidophosphonothioate; *N*-diethoxyphosphinothioylphthalimide   **CA name** O,O-diethyl (1,3-dihydro-1,3-dioxo-2*H*-isoindol-2-yl)phosphonothioate.   **M.f.** $C_{12}H_{14}NO_4PS$ **Code nos.** Dowco 199   **Selected tradenames** 'Plondrel'.   **Details** *PM6,* p. 222.

997   dithicrofos (common name).   **CAS RN** *[41219–31–2]* unstated stereochemistry. Insecticide evaluated by Hoechst AG.   **IUPAC name** S-(6-chloro-3,4-dihydro-= 2*H*-1-benzothi-in-4-yl) O,O-diethyl phosphorodithioate   **CA name** S-(6-chloro-= 3,4-dihydro-2*H*-1-benzothiopyran-4-yl) O,O-diethyl phosphorodithioate. **M.f.** $C_{13}H_{18}ClO_2PS_3$ **Code nos.** Hoe 19 510.

998   2-(1,3-dithiolan-2-yl)phenyl dimethylcarbamate (IUPAC/Chemical Abstracts name).   **CAS RN** *[21709–44–4].* Insecticide reported by F. Bachmann & J. B. Legge

(*J. Sci. Food Agric.*, Suppl., 1968, p. 39). Evaluated by Ciba AG (later Ciba-Geigy AG). **M.f.** $C_{12}H_{15}NO_2S_2$ **Code nos.** C 13 963. **Details** *PM2*, p. 209.

999   DKA-24 (development code). **CAS RN** *[97454–00–7]*. Herbicide safener reported by J. Nagy & K. Balogh (*Proc. Br. Crop Prot. Conf. - Weeds*, 1985, **1**, 107). Introduced by Eszakmagyarorszagi Vegyimuvek Chemical Works North Hungary (later SagroChem Co. Ltd). **IUPAC name** N^1,N^2-diallyl-N^2-dichloroacetylglycinamide **CA name** 2,2-dichloro-N-[2-oxo-2-(2-propenylamino)ethyl]-N-2-propenylacetamide. **M.f.** $C_{10}H_{14}Cl_2N_2O_2$ **Code nos.** DKA-24 (Eszakmagyarorszagi Vegyimuvek Chemical Works North Hungary). **Details** *PM10*, entry 255.

1000   DMPA (common name). **CAS RN** *[299–85–4]*. Herbicide introduced by Dow Chemical Co. (later DowElanco). **IUPAC name** O-2,4-dichlorophenyl O-methyl isopropylphosphoramidothioate **CA name** O-(2,4-dichlorophenyl) O-methyl (1-methylethyl)phosphoramidothioate. **M.f.** $C_{10}H_{14}Cl_2NO_2PS$ **Code nos.** K 22 023; Dowco 118; OMS 115; ENT 25 647. **Selected tradenames** 'Zytron'. **Details** *PM2*, p. 211.

1001   dodicin (common name). **CAS RN** *[6843–97–6]*. Fungicide and surface active agent. **IUPAC name** N-[2-(2-dodecylaminoethylamino)ethyl]glycine; 3,6,9-triaza= henicosanoic acid **CA name** N-[2-[[2-(dodecylamino)ethyl]amino]ethyl]glycine. **M.f.** $C_{18}H_{39}N_3O_2$.

1002   dofenapyn (common name). **CAS RN** *[42873–80–3]*. Acaricide discovered by Ciba-Geigy AG. **Common names** dofenapyn, dofénapyne **IUPAC name** 4-(pent-= 4-ynyloxy)phenyl phenyl ether **CA name** 1-(4-pentynyloxy)-4-phenoxybenzene. **M.f.** $C_{17}H_{16}O_2$ **Code nos.** CGA 29 170.

1003   drazoxolon (common name). **CAS RN** *[5707–69–7]*. Fungicide reported by M. J. Geoghegan (*Proc. Br. Insectic. Fungic. Conf., 4th*, 1967, **1**, 451). Introduced by ICI Plant Protection Division (later Zeneca Agrochemicals). **IUPAC name** 4-(2-chloro= phenylhydrazono)-3-methyl-1,2-oxazol-5(4H)-one; 4-(2-chlorophenylhydrazono)-= 3-methylisoxazol-5(4H)-one **CA name** 3-methyl-4,5-isoxazoledione 4-[(2-chloro= phenyl)hydrazone]. **M.f.** $C_{10}H_8ClN_3O_2$ **Code nos.** PP781 **Selected tradenames** 'Mil-Col'. **Details** *PM9*, entry 5500.

1004   DSP (common name). **CAS RN** *[3078–97–5]*. Insecticide introduced by Nippon Soda Co., Ltd. **IUPAC name** O-4-dimethylsulfamoylphenyl O,O-diethyl phosphorothioate; 4-diethoxyphosphinothioyloxy-N,N-dimethylbenzenesulfonamide **CA name** O-[4-[(dimethylamino)sulfonyl]phenyl] O,O-diethyl phosphorothioate. **M.f.** $C_{12}H_{20}NO_5PS_2$ **Code nos.** NK 0795 **Selected tradenames** 'Kaya-ace'.

1005   EBP (common name). **CAS RN** *[13286–32–3]*. Fungicide introduced by Kumiai Chemical Industry Co., Ltd. **IUPAC name** S-benzyl O,O-diethyl phosphorothioate **CA name** O,O-diethyl S-(phenylmethyl) phosphorothioate. **M.f.** $C_{11}H_{17}O_3PS$ **Selected tradenames** 'Kitazin'.

1006 eglinazine-ethyl; eglinazine (common name). **CAS RN** *[6616–80–4]*; (acid *[68228–19–3]*). The ethyl ester introduced as a herbicide by Nitrokémia Ipartelepek. **Common names** (for acid) eglinazine **IUPAC name** (for acid) N-(4-chloro-6-ethyl= amino-1,3,5-triazin-2-yl)glycine **CA name** (for acid) N-[4-chloro-6-(ethylamino)-= 1,3,5-triazin-2-yl]glycine. **M.f.** $C_9H_{14}ClN_5O_2$; (acid $C_7H_{10}ClN_5O_2$) **Selected tradenames** 'MG-06'. **Details** *PM9*, entry 5520.

1007 EI 1642 (development code). **CAS RN** *[16960–39–7]*. Insecticide evaluated by E. I. du Pont de Nemours and Co. **IUPAC name** S-methyl N-(carbamoyloxy)= thioacetimidate **CA name** methyl N-[(aminocarbonyl)oxy]ethanimidothioate. **M.f.** $C_4H_8N_2O_2S$ **Code nos.** EI 1642, ENT 27 411. **Details** *PM3*, p. 341.

1008 EL 177 (development code). **CAS RN** *[98477–07–7]*. Herbicide reported by H. E. Chamberlain *et al.* (*Proc. 1987 Br. Crop Prot. Conf. - Weeds*, **1**, 35). Introduced by Eli Lilly. **IUPAC name** 1-*tert*-butyl-5-cyano-N-methylpyrazole-4-carboxamide **CA name** 5-cyano-1-(1,1-dimethylethyl)-N-methyl-1H-pyrazole-4-carboxamide. **M.f.** $C_{10}H_{14}N_4O$ **Code nos.** EL-177; LYI 81 977. **Details** *PM9*, entry 1680.

1009 EMPC (common name). **CAS RN** *[18809–57–9]*. Insecticide introduced by Nippon Kayaku Co., Ltd. **IUPAC name** 4-ethylthiophenyl methylcarbamate **CA name** 4-(ethylthio)phenyl methylcarbamate. **M.f.** $C_{10}H_{13}NO_2S$ **Code nos.** NK-1 **Selected tradenames** 'Toxamate'.

1010 endothion (common name). **CAS RN** *[2778–04–3]*. Insecticide reported by F. Chaboussou & P. Ramadier (*Rev. Zool. Agric. Appl.*, 1957, **55**, 116). Introduced by Rhône-Poulenc Phytosanitaire and later by American Cyanamid Co. and FMC Corp. **IUPAC name** S-5-methoxy-4-oxo-4H-pyran-2-ylmethyl O,O-dimethyl phosphorothioate; 2-dimethoxyphosphinoylthiomethyl-5-methoxypyran-4-one **CA name** S-[(5-methoxy-4-oxo-4H-pyran-2-yl)methyl] O,O-dimethyl phosphorothioate. **M.f.** $C_9H_{13}O_6PS$ **Code nos.** 7175 RP (Rhône-Poulenc); AC 18 737 (American Cyanamid Co.); FMC 5767 (FMC Corp.); ENT 24 653. **Selected tradenames** 'Endocide'. **Details** *PM5*, p. 234.

1011 endrin (common name). **CAS RN** *[72–20–8]*. Developed by Shell International Chemical Co. after introduction by J. Hyman & Co. **Common names** endrin, endrine, nendrin **IUPAC name** (1R,4S,4aS,5S,6S,7R,8R,8aR)-1,2,3,4,10,10-hexachloro-= 1,4,4a,5,6,7,8,8a-octahydro-6,7-epoxy-1,4:5,8-dimethanonaphthalene; 1,2,3,4,10,10-hexachloro-6,7-epoxy-1,4,4a,5,6,7,8,8a-octahydro-*exo*-1,4-*exo*-= 5,8-dimethanonaphthalene **CA name** (1α,2β,2aβ,3α,6α,6aβ,7β,7aα)-= 3,4,5,6,9,9-hexachloro-1a,2,2a,3,6,6a,7,7a-octahydro-2,7:3,6-dimethanonaphth= [2,3-*b*]oxirene. **M.f.** $C_{12}H_8Cl_6O$ **Code nos.** Experimental Insecticide 269 (Hyman & Co.); OMS 197; ENT 17 251. **Details** *PM7*, entry 05560.

1012 ENT 17596 (ESA code). **CAS RN** *[126–15–8]*. Repellent action on cockroaches reported by L. D. Goodhue & C. Linnaid (*J. Econ. Entomol.*, 1952, **45**, 133). Introduced by Phillips Petroleum Co. and by McLaughlin Gormley King Co. **IUPAC name** 1,4,4a,5a,6,9,9a,9b-octahydrodibenzofuran-4a-carbaldehyde **CA name**

1,5a,6,9,9a,9b-hexahydro-4a(4*H*)-dibenzofurancarboxaldehyde. **M.f.** $C_{13}H_{16}O_2$
Other names butadiene-furfural copolymer; 2,3:4,5-bis(2-butylene)tetrahydrofurfural
Code nos. Phillips Repellent 11, ENT 17 596. **Selected tradenames** 'MGK Repellent
11'. **Details** *PM9*, entry 9100.

1013   ENT 92 (ESA code).   **CAS RN** *[115–31–1]*. Insecticide reported by R. L.
Pierpont (*Bull. Del. Univ. Agric. Exp. Sta.*, 1945, No. 253). Introduced by Hercules Inc.
(later Nor-Am.).   **IUPAC name** (1*R*,2*R*,4*R*)-born-2-yl thiocyanatoacetate   **CA name**
exo-1,7,7-trimethylbicyclo[2.2.1]hept-2-yl thiocyanatoacetate. **M.f.** $C_{13}H_{19}NO_2S$
Other names terpinyl thiocyanoacetate   **Code nos.** ENT 92.   **Selected tradenames**
'Thanite'.   **Details** *PM5*, p. 309.

1014   EPBP (common name).   **CAS RN** *[3792–59–4]*. Insecticide introduced by Nissan
Chemical Industries, Ltd.   **IUPAC name** *O*-2,4-dichlorophenyl *O*-ethyl phenyl=
phosphonothioate   **CA name** *O*-(2,4-dichlorophenyl) *O*-ethyl phenylphosphonothioate.
M.f. $C_{14}H_{13}Cl_2O_2PS$ **Code nos.** S-7   **Selected tradenames** 'S-Seven'.   **Details** *PM8*,
entry 04330.

1015   epofenonane (common name).   **CAS RN** *[57342–02–6]*. Insect growth
regulator evaluated by Hoffman-La Roche & Co.   **IUPAC name** 6,7-epoxy-3-ethyl-=
7-methylnonyl 4-ethylphenyl ether   **CA name** 2-ethyl-3-[3-ethyl-5-(4-ethylphenoxy)=
pentyl]-2-methyloxirane.   **M.f.** $C_{20}H_{32}O_2$ **Code nos.** Ro 10-3108.

1016   epronaz (common name).   **CAS RN** *[59026–08–3]*. Herbicide reported by L. G.
Copping & R. F. Brookes (*Proc. Br. Weed Control Conf., 12th*, 1974, **2**, p. 809). Evaluated
by Boots Co. Ltd (later Schering Agriculture).   **IUPAC name** *N*-ethyl-*N*-propyl-=
3-propylsulfonyl-1*H*-1,2,4-triazole-1-carboxamide; 1-(*N*-ethyl-*N*-propylcarbamoyl)-=
3-propylsulfonyl-1*H*-1,2,4-triazole   **CA name** *N*-ethyl-*N*-propyl-3-(propylsulfonyl)-=
1*H*-1,2,4-triazole-1-carboxamide.   **M.f.** $C_{11}H_{20}N_4O_3S$ **Code nos.** BTS 30 843.

1017   erbon (common name).   **CAS RN** *[136–25–4]*. Herbicide introduced by Dow
Chemical Co. (later DowElanco).   **IUPAC name** 2-(2,4,5-trichlorophenoxy)ethyl
2,2-dichloropropionate   **CA name** 2-(2,4,5-trichlorophenoxy)ethyl 2,2-dichloro=
propanoate.   **M.f.** $C_{11}H_9Cl_5O_3$ **Selected tradenames** 'Daron'.   **Details** *PM5*, p. 239.

1018   ESBP (common name).   **CAS RN** *[21722–85–0]*. Fungicide introduced by Nissan
Chemical Industries, Ltd.   **IUPAC name** *S*-benzyl *O*-ethyl phenylphosphonothioate
CA name *O*-ethyl *S*-(phenylmethyl) phenylphosphonothioate.   **M.f.** $C_{15}H_{17}O_2PS$
Selected tradenames 'Inezin'.

1019   etacelasil (common name).   **CAS RN** *[37894–46–5]*. Abscission agent reported
by J. Rufener & D. Pietà (*Riv. Ortoflorofruttic. Ital.*, 1974, **4**, 274). Introduced by Ciba-
Geigy AG.   **IUPAC name** 2-chloroethyltris(2-methoxyethoxy)silane   **CA name**
6-(2-chloroethyl)-6-(2-methoxyethoxy)-2,5,7,10-tetraoxa-6-silaundecane.
M.f. $C_{11}H_{25}ClO_6Si$ **Code nos.** CGA 13 586   **Selected tradenames** 'Alsol'.
Details *PM9*, entry 5640.

1020   etaconazole (common name).   **CAS RN** *[60207–93–4]* formerly *[71245–23–3]* unstated stereochemistry. Fungicide reported by T. Staub *et al.* (*Abstr. Int. Congr. Plant Protect.*, 1979, 310). Introduced by Ciba-Geigy AG as an agricultural fungicide, having been invented by Janssen Pharmaceutica.   **IUPAC name** ($\pm$)-1-[2-(2,4-dichloro= phenyl)-4-ethyl-1,3-dioxolan-2-ylmethyl]-1*H*-1,2,4-triazole   **CA name** 1-[[2-(2,4-dichlorophenyl)-4-ethyl-1,3-dioxolan-2-yl]methyl]-1*H*-1,2,4-triazole.   **M.f.** $C_{14}H_{15}Cl_2N_3O_2$ **Code nos.** CGA 64 251   **Selected tradenames** 'Sonax', 'Vangard'. **Details** *PM8*, entry 05650.

1021   etem (common name).   **CAS RN** *[33813–20–6]* (i); *[5782–83–3]* (ii). Fungicide originally believed to be (ii), later corrected to (i). Evaluated by Universal Crop Protection.   **Common names** etem for (i), ETM for (ii)   **IUPAC name** (i) 5,6-dihydro-(3*H*)-imidazo[2,1-*c*]-1,2,4-dithiazole-3-thione; (ii) 1,3,6-thiadiazepane-= 2,7-dithione; ethylenethiuram monosulfide   **CA name** (i) 5,6-dihydro-3*H*-imidazo= [2,1-*c*]-1,2,4-dithiazole-3-thione; (ii) hexahydro-1,3,6-thiadiazepine-2,7-dithione.   **M.f.** $C_4H_4N_2S_3$ (i), $C_4H_6N_2S_3$ (ii)   **Code nos.** UCP/21   **Selected tradenames** 'Vegita'. **Details** *PM6*, p. 239.

1022   ethidimuron (common name).   **CAS RN** *[30043–49–3]*. Herbicide reported by L. Eue *et al.* (*C. R. Journ. Etud. Herbic. Conf. COLUMA, 4th*, 1973, **1**, 14). Introduced by Bayer AG.   **IUPAC name** 1-(5-ethylsulfonyl-1,3,4-thiadiazol-2-yl)-1,3-dimethylurea   **CA name** *N*-[5-(ethylsulfonyl)-1,3,4-thiadiazol-2-yl]-*N*,*N*'-dimethylurea.   **M.f.** $C_7H_{12}N_4O_3S_2$ **Code nos.** MET 1486; BAY 107033   **Selected tradenames** 'Ustilan'.   **Details** *PM9*, entry 5690.

1023   ethiolate (common name).   **CAS RN** *[2941–55–1]*. Herbicide introduced by Gulf Oil Chemicals Co.   **IUPAC name** *S*-ethyl diethylthiocarbamate   **CA name** *S*-ethyl diethylcarbamothioate.   **M.f.** $C_7H_{15}NOS$ **Selected tradenames** 'Prefox'.   **Details** *PM4*, p. 246.

1024   ethoate-methyl (common name).   **CAS RN** *[116–01–8]*. Insecticide and acaricide reported by G. Lemetre *et al.* (*Not. Mal. Piante*, 1963, No. 64). Introduced by Bombrini Parodi-Delfino (later Snia Viscosa).   **Common names** ethoate-methyl, éthoate-méthyle   **IUPAC name** *S*-ethylcarbamoylmethyl *O*,*O*-dimethyl phosphorodithioate; 2-dimethoxyphosphinothioylthio-*N*-ethylacetamide   **CA name** *S*-[2-(ethylamino)-2-oxoethyl] *O*,*O*-dimethyl phosphorodithioate.   **M.f.** $C_6H_{14}NO_3PS_2$ **Code nos.** B/77; OMS 252; ENT 25 506.   **Selected tradenames** 'Fitios'.   **Details** *PM5*, p. 245.

1025   ethoxyquin (common name).   **CAS RN** *[91–53–2]*. Originally prepared by Knoevenagel (*Ber.*, 1921, **54**, 1722). Fungicide introduced by Monsanto Agriculture Co. **Common names** ethoxyquin, éthoxyquine, [polietoksichinolin]   **IUPAC name** 1,2-dihydro-2,2,4-trimethylquinolin-6-yl ether   **CA name** 6-ethoxy-1,2-dihydro-= 2,2,4-trimethylquinoline.   **M.f.** $C_{14}H_{19}NO$ **Selected tradenames** 'Deccoquin'.   **Details** *PM10*, entry 279.

1026   ethylene bis(trichloroacetate) (IUPAC name). **CAS RN** *[2514–53–6]*. Herbicide introduced by Hooker Chemical Corp. (and later by Nor-Am). **CA name** 1,2-ethanediyl bis(trichloroacetate). **M.f.** $C_6H_4Cl_6O_4$ **Other names** ethylene glycol bis(trichloroacetate) **Selected tradenames** 'Glytac'. **Details** *PM5*, p. 252.

1027   ethyl hexanediol (common name). **CAS RN** *[94–96–2]*. Insect repellent reported by P. Granett & H. L. Haynes (*J. Econ. Entomol.*, 1945, **38**, 671), by W. V. King (*ibid.*, 1951, **44**, 339) and by B. V. Travis & C. N. Smith (*ibid.*, p. 428). Introduced by Rutgers Co. and mainly used as a mixture with butopyronoxyl and dimethyl phthalate. **Common names** ethyl hexanediol, ethohexadiol **IUPAC name** 2-ethylhexane-1,3-diol **CA name** 2-ethyl-1,3-hexanediol. **M.f.** $C_8H_{18}O_2$ **Other names** ethyl hexylene glycol **Code nos.** Rutgers 6-12, ENT 375. **Details** *PM6*, p. 247.

1028   N-ethylmercurio-4-toluenesulfonanilide (IUPAC name). **CAS RN** *[517–16–8]*. Fungicide introduced by E. I. du Pont de Nemours and Co. **IUPAC name** N-ethylmercurio-= 4-toluenesulfonanilide; N-(ethylmercury)-p-toluenesulfonanilide **CA name** ethyl(4-methyl-N-phenylbenzenesulfonamidato-N)mercury. **M.f.** $C_{15}H_{17}HgNO_2S$ **Selected tradenames** 'Ceresan M', 'Granosan M'. **Details** *PM3*, p. 252.

1029   N-(2-ethyl-2H-pyrazol-3-yl)-N'-phenylurea (IUPAC name). **CAS RN** *[4058–90–6]* . Plant growth regulator introduced by Fahlberg-List to stimulate flower formation in glasshouse tomatoes. **CA name** N-(1-ethyl-1H-pyrazol-5-yl)-= N'-phenylurea. **M.f.** $C_{12}H_{14}N_4O$ **Other names** azoluron.

1030   etinofen (common name). **CAS RN** *[2544–94–7]*. Herbicide introduced by C. H. Boehringer Sohn (later Shell Agrar). **Common names** etinofen, étinofène **IUPAC name** α-ethoxy-4,6-dinitro-o-cresol **CA name** 2-(ethoxymethyl)-4,6-dinitrophenol. **M.f.** $C_9H_{10}N_2O_6$ **Selected tradenames** 'Dinethon'.

1031   etnipromid (common name). **CAS RN** *[76120–02–0]* unstated stereochemistry. Herbicide evaluated by Hoechst AG. **IUPAC name** (RS)-2-[5-(2,4-dichlorophenoxy)-2-nitrophenoxy]-N-ethylpropionamide **CA name** (±)-2-[5-(2,4-dichlorophenoxy)-2-nitrophenoxy]-N-ethylpropanamide. **M.f.** $C_{17}H_{16}Cl_2N_2O_5$ **Code nos.** Hoe 39 106.

1032   EXD (common name). **CAS RN** *[502–55–6]*. Herbicide reported by E. K. Alban & L. McCombs (*Proc. North Cent. Weed Control Conf.*, 1949, p. 81). Introduced by Roberts Chemicals Inc. and Monsanto Chemical Co. **IUPAC name** O,O-diethyl dithiobis(thioformate) **CA name** diethyl thioperoxydicarbonate; thioperoxy= dicarbonic acid ([(HO)C(S)]$_2$S$_2$) diethyl ester. **M.f.** $C_6H_{10}O_2S_4$ **Selected tradenames** 'Herbisan', 'Sulfasan'. **Details** *PM5*, p. 260.

1033   fenaminosulf (common name). **CAS RN** *[140–56–7]*. Fungicide reviewed by E. Urbschat (*Angew. Chem.*, 1960, **72**, 981). Introduced by Bayer AG. **Common names** fenaminosulf, phénaminosulf, DAPA **IUPAC name** sodium 4-dimethylamino= benzenediazosulfonate **CA name** sodium [4-(dimethylamino)phenyl]diazene=

sulfonate.   **M.f.** $C_8H_{10}N_3NaO_3S$ **Code nos.** Bayer 22 555; Bayer 5072   **Selected tradenames** 'Lesan', 'Dexon'.   **Details** *PM8*, entry 06000.

1034   fenapanil (common name).   **CAS RN** *[61019–78–1]*. Fungicide evaluated by Rohm & Haas Co.   **IUPAC name** ($\pm$)-2-(imidazol-1-ylmethyl)-2-phenylhexanenitrile; ($\pm$)-α-butyl-α-phenylimidazole-1-propanenitrile   **CA name** α-butyl-α-phenyl-= 1*H*-imidazole-1-propanenitrile.   **M.f.** $C_{16}H_{19}N_3$ **Code nos.** RH-2161   **Selected tradenames** 'Sisthane'.

1035   fenasulam (common name).   **CAS RN** *[78357–48–9]*. Herbicide evaluated by Shionogi and Co. Ltd.   **IUPAC name** methyl 4-[2-(4-chloro-o-tolyloxy)acetamido]= phenylsulfonylcarbamate; methyl (*N*-4-chloro-o-tolyloxyacetylsulfanilyl)carbamate   **CA name** methyl [[4-[[(4-chloro-2-methylphenoxy)acetyl]amino]phenyl]sulfonyl]= carbamate.   **M.f.** $C_{17}H_{17}ClN_2O_6S$ **Code nos.** SSH-60.

1036   fenazaflor (common name).   **CAS RN** *[14255–88–0]*. Acaricide reported by D. T. Saggers & M. L. Clark (*Nature (London)*, 1967, **215**, 275). Introduced by Fisons Pest Control Ltd (later Schering Agriculture).   **IUPAC name** phenyl 5,6-dichloro-= 2-trifluoromethylbenzimidazole-1-carboxylate   **CA name** phenyl 5,6-dichloro-= 2-(trifluoromethyl)-1*H*-benzimidazole-1-carboxylate.   **M.f.** $C_{15}H_7Cl_2F_3N_2O_2$ **Code nos.** NC 5016, OMS 1243.   **Selected tradenames** 'Lovozal'.   **Details** *PM4*, p. 267.

1037   fenchlorazole-ethyl; fenchlorazole (common name).   **CAS RN** *[103112–35–2]*; (acid *[103112–36–3]*). Herbicide safener reported by H. Bieringer *et al.* (*Proc. 1989 Br. Crop Prot. Conf. - Weeds*, **1**, 77). Fenclorazole-ethyl introduced by Hoechst AG (later AgrEvo GmbH).   **Common names** (for acid) fenchlorazole   **IUPAC name** ethyl 1-(2,4-dichlorophenyl)-5-trichloromethyl-1*H*-1,2,4-triazole-3-carboxylate   **CA name** ethyl 1-(2,4-dichlorophenyl)-5-(trichloromethyl)-1*H*-1,2,4-triazole-3-carboxylate.   **M.f.** $C_{12}H_8Cl_5N_3O_2$; (acid $C_{10}H_4Cl_5N_3O_2$)   **Code nos.** Hoe 070542; (acid Hoe 072829). **Details** *PM10*, entry 293.

1038   fenchlorphos (common name).   **CAS RN** *[299–84–3]*. Systemic insecticide introduced by Dow Chemical Co.   **Common names** fenchlorphos, ronnel   **IUPAC name** *O,O*-dimethyl *O*-2,4,5-trichlorophenyl phosphorothioate   **CA name** *O,O*-dimethyl *O*-(2,4,5-trichlorophenyl) phosphorothioate.   **M.f.** $C_8H_8Cl_3O_3PS$ **Code nos.** Dow ET-14; Dow ET-57; OMS 123; ENT 23 284.   **Selected tradenames** 'Nankor', 'Trolene', 'Korlan'.   **Details** *PM6*, p. 260.

1039   fenethacarb (common name).   **CAS RN** *[30087–47–9]*. Insecticide introduced by BASF AG.   **Common names** fenethacarb, phénétacarbe   **IUPAC name** 3,5-diethylphenyl methylcarbamate (I)   **CA name** (I).   **M.f.** $C_{12}H_{17}NO_2$ **Code nos.** BAS 235I.

1040   fenfluthrin (common name).   **CAS RN** *[75867–00–4]*. Insecticide reported by W. Behrenz & K. Naumann (*Pflanzenschutz-Nachr. (Engl. Ed.)*, 1982, **35**, 309). Evaluated by Bayer AG.   **Common names** fenfluthrin, fenfluthrine   **IUPAC name** 2,3,4,5,6-pentafluorobenzyl (1*R*)-*trans*-3-(2,2-dichlorovinyl)-2,2-dimethylcyclo=

propanecarboxylate; 2,3,4,5,6-pentafluorobenzyl (1R,3S)-3-(2,2-dichlorovinyl)-=
2,2-dimethylcyclopropanecarboxylate   CA name (1R-*trans*)-(pentafluorophenyl)methyl
3-(2,2-dichloroethenyl)-2,2-dimethylcyclopropanecarboxylate.   M.f. $C_{15}H_{11}Cl_2F_5O_2$
Code nos. NAK 1654, OMS 2013.   Selected tradenames 'Baynac'.

1041 fenitropan (common name).   CAS RN *[65934–94–3]* stated stereochemistry;
[65934–95–4] unstated stereochemistry. Fungicide reported by A. Kis-Tamás et al.
(*Proc. Br. Crop Prot. Conf. - Pest Dis.*, 1981, **1**, 29). Introduced by EGYT
Pharmacochemical Works.   Common names fenitropan, fénitropane   IUPAC name
(1RS,2RS)-2-nitro-1-phenyltrimethylene di(acetate)   CA name (1R*,2R*)-2-nitro-=
1-phenyl-1,3-propanediyl diacetate.   M.f. $C_{13}H_{15}NO_6$ Code nos. EGYT 2248
Selected tradenames 'Volparox'.   Details *PM9*, entry 6110.

1042 fenoprop; fenoprop-butotyl (common name).   CAS RN *[93–72–1]* unstated
stereochemistry. The plant-growth regulating properties of its salts described by M. E.
Synerholm & P. W. Zimmerman (*Contrib. Boyce Thompson Inst.*, 1945, **14**, 91). Its esters
introduced as herbicides by Dow Chemical Co. (later DowElanco) and its salts as
plant growth regulators by Amchem Products Inc. (later Rhône-Poulenc Agrochimie).
Common names fenoprop, silvex, 2,4,5-TP   IUPAC name (±)-2-(2,4,5-trichloro=
phenoxy)propionic acid   CA name (±)-2-(2,4,5-trichlorophenoxy)propanoic acid.
M.f. $C_9H_7Cl_3O_3$; (butotyl ester $C_{15}H_{19}Cl_3O_4$)   Other names 2,4,5-TP   Selected
tradenames 'Kuron', 'Fruitone T'.   Details *PM8*, entry 06140.

1043 fenoxacrim (common name).   CAS RN *[65400–98–8]*. Insecticide discovered by
Ciba-Geigy AG.   IUPAC name 3',4'-dichloro-1,2,3,4-tetrahydro-6-hydroxy-=
1,3-dimethyl-2,4-dioxopyrimidine-5-carboxanilide   CA name N-(3,4-dichlorophenyl)=
hexahydro-1,3-dimethyl-2,4,6-trioxo-5-pyrimidinecarboxamide (preferred tautomer).
M.f. $C_{13}H_{11}Cl_2N_3O_4$ Other names HHP.

1044 fenoxaprop-ethyl; fenoxaprop (common name).   CAS RN *[82110–72–3]*
fenoxaprop-ethyl (racemate); *[66441–23–4]* fenoxaprop-ethyl (unstated
stereochemistry); *[95617–09–7]* fenoxaprop (racemate); *[73519–55–8]* fenoxaprop
(unstated stereochemistry). Herbicide reported by H. Bieringer et al. (*Proc. Br. Crop
Prot. Conf. - Weeds*, 1982, **1**, 11). The ethyl ester introduced by Hoechst AG (later
AgrEvo GmbH).   Common names (for acid) fenoxaprop   IUPAC name ethyl
(±)-2-[4-(6-chloro-1,3-benzoxazol-2-yloxy)phenoxy]propionate; ethyl
(±)-2-[4-(6-chlorobenzoxazol-2-yloxy)phenoxy]propionate   CA name ethyl
(±)-2-[4-[(6-chloro-2-benzoxazolyl)oxy]phenoxy]propanoate.
M.f. $C_{18}H_{16}ClNO_5$; (acid $C_{16}H_{12}ClNO_5$)   Code nos. Hoe 033171; (acid Hoe 053022)
Selected tradenames 'Whip'.   Details *PM10*, entry 299.

1045 fenpirithrin (common name).   CAS RN *[68523–18–2]* (unstated
stereochemistry). Insecticide evaluated by Dow Chemical Co.   Common names
fenpirithrin, fenpirithrine   IUPAC name (RS)-cyano(6-phenoxy-2-pyridyl)methyl
(1RS)-*cis-trans*-3-(2,2-dichlorovinyl)-2,2-dimethylcyclopropanecarboxylate;
(RS)-cyano(6-phenoxy-2-pyridyl)methyl (1RS,3RS;1RS,3SR)-3-(2,2-dichlorovinyl)-=
2,2-dimethylcyclopropanecarboxylate   CA name cyano(6-phenoxy-2-pyridinyl)=

methyl 3-(2,2-dichloroethenyl)-2,2-dimethylcyclopropanecarboxylate. **M.f.** $C_{21}H_{18}Cl_2N_2O_3$ **Code nos.** Dowco 417 **Selected tradenames** 'Vivithrin'.

1046 fenridazon (common name). **CAS RN** *[68254–10–4]*. Plant growth regulator evaluated by Rohm & Haas Co. **IUPAC name** 1-(4-chlorophenyl)-1,4-dihydro-= 6-methyl-4-oxopyridazine-3-carboxylic acid **CA name** 1-(4-chlorophenyl)-= 1,4-dihydro-6-methyl-4-oxo-3-pyridazinecarboxylic acid. **M.f.** $C_{12}H_9ClN_2O_3$ **Code nos.** RH-0007.

1047 fenson (common name). **CAS RN** *[80–38–6]*. Introduced by Murphy Chemical Ltd (later DowElanco). **Common names** fenson, fénizon **IUPAC name** 4-chlorophenyl benzenesulfonate (I) **CA name** (I) **M.f.** $C_{12}H_9ClO_3S$ **Other names** CPBS; PCPBS **Selected tradenames** 'Murvesco'. **Details** *PM8,* entry 06240.

1048 fensulfothion (common name). **CAS RN** *[115–90–2]*. Nematicide reported by B. Homeyer (*Mitt. Biol. Bundesanst. Land-Forstwirtsch. Berlin-Dahlem*, 1967, **121**, 50). Introduced by Bayer AG. **IUPAC name** *O,O*-diethyl *O*-4-methylsulfinylphenyl phosphorothioate **CA name** *O,O*-diethyl *O*-[4-(methylsulfinyl)phenyl] phosphorothioate. **M.f.** $C_{11}H_{17}O_4PS_2$ **Code nos.** Bayer 25 141; S 767; OMS 37; ENT 24 945. **Selected tradenames** 'Dasanit', 'Terracur P'. **Details** *PM9,* entry 6250.

1049 fenteracol (common name). **CAS RN** *[2122–77–2]*. Herbicide introduced by Budapest Chemical Works. **IUPAC name** 2-(2,4,5-trichlorophenoxy)ethanol (I) **CA name** (I). **M.f.** $C_8H_7Cl_3O_2$ **Other names** TCPE **Selected tradenames** 'Klorinol'.

1050 fenthiaprop; fenthiaprop-ethyl (common name). **CAS RN** *[95721–12–3]* fenthiaprop (racemate); *[73519–50–3]* fenthiaprop (unstated stereochemistry); *[93921–16–5]* fenthiaprop-ethyl (racemate); *[66441–11–0]* fenthiaprop-ethyl (unstated stereochemistry). Herbicide reported by R. Handte *et al.* (*Proc. Br. Crop Prot. Conf. - Weeds*, 1982, **1**, 19). Ethyl ester, fenthiaprop-ethyl, introduced by Hoechst AG. **Common names** fenthiaprop, fentiaprop **IUPAC name** ($\pm$)-2-[4-(6-chloro-1,3-benzo= thiazol-2-yloxy)phenoxy]propionic acid; ($\pm$)-2-[4-(6-chlorobenzothiazol-2-yloxy)= phenoxy]propionic acid **CA name** ($\pm$)-2-[4-[(6-chloro-2-benzothiazolyl)oxy]phenoxy]= propanoic acid. **M.f.** $C_{16}H_{12}ClNO_4S$; (ethyl ester $C_{18}H_{16}ClNO_4S$) **Code nos.** Hoe 35 609 **Selected tradenames** 'Joker' (ethyl ester). **Details** *PM8,* entry 06270.

1051 fentrifanil (common name). **CAS RN** *[62441-54-7]*. Acaricide reported by N. Morton *et al.* (*Proc. Br. Crop Prot. Conf. - Pests Dis.*, 1977, **2**, 349). Evaluated by ICI Plant Protection Div. (later ICI Agrochemicals). **Common names** fentrifanil, hexafluoramin **IUPAC name** *N*-(6-chloro-α,α,α-trifluoro-*m*-tolyl)-α,α,α-trifluoro-4,6-dinitro-= *o*-toluidine; 2'-chloro-2,4-dinitro-5',6-bis(trifluoromethyl)diphenylamine **CA name** *N*-[2-chloro-5-(trifluoromethyl)phenyl]-2,4-dinitro-6-(trifluoromethyl)benzenamine. **M.f.** $C_{14}H_6ClF_6N_3O_4$ **Code nos.** PP199.

1052 flamprop (common name). **CAS RN** *[58667–63–3]*. Herbicidal properties of flamprop-methyl reported by E. Haddock *et al.*, (*Proc. Br. Weed Control Conf.*, 12th, 1974, **1**, 9); compared with analogues by B. Jeffcoat *et al.* (*Pestic. Sci.*, 1977, **8**, 1).

Flamprop-isopropyl and flamprop-methyl introduced by Shell Research Ltd (later American Cyanamid Co.); later replaced by the single stereoisomers (flamprop-M-= isopropyl and flamprop-M-methyl). **IUPAC name** N-benzoyl-N-(3-chloro-4-fluoro= phenyl)-DL-alanine **CA name** N-benzoyl-N-(3-chloro-4-fluorophenyl)-DL-alanine. **M.f.** $C_{16}H_{13}ClFNO_3$. **Details** *PM10*, entry 315.

1053 flamprop-isopropyl (common name). **CAS RN** *[52756–22–6]*. See entry 1052. **M.f.** $C_{19}H_{19}ClFNO_3$ **Code nos.** WL 29 672. **Details** *PM9*, entry 6370.

1054 flamprop-methyl (common name). **CAS RN** *[52756–25–9]*. See entry 1052. **IUPAC name** methyl N-benzoyl-N-(3-chloro-4-fluorophenyl)-DL-alaninate **CA name** N-benzoyl-N-(3-chloro-4-fluorophenyl)-DL-alanine methyl ester. **M.f.** $C_{17}H_{15}ClFNO_3$ **Code nos.** WL 29671 (Shell) **Selected tradenames** 'Mataven'. **Details** *PM10*, entry 315.

1055 flubenzimine (common name). **CAS RN** *[37893–02–0]* unstated stereochemistry. Acaricide reported by G. Zoebelein *et al.* (*Mitt. Biol. Bundesanst. Lund. Forstwirtsch., Berlin Dahlem*, 1979, **191**, 283; *Pflanzenschutz-Nachr. (Engl. Ed.)*, 1980, **33**, 169). Introduced by Bayer AG. **IUPAC name** $(2Z,4E,5Z)$-N^2,3-diphenyl-N^4,N^5-bis= (trifluoromethyl)-1,3-thiazolidine-2,4,5-triylidenetriamine **CA name** N-[3-phenyl-= 4,5-bis[(trifluoromethyl)imino]-2-thiazolidinylidene]benzenamine. **M.f.** $C_{17}H_{10}F_6N_4S$ **Code nos.** BAY SLJ 0312 **Selected tradenames** 'Cropotex'. **Details** *PM8*, entry 06410.

1056 flucofuron (common name). **CAS RN** *[370–50–3]*. Insecticide developed by Ciba-Geigy AG. **IUPAC name** 1,3-bis(4-chloro-α,α,α-trifluoro-m-tolyl)urea **CA name** N,N'-bis[4-chloro-3-(trifluoromethyl)phenyl]urea. **M.f.** $C_{15}H_8Cl_2F_6N_2O$ **Selected tradenames** 'Mitin N'. **Details** *PM9*, entry 6430.

1057 fluenetil (common name). **CAS RN** *[4301–50–2]*. Insecticide reported by P. de Pietri-Tonelli *et al.* (*Proc. Br. Insectic. Fungic. Conf., 3rd*, 1965, p. 478). Introduced by Montecatini Edison S.p.A. (later Agrimont S.p.A.). **Common names** fluenetil, fluénéthyl **IUPAC name** 2-fluoroethyl biphenyl-4-ylacetate **CA name** 2-fluoroethyl [1,1'-biphenyl]-4-acetate. **M.f.** $C_{16}H_{15}FO_2$ **Code nos.** M 2060 **Selected tradenames** 'Lambrol'; 'Lambrol EC'. **Details** *PM2*, p. 254.

1058 flufenican (common name). Herbicide evaluated by May & Baker Ltd (later Rhône-Poulenc Agrochimie). **IUPAC name** 2-(α,α,α-trifluoro-m-tolyloxy)nicotinanilide **CA name** N-phenyl-2-[3-(trifluoromethyl)phenoxy]-3-pyridinecarboximide. **M.f.** $C_{19}H_{13}F_3N_2O_2$.

1059 flumezin (common name). **CAS RN** *[25475–73–4]*. Herbicide evaluated by BASF AG. **Common names** flumezin, flumézine **IUPAC name** 2-methyl-= 4-(α,α,α-trifluoro-m-tolyl)-1,2,4-oxadazinane-3,5-dione; 2-methyl-4-(α,α,α-trifluoro-= m-tolyl)-2H-1,2,4-oxadiazine-3,5(4H,6H)-dione **CA name** 2-methyl-4-[3-(trifluoro= methyl)phenyl]-2H-1,2,4-oxadiazine-3,5(4H,6H)-dione. **M.f.** $C_{11}H_9F_3N_2O_3$ **Code nos.** BAS 348H.

1060 flumipropyn (common name).   CAS RN *[84478–52–4]*. Herbicide introduced by Sumitomo.   **IUPAC name** (±)-*N*-[4-chloro-2-fluoro-5-(1-methylprop-2-ynyl)oxy= phenyl]cyclohex-1-ene-1,2-dicarboximide   **CA name** (±)-2-[4-chloro-2-fluoro-= 5-[(1-methyl-2-propynyl)oxy]phenyl]-4,5,6,7-tetrahydro-1*H*-isoindole-1,3(2*H*)-dione. **M.f.** $C_{18}H_{15}ClFNO_3$ **Code nos.** S-23121 (Sumitomo).

1061 fluorbenside (common name).   CAS RN *[405–30–1]*. Acaricide reported by N. G. Clark et al. (*J. Sci. Food Agric.*, 1957, **8**, 566). Introduced by the Boots Co. Ltd Agriculture (later Schering Agriculture).   **IUPAC name** 4-chlorobenzyl 4-fluorophenyl sulfide   **CA name** 1-chloro-4-[[(4-fluorophenyl)thio]methyl]benzene.   **M.f.** $C_{13}H_{10}ClFS$ **Code nos.** HRS 924; RD 2454   **Selected tradenames** 'Fluorparacide', 'Fluorsulphacide'. **Details** *PM1*, p. 235.

1062 fluoridamid (common name).   CAS RN *[47000–42–0]*. Plant growth regulator introduced by 3M Co.   **Common names** fluoridamid, fluoridamide   **IUPAC name** 3′-(1,1,1-trifluoromethanesulfonamido)acet-*p*-toluidide   **CA name** *N*-[4-methyl-= 3-[[(trifluoromethyl)sulfonyl]amino]phenyl]acetamide.   **M.f.** $C_{10}H_{11}F_3N_2O_3S$ **Code nos.** MBR-6033   **Selected tradenames** 'Sustar'.

1063 fluorodifen (common name).   CAS RN *[15457–05–3]*. Herbicide reported by L. Ebner et al. (*Proc. Br. Weed Control Conf.*, 9th, 1968, **2**, 1026). Introduced by Ciba AG (later Ciba-Geigy AG).   **Common names** fluorodifen, fluorodifène   **IUPAC name** 4-nitrophenyl α,α,α-trifluoro-2-nitro-*p*-tolyl ether   **CA name** 2-nitro-1-(4-nitro= phenoxy)-4-(trifluoromethyl)benzene.   **M.f.** $C_{13}H_7F_3N_2O_5$ **Code nos.** C 6989 **Selected tradenames** 'Preforan'.   **Details** *PM8*, entry 06560.

1064 fluoromidine (common name).   CAS RN *[13577–71–4]*. Herbicide evaluated by Fisons Ltd (later Schering Agriculture).   **Common names** fluoromidine, fluromidine **IUPAC name** 6-chloro-2-trifluoromethyl-3*H*-imidazo[4,5-*b*]pyridine   **CA name** 6-chloro-2-(trifluoromethyl)-1*H*-imidazo[4,5-*b*]pyridine.   **M.f.** $C_7H_3ClF_3N_3$ **Code nos.** NC 4780.

1065 fluoronitrofen (common name).   CAS RN *[13738–63–1]*. Herbicide developed by Mitsui Toatsu Chemicals, Inc.   **Common names** fluoronitrofen, fluoronitrofène **IUPAC name** 2,4-dichloro-6-fluorophenyl 4-nitrophenyl ether   **CA name** 1,5-dichloro-3-fluoro-2-(4-nitrophenoxy)benzene.   **M.f.** $C_{12}H_6Cl_2FNO_3$ **Code nos.** Mo 500   **Selected tradenames** 'Mo 500'.   **Details** *PM9*, entry 6580.

1066 fluothiuron (common name).   CAS RN *[33439–45–1]*. Herbicide introduced by Bayer AG.   **IUPAC name** 3-[3-chloro-4-(chlorodifluoromethylthio)phenyl]-= 1,1-dimethylurea   **CA name** *N*′-[3-chloro-4-[(chlorodifluoromethyl)thio]phenyl]-= *N*,*N*-dimethylurea.   **M.f.** $C_{10}H_{10}Cl_2F_2N_2OS$ **Code nos.** BAY KUE 2079A   **Selected tradenames** 'Clearcide'.   **Details** *PM5*, p. 107.

1067 fluotrimazole (common name).   CAS RN *[31251–03–3]*. Fungicide reported by F. Grewe & K. H. Büchel (*Mitt. Biol. Bundesanst. Land-Forstwirtsch. Berlin-Dahlem*, 1973, **151**, 208). Introduced by Bayer AG.   **IUPAC name** 1-(3-trifluoromethyltrityl)-=

1H-1,2,4-triazole   **CA name** 1-[diphenyl[3-(trifluoromethyl)phenyl]methyl]-=
1H-1,2,4-triazole.   **M.f.** $C_{22}H_{16}F_3N_3$ **Code nos.** BAY BUE 0620   **Selected tradenames** 'Persulon'.   **Details** PM8, entry 06600.

1068   flupropadine; flupropadine hydrochloride (common name).   **CAS RN**
[81613–59–4]. Rodenticide reported by A. P. Buckle (J. Hyg., 1985, **95**, 505) and by
F. P. Rowe et al. (ibid., p. 513). Introduced by May & Baker Ltd (later Rhône-Poulenc
Agriculture).   **IUPAC name** 4-tert-butyl-1-[3-(α,α,α,α',α',α'-hexafluoro-3,5-xylyl)=
prop-2-ynyl]piperidine   **CA name** 1-[3-[3,5-bis(trifluoromethyl)phenyl]-2-propynyl]-=
4-(1,1-dimethylethyl)piperidine.   **M.f.** $C_{20}H_{23}F_6N$; (hydrochloride $C_{20}H_{24}ClF_6N$)
Code nos. M&B 36 892, OMS 3018.   **Details** PM9, entry 6620.

1069   fluvalinate (common name).   **CAS RN** [69409–94–5] fluvalinate. Insecticide
reported by C. A. Henrick et al. (Pestic. Sci., 1980, **11**, 224) and by R. J. Anderson
(J. Agric. Food Chem., 1985, **33**, 508). Fluvalinate introduced by Zoecon Corp. (later
Sandoz AG) and replaced by tau-fluvalinate (the D-valinate) [NOTE many recent
references in the literature have been made, in error, to fluvalinate rather than to
tau-fluvalinate].   **IUPAC name** (RS)-α-cyano-3-phenoxybenzyl N-(2-chloro-=
α,α,α-trifluoro-p-tolyl)-DL-valinate   **CA name** cyano(3-phenoxyphenyl)methyl
N-[2-chloro-4-(trifluoromethyl)phenyl]-DL-valinate.   **M.f.** $C_{26}H_{22}ClF_3N_2O_3$
Selected tradenames 'Mavrik'.

1070   FMC 1137 (development code).   **CAS RN** [2901–90–8] (i); [3031-21-8] (ii).
Insecticide and acaricide introduced by FMC Corp.   **IUPAC name** bis(diethoxy=
phosphinothioyl) disulfide (i) + bis(di-isopropoxyphosphinothioyl) disulfide (ii) 75:25
mixture   **CA name** bis(diethoxyphosphinothioyl) disulfide (i) + bis[di-(1-methyl=
ethoxy)phosphinothioyl] disulfide (ii).   **M.f.** $C_8H_{20}O_4P_2S_4$ (i), $C_{12}H_{28}O_4P_2S_4$ (ii)
Code nos. FMC 1137, ENT 23 584 (i + ii).   **Selected tradenames** 'Phostex'.

1071   FMC 19873 (development code).   **CAS RN** [40915–86–4]. Herbicide evaluated
by FMC Corp.   **IUPAC name** 6-tert-butyl-3-isopropyl[1,2]thiazolo[3,4-d]pyrimidin-=
4(5H)-one; 6-tert-butyl-4,5-dihydro-3-isopropylisothiazolo[3,4-d]pyrimidin-4-one   **CA
name** 6-(1,1-dimethylethyl)-3-(1-methylethyl)isothiazolo[3,4-d]pyrimidin-4(5H)-one.
M.f. $C_{12}H_{17}N_3OS$ **Code nos.** FMC 19873.   **Details** PM4, p. 66.

1072   FMC 21844 (development code).   **CAS RN** [35258–87–8]. Herbicide evaluated
by FMC Corp.   **IUPAC name** 6-tert-butyl-3-isopropyl-1,2-oxazolo[5,4-d]pyrimidin-=
4(5H)-one; 6-tert-butyl-4,5-dihydro-3-isopropylisoxazolo[5,4-d]pyrimidin-4-one
CA name 6-(1,1-dimethylethyl)-3-(1-methylethyl)isoxazolo[5,4-d]pyrimidin-4(5H)-one.
M.f. $C_{12}H_{17}N_3O_2$ **Code nos.** FMC 21844.   **Details** PM4, p. 68.

1073   FMC 21861 (development code).   **CAS RN** [35260–91–4]. Herbicide evaluated
by FMC Corp.   **IUPAC name** 6-tert-butyl-3-propyl-1,2-oxazolo[5,4-d]pyrimidin-=
4(5H)-one; 6-tert-butyl-4,5-dihydro-3-propylisoxazolo[5,4-d]pyrimidin-4-one
CA name 6-(1,1-dimethylethyl)-3-propylisoxazolo[5,4-d]pyrimidin-4(5H)-one.
M.f. $C_{12}H_{17}N_3O_2$ **Code nos.** FMC 21861.   **Details** PM3, p. 75.

1074 **FMC 23486** (development code). **CAS RN** *[38897–15–3]*. Herbicide reported by W. M. Dest *et al.* (*Proc. Northeast. Weed Sci. Soc.*, 1973, **27**, 31). Evaluated by FMC Corp. **IUPAC name** 6-*tert*-butyl-3-isopropyl[1,2]oxazolo[3,4-*d*]pyrimidin-4(5*H*)-one; 6-*tert*-butyl-4,5-dihydro-3-isopropylisoxazolo[3,4-*d*]pyrimidin-4-one **CA name** 6-(1,1-dimethylethyl)-3-(1-methylethyl)isoxazolo[3,4-*d*]pyrimidin-4(5*H*)-one. **M.f.** $C_{12}H_{17}N_3O_2$ **Code nos.** FMC 23486. **Details** *PM5*, p. 66.

1075 **FMC 25213** (development code). **CAS RN** *[41129–10–6]* (*cis*- isomer). Herbicide reported (*Ann. Meet. South. Weed Sci. Soc.*, *27th*, 1974) and evaluated by FMC Corp. **IUPAC name** 2-ethyl-5-methyl-1,3-dioxan-5-yl 2-methylbenzyl ether; 2-ethyl-5-methyl-5-(2-methylbenzyloxy)-1,3-dioxane **CA name** 2-ethyl-5-methyl-= 5-[(2-methylphenyl)methoxy]-1,3-dioxane. **M.f.** $C_{15}H_{22}O_3$ **Code nos.** FMC 25213. **Details** *PM5*, p. 254.

1076 **formparanate** (common name). **CAS RN** *[17702–57–7]*. Insecticide and acaricide evaluated by Union Carbide. **IUPAC name** 4-dimethylaminomethylene= amino-*m*-tolyl methylcarbamate **CA name** *N,N*-dimethyl-*N*-[2-methyl-4-[[(methyl= amino)carbonyl]oxy]phenyl]methanimidamide. **M.f.** $C_{12}H_{17}N_3O_2$ **Code nos.** UC 25 074.

1077 **fosmethilan** (common name). **CAS RN** *[83733–82–8]*. Insecticide reported by K. Sagi *et al.* (*Proc. 10th Int. Congr. Plant Prot.*, 1983, **1**, 384). **Common names** fosmethilan, fosméthilane **IUPAC name** *S*-[*N*-(2-chlorophenyl)butyramidomethyl] *O,O*-dimethyl phosphorodithioate; 2′-chloro-*N*-(dimethoxyphosphinothioyl= thiomethyl)butyranilide **CA name** *S*-[[(2-chlorophenyl)(1-oxobutyl)amino]methyl] *O,O*-dimethyl phosphorodithioate. **M.f.** $C_{13}H_{19}ClNO_3PS_2$ **Code nos.** NE-79 168. **Details** *PM9*, entry 6790.

1078 **fospirate** (common name). **CAS RN** *[5598–52–7]*. Insecticide introduced by Dow Chemical Co. **IUPAC name** dimethyl 3,5,6-trichloro-2-pyridyl phosphate **CA name** dimethyl 3,4,5-trichloro-2-pyridinyl phosphate. **M.f.** $C_7H_7Cl_3NO_4P$ **Code nos.** Dowco 217; OMS 1168; ENT 27 521.

1079 **fosthietan** (common name). **CAS RN** *[21548–32–3]*. Insecticide and nematicide described by W. K. Whitney & J. L. Aston (*Proc. Br. Insectic. Fungic. Conf. 8th*, 1975, **2**, 625). Introduced by American Cyanamid Co. **IUPAC name** diethyl 1,3-dithietan-= 2-ylidenephosphoramidate (I) **CA name** (I); 2-(diethoxyphosphinylimino)-= 1,3-dithietane. **M.f.** $C_6H_{12}NO_3PS_2$ **Code nos.** AC 64 475 **Selected tradenames** 'Nem-a-tak', 'Acconem', 'Geofos'. **Details** *PM7*, entry 06810.

1080 **furcarbanil** (common name). **CAS RN** *[28562–70–1]*. Fungicide evaluated by BASF AG. **IUPAC name** 2,5-dimethyl-3-furanilide **CA name** 2,5-dimethyl-*N*-phenyl-= 3-furancarboxamide. **M.f.** $C_{13}H_{13}NO_2$ **Code nos.** BAS 319F.

1081 **furconazole** (common name). **CAS RN** *[112839–33–5]* unstated stereochemistry. Fungicide reported by Rhône-Poulenc Agrochimie. **IUPAC name** (2*RS*,5*RS*;2*RS*,5*SR*)-5-(2,4-dichlorophenyl)tetrahydro-5-(1*H*-1,2,4-triazol-1-ylmethyl)-=

2-furyl 2,2,2-trifluoroethyl ether   **CA name** 1-[[2-(2,4-dichlorophenyl)tetrahydro-=
5-(2,2,2-trifluoroethoxy)-2-furanyl]methyl]-1H-1,2,4-triazole.   **M.f.** $C_{15}H_{14}Cl_2F_3N_3O_2$
Code nos. LS 840 608 (Rhône-Poulenc).   **Details** *PM9*, entry 6863.

1082   furconazole-cis (common name).   **CAS RN** *[112839–32–4]*. Fungicide reported
by B. Zech *et al.* (*Proc. 1988 Brighton Crop Prot. Conf. - Pests Dis.*, **2**, 503). Introduced by
Rhône-Poulenc Agrochimie.   **IUPAC name** (2*RS*,5*RS*)-5-(2,4-dichlorophenyl)=
tetrahydro-5-(1H-1,2,4-triazol-1-ylmethyl)-2-furyl 2,2,2-trifluoroethyl ether   **CA name**
cis-1-[[2-(2,4-dichlorophenyl)tetrahydro-5-(2,2,2-trifluoroethoxy)-2-furanyl]methyl]-=
1H-1,2,4-triazole.   **M.f.** $C_{15}H_{14}Cl_2F_3N_3O_2$ **Code nos.** LS 840 606 (Rhône-Poulenc).
Details *PM9*, entry 6865.

1083   furethrin (trivial name).   **CAS RN** *[17080–02–3]* (formerly *[7076–49–5]*).
Insecticide (M. Matsui *et al.*, *J. Am. Chem. Soc.*, 1952, **74**, 2181). The ester mixture from
the (*RS*)-alcohol and the (1*R*)-*trans*- acid is a more potent insecticide than the complete
mixture of stereoisomers (W. A. Gersdorff & N. Mitlin, *J. Econ. Entomol.*, 1952, **45**,
849).   **IUPAC name** (*RS*)-3-furfuryl-2-methyl-4-oxocyclopent-2-enyl
(1*RS*,2*RS*;1*RS*,2*SR*)-2,2-dimethyl-3-(2-methylprop-1-enyl)cyclopropanecarboxylate;
Roth: (*RS*)-3-furfuryl-2-methyl-4-oxocyclopent-2-enyl (1*RS*)-*cis*-*trans*-2,2-dimethyl-=
3-(2-methylprop-1-enyl)cyclopropanecarboxylate; 3-furfuryl-2-methyl-4-oxycyclo=
pent-2-enyl (±)-*cis*-*trans*-chrysanthemate   **CA name** (±)-3-(2-furanylmethyl)-=
2-methyl-4-oxo-2-cyclopenten-1-yl 2,2-dimethyl-3-(2-methyl-1-propenyl)cyclopropane=
carboxylate.   **M.f.** $C_{21}H_{26}O_4$ **Other names** furethrin.   **Details** *PM2*, p. 264.

1084   furmecyclox (common name).   **CAS RN** *[60568–05–0]*. Fungicide reported by
E.-H. Pommer & B. Zeeh (*Pestic. Sci.*, 1977, **8**, 320). Introduced by BASF AG.   **IUPAC
name** methyl *N*-cyclohexyl-2,5-dimethyl-3-furohydroxamate; methyl *N*-cyclohexyl-=
2,5-dimethylfuran-3-carbohydroxamate   **CA name** *N*-cyclohexyl-*N*-methoxy-=
2,5-dimethyl-3-furancarboxamide.   **M.f.** $C_{14}H_{21}NO_3$ **Code nos.** BAS 389F
Selected tradenames 'Campogran', 'Xyligen B'.   **Details** *PM8*, entry 06880.

1085   furophanate (common name).   **CAS RN** *[53878–17–4]*. Fungicide evaluated by
Rohm & Haas Co.   **IUPAC name** methyl 4-(2-furfurylideneaminophenyl)-3-thio=
allophanate   **CA name** methyl [[[2-[(2-furanylmethylene)amino]phenyl]amino]=
thioxomethyl]carbamate.   **M.f.** $C_{14}H_{13}N_3O_3S$ **Code nos.** RH-3928.

1086   furyloxyfen (common name).   **CAS RN** *[80020–41–3]* unstated
stereochemistry. Herbicide developed by Mitsui Toatsu Chemicals, Inc.   **Common
names** furyloxyfen, furyloxyfène   **IUPAC name** (±)-5-(2-chloro-α,α,α-trifluoro-=
p-tolyloxy)-2-nitrophenyl tetrahydro-3-furyl ether   **CA name** (±)-3-[5-[2-chloro-4-=
(trifluoromethoxy)phenoxy]-2-nitrophenoxy]tetrahydrofuran.   **M.f.** $C_{17}H_{13}ClF_3NO_5$
Code nos. MT-124 (Mitsui).   **Details** *PM9*, entry 6895.

1087   gliotoxin (antibiotic name).   **CAS RN** *[67–99–2]*. Antifungal antibiotic isolated
by R. Weindling & O. H. Emerson (*Phytopathology*, 1936, **26**, 1068) and structure
established by M. R. Bell (*J. Am. Chem. Soc.*, 1958, **80**, 1001). Instability has limited use in
crop protection.   **IUPAC name** (3*R*,5a*S*,6*S*,10a*R*)-2,3,5a,6,10,10-hexahydro-=

6-hydroxy-3-hydroxymethyl-2-methyl-3,10a-epidithiopyrazino[1,2-a]indole-1,4-dione
CA name [3R-(3α,5aβ,6β,10aα)]-2,3,5a,6-tetrahydro-6-hydroxy-3-(hydroxymethyl)-=
2-methyl-10H-3,10a-epidithiopyrazino[1,2-a]indole-1,4-dione.   **M.f.** $C_{13}H_{14}N_2O_4S_2$
Details *PM4*, p. 287.

1088   glyodin (common name).   **CAS RN** *[556–22–9]*. Fungicide reported by R. H.
Wellman & S. E. A. McCallan (*Contrib. Boyce Thompson Inst.*, 1946, **14**, 151). Introduced
by Union Carbide Corp.   **IUPAC name** 2-heptadecyl-2-imidazoline acetate; acetic acid -
2-heptadecyl-2-imidazoline (1:1)   **CA name** 2-heptadecyl-4,5-dihydro-1H-imidazolyl
monoacetate.   **M.f.** $C_{22}H_{44}N_2O_2$ **Code nos.** Crag Fruit Fungicide 341   **Selected
tradenames** 'Crag Fungicide 341'.   **Details** *PM1*, p. 245.

1089   glyphosine (common name).   **CAS RN** *[2439–99–8]*. Increase on carbohydrate
deposition in sugar cane described by C. A. Porter & L. E. Ahlrichs (*Hawaii Sugar Tech.
Rep. 1971*, 1972, **30**, 71). Plant growth regulator introduced by Monsanto Co.   **IUPAC
name** N,N-bis(phosphonomethyl)glycine (I)   **CA name** (I).   **M.f.** $C_4H_{11}NO_8P_2$ **Code
nos.** CP 41 845   **Selected tradenames** 'Polaris'.   **Details** *PM7*, entry 06960.

1090   griseofulvin (common name).   **CAS RN** *[126–07–8]*. Antifungal antibiotic
isolated by A. E. Oxford et al. (*Biochem. J.*, 1939, **33**, 240), structure elucidated by J. F.
Grove et al. (*Chem. Ind. (London)*, 1955, p. 160; *J. Chem. Soc.*, 1956, p. 3949; *Q. Rev.
Chem. Soc.*, 1963, **17**, 1). Introduced by Glaxo Ltd, now restricted to veterinary and
anthelmintic uses.   **Common names** griseofulvin, griséofulvine   **IUPAC name**
7-chloro-2',4,6-trimethoxy-6'-methylspiro[benzofuran-2(3H),1'-cyclohex-2'-ene]-=
3,4'-dione; 7-chloro-2',4,6-trimethoxybenzofuran-2(3H)-spiro-1'-cyclohex-2'-ene-=
3,4'-dione; 7-chloro-4,6-dimethoxycoumaran-3-one-2-spiro-1'-(2'-methoxy-=
6'-methylcyclohex-2'-en-4'-one)   **CA name** (1'S)-trans-7-chloro-2',4,6-trimethoxy-=
6'-methylspiro[benzofuran-2(3H),1'-[2]cyclohexene]-3,4'-dione.   **M.f.** $C_{17}H_{17}ClO_6$
Other names "curling factor".   **Details** *PM5*, p. 290.

1091   halacrinate (common name).   **CAS RN** *[34462–96–9]*. Fungicide reported by
J. M. Smith et al. (*Proc. Br. Insectic. Fungic. Conf., 8th*, 1975, **2**, 421). Evaluated by Ciba-
Geigy AG.   **IUPAC name** 7-bromo-5-chloro-8-quinolyl acrylate   **CA name** 7-bromo-=
5-chloro-8-quinolinyl 2-propenoate.   **M.f.** $C_{12}H_7BrClNO_2$ **Code nos.** CGA 30 599
Selected tradenames 'Tilt'.   **Details** *PM5*, p. 292.

1092   halosafen (common name).   **CAS RN** *[77227–69–1]*. Herbicide evaluated by ICI
Agrochemicals.   **Common names** halosafen, halosafène   **IUPAC name** 5-(2-chloro-=
α,α,α,6-tetrafluoro-p-tolyloxy)-N-ethylsulfonyl-2-nitrobenzamide   **CA name**
5-[2-chloro-6-fluoro-4-(trifluoromethyl)phenoxy]-N-(ethylsulfonyl)-2-nitrobenzamide.
M.f. $C_{16}H_{11}ClF_4N_2O_6S$.

1093   haloxydine (common name).   **CAS RN** *[2693–61–0]*. Herbicide evaluated by ICI
Plant Protection Division (later ICI Agrochemicals).   **IUPAC name** 3,5-dichloro-=
2,6-difluoropyridin-4-ol   **CA name** 3,5-dichloro-2,6-difluoro-4-pyridinol.
M.f. $C_5HCl_2F_2NO$ **Code nos.** PP493.

1094 **2-(2-heptadecyl-2-imidazolin-1-yl)ethanol** (IUPAC name). **CAS RN** *[95–19–2].* Fungicide evaluated by Union Carbide Corp. (later Rhône-Poulenc Agrochimie). **IUPAC name** 2-(2-heptadecyl-2-imidazolin-1-yl)ethanol; 2-heptadecyl-1-(2-hydroxy= ethyl)imidazoline **CA name** 2-heptadecyl-4,5-dihydro-1*H*-imidazole-1-ethanol. **M.f.** $C_{22}H_{44}N_2O$ **Code nos.** Fungicide 337.

1095 **heptopargil** (common name). **CAS RN** *[73886–28–9].* Plant growth regulator reported by A. Kis-Tamás et al. (*Proc. Br. Crop Prot. Conf. - Weeds*, 1980, **I**, 173). Introduced by EGYT Pharmacochemical Works. **IUPAC name** (*E*)-(1*RS*,4*RS*)-bornan-= 2-one *O*-prop-2-ynyloxime **CA name** (±)-1,7,7-trimethylbicyclo[2.2.1]heptan-2-one *O*-2-propynyloxime. **M.f.** $C_{13}H_{19}NO$ **Code nos.** EGYT 2250 **Selected tradenames** 'Limbolid'. **Details** *PM9*, entry 7050.

1096 **Hercules 3944** (development code). **CAS RN** *[2425–05–0].* Fungicide reported by W. R. Sitterly (*Plant Dis. Rep.*, 1961, **45**, 200). Introduced by Hercules Inc. **IUPAC name** 5-chloro-4-phenyl-1,2-dithiol-3-one **CA name** 5-chloro-4-phenyl-= 3*H*-1,2-dithiol-3-one. **M.f.** $C_9H_5ClOS_2$ **Code nos.** Hercules 3944. **Details** *PM3*, p. 115.

1097 **hexachloroacetone** (IUPAC name). **CAS RN** *[116–16–5].* Herbicide introduced by Allied Chemical Corp. Agricultural Division (later Hopkins Chemical Co.). **Common names** HCA, hexachloroacetone, hexachloracetone **CA name** 1,1,1,3,3,3-hexa= chloro-2-propanone. **M.f.** C_3Cl_6O **Code nos.** GC 1106 **Selected tradenames** 'Urox 379'. **Details** *PM5*, p. 296.

1098 **hexadecyl cyclopropanecarboxylate** (IUPAC/Chemical Abstracts name). **CAS RN** *[54460–46–7].* Acaricide evaluated by Zoecon Corp. (later Sandoz AG). **M.f.** $C_{20}H_{38}O_2$ **Other names** [cycloprate] (rejected common name proposal) **Code nos.** ZR 856 **Selected tradenames** 'Zardex'.

1099 **hexafluoroacetone trihydrate** (IUPAC name). **CAS RN** *[993–58–8]* (i); *[677–71–4]* (ii). Herbicide introduced by Allied Chemical Corp. Agricultural Division (later Hopkins Agricultural Chemical Co.). **IUPAC name** (i) hexafluoroacetone trihydrate; (ii) 1,1,1,3,3,3-hexafluoropropane-2,2-diol dihydrate **CA name** (i) 1,1,1,3,3,3-hexa= fluoro-2-propanone; (ii) 1,1,1,3,3,3-hexafluoro-2,2-propanediol. **M.f.** $C_3H_6F_6O_4$ **Code nos.** GC 7887. **Details** *PM4*, p. 295.

1100 **hexaflurate** (common name). **CAS RN** *[17029–22–0].* First prepared by M. C. Marignac (*Ann. Chem. Pharm.*, 1867, **145**, 237), evaluated as a herbicide by Pennwalt Corp. and Dow Chemical Co. **IUPAC name** potassium hexafluoroarsenate **CA name** potassium hexafluoroarsenate(1-). **M.f.** AsF_6K **Selected tradenames** 'Nopalmate'. **Details** *PM5*, p. 298.

1101 **hexylthiofos** (common name). **CAS RN** *[41495–67–4]* unstated stereochemistry. Fungicide evaluated by Bayer AG. **IUPAC name** *O*-cyclohexyl *O*,*S*-diethyl phosphorothioate (I) **CA name** (I). **M.f.** $C_{10}H_{21}O_3PS$ **Code nos.** NTN 3318.

1102 holosulf (common name). **CAS RN** *[21780–04–1]*. Plant growth regulator evaluated by Bayer AG. **IUPAC name** 2-chloroethanesulfinic acid (I) **CA name** (I). **M.f.** $C_2H_5ClO_2S$ **Code nos.** HOL 1302.

1103 1-hydroxy-1H-pyridine-2-thione (IUPAC name). **CAS RN** *[1121–30–8]* (i); *[1121–31–9]* (ii); *[32255–90–6]* manganese(II) derivative; *[13463–41–7]* zinc derivative. Compounds reported by E. Shaw et al. (J. Am. Chem. Soc., 1950, **72**, 4362). Various metal salts investigated as fungicides (M. Szkolnik & J. M. Hamilton, Plant Dis. Rep., 1957, **41**, 289, 293, 301) and evaluated by the Olin Mathieson Chemical Corp. **IUPAC name** 1-hydroxy-1H-pyridine-2-thione (i); tautomeric with pyridine-2-thiol 1-oxide (ii) **CA name** 1-hydroxy-2(1H)-pyridinethione (i); 2-pyridinethiol 1-oxide (ii). **M.f.** C_5H_5NOS **Selected tradenames** 'Omadine', 'Omadine OM 1565', 'Omadine OM 1564', 'Omadine OM 1563' (zinc derivative). **Details** PM2, p. 395.

1104 hyquincarb (common name). **CAS RN** *[56716–21–3]*. Insecticide reported by K.-D. Bock (Proc. Br. Crop. Prot. Conf. - Pests Dis., 1977, **3**, 1017). Evaluated by Hoechst AG. **IUPAC name** 5,6,7,8-tetrahydro-2-methyl-4-quinolyl dimethylcarbamate **CA name** 5,6,7,8-tetrahydro-2-methyl-4-quinolinyl dimethylcarbamate. **M.f.** $C_{13}H_{18}N_2O_2$ **Code nos.** Hoe 25 682.

1105 2-imidazolidone (IUPAC name). **CAS RN** *[120–93–4]*. Insect growth regulator reported by H. G. Simkover (J. Econ. Entomol., 1964, **57**, 574). Evaluated by Shell Development Co. **IUPAC name** 2-imidazolidone; imidazolidin-2-one **CA name** 2-imidazolidinone. **M.f.** $C_3H_6N_2O$ **Other names** ethyleneurea. **Details** PM1, p. 214.

1106 iodobonil (common name). **CAS RN** *[25671–45–8]*. Herbicide evaluated by Boehringer & Sohn (later Shell Agrar GmbH). **IUPAC name** allyl 4-cyano-= 2,5-di-iodophenyl carbonate; 4-(allyloxycarbonyloxy)-3,5-di-iodobenzonitrile **CA name** 4-cyano-2,6-di-iodophenyl 2-propenyl carbonate. **M.f.** $C_{11}H_7I_2NO_3$.

1107 ipazine (common name). **CAS RN** *[1912–25–0]*. Herbicide evaluated by J. R. Geigy S.A. (later Ciba-Geigy AG). **IUPAC name** 6-chloro-N^2,N^2-diethyl-= N^4-isopropyl-1,3,5-triazine-2,4-diamine; 2-chloro-4-diethylamino-6-isopropylamino-= 1,3,5-triazine **CA name** 6-chloro-N,N-diethyl-N'-(1-methylethyl)-1,3,5-triazine-= 2,4-diamine. **M.f.** $C_{10}H_{18}ClN_5$ **Code nos.** G 30 031 **Selected tradenames** 'Gesabal'.

1108 iprymidam (common name). **CAS RN** *[30182–24–2]*. Herbicide evaluated by Sandoz AG. **IUPAC name** 6-chloro-N^4-isopropylpyrimidine-2,4-diamine **CA name** 6-chloro-N^4-(1-methylethyl)-2,4-pyrimidinediamine. **M.f.** $C_7H_{11}ClN_4$ **Code nos.** SAN 52 123H.

1109 IPSP (common name). **CAS RN** *[5827–05–4]* . Insecticide reported by D. Murrayama et al. (Mem. Fac. Agric. Hokkaido Univ., 1966, **6**, 73). Introduced by Hokko Chemical Industry Co., Ltd. **IUPAC name** S-ethylsulfinylmethyl O,O-di-isopropyl phosphorodithioate **CA name** S-[(ethylsulfinyl)methyl] O,O-bis(1-methylethyl) phosphorodithioate. **M.f.** $C_9H_{21}O_3PS_3$ **Code nos.** PSP-204 (Hokko) **Selected tradenames** 'Aphidan'. **Details** PM8, entry 05910.

1110   isamidofos (common name).   **CAS RN** *[66602–87–7]*. Nematicide evaluated by Hoechst AG.   **IUPAC name** *O*-ethyl *S*-(*N*-methylcarbaniloylmethyl) *N*-isopropyl= phosphoramidothioate   **CA name** *O*-ethyl *S*-[2-(methylphenylamino)-2-oxoethyl] phosphorothioate.   **M.f.** $C_{14}H_{23}N_2O_3PS$ **Code nos.** Hoe 36 275.

1111   isobenzan (common name).   **CAS RN** *[297–78–9]*. Insecticide evaluated by Shell International Chemical Co. Manufactured in the Netherlands from 1958 to 1965. **Common names** isobenzan, telodrin   **IUPAC name** 1,3,4,5,6,7,8,8-octachloro-= 1,3,3a,4,7,7a-hexahydro-4,7-methanoisobenzofuran (I)   **CA name** (I).   **M.f.** $C_9H_4Cl_8O$ **Code nos.** SD 4402; OMS 206; ENT 25 545.   **Selected tradenames** 'Telodrin'.

1112   isocarbamid (common name).   **CAS RN** *[30979–48–7]*. Herbicide reported by L. Eue *et al.* (*Proc. Br. Weed Control Conf., 10th*, 1970, **2**, 610). Introduced by Bayer AG. **Common names** isocarbamid, isocarbamide   **IUPAC name** *N*-isobutyl-2-oxo= imidazolidine-1-carboxamide   **CA name** *N*-(2-methylpropyl)-2-oxo-1-imidazolidine= carboxamide.   **M.f.** $C_8H_{15}N_3O_2$ **Code nos.** BAY MNF 0166   **Selected tradenames** 'Merpelan AZ', 'Terratop'.   **Details** *PM7*, entry 07380.

1113   isocil (common name).   **CAS RN** *[314–42–1]*. Herbicide reported by H. C. Bucha *et al.* (*Science*, 1962, **137**, 537). Introduced by E. I. du Pont de Nemours and Co. **Common names** isocil, isoprocil   **IUPAC name** 5-bromo-3-isopropyl-6-methyluracil **CA name** 5-bromo-6-methyl-3-(1-methylethyl)-2,4(1*H*,3*H*)-pyrimidinedione.   **M.f.** $C_8H_{11}BrN_2O_2$ **Code nos.** Du Pont Herbicide 82   **Selected tradenames** 'Hyvar'. **Details** *PM1*, p. 260.

1114   isodrin (common name).   **CAS RN** *[465–73–6]*. Insecticide evaluated by Shell International Chemical Co.   **Common names** isodrin, isodrine   **IUPAC name** (1*R*,4*S*,5*R*,8*S*)-1,2,3,4,10,10-hexachloro-1,4,4a,5,8,8a-hexahydro-1,4:5,8-dimethano= naphthalene   **CA name** (1α,4α,4aβ,5β,8β,8aβ)-1,2,3,4,10,10-hexachloro-= 1,4,4a,5,8,8a-hexahydro-1,4:5,8-dimethanonaphthalene.   **M.f.** $C_{12}H_8Cl_6$ **Code nos.** Compound 711, OMS 198.

1115   isolane (common name).   **CAS RN** *[119–38–0]*. Insecticide reported by H. Gysin (*Proc. 3rd Int. Congr. Phytopharm., Paris*, 1952). Introduced by J. R. Geigy S.A. (later Ciba-Geigy AG).   **IUPAC name** 1-isopropyl-3-methylpyrazol-5-yl dimethylcarbamate **CA name** 3-methyl-1-(1-methylethyl)-1*H*-pyrazol-5-yl dimethylcarbamate.   **M.f.** $C_{10}H_{17}N_3O_2$ **Code nos.** G 23 611; OMS 62; ENT 19 060.   **Selected tradenames** 'Isolan'.   **Details** *PM1*, p. 262.

1116   isomethiozin (common name).   **CAS RN** *[57052–04–7]*. Herbicide reported by H. Hack (*Mitt. Biol. Bundesanst. Land- Fortwirtsch. Berlin-Dahlem*, 1975, **165**, 179; *Pflanzensuch-Nachr. (Engl. Ed.)*, 1975, **28**, 241). Introduced by Bayer AG.   **Common names** isomethiozin, isométhiozine   **IUPAC name** 6-*tert*-butyl-4-isobutylideneamino-= 3-methylthio-1,2,4-triazin-5(4*H*)-one   **CA name** 6-(1,1-dimethylethyl)-4-[(2-methyl= propylidene)amino]-3-(methylthio)-1,2,4-triazin-5(4*H*)-one.   **M.f.** $C_{12}H_{20}N_4OS$ **Code nos.** BAY DIC 1577   **Selected tradenames** 'Tantizon'; 'Tantizon Combi', 'Tantizon DP'.   **Details** *PM6*, p. 310.

1117 **isonoruron** (common name). **CAS RN** *[28805–78–9]* formerly *[28346–65–8]* (i + ii). Herbicide reported by A. Fischer (*Abstr. Int. Congr. Plant Prot.*, 1967, p. 446). Introduced by BASF AG. **IUPAC name** 1,1-dimethyl-3-(perhydro-4,7-methanoinden-=1-yl)urea (i) and 1,1-dimethyl-3-(perhydro-4,7-methanoinden-2-yl)urea (ii) **CA name** *N,N*-dimethyl-*N'*-(2,3,3a,4,5,6,7,7a-octahydro-4,7-methano-1*H*-inden-1-yl)urea (i) and *N,N*-dimethyl-*N'*-(2,3,3a,4,5,6,7,7a-octahydro-4,7-methano-1*H*-inden-2-yl)urea (ii); *N,N*-dimethyl-*N'*-[octahydro-4,7-methanoinden-1(or 2)-yl]urea (i + ii). **M.f.** $C_{13}H_{22}N_2O$ **Selected tradenames** 'Basanor', 'Basfitox'. **Details** *PM5*, p. 313.

1118 **isopamphos** (trivial name). Fungicide developed by Borsod. **IUPAC name** 3-nonyloxypropylammonium methylphosphonate. **M.f.** $C_{13}H_{32}NO_4P$ **Other names** isopamphos; izopamfos **Selected tradenames** 'BF-51'. **Details** *PM10*, entry 413.

1119 **isopolinate** (common name). **CAS RN** *[3134–70–1]*. Herbicide evaluated by Chugai Pharmaceutical Co. Ltd. **IUPAC name** *S*-isopropyl perhydroazepine-=1-carbothioate **CA name** (*S*)-(1-methylethyl) hexahydro-1*H*-azepine-1-carbothioate. **M.f.** $C_{10}H_{19}NOS$ **Code nos.** R-4574; CH-83.

1120 **isopropalin** (common name). **CAS RN** *[33820–53–0]*. Herbicide reported by L. R. Guse (*Proc. North Cent. Weed Control Conf.*, 1969, p. 44; G. J. Shoop, *ibid.*, p. 19). Introduced in USA (1972) by Eli Lilly & Co. (later DowElanco). **Common names** isopropalin, isopropaline **IUPAC name** 4-isopropyl-2,6-dinitro-*N,N*-dipropylaniline **CA name** 4-(1-methylethyl)-2,6-dinitro-*N,N*-dipropylbenzenamine. **M.f.** $C_{15}H_{23}N_3O_4$ **Code nos.** EL-179 **Selected tradenames** 'Paarlan'. **Details** *PM9*, entry 7450.

1121 **isopyrimol** (common name). **CAS RN** *[55283–69–7]*. Plant growth regulator evaluated by Eli Lilly. **IUPAC name** 1-(4-chlorophenyl)-2-methyl-1-pyrimidin-5-yl=propan-1-ol **CA name** α-(4-chlorophenyl)-α-(1-methylethyl)-5-pyrimidinemethanol. **M.f.** $C_{14}H_{15}ClN_2O$.

1122 **isothioate** (common name). **CAS RN** *[36614–38–7]*. Introduced by Nihon Nohyaku Co., Ltd. **IUPAC name** *S*-2-isopropylthioethyl *O,O*-dimethyl phosphorodithioate **CA name** *O,O*-dimethyl *S*-[2-[(1-methylethyl)thio]ethyl] phosphorodithioate. **M.f.** $C_7H_{17}O_2PS_3$ **Selected tradenames** 'Hosdon'. **Details** *PM8*, entry 07540.

1123 **isovaledione** (common name). Fungicide evaluated by Sumitomo Chemical Co., Ltd. **IUPAC name** 3-(3,5-dichlorophenyl)-1-isovalerylhydantoin **CA name** 3-(3,5-dichlorophenyl)-1-(3-methyl-1-oxobutyl)-2,4-imidazolidinedione. **M.f.** $C_{14}H_{14}Cl_2N_2O_3$ **Code nos.** S-9373.

1124 **2-isovalerylindan-1,3-dione** (IUPAC name). **CAS RN** *[83–28–3]*. Insecticide reported by L. B. Kilgore *et al.* (*Ind. Eng. Chem.*, 1942, **34**, 494). Introduced by Kilgore Chemical Co. **CA name** 2-(3-methyl-1-oxobutyl)-1*H*-indene-1,3(2*H*)-dione. **M.f.** $C_{14}H_{14}O_3$ **Selected tradenames** 'Valone'. **Details** *PM5*, p. 318.

SUPERSEDED ENTRIES

1125   isoxapyrifop (common name).   **CAS RN** *[87757–18–4]*. Herbicide reported by H. Ohyama *et al.* (*Proc. Br. Crop Prot. - Weeds,* 1989, **1**, 59). Introduced by Hokko Chemical Industry Co., Ltd.   **IUPAC name** (*RS*)-2-[2-(4-(3,5-dichloro-2-pyridyloxy)= phenoxy)propionyl]isoxazolidine   **CA name** (±)-2-[2-[4-[(3,5-dichloro-2-pyridinyl)= oxy]phenoxy]-1-oxopropyl]isoxazolidine.   **M.f.** $C_{17}H_{16}Cl_2N_2O_4$ **Code nos.** HOK-1566; HOK-868 (both Hokko); RH-0898 (Rohm & Haas).   **Details** *PM10,* entry 419.

1126   jodfenphos (common name).   **CAS RN** *[18181–70–9]*. Insecticide and acaricide reported by B. C. Haddow & T. G. Marks (*Proc. Br. Insectic. Fungic. Conf., 5th,* 1969, **2,** 531). Introduced by Ciba AG (later Ciba-Geigy AG).   **Common names** iodofenphos, jodfenphos   **IUPAC name** O-2,5-dichloro-4-iodophenyl O,O-dimethyl phosphorothioate   **CA name** O-(2,5-dichloro-4-iodophenyl) O,O-dimethyl phosphorothioate.   **M.f.** $C_8H_8Cl_2IO_3PS$ **Code nos.** C 9491, OMS 1211.   **Selected tradenames** 'Elocril', 'Nuvanol N', 'Waspex', 'Alfracon'.   **Details** *PM9,* entry 7595.

1127   karbutilate (common name).   **CAS RN** *[4849–32–5]*. Herbicide reported by J. H. Dawson (*Bull. Wash. Agric. Exp. Stn.,* 1967, No. 691). Introduced by Agricultural Division of FMC Corp. and later by Ciba-Geigy AG.   **IUPAC name** 3-(3,3-dimethyl= ureido)phenyl *tert*-butylcarbamate   **CA name** 3-[[(dimethylamino)carbonyl]amino]= phenyl (1,1-dimethylethyl)carbamate.   **M.f.** $C_{14}H_{21}N_3O_3$ **Code nos.** FMC 11 092; CGA 61 837   **Selected tradenames** 'Tandex'.   **Details** *PM9,* entry 7600.

1128   kelevan (common name).   **CAS RN** *[4234–79–1]*. Insecticide reported by E. E. Gilbert *et al.* (*J. Agric. Food Chem.,* 1966, **14**, 111); development reviewed by H. Maier-Bode (*Residue Rev.,* 1976, **63,** 45). Invented by the Allied Chemical Corp. and developed by C. F. Spiess & Sohn.   **Common names** kelevan, kélévane   **IUPAC name** ethyl 5-(1,2,3,4,6,7,8,9,10,10-decachloro-5-hydroxypentacyclo[5.3.0.0^{2,6}.0^{3,9}.0^{4,8}]dec-= 5-yl)-4-oxovalerate; ethyl 5-(1,2,4,5,6,7,8,8,9,10-decachloro-3-hydroxypentacyclo= [5.3.0^{2,6}.0^{4,10}.0^{5,9}]dec-3-yl)-4-oxovalerate   **CA name** ethyl 1,1a,3,3a,4,5,5,5a,5b,6-= decachlorooctahydro-α-2-hydroxy-γ-oxo-1,3,4-metheno-1*H*-cyclobuta[*cd*]pentalene-= 2-pentanoate.   **M.f.** $C_{17}H_{12}Cl_{10}O_4$ **Code nos.** GC-9160 (Allied Chemical Corp.)   **Selected tradenames** 'Despirol'.

1129   kinoprene (common name).   **CAS RN** *[42588–37–4]*. Introduced by Zoecon Corp. (later Sandoz AG).   **IUPAC name** prop-2-ynyl (±)-(*E,E*)-3,7,11-trimethyl= dodeca-2,4-dienoate   **CA name** (*E,E*)-2-propynyl 3,7,11-trimethyl-2,4-dodecadienoate.   **M.f.** $C_{18}H_{28}O_2$ **Code nos.** ZR 777   **Selected tradenames** 'Enstar'.   **Details** *PM7,* entry 07630.

1130   leptophos (common name).   **CAS RN** *[21609–90–5]*. Insecticide reported by A. K. Azab (*Proc. Br. Insectic. Fungic. Conf., 5th,* 1969, **2**, 550). Introduced by Velsicol Chemical Corp.   **Common names** leptophos, MBCP   **IUPAC name** O-4-bromo-= 2,5-dichlorophenyl O-methyl phenylphosphonothioate   **CA name** O-(4-bromo-= 2,5-dichlorophenyl) O-methyl phenylphosphonothioate.   **M.f.** $C_{13}H_{10}BrCl_2O_2PS$ **Code nos.** VCS 506, OMS 1438.   **Selected tradenames** 'Abar', 'Phosvel'.   **Details** *PM6,* p. 318.

1131   lirimfos (common name).   **CAS RN** *[38260–63–8]*. Insecticide evaluated by Sandoz AG.   **IUPAC name** *O*-6-ethoxy-2-isopropylpyrimidin-4-yl *O,O*-dimethyl phosphorothioate   **CA name** *O*-[6-ethoxy-2-(1-methylethyl)-4-pyrimidinyl] *O,O*-dimethyl phosphorothioate.   **M.f.** $C_{11}H_{19}N_2O_4PS$ **Code nos.** SAN 2011.

1132   LS830556 (development code).   **CAS RN** *[98565–18–5]*. Herbicide introduced by Rhône-Poulenc Agrochimie.   **IUPAC name** mesyl(methyl)carbamoylmethylamino= methylphosphonic acid   **CA name** [[[2-methyl(methylsulfonyl)amino]-2-oxoethyl]= amino]methylphosphonic acid.   **M.f.** $C_5H_{13}N_2O_6PS$ **Code nos.** LS830 556 (Rhône-Poulenc).   **Details** *PM9*, entry 7985.

1133   lythidathion (common name).   **CAS RN** *[2669–32–1]*. Insecticide evaluated by J. R. Gelgy SA (later Ciba-Geigy AG).   **IUPAC name** *S*-5-ethoxy-2,3-dihydro-2-oxo-= 1,3,4-thiadiazol-3-ylmethyl *O,O*-dimethyl phosphorodithioate; 3-dimethoxyphosphino= thioylthiomethyl-5-ethoxy-1,3,4-thiadiazol-2(3*H*)-one   **CA name** *S*-[5-ethoxy-2-oxo-= 1,3,4-thiadiazol-3(2*H*)-ylmethyl] *O,O*-dimethyl phosphorodithioate. **M.f.** $C_7H_{13}N_2O_4PS_3$ **Code nos.** GS 12 968 (Ciba-Geigy AG); NC 2962 (FBC Limited, now Schering Agriculture).

1134   malonoben (common name).   **CAS RN** *[10537–47–0]*. Acaricide reported by H. Fukashi *et al.* (*Agric. Biol. Chem.*, 1971, **35**, 2003). Evaluated by Gulf Oil Chemicals Co. **Common names** malonoben, malonobène   **IUPAC name** 2-(3,5-di-*tert*-butyl-= 4-hydroxybenzylidene)malononitrile   **CA name** [[3,5-bis(1,1-dimethylethyl)-= 4-hydroxyphenyl]methylene]propanedinitrile. **M.f.** $C_{18}H_{22}N_2O$ **Code nos.** GCP-5126; S 15 126; ENT 27 190.   **Selected tradenames** 'GCP-5126-2EC', 'GPC-5126-4L', 'GCP-5126-50W'.   **Details** *PM6*, p. 323.

1135   mazidox (common name).   **CAS RN** *[7219–78–5]*. Insecticide evaluated by Pest Control Ltd (became Schering Agrochemicals).   **IUPAC name** tetramethyl= phosphorodiamidic azide (I); tetramethylazidophosphonic diamide   **CA name** (I). **M.f.** $C_4H_{12}N_5OP$ **Code nos.** NC 7.

1136   mebenil (common name).   **CAS RN** *[7055–03–0]*. Fungicide reported by E.-H. Pommer and J. Kradel (*Proc. Br. Insectic. Fungic. Conf.*, *5th*, 1969, **2**, 563). Developed by BASF AG.   **IUPAC name** *o*-toluanilide   **CA name** 2-methyl-*N*-phenylbenzamide.   **M.f.** $C_{14}H_{13}NO$ **Code nos.** BAS 305F; BAS 3050F; BAS 3053F.   **Details** *PM5*, p. 332.

1137   mecarbinzid (common name).   **CAS RN** *[27386–64–7]*. Fungicide evaluated by BASF AG.   **Common names** mecarbinzid, mecarbinzide   **IUPAC name** methyl 1-(2-methylthioethylcarbamoyl)benzimidazol-2-ylcarbamate   **CA name** methyl 1-[[[2-(methylthio)ethyl]amino]carbonyl]-1*H*-benzimidazol-2-ylcarbamate.   **M.f.** $C_{13}H_{16}N_4O_3S$ **Code nos.** BAS 3201F.

1138   mecarphon (common name).   **CAS RN** *[29173–31–7]*. Insecticide reported by M. Pianka & W. S. Catling (*Proc. Int. Congr. Entomol.*, *13th*, Moscow, 1968, **2**, 263). Introduced by Murphy Chemical Co. (later DowElanco).   **IUPAC name** methyl [methoxy(methyl)phosphinothioylthio]acetyl(methyl)carbamate; *S*-(*N*-methoxy=

carbonyl-*N*-methylcarbamoylmethyl) *O*-methyl methylphosphonodithioate   **CA name** methyl 3,7-dimethyl-6-oxo-2-oxa-4-thia-7-aza-3-phosphaoctan-8-oate 3-sulfide; methyl [[methoxy(methylphosphinothioyl)thio]acetyl]methylcarbamate (former 9Cl).   **M.f.** $C_7H_{14}NO_4PS_2$ **Code nos.** MC 2420, OMS 1478.   **Details** *PM4*, p. 329.

1139   medinoterb; medinoterb acetate (common name).   **CAS RN** *[3996–59–6]* medinoterb; *[2487–01–6]* medinoterb acetate. Herbicide reported by G. A. Emery *et al.* (*Proc. 2nd EWRC Symp. New Herbic.*, 1965, p. 141). Introduced by Murphy Chemical Co. (later DowElanco).   **Common names** medinoterb, médinoterbe   **IUPAC name** 6-*tert*-butyl-2,4-dinitro-*m*-cresol; 6-*tert*-butyl-3-methyl-2,4-dinitrophenol   **CA name** 6-(1,1-dimethylethyl)-3-methyl-2,4-dinitrophenol.   **M.f.** $C_{11}H_{14}N_2O_5$; (medinoterb acetate $C_{13}H_{16}N_2O_6$)   **Code nos.** P 1488; MC 1488.   **Details** *PM5*, p. 335.

1140   menazon (common name).   **CAS RN** *[78–57–9]*. Insecticide reported by A. Calderbank (*Chem. Ind. (London)*, 1961, p. 630). Introduced by ICI Plant Protection Division (later ICI Agrochemicals).   **Common names** menazon, azidithion   **IUPAC name** *S*-4,6-diamino-1,3,5,-triazin-2-ylmethyl *O,O*-dimethyl phosphorodithioate   **CA name** *S*-(4,6-diamino-1,3,5-triazin-2-yl)methyl *O,O*-dimethyl phosphorodithioate.   **M.f.** $C_6H_{12}N_5O_2PS_2$ **Code nos.** PP175; OMS 503; ENT 25 760.   **Selected tradenames** 'Sayfos', 'Saphicol', 'Saphizon', 'SAlsan'; 'Abol X'.   **Details** *PM6*, p. 331.

1141   mephosfolan (common name).   **CAS RN** *[950–10–7]*. Insecticide and acaricide introduced by American Cyanamid Co.   **Common names** mephosfolan, méphospholan   **IUPAC name** diethyl 4-methyl-1,3-dithiolan-2-ylidenephosphoramidate; 2-(diethoxy= phosphinylimino)-4-methyl-1,3-dithiolane   **CA name** diethyl (4-methyl-1,3-dithiolan-= 2-ylidene)phosphoramidate.   **M.f.** $C_8H_{16}NO_3PS_2$ **Code nos.** EI 47 470 (Cyanamid), ENT 25 991.   **Selected tradenames** 'Cytrolane'.   **Details** *PM10*, entry 446.

1142   mesoprazine (common name).   **CAS RN** *[1824–09–5]*. Herbicide evaluated by J. R. Geigy SA (later Ciba-Geigy AG).   **IUPAC name** 6-chloro-N^2-isopropyl-N^4-= (3-methoxypropyl)-1,3,5-triazine-2,4-diamine   **CA name** 6-chloro-*N*-(3-methoxy= propyl)-*N'*-(1-methylethyl)-1,3,5-triazine-2,4-diamine.   **M.f.** $C_{10}H_{18}ClN_5O$ **Code nos.** G 34 698; CGA 4999.

1143   mesulfenfos (common name).   **CAS RN** *[3761–41–9]*. Insecticide evaluated by Bayer AG.   **IUPAC name** *O,O*-dimethyl *O*-4-methylsulfinyl-*m*-tolyl phosphorothioate   **CA name** *O,O*-dimethyl *O*-[3-methyl-4-(methylsulfinyl)phenyl] phosphorothioate.   **M.f.** $C_{10}H_{15}O_4PS_2$ **Code nos.** BAY S 2281.

1144   *Metarhizium anisopliae* (scientific name).   Soil insecticide evaluated by Bayer AG, Ecoscience and other companies. **Var.** Esc-1 (Ecoscience).   **Code nos.** BIO 1020 (Bayer).

1145   *Metaseiulus occidentalis* (scientific name).   Biological agent for control of spider mites, introduced by Koppert and other companies.   **Selected tradenames** 'Spidex-O'. **Details** *PM10*, entry 457.

1336   *Superseded entries – 1145*

1146 **metazoxolon** (common name). **CAS RN** *[5707–73–3]*. Fungicide reported by T. J. Purnell (*Proc. Br. Crop Prot. Conf., 7th*, 1973, **2**, 603). Evaluated by ICI Agrochemicals. **IUPAC name** 4-(3-chlorophenylhydrazono)-3-methyl-1,2-oxazol-5(4*H*)-one; 4-(3-chlorophenylhydrazono)-3-methylisoxazol-5(4*H*)-one **CA name** 3-methyl-= 4,5-isoxazoledione 4-[(3-chlorophenyl)hydrazone]. **M.f.** $C_{10}H_8ClN_3O_2$ **Code nos.** PP395.

1147 **metflurazon** (common name). **CAS RN** *[23576–23–0]*. Herbicide evaluated by Sandoz AG. **Common names** metflurazon, metflurazone **IUPAC name** 4-chloro-= 5-dimethylamino-2-(α,α,α-trifluoro-*m*-tolyl)pyridazin-3(2*H*)-one **CA name** 4-chloro-= 5-(dimethylamino)-2-[(3-trifluoromethyl)phenyl]-3(2*H*)-pyridazinone. **M.f.** $C_{13}H_{11}ClF_3N_3O$ **Code nos.** SAN 6706H.

1148 **methalpropalin** (common name). **CAS RN** *[57801–46–4]*. Herbicide evaluated by Eli Lilly & Co. **Common names** methalpropalin, méthalpropaline **IUPAC name** α,α,α-trifluoro-*N*-(2-methylallyl)-2,6-dinitro-*N*-propyl-*p*-toluidine **CA name** *N*-(2-methyl-2-propenyl)-2,6-dinitro-*N*-propyl-4-(trifluoromethyl)benzenamine. **M.f.** $C_{14}H_{16}F_3N_3O_4$.

1149 **methanesulfonyl fluoride** (IUPAC/Chemical Abstracts name). **CAS RN** *[558–25–8]*. Fumigant insecticide marketed by Bayer AG. **M.f.** CH_3FO_2S **Other names** MSF **Selected tradenames** 'Fumette'.

1150 **methazole** (common name). **CAS RN** *[20354–26–1]*. Herbicide reported by W. Furness (*Proc. Int. Congr. Plant Prot., 7th,* Paris, 1970, p. 314). Introduced by Velsicol Chemical Corp. and later manufactured and marketed by Sandoz AG. **IUPAC name** 2-(3,4-dichlorophenyl)-4-methyl-1,2,4-oxadiazolidine-3,5-dione **CA name** 2-(3,4-dichlorophenyl)-4-methyl-1,2,4-oxadiazolidine-3,5-dione. **M.f.** $C_9H_6Cl_2N_2O_3$ **Code nos.** VCS-438 (Sandoz) **Selected tradenames** 'Probe'. **Details** *PM9*, entry 8110.

1151 **methfuroxam** (common name). **CAS RN** *[28730–17–8]*. Fungicide reported by K. T. Alcock (*Plant Dis. Reptr.*, 1978, **62**, 854). Introduced in Germany by Uniroyal Chemical Co., Inc. **Common names** methfuroxam, méthfuroxame **IUPAC name** 2,4,5-trimethyl-3-furanilide **CA name** 2,4,5-trimethyl-*N*-phenyl-3-furancarboxamide. **M.f.** $C_{14}H_{15}NO_2$ **Code nos.** H719 (Uniroyal). **Details** *PM9*, entry 8120.

1152 **methiobencarb** (common name). **CAS RN** *[18357–78–3]*. Herbicide evaluated by Bayer AG. **Common names** methiobencarb, méthiobencarbe **IUPAC name** *S*-4-methoxybenzyl diethylthiocarbamate **CA name** *S*-[(4-methoxyphenyl)methyl] diethylcarbamothioate. **M.f.** $C_{13}H_{19}NO_2S$ **Code nos.** NTN 5810.

1153 **methiuron** (common name). **CAS RN** *[21540–35–2]*. Herbicide introduced by Yorkshire Tar Distillers Ltd. **IUPAC name** 1,1-dimethyl-3-*m*-tolyl-2-thiourea **CA name** *N,N*-dimethyl-*N'*-(3-methylphenyl)thiourea. **M.f.** $C_{10}H_{14}N_2S$ **Code nos.** MH 090. **Details** *PM2*, p. 306.

1154 **methocrotophos** (common name). **CAS RN** *[25601–84–7]*. Insecticide evaluated by Ciba AG (later Ciba-Geigy AG). **Common names** methocrotophos, métocrotophos   **IUPAC name** (*E*)-2-(*N*-methoxy-*N*-methylcarbamoyl)-1-methylvinyl dimethyl phosphate; 3-dimethoxyphosphinoyloxy-*N*-methoxy-*N*-methylisocrotonamide **CA name** (*E*)-3-(methoxymethylamino)-1-methyl-3-oxo-1-propenyl dimethyl phosphate.   **M.f.** $C_8H_{16}NO_6P$ **Code nos.** C 2307.

1155   methometon (common name).   **CAS RN** *[1771–07–9]*. Herbicide evaluated by J. R. Geigy SA (later Ciba-Geigy AG).   **Common names** methometon, métométon **IUPAC name** 6-methoxy-N^2,N^4-bis(3-methoxypropyl)-1,3,5-triazine-2,4-diamine **CA name** 6-methoxy-*N*,*N'*-bis(3-methoxypropyl)-1,3,5-triazine-2,4-diamine.   **M.f.** $C_{12}H_{23}N_5O_3$ **Code nos.** G 34 690.

1156   methoprotryne (common name).   **CAS RN** *[841–06–5]*. Herbicide reported by A. Gast *et al.* (*Z. Pflanzenkr. Pflanzenpathol. Pflanzenschutz*, 1965, **72**, 325). Introduced by J. R. Geigy S.A. (later Ciba-Geigy AG).   **Common names** methoprotryne, métoprotryne   **IUPAC name** N^2-isopropyl-N^4-(3-methoxypropyl)-6-methylthio-= 1,3,5-triazine-2,4-diamine   **CA name** *N*-(3-methoxypropyl)-*N'*-(1-methylethyl)-= 6-(methylthio)-1,3,5-triazine-2,4-diamine.   **M.f.** $C_{11}H_{21}N_5OS$ **Code nos.** G 36 393 (Ciba)   **Selected tradenames** 'Gesaran'.   **Details** *PM9*, entry 8210.

1157   methoquin-butyl (common name).   **CAS RN** *[19764–43–3]*. Insecticide. **Common names** methoquin-butyl, méthoquine-butyl   **IUPAC name** butyl 3-methyl= quinoline-4-carboxylate   **CA name** butyl 3-methyl-4-quinolinecarboxylate.   **M.f.** $C_{15}H_{17}NO_2$ **Other names** BMC.

1158   2-methoxyethylmercury acetate (IUPAC name).   **CAS RN** *[151–38–2]*. Fungicide introduced by Kenogard VT AB.   **CA name** (acetato-*O*)(2-methoxyethyl)= mercury.   **M.f.** $C_5H_{10}HgO_3$ **Selected tradenames** 'Panogen'.

1159   2-methoxyethylmercury chloride (IUPAC name).   **CAS RN** *[123–88–6]*. Fungicide introduced by I. G. Farbenindustrie AG (W. Bonrath, *Nachr. Schaedlingsbekaempfung I. G. Farbenindustrie (Leverkusen)*, 1935, **10**, 23) (later Bayer AG).   **CA name** chloro(2-methoxyethyl)mercury.   **M.f.** C_3H_7ClHgO **Selected tradenames** 'Agallol', 'Aretan', 'Ceresan Universal Liquid Seed Treatment'.   **Details** *PM7*, entry 08260.

1160   2-methoxyethylmercury silicate (chemical name).   **CAS RN** *[64491–92–5]*. Introduced by I. G. Farbenindustrie AG (later Bayer AG, who no longer manufacture or market it).   **IUPAC name** The compound is an organomercury silicate but there is some doubt about whether it is an orthosilicate or a metasilicate and therefore about its true molecular structure   **CA name** hydrogen [metasilicato(2-)-*O*](2-methoxyethyl)= mercurate(1-).   **Selected tradenames** 'Ceresan Universal Dry Seed Treatment', 'Soprasan'.   **Details** *PM8*, entry 08270.

1161   methoxyphenone (common name).   **CAS RN** *[41295–28–7]*. Herbicide reported (*Proc. Asian-Pacific Weed Sci. Soc. Conf.*, 4th, 1973, p. 215). Introduced by Nippon Kayaku Co., Ltd.   **IUPAC name** 4-methoxy-3,3'-dimethylbenzophenone

CA name (4-methoxy-3-methylphenyl)(3-methylphenyl)methanone. M.f. $C_{16}H_{16}O_2$
Code nos. NK-049   Selected tradenames 'Kayametone'.   Details *PM8*, entry 08250.

1162   methylarsenic sulfide (IUPAC name).   CAS RN *[2533-82-6]*. Fungicidal
properties of this chemical, which was reported in 1858 (von Bayer, *Ann. Chem. Pharm.*,
1858, **107**, 279), described in 1931. Introduced by Bayer AG (1958).   CA name
methylthioxoarsine.   M.f. CH_3AsS Other names MAS   Selected tradenames
'Rhizoctol', 'Urbasulf', 'Rhizoctol combi', 'Rhizoctol slurry'.   Details *PM4*, p. 349.

1163   methylmercury dicyandiamide (common name).   CAS RN *[502-39-6]*.
Fungicide introduced in Sweden in 1938, and in USA. by Panogen Inc.   IUPAC name
1-cyano-3-(methylmercurio)guanidine; (methylmercurio)guanidinocarbonitrile   CA
name (cyanoguanidinato-*N'*)methylmercury.   M.f. $C_3H_6HgN_4$ Selected tradenames
'Panogen'.   Details *PM5*, p. 361.

1164   3-methyl-1-phenylpyrazol-5-yl dimethylcarbamate (IUPAC name).   CAS RN
[87-47-8]. Insecticide introduced by J. R. Geigy AG (later Ciba-Geigy AG).   CA name
3-methyl-1-phenyl-1*H*-pyrazol-5-yl dimethylcarbamate.   M.f. $C_{13}H_{15}N_3O_2$ Code nos.
G 22 008; OMS 20; ENT 17 588.   Selected tradenames 'Pyrolan'.

1165   2-methyl(prop-2-ynyl)aminophenyl methylcarbamate (IUPAC name).   CAS RN
[23504-07-6]. Insecticide evaluated by Ciba AG (later Ciba-Geigy AG).   CA name
2-(methyl-2-propynylamino)phenyl methylcarbamate.   M.f. $C_{12}H_{14}N_2O_2$ Code nos.
C 17 018.   Details *PM3*, p. 339.

1166   4-methyl(prop-2-ynyl)amino-3,5-xylyl methylcarbamate (IUPAC name).   CAS
RN *[23623-49-6]*. Insecticide evaluated by Ciba AG (later Ciba-Geigy AG).   CA name
3,5-dimethyl-4-(methyl-2-propynylamino)phenyl methylcarbamate.   M.f. $C_{14}H_{18}N_2O_2$
Code nos. C 20 132.   Details *PM2*, p. 320.

1167   5-methyl-6-thioxo-1,3,5-thiadiazinan-3-ylacetic acid (IUPAC name).   CAS RN
[3655-88-7]. Nematicide introduced by Bayer AG.   IUPAC name 5-methyl-=
6-thioxo-1,3,5-thiadiazinan-3-ylacetic acid; tetrahydro-5-methyl-6-thioxo-=
2*H*-1,3,5-thiadiazin-3-ylacetic acid   CA name dihydro-5-methyl-6-thioxo-=
2*H*-1,3,5-thiadiazine-3(4*H*)-acetic acid.   M.f. $C_6H_{10}N_2O_2S_2$ Selected tradenames
'Terracur'.   Details *PM1*, p. 71.

1168   metoxadiazone (common name).   CAS RN *[60589-06-2]*. Insecticide reported
by D. Ambrosi et al. (*Proc. 1979 Br. Crop Prot. Conf. - Pests Dis.*, **2**, 533). Discovered by
Rhône-Poulenc Agrochimie and developed under license by Sumitomo Chemical Co.,
Ltd.   IUPAC name 5-methoxy-3-(2-methoxyphenyl)-1,3,4-oxadiazol-2(3*H*)-one (I)   CA
name (I).   M.f. $C_{10}H_{10}N_2O_4$ Code nos. 32 861 RP   Selected tradenames 'Elemic',
'Valsan', 'Wiper-jet'.

1169   metsulfovax (common name).   CAS RN *[21452-18-6]*. Fungicide evaluated by
Uniroyal Inc.   IUPAC name 2,4-dimethyl-1,3-thiazole-5-carboxanilide; 2,4-dimethyl=
thiazole-5-carboxanilide   CA name 2,4-dimethyl-*N*-phenyl-5-thiazolecarboxamide.

M.f. $C_{12}H_{12}N_2OS$ **Code nos.** G696   **Selected tradenames** 'Provax'.
Details *PM9*, entry 8582.

1170   mexacarbate (common name).   **CAS RN** *[315–18–4]*. Insecticide introduced by
Dow Chemical Co. (later DowElanco).   **IUPAC name** 4-dimethylamino-3,5-xylyl
methylcarbamate   **CA name** 4-(dimethylamino)-3,5-dimethylphenyl methylcarbamate.
M.f. $C_{12}H_{18}N_2O_2$ **Code nos.** Dowco 139; OMS 47; OMS 639; ENT 25 766.   **Selected
tradenames** 'Zectran'.   **Details** *PM4*, p. 359.

1171   MG 191 (development code).   Herbicide safener developed by Nitrokémia
Ipartelepek.   **IUPAC name** 2-dichloromethyl-2-methyl-1,3-dioxolane.   **M.f.** $C_5H_8Cl_2O_2$
Code nos. MG 191.

1172   milneb (common name).   **CAS RN** *[3773–49–7]*. Fungicide evaluated by E. I. du
Pont de Nemours and Co.   **Common names** milneb, thiadiazine   **IUPAC name**
4,4',6,6'-tetramethyl-3,3'-ethylenedi-1,3,5-thiadiazinane-2-thione; 4,4',6,6'-tetra=
methyl-3,3'-ethylenebis(tetrahydro-1*H*-1,3,5-thiadiazine-2-thione)   **CA name**
3,3'-(1,2-ethanediyl)bis[tetrahydro-4,6-dimethyl-2*H*-1,3,5-thiadiazine-2-thione].   **M.f.**
$C_{12}H_{22}N_4S_4$ **Code nos.** Experimental Fungicide 328   **Selected tradenames** 'Banlate'.

1173   mipafox (common name).   **CAS RN** *[371–86–8]*. Insecticide and acaricide
introduced by Pest Control Ltd (later Schering Agrochemicals).   **IUPAC name**
N,N'-di-isopropylphosphorodiamidic fluoride   **CA name** *N,N'*-bis(1-methylethyl)=
phosphorodiamidic fluoride.   **M.f.** $C_6H_{16}FN_2OP$ **Code nos.** Pestox 15   **Selected
tradenames** 'Isopestox'.

1174   mirex (common name).   **CAS RN** *[2385–85–5]*. Insecticide introduced by Allied
Chemical Corp. Agricultural Div. (later Hopkins Ltd).   **IUPAC name** dodecachloro=
pentacyclo[5.3.0.0^{2,6}.0^{3,9}.0^{4,8}]decane; perchloropentacyclo[5.3.0.0^{2,6}.0^{3,9}.0^{4,8}]decane;
dodecachloropentacyclo[5.2.1.0^{2,6}.0^{3,9}.0^{5,8}]decane   **CA name**
1,1a,2,2,3,3a,4,5,5,5a,5b,6-dodecachlorooctahydro-1,3,4-metheno-1*H*-cyclobuta=
[*cd*]pentalene.   **M.f.** $C_{10}Cl_{12}$ **Other names** perchlordécone   **Code nos.** GC 1283
(Allied Chemical Corp.), ENT 25 719.   **Selected tradenames** 'Mirex'.   **Details** *PM5*,
p. 368.

1175   MNFA (common name).   **CAS RN** *[5903–13–9]*. Acaricide introduced by
Nippon Soda Co., Ltd.   **IUPAC name** 2-fluoro-*N*-methyl-*N*-1-naphthylacetamide
CA name 2-fluoro-*N*-methyl-*N*-(1-naphthalenyl)acetamide.   **M.f.** $C_{13}H_{12}FNO$
Code nos. NA-26, ENT 27 403.   **Selected tradenames** 'Nissol'.

1176   MON 4620 (development code).   **CAS RN** *[40164–67–8]*. Plant growth
regulator introduced by Monsanto Agriculture Co.   **IUPAC name** *N*-acetamidomethyl-=
2-chloro-2',6'-diethylacetanilide   **CA name** *N*-[(acetylamino)methyl]-2-chloro-=
N-(2,6-diethylphenyl)acetamide.   **M.f.** $C_{15}H_{21}ClN_2O_2$ **Other names** [amidochlor]
(rejected common name proposal)   **Code nos.** MON 4620; CP-76 963   **Selected
tradenames** 'Limit'.

1177   monalide (common name).   CAS RN *[7287–36–7]*. Herbicide reported by F. Arndt (Z. *Pflanzenkr. Pflanzenpathol. Pflanzenschutz*, 1965, Sonderheft III, p. 277). Introduced by Schering AG (later AgrEvo GmbH).   **IUPAC name** 4′-chloro-= 2,2-dimethylvaleranilide; 4′-chloro-α,α-dimethylvaleranilide   **CA name** N-(4-chloro= phenyl)-2,2-dimethylpentanamide.   **M.f.** $C_{13}H_{18}ClNO$ **Code nos.** Schering 35 830 (AgrEvo)   **Selected tradenames** 'Potablan'.   **Details** *PM9*, entry 8650.

1178   monisouron (common name).   CAS RN *[55807–46–0]*. Herbicide introduced by Shionogi & Co. Ltd.   **IUPAC name** 1-(5-*tert*-butyl-1,2-oxazol-3-yl)-3-methylurea; 1-(5-*tert*-butylisoxazol-3-yl)-3-methylurea   **CA name** N-[5-(1,1-dimethylethyl)-= 3-isoxazolyl]-N′-methylurea.   **M.f.** $C_9H_{15}N_3O_2$ **Code nos.** SSH-41.

1179   monuron; monuron-TCA (common name).   **CAS RN** *[150–68–5]* monuron (i); *[140–41–0]* monuron-TCA (ii). Herbicidal activity of monuron reported by H. C. Bucha & C. W. Todd (*Science*, 1951, **114**, 493). Introduced by E. I. du Pont de Nemours and Co. (who no longer manufacture or market it) and as the trichloroacetate by Allied Chemical Corp. Agricultural Division (later Hopkins Agricultural Chemical Co.). **Common names** monuron, [chlorfenidim], CMU, monuron-TCA for the trichloroacetate salt (ii)   **IUPAC name** 3-(4-chlorophenyl)-1,1-dimethylurea (i); 3-(4-chlorophenyl)-1,1-dimethyluronium trichloroacetate (ii)   **CA name** N′-(4-chlorophenyl)-N,N-dimethylurea (i); trichloroacetic acid compound with N′-(4-chlorophenyl)-N,N-dimethylurea (1:1) (ii); 3-(p-chlorophenyl)-= 1,1-dimethylurea (i).   **M.f.** $C_9H_{11}ClN_2O$; (monuron-TCA $C_{11}H_{12}Cl_4N_2O_3$)   **Code nos.** GC-2996 (Allied) for (ii)   **Selected tradenames** 'Telvar'.   **Details** *PM8*, entry 08680.

1180   morfamquat; morfamquat dichloride (common name).   CAS RN *[4636–83–3]* morfamquat dichloride. Dichloride introduced as herbicide (H. M. Fox, *Proc. Br. Weed Control Conf., 7th*, 1964, 29; H. M. Fox & C. R. Beech, *ibid.*, p. 108) by ICI Plant Protection Div. (later ICI Agrochemicals).   **IUPAC name** 1,1′-bis(3,5-dimethyl= morpholinocarbonylmethyl)-4,4′-bipyridinium; 1,1′-bis(3,5-dimethylmorpholino= carbonylmethyl)-4,4′-bipyridyldiylium   **CA name** 1,1′-bis[2-(3,5-dimethyl-= 4-morpholinyl)-2-oxoethyl]-4,4′-bipyridinium.   **M.f.** $C_{26}H_{36}N_4O_4$; (morfamquat dichloride $C_{26}H_{36}Cl_2N_4O_4$)   **Code nos.** PP745   **Selected tradenames** 'Morfoxone'. **Details** *PM3*, p. 355.

1181   morphothion (common name).   CAS RN *[144–41–2]*. Systemic insecticide introduced by Sandoz AG.   **IUPAC name** O,O-dimethyl S-morpholinocarbonylmethyl phosphorodithioate   **CA name** O,O-dimethyl S-[2-(4-morpholinyl)-2-oxoethyl] phosphorodithioate.   **M.f.** $C_8H_{16}NO_4PS_2$ **Selected tradenames** 'Ekatin M'.

1182   mucochloric anhydride (trivial name).   CAS RN *[4412–09–3]*. Fungicide reported by A. E. Rich (*Plant Dis. Rep.*, 1960, **44**, 306). Evaluated by Allied Chemical Corp. Agrochemical Div. (later Hopkins Ltd).   **IUPAC name** 2,2′,3,3′-tetrachloro-= 4,4′-oxydibut-2-en-4-olide; 5,5′-oxybis(3,4-dichloro-5H-furan-2-one)   **CA name** 5,5′-oxybis[3,4-dichloro-2(5H)-furanone].   **M.f.** $C_8H_2Cl_4O_5$ **Other names** mucochloric anhydride   **Code nos.** GC 2466.   **Details** *PM2*, p. 334.

1183　myclozolin (common name). **CAS RN** *[54864–61–8]*. Fungicide reported by E.-H. Pommer & B. Zeeh (*Meded. Fac. Landbouwwet. Rijksuniv. Gent*, 1982, **47**, 935). Introduced by BASF AG. **Common names** myclozolin, myclozoline **IUPAC name** (*RS*)-3-(3,5-dichlorophenyl)-5-methoxymethyl-5-methyl-1,3-oxazolidine-2,4-dione **CA name** 3-(3,5-dichlorophenyl)-5-(methoxymethyl)-5-methyl-2,4-oxazolidinedione. **M.f.** $C_{12}H_{11}Cl_2NO_4$ **Code nos.** BAS 436F. **Details** *PM7*, entry 08730.

1184　naphthalene (IUPAC/Chemical Abstracts name). **CAS RN** *[91–20–3]*. Has long been used as a household fumigant against clothes moths. **Common names** naphthalene, naphtaléne. **M.f.** $C_{10}H_8$. **Details** *PM9*, entry 8760.

1185　naphthalic anhydride (common name). **CAS RN** *[81–84–5]*. Enhancement of the selectivity of thiocarbamate herbicides to maize reported by O. L. Hoffman (*Weed Sci. Soc. Am., Abstr.*, 1969, No. 12) and reviewed by G. R. Stephenson & F. Y. Chang (*Chemistry and Mode of Action of Herbicide Antidotes*, p. 35). Introduced by Gulf Oil Corp. **IUPAC name** naphthalene-1,8-dicarboxylic anhydride **CA name** 1*H*,3*H*-naphtho= [1,8-*cd*]pyran-1,3-dione. **M.f.** $C_{12}H_6O_3$ **Selected tradenames** 'Protect'. **Details** *PM8*, entry 08770.

1186　NC-170 (development code). Insect growth regulator reported by T. Miyake *et al.* (*Proc. 1988 Brighton Crop Prot. Conf. - Pests Dis.*, **2**, 535). Evaluated by Nissan Chemical Industries, Ltd. **IUPAC name** 4-chloro-5-(6-chloro-3-pyridylmethoxy)-= 2-(3,4-dichlorophenyl)pyridazin-3(2*H*)-one **CA name** 4-chloro-5-[(6-chloro-= 3-pyridinyl)methoxy]-2-(3,4-dichlorophenyl)-3(2*H*)-pyridazinone. **M.f.** $C_{16}H_9Cl_4N_3O_2$ **Code nos.** NC-170.

1187　NC-330 (development code). Herbicide reported by T. Nawamaki *et al.* (*Proc. Br. Crop Prot. Conf. - Weeds*, 1991, **1**, 45). Developed by Nissan Chemical Industries, Ltd. **IUPAC name** methyl 5-(4,6-dimethylpyrimidin-2-ylcarbamoylsulfamoyl)-1-(2-pyridyl)= pyrazole-4-carboxylate. **M.f.** $C_{17}H_{17}N_7O_5S$ **Code nos.** NC-330; NCI-851013 (both Nissan). **Details** *PM10*, entry 500.

1188　*Neodiprion Sertifer* NPV (scientific name). Use of this virus to control *N. sertifer* (European pine sawfly) larvae was pioneered by the USDA Forest Service at Hamden, Connecticut, by the Natural Environment Research Council Institute of Virology, Oxford, and the Forestry Commission in conjunction with Microbial Resources Ltd. **Scientific name** *Neodiprion Sertifer* Nuclear Polyhedrosis Virus. **Selected tradenames** 'Virox'. **Details** *PM9*, entry 8865.

1189　nifluridide (common name). **CAS RN** *[61444–62–0]*. Insecticide evaluated by Eli Lilly & Co. **IUPAC name** 6'-amino-α,α,α,2,2,3,3-heptafluoro-5'-nitropropion-= *m*-toluidide **CA name** *N*-[2-amino-3-nitro-5-(trifluoromethyl)phenyl]-= 2,2,3,3-tetrafluoropropanamide. **M.f.** $C_{10}H_6F_7N_3O_3$ **Code nos.** EL-468.

1190　nipyraclofen (common name). **CAS RN** *[99662–11–0]*. Herbicide discovered by Bayer AG. **IUPAC name** 1-(2,6-dichloro-α,α,α-trifluoro-*p*-tolyl)-4-nitropyrazol-= 5-ylamine **CA name** 1-[2,6-dichloro-4-(trifluoromethyl)phenyl]-4-nitro-1*H*-pyrazol-=

5-ylamine. **M.f.** $C_{10}H_5Cl_2F_3N_4O_2$ **Code nos.** SLA 3992 (Bayer). **Details** *PM9*, entry 8915.

1191 nitralin (common name). **CAS RN** *[4726–14–1]*. Herbicide reported by J. B. Regan *et al.* (*Proc. Northeast. Weed Control Conf.*, 1966, p. 36). Introduced by Shell Research Ltd. **IUPAC name** 4-methylsulfonyl-2,6-dinitro-*N,N*-dipropylaniline **CA name** 4-(methylsulfonyl)-2,6-dinitro-*N,N*-dipropylbenzenamine. **M.f.** $C_{13}H_{19}N_3O_6S$ **Code nos.** SD 11 831 **Selected tradenames** 'Planavin'. **Details** *PM8*, entry 08920.

1192 nitrilacarb; nitrilacarb 1:1 zinc chloride complex (common name). **CAS RN** *[29672–19–3]* nitrilacarb; *[58270–08–9]* formerly *[61332–32–9]* 1:1 complex with zinc chloride. Insecticide reported by W. K. Whitney & J. L. Aston (*Proc. Br. Insectic. Fungic. Conf., 8th*, 1975, **2**, 633). The 1:1 complex with zinc chloride introduced by American Cyanamid Co. **Common names** nitrilacarb, nitrilacarbe **IUPAC name** 4,4-dimethyl-= 5-(methylcarbamoyloxyimino)valeronitrile; 4,4-dimethyl-5-(methylcarbamoyloxy= imino)pentanenitrile **CA name** 4,4-dimethyl-5-[[[(methylamino)carbonyl]oxy]imino]= pentanenitrile. **M.f.** $C_9H_{15}N_3O_2$; (zinc chloride complex $C_9H_{15}Cl_2N_3O_2Zn$) **Code nos.** AC 82 258; CL 72 613 (nitrilacarb) **Selected tradenames** 'Accotril'. **Details** *PM6*, p. 385.

1193 nitrofen (common name). **CAS RN** *[1836–75–5]*. Herbicide introduced by Rohm & Haas Co. **Common names** nitrofen, nitrofène, niclofen, NIP **IUPAC name** 2,4-dichlorophenyl 4-nitrophenyl ether **CA name** 2,4-dichloro-1-(4-nitrophenoxy)= benzene. **M.f.** $C_{12}H_7Cl_2NO_3$ **Code nos.** FW-925 **Selected tradenames** 'Tok', 'Tokkorn'. **Details** *PM8*, entry 08950.

1194 nitrofluorfen (common name). **CAS RN** *[42874–01–1]*. Herbicide evaluated by Rohm & Haas Co. **Common names** nitrofluorfen, nitrofluorfène **IUPAC name** 2-chloro-α,α,α-trifluoro-*p*-tolyl 4-nitrophenyl ether **CA name** 2-chloro-1-(4-nitro= phenoxy)-4-(trifluoromethyl)benzene. **M.f.** $C_{13}H_7ClF_3NO_3$ **Code nos.** RH-2512.

1195 *N*-3-nitrophenylitaconimide (IUPAC name). **CAS RN** *[4137–12–6]*. Fungicide reported by B. von Schmeling (*Phytopathology*, 1962, **52**, 819). Evaluated by Uniroyal Inc. **CA name** 3-methylene-*N*-(3-nitrophenyl)pyrrolidine-2,5-dione. **M.f.** $C_{11}H_8N_2O_4$ **Code nos.** B 720.

1196 4-(2-nitroprop-1-enyl)phenyl thiocyanate (IUPAC name). **CAS RN** *[950–00–5]*. Fungicide introduced by Nippon Kayaku Co., Ltd. **Common names** nitrostyrene **CA name** 4-(2-nitro-1-propenyl)phenyl thiocyanate. **M.f.** $C_{10}H_8N_2O_2S$ **Selected tradenames** 'Styrocide'.

1197 norbormide (common name). **CAS RN** *[991–42–4]*. Rodenticide reported by A. P. Roszkowski *et al.* (*Science*, 1964, **144**, 412). Introduced by the McNeil Laboratories Inc. **Common names** norbormide, nobormide **IUPAC name** 5-(α-hydroxy-α-2-= pyridylbenzyl)-7-(α-2-pyridylbenzylidene)-8,9,10-trinorborn-5-ene-2,3-dicarboximide; 5-(α-hydroxy-α-2-pyridylbenzyl)-7-(α-2-pyridylbenzylidene)bicyclo[2.2.1]hept-5-ene-= 2,3-dicarboximide **CA name** 3a,4,7,7a-tetrahydro-5-(hydroxyphenyl-2-pyridinyl=

methyl)-8-(phenyl-2-pyridinylmethylene)-4,7-methano-1H-isoindole-1,3(2H)-dione.
M.f. $C_{33}H_{25}N_3O_3$ **Code nos.** McN-1025, ENT 51 762.   **Selected tradenames** 'Shoxin', 'Raticate'.   **Details** *PM8,* entry 09020.

1198   nornicotine (traditional name).   **CAS RN** *[494–97–3].* Insecticide.   **IUPAC name** 3-(pyrrolidin-2-yl)pyridine   **CA name** (S)-3-(2-pyrrolidinyl)pyridine.   **M.f.** $C_9H_{12}N_2$.

1199   noruron (common name).   **CAS RN** *[18530–56–8]* 3aα,4α,5α,7α,7aα-isomer; *[2163–79–3]* unstated stereochemistry. Herbicide reported by G. A. Buntin *et al.* (*Abstr. Am. Chem. Soc. Meeting,* Los Angeles, 1963). Introduced by Hercules Inc. Agrochemical Div. (and later Nor-Am.).   **Common names** noruron, norea   **IUPAC name** 1,1-dimethyl-= 3-(perhydro-4,7-methanoinden-5-yl)urea; 3-(hexahydro-4,7-methanoindan-5-yl)-= 1,1-dimethylurea   **CA name** 3aα,4α,5α,7α,7aα-N,N-dimethyl-N'-(octahydro-= 4,7-methano-1H-inden-5-yl)urea.   **M.f.** $C_{13}H_{22}N_2O$ **Code nos.** Hercules 7531   **Selected tradenames** 'Herban'.   **Details** *PM3,* p. 368.

1200   OCH (common name).   **CAS RN** *[4024–81–1].* Herbicide and fungicide introduced by Goodrich Chemical Co.   **IUPAC name** octachlorocyclohex-2-en-1-one   **CA name** 2,3,4,4,5,5,6,6-octachloro-2-cyclohexen-1-one.   **M.f.** C_6Cl_8O **Selected tradenames** 'Oktone'.

1201   OCS 21693 (development code).   **CAS RN** *[14419–01–3].* Herbicide reported by W. Furness (*Symp. New Herbic., 3rd,* 1969, p. 261). Introduced by Velsicol Corp. (later Sandoz AG).   **IUPAC name** methyl 2,3,5,6-tetrachloro-N-methoxy-N-methyl= terephthalamate   **CA name** methyl 2,3,5,6-tetrachloro-4-[[methoxy(methyl)amino]= carbonyl]benzoate.   **M.f.** $C_{11}H_9Cl_4NO_4$ **Code nos.** OCS 21693.   **Details** *PM4,* p. 354.

1202   oxapyrazon; oxapyrazon-sodium; oxapyrazon-(2-hydroxyethyldimethyl= ammonium) (common name).   **CAS RN** *[4489–31–0]* oxapyrazon; *[25316–56–7]* oxapyrazon-sodium; *[25316–57–8]* oxapyrazon-(2-hydroxyethyldimethylammonium). Herbicide developed by BASF AG.   **Common names** oxapyrazon, oxapyrazone, [oxapyrazon] (for 2-hydroxyethyldimethylammonium salt)   **IUPAC name** 5-bromo-= 1,6-dihydro-6-oxo-1-phenylpyridazin-4-yloxamic acid   **CA name** [(5-bromo-= 1,6-dihydro-6-oxo-1-phenyl-4-pyridazinyl)amino]oxoacetic acid.   **M.f.** $C_{12}H_8BrN_3O_4$; (sodium salt $C_{12}H_7BrN_3NaO_4$; 2-hydroxyethyldimethylammonium salt $C_{16}H_{19}BrN_4O_5$) **Code nos.** BAS 3380H (oxapyrazon-sodium); BAS 350H (oxapyrazon-(2-hydroxyethyl= dimethylammonium)).

1203   oxydeprofos (common name).   **CAS RN** *[2674–91–1].* Insecticide described by G. Schrader (*Die Entwicklung neuer insektizider Phosphorsäure-Ester,* 3rd Ed.). Introduced by Bayer AG.   **Common names** oxydeprofos, ESP   **IUPAC name** S-2-ethylsulfinyl-= 1-methylethyl O,O-dimethyl phosphorothioate   **CA name** S-[2-(ethylsulfinyl)-= 1-methylethyl] O,O-dimethyl phosphorothioate.   **M.f.** $C_7H_{17}O_4PS_2$ **Code nos.** Bayer 23 655; S 410; ENT 25 647.   **Selected tradenames** 'Metasystox S', 'Estox'. **Details** *PM7,* entry 09250.

1204 oxydisulfoton (common name). **CAS RN** *[2497–07–6].* Insecticide and acaricide introduced by Bayer AG. **IUPAC name** *O,O*-diethyl *S*-2-ethylsulfinylethyl phosphorodithioate **CA name** *O,O*-diethyl *S*-[2-(ethylsulfinyl)ethyl]phosphoro= dithioate. **M.f.** $C_8H_{19}O_3PS_3$ **Code nos.** Bayer 23 323; S 309; L 16/184 **Selected tradenames** 'Disyston-S'. **Details** *PM4,* p. 389.

1205 oxytetracycline (common name). **CAS RN** *[79–57–2].* Antibacterial antibiotic, still used in veterinary and human medicine, formerly also used against some plant pathogens. **Common names** oxytetracycline, ⌊terramicin⌋, ⌊terramitsin⌋ **IUPAC name** (4*S*,4a*R*,5*S*,5a*R*,6*S*,12a*S*)-4-dimethylamino-1,4,4a,5,5a,6,11,12a-octahydro-= 3,5,6,10,12,12a-hexahydroxy-6-methyl-1,11-dioxonaphthacene-2-carboxamide **CA name** [4*S*-(4α,4aα,5α,5aα,6β,12aα)]-4-(dimethylamino)-1,4,4a,5,5a,6,11,12a-octa= hydro-3,5,6,10,12,12a-hexahydroxy-6-methyl-1,11-dioxo-2-naphthacenecarboxamide. **M.f.** $C_{22}H_{24}N_2O_9$.

1206 parafluron (common name). **CAS RN** *[7159–99–1].* Herbicide introduced by Ciba AG (later Ciba-Geigy AG). **IUPAC name** 1,1-dimethyl-3-(α,α,α-trifluoro-= *p*-tolyl)urea **CA name** *N,N*-dimethyl-*N'*-[4-(trifluoromethyl)phenyl]urea. **M.f.** $C_{10}H_{11}F_3N_2O$ **Code nos.** C 15 935.

1207 perfluidone (common name). **CAS RN** *[37924–13–3].* Herbicide reported by W. A. Gentner (*Agric. Res. (Wash. D.C.),* 1971, 20(2), 5). Introduced by 3M Company. **IUPAC name** 1,1,1-trifluoro-*N*-(4-phenylsulfonyl-*o*-tolyl)methanesulfonamide; 1,1,1-trifluoro-2'-methyl-4'-(phenylsulfonyl)methanesulfonanilide **CA name** 1,1,1-trifluoro-*N*-[2-methyl-4-(phenylsulfonyl)phenyl]methanesulfonamide. **M.f.** $C_{14}H_{12}F_3NO_4S_2$ **Code nos.** MBR-8251 **Selected tradenames** 'Destun'. **Details** *PM9,* entry 9440.

1208 PH 60-38 (development code). **CAS RN** *[35409–97–3].* Insect growth regulator evaluated by Philips Duphar (later Uniroyal Chemical Co., Inc). **IUPAC name** 1-(4-chlorophenyl)-3-(2,6-dichlorobenzoyl)urea **CA name** 2,6-dichloro-= *N*-[[(4-chlorophenyl)amino]carbonyl]benzamide. **M.f.** $C_{14}H_9Cl_3N_2O_2$ **Code nos.** PH 60-38.

1209 phenisopham (common name). **CAS RN** *[57375–63–0].* Herbicide reported in *Schering AG Tech. Inf.* 1977. Introduced by Schering AG. **Common names** phenisopham, phénisophame **IUPAC name** isopropyl 3-[ethyl(phenyl)carbamoyloxy]carbanilate; 3-(isopropoxycarbonylamino)phenyl *N*-ethylcarbanilate; isopropyl 3-(*N*-ethyl-= *N*-phenylcarbamoyloxy)phenylcarbamate **CA name** 3-[[(1-methylethoxy)carbonyl]= amino]phenyl ethylphenylcarbamate. **M.f.** $C_{19}H_{22}N_2O_4$ **Code nos.** SN 58 132 **Selected tradenames** 'Diconal', 'Verdinal'. **Details** *PM8,* entry 09480.

1210 phenkapton (common name). **CAS RN** *[2275–14–1].* Insecticide and acaricide introduced by J. R. Geigy S.A. (later Ciba-Geigy AG). **Common names** phenkapton, CMP **IUPAC name** *S*-2,5-dichlorophenylthiomethyl *O,O*-diethyl phosphorodithioate **CA name** *S*-[[(2,5-dichlorophenyl)thio]methyl] *O,O*-diethyl phosphorodithioate. **M.f.** $C_{11}H_{15}Cl_2O_2PS_3$ **Code nos.** G 28 029, ENT 25 585.

1211 phenmedipham-ethyl (common name). **CAS RN** *[13684–44–1]*. Herbicide evaluated by Schering AG. **Common names** phenmedipham-ethyl, phenmédiphame-éthyl **IUPAC name** 3-ethoxycarbonylaminophenyl 3′-methylcarbanilate; ethyl 3′-(3′-methylcarbaniloyloxy)carbanilate **CA name** 3-[(ethoxycarbonyl)amino]= phenyl (3-methylphenyl)carbamate. **M.f.** $C_{17}H_{18}N_2O_4$ **Code nos.** SN 38 574.

1212 phenobenzuron (common name). **CAS RN** *[3134–12–1]*. Herbicidal activity reported by P. Poignant *et al.* (*Symp. New Herbic., 2nd*, 1965, p. 1). Introduced by Pechiney-Progil (later Rhône-Poulenc Agrochimie). **IUPAC name** 1-benzoyl-= 1-(3,4-dichlorophenyl)-3,3-dimethylurea **CA name** *N*-(3,4-dichlorophenyl)-= *N*-[(dimethylamino)carbonyl]benzamide. **M.f.** $C_{16}H_{14}Cl_2N_2O_2$ **Code nos.** PP 65-25 **Selected tradenames** 'Benzomarc'. **Details** *PM5*, p. 409.

1213 2-phenyl-4H-3,1-benzoxazin-4-one (IUPAC/Chemical Abstracts name). **CAS RN** *[1022–46–4]*. Herbicide reported by A. Fischer (*Meded. Landbouwhogesch. Opzoekingsstn. Staat Gent*, 1965, **30**, 163). Introduced by BASF AG. **M.f.** $C_{14}H_9NO_2$ **Code nos.** H-170 **Selected tradenames** 'Linurotox'. **Details** *PM5*, p. 412.

1214 phenylmercury dimethyldithiocarbamate (IUPAC name). **CAS RN** *[32407–99–1]*. Fungicide introduced by Berk Chemicals (later Steetley Chemicals Ltd). **CA name** (dimethylcarbamodithioato-*S,S′*)phenylmercury. **M.f.** $C_9H_{11}HgNS_2$ **Selected tradenames** 'Phelam'. **Details** *PM6*, p. 417.

1215 phenylmercury nitrate (common name). **CAS RN** *[8003–05–2]* formerly *[6059–33–1]*. Fungicide introduced by Steetley Ltd. **Common names** phenylmercury nitrate, nitrate de phénylmercure **IUPAC name** phenylmercury hydroxide with phenylmercury nitrate; phenylmercury nitrate (basic) **CA name** hydroxy(nitrato)= diphenyldimercury. **M.f.** $C_{12}H_{11}Hg_2NO_4$ **Selected tradenames** 'Harvesan Plus', 'Murcicide'. **Details** *PM6*, p. 418.

1216 phosacetim (common name). **CAS RN** *[4104–14–7]*. Rodenticide reported by V. B. Richens (*Pest Control*, 1967, **35**(9), 28). Introduced by Bayer AG. **Common names** phosacetim, phosacétime **IUPAC name** *O,O*-bis(4-chlorophenyl) *N*-acetimidoyl= phosphoramidothioate **CA name** *O,O*-bis(4-chlorophenyl) (1-iminoethyl)= phosphoramidothioate. **M.f.** $C_{14}H_{13}Cl_2N_2O_2PS$ **Code nos.** Bayer 38 819 **Selected tradenames** 'Gophacide'. **Details** *PM3*, p. 394.

1217 phosdiphen (common name). **CAS RN** *[36519–00–3]*. Fungicide reported by M. Hamada *et al.* (*Ann. Phytopathol. Soc. Japan*, 1971, **37**, 365). Introduced by Hokko Chemical Industry Co., Ltd. and Mitsui Toatsu Chemicals Inc. **IUPAC name** bis(2,4-dichlorophenyl) ethyl phosphate **CA name** bis(2,4-dichlorophenyl) ethyl phosphate. **M.f.** $C_{14}H_{11}Cl_4O_4P$ **Code nos.** MTO-460 (Mitsui Toatsu). **Details** *PM10*, entry 551.

1218 phosfolan (common name). **CAS RN** *[947–02–4]*. Insecticide introduced by American Cyanamid Co. **Common names** phosfolan, phospholan **IUPAC name** diethyl 1,3-dithiolan-2-ylidenephosphoramidate; 2-(diethoxyphosphinylimino)-1,3-dithiolane

CA name diethyl 1,3-dithiolan-2-ylidenephosphoramidate. **M.f.** $C_7H_{14}NO_3PS_2$ **Code nos.** EI 47 031 (Cyanamid); OMS 646; ENT 25 830. **Selected tradenames** 'Cyolane', 'Cyolan', 'Cyalane', 'Cylan'. **Details** *PM9*, entry 9670.

1219 phosnichlor (common name). Insecticide evaluated by Bayer AG. **Common names** phosnichlor, nichlorfos **IUPAC name** *O*-4-chloro-3-nitrophenyl *O*,*O*-dimethyl phosphorothioate **CA name** *O*-(4-chloro-3-nitrophenyl) *O*,*O*-dimethyl phosphorothioate. **M.f.** $C_8H_9ClNO_5PS$.

1220 phoxim-methyl (common name). **CAS RN** *[14816–16–1]*. Insecticide evaluated by Bayer AG. **Common names** phoxim-methyl, phoxime-méthyl **IUPAC name** *O*,*O*-dimethyl α-cyanobenzylineneamino-oxyphosphonothioate; dimethoxyphosphino= thioyloxyimino(phenyl)acetonitrile **CA name** 3-methoxy-6-phenyl-2,4-dioxa-5-aza-= 3-phosphapept-5-ene-7-nitrile 3-sulfide; α-[[(dimethoxyphosphinothioyl)oxy]imino]= benzeneacetonitrile (ex 9CI). **M.f.** $C_{10}H_{11}N_2O_3PS$ **Code nos.** SRA 7760.

1221 piproctanyl bromide (common name). **CAS RN** *[56717–11–4]*; (piproctanyl *[69309–47–3]*). Plant growth regulator reported by G. A. Hüppi *et al.* (*Experientia*, 1976, **32**, 37). The bromide introduced by Dr. R. Maag Ltd. **Common names** (for cation) piproctanyl **IUPAC name** 1-allyl-1-(3,7-dimethyloctyl)piperidinium bromide **CA name** 1-(3,7-dimethyloctyl)-1-(2-propenyl)piperidinium bromide. **M.f.** $C_{18}H_{36}BrN$; (piproctanyl $C_{18}H_{36}N$) **Code nos.** Ro 06-0761/000 (Maag); ACR-1222 **Selected tradenames** 'Alden', 'Stemtrol'. **Details** *PM9*, entry 9820.

1222 piprotal (common name). **CAS RN** *[5281–13–0]*. Synergist for pyrethrum reported by L. O. Hopkins & D. R. Maciver (*Pyrethrum Post*, 1965, **8**(2), 3). Introduced by McLaughlin Gormley King Co. **IUPAC name** 5-[bis[2-(2-butoxyethoxy)ethoxy]= methyl]-1,3-benzodioxole (I); 1-bis[2-(2-butoxyethoxy)ethoxy]methyl-3,4-methylene= dioxybenzene **CA name** (I). **M.f.** $C_{24}H_{40}O_8$ **Code nos.** OMS 1161. **Selected tradenames** 'Tropital'. **Details** *PM4*, p. 166.

1223 pirimetaphos (common name). **CAS RN** *[31377–69–2]*. Insecticide evaluated by Sandoz AG. **Common names** pirimetaphos, pyrimétaphos **IUPAC name** 2-diethylamino-6-methylpyrimidin-4-yl methyl methylphosphoramidate **CA name** 2-(diethylamino)-6-methyl-4-pyrimidinyl methyl methylphosphoramidate. **M.f.** $C_{11}H_{21}N_4O_3P$ **Code nos.** SAN I 52 135, OMS 1502.

1224 polychlorodicyclopentadiene isomers (IUPAC name). **CAS RN** *[8029–29–6]*. Insecticide and herbicide introduced by Velsicol Chemical Corp. **Selected tradenames** 'Bandane'.

1225 polychloroterpenes (traditional name). **CAS RN** *[8001–50–1]*. Insecticidal and acaricidal activity of this reaction mixture described by D. L. Kent *et al.* (*Soap Chem. Spec.*, 1953, **19**(6), 157). Introduced by Goodrich Chemical Co. **IUPAC name** heptachloro-2,2-dimethyl-3-methylenenorbornane **CA name** chlorinated mixed terpenes. **Other names** terpene polychlorinates **Code nos.** Compound 3961, ENT 19 442. **Selected tradenames** 'Strobane'. **Details** *PM1*, p. 410.

1226 potassium cyanate (IUPAC/Chemical Abstracts name).   CAS RN *[590–28–3]*. Introduced as herbicide by American Cyanamid Co.   **IUPAC name** potassium cyanate (I)   **CA name** (I).   **M.f.** CKNO   **Selected tradenames** 'Aero' cyanate.   **Details** *PM5*, p. 432.

1227 primidophos (common name).   CAS RN *[39247–96–6]*. Insecticide evaluated by ICI Plant Protection Division (later ICI Agrochemicals).   **Common names** primidophos, prymidophos   **IUPAC name** *O,O*-diethyl *O*-(2-*N*-ethylacetamido-= 6-methylpyrimidin-4-yl) phosphorothioate; *N*-(4-diethoxyphosphinothioyloxy-= 6-methylpyrimidin-2-yl)-*N*-ethylacetamide   **CA name** *O*-[2-(acetylethylamino)-= 6-methyl-4-pyrimidinyl] *O,O*-diethyl phosphorothioate.   **M.f.** $C_{13}H_{22}N_3O_4PS$ **Code nos.** PP484.

1228 proclonol (common name).   CAS RN *[14088–71–2]*. Acaricide introduced by Janssen Pharmaceutica.   **IUPAC name** 4,4′-dichloro-α-cyclopropylbenzhydrol   **CA name** 4-chloro-α-(4-chlorophenyl)-α-cyclopropylbenzenemethanol.   **M.f.** $C_{16}H_{14}Cl_2O$.

1229 procyazine (common name).   CAS RN *[32889–48–8]*. Herbicide introduced by Ciba-Geigy AG.   **IUPAC name** 2-(4-chloro-6-cyclopropylamino-1,3,5-triazin-= 2-ylamino)-2-methylpropionitrile   **CA name** 2-[[4-chloro-6-(cyclopropylamino)-= 1,3,5-triazin-2-yl]amino]-2-methylpropanenitrile.   **M.f.** $C_{10}H_{13}ClN_6$ **Code nos.** CGA 18 762.

1230 profluralin (common name).   CAS RN *[26399–36–0]*. Herbicide reported by T. D. Taylor *et al.* (*Annu. Meet. Weed Sci. Soc. Am.*, 1973, Abstr. 169). Introduced by Ciba-Geigy AG.   **Common names** profluralin, profluraline   **IUPAC name** *N*-(cyclopropyl= methyl)-α,α,α-trifluoro-2,6-dinitro-*N*-propyl-*p*-toluidine; *N*-cyclopropylmethyl-= 2,6-dinitro-*N*-propyl-4-trifluoromethylaniline   **CA name** *N*-(cyclopropylmethyl)-= 2,6-dinitro-*N*-propyl-4-(trifluoromethyl)benzenamine.   **M.f.** $C_{14}H_{16}F_3N_3O_4$ **Code nos.** CGA 10 832   **Selected tradenames** 'Tolban', 'Pregard'.   **Details** *PM9*, entry 10050.

1231 proglinazine-ethyl; proglinazine (common name).   CAS RN *[68228–18–2]*; (acid *[68228–20–6]*). The ethyl ester introduced as a herbicide by Nitrokémia Ipartelepek.   **Common names** (for the acid) proglinazine   **IUPAC name** (for the acid) *N*-(4-chloro-= 6-isopropylamino-1,3,5-triazin-2-yl)glycine   **CA name** (for the acid) *N*-[4-chloro-= 6-[(1-methylethyl)amino]-1,3,5-triazin-2-yl]glycine.   **M.f.** $C_{10}H_{16}ClN_5O_2$ **Code nos.** (acid) MG-07 (Nitrokémia).   **Details** *PM9*, entry 10060.

1232 promacyl (common name).   CAS RN *[34264–24–9]*. Insecticide and ixodicide introduced by ICI Australia.   **IUPAC name** 5-methyl-*m*-cumenyl butyryl(methyl)= carbamate   **CA name** 3-methyl-5-(1-methylethyl)phenyl methyl(1-oxobutyl)carbamate.   **M.f.** $C_{16}H_{23}NO_3$ **Other names** promecarb A   **Code nos.** CRC 7320   **Selected tradenames** 'Promicide'.   **Details** *PM9*, entry 8370.

1233 promecarb (common name).   CAS RN *[2631–37–0]*. Insecticide reported by A. Formigoni & G. P. Bellini (*Congr. Int. degli Antiparassitari, Naples*, 1965) and by A. Jäger (*Z. Angew. Entomol.*, 1966, **58**, 188). Introduced by Schering AG (later AgrEvo GmbH).   **Common names** promecarb, promécarbe   **IUPAC name** 5-methyl *m*-cumenyl

methylcarbamate; 3-isopropyl-5-methylphenyl methylcarbamate   **CA name** 3-methyl-=
5-(1-methylethyl)phenyl methylcarbamate.   **M.f.** $C_{12}H_{17}NO_2$ **Code nos.** SN 34 615;
OMS 716; ENT 27 300a.   **Selected tradenames** 'Carbamult'.   **Details** *PM9*, entry 10070.

1234   **propyl 3-*tert*-butylphenoxyacetate** (IUPAC name).   **CAS RN** *[66227–09–6]*.
Plant growth-regulating properties reported by C. J. Hibbitt & J. A. Hardisty (*Meded.
Fac. Landbouwwet. Rijksuniv. Gent*, 1979, **44**, 835). Introduced by May & Baker Ltd (later
Rhône-Poulenc Agrochimie).   **CA name** propyl [3-(1,1-dimethylethyl)phenoxy]acetate.
M.f. $C_{15}H_{22}O_3$ **Code nos.** M&B 25 105   **Selected tradenames** 'M&B 25-105'.   **Details**
PM8, entry 10200.

1235   **propyl isome** (common name).   **CAS RN** *[83–59–0]*. Synergist for pyrethrum
reported by M. E. Synerholm & A. Hartzell (*Contr. Boyce Thompson Inst.*, 1945, **14**, 79).
Introduced by S. B. Penick & Co.   **IUPAC name** dipropyl 5,6,7,8-tetrahydro-7-=
methylnaphtho[2,3-*d*]-1,3-dioxole-5,6-dicarboxylate (I); dipropyl 1,2,3,4-tetrahydro-=
3-methyl-6,7-methylenedioxynaphthalene-1,2-dicarboxylate   **CA name** (I).   **M.f.**
$C_{20}H_{26}O_6$ **Other names** dipropyl maleate isosafrole condensate.   **Details** *PM5*, p. 446.

1236   **prosulfalin** (common name).   **CAS RN** *[51528–03–1]*. Herbicide evaluated by Eli
Lilly & Co.   **Common names** prosulfalin, prosulfaline   **IUPAC name** *N*-(4-dipropyl=
amino-3,5-dinitrophenylsulfonyl)-*S,S*-dimethylsulfimide   **CA name** *N*-[[4-(dipropyl=
amino)-3,5-dinitrophenyl]sulfonyl]-*S,S*-dimethylsulfilimine.   **M.f.** $C_{14}H_{22}N_4O_6S_2$.

1237   **prothidathion** (common name).   Acaricide evaluated by J. R. Geigy S.A. (later
Ciba-Geigy AG).   **IUPAC name** *S*-2,3-dihydro-5-isopropoxy-2-oxo-1,3,4-thiadiazol-=
3-ylmethyl *O,O*-diethyl phosphorodithioate; 3-diethoxyphosphinothioylthiomethyl-=
5-isopropoxy-1,3,4-thiadiazol-2(3*H*)-one   **CA name** *O,O*-diethyl *S*-[[5-(1-methyl=
ethoxy)-2-oxo-1,3,4-thiadiazol-3(2*H*)-yl]methyl] phosphorodithioate.   **M.f.**
$C_{10}H_{19}N_2O_4PS_3$ **Code nos.** GS 13 010.

1238   **prothiocarb; prothiocarb hydrochloride** (common name).   **CAS RN**
[19622–08–3] prothiocarb; *[19622–19–6]* hydrochloride. Fungicide reported by
M. G. Bastiaansen et al. (*Meded. Fac. Landbouwwet. Rijksuniv. Gent*, 1974, **39**, 1019).
The hydrochloride introduced by Schering AG.   **Common names** prothiocarb,
prothiocarbe   **IUPAC name** *S*-ethyl (3-dimethylaminopropyl)thiocarbamate   **CA name**
S-ethyl [3-(dimethylamino)propyl]carbamothioate.   **M.f.** $C_8H_{18}N_2OS$;
(hydrochloride $C_8H_{19}ClN_2OS$)   **Code nos.** SN 41 703   **Selected tradenames**
'Previcur' (hydrochloride), 'Dynone'.   **Details** *PM7*, entry 10270.

1239   **prothoate** (common name).   **CAS RN** *[2275–18–5]*. Insecticide reported (*Ital.
Agric.*, 1955, **99**, 747). Discovered by American Cyanamid Co. and later introduced by
Montecatini S.p.A. (later Agrimont S.p.A.).   **IUPAC name** *O,O*-diethyl *S*-isopropyl=
carbamoylmethyl phosphorodithioate   **CA name** *O,O*-diethyl *S*-[2-[(1-methylethyl)=
amino]-2-oxoethyl] phosphorodithioate.   **M.f.** $C_9H_{20}NO_3PS_2$ **Code nos.**
E.I. 18 682 (Cyanamid); L 343 (Montedison); ENT 24 652.   **Selected tradenames** 'Fac'.
Details *PM9*, entry 10290.

1240 proxan; proxan-sodium (common name). **CAS RN** *[108–25–8]* proxan; *[140–93–2]* proxan-sodium. Herbicide reported by L. L. Baumgartner & B. Wolf (*Contrib. Boyce Thompson Inst.*, 1949, **15**, 403). Introduced by Goodrich Chemical Co. **Common names** proxan, proxane, IPX **IUPAC name** O-isopropyl hydrogen dithiocarbonate **CA name** O-(1-methylethyl) hydrogen carbonodithioate. **M.f.** $C_4H_8OS_2$; (sodium salt $C_4H_7NaOS_2$) **Selected tradenames** 'Good-rite n.i.x.'. **Details** *PM1*, p. 360.

1241 prynachlor (common name). **CAS RN** *[21267–72–1]*. Herbicide developed by BASF AG. **Common names** prynachlor, prynachlore **IUPAC name** 2-chloro-= N-(1-methylprop-2-ynyl)acetanilide **CA name** 2-chloro-N-(1-methyl-2-propynyl)-= N-phenylacetamide. **M.f.** $C_{12}H_{12}CINO$ **Code nos.** BAS 290H **Selected tradenames** 'Basamaize'.

1242 pydanon (common name). **CAS RN** *[22571–07–9]* unstated stereochemistry. Plant growth regulator developed by C. F. Spiess & Sohn. **IUPAC name** (±)-hexahydro-4-hydroxy-3,6-dioxopyridazin-4-ylacetic acid **CA name** hexahydro-= 4-hydroxy-3,6-dioxo-4-pyridazineacetic acid. **M.f.** $C_6H_8N_2O_5$ **Code nos.** H 1244.

1243 pyracarbolid (common name). **CAS RN** *[24691–76–7]*. Fungicide reported by H. Stingle *et al.* (*Int. Congr. Plant Prot., 7th, Paris.*, 1970, p. 205 (Abstr.)) and B. Jank & F. Grossman (*Pestic. Sci.*, 1971, **2**, 43). Introduced by Hoechst AG. **Common names** pyracarbolid, pyracarbolide **IUPAC name** 3,4-dihydro-6-methyl-2H-pyran-5-carboxanilide **CA name** 3,4-dihydro-6-methyl-N-phenyl-2H-pyran-5-carboxamide. **M.f.** $C_{13}H_{15}NO_2$ **Code nos.** Hoe 13 764; formerly [Hoe 02 989]; [Hoe 6052]; [Hoe 6053] **Selected tradenames** 'Sicarol'. **Details** *PM7*, entry 10330.

1244 pyresmethrin (common name). **CAS RN** *[24624–58–6]* formerly *[56194–68–4]* and *[20425–39–2]*. Insecticide discovered by M. Elliott *et al.* **Common names** pyresmethrin, pyresméthrine **IUPAC name** 5-benzyl-3-furyl= methyl (E)-(1R)-trans-3-(2-methoxycarbonylprop-1-enyl)-2,2-dimethylcyclopropane= carboxylate; 5-benzyl-3-furylmethyl (E)-(1R,3R)-3-(2-methoxycarbonylprop-1-enyl)-= 2,2-dimethylcyclopropanecarboxylate **CA name** [1R-[1α,3β(E)]]-[5-(phenylmethyl)-= 3-furanyl]methyl 3-(3-methoxy-2-methyl-3-oxo-1-propenyl)-2,2-dimethyl= cyclopropanecarboxylate. **M.f.** $C_{23}H_{26}O_5$ **Code nos.** NRDC 106.

1245 pyriclor (common name). **CAS RN** *[1970–40–7]*. Herbicide reported by M. J. Huraux & H. M. Lawson (*Symp. New Herbic., 2nd*, 1965, p. 269). Introduced by Dow Chemical Co. (later DowElanco). **IUPAC name** 2,3,5-trichloropyridin-4-ol **CA name** 2,3,5-trichloro-4-pyridinol. **M.f.** $C_5H_2Cl_3NO$ **Selected tradenames** 'Daxtron'. **Details** *PM1*, p. 363.

1246 pyridinitril (common name). **CAS RN** *[1086–02–8]*. Fungicide reported by G. Mohr *et al.* (*Meded. Rijksfac. Landbouwwet. Gent*, 1968, **33**, 1293). Introduced by E. Merck (later Shell Agrar GmbH). **Common names** pyridinitril, pyridinitrile, DDPP **IUPAC name** 2,6-dichloro-4-phenylpyridine-3,5-dicarbonitrile **CA name** 2,6-dichloro-=

4-phenyl-3,5-pyridinedicarbonitrile.   **M.f.** $C_{13}H_5Cl_2N_3$ **Code nos.** IT 3296   **Selected tradenames** 'Ciluan'.   **Details** *PM5*, p. 454.

1247   pyrimitate (common name).   **CAS RN** *[5221–49–8]*. Veterinary insecticide and acaricide introduced by ICI Pharmaceuticals Division.   **Common names** pyrimitate, [pyrimithate]   **IUPAC name** *O*-2-dimethylamino-6-methylpyrimidin-4-yl *O,O*-diethyl phosphorothioate   **CA name** *O*-[2-(dimethylamino)-6-methyl-4-pyrimidinyl] *O,O*-diethyl phosphorothioate.   **M.f.** $C_{11}H_{20}N_3O_3PS$ **Code nos.** ICI 29 661   **Selected tradenames** 'Diothyl'.

1248   pyrinuron (common name).   **CAS RN** *[53558–25–1]*. Rodenticide evaluated by Rohm & Haas Co.   **Common names** pyrinuron, piriminil   **IUPAC name** 1-(4-nitrophenyl)-= 3-(3-pyridylmethyl)urea   **CA name** *N*-(4-nitrophenyl)-*N'*-(3-pyridinylmethyl)urea. **M.f.** $C_{13}H_{12}N_4O_3$ **Code nos.** RH-787   **Selected tradenames** 'Vacor'.

1249   pyroxychlor (common name).   **CAS RN** *[7159–34–4]*. Fungicide evaluated by Dow Chemical Co. (later DowElanco).   **Common names** pyroxychlor, pyroxychlore **IUPAC name** 2-chloro-6-methoxy-4-trichloromethylpyridine; 6-chloro-4-trichloro= methyl-2-pyridyl methyl ether   **CA name** 2-chloro-6-methoxy-4-(trichloromethyl)= pyridine.   **M.f.** $C_7H_5Cl_4NO$ **Code nos.** Dowco 269   **Selected tradenames** 'Nurelle', 'Lorvek'.

1250   pyroxyfur (common name).   **CAS RN** *[70166–48–2]*. Fungicide evaluated by Dow Chemical Co. (later DowElanco).   **IUPAC name** 6-chloro-4-trichloromethyl-= 2-pyridyl furfuryl ether   **CA name** 2-chloro-6-(2-furanylmethoxy)-4-(trichloro= methyl)pyridine.   **M.f.** $C_{11}H_7Cl_4NO_2$ **Code nos.** Dowco 444   **Selected tradenames** 'Grandstand'.

1251   *N*-pyrrolidinosuccinamic acid (IUPAC name).   **CAS RN** *[23744–05–0]*. Plant growth regulator evaluated by Uniroyal Inc.   **CA name** 4-oxo-4-(1-pyrrolidinyl= amino)butanoic acid.   **M.f.** $C_8H_{14}N_2O_3$ **Code nos.** F 529   **Selected tradenames** 'F-Five'.

1252   quinacetol; quinacetol sulfate (common name).   **CAS RN** *[57130–91–3]* quinacetol sulfate; *[2598–31–4]* quinacetol. Fungicide reported by S. D. Hocombe *et al.* (*Proc. Br. Insectic. Fungic. Conf., 7th.,* 1973, **1,** 365). Quinacetol sulfate introduced by Ciba-Geigy AG.   **IUPAC name** 1-(8-hydroxyquinolin-5-yl)ethanone; 5-acetyl-= 8-hydroxyquinoline   **CA name** 1-(8-hydroxy-5-quinolinyl)ethanone.   **M.f.** $C_{11}H_9NO_2$; (quinacetol sulfate $C_{22}H_{20}N_2O_8S$)   **Code nos.** G 20 072   **Selected tradenames** 'Fongoren', 'Risoter'.   **Details** *PM5*, p. 455.

1253   quinalphos-methyl (common name).   **CAS RN** *[13593–08–3]*. Insecticide evaluated by Sandoz AG.   **Common names** quinalphos-methyl, chinalphos-méthyl **IUPAC name** *O,O*-dimethyl *O*-quinoxalin-2-yl phosphorothioate   **CA name** *O,O*-dimethyl *O*-2-quinoxalinyl phosphorothioate.   **M.f.** $C_{10}H_{11}N_2O_3PS$ **Code nos.** SAN 52 056I.

1254 quinazamid (common name). Fungicide evaluated by Boots Co. Ltd (later Schering Agrochemicals). **Common names** quinazamid, quinazamide **IUPAC name** *p*-benzoquinone monosemicarbazone **CA name** 2-(4-oxo-2,5-cyclohexadien-= 1-ylidene)hydrazinecarboxamide. **M.f.** $C_7H_7N_3O_2$ **Code nos.** RD 8684; BTS 8684.

1255 quinconazole (common name). **CAS RN** *[103970–75–8]*. Fungicide reported by C. R. Leake (*Proc. 1988 Brighton Crop Prot. Conf - Pests Dis.*, **1**, 343). Evaluated by Schering AG. **IUPAC name** 3-(2,4-dichlorophenyl)-2-(1*H*-1,2,4-triazol-1-yl)= quinazolin-4(3*H*)-one **CA name** 3-(2,4-dichlorophenyl)-2-(1*H*-1,2,4-triazol-1-yl)-= 4(3*H*)-quinazolinone. **M.f.** $C_{16}H_9Cl_2N_5O$ **Code nos.** SN 539 865.

1256 quinonamid (common name). **CAS RN** *[27541–88–4]*. Algicide reported by P. Hartz *et al.* (*Meded. Fac. Landbouwwet. Rijksuniv. Gent,* 1972, **37**, 699). Introduced by Hoechst AG. **Common names** quinonamid, quinonamide **IUPAC name** 2,2-dichloro-= *N*-(3-chloro-1,4-naphthoquinon-2-yl)acetamide **CA name** 2,2-dichloro-*N*-(3-chloro-= 1,4-dihydro-1,4-dioxo-2-naphthalenyl)acetamide. **M.f.** $C_{12}H_6Cl_3NO_3$ **Code nos.** Hoe 02 997 **Selected tradenames** 'Alginex', 'Nosprasit'. **Details** *PM7*, entry 10530.

1257 quinothion (common name). **CAS RN** *[22439–40–3]*. Insecticide. **IUPAC name** *O,O*-diethyl *O*-2-methylquinolin-4-yl phosphorothioate **CA name** *O,O*-diethyl *O*-(2-methyl-4-quinolinyl) phosphorothioate. **M.f.** $C_{14}H_{18}NO_3PS$.

1258 quintiofos (common name). **CAS RN** *[1776–83–6]*. Insecticide introduced by Bayer AG. **IUPAC name** *O*-ethyl *O*-8-quinolyl phenylphosphonothioate **CA name** *O*-ethyl *O*-8-quinolinyl phenylphosphonothioate. **M.f.** $C_{17}H_{16}NO_2PS$ **Code nos.** BAY 9037 **Selected tradenames** 'Bacdip'.

1259 R-1492 (development code). **CAS RN** *[953–17–3]*. Insecticide and acaricide reported by J. A. Harding (*J. Econ. Entomol.*, 1959, **52**, 1219). Introduced by Stauffer Chemical Co. (later ICI Americas). **IUPAC name** *S*-4-chlorophenylthiomethyl *O,O*-dimethyl phosphorodithioate **CA name** *S*-[[(4-chlorophenyl)thio]methyl] *O,O*-dimethyl phosphorodithioate. **M.f.** $C_9H_{12}ClO_2PS_3$ **Other names** methylcarbophenothione **Code nos.** R-1492; OMS 497; ENT 25 599. **Selected tradenames** 'Methyl Trithion', 'Tri-Me'. **Details** *PM6*, p. 111.

1260 RA-17 (development code). **CAS RN** *[105084–66–0]*. Acaricide reported by K. Balogh & G. Tarpai (*6th Intern. Congr. Pestic. Chem.*, 1986). Introduced in Hungary (1987) by Északmagyarországi Vegyimüvek. **IUPAC name** N^2-diethoxyphosphinothioyl-= N^2-ethyl-N^1,N^1-dipropylglycinamide **CA name** *O,O*-diethyl [2-(dipropylamino)-= 2-oxoethyl]ethylphosphoroamidothioate. **M.f.** $C_{14}H_{31}N_2O_3PS$ **Other names** phosglycin **Code nos.** RA-17 (Északmagyarországi Vegyimüvek) **Selected tradenames** 'Alkatox'. **Details** *PM9*, entry 4618.

1261 rabenzazole (common name). **CAS RN** *[40341–04–6]*. Fungicide reported by W. Specht & M. Tillkes (*Pflanzenschutz-Nachr. (Engl. Ed.)*, 1980, **33**, 61). Introduced by Bayer AG. **IUPAC name** 2-(3,5-dimethylpyrazol-1-yl)-1*H*-benzimidazole;

2-(3,5-dimethylpyrazol-1-yl)benzimidazole **CA name** 2-(3,5-dimethyl-1*H*-pyrazol-=
1-yl)-1*H*-benzimidazole. **M.f.** $C_{12}H_{12}N_4$ **Selected tradenames** 'Ciriom', 'Ciriom F'.

1262 rhodethanil (common name). Herbicide evaluated by Bayer AG. **Common
names** rhodethanil, rodéthanil **IUPAC name** 3-chloro-4-ethylaminophenyl thiocyanate
CA name 3-chloro-4-(ethylamino)phenyl thiocyanate. **M.f.** $C_9H_9ClN_2S$ **Code nos.**
BAY 5396 b H; BAY 53 427.

1263 RU 25475 (development code). **CAS RN** *[66841–26–7]* unstated
stereochemistry. Insecticide evaluated by Roussel Uclaf. **IUPAC name** (*S*)-α-cyano-=
3-phenoxybenzyl (*1R,3S*)-[(*RS*)-1,2-dibromo-2,2-dichloroethyl]-2,2-dimethyl=
cyclopropanecarboxylate; *Roth*: (*S*)-α-cyano-3-phenoxybenzyl (*1R*)-*trans*-3-=
((*RS*)-1,2-dibromo-2,2-dichloroethyl)-2,2-dimethylcyclopropanecarboxylate **CA name**
cyano(3-phenoxyphenyl)methyl 3-(1,2-dibromo-2,2-dichloroethyl)-2,2-dimethyl=
cyclopropanecarboxylate. **M.f.** $C_{22}H_{19}Br_2Cl_2NO_3$ **Other names** tralocythrin
(unaccepted common name) **Code nos.** RU 25475.

1264 ryanodine (traditional name). **CAS RN** *[15662–33–6]* formerly *[15800–60–9]*
(ryanodine); *[8047–13–0]* (ryania). Insecticidal activity of ryania reported by B. E.
Pepper & L. A. Carruth (*J. Econ. Entomol.*, 1945, **38**, 59). Introduced by S. B. Penick &
Co. **IUPAC name** (2*S,3S,4R,4aS,5S,5aS,8S,9R,9aR,9bR*)-2,3,4a,5a,9,9b-hexahydro-=
3-isopropyl-2a,5,8-trimethylperhydro-2,5-methanobenzo[1,2]pentaleno[1,6-*bc*]furan-=
4-yl pyrrole-2-carboxylate **CA name** [3*S*-(3α,4β,4aS*,6aβ,6aα,7a,8β,8aα,8bβ,9β,9aα)]-=
octahydro-3,6a,9-trimethyl-7-(1-methylethyl)-6,9-methanobenzo[1,2]pentaleno=
[1,6-*bc*]furan-4,6,7,8,8a,8b,9a-heptyl 3-(1*H*)-pyrrole-2-carboxylate. **M.f.** $C_{25}H_{35}NO_9$
Selected tradenames 'Ryno-tox'. **Details** *PM6*, p. 469.

1265 sabadilla (traditional name). Methods to enhance the insecticidal activity of
veratrine, a mixture of alkaloids from the seeds of *Schoenocaulon officinale*, were
reported by T. C. Allen *et al.* (*J. Econ. Entomol.*, 1944, **37**. 400). **Other names** cevadilla.
Details *PM5*, p. 464.

1266 salicylanilide (IUPAC name). **CAS RN** *[87–17–2]*. Fungicide reported by R. G.
Fargher *et al.*, (*Mem. Shirley Inst.*, 1930, **9**, 37). Introduced by ICI Limited. **CA name**
2-hydroxy-*N*-phenylbenzamide. **M.f.** $C_{13}H_{11}NO_2$ **Selected tradenames** 'Shirlan'.
Details *PM5*, p. 465.

1267 schradan (common name). **CAS RN** *[152–16–9]*. Systemic insecticide
discovered by G. Schrader & H. Küenthal (cited by G. Schrader, *Die Entwicklung neuer
insektizider Phosphorsaure-Ester*, 3rd Ed., p. 88). Introduced by Pest Control Ltd (later
Schering Agriculture). **Common names** schradan, schradane **IUPAC name**
octamethylpyrophosphoric tetra-amide **CA name** octamethyldiphosphoramide. **M.f.**
$C_8H_{24}N_4O_3P_2$ **Other names** OMPA **Selected tradenames** 'Pestox III', 'Pestox 3',
'Sytam'. **Details** *PM6*, p. 470.

1268 scilliroside (common name). **CAS RN** *[507–60–8]*. Rodenticide. Scilliroside, the
toxic principle of red squill, was isolated and identified by A. Stoll & J. Renz (*Helv. Chim.*

Acta, 1942, **25**, 377; 1943, **26**, 648). Chemistry and toxicology of cardiac glycosides reviewed by A. Stoll (*Experientia*, 1954, **10**, 282). The two varieties of *U. maritima*, red squill and white squill, contain glycosides but only the former was used against rats. **IUPAC name** 3β-(β-D-glucopyranosyloxy)-17β-(2-oxo-2H-pyran-5-yl)-14β-androst-= 4-ene-6β,8,14-triol 6-acetate **CA name** (3β,6β)-6-acetyloxy-3-(β-D-glucopyranosyloxy)-= 8,14-dihydroxybufa-4,20,22-trienolide. **M.f.** $C_{32}H_{44}O_{12}$. **Details** *PM9*, entry 10665.

1269 sebuthylazine (common name). **CAS RN** *[7286–69–3]*. Herbicide introduced by J. R. Geigy S.A. (later Ciba-Geigy AG). **IUPAC name** N^2-sec-butyl-6-chloro-= N^4-ethyl-1,3,5-triazine-2,4-diamine; 2-sec-butylamino-4-chloro-6-ethylamino-= 1,3,5-triazine **CA name** 6-chloro-N-ethyl-N'-(1-methylpropyl)-1,3,5-triazine-= 2,4-diamine. **M.f.** $C_9H_{16}ClN_5$ **Code nos.** GS 13 528 **Selected tradenames** 'Vorox'.

1270 secbumeton (common name). **CAS RN** *[26259–45–0]*. Herbicide reported by A. Gast & E. Fankhauser (*Proc. Br. Weed Control Conf., 8th*, 1966, p. 485). Introduced by J. R. Geigy S.A. (later Ciba-Geigy AG). **IUPAC name** N^2-sec-butyl-N^4-ethyl-= 6-methoxy-1,3,5-triazine-2,4-diamine; 2-sec-butylamino-4-ethylamino-6-methoxy-= 1,3,5-triazine **CA name** N-ethyl-6-methoxy-N'-(1-methylpropyl)-1,3,5-triazine-= 2,4-diamine. **M.f.** $C_{10}H_{19}N_5O$ **Code nos.** GS 14 254 **Selected tradenames** 'Etazine', 'Sumitol', 'Etazine 3585', 'Etazine 3947', 'Primatol 3588'. **Details** *PM8*, entry 10680.

1271 sesamex (common name). **CAS RN** *[51–14–9]*. Synergist for pyrethroid insecticides reported by M. Beroza (*J. Agric Food Chem.*, 1956, **4**, 49). Introduced by Shulton Inc. **IUPAC name** 5-[1-[2-(2-ethoxyethoxy)ethoxy]ethoxy]-1,3-benzodioxole (I); 2-(1,3-benzodioxol-5-yloxy)-3,6,9-trioxaundecane **CA name** (I). **M.f.** $C_{15}H_{22}O_6$ **Code nos.** ENT 20 871. **Selected tradenames** 'Sesoxane'. **Details** *PM6*, p. 472.

1272 sesasmolin (common name). **CAS RN** *[526–07–8]*. Synergistic activity with pyrethrum of sesame oil (from *Sesamum indicum*) reported by C. Eagleson (*Soap, Chem. Spec.*, 1942, **18**(12), 125) and traced to sesasmolin by H. L. Haller *et al.* (*J. Org. Chem.*, 1942, **7**, 183; *J. Econ. Entomol.*, 1942, **35**, 247) **IUPAC name** 1,3-benzodioxol-5-yl (1R,3aR,4S,6aR)-4-(1,3-benzodioxol-5-yl)perhydrofuro[3,4-c]furan-1-yl ether; 1-(1,3-benzodioxol-5-yl)-4-(1,3-benzodioxol-5-yloxy)tetrahydrofuro[3,4-c]furan; 2-(1,3-benzodioxol-5-yl)-6-(1,3-benzodioxol-5-yloxy)-3,7-dioxabicyclo[3.3.0]octane **CA name** [1S-(1α,3aα,4α,6aα)]-5-[4-(1,3-benzodioxol-5-yloxy)tetrahydro-= 1H,3H-furo[3,4-c]furan-1-yl]-1,3-benzodioxole. **M.f.** $C_{20}H_{18}O_7$. **Details** the related sesamin, **not** sesasmolin, described in *PM5*, p. 469.

1273 simeton (common name). **CAS RN** *[673–04–1]*. Herbicide developed by J. R. Geigy S. A. (later Ciba-Geigy AG). **IUPAC name** N^2,N^4-diethyl-6-methoxy-= 1,3,5-triazine-2,4-diamine; 2,4-bis(ethylamino)-6-methoxy-1,3,5-triazine **CA name** N,N'-diethyl-6-methoxy-1,3,5-triazine-2,4-diamine. **M.f.** $C_8H_{15}N_5O$ **Code nos.** G 30 044.

1274 SMY 1500 (development code). **CAS RN** *[64529–56–2]*. Herbicide reported by H. Hack & L. Eue (*Proc. 1985 Br. Crop Prot. Conf. - Weeds*, **1**, 35). Introduced in Israel (1989) by Bayer AG. **IUPAC name** 4-amino-6-tert-butyl-3-ethylthio-1,2,4-triazin-=

5(4H)-one   **CA name** 4-amino-6-(1,1-dimethylethyl)-3-(ethylthio)-1,2,4-triazin-=
5(4H)-one.   **M.f.** $C_9H_{16}N_4OS$ **Other names** [ethiozin] (rejected common name
proposal)   **Code nos.** SMY 1500 (Bayer)   **Selected tradenames** 'Tycor'.   **Details** PM9,
entry 0265.

1275   SN 72129 (development code).   **CAS RN** [77768–58–2] unstated
stereochemistry. Insecticide reported by H. Joppien et al. (Proc. 10th. Int. Congr. Plant
Prot., 1983, **1**, 392). Evaluated by Schering AG.   **IUPAC name** (E)-2-chlorobenzoyl=
(2,3-dihydro-4-phenyl-1,3-thiazol-2-ylidene)acetonitrile   **CA name** (E)-2-chloro-=
β-oxo-α-(4-phenyl-2(3H)-thiazolylidene)benzenepropanenitrile.   **M.f.** $C_{18}H_{11}ClN_2OS$
Other names thiapronil (unadopted proposed common name)   **Code nos.** SN 72129.

1276   sodium (Z)-3-chloroacrylate (IUPAC name).   **CAS RN** [4312–97–4]. Herbicide
reported by R. A. Herrett & A. N. Kurtz (Science, 1963, **141**, 1192). Introduced as a
defoliant by Union Carbide Corp. (later Rhône-Poulenc Ag.).   **IUPAC name** sodium
(Z)-3-chloroacrylate   **CA name** sodium (Z)-3-chloro-2-propenoate.   **M.f.** $C_3H_2ClNaO_2$
Other names sodium cis-3-chloroacrylate   **Code nos.** UC 20 299   **Selected tradenames**
'Prep'.   **Details** PM1, p. 386.

1277   sodium selenate (IUPAC name).   **CAS RN** [13410–01–0]. Introduced as an
insecticide.   **CA name** disodium selenate.   **M.f.** Na_2O_4Se.

1278   sophamide (common name).   Insecticide evaluated by Murphy Chemical Co.
IUPAC name S-methoxymethylcarbamoylmethyl O,O-dimethyl phosphorodithioate;
2-dimethoxyphosphinothioylthio-N-methoxymethylacetamide   **CA name**
S-[2-[(methoxymethyl)amino]-2-oxoethyl] O,O-dimethyl phosphorodithioate.
M.f. $C_6H_{14}NO_4PS_2$ **Code nos.** MC-62.

1279   SSF-109 (development code).   **CAS RN** [129586–32–9]. Broad-spectrum
fungicide introduced by Shionogi and Co. Ltd.   **CA name** (±)-cis-1-(4-chlorophenyl)-=
2-(1H-1,2,4-triazol-1-yl)cycloheptanol.   **M.f.** $C_{15}H_{18}ClN_3O$ **Code nos.** SSF-109
(Shionogi).   **Details** PM10, entry 633.

1280   SSI-121 (development code).   Acaricide evaluated by Shionogi and Co. Ltd.
M.f. $C_{22}H_{33}ClSiSn$ **Code nos.** SSI-121 (Shionogi).   **Details** PM10, entry 634.

1281   sulfallate (common name).   **CAS RN** [95–06–7]. Herbicide introduced by the
Monsanto Co.   **Common names** sulfallate, CDEC   **IUPAC name** 2-chloroallyl
diethyldithiocarbamate   **CA name** 2-chloro-2-propenyl diethylcarbamodithioate.
M.f. $C_8H_{14}ClNS_2$ **Code nos.** CP 4742   **Selected tradenames** 'Vegadex'.   **Details** PM6,
p. 483.

1282   sulfaquinoxaline.   **CAS RN** [59–40–5]. Rodenticide.   **CA name**
4-amino-N-2-quinoxalinylbenzenesulfonamide.   **M.f.** $C_{14}H_{12}N_4O_2S$.

1283   sulfoxide (common name).   **CAS RN** [120–62–7]. Synergist for pyrethroid
insecticides reported by M. E. Synerholm et al. (Contrib. Boyce Thompson Inst., 1947, **15**,

SUPERSEDED
ENTRIES

35). Introduced by S. B. Penick & Co.   **IUPAC name** 2-(1,3-benzodioxol-5-yl)ethyl octyl sulfoxide; 1-methyl-2-(3,4-methylenedioxyphenyl)ethyl octyl sulfoxide   **CA name** 5-[2-(octylsulfinyl)propyl]-1,3-benzodioxole.   **M.f.** $C_{18}H_{28}O_3S$ **Code nos.** ENT 16 634.   **Selected tradenames** 'Sulfocide'.   **Details** *PM6*, p. 485.

1284   sulglycapin (common name).   **CAS RN** *[51068–60–1]*. Herbicide reported by E. Eysell et al. (*Proc. Br. Crop Prot. Conf. - Weeds*, 1976, **2**, 709). Evaluated by BASF AG. **IUPAC name** azepan-1-ylcarbonylmethyl methylsulfamate; perhydroazepin-1-yl= carbonylmethyl methylsulfamate   **CA name** 2-(hexahydro-1*H*-azepin-1-yl)-2-oxoethyl methylsulfamate.   **M.f.** $C_9H_{18}N_2O_4S$ **Code nos.** BAS 461H; BAS 46 100H.

1285   sultropen (common name).   **CAS RN** *[963–22–4]*. Fungicide evaluated by Boots Co. Ltd (later Schering Agrochemicals).   **Common names** sultropen, sultropène   **IUPAC name** 2,4-dinitrophenyl pentyl sulfone   **CA name** 2,4-dinitro-1-(pentylsulfonyl)= benzene.   **M.f.** $C_{11}H_{14}N_2O_6S$ **Code nos.** RD 7901; BTS 7901.

1286   swep (common name).   **CAS RN** *[1918–18–9]*. Herbicide reported by H. R. Hudgins (*Proc. South. Weed Control Conf.*, 1963, **16**, 118). Introduced by FMC Corp. **Common names** swep, MCC   **IUPAC name** methyl 3,4-dichlorocarbanilate   **CA name** methyl (3,4-dichlorophenyl)carbamate.   **M.f.** $C_8H_7Cl_2NO_2$ **Code nos.** FMC 2995.   **Details** *PM5*, p. 486.

1287   2,4,5-T (common name).   **CAS RN** *[93–76–5]* 2,4,5-T; *[3813–14–7]* 2,4,5-T-= trolamine; *[57213–69–1]* 2,4,5-T-triethylammonium. Herbicide reported by C. L. Hamner & H. B. Tukey (*Science*, 1944, **100**, 154). Introduced by Amchem Products Inc. (later Rhône-Poulenc Agrochimie).   **IUPAC name** (2,4,5-trichlorophenoxy)acetic acid **CA name** (2,4,5-trichlorophenoxy)acetic acid.   **M.f.** $C_8H_5Cl_3O_3$ **Selected tradenames** 'Sylvoxone 850' (isoctyl ester).   **Details** *PM9*, entry 11120.

1288   tazimcarb (common name).   **CAS RN** *[40085–57–2]*, formerly *[62113–03–5]*. Insecticide and molluscicide evaluated by ICI Agrochemicals.   **Common names** tazimcarb, tazimcarbe   **IUPAC name** N-methyl-1-(3,5,5-trimethyl-4-oxo-= 1,3-thiazolidin-2-ylideneamino-oxy)formamide; 3,5,5-trimethyl-2-methylcarbamoyl= oxyimino-1,3-thiazolidin-4-one   **CA name** 3,5,5-trimethyl-2,4-thiazolidinedione 2-[*O*-[(methylamino)carbonyl]oxime]; formerly 2-[[[(aminocarbonyl)oxy]methyl]= imino]-3,5,5-trimethyl-4-thiazolidinone.   **M.f.** $C_8H_{13}N_3O_3S$ **Code nos.** PP505.

1289   2,4,5-TB (common name).   **CAS RN** *[93–80–1]*. Herbicide.   **IUPAC name** 4-(2,4,5-trichlorophenoxy)butyric acid   **CA name** 4-(2,4,5-trichlorophenoxy)butanoic acid.   **M.f.** $C_{10}H_9Cl_3O_3$.

1290   TDE (common name).   **CAS RN** *[72–54–8]*. Insecticide reported by P. Läuger et al. (*Helv. Chim. Acta*, 1944, **27**, 892). Introduced by Rohm & Haas Co. (found in nature as a degradation product of DDT).   **IUPAC name** 1,1-dichloro-2,2-bis(4-chloro= phenyl)ethane   **CA name** 1,1'-(2,2-dichloroethylidene)bis[4-chlorobenzene]. **M.f.** $C_{14}H_{10}Cl_4$ **Other names** DDD   **Code nos.** OMS 1078, ENT 4225.   **Selected tradenames** 'Rhothane'.   **Details** *PM3*, p. 457.

1291 tecoram (common name). Fungicide evaluated by Aagrunol NV. **IUPAC name** N',N',N''',N'''-tetramethyl-N,N''-ethylenedi(thiuram disulfide) **CA name** N,N,N',N'-tetramethyl-4,9-dithioxo-2,3,10,11-tetrathia-5,8-diazadodecanedithioamide. **M.f.** $C_{10}H_{18}N_4S_8$ **Selected tradenames** 'Azosan'.

1292 TEPP (common name). **CAS RN** *[107–49–3]*. Aphicide discovered in 1938 by G. Schrader & H. Kükenthal (cited by G. Schrader, *Die Entwickling neuer insektizider Phosphorsaure-Ester*, 3rd Ed.). In 1943 I. G. Farbenindustrie AG introduced a derivative then thought to be hexaethyl tetraphosphate (known as HETP) but since shown to contain TEPP as the main insecticidal component. **Common names** TEPP, ethyl pyrophosphate **IUPAC name** tetraethyl pyrophosphate **CA name** tetraethyl diphosphate. **M.f.** $C_8H_{20}O_7P_2$ **Code nos.** ENT 18 771. **Selected tradenames** 'Nifos T', 'Vapotone'. **Details** *PM8*, entry 11250.

1293 terallethrin (common name). **CAS RN** *[15589–31–8]* unstated stereochemistry. Insecticide introduced by Sumitomo Chemical Co., Ltd. **Common names** terallethrin, téralléthrine **IUPAC name** (RS)-3-allyl-2-methyl-4-oxocyclopent-= 2-enyl 2,2,3,3-tetramethylcyclopropanecarboxylate **CA name** $(\pm)$-2-methyl-4-oxo-=

3-(2-propenyl)-2-cyclopenten-1-yl 2,2,3,3-tetramethylcyclopropanecarboxylate. **M.f.** $C_{17}H_{24}O_3$ **Code nos.** M-108.

1294 terbucarb (common name). **CAS RN** *[1918–11–2]*. Herbicide reported by A. H. Haubein & J. R. Hansen, (*J. Agric. Food Chem.*, 1965, **13**, 555). Introduced by Hercules Inc. (later Nor-Am.). **Common names** terbucarb, terbucarbe, terbutol, MBPMC **IUPAC name** 2,6-di-*tert*-butyl-*p*-tolyl methylcarbamate **CA name** 2,6-bis-= (1,1-dimethylethyl)-4-methylphenyl methylcarbamate. **M.f.** $C_{17}H_{27}NO_2$ **Code nos.** Hercules 9573 **Selected tradenames** 'Azak'. **Details** *PM4*, p. 476.

1295 terbuchlor (common name). **CAS RN** *[4212–93–5]*. Herbicide evaluated by Monsanto Co. **Common names** terbuchlor, terbuchlore **IUPAC name** N-butoxy= methyl-6'-*tert*-butyl-2-chloroacet-*o*-toluidide **CA name** N-(butoxymethyl)-2-chloro-= N-[2-(1,1-dimethylethyl)-6-methylphenyl]acetamide. **M.f.** $C_{18}H_{28}ClNO_2$ **Code nos.** MON 0358; CP 46 358.

1296 tetcyclacis (common name). **CAS RN** *[77788–21–7]*. Plant growth regulator reported by J. Jung et al. (*Z. Acker-Pflanzenbau*, 1980, **149**, 128). Introduced by BASF AG. **IUPAC name** *rel*-(1R,2R,6S,7R,8R,11S)-5-(4-chlorophenyl)-3,4,5,9,10-penta-= azatetracyclo[5.4.1.0^{2,6}.0^{8,11}]dodeca-3,9-diene **CA name** (3aα,4β,4aα,6aα,7β,7aα)-= 1-(4-chlorophenyl)-3a,4,4a,6a,7,7a-hexahydro-4,7-methano-1H-[1,2]diazeto[3,4-f]= benzotriazole. **M.f.** $C_{13}H_{12}ClN_5$ **Code nos.** BAS 106W **Selected tradenames** 'Ken byo'. **Details** *PM7*, entry 11340.

1297 tetrachlorothiophene (IUPAC/Chemical Abstracts name). **CAS RN** *[6012–97–1]*. Nematicide developed by Pennwalt Corp. **M.f.** C_4Cl_4S **Other names** TCTP (also used for chlorthal-dimethyl) **Code nos.** ENT 25 764. **Selected tradenames** 'Penphene'. **Details** *PM3*, p. 464.

1298 tetrafluron (common name). **CAS RN** *[27954–37–6]* formerly *[34766–29–5]* and *[36510–94–8]*. Herbicide evaluated by Hoechst AG. **IUPAC name** 1,1-dimethyl-= 3-[3-(1,1,2,2-tetrafluoroethoxy)phenyl]urea   **CA name** N,N-dimethyl-= N'-[3-(1,1,2,2-tetrafluoroethoxy)phenyl]urea.   **M.f.** $C_{11}H_{12}F_4N_2O_2$ **Code nos.** Hoe 2991.

1299 O,O,O',O'-tetrapropyl dithiopyrophosphate (IUPAC name). **CAS RN** *[3244–90–4]*. Insecticide reported by A. D. F. Toy (*J. Am. Chem. Soc.*, 1951, **73**, 4670). Introduced by Stauffer Chemical Co. (later ICI Americas). **CA name** tetrapropyl thiodiphosphate; thiodiphosphoric acid ([(HO)$_2$P(S)]$_2$O) tetrapropyl ester. **M.f.** $C_{12}H_{28}O_5P_2S_2$ **Code nos.** ASP-51, ENT 16 894 **Selected tradenames** 'Aspon'. **Details** *PM8*, entry 11470.

1300 tetrasul (common name). **CAS RN** *[2227–13–6]*. Acaricide reported by J. Meltzer & F. C. Dietvoorts (*Proc. Int. Congr. Crop Prot.*, 4th, 1957, **1**, 669). Introduced by N. V. Philips-Roxane (later Uniroyal Chemical Co., Inc). **Common names** tetrasul, tetradisul, diphenylsulphide **IUPAC name** 4-chlorophenyl 2,4,5-trichlorophenyl sulfide; 2,4,4′,5-tetrachlorodiphenyl sulfide   **CA name** 1,2,4-trichloro-5-[4-(chlorophenyl)= thio]benzene. **M.f.** $C_{12}H_6Cl_4S$ **Code nos.** V-101; OMS 755; ENT 27 115. **Selected tradenames** 'Animert V101'. **Details** *PM8*, entry 11480.

1301 thiadifluor (common name). Fungicide introduced by Bayer AG. **IUPAC name** 3-(4-chlorophenyl)-N^2-methyl-N^4,N^5-bis(trifluoromethyl)-1,3-thiazolidine-= 2,4,5-triylidenetriamine   **CA name** N,N'-[3-(4-chlorophenyl)-2-(methylimino)-= 4,5-thiazolidinediylidene]bis[1,1,1-trifluoromethanamine]. **M.f.** $C_{12}H_7F_6N_4S$ **Code nos.** SLJ 4027a.

1302 thiazafluron (common name). **CAS RN** *[25366–23–8]*. Herbicide reported by G. Müller et al. (*C. R. Journ. Etud. Herbic. Conf. COLUMA*, 7th, 1973, p. 32). Introduced by Ciba-Geigy AG. **Common names** thiazafluron, thiazfluron **IUPAC name** 1,3-dimethyl-1-(5-trifluoromethyl-1,3,4-thiadiazol-2-yl)urea **CA name** N,N'-dimethyl-N-[5-(trifluoromethyl)-1,3,4-thiadiazol-2-yl]urea. **M.f.** $C_6H_7F_3N_4OS$ **Code nos.** GS 29 696 **Selected tradenames** 'Erbotan'. **Details** *PM9*, entry 11520.

1303 thicrofos (common name). **CAS RN** *[41219–32–3]* unstated stereochemistry. Insecticide evaluated by Hoechst AG. **IUPAC name** S-(6-chloro-3,4-dihydro-= 2H-1-benzothi-in-4-yl) O,O-diethyl phosphorothioate **CA name** S-(6-chloro-= 3,4-dihydro-2H-1-benzothiopyran-4-yl) O,O-diethyl phosphorothioate. **M.f.** $C_{13}H_{18}ClO_3PS_2$ **Code nos.** Hoe 20 906.

1304 thicyofen (common name). **CAS RN** *[116170–30–0]* unstated stereochemistry. Fungicide reported by T. W. Hofman et al. (*Proc. 1990 Brighton Crop Prot. Conf. - Pests Dis.*, 2, 431). Introduced by Duphar B.V. (later Uniroyal Chemical Co., Inc). **IUPAC name** (±)-3-chloro-5-ethylsulfinylthiophene-2,4-dicarbonitrile **CA name** (±)-3-chloro-5-(ethylsulfinyl)-2,4-thiophenedicarbonitrile. **M.f.** $C_8H_5ClN_2OS_2$ **Code nos.** PH 51-07; DU 510 311 (Duphar). **Details** *PM9*, entry 11535.

1305   thidiazimin (common name).   **CAS RN** *[123249–43–4]*. Herbicide, reported by R. Weiler *et al.*, (*Proc. Br. Crop Prot. Conf.* - Weeds, 1993, **1**, 29). Evaluated by Schering AG.   **Common names** thidiazimin, thidiazimine   **IUPAC name** (Z)-6-(6,7-dihydro-= 6,6-dimethyl-3H,5H-pyrrolo[2,1-c][1,2,4]thiadiazol-3-ylideneamino)-7-fluoro-= 4-(2-propynyl)-2H-1,4-benzoxazin-3(4H)-one   **CA name** 6-[(6,7-dihydro-= 6,6-dimethyl-3H,5H-pyrrolo[2,1-c][1,2,4]thiadiazol-3-ylidene)amino]-7-fluoro-= 4-(2-propynyl)-2H-1,4-benzoxazin-3(4H)-one.   **M.f.** $C_{18}H_{17}FN_4O_2S$ **Code nos.** SN 124 085.

1306   thiocarboxime (common name).   **CAS RN** *[25171–63–5]*. Insecticide and acaricide evaluated by Shell Research Ltd.   **IUPAC name** 3-[1-(methylcarbamoyl= oxyimino)ethylthio]propionitrile   **CA name** 2-cyanoethyl N-[[(methylamino)= carbonyl]oxy]ethanimidothioate.   **M.f.** $C_7H_{11}N_3O_2S$ **Code nos.** WL 21 959   **Selected tradenames** 'Talcord'.

1307   thiochlorfenphim (common name). Fungicide evaluated by Bayer AG.   **Common names** thiochlorfenphim, thiochlorphenphim, thiochlorphenphime   **IUPAC name** N-(4-chlorophenylthiomethyl)phthalimide   **CA name** 2-[[(4-chlorophenyl)thio]= methyl]-1H-isoindole-1,3(2H)-dione.   **M.f.** $C_{15}H_{10}ClNO_2S$ **Code nos.** BAY 66 109; SRA 3208.

1308   2-thiocyanatoethyl laurate (IUPAC name).   **CAS RN** *[301–11–1]*. Insecticide developed by Rohm & Haas Co.   **CA name** 2-thiocyanatoethyl dodecanoate.   **M.f.** $C_{15}H_{27}NO_2S$ **Selected tradenames** 'Lethane 60'.

1309   thionazin (common name).   **CAS RN** *[297–97–2]*. Nematicide introduced by American Cyanamid Co.   **Common names** thionazin, thionazine   **IUPAC name** O,O-diethyl O-pyrazin-2-yl phosphorothioate   **CA name** O,O-diethyl O-pyrazinyl phosphorothioate.   **M.f.** $C_8H_{13}N_2O_3PS$ **Code nos.** Experimental Nematicide 18 133, ENT 25 580.   **Selected tradenames** 'Nemafos', 'Zinophos', 'Cynem', 'Bulb Dip'.   **Details** *PM7*, entry 11650.

1310   thiophanate (common name).   **CAS RN** *[23564–06–9]*. Fungicide reported by K. Ishii (*Abstr. Int. Congr. Plant Prot., 7th, Paris*, 1970, p. 200); reviewed (*idem, Jpn. Pestic. Inf.*, 1971, No. 7, p. 27). Introduced by Nippon Soda Co., Ltd.   **Common names** thiophanate, thiophanate-éthyl   **IUPAC name** diethyl 4,4'-(o-phenylene)bis(3-thio= allophanate)   **CA name** diethyl [1,2-phenylenebis(iminocarbonothioyl)]bis[carbamate].   **M.f.** $C_{14}H_{18}N_4O_4S_2$ **Code nos.** NF 35   **Selected tradenames** 'Topsin', 'Cercobin', 'Nemafax'.   **Details** *PM8*, entry 11660.

1311   thioquinox (common name).   **CAS RN** *[93–75–4]*. Acaricide and fungicide reported by G. Unterstenhöfer (*Hoefchen-Briefe, (Engl. Ed.)*, 1960, **13**, 207) and by K. Sasse (*ibid.*, p. 197). Introduced by Bayer AG.   **IUPAC name** 1,3-dithiolo[4,5-b]= quinoxaline-2-thione (I); quinoxaline-2,3-diyl trithiocarbonate; 2-thioxo-1,3-dithiolo= [4,5-b]quinoxaline   **CA name** (I).   **M.f.** $C_9H_4N_2S_3$ **Code nos.** Bayer 30 686; Ss 1451; ENT 25 579.   **Selected tradenames** 'Eradex'.   **Details** *PM4*, p. 491.

1312 tioclorim (common name). CAS RN *[68925–41–7]*. Herbicide discovered by Produits Chimiques Ugine Kuhlmann. **IUPAC name** 6-chloro-5-(methylthio)= pyrimidine-2,4-diamine **CA name** 6-chloro-5-(methylthio)-2,4-pyrimidinediamine. **M.f.** $C_5H_7ClN_4S$ **Code nos.** UK-J 1506 (Produits Chimiques). **Details** *PM9*, entry 11710.

1313 tioxymid (common name). CAS RN *[70751–94–9]*. Fungicide introduced by Shionogi & Co. Ltd. **Common names** tioxymid, tioxymide **IUPAC name** 5-isothiocyanato-2-methoxy-*N,N*-dimethyl-*m*-toluamide **CA name** 5-isothiocyanato-= 2-methoxy-*N,N*,3-trimethylbenzamide. **M.f.** $C_{12}H_{14}N_2O_2S$ **Code nos.** SAF-787; 710 352-S.

1314 transpermethrin (common name). CAS RN *[52341–32–9]*. Insecticide discovered by M. Elliott *et al.* **Common names** transpermethrin, transperméthrine **IUPAC name** 3-phenoxybenzyl (1*RS*)-*trans*-3-(2,2-dichlorovinyl)-2,2-dimethyl= cyclopropanecarboxylate; 3-phenoxybenzyl (1*RS*,3*SR*)-3-(2,2-dichlorovinyl)-= 2,2-dimethylcyclopropanecarboxylate **CA name** *trans*-($\pm$)-(3-phenoxyphenyl)methyl 3-(2,2-dichloroethenyl)-2,2-dimethylcyclopropanecarboxylate. **M.f.** $C_{21}H_{20}Cl_2O_3$ **Code nos.** NRDC 146.

1315 triamiphos (common name). CAS RN *[1031–47–6]*. Acaricide, fungicide and insecticide reported by B. G. van den Bos *et al.* (*Rec. Trav. Chim. Pays-Bas*, 1960, **79**, 807). Introduced by Philips-Duphar B.V. (later Uniroyal Chemical Co., Inc). **IUPAC name** 5-amino-3-phenyl-1*H*-1,2,4-triazol-1-yl-*N,N,N',N'*-tetramethylphosphonic diamide; *P*-amino-3-phenyl-1*H*-1,2,4-triazol-1-yl-*N,N,N',N'*-tetramethylphosphonic diamide; 5-amino-1-(bisdimethylaminophosphinyl)-3-phenyl-1*H*-1,2,4-triazole **CA name** *P*-(5-amino-3-phenyl-1*H*-1,2,4-triazol-1-yl)-*N,N,N',N'*-tetramethylphosphonic diamide. **M.f.** $C_{12}H_{19}N_6OP$ **Code nos.** WP 155, ENT 27 223. **Selected tradenames** 'Wepsyn 155'. **Details** *PM6*, p. 525.

1316 triapenthenol (common name). CAS RN *[76608–88–3]* stereochemistry (*E*). Plant growth regulator reported at 44th Deutsche Pflanzenchutz-Tag (1984), K. Lürssen & W. Reiser (*Proc. 1985 Br. Crop Prot. Conf. - Weeds*, **1**, 121). Introduced in Belgium and France (1989) by Bayer AG. **IUPAC name** (*E*)-(*RS*)-1-cyclohexyl-4,4-dimethyl-= 2-(1*H*-1,2,4-triazol-1-yl)pent-1-en-1-ol **CA name** (*E*)-($\pm$)-β-(cyclohexylmethylene)-= α-(1,1-dimethylethyl)-1*H*-1,2,4-triazole-1-ethanol. **M.f.** $C_{15}H_{25}N_3O$ **Code nos.** BAY RSW 0411 **Selected tradenames** 'Baronet'. **Details** *PM9*, entry 11853.

1317 triarathene (common name). CAS RN *[65691–00–1]*. Acaricide reported by D. I. Relyea *et al.* (*Proc. Int. Congr. Plant Prot., 10th*, 1983, **1**, 355). Evaluated by Uniroyal Inc. **IUPAC name** 5-(4-chlorophenyl)-2,3-diphenylthiophene (I) **CA name** (I). **M.f.** $C_{22}H_{15}ClS$ **Code nos.** UBI-T930 **Selected tradenames** 'Micromite'.

1318 triarimol (common name). CAS RN *[26766–27–8]*. Fungicide reported by J. V. Gramlich *et al.* (*Proc. Br. Insectic. Fungic. Conf., 5th*, 1969, **2**, 576). Evaluated by Eli Lilly & Co. **IUPAC name** 2,4-dichloro-α-(pyrimidin-5-yl)benzhydryl alcohol **CA name**

α-(2,4-dichlorophenyl)-α-phenyl-5-pyrimidinemethanol. **M.f.** $C_{17}H_{12}Cl_2N_2O$ **Code nos.** EL 273 **Selected tradenames** 'Trimidal'. **Details** *PM3*, p. 481.

1319 triazbutil (common name). **CAS RN** *[16227–10–4]*. Fungicide evaluated by Rohm & Haas Co. **IUPAC name** 4-butyl-4*H*-1,2,4-triazole (I) **CA name** (I). **M.f.** $C_6H_{11}N_3$ **Code nos.** RH-124 **Selected tradenames** 'Indar'.

1320 tributyl phosphorotrithioite (IUPAC/Chemical Abstracts name). **CAS RN** *[150–50–5]*. Introduced by the Mobil Chemical Co. and later by Rhône-Poulenc Agrochimie. **M.f.** $C_{12}H_{27}PS_3$ **Other names** merphos **Selected tradenames** 'Folex'. **Details** *PM8*, entry 11910.

1321 tricamba (common name). **CAS RN** *[2307–49–5]*. Herbicide developed by Velsicol Corp. (later Sandoz AG). **IUPAC name** 3,5,6-trichloro-*o*-anisic acid **CA name** 2,3,5-trichloro-6-methoxybenzoic acid. **M.f.** $C_8H_5Cl_3O_3$ **Selected tradenames** 'Banvel T'.

1322 trichlamide (common name). **CAS RN** *[70193–21–4]*. Fungicide reported by T. Ohmori *et al.* (*Plant Dis.*, 1986, **70**, 51). Introduced in Japan (1985) by Nippon Kayaku Co., Ltd. **IUPAC name** (*RS*)-*N*-(1-butoxy-2,2,2-trichloroethyl)salicylamide **CA name** (±)-*N*-(1-butoxy-2,2,2-trichloroethyl)-2-hydroxybenzamide. **M.f.** $C_{13}H_{16}Cl_3NO_3$ **Code nos.** NK-483 (Nippon Kayaku) **Selected tradenames** 'Hataclean'. **Details** *PM9*, entry 11922.

1323 4,5,7-trichloro-2,1,3-benzothiadiazole (IUPAC/Chemical Abstracts name). **CAS RN** *[1982–55–4]*. Herbicide reported by J. Dams *et al.* (*Proc. Br. Weed Control Conf., 7th*, 1964, p. 1091). Evaluated by Philips-Duphar B.V. (later Uniroyal Chemical Co., Inc). **M.f.** $C_6HCl_3N_2S$ **Code nos.** PH 40-21. **Details** *PM3*, p. 483.

1324 trichlorobenzyl chloride (IUPAC name). **CAS RN** *[1344–32–7]* formerly *[25429-36-1]*. Herbicide introduced by Monsanto Co. **CA name** trichloro= (chloromethyl)benzene. **M.f.** $C_7H_4Cl_4$ **Other names** TCBC **Selected tradenames** 'Randox-T'. **Details** *PM4*, p. 500.

1325 2,2,2-trichloro-1-(3,4-dichlorophenyl)ethyl acetate (IUPAC name). **CAS RN** *[21757–82–4]*. Insecticide reported by W. Behrenz *et al.* (*Pflanzenschutz-Nachr. (Engl. Ed.)*, 1977, **30**, 237). Introduced by Bayer AG. **CA name** 3,4-dichloro-α-(trichloro= methyl)benzenemethyl acetate. **M.f.** $C_{10}H_7Cl_5O_2$ **Other names** [benzethazet]; [plifenate] (rejected common name proposals) **Code nos.** BAY MEB 6046 **Selected tradenames** 'Baygon MEB'. **Details** *PM8*, entry 11960.

1326 trichloronat (common name). **CAS RN** *[327–98–0]*. Insecticide reported by R. O. Drummond (*J. Econ. Entomol.*, 1963, **56**, 831). Introduced by Bayer AG. **Common names** trichloronat, trichloronate **IUPAC name** O-ethyl O-2,4,5-trichloro= phenyl ethylphosphonothioate **CA name** O-ethyl O-(2,4,5-trichlorophenyl) ethylphosphonothioate. **M.f.** $C_{10}H_{12}Cl_3O_2PS$ **Code nos.** Bayer 37 289; S 4400; OMS 412; OMS 578; ENT 25 712. **Selected tradenames** 'Agrisil', 'Agritox', 'Phytosol'. **Details** *PM8*, entry 11980.

1327 tridiphane (common name). **CAS RN** *[58138–08–2]*. Herbicide reported by E. S. Saunders *et al.* (*Proc. North Cent. Weed Control Conf.*, 1981, p. 133). Introduced by Dow Chemical Co. (later DowElanco). **IUPAC name** (*RS*)-2-(3,5-dichlorophenyl)-= 2-(2,2,2-trichloroethyl)oxirane   **CA name** (±)-2-(3,5-dichlorophenyl)-= 2-(2,2,2-trichloroethyl)oxirane.   **M.f.** $C_{10}H_7Cl_5O$ **Code nos.** Dowco 356 **Selected tradenames** 'Nelpon', 'Tandem'.   **Details** *PM8*, entry 12050.

1328 trifenmorph (common name). **CAS RN** *[1420–06–0]*. Molluscicide reported by C. B. C. Boyce *et al.* (*Nature (London)*, 1966, **210**, 1140). Introduced by Shell Research Ltd.   **Common names** trifenmorph, triphenmorphe   **IUPAC name** 4-tritylmorpholine (I); 4-(triphenylmethyl)morpholine (II)   **CA name** (II).   **M.f.** $C_{23}H_{23}NO$ **Code nos.** WL 8008   **Selected tradenames** 'Frescon'.   **Details** *PM6*, p. 536.

1329 trifenofos (common name).   **CAS RN** *[38524–82–2]*. Acaricide and insecticide introduced by Rohm & Haas Co.   **IUPAC name** *O*-ethyl *S*-propyl *O*-2,4,6-trichlorophenyl phosphorothioate   **CA name** *O*-ethyl *S*-propyl *O*-(2,4,6-trichlorophenyl) phosphorothioate.   **M.f.** $C_{11}H_{14}Cl_3O_3PS$ **Code nos.** RH-218.

1330 trifop; trifop-methyl (common name).   **CAS RN** *[58594–74–4]* trifop (unstated stereochemistry); *[59011–30–2]* trifop (racemate); *[58594–77–7]* trifop-methyl (unstated stereochemistry); *[59011–33–5]* trifop-methyl (racemate). Herbicide evaluated by Hoechst AG.   **IUPAC name** (*RS*)-2-[4-(α,α,α-trifluoro-*p*-tolyloxy)= phenoxy]propionic acid   **CA name** 2-[4-[4-(trifluoromethyl)phenoxy]phenoxy]= propanoic acid.   **M.f.** $C_{16}H_{13}F_3O_4$; (methyl ester $C_{17}H_{15}F_3O_4$)   **Code nos.** Hoe 29 152 (trifop-methyl).

1331 trifopsime (common name).   **CAS RN** *[72131–76–1]*. Herbicide evaluated by Hofmann-La Roche Ltd.   **IUPAC name** acetone (*R*)-*O*-[2-[4-(α,α,α-trifluoro-= *p*-tolyloxy)phenoxy]propionyl]oxime   **CA name** (*R*)-2-propanone *O*-[1-oxo-= 2-[4-[4-(trifluoromethyl)phenoxy]phenoxy]propyl]oxime.   **M.f.** $C_{19}H_{18}F_3NO_4$ **Code nos.** Ro 13-8895.

1332 trimeturon (common name).   **CAS RN** *[3050–27–9]*. Herbicide evaluated by Bayer AG.   **IUPAC name** 3-(4-chlorophenyl)-1,1,2-trimethylisourea   **CA name** methyl *N*′-(4-chlorophenyl)-*N,N*-dimethylcarbamimidate.   **M.f.** $C_{10}H_{13}ClN_2O$ **Code nos.** BAY 40 557.

1333 triprene (common name).   **CAS RN** *[40596–80–3]*. Insect growth regulator introduced by Zoecon Corp. (later Sandoz AG).   **IUPAC name** *S*-ethyl (*E,E*)-(*RS*)-11-methoxy-3,7,11-trimethyldodeca-2,4-dienethioate   **CA name** (*E,E*)-*S*-ethyl 11-methoxy-3,7,11-trimethyl-2,4-dodecadienethioate.   **M.f.** $C_{18}H_{32}O_2S$ **Code nos.** ZR 519   **Selected tradenames** 'Altorick'.

1334 tripropindan (common name).   **CAS RN** *[6682–77–5]*. Herbicide evaluated by Roche.   **Common names** tripropindan, tripropindane   **IUPAC name** 1-(6-isopropyl-1,1,4-trimethylindan-5-yl)propan-1-one   **CA name** 1-[2,3-dihydro-=

1,1,4-trimethyl-6-(1-methylethyl)-1*H*-inden-5-yl]-1-propanone.   **M.f.** $C_{18}H_{26}O$
Code nos. Ro 7-0668.

1335   tritac (common name).   **CAS RN** *[1861–44–5]*. Herbicide introduced by
Hooker Chemical Corp.   **IUPAC name** 1-(2,3,6-trichlorobenzyloxy)propan-2-ol
CA name 1-[(2,3,6-trichlorophenyl)methoxy]-2-propanol.   **M.f.** $C_{10}H_{11}Cl_3O_2$
Selected tradenames 'Tritac'.

1336   UBI-S734 (development code).   **CAS RN** *[60263–88–9]*. Herbicide evaluated by
Uniroyal Chemical Co., Inc.   **IUPAC name** 2-pyridyl 1-(2,5-xylyl)ethyl sulfone 1-oxide
CA name 2-[[1-(2,5-dimethylphenyl)ethyl]sulfonyl]pyridine 1-oxide.   **M.f.** $C_{15}H_{17}NO_3S$
Code nos. UBI-S734.

1337   urbacid (common name).   **CAS RN** *[2445–07–0]*. Fungicide introduced by Bayer
AG.   **IUPAC name** methylarsinediyl bis(dimethyldithiocarbamate)   **CA name**
dimethylcarbamodithioic acid bis(anhydrosulfide) with methylarsonodithious acid.   **M.f.**
$C_7H_{15}AsN_2S_4$ **Selected tradenames** 'Urbacid', 'Monzet', 'Tuzet'.   **Details** *PM5*, p. 356.

1338   WL 9385 (development code).   **CAS RN** *[2854–70–8]*. Herbicide reported by
R. A. Abbott & G. E. Barnsley (*J. Sci. Food Agric.*, 1968, **19**, 16). Evaluated by Shell
Research Ltd.   **IUPAC name** 2-azido-4-*tert*-butylamino-6-ethylamino-1,3,5-triazine;
6-azido-N^2-*tert*-butyl-N^4-ethyl-1,3,5-triazine-2,4-diamine   **CA name** 6-azido-=
N-(1,1-dimethylethyl)-N'-ethyl-1,3,5-triazine-2,4-diamine.   **M.f.** $C_9H_{16}N_8$
Code nos. WL 9385.

1339   XRD-563 (development code).   Reported by W. Arnold *et al.*, (*Proc. Br. Crop
Prot. Conf. - Pests Dis.*, 1992, **1**, 443). Fungicide evaluated by DowElanco.   **IUPAC name**
2,6-dichloro-N-(4-trifluoromethylbenzyl)benzamide.   **M.f.** $C_{15}H_{10}Cl_2F_3NO$
Code nos. XRD-563.

1340   xylachlor (common name).   **CAS RN** *[63114–77–2]*. Herbicide evaluated by
American Cyanamid Co.   **IUPAC name** 2-chloro-N-isopropylacet-2',3'-xylidide
CA name 2-chloro-N-(2,3-dimethylphenyl)-N-(1-methylethyl)acetamide.   **M.f.**
$C_{13}H_{18}ClNO$ **Code nos.** AC 206 784   **Selected tradenames** 'Combat'.

1341   zarilamid (common name).   **CAS RN** *[84527–51–5]* unstated stereochemistry.
Fungicide reported by S. P. Heaney *et al.* (*Proc. 1988 Brighton Crop Prot. Conf. - Pests Dis.*,
2, 551). Evaluated by ICI Agrochemicals.   **IUPAC name** (*RS*)-4-chloro-N-[cyano=
(ethoxy)methyl]benzamide   **CA name** ($\pm$)-4-chloro-N-(cyanoethoxymethyl)=
benzamide.   **M.f.** $C_{11}H_{11}ClN_2O_2$ **Code nos.** ICIA0001; PP001.

1342   zolaprofos (common name).   **CAS RN** *[63771–69–7]*. Insecticide evaluated by
BASF AG.   **IUPAC name** O-ethyl S-3-methyl-1,2-oxazol-5-ylmethyl S-propyl
phosphorodithioate; O-ethyl S-(3-methylisoxazol-5-ylmethyl) S-propyl
phosphorodithioate   **CA name** O-ethyl S-[(3-methyl-5-isoxazolyl)methyl] S-propyl
phosphorodithioate.   **M.f.** $C_{10}H_{18}NO_3PS_2$ **Code nos.** BAS 2681.

GLOSSARY

Latin–English

Names include those target species referred to in the Main Entries. For each name which is identified in the first column at the Genus level (i.e. names in italics), the third column gives: for fungi and bacteria, the Order; for insects and vertebrates, the Order then Family; for plants, the Family. The first column also includes some Families and Orders, with a corresponding higher level indicated in the third column.

Latin	English	Family and/or Order
Abutilon theophrasti	Velvetleaf	Malvaceae
Acari	Mites	Arachnida
Acaridae	Acarid mites	Acari
Acarina See *Acari*		
Acarus spp.	Flour mites; grain mites	Acari: Acaridae
Acrididae	Grasshoppers & locusts	Saltatoria
Actinomycetales	Filamentous bacteria	
Aculops spp.	Mites	Acari: Eriophyidae
Aculus schlechtendali	Rust mite, apple	Acari: Eriophyidae
Aculus spp.	Rust mites	Acari: Eriophyidae
Adoxophyes orana	Summer fruit tortrix moth	Lepidoptera: Tortricidae
Aedes aegypti	Yellow fever mosquito	Diptera: Culicidae
Aeschynomene spp.	Joint vetches	Leguminosae
Agaricales	Mushrooms, etc.	
Agriotes spp.	Wireworms	Coleoptera: Elateridae
Agromyza spp.	Leaf miners	Diptera: Agromyzidae
Agropyron repens See *Elymus repens*		
Agrostis gigantea	Black bent	Gramineae
Agrostis stolonifera	Creeping bent	Gramineae
Agrotis segetum	Turnip moth	Lepidoptera: Noctuidae
Agrotis spp.	Cutworms	Lepidoptera: Noctuidae
Alabama argillacea	Cotton leaf worm	Lepidoptera: Noctuidae
Albugo candida	White blister	Peronosporales
Aleurothrixus floccosus	A whitefly	Homoptera: Aleyrodidae
Aleyrodidae	Whiteflies	Homoptera
Alisma lanceolatum	Water plantain, narrow leaved	Alismataceae
Alisma plantago-aquatica	Water plantain	Alismataceae
Alisma spp.	Water plantains	Alismataceae
Alopecurus myosuroides	Black-grass	Gramineae
Alphitobius spp.	Mealworms	Coleoptera: Tenebrionidae
Alternaria brassicae	Dark leaf spot, brassicas	Hyphales
Alternaria brassicicola	Dark leaf spot, brassicas	Hyphales
Alternaria dauci	Carrot leaf blight	Hyphales

Latin	English	Family and/or Order
Alternaria spp.	Leaf spots, various	Hyphales
Amaranthus retroflexus	Redroot pigweed; common amaranth	Amaranthaceae
Amaranthus spp.	Amaranths	Amaranthaceae
Amblyomma spp.	Ticks	Acari: Ixodidae
Amblyseius andersoni	A predatory mite	Acari: Phytoseiidae
Amblyseius cucumeris	A predatory mite	Acari: Phytoseiidae
Amblyseius finlandicus	Fruit tree red spider mite predator	Acari: Phytoseiidae
Ambrosia artemisifolia	Ragweed, common	Compositae
Anobium punctatum	Furniture beetle	Coleoptera: Anobiidae
Anopheles spp.	Anopheles mosquitos	Diptera: Culicidae
Anoplura	Sucking lice	Insecta
Anthemis cotula	Stinking mayweed; stinking chamomile	Compositae
Anthemis spp.	Chamomiles (= some mayweeds)	Compositae
Anthomyiidae, *Delia* spp. (= some *Hylemya* spp.) and others	Root flies	Diptera: Anthomyiidae
Anthonomus grandis	Boll weevil	Coleoptera: Curculionidae
Anthonomus pomorum	Apple blossom weevil	Coleoptera: Curculionidae
Anthrenus spp.	Carpet beetles	Coleoptera: Dermestidae
Anticarsia gemmatalis	Soya bean looper	Lepidoptera: Noctuidae
Aonidiella aurantii	Californian red scale	Homoptera: Diaspididae
Apera spica-venti	Loose silky-bent	Gramineae
Aphanomyces cochlioides	Blackleg, beet crops	Saprolegniales
Aphanomyces spp.	Foot rot, root rot, various hosts	Saprolegniales
Aphididae	Aphids	Homoptera
Aphidius spp.	Aphid parasitoid wasps	Hymenoptera: Braconidae
Aphis citricola	Citrus aphid	Homoptera: Aphididae
Aphis fabae	Black bean aphid	Homoptera: Aphididae
Aphis gossypii	Melon and cotton aphid	Homoptera: Aphididae
Aphyllophorales		Basidiomycotina
Apodemus sylvaticus	Long-tailed field mouse	Rotentia: Muridae
Arrhenatherum elatius	False oat-grass	Gramineae
Arrhenatherum elatius var. bulbosum	Onion couch	Gramineae
Artemisia vulgaris	Mugwort; wormwood	Compositae
Arvicanthis niloticus	Field rat	Rotentia: Muridae
Ascochyta chrysanthemi See *Didymella ligulicola*		
Ascochyta fabae	Leaf spot, beans	Sphaeropsidales
Ascochyta pinodes	Leaf and pod spot, peas	Sphaeropsidales
Ascochyta pisi	Leaf and pod spot, peas	Sphaeropsidales
Ascochyta spp.	Leaf spots, various hosts	Sphaeropsidales

Latin	English	Family and/or Order
Ascomycotina	Fungi, sexually produced spores in sacs	
Aspergillus spp.	Storage fungi	Hyphales
Athous spp.	Garden wireworms	Coleoptera: Elateridae
Atomaria linearis	Pygmy beetle	Coleoptera: Cryptophagidae
Atriplex patula	Common orache	Chenopodiaceae
Aulacorthum solani	Glasshouse potato aphid	Homoptera: Aphididae
Avena barbata	Bearded oat	Gramineae
Avena fatua	Wild oat	Gramineae
Avena spp.	Oats (wild and cultivated)	Gramineae
Avena sterilis	Oat, sterile	Gramineae
Avena sterilis ssp. ludoviciana (= A. ludoviciana)	Wild oat, winter	Gramineae
Bandicota benghalensis	Lesser bandicoot mole rat	Rotentia: Muridae
Bandicota spp.	Bandicoot rats	Rotentia: Muridae
Basidiomycotina	Fungi, spores produced exogenously in basidia	
Begonia elatior	Begonia	Begoniaceae
Bemisia spp.	Whiteflies	Homoptera: Aleyrodidae
Bemisia tabaci	Tobacco whitefly	Homoptera: Aleyrodidae
Bilderdykia convolvulus See Fallopia convolvulus		
Bipolaris stenospila	Brown stripe, sugar cane	Hyphales
Blatta orientalis	Common cockroach	Dictyoptera: Blattidae
Blattella germanica	German cockroach	Dictyoptera: Blattidae
Blattella spp.	Cockroaches	Dictyoptera: Blattidae
Blissus leucopterus	Chinch bug	Lygaeidae
Blumeriella jaapii	Coccomycosis, Cherry leaf spot	Helotiales
Boophilus microplus	Ticks	Acari: Ixodidae
Botryosphaeria obtusa (= Physalospora obtusa)	Leaf spot and black rot, apple	Dothidiales
Botrytis allii	Neck rot, onions	Hyphales
Botrytis cinerea	Fruit rot, various hosts	Hyphales
Bovicola bovis	Cattle biting louse	Phthiraptera: Trichodectidae
Bovicola ovis	Sheep biting louse	Phthiraptera: Trichodectidae
Brachiaria mutica (= Panicum purpurascens)	Buffalograss	Gramineae
Brassica napus	Rape	Cruciferae
Bremia lactucae	Downy mildew, lettuce	Peronosporales
Brevipalpus phoenicis	Red crevice tea mite	Acari: Tenuipalpidae
Bromus sterilis	Barren brome	Gramineae
Bryobia praetiosa	Clover bryobia mite	Acari: Tetranychidae
Bryobia ribis	Gooseberry bryobia mite	Acari: Tetranychidae

Latin	English	Family and/or Order
Bryophyta	Mosses and liverworts	Bryophyta
Bucculatrix thurberiella	Cotton leaf perforator	Lepidoptera: Lyonetiidae
Butomus umbellatus	Rush, flowering	Butomaceae
Caloptilia theivora	Tea leaf roller	Lepidoptera: Gracillariidae
Calepitrimerus spp.	Mite	Acari: Eriophyidae
Calystegia sepium ssp. *sepium*	Bindweed, large	Convolvulaceae
Camponotus spp.	Carpenter ants	Hymenoptera: Formicidae
Candida spp.	Parasitic yeasts	Endomycetales
Canis familiaris	Dog	Carnivora: Canidae
Capsella bursa-pastoris	Shepherd's purse	Cruciferae
Carduus spp.	Thistles	Compositae
Carex spp.	Sedges	Cyperaceae
Cassia obtusifolia	Sickle pod	Leguminosae (Fabaceae)
Cecidomyiidae	Gall midges and predacious midges	Diptera
Cecidophyopsis ribis	Blackcurrant gall-mite	Acari: Eriophyidae
Centaurea spp.	Knapweeds	Compositae
Ceratitis capitata	Mediterranean fruit fly	Diptera
Ceratocystis spp.	Dutch-elm disease	Microascales
Ceratobasidium cereale	Sharp eyespot, cereals	Tulasnellales
Ceratodon purpureus	Type of moss	Bryophyta
Ceratophyllum demersum	Hornweed, common	Ceratophyllaceae
Ceratopogonidae	Biting midges	Diptera
Cercospora beticola	Leaf spot, beet crops	Hyphales
Cercospora spp.	Leaf spots, various	Hyphales
Cercospora zonata	Cercospora leaf spot, beans	Hyphales
Cercosporella herpotrichoides See *Pseudocercosporella herpotrichoides*		
Cercosporidium spp. (includes *C. sojinum* = *Cercospora sojina*)	Frog eye, leaf spot, soya	Hyphales
Ceutorhynchus assimilis	Cabbage seed weevil	Coleoptera: Curculionidae
Ceutorhynchus pleurostigmata	Turnip gall weevil	Coleoptera: Curculionidae
Ceutorhynchus quadridens	Cabbage stem weevil	Coleoptera: Curculionidae
Ceutorhynchus spp.	Brassica gall and stem weevils	Coleoptera: Curculionidae
Chaetocnema concinna	Mangold flea beetle	Coleoptera: Chrysomelidae
Chaetocnema spp.	Flea beetles	Coleoptera: Chrysomelidae
Chamomilla spp.	Mayweeds (some)	Compositae
Chenopodium album	Fat hen	Chenopodiaceae
Cheyletiella parasitivorax	Rabbit fur mite	Acari: Cheyletidae
Chilo plejadellus	Rice stem-borer	Lepidoptera: Pyralidae
Chilo spp.	Stem borers	Lepidoptera: Pyralidae
Chilo suppressalis	Rice stalk borer	Lepidoptera: Pyralidae

Latin	English	Family and/or Order
Chorioptes spp.	Mange mites	Acari: Psoroptidae
Chromatomyia syngenesiae See *Phytomyza syngenesiae*		
Chromolaena odorata	Siam weed	Compositae
Chrysanthemum segetum	Corn marigold	Compositae
Chrysomelidae	Chrysomelid beetles	Coleoptera
Chrysopa carnea See *Chrysoperla carnea*		
Chrysoperla carnea	Pearly green lacewing	Neuroptera: Chrysopidae
Chrysops spp.	Deer flies	Diptera: Tabanidae
Chrysoteuchia caliginosellus (= *Crambus caliginosellus*)	Grass moth	Lepidoptera: Pyralidae
Cicadellidae	Leafhoppers	Homoptera
Cimex lectularius	Bed bug	Heteroptera: Cimicidae
Cirsium arvense	Thistle, creeping	Compositae
Cladosporium carpophilum See *Stigmina carpophila*		
Cladosporium fulvum See *Fulvia fulva*		
Cladosporium spp.	Black mould; sooty mould	Hyphales
Clasterosporium carpophilum See *Stigmina carpophila*		
Clavibacter michiganensis	Tomato canker	Eubacteriales
Cnaphalocrocis medinalis	Rice leaf roller	Lepidoptera: Pyralidae
Cnemidocoptes spp.	Bird skin mites	Acari: Sarcoptidae
Coccidae	Scale insects	Homoptera
Coccomyces hiemalis See *Blumeriella jaapii*		
Coccus hesperidum	Brown soft scale	Homoptera: Coccidae
Coccus spp.	Scale insects	Homoptera: Coccidae
Cochliobolus miyabeanus	Brown spot, rice	Dothidiales
Cochliobolus sativus	Foot rot, root rot, cereals, grasses	Dothidiales
Coleoptera	Beetles	Insecta
Colletotrichum atramentarium See *C. coccodes*		
Colletotrichum coccodes	Root rot, tomato	Melanconiales
Colletotrichum coffeanum See *Glomerella cingulata*		
Colletotrichum gloeosporoides See *Glomorella cingulata*		
Colletotrichum lindemuthianum	Anthracnose, french beans	Melanconiales

Latin	English	Family and/or Order
Colletotrichum spp.	Anthracnose, various root rot and leaf curl diseases	Melanconiales
Commelina spp.	Dayflower; wandering Jew	Commelinaceae
Comstockaspis perniciosus	San José scale	Homoptera: Coccidae
Convolvulus arvensis	Field bindweed	Convolvulaceae
Coptotermes formosanus	Formosan termite	Isoptera: Rhinotermitidae
Coptotermes spp.	Termites	Isoptera: Rhinotermitidae
Corticium cerealis See Ceratobasidium cereale		
Corticium fuciforme See Laetisaria fuciformis		
Corticium sasakii See Pellicularia sasakii		
Corynebacterium michiganense See Clavibacter michiganensis		
Corynespora melonis	Leaf spot, melon	Hyphales
Cosmopolites sordidus	Banana root borer; banana weevil	Coleoptera: Curculionidae
Crambus caliginosellus	Grass moth	Lepidoptera: Pyralidae
Cricetus spp.	Crickets	Saltatoria: Gryllidae
Cronartium ribicola	Blackcurrant rust	Uredinales
Cryptolestes spp.	Grain beetles	Coleoptera: Cucujidae
Ctenocephalides canis	Dog flea	Siphonaptera: Pulicidae
Ctenocephalides felis	Cat flea	Siphonaptera: Pulicidae
Cucujidae	Flour beetles	Coleoptera
Culex fatigans (= C. quinquefasciatus)	House mosquito	Diptera
Culex quinquefasciatus See C. fatigans		
Culex spp.	Mosquitoes	Diptera
Culicidae	Mosquitoes	Diptera
Curculionidae	Weevils	Coleoptera
Cydia pomonella	Codling moth	Lepidoptera: Tortricidae
Cynodon dactylon	Bermuda grass	Gramineae
Cynodon spp.	Bermuda grass, star grasses	Gramineae
Cynomys spp.	Prairie dogs	Rodentia: Sciuridae
Cyperus brevifolius	Kyllinga, green	Cyperaceae
Cyperus difformis	Umbrella plant	Cyperaceae
Cyperus esculentus	Yellow nutsedge	Cyperaceae
Cyperus rotundus	Nutgrass	Cyperaceae
Cyperus serotinus	Late flowering cyperus	Cyperaceae
Cyperus spp.	Nutsedges	Cyperaceae
Dacnusa sibirica	Chrysanthemum leaf miner parasitoid	Hymenoptera: Braconidae
Dacus oleae	Olive fruit fly	Diptera: Tephritidae

Latin	English	Family and/or Order
Dacus spp.	Fruit flies	Diptera: Tephritidae
Damalinia bovis See *Bovicola bovis*		
Damalinia ovis See *Bovicola ovis*		
Damalinia spp. See *Bovicola* spp.		
Datura stramonium	Jimson weed, thorn apple	Solanaceae
Decoceras spp.	e.g. field slug	Mollusca: Gastropoda
Delia brassicae See *D. radicum*		
Delia coarctata	Wheat bulb fly	Diptera: Anthomyiidae
Delia radicum	Cabbage root fly	Diptera: Anthomyiidae
Delia spp. (= some *Hylemya* spp.)	Root flies	Diptera: Anthomyiidae
Demodex spp.	Follicle mites	Acari: Demodicidae
Dermanyssus gallinae	Chicken mite	Acari: Dermanyssidae
Dermaptera	Earwigs	Insecta
Desmodium tortuosum	Florida beggarweed	Leguminosae
Deuteromycetes	Fungi with no known sexual stage, or asexual stages of other fungi	Deuteromycotina (= Fungi Imperfecti)
Diabrotica spp.	Corn rootworms	Coleoptera: Chrysomelidae
Diaporthales		Ascomycotina
Diaporthe citri	Melanosis, citrus	Diaporthales
Diaporthe helianthi	Leaf spot and stem canker, sunflowers	Diaporthales
Diaporthe spp.	Includes stem canker fungi, various hosts	Diaporthales
Diaspidae (and others)	Scale insects	Homoptera
Diatraea saccharalis	Maize stalk borer	Lepidoptera: Pyralidae
Didesmococcus brevipes	Scale insect	Homoptera: Coccidae
Didymella applanata	Spur blight, cane fruit	Dothidiales
Didymella chrysanthemi See *Didymella ligulicola*		
Didymella ligulicola	Ray blight, chrysanthemum	Dothidiales
Digitaria adscendens (= *D. ciliaris*)	Crabgrass, tropical	Gramineae
Digitaria ciliaris See *Digitaria adscendens*		
Digitaria sanguinalis	Crabgrass	Gramineae
Digitaria spp.	Crabgrasses	Gramineae
Diplocarpon earliana	Leaf scorch, strawberry	Helotiales
Diplocarpon rosae	Blackspot, roses	Helotiales
Diplodia spp.	Stalk rots, various hosts	Sphaeropsidales
Diplopoda	Millepedes	Myriapoda
Diprion spp.	Sawflies	Hymenoptera: Diprionidae

REFERENCE

Latin	English	Family and/or Order
Diptera	Flies	Insecta
Distantiella theobroma	Cocoa capsid	Heteroptera: Miridae
Ditylenchus dipsaci	Stem nematode	Nematoda: Tylenchidae
Dothidiales		Ascomycotina
Drechslera graminea See *Pyrenophora graminea*		
Drepanopeziza ribis See *Pseudopeziza ribis*		
Drosophila spp.	Fruit flies	Diptera: Drosophilidae
Drosophilidae	Fruit flies	Diptera
Earias spp.	Spiny bollworms	Lepidoptera: Noctuidae
Echinochloa colonum	Barnyard grass, awnless	Gramineae
Echinochloa crus-galli	Barnyard grass	Gramineae
Echinochloa oryzicola (= *E. oryzoides*)	Cockspur, rice	Gramineae
Echinochloa spp.	Barnyard grasses	Gramineae
Eichhornia crassipes	Water hyacinth	Pontederiaceae
Elateridae	Click beetles; wireworms	Coleoptera
Eleocharis acicularis	Spike rush	Cyperaceae
Eleusine indica	Goosegrass	Gramineae
Elodea canadensis	Water weed; Canadian pondweed	Hydrocharitaceae
Elsinoe fawcettii	Scab, citrus	Dothidiales
Elymus repens	Common couch; quackgrass	Gramineae
Empoasca fabae	Green leafhopper	Homoptera: Cicadellidae
Empoasca spp.	Cotton leafhoppers	Homoptera: Cicadellidae
Encarsia formosa	Glasshouse whitefly parasitoid	Hymenoptera: Aphelinidae
Endomycetales	Yeasts	Ascomycotina
Eotetranychus spp.	Tetranychid mites	Acari: Tetranychidae
Ephestia elutella	Warehouse moth	Lepidoptera: Pyralidae
Epilachna spp.	Bean beetles	Coleoptera: Coccinellidae
Epilachna varivestis	Mexican bean beetle	Coleoptera: Coccinellidae
Epitrimerus pyri	Pear rust mite	Acari: Eriophyidae
Epitrix hirtipennis	Tobacco flea beetle	Coleoptera: Chrysomelidae
Eriophyes spp.	Mite	Acari: Eriophyidae
Eriophyidae	Eriophyid mites	Acari
Eriosoma lanigerum	Woolly aphid	Homoptera: Pemphigidae
Erwinia carotovora	Bacterial rot, celery; basal stem rot, cucurbits; blackleg, potatoes	Eubacteriales
Erysiphaceae		Erysiphales
Erysiphales	Powdery mildews	Ascomycotina
Erysiphe betae	Powdery mildew, beet crops	Erysiphales
Erysiphe cichoracearum	Powdery mildew, cucurbits	Erysiphales

Latin	English	Family and/or Order
Erysiphe graminis	Powdery mildew, cereals, grasses	Erysiphales
Erysiphe spp.	Powdery mildew, various hosts	Erysiphales
Eubacteriales	Cellular bacteria	
Eupatorium odoratum (= Chromolaena odorata)	Siam weed	Compositae
Euphorbia maculata	Spotted spurge	Euphorbiaceae
Eupterycyba jucunda	Potato leafhopper	Homoptera: Cicadellidae
Eutetranychus banksi	Texas citrus mite	Acari: Tetranychidae
Eutetranychus spp.	Tetranychid mites	Acari: Tetranychidae
Euxoa spp.	Cutworms, dart moths	Lepidoptera: Noctuidae
Exobasidium vexans	Blister blight, tea	Exobasidiales
Fallopia convolvulus	Black bindweed	Polygonaceae
Fannia canicularis	Lesser house fly	Diptera: Muscidae
Fimbristylis spp.	Fringe rushes	Cyperaceae
Fomes annosus See *Heterobasidion annosum*		
Formicidae	Ants	Hymenoptera
Frankliniella occidentalis	Western flower thrips	Thysanoptera: Thripidae
Fuchsia hybrida	Fuchsia	Onagraceae
Fulvia fulva	Leaf mould, tomato	Hyphales
Fulvia spp.	Leaf moulds	Hyphales
Fusarium coeruleum	Dry rot, post- harvest rot	Hyphales
Fusarium culmorum	Fusarium foot and root rots, various hosts	Hyphales
Fusarium graminearum See *Gibberella zeae*		
Fusarium moniliforme See *Gibberella fujikuroi*		
Fusarium nivale See *Microdochium nivalis*		
Fusarium oxysporum	Fusarium wilt, various hosts	Hyphales
Fusarium solani var. *coeruleum* See *F. coeruleum*		
Fusarium spp.	Rots, ear blights and wilts, various hosts (Imperfect fungi with perfect stages in various genera).	Hyphales
Galium aparine	Cleavers	Rubiaceae
Ganoderma spp.	White rot, timber	Aphyllophorales
Gastropoda	Slugs and snails	Mollusca
Geotrichum candidum	Rubbery rot, potatoes	Hyphales
Geranium spp.	Crane's bills	Geraniaceae

Latin	English	Family and/or Order
Gibberella fujikuroi	Banana black heart, cotton boll rot, maize stalk rot	Hypocreales
Gibberella spp. (= various *Fusarium* spp.)	Scab, cereals; brown foot rot and ear blight and other cereal diseases	Hypocreales
Gibberella zeae	Scab, cereals	Hypocreales
Globodera spp.	Potato cyst nematodes	Nematoda: Heteroderidae
Gloeodes pomigena	Sooty blotch, apple pear and citrus	Sphaeropsidales
Gloeosporium fructigenum See *Glomerella cingulata*		
Gloeosporium spp.	Gloeosporium rot, apples	Deuteromycotina
Glomerella cingulata	Gloeosporium rot, apples	Polystigmatales
Glossina spp.	Tsetse flies	Diptera: Glossinidae
Gryllidae	True crickets	Saltatoria
Gryllotalpa spp.	Mole crickets	Saltatoria
Guignardia bidwellii	Black rot, grapevines	Dothidiales
Gymnosporangium fuscum	Pear rust	Uredinales
Gymnosporangium spp.	Leaf scorch, apples; rust, various hosts	Uredinales
Haematobia irritans	Horn fly	Diptera: Muscidae
Haematobia irritans exigua	Buffalo fly	Diptera: Muscidae
Haematopinus eurysternus	Short-nose cattle louse	Phthiraptera: Haematopinidae
Haematopinus quadripertusus	Cattle tail louse	Phthiraptera: Haematopinidae
Hedera helix	Ivy	Araliaceae
Helianthus annuus	Sunflower	Compositae
Helicotylenchus spp.	Spiral nematodes	Nematoda: Tylenchidae
Helicoverpa armigera	Old World bollworm	Lepidoptera: Noctuidae
Helicoverpa assulta	Oriental tobacco budworm	Lepidoptera: Noctuidae
Helicoverpa zea	American bollworm	Lepidoptera: Noctuidae
Heliothis armigera See *Helicoverpa armigera*		
Heliothis assulta See *Helicoverpa assulta*		
Heliothis virescens	Tobacco budworm	Lepidoptera: Noctuidae
Heliothis zea See *Helicoverpa zea*		
Helminthosporium oryzae	Helminthosporium blight, rice	Hyphales
Helminthosporium solani	Silver scurf, potatoes	Hyphales
Helminthosporium turcicum	Northern leaf blight, maize	Hyphales
Helotiales		Ascomycotina
Hemileia vastatrix	Coffee rust	Uredinales
Hemitarsonemus latus See *Polyphagotarsonemus latus*		

Latin	English	Family and/or Order
Heterobasidion annosum	Butt rot, conifers	Aphyllophorales
Heterodera cruciferae	Brassica cyst nematode	Nematoda: Heteroderidae
Heterodera goettingiana	Pea cyst nematode	Nematoda: Heteroderidae
Heterodera schachtii	Beet cyst nematode	Nematoda: Heteroderidae
Heterodera spp.	Lemon-shaped cyst nematodes	Nematoda: Heteroderidae
Heteroderidae	Cyst nematodes	Nematoda
Heteropeza pygmaea	Mushroom cecid	Diptera: Cecidomyiidae
Heteroptera	Bugs	Hemiptera
Homona spp.	Moths	Lepidoptera: Tortricidae
Homoptera	Aphids, hoppers, etc.	Hemiptera
Hydrilla verticillata	Elodea, Florida	Hydrocharitaceae
Hylemya spp. See Delia spp.		
Hylotrupes bajulus	House longhorn beetle	Coleoptera: Cerambycidae
Hymenoptera	Ants; bees; wasps; sawflies	Insecta
Hyphales		Deuteromycotina
Hypocreales		Ascomycotina
Hypoderma bovis	Ox warble fly	Diptera: Oestridae
Hypoderma spp.	Warble flies	Diptera: Oestridae
Ipomoea hederacea	Morning glory, ivyleaf	Convolvulaceae
Ipomoea purpurea	Morning glory, tall	Convolvulaceae
Iridomyrmex humilis	Argentine ant	Hymenoptera: Formicidae
Ischaemum rugosum	Saramatta grass	Gramineae
Isopoda	Slaters; woodlice	Crustacea
Isoptera	Termites	Insecta
Ixodes holocyclus	Paralysis tick	Acari: Ixodidae
Ixodes ricinus	Sheep tick	Acari: Ixodidae
Ixodes spp.	Ticks	Acari: Ixodidae
Ixodidae	Ixodid ticks	Acari
Juncus maritimus	Sea-rush	Juncaceae
Jussiaea diffusa See Ludwigia peploides		
Jussiaea spp.	Water primroses	Onagraceae
Keiferia lycopersicella	Tomato pinworm	Lepidoptera: Gelechiidae
Knemidokoptes spp. See Cnemidocoptes spp.		
Kochia scoparia	Mock cypress	Chenopodiaceae
Laetisaria fuciformis	Red thread, turf	Aphyllophorales
Lamium purpureum	Red dead-nettle	Labiatae
Laodelphax striatella	Small brown planthopper	Homoptera: Delphacidae
Lapsana communis	Nipplewort	Compositae
Lasioderma serricorne	Cigarette beetle	Coleoptera: Anobiidae
Lepidoptera	Butterflies; moths	Insecta
Leptinotarsa decemlineata	Colorado beetle	Coleoptera: Chrysomelidae
Leptochloa chinensis	Sprangletop, red	Gramineae

Latin	English	Family and/or Order
Leptochloa fascicularis (= *Diplachne fascicularis*)	Sprangletop, bearded	Gramineae
Leptochloa spp.	Sprangletop grasses	Gramineae
Leptosphaeria nodorum (= *Septoria nodorum*)	Glume blotch, wheat	Dothidiales
Leucoptera malifoliella	Pear leaf blister moth	Lepidoptera: Lyonetiidae
Leucoptera scitella See *L. malifoliella*		
Leucoptera spp.	Leaf-mining moths	Lepidoptera: Lyonetiidae
Leveillula spp.	Powdery mildew, (includes *L. taurica*, peppers)	Erysiphales
Lindernia procumbens	Pimpernel, false	Scrophulariaceae
Linognathus ovillus	Sheep sucking louse	Phthiraptera: Linognathidae
Linognathus spp.	Sucking lice	Phthiraptera: Linognathidae
Linognathus vituli	Long-nosed cattle louse	Phthiraptera: Linognathidae
Liriomyza bryoniae	Tomato leaf miner	Diptera: Agromyzidae
Liriomyza huidobrensis	South American leaf miner	Diptera: Agromyzidae
Liriomyza spp.	Leaf miners	Diptera: Agromyzidae
Liriomyza trifolii	American serpentine leaf miner	Diptera: Agromyzidae
Lissorhoptrusoryzophilus	Rice water weevil	Coleoptera: Curculionidae
Lithocolletis spp. See *Phyllonorycter* spp.		
Lobesia botrana	European vine moth	Lepidoptera: Tortricidae
Lolium multiflorum	Ryegrass, italian	Gramineae
Lolium perenne	Ryegrass, perennial	Gramineae
Lolium rigidum	Wimmera ryegrass; annual ryegrass	Gramineae
Lolium spp.	Ryegrasses	Gramineae
Longidorus spp.	Needle nematodes	Nematoda
Lucilia cuprina	Australian sheep blowfly	Diptera: Calliphoridae
Lucilia sericata	Sheep maggot fly	Diptera: Calliphoridae
Ludwigia peploides	Water purslane	Onagraceae
Lycoriella auripila	Mushroom sciarid	Diptera: Sciaridae
Lymantria dispar	Gypsy moth	Lepidoptera: Lymantriidae
Lymantria spp.	e.g. gypsy moth	Lepidoptera: Lymantriidae
Lymnaea spp.	Fresh water snails	Mollusca: Gastropoda
Lyonetia clerkella	Apple leaf miner	Lepidoptera: Lyonetiidae
Macrosiphum euphorbiae	Potato aphid	Homoptera: Aphididae
Macrosiphum rosae	Rose aphid	Homoptera: Aphididae
Marasmius oreades and other species	Fairy rings	Agaricales
Margarodidae	Scale insects	Homoptera
Marsilea spp.	Four-leaved water clover	Marsileaceae
Marssonina potentillae ssp. *fragariae* See *Diplocarpon earliana*		

Latin	English	Family and/or Order
Marssonina spp.	Leaf blotches etc., various hosts	Melanconiales
Mastigomycotina	Primitive fungi ('Phycomycetes') producing motile spores	
Matricaria perforata (= *M. inodora*)	Scentless mayweed	Compositae
Matricaria spp.	Mayweeds (some)	Compositae
Megaselia spp.	Scuttle flies	Diptera: Phoridae
Melanconiales		Deuteromycotina
Meligethes aeneus	Pollen beetle	Coleoptera: Nitidulidae
Meligethes spp.	Blossom or pollen beetles	Coleoptera: Nitidulidae
Meloidogyne incognita	Southern root-knot nematode	Nematoda
Meloidogyne spp.	Root-knot nematodes	Nematoda
Melolontha melolontha	Cockchafer	Coleoptera: Scarabaeidae
Melolontha spp.	Cockchafers	Coleoptera: Scarabaeidae
Melophagus ovinus	Sheep ked	Diptera: Hippoboscidae
Mesocricetus auratus	Golden hamster	Rodentia: Cricetidae
Microdochium nivalis	Snow mould, grasses, cereals	Hyphales
Microthyriella rubi		Schizothyrium pomi
Microtus agrestis	Field vole	Rotentia: Muridae
Microtus montebelli	Japanese field vole	Rotentia: Muridae
Miridae	Capsid bugs	Heteroptera
Mollusca	Slugs and snails	
Monilia roreri	Pod rot, cocoa	Hyphales
Monilia laxa See *Sclerotinia fructigena*		
Monilia spp.	Various rots	Hyphales
Monilinia laxa See *Sclerotinia fructigena*		
Monilinia mali	Monilinia leaf blight, apple	Helotiales
Monilinia spp. See *Sclerotinia* spp.		
Monochoria vaginalis	Pickerel weed	Pontederiaceae
Monographella nivalis	Rice leaf scald	Sphaeriales
Monomorium pharaonis	Pharaoh's ant	Hymenoptera: Formicidae
Monomorium spp.	Seed-eating ants	Hymenoptera: Formicidae
Mucor spp.	Fruit rot, strawberries	Mucorales
Mucorales		Zygomycotina
Muridae	Mice and rats	Rotentia
Mus domesticus	House mouse	Rotentia: Muridae
Mus musculus	House mouse	Rotentia: Muridae
Mus spp.	Mice	Rotentia: Muridae
Musca domestica	House fly	Diptera: Muscidae
Musca spp.	Muscid flies	Diptera: Muscidae

Latin	English	Family and/or Order
Musca vetustissima	Australian bush fly	Diptera: Muscidae
Muscidae	House flies	Diptera
Mycogone perniciosa	White mould, mushrooms	Hyphales
Mycosphaerella arachidis	Brown spot, peanut	Dothidiales
Mycosphaerella brassicicola	Ring-spot, brassicas	Dothidiales
Mycosphaerella fijiensis	Black leaf streak, banana	Dothidiales
Mycosphaerella fragariae	White leaf spot, strawberry	Dothidiales
Mycosphaerella graminicola	Septoria leaf spot, wheat	Dothidiales
Mycosphaerella musicola	Banana leaf spot, sigatoka	Dothidiales
Mycosphaerella pinodes See *Ascochyta pinodes*		
Mycosphaerella pomi	Brooks spot, apple	Dothidiales
Mycosphaerella spp.	Leaf spot diseases, various hosts	Dothidiales
Myzus persicae	Peach-potato aphid	Homoptera
Najas guadalupensis	Moss, aquatic	Najadaceae
Nectria galligena	Canker, apple, pear	Hypocreales
Nematoda	Nematodes	
Neodiprion sertifer	European pine sawfly	Hymenoptera: Diprionidae
Nephotettix cincticeps	Green rice leafhopper	Homoptera: Cicadellidae
Nephotettix impicticepts	Green rice leafhopper	Homoptera: Cicadellidae
Nephotettix nigropictus	Tropical green rice leafhopper	Homoptera: Cicadellidae
Nephotettix spp.	Green leafhoppers	Homoptera: Cicadellidae
Nilaparvata lugens	Rice brown planthopper	Homoptera: Delphacidae
Nilaparvata spp.	Planthoppers	Homoptera: Delphacidae
Nitrosomonas spp.	N-fixing bacteria	Bacteria
Noctua pronuba	Yellow underwing moth; cutworm	Lepidoptera: Noctuidae
Noctua spp.	Cutworms	Lepidoptera: Noctuidae
Noctuidae	Noctuid moths	Lepidoptera
Notoedres spp.	Mange mites	Acari: Psoroptidae
Oidium hevea	Powdery mildew	Erysiphales
Oomycetes		Mastigomycotina
Oospora lactis See *Geotrichum candidum*		
Oospora pustulans See *Polyscytalum pustulans*		
Opomyza florum	Yellow cereal fly	Diptera: Opomyzidae
Opomyza spp.	Grass and cereal flies	Diptera: Opomyzidae
Opuntia spp.	Prickly pear cacti	Cactaceae
Orthoptera See *Saltatoria*		
Oryctolagus cuniculus	Rabbit	Leporidae: Lagomorpha
Oryzaephilus surinamensis	Saw-toothed grain beetle	Coleoptera: Cucujidae
Oscinella frit	Frit fly	Diptera: Chloropidae
Ostrinia nubilalis	European corn borer	Lepidoptera: Pyralidae
Otiorhynchus sulcatus	Vine weevil	Coleoptera: Curculionidae

Latin	English	Family and/or Order
Otodectes spp.	Ear-mange mites	Acari: Psoroptidae
Oulema melanopus	Cereal leaf beetle	Coleoptera: Chrysomelidae
Oulema oryzae	Rice leaf beetle	Coleoptera: Chrysomelidae
Paecilomyces spp.	Saprophytic fungi	Hyphales
Panicum dichotomiflorum	Fall panicum; smooth witchgrass	Gramineae
Panicum purpurascens See *Brachiaria mutica*		
Panicum spp.	Panic grasses	Gramineae
Panicum texanum	Millet, Texas	Gramineae
Panonychus citri	Citrus red mite	Acari: Tetranychidae
Panonychus spp.	Red spider mites	Acari: Tetranychidae
Panonychus ulmi	Fruit tree red spider mite	Acari: Tetranychidae
Papaver spp.	Poppies	Papaveraceae
Pectinophora gossypiella	Pink bollworm	Lepidoptera: Gelechiidae
Pediculus capitis	Head louse	Phthiraptera: Pediculidae
Pediculus humanus	Human body louse	Phthiraptera: Pediculidae
Pegomya betae See *P. hyoscamni*		
Pegomya hyoscamni	Beet leaf-miner; mangold fly	Diptera: Anthomyiidae
Pellicularia sasakii	Rice sheath blight	Tulasnellales
Pellicularia spp.	Rots, damping off, etc. various hosts	Tulasnellales
Penicillium digitatum	Green mould, citrus	Hyphales
Penicillium italicum	Blue mould, citrus	Hyphales
Penicillium spp.	Penicillium rots	Hyphales
Periplaneta americana	American cockroach	Dictyoptera: Blattidae
Peronosclerospora spp.	Downy mildew, sorghum	Peronosporales
Peronospora parasitica	Downy mildew, brassicae	Peronosporales
Peronospora spp.	Downy mildews	Peronosporales
Peronospora tabacina (= *Plasmopara tabacina*)	Blue mould, tobacco	Peronosporales
Peronosporaceae See *Peronosporales*		
Peronosporales	Downy mildews, etc.	Oomycetes
Petunia spp.	Petunia	Solanaceae
Phakopsora pachyrhizi	Rust, soya	Uredinales
Phalaris paradoxa	Canary grass, awned	Gramineae
Phalaris spp.	Canary grasses	Gramineae
Pheidole megacephala	Pheidole ants	Hymenoptera: Formicidae
Phoma exigua var. *foveata*	Gangrene, potatoes	Deuteromycotina
Phoma spp.	Root and stem rots, various	Deuteromycotina
Phomopsis citri See *Diaporthe citri*		
Phomopsis helianthi See *Diaporthe helianthi*		

Latin	English	Family and/or Order
Phomopsis spp. See *Diaporthe* spp.		
Phomopsis viticola	Dead arm, grape vines	Sphaeropsidales
Phoridae	Scuttle flies	Diptera
Phorodon humuli	Damson-hop aphid	Homoptera: Aphididae
Phragmidium mucronatum	Rust, roses	Uredinales
Phthorimaea operculella	Potato moth	Lepidoptera: Gelechiidae
'Phycomycetes'	Primitive fungi with coenocytic mycelium; includes the Divisions Mastigomycotina and Zygomycotina.	
Phyllactinia spp.	Powdery mildew, various hosts	Erysiphales
Phyllocoptes spp.	Mites	Acari: Eriophyidae
Phyllocoptruta oleivora	Citrus rust mite	Acari: Eriophyidae
Phyllocoptruta spp.	Rust mites	Acari: Eriophyidae
Phyllonorycter blancardella	Apple leaf miner	Lepidoptera: Gracillariidae
Phyllonorycter spp.	Leaf mining moths	Lepidoptera: Gracillariidae
Phyllotreta spp.	Flea beetles	Coleoptera: Chrysomelidae
Phyllotreta striolata	Flea beetle	Coleoptera: Chrysomelidae
Physalospora obtusa See *Botryosphaeria obtusa*		
Phytomyza spp.	Leaf miners	Diptera: Agromyzidae
Phytomyza syngenesiae	Chrysanthemum leaf miner	Diptera: Agromyzidae
Phytophthora cactorum	Collar rot, crown rot, apple	Peronosporales
Phytophthora capsici	Blight, capsicums	Peronosporales
Phytophthora fragariae	Red core, strawberry	Peronosporales
Phytophthora infestans	Blight, potato, tomato	Peronosporales
Phytophthora megasperma	Root rot, brassicas	Peronosporales
Phytophthora palmivora	Rot, various crops	Peronosporales
Phytophthora spp.	Blight, damping off, foot-rot, various hosts	Peronosporales
Phytoseiulus persimilis	Two-spotted spider mite predator	Acari: Phytoseiidae
Pieris brassicae	Large white butterfly	Lepidoptera: Pieridae
Pieris rapae	Small white butterfly	Lepidoptera: Pieridae
Pieris spp.	Cabbage white butterflies	Lepidoptera: Pieridae
Pistia stratiotes	Water duckweed	Araceae
Planococcus citri	Citrus mealybug	Homoptera: Pseudococcidae
Plantago spp.	Plantains	Plantaginaceae
Plasmodiophora brassicae	Clubroot, brassicas	Plasmodiophorales
Plasmodiophorales See *Plasmodiophoromycetes*		

Latin	English	Family and/or Order
Plasmodiophoromycetes	Parasitic members of the Myxomycota – a group with affinities with both primitive fungi and primitive animals.	
Plasmopara spp.	Downy mildews, various hosts	Peronosporales
Plasmopara tabacina See *Peronospora tabacina*		
Plasmopara viticola	Downy mildew, grapevine	Peronosporales
Plusia spp.	e.g. silvery moth	Lepidoptera: Noctuidae
Plutella xylostella	Diamond-back moth	Lepidoptera: Yponomeutidae
Poa annua	Meadow grass, annual	Gramineae
Poa spp.	Meadow grasses	Gramineae
Poa trivialis	Meadow-grass, rough	Gramineae
Podosphaera leucotricha	Powdery mildew, apple	Erysiphales
Podosphaera spp.	Powdery mildew, various hosts	Erysiphales
Pogonomyrmex spp.	Harvester ants	Hymenoptera: Formicidae
Polychrosis botrana See *Lobesia botrana*		
Polygonum aviculare	Knot grass	Polygonaceae
Polygonum convolvulus See *Fallopia convolvulus*		
Polygonum cuspidatum See *Reynoutria japonica*		
Polygonum lapathifolium	Pale persicaria	Polygonaceae
Polygonum persicaria	Redshank; persicaria; smartweed	Polygonaceae
Polygonum sachalinense See *Reynoutria sachalinensis*		
Polygonum spp.	Knotweeds	Polygonaceae
Polymyxa betae	Fungal vector or rhizomania virus	Plasmodiophorales
Polyphagotarsonemus latus	Broad mite	Acari: Tarsonemidae
Polyscytalum pustulans	Skin spot, potatoes	Hyphales
Polystigmatales		Ascomycetales
Polytrichum juniperinum	Moss	Musci
Portulaca oleracea	Purslane	Portulacaceae
Portulaca spp.	Purslane	Portulacaceae
Potamogeton distinctus	Pondweed, American	Potamogetonaceae
Potamogeton spp.	Pondweeds	Potamogetonaceae
Pratylenchus spp.	Root-lesion nematodes	Nematoda: Hoplolaimidae
Pseudocercosporellacapsellae	White leaf spot, oilseed rape	Hyphales

Latin	English	Family and/or Order
Pseudocercosporella herpotrichoides	Eye-spot, cereals	Hyphales
Pseudococcidae	Mealybugs	Homoptera
Pseudococcus spp.	Mealybugs	Homoptera: Pseudococcidae
Pseudomonas glumae	Bacterial grain rot, rice	Eubacteriales
Pseudomonas lachrymans	Angular leaf spot, cucurbits	Eubacteriales
Pseudomonas mors-prunorum	Bacteral canker, prunus	Eubacteriales
Pseudomonas phaseolicola	Halo blight, beans	Eubacteriales
Pseudomonas spp.	Bacterial blights and leaf spots, various hosts	Eubacteriales
Pseudomonas syringae pv. *lachrymans* See *P. lachrymans*		
Pseudomonas syringae pv. *mors prunorum* See *P. mors-prunorum*		
Pseudoperonospora cubensis	Downy mildew, cucurbits	Peronosporales
Pseudoperonospora humuli	Downy mildew, hops	Peronosporales
Pseudopeziza ribis	Leaf spot, currants, gooseberry	Helotiales
Psila rosae	Carrot fly	Diptera: Psilidae
Psocoptera	Booklice	Insecta
Psoregates spp.	Mange mites	Acari: Psoroptidae
Psoroptes equi	Psoroptic mange mite	Acari: Psoroptidae
Psoroptes ovis See *P. equi*		
Psoroptes spp.	Psoroptid mange mites	Acari: Psoroptidae
Psychodidae	Moth flies	Diptera
Psylla spp.	Psyllids	Homoptera: Psyllidae
Psyllidae	Psyllids	Homoptera
Pteridium aquilinum	Bracken	Filicales
Puccinia chrysanthemi	Brown rust, chrysanthemum	Uredinales
Puccinia graminis	Black stem rust, grasses	Uredinales
Puccinia hordei	Brown rust, barley	Uredinales
Puccinia recondita	Brown rust, wheat	Uredinales
Puccinia spp.	Rust, various hosts	Uredinales
Puccinia striiformis	Yellow rust, cereals	Uredinales
Pulicidae	Fleas	Siphonaptera
Pyralidae	Pyralid moths	Lepidoptera
Pyrenopeziza brassicae	Light leaf spot, brassicas	Helotiales
Pyrenophora graminea	Leaf stripe, barley	Dothidiales
Pyrenophora teres	Net blotch, barley	Dothidiales
Pyrenophora tritici-repentis	Tan spot, wheat	Dothidiales
Pyricularia oryzae	Blast, leaf blast, rice	Hyphales
Pythium spp.	Root rots, various	Peronosporales
Quelea quelea	Red-billed quelea	Passeriformes: Ploceidae

Latin	English	Family and/or Order
Radopholus similis	Burrowing nematode	Nematoda: Tylenchidae
Ramularia beticola	Leaf spot, beet crops	Hyphales
Ramularia spp.	Leaf spots, various	Hyphales
Ranunculus spp.	Buttercups	Ranunculaceae
Raphanus raphanistrum	Wild radish; runch	Cruciferae
Rattus argentiventer	Field rat	Rotentia: Muridae
Rattus hawaiiensis	Pacific rat	Rotentia: Muridae
Rattus norvegicus	Brown rat	Rotentia: Muridae
Rattus rattus	Black rat	Rotentia: Muridae
Rattus rattus diardii	Field rat	Rotentia: Muridae
Rattus tiomanicus	Palm rat	Rotentia: Muridae
Reynoutria japonica (= Polygonum cuspidatum)	Japanese knotweed	Polygonaceae
Reynoutria sachalinensis (= Polygonum sachalinense)	Giant knotweed	Polygonaceae
Rhinotermitidae	Termites	Isoptera
Rhipicephalus spp.	Ticks	Acari: Ixodidae
Rhizoctonia solani (= Thanetophorus cucumeris)	Damping off; root rots, various	Stereales
Rhizoctonia spp.	Foot rot, root rot, various hosts	Stereales
Rhizoglyphus callae, R. robini	Bulb mites	Acari: Acaridae
Rhizoglyphus echinopus See R. callae, R. robini		
Rhizopertha dominica See Rhyzopertha dominica		
Rhizopus spp.	Post-harvest rots	Mucorales
Rhododendron ponticum	Rhododendron	Ericaceae
Rhynchosporium secalis	Leaf blotch, barley and rye	Hyphales
Rhynchosporium spp.	Leaf spots, grasses	Hyphales
Rhyzopertha dominica	Lesser grain borer	Coleoptora: Bostrichidae
Rubus spp.	Brambles	Rosaceae
Rumex spp.	Docks and sorrels	Polygonaceae
Sagittaria sagittifolia	Arrowhead	Alismataceae
Sahlbergella singularis	Cocoa capsid	Heteroptera: Miridae
Saissetia coffeae	Hemispherical scale, helmet scale	Homoptera: Coccidae
Saissetia oleae	Mediterranean black scale, black olive scale	Homoptera: Coccidae
Salsola kali	Russian thistle	Chenopodiaceae
Saltatoria	Crickets, grasshoppers, etc.	Insecta
Saprolegniales		Oomycetes
Sarcoptes scabiei	Itch mite	Acari: Sarcoptidae
Sarcoptes spp.	Scabies mites, etc.	Acari: Sarcoptidae
Scerophthora macrospora	Downy mildew, cereals	Peronosporales
Sciaridae	Fungus gnats, sciarid flies	Diptera
Scirpus juncoides	Japanese bulrush	Cyperaceae

REFERENCE

Latin	English	Family and/or Order
Scirpus maritimus	Sea club-rush	Cyperaceae
Scirpus mucronatus	Roughseed bulrush	Cyperaceae
Scirpus spp.	Club-rushes	Cyperaceae
Scerophthora spp.	Downy mildew, wheat	Peronosporales
Schizothyrium pomi	Fly speck disease, apple	Dothidiales
Sclerospora spp.	Downy mildews, e.g. on pearl millet	Peronosporales
Sclerotinia fructicola	Brown rot, top fruit	Helotiales
Sclerotinia fructigena	Brown rot, apple, pear, plum	Helotiales
Sclerotinia homeocarpa	Dollar spot, turf	Helotiales
Sclerotinia laxa	Blossom wilt, apple, plum	Helotiales
Sclerotinia sclerotiorum	Rots of stems, storage organs, etc., various crops	Helotiales
Sclerotinia spp.	Sclerotinia rots, various hosts	Helotiales
Sclerotium cepivorum	White rot, onion	Agonomycetales
Sclerotium rolfsii	Rots, various hosts	Agonomycetales
Sclerotium spp.	Post-harvest rots, various hosts	Agonomycetales
Septoria nodorum See Leptosphaeria nodorum		
Septoria spp.	Leaf and glume spots, various hosts	Sphaeropsidales
Septoria tritici See Mycosphaerella graminicola		
Sesbania exaltata	Hemp sesbania	Leguminosae
Setaria faberi	Foxtail, giant	Gramineae
Setaria glauca (= S. lutescens)	Foxtail, yellow	Gramineae
Setaria spp.	Foxtail grasses	Gramineae
Setaria viridis	Foxtail, green	Gramineae
Sida spinosa	Spiny sida	Malvaceae
Sigmodon hispidus	Cotton rat	Rodentia: Cricetidae
Simuliidae	Black flies	Diptera
Simulium spp.	Black flies	Diptera: Simuliidae
Sinapis alba	White mustard	Cruciferae
Sinapis arvensis	Charlock	Cruciferae
Siphonaptera	Fleas	Insecta
Siphunculata See Anoplura		
Sitona spp.	Pea and bean weevils	Coleoptera: Curculionidae
Sitophilus oryzae	Rice weevil	Coleoptera: Curculionidae
Sitophilus zeamais	Maize weevil; rice weevil	Coleoptera: Curculionidae
Sitotroga cerealella	Angoumois grain moth	Lepidoptera: Gelechiidae
Sogatella furcifera	White-backed planthopper	Homoptera: Delphacidae

Latin	English	Family and/or Order
Solanum nigrum	Black nightshade	Solanaceae
Solenopotes capillatus	Blue cattle louse	Phthiraptera: Haematopinidae
Solenopsis spp.	Fire ants	Hymenoptera: Formicidae
Sonchus oleraceus	Smooth sowthistle	Compositae
Sorghum almum	Columbus grass	Gramineae
Sorghum bicolor	Shattercane	Gramineae
Sorghum halepense	Johnson grass	Gramineae
Sorghum spp.	Sorghum grasses	Gramineae
Sparganium erectum	Branched bur-reed	Sparganiaceae
Spergula arvensis	Corn spurrey	Caryophyllaceae
Sphacelotheca reiliana	Head smut, maize	Ustilaginales
Spheariales		Ascomycotina
Sphaeropsidales		Deuteromycotina
Sphaerotheca fuliginea	Powdery mildew, cucurbits	Erysiphales
Sphaerotheca pannosa	Powdery mildew, rose	Erysiphales
Sphaerotheca spp.	Powdery mildew, various hosts	Erysiphales
Sphenoclea zeylanica	Gooseweed	Sphenocleaceae
Spodoptera exigua	Beet army worm; lesser armyworm	Lepidoptera: Noctuidae
Spodoptera littoralis	Egyptian cotton leafworm	Lepidoptera: Noctuidae
Spodoptera spp.	Army worms	Lepidoptera: Noctuidae
Stellaria media	Common chickweed	Caryophyllaceae
Steneotarsonemus laticeps	Bulb scale mite	Acari: Tarsonemidae
Stethorus punctum	Minute black ladybird	Coleoptera: Coccinellidae
Stigmina carpophila	Shothole, prunus	Hyphales
Stomoxys calcitrans	Stable fly	Diptera: Muscidae
Streptomyces scabies	Common scab, potato, beet	Actinomycetales
Supella longipalpa	Brown-banded cockroach	Dictyoptera: Blattidae
Sus scrofa	Wild pig	Artiodactyla: Suidae
Symphyla spp.	Symphilids	Myriapoda
Tabanus spp.	Horse flies	Diptera: Tabanidae
Talpa spp.	Moles	Insectivora: Talpidae
Tanymecus pallidus	Beet leaf weevil	Coleoptera: Curculionidae
Taphrina deformans	Peach leaf-curl	Taphrinales
Tarsonemus spp. (= *Phytonemus*, in part)	Tarsonemid mites	Acari: Tarsonemidae
Taxus baccata	Yew	Taxaceae
Tenebrio spp.	Mealworms	Coleoptera: Tenebrionidae
Tenebrionidae	Mealworms	Coleoptera
Tephritidae	Large fruit flies	Diptera
Tetanops myopaeformis	Sugar beet root maggot	Diptera: Otitidae
Tetranychidae	Spider mites	Acari
Tetranychus spp.	Spider mites	Acari: Tetranychidae
Tetranychus cinnabarinus	Carmine spider mite	Acari: Tetranychidae

Latin	English	Family and/or Order
Tetranychus mcdanieli	McDaniel's spider mite	Acari: Tetranychidae
Tetranychus urticae	Two-spotted spider mite	Acari: Tetranychidae
Thanetophorus cucumeris (= *Rhizoctonia solani)*	Damping-off disease	Stereales
Thaumetopoea pityocampa	Pine processionary caterpillar	Lepidoptera: Thaumetopoeidae
Thielaviopsis spp.	Black root rot, tobacco	Deuteromycotina
Thripidae	Thrips	Thysanoptera
Thrips fuscipennis	Rose thrips	Thysanoptera: Thripidae
Thrips spp.	Thrips	Thysanoptera: Thripidae
Thrips tabaci	Onion thrips	Thysanoptera: Thripidae
Thysanoptera	Thrips	Insecta
Tilletia caries	Bunt, stinking smut	Ustilaginales
Tilletia spp.	Smut, various hosts	Ustilaginales
Tinea spp.	Clothes moths	Lepidoptera: Tineidae
Tineola spp.	Clothes moths	Lepidoptera: Tineidae
Tingidae	Lace bugs	Heteroptera
Tipula spp.	Crane flies; leatherjackets	Diptera: Tipulidae
Tortricidae	Tortrix moths	Lepidoptera
Tortrix spp.	Tortrix moths	Lepidoptera: Tortricidae
Tranzschelia discolor See *Tranzchelia pruni-spinosi*		
Tranzchelia pruni-spinosi	Plum rust	Uredinales
Trialeuroides vaporariorum	Glasshouse whitefly	Homoptera
Tribolium castaneum	Rust-red flour beetle	Coleoptera: Tenebrionidae
Tribolium confusum	Confused flour beetle	Coleoptera: Tenebrionidae
Tribolium spp.	Flour beetles	Coleoptera: Tenebrionidae
Trichodorus spp.	Stubby-root nematodes	Nematoda
Trichoplusia ni	Cabbage looper	Lepidoptera: Noctuidae
Tripleurospermum *maritimum (= Matricaria* *inodora, = M. perforata)*	Mayweed, scentless	
Trogoderma granarium	Khapra beetle	Coleoptera: Dermestidae
Tulasnellales		Basidiomycotina
Tunga penetrans	Chigoe flea	Siphonaptera: Pulicidae
Typha spp.	Bullrushes	Typhaceae
Typhula incarnata	Snow rot, cereals	Aphyllophorales
Typhlodromus pyri	Fruit tree red spider mite predator	Acari: Phytoseiidae
Uncinula necator	Powdery mildew, grapevines	Erysiphales
Uredinales	Rust fungi	Basidiomycotina
Urocystis spp.	Leaf smuts, various hosts	Ustilaginales
Uromyces betae	Rust, beet crops	Uredinales
Uromyces spp.	Rusts, various crops	Uredinales
Urtica dioica	Nettle, common	Urticaceae
Urtica spp.	Nettles	Urticaceae

Latin	English	Family and/or Order
Urtica urens	Nettle, small	Urticaceae
Ustilaginales	Smut fungi	Basidiomycotina
Ustilago nuda	Loose smut, barley, wheat	Ustilaginales
Ustilago spp.	Smut diseases, various hosts	Ustilaginales
Utricularia spp.	Bladderworts	Lentibulariaceae
Valsa ceratosperma	Valsa canker of apple	Diaporthales
Varroa jacobsoni	A parasite of honey bees	Acari: Laelaptidae
Vasates spp.	Mites	Acari: Eriophyidae
Venturia inaequalis	Scab, apples	Dothidiales
Venturia pirina	Scab, pears	Dothidiales
Veronica filiformis	Speedwell, slender	Scrophulariaceae
Veronica hederifolia	Speedwell, ivy-leaved	Scrophulariaceae
Veronica persica	Speedwell, common field	Scrophulariaceae
Veronica spp.	Speedwells	Scrophulariaceae
Verticillium fungicola	Dry bubble, mushrooms	Hyphales
Verticillium spp.	Verticillium wilt, various hosts	Hyphales
Vespa spp. See Vespula spp.		
Vespula spp.	Social wasps	Hymenoptera: Vespidae
Viola arvensis	Field pansy	Violaceae
Viola spp.	Wild pansies	Violaceae
Xanthium pennsylvanicum	Cocklebur	Compositae
Xanthium strumarium	Rough Cocklebur	Compositae
Xanthomonas campestris pv. citri See X. citri		
Xanthomonas campestris pv. malvacearum See X. malvacearum		
Xanthomonas campestris pv. oryzae See X. oryzae		
Xanthomonas citri	Citrus canker	Eubacteriales
Xanthomonas malvacearum	Bacteriosis, cotton	Eubacteriales
Xanthomonas oryzae	Leaf blight, rice	Eubacteriales
Xanthomonas spp.	Bacterial leaf spots, various hosts	Eubacteriales
Xestobium rufovillosum	Death watch beetle	Coleoptera: Dermestidae
Zygomycotina	Primitive fungi ('Phycomycetes') which do not produce motile spores. Sexually produced spores are non-motile zygospores.	
Zygopiala jamaicensis	Greasy blotch, carnation	Spaeropsidales

GLOSSARY

English–Latin

Species names were supplied by manufacturers in a mixture of Latin and English and, in many cases, these have been incorporated unchanged into the Main Entries. This glossary is intended to help the reader to identify the Latin name in cases where the English name may be unfamiliar.

However, it has been produced simply by inverting the Latin–English glossary, with limited subsequent editing. The reader should recognise that English-name and Latin-name groups of species are not congruent. For example, all *Peronospora* are downy mildews, but not all downy mildews are *Peronospora*; consequently, there are entries for various downy mildews, giving different Latin names.

In searching for English names, alternative forms of the name should be considered, especially with names containing an adjectival component.

Where these two problems operate together, this glossary needs to be used with particular care.

English	Latin	Family and/or Order
Amaranths	*Amaranthus* spp.	Amaranthaceae
American bollworm	*Helicoverpa zea*	Lepidoptera: Noctuidae
American cockroach	*Periplaneta americana*	Dictyoptera: Blattidae
American serpentine leaf miner	*Liriomyza trifolii*	Diptera: Agromyzidae
Angoumois grain moth	*Sitotroga cerealella*	Lepidoptera: Gelechiidae
Angular leaf spot, cucurbits	*Pseudomonas lachrymans*	Eubacteriales
Annual ryegrass	*Lolium rigidum*	Gramineae
Anopheles mosquitos	*Anopheles* spp.	Diptera: Culicidae
Anthracnose, french beans	*Colletotrichum lindemuthianum*	Melanconiales
Anthracnose, various root rot and leaf curl diseases	*Colletotrichum* spp.	Melanconiales
Ants	*Formicidae*	Hymenoptera
Aphid parasitoid wasps	*Aphidius* spp.	Hymenoptera: Braconidae
Apple blossom weevil	*Anthonomus pomorum*	Coleoptera: Curculionidae
Apple leaf miner	*Phyllonorycter blancardella*	Lepidoptera: Gracillariidae
Apple leaf miner	*Lyonetia clerkella*	Lepidoptera: Lyonetiidae
Argentine ant	*Iridomyrmex humilis*	Hymenoptera: Formicidae
Army worms	*Spodoptera* spp.	Lepidoptera: Noctuidae
Arrowhead	*Sagittaria sagittifolia*	Alismataceae
Australian bush fly	*Musca vetustissima*	Diptera: Muscidae
Australian sheep blowfly	*Lucilia cuprina*	Diptera: Calliphoridae
Bacteral canker, prunus	*Pseudomonas mors-prunorum*	Eubacteriales
Bacterial blights and leaf spots, various hosts	*Pseudomonas* spp.	Eubacteriales

English	Latin	Family and/or Order
Bacterial grain rot, rice	*Pseudomonas glumae*	Eubacteriales
Bacterial leaf spots, various hosts	*Xanthomonas* spp.	Eubacteriales
Bacterial rot, celery	*Erwinia carotovora*	Eubacteriales
Bacteriosis, cotton	*Xanthomonas malvacearum*	Eubacteriales
Banana black heart	*Gibberella fujikuroi*	Hypocreales
Banana leaf spot, sigatoka	*Mycosphaerella musicola*	Dothidiales
Banana root borer	*Cosmopolites sordidus*	Coleoptera: Curculionidae
Banana weevil	*Cosmopolites sordidus*	Coleoptera: Curculionidae
Bandicoot rats	*Bandicota* spp.	Rotentia: Muridae
Barnyard grass	*Echinochloa crus-galli*	Gramineae
Barnyard grass, awnless	*Echinochloa colonum*	Gramineae
Barren brome	*Bromus sterilis*	Gramineae
Basal stem rot, cucurbits	*Erwinia carotovora*	Eubacteriales
Bean beetles	*Epilachna* spp.	Coleoptera: Coccinellidae
Bean weevils	*Sitona* spp.	Coleoptera: Curculionidae
Bearded oat	*Avena barbata*	Gramineae
Bed bug	*Cimex lectularius*	Heteroptera: Cimicidae
Beet army worm	*Spodoptera exigua*	Lepidoptera: Noctuidae
Beet cyst nematode	*Heterodera schachtii*	Nematoda: Heteroderidae
Beet leaf-miner	*Pegomya hyoscamni*	Diptera: Anthomyiidae
Beet leaf weevil	*Tanymecus pallidus*	Coleoptera: Curculionidae
Begonia	*Begonia elatior*	Begoniaceae
Bermuda grass	*Cynodon dactylon*	Gramineae
Bindweed, large	*Calystegia sepium* ssp. *sepium*	Convolvulaceae
Bird skin mites	*Cnemidocoptes* spp.	Acari: Sarcoptidae
Biting midges	*Ceratopogonidae*	Diptera
Black bean aphid	*Aphis fabae*	Homoptera: Aphididae
Black bent	*Agrostis gigantea*	Gramineae
Black bindweed	*Fallopia convolvulus*	Polygonaceae
Black flies	*Simulium* spp.	Diptera: Simuliidae
Black-grass	*Alopecurus myosuroides*	Gramineae
Black leaf streak, banana	*Mycosphaerella fijiensis*	Dothidiales
Black mould	*Cladosporium* spp.	Hyphales
Black nightshade	*Solanum nigrum*	Solanaceae
Black olive scale	*Saissetia oleae*	Homoptera: Coccidae
Black rat	*Rattus rattus*	Rotentia: Muridae
Black root rot, tobacco	*Thielaviopsis* spp.	Deuteromycotina
Black rot, apple	*Botryosphaeria obtusa* (= *Physalospora obtusa*)	Dothidiales
Black rot, grapevines	*Guignardia bidwellii*	Dothidiales
Black stem rust, grasses	*Puccinia graminis*	Uredinales
Blackcurrant gall-mite	*Cecidophyopsis ribis*	Acari: Eriophyidae
Blackcurrant rust	*Cronartium ribicola*	Uredinales
Blackleg, beet crops	*Aphanomyces cochlioides*	Saprolegniales

English	Latin	Family and/or Order
Blackleg, potatoes	*Erwinia carotovora*	Eubacteriales
Blackspot, roses	*Diplocarpon rosae*	Helotiales
Bladderworts	*Utricularia* spp.	Lentibulariaceae
Blast, rice	*Pyricularia oryzae*	Hyphales
Blight, capsicums	*Phytophthora capsici*	Peronosporales
Blight, potato	*Phytophthora infestans*	Peronosporales
Blight, tomato	*Phytophthora infestans*	Peronosporales
Blister blight, tea	*Exobasidium vexans*	Exobasidiales
Blossom or pollen beetles	*Meligethes* spp.	Coleoptera: Nitidulidae
Blossom wilt, apple, plum	*Sclerotinia laxa*	Helotiales
Blue cattle louse	*Solenopotes capillatus*	Phthiraptera: Haematopinidae
Blue mould, citrus	*Penicillium italicum*	Hyphales
Blue mould, tobacco	*Peronospora tabacina* (= *Plasmopara tabacina*)	Peronosporales
Boll weevil	*Anthonomus grandis*	Coleoptera: Curculionidae
Booklice	*Psocoptera*	Insecta
Bracken	*Pteridium aquilinum*	Filicales
Brambles	*Rubus* spp.	Rosaceae
Branched bur-reed	*Sparganium erectum*	Sparganiaceae
Brassica cyst nematode	*Heterodera cruciferae*	Nematoda: Heteroderidae
Brassica gall and stem weevils	*Ceutorhynchus* spp.	Coleoptera: Curculionidae
Broad mite	*Polyphagotarsonemus latus*	Acari: Tarsonemidae
Brooks spot, apple	*Mycosphaerella pomi*	Dothidiales
Brown-banded cockroach	*Supella longipalpa*	Dictyoptera: Blattidae
Brown foot rot, cereals	*Gibberella* spp. (= various *Fusarium* spp.)	Hypocreales
Brown rat	*Rattus norvegicus*	Rotentia: Muridae
Brown rot, apple, pear, plum	*Sclerotinia fructigena, Sclerotinia fructicola*	Helotiales
Brown rust, barley	*Puccinia hordei*	Uredinales
Brown rust, chrysanthemum	*Puccinia chrysanthemi*	Uredinales
Brown rust, wheat	*Puccinia recondita*	Uredinales
Brown soft scale	*Coccus hesperidum*	Homoptera: Coccidae
Brown spot, peanut	*Mycosphaerella arachidis*	Dothidiales
Brown spot, rice	*Cochliobolus miyabeanus*	Dothidiales
Brown stripe, sugar cane	*Bipolaris stenospila*	Hyphales
Buffalo fly	*Haematobia irritans exigua*	Diptera: Muscidae
Buffalograss	*Brachiaria mutica* (= *Panicum purpurascens*)	Gramineae
Bugs	*Heteroptera*	Hemiptera
Bulb mites	*Rhizoglyphus callae, R. robini*	Acari: Acaridae
Bulb scale mite	*Steneotarsonemus laticeps*	Acari: Tarsonemidae
Bullrushes	*Typha* spp.	Typhaceae
Bunt, stinking smut	*Tilletia caries*	Ustilaginales

1390

English	Latin	Family and/or Order
Burrowing nematode	*Radopholus similis*	Nematoda: Tylenchidae
Butt rot, conifers	*Heterobasidion annosum*	Aphyllophorales
Buttercups	*Ranunculus* spp.	Ranunculaceae
Cabbage looper	*Trichoplusia ni*	Lepidoptera: Noctuidae
Cabbage root fly	*Delia radicum*	Diptera: Anthomyiidae
Cabbage seed weevil	*Ceutorhynchus assimilis*	Coleoptera: Curculionidae
Cabbage stem weevil	*Ceutorhynchus quadridens*	Coleoptera: Curculionidae
Cabbage white butterflies	*Pieris* spp.	Lepidoptera: Pieridae
Californian red scale	*Aonidiella aurantii*	Homoptera: Diaspididae
Canadian pondweed	*Elodea canadensis*	Hydrocharitaceae
Canary grass, awned	*Phalaris paradoxa*	Gramineae
Canary grasses	*Phalaris* spp.	Gramineae
Canker, apple, pear	*Nectria galligena*	Nectriaceae
Capsid bugs	*Miridae*	Heteroptera
Carmine spider mite	*Tetranychus cinnabarinus*	Acari: Tetranychidae
Carpenter ants	*Camponotus* spp.	Hymenoptera: Formicidae
Carpet beetles	*Anthrenus* spp.	Coleoptera: Dermestidae
Carrot fly	*Psila rosae*	Diptera: Psilidae
Carrot leaf blight	*Alternaria dauci*	Hyphales
Cat flea	*Ctenocephalides felis*	Siphonaptera: Pulicidae
Cattle biting louse	*Bovicola bovis*	Phthiraptera: Trichodectidae
Cattle tail louse	*Haematopinus quadripertusus*	Phthiraptera: Haematopinidae
Cereal leaf beetle	*Oulema melanopus*	Coleoptera: Chrysomelidae
Chamomiles	*Anthemis* spp.	Compositae
Charlock	*Sinapis arvensis*	Cruciferae
Cherry leaf spot	*Blumeriella jaapii*	Helotiales
Chicken mite	*Dermanyssus gallinae*	Acari: Dermanyssidae
Chigoe flea	*Tunga penetrans*	Siphonaptera: Pulicidae
Chinch bug	*Blissus leucopterus*	Lygaeidae
Chrysanthemum leaf miner	*Phytomyza syngenesiae*	Diptera: Agromyzidae
Chrysanthemum leaf miner parasitoid	*Dacnusa sibirica*	Hymenoptera: Braconidae
Chrysomelid beetles	*Chrysomelidae*	Coleoptera
Cigarette beetle	*Lasioderma serricorne*	Coleoptera: Anobiidae
Citrus aphid	*Aphis citricola*	Homoptera: Aphididae
Citrus canker	*Xanthomonas citri*	Eubacteriales
Citrus mealybug	*Planococcus citri*	Homoptera: Pseudococcidae
Citrus red mite	*Panonychus citri*	Acari: Tetranychidae
Citrus rust mite	*Phyllocoptruta oleivora*	Acari: Eriophyidae
Cleavers	*Galium aparine*	Rubiaceae
Click beetles	*Elateridae*	Coleoptera
Clothes moths	*Tinea* spp.	Lepidoptera: Tineidae
Clothes moths	*Tineola* spp.	Lepidoptera: Tineidae
Clover bryobia mite	*Bryobia praetiosa*	Acari: Tetranychidae

English	Latin	Family and/or Order
Club-rushes	*Scirpus* spp.	Cyperaceae
Clubroot, brassicas	*Plasmodiophora brassicae*	Plasmodiophorales
Coccomycosis	*Blumeriella jaapii*	Helotiales
Cockchafer	*Melolontha melolontha*	Coleoptera: Scarabaeidae
Cocklebur	*Xanthium pennsylvanicum*	Compositae
Cockroaches	*Blattella* spp.	Dictyoptera: Blattidae
Cockspur, rice	*Echinochloa oryzicola (= E. oryzoides)*	Gramineae
Cocoa capsid	*Sahlbergella singularis*	Heteroptera: Miridae
Cocoa capsid	*Distantiella theobroma*	Heteroptera: Miridae
Codling moth	*Cydia pomonella*	Lepidoptera: Tortricidae
Coffee rust	*Hemileia vastatrix*	Uredinales
Collar rot, apple	*Phytophthora cactorum*	Peronosporales
Colorado beetle	*Leptinotarsa decemlineata*	Coleoptera: Chrysomelidae
Columbus grass	*Sorghum almum*	Gramineae
Common amaranth	*Amaranthus retroflexus*	Amaranthaceae
Common chickweed	*Stellaria media*	Caryophyllaceae
Common cockroach	*Blatta orientalis*	Dictyoptera: Blattidae
Common couch	*Elymus repens*	Gramineae
Common orache	*Atriplex patula*	Chenopodiaceae
Common scab, potato, beet	*Streptomyces scabies*	Actinomycetales
Confused flour beetle	*Tribolium confusum*	Coleoptera: Tenebrionidae
Corn marigold	*Chrysanthemum segetum*	Compositae
Corn rootworms	*Diabrotica* spp.	Coleoptera: Chrysomelidae
Corn spurrey	*Spergula arvensis*	Caryophyllaceae
Cotton boll rot	*Gibberella fujikuroi*	Hypocreales
Cotton leaf perforator	*Bucculatrix thurberiella*	Lepidoptera: Lyonetiidae
Cotton leaf worm	*Alabama argillacea*	Lepidoptera: Noctuidae
Cotton leafhoppers	*Empoasca* spp.	Homoptera: Cicadellidae
Cotton rat	*Sigmodon hispidus*	Rodentia: Cricetidae
Crabgrass	*Digitaria sanguinalis*	Gramineae
Crabgrass, tropical	*Digitaria adscendens (= D. ciliaris)*	Gramineae
Crane flies	*Tipula* spp.	Diptera: Tipulidae
Crane's bills	*Geranium* spp.	Geraniaceae
Creeping bent	*Agrostis stolonifera*	Gramineae
Crickets	*Cricetus* spp.	Saltatoria: Gryllidae
Crown rot, apple	*Phytophthora cactorum*	Peronosporales
Cutworm	*Noctua pronuba*	Lepidoptera: Noctuidae
Cutworms	*Agrotis* spp., *Euxoa* spp., *Noctua* spp.	Lepidoptera: Noctuidae
Cyst nematodes	*Heteroderidae*	Nematoda
Damping off, various hosts	*Pellicularia* spp.	Tulasnellales
Damping off, various hosts	*Phytophthora* spp.	Peronosporales
Damson-hop aphid	*Phorodon humuli*	Homoptera: Aphididae

English	Latin	Family and/or Order
Dark leaf spot, brassicas	*Alternaria brassicae, Alternaria brassicicola*	Hyphales
Dart moths	*Euxoa* spp.	Lepidoptera: Noctuidae
Dayflower	*Commelina* spp.	Commelinaceae
Dead arm, grape vines	*Phomopsis viticola*	Sphaeropsidales
Death watch beetle	*Xestobium rufovillosum*	Coleoptera: Dermestidae
Deer flies	*Chrysops* spp.	Diptera: Tabanidae
Diamond-back moth	*Plutella xylostella*	Lepidoptera: Yponomeutidae
Docks and sorrels	*Rumex* spp.	Polygonaceae
Dog	*Canis familiaris*	Carnivora: Canidae
Dog flea	*Ctenocephalides canis*	Siphonaptera: Pulicidae
Dollar spot, turf	*Sclerotinia homeocarpa*	Helotiales
Downy mildew, brassicae	*Peronospora parasitica*	Peronosporales
Downy mildew, cereals	*Scerophthora macrospora*	Peronosporales
Downy mildew, cucurbits	*Pseudoperonospora cubensis*	Peronosporales
Downy mildew, grapevine	*Plasmopara viticola*	Peronosporales
Downy mildew, hops	*Pseudoperonospora humuli*	Peronosporales
Downy mildew, lettuce	*Bremia lactucae*	Peronosporales
Downy mildew, sorghum	*Peronosclerospora* spp.	Peronosporales
Downy mildew, wheat	*Scerophthora* spp.	Peronosporales
Dry bubble, mushrooms	*Verticillium fungicola*	Hyphales
Dry rot	*Fusarium coeruleum*	Hyphales
Dutch-elm disease	*Ceratocystis* spp.	Microasaceae
Ear blight, cereals	*Gibberella* spp. (= various *Fusarium* spp.)	Hypocreales
Ear blights, various hosts (Imperfect fungi with perfect stages in various genera)	*Fusarium* spp.	Hyphales
Ear-mange mites	*Otodectes* spp.	Acari: Psoroptidae
Earwigs	*Dermaptera*	Insecta
Egyptian cotton leafworm	*Spodoptera littoralis*	Lepidoptera: Noctuidae
Elodea, Florida	*Hydrilla verticillata*	Hydrocharitaceae
Eriophyid mites	*Eriophyidae*	Acari
European corn borer	*Ostrinia nubilalis*	Lepidoptera: Pyralidae
European pine sawfly	*Neodiprion sertifer*	Hymenoptera: Diprionidae
European vine moth	*Lobesia botrana*	Lepidoptera: Tortricidae
Eye-spot, cereals	*Pseudocercosporella herpotrichoides*	Hyphales
Fairy rings	*Marasmius oreades* and other species	Agaricaceae
Fall panicum	*Panicum dichotomiflorum*	Gramineae
False oat-grass	*Arrhenatherum elatius*	Gramineae
Fat hen	*Chenopodium album*	Chenopodiaceae
Field bindweed	*Convolvulus arvensis*	Convolvulaceae

REFERENCE

English	Latin	Family and/or Order
Field pansy	*Viola arvensis*	Violaceae
Field rat	*Arvicanthis niloticus, Rattus argentiventer, Rattus rattus diardii*	Rotentia: Muridae
Field vole	*Microtus agrestis*	Rotentia: Muridae
Filamentous bacteria	*Actinomycetales*	
Fire ants	*Solenopsis* spp.	Hymenoptera: Formicidae
Flea beetle	*Phyllotreta striolata*	Coleoptera: Chrysomelidae
Flea beetles	*Chaetocnema* spp., *Phyllotreta* spp.	Coleoptera: Chrysomelidae
Fleas	*Pulicidae*	Siphonaptera
Flies	*Diptera*	Insecta
Florida beggarweed	*Desmodium tortuosum*	Leguminosae
Flour beetles	*Cucujidae*	Coleoptera
Flour beetles	*Tribolium* spp.	Coleoptera: Tenebrionidae
Flour mites	*Acarus* spp.	Acari: Acaridae
Fly speck disease, apple	*Schizothyrium pomi*	Dothidiales
Follicle mites	*Demodex* spp.	Acari: Demodicidae
Foot rot, cereals, grasses	*Cochliobolus sativus*	Dothidiales
Foot rot, various hosts	*Aphanomyces* spp.	Saprolegniales
Foot rot, various hosts	*Phytophthora* spp.	Peronosporales
Foot rot, various hosts	*Rhizoctonia* spp.	Stereales
Formosan termite	*Coptotermes formosanus*	Isoptera: Rhinotermitidae
Four-leaved water clover	*Marsilea* spp.	Marsileaceae
Foxtail grasses	*Setaria* spp.	Gramineae
Foxtail, giant	*Setaria faberi*	Gramineae
Foxtail, green	*Setaria viridis*	Gramineae
Foxtail, yellow	*Setaria glauca* (= *S. lutescens*)	Gramineae
Fresh water snails	*Lymnaea* spp.	Mollusca: Gastropoda
Fringe rushes	*Fimbristylis* spp.	Cyperaceae
Frit fly	*Oscinella frit*	Diptera: Chloropidae
Frog eye, soya	*Cercosporidium* spp. (includes *C. sojinum* = *Cercospora sojina*)	Hyphales
Fruit flies	*Dacus* spp.	Diptera: Tephritidae
Fruit flies	*Drosophila* spp.	Diptera: Drosophilidae
Fruit rot, strawberries	*Mucor* spp.	Mucorales
Fruit rot, various hosts	*Botrytis cinerea*	Hyphales
Fruit tree red spider mite	*Panonychus ulmi*	Acari: Tetranychidae
Fruit tree red spider mite predator	*Amblyseius finlandicus*	Acari: Phytoseiidae
Fruit tree red spider mite predator	*Typhlodromus pyri*	Acari: Phytoseiidae
Fuchsia	*Fuchsia hybrida*	Onagraceae
Fungal virus vector	*Polymyxa betae*	Plasmodiophorales

English	Latin	Family and/or Order
Fungi which produce no spores	*Agonomycetales*	Deuteromycotina
Fungi with no known sexual stage, or asexual stages of other fungi	*Deuteromycetes*	Deuteromycotina (= Fungi Imperfecti)
Fungi, sexually produced spores in sacs	*Ascomycotina*	
Fungi, spores produced exogenously in basidia	*Basidiomycotina*	
Fungus gnats, sciarid flies	*Sciaridae*	Diptera
Furniture beetle	*Anobium punctatum*	Coleoptera: Anobiidae
Fusarium foot and root rots, various hosts	*Fusarium culmorum*	Hyphales
Fusarium wilt, various hosts	*Fusarium oxysporum*	Hyphales
Gall midges	*Cecidomyiidae*	Diptera
Gangrene, potatoes	*Phoma exigua* var. *foveata*	Deuteromycotina
Garden wireworms	*Athous* spp.	Coleoptera: Elateridae
German cockroach	*Blattella germanica*	Dictyoptera: Blattidae
Giant knotweed	*Reynoutria sachalinensis* (= *Polygonum sachalinense*)	Polygonaceae
Glasshouse potato aphid	*Aulacorthum solani*	Homoptera: Aphididae
Glasshouse whitefly	*Trialeuroides vaporariorum*	Homoptera
Glasshouse whitefly parasitoid	*Encarsia formosa*	Hymenoptera: Aphelinidae
Gloeosporium rot, apples	*Glomerella cingulata*	Polystigmatales
Gloeosporium rot, apples	*Gloeosporium* spp.	Deuteromycotina
Glume blotch, wheat	*Leptosphaeria nodorum* (= *Septoria nodorum*)	Dothidiales
Glume spots, various hosts	*Septoria* spp.	Sphaeropsidales
Golden hamster	*Mesocricetus auratus*	Rodentia: Cricetidae
Gooseberry bryobia mite	*Bryobia ribis*	Acari: Tetranychidae
Goosegrass	*Eleusine indica*	Gramineae
Gooseweed	*Sphenoclea zeylanica*	Sphenocleaceae
Grain beetles	*Cryptolestes* spp.	Coleoptera: Cucujidae
Grain mites	*Acarus* spp.	Acari: Acaridae
Grass and cereal flies	*Opomyza* spp.	Diptera: Opomyzidae
Grass moth	*Chrysoteuchia caliginosellus* (= *Crambus caliginosellus*)	Lepidoptera: Pyralidae
Grasshoppers	*Acrididae*	Saltatoria
Greasy blotch, carnation	*Zygopiala jamaicensis*	Spaeropsidales
Green leafhopper	*Empoasca fabae*	Homoptera: Cicadellidae
Green leafhoppers	*Nephotettix* spp.	Homoptera: Cicadellidae
Green mould, citrus	*Penicillium digitatum*	Hyphales
Green rice leafhopper	*Nephotettix impicticepts*	Homoptera: Cicadellidae
Green rice leafhopper	*Nephotettix cincticeps*	Homoptera: Cicadellidae
Gypsy moth	*Lymantria dispar*	Lepidoptera: Lymantriidae

REFERENCE

English	Latin	Family and/or Order
Halo blight, beans	*Pseudomonas phaseolicola*	Eubacteriales
Harvester ants	*Pogonomyrmex* spp.	Hymenoptera: Formicidae
Head louse	*Pediculus capitis*	Phthiraptera: Pediculidae
Head smut, maize	*Sphacelotheca reiliana*	Ustilaginales
Helmet scale	*Saissetia coffeae*	Homoptera: Coccidae
Helminthosporium blight, rice	*Helminthosporium oryzae*	Hyphales
Hemispherical scale	*Saissetia coffeae*	Homoptera: Coccidae
Hemp sesbania	*Sesbania exaltata*	Leguminosae
Horn fly	*Haematobia irritans*	Diptera: Muscidae
Hornweed, common	*Ceratophyllum demersum*	Ceratophyllaceae
Horse flies	*Tabanus* spp.	Diptera: Tabanidae
House fly	*Musca domestica*	Diptera: Muscidae
House longhorn beetle	*Hylotrupes bajulus*	Coleoptera: Cerambycidae
House mosquito	*Culex fatigans* (= *C. quinquefasciatus*)	Diptera
House mouse	*Mus domesticus, Mus musculus*	Rotentia: Muridae
Human body louse	*Pediculus humanus*	Phthiraptera: Pediculidae
Itch mite	*Sarcoptes scabiei*	Acari: Sarcoptidae
Ivy	*Hedera helix*	Araliaceae
Ixodid ticks	*Ixodidae*	Acari
Japanese bulrush	*Scirpus juncoides*	Cyperaceae
Japanese field vole	*Microtus montebelli*	Rotentia: Muridae
Japanese knotweed	*Reynoutria japonica* (= *Polygonum cuspidatum*)	Polygonaceae
Jimson weed	*Datura stramonium*	Solanaceae
Johnson grass	*Sorghum halepense*	Gramineae
Joint vetches	*Aeschynomene* spp.	Leguminosae
Khapra beetle	*Trogoderma granarium*	Coleoptera: Dermestidae
Knapweeds	*Centaurea* spp.	Compositae
Knot grass	*Polygonum aviculare*	Polygonaceae
Knotweeds	*Polygonum* spp.	Polygonaceae
Kyllinga, green	*Cyperus brevifolius*	Cyperaceae
Lace bugs	*Tingidae*	Heteroptera
Large fruit flies	*Tephritidae*	Diptera
Large white butterfly	*Pieris brassicae*	Lepidoptera: Pieridae
Late flowering cyperus	*Cyperus serotinus*	Cyperaceae
Leaf and pod spot, peas	*Ascochyta pinodes, Ascochyta pisi*	Sphaeropsidales
Leaf spots, various hosts	*Septoria* spp.	Sphaeropsidales
Leaf blast, rice	*Pyricularia oryzae*	Hyphales
Leaf blight, rice	*Xanthomonas oryzae*	Eubacteriales
Leaf blotch, barley and rye	*Rhynchosporium secalis*	Hyphales
Leaf blotches, etc., various hosts	*Marssonina* spp.	Melanconiales

English	Latin	Family and/or Order
Leaf miners	*Agromyza* spp., *Liriomyza* spp., *Phytomyza* spp.	Diptera: Agromyzidae
Leaf-mining moths	*Leucoptera* spp.	Lepidoptera: Lyonetiidae
Leaf-mining moths	*Phyllonorycter* spp.	Lepidoptera: Gracillariidae
Leaf mould, tomato	*Fulvia fulva*	Hyphales
Leaf scorch, apples	*Gymnosporangium* spp.	Uredinales
Leaf scorch, strawberry	*Diplocarpon earliana*	Helotiales
Leaf smuts, various hosts	*Urocystis* spp.	Ustilaginales
Leaf spot, apple	*Botryosphaeria obtusa* (= *Physalospora obtusa*)	Dothidiales
Leaf spots, various hosts	*Mycosphaerella* spp.	Dothidiales
Leaf spot, beans	*Ascochyta fabae*	Sphaeropsidales
Leaf spot, beet crops	*Cercospora beticola, Ramularia beticola*	Hyphales
Leaf spot, currants, gooseberry	*Pseudopeziza ribis*	Helotiales
Leaf spot, melon	*Corynespora melonis*	Hyphales
Leaf spot, soya	*Cercosporidium* spp. (includes *C. sojinum* = *Cercospora sojina*)	Hyphales
Leaf spot, sunflowers	*Diaporthe helianthi*	Diaporthales
Leaf spots, grasses	*Rhynchosporium* spp.	Hyphales
Leaf spots, various hosts	*Alternaria* spp., *Cercospora* spp., *Ramularia* spp.	Hyphales
Leaf spots, various hosts	*Ascochyta* spp., *Septoria* spp.	Sphaeropsidales
Leaf stripe, barley	*Pyrenophora graminea*	Dothidiales
Leafhoppers	*Cicadellidae*	Homoptera
Leatherjackets	*Tipula* spp.	Diptera: Tipulidae
Lemon-shaped cyst nematodes	*Heterodera* spp.	Nematoda: Heteroderidae
Lesser armyworm	*Spodoptera exigua*	Lepidoptera: Noctuidae
Lesser bandicoot mole rat	*Bandicota benghalensis*	Rotentia: Muridae
Lesser grain borer	*Rhyzopertha dominica*	Coleoptora: Bostrichidae
Lesser house fly	*Fannia canicularis*	Diptera: Muscidae
Light leaf spot, brassicas	*Pyrenopeziza brassicae*	Helotiales
Liverworts	*Bryophyta*	Bryophyta
Locusts	*Acrididae*	Saltatoria
Long-nosed cattle louse	*Linognathus vituli*	Phthiraptera: Linognathidae
Long-tailed field mouse	*Apodemus sylvaticus*	Rotentia: Muridae
Loose silky-bent	*Apera spica-venti*	Gramineae
Loose smut, barley, wheat	*Ustilago nuda*	Ustilaginales
Maize stalk borer	*Diatraea saccharalis*	Lepidoptera: Pyralidae
Maize stalk rot	*Gibberella fujikuroi*	Hypocreales
Maize weevil	*Sitophilus zeamais*	Coleoptera: Curculionidae
Mange mites	*Chorioptes* spp., *Notoedres* spp., *Psoregates* spp.	Acari: Psoroptidae
Mangold flea beetle	*Chaetocnema concinna*	Coleoptera: Chrysomelidae

REFERENCE

English	Latin	Family and/or Order
Mangold fly	*Pegomya hyoscamni*	Diptera: Anthomyiidae
Mayweed, scentless	*Tripleurospermum maritimum (= Matricaria inodora, = M. perforata)*	
Mayweeds	*Chamomilla* spp., *Matricaria* spp.	Compositae
McDaniel's spider mite	*Tetranychus mcdanieli*	Acari: Tetranychidae
Meadow grass, annual	*Poa annua*	Gramineae
Meadow-grass, rough	*Poa trivialis*	Gramineae
Mealworms	*Alphitobius* spp., *Tenebrio* spp.	Coleoptera: Tenebrionidae
Mealybugs	*Pseudococcus* spp.	Homoptera: Pseudococcidae
Mediterranean black scale	*Saissetia oleae*	Homoptera: Coccidae
Mediterranean fruit fly	*Ceratitis capitata*	Diptera
Melanosis, citrus	*Diaporthe citri*	Diaporthales
Melon and cotton aphid	*Aphis gossypii*	Homoptera: Aphididae
Mexican bean beetle	*Epilachna varivestis*	Coleoptera: Coccinellidae
Mice	*Mus* spp.	Rotentia: Muridae
Millepedes	*Diplopoda*	Myriapoda
Millet, Texas	*Panicum texanum*	Gramineae
Minute black ladybird	*Stethorus punctum*	Coleoptera: Coccinellidae
Mites	*Aculops* spp., *Calepitrimerus* spp., *Eriophyes* spp., *Phyllocoptes* spp., *Vasates* spp.	Acari: Eriophyidae
Mock cypress	*Kochia scoparia*	Chenopodiaceae
Mole crickets	*Gryllotalpa* spp.	Saltatoria
Moles	*Talpa* spp.	Insectivora: Talpidae
Monilinia leaf blight, apple	*Monilinia mali*	Helotiales
Morning glory, ivyleaf	*Ipomoea hederacea*	Convolvulaceae
Morning glory, tall	*Ipomoea purpurea*	Convolvulaceae
Mosquitoes	*Culex* spp.	Diptera
Moss	*Polytrichum juniperinum*	Musci
Moss, aquatic	*Najas guadalupensis*	Najadaceae
Mosses	*Bryophyta*	Bryophyta
Moth flies	*Psychodidae*	Diptera
Mugwort; wormwood	*Artemisia vulgaris*	Compositae
Muscid flies	*Musca* spp.	Diptera: Muscidae
Mushroom cecid	*Heteropeza pygmaea*	Diptera: Cecidomyiidae
Mushroom sciarid	*Lycoriella auripila*	Diptera: Sciaridae
Mushrooms, etc.	*Agaricales*	
N-fixing bacteria	*Nitrosomonas* spp.	Bacteria
Neck rot, onions	*Botrytis allii*	Hyphales
Needle nematodes	*Longidorus* spp.	Nematoda
Nematodes	*Nematoda*	
Net blotch, barley	*Pyrenophora teres*	Dothidiales

English	Latin	Family and/or Order
Nettle, common	*Urtica dioica*	Urticaceae
Nettle, small	*Urtica urens*	Urticaceae
Nettles	*Urtica* spp.	Urticaceae
Nipplewort	*Lapsana communis*	Compositae
Noctuid moths	*Noctuidae*	Lepidoptera
Northern leaf blight, maize	*Helminthosporium turcicum*	Hyphales
Nutgrass	*Cyperus rotundus*	Cyperaceae
Nutsedges	*Cyperus* spp.	Cyperaceae
Oat, sterile	*Avena sterilis*	Gramineae
Oats (wild and cultivated)	*Avena* spp.	Gramineae
Old World bollworm	*Helicoverpa armigera*	Lepidoptera: Noctuidae
Olive fruit fly	*Dacus oleae*	Diptera: Tephritidae
Onion couch	*Arrhenatherum elatius* var. *bulbosum*	Gramineae
Onion thrips	*Thrips tabaci*	Thysanoptera: Thripidae
Oriental tobacco budworm	*Helicoverpa assulta*	Lepidoptera: Noctuidae
Ox warble fly	*Hypoderma bovis*	Diptera: Oestridae
Pacific rat	*Rattus hawaiiensis*	Rotentia: Muridae
Pale persicaria	*Polygonum lapathifolium*	Polygonaceae
Palm rat	*Rattus tiomanicus*	Rotentia: Muridae
Panic grasses	*Panicum* spp.	Gramineae
Paralysis tick	*Ixodes holocyclus*	Acari: Ixodidae
Parasitic yeasts	*Candida* spp.	Endomycetales
Pea cyst nematode	*Heterodera goettingiana*	Nematoda: Heteroderidae
Pea weevils	*Sitona* spp.	Coleoptera: Curculionidae
Peach leaf-curl	*Taphrina deformans*	Taphrinales
Peach-potato aphid	*Myzus persicae*	Homoptera
Pear leaf blister moth	*Leucoptera malifoliella*	Lepidoptera: Lyonetiidae
Pear rust	*Gymnosporangium fuscum*	Uredinales
Pear rust mite	*Epitrimerus pyri*	Acari: Eriophyidae
Pearly green lacewing	*Chrysoperla carnea*	Neuroptera: Chrysopidae
Penicillium rots	*Penicillium* spp.	Hyphales
Persicaria	*Polygonum persicaria*	Polygonaceae
Petunia	*Petunia* spp.	Solanaceae
Pharaoh's ant	*Monomorium pharaonis*	Hymenoptera: Formicidae
Pheidole ants	*Pheidole megacephala*	Hymenoptera: Formicidae
Pickerel weed	*Monochoria vaginalis*	Pontederiaceae
Pimpernel, false	*Lindernia procumbens*	Scrophulariaceae
Pine processionary caterpillar	*Thaumetopoea pityocampa*	Lepidoptera: Thaumetopoeidae
Pink bollworm	*Pectinophora gossypiella*	Lepidoptera: Gelechiidae
Plantains	*Plantago* spp.	Plantaginaceae
Planthoppers	*Nilaparvata* spp.	Homoptera: Delphacidae
Plum rust	*Tranzchelia pruni-spinosi*	Uredinales
Pod rot, cocoa	*Monilia roreri*	Hyphales
Pollen beetle	*Meligethes aeneus*	Coleoptera: Nitidulidae
Pondweed, American	*Potamogeton distinctus*	Potamogetonaceae

REFERENCE

English	Latin	Family and/or Order
Pondweeds	*Potamogeton* spp.	Potamogetonaceae
Poppies	*Papaver* spp.	Papaveraceae
Post-harvest rot	*Fusarium coeruleum*	Hyphales
Post-harvest rots	*Rhizopus* spp.	Mucorales
Post-harvest rots	*Sclerotium* spp.	Agonomycetales
Potato aphid	*Macrosiphum euphorbiae*	Homoptera: Aphididae
Potato cyst nematodes	*Globodera* spp.	Nematoda: Heteroderidae
Potato leafhopper	*Eupterycyba jucunda*	Homoptera: Cicadellidae
Potato moth	*Phthorimaea operculella*	Lepidoptera: Gelechiidae
Powdery mildew	*Oidium hevea*	Erysiphales
Powdery mildew	*Leveillula* spp.	Erysiphales
Powdery mildew, apple	*Podosphaera leucotricha*	Erysiphales
Powdery mildew, beet crops	*Erysiphe betae*	Erysiphales
Powdery mildew, cereals, grasses	*Erysiphe graminis*	Erysiphales
Powdery mildew, cucurbits	*Erysiphe cichoracearum*	Erysiphales
Powdery mildew, cucurbits	*Sphaerotheca fuliginea*	Erysiphales
Powdery mildew, grape-vines	*Uncinula necator*	Erysiphales
Powdery mildew, rose	*Sphaerotheca pannosa*	Erysiphales
Powdery mildew, various hosts	*Phyllactinia* spp.	Erysiphales
Prairie dogs	*Cynomys* spp.	Rodentia: Sciuridae
Predacious midges	*Cecidomyiidae*	Diptera
Prickly pear cacti	*Opuntia* spp.	Cactaceae
Primitive fungi (Phycomycetes) producing motile spores	*Mastigomycotina*	
Primitive fungi (Phycomycetes) which do not produce motile spores. Sexually produced spores are non-motile zygospores.	*Zygomycotina*	
Psoroptic mange mite	*Psoroptes equi*	Acari: Psoroptidae
Psyllids	*Psylla* spp.	Homoptera: Psyllidae
Purslane	*Portulaca oleracea*	Portulacaceae
Pygmy beetle	*Atomaria linearis*	Coleoptera: Cryptophagidae
Pyralid moths	*Pyralidae*	Lepidoptera
Quackgrass	*Elymus repens*	Gramineae
Rabbit	*Oryctolagus cuniculus*	Leporidae: Lagomorpha
Rabbit fur mite	*Cheyletiella parasitivorax*	Acari: Cheyletidae
Ragweed, common	*Ambrosia artemisifolia*	Compositae
Rape	*Brassica napus*	Cruciferae
Ray blight, chrysanthemum	*Didymella ligulicola*	Dothidiales

1400

English	Latin	Family and/or Order
Red-billed quelea	*Quelea quelea*	Passeriformes: Ploceidae
Red core, strawberry	*Phytophthora fragariae*	Peronosporales
Red crevice tea mite	*Brevipalpus phoenicis*	Acari: Tenuipalpidae
Red dead-nettle	*Lamium purpureum*	Labiatae
Red spider mites	*Panonychus* spp.	Acari: Tetranychidae
Red thread, turf	*Laetisaria fuciformis*	Aphyllophorales
Redroot pigweed	*Amaranthus retroflexus*	Amaranthaceae
Redshank	*Polygonum persicaria*	Polygonaceae
Rhizomania virus	*Polymyxa betae*	Plasmodiophorales
Rhododendron	*Rhododendron ponticum*	Ericaceae
Rice brown planthopper	*Nilaparvata lugens*	Homoptera: Delphacidae
Rice leaf beetle	*Oulema oryzae*	Coleoptera: Chrysomelidae
Rice leaf roller	*Cnaphalocrocis medinalis*	Lepidoptera: Pyralidae
Rice leaf scald	*Monographella nivalis*	Sphaeriales
Rice sheath blight	*Pellicularia sasakii*	Tulasnellales
Rice stalk borer	*Chilo suppressalis*	Lepidoptera: Pyralidae
Rice stem borer	*Chilo plejadellus*	Lepidoptera: Pyralidae
Rice water weevil	*Lissorhoptrusoryzophilus*	Coleoptera: Curculionidae
Rice weevil	*Sitophilus oryzae*	Coleoptera: Curculionidae
Ring-spot, brassicas	*Mycosphaerella brassicicola*	Dothidiales
Root flies	Anthomyiidae, *Delia* spp. (= some *Hylemya* spp.) and others	Diptera: Anthomyiidae
Root-knot nematodes	*Meloidogyne* spp.	Nematoda
Root-lesion nematodes	*Pratylenchus* spp.	Nematoda: Hoplolaimidae
Root rot, brassicas	*Phytophthora megasperma*	Peronosporales
Root rot, cereals, grasses	*Cochliobolus sativus*	Dothidiales
Root rot, tomato	*Colletotrichum coccodes*	Melanconiales
Root rots, various hosts	*Aphanomyces* spp.	Saprolegniales
Root rots, various hosts	*Phoma* spp.	Deuteromycotina
Root rots, various hosts	*Pythium* spp.	Peronosporales
Root rots, various hosts	*Rhizoctonia* spp.	Stereales
Rose aphid	*Macrosiphum rosae*	Homoptera: Aphididae
Rose thrips	*Thrips fuscipennis*	Thysanoptera: Thripidae
Rot, various crops	*Phytophthora palmivora*	Peronosporales
Rots of stems, storage organs, etc., various crops	*Sclerotinia sclerotiorum*	Helotiales
Rots, various hosts (Imperfect fungi with perfect stages in various genera)	*Fusarium* spp.	Hyphales
Rots, various hosts	*Pellicularia* spp.	Tulasnellales
Rots, various hosts	*Sclerotium rolfsii*	Agonomycetales
Rough Cocklebur	*Xanthium strumarium*	Compositae
Roughseed bulrush	*Scirpus mucronatus*	Cyperaceae
Rubbery rot, potatoes	*Geotrichum candidum*	Hyphales

English	Latin	Family and/or Order
Runch	*Raphanus raphanistrum*	Cruciferae
Rush, flowering	*Butomus umbellatus*	Butomaceae
Russian thistle	*Salsola kali*	Chenopodiaceae
Rust, beet crops	*Uromyces betae*	Uredinales
Rust fungi	*Uredinales*	Basidiomycotina
Rust mite, apple	*Aculus schlechtendali*	Acari: Eriophyidae
Rust mites	*Aculus spp.*	Acari: Eriophyidae
Rust mites	*Phyllocoptruta spp.*	Acari: Eriophyidae
Rust-red flour beetle	*Tribolium castaneum*	Coleoptera: Tenebrionidae
Rust, roses	*Phragmidium mucronatum*	Uredinales
Rust, soya	*Phakopsora pachyrhizi*	Uredinales
Rust, various hosts	*Puccinia spp., Uromyces spp.*	Uredinales
Ryegrass, italian	*Lolium multiflorum*	Gramineae
Ryegrass, perennial	*Lolium perenne*	Gramineae
Ryegrasses	*Lolium spp.*	Gramineae
San José scale	*Comstockaspis perniciosus*	Homoptera: Coccidae
Saprophytic fungi	*Paecilomyces spp.*	Hyphales
Saramatta grass	*Ischaemum rugosum*	Gramineae
Saw-toothed grain beetle	*Oryzaephilus surinamensis*	Coleoptera: Cucujidae
Sawflies	*Diprion spp.*	Hymenoptera: Diprionidae
Scab, apples	*Venturia inaequalis*	Dothidiales
Scab, cereals	*Gibberella spp. (= various Fusarium spp.)*	Hypocreales
Scab, citrus	*Elsinoe fawcettii*	Dothidiales
Scab, pears	*Venturia pirina*	Dothidiales
Scabies mites, etc.	*Sarcoptes spp.*	Acari: Sarcoptidae
Scale insect	*Didesmococcus brevipes*	Homoptera: Coccidae
Scale insects	*Coccus spp.*	Homoptera: Coccidae
Scale insects	*Coccidae, Diaspidae, Margarodidae*	Homoptera
Scentless mayweed	*Matricaria perforata (= M. inodora)*	Compositae
Sclerotinia rots, various hosts	*Sclerotinia spp.*	Helotiales
Scuttle flies	*Megaselia spp.*	Diptera: Phoridae
Sea club-rush	*Scirpus maritimus*	Cyperaceae
Sea-rush	*Juncus maritimus*	Juncaceae
Sedges	*Carex spp.*	Cyperaceae
Seed-eating ants	*Monomorium spp.*	Hymenoptera: Formicidae
Septoria leaf spot, wheat	*Mycosphaerella graminicola*	Dothidiales
Sharp eyespot, cereals	*Ceratobasidium cereale*	Tulasnellales
Shattercane	*Sorghum bicolor*	Gramineae
Sheep biting louse	*Bovicola ovis*	Phthiraptera: Trichodectidae
Sheep ked	*Melophagus ovinus*	Diptera: Hippoboscidae
Sheep maggot fly	*Lucilia sericata*	Diptera: Calliphoridae
Sheep sucking louse	*Linognathus ovillus*	Phthiraptera: Linognathidae

English	Latin	Family and/or Order
Sheep tick	*Ixodes ricinus*	Acari: Ixodidae
Shepherd's purse	*Capsella bursa-pastoris*	Cruciferae
Short-nose cattle louse	*Haematopinus eurysternus*	Phthiraptera: Haematopinidae
Shothole, prunus	*Stigmina carpophila*	Hyphales
Siam weed	*Eupatorium odoratum (= Chromolaena odorata)*	Compositae
Sickle pod	*Cassia obtusifolia*	Leguminosae (Fabaceae)
Silver scurf, potatoes	*Helminthosporium solani*	Hyphales
Skin spot, potatoes	*Polyscytalum pustulans*	Hyphales
Slaters	*Isopoda*	Crustacea
Slugs	*Gastropoda*	Mollusca
Small brown planthopper	*Laodelphax striatella*	Homoptera: Delphacidae
Small white butterfly	*Pieris rapae*	Lepidoptera: Pieridae
Smartweed	*Polygonum persicaria*	Polygonaceae
Smooth sowthistle	*Sonchus oleraceus*	Compositae
Smooth witchgrass	*Panicum dichotomiflorum*	Gramineae
Smut diseases, various hosts	*Ustilago* spp.	Ustilaginales
Smut, various hosts	*Tilletia* spp.	Ustilaginales
Snails	*Gastropoda*	Mollusca
Snow mould, grasses, cereals	*Microdochium nivalis*	Hyphales
Snow rot, cereals	*Typhula incarnata*	Aphyllophorales
Social wasps	*Vespula* spp.	Hymenoptera: Vespidae
Sooty blotch, apple pear and citrus	*Gloeodes pomigena*	Sphaeropsidales
Sooty mould	*Cladosporium* spp.	Hyphales
Sorghum grasses	*Sorghum* spp.	Gramineae
South American leaf miner	*Liriomyza huidobrensis*	Diptera: Agromyzidae
Southern root-knot nematode	*Meloidogyne incognita*	Nematoda
Soya bean looper	*Anticarsia gemmatalis*	Lepidoptera: Noctuidae
Speedwell, common field	*Veronica persica*	Scrophulariaceae
Speedwell, ivy-leaved	*Veronica hederifolia*	Scrophulariaceae
Speedwell, slender	*Veronica filiformis*	Scrophulariaceae
Spider mites	*Tetranychus* spp.	Acari: Tetranychidae
Spike rush	*Eleocharis acicularis*	Cyperaceae
Spiny bollworms	*Earias* spp.	Lepidoptera: Noctuidae
Spiny sida	*Sida spinosa*	Malvaceae
Spiral nematodes	*Helicotylenchus* spp.	Nematoda: Tylenchidae
Spotted spurge	*Euphorbia maculata*	Euphorbiaceae
Sprangletop, bearded	*Leptochloa fascicularis (= Diplachne fascicularis)*	Gramineae
Sprangletop, red	*Leptochloa chinensis*	Gramineae
Spur blight, cane fruit	*Didymella applanata*	Dothidiales
Stable fly	*Stomoxys calcitrans*	Diptera: Muscidae

English	Latin	Family and/or Order
Stalk rots, various hosts	*Diplodia* spp.	Sphaeropsidales
Stem borers	*Chilo* spp.	Lepidoptera: Pyralidae
Stem canker, sunflowers	*Diaporthe helianthi*	Diaporthales
Stem nematode	*Ditylenchus dipsaci*	Nematoda: Tylenchidae
Stem rots, various hosts	*Phoma* spp.	Deuteromycotina
Stinking chamomile	*Anthemis cotula*	Compositae
Stinking mayweed	*Anthemis cotula*	Compositae
Storage fungi	*Aspergillus* spp.	Hyphales
Stubby-root nematodes	*Trichodorus* spp.	Nematoda
Sucking lice	*Linognathus* spp.	Phthiraptera: Linognathidae
Sugar beet root maggot	*Tetanops myopaeformis*	Diptera: Otitidae
Summer fruit tortrix moth	*Adoxophyes orana*	Lepidoptera: Tortricidae
Sunflower	*Helianthus annuus*	Compositae
Symphilids	*Symphyla* spp.	Myriapoda
Tan spot, wheat	*Pyrenophora tritici-repentis*	Dothidiales
Tarsonemid mites	*Tarsonemus* spp. (= *Phytonemus*, in part)	Acari: Tarsonemidae
Tea leaf roller	*Caloptilia theivora*	Lepidoptera: Gracillariidae
Termites	*Coptotermes* spp.	Isoptera: Rhinotermitidae
Tetranychid mites	*Eotetranychus* spp., *Eutetranychus* spp.	Acari: Tetranychidae
Texas citrus mite	*Eutetranychus banksi*	Acari: Tetranychidae
Thistle, creeping	*Cirsium arvense*	Compositae
Thistles	*Carduus* spp.	Compositae
Thorn apple	*Datura stramonium*	Solanaceae
Thrips	*Thrips* spp.	Thysanoptera: Thripidae
Ticks	*Amblyomma* spp.	Acari: Ixodidae
Ticks	*Boophilus microplus*	Acari: Ixodidae
Ticks	*Ixodes* spp.	Acari: Ixodidae
Ticks	*Rhipicephalus* spp.	Acari: Ixodidae
Tobacco budworm	*Heliothis virescens*	Lepidoptera: Noctuidae
Tobacco flea beetle	*Epitrix hirtipennis*	Coleoptera: Chrysomelidae
Tobacco whitefly	*Bemisia tabaci*	Homoptera: Aleyrodidae
Tomato canker	*Clavibacter michiganensis*	Eubacteriales
Tomato leaf miner	*Liriomyza bryoniae*	Diptera: Agromyzidae
Tomato pinworm	*Keiferia lycopersicella*	Lepidoptera: Gelechiidae
Tortrix moths	*Tortrix* spp.	Lepidoptera: Tortricidae
Tropical green rice leafhopper	*Nephotettix nigropictus*	Homoptera: Cicadellidae
True crickets	*Gryllidae*	Saltatoria
Tsetse flies	*Glossina* spp.	Diptera: Glossinidae
Turnip gall weevil	*Ceutorhynchus pleurostigmata*	Coleoptera: Curculionidae
Turnip moth	*Agrotis segetum*	Lepidoptera: Noctuidae
Two-spotted spider mite	*Tetranychus urticae*	Acari: Tetranychidae
Two-spotted spider mite predator	*Phytoseiulus persimilis*	Acari: Phytoseiidae

English	Latin	Family and/or Order
Umbrella plant	*Cyperus difformis*	Cyperaceae
Valsa canker of apple	*Valsa ceratosperma*	Diaporthales
Velvetleaf	*Abutilon theophrasti*	Malvaceae
Verticillium wilt, various hosts	*Verticillium* spp.	Hyphales
Vine weevil	*Otiorhynchus sulcatus*	Coleoptera: Curculionidae
Wandering Jew	*Commelina* spp.	Commelinaceae
Warble flies	*Hypoderma* spp.	Diptera: Oestridae
Warehouse moth	*Ephestia elutella*	Lepidoptera: Pyralidae
Water duckweed	*Pistia stratiotes*	Araceae
Water hyacinth	*Eichhornia crassipes*	Pontederiaceae
Water plantain	*Alisma plantago-aquatica*	Alismataceae
Water plantain, narrow leaved	*Alisma lanceolatum*	Alismataceae
Water primroses	*Jussiaea* spp.	Onagraceae
Water purslane	*Ludwigia peploides*	Onagraceae
Water weed	*Elodea canadensis*	Hydrocharitaceae
Weevils	*Curculionidae*	Coleoptera
Western flower thrips	*Frankliniella occidentalis*	Thysanoptera: Thripidae
Wheat bulb fly	*Delia coarctata*	Diptera: Anthomyiidae
White-backed planthopper	*Sogatella furcifera*	Homoptera: Delphacidae
White blister	*Albugo candida*	Peronosporales
White leaf spot, oilseed rape	*Pseudocercosporellacapsellae*	Hyphales
White leaf spot, strawberry	*Mycosphaerella fragariae*	Dothidiales
White mould, mushrooms	*Mycogone perniciosa*	Hyphales
White mustard	*Sinapis alba*	Cruciferae
White rot, onion	*Sclerotium cepivorum*	Agonomycetales
White rot, timber	*Ganoderma* spp.	Ganodermataceae
Whiteflies	*Bemisia* spp.	Homoptera: Aleyrodidae
Wild oat	*Avena fatua*	Gramineae
Wild oat, winter	*Avena sterilis* ssp. *ludoviciana* (= *A. ludoviciana*)	Gramineae
Wild pansies	*Viola* spp.	Violaceae
Wild pig	*Sus scrofa*	Artiodactyla: Suidae
Wild radish	*Raphanus raphanistrum*	Cruciferae
Wilts, various hosts (Imperfect fungi with perfect stages in various genera)	*Fusarium* spp.	Hyphales
Wimmera ryegrass	*Lolium rigidum*	Gramineae
Wireworms	*Agriotes* spp.	Coleoptera: Elateridae
Woodlice	*Isopoda*	Crustacea
Woolly aphid	*Eriosoma lanigerum*	Homoptera: Pemphigidae
Yeasts	*Endomycetales*	Ascomycotina
Yellow cereal fly	*Opomyza florum*	Diptera: Opomyzidae
Yellow fever mosquito	*Aedes aegypti*	Diptera: Culicidae

English	Latin	Family and/or Order
Yellow nutsedge	*Cyperus esculentus*	Cyperaceae
Yellow rust, cereals	*Puccinia striiformis*	Uredinales
Yellow underwing moth	*Noctua pronuba*	Lepidoptera: Noctuidae
Yew	*Taxus baccata*	Taxaceae

DIRECTORY OF COMPANIES

Parts of names given in bold represent the short form of the company name which is used in the text of Main Entries.

A H **Marks** & Co., Ltd
Wyke Lane, Wyke, Bradford, W. Yorks
BD12 9EJ, UK.
Tel: 44 1274 691234
Fax: 44 1274 691176

Abbott Laboratories
Chemical & Agricultural Products
Division, 1401 Sheridan Rd.,
North Chicago IL 60064, USA.
Tel: 1 800 323 9597
Fax: 1 847 937 3679

ACF Chemiefarma N.V.
Straatweg 2, Postbus 5,
3600 AA Maarssen, Netherlands.

Ag Pesticides (Pvt.) Ltd
18-20 Naval Fleet Club, Inverarity Rd.,
Saddar Karachi 75530, Pakistan.
Tel: 92 21 7781626/778
Fax: 92 21 7781635

Agan Chemical Manufacturers Ltd
Northern Industrial Zone, P.O. Box 262,
Ashdod 77102, Israel.
Tel: 972 8 8515211
Fax: 972 8 8522806

AGC MicroBio Ltd
See MicroBio

AgrEvo
See Hoechst Schering AgrEvo

AgrEvo (France)
Usine de Saint Marcel, 13011 Marseille,
France.
Tel: 33 66 59 2041
Fax: 33 66 58 5003

AgrEvo UK Ltd Environmental Health
Hauxton, Cambridge CB2 5HU
UK.
Tel: 44 1223 252638
Fax: 44 1223 252639

AgrEvo Environmental Health Inc.
95 Chestnut Ridge Rd.,
Montvale NJ 07645, USA.
Tel: 1 201 307 3281
Fax: 1 201 307 3281

AgrEvo USA Co.
Little Falls Centre One,
2711 Centerville Rd.,
Wilmington DE 19808, USA.
Tel: 1 302 892 3000
Fax: 1 302 892 3013

Agri-Pharm International Inc.
See Cyclo

Agrichem BV
4900 AG Oosterhout, Netherlands.

Agridyne Technologies Inc.
See Thermo Trilogy

Agriphar
See Chimac-Agriphar

Agro-Kanesho Co., Ltd
Akasaka Shasta-East 7th Fl.,
4-2-19 Akasaka, Minato-Ku, Tokyo, Japan.
Tel: 81 3 5570 4711
Fax: 81 3 5570 4708

Agrokémia Sellye Rt.
7960 Sellye, Sósvertikei u., Hungary.
Tel: 36 1 342 5392
Fax: 36 1 342 1524

Agrolinz Melamin
See Nufarm GmbH

Agropharm Ltd
Buckingham House, Church Road, Penn,
High Wycombe, Bucks HP10 8LN, UK.
Tel: 44 1494 816575
Fax: 44 1494 816578

Aimco Pesticides Ltd
Akhand Jyoti, Block 1 & 3, 8th Road,
Santacruz (East), P.O. Box 6822,
Mumbai 400 055, India.
Tel: 91 22 618 3042
Fax: 91 22 611 6736

Akzo Nobel B.V.
See Nufarm B.V.

Alkaloida Chemical Co., Ltd
Kabay Janos u 29, H-4440 Tiszavasvari,
Hungary.
Tel: 36 42 372511
Fax: 36 42 3725328

All India Medical Corp.
See Aimco

Allied Colloids
Agriculture Division, P O Box 38,
Low Moor, Bradford, BD12 0JZ, UK.
Tel: 44 1274 417000
Fax: 44 1274 606499

American **Cyanamid** Co.
Agricultural Research Division,
P.O. Box 400, Princeton, NJ 08543 0400,
USA.
Tel: 1 609 716 2000
Fax: 1 609 275 3523

Amvac Chemical Corp.
4695 MacArthur Court, Site 1250,
Newport Beach, CA 92660, USA.
Tel: 1 714 260 1212
Fax: 1 714 260 1213

Ancom Berhad
Persiaran Selangor, 40000 Shah Alam,
Selangor Darul, Ehsan, Malaysia.
Fax: 60 3 7369818

Andermatt Biocontrol AG
Unterdorf, CH-6146 Grossdietwil,
Switzerland.
Tel: 41 62 927 2840
Fax: 41 62 927 2123

Ankur Agro-Chem Ltd
512/513 'MIDAS', Sahar Plaza Complex,
J. B. Nagar, M. V. Road, Andheri (East),
Mumbai 400 059, India.
Tel: 91 22 8213508
Fax: 91 22 8213509

Applied Bio-nomics Ltd
11074 W Saanich Rd, Sidney, BC,
V8L 5P5, Canada.
Tel: 1 604 656 2123
Fax: 1 604 656 3844

AQ Group
See Westrade Guatemala

Aragonesas Agro, S.A.
Po Recoletos 27, 28004 Madrid, Spain.
Tel: 34 5853800
Fax: 34 5852310

Arbico Inc.
P.O. Box 4247, Tucson, AZ 85738-1247,
USA.
Tel: 1 520 825 9785
Fax: 1 520 825 2038

Asahi Chemical Mnfg. Co., Ltd
500 Takayasu, Ikaruga-cho, Ikoma-gun,
Nara Pref., Japan.
Tel: 81 7457 4 1131
Fax: 81 7457 4 1961

Atabay Agrochemicals & Veterinary
Products
Acibadem Koftuncu Sok No. 1, Kadikoy,
81010 Istanbul, Turkey.
Tel: 90 216 326 6965
Fax: 90 216 340 1377

Atlas Crop Protection Ltd
Moorend House, Moorend Lane,
Dewsbury, W. Yorks WF13 4QQ, UK.
Tel: 44 1924 411138
Fax: 44 1924 411139

Atomergic Chemetals Corp.
222 Sherwood Avenue, Farmingdale,
NY 11735-1718, USA.
Tel: 1 516 694 9000
Fax: 1 516 694 9177

Atul Ltd
Agro & Pharma Division, Atul 396 020,
Gujarat, India.
Tel: 91 2632 33 261 5
Fax: 91 2632 33024, 33619

Baker Performance Chemicals Inc.
3920 Essex Lane, P.O. Box 27714,
Houston, TX 77227-7714, USA.
Tel: 1 713 599 7400
Fax: 1 713 599 7585

Barclay Chemicals Mfg. Ltd
Lilmar Industrial Estate, Santry, Dublin 9,
Ireland.
Tel: 3531 842 5755
Fax: 3531 842 5413

BASF AG
Postfach 120, D-67114 Limburgerhof,
Germany.
Tel: 49 621 600
Fax: 49 621 60 425 25

Bayer AG
D-51368 Leverkusen, Germany.
Tel: 49 2173 38 0
Fax: 49 2173 38 3564

Bayer Corp.
Agriculture Div., 8400 Hawthorn Rd.,
P.O. Box 4913, Kansas City,
MO 64120 0013, USA.
Tel: 1 816 242 2000
Fax: 1 816 242 2738

BCP
See Biological Crop Protection

Becker Microbial Products Inc.
9464 NW 11th St., Plantation, FL 33322,
USA.
Tel: 1 954 474 7590
Fax: 1 954 474 2463

Beneficial Insectary
14751 Oak Run Rd, Oak Run, CA 96069,
USA.
Tel: 1 916 472 3715
Fax: 1 916 472 3523

Bharat Pulverising Mills Ltd
19 Shriniketan, P.O. Box 11481,
14 M. Karve Road, 2nd Marine Cross,
Churchgate, Bombay 400 020, India.
Tel: 91 22 2032877/203
Fax: 91 22 2062751

Bio Collect
5841 Crittenden St, Oakland, CA 94601,
USA.
Tel: 1 510 436 8052
Fax: 1 510 532 0288

Bio Protection Pty. Ltd
P.O. Box 35, Warwick, Queensland 4370,
Australia.
Tel: 61 76 661590
Fax: 61 76 661639

Biobest
Ilse Velden 18, B-2260 Westerlo, Belgium.
Tel: 32 14 231701
Fax: 32 14 231831

Biocontrol Ltd
PO Box 515, Warwick, Queensland 4370,
Australia.
Tel: 61 76 61 4488
Fax: 61 76 61 7211

Biofac Inc.
PO Box 87, Mathis, TX 78368, USA.
Tel: 1 512 547 3259
Fax: 1 512 517 9660

BioLogic
Springtown Rd, PO Box 177, Willow Hill,
PA 17271, USA.
Tel: 1 717 349 2789
Fax: 1 717 349 2922

Biological Crop Protection
Occupation Rd, Wye, Ashford, Kent
TN25 5AH, UK.
Tel: 44 1233 813240
Fax: 44 1233 813383

BioSafer
99/220 Tessabansongkraoh Rd, Ladyao,
Jatujak, Bangkok 10900, Thailand.
Tel: 662 9543120 6
Fax: 662 9543128, 5802178

Biosys
See Thermo Trilogy

Borax Français
B.P. 8, 78102 St Germain en Laye Cedex,
France.

BPCI
See Baker Performance Chemicals Inc.

Buckman Laboratories Inc.
1256 N McLean Blvd., Memphis,
TN 38108, USA.
Tel: 1 901 278 0330
Fax: 1 901 276 5970

Budapest Chemical Works
See Budapesti Vegyimüvek

Budapesti Vegyimüvek Rt.
1476 Budapest, Pf.236, Hungary.

Bugs for Bugs
28 Orton St, Mundubbera 4626, Australia.
Tel: 61 71 654576
Fax: 61 71 654626

Burlington Bio-Medical & Scientific Corp.
222 Sherwood Avenue, Farmingdale
NY 11735, USA.
Tel: 1 516 694 9000
Fax: 1 516 694 9177

Caffaro S.p.A.
Via Friuli 55, 20031 Cesano Maderno,
Milano, Italy.
Tel: 39 362 51 4266
Fax: 39 362 51 4454

Cedar Chemical Corp.
5100 Poplar Ave., Suite 2414, Memphis,
TN 38137, USA.
Tel: 1 901 685 5348
Fax: 1 901 684 5398

Cequisa
Muntaner 322 1°, 08021 Barcelona, Spain.
Tel: 34 3 200 0322
Fax: 34 3 200 5648

CFPI
28 Boulevard Camelinat, BP 75,
92233 Gennevilliers, France.
Tel: 33 1 40 85 5050
Fax: 33 1 47 92 2545

Chemia S.p.A.
Via Statale 327, 44040 Dosso (Ferrara),
Italy.
Tel: 39 532 848477
Fax: 39 532 848383

Cheminova Agro A/S
PO Box 9, 7620 Lemvig, Denmark.
Tel: 45 97 83 4100
Fax: 45 97 83 455

Chemol Trading Ltd Co.
H-1439, Budapest 70, P.O.B. 696,
Hungary.
Tel: 36 1 269 8521
Fax: 36 1 269 8590

Chiltern Farm Chemicals Ltd
11 High St, Thornborough, Buckingham,
MK18 2DF, UK.
Tel: 44 1280 822400
Fax: 44 1223 835211

Chimac-Agriphar S.A.
Rue De Renory 26, 4102 Ougrée
(Seraing), Belgium.
Tel: 32 43 30 1711
Fax: 32 43 30 1749

Chinoin
AgChem Business Unit, H-1780,
P.O. Box 49, Budapest, Hungary.
Tel: 36 1 226 1637
Fax: 36 1 226 1654

Chugai Pharmaceutical Co., Ltd
See Eikou Kasei

Chunhu
Xingang, Lucheng Town, North District,
Yueyang City, Hunan, 414013, China.
Tel: 86 730 8461060
Fax: 86 730 8461061

Ciba-Geigy
See Novartis

Ciba Bunting
See Novartis BCM

Crop Genetics International
See Thermo Trilogy

Crystal Chemical Inter-America
10303 NW Freeway, Suite 512, Houston,
TX 77092, USA.
Tel: 1 713 956 6196
Fax: 1 713 956 6835

Cyclo International S. de R. L. de C. V.
Calle Laurel No. 10, Col. Basso, Rosarito,
BC 22710, Mexico.
Tel: 52 661 20436
Fax: 52 661 21976

Daikin Industries, Ltd
Planning Dept, 4-12 Nakazaki-Nishi,
2-chome, Kita-ku, Osaka 530, Japan.

Dainippon Ink & Chemicals Inc.
7-20 Nihonbashi 3-chome, Chuo-ku,
Tokyo 103, Japan.
Tel: 81 3 3272 4511
Fax: 81 3 3281 8589

Defensa S.A.
Industria de Defensivos Agricolas S,
Rua Padra Chagas, 79 - 7th Fl.,
90570-080 Porto Alegre RS, Brazil.
Tel: 55 51 346 2121
Fax: 55 51 346 1844

Degesch GmbH
See Detia Degesch

DELICIA GmbH Delitzsch
Duebener Str. 137, 04509 Delitzsch,
Germany.
Tel: 49 34202 65 300
Fax: 49 34202 65 309

Denka International B.V.
Hanzeweg 1, 3771 NG Barneveld,
Netherlands.
Tel: 31 342 412174
Fax: 31 342 490587

Detia Degesch GmbH
Dr. Werner Freyberg Str.,
69510 Laudenbach, Germany.
Tel: 49 6201 708 0
Fax: 49 6201 708 402

Detia Freyberg GmbH
See Detia Degesch

Dhanuka Pesticides Ltd
Rajendra Mansion, 19-A, Ansari Rd,
Darya Gani, New Delhi 110 002, India.
Tel: 91 11 3261771
Fax: 91 11 3265805

Diachem S.p.A.
via Tonale 15, 24061 Albano S.
Alessandro, Bergamo 24061, Italy.
Tel: 39 35 581120
Fax: 39 35 581357

DowElanco
9330 Zionsville Rd., Indianapolis,
IN 46268-1054, USA.
Tel: 1 317 337 4974
Fax: 1 317 337 7344

Drexel Chemical Co.
1700 Channel Ave., Memphis, TN 38113,
USA.
Tel: 1 901 774 4370
Fax: 1 901 774 4666

Du Pont
See E. I. du Pont de Nemours

E. I. **du Pont** de Nemours
Du Pont Agricultural Products, Walker's
Mill, Barley Mill Plaza, Wilmington,
DE 19880, USA.
Tel: 1 800 441 7515
Fax: 1 302 992 6470

Eastman Chemical Co.
P.O. Box 431, Kingsport, TN 37662, USA.
Fax: 1 423 224 0648

Ecogen Inc.
2005 Cabot Blvd West, P.O. Box 3023,
Langhorne, PA 19047-1810, USA.
Tel: 1 215 757 1590
Fax: 1 215 757 2956

Ecoscience Corp.
10 Alvin Court, East Brunswick,
NJ 08816, USA.

Efthymiadis
See K. & N. Efthymiadis

Eikou Kasei Co., Ltd
Agrochemicals Division, Violet Akihabara
Bldg, 18-1 Kanda Matsunaga-cho,
Chiyoda-ku, Tokyo 101, Japan.
Tel: 81 3 5256 3861/2
Fax: 81 3 5256 3864

Elf Atochem Agri S.A.
1, rue des Frères Lumiére, B.P. 9,
78373 Plaisir Cedex, France.
Tel: 33 1 30 81 73 00
Fax: 33 1 30 81 72 50

Elf Atochem N America Inc.
2000 Market St, 21st Floor, Philadelphia,
PA 19103-3222, USA.
Tel: 1 215 419 7219
Fax: 1 215 419 7243

Embetec
See Rhône-Poulenc

ÉMV Fa.
3792 Sajóbábony, Hungary.

Endura S.p.A.
Viale Pietramellara 5, 40121 Bologna,
Italy.
Tel: 39 51 558444
Fax: 39 51 557255

English Woodlands Biocontrol
Hoyle Depot, Graffham, Petworth,
W. Sussex GU28 0LR, UK.
Tel: 44 1798 867574
Fax: 44 1798 867574

Excel Industries Ltd
184-87 Swami Vivekanand Rd, Jogeshwari
(West), Bombay 400 102, India.
Tel: 91 22 6288258
Fax: 91 22 6203657

Fair Products, Inc.
P.O. Box 386, Cary, NC 27512-0386,
USA.
Tel: 1 919 467 8352
Fax: 1 919 467 9142

Fargro Ltd
Toddington Lane, Littlehampton,
W. Sussex BN17 7PP, UK.
Tel: 44 1903 721591
Fax: 44 1903 730737

FCC
Farmers Crop Chemicals Ltd,
Thorn Farm, Evesham Rd, Inkberrow,
Worcs. WR7 4LJ, UK.
Tel: 44 1386 793401
Fax: 44 1386 793184

Feinchemie Schwebda
Strassburger Strasse 5,
D-37269 Eschwege, Germany.
Tel: 49 5651 92370
Fax: 49 5651 22442

Ficom Organics Ltd
Export Dept, 162 Maker Chamber III,
Nariman Point, Bombay 400 021, India.
Tel: 91 22 2855481
Fax: 91 22 204 3961

Fine Agrochemicals Ltd
3 The Bull Ring, Worcester WR2 5AA,
UK.
Tel: 44 1905 748444
Fax: 44 1905 748440

Fitoquimica Produtos Para a Agricultura,
Lda.
E.N. 3-Cartaxo-Lisbon KM 1, P.O. Box 75,
2070 Cartaxo, Portugal.
Tel: 351 43 779084
Fax: 351 43 770902

Florin
Szeged, Kenyárgyár u. 5, Hungary.

FMC Corp.
Agricultural Products Group,
1735 Market St, Philadelphia PA 19103,
USA.
Tel: 1 215 299 6661
Fax: 1 215 299 6256

Fujisawa Pharmaceutical
Chemicals Group, 3-4-6 Nihonbashi
Honcho, Chuo-ku, Tokyo 103, Japan.
Tel: 81 3 3279 0882
Fax: 81 3 3241 5805

Gedeon Richter Rt.
1103 Budapest, Gyömröi út 19-21,
Hungary.

General Quimica, S.A.
Apartado 13, 09200 Miranda de Ebro,
Burgos, Spain.
Tel: 34 96 351 34 26
Fax: 34 96 351 03 57

Gharda Chemicals Ltd
B-27-29, MIDC, Dombivli (E), 421 203,
Dist. Thane, India.
Tel: 91 251 471215
Fax: 91 251 472777

Gilmore Marketing & Development, Inc.
152 Collins St, Memphis, TN 38112, USA.
Tel: 1 901 323 5870
Fax: 1 901 454 0295

Gist-Brocades B.V.
Wateringseweg 1, Postbus 1,
2600 MA Delft, Netherlands.
Tel: 31 15 2799111
Fax: 31 15 2793200

Gowan Co.
P.O. Box 5569, Yuma AZ 85366, USA.
Tel: 1 520 783 8844
Fax: 1 520 343 9255

Grace
See Thermo Trilogy

Great Lakes Chemical Corp.
P.O. Box 2200, One Great Lakes Blvd.,
West Lafayette, IN 47906-0200, USA.
Tel: 1 317 497 6100
Fax: 1 317 497 6123

Griffin Corp.
P.O. Box 1847, Rocky Ford Road,
Valdosta, GA 31603, USA.
Tel: 1 912 249 5203
Fax: 1 912 244 5978

Gustafson Inc.
P.O. Box 660065, Dallas, TX 75266-0065,
USA.
Tel: 1 214 985 8877
Fax: 1 214 867 0816

HACCO Inc.
P.O. Box 7190, Madison, WI 53707-7190,
USA.
Tel: 1 608 221 6200
Fax: 1 608 221 6208

Hanwha Corp.
1 Changgyo-Dong, Chung-ku,
Seoul 100797, Korea.
Tel: 82 2 729 1820
Fax: 82 2 729 1850

Hawkesbury Integrated Pest Management
Service
PO Box 436, Richmond, NSW 2753,
Australia.
Tel: 61 45 701331
Fax: 61 45 701314

Headland Agrochemicals Ltd
Norfolk House, Gt. Chesterford Court,
Gt. Chesterford, Essex CB10 1PF, UK.
Tel: 44 1799 530146
Fax: 44 1799 530229

Herbex Produtos Quimicos, Lda.
Estrada de Albarraque, 2710 Sintra,
Portugal.
Tel: 351 1 915 81 35/36
Fax: 351 1 915 00 21

Hico Products Ltd
P.B. 16483, 771 Pandit Satavlekar Marg.,
Mahim, Bombay 400 016, India.
Tel: 91 22 4377231
Fax: 91 22 4221526

High Kite Ltd
7th Floor, 11-15 Chatham Road South,
Tsimshatsui, Kowloon, Hong Kong.
Tel: 852 23 68 71 17
Fax: 852 23 66 81 82

Hindustan Insecticides Ltd
Scope Complex Core 6, 2nd Floor,
7 Lodi Road, New Delhi 1100 03, India.
Tel: 91 11 436 2165/19
Fax: 91 11 436 2116

Hodogaya Chemical Co., Ltd
66-2 Horikawa-cho, Saiwai-ku, Kawasaki,
Kanagawa 210, Japan.
Tel: 81 44 549 6600
Fax: 81 44 549 6630

Hoechst Roussel Vet.
P.O. Box 2500, Rt. 202-206, Somerville,
NJ 08876-1258, USA.
Tel: 1 908 231 2000
Fax: 1 908 231 4462

Hoechst Schering AgrEvo GmbH
D-65926, Frankfurt am Main, Germany.
Tel: 49 69 305 6699

Hoechst Veterinar GmbH
PGE Antiparasitica, Rheingaustrasse 190,
Postfach 3311, 65023 Wiesbaden,
Germany.
Tel: 49 611 9627 977
Fax: 49 611 9627 896

Hokko Chemical Industry Co., Ltd
Central Research Laboratories,
2165 Toda, Atsugi, Kanagawa 243, Japan.
Tel: 81 462 28 5881
Fax: 81 462 28 0164

Hortichem Ltd
14 Edison Rd, Churchfields Industrial
Estate, Salisbury, Wilts SP2 7NU, UK.
Tel: 44 1722 320133
Fax: 44 1722 326799

Hubei Sanonda Co., Ltd
1 East Beijing Road, Shashi,
Hubei 434001, China.
Tel: 86 716 8316975
Fax: 86 716 8315265

Hui Kwang Chemical Co., Ltd
17-10 Ling Tzyy Lin, Matou, Tainan,
Tainan Hsien, Taiwan, Rep. of China.
Tel: 886 6 5702181
Fax: 886 6 5700065

Idemitsu Kosan Co., Ltd
1280 Kami-Izumi, Sodegaura,
Chiba 299-02, Japan.

Ihara Chemical Industry Co., Ltd
1-4-26 Ikenohata, Taito-ku, Tokyo 110,
Japan.
Tel: 81 3 3822 5223
Fax: 81 3 3828 9887

India Pesticides Ltd
Water Works Road, Aishbagh,
Lucknow 226 004 (U.P.), India.
Tel: 91 522 269194
Fax: 91 522 269873

Industrie Chimiche **Caffaro** S.p.A.
See Caffaro

Ingenieria Industrial S.A. de C.V.
Ave. Coyoacan 1878-403,
Colonia del Valle 03100, Mexico D.F.,
Mexico.
Tel: 52 5 524 8369
Fax: 52 5 524 8270

Inquinosa
Hermanos Escartin, 7,
28224 Pozuelo de Alarcon, Madrid, Spain.
Tel: 34 1 351 1938
Fax: 34 1 351 1830

Inquiport, S.A.
Avda. Los Pioneros, Salida Hacia Guanare,
Local Inquiport, Araure Edo Portuguesa,
Venezuela.
Tel: 58 55 21 2332
Fax: 58 55 21 2330

Integrated Pest Management p/l
See Bugs for Bugs

Intrachem (International) S.A.
34 Quai de Cologny, Cologny,
CH-1223 Geneva, Switzerland.
Tel: 41 22 736 78 87
Fax: 41 22 736 24 10

IPESA S.A.
Joaquin V Gonzalez 4977, CP 1419,
Buenos Aires, Argentina.
Tel: 54 1 501 6800
Fax: 54 1 502 0305

IPM Laboratories Inc.
Main St, Locke, NY 13092-0300, USA.
Tel: 1 315 497 2063
Fax: 1 315 497 3129

Isagro S.p.A.
Centro Direzionale Milano Oltre,
Palazzo Raffaello, Via Cassanese 224,
20090 Milano, Italy.
Tel: 39 2 26996 425
Fax: 39 2 26996 287

Ishihara Sangyo Kaisha, Ltd
3-1, Nishi-Shibukawa 2-chome, Kusatsu,
Shiga 525, Japan.
Tel: 81 775 62 8338
Fax: 81 775 62 9506

ISK Biosciences Corp.
5966 Heisley Road, P O Box 8000,
Mentor, Ohio 44061-8000, USA.
Tel: 1 216 357 4100
Fax: 1 216 354 9506

J. J. **Mauget** Co.
2810 North Figueroa St, Los Angeles,
CA 90065, USA.

J. R. **Simplot** Co.
Minerals and Chemicals Group,
Group Headquarters, P.O. Box 912,
Pocatello, ID 83204, USA.
Tel: 1 208 238 2700
Fax: 1 208 238 2760

Janssen Pharmaceutica
Plant Protection Division, Turnhoutseweg
30, B-2340 Beerse, Belgium.
Tel: 32 14 60 2527
Fax: 32 14 60 2841

Japan Tobacco Inc.
Agribusiness Division,
4-12-62 Higashi-Shinagawa, Shinagawa-ku,
Tokyo 140, Japan.
Tel: 81 3 3474 3111
Fax: 81 3 5479 0360

Jewnin-Joffe Industry Ltd
Shalom Tower, 22nd Floor,
P.O. Box 29511, Tel-Aviv 61294, Israel.
Tel: 972 3 517 0034
Fax: 972 3 510 0050

Jiangsu Eternal Union Corp.
178 Huancheng Nan Road,
Jiangyin 214400, Jiangsu, China.
Tel: 86 510 6891042
Fax: 86 510 6891233

Jin Hung Fine Chemicals Co., Ltd
#708 Sinsong Bldg., 25-4 Yeoido-dong,
Yeongdeungpo-ku, Seoul, 150-010, Korea.
Fax: 82 2 782 7266

Jingma Chemicals Ltd
No. 104 Yu Gu Road, Hangzhou, China.
Tel: 86 571 7970282
Fax: 86 571 7980702

K. & N. **Efthymiadis** S.A.
P.O. Box 48, 570 22 Sindos Industrial
Area, Thessaloniki, Macedonia, Greece.
Tel: 30 31 798403
Fax: 30 31 798423

Kaken Pharmaceutical Co., Ltd
4-18-4 Nihonbashi-honcho, Chuo-ku,
Tokyo 103, Japan.
Tel: 81 3 3231 1223
Fax: 81 3 3270 5360

Kemira Agro Oy
P.O. Box 330, FIN-00101 Helsinki 10,
Finland.
Tel: 358 9 132 1561
Fax: 358 9 132 1384

Kemira Fine Chemicals Oy
P.O. Box 330, FIN-00101 Helsinki 10,
Finland.
Tel: 358 9 132 1612
Fax: 358 9 132 1624

Kerr-McGee Chemical Corp.
P.O. Box 25861, 123 Robert S. Kerr Ave.,
Oklahoma City, OK 73125, USA.
Tel: 1 405 270 1313
Fax: 1 405 270 3123

Khatau Junker Ltd
See Rallis

Killgerm Chemicals Ltd
115 Wakefield Rd, Flushdyke, Ossett,
W. Yorks WF5 9BW, UK.
Tel: 44 1924 265090
Fax: 44 1924 265033

Kincaid Enterprises, Inc.
P.O. Box 549, Nitro, WV 25143, USA.
Tel: 1 304 755 3377
Fax: 1 304 755 4547

KMG-Bernuth
10611 Harwin, Suite 402
Houston, TX 77306, USA.
Tel: 1 713 988 9252
Fax: 1 713 988 9298

Kohjin Co., Ltd
1-1 Shimbashi 1-chome, Minato-ku,
Tokyo, Japan.

Koppert BV
Veilingweg 17, PO Box 155, 2650 AD
Berkel en Rodenrijs, Netherlands.
Tel: 31 10 5140444
Fax: 31 10 5115203

Krishi Rasayan (Bihar)
FMC Fortuna, Block No. A11, 4th Fl.,
234/3A Acharya Jagadish Chandra Bose
Rd, Calcutta 700 020, India.
Tel: 91 33 247 5719/37
Fax: 91 33 247 1436

Kubota Corp.
1-2-47 Shikitsuhigashi, Naniwa-ku,
Osaka 556-91. Japan.
Tel: 81 6 648 2111
Fax: 81 6 648 3826

Kumiai Chemical Industry Co., Ltd
4-26 Ikenohata 1-chome, Taitoh-ku,
Tokyo 110, Japan.
Tel: 81 3 3822 5165
Fax: 81 3 3822 5005

Kunafin
Rte 1, Box 39, Quemado, TX 78877, USA.
Tel: 1 800 832 1113
Fax: 1 512 757 1468

Kuo Ching Chemical Co., Ltd
No. 53, Chung Ming Rd. 4F, Taichung,
Taiwan, Rep. of China.
Tel: 886 4 326 5021
Fax: 886 4 325 5197

Kureha Chemical Industry Co., Ltd
Nishiki Research Laboratories, 16 Ochiai,
Nishiki, Iwaki 974, Japan.
Tel: 81 246 63 5111
Fax: 81 246 63 7356

La Cornubia S.A.
B.P. 55, 33016 Bordeaux Cedex, France.
Tel: 33 557 77 55 00
Fax: 33 556 32 50 13

La Quinoleine
See Novartis

Lainco S.A.
P.O. Box 73 - Rubi,
08191 Rubi (Barcelona), Spain.
Tel: 34 3 588 5050
Fax: 34 3 588 4592

Lipha, Lyonnaise Industrielle
Pharmaceutique s.a.
34 rue Saint-Romain,
F 69379 Lyon Cedex 08, France.
Tel: 33 7877 1818
Fax: 33 7877 1939

Lonza Ltd
Muenchensteinerstrasse 38,
CH-4002 Basel, Switzerland.
Tel: 41 61 316 8111
Fax: 41 61 316 8733

Luxan B.V.
P.O. Box 9, Industrieweg 2,
6660 AA Elst (Gld), Netherlands.
Tel: 31 481 360811
Fax: 31 481 372479

Luxembourg Industries (Pamol) Ltd
27 Hamered St., P.O. Box 13,
Tel Aviv 61000, Israel.
Tel: 972 3 5103388
Fax: 972 3 5100474

M&R Durango Inc.
PO Box 886, Bayfield, CO 81122, USA.
Tel: 1 303 259 3521
Fax: 1 303 259 3857

Makhteshim-Agan
See Makhteshim *and* Agan

Makhteshim Chemical Works Ltd
P.O. Box 60, 84100 Beer-Sheva, Israel.
Tel: 972 7 6296611
Fax: 972 7 6280304, 6280364

Mallinckrodt Inc.
675 McDonnell Boulevard, St Louis,
MO 63134, USA.

Marks
See A. H. Marks

Mauget
See J. J. Mauget

McLaughlin Gormley King Co.
8810 Tenth Avenue N, Minneapolis
MN 55427, USA.
Tel: 1 612 544 0341
Fax: 1 612 544 6437

Meiji Seika Kaisha Ltd
Agrochemical Department,
2-4-16 Kyobashi, 2-chome,
Chuo-ku, Tokyo 104, Japan.
Tel: 81 3 3272 6511
Fax: 81 3 3281 4058

Merck & Co., Inc.
Merck AgVet Div., P.O. Box 2000,
Rahway, NJ 07065, USA.
Tel: 1 908 855 4277
Fax: 1 908 855 6480

MGK
See McLaughlin Gormley King

Mico Farm Chemicals Ltd
Viscose Towers, 1078 Avanashi Rd,
Coimbatore 641 018, India.
Tel: 91 422 218020
Fax: 91 422 218030

Microbio Ltd
Dales Manor Business Park,
Babraham Road, Sawston,
Cambridge CB2 4LJ, UK.
Tel: 44 1223 8308608
Fax: 44 1223 830861

Mikasa Chemical Industry Co., Ltd
See Yashima

Miles Inc.
See Bayer

Miller Chemical & Fertilizer Corp.
P.O. Box 333, 120 Radio Road, Hanover,
PA 17331, USA.
Tel: 1 717 632 8921
Fax: 1 717 632 9638

Mirfield Sales Services Ltd
Moorend House, Moorend Lane,
Dewsbury, W. Yorks WF13 4QQ, UK.
Tel: 44 1924 409782
Fax: 44 1924 410792

Mitchell Cotts Chemicals Ltd
PO Box 6, Steanard Lane, Mirfield,
W. Yorks WF14 8QB, UK.
Tel: 44 1924 493861
Fax: 44 1924 490972

Mitsu Industries Ltd
2803/2 Third Phase, GIDC,
Vapi-396 195 (Gujarat), India.
Tel: 91 2638 30782
Fax: 91 2638 30781

Mitsubishi Chemical Corp.
Mitsubishi Building, 5-2 Marunouchi
2-chome, Chiyoda-ku, Tokyo 100, Japan.

Mitsui Toatsu Chemicals, Inc.
Fine Chemicals Division,
2-5, Kasumigaseki 3-chome,
Chiyoda-ku, Tokyo 100, Japan.
Tel: 81 3 3592 4852
Fax: 81 3 3592 4282

Monsanto Co.
Crop Protection, 800 N. Lindbergh Blvd.,
St. Louis, MO 63167, USA
Tel: 1 314 694 3540
Fax: 1 314 694 2306

MotomCo, Ltd
29. N. Fort Harrison Ave., Clearwater,
FL 34615, USA.
Tel: 1 608 244 2904
Fax: 1 813 447 5141

MTM Agrochemicals Ltd
See United Phosphorus

Mycogen Corp.
5501 Oberlin Drive, San Diego,
CA 92121, USA.
Tel: 1 619 453 8030
Fax: 1 619 453 9089

Mycotech Corp.
630 Utah Avenue, Butte, MT 59701, USA.
Tel: 1 406 782 2386

N.P.P.
Route d'Artix, B.P. 80, 64150 Nogueres,
France.
Tel: 33 559 60 92 92
Fax: 33 559 60 92 19

Nagarjuna Agrichem Ltd
Auto Plaza, First floor, Road No. 3,
Banjara Hills, Hyderabad 500 034, India.
Tel: 91 40 318217
Fax: 91 40 319234

Natural Plant Protection
See N.P.P.

Nature's Alternative Insectary Ltd
Box 19, Dawson Rd, Nanoose Bay, BC,
V0R 2R0, Canada.
Tel: 1 604 468 7912
Fax: 1 604 468 7912

Neudorff
See W. Neudorff

New Chemi Industries Ltd
33/3rd Floor, Maker Chambers VI,
220 Nariman Point, Bombay-400 021,
India.
Tel: 91 287 1173
Fax: 91 22 2870923

Nicobrand
See The Nicobrand Company

Nihon Bayer Agrochem K.K.
P.O. Box 157, Nihonbashi, Tokyo, Japan.
Tel: 81 3 3280 9894
Fax: 81 3 3280 9906

Nihon Nohyaku Co., Ltd
8th Floor Eitaro Building, 2-5 Nihonbashi
1-chome, Chuo-ku, Tokyo 103, Japan.
Tel: 81 3 3278 0461
Fax: 81 3 3281 2443

Niklor Chemical Co., Inc.
2060 E. 220th St., Long Beach, CA 90810,
USA.
Tel: 1 310 830 2253
Fax: 1 310 830 2835

Nippon Kayaku Co., Ltd
Agrochemicals Division, 11-2 Fujimi
1-chome, Chiyoda-ku, Tokyo, Japan.
Tel: 81 3 3237 5221
Fax: 81 3 3237 5089

Nippon Soda Co., Ltd
Agrochemicals Division, 2-1 Ohtemachi,
2-chome, Chiyoda-ku, Tokyo 100, Japan.
Tel: 81 3 3245 6266
Fax: 81 3 3245 6289

Nissan Chemical Industries Ltd
Kowa-Hitotsubashi Building,
7-1, 3-chome, Kanda-nishiki-cho,
Chiyoda-ku, Tokyo 101, Japan.
Tel: 81 3 3296 8151
Fax: 81 3 3296 8016

Nitrokémia Rt.
8184 Füzfögyártelep, Pf. 45, Hungary.
Tel: 36 88 352 011
Fax: 36 88 3 51002

NOR-AM
See AgrEvo USA

NORDOX Industrier AS
Østensjøveien 13, N-0661 Oslo, Norway.
Tel: 47 22 659090
Fax: 47 22 641208

Novartis BCM
Aldham Business Centre, New Road,
Aldham, Colchester CO6 3PN, UK.
Tel: 44 1206 243200
Fax: 44 1206 243209

Novartis Crop Protection AG
CH-4002, Basel, Switzerland.
Tel: 41 61 697 1111
Fax: 41 61 324 8001

Novo Nordisk
See Abbott

NPP
See N.P.P.

Nufarm B.V.
Welplaatweg 12, Rotterdam Botlek 3197,
Netherlands.
Tel: 31 10 438 9545
Fax: 31 10 472 2826

Nufarm GmbH & Co. KG
Postfach 21, St Peter Str. 25, A-4021 Linz,
Austria.
Tel: 43 70 6918 2006
Fax: 43 70 6918 2004

Nufarm Ltd
102-105 Pipe Rd., Laverton North,
Victoria 3026, Australia.
Tel: 61 39 282 1000
Fax: 61 39 282 1001

Nufarm UK Ltd
Crabtree Manorway North, Belvedere,
Kent DA17 6BQ, UK.
Tel: 44 181 311 7000
Fax: 44 181 310 5636

Otsuka Chemical Co., Ltd
Agrochemicals Development Department,
3-2-27 Ote-dori, Chuo-ku, Osaka 540, Japan.
Tel: 81 6 943 7711
Fax: 81 6 943 7703

Oxon Italia S.p.A.
Via Sempione 195, Pero (Milano) 20016,
Italy.
Tel: 39 2 353 781
Fax: 39 2 339 10759

Pace International LP
P.O. Box 519, 500 7th Ave. S., Kirkland,
WA 98083, USA.
Tel: 1 206 827 8711
Fax: 1 206 822 8261

Pacific Biocontrol Corp.
400 E. Evergreen Blvd., #306, Vancouver,
WA 98660, USA.
Tel: 1 206 693 2866
Fax: 1 206 693 3088

Pamol
See Luxembourg

Pan Britannica Industries Ltd
Britannica House, Waltham Cross, Herts
EN8 7DY, UK.
Tel: 44 1992 623691
Fax: 44 1992 626452

PBI
See Pan Britannica

PBI/Gordon Corp.
1217 W. 12th St, Kansas City, MO 64101,
USA.
Tel: 1 816 421 4070
Fax: 1 816 474 0462

Pepro
See Rhône-Poulenc

Perifleur Products Ltd
Hangleton Lane, Ferring, Worthing,
West Sussex BN12 6PP, UK.

Pestcon Systems, Inc.
5511 Capital Center Dr., Suite 302,
Raleigh, NC 27606, USA.
Tel: 1 919 859 2500
Fax: 1 919 859 2155

Pesticides India
P.O. Box 20 Udaisagar Rd.,
Udaipur 313 001, Rajasthan, India.
Tel: 91 294 414751
Fax: 91 294 524946

Phelps Dodge Refining Corp.
Specialty Metal & Chemical Sales,
P.O. Box 20001, El Paso, TX 79998, USA.
Tel: 1 800 223 8567
Fax: 1 915 775 8350

Phibro-Tech, Inc.
1 Parker Plaza, Ft. Lee, NJ 07024, USA.
Tel: 1 201 944 6000
Fax: 1 201 944 7916

Philagro France
Telebase, Batiment 2, rue Claude Chappe,
69370 Saint-Didier-au-Mont-d'Or, France.
Tel: 33 78 64 32 14
Fax: 33 72 53 04 58

Pilarquim Corp.
P.O. Box 7-777, Taipei, Taiwan,
Rep. of China.
Tel: 886 2 362 2222
Fax: 886 2 363 0000

Planters Products, Inc.
Planters Products Bldg., Esteban St.,
Legaspi Village, Makati Metro, Manila,
Philippines.
Tel: 632 818 2119
Fax: 632 816 4388

PPG Industries
One PPG Place, Pittsburgh, PA 15272,
USA.
Tel: 1 412 434 3131

Prentiss Incorporated
21 Vernon St CB 2000, Floral Park,
NY 11001, USA.
Tel: 1 516 326 1919
Fax: 1 516 326 2312

Probelte, S.A.
P.O. Box 4579, Murcia 30080, Spain.
Tel: 34 68 307250
Fax: 34 68 305432

Procida
See AgrEvo (France)

Productos Fitosanitarios Proficol
See Proficol El Carmen S.A.

Productos OSA
Avenida de Mayo 1161 - 1° Piso,
1085 Buenos Aires, Argentina.
Tel: 54 1 325 6481
Fax: 54 1 383 8139

Proficol El Carmen S.A.
P.O. Box 92126, Calle 85, No. 9-65,
Santafe de Bogota, Colombia.
Tel: 57 1 2579100/2181
Fax: 57 1 2187168

Pyrethrum Board of Kenya
P.O. Box 591, Nakuru, Kenya.
Tel: 254 9037211 567
Fax: 254 903745 274

Q.E.A.C.A. S.A.
Av Madero 942 - 5° Piso,
1106 Buenos Aires, Argentina.
Tel: 54 1 310 1363
Fax: 54 1 313 7571

Quadrangle Agrochemicals
Bishop Monkton, Harrogate, N. Yorks
HG3 3QQ, UK.
Tel: 44 1765 677116
Fax: 44 1765 677168

Rallis India Ltd
Agrochemical Res. Station,
21/22 Peenya Industrial Area,
P.O. Box 5813, Bangalore 560 058,
Karnataka, India.
Tel: 91 8394959
Fax: 91 80 8394015

Ramcides
See Sree Ramcides

Raschig AG
Mundenheimer Strasse 100, Postfach
211128, D-67 Ludwigshafen/Rhein,
Germany.

Reanal Fine Chemical Co., Rt.
1147 Budapest, Telepes u.53, Hungary.

Rhône-Poulenc Secteur Agro
14/20 Rue Pierre Baizet, B.P. 9163,
F-69263 Lyon Cedex 09, France.
Tel: 33 472 29 25 25
Fax: 33 472 29 27 99

Rhône-Poulenc Yuka Agro KK
No. 16 Kowa Building, Annex 9-20,
Akasaka 1-chome, Minato-ku, Tokyo 107,
Japan.
Tel: 81 3 5570 6061
Fax: 81 3 5570 6070

Riken Green Co., Ltd
NDK Lotus Bldg 12-20, Ueno 2-chome,
Taitoh-ku, Tokyo 110, Japan.
Tel: 81 3 3833 6321
Fax: 81 3 3833 6325

Rincon-Vitova Insectaries Inc.
P.O. Box 1555, Ventura CA 93002, USA.
Tel: 1 805 643 5407
Fax: 1 805 643 6267

Rohm & Haas Co.
100 Independence Mall West, Philadelphia
PA 19106, USA.
Tel: 1 215 592 3000
Fax: 1 215 592 2797

RohMid L.L.C.
One Campus Drive, Parsippany,
NJ 07054-4492, USA.
Tel: 1 800 545 9525
Fax: 1 201 831 3858

Rotam Agrochemical Co., Ltd
7/F, Cheung Tat Centre,
18 Cheung Lee St., Chai Wan, Hong Kong.
Tel: 852 2896 5608
Fax: 852 2558 6577

Roussel Uclaf Corp.
See AgrEvo Environmental Health Inc.

Sagrochem
See EMV

Sanachem (Pty) Ltd
Old Mill Site, Canelands, P.O. Box 1454,
Durban 4000, South Africa.
Tel: 27 0322 301111
Fax: 27 0322 331278

Sandoz Agro Ltd
See Novartis

Sanex Inc.
5300 Harvester Rd., Burlington,
Ont. L7L 5N5, Canada.
Tel: 1 905 639 7535
Fax: 1 905 639 3488

Sankyo Co., Ltd
Agrochemicals Division, 7-12 Ginza
2-chome, Chuo-ku, Tokyo 104, Japan.
Tel: 81 3 3562 7524
Fax: 81 3 3562 7525

Sanonda
See Hubei Sanonda

Sautter & Stepper GmbH
Rosenstr. 19, D-72119 Ammerbuch 5,
Altingen, Germany.
Tel: 49 7032 75501
Fax: 49 7032 74199

SCC
Eckelsheimer Str. 37, D-55597,
Woellstein, Germany.
Tel: 49 6703 9344 0
Fax: 49 6703 9344 44

SDS Biotech K.K.
12-7 Higashi Shimbashi 2-chome,
Minato-ku, Tokyo 105, Japan.
Tel: 81 3 3436 3812
Fax: 81 3 3436 0989

Searle (India) Ltd
21 D. Sukhadvala Marg, P.O. Box 233,
Mubai 400 001, India.
Tel: 91 2047731
Fax: 91 22 2047009

SePRO Corp.
11550 North Meridian Street, Suite 180,
Carmel, Indiana 46032, USA.
Tel: 1 317 580 8282
Fax: 1 317 580 8280

Shanghai Zhong Xi Pharmaceutical Co.,
Ltd
1515 Jiao Tong Road, Shanghai 200065,
China.
Tel: 86 21 56081348
Fax: 86 21 56083040

Shaw Wallace & Co., Ltd
154 Thambu Chetty St, P.O. Box 14,
Madras 600 001, India.
Tel: 91 44 5340021
Fax: 91 44 5341804

Shenzhen Jiangshan Commerce & Industry
9/ F. Real Estate Bldg., Renmin South Rd,
Shenzhen, China.
Tel: 86 755 2320413
Fax: 86 755 2338993

Shin-Etsu Chemical Co., Ltd
Fine Chemicals Dept., 2-6-1 Ohtemachi,
Chiyoda-ku, Tokyo 100, Japan.
Tel: 81 3 3246 5280
Fax: 81 3 3246 5371

Shinung Corp.
See Sinon

Shionogi & Co., Ltd
1-8 Doshomachi 3-chome, Chuo-ku,
Osaka 541, Japan.
Tel: 81 6 202 2161
Fax: 81 6 202 1318

Siapa S.p.A.
via Yser 16, 00198 Roma, Italy.
Tel: 39 6 84345
Fax: 39 6 855 1242

Simplot
See J. R. Simplot

Sinon Corp.
45 Wu Chuan Center St, Taichung,
Taiwan, Rep. of China.
Tel: 886 4 6934261
Fax: 886 4 6934265

Sipcam Phyteurop
Immeuble Courcellor II, 35 rue d'Alsace,
92531 Levallois-Perret Cedex, France.
Tel: 33 1 47 59 77

SKW Trostberg AG
Postfach 1262, D-83303 Trostberg,
Germany.
Tel: 49 8621 86 2833
Fax: 49 8621 86 2252

Solvay Duphar B.V.
See Uniroyal

Sorex Ltd
St Michael's Industrial Est., Hale Rd,
Widnes, Cheshire WA8 8TJ, UK.
Tel: 44 51 420 7151
Fax: 44 51 495 1163

Source Technology Biologicals, Inc.
3355 Hiawatha Ave., Suite 222,
Minneapolis, MN 55406, USA.
Tel: 1 612 724 7102
Fax: 1 612 724 1642

Sree Ramcides Chemicals Pvt. Ltd
Aishwarya Complex, 2nd Floor,
P B No 1013, 4 Doraisamy Rd, T Nagar,
Chennai, Madras 600 017, India.
Tel: 91 44 4345770
Fax: 91 44 4347569

Staehler Agrochemie GmbH & Co. KG
Postfach 2047, Stader Elbstrasse,
2160 Stade, Germany.

Stefes
See Hoechst Schering AgrEvo

Stoller Enterprises Inc.
8580 Katy Freeway, Suite 200, Houston,
TX 77024, USA.
Tel: 1 713 461 1493
Fax: 1 713 461 4467

Sumitomo Chemical Company, Ltd
5-33 Kitahama 4-chome, Chuo-ku,
Osaka 541, Japan.
Tel: 81 6 220 3683
Fax: 81 6 220 3342

Sundat (S) Pte. Ltd
26 Gul Crescent, Singapore 629532,
Singapore.
Tel: 65 8612460
Fax: 65 8620287

Synexus
Avenue de Tervuren 270 272,
1150 Brussels, Belgium.
Tel: 32 2 776 4111
Fax: 32 2 776 4385

Synthesia a.s.
Semtin, 532 17 Pardubice,
Czech Republic.
Tel: 42 6821111
Fax: 42 46465

Tagros Chemicals India Ltd
Jhaver Centre, Rajah Annamalai Building,
IV Floor, No. 19, Marshalls Rd, Egmore,
Madras 600 008, India.
Tel: 91 44 8557841
Fax: 91 44 8557873

Taiwan Tainan Giant Industrial Co., Ltd
3F No 53, Chung Ming Rd., Taichung,
Taiwan, Rep. of China.
Tel: 886 4 320 9988
Fax: 886 4 321 2824

Takasago International (Nederland) B.V.
Gooimeer 9, NL 1411 DD, Naarden,
Netherlands.
Tel: 31 35 6954233
Fax: 31 35 6954239

Takeda Chemical Industries Ltd
Agro Company, 12-10 Nihonbashi
2-chome, Chuo-ku, Tokyo 103, Japan.
Fax: 81 3 3278 2750

Tamogan Chemical Ltd
P.O. Box 2438, Tel-Aviv 61024, Israel.
Tel: 972 3 9265272
Fax: 972 3 926521

Tekchem S.A. de C.V.
Ave. Jalisco 180, 5 PISO Tacubaya,
Mexico City D.F. 11870, Mexico.
Tel: 52 5 272 2221
Fax: 52 5 272 2950

The **Nicobrand** Company
189 Castleroe Rd, Coleraine,
Northern Ireland, BT51 3RP, UK.
Tel: 44 1265 868733
Fax: 44 1265 868735

Thermo Trilogy Corp.
7500 Grace Dr., Columbia, MD 21044,
USA.
Tel: 1 410 531 4711
Fax: 1 410 531 4780

Tifa (C.I.) Ltd
Tifa Square, Millington, NJ 07946, USA.
Tel: 1 908 647 2517
Fax: 1 908 647 7338

Tokuyama Corp.
Specialty Chemicals Business Div.,
3-1 Shibuya 3-chome, Shibuya-ku,
Tokyo 105, Japan.
Tel: 81 3 3499 8937
Fax: 81 3 3499 8967

Tomen Agro Inc.
100 First St, Suite 1610, San Francisco,
CA 94105, USA.
Tel: 1 415 536 3480
Fax: 1 415 284 9883

Tomono Agrica Co., Ltd
2-12-25 Kasuga, Shizuoka City 420, Japan.
Tel: 81 54 254 6261
Fax: 81 54 254 6263

Tripart Farm Chemicals Ltd
The Beeches, 42 West St,
Godmanchester, Huntingdon, Cambs
PE18 8HH, UK.
Tel: 44 1480 445700
Fax: 44 1480 435102

Troy Biosciences Inc.
2620 N 37th Dr., Phoenix, AZ 85009,
USA.
Tel: 1 602 233 9047
Fax: 1 602 254 7989

Truchem Ltd
Brook House, 30 Larwood Grove,
Sherwood, Nottingham, NG5 3JD, UK.
Tel: 44 115 926 0762
Fax: 44 115 967 1153

Ube Industries Ltd
Ube Research Lab, 1978-5 Kogushi,
Ube City, Yamaguchi Pref 755, Japan.
Tel: 81 836 31 6438
Fax: 81 836 31 6282

UCB Chemicals
Pantserschipstraat 207, B-9000 Gent,
Belgium.
Tel: 32 9 254 14 11
Fax: 32 9 254 14 10

Unicrop
Universal Crop Protection Ltd,
Park House, Maidenhead Rd, Cookham,
Berks SL6 9DS, UK.
Tel: 44 1628 526083
Fax: 44 1628 810457

Uniroyal Chemical Ltd
Benson Rd., Middlebury, CT 06749, USA.
Tel: 1 203 573 2000
Fax: 1 203 573 3394

United Phosphorus Ltd
Readymoney Terrace, 167 Dr. Annie
Besant Rd., Worli, Bombay Maharashtra
400 018, India.
Tel: 91 22 493 0681/49
Fax: 91 22 493 7331

Urania Agrochem GmbH
P.O. Box 106220, D-20042 Hamburg,
Germany.
Tel: 49 40 236 52322
Fax: 49 40 236 52255

US Borax Inc.
26877 Tourney Rd., Valencia, CA 91355,
USA.
Tel: 1 805 287 5400
Fax: 1 805 287 5455

Valent U.S.A. Corp.
1333 N. California Blvd., Suite 600,
Walnut Creek, CA 94596-8025, USA.
Tel: 1 510 256 2700
Fax: 1 510 256 2844

Velpol, S.A. de C.V.
See Tekchem

Velsicol Chemical Corp.
10400 West Higgins Rd, Rosemont,
IL 60018, USA.
Tel: 1 847 298 9000
Fax: 1 847 298 9015

Vischim
See Sipcam

Vitax Ltd
Owen St, Coalville, Leics LE6 2DE, UK.
Tel: 44 1530 510060
Fax: 44 1530 510299

Voltas Ltd
Chemicals & Agro Products,
616 Swapnalok Complex,
Sarojini Devi Rd., Secunderabad 500 003,
Andhra Pradesh 500 003, India.
Tel: 91 040 813495
Fax: 91 040 810050

W. **Neudorff** GmbH KG
Abt. Nutzorgardsmen, An der Muhle 3,
Postfach 1209, D-31857 Emmerthal,
Germany.
Tel: 49 5155 62460
Fax: 49 5155 62457

Wacker-Chemie GmbH
Prinzregentenstrasse 22, 8000 Munchen
83, Hanns-Seidel-Platz 4, Germany.
Fax: 49 89 6279 1772

Westrade Guatemala S.A.
Avenida La Reforma 13-70,
Zona 9-01009, Edificio Real Reforma,
Guatemala City, Guatemala, C.A.
Fax: 502 3315608

Wilbur-Ellis Co.
191 W. Shaw Ave., Suite 107, Fresno,
CA 93704-2876, USA.
Tel: 1 209 226 1934
Fax: 1 209 226 7630

Wockhardt Ltd
Readymoney Terrace, 167 Dr. Annie
Besant Rd, Bombay 400 018, India.
Tel: 91 492 66 44, 495 11 16
Fax: 91 22 495 22 95, 495 30 90

Wujiang Luosen Chemicals Co., Ltd
North End, Tong Luo Town,
Wujing City 215327, Jiangsu Province,
China.

Wujin 3rd Veterinary Pharmaceutical
Wujin Huaxia Agrochemical Factory,
Xiaxi Town, Wujin, Jingsu, China.
Tel: 86 519 3581014
Fax: 86 519 3581236

Yashima Chemical Industry Co., Ltd
YTT Bldg., 757-1 Futako, Takatsu-ku,
Kawasaki-Shi, Kanagawa 213, Japan.
Tel: 81 44 813 4206
Fax: 81 44 813 5299

Young IL Chemical Co., Ltd
1338-20, Seocho-dong, Seocho-ku,
Seoul 137-072, Korea.
Tel: 82 2 555 0471
Fax: 82 2 553 2639

Zeneca Agrochemicals
Fernhurst, Haslemere, Surrey GU27 3JE,
UK.
Tel: 44 1428 644061
Fax: 44 1428 652922

Zhong-Xi
See Shanghai Zhong-Xi

ABBREVIATIONS AND CODES

1. Common names – recommended names for ions and radicals
2. GCPF (formerly GIFAP) formulation codes
3. WIPO country codes for patents
4. WHO and EPA Toxicity Classification
5. EC Risk Symbols and Phrases
6. General abbreviations

1. Common names - recommended names for ions and radicals

ISO/TC 81 and BSI have adopted abbreviations for several ions and radicals. Those used in this edition of *The Pesticide Manual* are listed below.

Abbreviation	IUPAC name	Structure
butotyl	2-butoxyethyl	$CH_3CH_2CH_2CH_2OCH_2CH_2-$
diolamine	bis(2-hydroxyethyl)ammonium	$(HOCH_2CH_2)_2N^+H_2$
etotyl	2-ethoxyethyl	$CH_3CH_2OCH_2CH_2-$
meptyl	1-methylheptyl	$CH_3(CH_2)_5CH(CH_3)-$
metilsulfate	methyl sulfate	$CH_3OSO_2O^-$
mexyl	1-methylhexyl	$CH_3(CH_2)_4CH(CH_3)-$
olamine	2-hydroxyethylammonium	$HOCH_2CH_2N^+H_3$
tefuryl	(±)-tetrahydrofurfuryl	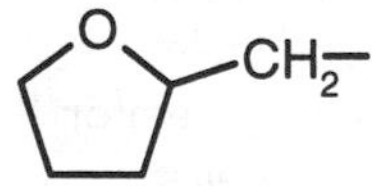
trimesium	trimethylsulfonium	$(CH_3)_3S^+$
trolamine	tris(2-hydroxyethyl)ammmonium	$(HOCH_2CH_2)_3N^+H$

2. GCPF (formerly GIFAP) formulation codes

The following standard codes are used. For further details, see *Catalogue of Pesticide Formulation Types and International Coding System,* Technical monograph No. 2, February 1989, ref. MT02E, (and addendum dated 20 November 1994), GCPF, Brussels.

CODE	TERM	CODE	TERM
AB	Grain bait	KN	Cold fogging concentrate
AE	Aerosol dispenser	KP	Combi-pack solid/solid
AI	Active ingredient	LA	Lacquer
AL	Other liquids to be applied undiluted	LS	Solution for seed treatment
AP	Other powder	MG	Microgranule
BB	Block bait	OF	Oil miscible flowable concent
BR	Briquette		miscible suspension)
CB	Bait concentrate	OL	Oil miscible liquid
CG	Encapsulated granule	OP	Oil dispersible powder
CS	Capsule suspension	PA	Paste
DC	Dispersible concentrate	PB	Plate bait
DP	Dispersible powder	PC	Gel or paste concentrate
DS	Powder for dry seed treatment	PO	Pour-on
EC	Emulsifiable concentrate	PR	Plant rodlet
ED	Electrochargeable liquid	PS	Seed coated with a pesticide
EG*	Emulsifiable granule	RB	Bait (ready for use)
EO	Emulsion, water in oil	SA	Spot-on
ES	Emulsion for seed treatment	SB	Scrap bait
EW	Emulsion, oil in water	SC	Suspension concentrate (= flc
FD	Smoke tin		concentrate)
FG	Fine granule	SE	Suspo-emulsion
FK	Smoke candle	SG	Water soluble granules
FP	Smoke cartridge	SL	Soluble concentrate
FR	Smoke rodlet	SO	Spreading oil
FS	Flowable concentrate for seed treatment	SP	Water soluble powder
		SS	Water soluble powder for see treatment
FT	Smoke tablet		
FU	Smoke generator	SU	Ultra-low volume (ULV) suspe
FW	Smoke pellet	TB	Tablet
GA	Gas	TC	Technical material
GB	Granular bait	TK	Technical concentrate
GE	Gas generating product	TP	Tracking powder
GC	Macrogranule	UL	Ultra-low volume (ULV) liquid
GL	Emulsifiable gel	VP	Vapour releasing product
GP	Flo-dust	WG	Water dispersible granules
GR	Granule	WP	Wettable powder
GS	Grease	WS	Water dispersible powder for treatment
GW	Water soluble gel		
HN	Hot fogging concentrate	XX	Others
KK	Combi-pack solid/liquid		
KL	Combi-pack liquid/liquid	* Proposed code, not adopted by GCP

1426

3. WIPO country codes for patents

The World Intellectual Property Organisation (WIPO) uses standard two letter country codes for patents. The following have been used in this edition of *The Pesticide Manual*.

AT - Austria   AU - Australia
BE - Belgium   CA - Canada
CH - Switzerland   DE - Germany
EP - European Patent Organisation
FR - France   GB - United Kingdom
HU - Hungary   IT - Italy
JP - Japan   NL - Netherlands
SE - Sweden   US - USA

4. WHO and EPA Toxicity Classification

WHO toxicity classification

The World Health Organisation classification for estimating the acute toxicity of pesticides.

	Class	LD_{50} for the rat (mg/kg b.w.) Oral Solids	Liquids	Dermal Solids	Liquids
Extremely hazardous	Ia	≤ 5	≤ 20	≤ 10	≤ 40
Highly hazardous	Ib	5-50	20-200	10-100	40-400
Moderately hazardous	II	50-500	200-2000	100-1000	400-4000
Slightly hazardous	III	≥ 501	≥ 2001	≥ 1001	≥ 4001
Product unlikely to present acute hazard in normal use	Table 5	≥ 2000	≥ 3000	–	–
Not classified; believed obsolete	Table 6				
Fumigants not classified under WHO	Table 7				

EPA toxicity classification

Class	Oral LD$_{50}$ (mg/kg)	Acute toxicity to rat Dermal LD$_{50}$ (mg/kg)	Inhalation LC$_{50}$ (mg/l)	Eye effects	Skin effect
I	≤50	≤200	≤0.2	Corrosive; corneal opacity not reversible within 7 days.	Corrosive
II	50-500	200-2000	0.2-2.0	Corneal opacity reversible within 7 days; irritation persisting for 7 days.	Severe irritation at 72 hours.
III	500-5000	2000-20 000	2.0-20	No corneal opacity; irritation reversible within 7 days.	Moderate irritation at 72 hour
IV	≥5000	≥20 000	≥20	No irritation.	Mild or slig irritation at 72 hours.

5. EC Risk Symbols and Phrases

For a brief explanation, see Notes to the Sample Entry, p. xvi. The following appear in this of *The Pesticide Manual*. For a full list, and for fuller details see "*Approved guide to the Classi and Labelling of substances and preparations dangerous for Supply* (2nd Ed.)"

RISK SYMBOLS:

C   - Corrosive
O   - Oxidising
F   - Highly flammable
F+   - Extremely flammable
T   - Toxic
T+   - Very toxic
Xi   - Irritant
Xn   - Sensitising

ASSOCIATED RISK PHRASES

9	Explosive when mixed with combustible material
10	Flammable
11	Highly flammable
12	Extremely flammable
20	Harmful by inhalation
21	Harmful in contact with skin
22	Harmful if swallowed
23	Toxic by inhalation
24	Toxic in contact with skin
25	Toxic if swallowed
26	Very toxic by inhalation
27	Very toxic in contact with skin
28	Very toxic if swallowed
31	Contact with acids liberates toxic gas
32	Contact with acids liberates very toxic gas
33	Danger of cumulative effects
34	Causes burns
35	Causes severe burns
36	Irritating to the eyes
37	Irritating to the respiratory system
38	Irritating to the skin
40	Possible risk of irreversible effects
41	Risk of serious damage to eyes
43	May cause sensitisation by skin contact
44	Risk of explosion if heated under confinement
45	May cause cancer
48	Danger of serious damage to health by prolonged exposure
61	May cause harm to the unborn child
63	Possible risk of harm to the unborn child

COMBINATION OF PARTICULAR RISKS

15/29	Contact with water liberates toxic, extremely flammable gas
20/21	Harmful by inhalation and in contact with skin
20/21/22	Harmful by inhalation, in contact with skin and if swallowed
20/22	Harmful by inhalation and if swallowed
21/22	Harmful in contact with skin and if swallowed
23/24/25	Toxic by inhalation, in contact with skin, and if swallowed
23/25	Toxic by inhalation and if swallowed
24/25	Toxic in contact with skin and if swallowed
26/27	Very toxic by inhalation and in contact with skin
26/27/28	Very toxic by inhalation, in contact with skin and if swallowed
26/28	Very toxic by inhalation and if swallowed

27/28	Very toxic in contact with skin and if swallowed
36/37	Irritating to eyes and respiratory system
36/37/38	Irritating to eyes, respiratory system and skin
36/38	Irritating to eyes and skin
37/38	Irritating to respiratory system and skin
40/22	Harmful: possible risk of irreversible effects if swallowed
48/22	Harmful: danger of serious damage to health by prolonged exposure if swa
48/23/24/25	Toxic: danger of serious damage to health by prolonged exposure through inhalation, in contact with skin and if swallowed
48/24/25	Toxic: danger of serious damage to health by prolonged exposure in conta skin and if swallowed
48/25	Toxic: danger of serious damage to health by prolonged exposure if swallo

6. General abbreviations

The following abbreviations have been used, some being SI units

a	acre
ACS	American Chemical Society
ADI	acceptable daily intake
a.e.	acid equivalent (active ingredient expressed in terms of parent acid)
AG	Aktiengesellschaft (Company)
a.i.	active ingredient
ALC_{50}	approximate concentration required to kill 50% of test organisms
als	acetolactate synthase
ANPP	Association Nationale pour la Protection des Plantes
ANSI	American National Standards Institute
AOAC	Association of Official Analytical Chemists
AOAC Methods	Official Methods of Analysis of The Association of Official Analytical Che
BAN	British Approved Name (by British Pharmacopoeia Commission)
BBA	Biologische Bundesanstalt Abteilung
BCPC	British Crop Protection Council
BIOS	British Intelligence Objective Sub-Committee
b.p.	boiling point at stated pressure
BS	British Standard
BSI	British Standards Institution
B.V.	Beperkt Vennootschap (Limited)
b.w.	body weight
c.	circa (about)
C.A.	Chemical Abstracts
calc.	calculated

CAS RN	Chemical Abstracts Services Registry Number
CBI	carotenoid biosynthesis inhibitor
cf	compare
cfu	colony forming units
CHO	Chinese hamster ovary
9CI	9th collective index period (1972-1976) of *Chemical Abstracts*
CIPAC	Collaborative International Pesticides Analytical Council Limited
Co.	Company
COLUMA	Comite de Lutte Contre les Mauvaises Herbes
concn.	concentration
Corp.	Corporation
cSt	centiStokes
cwt	hundredweight
d	day(s)
decomp.	with decomposition
DMF	*N,N*-dimethylformamide
DMSO	dimethyl sulfoxide
dt	decitonne
DT_{50}	time for 50% loss; half-life
E_bC_{50}	median effective concentration (biomass, e.g. of algae)
EC_{50}	median effective concentration
E_rC_{50}	median effective concentration (growth rate, e.g. of algae)
ECD	electron-capture detector
ed.	editor
Ed.	edition
ED	electrochargeable liquid
e.g.	for example
EHC	Environmental Health Criteria (see the Guide to use of the Main Entries)
E-ISO	ISO name (English spelling)
EPA	Environmental Protection Agency (of USA)
EPPO	European and Mediterranean Plant Protection Organisation
ESA	Entomological Society of America
est.	estimated
et al.	and others (authors)
EU	European Union
EWRC	European Weed Research Council (pre-1975)
EWRS	European Weed Research Society (since 1975)
f	femto, multiplier (10^{-15}) for SI units
FAO	Food and Agricultural Organisation (of the United Nations)
FID	flame-ionisation detector
F-ISO	ISO name (French spelling)

f.p.	freezing point
FPD	flame photometric detector
FTP	flame thermionic detector
g	gram (hence also ng, μg, mg, kg, *etc.*)
gc	gas chromatography
gc-ms	combined gas chromatography-mass spectrometry
GCPF	Global Crop Protection Federation (formerly GIFAP)
GIFAP	Groupement International des Associations Nationales de Fabricants de Products Agrochimiques (now known as GCPF)
glc	gas-liquid chromatography
GUS	groundwater ubiquity score
GV	granulosis virus
h	hour(s)
ha	hectare(s) (10^4 m^2)
hl	hectolitre (100 l)
HGPRT	hypoxanthine-guanine phosphoribosyltransferase (enzyme involved in a cell culture test for mutagens)
HMSO	Her Majesty's Stationary Office (UK)
hplc	high performance liquid chromatography
IARC	International Agency for Research on Cancer
ibid.	in the journal last mentioned
idem	by the author(s) last mentioned
i.e.	that is
Inc.	Incorporated
INN	International Nonproprietary Name (by WHO)
INRA	Institut National de la Recherche Agronomique
IOBC	International Organisation for Biological Control
i.p.	intraperitoneal
IPM	integrated pest management
i.r.	infrared
ISO	International Organization for Standardization
i.u.	international unit (measure of activity of micro-organisms)
i.v.	intravenous
IUPAC	International Union of Pure and Applied Chemistry
J	Joules
JMAF	Japanese Ministry for Agriculture, Forestry and Fisheries (*formerly* Japanese Ministry for Agriculture and Forestry)
JMPR	Joint meeting of the FAO Panel of Experts on Pesticide Residues and the Environment and the WHO Expert Group on Pesticide Residues
j.v.	joint venture

k	kilo, multiplier (1000) for SI units
K_d	soil parameter; see **Soil\Environment** in the Guide to use of the Main Entries
K_{oc}	soil parameter; see **Soil\Environment** in the Guide to use of the Main Entries
K_{om}	soil parameter; see **Soil\Environment** in the Guide to use of the Main Entries
K_{ow}	distribution coefficient between *n*-octanol and water
kg	kilogram(s)
kPa	kilopascal (1000 Pa)
l	litre (hence also ml, etc.)
lb/a	pounds per acre
lc	liquid chromatography
LC_{50}	concentration required to kill 50% of test organisms
L:D	light:dark
LD_{50}	dose required to kill 50% of test organisms
LOEC	lowest observed effect concentration
Ltd	Limited
m	metre (hence also nm, mm, etc.)
m	milli, multiplier (10^{-3}) for SI units
M	mega, multiplier (10^6) for SI units
mo	(in NOEL) month(s)
MAFF	Ministry of Agriculture Fisheries and Food (England and Wales)
MATC	maximum acceptable toxicant concentration
MCD	microcoulometric detector
mg	milligram(s), (0.001 g)
MHC	moisture-holding capacity
MIBK	methyl isobutyl ketone
mm	millimetre(s), (0.001 m)
m/m	proportion by mass
mmHg	pressure equivalent to 1 mm of mercury (133.3 Pa)
m.p.	melting point
mPa	millipascal, (0.001 Pa)
m/V	mass per volume
MWC	maximum water content
n	nano, multiplier (10^{-9}) for SI units
N	Newtons (hence also mN, etc.)
ng	nanogram, (10^{-9} g)
nm	nanometre, (10^{-9} m)
mmr	nuclear magnetic resonance
NOAEL	no observed adverse effect level
NOEC	no observed effect concentration
NOEL	no observed effect level
nPa	nanopascal, (10^{-9} Pa)
NPD	nitrogen-phosphorus detector

NPV	nuclear polyhedrosis virus
NRDC	National Research and Development Corporation (former, of UK)
N.V.	Naamloze Vennootschap (Limited)
o.c.	organic carbon
o.m.	organic matter
OMS	Organisation Mondiale de la Sante (WHO)
p	pico, multiplier (10^{-12}) for SI units
P	same as K_{ow}
Pa	pascal (SI unit for pressure)
pa	provisionally approved (of ISO names, see Guide to use of the Main Entrie
pF	$-\log_{10}$ water pressure (in cm) in soil
pH	$-\log_{10}$ hydrogen ion concentration
pK_a	$-\log_{10}$ acid dissociation constant
PIB	polyhedral inclusion body
PIC	Prior Informed Consent (see the Guide to use of the Main Entries)
PLC	Public Limited Company
PMn	*The Pesticide Manual, nth edition*
post-em.	after emergence
ppb	parts per billion
ppi	pre-plant incorporated
ppm	parts per million
PRC	Peoples Republic of China
pre-em.	before emergence
PTDI	Provisional tolerable daily intake
q.v.	*quod vide* (which see)
r.h.	relative humidity
respy.	respectively
RINN	Recommended International Non-proprietary Name
Roth:	Rothamsted (re: pyrethroid nomenclature system)
rp-tlc	reversed phase thin layer chromatography
rplc	reversed phase chromatography
s	second(s)
S.A.	Société Anonyme (Company)
S-FPD	Sulfur-specific flame Photometric Detector
SI	International System of Units
sp.	species (singular)
S.p.A.	Societe par Actions (Company)
spp.	species (plural)

t	tonne, 1000 kg
T.C.	total carbon (= mineral C + organic C)
TCD	thermal conductivity detector
tech.	technical grade
TID	thermionic detector
tlc	thin-layer chromatography
TLm	median tolerance limit
UK	United Kingdom
UNEP	United Nations Environment Programme
USDA	United States Department of Agriculture
u.v.	ultraviolet
U.V.	ultraviolet (at the beginning of a sentence)
v.p.	vapour pressure
w	week(s)
W	watt(s)
WHO	World Health Organisation (of the United Nations) = OMS
WIPO	World Intellectual Property Organisation
WSSA	Weed Science Society of America
wt.	weight
y	year(s)
$[\alpha]_D^t$	specific rotation (degrees) for sodium D lines at temperature t °C
μ	micro, multiplier (10^{-6}) for SI units
>	greater than
$\geq$	greater than or equal to
<	less than
$\leq$	less than or equal to

BIBLIOGRAPHY

Titles are listed in alphabetical order. Sources referred to in the text are indicated here by emboldening of the abbreviated title used in the text, followed by the full title.

'Analytical Methods for Pesticides, Plant Growth Regulators and Food Additives', Series editor, G. Zweig. New York: Academic Press. Volumes **2–4** (1964)

'Analytical Methods for Pesticides and Plant Growth Regulators', Series editor, G. Zweig. New York: Academic Press. Volumes **5–16** (1967-1988).

AOAC Methods - 'Official Methods of Analysis of the Association of Official Analytical Chemists' (16th Ed.), P. Cunniff, ed. AOAC International, Arlington, Virginia (1995). ISBN 0-935584-54-4.

'Approved guide to the Classification and Labelling of substances and preparations dangerous for Supply' (2nd Ed.), Health and Safety Commission (UK) (1994). ISBN 0-7176-0860-3.

'Approved Supply List' (2nd Ed.), Health and Safety Commission (UK) (1994). ISBN 0-7176-0858-1. [See also 'Draft Proposals for Chemicals', below]

'Attractants and Pheromones of Noxious Insects', IOBC, West Palearctic Regional Section (1984/VII/1), A. K. Minks.

Catalogue of Pesticide Formulation Types and International Coding System, Technical monograph No. 2, February 1989, ref. MT02E, (and addendum dated 20 November 1994), GCPF, Brussels.

CIPAC Handbook – CIPAC Handbook E, 'Analysis of Technical and Formulated Pesticides', A. Martijn & W. Dobrat, eds., Collaborative International Pesticides Analytical Council Ltd (1992).
    CIPAC Handbook F, 'Physico-chemical Methods for Technical and Formulated Pesticides', W. Dobrat & A. Martijn, eds., CIPAC (1994).
    CIPAC Handbook G, 'Analysis of Technical and Formulated Pesticides', W. Dobrat & A. Martijn, eds., CIPAC (1995).
    and earlier titles in the series.

Comp. Anal. Profiles – 'Comprehensive Analytical Profiles of Important Pesticides', in the series *Modern Methods of Pesticide Analysis*, J. Sherma & T. Cairns, eds., CRC Press (1993). ISBN 0-8493-7992-X.

REFERENCE

'Dictionary of Organic Compounds' (5th Ed.), J. Buckingham, ed., Chapman & Hall (1982). ISBN 0-412-17000-0.

'Die Entwicklung neuer insektizider Phosphorsäure-Ester' (3rd Ed.), G. Schrader. Weinheim: Verlag Chemie (1963).

'Draft Proposals for Chemicals (Hazard Information and Packaging for Supply) (Amendment) Regulations 1996', issued by Health and Safety Commission (UK).

'Farm Chemicals Handbook', Meister (1977).

'Glyphosate, A Unique Global Herbicide', J. E. Franz et al., ACS Monograph 189 (1996). ISBN 0-8412-3458-2.

'Herbicides Inhibiting Branched-Chain Amino Acid Biosynthesis - Recent Developments', J. Stetter, ed., Vol. 10 in the series 'Chemistry of Plant Protection', Springer-Verlag (1994). ISBN 0-387-58181-2.

'The Imidazolinone Herbicides', D. L. Shaner & S. L. O'Connor, eds., CRC (1991). ISBN 0-8493-5763-2.

'Insect Pheromones', IOBC wprs Bulletin 16 (10) (1993), L. J. McVeigh et al., eds.

'Insect Pheromones and Other Behaviour-Modifying Chemicals', R. L. Ridgway, M. Inscoe & H. Arn, eds. BCPC Monograph No. 51 (1992).

Insecticides, Vol. 5 in 'Progress in Pesticide Biochemistry and Toxicology', D. H. Hutson & T. R. Roberts, eds., Wiley (1985).

'Inventory of pesticide toxicity evaluations, use recommendations etc. made by FAO, IARC, IPCS, IRPTC, JMPR, UNEP and WHO' (4th annotated Ed.), G. Ekström, ed., The Swedish National Chemicals Inspectorate, P.O. Box 1384, Solna, Sweden.

'The Japanese Agrochemical Industry' (DS 99) (Agrow Reports Series), S. Watkins; PJB Publications Ltd, Richmond, Surrey, UK.

Merck Index (12th Ed.), S. Budavari, ed., Merck & Co, Whitehorse Station, NJ, USA (1996).

'Modern Selective Fungicides', (2nd Ed.), H. Lyr, ed., Fischer Verlag (1995). ISBN 3-334-60455-1.

'Nomenclature of Organic Chemistry', Sections A, B, C, D, E, F and H. International Union of Pure and Applied Chemistry. Oxford: Pergamon (1979).

'Nomenclature of Organic Compounds', Recommendations 1993; (supplement to preceding item). International Union of Pure and Applied Chemistry. Blackwell Scientific (1993).

'The Pesticide Index' (3rd Ed.), I. G. Copping et al., eds., BCPC and The Royal Society of Chemistry (1995). ISBN 0-948404-88-4.

'Semiochemicals in Crop Protection' (DS 69) (Agrow Reports Series), A. Smith. PJB Publications Ltd, Richmond, Surrey, UK.

'Prior Informed Consent: A Guide to its Working', GIFAP document D/1990/2537/13, Dec 1990.

'Target Sites for Herbicide Action', R. C. Kirkwood, ed., in the series 'Topics in Applied Chemistry'. Plenum (1991). ISBN 0-306-43846-1.

'The UK Pesticide Guide', R. Whitehead, ed., CABI and BCPC (1997). ISBN 0-85199-205-6. And earlier volumes in the series.

'The WHO Recommended Classification of Pesticides by Hazard and Guidelines to Classification 1996-1997', IPCS, WHO/PCS/96.3 and earlier editions of this series. Available from WHO, Geneva.

REFERENCE

INDEX 1

Chemical Abstracts Service Registry Numbers

A Chemical Abstracts Service Registry Number (CAS RN) comprises between one and seven digits followed by a hyphen, two further digits, a further hyphen and finally a check digit. There is normally no chemical relationship between compounds with consecutive numbers.

This Index contains the Registry Numbers in numerical order (e.g. *85-00-7, 2764-72-9, 6385-62-2, 15263-52-2, 15263-53-3, etc.*).

All references are to Entry Numbers; unqualified numbers (e.g. 230) refer to Main Entries whilst references to superseded materials are preceded by an 'S' (e.g. S768).

Chemical Abstracts Service normally allocates one Registry Number to every chemical or all mixtures of the same components (irrespective of the ratio for these components). Occasionally a pesticide is first reported under a code number (without the chemical name) and later as the name (without the code number), and so receives more than one Registry Number. In due course, only one Registry Number for a compound is continued, but the 'discontinued' numbers are retained for searching the original abstracts on-line.

Individual Registry Numbers are used for related compounds (e.g. *[15263-53-3]* for cartap, *[22042-59-7]* for cartap hydrochloride (unstated ratio), *[15263-52-2]* cartap monohydrochloride; *[2764-72-9]* for diquat ion, *[85-00-7]* for diquat dibromide, *[6385-62-2]* for diquat dibromide monohydrate).

Whenever possible, stereoisomerism of pesticidal molecules is fully defined by their IUPAC names in the standards issued by the International Organization for Standardization and the British Standards Institution. In a few cases, however, the stereochemistry of some molecules is not fully defined by the names employed by the Chemical Abstracts Service. Thus, exceptionally, a given Registry Number has been applied to two (or more) common names where the difference is confined to the stereochemistry of the molecules concerned. Also, different Registry Numbers may be used for a known racemate and for the analogous compound with 'unstated stereochemistry' (though this is usually a racemate or a mixture of all possible stereoisomers).

CAS RN	Entry no.	CAS RN	Entry no.
133-07-3	364	309-00-2	S763
133-32-4	424	311-47-7	S879
133-90-4	119	314-40-9	82
134-31-6	407	314-42-1	S1113
136-25-4	S1017	315-18-4	S1170
136-63-0	S830	317-83-9	S979
136-78-7	S995	319-84-6	392
137-26-8	712	319-85-7	392
137-30-4	758	319-86-8	392
137-42-8	473	327-98-0	S1326
137-98-4	17	330-54-1	260
139-40-2	604	330-55-2	445
140-41-0	S1179	333-41-5	209
140-56-7	S1033	338-45-4	499
140-57-8	S779	370-50-3	S1056
140-93-2	S1240	371-86-8	S1173
141-03-7	S940	405-30-1	S1061
141-66-2	223	420-04-2	167
142-59-6	506	465-73-6	S1114
143-33-9	405	467-69-6	351
143-50-0	S856	468-44-0	380
144-21-8	486	470-90-6	124
144-41-2	S1181	485-31-4	S804
144-54-7	473	494-52-0	S775
145-73-3	271	494-97-3	S1198
148-79-8	701	495-73-8	S795
149-26-8	S995	500-28-7	S790
150-50-5	S1320	502-39-6	S1163
150-68-5	S1179	502-55-6	S1032
151-38-2	S1158	507-60-8	S1268
152-16-9	S1267	510-15-6	S869
156-62-7	S835	510-75-8	380
297-78-9	S1111	513-77-9	S788
297-97-2	S1309	517-16-8	S1028
297-99-4	573	520-45-6	S928
298-00-0	552	526-07-8	S1272
298-01-1	499	532-34-3	S829
298-02-2	570	533-74-4	198
298-03-3	S931	534-52-1	261
298-04-4	257	535-89-7	S902
299-45-6	S965	542-75-6	215
299-84-3	S1038	555-37-3	516
299-85-4	S1000	555-89-5	S925
299-86-5	S904	556-22-9	S1088
300-76-5	507	556-61-6	489
301-11-1	S1308	558-25-8	S1149
301-12-2	546	563-12-2	283

INDEXES

CAS RN	Entry no.	CAS RN	Entry no.
1610-18-0	596	2032-65-7	482
1646-88-4	S762	2074-50-2	550
1689-83-4	425	2078-42-4	676
1689-84-5	88	2079-00-7	78
1689-99-2	88	2104-64-5	273
1698-60-8	126	2104-96-3	S816
1702-17-6	153	2122-70-5	510
1715-40-8	S815	2122-77-2	S1049
1746-81-2	503	2136-79-0	140
1754-58-1	S936	2157-98-4	502
1771-07-9	S1155	2163-69-1	S916
1773-37-1	S840	2163-79-3	S1199
1776-83-6	S1258	2163-80-6	486
1824-09-5	S1142	2164-08-1	443
1836-75-5	S1193	2164-09-2	S850
1836-77-7	S868	2164-17-2	342
1861-32-1	140	2212-67-1	501
1861-40-1	57	2227-13-6	S1300
1861-44-5	S1335	2227-17-0	225
1891-95-8	S882	2274-67-1	246
1897-45-6	133	2274-74-0	S864
1910-42-5	550	2275-14-1	S1210
1912-24-9	34	2275-18-5	S1239
1912-25-0	S1107	2275-23-2	748
1912-26-1	737	2282-34-0	S820
1918-00-9	210	2300-66-5	210
1918-02-1	578	2302-17-2	33
1918-11-2	S1294	2303-16-4	S935
1918-13-4	141	2303-17-5	722
1918-16-7	598	2307-49-5	S1321
1918-18-9	S1286	2307-68-8	559
1928-43-4	192	2310-17-0	571
1929-73-3	192	2312-35-8	603
1929-77-7	749	2314-09-2	351
1929-82-4	523	2385-85-5	S1174
1929-88-0	S803	2401-85-6	S870
1966-58-1	S947	2425-05-0	S1096
1967-16-4	S855	2425-06-1	104
1970-40-7	S1245	2425-10-7	756
1982-47-4	S880	2425-66-3	S874
1982-49-6	649	2439-00-1	S859
1982-55-4	S1323	2439-01-2	116
1982-69-0	210	2439-10-3	265
2008-39-1	192	2439-99-8	S1089
2008-41-5	100	2445-07-0	S1337
2027-47-6	532	2463-84-5	S942
2032-59-9	S771	2464-37-1	S867

INDEXES

CAS RN	Entry no.	CAS RN	Entry no.
4104-14-7	S1216	5902-95-4	486
4137-12-6	S1195	5903-13-9	S1175
4147-51-7	S993	5915-41-3	692
4151-50-2	666	6012-83-5	S807
4212-93-5	S1295	6012-92-6	S875
4234-79-1	S1128	6012-97-1	S1297
4301-50-2	S1057	6013-05-4	S911
4312-97-4	S1276	6059-33-1	S1215
4412-09-3	S1182	6062-26-6	457
4466-14-2	622	6073-72-9	S988
4482-55-7	318	6108-10-7	392
4489-31-0	S1202	6164-98-3	S857
4602-84-0	298	6365-83-9	251
4636-83-3	S1180	6385-62-2	256
4658-28-0	S783	6386-63-6	156
4685-14-7	550	6392-46-7	S765
4726-14-1	S1191	6465-51-6	S980
4727-29-1	568	6465-60-7	S980
4824-78-6	S817	6597-78-0	S941
4849-32-5	S1127	6616-80-4	S1006
5103-71-9	121	6682-77-5	S1334
5103-74-2	121	6734-80-1	473
5131-24-8	S996	6753-24-8	199
5221-49-8	S1247	6843-97-6	S1001
5221-53-4	242	6923-22-4	502
5234-68-4	111	6980-18-3	438
5251-93-4	S797	6988-21-2	S990
5259-88-1	545	7003-89-6	129
5281-13-0	S1222	7055-03-0	S1136
5328-04-1	S895	7076-49-5	S1083
5386-57-2	S982	7076-53-1	S958
5386-77-6	S987	7076-63-3	S909
5543-58-8	753	7085-19-0	459
5598-13-0	138	7122-04-5	S974
5598-52-7	S1078	7159-34-4	S1249
5707-69-7	S1003	7159-99-1	S1206
5707-73-3	S1146	7212-44-4	298
5742-19-8	192	7219-78-5	S1135
5782-83-3	S1021	7257-41-2	S983
5825-79-6	S881	7286-69-3	S1269
5826-91-5	S966	7287-19-6	597
5827-05-4	S1109	7287-36-7	S1177
5834-94-6	S784	7292-16-2	601
5834-96-8	S786	7345-69-9	386
5836-10-2	S878	7487-94-7	467
5836-29-3	162	7546-30-7	469
5902-51-2	689	7547-66-2	216

CAS RN	Entry no.	CAS RN	Entry no.
13952-84-6	99	17702-57-7	S1076
14088-71-2	S1228	17756-81-9	S845
14214-32-5	S969	17781-16-7	109
14255-88-0	S1036	17804-35-2	60
14419-01-3	S1201	18181-70-9	S1126
14437-17-3	S862	18181-80-1	87
14484-64-1	320	18221-59-5	460
14491-59-9	S901	18250-63-0	223
14698-29-4	544	18357-78-3	S1152
14798-36-8	118	18467-77-1	235
14816-16-1	S1220	18530-56-8	S1199
14816-18-3	575	18691-97-9	477
14816-20-7	S883	18708-86-6	124
15096-52-3	164	18708-87-7	124
15165-67-0	217	18809-57-9	S1009
15263-52-2	113	18854-01-8	437
15263-53-3	113	19000-52-3	S980
15299-99-7	513	19000-58-9	S980
15302-91-7	465	19044-88-3	536
15310-01-7	S793	19360-02-2	S766
15457-05-3	S1063	19396-06-6	586
15521-65-0	517	19408-46-9	438
15545-48-9	134	19480-40-1	199
15589-31-8	S1293	19493-94-8	S760
15652-38-7	S926	19622-08-3	S1238
15662-33-6	S1264	19622-19-6	S1238
15800-60-9	S1264	19666-30-9	539
15845-66-6	372	19670-15-6	270
15879-93-3	118	19691-80-6	S781
15972-60-8	14	19750-95-9	S857
16118-49-3	108	19764-43-3	S1157
16202-91-8	S760	19937-59-8	496
16227-10-4	S1319	20354-26-1	S1150
16376-36-6	118	20425-39-2	S1244
16484-77-8	460	20427-59-2	158
16606-64-7	S774	20711-10-8	696
16672-87-0	281	20782-58-5	S858
16752-77-5	483	20856-57-9	S848
16807-48-0	S928	20859-73-8	574
16893-85-9	656	21087-64-9	497
16960-39-7	S1007	21267-72-1	S1241
17029-22-0	S1100	21452-18-6	S1169
17040-19-6	S932	21540-35-2	S1153
17080-02-3	S1083	21548-32-3	S1079
17109-49-8	267	21609-90-5	S1130
17439-94-0	271	21634-96-8	S946
17606-31-4	64	21689-84-9	S910

CAS RN	Entry no.	CAS RN	Entry no.
28249-77-6	706	33245-39-5	331
28346-65-8	S1117	33271-65-7	S908
28434-00-6	73	33399-00-7	S811
28434-01-7	74	33439-45-1	S1066
28559-00-4	S918	33629-47-9	97
28562-70-1	S1080	33693-04-8	691
28631-35-8	216	33813-20-6	S1021
28730-17-8	S1151	33820-53-0	S1120
28772-56-7	83	33956-49-9	157
28805-78-9	S1117	34014-18-1	683
28874-46-6	271	34123-59-6	433
28956-64-1	S796	34205-21-5	236
29091-05-2	248	34256-82-1	6
29091-21-2	593	34264-24-9	S1232
29104-30-1	67	34407-87-9	S971
29173-31-7	S1138	34462-96-9	S1091
29232-93-7	585	34494-03-6	383
29672-19-3	S1192	34622-58-7	534
29973-13-5	282	34643-46-4	614
30043-49-3	S1022	34681-10-2	95
30043-55-1	S827	34681-23-7	96
30087-47-9	S1039	34766-29-5	S1298
30182-24-2	S1108	34849-42-8	S914
30560-19-1	4	35256-85-0	682
30864-28-9	478	35256-86-1	S800
30979-48-7	S1112	35258-87-8	S1072
31182-61-3	74	35260-91-4	S1073
31218-83-4	605	35367-38-5	231
31251-03-3	S1067	35400-43-2	673
31377-69-2	S1223	35409-97-3	S1208
31530-30-0	608	35554-44-0	410
31717-87-0	264	35575-96-3	38
31848-11-0	S838	35597-43-4	71
31895-21-3	707	35597-44-5	382
31895-22-4	707	35764-59-1	74
32255-90-6	S1103	35842-17-2	264
32407-99-1	S1214	35857-62-6	504
32534-96-6	S980	36001-88-4	S772
32535-08-3	S980	36145-08-1	S884
32791-87-0	156	36335-67-8	94
32809-16-8	592	36510-94-8	S1298
32861-85-1	117	36519-00-3	S1217
32889-48-8	S1229	36614-38-7	S1122
33089-61-1	22	36734-19-7	428
33189-72-9	696	36756-79-3	713
33213-65-9	270	37248-47-8	747
33213-66-0	270	37273-85-1	86

INDEXES

CAS RN	Entry no.	CAS RN	Entry no.
51877-74-8	S805	56194-68-4	S1244
51971-67-6	3	56425-91-3	355
52207-99-5	384	56716-21-3	S1104
52315-07-8	183	56717-11-4	S1221
52315-07-8	184	57018-04-9	714
52315-07-8	187	57052-04-7	S1116
52341-32-9	S1314	57130-91-3	S1252
52508-35-7	235	57213-69-1	S1287
52570-16-8	512	57342-02-6	S1015
52645-53-1	561	57353-42-1	324
52756-22-6	S1052	57369-32-1	633
52756-22-6	S1053	57375-63-0	S1209
52756-25-9	S1054	57608-04-5	561
52820-00-5	204	57646-30-7	375
52888-80-9	612	57754-85-5	153
52918-63-5	204	57801-46-4	S1148
53042-79-8	384	57837-19-1	470
53112-28-0	628	57960-19-7	12
53369-07-6	382	57966-95-7	182
53558-25-1	S1248	57973-67-8	324
53780-34-0	463	58011-68-0	618
53878-17-4	S1085	58041-19-3	S970
53988-93-5	451	58138-08-2	S1327
54364-63-5	263	58270-08-9	S1192
54381-26-9	S770	58580-13-5	S811
54406-48-3	268	58580-14-6	S811
54460-46-7	S1098	58594-74-4	S1330
54593-83-8	122	58594-77-7	S1330
54864-61-8	S1183	58667-63-3	S1052
55179-31-2	77	58769-20-3	646
55219-65-3	721	58810-48-3	531
55283-68-6	279	58842-20-9	522
55283-69-7	S1121	58858-18-7	126
55285-14-8	110	59011-30-2	S1330
55290-64-7	241	59011-33-5	S1330
55335-06-3	733	59026-08-3	S1016
55511-98-3	S825	59404-06-7	S862
55512-33-9	626	59604-10-3	S862
55634-91-8	18	59604-11-4	S862
55774-32-8	263	59621-49-7	S893
55807-46-0	S1178	59669-26-0	708
55808-13-4	S844	59682-52-9	371
55814-41-0	466	59756-60-4	352
55861-78-4	434	60074-25-1	S851
56073-07-5	227	60168-88-9	300
56073-10-0	81	60207-31-0	35
56141-00-5	S836	60207-90-1	607

INDEXES

CAS RN	Entry no.	CAS RN	Entry no.
68523-18-2	S1045	72619-32-0	390
68672-17-3	S794	72850-64-7	350
68694-11-1	738	72963-72-5	420
68721-60-8	S819	73250-68-7	461
68925-41-7	S1312	73519-50-3	S1050
69148-12-5	S891	73519-55-8	S1044
69309-47-3	S1221	73790-28-0	410
69327-76-0	92	73886-28-9	S1095
69335-91-7	327	74051-80-2	648
69377-81-7	354	74070-46-5	8
69409-94-5	362	74115-24-5	149
69409-94-5	S1069	74222-97-2	667
69581-33-5	S922	74223-56-6	667
69770-45-2	337	74223-64-6	498
69806-34-4	390	74267-69-9	S851
69806-40-2	390	74310-70-6	S851
69806-50-4	327	74712-19-9	85
69865-47-0	183	74738-17-3	311
70124-77-5	333	74782-23-3	537
70166-48-2	S1250	75021-72-6	219
70193-21-4	S1322	75202-10-7	520
70585-36-3	77	75736-33-3	S960
70585-38-5	77	75867-00-4	S1040
70630-17-0	471	76120-02-0	S1031
70751-94-9	S1313	76280-91-6	684
71048-99-2	71	76578-12-6	640
71101-05-8	S968	76608-88-3	S1316
71245-23-3	S1020	76636-10-7	S768
71283-65-3	219	76674-21-0	361
71283-80-2	309	76714-83-5	746
71422-67-8	125	76714-88-0	247
71441-80-0	648	76738-62-0	548
71561-11-0	621	77107-49-4	S851
71611-31-9	333	77182-82-2	382
71626-11-4	54	77227-69-1	S1092
71697-59-1	186	77458-01-6	616
71751-41-2	1	77501-60-1	344
72131-76-1	S1331	77501-63-4	442
72178-02-0	365	77501-90-7	344
72204-43-4	185	77732-09-3	540
72269-48-8	735	77768-58-2	S1275
72280-52-5	S851	77788-21-7	S1296
72391-46-9	142	78168-93-1	S767
72459-58-6	727	78357-48-9	S1035
72490-01-8	310	78370-21-5	S917
72492-54-7	S851	78587-05-0	401
72492-55-8	S851	79127-80-3	310

INDEXES

CAS RN	Entry no.	CAS RN	Entry no.
94051-08-8	641	103112-36-3	S1037
94125-34-5	613	103361-09-7	341
94361-06-5	189	103833-18-7	9
94361-07-6	189	103970-75-8	S1255
94593-91-6	145	104030-54-8	440
94596-45-9	460	104040-78-0	325
95266-40-3	744	104098-48-8	2
95485-99-9	101	104098-49-9	2
95480-33-4	S894	104273-73-6	744
95617-09-7	S1044	104459-82-7	13
95721-12-3	S1050	104653-34-1	230
95737-68-1	631	105024-66-6	650
96182-53-5	681	105084-66-0	S1260
96489-71-3	624	105512-06-9	147
96491-05-3	700	105779-78-0	629
96525-23-4	356	106040-48-6	728
97454-00-7	S999	106325-08-0	48
97780-06-8	280	106917-52-6	358
97886-45-8	259	107534-96-3	678
98389-04-9	620	107713-58-6	336
98477-07-7	S1008	108173-90-6	385
98565-18-5	S1132	109293-97-2	50
98730-04-2	61	109293-98-3	50
98886-44-3	373	110235-47-7	464
98967-40-9	339	110488-70-5	244
99105-77-8	664	110956-75-7	560
99129-21-2	146	111353-84-5	280
99257-43-9	419	111479-05-1	602
99283-00-8	127	111578-32-6	491
99283-01-9	62	111812-58-9	315
99387-89-0	738	111872-58-3	387
99434-58-9	174	111991-09-4	519
99607-70-2	154	112143-82-5	724
99662-11-0	S1190	112226-61-6	388
100646-51-3	641	112281-77-3	695
100728-84-5	411	112410-23-8	679
100760-10-9	640	112636-83-6	224
100784-20-1	389	112839-32-4	S1082
101007-06-1	9	112839-33-5	S1081
101012-85-5	460	112860-04-5	409
101200-48-0	728	113036-87-6	589
101205-02-1	174	113036-88-7	332
101463-69-8	335	113096-99-4	189
101903-30-4	554	113136-77-9	170
102851-06-9	362	113158-40-0	309
103055-07-8	446	114311-32-9	412
103112-35-2	S1037	114369-43-6	302

INDEXES

INDEX 2

Molecular Formulae

These entries include both Main and Superseded Entries. Ions (e.g. diquat, $C_{12}H_{12}N_2$) and salts (e.g. diquat dibromide, $C_{12}H_{12}Br_2N_2$) are listed as such, so as to agree with the formulae in the text.

All references are to Entry Numbers: unqualified numbers (e.g. 230) refer to Main Entries, whilst references to superseded materials are preceded by an 'S' (e.g. S768).

Mol. Formula	Entry no.
AlF_6Na_3	164
AlP	574
AsF_6K	S1100
As_2O_3	S780
$B_4H_{20}Na_2O_{17}$	79
BaS_x	S789
$CBaO_3$	S788
$CCaN_2$	S835
CCl_3NO_2	132
CCl_4	S842
CH_2N_2	167
CH_2O	368
$CH_3AsNa_2O_3$	486
CH_3AsS	S1162
CH_3Br	487
CH_3FO_2S	S1149
CH_4AsNaO_3	486
CH_5AsO_3	486
CHN	405
$CKNO$	S1226
CNa_2S_4	386
$CNNa$	405
CS_2	S841
C_2CaN_2	405
$C_2Cl_3NaO_2$	677
$C_2HCl_3O_2$	677
$C_2H_2ClNaO_2$	130
$C_2H_2FNaO_2$	655
$C_2H_3ClO_2$	130
$C_2H_3Cl_2NO_2$	S950
C_2H_3NS	489
C_2H_4BrCl	S813
$C_2H_4Br_2$	289
$C_2H_4Cl_2$	290
C_2H_4FNO	343
$C_2H_4NNaS_2$	473
$C_2H_4N_4$	23
$(C_2H_4O)_x$	472
$C_2H_5ClO_2S$	S1102
$C_2H_5NS_2$	473
$C_2H_6AsNaO_2$	245
$C_2H_6ClO_3P$	281
$C_2H_6Hg_2O_4S$	S808
$C_2H_7AsO_2$	245
$C_2H_7O_3P$	372
$C_2H_8NO_2PS$	479
$C_2H_8N_2O$	404
C_3Cl_6O	S1097
$C_3HF_4NaO_2$	347
$C_3H_2ClNaO_2$	S1276

Mol. Formula	Entry no.
$C_3H_2Cl_2F_4O_2$	S956
$C_3H_2F_4O_2$	347
$C_3H_3Cl_2NaO_2$	196
$C_3H_3Cl_3O_2$	S877
C_3H_3N	S761
C_3H_4BrCl	S814
$C_3H_4Cl_2$	215
$C_3H_4Cl_2O_2$	196
C_3H_4O	10
$C_3H_5Br_2Cl$	S937
$C_3H_6BrNO_4$	90
$C_3H_6ClNO_2$	S874
$C_3H_6Cl_2$	S954
$C_3H_6Cl_2Cu_3NS_2$	S909
$C_3H_6F_6O_4$	S1099
$C_3H_6HgN_4$	S1163
$C_3H_6N_2O$	S1105
C_3H_7ClHgO	S1159
$C_3H_7NNaO_5P$	383
$C_3H_8NO_4P$	371
$C_3H_8NO_5P$	383
$C_3H_{10}NO_3P$	S774
$C_3H_{10}N_6S_2$	S809
$C_3H_{11}N_2O_4P$	371
$C_3H_{11}N_2O_5P$	383
C_4Cl_4S	S1297
$C_4H_3KN_2O_2$	449
$C_4H_4N_2O_2$	449
$C_4H_4N_2S_3$	S1021
$C_4H_5NO_2$	408
$C_4H_6Cl_2O_2S$	S957
$C_4H_6MnN_2S_4$	453
$C_4H_6N_2Na_2S_4$	506
$C_4H_6N_2S_3$	S1021
$C_4H_6N_2S_4Zn$	757
$C_4H_6O_2S_4$	S977
$C_4H_7Br_2Cl_2O_4P$	507
$C_4H_7Cl_2O_4P$	218
$C_4H_7NaOS_2$	S1240
$C_4H_8Cl_2O$	S806
$C_4H_8Cl_3O_4P$	730
$C_4H_8N_2O_2S$	S1007
$C_4H_8OS_2$	S1240
$C_4H_{10}NO_3PS$	4
$C_4H_{11}N$	99
$C_4H_{11}NO_8P_2$	S1089
$C_4H_{12}FN_2OP$	S972
$C_4H_{12}N_5OP$	S1135
$C_5HCl_2F_2NO$	S1093
$C_5HCl_5O_3$	S766

INDEXES

Index 2 - Molecular Formulae

INDEXES

Mol. Formula	Entry no.
$C_{13}H_{15}N_3O_2$	S1164
$C_{13}H_{15}N_3O_3$	413
$C_{13}H_{16}Cl_3NO_3$	S1322
$C_{13}H_{16}Cl_3NO_4$	733
$C_{13}H_{16}F_3N_3O_4$	57
$C_{13}H_{16}F_3N_3O_4$	740
$C_{13}H_{16}NO_4PS$	437
$C_{13}H_{16}N_2O_6$	S1139
$C_{13}H_{16}N_2O_7$	S988
$C_{13}H_{16}N_4O_3S$	S1137
$C_{13}H_{16}N_{10}O_5S$	39
$C_{13}H_{16}O_2$	S1012
$C_{13}H_{16}O_5$	744
$C_{13}H_{17}ClN_2O_4$	S885
$C_{13}H_{17}ClO_3$	457
$C_{13}H_{17}F_3N_4O_4$	593
$C_{13}H_{17}NO_4$	S974
$C_{13}H_{18}ClNO$	559
$C_{13}H_{18}ClNO$	S1177
$C_{13}H_{18}ClNO$	S1340
$C_{13}H_{18}ClNO_2$	238
$C_{13}H_{18}ClO_2PS_3$	S997
$C_{13}H_{18}ClO_3PS_2$	S1303
$C_{13}H_{18}N_2O_2$	443
$C_{13}H_{18}N_2O_2$	S1104
$C_{13}H_{18}O_5S$	285
$C_{13}H_{19}ClNO_3PS_2$	28
$C_{13}H_{19}ClNO_3PS_2$	S1077
$C_{13}H_{19}ClN_2$	S871
$C_{13}H_{19}Cl_3N_2O_3$	733
$C_{13}H_{19}NO$	S1095
$C_{13}H_{19}NO_2$	S820
$C_{13}H_{19}NO_2$	S914
$C_{13}H_{19}NO_2S$	308
$C_{13}H_{19}NO_2S$	S1013
$C_{13}H_{19}NO_2S$	S1152
$C_{13}H_{19}N_3O_4$	557
$C_{13}H_{19}N_3O_6S$	S1191
$C_{13}H_{21}N_2O_4PS$	94
$C_{13}H_{21}O_3PS$	427
$C_{13}H_{21}O_4PS$	601
$C_{13}H_{22}NO_3PS$	299
$C_{13}H_{22}N_2O$	S1117
$C_{13}H_{22}N_2O$	S1199
$C_{13}H_{22}N_3O_4PS$	S1227
$C_{13}H_{22}N_4O_3S$	724
$C_{13}H_{23}N_2O_3PS$	681
$C_{13}H_{24}N_3O_3PS$	584
$C_{13}H_{24}N_4O_3S$	91
$C_{13}H_{29}N_3$	265

Mol. Formula	Entry no.
$C_{13}H_{32}NO_4P$	S1118
$C_{14}H_4N_2O_2S_2$	258
$C_{14}H_5Cl_6NO_3$	684
$C_{14}H_6ClF_3NNaO_5$	7
$C_{14}H_6ClF_6N_3O_4$	S1051
$C_{14}H_6Cl_2F_4N_2O_2$	686
$C_{14}H_7Br_3F_3N_3O_4$	84
$C_{14}H_7ClF_2N_4$	674
$C_{14}H_7ClF_3NO_5$	7
$C_{14}H_8Cl_2N_4$	149
$C_{14}H_8Cl_2O_3$	S946
$C_{14}H_8O_2$	29
$C_{14}H_9ClF_2N_2O_2$	231
$C_{14}H_9ClO_2$	S866
$C_{14}H_9ClO_3$	S867
$C_{14}H_9Cl_2NO_5$	69
$C_{14}H_9Cl_3N_2O_2$	S1208
$C_{14}H_9Cl_5$	201
$C_{14}H_9Cl_5O$	222
$C_{14}H_9NO_2$	S1213
$C_{14}H_{10}Cl_4$	S1290
$C_{14}H_{10}F_4N_4O_7S$	589
$C_{14}H_{10}O_3$	351
$C_{14}H_{11}ClFN_5O_5S$	155
$C_{14}H_{11}ClFNO_4$	147
$C_{14}H_{11}Cl_2NO_4$	S851
$C_{14}H_{11}Cl_4O_4P$	S1217
$C_{14}H_{11}NO_3$	568
$C_{14}H_{12}Cl_2N_2O$	627
$C_{14}H_{12}Cl_2O$	S861
$C_{14}H_{12}F_3NO_4S_2$	S1207
$C_{14}H_{12}F_3N_5O_7S$	348
$C_{14}H_{12}N_4O_2S$	S1282
$C_{14}H_{13}ClN_6O_5S$	416
$C_{14}H_{13}ClO_5S$	664
$C_{14}H_{13}Cl_2N_2O_2PS$	S1216
$C_{14}H_{13}Cl_2N_5O_4S$	495
$C_{14}H_{13}Cl_2O_2PS$	S1014
$C_{14}H_{13}F_4N_3O_2S$	51
$C_{14}H_{13}NO$	S1136
$C_{14}H_{13}N_3$	464
$C_{14}H_{13}N_3O_3S$	S1085
$C_{14}H_{14}ClNO_3$	S922
$C_{14}H_{14}ClN_2O_3PS$	S786
$C_{14}H_{14}Cl_2N_2O$	410
$C_{14}H_{14}Cl_2N_2O_3$	S1123
$C_{14}H_{14}NO_4PS$	273
$C_{14}H_{14}N_4O_5S$	667
$C_{14}H_{14}O_2$	510
$C_{14}H_{14}O_3$	579

INDEXES

Mol. Formula	Entry no.
$C_{15}H_{15}Cl_2N_3O_5$	518
$C_{15}H_{15}F_3N_2O_2$	355
$C_{15}H_{16}ClF_2N_3O$	233
$C_{15}H_{16}Cl_3N_3O_2$	591
$C_{15}H_{16}F_3N_5O_4S$	613
$C_{15}H_{16}F_5NO_2S_2$	259
$C_{15}H_{16}N_2O_2$	27
$C_{15}H_{16}N_4O_5S$	667
$C_{15}H_{16}N_4O_7S$	62
$C_{15}H_{17}Br_2NO_2$	88
$C_{15}H_{17}ClN_2O_2$	S890
$C_{15}H_{17}ClN_4$	505
$C_{15}H_{17}Cl_2N_3O$	247
$C_{15}H_{17}Cl_2N_3O_2$	607
$C_{15}H_{17}HgNO_2S$	S1028
$C_{15}H_{17}I_2NO_2$	425
$C_{15}H_{17}NO_2$	S1157
$C_{15}H_{17}NO_3S$	S1336
$C_{15}H_{17}N_5O_6S$	728
$C_{15}H_{17}O_2PS$	S1018
$C_{15}H_{18}ClNO_4$	S846
$C_{15}H_{18}ClN_3O$	189
$C_{15}H_{18}ClN_3O$	746
$C_{15}H_{18}ClN_3O$	S1279
$C_{15}H_{18}Cl_2N_2O_3$	539
$C_{15}H_{18}Cl_3NO$	440
$C_{15}H_{18}N_2O_3$	411
$C_{15}H_{18}N_2O_6$	S804
$C_{15}H_{18}N_4$	321
$C_{15}H_{18}N_4O_7S$	287
$C_{15}H_{18}N_6O_6S$	280
$C_{15}H_{18}N_6O_6S$	519
$C_{15}H_{19}ClN_4O_3$	236
$C_{15}H_{19}Cl_2N_3O$	S960
$C_{15}H_{19}Cl_3O_4$	S1042
$C_{15}H_{19}N_3O_3$	415
$C_{15}H_{19}N_3O_4$	412
$C_{15}H_{19}N_5O_7S$	145
$C_{15}H_{20}ClN_3O$	548
$C_{15}H_{20}Cl_2O_4$	216
$C_{15}H_{20}N_2O_7$	S982
$C_{15}H_{21}ClN_2O_2$	S1176
$C_{15}H_{21}ClO_4$	455
$C_{15}H_{21}Cl_2FN_2O_3$	354
$C_{15}H_{21}NO_4$	470
$C_{15}H_{21}NO_4$	471
$C_{15}H_{21}NOS$	237
$C_{15}H_{22}BrNO$	85
$C_{15}H_{22}ClNO_2$	493
$C_{15}H_{22}ClNO_2$	609

Mol. Formula	Entry no.
$C_{15}H_{22}ClNO_2$	S929
$C_{15}H_{22}O_3$	S1075
$C_{15}H_{22}O_3$	S1234
$C_{15}H_{22}O_6$	S1271
$C_{15}H_{23}ClO_4S$	S779
$C_{15}H_{23}NO$	682
$C_{15}H_{23}NO_4$	S915
$C_{15}H_{23}NOS$	278
$C_{15}H_{23}N_3O_4$	S1120
$C_{15}H_{24}NO_4PS$	430
$C_{15}H_{25}N_3O$	S1316
$C_{15}H_{26}O$	298
$C_{15}H_{27}NO_2S$	S1308
$C_{15}H_{28}O_2$	735
$C_{15}H_{33}N_3O_2$	265
$C_{16}H_8Cl_2FN_5O$	349
$C_{16}H_8Cl_2F_6N_2O_3$	399
$C_{16}H_9ClF_3NO_7$	344
$C_{16}H_9Cl_2N_5O$	S1255
$C_{16}H_9Cl_4N_3O_2$	S1186
$C_{16}H_{11}ClF_4N_2O_6S$	S1092
$C_{16}H_{12}ClF_4N_3O_4$	338
$C_{16}H_{12}ClNO_4S$	S1050
$C_{16}H_{12}ClNO_5$	309
$C_{16}H_{12}ClNO_5$	S1044
$C_{16}H_{13}ClFNO_3$	324
$C_{16}H_{13}ClFNO_3$	S1052
$C_{16}H_{13}ClFNO_5$	340
$C_{16}H_{13}ClF_3NO_4$	390
$C_{16}H_{13}Cl_2NO_3$	S802
$C_{16}H_{13}F_2N_3O$	361
$C_{16}H_{13}F_3O_4$	S1330
$C_{16}H_{14}Cl_2N_2O_2$	S1212
$C_{16}H_{14}Cl_2O$	S1228
$C_{16}H_{14}Cl_2O_3$	S869
$C_{16}H_{14}Cl_2O_4$	219
$C_{16}H_{14}N_2O_2S$	461
$C_{16}H_{14}OS$	S970
$C_{16}H_{15}Cl_2NO_2$	151
$C_{16}H_{15}Cl_2NO_2$	S905
$C_{16}H_{15}Cl_2NO_3$	291
$C_{16}H_{15}Cl_3O_2$	485
$C_{16}H_{15}Cl_4O_5P$	S933
$C_{16}H_{15}FO_2$	S1057
$C_{16}H_{15}F_2N_3Si$	357
$C_{16}H_{16}Cl_2NO_2$	580
$C_{16}H_{16}Cl_2O_2$	S807
$C_{16}H_{16}NO_5PS$	S791
$C_{16}H_{16}N_2O_3$	658
$C_{16}H_{16}N_2O_4$	206

INDEXES

Index 2 - Molecular Formulae

Mol. Formula	Entry no.
$C_{24}H_{25}NO_3$	188
$C_{24}H_{27}N_3O_4$	315
$C_{24}H_{32}O_4$	12
$C_{24}H_{33}NO_4$	98
$C_{24}H_{37}N_3O_5$	S979
$C_{24}H_{40}O_8$	S1222
$C_{24}H_{53}N_7O_6$	419
$C_{14}H_{34}OSn_2$	S810
$C_{25}H_{20}ClF_2N_3O_3$	332
$C_{25}H_{22}ClNO_3$	277
$C_{25}H_{22}ClNO_3$	319
$C_{25}H_{24}F_6N_4$	403
$C_{25}H_{28}O_3$	292
$C_{25}H_{29}FO_2Si$	650
$C_{25}H_{35}NO_9$	S1264
$C_{26}H_{18}CuO_6$	S895
$C_{26}H_{21}Cl_2NO_4$	172
$C_{26}H_{21}F_6NO_5$	9
$C_{26}H_{22}BrF_2NO_4$	759
$C_{26}H_{22}ClF_3N_2O_3$	362
$C_{26}H_{22}ClF_3N_2O_3$	S1069
$C_{26}H_{23}Cl_4NO_6$	S836
$C_{26}H_{23}F_2NO_4$	333
$C_{26}H_{36}Cl_2N_4O_4$	S1180
$C_{26}H_{36}N_4O_4$	S1180
$C_{27}H_{44}O$	752
$C_{28}H_{22}Cl_2FNO_3$	337
$C_{28}H_{44}O$	276
$C_{30}H_{23}BrO_4$	83
$C_{31}H_{23}BrO_2S$	230
$C_{31}H_{23}BrO_3$	81
$C_{31}H_{24}O_3$	227
$C_{31}H_{44}O_7$	500
$C_{32}H_{44}O_{12}$	S1268
$C_{32}H_{46}O_7$	500

Mol. Formula	Entry no.
$C_{33}H_{25}F_3O_4$	326
$C_{33}H_{25}N_3O_3$	S1197
$C_{33}H_{47}NO_{13}$	515
$C_{35}H_{44}O_{16}$	36
$C_{41}H_{65}NO_{10}$	754
$C_{42}H_{67}NO_{10}$	754
$C_{42}H_{84}N_{14}O_{36}S_3$	661
$C_{46}H_{51}BrClPSn$	S926
$C_{47}H_{70}O_{14}$	1
$C_{48}H_{72}O_{14}$	1
$C_{60}H_{78}OSn_2$	303
$C_{72}H_{131}N_7O_9S_3$	419
$C_{84}H_{111}FeN_{12}$	S792
CaS_x	103
$ClCu_2H_3O_3$	159
$ClNaO_3$	653
Cl_2Hg	467
Cl_2Hg_2	469
CuH_2O_2	158
$CuH_{10}N_4O_8S_2$	S908
$CuH_{10}O_9S$	160
Cu_2O	166
FNa	654
F_2O_2S	672
F_6Na_2Si	656
FeO_4S	322
H_2O_4S	671
H_3P	574
$H_6N_2O_3S$	24
HgO	468
Mg_3P_2	574
Na_2O_4Se	S1277
P_2Zn_3	574
S_x	670

INDEXES

INDEX 3

Code Numbers

This index includes official and development codes. A description of these code types is given in the Guide to using the Main Entries page x.

The index is arranged in strictly alphabetical order; the numerical components of a code are not in order of magnitude.

Hyphens and spaces, which tend not to be consistently displayed in development codes, are ignored for sorting purposes, and are not printed in this index.

Thus AC 189635 (printed as AC189635) precedes AC 3911 (printed as AC3911).

All references are to Entry Numbers; unqualified numbers (e.g. 230) refer to Main Entries, whilst references to superseded materials are preceded by an 'S' (e.g. S768).

Code	Entry no.	Code	Entry no.
Bayer10756	S931	BayerL13/59	730
Bayer15080	S795	BayerSAS2074	116
Bayer15922	730	BAYFCR1272	176
Bayer16259	40	BAYFOE5043	51
Bayer17147	41	BAYKUE2079A	S1066
Bayer18436	205	BAYKWG0519	721
Bayer19149	218	BAYKWG0599	77
Bayer19639	257	BAYMAT7484	681
Bayer20315	S932	BAYMEB6046	S1325
Bayer21/199	161	BAYMEB6401	S890
Bayer21097	546	BAYMEB6447	720
Bayer22/190	S790	BAYMNF0166	S1112
Bayer22408	S791	BAYNTN6867	S772
Bayer22555	S1033	BAYRSW0411	S1316
Bayer23129	710	BAYS2281	S1143
Bayer23323	S1204	BAYSLJ0312	S1055
Bayer23655	S1203	BAYSRA12869	430
Bayer25/154	205	BAYSRA7502	575
Bayer25141	S1048	BAYSRA7747	S883
Bayer25648	518	BAYSSH0860	S767
Bayer29493	316	BAYV16045	337
Bayer30130	600	BAYVq1950	337
Bayer30686	S1311	BcS3	78
Bayer32394	S792	BIO1020	S1144
Bayer33172	374	BOQ5812315	610
Bayer36205	116	BTS14639	S822
Bayer37289	S1326	BTS27419	22
Bayer37344	482	BTS30843	S1016
Bayer38819	S1216	BTS40542	591
Bayer39007	610	BTS44584	S819
Bayer41367c	307	BTS7901	S1285
Bayer41831	306	BTS8684	S1254
Bayer44646	S771	BX112	595
Bayer45432	533		
Bayer46131	608	C076	1
Bayer47531	212	C10015	S974
Bayer50282	S765	C13963	S998
Bayer5072	S1033	C1414	502
Bayer60618	S803	C15935	S1206
Bayer70533	S862	C17018	S1165
Bayer71628	479	C177	218
Bayer73	518	C18244	S949
Bayer74283	477	C18898	239
Bayer77488	575	C19490	582
Bayer78418	267	C1983	S880
Bayer79770	S848	C20132	S1166
Bayer94337	497	C2059	342
Bayer96610	576	C2242	134
BayerE393	669	C2307	S1154
BayerKHE0145	431	C3	494

INDEXES

Code	Entry no.	Code	Entry no.
CL304415	S889	D2341	193
CL336379	244	D735	111
CL38023	297	DAC893	140
CL72613	S1192	DAM18654	S918
CL7521	265	DDCNi	517
CL84777	229	DE105	754
CL900050	724	DE436	301
CL901444	324	DE473	399
CL901445	324	DE498	339
CL94377	3	DE511	495
CME107	258	DE535	390
CME127	8	DE795	638
CME134	686	DEH112	178
CME151	244	DF125	419
CN112936	593	DF250	419
CN113183	367	DIC1468	497
Co6054	S952	DIC3202	S768
Co755	95	DKA24	S999
Co859	96	DN1	S979
Compound118	S763	DN111	S979
Compound3961	S1225	DN289	S985
Compound497	S962	DNC302	641
Compound711	S1114	DO14	603
CP15336	S935	DOE81680H	354
CP17029	S900	DOW43300H	354
CP23426	722	DOW43304H	354
CP31393	598	Dowco118	S1000
CP41845	S1089	Dowco132	S904
CP46358	S1295	Dowco139	S1170
CP4742	S1281	Dowco163	523
CP50144	14	Dowco169	S936
CP52223	S929	Dowco179	137
CP53619	93	Dowco199	S996
CP6343	S764	Dowco213	181
CP67573	383	Dowco214	138
CP76963	S1176	Dowco217	S1078
CR1693	250	Dowco233	733
CragFruitFungicide341	S1088	Dowco269	S1249
CragFungicide658	S896	Dowco290	153
CragFungicide974	198	Dowco356	S1327
CragHerbicide2	S945	Dowco417	S1045
CRC7320	S1232	Dowco429	S823
CS645A	S905	Dowco433	354
CS674A	S905	Dowco433MHE	354
CS708	S905	Dowco444	S1250
CS847	S787	Dowco453	390
CS8890	S959	Dowco453EE	390
CUR616	S860	Dowco453ME	390
		DowET14	S1038
D1221	109	DowET57	S1038

Code	Entry no.	Code	Entry no.
ENT15108	551	ENT23648	222
ENT15152	394	ENT23708	S843
ENT15349	289	ENT23737	697
ENT154	261	ENT23969	106
ENT15406	S954	ENT23979	270
ENT157	S979	ENT24042	570
ENT15949	S763	ENT24105	283
ENT16225	S962	ENT24182	223
ENT16273	669	ENT24650	243
ENT16275	72	ENT24652	S1239
ENT16358	S863	ENT24653	S1010
ENT16391	S856	ENT24717	S903
ENT16519	S779	ENT24723	S967
ENT1656	290	ENT24727	250
ENT16634	S1283	ENT24728	S975
ENT16894	S1299	ENT24945	S1048
ENT17034	448	ENT24964	546
ENT17035	S942	ENT24969	124
ENT1716	485	ENT24970	S791
ENT17251	S1011	ENT24979	S810
ENT17292	552	ENT24980X	S773
ENT17295	S931	ENT24988	507
ENT17298	273	ENT25208	317
ENT17470	S943	ENT25445	23
ENT17510	17	ENT25500	S907
ENT17588	S1164	ENT25506	S1024
ENT17596	S1012	ENT25515	573
ENT17957	161	ENT25540	316
ENT18060	136	ENT25545	S1111
ENT18066	S905	ENT25579	S1311
ENT18596	S869	ENT25580	S1309
ENT18771	S1292	ENT25585	S1210
ENT19060	S1115	ENT25599	S1259
ENT19109	S972	ENT25606	116
ENT19442	S1225	ENT25647	S1000
ENT19507	209	ENT25647	S1203
ENT19763	730	ENT25670	431
ENT20696	S853	ENT25671	610
ENT20738	218	ENT25705	572
ENT20852	88	ENT25712	S1326
ENT20852	S828	ENT25715	306
ENT20871	S1271	ENT25718	225
ENT20993	S773	ENT25719	S1174
ENT22014	40	ENT25726	482
ENT22374	499	ENT25760	S1140
ENT22879	S991	ENT25764	S1297
ENT23233	41	ENT25766	S1170
ENT23284	S1038	ENT25784	S771
ENT23347	257	ENT25793	S804
ENT23584	S1070	ENT25830	S1218

INDEXES

Index 3 - Code Numbers

Code	Entry no.	Code	Entry no.
H321	482	Hoe20906	S1303
H468T	618	Hoe22870	S893
H52143	526	Hoe25682	S1104
H69	755	Hoe26150	S985
H719	S1151	Hoe29152	S1330
H722	S901	Hoe2991	S1298
H7236	327	Hoe30374	28
H95	S833	Hoe35609	S1050
H9789	526	Hoe36275	S1110
HAG107	718	Hoe39106	S1031
HC1281	676	Hoe404	287
HC252	391	Hoe498	650
Hercules14503	S934	HOE55502A	315
Hercules22234	S964	Hoe6052	S1243
Hercules3944	S1096	Hoe6053	S1243
Hercules3956	S837	HOK1566	S1125
Hercules426	S854	HOK7501	456
Hercules7531	S1199	HOK868	S1125
Hercules9573	S1294	HOL1302	S1102
HerculesAC258	S991	HOX1901	282
HF6305	417	HOX2709	S884
HF8505	417	HRS860	S853
Hoe002747	503	HRS924	S1061
Hoe002782	317	HW52	291
Hoe002960	726	HWG1608	678
Hoe017411	107		
Hoe021079	219	ICI146814	179
Hoe023408	219	ICI29661	S1247
Hoe02671	270	ICIA0001	S1341
Hoe02784	S804	ICIA0005	328
Hoe02810	445	ICIA0009	327
Hoe02824	317	ICIA0051	664
Hoe02873	619	ICIA0192	329
Hoe02904	S986	ICIA0224	383
Hoe02982	395	ICIA0321	180
Hoe02989	S1243	ICIA0500	98
Hoe02997	S1256	ICIA0523	398
Hoe033171	S1044	ICIA0574	612
Hoe039866	382	ICIA0604	717
Hoe046360	309	ICIA0754	148
Hoe070542	S1037	ICIA0858	409
Hoe075032	21	ICIA2957	278
HOE076003	9	ICIA5504	43
Hoe084498	650	ICIA5682	336
Hoe088406	309	IDH1105	725
Hoe095404	287	IK141	200
Hoe107892	462	IKF1216	329
Hoe13764	S1243	IKI1145	373
Hoe16410	433	IKI7899	125
Hoe19510	S997	IMC3950	706

INDEXES

Index 3 - Code Numbers

Code	Entry no.	Code	Entry no.
OMS174	S839	OMS253	223
OMS18	S962	OMS29	106
OMS1800	74	OMS3002	292
OMS1804	231	OMS3004	184
OMS1806	295	OMS3010	310
OMS1809	564	OMS3011	64
OMS1810	564	OMS3018	S1068
OMS1820	22	OMS3020	84
OMS1821	561	OMS3021	180
OMS1823	S805	OMS3022	110
OMS1825	38	OMS3023	277
OMS1845	395	OMS3024	70
OMS186	41	OMS3026	708
OMS187	S973	OMS3029	181
OMS19	551	OMS3032	188
OMS193	394	OMS3033	587
OMS194	S763	OMS3034	72
OMS197	S1011	OMS3035	699
OMS198	S1114	OMS3040	616
OMS1998	204	OMS3041	332
OMS1999	312	OMS3043	74
OMS2	316	OMS3044	72
OMS20	S1164	OMS3045	73
OMS2000	319	OMS3046	73
OMS2002	183	OMS3047	326
OMS2004	594	OMS3048	718
OMS2005	478	OMS3049	172
OMS2006	614	OMS3051	177
OMS2007	333	OMS3053	230
OMS2011	179	OMS313	307
OMS2012	176	OMS32	431
OMS2013	S1040	OMS33	610
OMS2014	191	OMS37	S1048
OMS2015	739	OMS410	366
OMS202	S815	OMS412	S1326
OMS204	270	OMS43	306
OMS205	270	OMS466	485
OMS206	S1111	OMS467	S905
OMS211	S854	OMS468	17
OMS213	552	OMS469	209
OMS214	S942	OMS47	S1170
OMS217	S790	OMS479	S976
OMS219	273	OMS485	161
OMS223	306	OMS497	S1259
OMS226	169	OMS503	S1140
OMS227	S820	OMS570	270
OMS232	572	OMS571	S804
OMS239	S903	OMS578	S1326
OMS244	S843	OMS584	297
OMS252	S1024	OMS595	694

INDEXES

Index 3 - Code Numbers

Code	Entry no.	Code	Entry no.
UC80502	708	WL43467	183
UCP/21	S1021	WL43479	561
UHF8615	554	WL43775	319
UKJ1506	S1312	WL5792	141
UMP488	491	WL63611	S910
UR0003	554	WL8008	S1328
USB3153	593	WL85871	184
USB3584	248	WL9385	S1338
USR604	S944	WL95481	144
		WP155	S1315
V101	S1300	WVII/117	374
V18	697		
V23031	340	X52	117
V53482	341	XDE105	754
VC13Nemacide	S943	XDE436	301
VC9104	286	XDE511	495
VCS438	S1150	XDE537	178
VCS506	S1130	XDE565	155
Vel4207	S836	XE1019	746
Vel5026	S825	XE773	S968
Velsicol104	394	XE779	247
Velsicol1068	121	XRD429	S823
Velsicol58CS11	210	XRD4331MHE	354
		XRD473	399
VP1940	317	XRD498	339
W4189	139	XRD511	495
WBA8119	81	XRD563	S1339
WL108366	326		
WL115110	335	YI5301	293
WL127294	244		
WL136184	476	ZK100309	628
WL145158	724	ZK10970	369
WL147281	476	ZK14494	206
WL17731	S802	ZK15320	563
WL19805	168	ZK49913	285
WL208304	122	Zk66752	599
WL21959	S1306	ZR512	406
WL22361	305	ZR515	484
WL29671	S1054	ZR519	S1333
WL29672	S1053	ZR777	S1129
WL43423	324	ZR856	S1098
WL43425	324		

INDEXES

INDEX 4

Common, Trade and Other Names

This index includes all names with the exceptions of (i) all but a few simple chemical names (chemical names appear in a separate Index 5) and (ii) development and official codes (which appear in Index 3).

Names with a numeric component may appear twice, in places which respectively do and do not take account of the digits. For example 1,3-dichloropropene appears near the beginning of the index (numbers being placed before letters) and in the place appropriate for 'dichloropropene'. Except near the beginning of the index, where numbers are also observed in determining sort order, the index ignores hyphens and spaces for sorting purposes.

All references are to the Entry number; unqualified numbers (e.g. 230) refer to Main Entries whilst references to Superseded materials are preceded by an 'S' (e.g. S798).

Italic print is used for index terms which are scientific names for living organisms (e.g. *Bacillus thuringiensis*).

Trade names are enclosed within single quotation marks (e.g. 'Ambush'). Their use, even without this indication, does not imply that they are not protected by law. No attempt has been made to distinguish between registered and unregistered trade marks.

Names appearing in this index may be obsolete; this includes common names that have been replaced and trade names whose (original) application has been discontinued. No indication that a name is obsolete is given here, but obsolete names which appear in Main Entries do so within brackets, e.g. [arprocarb].

Note: Readers should bear in mind national variations in spelling such as those between 'f' and 'ph', and 't' and 'th'.

Index 4 - General Names

Name	Entry no.	Name	Entry no.
'Acifon'	41	AIA	423
aclonifen	8	'Aim'	125
aclonifène	8	'Aimcosystox'	546
ACN	637	'Aimcozeb'	452
ACNQ	637	'Aimcozim'	107
'Acrex'	249	'Aimphon'	573
'Acricid'	S804	'Aimsan'	565
acrinathrin	9	'Aimthene'	4
acrinathrine	9	'Aimthyl'	711
'Acrobat'	244	'Akar'	S869
'Acrobe'	46	aker-tuba	645
acrolein	10	'Aktikon'	34
acrylonitrile	S761	'Aktuan'	258
'Act'	461	alachlor	14
'Act'	620	alachlore	14
'Actellic'	585	'Alanap'	514
'Actellifog'	585	'Alanap-L'	514
'Acti-Aid'	S915	'Alanex'	14
'Acti-dione'	S915	alanycarb	15
'Action'	359	'Alar'	197
'Actipron'	562	'Albrass'	598
'Activol'	379	aldéhyde formique	368
'Actril'	425	'Alden'	S1221
'Actril M'	459	aldicarb	16
'Actril M75'	455	aldicarbe	16
'Addstem'	107	aldoxycarb	S762
adjuvant oils	562	aldoxycarbe	S762
'Admire'	418	aldrin	S763
'Adol'	443	aldrine	S763
Adoxophyes orana granulosis virus	11	'Alegro'	563
Adoxophyes orana GV	11	'Alfacron'	38
'Aero' cyanate	S1226	'Alfadex'	622
'Afalon'	445	alfoxylate	184
'Affinity'	112	'Alfracon'	S1126
'Afidanil'	270	'Alginex'	S1256
'Afugan'	619	alidochlore	S764
'Agallol'	S1159	'Aliette'	372
'Agil'	602	'Alios'	745
'Agree'	46	'Alipur'	S916
'Agreen'	620	'Alirox'	274
'Agricorn'	455	'Alkatox'	S1260
'Agricorn D'	192	'Allegro'	439
'Agrimycin 17'	661	allethrin	17
'Agrisil'	S1326	*d*-allethrin	72
'Agritox'	455	*d-trans*-allethrin	72
'Agritox'	S1326	alléthrine	17
'Agrofuran'	109	allidochlor	S764
'Agromil'	137	'Allisan'	221
'Agrotrina'	183	alloxydim	18
'Agroxone'	455	alloxydim-sodium	18
'Agtoxin'	574	'Ally'	498
'Agzinphos'	574	allyxycarb	S765

Name	Entry no.	Name	Entry no.
allyxycarbe	S765	amiprofos-methyl	S772
'Alodan'	S854	'Amistar'	43
'Alon'	433	'Amitan'	757
alorac	S766	amiton	S773
alpha-cypermethrin	184	amiton hydrogen oxalate	S773
alpha-cyperméthrine	184	amitraz	22
alphachloralose	118	amitraze	22
'Alphadhan'	184	amitrole	23
'Alphaguard'	184	'Amizina'	651
alphamethrin	184	ammonium sulfamate	24
'Alpha Raxil CA'	29	ammonium sulphamate	24
'Al-Phos'	574	Ampelomyces quisqualis	25
'Alsol'	S1019	ampropylfos	S774
'Alsystin'	739	AMS	24
'Alto'	189	anabasine	S775
'Altorick'	S1333	Anagrus atomus	26
'Alugan'	S815	'Anchor'	540
aluminium phosphide	574	ancymidol	27
aluminum phosphide	574	ancymidole	27
'Alzodef'	167	'Andalin'	332
amben	119	'Anelda'	100
'Ambly-line cu CRS'	19	anilazine	S776
'Ambly-line d'	19	anilofos	28
Amblyseius californicus	19	'Aniloguard'	28
Amblyseius cucumeris	19	'Animert V101'	S1300
Amblyseius degenerans	19	anisuron	S777
Amblyseius mckenziei	19	'Aniten'	351
Amblyseius spp.	19	'Anniverse'	387
Amblyseius tardi	577	'Ansar 8100'	486
'Ambush'	561	'Antak'	203
'Amdro'	403	'Anthio'	370
'Amesip'	20	anthraquinone	29
'Ametrex'	20	antiphen	214
ametridione	S767	'Antor'	S964
ametryn	20	'Antracol'	608
ametryne	20	antu	S778
'Amexine'	97	'Anvil'	398
'Amiben'	119	'Apache'	101
amibuzin	S768	'Apachlor'	124
amibuzine	S768	'Apadrin'	502
amidiphos	S769	APC	S765
amidithion	S769	Aphelinus abdominalis	30
amidochlor	S1176	'Aphibank'	31
amidocyanogen	167	'Aphidamia'	402
'Amidos'	218	'Aphidan'	S1109
amidosulfuron	21	'Aphidend'	32
amidothioate	S770	Aphidius colemani	31
'Amid-Thin'	509	Aphidius platensis	31
aminocarb	S771	Aphidius transcaspicus	31
aminocarbe	S771	Aphidoletes aphidimyza	32
aminotriazole	23	'Aphido-line a'	32
'Amiphos'	S924	'Aphipar'	31

INDEXES

1505

Index 4 - General Names

Name	Entry no.	Name	Entry no.
benomyl	60	'Betasan'	63
'Benor'	60	'Betel'	53
benoxacor	61	bethrodine	57
benquinox	S795	'Betozon'	126
bensulfuron	62	'BF-51'	S1118
bensulfuron-methyl	62	BHC	392
bensulide	63	bialaphos	71
bensultap	64	'Bicep'	34
'Bensumac'	63	'Bidisin'	S862
'Benta'	65	'Bidrin'	223
bentaluron	S796	bifenazate	193
bentazon	65	bifenox	69
bentazone	65	bifenthrin	70
benthiocarb	706	bifenthrine	70
benzadox	S797	'Bigturn'	706
benzadox-ammonium	S797	bilanafos	71
benzamacril	S798	bilanafos-sodium	71
benzamacril-isobutyl	S798	'Billy'	620
benzamizole	435	binapacryl	S804
benzamorf	S799	bioallethrin	72
benzamorphe	S799	S-bioallethrin	73
benzene hexachloride	392	'Bioallethrine'	72
benzethazet	S1325	bioallethrin S-cyclopentenyl isomer	73
benzipram	S800	'Biobit'	46
benziprame	S800	'Biocel Sg 85'	558
benzoepin	270	'Bion'	114
benzofenap	66	biopermethrin	S805
benzofluor	S794	bioperméthrine	S805
benzofos	571	bioresmethrin	74
'Benzomarc'	S1212	bioresméthrine	74
benzomate	67	'BioSafe-N'	659
benzoximate	67	'BioVector'	659
benzoylprop	S802	'Biovector 335'	659
benzoylprop-ethyl	S802	'BiPC'	S855
benzphos	571	biphenyl	75
benzthiazuron	S803	'Birlane'	124
6-benzylaminopurine	68	'Biruku'	293
'Berelex'	379	bisclofentezin	149
'Beret'	311	bis ether	S806
'Bestguard'	521	bis oxide	S810
'Bestox'	184	bispyribac-sodium	76
'Bestseller'	184	bisthiosemi	S809
'Beta'	563	bitertanol	77
beta-cyfluthrin	177	'Bladafum'	669
beta-cyfluthrine	177	'Bladex'	168
beta-cypermethrin	185	'Blagal'	455
'Betanal'	563	'Bla-S'	78
'Betanal AM'	206	'Blasin'	321
'Betanal Progress'	285	'Blasin'	576
'Betanal Tandem'	285	blasticidin-S	78
'Betapal'	511	'Blazer'	7
'Betapost'	563	'Blitox'	159

1508

Name	Entry no.	Name	Entry no.
blue copperas	160	bromophos-ethyl	S817
blue stone	160	bromopropylate	87
'Blue Viking'	160	'Bromotril'	88
blue vitriol	160	'Bromox'	88
BMC	107	'Bromoxan'	88
BMC	S1157	bromoxynil	88
'B-Nine'	197	bromoxynil heptanoate	88
'Bo-Ana'	297	bromoxynil octanoate	88
BOH	404	bromoxynil-potassium	88
'Bolero'	706	brompyrazon	S818
'Bollgard'	46	brompyrazone	S818
'Bolstar'	673	bromuconazole	89
'Boltage'	616	bromure de méthyle	487
'Bombardier'	133	bronopol	90
'Bonzi'	548	'Bronotak'	90
'Boral'	665	broprodifacoum	83
borax	79	brown oil of vitriol	671
Bordeaux mixture	80	BRP	507
'BotaniGard' ES	52	Bt	46
'Botran'	221	Btk	46
BOV	671	'Bucarpolate'	S830
'Boxer'	612	'Buctril'	88
BPMC	307	'Bueno'	486
BPPS	603	bufencarb	S820
'Brassicol'	639	bufencarbe	S820
'Bravo'	133	'Bulan'	S905
'Brestan'	317	'Bulb Dip'	S1309
'Brestan'	453	'Bulldock'	177
'Brestan Flow'	317	'Bullet'	137
'Brio'	29	buminafos	S821
'Brio'	727	'Bumper'	607
'Broadstrike'	339	bupirimate	91
brodifacoum	81	buprofezin	92
bromacil	82	buprofézine	92
bromadiolone	83	'Bushwacker'	683
bromchlophos	507	butacarb	S822
bromethalin	84	butacarbe	S822
brométhaline	84	butachlor	93
'Bromex'	507	'Butacide'	581
bromfenvinfos	S811	butadiene-furfural copolymer	S1012
'Brominal'	88	butam	682
'Brominal Plus'	455	butamifos	94
bromobonil	S812	'Butanex'	93
bromobutide	85	'Butataf'	93
bromoconazole	89	butathiofos	S823
bromocyclen	S815	butenachlor	S824
bromocyclène	S815	buthidazole	S825
'Bromodan'	S815	buthiobate	S826
bromofenoxim	86	buthiuron	S827
'Brom-O-Gas'	487	butilate	100
bromophénoxime	86	butilchlorofos	S828
bromophos	S816	'Butisan S'	475

INDEXES

INDEXES

Index 4 - General Names

INDEXES

INDEXES

Name	Entry no.	Name	Entry no.
'Furacarb'	109	'GCP-5126-50W'	S1134
'Furacon'	58	'Gemstar'	393
'Furadan'	109	'Gencor'	406
furalaxyl	375	'Genesis'	148
furathiocarb	376	'Genite'	S951
furcarbanil	S1080	'Genitol'	S951
furconazole	S1081	'Geofos'	S1079
furconazole-cis	S1082	'Geonter'	689
furethrin	S1083	'Gesabal'	S1107
furidazol	374	'Gesagard'	597
furidazole	374	'Gesamil'	604
furilazole	377	'Gesapax'	20
furmecyclox	S1084	'Gesaprim'	34
furophanate	S1085	'Gesaran'	S1156
'Fury'	187	'Gesatop'	651
furyloxyfen	S1086	'Gestatamin'	S782
furyloxyfène	S1086	'Giant'	479
'Fusaclean G'	378	'GIB'	379
'Fusaclean L'	378	gibberellic acid	379
'Fusarex'	685	gibberellin A3	379
Fusarium oxysporum	378	gibberellin A4	380
'Fusilade'	327	gibberellin A4 with gibberellin A7	380
'Fusilade'	328	gibberellin A7	380
'Fyfanon'	448	'Gibbex'	379
		'Gibrel'	379
GA3	379	'Gladiator'	126
'Galaxy'	65	Glasshouse whitefly parasite	269
'Galaxy'	7	'Glean'	139
'Galben'	54	Gliocladium virens	381
'Galecron'	S857	'Glion'	383
'Galigan'	547	gliotoxin	S1087
'Gallant'	390	'Glistar'	383
'Gallant Super'	390	glucochloral	118
'Galleon'	664	glucochloralose	118
'Galtak'	55	glufosinate	382
'Gambit'	311	glufosinate-ammonium	382
gamma benzene hexachloride	392	'Glycel'	383
gamma-BHC	392	glycophene	428
'Gamma-Col'	392	glyodin	S1088
gamma-HCH	392	'Glyphogan'	383
gamma-HKhTsH	392	glyphosate	383
'Gardona'	694	glyphosate-ammonium	383
'Gardoprim'	692	glyphosate-isopropylammonium	383
'Garlon'	733	glyphosate-sodium	383
'Garlon 3A'	733	glyphosate-trimesium	383
'Garvox'	56	glyphosine	S1089
'Gatnon'	S803	'Glyphotox'	383
'Gaucho'	418	'Glytac'	S1026
'Gaucho Ble'	29	'Gnatrol'	46
'Gaucho Orge'	727	'Goal'	547
'Gavelan'	636	'Gokilaht'	188
'GCP-5126-2EC'	S1134	'Goltix'	474

Name	Entry no.	Name	Entry no.
hexazinone	400	'Illoxan'	219
'Hexygon'	401	imazalil	410
hexylthiocarbam	171	imazalil sulfate	410
hexylthiofos	S1101	imazameth	2
hexythiazox	401	imazamethabenz	411
HHDN	S763	imazamethabenz-methyl	411
HHP	S1043	imazamox	412
'Hinge'	107	imazapic	2
'Hinochloa'	461	imazapyr	413
'Hinorabcide'	267	imazaquin	414
'Hinosan'	267	imazaquin-ammonium	414
Hippodamia convergens	402	imazaquine	414
HKhTsH	392	imazethapyr	415
'Hodocide'	291	imazosulfuron	416
'Hoegrass'	219	imibenconazole	417
'Hoelon'	219	imidacloprid	418
holosulf	S1102	'Imidan'	572
'Horizon'	678	iminoctadine	419
'Hormocel'	129	iminoctadine triacetate	419
'Hosdon'	S1122	iminoctadine tris	419
'Hostaquick'	395	imiprothrin	420
'Hostathion'	726	'Impact'	361
'Hubelux'	634	'Imugan'	S848
'Hungazin'	34	inabenfide	421
'Hunter'	690	'Indalone'	S829
'Hurler'	354	indanofan	422
hydramethylnon	403	'Indar'	302
hydraméthylnone	403	'Indar'	S1319
hydrazide maléique	449	indol-3-ylacetic acid	423
2-hydrazinoethanol	404	4-indol-3-ylbutyric acid	424
hydrocyanic acid	405	indoxacarb	266
hydrogen cyanamide	167	'Inezin'	S1018
hydrogen cyanide	405	'Inovat'	572
hydrogen phosphide	574	'Inovitan'	572
'Hydrol'	S765	'Insegar'	310
hydroprene	406	'Investt'	496
2-hydroxybiphenyl	567	iodobonil	S1106
hydroxyisoxazole	408	iodofenphos	S1126
8-hydroxyquinoline sulfate	407	'Iotril'	425
hymexazol	408	ioxynil	425
hyquincarb	S1104	ioxynil octanoate	425
'Hyvar'	S1113	ioxynil-sodium	425
'Hyvar X'	82	ipazine	S1107
'Hyvar X-L'	82	IPC	136
		IPC	606
IAA	423	ipconazole	426
IBA	424	'Ipersan'	740
'Ibertox'	261	Ipheseius degenerans	19
IBP	427	iprobenfos	427
'Icon'	180	iprodione	428
IFK	606	'Iprofile'	433
'Igran'	693	iprymidam	S1108

1527

INDEXES

Name	Entry no.	Name	Entry no.
N-pyrrolidinosuccinamic acid	S1251	'Oribright'	658
'N-Serve'	523	Oriental fruit moth pheromone	262
nuarimol	528	'Oriflam'	23
'Nudrin'	483	'Ori-line'	535
'Nurelle'	S1249	'Orion'	15
'Nustar'	357	*Orius albidepennis*	535
'Nuvacron'	502	*Orius insidiousus*	535
'Nuvan'	218	*Orius laevigatus*	535
'Nuvanol N'	S1126	*Orius majusculus*	535
		Orius spp.	535
'Occidor'	107	'Orthene'	4
OCH	S1200	orthobencarb	534
'Octachlor'	121	orthophenylphenol	567
'Octalene'	S763	'Ortran'	4
'Octalox'	S962	'Ortus'	315
octhilinone	529	oryzalin	536
N-octylbicycloheptenedicarboximide	272	'Oryzemate'	590
'Odyssey'	412	'Osadan'	303
'Oftanol'	430	'Osaquat'	550
'Ofunack'	625	'Osbac'	307
'Ofunack M'	494	'Oscar'	680
ofurace	531	'Ossirame'	159
'Ohric'	S953	'Ostrinil'	52
'Okishindo'	543	'Otinem S'	659
'Oktone'	S1200	'Oust'	667
'Oleate'	532	'Outflank'	561
oleic acid	532	'Outfox'	S920
'Olitref'	740	'Ovation'	346
'Olymp'	357	ovatran	S863
'Omadine'	S1103	'Overtop'	415
'Omadine disulfide'	S994	ovex	S863
'Omadine-DS'	S994	'Ovitox'	S863
'Omadine OM 1563'	S1103	'Ovotran'	S863
'Omadine OM 1564'	S1103	oxabetrinil	537
'Omadine OM 1565'	S1103	oxadiargyl	538
'Omazene'	S908	oxadiazon	539
'Omega'	641	oxadixyl	540
omethoate	533	oxamil	541
'Omexan'	137	oxamyl	541
'Omite'	603	oxapyrazon	S1202
OMPA	S1267	oxapyrazone	S1202
OMU	S916	oxapyrazon-sodium	S1202
'Oncol'	58	oxasulfuron	542
'Onebest'	700	'Oxatin'	111
'Onecide'	327	oxine-copper	543
'Optica'	460	oxine-Cu	543
'Opus'	48	oxine-cuivre	543
orbencarb	534	oxolinic acid	544
'Ordoval'	401	'Oxotin'	181
'Ordram'	501	oxycarboxin	545
'Oreste'	625	oxycarboxine	545
organic nickel	517	oxychlorure de cuivre	159

INDEXES

INDEXES

Name	Entry no.	Name	Entry no.
sesasmolin	S1272	'Sirius'	620
sesone	S995	'Sisthane'	S1034
'Sesoxane'	S1271	'Sitofex'	367
sethoxydim	648	'Sivel'	210
séthoxydime	648	'Skeetal'	46
'Setoff'	145	'Skipper'	708
'Sevin'	106	'Slam C'	S786
'Shaminliulin'	306	SMA	130
'Shandon'	189	'Smash'	634
'Shaphos'	574	SMCA	130
'Shasicidin'	319	SMDC	473
'Sheriff'	641	'Snip'	S976
'Sherpa'	183	S.N.P.	551
'Shibagen'	325	'Sochlor'	653
'Shirahagen-S'	684	sodium aminofluoride	164
'Shirlan'	329	sodium chlorate	653
'Shirlan'	S1266	sodium chloroacetate	130
'Sholay'	14	sodium cis-3-chloroacrylate	S1276
'Short-keep'	76	sodium cyanide	405
'Showrone'	195	sodium dimethylarsinate	245
'Shoxin'	S1197	sodium fluoride	654
'Siacarb'	706	sodium fluoroacetate	655
'Sialite'	250	sodium fluorosilicate	656
'Sibatito'	416	sodium fluosilicate	656
'Sibutol'	29	sodium hexafluorosilicate	656
'Sibutol'	77	sodium pentachlorophenoxide	558
'Sibutol'	374	sodium 2-phenylphenoxide	567
'Sicarol'	S1243	sodium selenate	S1277
siduron	649	sodium silicofluoride	656
'Sierra'	281	'Sofit'	304
sigma-cypermethrin	186	'Sofit'	588
silafluofen	650	'SoilGard'	381
silaneophan	650	'Solacol'	747
'Silatop'	650	solan	559
'Silbenil'	211	'Solbar'	S789
'Silrifos'	137	'Solfac'	176
'Silvacur'	678	'Solicam'	526
'Silvacur'	721	'Solnet'	588
silvex	S1042	'Solvirex'	257
'Simanex'	651	'Sonalan'	279
'Simatylone LA'	651	'Sonar'	352
simazine	651	'Sonax'	S1020
simeton	S1273	'Sonis'	744
simetryn	652	sophamide	S1278
simetryne	652	'Soprasan'	S1160
'Sinal'	495	'Sorexa CD'	276
'Sinbar'	689	'Sorex Brodifacoum Rat & Mouse Bait'	81
'Sinfluran'	740	soufre	670
'Sinox General'	S984	'Spannit'	137
'Siolcid'	445	'Spanone'	S857
'Sipcatin'	181	'Spark'	726
'Sirbon'	387	'Sparkstar'	239

INDEXES

1543

INDEXES

Index 4 – General Names

Name	Entry no.	Name	Entry no.
Verticillium lecanii	750	'White Star'	448
'Vetrazin'	191	'Win'	440
'Victenon'	64	'Win Admire'	440
'Vidden D'	S955	'Winner'	328
'Vigil'	S960	'Wiper-jet'	S1168
'Vikane'	672	'Wolf-Ace'	62
'Vincit'	361	'Wolf-Ace'	461
vinclozolin	751		
vinclozoline	751	'XenTari'	46
'Vinicur'	S922	'X-Gnat'	659
vinyl cyanide	S761	'XL-2G'	57
'Virox'	S1188	'XL All Nicotine'	520
'Visclor'	133	XMC	755
'Visor'	702	xylachlor	S1340
'Vista'	349	'Xyligen B'	S1084
'Vital'	4	xylylcarb	756
Vitamin D2	276		
vitamin D3	752	yellow oxide of mercury	468
'Vitavax'	111	'Yukahope'	151
'Vitex'	757	'Yukamate'	237
'Viviful'	595	'Yukawide'	66
'Vivithrin'	S1045		
'Volaton'	575	'Zabara'	440
'Volcano'	666	'Zardex'	S1098
'Volcyper'	183	zarilamid	S1341
'Volparox'	S1041	'Zark'	62
'Volphor'	570	'Zark'	461
'Voltage'	616	'Zawa'	574
'Volzim'	107	'Z-Bordeaux'	80
'Vondozeb'	452	'Zebra'	452
'Vorox'	S1269	'Zectran'	S1170
'Vydate'	541	zeidane	201
		'Zeldox'	401
'Warbexol'	297	'Zeltoxone'	740
'Warefog'	136	'Zero One'	456
warfarin	753	zeta-cypermethrin	187
warfarine	753	zinc naphthenate	508
'Warrant'	570	zinc phosphide	574
'Warrior'	180	'Zinc-Tox'	574
'Waspex'	S1126	zineb	757
'Wechser'	560	zinèbe	757
'Weedazol'	23	'Zinophos'	S1309
'Weedless'	550	ziram	758
'Weedone'	216	zirame	758
'Weedtox'	192	zolaprofos	S1342
'Wepsyn 155'	S1315	'Zolone'	571
'Whip'	S1044	zoocoumarin	753
'Whip Super'	309	'Zorial'	526
white arsenic	S780	ZXI 8901	759
white halo fungus	750	'Zytron'	S1000
white oils	562		

INDEX 5

Chemical Names

Chemical names (IUPAC and *Chemical Abstracts*) which are less than about 90 characters long are given in this index.

All references are to Entry Numbers; unqualified numbers (e.g. 230) refer to Main Entries, whilst references to superseded materials are preceded by an 'S' (e.g. S768).

The order in this index is strictly alphabetical; spaces and all punctuation marks between words and accents are ignored. Letter locants [*o-*, *m-*, *sec-*, *tert-*, *N-*, *O-*, *S-*, etc.] and stereochemical descriptors [*cis-*, *trans-*, *(R)-*, *(S)-*, *(E)-*, *(Z)-*, *endo-*, etc.] are also ignored in deciding order, as are Greek letters. Numerals are used, however, to separate terms which would otherwise be identical.

In many cases, where the IUPAC and *Chemical Abstracts* names are closely similar, and would occur consecutively, only one name has been included.

Chemical Name	Entry no.

Chemical Name	Entry no.
1,2,3-benzothiadiazole-7-carbothioic acid S-methyl ester	114
1-(1,3-benzothiazol-2-yl)-1,3-dimethylurea	477
1-benzothiazol-2-yl-1,3-dimethylurea	477
N-2-benzothiazolyl-N,N'-dimethylurea	477
1-(1,3-benzothiazol-2-yl)-3-isopropylurea	S796
1-benzothiazol-2-yl-3-isopropylurea	S796
N-2-benzothiazolyl-N'-(1-methylethyl)urea	S796
1-(1,3-benzothiazol-2-yl)-3-methylurea	S803
1-benzothiazol-2-yl-3-methylurea	S803
N-2-benzothiazolyl-N'-methylurea	S803
2-(1,3-benzothiazol-2-yloxy)-N-methylacetanilide	461
2-benzothiazol-2-yloxy-N-methylacetanilide	461
2-(2-benzothiazolyloxy)-N-methyl-N-phenylacetamide	461
[(benzoylamino)oxy]acetic acid	S797
N-benzoyl-N-(3-chloro-4-fluorophenyl)-DL-alanine	S1052
N-benzoyl-N-(3-chloro-4-fluorophenyl)-D-alanine	324
N-benzoyl-N-(3-chloro-4-fluorophenyl)-DL-alanine methyl ester	S1054
1-benzoyl-1-(3,4-dichlorophenyl)-3,3-dimethylurea	S1212
6-benzyladenine	68
benzyl 2-chloro-4-trifluoromethyl-1,3-thiazole-5-carboxylate	350
S-benzyl di-sec-butylthiocarbamate	713
S-benzyl O,O-diethyl phosphorothioate	S1005
S-benzyl O,O-di-isopropyl phosphorothioate	427
S-benzyl 1,2-dimethylpropyl(ethyl)thiocarbamate	278
S-benzyl dipropylthiocarbamate	612
S-benzyl O-ethyl phenylphosphonothioate	S1018
5-benzyl-3-furylmethyl (+)-trans-chrysanthemate	74
5-benzyl-3-furylmethyl (±)-cis-trans-chrysanthemate	642
5-benzyl-3-furylmethyl (1R,3R)-2,2-dimethyl-3-(2-methylprop-1-enyl)cyclopropanecarboxylate	74
N-benzyl-N-isopropyl-3,5-dimethylbenzamide	S800
N-benzyl-N-isopropylpivalamide	682
3-[benzyl(methyl)amino]-2-cyanoacrylic acid	S798
1,1'-biphenyl	75
biphenyl	75
[1,1'-biphenyl]-2-ol	567
biphenyl-2-ol	567
β-([1,1'-biphenyl]-4-yloxy)-α-(1,1-dimethylethyl)-1H-1,2,4-triazole-1-ethanol	77
1-(biphenyl-4-yloxy)-3,3-dimethyl-1-(1H-1,2,4-triazol-1-yl)butan-2-ol	77
3-(3-biphenyl-4-yl-1,2,3,4-tetrahydro-1-naphthyl)-4-hydroxycoumarin	227
5-[bis[2-(2-butoxyethoxy)ethoxy]methyl]-1,3-benzodioxole	S1222
1-bis[2-(2-butoxyethoxy)ethoxy]methyl-3,4-methylenedioxybenzene	S1222
bis(2-chloroethyl) ether	S806
bis(2-chloro-1-methylethyl) ether	200
bis(4-chlorophenoxy)methane	S925
O,O-bis(4-chlorophenyl) N-acetimidoylphosphoramidothioate	S1216
1,1-bis(4-chlorophenyl)ethanol	S861
1,1-bis(4-chlorophenyl)-2-ethoxyethanol	S807
O,O-bis(4-chlorophenyl) (1-iminoethyl)phosphoramidothioate	S1216

Chemical Name	Entry no.
1,3-bis(2,2,2-trichloro-1-hydroxyethyl)urea	S945
1-[3-[3,5-bis(trifluoromethyl)phenyl]-2-propynyl]-4-(1,1-dimethylethyl)piperidine	S1068
bis[tris(2-methyl-2-phenylpropyl)tin] oxide	303
borax	79
Bordeaux mixture	80
(E)-(1RS,4RS)-bornan-2-one O-prop-2-ynyloxime	S1095
(1R,2R,4R)-born-2-yl thiocyanatoacetate	S1013
3-[3-(4'-bromobiphenyl-4-yl)-3-hydroxy-1-phenylpropyl]-4-hydroxycoumarin	83
3-[3-(4'-bromobiphenyl-4-yl)-1,2,3,4-tetrahydro-1-naphthyl]-4-hydroxycoumarin	81
5-bromo-3-sec-butyl-6-methyluracil	82
1-bromo-2-chloroethane	S813
4-bromo-2-(4-chlorophenyl)-1-(ethoxymethyl)-5-(trifluoromethyl)-1H-pyrrole-3-carbonitrile	123
O-(4-bromo-2-chlorophenyl) O-ethyl S-propyl phosphorothioate	594
3-(4-bromo-3-chlorophenyl)-1-methoxy-1-methylurea	120
3-bromo-1-chloroprop-1-ene	S814
7-bromo-5-chloro-8-quinolinyl 2-propenoate	S1091
7-bromo-5-chloro-8-quinolyl acrylate	S1091
O-(4-bromo-2,5-dichlorophenyl) O,O-diethyl phosphorothioate	S817
O-(4-bromo-2,5-dichlorophenyl) O,O-dimethyl phosphorothioate	S816
2-bromo-1-(2,4-dichlorophenyl)ethenyl diethyl phosphate	S811
O-(4-bromo-2,5-dichlorophenyl) O-methyl phenylphosphonothioate	S1130
1-[[4-bromo-2-(2,4-dichlorophenyl)tetrahydro-2-furanyl]methyl]-1H-1,2,4-triazole	89
1-[(2RS,4RS:2RS,4SR)-4-bromo-2-(2,4-dichlorophenyl)tetrahydrofurfuryl]-1H-1,2,4-triazole	89
2-bromo-1-(2,4-dichlorophenyl)vinyl diethyl phosphate	S811
1-[[2-[4-(bromodifluoromethoxy)phenyl]-2-methylpropoxy]methyl]-3-phenoxybenzene	387
2-(4-bromodifluoromethoxyphenyl)-2-methylpropyl 3-phenoxybenzyl ether	387
[(5-bromo-1,6-dihydro-6-oxo-1-phenyl-4-pyridazinyl)amino]oxoacetic acid	S1202
5-bromo-1,6-dihydro-6-oxo-1-phenylpyridazin-4-yloxamic acid	S1202
2-bromo-N-(α,α-dimethylbenzyl)-3,3-dimethylbutyramide	85
2-bromo-3,3-dimethyl-N-(1-methyl-1-phenylethyl)butanamide	85
2-bromo-3,3-dimethyl-N-(1-methyl-1-phenylethyl)butyramide	85
5-bromo-3-isopropyl-6-methyluracil	S1113
bromomethane	487
5-bromomethyl-1,2,3,4,7,7-hexachlorobicyclo[2.2.1]hept-2-ene	S815
5-bromomethyl-1,2,3,4,7,7-hexachloro-8,9,10-trinorborn-2-ene	S815
5-bromo-6-methyl-3-(1-methylethyl)-2,4(1H,3H)-pyrimidinedione	S1113
5-bromo-6-methyl-3-(1-methylpropyl)-2,4(1H,3H)-pyrimidinedione	82
2-bromo-2-nitro-1,3-propanediol	90
2-bromo-2-nitropropane-1,3-diol	90
3-(4-bromophenyl)-1-methoxy-1-methylurea	492
2-butanamine	99
butanedioic acid mono(2,2-dimethylhydrazide)	197
(Z)-N-but-2-enyloxymethyl-2-chloro-2',6'-diethylacetanilide	S824
5-[2-(2-butoxyethoxy)ethoxymethyl]-6-propyl-1,3-benzodioxole	581
2-(2-butoxyethoxy)ethyl 1,3-benzodioxole-5-carboxylate	S830
2-(2-butoxyethoxy)ethyl piperonylate	S830
2-(2-butoxyethoxy)ethyl 6-propylpiperonyl ether	581
2-(2-butoxyethoxy)ethyl thiocyanate	S831

Chemical Name	Entry no.
N'-(4-chloro-2-methylphenyl)-*N*,*N*-dimethylthiourea	S872
N'-(3-chloro-4-methylphenyl)-*N*,*N*-dimethylurea	134
N-(3-chloro-4-methylphenyl)-2-methylpentanamide	559
2-chloro-*N*-(1-methylprop-2-ynyl)acetanilide	S1241
2-chloro-*N*-(1-methyl-2-propynyl)-*N*-phenylacetamide	S1241
O-(3-chloro-7-methylpyrazolo[1,5-*a*]pyrimidin-2-yl) *O*,*O*-diethyl phosphorothioate	S884
7-chloro-3-methylquinoline-8-carboxylic acid	636
2-[2-chloro-1-(methylsulfonyl)benzoyl]-1,3-cyclohexanedione	664
O-[2-chloro-4-(methylthio)phenyl] *O*-methyl ethylphosphoramidothioate	S770
6-chloro-5-(methylthio)pyrimidine-2,4-diamine	S1312
3'-chloro-2-methylvalero-*p*-toluidide	559
2-chloro-6-nitro-3-phenoxyaniline	8
2-chloro-1-(4-nitrophenoxy)-4-(trifluoromethyl)benzene	S1194
O-(2-chloro-4-nitrophenyl) *O*,*O*-dimethyl phosphorothioate	S942
O-(3-chloro-4-nitrophenyl) *O*,*O*-dimethyl phosphorothioate	S790
O-(4-chloro-3-nitrophenyl) *O*,*O*-dimethyl phosphorothioate	S1219
1-chloro-2-nitropropane	S874
4-chloro-2-oxo-3(2*H*)-benzothiazoleacetic acid	55
4-chloro-2-oxobenzothiazolin-3-ylacetic acid	55
S-[(6-chloro-2-oxo-3(2*H*)-benzoxazolyl)methyl] *O*,*O*-diethyl phosphorodithioate	571
S-[(6-chloro-2-oxooxazolo[4,5-*b*]pyridin-3(2*H*)-yl)methyl] *O*,*O*-dimethyl phosphorothioate	38
(*E*)-2-chloro-β-oxo-α-(4-phenyl-2(3*H*)-thiazolylidene)benzenepropanenitrile	S1275
(4-chlorophenoxy)acetic acid	163
4-chlorophenoxyacetic acid	163
3-(4-chlorophenoxy)benzyl (*RS*)-2-(4-ethoxyphenyl)-3,3,3-trifluoropropyl ether	336
β-(4-chlorophenoxy)-α-(1,1-dimethylethyl)-1*H*-1,2,4-triazole-1-ethanol	721
(1*RS*,2*RS*;1*RS*,2*SR*)-1-(4-chlorophenoxy)-3,3-dimethyl-1-(1*H*-1,2,4-triazol-1-yl)butan-2-ol	721
1-(4-chlorophenoxy)-3,3-dimethyl-1-(1*H*-1,2,4-triazol-1-yl)-2-butanone	720
1-(4-chlorophenoxy)-3,3-dimethyl-1-(1*H*-1,2,4-triazol-1-yl)butan-2-one	720
1-(4-chlorophenoxy)-3-[[2-(4-ethoxyphenyl)-3,3,3-trifluoropropoxy]methyl]benzene	336
1-(4-chlorophenoxy)-1-(1*H*-imidazol-1-yl)-3,3-dimethyl-2-butanone	S890
1-(4-chlorophenoxy)-1-(imidazol-1-yl)-3,3-dimethylbutanone	S890
(±)-2-[4-(4-chlorophenoxy)phenoxy]propanoic acid	S893
(±)-2-[4-(4-chlorophenoxy)phenoxy]propionic acid	S893
3-[4-(4-chlorophenoxy)phenyl]-1,1-dimethylurea	S880
N'-[4-(4-chlorophenoxy)phenyl]-*N*,*N*-dimethylurea	S880
(±)-2-(3-chlorophenoxy)propanoic acid	152
(±)-2-(3-chlorophenoxy)propionic acid	152
N-[[[(4-chlorophenyl)amino]carbonyl]-2,6-difluorobenzamide	231
O-[4-[(4-chlorophenyl)azo]phenyl] *O*,*O*-dimethyl phosphorothioate	S786
O-4-(4-chlorophenylazo)phenyl *O*,*O*-dimethyl phosphorothioate	S786
4-chlorophenyl benzenesulfonate	S1047
2-(2-chlorophenyl)-1*H*-benzimidazole	S860
2-(2-chlorophenyl)benzimidazole	S860
S-[*N*-(2-chlorophenyl)butyramidomethyl] *O*,*O*-dimethyl phosphorodithioate	S1077
4-chlorophenyl 4-chlorobenzenesulfonate	S863
(±)-α-(2-chlorophenyl)-α-(4-chlorophenyl)-5-pyrimidinemethanol	300
(4*RS*,5*RS*)-5-(4-chlorophenyl)-*N*-cyclohexyl-4-methyl-2-oxo-1,3-thiazolidine-3-carboxamide	401

Chemical Name	Entry no.
trans-5-(4-chlorophenyl)-N-cyclohexyl-4-methyl-2-oxo-3-thiazolidinecarboxamide	401
(±)-α-[N-(3-chlorophenyl)cyclopropanecarboxamido]-γ-butyrolactone	S922
α-(4-chlorophenyl)-α-(1-cyclopropylethyl)-1H-1,2,4-triazol-1-ethanol	189
(2RS,3RS;2RS,3SR)-2-(4-chlorophenyl)-3-cyclopropyl-1-(1H-1,2,4-triazol-1-yl)butan-2-ol	189
1-(4-chlorophenyl)-3-(2,6-dichlorobenzoyl)urea	S1208
2-(2-chlorophenyl)-2-(diethoxyphosphinothioyloxyimino)acetonitrile	S883
1-(4-chlorophenyl)-3-(2,6-difluorobenzoyl)urea	231
3-(2-chlorophenyl)-6-(2,6-difluorophenyl)-1,2,4,5-tetrazine	674
1-(4-chlorophenyl)-1,4-dihydro-6-methyl-4-oxopyridazine-3-carboxylic acid	S1046
(E,Z)-4-[3-(4-chlorophenyl)-3-(3,4-dimethoxyphenyl)acryloyl]morpholine	244
(E,Z)-4-[3-(4-chlorophenyl)-3-(3,4-dimethoxyphenyl)-1-oxo-2-propenyl]morpholine	244
N-(4-chlorophenyl)-2,2-dimethylpentanamide	S1177
(RS)-1-p-chlorophenyl-4,4-dimethyl-3-(1H-1,2,4-triazol-1-ylmethyl)pentan-3-ol	678
(2RS,3RS)-1-(4-chlorophenyl)-4,4-dimethyl-2-(1H-1,2,4-triazol-1-yl)pentan-3-ol	548
(E)-(RS)-1-(4-chlorophenyl)-4,4-dimethyl-2-(1H-1,2,4-triazol-1-yl)pent-1-en-3-ol	746
(E)-(S)-1-(4-chlorophenyl)-4,4-dimethyl-2-(1H-1,2,4-triazol-1-yl)pent-1-en-3-ol	746
3-(4-chlorophenyl)-1,1-dimethylurea	S1179
3-(4-chlorophenyl)-1,1-dimethyluronium trichloroacetate	S1179
5-(4-chlorophenyl)-2,3-diphenylthiophene	S1317
5-chloro-4-phenyl-1,2-dithiol-3-one	S1096
(2RS,3SR)-1-[3-(2-chlorophenyl)-2,3-epoxy-2-(4-fluorophenyl)propyl]-1H-1,2,4-triazole	48
(RS)-2-[2-(3-chlorophenyl)-2,3-epoxypropyl]-2-ethylindan-1,3-dione	422
7-(2-chlorophenyl)-4-ethoxy-3,5-dioxa-6-aza-4-phosphaoct-6-ene-8-nitrile 4-sulfide	S883
2-(4-chlorophenyl)-3-ethyl-2,5-dihydro-5-oxopyridazine-4-carboxylic acid	148
(±)-α-[2-(4-chlorophenyl)ethyl]-α-(1,1-dimethylethyl)-1H-1,2,4-triazole-1-ethanol	678
α-[2-(4-chlorophenyl)ethyl]-α-phenyl-1H-1,2,4-triazole-1-propanenitrile	302
cis-1-[[3-(2-chlorophenyl)-2-(4-fluorophenyl)oxiranyl]methyl]-1H-1,2,4-triazole	48
(±)-α-(2-chlorophenyl)-α-(4-fluorophenyl)-5-pyrimidinemethanol	528
4-(2-chlorophenylhydrazono)-3-methyl-1,2-oxazol-5(4H)-one	S1003
4-(3-chlorophenylhydrazono)-3-methyl-1,2-oxazol-5(4H)-one	S1146
3-(4-chlorophenyl)-1-methoxy-1-methylurea	503
N-[(4-chlorophenyl)methyl]-N-cyclopentyl-N'-phenylurea	556
(4-chlorophenyl)methyl N-(2,4-dichlorophenyl)-1H-1,2,4-triazole-1-ethanimidothioate	417
S-[(2-chlorophenyl)methyl] diethylcarbamothioate	534
S-[(4-chlorophenyl)methyl] diethylcarbamothioate	706
(R*,R*)-(±)-β-[(4-chlorophenyl)methyl]-α-(1,1-dimethylethyl)-1H-1,2,4-triazole-1-ethanol	548
2-[(4-chlorophenyl)methyl]-4,4-dimethyl-3-isoxazolidinone	150
5-[(4-chlorophenyl)methyl]-2,2-dimethyl-1-(1H-1,2,4-triazol-1-ylmethyl)cyclopentanol	476
(E)-(±)-β-[(4-chlorophenyl)methylene]-α-(1,1-dimethylethyl)-1H-1,2,4-triazole-1-ethanol	746
5-[(4-chlorophenyl)methylene]-2,2-dimethyl-1-(1H-1,2,4-triazol-1-ylmethyl)cyclopentanol	745
S-[2-[(4-chlorophenyl)(1-methylethyl)amino]-2-oxoethyl] O,O-dimethyl phosphorodithioate	28
α-(4-chlorophenyl)-α-(1-methylethyl)-5-pyrimidinemethanol	S1121
2-[(4-chlorophenyl)methyl]-5-(1-methylethyl)-1-(1H-1,2,4-triazol-1-ylmethyl)cyclopentanol	426
3-(4-chlorophenyl)-1-methyl-1-(1-methylprop-2-ynyl)urea	S833
1-(4-chlorophenyl)-2-methyl-1-pyrimidin-5-ylpropan-1-ol	S1121
3-(4-chlorophenyl)-5-methylrhodanine	S875
3-(4-chlorophenyl)-5-methyl-2-thioxo-1,3-thiazolidin-4-one	S875
2-[[2-(3-chlorophenyl)oxiranyl]methyl]-2-ethyl-1H-indene-1,3(2H)-dione	422

INDEXES

Chemical Name	Entry no.
2,4-dichloro-6-fluorophenyl 4-nitrophenyl ether	S1065
3,4-dichloro-1-(4-fluorophenyl)-1*H*-pyrrole-2,5-dione	345
(*RS*)-1-[2,5-dichloro-4-(1,1,2,3,3,3-hexafluoropropoxy)phenyl]-3-(2,6-difluorobenzoyl)urea	446
3,5-dichloro-4-hydroxybenzonitrile	S882
2,7-dichloro-9-hydroxy-9*H*-fluorene-9-carboxylic acid	S946
2,7-dichloro-9-hydroxyfluorene-9-carboxylic acid	S946
C (2,5-dichloro-4-iodophenyl) *O,O*-dimethyl phosphorothioate	S1126
O-(2,5-dichloro-4-iodophenyl) *O*-ethyl ethylphosphonothioate	S949
3,6-dichloro-2-methoxybenzoic acid	210
2,4-dichloro-1-(3-methoxy-4-nitrophenoxy)benzene	117
3′,4′-dichloro-2-methylacrylanilide	S850
4,4′-dichloro-2,2′-methylenediphenol	214
3-[2,4-dichloro-5-(1-methylethoxy)phenyl]-5-(1,1-dimethylethyl)-1,3,4-oxadiazol-2(3*H*)-one	539
2-dichloromethyl-2-methyl-1,3-dioxolane	S1171
(±)-2-(2,4-dichloro-3-methylphenoxy)-*N*-phenylpropanamide	151
N-(2,6-dichloro-3-methylphenyl)-5,7-dimethoxy[1,2,4]triazolo[1,5-*a*]pyrimidine-2-sulfonamide	495
O-(2,6-dichloro-4-methylphenyl) *O,O*-dimethyl phosphorothioate	714
6-(3,5-dichloro-4-methylphenyl)pyridazin-3(2*H*)-one	220
O-[2,5-dichloro-4-(methylthio)phenyl] *O,O*-diethyl phosphorothioate	S887
O-[4,5-dichloro-2-(methylthio)phenyl] *O,O*-diethyl phosphorothioate	S887
O-[2,4-dichloro-5-(methylthio)phenyl] *O,O*-diethyl phosphorothioate	S887
2,3-dichloro-1,4-naphthalenedione	S944
2,3-dichloro-1,4-naphthoquinone	S944
2,6-dichloro-4-nitroaniline	221
1,1-dichloro-1-nitroethane	S950
2,4-dichloro-1-(4-nitrophenoxy)benzene	S1193
2′,5-dichloro-4′-nitrosalicylanilide	518
2,2-dichloro-*N*-[2-oxo-2-(2-propenylamino)ethyl]-*N*-2-propenylacetamide	S999
(2,4-dichlorophenoxy)acetic acid	192
4-(2,4-dichlorophenoxy)butanoic acid	199
4-(2,4-dichlorophenoxy)butyric acid	199
2-(2,4-dichlorophenoxy)ethyl hydrogen sulfate	S995
4-(2,4-dichlorophenoxy)-2-methoxy-1-nitrobenzene	117
5-(2,4-dichlorophenoxy)-2-nitroanisole	117
(*RS*)-2-[5-(2,4-dichlorophenoxy)-2-nitrophenoxy]-*N*-ethylpropionamide	S1031
(*RS*)-2-[4-(2,4-dichlorophenoxy)phenoxy]propionic acid	219
(*R*)-2-(2,4-dichlorophenoxy)propionic acid	217
(*RS*)-2-(2,4-dichlorophenoxy)propionic acid	216
1-[[[(2,4-dichlorophenyl)amino]carbonyl]cyclopropanecarboxylic acid	170
2,4-dichlorophenyl benzenesulfonate	S951
N-(3,4-dichlorophenyl)cyclopropanecarboxamide	S923
O-(2,4-dichlorophenyl) *O,O*-diethyl phosphorothioate	S943
N-(3,4-dichlorophenyl)-*N*-[(dimethylamino)carbonyl]benzamide	S1212
N-(3,4-dichlorophenyl)-*N*-[(dimethylamino)carbonyl]-4-methoxybenzamide	S777
3-(3,5-dichlorophenyl)-1,5-dimethyl-3-azabicyclo[3.1.0]hexane-2,4-dione	592
N-(3,5-dichlorophenyl)-1,2-dimethylcyclopropane-1,2-dicarboximide	592
3-(3,5-dichlorophenyl)-5,5-dimethyl-1,3-oxazolidine-2,4-dione	S959
(2*RS*,3*RS*)-1-(2,4-dichlorophenyl)-4,4-dimethyl-2-(1*H*-1,2,4-triazol-1-yl)pentan-3-ol	S960

Chemical Name	Entry no.
O,O-dimethyl O-4-methylthio-m-tolyl phosphorothioate	316
O,O-dimethyl S-morpholinocarbonylmethyl phosphorodithioate	S1181
O,O-dimethyl S-[2-(4-morpholinyl)-2-oxoethyl] phosphorodithioate	S1181
dimethyl–sulfatodimercury	S808
O,O-dimethyl O-(4-nitrophenyl) phosphorothioate	552
O,O-dimethyl O-4-nitro-m-tolyl phosphorothioate	306
3aα,4α,5α,7α,7aα-N,N-dimethyl-N′-(octahydro-4,7-methano-1H-inden-5-yl)urea	S1199
N,N-dimethyl-N′-(2,3,3a,4,5,6,7,7a-octahydro-4,7-methano-1H-inden-1-yl)urea	S1117
N,N-dimethyl-N′-(2,3,3a,4,5,6,7,7a-octahydro-4,7-methano-1H-inden-2-yl)urea	S1117
1-(3,7-dimethyloctyl)-1-(2-propenyl)piperidinium bromide	S1221
O,O-dimethyl S-[(4-oxo-1,2,3-benzotriazin-3(4H)-yl)methyl] phosphorodithioate	41
5,5-dimethyl-3-oxocyclohex-1-enyl dimethylcarbamate	S975
3-[(2R)-2-[(1S,3S,5S)-3,5-dimethyl-2-oxocyclohexyl]-2-hydroxyethyl]glutarimide	S915
4-[(2R)-2-[(1S,3S,5S)-(3,5-dimethyl-2-oxocyclohexyl)]-2-hydroxyethyl]piperidine-2,6-dione	S915
2-(2,2-dimethyl-1-oxopropyl)-1H-indene-1,3(2H)-dione	579
1,1-dimethyl-3-(perhydro-4,7-methanoinden-1-yl)urea	S1117
1,1-dimethyl-3-(perhydro-4,7-methanoinden-2-yl)urea	S1117
1,1-dimethyl-3-(perhydro-4,7-methanoinden-5-yl)urea	S1199
N,N-dimethyl-α-phenylbenzeneacetamide	254
N′-(2,4-dimethylphenyl)-N-[[(2,4-dimethylphenyl)imino]methyl]-N-methylmethanimidamide	22
dimethyl [1,2-phenylenebis(iminocarbonothioyl)]bis[carbamate]	711
dimethyl 4,4′-(o-phenylene)bis(3-thioallophanate)	711
2-[[1-(2,5-dimethylphenyl)ethyl]sulfonyl]pyridine 1-oxide	S1336
2,5-dimethyl-N-phenyl-3-furancarboxamide	S1080
N-(2,6-dimethylphenyl)-N-(2-furanylcarbonyl)-DL-alanine methyl ester	375
N-(2,6-dimethylphenyl)-2-methoxy-N-(2-oxo-3-oxazolidinyl)acetamide	540
3,4-dimethylphenyl methylcarbamate	756
3,5-dimethylphenyl methylcarbamate	755
4,6-dimethyl-N-phenyl-2-pyrimidinamine	628
2,4-dimethyl-N-phenyl-5-thiazolecarboxamide	S1169
1,1-dimethyl-3-phenylurea	318
1,1-dimethyl-3-phenyluronium trichloroacetate	318
O,S-dimethyl phosphoramidothioate	479
O,O-dimethyl S-phthalimidomethyl phosphorodithioate	572
1,1-dimethylpiperidinium	465
dimethyl(4-piperidinocarbonyloxy-2,5-xylyl)sulfonium toluene-4-sulfonate	S819
N-(1,2-dimethylpropyl)-N′-ethyl-6-(methylthio)-1,3,5-triazine-2,4-diamine	239
2-(3,5-dimethyl-1H-pyrazol-1-yl)-1H-benzimidazole	S1261
(Z)-4,6-dimethyl-2(1H)-pyrimidinone [1-(2-methylphenyl)ethylidene]hydrazone	321
2-[[[[(4,6-dimethyl-2-pyrimidinyl)amino]carbonyl]amino]sulfonyl]benzoic acid	667
N-(4,6-dimethylpyrimidin-2-yl)aniline	628
2-(4,6-dimethylpyrimidin-2-ylcarbamoylsulfamoyl)benzoic acid	667
2-[3-(4,6-dimethylpyrimidin-2-yl)ureidosulfonyl]benzoic acid	667
O,O-dimethyl O-quinoxalin-2-yl phosphorothioate	S1253
O-4-dimethylsulfamoylphenyl O,O-diethyl phosphorothioate	S1004
O-4-dimethylsulfamoylphenyl O,O-dimethyl phosphorothioate	297
dimethyl 2,3,5,6-tetrachloro-1,4-benzenedicarboxylate	140
1,1-dimethyl-3-[3-(1,1,2,2-tetrafluoroethoxy)phenyl]urea	S1298

Chemical Name	Entry no.
2-[1-(ethoxyimino)propyl]-3-hydroxy-5-(2,4,6-trimethylphenyl)-2-cyclohexen-1-one	717
O-6-ethoxy-2-isopropylpyrimidin-4-yl O,O-dimethyl phosphorothioate	S1131
2-(ethoxymethyl)-4,6-dinitrophenol	S1030
O-[6-ethoxy-2-(1-methylethyl)-4-pyrimidinyl] O,O-dimethyl phosphorothioate	S1131
S-[5-ethoxy-2-oxo-1,3,4-thiadiazol-3(2H)-ylmethyl] O,O-dimethyl phosphorodithioate	S1133
(4-ethoxyphenyl)[3-(4-fluoro-3-phenoxyphenyl)propyl](dimethyl)silane	650
1-[[2-(4-ethoxyphenyl)-2-methylpropoxy]methyl]-3-phenoxybenzene	292
2-(4-ethoxyphenyl)-2-methylpropyl 3-phenoxybenzyl ether	292
5-ethoxy-3-(trichloromethyl)-1,2,4-thiadiazole	294
2-ethylamino-4-isopropylamino-6-methoxy-1,3,5-triazine	S782
2-[[4-(ethylamino)-6-(methylthio)-1,3,5-triazin-2-yl]amino]-2-methylpropanenitrile	S910
S-[2-(ethylamino)-2-oxoethyl] O,O-dimethyl phosphorodithioate	S1024
S-ethyl azepane-1-carbothioate	501
ethyl O-benzoyl-3-chloro-2,6-dimethoxybenzohydroximate	67
S-ethyl bis(2-methylpropyl)carbamothioate	100
O-ethyl S,S-bis(1-methylpropyl) phosphorodithioate	101
ethyl (3-tert-butyl-1-dimethylcarbamoyl-1H-1,2,4-triazol-5-ylthio)acetate	724
(R)-1-(ethylcarbamoyl)ethyl carbanilate	108
S-ethylcarbamoylmethyl O,O-dimethyl phosphorodithioate	S1024
ethyl (R)-2-[4-[(6-chloro-2-benzoxazolyl)oxy]phenoxy]propanoate	309
ethyl (±)-2-[4-(6-chloro-1,3-benzoxazol-2-yloxy)phenoxy]propionate	S1044
ethyl 2-chloro-5-(4-chloro-5-difluoromethoxy-1-methylpyrazol-3-yl)-4-fluorophenoxyacetate	617
ethyl 4-chloro-α-(4-chlorophenyl)-α-hydroxybenzeneacetate	S869
ethyl (S)-2-chloro-5-[2-chloro-4-(trifluoromethyl)phenoxy]benzoate	391
ethyl O-[2-chloro-5-(2-chloro-α,α,α-trifluoro-p-tolyloxy)benzoyl]-L-lactate	391
ethyl 5-chloro-1H-3-indazole-3-acetate	288
ethyl 2-[[[[(4-chloro-6-methoxypyrimidin-2-yl)amino]carbonyl]amino]sulfonyl]benzoate	127
ethyl 4-(4-chloro-2-methylphenoxy)butanoate	457
S-ethyl (4-chloro-2-methylphenoxy)ethanethioate	456
ethyl (R)-2-[4-(6-chloroquinoxalin-2-yloxy)phenoxy]propionate	641
ethyl (RS)-2-[4-(6-chloroquinoxalin-2-yloxy)phenoxy]propionate	640
ethyl 4-(4-chloro-o-tolyloxy)butyrate	457
S-ethyl 4-chloro-o-tolyloxythioacetate	456
ethyl O-[5-(2-chloro-α,α,α-trifluoro-p-tolyloxy)-2-nitrobenzoyl]-DL-lactate	442
S-ethyl cyclohexyl(ethyl)thiocarbamate	171
ethyl 4-(cyclopropylhydroxymethylene)-3,5-dioxocyclohexanecarboxylate	744
ethyl 4,4'-dichlorobenzilate	S869
ethyl (±)-3-(3,5-dichlorophenyl)-5-methyl-2,4-dioxo-oxazolidine-5-carboxylate	142
ethyl 1-(2,4-dichlorophenyl)-5-(trichloromethyl)-1H-1,2,4-triazole-3-carboxylate	S1037
ethyl 2-[(diethoxyphosphinothioyl)oxy]-5-methylpyrazolo[1,5-a]pyrimidine-6-carboxylate	619
ethyl (diethoxyphosphinothioylthio)acetyl(methyl)carbamate	458
ethyl N-(diethoxythiophosphorylthio)acetyl-N-methylcarbamate	458
S-ethyl diethylcarbamothioate	S1023
S-ethyl diethylthiocarbamate	S1023
5-ethyl-5,8-dihydro-8-oxo[1,3]dioxolo[4,5-g]quinoline-7-carboxylic acid	544
S-ethyl di-isobutylthiocarbamate	100
ethyl α-[(dimethoxyphosphinothioyl)thio]benzeneacetate	565
ethyl 5-(4,6-dimethoxypyrimidin-2-ylcarbamoylsulfamoyl)-1-methylpyrazole-4-carboxylate	620

Chemical Name	Entry no.
ethyl [[1-[(dimethylamino)carbonyl]-3-(1,1-dimethylethyl)-1*H*-1,2,4-triazol-5-yl]thio]acetate	724
S-ethyl [3-(dimethylamino)propyl]carbamothioate	S1238
S-ethyl (3-dimethylaminopropyl)thiocarbamate	S1238
(2*RS*,4*RS*)-4-(2-ethyl-1,3-dioxolan-4-ylmethoxy)phenyl phenyl ether	252
(2*RS*,4*SR*)-4-(2-ethyl-1,3-dioxolan-4-ylmethoxy)phenyl phenyl ether	252
O-ethyl S,S-diphenyl phosphorodithioate	267
S-ethyl dipropylcarbamothioate	274
O-ethyl S,S-dipropyl phosphorodithioate	286
S-ethyl dipropylthiocarbamate	274
1,1'-ethylene-2,2'-bipyridyldiylium	256
ethylenebis(dithiocarbamate) mixed metal complex	451
ethylene bis(trichloroacetate)	S1026
ethylenethiuram monosulfide	S1021
ethyl 6-ethoxy-2-methyl-3-oxo-7-oxa-5-thia-2-aza-6-phosphanonanoate 6-sulfide	458
2-ethyl-3-[3-ethyl-5-(4-ethylphenoxy)pentyl]-2-methyloxirane	S1015
N-ethyl-1,1,2,2,3,3,4,4,5,5,6,6,7,7,8,8,8-heptadecafluoro-1-octanesulfonamide	666
S-ethyl hexahydro-1*H*-azepine-1-carbothioate	501
2-ethylhexane-1,3-diol	S1027
N-(2-ethylhexyl)bicyclo[2.2.1]hept-5-ene-2,3-dicarboximide	272
2-(1-ethylhexyl)-4,6-dinitrophenyl methyl carbonate	S980
4-(1-ethylhexyl)-2,6-dinitrophenyl methyl carbonate	S980
2-(2-ethylhexyl)-3a,4,7,7a-tetrahydro-4,7-methano-1*H*-isoindole-1,3(2*H*)-dione	272
N-(2-ethylhexyl)-8,9,10-trinorborn-5-ene-2,3-dicarboximide	272
ethyl hydrogen (aminocarbonyl)phosphonate	371
ethyl hydrogen carbamoylphosphonate	371
ethyl hydrogen phosphonate	372
O-ethyl O-2-isopropoxycarbonylphenyl N-isopropylphosphoramidothioate	430
N^2-ethyl-N^4-isopropyl-6-methoxy-1,3,5-triazine-2,4-diamine	S782
(*RS*)-5-ethyl-2-(4-isopropyl-4-methyl-5-oxo-2-imidazolin-2-yl)nicotinic acid	415
N^2-ethyl-N^4-isopropyl-6-methylthio-1,3,5-triazine-2,4-diamine	20
N-ethylmercurio-4-toluenesulfonanilide	S1028
N-ethyl-6-methoxy-N'-(1-methylethyl)-1,3,5-triazine-2,4-diamine	S782
N-ethyl-6-methoxy-N'-(1-methylpropyl)-1,3,5-triazine-2,4-diamine	S1270
S-ethyl (*E,E*)-(*RS*)-11-methoxy-3,7,11-trimethyldodeca-2,4-dienethioate	S1333
O-ethyl S-(N-methylcarbaniloylmethyl) N-isopropylphosphoramidothioate	S1110
ethyl 3'-(3'-methylcarbaniloyloxy)carbanilate	S1211
2-ethyl-5-methyl-1,3-dioxan-5-yl 2-methylbenzyl ether	S1075
N-ethyl-N'-(1-methylethyl)-6-(methylthio)-1,3,5-triazine-2,4-diamine	20
O-ethyl S-(3-methylisoxazol-5-ylmethyl) S-propyl phosphorodithioate	S1342
2-ethyl-5-methyl-5-(2-methylbenzyloxy)-1,3-dioxane	S1075
2-ethyl-5-methyl-5-[(2-methylphenyl)methoxy]-1,3-dioxane	S1075
ethyl 3-methyl-4-(methylthio)phenyl (1-methylethyl)phosphoramidate	299
O-ethyl O-(5-methyl-2-nitrophenyl) (1-methylpropyl)phosphoramidothioate	94
O-ethyl S-3-methyl-1,2-oxazol-5-ylmethyl S-propyl phosphorodithioate	S1342
O-ethyl S-[2-(methylphenylamino)-2-oxoethyl] phosphorothioate	S1110
ethyl(4-methyl-N-phenylbenzenesulfonamidato-N)mercury	S1028
N-ethyl-N-(2-methyl-2-propenyl)-2,6-dinitro-4-(trifluoromethyl)benzenamine	279
N-[3-(1-ethyl-1-methylpropyl)isoxazol-5-yl]-2,6-dimethoxybenzamide	435

Chemical Name	Entry no.

INDEXES

INDEXES

Chemical Name	Entry no.
3-methyl-2-(1-methylethyl)-4,6-dinitrophenol	S983
exo-(±)-1-methyl-4-(1-methylethyl)-2-[(2-methylphenyl)methoxy]-7-oxabicyclo[2.2.1]heptane	144
N-methyl-N'-(1-methylethyl)-6-(methylthio)-1,3,5-triazine-2,4-diamine	207
3-methyl-5-(1-methylethyl)phenyl methylcarbamate	S1233
3-methyl-5-(1-methylethyl)phenyl methyl(1-oxobutyl)carbamate	S1232
3-methyl-1-(1-methylethyl)-1H-pyrazol-5-yl dimethylcarbamate	S1115
S-methyl O-2-(1-methylheptyl)-4,6-dinitrophenyl thiocarbonate	S987
O-methyl O-(4-methyl-2-nitrophenyl) (1-methylethyl)phosphoramidothioate	S772
1-methyl-3-(1-methyl-1-phenylethyl)-1-phenylurea	488
[[[2-methyl(methylsulfonyl)amino]-2-oxoethyl]amino]methylphosphonic acid	S1132
2-methyl-2-(methylsulfonyl)propanal O-[(methylamino)carbonyl]oxime	S762
2-methyl-2-methylsulfonylpropionaldehyde O-methylcarbamoyloxime	S762
methyl 1-[[[2-(methylthio)ethyl]amino]carbonyl]-1H-benzimidazol-2-ylcarbamate	S1137
methyl 1-(2-methylthioethylcarbamoyl)benzimidazol-2-ylcarbamate	S1137
2-methyl-2-(methylthio)propanal O-[(methylamino)carbonyl]oxime	16
2-methyl-2-(methylthio)propionaldehyde O-methylcarbamoyloxime	16
O-methyl O-2-nitro-p-tolyl isopropylphosphoramidothioate	S772
2-(3-methyl-1-oxobutyl)-1H-indene-1,3(2H)-dione	S1124
2-[1-methyl-2-(4-phenoxyphenoxy)ethoxy]pyridine	631
5-methyl-5-(4-phenoxyphenyl)-3-(phenylamino)-2,4-oxazolidinedione	296
3-(2-methylphenoxy)pyridazine	S901
methyl N-phenylacetyl-N-2,6-xylyl-DL-alaninate	54
2-methyl-N-phenylbenzamide	S1136
S-1-methyl-1-phenylethyl piperidine-1-carbothioate	237
1-(1-methyl-1-phenylethyl)-3-p-tolylurea	195
2-methyl-N-phenyl-3-furancarboxamide	305
3-methylphenyl methylcarbamate	494
N-(4-methylphenyl)-N'-(1-methyl-1-phenylethyl)urea	195
4-methyl-N-phenyl-6-(1-propynyl)-2-pyrimidinamine	464
3-methyl-1-phenyl-1H-pyrazol-5-yl dimethylcarbamate	S1164
1-methyl-4-phenylpyridinium	S919
1-methyl-3-phenyl-5-[3-(trifluoromethyl)phenyl]-4(1H)-pyridinone	352
S-2-methylpiperidinocarbonylmethyl O,O-dipropyl phosphorodithioate	582
3-(2-methylpiperidino)propyl 3,4-dichlorobenzoate	580
S-[2-(2-methyl-1-piperidinyl)-2-oxoethyl] O,O-dipropyl phosphorodithioate	582
N-(2-methyl-2-propenyl)-2,6-dinitro-N-propyl-4-(trifluoromethyl)benzenamine	S1148
2-(1-methylpropyl)-4,6-dinitrophenol	S985
2-(1-methylpropyl)-4,6-dinitrophenyl acetate	S986
2-(1-methylpropyl)-4,6-dinitrophenyl 3-methyl-2-butenoate	S804
N-(2-methylpropyl)-2-oxo-1-imidazolidinecarboxamide	S1112
2-(1-methylpropyl)phenyl methylcarbamate	307
2-methyl(prop-2-ynyl)aminophenyl methylcarbamate	S1165
4-methyl(prop-2-ynyl)amino-3,5-xylyl methylcarbamate	S1166
1-methylprop-2-ynyl 3-chlorocarbanilate	S855
1-methyl-2-propynyl (3-chlorophenyl)carbamate	S855
N-(4-methyl-6-prop-1-ynylpyrimidin-2-yl)aniline	464
(RS)-2-methyl-1-pyrimidin-5-yl-1-(4-trifluoromethoxyphenyl)propan-1-ol	355
(S)-3-(1-methylpyrrolidin-2-yl)pyridine	520

INDEXES

Chemical Name	Entry no.
1-{N-propyl-N-[2-(2,4,6-trichlorophenoxy)ethyl]}carbamoylimidazole	591
N-propyl-N-[2-(2,4,6-trichlorophenoxy)ethyl]-1H-imidazole-1-carboxamide	591
prop-2-ynyl (R)-2-[4-(5-chloro-3-fluoropyridin-2-yloxy)phenoxy]propionate	147
prop-2-ynyl (±)-(E,E)-3,7,11-trimethyldodeca-2,4-dienoate	S1129
pyridazin-3-yl o-tolyl ether	S901
pyridine-2-thiol 1-oxide	S1103
2-pyridyl 1-(2,5-xylyl)ethyl sulfone 1-oxide	S1336
N-pyrrolidinosuccinamic acid	S1251
3-(pyrrolidin-2-yl)pyridine	S1198
8-quinolinol sulfate	407
quinoxaline-2,3-diyl trithiocarbonate	S1311
salicylanilide	S1266
(3β,5Z,7E)-9,10-secocholesta-5,7,10(19)-trien-3-ol	752
(5Z,7E,22E)-(3S)-9,10-secoergosta-5,7,10(19),22-tetraen-3-ol	276
sodium 2,6-bis(4,6-dimethoxypyrimidin-2-yloxy)benzoate	76
sodium chlorate	653
sodium (Z)-3-chloroacrylate	S1276
sodium 5-chloro-2-[4-chloro-2-[3-(3,4-dichlorophenyl)ureido]phenoxy]benzenesulfonate	663
sodium 2-chloro-6-(4,6-dimethoxypyrimidin-2-ylthio)benzoate	632
sodium [4-chloro-2-(hydroxymethyl)phenoxy]acetate	156
sodium 4-chloro-α-hydroxy-o-tolyloxyacetate	156
sodium (Z)-3-chloro-2-propenoate	S1276
sodium 5-(2-chloro-α,α,α-trifluoro-p-tolyloxy)-2-nitrobenzoate	7
sodium 4-dimethylaminobenzenediazosulfonate	S1033
sodium fluoride	654
sodium fluoroacetate	655
sodium fluorosilicate	656
sodium hydrogen methylarsonate	486
sodium pentachlorophenate	558
sodium pentachlorophenoxide	558
sodium selenate	S1277
sodium tetrathio(peroxocarbonate)	386
sodium trichloroacetate	677
strychnidin-10-one	662
1,1'-sulfonylbis[benzene]	S992
sulfur	670
sulfuric acid	671
sulfuryl fluoride	672
2,4,5,6-tetrachloro-1,3-benzenedicarbonitrile	133
2,3,5,6-tetrachloro-1,4-benzenedicarboxylic acid	140
2,3,5,6-tetrachloro-1,4-benzoquinone	S849
2,3,5,6-tetrachloro-2,5-cyclohexadiene-1,4-dione	S849
2,3,4,5-tetrachloro-6-[[(2,3-dichlorophenyl)amino]carbonyl]benzoic acid	684
3,4,5,6-tetrachloro-N-(2,3-dichlorophenyl)phthalamic acid	684
2,4,4',5-tetrachlorodiphenyl sulfide	S1300
N-(1,1,2,2-tetrachloroethylthio)cyclohex-4-ene-1,2-dicarboximide	104
4,5,6,7-tetrachloro-1(3H)-isobenzofuranone	576
tetrachloroisophthalonitrile	133

Chemical Name	Entry no.
2-thioxo-1,3-dithiolo[4,5-*b*]quinoxaline	S1311
o-toluanilide	S1136
m-tolyl methylcarbamate	494
N-*m*-tolylphthalamic acid	716
toxaphene	S837
3,6,9-triazahenicosanoic acid	S1001
1*H*-1,2,4-triazol-3-amine	23
1*H*-1,2,4-triazol-3-ylamine	23
tributyl(2,4-dichlorobenzyl)phosphonium	135
tributyl[(2,4-dichlorophenyl)methyl]phosphonium	135
S,S,S-tributyl phosphorotrithioate	729
tributyl phosphorotrithioite	S1320
trichloroacetic acid	677
trichloroacetic acid compound with *N*'-(4-chlorophenyl)-*N,N*-dimethylurea	S1179
trichloroacetic acid compound with *N,N*-dimethyl-*N*-phenylurea	318
S-2,3,3-trichloroallyl di-isopropyl(thiocarbamate)	722
S-2,3,3-trichloroallyl di-isopropylthiocarbamate	722
3,5,6-trichloro-*o*-anisic acid	S1321
2,3,6-trichlorobenzeneacetic acid	S859
2,3,6-trichlorobenzoic acid	676
4,5,7-trichloro-2,1,3-benzothiadiazole	S1323
trichlorobenzyl chloride	S1324
1-(2,3,6-trichlorobenzyloxy)propan-2-ol	S1335
1,1,1-trichloro-2,2-bis(4-chlorophenyl)ethane	201
2,2,2-trichloro-1,1-bis(4-chlorophenyl)ethanol	222
1,1,1-trichloro-2,2-bis(4-methoxyphenyl)ethane	485
trichloro(chloromethyl)benzene	S1324
1,2,4-trichloro-5-[(4-chlorophenyl)sulfonyl]benzene	697
1,2,4-trichloro-5-[4-(chlorophenyl)thio]benzene	S1300
N-[2,2,2-trichloro-1-(3,4-dichloroanilino)ethyl]formamide	S848
N-[2,2,2-trichloro-1-[(3,4-dichlorophenyl)amino]ethyl]formamide	S848
1,1,1-trichloro-di-(4-chlorophenyl)ethane	201
2,2,2-trichloro-1-(3,4-dichlorophenyl)ethyl acetate	S1325
2,2,2-trichloro-1-(dimethoxyphosphinyl)ethyl butanoate	S828
1,1'-(2,2,2-trichloroethylidene)bis[4-chlorobenzene]	201
1,1'-(2,2,2-trichloroethylidene)bis[4-methoxybenzene]	485
(*R*)-1,2-*O*-(2,2,2-trichloroethylidene)-α-D-glucofuranose	118
N-(trichloromethanesulfenyl)phthalimide	364
2,3,5-trichloro-6-methoxybenzoic acid	S1321
N-(trichloromethylthio)cyclohex-4-ene-1,2-dicarboximide	105
2-[(trichloromethyl)thio]-1*H*-isoindole-1,3(2*H*)-dione	364
N-(trichloromethylthio)phthalimide	364
trichloronitromethane	132
1,3,5-trichloro-2-(4-nitrophenoxy)benzene	S868
(2,4,5-trichlorophenoxy)acetic acid	S1287
4-(2,4,5-trichlorophenoxy)butyric acid	S1289
2-(2,4,5-trichlorophenoxy)ethanol	S1049
2-(2,4,5-trichlorophenoxy)ethyl 2,2-dichloropropionate	S1017

INDEXES

INDEX 6

Classes

For a description of classes, see the Guide to using the Main Entries page x.

Note that similar class names may refer to profoundly different types of activity. For example, 'organophosphorus' is used to describe the phosphate ester insecticides, whilst 'organophosphate ester' is used for a group of fungicides, and 'organophosphorus herbicide' for a group of herbicides.

Unlike the other indexes, references are to entry names. Only Main Entries are included.

INDEX

Class	Name	Class	Name
1,2,4-triazinone	hexazinone		metiram
	metamitron		nabam
	metribuzin		propineb
1,3,5-triazine	ametryn		zineb
	atrazine	amide	isoxaben
	cyanazine		propyzamide
	desmetryn		tebutam
	dimethametryn	amidine	amitraz
	prometon	aminoglycoside antibiotic	kasugamycin
	prometryn	anilide	pentanochlor
	propazine		propanil
	simazine	anilinopyrimidine	cyprodinil
	simetryn		mepanipyrim
	terbumeton		pyrimethanil
	terbuthylazine	antibiotic	blasticidin-S
	terbutryn		polyoxins
	trietazine		polyoxin B
2-(4-aryloxyphenoxy)propionic acid		aromatic hydrocarbon	biphenyl
	clodinafop-propargyl	aromatic hydrocarbon derivative	quintozene
	cyhalofop-butyl	arylalanine	flamprop-M
	diclofop-methyl	aryloxyalkanoic acid	clomeprop
	fenoxaprop-P-ethyl		cloprop
	fluazifop-butyl		cloxyfonac
	fluazifop-P-butyl		4-CPA
	haloxyfop		2,4-D
	propaquizafop		2,4-DB
	quizalofop		dichlorprop
	quizalofop-P		dichlorprop-P
2,6-dinitroaniline	benfluralin		fluroxypyr
	butralin		MCPA
	dinitramine		MCPA-thioethyl
	ethalfluralin		MCPB
	fluazinam		mecoprop
	fluchloralin		mecoprop-P
	flumetralin		triclopyr
	oryzalin	aryloxyphenoxypropionic acid	
	pendimethalin	See 2-(4-aryloxyphenoxy)propionic acid	
	prodiamine	auxin	indol-3-ylacetic acid
	trifluralin		4-indol-3-ylbutyric acid
2-dimethylaminopropane-1,3-dithiol		avermectin	abamectin
	bensultap	azole	azaconazole
	cartap		BAS 480F
2-dimethylaminopropane-1,3-dithiol analogue	thiocyclam		bitertanol
			bromuconazole
Acari	Amblyseius spp.		cyproconazole
	Phytoseiulus persimilis		difenoconazole
alkanamide	diphenamid		diniconazole
	naproanilide		fenbuconazole
	napropamide		fluquinconazole
alkylenebis(dithiocarbamate)	mancopper		flusilazole
	mancozeb		flutriafol
	maneb		hexaconazole

Class	Name	Class	Name
	imazalil		phenmedipham
	imibenconazole	carbamate	bendiocarb
	ipconazole		benfuracarb
	metconazole		carbaryl
	myclobutanil		carbetamide
	paclobutrazol		carbofuran
	pefurazoate		carbosulfan
	penconazole		chlorpropham
	prochloraz		ethiofencarb
	propiconazole		fenobucarb
	tebuconazole		formetanate
	tetraconazole		furathiocarb
	triadimefon		isoprocarb
	triadimenol		methiocarb
	triflumizole		metolcarb
	triticonazole		pirimicarb
	uniconazole		propamocarb hydrochloride
azole analogue	triforine		propham
azomethine	pymetrozine		propoxur
bacterium	*Bacillus sphaericus*		trimethacarb
	Bacillus subtilis		XMC
	Bacillus thuringiensis		xylylcarb
	Streptomyces griseoviridis	carboxamide	carboxin
benzilate	bromopropylate		fenfuram
benzimidazole	benomyl		flutolanil
	carbendazim		mepronil
	debacarb		oxycarboxin
	fuberidazole	chloroacetanilide	acetochlor
	thiabendazole		alachlor
benzimidazole precursor	thiophanate-methyl		butachlor
benzofuranyl alkanesulfonate	benfuresate		dimethachlor
	ethofumesate		metazachlor
benzoic acid	chlorthal-dimethyl		metolachlor
benzoic acid (auxin)	chloramben		pretilachlor
	dicamba		propachlor
	2,3,6-TBA		propisochlor
benzonitrile	chlorthiamid		thenylchlor
	dichlobenil	chloroamide	dichlormid
benzoxazine	benoxacor	chlorophenol	dichlorophen
benzoylurea	chlorfluazuron	Coleoptera	*Cryptolaemus montrouzieri*
	diflubenzuron		*Hippodamia convergens*
	fluazuron	coumarin anticoagulant	brodifacoum
	flucycloxuron		bromadiolone
	flufenoxuron		coumatetralyl
	hexaflumuron		difenacoum
	lufenuron		flocoumafen
	novaluron		warfarin
	teflubenzuron	coumarin anticoagulant analogue	difethialone
	triflumuron	cyclodiene organochlorine	chlordane
bipyridylium	diquat dibromide		endosulfan
	paraquat dichloride		heptachlor
bis-carbamate	desmedipham	cyclohexanedione oxime	alloxydim

Class	Name
	clethodim
	cycloxydim
	sethoxydim
	tralkoxydim
cytokinin	6-benzylaminopurine
diacylhydrazine	halofenozide
	RH-2485
	tebufenozide
dicarboximide	chlozolinate
	iprodione
	procymidone
	vinclozolin
dimethylaminopropane-1,3-dithiol	
See 2-dimethylaminopropane-1,3-dithiol	
dimethyldithiocarbamate	ferbam
	thiram
	ziram
dinitroaniline	See 2,6-dinitroaniline
dinitrophenol	dinoterb
	DNOC
dinitrophenol derivative	dinocap
diphenyl ether	acifluorfen
	aclonifen
	bifenox
	chlomethoxyfen
	diofenolan
	fluoroglycofen-ethyl
	fomesafen
	HC-252
	lactofen
	oxyfluorfen
Diptera	Aphidoletes aphidimyza
dithiocarbamate	
	nickel bis(dimethyldithiocarbamate)
DMI fungicide	fenarimol
entomopathogenic fungus	Beauveria bassiana
	Beauveria brongniartii
	Paecilomyces fumosoroseus
	Verticillium lecanii
ethylene generator	ethephon
fatty acid	oleic acid (fatty acids)
fungus	Ampelomyces quisqualis
	Fusarium oxysporum
	Gliocladium virens
	Phlebiopsis gigantea
	Trichoderma harzianum
granulosis virus	Adoxophyes orana GV
	Cydia pomonella GV
guanidine	dodine
	guazatine
	iminoctadine
halogenated alkanoic acid	dalapon

Class	Name
	flupropanate
	TCA-sodium
Hemiptera	Macrolophus caliginosus
	Orius spp.
hydroxybenzonitrile	bromoxynil
	ioxynil
hydroxybenzonitrile precursor	
	bromofenoxim
Hymenoptera	Anagrus atomus
	Aphelinus abdominalis
	Aphidius colemani
	Dacnusa sibirica
	Diglyphus isaea
	Encarsia formosa
	Eretmocerus californicus
	Leptomastix dactylopii
	Trichogramma spp.
imidazolinone	AC 263,222
	imazamethabenz-methyl
	imazamox
	imazapyr
	imazaquin
	imazethapyr
indandione anticoagulant	chlorophacinone
	diphacinone
	pindone
inorganic	Bordeaux mixture
	copper hydroxide
	copper oxychloride
	cuprous oxide
	mercuric oxide
inorganic fluoride	sodium hexafluorosilicate
juvenile hormone mimic	pyriproxyfen
	hydroprene
	methoprene
methyl isothiocyanate precursor	dazomet
	metam
morphactin	flurenol
morpholine	dodemorph
	fenpropimorph
	tridemorph
morpholine analogue (piperidine derivative)	fenpropidin
N-trihalomethylthio	captafol
	captan
	dichlofluanid
	folpet
	tolylfluanid
natural pyrethrin	pyrethrins (pyrethrum)
nematode	
Heterorhabditis bacteriophora and H. megidis	
Steinernema spp.	

INDEXES

INDEXES

Class	Name	Class	Name
	bensulfuron-methyl		pyributicarb
	chlorimuron-ethyl		thiobencarb
	chlorsulfuron		tiocarbazil
	cinosulfuron		tri-allate
	ethametsulfuron-methyl		vernolate
	ethoxysulfuron	**triazine**	See 1,3,5-triazine
	flazasulfuron	**triazinone**	See 1,2,4-triazinone
	flupyrsulfuron-methyl-sodium	**triazole**	amitrole
	halosulfuron-methyl	**triazolopyrimidine sulfonanilide**	
	imazosulfuron		cloransulam-methyl
	metsulfuron-methyl		flumetsulam
	nicosulfuron		metosulam
	oxasulfuron	**triketone**	sulcotrione
	primisulfuron-methyl	**uracil**	bromacil
	prosulfuron		lenacil
	pyrazosulfuron-ethyl		terbacil
	rimsulfuron	**urea**	chlorbromuron
	sulfometuron-methyl		chlorotoluron
	sulfosulfuron		dimefuron
	thifensulfuron-methyl		diuron
	triasulfuron		fenuron
	tribenuron-methyl		fluometuron
	triflusulfuron-methyl		isoproturon
synthetic auxin	2-(1-naphthyl)acetamide		isouron
	2-(1-naphthyl)acetic acid		linuron
	(2-naphthyloxy)acetic acid		methabenzthiazuron
tetrazine	clofentezine		metobenzuron
	SZI-121		metobromuron
thiocarbamate	butylate		metoxuron
	cycloate		monolinuron
	dimepiperate		neburon
	EPTC		siduron
	esprocarb		tebuthiuron
	molinate		
	orbencarb	**virus**	See granulosis virus *and* nuclear
	pebulate		polyhedrosis virus
	prosulfocarb	**wasp**	See Hymenoptera